ASTRONOMY AND ASTROPHYSICS ABSTRACTS

A Publication of the Astronomisches Rechen-Institut Heidelberg
Produced in Cooperation
with the Fachinformationszentrum Karlsruhe

Astronomy and Astrophysics Abstracts is Prepared
Under the Auspices of the International Astronomical Union

Volume 54 A
Literature 1991, Part 2

Edited by
G. Burkhardt U. Esser H. Hefele I. Heinrich W. Hofmann
D. Krahn V. R. Matas L. D. Schmadel R. Wielen G. Zech

Springer-Verlag Berlin Heidelberg GmbH

Astronomisches Rechen-Institut, Mönchhofstraße 12–14,
W-6900 Heidelberg 1, Fed. Rep. of Germany
Telex: 461 336 ARIHD D
Telefax: 0 62 21-40 52 97
Telephone: 0 62 21-40 50
Director: Prof. Dr. Roland Wielen

Astronomy and Astrophysics Abstracts
Department Head: Dr. Lutz D. Schmadel
Editors-in-Chief: Inge Heinrich, Dr. Lutz D. Schmadel

ISBN 978-3-662-12378-2 ISBN 978-3-662-12376-8 (eBook)
DOI 10.1007/978-3-662-12376-8

Originally published by Springer-Verlag Berlin Heidelberg New York in 1992
Softcover reprint of the hardcover 1st edition 1992

Library of Congress Catalog Card Number 72-104650.

Media conversion: Daten- und Lichtsatz-Service, Würzburg.

55/3140-5 4 3 2 1 0 – Printed on acid-free paper

Preface

Astronomy and Astrophysics Abstracts aims to present a comprehensive documentation of the literature concerning all aspects of astronomy, astrophysics, and their neighbouring fields. It is devoted to the recording, summarizing, and indexing of the relevant publications throughout the world. *Astronomy and Astrophysics Abstracts* is prepared by a special department of the *Astronomisches Rechen-Institut* in close cooperation with the *Fachinformationszentrum Karlsruhe*. The work is done under the auspices of the *International Astronomical Union*.

Volume 54 records literature published in 1991 and received before 30 April 1992. Some older documents which we received late and which are not surveyed in earlier volumes are included too. The input work was shared between the Fachinformationszentrum Karlsruhe and the Astronomisches Rechen-Institut. We thank our Karlsruhe colleagues Ms. Ingrid Ernst and Mr. Wolfgang Lück for their willing and effective cooperation. We acknowledge with thanks contributions from our colleagues all over the world. Last but not least we express our gratitude to all the organizations, observatories, and publishers that provide us with complimentary copies of their publications.

The recording process and corrections were carried out by our technical staff members Ms. Helga Ballmann, Ms. Monika Kohl, and Ms. Sylvia Matyssek. It is a pleasure to thank them all for their diligence.

Heidelberg, July 1992 **The Editors**

Contents

PART A

PART B

Stars

Interstellar Matter, Nebulae

Radio Sources, X-ray Sources, Cosmic Rays

Stellar Systems, Galaxy, Extragalactic Objects, Cosmology

Introduction

Astronomical Bibliographies

Astronomy and Astrophysics Abstracts started documentation and abstracting work in 1969 as the direct successor of the Astronomischer Jahresbericht. For information on astronomical literature before this date consultation of one of the following bibliographies is suggested:

(1) J. J. de Lalande, Bibliographie Astronomique, Paris 1803 (this work covers the time from 480 B. C. to the year 1803, VIII + 966 pages).

(2) J. C. Houzeau, A. Lancaster, Bibliographie générale de l'astronomie, Volume I (in two parts), Bruxelles 1887, 1889, Volume II, Bruxelles 1882. The complete title of Volume II is "Bibliographie générale de l'astronomie ou catalogue méthodique des ouvrages, des mémoires et des observations astronomiques, publiés depuis l'origine de l'imprimerie jusqu'en 1880". A new edition of these volumes was prepared by D. W. Dewhirst in 1964.

(3) Bibliography of Astronomy, 1881–1898. The literature of this period was recorded on standard slips by the Observatoire Royal de Belgique. From the material (some 52,000 items) a microfilm version was produced by University Microfilms Limited, Tylers Green, High Wycombe, Buckinghamshire, England, in 1970.

(4) Astronomischer Jahresbericht, 1899 gegründet von Walter Wislicenus, herausgegeben vom Astronomischen Rechen-Institut in Heidelberg (formerly in Berlin), Verlag W. de Gruyter, Berlin. For the period from 1899 to 1968 sixty-eight volumes were published, each of which, in general, covers the literature of one year.

Concept of *Astronomy and Astrophysics Abstracts*

This abstracting service aims to present a comprehensive documentation of the literature in all fields of astronomy and astrophysics and their border fields. It appears in semi-annual volumes. Two of these volumes cover the literature of one calendar year. Every effort will be made to ensure that the average time interval between the receiving date of the original documents and publication of the abstracts will not exceed eight months. This time interval is near to that achieved by monthly abstracting journals, compared to which our system of accumulation of information over six months offers the advantage of greater convenience for the user.

The main characteristics of the concept of *Astronomy and Astrophysics Abstracts* may be summarized as follows:

(1) The subdivision of astronomy and its border fields into subject categories is facilitated by the fact that the astronomical objects appear to be particularly well suited for the formation of categories. It may be assumed that such subdivisions can be maintained for a long period. Experience shows, however, that progress in research might imply minor changes in the classification scheme.

(2) Each paper has been classified into one of 106 numbered subject categories and given a serial number within the category. In this way each item is numbered by six figures: the first three indicate the number of the category, the following three the serial number within the category. Reference to an abstract in Volume 1 is indicated by "01" before the number of the category; for example: 01.074.028, denotes Volume 1, category 074, abstract 028. A paper might be classified into more than one category. In this case, its abstract is placed only in one category, whereas in the other categories only cross references are given. These are listed at the end of each category.

(3) Authors' abstracts are used whenever possible. Popular articles are not abstracted.

(4) If possible, titles of papers and abstracts are given in English. A special reference is made to titles which we have not taken in the original language.

The whole material was recorded by means of a local AT network. All text recording programs and other data processing software were developed by Multicom GmbH, Gröbenzell, F. R. Germany and by our staff members as well. The index computations were carried out on the IBM 3090-180 computer of the University of Heidelberg.

Classification Systems

The two most common and widely used classification systems in astronomy and astrophysics are given by Class 9 of the revised edition of the International Classification System for Physics, published by the International Council of Scientific Unions Abstracting Board (Second edition 1978. ICSU-AB, 17 Rue Mirabeau, 75017 Paris, France, ISSN 0305-9618), and the *Astronomy and Astrophysics Abstracts* classification. In order to facilitate literature searches, we introduce a concordance relation between these two very different systems. This solution is only a unilateral one. Starting from the third hierarchical level of the PHYS-Classification Scheme 1987, the appropriate *Astronomy and Astrophysics Abstracts* chapter numbers are listed. This cannot imply an identical content of the respective chapters in both systems. In many cases there is only a rather partial concordance, and therefore the *Astronomy and Astrophysics Abstracts* numbers are enclosed in parentheses.

Transliteration Scheme for the Russian Alphabet

The transliteration of the Russian alphabet in use in *Astronomy and Astrophysics Abstracts* is presented here.

А	а	a	П	п	p
Б	б	b	Р	р	r
В	в	v	С	с	s
Г	г	g	Т	т	t
Д	д	d	У	у	u
Е	е	e	Ф	ф	f
Ё	ё	e	Х	х	kh
Ж	ж	zh	Ц	ц	ts
З	з	z	Ч	ч	ch
И	и	i	Ш	ш	sh
Й	й	j	Щ	щ	shch
К	к	k	Ы	ы	y
Л	л	l	Ь	ь	'
М	м	m	Э	э	eh
Н	н	n	Ю	ю	yu
О	о	o	Я	я	ya

This transliteration was recommended by the Abstracting Board of the International Council of Scientific Unions in 1969. It corresponds essentially to the transliteration proposed by the Academy of Sciences, Moscow. In this case the letters can be read and printed by usual data processing machines. If the names of Russian authors in the literature are transliterated in a different scheme, we present the names as they are given in the references cited and in addition in brackets according to our transliteration table.

Sources of Information

The majority of sources of information for this volume is given in category **001 Periodicals** and in category **008 Observatories, Institutes.** It may be noted that the titles of the periodicals are given in the original languages, and that Russian titles have been transliterated applying the transliteration scheme given above. Category 008 records publication series of observatories and astronomical institutes. Titles of the periodicals have been given following the recommendations of the "International List of Periodical Title Word Abbreviations" and its additions (see also **Abbreviations**). In most cases they permit recognition of the full title without recourse to the key in category 001. If other secondary sources have been consulted, we cite these papers and give reference to the respective services.

The total number of papers (some do not give names of authors) recorded in this volume amounts to 11,082.

Author, Subject, and Object Indexes

The subject category and the serial number have been used as a reference in all three indexes. These references are more precise than page references and offer considerable advantages in indexing by means of computers.

The author index of this volume contains 14,186 names.

We consider the subject index as an approximation to an optimal index covering all fields of astronomy and astrophysics. At present, the *Astronomy and Astrophysics Abstracts Vocabulary* 29.92, containing 2,438 key words, is in use. This is done not only for the users' convenience, but also with the intention to propose the use of special key words to authors and publishers.

While each volume is scheduled to contain an author index, a subject index and an object index, the magnetic tapes containing the index information will be used to produce separate index volumes (authors, subjects, and objects) at intervals of five years.

Beginning with Vol. 39 *Astronomy and Astrophysics Abstracts* will additionally contain an object index providing a further key to the documents abstracted. For detailed information concerning selection, standardization, sorting, etc, see introduction to the object index.

The sorting program for the indexes is based on the IBM SORT/MERGE Program. This program sorts blank before hyphen and before letters. Apostrophes are ignored by a special routine.

The users are requested to inform us on spelling errors within the author index in order to assist us in eliminating mistakes in future cumulative indexes.

Information Retrieval

Most of the information given in this volume of *Astronomy and Astrophysics Abstracts* is incorporated into the data base PHYS of the Fachinformationszentrum Karlsruhe. PHYS is on-line retrievable via STN International. Inquiries about PHYS should be directly addressed to:

STN International
c/o Fachinformationszentrum Karlsruhe
P.O. Box 2465
D-7500 Karlsruhe 1
F. R. Germany

Telex: 17 724 710+
Telefax: 0 72 47-80 86 66
Teletex: 72 47 10-FIZKA
Telephone: 0 72 47-80 85 55

The Astronomisches Rechen-Institut Heidelberg does not distribute the astronomical bibliography in any on-line or off-line form.

Astronomy and Astrophysics Abstracts is prepared at the *Astronomisches Rechen-Institut,* Heidelberg under the auspices of the *International Astronomical Union* on a non-profit basis. The editors urge publishers of literature related to astronomy and astrophysics to provide our service in due time with complimentary copies of their material.

Publications should be mailed to:

Astronomy and Astrophysics Abstracts
Astronomisches Rechen-Institut
Moenchhofstrasse 12–14
W-6900 Heidelberg 1
Fed. Rep. of Germany
Telex: 461 336 ARIHD D
Telefax: 0 62 21-40 52 97
Telephone: 0 62 21-4050

Concordance Relation

between the PHYS-Classification Scheme 1987
and the *Astronomy and Astrophysics Abstracts* Classification Scheme

PHYS Classification Scheme		AAA Classification Scheme
0 General		
0100	**Communication, education, history, and philosophy**	
0130	Physics literature and publications	002, 003, 012, 046
0140	Education	014
0150	Educational aids	014
0160	Biographical, historical, and personal notes	004–007
0190	Other topics of general interest	015
9 Geophysics, Astronomy, and Astrophysics		
9100	**Solid Earth physics**	
9110	Geodesy and gravity	044, 045, 081
9125	Geomagnetism and paleomagnetism; geoelectricity	084 (081)
9135	Earth's interior structure and properties	081
9190	Other topics in solid Earth physics	081
9400	**Aeronomy**	
9410	Physics of the neutral atmosphere	082 (063, 084)
9420	Physics of the ionosphere	083 (062, 084)
9430	Physics of the magnetosphere	084 (062, 106)
9440	Cosmic-ray interactions with the Earth	078, 144 (085, 105)
9500	**Fundamental astronomy and astrophysics, instrumentation, techniques, and astronomical observations**	
9510	Fundamental astronomy	041–043 (052, 079, 095, 096)
9520	Historical and archaeoastronomy	004

PHYS Classification Scheme		AAA Classification Scheme
9530	Fundamental aspects of astrophysics	022, 061–063, 066
9540	Artificial Earth satellites	051, 053
9545	Observatories	008, 009
9555	Astronomical and space-research instrumentation	032–035 (031, 036, 053)
9575	Observation and reduction techniques	036 (021)
9580	Catalogues, atlases, etc.	002 (046)
9600	**Solar system**	
9610	General, solar nebula, and cosmogony	107 (091)
9620	Moon	094
9630	Planets and satellites (excluding the Moon)	091–093, 097, 099–101
9640	Cosmic rays	078, 144 (106)
9650	Interplanetary space	074, 098, 102–106
9660	Solar physics	071–080
9700	**Stars**	
9710	Stellar characteristics and properties	064, 065, 111–116, 131, (061, 062, 063)
9720	Normal stars (by class); general or individual	113–116, 121, 126
9730	Variable and peculiar stars (including novae)	112, 116, 117, 122–124
9760	Late stages of stellar evolution (including black holes)	067, 125, 126 (065)
9780	Binary and multiple stars (including extrasolar planetary systems)	117–120
9800	**Stellar systems; galactic and extragalactic objects and systems; the Universe**	
9810	Stellar dynamics	151
9820	Stellar clusters and associations	152–154
9840	Interstellar matter and nebulae	125, 131–134
9850	Galaxies, extragalactic objects and systems	155–160 (066, 161)
9870	Other objects and background radiations of unknown origin or distances	133, 141–144, 161
9880	Cosmology	161 (061, 066)

Abbreviations

Abbreviations used in *Astronomy and Astrophysics Abstracts* are primarily based on the 'International List of Periodical Title Word Abbreviations', prepared for the UNISIST/ICSU-AB Working Group on Bibliographic Descriptions (1970).

A.A.B.	Associazione Astrofili Bolognesi
Aarg.	Aargang
AAS	American Astronomical Society
AAVSO	American Association of Variable Star Observers
Abh.	Abhandlung –
Abstr.	Abstract –
Abt.	Abteilung
Acad.	Academi –, Academy
Accad.	Accademi –
Act.	Active, Activit –
Adm.	Administr –
Adv.	Advanc –
Aehron.	Aehronomi –
Aeron.	Aeronom –
Aeronaut.	Aeronauti –
Aerosp.	Aerospace
Afr.	Africa –
AG	Astronomische Gesellschaft
AIAA	American Institute of Aeronautics and Astronautics
AJB	Astronomischer Jahresbericht
Akad.	Akadem –
AL	Alabama
Alm.	Almanac –
Am.	America –, Amerika –, Amerique –
Amat.	Amateur –
Amst.	Amsterdam
An.	Anais, Anale –, Anali –, Anals
Anal.	Analis –, Analit –, Analys –, Analyt –
Angew.	Angewandt –
Ann.	Annaes, Annal –
Annu.	Annu –
Anst.	Anstalt
Anu.	Anual –, Anuar –
Anz.	Anzeiger
Appl.	Applied
AR	Arkansas
Arb.	Arbeit
Arch.	Archiv –
Årg.	Årgang
Argent	Argentin –
Ark.	Arkiv –
Arkh.	Arkhiv –
Artif.	Artifici –
ASA	Astronomical Society of Australia
Asoc.	Asocia –
ASP	Astronomical Society of the Pacific
ASSA	Astronomical Society of Southern Africa
Assem.	Assembl –
Assoc.	Associ –
Assoz.	Assozi –
Astrofis.	Astrofisic –
Astrofiz.	Astrofizi –
Astrometr.	Astrometr –
Astron.	Astronom –
Astronaut.	Astronauti –, Astronauty –
Astrophys.	Astrophys –
ASV	Astronomical Society of Victoria
ASWA	Astronomical Society of Western Australia
At.	Atom –
Atmos.	Atmosf –, Atmosph –
Aust.	Australi –
AZ	Arizona
BAA	British Astronomical Association
Barc.	Barcelona
Bayer.	Bayerisch –
Beitr.	Beitrag, Beiträge
Belg.	Belge –, Belgi –
Beob.	Beobacht –
Beogr.	Beograd –
Ber.	Bericht –
Bibl.	Bibliot –
Bibliogr.	Bibliograf –, Bibliograph –
BIH	Bureau International de l'Heure
Bimest.	Bimestr –
Bl.	Blatt, Blätter
Bol.	Boletin
Boll.	Bolletino
Br.	British
Bras.	Brasil –
Brun.	Brunens –
Bruss.	Brussel, s
Brux.	Bruxelles
Bul.	Buleten –, Buletin –, Bulten
Bulg.	Bulgar –
Bull.	Bulletin –, Bullettino
Bur.	Bureau –
Byul.	Byuleten –, Byuletin –
Byull.	Byulleten –
C. R.	Comptes Rendus
CA	California
Cah.	Cahier –
Camb.	Cambridge
Can.	Canadi –, Canada
Carol.	Carolina –
Cas.	Casopis
Cat.	Catalog –
Celest.	Celestial
Cent.	Center, Central, Centrale, Centrally, Centre
Cercet.	Cercetari
Cesk.	Ceskoslov –
Chem.	Chemi –
Chim.	Chimi –
Chin.	Chinese
Chron.	Chronic –, Chronik, Chronique
Chronom.	Chronometr –
Cie.	Compagnie
Cienc.	Ciencia –
Cient.	Cientific –
Circ.	Circolar –, Circolo, Circolaire –, Circular –, Circulo
Cirk.	Cirkulaer –
Cl.	Clasa, Classe –
CO	Colorado
Co.	Companies, Company
Coll.	College
Collect.	Collect –
Colloq.	Colloqui –

Comet.	Cometary
Commentat.	Commentat –
Commun.	Communica –
Comput.	Computation, Computer –, Computing
Comun.	Comunica –
Conf.	Conferen –
Congr.	Congres –
Contract.	Contract –
Contrib.	Contribu –
Copenh.	Copenhagen
Cosmochim.	Cosmochimi –
COSPAR	Committee on Space Research
Crystallogr.	Crystallograph –
CSIRO	Commonwealth Scientific and Industrial Research Organization
CT	Connecticut
Cult.	Cultur –, Cultuur
Curr.	Current
Czech.	Czechoslovak –
DC	District of Columbia
DE	Delaware
Dep.	Departament, Département, Department
Dev.	Development –, Développement –
Dig.	Digest
Dir.	Director –
Diss.	Disserta –
Div.	Divis –
Doc.	Document –
Dok.	Dokument –
Dokl.	Doklad –
Dom.	Dominion
Dtsch.	Deutsch
Ed.	Edit –
Edinb.	Edinburgh
Ehksp.	Ehksperiment –
Eidg.	Eidgenössisch –
Eksp.	Eksperiment –
Electron.	Electroni –
Eng.	Engineer –
Environ.	Environment –
Equip.	Equipement, Equipment
Ergeb.	Ergebnis –
ESA	European Space Agency
ESO	European Southern Observatory
Espec.	Especial –
ESRO	European Space Research Organization
Eur.	Europ –
Eval.	Evaluation –
Exp.	Experiment –
Extraterr.	Extraterrestr –
Fac.	Facolt –, Faculd –, Facult –
Fak.	Fakult
Fasc.	Fascicul –
Fenn.	Fenni –
Finn.	Finni –
Fis.	Fisic –, Fisik –
Fiz.	Fizic –, Fizik –, Fizyk –
FL	Florida
Fluid.	Fluidi –
Fond.	Fondation –, Fondazione
Fortschr.	Fortschritt –
Fotogr.	Fotograf –
Found.	Foundation –
Fr.	Français –
Freq.	Frequen –
Fundam.	Fundamenta –
Fys.	Fysik –, Fysisch, Fysisk –
Fyz.	Fyzik –
G.	Giornale
GA	Georgia
Gaz.	Gazeta, Gazette
Gazz.	Gazzetta
Gen.	General
Geochem.	Geochem –
Geochim.	Geochim –
Geod.	Geodaes –, Geodaet –, Geodes –, Geodet –, Geodez –
Geofis.	Geofis –
Geofiz.	Geofiz –
Geofys.	Geofys –
Geogr.	Geograf –, Geograph –
Geokhim.	Geokhim –
Geol.	Geolog –, Geolosk –
Geomagn.	Geomagneti –
Geophys.	Geophys –
Ges.	Gesellschaft
Gesch.	Geschichte
Gl.	Glavno –
Glas.	Glasnik
Gos.	Gosudarst –
Gov.	Government –
Grenzgeb.	Grenzgebiet –
GSFC	Goddard Space Flight Center
H.M.	Her Majesty's, His Majesty's
Hamb.	Hamburg
Handb.	Handbook, Handbuch
Heidelb.	Heidelberg –
Helv.	Helveti –
Her.	Herald –
HI	Hawaii
Hist.	History
Hochsch.	Hochschule
Hoegsk.	Hoegskol –
HR-diagram	Hertzsprung-Russell diagram
Hung.	Hungar –
Hydrogr.	Hydrograf –, Hydrograph –
IA	Iowa
IAF	International Astronautical Federation
IAU	International Astronomical Union
IBM	International Business Machines Corporation
ICSTI	International Council for Scientific and Technical Information
ICSU	International Council of Scientific Unions
ID	Idaho
IEEE	Institute of Electrical and Electronics Engineers
IL	Illinois
IN	Indiana
Inc.	Incorporated
Ind.	Industr –
Inf.	Informat –, Informaz –, Informe –
Ing.	Ingenieur
INIS	International Nuclear Information System
INSPEC	International Information Services for the Physics and Engineering Communities
Inst.	Institut –, Instytut –
Instrum.	Instrument –
Int.	International, Internazional –
Intell.	Intelligenc –
Inter.	Intérieur –, Interior
Interplanet.	Interplanetary
Intez.	Intezet –
Invest.	Investiga –
Ionos.	Ionosfer –, Ionospher –
Ir.	Irish
Iskusstv.	Iskusstvenn –
Isr.	Israel –
Issled.	Issledovan –
Ist.	Istitut
Ital.	Itali –
Izd.	Izdatel –
Izv.	Izvesti –

J.	Joernaal –, Jornal –, Journal –
Jaarb.	Jaarboek –
Jahrb.	Jahrbuch, Jahrbücher
Jahresber.	Jahresbericht –
Jahresschr.	Jahresschrift
Jahrg.	Jahrgang
JPL	Jet Propulsion Laboratory
Jpn.	Japan –
K.	Königlich –, Koninkljik –, Kunglig –
Kartogr.	Kartograf –
Kernforsch.	Kernforschung
Kernphys.	Kernphysik –
Khem.	Khemyi –
Khim.	Khimi –
Kim.	Kimija –, Kimya
Kl.	Klass –
Kolloq.	Kolloquium –
Komet.	Kometnyj
Komm.	Kommission –
Konf.	Konfer –
Kongr.	Kongress
Kosm.	Kosmich –
Kosmog.	Kosmogon –
Kozp.	Kozponti
KPNO	Kitt Peak National Observatory
KS	Kansas
KY	Kentucky
LA	Louisiana
Lab.	Laborato –
LEST	Large European Solar Telescope
Lett.	Letter –, Lettra, Lettre
Libr.	Librair –, Librar –
MA	Massachusetts
Madr.	Madrid
Mag.	Magasin, Magazin –
Magn.	Magneti –, Magnitn –
Mar.	Marin –
Mat.	Matemaat –, Matemat –
Mater.	Material –
Math.	Mathemat –
MD	Maryland
ME	Maine
Meas.	Measur –
Mec.	Mecani –
Mech.	Mechani –
Medd.	Meddelande –, Meddelelse
Meded.	Mededeeling, Mededeling –
Mekh.	Mekhani –
Mem.	Memento –, Memoir –, Memori –, Memory –, Memuary
Memo.	Memorand –
Mens.	Mensile, Mensual –, Mensuel –
Messtech.	Messtechni –
Meteorol.	Meteorolog –
Mex.	Mexic –
MI	Michigan
Micromec.	Micromecaniq –
Miner.	Mineral, Minerale –, Minerali –
Mineral.	Mineralog –
MIT	Massachusetts Institute of Technology
Mitt.	Mitteilung –
MN	Minnesota
MO	Missouri
Mod.	Modern –
Mol.	Molecul –, Molekul –
Mon.	Monat, Monatlich –, Month –
Monogr.	Monograph –
MPI	Max-Planck-Institut
MS	Mississippi
MT	Montana
Mt.	Mount
Muench.	Muenchen
Mus.	Museum
N.Z.	New Zealand
Nablyud.	Nablyudeni –
Nac.	Nacion –
Nachr.	Nachricht –
NASA	National Aeronautics and Space Administration
Nat.	Natur –
Natl.	National –
Naturforsch.	Naturforsch –
Naturwiss.	Naturwissenschaft –
Natuurkd.	Natuurkunde
Nauchn.	Nauchny –
Nauk.	Nauka, Naukite, Naukov –, Naukow –
Naut.	Nautic –
Nav.	Naval –
Navig.	Navigat –
Naz.	Nazion –
NC	North Carolina
ND	North Dakota
NE	Nebraska
Ned.	Nederland –
Newsl.	Newsletter –
NH	New Hampshire
NJ	New Jersey
NM	New Mexico
Not.	Notationes, Notic –, Notise, Notizi –
Nouv.	Nouveau –, Nouvell –
Nov.	Novoe
Nucl.	Nucléaire –, Nuclear –, Nucl –
Nukl.	Nukle –
Numer.	Numeri –
NV	Nevada
NY	New York
O-va	Obshchestva
O-vo	Obshchestvo
Obs.	Observ –
Österr.	Österreich –
Off.	Offic –
OH	Ohio
OK	Oklahoma
Opt.	Optic –, Optik –, Optique
OR	Oregon
Oss.	Osserva –
PA	Pennsylvania
Pac.	Pacific
Paleontol.	Paleontolog –
Pap.	Paper –, Papier
Part.	Particle
Pekin.	Pekinens –
Perem.	Peremenn –
Period.	Periodi –
Petrol.	Petrolog –
Philos.	Philosoph –
Photogr.	Photograf –, Photograph –
Photogramm.	Photogrammetr –
Photom.	Photometr –
Phys.	Physic –, Physik –, Physique –, Physisch –
Pict.	Picture –
Planet.	Planetary
Pol.	Polish, Polon –
Pr.	Prac –
Prelim.	Prelimin –
Prepr.	Preprint
Prib.	Pribor –
Prikl.	Prikladn –
Prilozh.	Prilozhen –
Prir.	Prirodn –
Prirodoved.	Prirodoved –

Probl.	Problem –
Proc.	Proceedings
Prod.	Prodott –, Produc –, Produkt,
Prog.	Progres –
Propag.	Propagation
Prospect.	Prospecting
Prov.	Provinc –, Provints –, Provinz –
Pubbl.	Pubblicazion –
Publ.	Publicac –, Publicas –, Publicat –, Publikas –, Publikat –
Q.	Quarterly
Quant.	Quantit –
R.	Royal
Radiat.	Radiati –
Radioact.	Radioactiv –, Radioaktiv –
Radioisot.	Radioisotop –
Rap.	Raport –
Rapp.	Rapport –
RAS	Royal Astronomical Society
Rec.	Record –
Rech.	Recherche –
Ref.	Referat –, Reference –, Referieren
Relat.	Related, Relation –
Relativ.	Relativit –
Rend.	Rendicont –
Rep.	Report –
Repr.	Reprint –
Repub.	Republi –
Res.	Research –
Result.	Resultad –, Resultat –
Rev.	Review –, Revisio, Revista, Revue –
Rezul't.	Rezul'tat –
RI	Rhode Island
Ric.	Ricerca, Ricerche
Riv.	Rivist –
Roum.	Roumain –
Rundsch.	Rundschau
S. Afr.	South Africa
SAF	Société Astronomique de France
SAI	Società Astronomica Italiana
Samml.	Sammlung –
SAO	Smithsonian Astrophysical Observatory
SAS	Société Astronomique de Suisse
Satell.	Satellite
Sb.	Sbornik –
SC	South Carolina
Scand.	Scandinavi –
Sch.	Schul –
Schr.	Schrift –
Schriftenr.	Schriftenreihe
Schweiz.	Schweizer –
Sci.	Scienc –, Scient –, Scienz –
Scr.	Scripta, Scritt –
SD	South Dakota
Secc.	Seccion –
Sect.	Secti –
Sekc.	Sekci –, Sekcj –
Sekt.	Sektion –, Sektor –
Sekts.	Sektsi –
Sel.	Seleccion –, Select –, Selek –, Selezione
Selsk.	Selskab –, Selskap –
Semin.	Séminair –, Seminar –
Sep.	Separat –
Ser.	Seria –, Serie –, Seriya
Serv.	Servic –, Serviz –
Sess.	Sessi –
Signal.	Signalétique –
Simp.	Simpoz –
Sin.	Sinica
Sitzungsber.	Sitzungsbericht –
Skr.	Skrift –
Smithson.	Smithsonian
Soc.	Sociedad –, Societ –
Sol.	Solar
Soln.	Solnechn –
Sonderdr.	Sonderdruck –
Soobshch.	Soobshchen –
South.	Southern
Spacecr.	Spacecraft
Spat.	Spatial –
Spec.	Special –
Spectrosc.	Spectroscop –
Spectrosk.	Spectroskop –
Spets.	Spetsial –
Spez.	Spezial –, Speziell –
SSR	Sovetskaya Sotsialisticheskaya Respublika
SSSR	Soyuz Sovetskikh Sotsialisticheskikh Respublik
St.	Saint –, Sankt –, Sant –
– St.	– Straße, Street
Stand.	Standard –, Standart –
Sternw.	Sternwarte –
Stiint.	Stiintific –
Stn.	Station, Stazione
Stud.	Studia, Studie –, Studii
Supl.	Suplement –, Supliment –
Suppl.	Supplement –
Surv.	Survey –
Syd.	Sydney
Symp.	Sympos –, Sympoz –
Syst.	System –
Sz.	Szemle
Teach.	Teacher –, Teaching
Tec.	Tecni –
Tech.	Techni –
Technol.	Technolog –
Tecnol.	Tecnolog –
Teh.	Tehnic –, Tehnika, Tehnisk –
Tehnol.	Tehnolog –, Tehnolosk –
Tek.	Tekni –
Tekh.	Tekhni –
Tekhnol.	Tekhnolog –
Teknol.	Teknolog –
Telesc.	Telescop –
Telev.	Television –
Teor.	Teoret –, Teori –
Terr.	Terrestr –
Test.	Testing
TH	Technische Hochschule
Theor.	Theoret –, Theori –
Tidschr.	Tidschrift –
Tidskr.	Tidskrift –
Tidsskr.	Tidsskrift –
TN	Tennessee
Top.	Topic –
Torun.	Torunensis
Tr.	Trudy
Trans.	Transactions, Transazione
Tsentr.	Tsentral –
Tsirk.	Tsirkulyar –
TU	Technical University
TX	Texas
Uch.	Uchen –
Uchebn.	Uchebn –
UK	United Kingdom
Umsch.	Umschau
UN	United Nations
Univ.	Universidad –, Universit –, Univerzitet
Ups.	Upsaliens –

US	United States
USA	United States of America
USSR	Union of Soviet Socialist Republics
UT	Utah
Utr.	Utrecht
VA	Virginia
Var.	Various
Ver.	Verein –, Verenig –
Veränderl.	Veränderlich –
Verh.	Verhandl –
Vermess.	Vermessung –
Vermessungswes.	Vermessungswesen
Veröff.	Veröffentlich –
Vesn.	Vesnik
Vestn.	Vestnik
Vetensk.	Vetenskap –
Vgl.	Vergleich –
Vidensk.	Videnskab –, Videnskap
Vierteljahresschr.	Vierteljahresschrift –
Vierteljahrsschr.	Vierteljahrsschrift
VLB	Very Long Baseline
Volcanol.	Volcanolog –
Vopr.	Vopros –
Vortr.	Vorträge
Vses.	Vsesoyuzn –
VT	Vermont
Vyp.	Vypusk –
Vyssh.	Vyssh –
Vyzk.	Vyzkum –
WA	Washington
West.	Western
Wet.	Wetenschap –, Wetenskap –
WI	Wisconsin
Wiss.	Wissenschaft –
WV	West Virginia
WY	Wyoming
Yad.	Yadern –
Z.	Zeitschrift –
ZA	Zero Age
ZAED	Zentralstelle für Atomkernenergie-Dokumentation
Zap.	Zapisk –, Zapyisk –
Zaved.	Zaveden –
Zent.	Zentral
Zentralbl.	Zentralblatt
Zesz.	Zeszyt
Zh.	Zhurnal –
Zirk.	Zirkular

Periodicals, Proceedings, Books, Activities

001 Periodicals

AAO Newsl.
AAO Newsletter. Anglo–Australian Observatory, PO Box 296, Epping, NSW 2121 (Australia). Quarterly.

Acta Astron.
Acta Astronomica. Polska Akademia Nauk, Komitet Astronomii. Państwowe Wydawnictwo Naukowe, Warszawa–Kraków (Poland). Quarterly. Subscription address: Ars Polona, 00–068 Warszawa, Krakowskie Przedmieście 7, Poland. ISSN 0001-5237, AASWAM.

Acta Astron. Sin.
Acta Astronomica Sinica. Purple Mountain Observatory, Academia Sinica, Nanjing (China). Quarterly. Selected English translation in Chinese Astronomy and Astrophysics. ISSN 0001-5245, TIWHAO.

Acta Astrophys. Sin.
Acta Astrophysica Sinica. Academia Sinica. Beijing Astronomical Obervatory, Beijing (China). Subscription address: Guozi Shudian, PO Box 399, Beijing, China. Selected English translation in Chinese Astronomy and Astrophysics. ISSN 0253–2379.

Acta Geophys. Sin.
Acta Geophysica Sinica. Department of Geophysical Research, Academia Sinica, Beijing (China). Bimonthly. Subscription address: Guozi Shudian, PO Box 399, Beijing, China. ISSN 0001–5733, AHGSBY.

Acta Opt. Sin.
Acta Optica Sinica. Monthly. ISSN 0253–2239.

Acta Phys. Hung.
Acta Physica Hungarica. Akademiai Kiado, Publishing House of the Hungarian Academy of Sciences, H–1054 Budapest, Alkotmany U. 21 (Hungary). Quarterly. Formerly: Acta Physica Academiae Scientarum Hungaricae. ISSN 0001–6705, APHUE2.

Acta Phys. Pol., A
Acta Physica Polonica, Series A (General Physics, Atomic and Molecular Spectroscopy, Applied Physics). Polska Akademia Nauk (PAN), Warsaw (Poland). Bimonthly. ISSN 0587–4246, ATPLB6.

Acta Phys. Pol., B
Acta Physica Polonica, Series B (Elementary Particle Physics, Nuclear Physics, Theory of Relativity, Field Theory). Polska Akademia Nauk (PAN), Warsaw (Poland). Monthly. ISSN 0587–4254, APOBBB.

Acta Phys. Sin.
Acta Physica Sinica. Academia Sinica, Beijing (China). Monthly. ISSN 0372–736X, WLHPAR.

Adv. Space Res.
Advances in Space Research. Pergamon Press, Oxford (UK). Irregularly. The official journal of the Committee on Space Research (COSPAR). ISSN 0273–1177, ASRSDW.

AIP Conf. Proc.
AIP Conference Proceedings. American Institute of Physics, New York, NY (USA). Irregularly. ISSN 0094–243X, APCPCS.

Am. Assoc. Variable Star Obs. Bull.
The American Association of Variable Star Observers Bulletin. The American Association of Variable Star Observers, 25 Birch Street, Cambridge, MA 02138 (USA). Irregularly. ISSN 0097–5265, BAAODO.

Am. Assoc. Variable Star Obs. Circ.
American Association of Variable Star Observers Circular. American Association of Variable Star Observers, Cambridge, MA (USA). Irregularly. Subscription address: AAVSO, C.E. Scovil, Stamford Observatory, 39 Scofieldtown Rd., Stamford, CT 06903, USA. ISSN 0197–2979, AACID5.

Am. J. Phys.
American Journal of Physics. American Institue of Physics, New York, NY (USA). Monthly. ISSN 0002–9505, AJPIAS.

Ann. Geophys.
Annales Geophysicae (Upper Atmosphere and Space Sciences). European Geophysical Society. Gauthier–Villars, Montrouge (France). Bimonthly. Formerly: Annales Geophysicae, Series A. Subscription address: C.D.R. Centrale des Revues, 11 rue Gossin, F–92543. Montrouge Cedex, France.

Ann. Inst. Henri Poincaré, Phys. Théor.
Annales de l'Institut Henri Poincaré, Physique Théorique. Gauthier–Villars, Montrouge (France). Quarterly. ISSN 0246–0211, AHPAAO.

Ann. Phys. (Leipzig)
Annalen der Physik (Leipzig). Barth, Leipzig (Germany). 8 issues per year. ISSN 0003–3804, ANPYAZ.

Ann. Phys. (N.Y.)
Annals of Physics (New York). Academic Press, New York, NY (USA). 15 issues per year. ISSN 0003–4916, APNYA6.

Ann. Phys. (Paris) Colloq.
Annales de Physique (Paris), Colloque. Les Editions de Physique, Les Ulis (France). ISSN 0003–4169, ANPHAJ.

Ann. Shanghai Obs., Acad. Sin.
Annals of Shanghai Observatory, Academia Sinica. Shanghai Scientific and Technical Publishers, Shanghai, Rei Jing Er Street 450 (China). Irregularly.

Annu. Rev. Astron. Astrophys.
Annual Review of Astronomy and Astrophysics. Annual Reviews Inc., 4139 El Camino Way, Palo Alto, CA 94303–0897 (USA). Annual. ISSN 0066–4146, ARAAAJ.

Appl. Opt.
Applied Optics. Optical Society of America. American Institute of Physics, New York, NY (USA). Semimonthly. ISSN 0003–6935, APOPAI.

Appl. Phys. Lett.
Applied Physics Letters. American Institute of Physics, New York, NY (USA). Semimonthly. ISSN 0003–6951, APPLAB.

Appl. Phys., B
Applied Physics, B (Photophysics and Laser Chemistry). Springer, Berlin (Germany, F.R.). Monthly. Formerly: Applied Physics. ISSN 0721–7269, APPCDL.

Appl. Spectrosc.
Applied Spectroscopy. Society for Applied Spectroscopy. Williams and Wilkins, Baltimore, MD (USA). Bimonthly. ISSN 0003–7028, APSPA4.

Appl. Spectrosc. Rev.
Applied Spectroscopy Reviews. Marcel Dekker, 270 Madison Avenue, New York, NY 10016 (USA). Quarterly. ISSN 0570–4928, APSRBB.

Arch. Hist. Exact Sci.
Archive for History of Exact Sciences. Springer, Berlin (Germany, F.R.). ISSN 0003–9519, AHESAN.

Artif. Satell.
Artificial Satellites. Polish Academy of Sciences, Space Research Centre. Państwowe Wydawnictwo Naukowe, Warszawa–Łódź (Poland). Subscription address: Ars Polona, 00–068 Warszawa, Krakowskie Przedmieście 7, Poland. ISSN 0208–841X.

Astrofiz. Issled. Izv. Spets. Astrofiz. Obs.
Astrofizicheskie Issledovaniya. Izvestiya Spetsial'noj Astrofizicheskoj Observatorii. Akademiya Nauk SSSR. Nauka, Leningrad (USSR). Irregularly. English translation in Bulletin of the Special Astrophysical Observatory – North Caucasus. ISSN 0320–9318, AILEDQ.

Astrofizika
Astrofizika. Izdatel'stvo Akademii Nauk Armyanskoj SSR, Erevan (USSR). Bimonthly. English translation in Astrophysics. ISSN 0571–7132, ASTKBG.

Astron. Astrophys.
Astronomy and Astrophysics. Springer, Berlin (Germany, F.R.). Semimonthly. ISSN 0004–6361, AAEJAF.

Astron. Astrophys. Rev.
The Astronomy and Astrophysics Review. Springer, Berlin (Germany, F.R.). Irregularly. ISSN 0935–4956, AASREB.

Astron. Astrophys., Suppl. Ser.
Astronomy and Astrophysics, Supplement Series. Les Editions de Physique, Z.I. de Courtabœuf, B.P. 112, F-91944 Les Ulis Cedex (France). Monthly. ISSN 0365–0138, AAESBO.

Astron. Her.
The Astronomical Herald. Astronomical Society of Japan, Tokyo (Japan). Monthly. ISSN 0374–2466, TGEPAC.

Astron. Inst. Univ. Brno, Contrib.
Astronomical Institute of the University Brno, Contributions. Astronomical Institute, University Brno, Kotlářská 2, Brno (Czechoslovakia). Irregularly.

Astron. J.
The Astronomical Journal. American Astronomical Society. American Institute of Physics, New York, N.Y. (USA). Monthly. ISSN 0004–6256, ANJOAA.

Astron. Nachr.
Astronomische Nachrichten. Akademie Verl., Berlin (Germany). Bimonthly. ISSN 0004–6337, ASNAAN.

Astron. Now
Astronomy Now. Intra Press, London (UK). Monthly. ISSN 0951–9726.

Astron. Q.
The Astronomy Quarterly. Pergamon Press, New York, NY (USA). Quarterly. ISSN 0364–9229.

Astron. Rechen-Inst. Heidelb., Mitt., Ser. A
Astronomisches Rechen-Institut Heidelberg, Mitteilungen Serie A. Astronomisches Rechen-Institut, Mönchhofstraße 12–14, D–6900 Heidelberg 1 (Germany, F.R.). Irregularly. ISSN 0440–5951, ARIMBS.

Astron. Rechen-Inst., Heidelb., Mitt., Ser. B
Astronomisches Rechen-Institut Heidelberg, Mitteilungen, Serie B. Astronomisches Rechen-Institut, Mönchhofstraße 12–14, D–6900 Heidelberg (Germany, F.R.). Irregularly.

Astron. Sch.
Astronomie in der Schule. Friedrich Verlag, Velber (Germany). Bimonthly. ISSN 0004–6310.

Astron. Soc. Pac. Conf. Ser.
Astronomical Society of the Pacific Conference Series. Brigham Young University Print Services, Provo, UT (USA). Astronomical Society of the Pacific (ASP), San Francisco, CA (USA).

Astron. Tidsskr.
Astronomisk Tidsskrift. Astronomisk Selskab, København (Denmark). Quarterly. ISSN 0004–6345, ANTKBF.

Astron. Tsirk.
Astronomicheskij Tsirkulyar. Astronomicheskij Sovet AN SSSR. Byuro Astronomicheskikh Soobshchenij, Moskva (USSR). Monthly. ISSN 0236–2457.

Astron. Vestn.
Astronomicheskij Vestnik. Issledovaniya Solnechnoj Sistemy. Nauka, Moskva (USSR). Bimonthly. English translation in Solar System Research. ISSN 0320–930X, ASVEA7.

Astron. Zh.
Astronomicheskij Zhurnal. Akademiya Nauk SSSR. Izdatel'stvo Nauka, Moskva (USSR). Bimonthly. English translation in Soviet Astronomy. ISSN 0004–6299, ASZHA2.

Astronomia UAI
Astronomia UAI. Unione Astrofili Italiani. Università degli Studi di Padova, Dipartimento di Astronomia, Vicolo dell'Osservatorio, 5, I–35122 Padova (Italy). Bimonthly. Periodico bimestrale dell' Unione Astrofili Italiani. ISSN 0392–2308.

Astronomie
L'Astronomie. Société Astronomique de France, 3, rue Beethoven, F–75016 Paris (France). Monthly. ISSN 0004–6302, ATROAO.

Astronomy
Astronomy. Kalmbach Publishing Co., 1027 N. 7th Street, Milwaukee, WI 53233 (USA). Monthly. ISSN 0091–6358, ASTRD5.

Astrophys. J.
The Astrophysical Journal. American Astronomical Society. University of Chicago Press, Chicago, IL (USA). Semimonthly. ISSN 0004–637X, ASJOAB.

Astrophys. J., Lett.
The Astrophysical Journal, Letters. American Astronomical Society. University of Chicago Press, Chicago, IL (USA). Semimonthly. Formerly: Astrophysical Journal, Letters to the Editor. ISSN 0004–637X, ASJOAB.

Astrophys. J., Suppl. Ser.
The Astrophysical Journal, Supplement Series. American Astronomical Society. University of Chicago Press, Chicago, IL (USA). Monthly. ISSN 0067–0049, APJSA2.

Astrophys. Lett. Commun.
Astrophysical Letters and Communications. Gordon and Breach, New York, NY (USA). Quarterly. Formerly: Astrophysical Letters. ISSN 0888–6512, ALECE7.

Astrophys. Space Sci.
Astrophysics and Space Science. Kluwer Academic Publ., Dordrecht (Netherlands). Semimonthly. ISSN 0004–640X, APSSBE.

Astrophys. Space Sci. Libr.
Astrophysics and Space Science Library. Kluwer Academic Publ., Dordrecht (Netherlands).

Astrophysics
Astrophysics. Consultants Bureau, New York, NY (USA). Bimonthly. A cover–to–cover translation of Astrofizika. ISSN 0571–7256, ATPYAA.

At. Data Nucl. Data Tables
Atomic Data and Nuclear Data Tables. Academic Press, New York, NY (USA). Bimonthly. ISSN 0092–640X, ADNDAT.

Atmos. Res.
Atmospheric Research. Elsevier, Amsterdam (Netherlands). Quarterly. ISSN 0169–8095, ATREEW.

Aust. J. Astron.
Australian Journal of Astronomy. Astral Press, Wembley, W.A. 6014 (Australia). Semiannual. ISSN 0814–5628.

Aust. J. Phys.
Australian Journal of Physics. Commonwealth Scientific and Industrial Research Organization (CSIRO), 314 Albert Street, East Melbourne, Victoria 3002 (Australia). Bimonthly. ISSN 0004–9506, AUJPAS.

Babes–Bolyai Univ., Fac. Math. Phys., Res. Semin.
Babes–Bolyai University, Faculty of Mathematics and Physics, Research Seminars. Babes–Bolyai University, Cluj–Napoca (Romania).

BAV Mitt.
BAV Mitteilungen. Berliner Arbeitsgemeinschaft für Veränderliche Sterne e. V., Munsterdamm 90, D–1000 Berlin 41 (Germany, F.R.).

BAV Rundbrief
BAV Rundbrief. Berliner Arbeitsgemeinschaft für Veränderliche Sterne e.V., Munsterdamm 90, D–1000 Berlin 41 (Germany, F.R.). ISSN 0405–5497.

BBSAG Bull.
Bedeckungsveränderlichen Beobachter der Schweizerischen Astronomischen Gesellschaft Bulletin. Subscription address: N. Hasler, Huzlenstraße 3, CH–8604 Volketswil, Switzerland.

Bibliogr. Program Notes Close Bin.
Bibliography and Program Notes on Close Binaries. International Astronomical Union. Department of Geoscience, National Defence Academy, Yokosuka (Japan). Irregularly. ISSN 1017–0634.

Bild Wiss.
Bild der Wissenschaft. Deutsche Verlagsanstalt, Stuttgart (Germany, F.R.). Monthly. ISSN 0006–2375, BIWIAX.

BIPM Circ. T
Bureau International des Poids et Mesures (BIPM), Circular T. Bureau International des Poids et Mesures (BIPM), Section du temps, Pavillion de Breteuil, F–92312 Sèvres CEDEX (France). Monthly. ISSN 1143–1393.

Bol. Acad. Nac. Cienc. Cordoba, Argent.
Boletin de la Academia Nacional de Ciencias, Cordoba, Argentina. Academia Nacional de Ciencias, Cordoba (Argentina).

Bol. Asoc. Argent. Astron.
Boletin de la Asociación Argentina de Astronomía. Acociación Argentina de Astronomía, San Juan (Argentina). Irregularly. ISSN 0571–3285.

Boundary–Layer Meteorol.
Boundary–Layer Meteorology. Kluwer Academic Publ., Dordrecht (Netherlands). Monthly. ISSN 0006–8314, BLMEBR.

Br. Astron. Assoc. Circ.
British Astronomical Association Circular. British Astronomical Association, Burlington House, Piccadilly, London W1V 9AG (UK). ISSN 0264–4185.

Bull. Am. Astron. Soc.
Bulletin of the American Astronomical Society. American Astronomical Society. American Institute of Physics, New York, NY (USA). Quarterly. ISSN 0002–7537, ASLBBH.

Bull. Assoc. Fr. Obs. Etoiles Variables
Bulletin de l'Association Française des Observateurs d'Etoiles Variables. Association Française des Observateurs d'Etoiles Variables, Observatoire de Strasbourg, 11, rue de l'Université, F–67000 Strasbourg (France). Quarterly. ISSN 0153–9949.

Bull. Astron. Inst. Czech.
Bulletin of the Astronomical Institutes of Czechoslovakia. Academia, Publishing House of the Czechoslovak Academy of Sciences, Vodičkova 40, 11229 Praha 1 (Czechoslovakia). Bimonthly. ISSN 0004–6248, BAICAP.

Bull. Astron. Soc. India
Bulletin of the Astronomical Society of India. Astronomical Society of India, Osmania University, Hyderabad 500 007 (India). Quarterly. ISSN 0304–9523, BANID3.

Bull. Crimean Astrophys. Obs.
Bulletin of the Crimean Astrophysical Observatory. Allerton Press, New York, NY (USA). Annual. English translation of Izvestiya Krymskoj Astrofizicheskoj Observatorii. ISSN 0190–2717, BCAOD4.

Bull. Inf. Cent. Données Stellaires
Bulletin d'Information du Centre de Données Stellaires. Observatoire de Strasbourg, 11, rue de l'Université, F–67000 Strasbourg (France). Semiannual. ISSN 0242–6536, BICSDY.

Bull. Spec. Astrophys. Obs. – North Caucasus
Bulletin of the Special Astrophysical Observatory – North Caucasus. Allerton Press, New York, NY (USA). Irregularly. A cover–to–cover translation of Astrofizicheskie Issledovaniya – Izvestiya Spetsial'noj Astrofizicheskoj Observatorii. ISSN 0190–2709, BSACDC.

C. R. Acad. Sci., Sér. Gén., Vie Sci.
Comptes Rendus de l'Académie des Sciences, Série Générale, La Vie des Sciences. Académie des Sciences, Paris (France). Gauthier-Villars, Montrouge (France). Annual. ISSN 0762-0969.

C. R. Acad. Sci., Sér. II
Comptes Rendus de l'Académie des Sciences, Série II: Mécanique, Physique, Chimie, Sciences de la Terre et de l'Univers. Académie des Sciences, Paris (France). Gauthier-Villars, Montrouge (France). Weekly. ISSN 0764-4450, CRAMED.

Can. J. Phys.
Canadian Journal of Physics. National Research Council of Canada, Ottawa, K1A 0R6, Ontario, (Canada). Monthly. ISSN 0008-4204, CJPHAD.

Celest. Mech. Dyn. Astron.
Celestial Mechanics and Dynamical Astronomy. Kluwer Academic Publ., Dordrecht (Netherlands). Quarterly. ISSN 0923-2958, CLMCAV.

Chaos Solitons Fractals
Chaos, Solitons and Fractals. Pergamon, Oxford (UK). ISSN 0960-0779.

Chem. Phys. Lett.
Chemical Physics Letters. North-Holland, Amsterdam (Netherlands). Semimonthly. ISSN 0009-2614, CHPLBC.

Chin. Astron. Astrophys.
Chinese Astronomy and Astrophysics. Pergamon Press, Oxford (UK). Quarterly. A selected translation from the current issues of Acta Astronomica Sinica, Acta Astrophysica Sinica, and Chinese Journal of Space Sciences. ISSN 0275-1062.

Chin. J. Phys.
Chinese Journal of Physics. Physical Society of the Republic of China, Physics Department, National Taiwan University, Taipei, Taiwan, Republic of China. Irregularly. ISSN 0577-9073, CJOPAW.

Chin. J. Space Sci.
Chinese Journal of Space Science. Quarterly. ISSN 0254-6124.

Chin. Phys.
Chinese Physics. American Institute of Physics, New York, NY (USA). Quarterly. English translation of Wuli Xuebao. ISSN 0273-429X, CHPHD2.

Chin. Phys. Lett.
Chinese Physics Letters. Science Press, 137 Chaoyangmennei Street, Beijing (China). Monthly. ISSN 0256-307X, CPLEEU.

Chin. Sci. Bull.
Chinese Science Bulletin.

Ciel
Le Ciel. Société Astronomique de Liège, B-4200 Cointe-Liège (Belgium). Monthly. Bulletin de la Société Astronomique de Liège. ISSN 0771-3010.

Ciel Terre
Ciel et Terre. Société Royale Belge d'Astronomie, de Météorologie et de Physique du Globe, Avenue Circulaire 3, 1180 Bruxelles (Belgium). Bimonthly. Bulletin de la Société Royale Belge d'Astronomie, de Météorologie et de Physique du Globe. ISSN 0009-6709, CIELAV.

Circ. Inf. – IAU Comm. Etoiles Doubles
Circulaire d'Information. Union Astronomique Internationale, Commission des Etoiles Doubles. Observatoire de la Côte d'Azur, B.P. 139, F-06003 Nice (France). 3 Nos. per year.

Classical Quantum Gravity
Classical and Quantum Gravity. Institute of Physics, London (UK). Bimonthly. ISSN 0264-9381, CQGRDG.

Commentat. Phys.-Math. Chem.-Med.
Commentationes Physico-Mathematicae et Chemico-Medicae. The Finnish Society of Sciences and Letters, Mariankatu 5, SF-00170 Helsinki (Finland). Irregularly. ISSN 0788-5717, CPHMAU.

Comments Astrophys.
Comments on Astrophysics. Comments on Modern Physics: Part C. Gordon and Breach, New York, NY (USA). Bimonthly. ISSN 0146-2970, COASB7.

Commun. Math. Phys.
Communications in Mathematical Physics. Springer, Berlin (Germany, F.R.) Semimonthly. ISSN 0010-3616, CMPHAY.

Comput. Phys.
Computers in Physics. American Institute of Physics, New York, N.Y. (USA). Bimonthly. ISSN 0894-1866, CPHYE2.

Comput. Phys. Commun.
Computer Physics Communications. North-Holland, Amsterdam (Netherlands). Monthly. ISSN 0010-4655, CPHCBZ.

Contemp. Phys.
Contemporary Physics. Taylor and Francis, London (UK). Bimonthly. ISSN 0010-7514, CTPHAF.

Contrib. Atmos. Phys.
Contributions to Atmospheric Physics. Vieweg, Wiesbaden (Germany, F.R.). Quarterly. ISSN 0005-8173, BPYAAY.

Contrib. Dep. Astron., Univ. Tokyo
Contributions from the Department of Astronomy, University of Tokyo. Department of Astronomy, University of Tokyo, Bunkyo-ku, Tokyo 113 (Japan). Irregularly. ISSN 0563-8038, TDRT-A.

Contrib. Plasma Phys.
Contributions to Plasma Physics. Akademie-Verl., Berlin (Germany) Bimonthly. Formerly: Beiträge aus der Plasma Physik. ISSN 0863-1042, BPPHAA.

Cosmic Res.
Cosmic Research. Consultants Bureau, New York, NY (USA). Bimonthly. A cover-to-cover translation of Kosmicheskie Issledovaniya. ISSN 0010-9525, CSCRA7.

COSPAR Colloq. Ser.
COSPAR Colloquia Series. Pergamon Press, Oxford (UK). Irregularly.

Dtsch. Geod. Komm. Bayer. Akad. Wiss., Reihe C
Deutsche Geodätische Kommission bei der Bayerischen Akademie der Wissenschaften, Reihe C: Dissertationen. Beck, München (Germany, F.R.). Irregularly. ISSN 0065-5325.

Earth Planet. Sci. Lett.
Earth and Planetary Science Letters. Elsevier, Amsterdam (Netherlands). Irregularly. ISSN 0012-821X, EPSLA2.

Earth Space
Earth in Space. American Geophysical Union, Washington, DC (USA). 9 issues per year (monthly Sep – May). A magazine for teachers and students of science. ISSN 1040-3124.

Earth, Moon, Planets
Earth, Moon, and Planets. Kluwer Academic Publ., Dordrecht (Netherlands). Monthly. Formerly: The Moon and the Planets. ISSN 0167-9295, EMPLD3.

Earth-Sci. Rev.
Earth-Science Reviews. Elsevier, Amsterdam (Netherlands). Irregularly. ISSN 0012-8252, ESREAV.

Electron. Lett.
Electronics Letters. Institution of Electrical Engineers, Savoy Place, London WC2R 0BL (UK). Semimonthly. ISSN 0013-5194, ELLEAK.

ESA Bull.
ESA Bulletin. ESA Publications Division, c/o ESTEC, Noordwijk (Netherlands). Quarterly. ISSN 0376-4265, ESABD8.

ESA IUE Newsl.
ESA IUE Newsletter. The ESA IUE Observatory, Apartado 54065, 28080 Madrid (Spain). ISSN 1011-0100.

ESA J.
ESA Journal. ESA Publications Division, ESTEC, PO Box 299, 2200 AG Noordwijk (Netherlands). Quarterly. ISSN 0379-2285, ESAJDW.

ESO Conf. Workshop Proc.
ESO Conference and Workshop Proceedings. European Southern Observatory, Karl-Schwarzschild-Straße 2, D-8046 Garching bei München (Germany, F.R.). Irregularly.

ESO Sci. Rep.
ESO Scientific Report. European Southern Observatory, Karl-Schwarzschild-Straße 2, D-8046 Garching bei München (Germany, F.R.). Irregularly.

Eur. J. Phys.
European Journal of Physics. Institute of Physics, London (UK). Quarterly. ISSN 0143-0807, EJPHD4.

Europhys. Lett.
Europhysics Letters. European Physical Society, PO Box 69, CH-1213 Petit-Lancy 2 (Switzerland). Semimonthly. ISSN 0302-072X, EULEEJ.

Europhys. News
Europhysics News. European Physical Society, PO Box 69, CH-1213 Petit-Lancy 2 (Switzerland). Monthly. ISSN 0531-7479, EUPNAS.

Exp. Astron.
Experimental Astronomy. Kluwer Academic Publ., Dordrecht (Netherlands). Bimonthly. A supplement to Astrophysics and Space Science. ISSN 0922-6435.

Fizika
Fizika. Mladost Export-Import, 41000 Zagreb, Ilica 30 (Yugoslavia). Quarterly. ISSN 0015-3206, FZKAAA.

Fortschr. Phys.
Fortschritte der Physik. Akademie-Verl., Berlin (Germany). Monthly. ISSN 0015-8208, FPYKAG.

Found. Phys.
Foundations of Physics. Plenum Publishing Corporation, New York, NY (USA). Monthly. ISSN 0015-9018, FNDPA4.

Found. Phys. Lett.
Foundations of Physics Letters. Plenum Press, New York, NY (USA). Bimonthly. ISSN 0894-9875, FPLEET.

G. Astron.
Giornale di Astronomia. Società Astronomica Italiana, Largo E. Fermi, 5-50125 Firenze (Italy). Quarterly. ISSN 0390-1106.

Gemini
Gemini. Newsletter of the Royal Greenwich Observatory. Royal Greenwich Observatory, Madingley Road, Cambridge, CB3 0EZ (UK). Quarterly. ISSN 0960-670X.

Gen. Relativ. Gravitation
General Relativity and Gravitation. Plenum Publishing Corporation, New York, NY (USA). Monthly. ISSN 0001-7701, GRGVA8.

Geochim. Cosmochim. Acta
Geochimica et Cosmochimica Acta. Pergamon Press, New York (USA). Monthly. Journal of the Geochemical Society and the Meteoritical Society. ISSN 0016-7037, GCACAK.

Geomagn. Aehron.
Geomagnetizm i Aehronomiya. Akademiya Nauk SSSR. Izdatel'stvo Nauka, Moskva (USSR). Bimonthly. English translation in Geomagnetism and Aeronomy. ISSN 0016-7940, GEAEA6.

Geomagn. Aeron.
Geomagnetism and Aeronomy. American Geophysical Union, 2000 Florida Avenue, N.W., Washington, DC 20009 (USA). Bimonthly. Cover-to-cover translation of Geomagnetizm i Aehronomiya. ISSN 0016-7932, GMARAX.

Geophys. Astrophys. Fluid Dyn.
Geophysical and Astrophysical Fluid Dynamics. Gordon and Breach, New York, NY (USA). Irregularly. ISSN 0309-1929, GAFDD3.

Geophys. J. Int.
Geophysical Journal International. Royal Astronomical Society, Deutsche Geophysikalische Gesellschaft, European Geophysical Society. Blackwell, Oxford (UK). Monthly. Formed by merger of Geophys. J. R. Astron. Soc., J. Geophys, and Ann. Geophys., Ser. B. ISSN 0955-419X.

Geophys. Res. Lett.
Geophysical Research Letters. American Geophysical Union, Washington, DC (USA). Monthly. ISSN 0094-8276, GPRLAJ.

Geophysics
Geophysics. Society of Exploration Geophysicists, PO Box 3098, Tulsa, OK 74101 (USA). Monthly. ISSN 0016-8033, GPYSA7.

GEOS Circ.
GEOS Circular. GEOS (Groupe d'Etude et d'Observation Stellaire), 3, promenade Venezia, F-78000 Versailles (France). Irregularly. ISSN 1148-6252.

Glob. Planet. Change
Global and Planetary Change. Elsevier, Amsterdam (Netherlands). Irregularly. ISSN 0921-8181, GPCHE4.

Hadronic J.
Hadronic Journal. Hadronic Press, Nonantum, MA 02195 (USA). Bimonthly. ISSN 0162-5519, HAJODX.

Heavens
The Heavens. Oriental Astronomical Association, Ōtsu-shi, Shiga-ken (Japan). Monthly. In Japanese. ISSN 0287-6906.

Helsinki Univ. Technol., Metsähovi Radio Res. Stn., Rep. Ser. A
Helsinki University of Technology, Metsähovi Radio Research Station, Report Series A. Metsähovi Radio Research Station, Kylmälä (Finland). ISSN 0783-8751.

Helv. Phys. Acta
Helvetica Physica Acta. Schweizerische Physikalische Gesellschaft. Birkhäuser, Basel (Switzerland). Bimonthly. ISSN 0018-0238, HPACAK.

High Energ. Phys. Nucl. Phys.
High Energy Physics and Nuclear Physics. Science Press, Beijing (China). Bimonthly. Formerly: Physica Energiae Fortis et Physica Nuclearis. ISSN 0254-3052, KNWLD9.

Hyperfine Interact.
Hyperfine Interactions. Baltzer, Basel (Switzerland). Monthly. ISSN 0304-3843, HYINDN.

I.A.P.P.P. Commun.
International Amateur–Professional Photoelectric Photometry Communication. Dyer Observatory, Vanderbilt University, Nashville, TN 37235 (USA). Quarterly. ISSN 0886-6961.

IAU Circ.
International Astronomical Union, Circular. International Astronomical Union. Central Bureau for Astronomical Telegrams, Smithsonian Astrophysical Observatory, 60 Garden Street, Cambridge, MA 02138 (USA). Irregularly. ISSN 0081-0304, IANUAB.

Icarus
Icarus. Academic Press, 1 East First Street, Duluth, MN 55802 (USA). Monthly. ISSN 0019-1035, ICRSA5.

IEEE Trans. Magn.
IEEE Transactions on Magnetics. The Institute of Electrical and Electronics Engineers, Piscataway, NJ (USA). Bimonthly. ISSN 0018-9464, IEMGAQ.

IEEE Trans. Nucl. Sci.
IEEE Transactions on Nuclear Science. The Institute of Electrical and Electronics Engineers, 345 East 47th Street, New York, NY 10017 (USA). Bimonthly. ISSN 0018-9499, IETNAE.

IERS Bull., A
IERS Bulletin. A. International Earth Rotation Service, U.S. Naval Observatory, Washington, DC (USA). Weekly. ISSN 1048-6127.

IERS Bull., B
IERS Bulletin. B. International Earth Rotation Service, Central Bureau of IERS, Observatoire de Paris, Paris (France). Monthly.

IERS Tech. Note
IERS (International Earth Rotation Service) Technical Note. Central Bureau of IERS, Observatoire de Paris, Paris (France). Irregularly.

Indian J. Phys., Part B
Indian Journal of Physics, Part B. Indian Association for the Cultivation of Science, 2 & 3 Raja Subodh Chandra Mallik Road, Calcutta 700 032 (India). Bimonthly. ISSN 0019-5480, IJPYBT.

Indian J. Pure Appl. Phys.
Indian Journal of Pure and Applied Physics. Council of Scientific and Industrial Research, Publications and Information Directorate, Hillside Road, New Delhi 110 012 (India). Monthly. ISSN 0019-5596, IJOPAU.

Indian J. Radio Space Phys.
Indian Journal of Radio and Space Physics. Council of Scientific and Industrial Research, Publications and Information Directorate, Hillside Road, New Delhi 110 012 (India). Bimonthly. ISSN 0367-8393, IJRSAK.

Inf. Bull. Variable Stars
Information Bulletin on Variable Stars. Commission 27 of the IAU. Konkoly Observatory, Budapest (Hungary). Irregularly. ISSN 0374-0676.

Infrared Phys.
Infrared Physics. Pergamon Press, Oxford (UK). Bimonthly. ISSN 0020-0891, INFPAD.

Int. Assoc. Geod. Symp.
International Association of Geodesy Symposia. Springer, New York, NY (USA). Irregularly.

Int. Astron. Union Colloq.
International Astronomical Union Colloquium. Various publishers.

Int. Astron. Union Symp.
International Astronomical Union Symposium. Kluwer Academic Publ., Dordrecht (Netherlands).

Int. Comet Q.
The International Comet Quarterly. Smithsonian Astrophysical Observatory, 60 Garden Street, Cambridge, MA 02138, (USA). Quarterly. ISSN 0736-6922, ICOQDL.

Int. J. Infrared Millimeter Waves
International Journal of Infrared and Millimeter Waves. Plenum Publishing Corporation, New York, NY (USA). Monthly. ISSN 0195-9271, IJIWDO.

Int. J. Mod. Phys. A
International Journal of Modern Physics A. World Scientific, Singapore (Singapore). 28 issues per year. ISSN 0217-751X.

Inverse Probl.
Inverse Problems. Institute of Physics, Bristol (UK). Quarterly. ISSN 0266-5611, INPEEY.

Ir. Astron. J.
The Irish Astronomical Journal. Armagh Observatory, Armagh BT61 9DG (UK). Semiannual. ISSN 0021-1052, IRAJAW.

IRIS Bull. A
IRIS Bulletin A. Subcommission International Radio Interferometric Surveying (IRIS) Steering Committee. National Geodetic Survey, National Oceanic and Atmospheric Administration, N/CG114, Rockville, MD 20852 (USA). Monthly. ISSN 1016-8648.

Isis
Isis. Journal of the History of Science Society. Department of the History of Medicine, University of Wisconsin, Madison, WI (USA). Quarterly. An international review devoted to the history of science and its cultural influences. ISSN 0021-1753.

Istor.–Astron. Issled.
Istoriko–Astronomicheskie Issledovaniya. Nauka, Moskva (Russia). Irregularly. Research in the History of Astronomy. ISSN 0234-8632.

Izv. Acad. Sci. USSR, Phys. Solid Earth
Izvestiya Academy of Sciences USSR, Physics of the Solid Earth. American Geophysical Union (AGU), Washington, DC (USA). Monthly. Cover–to–cover translation of Izv. Akad. Nauk SSSR, Fiz. Zemli. ISSN 0001-4354, IPSEBQ.

Izv. Akad. Nauk SSSR, Fiz. Atmos. Okeana
Izvestiya Akademii Nauk SSSR, Fizika Atmosfery i Okeana. AN SSSR, Moscow (USSR). Monthly. English translation in Izvestiya, Academy of Sciences USSR, Atmospheric and Oceanic Physics. ISSN 0002-3515, IVNMAW.

Izv. Astron. Obs. Ehngel'gardt.
Izvestiya Astronomicheskoj Observatorii Ehngel'gardta. Izdatel'stvo Kazanskogo Universiteta, Kazan (USSR). Irregularly. ISSN 0321-1762.

Izv. Vyssh. Uchebn. Zaved., Radiofiz.
Izvestiya Vysshikh Uchebnykh Zavedenij, Radiofizika. Gor'kovskij Universitet, Gor'kij, ul. Lyadova 25 (USSR) Monthly. English translation in Radiophysics and Quantum Electronics. ISSN 0021-3462, IVYRAY.

J. Acoust. Soc. Am.
Journal of the Acoustical Society of America. American Institute of Physics, New York, N.Y. (USA). Monthly. ISSN 0001-4966, JASMAN.

J. Appl. Phys.
Journal of Applied Physics. American Institute of Physics, New York, NY (USA). Semimonthly. ISSN 0021-8979, JAPIAU.

J. Astron. Fr.
Journal des Astronomes Français. Société Française des Spécialistes d'Astronomie. Edité à l'Observatoire de Midi-Pyrénées, 14, avenue Ed. Belin, F-31400 Toulouse (France). Bulletin de la Société Française des Spécialistes d'Astronomie. ISSN 0181-9429.

J. Astrophys. Astron.
Journal of Astrophysics and Astronomy. Indian Academy of Sciences, PO Box 8005, Bangalore 560080 (India). Quarterly. ISSN 0250-6335, JASRD7.

J. Atmos. Sci.
Journal of the Atmospheric Sciences. American Meteorological Society, 45 Beacon Street, Bosten, MA 02108 (USA). Semimonthly. ISSN 0022-4928, JAHSAK.

J. Atmos. Terr. Phys.
Journal of Atmospheric and Terrestrial Physics. Pergamon Press, Oxford (UK). Bimonthly. ISSN 0021-9169, JATPA3.

J. Beijing Normal Univ. Nat. Sci.
Journal of the Beijing Normal University Natural Sciences. Quarterly. In Chinese.

J. Br. Astron. Assoc.
Journal of the British Astronomical Association. The British Astronomical Association, Burlington House, Picadilly, London W1V 9AG (UK). Bimonthly. ISSN 0007-0297, JBAAA6.

J. Br. Interplanet. Soc.
Journal of the British Interplanetary Society. The British Interplanetary Society, 27/29 South Lambeth Road, London SW8 1SZ (UK). Monthly. ISSN 0007-084X, JBISAW.

J. Chem. Phys.
Journal of Chemical Physics. American Institute of Physics, New York, NY (USA). Monthly. ISSN 0021-9606, JCPSA6.

J. Fluid Mech.
Journal of Fluid Mechanics. Cambridge University Press, London (UK). Monthly. ISSN 0022-1120, JFLSA7.

J. Geophys. Res.
Journal of Geophysical Research, Part A: Space Physics. American Geophysical Union, Washington, DC (USA). Monthly. ISSN 0148-0227, JGREA2.

J. Geophys. Res.
Journal of Geophysical Research, Part B: Solid Planets. American Geophysical Union, Washington, DC (USA). Monthly. ISSN 0148-0227, JJGBDO.

J. Geophys. Res.
Journal of Geophysical Research, Part C: Oceans and Atmospheres. American Geophysical Union, Washington, DC (USA). Monthly. ISSN 0148-0227, JJGADR.

J. Geophys. Res.
Journal of Geophysical Research, Part D: Atmospheres. American Geophysical Union, Washington, DC (USA). Monthly. ISSN 0148-0227, JGREA2.

J. Geophys. Res.
Journal of Geophysical Research, Part E: Planets American Geophysical Union, Washington, DC (USA). Monthly. ISSN 0148-0227, JGREA2.

J. Hist. Astron.
Journal for the History of Astronomy. Science History Publications Ltd., Halfpenny Furze, Mill Lane, Chalfont St Giles, Buckinghamshire, HP8 4NR (UK). Quarterly. ISSN 0021-8286, JHSAA2.

J. Korean Astron. Soc.
The Journal of the Korean Astronomical Society. Department of Earth Science Education, Seoul National University, Seoul (Republic of Korea). Semiannual. ISSN 0253-3065.

J. Magn. Magn. Mater.
Journal of Magnetism and Magnetic Materials. North-Holland, Amsterdam (Netherlands) Monthly. ISSN 0304-8853, JMMMDC.

J. Mater. Sci.
Journal of Materials Science. Chapman and Hall, London (UK). Monthly. ISSN 0022-2461, JMTSAS.

J. Math. Phys.
Journal of Mathematical Physics. American Institute of Physics, New York, NY (USA). Monthly. ISSN 0022-2488, JMAPAQ.

J. Mod. Opt.
Journal of Modern Optics. Taylor and Francis, London (UK). Monthly. Formerly: Optica Acta. ISSN 0950-0340.

J. Mol. Spectrosc.
Journal of Molecular Spectroscopy. Academic Press, New York, NY (USA) Monthly. ISSN 0022-2852, JMOSA3.

J. Nanjing Univ.
Journal of Nanjing University. (Natural Sciences Edition). Nanjing University, Nanjing (China). Quarterly. Nanjing Daxue Xuebao. ISSN 0465-7926.

J. Non-Cryst. Solids
Journal of Non-Crystalline Solids. North-Holland, Amsterdam (Netherlands). Irregularly. ISSN 0022-3093, JNCSBJ.

J. Opt. (Paris)
Journal of Optics. Masson, Paris (France). Bimonthly. Formerly: Nouvelle Revue d'Optique. ISSN 0150-536X, JOOPDB.

J. Opt. Soc. Am. A
Journal of the Optical Society of America A. (Optics and Image Science). Optical Society of America. American Institute of Physics, New York, NY (USA) 13 issues per year. Formerly: Journal of the Optical Society of America. ISSN 0740-3232, JOAOD6.

J. Opt. Soc. Am. B
Journal of the Optical Society of America B (Optical Physics). Optical Society of America. American Institute of Physics, New York, NY (USA). 13 issues per year. Formerly: Journal of the Optical Society of America. ISSN 0740-3224, JOBPDE.

J. Phys. A
Journal of Physics A (Mathematical and General Physics). Institute of Physics, London (UK). 18 issues per year. ISSN 0305–4470, JPHAC5.

J. Phys. B
Journal of Physics B (Atomic, Molecular and Optical Physics). Institute of Physics, London (UK). Semimonthly. Formerly: Journal of Physics, B (London). Atomic and Molecular Physics. ISSN 0935–4075, JPAPEH.

J. Phys. D
Journal of Physics D (Applied Physics). Institute of Physics, London (UK). Monthly. ISSN 0022–3727, JPAPBE.

J. Phys. Earth
Journal of Physics of the Earth. University of Tokyo Press, Tokyo (Japan). Subscription address: Japan Publications Trading Company, C.P.O. 722, Tokyo, Japan. ISSN 0022–3743, JPHEAF.

J. Phys. G
Journal of Physics G (Nuclear Physics). Institute of Physics, London (UK). Monthly. ISSN 0305–4616, JPHGBM.

J. Phys. Soc. Jpn.
Journal of the Physical Society of Japan. Physical Society of Japan. Room 211, Kikai Shinko Building, 3–5–8 Shiba Koen, Minato–ku, Tokyo 105 (Japan) Monthly. ISSN 0031–9015, JUPSAU.

J. Plasma Phys.
Journal of Plasma Physics. Cambridge University Press, London (UK) Bimonthly. ISSN 0022–3778, JPLPBZ.

J. Quant. Spectrosc. Radiat. Transfer
Journal of Quantitative Spectroscopy and Radiative Transfer. Pergamon Press, Oxford (UK). Monthly. ISSN 0022–4073, JQSRAE.

J. R. Astron. Soc. Can.
The Journal of the Royal Astronomical Society of Canada. The Royal Astronomical Society of Canada, 136 Dupont Street, Toronto, Ontario M5R 1V2 (Canada). Bimonthly. ISSN 0035–872X, JRASA2.

J. Stat. Phys.
Journal of Statistical Physics. Plenum Publishing Corporation, New York, NY (USA) Semiannual. ISSN 0022–4715, JSTPBS.

J. Vac. Sci. Technol., A
Journal of Vacuum Science and Technology, A: Vacuums Surfaces and Films. American Institute of Physics, New York, NY (USA). Bimonthly. ISSN 0734–2101, JVTAD6.

Jenaer Rundsch.
Jenaer Rundschau. Jenoptik Jena GmbH, Verlag Technik, Berlin (Germany). Quarterly. ISSN 0368-203X, JERUA2.

JETP Lett.
JETP Letters. American Institute of Physics, New York, NY (USA). Monthly. Cover–to–cover translation of Pis'ma v Zhurnal Ehksperimental'noj i Teoreticheskoj Fiziki. ISSN 0021–3640, JTPLA2.

Johns Hopkins APL Tech. Dig.
Johns Hopkins APL Technical Digest. The Johns Hopkins University Applied Physics Laboratory, Johns Hopkins Road, Laurel, Md. 20707 (USA). Quarterly. ISSN 0270–5214, APLDAZ.

Jpn. J. Appl. Phys., Part 1
Japanese Journal of Applied Physics, Part 1 (Regular Papers and Short Notes). Publication Office, Daini Toyokaiji Building, 24–8, Shinbashi 4–chome, Minato–ku, Tokyo 105 (Japan). Monthly. Formerly: Japanese Journal of Applied Physics. ISSN 0021–2922, JAPNDE.

Jpn. J. Appl. Phys., Part 2
Japanese Journal of Applied Physics, Part 2 (Letters). Publication Office, Daini Toyokaiji Building, 24–8, Shinbashi 4–chome, Minato–ku, Tokyo 105 (Japan). Monthly. Formerly: Japanese Journal of Applied Physics. ISSN 0021–4922, JAPLDB.

Kexue Tongbao
Kexue Tongbao (Science Bulletin). Academia Sinica. Science Press, No. 137, Chaoyangmennei Street, Beijing (China). Monthly. ISSN 0023–074X, KETOD8.

Kinematics Phys. Celest. Bodies
Kinematics and Physics of Celestial Bodies. Allerton Press, New York, NY (USA). A cover–to–cover translation of Kinematika i Fizika Nebesnykh Tel. ISSN 0884–5913.

Kinematika Fiz. Nebesn. Tel
Kinematika i Fizika Nebesnykh Tel. Akademiya Nauk Ukrainskoj SSR. Otdelenie Fiziki i Astronomii. Glavnaya Astronomicheskaya Observatoriya Akademii Nauk USSR. Naukova Dumka, Kiev (USSR). Bimonthly. English translation in Kinematics Phys. Celest. Bodies. ISSN 0233–7665.

Kosm. Issled.
Kosmicheskie Issledovaniya. Akademiya Nauk SSSR. Nauka, Moskva (USSR). Bimonthly. English translation in Cosmic Research. ISSN 0023–4206.

KPM
Kometen Planetoiden Meteore. 3 issues per year. Subscription address: Michael Möller, Steiluferallee 7, D–2408 Timmendorfer Strand (Germany F.R.). ISSN 0930–102X.

Laser Optoelektron.
Laser und Optoelektronik. AT–Fachverl., Stuttgart (Germany, F.R.). Quarterly. ISSN 0722–9003, LAOPD3.

Laser Part. Beams
Laser and Particle Beams. Cambridge University Press, New York, NY (USA). Quarterly. ISSN 0263–0346, LPBEDA.

Lect. Notes Earth Sci.
Lecture Notes in Earth Sciences. Springer, Berlin (Germany, F.R.).

Lect. Notes Phys.
Lecture Notes in Physics. Springer, Berlin (Germany, F.R.). Irregularly.

Lect. Notes Phys., New Ser. m, Monogr.
Lecture Notes in Physics, New Series m: Monographs. Springer, Berlin (Germany). Irregularly.

Lett. Math. Phys.
Letters in Mathematical Physics. Kluwer, Dordrecht (Netherlands). Quarterly. ISSN 0377–9017, LMPDHQ.

Manuscr. Geod.
Manuscripta Geodaetica. Springer, Berlin (Germany, F.R.) Bimonthly. ISSN 0340–8825, MANGEH.

Mater. Lett.
Materials Letters. North–Holland, Amsterdam (Netherlands). Monthly. ISSN 0167–577X, MLETDJ.

Mater. Res. Bull.
Materials Research Bulletin. Pergamon Press, Elmsford, NY (USA). Monthly. ISSN 0025–5408, MRBUAC.

Meas. Sci. Technol.
Measurement Science and Technology. Institute of Physics (IOP), London (UK). Monthly. Formerly: J. Phys. E (London). Sci. Instrum. ISSN 0957–0233, MSTCEP.

Mem. Soc. Astron. Ital.
Memorie della Società Astronomica Italiana. Osservatorio Astrofisica di Arcetri, Largo Fermi, 5, I–50125 Firenze (Italy). Quarterly. ISSN 0037–8720, MSATAB.

Mercury
Mercury. Astronomical Society of the Pacific, 390 Ashton Avenue, San Francisco, CA 94112 (USA). Bimonthly. The Journal of the Astronomical Society of the Pacific. ISSN 0047–6773, MRCYAT.

Messenger
The Messenger – El Mensajero. European Southern Observatory, Karl-Schwarzschild-Straße 2, D–8046 Garching bei München (Germany, F.R.). Quarterly. ISSN 0722–6691, MESSE4.

Meteoritics
Meteoritics. Meteoritical Society at the Institute of Geophysics and Planetary Physics. University of California, Los Angeles, CA 90024–1567 (USA). Quarterly. ISSN 0026–1114, MERTAW.

Meteorol. Atmos. Phys.
Meteorology and Atmospheric Physics. Springer, Wien (Austria). Irregularly. ISSN 0177–8971, MAPHEU.

Minor Planet Bull.
The Minor Planet Bulletin. Association of Lunar and Planetary Observers. Editorial Office: R. P. Binzel, MIT 54–426, Cambridge, MA 02139 (USA). Quarterly. Bulletin of the Minor Planet Section of the Association of Lunar and Planetary Observers. ISSN 1052–8091.

Minor Planet Circ.
The Minor Planet Circulars/Minor Planets and Comets. International Astronomical Union. Minor Planet Center, Smithsonian Astrophysical Observatory, 60 Garden Street, Cambridge, MA 02138, (USA). Published usually at the date of full moon. ISSN 0736–6884, MPCIB2.

Mitt. Lohrmann–Obs., Tech. Univ. Dresden
Mitteilungen des Lohrmann–Observatoriums der Technischen Universität Dresden. Lohrmann–Observatorium, Technische Universität Dresden, O–8027 Dresden, Mommsenstraße 13 (Germany). Irregularly.

Mitt. Sternw. Sonneberg
Mitteilungen der Sternwarte zu Sonneberg. Sternwarte Sonneberg, Sonneberg (Germany). Irregularly. ISSN 0373–6202.

Mitt. Univ.–Sternw. Jena
Mitteilungen der Universitäts–Sternwarte zu Jena. Universitäts–Sternwarte Jena, Schillergäßchen 2, O–6900 Jena (Germany). Irregularly. ISSN 0863–5560.

Mitt. Veränderliche Sterne
Mitteilungen über Veränderliche Sterne. Sternwarte Sonneberg, Sonneberg (Germany). Irregularly. ISSN 0232–7554.

Mod. Phys. Lett. A
Modern Physics Letters A. World Scientific, Singapore (Singapore). 32 issues per year. ISSN 0217–7323.

Mon. Not. R. Astron. Soc.
Monthly Notices of the Royal Astronomical Society. Royal Astronomical Society. Blackwell, London (UK). Semimonthly. ISSN 0035–8711, MNRAA4.

Mon. Notes Astron. Soc. S. Afr.
Monthly Notes of the Astronomical Society of Southern Africa. South African Astronomical Observatory, PO Box 9, Observatory, 7935 Cape (South Africa). Bimonthly. ISSN 0024–8266, MASAAK.

Nachr. Olbers–Ges. Bremen
Nachrichten der Olbers–Gesellschaft Bremen. Olbers–Gesellschaft Bremen e.V., Bremen (Germany, F.R.). Quarterly.

Natl. Astron. Obs. (Jpn.), Repr.
National Astronomical Observatory (Japan), Reprint. National Astronomical Observatory, Mitaka, Tokyo 181 (Japan). Formerly entitled Tokyo Astron. Obs. Repr. ISSN 0915–0021.

NATO ASI Ser., Ser. C, Math. Phys. Sci.
NATO Advanced Science Institutes Series. Series C: Mathematical and Physical Sciences. Kluwer Academic Publ., Dordrecht (Netherlands). Irregularly.

Nature
Nature. Macmillan Magazines Ltd., London (UK). Weekly. Subscription address: Nature, Subscription Dept., Brunel Road, Basingstoke, Hants RG21 2XS, UK. ISSN 0028–0836, NATUAS.

Naturwiss. Rundsch.
Naturwissenschaftliche Rundschau. Wiss. Verlagsges., Stuttgart (Germany, F.R.). Monthly. ISSN 0028–1050, NARSAC.

Naturwissenschaften
Die Naturwissenschaften. Springer, Heidelberg (Germany, F.R.). Monthly. ISSN 0028–1042, NATWAY.

News Lett. Astron. Soc. N.Y.
News Letter of the Astronomical Society of New York. Astronomical Society of New York, 1125 Oxford Place, Schenectady, NY 12308 (USA). Semiannual. ISSN 0148–9992.

Nizamiah Rangapur Obs. Dep. Astron., Osmania Univ., Repr.
Nizamiah & Rangapur Observatories and Department of Astronomy, Osmania University, Reprint. Astronomy Department, Osmania University, Hyderabad–7 (India). Irregularly. ISSN –.

Nucl. Instrum. Methods Phys. Res., Sect. A
Nuclear Instruments and Methods in Physics Research, Section A (Accelerators, Spectrometers, Detectors and Associated Equipment). North–Holland, Amsterdam (Netherlands). Bimonthly. Formerly: Nuclear Instruments and Methods in Physics Research. ISSN 0168–9002, NIMRD9.

Nucl. Instrum. Methods Phys. Res., Sect. B
Nuclear Instruments and Methods in Physics Research, Section B (Beam Interactions with Materials and Atoms). North–Holland, Amsterdam (Netherlands). Monthly. Formerly: Nuclear Instruments and Methods in Physics Research. ISSN 0168–583X, NIMAER.

Nucl. Phys. A
Nuclear Physics A. North–Holland, Amsterdam (Netherlands). Weekly. ISSN 0375–9474, NUPABL.

Nucl. Phys. B, Part. Phys.
Nuclear Physics B, Particle Physics. North–Holland, Amsterdam (Netherlands). Weekly. Formerly: Nuclear Physics, B. ISSN 0550–3213, NUPBBO.

Nucl. Phys. B, Proc. Suppl.
Nuclear Physics B, Proceedings Supplements. North-Holland, Amsterdam (Netherlands). Irregularly. ISSN 0920-5632, NPBSE7.

Nuovo Cimento A
Il Nuovo Cimento della Società Italiana di Fisica A (Nuclei, Particles and Fields). Società Italiana di Fisica. Editrice Compositori, viale XII Giugno, 1, I-40124 Bologna (Italy). 20 issues per year. ISSN 0369-3546, NIFAAM.

Nuovo Cimento B
Il Nuovo Cimento della Società Italiana di Fisica B (General Physics, Relativity, Astronomy and Plasmas). Società Italiana di Fisica. Editrice Compositori, viale XII Giugno, 1, I-40124 Bologna (Italy). Monthly. ISSN 0369-3554, NIFBAP.

Nuovo Cimento C
Il Nuovo Cimento della Società Italiana di Fisica C (Geophysics and Space Physics). Società Italiana di Fisica. Editrice Compositori, viale XII Giugno, 1, I-40124 Bologna (Italy). Bimonthly. ISSN 0390-5551, NIFCAS.

Nuovo Cimento D
Il Nuovo Cimento della Società Italiana di Fisica D (Condensed Matter, Atomic, Molecular and Chemical Physics, Biophysics). Società Italiana di Fisica. Editrice Compositori, viale XII Giugno, 1, I-40124 Bologna (Italy). Bimonthly. ISSN 0392-6737, NCSDDN.

OAA Comput. Sect. Circ.
Oriental Astronomical Association Computing Section Circular. S. Nakano, PO Box No. 32, Sumoto Post Office, Hyogo-Ken, 656-91 (Japan). Irregularly.

Obs. Trav.
Observations et Travaux. Société Astronomique de France, 3, rue Beethoven, F-75016 Paris (France). Quarterly. ISSN 0769-0878.

Observatory
The Observatory. Rutherford Appleton Laboratory, Chilton, Didcot, Oxon, OX11 0QX (UK). Bimonthly. ISSN 0029-7704, OBSEAR.

Occultation Newsl.
Occultation Newsletter. International Occultation Timing Association (IOTA), 6N106 White Lake Lane, St. Charles, IL 60175 (USA). Quarterly. ISSN 0737-6766.

Österr. Z. Vermessungswes. Photogramm.
Österreichische Zeitschrift für Vermessungswesen und Photogrammetrie. Österreichischer Verein für Vermessungswesen und Photogrammetrie, Wien (Austria). Quarterly. ISSN 1017-2963.

Opt. Commun.
Optics Communications. North-Holland, Amsterdam (Netherlands) Biweekly. ISSN 0030-4018, OPCOB8.

Opt. Eng.
Optical Engineering. Society of Photo-Optical Instrumentation Engineers, Bellingham, WA (USA). Monthly. ISSN 0091-3286, OPEGAR.

Opt. Laser Technol.
Optics and Laser Technology. IPC Science and Technology Press, Guildford (UK). Bimonthly. ISSN 0030-3992, OLTCAS.

Opt. Lett.
Optics Letters. American Institute of Physics, New York, NY (USA). Monthly. ISSN 0146-9592, OPLEDP.

Opt. Photonics News
Optics and Photonics News. Optical Society of America, Washington, DC (USA). Monthly. ISSN 1047-6938, OPPHRI.

Opt. Pura Apl.
Optica Pura y Aplicada. Sociedad Espanola de Optica, Madrid (Spain). 3 issues per year. ISSN 0030-3917, OPAPAY.

Opt. Spectrosc.
Optics and Spectroscopy. American Institute of Physics, New York, NY (USA). Monthly. Cover-to-cover translation of Optica i Spektroskopiya. ISSN 0030-400X, OPSUA3.

Optik
Optik. Wiss. Verlagsges., Stuttgart (Germany, F.R.). Quarterly. ISSN 0030-4026, OTIKAJ.

Origins Life Evol. Biosphere
Origins of Life and Evolution of the Biosphere. Kluwer Academic Publ., Dordrecht (Netherlands). Bimonthly. The Journal of the International Society for the Study of the Origin of Life. ISSN 0169-6149, OGLFAU.

Orion
Orion. Zeitschrift der Schweizerischen Astronomischen Gesellschaft. Schweizerische Astronomische Gesellschaft, Chemin du Marais-Long, 10, CH-1217 Meyrin (Switzerland). Bimonthly. ISSN 0030-557X, ORIOAH.

Orione
Orione. Il Castello, Via C. Ravizza 14-16, I-20149 Milano (Italy). Bimonthly. ISSN 0393-0858.

Phys. Bl.
Physikalische Blätter. Physik-Verlag, Postfach 1260/1280, D-6940 Weinheim (Germany, F.R.). Monthly. ISSN 0031-9279, PHBLAG.

Phys. Chem. Miner.
Physics and Chemistry of Minerals. Springer, Berlin (Germany, F.R.). Irregularly. ISSN 0342-1791, PCMIDU.

Phys. Earth Planet. Inter.
Physics of the Earth and Planetary Interiors. Elsevier, Amsterdam (Netherlands). Quarterly. ISSN 0031-9201, PEPIAM.

Phys. Fluids, A
Physics of Fluids, A. American Institute of Physics, New York, NY (USA). Monthly. ISSN 0031-9171, PFLDAS.

Phys. Fluids, B
Physics of Fluids, B. American Institute of Physics, New York, NY (USA). Monthly. ISSN 0031-9171, PFLDAS.

Phys. Lett. A
Physics Letters A (General, Atomic, and Solid State Physics). North-Holland, Amsterdam (Netherlands). Weekly. ISSN 0375-9601, PYLAAG.

Phys. Lett. B
Physics Letters B (Nuclear, Elementary Particle, and High-Energy Physics). North-Holland, Amsterdam (Netherlands). Weekly. ISSN 0370-2693, PYLBAJ.

Phys. Rep.
Physics Reports. North-Holland, Amsterdam (Netherlands). Weekly. Review Section of Physics Letters, Section C. ISSN 0370-1573, PRPLCM.

Phys. Rev. A
Physical Review A (General Physics). American Physical Society. American Institute of Physics, New York, NY (USA). Biweekly. ISSN 0556-2791, PLRAAN.

Phys. Rev. B
Physical Review B (Condensed Matter). American Physical Society. American Institute of Physics, New York, NY (USA). 36 issues per year. ISSN 0556–2805, PRBMDO.

Phys. Rev. C
Physical Review C (Nuclear Physics). American Physical Society. American Institute of Physics, New York, NY (USA). Bimonthly. ISSN 0556–2813, PRVCAN.

Phys. Rev. D
Physical Review D (Particles and Fields). American Physical Society. American Institute of Physics, New York, NY (USA). Biweekly. ISSN 0556–2821, PRVDAQ.

Phys. Rev. Lett.
Physical Review Letters. American Physical Society. American Institute of Physics, New York, NY (USA). Weekly. ISSN 0031–9007, PRLTAO.

Phys. Scr.
Physica Scripta. The Royal Swedish Academy of Sciences, Box 50005, S–10405 Stockholm, (Sweden). Monthly. ISSN 0031–8949, PHSTBO.

Phys. Scr. T
Physica Scripta T. The Royal Swedish Academy of Sciences, Box 50005, S–10405 Stockholm (Sweden). Irregularly. ISSN 0281–1847, PHSTBO.

Phys. Teach.
Physics Teacher. American Institute of Physics, New York, NY (USA). 9 issues per year. ISSN 0031–921X, PHTEAH.

Phys. Today
Physics Today. American Institute of Physics, New York, NY (USA). Monthly. ISSN 0031–9228, PHTOAD.

Phys. Unserer Zeit
Physik in Unserer Zeit. Verlag Chemie, Postfach 1260/1280, D–6940 Weinheim (Germany, F.R.). Bimonthly. ISSN 0031–9252, PHUZAH.

Physica A
Physica A (Theoretical and Statistical Physics). North–Holland, Amsterdam (Netherlands). Monthly. ISSN 0378–4371, PHYADX.

Physica C
Physica C (Superconductivity). North–Holland, Amsterdam (Netherlands). Monthly. Formerly: Physica B, C. ISSN 0921–4534, PHYCE6.

Pis'ma Astron. Zh.
Pis'ma v Astronomicheskij Zhurnal. Akademiya Nauk SSSR. Nauka, Moskva (USSR). Monthly. English translation in Soviet Astronomy Letters. ISSN 0320–0108, PAZHDA.

Planet. Space Sci.
Planetary and Space Science. Pergamon Press, Oxford (UK). Monthly. ISSN 0032–0633, PLASSE.

Plasma Phys. Controlled Fusion
Plasma Physics and Controlled Fusion. Pergamon Press, Oxford (UK). Monthly. Formerly: Plasma Physics. ISSN 0741–3335, PLPHBZ.

Port. Phys.
Portugaliae Physica. Sociedade Portuguesa de Fisica, Avenida de Republica, 37–40, P–1000 Lisboa (Portugal). Quarterly. ISSN 0048–4903, POPYA4.

Postepy Astron.
Postepy Astronomii. Czasopismo Póswiecone Upowszechnianiu Wiedzy Astronomicznej. Polskie Towarzystwo Astronomiczne, Warszawa. Pánstwowe Wydawnictwo Naukowe, Warszawa (Poland). Quarterly., PYAIAJ.

Pract. Astron. Handb.
Practical Astronomy Handbooks. Cambridge University Press, Cambridge (UK).

Pramāna
Pramāna. Indian Academy of Sciences, Bangalore 560 006 (India). Bimonthly. ISSN 0304–4289, PRAMCI.

Priroda
Priroda. Nauka, Moskva (USSR). Monthly. ISSN 0032–874X.

Proc. Astron. Soc. Aust.
Proceedings of the Astronomical Society of Australia. CSIRO Division of Radiophysics, Box 76, P.O., Epping, N.S.W. 2121 (Australia). Semiannual. ISSN 0066–9997, AAUPBC.

Proc. IEEE
Proceedings of the IEEE. The Institute of Electrical and Electronics Engineers, 345 East 47th Street, New York, NY 10017–2394 (USA). Monthly. Subscriptions: 445 Hoes Lane, Piscataway, NJ 08854–4150, USA. ISSN 0018–9219.

Proc. Indian Acad. Sci., Earth Planet. Sci.
Proceedings of the Indian Academy of Sciences, Earth and Planetary Sciences. Indian Academy of Sciences, Bangalore 560 080 (India). Irregularly. ISSN 0370–0089, PIESDS.

Prog. Astron.
Progress in Astronomy. Knowledge Press, 650 Gubei Road, Shanghai (China). Quarterly. Distributed by Shanghai Distribution Office of Xinhua Bookstore. ISSN 1000–8349.

Prog. Theor. Phys.
Progress of Theoretical Physics. Kyoto Univ. (Japan). Research Inst. for Fundamental Physics. Monthly. ISSN 0033–068X, PTPKAV.

Publ. Astron. Inst. Czech. Acad. Sci.
Publications of the Astronomical Institute of the Czechoslovak Academy of Sciences. Astronomical Institute of the Czechoslovak Academy of Sciences. 251 65 Ondřejov (Czechoslovakia). Irregularly.

Publ. Astron. Soc. Jpn.
Publications of the Astronomical Society of Japan. Astronomical Society of Japan. National Astronomical Observatory, Mitaka–shi, Tokyo 181 (Japan). Bimonthly. ISSN 0004–6264, PASJA.

Publ. Astron. Soc. Pac.
Publications of the Astronomical Society of the Pacific. Astronomical Society of the Pacific, 390 Ashton Avenue, San Francisco, CA 94122 (USA). Monthly. ISSN 0004–6280, PASPAU.

Publ. Beijing Astron. Obs.
Publications of the Beijing Astronomical Observatory. Beijing Astronomical Observatory, Academica Sinica, Beijing (China). Irregularly.

Publ. Ege Univ. Obs.
Publications of Ege University Observatory. Ege University Observatory, Bornova, Izmir (Turkey). Irregularly.

Publ. Natl. Astron. Obs. Jpn.
Publications of the National Astronomical Observatory of Japan. National Astronomical Observatory of Japan, Mitaka, Tokyo (Japan). Irregularly. ISSN 0915–3640.

Publ. Obs. Ebro, Misc.
Publicaciones del Observatorio del Ebro, Miscelánea. Consejo Superior de Investigaciones Cientificas, Roquetes (Spain). Irregularly.

Publ. Purple Mt. Obs.
Publications of Purple Mountain Observatory. Purple Mountain Observatory, Academia Sinica, Nanking (China). Quarterly. ISSN 1000–3681.

Publ. Shaanxi Astron. Obs.
Publications of the Shaanxi Astronomical Observatory. Shaanxi Astronomical Observatory, Academia Sinica, Lintong, Xian (China). Irregularly. ISSN 1001–1544.

Publ. Yunnan Obs.
Publications of Yunnan Observatory. Yunnan Observatory, Academia Sinica, PO Box 110, Kunming, Yunnan Province (China). Quarterly.

Pure Appl. Geophys.
Pure and Applied Geophysics. Birkhäuser, Basel (Switzerland). Bimonthly. ISSN 0033–4553, PAGYAV.

Q. J. R. Astron. Soc.
The Quarterly Journal of the Royal Astronomical Society. Royal Astronomical Society. Blackwell, Oxford (UK). Quarterly. ISSN 0035–8738, QJRAAK.

Q. J. R. Meteorol. Soc.
Quarterly Journal of the Royal Meteorological Society. Royal Meteorological Society, Bracknell (UK). Quarterly. ISSN 0035–9009, QJRMAM.

Radiant
Radiant. Dutch Meteor Society, Lederkarper 4, 2318 NB Leiden (Netherlands). Bimonthly. Journal of the Dutch Meteor Society. ISSN 0925–8566.

Radio Sci.
Radio Science. American Geophysical Union, 2000 Florida Avenue, N.W., Washington, DC 20009 (USA). Bimonthly. ISSN 0048–6604, RASCAD.

Recherche
Recherche. Société d'Editions Scientifiques, 57, rue de Seine, F–75006 Paris (France). 11 issues per year. ISSN 0029–5671, RCCHBV.

Rep. Prog. Phys.
Reports on Progress in Physics. Institute of Physics, Bristol (UK). Monthly. ISSN 0034–4885, RPPHAG.

Rev. Astron.
Revista Astronomica. Asociación Argentina Amigos de la Astronomia, Patricias Argentinas 550, 1405 Buenos Aires (Argentina). ISSN 0044–9253.

Rev. Bras. Fis.
Revista Brasileira de Fisica. Sociedade Brasileira de Fisica, C.P. 20553, São Paulo (Brazil). Quarterly. ISSN 0374–4922, RBFSA3.

Rev. Geophys.
Reviews of Geophysics. American Geophysical Union, 2000 Florida Avenue, N.W., Washington, DC 20009 (USA). Four times a year. ISSN 8755–1209, RGSPBL.

Rev. Mex. Astron. Astrofis.
Revista Mexicana de Astronomia y Astrofisica. Instituto de Astronomia, Universidad Nacional Autónoma de México, Apartado Postal 70–264, México 20, D.F. (México). Semiannual. ISSN 0185–1101, RMAAD4.

Rev. Mex. Fis.
Revista Mexicana de Fisica. Sociedad Mexicana de Fisica, Apartado Postal 20–364, México 20, D.F. (México). Quarterly. ISSN 0035–001X, RMXFAT.

Rev. Mod. Astron.
Reviews in Modern Astronomy. Springer-Verlag, Berlin (Germany, F.R.). Irregularly. Series of publications of the Astronomische Gesellschaft (AG).

Rev. Roum. Phys.
Revue Roumaine de Physique. Academie Republicii Populare Romine, B.P. 134–135, Bucuresti (Romania). 10 issues per year. ISSN 0035–4090, RRPQAN.

Rev. Sci. Instrum.
Review of Scientific Instruments. American Institute of Physics, New York, NY (USA). Monthly. ISSN 0034–6748, RSINAK.

SAAO Newsl.
SAAO Newsletter. South African Astronomical Observatory, PO Box 9, Observatory 7935, Cape (South Africa).

Sci. Am.
Scientific American. Scientific American Inc., 415 Madison Avenue, New York, NY 10017 (USA). Monthly. ISSN 0036–8733, SCAMAC.

Sci. China, Ser. A
Science in China, Series A: Mathematics, Physics, Astronomy & Technological Sciences. Science Press, Beijing (China). Monthly. (Scientia Sinica). ISSN 0253–5831, SSINAV.

Sci. Rep. Fukushima Univ.
Science Reports of the Fukushima University. Faculty of Education, Fukushima University, Fukushima (Japan). ISSN 0387–0855.

Sci. Rep. Tôhoku Univ., Eighth Ser.
The Science Reports of the Tôhoku University, Eighth Series (Physics and Astronomy). Faculty of Science, Tôhoku University, Sendai (Japan). Irregularly. ISSN 0388–5607.

Science
Science. American Association for the Advancement of Science, 1333 H Street, NW, Washington, DC 20005 (USA). Weekly. ISSN 0036–8075, SCIEAS.

Scientometrics
Scientometrics. Elsevier, Amsterdam (Netherlands).

SID Tech. Bull.
SID Technical Bulletin. The American Association of Variable Star Observers. A. J. Stokes, PO Box 398, Hudson, OH 44236 (USA). Quarterly.

Sidereal Times
The Sidereal Times. Astronomical Society of Western Australia, PO Box 421, Subiaco, W.A. 6008 (Australia).

Sitzungsber., Österr. Akad. Wiss., Math.–Naturwiss. Kl., Abt. II
Sitzungsberichte, Österreichische Akademie der Wissenschaften, Mathematisch-Naturwissenschaftliche Klasse, Abteilung II (Mathematische, Physikalische und Technische Wissenschaften). Springer, Wien (Austria). Irregularly. ISSN 0029–8816.

Sky Telesc.
Sky and Telescope. Sky Publ., Cambridge, MA (USA). Monthly. ISSN 0037–6604, SKTEA3.

Sol. Bull. (AAVSO)
Solar Bulletin (American Association of Variable Star Observers). The American Association of Variable Star Observers – Solar Division. Peter O. Taylor, PO Box 5685, Athens, GA 30604 (USA). Monthly. ISSN 0251–8480.

Sol. Phys.
Solar Physics. Kluwer Academic Publ., Dordrecht (Netherlands). Irregularly. ISSN 0038–0938, SLPHAX.

Sol. Radio Data
Solar Radio Data. Observatorium für solare Radioastronomie. Akademie der Wissenschaften, Zentralinstitut für Astrophysik, 1501 Tremsdorf, Observatorium (Germany). Quarterly.

Sol. Syst. Res.
Solar System Research. Consultants Bureau, New York, NY (USA). Quarterly. Cover–to–cover translation of Astronomicheskij Vestnik. ISSN 0038–0946, SSYRAL.

Soln. Dannye, Byull.
Solnechnye Dannye, Byulleten'. Akademiya Nauk SSSR, Astronomicheskij Sovet i Glavnaya Astronomicheskaya Observatoriya. Nauka, Leningradskoe Otdelenie, Leningrad (USSR). Monthly. ISSN 0552–5829.

Sonne
Sonne. Wilhelm–Foerster–Sternwarte, Munsterdamm 90, D–1000 Berlin 41 (Germany, F.R.). Quarterly. Mitteilungsblatt der Amateursonnenbeobachter. ISSN 0721–0094.

South. Stars
Southern Stars. Royal Astronomical Society of New Zealand, PO Box 3181, Wellington, (New Zealand). Quarterly. ISSN 0049–1640, SNSRBP.

Sov. Astron.
Soviet Astronomy. American Institute of Physics, New York, NY (USA). Bimonthly. Cover–to–cover translation of Astronomicheskij Zhurnal. ISSN 0038–5301, SAAJAN.

Sov. Astron. Lett.
Soviet Astronomy Letters. American Institute of Physics, New York, NY (USA). Bimonthly. Cover–to–cover translation of Pis'ma v Astronomicheskij Zhurnal. ISSN 0360–0327, SALEDW.

Sov. J. Nucl. Phys.
Soviet Journal of Nuclear Physics. American Institute of Physics, New York, NY (USA). Bimonthly. A cover–to–cover translation of Yadernaya Fizika. ISSN 0038–5506, SJNCAS.

Sov. J. Opt. Technol.
Soviet Journal of Optical Technology. American Institute of Physics, New York, NY (USA). Monthly. Cover–to–cover translation of Optiko–Mekhanicheskaya Promyshlennost'. ISSN 0038–5514, SJOTBH.

Sov. J. Part. Nucl.
Soviet Journal of Particles and Nuclei. American Institute of Physics, New York, NY (USA). Bimonthly. Cover–to–cover translation of Fizika Ehlementarnykh Chastits i Atomnogo Yadra. ISSN 0090–4759, SJPNA3.

Sov. J. Quantum Electron.
Soviet Journal of Quantum Electronics. American Inst. of Physics (AIP), New York, NY (USA). Monthly. Cover–to–cover translation of Kvantovaya Ehlektronika (Moskva). ISSN 0049–1748, SJQEAF.

Sov. Phys. – Dokl.
Soviet Physics – Doklady. American Institute of Physics, New York, NY (USA). Monthly. Cover–to–cover translation of Doklady Akademii Nauk SSSR. ISSN 0038–5689, SPHDAO.

Sov. Phys. – JETP
Soviet Physics – JETP. American Institute of Physics, New York, NY (USA). Monthly. Cover–to–cover translation of Zhurnal Ehksperimental'noj i Teoreticheskoj Fiziki. ISSN 0038–5646, SPHJAR.

Sov. Phys. – Usp.
Soviet Physics – Uspekhi. American Institute of Physics, New York, NY (USA). Monthly. Cover–to–cover translation of Uspekhi Fizicheskikh Nauk. ISSN 0038–5670, SOPUAP.

Space Sci. Rev.
Space Science Reviews. Kluwer Academic Publ., Dordrecht (Netherlands). Quarterly. ISSN 0038–6308, SPSRA4.

Space Telesc. Sci. Inst. Symp. Ser.
Space Telescope Science Institute Symposium Series. Cambridge University Press, Cambridge (UK). Irregularly.

Space Telesc. Sci. Inst., Newsl.
Space Telescope Science Institute, Newsletter. Space Telescope Science Institute, 3700 San Martin Drive, Baltimore, MD 21218 (USA). Quarterly. ISSN 1055–6524.

Spaceflight
Spaceflight. The British Interplanetary Society Ltd., 27/29 South Lambeth Road, London SW8 1SZ (UK). Monthly. The international magazine of space and astronautics. ISSN 0038–6340, SPFLAN.

Spektrum Wiss.
Spektrum der Wissenschaft. VCH Verlagsgesellschaft, Weinheim (Germany, F.R.). Monthly. ISSN 0170–2971, SPWIDY.

ST–ECF Newsl.
Space Telescope–European Coordinating Facility Newsletter. Space Telescope–European Coordinating Facility, European Southern Observatory, Karl–Schwarzschild–Str. 2, D–8046 Garching bei München (Germany, F.R.). ISSN 0258–8080.

Sterne
Die Sterne. Barth, Leipzig (Germany). Bimonthly. ISSN 0039–1255, STNEA2.

Sterne Weltraum
Sterne und Weltraum. Verlag Sterne und Weltraum Dr. Vehrenberg GmbH, Portiastraße 10, D–8000 München 90 (Germany, F.R). Monthly. ISSN 0039–1263, STUWAN.

Sternenbote
Der Sternenbote. Astronomisches Büro, Hasenwartgasse 32, A–1238 Wien (Austria). Monthly. Österreichische Astronomische Monatsschrift. ISSN 0039–1271.

Strolling Astron.
The Strolling Astronomer. The Journal of the Association of Lunar and Planetary Observers. A.L.P.O., PO Box 16131, San Francisco CA 94116 (USA). ISSN 0039–2502, STASAD.

Stud. Cercet. Fiz.
Studii si Cercetari de Fizica. Academia Republicii Populare Romine, PO Box 134–5, Calca Victoriei 126, Bucuresti (Romania). 10 issues per year. ISSN 0039–3940, SCEFAB.

Surv. Geophys.
Surveys in Geophysics. An International Review Journal of Geophysics and Planetary Sciences. Kluwer Academic Publ., Dordrecht (Netherlands). Quarterly. ISSN 0169–3298, SUGEEC.

Surv. High Energy Phys.
Surveys in High Energy Physics. Harwood Academic Publishers, Chur, Switzerland. Quarterly. ISSN 0142–2413, SHEPDB.

Tectonophysics
Tectonophysics. Elsevier, Amsterdam (Netherlands). Irregularly. ISSN 0040–1951, TCTOAM.

Telesc. Naz. Galileo, Tech. Rep.
Telescopio Nazionale Galileo, Technical Report. Osservatorio Astronomico di Padova–Asiago, Asiago (Italy). Irregularly.

Tellus, Ser. A
Tellus, Series A. Munksgaard, Copenhagen (Denmark). 5 issues per year. Dynamic Meteorology and Oceanography. ISSN 0280–6495, TSAOD8.

Tellus, Ser. B
Tellus, Series B. Munksgaard, Copenhagen (Denmark). 5 issues per year. Chemical and Physical Meteorology. ISSN 0280–6509, TSBMD7.

Teor. Mat. Fiz.
Teoreticheskaya i Matematicheskaya Fizika. AN SSSR, Moscow (USSR). Inst. Fiziki. 3 issues per year. English translation in Theoretical and Mathematical Physics. ISSN 0564–6162, TMFZAL.

Tokyo Meteor Network Rep.
Tokyo Meteor Network Report. Tokyo Meteor Network, 1–27–5, Daisawa, Setagaya-ku, Tokyo 155 (Japan). Irregularly.

Tr. Gos. Astron. Inst. im. Shternberga
Trudy Gosudarstvennogo Astronomicheskogo Instituta im. P. K. Shternberga. Izdatel'stvo Moskovskogo Universiteta, Moskva (USSR). Irregularly. ISSN 0371–6791.

Transp. Theory Stat. Phys.
Transport Theory and Statistical Physics. Marcel Dekker, 270 Madison Avenue, New York, NY 10016 (USA). Irregularly. ISSN 0041–1450, TTSPB4.

Ukr. Fiz. Zh.
Ukrainskij Fizicheskij Zhurnal. Naukova Dumka, Kiev (Ukrainian SSR). Monthly. ISSN 0503–1265, UFIZAW.

Universe Classroom
The Universe in the Classroom. Astronomical Society of the Pacific, 390 Ashton Ave., San Francisco, CA 94112 (USA). A newsletter on teaching astronomy.

Universo
Universo. Liga Ibero-Americana de Astronomía (LIADA), Apartado 700, Mérida 5101–A (Venezuela). 3 issues per year. ISSN 0012–9820.

Urania
Urania. Zakład Narodowy imienia Ossolińskich, Wydawnictwo Polskiej Akademii Nauk, Kraków (Poland). Monthly. Miesiecznik Polskiego Towarzystwa Miłośników Astronomii. ISSN 0042–0794.

U.S. Nav. Obs., Ser. 4
United States Naval Observatory, Series 4, Daily time differences. U.S. Naval Observatory, Washington D.C. 20392 (USA). Weekly.

Usp. Fiz. Nauk
Uspekhi Fizicheskikh Nauk. Mezhdunarodnaya Kniga, Moskva (USSR). Monthly. English translation in Soviet Physics – Uspekhi. ISSN 0042–1294, UFNAAG.

Van Vleck Obs. Contrib.
Van Vleck Observatory Contribution. Van Vleck Observatory, Middletown, CT (USA). Irregularly.

Vasiona
Vasiona. Astronomsko Drushtvo "Rudjer Bošković", Beograd (Yugoslavia). Quarterly. In Serbo-Croat. Revue d'Astronomie. ISSN 0506–4295.

Vatican Obs. Publ.
Vatican Observatory Publications. Specola Vaticana, 1-00120 Città del Vaticano. Irregularly. ISSN 0083–5293.

Verh. Dtsch. Phys. Ges.
Verhandlungen der Deutschen Physikalischen Gesellschaft. VCH-Verlagsgesellschaft, Weinheim (Germany, F.R.). 7 issues per year. ISSN 0420–0195, VDPEAZ.

Veröff. Bayer. Komm. Int. Erdmessung, Bayer. Akad. Wiss., Astron.–Geod. Arb.
Veröffentlichungen der Bayerischen Kommission für die Internationale Erdmessung der Bayerischen Akademie der Wissenschaften. Verlag der Bayerischen Akademie der Wissenschaften, München (Germany). Irregularly. ISSN 0340–7691.

Vestn. Leningr. Univ., Mat. Mekh. Astron.
Vestnik Leningradskogo Universiteta, Matematika, Mekhanika, Astronomiya. Leningradskij Universitet, Universitetskaya Nab. 7/9 Leningrad B-164 (USSR). Irregularly. ISSN 0024–0850, VMMAA3.

Vistas Astron.
Vistas in Astronomy. Pergamon Press, Oxford (UK). Quarterly. ISSN 0083–6656, VASTA6.

WGN
WGN. Werkgroepnieuws – meteoren. The International Meteor Organization. P. Roggemans, Pijnboomstraat 25, B-2800 Mechelen (Belgium). Bimonthly. The Journal of the International Meteor Organization. ISSN 1016–3115.

Yamamoto Circ.
Yamamoto Circular. Yamamoto Observatory, 289 Kamitanakami-Kiryutyo, Otu, Sigaken, 520–21 (Japan). Irregularly.

Z. Naturforsch., A
Zeitschrift für Naturforschung, Section A. A Journal of Physical Sciences. Verlag der Zeitschrift für Naturforschung, PO Box 2645, D-7400 Tübingen (Germany, F.R). Monthly. ISSN 0932–0784, ZNASEI.

Z. Phys., A
Zeitschrift für Physik, A (Atomic Nuclei). Springer, Berlin (Germany, F.R.). Monthly. ISSN 0930–1151, ZAANEE.

Z. Phys., B
Zeitschrift für Physik, B (Condensed Matter). Springer, Berlin (Germany, F.R.). Monthly. ISSN 0722–3277, ZPCMDN.

Z. Phys., C
Zeitschrift für Physik, C (Particles and Fields). Springer, Berlin (Germany, F.R.). Monthly. ISSN 0170–9739, ZPCFD2.

Z. Vermessungswes.
Zeitschrift für Vermessungswesen. Deutscher Verein für Vermessungswesen e.V.

Zemlya Vselennaya
Zemlya i Vselennaya. Nauka, Moskva (USSR). Bimonthly. Nauchno-Populyarnyj Zhurnal Akademii Nauk SSSR. ISSN 0044–3948.

Zh. Ehksp. Teor. Fiz.
Zhurnal Ehksperimental'noj i Teoreticheskoj Fiziki. Akademiya Nauk SSSR, Leninskij Prosp. 14, Moskva (USSR). Bimonthly. English translation in Soviet Physics – JETP. ISSN 0044–4510, ZETFA7.

Journals Abstracted Completely

A selected number of journals listed in category 001 (periodicals) are central to the subject scope of *Astronomy and Astrophysics Abstracts*. Depending on their relevance, almost all papers of the journals listed below are abstracted in our service.

Acta Astron.
Acta Astron. Sin.
Acta Astrophys. Sin.
Acta Cosmologica
Am. Assoc. Variable Star Obs. Bull.
Annu. Rev. Astron. Astrophys.
Archaeoastronomy (U.K.)
Astrofizika
Astron. Astrophys.
Astron. Astrophys. Rev.
Astron. Astrophys., Suppl. Ser.
Astron. Ges., Abstr. Ser.
Astron. J.
Astron. Nachr.
Astron. Q.
Astron. Tidsskr.
Astron. Tsirk.
Astron. Vestn.
Astron. Zh.
Astronomia UAI
Astronomie
Astrophys. J.
Astrophys. J., Lett.
Astrophys. J., Suppl. Ser.
Astrophys. Lett. Commun.
Astrophys. Space Sci.
Aust. J. Astron.

BAV Rundbrief
BBSAG Bull.
Bol. Asoc. Argent. Astron.
Br. Astron. Assoc. Circ.
Bull. Am. Astron. Soc.
Bull. Assoc. Fr. Obs. Etoiles Variables
Bull. Astron. Inst. Czech.
Bull. Astron. Soc. India
Bull. Inf. Cent. Données Stellaires

Celest. Mech. Dyn. Astron.
Chin. Astron. Astrophys.
Ciel
Ciel Terre
Circ. Inf.
Comments Astrophys.

Earth, Moon, Planets
Exp. Astron.

Fundam. Cosmic Phys.

G. Astron.
Gen. Relativ. Gravitation
GEOS Circ.

I.A.P.P.P. Commun.
IAU Circ.
Icarus
Inf. Bull. Variable Stars
Int. Comet Q.
Ir. Astron. J.
Issled. Solntsa Krasnykh Zvezd
Itogi Nauki Tekh., Ser. Astron.

J. Am. Assoc. Variable Star Obs.
J. Astrophys. Astron.
J. Br. Astron. Assoc.
J. Hist. Astron.
J. Korean Astron. Soc.
J. R. Astron. Soc. Can.

Kinematika Fiz. Nebesn. Tel
Komet. Tsirk.
Komety Meteory

Mem. Soc. Astron. Ital.
Mercury
Messenger
Meteoritics
Meteoritika
Minor Planet Bull.
Minor Planet Circ.
Mitt. Astron. Ges.
Mon. Not. R. Astron. Soc.
Mon. Notes Astron. Soc. S.Afr.

Nablyud. Iskusstv. Nebesn. Tel
Nauchn. Inf.
News Lett. Astron. Soc. N.Y.

Observatory
Occultation Newsl.
Orion
Orione

Perem. Zvezdy
Perem. Zvezdy, Prilozh.
Pis'ma Astron. Zh.
Postępy Astron.
Probl. Kosm. Fiz.
Proc. Astron. Soc. Aust.
Prog. Astron.
Publ. Astron. Soc. Jpn.
Publ. Astron. Soc. Pac.
Publ. Variable Star Sect., R. Astron. Soc. N.Z.

Q. J. R. Astron. Soc.

Rev. Astron.
Rev. Mex. Astron. Astrofis.
Rev. Mod. Astron.

Sky Telesc.
Sol. Phys.
Soln. Dannye Byull.
South. Stars
Sov. Sci. Rev., Sect. E
Space Sci. Rev.
Sterne
Sterne Weltraum
Strolling Astron.

Trans. IAU

Vasiona
Vestn. Kiev. Univ. Astron.
Vistas Astron.

Yamamoto Circ.

Zenit
Zvaigžnota Debess

Publications of Observatories and Astronomical Institutes

Reports, communications, publications, and numbered series of reprints of observatories and astronomical institutes (listed in category 001) which are scanned completely in our service are listed below.

AAO Newsl.
Abastumanskaya Astrofiz. Obs., Byull.
Anglo-Aust. Telesc., Annu. Rep.
Ann. Shanghai Obs., Acad. Sin.
Annu. Rep. BIPM Time Sect.
Annu. Rep. IERS
Archenhold-Sternw. Berlin-Treptow, Sonderdr.
Archenhold-Sternw. Berlin-Treptow, Vortr. Schr.
Astrofiz. Issled. Izv. Spets. Astrofiz. Obs.
Astron. Bull., Carter Obs.
Astron. Inst. Univ. Brno, Contrib.
Astron. Inst. Univ. Brno, Publ.
Astron. Pap.
Astron. Rechen-Inst. Heidelb., Mitt., Ser. A
Astron. Rechen-Inst. Heidelb., Mitt., Ser. B

BAV Mitt.
BIPM Circ. T
Bol. Astron. Obs. Madr.
Bol. Obs. Astron. Quito
Bol. Obs. Ebro
Bol. R. Inst. Obs. Armada
Boyden Obs., Occas. Publ.
Boyden Obs., Repr.
Bull. Astron.
Bull. Crimean Astrophys. Obs.
Bull. Inf. Cent. Données Stellaires
Bull. Obs. Astron. Belgr.
Bull. Spec. Astrophys. Obs. – North Caucasus
Byull. Inst. Astrofiz.
Byull. Inst. Teor. Astron.

Carter Obs., Repr. Ser.
Cartes Synoptiques
Cent. Astron. Sci. Spat., Obs. Sol.
Circ. Czech. Obs., Time Latitude
Circ. Stn. Astron. Int. Latitudine, Carloforte-Cagliari, Ser. A
Commun. Konkoly Obs.
Commun. Univ. Lond. Obs.
Contrib. Astron. Obs. Skalnaté Pleso
Contrib. Bosscha Obs.
Contrib. Dept. Astron., Univ. Tokyo
Contrib. Inst. Argent. Radioastron.
Contrib. Nicholas Copernicus Obs. Planetarium Brno
Contrib. Nizamiah Japal-Rangapur Obs.
Contrib. Van Vleck Obs.

Data Rep. Hydrogr. Obs., Ser. Astron. Geod.
Dept. Radio Space Sci., Onsala Space Obs., Res. Rep.
Dtsch. Hydrogr. Inst. Hamb., Jahresber.

ESO Annu. Rep.
ESO Conf. Workshop Proc.
ESO Sci. Rep.
ESO Tech. Rep.

Gemini

Hvar Obs. Bull.

IERS Bull. A
IERS Bull. B
IERS Tech. Notes
Inst. Astrofis. Canarias, Mem.
Inst. Astron. Astrophys. Tech. Univ. Berlin, Mitt.
Inst. Astron. Geod., Univ. Madr., Publ.
Inst. Astron., Univ. Camb., Annu. Rep.
Inst. Astrophys., Univ. Liège, Annu. Rep.
Izv. Astron. Ehngel'gardt. Obs.
Izv. Glav. Astron. Obs. Pulkovo
Izv. Krym. Astrofiz. Obs.

Kapteyn Astron. Inst., Annu. Rep.
Kodaikanal Obs. Bull.

Latitude Circ.
Lick Obs. Bull.

Max-Planck-Inst. Kernphys., Jahresber.
Messenger
Mitt. Archenhold-Sternw. Berlin-Treptow
Mitt. Geod. Inst. Rheinischen Friedrich-Wilhelms-Univ. Bonn
Mitt. Karl-Schwarzschild Obs. Tautenburg
Mitt. Lohrmann-Obs. Tech. Univ. Dresden
Mitt. Satell.-Beobachtungsstn. Zimmerwald
Mitt. Sonnenobs. Kanzelhöhe
Mitt. Sternw. Sonneberg
Mitt. Universitätssternw. Graz
Mitt. Univ.-Sternw. Jena
Mitt. Veränderliche Sterne
Mitt. Zentralinst. Phys. Erde
MPE Rep.
MPE Repr.

Natl. Astron. Obs. (Jpn.), Repr.
Natl. Radio Astron. Obs., Repr., Ser. A
Natl. Radio Astron. Obs., Repr., Ser. B
Nizamiah Rangapur Obs. Dept. Astron. Osmania Univ., Repr.
NRAO Workshops

Obs. Astron. Córdoba, Tirada Aparte
Obs. R. Belg., Commun., Sér. A
Obs. R. Belg., Commun., Sér. B
Occas. Rep. R. Obs., Edinb.
Oss. Astrofis. Catania, Pubbl.

Pubbl. Stn. Astron. Int. Latitudine Carloforte-Cagliari, Nuov. Ser.
Publ. Astron. Dept. Eötvös Univ.
Publ. Astron. Inst. Czech. Acad. Sci.
Publ. Astron. Opservatorije Beogr.
Publ. Astrophys. Obs. Potsdam
Publ. Beijing Astron. Obs.
Publ. Debrecen Heliophys. Obs.
Publ. Debrecen Heliophys. Obs., Heliogr. Ser.
Publ. Dep. Astron. Univ. Beogr.
Publ. Dom. Astrophys. Obs.
Publ. Ege Univ. Obs.
Publ. Istanbul Univ. Obs.
Publ. Natl. Astron. Obs. Jpn.
Publ. Obs. Astron. Strasbourg, Sér. Astron. Sci. Hum.
Publ. Obs. Ebro
Publ. Purple Mt. Obs.
Publ. Shaanxi Astron. Obs.
Publ. Spéc. Cent. Données Stellaires
Publ. U.S. Nav. Obs., Second Ser.
Publ. Warner Swasey Obs.
Publ. Yunnan Obs.

Q. Bull. Sol. Act.

R. Greenwich Obs., Annu. Rep.
R. Greenwich Obs., Bull.
R. Obs. Edinb., Res. Facilities
Rep. Natl. Astron. Obs. Jpn.
Rep. Ser., Dept. Phys. Sci., Univ. Turku
Rutherford Appleton Lab., Rep.

SAAO Newsl.
S.Afr. Astron. Obs., Annu. Rep.
S.Afr. Astron. Obs., Circ.
Sendai Astron. Rap.
Sol. Maps Act.
Sol. Radio Data
Soobshch. Byurakan. Obs.
Soobshch. Spets. Astrofiz. Obs.
Space Telesc. Sci. Inst., Newsl.
ST-ECF Newsl.

Tartu Astrofüüs. Obs. Publ.
Tartu Astrofüüs. Obs. Teated
Telesc. Naz. Galileo, Tech. Rep. (Asiago)
Theor. Pap.
Time Freq. Serv., Bull.
Time Serv. Bull.
Tr. Astrofiz. Inst. (Alma-Ata)
Tr. Astron. Obs. (Leningrad)
Tr. Glav. Astron. Obs. Pulkovo, Ser. 2
Tr. Gos. Astron. Inst. im. Shternberga
Tr. Inst. Teor. Astron., Leningrad
Tr. Kazan. Gorod. Astron. Obs.
Tr. Tashkent. Astron. Obs.
Tsirk. Astron. Inst. (Tashkent)
Tsirk. Astron. Obs. L'vov
Tsirk. Shemakh. Astrofiz. Obs.
Turku Univ. Obs., Informeto
Turku Univ. Obs., Informo

Univ. Tex., Monogr. Astron.
Univ. Tex., Publ. Astron.
Upps. Astron. Obs. Ann.
Upps. Astron. Obs., Rep.
U.S. Nav. Obs., Circ.
U.S. Nav. Obs., Time Serv. Publ., Ser. 4
U.S. Nav. Obs., Time Serv. Publ., Ser. 14
Uttar Pradesh State Obs., Repr.

Vatican Obs. Publ.
Vatican Obs. Publ., Spec. Ser., Studi Galileiani
Veröff. Archenhold-Sternw. Berlin-Treptow
Veröff. Astron. Rechen-Inst. Heidelb.
Veröff. Remeis-Sternw. Bamberg
Veröff. Sternw. Pulsnitz
Veröff. Sternw. Sonneberg
Veröff. Wilhelm-Foerster-Sternw.
Vilniaus Astron. Obs. Biul.

002 Bibliographical Publications, Documentation, Catalogues, Data Bases

002.001 **La base de données extragalactiques de Lyon–Meudon revisitée.**
G. Paturel, A. M. Garcia.
Bull. Inf. Cent. Données Stellaires, No. 38, p. 3 – 6 (Dec 1990). – See Abstr. 012.001 for the main entry.
After a short historical recall, a description of the Lyon–Meudon Extragalactic Database is given, highlighting the basic principle that only raw data have to be stored. The different ways of distribution are summarized.

002.002 **The NASA/IPAC extragalactic database.**
G. Helou.
Bull. Inf. Cent. Données Stellaires, No. 38, p. 7 – 8 (Dec 1990). – See Abstr. 012.001 for the main entry.
The current version of NASA/IPAC extragalactic database provides positions, names, and basic data for extragalactic objects, as well as related bibliographic references, and notes from catalogs and other publications. This version also offers for browsing the abstracts of recent articles of extragalactic interest from the major journals.

002.003 **La situation de l'extragalactique au CDS (Juin 1990).**
P. Dubois.
Bull. Inf. Cent. Données Stellaires, No. 38, p. 9 – 12 (Dec 1990). – See Abstr. 012.001 for the main entry.
There exist three astronomical databases with extragalactic data: NED at IPAC in the United–States, the database of Lyon–Meudon, SIMBAD at the CDS in France. A comparison is made between these databases in order to examine the possible collaborations between them.

002.004 **Flamsteed stars not contained in the Yale "Catalogue of Bright Stars".**
W. P. Bidelman.
Bull. Inf. Cent. Données Stellaires, No. 38, p. 13 – 17 (Dec 1990).

002.005 **An updated list of keywords for indexed access to the list of CDS catalogues.**
O. Malkov, F. Bonnarel.
Bull. Inf. Cent. Données Stellaires, No. 38, p. 21 – 28 (Dec 1990).

002.006 **Recent aspects of the bibliography of astronomical objects.**
F. Ochsenbein, P. Dubois.
Bull. Inf. Cent. Données Stellaires, No. 38, p. 29 – 34 (Dec 1990).
This paper presents a brief statistical summary of the distribution of quotations about astronomical objects found in the literature over the period 1985 – 1989.A table lists the most "popular" objects over this period; the evolution over the period covered of some objects listed in this table is discussed. The contribution of such popular objects to the general bibliography is quantified.

002.007 **Computerizing and updating the "First dictionary of the nomenclature of celestial objects". I. New presentation and files.**
S. Borde, M.–C. Lortet.
Bull. Inf. Cent. Données Stellaires, No. 38, p. 35 – 39 (Dec 1990).
The dictionary of the nomenclature of celestial objects and its first supplement are presently being updated and computerized as VAX files. New data collected since 1986 are immediatly available.

002.008 **Computerizing and updating the "First dictionary of the nomenclature of celestial objects". II. New data collection since 1986.**
S. Borde, M.–C. Lortet.
Bull. Inf. Cent. Données Stellaires, No. 38, p. 41 – 67 (Dec 1990).
The authors present the new publications collected from 1986 to the end of Feb 1990 for the dictionary of nomenclature of celestial objects.

002.009 **The need for a radio astronomical data base: first results of a campaign.**
H. Andernach.
Bull. Inf. Cent. Données Stellaires, No. 38, p. 69 – 93 (Dec 1990).
The author is preparing a bibliography of papers on flux density data of radio sources in general, which estimates that the number of published flux densities is approaching 400,000 by mid–1990, and this number is growing at ever higher speed. A proposal for creating a publicly accessible pool for data on radio sources is described and has been distributed to about 200 individual radio astronomers and a few observatories. The results of this campaign are presented and future possibilities to achieve this goal are outlined.

002.010 **Catalogue of Lambda Bootis candidates.**
P. Renson, R. Faraggiana, C. Boehm.
Bull. Inf. Cent. Données Stellaires, No. 38, p. 137 – 149 (Dec 1990).
The table presented is a compilation of all the stars that, according to the literature, may have been considered at least once as probable λ Boo stars. 101 objects are listed, but some of them are probably not real λ Boo stars. The catalogue is intended as a basis for the selection of stars which present spectral anomalies similar to those of λ Boo itself.

002.011 **The catalogue of multiple and double stars of T–associations in a machine–readable form.**
G. N. Salukvadze.
Bull. Inf. Cent. Données Stellaires, No. 38, p. 165 – 170 (Dec 1990).

002.012 **The Catalogue of Trapezium–type multiple systems in machine–readable form.**
G. N. Salukvadze.
Bull. Inf. Cent. Données Stellaires, No. 38, p. 171 – 173 (Dec 1990).

002.013 **Close binaries observed polarimetrically.**
R. H. Koch.
Bull. Inf. Cent. Données Stellaires, No. 38, p. 175 – 193 (Dec 1990).

002.014 **The Washington Double Star (WDS) catalogs.**
C. E. Worley, G. G. Douglass.
Bull. Inf. Cent. Données Stellaires, No. 38, p. 195 – 196 (Dec 1990).

002.015 **Introduction to PPM–South. Preliminary version (May 1990).**
U. Bastian, S. Roeser.
Bull. Inf. Cent. Données Stellaires, No. 38, p. 197 – 207 (Dec 1990).
This is a short explanation of the preliminary version of the southern part of PPM catalogue. It shortly describes the way it is constructed and the contents of the catalogue on magnetic tape.

002.016 **Third list of corrections to the identifications of IRAS sources in Astron. & Astrophys. Suppl. 65, 607 and Astron. J. 98, 931.**
W. P. Bidelman.
Bull. Inf. Cent. Données Stellaires, No. 38, p. 215 - 216 (Dec 1990).

002.017 **Star catalogs and files available at the Stellar Data Center. Additions.**
Bull. Inf. Cent. Données Stellaires, No. 38, p. 231 - 233 (Dec 1990).

002.018 **Catalogs currently available on microfiches [from the Stellar Data Center].**
Bull. Inf. Cent. Données Stellaires, No. 38, p. 235 - 241 (Dec 1990).

002.019 **PPM: der Nachfolger des SAO-Katalogs.**
U. Bastian, S. Röser.
Sterne Weltraum, Jahrg. 30, Nr. 10, p. 592 - 596 (Oct 1991).

002.020 **A database of Ap stars.**
P. Renson, D. Kobi, P. North.
Astron. Astrophys., Suppl. Ser., Vol. 89, No. 1, p. 61 - 76 (Jul 1991).
A database for 3195 Ap stars, which can be used on an IBM-compatible personal computer, is described. It is essentially based on Renson's general catalogue of Ap and Am stars, but contains much additional information. It operates within the Dbase III Plus system.

002.021 **The IUE Final Archive - II.**
A. Talavera, J. D. Ponz.
ESA IUE Newsl., No. 38, p. 38 - 40 (Aug 1991).

002.022 **Index der KPM-Ausgaben 1 bis 18.**
J. Jahn.
KPM, Jahrg. 6, No. 18, p. 4 - 11 (Dec 1991).

002.023 **The complete Hard X-Ray Burst Spectrometer event list, 1980 - 1989.**
B. R. Dennis, L. E. Orwig, G. S. Kennard, G. J. Labow, R. A. Schwartz, A. R. Shaver, A. K. Tolbert.
National Aeronautics and Space Administration, Washington, DC (USA), NASA-TM—4332. 233 p. (Dec 1991).
This event list is a comprehensive reference for all hard X-ray bursts detected with the Hard X-Ray Burst Spectrometer on the Solar Maximum Mission from the time of launch on February 14, 1980 to the end of the mission in December 1989. 12,776 events were detected in the energy range ~30 to ~600 keV with the vast majority being solar flares. This list includes the start time, peak time, duration, and peak rate of each event.

002.024 **An atlas of optical counterparts of low mass X-ray binaries. I. The northern sample.**
M. Gottwald, H. Steinle, U. Graser, W. Pietsch.
Astron. Astrophys., Suppl. Ser., Vol. 89, No. 2, p. 367 - 388 (Aug 1991). Also published as report MPE—195(prepr.), 48. p. (Nov 1990).
The authors present the first part of an atlas of optical counterparts of low mass X-ray binaries. It comprises those sources that can be observed from a site at moderate northern latitudes. The declination threshold is $\delta > -22°$ so that the sample of low mass X-ray binaries is referred to as the Northern Sample. In total this part of the atlas includes 19 identified and 9 unidentified objects. The latter define the list of bright X-ray sources being possible low mass binary candidates but lacking any obvious optical counterpart. The information presented in this atlas is compiled from various papers from the past two decades. Where appropriate, the most recent X-ray source positions and error radii are quoted. Some of them have not been published before or were not reported in the context of optical identifications. The maps of the atlas were obtained with a CCD camera at the focus of the 1.23 m, the 2.2 m and the 3.5 m telescope at the German-Spanish Astronomical Center/Calar Alto, as uniformly as possible, for the complete Northern Sample.

002.025 **UBV photometry of stars whose positions are accurately known. VI.**
T. Oja.
Astron. Astrophys., Suppl. Ser., Vol. 89, No. 2, p. 415 - 419 (Aug 1991).
UBV photometry is presented for about 1000 stars of the Bright Star Catalogue and the faint extension of the FK5.

002.026 **General catalogue of Ap and Am stars.**
P. Renson, M. Gerbaldi, F. A. Catalano.
Astron. Astrophys., Suppl. Ser., Vol. 89, No. 3, p. 429 - 434 (Sep 1991).
A catalogue of all the Ap stars and Am stars ever discovered is presented. It contains 6684 objects. However some of them, marked with a / on the left of the corresponding line, have been wrongly considered as Ap or Am stars. For many other ones, with a ? on the left of their line, it is not yet well established that they belong to this group of stars. For each object the various identificators and the main observational data (magnitude, colour, spectral type, duplicity, periodicity, magnetic field, projected rotational velocity, ...) are given. This catalogue is available on magnetic tape from the Centre de Données Stellaires (Strasbourg Observatory, France).

002.027 **The interstellar lines catalogue.**
B. García.
Astron. Astrophys., Suppl. Ser., Vol. 89, No. 3, p. 469 - 527 (Sep 1991).
A bibliographical catalogue of the interstellar lines has been prepared (2003 stars in our Galaxy, 86 stars in the Magellanic Clouds and 41 extragalactic objects, and 371 bibliographical references. The catalogue is available at the CDS on magnetic tape.

002.028 **Positional reference stars in the Magellanic Clouds.**
J. Périé, L. Prévot, M. Rousseau, Y. Peyrin, A. Robin.
Astron. Astrophys., Suppl. Ser., Vol. 90, No. 1, p. 1 - 17 (Oct 1991).
From a total of 27 multicolour plates obtained with the ESO Schmidt telescope in the direction of the Magellanic Clouds, the authors have determined the equatorial coordinates of 926 stars (mainly of galactic origin) at the mean epoch T = 1978.4 with an overall accuracy characterized by the mean values of the O-C quantities, $S_a = 0.35''$ and $S_d = 0.38''$, calculated from the coordinates of the Perth reference stars. These values are larger than the accuracy expected for primary standard stars; however they allow to consider the new positions as those of reliable secondary standard stars; the published positions correspond, in any case, to an unquestionable improvement of the quality of the coordinates provided in the current catalogues. This work represents an "astrometric step" in the starting of a "Durchmusterung" of the Magellanic Clouds organized by de Boer (1988, 1989).

002.029 **50 volumes of Astronomy and Astrophysics Abstracts: an interesting statistic.**
A. G. D. Philip.
Bull. Am. Astron. Soc., Vol. 23, No. 2, p. 876 (1991). Abstract. - See Abstr. 012.008 for the main entry.

002.030 **A directory of astronomy, space sciences and related organizations of the world.**
A. Heck.
Astron. Astrophys., Suppl. Ser., Vol. 90, No. 1, p. 103 - 104 (Oct 1991).
A directory of astronomy, space sciences and related organizations of the world (ASpScROW 1991) has been compiled. It is made available as CDS Publication n° 16 through the Strasbourg Data Centre.

002.031 **Santiago declination catalogue. II. A declination catalogue of 493 FK5 stars (equinox J2000.0).**
G. Carrasco, P. Loyola.
Astron. Astrophys., Suppl. Ser., Vol. 90, No. 1, p. 109 – 120 (Oct 1991).
The positions in declination of 493 FK5 stars, observed with the Repsold Meridian Circle at Cerro Calán National Astronomical Observatory, during the period 1979 to 1983, are given. The average of the mean square error, for the whole catalogue, of one observation is $\bar{\varepsilon}_\delta = \pm 0\overset{''}{.}08$. The epoch of observation of the catalogue is 1981.3.

002.032 **Kinematical observations of ordinary spiral galaxies: a bibliographical compilation.**
R. L. M. Corradi, M. Capaccioli.
Astron. Astrophys., Suppl. Ser., Vol. 90, No. 1, p. 121 – 159 (Oct 1991).
The authors present a collection of data extracted from 280 papers reporting observations of the kinematics of 245 non-barred spiral galaxies. Information is provided on the type of observations (instruments, spectral lines used, etc.) and the derived geometrical and kinematical parameters of the galaxies (major axis position angle, inclination, heliocentric systemic velocity, maximum extension of the kinematical measurements, etc...). In addition, whenever possible, a "mean" rotation curve has been considered, from which was derived the maximum rotational velocity of the galaxy and a parameter describing the essential shape of the rotation curve within r_{25}. Histograms illustrating the distribution of morphological types, inclinations, extensions of the kinematical measurements, and maximum rotational velocities account for the statistical properties of this sample of spiral galaxies.

002.033 **An atlas of normal and active galaxies from IUE archive.**
D. Calzetti, A. L. Kinney, R. C. Bohlin, N. Panagia, R. Wyse.
Bull. Am. Astron. Soc., Vol. 23, No. 2, p. 893 – 894 (1991). Abstract. – See Abstr. 012.008 for the main entry.

002.034 **A database of IUE spectra of active galactic nuclei.**
G. Pike, J. Saken, R. Edelson, A. Koratkar, A. Kinney.
Bull. Am. Astron. Soc., Vol. 23, No. 2, p. 894 (1991). Abstract. – See Abstr. 012.008 for the main entry.

002.035 **A database of IUE spectra of active galactic nuclei – the results for quasars.**
A. Koratkar, A. Kinney, R. Bohlin, R. Edelson, J. Saken, G. Pike.
Bull. Am. Astron. Soc., Vol. 23, No. 2, p. 894 (1991). Abstract. – See Abstr. 012.008 for the main entry.

002.036 **The NSSDC Data Archiving and Distribution Service (NDADS).**
J. E. Gass, M. E. Van Steenberg, J. Behnke, J. L. Green.
Bull. Am. Astron. Soc., Vol. 23, No. 2, p. 907 – 908 (1991). Abstract. – See Abstr. 012.008 for the main entry.

002.037 **Astronomical Data Center CD–ROM: selected astronomical catalogs, Volume 1.**
S. E. Gessner, L. E. Brotzman, J. M. Mead, M. E. Van Steenberg.
Bull. Am. Astron. Soc., Vol. 23, No. 2, p. 908 (1991). Abstract. – See Abstr. 012.008 for the main entry.

002.038 **Second generation IRAS sky intensity product.**
S. Wheelock, J. Chillemi, N. Gautier, D. Gregorich, D. Kester, H. McCallon, C. Oken, J. White, T. Chester.
Bull. Am. Astron. Soc., Vol. 23, No. 2, p. 908 (1991). Abstract. – See Abstr. 012.008 for the main entry.

002.039 **The Second HST Guide Star Photometric Catalog.**
L. Siciliano, M. Postman, M. Shara, D. Rehner, C. Sturch.
Bull. Am. Astron. Soc., Vol. 23, No. 2, p. 908 (1991). Abstract. – See Abstr. 012.008 for the main entry.

002.040 **Catalogue of blue stars between the Magellanic Clouds.**
S. Demers, M. J. Irwin.
Astron. Astrophys., Suppl. Ser., Vol. 91, No. 1, p. 171 – 182 (Nov 1991).
The authors present a list of 1679 blue stars located in a region of the southern sky between the Magellanic Clouds. Equatorial coordinates and photographic V and B–V are given for stars brighter than V = 18.5. The instrumental magnitudes obtained from UKST IIIa–J and IIIa–F plates are calibrated with nearly two hundred stars with CCD B and V observations. This list can be used for follow–up identification, spatial distribution and other studies of young OB stars belonging to the Magellanic Clouds.

002.041 **An Einstein Observatory catalog of B–type stars.**
F. R. Harnden Jr., F. Grillo, S. Sciortino, G. Micela, G. S. Vaiana.
Bull. Am. Astron. Soc., Vol. 23, No. 2, p. 913 (1991). Abstract. – See Abstr. 012.008 for the main entry.

002.042 **The UVSP data archive.**
J. B. Gurman.
Bull. Am. Astron. Soc., Vol. 23, No. 2, p. 925 (1991). Abstract. – See Abstr. 012.008 for the main entry.

002.043 **The archive data set for the gamma–ray spectrometer on SMM.**
K. Levenson.
Bull. Am. Astron. Soc., Vol. 23, No. 2, p. 925 (1991). Abstract. – See Abstr. 012.008 for the main entry.

002.044 **Publication practices in various sciences.**
H. A. Abt.
Bull. Am. Astron. Soc., Vol. 23, No. 2, p. 961 – 962 (1991). Abstract. – See Abstr. 012.008 for the main entry.

002.045 **Index to Nos. 1 – 38 of the CDS Bulletin.**
C. Jaschek.
Bull. Inf. Cent. Données Stellaires, No. 39, p. 1 – 93 (1991).

002.046 **Determination of positions and proper motions of stars from the plates of the northern sky photographic survey program.**
A. I. Yatsenko.
Kinematika Fiz. Nebesn. Tel, Tom 7, No. 4, p. 56 – 60 (Jul–Aug 1991). In Russian. English translation in Kinematics Phys. Celest. Bodies, Vol. 7, No. 4.
The method of reduction of plates measurements of the northern sky photographic survey programs for the determination of positions and proper motions of stars has been described. The systematic differences of positions and proper motions of stars (AGK3–PPM) have been investigated.

002.047 **Catalogue of proper motions of stars in the North Galactic pole region.**
L. K. Pakulyak.
Kinematika Fiz. Nebesn. Tel, Tom 7, No. 4, p. 61 – 66 (Jul–Aug 1991). In Russian. English translation in Kinematics Phys. Celest. Bodies, Vol. 7, No. 4.
The reduction of 43 plates from the Oxford and Paris sections of the Astrographic Catalogue resulted in the catalogue of 3304 stars for the North Galactic pole region with positional error of 0.2″ and the magnitude error in the range of $0.1 - 0.4^m$. Proper motions of 1561 stars were determined with that catalogue and the previous standard region catalogue the difference of epochs being $\sim$100 years. The proper motion error is of 0.003″/year for both coordinates. The proper motions magnitude equation was investigated and accounted for.

002.048 **Spectroscopic investigations of stellar objects from the Second Byurakan Sky Survey. II. The fields centered on $\alpha = 09^h50^m$, $\delta = +55°00'$ and $\alpha = 11^h30^m$, $\delta = +59°00'$.**
D. A. Stepanyan, V. A. Lipovetskij, A. I. Shapovalova, L. K. Erastova, V. O. Chavushyan.
Astrophysics, Vol. 33, No. 2, p. 411 – 418 (Sep – Oct 1990). English translation of Astrofizika, Tom 33, Vyp. 2, p. 199 – 212 (Oct 1990).
Data are given on 49 stellar objects from two fields of the SBS Survey with coordinates of their centers $\alpha = 09^h50^m$, $\delta = +55°00'$ and $\alpha = 11^h30^m$, $\delta = +59°00'$. The authors have found 21 QSOs, four Seyfert galaxies, and three emission galaxies. The remaining objects are white dwarfs and hot subdwarfs of various types. Three continuum objects have been found. They give the basic parameters of the emission lines and other data for all quasistellar objects and also scans of the overwhelming majority of the QSOs. Scans of some typical white dwarfs and subdwarfs are also given.

002.049 **Bibliography and program notes on close binaries.**
A. Yamasaki, O. Demircan, H. Drechsel, T. J. Herczeg, V. G. Karetnikov, C. D. Scarfe, M. Vetesnik, H. L. Malasan (eds.).
Bibliogr. Program Notes Close Bin., No. 53.
International Astronomical Union. Commission 42. 26 p. (Jan 1992).
Material published by September 15, 1991.

002.050 **First Byurakan Spectral Sky Survey. Blue stellar objects. II. Region $\delta = +39°$.**
G. V. Abramyan, V. A. Lipovetskij, A. M. Mikaelyan, D. A. Stepanyan.
Astrophysics, Vol. 33, No. 2, p. 418 – 426 (Sep – Oct 1990). English translation of Astrofizika, Tom 33, Vyp. 2, p. 213 – 221 (Oct 1990).
This paper gives the second list of blue stellar objects of the second part of the First Byurakan Spectral Sky Survey. The objects are situated in the strip $+37° \leqslant \delta \leqslant +41°$, $13^h30^m \leqslant \alpha \leqslant 18^h30^m$ and $21^h30^m \leqslant \alpha \leqslant 24^h00^m$. The list contains data on 103 blue stellar objects, of which 64 have been discovered for the first time. A preliminary classification of the objects has been made. The statistics of the objects in the $\delta = +39°$ region has been established, and the list has been compared with the PG and Kiso surveys in this part of the sky. A statistical correlation between the preliminary classification of the authors and the widely adopted spectral classification has been established.

002.051 **Solar variability data on CD–ROM NGDC–05.**
H. E. Coffey, D. C. Wilkinson, J. A. McKinnon, C. C. Abston, J. H. Allen.
Bull. Am. Astron. Soc., Vol. 23, No. 2, p. 1054 (1991). Abstract. – See Abstr. 012.012 for the main entry.

002.052 **The Solar Maximum Mission data analysis center.**
J. B. Gurman.
Bull. Am. Astron. Soc., Vol. 23, No. 2, p. 1063 (1991). Abstract. – See Abstr. 012.012 for the main entry.

002.053 **The archive data set for the Gamma–Ray Spectrometer on SMM.**
K. Levenson.
Bull. Am. Astron. Soc., Vol. 23, No. 2, p. 1063 (1991). Abstract. – See Abstr. 012.012 for the main entry.

002.054 **A study of the publishing activity of astronomers since 1969.**
E. Davoust, L. D. Schmadel.
Scientometrics, Vol. 22, No. 1, p. 9 – 39 (1991).
This is a statistical analysis of the publishing activity of astronomers worldwide, as measured by the number of papers and monographs published in the period 1969–87. Both the astronomical literature and the number of authors publishing in the field are increasing rapidly. The distribution of productivity among astronomers is followed in time, thus revealing the evolution of research methods and publication strategies in the past 19 years. Fourteen "superproductive" astronomers, who published over 150 papers in 15 years, and the subsample of French astronomers active in 1986 are investigated separately.

002.055 **The revised catalogue of DC–dwarfs (a variant of the year 1987).**
G. M. Beskin, S. N. Mitronova.
Astrofiz. Issled. Izv. Spets. Astrofiz. Obs., Tom 31, p. 36 – 79 (1991). In Russian. English translation in Bull. Spec. Astrophys. Obs. – North Caucasus.
Using the literature data a list of 123 objects with apparent proper motions, the spectra of which have no lines, is compiled. Their coordinates, kinematic, photometric and spectral characteristics are presented. The analysis of parameters of objects from Nesterenko's and Chuprina's list (1976) with detected lines has shown that DC–dwarfs are mostly white dwarfs of different spectral classes with weakened lines, but not cold DB–dwarfs, as it was suspected earlier. The study of *UBV*–characteristics and photographic data allowed us to select 24 DC–dwarfs with spectra different from pure black–hole ones and which are, probably, variable on time–scales of months–years. These objects are of interest as probable black holes and hence are first candidates for their superrapid brightness variability to be searched.

002.056 **Compiled list of clusters of galaxies with measured redshifts.**
V. S. Lebedev, I. A. Lebedeva.
Astrofiz. Issled. Izv. Spets. Astrofiz. Obs., Tom 31, p. 91 – 127 (1991). In Russian. English translation in Bull. Spec. Astrophys. Obs. – North Caucasus.
The data for 1943 clusters of galaxies with measured radial velocities are presented.

002.057 **Specifications concerning names, designations, and nomenclature for astronomical radiation sources outside the Solar System. A message from the Working Group on Designations from Commission 5 of the International Astronomical Union.**
F. Spite.
Astrophys. Space Sci., Vol. 183, No. 1, p. 163 – 165 (Sep 1991).

002.058 **Second general catalogue of stars observed with the photoelectric astrolabes (GCPA2).**
Lu Lizhi.
Publ. Beijing Astron. Obs., No. 17, p. 1 – 58 (Jun 1991).
On the basis of the data observed with the photoelectric astrolabes in China during the period from 1982 to 1990, the second general catalogue of stars has been compiled. With magnitudes ranging from $0^m\!.1$ to $7^m\!.3$, the catalogue consists of 2606 stars from $\delta = -3°$ to $\delta = 69°$, in which 980 FK5 stars and 1626 GC stars are included. The mean precisions of 2524 $\Delta\alpha$'s and 1748 $\Delta\alpha$'s are ± 3.8 ms and $\pm 0''\!.064$, respectively. The mean epoch of the catalogue is 1988.0. Finally, systematic corrections of (CAT–FK5) are given.

002.059 **Fourth preliminary catalogue of stars observed with the photoelectric astrolabe (PACP4).**
Lu Lizhi.
Publ. Beijing Astron. Obs., No. 17, p. 59 – 104 (Jun 1991).
Using the data observed with the photoelectric astrolabe Mark II of the Beijing Astronomical Observatory during the period from 1984 – 1990, Fourth Preliminary Catalogue of Stars (PACP4) has been compiled. It consists of 1969 stars included 650 FK5 stars and 1319 GC stars. The mean precisions of $\Delta\alpha$'s and $\Delta\delta\alpha$'s are ± 4.1 ms and $\pm 0''\!.067$, respectively. The mean epoch is 1986.8. Finally, systematic corrections of (PACP4–FK5) are given.

002.060 **Third preliminary catalogue of stars' right ascension observed with the photoelectric transit instrument (PPCP3).**
Lu Lizhi.
Publ. Beijing Astron. Obs., No. 17, p. 105 – 119 (Jun 1991).
On the basis of the data observed with the Photoelectric Transit Instrument of the Beijing Astronomical Observatory during the period from 1980 through 1984, the third preliminary catalogue of stars on right ascension observed with the Photoelectric Transit Instrument (PPCP3) has been compiled. The relations between the residuals and the magnitude, and the spectral type of FK4 stars are discussed. There are 441 stars in this catalogue. The mean precisions of $\Delta\alpha$'s is ± 2.3 ms. Finally, systematic corrections of (PPCP3–FK4) are given.

002.061 **Fourth preliminary catalogue of stars' right ascension observed with the photoelectric transit instrument (PPCP4).**
Wang Ray, Lu Lizhi.
Publ. Beijing Astron. Obs., No. 17, p. 120 – 132 (Jun 1991).
On the photon counter data observed with Photoelectric Transit Instrument of Beijing Astronomical Observatory during the period from 1988 to 1990, corrections of right ascensions of 324 stars are given. The mean precision is 4.7 ms. The mean observational epoch is 1989.7. The relations of residuals with magnitudes and spectral types are discussed and given respectively. Finally, the systematic corrections (PPCP4–FK4) are also analysed.

002.062 **Automated information system for the Lunar Nomenclature data base.**
V. V. Shevchenko, S. G. Pugacheva, K. I. Dekhtyareva, T. P. Skobeleva.
Astron. Vestn., Tom 25, No. 5, p. 578 – 592 (Sep – Oct 1991). In Russian. English translation in Solar Syst. Res., Vol. 25, No. 5.
The whole complex information about toponymy, morphology, selenographical coordinates and dimensions of the lunar objects was included in the Lunar Nomenclature data base. The terminology used for lunar names is approved by the IAU. The Lunar Nomenclature data base is constructed in the form of logical structure of the relational type. The paper reports the results of statistics of the data base automated system. The data of the Lunar Nomenclature were systematized in the following categories: type of lunar relief, origin and year awarding of the name, bibliographical information about famous scientists which surnames were immortalized in the names of lunar objects. The map of density of named craters was drawn for various diameters and analysis of a load of named lunar craters was implemented for maps scale to be 1:5000000, 1:1000000, 1:250000. The number of named lunar objects is 1920.

002.063 **The Second Palomar Sky Survey.**
I. N. Reid, C. Brewer, R. J. Brucato, W. R. McKinley, A. Maury, D. Mendenhall, J. R. Mould, J. Mueller, G. Neugebauer, J. Phinney, W. L. W. Sargent, J. Schombert, R. Thicksten.
Publ. Astron. Soc. Pac., Vol. 103, No. 665, p. 661 – 674 (Jul 1991).
The authors describe the main characteristics of the Second Palomar Observatory Sky Survey, currently being taken using the Oschin telescope on Palomar Mountain. The limiting magnitudes of the POSS II plates are directly comparable with those of the SERC/ESO southern surveys, at $B_J = 22.5$ mag, $R_c = 20.8$ mag, and $I_c = 19.5$ mag. They discuss the main modifications made to the telescope and the photographic methods employed in the course of the survey and compare the POSS I and POSS II plates.

002.064 **Comparison of the Catalogues KCB and FK5.**
G. M. Blank, N. S. Blinov, I. I. Vdovenko, E. N. Fedoseev.
Astron. Tsirk., No. 1543, p. 39 (Mar – Apr 1990). In Russian.
Differences of the coordinates of stars in KCB and FK5 Catalogues are analyzed with epochs 1960 and 2000.

002.065 **Early-type emission-line stars with large infrared excesses.**
Dong Yisun, Hu Jingyao.
Chin. Astron. Astrophys., Vol. 15, No. 3, p. 275 – 283 (Sep 1991). English translation of Acta Astrophys. Sin., Vol. 11, No. 2, p. 172 – 180 (Apr 1991). See Abstr. 53.002.056.
The authors give a catalogue of early-type emission-line stars with V–[25] $\geqslant 8$, obtained by cross-identification between Wackerling's catalogue and the IRAS catalogue of point sources. A study of the distribution in space of the stars show that these stars belong to the extreme Population I, thus the present compilation provides a rather complete sample for further study of the evolution of pre- and post-main sequence stars of medium and high masses.

002.066 **On the "Ekat" catalogue and prospects of its application in photographic observations of minor planets.**
D. D. Polojentsev (*D. D. Polozhentsev*), Kh. I. Potter, L. I. Yagudin, R. F. Zalles.
1. Spain–USSR Workshop on Positional Astronomy and Celestial Mechanics, p. 63 – 66 (1991). – See Abstr. 012.025 for the main entry.
Precise photographic observations of minor planets that are being made with the final aim of improving the parameter of the fundamental catalogue require a good astrometric tie to the existing fundamental system. The catalogue "Ekat" (Equatorial Catalogue) being compiled at present by the astronomers of the Pulkovo Observatory and Observatorio Astronomico Boliviano–Sovietico can help to solve this problem. The observations are made with the expedition astrograph (D = 23 cm, F = 230 cm, field 5° x 5°) at Tarija, Bolivia and Ordubad, Azerbajan, USSR. More than 60% of observations have been performed from the declination +10 in the southward zones. The measurements with the automatic measuring machine "Fantasia" and their reduction will be made at Pulkovo. The catalogue will contain about 1 million stars down to 12 mag in the declination zone from –20° to +20°. The "Ekat" catalogue will enable to use long focus astrograph with the field up to 0°.5 for observations of minor planets, which will considerably improve the precision of their coordinate determinations.

002.067 **A search for steep-spectrum sources in the UTRAO, GB and RC catalogues.**
P. Reich, N. S. Soboleva, A. V. Temirova, E. Fürst, N. E. Gol'neva.
Astron. Zh., Tom 68, Vyp. 4, p. 681 – 694 (Jul–Aug 1991). In Russian. English translation in Sov. Astron., Vol. 35, No. 4.
A comparison of three high-sensitivity catalogues at meter and centimeter wavelengths, UTRAO (365 MHz), GB (4.85 GHz) and RC (3.94 GHz) is made. As a result, spectral index distributions are constructed for $\sim$2000 common objects in the declination strip $5 \pm 2°$, and a list of radio sources, containing about 300 objects with steep spectra ($\alpha > 1.1$) is compiled. It is shown that there is a real distinction between the spectral index distributions for point-like and double sources; there is also a difference in the values of the mean spectral index for these classes of objects. A systematic shift between the coordinates of the UTRAO and GB catalogues: $\approx 1.3^s$ in right ascension and –14″ in declination.

002.068 **Some results of determining the SAO catalogue systematic errors from observations of minor planets.**
Yu. V. Batrakov, E. N. Makarova, V. A. Shor, A. S. Vashkevich.
1. Spain–USSR Workshop on Positional Astronomy and Celestial Mechanics, p. 133 – 136 (1991). – See Abstr. 012.025 for the main entry.
The values of parameters which represent the systematic errors of the SAO catalogue have been obtained using positional observations of minor planets (1) Ceres and (2) Pallas. The results are tentative and further improvement is needed.

002.069 **New reference catalogues for photographic observations of minor planets.**
L. I. Yagudin.
1. Spain–USSR Workshop on Positional Astronomy and Celestial Mechanics, p. 137 (1991). Abstract. – See Abstr. 012.025 for the main entry.

002.070 **An extragalactic data base. IV. Errors and misprints in catalogues of galaxies.**
G. Paturel, C. Petit, N. Kogoshvili, P. Dubois, L. Bottinelli, P. Fouqué, R. Garnier, L. Gouguenheim.
Astron. Astrophys., Suppl. Ser., Vol. 91, No. 3, p. 371 – 397 (Dec 1991).
This paper collects more than 1500 errors or misprints which plague proper identification of galaxies in the following catalogues: NGC, IC, MCG, UGC, CGCG, ESO, MK, SGC, KARA, FAIR, RC2, ZW–compact, UGCA, TULLY, NGC2000, ARAK, VCC, DDO, VV, SBS, PGC. The Lyon–Meudon Extragalactic Database is consistent with the corrections which are collected. Some difficult cases are discussed individually.

002.071 **The activities of the NSSDC and ADC in the area of archiving photometric data.**
W. H. Warren Jr.
Meeting on Precision Photometry: Astrophysics of the Galaxy, p. 275 – 282 (1991). – See Abstr. 012.027 for the main entry.

002.072 **Sky Catalogue 2000.0. Volume 1: Stars to magnitude 8.0.**
A. Hirshfeld, R. W. Sinnott, F. Ochsenbein.
Cambridge University Press, Cambridge (UK). 2nd edition 700 p. (1991). ISBN 0–521–41743–0. Sky Publishing Corp., Cambridge, MA (USA).
This wholly revised second edition of Sky Catalogue 2000.0, Vol.1, first published in 1982, contains 50,071 stars which is almost 11% more than the 45,269 of the first edition. Substantial improvements have been made in several of the data columns. The enhancements have become possible through the extensive use of the SIMBAD data base and this catalogue is the largest printed subset of SIMBAD currently available. The catalogue summarizes the essential characteristics of the 50,071 stars of visual magnitude 8.0 and brighter. Positional data are given for the new standard epoch 2000.0; the modern UBV photometric system has been adopted to express stellar magnitudes and colors; whenever possible, complete spectral–type information is listed, including luminosity class and peculiarity code; distances are given along with the methods by which they were derived; multiple and variable stars have been flagged. In addition, each star has reference numbers from SAO and HD catalogues.

002.073 **Erratum: "An early TYCHO astrometric catalogue" [Astrophys. Space Sci., Vol. 177, Nos. 1 – 2, p. 109 – 114 (Mar 1991)].**
E. Høg, G. Bässgen, U. Bastian, D. Egret, J. L. Halbwachs.
Astrophys. Space Sci., Vol. 184, No. 2, p. 331 (Oct 1991). See Abstr. 53.002.061.

002.074 **Search for steep–spectrum sources in the UTRAO, GB, and RC catalogs.**
P. Reich, N. S. Soboleva, A. V. Temirova, E. Fürst, N. E. Gol'neva.
Sov. Astron., Vol. 35, No. 4, p. 335 – 341 (Jul 1991). Current Physics Microform No.: 9203X0521. English translation of Astron. Zh., Tom 68, Vyp. 4, p. 681 – 694 (1991).
Three high–sensitivity catalogs at meter and centimeter wavelengths are compared: the UTRAO (365 MHz), GB (4.85 GHz), and RC (3.94 GHz) catalogs. As a result, distributions of spectral indices are constructed for ~2000 common objects in the declination band $5 \pm 2°$ and a list of radio sources containing about 300 objects with a steep spectral index ($\alpha > 1.1$) is compiled. It is shown that the distributions of spectral indices for point–like and double objects actually differ, as do the mean values of spectral index for these classes of objects. A systematic offset is found between coordinates in the UTRAO and GB catalogs: $\sim 1.3^s$ in right ascension and –14″ in declination.

002.075 **New high density reference star catalogues, PPM and ACRS.**
M. Miyamoto.
Astron. Her., Vol. 84, No. 7, p. 224 – 228 (1991). In Japanese.

002.076 **The Einstein Observatory Extended Medium–Sensitivity Survey. II. The optical identifications.**
J. T. Stocke, S. L. Morris, I. M. Gioia, T. Maccacaro, R. Schild, A. Wolter, T. A. Fleming, J. P. Henry.
Astrophys. J., Suppl. Ser., Vol. 76, No. 3, p. 813 – 874 (Jul 1991).
The optical identifications of the Einstein Extended Medium Sensitivity Survey (EMSS) are presented. At present over 96% of the 835 X–ray sources in the survey have been successfully identified in the following proportions: active galactic nuclei (QSOs, quasars, and Seyferts), 51.1%; BL Lacertae objects, 4.3%; clusters of galaxies, 12.2%; normal galaxies, 2.1%; cooling flow galaxies, 0.6%; Galactic stars, 25.8%; and unidentified, 3.9%. All of the EMSS sources are identified with previously known classes of X–ray emitters except for five sources now identified with cooling flow galaxies whose novel attributes are described. The Galactic stars are primarily normal dwarf stars but also include RS CVn and W UMa binary systems, two FK Comae stars, five cataclysmic variables, one white dwarf, and one X–ray binary candidate. The extragalactic sources are predominantly AGNs, which range from low–z, low–luminosity Seyfert galaxies to high–z QSOs ($z > 2$), but also include distant clusters of galaxies, BL Lac objects and nearby, normal galaxies. Most of the individual optical counterparts are previously unknown objects and so constitute large statistical samples independent of previous selection methods. The details of the optical identification process are described including the observational criteria used to establish what constitutes a plausible optical counterpart.

002.077 **The Case Low–Dispersion Northern Sky Survey. XII. A region in southern Canes Venatici.**
P. Pesch, N. Sanduleak, C. B. Stephenson.
Astrophys. J., Suppl. Ser., Vol. 76, No. 4, p. 1043 – 1053 (Aug 1991). With plates 43 – 52.
Positions, estimated magnitudes, and finding charts (when necessary) are provided for 194 blue and/or emission–line galaxies, 15 H II regions in seven galaxies, 157 unresolved blue and/or emission–line objects, including QSO candidates, and 32 known and suspected blue stars in a 107 deg^2 region in Canes Venatici encompassed by $12^h15^m < R.A. < 14^h00^m$ and $+33° < Decl. < +38°$ (1950). The objects, whose blue magnitudes are mostly within the range 15 – 18, were identified on low–dispersion objective–prism plates taken with the Burrell Schmidt telescope at Kitt Peak.

002.078 **Catalogue of magnitudes and colour indices of stars in direction of nebula M20.**
I. A. Dubyago.
Izv. Astron. Obs. Ehngel'gardt., Vyp. 54, p. 53 – 78 (1989). In Russian.
1000 stars were measured in UBV system in a field of 1 square degree in the direction of $l = 7°$. A number of stars, V magnitudes, (B–V) and (U–B) colour indices, identification maps are given.

002.079 **Bibliography (April – June 1989).**
E. B. Carling, Z. Kopal, F. B. Waranius (eds.).
Earth, Moon, Planets, Vol. 53, No. 3, p. 239 – 284 (Jun 1991).
Contents: 1. The Moon. 2. The planets. 3. Jupiter. 4. Satellites of Jupiter. 5. Mars. 6. Satellites of Mars. 7. Mercury. 8. Neptune. 9. Satellites of Neptune. 10. Pluto. 11. Satellites of Pluto. 12. Saturn. 13. Satellites of Saturn. 14. Uranus. 15. Satellites of Uranus. 16. Venus.

002.080 **Bibliography (April – June 1989).**
E. B. Carling, Z. Kopal, F. B. Waranius (eds.).
Earth, Moon, Planets, Vol. 54, No. 1, p. 67 – 87 (Jul 1991).
Contents: 18. Space utilization. 19. Asteroids. 20. Comets.

002.081 **Bibliography (April – June 1989).**
E. B. Carling, Z. Kopal, F. B. Waranius (eds.).
Earth, Moon, Planets, Vol. 54, No. 2, p. 151 – 192 (Aug 1991).
Contents: 20. Comets (continued). 21. Meteorites. 22. Miscellaneous.

002.082 **Bibliography (July – September 1989).**
E. B. Carling, Z. Kopal, F. B. Waranius (eds.).
Earth, Moon, Planets, Vol. 55, No. 1, p. 75 – 96 (Oct 1991).
Contents: 1. The Moon (including aspects of the Earth–Moon system). 2. Planets (articles about more than one body). 3. Jupiter.

002.083 **Catalogue of physical and kinematical parameters for main sequence F stars within 80 parsecs of the Sun.**
V. A. Marsakov, Yu. G. Shevelev.
Astron. Tsirk., No. 1545, p. 11 – 12 (Aug – Sep 1990). In Russian.
The effective temperature, log g, [Fe/H], distance, tangential velocity, space velocity components and elements of orbit for about 6000 F stars are calculated on the basis of uvbyβ photometry, proper motions and radial velocity from the literature.

002.084 **Comparison of star proper motions in the PPM (Heidelberg) and SSKASD (Kiev) catalogues in regions of the sky with galaxies.**
S. P. Rybka.
Kinematics Phys. Celest. Bodies, Vol. 7, No. 2, p. 51 – 54 (1991). English translation of Kinematika Fiz. Nebesn. Tel, Tom 7, No. 2, p. 58 – 61 (1991).
Comparison of the proper motions of stars in the PPM and SSKASD catalogues yields precession–constant corrections: $\Delta k = -0.0035 \pm 0.0008''$/yr and $\Delta n = -0.0008 \pm 0.00008''$/yr. The systematic differences between these catalogues are represented by a series of orthogonal functions that are products of the terms of a Fourier series and Hermite and Legendre polynomials.

002.085 **Spectroscopic and photometric investigation of objects in selected fields with the 6–m telescope.**
V. L. Afanas'ev, V. V. Vlasyuk, S. N. Dodonov, H. Lorenz, V. Yu. Terebizh.
Bull. Spec. Astrophys. Obs. – North Caucasus, Vol. 32, p. 23 – 50 (1990). English translation of Astrofiz. Issled. Izv. Spets. Astrofiz. Obs., Tom 32, p. 31 – 65 (1990).
A description of a photometric and spectroscopic investigation of objects of fields SA68 ($00^h14^m51.03^s$, $+15°36'19.1''$), M82 ($9^h53^m28.23^s$, $+69°52'12.7''$), and SA57 ($13^h06^m03.69^s$, $+29°36'31.8''$) and reductions of the data are given. The photometric samples of the objects are complete to 21.75^m in B for SA68, to 22.25^m in B for field M82, and to 23^m in B for field SA57. Spectroscopic investigations of starlike objects have been performed for 64% of objects of field SA68, for 63% of objects of field M82, and for 56% of objects of field SA57. Photometric and spectroscopic catalogues of data on objects of the fields are given.

002.086 **Photometric investigation of a sample of objects complete to B = 21^m in the field 9^h40^m, +50°.**
V. L. Afanas'ev, V. V. Vlasyuk, S. N. Dodonov, O. N. Spiridonova, G. Richter, S. B. Novikov, O. I. Bugaenko, Yu. A. Shokin, V. V. Shul'ga.
Bull. Spec. Astrophys. Obs. – North Caucasus, Vol. 32, p. 57 – 88 (1990). English translation of Astrofiz. Issled. Izv. Spets. Astrofiz. Obs., Tom 32, p. 73 – 105 (1990).
A photometric investigation of a sample of objects complete to $B = 21^m$ in a 1 sq. degree field with center coordinates 9^h40^m, $+50°$ was carried out on the basis of photographs obtained on the 1–m telescope on Mt. Maidanak. The number of objects in the sample is 2222. The method of observation and data processing is described. The results of the photometric investigation are gathered into a catalogue. The sample obtained is the basic one for a subsequent spectroscopic investigation of the objects.

002.087 **The Asiago supernovae catalog.**
R. Barbon, E. Cappellaro, M. Turatto.
10. Santa Cruz Summer Workshop in Astronomy and Astrophysics: Supernovae, p. 720 – 723 (1991). – See Abstr. 012.052 for the main entry.

002.088 **The first Byurakan Spectral Sky Survey. Blue stellar objects. III. Zone $\delta = +43°$.**
G. V. Abramyan, V. A. Lipovetskij, A. M. Mikaelyan, D. A. Stepanyan.
Astrofizika, Tom 33, Vyp. 3, p. 345 – 350 (Dec 1990). In Russian. English translation in Astrophysics, Vol. 33, No. 3. With 7 plates.
The third list of blue stellar objects of the second part of the First Byurakan Spectral Sky Survey is given. The objects are located in the zone $+41° \leqslant \delta \leqslant +45°$, $0^h \leqslant \alpha \leqslant 3^h40^m$ and $6^h \leqslant \alpha \leqslant 13^h30^m$. The list contains data for 120 blue stellar objects among which 79 are new. A preliminary classification of objects is made.

002.089 **Analysis of observational catalogues and formation of the bright stars compiled catalogue system.**
N. D. Kanivets, A. A. Molotaj, V. V. Tel'nyuk–Adamchuk, N. A. Chernega.
International Seminar on Problems of Astrometry and Space Geodynamics, p. 55 – 66 (1991). In Russian. – See Abstr. 012.049 for the main entry.
The characteristics are presented of 18 differential catalogues used to form the Kiev compiled catalogue of positions of 5115 bright stars of the whole sky. An analytical comparison of individual catalogues with the first version of the compiled one has been carried out. Systematic catalogues weights resulting from these comparisons are taken as the basis of the compiled catalogue system formation.

002.090 **Spectral investigations of the Second Byurakan Sky Survey objects. Stellar objects. III. Fields $\alpha = 12^h22^m$, $\delta = +55°00'$ and $\alpha = 15^h30^m$, $\delta = +59°00'$.**
D. A. Stepanyan, V. A. Lipovetskij, A. I. Shapovalova, L. K. Erastova, V. O. Chavushyan.
Astrofizika, Tom 33, Vyp. 3, p. 351 – 362 (Dec 1990). In Russian. English translation in Astrophysics, Vol. 33, No. 3.
The data for 53 stellar objects from the Second Byurakan Survey (SBS) are presented. 25 QSO, two Sy1, one Sy2 and two emission–line galaxies are found. The remainder objects are white dwarfs and hot subdwarfs. Five continuum objects are found. The main parameters of emission lines and the other data for all QSO's and the scans of the main part of QSO's are presented. Some scans of typical white dwarfs and subdwarfs are also given.

002.091 **A Be star atlas of far UV and optical high–resolution spectra.**
V. Doazan, G. Sedmak, M. Barylak, L. Rusconi.
European Space Agency, Paris (France), ESA–SP—1147. 382 p. (Nov 1991). ISBN 92–9092–025–4. Price Dfl. 75.00.
The aim of this Atlas is to provide an overview of the spectral characteristics of Be stars in both the far UV and optical wavelength ranges and of their variable behaviour when data are available. Be stars have been frequently observed with the International Ultraviolet Explorer (IUE) (Boggess et al., 1978). The IUE archives contain an enormous amount of information on these objects. After one decade, in 1988, more than 2000 high resolution spectra obtained in the short and long wavelength ranges for about 190 Be stars were stored in the IUE Data Bank. Although a large number of these observations have been published in several publications (see the ESA and NASA Newsletter bibliography), these data were reported for different specific studies. The authors felt that it was time to collect all these observations and present a selection of them in a condensed form in an Atlas. In order to complete these far UV data they also

provide for the majority of these stars line–profiles of Hα and/or Hβ obtained for the long term Be star program of V.D. and collaborators.

002.092 **Bibliography (July – September 1989).**
E. B. Carling, Z. Kopal, F. B. Waranius (eds.).
Earth, Moon, Planets, Vol. 55, No. 3, p. 233 – 299 (Dec 1991).
Contents: 4. Satellites of Jupiter. 5. Mars. 6. Satellites of Mars. 8. Neptune. 9. Satellites of Neptune. 10. Pluto. 12. Saturn. 13. Satellites of Saturn. 14. Uranus. 15. Satellites of Uranus. 16. Venus. 18. Space utilization. 19. Asteroids. 20. Comets. 21. Meteorites. 22. Miscellaneous.

002.093 **Analysis of papers published from the Isaac Newton Group of Telescopes 1984 – 88.**
W. L. Martin, P. G. Murdin, K. P. Tritton.
Astrophys. Lett. Commun., Vol. 28, No. 2–4, p. 69 – 71 (Sep 1991). – See Abstr. 012.072 for the main entry.
A bibliography of papers published with results from the Isaac Newton Group of Telescopes on La Palma has been analysed to show trends on usage.

002.094 **The Case Low–Dispersion Northern Sky Survey.**
P. Pesch.
Objective–prism and other surveys – a meeting in memory of Nicholas Sanduleak, p. 3 – 16 (1991). – See Abstr. 012.074 for the main entry.
The Case Low–Dispersion Northern Sky Survey became N. Sanduleak's primary scientific activity for the last ten years of his life. Contents: 1. Background. 2. Description of the survey. 3. Current status of the survey. 4. Future plans.

002.095 **A catalogue of known EUV sources.**
B. Y. Welsh, P. W. Vedder.
1. Berkeley Colloquium on Extreme Ultraviolet Astronomy, p. 483 – 484 (1991). – See Abstr. 012.045 for the main entry.

002.096 **Surveys for cool supergiants in the Galaxy.**
D. J. MacConnell.
Objective–prism and other surveys – a meeting in memory of Nicholas Sanduleak, p. 17 – 26 (1991). – See Abstr. 012.074 for the main entry.
Most of the roughly two hundred K–M supergiants known presently in the Galaxy were found on wide–angle photographs taken through objective prisms. These stars are important signposts for galactic structure studies and for evolutionary studies of massive stars. They are important sources of mass loss and are often surrounded by extensive dust shells, and, the case of Sanduleak –69°202 to the contrary notwithstanding, they are expected to be the immediate precursors of most type II supernovae. The author discusses earlier surveys, the present large one, and makes some comments about a list of unpublished candidates of Nick Sanduleak.

002.097 **Planetary nebulae. A practical guide and handbook for amateur astronomers.**
S. J. Hynes.
Willmann–Bell, Richmond, VA (USA). 270 p. (1991). ISBN 0–943396–30–1. Price US$ 24.95.
This handbook on planetary nebulae includes 253 finder charts and a catalogue of 1340 objects.
Contents: 1. The discovery of planetary nebulae. 2. The astrophysics of planetary nebulae. 3. Planetary nebula mimics. 4. Amateur observation of planetary nebulae. 5. Catalogue and atlas of planetary nebulae.

002.098 **Lunar tables and programs from 4000 B.C. to A.D. 8000.**
M. Chapront–Touzé, J. Chapront.
Willmann–Bell, Richmond, VA (USA). 170 p. (1991). ISBN 0–943396–33–6. Price US$ 19.95.
This book contains two sets of tables for the geocentric motion of the Moon. The first set provides time–dependent expansions of the longitude, latitude, and radius vector of the Moon, referred to the mean ecliptic and equinox of date. These tables include a large number of terms in order to ensure a sufficient precision in the present period for most users. The second set of tables provide expansions of the semimajor axis, eccentricity, sine of half the inclination, longitude of perigee, longitude of node, and mean longitude, referred to the mean ecliptic and equinox of date. These tables are intended for the user who does not need high precision. Microcomputer program listings in FORTRAN, BASIC and PASCAL are provided to implement the tables and formulae presented in the book.

002.099 **The space distribution of supergiants and other stars in the Michigan Spectral Catalogue, Volumes 1 – 4.**
N. Houk.
Objective–prism and other surveys – a meeting in memory of Nicholas Sanduleak, p. 27 – 32 (1991). – See Abstr. 012.074 for the main entry.
MacSpin software is used to display the space distribution of stars south of $\delta = -12°$ in the Michigan Spectral Catalogues. B, A and F – M supergiants are shown separately as well as combined. In contrast, F and G dwarf space distributions are also displayed.

002.100 **Photographic atlas for supernovae search. Vol. II: Declinations –20° to +20°.**
M. Lopez Alvarez.
Editorial Argenta Sarlep, Buenos Aires (Argentina). 433 p. (1990). ISBN 950–9332–60–7. Price US$ 39.95.
This atlas presents photos of all important galaxies in the equatorial belt, within the reach of modest amateur telescopes. On each plate, local stars up to magnitude 15.0 to 15.5 are included, in order to identify any new star appearing which will be suspected as a supernova.

002.101 **Thin prism survey for F and G stars at the south galactic pole.**
P. K. Lu.
Objective–prism and other surveys – a meeting in memory of Nicholas Sanduleak, p. 33 – 44 (1991). – See Abstr. 012.074 for the main entry.
Thin prism spectra with a dispersion of 1360 Å/mm were used to search for F and G stars in a study to determine the galactic gravitational force K(z). Prism plates covering approximately 100 sq. degrees centered on the SGP were obtained using the Michigan Curtis Schmidt telescope at CTIO. A catalog of identifications and positions for 3154 F and G star candidates in a 40 square degree area is now completed. The positions are based on PDS scans of a 6°.3 × 6°.3 UK Schmidt direct film centered on the SGP. The new thin–prism spectra yield a set of uniformly–selected F and G stars as a tracer population to a limiting magnitude of 16.0.

002.102 **A catalog of absorption in the spectra of quasi–stellar objects.**
V. Junkkarinen, A. Hewitt, G. Burbidge.
Astrophys. J., Suppl. Ser., Vol. 77, No. 2, p. 203 – 254 (Oct 1991).
This paper contains a compilation of basic information on QSO absorption systems. A total of 724 QSOs are listed in three tables. Table 1 contains 353 QSOs that have absorption–line systems with measured redshifts. The numerous Lyα forest systems based only on narrow hydrogen lines are excluded from this table. Absorption systems with multiple components are listed only once if all of the components fall within a span of 500 km s^{-1}. A total of 862 systems with narrow metal lines, which are considered to be reliably identified, are listed in Table 1. Table 1 contains an additional 260 systems consisting of less certain systems, damped Lyα systems, and Lyman limit systems that have no detected narrow metal lines. In Table 2 the authors list 146 broad absorption line QSOs. In Table 3 the authors give an inhomogeneous list of 240 QSOs in which absorption systems have been seen but not measured. Also shown are histograms of emission redshift, absorption redshift, and $(1+z_{em})/(1+z_{abs})$ for the reliably identified narrow metal line

systems. Several selection effects are apparent in this collection of data.

002.103 **A compilation of redshifts and velocity dispersions for Abell clusters (Epoch 1991.2).**
M. F. Struble, H. J. Rood.
Astrophys. J., Suppl. Ser., Vol. 77, No. 3, p. 363 – 377 (Nov 1991).
The authors present a list of redshifts for 758 Abell clusters, and velocity dispersions for 121, published as of 1991 March. The authors present another list of 33 Abell clusters with published redshifts, most of which are probably redshifts of foreground or background galaxies superposed on, or near, the Abell clusters. Over the past 4 years, the published number of redshifts of member galaxies in clusters increased by a factor of 2.0, but the number of cluster redshifts increased by only a factor of 1.3, while the number of cluster velocity dispersions derived from 50 or more redshifts of member galaxies increased by a factor of 3.8 (from 6 to 23).

002.104 **Highly reddened, probable supergiants found in the recent Milky Way infrared survey at the Warner and Swasey Observatory.**
C. B. Stephenson.
Objective–prism and other surveys – a meeting in memory of Nicholas Sanduleak, p. 155 (1991). Abstract. – See Abstr. 012.074 for the main entry.

002.105 **Spectroscopy of highly evolved stars in the Case Low Dispersion Northern Sky Survey.**
E. M. Sion, R. M. Wagner, S. G. Starrfield, J. Liebert.
Objective–prism and other surveys – a meeting in memory of Nicholas Sanduleak, p. 163 – 166 (1991). – See Abstr. 012.074 for the main entry.

002.106 **First Byurakan Spectral Sky Survey. Blue stellar objects. III. Region $\delta = +43°$.**
G. V. Abramyan, V. A. Lipovetskij, A. M. Mikaelyan, D. A. Stepanyan.
Astrophysics, Vol. 33, No. 3, p. 493 – 500 (Nov – Dec 1990). English translation of Astrofizika, Tom 33, Vyp 3, p. 345 – 350 (Dec 1990).
The third list of blue stellar objects of the second part of the First Byurakan Spectral Sky Survey is given. The objects are situated in the strip $+41° \leqslant \delta \leqslant +45°$, $0^h \leqslant \alpha \leqslant 3^h40^m$, and $6^h \leqslant \alpha \leqslant 13^h30^m$. The list contains data on 120 blue stellar objects, of which 79 have been discovered for the first time. A preliminary classification of the objects is given.

002.107 **Spectroscopic investigations of stellar objects from the Second Byurakan Sky Survey. III. The fields centered on $\alpha = 12^h22^m$, $\delta = +55°00'$ and $\alpha = 15^h30^m$, $\delta = +59°00'$.**
D. A. Stepanyan, V. A. Lipovetskij, A. I. Shapovalova, L. K. Erastova, V. O. Chavushyan.
Astrophysics, Vol. 33, No. 3, p. 501 – 508 (Nov – Dec 1990). English translation of Astrofizika, Tom 33, Vyp 3, p. 351 – 362 (Dec 1990).
Data are given on 53 stellar objects from two fields of the SBS survey with coordinates of their centers $\alpha = 12^h22^m$, $\delta = +55°00'$ and $\alpha = 15^h30^m$, $\delta = +59°00'$. The investigations have resulted in the discovery of 25 QSOs, two type 1 Seyfert galaxies, one type 2 Seyfert galaxy, and two emission galaxies. The remaining objects are white dwarfs and hot subdwarfs of various types. Five continuum objects were found. The main parameters of the emission lines and other data are given for all the quasi–stellar objets, and also scans of the overwhelming majority of the QSOs. Scans of some typical white dwarfs and subdwarfs are also given.

002.108 **PKSCAT90: the southern radio source database.**
R. E. Otrupcek, A. E. Wright.
Proc. Astron. Soc. Aust., Vol. 9, No. 1, p. 170 (1991). – See Abstr. 012.078 for the main entry.
The PKSCAT90 database consists of radio and optical data for 8264 radio sources. It covers all the sky south of declination +27°. Most of the galactic plane and the Magellanic Cloud regions are excluded from this catalogue but have been the subject of other specialised surveys. A few data errors in the initial PKSCAT90 version 1.00 released in March 1990 have been corrected in the current version 1.01 edition.

002.109 **The real stuff.**
A. Heck.
Applying fractals in astronomy, p. 201 – 210 (1991). – See Abstr. 003.023 for the main entry.
The present situation in the field of astronomical databases and archives is discussed. Shortcomings and possible future trends are mentioned, as well as some considerations on a future ideal environment for direct home applications on real and remotely located astronomical data.

002.110 **Catalogue of proper motions in the declination of stars of the Moscow zenith zone.**
L. P. Gribko.
Tr. Gos. Astron. Inst. im. Shternberga, Tom 62, p. 243 – 252 (1991). In Russian.
The present article presents the catalogue of proper motions in the declination of 426 stars of the Moscow zenith zone. The mean value of the mean square error of the proper motion of the individual star makes up $\pm 0.003''$.

002.111 **A concise dictionary of astronomy.**
J. Mitton.
Oxford University Press, Oxford (UK). 430 p. (1991). ISBN 0–19–853967–3. Price £ 12.95.
Containig over 2,400 entries, this wide–ranging dictionary encompasses all aspects of classical and modern astronomy.

002.112 **A catalogue of quasars and active nuclei. (5th edition).**
M.-P. Véron–Cetty, P. Véron.
ESO Sci. Rep., No. 10.
European Southern Observatory, Garching (Germany, F.R.). 199 p. (Oct 1991).
Because of the fast increase in the number of published quasars, the authors have prepared an updated version of their catalogue of quasars and active nuclei (Véron–Cetty and Véron, 1984, 1985, 1987, 1989) which now contains 6225 quasars and 1540 active galaxies (of which 575 are Seyfert 1), compared with 4169 quasars and 1108 Seyfert and related galaxies in the fourth edition. Like the fourth edition, it includes positions, redshift, photometry (U, B, V) and 6 and 11 cm flux densities. Now 162 BL Lac objects (117 in the fourth edition) are listed.

002.113 **Catalogue of stars measured in the Geneva Observatory photometric system (Rufener 1988). Documentation for the machine–readable version.**
W. H. Warren Jr.
National Aeronautics and Space Administration (NASA), Greenbelt, MD (USA). National Space Science Data Center (NSSDC) / World Data Center A for Rockets and Satellites, NSSDC/WDC–A–R/S—91–17. 22 p. (Jul 1991).
The machine–readable version of the catalog, as it is currently being distributed from the astronomical data centers, is described. The catalog is a compilation of photoelectric measurements made in the Geneva Observatory seven–color photometric system and it contains data for 29,397 stars. The data include star identifications (name, HR, HD, and DM number), equatorial coordinates, the normalized magnitudes U, V, $B1$, $B2$, $V1$, G, the color indices $B1$–$B2$, $B2$–$V1$, and $V1$–G, the system parameters d, Δ, g, m_2, and the magnitude in the UBV system. The numbers of measurements and standard deviations of the means are included separately for the colors and the V magnitude.

002.114 **Third Reference Catalogue of Bright Galaxies (de Vaucouleurs, de Vaucouleurs, Corwin, Buta, Paturel, and Fouqué 1991). Documentation for the machine-readable version.**
N. G. Roman, G. de Vaucouleurs, A. de Vaucouleurs, H. G. Corwin Jr., R. J. Buta, G. Paturel, P. Fouqué.
National Aeronautics and Space Administration (NASA), Greenbelt, MD (USA). National Space Science Data Center (NSSDC) / World Data Center A for Rockets and Satellites, NSSDC/WDC-A-R/S—91-18. 37 p. (Jul 1991).
The University of Texas has prepared a third edition of its catalogue of bright galaxies. This not only contains many more entries than the second edition (23,023) but substantially more information for each entry. This document describes the machine-readable version of that catalogue and provides some information about the contents.

002.115 **Sydney Southern Star Catalogue (King and Lomb). Documentation for the machine-readable version.**
N. G. Roman, D. S. King, N. R. Lomb.
National Aeronautics and Space Administration (NASA), Greenbelt, MD (USA). National Space Science Data Center (NSSDC) / World Data Center A for Rockets and Satellites, NSSDC/WDC-A-R/S—91-02. 20 p. (Aug 1991).
The machine-readable version of the *Sydney Southern Star Catalogue*, as produced at the Sydney Observatory, is described. This catalogue principally covers the range between −51° and −63°30′ and contains positions and proper motions for 29,926 stars. 3244 faint Astrographic stars were added to supplement the stars at fainter magnitudes. Positions in standard form (hours, minutes, and seconds of time for the right ascensions and degrees, minutes, and seconds of arc for the declinations) have been added to the machine readable version of this catalogue.

002.116 **NSSDC Data Listing.**
R. Horowitz (comp.).
National Aeronautics and Space Administration (NASA), Greenbelt, MD (USA). National Space Science Data Center (NSSDC) / World Data Center A for Rockets and Satellites, NSSDC/WDC-A-R/S—91-19. 129 p. (Aug 1991).
This document is intended to identify, in a highly summarized way, data available from the National Space Science Data Center (NSSDC). Most data are maintained as offline data sets (on magnetic tape or as film/print products of various sizes, or as disks) gathered from individual instruments carried on spacecraft; these comprise the Satellite Data Listing. Descriptive names, time spans, data form, and quantity of these data sets are identified in the listing, which is sorted alphabetically – first by spacecraft name and then by the principal investigator's or team leader's last name. Several data sets held at the NSSDC, not associated with individual spaceflight instruments, are identified in separate listings following the Satellite Data Listing. These data sets make up the Supplementary Data Listings and include composite spacecraft data sets, ground-based data, models, and computer routines.

002.117 **A bibliography of planetary geology and geophysics principal investigators and their associates, 1990 – 1991.**
National Aeronautics and Space Administration, Washington, DC (USA). NASA Office of Space Science and Applications, NASA-TM—4299. 50 p. (Oct 1991).
This document, a companion piece to NASA-TM—4300 (53.091.021), is a compilation of selected bibliographic data specifically relating to recent publications submitted by principal investigators and their associates, supported through the NASA Office of Space Science and Applications, Solar System Exploration Division, Planetary Geology and Geophysics Program.

002.118 **The global paleomagnetic database. Design, installation and use with ORACLE.**
J. Lock, M. W. McElhinny.
Surv. Geophys., Vol. 12, No. 4/5, p. 317 – 491 (Jul 1991).
The Global Paleomagnetic Database is the result of an international project sponsored by Division I of IAGA and funded through contributions from nine countries. It has been developed using the ORACLE Relational Database Management System as agreed at the first Workshop on databases in paleomagnetism and rock magnetism held in Czechoslovakia in June/July 1988. All published data up to the end of 1988 are included in the database and updates will occur every four years through the IAGA Division I Working Groups in Paleomagnetism and Rock Magnetism. This document outlines the structure of the database and provides details for installing and maintaining both the ORACLE software and the Global Database on a Personal Computer under MS-DOS. Examples of querying the database using SQL (Structured Query Language), now designated the official international language for database systems, are given.

002.119 **Radial velocities of dwarf stars in the solar neighborhood.**
A. A. Tokovinin.
Catalogue of stellar radial velocities. Catalogue of proper motions, p. 4 – 28 (1990). In Russian. – See Abstr. 003.032 for the main entry.
A catalogue of radial velocity observations of nearby K, M dwarfs is given. The velocities were measured by the correlation technique, with accuracy of 0.3 – 0.6 km/s. The method of observations and data reduction are briefly described. The observational program is outlined. The catalogue contains 1197 observations of 247 objects. The mean velocity, the mean width and the mean contrast of the correlation dip are calculated for each star.

002.120 **Catalogue of radial velocities of late-type stars.**
A. S. Rastorguev, A. M. Valitova, E. V. Glushkova, Yu. V. Kulagin, M. V. Pavlov, N. N. Samus'.
Catalogue of stellar radial velocities. Catalogue of proper motions, p. 29 – 69 (1990). In Russian. – See Abstr. 003.032 for the main entry.
The precise radial velocities of F5–M5 stars observed by a CORAVEL-type velocity spectrometer are given. The catalogue includes more than 1400 individual measurements for 1200 stars (high-luminosity stars, some cepheids, late-type stars in Kapteyn areas). The observations were made in 1987 – 1989.

002.121 **Catalogue of proper motions of high-luminosity O and B stars.**
A. M. Valitova, D. K. Karimova, E. D. Pavlovskaya.
Catalogue of stellar radial velocities. Catalogue of proper motions, p. 70 – 91 (1990). In Russian. – See Abstr. 003.032 for the main entry.
The proper motions of high-luminosity stars of O and B spectral types are given. Proper motions and coordinates are derived in the FK4 system for the equinox 1950.0. The FK4 proper motions contain the corrections for the new constants of precession.

002.122 **The IUE archive at the Canadian Astronomical Data Centre.**
D. R. Crabtree, D. Durand, G. Farrell, S. Gaudet.
J. R. Astron. Soc. Can., Vol. 85, No. 4, p. 217 (Aug 1991). Abstract. – See Abstr. 012.091 for the main entry.

002.123 **The precise positions and proper motions of 1980 GC stars.**
Li Zhenxing, Lu Chunlin.
Publ. Purple Mt. Obs., Vol. 10, No. 4, p. 288 – 351 (Dec 1991).
In this paper, the positions and proper motions of 1980 GC stars are determined using 9 catalogs. The procedure of compilation is described. The results are discussed.

002.124 **Establishment and query of text database for the study of radio astronomy.**
Zhuang Xuejuan.
Ann. Shanghai Obs., Acad. Sin., No. 12, p. 179 – 184 (1991). In Chinese.

002.125 **Tentative idea and practice of database of astrometry and geodynamics at Shanghai Observatory.**
Zhuang Xuejuan, Cheng Youfen, Li Zhengxin.
Ann. Shanghai Obs., Acad. Sin., No. 12, p. 185 – 187 (1991). In Chinese.

002.126 **Sur les publications des astronomes.**
C. Jaschek.
J. Astron. Fr., No. 40, p. 3 – 5 (Jul 1991).

002.127 **Comparison of catalogues through asteroid observations.**
J. G. Sanguin, R. L. Branham Jr.
Bol. Asoc. Argent. Astron., No. 36, p. 201 – 208 (1990). In Spanish. – See Abstr. 012.095 for the main entry.
The authors have obtained 525 observations, made by four observatories between 1965 and 1989, of 16 minor planets to study the behavior of the FK5 catalog near −30°.

002.128 **Black holes. A selected bibliography.**
A. Fraknoi.
Mercury, Vol. 20, No. 4, p. 109 – 110 (Jul – Aug 1991).

002.129 **The core journals and the literatures ageing law on Chinese astronomical research.**
Pang Xiaojun, Xu Lei, Xu Mianqin, Zhou Shihu, Ke Darong.
Ann. Shanghai Obs., Acad. Sin., No. 12, p. 191 – 199 (1991). In Chinese.

002.130 **Spectral investigations of the Second Byurakan Sky Survey objects. Stellar objects. IV.**
D. A. Stepanyan, V. A. Lipovetskij, V. O. Chavushyan, L. K. Erastova, A. I. Shapovalova.
Astrofizika, Tom 34, Vyp. 1, p. 5 – 12 (Feb 1991). In Russian. With 2 plates. English translation in Astrophysics, Vol. 34, No. 1.
The data for 26 quasistellar objects from the Second Byurakan Survey (SBS) located on the northern sky in the field $8^h < \alpha < 17^h$ and $+49° < \delta < +61°$ are given. The main parameters of emission lines and the other data for all QSO's and the scans of the 20 QSO's are presented.

002.131 **The First Byurakan Spectral Sky Survey. Blue stellar objects. IV. Zone $+41° \leqslant \delta \leqslant +45°$.**
G. V. Abramyan, V. A. Lipovetskij, A. M. Mikaelyan, D. A. Stepanyan.
Astrofizika, Tom 34, Vyp. 1, p. 13 – 20 (Feb 1991). In Russian. With 6 plates. English translation in Astrophysics, Vol. 34, No. 1.
The fourth list of blue stellar objects of the second part of the First Byurakan Spectral Sky Survey is given. The objects are situated in the zone with $+41° \leqslant \delta \leqslant +45°$, $13^h30^m \leqslant \alpha \leqslant 19^h10^m$, and $21^h50^m \leqslant \alpha \leqslant 24^h00^m$. The list contains data for 106 blue stellar objects among which 64 are new. A preliminary classification of objects is made. A comparison of the authors' preliminary classification with the generally used spectral classification is carried out and subsamples of FBS objects are singled out for searching QSO, WD and sd.

002.132 **A catalog of flare stars in Orion nebula region.**
R. Sh. Natsvlishvili.
Astrofizika, Tom 34, Vyp. 1, p. 107 – 139 (Feb 1991). In Russian. With 18 plates. English translation in Astrophysics, Vol. 34, No. 1.
The catalog and identification charts for 491 flare stars in the Orion nebula region discovered up to 1986 are presented. The results of a preliminary analysis of the data on these stars are given.

002.133 **The Second Byurakan Sky Survey. Spectra of galaxies. IV.**
D. A. Stepanyan, V. A. Lipovetskij, L. K. Erastova, A. I. Shapovalova.
Astrofizika, Tom 34, Vyp. 2, p. 205 – 211 (Apr 1991). In Russian. English translation in Astrophysics, Vol. 34, No. 2.
The results of spectroscopic observations of 32 galaxies from the Second Byurakan Survey are presented. The observations have been carried out with the 6 m telescope of SAO AS USSR and 2.6 m telescope of BAO AS of Armenia. Emission lines are found in the spectra of most of the investigated galaxies. The redshifts and luminosities of all galaxies have been determined. It is shown that SBS 1212 + 558 and SBS 1213 + 549 A are Sy1 galaxies and SBS 1220 + 544 is probably a Sy galaxy too. The galaxy SBS 0948 + 532 was classified as a BCDG.

002.134 **Spacecraft tables 1957 – 1990.**
J. Heyman.
Univelt, San Diego, CA (USA). 251 p. (1991). ISBN 0–912183–07–1. Price US$ 60.00.
This book presents a record of all space launches of all space-active countries from 1957 to 1990. In each case, the name and number designation, launch date, reentry date (when applicable) are given. Launches are organized by country, and by type of mission. If known, the purpose of the mission is indicated. The countries represented include the U.S., U.S.S.R., Argentina, Australia, Brazil, Bulgaria, Canada, China, France, Germany, Hong Kong, India, Indonesia, Israel, Italy, Japan, Luxembourg, Mexico, The Netherlands, Pakistan, Spain, Sweden, United Kingdom. Among organizations covered are NASA, Amsat, Intelsat, ESA, Interkosmos, NATO, Inmarsat, Entelsat, Arabsat.

002.135 **Stjernenes navn.**
T. S. Ringnes.
Astron. Tidsskr., Årg. 24, Nr. 4, p. 161 – 164 (1991).

On the occasion of the 100th anniversary of the "Carte du Ciel" and the "Astrographic Catalogue".
See Abstr. 004.048.

Christoph Clavius. Ein Namen– und Schriftenverzeichnis zu seinen *Opera mathematica*.
See Abstr. 005.025.

Objective–prism and other surveys. Proceedings.
See Abstr. 012.074.

The programme FON–OBZOR–S.
See Abstr. 013.066.

The programme MEGA: results and outlook.
See Abstr. 013.067.

The RGO plate library and measuring machine facilities.
See Abstr. 034.087.

ESA's on–line directory of space and earth–science data.
See Abstr. 036.014.

On the derivation of selection functions from redshift survey data.
See Abstr. 036.132.

First results of the photographic survey program of the northern sky.
See Abstr. 036.167.

Creation of the star position catalogue by five CdC plate–block processing.
See Abstr. 036.169.

The catalogue of geodetical stars observed at Yunnan Observatory.
See Abstr. 041.008.

The results and prospects of use of minor planets observations for improvement of the systems of the fundamental catalogues.
See Abstr. 041.011.

On the influence of atmospheric dispersion on the photographic positions of stars.
See Abstr. 041.022.

Transformation from FK4 system to FK5 system.
See Abstr. 043.014.

Keeping track of comets and minor planets.
See Abstr. 098.026.

Standard stars for the Alma–Ata catalog. Discussion.
See Abstr. 113.005.

JHKLM standard stars in the ESO system.
See Abstr. 113.015.

The extension of the Geneva catalogue.
See Abstr. 113.016.

BVR photoelectric sequences for selected fields in the Palomar Sky Survey and the magnitude–diameter relation.
See Abstr. 113.030.

Photometric provision for the photographic survey of the north sky and MEGA programmes.
See Abstr. 113.037.

The standards of stellar magnitudes and colours in the UBVR system in programme MEGA areas: observational results.
See Abstr. 113.038.

A standard stellar spectral sequence in the red/near–infrared: classes K5 to M9.
See Abstr. 114.094.

Speckle–interferometric observations of stars from the Gliese catalogue.
See Abstr. 118.005.

Supernova studies. VII. An atlas of light curves of supernovae type I.
See Abstr. 125.004.

A spectroscopic glance at the CfA supernovae atlas.
See Abstr. 125.065.

Catalog of SN 1987A polarimetry corrected for interstellar polarization.
See Abstr. 125.239.

Distribution and origin of high–velocity clouds. III. Clouds, complexes and populations.
See Abstr. 131.097.

Atlas of extinction curves derived from ultraviolet spectra of the TD–1 satellite.
See Abstr. 131.322.

The EXOSAT GSPC iron line catalogue.
See Abstr. 142.059.

Liste des étoiles Ap et Am dans les amas ouverts (Edition révisée).
See Abstr. 153.002.

A photometric study of the bright cloud B in Sagittarius. VII. 1165 new variable stars and 65 diffuse objects.
See Abstr. 155.020.

A few statistics from the catalog of polar–ring galaxies.
See Abstr. 157.192.

The QMW IRAS galaxy catalogue: a highly complete and reliable IRAS 60–μm galaxy catalogue.
See Abstr. 157.302.

The large, bright QSO survey, a digital objective–prism survey for quasars.
See Abstr. 159.114.

A catalog of galaxies in the Dorado cloud.
See Abstr. 160.022.

Statistical properties of X–ray clusters of galaxies.
See Abstr. 160.066.

Correlation functions of galaxies in the ESO–Uppsala catalog.
See Abstr. 161.211.

Objective determination of the large–scale distribution of galaxies.
See Abstr. 161.552.

003 Books

003.001 **The new cosmos.**
A. Unsöld, B. Baschek.
Springer, Berlin (Germany). 4.completely rev. ed 454 p. (1991). ISBN 3–540–52593–9. Price DM 118.00. ISBN 0–387–52593–9 (USA)

This book, the fourth, completely revised edition, is an English translation, by W. D. Brewer, of the German original (5th edition) "*Der neue Kosmos*", published in 1991. (See 003.008). Astronomy, astrophysics and space research have developed extensively and rapidly in the last few decades. The new opportunities for observation afforded by space travel, the development of high–sensitivity light detectors and the use of powerful computers have revealed new aspects of the fascinating world of galaxies and quasars, stars and planets. This edition of "The new cosmos" bears witness to this explosive development. It provides a comprehensive but concise introduction to all of astronomy and astrophysics. It stresses observations and theoretical principles equally, requiring of the reader only basic mathematical and scientific background knowledge.
Contents: 1. Introduction. 2. Classical astronomy. The solar system. 3. Astronomical and astrophysical instruments. 4. The Sun and stars. Astrophysics of individual stars. 5. Stellar systems: the Milky Way and other galaxies. Cosmogony and cosmology. 6. The origin of the solar system. The evolution of the Earth and of life.

003.002 **Physics of the inner heliosphere 2. Particles, waves and turbulence.**
R. Schwenn, E. Marsch (eds.).
Phys. Chem. Space., Vol. 21.
Springer, Berlin (Germany). 363 p. (1991). ISBN 3–540–52083–X. Price DM 178.00. ISBN 0–387–52083–X (USA).

Physics of the inner heliosphere gives for the first time a comprehensive and complete summary of our knowledge of the inner solar system. Using data collected over more than 11 years

by the Helios twin solar probes, one of the most successful ventures in unmanned space exploration, the authors have compiled 11 extensive reviews of the physical processes of the inner heliosphere and their relation to the solar atmosphere. Researchers and advanced students in space and plasma physics, astronomy, and solar physics will be surprised to see just how closely the heliosphere is tied to, and how sensitively it depends on, the Sun. Volume 2 deals with particles, waves, and turbulence, with chapters on: magnetic clouds, interplanetary slow shocks, kinetic physics of the solar wind plasma, waves and instabilities, MHD turbulence, energetic particles in the inner solar system.
The individual contributions within the subject scope of AAA are included in their corresponding categories – see abstracts 062.035, 074.012, 074.013, 078.004, 106.013, 106.014.

003.003 **Gauge field theories. An introduction with applications.**
M. Guidry.
Wiley, New York, NY (USA). 620 p. (1991). ISBN 0-471-63117-5.
The book is divided into three parts. Part I is an introduction to the general principles of relativistic quantum field theory. In Part II the essential ingredients of gauge fields for the weak, electromagnetic, and strong interactions are introduced. In Part III, the attention to the interface between modern elementary particle physics and "applied disciplines" such as nuclear physics, astrophysics, and cosmology is shifted. In the last two chapters topics that are of particular concern to nuclear physicists and cosmologists are addressed. The quark-gluon plasma that presumably characterized the universe microseconds after the big bang and that may be produced soon in laboratory experiments is considered, and one examines the part that gauge theories may have taken in producing some of the essential characteristics of our universe.

003.004 **Our universes.**
D. H. Wilkinson.
George B. Pegram Lect. Ser.
Columbia University Press, New York, NY (USA). 224 p. (1991). ISBN 0-231-07184-1. Price US$ 45.00.
This book on modern astrophysical cosmology is the author's look at the relationship between the physical universe (if such a thing exists) and our perception of it. He journeys to the frontier where science and philosophy meet.
Contents: 1. Our nature. 2. Our intellectual tools. 3. The big bang. 4. The goodness of fit.

003.005 **The Arabs and the stars. Texts and traditions on the fixed stars, and their influence in medieval Europe.**
P. Kunitzsch.
Collect. Stud. Ser., Vol. 307.
Variorum Reprints, Northampton (UK). 350 p. (1989). ISBN 0-86078-255-7. Price £ 47.50.
The important role played by "the Arabs" in the transmission and development of the sciences is well known. This volume presents a selection of 29 articles dedicated mainly to the knowledge of the fixed stars. The articles are reproduced in their original form (adapted in size to the page format here used), but taking the opportunity to correct various misprints and minor errors on the pages themselves.

003.006 **POSTPRINCIPIA. Gravitation for physicists and astronomers.**
P. Rastall.
World Scientific, Singapore (Singapore). 244 p. (1991). ISBN 981-02-0778-6. Price £ 22.00.
Contents: 1. Newtonian gravity. 2. Static gravitational fields. 3. Particle mechanics. 4. Weight and energy. 5. The postnewtonian theory. 6. The postnewtonian metric. 7. Postnewtonian fluid mechanics. 8. Motion of the centre of mass. 9. Exact theories of gravity.

003.007 **Differential geodesy.**
M. Hotine.
Springer, Berlin (Germany). 195 p. (1991). ISBN 3-540-53799-6. Price DM 174.00. ISBN 0-387-53799-6 (USA).
Apart from Hotine's work on mathematical geodesy, several previously unpublished reports are collected in this monograph, supplemented by editorial comments on these contributions and a complete bibliography of Hotine.
Contents: Foreword (*C. A. Whitten, B. H. Chovitz*). Editorial introduction (*J. Zund*). 1. Trends in mathematical geodesy (1964). 2. Adjustment of triangulation in space (1956). 3. Metrical properties of the Earth's gravitational field (1957). 4. Geodetic coordinate systems (1957). 5. A primer on non-classical geodesy (1959). 6. The third dimension in geodesy (1960). 7. Harmonic functions (1962). 8. Downward continuation of the gravitational potential (1967). 9. Curvature corrections in electronic distance measurements (1967). Bibliography of Hotine (*J. Nolton*). Hotine's *Mathematical geodesy* (*B. H. Chovitz*). Martin Hotine: friend and pro (*C. A. Whitten*).

003.008 **Reviews in Modern Astronomy 4.**
G. Klare (ed.).
Springer, Berlin (Germany). 294 p. (1991). ISBN 3-540-54245-0. Price DM 98.00. ISBN 0-387-54245-0 (USA).
The fourth yearbook of the Astronomische Gesellschaft (AG) contains 18 contributions, which were presented during the fall meeting of the AG at Freiburg/Breisgau in September 1990, on solar physics, solar system problems (such as dust rings around planets, and the solar wind), the first results of the Hipparcos and ROSAT space missions, protostellar and protoplanetary evolution, the supernova SN 1987A in the LMC, and dynamos in galaxies. Eugene Parker's Schwarzschild Lecture is dedicated to the topic "Convection, spontaneous discontinuities, and stellar winds and X-ray emission".
The individual contributions within the subject scope of AAA are included in their corresponding categories – see abstracts 034.068, 036.115, 051.030, 051.031, 062.097, 062.098, 062.099, 062.100, 064.060, 071.035, 072.086, 074.074, 074.075, 075.034, 091.012, 098.037, 116.022, 125.181.

003.009 **Der neue Kosmos.**
A. Unsöld, B. Baschek.
Springer, Berlin (Germany). 5.überarb.u.erw. ed. 466 p. (1991). ISBN 3-540-53757-0. Price DM 98.00.
Contents: 1. Einführung. 2. Klassische Astronomie. Das Planetensystem. 3. Astronomische and astrophysikalische Instrumente. 4. Sonne und Sterne. Astrophysik des einzelnen Sterns. 5. Sternsysteme: Milchstraße und Galaxien. Kosmogonie und Kosmologie. 6. Entstehung des Planetensystems. Entwicklung der Erde und des Lebens.
For the English translation of this fifth, revised and enlarged German edition see 003.001.

003.010 **Annual Review of Astronomy and Astrophysics. Volume 29.**
G. Burbidge, D. Layzer, A. Sandage (eds.).
Annu. Rev. Astron. Astrophys., Vol. 29, 711 p. (1991). Price US$ 53.00 (USA, Canada), US$ 58.00 (elsewhere).

003.011 **Elementary particles.**
I. S. Hughes.
Cambridge University Press, Cambridge (UK). 3. ed. 453 p. (1991). ISBN 0-521-40739-7.
This is the third edition of a text that is already well established as one of the standard undergraduate books on the subject of elementary particle physics. The author has updated the whole text in line with current particle nomenclature and has added material to cover important new developments. There is also a completely new major chapter on particle physics and cosmology.

003.012 **Paläomagnetismus und Archäomagnetismus.**
H. C. Soffel.
Springer, Berlin (Germany). 287 p. (1991). ISBN 3-540-53890-9. Price DM 78.00. ISBN 0-387-53890-9 (USA).

This introduction textbook deals with the methods of the field of paleomagnetism in a comprehensive manner. The geoscientific oriented application is based on the foundations of rock magnetism, as far as they are important far paleomagnetic working. The reader will become familiar with the various laboratory and measuring procedures, the possibilities of application and the limits of the method, to find his own estimation of the applicability. Theoretical foundations are connected with geological and archeological questions which reach from investigations of single rock samples or objects found till to the discussion of the continental drift. An extensive bibliography is added.

003.013 **The YOHKOH (SOLAR-A) mission. A series of invited contributions.**
Z. Švestka, Y. Uchida (eds.).
Kluwer, Dordrecht (Netherlands). 115 p. (1991). ISBN 0-7923-1532-4. Price Dfl. 160.00, US$ 89.00, £ 55.00. Reprinted from Sol. Phys., Vol. 136, No. 1, p. i – v, 1 – 110 (Nov 1991).

The SOLAR-A spacecraft was launched by the Institute of Space and Astronautical Science, Japan (ISAS) in August 1991, and, following an ISAS tradition, was renamed YOHKOH. This mission is dedicated principally to the study of solar flares, especially of high-energy phenomena observed in the X- and gamma-ray ranges. With a coordinated set of instruments including hard X-ray and soft X-ray imaging telescopes as well as spectrometers with advanced capabilities, it will reveal many new aspects of flares and help better understand their physics, supporting international collaborations with ground-based observatories as well as theoretical investigations. An overview of this mission, including the satellite, its scientific instruments, and its operation, is given.

003.014 **From the hill: the story of Lowell Observatory.**
R. Houk.
Northland Printing, Flagstaff, AZ (USA). 56 p. (1991).

003.015 **The aurora. Sun-Earth interactions.**
N. Bone.
Ellis Horwood Lib. Space Sci. Space Technol., Ser. Astron.
Ellis Horwood, New York, NY (USA). 156 p. (1991). ISBN 0-13-051392-X. Price £ 38.00.

Contents: 1. Introduction. 2. The aurora in history. 3. Scientific investigations of the aurora. 4. The active Sun. 5. The Earth's magnetosphere. 6. The polar aurorae. 7. The mid-latitude aurora. 8. Other effects. 9. Related high-atmosphere phenomena.

003.016 **The new solar system.**
J. K. Beatty, A. Chaikin (eds.).
Cambridge University Press, Cambridge (UK). 3. ed. 334 p. (1990). ISBN 0-521-36162-1. ISBN 0-521-36965-7 (paper). Sky Publishing Corporation, Cambridge, MA (USA), ISBN 0-933346-55-7, ISBN 0-933346-56-5 (paper).

This book is the third edition of "*The new solar system*", first published in 1981 (see 30.003.026) and updated in 1982 (2. ed. see 34.003.031).

Contents:
1. The golden age of solar-system exploration (*N. W. Hinners*). 2. The Sun (*R. W. Noyes*). 3. Magnetospheres, cosmic rays, and the interplanetary medium (*J. A. Van Allen*). 4. The Moon (*P. D. Spudis*). 5. Mars (*M. H. Carr*). 6. Planet Earth (*D. L. Anderson*). 7. Surfaces of the terrestrial planets (*J. W. Head III*). 8. Atmospheres of the terrestrial planets (*J. B. Pollack*). 9. The Voyager encounters (*B. A. Smith*). 10. Interiors of the giant planets (*W. B. Hubbard*). 11. Atmospheres of the giant planets (*A. P. Ingersoll*). 12. Planetary rings (*J. A. Burns*). 13. The Galilean satellites (*T. V. Johnson*). 14. Titan (*T. Owen*). 15. Icy bodies of the outer solar system (*D. P. Cruikshank, D. Morrison*). 16. The Halley encounters (*R. Reinhard*). 17. Comets (*J. C. Brandt*). 18. Asteroids (*C. R. Chapman*). 19. Meteorites (*J. A. Wood*). 20. Small bodies and their origins (*W. K. Hartmann*). 21. The collision of solid bodies (*E. M. Shoemaker, C. S. Shoemaker*). 22. Life in the new solar system? (*G. A. Soffen*). 23. Putting it all together (*J. S. Lewis*).

003.017 **Istoriko-Astronomicheskie Issledovaniya (XXI). Minuvshee – sovremennost' – prognozy.** ***(Research in the History of Astronomy (XXI). Past – presence – future.)***
A. A. Gurshtejn (ed.).
Nauka, Moskva (USSR). 416 p. (1989). ISBN 5-02-014065-1. Price 3 Rbl. 60 Kop.
The individual contributions within the subject scope of AAA are included in their corresponding categories – see abstracts 004.046, 004.047, 004.048, 004.049, 004.050, 004.051, 004.056, 004.057, 004.058, 004.059, 004.060, 004.061, 004.064, 004.065, 004.066, 005.019, 005.020, 005.021, 005.023, 005.024, 009.007, 011.018, 011.019, 015.019, 051.067.

003.018 **Yarkostnye kharakteristiki infrakrasnogo i vidimogo izlucheniya lunnej poverkhnosti.** ***(Infrared and visible brightness characteristics of the lunar surface.)***
V. V. Shevchenko, S. G. Pugacheva, K. I. Dekhtyareva.
Gos. Astron. Inst. im. P. K. Shternberga.
Izdatel'stvo Moskovskogo Universiteta, Moskva (USSR). 183 p. (1989). ISBN 5-211-01720-X. Price 75 Kop.

Contents: 1. Introduction. 2. A selenographic technique for a systematic affiliation and identification of catalogue reference points at optical and IR wavelengths. 3. Selenographic and rectangular coordinates of lunar surface sections. 4. Catalogue of "Infrared and visible brightness characteristics of lunar surface sections". 5. Contour maps of thermal anomalies of the visible hemisphere of the Moon.

003.019 **The meteorite and tektite collector's handbook. A practical guide to their acquisition, preservation and display.**
P. M. Bagnall.
Willmann-Bell, Richmond, VA (USA). 170 p. (1991). ISBN 0-943396-31-X. Price US$ 24.95.

The book starts with a brief review of the history of meteorite collecting. This is followed by details of how meteorites are priced and sold and three chapters dealing with compositions, structure, and classification from a practical viewpoint. The chapter, "The finer points of meteorite collecting", deals with testing, preservation, and exhibition techniques and suggests ways in which the meteoritophile can derive the greatest rewards from collecting. The final chapter covers tektites.

003.020 **Astronomical algorithms.**
J. Meeus.
Willmann-Bell, Richmond, VA (USA). 435 p. (1991). ISBN 0-943396-35-2. Price US$ 24.95 and a companion diskette US$ 24.95.

The present work is a greatly revised version of "*Astronomical formulae for calculators*" published in 1978. It is, in fact, a completely new book. The subjects have been expanded and the content has been improved. Changes were needed to take into account new resolutions of the International Astronomical Union, particularly the adoption of the new standard epoch J2000.0, while moreover one profited by the new planetary and lunar theories constructed at the Bureau des Longitudes of Paris. Astronomical algorithms intends to be a guide for the (professional or amateur) astronomer who wants to do calculations. This book is not a general textbook on astronomy. The reader will find no theoretical derivations. Except in a few rare cases, no programs are given in this book. The few programs the author gives are in standard BASIC. This book is restricted to the "classical", mathematical astronomy, although a few astronomy oriented mathematical techniques are dealt with, such as interpolation, fitting curves and sorting data. But astrophysics is not considered at all.

003.021 **Methods of orbit determination for the microcomputer.**
D. L. Boulet.
Willmann-Bell, Richmond, VA (USA). 578 p. (1991). ISBN 0-943396-34-4. Price US$ 24.95.
This book describes how the principles of celestial mechanics may be applied to determine the orbits of planets, comets, and Earth satellites. More specifically, it shows how a dedicated novice can learn, by first-hand experience: how orbital motion conforms to Newtonian physics, how a set of orbital elements can be translated into quantities which can be compared with observations, and how a record of observed motion can be used to determine an orbit from scratch or improve a preliminary orbit. This is a how-to-do-it book. Even though the derivations of many important relationships are described in some detail, the emphasis throughout is on practical applications.
Contents: 1. Fundamentals of orbital motion. 2. Time and position. 3. The two-body problem. 4. Orbit geometry. 5. Ephemeris generation. 6. Special perturbations. 7. Applied numerical methods. 8. Preliminary orbit data. 9. The method of Laplace. 10. The method of Gauß. 11. The method of Olbers. 12. Orbit improvement.

003.022 **Numerical modeling in applied physics and astrophysics.**
R. L. Bowers, J. R. Wilson.
Jones and Bartlett Publishers, Boston, MA (USA). 718 p. (1991). ISBN 0-86720-123-1. Price US$ 67.60.
This book is intended to be a self-contained introductory presentation of selected, practical computational methods which have been extensively and successfully applied to a wide range of astrophysical problems.
Contents: 1. Astrophysics on a grid. 2. Overview of finite difference methods. 3. Properties of matter. 4. Lagrangian hydrodynamics. 5. Eulerian hydrodynamics. 6. Radiative transport. 7. Radiation-matter coupling. 8. Magnetohydrodynamics. 9. Gravitation. 10. Thermonuclear and neutrino reactions.

003.023 **Applying fractals in astronomy.**
A. Heck, J. M. Perdang (eds.).
Lect. Notes Phys., New Ser. m, Monogr., Vol. m 3.
Springer, Berlin (Germany). 219 p. (1991). ISBN 3-540-54353-8. Price DM 46.00. ISBN 0-387-54353-8 (USA).
The nine review articles in this book give a fairly complete survey of the applications of fractals to astronomical and astrophysical problems. The book is self-contained and written with the particular aim of teaching graduate students how to apply fractal geometry to new problems. After a general introduction to the basic concepts, applications are given to astrophysical dynamical systems, to strange attractors of stellar pulsations, to the solar granular pattern, and to both the large-scale structure of the Universe – a central theme of book – and to astronomical data analysis.
The individual contributions within the subject scope of AAA are included in their corresponding categories – see abstracts 002.109, 021.041, 062.188, 065.117, 066.203, 161.602, 161.603, 161.604, 161.607.

003.024 **Seeing the solar system. Telescopic projects, activities, and explorations in astronomy.**
F. Schaaf.
Wiley Sci. Ed.
Wiley, New York, NY (USA). 223 p. (1991). ISBN 0-471-53070-0. Price US$ 29.95. ISBN 0-471-53071-9, Price US$ 14.95 (paper).
This entertaining handbook teaches the art and science of astronomical observation using nothing more than a basic telescope. It includes easy telescopic astronomy projects that cover the planets and their moons, asteroids, comets and meteors, the Sun and much more.

003.025 **Unusual telescopes.**
P. L. Manly.
Cambridge University Press, Cambridge (UK). 238 p. (1991). ISBN 0-521-38200-9. Price £ 19.95, US$ 39.95.
The author surveys more than 150 unusual telescope designs. These are telescopes built by amateur and professional astronomers to suit some special need. There is an inflatable telescope and one with a liquid mirror. Every so often a neglected design comes back into fashion: the largest telescopes now under construction use the alt-azimuth design that was ignored for over a century and liquid mirror telescopes can be used for zenithal astronomy. The author shows why a particular engineering approach makes each telescope unique and explains the rationale behind the design. The effects on telescope performance are discussed where possible.
Contents: 1. Optics. 2. Telescope mounts. 3. Telescope and mount materials. 4. Strange drivers. 5. Moving the whole telescope. 6. The moving eyepiece. 7. The stationary eyepiece. 8. Limits. 9. Whimsy.

003.026 **Observing the Sun.**
P. O. Taylor.
Pract. Astron. Handb., Vol. 3.
Cambridge University Press, Cambridge (UK). 172 p. (1991). ISBN 0-521-40110-0. Price £ 17.95, US$ 29.95.
The author explains in a clear and practical way everything that a telescope user needs to know in order to make solar observations. The book deals with the following topics: historical background, choice of equipment for the safe conduct of solar observations, observations of sunspots and eclipses, and reporting observations. New techniques, such as the electronic recording of solar flares, are included.

003.027 **The Sun in time.**
C. P. Sonett, M. S. Giampapa, M. S. Matthews (eds.).
Space Sci. Ser.
University of Arizona Press, Tucson, AZ (USA). 1003 p. (1991). ISBN 0-8165-1297-3. Price US$ 60.00.
The authors try to achieve an improved understanding of our nearest star, the Sun. They have adopted a highly interdisciplinary approach that utilizes the data and analytical tools from a broad range of fields of active research. The result is encompassed in 36 chapters that, in total, trace the past evolution of the Sun and provide a review of our current understanding of both its structure and its role in the origin and evolution of the solar system. Contents: 1. The Sun. 2. Energetic particles. 3. Isotopes. 4. The Sun and climate. 5. The early Sun. 6. Solar-like stars.
The individual contributions within the subject scope of AAA are included in their corresponding categories – see abstracts 004.062, 072.129, 074.109, 078.021, 078.022, 078.023, 078.024, 078.025, 078.026, 080.115, 080.116, 080.117, 080.118, 085.051, 085.052, 085.053, 085.054, 085.055, 085.056, 105.048, 106.042, 107.019, 107.020, 107.021, 107.022, 107.023, 115.033, 115.034, 116.086, 116.087, 121.093, 121.094, 144.103, 144.104, 144.105, 144.106.

003.028 **Solar interior and atmosphere.**
A. N. Cox, W. C. Livingston, M. S. Matthews (eds.).
Space Sci. Ser.
University of Arizona Press, Tucson, AZ (USA). 1430 p. (1991). ISBN 0-8165-1229-9. Price US$ 65.00.
The chapters of this volume are based on talks given at a meeting held at Tucson, AZ (USA), 15 - 18 Nov 1988.
This textbook for graduate students and solar and stellar astrophysicists studying in Sun is the most comprehensive of many recent books on this subject. Contents: 1. The solar interior. 2. What surface oscillations reveal about the inside. 3. What surface radiation reveals about the inside. 4. Expanding magnetic fields and the atmosphere. 5. Our Sun, other suns, other stars. 6. Basic data.
The individual contributions within the subject scope of AAA are included in their corresponding categories – see abstracts 036.276, 062.209, 062.210, 063.110, 065.125, 071.054, 071.055, 071.056, 071.057, 071.058, 072.131, 072.132, 073.164, 073.165, 073.166, 074.113, 074.114, 075.063, 077.073, 080.120, 080.121, 080.122, 080.123, 080.124, 080.125, 080.126, 080.127, 080.128,

080.129, 080.130, 080.131, 080.132, 080.133, 080.134, 080.135, 080.136, 080.137, 080.138, 115.036.

003.029 **The almighty chance.**
Ya. B. Zel'dovich, A. A. Ruzmaikin (*A. A. Ruzmajkin*), D. D. Sokoloff (*D. D. Sokolov*).
World Scientific Lect. Notes Phys., Vol. 20.
World Scientific, Singapore (Singapore). 327 p. (1990). ISBN 9971-50-916-4. Price £ 39.00. ISBN 9971-50-917-2 (paper).
Contents: 1. Introduction. 2. The chance on stage. 3. Diffusion. 4. The Brownian motion. 5. Fractals and dimensions. 6. Percolation and random boundaries. 7. Random hydrodynamic motions. 8. General conception of intermittency. 9. Magnetic field in a flow of conductive fluid. 10. The casual universe.

003.030 **Neutrino physics.**
K. Winter (ed.).
Camb. Monogr. Part. Phys. Nucl. Phys. Cosmol., Vol. 1.
Cambridge University Press, Cambridge (UK). 684 p. (1991). ISBN 0-521-36452-3.
This book presents an up-to-date, coherent and comprehensive overview of modern neutrino physics. Neutrino physics began as a branch of nuclear physics, went on to become an important field of particle physics and has more recently become of fundamental importance in astrophysics and cosmology. This book covers all the major areas of current interest. It begins with a chapter charting the history of the subject through a careful selection of articles by authors such as W. Pauli, B. Pontecorvo and M. Schwartz. This is followed by separate chapters discussing the intrinsic properties of neutrinos, the theory of the interaction of neutrinos with matter, experimental investigations of the weak interaction in neutrino processes, the theory and supporting experiment for the basic properties of the interaction of neutrinos with fermions, and on astrophysics and cosmology.

003.031 **Bausteine des Universums. Quarks und Leptonen.**
J. Ranft.
Deutscher Verlag der Wissenschaften, Berlin (Germany). 191 p. (1991). ISBN 3-326-00520-2. Price DM 42.00.
Contents: 1. Die Physik des 20. Jahrhunderts und der "gesunde Menschenverstand". 2. Symmetrien und Erhaltungssätze in der Physik. 3. Die vier Kräfte der Physik. 4. Elementarteilchen, von den vier Elementen der Griechen bis zu Quarks und Leptonen. 5. Quarks, existieren sie wirklich? 6. Das Jahr 1974, die neuen schweren Teilchen. 7. Die starke Wechselwirkung der Quarks und Gluonen. 8. Haben Neutrinos eine Masse? 9. Sturz der Spiegelsymmetrie und der Zeitumkehrsymmetrie. 10. Die schwache Wechselwirkung der Elementarteilchen. 11. Die elektroschwache Wechselwirkung, eine Folge brillanter Ideen und brillanter Experimente. 12. Protonenzerfall und magnetische Monopole. 13. Gibt es eine Superkraft? Neue Fragen und unkonventionelle Ideen. 14. Hochenergiebeschleuniger, eine Folge unerahnter Erfindungen. 15. Die Superkollider der Zukunft. 16. Die heutigen Universaldetektoren der Hochenergiephysik. 17. Elementarteilchenphysik und der frühe Kosmos.

003.032 **Katalog izmerenij luchevykh skorostej zvezd. Katalog sobstvennykh dvizhenij.** ***(Catalogue of stellar radial velocities. Catalogue of proper motions.)***
A. A. Tokovinin (ed.).
Izdatel'stvo Moskovskogo Universiteta, Moskva (USSR). 92 p. (1990). ISBN 5-211-01809-5. Price 30 Kop.
The individual contributions within the subject scope of AAA are included in their corresponding categories – see abstracts 002.119, 002.120, 002.121.

003.033 **Mars and the development of life.**
A. Hansson.
Ellis Horwood Lib. Space Sci. Space Technol., Ser. Space Life Sci.
Ellis Horwood, New York, NY (USA). 144 p. (1991). ISBN 0-13-559022-1. Price US$ 69.95. ISBN 0-13-561580-1 (paper).
Opening with a look at the biophysical perspective of life on Mars, the book continues with a discussion of the presence of iron on Mars. The search for signals of life in meteorites, in Antarctica and on Mars is comprehensively covered. The detection of life (in the form of clay) on Mars leads to a review of the hypothesis of the existence of clay-life on Mars – the author discusses several interesting points which have arisen from the Viking results – and on transforming Mars. The contamination of Mars and the technology of the returning samples is also discussed. A look at the planning of Mars activities since the Phobos 88 mission is given.

003.034 **The hidden universe.**
R. J. Tayler.
Ellis Horwood Lib. Space Sci. Space Technol., Ser. Astron.
Ellis Horwood, New York, NY (USA). 213 p. (1991). ISBN 0-13-388711-1. Price US$ 69.95. ISBN 0-13-388133-4 (paper).
The author discusses the evidence that there may be much more matter in the universe than is immediately visible. This hidden matter only becomes apparent because of the gravitational influence that it exerts on other bodies in the universe. After a brief introduction the author discusses observations of the principal constituents of the universe, stars and galaxies. He moves on to review the evidence that both galaxies and clusters of galaxies are more massive than they appear, and provides cosmological evidence that there may be a large amount of hidden matter. This matter may be ordinary matter but it could also be exotic elementary particles. The present balance of evidence is that much of it is of the latter form, and the author discusses what influence the particles would have and whether they might be detected by terrestrial experiments.

003.035 **Planetary mapping.**
R. Greeley, R. M. Batson (eds.).
Camb. Planet. Sci. Ser., Vol. 6.
Cambridge University Press, Cambridge (UK). 307 p. (1990). ISBN 0-521-30774-0. Price £ 45.00, US$ 70.00.
In this book, the key elements involved in making maps of planets and satellites are presented, problems and their solutions are discussed, and prospects for the future are outlined.
Contents: 1. Introduction (*R. Greeley, R. M. Batson*). 2. History of planetary cartography (*R. M. Batson, E. A. Whitaker, D. E. Wilhelms*). 3. Cartography (*R. M. Batson*). 4. Planetary nomenclature (*M. E. Strobell, H. Masursky*). 5. Geodetic control (*M. E. Davies*). 6. Topographic mapping (*S. S. C. Wu, F. J. Doyle*). 7. Geologic mapping (*D. E. Wilhelms*). Appendix I: Map formats and projections used in planetary cartography (*R. M. Batson*). Appendix II: Halftone processes for planetary maps (*J. L. Inge*). Appendix III: Digital planetary cartography (*R. M. Batson*).

003.036 **Planetary geosciences – 1988.**
M. Zuber, O. James, G. MacPherson, J. Plescia (eds.).
National Aeronautics and Space Administration, Washington, DC (USA), NASA-SP—498. 114 p. (1989).
This document is intended to provide an overview of some of the more significant scientific findings and discoveries made in 1988 by scientists supported by the Planetary Geosciences Programs. These scientists have provided short articles describing their research and its importance in the context of the larger problems of planetary geoscience.

Contents: 1. Inner solar system. 2. Outer solar system. 3. Lunar samples. 4. Meteorites and cosmic dust.

003.037 **Das Projekt "Satelliten- und Inertialgeodäsie" der Bayerischen Kommission für die Internationale Erdmessung.**
R. Sigl, W. Ehrnsperger, W. Wende, G. Boedecker, K. H. Neumayer.
Veröff. Bayer. Komm. Int. Erdmessung, Bayer. Akad. Wiss., Astron.-Geod. Arb., Nr. 51.
Verlag der Bayerischen Akademie der Wissenschaften, München (Germany, F.R.). 71 p. (1991). ISBN 3-7696-9794-4.
Contents: 1. Zur Entwicklung des Projektes "Satelliten- und Inertialgeodäsie" (*R. Sigl*). 2. EUREF - Erstellung eines fundamentalen 3-D Festpunktfeldes in Europa (*W. Ehrnsperger*). 3. Genauigkeitssteigerung in der geodätischen Astronomie durch den FK5 Sternkatalog. Neuauswertung von astronomisch bestimmten Längendifferenzen im Europäischen Längennetz (ELN) (*W. Wende*). 4. Status des Projektes "Inertialschwere" zur Bestimmung des Schwerevektors aus Inertial- und GPS-Messungen (*G. Boedecker*). 5. Stochastische Modelle für Satelliten- und Empfängeruhren sowie für das Gravitationsfeld (*K. H. Neumayer*).

The individual contributions within the subject scope of AAA are included in their corresponding categories - see abstracts 045.021.

003.038 **Nuclei in the cosmos.**
H. Oberhummer (ed.).
Grad. Texts Contemp. Phys.
Springer, Berlin (Germany). 248 p. (1991). ISBN 3-540-54198-5. Price DM 88.00. ISBN 0-387-54198-5 (USA).
This review volume provides an overview of the current status of nuclear astrophysics. Experts in their respective fields provide in ten contributions detailed surveys of our understanding of different astrophysical environments and their connection to nuclear processes. Special emphasis is given to the interdisciplinary nature of the field: astronomy, nuclear physics, astrophysics, and elementary particle physics are equally involved. The topics of the book range from the collection of observational facts with astronomical methods to experimental work and theoretical models to describe the nuclear processes involved in different astrophysical scenarios. Astrophysical models and calculations simulate how nuclear processes contribute to driving the evolution of stars, interstellar matter, and the whole universe.
The individual contributions within the subject scope of AAA are included in their corresponding categories - see abstracts 022.176, 022.177, 061.127, 061.128, 061.129, 061.130, 144.118, 157.352, 161.677, 161.678.

004 History of Astronomy

004.001 **The astronomer's tale.**
J. Lester.
Astron. Now, Vol. 5, No. 7, p. 20 - 21 (Jul 1991).

004.002 **Die beiden Observatorien in Potsdam und Babelsberg. Ihre Geschichte und Bedeutung, ihr heutiger Zustand.**
P. Müller.
Sterne Weltraum, Jahrg. 30, Nr. 8-9, p. (Aug-Sep 1991).

004.003 **Astronomy then and now.**
B. Jones.
Astron. Now, Vol. 5, No. 8, p. 33 - 38 (Aug 1991).
The last 80 years have seen our understanding of the universe in which we live grow in leaps and bounds but where would you start explaining everything? The author gives a few ideas.

004.004 **An unusual observatory.**
G. Young.
Astron. Now, Vol. 5, No. 8, p. 42 - 43 (Aug 1991).

004.005 **Der "Zij Muhammad Shahi". Die Kalenderreform Jai Singhs II.**
B. du Mont.
Sterne Weltraum, Jahrg. 30, Nr. 12, p. 732 - 735 (Dec 1991).

004.006 **Tycho's new star.**
I. Seymour.
Astron. Now, Vol. 5, No. 11, p. 46, 48 (Nov 1991).

004.007 **The system of formulating the calendar by observing the Beidou stars.**
Li Yong.
J. Nanjing Univ., Vol. 27, No. 4, p. 653 - 659 (Oct 1991). In Chinese.

004.008 **Physicists and radiometers: the early history of infrared astronomy (1860 - 1940).**
R. S. Brashear.
Bull. Am. Astron. Soc., Vol. 23, No. 2, p. 871 (1991). Abstract. - See Abstr. 012.008 for the main entry.

004.009 **The first solar neutrino conference: Irvine 1972.**
V. Trimble.
Bull. Am. Astron. Soc., Vol. 23, No. 2, p. 871 (1991). Abstract. - See Abstr. 012.008 for the main entry.

004.010 **Search for gravitational radiation.**
J. Weber.
Bull. Am. Astron. Soc., Vol. 23, No. 2, p. 871 (1991). Abstract. - See Abstr. 012.008 for the main entry.

004.011 **The appointment of a physicist as director of the astronomical center of the world.**
D. E. Osterbrock.
Bull. Am. Astron. Soc., Vol. 23, No. 2, p. 872 (1991). Abstract. - See Abstr. 012.008 for the main entry.

004.012 **Moon echoes and Magellanic Clouds.**
F. J. Kerr.
Bull. Am. Astron. Soc., Vol. 23, No. 2, p. 872 (1991). Abstract. - See Abstr. 012.008 for the main entry.

004.013 **The establishment of radio astronomy in Canada: instrumentation pull or programmatic push?**
R. A. Jarrell.
Bull. Am. Astron. Soc., Vol. 23, No. 2, p. 872 (1991). Abstract. - See Abstr. 012.008 for the main entry.

004.014 **The reception given to radio astronomy by optical astronomers in the decade following World War II.**
W. T. Sullivan.
Bull. Am. Astron. Soc., Vol. 23, No. 2, p. 872 (1991). Abstract. – See Abstr. 012.008 for the main entry.

004.015 **Project Ozma: the opening of the electromagnetic spectrum for SETI.**
S. J. Dick.
Bull. Am. Astron. Soc., Vol. 23, No. 2, p. 872 (1991). Abstract. – See Abstr. 012.008 for the main entry.

004.016 **Inclusive counting as the source of the misunderstanding about the luni–solar calendar of Geminus prior to the octaeteris.**
T. D. Worthen.
Bull. Am. Astron. Soc., Vol. 23, No. 2, p. 898 (1991). Abstract. – See Abstr. 012.008 for the main entry.

004.017 **Sacrobosco in Ulster.**
A. S. Drennan.
Ir. Astron. J., Vol. 20, No. 2, p. 78 – 83 (Sep 1991).

004.018 **The Geminid meteor shower: a possible pre–discovery observation in 1834.**
L. P. Connolly, D. Hoffleit.
Bull. Am. Astron. Soc., Vol. 23, No. 2, p. 941 (1991). Abstract. – See Abstr. 012.008 for the main entry.

004.019 **The stationary radiant debate revisited.**
M. Beech.
Q. J. R. Astron. Soc., Vol. 32, No. 3, p. 245 – 264 (Sep 1991).
Under the direction and influence of essentially one man, William F. Denning, the stationary radiant debate raged for some 50 yrs, between *c.* 1880 and *c.* 1930, in the field of meteor astronomy. This article reviews the debate, looks to the people involved, and attempts to offer a clear explanation of the observational principles that lead to the erroneous identification of stationary radiants.

004.020 **Humboldt and the Earth's magnetic field.**
S. R. C. Malin, D. R. Barraclough.
Q. J. R. Astron. Soc., Vol. 32, No. 3, p. 279 – 293 (Sep 1991).
About 150 years ago, many magnetic observatories were established to monitor the variations of the geomagnetic field with time. In most instances the foundation was a result of the initiative or influence of Alexander von Humboldt. In this paper the authors review Humboldt's contribution to geomagnetism, as an observer, an organizer of geomagnetic studies, an encourager of other scientists, and an analyst.

004.021 **La gran meridiana de Gian Domenico Cassini.**
M. Quadrelli.
Rev. Astron., No. 246, p. 5 – 8 (Jun 1991).

004.022 **150 Jahre Bessel–Ellipsoid, 1841 – 1991.**
J. Zeger.
Österr. Z. Vermessungswes. Photogramm., Jahrg. 79, Heft 4, p. 337 – 340 (1991).

004.023 **Comments on the astronomical ideas of Zhou Bi Classic on Mathematics.**
Guo Shengchi.
Acta Astron. Sin., Vol. 32, No. 2, p. 208 – 214 (Jun 1991). In Chinese.
There are intact astronomical ideas in the Zhou Bi Classic on Mathematics. In this paper they are commented as objectively as possible. Being not as advanced as the theory of Sphere–heavens, the second theory of Canopy–heavens shows the progress of the knowledge about the universe. Under the guidance of this theory, people got a further understanding of the origins of the day, night, and season and established a rational idea of orientation as well as the primary idea of the local time. The inference of the arctic regions and equatorial areas is close to reality too. The method of determination of the solar diameter is also commented as objectively as possible.

004.024 **The astronomical tables of Jacob ben David Bonjorn.**
J. Chabás.
Arch. Hist. Exact Sci., Vol. 42, No. 4, p. 279 – 314 (16 Sep 1991).

004.025 **De l'influence de la gravitation sur la propagation de la lumière en théorie newtonienne. L'archéologie des trous noirs.**
J. Eisenstaedt.
Arch. Hist. Exact Sci., Vol. 42, No. 4, p. 315 – 386 (16 Sep 1991).

004.026 **The introduction of dated observations and precise measurement in Greek astronomy.**
B. R. Goldstein, A. C. Bowen.
Arch. Hist. Exact Sci., Vol. 43, No. 2, p. 93 – 132 (29 Oct 1991).

004.027 **Ptolemy's maps of earth and the heavens: a new interpretation.**
J. L. Berggren.
Arch. Hist. Exact Sci., Vol. 43, No. 2, p. 133 – 144 (29 Oct 1991).

004.028 **Pourquoi la nuit est–elle noire? Le paradoxe de Chéseaux–Olbers (1re partie).**
C. Nitschelm.
Orion, Jahrg. 49, Nr. 245, p. 138 – 142 (Aug 1991).

004.029 **Pourquoi la nuit est–elle noire? Le paradoxe de Chéseaux–Olbers (2e partie).**
C. Nitschelm.
Orion, Jahrg. 49, Nr. 246, p. 172 – 175 (Oct 1991).

004.030 **Der Stern von Bethlehem.**
W. Lüthi.
Orion, Jahrg. 49, Nr. 247, p. 235 – 238 (Dec 1991).

004.031 **La fondation de l'observatoire et les débuts de l'astronomie au Pic du Midi (1852 – 1947). Introduction. Première partie: La fondation de l'observatoire.**
J.–M. Malherbe, C. Coutard, T. Roudier.
Astronomie, p. 6 – 9 (Oct 1991).

004.032 **La fondation de l'observatoire et les débuts de l'astronomie au Pic du Midi (1852 – 1947). Deuxième partie: L'introduction de l'astronomie.**
J.–M. Malherbe, C. Coutard, T. Roudier.
Astronomie, p. 1 – 3 (Nov 1991).

004.033 **La fondation de l'observatoire et les débuts de l'astronomie au Pic du Midi (1852 – 1947). Dernière partie: Le règne de la coupole Baillaud.**
J.–M. Malherbe, C. Coutard, T. Roudier.
Astronomie, p. 1 – 9 (Dec 1991).

004.034 **"Almagest" and the "new chronology".**
Yu. N. Efremov.
Priroda, No. 7, p. 94 – 100 (Jul 1991). In Russian.
The recent attempts of certain Soviet mathematicians to supply a new (later) date for one of the greatest creations of the ancients – Ptolemy's "Almagest" – are, probably, groundless despite their sensational nature.

004.035 **The Cracow period of Copernicus' studies.**
S. R. Brzostkiewicz, D. Górnicza.
Urania, Rok 62, Nr. 9, p. 234 – 241 (Sep 1991). In Polish.

004.036 **Who was Jean Philippe Louis de Chéseaux's?**
S. Gil.
Urania, Rok 62, Nr. 9, p. 243 – 245 (Sep 1991). In Polish.

004.037 **Les astronomes et la naissance de la statistique.**
F. Mignard.
Journées 1991: Systèmes de référence spatio–temporels, p. 59 – 65 (1991). – See Abstr. 012.048 for the main entry.
In the astronomical daily work, statistical technics are mainly used through the tool–box approach, because the statistical know–how is derived from an independent branch of science. Such a situation has not always been the prevailing one and in the past astronomers played a significant role in the emergence of the statistical concepts, both in relation with their activity of star gazers, but also by applying their ability of number crunchers to a fledging quantitative social science that was in the making in the 17th and 18th century. The main aspects of this interaction are investigated in this paper.

004.038 **Tobias Mayer et le traitement des observations.**
F. Mignard, B. Morando, G. Theureau.
Journées 1991: Systèmes de référence spatio–temporels, p. 74 – 79 (1991). – See Abstr. 012.048 for the main entry.
In order to determine the parameters of the rotation of the Moon Tobias Mayer in 1750 made measurements of the position of some craters on the surface. He obtained 27 linear equations for the three unknowns. Instead of choosing arbitrarily three equations, he devised a clever method using all equations and obtained as good a solution as if he had used the method of least squares which was not yet known.

004.039 **The origins of the reflecting telescope.**
C. A. Ronan.
J. Br. Astron. Assoc., Vol. 101, No. 6, p. 335 – 342 (Dec 1991).
This paper discusses the origins of the reflecting telescope. A case is made for the reflector's origin in England during the last decades of the sixteenth century.

004.040 **The 8 1/4–inch Clark refractor of the Temple Observatory, Rugby.**
R. A. Marriott.
J. Br. Astron. Assoc., Vol. 101, No. 6, p. 343 – 350 (Dec 1991).
During the 1850s W. R. Dawes purchased five object glasses from the American opticians Alvan Clark and Sons. The only one to have been in continuous use since then was the last of them, in a complete 8 1/4–inch refractor.

004.041 **The Lilienthal tragedy.**
R. Baum.
J. Br. Astron. Assoc., Vol. 101, No. 6, p. 369 – 370 (Dec 1991).

004.042 **The ancient history of the New Plymouth 6 inch refractor.**
W. Orchiston.
South. Stars, Vol. 34, No. 4, p. 277 – 284 (Sep 1991).
This paper examines the trans–Tasman associations – the "ancient history" – of the New Plymouth refractor, and the major achievements of its manufacturers.

004.043 **Feeling the heat: the early years of infrared astronomy.**
R. S. Brashear.
Publ. Astron. Soc. Pac., Vol. 103, No. 668, p. 1117 (Oct 1991). Abstract. Paper presented at the history of astronomy sessions of the Astronomical Society of the Pacific Summer Meeting, Laramie, WY (USA), 25 – 27 Jun 1991.

004.044 **Steps toward the velocity–distance relation of extragalactic nebulae.**
H. W. Duerbeck, W. C. Seitter.
Publ. Astron. Soc. Pac., Vol. 103, No. 668, p. 1117 – 1118 (Oct 1991). Abstract. Paper presented at the history of astronomy sessions of the Astronomical Society of the Pacific Summer Meeting, Laramie, WY (USA), 25 – 27 Jun 1991.

004.045 **The adaptation of Babylonian methods in Greek numerical astronomy.**
A. Jones.
Isis, Vol. 82, No. 313, p. 441 – 453 (Sep 1991).

004.046 **On the history of investigations of nonstable stars in the USSR.**
V. G. Gorbatskij.
Istor.–Astron. Issled., Tom 21, p. 77 – 86 (1989). In Russian. – See Abstr. 003.017 for the main entry.

004.047 **A. L. Chizhevskij's views on solar–terrestrial relations.**
N. I. Moiseeva.
Istor.–Astron. Issled., Tom 21, p. 87 – 99 (1989). In Russian. – See Abstr. 003.017 for the main entry.

004.048 **On the occasion of the 100th anniversary of the "Carte du Ciel" and the "Astrographic Catalogue".**
I. G. Kolchinskij.
Istor.–Astron. Issled., Tom 21, p. 100 – 117 (1989). In Russian. – See Abstr. 003.017 for the main entry.

004.049 **O. W. Struve's Levelling Project in European Russia.**
M. V. Kuznetsov.
Istor.–Astron. Issled., Tom 21, p. 118 – 132 (1989). In Russian. – See Abstr. 003.017 for the main entry.
This paper reports on the contribution made by the Pulkovo Observatory to the development of high–precision geodetic work in Russia.

004.050 **On the problem of dissemination of heliocentric ideas in Russia.**
A. T. Grigor'yan.
Istor.–Astron. Issled., Tom 21, p. 133 – 143 (1989). In Russian. – See Abstr. 003.017 for the main entry.

004.051 **Galileo Galilei's "Two Books".**
E. G. Grishko.
Istor.–Astron. Issled., Tom 21, p. 144 – 154 (1989). In Russian. – See Abstr. 003.017 for the main entry.

004.052 **Amateur–professional co–operation in astronomy education: a nineteenth century Australian model.**
W. Orchiston.
Proc. Astron. Soc. Aust., Vol. 9, No. 1, p. 186 – 188 (1991). – See Abstr. 012.078 for the main entry.
During the second half of the nineteenth century Australian amateur and professional astronomers worked together in actively promoting astronomy among the general public. This collaboration was a product of the cordial amateur–professional relations that permeated pre–astrophysics Australian astronomy.

004.053 **Nineteenth century investigations of periodicities in the Jovian atmosphere.**
T. A. Hockey.
Vistas Astron., Vol. 34, Part 4, p. 409 – 414 (1991).

004.054 **The Hubble–van Maanen conflict over internal motions in spiral nebulae: yet more new information on an already old topic.**
R. W. Brashear, N. S. Hetherington.
Vistas Astron., Vol. 34, Part 4, p. 415 – 423 (1991).

004.055 **The Herschel–Denning correspondence: 1871 – 1900.**
M. Beech.
Vistas Astron., Vol. 34, Part 4, p. 425 – 447 (1991).

004.056 **Jean Buridan and his contribution to progress in astronomy.**
I. V. Lupandin.
Istor.–Astron. Issled., Tom 21, p. 155 – 163 (1989). In Russian. – See Abstr. 003.017 for the main entry.
The author presents a translation from Latin of, and comments on astronomy–related statements of Buridan (~1295 – ~1358), a famous French scholar, rector of the Paris University in the XIV century, which show quite a new picture of the niveau of knowledge of astronomy in medieval Europe.

004.057 **On the scientific heritage of the astronomer and mathematician of the Xth century Kushjar ibn Labban Djili (~940 – ~1010).**
Kh. F. Abdulla-Zade, M. M. Rozhanskaya.
Istor.-Astron. Issled., Tom 21, p. 164 – 174 (1989). In Russian. – See Abstr. 003.017 for the main entry.
This article deals with medieval astronomy in Islamic countries.

004.058 **Determination of the epoch of the "Almagest" star catalogue using an analysis of proper motions of stars. (On the problem of authorship of Ptolemy's Star Catalogue).**
Yu. N. Efremov, E. D. Pavlovskaya.
Istor.-Astron. Issled., Tom 21, p. 175 – 192 (1989). In Russian. – See Abstr. 003.017 for the main entry.
The authors have worked out a new and unusual approach to dating of old star catalogues, and give a description of their method.

004.059 **Incarnation of the Universe. (Archeological relics as objects of paleoastronomy).**
I. L. Kyzlasov.
Istor.-Astron. Issled., Tom 21, p. 193 – 212 (1989). In Russian. – See Abstr. 003.017 for the main entry.
The author introduces some quite new ideas on archeological relics being objects of paleoastronomy. His concept is of fundamental value for solution of problems of science genesis.

004.060 **Astrophysics' contribution to the study of comets in the XIX century.**
D. B. Herrmann.
Istor.-Astron. Issled., Tom 21, p. 229 – 238 (1989). In Russian. – See Abstr. 003.017 for the main entry.

004.061 **Medieval observations of comets in Latvia.**
J. M. Kletnieks.
Istor.-Astron. Issled., Tom 21, p. 239 – 250 (1989). In Russian. – See Abstr. 003.017 for the main entry.

004.062 **Sunspots: the historical background.**
G. V. Coyne.
The Sun in time, p. 3 – 10 (1991). – See Abstr. 003.027 for the main entry.
After a brief review of the earliest observations of sunspots, this chapter concentrates on the controversies of the 16th and 17th centuries concerning the nature of sunspots and evaluates these controversies in light of the traditions inherited from Aristotle and against the background of the conflict between heliocentric and geocentric cosmologies. The role played by the early study of sunspots in the development of the scientific method is also discussed.

004.063 **Hetu'u Rapanui: the archaeoastronomy of Easter Island.**
W. Liller.
Liller Symposium: Asteroids to quasars, p. 267 – 286 (1991). – See Abstr. 012.087 for the main entry.

004.064 **Accomplishments and alarms of Pulkovo, some pages of reminescenses.**
M. N. Gnevyshev.
Istor.-Astron. Issled., Tom 21, p. 342 – 368 (1989). In Russian. – See Abstr. 003.017 for the main entry.

004.065 **Survived. The Pulkovo Observatory in the years 1938 – 1954.**
T. V. Krat.
Istor.-Astron. Issled., Tom 21, p. 369 – 386 (1989). In Russian. – See Abstr. 003.017 for the main entry.

004.066 **From Pulkovo's history.**
D. E. Shchegolev.
Istor.-Astron. Issled., Tom 21, p. 387 – 391 (1989). In Russian. – See Abstr. 003.017 for the main entry.

004.067 **Medicine wheels in Western Canada.**
D. Hladiuk.
J. R. Astron. Soc. Can., Vol. 85, No. 4, p. 174 (Aug 1991). Abstract. – See Abstr. 012.090 for the main entry.

004.068 **Airy correspondence held at the National Archives of Canada.**
J. E. Kennedy.
J. R. Astron. Soc. Can., Vol. 85, No. 4, p. 175 (Aug 1991). Abstract. – See Abstr. 012.090 for the main entry.

004.069 **Skywatchers of the Salt River Valley Hohokam.**
B. Mixon, R. E. White.
Astron. Q., Vol. 8, No. 4, p. 245 – 259 (1991).

004.070 **The Venus table of the Dresden codex and the movements of the planet Venus.**
B. Wells.
J. R. Astron. Soc. Can., Vol. 85, No. 6, p. 291 – 315 (Dec 1991).
The movements of the planet Venus in the 10th century A.D. are compared with the record of the Maya scribes in the Dresden codex. The manner in which the Venus table may have been used by the Maya is examined in detail. Orbital patterns of Venus are related to Maya glyphic representations of Venus.

004.071 **Un observatoire privé à Amiens dans les années 1930 – 1940.**
A. Le Boeuffle.
Obs. Trav., No. 28, p. 24 – 25 (1991).

004.072 **A century of national time.**
S. Debarbat.
Recherche, Vol. 22, No. 234, p. 974 – 975 (Jul – Aug 1991). In French.

004.073 **Epicurean cosmology: a "steady-state universe".**
O. Longo.
Mem. Soc. Astron. Ital., Vol. 62, No. 3, p. 433 – 441 (1991). – See Abstr. 012.093 for the main entry.
The aim of this essay is to re-establish a meaningful relationship and contact between ancient and modern science, starting from Epicurean cosmology and moving in the direction recently suggested by V. Goldschmidt: to consider ancient science by an immediate, direct reading of the texts, evaluating them according to our own standards and within the scheme of a genuine *scientia perennis*.

004.074 **The infinity theme in Galilei.**
N. Badaloni.
Mem. Soc. Astron. Ital., Vol. 62, No. 3, p. 443 – 454 (1991). – See Abstr. 012.093 for the main entry.
The author demonstrates: 1) the existence of a strong temptation in Galilei to interpret the fixed stars as a series of worlds somehow dangerously similar to the one suggested by Bruno who followed the tracks of Lucretius tradition; 2) Galilei's conception of infinity is an open scenery built up on the base of an encounter with the external world, that is with observation experiences confirmed by fortunate joints with other experiences and conclusions; 3) the implications of this new conception what considers not only the earth but the whole solar system as the center of the world, in spite of the obstacles and the oppositions they met, drive gradually Galilei to show the presence of a peculiar form of infinitude in the same natural and artificial machines he is conceiving, as if, in doing so, he were to take his revenge for the rejection imposed to him by the astronomical infinity.

004.075 **Classic and non–intentional consequences of some Newtonian problems.**
E. Bellone.
Mem. Soc. Astron. Ital., Vol. 62, No. 3, p. 477 – 485 (1991). – See Abstr. 012.093 for the main entry.
The author examines the consequences of Newtonian problems concerning non–gravitational interactions (forces of attraction and repulsion inversely proportional to the nth power of the distance), the propagation of gravity in a medium (outlined in Newtonian's third letter to Bentley), the deflection of light in the vicinity of masses – stars – (presented as a problem in Query 1 of the treatise on optics), and the general theme of the death of the Universe due to the fact that motion is much more apt to be lost than got, and is always upon the decay, as started in Query 31 of the same treatise.

004.076 **Gill, Grubb and the astrographic telescopes – a correspondence.**
I. S. Glass.
Mon. Notes Astron. Soc. S. Afr., Vol. 50, Nos. 11 – 12, p. 121 – 136 (Dec 1991).

004.077 **Along the Milky Way.**
E. C. Krupp.
Mercury, Vol. 20, No. 6, p. 162 – 174 (Nov – Dec 1991).

004.078 **L'activité solaire et l'activité aurorale au XVIIe siècle.**
J.–P. Legrand, M. Le Goff, C. Mazaudier, W. Schröder.
C. R. Acad. Sci., Sér. Gén., Vie Sci., Tome 8, No. 3, p. 181 – 219 (1991).

004.079 **Precision of medieval Islamic eclipse measurements.**
F. R. Stephenson, S. S. Said.
J. Hist. Astron., Vol. 22, Part 3, p. 195 – 207 (Aug 1991).

004.080 **The United States Naval astronomical expedition (1849 – 52) for the solar parallax.**
W. W. Huffman.
J. Hist. Astron., Vol. 22, Part 3, p. 208 – 220 (Aug 1991).

004.081 **The use of astronomical tables by Albertus Magnus.**
B. B. Price.
J. Hist. Astron., Vol. 22, Part 3, p. 221 – 240 (Aug 1991).

004.082 **The Rosse spirals.**
D. W. Dewhirst, M. Hoskin.
J. Hist. Astron., Vol. 22, Part 4, p. 257 – 266 (Nov 1991).

004.083 **The Kenya expedition of Leiden Observatory.**
J. K. Katgert–Merkelijn.
J. Hist. Astron., Vol. 22, Part 4, p. 267 – 296 (Nov 1991).

004.084 **Accuracy of eclipse observations recorded in medieval Arabic chronicles.**
S. S. Said, F. R. Stephenson.
J. Hist. Astron., Vol. 22, Part 4, p. 297 – 310 (Nov 1991).

004.085 **Historical note on the discovery of cosmological background radiation.**
A. Le Floch, F. Bretenaker.
C. R. Acad. Sci., Sér. II, Tome 313, No. 1, p. 35 – 40 (4 Jul 1991).

004.086 **La cosmologia nella cultura del '900: appunti introduttivi.**
A. Masani.
G. Astron., Vol. 17, N. 1 – 2, p. 3 – 6 (Mar – Jun 1991). Paper presented at "La cosmologia nella cultura dell '900". Conferenza Internazionale nel centenario della nascita di Hubble, Torino (Italy), 21 – 23 Nov 1990.

004.087 **Traek of Astronomisk Selskabs historie gennem 75 år.**
L. Moustgaard.
Astron. Tidsskr., Årg. 24, Nr. 3, p. 97 – 111 (1991).

004.088 **Tre tyske obsdervatorier for 200 år siden.**
H. Meyer.
Astron. Tidsskr., Årg. 24, Nr. 4, p. 165 – 167 (1991).

004.089 **L'area megalitica di Aosta e gli allineamenti lunari.**
G. Romano.
Orione, Vol. 11, N. 4, p. 18 – 20 (Jul – Aug 1991).

004.090 **Le donne e l'astronomia fra XVII e XVIII secolo.**
M. G. Pancaldi.
Orione, Vol. 11, N. 4, p. 30 – 43 (Jul – Aug 1991).

004.091 **Il principio di Weyl, passo essenziale verso la precisazione della nozione di modello di Universo.**
S. Bergia.
G. Astron., Vol. 17, N. 1 – 2, p. 48 – 53 (Mar – Jun 1991). Paper presented at "La cosmologia nella cultura dell '900". Conferenza Internazionale nel centenario della nascita di Hubble, Torino (Italy), 21 – 23 Nov 1990.

004.092 **Illuminating incidents in antipodean astronomy: section A of the Royal Society of New South Wales.**
W. Orchiston, R. Bhathal.
Aust. J. Astron., Vol. 4, No. 2, p. 49 – 67 (Oct 1991).

Stjernenes navn.
See Abstr. 002.135.

The Arabs and the stars. Texts and traditions on the fixed stars, and their influence in medieval Europe.
See Abstr. 003.005.

Research in the History of Astronomy (XXI). Past – presence – future.
See Abstr. 003.017.

The McDonald Observatory and I way back then.
See Abstr. 005.018.

Johann Franz Encke, 1791 – 1865.
See Abstr. 005.028.

50 years of McDonald Observatory.
See Abstr. 009.009.

Drawing down the Moon.
See Abstr. 015.004.

Göttergeschichten am Himmel.
See Abstr. 015.012.

Infinity, finite and error.
See Abstr. 015.027.

How the world ceased to be eternal: Christian objections to pagan cosmology.
See Abstr. 015.032.

The arrow of time.
See Abstr. 015.034.

Il dio dei teologi e il dio dei cosmologi. Apologia di Bellarmino.
See Abstr. 015.044.

Filosofia e cosmologia: alcune considerazioni sul problema del realismo.
See Abstr. 015.046.

Qualche aspetto filosofico delle rivoluzioni in cosmologia.
See Abstr. 015.048.

On the Lehmann–Filhés theorem as a consequence of one theorem of Ostrogradsky.
See Abstr. 021.038.

The Star of Bethlehem – a comet in 5 BC – and the date of the birth of Christ.
See Abstr. 046.007.

The selection of superior epoch in Chinese calendars during the period from East Han to Liu Song dynasty.
See Abstr. 046.015.

The USSR Academy of Sciences and space research.
See Abstr. 051.067.

Un antico incostante sole. I.
See Abstr. 072.149.

At the "sources" of the contemporary theory of the origin of the Earth.
See Abstr. 107.010.

Interpretazione di H. Poincaré della ipotesi cosmogonica di P. S. Laplace sull'origine del sistema solare.
See Abstr. 107.029.

005 Biography

005.001 **Gustav Witt zum 125. Geburtstag.**
K.-H. Tiemann.
Sterne Weltraum, Jahrg. 30, Nr. 12, p. 722 (Dec 1991).

005.002 **The pit and the pendulum.**
A. Chapman.
Astron. Now, Vol. 5, No. 12, p. 18 – 21 (Dec 1991).
George Airy, the seventh Astronomer Royal, was one of the more controversial astronomers to hold this office. The author looks back at one experiment which spanned a large part of his career the quest to determine gravity and thus provide astronomers with the density of planet Earth.

005.003 **Plasma waves at collisionless shocks in space: the observations of Frederic L. Scarf.**
S. Moses, C. Kennel.
Adv. Space Res., Vol. 11, No. 9, p. 3 – 14 (1991). – See Abstr. 012.006 for the main entry.
Collisionless shock have been found upstream of every planet and comet visited by spacecraft and in interplanetary space between the orbits of Mercury and Neptune. The study of these obiquitous phenomena played an important part in the career of one of the most prominent founders of space plasma physics, Frederic L. Scarf. The detection of electrostatic turbulence associated with collisionless shocks was an early goal for the plasma wave instruments developed by Scarf and his co-workers, and its discovery emphasized the importance of the application of plasma kinetic theory to the space environment. Later in Scarf's career, detection of shocks became routine, and understanding their structure was the scientific challenge.

005.004 **William H. Wright's early planetary photography in the near infrared and near ultraviolet.**
J. S. Tenn.
Bull. Am. Astron. Soc., Vol. 23, No. 2, p. 873 (1991). Abstract. – See Abstr. 012.008 for the main entry.

005.005 **Companions in astronomy: Margaret Lindsay Huggins and Agnes Mary Clerke.**
M. T. Bruck.
Ir. Astron. J., Vol. 20, No. 2, p. 70 – 77 (Sep 1991).

005.006 **Karl Schwarzschild: Zu seinem 75. Todestag.**
D. B. Herrmann.
Astron. Sch., Jahrg. 28, Heft 4, p. 10 – 11 (Aug 1991).

005.007 **Johann Franz Encke – zu seinem 200. Geburtstag.**
D. Frisch, E. Graiche, J. Sendel.
Astron. Sch., Jahrg. 28, Heft 5, p. 13 – 15 (Oct 1991).

005.008 **Otto Yul'evich Schmidt, on the occasion of the 100th anniversary of his birthday, 30 September 1991.**
Astron. Vestn., Tom 25, No. 6, p. 723 – 724 (Nov – Dec 1991). In Russian. English translation in Sol. Syst. Res., Vol. 25, No. 6.

005.009 **Er suchte die Sterne – und blieb auf der Erde: Leben und Werk des astronomischen Volksschriftstellers Bruno Hans Bürgel (Teil 2).**
A. Zenkert.
Orion, Jahrg. 49, Nr. 245, p. 136 – 138 (Aug 1991).

005.010 **Charles Babbage (1791 – 1871).**
C. Parker.
Gemini, No. 34, p. 26 – 29 (Dec 1991).

005.011 **The thorns and stars of Friedrich Houtermans (22 January 1903 – 1 March 1966).**
I. B. Khriplovich.
Priroda, No. 7, p. 86 – 91 (Jul 1991). In Russian.
This is a story of a strange fate of a talented German physicist who was the first to prove that thermonuclear reactions supplied stars with energy.

005.012 **Lick Observatory Sir Galahad: Arthur B. Wyse.**
D. E. Osterbrock.
Publ. Astron. Soc. Pac., Vol. 103, No. 668, p. 1118 (Oct 1991). Abstract. Paper presented at the history of astronomy sessions of the Astronomical Society of the Pacific Summer Meeting, Laramie, WY (USA), 25 – 27 Jun 1991.

005.013 **Samuil Aronovich Kaplan (1921 – 1978), on the occasion of the 70th birthday.**
V. N. Tsytovich, I. M. Ojringel', E. B. Klejman.
Zemlya Vselennaya, No. 4, p. 46 – 48 (Jul – Aug 1991). In Russian.

005.014 **Solomon Borisovich Pikel'ner (1921 – 1975), on the occasion of his 70th birthday.**
K. V. Bychkov.
Zemlya Vselennaya, No. 4, p. 49 – 51 (Jul – Aug 1991). In Russian.

005.015 **Restoring pages of history. Daniil Osipovich Svyatskij (1881 – 1940).**
V. A. Bronshtehn.
Zemlya Vselennaya, No. 4, p. 68 – 74 (Jul – Aug 1991). In Russian.

005.016 **Restoring pages of history. Nina Mikhajlovna Staude (1888 – 1980).**
V. A. Bronshtehn.
Zemlya Vselennaya, No. 5, p. 68 – 74 (Sep – Oct 1991). In Russian.

005.017 **Iosif Samuilovich Shklovskij (1 July 1916 – 3 March 1985), on the occasion of his 75th birthday.**
V. G. Kurt, I. N. Minin.
Zemlya Vselennaya, No. 6, p. 37 – 41 (Nov – Dec 1991). In Russian.

005.018 **The McDonald Observatory and I way back then.**
D. M. Popper.
Symposium in Celebration of the 50. Anniversary of McDonald Observatory of the University of Texas at Austin: Frontiers of stellar evolution, p. 19 – 22 (1991). – See Abstr. 012.085 for the main entry.

005.019 **In memory of Boris Petrovich Gerasimovich (1889 – 1937).**
V. K. Abalakin.
Istor.-Astron. Issled., Tom 21, p. 251 – 252 (1989). In Russian. – See Abstr. 003.017 for the main entry.
B. P. Gerasimovich, director of the Pulkovo Observatory perished tragically in the time of Stalin's repressions at the end of the 1930's.

005.020 **Life and scientific work of Boris Petrovich Gerasimovich, on the occasion of the 100th anniversary of his birthday (1889 – 1937).**
A. I. Eremeeva.
Istor.-Astron. Issled., Tom 21, p. 253 – 301 (1989). In Russian. – See Abstr. 003.017 for the main entry.

005.021 **Vasilij Grigor'evich Fesenkov – an outstanding astrophysicist, on the occasion of the 100th anniversary of his birthday (1889 – 1972).**
N. B. Divari.
Istor.-Astron. Issled., Tom 21, p. 302 – 326 (1989). In Russian. – See Abstr. 003.017 for the main entry.

005.022 **Robert Wheeler Willson: his life and legacy.**
B. L. Welther.
Liller Symposium: Asteroids to quasars, p. 255 – 261 (1991). – See Abstr. 012.087 for the main entry.
For the 20 years between 1962 and 1982, William Liller held at the Robert Wheeler Willson Professorship of Applied Astronomy at Harvard University. Therefore, the Liller Symposium presents an appropriate occasion to remember Willson and his contribution to astronomy at Harvard.

005.023 **E. P. Fedorov and his school of sciences (1909 – 1986).**
A. A. Korsun'.
Istor.-Astron. Issled., Tom 21, p. 327 – 341 (1989). In Russian. – See Abstr. 003.017 for the main entry.

005.024 **On the occasion of the 100th anniversary of A. A. Friedmann's (1888 – 1925) birth.**
G. E. Gorelik.
Istor.-Astron. Issled., Tom 21, p. 397 – 398 (1989). In Russian. – See Abstr. 003.017 for the main entry.

005.025 **Christoph Clavius. Ein Namen- und Schriftenverzeichnis zu seinen *Opera mathematica*.**
E. Knobloch.
Bollettino di Storia delle Scienze Matematiche, Vol. 10, Fasc. 2, p. 135 – 189 (1990).

005.026 **Edwin Powell Hubble and cosmology.**
R. Kippenhahn.
Mem. Soc. Astron. Ital., Vol. 62, No. 3, p. 533 – 552 (1991). – See Abstr. 012.094 for the main entry.
The author discusses some of Hubble's contributions to cosmology, expecially using the method of pulsating stars as distance indicators.

005.027 **The happiest thought of Einstein's life.**
M. Heller.
Astron. Q., Vol. 8, No. 3, p. 177 – 180 (1991).

005.028 **Johann Franz Encke, 1791 – 1865.**
A. H. Batten.
J. R. Astron. Soc. Can., Vol. 85, No. 6, p. 316 – 323 (Dec 1991).
A brief account is given of the life and work of Johann Encke, born on September 23, 1791, with special reference to his studies of comets and celestial mechanics, his work on the Berliner Astronomisches Jahrbuch, and his role in the discovery of Neptune.

005.029 **Beatrice Muriel Hill Tinsley: una vita per la scienza.**
M. D. Guarnieri, M. G. Pancaldi Stagni.
Orione, Vol. 11, N. 2, p. 28 – 33 (Mar – Apr 1991).

005.030 **God's turnstile: the work of John Wheeler and Stephen Hawking.**
D. Overbye.
Mercury, Vol. 20, No. 4, p. 98 – 108 (Jul – Aug 1991).

005.031 **Henri Poincaré: the ninth Bruce medalist.**
J. S. Tenn.
Mercury, Vol. 20, No. 4, p. 111 – 112 (Jul – Aug 1991).

005.032 **Jacobus Cornelius Kapteyn: the tenth Bruce medalist.**
J. S. Tenn.
Mercury, Vol. 20, No. 5, p. 145 – 147, 159 (Sep – Oct 1991).

005.033 **Oskar Backlund: the eleventh Bruce medalist.**
J. S. Tenn.
Mercury, Vol. 20, No. 6, p. 175, 178 – 179 (Nov – Dec 1991).

005.034 **En succesrig dansk amatørastronom, E. V. Petersen.**
P. B. Darnell.
Astron. Tidsskr., Årg. 24, Nr. 3, p. 112 – 114 (1991).

005.035 **Hans Moritz Graf von Brühl und seine Leistungen im Zusammenhang mit dem Übergang zu den astronomischen Meßinstrumenten mit Vollkreisen Ende des 18. Jahrhunderts.**
K.-D. Herbst.
Sterne, Band 67, Heft 6, p. 335 – 342 (1991).

005.036 **I contributi astronomici di Jean-Baptiste Le Rond D'Alembert.**
S. D'Amico.
Orione, Vol. 11, N. 2, p. 22 – 27 (Mar – Apr 1991).

Research in the History of Astronomy (XXI). Past – presence – future.
See Abstr. 003.017.

Tycho's new star.
See Abstr. 004.006.

Tobias Mayer et le traitement des observations.
See Abstr. 004.038.

The Lilienthal tragedy.
See Abstr. 004.041.

Illuminating incidents in antipodean astronomy: section A of the Royal Society of New South Wales.
See Abstr. 004.092.

50 years of McDonald Observatory.
See Abstr. 009.009.

Elementary particles and the universe. Essays in honor of Murray Gell–Mann.
See Abstr. 012.007.

A. A. Friedmann: centenary volume. Proceedings.
See Abstr. 012.114.

On Bengt Strömgren's work in photometry.
See Abstr. 013.053.

The interplanetary plasma.
See Abstr. 013.060.

40 years numerical stellar atmospheres. Concluding remarks and personal considerations.
See Abstr. 064.117.

Minor planet 3900 Knežević.
See Abstr. 098.086.

006 Personal Notes

006.001 **Fred C. Adams received the Robert J. Trumpler Award 1991 of the Astronomical Society of the Pacific.**
N. Morrison.
Mercury, Vol. 20, No. 6, p. 184 – 186 (Nov – Dec 1991).

006.002 **Richard Berry and the staff of Astronomy Magazine received the Klumpke–Roberts Award 1991 of the Astronomical Society of the Pacific.**
N. Morrison.
Mercury, Vol. 20, No. 6, p. 189 – 190 (Nov – Dec 1991).

006.003 **Address by the President (Professor K. A. Pounds) on the presentation of the Chapman Medal to Professor S. W. H. Cowley on Friday 1991 March 8.**
K. A. Pounds.
Q. J. R. Astron. Soc., Vol. 32, No. 3, p. 215 – 216 (Sep 1991).

006.004 **The address by the President (Professor K. A. Pounds) on the presentation of the Eddington Medal to Professor I. Iben Jr. on Friday 1991 May 10.**
K. A. Pounds.
Q. J. R. Astron. Soc., Vol. 32, No. 4, p. 353 – 354 (Dec 1991).

006.005 **Otto Kippes received the Amateur Achievement Award 1991 of the Astronomical Society of the Pacific.**
N. Morrison.
Mercury, Vol. 20, No. 6, p. 188 – 189 (Nov – Dec 1991).

006.006 **Robert D. McClure, René Racine and the Instrumentation Group of the Dominion Astrophysical Observatory received the Muhlmann Prize 1991 of the Astronomical Society of the Pacific.**
N. Morrison.
Mercury, Vol. 20, No. 6, p. 186 – 188 (Nov – Dec 1991).

006.007 **Donald E. Osterbrock received the Bruce Medal 1991 of the Astronomical Society of the Pacific.**
N. Morrison.
Mercury, Vol. 20, No. 6, p. 182 – 184 (Nov – Dec 1991).

006.008 **René Racine, Robert D. McClure and the Instrumentation Group of the Dominion Astrophysical Observatory received the Muhlmann Prize 1991 of the Astronomical Society of the Pacific.**
N. Morrison.
Mercury, Vol. 20, No. 6, p. 186 – 188 (Nov – Dec 1991).

006.009 **Allan R. Sandage received the 1991 Crafoord Prize from the Swedish Academy of Sciences.**
Phys. Today, Vol. 44, No. 12, p. 91 (Dec 1991).

006.010 **The address by the President (Professor K. A. Pounds) on the presentation of the Gold Medal to Professor G. J. Wasserburg on Friday 1991 May 10.**
K. A. Pounds.
Q. J. R. Astron. Soc., Vol. 32, No. 4, p. 351 – 352 (Dec 1991).

007 Obituaries

007.001 **Eloge: Eric John Aiton, 1920 – 21 February 1991.**
J. B. Brackenridge.
Isis, Vol. 82, No. 314, p. 689 – 691 (Dec 1991).

007.002 **Pavel Andrle (1 September 1936 – 28 November 1991).**
L. Perek, J. Ruprecht.
Bull. Astron. Inst. Czech., Vol. 42, No. 6, p. 408 (Nov 1991).

007.003 **In memoriam: Oleg V. Dobrovolsky (12 December 1914 – 12 December 1989).**
M. Ya. Marov, S. I. Ibadov.
Icarus, Vol. 94, No. 2, p. 259 (Dec 1991).

007.004 **Henry Rainsford Hulme, 9 August 1908 – 8 January 1991.**
R. J. Tayler.
Q. J. R. Astron. Soc., Vol. 32, No. 3, p. 313 (Sep 1991).

007.005 **In memoriam: Peter Mackenzie Millman (10 August 1906 – 11 December 1990).**
I. Halliday.
Icarus, Vol. 93, No. 2, p. 181 – 182 (Oct 1990).

007.006 **Eloge: Otto Neugebauer, 26 May 1899 – 19 February 1990.**
D. Pingree.
Isis, Vol. 82, No. 311, p. 87 – 88 (Mar 1991).

007.007 **Michael Victor Penston, 17 February 1943 – 23 December 1990.**
B. Pagel.
Q. J. R. Astron. Soc., Vol. 32, No. 3, p. 314 – 318 (Sep 1991).

007.008 **S. Plakidis (22 May 1893 – 30 January 1991).**
G. Contopoulos.
Q. J. R. Astron. Soc., Vol. 32, No. 4, p. 483 – 484 (Dec 1991).

007.009 **P. D. Stern (1913 – 31 January 1991).**
D. L. Schrader.
Q. J. R. Astron. Soc., Vol. 32, No. 4, p. 485 – 486 (Dec 1991).

007.010 **Katsuo Tanaka (1943 – 1990).**
E. Hiei, T. Sakurai.
Sol. Phys., Vol. 136, No. 1, p. iv – v (Nov 1991). Special issue: The YOHKOH (SOLAR-A) mission. For the hardcover edition see 003.013.

007.011 **In memoriam Katsuo Tanaka.**
H. Zirin, T. Sakurai.
International SOLAR-A Science Meeting: Flare physics in solar activity maximum 22, p. 5 – 8 (1991). – See Abstr. 012.104 for the main entry.

007.012 **Eloge: Victor E. Thoren, 13 May 1935 – 9 March 1991.**
O. Gingerich.
Isis, Vol. 82, No. 314, p. 693 – 694 (Dec 1991).

007.013 **Victor E. Thoren, 13 May 1935 – 9 March 1991.**
R. S. Westfall.
J. Hist. Astron., Vol. 22, Part 3, p. 253 – 254 (Aug 1991).

007.014 **Betty Louise Turtle (1941 – 1990).**
J. W. V. Storey, D. J. Faulkner.
Proc. Astron. Soc. Aust., Vol. 9, No. 1, p. 6 – 7 (1991). – See Abstr. 012.078 for the main entry.

007.015 **Erinnerungen an Hans Vehrenberg, 6. März 1910 – 2. August 1991.**
G. D. Roth.
Sterne Weltraum, Jahrg. 30, Nr. 12, p. 766 – 768 (Dec 1991).

007.016 **Basilis Xanthopoulos died 27 November 1990.**
7. Latin American Symposium on Relativity and Gravitation, p. VII (1991). – See Abstr. 012.005 for the main entry.

008 Publications of Observatories, Institutes

Reports, communications, and publications of observatories and astronomical institutes are recorded in this section; included are numbered series of reprints. Whenever possible, the numbers of the abstracts referring to the publications are given. The places of observatories and institutes are listed in alphabetical order. If only the formal name of an observatory or institute is known, its place can be found in the following index list.

Aerospace Corporation	**Los Angeles,** CA (USA)
Allegheny Observatory	**Pittsburgh,** PA (USA)
Anglo-Australian Observatory	**Epping** (Australia)
Arcetri Astrophysical Observatory	**Florence** (Italy)
Archenhold Observatory	**Berlin** (Germany)
Argentine Institute of Radio Astronomy	**Villa Elisa** (Argentina)
Arizona State University	**Tempe,** AZ (USA)
Astronomical Institute of the Czechoslovak Academy of Sciences	**Ondřejov** (Czechoslovakia)
Astronomisches Rechen-Institut	**Heidelberg** (Germany)
AT&T Bell Laboratories	**Holmdel,** NJ (USA)
Babelsberg Observatory	**Potsdam** (Germany)
Bartol Research Foundation	**Newark,** DE (USA)
Battelle Memorial Institute	**Richland,** WA (USA)
Behlen Observatory	**Lincoln,** NB (USA)
Bergedorf Observatory	**Hamburg** (Germany)
Blindern Institute of Theoretical Astrophysics	**Oslo** (Norway)
Bosscha Observatory	**Lembang** (Indonesia)
Boyden Observatory	**Bloemfontein** (South Africa)
Braeside Observatory	**Flagstaff,** AZ (USA)
Bundesamt für Seeschiffahrt und Hydrographie	**Hamburg** (Germany)
Bureau Central de l'IERS	**Paris** (France)
Bureau International des Poids et Mesures, Time Section	**Sèvres** (France)
C. E. Kenneth Mees Observatory	**Rochester,** NY (USA)
California Institute of Technology	**Pasadena,** CA (USA)
Cantonal Observatory	**Neuchâtel** (Switzerland)
Capodimonte Astronomical Observatory	**Naples** (Italy)
Carlsberg Automatic Meridian Circle	**La Palma** (Spain)
Carnegie Institution of Washington, The Observatories	**Pasadena,** CA (USA)
Carter Observatory	**Wellington** (New Zealand)
Center for Astrophysics	**Cambridge,** MA (USA)
Center for High Angular Resolution Astronomy, Georgia State University	**Atlanta,** GA (USA)
Centre de Données Stellaires	**Strasbourg** (France)
Centro Astronomico Hispano-Aleman	**Calar Alto** (Spain)
Centro de Investigación de Astronomia (CIDA)	**Mérida** (Venezuela)
Cerro Calan National Astronomical Observatory	**Santiago** (Chile)
Cerro Tololo Interamerican Observatory	**La Serena** (Chile)
Charles University Astronomical Institute	**Prague** (Czechoslovakia)
Climenhaga Observatory	**Victoria** (Canada)
Clyde W. Tombaugh Observatory	**Lawrence,** KS (USA)
Cointe Observatory	**Liège** (Belgium)
Collurania Astronomical Observatory	**Teramo** (Italy)
Computer Sciences Corporation	**Beltsville,** MD (USA)
Cornell University	**Ithaca,** NY (USA)
David Dunlap Observatory	**Richmond Hill** (Canada)
Dearborn Observatory	**Evanston,** IL (USA)
Department of Astronomy and Space Sciences, Punjabi University	**Patiala** (India)
Department of Mathematics, University of Poona	**Poona** (India)
Dominion Astrophysical Observatory	**Victoria** (Canada)
Dominion Radio Astrophysical Observatory	**Penticton** (Canada)
Dudley Observatory	**Schenectady,** NY (USA)
Dunsink Observatory	**Dublin** (Ireland)
Dyer Observatory	**Nashville,** TN (USA)
Ebro Observatory	**Roquetas** (Spain)
Effelsberg Radio Observatory	**Bonn** (Germany)
Ege University Observatory	**Izmir** (Turkey)
Electronics Research Laboratory	**Göteborg** (Sweden)
Engelhardt Observatory	**Kazan'** (Russia)
Erwin W. Fick Observatory	**Ames,** IA (USA)

European Southern Observatory	**Garching** (Germany)
European Southern Observatory	**La Silla** (Chile)
Felix Aguilar Observatory	**San Juan** (Argentina)
Figl Observatory	**Vienna** (Austria)
Five College Astronomy Department	**Amherst,** MA (USA)
Floirac Observatory	**Bordeaux** (France)
Florida State University Radio Observatory	**Tallahassee,** FL (USA)
Flower and Cook Observatory	**Malvern,** PA (USA)
Franko State University Astronomical Observatory	**L'vov** (Ukraine)
Georgia State University	**Atlanta,** GA (USA)
Goddard Space Flight Center	**Greenbelt,** MD (USA)
Goethe Link Observatory	**Bloomington,** IN (USA)
Griffith Observatory	**Los Angeles,** CA (USA)
Hartebeesthoek Radio Astronomy Observatory	**Johannesburg** (South Africa)
Haute Provence Observatory	**Saint Michel l'Observatoire** (France)
Haystack Observatory	**Westford,** MA (USA)
Heinrich-Hertz-Institut	**Berlin** (Germany)
Heliophysical Observatory	**Debrecen** (Hungary)
Herzberg Institute of Astrophysics	**Ottawa** (Canada)
High Altitude Observatory	**Boulder,** CO (USA)
H.M. Nautical Almanac Office	**Cambridge,** (UK)
Hoher List Observatory	**Bonn** (Germany)
Hopkins Observatory	**Williamstown,** MA (USA)
Hvar Observatory	**Zagreb** (Croatia)
IBM Thomas J. Watson Research Center	**Yorktown Heights,** NY (USA)
Indian Institute of Astrophysics	**Bangalore** (India)
Infrared Telescope Facility	**Honolulu,** HI (USA)
Institut Astrofiziki	**Dushanbe** (Tajikistan)
Instituto de Astrofisica de Canarias	**La Laguna** (Spain)
International Earth Rotation Service	**Paris** (France)
International Latitude Observatory	**Carloforte-Cagliari** (Italy)
IUE Observatory	**Villafranca** (Spain)
James Clerk Maxwell Telescope	**Hilo,** HI (USA)
Jet Propulsion Laboratory	**Pasadena,** CA (USA)
Jodrell Bank Radio Observatory	**Manchester** (UK)
Judson B. Coit Observatory	**Boston,** MA (USA)
Kandilli Observatory	**Istanbul** (Turkey)
Kanzelhöhe Solar Observatory	**Graz** (Austria)
Kapteyn Astronomical Laboratory	**Groningen** (Netherlands)
Karl Schwarzschild Observatory	**Tautenburg** (Germany)
Kiepenheuer Institut	**Freiburg** (Germany)
Kiso Observatory	**Tokyo** (Japan)
Kitt Peak National Observatory	**Tucson,** AZ (USA)
Kodaikanal Observatory	**Bangalore** (India)
Königstuhl Observatory	**Heidelberg** (Germany)
Konkoly Observatory	**Budapest** (Hungary)
Korean National Astronomical Observatory	**Seoul** (Korea)
Kwasan and Hida Observatories	**Kyoto** (Japan)
Laboratory for High Energy Astrophysics	**Greenbelt,** MD (USA)
Las Campanas Observatory	**Pasadena,** CA (USA)
Latitude Station of the Polish Academy of Sciences	**Borowiec** (Poland)
Lawrence Livermore National Laboratory	**Livermore,** CA (USA)
Leander McCormick Observatory	**Charlottesville,** VA (USA)
Lick Observatory	**Santa Cruz,** CA (USA)
Lindheimer Astronomical Research Center	**Evanston,** IL (USA)
Lockheed Palo Alto Research Laboratory	**Palo Alto,** CA (USA)
Lockheed Solar Observatory	**Saugus,** CA (USA)
Lohrmann Observatory	**Dresden** (Germany)
Louisiana State University Observatory	**Baton Rouge,** LA (USA)
Lowell Observatory	**Flagstaff,** AZ (USA)
Lunar and Planetary Laboratory	**Tucson,** AZ (USA)
Main Astronomical Observatory of the USSR Academy of Sciences	**Pulkovo** (Russia)
Max-Planck-Institut für Astronomie	**Calar Alto** (Spain)
Max-Planck-Institut für Astronomie	**Heidelberg** (Germany)
Max-Planck-Institut für Kernphysik	**Heidelberg** (Germany)
Max-Planck-Institut für Physik und Astrophysik, Institut für Astrophysik	**Garching** (Germany)

Max-Planck-Institut für Physik und Astrophysik, Institut für Extraterrestrische Physik	**Garching** (Germany)
Max-Planck-Institut für Radioastronomie	**Bonn** (Germany)
McDonald Observatory	**Fort Davis,** TX (USA)
McDonnell Center for the Space Sciences	**St. Louis,** MO (USA)
Meudon Observatory	**Paris** (France)
Michigan State University Observatory	**East Lansing,** MI (USA)
Millstone Hill Radar Observatory	**Westford,** MA (USA)
Minor Planet Center	**Cambridge,** MA (USA)
Molonglo Radio Observatory	**Sydney** (Australia)
Monterey Institute for Research in Astronomy	**Carmel Valley,** CA (USA)
Mount Hamilton	**Santa Cruz,** CA (USA)
Mount John University Observatory	**Lake Tekapo** (New Zealand)
Mount Palomar Observatory	**Pasadena,** CA (USA)
Mount Stromlo Observatory	**Canberra** (Australia)
Mount Wilson Observatory	**Pasadena,** CA (USA)
Mullard Radio Astronomy Observatory	**Cambridge** (UK)
Mullard Space Science Laboratory	**London** (UK)
Multiple Mirror Telescope Observatory	**Tucson,** AZ (USA)
N. Copernicus Astronomical Center	**Warsaw** (Poland)
N. Copernicus University Observatory	**Torun** (Poland)
NASA Headquarters	**Washington,** DC (USA)
National Astronomical Observatory	**Tokyo** (Japan)
National Astronomy and Ionosphere Center	**Ithaca,** NY (USA)
National Bureau of Standards	**Washington,** DC (USA)
National Radio Astronomy Observatory	**Charlottesville,** VA (USA)
National Radio Astronomy Observatory	**Green Bank,** WV (USA)
National Radio Astronomy Observatory	**Socorro,** NM (USA)
National Solar Observatory	**Sunspot,** NM (USA)
New Mexico State University Observatory	**Las Cruces,** NM (USA)
Nicholas Copernicus Observatory	**Brno** (Czechoslovakia)
Nizamiah and Japal-Rangapur Observatories	**Hyderabad** (India)
Nuffield Radio Astronomy Laboratories	**Manchester** (UK)
Observatorio del Ebro	**Roquetas** (Spain)
Ole Roemer Observatory	**Aarhus** (Denmark)
Onsala Space Observatory	**Göteborg** (Sweden)
Osservatorio Astronomico di Brera	**Milano-Merate** (Italy)
Owens Valley Radio Observatory	**Big Pine,** CA (USA)
Pennsylvania State University	**University Park,** PA (USA)
Perkins Observatory	**Delaware,** OH (USA)
Perth Observatory	**Bickley** (Australia)
Pic du Midi Observatory	**Toulouse** (France)
Pino Torinese Observatory	**Turin** (Italy)
Planetary Science Institute	**Tucson,** AZ (USA)
Purple Mountain Observatory	**Nanking** (China)
Raman Research Institute	**Bangalore** (India)
Rattlesnake Mountain Observatory	**Richland,** WA (USA)
Real Instituto y Observatorio de la Armada	**San Fernando** (Spain)
Remeis Observatory	**Bamberg** (Germany)
Rensselaer Observatory	**Troy,** NY (USA)
Ritter Astrophysical Research Center	**Toledo,** OH (USA)
Rosemary Hill Observatory	**Bronson,** FL (USA)
Rothney Astrophysical Observatory	**Calgary** (Canada)
Royal Greenwich Observatory	**Cambridge** (UK)
Royal Observatory	**Edinburgh** (UK)
Royal Observatory of Belgium	**Uccle** (Belgium)
Rutgers University	**Piscataway,** NJ (USA)
Rutherford Appleton Laboratory	**Chilton** (UK)
Sagamore Hill Radio Observatory	**Hanscom,** MA (USA)
Saltsjöbaden Observatory	**Stockholm** (Sweden)
San Vittore Observatory	**Bologna** (Italy)
Satelliten-Beobachtungsstation	**Zimmerwald** (Switzerland)
Shaanxi Astronomical Observatory	**Lintong** (China)
Smithsonian Astrophysical Observatory	**Cambridge,** MA (USA)
Sonnenborgh Observatory	**Utrecht** (Netherlands)
Sonoma State University	**Rohnert Park,** CA (USA)
South African Astronomical Observatory	**Cape Town** (South Africa)
Space Telescope Science Institute	**Baltimore,** MD (USA)
Special Astrophysical Observatory	**Zelenchukskaya** (Russia)
Sproul Observatory	**Swarthmore,** PA (USA)

Stanford Center for Radar Astronomy	**Menlo Park,** CA (USA)
Stellar Data Center	**Strasbourg** (France)
Sternberg State Astronomical Institute	**Moscow** (Russia)
Steward Observatory	**Tucson,** AZ (USA)
Stockert Radio Observatory	**Bonn** (Germany)
Struve Astrophysical Observatory	**Tartu** (Estonia)
Table Mountain Observatory	**Wrightwood,** CA (USA)
Tata Institute of Fundamental Research	**Bombay** (India)
Tôhoku University Observatory	**Sendai** (Japan)
United Kingdom Infrared Telescope (UKIRT)	**Hilo,** HI (USA)
University of Sussex	**Brighton** (UK)
U.S. Naval Observatory	**Washington,** DC (USA)
UT Radio Astronomy Observatory	**Marfa,** TX (USA)
Uttar Pradesh State Observatory	**Naini Tal** (India)
Van Vleck Observatory	**Middletown,** CT (USA)
Vassar College Observatory	**Poughkeepsie,** NY (USA)
Vatican Observatory	**Castel Gandolfo** (Vatican City)
W. M. Keck Observatory	**Berkeley,** CA (USA)
W. M. Keck Observatory	**Kamuela,** HI (USA)
Warner and Swasey Observatory	**Cleveland,** OH (USA)
Washburn Observatory	**Madison,** WI (USA)
Wesleyan Radio Observatory	**Delaware,** OH (USA)
Western Ontario University Observatory	**London** (Canada)
Whipple Observatory (MMT)	**Tucson,** AZ (USA)
Wilhelm Foerster Observatory	**Berlin** (Germany)
William Marsh Rice University, Department of Space Physics and Astronomy	**Houston,** TX (USA)
Wyoming Infrared Observatory	**Laramie,** WY (USA)
Yale Astronomy Department	**New Haven,** CT (USA)
Yale University Astronomical Observatory	**New Haven,** CT (USA)
Yerkes Observatory	**Chicago,** IL (USA)
Yunnang Obs.	**Kunming** (China)
Zentralinstitut für Physik der Erde	**Potsdam** (Germany)

Ahmedabad (India)

008.001 **Physical Research Laboratory. Report for the period 1988 April 1 – 1989 March 31.**
Bull. Astron. Soc. India, Vol. 19, Nos. 1 – 2, p. 123 – 137 (Mar – Jun 1991).

008.002 **Physical Research Laboratory. Report for the period 1989 April 1 – 1990 March 31.**
Bull. Astron. Soc. India, Vol. 19, Nos. 1 – 2, p. 139 – 153 (Mar – Jun 1991).

Asiago (Italy)

008.003 **Telescopio Nazionale Galileo, Technical Report.**
Nos. 5 (54.034.137), 6 (54.032.024), 7 (54.032.025), 8 (54.031.073), 9 (54.032.026).

Austin, TX (USA)

008.004 **GaussFIT, University of Texas at Austin.**
W. Jefferys, B. McArthur, J. McCartney.
Bull. Am. Astron. Soc., Vol. 23, No. 2, p. 997 – 999 (1991). AAS Working Group on Astronomical Software. Annual and technical reports for astronomical software facilities.

008.005 **Pickles, University of Texas at Austin.**
J. McCartney, W. Jefferys, B. McArthur.
Bull. Am. Astron. Soc., Vol. 23, No. 2, p. 1005 – 1006 (1991). AAS Working Group on Astronomical Software. Annual and technical reports for astronomical software facilities.

Baltimore, MD (USA)

008.006 **Space Telescope Science Institute, Newsletter.**
Vol. 8, No. 3 (Nov 1991).

008.007 **Space Telescope Science Institute Symposium Series.**
Vols. 4 (54.012.124), 5 (54.012.062).

Brighton (UK)

008.009 **Astronomy Centre, University of Sussex. Report for the period 1990 January 1 to 1990 December 31.**
R. J. Tayler.
Q. J. R. Astron. Soc., Vol. 32, No. 4, p. 449 – 458 (Dec 1991).

Brno (Czechoslovakia)

008.010 **Astronomical Institute of the University Brno, Contributions.**
Nos. 28 (54.124.039), 29 (54.014.064), 30 (54.014.065).

Cambridge (UK)

008.011 **Institute of Astronomy, University of Cambridge. Annual report Oct 1990 – Sep 1991.**
M. J. Rees.
Cambridge Univ. (UK). Inst. of Astronomy, 55 p. (1991).

008.012 **Gemini. Newsletter of the Royal Greenwich Observatory.**
Nos. 33 (Sep 1991), 34 (Dec 1991).

Cambridge, MA (USA)

008.013 **PROS, Smithsonian Astrophysical Observatory. Report for the period Feb 1990 – Jan 1991.**
D. M. Worrall.
Bull. Am. Astron. Soc., Vol. 23, No. 2, p. 1007 – 1008 (1991). AAS Working Group on Astronomical Software. Annual and technical reports for astronomical software facilities.

008.014 **International Astronomical Union, Circular.**
Nos. 5296 – 5419 (Jul – Dec 1991).

008.015 **The Minor Planet Circulars/Minor Planets and Comets.**
Nos. 18485 – 19346 (Aug – Nov 1991).

Cape Town (South Africa)

008.016 **SAAO Newsletter.**
No. 18 (Oct 1991).

Castel Gandolfo (Vatican City)

008.017 **Vatican Observatory Publications.**
Vol. 3, No. 2 (1992) (54.008.018).

008.018 **Vatican Observatory. Annual report 1991.**
G. V. Coyne.
Vatican Obs. Publ., Vol. 3, No. 2, p. 21 – 45 (1992).

Crimea (Ukraine)

008.019 **Bulletin of the Crimean Astrophysical Observatory.**
Vol. 81 (1990).

Dresden (Germany)

008.020 **Mitteilungen des Lohrmann–Observatoriums der Technischen Universität Dresden.**
Nr. 61 (54.053.005), 62 (54.036.066), 63 (53.008.053), 64 (51.041.030), 65 (54.041.025).

Epping (Australia)

008.021 **AAO Newsletter.**
Nos. 58 (Jul 1991), 59 (Oct 1991).

008.022 **Anglo–Australian Observatory. Report of the Anglo–Australian Telescope Board, 1 July 1990 to 30 June 1991.**
Australian Government Publishing Service, Canberra (Australia), 96 p. (1991). ISSN 0728–6554.

Garching (Germany)

008.023 **MPE Reports.**
No. 231 (54.035.009).

008.024 **MPA Reports.**
MPA—593 (54.161.149), MPA—620 (54.160.037).

008.025 **MIDAS, Image Processing Group, European Southern Observatory. Report for the period Mar 1990 – Mar 1991.**
Bull. Am. Astron. Soc., Vol. 23, No. 2, p. 1004 (1991). AAS Working Group on Astronomical Software. Annual and technical reports for astronomical software facilities.

008.026 **ESO Conference and Workshop Proceedings.**
Nos. 36 (54.012.092), 37 (54.012.075), 38 (54.012.061).

008.027 **ESO Scientific Report.**
No. 10 (54.002.112).

008.028 **Space Telescope–European Coordinating Facility Newsletter.**
No. 16 (Aug 1991).

008.029 **The Messenger – El Mensajero.**
Nos. 65 (Sep 1991), 66 (Dec 1991).

Graz (Austria)

008.030 **Mitteilungen des Sonnenobservatoriums Kanzelhöhe.**
Nr. 56 (51.080.049), 57 (51.071.005), 58 (52.072.089), 59 (52.072.107), 60 (52.080.071), 61 (52.071.013), 62 (52.072.085), 63 (52.071.030), 64 (54.044.046), 65 (52.034.057), 66 (50.073.089), 67 (52.077.037), 68 (52.073.116), 69 (52.072.005), 70 (52.071.050), 71 (53.071.026), 72 (54.071.022), 73 (52.071.005), 74 (54.072.064), 75 (54.071.043), 76 (54.071.045), 77 (54.082.067).

Greenbelt, MD (USA)

008.031 **FITS, IAU FITS Working Group. Report for the period Sep 1987 – Mar 1991.**
B. M. Schlesinger, P. Grosbøl, D. Wells.
Bull. Am. Astron. Soc., Vol. 23, No. 2, p. 993 – 994 (1991). AAS Working Group on Astronomical Software. Annual and technical reports for astronomical software facilities.

008.032 **FITS Support Office, National Space Science Data Center. Report for the period 1 Mar 1990 – 15 Feb 1991.**
B. M. Schlesinger, D. M. Sawyer.
Bull. Am. Astron. Soc., Vol. 23, No. 2, p. 995 – 996 (1991). AAS Working Group on Astronomical Software. Annual and technical reports for astronomical software facilities.

008.033 **National Aeronautics and Space Administration (NASA), Greenbelt, MD (USA). National Space Science Data Center (NSSDC) / World Data Center A for Rockets and Satellites**
NSSDC/WDC-A-R/S—91-02 (54.002.115), 91-17 (54.002.113), 91-18 (54.002.114), 91-19 (54.002.116).

Heidelberg (Germany)

008.034 **Apparent Places of Fundamental Stars 1993.**
(54.046.009).

008.035 **Astronomisches Rechen-Institut Heidelberg, Mitteilungen Serie A.**
Nr. 230 (54.151.059), 232 (54.151.091).

008.036 **Astronomisches Rechen-Institut Heidelberg, Mitteilungen, Serie B.**
Nr. 176 (54.036.151), 177 (54.153.029), 178 (53.041.010).

Helsinki (Finland)

008.037 **Helsinki University of Technology, Metsähovi Radio Research Station, Report Series A.**
No. 7 (1991) (54.077.075).

Hilo, HI (USA)

008.038 **United Kingdom Infrared Telescope. Annual report 1990.**
Royal Observatory, Edinburgh (UK); Joint Astronomy Centre, Hilo, HI (USA), 55 p. (1991).

Hyderabad (India)

008.039 **Nizamiah & Rangapur Observatories and Department of Astronomy, Osmania University, Reprint.**
Nos. 144 (53.117.322), 145 (53.122.141), 146 (53.114.131).

Izmir (Turkey)

008.040 **Publications of Ege University Observatory.**
No. 28 (1988-1989) (45.119.034, 45.002.004, 45.119.012, 45.117.218, 45.119.045, 46.119.017, 45.117.258, 46.119.065, 49.119.005, 49.082.001, 49.119.036, 50.119.042, 50.119.043, 49.117.294).

Jena (Germany)

008.041 **Mitteilungen der Universitäts-Sternwarte zu Jena.**
Nr. 175 (43.153.017), 176 (44.103.028), 177 (49.036.049), 178 (44.117.022), 179 (50.153.035), 180 (50.022.038), 181 (50.112.104), 182 (50.131.391), 183 (50.131.378), 184 (50.131.392), 185 (50.131.399).
This series discontinues with Nr. 185.

Kazan' (Russia)

008.042 **Izvestiya Astronomicheskoj Observatorii Ehngel'gardta.**
Vyp. 54 (1989), 55 (1990).

Kunming (China)

008.043 **Publications of Yunnan Observatory.**
Nos. 4 (1990) (54.012.047), 1 - 4 (1991).

Lintong (China)

008.044 **Publications of the Shaanxi Astronomical Observatory.**
Vol. 14, No. 2 (Dec 1991).

London (UK)

008.045 **Mullard Space Science Laboratory, University College London. Report for the period 1989 August 1 to 1991 March 31.**
J. L. Culhane.
Q. J. R. Astron. Soc., Vol. 32, No. 4, p. 459 - 469 (Dec 1991).

Manchester (UK)

008.046 **Astrophysics Group, Mathematics Department, UMIST, Manchester. Report for the period 1989 January 1 to 1990 December 31.**
D. A. Williams.
Q. J. R. Astron. Soc., Vol. 32, No. 3, p. 305 - 312 (Sep 1991).

008.047 **University of Manchester, Nuffield Radio Astronomy Laboratories, Jodrell Bank. Report of the researches in radio astronomy at Jodrell Bank for the year ending 1990 December 31.**
R. D. Davies.
Q. J. R. Astron. Soc., Vol. 32, No. 4, p. 429 - 447 (Dec 1991).

Middletown, CT (USA)

008.048 **Van Vleck Observatory Contribution.**
No. 11 (54.012.027), 12 (54.012.074).

Naini Tal (India)

008.049 **Uttar Pradesh State Observatory. Report for the period 1987 January 1 - 1989 December 31.**
Bull. Astron. Soc. India, Vol. 19, Nos. 1 - 2, p. 155 - 161 (Mar - Jun 1991).

Nanking (China)

008.050 **Publications of Purple Mountain Observatory.**
Vol. 10, Nos. 2 (Jun 1991), 4 (Dec 1991).

Neuchâtel (Switzerland)

008.051 **Rapport de l'Observatoire cantonal de Neuchâtel. Exercice 1991.**
G. Busca, G. Fischer.
Observatoire cantonal, Neuchâtel (Switzerland), 32 p. (1992).

Ondřejov (Czechoslovakia)

008.052 **Publications of the Astronomical Institute of the Czechoslovak Academy of Sciences.**
No. 78 (54.077.069).

Paris (France)

008.053 **IERS Bulletin. B.**
Nos. 41 – 46 (Jul – Dec 1991) (54.044.002).

008.054 **Explanatory supplement to IERS Bulletins A and B.**
Observatoire de Paris (France). Bureau Central de l'IERS, 7 p. (Mar 1992).

008.055 **IERS (International Earth Rotation Service) Technical Note.**
Nos. 7 (Dec 1991) (54.043.011), 8 (Oct 1991) (54.044.044), 9 (Dec 1991) (54.043.012).

Pasadena, CA (USA)

008.056 **Caltech VLBI Analysis Programs, California Institute of Technology.**
T. J. Pearson.
Bull. Am. Astron. Soc., Vol. 23, No. 2, p. 991 – 992 (1991). AAS Working Group on Astronomical Software. Annual and technical reports for astronomical software facilities.

Peking (China)

008.057 **Publications of the Beijing Astronomical Observatory.**
No. 17 (Jun 1991).

Poona (India)

008.058 **Khagol. The IUCAA Bulletin.**
Nos. 7 (Jul 1991), 8 (Oct 1991).

Potsdam (Germany)

008.059 **Solar Radio Data.**
Parts III/91 (54.077.027), IV/91 (54.077.028).
Part IV/91 is the final issue of this series.

Preston (UK)

008.060 **Lancashire Polytechnic Department of Physics and Astronomy. Report for the period 1989 October to 1990 September.**
E. I. Robson.
Q. J. R. Astron. Soc., Vol. 32, No. 4, p. 471 – 480 (Dec 1991).

Roquetas (Spain)

008.061 **Publicaciones del Observatorio del Ebro, Miscelánea.**
No. 39 (54.084.038).

Saint Michel l'Observatoire (France)

008.062 **La Lettre de l'OHP.**
Observatoire de Haute-Provence, St. Michel l'Observatoire (France)
No. 7 (Oct 1991).

Sendai (Japan)

008.063 **Annual report of research activities of the Astronomy Group, Tôhoku University.**
Sci. Rep. Tôhoku Univ., Eighth Ser., Vol. 12, No. 1, p. 105 – 110 (Jul 1991).

Sèvres (France)

008.064 **Bureau International des Poids et Mesures (BIPM), Circular T.**
Nos. 41 – 46 (Jul – Dec 1991) (54.044.045).

Shanghai (China)

008.065 **Annals of Shanghai Observatory, Academia Sinica.**
No. 12 (1991).

Sonneberg (Germany)

008.066 **Mitteilungen der Sternwarte zu Sonneberg.**
Nr. 99 (54.122.029).

008.067 **Mitteilungen über Veränderliche Sterne.**
Band 12, Heft 6 (Aug 1991), Heft 7 (Oct 1991).

Strasbourg (France)

008.068 **Bulletin d'Information du Centre de Données Stellaires.**
Nos. 38 (Dec 1990), 39 (1991) (Index to Nos. 1 – 38).

Tartu (Estonia)

008.069 **Chronicle (1990).**
H. Ibrus.
Calendar of the Tartu Astronomical Observatory for the year 1992, p. 39 – 43 (1991). In Estonian. See Abstr. 046.016.

Tokyo (Japan)

008.070 **National Astronomical Observatory (Japan), Reprint.**
Nos. 110 (54.131.352), 111 (54.131.100), 113 (54.160.013), 114 (54.031.026), 115 (54.142.026), 116 (54.131.323), 118 (54.073.111), 119 (53.051.046), 120 (54.014.052), 121 (54.082.060), 122 (54.159.185), 124, 125 (54.161.366), 126 (54.042.072), 127 (54.131.280), 128 (53.013.060), 132 (54.072.121), 136 (54.033.017).

008.071 **Contributions from the Department of Astronomy, University of Tokyo.**
Nos. 386 (54.064.201), 387 (54.064.202).

Torino (Italy)

008.072 **Publications of the Osservatorio Astronomico di Torino.**
Nos. 483 (50.117.091), 484 (50.098.039), 485 (49.117.293), 486 (50.098.098), 487, 488 (50.008.089), 489 (50.101.129), 490, 491, 492 (50.098.052), 493 (51.022.045), 494 (51.098.171), 495 (51.098.146), 496 (51.098.163), 497 (51.098.189), 498 (51.022.146), 499 (51.098.162), 500 (51.131.077), 501 (51.131.417), 502 (51.131.415), 503 (51.116.040), 504 (49.063.007), 505 (49.158.162), 506 (49.063.073), 507 (50.063.090), 508 (50.134.124), 509, 510 (51.126.043), 511 (50.064.022), 512 (51.011.004), 513 (51.098.042), 514 (51.036.002), 515 (51.100.013), 516 (49.066.299), 517 (51.065.059), 518 (51.098.092), 519 (51.064.061), 520 (51.062.134), 521 (51.117.350), 522 (51.131.415), 523 (51.131.417), 524 (52.125.007), 525 (51.041.038), 526 (51.041.026), 527 (51.098.191), 528 (51.062.043), 529 (51.062.019), 530 (52.041.014), 531 (52.098.123), 532 (52.098.122), 533 (52.043.001), 534 (52.036.230), 535 (52.036.017), 536 (51.041.054), 537 (52.158.163), 538 (52.098.025), 539 (53.117.015), 540 (52.159.058), 541 (52.062.221), 542 (52.121.060), 543 (52.098.100), 544 (52.098.103), 545 (52.043.001), 546 (51.158.171), 547 (52.063.048), 548 (53.063.034), 549, 550 (52.117.227), 551 (51.117.350), 552 (51.131.417), 553, 554, 555 (52.065.109) 556 (53.065.097), 557 (53.036.056), 558 (52.061.076), 559 (53.117.084), 560 (53.101.015), 561 (54.064.151), 562 (53.134.063), 563 (54.158.131), 564 (54.158.308), 565 (50.158.321), 566 (52.063.120), 567 (52.063.119), 568, 569 (54.117.078), 570 (54.153.036), 571 (53.158.149), 572, 573 (53.065.047), 574 (53.065.171), 575 (53.065.172), 576 (53.041.041), 577 (54.121.039), 578, 579, 580, 581, 582 (52.061.122), 583 (52.061.134), 584 (53.061.138), 585 (53.061.139), 586 (53.061.140), 587 (53.061.147), 588, 589 (54.066.140), 590 (54.158.249), 591 (53.065.165), 592 (54.064.053), 593 (54.157.092), 594 (54.063.101).

Tucson, AZ (USA)

008.073 **GONG, National Solar Observatory.**
J. A. Pintar.
Bull. Am. Astron. Soc., Vol. 23, No. 2, p. 1000 – 1003 (1991). AAS Working Group on Astronomical Software. Annual and technical reports for astronomical software facilities.
This report reviews the current status of the development of the data management system for the GONG helioseismology project.

Washington, DC (USA)

008.074 **IERS Bulletin. A.**
Vol. 4, Nos. 27 – 52 (Jul – Dec 1991) (54.044.007).

008.075 **United States Naval Observatory, Series 4, Daily time differences.**
Nos. 1274 – 1299 (Jul – Dec 1991) (54.044.008).

Zelenchukskaya (Russia)

008.076 **Astrofizicheskie Issledovaniya. Izvestiya Spetsial'noj Astrofizicheskoj Observatorii.**
Tom 31 (1991).

008.077 **Chronicle.**
Astrofiz. Issled. Izv. Spets. Astrofiz. Obs., Tom 31, p. 180 – 181 (1991). In Russian.

008.078 **Bulletin of the Special Astrophysical Observatory – North Caucasus.**
Vol. 32 (1990).

009 Notes on Observatories, Planetaria, Exhibitions

009.001 **High resolution imaging from Mauna Kea.**
R. D. McClure.
Bull. Am. Astron. Soc., Vol. 23, No. 2, p. 898 (1991). Abstract. – See Abstr. 012.008 for the main entry.

009.002 **Bericht uit Flagstaff.**
P. Jenniskens.
Radiant, Jaarg. 13, Nr. 4, p. 86 – 87 (Aug 1991).

009.003 **Als Neuling auf dem Calar Alto.**
W. Pfau.
Sterne, Band 67, Heft 4, p. 185 – 192 (1991).

009.004 **Geschichte der Universitätssternwarte Leipzig.**
J. Helfricht, J. Classen.
Sterne, Band 67, Heft 4, p. 201 – 213 (1991).

009.005 **Kleines Planetarium unter einer Sternwartenkuppel.**
H. Mucke.
Astron. Sch., Jahrg. 28, Heft 4, p. 25 – 27 (Aug 1991).

009.006 **The Canarian Observatories.**
F. Sánchez, J. V. Wall.
Astrophys. Lett. Commun., Vol. 28, No. 2–4, p. 47 – 67 (Sep 1991). – See Abstr. 012.072 for the main entry.
This is an overview of the present state of the Observatories of the Instituto de Astrofisica de Canarias (IAC): the Observatorio del Teide at Izaña on the island of Tenerife, and the Observatorio del Roque de Los Muchachos on the island of La Palma. The purpose is to provide the context for results obtained at these Observatories presented in the companion papers in this celebratory issue.

009.007 **The second birth of Pulkovo.**
A. N. Dadaev.
Istor.–Astron. Issled., Tom 21, p. 17 – 49 (1989). In Russian. – See Abstr. 003.017 for the main entry.
In 1989 the Pulkovo Observatory celebrated the 150th anniversary. During the Second World War the Observatory was destroyed and in March 1945 it was decided to reconstruct the Main Astronomical Observatory of the Academy of Sciences of the USSR. The author is one of the eldest scholars of the Observatory, he reports in detail on the past and present-day situation of Pulkovo.

009.008 **Astronomy in the city.**
J. A. Perwick, G. K. Urquhart, W. S. G. Walker.
Proc. Astron. Soc. Aust., Vol. 9, No. 1, p. 177 (1991). – See Abstr. 012.078 for the main entry.
Auckland Observatory is a public observatory with a strong research interest. Its activities are reviewed and described. The important role which public access to current astronomy plays is mentioned.

009.009 **50 years of McDonald Observatory.**
H. J. Smith.
Symposium in Celebration of the 50. Anniversary of McDonald Observatory of the University of Texas at Austin: Frontiers of stellar evolution, p. 1 - 17 (1991). - See Abstr. 012.085 for the main entry.
The roots of McDonald Observatory go back well over a century, to the great Struve astronomical dynasty in Russia and to the early settlement of Texas. A brief tracing of this history leads to the founding of the Observatory by Otto Struve and to the first 25 years of its development into one of the world's major observatories during the initial direction and operation. The second quarter-century of substantial further growth is outlined still mnore briefly in this narrative account which stresses some of the personal contributions which have made the Observatory what it is today.

009.010 **Starry, starry night: observing on Kitt Peak.**
R. G. Teske.
Mercury, Vol. 20, No. 4, p. 115 - 119 (Jul - Aug 1991).

009.011 **Will La Silla succumb to the VLT?**
S. Cristiani.
Messenger, No. 65, p. 2 - 4 (Sep 1991).

009.012 **Nanango Astronomical Observatory establishment and function.**
J. Van Vegchel.
South. Stars, Vol. 34, No. 3, p. 37 - 38 (18 Jun 1991). - See Abstr. 012.120 for the main entry.
The aims and objectives of the Nanango Astronomical Observatory are presented, together with a listing of available equipment and facilities.

009.013 **A short history of the U.S. Naval Observatory.**
T. J. Rafferty.
South. Stars, Vol. 34, No. 3, p. 57 - 58 (18 Jun 1991). - See Abstr. 012.120 for the main entry.

009.014 **Origins of the Molonglo Radio Observatory - The Cross and the MOST.**
B. Y. Mills.
Aust. J. Phys., Vol. 44, No. 6, p. 719 - 727 (1991). Paper presented at the Molonglo Observatory 25th Anniversary Symposium, Univ. Sydney, 22 - 23 Nov 1990.
The history of the Molonglo Radio Observatory is reviewed from initial steps in 1990 through the construction of the Cross and its conversion to the MOST. A selection of early scientific programs carried out with the Cross is briefly described.

009.015 **L'Osservatorio di Roque de los Muchachos.**
W. Ferreri.
Orione, Vol. 11, N. 4, p. 10 - 17 (Jul - Aug 1991).

Analysis of papers published from the Isaac Newton Group of Telescopes 1984 - 88.
See Abstr. 002.093.

From the hill: the story of Lowell Observatory.
See Abstr. 003.014.

Die beiden Observatorien in Potsdam und Babelsberg. Ihre Geschichte und Bedeutung, ihr heutiger Zustand.
See Abstr. 004.002.

La fondation de l'observatoire et les débuts de l'astronomie au Pic du Midi (1852 - 1947). Introduction. Première partie: La fondation de l'observatoire.
See Abstr. 004.031.

La fondation de l'observatoire et les débuts de l'astronomie au Pic du Midi (1852 - 1947). Deuxième partie: L'introduction de l'astronomie.
See Abstr. 004.032.

La fondation de l'observatoire et les débuts de l'astronomie au Pic du Midi (1852 - 1947). Dernière partie: Le règne de la coupole Baillaud.
See Abstr. 004.033.

Accomplishments and alarms of Pulkovo, some pages of reminescenses.
See Abstr. 004.064.

Survived. The Pulkovo Observatory in the years 1938 - 1954.
See Abstr. 004.065.

From Pulkovo's history.
See Abstr. 004.066.

Un observatoire privé à Amiens dans les années 1930 - 1940.
See Abstr. 004.071.

The Kenya expedition of Leiden Observatory.
See Abstr. 004.083.

High-resolution imaging in La Palma.
See Abstr. 013.054.

The construction of the East Asia Meridina Chain of geomagnetic observatories.
See Abstr. 013.064.

Stellar astronomy with the Isaac Newton Group telescopes.
See Abstr. 013.073.

Status of the Perth Observatory automated supernova search program.
See Abstr. 013.091.

Progress on light pollution control at the David Dunlap Observatory.
See Abstr. 013.100.

Das Sonnenobservatorium Einsteinturm in Potsdam. Erforschung solarer Magnetfelder und der Physik von Sonnenflecken.
See Abstr. 075.001.

010 Societies, Associations, Organizations

American Association of Variable Star Observers (AAVSO)

010.001 **American Association of Variable Star Observers Circular.**
Nos. 249 – 254 (Jul – Dec 1991).

010.002 **SID Technical Bulletin.**
Vol. 2, No. 3 (Jul 1991).

010.003 **Solar Bulletin (American Association of Variable Star Observers).**
Vol. 47, Nos. 7 – 12 (Jul – Dec 1991).

010.004 **The American Association of Variable Star Observers Bulletin.**
No. 55 (123.027).

American Astronomical Society (AAS)

010.005 **Bulletin of the American Astronomical Society.**
Vol. 23, No. 2 (1991) (012.008, 012.012).

Asociación Argentina de Astronomía

010.006 **Boletin de la Asociación Argentina de Astronomía.**
No. 36 (1990) (012.095).

Association Française des Observateurs d'Etoiles Variables

010.007 **Activité de l'AFOEV en 1990 (2ème partie).**
E. Schweitzer.
Bull. Assoc. Fr. Obs. Etoiles Variables, No. 57, p. 1 – 16 (Sep 1991).
Contents: 1. Variables du type UG. 2. SS Cygni. 3. Variables du type RCB. 4. Variables du type Z And. 5. Novae, supernovae et assimilées. 6. Etoiles particulières.

010.008 **La vie de l'Association.**
E. Schweitzer.
Bull. Assoc. Fr. Obs. Etoiles Variables, No. 57, p. 23 (Sep 1991).
Further information is given in No. 58, p. 8 (Dec 1991).

010.009 **Bulletin de l'Association Française des Observateurs d'Etoiles Variables.**
Nos. 57 (Sep 1991), 58 (Dec 1991).

Association of Lunar and Planetary Observers (A.L.P.O.)

010.010 **The Minor Planet Bulletin.**
Vol. 18, Nos. 3 (Jul – Sep 1991), 4 (Oct – Dec 1991).

010.011 **The Strolling Astronomer. The Journal of the Association of Lunar and Planetary Observers.**
Vol. 35, Nos. 3 (Sep 1991), 4 (Dec 1991).

Astronomical Society of Australia (ASA)

010.012 **Proceedings of the Astronomical Society of Australia.**
Vol. 9, No. 1 (1991) (012.078).

Astronomical Society of India

010.013 **Bulletin of the Astronomical Society of India.**
Vol. 19, Nos. 1 – 2 (Mar – Jun 1991).

Astronomical Society of Japan

010.014 **The Astronomical Herald.**
Vol. 84, Nos. 7 – 12 (1991). In Japanese.

010.015 **Publications of the Astronomical Society of Japan.**
Vol. 43, Nos. 4 – 6 (1991).

Astronomical Society of New York

010.016 **News Letter of the Astronomical Society of New York.**
Vol. 3, No. 9 (Feb 1991) (012.119).

Astronomical Society of Southern Africa (ASSA)

010.017 **Monthly Notes of the Astronomical Society of Southern Africa.**
Vol. 50, Nos. 7/8 (Aug 1991), 9/10 (Oct 1991), 11/12 (Dec 1991).

010.018 **Proceedings of the Annual General Meeting 1991.**
Mon. Notes Astron. Soc. S. Afr., Vol. 50, Nos. 9 – 10, p. 84 – 102 (Oct 1991).

010.019 **Centre reports: Cape Centre, Harare Centre, Natal Centre, Natal Midland Centre, Pretoria Centre, Transvaal Centre.**
J. Davison, M. J. R. Begbie, E. Calvert, B. Lawrence, M. Poll, T. Hilton.
Mon. Notes Astron. Soc. S. Afr., Vol. 50, Nos. 11 – 12, p. 110 – 118 (Dec 1991).

Astronomical Society of the Pacific (ASP)

010.020 **Publications of the Astronomical Society of the Pacific.**
Vol. 103, Nos. 665 (Jul 1991), 666 (Aug 1991) (012.034), 667 (Sep 1991), 668 (Oct 1991), 669 (Nov 1991), 670 (Dec 1991).

010.021 **Astronomical Society of the Pacific Conference Series.**
Vol. 13 (012.050), 16 (012.051), 19 (012.089), 20 (012.085), 21 (012.084).

010.022 **The Universe in the Classroom.**
No. 19 (Win 1991–1992).

010.023 **Mercury.**
Vol. 20, Nos. 4 (Jul – Aug 1991), 5 (Sep – Oct 1991), 6 (Nov – Dec 1991).

Astronomical Society of Western Australia (ASWA)

010.024 **The Sidereal Times.**
Vol. 42, Nos. 1 (Aug - Sep 1991), 2 (Oct - Nov 1991), 3 (Dec 1991 - Jan 1992).

Astronomische Gesellschaft (AG)

010.025 **Reviews in Modern Astronomy.**
Vol. 4 (1991) (003.008).

Berliner Arbeitsgemeinschaft für Veränderliche Sterne e.V. (BAV)

010.026 **BAV Rundbrief.**
40. Jahrg., Nr. 3/4 (1991).

010.027 **BAV Mitteilungen.**
Nr. 59 (54.123.004).

British Astronomical Association (BAA)

010.028 **Report of the council for the session 1990 August 1 to 1991 July 31.**
J. Br. Astron. Assoc., Vol. 101, No. 5, p. 301 - 313 (Oct 1991).

010.029 **Journal of the British Astronomical Association.**
Vol. 101, Nos. 4 (Aug 1991), 5 (Oct 1991), 6 (Dec 1991).

010.030 **Meeting reports.**
J. Br. Astron. Assoc., Vol. 101, No. 4, p. 248 - 251 (Aug 1991).
Further informations are given in No. 5, p. 315 - 318 (Oct 1991).

010.031 **The British Astronomical Association Newsletter.**
Nos. 49 (Aug 1991), 50 (Oct 1991), 51 (Dec 1991).

010.032 **British Astronomical Association Circular.**
Nos. 710 (2 Aug 1991), 711 (4 Oct 1991).

British Interplanetary Society (BIS)

010.033 **Journal of the British Interplanetary Society.**
Vol. 44, Nos. 7 - 12 (Jul - Dec 1991).

010.034 **Spaceflight.**
Vol. 33, Nos. 7 - 12 (Jul - Dec 1991).

European Space Agency (ESA)

010.035 **ESA IUE Newsletter.**
No. 38 (Aug 1991).

010.036 **ESA Bulletin.**
Nos. 67 (Aug 1991), 68 (Nov 1991).

010.037 **ESA Special Publication.**
ESA-SP—1135 (036.203), ESA-SP—1147 (002.091).

010.038 **ESA Journal.**
Vol. 15, No. 3/4 (1991).

International Amateur–Professional Photoelectric Photometry

010.039 **International Amateur–Professional Photoelectric Photometry Communication.**
Nos. 45 (Sep - Nov 1991), 46 (Dec 1991 - Feb 1992).

010.040 **IAPPP: the next generation.**
T. D. Oswalt.
I.A.P.P.P. Commun., No. 45, p. 59 (Sep - Nov 1991). See Abstr. 012.002 for the main entry.

International Astronomical Union (IAU)

010.041 **Information Bulletin on Variable Stars. Commission 27 of the IAU.**
Nos. 3631 - 3692 (Jul - Dec 1991).

010.042 **Bibliography and Program Notes on Close Binaries.**
No. 53 (002.049).

010.043 **International Astronomical Union Colloquium.**
Nos. 123 (012.038), 129 (012.046), 130 (012.024).

010.044 **International Astronomical Union Symposium.**
No. 147 (012.043).

010.045 **International Astronomical Union, Circular.**
Nos. 5296 - 5419 (Jul - Dec 1991).

010.046 **The Minor Planet Circulars/Minor Planets and Comets.**
Nos. 18485 - 19346 (Aug - Nov 1991).

010.047 **IAU Working Group on Wide-Field Imaging, Newsletter.**
No. 1 (Jan 1991) (034.138).

010.048 **Circulaire d'Information. Union Astronomique Internationale, Commission des Etoiles Doubles.**
No. 115 (Oct 1991).

International Occultation Timing Association (IOTA)

010.049 **Occultation Newsletter.**
Vol. 5, Nos. 4 (Aug 1991), 5 (Nov 1991), 6 (Dec 1991).

Korean Astronomical Society

010.050 **The Journal of the Korean Astronomical Society.**
Vol. 24, No. 2 (Dec 1991).

Meteoritical Society

010.051 **Meteoritics.**
Vol. 26, No. 3 (Sep 1991).

010.052 **Geochimica et Cosmochimica Acta.**
Vol. 55, Nos. 7 - 12 (Jul - Dec 1991).

Oriental Astronomical Association

010.053 **The Heavens.**
Vol. 72, Nos. 7 - 12 (Jul - Dec 1991). In Japanese.

Royal Astronomical Society (RAS)

010.054 **Summary of the RAS Specialist Discussion Meeting on 150 Years of Magnetic Observatories: Recent Researches on World Data, London (UK), 11 Jan 1991.**
D. Barraclough.
Observatory, Vol. 111, No. 1103, p. 148 – 154 (Aug 1991).

010.055 **Meetings of the Society.**
Observatory, Vol. 111, No. 1103, p. 145 – 148 (Aug 1991).
Further information is given in No. 1105, p. 261 – 280 (Dec 1991).

010.056 **RAS G/MIST Discussion Meeting on Radar Studies of the Upper Atmosphere and the Ionosphere, London (UK), 12 Oct 1990.**
J. P. Morelli, S. W. H. Cowley, T. K. Yeoman, I. W. McCrea.
Q. J. R. Astron. Soc., Vol. 32, No. 3, p. 295 – 300 (Sep 1991).

010.057 **Reports of meetings.**
Q. J. R. Astron. Soc., Vol. 32, No. 3, p. 319 – 348 (Sep 1991).

010.058 **The Quarterly Journal of the Royal Astronomical Society.**
Vol. 32, Nos. 3 (Sep 1991), 4 (Dec 1991).

010.059 **RAS Discussion Meeting on "The Common-User Principle: Advantages and Disadvantages", London (UK), 9 Dec 1988.**
Observatory, Vol. 111, No. 1105, p. 281 – 290 (Dec 1991).

010.060 **Monthly Notices of the Royal Astronomical Society.**
Vol. 251, Nos. 1 – 4, Vol. 252, Nos. 1 – 4, Vol. 253, Nos. 1 – 4 (1 Jul – 15 Dec 1991).

Royal Astronomical Society of Canada

010.061 **Royal Astronomical Society of Canada Bulletin.**
Vol. 1, Nos. 4 (Aug 1991), 5 (Oct 1991), 6 (Dec 1991).

010.062 **The Journal of the Royal Astronomical Society of Canada.**
Vol. 85, Nos. 4 (Aug 1991), 5 (Oct 1991), 6 (Dec 1991).

Royal Astronomical Society of New Zealand

010.063 **Southern Stars.**
Vol. 34, Nos. 3 (Jun 1991), 4 (Sep 1991), 5 (Dec 1991).

Schweizerische Astronomische Gesellschaft (SAG)

010.064 **Orion. Zeitschrift der Schweizerischen Astronomischen Gesellschaft.**
49. Jahrg., Nr. 245 (Aug 1991), 246 (Oct 1991), 247 (Dec 1991).

010.065 **Mitteilungen.**
Orion, Jahrg. 49, Nr. 245, p. 149 – 152 (Aug 1991).
Further information is given in Nr. 246, p. 187 – 196 (Oct 1991), Nr. 247, p. 239 – 246 (Dec 1991).

010.066 **Bedeckungsveränderlichen Beobachter der Schweizerischen Astronomischen Gesellschaft Bulletin.**
No. 98 (15 Sep 1991).

Società Astronomica Italiana (S.A.It.)

010.067 **Memorie della Società Astronomica Italiana.**
Vol. 62, Nos. 1 (012.033), 2, 3 (012.093, 012.094) (1991).

010.068 **Giornale di Astronomia.**
Vol. 17, N. 1 – 2 (Mar – Jun 1991), N. 3 – 4 (Sep – Dec 1991).

Société Astronomique de France

010.069 **L'Astronomie.**
Juillet – décembre (1991).

010.070 **Observations et Travaux.**
Nos. 27, 28 (1991).

Société Astronomique de Liège

010.071 **Le Ciel.**
Vol. 53, septembre – décembre (1991).

Société Française des Spécialistes d'Astronomie (SFSA)

010.072 **Journal des Astronomes Français.**
Nos. 39 (012.099), 40 (Jul 1991).

Société Royale Belge d'Astronomie

010.073 **Ciel et Terre.**
Vol. 107, Nos. 4 (Jul – Aug 1991), 5 (Sep – Oct 1991), 6 (Nov – Dec 1991).

Vereinigung der Sternfreunde e.V. (VdS)

010.074 **VdS Nachrichten für Sternfreunde.**
Sterne Weltraum, Jahrg. 30, Nr. 7, p. 472 – 475 (Jul 1991).
Further informations are given in Nr. 8 – 9, p. 556 – 559 (Aug – Sep 1991), Nr. 10, p. 624 – 627 (Oct 1991), Nr. 11, p. 698 – 701 (Nov 1991), Nr. 12, p. 776 – 779 (Dec 1991).

010.075 **Sonne.**
Jahrg. 15, Nr. 58 (Aug 1991), 59 (Sep 1991), 60 (Dec 1991).

Vsesoyuznoe Astronomo-Geodezicheskoe Obshchestvo (VAGO)

010.076 **Plenary session of the Central Council of VAGO in Simferopol, 1 – 3 April 1991.**
N. N. Spasskij.
Zemlya Vselennaya, No. 5, p. 105 – 106 (Sep – Oct 1991). In Russian.

011 Reports on Colloquia, Congresses, Meetings, Symposia, Expeditions

011.001 **International scientific seminar on problems of astrometry and geodynasmics, held at the Main Astronomical Observatory of the Ukrainian Academy of Sciences, 10 - 12 September 1990.**
N. T. Mironov, N. V. Kharchenko.
Kinematika Fiz. Nebesn. Tel, Tom 7, No. 4, p. 91 - 92 (Jul-Aug 1991). In Russian.

011.002 **Conference of the Working Group "Spectrophotometric and photometric standards and catalogues", held at the Main Astronomical Observatory of the Ukrainian Academy of Sciences, 29 - 31 October 1990.**
A. V. Kharitonov.
Kinematika Fiz. Nebesn. Tel, Tom 7, No. 4, p. 93 - 96 (Jul-Aug 1991). In Russian.

011.003 **Third Annual Meeting of the International Dark-Sky Association (IDA), Inc.**
T. Hunter.
I.A.P.P.P. Commun., No. 46, p. 15 - 16 (Dec 1991 - Feb 1992).

011.004 **Bioastronomy - the exploration broadens: the third international symposium on bioastronomy.**
M. D. Papagiannis.
J. Br. Interplanet. Soc., Vol. 44, No. 8, p. 405 - 408 (Aug 1991).
The third International Symposium on Bioastronomy was held in Val Cenis in the French Alps, June 18 - 23, 1990.

011.005 **Autumn MIST meeting, 1990.**
M. J. Laird.
Q. J. R. Astron. Soc., Vol. 32, No. 4, p. 423 - 428 (Dec 1991).
The autumn MIST meeting was held on 1990 Nov 30 at the Society of Antiquaries, Burlington House.

011.006 **Gamma ray line astrophysics. Report on an international symposium, Saclay, France, December 10 - 13, 1990.**
R. Ramaty, N. Prantzos.
Comments Astrophys., Vol. 15, No. 5, p. 301 - 319 (Sep 1991).
Contents: 1. Galactic positron annihilation radiation. 2. Galactic ^{26}Al line emission. 3. Supernovae and SN 1987A. 4. The Crab nebula and pulsar. 5. Gamma ray bursts. 6. Solar flares.

011.007 **Conference on active nuclei of galaxies held at the Crimean Astrophysical Observatory on 30 October - 2 November 1990.**
V. I. Pronik.
Astron. Zh., Tom 68, Vyp. 6, p. 1332 - 1334 (Nov-Dec 1991). In Russian. English translation in Sov. Astron., Vol. 35, No. 6.

011.008 **The XXIst Congress of the International Astronomical Union.**
J. M. Kreiner.
Urania, Rok 62, Nr. 12, p. 322 - 328 (Dec 1991). In Polish.

011.009 **CAS Symposium No. 106 and No. 107: Connections of reference systems and techniques of photographic astrometry, Yichang (China), 16 - 19 Nov 1990.**
Xia Yifei.
Prog. Astron., Vol. 9, No. 2, p. 161 (Jun 1991). In Chinese.

011.010 **CAS Symposium No. 110: Numerical investigation in dynamical astronomy, Yichang (China), 12 - 14 Nov 1990**
He Miaofu.
Prog. Astron., Vol. 9, No. 2, p. 161 - 162 (Jun 1991). In Chinese.

011.011 **CAS Symposium No. 118: The 10th Meeting of Astronomical Instrumentation, Xinanjiang Zejiang (China), 3 - 6 Nov 1990.**
Li Ting, Zhang Youyi.
Prog. Astron., Vol. 9, No. 2, p. 162 - 163 (Jun 1991). In Chinese.

011.012 **The National Symposium of the Dynamics of Solar Active Regions, Dali (China), 5 - 11 Nov 1990.**
Xuan Jiayu.
Prog. Astron., Vol. 9, No. 2, p. 163 (Jun 1991). In Chinese.

011.013 **CAS Symposium No. 111, 112, 113, 114: The first symposium in memory of Zhang Heng - a joint workshop on astrophysics in China, Lintong, Shaanxi (China), 23 - 29 Aug 1990.**
Peng Qiuhe.
Prog. Astron., Vol. 9, No. 3, p. 250 (Sep 1991). In Chinese.

011.014 **CAS Symposium No. 119: The MHD of solar active regions and redshift precursor of flares, Xiamen (China), 4 - 8 Jan 1991.**
Wu Meixia.
Prog. Astron., Vol. 9, No. 3, p. 250 - 251 (Sep 1991). In Chinese.

011.015 **CAS Symposium No. 115: The 4th Symposium of Precise Time and Time Dissemination, Guilin (China), 15 - 19 Nov 1990.**
Wu Guichen.
Prog. Astron., Vol. 9, No. 3, p. 251 - 252 (Sep 1991). In Chinese.

011.016 **Transportation system for a lunar base.**
B. I. Sotnikov, G. M. Bajdal, G. A. Sizentsev.
Zemlya Vselennaya, No. 5, p. 19 - 25 (Sep - Oct 1991). In Russian.
The authors report on a conference held at the Sternberg Astronomical Institute in Moscow, 5 - 8 February 1991, concerning the problems of establishing a base on the Moon for scientific investigation.

011.017 **Riverside 1991.**
W. P. Clarke.
Vasiona, Année 39, No. 5, p. 74 - 75 (1991). In Croatian.
Twenty Third Annual Riverside Telescope Makers Conference was held from May 24 to 27 in Big Bear, California. More then 1700 amateurs from across US converged on the world largest star party. The article covers a short presentation of development in construction of telescopes, talks and activities during the conference.

011.018 **Assembly of the Section on History of Astronomy in Riga, 9 - 11 February 1988.**
M. V. Kuznetsov.
Istor.-Astron. Issled., Tom 21, p. 392 - 393 (1989). In Russian. - See Abstr. 003.017 for the main entry.

011.019 **Bulgaria: the First National Symposium on Archaeoastronomy, 22 - 24 November 1988.**
Ts. Radoslavova.
Istor.-Astron. Issled., Tom 21, p. 394 - 396 (1989). In Russian. - See Abstr. 003.017 for the main entry.

011.020 **The role of impacts in the history of the early Earth.**
B. M. French.
Earth Space, Vol. 4, No. 1, p. 5 - 6 (Sep 1991).
The author reports on the meeting "Meteorite impact and the early Earth", Perth (Australia), Sep 1990, which presents a

collection of evidence to support the idea that large extraterrestrial impacts played a major role in the development of the early Earth.

011.021 **The tenth meeting of the European Symposium on Occultation Projects.**
D. W. Dunham.
Occultation Newsl., Vol. 5, No. 5, p. 120 – 122 (Nov 1991).

011.022 **Impressions from the XXI IAU General Assembly.**
D. Baade, F. Merkle, P. Traat, C. Sterken, P. Grosbøl, M. Rosa, C. Madsen, R. Warmels, H. Van Der Laan.
Messenger, No. 65, p. 16 – 24 (Sep 1991).

011.023 **Mini-workshop on large-size CCDs at ESO.**
S. D'Odorico, T. Ducros, O. Iwert, R. Reiss.
Messenger, No. 65, p. 43 – 45 (Sep 1991).

011.024 **Symposium on problems of the theory of radiative transfer, held in Leningrad, 22 – 24 October 1990.**
A. K. Kolesov.
Astrofizika, Tom 34, Vyp. 1, p. 141 – 144 (Feb 1991). In Russian. English translation in Astrophysics, Vol. 34, No. 1.

011.025 **A report on the second ESO Conference on High Resolution Imaging by Interferometry.**
J. M. Beckers, F. Merkle.
Messenger, No. 66, p. 5 – 10 (Dec 1991).

011.026 **Universums begynnelse bland de jämtländska fjällen.**
U. Torkelsson.
Astron. Tidsskr., Årg. 24, Nr. 4, p. 168 – 178 (1991).

011.027 **Report on the 21st General Assembly of the IAU, Buenos Aires, 1991.**
E. Budding.
South. Stars, Vol. 34, No. 5, p. 327 – 335 (Dec 1991).

Specifications concerning names, designations, and nomenclature for astronomical radiation sources outside the Solar System. A message from the Working Group on Designations from Commission 5 of the International Astronomical Union.
See Abstr. 002.057.

On the occasion of the 100th anniversary of A. A. Friedmann's (1888 – 1925) birth.
See Abstr. 005.024.

On the studies of astronomical photography problems and information on the proposed new photomaterials for astronomical photography (1987)
See Abstr. 034.134.

012 Proceedings of Colloquia, Congresses, Meetings, Symposia

012.001 **Grandes formations extragalactiques. Comptes rendus de la Réunion du Conseil Scientifique du CDS, Lyon (France), 15 May 1990.**
Bull. Inf. Cent. Données Stellaires, No. 38, p. 1 – 12 (Dec 1990).
The individual contributions within the subject scope of AAA are included in their corresponding categories – see abstracts 002.001, 002.002, 002.003, 036.002.

012.002 **The IAPPP – 12th Annual Fairborn Symposium. Proceedings. The IAPPP – 12th Annual Fairborn Symposium, Las Vegas, NV (USA), 28 Feb – 3 Mar 1991.**
R. M. Genet (ed.).
I.A.P.P.P. Commun., No. 45, p. 1 – 85 (Sep – Nov 1991).
The individual contributions within the subject scope of AAA are included in their corresponding categories – see abstracts 013.002, 013.003, 013.004, 032.001, 032.002, 032.003, 036.003, 036.004, 036.006, 036.007, 115.002, 116.001, 117.005, 117.006, 117.007, 119.003, 119.004, 123.002.

012.003 **The restoration of HST images and spectra. Proceedings. Workshop on the Restoration of HST Images and Spectra, Baltimore, MD (USA), 20 – 21 Aug 1990.**
R. L. White, R. J. Allen (eds.).
Space Telescope Science Institute, Baltimore, MD (USA); National Aeronautics and Space Administration, Washington, DC (USA). 146 p. (1991).
The papers in these proceedings focus entirely on computational methods for restoring the resolution of HST images and spectra.
The individual contributions within the subject scope of AAA are included in their corresponding categories – see abstracts 036.018, 036.019, 036.020, 036.021, 036.022, 036.023, 036.025, 036.026, 036.027, 036.028, 036.029, 036.030, 036.031, 036.032, 036.033, 036.034, 036.035, 036.036, 036.037, 036.038.

012.004 **Quantum gravity. Proceedings. 5. Seminar on Quantum Gravity, Moscow (USSR), 28 May – 1 Jun 1990.**
M. A. Markov, V. A. Berezin, V. P. Frolov (eds.).
World Scientific, Singapore (Singapore). 712 p. (1991). ISBN 981-02-0440-X. Price £ 50.00.
The aim of the Seminar was to discuss the most important problems of modern quantum gravity, viz.: (1) quantum gravity: the state of art; (2) quantum black hole physics; (3) quantum effects in the very early universe, including inflationary cosmology, vacuum phase transitions and cosmic strings.
The individual contributions within the subject scope of AAA are included in their corresponding categories – see abstracts 061.020, 061.021, 061.022, 061.023, 061.024, 061.025, 066.009, 066.010, 066.012, 066.013, 066.014, 066.015, 066.016, 066.017, 066.018, 066.019, 066.020, 066.021, 066.022, 066.023, 066.024, 066.025, 066.026, 066.032, 067.018, 067.019, 067.020, 161.044, 161.061, 161.062, 161.063, 161.064, 161.065, 161.072, 161.074, 161.075, 161.076, 161.077, 161.078, 161.079, 161.080, 161.081, 161.089, 161.090.

012.005 **Relativity and gravitation: classical and quantum. Proceedings. 7. Latin American Symposium on Relativity and Gravitation, Cocoyoc (Mexico), 2 – 8 Dec 1990.**
J. C. D'Olivo, E. Nahmad-Achar, M. Rosenbaum, M. P. Ryan Jr., L. F. Urrutia, F. Zertuche (eds.).
World Scientific, Singapore (Singapore). 562 p. (1991). ISBN 981-02-0715-8. Price £ 39.00.
Contents: 1. Plenary lectures. 2. Formal developments. 3. Exact solutions. 4. Radiative spacetimes. 5. Relativistic astrophysics. 6. Cosmology (classical and quantum). 7. 2+1 gravity and quantum gravity. 8. Other theories of gravitation. 9. Field theory (classical and quantum).
The individual contributions within the subject scope of AAA are included in their corresponding categories – see abstracts 021.007, 021.008, 061.027, 061.028, 061.029, 061.030, 061.031,

061.032, 066.037, 066.038, 066.039, 066.040, 066.041, 066.042, 066.043, 066.044, 066.045, 066.046, 066.047, 066.048, 066.049, 066.050, 066.051, 066.052, 066.053, 066.054, 066.055, 066.056, 066.057, 066.058, 066.059, 066.060, 066.062, 066.063, 066.064, 066.065, 066.066, 066.067, 066.068, 066.069, 066.071, 066.072, 067.022, 067.024, 067.025, 122.066, 151.009, 161.091, 161.092, 161.096, 161.098, 161.102, 161.103, 161.105, 161.107, 161.108, 161.109, 161.111, 161.112, 161.115, 161.116, 161.117, 161.118.

012.006 **Planetary magnetospheric physics I. Proceedings. Symposium 6 and Topical Meeting of the COSPAR Interdisciplinary Scientific Commission D (Meetings D2, D4) of the COSPAR 28. Plenary Meeting, The Hague (Netherlands), 25 Jun – 6 Jul 1990.**
R. J. L. Grard, J. Büchner, M. Scholer, D. Burgess (eds.).
Adv. Space Res., Vol. 11, No. 9, 296 p. (1991). ISBN 0-08-041834-1.
Contents: 1. Role of plasma waves in the physics of planetary environments. 2. Particle populations in magnetospheric current sheets. 3. Quasi-parallel and parallel shocks.
The individual contributions within the subject scope of AAA are included in their corresponding categories – see abstracts 005.003, 022.014, 022.015, 022.016, 022.017, 022.018, 035.007, 062.021, 062.022, 062.023, 062.024, 062.025, 062.026, 062.027, 062.028, 062.029, 062.030, 074.011, 084.009, 084.010, 084.011, 084.012, 084.013, 084.014, 084.015, 084.016, 084.017, 084.018, 084.019, 084.020, 084.021, 084.022, 084.023, 084.024, 084.025, 084.026, 084.027, 084.028, 084.029, 084.030, 093.004, 097.005, 102.004, 103.074, 103.084, 103.085, 103.086, 106.011, 106.012.

012.007 **Elementary particles and the universe. Essays in honor of Murray Gell-Mann. Symposium in celebration of Murray Gell-Mann's 60th birthday "Where are our efforts leading?", Pasadena, CA (USA), 27 – 28 Jan 1989.**
J. H. Schwarz (ed.).
Cambridge University Press, Cambridge (UK). 219 p. (1991). ISBN 0-521-41253-6. Price £ 24.95, US$ 49.95.
Professor Murray Gell-Mann is one of the most influential and brilliant scientists of the twentieth century. His work on symmetries, including the invention of the "quark", in the 1950s and early 1960s has provided a foundation for much of modern particle physics and was recognised by the award of the Nobel Prize for Physics in 1969. This book is a collection of research articles to celebrate Gell-Mann's 60th birthday, in September 1989. The main body of contributions are concerned with theoretical particle physics and its applications to cosmology.

012.008 **178. Meeting of the American Astronomical Society (AAS), Seattle, WA (USA), 26 – 30 May 1991. Abstracts of presented papers.**
Bull. Am. Astron. Soc., Vol. 23, No. 2, p. 843 – 988 (1991).
The individual contributions within the subject scope of AAA are included in their corresponding categories – see abstracts 002.029, 002.033, 002.034, 002.035, 002.036, 002.037, 002.038, 002.039, 002.041, 002.042, 002.043, 002.044, 004.008, 004.009, 004.010, 004.011, 004.012, 004.013, 004.014, 004.015, 004.016, 004.018, 005.004, 009.001, 013.009, 013.010, 013.011, 013.018, 014.023, 014.024, 014.025, 014.026, 014.028, 014.029, 014.030, 014.033, 014.034, 014.035, 014.036, 014.037, 015.007, 015.008, 022.023, 022.028, 022.029, 022.030, 031.005, 031.006, 031.007, 031.008, 031.009, 032.004, 032.005, 032.008, 034.010, 034.011, 034.012, 034.013, 034.014, 034.015, 034.017, 034.021, 035.010, 035.011, 035.012, 035.013, 035.014, 035.017, 035.018, 035.019, 035.020, 035.021, 036.039, 036.040, 036.042, 036.043, 036.044, 036.045, 036.046, 036.047, 036.048, 036.049, 036.050, 036.051, 036.052, 036.053, 036.054, 036.055, 036.057, 036.058, 041.007, 051.002, 051.003, 051.004, 051.005, 051.006, 051.007, 051.008, 051.009, 051.010, 051.011, 051.012, 061.033, 062.037, 062.038, 062.039, 062.040, 062.041, 063.013, 063.015, 064.011, 064.012, 064.013, 064.014, 064.015, 064.016, 064.017, 064.018, 064.019, 065.016, 065.017, 065.018, 066.070, 066.074, 067.023, 067.026, 067.027, 067.028, 071.003, 071.005, 072.008, 072.012, 073.008, 073.009, 073.010, 074.014, 074.016, 074.017, 074.018, 074.024, 076.001, 080.007, 080.008, 080.009, 080.012, 082.010, 082.011, 082.012, 091.007, 093.005, 099.005, 099.006, 099.009, 100.002, 101.006, 102.008, 102.009, 103.033, 103.035, 103.038, 103.088, 106.015, 111.001, 112.008, 112.010, 112.011, 112.012, 112.013, 112.014, 112.015, 112.016, 112.017, 112.018, 112.019, 112.020, 112.021, 112.022, 112.023, 114.009, 114.010, 114.011, 114.014, 114.015, 114.016, 114.017, 114.018, 114.019, 115.009, 115.010, 115.011, 116.003, 116.005, 116.006, 116.007, 116.008, 116.009, 117.015, 117.017, 117.018, 117.019, 117.020, 117.021, 117.022, 117.023, 117.024, 117.026, 117.027, 117.029, 117.030, 117.031, 117.032, 117.033, 117.034, 117.035, 117.036, 117.037, 117.038, 118.006, 119.008, 119.009, 119.010, 119.011, 119.012, 119.013, 120.003, 120.004, 121.004, 121.005, 121.006, 121.009, 121.011, 121.012, 122.008, 122.009, 122.010, 122.011, 122.012, 122.013, 122.014, 122.015, 122.017, 122.018, 122.021, 122.022, 122.024, 124.023, 125.006, 125.007, 125.008, 125.009, 125.010, 125.011, 125.012, 125.013, 125.014, 125.015, 125.016, 125.017, 125.018, 125.019, 125.175, 125.176, 125.177, 126.006, 126.009, 126.013, 131.016, 131.017, 131.018, 131.019, 131.020, 131.021, 131.022, 131.023, 131.024, 131.025, 131.026, 131.027, 131.028, 131.029, 131.030, 131.031, 131.032, 131.033, 131.034, 131.036, 131.037, 131.038, 131.040, 131.043, 131.044, 131.045, 131.046, 131.049, 131.050, 131.051, 131.052, 132.003, 132.004, 132.005, 132.006, 132.007, 132.008, 134.008, 134.009, 134.010, 134.011, 134.012, 134.013, 134.014, 134.015, 142.004, 142.005, 142.006, 142.007, 142.008, 142.009, 142.010, 142.011, 143.010, 143.011, 143.012, 144.015, 144.017, 151.014, 151.017, 151.018, 152.001, 152.002, 152.003, 152.004, 153.009, 153.010, 153.011, 154.005, 154.006, 154.007, 154.008, 154.009, 154.010, 154.011, 154.012, 154.013, 154.014, 154.015, 154.016, 154.017, 155.012, 155.013, 155.014, 155.015, 155.016, 155.017, 155.018, 155.019, 155.023, 155.024, 155.025, 155.026, 155.027, 155.028, 156.003, 156.004, 156.007, 156.008, 156.009, 157.023, 157.024, 157.025, 157.026, 157.027, 157.028, 157.029, 157.030, 157.031, 157.032, 157.035, 157.036, 157.037, 157.038, 157.039, 157.040, 157.041, 157.042, 157.043, 157.047, 157.048, 157.049, 157.052, 157.053, 157.054, 157.055, 157.056, 157.057, 157.058, 157.059, 157.060, 157.061, 157.064, 157.065, 157.066, 157.067, 157.068, 157.069, 157.070, 157.071, 157.072, 157.073, 157.074, 157.075, 158.019, 158.020, 158.021, 158.022, 158.023, 158.024, 158.025, 158.026, 158.028, 158.029, 158.030, 158.031, 158.032, 158.033, 158.034, 158.035, 158.036, 158.037, 158.038, 158.039, 158.040, 158.044, 158.045, 158.046, 158.047, 158.048, 158.049, 158.050, 159.009, 159.010, 159.011, 159.013, 159.015, 159.016, 159.017, 159.018, 159.019, 159.020, 159.024, 159.025, 159.026, 160.015, 160.016, 160.017, 160.018, 160.019, 160.020, 160.021, 160.022, 160.023, 160.024, 160.025, 160.026, 160.027, 160.028, 160.029, 161.082, 161.083, 161.084, 161.085, 161.086, 161.087, 161.088, 161.110, 161.119, 161.120, 161.121, 161.122, 161.123, 161.124, 161.125, 161.592.

012.009 **National Large Telescope Workshop. Proceedings. National Large Telescope Workshop: Scientific requirements and requirement of site, Bangalore (India), 19 – 21 Oct 1989.**
M. Parthasarathy (ed.).
Government of India, Ministry of Science and Technology, New Delhi. 158 p. (1991).
In India there is a growing need to build a large telescope which will enable the Indian astronomers to carry out front rank research in several fields of astrophysics. Various aspects of selecting a good site for optical and IR astronomy are discussed.
The individual contributions within the subject scope of AAA are included in their corresponding categories – see abstracts 013.019, 013.020, 013.021, 013.022, 013.023, 013.024, 013.025, 013.026, 013.027, 013.028, 013.029, 013.030, 013.031, 013.032, 013.033, 082.015, 082.016, 082.017, 082.018, 082.019, 082.020, 082.021, 082.022, 082.023, 157.076, 158.052.

012.010 **Particle physics. Proceedings. NATO Institut d'Etudes Scientifiques sur Physique des Particules, Cargèse (France), 18 Jul – 4 Aug 1989.**
M. Lévy, J. L. Basdevant, M. Jacob, D. Speiser, J. Weyers, R. Gastmans (eds.).
NATO ASI Ser., Ser. B, Phys., Vol. 223.
Plenum Press, New York, NY (USA). 369 p. (1990). ISBN 0-306-43601-9.
The 1989 Cargèse Summer Institute on Particle Physics centered on the following topics: new experimental results –

strings, superstrings and conformal field theory – latter approximations.
The individual contributions within the subject scope of AAA are included in their corresponding categories – see abstracts 067.030.

012.011 **Mechanisms of chromospheric and coronal heating. Proceedings. International Conference on Mechanisms of Chromospheric and Coronal Heating, Heidelberg (Germany, F.R.), 5 – 8 Jun 1990.**
P. Ulmschneider, E. R. Priest, R. Rosner (eds.).
Springer, Berlin (Germany, F.R.). 663 p. (1991). ISBN 3–540–53436–9. Price DM 128.00. ISBN 0–387–53436–9 (USA).
This book is a comprehensive inventory of the proposed chromospheric and coronal heating theories. The proposed heating processes are critically compared and the observational evidence for the various mechanisms is reviewed.
The individual contributions within the subject scope of AAA are included in their corresponding categories – see abstracts 035.025, 051.013, 062.059, 062.060, 062.061, 062.062, 062.063, 062.064, 062.065, 062.066, 062.069, 062.070, 062.071, 062.072, 062.073, 062.074, 062.075, 062.076, 062.077, 062.078, 062.079, 062.080, 062.081, 064.020, 064.022, 064.023, 064.024, 064.025, 064.026, 064.027, 064.028, 064.029, 064.030, 064.032, 064.033, 064.034, 064.035, 064.036, 064.037, 064.038, 064.039, 064.040, 064.041, 064.042, 064.043, 064.044, 064.045, 064.046, 064.047, 064.048, 071.008, 071.009, 072.023, 073.018, 073.019, 073.020, 073.021, 073.022, 073.023, 073.024, 073.025, 073.026, 073.046, 073.057, 073.058, 073.075, 074.027, 074.028, 074.029, 074.030, 074.031, 074.032, 074.033, 074.034, 074.035, 074.036, 074.037, 074.039, 074.046, 074.047, 074.048, 074.049, 074.050, 074.051, 074.064, 074.066, 074.067, 074.068, 075.004, 075.005, 075.020, 076.004, 077.009, 077.010, 080.017, 080.018, 080.026, 080.027, 080.028, 080.035, 080.044, 080.045, 080.046, 080.047, 112.025, 116.015.

012.012 **21. Meeting of the Solar Physics Division of the American Astronomical Society (AAS), Huntsville, AL (USA), 9 – 11 Apr 1991. Abstracts of presented papers.**
Bull. Am. Astron. Soc., Vol. 23, No. 2, p. 1009 – 1078 (1991).
The individual contributions within the subject scope of AAA are included in their corresponding categories – see abstracts 002.051, 002.052, 002.053, 013.034, 013.035, 013.036, 013.037, 013.038, 013.039, 013.040, 013.041, 013.042, 013.043, 031.013, 034.029, 034.031, 034.032, 034.033, 034.035, 035.026, 035.027, 035.028, 035.029, 035.030, 035.031, 035.033, 035.034, 035.035, 036.067, 036.068, 036.069, 036.070, 036.071, 036.072, 036.073, 036.074, 051.014, 051.015, 051.017, 051.018, 062.048, 062.049, 062.050, 062.051, 062.052, 062.053, 062.054, 062.055, 062.056, 062.057, 062.067, 062.068, 063.018, 063.020, 071.010, 071.011, 071.012, 071.013, 071.014, 071.015, 071.016, 071.017, 071.018, 071.019, 071.020, 072.024, 072.025, 072.026, 072.027, 072.028, 072.029, 072.030, 072.031, 072.032, 072.033, 072.034, 072.035, 072.036, 072.037, 072.038, 072.039, 072.040, 072.041, 072.042, 072.043, 072.044, 073.027, 073.028, 073.029, 073.030, 073.031, 073.032, 073.033, 073.034, 073.035, 073.036, 073.038, 073.039, 073.040, 073.041, 073.042, 073.043, 073.044, 073.045, 073.048, 073.049, 073.050, 073.051, 073.052, 073.053, 073.054, 073.055, 073.056, 073.059, 073.060, 073.061, 073.062, 073.063, 073.064, 073.065, 073.066, 073.067, 073.068, 073.069, 073.070, 073.071, 073.072, 073.073, 073.074, 074.038, 074.040, 074.041, 074.042, 074.043, 074.044, 074.045, 074.052, 074.053, 074.054, 074.055, 074.056, 074.057, 074.058, 074.059, 074.060, 074.061, 074.062, 074.063, 075.006, 075.007, 075.008, 075.009, 075.010, 075.011, 075.012, 075.013, 075.014, 075.015, 075.016, 075.017, 075.018, 075.019, 075.021, 075.022, 075.023, 076.002, 076.003, 076.005, 076.006, 076.007, 076.008, 076.009, 076.010, 076.011, 076.012, 076.013, 076.014, 076.015, 076.016, 076.017, 077.008, 077.011, 077.012, 077.013, 077.014, 077.015, 077.016, 077.017, 077.018, 077.019, 077.020, 077.021, 079.008, 080.019, 080.020, 080.021, 080.022, 080.023, 080.024, 080.025, 080.029, 080.030, 080.031, 080.032, 080.033, 080.034, 080.036, 080.037, 080.038, 080.039, 080.040, 080.041, 080.042, 080.043, 085.014, 116.016.

012.013 **Enlarged space and ground data base for ionospheric modelling. Proceedings. Topical Meeting of the COSPAR Interdisciplinary Scientific Commission C (Meeting C5) of the COSPAR 28. Plenary Meeting, The Hague (Netherlands), 25 Jun – 6 Jul 1990.**
K. Rawer, W. R. Piggott (eds.).
Adv. Space Res., Vol. 11, No. 10, 212 p. (1991). ISBN 0–08–041166–5.
Contents: 1. Mapping of ionospheric parameters. 2. Electron density profiles. 3. Plasma temperatures. 4. Ion composition.
The individual contributions within the subject scope of AAA are included in their corresponding categories – see abstracts 083.007, 083.008, 083.009, 083.010, 083.011, 083.012, 083.013, 083.014, 083.015, 083.016, 083.017, 083.018.

012.014 **International Lens Design Conference. Proceedings. OSA International Lens Design Conference, Monterey, CA (USA), 11 – 14 Jun 1990.**
G. N. Lawrence (ed.).
SPIE Proc. Ser., Vol. 1354.
SPIE, Bellingham, WA (USA). 772 p. (1990). ISBN 0–8194–0415–2.
The individual contributions within the subject scope of AAA are included in their corresponding categories – see abstracts 021.010, 021.011, 021.012, 021.013, 021.014, 021.015, 031.027, 031.028, 031.029, 031.030, 031.031, 031.032, 031.033, 031.034, 031.035, 031.036, 031.037, 031.038, 031.039, 031.040, 031.041, 031.042, 031.043, 031.044, 031.045.

012.015 **Optical fabrication and testing. Proceedings. SPIE Conference on Optical Fabrication and Testing as Part of the Asia Pacific Conference on Optical Technology (APCOT '90), Singapore (Singapore), 22 – 27 Oct 1990.**
M. Lorenzen, D. R. Campbell, C. W. Johnson (eds.).
SPIE Proc. Ser., Vol. 1400.
SPIE, Bellingham, WA (USA). 201 p. (1991). ISBN 0–8194–0483–7.
The individual contributions within the subject scope of AAA are included in their corresponding categories – see abstracts 031.020, 031.021.

012.016 **The physics of solar flares. Proceedings. Royal Society Discussion Meeting: The physics of solar flares, London (UK), 13 – 14 Mar 1991.**
J. L. Culhane, C. Jordan (eds.).
Royal Society, London (UK). 175 p. (1991). ISBN 0–85403–444–7. Price £ 19.50 (UK), £ 21.00 (overseas).
First published in Philos. Trans. R. Soc. London, Ser A, Vol. 336, No. 1643, p. 321 – 495 (1991).
The purpose of this volume is to review the contribution of the three spacecraft (P78–1, SMM, and Hinotori) to the study of solar flares. It also includes discussions of the current theoretical basis of some aspects of solar flares (including magnetic energy conversion).
The individual contributions within the subject scope of AAA are included in their corresponding categories – see abstracts 022.053, 051.028, 073.080, 073.081, 073.082, 073.083, 073.084, 073.085, 073.086, 073.090, 073.091, 073.092, 073.093, 074.072.

012.017 **The first year of HST observations. Proceedings. Workshop on the First Year of HST Observations, Baltimore, MD (USA), 14 – 16 May 1991.**
A. L. Kinney, J. C. Blades (eds.).
Space Telescope Science Institute, Baltimore, MD (USA); National Aeronautics and Space Administration, Washington, DC (USA). 304 p. (1991).
The goals of the meeting were to gather the collective experience of the scientists who were instrumental in preparing for the HST mission and analyzing the early science and calibration data, and to inform both the HST community and prospective observers of the scientific potential of the Observatory, even with its aberrated optics. The presentations included topics such as planets, hot stars, cool stars, chemically peculiar

stars, stellar systems, galactic phenomenon, gravitationally lensed quasars, and the absorption systems observed in nearby quasars. Also presented are papers on the spacecraft performance, including the improved understanding of the Optical Telescope Assembly and the resultant imaging quality. Posters on the planning and scheduling system, which are included as appendices in this volume, show how science observations are executed with the telescope.
The individual contributions within the subject scope of AAA are included in their corresponding categories – see abstracts 035.040, 035.041, 035.042, 035.043, 035.044, 035.045, 035.047, 035.048, 035.049, 035.050, 035.051, 036.085, 036.086, 036.087, 036.096, 036.097, 036.098, 051.019, 051.020, 051.021, 051.022, 051.023, 051.024, 051.025, 051.026, 051.027, 112.031, 114.031, 114.032, 114.033, 131.067, 154.023, 156.011, 157.100, 158.065, 159.033, 159.034, 159.035, 159.036, 159.037, 159.038, 159.040.

012.018 **Spectal line shapes. Volume 6. Proceedings. 10. International Conference on Spectral Line Shapes (ICSLS–10), Austin, TX (USA), 25 – 29 Jun 1990.**
L. Frommhold, J. W. Keto (eds.).
AIP Conf. Proc., No. 216.
American Institute of Physics, New York, NY (USA). 643 p. (Dec 1990). ISBN 0–88318–791–4.
These proceedings contain the 26 invited presentations and abstracts for the 90 contributed papers.
Contents: 1. Plasma broadening. 2. Laser produced dense plasmas. 3. Neutral broadening – atomic transitions. 4. Ulracold atoms. 5. Collisional redistribution. 6. Photon–assisted collisions. 7. Reaction dynamics. 8. Coherent and nonlinear processes. 9. Neutral broadening – molecular transitions. 10. Dimers and excimers. 11. Collision–induced absorption. 12. Collision–induced light scattering. 13. Terrestrial and astrophysics.
The individual contributions within the subject scope of AAA are included in their corresponding categories – see abstracts 013.045, 022.043, 022.044, 022.045, 022.046, 022.047, 022.048, 022.049, 126.019.

012.019 **Solar and galactic cosmic rays. Proceedings. 12. European Cosmic Ray Symposium, Nottingham (UK), 15 – 21 Jul 1990.**
P. R. Blake, W. F. Nash (eds.).
Nucl. Phys. B, Proc. Suppl., Vol. 22B, 232 p. (Jul 1991).
The individual contributions within the subject scope of AAA are included in their corresponding categories – see abstracts 022.042, 078.006, 143.016, 143.017, 144.022, 144.023, 144.024, 144.025, 144.026, 144.027, 144.028, 144.029, 144.030, 161.155.

012.020 **High energy gamma–ray astronomy. Proceedings. International Conference on High Energy Gamma–Ray Astronomy (ICHEGA), Ann Arbor, MI (USA), 2 – 5 Oct 1990.**
J. Matthews (ed.).
AIP Conf. Proc., No. 220.
American Institute of Physics, New York, NY (USA). 353 p. (Apr 1991). ISBN 0–88318–812–0. Price US$ 85.00.
High energy γ–ray astronomy has undergone an astonishing growth in the last decade. The purpose of the conference was to present a forum for a unified discussion of all aspects of γ–ray astronomy. It provided a unique opportunity for the diverse community of observers to coordinate their efforts across the boundaries of different energy regimes. The focus of the meeting was γ–ray astronomy as a single discipline.
The individual contributions within the subject scope of AAA are included in their corresponding categories – see abstracts 034.046, 034.047, 034.049, 034.051, 034.052, 034.053, 034.054, 034.055, 034.057, 034.058, 034.060, 034.061, 034.062, 034.063, 034.064, 035.058, 035.060, 036.102, 036.103, 036.105, 036.106, 036.108, 061.047, 067.047, 080.055, 143.018, 143.019, 143.020, 143.021, 143.022, 143.023, 143.024, 144.031, 144.032, 144.033, 144.034, 144.035, 144.036, 144.037, 144.038, 144.039, 144.040, 144.041, 144.042, 144.043, 144.044, 144.046, 144.047, 144.048, 144.050, 144.051, 155.033, 156.012, 159.041, 161.158.

012.021 **From the solar core to the Earth's magnetosphere. Extended abstracts of reports. All–Union Meeting on the Quasibiennial Solar Variations, Pushchino (USSR), 25 – 27 Feb 1991.**
Soln. Dannye, Byull., No. 2/91, p. 89 – 102 (Feb 1991).
Further abstracts of this meeting are published in No. 3/91, p. 97 – 106 (Mar 1991).
The individual contributions within the subject scope of AAA are included in their corresponding categories – see abstracts 021.046, 072.068, 072.069, 072.070, 072.071, 072.072, 072.073, 072.074, 072.075, 072.077, 072.078, 075.030, 075.031, 075.032, 077.031, 078.007, 080.056, 084.036, 085.023, 085.024, 085.025, 085.026, 144.052.

012.022 **Current problems in particle physics and cosmology. Proceedings. Workshop on Current Problems in Particle Physics and Cosmology, Ioannina (Greece), 25 – 29 Jun 1990.**
K. Tamvakis, C. E. Vayonakis, J. D. Vergados (eds.).
Nucl. Phys. B, Proc. Suppl., Vol. 22A, 150 p. (Jul 1991).
The individual contributions within the subject scope of AAA are included in their corresponding categories – see abstracts 066.083.

012.023 **Quantum mechanics in curved space–time. Proceedings. NATO Advanced Research Workshop on Quantum Mechanics in Curved Space–Time, Erice (Italy), 2 – 12 May 1989.**
J. Audretsch, V. de Sabbata (eds.).
NATO ASI Ser., Ser. B, Phys., Vol. 230.
Plenum Press, New York, NY (USA). 566 p. (1990). ISBN 0–306–43661–2.
The individual contributions within the subject scope of AAA are included in their corresponding categories – see abstracts 066.084, 066.085, 066.086, 067.049, 067.050, 067.051, 067.052, 161.161, 161.162, 161.163.

012.024 **The Sun and cool stars: activity, magnetism, dynamos. Proceedings. IAU Colloquium No. 130: The Sun and cool stars – activity, magnetism, dynamos, Helsinki (Finland), 17 – 20 Jul 1990.**
I. Tuominen, D. Moss, G. Rüdiger (eds.).
Lect. Notes Phys., Vol. 380.
Springer, Berlin (Germany). 540 p. (1991). ISBN 3–540–53955–7. Price DM 106.00. ISBN 0–387–53955–7 (USA).
The contributions in this volume report recent studies on the Sun and late–type stars. Particular emphasis is placed on observations that are relevant to the question of large–scale magnetic activity, and also on the theoretical (dynamo) models for such activity. Experimental papers deal with surface imaging techniques. Explicit computer simulations of hydromagnetic turbulence give insight into the magnetic topology and associatied fluid motions, especially near the base of the convection zone. In addition, mean–field dynamo models are presented.
The individual contributions within the subject scope of AAA are included in their corresponding categories – see abstracts 036.117, 036.118, 036.119, 036.120, 062.101, 062.102, 062.103, 062.104, 062.105, 062.106, 062.107, 062.108, 062.109, 064.061, 064.062, 064.063, 065.035, 065.036, 065.037, 065.038, 065.039, 065.040, 065.041, 065.042, 065.044, 065.045, 065.046, 071.037, 071.038, 071.039, 072.087, 072.088, 072.089, 074.077, 075.035, 075.036, 075.037, 075.038, 075.039, 075.040, 075.041, 075.042, 075.043, 077.036, 080.061, 080.062, 080.063, 080.064, 080.065, 080.066, 080.067, 080.068, 080.069, 080.070, 080.071, 080.072, 080.073, 080.075, 112.041, 113.007, 114.039, 114.040, 115.015, 115.016, 116.023, 116.024, 116.025, 116.026, 116.027, 116.028, 116.029, 116.030, 116.031, 116.032, 116.033, 116.034, 116.035, 116.036, 116.038, 116.039, 116.040, 116.041, 117.070, 117.071, 117.072, 117.073, 117.074, 117.075, 117.078, 117.079, 117.080, 117.082, 119.020, 120.013, 121.019, 121.020, 122.040.

012.025 **First Spain–USSR Workshop on Positional Astronomy and Celestial Mechanics. Proceedings. 1. Spain–USSR Workshop on Positional Astronomy and Celestial Mechanics, Valencia (Spain), 11 – 15 Mar 1991.**
A. López García, R. F. López Machí, A. G. Sokolsky (*A. G. Sokol'skij*) (eds.).
Universitat de València (Spain). Observatorio Astronomico. 158 p. (1991). ISBN 84–370–0814–X.

This Workshop was aimed at the following goals: (1) mutual exchange of information about scientific activity at the institutions which signed the agreement on scientific cooperation between the Astronomical Observatory of the University of Valencia and the Institute of Theoretical Astronomy of the USSR Academy of Sciences, Leningrad; (2) to correct the form and the contents of future work to be done in cooperation and to elaborate the clear prospects for the long–time cooperation.
The individual contributions within the subject scope of AAA are included in their corresponding categories – see abstracts 002.066, 002.068, 002.069, 013.051, 022.063, 036.121, 041.010, 041.011, 042.022, 042.024, 042.025, 042.026, 042.027, 042.028, 042.029, 042.030, 042.031, 096.002, 097.011, 098.038, 098.041, 098.046, 102.013, 103.068.

012.026 **High energy astrophysics: compact stars and active galaxies. Proceedings. 3. Chinese Academy of Sciences and Max–Planck Society Workshop: High energy astrophysics: compact stars and active galaxies, Huangshan (China), 19 – 23 Oct 1990.**
Li Qibin (ed.).
World Scientific, Singapore (Singapore). 356 p. (1991). ISBN 981–02–0697–6. Price £ 37.00.

Contents: 1. Accretion disk physics. 2. Ejection and streams. 3. X–ray and γ–ray sources, cataclysmic variables and SNRs. 4. Active galaxies. 5. Large scale structure of the Universe. 6. Development in theoretical astrophysics.
The individual contributions within the subject scope of AAA are included in their corresponding categories – see abstracts 062.122, 062.126, 062.127, 064.071, 064.072, 064.078, 065.058, 067.077, 067.078, 067.079, 067.084, 113.021, 117.086, 117.087, 117.088, 117.090, 117.091, 117.094, 125.037, 126.026, 132.023, 143.030, 151.047, 151.065, 155.053, 156.017, 157.141, 158.081, 158.090, 158.091, 158.092, 158.093, 158.094, 158.095, 158.096, 158.098, 159.052, 159.053, 159.054, 159.055, 159.056, 160.060, 161.207, 161.211, 161.213, 161.214, 161.215, 161.216, 161.217, 161.218, 161.219, 161.220, 161.221.

012.027 **Precision photometry: astrophysics of the Galaxy. Proceedings. Meeting on Precision Photometry: Astrophysics of the Galaxy, Schenectady, NY (USA), 3 – 5 Oct 1990.**
A. G. D. Philip, A. R. Upgren, K. A. Janes (eds.).
Van Vleck Obs. Contrib., No. 11.
L. Davis Press, Schenectady, NY (USA). 400 p. (1991). ISBN 0–933485–13–1.
The individual contributions within the subject scope of AAA are included in their corresponding categories – see abstracts 002.071, 013.053, 013.057, 021.017, 034.078, 035.087, 036.122, 036.123, 036.124, 036.125, 036.126, 036.127, 064.066, 111.006, 113.009, 113.010, 113.011, 113.012, 113.014, 113.016, 113.017, 113.018, 113.019, 113.020, 113.047, 114.045, 114.046, 153.021, 153.023, 153.024, 153.025, 154.033, 154.034, 154.035, 154.036, 154.037, 154.038, 155.043, 155.044, 155.045, 155.046, 155.047, 155.048.

012.028 **Dynamics of disc galaxies. Proceedings. Conference on Dynamics of Disc Galaxies, Varberg Castle (Sweden), 25 – 30 May 1991.**
B. Sundelius (ed.).
Göteborg Univ. (Sweden). Dept. of Astronomy/Astrophysics; Chalmers Univ. of Tech., Göteborg (Sweden). 415 p. (1991). ISBN 91–7032–630–4. Price SEK 200.00, US$ 35.00, £ 19.00.

The conference brought together present ideas on the dynamical aspects of the structure and evolution of galactic discs.
The individual contributions within the subject scope of AAA are included in their corresponding categories – see abstracts 021.016, 065.050, 151.032, 151.033, 151.034, 151.035, 151.037, 151.038, 151.039, 151.040, 151.041, 151.042, 151.043, 151.044, 151.049, 151.050, 151.051, 151.052, 151.053, 151.054, 151.055, 151.056, 151.057, 151.058, 151.059, 151.062, 151.063, 151.064, 155.038, 155.039, 155.040, 155.041, 155.042, 155.050, 157.111, 157.112, 157.113, 157.114, 157.116, 157.117, 157.120, 157.123, 157.124, 157.125, 157.127, 157.128, 157.129, 157.133, 157.136, 157.140, 157.142, 160.059.

012.029 **Advanced laser concepts and applications. Proceedings. Symposium on Advanced Laser Concepts and Applications as Part of the 4. International Congress on Optical Science and Engineering (ECO–4), The Hague (Netherlands), 12 – 13 Mar 1991.**
S. Singer (ed.).
SPIE Proc. Ser., Vol. 1501.
SPIE, Bellingham, WA (USA). 215 p. (1991). ISBN 0–8194–0610–4.

Contents: 1. New concepts and advances in solid–state lasers. 2. Detectors and diagnostics. 3. New laser concepts. 4. Free–electron lasers and optics. 5. Applications (which ranged from drilling to spectroscopy to astronomy to advanced photolithography).
The individual contributions within the subject scope of AAA are included in their corresponding categories – see abstracts 034.075.

012.030 **Space astronomical telescopes and instruments. Proceedings. Conference on Space Astronomical Telescopes and Instruments as Part of SPIE's Technical Symposia on Optical Engineering and Photonics in Aerospace Sensing (OE/Aerospace Sensing '91), Orlando, FL (USA), 1 – 4 Apr 1991.**
P. Y. Bely, J. B. Breckinridge (eds.).
SPIE Proc. Ser., Vol. 1494.
SPIE, Bellingham, WA (USA). 545 p. (1991). ISBN 0–8194–0603–1. Price US$ 87.00.

This conference provided an excellent forum for the recent progress being made in space UV/visible/IR telescopes and instruments. Session 1: "Lessons from the Hubble Space Telescope", describes the status of the observatory after almost a year of operation, and presented the plans for correcting the problem of spherical aberration during the repair mission scheduled for 1993. The papers presented at Session 2: "Telescopes on the Moon", testifies to the enthusiasm generated by the proposals to build astronomical instruments on the Moon. The space environment enables new scientific measurements not only in wavelength regions unknown to ground–based astronomers, but also in the new ultralow scattered–light regimes and at spatial resolutions previously unavailable. State–of–the–art papers on new telescope design, optomechanical structures, integrated optics modeling, optical materials, wavefront sensing and control, and sensor optics were presented in Session 3 within the framework of new space science mission objectives. Finally, Session 4 explored current techniques and ideas on mirror fabrication and active control for space application.
The individual contributions within the subject scope of AAA are included in their corresponding categories – see abstracts 031.049, 031.050, 031.051, 032.016, 034.076, 034.079, 034.080, 034.081, 034.082, 034.083, 035.064, 035.065, 035.066, 035.067, 035.068, 035.069, 035.070, 035.071, 035.072, 035.073, 035.074, 035.075, 035.076, 035.077, 035.078, 035.079, 035.080, 035.081, 035.082, 035.083, 035.084, 035.085, 035.086, 035.088, 035.089, 035.090, 035.091, 035.092, 035.093, 035.094, 035.095, 035.096, 035.097, 035.098, 035.099, 035.100, 035.101.

012.031 **Electron image tubes and image intensifiers II. Proceedings. 2. Conference on Image Tubes and Image Intensifiers, San Jose, CA (USA), 28 Feb – 1 Mar 1991.**
I. P. Csorba (ed.).
SPIE Proc. Ser., Vol. 1449.
SPIE, Bellingham, WA (USA). 202 p. (1991). ISBN 0–8194–0548–5.

This conference reviewed the recent progress achieved in UV image intensification, high–speed gated image intensification, state–of–the–art materials and devices technology, X–ray image intensification, high-resolution low-light-level astronomical image detection, high-definition television (HDTV), and IR imaging and computerized image detection methods.

012.032 **The Armagh Meeting of the Royal Astronomical Society, Armagh (UK), 3 – 5 Apr 1991.**
Vistas Astron., Vol. 34, Part 3, p. 187 – 368 (1991).
The individual contributions within the subject scope of AAA are included in their corresponding categories – see abstracts 013.054, 013.055, 013.056, 071.042, 073.105, 073.106, 073.107, 112.043, 114.043, 116.043, 116.044, 156.014, 157.115.

012.033 **The problem of lithium. Proceedings. European Workshop on the Problem of Lithium in Honour of Evry Schatzman, Monte Porzio, Rome (Italy), 12 – 14 Sep 1990.**
F. D'Antona (ed.).
Mem. Soc. Astron. Ital., Vol. 62, No. 1, p. 1 – 195 (1991).
In this meeting observations and theory of lithium in stars have been discussed with the aims of focusing the patterns of lithium destruction and putting into evidence the possible controversial indications of lithium production in the course of the galactic lifetime.
The individual contributions within the subject scope of AAA are included in their corresponding categories – see abstracts 065.051, 065.052, 065.053, 065.054, 065.055, 065.056, 065.057, 114.047, 114.048, 114.049, 114.050, 114.051, 114.052, 114.053, 114.054, 114.055, 114.057, 114.058, 114.059.

012.034 **Plasmas in astrophysics. Proceedings. 3. Texas-Mexico Conference on Astrophysics: Plasmas in astrophysics, Houston, TX (USA), 14 – 15 Feb 1991.**
C. R. O'Dell (ed.).
Publ. Astron. Soc. Pac., Vol. 103, No. 666, p. 759 – 927 (Aug 1991).
The meeting was organized into four topical sessions: 1. Compact objects. 2. Supernova remnants. 3. Emission nebulae. 4. Active galactic nuclei.
The individual contributions within the subject scope of AAA are included in their corresponding categories – see abstracts 062.123, 067.080, 067.081, 067.082, 067.083, 125.032, 125.033, 125.034, 125.035, 125.036, 131.098, 132.016, 132.017, 132.018, 132.019, 132.020, 132.021, 132.022, 134.037, 134.038, 134.039, 142.019, 155.049, 155.054, 157.134, 157.135, 157.137, 157.138, 158.085, 158.086, 158.087, 158.088, 160.057, 160.058.

012.035 **Determination of the geoid – present and future. Proceedings. IAG Symposium: Determination of the geoid, Milan (Italy), 11 – 13 Jun 1990.**
R. H. Rapp, F. Sansò (eds.).
Int. Assoc. Geod. Symp., No. 106.
Springer, New York, NY (USA). 511 p. (1991). ISBN 0-387-97470-9. ISBN 3-540-97470-9 (Germany), Price DM 128.00.
Contents: 1. Global geopotential models: present and future. 2. The role of topography in geoid computations: the marine geoid. 3. The geoid and GPS. 4. State of computation of national or regional geoids. 5. Theoretical improvements in the determination of the geoid. 6. Software and data improvement for geoid computations. 7. GEOMED: an international project for the determination of the geoid in the Mediterranean.
The individual contributions within the subject scope of AAA are included in their corresponding categories – see abstracts 021.019, 021.020, 021.021, 021.022, 021.023, 021.024, 021.025, 021.026, 045.009, 081.018, 081.019, 081.020, 081.021, 081.022, 081.023, 081.024, 081.025, 081.026, 081.027, 081.029, 081.030, 081.031, 081.032, 081.033, 081.034, 081.035.

012.036 **Quantum cosmology and baby universes. Proceedings. 7. Jerusalem Winter School for Theoretical Physics: Quantum cosmology and baby universes, Jerusalem (Israel), 27 Dec 1989 – 4 Jan 1990.**
S. Coleman, J. B. Hartle, T. Piran, S. Weinberg (eds.).
World Scientific, Singapore (Singapore). 367 p. (1991). ISBN 981-02-0345-4. Price £ 36.00.
Recently there has been a flurry of activity in quantum gravity, inspired both by the importation of ideas from string theory and by speculations about connections between topology-changing quantum fluctuations and the vanishing of the cosmological constant. These developments were among the topics of the Seventh Jerusalem Winter School 1989/1990.
The individual contributions within the subject scope of AAA are included in their corresponding categories – see abstracts 036.148, 066.115, 066.116, 161.208, 161.209, 161.210, 161.212.

012.037 **Astroparticle physics. Proceedings. International School of Astroparticle Physics, Houston, TX (USA), 6 – 12 Jan 1991.**
D. V. Nanopoulos (ed.).
World Scientific, Singapore (Singapore). 463 p. (1991). ISBN 981-02-0582-1. Price £ 65.00. ISBN 981-02-0583-X (paper).
The individual contributions within the subject scope of AAA are included in their corresponding categories – see abstracts 051.032, 051.033, 061.063, 061.064, 061.065, 080.085, 161.222, 161.223, 161.226, 161.227, 161.228, 161.234, 161.238, 161.239, 161.240, 161.241, 161.243, 161.244.

012.038 **Observatories in Earth orbit and beyond. Proceedings. IAU Colloquium No. 123: Observatories in Earth orbit and beyond, Greenbelt, MD (USA), 24 – 27 Apr 1990.**
Y. Kondo (ed.).
Astrophys. Space Sci. Libr., Vol. 166.
Kluwer Academic Publ., Dordrecht (Netherlands). 584 p. (1990). ISBN 0-7923-1133-7. Price Dfl. 140.00, US$ 69.00, £ 48.00.
The Colloquium featured invited reviews of virtually all current and approved future projects in space astronomy, over the entire electromagnetic spectrum from gamma-ray to radio, of all major space faring nations of the world. Also included are reviews and panel discussions of current and future launch systems, advantages and disadvantages of various orbits and sites in space for astronomical observatories, and major unsolved problems of astronomy. A number of contributed papers were also given at the meeting and are included in the current proceedings.
The individual contributions within the subject scope of AAA are included in their corresponding categories – see abstracts 031.053, 035.102, 035.103, 035.104, 035.105, 035.106, 035.107, 035.108, 035.109, 035.110, 035.111, 035.112, 035.113, 035.114, 035.115, 035.116, 035.117, 035.118, 035.119, 035.120, 035.121, 035.122, 035.123, 035.124, 035.125, 035.126, 035.127, 035.128, 035.129, 035.130, 035.131, 035.132, 035.133, 035.134, 035.135, 035.136, 035.137, 035.138, 035.139, 035.140, 035.141, 035.142, 035.143, 035.144, 035.145, 035.146, 035.147, 035.148, 035.149, 035.150, 035.151, 035.152, 035.153, 035.154, 035.155, 036.134, 036.137, 036.138, 036.139, 036.140, 036.141, 036.142, 036.143, 042.044, 051.034, 051.035, 051.036, 051.037, 051.038, 051.039, 051.040, 051.041, 062.141, 066.119, 142.024, 156.022, 158.144, 161.255.

012.039 **The environmental model of Mars. Proceedings. 2. COSPAR Colloquium: The environmental model of Mars, Sopron (Hungary), 22 – 26 Jan 1990.**
K. Szegö (ed.).
COSPAR Colloq. Ser., Vol. 2.
Pergamon Press, Oxford (UK). 167 p. (1991). ISBN 0-08-040787-0. Price US$ 30.00.
The main aim of this volume is to bring together information from scientists from all the agencies and countries where missions to Mars or instruments for these missions are planned. It also enables the discussion and updating of the current model for the Mars environment. The exploration of planet Mars will be the focal point of the planetary missions in the coming years. Among others, the investigation of the surface and the upper layers of the soil is of primary interest. So it can be safely predicted that the planet will be populated soon by penetrators, small ground stations, rovers, balloons will fly around in the atmosphere, etc.
The individual contributions within the subject scope of AAA are included in their corresponding categories – see abstracts 022.088, 035.156, 036.147, 051.042, 051.043, 051.044, 051.045, 051.046, 097.022, 097.023, 097.024, 097.025, 097.026, 097.027,

097.028, 097.029, 097.030, 097.031, 097.032, 097.033, 097.034, 097.035, 097.036, 105.032.

012.040 **Variability of active galactic nuclei. Proceedings. Georgia State University Conference on Variability of Active Galactic Nuclei, Atlanta, GA (USA), 2 – 4 May 1990.**
H. R. Miller, P. J. Wiita (eds.).
Cambridge University Press, Cambridge, NY (USA). 400 p. (1991). ISBN 0–521–41295–1. Price £ 35.00, US$ 49.50.
Active galactic nuclei (AGNs) continue to receive a great deal of attention from observational astronomers and theorists. This book collects together the results presented at Georgia State University in May 1990. It is clear that careful analysis of variability over many wave bands is likely to give clues as to the nature of the central engine in AGNs. The contributions are grouped in themes: multifrequency variability, UV–optical–IR variability, radio variability, and X–ray variability. One of the most exciting results reported is that many blazars exhibit micro–variability. There is an important group of papers on NGC 5548 and other Seyfert galaxies. New results are also announced for many quasars, Seyfert galaxies, BL Lacertae objects and radio galaxies. The data are interspersed with theoretical insights. The book is a valuable summary of the variability observed in a wide variety of AGNs, and the substantial constraints now imposed on models for the central engine and the production of variability.
The individual contributions within the subject scope of AAA are included in their corresponding categories – see abstracts 036.144, 036.145, 036.146, 064.088, 158.103, 158.104, 158.105, 158.106, 158.107, 158.108, 158.109, 158.110, 158.111, 158.112, 158.113, 158.114, 158.115, 158.116, 158.117, 158.118, 158.119, 158.120, 158.121, 158.122, 158.123, 158.124, 158.125, 158.126, 158.127, 158.128, 158.129, 158.130, 158.131, 158.133, 158.134, 158.135, 158.136, 158.137, 158.138, 158.139, 158.140, 158.141, 158.142, 158.143, 158.145, 158.146, 158.147, 158.148, 158.149, 158.150, 158.151, 159.058, 159.059, 159.060, 159.061, 159.062, 159.063, 159.064, 159.065, 159.066.

012.041 **The interpretation of modern synthesis observations of spiral galaxies. Proceedings. Workshop on the Interpretation of Modern Synthesis Observations of Spiral Galaxies, Albuquerque, NM (USA), 12 – 15 Mar 1990.**
N. Duric, P. C. Crane (eds.).
Astron. Soc. Pac. Conf. Ser., Vol. 18.
Astronomical Society of the Pacific, San Francisco, CA (USA). 292 p. (1991). ISBN 0–937707–37–6. Price US$ 32.00.
Contents: 1. Cosmic rays and magnetic fields. 2. Discrete radio sources in spiral galaxies. 3. Statistical studies. 4. Spectral–line studies. 5. Non–spiral galaxies. 6. Nuclear activity and active galaxies.
The individual contributions within the subject scope of AAA are included in their corresponding categories – see abstracts 061.067, 125.048, 144.060, 144.061, 144.062, 155.068, 155.069, 157.171, 157.172, 157.173, 157.174, 157.175, 157.176, 157.177, 157.178, 157.179, 157.180, 157.181, 157.182, 157.183, 157.184, 157.185, 157.186, 157.187, 158.158, 158.159, 158.160, 158.161, 160.074.

012.042 **The physics of star formation and early stellar evolution. Proceedings. NATO Advanced Study Institute on The Physics of Star Formation and Early Stellar Evolution, Agia Pelagia, Crete (Greece), 27 May – 8 Jun 1990.**
C. J. Lada, N. D. Kylafis (eds.).
NATO ASI Ser., Ser. C, Math. Phys. Sci., Vol. 342.
Kluwer Academic Publ., Dordrecht (Netherlands). 777 p. (1991). ISBN 0–7923–1349–6. Price Dfl. 350.00, US$ 194.00, £ 117.00.
These proceedings provides simultaneously a broad and systematic overview of, as well as a rigorous introduction to, the fundamental physics and astronomy at the heart of modern research in star formation and early evolution.
The individual contributions within the subject scope of AAA are included in their corresponding categories – see abstracts 062.142, 063.036, 064.090, 064.091, 064.092, 121.030, 121.032, 121.033, 121.034, 131.204, 131.205, 131.206, 131.207, 131.208, 131.209, 131.210, 131.211, 131.212, 131.213, 131.214, 131.215, 131.216.

012.043 **Fragmentation of molecular clouds and star formation. Proceedings. IAU Symposium No. 147: Fragmentation of molecular clouds and star formation, Grenoble (France), 11 – 16 Jun 1990.**
E. Falgarone, F. Boulanger, G. Duvert (eds.).
Int. Astron. Union, Symp., No. 147.
Kluwer Academic Publ., Dordrecht (Netherlands). 580 p. (1991). ISBN 0–7923–1158–2. Price Dfl. 225.00, US$ 146.00, £ 77.00.
These proceedings are concerned with interstellar matter and molecular clouds: their large scale structure, velocity fields, magnetic fields, chemistry, shocks and instabilities, small scale structure, star formation, and early stages of stellar evolution.
The individual contributions within the subject scope of AAA are included in their corresponding categories – see abstracts 062.138, 062.140, 107.006, 112.058, 115.018, 121.026, 121.027, 121.028, 121.029, 131.114, 131.116, 131.117, 131.118, 131.119, 131.121, 131.122, 131.126, 131.128, 131.129, 131.130, 131.131, 131.132, 131.133, 131.134, 131.135, 131.136, 131.137, 131.138, 131.139, 131.140, 131.141, 131.143, 131.144, 131.145, 131.146, 131.147, 131.148, 131.149, 131.150, 131.151, 131.152, 131.153, 131.155, 131.156, 131.157, 131.158, 131.159, 131.160, 131.161, 131.162, 131.163, 131.164, 131.165, 131.166, 131.167, 131.168, 131.169, 131.170, 131.171, 131.172, 131.173, 131.174, 131.175, 131.176, 131.177, 131.178, 131.179, 131.180, 131.181, 131.182, 131.183, 131.184, 131.185, 131.186, 131.187, 131.188, 131.189, 131.190, 131.191, 131.192, 131.193, 131.194, 131.195, 131.196, 131.197, 131.198, 131.199, 131.200, 131.201, 131.202, 131.203, 132.026, 132.028, 157.162, 157.164, 161.252.

012.044 **White dwarfs. Proceedings. NATO Advanced Research Workshop – 7. European Workshop on White Dwarfs, Toulouse (France), 3 – 5 Sep 1990.**
G. Vauclair, E. Sion (eds.).
NATO ASI Ser., Ser. C, Math. Phys. Sci., Vol. 336.
Kluwer Academic Publ., Dordrecht (Netherlands). 485 p. (1991). ISBN 0–7923–1266–X. Price Dfl. 240.00, US$ 147.00, £ 82.00.
The 7th European Workshop on White Dwarfs took place in the year of the seventieth birthday of Prof. Evry Schatzman, whose pioneering work has been the inspiration for many of the new ideas in the white dwarf community. The Scientific Organizing Committee has agreed to dedicate the workshop to him on this occasion.
Contents: 1. Pre–white dwarf evolution and white dwarf cooling. 2. Asteroseismology of white dwarfs. 3. Atmospheres and envelopes. 4. White dwarfs in binaries.
The individual contributions within the subject scope of AAA are included in their corresponding categories – see abstracts 013.068, 013.070, 064.122, 064.123, 064.124, 064.125, 065.097, 065.098, 065.099, 065.100, 067.133, 114.086, 114.087, 117.168, 117.169, 117.170, 117.171, 117.172, 126.042, 126.043, 126.044, 126.045, 126.046, 126.047, 126.048, 126.049, 126.050, 126.051, 126.052, 126.053, 126.054, 126.055, 126.056, 126.057, 126.058, 126.059, 126.060, 126.061, 126.062, 126.063, 126.064, 126.065, 126.066, 134.055, 134.056, 155.087.

012.045 **Extreme ultraviolet astronomy. A selection of papers presented at the First Berkeley Colloquium on Extreme Ultraviolet Astronomy. 1. Berkeley Colloquium on Extreme Ultraviolet Astronomy, Berkeley, CA (USA), 19 – 20 Jan 1989.**
R. F. Malina, S. Bowyer (eds.).
Pergamon Press, New York, NY (USA). 532 p. (1991). ISBN 0–08–037302–X. Price £ 52.90.
Contents: 1. Overview of EUV observations. 2. Sources of EUV radiation. 3. Interstellar absorption. 4. Innovative EUV instrumentation. 5. Future EUV missions.
The individual contributions within the subject scope of AAA are included in their corresponding categories – see abstracts 002.095, 031.066, 031.067, 034.118, 035.189, 035.190, 035.191, 035.192, 035.193, 035.194, 035.195, 051.054, 051.055, 051.056, 051.057, 051.059, 051.060, 051.061, 051.062, 051.063, 051.064, 062.165, 063.092, 063.093, 064.130, 064.131, 064.132, 064.133, 067.155, 067.156, 073.151, 091.036, 099.048, 112.088, 112.089, 114.091, 117.177, 117.178, 117.180, 117.181, 117.182, 126.069, 126.070, 126.071, 131.292, 131.293, 131.294, 131.295, 142.031, 142.032, 159.090, 159.091.

012.046 **Structure and emission properties of accretion disks. Proceedings. 6. IAP Astrophysics Meeting – IAU Colloquium No. 129: Structure and emission properties of accretion disks, Paris (France), 2 – 6 Jul 1990.**
C. Bertout, S. Collin-Souffrin, J. P. Lasota, J. Tran Thanh Van (eds.).
Int. Astron. Union Colloq., No. 129.
Editions Frontières, Gif sur Yvette (France). 595 p. (1991). ISBN 2-86332-093-9. Price FF 480.00, US$ 80.00.
In recent years accretion disks have become an established part of astrophysics. They have proved to be an essential element of binary systems, and to play an equally vital role both in describing the formation of young stars and in constructing models of active galactic nuclei. Many problems of physics, and some specific difficulties, are shared by these three areas, even though differences of scale between them can attain several orders of magnitude. Therefore, this colloquium was organized to bring together all those who are interested in accretion disks. Contents: (A) Review and invited papers: 1. Observations of disks. 2. Physical processes in disks. 3. Disk structure and models. 4. Boundary layers and winds from disks. (B) Poster contributions.
The individual contributions within the subject scope of AAA are included in their corresponding categories – see abstracts 062.167, 062.168, 062.169, 062.170, 062.171, 062.172, 062.176, 062.180, 062.182, 063.098, 063.099, 063.100, 064.134, 064.135, 064.136, 064.137, 064.138, 064.140, 064.141, 064.142, 064.143, 064.145, 064.146, 064.147, 064.148, 064.149, 064.150, 064.151, 064.152, 064.153, 064.155, 064.156, 064.158, 064.159, 064.160, 064.161, 064.162, 064.163, 064.165, 064.166, 064.167, 064.168, 064.169, 064.170, 064.171, 067.161, 067.162, 067.163, 067.166, 067.167, 067.168, 067.169, 067.170, 067.171, 067.172, 067.173, 067.175, 067.176, 067.177, 067.180, 067.183, 067.184, 067.185, 067.187, 067.188, 067.189, 067.191, 067.192, 067.193, 067.194, 107.016, 112.091, 112.093, 112.097, 117.185, 117.186, 117.192, 117.193, 117.196, 117.199, 117.200, 117.201, 117.202, 117.204, 117.205, 117.206, 117.207, 117.208, 117.209, 117.210, 117.215, 117.219, 117.221, 117.222, 117.224, 117.227, 117.229, 117.230, 120.027, 121.066, 121.067, 121.068, 121.069, 121.070, 121.072, 121.073, 121.075, 121.076, 121.077, 121.078, 121.079, 121.080, 121.081, 124.005, 124.015, 131.309, 143.044, 158.221, 158.236, 158.237, 158.239, 158.240, 158.242, 158.246, 158.247, 158.251, 158.252, 158.253, 158.254, 158.257, 158.259, 158.260, 158.262, 158.263, 158.265, 158.266, 158.267, 158.269, 158.270, 158.271, 158.272, 158.273, 159.107, 159.111, 159.115.

012.047 **22 MAX. Global character research of solar–terrestrial system. Special issue III. Proceedings. Workshop on major solar–geophysical events in March and August 1989, Beijing (China), 23 – 26 Apr 1990.**
Hu Wenrui, Zhang Bairong, Shen Haizhang, Li Xixian (eds.).
Publ. Yunnan Obs., No. 4, 318 p. (Dec 1990). With 25 plates.
The individual contributions within the subject scope of AAA are included in their corresponding categories – see abstracts 013.061, 013.062, 013.063, 013.064, 036.150, 071.046, 072.101, 072.102, 072.103, 072.106, 072.107, 072.108, 072.109, 072.110, 072.123, 073.119, 073.120, 073.121, 073.122, 073.123, 073.124, 073.125, 073.127, 073.130, 073.131, 073.132, 073.133, 073.134, 073.135, 073.136, 073.137, 073.138, 075.054, 075.056, 076.027, 077.049, 077.050, 077.051, 077.052, 077.053, 077.054, 077.055, 077.056, 077.057, 077.058, 077.059, 077.063, 078.014, 078.015, 078.016, 083.023, 083.024, 083.025, 083.026, 083.027, 083.031, 083.032, 083.033, 083.034, 083.035, 084.051, 085.033, 085.034, 085.035, 085.036, 085.037, 085.038, 085.039, 085.042, 085.043, 085.044, 085.045, 085.046, 085.047, 085.048, 085.049.

012.048 **Systèmes de référence spatio–temporels: métrologie et astrométrie. Journées 1991: Systèmes de référence spatio–temporels, Paris (France), 2 – 4 Jun 1991.**
N. Capitaine (ed.).
Observatoire de Paris (France). Dept. d'Astronomie Fondamentale. 239 p. (1991). ISBN 2-901057-23-3.
Contents: 1. Métrologie des échelles de temps. 2. Méthodes statistiques et traitement numérique. 3. Mesures et erreurs. 4. Earth rotation in the Hipparcos reference frame.
The individual contributions within the subject scope of AAA are included in their corresponding categories – see abstracts 004.037, 004.038, 021.029, 021.032, 021.033, 021.034, 021.035, 032.020, 034.096, 034.097, 036.157, 036.163, 036.166, 036.168, 041.014, 041.015, 041.016, 041.017, 041.019, 043.008, 043.009, 043.010, 044.025, 044.026, 044.027, 044.028, 044.029, 044.031, 044.032, 044.033, 044.035, 044.036, 044.037, 066.161, 091.026, 097.041, 111.010.

012.049 **Problemy astrometrii i kosmicheskoj geodinamiki. Materialy. *(Problems of astrometry and space geodynamics. Proceedings.)* International Seminar on Problems of Astrometry and Space Geodynamics, Kiev (USSR), 10 – 12 Sep 1990.**
N. V. Kharchenko (ed.).
Naukova Dumka, Kiev (USSR). 136 p. (1991). ISBN 5-12-003339-3. Price 3 Rbl.
The individual contributions within the subject scope of AAA are included in their corresponding categories – see abstracts 002.089, 013.066, 013.067, 034.106, 036.167, 036.169, 041.018, 041.020, 041.021, 041.022, 044.038, 044.039, 045.012, 081.041, 082.052, 111.012, 113.037, 113.038, 113.039, 153.066, 155.081.

012.050 **The formation and evolution of star clusters. Proceedings. Symposium on the Formation and Evolution of Star Clusters as Part of the 102. Summer Scientific Meeting of the Astronomical Society of the Pacific, Boston, MA (USA), 14 – 15 Jul 1990.**
K. Janes (ed.).
Astron. Soc. Pac. Conf. Ser., Vol. 13.
Astronomical Society of the Pacific, San Francisco, CA (USA). 631 p. (1991). ISBN 0-937707-32-5. Price US$ 28.80.
Contents: 1. Have star and cluster formation processes changed with time? 2. Stellar evolution: new approaches to unresolved problems. 3. Clusters as probes of dynamical evolution. 4. Clusters as tracers of the chemical evolution of galaxies.
The individual contributions within the subject scope of AAA are included in their corresponding categories – see abstracts 021.030, 065.069, 065.070, 113.031, 114.081, 114.082, 114.083, 114.084, 119.040, 120.020, 120.022, 120.023, 122.074, 122.075, 126.034, 131.223, 131.224, 131.225, 131.226, 131.227, 131.228, 131.229, 132.029, 132.030, 151.091, 151.093, 151.094, 151.095, 151.096, 152.007, 152.008, 152.009, 153.040, 153.041, 153.042, 153.043, 153.044, 153.045, 153.046, 153.047, 153.048, 153.059, 153.060, 153.061, 153.062, 153.064, 153.067, 153.068, 153.069, 153.070, 154.045, 154.046, 154.047, 154.048, 154.049, 154.050, 154.051, 154.052, 154.053, 154.054, 154.056, 154.057, 154.058, 154.059, 154.060, 154.061, 154.062, 154.063, 154.064, 154.073, 154.074, 154.075, 154.076, 154.077, 154.078, 154.079, 154.080, 154.081, 154.082, 154.083, 154.084, 154.085, 154.086, 154.087, 154.088, 154.089, 154.090, 154.091, 154.092, 154.093, 154.094, 154.095, 154.096, 154.097, 154.098, 154.099, 154.100, 154.104, 154.105, 154.106, 154.107, 154.108, 154.109, 154.110, 154.111, 154.112, 154.113, 154.114, 154.115, 154.116, 154.117, 154.118, 154.119, 154.120, 154.121, 154.122, 154.123, 155.082, 155.083, 155.085.

012.051 **Skylines. Proceedings. 3. Haystack Observatory Conference on the Interstellar Medium: Skylines – atoms, ions and molecules – new results in spectral line astrophysics, Westford, MA (USA), 23 – 25 May 1990.**
A. D. Haschick, P. T. P. Ho (eds.).
Astron. Soc. Pac. Conf. Ser., Vol 16.
Astronomical Society of the Pacific, San Francisco, CA (USA). 491 p. (1991). ISBN 0-937707-35-X. Price US$ 32.00.
The scientific sessions of this meeting consist of 24 invited talks, 20 contributed talks and 28 poster papers on the topics: 1. Protogalaxies, galaxies, megamasers. 2. Giant molecular clouds, massive stars, protostars. 3. Low mass cores, outflows, Herbig–Haro objects. 4. New molecules, chemistry, magnetic fields. 5. Instrumentation.
The individual contributions within the subject scope of AAA are included in their corresponding categories – see abstracts 013.069, 022.095, 033.014, 033.015, 033.016, 035.160, 051.051, 061.085, 062.150, 064.121, 112.075, 112.076, 112.077, 112.078,

112.081, 112.082, 112.083, 112.084, 121.049, 121.050, 121.051, 121.052, 121.053, 121.054, 121.055, 121.056, 121.057, 121.058, 121.059, 122.073, 122.076, 131.253, 131.254, 131.255, 131.256, 131.258, 131.259, 131.262, 131.263, 131.264, 131.265, 131.266, 131.267, 131.268, 131.269, 131.270, 131.271, 131.272, 131.276, 131.277, 131.278, 131.281, 131.282, 132.032, 132.033, 132.034, 155.079, 155.080, 155.084, 157.220, 157.221, 157.222, 157.223, 157.224, 157.225, 158.184.

012.052 **Supernovae. Proceedings. 10. Santa Cruz Summer Workshop in Astronomy and Astrophysics: Supernovae, Santa Cruz, CA (USA), 9 – 21 Jul 1989.**
S. E. Woosley (ed.).
Springer, New York, NY (USA). 812 p. (1991). ISBN 0–387–97071–1. Price DM 84.00.
This book presents the state of supernova studies at the beginning of the 1990s as reported at a meeting in Jul 1989. The 113 papers contained in this volume report on all aspects of the field – observations at wavelengths from radio through gamma–rays, bolometric light curves and spectra, neutrino observations, the theory of stellar explosions, multidimensional models for mixing, nucleosynthesis calculations, synthetic spectral modeling, presupernova evolution, supernova remnants, supernova rates, supernovae as standard candles, the interaction of supernovae with their surroundings, and the most comprehensive and up–to–date treatment of SN 1987A currently available.
The individual contributions within the subject scope of AAA are included in their corresponding categories – see abstracts 002.087, 013.065, 035.046, 036.162, 061.069, 061.070, 061.071, 061.072, 061.073, 061.074, 065.087, 065.088, 065.089, 066.159, 067.092, 067.093, 067.097, 117.148, 117.152, 117.154, 124.003, 125.051, 125.052, 125.053, 125.054, 125.055, 125.056, 125.057, 125.058, 125.059, 125.060, 125.061, 125.062, 125.063, 125.064, 125.065, 125.066, 125.067, 125.068, 125.069, 125.070, 125.071, 125.072, 125.073, 125.074, 125.075, 125.076, 125.077, 125.078, 125.079, 125.080, 125.081, 125.082, 125.083, 125.084, 125.085, 125.086, 125.087, 125.088, 125.089, 125.164, 125.168, 125.170, 125.188, 125.189, 125.190, 125.191, 125.192, 125.193, 125.194, 125.195, 125.196, 125.197, 125.198, 125.199, 125.200, 125.201, 125.202, 125.203, 125.204, 125.205, 125.206, 125.207, 125.208, 125.209, 125.210, 125.211, 125.212, 125.213, 125.214, 125.215, 125.216, 125.218, 125.219, 125.220, 125.221, 125.223, 125.224, 125.225, 125.226, 125.227, 125.228, 125.229, 125.230, 125.303, 126.037, 126.039, 131.246, 131.247, 156.026, 156.028, 157.212, 158.180, 161.285.

012.053 **Cosmic rays, supernovae and the interstellar medium. Proceedings. 7. Course of the International School of Cosmic–Ray Astrophysics: Cosmic rays, supernovae, and the interstellar medium – A NATO Advanced Study Institute, Erice (Italy), 26 Jul – 5 Aug 1990.**
M. M. Shapiro, R. Silberberg, J. P. Wefel (eds.).
NATO ASI Ser., Ser. C, Math. Phys. Sci., Vol. 337.
Kluwer Academic Publ., Dordrecht (Netherlands). 359 p. (1991). ISBN 0–7923–1278–3. Price Dfl. 175.00, US$ 99.00, £ 59.00.
Contents: 1. Cosmic rays and the interstellar medium. 2. Supernovae, acceleration, propagation, and source composition. 3. Gamma rays and their role as cosmic ray tracers. 4. Ultra–energetic cosmic nuclei.
The individual contributions within the subject scope of AAA are included in their corresponding categories – see abstracts 061.081, 061.082, 125.092, 125.093, 125.094, 125.232, 131.273, 131.274, 131.275, 143.037, 143.038, 144.070, 144.071, 144.072, 144.073, 144.074, 144.075, 144.076, 144.077, 144.078, 157.230.

012.054 **Warped disks and inclined rings around galaxies. Proceedings. Workshop on Warped Disks and Inclined Rings Around Galaxies, Pittsburgh, PA (USA), 30 May – 1 Jun 1990.**
S. Casertano, P. D. Sackett, F. H. Briggs (eds.).
Cambridge University Press, Cambridge (UK). 308 p. (1991). ISBN 0–521–40184–4. Price £ 35.00, US$ 54.50.
This volume contains 27 papers describing observational results, statistical interpretations, numerical simulations, and theoretical models aimed at better understanding the mechanisms that cause stars and gas to organize into complex warps and inclined rings that are now being discovered around galaxies in increasing numbers. Warped and polar–ringed galaxies are examined in detail, including: their origin, evolution, and ultimate fate, the frequency of their occurrence, their morphology, kinematics, and dynamics, and their relationship to larger questions of galaxy interactions and the nature of dark matter.
The individual contributions within the subject scope of AAA are included in their corresponding categories – see abstracts 151.072, 151.073, 151.074, 151.075, 151.077, 151.078, 151.079, 151.080, 151.081, 151.082, 151.083, 151.084, 151.085, 155.071, 157.189, 157.190, 157.191, 157.192, 157.193, 157.201, 157.202, 157.203, 157.204, 158.167, 158.168, 160.075, 160.077.

012.055 **Angular momentum evolution of young stars. Proceedings. NATO Advanced Research Workshop on Angular Momentum Evolution of Young Stars, Noto (Italy), 17 – 21 Sep 1990.**
S. Catalano, J. R. Stauffer (eds.).
NATO ASI Ser., Ser. C, Math. Phys. Sci., Vol. 340.
Kluwer Academic Publ., Dordrecht (Netherlands). 481 p. (1991). ISBN 0–7923–1316–X. Price Dfl. 240.00, US$ 134.00, £ 80.00.
At this workshop observers provided new fundamental data on the rotation of T Tauri stars, low mass stars in young clusters, and post–main sequence stars. Additional papers dealt with how new observational data on protostellar disks, lithium abundances in young stars and magnetic field measurements might provide useful constraints on models of rotational velocity evolution. These observational reports provided the basis for a number of theoretical papers which attempted to compare the predictions of angular momentum loss via stellar winds with the observations.
The individual contributions within the subject scope of AAA are included in their corresponding categories – see abstracts 051.048, 051.049, 065.072, 065.073, 065.074, 065.075, 065.078, 065.079, 065.080, 065.081, 065.082, 065.083, 065.085, 065.086, 080.094, 080.095, 080.096, 115.020, 116.062, 116.063, 116.064, 116.065, 116.066, 116.067, 116.068, 116.069, 116.070, 116.071, 116.072, 121.036, 121.037, 121.038, 121.039, 131.233, 131.234, 153.057, 153.058.

012.056 **Stellar atmospheres: beyond classical models. Proceedings. NATO Advanced Research Workshop on Stellar Atmospheres: Beyond Classical Models, Trieste (Italy), 3 – 7 Sep 1990.**
L. Crivellari, I. Hubeny, D. G. Hummer (eds.).
NATO ASI Ser., Ser. C, Math. Phys. Sci., Vol. 341.
Kluwer Academic Publ., Dordrecht (Netherlands). 475 p. (1991). ISBN 0–7923–1343–7. Price Dfl. 230.00, US$ 129.00, £ 77.00.
This workshop was decided to center the discussion around three closely related topics: 1. New computational algorithms. 2. The degree of physical sophistication that can now be achieved. 3. Assessment of the reliability of spectroscopic diagnostics based on theoretical models.
The individual contributions within the subject scope of AAA are included in their corresponding categories – see abstracts 063.042, 063.043, 063.044, 063.045, 063.046, 063.047, 063.048, 063.049, 063.050, 063.051, 063.052, 063.057, 063.058, 063.059, 063.060, 063.061, 063.063, 064.095, 064.096, 064.097, 064.099, 064.100, 064.101, 064.102, 064.103, 064.104, 064.105, 064.106, 064.107, 064.108, 064.109, 064.110, 064.111, 064.112, 064.113, 064.114, 064.115, 064.116, 064.117, 067.099, 071.051, 072.113, 114.075.

012.057 **Neutron stars: theory and observation. Proceedings. NATO Advanced Study Institute on Neutron Stars: An Interdisciplinary Field, Agia Pelagia, Crete (Greece), 3 – 14 Sep 1990.**
J. Ventura, D. Pines (eds.).
NATO ASI Ser., Ser. C, Math. Phys. Sci., Vol. 344.
Kluwer Academic Publ., Dordrecht (Netherlands). 603 p. (1991). ISBN 0–7923–1397–6. Price Dfl. 270.00, US$ 146.00, £ 92.00.
Contents: 1. Inside neutron stars. 2. Formation and evolution. 3. Observations on binaries and QPO's. 4. Pulsar and X–ray emission. 5. Bursts and TeV emission.

The individual contributions within the subject scope of AAA are included in their corresponding categories – see abstracts 067.104, 067.105, 067.106, 067.107, 067.108, 067.109, 067.110, 067.111, 067.112, 067.113, 067.114, 067.115, 067.116, 067.117, 067.118, 067.120, 067.121, 067.122, 067.123, 067.124, 067.125, 067.127, 067.128, 117.158, 117.159, 117.161, 117.162, 117.163, 143.036.

012.058 **Charge-coupled devices and solid state optical sensors. SPIE's Symposium on Charge-Coupled Devices and Solid State Optical Sensors, Santa Clara, CA (USA), 12 – 14 Feb 1990.**
M. M. Blouke (ed.).
Proc. Soc. Photo-Opt. Instrum. Eng., Vol. 1242.
SPIE, Bellingham, WA (USA). 272 p. (1990). ISBN 0-8194-0289-3.

Solid state optical sensors in general and CCDs in particular have established themselves as the detectors of choice for many applications. The advantages of low power, low noise, high resolution, high geometric fidelity, unparalleled sensitivity to a broad range of wavelengths, and extremely high quantum efficiency over this range have led to a number of revolutionary uses. It was the intent of this conference to update the optics community on recent developments in the area of high-performance devices and to explore some new applications. The conference covered topics from the uses of CCDs in various systems to focal-plane processing and a variety of device architectures.
The individual contributions within the subject scope of AAA are included in their corresponding categories – see abstracts 034.105.

012.059 **Charge-coupled devices and solid state optical sensors II. Proceedings. Conference on Charge-Coupled Devices and Solid State Optical Sensors II as Part of the SPIE/IS and T Symposium on Electronic Imaging Science and Technology, San Jose, CA (USA), 25 – 27 Feb 1991.**
M. M. Blouke (ed.).
SPIE Proc. Ser., Vol. 1447.
SPIE, Bellingham, WA (USA). 328 p. (1991). ISBN 0-8194-0546-9.

These proceedings deal with the state of the art in CCDs and optical sensors. The goal of the first session was to acquaint the user community with the large range of applications that solid state sensors are now finding. The subject areas span from astronomy to chemical analysis to computed tomography to endoscopy. The second session was concerned with radiation damage in CCDs. Thinning technology was covered in the third session. New imager technologies and new devices were the subject of the fourth session. The fifth and final session covered the subject of new megapixel CCDs and arrays.
The individual contributions within the subject scope of AAA are included in their corresponding categories – see abstracts 034.108, 034.109, 034.110, 034.111, 035.163, 035.164.

012.060 **Primordial nucleosynthesis. Proceedings. Workshop on Primordial Nucleosynthesis, Chapel Hill, NC (USA), 6 – 8 Oct 1989.**
W. J. Thompson, B. W. Carney, H. J. Karwowski (eds.).
World Scientific, Singapore (Singapore). 129 p. (1990). ISBN 981-02-0149-4. Price £ 28.00.

The workshop focussed on the epoch beginning with hadronization and lasting until completion of the initial synthesis of the very light nuclei through ^{7}Li and ^{9}Be, that is, from about 1 second to 1000 seconds after the big bang.
The individual contributions within the subject scope of AAA are included in their corresponding categories – see abstracts 061.086, 061.087, 061.088, 061.089, 161.303, 161.304, 161.305, 161.306.

012.061 **3rd ESO/ST-ECF Data Analysis Workshop. Proceedings. 3. ESO/ST-ECF Data Analysis Workshop, Garching (Germany), 22 – 23 Apr 1991.**
P. J. Grosbøl, R. H. Warmels (eds.).
ESO Conf. Workshop Proc., No. 38.
ESO, Garching (Germany). 242 p. (1991). ISBN 3-923524-39-0. Price DM 30.00.

Each year the Data Analysis Workshop centers on a specific topic, this year being direct imaging data. The sessions included such topics as digital filters, deconvolution and decomposition.
The individual contributions within the subject scope of AAA are included in their corresponding categories – see abstracts 034.113, 036.175, 036.176, 036.177, 036.178, 036.179, 036.180, 036.181, 036.182, 036.183, 036.184, 036.185, 036.186, 036.187, 036.188, 036.189, 036.190, 036.191, 036.192, 036.193, 036.194, 036.195, 036.196, 036.197, 036.198, 036.199, 036.200, 036.201, 036.202, 103.102, 158.194.

012.062 **Massive stars in starbursts. Proceedings. Massive Stars in Starbursts Meeting, Baltimore, MD (USA), 15 – 17 May 1990.**
C. Leitherer, N. R. Walborn, T. M. Heckman, C. A. Norman (eds.).
Space Telesc. Sci. Inst. Symp. Ser., Vol. 5.
Cambridge University Press, Cambridge (UK). 348 p. (1991). ISBN 0-521-40465-7. Price £ 35.00, US$ 49.50.

This book reviews the importance of massive stars in several areas of astrophysics. Massive stars are objects that are 10 – 100 times the mass of our Sun. Above ten solar masses, loss through stellar winds begins to have a major impact on the evolution of a star. The upper limit of 100 solar masses is derived from observations. Significant progress has now been achieved in massive star research. New models, along with high quality observations, have improved our understanding of the formation, structure, atmosphere and evolution of these massive objects. They are formed in violent bursts of star formation and are probably related to the phenomena observed in active galactic nuclei. The workshop at the Space Telescope Science Institute examined the interplay between the astrophysics of massive stars and their location in extragalactic starburst regions. The workshop was conceived not merely as a review of the pertinent fields but as a dynamical contribution toward future research.
The individual contributions within the subject scope of AAA are included in their corresponding categories – see abstracts 064.126, 114.089, 115.024, 115.025, 115.026, 115.027, 131.283, 131.284, 132.036, 157.233, 157.234, 157.235, 157.236, 157.237, 157.238, 157.239, 157.240, 157.241.

012.063 **Chapman Conference on Venus and Mars: Atmospheres, Ionospheres, and Solar Wind Interactions, Balatonfüred (Hungary), 4 – 8 Jun 1990.**
J. G. Luhmann (ed.).
J. Geophys. Res., Vol. 96, No. A7, p. 11019 – 11290 (1 Jul 1991).
The individual contributions within the subject scope of AAA are included in their corresponding categories – see abstracts 093.043, 093.044, 093.045, 093.046, 093.047, 093.048, 093.049, 093.050, 093.051, 093.052, 093.053, 093.054, 093.055, 093.056, 097.049, 097.051, 097.052, 097.053, 097.054, 097.055, 097.056, 097.057, 097.058, 097.059, 097.060.

012.064 **Atomic physics of highly charged ions. Proceedings. 5. Internationale Konferenz über die Physik hochgeladener Ionen, Gießen (Germany), 10 – 14 Sep 1990.**
E. Salzborn, A. Müller, P. H. Mokler (eds.).
Springer, Berlin (Germany). 380 p. (1991). ISBN 3-540-54065-2.
This book originally appeared as a supplement to Vol. 21 (1991) of the journal Z. Phys., D.

The present status of scientific work involving highly ionized atoms is reviewed in this proceedings volume, which incorporates the progress made over the last two years. More than 20 invited articles cover the whole area of active research in this field, including applications in plasmas, ion-surface interactions, precision spectroscopy, electron-ion recombination and ionization, ion-atom collisions, storage-ring physics, and ion sources.

Nearly 100 contributed papers deal with the latest theoretical and experimental achievements, presenting the reader with the current state of the art in highly charged ion physics.
The individual contributions within the subject scope of AAA are included in their corresponding categories – see abstracts 062.155, 076.025.

012.065 **EUV, X-ray, and gamma-ray instrumentation for astronomy II. Proceedings. SPIE Conference on EUV, X-Ray, and Gamma-Ray Instrumentation for Astronomy II as Part of SPIE International Symposium on Optical Applied Science and Engineering, San Diego, CA (USA), 24 – 26 Jul 1991.**
O. H. Siegmund, R. E. Rothschild (eds.).
SPIE Proc. Ser., Vol. 1549.
SPIE, Bellingham, WA (USA). 449 p. (1991). ISBN 0-8194-0677-5.
The topics represented in these conference proceedings are detailed as follows: Session 1: Proportional counter detectors, covers the development of proportional counter techniques and counter windows. Applications in several satellite missions, specifically for imaging and energy resolution, are discussed. Sessions 2 and 3: Microchannel-plate detectors, summarize the response characteristics of microchannel plates at high input rates. MAMA image readout schemes and position-encording electronics are also discussed. Session 4: X-ray and gamma-ray spectroscopy and imaging techniques, details topics including crystal diffractions, HEXTE instrument design, and polymer supports for thin windows. Sessions 5, 7, and 9: Spaceborne experiments and missions, encompass a considerable range of spaceflight instrumentation. This includes new mission concepts, development of sounding rocket payloads, novel window technologies, interferometric techniques, and instrumentation for major astrophysics experiments. Sessions 6 and 8: Solid state and cryogenic detectors, describe work with CCDs, superconducting tunneling junctions, and Si and GaAs detectors for use in detection of X rays and particles. Applications on the XMM and spectrum-x-gamma satellites are discussed.
The individual contributions within the subject scope of AAA are included in their corresponding categories – see abstracts 034.114, 035.166, 035.167, 035.168, 035.169, 035.170, 035.171, 035.172, 035.173, 035.174, 035.175, 035.176, 035.177, 035.178, 035.179, 035.180, 035.181, 035.182, 035.183.

012.066 **Reflective and refractive optical materials for Earth and space applications. Proceedings. Conference on Reflective and Refractive Optical Materials for Earth and Space Applications as Part of SPIE's Technical Symposia on Optical Engineering and Photonics in Aerospace Sensing (OE/Aerospace Sensing '91), Orlando, FL (USA), 4 – 5 Apr 1991.**
M. J. Riedl, R. R. Hale, T. B. Parsonage (eds.).
SPIE Proc. Ser., Vol. 1485.
SPIE, Bellingham, WA (USA). 290 p. (1991). ISBN 0-8194-0594-9.
The individual contributions within the subject scope of AAA are included in their corresponding categories – see abstracts 031.060, 031.061, 031.062, 031.063, 031.064.

012.067 **Theoretical problems in space and fusion plasmas. Proceedings. 2. International Workshop on Plasma Physics: Plasma Theoretical Problems in Astro- and Fusion Physics, Pichl (Austria), 1 – 3 Mar 1990.**
H. K. Biernat, S. J. Bauer, M. Heindler (eds.).
Verl. der Österreichischen Akademie der Wissenschaften, Graz (Austria). 292 p. (1991). ISBN 3-7001-1875-9.
The individual contributions within the subject scope of AAA are included in their corresponding categories – see abstracts 062.156, 062.157, 080.101, 083.030, 084.058, 084.059, 084.060, 084.061, 084.062, 091.032, 091.033, 099.044, 161.475, 161.476, 161.477.

012.068 **Strange quark matter in physics and astrophysics. Proceedings. International Workshop on Strange Quark Matter in Physics and Astrophysics, Aarhus (Denmark), 20 – 24 May 1991.**
J. Madsen, P. Haensel (eds.).
Nucl. Phys. B, Proc. Suppl., Vol. 24B, 310 p. (Dec 1991).
Contents: 1. Physical properties of strange quark matter. 2. The quark-hadron phase transition. 3. Strange quark matter in cosmology. 4. Strange stars. 5. Experimental searches for strange quark matter.
The individual contributions within the subject scope of AAA are included in their corresponding categories – see abstracts 067.140, 067.141, 067.142, 067.143, 067.144, 067.145, 067.146, 067.147, 067.148, 067.149, 067.150, 067.151, 067.236, 067.237, 067.238, 067.239, 144.082, 144.083, 144.084, 161.329, 161.330, 161.331, 161.332, 161.333, 161.334.

012.069 **After the first three minutes. Proceedings. After the first three minutes – an astrophysics workshop, College Park, MD (USA), 15 – 17 Oct 1990.**
S. S. Holt, C. L. Bennett, V. Trimble (eds.).
AIP Conf. Proc., No. 222.
American Institute of Physics, New York, NY (USA). 644 p. (1991). ISBN 0-88318-828-7. Price US$ 85.00.
Contents: 1. Introductory overviews. 2. Spectrum of the cosmic microwave background. 3. Isotropy of the cosmic microwave background. 4. The IR background: early star and galaxy formation. 5. Large scale structure from galaxy distributions. 6. Large scale structure and the intergalactic medium. 7. (Mostly) velocity-dependent evidence for dark matter. 8. Velocity-independent evidence for dark matter. 9. Dark matter candidates and alternatives. 10. Summary overviews.
The individual contributions within the subject scope of AAA are included in their corresponding categories – see abstracts 034.119, 034.120, 034.121, 051.065, 051.066, 106.039, 151.103, 151.104, 154.131, 155.102, 155.103, 157.259, 157.260, 157.261, 157.262, 159.109, 160.101, 160.102, 160.103, 160.104, 160.105, 160.106, 160.107, 160.108, 161.343, 161.344, 161.345, 161.346, 161.347, 161.348, 161.349, 161.350, 161.351, 161.352, 161.353, 161.355, 161.356, 161.357, 161.358, 161.359, 161.361, 161.362, 161.363, 161.365, 161.374, 161.375, 161.376, 161.377, 161.378, 161.379, 161.380, 161.381, 161.382, 161.383, 161.384, 161.385, 161.387, 161.388, 161.389, 161.391, 161.392, 161.393, 161.395, 161.396, 161.397, 161.398, 161.399, 161.401, 161.402, 161.403, 161.404, 161.405, 161.406, 161.407, 161.408, 161.409, 161.410, 161.411, 161.412, 161.413, 161.414, 161.415.

012.070 **Variability of active galaxies. Proceedings. Workshop on Variability of Active Galaxies, Heidelberg (Germany, F.R.), 3 – 5 Sep 1990.**
W. J. Duschl, S. J. Wagner, M. Camenzind (eds.).
Lect. Notes Phys., Vol. 377.
Springer, Berlin (Germany, F.R.). 324 p. (1991). ISBN 3-540-53860-7. Price DM 69.00. ISBN 0-387-53860-7 (USA).
The enormous amounts of energy radiated from the active nuclei of galaxies vary on short time scales, and the emission regions are difficult to observe. To provide a complete understanding of these phenomena a wide variety of studies is presented in this volume. The contributions are broadly divided between line and continuum variability, with observational results, methodological approaches, and theoretical models accompanying each. The final part is devoted to the important aspect of propagation-induced variability.
The individual contributions within the subject scope of AAA are included in their corresponding categories – see abstracts 063.094, 063.095, 063.096, 067.159, 067.160, 158.209, 158.210, 158.211, 158.212, 158.213, 158.215, 158.216, 158.217, 158.218, 158.219, 158.222, 158.223, 158.224, 158.225, 158.226, 158.227, 158.228, 158.229, 158.230, 158.231, 158.232, 158.233, 158.234, 158.235, 159.094, 159.095, 159.096, 159.097, 159.098, 159.099, 159.100, 159.101, 159.102.

012.071 **Primordial nucleosynthesis and evolution of early universe. Proceedings. IUPAP Conference on Primordial Nucleosynthesis and Evolution of Early Universe, Tokyo (Japan), 4 – 8 Sep 1990.**
K. Sato, J. Audouze (eds.).
Astrophys. Space Sci. Libr., Vol. 169.
Kluwer Academic Publ., Dordrecht (Netherlands). 647 p. (1991). ISBN 0–7923–1193–0. Price Dfl. 240.00, US$ 129.00, £ 82.00.
The most important probe of the early stages of the Universe is primordial nucleosynthesis. Light–element abundances depend critically upon the temperature and the density evolution in the early Universe. Recent developments show that the synthesis of elements in the early Universe is also an indicator of inhomogeneities at that time. This gives an immediate connection to observational cosmology. The link between large scale structures and their origin may be found in the synthesis of elements in the very early stages of our Universe. The dark matter "problem" might also find clues for solution in the early nucleosynthesis. The same can be said for the uniform cosmic background found by COBE. These proceedings cover the several topics in cosmology in a judicious and expert manner.
The individual contributions within the subject scope of AAA are included in their corresponding categories – see abstracts 061.104, 061.105, 061.106, 061.107, 061.108, 066.192, 067.195, 080.108, 142.045, 151.117, 157.281, 157.284, 158.268, 159.117, 160.121, 160.122, 160.126, 161.421, 161.422, 161.423, 161.424, 161.425, 161.426, 161.427, 161.428, 161.429, 161.436, 161.437, 161.438, 161.439, 161.440, 161.441, 161.442, 161.443, 161.444, 161.445, 161.446, 161.447, 161.448, 161.449, 161.451, 161.452, 161.453, 161.454, 161.455, 161.456, 161.457, 161.458, 161.459, 161.460, 161.461, 161.462, 161.463, 161.465, 161.466, 161.467, 161.468, 161.481, 161.485, 161.487, 161.488, 161.490, 161.491, 161.492, 161.493, 161.494, 161.495, 161.496, 161.497, 161.498, 161.499, 161.500, 161.501, 161.502, 161.503, 161.504, 161.505, 161.506, 161.507, 161.508, 161.509, 161.510, 161.511, 161.512, 161.528, 161.529, 161.530, 161.531, 161.532, 161.533, 161.535, 161.536, 161.538, 161.540, 161.542, 161.545, 161.546, 161.547, 161.548, 161.549, 161.550, 161.551.

012.072 **Meeting to commemorate the 10th anniversary of the signature of the international agreements in astrophysics, La Laguna, Tenerife (Spain), 9 Jun 1989.**
Astrophys. Lett. Commun., Vol. 28, No. 2–4, p. 47 – 215 (Sep 1991).
The individual contributions within the subject scope of AAA are included in their corresponding categories – see abstracts 002.093, 009.006, 013.073, 013.074, 013.075, 013.076, 013.077, 013.078, 013.079, 013.080, 032.023.

012.073 **Laboratory research for planetary atmospheres. A selection of papers. 2. International Conference on Laboratory Research for Planetary Atmospheres (ICLRPA), Charlottesville, VA (USA), 21 Oct 1990.**
K. Fox (ed.).
J. Geophys. Res., Vol. 96, No. E2, p. 17453 – 17572 (25 Sep 1991).
The individual contributions within the subject scope of AAA are included in their corresponding categories – see abstracts 022.110, 022.111, 022.112, 022.113, 022.114, 022.115, 022.116, 022.117, 022.118, 022.119, 022.120, 022.121, 022.122, 022.123, 022.124, 022.125, 022.126, 022.127, 022.128, 099.049.

012.074 **Objective–prism and other surveys. Proceedings. Objective–prism and other surveys – a meeting in memory of Nicholas Sanduleak, Middletown, CT (USA), 9 – 11 May 1991.**
A. G. D. Philip, A. R. Upgren (eds.).
Van Vleck Obs. Contrib., No. 12.
L. Davis Press, Schenectady, NY (USA). 214 p. (1991). ISBN 0–933485–15–8.
The individual contributions within the subject scope of AAA are included in their corresponding categories – see abstracts 002.094, 002.096, 002.099, 002.101, 002.104, 002.105, 036.211, 113.042, 114.097, 114.099, 114.100, 115.030, 125.262, 132.040, 134.062, 155.097, 155.107, 155.108, 155.110, 156.036, 159.114.

012.075 **Supernova 1987A and other supernovae. Proceedings. ESO/EIPC Workshop on Supernova 1987A and other Supernovae, Marciana Marina, Isola d'Elba (Italy), 17 – 22 Sep 1990.**
I. J. Danziger, K. Kjär (eds.).
ESO Conf. Workshop Proc., No. 37.
ESO, Garching (Germany). 769 p. (Jun 1991). ISBN 3–923524–38–2. Price DM 80.00.
Contents: 1. Progenitors. 2. Models – core collapse, neutrinos, mixing, nucleosynthesis. 3. Spectra, light curves. 4. Envelope models. 5. γ– and X–ray emission. 6. Pulsars, compact objects and late–time energy deposition. 7. Dust and echoes. 8. Circumstellar shells. 9. Radio. 10. Supernova remnants. 11. Supernova rates and chemical evolution. 12. Future campaigns and prospects.
The individual contributions within the subject scope of AAA are included in their corresponding categories – see abstracts 013.092, 034.122, 061.109, 065.109, 065.111, 080.110, 125.099, 125.100, 125.101, 125.102, 125.103, 125.104, 125.105, 125.106, 125.107, 125.108, 125.109, 125.110, 125.112, 125.113, 125.114, 125.115, 125.116, 125.117, 125.118, 125.119, 125.120, 125.121, 125.122, 125.123, 125.124, 125.125, 125.126, 125.127, 125.129, 125.130, 125.131, 125.132, 125.133, 125.134, 125.135, 125.136, 125.165, 125.171, 125.240, 125.241, 125.242, 125.243, 125.244, 125.245, 125.246, 125.247, 125.248, 125.249, 125.250, 125.251, 125.252, 125.253, 125.254, 125.255, 125.256, 125.257, 125.258, 125.259, 125.260, 125.261, 125.263, 125.265, 125.266, 125.267, 125.268, 125.269, 125.270, 125.271, 125.272, 125.273, 125.275, 125.276, 125.277, 125.278, 125.279, 125.280, 125.281, 125.305, 157.287, 157.291, 161.556.

012.076 **Observational tests of cosmological inflation. Proceedings. NATO Advanced Research Workshop on Observational Tests of Inflation, Durham (UK), 10 – 14 Dec 1990.**
T. Shanks, A. J. Banday, R. S. Ellis, C. S. Frenk, A. W. Wolfendale (eds.).
NATO ASI Ser., Ser. C, Math. Phys. Sci., Vol. 348.
Kluwer Academic Publ., Dordrecht (Netherlands). 503 p. (1991). ISBN 0–7923–1431–X. Price Dfl. 240.00, US$ 134.00, £ 81.00.
The aim of the workshop was to discuss the current status of the inflation theory and its observational predictions.
Contents: 1. Inflation theory. 2. Cosmological timescale test. 3. High redshift tests of Ω_0. 4. Galaxy clustering, Ω_0 and the primordial spectrum. 5. Microwave background anisotropy.
The individual contributions within the subject scope of AAA are included in their corresponding categories – see abstracts 151.121, 154.140, 154.141, 157.279, 157.295, 158.280, 159.118, 159.123, 159.124, 160.118, 160.119, 160.120, 160.127, 161.470, 161.471, 161.472, 161.473, 161.474, 161.478, 161.479, 161.480, 161.482, 161.483, 161.484, 161.486, 161.515, 161.516, 161.517, 161.518, 161.519, 161.520, 161.522, 161.523, 161.524, 161.525, 161.526, 161.527, 161.534, 161.537, 161.539, 161.541, 161.543, 161.544, 161.557, 161.558, 161.559, 161.560, 161.561, 161.562, 161.563, 161.564, 161.565, 161.566, 161.567, 161.569.

012.077 **Uranus. Invited reviews. Uranus Colloquium, Pasadena, CA (USA), 28 Jun – 1 Jul 1988.**
J. T. Bergstralh, E. D. Miner, M. S. Matthews (eds.).
Space Sci. Ser.
University of Arizona Press, Tucson, AZ (USA). 1088 p. (1991). ISBN 0–8165–1208–6. Price US$ 65.00.
The twenty invited reviews of this Uranus Colloquium form the core of the chapters in this book. They emphasize results from the Voyager encounter.
Contents: 1. Introduction: The Uranian system – an overview. 2. Interior. 3. Atmosphere. 4. Rings. 5. Satellites. 6. Magnetosphere.
The individual contributions within the subject scope of AAA are included in their corresponding categories – see abstracts 101.096, 101.097, 101.098, 101.099, 101.100, 101.101, 101.102,

101.103, 101.104, 101.105, 101.106, 101.107, 101.108, 101.109, 101.110, 101.111, 101.112, 101.113, 101.114, 101.115, 101.116.

012.078 **5. Asian-Pacific Regional Astronomy Meeting of the IAU, Sydney (Australia), 16 – 20 Jul 1990. Proceedings.**
M. C. B. Ashley, J. L. Caswell, W. J. Couch, R. W. Hunstead, K. M. Proust (eds.).
Proc. Astron. Soc. Aust., Vol. 9, No. 1, p. 198 (1991).
These proceedings are dedicated to the memory of Louise Webster (Turtle).
The aim of the meeting was to provide a forum for discussion of a broad range of topics within astronomy and an opportunity to foster scientific co-operation and exchange within the Asian-Pacific region. The program was organised into the following scientific sessions: (1) extragalactic astronomy, (2) galactic structure and the interstellar medium, (3) comets and the solar system, (4) solar astronomy, (5) stellar evolution, (6) supernovae, especially SN 1987A, (7) instrumentation and space observatories, and (8) education in astronomy. In addition, a panel discussion was held on the role of prediction and theory in science and the relevance for the media and public, centering on the results of the Voyager-Neptune encounter.
The individual contributions within the subject scope of AAA are included in their corresponding categories – see abstracts 002.108, 004.052, 009.008, 013.085, 013.086, 013.087, 013.088, 013.089, 013.090, 013.091, 014.051, 014.052, 014.053, 014.054, 014.055, 014.056, 014.057, 014.058, 014.059, 014.060, 014.061, 014.062, 015.015, 015.016, 015.018, 034.130, 034.131, 034.132, 034.133, 036.224, 036.226, 036.227, 051.068, 063.103, 065.115, 072.125, 080.109, 082.060, 098.087, 098.088, 112.100, 112.101, 112.102, 114.103, 117.241, 119.046, 121.085, 121.086, 121.087, 122.092, 122.093, 125.128, 125.141, 125.282, 125.289, 125.290, 125.294, 125.295, 126.082, 131.331, 131.332, 131.334, 131.335, 134.065, 142.051, 144.092, 151.118, 154.144, 154.145, 155.118, 155.119, 156.037, 156.038, 157.292, 157.294, 157.296, 158.279, 158.281, 159.121, 159.122, 160.129, 161.552, 161.586.

012.079 **Frontier objects in astrophysics and particle physics. Proceedings. Vulcano Workshop 1988: Frontier objects in astrophysics and particle physics, Vulcano (Italy), 23 – 27 May 1988.**
F. Giovannelli, G. Mannocchi (eds.).
Soc. Ital. Fis., Atti Conf., Vol. 19.
Italian Physical Society, Bologna (Italy). 443 p. (1989). ISBN 88-7794-022-0. Price £ 90.00.
At the 1988 Vulcano Workshop the physical processes occurring in cosmic frontier objects are discussed.
The individual contributions within the subject scope of AAA are included in their corresponding categories – see abstracts 022.143, 034.124, 034.125, 034.126, 034.129, 035.197, 064.173, 080.111, 112.099, 117.235, 117.236, 117.238, 117.239, 125.137, 125.138, 125.139, 125.140, 125.283, 125.284, 125.285, 125.286, 125.287, 125.288, 125.291, 125.292, 125.293, 125.296, 142.047, 142.048, 142.049, 142.050, 143.049, 144.095, 144.096, 144.097, 144.098, 153.076.

012.080 **The birth and early evolution of our universe. Proceedings. Nobel Symposium 79, Gräftåvallen (Sweden), 11 – 16 Jun 1990.**
J. S. Nilsson, B. Gustafsson, B.-S. Skagerstam (eds.).
Phys. Scr. T, Vol. T36, 301 p. (1991). ISBN 91-87308-76-2. Also available as hardbound edition from World Scientific, Singapore (Singapore), ISBN 981-02-0357-8, Price £ 52.00.
Contents: 1. Cosmic background radiation – primordial nucleosynthesis – inflation. 2. Large scale structure. 3. Cosmic strings and textures – dark matter – standard model physics. 4. Dark matter – standard model physics. 5. Quantum cosmology.
The individual contributions within the subject scope of AAA are included in their corresponding categories – see abstracts 061.110, 061.111, 061.112, 067.206, 067.207, 161.574, 161.575, 161.576, 161.577, 161.578, 161.579, 161.580, 161.581, 161.582, 161.583, 161.584, 161.585, 161.588, 161.589, 161.590, 161.591, 161.593, 161.594, 161.595, 161.596, 161.597, 161.598, 161.599, 161.600, 161.601.

012.081 **Maximum entropy and Bayesian methods. Proceedings. 10. International Workshop on Maximum Entropy and Bayesian Methods, MaxEnt 90, Laramie, WY (USA), 30 Jul – 3 Aug 1990.**
W. T. Grandy Jr., L. H. Schick (eds.).
Fundam. Theor. Phys., Vol. 43.
Kluwer Academic Publ., Dordrecht (Netherlands). 383 p. (1991). ISBN 0-7923-1140-X. Price Dfl. 170.00, US$ 99.00, £ 59.00.
This volume contains a selection of papers presented at the Tenth Annual Workshop on Maximum Entropy and Bayesian Methods. The thirty-six papers included cover a wide range of applications in areas such as economics and econometrics, astronomy and astrophysics, general physics, complex systems, image reconstruction, and probability and mathematics. Together they give an excellent state-of-the-art overview of fundamental methods of data analysis.
The individual contributions within the subject scope of AAA are included in their corresponding categories – see abstracts 036.215, 036.216, 036.217, 036.218, 036.219, 036.220, 036.221, 036.222.

012.082 **22. EGAS Conference of the European Group for Atomic Spectroscopy, Uppsala (Sweden), 10 – 13 Jul 1990. Proceedings.**
A. Arnesen, R. Hallin (eds.).
Phys. Scr. T, Vol. T34, 109 p. (1991). ISBN 91-87308-72-X.
This volume contains the invited lectures of this conference on pure and applied atomic and molecular spectroscopy and related topics.
The individual contributions within the subject scope of AAA are included in their corresponding categories – see abstracts 022.141.

012.083 **ISPP-7. Controlled active global experiments (CAGE). Proceedings. International Workshop on Controlled Active Global Experiments (CAGE), Varenna (Italy), 5 – 12 Sep 1990.**
E. Sindoni, A. Y. Wong (eds.).
Editrice Compositori, Bologna (Italy). 400 p. (1991). ISBN 88-7794-036-0.
Two phenomena were the major topics during the workshop – ozone depletion and global warming.
The individual contributions within the subject scope of AAA are included in their corresponding categories – see abstracts 144.094.

012.084 **The space distribution of quasars. Proceedings. Workshop on Space Distribution of Quasars as Part of the 75. Anniversary Celebrations of the National Research Council of Canada (NRCC), Victoria (Canada), 3 – 5 Jun 1991.**
D. Crampton (ed.).
Astron. Soc. Pac. Conf. Ser., Vol. 21.
Astronomical Society of the Pacific, San Francisco, CA (USA). 426 p. (1991). ISBN 0-937707-40-6. Price US$ 32.00.
The aims of the workshop were to bring together the results of various quasar surveys underway (at all wavelengths), to compare them with the large scale structure picture emerging from galaxy surveys, and to confront them with recent theories on the origin and evolution of structure in the universe.
The individual contributions within the subject scope of AAA are included in their corresponding categories – see abstracts 142.053, 142.054, 158.284, 158.285, 158.286, 158.287, 158.290, 158.291, 158.292, 158.293, 159.125, 159.126, 159.127, 159.128, 159.129, 159.130, 159.131, 159.132, 159.133, 159.134, 159.135, 159.136, 159.137, 159.138, 159.139, 159.140, 159.141, 159.142, 159.145, 159.146, 159.147, 159.148, 159.149, 159.150, 159.151, 159.152, 159.153, 159.154, 159.155, 159.156, 159.157, 159.158, 159.159, 159.160, 159.161, 159.162, 159.163, 159.164, 159.165, 159.166, 159.167, 159.168, 159.169, 159.170, 161.618.

012.085 **Frontiers of stellar evolution. Proceedings. Symposium in Celebration of the 50. Anniversary of McDonald Observatory of the University of Texas at Austin: Frontiers of stellar evolution, Alpine, TX (USA), 24 – 28 Jul 1989.**
D. L. Lambert (ed.).
Astron. Soc. Pac. Conf. Ser., Vol. 20.
Astronomical Society of the Pacific, San Francisco, CA (USA). 630 p. (1991). ISBN 0-937707-39-2. Price US$ 32.00.
In 1989, the 82-inch reflector, known as the Otto Struve telescope, completed 50 years of astronomical observations. To celebrate this anniversary, the W. J. McDonald Observatory and the Department of Astronomy of the University of Texas at Austin decided to hold a symposium on "Frontiers of stellar evolution". This volume contains all but three of the review papers presented at the symposium.
Contents: 1. Historical perspectives. 2. Formation of stars. 3. Evolution of single stars. 4. Deaths and transfiguration. 5. Interacting binary stars. 6. Applications to galactic evolution. 7. Abstracts of poster papers.
The individual contributions within the subject scope of AAA are included in their corresponding categories – see abstracts 005.018, 009.009, 034.135, 036.229, 064.177, 065.118, 065.119, 065.121, 065.122, 067.212, 067.217, 080.113, 112.103, 112.104, 112.105, 112.106, 112.107, 112.108, 114.106, 114.107, 114.108, 114.109, 114.110, 114.111, 115.032, 116.083, 117.245, 117.246, 117.247, 117.248, 117.249, 117.252, 117.254, 119.050, 124.006, 124.007, 125.142, 125.297, 126.084, 126.085, 126.086, 131.340, 131.341, 131.345, 134.069, 134.070, 134.071, 134.072, 152.010, 153.078, 154.146, 154.147, 154.148, 154.149, 154.150, 155.120, 155.121, 155.122, 156.039, 156.040, 156.041, 158.289.

012.086 **Advances in solar system magnetohydrodynamics. Proceedings. Symposium on Geophysical and Astrophysical MHD as Part of the British Theoretical Mechanics Colloquium: Advances in solar system magnetohydrodynamics, St. Andrews (UK), 2 – 6 Apr 1990.**
E. R. Priest, A. W. Hood (eds.).
Cambridge University Press, Cambridge (UK). 450 p. (1991). ISBN 0-521-40325-1. Price £ 35.00, US$ 59.50.
This book summarises the present state of understanding in many of the main branches of solar system MHD. Several chapters are devoted to applications, namely to solar coronal heating, solar prominences, solar flares, and MHD turbulence in the solar wind. The Earth's magnetosphere is covered as well as astrophysical disks and jets.
The individual contributions within the subject scope of AAA are included in their corresponding categories – see abstracts 062.191, 062.192, 062.193, 062.194, 062.195, 062.196, 062.197, 062.198, 062.199, 062.200, 062.201, 062.202, 062.203, 073.157, 073.158, 074.106, 084.049.

012.087 **Asteroids to quasars. A symposium honouring William Liller. Proceedings. Liller Symposium: Asteroids to quasars, Cambridge, MA (USA), 6 Jun 1987.**
P. M. Lugger (ed.).
Cambridge University Press, Cambridge (UK). 332 p. (1991). ISBN 0-521-35321-2.
This volume contains insightful reviews of the current state of a broad range of frontier astrophysical areas, with an emphasis on observational studies. Particular topics include the nature of asteroids, application of stellar occultations to solar system studies, solar eclipse expeditions, planetary nebulae, identification of galactic X-ray sources, ages of globular star clusters, the globular star cluster system of the Andromeda Galaxy, star formation and spiral structure, the dynamical evolution of clusters of galaxies, the identification of extragalactic X-ray sources with quasars, and the archaeoastronomy of Easter Island. The reviews are of special interest to professional astronomers and students of astronomy, and are also accessible to the interested non-specialist reader.
The individual contributions within the subject scope of AAA are included in their corresponding categories – see abstracts 004.063, 005.022, 015.020, 036.231, 036.233, 079.003, 096.003, 098.092, 114.112, 117.258, 134.073, 134.074, 142.055, 142.056, 154.151, 157.307, 157.308, 157.309, 160.137, 160.138.

012.088 **Iron line diagnostics in X-ray sources. Proceedings. Workshop on Iron Line Diagnostics in X-Ray Sources, Varenna, Como (Italy), 9 – 12 Oct 1990.**
A. Treves, G. C. Perola, L. Stella (eds.).
Lect. Notes Phys., Vol. 385.
Springer, Berlin (Germany). 330 p. (1991). ISBN 3-540-54354-6. Price DM 69.00. ISBN 0-387-54354-6 (USA).
Contents: 1. Thin plasma: SN remnants, stellar coronae and clusters. 2. X-ray binaries. 3. Active galactic nuclei. 4. Future missions.
The individual contributions within the subject scope of AAA are included in their corresponding categories – see abstracts 035.206, 035.207, 035.208, 051.071, 051.072, 051.073, 051.074, 064.179, 067.228, 067.229, 112.111, 112.112, 117.259, 117.260, 117.261, 117.262, 117.263, 117.264, 117.265, 117.267, 117.268, 117.270, 117.271, 121.092, 125.144, 125.145, 125.146, 142.057, 142.058, 142.059, 155.124, 158.300, 158.301, 158.303, 158.304, 158.305, 158.306, 158.308, 158.309, 158.310, 158.311, 158.317, 158.318, 158.319, 160.146, 160.147, 160.148, 160.149.

012.089 **Radio interferometry: theory, techniques, and applications. Proceedings. IAU Colloquium No. 131: Radio interferometry – theory, techniques, and applications, Socorro, NM (USA), 8 – 12 Oct 1990.**
T. J. Cornwell, R. A. Perley (eds.).
Astron. Soc. Pac. Conf. Ser., Vol. 19.
Astronomical Society of the Pacific, San Francisco, CA (USA). 489 p. (1991). ISBN 0-937707-38-4. Price US$ 32.00.
IAU Colloquium No. 131 was held at the tenth anniversary of the dedication of the VLA and discusses the present and future of radio interferometry.
Contents: 1. Theory of interferometry. 2. Hardware for interferometry. 3. Space-based VLBI. 4. Seeing. 5. Computing. 6. History. 7. Imaging. 8. VLBI imaging. 9. Other fields. 10. Interferometric telescopes.
The individual contributions within the subject scope of AAA are included in their corresponding categories – see abstracts 032.027, 033.021, 033.022, 033.023, 033.024, 033.025, 033.026, 033.027, 033.028, 033.029, 033.030, 033.031, 033.032, 033.033, 033.034, 033.035, 033.036, 033.037, 033.038, 033.039, 033.040, 033.041, 033.042, 033.043, 033.044, 033.045, 033.047, 033.048, 033.049, 033.050, 033.051, 033.052, 033.053, 034.136, 035.200, 035.201, 035.202, 035.204, 035.205, 036.234, 036.235, 036.236, 036.237, 036.238, 036.239, 036.240, 036.241, 036.242, 036.243, 036.244, 036.245, 036.246, 036.247, 036.248, 036.249, 036.250, 036.251, 036.252, 036.253, 036.254, 036.255, 036.256, 036.257, 036.258, 036.259, 036.260, 036.261, 036.262, 036.263, 036.264, 036.265, 036.266, 036.267, 036.268, 036.269, 036.270, 036.271, 036.272, 036.273, 036.274, 036.275, 036.277, 041.024, 082.065, 131.347, 131.351, 141.012, 158.302.

012.090 **Abstracts of contributed papers presented at the General Assembly of the Royal Astronomical Society of Canada, Vancouver, BC (Canada), 17 – 20 May 1991.**
J. R. Astron. Soc. Can., Vol. 85, No. 4, p. 173 – 176 (Aug 1991).
The individual contributions within the subject scope of AAA are included in their corresponding categories – see abstracts 004.067, 004.068, 013.095, 014.066, 014.067, 014.068, 072.133, 079.011, 081.049, 098.095, 112.127, 112.132.

012.091 **Abstracts of contributed papers presented at the 22. Annual Meeting of the Canadian Astronomical Society, Toronto (Canada), 10 – 12 Jun 1991.**
J. R. Astron. Soc. Can., Vol. 85, No. 4, p. 178 – 222 (Aug 1991).
The individual contributions within the subject scope of AAA are included in their corresponding categories – see abstracts 002.122, 013.100, 014.069, 014.070, 014.071, 014.072, 022.153, 022.154, 031.077, 032.031, 032.032, 032.033, 033.054, 034.140, 034.141, 034.157, 035.210, 035.211, 035.215, 036.282, 036.283, 036.284, 036.285, 036.286, 036.287, 036.288, 036.289, 062.224, 063.115, 063.118, 064.192, 065.143, 067.244, 071.059, 075.064,

091.048, 098.097, 100.027, 100.028, 107.024, 112.146, 112.147, 112.151, 112.152, 113.046, 114.124, 114.125, 114.126, 114.127, 114.128, 114.129, 117.280, 117.281, 117.282, 117.283, 117.284, 117.285, 118.027, 121.104, 122.104, 122.106, 122.107, 125.151, 125.153, 125.154, 131.367, 131.370, 131.371, 131.372, 131.373, 131.374, 131.375, 131.376, 131.377, 131.378, 131.379, 132.042, 132.044, 132.045, 132.046, 132.047, 134.077, 134.078, 141.013, 151.137, 151.138, 151.139, 152.012, 153.083, 153.085, 153.086, 154.156, 154.157, 154.158, 154.159, 154.160, 157.324, 157.325, 157.326, 157.327, 157.328, 157.329, 157.330, 157.331, 157.332, 157.333, 157.334, 157.335, 157.336, 157.338, 157.339, 158.326, 158.327, 158.328, 158.331, 159.177, 160.160, 161.650, 161.651, 161.652, 161.655, 161.656.

012.092 **Rapid variability of OB–stars: nature and diagnostic value. ESO Workshop on Rapid Variability of OB–Stars: Nature and Diagnostic Value, Garching (Germany), 15 – 17 Oct 1990.**
D. Baade (ed.).
ESO Conf. Workshop Proc., No. 36.
European Southern Observatory, Garching (Germany). 390 p. (1991). ISBN 3–923524–37–4. Price DM 45.00.

In the past decade, rapid photospheric variability has been recognized as the non–standard property that perhaps is the most common one among early–type stars. These proceedings offer an unusually complete overview of the existing observations. They are equally complete in their reflectance of the presently considered models. Because the simple definition 'on a rotational time scale' of the qualifier 'rapid' used in the title is very adequate for many stars, modulation is a strong contender also as a general model. The model that can be made to formally reproduce the widest range of observations is nonradial pulsation which, therefore, has earned itself the somewhat ambiguous reputation as a model for everything. An attraction of this model is that it would give the possibility to infer also structural and evolutionary quantities. It was the second purpose of the workshop to offer at least a glimpse of this potential.
Contents: 1. Observations of line–profile variability. 2. Light variations. 3. Possibly related line–profile variations in other stars. 4. Connection between rapid photospheric variability and circumstellar activity. 5. Ad hoc models. 6. Theory of nonradial pulsation. 7. Atmospheric diagnostics. 8. Other diagnostic considerations. 9. Workshop summaries.
The individual contributions within the subject scope of AAA are included in their corresponding categories – see abstracts 036.278, 036.279, 063.116, 063.117, 064.180, 064.181, 064.184, 064.185, 064.186, 065.129, 065.130, 065.131, 065.132, 065.133, 065.134, 065.135, 112.115, 112.116, 112.117, 112.118, 112.119, 112.120, 112.121, 112.122, 112.123, 112.124, 112.125, 112.128, 112.129, 112.130, 112.131, 112.136, 112.137, 112.138, 112.139, 112.140, 112.141, 112.142, 112.144, 113.044, 113.045, 114.117, 114.118, 114.119, 114.120, 114.121, 114.122, 116.088, 116.090, 116.091, 116.093, 116.094, 116.095, 117.272, 122.098, 122.099, 122.100, 122.102, 122.103, 153.080.

012.093 **Kosmos – cosmology today between science and philosophy. Proceedings. 1. Venice Conference on Cosmology and Philosophy, Venice (Italy), 8 – 9 May 1987**
F. Bertola, M. Calvani, U. Curi (eds.).
Mem. Soc. Astron. Ital., Vol. 62, No. 3, p. 411 – 530 (1991).
The individual contributions within the subject scope of AAA are included in their corresponding categories – see abstracts 004.073, 004.074, 004.075, 015.022, 015.026, 015.027, 015.028, 015.029, 015.030, 015.031, 161.660, 161.661.

012.094 **Origin and evolution of the Universe. A Hubble Centenary Conference. Proceedings. 3. Venice Conference on Cosmology and Philosophy, Venice (Italy), 15 – 16 Dec 1989.**
F. Bertola, M. Calvani, U. Curi (eds.).
Mem. Soc. Astron. Ital., Vol. 62, No. 3, p. 531 – 704 (1991).
The individual contributions within the subject scope of AAA are included in their corresponding categories – see abstracts 005.026, 015.032, 015.033, 015.034, 015.035, 015.036, 015.037, 161.662, 161.663.

012.095 **35. Reunión Anual de la Asociación Argentina de Astronomia, San Juan (Argentina), 18 – 22 Sep 1990.**
Bol. Asoc. Argent. Astron., No. 36, 480 p. (1990).
Contents: 1. Stellar structure and evolution. 2. Galactic structure and stellar dynamics. 3. Interstellar material. 4. Galaxies and cosmology. 5. Solar system. 6. Astrometry. 7. Instrumental and reduction techniques. 8. Abstracts.
The individual contributions within the subject scope of AAA are included in their corresponding categories – see abstracts 002.127, 013.107, 014.076, 021.051, 031.081, 032.046, 034.165, 034.166, 034.167, 036.300, 036.311, 036.312, 036.313, 036.314, 041.026, 043.006, 043.016, 043.017, 044.055, 044.056, 051.087, 064.193, 076.045, 076.046, 082.071, 092.006, 098.103, 102.032, 112.157, 112.158, 113.048, 114.137, 114.138, 114.139, 116.097, 116.098, 117.293, 117.294, 117.295, 131.383, 131.384, 131.385, 131.386, 131.388, 134.080, 151.142, 153.087, 154.154, 154.163, 155.129, 156.043, 157.340, 157.341, 157.342, 157.343, 160.163, 160.164, 161.669.

012.096 **Challenges to theories of the structure of moderate–mass stars. Proceedings. Conference on Challenges to Theories of the Structure of Moderate–Mass Stars, Santa Barbara, CA (USA), 19 – 22 Jun 1990.**
D. Gough, J. Toomre (eds.).
Lect. Notes Phys., Vol. 388.
Springer, Berlin (Germany). 421 p. (1991). ISBN 3–540–54420–8. Price DM 83.00. ISBN 0–387–54420–8 (USA).

The above conference reviewed the achievements of the subject, paying particular attention to work that had been done at the Institute for Theoretical Physics at the University of California, Santa Barbara. The Proceedings summarize the latest advances in helioseismology, including the measurement of the internal stratification of the sun, the assessment of the sensitivity of oscillations to uncertainties in fundamental physics, and the study of the angular velocity and the dynamics of the convection zone.
The individual contributions within the subject scope of AAA are included in their corresponding categories – see abstracts 013.096, 013.102, 021.045, 036.290, 062.211, 062.212, 062.213, 062.214, 062.215, 062.216, 062.217, 062.218, 064.191, 065.141, 065.144, 065.145, 065.146, 065.148, 065.149, 065.150, 065.151, 065.152, 071.060, 071.061, 080.142, 080.143, 080.144, 080.145, 080.146, 080.147, 080.148, 080.149, 080.150, 080.151, 080.152, 080.154, 080.155, 080.156, 080.157, 080.158, 080.159, 080.160, 080.161, 080.162, 114.130, 122.109.

012.097 **The physical universe: the interface between cosmology, astrophysics and particle physics. Proceedings. 12. Autumn School of Physics. The physical universe – the interface between cosmology, astrophysics and particle physics, Lisbon (Portugal), 1 – 5 Oct 1990.**
J. D. Barrow, A. B. Henriques, M. T. V. T. Lago, M. S. Longair (eds.).
Lect. Notes Phys., Vol. 383.
Springer, Berlin (Germany). 320 p. (1991). ISBN 3–540–54293–0. Price DM 69.00.
The individual contributions within the subject scope of AAA are included in their corresponding categories – see abstracts 035.212, 036.281, 061.122, 066.217, 067.240, 067.241, 161.641, 161.642, 161.643, 161.644, 161.645, 161.646, 161.647, 161.648, 161.649.

012.098 **Gravitational astronomy: instrument design and astrophysical prospects. Proceedings. Elizabeth and Frederick White Research Conference: Gravitational astronomy – instrument design and astrophysical prospects, Canberra (Australia), 24 – 26 Sep 1990.**
D. E. McClelland, H.–A. Bachor (eds.).
World Scientific, Singapore (Singapore). 463 p. (1991). ISBN 981–02–0688–7. Price £ 39.00.

The scientific program of the conference included both general theoretical presentations and progress reports from major research groups involved in the search for gravitational waves.

The individual contributions within the subject scope of AAA are included in their corresponding categories – see abstracts 013.097, 034.142, 034.143, 034.144, 034.145, 034.146, 034.147, 034.148, 034.149, 034.150, 034.151, 034.152.

012.099 **Etoiles à enveloppes vues à haute résolution angulaire. Journées d'Etudes Scientifiques, Nice (France), 24 – 25 Oct 1990.**
J. Astron. Fr., No. 39, p. 2 – 26 (Jul 1991).
The individual contributions within the subject scope of AAA are included in their corresponding categories – see abstracts 031.080, 032.043, 032.044, 036.305, 036.306, 036.307, 036.308, 112.149, 112.154, 112.155, 112.156, 115.038, 121.105, 122.111.

012.100 **High-pressure equations of state: theory and applications. Proceedings. International School of Physics 'Enrico Fermi'. Course 113: High-pressure equations of state: theory and applications, Varenna (Italy), 25 Jul – 4 Aug 1989.**
S. Eliezer, R. A. Ricci (eds.).
North-Holland, Amsterdam (Netherlands). 579 p. (1991). ISBN 0-444-89259-1.
The individual contributions within the subject scope of AAA are included in their corresponding categories – see abstracts 062.221, 062.222, 062.223, 065.147, 091.052.

012.101 **Problems of fundamental modern physics, II. Proceedings. 5. Winter School on Hadronic Physics: Problems of fundamental modern physics II, Folgaria (Italy), 5 – 10 Feb 1990.**
R. Cherubini, P. Dalpiaz, B. Minetti (eds.).
World Scientific, Singapore (Singapore). 532 p. (1991). ISBN 981-02-0295-4.
The individual contributions within the subject scope of AAA are included in their corresponding categories – see abstracts 034.153, 034.154, 034.155, 034.156, 066.222, 161.658.

012.102 **Techniques and concepts of high-energy physics 6. Proceedings. NATO Advanced Study Institute on Techniques and Concepts of High Energy Physics, St. Croix, VI (USA), 14 – 25 Jun 1990.**
T. Ferbel (ed.).
NATO ASI Ser., Ser. B, Phys., Vol. 275.
Plenum Press, New York, NY (USA). 447 p. (1991). ISBN 0-306-44043-1.
These proceedings contain the lectures which concern gauge theories and unification, the connection between particle physics and cosmology, experimental aspects of hadron-collider physics, meson spectroscopy, the physics of accelerators, calorimetry in high energy physics, noise limits in charge measurements, and silicon JFET's in low-noise circuits.
The individual contributions within the subject scope of AAA are included in their corresponding categories – see abstracts 161.659.

012.103 **Relativistic hadrons in cosmic compact objects. Proceedings. Workshop on Relativistic Hadrons in Cosmic Compact Objects, Koninki/Suhora (Poland), 9 – 11 Oct 1990.**
A. A. Zdziarski, M. Sikora (eds.).
Lect. Notes Phys., Vol. 391.
Springer, Berlin (Germany). 194 p. (1991). ISBN 3-540-54789-4. Price DM 59.00. ISBN 0-387-54789-4 (USA).
One of an array of problems discussed at the workshop was the question whether a large fraction of the energy produced by active galactic nuclei is channeled into a population of relativistic protons. It was pointed out by several speakers that reactions involving such protons would lead to an electron-positron pair cascade which could be responsible for the observed nonthermal continuum radiation. Neutrons, which are a byproduct of pion-producing reactions of relativistic protons, can escape from a central region in which they are produced and decay at distances of several light years from the central black hole in an active nucleus. This has dramatic consequences for the dynamics of accretion and the emission of very high energy γ-rays. Direct evidence of relativistic hadron production in active galactic nuclei could come from detection of TeV neutrinos by DUMAND-type detectors.
The individual contributions within the subject scope of AAA are included in their corresponding categories – see abstracts 062.226, 062.227, 062.228, 062.229, 062.230, 062.231, 062.232, 062.233, 062.234, 063.120, 067.251, 067.252, 067.253, 067.254, 067.255, 067.256, 067.257, 067.258, 067.259, 067.260, 117.289, 143.055, 143.056.

012.104 **Flare physics in solar activity maximum 22. Proceedings. International SOLAR-A Science Meeting: Flare physics in solar activity maximum 22, Tokyo (Japan), 23 – 26 Oct 1990.**
Y. Uchida, R. C. Canfield, T. Watanabe, E. Hiei (eds.).
Lect. Notes Phys., Vol. 387.
Springer, Berlin (Germany). 370 p. (1991). ISBN 3-540-54383-X. Price DM 76.00. ISBN 0-387-54383-X (USA).
These proceedings are dedicated to the memory of the late Professor Katsuo Tanaka. The stress of the conference was put on (i) making clearer through discussions what could and should be done using SOLAR-A, thus helping the development of analysis software, and (ii) seeking possibilities for collaborations with ground-based optical and radio observers.
Contents: 1. Introduction and memoriam for the late Professor Katsuo Tanaka. 2. Experiments on the SOLAR-A satellite. 3. Magnetic flux emergence and preflare states. 4. Acceleration and propagation of high energy particles in flares. 5. Heating and flows in flares. 6. Physical picture of flares. 7. Global coronal structure and solar activity. 8. Collaborative ground-based observations with SOLAR-A.
The individual contributions within the subject scope of AAA are included in their corresponding categories – see abstracts 013.103, 032.035, 033.055, 035.216, 035.217, 035.218, 035.219, 036.293, 051.078, 051.079, 051.080, 051.081, 051.082, 062.235, 062.236, 072.137, 073.171, 073.172, 073.173, 073.175, 073.176, 073.177, 073.178, 073.179, 073.180, 073.181, 073.183, 073.184, 073.185, 073.186, 074.121, 074.122, 074.123, 074.124, 074.125, 074.126, 075.066, 075.067, 075.068, 076.028, 076.029, 076.030, 076.031, 076.032, 076.033, 076.034, 076.035, 076.036, 076.037, 076.038, 076.039, 076.040, 076.041, 076.042, 076.043, 076.044, 077.077, 077.078, 078.029, 080.163.

012.105 **High energy physics and cosmology. Proceedings. 1990 Summer School in High Energy Physics and Cosmology, Miramare, Trieste (Italy), 18 Jun – 28 Jul 1990.**
J. C. Pati, S. Randjbar-Daemi, E. Sezgin, Q. Shafi (eds.).
ICTP Ser. Theor. Phys., Vol. 7.
World Scientific, Singapore (Singapore). 728 p. (1991). ISBN 981-02-0469-8. Price £ 50.00.
The International Centre for Theoretical Physics in Trieste has been running a Summer Workshop/School in High Energy Physics and Cosmology since 1981. The primary goal of these workshop/schools is to bring together active physicists from around the world, particularly from developing countries, and to provide a platform where the frontier developments in the field are presented. The 1990 Summer School was devoted to superstrings and conformal field theories, and supersymmetry, phenomenology and cosmology.
The individual contributions within the subject scope of AAA are included in their corresponding categories – see abstracts 066.223, 066.224, 161.664, 161.665, 161.666, 161.667.

012.106 **Solid-state astrophysics. Proceedings. International School of Physics 'Enrico Fermi'. Course 111: Solid-state astrophysics, Varenna (Italy), 27 Jun – 7 Jul 1989.**
E. Bussoletti, G. Strazzulla (eds.).
North-Holland, Amsterdam (Netherlands). 442 p. (1991). ISBN 0-444-89218-4. Price Dfl. 290.00, US$ 148.50.
Solid-state astrophysics is the research field having as object of interest solid bodies in space, their processing and laboratory simulations. The need to have a school mainly came out of the data obtained in the last 10 years by various groups. The results

have been based on an interdisciplinary interaction between astronomers, physicists and chemists. The school covered a wide area of both material processing in space and in the laboratory, as well as the properties of grains, comets, meteorites, planets and satellites, with a view to past and future space missions. The effects of irradiation by UV and ions were presented in detail. Vibrational spectroscopy of relevant materials as well as of astronomical sources has been fully reviewed.
The individual contributions within the subject scope of AAA are included in their corresponding categories – see abstracts 022.159, 022.160, 022.161, 022.162, 022.163, 022.164, 022.165, 022.166, 022.168, 022.169, 051.083, 051.084, 091.054, 098.101, 100.033, 131.380, 131.381, 131.382.

012.107 **Astrophysics at FUV and EUV wavelengths. Proceedings. Topical Meeting of the COSPAR Interdisciplinary Scientific Commission E (Meeting E3) of the COSPAR 28. Plenary Meeting, The Hague (Netherlands), 25 Jun – 6 Jul 1990.**
B. Aschenbach (ed.).
Adv. Space Res., Vol. 11, No. 11, 234 p. (1991). ISBN 0–08–041841–4.
The objective of the meeting was to provide a comprehensive overview about current astrophysical research in the wavelength region from about 100 – 1200 Å, both from a theoretical point of view and instrumental. A key aspect was certainly spectroscopy, basically because the FUV/EUV wavelength band is so rich in spectral lines. Very recent results from air– and space–borne experiments were reported and new space missions presented and discussed.
The individual contributions within the subject scope of AAA are included in their corresponding categories – see abstracts 035.220, 035.221, 035.222, 035.223, 035.224, 035.225, 035.226, 035.227, 035.228, 035.229, 035.230, 036.316, 112.153, 114.132, 114.133, 114.134, 114.135, 114.136, 117.292, 117.297, 126.092, 131.389, 131.390, 131.391, 142.062, 154.164, 157.344, 159.179.

012.108 **13. Meeting of the Astronomical Society of India, Srinagar (India), 21 – 24 Jun 1989. Abstracts of papers presented.**
Bull. Astron. Soc. India, Vol. 19, Nos. 1 – 2, p. 75 – 95 (Mar – Jun 1991).
The individual contributions within the subject scope of AAA are included in their corresponding categories – see abstracts 015.039, 022.170, 032.047, 032.048, 034.168, 034.169, 034.170, 036.319, 042.080, 063.125, 065.155, 067.261, 067.268, 067.269, 071.063, 074.127, 074.128, 102.033, 103.112, 103.113, 112.163, 117.300, 117.301, 117.302, 117.303, 117.304, 119.056, 120.029, 121.107, 125.156, 125.157, 126.093, 131.392, 131.393, 131.394, 133.007, 134.081, 143.058, 143.059, 143.063, 143.064, 143.065, 143.066, 143.067, 144.115, 144.116, 156.044, 157.345, 159.180, 159.181, 160.162, 161.670, 161.671.

012.109 **Particles, strings and supernovae. Volumes 1, 2. Proceedings. 1988 Theoretical Advanced Study Institute in Elementary Particle Physics (TASI–88): Particles, strings and supernovae, Providence, RI (USA), 6 Jun – 2 Jul 1988.**
A. Jevicki, C.–I. Tan (eds.).
World Scientific, Singapore (Singapore). 1086 p. (1989). ISBN 9971–50–792–7. Price £ 56.00. ISBN 9971–50–793–5 (paper).
The central theme of TASI–88 was "particles, strings, and supernovae", with primary focus on recent developments. In addition to introductory lectures on standard particle physics and on strings and superstrings, special emphasis was placed on non–perturbative aspects of quantum field theories. It has become increasingly clear that non–accelerator events promise to emerge as an important branch of experimental particle physics. The rapidly increasing information about structure in the universe on very large scales can provide a clue to and a constraint on high energy particle theories. Lectures on cosmology and neutrino–astrophysics were also provided.

012.110 **Data analysis in astronomy. III. Proceedings. 3. International Workshop on Data Analysis in Astronomy, Erice, Sicily (Italy), 20 – 27 Jun 1988.**
V. Di Gesù, L. Scarsi, P. Crane, J. H. Friedman, S. Levialdi, M. C. Maccarone (eds.).
Ettore Majorana Int. Sci. Ser., Phys. Sci., Vol. 40.
Plenum Press, New York, NY (USA). 428 p. (1989). ISBN 0–306–43158–0. Price US$ 95.00 (North America), US$ 114.00 (elsewhere).
This workshop was intended to provide an overview on the state of the art in the data analysis methodologies and tools in the new frontiers of astrophysics (γ–astronomy, neutrino astronomy, gravitational waves, background radiation and extreme cosmic ray energy spectrum). The book is organized in two sessions: 1. Data analysis methods and tools. 2. New frontiers in astronomy.
The individual contributions within the subject scope of AAA are included in their corresponding categories – see abstracts 034.176, 036.323, 036.325, 036.326, 036.328, 036.329, 036.330, 036.331, 036.332, 036.333, 036.334, 036.335, 036.336, 036.337, 036.338, 036.339, 036.340, 036.341, 036.342, 036.343, 036.344, 036.347, 036.348, 036.349, 036.350, 036.351, 036.352, 036.353, 036.354, 036.355, 036.356, 036.357, 159.183, 161.673.

012.111 **Heavy ion physics and nuclear astrophysical problems. Proceedings. International Symposium on Heavy Ion Physics and Nuclear Astrophysical Problems, Tokyo (Japan), 21 – 23 Jul 1988.**
S. Kubono, M. Ishihara, T. Nomura (eds.).
World Scientific, Singapore (Singapore). 437 p. (1989). ISBN 9971–50–751–X. Price £ 44.00.
The symposium has focussed on current topics and future trends in intermediate–energy heavy ion physics. Particular emphasis is placed on problems associated with nuclei far from the stability line and nuclear astrophysical problems relevant to unstable nuclei. These subjects are tied up with new research opportunities to be offered by the recently completed K = 50 cyclotron at Riken and a radioactive beam factory.

012.112 **The Fifth Marcel Grossmann Meeting on Recent Developments in Theoretical and Experimental General Relativity, Gravitation and Relativistic Field Theories. Parts A, B. Proceedings. 5. Marcel Grossmann Meeting on General Relativity, Perth (Australia), 8 – 13 Aug 1988.**
D. G. Blair, M. J. Buckingham (eds.).
World Scientific, Singapore (Singapore). 1961 p. (1989). ISBN 9971–50–784–6. Price £ 96.00.
Contents: Part A.
1. Rigorous and exact.
Classical general relativity: highly non–linear behaviour. Spin, geometry and topology. Approximation methods. Exact solutions. Black hole physics. Alternative theories and torsion.
2. Quantum gravity.
Critical accelerations. Quantum gravity. String theories. Cosmic strings, superstrings and supergravity. Quantum cosmology: wavefunction of the universe. Quantum cosmology.
3. Cosmology.
Early cosmology and quantum field theory. Supersymmetry, multidimensional cosmology and Kaluza–Klein theory. Theoretical cosmology. Large–scale structure of the universe. Dark matter.
Part B.
4. Mathematical astrophysics.
Algebraic computing. Numerical relativity. Astrophysics of collapsed objects. Self gravitating systems. History of general relativity.
5. Observational astrophysics.
Sources of gravitational radiation. Relativistic astrophysics. Supernovae. Observation of collapsed objects. Cosmic background.
5. Precision experiments.
The fifth force. Measuring the gravitational interaction in precision space experiments. Resonant bar antennas. Laser interferometer antennas. Detection of gravitational radiation.

Quantum technology for gravitational radiation detection. Precision clocks in general relativity.

012.113 **Big bang, active galactic nuclei and supernovae. Proceedings. 20. Yamada Conference: Big bang, active galactic nuclei and supernovae, Tokyo (Japan), 28 Mar – 1 Apr 1988.**
S. Hayakawa, K. Sato (eds.).
Universal Academy Press, Tokyo (Japan). 633 p. (1989). ISBN 4-946443-05-3. Price Yen 12000.00.

The conference covered fields which are thought to play key roles in the evolution of the universe. The articles contained are arranged roughly in chronological order, from the creation of the universe to the death of stars as supernovae, and intend to give a coherent scenario of the evolution of the universe, although conflicting opinions are expressed sometimes by different authors. The book begins with the early universe to which particle physics is relevant, describes the formation of structure of various scales in chapters of galaxy formation and background radiation, and discusses active galactic nuclei, including quasars which not only play an important role in the galaxy formation but also serve as light sources to probe the deep space in the universe. The book also includes supernovae and neutrino astronomy whose progress is stimulated by the detection of a neutrino burst from SN 1987A.

012.114 **A. A. Friedmann: centenary volume. Proceedings. Friedmann Centenary Conference, Leningrad (USSR), 22 – 26 Jun 1988.**
M. A. Markov, V. A. Berezin, V. F. Mukhanov (eds.).
World Scientific, Singapore (Singapore). 352 p. (1990). ISBN 9971-50-899-0. Price £ 41.00, US$ 64.00.

In June 1988 Alexander Alexandrovich Friedmann would be 100. Much has been done by him for the development of hydrodynamics and theoretical meteorology. But first of all his name is a symbol of the completely new epoch in cosmology. Friedmannian cosmological models appeared to be a basis for modern cosmology including investigations on the nucleosynthesis in a hot universe, galaxy formation theory, explanation of the baryon asymmetry of the universe and so on.

012.115 **General relativity and gravitational physics. Proceedings. 8. Italian Conference on General Relativity and Gravitational Physics, Cavalese, Trento (Italy), 30 Aug – 3 Sep 1988.**
M. Cerdonio, R. Cianci, M. Francaviglia, M. Toller (eds.).
World Scientific, Singapore (Singapore). 648 p. (1989). ISBN 9971-50-844-3. Price £ 54.00, US$ 86.00.

These proceedings contain all the invited talks and most of the contributed talks presented at the conference. The contributed papers were subdivided in five workshops according to the following main themes:
A. General relativity and classical theories of gravitation. B. Relativistic astrophysics and cosmology. C. Experimental and observational gravitation. D. Supergravity and string theories. E. Quantum gravity.

012.116 **Tests of fundamental laws in physics. Proceedings. 24. Rencontre de Moriond, 9. Moriond Workshop: Tests of fundamental laws in physics, Les Arcs, Savoie (France), 21 – 28 Jan 1989.**
O. Fackler, J. Trân Thanh Vân (eds.).
Editions Frontières, Gif-sur-Yvette (France). 678 p. (1989). ISBN 2-86332-064-5. Price FF 500.00.

The subject of this workshop focuses on the tests of fundamental laws in physics and on the search for new and exotic phenomena. Special emphasis is given to the new narrow e^+e^- resonances at low energies, to the possibility of a fifth force, and to recent developments in neutrino physics.

012.117 **Geoscience and a lunar base. A comprehensive plan for lunar exploration. Proceedings. Workshop on Geoscience from a Lunar Base, Houston, TX (USA), 25 – 26 Aug 1988.**
G. J. Taylor, P. D. Spudis (eds.).
National Aeronautics and Space Administration, Washington, DC (USA), NASA-CP—3070. 78 p. (Apr 1990).

These proceedings describe a comprehensive plan for the geologic exploration of the Moon. The document begins by explaining the scientific importance of studying the Moon and outlines the many unsolved problems in lunar science. Subsequent chapters detail different, complementary approaches to geologic studies: global surveys, including orbiting spacecraft such as Lunar Observer and installation of a global geophysical network; reconnaissance sample return mission, by either automated rovers or landers, or by piloted forays; detailed field studies, which involve astronauts and teleoperated robotic field geologists. The document then develops a flexible scenario for exploration and sketches the technological developments needed to carry out the exploration scenario.

012.118 **Les méthodes modernes de la mécanique céleste. Théorie des perturbations et chaos intrinsèque. Comptes rendus. *(Modern methods in celestial mechanics. Proceedings.)* 13. Ecole de Printemps d'Astrophysique de Goutelas: Les méthodes modernes de la mécanique céleste, Château de Goutelas (France), 24 – 29 Apr 1989.**
D. Benest, C. Froeschlé (eds.).
Editions Frontières, Gif-sur-Yvette (France). 516 p. (1990). ISBN 2-86332-091-2. Price FF 380.00.
The individual contributions within the subject scope of AAA are included in their corresponding categories – see abstracts 021.053, 021.055, 021.056, 022.172, 042.084, 042.085, 042.086, 042.087, 042.088, 042.089, 042.090, 042.094, 042.095, 042.096, 091.057.

012.119 **Fall meeting of the Astronomical Society of New York, Troy, NY (USA), 3 Nov 1990. Abstracts of papers presented.**
A. G. D. Philip (ed.).
News Lett. Astron. Soc. N.Y., Vol. 3, No. 9, 37 p. (Feb 1991).
The individual contributions within the subject scope of AAA are included in their corresponding categories – see abstracts 034.186, 100.037, 112.161, 112.162, 126.095, 131.402, 131.403, 131.404, 131.405, 133.009, 144.117, 158.341, 159.184.

012.120 **Toward the global APT network. Proceedings. 3. New Zealand Conference on Photoelectric Photometry, Blenheim (New Zealand), 9 – 12 Mar 1989.**
E. Budding, J. Richard (eds.).
South. Stars, Vol. 34, No. 3, 216 p. (18 Jun 1991).

Contents: 1. Surveying the subject. 2. Automatic Photoelectric Telescopes (ATPs). 3. Related procedures and instruments. 4. CCD technology. 5. Multiwavelength facilities.
The individual contributions within the subject scope of AAA are included in their corresponding categories – see abstracts 009.012, 009.013, 032.053, 032.054, 032.055, 034.183, 034.184, 034.185, 034.187, 034.188, 034.189, 036.361, 036.363, 036.364, 036.367, 036.368, 036.369, 117.306, 117.309, 117.310, 119.058, 119.059, 122.117, 122.118, 122.120, 122.122, 122.123, 125.163, 125.301, 143.062, 158.340.

012.121 **Theory of accretion disks. Proceedings. NATO Advanced Research Workshop on Theory of Accretion Disks, Garching (Germany), 6 – 10 Mar 1989.**
F. Meyer, W. J. Duschl, J. Frank, E. Meyer-Hofmeister (eds.).
NATO ASI Ser., Ser. C, Math. Phys. Sci., Vol. 290.
Kluwer Academic Publ., Dordrecht (Netherlands). 491 p. (1989). ISBN 0-7923-0453-5. Price Dfl. 240.00, £ 82.00.

Accretion of material on to compact objects is an ubiquitous mechanism powering very diverse astrophysical sources ranging in size and luminosity by many orders of magnitude. The formation of an *accretion disk* which allows angular momentum re-distribution and converts potential energy into radiation with an efficiency which can be higher than the nuclear burning yield is discussed for systems ranging in size from quasars and active galactic nuclei to accretion disks around forming stars and the

early solar system and to compact binaries such as cataclysmic variables and low-mass X-ray binaries. Other objects that should be mentioned in this context are SS433, the black hole binary candidates, and possibly gamma-ray burst sources. Observations of these systems have provided important constraints for theoretical accretion disk models on widely differing scales, luminosities, mass-transfer rates and physical environments.
The individual contributions within the subject scope of AAA are included in their corresponding categories - see abstracts 062.244, 062.246, 062.249, 062.261, 062.263, 062.264, 062.265, 062.266, 062.267, 062.268, 062.269, 063.127, 063.128, 063.129, 064.195, 064.196, 064.197, 064.198, 064.199, 064.200, 064.201, 064.202, 064.203, 064.204, 064.205, 064.206, 064.207, 067.267, 067.271, 067.272, 117.308, 117.312, 117.313, 117.314, 131.401, 142.064, 158.336, 158.337, 158.338, 158.339, 158.343.

012.122 **Anomalous nuclear effects in deuterium/solid systems. Proceedings. Workshop on Anomalous Nuclear Effects in Deuterium/Solid Systems, Provo, UT (USA), 1990.**
S. E. Jones, F. Scaramuzzi, D. Worledge.
AIP Conf. Proc., No. 228.
American Institute of Physics, New York, NY (USA). 961 p. (1991). ISBN 0-88318-833-3.
Nearly seventy papers were presented, with discussions divided into five broad categories: 1. Neutron emission studies. 2. Charged particle detection. 3. Tritium/helium studies. 4. Fusion in the planets. 5. Theory.
The individual contributions within the subject scope of AAA are included in their corresponding categories - see abstracts 081.059, 081.060, 099.075, 099.076.

012.123 **Boron-rich solids. Proceedings. 10. International Symposium on Boron, Borides and Related Compounds, Albuquerque, NM (USA), 27 - 30 Aug 1990.**
D. Emin, T. L. Aselage, A. C. Switendick, B. Morosin, C. L. Beckel (eds.).
AIP Conf. Proc., No. 231.
American Institute of Physics, New York, NY (USA). 692 p. (1991). ISBN 0-88318-793-4.
The individual contributions within the subject scope of AAA are included in their corresponding categories - see abstracts 161.674.

012.124 **Clusters of galaxies. Proceedings. Clusters of Galaxies Meeting, Baltimore, MD (USA), 15 - 17 May 1989.**
W. R. Oegerle, M. J. Fitchett, L. Danly (eds.).
Space Telesc. Sci. Inst., Symp. Ser., Vol. 4.
Cambridge University Press, Cambridge (UK). 408 p. (1990). ISBN 0-521-38462-1. Price £ 35.00. US$ 54.50.
Collected in this volume are the review papers from the Space Telescope Science Institute Symposium on Clusters of Galaxies in May 1989. Subjects covered include the existence and importance of subclustering, models of the evolution of clusters and the intracluster medium, the effect of the cluster environment on galaxies, observations of high redshift clusters, and the use of clusters as tracers of large-scale structure.
The individual contributions within the subject scope of AAA are included in their corresponding categories - see abstracts 160.172, 160.173, 160.174, 160.175, 160.176, 160.177, 160.178, 160.179, 160.180, 160.181, 160.182, 160.183, 160.184, 160.185, 161.675.

012.125 **2. Symposium on Reversals, Secular Variation and Dynamo Theory, Santa Fe, NM (USA), 6 - 10 Aug 1990.**
Geophys. Astrophys. Fluid Dyn., Vol. 60, No. 1-4, p. 89 - 436 (Nov 1991).
The individual contributions within the subject scope of AAA are included in their corresponding categories - see abstracts 044.058, 062.238, 062.239, 062.240, 062.241, 062.243, 062.245, 081.061, 081.062, 081.063, 081.064, 084.073, 084.074, 084.075, 084.077, 084.078, 084.079, 084.080, 084.081, 084.082, 084.083, 084.084, 084.085, 084.086, 084.087, 084.088.

012.126 **T. G. Cowling Memorial Symposium on Solar System Plasmas as Part of the 16. General Assembly of the European Geophysical Society (EGS), Wiesbaden (Germany), 22 - 26 Apr 1991.**
Geophys. Astrophys. Fluid Dyn., Vol. 62, No. 1-4, p. 15 - 281 (Dec 1991).
The individual contributions within the subject scope of AAA are included in their corresponding categories - see abstracts 062.254, 062.255, 062.256, 072.147, 073.187, 074.130, 075.070, 080.166, 080.167, 080.168, 080.169, 080.170, 084.089, 084.090, 084.091.

012.127 **Memorias de la Reunión Astronómica Sur-Andina, Santiago (Chile), 18 - 19 Mar 1991. Abstracts.**
D. Geisler, L. Infante (eds.).
Rev. Mex. Astron. Astrofis., Vol. 22, No. 2, p. 299 - 310 (Oct 1991).
The individual contributions within the subject scope of AAA are included in their corresponding categories - see abstracts 013.109, 013.110, 063.126, 111.017, 112.164, 112.165, 113.049, 117.315, 124.009, 124.027, 154.161, 157.354, 157.355, 160.167, 160.168, 160.169, 160.170, 160.171, 161.681, 161.682, 161.683, 161.686.

013 Reports on Astronomy in Various Countries and Particular Fields

013.001 **Some problems of the investigation of the Sun within the framework of the STEP project.**
G. V. Kuklin, V. N. Obridko.
Geomagn. Aeron., Vol. 30, No. 4, p. 449 - 454 (Feb 1990). Paper presented at the 5. Symposium on Solar Terrestrial Physics, Samarkand (USSR), 2 - 6 Oct 1989. English translation of Geomagn. Aehron., Tom 30, No. 4, p. 529 - 537 (1990).
The main problems of investigation of the Sun within the framework of the international program Energy in the Sun-Earth System (STEP) are formulated. The problems of energy transport from the inner layers outwards right to the corona and heliosphere are discussed. The problems of generation of magnetic fields and helioseismological methods of studying the inner structure of the Sun are discussed. Special attention is devoted to the interrelation of local and global fields.

013.002 **Project SARV (Small-amplitude red variable).**
J. R. Percy.
I.A.P.P.P. Commun., No. 45, p. 24 (Sep - Nov 1991). Abstract. - See Abstr. 012.002 for the main entry.

013.003 **Scientific results from the AAVSO photometry program.**
J. R. Percy.
I.A.P.P.P. Commun., No. 45, p. 33 – 36 (Sep – Nov 1991). – See Abstr. 012.002 for the main entry.

013.004 **The role of the amateur photometrist in the age of robotic observatories.**
R. D. Lines.
I.A.P.P.P. Commun., No. 45, p. 52 (Sep – Nov 1991). Abstract. – See Abstr. 012.002 for the main entry.

013.005 **Astronomy under threat.**
A. Wolfendale.
Astron. Now, Vol. 5, No. 7, p. 24 – 25 (Jul 1991).

013.006 **Optik für Röntgenstrahlen. Entwicklungen in der ČSFR.**
R. Hudec.
Sterne Weltraum, Jahrg. 30, Nr. 10, p. 580 – 582, 584 – 586 (Oct 1991).

013.007 **Astronomie in Australien.**
G. W. E. Beekman.
Sterne Weltraum, Jahrg. 30, Nr. 10, p. 587 – 591 (Oct 1991).

013.008 **Report on the Rosat–IUE All Sky Survey (RIASS) program.**
D. de Martino, W. Wamsteker, J. M. Mas Hesse, J. Bonnell, C. Imhoff.
ESA IUE Newsl., No. 38, p. 10 – 26 (Aug 1991).

013.009 **The decade of discovery.**
J. N. Bahcall.
Bull. Am. Astron. Soc., Vol. 23, No. 2, p. 904 (1991). Abstract. – See Abstr. 012.008 for the main entry.

013.010 **Astronomy from space: the 1990's and beyond.**
C. Beichman.
Bull. Am. Astron. Soc., Vol. 23, No. 2, p. 904 (1991). Abstract. – See Abstr. 012.008 for the main entry.

013.011 **The AASC report: the ground–based program.**
S. C. Wolff.
Bull. Am. Astron. Soc., Vol. 23, No. 2, p. 904 (1991). Abstract. – See Abstr. 012.008 for the main entry.

013.012 **New frontiers in ground–based optical astronomy.**
S. E. Strom.
Sky Telesc., Vol. 82, No. 1, p. 18 – 23 (Jul 1991).
Observatories in space have shown the cosmos at invisible wavelengths and revealed unimagined phenomena. Powerful new ground–based telescopes are needed to follow up these discoveries with visible and infrared observations.

013.013 **Infrared astronomy: pixels to spare.**
M. McCaughrean.
Sky Telesc., Vol. 82, No. 1, p. 31 – 35 (Jul 1991).
Large–format infrared cameras are producing spectacular high–resolution images of everything from nearby satellites to distant galaxies.

013.014 **Infrared astronomy takes center stage.**
F. C. Gillett, I. Gatley, D. Hollenbach.
Sky Telesc., Vol. 82, No. 2, p. 148 – 153 (Aug 1991).
The 1990's promise to be the decade of infrared astronomy, as astronomers plan to build major new infrared telescopes for use on the ground, in the air, and in space.

013.015 **Radio astronomy: the next decade.**
K. I. Kellermann.
Sky Telesc., Vol. 82, No. 3, p. 247 – 253 (Sep 1991).
With a new generation of highly capable radio telescopes poised to begin work in the 1990's, radio astronomers hope to top the significant discoveries they have made in past decades in virtually every subfield in astronomy.

013.016 **Astronomy in the mind and the lab.**
D. N. Schramm, C. F. McKee.
Sky Telesc., Vol. 82, No. 4, p. 352 – 357 (Oct 1991).
Theoretical studies and laboratory experiments will play a key role in driving astronomy forward in the 1990's.

013.017 **Radio astronomy's crumbling showpiece.**
M. M. Waldrop.
Science, Vol. 253, No. 5017, p. 268 – 269 (19 Jul 1991).

013.018 **The EUVE guest observer program.**
C. A. Christian.
Bull. Am. Astron. Soc., Vol. 23, No. 2, p. 972 (1991). Abstract. – See Abstr. 012.008 for the main entry.

013.019 **On doing cosmology with a large optical telescope.**
J. V. Narlikar.
National Large Telescope Workshop: Scientific requirements and requirement of site, p. 6 – 10 (1991). – See Abstr. 012.009 for the main entry.
This paper highlights the interesting problems in cosmology that require observational inputs normally available from optical telescopes in the 4 – 6 m class.

013.020 **Galactic bulges and discs.**
T. P. Prabhu.
National Large Telescope Workshop: Scientific requirements and requirement of site, p. 11 – 16 (1991). – See Abstr. 012.009 for the main entry.
Last two decades have witnessed considerable improvement in our understanding of formation, structure and dynamical evolution of elliptical and spiral galaxies, largely due to detailed observations obtained using existing large optical telescopes and theoretical models inspired by these observations. The present state of knowledge is summarized here. Need is currently felt for high spatial and spectral resolution observations.

013.021 **Some aspects of quasars.**
C. Sivaram.
National Large Telescope Workshop: Scientific requirements and requirement of site, p. 17 – 20 (1991). – See Abstr. 012.009 for the main entry.

013.022 **Polarization studies of AGN using a large telescope.**
M. R. Deshpande.
National Large Telescope Workshop: Scientific requirements and requirement of site, p. 21 – 24 (1991). – See Abstr. 012.009 for the main entry.
The author discusses the astrophysical problems that can be taken up by a large telescope. He confines his discussions to polarization studies of AGN's, which encompass quasars, QSO's, Seyfert galaxies and BL Lac objects.

013.023 **Star forming regions in galaxies.**
Y. D. Mayya.
National Large Telescope Workshop: Scientific requirements and requirement of site, p. 34 – 36 (1991). – See Abstr. 012.009 for the main entry.
Detailed mapping of the velocity fields in the central regions of galaxies will greatly help in understanding several problems involving enhanced star formation in galaxies.

013.024 **Quasar astronomy with a large optical telescope.**
U. C. Joshi.
National Large Telescope Workshop: Scientific requirements and requirement of site, p. 47 – 53 (1991). – See Abstr. 012.009 for the main entry.
A few areas of interest in extragalactic astronomy which demand high resolution spectroscopy are discussed.

013.025 **Novae and supernovae: outburst and quiescence.**
G. C. Anupama.
National Large Telescope Workshop: Scientific requirements and requirement of site, p. 54 – 60 (1991). – See Abstr. 012.009 for the main entry.
Studies of novae and supernovae are of prime interest in the understanding of stellar evolution and nucleosynthesis, chemical evolution of galaxies, aspects of binary systems such as mass transfer and accretion onto evolved compact objects. These objects can also be used to determine distances to galaxies. Some of the observations needed in understanding these systems are discussed.

013.026 **Stellar abundance studies with a large telescope.**
M. Parthasarathy.
National Large Telescope Workshop: Scientific requirements and requirement of site, p. 61 – 66 (1991). – See Abstr. 012.009 for the main entry.
To determine accurate element abundances requires high resolution and high signal to noise ratio spectra of stars in the galactic halo, globular clusters, galactic disc and bulge and galactic plane. We need to obtain spectra of stars of different masses, luminosities and evolutionary stages. This in turn requires reaching very fainter magnitudes, and higher resolutions.

013.027 **X–ray astronomy and optical studies.**
M. V. K. Apparao.
National Large Telescope Workshop: Scientific requirements and requirement of site, p. 67 – 70 (1991). – See Abstr. 012.009 for the main entry.
Almost every class of objects in the sky is found to be emitting X–rays at some level and optical studies of each of these classes are interesting and necessary.

013.028 **Study of galactic and extra galactic star clusters with a large ground based optical telescope.**
R. Sagar.
National Large Telescope Workshop: Scientific requirements and requirement of site, p. 71 –78 (1991). – See Abstr. 012.009 for the main entry.

013.029 **Infrared observations with a large telescope.**
M. R. Deshpande.
National Large Telescope Workshop: Scientific requirements and requirement of site, p. 79 – 82 (1991). – See Abstr. 012.009 for the main entry.
The author emphasises the importance of infrared work. He also discusses some of the key astrophysical problems which need a large telescope.

013.030 **Study of planetary nebulae: need for a large optical telescope.**
N. Kameswara Rao.
National Large Telescope Workshop: Scientific requirements and requirement of site, p. 83 – 86 (1991). – See Abstr. 012.009 for the main entry.
The author presents some of the relevant areas in planetary nebulae research which need a large telescope.

013.031 **High resolution spectroscopy with large aperture telescopes.**
J. N. Desai.
National Large Telescope Workshop: Scientific requirements and requirement of site, p. 87 – 88 (1991). – See Abstr. 012.009 for the main entry.

013.032 **High resolution spectroscopy of T Tauri stars.**
M. V. Mekkaden.
National Large Telescope Workshop: Scientific requirements and requirement of site, p. 89 – 90 (1991). – See Abstr. 012.009 for the main entry.

013.033 **A few research problems for large telescopes.**
S. C. Joshi.
National Large Telescope Workshop: Scientific requirements and requirement of site, p. 91 – 95 (1991). – See Abstr. 012.009 for the main entry.

013.034 **The Solar Radio Astronomy Programme at the National Research Council of Canada.**
K. F. Tapping.
Bull. Am. Astron. Soc., Vol. 23, No. 2, p. 1039 (1991). Abstract. – See Abstr. 012.012 for the main entry.

013.035 **Objectives and scope of RISE.**
P. Foukal.
Bull. Am. Astron. Soc., Vol. 23, No. 2, p. 1040 (1991). Abstract. – See Abstr. 012.012 for the main entry.

013.036 **RISE telescope concept for CaK imaging.**
W. Livingston.
Bull. Am. Astron. Soc., Vol. 23, No. 2, p. 1040 (1991). Abstract. – See Abstr. 012.012 for the main entry.

013.037 **Some theoretical topics for RISE.**
T. Holzer.
Bull. Am. Astron. Soc., Vol. 23, No. 2, p. 1040 (1991). Abstract. – See Abstr. 012.012 for the main entry.

013.038 **SunRISE at the NSF.**
R. A. Goldberg.
Bull. Am. Astron. Soc., Vol. 23, No. 2, p. 1040 (1991). Abstract. – See Abstr. 012.012 for the main entry.

013.039 **RISE–related programs at NASA.**
W. J. Wagner.
Bull. Am. Astron. Soc., Vol. 23, No. 2, p. 1041 (1991). Abstract. – See Abstr. 012.012 for the main entry.

013.040 **NOAA's program of research on solar influences in global change.**
E. Hildner.
Bull. Am. Astron. Soc., Vol. 23, No. 2, p. 1041 (1991). Abstract. – See Abstr. 012.012 for the main entry.

013.041 **RISE and the NRC's Committee on Global Change. Working Group on Solar Influences on Global Change.**
J. L. Lean.
Bull. Am. Astron. Soc., Vol. 23, No. 2, p. 1041 (1991). Abstract. – See Abstr. 012.012 for the main entry.

013.042 **The MSFC solar GRO guest investigation.**
M. J. Hagyard, G. A. Gary, R. L. Moore.
Bull. Am. Astron. Soc., Vol. 23, No. 2, p. 1073 (1991). Abstract. – See Abstr. 012.012 for the main entry.

013.043 **The distribution of BATSE quick–look flare data through the SMM Data Analysis Center.**
R. A. Schwartz, B. R. Dennis, L. E. Orwig, A. L. Kiplinger.
Bull. Am. Astron. Soc., Vol. 23, No. 2, p. 1074 (1991). Abstract. – See Abstr. 012.012 for the main entry.

013.044 **Photoelectric photometry in Romania.**
G. Oprescu.
I.A.P.P.P. Commun., No. 46, p. 41 (Dec 1991 – Feb 1992).

013.045 **Infared spectroscopy as a tool to study earth's atmosphere.**
J. M. Flaud.
AIP Conf. Proc., No. 216, p. 523 – 541 (Dec 1990). – See Abstr. 012.018 for the main entry.

013.046 **Borowiec laser station 1986 – 1990.**
S. Schillak.
Artif. Satell., Vol. 26, No. 1, p. 13 – 18 (1991).
The activity of Borowiec satellite laser station in period 1986 – 1990 is presented. Main technical parameters of the system in this time and short description of programs are also included.

013.047 **Time Service of Borowiec Astronomical Latitude Observatory in the years 1985 – 1990.**
P. Fraczyk, J. Nawrocki.
Artif. Satell., Vol. 26, No. 1, p. 139 – 174 (1991).
Activity of the Observatory Time Service in the field of time scale generation and new time & frequency techniques development is presented.

013.048 **Satellite dynamics research in Poznań University Observatory in 1986 – 1990.**
E. Wnuk.
Artif. Satell., Vol. 26, No. 1, p. 221 – 229 (1991).

013.049 **Tidal research in geodynamics at Warsaw Space Research Centre (1985 – 1990).**
T. Chojnicki, M. Kaczorowski.
Artif. Satell., Vol. 26, No. 1, p. 291 – 299 (1991).

013.050 **The BIPM and the accurate measurement of time.**
T. J. Quinn.
Proc. IEEE, Vol. 79, No. 7, p. 894 – 905 (Jul 1991). Special issue: Time and frequency. J. Jespersen, D. W. Hanson (eds.).
Accurate measurements in science and technology require world–wide agreement on standards and units. The role of the Bureau International des Poids et Mesures (BIPM) in assuming the uniformity and reliability of the world's measurement system is described. The development of atomic time scales is outlined with particular reference to the Bureau International de l'Heure and the origins of TAI and UTC. The article ends with a brief description of the work currently being carried out at the BIPM.

013.051 **Soviet programs "Jupiter" and "Ceres" for improving orbits of bodies of solar system.**
A. G. Sokolsky (*A. G. Sokol'skij*).
1. Spain–USSR Workshop on Positional Astronomy and Celestial Mechanics, p. 49 – 50 (1991). – See Abstr. 012.025 for the main entry.
Soviet scientific programs "Jupiter" and "Ceres" are being accomplished according to the decision of the USSR Academy of Sciences within the framework of the "Astronomy" global program of fundamental studies. It is supposed that the obtained results will be used in many fundamental and application programs in astronomy, cosmonautics, geodesy, navigation, etc. The period of accomplishment of the projects is 1990 – 1992.

013.052 **Interference and radioastronomy.**
A. R. Thompson, T. E. Gergely, P. A. Vanden Bout.
Phys. Today, Vol. 44, No. 11, p. 41 – 49 (Nov 1991). Current Physics Microform No.: 9201F2137.
The radioastronomer's struggle against a growing flood of interfering sources, from garage door openers to digital audio broadcast satellites, must be fought in the technical and political arenas.

013.053 **On Bengt Strömgren's work in photometry.**
J. Knude.
Meeting on Precision Photometry: Astrophysics of the Galaxy, p. 3 – 12 (1991). – See Abstr. 012.027 for the main entry.

013.054 **High–resolution imaging in La Palma.**
R. M. Redfern.
Vistas Astron., Vol. 34, Part 3, p. 201 – 233 (1991). – See Abstr. 012.032 for the main entry.
Currently, in the La Palma Observatory, there are a number of instrumentation research groups actively seeking to achieve high resolution images. This article briefly reviews (a) some basic results of seeing theory, (b) seeing measurements in La Palma, (c) methods of high resolution imaging (HRI) and current work in HRI in La Palma, and (d) the future for HRI instrumentation in La Palma, (and elsewhere).

013.055 **The EUV background, the interstellar medium, and ROSAT.**
H. J. Walker, A. W. Harris, T. J. Sumner, G. C. Brebner, M. A. Barstow, R. Willingale.
Vistas Astron., Vol. 34, Part 3, p. 281 – 290 (1991). – See Abstr. 012.032 for the main entry.
This is a review of work done by several people in the UK Interstellar Medium/Diffuse Emission ROSAT Special Interest Group.

013.056 **First results from the ROSAT Wide Field Camera source identification programme.**
K. O. Mason, G. Branduardi–Raymont, G. E. Bromage, D. Buckley, P. A. Charles, B. J. M. Hassall, M. R. S. Hawkins, S. Hodgkin, C. D. Pike, C. M. Jomaron, D. H. P. Jones, I. McHardy, T. T. Naylor, T. J. Ponman, M. G. Watson.
Vistas Astron., Vol. 34, Part 3, p. 343 – 351 (1991). – See Abstr. 012.032 for the main entry.
This progress report summarises the first weeks of observing on the INT and at SAAO 1990/91. More than 180 WFC sources have been observed optically, and over 90 positive identifications made.

013.057 **Summary paper: precision photometry, past, present, future.**
K. A. Janes.
Meeting on Precision Photometry: Astrophysics of the Galaxy, p. 283 – 289 (1991). – See Abstr. 012.027 for the main entry.

013.058 **Coordinated multi–site photometry of southern rapidly oscillating Ap stars. I: The 1992 campaign on HD 84041 – a call for collaborators.**
P. Martinez, D. W. Kurtz.
Inf. Bull. Variable Stars, No. 3634, p. 1 – 4 (9 Jul 1991).

013.059 **Observational project on gravitational microlensing.**
M. Kubiak, A. Udalski.
Postepy Astron., Tom 39, Zesz. 4, p. 148 – 150 (1991). In Polish.

013.060 **The interplanetary plasma.**
B. Rossi.
Annu. Rev. Astron. Astrophys., Vol. 29, p. 1 – 8 (1991). – See Abstr. 003.010 for the main entry.
The author felt motivated to initiate a study of interplanetary plasma with his group at MIT. Thus was born a research program that continues to this day, and that has been one of the major research activities of the Institute.

013.061 **The solar–terrestrial sciences in ninety's.**
Hu Wenrui.
Publ. Yunnan Obs., No. 4, p. 3 – 8 (Dec 1990). – See Abstr. 012.047 for the main entry.

013.062 **A brief review, recent development in interplanetary physics and magnetospheric physics.**
Fu Zhufeng.
Publ. Yunnan Obs., No. 4, p. 14 – 15 (Dec 1990). – See Abstr. 012.047 for the main entry.

013.063 **The progress in the research on radio rapid fluctuation in solar flares for the last year.**
Fu Qijun.
Publ. Yunnan Obs., No. 4, p. 234 – 240 (Dec 1990). – See Abstr. 012.047 for the main entry.

013.064 **The construction of the East Asia Meridina Chain of geomagnetic observatories.**
Liu Changfa, Wang Juyi, Zhang Weixi, Dou Shaopu, Zhang Ping.
Publ. Yunnan Obs., No. 4, p. 253 – 257 (Dec 1990). – See Abstr. 012.047 for the main entry.

013.065 **Progress and new directions for the Berkeley supernova search.**
S. Perlmutter, H. J. Marvin, R. A. Muller, C. R. Pennypacker, T. P. Sasseen, C. K. Smith, Wang Liping.
10. Santa Cruz Summer Workshop in Astronomy and Astrophysics: Supernovae, p. 727 – 730 (1991). – See Abstr. 012.052 for the main entry.

013.066 **The programme FON–OBZOR–S.**
V. S. Kislyuk.
International Seminar on Problems of Astrometry and Space Geodynamics, p. 3 – 11 (1991). In Russian. – See Abstr. 012.049 for the main entry.
The current status of the photographic survey programme of the northern sky as well as the problem of the creation of the northern part of the survey astrometrical catalogue using the FON and Carte du Ciel observations are discussed.

013.067 **The programme MEGA: results and outlook.**
N. V. Kharchenko.
International Seminar on Problems of Astrometry and Space Geodynamics, p. 15 – 23 (1991). In Russian. – See Abstr. 012.049 for the main entry.
Problems and observational data of the complex programme of investigation of the main meridional section of the Galaxy (MEGA) are discussed. Based on data of the programme MEGA catalogue, containing astrometric and some astrophysical stellar characteristics, stellar kinematics in the Galaxy meridional section, the structure and evolutionary peculiarities of centroids in the solar neighborhood are investigated.

013.068 **Asteroseismology of white dwarf stars with the Whole Earth Telescope.**
D. E. Winget.
NATO Advanced Research Workshop – 7. European Workshop on White Dwarfs, p. 129 – 141 (1991). – See Abstr. 012.044 for the main entry.
The Whole Earth Telescope (WET) project is a cooperative effort to obtain global time–series photometry of compact variable stars. The aim of WET is to obtain essentially continuous light curves, from which high–resolution, high signal–to–noise power–spectra are computed. These enable to apply the powerful theoretical machinery of asteroseismology to determine many of the fundamental parameters of white dwarf stars: rotation rates, magnetic field strengths, total mass, compositional stratification of the envelope, core composition, and more. The purpose of this review is to provide a brief update on the WET project. The author presents a list of principle targets for which ixtensive data sets currently exist, or are planned (as of May 1991). All of the data on each target is in the public domain 18 months after a run, and possibly earlier (at the discretion of the principle investigator for the specific target). The author discusses the results for the DOV star PG 1159–035 as a case study, and finally attempts to chart some of the future directions we can expect the WET project to take.

013.069 **Summary remarks on star formation.**
R. B. Larson.
3. Haystack Observatory Conference on the Interstellar Medium: Skylines – atoms, ions and molecules – new results in spectral line astrophysics, p. 457 – 461 (1991). – See Abstr. 012.051 for the main entry.

013.070 **Space Telescope observations of white dwarf stars.**
H. L. Shipman.
NATO Advanced Research Workshop – 7. European Workshop on White Dwarfs, p. 369 – 378 (1991). – See Abstr. 012.044 for the main entry.
The Hubble Space Telescope will present an opportunity for a dramatic improvement in the understanding of white dwarf stars. A collaboration of reasearchers has obtained observing time to study white dwarf stars. This paper reviews the proposed observations, the genesis and history of the collaboration, and the current status of this project in light of the telescope's shortcomings.

013.071 **Correlation between the Maryland and Rome gravitational–wave detectors and the Mont Blanc, Kamioka and IMB particle detectors during SN 1987A.**
M. Aglietta, A. Castellina, W. Fulgione, G. Trinchero, S. Vernetto, P. Astone, G. Badino, G. Bologna, M. Bassan, E. Coccia, I. Modena, P. Bonifazi, M. G. Castellano, M. Visco, C. Castagnoli, P. Galeotti, O. Saavedra, C. Cosmelli, S. Frasca, G. V. Pallottino, G. Pizzella, P. Rapagnani, F. Ricci, E. Majorana, D. Gretz, J. Weber, G. Wilmot.
Nuovo Cimento B, Vol. 106, No. 11, p. 1257 – 1269 (Nov 1991).
Following a previously found correlation between the gravitational–wave detectors and the Mont Blanc particle detector, the authors have searched for a similar correlation between the data of the experiments mentioned in the title. They have found that both the Kamioka and the IMB data have a correlation with the gravitational–wave data that occurs with the same characteristics and at the same time of that already found with Mont Blanc. This correlation extends for a period of one or two hours centred at the hour 2:45 UT of 23 February 1987. It shows that the particle detector signals are delayed with respect to the gravitational–wave detector signals by (1.2 ± 0.5)s. The probability that the additional correlation due to Kamioka and IMB is only accidental is estimated of the order of 10^{-3} or 10^{-4}.

013.072 **Growing interests in the high S/N spectroscopy of stellar physics.**
Zhai Disheng, Jiang Shiyang.
Prog. Astron., Vol. 9, No. 2, p. 115 – 123 (Jun 1991). In Chinese.
Great developments have been made in the field of high S/N spectroscopy on stellar physics in recent 10 years. This paper give a short review on the important advances of observational approach to better understanding the chemical abundances and physical structure of stars by high S/N spectroscopy. The authors give prospects for research work with the Chinese 2.16 m telescope with a fiber–feeding echelle coudé spectrograph.

013.073 **Stellar astronomy with the Isaac Newton Group telescopes.**
P. A. Charles.
Astrophys. Lett. Commun., Vol. 28, No. 2–4, p. 73 – 99 (Sep 1991). – See Abstr. 012.072 for the main entry.

013.074 **Solar research at the IAC.**
M. Vazquez.
Astrophys. Lett. Commun., Vol. 28, No. 2–4, p. 113 – 129 (Sep 1991). – See Abstr. 012.072 for the main entry.
The main aspects of solar physics research at the IAC are presented. Two main branches currently exist: a) Helioseismology, which studies the physical conditions in the solar interior, using measurements of the solar oscillations in radial velocity and brightness, and b) Structure and dynamics of the solar atmosphere, whose main aim is observation, mainly with high spatial resolution, of the different photospheric structures and their interpretation using MHD and radiative–transfer diagnostics.

013.075 **Stellar astrophysics at the IAC from 1979 to 1989.**
R. Rebolo, S. Arribas.
Astrophys. Lett. Commun., Vol. 28, No. 2–4, p. 131 – 146 (Sep 1991). – See Abstr. 012.072 for the main entry.
The authors summarize the work carried out during the last ten years at the IAC in the field of stellar astrophysics. The paper

considers three areas: stellar interiors, atmospheres and envelopes, stellar evolution and stars in our galaxy. The main results obtained are briefly summarized.

013.076 **Galaxies as seen by the ING.**
D. Carter.
Astrophys. Lett. Commun., Vol. 28, No 2–4, p. 147 – 156 (Sep 1991). – See Abstr. 012.072 for the main entry.
The author presents a short review of extragalactic astronomy as performed by the telescopes of the Isaac Newton Group. As such it can be neither complete nor detailed. He has chosen five problems which seem to him to be the most important that we face at present. For each problem he looks at the work that has been done with the ING telescopes and then concentrate on the future to try to predict what important observations will be done with the ING telescopes, with existing equipment and with that planned for the next five years.

013.077 **Solar research and programs of the DFG at Observatorio del Teide.**
W. Mattig.
Astrophys. Lett. Commun., Vol. 28, No. 2–4, p. 157 – 169 (Sep 1991). – See Abstr. 012.072 for the main entry.
Three German instruments are now in operation at the Observatorio del Teide in Izaña (Tenerife). The Vacuum–Newton–Telescope (VNT), built for the site–test campaign, is still in use. The Vacuum–Gregory–Coudé–Telescope (GCT) started in 1987, and the Vacuum–Tower–Telescope (VTT) in 1989 with regular observations. The GCT and VTT are equipped with high resolution spectrographs, the latter with an echelle grating. The results obtained with the VNT and GCT are mostly concentrated on the fine structures of the photosphere and the sunspots including the magnetic fields. Data can be obtained with the highest spatial resolution, up to theoretical instrumental limits. The point spread function of the VTT, measured by observing a star image, has a FWHM of 0.3 arc sec. The paper summarizes the scientific work at Izaña over the last few years.

013.078 **Infrared observations in the Canary Islands.**
M. J. Selby, A. Mampaso.
Astrophys. Lett. Commun., Vol. 28, No. 2–4, p. 171 – 193 (Sep 1991). – See Abstr. 012.072 for the main entry.
The infrared observational activities in the Canaries are described. These span nearly twenty years and are shown to be very varied in nature and are mainly concentrated in the near infrared 1 – 5 μm. A description of the telescopes is followed by an assessment of the quality of the infrared sky on the high dry mountain sites. It indicates that the Canaries provide the best infrared site close to Europe.

013.079 **Ultra high energy gamma ray astronomy at the Observatorio del Roque de los Muchachos on La Palma.**
M. Bott–Bodenhausen.
Astrophys. Lett. Commun., Vol. 28, No. 2–4, p. 195 – 202 (Sep 1991). – See Abstr. 012.072 for the main entry.
A short review on observations of high energy gamma rays in the energy range between 10^{10}eV and 10^{17}eV is presented. The implications of the existence of high energy gamma ray emissions from cosmic point sources for elementary particle physics and astrophysics are discussed. The High Energy Gamma Ray Array (HEGRA) is described as installed on the Roque de los Muchachos on La Palma for the detection of cosmic particle induced extensive air showers.

013.080 **Extragalactic research at the Instituto de Astrofisica de Canarias.**
E. Pérez, J. M. Vilchez, A. Varela, J. M. Rodríguez–Espinosa, M. Prieto, I. Pérez–Fournon, C. Muñoz–Tuñón, E. Mediavilla, M. R. Kidger, J. Cepa, J. E. Beckman, R. A. Watson.
Astrophys. Lett. Commun., Vol. 28, No. 2–4, p. 203 – 215 (Sep 1991). – See Abstr. 012.072 for the main entry.
The authors give a general overview of the extragalactic research projects presently being carried out at the Instituto de Astrofisica de Canarias (IAC). The interests in this field of astronomical research range widely, from normal galaxies through the most active galaxies and quasars as well as cosmology, and from purely kinematical studies to the more comprehensive ones including dynamical and chemical evolution.

013.081 **Neutrino astronomy: Putting a project on ice.**
D. Kazanas.
Nature, Vol. 353, No. 6342, p. 298 – 299 (26 Sep 1991).

013.082 **Observation of muons using the polar ice cap as a Cerenkov detector.**
D. M. Lowder, T. Miller, P. B. Price, A. Westphal, S. W. Barwick, F. Halzen, R. Morse.
Nature, Vol. 353, No. 6342, p. 331 – 333 (26 Sep 1991). Letter–to–the–editor.
Detection of the small flux of extraterrestrial neutrinos expected at energies above 1 TeV, and identification of their astrophysical point sources, will require neutrino telescopes with effective areas measured in square kilometres. Such a device can be built only by using some naturally occurring detecting medium of enormous extent: deep Antarctic ice is a strong candidate. A neutrino telescope could be constructed by drilling holes in the ice with hot water into which photomultiplier tubes could be placed to a depth of 1 km. Neutrinos would be recorded, as in underground neutrino detectors using water as the medium, by the observation of Cerenkov radiation from secondary muons. The authors have begun the AMANDA (Antarctic Muon and Neutrino Detector Array) project to test this idea, and here they describe a pilot experiment using photomultiplier tubes placed into Arctic ice in Greenland. Cerenkov radiation from muons was detected, and a comparison of count rate with the expected muon flux indicates that the ice is very transparent, with an absorption length greater than 18 m.

013.083 **Astronomy: When big is beautiful.**
S. R. Kulkarni, M. Shao, C. A. Haniff.
Nature, Vol. 352, No. 6334, p. 383 – 384 (1 Aug 1991).
The authors report on projects concerning stellar interferometers.

013.084 **Gamma–ray bursts: Astronomy versus astrophysics.**
D. Lindley.
Nature, Vol. 354, No. 6348, p. 20 – 21 (7 Nov 1991).

013.085 **The Japanese National Large Telescope.**
T. Kogure, K. Kodaira.
Proc. Astron. Soc. Aust., Vol. 9, No. 1, p. 52 – 54 (1991). – See Abstr. 012.078 for the main entry.
The Japanese National Large Telescope is an 8–metre class optical–infrared reflector with a monolithic thin meniscus mirror, to be constructed at the Mauna Kea summit, Hawaii. The JNLT will be characterised by high quality performance in the optical and infrared regions, achieved by adopting new technologies such as active mirror support, fast optics and a thermally controlled dome. In particular, high infrared qualities are regarded as the most important charactristics among various design goals. The JNLT project is now close to the final study phase before construction. This paper reviews the scientific motivations and the special technical features of the JNLT. Finally, the promotion of international collaboration around the JNLT is emphasised.

013.086 **Prospects for an Antarctic Observatory.**
P. R. Gillingham.
Proc. Astron. Soc. Aust., Vol. 9, No. 1, p. 55 – 56 (1991). – See Abstr. 012.078 for the main entry.
Following a 1989 USA conference on astrophysics from Antarctica, a number of proposals have been advanced to exploit the great astronomical potential of this region. Near the highest point of the ice plateau (where the pressure altitude is about 5000 m), the extremely dry cold atmosphere will allow effective observation through many wavelength ranges currently inaccessible from the ground. There is also reason to believe that better seeing than is otherwise obtainable from the Earth's surface may

be achievable. Astronomers interested in future Antarctic observing should co–operate in presenting cases to their governments for support facilities there.

013.087 **Space astrophysics in Japan.**
K. Makishima.
Proc. Astron. Soc. Aust., Vol. 9, No. 1, p. 57 – 59 (1991). – See Abstr. 012.078 for the main entry.
This paper describes some of the space science programs currently underway and planned in Japan.

013.088 **Newly installed radio and optical telescopes in China.**
Wang Jingsheng.
Proc. Astron. Soc. Aust., Vol. 9, No. 1, p. 60 – 61 (1991). – See Abstr. 012.078 for the main entry.
New radio and optical telescopes installed in recent years in China are summarised. These include the 2.16–m optical telescope, the solar magnetic field telescope, the Miyun synthesis radio telescope, the 1.26–m infrared telescope (Beijing Astronomical Observatory), the 25–m radio telescope as the first station of China's VLBI network, the 1.56–m astrometric telescope (Shanghai Observatory), and the 13.7–m millimetre wave radio telescope.

013.089 **Asian–Pacific co–operation in optical astronomy.**
R. Cannon.
Proc. Astron. Soc. Aust., Vol. 9, No. 1, p. 62 – 63 (1991). – See Abstr. 012.078 for the main entry.
This paper discusses opportunities for collaboration in astronomy among the countries of the Asian–Pacific region.

013.090 **New Zealand astronomy in the 1990s.**
P. L. Cottrell.
Proc. Astron. Soc. Aust., Vol. 9, No. 1, p. 64 – 68 (1991). – See Abstr. 012.078 for the main entry.
There has been a dramatic increase in astronomical research output in New Zealand over the last decade. A brief overview of the existing New Zealand astronomical scene is followed by discussion of a number of new initiatives being proposed. In addition, for programs which require a long timebase of observations, extreme southerly latitudes or longitudinal coverage, New Zealand could provide a unique opportunity.

013.091 **Status of the Perth Observatory automated supernova search program.**
A. J. Williams, D. G. Blair, R. Burman, M. Evans, R. Martin, P. V. Birch, M. P. Candy, A. Verveer, R. Koch, M. J. Lynch, X. Dai, M. G. Zadnik, D. Foster.
Proc. Astron. Soc. Aust., Vol. 9, No. 1, p. 84 – 85 (1991). – See Abstr. 012.078 for the main entry.
The Perth Astronomy Research Group (PARG), consisting of members from Curtin University of Technology, Perth Observatory and the University of Western Australia, is in the process of developing an automated supernova search system, using the 61–cm Lowell–Perth reflector, a CCD camera and an 80386–based computer for image analysis. Computer control of the telescope and dome, a liquid–nitrogen–cooled CCD camera, and modified VISTA image analysis software will be completed in late 1990, allowing initial semi–automatic searching of external galaxies, together with CCD photometry of flare stars and newly discovered supernovae. Full–scale automation will be introduced subsequently, in collaboration with the Berkeley group. This paper describes the project, and reports on its current status.

013.092 **The ESO supernova program.**
E. Cappellaro, P. Bouchet, M. Della Valle, J. Danziger, C. Fransson, C. Gouiffes, L. Lucy, P. Mazzali, M. Philips, M. Turatto.
ESO/EIPC Workshop on Supernova 1987A and other Supernovae, p. 725 – 730 (Jun 1991). – See Abstr. 012.075 for the main entry.

013.093 **Strategy for the detection and study of other planetary systems and extrasolar planetary materials: 1990 – 2000.**
National Research Council, Washington, DC (USA). Commission on Physical Sciences, Mathematics, and Applications. Space Studies Board. Committee on Planetary and Lunar Exploration, National Academy Press, Washington, DC (USA). 94 p. (1990). ISBN 0–309–04193–7. Price US$ 15.00.
Contents: 1. Introduction. 2. Basic scientific goals. 3. Present understanding of the origin of planetary systems. 4. Observations of extrasolar planetary and protoplanetary material. 5. Observational requirements for identification of extrasolar–system planets. 6. Future investigations of precursor planetary systems. 7. Relation to other astronomical and astrophysical studies. 8. Recommendations.

013.094 **Scanning with charge–coupled devices.**
T. Gehrels.
Space Sci. Rev., Vol. 58, No. 3/4, p. 347 – 375 (Nov 1991).
The 0.91–m "Spacewatch Telescope" of the Steward Observatory of the University of Arizona on Kitt Peak is dedicated to scanning with charge–coupled devices (CCDs) during the dark half of the month. The author explored six modes of using CCDs for searches of gamma–ray bursters, debris in geosynchronous space, satellites of asteroids, brown dwarfs, the tenth planet, comets, cometesimals, and various types of asteroids. In the process, the author gained experience with cosmic rays and artifacts in CCD observations. Each of these topics is described. The author especially notes that the existence of cometesimals has not been confirmed by the Spacewatch Telescope, contrary to reports published by others.

013.095 **Light pollution: "light" at the end of the tunnel?**
R. Lewis.
J. R. Astron. Soc. Can., Vol. 85, No. 4, p. 175 (Aug 1991). Abstract. – See Abstr. 012.090 for the main entry.

013.096 **Program at ITP: helioseismology – probing the interior of a star.**
D. Gough, J. Toomre.
Conference on Challenges to Theories of the Structure of Moderate–Mass Stars, p. 1 – 9 (1991). – See Abstr. 012.096 for the main entry.
The research program in helioseismology carried out at the Institute for Theoretical Physics (ITP) at the University of California, Santa Barbara during the six–month interval from January to June 1990 involved 61 scientists of diverse disciplines: theorists, observers and instrumentalists in physics, astrophysics and geophysics. The main topics of research and joint discussion included mode excitation and decay, the internal structure of the Sun and its sensitivity to the physics of the equation of state and opacity, seismological inverse procedures, and solar rotation and convection–zone dynamics.

013.097 **Prospects of experimental gravitational wave research in Moscow State University.**
V. N. Rudenko.
Elizabeth and Frederick White Research Conference: Gravitational astronomy – instrument design and astrophysical prospects, p. 211 – 216 (1991). – See Abstr. 012.098 for the main entry.
A brief outline of the current status of and future prospects for gravitational wave research in the USSR is presented.

013.098 **Favourites of astronomers.**
Calendar of the Tartu Astronomical Observatory for the year, 1992.
p. 30 (1991). In Estonian. See Abstr. 046.016.

013.099 **The progress of solar–terrestrial sciences in China.**
Hu Wenrui, Fu Zhufeng.
Chin. Sci. Bull., Vol. 36, No. 17, p. 1409 – 1418 (Sep 1991).
The solar–terrestrial sciences involve solar physics, interplanetary physics, magnetospheric physics, ionospheric physics, thermospheric and atmospheric physics and chemistry, and the related field of Earth sciences. However, more emphasis will be laid on the correlations between different regions of the solar–terrestrial system, and the cause and effect connection of the related processes.

013.100 **Progress on light pollution control at the David Dunlap Observatory.**
C. T. Bolton.
J. R. Astron. Soc. Can., Vol. 85, No. 4, p. 216 – 217 (Aug 1991). Abstract. – See Abstr. 012.091 for the main entry.

013.101 **Rendez nous notre ciel nocturne! (Chapitre 2).**
B. Rousseau.
Obs. Trav., No. 28, p. 2 – 4 (1991).

013.102 **Challenges to theories of the structure of moderate mass stars.**
I. W. Roxburgh.
Conference on Challenges to Theories of the Structure of Moderate-Mass Stars, p. 411 – 414 (1991). – See Abstr. 012.096 for the main entry.
Closing remarks to the conference held in Santa Barbara, CA (USA), 19 – 22 June 1990.

013.103 **Max '91/FLARES 22.**
B. R. Dennis.
International SOLAR-A Science Meeting: Flare physics in solar activity maximum 22, p. 305 – 311 (1991). – See Abstr. 012.104 for the main entry.
The purpose of this paper is to describe in some detail the various aspects of the Max '91 program, the observational capabilities, the campaign mode of operating, and the coordination and communications facilities.

013.104 **Relazione della commissione UAI sull'inquinamento luminoso: attività svolta.**
P. Cinzano.
Astronomia UAI, N. 10, p. 23 – 24 (Nov – Dec 1991).

013.105 **Bilan et perspectives du programme VSNA en astronomie.**
D. Alloin, P. Léna.
J. Astron. Fr., No. 40, p. 5 – 10 (Jul 1991).

013.106 **A quinquennial observational report of A.L.P.O. remote planets section for the years 1985 – 1989.**
R. G. Hodgson.
Strolling Astron., Vol. 35, No. 3, p. 97 – 99 (Sep 1991).

013.107 **Radio–stars and SAO–stars observations with the Repsold Meridian Circle of the "Felix Aguilar" Astronomical Observatory.**
R. A. Carestia, W. L. Castro, M. Gallego, C. C. Mallamaci, L. F. Marmolejo, J. L. Navarro, J. A. Perez, J. A. Sedeño.
Bol. Asoc. Argent. Astron., No. 36, p. 245 – 247 (1990). In Spanish. – See Abstr. 012.095 for the main entry.
General information is supplied about the present conditions of radio–stars and SAO–stars observation program, which are being carried out at the "Felix Aguilar" Astronomical Observatory, with the Repsold Meridian Circle.

013.108 **VLBI in Australia – a review.**
D. L. Jauncey.
Aust. J. Phys., Vol. 44, No. 6, p. 785 – 803 (1991). Paper presented at the Monglolo Observatory 25th Anniversary Symposium, Univ. Sydney, 22 – 23 Nov 1990.
After two decades of Australian VLBI, high-resolution radio astronomy continues to be an active and fruitful research field. The status of Australian VLBI programs in astrophysics, astrometry and geodesy is reviewed and likely future developments are outlined.

013.109 **Dispersión de velocidades en galaxias elípticas.**
J. H. Calderón, S. Paolantonio, M. V. Alonso, P. S. Pellegrini.
Rev. Mex. Astron. Astrofis., Vol. 22, No. 2, p. 304 (Oct 1991). Abstract. – See Abstr. 012.127 for the main entry.

013.110 **Corrimiento al rojo de galaxias australes.**
J. H. Calderón, P. S. Pellegrini, C. Willmer, L. N. da Costa.
Rev. Mex. Astron. Astrofis., Vol. 22, No. 2, p. 304 (Oct 1991). Abstract. – See Abstr. 012.127 for the main entry.

013.111 **The international project CANGAROO at Woomera.**
J. R. Patterson.
Aust. J. Astron., Vol. 4, No. 2, p. 37 – 41 (Oct 1991).
Very high energy gamma ray astronomy is at a crucial stage of development. While present results on several sources are encouraging, they are not considered conclusive by other astronomers, except for the Crab. Project CANGAROO is an international collaboration between The University of Adelaide and five Japanese institutes. A 3.8–m astronomical telescope is now located 100 m east of BIGRAT, the Adelaide telescope. Both will produce stereo images using small fast photomultipliers to distinguish between gamma ray– and cosmic ray–initiated Cerenkov showers.

Traek of Astronomisk Selskabs historie gennem 75 år.
See Abstr. 004.087.

Following the Sun.
See Abstr. 014.016.

Time synchronization in China using current techniques.
See Abstr. 034.177.

New reduction of Earth rotation parameters and Hipparcos reference frame.
See Abstr. 044.038.

Common mechanisms in the solar–terrestrial system – part of the STEP international program (a review).
See Abstr. 085.003.

The AAT SN 1987A data archive.
See Abstr. 125.295.

Gravitational microlensing and the Hamburg quasar monitoring program.
See Abstr. 159.102.

014 Teaching in Astronomy

014.001 **Cepheid variables: a spectroscopic study, phase III.**
B. W. Thomas.
I.A.P.P.P. Commun., No. 45, p. 104 – 107 (Sep – Nov 1991).

014.002 **Snapping the planets electronically.**
R. Berry.
Astronomy, Vol. 19, No. 8, p. 54 – 55 (Aug 1991).

014.003 **Glorious Universe.**
R. Burnham, A. Schaller.
Astronomy, Vol. 19, No. 10, p. 30 – 43 (Oct 1991).

014.004 **Study of the effect of relativistic time dilation on cosmic ray muon flux – an undergraduate modern physics experiment.**
N. Easwar, D. A. MacIntire.
Am. J. Phys., Vol. 59, No. 7, p. 589 – 592 (Jul 1991). Current Physics Microform No.: 9107A1853.

An experiment to study the effect of relativistic time dilation on secondary muon fluxes observed at different altitudes is described in this article. Muons, produced as secondary particles from the interaction of primary cosmic rays with the upper atmosphere, form a natural and abundant source of subatomic "clocks" moving at very high speeds. The measured muon flux on a mountain relative to that measured at sea level can be compared to predictions from calculations that take into account the relativistic time dilation in the muon frame. Situations under which such an experiment can be successfully performed are explored. This experiment has been developed as a module in an undergraduate advanced laboratory course.

014.005 **Sonnenbeobachtung mit der Lochkamera.**
K.-L. Bath.
Sterne Weltraum, Jahrg. 30, Nr. 7, p. 444 – 446 (Jul 1991).

014.006 **Our nearest star.**
I. Nicolson.
Astron. Now, Vol. 5, No. 7, p. 16 – 17 (Jul 1991).

014.007 **Sterne modellieren im PC. Astrophysik für den Amateur.**
C. Wolf.
Sterne Weltraum, Jahrg. 30, Nr. 8–9, p. 510 – 516 (Aug–Sep 1991).

014.008 **Das Sonnenplanetarium nach Eudoxos. Ein Modell zur Darstellung der scheinbaren Bewegung der Sonne um die Erde.**
H. Paulus.
Sterne Weltraum, Jahrg. 30, Nr. 8–9, p. 517 – 518 (Aug–Sep 1991).

014.009 **Planetesimals.**
I. Nicolson.
Astron. Now, Vol. 5, No. 8, p. 16 – 17 (Aug 1991).

014.010 **Optical telescopes.**
I. Nicolson.
Astron. Now, Vol. 5, No. 9, p. 16 – 17 (Sep 1991).

014.011 **Astronomi i østtyske skoler.**
K. Lindner.
Astron. Tidsskr., Årg. 24, Nr. 3, p. 148 – 150 (1991). Translated by K. Johansen.

014.012 **Quasare in der Mittelstufe. Was Zehntklässler schon immer über PG 0026 + 129 wissen wollten.**
F. W. Volck.
Sterne Weltraum, Jahrg. 30, Nr. 10, p. 598 – 599 (Oct 1991).

014.013 **Light and the spectrum.**
I. Nicolson.
Astron. Now, Vol. 5, No. 10, p. 16 – 17 (Oct 1991).

014.014 **Through rose coloured glasses.**
D. Trombino.
Astron. Now, Vol. 5, No. 10, p. 22 – 25 (Oct 1991).

014.015 **Stars of many kinds.**
I. Nicolson.
Astron. Now, Vol. 5, No. 11, p. 16 – 17 (Nov 1991).

014.016 **Following the Sun.**
N. Morgan.
Astron. Now, Vol. 5, No. 11, p. 42 – 43 (Nov 1991).

014.017 **Algol für Schulklassen.**
U. Bastian.
Sterne Weltraum, Jahrg. 30, Nr. 12, p. 739 – 741 (Dec 1991).

014.018 **Astrometrische Beobachtungen von Kometen.**
E. Meyer, H. Raab.
Sterne Weltraum, Jahrg. 30, Nr. 12, p. 772 – 773 (Dec 1991).

014.019 **Life and times of the stars.**
I. Nicolson.
Astron. Now, Vol. 5, No. 12, p. 16 – 17 (Dec 1991).

014.020 **The orbit of Mercury.**
M. B. Steward.
Phys. Teach., Vol. 29, No. 6, p. 346 – 347 (Sep 1991). Abstract. Current Physics Microform No.: 9111F1704.

014.021 **Daytime observations for astronomy students I and II.**
N. L. Lark.
Phys. Teach., Vol. 29, No. 7, p. 459 – 460 (Oct 1991). Abstract. Current Physics Microform No.: 9111F1812.

014.022 **Daytime observations for astronomy students III.**
J. Hickman, N. L. Lark.
Phys. Teach., Vol. 29, No. 8, p. 500 – 501 (Nov 1991). Abstract. Current Physics Microform No.: 9201D2194.

014.023 **Integrating observations into astronomy education.**
D. B. Friend.
Bull. Am. Astron. Soc., Vol. 23, No. 2, p. 925 – 926 (1991). Abstract. – See Abstr. 012.008 for the main entry.

014.024 **Visualization and geometrical transformation in elementary astronomy: 2500 years of computer–aided instruction.**
D. N. Brown, J. Evans.
Bull. Am. Astron. Soc., Vol. 23, No. 2, p. 926 (1991). Abstract. – See Abstr. 012.008 for the main entry.

014.025 **Interactive videotape with HyperCard.**
J. W. Allen.
Bull. Am. Astron. Soc., Vol. 23, No. 2, p. 926 (1991). Abstract. – See Abstr. 012.008 for the main entry.

014.026 **The science requirement blues: astronomy as a tonic.**
R. B. Partridge.
Bull. Am. Astron. Soc., Vol. 23, No. 2, p. 929 – 930 (1991). Abstract. – See Abstr. 012.008 for the main entry.

014.027 **Astronomical computing: North America's solar eclipse.**
R. W. Sinnott.
Sky Telesc., Vol. 82, No. 1, p. 71, 73 – 74 (Jul 1991).

014.028 **Concepts of the cosmos.**
G. S. Mumford.
Bull. Am. Astron. Soc., Vol. 23, No. 2, p. 930 (1991). Abstract. – See Abstr. 012.008 for the main entry.

014.029 **The relevance of astronomy in a liberal arts education.**
K. Mukherjee.
Bull. Am. Astron. Soc., Vol. 23, No. 2, p. 930 (1991). Abstract. – See Abstr. 012.008 for the main entry.

014.030 **Access to contemporary research results for teachers of astronomy courses on the school and university levels.**
J. M. Pasachoff.
Bull. Am. Astron. Soc., Vol. 23, No. 2, p. 930 (1991). Abstract. – See Abstr. 012.008 for the main entry.

014.031 **Astronomical computing: Indexes for star charts.**
T. Hunter.
Sky Telesc., Vol. 82, No. 2, p. 182 (Aug 1991).

014.032 **Astronomical computing: a spreadsheet for polar alignment.**
G. C. Baldwin.
Sky Telesc., Vol. 82, No. 3, p. 299 – 301 (Sep 1991).

014.033 **The Martian moon dilemma: implications of a study of spatial reasoning ability of students taking introductory college astronomy.**
D. Schatz.
Bull. Am. Astron. Soc., Vol. 23, No. 2, p. 935 (1991). Abstract. – See Abstr. 012.008 for the main entry.

014.034 **Methods of astronomy: an activities class.**
M. K. Hemenway.
Bull. Am. Astron. Soc., Vol. 23, No. 2, p. 935 (1991). Abstract. – See Abstr. 012.008 for the main entry.

014.035 **An undergraduate astronomy course for non–major honors students.**
D. G. Iadevaia.
Bull. Am. Astron. Soc., Vol. 23, No. 2, p. 935 (1991). Abstract. – See Abstr. 012.008 for the main entry.

014.036 **The UC Santa Barbara remote access telescope project.**
P. Lubin.
Bull. Am. Astron. Soc., Vol. 23, No. 2, p. 935 (1991). Abstract. – See Abstr. 012.008 for the main entry.

014.037 **Do atoms really "emit" absorption lines?**
K. Brecher.
Bull. Am. Astron. Soc., Vol. 23, No. 2, p. 936 (1991). Abstract. – See Abstr. 012.008 for the main entry.

014.038 **Visual limiting magnitude determination charts.**
A. McBeath.
J. Br. Astron. Assoc., Vol. 101, No. 4, p. 213 – 218 (Aug 1991).

A set of 14 small, epoch 2000.0, star charts, suitable for determining naked–eye limiting magnitudes down to +6.5 mag in various areas of the sky and from most latitudes throughout the world, is presented.

014.039 **The visibility of the aurora borealis.**
R. J. Livesey.
J. Br. Astron. Assoc., Vol. 101, No. 4, p. 233 – 234 (Aug 1991).

The Aurora Section regularly receives enquiries from the public as to where and when they might go and see the aurora borealis. This paper looks at the currently available observing statistics in an attempt to answer this question.

014.040 **Perturbations – a simple approach.**
D. Tattersfield.
J. Br. Astron. Assoc., Vol. 101, No. 5, p. 292 – 293 (Oct 1991).

014.041 **Research in the high school classroom: an informal account of a new way to learn science.**
J. Lockwood.
I.A.P.P.P. Commun., No. 46, p. 57 – 62 (Dec 1991 – Feb 1992).

014.042 **Sterngeburt.**
H. Zimmermann.
Astron. Sch., Jahrg. 28, Heft 4, p. 4 – 5 (Aug 1991).

014.043 **Computer in der astronomischen Bildung.**
J. Lichtenfeld.
Astron. Sch., Jahrg. 28, Heft 4, p. 15 – 16 (Aug 1991).

014.044 **Zum Anblick der Sonne von den Planeten des Sonnensystems aus.**
H. J. Nitschmann.
Astron. Sch., Jahrg. 28, Heft 4, p. 17 – 18 (Aug 1991).

014.045 **Sonnenbeobachtung mit Schülern der Sekundarstufe I.**
P. Kriesel.
Astron. Sch., Jahrg. 28, Heft 4, p. 19 – 21 (Aug 1991).

014.046 **Zur Veranschaulichung astronomischer Größen.**
J. Naumann.
Astron. Sch., Jahrg. 28, Heft 5, p. 11 – 12 (Oct 1991).

014.047 **Langzeitbeobachtungen des Mondes.**
P. Kriesel.
Astron. Sch., Jahrg. 28, Heft 5, p. 21 – 26 (Oct 1991).

014.048 **Der Radiohimmel auch für Amateure.**
K. Hopf, W. Wolfram, F. Fischer.
Astron. Sch., Jahrg. 28, Heft 5, p. 32 – 33 (Oct 1991).

014.049 **The Colorado scale–model solar system.**
J. O. Bennett, T. R. Ayres, K. B. Center, M. F. Carter, R. S. Bass.
Phys. Teach., Vol. 29, No. 6, p. 371 – 374 (Sep 1991). Abstract. Current Physics Microform No.: 9111F1729.

014.050 **A demonstration of the apparent ellipticity of the rising or setting Sun.**
Cui Jinsong.
Phys. Teach., Vol. 29, No. 9, p. 566 (Dec 1991). Abstract. Current Physics Microform No.: 9201E2166.

014.051 **Science education in a planetarium.**
M. Othman.
Proc. Astron. Soc. Aust., Vol. 9, No. 1, p. 69 – 71 (1991). – See Abstr. 012.078 for the main entry.

This paper discusses the role of the planetarium as an indispensable means of teaching astronomy and as an increasingly popular way of teaching science.

014.052 **Proposed structure of education in astronomy.**
S. Isobe.
Proc. Astron. Soc. Aust., Vol. 9, No. 1, p. 72 – 75 (1991). – See Abstr. 012.078 for the main entry.

This paper discusses the importance of having a structured approach to educating the general public in astronomy. It outlines the role schools, planetaria and national observatories must play in this process. It also addresses the question of how astronomical events are best exploited to active the public's interest in astronomy.

014.053 **Edinburgh astronomy and educational teaching packages.**
A. Savage, S. Tritton, M. Brück.
Proc. Astron. Soc. Aust., Vol. 9, No. 1, p. 155 (1991). – See Abstr. 012.078 for the main entry.

Film copies of original astronomical plates taken with the 1.2–m UK Schmidt Telescope at Siding Spring Observatory, New South Wales have been used for several years at Edinburgh

University. Two teaching packages are intended for undergraduate use; the educational packages are mainly designed as visual aids for colleges, schools and amateur groups.

014.054 **Tertiary astronomy programs in New Zealand.**
P. L. Cottrell.
Proc. Astron. Soc. Aust., Vol. 9, No. 1, p. 172 – 173 (1991). – See Abstr. 012.078 for the main entry.
A number of New Zealand universities offer astronomy courses at different undergraduate levels. The courses are taught in Physics or Mathematics Departments. Some are only sub-courses within a physics major rather than a dedicated astronomy course. The author details these and shows material relating to the course content and some of the texts used. Postgraduate programs in astronomy are also offered, principally at the University of Canterbury but also at the University of Auckland and Victoria University of Wellington.

014.055 **The Carter Observatory education program.**
F. Andrews.
Proc. Astron. Soc. Aust., Vol. 9, No. 1, p. 174 (1991). – See Abstr. 012.078 for the main entry.
Over the past ten years a comprehensive educational program has been developed at Carter Observatory which covers every level from pre–school to PhD. A great deal of time also is expended in a varied adult education program. Every effort has been made to co–ordinate the various parts of the program in such a way that people will be encouraged to take a productive interest in astronomy.

014.056 **China's postgraduate education in astronomy.**
Liu Lin.
Proc. Astron. Soc. Aust., Vol. 9, No. 1, p. 178 – 179 (1991). – See Abstr. 012.078 for the main entry.
This paper summarises the opportunities available for postgraduate education in astronomy and related disciplines in China.

014.057 **Problems with teaching astronomy in Japan.**
R. Miyawaki.
Proc. Astron. Soc. Aust., Vol. 9, No. 1, p. 184 (1991). – See Abstr. 012.078 for the main entry.
By 1989 a new curiculum in Japanese elementary and secondary schools had been devised and started. The author reports on the contents of the new curriculum and point out some problems in teaching astronomy in Japan identified from the results of recent research in science education. Recent research shows that it is important to know how children's ideas and misconceptions are constructed and what role the philosophy of science may play in shaping them.

014.058 **Education in astronomy: some suggestions.**
G. L. Kalra.
Proc. Astron. Soc. Aust., Vol. 9, No. 1, p. 185 (1991). – See Abstr. 012.078 for the main entry.
Astronomy is one of the most ancient disciplines of knowledge which has inspired a great deal of human thought. Its study has triggered developments in many fields. Bohr's atomic model and our idea about the thermonuclear fusion are two of the many examples which have significantly influenced (and will continue to influence) the course of human civilisation.

014.059 **Astronomy and astrophysics: a targeted approach.**
W. J. Zealey.
Proc. Astron. Soc. Aust., Vol. 9, No. 1, p. 189 (1991). – See Abstr. 012.078 for the main entry.
Astronomy and astrophysics subjects at the University of Wollongong are seen as a vehicle for emphasising and clarifying concepts in the physics degree course. They also serve to introduce both undergraduate and postgraduate students to skills in computing and instrumentation not encountered in traditional subjects.

014.060 **IUCAA's role in astronomy education in India.**
N. Dadhich.
Proc. Astron. Soc. Aust., Vol. 9, No. 1, p. 190 (1991). – See Abstr. 012.078 for the main entry.
This paper summarises the function of the Inter–University Centre for Astronomy and Astrophysics in India.

014.061 **The bulletin "Teaching of Astronomy".**
S. Isobe.
Proc. Astron. Soc. Aust., Vol. 9, No. 1, p. 191 (1991). – See Abstr. 012.078 for the main entry.
The paper briefly describes a new publication in Japan, the bulletin "Teaching of Astronomy in the Asian–Pacific Region". The author is the Chairman of the Working Committee of Teaching of Astronomy in the Asian–Pacific region of the IAU.

014.062 **Astronomy education in a multi–lingual rural society.**
S. Ramadurai.
Proc. Astron. Soc. Aust., Vol. 9, No. 1, p. 192 (1991). – See Abstr. 012.078 for the main entry.
This paper examines the importance of astronomy education for a rural society in a third world country. Some conclusions are drawn about methods of introducing astronomy to school students and, through them, the entire village. The conclusions are based on the author's direct experience in addressing high school students in several rural towns in India.

014.063 **The search for planets around other stars.**
Universe Classroom, No. 19, p. 1 – 3 (Win 1991–1992).

014.064 **Knowledge of pupils in astrophysics at secondary schools in Czechoslovakia.**
V. Štefl.
Astron. Inst. Univ. Brno, Contrib., No. 29.
Universita J. E. Purkyne, Brno (Czechoslovakia). Astronomical Institute. 4 p. (1991). Reprinted from Scripta – Physics, Vol. 21, p. 35 – 38 (1991).
This paper deals with a detailed analysis of the knowledge and skills of pupils in astrophysics on the basis of the results of a didactic test.

014.065 **Comparative research of pupils' knowledge of astrophysics at secondary schools in Czechoslovakia and Poland.**
V. Štefl, J. Domański.
Astron. Inst. Univ. Brno, Contrib., No. 30.
Universita J. E. Purkyne, Brno (Czechoslovakia). Astronomical Institute. 10 p. (1991). Reprinted from Scripta – Physics, Vol. 21, p. 39 – 48 (1991).
The paper presents results of a didactic test of astrophysics, that had 20 multiple–choice items. The test was parallel set in the school year 1988/89 at five secondary schools in Czechoslovakia and at three lyceums in Poland.

014.066 **Teaching astronomy by distance education.**
M. Connors.
J. R. Astron. Soc. Can., Vol. 85, No. 4, p. 173 (Aug 1991). Abstract. – See Abstr. 012.090 for the main entry.

014.067 **Longitude from lunar distances.**
T. Hicks.
J. R. Astron. Soc. Can., Vol. 85, No. 4, p. 173 – 174 (Aug 1991). Abstract. – See Abstr. 012.090 for the main entry.

014.068 **Computer simulations of Mars and stars.**
J. R. Naiden.
J. R. Astron. Soc. Can., Vol. 85, No. 4, p. 175 – 176 (Aug 1991). Abstract. – See Abstr. 012.090 for the main entry.

014.069 **Attracting and training the next generation of Canadian astronomers.**
J. R. Percy.
J. R. Astron. Soc. Can., Vol. 85, No. 4, p. 221 (Aug 1991). Abstract. – See Abstr. 012.091 for the main entry.

014.070 **Animating the Hertzsprung–Russell diagram.**
J. Pocock, M. L. McCall.
J. R. Astron. Soc. Can., Vol. 85, No. 4, p. 221 (Aug 1991). Abstract. – See Abstr. 012.091 for the main entry.

014.071 **New tools for introducing astronomy to secondary school students.**
T. D. Stiff.
J. R. Astron. Soc. Can., Vol. 85, No. 4, p. 221 – 222 (Aug 1991). Abstract. – See Abstr. 012.091 for the main entry.

014.072 **Introducing the sky to students in astronomy survey courses.**
D. G. Turner.
J. R. Astron. Soc. Can., Vol. 85, No. 4, p. 222 (Aug 1991). Abstract. – See Abstr. 012.091 for the main entry.

014.073 **Le stelle doppie.**
P. Tempesti.
Astronomia UAI, N. 10, p. 5 – 10 (Nov – Dec 1991).
In the first part of this article on double stars, the author recalls the Kepler laws and the newtonian gravitational law to explain the orbit or these stars. Then he introduces the concept of absolute and relative orbital motion and the features of relative orbit.

014.074 **Acquisizione digitale ed elaborazione numerica di immagini planetarie.**
A. Leo, G. Quarra Sacco, D. Sarocchi.
Astronomia UAI, N. 10, p. 11 – 15 (Nov – Dec 1991).
The problems of planetary digital images collection are examined. The acquisition with a CCD camera and a cooled CCD are discussed together with the recorder system.

014.075 **Proposition d'une maitrise de "Sciences de l'univers" à l'Université de Toulouse.**
S. Vauclair.
J. Astron. Fr., No. 40, p. 11 – 13 (Jul 1991).

014.076 **About the conceptualization of the Earth–Sun–Moon system in six to eleven years old children.**
H. Tignanelli, M. Espíndola, F. Suarez, S. Durilén.
Bol. Asoc. Argent. Astron., No. 36, p. 209 – 224 (1990). In Spanish. – See Abstr. 012.095 for the main entry.
The authors show the methodology employed to gather and study the material that allow one to know the degree of conception of the sun–earth–moon system, in a group of children from 6 to 11 years. This method of investigation is in agreement with those used in similar studies made in Australia by B. Jones.

014.077 **Getting started: telescope selection.**
H. D. Jamieson.
Strolling Astron., Vol. 35, No. 4, p. 181 – 183 (Dec 1991).

014.078 **The Ooty summer training program 1990.**
D. J. Saikia, C. J. Salter, D. G. Banhatti, T. Ghosh, P. Gothoskar, P. K. Manoharan.
Bull. Astron. Soc. India, Vol. 19, Nos. 1 – 2, p. 109 – 122 (Mar – Jun 1991).

Amateur–professional co–operation in astronomy education: a nineteenth century Australian model.
See Abstr. 004.052.

Ein Präzisionsfernrohr für Kleinsternwarten: 100/1000 von Jenoptik Carl Zeiss Jena.
See Abstr. 032.006.

015 Miscellanea (Philosophical Aspects, Extraterrestrial Life, etc.)

015.001 **Demise of the dinosaurs: a mystery solved?**
R. S. Dietz.
Astronomy, Vol. 19, No. 7, p. 30 – 37 (Jul 1991).
Scientists are looking very closely at a buried structure in Mexico's Yucatan peninsula. It may well be the "smoking gun" of the asteroid impact that killed off the dinosaurs.

015.002 **Night views.**
S. Odenwald.
Astronomy, Vol. 19, No. 8, p. 34 – 37 (Aug 1991).
Archaeologists and astronomers, among all scientists, are kindred spirits whose work encompasses the two great extremes of our human limitations: time and space.

015.003 **Leben auf Planeten anderer Sonnensysteme?**
R. W. Schlögl.
Sterne Weltraum, Jahrg. 30, Nr. 7, p. 426 – 428 (Jul 1991).

015.004 **Drawing down the Moon.**
I. Seymour.
Astron. Now, Vol. 5, No. 7, p. 18 (Jul 1991).
The collapse of Aristotelian cosmology during the 16th century transformed our view of the Universe. The Moon and planets were no longer beyond the laws of earthly physics; they were worlds like our own.

015.005 **Catastrophism now.**
S. V. M. Clube, W. M. Napier.
Astron. Now, Vol. 5, No. 8, p. 46 – 49 (Aug 1991).
The idea that evolution advances by "astronomical punctuation" due to large impacts has been very popular in recent years. The latest finding, the dinosaur extinction is most likely to be associated with the more continuous débris of a giant comet alters our perception of "punctuated equilibrium". Thus our present environment probably also includes a giant comet in an advanced stage of disintegration and may therefore be much more hazardous than previously realised, not least through impacts, ozone depletion and sudden changes in global temperature.

015.006 **Needles in the cosmic haystack. The search for extraterrestrial intelligence (SETI).**
B. Rose.
Astron. Now, Vol. 5, No. 11, p. 19 – 24 (Nov 1991).

015.007 **Dust and light pollution.**
R. H. Garstang.
Bull. Am. Astron. Soc., Vol. 23, No. 2, p. 876 (1991). Abstract. – See Abstr. 012.008 for the main entry.

015.008 **Confessions of an astronomy textbook writer.**
M. Zeilik.
Bull. Am. Astron. Soc., Vol. 23, No. 2, p. 930 (1991). Abstract. – See Abstr. 012.008 for the main entry.

015.009 **S & T looks back... and ahead.**
T. Ferris, P. Morrison, M. Collins, C. Sagan, R. Jastrow, A. C. Clarke.
Sky Telesc., Vol. 82, No. 5, p. 464 – 471 (Nov 1991).

015.010 **La búsqueda de vida extraterrestre. Parte 1.**
A. Ballardini.
Rev. Astron., No. 246, p. 2 – 5 (Jun 1991).

015.011 **The astrological material of Zuozhuan and the ancient Chinese astrology.**
Li Yong.
Acta Astron. Sin., Vol. 32, No. 2, p. 215 – 221 (Jun 1991). In Chinese.

015.012 **Göttergeschichten am Himmel.**
K. Bargfried.
Astron. Sch., Jahrg. 28, Heft 5, p. 27 – 29 (Oct 1991).

015.013 **A list of near–ecliptical sun–like stars for the "ZODIAC" SETI–program.**
L. N. Filippova.
Astron. Tsirk., No. 1544, p. 37 – 38 (May – Jul 1990).

015.014 **The fight without defeat: the 1951 conference on planetary cosmogony.**
A. E. Levin.
Priroda, No. 9, p. 99 – 107 (Sep 1991). In Russian.

015.015 **Astronomy and human life.**
S. Ramadurai.
Proc. Astron. Soc. Aust., Vol. 9, No. 1, p. 76 – 77 (1991). – See Abstr. 012.078 for the main entry.
The role of astronomy in the cultural exchanges between India and China through several centuries is taken as an example of astronomy shaping human life. Then the tradition of astronomy in the development of scientific thought is examined. The contribution of astronomy in shaping the post–Renaissance society is briefly introduced. It is concluded that the recent developments of ideas in cosmology throw light on the intrinsic nature and limitations of the scientist as a human being.

015.016 **Science in the public eye.**
P. Pockley.
Proc. Astron. Soc. Aust., Vol. 9, No. 1, p. 78 – 79 (1991). – See Abstr. 012.078 for the main entry.

015.017 **Legal aspects of the search for extraterrestrial civilizations.**
G. V. Sil'vestrov.
Zemlya Vselennaya, No. 4, p. 75 – 79 (Jul – Aug 1991). In Russian.

015.018 **Communicating space science to the public.**
K. A. Dougherty, M. L. James.
Proc. Astron. Soc. Aust., Vol. 9, No. 1, p. 175 – 176 (1991). – See Abstr. 012.078 for the main entry.
A version of this paper was originally presented at the Fifth National Space Engineering Symposium at Canberra in November 1989. A copy of the full paper will be found in the text of the Proceedings. It is presented here because the authors believe that the issues raised and strategies suggested are as relevant to the astronomical community as they are to other areas of space science and technology.

015.019 **On the problem of the method for scientific reconstruction of the history of exact sciences.**
V. M. Rozin.
Istor.–Astron. Issled., Tom 21, p. 213 – 228 (1989). In Russian. – See Abstr. 003.017 for the main entry.

015.020 **The great mnemonics contest.**
O. Gingerich.
Liller Symposium: Asteroids to quasars, p. 263 – 265 (1991). – See Abstr. 012.087 for the main entry.

015.021 **Tides in a space colony.**
J. Fukue.
Publ. Astron. Soc. Jpn., Vol. 43, No. 4, p. 657 – 660 (1991).
The author examines the tides of a cylindrical ocean in a space colony. When the rotation axis of the colony is perpendicular to the direction to the Earth, the shape of the cylindrical ocean becomes an ellipse whose major axis is perpendicular to the direction to the Earth. On the other hand, in the case of tides in an ocean of the Earth, the major axis is parallel to the direction to the Moon. In both cases, however, a full tide takes place at a point under the body which exerts the tidal force.

015.022 **Cosmology at the crossroad: science or philosophy?**
U. Curi.
Mem. Soc. Astron. Ital., Vol. 62, No. 3, p. 419 – 425 (1991). – See Abstr. 012.093 for the main entry.

015.023 **Representations of astronomers in literature.**
R. D. Haynes.
Astron. Q., Vol. 8, No. 3, p. 131 – 176 (1991).

015.024 **The teaching of astronomy in Platon's *Republic*.**
A. Sinachopoulos, D. Sinachopoulos.
Astron. Q., Vol. 8, No. 3, p. 181 – 187 (1991).

015.025 **Laplace's demon in the relativistic universe.**
M. Heller.
Astron. Q., Vol. 8, No. 4, p. 219 – 243 (1991).

015.026 **Cosmology and epistemology.**
L. Gratton.
Mem. Soc. Astron. Ital., Vol. 62, No. 3, p. 427 – 432 (1991). – See Abstr. 012.093 for the main entry.
After recalling Hume's criticism of Berkeley's philosophy, the author expresses the view that the more reasonable way to understand the meaning of science (and particularly of cosmology) is that of the great Swiss psychologist Jean Piaget's "*Genetic epistemology*" which moves from the observation of how the fundamental concepts of space, time and causality form and develop in the minds of single individuals from childhood to adult age.

015.027 **Infinity, finite and error.**
P. Zellini.
Mem. Soc. Astron. Ital., Vol. 62, No. 3, p. 455 – 463 (1991). – See Abstr. 012.093 for the main entry.
Philosophical, mathematical, and historical aspects are discussed.

015.028 **How can philosophy cope with changing fashions in science.**
H. Arp.
Mem. Soc. Astron. Ital., Vol. 62, No. 3, p. 465 – 469 (1991). – See Abstr. 012.093 for the main entry.
The author believes that science would give us the facts which philosophy would link together in a system meaningful to human beings.

015.029 **Boundaries of the world and ubiquity of the mind.**
C. Sini.
Mem. Soc. Astron. Ital., Vol. 62, No. 3, p. 471 – 475 (1991). – See Abstr. 012.093 for the main entry.
Philosophical aspects of cosmological thinking are traced.

015.030 **1980's cosmology and bioastronomy: an Odyssean perspective.**
J. Heidmann.
Mem. Soc. Astron. Ital., Vol. 62, No. 3, p. 487 – 497 (1991). – See Abstr. 012.093 for the main entry.
Cosmology is the study of the universe in its entirety, on a grand space and time scale. Bioastronomy is the endeavour to document the emergence and evolution of life from a cosmological standpoint. Both have a major philosophical impact because the cosmos is our home, our haven, substratum. Our fate is linked to its own through a fundamental symbiosis.

015.031 **Some remarks on cosmology and biology.**
F. Hoyle.
Mem. Soc. Astron. Ital., Vol. 62, No. 3, p. 513 – 518 (1991). – See Abstr. 012.093 for the main entry.
Cosmological theories are based almost wholly on only three facts: the redshifts of galaxies, the existence of the so–called microwave background and the synthesis of three light nuclei. Even though these facts are powerful and startling, a small number of facts alone cannot determine the structure of the entire universe uniquely. The author briefly touches upon microorganisms, fossils, and the *anthropic principle*.

015.032 **How the world ceased to be eternal: Christian objections to pagan cosmology.**
R. Bodei.
Mem. Soc. Astron. Ital., Vol. 62, No. 3, p. 615 – 629 (1991). – See Abstr. 012.094 for the main entry.

015.033 **The cosmic experiment.**
L. Gratton.
Mem. Soc. Astron. Ital., Vol. 62, No. 3, p. 631 – 639 (1991). – See Abstr. 012.094 for the main entry.
The author tries to explore which is or may be the position of cosmology among the various branches of science, but somebody denies that cosmology may be called a branch of science. This of course depends on what we agree to call "science" and on which kind of activity is cosmology.

015.034 **The arrow of time.**
P. Davies.
Mem. Soc. Astron. Ital., Vol. 62, No. 3, p. 641 – 652 (1991). – See Abstr. 012.094 for the main entry.
It has often been said that the history of philosophy is a series of footnotes to Plato. The essence of the problem addressed here – the nature and origin of the arrow of time – can indeed be traced back to some of the deepest examples of Platonic thought.

015.035 **Will the Universe expand for ever?**
M. Rees.
Mem. Soc. Astron. Ital., Vol. 62, No. 3, p. 653 – 673 (1991). – See Abstr. 012.094 for the main entry.
The author's theme is the cosmic evolution in the long–term future. Will the Universe go on expanding forever, or will it recollapse? Nobody knows. But the author tries to argue that it is not foolish nor presumptuous for us to ask such a cosmic question, and that research programmes now in progress may offer an answer within the next decade.

015.036 **The concept of the Universe and the problem of reality.**
V. Weidemann.
Mem. Soc. Astron. Ital., Vol. 62, No. 3, p. 675 – 688 (1991). – See Abstr. 012.094 for the main entry.
Contents: 1. Evolutionary Universe. 2. The inflationary Universe. 3. Some critical remarks. 4. From the Universe to universes. 5. Is cosmology a science? 6. The origin of the Universe – the mystery of time. 7. The problem of reality.

015.037 **Origin and evolution of the Universe. Final panel discussions I – III.**
J. Heidmann, D. W. Sciama, J. Casanovas.
Mem. Soc. Astron. Ital., Vol. 62, No. 3, p. 689 – 704 (1991). – See Abstr. 012.094 for the main entry.

015.038 **Remarques sur la confrontation entre la démarche scientifique et les positions spiritualistes.**
J. Schneider.
J. Astron. Fr., No. 40, p. 13 – 16 (Jul 1991).

015.039 **The Moon illusion and the height of the sky.**
K. D. Abhyankar.
Bull. Astron. Soc. India, Vol. 19, Nos. 1 – 2, p. 94 – 95 (Mar – Jun 1991). Abstract. – See Abstr. 012.108 for the main entry.

015.040 **Les cultes solaires.**
J. Duchesne–Guillemin.
Ciel, Vol. 53, p. 309 – 319 (Nov 1991). In French.

015.041 **ETI, SETI und die Öffentlichkeit heute.**
R. Pinotti.
Sterne, Band 67, Heft 5, p. 271 – 278 (1991).

015.042 **Cosmologia e teologia. Le vie della loro mutua interazione.**
W. R. Stoeger.
G. Astron., Vol. 17, N. 1 – 2, p. 15 – 19 (Mar – Jun 1991). Paper presented at "La cosmologia nella cultura dell '900". Conferenza Internazionale nel centenario della nascita di Hubble, Torino (Italy), 21 – 23 Nov 1990.

015.043 **La teologia di fronte alla cosmologia.**
J. Casanovas.
G. Astron., Vol. 17, N. 1 – 2, p. 20 – 23 (Mar – Jun 1991). Paper presented at "La cosmologia nella cultura dell '900". Conferenza Internazionale nel centenario della nascita di Hubble, Torino (Italy), 21 – 23 Nov 1990.

015.044 **Il dio dei teologi e il dio dei cosmologi. Apologia di Bellarmino.**
M. Pera.
G. Astron., Vol. 17, N. 1 – 2, p. 31 – 39 (Mar – Jun 1991). Paper presented at "La cosmologia nella cultura dell '900". Conferenza Internazionale nel centenario della nascita di Hubble, Torino (Italy), 21 – 23 Nov 1990.

015.045 **Cosmologia ed evoluzione.**
C. Sini.
G. Astron., Vol. 17, N. 1 – 2, p. 39 – 41 (Mar – Jun 1991). Paper presented at "La cosmologia nella cultura dell '900". Conferenza Internazionale nel centenario della nascita di Hubble, Torino (Italy), 21 – 23 Nov 1990.

015.046 **Filosofia e cosmologia: alcune considerazioni sul problema del realismo.**
F. Minazzi.
G. Astron., Vol. 17, N. 1 – 2, p. 41 – 48 (Mar – Jun 1991). Paper presented at "La cosmologia nella cultura dell '900". Conferenza Internazionale nel centenario della nascita di Hubble, Torino (Italy), 21 – 23 Nov 1990.

015.047 **Macchine del tempo cosmiche.**
F. de Felice.
G. Astron., Vol. 17, N. 1 – 2, p. 53 – 54 (Mar – Jun 1991). Paper presented at "La cosmologia nella cultura dell '900". Conferenza Internazionale nel centenario della nascita di Hubble, Torino (Italy), 21 – 23 Nov 1990.

015.048 **Qualche aspetto filosofico delle rivoluzioni in cosmologia.**
P. Budinich.
G. Astron., Vol. 17, N. 3 – 4, p. 66 – 73 (Sep – Dec 1991). Paper presented at "La cosmologia nella cultura dell '900". Conferenza Internazionale nel centenario della nascita di Hubble, Torino (Italy), 21 – 23 Nov 1990.

015.049 **MegaSETI.**
J. Heidmann.
G. Astron., Vol. 17, N. 3 – 4, p. 74 – 77 (Sep – Dec 1991). Paper presented at "La cosmologia nella cultura dell '900". Conferenza Internazionale nel centenario della nascita di Hubble, Torino (Italy), 21 – 23 Nov 1990.

015.050 **Cosmologia e domande ultime: commenti sul confronto fra evoluzione del cosmo e teologia cristiana della creazione.**
G. Tanzella-Nitti.
G. Astron., Vol. 17, N. 3 – 4, p. 93 – 98 (Sep – Dec 1991). Paper presented at "La cosmologia nella cultura dell '900". Conferenza Internazionale nel centenario della nascita di Hubble, Torino (Italy), 21 – 23 Nov 1990.

015.051 **Oggi la cosmologia è una scienza?**
L. Geymonat.
G. Astron., Vol. 17, N. 1 – 2, p. 14 – 15 (Mar – Jun 1991). In Italian. Paper presented at "La cosmologia nella cultura dell '900". Conferenza Internazionale nel centenario della nascita di Hubble, Torino (Italy), 21 – 23 Nov 1990.

Our universes.
See Abstr. 003.004.

Mars and the development of life.
See Abstr. 003.033.

Project Ozma: the opening of the electromagnetic spectrum for SETI.
See Abstr. 004.015.

Kosmos – cosmology today between science and philosophy. Proceedings. 1. Venice Conference on Cosmology and Philosophy, Venice (Italy), 8 – 9 May 1987
See Abstr. 012.093.

Origin and evolution of the Universe. A Hubble Centenary Conference. Proceedings. 3. Venice Conference on Cosmology and Philosophy, Venice (Italy), 15 – 16 Dec 1989.
See Abstr. 012.094.

Computational problems and signal processing in SETI.
See Abstr. 036.217.

A simple algorithm for simulating the solar ultraviolet radiation at the Earth's surface: an application in determining the minimum erythema dose.
See Abstr. 085.031.

The distribution of solar flares and probable relations to biological effects.
See Abstr. 085.032.

Making Mars habitable.
See Abstr. 097.071.

Major themes in the relation between philosophy and cosmology.
See Abstr. 161.662.

Applied Mathematics, Physics

021 Mathematical Papers Related to Astronomy and Astrophysics, Computing

021.001 **Some further reduction formulas for certain classes of generalized multiple hypergeometric series arising in physical, astrophysical, and quantum chemical applications.**
H. M. Srivastava.
Astrophys. Space Sci., Vol. 181, No. 2, p. 195 – 202 (Jul 1991).
The object of this paper is to derive a further reduction formula for the multivariable hypergeometric function considered recently by A. V. Niukkanen and H. M. Srivastava from substantially more general identities involving multiple series with essentially arbitrary terms. Some interesting connections of the results considered here with those given in the literature, and some indication of their applicability, are also provided.

021.002 **A process of extrapolation of polar coordinates with data in changeable length.**
Xiao Naiyuan, Xia Yifei, Ding Yuerong.
J. Nanjing Univ., Vol. 27, No. 3, p. 456 – 466 (Jul 1991). In Chinese.
A method of separating polar coordinates into several components has been devised. According to this method polar coordinates can be extrapolated by means of data in changeable length. Based on a wealth of testing calculations an empirical criterion has been concluded for data length. With this criterion and by the method to extrapolate polar coordinates, the extrapolation precision of between 0.″017 and 0.″030 can be obtained in most of the cases.

021.003 **Linear and circular digital spectral analysis of serial data.**
R. B. Stothers.
Astrophys. J., Vol. 375, No. 1, p. 423 – 426 (1 Jul 1991).
Two methods of digital spectral analysis of unevenly sampled data are developed and illustrated here. One method uses a linear function of time (or space), the other uses circular functions. The circular method turns out to be essentially equivalent to a least–squares sine–wave analysis. The linear, anharmonic method uses only the field of real numbers and elementary algebraic operations, and hence it can be made computationally very fast and accurate. Both methods are very general, properly handling all kinds of time series ranging from simple series consisting only of the times of events to complicated series consisting of pulses with long duty cycles. The two methods are here applied to the analysis of annual mean relative sunspot numbers.

021.004 **Accretional evolution of a planetesimal swarm. 1. A new simulation.**
D. Spaute, S. J. Weidenschilling, D. R. Davis, F. Marzari.
Icarus, Vol. 92, No. 1, p. 147 – 164 (Jul 1991).
The authors describe a new simulation for planetary accretion which treats simultaneous accretion in many interacting heliocentric distance zones, rather than the single one characteristic of earlier models. The planetesimals are characterized by keplerian elements rather than a single "random velocity". This numerical code can follow the evolution of both the size distribution and the orbit element distribution of a planetesimal swarm from any initial distributions of sizes and orbits. The authors find that their accretion algorithm yields good agreement with the analytic solutions.

021.005 **Tidal disruption of inviscid planetesimals.**
A. P. Boss, A. G. W. Cameron, W. Benz.
Icarus, Vol. 92, No. 1, p. 165 – 178 (Jul 1991).
The authors develop a simple analytical model including self–gravity that yields a criterion implying tidal disruption is strongly dependent on the mass of planetesimal. They then present and compare numerical, 3D simulations of tidal disruption, based on the smoothed particle hydrodynamics (SPH) code. In part because each particle is represented by a carefully chosen spatial function, the SPH method offers a considerably more sophisticated method for representing continuous bodies than the hard–particle approach used by previous workers. The SPH code is also better suited to studying the tidal disruption problem than the finite difference code employed by Mizuno and Boss, because of the superior ability of the SPH code in following arbitrary deformations of two or more bodies and in calculating the orbits of objects on ballistic trajectories.

021.006 **The origin of the Moon and the single impact hypothesis. IV.**
A. G. W. Cameron, W. Benz.
Icarus, Vol. 92, No. 2, p. 204 – 216 (Aug 1991).
Previous papers in this series have described the smoothed particle hydrodynamics method which the authors have employed to explore the conditions in which a major planetary collision may have been responsible for the formation of the Moon. In this paper they have run 41 simulations of the planetary collisions; these constitute an exploration of some of the relevant parameter space. The results lead to a giant blowoff hypothesis. Not only can the primitive atmosphere escape, but so can large amounts of mantle material which flows over the surface of the orbiting disk and blows off in the forward direction. This greatly extends the range of interesting parameters that can be seriously considered in the Giant Impact hypothesis.

021.007 **Determining isometries of space–times using computer algebra.**
M. E. Araujo, T. Dray, J. E. F. Skea.
7. Latin American Symposium on Relativity and Gravitation, p. 264 – 270 (1991). – See Abstr. 012.005 for the main entry.
An algorithm is given for determining the isometry group of an arbitrary spacetime (in four dimensions). An example is given and the partial implementation of this algorithm using the symbolic manipulation package CLASSI is discussed.

021.008 **Helmholtz conditions for differential equations of any order.**
F. Pardo.
7. Latin American Symposium on Relativity and Gravitation, p. 512 – 518 (1991). – See Abstr. 012.005 for the main entry.
The Inverse Problem of the variational calculus is studied in its nonrestricted form for differential equations of any order. The Helmholtz conditions are written out explicitly, in such a way that they can be considered as a set of conditions over the

existence of a non–singular matrix, that can be interpreted as an integrating factor for the existence of an Action Integral.

021.009 **On the evaluation of the Fermi–Dirac integrals.**
R. P. Sagar.
Astrophys. J., Vol. 376, No. 1, p. 364 – 366 (20 Jul 1991).
The author illustrates how Gaussian quadratures obtained from the weight functions, $w(x) = x^k/(e^x+1)$, for $k = -1/2, 1/2, 3/2, 5/2$, on the interval $[0, \infty]$, may be used to accurately determine Fermi–Dirac integrals. The author gives numerical examples and lists the recursion coefficients and the 24 point quadrature rule for $k = -1/2$.

021.010 **Overview of CODE V™ development.**
T. I. Harris.
OSA International Lens Design Conference, p. 104 – 111 (1990). – See Abstr. 012.014 for the main entry.
This paper is part of a session that is aimed at briefly describing some of today's optical design software packages, with emphasis on the program's philosophy and technology. CODE V is the ongoing result of a development process that began in the 1960's; it is now the result of many people's efforts. This paper summarizes the roots of the program, some of its history, dominant philosophies and technologies that have contributed to its usefulness and some that drive its continued development.

021.011 **SYNOPSYS™ – a lens design computer program package.**
D. C. Dilworth.
OSA International Lens Design Conference, p. 112 – 115 (1990). – See Abstr. 012.014 for the main entry.
Originally conceived at a time when lens design was carried out on Marchand calculators, SYNOPSYS was one of the earliest and is now one of the largest design programs. SYNOPSYS was for many years the personal design tool of its author. Since it was then the only tool available, it was necessary to develop a comprehensive set of capabilities requiring much innovation and experimentation.

021.012 **Optimization of the optical transfer function.**
M. Kidger, P. Benham.
OSA International Lens Design Conference, p. 92 – 96 (1990). – See Abstr. 012.014 for the main entry.
A technique for optimizing the real part of the optical transfer function (OTF) is described, with an example illustrating the results to be obtained.

021.013 **Practical use of generalized simulated annealing optimization on microcomputers.**
G. K. Hearn.
OSA International Lens Design Conference, p. 186 – 191 (1990). – See Abstr. 012.014 for the main entry.
The practical use of Generalized Simulated Annealing (GSA) optimization is explored to determine additional benefits beyond the global minima location capability. In particular, a means of using the strengths of both GSA and Damped Least Squares (DLS) optimization techniques is examined in order to increase the amount of usable information and to enhance usefulness in a microcomputer environment.

021.014 **Quadric surfaces: some derivations and applications.**
E. W. Larkin.
OSA International Lens Design Conference, p. 222 – 231 (1990). – See Abstr. 012.014 for the main entry.
This paper examines some derivations and applications related to quadric surfaces. The classic lens design conicoids (planes, spheres, ellipsoids, paraboloids, and hyperboloids) are special cases of quadric surfaces. Facility with quadric surfaces has applications in optical engineering and analysis, aspheric shape specification, and reduction of coordinate measuring machine data. A numeric example supports all derivations and applications.

021.015 **A small–computer program for optical design and analysis written in 'C'.**
L. H. J. F. Beckmann.
OSA International Lens Design Conference, p. 254 – 261 (1990). – See Abstr. 012.014 for the main entry.

021.016 **Exact solutions for Hamiltonian systems using integer arithmetic.**
D. J. D. Earn, S. Tremaine.
Conference on Dynamics of Disc Galaxies, p. 137 – 141 (1991). – See Abstr. 012.028 for the main entry.
The authors present a method for modifying a traditional integration scheme so that if the original system is Hamiltonian then the computed system is also exactly Hamiltonian. The effect of this method is to provide the exact solution for a Hamiltonian system that is slightly different from the system one really wishes to integrate. The method is applicable to an important class of problems in Hamiltonian dynamics. In particular, it can be used to perform exact integrations of Hamiltonian systems that closely approximate the gravitational N–body problem.

021.017 **Iterated function systems and the dynamical representation of natural structures.**
M. Frame.
Meeting on Precision Photometry: Astrophysics of the Galaxy, p. 351 – 360 (1991). – See Abstr. 012.027 for the main entry.

021.018 **Experimental test of a new algorithm estimating a process with perturbations that are impossible to simulate.**
M. L. Lidov, L. M. Bakuma.
Cosmic Res., Vol. 29, No. 1, p. 105 – 116 (Jul 1991). English translation of Kosm. Issled., Tom 29, Yyp. 1, p. 115 – 126 (1991).

021.019 **Nonspherical boundary value problem by pseudodifferential operators.**
W. Keller.
IAG Symposium: Determination of the geoid, p. 299 – 310 (1991). – See Abstr. 012.035 for the main entry.

021.020 **Coerciveness and geometry in the linear gravimetric boundary value problem – reduced version.**
P. Holota.
IAG Symposium: Determination of the geoid, p. 339 – 348 (1991). – See Abstr. 012.035 for the main entry.

021.021 **Holes in boundary and out–of–boundary data.**
F. Sacerdote, F. Sansò.
IAG Symposium: Determination of the geoid, p. 349 – 356 (1991). – See Abstr. 012.035 for the main entry.

021.022 **A new direct solution of the transformation problem of Cartesian into ellipsoidal coordinates.**
M. Lapaine.
IAG Symposium: Determination of the geoid, p. 395 – 404 (1991). – See Abstr. 012.035 for the main entry.
The author found that from a geometrical point of view, this problem consists of finding the intersection point at the reference ellipsoid by orthogonal (Helmert's) projection of a space point to the ellipsoid surface. It is shown that this problem is equivalent to the problem of finding the roots of a fourth degree algebraic equation. By means of certain unconventional parametric representation of the ellipsoid, the author shows that one can easily and accurately calculate just the needed solution of the corresponding fourth degree equation.

021.023 **Stokes formula using Fast Fourier Techniques.**
G. Strang van Hees.
IAG Symposium: Determination of the geoid, p. 405 – 408 (1991). – See Abstr. 012.035 for the main entry.
The full paper is published in Manuscripta Geodaetica, Vol. 15, p. 235 – 239 (1990).

021.024 **Simulation of regional gravity field recovery from satellite gravity gradiometer data using collocation and FFT.**
D. Arabelos, C. C. Tscherning.
IAG Symposium: Determination of the geoid, p. 409 (1991). Abstract. – See Abstr. 012.035 for the main entry.

021.025 **The GEMT1 variance–covariance matrix, its characteristics and applications in geoid computations.**
M. van Gelderen, R. H. N. Haagmans.
IAG Symposium: Determination of the geoid, p. 410 – 421 (1991). – See Abstr. 012.035 for the main entry.

021.026 **Fast collocation.**
R. Barzaghi, G. Bottoni.
IAG Symposium: Determination of the geoid, p. 442 – 443 (1991). Abstract. – See Abstr. 012.035 for the main entry.

021.027 **Prediction of solar activity with a neural network and its effect on orbit prediction.**
K. E. Williams.
Johns Hopkins APL Tech. Dig., Vol. 12, No. 4, p. 310 – 317 (Oct – Dec 1991).

The variability of solar activity over time demonstrates highly complex behavior. Neural networks provide useful models for predicting that behavior. Accurate predictions of solar activity are essential for satellite tracking and constellation management, orbit lifetime studies, and mission planning for long–term orbiting platforms.

021.028 **Determination of the period of a variable signal from the times of characteristic events.**
I. L. Andronov.
Kinematics Phys. Celest. Bodies, Vol. 7, No. 2, p. 70 – 78 (1991). English translation of Kinematika Fiz. Nebesn. Tel, Tom 7, No. 2, p. 78 – 86 (1991).

A method considered for determination of the period of a variable signal from the times of "characteristic" points on the phase curve makes it possible to abandon calculation of the "continuous" periodogram in favor of a local–extreme search, which saves a great deal of machine time. Statistical periodograms are investigated using orthogonal Chebyshev functions. It is shown that the classical statistical criteria that are usually used when the values of the function to be smoothed are normally distributed cannot be used to estimate the significance of peaks on the periodogram because the phases are uniformly distributed when the trial frequency is incorrect. The algorithm is illustrated in the case of the eclipsing cataclysmic variable PG 0818 + 513 and a model random distribution of the times.

021.029 **Elements of a condition theory for the computational analysis of algorithms.**
F. Chatelin, V. Frayssé.
Journées 1991: Systèmes de référence spatio–temporels, p. 66 – 73 (1991). – See Abstr. 012.048 for the main entry.

021.030 **Techniques for N–body codes on the connection machine.**
P. Hertz, S. L. W. McMillan.
Symposium on the Formation and Evolution of Star Clusters as Part of the 102. Summer Scientific Meeting of the Astronomical Society of the Pacific, p. 401 – 403 (1991). – See Abstr. 012.050 for the main entry.

Direct summation N–body codes are necessary to understand certain dynamical issues in star cluster evolution, yet current codes run prohibitively slowly on serial machines for large N. Massively parallel SIMD computers allow significantly larger systems to be studied. The authors have implemented a simple Aarseth direct summation N–body code, for which the CPU time scales as O(N logN) for large N. For comparison, most high–precision serial codes scale asymptotically as $O(N^2)$.

021.031 **Two–point correlation function and estimate of its confidence intervals.**
V. S. Lebedev, I. A. Lebedeva.
Bull. Spec. Astrophys. Obs. – North Caucasus, Vol. 32, p. 120 – 127 (1990). English translation of Astrofiz. Issled. Izv. Spets. Astrofiz. Obs., Tom 32, p. 147 – 156 (1990).

A method of calculating the confidence intervals for numerical estimates of the values of correlation functions is described. Algorithms and examples of their use are given.

021.032 **Composition des mesures ou des données tabulées, composition des erreurs.**
P. Bacchus.
Journées 1991: Systèmes de référence spatio–temporels, p. 80 – 85 (1991). – See Abstr. 012.048 for the main entry.

This paper investigates how values and errors are composed when computing the value of a function from Tables. Two different forms of Tables are considered corresponding to the tabulated values of the function for given values of its variable or to the tabulated values of the coefficients of the development of the function into Chebyshev polynomials (as used in the "Connaissance des Temps") for a given interval of its variable.

021.033 **Collinearity in the linear model. Statistical and numerical alternatives to least squares. Proximal determination, ridge and bayesian estimators.**
M. L. Bougeard, C. Michelot.
Journées 1991: Systèmes de référence spatio–temporels, p. 86 – 91 (1991). – See Abstr. 012.048 for the main entry.

021.034 **Problèmes de convergence d'algorithmes dans la détermination de trajectoires perturbées.**
A. Bec–Borsenberger, M. L. Bougeard.
Journées 1991: Systèmes de référence spatio–temporels, p. 103 – 106 (1991). – See Abstr. 012.048 for the main entry.

The problem of the robustness of the algorithms used is studied in the case of the determination of trajectories of objects submitted to perturbations. The algorithm used is presented here and the numerical resolution is given; the notions of stability and robustness are distinguished and a procedure to obtain a robust solution is detailed.

021.035 **Erreurs et fautes dans les logiciels.**
J. Arsac.
Journées 1991: Systèmes de référence spatio–temporels, p. 125 – 133 (1991). – See Abstr. 012.048 for the main entry.

A software package may give wrong results for two reasons: computations are made with a limited number of digits, programs do contain wrong statements. Round–off errors may occur even with a small number of operations. A statistical approach of these errors leads to a way to estimate the accuracy of numerical results.

021.036 **Constrained realizations of Gaussian fields: a simple algorithm.**
Y. Hoffman, E. Ribak.
Astrophys. J., Lett., Vol. 380, No. 1, p. L5 – L8 (10 Oct 1991).

The authors present a straightforward method for the construction of constrained realizations of Gaussian fields. Consider a Gaussian random field and its ensemble mean field given a set of constraints. The residual of the field from its mean is statistically independent of the actual numerical value of the constraints. The algorithm is based on a simple construction of this residual field, which is then added to the analytically calculated mean field. This algorithm is exact and involves no iterations. The computational effort involves only setting one random realization in the inverse k–space, its FFT to real space, and the calculation of the mean field (given the set of constraints) over all grid points. Furthermore, it can be applied to a large number of constraints of arbitrary amplitudes and positions. The algorithm facilitates the generation of the initial conditions for N–body simulations which obey a few hundreds of constraints imposed by the observable universe.

021.037 **The computer simulation of space.**
N. Trajković.
Vasiona, Année 39, No. 5, p. 83 – 84 (1991). In Croatian.
The program for simulation of the space as a function of the mass distribution is given, written in SUPER BASIC.

021.038 **On the Lehmann–Filhés theorem as a consequence of one theorem of Ostrogradsky.**
G. I. Shirmin.
Tr. Gos. Astron. Inst. im. Shternberga, Tom 62, p. 40 – 65 (1991). In Russian.
It is shown that the so–called Lehmann–Filhés theorem playing the role of a groundwork in the "method of incomplete integral" of dynamical problems solving is a simple consequence of one theorem of Russian academician M. V. Ostrogradsky. The appropriate proof of that is contained in his famous paper "The Memoir on Differential Equations of the isoperimetric Problem" having been reported to the Russian Academy of Sciences on the 29–th of November 1848. So it is prooved that on the basis of the above mentioned method the priority belongs to M. V. Ostrogradsky who published his results more than half a century earlier than the same was made by Lehmann–Filhés (1904).

021.039 **On computing Laplace's coefficients and their derivatives.**
I. A. Gerasimov, E. L. Vinnikov.
Tr. Gos. Astron. Inst. im. Shternberga, Tom 62, p. 66 – 83 (1991). In Russian.
The algorithm of computing Laplace's coefficients and their derivatives is proposed with application of recurrent relations. The A.G.M.–method is used for the calculation of values $L_0^{(0)}$, $L_0^{(1)}$. The FORTRAN–program corresponding to the algorithm is given. The precision control was provided with numerical integrating by Simpsons method. The behavior of Laplace's coefficients and their third derivatives whith varying indices K, n for fixed values of the α–parameter is presented graphically.

021.040 **Calculation and inversion of the Weierstrass γ–function in limiting cases.**
I. A. Gerasimov, B. R. Mushailov.
Tr. Gos. Astron. Inst. im. Shternberga, Tom 62, p. 84 – 90 (1991). In Russian.
The complete algorithm of calculation and inversion of the Weierstrass γ–function with the real invariants is presented. Limiting cases of practical interest are discussed.

021.041 **Geometry and dynamics of fractal sets.**
J. M. Perdang.
Applying fractals in astronomy, p. 1 – 41 (1991). – See Abstr. 003.023 for the main entry.
The most relevant tools of fractal geometry are introduced and several classes of mathematical frameworks which generate fractal configurations dynamically (in spacetime as well as in parameter spaces) are examined.

021.042 **Treecode with a special–purpose processor.**
J. Makino.
Publ. Astron. Soc. Jpn., Vol. 43, No. 4, p. 621 – 638 (1991).
The author describes an implementation of the modified Barnes–Hut tree algorithm for a gravitational N–body calculation on a GRAPE (GRAvity PipE) backend processor. GRAPE is a special–purpose computer for N–body calculations. It receives the positions and masses of particles from a host computer and then calculates the gravitational force at each coordinate specified by the host. To use this GRAPE processor with the hierarchical tree algorithm, the host computer must maintain a list of all nodes that exert force on a particle. If one creates this list for each particle of the system at each timestep, the number of floating–point operations on the host and that on GRAPE would become comparable, and the increased speed obtained by using GRAPE would be small. In their modified algorithm, the author creates a list of nodes for many particles. Thus, the amount of the work required of the host is significantly reduced. This algorithm was originally developed by Barnes in order to vectorize the force calculation on a Cyber 205. With this algorithm, the computing time of the force calculation becomes comparable to that of the tree construction, if the GRAPE backend processor is sufficiently fast. The obtained speed–up factor is 30 to 50 for a RISC–based host computer and GRAPE–1A with a peak speed of 240 Mflops.

021.043 **DREAM: a special–purpose computer for large scale mesh calculations.**
J. Makino, T. Ebisuzaki, S. K. Okumura.
Publ. Astron. Soc. Jpn., Vol. 43, No. 4, p. 639 – 648 (1991).
The authors present the concept of DREAM (Disk REsource Array Machine), a computer designed for large–scale simulations using finite–difference grids. At present, such simulations are usually performed on vector processors, on which the size of the calculation is limited not by the speed of computation but by the size of the memory. The DREAM system uses an array of magnetic disks as the main memory, instead of silicon solid–state memory chips. Since the price per bit of a magnetic disk is about 1/100 that of a silicon memory, one can construct a machine with a 100–times larger memory for the same cost. The access time of a disk is about 10^5 times longer than that of a silicon memory. In the case of finite–difference calculations using regular grids, however, this access time becomes negligible, since the memory is accessed only in long vectors. Using currently available technology, a DREAM system with 2 Gbytes of disk space and effective speed of 40 Mflops costs around 3 million Japanese yen. A system with a 200 Gbytes disk and a 10 Gflops speed would cost a few hundred million yen. The authors also discuss the reliability of the hardware and the possibility of applying DREAM to other fields, such as the reduction of large amount of observational data.

021.044 **On the breakdown point of *M*–estimators with the convex object functions.**
Jia Peizhang.
Acta Astron. Sin., Vol. 32, No. 4, p. 429 – 435 (Dec 1991). In Chinese.
The breakdown point of M–estimators with the convex object functions is discussed. As an example, the breakdown point of the linear model of a polynomial with degree two is derived and it is only 15%.

021.045 **On the 2–dimensional rotational inversion problem.**
J. Schou.
Conference on Challenges to Theories of the Structure of Moderate–Mass Stars, p. 93 – 100 (1991). – See Abstr. 012.096 for the main entry.
Inspired by the GONG hound and hare exercise, where 2–dimensional inversions are being performed, the author has started writing a set of programs to do both the forward and inverse problem, in both the 1–dimensional case (with possible expansion in latitude) and the full 2–dimensional case. Some preliminary results of running these programs are presented here, together with an outline of the whole package.

021.046 **Search for periodicity in time series through prediction.**
A. S. Grebinskij, M. A. Polyakov.
Soln. Dannye, Byull., No. 3/91, p. 104 (Mar 1991). Abstract. – See Abstr. 012.021 for the main entry.

021.047 **On the period analysis of observable series.**
Zhao Ming.
Ann. Shanghai Obs., Acad. Sin., No. 12, p. 19 – 25 (1991). In Chinese.

021.048 **Expanding function for PDP–11/24 computer system.**
Tan Xiaoyu, Sun Yongyuan.
Ann. Shanghai Obs., Acad. Sin., No. 12, p. 188 – 190 (1991). In Chinese.

021.049 **Free–PLOT graphics software system.**
Wu Guangjie.
Publ. Yunnan Obs., No. 3, p. 41 – 47 (1991). In Chinese.
Based on PGPLOT Graphics Subroutine Library a new graphics software system Free–PLOT is developed on VAX 8350 in Yunnan Observatory. It includes nearly all advantages of PGPLOT. In addition, since it has some special graphical symbols, some basic geometric curves and powerful analytic geometric calculation functions, it could be as a preliminary tools of CAD to make circuit diagrams, mechanics diagrams, optics diagrams, etc.

021.050 **Mutual calling between C and FORTRAN program.**
Wu Junqin.
Publ. Yunnan Obs., No. 3, p. 48 – 52 (1991). In Chinese.
An interface method between programs written in C and FORTRAN computer languages is introduced. Using this method, both C and FORTRAN programs can call the processes or functions written by the other computer language simply. When developing a system software package for a special function, one can comprehensively utilize the superiority of two computer languages.

021.051 **About the problem of generating three–dimensional pseudo–random points.**
D. D. Carpintero.
Bol. Asoc. Argent. Astron., No. 36, p. 166 – 174 (1990). In Spanish. – See Abstr. 012.095 for the main entry.
The author demonstrates that a popular pseudo–random number generator is not adequate in some circumstances to generate n–dimensional random points, $n > 2$. This problem is particularly noxious when direction cosines are generated. He proposes several soultions, among them a good generator that satisfies all statistical criteria.

021.052 **GRAPE–1A: special–purpose computer for *N*–body simulation with a tree code.**
T. Fukushige, T. Ito, J. Makino, T. Ebisuzaki, D. Sugimoto, M. Umemura.
Publ. Astron. Soc. Jpn., Vol. 43, No. 6, p. 841 – 858 (1991).
The authors have designed and built GRAPE (GRAvity PipE)–1A, a special–purpose computer for *N*–body simulations using the $O(N \log N)$ tree code. GRAPE–1A calculates the gravitational force between particles. A host computer, which is connected to GRAPE–1A, performs all other calculations. GRAPE–1A can complete either a simulation of interacting galaxies ($N = 10^5$, 5000 timesteps) or a cosmological simulation ($N = 10^6$, 500 timesteps) within one week. GRAPE–1A can be also applied to particle–based hydrodynamical simulations such as smoothed–particle hydrodynamics (SPH). The performance of GRAPE–1A is comparable to that of a supercomputer with a peak–speed of 500 Mflops for any type of the direct summation, the modified tree algorithm and SPH.

021.053 **Manipulation des séries.**
J. Laskar.
13. Ecole de Printemps d'Astrophysique de Goutelas: Les méthodes modernes de la mécanique céleste, p. 89 – 107 (1990). – See Abstr. 012.118 for the main entry.
Contents: 0. Introduction. 1. Manipuler quoi?. 2. Séries à plusieurs variables. 3. Différentes représentations. 4. Algorithmes. 5. Conclusion.

021.054 **Symplectic integrators for long–term integrations in celestial mechanics.**
B. Gladman, M. Duncan, J. Candy.
Celest. Mech. Dyn. Astron., Vol. 52, No. 3, p. 221 – 240 (1991).
The authors discuss the use of symplectic integration algorithms in long–term integrations in the field of celestial mechanics. The method's advantages and disadvantages (with respect to more common integration methods) are discussed. The numerical performance of the algorithms is evaluated using the 2–body and circular restricted 3–body problems. Symplectic integration methods have the advantages of linear phase error growth in the 2–body problem (unlike most other methods), good conservation of the integrals of the motion, good performance for moderately eccentric orbits, and ease of use. Its disadvantages include a relatively large number of force evaulations and an inability to continuously vary the step size.

021.055 **On the analytical and numerical approximation of invariant manifolds.**
C. Simo.
13. Ecole de Printemps d'Astrophysique de Goutelas: Les méthodes modernes de la mécanique céleste, p. 285 – 329 (1990). – See Abstr. 012.118 for the main entry.
Contents: 1. The continuation method. 2. Analysis of bifurcations in the continuation method. 3. Fixed points and their stability. 4. The numerical integration of ordinary differential equations. 5. The variational equations and the differentials of the Poincaré map. 6. The computation of periodic orbits and their stability. 7. On the computation of Lyapunov elements. 8. Computation of quasi periodic orbits. 9. Normal forms. 10. Analytical approximation of invariant manifolds. 11. Numerical approximation of invariant manifolds. 12. Computation of homoclinic and heteroclinic point. Global bifurcation curves.

021.056 **The Connection Machine: principle and application in astronomy and astrophysics.**
H. Scholl.
13. Ecole de Printemps d'Astrophysique de Goutelas: Les méthodes modernes de la mécanique céleste, p. 497 – 507 (1990). – See Abstr. 012.118 for the main entry.
Contents: 1. Introduction. 2. The principle of the Connection Machine. 3. The N–body problem. 4. Monte–Carlo methods in radiative transfer.

021.057 **A modified Aarseth code for GRAPE and vector processors.**
J. Makino.
Publ. Astron. Soc. Jpn., Vol. 43, No. 6, p. 859 – 876 (1991).
The author discusses the performance of a hierarchical timestep algorithm, which is Aarseth's individual timestep algorithm for *N*–body problems modified for use with GRAPE hardware and/or vector processors.

Astronomical algorithms.
See Abstr. 003.020.

Methods of orbit determination for the microcomputer.
See Abstr. 003.021.

Numerical modeling in applied physics and astrophysics.
See Abstr. 003.022.

Applying fractals in astronomy.
See Abstr. 003.023.

Infinity, finite and error.
See Abstr. 015.027.

Optimization using the OSLO and Super–OSLO programs.
See Abstr. 031.027.

Optical design with physical optics using GLAD.
See Abstr. 031.028.

Search algorithm for weak periodicities in definite time intervals.
See Abstr. 036.084.

Non–linear filters based on ordered statistics.
See Abstr. 036.182.

Image analysis context development in ESO–MIDAS environment at the IFCAI–CNR.
See Abstr. 036.184.

On the Theil's fitting method.
See Abstr. 036.199.

Comparison between different isophote–fitting programs.
See Abstr. 036.202.

Object detection and analysis in digitised images: algorithmic components and software structure.
See Abstr. 036.335.

STSDAS: the Space Telescope Science Data Analysis System.
See Abstr. 036.337.

An expert assistant supporting Hubble Space Telescope proposal preparation.
See Abstr. 036.349.

Comments on the symmetric difference quotient and its applications to the correction of orbits by R. A. Serafin and E. Wnuk.
See Abstr. 042.001.

A general scheme of Newtonian interactions in systems of gravitating masses.
See Abstr. 042.013.

Modern problems of computerization of celestial mechanics.
See Abstr. 042.022.

The influence of orbital eccentricity of a disturbing body on chaotic region.
See Abstr. 042.023.

Present status of a new algorithm for orbital elements improvement.
See Abstr. 042.030.

The "CELESTA" specialized system for analytical support of local methods of non–linear analysis in celestial mechanics.
See Abstr. 042.031.

The Lissajous transformation. I. Basics.
See Abstr. 042.039.

The Lissajous transformation. II. Normalization.
See Abstr. 042.040.

The Lissajous transformation. III. Parametric bifurcations.
See Abstr. 042.041.

The Lissajous transformation. IV. Delaunay and Lissajous variables.
See Abstr. 042.042.

On the representativity of initial conditions of triple systems.
See Abstr. 042.048.

Solving Kepler's equation with high efficiency and accuracy.
See Abstr. 042.060.

Periodic motion around the triangular equilibrium points of the photogravitational restricted problem of three bodies.
See Abstr. 042.061.

Three–dimensional bifurcations of periodic solutions around the triangular equilibrium points of the restricted three–body problem.
See Abstr. 042.062.

A numerical investigation of the one–dimensional newtonian three–body problem. III. Mass dependence in the stability of motion.
See Abstr. 042.064.

Mean values of particular functions in the elliptic motion.
See Abstr. 042.066.

Second–species solutions in the circular and elliptic restricted three–body problem. I. Existence and asymptotic approximation.
See Abstr. 042.067.

Second–species solutions in the circular and elliptic restricted three–body problem. II. Numerical explorations.
See Abstr. 042.068.

A general timing conduction for consecutive collision orbits in the limiting case $\mu = 0$ of the elliptic restricted problem.
See Abstr. 042.069.

About stability of libration points in the restricted photogravitational three body problem.
See Abstr. 042.070.

Kepler's Equation and Newton's method.
See Abstr. 042.071.

The influence of the orbital eccentricity of a primary on the chaotic region.
See Abstr. 042.073.

Le développement de la fonction perturbatrice.
See Abstr. 042.085.

Averaging Hamiltonian systems.
See Abstr. 042.088.

1. Action–angle variables. 2. Semi–numerical perturbation method.
See Abstr. 042.089.

Expériences numériques et mécanique céleste.
See Abstr. 042.094.

A complex analytic solution for the attitude motion of a near–symmetric rigid body under body–fixed torques.
See Abstr. 052.005.

Motion of artificial satellites in the set of Eulerian redundant parameters. II.
See Abstr. 052.017.

Towards a theory of wave transport.
See Abstr. 062.217.

A direct numerical approach to the Chandrasekhar's *H*–functions for arbitrary characteristic functions.
See Abstr. 063.019.

Recent advances in computational methods.
See Abstr. 063.042.

Acceleration of convergence.
See Abstr. 063.043.

Global and local methods for 1–D problems; implementation aspects and CPU–time and memory scalings.
See Abstr. 063.046.

2–D axisymmetric line transport.
See Abstr. 063.047.

Comparing Victoria and Yale evolutionary tracks.
See Abstr. 065.148.

Superalgebras of Duffin–Kemmer type.
See Abstr. 066.046.

Superholomorphic structures, moment maps and super self dual Yang Mills connections over super Riemann surfaces.
See Abstr. 066.047.

Accuracy tests for the computation of solar models.
See Abstr. 080.145.

The NASA environmental models of Mars.
See Abstr. 097.023.

Monte–Carlo simulation of open cluster colour–magnitude diagrams.
See Abstr. 153.083.

The three–point correlation function for Abell clusters: observations and simulations.
See Abstr. 160.153.

The two–point correlation functions applied to systems with or without a mean density.
See Abstr. 161.231.

022 Physical Papers Related to Astronomy and Astrophysics

022.001 **Molecular calcium in interstellar space and circumstellar envelopes? A possible candidate: CaC_2.**
P. Redondo, Y. Ellinger, G. Berthier, R. Cimiraglia.
Astron. Astrophys., Vol. 247, No. 1, p. L1 – L4 (Jul 1991). Letter–to–the–editor.

A theoretical study of the possible structures of calcium dicarbide has been carried out at the SCF–CI level using the perturbative MP2 and multireference MP2 formalisms to account for electronic correlation. The ground state of the system has a triangular geometry of C_{2v} symmetry. The first linear structure is found 8 Kcal/mol higher. The strong dipole moment of the triangular form, 13.3 Debye, reflects the ionic character $Ca^+C_2^-$ of the dominant configuration and makes this molecule a promising candidate for radio detection in circumstellar envelopes and interstellar space.

022.002 **Laboratory rotational spectrum of SiC_2 in the v_3 vibrationally excited state.**
M. Bogey, C. Demuynck, J. L. Destombes, A. D. Walters.
Astron. Astrophys., Vol. 247, No. 1, p. L13 – L16 (Jul 1991). Letter–to–the–editor.

The millimeter wave spectrum of SiC_2 in its antisymmetric v_3 mode was observed in a glow discharge, using a mixture of SiH_4, C_2H_2 and Ar. A total of 95 lines, characterised by $8 \leqslant J \leqslant 21$ and $1 \leqslant K_a \leqslant 13$ were measured between 190 and 470 GHz, and a set of molecular constants was derived. Evidence for large amplitude motion was found from the comparison with the ground state molecular constants. The frequencies of the lines of astrophysical interest were accurately predicted.

022.003 **Optical reflection spectroscopy of mineral and ice surfaces in comet simulation experiments.**
E. Langenscheidt, K. Roessler.
Fresenius J. Anal. Chem., Vol. 341, No. 1/2, p. 95 – 100 (Sep 1991). Paper presented at the 6. Working Conference on Applied Surface Analysis, Kaiserslautern (Germany), 9 – 12 Jul 1990.

Optical and photoacoustic absorption and emission spectra in the region of 25000 to 400 cm^{-1} of surfaces of minerals and ice–mineral mixtures were measured. The experiments were undertaken to better interpret the results of laboratory simulation of cometary nuclei (KOSI), in particular the determination of the albedo. The work compares the different sampling techniques in IR spectroscopy and light scattering.

022.004 **The coal model for the unidentified infrared bands. II. The thermal emission mechanism.**
R. Papoular, C. Reynaud, I. Nenner.
Astron. Astrophys., Vol. 247, No. 1, p. 215 – 225 (Jul 1991).

New spectroscopic data are presented in support of the coal model for the composition and structure of the Unidentified Infrared Band (UIB) emitters; these data were obtained by diffuse–reflectance, Fourier–transform spectroscopy of two high ranking coals of different origins in the range 2.5–20 μm. Pursuing the elaboration of the model, the authors then address the problem of the emission mechanism and consider, in particular, coal emission in thermal equilibrium. This requires a knowledge of the spectral absorptivity of coal at various temperatures and in vacuum. For this purpose, the authors used an original technique based on the measurement of the amplitude and phase of the sample emission (between 1.5 and 4.5 μm) under irradiation by a chopped beam of UV/visible light. Absorptivity curves were obtained in the range ~300 to ~600K, from which spectral emissions were deduced; these compare reasonably well with the spectra of various astronomical sources, showing that this is the right range of grain temperatures. It is then possible to estimate the input energy flux density required to maintain a grain of size d(Å) at a temperature T. All these properties properly fit the available data on UIB's. The size distribution of coal–like particles in space is expected to extend down to the smallest structural entity in coal, 10–20 Å in size. In this limit, transient heating by single UV photons may compete with the thermal equilibrium mechanism in regions of weak ISRF.

022.005 **A study of the Ne II $2p^43s$–$2p^43p$ transition probabilities.**
M. L. Burshtein, V. Vujnović.
Astron. Astrophys., Vol. 247, No. 1, p. 252 – 258 (Jul 1991).

The lifetimes of Ne II levels $2p^43p$ were used in combination with the multiplet branching ratios to obtain the absolute transition probabilities of the $2p^43s$–$2p^43p$ transition array.

022.006 **Fullerite. Neue Formen des Kohlenstoffs von möglicher interstellarer Bedeutung.**
W. Krätschmer.
Sterne Weltraum, Jahrg. 30, Nr. 12, p. 736 – 738 (Dec 1991).

022.007 **Ab initio calculations on HC^+_5, a cation of interest to astrochemistry.**
P. Botschwina.
J. Chem. Phys., Vol. 95, No. 6, p. 4360 – 4365 (15 Sep 1991). Current Physics Microform No.: 9110B0728.

HC^+_5, a cation of astrochemical interest, has been studied by the coupled electron pair approximation (CEPA) using a basis set of 111 contracted Gaussian–type orbitals (cGTOs). It has a large equilibrium dipole moment of –2.88 D with the positive end at the proton site and is thus a suitable candidate for future investigation by radioastronomy or microwave spectroscopy. Most promising for a study in the infrared region of the spectrum is the Fermi resonance doublet v_2/v_4+v_5 with band origins calculated at 2199 and 2174 cm^{-1}. Using larger basis sets of 146 and 160 cGTOs, the equilibrium proton affinity of C_5 is calculated to be 887.7 and 886.6 kJ mol^{-1}, respectively. For the proton affinity at 298K, a value of 860 ± 5 kJ mol^{-1} is recommended.

022.008 **Fourier transform spectroscopy of the CS $d^3\Delta_i$–$a^3\Pi_r$ system: indirect measurement of the $^1\Sigma^-$ valence state.**
J. I. Choe, Y. M. Rho, S. M. Lee, A. C. Floch le, S. G. Kukolich.
J. Mol. Spectrosc., Vol. 149, No. 1, p. 185 – 213 (Sep 1991).
The electronic emission spectrum of CS in the range of 9500 – 17500 cm^{-1} was recorded with 0.03 cm^{-1} resolution using the 1–m Fourier transform spectrometer associated with the McMath solar telescope at Kitt Peak National Observatory, with an electric discharge in a CS_2 and helium mixture.

022.009 **E1 transitions of astrophysical interest in neutral oxygen.**
A. Hibbert, E. Biémont, M. Godefroid, N. Vaeck.
J. Phys. B, Vol. 24, No. 18, p. 3943 – 3958 (28 Sep 1991).
Atomic transition rates and f values have been calculated in the triplet and quintet systems and for some intercombination lines of neutral oxygen. Configuration interaction was included in the calculations performed with the CIV3 code of Hibbert for all the transitions connecting the n = 3 and n = 4 energy levels. In order to reproduce the observed energy splittings between the energy states, small empirical adjustments were introduced to the diagonal matrix elements. Comparisons with previously published f values and radiative lifetimes are discussed and the subsequent astrophysical applications of the results are briefly mentioned.

022.010 **Interferometric gravitational wave detector using the transmission light on the Fabry–Perot cavity.**
N. Mio, M. Ohashi.
Jpn. J. Appl. Phys., Part 2, Vol. 30, No. 10B, p. 1833 – 1835 (15 Oct 1991).
The authors show the feasibility of an interferometric gravitational wave detector using the transmission light of the Fabry–Perot cavity, which eliminates the effect due to the frequency noise of the light source. The characteristics of the detector are discussed.

022.011 **The use of the reaction $^9Be(\alpha, \gamma n)^{12}C$ to determine the energy scale and the efficiency at low energies of stellar collapse neutrino detectors.**
A. Baldini, C. Bemporad, F. Cei, G. Giannini, R. Pazzi, M. Grassi.
Nucl. Instrum. Methods Phys. Res., Sect. A, Vol. 305, No. 2, p. 475 – 483 (20 Jul 1991).
Experiments for the detection of gravitational stellar collapse $\bar{\nu}_e$–events are most frequently based on the reaction: $\bar{\nu}_e + p \rightarrow n + e^+$, followed by delayed neutron capture in hydrogen $n + p \rightarrow \gamma + d$, $E_\gamma = 2.2$ MeV. Data obtained by a low intensity Am/Be source (as n and γ-ray emitter), in association with MACRO scintillation counters, are being used to determine the capability of detecting delayed neutron capture from cascade n, γ-ray emissions. The $^9Be(\alpha, \gamma n)^{12}C$ reaction is a valuable tool for setting the optimum 'stellar gravitational collapse trigger' working conditions both for the detection of the primary $\bar{\nu}_e$–events and for the detection of the delayed neutron captures. The use of the Am/Be source allowed several experimental checks of Monte Carlo calculated detection efficiencies.

022.012 **Stark broadening of spectral lines of multicharged ions of astrophysical interest. I. C IV lines.**
M. S. Dimitrijević, S. Sahal–Bréchot, V. Bommier.
Astron. Astrophys., Suppl. Ser., Vol. 89, No. 3, p. 581 – 590 (Sep 1991).
Using a semi–classical approach, the authors have calculated electron–, proton–, and ionized helium–impact line widths and shifts for 31 C IV multiplets. The obtained results have been compared with available experimental data. The comprehensive set of obtained data has also been used for the investigation of Stark broadening parameter regularities within the spectral series.

022.013 **Stark broadening of spectral lines of multicharged ions of astrophysical interest. II. Si IV lines.**
M. S. Dimitrijević, S. Sahal–Bréchot, V. Bommier.
Astron. Astrophys., Suppl. Ser., Vol. 89, No. 3, p. 591 – 598 (Sep 1991).
Using a semi–classical approach, the authors have calculated electron–, proton–, and ionized helium–impact line widths and shifts for 40 Si IV multiplets. The comprehensive set of obtained data has also been used for investigation of Stark broadening parameter regularities within spectral series.

022.014 **Hybrid simulations of rotational discontinuities.**
C. C. Goodrich, P. J. Cargill.
Adv. Space Res., Vol. 11, No. 9, p. 107 – 110 (1991). – See Abstr. 012.006 for the main entry.

022.015 **On the structure of rotational discontinuities with large phase angles.**
P. Richter, M. Scholer.
Adv. Space Res., Vol. 11, No. 9, p. 111 – 115 (1991). – See Abstr. 012.006 for the main entry.
Using hybrid simulations (particle ions and electron fluid) the authors have investigated the stability of rotational discontinuities. A density increase builds up at the position of the rotational discontinuity. The final density structure of the rotational discontinuity develops independently of the initial phase angle and of the half width. Only minor differences occur for different senses of rotation, i.e. for a rotational discontinuity exhibiting an ion sense or an electron sense of rotation, respectively. However the structure of the magnetic field strongly depends on the phase as shown by simulations of rotational discontinuities with 270° phase angle.

022.016 **Particle simulations of collisionless magnetic reconnection in magnetospheric current sheets.**
D. Q. Ding, L. C. Lee.
Adv. Space Res., Vol. 11, No. 9, p. 117 – 127 (1991). – See Abstr. 012.006 for the main entry.
The driven collisionless magnetic reconnection processes at the dayside magnetopause and in the magnetotail are investigated through 2 1/2–dimensional particle simulations. A review of simulation studies of driven collisionless magnetic reconnection and slow shocks in the magnetospheric current sheets is presented.

022.017 **Numerical simulation of flux transfer events at the magnetopause: electron dynamics.**
A. Otto, U. Arendt.
Adv. Space Res., Vol. 11, No. 9, p. 129 – 132 (1991). – See Abstr. 012.006 for the main entry.
Magnetic Flux Transfer Events (FTE's) at the magnetopause are expected to play an important role in the interaction between solar wind and magnetospheric plasma. Among a large number of observational signatures, typical properties of FTE's in the magnetosheath are the intermittent appearance of a particle population of hot plasma and a bipolar variation of the magnetic field component normal to the magnetopause. This led to the interpretation that FTE's might be the result of localized magnetic reconnection which allows for energetic magnetospheric particles to enter the magnetosheath. In the framework of resistive MHD computations the authors investigate a three–dimensional magnetic reconnection process which is caused by a locally enhanced nonidealness (resistivity). It is shown that the dynamical evolution leads to structures which exhibit interesting properties of FTE's, where test particles are used as a diagnostic tool.

022.018 **Ion acceleration in a current sheet with magnetic islands.**
A. T. Altyntsev, N. V. Lebedev, N. A. Strokin.
Adv. Space Res., Vol. 11, No. 9, p. 189 – 191 (1991). – See Abstr. 012.006 for the main entry.
This paper reports on results of an experimental study of proton acceleration processes in a neutral current sheet. An effect of resonance acceleration during the interaction of particles with

a macroscopic potential jump in the sheet was detected. Results of laboratory experiments are compared with measurements in the geomagnetic tail.

022.019 **Heating by optical absorption and the performance of interferometric gravitational–wave detectors.**
W. Winkler, K. Danzmann, A. Rüdiger, R. Schilling.
Phys. Rev. A, Vol. 44, No. 11, p. 7022 – 7036 (1 Dec 1991). Current Physics Microform No.: 9201F0636.
The ultimate sensitivity of interferometric gravitational–wave detectors requires extremely high light powers sensing the separation of test masses. Absorption of light at the optical components causes wave–front distortions via the thermally deformed substrates or via thermal lensing. The performance of thermally distorted interferometers is treated quantitatively, and the two schemes for increasing the optical path, the delay–line and Fabry–Pérot methods, are compared.

022.020 **Velocity distribution of fragments formed in a simulated collisional disruption.**
A. Nakamura, A. Fujiwara.
Icarus, Vol. 92, No. 1, p. 132 – 146 (Jul 1991).
The velocity distribution of fragments resulting from catastrophic disruption was determined experimentally which is of great importance to understand the collisional evolution of planetary bodies.

022.021 **Modeling of collision–induced infrared absorption spectra of H_2–H_2 pairs in the fundamental band at temperatures from 20 to 300K.**
A. Borysow.
Icarus, Vol. 92, No. 2, p. 273 – 279 (Aug 1991).
A numerical method is presented which generates the rotovibrational (free–free) collision–induced absorption (RV CIA) spectra of H_2–H_2 pairs in the fundamental band of hydrogen, at temperatures from 20 to 300K. Computed synthesized spectra are found to be in close agreement with all presently known CIA spectra of H_2–H_2 and reproduce the results of the quantum mechanical computations to within a few percent.

022.022 **Atlas of the absorption/dissociation spectra of CO and its isotopes between 91.2 nm and 115.2 nm.**
M. Eidelsberg, J. J. Benayoun, Y. Viala, F. Rostas.
Astron. Astrophys., Suppl. Ser., Vol. 90, No. 2, p. 231 – 282 (Oct 1991).
Based on the data on the CO molecule recently obtained by Eidelsberg and Rostas (1990), this paper presents a complete atlas of the UV absorption lines of the four main isotopes: $^{12}C^{16}O$, $^{13}C^{16}O$, $^{12}C^{18}O$ and $^{13}C^{18}O$ in the wavelength range 91.2 – 115.2 nm. The results are displayed in tables which contain the parameters of the absorption bands as well as the wavenumber and the wavelength of each line either observed or computed from the rotational constants of the upper electronic Rydberg state. Besides these spectroscopic data, for each transition, the absorption oscillator strength, the lifetime of the upper electronic state and its dissociation probability are also given. The set of data, which can be useful to compute the photodissociation rate and the abundance of CO in various astrophysical environments is available upon request, on computer compatible media. The synthetic absorption spectrum of $^{12}C^{16}O$ in the range 91.2 – 115.2 nm, calculated for a temperature of 50K by assuming a Boltzmann distribution of the rotational levels, is presented in a graphic form.

022.023 **Collisional excitation of molecular hydrogen by hydrogen atoms.**
P. G. Martin, M. E. Mandy.
Bull. Am. Astron. Soc., Vol. 23, No. 2, p. 888 (1991). Abstract. – See Abstr. 012.008 for the main entry.

022.024 **Calculated frequencies for the CO ground state transitions.**
A. Le Floch.
Astron. Astrophys., Suppl. Ser., Vol. 90, No. 3, p. 513 – 540 (Nov 1991).
Calculated frequencies of transitions dealing with the ground state of CO have been obtained from a new set of frequency Dunham parameters. These data concern numerous bands belonging to the four sequences $\Delta V = 0$ to 3.

022.025 **Ion equilibrium for minor components in a thin plasma.**
M. Landini, B. C. Monsignori Fossi.
Astron. Astrophys., Suppl. Ser., Vol. 91, No. 1, p. 183 – 196 (Nov 1991).
The ionization equilibrium of some cosmically less abundant elements (F, P, Cl, K, Ti, Cr, Mn and Co) is computed. The computation is performed using the Shull and Van Steenberg (1982) approximations for the ionization, radiative, dielectronic and autoionization rates. The fitting coefficients for the rates are given and the ion abundances are supplied for temperatures between 10^5 and 10^8K.

022.026 **Erratum: "A treatment of opacity and emission suitable for media of low density and temperature. II. Molecular hydrogen" [Astron. Astrophys., Suppl. Ser., Vol. 88, No. 3, p. 451 – 470 (Jun 1991)].**
R. Capuzzo–Dolcetta, A. Di Fazio, F. Palla.
Astron. Astrophys., Suppl. Ser., Vol. 91, No. 1, p. 197 (Nov 1991). See Abstr.53.022.085.

022.027 **A synthetic telluric spectrum in the wavelength region surrounding the D_1 and D_2 lines of sodium.**
I. Lundström, A. Ardeberg, E. Maurice, H. Lindgren.
Astron. Astrophys., Suppl. Ser., Vol. 91, No. 2, p. 199 – 208 (Dec 1991).
The telluric spectrum has been studied in the wavelength region surrounding the D_1 and D_2 lines of sodium. From spectrograms at high spectral resolution, detailed measurements have been made of 222 lines, mostly identified as due to atmospheric water vapour. It is obvious, that the strength and pronounced variability of the telluric lines make it very difficult to obtain meaningful data for lines of interstellar sodium, especially fainter components. In particular, spectrogram division for elimination of the effect of telluric lines seems highly doubtful, if weak components of interstellar lines are to be measured. However, from spectrograms with high signal–to–noise, ratio, strong telluric and faint or absent stellar and interstellar lines, the authors have constructed a synthetic spectrum of telluric lines from 586.9 nm to 591.7 nm. Individual line positions for stronger telluric lines are accurate to around 7×10^{-5}nm, corresponding to a radial–velocity precision around 40 m s^{-1}. Individual relative equivalent widths for the same lines are accurate to around seven percent. This means that, with only a few resolved telluric lines measureable, the synthetic relative telluric spectrum can be converted into an absolute spectrum, reproduced to rather high accuracy. Subsequent elimination of the telluric spectrum is very accurate, allowing for safe recuperation of interstellar lines, also those of only marginal strength. At the same time, one can show that use of the telluric spectrum is excellent for the purpose of high precision radial velocities. When well exposed, the telluric spectrum gives a zero point in radial velocity with a precision around 20 m s^{-1} for spectrograms with resolution around 10^5.

022.028 **Lab spectra of MgTBP and H_2TBP: assignment of 79 diffuse interstellar bands.**
F. M. Johnson.
Bull. Am. Astron. Soc., Vol. 23, No. 2, p. 933 (1991). Abstract. – See Abstr. 012.008 for the main entry.

022.029 **Off-center point explosion in a radially stratified medium – Kompaneets approximation.**
D. G. Korycansky.
Bull. Am. Astron. Soc., Vol. 23, No. 2, p. 940 (1991). Abstract. – See Abstr. 012.008 for the main entry.

022.030 **Absolute electron excitation cross sections in ions using energy-loss and merged-beams methods.**
A. Chutjian, S. J. Smith, K.-F. Man, R. J. Mawhorter, I. D. Williams.
Bull. Am. Astron. Soc., Vol. 23, No. 2, p. 975 – 976 (1991). Abstract. – See Abstr. 012.008 for the main entry.

022.031 **Calculations of the $b_n \times \beta_n$ deviation coefficient for carbon at low temperatures.**
V. O. Ponomarev.
Astron. Tsirk., No. 1549, p. 3 – 4 (May – Aug 1991).
The author presents calculations of the $b_n \times \beta_n$ deviation coefficient for carbon taking into account radiationless electron capture and collisional stabilization of a highly-excited state at low temperatures and electron densities which are characteristic for clouds in the Cas A direction.

022.032 **Electron collisional rates for atomic hydrogen, revisited.**
E. S. Chang, E. H. Avrett, R. Loeser.
Astron. Astrophys., Vol. 247, No. 2, p. 580 – 583 (Jul 1991).
The authors show that there are significant inconsistencies in the compilation of electron impact transition rate coefficients for atomic hydrogen by Giovanardi et al. (1987) and Giovanardi and Palla (1989). In particular, the rates of the *n-n'* transitions 1-4, 2-4, 3-4, 3-5, 4-5, and 4-6 are not consistent with neighboring transitions rates, and qualitatively disagree with the rates of Johnson (1972).

022.033 **New measurements of H_3^+ and HCO^+ dissociative recombination rate coefficient.**
A. Canosa, B. R. Rowe, J. B. A. Mitchell, J. C. Gomet, C. Rebrion.
Astron. Astrophys., Vol. 248, No. 1, p. L19 – L21 (Aug 1991). Letter-to-the-editor.
Dissociative recombination rate coefficients of H_3^+, with low vibrational excitation, and HCO^+ have been measured by means of a new flowing afterglow apparatus at 300K. For H_3^+, it has been obtained $\alpha(H_3^+(v \leqslant 2)) = 1.5 \times 10^{-7} cm^3 s^{-1}$ and $\alpha(H_3^+(v=0)) = 1.1 \times 10^{-7} cm^3 s^{-1}$. The problem of the vibrational state of H_3^+ is pointed out. For HCO^+, the rate coefficient has been measured to be $\alpha(HCO^+) = 2.4 \times 10^{-7} cm^3 s^{-1}$. It is stressed that this results are in agreement with the recent experiment of Amano, but are in conflict with the FALP study of Adams and Smith.

022.034 **Accurate oscillator strengths of astrophysical interest for neutral oxygen.**
E. Biémont, A. Hibbert, M. Godefroid, N. Vaeck, B. C. Fawcett.
Astrophys. J., Vol. 375, No. 2, p. 818 – 822 (10 Jul 1991).
Knowledge of accurate oscillator strenghts in neutral oxygen is of crucial importance in astrophysics. The accurate determination of the solar abundance of O I, which remains the primary source for the entry in cosmic abundance tables, requires the use of E1 transitions appearing in the solar photospheric spectrum. A new accurate scale of oscillator strengths has been obtained for the $2p^3(^4S)nl$ – $2p^3(^4S)n'l'$ transitions of O I of astrophysical interest. Configuration interaction effects have been considered in detail. The results have been used to improve the solar abundance determination for oxygen. The new result is $A(O) = 8.86 \pm 0.04$ (in the usual logarithmic scale).

022.035 **Association of CH_3^+ and H_2: conflicting results.**
D. R. Bates.
Astrophys. J., Vol. 375, No. 2, p. 833 – 842 (10 Jul 1991).
Earlier calculations on association in $CH_3^+ + H_2$ and $CD_3^+ + D_2$ collisions are extended and refined. It is concluded that there are serious discrepancies between, on the one hand, recent measurements made with an 80K low-pressure ion trap and, on the other hand, theory and measurements made with the SIFT apparatus at 80K, 0.4 Torr, and a 13K ultra-low-pressure ion trap. No explanation is offered.

022.036 **Theoretical study of the spectroscopy of FeH^+.**
S. R. Langhoff, C. W. Bauschlicher Jr.
Astrophys. J., Vol. 375, No. 2, p. 843 – 845 (10 Jul 1991).
High-level theoretical calculations are presented for selected triplet, quintet, and septet states of FeH^+. The ordering of the quintet states arising from the ^6D ground state of Fe^+ is definitively determined to be $X^5\Delta$, $A^5\Pi$, and $B^5\Sigma^+$, which is consistent with Hund's rules and mixing arguments. The $A^5\Pi - X^5\Delta$ separation is $\sim 600\ cm^{-1}$, whereas the lowest triplet state ($a^3\Sigma^-$) lies $\sim 10{,}000\ cm^{-1}$ above the ground state. The dissociation energy of FeH^+ is determined to be 2.29 eV, in good agreement with previous theoretical studies and experiment. The excited quintet states of Fe^+ are repulsive in the Franck-Condon region of the $X^5\Delta$ state. Thus FeH^+ is probably photodissociated in stellar atmospheres, which may explain why it has not been observed, despite having a high predicted abundance based on thermodynamic arguments.

022.037 **Search for anomalous spin-dependent forces using stored ion spectroscopy.**
D. J. Wineland, J. J. Bollinger, D. J. Heinzen, W. M. Itano, M. G. Raizen.
Phys. Rev. Lett., Vol. 67, No. 13, p. 1735 – 1738 (23 Sep 1991). Current Physics Microform No.: 9111C2109.
Resonances in atomic ions can be used to search for new, weak, spin-dependent interactions. Upper limits on anomalous dipole-monopole and dipole-dipole couplings for the neutron and electron are determined by examining hyperfine resonances in stored $^9Be^+$ ions. These experiments also place strict limits on anomalous weights of spinning gyroscopes.

022.038 **Clathrate hydrate formation in amorphous cometary ice analogs in vacuo.**
D. Blake, L. Allamandola, S. Sandford, D. Hudgins, F. Freund.
Science, Vol. 254, No. 5031, p. 548 – 551 (25 Oct 1991).
The presence of clathrate hydrates in cometary ice has been suggested to account for anomalous gas release at large radial distances from the sun as well as the retention of volatiles in comets to elevated temperatures. However, how clathrate hydrates can form in low-pressure environments, such as in cold interstellar molecular clouds, in the outer reaches of the early solar nebula, or in cometary ices, has been poorly understood. Experiments performed with the use of a modified electron microscope demonstrate that during the warming of vapor-deposited amorphous ices in vacuo, clathrate hydrates can form by rearrangements in the solid state.

022.039 **Iron and chromium absorption bands in the spectra of terrestrial pyroxenes: application to mineralogic remote sensing of asteroid surfaces.**
D. I. Shestopalov, L. F. Golubeva, M. N. Taran, V. M. Khomenko.
Astron. Vestn., Tom 25, No. 4, p. 442 – 452 (Jul – Aug 1991). In Russian. English translation in Sol. Syst. Res., Vol. 25.

022.040 **Fe I oscillator strengths of lines of astrophysical interest.**
A. Bard, A. Kock, M. Kock.
Astron. Astrophys., Vol. 248, No. 1, p. 315 – 322 (Aug 1991).
Oscillator strengths of 114 Fe I lines in the wavelength range 260-1170 nm are presented which were obtained by means of emission measurements on a high-current hollow-cathode and by using a high-resolution Fourier-transform spectrometer, in conjunction with literature data of lifetime that are considered reliable. In order to comprise a large number of lines of astrophysical interest the authors have included some branching fractions from Bridges & Kornblith; May et al.; and Kock et al. in this analysis. As a result the uncertainty of the *f*-values typically is between 7 and 12%, and it is 33% at the most. The data are

compared with literature data, especially with those of the Oxford group.

022.041 **Can phase space theory reproduce experimental neutral product branching ratios for dissociative recombination reactions?**
E. T. Galloway, E. Herbst.
Astrophys. J., Vol. 376, No. 2, p. 531 – 539 (1 Aug 1991).
A formulation of the statistical phase space theory has been used to obtain neutral product branching ratios for tetratomic ion–electron dissociative recombination reactions. The results have been compared with recent experimental results for the H_3O^+ + e, $NNOH^+$ + e, $HOCO^+$ + e, and H_3S^+ + e reactions. Two conclusions have been reached. First, some dissociative recombination reactions can be described by a phase space theory that contains neutral product distributions with limited rotational excitation. Second, phase space theory does not, however, currently possess general predictive power, free from parameterization, for tetratomic ion–electron dissociative recombination reactions.

022.042 **Reminiscences on experimental work with various particle detectors.**
A. L. Hodson.
Nucl. Phys., B Proc. Suppl., Vol. 22B, p. 178 – 196 (Jul 1991). – See Abstr. 012.019 for the main entry.
This paper contains some chamber and discharge chamber photographs of phenomena in particle and cosmic ray physics. Among those reproduced are hitherto unpublished photographs of unusual near–horizontal showers, of cores and multicores in extensive air showers, and some illustrating the discharge mechanism of proportional and Geiger counters and an early attempt at pulsing a spark chamber. The accompanying text includes comments, based on the photographs, which are relevant to the performance of several types of particle detector.

022.043 **Astrophysical aspects of neutral atom line broadening.**
E. L. Lewis.
AIP Conf. Proc., No. 216, p. 541 – 562 (Dec 1990). Current Physics Microform No.: 9101E0541. – See Abstr. 012.018 for the main entry.
Recent work on aspects of line shapes which are relevant to astrophyscial spectra are discussed, Namely (1) the impact broadening of strong resonance lines by atomic hydrogen, (2) the effects on line profiles of collisional redistribution of radiation in the impact region and (3) the influence of velocity changing collisions on impact region redistribution.

022.044 **Spectroscopy in mega–Gauss fields: theory.**
G. Wunner.
AIP Conf. Proc., No. 216, p. 563 – 573 (Dec 1990). Current Physics Microform No.: 9101E0563. – See Abstr. 012.018 for the main entry.
The author reports on how the recent numerical solution of the problem of a hydrogen atom in a magnetic field of arbitrary strength – one of the remaining "unsolved elementary problems in atomic physics" (Kleppner et al.) – has helped in identifying the mysterious line features unexplained for almost five decades in the optical spectrum of the white dwarf star Grw +70°8247 as due to hydrogen atoms exposed to a magnetic field of ~320 million Gauss, and, at the same time, has laid the foundation for the quantitative spectroscopy of these compact cosmic objects in the magnetic field range up to $\sim 10^3$mega–Gauss.

022.045 **The Lyman α line wing.**
N. Allard, J. Kielkopf.
AIP Conf. Proc., No. 216, p. 604 (Dec 1990). Current Physics Microform No.: 9101E0604. – See Abstr. 012.018 for the main entry.
Resonance broadening in the far wing of Lyman α is analyzed. The extreme far wing profiles for hydrogen–rich DA white dwarf stars are estimated.

022.046 **Simple convergent formula for estimating Stark widths of neutral atom lines.**
M. S. Dimitrijevíc, N. Ben Nessib.
AIP Conf. Proc., No. 216, p. 605 – 606 (Dec 1990). Current Physics Microform No.: 9101E0605. – See Abstr. 012.018 for the main entry.
A simple analytical expression for estimating Stark widths of neutral atoms has been derived from a convergent theory that avoids the computation of the minimum impact parameter. The formula obtained was applied to He I lines. The results are compared to other calculations.

022.047 **Hydrogen Stark profiles for astrophysics.**
C. Stehlé.
AIP Conf. Proc., No. 216, p. 607 – 608 (Dec 1990). Current Physics Microform No.: 9101E0607. – See Abstr. 012.018 for the main entry.
A large number of hydrogen line profiles are calculated in the framework of the model microfield method with specific modification for electronic contributions. Profiles of the Lyman, Balmer and Paschen series have been obtained from the center to the Holtsmark wings of the lines.

022.048 **Stark parameter regularities along the periodic table.**
J. Purić, M. Ćuk, S. Djeniže, A. Lesage.
AIP Conf. Proc., No. 216, p. 609 – 610 (Dec 1990). Current Physics Microform No.: 9101E0609. – See Abstr. 012.018 for the main entry.

022.049 **Swan band pressure shifts in DQ white dwarfs.**
G. L. Hammond.
AIP Conf. Proc., No. 216, p. 611 – 612 (1 Dec 1990). Current Physics Microform No.: 9101E0611. – See Abstr. 012.018 for the main entry.

022.050 **A discussion of the formula in the calculation of the column density at second level of hydrogen atoms.**
Li Kejun.
Publ. Yunnan Obs., No. 2, p. 19 – 23 (Jun 1991). In Chinese.

022.051 **Electron transport in atmospheric gases I. Integral characteristics.**
T. I. Sergienko, V. E. Ivanov.
Geomagn. Aehron., Tom 31, No. 4, p. 635 – 642 (Jul – Aug 1991). In Russian. English translation in Geomagn. Aeron., Vol. 31, No. 4.

022.052 **Electron transport in atmospheric gases. II. Energy spectra.**
V. E. Ivanov, B. V. Kozelov, T. I. Sergienko.
Geomagn. Aehron., Tom 31, No. 4, p. 643 – 648 (Jul – Aug 1991). In Russian. English translation in Geomagn. Aeron., Vol. 31, No. 4.

022.053 **Atomic physics calculations relevant to solar flare spectra.**
H. E. Mason.
Royal Society Discussion Meeting: The physics of solar flares, p. 471 – 484 (1991). – See Abstr. 012.016 for the main entry.
Solar flare spectra in the ultraviolet and X–ray wavelength regions are rich in emission lines from highly ionized ions, formed at temperatures around 10^7K. These lines can be used as valuable diagnostics for probing the physical conditions in solar flares. Such analyses require accurate atomic data for excitation, ionization and recombination processes. The author presents a review of work which has already been carried out, in particular for the Solar Maximum Mission observations.

022.054 **Electron transport in atmospheric gases. III. Angular distributions.**
B. V. Kozelov, V. E. Ivanov, T. I. Sergienko.
Geomagn. Aehron., Tom 31, No. 5, p. 940 – 943 (Sep – Oct 1991). In Russian. English translation in Geomagn. Aeron., Vol. 31, No. 5.

022.055 **Oscillator strengths for Y III and Zr III in the IUE region.**
A. Redfors.
Astron. Astrophys., Vol. 249, No. 2, p. 589 – 593 (Sep 1991).
Oscillator strengths for Y III and Zr III in the region 1190–3100 Å have been calculated for transitions from all levels below 13 eV. The presence of these spectra in IUE observations of chemically peculiar stars is discussed.

022.056 **Determination of major, minor and trace constituents in meteorites by inductively coupled plasma–atomic emission spectroscopy.**
Zhang Zhanzia, B. R. Bear, V. A. Fassel.
Spectrochim. Acta, Part B, Vol. 46, No. 8, p. 1171 – 1183 (1991).
The ICP–AES determination of major, minor and trace constituents in meteorites is described. The accuracy of the determinations was assessed by: (a) comparing the results with those obtained by neutron activation and X–ray fluorescence and (b) analysing several geological reference samples. The results showed good agreement. The relative standard deviation of the determination of the major elements ranged from 0.5 to 2% and of the minor and trace elements from 0.5 to 8%.

022.057 **The initial results of researching in X–band YBCO high T_c superconducting cavity.**
Su Jinyuan, Kou Wenwei, Luo Jiazeng, Xin Xuanjiao, Jiang Shunda, Zhao Meiyu, Li Chengen, Jin Hangyun.
Ann. Shanghai Obs., Acad. Sin., No. 12, p. 140 – 145 (1991). In Chinese.

022.058 **A critical assessment of electron excitation between the $2s^22p^2$ and $2s2p^3$ configurations of Mg^{+6}.**
A. Burgess, H. E. Mason, J. A. Tully.
Astrophys. J., Vol. 376, No. 2, p. 803 – 808 (1 Aug 1991).
Using the programs RMATRX and DSTWAV, which are based, respectively, on the close coupling and distorted wave approximations, the authors compute and compare partial collision strengths for the optically allowed transitions between the $2s^22p^2$ and $2s2p^3$ configurations of Mg^{+6}. From this investigation the authors conclude that the differences between the published results of Mason and Bhatia (DSTWAV) and Aggarwal (RMATRX) arise not simply because they used different target wave functions, but also because the method used by Aggarwal for treating the outer region in his close coupling calculation is inadequate for optically allowed transitions.

022.059 **(^{6}Li,^{6}He) measurements as an alternative calibration for solar neutrino detectors.**
E. Aschenauer, H. Dennert, W. Eyrich, A. Lehmann, M. Moosburger, H. Wirth, H. J. Gils, H. Rebel, S. Zagromski.
Phys. Rev. C, Vol. 44, No. 6, p. 2771 – 2775 (Dec 1991). Current Physics Microform No.: 9203A1283.
The (^{6}Li,^{6}He) reaction was studied on the nuclei ^{37}Cl and ^{71}Ga at E_{6Li} = 156 MeV at extreme forward angles including zero degree. Gamow–Teller strength and the corresponding B(GT) values were extracted. It is shown that these measurements provide an alternative method to calibrate solar neutrino detectors.

022.060 **Optical coatings on KCl windows for CO_2 lasers.**
R. Medianu, S. V. Nistor.
Stud. Cercet. Fiz., Vol. 42, No. 7, p. 687 – 691 (1990).
The surface of polished windows, made from ultratransparent KCl by IRA technique, has been studied by both optical microscopy and spectroscopy during various stages of coatings with a partial reflecting layer of BaF_2–ZnSe. It has been found that the adherence of the coating seems to be associated to the presence on the substrate of organic molecules resulting from the diffusion oil.

022.061 **Simularea numerica a proceselor microfizice din ceata.**
S. Stefan, D. Borsan.
Stud. Cercet. Fiz., Vol. 42, No. 8–10, p. 747 – 754 (1990).
Fog as a polydisperse system may go spontaneously through different microphysical processes such as growth by condensation and evaporation, coagulation and gravitational settling. The paper provides a numerical model of the time evolution of the dimension spectrum of fog droplets, due to the joint effects of evaporation and gravitational settling as laser radiation passes through the fog. The results point to the importance of original conditions (temperature, relative humidity) in the determination of the optical path in the fog. Microphysical processes occurring in laboratory fog are also useful for studying both the motion and dissipation of natural fog.

022.062 **Theoretical studies of the absorption spectra of polycyclic aromatic hydrocarbons.**
S. Canuto, M. C. Zerner, G. H. F. Diercksen.
Astrophys. J., Vol. 377, No. 1, p. 150 – 157 (10 Aug 1991).
The authors examine the UV – visible spectra of graphite–like fragments such as tetracene, anthanthrene, and coronene, as well as their dimers and trimers using the intermediate neglect of differential overlap quantum chemical technique for comparison with the observed interstellar extinction believed to stem from graphite fragments. In general, the calculated results are in good agreement with experiments for the monomeric species which have well characterized spectra. The low–resolution interstellar extinction, however, is reproduced by several of the model systems, and in particular by a superposition of all three monomers as well as the dimers and trimers of this study at their calculated interplanar distances.

022.063 **On the problem of movement of a gyroscope in Cardan suspension, studied in a set of redundant variables.**
R. Cid Palacios, M. C. Fernandez Quiñones.
1. Spain–USSR Workshop on Positional Astronomy and Celestial Mechanics, p. 125 – 131 (1991). – See Abstr. 012.025 for the main entry.

022.064 **The absorption coefficient of the liquid N_2 2.15–μm band and application to Triton.**
W. M. Grundy, U. Fink.
Icarus, Vol. 93, No. 1, p. 169 – 173 (Sep 1991).
The authors have measured the temperature dependence of the absorption coefficient and integrated absorption of the liquid nitrogen 2.15–μm 2–0 collision–induced band. The integrated absorption of the liquid is observed to be smaller than that of the gas, and decreases slightly with decreasing temperature. The band gets sharper with decreasing temperature with a half–width proportional to the square root of the temperature. If this temperature–dependent behavior is extrapolated to the nominal surface temperature of Triton, a new estimate of the grain size of N_2 ice on Triton's south polar cap can be made.

022.065 **Simulation of the spectral content of the 7.8–μm emission from Jupiter at high resolution.**
R. Halthore, D. Kratz.
Icarus, Vol. 93, No. 1, p. 174 – 178 (Sep 1991).
Jovian emission due to $^{12}CH_4$, $^{13}CH_4$, and CH_3D in the 1100 to 1400 cm^{-1} region is simulated at high spectral resolution as seen from the Earth from an altitude of 4 km above sea level. Due to absorption in the Earth's atmosphere, strong lines do not contribute significantly to the total intensity in this region; it is the weak to marginally strong lines that contribute most. Polar hot spots and other midlatitude features observed in broadband observations are conclusively shown to be stratospheric effects. Signal enhancement in the broadband observations is obtained at zenith angles close to zero and increased doppler shifts; however, the magnitude in the latter case is a sensitive function of the stratospheric temperature.

022.066 **Submarine measurement of the Newtonian gravitational constant.**
M. A. Zumberge, J. A. Hildebrand, J. M. Stevenson, R. L. Parker, A. D. Chave, M. E. Ander, F. N. Spiess.
Phys. Rev. Lett., Vol. 67, No. 22, p. 3051 – 3054 (25 Nov 1991).
Current Physics Microform No.: 9201B1275.
The authors have measured the Newtonian gravitational constant using the ocean as an attracting mass and a research submersible as a platform for gravity measurements. Gravitational acceleration was measured along four continuous profiles to depths of 5000 m with a resolution of 0.1 mGal. These data, combined with satellite altimetry, sea surface and seafloor gravity measurements, and seafloor bathymetry, yield an estimate of $G = (6.677 \pm 0.013) \times 10^{-11}$ $m^3 s^{-2} kg^{-1}$; the fractional uncertainty is 2 parts in 1000. Within this accuracy, the submarine value for G is consistent with laboratory determinations.

022.067 **Compilation of atomic oscillator strengths for carbon, nitrogen and oxygen ions. II. Addendum for the beryllium isoelectronic sequence.**
N. Allard, M.-C. Artru, T. Lanz, M. Le Dourneuf.
Astron. Astrophys., Suppl. Ser., Vol. 91, No. 3, p. 399 – 408 (Dec 1991).
This addendum complements the recent compilation of oscillator strengths for C III, N IV and O V, by Allard et al. (1990). It concerns high-excitation multiplets of C III and N IV. Adopted gf-values are proposed, following the same scheme of averaging. Additional references are given to recent theoretical results. In particular the authors present a comparative analysis of the data of Nussbaumer and Storey (1984) which were not previously considered.

022.068 **Shape and clustering effects on the optical properties of amorphous carbon.**
F. Rouleau, P. G. Martin.
Astrophys. J., Vol. 377, No. 2, p. 526 – 540 (20 Aug 1991).
Optical constants for some samples of amorphous carbon are presented for the energy range 4.1×10^{-3} to 3.5×10^3eV, using a self-consistent method which combines extinction data in the infrared with other measurements of optical constants at higher energy. By construction, the optical constants satisfy the Kramers-Kronig relations. The effects of shape and clustering on the optical constants are examined and found to be significant. The CDE shape distribution mimics qualitatively the effects of compact clusters. Other data showing how the optical properties depend on the degree of hydrogenation and on the type of bonding in the amorphous carbon are discussed. Further laboratory measurements are suggested. The relevance of these optical constants to circumstellar and interstellar grains is reviewed.

022.069 **Laboratory simulation, a tool for comet research.**
E. Grün, H. Kochan, K. J. Seidensticker.
Geophys. Res. Lett., Vol. 18, No. 2, p. 245 – 248 (Feb 1991).
Simulation experiments are performed in order to study processes which could lead to a better understanding of cometary nuclei. During the KOSI (Kometensimulation) experiments decimeter-sized samples of porous H_2O and CO_2 ice-dust mixtures are irradiated with light. Phenomena like heat transport into the sample, compositional and structural modifications of the sample and release of dust and gas are studied which occur within centimeter to decimeter distance from the sample surface. Techniques to handle and analyse low temperature ice samples are developed. So far, six KOSI experiments have been performed and about six more are expected to be carried out within the next three years. This paper describes the approach taken and techniques used in the KOSI experiments. The authors discuss the relevance of some initial results obtained for cometary research.

022.070 **KOSI?**
H. U. Keller, W. J. Markiewicz.
Geophys. Res. Lett., Vol. 18, No. 2, p. 249 – 252 (Feb 1991).
KOSI experiments are discussed from the point of view of their possible applications to cometary sciences. Scales and geometries of KOSI and comets are compared, making it obvious that laboratory simulation of any global cometary features is impossible. To indicate the difficulties in process simulation, the authors discuss in some detail the properties of the gas flow generated by the sublimating ice/dust samples, the interaction of the gas with the emitted grains and heat transport within the sample. Some emphasis is laid on the gas-grain interaction showing that large grains have to be very fluffy (fractal) if gas drag is responsible for their emission.

022.071 **Visible and near IR albedo measurements of ice/dust mixtures.**
A. Oehler, G. Neukum.
Geophys. Res. Lett., Vol. 18, No. 2, p. 253 – 256 (Feb 1991).
The surface energy budget is an important input parameter for modelling the dynamical processes of the KOSI samples. In order to calculate the absorbed fraction of the irradiation the directional-hemispherical albedo of the samples must be known. The albedo is calculated from the spectral radiance factor in the wavelength range from 0.36 μm to 2.5 μm. All samples were analysed before and after the experiments. The spectra of the experiments KOSI 3, 4 and 6 have been measured under three different illumination geometries. During the KOSI sublimation experiments a dust mantle 1 to 5 millimeter thick forms on top of the samples. This changes the spectrophotometric properties of the sample surfaces in terms of scattering characteristics and water ice absorptions at 1.5 μm and 2.0 μm. Water ice absorptions observed after the experiment may be partly due to frost contamination.

022.072 **Energy balance of the KOSI 4 experiment.**
E. Grün, J. Benkhoff, R. Heidrich, P. Hesselbarth, H. Kohl, E. Kührt.
Geophys. Res. Lett., Vol. 18, No. 2, p. 257 – 260 (Feb 1991).
A detailed energy analysis of an ice-dust-sample under interplanetary space conditions is presented. It is based on measurements obtained during a comet simulation (KOSI) experiment. Results of the analysis show that visible and thermal reradiation from the sample surface, the sublimation energy of the ices and the increase of internal heat are the most important contributions to the energy balance. At the beginning of irradiation a relatively high ice sublimation rate is observed. Dust particles remaining on the surface, build up a nonvolatile mantle, producing an obstacle for heat flow inwards and gas flow outwards. Consequences for cometary nuclei models are discussed.

022.073 **Thermal histories of the KOSI samples.**
J. Benkhoff, T. Spohn.
Geophys. Res. Lett., Vol. 18, No. 2, p. 261 – 264 (Feb 1991).
Thermal histories of (H_2O, CO_2, CH_3OH) porous ice/dust samples during KOSI sublimation experiments are presented and interpreted.

022.074 **The effect of non-volatile porous layers on temperature and vapor pressure of underlying ice.**
N. I. Kömle, G. Steiner, M. Baguhl, H. Kohl, H. Kochan, K. Thiel.
Geophys. Res. Lett., Vol. 18, No. 2, p. 265 – 268 (Feb 1991).
The space missions to comet Halley have confirmed the view that on large parts of the surface of a cometary nucelus the ice is buried below a non-volatile dust layer or even a sintered cohesive crust. In order to obtain a better understanding of such ice/crust systems, the thermal behavior of ice samples covered by non-volatile porous material under illumination by an artificial "sun" was studied by means of laboratory experiments. Both loose dust mantles and cohesive porous crusts were used as covers. For comparison an experiment with uncovered ice was also performed. The authors recorded temperature and gas pressure close to the surface of the porous ice and it was found that due to the suppression of gas outflow the ice temperatures in the covered ice became always higher than in the "free sublimation" experiment. Furthermore, a pressure build-up of several Pa was observed in the covered ice close to the surface. In some experiments the

measured pressures remained significantly below the saturation vapor pressure which corresponds to the measured temperature at the same depth, indicating strong deviations from equilibrium conditions.

022.075 **Gas release from ice/dust mixtures.**
P. Hesselbarth, D. Krankowsky, P. Lämmerzahl, K. Mauersberger, A. Winkler, P. Hsiung, K. Rössler.
Geophys. Res. Lett., Vol. 18, No. 2, p. 269 – 272 (Feb 1991).
Gas fluxes from insolated ice/dust mixtures were measured by pressure gauges and a mass spectrometer placed one meter in front of the sample. As soon as the insolation of about one solar constant was started a fast rise in H_2O and CO_2 emissions from the fresh surface was observed. Thereafter gas fluxes slowly decreased. When intermittent dark periods occurred, water emission dropped rapidly while CO_2 fluxes were measured for hours. Results from labeled isotope layers gave information on sublimation depth and temperature and showed that sublimated CO_2 migrated also toward deeper parts of the sample where it recondensed, producing an enrichment. A simple model is presented which describes the redistribution of CO_2 within the sample consistent with the gas phase and solid phase measurements of CO_2 abundances.

022.076 **KOSI: gas drag derived from ice/dust–particle trajectories.**
H. Kochan, W. J. Markiewicz, H. U. Keller.
Geophys. Res. Lett., Vol. 18, No. 2, p. 273 – 276 (Feb 1991).
Trajectories of the emitted ice/dust particles were successfully monitored by a high speed shutter video camera during the large scale experiment KOSI–5 (Nov. 1989) performed in the DLR Space Simulator. These new results complement and quantify the high resolution observations of the surface obtained previously in the small chamber. The particles are coupled rather than loosely deposited on the surface. The authors conclude that the particle dynamics is dominated by the acceleration during their flight. This acceleration is caused by the momentum transfer from the gas sublimating from the sample. The net observed acceleration in the direction normal to the surface is in the range 10–20 m/sec^2. The ice/dust particles appear to be entrained into the gas flow, rather than ejected from the surface.

022.077 **Measurement of the volatile component in particles emitted from an ice/dust mixture.**
K. Mauersberger, H.-J. Michel, D. Krankowsky, P. Lämmerzahl, P. Hesselbarth.
Geophys. Res. Lett., Vol. 18, No. 2, p. 277 – 280 (Feb 1991).
A set of novel instruments has been developed to measure the volatile component of solid particles emitted from the surface of an isolated ice/dust mixture. This paper reports first results of the experiments.

022.078 **Dust emission of mineral/ice mixtures: residue structure and dynamical parameters.**
K. Thiel, G. Kölzer, H. Kohl.
Geophys. Res. Lett., Vol. 18, No. 2, p. 281 – 284 (Feb 1991).
In a series of experiments the particle emission from ice/dust mixtures under insolation at space conditions was investigated using an array of particle collectors and Piezo impact detectors. The dust data provide an estimate of the ratio of ice containing to ice free particles for different size ranges, indicating a decrease in the frequency of ice free particles with increasing particle size. The measurement of the main dynamical parameters of the particle emission reveals a significant interrelation of particle size, starting angle and velocity of the emission process. The near surface particle stream exhibits an internal structure of a low–speed central core of large grains superimposed by a high–speed background of small grains. Possible implications for the dynamics of a dust jet emanating from a comet nucleus are discussed.

022.079 **Cometary analogue material: preparation, composition, and thin section petrography.**
D. Stöffler, H. Düren, J. Knölker, R. Hische, A. Bischoff.
Geophys. Res. Lett., Vol. 18, No. 2, p. 285 – 288 (Feb 1991).
Cometary analogue material has been synthesized in kilogram quantities by spraying suspensions of silicate mineral dust with a median grain size of 4–8 μm, and carbon (soot of $\sim$25 nm grain size) into liquid nitrogen under controlled conditions. The samples with densities between 0.3 and 0.7 g/cm^3 were used for sublimation experiments and for uniaxial compression tests. In one case fragments of consolidated phonolitic tuff (0.6–2.5 and 2.5–10 nm size fractions) were admixed mechanically. By modifying existing technique polished sections and thin sections were produced from these samples for optical microscopy. Changes in the intergrowth characteristics, the grain size of the ice crystals, and the porosity were observed in the irradiated and compressed samples. They are due to sublimation–condensation–recrystallization and compaction–recrystallization effects, respectively.

022.080 **Comet simulation experiments as a prelude to the CRAF and ROSETTA missions.**
A. Bar-Nun.
Geophys. Res. Lett., Vol. 18, No. 2, p. 289 – 291 (Feb 1991).
Some of the achievements of small scale comet simulation experiments and modeling are described, which led to a need of large scale simulation experiments such as KOSI. As part of the KOSI effort, a new machine is being built at Tel Aviv University to produce large samples of gas–laden amorphous water ice. These large scale experiments are a preparation for the coming CRAF and ROSETTA missions to comets.

022.081 **Dissociative recombination of the ground state of $N^+{}_2$.**
S. L. Guberman.
Geophys. Res. Lett., Vol. 18, No. 6, p. 1051 – 1054 (Jun 1991).
Large scale theoretical quantum chemical calculations are reported for the dissociative recombination with an electron of the lowest vibrational level of the ground state of the nitrogen molecular ion. The calculated dissociative recombination rate is $1.6x10^{-7}x(T_e/300)^{-.37}cm^3/sec$ for $100 < T_e < 1000K$ and is in excellent agreement with prior microwave afterglow experiments but disagrees with recent merged beam results. The dissociation limit yields mostly $^4S + {}^2D$ atoms in disagreement with plasma flow tube results. The dominant mechanism for dissociative recombination imparts sufficient energy to the product atoms to allow for escape from the Martian atmosphere.

022.082 **Theoretical determination of the intermolecular potentials of the C_2–H_2 system. Application to the collisional deexcitation of C_2 in collisions with H_2.**
H. Lavendy, J. M. Robbe, G. Chambaud, B. Levy, E. Roueff.
Astron. Astrophys., Vol. 251, No. 1, p. 365 – 368 (Nov 1991).
Rotational deexcitation cross–sections are calculated for C_2 ground state by collisions with H_2 in an IOS treatment after the relevant potential surfaces have been determined. The corresponding rates are then determined for temperatures relevant to interstellar clouds.

022.083 **Collisional excitation of formaldehyde in "hot" interstellar molecular regions.**
S. Green.
Astrophys. J., Suppl. Ser., Vol. 76, No. 3, p. 979 – 983 (Jul 1991).
Earlier calculations for the rotational excitation of formaldehyde by collisions with He atoms are extended to include the lowest 81 rotational levels and kinetic temperatures to 300K.

022.084 **Forbidden lines of np^q ions. II. Line intensities.**
J. P. Lynch, M. Kafatos.
Astrophys. J., Suppl. Ser., Vol. 76, No. 4, p. 1169 – 1191 (Aug 1991).
Ground state forbidden transitions of np^q ions of C, N, O, Ne, Mg, Si, S, and Fe can provide important information on the state of cosmic ionized gases. Wavelengths of these lines are in the far– and near–UV, visible, and near– and far–IR regions of the

spectrum. The line intensity ratios of particular transitions in q = 2, 4 ions can provide information on the temperature of the gas and in q = 3 ions information on the density of the gas. The authors have tabulated the line intensities of 95 transitions of these ions. These data can be used for calculations of the absolute line intensities, if the ionic abundances are known. Graphical presentations for 25 of the most interesting cases are given. These calculations are important for studies of the solar transition region, ionized nebulae, circumstellar shells, supernova remnants, interstellar bubbles produced by stellar winds, and emission regions in active galactic nuclei.

022.085 **Atomic data for resonance absorption lines. I. Wavelengths longward of the Lyman limit.**
D. C. Morton.
Astrophys. J., Suppl. Ser., Vol. 77, No. 1, p. 119 – 202 (Sep 1991).
This compilation presents wavelengths, statistical weights, and excitation energies for 2031 atomic spectral lines in 658 multiplets whose lower energy levels lie below 2217 cm^{-1} or 0.275 eV. Transition probabilities and oscillator strengths (f–values) are included for most of the lines, and radiation damping constants for many of the stronger ones. This list of the resonance lines contains the transitions likely to be observed in absorption in interstellar gas clouds, circumstellar shells, stellar winds, QSO winds, and clouds on the sight lines to QSOs, where neither the particle density nor the radiation density is large enough to populate the higher levels significantly. The tables cover transitions of all the ion stages of all elements from hydrogen to germanium which have permitted or intersystem lines that originate from the ground level or excited fine–structure levels of the ground term at wavelengths longward of the Lyman limit at 911.75 Å. Forbidden lines and molecular lines are excluded. Many weak lines are included for the more abundant elements, so that to the extent that the available estimates of f–values permit, the tables are complete for all interstellar atomic absorption lines with $Z \leqslant 32$ that are likely to be detected by the Hubble Space Telescope.

022.086 **Laboratory absorption spectra of molecules at interstellar cloud temperatures: first measurements on CO at ~97 nm.**
P. L. Smith, G. Stark, K. Yoshino, K. Ito, M. H. Stevens.
Astron. Astrophys., Vol. 252, No. 1, p. L13 – L15 (Dec 1991). Letter–to–the–editor.
In the 91–100 nm spectral region, where absorption of photons by interstellar CO usually leads to dissociation, laboratory spectra obtained at ~295K show that most CO bands are both overlapped and perturbed. Reliable band oscillator strengths cannot be extracted from such spectra. As a consequence, synthetic extreme–UV absorption spectra for CO at the low temperatures that prevail in interstellar clouds are uncertain. The authors have used a supersonic expansion technique to cool CO to ~30K and have studied, with high spectral resolution, three bands in the 97 nm region. The measured spectrum at 30K is in reasonable agreement with some published modelled spectra, but the ratios of integrated cross sections are somewhat different from those determined from low resolution spectra obtained at 295K, in which the bands are blended. The significance of the low temperature cross section measurements for interstellar and circumstellar chemistry remains to be investigated.

022.087 **Pure rotational spectrum of hydrogen deuteride by far–infrared Fourier transform spectroscopy.**
L. Ulivi, P. De Natale, M. Inguscio.
Astrophys. J., Lett., Vol. 378, No. 1, p. L29 – L31 (1 Sep 1991).
The far–infrared spectrum of HD from 170 to 350 cm^{-1} is measured with a frequency accuracy of a few parts in 10^7 by means of Fourier transform spectroscopy. A general fit including other existing data provides values for B_0, D_0, and H_0 ground–state rotational constants with almost two orders of magnitude improved accuracy.

022.088 **A technique for measuring the electric properties of planetary grounds.**
R. Grard, B. van Deenen, A. Tabbagh.
2. COSPAR Colloquium: The environmental model of Mars, p. 63 – 67 (1991). – See Abstr. 012.039 for the main entry.
An original method for measuring the resistivity and dielectric constant of planetary grounds is described. The instrument can be integrated on a lander or a penetrator, or carried by a rover. The working principle of the instrument is first recapitulated and experimental results obtained with two different prototypes operated in entirely different environments are presented.

022.089 **Experimental extinction properties of granular mixtures of silicon carbide and amorphous carbon.**
V. Orofino, V. Mennella, A. Blanco, E. Bussoletti, L. Colangeli, S. Fonti.
Astron. Astrophys., Vol. 252, No. 1, p. 315 – 319 (Dec 1991).
The authors report laboratory data on the infrared extinction properties of mixtures composed of amorphous carbon (AC) and silicon carbide (SiC) submicron grains, relevant to the study of dust envelopes around C–rich stars. They have experimentally verified that the extinction spectra of such mixtures can be conveniently evaluated simply by using those derived for the single components. Extrapolation of laboratory data to vacuum has been done starting from infrared spectra of grains embedded in different matrices [(postassium bromide (KBr) and cesium iodide (CsI)]. A first application of the data to the infrared spectrum of a typical IR source GL 799 suggests the presence of AC and SiC (both α and β) particles in its shell. The relative amounts of the three components are in agreement with the elemental abundances in the envelopes around carbon–rich stars.

022.090 **The 3.1 μm feature in ion–irradiated water ice.**
G. A. Baratta, G. Leto, F. Spinella, G. Strazzulla, G. Foti.
Astron. Astrophys., Vol. 252, No. 1, p. 421 – 424 (Dec 1991).
The authors present new experimental data demonstrating the transition from crystalline to amorphous water ice induced by keV ion–irradiation at temperatures between 10 and 80K. The technique used to monitor the structural changes is in situ IR spectroscopy. Some astrophysical applications of the results show that such a process has to be considered when attempts to fit the astronomical spectra to those obtained in the laboratory are done and especially when one attempts to deduce, from the shape of the observed ice band, information on the physical status (e.g. temperature history) of the responsible dust.

022.091 **"Polarization shadows" on the negative branch of the polarization curve.**
V. V. Titov.
Astron. Tsirk., No. 1545, p. 33 – 34 (Aug – Sep 1990). In Russian.
The results of a study of polarization curves' dependence on the phase angle are given. The effect of "polarization shadows" was detected on the negative branch of the polarization curve. The "polarization shadows" are considered to be, probably, properties of the particles surface.

022.092 **Radiative livetimes in Ti I.**
J. E. Lawler.
Astron. Astrophys., Vol. 252, No. 2, p. 853 (Dec 1991).
Radiative lifetimes for 12 of the 161 levels of Ti I reported recently by Salih and Lawler were found to be slightly in error due to misalignment of the fluorescence collection system in the laser induced fluorescence experiment. Improved lifetimes for these long lived levels are reported here. The renormalization of the Oxford Ti I oscillator strengths and revised value for the solar abundance of Ti are not affected by these new results.

022.093 **Millimeter– and submillimeter–wave spectrum of highly excited states of water.**
J. C. Pearson, T. Anderson, E. Herbst, F. C. De Lucia, P. Helminger.
Astrophys. J., Lett., Vol. 379, No. 1, p. L41 – L43 (20 Sep 1991).
To facilitate studies of water in the interstellar medium and late–type stars, the authors have measured the frequencies of 30

new millimeter– and submillimeter–wave transitions of $H_2{}^{16}O$ which lie between 100 GHz and 600 GHz. This represents almost a doubling of the number of water lines that have been observed in the laboratory in this spectral region at high resolution. All of the newly observed lines are highly excited, lying between 2400 and 4200 cm^{-1} above the ground level. Some of these have large excitation energies because of their high rotational states and others because they lie in excited vibrational states.

022.094 **Dielectronic recombination rate coefficients for S^{q+} (q = 1 – 5) ions.**
N. R. Badnell.
Astrophys. J., Vol. 379, No. 1, p. 356 – 358 (20 Sep 1991).
The author has calculated the total dielectronic recombination rate coefficient for S^{q+} (q = 1 – 5) ions, in the low–density limit. The results are tabulated over the temperature range $T = 3\times10^4 - 10^6$K. At electron temperatures ($\sim10^5$K) and densities ($\leqslant10^4 cm^{-3}$) applicable to the Io plasma torus, these results differ substantially from existing data.

022.095 **Laboratory and astronomical spectroscopy of new interstellar molecules.**
S. Saito.
3. Haystack Observatory Conference on the Interstellar Medium: Skylines – atoms, ions and molecules – new results in spectral line astrophysics, p. 349 – 361 (1991). – See Abstr. 012.051 for the main entry.
A high–sensitivity millimeter–and submillimeter–wave spectrometer has been developed and used to study more than 25 new transient molecules, which are potential candidates for interstellar molecules. Various collaborations between laboratory spectroscopy and radio astronomy have led to identification of seven transient species to new interstellar molecules. The discussion has been directed at the spectroscopically and astronomically important aspects of recent studies: C_6H, CCS, C_3S and CH_2CN.

022.096 **Sticking probability of H and D atoms on amorphous ice: a computational study.**
V. Buch, Q. Zhang.
Astrophys. J., Vol. 379, No. 2, p. 647 – 652 (1 Oct 1991).
Sticking of H and D atoms on amorphous ice was studied computationally, using the molecular dynamics simulation technique. A simple formula $S_T = [k_BT/E_0)+1]^{-2}$ was obtained for the sticking probability at $T_{gas} \lesssim 300$K, with $E_0 = 200$K for the D atom, and $E_0 = 102$K for the H atom. The formula is suggested as a reasonable estimate for sticking on dust grains in general – provided that most grains are covered by at least a few layers of weakly bonded disordered molecular material.

022.097 **Improvements in the sensitivity of screen–type X–ray films for cascade showers and the charge resolution for heavy ions.**
T. Kobayashi, M. Fujii, J. Nishimura, T. Taira, H. Aizu, Y. Komori.
Nuovo Cimento A, Vol. 104, No. 12, p. 1753 – 1770 (Dec 1991).
The characteristics of the screen–type X–ray film, and its applications to observations of cosmic–ray electrons and heavy primaries are described in this paper.

022.098 **Determination of absolute strengths of N_2 quadrupole lines from high–resolution ground–based IR solar observations.**
P. Demoulin, C. B. Farmer, C. P. Rinsland, R. Zander.
J. Geophys. Res., Vol. 96, No. D7, p. 13003 – 13008 (20 Jul 1991).
High–resolution, high signal–to–noise ratio, solar absorption spectra recorded with a Fourier transform spectrometer at the International Scientific Station of the Jungfraujoch, Switzerland, have been analyzed to determine the strengths of several lines belonging to the S branch of the N_2 (1–0) electric quadrupole vibration–rotation band centered at 2329.9168 cm^{-1}. The method applied was based on equivalent width measurements of lines observed over a broad range of air masses; extrapolation of these measurements to zero air mass gave the line strengths for the transitions S_7 to S_{10}, independent of half widths, an ambiguity unavoidable with the use of curve–fitting techniques. The resulting absolute accuracies of the line strengths derived here, estimated to be better than $\pm2.5\%$ for S_8, $\pm2.6\%$ for S_{10}, $\pm3.4\%$ for S_9, and $\pm5.1\%$ for S_7, are due largely to the high quality and quantity of the spectra retained in this analysis and the accuracy with which the observation conditions are known. An important application of the improved values for these N_2 transitions, which have low initial ground state energies, is the direct determination of the line–of–sight atmospheric air masses associated with remotely sensed infrared spectroscopic observations. Positions of the N_2 transitions studied here have further been redetermined with an absolute accuracy better than 0.0002 cm^{-1}.

022.099 **On the state of the emitter of the 3.3 micron unidentified infrared band: absorption spectroscopy of polycyclic aromatic hydrocarbon species.**
G. C. Flickinger, T. J. Wdowiak, P. L. Gómez.
Astrophys. J., Lett., Vol. 380, No. 1, p. L43 – L46 (10 Oct 1991).
Experiments directed toward an understanding of the unidentified infrared bands have resulted in infrared absorption spectra of the C–H stretch feature of various polycyclic aromatic hydrocarbon (PAH) molecules while a vapor at elevated temperatures. These experiments demonstrate nonpermanent temperature–dependent changes in the character of the feature similar to what has been seen in previous experiments performed with PAHs encapsulated in a KBr matrix. However, there is a distinct difference in peak absorption wavelength of the vapor phase feature from the feature of the PAH in the condensed state. The C–H stretch peak absorption wavelength distinguishes four classes of PAH at elevated temperature: condensed compact, condensed noncompact, vapor phase compact, and vapor phase noncompact PAH species. The compact PAH coronene has a C–H stretch peak wavelength in the condensed state at elevated temperature of 3.290 μm, whereas it displays an elevated temperature vapor phase spectrum with peak wavelength at 3.276 μm. All vapor phase PAH species studied consistently show peak absorption shortward (0.01 – 0.02 μm) of the peak exhibited by the species in the condensed state. This result suggests that much of the interstellar PAHs may not be isolated "free–flying" molecules but exist, rather, in a condensed form.

022.100 **Stark broadening of Li(I) lines.**
M. S. Dimitrijević, S. Sahal–Bréchot.
J. Quant. Spectrosc. Radiat. Transfer, Vol. 46, No. 1, p. 41 – 53 (Jul 1991).
A semiclassical approach has been used to evaluate electron–, proton– and ionized helium–impact line widths and shifts for 61 neutral lithium multiplets. Calculations have been performed as a function of temperature and at a perturber density of $10^{13} cm^{-3}$. The results have been used to investigate Stark–broadening–parameter regularities within spectral series.

022.101 **Larmor precession in Zeeman spectra of the Cs–D_2 line.**
I. Hirano.
J. Quant. Spectrosc. Radiat. Transfer, Vol. 46, No. 3, p. 151 – 158 (Sep 1991).
In the dilute gas phase of Cs metal, the Larmor precession is found to modulate the optical coherence in a density–matrix formalism of the magneto–optics effect. Through spectral analysis, it is proved that in addition to the actual level splittings, there exists in the Cs Zeeman spectra at 0.015 T a frequency shift of a few hundred megahertz which far exceeds shifts due to atomic collisions. The cause of the shift is attributed to the Larmore precession in a magnetic field.

022.102 **Influence of a d.c. magnetic field on the He I 447.1 nm line in pure helium plasmas**
Z. Mijatović, M. Pavlov, S. Djurović, B. Vujičić, R. Kobilarov.
J. Quant. Spectrosc. Radiat. Transfer, Vol. 46, No. 6, p. 557 – 563 (Dec 1991).
The He I 4471 Å line with its forbidden component has been studied experimentally in pure helium plasmas in the presence

and in the absence of a magnetic field with a flux density of 0.5 T. Profiles were recorded parallel and perpendicularly to the magnetic field lines for an electron–density range from 4×10^{22} to $9.2 \times 10^{22} m^{-3}$ and a temperature range between 20,000 and 30,000K. The magnetic field influence only the peak separation between the allowed and forbidden components of the line.

022.103 **Stark broadening regularities along a homologous sequence of the second group in the periodic system.**
S. Djeniže, A. Srećković, J. Labat.
J. Quant. Spectrosc. Radiat. Transfer, Vol. 46, No. 5, p. 433 – 437 (Nov 1991).
The existence of a Stark–width dependence on the upper–level ionization potential has been demonstrated for the homologous sequence of the second group of the elements in the periodic system for three types of transitions in the case of singly–ionized atoms. Equality has been noted for the reduced Stark widths of ionic spectral lines in the A and B subgroups of the fourth, fifth and sixth periods. Using the observed trends, the authors have predicted Stark widths for the Be II, Sr II and Ra II spectral lines that have not yet been measured or calculated.

022.104 **Investigation of the continuum emission of krypton plasmas.**
S. Manola, S. Dević, A. Lesage.
J. Quant. Spectrosc. Radiat. Transfer, Vol. 46, No. 6, p. 577 – 582 (Dec 1991).
Spectral continuum emission coefficients of krypton over the wavelength range 370 – 670 nm have been measured in a plasma produced in a shock tube. The ξ factor was derived after subtraction of line emission. The plasma electron density was measured with a 3.39 μm Michelson laser interferometer. Dynamic plasma pressure measurements permitted calculation of neutral contributions and the index of refraction variation. The measured ξ factors agree with other experimental data within the experimental errors. The dependence of ξ on wavelength and electron density in the range $3 - 10 \times 10^{16} cm^{-3}$ is discussed.

022.105 **Stark broadening of the He I 4471 Å line and its forbidden component at high electron densities.**
N. I. Uzelac, I. Stefanović, N. Konjević.
J. Quant. Spectrosc. Radiat. Transfer, Vol. 46, No. 5, p. 447 – 453 (Nov 1991).
Results are reported of measurements on the neutral helium 4471 Å line and its forbidden component at 4470 Å in the plasma of a high–pressure pulsed discharge at an electron temperature of about 2.5 eV and at electron densities ranging from 0.6×10^{17} to $3.4 \times 10^{17} cm^{-3}$. Good agreement is found with other low–electron–temperature experiments at about 1.5 eV. Differences between the results obtained in these studies and other experimental data show, in some cases, systematic discrepancies with the results derived from available calculations.

022.106 **Transition probabilities of *6p–ns* (*n* = 7, 8, 9, 10, 11, 12, 13) lines of Xe(I).**
C. Peraza, P. Martin, J. Campos.
J. Quant. Spectrosc. Radiat. Transfer, Vol. 46, No. 5, p. 455 – 458 (Nov 1991).
Relative transition probabilities for 60 lines belonging to the *6p–ns* (n = 7, 8, 9, 10, 11, 12, 13) transition arrays of Xe(I) have been determined from emission line–intensity measurements of an optically–thin light source. Transition probabilities were placed on an absolute scale by using line–strength sum rules. Comparison is made with theoretical results.

022.107 **Core–excited doublet and quartet states in the sodium isoelectronic sequence.**
T. M. El–Sherbini, A. A. Farrag.
J. Quant. Spectrosc. Radiat. Transfer, Vol. 46, No. 5, p. 473 – 475 (Nov 1991).
The authors have calculated the energies of doublet and quartet states of some $2s2p^6nln'l'$ configurations in Na I and Na–like ions using the Hartree–Fock method. Good agreement has been obtained between the authors' results and previously derived data using photoabsorption measurements.

022.108 **Collisional rates and cooling within atomic hydrogen plasmas.**
T. T. Scholz, H. R. J. Walters.
Astrophys. J., Vol. 380, No. 1, p. 302 – 306 (10 Oct 1991).
Accurate collisional rates for electron impact ionization of the ground state of atomic hydrogen are evaluated by convolving the experimental cross section with a Maxwellian distribution of electronic energies. An extension of previous work on 1s – 2s and 1s – 2p scattering rates is also reported. These results are supplemented by semiempirical collisional rates for other excitation processes to yield the cooling rate for optically thin hydrogen plasmas. The numerical results are evaluated for temperatures ranging from 2×10^3K to 1×10^8K, and are fitted to simple analytic formulae designed to be easily incorporated into computer programs.

022.109 **Temperature–sensitive line ratios of the lithium–like ion O VI.**
E. Haug.
Sol. Phys., Vol. 136, No. 1, p. 111 – 116 (Nov 1991).
The intensity ratios of emission lines from the lithium–like ion O VI are calculated as a function of electron temperature using recent computations of collision strengths by Zhang, Sampson, and Fontes (1990). The line ratios are strongly dependent on temperature and are appropriate as a diagnostic of hot thin plasmas such as in the solar transition region and corona.

022.110 **Collisional cooling as an environment for planetary research.**
T. M. Goyette, F. C. de Lucia, D. R. Willey.
J. Geophys. Res., Vol. 96, No. E2, p. 17455 – 17462 (25 Sep 1991). – See Abstr. 012.073 for the main entry.
A new experimental technique which provides a laboratory environment much like that of the outer planets is described. More specifically this technique, collisional cooling, allows the study of the collisional properties of molecular species at temperatures which in ordinary spectroscopic cells would result in vanishingly small vapor pressures due to condensation. In this paper the basic theory of collisions is put forth, the experimental techniques described, and the results of pressure broadening studies for a number of species described.

022.111 **Laboratory measurement of the millimeter wave properties of liquid sulfuric acid (H_2SO_4).**
A. K. Fahd, P. G. Steffes.
J. Geophys. Res., Vol. 96, No. E2, p. 17470 – 17476 (25 Sep 1991). – See Abstr. 012.073 for the main entry.
Recent observations of the millimeter wave (2.6 mm) emission from Venus have shown significant variations in its continuum flux emission (de Pater et al., 1991). Some of this change in emission may be attributed to variability in the abundance of Venus cloud constituents, specifically H_2SO_4 condensates. In order to evaluate this hypothesis, a knowledge of the dielectric properties of liquid sulfuric acid at millimeter wave frequencies ($30 < f < 300$ GHz) is needed. The dielectric constants can then be used to determine the expected absorptivity from such a condensate and hence the effect of H_2SO_4 droplets on the millimeter wave emission from Venus. Therefore the authors have performed laboratory measurements of the complex dielectric constant of liquid sulfuric acid between 30–40 and 90–100 GHz. The methodology and the results are reported in this paper along with a calculation of the absorption properties of H_2SO_4 droplets under Venus–like conditions. In addition, this paper discusses the effects of these condensates on the variation in the flux emission of Venus and compares the effect of H_2SO_4 droplets with other absorbers in the atmosphere of Venus such as gaseous SO_2. The authors conclude that this condensate does affect the brightness temperature of Venus and its effect cannot be omitted in future modeling of the atmosphere of Venus.

022.112 **Optical constants of solid methane and ethane from 10,000 to 450 cm^{-1}.**
J. Pearl, M. Ngoh, M. Ospina, R. Khanna.
J. Geophys. Res., Vol. 96, No. E2, p. 17476 - 17482 (25 Sep 1991). – See Abstr. 012.073 for the main entry.
The authors made thin film transmission measurements of solid CH_4 (phases I and II) and C_2H_6 (phase II). From these, the complex indices of refraction at near- and mid-infrared wavelengths were determined by using a combined least squares and Kramers-Kronig analysis.

022.113 **Methane spectral line widths and shifts, and dependences on physical parameters.**
K. Fox, D. T. Quillen, D. E. Jennings, J. Wagner, C. Plymate.
J. Geophys. Res., Vol. 96, No. E2, p. 17483 - 17488 (25 Sep 1991). – See Abstr. 012.073 for the main entry.
High-resolution measurements of widths and shifts of vibrational-rotational lines in infrared spectra of gaseous methane have been made in the laboratory. This research introduced at ambient temperatures the rare gases helium, neon, argon, krypton, and xenon and the diatomics hydrogen, nitrogen, oxygen, and carbon monoxide at pressures from 100 to 700 torr. Detailed quantitative studies of correlations with masses, polarizabilities and quadrupole moments suggest that there may be some functional dependences on these physical parameters.

022.114 **Laboratory spectra of ^{13}C ethane.**
J. Kurtz, D. C. Reuter, D. E. Jennings, J. J. Hillman.
J. Geophys. Res., Vol. 96, No. E2, p. 17489 - 17492 (25 Sep 1991). – See Abstr. 012.073 for the main entry.
The laboratory infrared spectrum of ^{13}C monosubstituted ethane has been obtained at high resolution (0.0025 cm^{-1}) using the McMath Fourier transform spectrometer at Kitt Peak National Observatory in May 1990. A preliminary analysis of the $\nu_{12}\,{}^{r}Q_0$ branch (substituted species) suggests that its intensity is 1.15 ± 0.05 times stronger than the equivalent ν_9 branch in the normal $^{12}C_2H_6$ species. This result leads to a correction of a previously published estimate for the $^{12}C/^{13}C$ ratio in the atmosphere of Jupiter from ~ 94 to ~ 106.

022.115 **Herman-Wallis factors in the C_2H_2 ν_5 infrared fundamental near 14 μm.**
W. C. Maguire, J. J. Hillman, M. Weber, W. E. Blass.
J. Geophys. Res., Vol. 96, No. E2, p. 17493 - 17496 (25 Sep 1991). – See Abstr. 012.073 for the main entry.
The presence of acetylene has been confirmed for some time in the atmospheres of the outer planets Jupiter, Saturn, Neptune, and Saturn's satellite Titan. For these atmospheres, the determination of C_2H_2 abundances using its strong ν_5 fundamental requires laboratory line position and intensity measurements. The 1-meter Fourier transform spectrometer at McMath solar telescope of Kitt Peak National Observatory was used to measure C_2H_2 at an unapodized spectral resolution of 0.0025 cm^{-1}. Synthetic spectra are generated by convolving a Voigt line shape with an instrument function and varying intensity parameters by means of nonlinear least squares technique. Intensities of 37 ν_5 lines spanning P18 through R20 were measured using 0.123 torr of gas in a 1-cm cell. A Herman-Wallis intensity correction parameter $A^{pr}{}_1 = 1.3(4) \times 10^{-3}$ has been derived using a least squares linear fit.

022.116 **Measurements of pure CO_2 absorption at high densities near 2.3 μm.**
C. Brodbeck, Nguyen Van-Thanh, J. P. Bouanich, C. Boulet, A. Jean-Louis, B. Bezard, C. de Bergh.
J. Geophys. Res., Vol. 96, No. E2, p. 17497 - 17500 (25 Sep 1991). – See Abstr. 012.073 for the main entry.
Thermal emission from the deep atmosphere of Venus can be detected on the nightside around 2.3 μm. The analysis of this radiation requires a reliable knowledge of the absorption in the far wings of the nearby allowed CO_2 bands and of the absorption due to collision-induced bands. The authors measured absorption coefficients for pure CO_2 at pressures varying from 30 to 60 bars in the frequency range 3910-4570 cm^{-1} at 297.5K. Values between 1.0 and $1.6 \times 10^{-7} cm^{-1} amagat^{-2}$ are found in the 4100-4500 cm^{-1} interval where emission from the Venus nightside occurs. The comparison of experimental results with synthetic spectra calculated from a line by line code demonstrates that the Lorentzian line shape strongly overestimates the observed absorption, whereas the use of a χ factor extrapolated from the 3800-4000 cm^{-1} region does not provide enough opacity.

022.117 **Pressure-induced absorption in the infrared: a data base for the modelling of planetary atmospheres.**
A. Borysow, L. Frommhold.
J. Geophys. Res., Vol. 96, No. E2, p. 17501 - 17506 (25 Sep 1991). – See Abstr. 012.073 for the main entry.
Recent progress related to the modeling of various pressure-induced absorption bands of binary molecular systems, involving dissimilar and like pairs of molecules such as H_2, He, N_2, CH_4, etc., is reported. The notes include references to the existing efforts based on quantum line shape computations that reproduce the laboratory measurements of pressure-induced absorption closely. The results may be considered a most reliable data base permitting an evaluation of pressure-induced absorption as function of frequency, temperature, and gas density for various binary molecular systems and spectral bands in the infrared region of the spectrum.

022.118 **New measurements of the 6190-Å band of methane.**
M. E. Mickelson, L. E. Larson, A. Schubert.
J. Geophys. Res., Vol. 96, No. E2, p. 17507 - 17512 (25 Sep 1991). – See Abstr. 012.073 for the main entry.
New laboratory measurements have been made of the absorption coefficient of the visible methane band at 6190 Å. Data were obtained using a tunable dye laser system operating with a line width of 0.067 cm^{-1} (0.025 Å). Spectra were recorded at approximately 1-Å intervals with the beam coupled to a 22-mm base length White-type absorption cell adjusted for an optical path of 1.584 km and filled to a density of 0.884 amagats. These room temperature results are compared with previous low-resolution measurements in the spectral region from 6000 to 6400 Å.

022.119 **A laboratory facility for mid-UV absorption spectroscopy of molecular compounds for planetary atmospheres.**
P. Bruston, F. Raulin, H. Poncet.
J. Geophys. Res., Vol. 96, No. E2, p. 17513 - 17517 (25 Sep 1991). – See Abstr. 012.073 for the main entry.
Available laboratory UV spectroscopic data do not usually fulfill all the conditions which are required for reducing observational data of planetary atmospheres: absolute wavelength calibration, wide wavelength range including the mid-UV, spectral resolution, absolute determination of the absorption coefficients, data at various temperatures. Faced with this fact in their research program on the structure and composition of Titan atmosphere, The authors have developed a suitable laboratory facility. They describe this experimental facility: it gives absolute absorption coefficients at 0.1 Å resolution in the range 185-350 nm with sample temperatures adjustable between -50°C and +50°C. The future program to be implemented with this facility includes studies on nitriles and polyacetylenes.

022.120 **High-resolution, low-temperature photoabsorption cross sections of C_2H_2, PH_3, AsH_3, and GeH_4, with application to Saturn's atmosphere.**
F. Chen, D. L. Judge, C. Y. R. Wu, J. Caldwell, H. P. White, R. Wagener.
J. Geophys. Res., Vol. 96, No. E2, p. 17519 - 17528 (25 Sep 1991). – See Abstr. 012.073 for the main entry.
Using synchrotron radiation as a continuum background light source, the VUV absorption cross sections of (1) C_2H_2 and PH_3 have been obtained under physical conditions approximating the stratosphere of Saturn and (2) AsH_3 and GeH_4 have been obtained under room temperature conditions. The authors have combined these new laboratory measurements with International Ultraviolet Explorer observations of the albedo of Saturn, for which improved data reduction techniques have been used, to

produce new models for that atmosphere. When the effects of C_2H_2 absorption are accounted for, additional absorption by other molecules is required. The best fitting model also includes absorption by H_2O, CH_4, C_2H_6, PH_3, and GeH_4. The abundance of H_2 is reduced by a factor of 2 as a direct consequence of the new low-temperature cross section data of C_2H_2. The model abundance distributions of PH_3 and GeH_4 on Saturn decrease with increasing altitude, suggesting a possible consequence of UV photolysis.

022.121 **High-resolution, VUV (147–201 nm) photoabsorption cross sections for C_2H_2 at 195 and 295K.**
P. L. Smith, K. Yoshino, W. H. Parkinson, K. Ito, G. Stark.
J. Geophys. Res., Vol. 96, No. E2, p. 17529 – 17534 (25 Sep 1991). – See Abstr. 012.073 for the main entry.

Acetylene, C_2H_2, plays an important role in the photochemistry of the atmospheres of the giant planets, in part because of its strong absorption features in the 145–201 nm region where there is significant solar flux. Accurate photoabsorption cross section data at the temperatures of planetary atmospheres are required for interpretation of observations and for reliable photochemical models. The authors have measured the photoabsorption cross sections of acetylene at 195 and 295K in the wavelength range 147–201 nm. (The 295K measurements extend to 137 nm.) The high-resolution spectrometers employed had instrumental bandwidths smaller than those used in previous measurements and therefore produced cross sections that are less influenced by instrumental effects. The results indicate that hot bands make small but significant contributions to the room temperature absorption spectrum of C_2H_2. Photochemical models of planetary atmospheres should employ cross sections measured at the temperatures of these atmospheres.

022.122 **Alteration of the UV-visible reflectance spectra of H_2O ice by ion bombardment.**
N. J. Sack, J. W. Boring, R. E. Johnson, R. A. Baragiola, M. Shi.
J. Geophys. Res., Vol. 96, No. E2, p. 17535 – 17539 (25 Sep 1991). – See Abstr. 012.073 for the main entry.

Satellites in the Jovian and Saturnian system exhibit differences in reflectivity between their "leading" and "trailing" surfaces which can affect the local vapor pressure. Since these differences are thought to be due to differences in the flux of bombarding magnetospheric ions, the authors studied the influence of ion impact on the UV-visible reflectance of water ice surfaces (20–90K) by keV ion bombardment. An observed decrease in reflectance in the UV is attributed to rearrangement processes that affect the physical microstructure and surface "roughness". The ratio in reflectance of bombarded to freshly deposited films is compared to the ratio of the reflectance of the leading and trailing hemispheres for Europa and Ganymede.

022.123 **Studies of proton irradiated H_2O+CO_2 and H_2O+CO ices and analysis of synthesized molecules.**
M. H. Moore, R. Khanna, B. Donn.
J. Geophys. Res., Vol. 96, No. E2, p. 17541 – 17546 (25 Sep 1991). – See Abstr. 012.073 for the main entry.

Infrared spectra of H_2O+CO_2 and H_2+CO ices before and after proton irradiation showed that a major reaction in both mixtures was the interconversion of $CO_2 \rightleftharpoons CO$. Radiation synthesized organic compounds such as carbonic acid were identified in the H_2O+CO_2 ice. Different chemical pathways dominate in the H_2O+CO ice in which formaldehyde, methanol, ethanol, and methane were identified. Sublimed material was also analyzed using a mass spectrometer. Implications of these results are discussed in references to comets.

022.124 **Ion irradiation experiments relevant to cometary physics.**
G. Strazzulla, G. Leto, G. A. Baratta, F. Spinella.
J. Geophys. Res., Vol. 96, No. E2, p. 17547 – 17552 (25 Sep 1991). – See Abstr. 012.073 for the main entry.

During their stay in the Oort cloud (4.6×10^9years), comets are subject to the flux of galactic cosmic rays. Experimental results on the chemical and physical changes induced by ion and electron irradiation of relevant materials have been recently reviewed with a view to their relevance for cometary astrophysics. In particular, the ability of ion irradiation of simple carbon-containing ices to produce complex refractory organic materials has been discussed. Here the authors discuss further experiments carried out in their laboratory whose results are relevant to cometary physics.

022.125 **Irradiation effects in a comet's outer layers.**
R. E. Johnson.
J. Geophys. Res., Vol. 96, No. E2, p. 17553 – 17557 (25 Sep 1991). – See Abstr. 012.073 for the main entry.

Laboratory data and cosmic ray doses are used to understand the irradiation-induced formation and destruction of organics in the outer layers (mantle) of a comet in the Oort cloud and for determining the fate of volatiles in the mantle. This is used to estimate the change in the orthowater/parawater ratio with depth in the outer layers of a new comet and the average thickness of the refractory part of the mantle of a new comet. Because comets will have large-scale irregularities, regions of the mantle will be unstable, leading to enhanced activity on a new comet and producing "permanently" active regions.

022.126 **Electron impact excitation and dissociation of N_2 via the $b^1\Pi_u$ state.**
J. M. Ratliff, G. K. James, S. Trajmar, J. M. Ajello, D. E. Shemansky.
J. Geophys. Res., Vol. 96, No. E2, p. 17559 – 17561 (25 Sep 1991). – See Abstr. 012.073 for the main entry.

Electron impact excitation of the $b^1\Pi_u$ state in N_2 plays a prominent role in the dissociation of the molecule and thus in the production of atomic nitrogen in planetary atmospheres. Electron impact excitation cross sections combined with electron-impact-induced fluorescence measurements can yield the corresponding dissociation cross sections. Serious discrepancies exist among excitation cross sections reported in the literature. To clarify the situation, the authors measured these cross sections at two impact energies using electron energy loss spectroscopy. The new results are in agreement with recent values deduced from optical measurements and fall midway between previous results which are too high or low by factors of 2.

022.127 **Cross sections for direct and dissociative ionization of NH_3 and CS_2 by electron impact.**
M. V. V. S. Rao, S. K. Srivastava.
J. Geophys. Res., Vol. 96, No. E2, p. 17563 – 17568 (25 Sep 1991). – See Abstr. 012.073 for the main entry.

Cross sections for the production of positive ions by electron impact on NH_3 and CS_2 have been measured using a crossed electron beam-molecular beam collision geometry. The present results on NH_3 are considerably different from previous measurements which disagree among themselves. Except for the formation of CS_2, cross sections for other fragment ions resulting from CS_2 have not been measured in the past. Threshold energies for the appearance of various positive ions are also reported. It is estimated that the present results are accurate to within 15%.

022.128 **Implications of C_2H photochemistry on the modeling of C_2 distributions in comets.**
W. M. Jackson, Y. Bao, R. S. Urdahl.
J. Geophys. Res., Vol. 96, No. E2, p. 17569 – 17572 (25 Sep 1991). – See Abstr. 012.073 for the main entry.

Laboratory studies of the secondary photolysis of the C_2H radical are summarized and used to explain some discrepancies between models of C_2 emission in comets. These studies show that several states of the C_2 radicals produced in the photolysis of C_2H_2 at 193 nm have bimodal rotational distributions when plotted as a Boltzmann diagram. They also establish that the C_2 radicals are formed with varying degrees of vibrational excitation, so that if they are formed in a similar manner in comets, the C_2 radicals must start out with this initial vibrational excitation.

022.129 **Spectral signature of oxidized pyroxenes: implications to remote sensing of terrestrial planets.**
D'A. W. Straub, R. G. Burns, S. F. Pratt.
J. Geophys. Res., Vol. 96, No. E3, p. 18819 – 18830 (25 Oct 1991).
The Fe^{2+} ions in pyroxenes occurring in basalts on surfaces of terrestrial planets are vulnerable to oxidation, particularly in magma in contact with the atmosphere; this produces Fe^{3+}–bearing assemblages that may mask spectral features of remnant Fe^{2+} in pyroxenes in telescopic reflectance spectra of oxidized planetary surfaces. To assess such interferences, reflectance spectra (0.3–2.6 μm) were obtained for a variety of pyroxenes that had been heated in air at elevated temperatures.

022.130 **Thermal budget of multicomponent porous ices.**
G. Steiner, N. I. Kömle.
J. Geophys. Res., Vol. 96, No. E3, p. 18897 – 18902 (25 Oct 1991).
A relation for the temperature evolution of multicomponent porous ices is derived from the energy equation of multicomponent systems. It is demonstrated that the predominant energy transfer mechanisms can be incorporated into an effective thermal conductivity (λ_{eff}), which is strongly temperature dependent, and that the heat conduction equation can be written in a Fourier type form. The vapor of volatile ices significantly contributes to λ_{eff} by the transfer of latent heat of sublimation but otherwise hardly affects the energy balance. The general results are applied to porous water ice, porous H_2O–CO_2 ice and porous H_2O–CO ice. In all cases it is found that the major part of the deposited energy per unit volume is consumed for heating of the ice. Depending on the material parameters and temperature, energy is transferred mainly by solid state conduction or by the transport of latent heat due to sublimation–condensation processes.

022.131 **Theoretical He I line intensities in low–density plasmas.**
D. P. Smits.
Mon. Not. R. Astron. Soc., Vol. 251, No. 2, p. 316 – 317 (15 Jul 1991). With microfiche MN 251/10.
A capture–collision–cascade model has been developed for calculating recombination spectra of He I in nebular conditions. The method of solution closely follows the treatment of Brocklehurst. The results of computations at temperatures from 312.5 to 20,000K and electron densities N_e of 10^2, 10^4 and $10^6 cm^{-3}$ are presented. The singlets have been treated in both case A and B conditions, the triplets in case B only. The He I results use newer data than the calculations of Brocklehurst and highlight an error in his work. These calculations can be used to interpret the line emissions from He I in the optical and infrared bands.

022.132 **Laboratory simulations of large scale fragmentation events.**
K. R. Housen, R. M. Schmidt, K. A. Holsapple.
Icarus, Vol. 94, No. 1, p. 180 – 190 (Nov 1991).
Experimental studies of catastrophic fragmentation are necessarily limited to centimeter–sized targets. This, coupled with uncertainties in the scaling relationships for fragmentation, have made it difficult to estimate the conditions required to collisionally disrupt large asteroids or planetary satellites. A fundamental uncertainty arises because the results for small scale experiments are dominated by the strength of the target, while sufficiently large bodies are dominated by the pressure due to gravitational forces. This paper summarizes the results of fragmentation experiments conducted at elevated external pressure. By applying the appropriate overpressure to a small target, an experiment matches the lithostatic compressive stress which exists in the interior of a larger body. As the overpressure is increased, the fragmentation transitions form a regime dominated by material strength to one dominated by the applied overpressure. An analogy is made to the transition from strength to gravity effects. The high pressure tests allow direct measurement of the energy per unit target mass required to catastrophically fragment a body subject to large compressive stresses. These specific energies compare well with those estimated from observations of the Themis, Eos, and Koronis families.

022.133 **Derivation of midinfrared (5 – 25 μm) optical constants of some silicates and palagonite.**
T. Roush, J. Pollack, J. Orenberg.
Icarus, Vol. 94, No. 1, p. 191 – 208 (Nov 1991).
The midinfrared 5 – 25 μm optical constants are presented for: (1) pyrophyllite; (2) kaolinite; (3) serpentine; (4) montmorillonite; (5) saponite; (6) palagonite; and (7) orthopyroxene. Comparison of the values derived here with those previously presented for serpentine, montmorillonite, and palagonite is generally quite good and discrepancies between values are probably due to either chemical differences between the actual samples or different techniques used to derive the values. For montmorillonite, saponite, and palagonite the authors were able to derive optical constants in the region of the H_2O–bending fundamental near 6 μm. They find that if a pellet of pure material can be produced with a mirror–like surface then the optical constants of clays and other noncohesive materials can be readily derived.

022.134 **Laboratory reflectance measurements of analogues to "dirty" ice surfaces on atmosphereless solar system bodies.**
J. R. Stephens, B. Å. S. Gustafson.
Icarus, Vol. 94, No. 1, p. 209 – 217 (Nov 1991).
This article presents laboratory measurements of bidirectional reflectances from particle/ice mixtures representing "dirty" ice surfaces on atmosphereless Solar System bodies. The reflectance was monitored as ice sublimed under vacuum from the mixtures leaving behind porous refractory particle layers up to a millimeter thick. Similar layers or "dust–mantles" may evolve on cometary surfaces and on icy asteroids. Before significant ice sublimation, a 10% (by mass) colloidal silica in water ice sample was the brightest with reflectances in the 0.20 – 0.30 range. A similar mixture with the addition of 1% carbon had only 0.03 to 0.06 reflectance. A mixture of 1% carbon in water ice (without silica) has an intermediate reflectance of 0.05 to 0.20. Only carbon in ice showed marked angular dependence with a broad increase toward the specular angle. The formation of porous refractory layers or mantles through sublimation is coupled with significant changes in reflectance and color. The processed carbon/ice sample is optically most similar to dark icy Solar System bodies including C type asteroids. The sample is more neutral in color than D type asteroids and comet Halley. The carbon/silica/ices samples, with comparatively high reflectance and blue reflectivity gradients, are unlike asteroids or comets despite containing a mixture of silicate and carbonaceous material, a composition expected of these bodies. Differences in reflectance and reflectivity gradient between the samples and solar system bodies may reflect differing physical processes during mantle formation as well as differences in chemistry of silicate and carbonaceous materials.

022.135 **A table of astronomically important ro–vibrational transitions for the H_3^+ molecular ion.**
L. Kao, T. Oka, S. Miller, J. Tennyson.
Astrophys. J., Suppl. Ser., Vol. 77, No. 2, p. 317 – 329 (Oct 1991).
A table of transition frequencies and Einstein A_{ij} coefficients for the astronomically important molecular ion H_3^+ is presented. The table consists of calculated frequencies and A_{ij} values, and frequencies measured in the laboratory, augmented by recent observations of the H_3^+ spectrum taken from the upper atmosphere of Jupiter. The frequencies given cover the L and K atmospheric windows.

022.136 **Effective collision strengths among fine–structure levels of Ca XV: Erratum.**
K. M. Aggarwal, K. A. Berrington, F. P. Keenan.
Astrophys. J., Suppl. Ser., Vol. 77, No. 3, p. 441 – 449 (Nov 1991). See Abstr. 51.022.068.
An error has been found in the calculation of Aggarwal, Berrington, and Keenan, made in 1990, due to a mistake in the

term coupling coefficients used to transform the collision strengths to intermediate coupling. The 190 fine-structure transitions among the $1s^22s^22p^2$, $1s^22s2p^3$ and $1s^22p^4$ configurations of Ca XV are therefore corrected and retabulated. Major differences between the R-matrix calculation of Aggarwal et al. and the distorted wave method of Dere et al. have now been resolved, though minor differences still remain. The effective collision strengths obtained after averaging the collision strengths over a Maxwellian distribution of electron velocities are also presented in the temperature range below 10^7K. The differences between the R-matrix and distorted wave effective collision strengths can reach one order of magnitude in some transitions, and these differences are largely attributed to resonances.

022.137 **New low-temperature effective collision strengths for transitions among the $1s^22s^22p^2$ terms of Ne V.**
D. J. Lennon, V. M. Burke.
Mon. Not. R. Astron. Soc., Vol. 251, No. 4, p. 628 - 631 (15 Aug 1991).
New collision strengths for transitions among the $1s^22s^22p^2$ terms of Ne V are calculated permitting effective collision strengths to be estimated up to temperatures of approximately 20000K. Results for transitions between the ground state $1s^22s^22p^2\ ^3P_J$ ($J = 0, 1, 2$) levels agree well with previous work by Aggarwal implying that difficulties in interpreting observed Ne V line ratios do not arise from inaccurate atomic data.

022.138 **Comparison of a calculated spectrum of $C_{60}H_{60}$ with the unidentified astronomical infrared emission features.**
A. Webster.
Nature, Vol. 352, No. 6334, p. 412 - 414 (1 Aug 1991). Letter-to-the-editor.
Infrared emission features consisting of narrow lines and broad plateaux, ranging from 3 μm in wavelength to over 12 μm, are seen in a wide variety of astronomical objects in which interstellar or circumstellar gas is illuminated by ultraviolet radiation from a star. Several candidates have been proposed as the source of this emission, but none is widely accepted. The author calculates the vibrational spectrum of the fullerane $C_{60}H_{60}$, the saturated hydride of the soccerball-shaped molecule C_{60}, using a force-field model. Six of the infrared active frequencies match unidentified emission lines to within 4%, and a seventh differs from an observed line by 8%. The calculation suggests why the astronomical feature at 7.7 μm is the strongest, and the observed variation of the 3.4 μm and 3.28 μm features with astrophysical environment is consistent with the idea that the former is attributable to stretching of the C-H bond in heavily hydrogenated fulleranes, whereas the latter is due to the same transition in lightly hydrogenated fulleranes.

022.139 **Atoms in very strong magnetic fields.**
M. C. Miller, D. Neuhauser.
Mon. Not. R. Astron. Soc., Vol. 253, No. 1, p. 107 - 122 (1 Nov 1991). With microfiche MN 253/1.
The X-ray spectra of neutron stars are expected to be determined by the opacities of atoms with atomic number $Z > 2$ in strong magnetic fields. The authors calculate the energy levels, wavefunctions and transition rates of hydrogen, helium, carbon, nitrogen and silicon in the very strong ($B > 4.7 \times 10^9 Z^2$G) magnetic fields expected in neutron stars. The wavefunctions are represented in terms of Landau states, and are calculated with a high-field multiconfigurational Hartree-Fock code. The authors compare their results for hydrogen with previous work and use their wavefunctions to compute bound-bound and bound-free oscillator strengths for heavier elements.

022.140 **Si II emission line diagnostics.**
P. G. Judge, K. G. Carpenter, G. M. Harper.
Mon. Not. R. Astron. Soc., Vol. 253, No. 1, p. 123 - 128 (1 Nov 1991).
Ratios of Si II UV emission lines in a variety of objects are compared with computations based upon recently published collisional data of Dufton and Kingston. The authors analyse a high-quality spectrum of α Tau (K5 III) obtained by Carpenter et al. using the G270M mode of the GHRS on the Hubble Space Telescope, in addition to high-dispersion data from Skylab and IUE. Agreement between observation and theory is satisfactory for most lines with the exception of ratios within the $3s^23p^2P^0 \rightarrow 3s3p^2\ ^4P$ (UV 0.01) multiplier near 2335 Å. A blend with a line of Ni II accounts partially for the discrepancies and solves a problem with Si II] emission measures. However, a further problem remains which suggests that the computed collision strengths for the transitions $3s^23p^2P^0{}_{1/2,3/2} \rightarrow 3s3p^2\ ^4P_{5/2}$ may be too large by a factor of about 1.5.

022.141 **The future of stellar spectroscopy and its dependence on YOU.**
B. Gustafsson.
Phys. Scr. T, Vol. T34, p. 14 - 19 (1991). - See Abstr. 012.082 for the main entry.
The use of atomic data in contemporary stellar spectroscopy is illustrated by a number of examples of recent applications. The needs for extended and improved data are stressed.

022.142 **Doppler effect in astronomy.**
Lj. Dokić.
Vasiona, Année 39, No. 5, p. 76 - 80 (1991). In Croatian.
The Doppler effect has been discussed in detail from the astronomical point of view.

022.143 **Electron-proton beam irradiations with matter and radiation.**
W. Bednarek, S. Karakuła, W. Tkaczyk, F. Giovannelli.
Vulcano Workshop 1988: Frontier objects in astrophysics and particle physics, p. 189 - 195 (1989). - See Abstr. 012.079 for the main entry.
The photon energy spectra from the electron-proton (e-p) beam interactions with matter and radiation were calculated. One obtains the photon energy spectra from inverse Compton scattering of an arbitrary background radiation by relativistic electron-beams and photon energy spectra from p + p interactions via π^0 decay for relativistic proton-beams. The results for selected beam Lorentz factors and selected angles of emission with respect to the beam axis are presented. The computed comptonized black body spectrum has been used succesfully to fit the spectrum of Cyg X-1 in the energy range greater than 1 MeV.

022.144 **The Si II intercombination multiplet in late-type stars.**
P. L. Dufton, F. P. Keenan, A. Hibbert, R. P. Stafford, P. B. Byrne, D. Agnew.
Mon. Not. R. Astron. Soc., Vol. 253, No. 3, p. 474 - 478 (1 Dec 1991).
New atomic data are used to calculate the relative intensities of transitions within the Si II $3s^23p^2P$-$3s3p^{24}P$ intercombination multiplet for plasma parameters appropriate to late-type stellar atmospheres. These ratios are found to be significantly different from those of Dufton and Kingston, principally due to changes in the radiative rates. A comparison with line ratios for the Sun (from Skylab S082B spectra) and late-type stars (from IUE data) indicates that although the new theoretical line ratios are in better agreement with observation, significant discrepancies still exist. Possible explanations for these discrepancies are briefly discussed.

022.145 **The connection of optical characteristics with the chemical composition of the Earth's and Moon's magnetic rock.**
O. I. Kvaratskhelia, V. Yu. Losev, V. V. Novikov.
Tr. Gos. Astron. Inst. im. Shternberga, Tom 62, p. 188 - 201 (1991). In Russian.
Albedo, colour index and polarization degree of the standards magmatic rock having equal size of particles have been investigated in the spectral diapason 419 - 704 nm. The connections of optical and chemical parameters of samples have been analysed. Correlations of the degree of polarization with the concentration of iron oxides, albedo with the concentration of silica and alumina, colour index with magnesia of the samples are exposed.

According to the results of measuring the degree of polarization of 100 lunar areas the status of the optical situation has been investigated. The connection of optical and chemical parameters for the lunar landing sites is under discussion now.

022.146 **New emission bands in the high resolution spectrum of molecular nitrogen between 107.7 and 124.2 nm.**
J.-Y. Roncin, J.-L. Subtil, F. Launay, K. Yoshino.
Planet. Space Sci., Vol. 39, No. 9, p. 1301 - 1304 (Sep 1991).
The vacuum ultraviolet emission spectrum of molecular nitrogen, produced by a low pressure discharge, is investigated at high spectral resolution from 107.7 to 124.2 nm. Band origins are reported for the 103 bands rotationally analyzed.

022.147 **Integral and differential cross sections for electron impact ionization.**
R. E. H. Clark, J. Abdallah Jr., J. B. Mann.
Astrophys. J., Vol. 381, No. 2, p. 597 - 600 (10 Nov 1991).
Calculated Couloumb-Born exchange cross sections for hydrogenic ions from 1*s* through 6*h* exist in the literature. In this paper, an analytic fit formula to these cross sections has been derived by fitting simultaneously to n, l, and u, where n is the principal quantum number, l is the orbital angular momentum, and u is the impact electron energy in threshold units. The resulting fit, which is easily integrated over a Maxwellian velocity distribution, reproduces the calculated cross sections to better than 10%. A prescription is given for applying the cross sections to nonhydrogenic ions. In addition, the differential cross sections with respect to the energy of the scattered electron have been fitted.

022.148 **Laboratory infrared spectra of CH_2D^+ and $HCCD^+$ and predicted microwave transitions.**
M. Rösslein, M.-F. Jagod, C. M. Gabrys, T. Oka.
Astrophys. J., Lett., Vol. 382, No. 1, p. L51 - L53 (20 Nov 1991).
In order to provide radio-astronomical probes for CH_3^+ and $HCCH^+$, which play crucial roles in the chemical evolution of molecular clouds, infrared vibration-rotation spectra of their monodeuterated species have been measured. Rotational transitions of CH_2D^+ and $HCCD^+$ are predicted.

022.149 **Hydrogenated amorphous carbon grains and the 2175 Å interstellar hump.**
A. Blanco, E. Bussoletti, L. Colangeli, S. Fonti, J. R. Stephens.
Astrophys. J., Lett., Vol. 382, No. 2, p. L97 - L99 (1 Dec 1991).
The authors present the first extinction spectra of carbon grains condensed in a hydrogen atmosphere to simulate the formation of dust particles in hydrogen-rich astrophysical environments. In order to assess the influence of hydrogen during grain formation unambiguously, experiments were performed in hydrogen-poor (Ar) and hydrogen-rich (H_2) atmospheres allowing the carbon vapor to condense homogeneously into grains. The experimental data confirm the theoretical predictions on the lack of any feature in the ultraviolet spectra of fully hydrogenated amorphous carbon grains and clearly show the relevance of the different atmospheres during the condensation process. These results support the idea that grains observed in the interstellar medium, which show a strong 2175 Å feature, have undergone some processing subsequent to their formation.

022.150 **Regularities of Stark parameters along the periodic table.**
J. Purić, M. Ćuk, M. S. Dimitrijević, A. Lesage.
Astrophys. J., Vol. 382, No. 1, p. 353 - 357 (20 Nov 1991).
Stark width and shift dependence on the upper-level ionization potential of the corresponding transitions have been investigated on the basis of semiclassical and semiempirical data. The simple formulas obtained from that dependence can be used to predict the Stark broadening and shift parameters of resonance $n_0s - n_0p$ and $n_0s - (n_0+1)p$ (n_0 for ground level) and off-resonance $[(n_0+1)s - (n_0+1)p$ lines of neutral atoms and ions. The predicted data are compared with the known experimental and theoretical values. The agreement is found to be within the quoted experimental and theoretical uncertainties. All data are given for an electron density $N = 10^{17} cm^{-3}$ and for electron temperatures T = 10,000, 20,000, 40,000, and 80,00K (for neutral atoms and singly charged ions); and for T = 20,000, 40,000, 80,000, and 160,000K (for doubly and multiply charged ions).

022.151 **Collision strengths and excitation rate coefficients for transitions in Fe XXI.**
K. M. Aggarwal.
Astrophys. J., Suppl. Ser., Vol. 77, No. 4, p. 677 - 696 (Dec 1991).
Using the Dirac-Fock R-matrix program, collision strengths have been computed for all transitions belonging to 20 fine-structure levels among the $1s^22s^22p^2$, $1s^22s2p^3$, and $1s^22p^4$ configurations of Fe XXI. All partial waves with $J \leqslant 29/2$ of both even and odd parities have been included, and collision strengths are tabulated at energies above thresholds in the range below 300 Ry. Resonances are highlighted in the threshold energy region, and effective collision strengths have been obtained after integrating collision strengths over a Maxwellian distribution of electron velocities. The results for effective collision strengths are tabulated for a series of electron temperatures below $\log T_e = 7.4$K, suitable for applications in astrophysical and laboratory plasmas. Comparisons have been made with the earlier available distorted wave data, and accuracy is assessed. The earlier collision strengths have been underestimated up to a factor of 3 for some transitions, whereas effective collision strengths have been underestimated up to an order of magnitude mainly due to the omission of resonances.

022.152 **3.3 micron emission from ultraviolet excitation of some aromatic molecules.**
J. Shan, M. Suto, L. C. Lee.
Astrophys. J., Vol. 383, No. 1, p. 459 - 465 (10 Dec 1991).
Infrared emissions are observed from photoexcitation of benzene, naphthalene, phenanthrene, anthracene, pyrene, and their methyl derivatives in the gas phase at 193 nm using an ArF laser as a light source. An emission band with peak at 3.3 μm is observed from all the molecules studied. This band is attributed to the aromatic C-H stretching vibrational modes. In addition to the 3.3 μm band, an emission band at 3.4 - 3.6 μm is observed from photoexcitation of the methyl derivatives, which is attributed to the C-H stretching vibrational modes of the CH_3 group. The observed spectra are compared with the IR emission bands observed in many astronomical objects. The current laboratory data lend support to a model in which PAH molecules are responsible for the "unidentified" interstellar IR emissions. IR emissions from the 193 nm excitation of several interstellar molecules other than PAHs are also observed, but they are quite different from the interstellar IR bands.

022.153 **Collisional excitation of molecular hydrogen by hydrogen atoms.**
P. G. Martin, M. E. Mandy.
J. R. Astron. Soc. Can., Vol. 85, No. 4, p. 200 - 201 (Aug 1991).
Abstract. - See Abstr. 012.091 for the main entry.

022.154 **Shape and clustering effects on the optical properties of amorphous carbon.**
F. Rouleau, P. G. Martin.
J. R. Astron. Soc. Can., Vol. 85, No. 4, p. 201 (Aug 1991).
Abstract. - See Abstr. 012.091 for the main entry.

022.155 **Angular momentum transfer in low velocity oblique impacts: implications for asteroids.**
M. Yanagisawa, J. Eluszkiewicz, T. J. Ahrens.
Icarus, Vol. 94, No. 2, p. 272 - 282 (Dec 1991).
The efficiency of angular momentum transfer ζ in low velocity oblique impacts was studied experimentally. ζ is defined as the fraction of incident angular momentum transferred to the rotation of the target. Plaster, mortar, cement, and one granite target were studied. Lead and aluminum projectiles were used. Only cratering impacts were considered. ζ was found to decrease with increasing incidence angle. This suggests that more energetic

impacts transfer angular momentum more efficiently. If asteroid spin rates are the result of mutual noncatastrophic collisions and the taxonomic classes are indicative of the bulk properties (C asteroids being weaker than S asteroids and the latter weaker than M objects), then the differences between the corresponding spin rates are predicted to be smaller on the basis of this results than what could be expected from a consideration of the relative strength and density only.

022.156 **Impact experiments. 3: Catastrophic fragmentation of aggregate targets and relation to asteroids.**
E. V. Ryan, W. K. Hartmann, D. R. Davis.
Icarus, Vol. 94, No. 2, p. 283 – 298 (Dec 1991).

The authors report results of the first collisional experiments using artificial aggregates as targets. The purpose was to study fragmentation of weakly bonded bodies consisting of stronger constituent particles. Examples of such bodies are early, loosely aggregated planetesimals or asteroid "rubble piles". The authors present results on impact strengths and size distributions from 18 collision experiments, as well as 53 drop tests and two pendulum impact tests. They find that weakly bonded aggregate bodies can have unexpectedly high impact strengths because of energy dissipation among the constituent particles. The cumulative size distributions of fragments larger than the constituent grains form power law segments similar to those formed by fragmentation of homogeneous targets, but peaks in the size distribution arise at the sizes of the small pebbles for the glued aggregate targets.

022.157 **An experimental study of the sublimation of water ice and the release of trapped gases.**
R. L. Hudson, B. Donn.
Icarus, Vol. 94, No. 2, p. 326 – 332 (Dec 1991).

The sublimation of Ar/H_2O, CO/H_2O, and CO_2/H_2O mixtures condensed at 23 – 30K was investigated. The release of the more volatile component of the ice mixture occurred in several discrete temperature regions. Gas release, as monitored by a mass spectrometer, was correlated with changes in the infrared spectrum of the solid ice. The results are in substantial agreement with those of Bar–Nun and co–workers, who studied ice sublimation using mass spectrometry alone, and Sandford and Allamandola, who used infrared spectroscopy alone. Several different concentrations were used for all gas/H_2O mixtures, and relative peak intensities during warming were found to depend on the initial gas/H_2O ratio. Some implications for cometary phenomena are given.

022.158 **$CH_4/NH_3/H_2O$ spark tholin: chemical analysis and interaction with Jovian aqueous clouds.**
G. D. McDonald, B. N. Khare, W. R. Thompson, C. Sagan.
Icarus, Vol. 94, No. 2, p. 354 – 367 (Dec 1991).

The organic solid (tholin) produced by spark discharge in a $CH_4+NH_3+H_2O$ atmosphere is investigated, along with the separable components of its water–soluble fraction. The chemistry of this material serves as a provisional model for the interaction of Jovian organic heteropolymers with the deep aqueous clouds of Jupiter.

022.159 **Ion–solid interaction: the experimental approach.**
G. Foti.
International School of Physics 'Enrico Fermi'. Course 111: Solid–state astrophysics, p. 73 – 90 (1991). – See Abstr. 012.106 for the main entry.

Solid–state matter in the Universe changes continuously under the action of an external agent, mainly UV photons and energetic particles flowing on its constituents. In the last decades the development of ground–based and space technologies have given, for the first time, the possibility to observe details attributable to the irradiation they have suffered for a long time. However, a detailed description of the effects due to energetic particles bombarding solids in astrophysics and their role to determine the evolutionary history is impossible at the present time, largely due to the lack of ad hoc thought laboratory experiments. To get insight into these intriguing problems for few years some experimental works have been performed in several laboratories around the world. In "simulation" experiments of this kind the important parameters to reproduce are mass and energy of the incoming ion, composition and physical state (pressure, temperature, density, etc.) of the target, dose rate and total ion irradiation dose.

022.160 **Spectroscopy of PAH molecules.**
L. B. D'Hendecourt, A. Léger, L. Verstraete, P. Ehrenfreund.
International School of Physics 'Enrico Fermi'. Course 111: Solid–state astrophysics, p. 91 – 105 (1991). – See Abstr. 012.106 for the main entry.

It is important to settle the identification of PAHs in the observed IR spectra on grounds as safe as possible. Léger and Puget (1984) have emphasized that they expected the interstellar PAHs to be a large family of species, whereas their spectroscopical comparison was done with a single molecule (coronene). The aim of the present lecture is to extend this comparison to other relevant PAHs and to describe, with precision, the emission mechanism. Few examples of small and large PAHs are reported.

022.161 **Irradiation of solids: theory.**
R. E. Johnson.
International School of Physics 'Enrico Fermi'. Course 111: Solid–state astrophysics, p. 129 – 168 (1991). – See Abstr. 012.106 for the main entry.

Contents: 1. Plasma environment. 2. Materials. 3. Interactions. 4. Material alterations. 5. Defects and sputtering: elastic collisions. 6. Electronically induced alterations and sputtering. 7. Applications. 8. Summary.

022.162 **Suprathermal chemistry in space.**
K. Roessler.
International School of Physics 'Enrico Fermi'. Course 111: Solid–state astrophysics, p. 197 – 266 (1991). – See Abstr. 012.106 for the main entry.

Ions or atoms which are injected with high energies into solids do not only create reactive defects and excited states but may undergo specific chemical reactions different from the classical thermal ones. They impart their kinetic and/or electronic energy to the reaction complex and form excited molecules or intermediate states which can stabilize by a variety of reaction channels. These reactions are called suprathermal or high energy or hot. The expression "hot–atom chemistry" applies in particular to solids where fast neutralization processes will change most of the ions injected into neutrals, at least at the end of their trajectory when chemical reactions are likely to occur. Hot–atom chemistry has received quite a lot of attention in nuclear chemistry and physics. Suprathermal chemistry is at present considered by many astrophyiscists an interesting but not very important side effect. The task of this lecture is to show that it constitutes a central aspect of chemistry in space, in particular for the problems of solid–state organic chemistry and general problems of prebiotic chemical evolution in cosmos.

022.163 **CO_2 profiles in cometary analogs.**
P. Hsiung, K. Roessler.
International School of Physics 'Enrico Fermi'. Course 111: Solid–state astrophysics, p. 267 – 277 (1991). – See Abstr. 012.106 for the main entry.

The cometary environment has been studied by five spacecrafts when they passed close to comet P/Halley in March 1986. A large amount of data has been collected about the cometary atmosphere and the composition of the dust grains emitted by the nucleus. A further information is still necessary to understand the dynamic processes involved in gas and dust emission from the nucleus, as well as the physical properties of its superficial region. The task of the comet simulation (KOSI experiments) is to produce cometary analog material in view of the latest recorded data, and to test theoretical models. This material consists mainly of an H_2O–mineral ice mixture (KOSI 1 and 2), with an addition of CO_2 as a representative for the volatiles (KOSI 3, November 1988, and KOSI 4, May 1989). The solar irradiation is simulated by a set of xenon lamps of variable intensity. This insolation stimulates dust grain ejection, heats the surface, and generates various transformations. The processes take place at a scale smaller than 10 cm down to 1 mm or even below (sublimation, formation and ablation of the crust, dust emission, etc.). Another important goal is the study of the interactive parameters (temperature, gas flux and dust emission, etc.). The study of surface changes such as chemical reactions, crust strength or porosity and petrography are expected to give clues to the dynamics of the processes involved during irradiation. Finally, the KOSI experiments are conceived to prepare the comet nucleus sample return mission Rosetta.

022.164 **Spectroscopy of solid materials simulating cosmic dust.**
T. J. Wdowiak.
International School of Physics 'Enrico Fermi'. Course 111: Solid-state astrophysics, p. 279 – 337 (1991). – See Abstr. 012.106 for the main entry.

Contents: 1. Introduction. 2. Conditions in the interstellar, circumstellar and interplanetary medium. 3. Sources of material of astrophysical interest. 4. Sample preparation and modification. 5. Sample containment. 6. Spectroscopic insights into astrophysical phenomena. 7. Future directions.

022.165 **Modification of solid by cyclotron beams.**
V. O. Naidenov (*V. O. Najdenov*).
International School of Physics 'Enrico Fermi'. Course 111: Solid-state astrophysics, p. 371 – 380 (1991). – See Abstr. 012.106 for the main entry.

A lot of investigations have been carried out on the modifications of solids by some radiative influences. Main attention was given to the study of the influence on the thin surface layers due to the low-energy ion beams (up to several hundred keV) and due to electrons up to several MeV. The investigations of the influence of gamma-rays and neutrons were also given much attention. The behaviour of solids after the irradiation by ion beams of several dozens MeV was not sufficiently studied. However, these beams may be of interest from both the physical and technical points of view because they are able to produce "finer" and "purer" effects. In fact, the tracks of protons and heavy ions are straight and do not overlap at small radiation doses. Their range can be changed by altering the particle energy and hence it is possible to produce radiation damages at the required depth. It is also possible to alter the type and density of the damages by using different types of ions. Sometimes it is possible to incorporate chemical impurities into a solid at the required depth. The direction of flux can be chosen considering different cuts of a crystal. There are some more advantages of high-energy ion beams. One of them is a good reproduction of results. The advantages are obvious in the processes which do not require a large dose (less than $10^{13}cm^{-2}$) for its production. The above condition is favourable for cyclotron operation. Below the results of studying GaAs and Si properties are presented and also a number of devices based on them are described.

022.166 **Molecule formation by fast particles in astronomical objects.**
V. Pirronello.
International School of Physics 'Enrico Fermi'. Course 111: Solid-state astrophysics, p. 381 – 390 (1991). – See Abstr. 012.106 for the main entry.

Contents: 1. Introduction. 2. Some selected experimental results. 3. Molecules produced in astronomical environments by the interaction among particles and solids. 4. Conclusions.

022.167 **On the theoretical calculation of line formation.**
Wu Guanqian.
Publ. Yunnan Obs., No. 4, p. 46 – 54 (1991). In Chinese.

022.168 **Experimental data on the nucleation and growth of solid particles from high-temperature gas mixtures.**
J. R. Stephens.
International School of Physics 'Enrico Fermi'. Course 111: Solid-state astrophysics, p. 391 – 402 (1991). – See Abstr. 012.106 for the main entry.

The author briefly reviews the chemistry of the gas mixtures that occur in astronomical environments, largely restricting himself to oxygen-rich gas mixtures in which the O/C atomic ratio is greater than 1. Under these conditions silicates are the predominant condensates and have been the most extensively studied. The author then reviews differences between the condensation behavior of a gas at equilibrium and the behavior as described by condensation theories. Methods of measuring the condensation of simple and complex gas mixtures are discussed. The author focuses then on the results of several experiments that he has performed on the condensation of gas mixtures of astronomical interest. Finally some general conclusions are drawn on the condensation behavior of complex gas mixtures of nearly cosmic abundance.

022.169 **Vibrational spectroscopy of ion-irradiated carbonaceous materials.**
G. Strazzulla, G. A. Baratta, A. Magazzù.
International School of Physics 'Enrico Fermi'. Course 111: Solid-state astrophysics, p. 403 – 421 (1991). – See Abstr. 012.106 for the main entry.

For the last 10 years many experimental results have been obtained on the chemical and physical changes induced by ion and electron irradiation of materials. Those results have been reviewed in recent literature with a view to their astrophysical relevance. Here the authors present IR and Raman spectra of organic materials both commercially available and newly produced in the laboratory and their changes due to fast-ion bombardment. The organics on/in the various astronomical objects are continuously modified by external radiation, mainly UV photons and energetic ions determining both their formation starting from simple carbon-containing ices and their further evolution. In particular, some physico-chemical effects induced by keV-MeV ions colliding on frozen gases and organic materials have been studied in some laboratories, among which in Catania.

022.170 **New molecules in space.**
M. Singh, J. P. Chaturvedi.
Bull. Astron. Soc. India, Vol. 19, Nos. 1 – 2, p. 78 (Mar – Jun 1991). Abstract. – See Abstr. 012.108 for the main entry.

022.171 **Time and frequency in fundamental metrology.**
B. W. Petley.
Proc. IEEE, Vol. 79, No. 7, p. 1070 – 1076 (Jul 1991). Special issue: Time and frequency. J. Jespersen, D. W. Hanson (eds.).

The role of time and frequency in a wide range of measurements is discussed, particularly those involving the International System of Units and fundamental physical constants.

022.172 **Mouvement du solide.**
F. Mignard.
13. Ecole de Printemps d'Astrophysique de Goutelas: Les méthodes modernes de la mécanique céleste, p. 433 – 496 (1990). – See Abstr. 012.118 for the main entry.

Contents: 1. Introduction. 2. Corps rigides et corps déformables. 3. Transformations orthogonales. 4. Description paramétrique des rotations. 5. Angles d'Euler. 6. Relations cinématiques. 7. Equations du mouvement.

022.173 **Absolute state-selected total cross sections for the $O^+(^4S)+CO_2$ reactions.**
G. D. Flesch, C. Y. Ng.
J. Geophys. Res., Vol. 96, No. A12, p. 21403 – 21405 (1 Dec 1991).

Absolute state-selected total cross sections for O_2^+, CO_2^+, CO^+, and C^+ resulting from the reaction $O^+(^4S)+CO_2$ have been measured in the center-of-mass collision energy ($E_{c.m.}$) range of 0.2–150 eV. The cross section for O_2^+ drops off rapidly with increasing $E_{c.m.}$ and becomes essentially zero at $E_{c.m.} > 3$ eV. The charge transfer CO_2^+ is overwhelmingly the major product ion at $E_{c.m.} > 0.5$ eV. The onsets measured for CO^+ and C^+ are consistent with the thermochemical thresholds for the formation of $CO^+ + 2\,O$ and $C^+ + 3\,O$, respectively.

022.174 **Abosolute state–selected total cross sections for the $O^+(^2D,^2P)+N_2$ reactions.**
G. D. Flesch, C. Y. Ng.
J. Geophys. Res., Vol. 96, No. A12, p. 21407 – 21410 (1 Dec 1991).
Absolute state–selected total cross sections for the reactions $O^+(^2D)$ (or $O^+(^2P)$)$+N_2 \rightarrow N_2^+ + O$, $NO^+ + N$, and $N^+ + NO$ (or $N+O$) have been measured in the center–of–mass (c.m.) collision energy ($E_{c.m.}$) range of 1.3–25 eV. Charge transfer is overwhelmingly the major product channel. The ratios of the charge transfer cross section due to $O^+(^2P)$ to that associated with $O^+(^2D)$ at the corresponding $E_{c.m.}$ are found to be in the range of 0.71–0.88. The cross sections for N^+ resulting from the $O^+(^2D,^2P)+N_2$ reactions are significantly greater than those fomed by the $O^+(^4S)+N_2$ reaction. The $E_{c.m.}$ dependences observed for the N^+ cross section indicate that the formation of $N^+ + NO$ from the collision of $O^+(^2P)+N_2$ is a viable process, whereas N^+ formed in the $O^+(^2D)+N_2$ reaction is mainly associated with the $N^+ + N + O$ channel.

022.175 **Partitioning of platinum group elements in the Fe–O–S system to 11 GPa and their fractionation in the mantle and meteorites.**
M. E. Fleet, R. G. Tronnes, W. E. Stone.
J. Geophys. Res., Vol. 96, No. B13, p. 21949 – 21958 (10 Dec 1991).
The partitioning of minor amounts of platinum group elements (PGE: Ru, Rh, Pd, Os, Ir, Pt) in the Fe–O–S system has been investigated to 11 GPa. The marked fractionation of Pt from Pd in the presence of coexisting alloy and sulfide liquid, reported earlier for the Fe–Ni–S system at low pressure, is extended to high pressure. The experiments were carried out at 1000°–1375°C and 4.5–11 GPa with graphite capsules using 18–mm octahedral pressure cells in a uniaxial split–sphere multianvil apparatus.

022.176 **Experimental determination of stellar reaction rates.**
C. Rolfs.
Nuclei in the cosmos, p. 1 – 7 (1991). – See Abstr. 003.038 for the main entry.
Contents: 1. Introduction. 2. Direct measurements. 3. Indirect approaches. 4. Reactions involving radioactive nuclides.

022.177 **Radioactive ion beams in nuclear astrophysics.**
J. Vervier.
Nuclei in the cosmos, p. 9 – 27 (1991). – See Abstr. 003.038 for the main entry.
Contents: 1. Introduction. 2. Interest of radioactive ion beams in nuclear astrophysics. 3. Production of radioactive ion beams. 4. Special experimental techniques in nuclear astrophysics using radioactive ion beams. 5. Conclusions.

The almighty chance.
See Abstr. 003.029.

Spectal line shapes. Volume 6. Proceedings.
See Abstr. 012.018.

Laboratory research for planetary atmospheres. A selection of papers. 2. International Conference on Laboratory Research for Planetary Atmospheres (ICLRPA), Charlottesville, VA (USA), 21 Oct 1990.
See Abstr. 012.073.

Solid–state astrophysics. Proceedings.
See Abstr. 012.106.

Attractor reconstruction and dimensional analysis of chaotic signals.
See Abstr. 036.323.

Stability in factor analyses with respect to small perturbations of the data and to departure from Gaussian law of the variables.
See Abstr. 036.326.

Modeling Fermi acceleration at parallel shocks with Monte Carlo techniques.
See Abstr. 062.030.

Role of highly–charged ions in astrophysical and laboratory plasmas.
See Abstr. 062.155.

Total and effective radiative recombination coefficients.
See Abstr. 063.033.

The $H+H^- \rightleftharpoons H_2(v,j)+e^-$ reaction: a consistent description of the associative detachment and dissociative attachment processes using the resonant scattering theory.
See Abstr. 063.053.

New opacity calculations.
See Abstr. 063.063.

A novel collisional radiative model with a numerical bottom and an analytical top.
See Abstr. 063.078.

Relative estimation of the mineral ages using uranium migration.
See Abstr. 081.013.

A note on the thermal component of the equation of state in solids.
See Abstr. 081.038.

Coherent backscatter model for the unusual radar reflectivity of icy satellites.
See Abstr. 091.038.

Modeling of Jupiter's millimeter wave emission utilizing laboratory measurements of ammonia (NH_3) opacity.
See Abstr. 099.049.

Elusive H_3^+ ion tracked down.
See Abstr. 099.052.

Methanol in space and in the laboratory.
See Abstr. 131.276.

Laboratory, observational and theoretical studies of interstellar ices.
See Abstr. 131.381.

The ratio of optical to near–infrared emission line strengths in S II as electron density diagnostics for planetary nebulae.
See Abstr. 134.052.

Astronomical Instruments and Techniques

031 Astronomical Optics

031.001 **Delayed elastic effects in the glass ceramics Zerodur and ULE at room temperature.**
J. W. Pepi, D. Golini.
Appl. Opt., Vol. 30, No. 22, p. 3087 – 3088 (1 Aug 1991). Current Physics Microform No.: 9109A0015.
The time–dependent phenomenon of delayed elasticity was studied at room temperature and is presented for Zerodur and ULE. The effect is believed to be related to the alkali oxide content of the glass and to the rearrangement of the ion groups within the structure during stress.

031.002 **Procedure for computer–controlled milling of accurate surfaces of revolution for millimeter and far–infrared mirrors.**
L. Emmons, R. de Zafra.
Appl. Opt., Vol. 30, No. 22, p. 3163 – 3165 (1 Aug 1991). Current Physics Microform No.: 9109A0091.
The authors discuss a simple method for milling accurate off–axis parabolic mirrors with a computer–controlled milling machine. For machines with a built–in circle–cutting routine, an exact paraboloid can be milled with few computer commands and without the use of the spherical or linear approximations that have been discussed in other mirror–cutting procedures in the literature. The method can be adapted easily to cut off–axis sections of elliptical or spherical mirrors.

031.003 **Ensquared power for obscured circular pupils with off–center imaging.**
L. M. Beyer, S. H. Cobb, L. C. Clune.
Appl. Opt., Vol. 30, No. 25, p. 3569 – 3574 (1 Sep 1991). Current Physics Microform No.: 9111A1635.
Power incident upon square detectors in the focal plane of diffraction–limited optical systems with obscured circular pupils is computed as a function of the line–of–sight position on the detector. The results are applied to an analysis of the dependence of the signal–to–noise ratio on detector size for obscuration ratios of 0, 0.25, and 0.5 and for line–of–sight position on the detector.

031.004 **Erster 8–Meter–Spiegelträger für das Very Large Telescope gegossen.**
D. Fischer.
Sterne Weltraum, Jahrg. 30, Nr. 7, p. 436 – 438 (Jul 1991).

031.005 **Hartmann testing of the 2.4–m Hiltner primary.**
P. L. Schechter, P. Mack.
Bull. Am. Astron. Soc., Vol. 23, No. 2, p. 873 (1991). Abstract. – See Abstr. 012.008 for the main entry.

031.006 **Design and demonstrated performance of adaptive optics system.**
G. Rousset.
Bull. Am. Astron. Soc., Vol. 23, No. 2, p. 895 (1991). Abstract. – See Abstr. 012.008 for the main entry.

031.007 **Adaptive optics experiments using synthetic beacons.**
C. A. Primmerman.
Bull. Am. Astron. Soc., Vol. 23, No. 2, p. 898 – 899 (1991). Abstract. – See Abstr. 012.008 for the main entry.

031.008 **The LASTAR adaptive optics system for infra–red astronomy.**
E. Kibblewhite.
Bull. Am. Astron. Soc., Vol. 23, No. 2, p. 899 (1991). Abstract. – See Abstr. 012.008 for the main entry.

031.009 **Adaptive optics including Rayleigh scattering experiments.**
L. A. Thompson.
Bull. Am. Astron. Soc., Vol. 23, No. 2, p. 899 (1991). Abstract. – See Abstr. 012.008 for the main entry.

031.010 **Efficiency of grazing incident optics: the spiral collimator.**
I. A. Artyukov, A. V. Vinogradov, I. Kozhevnikov.
Appl. Opt., Vol. 30, No. 28, p. 4154 – 4157 (1 Oct 1991). Current Physics Microform No.: 9111F1832.
Geometric properties and construction of the spiral collimator are discussed. The expression for the efficiency of power conversion of a divergent soft X–ray point source into a parallel beam is obtained. The efficiency of the spiral collimator is shown to be up to 30 – 90% in the whole soft X–ray spectral range. This efficiency is limited only by the fundamental reason: X–ray radiation absorption in the mirror coating of the spiral collimator. In this sense, the spiral collimator is more effective than any known X–ray optical collimation system with one or manifold reflections.

031.011 **X–ray imaging at the diffraction limit.**
E. L. Raab, D. M. Tennant, W. K. Waskiewicz, A. A. MacDowell, R. R. Freeman.
J. Opt. Soc. Am. A, Vol. 8, No. 10, p. 1614 – 1621 (Oct 1991). Current Physics Microform No.: 9111F0532.
The authors tested the ability of normal–incidence, multilayer–coated X–ray optics to produce diffraction–limited images. The imaging performance at 14 nm of a spherical mirror substrate coated with a single metal layer was compared with that of the same substrate coated with a Mo–Si multilayer. A Foucault knife–edge test was used to quantify the image aberrations. The knife–edge data were fitted with a model based on Fresnel diffraction theory. The authors found that the multilayer coating introduced a measurable but small figure error to the optic; the change in figure was less than $\lambda/16$ peak to valley. They conclude that a properly deposited multilayer coating will not degrade the imaging performance of an optical system designed to operate at the diffraction limit.

031.012 **Equivalence of two theories of beam tilt.**
B. R. Merrill.
J. Opt. Soc. Am. A, Vol. 8, No. 8, p. 1316 – 1318 (Aug 1991). Current Physics Microform No.: 9109F2065.
The author demonstrates the equivalence of two different problems of beam tilt or angle of arrival. One approach employs the optical analog of the Ehrenfest theorem from quantum physics to derive the angle of arrival for a beam wave. The other approach models the tilt based on the average angle of arrival over a telescope aperture.

031.013 **Cleaning of contaminated UV optics by low energy O^+.**
D. K. Prinz, C. W. Bowers, B. Lippey, R. Goldberg.
Bull. Am. Astron. Soc., Vol. 23, No. 2, p. 1038 (1991). Abstract. – See Abstr. 012.012 for the main entry.

031.014 **Fabrication and characterization of replicated and lacquercoated grazing incidence optics for X–ray astronomy.**
M. P. Ulmer, R. Haidle, R. Altkorn, P. Georgopoulos, B. Rodricks, P. Z. Takacs.
Opt. Eng., Vol. 30, No. 8, p. 1150 – 1154 (Aug 1991).
The authors discuss the fabrication and testing of electroformed replica Wolter I optics made from gold–coated lacquered mandrels. They also discuss testing of gold– and palladium–coated lacquered test flats. X–ray (5 keV for Wolter I mirror and 8 to 40 keV for test flats) and optical (Wyko NCP–1000 profiler) measurements were used to evaluate the mirrors.

031.015 **Analysis of constructions of silicon carbide mirrors.**
V. A. Alekseev, V. V. Bokov, V. N. Egorov, G. I. Eremeev, E. A. Narusbek, N. L. Tkachenko, A. I. Filippov.
Sov. J. Opt. Technol., Vol. 58, No. 7, p. 390 – 391 (Jul 1991). Current Physics Microform No.: 9201X2166.
This paper presents the results of a test of four types of constructions of large silicon carbide mirrors. The tests were performed using a method that makes it possible to characterize the stiffness of the mirror by the magnitude of the elastic deformation at the center of the operating surface due to the action of a uniformly distributed load under certain clamping conditions. The honeycomb (to reduce weight) construction of the mirror, covered by a plate on the rear side, makes it possible to increase the stiffness by a factor of 1.9 compared with a monolithic mirror of the same mass.

031.016 **Quantitative testing of optical components by schlieren method.**
A. M. Lyalikov.
Sov. J. Opt. Technol., Vol. 58, No. 8, p. 470 – 471 (Aug 1991). Current Physics Microform No.: 9202X0016.
This paper examines a new scheme for the testing of transparent objects by the schlieren method, making it possible to arrive at a quantitative estimate of the deflection angle of rays in optical inhomogeneities. The proposed scheme is certified for testing for optical inhomogeneities in glass substrates.

031.017 **Interrelationship of aberration properties of axisymmetrical aspherical surfaces.**
V. A. Zverev, S. M. Marchuk.
Sov. J. Opt. Technol., Vol. 58, No. 8, p. 496 – 498 (Aug 1991). Current Physics Microform No.: 9202X0042.
The interrelationship of the geometrical and aberration properties of traditional aspherical surfaces of rotation and aspherical surfaces of a new type, formed in a certain manner, is shown.

031.018 **The precise adjustment of optical system for horizontal transit circle.**
Li Zhigang, Li Huanxin.
Publ. Shaanxi Astron. Obs., Vol. 14, No. 1, p. 1 – 6 (Aug 1991). In Chinese.
The first meridian circle of China has been installed at CSAO in Apr 1990. The optical system has been adjusted accurately based on the primary installation. Because of the constructional characteristics of the horizontal meridian circle, there are some difficulties for the adjustment. However, the practice has verified that the method of approximating step by step is very successful. Now the optical system has reached the required level.

031.019 **The fine adjusting method for the optical system of the DCMT.**
Hu Qiqian, Li Depei, Li Zhigang.
Publ. Shaanxi Astron. Obs., Vol. 14, No. 1, p. 7 – 13 (Aug 1991). In Chinese.
The principle of the adjusting for the optical system of the Danish–Chinese Meridian Telescope is discussed. The method based on the auto–collimation system is proposed. Since the parts of this instrument are in a weak linkage between them, all the parameters for the DCMT have to be adjusted. The special program for adjustment of this instrument is needed to be planned, because the adjusting for any parts of this instrument will influence several parameter.

031.020 **New glasses for optics and optoelectronics.**
H. F. Morian.
SPIE Conference on Optical Fabrication and Testing as Part of the Asia Pacific Conference on Optical Technology (APCOT '90), p. 146 – 157 (1991). – See Abstr. 012.015 for the main entry.
Optical glasses, due to their high homogeneity regarding physical properties, and the ability to tailor key characteristics to meet the desired performance, find new applications in the wide fields of optics and optoelectronics. Several new developments are stressed.

031.021 **Application and machining of "Zerodur" for optical purposes.**
N. Reisert.
SPIE Conference on Optical Fabrication and Testing as Part of the Asia Pacific Conference on Optical Technology (APCOT '90), p. 171 – 178 (1991). – See Abstr. 012.015 for the main entry.
The paper aims to provide a general view of the most essential proporties of "Zerodur", its major fields of application, the manufacture and the machining in the form of grinding and polishing.

031.022 **Some parameters and characteristics of an off–axis paraboloid.**
D. Malacara.
Opt. Eng., Vol. 30, No. 9, p. 1277 – 1280 (Sep 1991).
A reflecting off–axis paraboloid is frequently used either to collimate the light from a point source or to concentrate in a point the light from a collimated beam. Because these paraboloids are expensive and difficult to manufacture, a spherical mirror is frequently used instead. The aberrations arising with this substitution are analyzed and evaluated.

031.023 **Ray misidentification errors in Hartmann–like wavefront sensors.**
J. M. Geary.
Opt. Eng., Vol. 30, No. 9, p. 1388 – 1393 (Sep 1991).
Hartmann wavefront sensors rely on an exact correlation between a transverse ray aberration measurement in the focal plane and the sampling hole position in the pupil identifying that ray. If this relationship breaks down, the reconstructed wavefronts will not correspond to the actual incident wavefronts. This occurs if rays yielding certain transverse errors are misidentified, e.g., if sampling is in a plane not conjugate to the pupil. A case study is made of the SHAPE wavefront sensor in which such ray misidentification can arise.

031.024 **Large–mirror testing facility at the National Optical Astronomy Observatories.**
L. D. Barr, V. Coudé du Foresto, J. Fox, G. A. Poczulp, J. Richardson, C. Roddier, F. Roddier.
Opt. Eng., Vol. 30, No. 9, p. 1405 – 1414 (Sep 1991).
A method for testing the surfaces of large mirrors has been developed to be used even when conditions of vibration and thermal turbulence in the light path cannot be eliminated. The full aperture of the mirror under test is examined by means of a scatterplate interferometer that has the property of being a quasi–common–path method, although any means for obtaining

interference fringes will do. The method uses a remotely operated CCD camera system to record the fringe pattern from the workpiece. The typical test is done with a camera exposure of about a millisecond to "freeze" the fringe pattern on the detector. Averaging up to 10 separate exposures effectively eliminates the turbulence effects. The method described provides the optician with complete numerical information and visual plots for the surface under test and the diffracted image the method will produce, all within a few minutes, to an accuracy of 0.01 μm measured peak-to-valley.

031.025 **Adaptive Optik.**
F. Merkle.
Phys. Unserer Zeit, Jahrg. 22, Nr. 6, p. 260 – 266 (Nov 1991).

031.026 **Evaluation of seeing on a 62-cm mirror.**
M. Iye, T. Noguchi, Y. Torii, Y. Mikami, H. Ando.
Publ. Astron. Soc. Pac., Vol. 103, No. 665, p. 712 – 722 (Jul 1991).

A 62-cm active optics telescope model with a Shack-Hartmann wavefront analyzer was used to measure the mirror seeing effect. The degradation of the imaging quality due to the generation of microthermal convection was quantitatively evaluated from diurnal monitoring measurements over 90 days and nights. The dependence of mirror seeing on the temperature difference between the mirror and the ambient air and the effect of flushing flow to blow away the microthermal turbulence were measured.

031.027 **Optimization using the OSLO and Super-OSLO programs.**
D. C. Sinclair.
OSA International Lens Design Conference, p. 116 – 119 (1990). – See Abstr. 012.014 for the main entry.

The OSLO series of optical design programs are intended to be "open" programs that can be easily adapted to a large variety of optical design tasks. Some techniques and tradeoffs used during the development of the programs will be described.

031.028 **Optical design with physical optics using GLAD.**
G. N. Lawrence.
OSA International Lens Design Conference, p. 126 – 135 (1990). – See Abstr. 012.014 for the main entry.

A physical optics code provides the optical designer a means to develop many types of systems which can not be treated properly by geometrical optics. Physical optics must consider a much broader range of effects and components and computational times are longer, but the sophistication of physical optics software is now comparable to that of geometrical optics programs and allow the optical designer to work on systems that go beyond "classical" geometrical optics.

031.029 **Nonlinearity and lens design.**
B. Tatian.
OSA International Lens Design Conference, p. 154 – 164 (1990). – See Abstr. 012.014 for the main entry.

This article addresses the impact of nonlinearity on the problem of designing lenses and the connection of Newton's method with ordinary differential equations, and describes a method for improving its convergence inspired by the total least squares problem.

031.030 **Rank down method for automatic lens design.**
H. Ooki.
OSA International Lens Design Conference, p. 171 – 176 (1990). – See Abstr. 012.014 for the main entry.

A new automatic lens design method using singular value decomposition (SVD) is proposed. This method enables the designer to reflect his own intention in optimization via the Rank Down Constant introduced in this paper. This characteristic also makes it possible to accelerate the optimization. An example showing the high performance of this method is demonstrated. Some useful applications of S.V.D. are also proposed.

031.031 **Calculation method of the Strehl Definition for decentral optical systems.**
Zhuang Songlin, Chen Huai'an.
OSA International Lens Design Conference, p. 252 – 253 (1990). – See Abstr. 012.014 for the main entry.

A new method is proposed to calculate the Strehl Definition for optical systems lacking rotational symmetry.

031.032 **Designs of two-glass apochromats and superachromats.**
R. I. Mercado.
OSA International Lens Design Conference, p. 262 – 272 (1990). – See Abstr. 012.014 for the main entry.

Superachromatic designs of lens systems utilizing two optical materials are described. Design examples comprising an infrared doublet, a hybrid refractive-diffractive lens, catadioptric systems and Gauss objectives are presented. The performance evaluations of these designs are also shown.

031.033 **A simple multidimensional quadratic extrapolation method for the correction of specific aberrations in lens systems.**
J. Maxwell, C. C. Hull.
OSA International Lens Design Conference, p. 277 – 285 (1990). – See Abstr. 012.014 for the main entry.

The need to achieve control of specific aberrations and to investigate the aberrational properties of lens systems, has led to the development of an interactive design tool based on multi-dimensional quadratic extrapolation. A quadratic approximation to the non-linearly varying aberrations allows the step length of this differential correction technique to be extended and the underlying relationships between groups of constructional parameters and aberrations are revealed. Examples are given.

031.034 **Design of two- and three-element diffractive telescopes.**
D. A. Buralli, G. M. Morris.
OSA International Lens Design Conference, p. 292 – 296 (1990). – See Abstr. 012.014 for the main entry.

Design procedures for simple two- and three-element diffractive telescopes are described. The basic configuration for the two-element design is obtained analytically by solving design equations to set the Seidel aberrations to target values. Computer optimization is used to complete the design of the doublet and triplet telescopes. It is shown that diffraction limited performance can be obtained from these diffractive systems.

031.035 **Diffraction performance calculations in lens design.**
D. Malacara.
OSA International Lens Design Conference, p. 2 – 14 (1990). – See Abstr. 012.014 for the main entry.

A review of the methods employed to compute the wavefront shape, the point spread function and the optical transfer function in lens evaluation programs is presented. One of the simplest methods to perform numerical calculations of the diffraction performance of optical systems is to divide the aperture in small squares, and then to consider the wavefront in each of these small squares to be flat and perpendicular to the ray direction in that region. This method however, presents some limitations, since the wavefront has to be flat within a fraction of the wavelength in that small square. This might not be the case if the wavefront is either too aberrated, or the defocusing is too large.

031.036 **Importance of dispersion tolerances in infrared lens design.**
R. J. Korniski, K. P. Thompson.
OSA International Lens Design Conference, p. 402 – 407 (1990). – See Abstr. 012.014 for the main entry.

The sensitivity of infrared optical designs to errors in dispersion is explored through the use of examples. The main cause for a high sensitivity to dispersion tolerances in achromatic designs is explained. Conditions for alerting the optical designer to the need for including dispersion tolerances in the design process as well as to suspect raw refractive-index data for an infrared material are provided.

031.037 **The design of high performance aplanatic achromats for the near ultraviolet waveband.**
T. I. J. Al-Baho, R. C. M. Learner, J. Maxwell.
OSA International Lens Design Conference, p. 417 – 428 (1990). – See Abstr. 012.014 for the main entry.
Due to an apparent conflict between transparency and dispersion, the near ultraviolet transmission of achromatic lenses is strongly dependant on glass choice. The design of achromats for this waveband is described and the choice of optical glass is reviewed in detail. Designs are given.

031.038 **Easily fabricated wide angle telescope.**
R. C. Owen.
OSA International Lens Design Conference, p. 430 – 433 (1990). – See Abstr. 012.014 for the main entry.
A novel configuration for a wide angle, flat field, all reflective, unobscured telescope, using two figured surfaces which are easily fabricated and aligned, is described.

031.039 **All–refractive telescope for next generation inter–satellite communication.**
H. J. Heimbeck.
OSA International Lens Design Conference, p. 434 – 437 (1990). – See Abstr. 012.014 for the main entry.
Telescopes are used for optical communication links between satellites. For high power laser sources an all–refractive telescope is preferred. A solution is presented meeting the demanding specifications.

031.040 **All–reflective phased array imaging telescopes.**
T. W. Stuhlinger.
OSA International Lens Design Conference, p. 438 – 446 (1990). – See Abstr. 012.014 for the main entry.
Optically phased telescope arrays allow high resolution, large effective–aperture imaging, using small optical components. A new, all–reflective, wide–field–of–view design is discussed.

031.041 **Design of array systems using shared symmetry.**
Miao Chenghsi.
OSA International Lens Design Conference, p. 447 – 456 (1990). – See Abstr. 012.014 for the main entry.
Multiple–array telescope systems with a minimum number of components and using decentered aspherical correctors that share the symmetry of the combined units are discussed. The theory and practice of designing such shared symmetry systems is described.

031.042 **Concentric systems for adaptation as spectrographs.**
L. N. Metz.
OSA International Lens Design Conference, p. 457 – 459 (1990). – See Abstr. 012.014 for the main entry.
A few unusual concentric optical configurations that are evolved from Dyson and Offner systems are presented for possible adaptation as spectrographs.

031.043 **Simultaneous alignment and multiple surface figure testing of optical system components via wavefront aberration measurement and reverse optimization.**
M. A. Lundgren, W. L. Wolfe.
OSA International Lens Design Conference, p. 533 – 539 (1990). – See Abstr. 012.014 for the main entry.
Simultaneous alignment and figure testing have been simulated by computer for components of a Cooke triplet and a two–Petzval telescope using end–to–end wavefront aberration measurement and computer model reverse optimization. Experimental confirmation is discussed.

031.044 **Ellipsoidal mirrors as modular elements in the design of off–axis optical systems.**
O. N. Stavroudis.
OSA International Lens Design Conference, p. 627 – 646 (1990). – See Abstr. 012.014 for the main entry.

031.045 **All–reflective four–element zoom telescope: design and analysis.**
R. B. Johnson, J. B. Hadaway, T. Burleson, B. Watts, E. D. Park.
OSA International Lens Design Conference, p. 669 – 675 (1990). – See Abstr. 012.014 for the main entry.
The design process for an all–reflective zoom telescope is presented. Special consideration is given to the development of the starting configuration and the subsequent optimization process. The non–traditional optimization route utilized as a result of the unusual pupil characteristics of such an all–reflective zoom system is examined. Results of the design are presented.

031.046 **Adaptive optics on a 3.6–m telescope: results and performance.**
F. Rigaut, G. Rousset, P. Kern, J. C. Fontanella, J. P. Gaffard, F. Merkle, P. Léna.
Astron. Astrophys., Vol. 250, No. 1, p. 280 – 290 (Oct 1991).
This paper reports the results of the first observations made on a large (3.6–m) telescope with the adaptive optics (AO) prototype, named COME–ON, conceived to explore the application of this technique for the European Very Large Telescope (VLT). The wavefront is studied, after its correction by the servo–system, using a Zernike polynomial expansion. The analysis of uncorrected and corrected images in the near infrared ($\leqslant 5\ \mu m$) leads to a detailed evaluation of the system performance in terms of improvement of angular resolution – reaching nearly the ideal diffraction profiles down to $2.2\ \mu m$ –, and Strehl ratio – approaching 0.6 to 0.8 at $3.8\ \mu m$. Sets of short exposure images allow to deduce some important parameters on the temporal effects of the correction, such as an enhancement of the speckle modulation transfer function. Finally, the authors propose some views on potential applications of adaptive optics to astronomy.

031.047 **Technological system for the automated shaping of large optical components.**
A. I. Grodnikov, V. A. Gorshkov.
Sov. J. Opt. Technol., Vol. 58, No. 9, p. 588 – 592 (Sep 1991). Current Physics Microform No.: 9202X0134.
The technological equipment of a system for the automated shaping of large optical components, guaranteeing the achievement of high–precision shaping of flat, spherical and aspherical surfaces, is examined. The specifications and structural features of the elements of a system of automated machines, support and basing mounts, interferometers of various types, and the software for the technological process are presented. The technological capabilities of the system are demonstrated for the example of finishing an aspherical mirror having a diameter of 1500 mm and a radius of curvature of $\sim$10,000 mm.

031.048 **Calculation of the parameters of a modified Hindle system for testing convex hyperbolic surfaces.**
A. G. Seregin, I. S. Potemin, T. N. Tul'eva.
Sov. J. Opt. Technol., Vol. 58, No. 9, p. 607 – 609 (Sep 1991). Published in summary form only. Current Physics Microform No.: 9202X0153.

031.049 **Optical design of an off–axis, low distortion UV telescope.**
A. H. Richardson.
Conference on Space Astronomical Telescopes and Instruments as Part of SPIE's Technical Symposia on Optical Engineering and Photonics in Aerospace Sensing (OE/Aerospace Sensing '91), p. 314 – 319 (1991). – See Abstr. 012.030 for the main entry.
An off–axis three–mirror telescope with a flat focal surface was designed. This 20 cm telescope has an f/2 primary and a f/8 tertiary focus. The focal surface has a diameter of 4 cm which subtents 1.3°. The telescope is designed to have very low distortion so that resolution would not be degraded during integration of images moving on the CCD detector at the same rate as that of the CCD readout.

031.050 **Metallic alternative to glass mirrors (active mirrors in aluminium). A review.**
J. P. Rozelot, J. M. Leblanc.
Conference on Space Astronomical Telescopes and Instruments as Part of SPIE's Technical Symposia on Optical Engineering and Photonics in Aerospace Sensing (OE/Aerospace Sensing '91), p. 481 – 490 (1991). – See Abstr. 012.030 for the main entry.
Glass mirrors for telescopes had their golden ages with the improvements of new glass ceramics (such as Zerodur), or with the developments of new techniques such as light-weighted blanks, honeycomb structures, etc. Metallic mirrors are manufactured at reduced cost, have no risks (brittleness) and may be delivered in short time and large size. All deformations (such as thermal, or even gravitational in case of planetary exploration) can be compensated through active optics, and the optical quality can be controlled even if unexpected incidents should occur. This presentation gives a review of all such existing new facilities, including the last results obtained with aluminium mirrors developed within the European EUREKA procedure (LAMA programme).

031.051 **A method of making ultralight primary mirrors.**
R. R. Zito.
Conference on Space Astronomical Telescopes and Instruments as Part of SPIE's Technical Symposia on Optical Engineering and Photonics in Aerospace Sensing (OE/Aerospace Sensing '91), p. 491 – 497 (1991). – See Abstr. 012.030 for the main entry.
Space vehicle weight limitations require space telescope components to be as light as possible. In this paper the fabrication of ultralight mirror segments is described. The mirrors are made from HTP ("High Thermal Performance" material), better known as a third generation derivative of the space shuttle heat shield tile. When one side of an HTP surface is coated with a suitable glassy substance, such as fused silica, fine mirrors can be made. Experiments have shown that a one centimeter thick specimen of HTP material will outgas from one atmosphere pressure down to the ambient pressure at low Earth orbit in eight minutes when the pumping rate is 200 litres/sec.

031.052 **Zentrieren, justieren, kollimieren. Teil 2.**
H. G. Ziegler.
Orion, Jahrg. 49, Nr. 247, p. 220 – 223 (Dec 1991).

031.053 **Project of a three reflection telescope for wide field ultraviolet observations.**
A. Amoretti, M. Badiali, A. Preite-Martinez, R. Viotti.
IAU Colloquium No. 123: Observatories in Earth orbit and beyond, p. 487 – 491 (1990). – See Abstr. 012.038 for the main entry.
The authors present a project of a three-reflection 1.5 m f/3 aplanatic wide-field telescope, in view of a small-satellite mission for the all-sky survey in the ultraviolet.

031.054 **Efficiency of compensation for phase front distortions in adaptive optical systems when measuring the average phase-front tilt.**
V. E. Kirakosyants, V. A. Loginov, V. V. Slonov,
A. O. Sulimov.
Sov. J. Quantum Electron., Vol. 21, No. 12, p. 1374 – 1377 (Dec 1991). Current Physics Microform No.: 9203X1820.
An analysis is made of the enhancement of the efficiency of detection or transmission of laser radiation achieved by compensation for large-scale distortions of the phase front. An algorithm for the compensation of phase distortions is determined taking into account a priori information on the statistical properties of the received signal and the noise characteristics of the meter used to measure the aperture-averaged tilt of the phase front. The efficiencies of optimal and suboptimal algorithms are compared.

031.055 **Differential adaptive optics for astronomical interferometry.**
J. M. Beckers.
Appl. Opt., Vol. 30, No. 34, p. 5010 – 5011 (1 Dec 1991). Current Physics Microform No.: 9203A1761.
Sometimes it suffices to make the wave-front distortions across individual telescopes equal rather than zero. This leads to the concept of differential adaptive optics.

031.056 **Use of minimum-energy modes for modal-active optics corrections of thin meniscus mirrors.**
L. Noethe.
J. Mod. Opt., Vol. 38, No. 6, p. 1043 – 1066 (Jun 1991).
For thin meniscus mirrors the modes which correct arbitrary wavefront aberrations in the most efficient way are derived. These modes are similar to the free-vibration modes of the mirror. The relationship between the two types of modes is shown. The most important features such as orthogonality and energy of the modes are discussed. This modal approach is applied to a few practical problems in connection with the primary mirrors of the Very Large Telescope of the European Southern Observatory.

031.057 **Interferometric testing of weak aspheric surfaces versus design specifications.**
V. Greco, G. Molesini, F. Quercioli, A. Novi.
Optik, Vol. 87, No. 4, p. 159 – 162 (Jun 1991).
The problem of testing aspheric surfaces with respect to design specifications is discussed. Conversion from aspherics formula to Zernike polynomials for the special case of rotational symmetry is presented. Application to interferometric testing of weak aspheric surfaces is illustrated experimentally.

031.058 **Preparation of metal substrates for deposition of optical coatings.**
A. F. Bol'shanin, S. V. Orlov.
Sov. J. Opt. Technol., Vol. 58, No. 10, p. 671 – 672 (Oct 1991). Current Physics Microform No.: 9204X0141.
A standard procedure is proposed for the cleaning of substrates made from soft metals and alloys (aluminum, copper) before depositing optical coatings, to ensure a high reproducibility of the results. Electron-beam heating of the substrates in order to modify the original surface is shown to be promising.

031.059 **Rapid optical fabrication with computer-controlled optical surfacing.**
R. A. Jones, W. J. Rupp.
Opt. Eng., Vol. 30, No. 12, p. 1962 – 1968 (Dec 1991). This paper is a revision of a paper presented at the SPIE Conference on Advanced Optical Manufacturing and Testing, Jul 1990, San Diego, CA (USA).
In the next decades, we foresee the need for the fabrication of many large aspheric mirrors for space and astronomy applications. Itek Optical Systems has developed computer-controlled optical surfacing for rapidly fabricating such optics. Recently, these operations have been improved with the development of a sheaf of new processes based on a greatly increased understanding of the mechanics of glass surfacing. These improvements have already produced a reduction of large optics surfacing time by a factor of four, and the technology points toward a possible further reduction by a factor of two.

031.060 **Catalog of infrared and cryo-optical properties for selected materials.**
J. B. Heaney, P. W. Alley, S. E. Bradley.
Conference on Reflective and Refractive Optical Materials for Earth and Space Applications as Part of SPIE's Technical Symposia on Optical Engineering and Photonics in Aerospace Sensing (OE/Aerospace Sensing '91), p. 140 – 159 (1991). – See Abstr. 012.066 for the main entry.
This report describes the status of an effort to catalog the refractive index, absorption, reflectance and transmittance of selected infrared transmitting materials over the wavelength

range from 1 μm to 1000 μm and at temperatures from 300K to about 2K.

031.061 **Design and optical performance of beryllium assessment mirrors.**
B. B. Thomas, L. C. Maxey, A. C. Miller.
Conference on Reflective and Refractive Optical Materials for Earth and Space Applications as Part of SPIE's Technical Symposia on Optical Engineering and Photonics in Aerospace Sensing (OE/Aerospace Sensing '91), p. 20 – 30 (1991). – See Abstr. 012.066 for the main entry.

Beryllium mirrors, approximately 19.8 cm × 8.8 cm in size, were designed, procured, and evaluated. The mirrors are off–axis rectangular segments of an f/1 paraboloid. These segments, which are the primary mirrors in a Cassegrainian telescope design, are lightweighted and incorporate a stress–free mounting technique which minimizes post–assembly alignment. Upon delivery, the mirrors were characterized to determine their compliance with the figure, scatter, reflectance, and dimensional specifications.

031.062 **Broad band anti–reflection coating designs for large aperture infrared windows.**
K. Balasubramanian, T. V. Le, K. H. Guenther, V. Kumar.
Conference on Reflective and Refractive Optical Materials for Earth and Space Applications as Part of SPIE's Technical Symposia on Optical Engineering and Photonics in Aerospace Sensing (OE/Aerospace Sensing '91), p. 245 – 253 (1991). – See Abstr. 012.066 for the main entry.

Effective antireflection coating designs for infrared windows are developed for the 3 – 5 μm and 8 – 12 μm bands. Dual band performance with these designs are optimized with computer aided techniques. A simple scheme to improve their oblique incidence performance is examined. Potential use of these designs in practical coatings without the use of radioactive materials is also discussed.

031.063 **Lessons learned in recent beryllium mirror fabrication.**
J. A. Wells, C. M. Lombard, G. B. Sloan, W. W. Moore, C. E. Martin.
Conference on Reflective and Refractive Optical Materials for Earth and Space Applications as Part of SPIE's Technical Symposia on Optical Engineering and Photonics in Aerospace Sensing (OE/Aerospace Sensing '91), p. 2 – 12 (1991). – See Abstr. 012.066 for the main entry.

The lessons learned in recent fabrication of beryllium mirrors could have a significant impact on how beryllium optics of the future are produced. This paper provides an overview of the latest techniques for beryllium optics fabrication and a comparison of the results achieved.

031.064 **Fabrication of a fast aspheric beryllium mirror.**
R. A. Paquin, G. J. Gardopée.
Conference on Reflective and Refractive Optical Materials for Earth and Space Applications as Part of SPIE's Technical Symposia on Optical Engineering and Photonics in Aerospace Sensing (OE/Aerospace Sensing '91), p. 39 – 45 (1991). – See Abstr. 012.066 for the main entry.

A program to fabricate a large, optically fast, aspheric lightweight Be mirror was initiated in order to demonstrate state–of–the–art technology. The mirror blank was fabricated as a 1.0–m diameter, f/0.58 ellipse directly from IP–70 grade power using near–net–shape hot isostatic pressing (HIPing) and a patented tooling approach that produced a closed back, honeycomb–cored mirror weighing less than 18–kg. Details of the mirror design and of the assembly for HIPing are given.

031.065 **Application of adaptive optics to astronomy.**
Zhao Peiqian, Zhou Bifang.
Prog. Astron., Vol. 9, No. 2, p. 151 – 160 (Jun 1991). In Chinese.

The basic principle of adaptive optics applied to astronomy, as well as the components and performance of an adaptive optics system are described systematically. The status of adaptive optics in astronomy is given. Finally, the prospect of adaptive optics used in astronomy is discussed.

031.066 **Characterization of multilayer–coated concave diffraction gratings in the EUV.**
J. Bixler, D. Dietrich, T. Barbee Jr.
1. Berkeley Colloquium on Extreme Ultraviolet Astronomy, p. 364 – 367 (1991). – See Abstr. 012.045 for the main entry.

The authors describe their work to date on how to characterize two multilayer–coated concave diffraction gratings, one conventionally ruled, the other holografically ruled. They have obtained photographic spectra with both gratings that indicate their good resolution properties and promising effieciencies.

031.067 **Recent advances in EUV astronomy with multilayers.**
B. M. Haisch, G. J. Rottman, T. E. Whittemore.
1. Berkeley Colloquium on Extreme Ultraviolet Astronomy, p. 368 – 379 (1991). – See Abstr. 012.045 for the main entry.

The purpose of this paper is to discuss some of the details of EUV multilayer optics for astronomy in the context of the design, fabrication, calibration, and flight performance of an experiment the authors built and flew on a rocket in March 1988 and of a follow–on imaging/spectroscopy experiment being designed for a later flight.

031.068 **New, fast, high–performance single mirror spectrograph cameras.**
C. G. Wynne.
Mon. Not. R. Astron. Soc., Vol. 252, No. 1, p. 171 – 176 (1 Sep 1991).

The considerations that led to a new form of single mirror spectrograph camera are discussed, and design data and performance figures given for two f/1.2 cameras, illustrating some aspects of versatility of these systems.

031.069 **Achromatized Bowen–Vaughan telescopes.**
C. G. Wynne.
Mon. Not. R. Astron. Soc., Vol. 253, No. 1, p. 160 – 166 (1 Nov 1991).

The Bowen–Vaughan telescope (1973) consisting of a Ritchey–Chrétien type mirror pair with a rear aspheric plate to correct the residual astigmatism of the mirrors, is limited in performance by the chromatic difference of astigmatism of the aspheric. The use of an aspheric doublet makes it possible to achieve subarcsec imagery over a wide spectral range over field diameters from 2° to 3°. The performance and limitations of a range of designs are described.

031.070 **Test of the optical quality of a telescope.**
Lj. Jovanović.
Vasiona, Année 39, No. 5, p. 75 – 76 (1991). In Serbo–Croatian.

A series of test, using observations of Jupiters, his satellites and the Moon is described.

031.071 **A flat array large telescope concept for use on the moon, earth, and in space.**
B. E. Woodgate.
Rev. Sci. Instrum., Vol. 62, No. 12, p. 2931 – 2935 (Dec 1991). Current Physics Microform No.: 9201B0075.

An astronomical optical telescope concept is described which can provide very large collecting areas, of order 1000 m^2. This is an order of magnitude larger than the new generation of telescopes now being designed and built. Multiple gimballed flat mirrors direct the beams from a celestial source into a single telescope of the same aperture as each flat mirror. Multiple images of the same source are formed at the telescope focal plane. A beam combiner collects these images and superimposes them into a single image, onto a detector or spectrograph aperture. This telescope could be used on the earth, the moon, or in space.

031.072 **Active optics and the New Technology Telescope (NTT): the key to improved optical quality at lower cost in large astronomical telescopes.**
R. N. Wilson.
Contemp. Phys., Vol. 32, No. 3, p. 157 – 172 (May–Jun 1991).
The first fully "active" telescope, the ESO NTT, produced its first astronomical images in Mar 1989. The spectacular results and the background work leading to them are described. The origins of the development of the active optics concept are reviewed and a description of the NTT system is given. Recent operational results and limitations to optical quality are discussed, as well as the steps taken to establish a fully automatic correction system. Finally, the optical efficiency of telescopes is considered in the light of the NTT experience and the implications for future ground– and space–based telescopes summarized.

031.073 **Moving M2 mirror without pointing offset.**
R. Ragazzoni, F. Bortoletto.
Telesc. Naz. Galileo, Tech. Rep., No. 8.
Osservatorio Astronomico di Padova–Asiago, Asiago (Italy). 16 p. (Sep 1991). To be published in Proc. SPIE, Vol. 1542, "Active and adaptive optical systems".
Moving the secondary mirror M2 to introduce an amount of decentering coma is one of the tasks of active optics. The authors show that this target is accomplished with high accuracy rotating the mirror around a point located near, but not exactly at the center of curvature of M2. Ray tracing results are compared to analytical ones in the case of the Italian Galileo telescope, that will be equipped with an high precision M2 driving device; the close matching with the analytical calculations is demonstrated.

031.074 **Of speckles, fringes, and an optical very large array.**
A. Labeyrie.
Comments Astrophys., Vol. 15, No. 6, p. 323 – 342 (Oct 1991).
The author reviews the history of speckle interferometry, some results obtained with this technique, and potential developments in the future. The paper is derived from the author's Tinsley Prize Lecture, presented to the American Astronomical Society in January 1991.

031.075 **Study of making a Schmidt corrector plate by vacuum deformation.**
Zhang Xincan, Li Depei.
Acta Astron. Sin., Vol. 32, No. 3, p. 304 – 308 (Sep 1991). In Chinese.
The vacuum deformation method and principles of making a Schmidt corrector plate are described. The theoretical analysis and the relevant formulae are also presented. The authors have successfully fabricated a Schmidt corrector plate (200 mm diam., $t/d = 1/40$) by means of the vacuum deformation. This paper describes how the plate was made and tested and shows the satisfactory results.

031.076 **Geometrical parameters in the Hartmann screen of aspherical mirrors.**
Xiong Yaoheng, Feng Hesheng.
Acta Astron. Sin., Vol. 32, No. 4, p. 409 – 414 (Dec 1991). In Chinese.
The optimum geometrical parameters of the square array screen in the two–dimensional Hartmann quantitative test of an aspherical mirror have been chosen. The results for the 1.2 m paraboloid mirror at the Yunnan Observatory are given.

031.077 **Liquid mirrors: a progress report.**
E. F. Borra, R. Content, L. Girard, J. Gauvin, P. Hickson, B. Gibson.
J. R. Astron. Soc. Can., Vol. 85, No. 4, p. 188 (Aug 1991). Abstract. – See Abstr. 012.091 for the main entry.

031.078 **Large–telescope focal–length reduction systems.**
G. M. Popov.
Bull. Crimean Astrophys. Obs., Vol. 81, p. 157 – 163 (1990). English translation of Izv. Krym. Astrofiz. Obs., Tom 81, p. 175 – 183 (1990). See Abstr. 53.031.040.
Existing focal–length converters are surveyed. A simplified mirror–lens system is described, which has low light losses. A method is given for designing it. A converter has been designed for the principal focus of a 6 m telescope. The system with the converter is intended for researching the spectra of faint sources. Forms of the design are described with internal or external focal–surface placing.

031.079 **Two–dimensional quantitative Hartmann test of the 1.2 m primary mirror and the 1.2 m telescope.**
Xiong Yaoheng, Feng Hesheng, Wang Wu, Jiang Chongguo, Zhang Yuncheng.
Publ. Yunnan Obs., No. 4, p. 23 – 34 (1991). In Chinese.
The two–dimensional quantitative Hartmann tests for the 1.2 m aspherical primary mirror and the 1.2 m alt azimuth telescope at the Yunnan Observatory are performed. The light energy concentration, the distribution of the spot diagram at the optimum focal plane of the 1.2 m primary mirror and the surface deviation distribution are obtained.

031.080 **Exploitation scientifique des données en interférométrie 2 télescopes.**
Y. Rabbia.
J. Astron. Fr., No. 39, p. 22 (Jul 1991). Abstract. – See Abstr. 012.099 for the main entry.

031.081 **Elementary software for optical design.**
J. Campitelli.
Bol. Asoc. Argent. Astron., No. 36, p. 334 – 341 (1990). In Spanish. – See Abstr. 012.095 for the main entry.
The main features of two programs are shown. One is interactive, and provides means of evolving a design by easy but complete evaluation and parameter change. The other is intended for documentation of an optimized design.

031.082 **The VLT adaptive optics programme.**
F. Merkle, N. Hubin, G. Gehring, F. Rigaut.
Messenger, No. 65, p. 13 – 14 (Sep 1991).

031.083 **Aluminium mirror technology at ESO: positive results obtained with 1.8–m test mirrors.**
P. Dierickx, F. Zigmann.
Messenger, No. 65, p. 66 – 68 (Sep 1991).

International Lens Design Conference. Proceedings.
See Abstr. 012.014.

Optical fabrication and testing. Proceedings.
See Abstr. 012.015.

Reflective and refractive optical materials for Earth and space applications. Proceedings.
See Abstr. 012.066.

Optik für Röntgenstrahlen. Entwicklungen in der ČSFR.
See Abstr. 013.006.

Overview of CODE V™ development.
See Abstr. 021.010.

SYNOPSYS™ – a lens design computer program package.
See Abstr. 021.011.

Optimization of the optical transfer function.
See Abstr. 021.012.

Practical use of generalized simulated annealing optimization on microcomputers.
See Abstr. 021.013.

Quadric surfaces: some derivations and applications.
See Abstr. 021.014.

A small–computer program for optical design and analysis written in 'C'.
See Abstr. 021.015.

Mirror, dome, and natural seeing at CFHT.
See Abstr. 032.017.

Primary mirror control system for the Galileo telescope.
See Abstr. 032.025.

Experimental demonstration of real time atmospheric compensation with adaptive optics employing laser guide stars.
See Abstr. 034.017.

Determination of the crystal elements' orientation arrangement in adjustable–wavelength birefringent filters.
See Abstr. 034.077.

A quadrant–CCD star tracker.
See Abstr. 034.105.

A fast focal–reducing camera for CCD imaging.
See Abstr. 034.115.

High–efficiency "superachromatic" polarimetry optics for use in optical astronomical spectrographs.
See Abstr. 034.116.

Design of a two–channel echelle spectrograph.
See Abstr. 034.137.

Hubble Space Telescope optical performance.
See Abstr. 035.040.

Fixing the Hubble Space Telescope.
See Abstr. 035.080.

Comparison of the angular resolution limit and SNR of the Space Hubble Telescope and the large ground–based telescopes.
See Abstr. 035.095.

Hubble Space Telescope optics – problems and solutions.
See Abstr. 035.096.

Direct imaging of nonsolar planets with infrared telescopes using apodized coronagraphs.
See Abstr. 036.001.

The HST spherical aberration and its effects on images.
See Abstr. 036.018.

Restoration of images degraded by telescope aberrations.
See Abstr. 036.086.

Imaging performance analysis of adaptive optical telescopes using laser guide stars.
See Abstr. 036.152.

Estimation of the accuracy of wavefront reconstruction by adaptive optical phase–conjugation systems.
See Abstr. 036.154.

Laser beams untwinkle stars.
See Abstr. 036.212.

Compensation of atmospheric optical distortion using a synthetic beacon.
See Abstr. 036.213.

Measurement of atmospheric wavefront distortion using scattered light from a laser guide–star.
See Abstr. 036.214.

High spatial resolution techniques.
See Abstr. 036.276.

Possibility of adaptive image correction in an early visual–information processing stage.
See Abstr. 036.360.

032 Astronomical Instruments

032.001 **Worms and belts.**
G. W. Henry, R. Nagarajan, M. R. Busby.
I.A.P.P.P. Commun., No. 45, p. 11 - 17 (Sep - Nov 1991). - See Abstr. 012.002 for the main entry.

032.002 **The TS-24 telescope simulator: detailed product description.**
C. Young, R. S. Schwartz.
I.A.P.P.P. Commun., No. 45, p. 75 - 79 (Sep - Nov 1991). - See Abstr. 012.002 for the main entry.
TS-24 is a computer program for the IBM or Macintosh computer that simulates the actual use of a 24" research telesocpe. TS-24 is a low cost alternative to actual scientific instrumentation for use in training students from grades 7 through college in the process of data collection, analysis and presentation.

032.003 **Progress report on the Berkeley Automatic Imaging Telescope (BAIT).**
M. W. Richmond.
I.A.P.P.P. Commun., No. 45, p. 80 - 85 (Sep - Nov 1991). - See Abstr. 012.002 for the main entry.

032.004 **A new type of telescope flow control – the unclosure.**
S. Limmongkol, R. E. Breidenthal.
Bull. Am. Astron. Soc., Vol. 23, No. 2, p. 875 (1991). Abstract. - See Abstr. 012.008 for the main entry.

032.005 **Telescope enclosure flow visualization.**
F. Forbes, W. Y. Wong, J. Baldwin, W. Siegmund, S. Limmongkol, C. Comfort.
Bull. Am. Astron. Soc., Vol. 23, No. 2, p. 875 - 876 (1991). Abstract. - See Abstr. 012.008 for the main entry.

032.006 **Ein Präzisionsfernrohr für Kleinsternwarten: 100/1000 von Jenoptik Carl Zeiss Jena.**
H. Mucke.
Sternenbote, Jahrg. 34, Nr. 9, p. 170 - 185 (1991).

032.007 **The instrumental polarization of a Gregory–Coudé telescope.**
J. Sánchez Almeida, V. Martínez Pillet, A. D. Wittmann.
Sol. Phys., Vol. 134, No. 1, p. 1 - 13 (Jul 1991).
The authors calculate a theoretical model of the polarization properties of a Gregory-Coudé telescope to predict the behaviour of the German Gregory-Coudé Telescope installed at the Observatorio del Teide (Spain). Measurements of the real effects produced by this telescope acting upon light of known polarization are compared with the model. The authors estimate an uncertainty in its predictions of about 10%, which is produced by the uncertainties of the (complex) refractive index of the metallic layers covering the mirrors. The way in which the plain telescope changes the Stokes' profiles is briefly considered.

032.008 **The resolution limits of large ground-based telescopes.**
T. S. McKechnie.
Bull. Am. Astron. Soc., Vol. 23, No. 2, p. 977 (1991). Abstract. - See Abstr. 012.008 for the main entry.

032.009 **Lateral support of very large telescope mirrors by edge forces only.**
G. Schwesinger.
J. Mod. Opt., Vol. 38, No. 8, p. 1507 - 1516 (Aug 1991).
Until now a diameter of about 4 m seemed to be the upper-size limit of telescope mirrors that still permitted cost-saving designs of lateral supports by edge forces alone. In some designs, the supporting edge forces comprised not only basic push-pull action normal to the edge but also a specific, moderate amount of tangential shear. However, this was a by-product of design economy rather than the result of understanding the potential of tangential support forces as a means of reducing mirror flexure systematically, down to residuals in the 1% region. The surprising possibility of extending the usefulness of pure edge supports is demonstrated by the example of the 8 m mirror of the ESO's VLT. Fitted with a lateral support at the outer edge alone, this thin mirror will exhibit a wavefront aberration with a calculated rms value of only 18 nm, without taking into account possible active control.

032.010 **Fernrohrantriebe mit Schrittmotoren (II).**
R. Gröbel.
BAV Rundbrief, Jahrg. 40, Nr. 3 - 4, p. 112 - 118 (1991).

032.011 **Design and accomplishment of the software for the control system of the 1.2-m horizontal telescope.**
Wang Wu, Wang Jinsuo, Jiang Chongguo.
Publ. Yunnan Obs., No. 1, p. 53 - 60 (Mar 1991). In Chinese.

032.012 **The stability of the horizontal axis of the DCMT.**
Li Zhigang, Li Huanxin, Fang Fusheng, Zhang Jing.
Publ. Shaanxi Astron. Obs., Vol. 14, No. 1, p. 23 - 32 (Aug 1991). In Chinese.
The stability of the horizontal axis of the DCMT is investigated. Using the self-calibration system of the DCMT, the error of the horizontal axis has been measured. With the visual micrometer the errors of the readings have been estimated to be 0".10 - 0".15. The observation data have been reduced, and the systematic error and the repeatability for the horizontal axis are estimated. The results of the test observation indicate that the stability of the horizontal axis of this instrument is fairly good.

032.013 **Summary on development of optical and mechanical construction for the Danish-Chinese Meridian Telescope.**
Hu Qiqian.
Publ. Shaanxi Astron. Obs., Vol. 14, No. 1, p. 33 - 36 (Aug 1991). In Chinese.
The difficult points in design and technology of the DCMT are discussed preliminarily based on the author's personal experience. The technical measures taken in solving these difficult points have been introduced briefly and some suggestions have been proposed.

032.014 **Test and analysis of the sway for the DCMT horizontal axis.**
Fang Fusheng, Li Huanxin, Liu Shaomin, Zhang Jing, Zhu Zi, Li Zhigang.
Publ. Shaanxi Astron. Obs., Vol. 14, No. 1, p. 37 - 43 (Aug 1991). In Chinese.
The sway of the horizontal axis for the DCMT has been analyzed and studied. The direction of the horizontal axis of the DCMT's is determined by the basic plane of the end of the big gear. The deviation from a ideal plane for the basic plane causes the sway of the horizontal axis. There are two kinds of effects of the sway on instrument precision: (1) causing sway of main mirror, (2) causing displacement of coordinate system of the micrometer. Since there exists an automatic collimation system of the DCMT, the observation accuracy is mainly influenced by the repeatability of the horizontal axis, so that the repeatability of the sway for the horizontal axis is discussed in this paper. The test and estimation show that the accuracy of repeatability of sway is about 0".18 for the horizontal axis of the DCMT.

032.015 **Test and analysis of the azimuth variation for DCMT base frame.**
Fang Fusheng, Liu Shaomin, Zhu Zi.
Publ. Shaanxi Astron. Obs., Vol. 14, No. 1, p. 44 - 49 (Aug 1991). In Chinese.
The variation of the azimuth of the DCMT base frame, related to the piers, with the temperature of the frame is discussed and analyzed. The construction of the base frame and the principle of

four points support are described. The test method for the variation of the azimuth of the base frame are introcuced, and the test results are presented.

032.016 **Proposed conversion of the McMath telescope to 4.0 meter aperture for solar observations in the IR.**
W. Livingston.
Conference on Space Astronomical Telescopes and Instruments as Part of SPIE's Technical Symposia on Optical Engineering and Photonics in Aerospace Sensing (OE/Aerospace Sensing '91), p. 498 – 502 (1991). – See Abstr. 012.030 for the main entry.
Located on a 2076 m summit in Arizona, the present all–reflective McMath optical system consists of a 2.0 m CERVIT flat mounted as a heliostat to follow the Sun, a 1.6 m 86.4 m focal length quartz concave positioned within an inclined underground tunnel, and a 1.5 m CERVIT flat which directs the image of different fixed instrument stations. The building is adequate to accommodate a 6.0 m tracking feed and a 4.0 m concave, resulting in an f/22 beam which is hot but manageable. A 4.0 m aperture is desirable for adequate flux and resolution at 12 μm where the diffraction limit is 0.75″. The telescope would be used at night by the solar/stellar community.

032.017 **Mirror, dome, and natural seeing at CFHT.**
R. Racine, D. Salmon, D. Cowley, J. Sovka.
Publ. Astron. Soc. Pac., Vol. 103, No. 667, p. 1020 – 1032 (Sep 1991).
A large and homogeneous set of image quality data, obtained with HRCam at the prime focus of the CFHT, was correlated with thermal sensor data to identify and quantify "local seeing" effects. The main findings are as follows. 1. Mirror seeing is an important source of image spread and amounts to FWHM = $0''.40/°C^{6/5}$ when the primary mirror is warmer than the air inside the dome. 2. Dome seeing is marginally significant, at the rate of $\sim 0''.1/°C^{6/5}$ of temperature difference between the air inside and outside the dome. 3. Optical aberrations from the CFHT primary mirror and from HRCam currently impose a limit of FWHM = 0″.38 to the image quality. 4. The median natural atmospheric seeing at the CFHT site on Mauna Kea is FWHM = $0''.43 \pm 0''.05$, the 10th and 90th percentiles being $\sim$0″.25 and $\sim$0″.7, respectively.

032.018 **Pointing models for the Calar Alto 2.2 m and 3.5 m telescopes.**
U. Graser, U. Hopp.
Astron. Astrophys., Vol. 251, No. 2, p. 737 – 742 (Nov 1991).
The authors discuss the pointing properties of the 2.2 m and the 3.5 m telescopes of Calar Alto Observatory. An empirical approach for the pointing models was chosen. The procedure for the development of these models is described. As result the authors get rms–accuracies between 2″.6 and 4″.0 for the Ritchey–Chretien and prime foci observations. The main parameters in the pointing models can be interpreted as encoder alignment errors, alignment errors of the mounting relative to the absolute reference system, and some flexure of the systems. Finally the authors report on the time dependencey of the pointing model and on tests of the tracking accuracy of the 3.5 m telescope.

032.019 **A proposal for a low instrumental polarization coudé telescope.**
V. Martínez Pillet, J. Sánchez Almeida.
Astron. Astrophys., Vol. 252, No. 2, p. 861 – 865 (Dec 1991).
The authors describe how any coudé telescope containing two flat mirrors with identical properties and incidence angles can be freed from its instrumental polarization. A half–wave plate positioned between the two mirrors with the proper orientation produces the desired effect. The present work discusses this very simple design and the accuracy it allows. In addition, the authors consider the feasibility of the technique applied to a particular telescope: the evacuated Gregory Coudé Telescope at the Observatorio del Teide.

032.020 **Mesures concernant la stabilité d'emplacement du cercle meridien de Bucarest.**
P. Popescu, M. Liculescu.
Journées 1991: Systèmes de référence spatio–temporels, p. 186 – 187 (1991). – See Abstr. 012.048 for the main entry.
This paper presents the variations of the instrumental parameters (collimation, azimuth and inclination) during the period 1986 – 1990; they reveal the stability in time and the precision of the Meridian Circle of Bucarest Observatory.

032.021 **Mirroring the cosmos.**
C. S. Powell.
Sci. Am., Vol. 265, No. 5, p. 80 – 89 (Nov 1991).
When the Hale Telescope was completed in 1947, many astronomers believed it would be the largest ever built. But thick, rigid mirrors are being replaced by multifaceted reflecting surfaces, lightweight honeycombs and flexible sheets of glass. Combined with optical technologies that cancel out atmospheric distortion, the new telescopes promise the clearest and brightest view yet of the heavens.

032.022 **Wind–tunnel study of an observatory dome with a circular aperture.**
G. G. Zilliac, E. W. Cliffton.
Publ. Astron. Soc. Pac., Vol. 103, No. 669, p. 1211 – 1217 (Nov 1991).
This report describes results of a wind–tunnel test of a new concept in observatory dome design, the fixed shutter dome, which is intended to protect telescopes from wind shake. The concept is an evolution of the split–sphere topology. In order to close larger apertures, the new design allows an upper spherical segment to rotate under a shell of slightly larger diameter which is fixed to a lower rotatable segment. This design is structurally more efficient than conventional designs. From an aerodynamic standpoint, the new dome configuration is similar in overall shape to conventional observatory domes, with the exception of the telescope viewing aperture. The new design consists of a circular aperture of reduced area in contrast to conventional domes with rectangular or slotted openings. Wind–tunnel results of a side–by–side comparison of the new dome with a conventional dome demonstrate that the mean and fluctuating velocity through the aperture and in the center of the new dome configuration are lower than those of conventional domes, thus reducing the likelihood of telescope flow–induced vibration.

032.023 **The Carlsberg Automatic Meridian Circle: La Palma.**
L. Helmer.
Astrophys. Lett. Commun., Vol. 28, No. 2–4, p. 101 – 112 (Sep 1991). – See Abstr. 012.072 for the main entry.
The "Carlsberg Automatic Meridian Circle" project on La Palma is presented and some of the results of the first five years of operation are discussed. The performance of the telescope under the excellent conditions on La Palma is tabled, and the vast improvements in accuracy and limiting magnitude due to the latest improvement in spring 1988 of the instrument are reported. A short description is given of the new transit micrometer, which yields accuracies in positions of 0.14 arcsec for a single observation. Mentioned also are the results of sampling meterological information on a continuous basis throughout the whole period of operation.

032.024 **The TNG – command architecture.**
A. Baruffolo, C. Bonoli, A. Ciani.
Telesc. Naz. Galileo, Tech. Rep., No. 6.
Osservatorio Astronomico di Padova–Asiago, Asiago (Italy). 12 p. (Jun 1991).
The authors describe the overall structure of the software dedicated to the control and operation of the Telescopio Nazionale Galileo (TNG). Special attention is dedicated to the commanding architecture implemented on the VMEs.

032.025 **Primary mirror control system for the Galileo telescope.**
F. Bortoletto, A. Baruffolo, C. Bonoli,
M. D'Alessandro, D. Fantinel, G. Giudici, R. Ragazzoni,
L. Salvadori, P. Vanini.
Telesc. Naz. Galileo, Tech. Rep., No. 7.
Osservatorio Astronomico di Padova–Asiago, Asiago (Italy). 15 p. (Sep 1991). To be published in Proc. SPIE, Vol. 1542, "Active and adaptive optical systems".

The Italian Galileo telescope (TNG) is in an advanced phase of construction. Among the various new technical aspects of this telescope the active optics system is now receiving special consideration. The optical and informatic groups are considering the definition of the control environment dedicated to the active-optics. A solution based on an array of interconnected 16bit transputers is described with the main requirements for the inter-communication and monitoring software.

032.026 **Galileo project – workstation software system. Architecture design document.**
A. Balestra, P. Marcucci, F. Pasian, M. Pucillo, R. Smareglia, C. Vuerli.
Telesc. Naz. Galileo, Tech. Rep., No. 9.
Osservatorio Astronomico di Padova–Asiago, Asiago (Italy). 37 p. (Dec 1991). = Pubbl. Oss. Astron. Trieste, No. 1428.

This document contains the architectural design of the workstation software system, which, together with the telescope software system, is part of the Galileo telescope control system. Special emphasis is put both on general issues like software standards, portability and modularity, and on more particular ones as remote control capabilities and fault tolerance. The document is divided into two main sections: the first describes the environment needed for the WSS to operate, including all off-line utilities dedicated to the housekeeping of the characteristics files. The other one concerns the structure of the on-line software, i.e. the monitoring processes and the user interface.

032.027 **Imaging optical interferometry.**
R. S. Simon, T. S. Brackett, K. J. Johnston,
D. Mozurkewich, K. W. Weiler, D. J. Hutter, J. T. Armstrong.
IAU Colloquium No. 131: Radio interferometry – theory, techniques, and applications, p. 358 – 367 (1991). – See Abstr. 012.089 for the main entry.

Interferometry at optical wavelengths is very similar to radio interferometry, once the fundamental differences in detectors are accounted for. The Mount Wilson Mark III optical interferometer has been used for optical interferometry of stars and stellar systems. Success with the Mark III has lead to the current program at the Naval Research Laboratory to build the Big Optical Array (BOA), which will be an imaging interferometer. Imaging simulations show that BOA will be able to produce images of complex stellar systems, with a resolution as fine as 0.2 mas.

032.028 **Improvement of type II photo–astrolabe and its preliminary results.**
Zhao Gang, Zhang Jianwei.
Prog. Astron., Vol. 9, No. 4, p. 326 (Dec 1991). In Chinese.

032.029 **Liquid mirror telescopes: history.**
B. K. Gibson.
J. R. Astron. Soc. Can., Vol. 85, No. 4, p. 158 – 171 (Aug 1991).

A history of Liquid Mirror Telescope (LMT) experimentation is provided. The concept can be traced to Ernesto Capocci of the Naples Observatory (1850). The recent work of Ermanno Borra and colleagues at Laval University has culminated with the realization of large diffraction–limited mirrors. An offshoot of the Laval laboratory research is the University of British Columbia/Laval fully–automated 2.7–metre *f*/1.9 LMT, a prototype deep–sky survey instrument.

032.030 **New largest optical telescope.**
M. Jõeveer.
Calendar of the Tartu Astronomical Observatory for the year, 1992.
p. 34 – 36 (1991). In Estonian. See Abstr. 046.016.

032.031 **UBC/Laval 2.7–metre liquid mirror telescope: status report.**
P. Hickson, B. K. Gibson, D. Hogg, E. F. Borra, R. Content, L. Girard.
J. R. Astron. Soc. Can., Vol. 85, No. 4, p. 218 (Aug 1991). Abstract. – See Abstr. 012.091 for the main entry.

032.032 **Expanded spectrographic capabilities of the DDO 1.88 m telescope.**
K. W. Kamper, C. T. Bolton.
J. R. Astron. Soc. Can., Vol. 85, No. 4, p. 218 – 219 (Aug 1991). Abstract. – See Abstr. 012.091 for the main entry.

032.033 **Automatic telescope control.**
P. Delaney, A. Prior, A. Ebbers, A. Rana.
J. R. Astron. Soc. Can., Vol. 85, No. 4, p. 220 – 221 (Aug 1991). Abstract. – See Abstr. 012.091 for the main entry.

032.034 **Observational examinations of some properties for the 1.56 m astrometric telescope of Shanghai Observatory.**
Wang Jiaji, Zhao Jianhong, Chen Li, Jiang Peifang, Chen Jin.
Ann. Shanghai Obs., Acad. Sin., No. 12, p. 31 – 39 (1991). In Chinese.

Some experimental astrometric plates were taken with the 1.56 m telescope of Shanghai Observatory during 1988 – 1989. Based on these plates, some properties of the telescope, such as imaging, magnitude limits, the focus length, and the position precision, are discussed.

032.035 **Solar Flare Telescope and 10–cm New Coronagraph.**
K. Ichimoto, T. Sakurai, A. Yamaguchi, K. Kumagai, Y. Nishino, Y. Suematsu, E. Hiei, T. Hirayama.
International SOLAR–A Science Meeting: Flare physics in solar activity maximum 22, p. 320 – 323 (1991). – See Abstr. 012.104 for the main entry.

Two new telescopes were built at the National Astronomical Observatory of Japan, i.e. the Solar Flare Telescope and the 10–cm New Coronograph. The Solar Flare Telescope was constructed to make observations of photospheric velocity fields, vector magnetic fields, and Hα and continuum images of active regions simultaneously. The 10–cm New Coronagraph, which was developed to make precise measurements of the coronal intensity in several wavelengths, has already been in operation for one year at the Norikura Solar Observatory. At present the accuracy of about 10^{-6} of the solar disk intensity is achieved in continuum light, but further improvement in the photometric accuracy remains to be done.

032.036 **The design and application of an automatic focusing device for the two–mirror optical system.**
Ding Hua, Wang Lanjuan, Gong Shoushen.
Ann. Shanghai Obs., Acad. Sin., No. 12, p. 107 – 113 (1991). In Chinese.

032.037 **An accurate test and the analysis of the tested results of the axial system of the 1.2 m alt–az telescope.**
Feng Hesheng, Xiong Yaoheng, Zhang Yuncheng, Wang Jinso.
Publ. Yunnan Obs., No. 4, p. 1 – 6 (1991). In Chinese.

The accurate tested results of the axes of the 1.2 m alt–az telescope at the Yunnan Observatory are given. The systematic erorrs are separated from the results and the mechanism of each systematic error is found.

032.038 **Test and analysis of the stability of friction drive of the vertical axis of the 1.2 m alt–az telescope.**
Feng Hesheng, Jiang Chongguo, Wang Wu, Wang Jinso, Zhang Yuncheng, Xiong Yaoheng, Yang Wanjiang, Niu Xin.
Publ. Yunnan Obs., No. 4, p. 7 – 13 (1991). In Chinese.
The test and analysis of the static and dynamic stability of friction drive of the vertical axis of the 1.2 m alt–az telescope at Yunnan Observatory are carried out. The actual velocities and accelerations to be needed for tracking the Moon and the Lageos satellite are simulated. The tested results show that the stability of friction drive system can meet the needs of the $\pm 1''$ tracking accuracy.

032.039 **Rigidity of friction drive of the vertical axis of the 1.2 m alt–az telescope.**
Feng Hesheng, Jiang Chongguo, Xiong Yaoheng, Zhang Yuncheng, Wang Wu, Wang Jinso.
Publ. Yunnan Obs., No. 4, p. 14 – 18 (1991). In Chinese.

032.040 **On the development of the control system of the 1.2 m horizontal telescope.**
Jiang Chongguo, He Miaochan, Yang Jing, Yang Wanjiang.
Publ. Yunnan Obs., No. 4, p. 19 – 22 (1991). In Chinese.

032.041 **Mounting system of the primary mirror of the 1.2 m alt–az telescope.**
Wang Yong, Jiang Xiaoru.
Publ. Yunnan Obs., No. 4, p. 35 – 39 (1991). In Chinese.

032.042 **The rotation problem of the 1.2 m alt–az telescopes field.**
Xiong Yaoheng, Feng Hesheng.
Publ. Yunnan Obs., No. 4, p. 40 – 45 (1991). In Chinese.
The formula for the object field rotation due to alt–az mounting is presented and the image field rotation due to plane mirrors system with the movement of azimuth and altitude at 1.2 m alt–az telescope is determined.

032.043 **Grand interferomètre à deux télescopes.**
D. Mourard.
J. Astron. Fr., No. 39, p. 15 – 16 (Jul 1991). – See Abstr. 012.099 for the main entry.

032.044 **Méthodes de haute résolution angulaire mono–pupille en infrarouge.**
C. Perrier.
J. Astron. Fr., No. 39, p. 16 – 20 (Jul 1991). – See Abstr. 012.099 for the main entry.

032.045 **Le télescope Bernard Lyot.**
J. Rösch.
J. Astron. Fr., No. 40, p. 2 – 3 (Jul 1991).

032.046 **Two Chinese new telescopes for astronomy.**
J. Cheng.
Bol. Asoc. Argent. Astron., No. 36, p. 301 – 310 (1990). In Spanish. – See Abstr. 012.095 for the main entry.
Recently, two telescope projects have been in progress in China. These are 13.7 m millimeter wavelength telescope and 1.05 m multi–focus optical telescope. Both telescopes are introduced in detail. The discussion includes design features, telescope characteristics and their performance.

032.047 **Classical astronomical instruments used in Survey of India for positional astronomy.**
V. K. Nagar, P. S. Bains, M. J. S. Grover, S. Singh.
Bull. Astron. Soc. India, Vol. 19, Nos. 1 – 2, p. 84 (Mar – Jun 1991). Abstract. – See Abstr. 012.108 for the main entry.

032.048 **Gamma–ray astronomy with a conventional optical telescope.**
C. L. Bhat, R. K. Kaul, H. S. Rawat, V. K. Senecha, R. C. Rannot, M. L. Sapru, A. K. Tickoo, H. Razdan.
Bull. Astron. Soc. India, Vol. 19, Nos. 1 – 2, p. 88 (Mar – Jun 1991). Abstract. – See Abstr. 012.108 for the main entry.

032.049 **A progress report on the VLT instrumentation plan.**
S. D'Odorico, J. Beckers, A. Moorwood.
Messenger, No. 65, p. 10 – 13 (Sep 1991).

032.050 **Siderostat mounting and its servo parameters.**
Xu Jiayan, Wang Zhengming, Xiao Jinhong, Wang Hongqi.
Publ. Shaanxi Astron. Obs., Vol. 14, No. 2, p. 106 – 110 (Dec 1991). In Chinese.
The authors discuss advantages of Alt–Azi mounted siderostats for an optical interferometer, and suggest an optimum optical configuration of siderostat–telescope, which could cover a large field of sky. The values of positions, speeds and accelerations for the normal of a siderostat were calculated for stars located at different zenith and azimuth.

032.051 **The stability of collimation error for the DCMT.**
Li Huanxin, Liu Shaomin, Fang Fusheng, Zhang Jing, Li Zhigang.
Publ. Shaanxi Astron. Obs., Vol. 14, No. 2, p. 117 – 121 (Dec 1991). In Chinese.
For the DCMT which is a kind of new instruments and is different from the traditional meridian circle, the collimation error is discussed. It is shown that this error depends on angle between S1 and S2 mirrors. The results about the change of collimation error with temperature are shown. The collimation error seems to be stable.

032.052 **Astrophysical functions of the reflector with diameter 1.05 m and its necessary auxiliary facilities.**
Jiang Shiyang.
Publ. Shaanxi Astron. Obs., Vol. 14, No. 2, p. 141 – 151 (Dec 1991). In Chinese.

032.053 **Designing at APT programme.**
E. Budding.
South. Stars, Vol. 34, No. 3, p. 61 – 66 (18 Jun 1991). – See Abstr. 012.120 for the main entry.
The factors involved in planning for an Automatic Photoelectric Telescope (APT) operation are reviewed.

032.054 **An intelligent APT.**
B. Warner.
South. Stars, Vol. 34, No. 3, p. 67 – 71 (18 Jun 1991). – See Abstr. 012.120 for the main entry.
How to add "intelligence" to the basic APT concept forms the subject of this brief review. This refers to improvements in field location, extending available magnitude ranges, constraint fixing, selecting and queuing observations. A few practical cases are considered for illustration.

032.055 **Lunar automatic photometric telescopes (LAPT's).**
M. Zeilik.
South. Stars, Vol. 34, No. 3, p. 73 – 76 (18 Jun 1991). – See Abstr. 012.120 for the main entry.
A viable proposal for a network of modest sized robust remote telescopes distributed on the moon's surface is presented together with some possible applications.

Unusual telescopes.
See Abstr. 003.025.

The origins of the reflecting telescope.
See Abstr. 004.039.

The 8 1/4–inch Clark refractor of the Temple Observatory, Rugby.
See Abstr. 004.040.

Gill, Grubb and the astrographic telescopes – a correspondence.
See Abstr. 004.076.

Will La Silla succumb to the VLT?
See Abstr. 009.011.

Toward the global APT network. Proceedings. 3. New Zealand Conference on Photoelectric Photometry, Blenheim (New Zealand), 9 – 12 Mar 1989.
See Abstr. 012.120.

Astronomy: When big is beautiful.
See Abstr. 013.083.

The Japanese National Large Telescope.
See Abstr. 013.085.

Newly installed radio and optical telescopes in China.
See Abstr. 013.088.

Optical telescopes.
See Abstr. 014.010.

Design and demonstrated performance of adaptive optics system.
See Abstr. 031.006.

The precise adjustment of optical system for horizontal transit circle.
See Abstr. 031.018.

The fine adjusting method for the optical system of the DCMT.
See Abstr. 031.019.

Evaluation of seeing on a 62-cm mirror.
See Abstr. 031.026.

Adaptive optics on a 3.6-m telescope: results and performance.
See Abstr. 031.046.

Active optics and the New Technology Telescope (NTT): the key to improved optical quality at lower cost in large astronomical telescopes.
See Abstr. 031.072.

Moving M2 mirror without pointing offset.
See Abstr. 031.073.

Of speckles, fringes, and an optical very large array.
See Abstr. 031.074.

Large-telescope focal-length reduction systems.
See Abstr. 031.078.

The VLT adaptive optics programme.
See Abstr. 031.082.

A proposal for an ATIS command set extension for CCD cameras.
See Abstr. 034.002.

The micrometers of the Carlsberg Automatic Meridian Circle.
See Abstr. 034.107.

Quantitative assessment of the science return from an orbiting, imaging, optical interferometer.
See Abstr. 035.204.

Telescope beam-profile diagnostics and the solar limb.
See Abstr. 036.024.

Absolute spectrophotometry of the Sun. III. SEF-1 telescope automation system.
See Abstr. 036.056.

Effect of a finite outer scale on the Zernike decomposition of atmospheric optical turbulence.
See Abstr. 082.024.

033 Radio Telescopes and Equipment

033.001 **Millimeter-wave Schottky mixer receivers for applications in antenna measurements and radio astronomy.**
A. O. Lehto.
Thesis (D.Sc.), Helsinki Univ. of Tech., Espoo (Finland). Radio Lab. Report-S--185, 18 p. (16 Aug 1990). ISBN 951-22-0271-9.

This thesis concentrates on whisker-contacted Schottky mixer receivers and on some of their applications. In this introductory part operation, noise sources, and construction of Schottky mixers and receivers are explained. An equivalent network of the embedding impedance seen by the mixer diode is presented in this paper. It is based on scaled model measurements and verified by comparing measured receiver noise performance with results given by computer simulations. Two room-temperature Schottky mixer receivers working at 110 GHz and 183 GHz have been used to solve certain problems in antenna measurements: the quality of anechoic chambers and absorber panels, and amplitude and phase patterns of two corrugated feed horns have been measured. Reflections in the quiet zone, which is the volume used for antenna measurements, were measured in three chambers. The phase patterns of the horns were measured by using a novel differential phase method which avoids problems associated with conventional phase measurement methods.

033.002 **The express-control method of pointing the RATAN-600 antenna pattern in height direction.**
O. A. Golubchina.
Astrofiz. Issled. Izv. Spets. Astrofiz. Obs., Tom 31, p. 144 - 148 (1991). In Russian. English translation in Bull. Spec. Astrophys. Obs. - North Caucasus.

The express-control method of pointing the RATAN-600 antenna pattern center in the height direction is suggested. The method enables to detect the height position of the antenna pattern center using one observation of the Sun, the azimuth of which does not equal to 0 at the moment of the observation. The accuracy of this method is higher than 1′. Results of the experiments are presented. The express-control method can be used for detection of radio-refraction in low heights, that is confirmed experimentally.

033.003 **Fast anti-jamming estimation of the noise-track level of a radiometer with absolute median deviation.**
B. L. Erukhimov.
Astrofiz. Issled. Izv. Spets. Astrofiz. Obs., Tom 31, p. 149 - 152 (1991). In Russian. English translation in Bull. Spec. Astrophys. Obs. - North Caucasus.

A real time method of anti-jamming estimation of the σ level of a radiometer is discussed. A known algorithm of the absolute median deviation is used. A comparison of the proposed method with a classical one is given on noise records contaminated by jamming typical for radioastronomical observations. The efficiency of the used estimation is shown.

033.004 **On the accuracy of coordinate measurements with the RATAN-600 radio telescope.**
M. G. Mingaliev, V. N. Chernenkov.
Astrofiz. Issled. Izv. Spets. Astrofiz. Obs., Tom 31, p. 153 - 162 (1991). In Russian. English translation in Bull. Spec. Astrophys. Obs. - North Caucasus.

The possible limitations of accuracy of coordinate measurements with the RATAN-600 radio telescope are discussed. The main source of position measurement errors deals with unstability of the electric axis and horn type antenna orientation (the accuracy of RA determiantion is about 4″). Using special methods for control of the electric axis and horn type antenna orientation makes it possible to decrease the accuracy of position measurements up to 0.3″ – 0.5″. The part of errors of coordinate measurements duc to the atmosphere events is discussed also.

033.005 **Scaling posture of the antenna of a solar radio telescope and the design of the control program.**
Yang Kaiping.
Publ. Yunnan Obs., No. 1, p. 40 – 44 (Mar 1991). In Chinese.
A method is proposed for the design of a control program in order to accomplish the automation of the control of the scaling posture of the antenna of the telescope; an example is also given.

033.006 **Two feeds with single common antenna operating at 15 and 21 cm wavelengths.**
Chen Guoqian, Xie Ruixiang, Wang Congyun.
Publ. Yunnan Obs., No. 2, p. 24 – 27 (Jun 1991). In Chinese.
The construction and principle of two feeds with single common antenna operating at 15 and 21 cm wavelengths are presented. Its designing parameters and the results of solar radio emission observation are also given.

033.007 **The analysis of the pointing and tracking accuracies of the 10 m radio telescope.**
Du Hong, Su Bumei.
Publ. Yunnan Obs., No. 2, p. 52 – 60 (Jun 1991). In Chinese.
The factors contributing to the pointing and tracking accuracies of the 10 m radio telescope at Yunnan Observatory are analyzed and discussed. Some methods of improving the pointing and tracking accuracies of the antenna system are given. The possibility of the pointing accuracy of 33″ at the 1.3 cm is expounded and proved.

033.008 **Make old masers radiate vigour of their youth.**
Zhai Zaocheng, Huang Hengxiang, Lin Chuanfu, Lu Jiafu, Jiang Guoxing, Luo Weihua.
Publ. Shaanxi Astron. Obs., Vol. 14, No. 1, p. 89 – 94 (Aug 1991). In Chinese.
Three old models of hydrogen masers which were developed at Shanghai Observatory have been reformed since 1985. Two years operations indicate that the stabilities reach 10^{-15} for periods of time beyond 100 sec, and the masers are reliable. This paper describes the techniques used for reforming of old masers. The performances and operations of the reformed masers in VLBI observations are also introduced.

033.009 **The calibration of celestial emission antenna temperature.**
Zhou Shurong.
Publ. Purple Mt. Obs., Vol. 10, No. 2, p. 136 – 143 (Jun 1991). In Chinese.
Two methods of the calibration of the celestial emission antenna temperature that has been calibrated for both the temperature at the end of the antenna and at the aperture plane of the antenna are presented. Emission characteristic and application of noise sources of the microwave blackbody is described which have been widely used to the calibration at the aperture plane of antenna and the calibration of celestial emission temperature.

033.010 **A high speed data collector and buffer.**
Ding Zugao.
Publ. Purple Mt. Obs., Vol. 10, No. 2, p. 151 – 156 (Jun 1991). In Chinese.
A high speed data collector and buffer applied to detect weak signal in strong noise is described in the paper. Its characteristics are high speed, long time buffer, and large capactity. It can communicate between a computer and various detectors (e.g. multi–channel filter, surface acoustic wave filter, acoustic optical spectrometer, and so on) and multiplex sensors.

033.011 **The Canadian geophysical long baseline interferometer.**
J. L. Yen, P. Leone, G. A. Watson, J. K. Zao, J. Popelar, W. T. Petrachenko, G. Feil, W. H. Cannon, P. Mathieu, P. Newby, H. Tan, R. D. Wietfeldt, J. A. Galt.
Radio Sci., Vol. 26, No. 1, p. 89 – 99 (Jan–Feb 1991).
The Canadian geophysical long baseline interferometer, a new very long baseline interferometer system, has been developed and put into operation. The system tracks source delay and fringe rotation during observation using a wave front clock. After the low–noise receiver the signals from radio sources are channelized and down converted to baseband, each baseband channel is one bit digitized at a rate of 12 mbar/s and then recorded on a single video cassette recorder for processing. The total bandwidth of the system can be easily increased by replication of baseband channels. During processing the tapes are played back in synchronism and the recovered astronomy data are processed in a simple correlator to obtain the source visibility. The use of the wave front clock allows bursts of wideband astronomy data to be sampled at a high rate at each station and then recorded at a lower rate for processing. Because bursts recorded at different stations at the same wave front clock epoch are emitted by the source at the same time, they are correlated.

033.012 **Interference rejection in Miyun Synthesis Radio Telescope.**
Yang Yipei, Zheng Yijia, Zhang Xizhen, Chen Hongsheng, Wu Xiaotao.
Chin. Astron. Astrophys., Vol. 15, No. 3, p. 364 (Sep 1991). Abstract. The full paper is published in Acta Astrophys. Sin., Vol. 11, No. 1, p. 73 – 78 (Jan 1991). See Abstr. 53.033.017.

033.013 **A microwave square–loop ferrite switch as a balancer in radiometers.**
V. R. Amirkhanyan, D. V. Musina.
Bull. Spec. Astrophys. Obs. – North Caucasus, Vol. 32, p. 166 – 168 (1990). English translation of Astrofiz. Issled. Izv. Spets. Astrofiz.. Obs., Tom 32, p. 207 – 210 (1990).
A microwave switch with ferrite material having a rectangular hysteresis loop was investigated. The possibility of using it as a balancer in a Dicke radiometer is shown. A simple control scheme is proposed.

033.014 **The Nobeyama Millimeter Array and its future plan.**
M. Ishiguro.
3. Haystack Observatory Conference on the Interstellar Medium: Skylines – atoms, ions and molecules – new results in spectral line astrophysics, p. 421 (1991). Abstract. – See Abstr. 012.051 for the main entry.

033.015 **Millimeter–wave receivers for radioastronomy.**
R. Blundell.
3. Haystack Observatory Conference on the Interstellar Medium: Skylines – atoms, ions and molecules – new results in spectral line astrophysics, p. 423 – 435 (1991). – See Abstr. 012.051 for the main entry.
The field of millimeter–wave radioastronomy has advanced rapidly since the discovery of the cosmic background radiation by Penzias and Wilson in 1965. One of the main reasons for the rapid advances in millimeter–wave radioastronomy has been the continual evolution of the sophisticated techniques required to produce low–noise receivers. Recently, a new generation of low–noise millimeter–wave heterodyne receivers has emerged which is of particular interest. This type of receiver makes use of the extremely sharp nonlinearity produced by the tunneling of single electrons between two superconductors. The present status in the development of such receivers for radioastronomical applications is reviewed and the possibilities of extending the new technology to higher frequencies are examined.

033.016 **The Haystack 3–mm upgrade.**
R. Barvainis, J. A. Ball, R. P. Ingalls, J. Antebi, M. Zarghamee.
3. Haystack Observatory Conference on the Interstellar Medium: Skylines – atoms, ions and molecules – new results in spectral line astrophysics, p. 437 – 438 (1991). – See Abstr. 012.051 for the main entry.

033.017 **Frequency selective surfaces for radio astronomy.**
Y. Irimajiri, T. Takano.
Exp. Astron., Vol. 2, No. 2, p. 123 – 136 (1991).
Frequency selective surfaces (FSSs) with the Jerusalem–cross array configuration have been developed for simultaneous observations of radio astronomical sources at the 40 GHz and 80 GHz bands. For supporting material, PET (polyethylene terephthalate: 200 × 220 mm large, 12 μm thick, permittivity 3.0) was used, and copper was used for the conducting array. The measured characteristics were in good agreement with the calculated ones. The fabricated FSS has low insertion losses of less than 0.2 dB and little distortion of polarization for each band. The characteristics of the developed FSS satisfy radio astronomical requirements.

033.018 **VSWR reduction for large millimeter–wave Cassegrain radiotelescopes.**
R. Padman, R. E. Hills.
Int. J. Infrared Millimeter Waves, Vol. 12, No. 6, p. 589 – 599 (Jun 1991).
Multiple reflections in large radiotelescopes used for astronomical spectroscopy cause characteristic modulations of the observed spectrum ("baseline ripple"). For a given mechanism, the magnitude of the effect depends primarily on the reflection coefficient, which for the most important paths is proportional to Λ. Although ripple is thus generally negligible at mm–wavelengths, there are some instances where it may still be significant, and a recent experiment at the 15 m James Clerk Maxwell Telescope was severely affected. The authors describe a technique which has proved effective in reducing the ripple to an acceptable level. The purpose of this research note is to draw this technique to the attention of the astronomical community. The authors calculate the reflection coefficient for a typical instrument and show that it can be greatly reduced over a very broad bandwidth by means of a tapered absorber in the centre of the secondary mirror. The graphical analysis of the tapered absorber also demonstrates that the scattering cones and circular absorbers widely used in radioastronomy do not in fact reduce the reflection coefficient significantly.

033.019 **Performance of the acousto–optical spectral line receivers at the Metsahovi 14 m radio telescope.**
L. Malkamäki, K. Mattila, A. Räisänen, A. Lehto, J. Peltonen.
Exp. Astron., Vol. 2, No. 3, p. 149 – 161 (1991).
The spectral line receiving system at the 13.7 m radio telescope at the Metsähovi Radio Research Station is described. A cooled low noise Schottky mixer receiver in 72 – 115 GHz range is employed. The back end system comprises three acousto–optical spectrometers (AOS) with bandwidths of 11 MHz, 100 MHz and 600 MHz. Active frequency stabilization has been realized for the spectrometers, and is described here in detail for the first time. A description of the receiver and back end system and the results of performance measurements are given. As a result of active stabilization an excellent frequency stability has been achieved, amounting to ± 0.005, ± 0.003 and ± 0.01 in units of resolution bandwidth for the 11, 100 and 600 MHz AOS, respectively. Special measures have been taken to eliminate background light and prevent interference fringe patterns. Tests of the noise performance of the three AOS show that cycle times of > 250 s are achieved.

033.020 **The far sidelobes and noise temperature of a small paraboloidal antenna used for radio astronomy.**
M. D. Anderson, T. L. Landecker, D. Routledge, J. F. Vaneldik.
Radio Sci., Vol. 26, No. 2, p. 353 – 361 (Mar–Apr 1991).
Measurements have been made of the radiation pattern of a symmetrical, prime–focus paraboloidal antenna which is used as a radio telescope at 1420 MHz. Of a measured total antenna temperature of 26.8K with the antenna pointing at the zenith, the following contributions can be accounted for: cosmic microwave background (2.7K), galactic emission (1.0K), atmospheric emission (2.0K), direct spillover from the ground into the feed (8.0K), leakage through the reflector mesh (5.9K), and diffraction around the reflector rim (0.6K).

033.021 **The design of high frequency antennas.**
J. Delannoy.
IAU Colloquium No. 131: Radio interferometry – theory, techniques, and applications, p. 15 – 25 (1991). – See Abstr. 012.089 for the main entry.
The conceptual guidelines in designing mm radio–telescope antennas are enlightened by some recent progress in their mechanical construction, and opto–radio–geometrical adjustment. This paper reviews concepts, illustrated in many existing solutions, including the new "phase–retrieval" holography for single dish adjustment (Morris 1982): 2 amplitude only maps at focus and out of focus, with enough dynamic range and signal to noise ratio, give surface errors within hours.

033.022 **Correlators for interferometry – today and tomorrow.**
Y. Chikada.
IAU Colloquium No. 131: Radio interferometry – theory, techniques, and applications, p. 26 – 32 (1991). – See Abstr. 012.089 for the main entry.
History of correlators for radio interferometer is briefly reviewed. Advantages and disadvantages of various methods are discussed and the most appropriate ways are given for various interferometers. Challenges to the next generation short–wavelength interferometer are also discussed.

033.023 **The multifrequency front end: a new type of front end for the Westerbork Synthesis Radio Telescope.**
G. H. Tan.
IAU Colloquium No. 131: Radio interferometry – theory, techniques, and applications, p. 42 – 46 (1991). – See Abstr. 012.089 for the main entry.
The Westerbork Synthesis Radio Telescope will be equipped with new front ends. These front ends will cover 8 frequency bands in the range from 250 MHz to 8.6 GHz. For the frequency bands above 1.2 GHz the sensitivity of the instrument will be drastically improved. Two independent local oscillator systems make it possible to observe in two frequency bands simultaneously.

033.024 **Hardware considerations for high–dynamic–range imaging.**
D. S. Bagri, A. R. Thompson.
IAU Colloquium No. 131: Radio interferometry – theory, techniques, and applications, p. 47 – 54 (1991). – See Abstr. 012.089 for the main entry.
The dynamic range of an image obtained using an interferometer array depends on the (interferometer) gain closure errors. This paper discusses hardware imperfections which cause closure errors and how these errors can be minimised.

033.025 **A pulse calibration system for the VLBA.**
A. R. Thompson, D. S. Bagri.
IAU Colloquium No. 131: Radio interferometry – theory, techniques, and applications, p. 55 – 59 (1991). – See Abstr. 012.089 for the main entry.
In pulse calibration of an interferometer a series of short pulses is inserted into the receiver input to produce a spectrum of equi–spaced lines. The use of these for calibration, and the proposed system for the VLBA, are described.

033.026 **Cryogenically–cooled, HEMT amplifiers and receivers in 1 – 50 GHz range: state–of–the–art.**
M. W. Pospieszalski.
IAU Colloquium No. 131: Radio interferometry – theory, techniques, and applications, p. 60 – 64 (1991). – See Abstr. 012.089 for the main entry.
A review of the recent developments in the design, construction and performance of cryogenically–coolable, high–electron–mobility transistor (HEMT, MODFET) amplifiers and their application in compact cryogenic receivers for radio astronomy applications is presented.

033.027 **Tape recorders for VLBI.**
A. E. E. Rogers.
IAU Colloquium No. 131: Radio interferometry – theory, techniques, and applications, p. 65 – 75 (1991). – See Abstr. 012.089 for the main entry.
Magnetic tape recorders developed for VLBI have taken advantage of the best available technology to achieve high data rates and large storage capacity. The Mark I, II and III VLBI recording systems have data rates of 0.72, 4 and 224 Mbits/sec respectively. The recorder developed for the VLBA has a data rate of 256 Mbits/sec and can store 5.5 Terabits of data on one tape.

033.028 **Direct imaging digital lens for transient radio source survey.**
T. Daishido, K. Asuma, K. Nishibori, J. Nakajima, M. Yano, E. Otobe, N. Watanabe, A. Tsuchiya, S. Iwase.
IAU Colloquium No. 131: Radio interferometry – theory, techniques, and applications, p. 86 – 89 (1991). – See Abstr. 012.089 for the main entry.
FFT based 2D Dygital Lens is discussed comparing with Fourier synthesis. The sensitivity of constructing Large Array will be 50 mJy. Precise manipulation of the radiation field is possible in it.

033.029 **A portable holographic antenna measurement system.**
D. B. Shaffer, J. M. Gipson.
IAU Colloquium No. 131: Radio interferometry – theory, techniques, and applications, p. 94 – 97 (1991). – See Abstr. 012.089 for the main entry.
Interferometrics has developed a portable holographic antenna measurement system. The system is comprised of a small interferometer which uses satellite signals to measure the complex beam pattern of the antenna under test. A PC controls all aspects of data collection and analysis. The system produces maps of surface profile errors and the illumination power in reflector antennas.

033.030 **Compatibility issues in VLBI.**
R. D. Wietfeldt, L. R. d'Addario.
IAU Colloquium No. 131: Radio interferometry – theory, techniques, and applications, p. 98 – 101 (1991). – See Abstr. 012.089 for the main entry.
New VLBI systems are under development in many countries in preparation for large dedicated arrays, including orbiting radio telescopes to be launched in the 1990s. In particular, several new recording systems and correlators are being constructed: the VLBA in the U.S., the S2 in Canada, and the K4/VSOP in Japan. This paper reviews the issues of technical compatibility among these systems in an effort to facilitate global VLBI experiments.

033.031 **Opacity at the Owens Valley Millimeter Array.**
S. L. Scott, D. P. Woody.
IAU Colloquium No. 131: Radio interferometry – theory, techniques, and applications, p. 119 – 123 (1991). – See Abstr. 012.089 for the main entry.
The authors have measured the opacity at the Owens Valley Millimeter Array with tip curves at frequencies from 95 to 248 GHz. These measurements have been interpreted using a model and then scaled to a standard frequency of 100 GHz. The unique aspect of this data is the wide range of frequencies covered and the use of receivers with different sideband gain ratios.

033.032 **Software systems for the Ryle Telescope.**
D. J. Titterington.
IAU Colloquium No. 131: Radio interferometry – theory, techniques, and applications, p. 128 – 131 (1991). – See Abstr. 012.089 for the main entry.
An overview of the software systems developed for the broad-banded Ryle Telescope is given.

033.033 **Reminiscences of early days of the VLA.**
D. S. Heeschen.
IAU Colloquium No. 131: Radio interferometry – theory, techniques, and applications, p. 150 – 159 (1991). – See Abstr. 012.089 for the main entry.
The author gives a short chronological overview of the VLA project with special emphasis to the very early days from about 1961 through 1966.

033.034 **The early technical development of the VLA.**
G. W. Swenson Jr.
IAU Colloquium No. 131: Radio interferometry – theory, techniques, and applications, p. 160 – 164 (1991). – See Abstr. 012.089 for the main entry.

033.035 **The design and construction of the VLA: the project manager's view.**
J. H. Lancaster.
IAU Colloquium No. 131: Radio interferometry – theory, techniques, and applications, p. 165 – 170 (1991). – See Abstr. 012.089 for the main entry.

033.036 **GMRT mapping strategies.**
A. P. Rao.
IAU Colloquium No. 131: Radio interferometry – theory, techniques, and applications, p. 212 – 217 (1991). – See Abstr. 012.089 for the main entry.
The Giant Metrewave Radio Telescope (GMRT) that is being built to operate at metre wavelenghts, will encounter a number of problems that are specific to synthesis imaging at low frequencies. Some of these problems are highlighted and possible solutions that are being considered, are discussed.

033.037 **Giant Metrewave Radio Telescope (GMRT).**
G. Swarup.
IAU Colloquium No. 131: Radio interferometry – theory, techniques, and applications, p. 376 – 379 (1991). – See Abstr. 012.089 for the main entry.
The Giant Metrewave Radio Telescope (GMRT) is being set up near Pune in India consisting of 30 numbers of 45–m diameter parabolic dishes. Twelve of these are being placed in a central array, about 1×1 km^2 in size, and the remaining eighteen along three 14–km long arms forming a Y–shaped array. GMRT will operate at six frequency bands near 38, 153, 233, 327, 611 and 1420 MHz. It is expected to be completed progressively by 1993, and will be the world's largest radio telescope operating at metre and decimetre wavelengths.

033.038 **The MERLIN – phase 2.**
P. N. Wilkinson.
IAU Colloquium No. 131: Radio interferometry – theory, techniques, and applications, p. 381 – 385 (1991). – See Abstr. 012.089 for the main entry.
The Jodrell Bank MERLIN is currently being upgraded to produce higher sensitivity and higher resolving power. The major capital item has been a new 32 m telescope located at MRAO Cambridge which will operate to at least 50 GHz. A brief outline of the upgraded MERLIN and its performance is given.

033.039 **Progress on Chinese VLBI Network project.**
Ye Shuhua, Wan Tongshan, Qian Zhihan.
IAU Colloquium No. 131: Radio interferometry – theory, techniques, and applications, p. 386 – 389 (1991). Abstract. – See Abstr. 012.089 for the main entry.

033.040 **The Very Long Baseline Array.**
P. J. Napier.
IAU Colloquium No. 131: Radio interferometry – theory, techniques, and applications, p. 390 – 394 (1991). – See Abstr. 012.089 for the main entry.
The Very Long Baseline Array, currently under construction, is a dedicated VLBI array consisting of ten 25 m diameter antennas and a 20 station correlator. The key design features that make it a flexible, general purpose instrument are reviewed and some

details are provided of the array layout, frequency coverage and sensitivity.

033.041 **Microwave background interferometry in Cambridge.**
M. E. Jones.
IAU Colloquium No. 131: Radio interferometry – theory, techniques, and applications, p. 395 – 399 (1991). – See Abstr. 012.089 for the main entry.

Interferometric methods of studying the Cosmic Microwave Background (CMB) have some distinct advantages over the switched–beam techniques which have mostly been used. However, most existing interferometers are not well suited to CMB observations, for a variety of reasons. A new instrument, the Ryle Telescope, has been developed in Cambridge which has a high temperature sensitivity (120 μK in 12 h) and the ability to integrate for several hundred hours on the same field. It will be used to study the CMB on angular scale of a few arcminutes, with particular emphasis on the Sunyaev–Zel'dovich effect. A second instrument to study the CMB on angular scales of tens of arcminutes, the Cosmic Anisotropy Telescope (CAT), is also being developed.

033.042 **The Owens Valley Millimeter Array.**
S. Padin, S. L. Scott, D. P. Woody.
IAU Colloquium No. 131: Radio interferometry – theory, techniques, and applications, p. 400 – 404 (1991). – See Abstr. 012.089 for the main entry.

The telescopes and signal processing systems of the Owens Valley Millimeter Array are briefly described.

033.043 **The Smithsonian Astrophysical Observatory submillimeter wavelength array.**
C. R. Masson.
IAU Colloquium No. 131: Radio interferometry – theory, techniques, and applications, p. 405 – 409 (1991). – See Abstr. 012.089 for the main entry.

The Smithsonian Astrophysical Observatory (SAO) is building a submillimeter–wavelength interferometer array. The telescope will be located at a high mountain site and will operate between 200 and 900 GHz. The array will consist of six movable 6 m antennas, on a Y–shaped track, with maximum baselines of 500 m, giving sub–arcsecond resolution. It will operate simultaneously in 2 different bands, and will have a versatile spectrometer.

033.044 **The millimeter array.**
R. L. Brown.
IAU Colloquium No. 131: Radio interferometry – theory, techniques, and applications, p. 410 – 414 (1991). – See Abstr. 012.089 for the main entry.

Contents: Genesis of the MMA. MMA overview. Imaging at millimeter wavelengths. The Millimeter Array Project. Construction costs and schedule.

033.045 **The proposed Radio Schmidt Telescope: the technical challenges.**
P. E. Dewdney, T. L. Landecker.
IAU Colloquium No. 131: Radio interferometry – theory, techniques, and applications, p. 415 – 419 (1991). – See Abstr. 012.089 for the main entry.

The concept of the Radio Schmidt Telescope is designed to fill a scientific requirement for a radio telescope which is very sensitive to diffuse, extended emission, but has much higher resolution than single–antenna telescopes. This telescope undertakes to utilize a new region of "parameter space" in the aperture synthesis technique. The emphasis of this article is on a process leading to the best implementation of the concept at reasonable cost while satisfying all the scientific objectives.

033.046 **Progress of VLBI techniques and the role of the Chinese VLBI network.**
Wan Tongshan.
Prog. Astron., Vol. 9, No. 4, p. 309 – 320 (Dec 1991). In Chinese.

The author presents a brief review of VLBI techniques and astrophysics and mentions the limitations of current VLBI observations, namely, sensitivity, dynamic range and time consuming in image processing. It is expected that this situation will be improved with the development of ground VLBI, space VLBI and millimeter VLBI. Also mentioned is the role of the Chinese VLBI network.

033.047 **The Hydrogen Array.**
P. N. Wilkinson.
IAU Colloquium No. 131: Radio interferometry – theory, techniques, and applications, p. 428 – 432 (1991). – See Abstr. 012.089 for the main entry.

The time is ripe for planning an array with a collecting area of 1 km^2 (14 times larger than Arecibo and 75 times larger than the VLA). In view of its major astronomical target the author has dubbed this concept "The Hydrogen Array", although 1 μJy continuum sources will also be reliably detected. The author presents some initial thoughts about the issues involved.

033.048 **Imaging with a homogeneous millimeter array.**
M. A. Holdaway.
IAU Colloquium No. 131: Radio interferometry – theory, techniques, and applications, p. 437 – 440 (1991). – See Abstr. 012.089 for the main entry.

Presented are the results of simulations of the effects of various errors on the imaging capabilities of the compact (72 m) configuration of a 39 element homogeneous millimeter array using the mosaic algorithm to image sources larger than the primary beam. Included in the errors investigated to date are systematic and random pointing errors, Gaussian receiver and atmospheric noise, and primary beam errors. Various methods of quantifying the quality of the reconstructed image are also discussed.

033.049 **Imaging characteristics of the proposed NRAO MMA, with and without a single large element, and a pointing correction algorithm.**
D. T. Emerson.
IAU Colloquium No. 131: Radio interferometry – theory, techniques, and applications, p. 441 – 444 (1991). – See Abstr. 012.089 for the main entry.

This is a continuation of earlier imaging studies for the MMA. Earlier work is confirmed, and the differences in imaging quality resulting from small and larger central antennas used for short–spacing information are investigated. An algorithm is presented which uses interferometer data to correct for an important component of central element pointing errors.

033.050 **Monitoring source variability with the Fast All Sky Telescope.**
K. J. Johnston, R. L. Fiedler, R. S. Simon.
IAU Colloquium No. 131: Radio interferometry – theory, techniques, and applications, p. 445 – 448 (1991). – See Abstr. 012.089 for the main entry.

The proposed Fast All Sky Telescope (FAST) is an interferometer which is intended to monitor the northern four–fifths of the celestial sphere every two days at 8.1 GHz and daily at 2.7 GHz. The design goal is to have a rms sensitivity of 10 mJy/beam at both frequencies. The array is planned to comprise 20 3–meter diameter antennas with a maximum baseline of 0.7 km.

033.051 **The IRAM Plateau de Bure Interferometer.**
R. Lucas.
IAU Colloquium No. 131: Radio interferometry – theory, techniques, and applications, p. 449 – 452 (1991). – See Abstr. 012.089 for the main entry.

The author gives a brief report on the status of the IRAM Interferometer which is now in operation on Plateau de Bure.

033.052 **The Canadian Geophysical LBI system: status and future plans.**
W. T. Petrachenko, P. Mathieu, J. Popelar, M. Daniels.
IAU Colloquium No. 131: Radio interferometry – theory, techniques, and applications, p. 453 – 456 (1991). – See Abstr. 012.089 for the main entry.

Emphasis in the design of the Canadian Geophysical LBI (CGLBI) system has been placed on economy, operational efficiency, and performance in geophysical applications. Unique features of the system include the use of modified VHS recorders to store digitized astronomical data and the capability to track delay and fringe rotation during observation. A prototype system has been completed and tested at L–band using the ARO–DRAO interferometer baseline. The prototype system converts a single fixed 12 Mhz wide IF channel to baseband.

033.053 **Optimal antenna configurations for the BIMA Array.**
N. Thatte, W. J. Welch.
IAU Colloquium No. 131: Radio interferometry – theory, techniques, and applications, p. 457 – 463 (1991). – See Abstr. 012.089 for the main entry.

The authors attempt to determine optimal configurations of six antennas for the proposed Berkeley Illinois Maryland Array. The instrumental response is used as a performance measure. A two dimensional sinc function is chosen as the ideal beam, supported by theoretical considerations. A mean square comparison with this ideal is used to perform optimizations for arrays of various resolutions, using the current station locations at Hat Creek. Configurations for larger fields of view are also discussed, employing two days of coverage with six antennas.

033.054 **The proposed Radio Schmidt Telescope.**
T. L. Landecker, P. E. Dewdney.
J. R. Astron. Soc. Can., Vol. 85, No. 4, p. 219 – 220 (Aug 1991). Abstract. – See Abstr. 012.091 for the main entry.

033.055 **Nobeyama Radioheliograph.**
S. Enome.
International SOLAR–A Science Meeting: Flare physics in solar activity maximum 22, p. 330 – 337 (1991). – See Abstr. 012.104 for the main entry.

The Nobeyama Radioheliograph is a two–year project. A brief progress report is presented.

033.056 **Shanghai VLBI system for astrophysics.**
Wan Tongshan, Liang Shiguang.
Ann. Shanghai Obs., Acad. Sin., No. 12, p. 95 – 106 (1991). In Chinese.

033.057 **Measurements of electric parameters of the 25–meter antenna.**
Huang Xinyong, Gao Yihua.
Ann. Shanghai Obs., Acad. Sin., No. 12, p. 127 – 133 (1991). In Chinese.

033.058 **The thermal control of the millimeter radio telescope.**
Wu Wenda.
Publ. Yunnan Obs., No. 3, p. 53 – 58 (1991). In Chinese.

The thermal effects of the radio telescope antenna are summrized and the method of thermal control for a uniformity of $\leqslant 0.5°C$ of the 30–m millimeter radio telescope is described.

033.059 **The MOST and other radio telescopes.**
J. G. Robertson.
Aust. J. Phys., Vol. 44, No. 6, p. 729 – 742 (1991). Paper presented at the Molonglo Observatory 25th Anniversary Symposium, Univ. Sydney, 22 – 23 Nov 1990.

The MOST performs rotational synthesis, in common with the Australia Telescope and other arrays overseas. However, the MOST's unique construction fits it into a complementary rather than a competitive niche, with its particular strength being rapid surveys of large areas of sky.

033.060 **Position and flux density calibration at Molonglo.**
R. W. Hunstead.
Aust. J. Phys., Vol. 44, No. 6, p. 743 – 757 (1991). Paper presented at the Molonglo Observatory 25th Anniversary Symposium, Univ. Sydney, 22 – 23 Nov 1990.

When the Molonglo Cross was brought into full operation its higher angular resolution and greater pointing accuracy posed special problems for position calibration. The development of a southern radio calibration grid is described, together with its extension for the MOST.

Origins of the Molonglo Radio Observatory – The Cross and the MOST.
See Abstr. 009.014.

Radio interferometry: theory, techniques, and applications. Proceedings.
See Abstr. 012.089.

Radio astronomy: the next decade.
See Abstr. 013.015.

Radio astronomy's crumbling showpiece.
See Abstr. 013.017.

Astronomy: When big is beautiful.
See Abstr. 013.083.

Newly installed radio and optical telescopes in China.
See Abstr. 013.088.

Der Radiohimmel auch für Amateure.
See Abstr. 014.048.

The Ooty summer training program 1990.
See Abstr. 014.078.

Two Chinese new telescopes for astronomy.
See Abstr. 032.046.

The use of optical fibres in radioastronomy.
See Abstr. 034.136.

High–resolution imaging spectroscopy at terahertz frequencies.
See Abstr. 035.155.

IVS – a second generation space VLBI mission.
See Abstr. 035.200.

Simulations of space VLBI.
See Abstr. 035.201.

Space ground interferometer.
See Abstr. 035.202.

Aperture synthesis imaging from the Moon.
See Abstr. 035.205.

Telescope beam–profile diagnostics and the solar limb.
See Abstr. 036.024.

Correcting atmospheric–induced phase errors of a synthetic aperture antenna array by time series modeling and Kalman filtering.
See Abstr. 036.061.

Interstellar scattering effects on the detection of narrow–band signals.
See Abstr. 036.080.

A local complex for data collection and control of radiometers with maintenance of a parallel process mode.
See Abstr. 036.161.

Continuous calibration of radio observations.
See Abstr. 036.165.

Self–noise in interferometers.
See Abstr. 036.234.

Source–noise in radio synthesis images.
See Abstr. 036.235.

Burst mode VLBI and pulsar applications.
See Abstr. 036.237.

Using sub–aperture illumination to solve the short–spacing problem.
See Abstr. 036.238.

Phase control in the digital lens.
See Abstr. 036.239.

Some highlights of interferometry in early radio astronomy.
See Abstr. 036.241.

The use of beam–sets in the analysis of wide–field maps from the CLFST.
See Abstr. 036.243.

Imaging with non–coplanar data.
See Abstr. 036.244.

Self–calibration of M.O.S.T. data.
See Abstr. 036.248.

High accuracy ($\ll 1\%$) polarisation measurements with the WSRT.
See Abstr. 036.249.

Imaging for the Australia Telescope: present needs and future strategies.
See Abstr. 036.253.

Phase calibrators for MERLIN.
See Abstr. 036.257.

Observational test of the homogeneous array concept: an X–band image of the Crab Nebula.
See Abstr. 036.258.

A test of radio interferometry by lunar reflection.
See Abstr. 036.277.

A software for radio source antenna temperature measurements.
See Abstr. 036.298.

Spectral observations of solar radio bursts in the range of 2.0–4.5 GHz by a new digital radio spectrometer.
See Abstr. 077.061.

Absolute calibration of solar radio flux density at 9375 MHz.
See Abstr. 077.070.

The NRAO Millimeter Array site survey.
See Abstr. 082.065.

85 - 90 GHz molecular line observations with a maser receiver at the CrAO 22–meter radio telescope.
See Abstr. 131.400.

Uncovering the secrets of star formation with millimeter arrays.
See Abstr. 131.403.

A Zenith survey of 1988 with RATAN–600 at 8.0 cm.
See Abstr. 141.002.

The 1988 8.0–cm zenith survey with the RATAN–600.
See Abstr. 141.007.

The first VLA observations of radio sources at 73.8 MHz.
See Abstr. 141.012.

Observations of microwave background anisotropy at Tenerife and Cambridge.
See Abstr. 161.558.

034 Auxiliary Instrumentation, Photographic Materials, Clocks

034.001 **High–resolution spectrometry with a masked Fabry–Perot interferometer.**
L. Crescentini.
Appl. Opt., Vol. 30, No. 22, p. 3114 - 3118 (1 Aug 1991). Current Physics Microform No.: 9109A0042.

Features of a real Fabry–Perot interferometer are severely affected by surface irregularities. A new configuration, based on appropriately masking the plates, is proposed. Theoretical results show an increase in the finesse of as much as a factor of 10 as well as an increase in the contrast factor of as much as a factor of 100, while the luminosity–finesse product is preserved. The case of nonnormal incidence is also considered, and the differences arising from that case are discussed.

034.002 **A proposal for an ATIS command set extension for CCD cameras.**
B. Neely, D. Epand.
I.A.P.P.P. Commun., No. 45, p. 92 - 103 (Sep - Nov 1991).

034.003 **Fieldtesting Kodak's Ektar films.**
C. Schur.
Astronomy, Vol. 19, No. 9, p. 70 - 75 (Sep 1991).

034.004 **A high data rate recorder for astronomy.**
H. F. Hinteregger, A. E. E. Rogers, R. J. Capallo, J. C. Webber, W. T. Petrachenko, H. Allen.
IEEE Trans. Magn., Vol. 27, No. 3, p. 3455 - 3465 (May 1991).

A magnetic tape recorder has been developed for the special requirements of radio astronomy and geodesy. These requirements include a high data rate, a high bit packing density, and long record times. The current version of this longitudinal recorder used by the very long baseline array (VLBA) records 5.5 Terabits on a 14–in diameter reel of inch–wide tape. A maximum record rate of 256 Mb/s is achieved in the VLBA configuration with one recorder operating at 4 ms and utilizing 32 of the heads in a single stack.

034.005 **Sonnenuhren unserer Zeit.**
A. Zenkert.
Sterne Weltraum, Jahrg. 30, Nr. 8–9, p. 538 - 540 (Aug–Sep 1991).

034.006 **Schrittmotor–Steuerung zur Fernrohr–Nachführung.**
M. Kempf, T. Koziel.
Sterne Weltraum, Jahrg. 30, Nr. 8–9, p. 546 – 547 (Aug–Sep 1991).

034.007 **Zur Kontraststeigerung astronomischer Photographien. Kompositverfahren, unscharfes Maskieren und andere Techniken.**
G. Rhemann.
Sterne Weltraum, Jahrg. 30, Nr. 12, p. 774 – 775 (Dec 1991).

034.008 **Autocollimation echelle–spectrometer of the 6–m telescope.**
V. G. Klochkova, V. E. Panchuk, V. P. Ryadchenko.
Pis'ma Astron. Zh., Tom 17, No. 7, p. 645 – 652 (Jul 1991). In Russian. English translation in Sov. Astron. Lett., Vol. 17, No. 4.
This is a report on putting into operation the autocollimation echelle–spectrometer (ESPAC) at the 6–meter telescope that can be used with a two dimensional photon–counting system or with CCD–image system. Obervations have shown that the stars of $V = 12^m$ can be observed with the spectrometer with spectral resoltuion $\Delta\lambda = 0.2$ Å and signal to noise ratio more than 40. Being operated with a photon–counting system it is capable to record 24 spectral orders; 27 is the number of orders in the CCD mode of operation.

034.009 **New stellar spectrometer on CCD–line.**
A. G. Zuev, G. V. Kir'yan, V. S. Korepanov, M. I. Rybakov, N. A. Sokolov.
Pis'ma Astron. Zh., Tom 17, No. 8, p. 754 – 767 (Aug 1991). In Russian. English translation in Sov. Astron. Lett., Vol. 17, No. 4.
A new stellar spectrometer for measuring the energy distribution in the spectra of stars in the wavelength range from 0.4 to 1.0 microns was made at the Central Astronomical Observatory of the USSR Academy of Sciences. Registration of the spectra is made with CCD–line. The optical scheme with cross–dispersion is used in the stellar spectrometer, which allows to observe the spectrum of the star in two orders of the grating: the first order has the wavelength range from 0.63 to 1.0 microns and the second one – from 0.4 to 0.67 microns. This spectrometer is controlled by a computer (DVK–3, MERA–660, SM–4) through the CAMAC standard.

034.010 **The Nevada CCD camera project: a tool for remote–access astronomy.**
G. Ebner, T. Colegrove, B. Johnson, J. Kliwer, A. Schultz.
Bull. Am. Astron. Soc., Vol. 23, No. 2, p. 874 (1991). Abstract. – See Abstr. 012.008 for the main entry.

034.011 **Real–time acquisition, analysis, and display of photometric data at the Gettysburg College Observatory.**
L. Marschall, C. Gauthier.
Bull. Am. Astron. Soc., Vol. 23, No. 2, p. 874 (1991). Abstract. – See Abstr. 012.008 for the main entry.

034.012 **The Maryland–Caltech imaging Fabry–Perot.**
S. N. Vogel, J. Roth, S. R. Kulkarni, J. J. Hester, R. Gruendl.
Bull. Am. Astron. Soc., Vol. 23, No. 2, p. 874 (1991). Abstract. – See Abstr. 012.008 for the main entry.

034.013 **An imaging Fourier transform spectrometer: tests and results.**
D. A. Simons, J. P. Maillard, L. L. Cowie.
Bull. Am. Astron. Soc., Vol. 23, No. 2, p. 875 (1991). Abstract. – See Abstr. 012.008 for the main entry.

034.014 **Astronomical observations at 10 and 20 microns with the NRL infrared camera.**
S. Odenwald, K. Shivanandan, H. Thronson, G. Lamb.
Bull. Am. Astron. Soc., Vol. 23, No. 2, p. 875 (1991). Abstract. – See Abstr. 012.008 for the main entry.

034.015 **The drive system of the Spacewatch CCD–scanning telescope.**
M. L. Perry, J. E. Frecker.
Bull. Am. Astron. Soc., Vol. 23, No. 2, p. 875 (1991). Abstract. – See Abstr. 012.008 for the main entry.

034.016 **A new CCD–camera for astronomical observations.**
V. Yu. Berezin, A. G. Zuev, G. V. Kir'yan, M. I. Rybakov, A. T. Khvilivitskij, I. V. Il'in, P. P. Petrov, I. S. Savanov, A. G. Shcherbakov.
Pis'ma Astron. Zh., Tom 17, No. 10, p. 953 – 960 (Oct 1991). In Russian. English translation in Sov. Astron. Lett., Vol. 17, No. 5.
A new Soviet CCD–camera is briefly described. High resolution spectra of stars taken at the 2.6–m Shajn reflector with this camera are presented. Comparison of the spectra taken with the new camera and with the camera CCD–2000 at the same telescope shows that the new camera is more sensitive within the whole spectral range from 3940 Å to 10830 Å.

034.017 **Experimental demonstration of real time atmospheric compensation with adaptive optics employing laser guide stars.**
R. Q. Fugate, C. H. Higgins, J. L. Wynia, W. J. Lange, A. C. Slavin, W. J. Wild, M. P. Jelonek, M. T. Donovan, S. J. Cusumano, J. M. Spinhirne, B. R. Boeke, J. F. Moroney, K. S. Nickerson, J. M. Anderson, R. E. Ruane, D. W. Swindle, R. A. Cleis.
Bull. Am. Astron. Soc., Vol. 23, No. 2, p. 898 (1991). Abstract. – See Abstr. 012.008 for the main entry.

034.018 **An autocollimation échelle spectrometer for the six–meter telescope.**
V. G. Klochkova, V. E. Panchuk, V. P. Ryadchenko.
Sov. Astron. Lett., Vol. 17, No. 4, p. 274 – 277 (Jul 1991). Current Physics Microform No.: 9202X0450. English translation of Pis'ma Astron. Zh., Tom 17, No. 7, p. 645 – 652 (Jul 1991).
The authors report on the installation of the ESPAC autocollimation échelle spectrometer on the 6–m telescope, intended for use with a panoramic photon counter of a CCD array. Observations on the 6–m telescope have shown that stars with $V = 12^m$ are accessible to the spectrometer with $\Delta\lambda = 0.2$ Å spectral resolution and a signal–to–noise ratio of 40 or better. Twenty–four spectral orders are recorded simultaneously during operation with the photon counter, and 27 orders are recorded during operation with the CCD array.

034.019 **A new stellar spectrometer based on a CCD linear array.**
A. G. Zuev, G. V. Kir'yan, V. S. Korepanov, M. I. Rybakov, N. A. Sokolov.
Sov. Astron. Lett., Vol. 17, No. 4, p. 319 – 324 (Jul 1991). Current Physics Microform No.: 9202X0495. English translation of Pis'ma Astron. Zh., Tom 17, No. 8, p. 754 – 767 (Aug 1991).
A new stellar spectrometer for measuring the energy distribution in a stellar spectrum in the wavelength range from 0.45 to 1.0 μm has been built at the Main Astronomical Observatory, USSR Academy of Sciences. The spectrum is recorded with a linear charge–coupled device. A cross–dispersion optical system is used in the spectrometer, making it possible to observe stellar spectra alternately in two grating orders: the first order covers 0.63 to 1.0 μm and the second order covers 0.4 to 0.67 μm. The spectrometer and data transfer are computer–controlled (DVK–3, MERA–660, SM–4) by a CAMAC system.

034.020 **An automated astronomical scanning complex for parallel and independent operation in two spectral bands.**
V. D. Vdovichenko, S. M. Gajsin, A. G. Lukichev, S. A. Mosina.
Kinematics Phys. Celest. Bodies, Vol. 7, No. 1, p. 71 – 76 (1991). English translation of Kinematika Fiz. Nebesn. Tel, Tom 7, No. 1, p. 78 – 85 (1991).
A new medium–resolution spectrometer for the 1.5–meter telescope is described. To improve efficiency and broaden the working band in the spectrum, the beam is split by a dichroic filter after reflection from the collimator and directed into two

spectral channels, each with its own diffraction grating and camera. The first ("blue") channel works in the spectral range 300 – 780 nm, and the second ("red") channel at 760 – 1100 nm. The FEU–39, –79, –136, and –83 photomultiplier radiation detectors used in these channels are operated in the photon-counting mode. The information in each channel is registered independently and concurrently. In addition to the spectral channels there is a reference photometric channel for atmospheric–transparency monitoring. A diaphragm unit is placed in the plane of the spectrometer entrance slit so that the photometric profiles of extended objects can be obtained by image scanning. All mechanical operations are performed with the aid of computer–controlled stepping motors and electromagnets.

034.021 **Optical gyroscopes applied to telescopes.**
H. W. Babcock.
Bull. Am. Astron. Soc., Vol. 23, No. 2, p. 977 (1991). Abstract. The full paper is published in Publ. Astron. Soc. Pac., Vol. 103, No. 663, p. 468 – 469 (May 1991). – See Abstr. 012.008 for the main entry.

034.022 **Neutrino and dark–matter detection at low temperature.**
L. Stodolsky.
Phys. Today, Vol. 44, No. 8, Pt. 1, p. 24 – 32 (Aug 1991). Current Physics Microform No.: 9109A1338.
The use of low–temperature techniques to see tiny energies in massive detectors may be opening a "low energy" frontier of particle physics.

034.023 **Deep–sky photography – Tmax or Technical Pan?**
C. R. Martys.
J. Br. Astron. Assoc., Vol. 101, No. 4, p. 223 – 229 (Aug 1991).
Baked Technical Pan 2415 film was compared with the new T–grain emulsions Tmax 100, Tmax 400 and Tmax 3200 (untreated) in order to evaluate the effectiveness of the new materials for long exposure deep–sky photography. A series of exposures was made of the M42 region of Orion with each film, and the processed negatives compared for speed, grain, and relative blue and red sensitivity.

034.024 **CCD imaging.**
M. Gavin.
J. Br. Astron. Assoc., Vol. 101, No. 4, p. 235 – 237 (Aug 1991).

034.025 **Spectrograph distortion in sunspot line profiles.**
V. Martínez Pillet, J. Sánchez Almeida.
Sol. Phys., Vol. 134, No. 2, p. 403 – 406 (Aug 1991). Letter–to–the–editor.
The authors show empirically how the effect of the instrumental polarization of the spectrograph can distort the shape of the intensity profiles in sunspots. In order to avoid the problems an analysis of the polarization of the light should be performed at the entrance slit of the spectrograph.

034.026 **The use of CCD for observations of large planets.**
L. A. Akimov, V. V. Korokhin, O. M. Starodubtseva, A. N. Aksenov.
Astron. Tsirk., No. 1549, p. 31 – 32 (May – Aug 1991). In Russian.
The photopolarimetric complex, which was made on the basic of a one–dimensional CCD and a personal computer, is briefly described. It was used for polarimetric observations of Jupiter in November 1989.

034.027 **Position offsets in curved–channel microchannel plate detectors.**
D. M. Hassler, G. J. Rottman, G. M. Lawrence.
Appl. Opt., Vol. 30, No. 25, p. 3575 – 3581 (1 Sep 1991). Current Physics Microform No.: 9111A1641.
Microchannel plate detectors are widely used for ultraviolet and extreme–ultraviolet observations. Curved–channel microchannel plates, or C plates, provide an electron gain of $\approx 10^6$ in a single plate because the curved channels prevent ion feedback. However, offsets between input and output in curved–channel microchannel plates produce slight distortions in the spatial or spectral scale of the detector, complicating the use of such detectors for high–resolution imaging and spectroscopy. The authors have examined the image distortion caused by nonconstant curvature. They find that a typical C plate with 25 μm diameter pores exhibits ± 25 μm image distortion across the plate.

034.028 **Influence of coating thickness on the performance of a Fabry–Perot interferometer.**
J. J. Monźon, L. L. Sánchez–Soto, E. Bernabeu.
Appl. Opt., Vol. 30, No. 28, p. 4126 – 4132 (1 Oct 1991). Current Physics Microform No.: 9111F1804.
The effects of the coating thickness on the physical performance of a Fabry–Perot interferometer (FP) are investigated. The FP is modeled as three media separated by two thin films and not merely by two interfaces. The authors show that the transmitted intensity obeys an Airy function, but not the reflected intensity because of the appearance of a complex factor accounting for the coupling between the reflected waves in the coatings. The Stokes relations are generalized for this model. The authors study the dependence of the phase lag of reflection in the coatings on the angle of incidence, which causes a shift in the position of the intensity maxima. They discuss as well the properties of the FP with an absorbing medium in the cavity, defining a merit function that optimizes the compromise between peak transmission and finesse.

034.029 **The Bartol/NASA/NSO high–degree helioseismometer.**
J. W. Harvey, F. Hill, T. L. Duvall Jr., H. Jones, S. M. Jefferies, M. A. Pomerantz.
Bull. Am. Astron. Soc., Vol. 23, No. 2, p. 1032 – 1033 (1991). Abstract. – See Abstr. 012.012 for the main entry.

034.030 **Performance limits of stationary Fourier spectrometers.**
M. L. Junttila, J. Kauppinen, E. Ikonen.
J. Opt. Soc. Am. A, Vol. 8, No. 9, p. 1457 – 1462 (Sep 1991). Current Physics Microform No.: 9111F0375.
Theoretical performance limits of stationary Fourier spectrometers without mechanical scanning are analyzed and compared with the performance of a scanning Fourier spectrometer. Spectrometers employing uncollimated light are most favorable. In amplitude–splitting interferometers the reduction in fringe visibility brought about by the extended source can be avoided and leads to high optical throughput in the corresponding spectrometer. In a stationary wave–front–splitting interferometer, realized without a beam splitter, the fringe contrast depends on the size of the source. The use of a slit source increases the optical throughput of source–size–limited spectrometers.

034.031 **The H–Lyman α coronagraph/polarimeter.**
S. Fineschi, R. B. Hoover, A. B. C. Walker Jr.
Bull. Am. Astron. Soc., Vol. 23, No. 2, p. 1038 (1991). Abstract. – See Abstr. 012.012 for the main entry.

034.032 **1024 x 1024 CCD cameras under development and operation at Lockheed PARL.**
I. Zayer, D. Duncan, C. Edwards, G. Kelly, M. Levay, M. Morrill, A. Title, T. Tarbell, J. Wolfson, W. Rosenberg.
Bull. Am. Astron. Soc., Vol. 23, No. 2, p. 1056 – 1057 (1991). Abstract. – See Abstr. 012.012 for the main entry.

034.033 **Narrow passband filter solar observations.**
L. A. Smaldone, G. Cauzzi, S. L. Keil.
Bull. Am. Astron. Soc., Vol. 23, No. 2, p. 1057 (1991). Abstract. – See Abstr. 012.012 for the main entry.

034.034 **Polarization pinhole interferometer.**
Wang Guiying, Zheng Yanling, Sun Aimin, Wu Shudong, Wang Zhijiang.
Opt. Lett., Vol. 16, No. 17, p. 1352 – 1354 (1 Sep 1991). Current Physics Microform No.: 9111A2238.
A polarization pinhole interferometer is based on the principles of pinhole diffraction, a common light path, and the

polarization technique in its design. It has many advantages, such as adjustable contrast of the fringes, nondestructive testing, and less sensitivity to environmental factors. The instrument can be used extensively in optical element testing and glue–layer testing. The accuracy of the fringe pattern analysis is controlled by the number of sampled points in the fringes. The analysis results of the closed fringe pattern are given.

034.035 **Precision visible and infrared solar photometry.**
Lin Haosheng, J. R. Kuhn.
Bull. Am. Astron. Soc., Vol. 23, No. 2, p. 1069 (1991). Abstract. – See Abstr. 012.012 for the main entry.

034.036 **Evaluation of the stability of the Optec SSP–3 photometer.**
R. Riordan.
I.A.P.P.P. Commun., No. 46, p. 29 – 34 (Dec 1991 – Feb 1992).

034.037 **A remote filter box controller.**
D. L. DuPuy.
I.A.P.P.P. Commun., No. 46, p. 35 – 39 (Dec 1991 – Feb 1992).

034.038 **Photometric filters for a CCD camera.**
P. Boltwood.
I.A.P.P.P. Commun., No. 46, p. 48 – 56 (Dec 1991 – Feb 1992).

034.039 **Solar 2D spectroscopy: a new MSDP instrument.**
P. Mein.
Astron. Astrophys., Vol. 248, No. 2, p. 669 – 676 (Aug 1991).
A new instrument using the multichannel subtractive double pass (MSDP) technique has been used with the large spectrograph of the Vacuum Tower Telescope of the Teide Observatory, as a result of cooperation between the Paris–Meudon Observatory and the Kiepenheuer Institute. The main capabilities are: simultaneous spectro–imaging in 2 line profiles (medium and strong lines), high spatial resolution in all the 2D field, no differential seeing effects at all wavelengths, best compromises between exposure time and spectral resolution for lines, fast change of wavelength, possibility of spectro–imaging with high spectral resolution by successive exposures (weak lines). Either 9 or 11 channels are recorded simultaneously in each line profile. The size of the field is typically 120×1000 or 60×1000 diffraction widths of the telescope image. The spectral resolving power is 10^5 in one exposure ($n \times 10^5$ in n exposures; $n < 10$). The optical design and the codes for data processing are reviewed. The performances of the MSDP and the other 2D spectroscopy techniques are briefly compared. Future prospects are discussed.

034.040 **Comparative characteristics of the Coudé–spectrograph of the Shemakha 2–m telescope and the main stellar spectrograph of the BTA.**
A. Kh. Rzaev, E. L. Chentsov.
Astrofiz. Issled. Izv. Spets. Astrofiz. Obs., Tom 31, p. 134 – 143 (1991). In Russian. English translation in Bull. Spec. Astrophys. Obs. – North Caucasus.
Spectrophotometric characteristics, penetrating capability and position accuracy are compared.

034.041 **Improved band filters for a millimeter photometer.**
G. A. Chuntonov, I. A. Maslov, V. A. Soglasnova.
Astrofiz. Issled. Izv. Spets. Astrofiz. Obs., Tom 31, p. 163 – 164 (1991). In Russian. English translation in Bull. Spec. Astrophys. Obs. – North Caucasus.
Filters for investigating the Syunyaev–Zel'dovich effect in the mean millimetre region have been designed and built.

034.042 **Digicon spectrophotometer. Description of light detectors and their characteristics.**
V. S. Rylov.
Astrofiz. Issled. Izv. Spets. Astrofiz. Obs., Tom 31, p. 165 – 179 (1991). In Russian. English translation in Bull. Spec. Astrophys. Obs. – North Caucasus.
Description and characteristics of all the elements of the Digicon spectrophotometer (DSP–40) are presented. In its construction the DSP–40 is attached to the fast spectrograph FS–160 of the 6–m telescope, or to the universal spectrograph UAGS (DDR), so the author presents their brief characteristics together with the characteristics of the 40–channel photon detector – Digicon. The procedure of Digicon focusing in the DSP–40 scheme, setting the discriminating threshold, identification of the comparison spectrum and calculation of the wavelength scale, and one of the possible ways of preliminary data reduction, including the transfer curve of the whole system is considered. In conclusion the results of testing the DSP–40 in the laboratory and at the telescope are presented.

034.043 **A thick CCD spectral observation system at the Yunnan Observatory.**
Zhao Zhaowang, Xu Jun, Tan Huisong.
Publ. Yunnan Obs., No. 1, p. 45 – 52 (Mar 1991). In Chinese.
The principal characteristics of the first thick CCD system for the observation of the spectra of stars at the Yunnan Observatory are briefly introduced. The structure of the electronic system and the software for the data acquisition and processing are also presented. In addition, some observed results of more than 200 stars obtained with the CCD system are given.

034.044 **A CCD detector attached to the lower latitude meridian circle. (II) The determination and related correction of a few constants of the detector.**
Zhang Jing, Peng Qingyu, Li Zhiming, Hu Xiaochun, Mao Wei.
Publ. Yunnan Obs., No. 1, p. 68 – 76 (Mar 1991). In Chinese.
The influences of the errors of making and mounting of the CCD chip on the nonorthogonality of the two measuring coordinate axes and the unequality of the scales of the two axes are described and a method for determining directly the errors on the LLMC is proposed.

034.045 **A modified observational software system for multi–waveband solar spectrograph.**
Chen Jingying, Feng Yongming.
Publ. Yunnan Obs., No. 2, p. 28 – 34 (Jun 1991). In Chinese.

034.046 **Was the Crab seen by the ASGAT telescope?**
V. Basiuk, P. Goret, N. Petrou, A. Raviart, A. Tabary, L. Treguer, R. Bazer–Bachi.
AIP Conf. Proc., No. 220, p. 65 – 69 (Apr 1991). Current Physics Microform No.: 9106B1199. – See Abstr. 012.020 for the main entry.
The ASGAT gamma ray telescope, located 1700 m a.s.l. in the French Pyrénées, was fully operational by the end of 1989. The Crab nebula was tracked for 22 hours in December '89 and January '90. Fast timing information from the 7 mirrors array was used to reconstruct the arrival direction of each event with an accuracy better than a degree. A special data analysis, designed to keep reliable control of the cosmic background, was tailored leading to the detection of a gamma ray signal from the Crab direction, with a flux of $3.4 \pm 1.5 \times 10^{-11} cm^{-2} s^{-1}$ above 500 GeV. The realiability of this 2.3σ result is discussed.

034.047 **Design and operation of the Chicago Air Shower Array.**
L. J. Rosenberg, J. W. Cronin, B. E. Fick, K. G. Gibbs, H. A. Krimm, N. C. Mascarenhas, T. A. McKay, D. Müller, B. J. Newport, R. A. Ong, M. E. Wiedenbeck, K. D. Green, J. Matthews, D. Nitz, D. Sinclair, J. C. van der Velde.
AIP Conf. Proc., No. 220, p. 111 – 121 (Apr 1991). Current Physics Microform No.: 9106B1245. – See Abstr. 012.020 for the main entry.
The Chicago Air Shower Array (CASA) is a large–area surface scintillator array designed to study PeV sources of cosmic rays. The complete detector will consist of 1089 detector stations, distributed on a square 15 m grid. The authors have operated an array of 49 stations for much of the 1989 calendar year, an array of 529 stations for much of 1990, and the balance of the 1089 stations will be operating in early 1991. This surface array, together with the University of Michigan underground muon

detectors (MIA), and the University of Utah atmospheric Čerenkov telescopes and Fly's Eye air fluorescence detector, constitute a uniquely powerful instrument, dubbed the Utah Michigan Chicago (UMC) experiment, for the study of PeV sources. The authors report the performance and current status of these detectors.

034.048 **High–performance servo positioning and data transmission for DCMT.**
Xu Xinqi.
Publ. Shaanxi Astron. Obs., Vol. 14, No. 1, p. 14 – 22 (Aug 1991). In Chinese.
The control strategy and the test for the servo positioning and data transmission empolyed in the full automatic horizontal meridian cirlce, which now is at its second engineering stage, are the main topic of this paper.

034.049 **The angular resolution of the HEGRA scintillation counter array at La Palma.**
A. Karle, HEGRA Collaboration.
AIP Conf. Proc., No. 220, p. 127 – 131 (Apr 1991). Current Physics Microform No.: 9106B1261. – See Abstr. 012.020 for the main entry.
The extended HEGRA–detector (High Energy Gamma Ray Array) at La Palma is in operation with 100 scintillation counters since August 1989. More than 3×10^7 events have been recorded at an energy above 0.7×10^{14}eV. The overall angular resolution of the array is better than 0.8°. The observation of shadowing of ultra high energy cosmic rays by the moon and the sun confirms both the predicted resolution and the correct alignment of the detector.

034.050 **Applications of operational amplifiers to photoelectric scanning micrometers.**
Xu Linsheng, Li Zhigang.
Publ. Shaanxi Astron. Obs., Vol. 14, No. 1, p. 58 – 64 (Aug 1991). In Chinese.
The successful application of the operational amplifiers to the photoelectric scanning micrometers is discussed. The gain accuracy related to the measurement accuracy as well as the effects of detuning and drafting are analyzed, so as to provide the basis for the reasonable choice of the operational amplifiers. Besides, the method of eliminating the interference from surrounding are introduced.

034.051 **Status of the THEMISTOCLE multi–mirror Cerenkov array for VHE/UHE source studies.**
L. Behr, S. Danagoulian, B. Dudelzak, P. Eschstruth, R. Riskalla, P. Roy, P. Espigat, G. Fontaine, C. Ghesquière, P. Schune, J. Vrana, R. George, F. Kovacs, Y. Pons, M. Rivoal, T. Socroun, A. M. Touchard, P. Baillon, B. Fabre, C. Meynadier.
AIP Conf. Proc., No. 220, p. 237 – 241 (Apr 1991). Current Physics Microform No.: 9106B1371. – See Abstr. 012.020 for the main entry.
The THEMISTOCLE collaboration has brought into operation an 18 telescope array designed to study the angular resolution that can be obtained through multiple, high precision timing measurements on the Cerenkov light front for primary gamma–ray energies above about 5 TeV. Detector characteristics, calibration techniques and preliminary hadronic shower analysis are presented.

034.052 **GRAND particle identification.**
J. Gress, Y. Lu, A. Anagnostopoulos, J. Kochocki, J. Olson, J. Poirier, S. Mikocki, A. Trzupek.
AIP Conf. Proc., No. 220, p. 242 – 245 (Apr 1991). Current Physics Microform No.: 9106B1376. – See Abstr. 012.020 for the main entry.
The method of using a thin absorber and tracking chambers to identify particles in air showers has been studied using a test setup in the GRAND air shower array. Dramatic differences between muons and electrons are seen in the fraction of particles which stop in the absorber and in the distribution of scattering angles for particles which penetrate the absorber. A simple algorithm is applied to 93% of two samples of data which were preidentified as muons and as electrons respectively. Of these, 96.2±0.2% of the muons and 95.1±0.6% of the electrons are correctly identified.

034.053 **A first optimization of the SINGAO air shower array.**
M. Ambrosio, G. C. Barbarino, B. Bartoli, D. Campana, J. W. Elbert, M. Iacovacci, F. Guarino, G. Osteria, V. Silvestrini.
AIP Conf. Proc., No. 220, p. 246 – 252 (Apr 1991). Current Physics Microform No.: 9106B1380. – See Abstr. 012.020 for the main entry.
In order to investigate discrete sources of ultra–high energy gamma rays, by studying extensive air showers, the authors require an apparatus with good resolution, both in angle and energy, as well as good sensitivity. They consider an array of n^2 detectors, uniformly placed on an area of ~10 km². The authors analyze the detectors by simulation experiments. Such an array can be used to perform gamma ray astronomy studies, for example in the SINGAO experiment, foreseen to be built in southern Italy.

034.054 **GRANITE – a new telescope for TeV gamma ray astronomy.**
D. I. Meyer, C. W. Akerlof, M. F. Cawley, D. J. Fegan, A. M. Hillas, R. C. Lamb, M. Lawrence, D. A. Lewis, P. Reynolds, M. Schubnell, T. C. Weekes.
AIP Conf. Proc., No. 220, p. 253 – 256 (Apr 1991). Current Physics Microform No.: 9106B1387. – See Abstr. 012.020 for the main entry.

034.055 **Status and performance of the AS array of the Tibet AS_γ experiment.**
Tibet AS_γ Collaboration, M. Amenomori, Z. W. Bai, Z. Cao, L. K. Ding, Z. Y. Feng, K. Hibino, N. Hotta, Q. Huang, A. X. Huo, H. Y. Jia, G. Z. Jiang, S. Q. Jiao, F. Kajino, K. Kasahara, Labaciren, D. M. Mei, L. Meng, Mimaciren, K. Mizutani, J. Mu, H. Nanjo, M. Nishizawa, Nusang, M. Ohnishi, I. Ohta, J. R. Ren, T. Saito, M. Sakata, Z. Z. Shi, M. Shibata, T. Shirai, X. X. Sun, H. Sugimoto, A. Tai, K. Taira, Y. H. Tan, N. Tateyama, S. Torii, H. Wang, J. X. Wang, C. Z. Wen, Y. Yamamoto, H. M. Yang, G. C. Yu, P. Yuan, T. Yuda, J. G. Zeng, C. H. Zhao, C. S. Zhang, H. M. Zhang, L. Zhang, M. Zhang, Zhasang, Zhaxiciren, W. D. Zhou.
AIP Conf. Proc., No. 220, p. 257 – 264 (Apr 1991). Current Physics Microform No.: 9106B1391. – See Abstr. 012.020 for the main entry.
The Tibet $AS\gamma$ experiment, which has started since January, 1990, is located at an altitude of 4300 m at Yangbaging in Tibet, China (90.5°E, 30.1°N). The air–shower array is composed of 49 scintillation counters for fast timing, each counter having an area of 0.5 m², in a grid pattern with a spacing of 15 m and 16 density detectors. The analysis of experimental data shows that the performance of array is realized as expected by the Monte Carlo simulation. The detection threshold energy for γ–rays is ~40 TeV; the peak energy 20 TeV. The angular resolution around 100 TeV is ~0.5°. The systematic error of arrival direction is considerably smaller than the angular resolution in the concerned energy region.

034.056 **The artificial satellite observation chronograph controlled by single chip microcomputer.**
Pan Guangrong, Tan Jufan, Ding Yuanjun.
Publ. Purple Mt. Obs., Vol. 10, No. 2, p. 144 – 150 (Jun 1991). In Chinese.
The instrument specifications, hardware structure, software design, and other characteristics of the chronograph mounting on a theodolite used for artificial satellite observation are presented. The instrument is a real time control system with a single chip microcomputer.

034.057 **The CANGAROO project.**
P. Edwards.
AIP Conf. Proc., No. 220, p. 265 – 269 (Apr 1991). Current Physics Microform No.: 9106B1399. – See Abstr. 012.020 for the main entry.
Detections of TeV gamma–rays from the Crab Nebula by the Whipple group have shown the effectiveness of the Čerenko imaging technique in discriminating against the cosmic ray background. "Stereoscopic" observations of atmospheric Čerenkov emission from air showers, using two telescopes separated by ~100 m, are expected to increase the sensitivity of this technique. The CANGAROO project, a collaboration between the University of Adelaide and a number of Japanese institutions, will employ both of these techniques for observations from Woomera, Australia.

034.058 **Gamma ray astronomy above 10^{17}eV with the high resolution Fly's Eye.**
D. B. Kieda, HiRes Collaboration.
AIP Conf. Proc., No. 220, p. 270 – 274 (Apr 1991). Current Physics Microform No.: 9106B1404. – See Abstr. 012.020 for the main entry.
Recent observations by the Fly's Eye, Akeno, and Yakutsk cosmic ray experiment have provided evidence for emission of neutral particles by Cygnus X-3 in the EHE regime ($E > 10^{17}$eV). A substantial improvement in detection area is needed to provide significant measurements. The authors describe the need for the high resolution Fly's Eye; a proposed large collection area cosmic ray detector capabilities for measurement of diffuse and point sources of γ rays. The deployment status of four prototype detection units is described.

034.059 **EUV/FUV response characteristics of photographic films for the multi–spectral solar telescope array.**
R. B. Hoover, A. B. C. Walker Jr., C. E. DeForest, M. J. Allen, J. F. Lindblom.
Opt. Eng., Vol. 30, No. 8, p. 1116 – 1124 (Aug 1991). Paper presented at SPIE Conference on X-Ray/EUV Optics for Astronomy, Microscopy, Polarimetry, and Projection Lithography as Part of SPIE's International Symposium on Optical and Optoelectronic Applied Science and Engineering, San Diego, CA (USA), 9 – 13 July 1990.
The Multi–Spectral Solar Telescope Array (MSSTA) is a sounding rocket–borne observatory designed to produce ultra-high–resolution full–disk images of the sun. The desire for ultrahigh–resolution (~0.1 to 0.3″) images of the solar disk and corona out to 1.5 $R_\odot$ demands an information storage capacity that at the present time can be met only by the highest quality photographic emulsions. The authors describe the performance and characteristics required of the MSSTA photographic films for solar observations in the soft X–ray/EUV and FUV wavelength regimes. They discuss the properties of the important new emulsions selected for flight.

034.060 **MILAGRO – a TeV air shower array.**
J. A. Goodman, C. Y. Chang, T. J. Haines, P.–W. Kwok, X.–Q. Lu, G. B. Yodh, R. L. Burman, C. M. Hoffman, E. C. Milner, D. E. Nagle, M. E. Potter, V. Sandberg, D. H. White, W.–P. Zhang, R. W. Ellsworth, S. Freedman, B. Fujikawa, D. Krakauer, R. Lamb, A. I. Mincer, R. Svoboda, T. Tumer, A. Zych, M. Cavalli, D. Coyne, D. Dorfan, D. Williams.
AIP Conf. Proc., No. 220, p. 275 – 284 (Apr 1991). Current Physics Microform No.: 9106B1409. – See Abstr. 012.020 for the main entry.
The authors describe a new proposed water–Čerenkov detector called MILAGRO, for observing cosmic gamma rays over a broad energy range of 1 – 1000 TeV. This detector will have an angular resolution better than 0.5° and an energy threshold near 1 TeV.

034.061 **DUMAND II status report.**
R. J. Wilkes, T. Aoki, R. Becker–Szendy, P. Bosetti, P. E. Boynton, H. Bradner, U. Camerini, J. Clem, V. Commichau, D. Dau, S. Dye, P. K. F. Grieder, T. Hayashino, E. Hazen, M. Jaworski, T. Kitamura, K. Kobayakawa, P. Koske, J. G. Learned, C. Ley, J. J. Lord, R. March, S. Matsuno, P. Minkowski, K. Mitsui, D. O'Connor, Y. Ohashi, A. Okada, V. Z. Peterson, J. Rathlev, A. Roberts, C. E. Roos, M. Sakuda, D. Samm, V. J. Stenger, S. Tanaka, S. Uehara, M. Webster, G. Wilkins, A. Yamaguchi, I. Yamamoto, K. K. Young.
AIP Conf. Proc., No. 220, p. 285 – 289 (Apr 1991). Current Physics Microform No.: 9106B1419. – See Abstr. 012.020 for the main entry.
The scientific goals, design, capabilities, and status of the DUMAND II detector system are described. In June, 1989, the construction of DUMAND II was recommended, funding began in 1990, and prototype development for various detector subsystems is under way. Current plans include deployment of the shore cable, junction box and three strings of optical detector modules in 1992, and expansion to the full 9–string configuration in 1993.

034.062 **AIROBICC – a new array of angle integrating Cerenkov counters for improved γ/hadron separation in extended air showers.**
M. Bott–Bodenhausen, I. Holl, A. Kabelschacht, A. Karle, E. Lorenz, R. Maier, M. Merck, M. Rozanska, P. Weisbach.
AIP Conf. Proc., No. 220, p. 305 – 309 (Apr 1991). Current Physics Microform No.: 9106B1439. – See Abstr. 012.020 for the main entry.
The authors propose to build an array of angle integrating air Cerenkov counters and to combine it with the HEGRA scintillator matrix at La Palma. The array will be sensitive to extended air showers above an energy of $\approx 5 \times 10^{13}$eV and up to a zenith angle of 35°. The incident shower direction will be determined from fast timing measurements with an angular resolution of 2 – 0.5 mrad. The combination of the Cerenkov– and scintillator matrix data will provide a powerful tool for γ/hadron discrimination. Together with the proposed muon detectors the authors expect to achieve rejection ratios of $\geqslant 10^4$ thus approaching the possibility to detect diffuse γ–radiation.

034.063 **Maximizing the sensitivity of a low threshold VHE gamma ray telescope by the use of neural nets and other methods.**
M. P. Kertzman, G. H. Sembroski.
AIP Conf. Proc., No. 220, p. 290 – 294 (Apr 1991). Current Physics Microform No.: 9106B1424. – See Abstr. 012.020 for the main entry.
This study shows that there are several methods available to increase the sensitivity of the Haleakala Gamma Ray Telescope.

034.064 **JASA: a prototype water–Čerenkov air–shower detector.**
D. Berley, C. Dion, J. A. Goodman, T. J. Haines, P. W. Kwok, M. J. Stark, R. C. Svoboda, H. Ferguson, C. M. Hoffman, E. Horch, R. W. Ellsworth, R. S. Delay, X.–Q. Lu, G. B. Yodh.
AIP Conf. Proc., No. 220, p. 310 – 315 (Apr 1991). Current Physics Microform No.: 9106B1444. – See Abstr. 012.020 for the main entry.
A small pilot experiment to examine the use of the water–Čerenkov technique for air shower detection was installed near the center of the CYGNUS air shower array. Preliminary results showing general agreement with simulations are presented. Thus, the technique promises to offer significant advances for VHE–UHE γ–ray astronomy.

034.065 **A star–and–sky chopping polarimeter: design and performance.**
S. K. Jain, G. Srinivasulu.
Opt. Eng., Vol. 30, No. 9, p. 1415 – 1421 (Sep 1991).
The authors have developed a star–and–sky chopping polarimeter for accurate measurements of linear polarization of starlight in the standard astronomical photometric U, B, V, R, and I bands. The instrumental polarization, as determined by observing the standard unpolarized stars, is 0.04%. It is possible

to use the instrument for the measurements of circular polarization as well. A microcomputer controls the various operations of the instrument, acquires the data, and does the on-line data reduction. This paper describes the design and performance of the polarimeter.

034.066 **Elaboration of an apparatus on the basis of CCD for meteor observations.**
S. Mukhamednazarov, I. S. Zejda, V. A. Shul'gin, N. L. Vlasov.
Astron. Tsirk., No. 1543, p. 31 – 32 (Mar – Apr 1990). In Russian.

An apparatus on the basis of a CCD-camera and video cassette recorder for meteor observations is elaborated. The first results have been obtained.

034.067 **On the increase of the efficiency of U-photoplates by means of preexposure in B-light.**
Ts. B. Georgiev.
Astron. Tsirk., No. 1543, p. 37 – 38 (Mar – Apr 1990). In Russian.

The efficiency of one 3.5 h U-plate is increased by about 3 times by means of 10 min preexposure in B-light. The change of the color system in the case of surface photometry is negligible.

034.068 **Simultaneous observations with a tunable filter and the Echelle Spectrograph of the Vacuum Tower Telescope at Tenerife.**
W. Schmidt.
Reviews in Modern Astronomy 4, p. 117 – 125 (1991). – See Abstr. 003.008 for the main entry.

The observing project described makes use of the great flexibility of the instrumentation of the Vacuum Tower Telescope at Tenerife. The most important characteristics of the telescope are described. The science observing programs were dedicated to the measurements of small scale magnetic structures and the velocity fields in both horizontal coordinates and at different heights in the solar atmosphere. Narrow band filtergrams in left and right handed circular polarized light with a field of view of about 120 arcseconds squared were recorded in order to construct longitudinal magnetograms while the spectrograph was equipped with a Stokes-V polarimeter for the measurement of Stokes-V profiles with high spectral resolution. Slit jaw reference pictures were taken in white light in order to locate the spectograph slit in the magnetograms.

034.069 **Atomic ion frequency standards.**
W. M. Itano.
Proc. IEEE, Vol. 79, No. 7, p. 936 – 942 (Jul 1991). Special issue: Time and frequency. J. Jespersen, D. W. Hanson (eds.).

A new class of atomic frequency standard, based on ions trapped by electromagnetic fields, is under development. Such standards have the potential of achieving higher frequency accuracy than currently available standards. They are also capable of very good frequency stability. The history and status of trapped-ion frequency standards are reviewed. Prospects for future standards are discussed.

034.070 **Laser-cooled neutral atom frequency standards.**
S. L. Rolston, W. D. Phillips.
Proc. IEEE, Vol. 79, No. 7, p. 943 – 951 (Jul 1991).

Recent advances in laser cooling, including temperatures as low a few microkelvin, and the demonstration of the ability to launch atoms while maintaining these low temperatures, now allows the realization of atomic fountain frequency standards. An analysis of a Cs microwave fountain predicts potential stability and accuracy improvements of a factor of 100 over present laboratory standards. Neutral atom frequency standards will have much better signal-to-noise ratios than ion standards, but suffer from larger systematic effects, making the two technologies complementary in many ways.

034.071 **Characterization of frequency stability in precision frequency sources.**
J. Rutman, F. L. Walls.
Proc. IEEE, Vol. 79, No. 7, p. 952 – 960 (Jul 1991). Special issue: Time and frequency. J. Jespersen, D. W. Hanson (eds.).

This paper presents a short review of the progress that has occurred over the past 25 – 30 years in both the theoretical and practical characterization of frequency stability of precision frequency sources. The emphasis is on the evolution of ideas and concepts for the characterization of random noise processes in such standards in the time domain and the Fourier frequency domain.

034.072 **Time synchronization using laser techniques.**
S. Leschiutta.
Proc. IEEE, Vol. 79, No. 7, p. 1001 – 1008 (Jul 1991). Special issue: Time and frequency. J. Jespersen, D. W. Hanson (eds.).

Pulsed lasers enable very accurate time comparisons between remote clocks. The reasons for investigating this technique are reviewed along with a discussion of laser timing experiments, proposed or performed, giving relevant results. The specific problems of timing via lasers are listed and the error sources are investigated.

034.073 **A new hydrogen-maser time and frequency standard for Chinese VLBI network.**
Zhai Zaocheng, Lin Chuanfu, Huang Hengxiang,
Jiang Guoxing, Luo Weihua, Zhang Weiqun, He Jianwei.
Acta Astron. Sin., Vol. 32, No. 2, p. 194 – 200 (Jun 1991). In Chinese.

A new kind of hydrogen-maser time and frequency standard was installed at the Sheshan 25-meter radio telescope in Shanghai, for very long baseline interferometry (VLBI) experiments. This paper describes its design feature and structure as well as the measured results of its performance. Its application and data in VLBI experiments are also included.

034.074 **The absolute calibration of the sensitometer at Shanghai Observatory.**
Yao Baoan, Jing Jiayun.
Acta Astron. Sin., Vol. 32, No. 2, p. 201 – 207 (Jun 1991). In Chinese.

The authors present a new sensitometer. The main difference between their sensitometer and that of others is that the test plate is in contact with the surface of the wedge instead of projecting the image of the wedge onto the emulsion of the plate. So the radiation measured by contacting the window of the radiometer to the surface of the wedge is not the real radiation received by the test plate, because there is a certain distance between the window and the detector of the radiometer, i.e. only the specular density of the wedge is used in this case. After designing the correct method, the authors have measured the diffuse density of the wedge through the interference filters and have shown that the densities so measured are actually equivalent to the diffuse density measured in the combined radiation. Therefore, the absolute calibration is simply reduced to the measurement of the incident radiation in front of the wedge and multiplication of it by the diffuse density of the wedge.

034.075 **FIR lasers as local oscillators in submillimeter astronomy.**
H. P. Röser, P. van der Wal.
Symposium on Advanced Laser Concepts and Applications as Part of the 4. International Congress on Optical Science and Engineering (ECO-4), p. 194 – 197 (1991). – See Abstr. 012.029 for the main entry.

Optically pumped molecular lasers have been developed for high resolution spectroscopy in the wavelength range from 100 to 500 μm with a resolving power of $>10^6$. With more than 1,000 discrete laser lines they cover the whole spectral range and offer high cw output power up to 50 mW, good signal to noise ratio and high spectral purity. This paper summarises the state of the art of optically pumped lasers and their use as local oscillator in a heterodyne spectrometer specially developed for airborne astronomy.

034.076 **The Filled Arm Fizeau Telescope (FFT).**
S. P. Synnott.
Conference on Space Astronomical Telescopes and Instruments as Part of SPIE's Technical Symposia on Optical Engineering and Photonics in Aerospace Sensing (OE/Aerospace Sensing '91), p. 334 – 343 (1991). – See Abstr. 012.030 for the main entry.

The design of a Mills Cross imaging interferometer is described in which the arms are fully filled with mirror segments of a Ritchey–Chretien primary. The optical design, structural configuration, thermal disturbances and vibrations, materials, controls and metrology issues and scientific capabilities are discussed, and technology needs identified.

034.077 **Determination of the crystal elements' orientation arrangement in adjustable–wavelength birefringent filters.**
Wang Jingshan, Ai Guoxiang, Deng Yuanyong.
Acta Astrophys. Sin., Vol. 11, No. 4, p. 383 – 388 (Oct 1991). In Chinese.

This paper discusses several principles of determination of the crystal elements' orientation arrangement in adjustable–wavelength birefringent filters. By using these principles, the authors determine the relations of the orientations among the birefringent crystal plates, the waveplates and the rotated 1/2 waveplates simply, clearly, and accurately without any complicated mathematical calculations.

034.078 **A CCD photometer for the Vassar College Observatory.**
F. R. Chromey, M. E. Hunt, M. Edelstein, E. Bonar.
Meeting on Precision Photometry: Astrophysics of the Galaxy, p. 293 – 297 (1991). – See Abstr. 012.027 for the main entry.

The authors have installed a CCD camera on the 0.4 meter reflector at the Vassar College Observatory. The hardware and software components installed are described and the photometric observations of standard stars used to evaluate the system are presented.

034.079 **Imaging pulse–counting detector systems for space ultraviolet astrophysics missions.**
J. G. Timothy.
Conference on Space Astronomical Telescopes and Instruments as Part of SPIE's Technical Symposia on Optical Engineering and Photonics in Aerospace Sensing (OE/Aerospace Sensing '91), p. 394 – 402 (1991). – See Abstr. 012.030 for the main entry.

A family of imaging, pulse–counting, photoelectric detector systems, the Multi–Anode Microchannel Arrays (MAMAs), are now under active development for use on a number of space ultraviolet astrophysics missions at far–ultraviolet (FUV) and extreme–ultraviolet (EUV) wavelengths between about 300 and 28 nm. The construction and performance characteristics of different MAMA detector systems are described, and techniques for improving the spatial resolution of the detector systems by the use of custom Application–Specific Integrated Circuits (ASICs) in the electronics are discussed.

034.080 **Further performance characteristics of a high sensitivity 64 × 64 element InSb hybrid focal plane array.**
R. C. Fischer, C. J. Martin, C. A. Niblack, H. A. Timlin, J. T. Wimmers.
Conference on Space Astronomical Telescopes and Instruments as Part of SPIE's Technical Symposia on Optical Engineering and Photonics in Aerospace Sensing (OE/Aerospace Sensing '91), p. 414 – 418 (1991). – See Abstr. 012.030 for the main entry.

Recent performance results for a two–dimensional InSb hybrid focal plane array are presented. The short wavelength (1 to 3 micron) response at liquid helium temperatures has been improved, making the device more useful for astronomy and other scientific applications.

034.081 **Si:Ga focal plane arrays for satellite and ground–based telescopes.**
P. Mottier, P. Agnese, P. O. Lagage.
Conference on Space Astronomical Telescopes and Instruments as Part of SPIE's Technical Symposia on Optical Engineering and Photonics in Aerospace Sensing (OE/Aerospace Sensing '91), p. 419 – 426 (1991). – See Abstr. 012.030 for the main entry.

Since 1984 infrared sensor devices have been developed in the CEA–LETI laboratories for astronomical observation in the 4 – 17 micron wavelength band using Si:Ga detectors. These devices are to equip ISOCAM, a camera which will operate from ISO (Infrared Space Observatory), the European satellite expected to be launched in May 1993, and C10μ, a French astronomical camera which will be based at the CFHT (Canadian French Hawai Telescope). In this paper, a brief description of the devices and their operating conditions, as well as main preliminary results are given.

034.082 **Microchannel–plate detectors for space–based astronomy.**
J. H. Crocker, C. R. Cox, K. A. Ray, A. Sen.
Conference on Space Astronomical Telescopes and Instruments as Part of SPIE's Technical Symposia on Optical Engineering and Photonics in Aerospace Sensing (OE/Aerospace Sensing '91), p. 434 – 439 (1991). – See Abstr. 012.030 for the main entry.

Photon–counting detectors such as the Ranicon have advantages over other detectors in space based astronomy in the area of time–resolved imaging and spectroscopy. The authors describe the details of a system to record positions and absolute arrival times of individual photons. They suggest the use of a Global Positioning Satellite system for accurate absolute timing when synchronizing observations from different observatories.

034.083 **The fabrication of toroidal and coma–corrected toroidal diffraction gratings from spherical master gratings using elastically–deformable substrates: a progress report.**
M. C. E. Huber, J. G. Timothy, J. S. Morgan, G. Lemaître, G. Tondello, G. Naletto.
Conference on Space Astronomical Telescopes and Instruments as Part of SPIE's Technical Symposia on Optical Engineering and Photonics in Aerospace Sensing (OE/Aerospace Sensing '91), p. 472 – 480 (1991). – See Abstr. 012.030 for the main entry.

The authors have developed a technique which permits toroidal, and coma–corrected toroidal, diffraction gratings to be replicated from spherical master gratings by the use of elastically–deformable substrates. Toroidal gratings correct for astigmatism, and, thus make it possible to construct stigmatic spectrometers that employ a single reflective diffraction grating. These spectrometers are particularly useful for the extreme–ultraviolet (EUV) wavelength range. The use of the toroidal gratings in some space missions is briefly described.

034.084 **The Mees CCD imaging spectrograph.**
M. J. Penn, D. L. Mickey, R. C. Canfield, B. J. LaBonte.
Sol. Phys., Vol. 135, No. 1, p. 163 – 177 (Sep 1991).

The Mees CCD (MCCD) instrument is an imaging spectroscopy device which uses the 25 cm coronagraph telescope and the 3.0 m coudé spectrograph at Mees Solar Observatory (MSO) on Haleakala, Maui. The instrument works with resolving power up to $R \approx 200000$ with significant throughput from $\lambda 3934$ Å (Ca II K) to $\lambda \approx 10000$ Å. A fast guiding active mirror stabilizes the image during observations. A rapidly writing magnetic tape storage system allows observations to be recorded at 256 kbytes s^{-1}. Currently, the MCCD is used for imaging spectroscopy of solar flares at $\lambda 6563$ Å (Hα), and velocity measurements of umbral oscillations; future plans include emission line studies of active region coronae, and photospheric studies of solar oscillations.

034.085 **An electrograph for measurement of macroscopic electric fields in prominences and flares.**
T. Moran, P. Foukal.
Sol. Phys., Vol. 135, No. 1, p. 179 – 191 (Sep 1991).
The authors describe an "electrograph" instrument designed for measurement of macroscopic electric fields in solar plasmas, using the polarization dependence of line width in Stark–broadened hydrogen Paschen emission lines. Observations of quiescent prominences and limb chromosphere provide upper limits of 5 – 10 V cm^{-1} for transverse macroscopic electric fields in these structures, averaged over an area of about 5 × 7 arcsec. Random thermal motions of hydrogen ions across magnetic field lines generate a quasi–static electric field, which should be distinguishable from pressure broadening in the intensely magnetized chromosphere over a sunspot, given an electrograph sensitivity a factor 2 – 3 better than that achieved here.

034.086 **Polarization characteristics of Jensch coelostat.**
M. L. Demidov.
Sol. Phys., Vol. 135, No. 1, p. 193 – 197 (Sep 1991).
The polarization characteristics of the Jensch coelostat, used at the Sayan Observatory, are presented. Results of calculations of the instrumental polarization (IP) are compared with observations. It is shown that structural features of the Jensch coelostat lead to a very important dependence of the Stokes's parameters, that characterize the IP, on solar coordinates.

034.087 **The RGO plate library and measuring machine facilities.**
M. Irwin, J. Pilkington, R. Catchpole.
Gemini, No. 33, p. 15 – 19 (Sep 1991).

034.088 **FAST on the WHT.**
R. Clegg, M. Ward, M. Blietz, M. Cameron, S. Drapatz, R. Genzel, A. Krabbe, V. Rotaciuc, P. van der Werf.
Gemini, No. 34, p. 1 – 4 (Dec 1991).

034.089 **More silicon acreage on the ING telescopes.**
P. Jordan, P. Oates.
Gemini, No. 34, p. 5 – 7 (Dec 1991).

034.090 **CCD Chips für die Astronomie.**
F. Möhle.
Orion, Jahrg. 49, Nr. 246, p. 209 – 211 (Oct 1991).

034.091 **Photometric systems of astrograph AFR–18 of the Engelhardt Observatory.**
L. A. Urasin, S. S. Tokhtas'ev, S. K. Fomin, I. E. Tselishchev.
Izv. Astron. Obs. Ehngel'gardt., Vyp. 54, p. 82 – 85 (1989). In Russian.
The photometric systems of the telescope AFR–18 in combinations of the photographic plates and filters are given. The equations of the correlation between these systems and UBV system were calculated.

034.092 **Deux nouvelles voix pour l'horloge parlante.**
Astronomie, p. 4 – 7 (Nov 1991).

034.093 **Computing facilities on La Palma.**
C. Benn.
Gemini, No. 34, p. 22 – 23 (Dec 1991).

034.094 **On an analytical representation of the characteristic curve of photomaterials.**
M. A. Mnatsakan'yan, R. A. Sarkis'yan, M. Sh. Karapet'yan.
Astron. Tsirk., No. 1544, p. 35 – 36 (May – Jul 1990). In Russian.

034.095 **Properties of fiber optics for application in astronomical interferometry.**
R. R. Rohloff, C. Leinert.
Appl. Opt., Vol. 30, No. 34, p. 5031 – 5036 (1 Dec 1991). Current Physics Microform No.: 9203A1782.
The authors tested properties of single–mode and multimode fiber optics that are relevant for use in a long–baseline astronomical interferometer, and give quantitative values for modulation, transmission, and stability of the fiber optics as measured in a Mach–Zehnder interferometer configuration. Only polarization–maintaining single–mode fibers gave satisfying results. Good preservation of modulation and ease of beam combining are strong features of these fibers, whereas the light losses that are due to imperfect coupling may present a serious problem.

034.096 **Construction d'une horloge primaire au Laboratoire Primaire du Temps et des Fréquences.**
P. Aynié, J. P. Berthet, S. Bouzid, A. Clairon, B. Dahmani, A. H. Gérard, P. Leclerc, M. Lours, D. Rovera, G. Santarelli, E. De Clercq.
Journées 1991: Systèmes de référence spatio–temporels, p. 29 – 36 (1991). – See Abstr. 012.048 for the main entry.
The recent advent of laser diodes in the near infrared opened the way to the realization of optically pumped cesium clocks. Such a clock is being constructed in the authors' laboratory. The authors give here the main characteristics of the prototype. Preliminary results are reported.

034.097 **A laser cooled cesium atomic fountain: towards a high performance clock.**
A. Clairon, C. Salomon, S. Guellati, W. D. Phillips.
Journées 1991: Systèmes de référence spatio–temporels, p. 37 (1991). – See Abstr. 012.048 for the main entry.
The authors report the first realization of Zacharias's proposal for observing a Ramsey resonance in an atomic fountain. With presently achieved atom density and temperature, one envisions cold cesium fountain clocks with unprecedented performances both in stability (10^{-16}/day) and accuracy (10^{-16}).

034.098 **An automated ocular micrometer for photographic observation of comets and asteroids.**
D. A. Rozhkovskij, D. I. Gorodetskij, G. I. Bublik, A. M. Sredinin.
Kinematics Phys. Celest. Bodies, Vol. 7, No. 2, p. 86 – 87 (1991). English translation of Kinematika Fiz. Nebesn. Tel, Tom 7, No. 2, p. 96 – 97 (1991).
A new guide–telescope accessory that relieves the observer of manual operations with ocular micrometers in photographic observations of comets and asteroids is briefly described.

034.099 **Image recording complex with the CCD array of the 6–m telescope.**
A. N. Borisenko, V. V. Vitkovskij, O. P. Zhelenkova, A. I. Kopylov, S. V. Markelov, V. P. Ryadchenko, V. S. Shergin.
Bull. Spec. Astrophys. Obs. – North Caucasus, Vol. 32, p. 128 – 134 (1990). English translation of Astrofiz. Issled. Izv. Spets. Astrofiz. Obs., Tom 32, p. 157 – 165 (1990).
The structure of the complex, its operating principles, and software are described. The results of laboratory investigations of the equipment with a type PhCCD–1M array and the first experience of using the complex in observations with the 6–m telescope are given.

034.100 **Improvement of the characteristics of the main stellar spectrograph of the 6–m telescope. Experience in operating with image slicers.**
I. F. Bikmaev, E. B. Gazhur.
Bull. Spec. Astrophys. Obs. – North Caucasus, Vol. 32, p. 146 – 152 (1990). English translation of Astrofiz. Issled. Izv. Spets. Astrofiz. Obs., Tom 32, p. 181 – 190 (1990).
Results of investigations and observations with the main stellar spectrograph of the 6–m telescope developed at the Special Astrophysical Observatory, USSR Academy of Sciences, are presented. It is shown that the use of the slicers increases the luminous efficiency of the spectrograph by a factor of 3.5 – 4.5 relative to the classical slit. The photometric accuracy of the

spectra and penetrating capacity of the spectrograph increase accordingly.

034.101 **High-dispersion of an echelle spectrometer for the coude focus of large telescopes.**
A. B. Dombrovskaya, G. G. Paritskaya.
Bull. Spec. Astrophys. Obs. - North Caucasus, Vol. 32, p. 153 - 156 (1990). English translation of Astrofiz. Issled. Izv. Spets. Astrofiz.. Obs., Tom 32, p. 191 - 196 (1990).
Various high-resolution spectrometer designs based on the use of a large echelle grating are examined.

034.102 **Improved microphotometer of the Rozhen Observatory.**
Yu. Stajkov, I. Pamukchiev, Ts. Georgiev, B. Bilkina.
Bull. Spec. Astrophys. Obs. - North Caucasus, Vol. 32, p. 157 - 160 (1990). English translation of Astrofiz. Issled. Izv. Spets. Astrofiz.. Obs., Tom 32, p. 197 - 200 (1990).
A simple improvement of the GII - GIII microphotometer (Germany) to broaden its capabilities in investigations of photographic emulsions and stellar photometry is reported. Results of tests and examples of use are given.

034.103 **Hypersensitization of emulsion IV-N by means of a solution of silver nitrate and benzotriazole.**
V. Tsintsarov, M. Panov, Ts. Georgiev.
Bull. Spec. Astrophys. Obs. - North Caucasus, Vol. 32, p. 161 - 162 (1990). English translation of Astrofiz. Issled. Izv. Spets. Astrofiz.. Obs., Tom 32, p. 201 - 203 (1990).
An earlier proposed method of hypersensitization for type I-N emulsion was successfully applied to type IV-N emulsion. By adding benzotriazole and reducing the immersion temperature to 5°C, the authors increased the sensitivity by a factor of 1.5 - 2 in comparison with the use of silver nitrate alone.

034.104 **Polarization characteristics of the Jensch coelostat.**
M. L. Demidov.
Kinematika Fiz. Nebesn. Tel, Tom 7, No. 6, p. 62 - 70 (Nov-Dec 1991). In Russian. English translation in Kinematics Phys. Celest. Bodies, Vol. 7, No. 6.
By considering an example of the coelostat of the solar telescope of operative predictions (STOP) at the Sayan observatory, the paper deals with polarization characteristics of the Jensch coelostats. The theory of coelostats of this design is briefly considered, and all relations are given, which are required for practical calculations of the conversion of Stokes parameters during reflection of light from coelostat mirrors, depending on declination and on the hour angle of the Sun. The significant dependence of the instrumental polarization (IP) on the Sun's coordinates as well as on the coelostat site latitudes is noted. At the same time the IP dependence on the adjustment errors of the coelostats is rather weak. The results of theoretical calculations of IP are illustrated by particular observations with the STOP.

034.105 **A quadrant-CCD star tracker.**
M. Clampin, S. T. Durrance, R. Barkhouser, D. A. Golimowski, A. Wald, W. G. Fastie, D. L. Heidtmann, M. M. Blouke, J. A. Westphal, J. R. Janesick, J. E. Gunn.
SPIE's Symposium on Charge-Coupled Devices and Solid State Optical Sensors, p. 217 - 222 (1990). - See Abstr. 012.058 for the main entry.
The authors discuss the characteristics of a quadrant-CCD developed by Tektronix for pointing and tracking applications. The device is discussed in the context of its application to the Johns Hopkins University adaptive optics program, where it is being used for the correction of stellar image motion resulting from atmospheric turbulence. The transfer function of the device is discussed and results are presented which demonstrate the quadrant-CCDs capability to provide offset signals to control a rapid guiding mirror in the Johns Hopkins instrument.

034.106 **Position observations of geostationary artificial satellites at the Main Astronomical Observatory of the Ukrainian Academy of Sciences and estimate for their precision.**
Yu. N. Ivashchenko, L. N. Kizyun, Yu. I. Safronov.
International Seminar on Problems of Astrometry and Space Geodynamics, p. 86 - 90 (1991). In Russian. - See Abstr. 012.049 for the main entry.
A description of a complex, equipped with a special photocassette, constructed at the Main Astronomical Observatory of the Ukrain Academy of Sciences for determination of high-precision positions of artificial and natural celestial bodies is given. The observations made with this special photocassette display a high precision.

034.107 **The micrometers of the Carlsberg Automatic Meridian Circle.**
L. Helmer, C. Fabricius, L. V. Morrison.
Exp. Astron., Vol. 2, No. 2, p. 85 - 99 (1991).
A new moving slit star micrometer has been developed for the 18 cm Carlsberg Automatic Meridian Circle on La Palma. A zenith mean error of 0".15 for one transit and a limiting magnitude of 14^m8 is obtained. It has a high efficiency and is well suited for improving the FK5 system and extending it to much fainter stars. A new collimator alignment micrometer and the necessary calibrations are also described.

034.108 **Megapixel CCD thinning/backside progress at SAIC.**
A. R. Schaefer, R. H. Varian, J. Cover, R. Larsen.
Conference on Charge-Coupled Devices and Solid State Optical Sensors II as Part of the SPIE/IS and T Symposium on Electronic Imaging Science and Technology, p. 165 - 176 (1991). - See Abstr. 012.059 for the main entry.
The authors will discuss the thinning efforts to date on SAIC 1024 × 1024 CCD arrays, including flattening mounts and anti-reflection coatings. Test and characterization results are presented for various chips which have been passivated using either flashgate or p^+ substrate remnant. Finally, preliminary results in two related areas are discussed: initial thinning efforts on SAIC/LORAL 18 μm 2048 × 2048 CCD's; and electron imaging with thinned chips in a Digicon tube.

034.109 **Performance tests of large CCDs.**
L. Robinson, W. Brown, K. Gilmore, R. Stover, Wei Mingzhi, J. C. Geary.
Conference on Charge-Coupled Devices and Solid State Optical Sensors II as Part of the SPIE/IS and T Symposium on Electronic Imaging Science and Technology, p. 214 - 228 (1991). - See Abstr. 012.059 for the main entry.
Results obtained in fabricating and testing of large CCD image sensors are reported. The emphasis is on high quantum efficiency, excellent charge transfer efficiency at low signal level, large pixel count, low readout noise, and very low dark current, since the main interest is in the use of the devices for optical astronomy where these parameters are most important. Test results for CCDs are presented. Experiments to demonstrate the feasibility of a reproducible biased-gate using transparent indium tin oxide as a conducting layer over a silicon oxide insulating layer are discussed. Quantum efficiency of bias-gate thinned CCDs is compared with results obtained from a phosphor-coated front-illuminated CCD.

034.110 **A 4096 × 4096 pixel CCD mosaic imager for astronomical applications.**
J. C. Geary, G. A. Luppino, R. J. Hlivak, R. Bredthauer, L. Robinson.
Conference on Charge-Coupled Devices and Solid State Optical Sensors II as Part of the SPIE/IS and T Symposium on Electronic Imaging Science and Technology, p. 264 - 273 (1991). - See Abstr. 012.059 for the main entry.
The authors have designed and successfully fabricated a large-format CCD imager to be used in a 2 × 2 mosaic array. Each quadrant is an independent imager of 2048 × 2048 15 μm pixels, designed to be edge-butted on two sides. The authors also

describe smaller 2688 × 512 15 μm pixel CCD imagers designed for spectrographic applications.

034.111 **Virtual phase charge–coupled device (CCD) imaging system for astronomy.**
A. T. Khvilivitzky, A. G. Zuev, M. I. Rybakov, G. V. Kiryan, V. Yu. Berezin.
Conference on Charge–Coupled Devices and Solid State Optical Sensors II as Part of the SPIE/IS and T Symposium on Electronic Imaging Science and Technology, p. 64 – 68 (1991). – See Abstr. 012.059 for the main entry.
The CCD imaging system for use at the ground telescopes was manufactured and tested. The two–level controller provides great flexibility in system operation. The results of system performance at the Crimea Observatory are presented.

034.112 **A Cassegrain echelle spectrograph.**
D. M. Hunten, W. K. Wells, R. A. Brown, N. M. Schneider, R. L. Hilliard.
Publ. Astron. Soc. Pac., Vol. 103, No. 669, p. 1187 – 1192 (Nov 1991).
The "LPL echelle" is a CCD spectrograph with a resolving power of 160,000 and enough sensitivity to detect the brighter airglow emissions in a few minutes. It uses an R3.2 grating in a novel out–of–plane mounting that permits any wavelength to be placed on the blaze. A single order is selected by a medium–band interference filter. Lower dispersions and broader wavelength coverage are obtained by replacing the echelle by a 600 line mm^{-1} grating or a Littrow prism.

034.113 **MONICA – the Wendelstein Observatory Monochromatic Imaging Camera.**
M. M. Roth, T. Soffner, J. W. Hoffmüller.
3. ESO/ST–ECF Data Analysis Workshop, p. 107 – 111 (1991). – See Abstr. 012.061 for the main entry.
Preliminary test results are presented for the newly developed CCD imaging system at Wendelstein Observatory. The authors present information on an associated photometric test facility, and a brief description of image processing software available at the institute.

034.114 **High–dynamic–range MCP structures.**
D. C. Slater, J. G. Timothy.
SPIE Conference on EUV, X–Ray, and Gamma–Ray Instrumentation for Astronomy II as Part of SPIE International Symposium on Optical Applied Science and Engineering, p. 68 – 80 (1991). – See Abstr. 012.065 for the main entry.
The authors report on the development of a new high–dynamic–range two–stage Multi–Anode Microchannel Array (MAMA) imaging tube designed for improved high count rate performance at FUV and EUV wavelengths. The new two–stage MAMA tube employs two 25–mm–diameter format MCPs placed in tandem with a small gap between the plates. The front (input) MCP is designed to be a low–gain converter plate that supports an opaque photocathode and converts the detected photons to electrons, while the second (output) MCP is of higher conductivity and thus maintains the overall gain of the multiplier at high count rates. The second MCP is mounted in proximity focus with a (224 × 960) – pixed fine–fine coincidence MAMA array for high–spatial–resolution imaging studies.

034.115 **A fast focal–reducing camera for CCD imaging.**
G. S. Aldering, G. D. Bothun.
Publ. Astron. Soc. Pac., Vol. 103, No. 670, p. 1296 – 1313 (Dec 1991).
This paper describes the design and performance of the OREAD focal–reducing camera, constructed for use at the Cassegrain focus of the Michigan–Dartmouth–MIT Observatory McGraw–Hill 1.3–meter telescope. OREAD is an all–refractive optical system providing a 50–mm collimated beam suitable for commercially available narrow–band filters or grisms. Its collimator and field lens are custom–made to decrease the severe coma which would otherwise be present at the extremities of its 30–arc–min field. When used with a CCD such as the TI 800 × 800, OREAD provides a field of view of 0.1 square degree of 1''.5 $pixel^{-1}$. The focal ratio of OREAD of f/1.63 allows sky–limited narrow–band images to be obtained with exposures of only moderate duration. In an Appendix the authors examine some limitations inherent in coarsely sampled data obtained with CCDs.

034.116 **High–efficiency "superachromatic" polarimetry optics for use in optical astronomical spectrographs.**
R. W. Goodrich.
Publ. Astron. Soc. Pac., Vol. 103, No. 670, p. 1314 – 1322 (Dec 1991).
A design for an optical spectropolarimeter consisting of a Pancharatnam half–wave plate and a modified Glan–Taylor polarizing beam splitter retrofitted to an existing astronomical spectrograph is described. The wave plate has a retardance of $180° \pm 2°$ over the entire design wavelength range from 3200 Å to 1.1 μm. The calcite polarizing beam splitter has a beam separation of one beam diameter, and a low length–to–width ratio of 1.035, making efficient use of the limited space available in most spectrographs. Cassegrain CCD spectrographs on the Mount Locke 2.7–m and Mount Palomar 5–m telescopes have been retrofitted with spectropolarimeters of the type described. The combination of these "superachromatic" polarimeter parts with the high–efficiency CCD detectors now used in these and many other astronomical spectrographs should lead to a great expansion of spectropolarimetry in many branches of astronomical research.

034.117 **CCD array for astronomical observations.**
V. Yu. Berezin, A. G. Zuev, G. V. Kir'yan, M. I. Rybakov, A. T. Khvilivitskij, I. V. Il'in, P. P. Petrov, I. S. Savanov, A. G. Shcherbakov.
Sov. Astron. Lett., Vol. 17, No. 5, p. 405 – 408 (Sep 1991). Current Physics Microform No.: 9204X1352. English translation of Pis'ma Astron. Zh., Tom 17, No. 10, p. 953 – 960 (1991).
The authors describe a Soviet–made CCD system based on a liquid nitrogen cooled virtual–phase array. High–dispersion stellar spectra obtained at the 2.6–m Shajn telescope are presented. The new system is more sensitive than the CCD–2000 system operating at the same telescope over the entire spectral range from 3940 to 10830 Å.

034.118 **A high efficiency imaging extreme ultraviolet spectrometer.**
J. S. Morgan, D. C. Slater, G. Tondello, G. Naletto, G. Lemaitre, J. G. Timothy, M. C. E. Huber, E. Jannitti, P. Nicolosi.
1. Berkeley Colloquium on Extreme Ultraviolet Astronomy, p. 380 – 389 (1991). – See Abstr. 012.045 for the main entry.
The authors describe the results of the first tests of an imaging EUV spectrometer employing a single concave toroidal diffraction grating and a two–dimensional pulse–counting MAMA detector system utilizied in an open–structure configuration.

034.119 **Development of 3–D UV and soft X–ray detectors based on energy sensitive superconducting elements.**
A. K. Drukier.
AIP Conf. Proc., No. 222, p. 475 – 478 (15 Apr 1991). Current Physics Microform No.: 9106E0475. – See Abstr. 012.069 for the main entry.

034.120 **Detecting halo weakly interacting massive particles with a cryogenic phonon detector.**
N. Wang, P. D. Barnes, A. Cummings, J. Emes, Y. Giraud–Héraud, E. E. Haller, A. Lange, R. Ross, B. Sadoulet, T. Shutt, C. Stubbs.
AIP Conf. Proc., No. 222, p. 471 – 474 (15 Apr 1991). Current Physics Microform No.: 9106E0471. – See Abstr. 012.069 for the main entry.
If dark matter in the halo is made of weakly interacting massive particles (WIMPs), it may be detected with a cryogenic phonon detector. Such a detector must have low energy threshold, high energy resolution, and, more importantly, good background

rejection capability. It has been proposed that a good background rejection ratio can be obtained by a simultaneous measurement of phonons and ionization. To test this experimentally, the authors have built a 60 g Ge detector and have measured simultaneous phonon and ionization signals, each with 4 keV energy resolution. They present a simple estimate of the rejection ratios and WIMP event rates that may be achieved by such a detector.

034.121 **Techniques for detection of cosmic diffuse neutrinos and anti-neutrinos.**
D. B. Cline.
AIP Conf. Proc., No. 222, p. 512 - 519 (15 Apr 1991). Current Physics Microform No.: 9106E0512. - See Abstr. 012.069 for the main entry.
The author first indicates the energy regions where diffuse neutrinos or anti-neutrinos are expected. Techniques for the detection of these neutrinos are described. The author focusses on the prospects for detecting neutrinos or anti-neutrinos in the window between the atmospheric and solar neutrino flux, from 10 - 30 MeV. He also describes methods to detect neutrinos in the 100 MeV to GeV region using a lunar neutrino telescope. A brief comment is made concerning the detection of the 1.9° relic neutrinos.

034.122 **Experimental limit to the galactic stellar core collapse rate.**
E. N. Alexeyev (*E. N. Alekseev*), L. N. Alexeyeva (*L. N. Alekseeva*), A. E. Chudakov, I. V. Krivosheina, V. I. Volchenko, V. N. Zakidyshev.
ESO/EIPC Workshop on Supernova 1987A and other Supernovae, p. 697 - 702 (Jun 1991). - See Abstr. 012.075 for the main entry.
The Baksan experiment on the search for stellar core collapse neutrino bursts is described. The search range of the telescope is 20 kpc. No signals which can be interpreted as a $\bar{\nu}_e$ burst were found during the total observational time.

034.123 **The first results of work with CCD detector ST-4.**
B. Dintinjana, H. Mikuž.
Vasiona, Année 39, No. 5, p. 83 (1991). In Croatian.
The description of the first observations with CCD in Idria is given.

034.124 **Search for neutrinos from collapsing stars at Mont Blanc.**
M. Aglietta, G. Badino, G. Bologna, C. Castagnoli, A. Castellina, V. L. Dadykin, W. Fulgione, P. Galeotti, F. F. Kalchukov, V. B. Kortchaguin, P. V. Kortchaguin, A. S. Malguin, V. G. Ryassny, O. G. Ryazskaya, O. Saavedra, V. P. Talockin, G. Trinchero, S. Vernetto, G. T. Zatsepin, V. F. Yakushev.
Vulcano Workshop 1988: Frontier objects in astrophysics and particle physics, p. 103 - 120 (1989). - See Abstr. 012.079 for the main entry.
The authors present the upgrading of the LSD experiment, presently running in the Mont Blanc Laboratory. The data recorded during the period when supernova 1987A exploded are analysed in detail. The research program of LSD-2, the same experiment as LSD but with an higher sensitivity to search for neutrino bursts from collapsing stars, is also discussed.

034.125 **A fast imaging photometer and the detection of optical flashes from gamma-ray burst sources.**
A. Piccioni, C. Bartolini, A. Guarnieri, G. Pizzichini, J. M. Poulsen, S. R. Rosellini.
Vulcano Workshop 1988: Frontier objects in astrophysics and particle physics, p. 217 - 224 (1989). - See Abstr. 012.079 for the main entry.
There is growing evidence that gamma-ray burst (GRB) sources emit also optical flashes, although up to now no coincidence between a GRB and an optical flash has been observed. The authors are setting up an instrument for continuous optical monitoring of GRB error boxes. It will have a trigger system for recording optical flashes with a pre-trigger memory and imaging capability. It will be used at the focus of a 62 cm telescope, the largest instrument until now employed for this purpose. In this way one shall have prompt detection and localization of any events, and a fairly good possibility of identifying spurious flashes.

034.126 **The South Pole Air Shower Experiment.**
A. M. Hillas.
Vulcano Workshop 1988: Frontier objects in astrophysics and particle physics, p. 309 - 312 (1989). - See Abstr. 012.079 for the main entry.
A scintillator array to detect air showers with a median energy of 140 TeV has been set up within 0.4 km of the south pole. The aim is to provide continuous monitoring of many X-ray sources in the southern sky, to look for ultra high energy gamma ray emission. The array is operating with high efficiency. This is illustrated in the case of SN 1987A, as data from the first 18 days of operation will give the most sensitive limit so far obtained for the 100 TeV emission from this object.

034.127 **An intensified C.C.D. camera for telescope guidance.**
G. Herpe, M. Chevreton, J. C. Robin, P. Prat, B. Servan, F. Gex.
Exp. Astron., Vol. 2, No. 3, p. 163 - 177 (1991).
This paper deals with the study and the fabrication of an intensified C.C.D. camera for guiding telescopes or for searching fields containing faint stars. It describes the sensor of the camera (by connecting the Thomson C.C.D. and the RTC image intensifier) and its three working modes. The first is the direct exploitation of the video signal sent to the TV monitor, in conformity with the C.C.I.R. standard. However, the study of stability and sensitivity of the intensifier lead to create a second working mode necessary for the correct detection of faint objects (by an "averaging filter" improving the signal to noise ratio). The last mode is the "integration" by the C.C.D. itself: it allows a maximum of three-seconds integration time (therefore the detection of fainter objets), with the help of a thermoelectric cooling system. Moreover, the tests made on the 1.52 meter telescope of the O.A.N. (Calar-Alto, Spain) have been successful and confirm the capabilities of this camera to replace the intensified TV tube cameras, like the super-isocon camera.

034.128 **Charge coupled devices (CCDs) in X-ray astronomy.**
D. H. Lumb, G. D. Berthiaume, D. N. Burrows, G. P. Garmire, J. A. Nousek.
Exp. Astron., Vol. 2, No. 3, p. 179 - 201 (1991).
The application of CCDs to X-ray imaging and spectroscopy for astronomy is described. The requirements which differ markedly from those of traditional optical applications are highlighted. Results of recent research programs to optimize CCD X-ray response are presented. It is shown that very high quantum efficiencies and Fano noise limited energy resolutions can be obtained. A number of issues related to the practical implementation of future instruments are reviewed.

034.129 **The CLUE experiment for γ ray astronomy.**
Vulcano Workshop 1988: Frontier objects in astrophysics and particle physics, p. 325 - 328 (1989). - See Abstr. 012.079 for the main entry.
The purpose of the Clue experiment is the study of energetic γ-rays coming from localized cosmic sources. The authors shall do this by selecting the light coming from the electromagnetic showers produced in the atmosphere. The detector is a finely grained gas chamber, sensitive to the middle region of the ultraviolect spectrum and capable of resolving single events.

034.130 **A Peltier-cooled CCD camera.**
B. D. Carter, M. C. B. Ashley.
Proc. Astron. Soc. Aust., Vol. 9, No. 1, p. 158 - 159 (1991). - See Abstr. 012.078 for the main entry.
The authors describe the application of Peltier effect cooling to charge coupled device (CCD) detectors. They are developing this technique to produce a CCD camera which requires low

maintenance, yet has sufficiently small dark–current for long exposure imaging. This camera will be used in an automated imaging telescope at Siding Spring Observatory. The design principles used to maximise cooling of the detector, and hence minimise dark–current, are discussed. A small dark–current can be obtained only if great care is taken to reduce or eliminate convective, conductive and radiative heating of the chip. In addition, a path of high thermal conductivity must be provided for the heat removed from the CCD. A recent laboratory test of the cooling system demonstrates that careful design can lead to sufficiently low CCD dark–current for many astronomical applications.

034.131 **A coudé auxiliary feed for the AAT: a focus for international co–operation in optical astronomy.**
P. Gillingham, M. Pettini, M. Bessell, L. Cram, K. Ogura.
Proc. Astron. Soc. Aust., Vol. 9, No. 1, p. 160 – 161 (1991). – See Abstr. 012.078 for the main entry.

In 1988, a very powerful and versatile coudé echelle spectrograph was installed at the Anglo–Australian Telescope. Since then it has been winning about 30% of AAT time and has been highly productive scientifically. Therefore, using a coudé auxiliary telescope (CAT) to give the spectrograph more use and to free the AAT for other work is a very attractive possibility, particularly if the CAT can give a sensitivity similar to that obtained with the 3.9–m telescope. While a conventional CAT with such performance would be very expensive, a much cheaper instrument should approach this goal, at least for high resolution spectroscopy. Even so, the cost could not be met in a reasonable time within the AAO's budget; hence, additional international collaboration is being sought.

034.132 **Extending integration time of a Philips type 56470 CCD camera for use in telescope offset guiding.**
C. Maclean, W. Tobin.
Proc. Astron. Soc. Aust., Vol. 9, No. 1, p. 167 – 169 (1991). – See Abstr. 012.078 for the main entry.

The authors have extended the integration time of a Philips type 56470 video camera to make it useful for visual offset guiding on 0.6–m and 1–m telescopes.

034.133 **Sapphire clocks in VLBI radioastronomy.**
A. N. Luiten, D. G. Blair.
Proc. Astron. Soc. Aust., Vol. 9, No. 1, p. 171 (1991). – See Abstr. 012.078 for the main entry.

This paper describes a system utilising a sapphire–loaded superconductor cavity to provide a high–stability oscillator signal for use in VLBI observations.

034.134 **On the studies of astronomical photography problems and information on the proposed new photomaterials for astronomical photography (1987)**
O. D. Dokuchaeva.
Tr. Gos. Astron. Inst. im. Shternberga, Tom 62, p. 268 – 273 (1991). In Russian.

Information on the meeting of the Working Group "Astrophotography" of the Astronomical Council of USSR Academy of Sciences is given. The meeting took place in the Astronomical Observatory in Odessa in 1987, April. The problem of new photomaterials, made in Kazan and in Pereslavl was discussed, the methods of hypersensitizing of materials by soaking and baking in hydrogen atmosphere and many other questions were examined in detail. 21 reports were presented by astronomers and by specialists in scientific photography.

034.135 **Observational advantages of the McDonald Observatory white–pupil 2D Coudé spectrometer.**
P. J. MacQueen, R. G. Tull.
Symposium in Celebration of the 50. Anniversary of McDonald Observatory of the University of Texas at Austin: Frontiers of stellar evolution, p. 608 (1991). Abstract. – See Abstr. 012.085 for the main entry.

034.136 **The use of optical fibres in radioastronomy.**
A. C. Young.
IAU Colloquium No. 131: Radio interferometry – theory, techniques, and applications, p. 33 – 41 (1991). – See Abstr. 012.089 for the main entry.

Optical fibres have been used extensively within the Australia Telescope to carry IF data, LO reference, timing signals and computer communications. This paper discusses the various options considered in the design of a system which should assist similar projects in making the best use of optical fibres.

034.137 **Design of a two–channel echelle spectrograph.**
R. K. Bhatia, R. Gratton, A. Iovino.
Telesc. Naz. Galileo, Tech. Rep., No. 5.
Osservatorio Astronomico di Padova–Asiago, Asiago (Italy). 22 p. (Jun 1991).

This paper discusses in detail the design considerations for a two–channel echelle, one channel optimized for the blue wavelength region, and the other for the red, for the 3.5 m telescope Galileo. The instrument, fed by fibres, will be placed in the basement of the telescope.

034.138 **IAU Working Group on Wide–Field Imaging.**
H. T. MacGillivray (ed.).
IAU Working Group on Wide–Field Imaging, Newsl., No. 1.
International Astronomical Union. Commission 9. 55 p. (Jan 1991).

Contents: 1. Introduction – The IAU Working Group on Wide–Field Imaging (*R. M. West*). 2. Reports from the Sub–Sections of the Working Group – a. Sky surveys and patrols (*R. M. West*). b. Photographic techniques (*D. F. Malin*). c. Digitization techniques (*H. T. MacGillivray*). d. Archival and retrieval of wide–field data (*B. Lasker*). 3. Meeting of the Organising Committee (*R. M. West*). 4. Wide–field plate archives (*M. Tsvetkov*). 5. Reproduction of the Palomar Observatory Sky Surveys (*R. J. Brucato*). 6. Status of the St ScI scan–distribution program (*B. Lasker*). 7. Pixel addition – pushing Schmidt plates to B = 25 (*M. R. S. Hawkins*). 8. Photometry from Estar film (*S. Phillipps, Q. Parker*). 9. ASCHOT – Astrophysical Schmidt Orbital Telescope (*H. Lorenz*). 10. The Hitchhiker parallel CCD camera (*J. Davies, M. Disney, S. Driver, I. Morgan, S. Phillipps*).

034.139 **Time Service and Astrolabe Station in Coimbra.**
I. Duque Simões.
Coimbra Univ. (Portugal). Observatório Astronómico. 25 p. (1990).

034.140 **The Montreal Infrared Camera (MONICA).**
R. Doyon, D. Nadeau, M. Riopel, A. F. J. Moffat, S. Demers, P. Bastien, H. Yee.
J. R. Astron. Soc. Can., Vol. 85, No. 4, p. 182 (Aug 1991). Abstract. – See Abstr. 012.091 for the main entry.

034.141 **A polarizing Fourier transform spectrometer for intermediate resolution spectroscopy at millimetre and sub–millimetre wavelengths with the James Clerk Maxwell telescope.**
D. A. Naylor, G. T. Tompkins, T. A. Clark, G. R. Davis, W. D. Duncan.
J. R. Astron. Soc. Can., Vol. 85, No. 4, p. 185 – 186 (Aug 1991). Abstract. – See Abstr. 012.091 for the main entry.

034.142 **Modelling of the Virgo interferometer.**
P. Hello, J. Y. Vinet.
Elizabeth and Frederick White Research Conference: Gravitational astronomy – instrument design and astrophysical prospects, p. 1 – 13 (1991). – See Abstr. 012.098 for the main entry.

The authors present a review of the modelling program of the VIRGO project. This program involves optics, thermal behaviour, mechanical behaviour, analysed either with numerical or analytical methods. They present first the results obtained with an optical simulator based on an original Fabry–Perot model in the field of mirrors specifications. The second topic is related to the thermal and mechanical behaviour of the cavity input mirrors illuminated by high power laser beams.

034.143 **Ultralow temperature resonant–mass gravitational radiation detectors: current status and future prospects.**
P. F. Michelson, T. Aldcroft, J. Chiang, D. DeBra, J. Henderson, L. Mann, D. McKenzie, F. McLoughlin, H. J. Paik, R. Penny, J. C. Price, T. Stevenson, B. Vaughan, C. Zhou.
Elizabeth and Frederick White Research Conference: Gravitational astronomy – instrument design and astrophysical prospects, p. 14 – 56 (1991). – See Abstr. 012.098 for the main entry.
The authors describe the Stanford ultralow temperature detector design and the technology. They also describe preliminary investigations that may eventually lead to the construction of resonant–mass detectors with sensitivities approaching $h \sim 10^{-22}$.

034.144 **Development of 10 m laser interferometer gravitational wave antenna (TENKO–10) at ISAS.**
N. Kawashima.
Elizabeth and Frederick White Research Conference: Gravitational astronomy – instrument design and astrophysical prospects, p. 57 – 70 (1991). – See Abstr. 012.098 for the main entry.
Development of 9.2 m 102 bounce laser interferometer gravitational wave antenna (TENKO–10) at ISAS is described. The objective is to establish the engineering background for the laser interferometer gravitational wave antenna in Japan so that one can construct a km class antenna for the detection of gravitational waves in future. A particular emphasis was on the achievement of a long term stable operation. An upgrade of the interferometer to 100 m (TENKO–100) is going to be started in 1991 and will be completed in 1994.

034.145 **Conceptual design of a Japanese interferometric gravitational wave detector.**
M.–K. Fujimoto, M. Ohashi, N. Mio, K. Tsubono.
Elizabeth and Frederick White Research Conference: Gravitational astronomy – instrument design and astrophysical prospects, p. 75 – 90 (1991). – See Abstr. 012.098 for the main entry.
A design study on the interferometric gravitational wave detector is described. The detector consists of two orthogonal 3 km FP arms which are illuminated by 1 kW infrared light from a LD pumped Nd:YAG laser.

034.146 **Interferometric detection of gravitational waves: the VIRGO project.**
C. Bradaschia, E. Calloni, M. Cobal, R. Del Fabbro, A. Di Virgilio, A. Giazotto, H. Kautzky, V. Montelatici, D. Passuello, W. Velloso, F. Barone, L. Di Fiore, L. Milano, G. Russo, S. Solimeno, L. Gammaitoni, E. Marchesoni, S. Santucci, A. Brillet, M. Davier, E. Durand, P. Hello, D. Ivanov, M. Lesimple, C. N. Man, A. Marraud, M. Pham Tu, J. Y. Vinet, C. Boccara, J. P. Roger, J. Colas, M. Girard, M. Yvert, L. Massonetr, B. Mours, I. Wingerter.
Elizabeth and Frederick White Research Conference: Gravitational astronomy – instrument design and astrophysical prospects, p. 110 – 135 (1991). – See Abstr. 012.098 for the main entry.
To detect gravitational waves in a frequency range between 10 Hz and few kHz a ground based antenna was designed by the French–Italian collaboration VIRGO. The antenna is a 3 km long interferometer with special suspensions for the optical components, planned to reduce the seismic noise in order to reach the low frequency region. New developments of this project are presented.

034.147 **Interferometer measurements using non–classical light.**
D. F. Walls, M. J. Holland.
Elizabeth and Frederick White Research Conference: Gravitational astronomy – instrument design and astrophysical prospects, p. 136 – 146 (1991). – See Abstr. 012.098 for the main entry.
It has been proposed that non–classical light may be useful in improving the sensitivity of optical interferometers. This has application in reducing the threshold for detection of gravitational radiation. The authors review measurements based on this principle, as well as ways of producing the non–classical light using lasers with output intensity fluctuations below the shot noise limit.

034.148 **A proposed large–scale laser interferometer gravitational wave detector in Australia.**
D. E. McClelland, H.–A. Bachor, D. G. Blair, J. Collett, R. J. Sandeman, P. J. Turner.
Elizabeth and Frederick White Research Conference: Gravitational astronomy – instrument design and astrophysical prospects, p. 147 – 171 (1991). – See Abstr. 012.098 for the main entry.
The authors present a brief outline of the major optical and mechanical design features of the proposed Australian International Gravitational wave Observatory (AIGO).

034.149 **The gravitational wave experiment of the Rome group. Status report for the resonant antennas Explorer and Nautilus.**
P. Astone, M. Bassan, P. Bonifazi, M. G. Castellano, G. Cavallari, E. Coccia, C. Cosmelli, S. Frasca, E. Majorana, I. Modena, G. V. Pallottino, G. Pizzella, P. Rapagnani, F. Ricci, M. Visco.
Elizabeth and Frederick White Research Conference: Gravitational astronomy – instrument design and astrophysical prospects, p. 189 – 206 (1991). – See Abstr. 012.098 for the main entry.
The basic features of the resonant antennas for the search of gravitational waves are reviewed. The status report for the Rome detectors, Explorer and Nautilus, is presented. Explorer is recording data since July 1990 with $T_{eff} = 13$ mK corresponding to $h_{min} \approx 9 \times 10^{-19}$. Nautilus has been completed and will be cooled early in 1991.

034.150 **Modern status and future plans of the Moscow University interferometer.**
V. K. Milyukov, V. Yu. Novichkov, V. N. Rudenko.
Elizabeth and Frederick White Research Conference: Gravitational astronomy – instrument design and astrophysical prospects, p. 217 – 221 (1991). – See Abstr. 012.098 for the main entry.
The gravitational measurements research group of Sternberg Astronomical Institute has constructed a long based (75 m) geophysical laser deformograph as a first step on the way to a laser interferometrical gravitational antenna. The preliminary tests have shown that the sensitivity of that set–up is better than $10^{-12}\mathrm{Hz}^{-1/2}$ for a relative deformation of end masses installed on the ground for a frequency above 300 Hz.

034.151 **Low frequency noise in gravitational wave interferometers.**
P. R. Saulson.
Elizabeth and Frederick White Research Conference: Gravitational astronomy – instrument design and astrophysical prospects, p. 248 – 270 (1991). – See Abstr. 012.098 for the main entry.
The ultimate low frequency performance of terrestrial gravitational wave detectors is limited by the uncertainty principle and by fluctuating gravitational fields from the noisy environment. The author presents the expected amplitudes of these noise sources.

034.152 **Some recent developments in interferometric gravitational–wave detection.**
B. Meers.
Elizabeth and Frederick White Research Conference: Gravitational astronomy – instrument design and astrophysical prospects, p. 281 – 298 (1991). – See Abstr. 012.098 for the main entry.
The author discusses some ideas on the importance of wavefront distortion in both standard and dual recycling, some thoughts on how narrowband interferometers might be used to detect gravitational radiation from coalescing compact binaries,

and he reports on the first experimental demonstration of dual recycling.

034.153 **Coincidences between gravitational wave bar detectors.**
G. Pizzella.
5. Winter School on Hadronic Physics: Problems of fundamental modern physics II, p. 300 – 324 (1991). – See Abstr. 012.101 for the main entry.
The basic features of the resonant antennas for the search of gravitational waves are reviewed. The correlations between the Rome and Maryland gravitational wave detectors and various particle detectors during the SN 1987A are shown.

034.154 **Fabry–Perot resonators for interferometric GW antennas.**
S. Solimeno, F. Barone, L. di Fiore, L. Milano.
5. Winter School on Hadronic Physics: Problems of fundamental modern physics II, p. 325 – 340 (1991). – See Abstr. 012.101 for the main entry.
The authors discuss some problems relative to the alignment of Fabry–Perot cavities used in long baseline interferometers. In particular, they derive analytic expressions of the response of a misaligned F–P resonator with oscillating mirrors by taking into account longitudinal and transverse optical modes. Finally they introduce a measure of the misalignment and analyze the onset of instabilities induced by the radiation pressure.

034.155 **VIRGO: a ground based interferometric antenna for gravitational wave detection above 10 Hz.**
C. Bradaschia, E. Calloni, M. Cobal, R. del Fabbro, A. di Virgilio, A. Giazotto, L. E. Holloway, H. Kautzky, B. Michelozzi, V. Montelatici, D. Passuello, W. Velloso.
5. Winter School on Hadronic Physics: Problems of fundamental modern physics II, p. 341 – 356 (1991). – See Abstr. 012.101 for the main entry.
A French–Italian collaboration plan to build a ground based antenna to detect gravitational waves in a wide frequency range from 10 Hz to a few kHz. The antenna is a high sensitivity interferometer with 3 km arms. Sensitivity in the 10 Hz region is achieved by means of ad–hoc suspensions for the optical components, in order to reduce the seismic noise.

034.156 **Experimental gravitation at very low temperatures.**
M. Cerdonio, P. Falferi, P. Fortini, R. Macchietto, A. Ortolan, G. A. Prodi, C. Ravanelli, B. Tiveron, J. P. Zendi, M. Bonaldi, A. Cavalleri, G. Durin, G. Fontana, A. Maddalosso, A. Maraner, R. Mezzena, R. Tommasini, S. Vitale.
5. Winter School on Hadronic Physics: Problems of fundamental modern physics II, p. 357 – 378 (1991). – See Abstr. 012.101 for the main entry.
Gravitational waves and "gravitomagnetic" components of general–relativistic gravitation have as yet escaped direct detection with laboratory experiments. The authors report on progress in their attempt to make feasible a ground based cryogenic experiment aimed to measure the Lense–Thirring–Schiff dragging of inertial frames due to the rotating earth, which is the main component of the local "gravitomagnetic" field. They present the goals of the experiment AURIGA, to be set up at LNL: an ultracryogenic, $T \approx 0.1K$, resonant antenna, $M \cong 2300$ kg, will be operated in continuous coincidence with the similar NAUTILUS at CERN to search for gravitational waves impulses from galaxies of the Local Group.

034.157 **A new CCD for the classification spectrograph at the University of Toronto Southern Observatory (UTSO) in Chile.**
R. F. Garrison, K. W. Kamper, A. Ridder, I. Shelton.
J. R. Astron. Soc. Can., Vol. 85, No. 4, p. 217 (Aug 1991). Abstract. – See Abstr. 012.091 for the main entry.

034.158 **Atomic hydrogen maser frequency stability measurement system.**
Luo Weihua, Gang Guoxing, He Jianwei, Zhang Weiqun.
Ann. Shanghai Obs., Acad. Sin., No. 12, p. 146 – 151 (1991). In Chinese.

034.159 **A phase–locked receive system used in cavity auto–tuning for hydrogen masers.**
He Jianwei, Luo Weihua.
Ann. Shanghai Obs., Acad. Sin., No. 12, p. 152 – 158 (1991). In Chinese.

034.160 **The evaluation of the velocity distribution by using beam optic parameter in a cesium beam frequency standard.**
Peng Jixing.
Ann. Shanghai Obs., Acad. Sin., No. 12, p. 159 – 164 (1991). In Chinese.

034.161 **The correlate analysis of the industry cesium clock rate change resulted from the ion pump parameter.**
Hu Jinlun.
Ann. Shanghai Obs., Acad. Sin., No. 12, p. 165 – 169 (1991). In Chinese.

034.162 **Microwave spectrum of optical system improvement of cesium beam frequency standard at Shanghai Observatory.**
Wang Guanzhong, Lu Zhongzhen.
Ann. Shanghai Obs., Acad. Sin., No. 12, p. 170 – 172 (1991). In Chinese.

034.163 **The comparison of optical character of the fiber with the optical transmission of the coudé spectrograph.**
Qin Songnian, Gao Cai.
Publ. Yunnan Obs., No. 3, p. 25 – 29 (1991). In Chinese.
The possibility and the advantage of replacing the optical system by the fiber is described. Optical losses of the energy of fiber transmission system with the optical transmission system are also compared.

034.164 **An automated drive for a gamma telescope.**
V. G. Shitov.
Bull. Crimean Astrophys. Obs., Vol. 81, p. 164 – 178 (1990). English translation of Izv. Krym. Astrofiz. Obs., Tom 81, p. 183 – 198 (1990). See Abstr. 53.034.046.
The automatic drive structure for a gamma telescope is considered. The dynamic characteristics are examined along with the stability and performance. Design principles for system modules are presented, the structures are described, and the major characteristics are examined.

034.165 **Video signal digitalizer system.**
P. Recabarren, J. Calderón, G. Giovanola, J. Sanchez, L. Espinosa.
Bol. Asoc. Argent. Astron., No. 36, p. 326 – 333 (1990). In Spanish. – See Abstr. 012.095 for the main entry.
The authors have developed an equipment designed to acquire and store standard TV camera images in order to be processed with personal computers. The unit permit the acquisition of 384×485 pixel video camera images in 40 milliseconds. The astronomical applications of the system are briefly discussed.

034.166 **A new automatic photopolarimeter for CASLEO: electronics of instrument.**
E. Martinez, J. L. Aballay, A. Marún, H. Ruartes.
Bol. Asoc. Argent. Astron., No. 36, p. 342 – 385 (1990). In Spanish. – See Abstr. 012.095 for the main entry.
A new photopolarimeter is under construction in CASLEO. The instrument will be operated automatically by a PC/XT computer from the control room. The computer configurates the instrument as photometer or polarimeter and controls the data aquisition and movements of the several mechanisms of the equipment. A description and the theory of operation of the

electronics of the instrument is presented. The hardware and the control software of the instrument are described in this article.

034.167 **Automation of CASLEO's Cassegrain spectrograph.**
E. Martínez, J. L. Aballay, A. Marún, H. Ruartes.
Bol. Asoc. Argent. Astron., No. 36, p. 386 – 396 (1990). In Spanish. – See Abstr. 012.095 for the main entry.
The Casleo's Cassegrain spectrograph will be made fully automatically. Fromn the control room the equipment will allow a remote and automatic operation. The offset guider will be included in the automation too. The user will be able to operate the instrument from a keyboard and a display or from a PC/XT computer.

034.168 **A high speed two–star photometer.**
G. V. Rao, V. A. Thomas, V. C. Babu, K. R. N. Kutty, B. N. Ashoka, S. Sugumar, K. C. Navada, S. Seetha, T. M. K. Marar, K. Kasturirangan, U. R. Rao, J. C. Bhattacharyya.
Bull. Astron. Soc. India, Vol. 19, Nos. 1 – 2, p. 85 (Mar – Jun 1991). Abstract. – See Abstr. 012.108 for the main entry.

034.169 **A portable photon counting system for fast photometry.**
R. Venkatesha, S. Seetha.
Bull. Astron. Soc. India, Vol. 19, Nos. 1 – 2, p. 85 (Mar – Jun 1991). Abstract. – See Abstr. 012.108 for the main entry.

034.170 **Echelle spectrograph of the Japal–Rangapur Observatory – design and fabrication details.**
C. H. Kumar.
Bull. Astron. Soc. India, Vol. 19, Nos. 1 – 2, p. 86 (Mar – Jun 1991). Abstract. – See Abstr. 012.108 for the main entry.

034.171 **The use of photography in astronomy: some thoughts about Schmidt telescope wide–field work.**
R. M. West.
Messenger, No. 65, p. 45 – 50 (Sep 1991).

034.172 **CASPEC's new look.**
L. Pasquini, A. Gilliotte.
Messenger, No. 65, p. 50 – 53 (Sep 1991).

034.173 **Multi–object spectroscopy with an automatic fibre positioning system in a one–degree field. First technical run of MEFOS at the prime focus of the 3.6–m telescope.**
R. Bellenger, M. Dreux, P. Felenbok, A. Fernandez, J. Guerin, R. Schmidt, G. Avila, S. D'Odorico, W. Eckert, G. Rupprecht.
Messenger, No. 65, p. 54 – 59 (Sep 1991).

034.174 **L'entraînement d'un télescope par un moteur pas à pas (3). Un diviseur de fréquence ultra–stable.**
J. Bourgeois.
Ciel Terre, Vol. 107, No. 5, p. 172 – 176 (Sep–Oct 1991). In French.

034.175 **Long–slit spectroscopy at La Silla: an annotated menu.**
W. W. Zeilinger.
Messenger, No. 66, p. 63 – 65 (Dec 1991).

034.176 **Data analysis and algorithms for g.w. antennas.**
G. V. Pallottino, G. Pizzella.
3. International Workshop on Data Analysis in Astronomy, p. 361 – 391 (1989). – See Abstr. 012.110 for the main entry.
The analysis of the sensitivity of a gravitational wave detector shows that the role of the algorithms used for the analysis of the data is comparable to that of the experimental apparatus. The authors discuss the response of an antenna to an incoming gravitational wave, deriving a mathematical model of the system, which is used for examining various algorithms of data analysis, including the optimum filter and the matched filter. A brief discussion of coincidence techniques, using two detectors, is also included.

034.177 **Time synchronization in China using current techniques.**
Miao Yongrui, Ye Shuhua, Jin Wenjing, Zhang Jintong.
Publ. Shaanxi Astron. Obs., Vol. 14, No. 2, p. 136 – 140 (Dec 1991).

034.178 **Theoretical and technical problems in design of DMTD (*dual mixer time difference*) system.**
Bian Yujing.
Publ. Shaanxi Astron. Obs., Vol. 14, No. 2, p. 156 – 163 (Dec 1991). In Chinese.

034.179 **Low noise 5.001 MHz frequency synthesizer.**
Wang Danni, Bian Yujing.
Publ. Shaanxi Astron. Obs., Vol. 14, No. 2, p. 164 – 167 (Dec 1991). In Chinese.

034.180 **Design of high performance beater.**
Chen Sufang, Bian Yujing.
Publ. Shaanxi Astron. Obs., Vol. 14, No. 2, p. 168 – 171 (Dec 1991). In Chinese.

034.181 **The application of the standard frequency in television signals to the synchro–broadcast.**
Liang Shuangyou.
Publ. Shaanxi Astron. Obs., Vol. 14, No. 2, p. 172 – 176 (Dec 1991). In Chinese.

034.182 **En ny generation CCD–detektorer til nordisk astronomi.**
J. Andersen, R. F. Nielsen, P. Nørregaard.
Astron. Tidsskr., Årg. 24, Nr. 3, p. 138 – 141 (1991).

034.183 **Photoelectric astronomy at Mt. Tamborine Observatory.**
A. A. Page, G. E. Neilsen.
South. Stars, Vol. 34, No. 3, p. 33 – 36 (18 Jun 1991). – See Abstr. 012.120 for the main entry.

034.184 **An automatic photopolarimeter for small telescopes.**
S. Tapia, R. A. Tucker, E. R. Craine.
South. Stars, Vol. 34, No. 3, p. 77 – 83 (18 Jun 1991). – See Abstr. 012.120 for the main entry.
Free of the observers' inherent inefficiency Automatic Photometric Telescopes (APT's) appear to be very close to an ideal level of performance in the collection of astronomical data; their limitations now depend largely on photometric atmospheric conditions. The efficiency of APT's can still be increased by replacing the photometer with a photopolarimeter.

034.185 **The JEAP (*Joint European Amateur Photometer*): the four star photometer and the future of amateur/professional collaboration.**
E. N. Walker.
South. Stars, Vol. 34, No. 3, p. 85 – 92 (18 Jun 1991). – See Abstr. 012.120 for the main entry.

034.186 **First light with the Vassar CCD.**
F. R. Chromey.
News Lett. Astron. Soc. N.Y., Vol. 3, No. 9, p. 21 (Feb 1991). Abstract. – See Abstr. 012.119 for the main entry.

034.187 **A CCD for Mount John.**
W. Tobin.
South. Stars, Vol. 34, No. 3, p. 155 – 158 (18 Jun 1991). – See Abstr. 012.120 for the main entry.

034.188 **A fast astrograph for CCD imaging devices.**
D. Beach.
South. Stars, Vol. 34, No. 3, p. 159 – 161 (18 Jun 1991). – See Abstr. 012.120 for the main entry.
A compact optical system has been invented for CCD imaging/photometry with maximum information throughput.

034.189 **A fast two–star photometer.**
D. J. Sullivan.
South. Stars, Vol. 34, No. 3, p. 201 – 208 (18 Jun 1991). – See Abstr. 012.120 for the main entry.
An outline is presented of the main features of a computer controlled, two channel photometric system, that is capable of high speed operation. Some examples of data obtained using the instrument are also presented.

034.190 **Meridiane del centro Italia.**
N. Severino.
Orione, Vol. 11, N. 4, p. 44 – 48 (Jul – Aug 1991).

034.191 **Photometric calibration of the CCD camera of 1–m telescope at VBO.**
Y. D. Mayya.
J. Astrophys. Astron., Vol. 12, No. 4, p. 319 – 331 (Dec 1991).
Calibration of the CCD camera of the 1–m telescope at the Vainu Bappu Observatory, Kavalur, to the BVR system is reported here based on the observations of stars in the "dipper asterism" in the open cluster M67 (NGC 2682). Transformations involving B and V have negligible colour terms, while those involving R are slightly colour dependent.

034.192 **Een middenformaat fish–eye camera geautomatiseerd.**
H. Betlem.
Radiant, Jaarg. 13, Nr. 6, p. 138 – 142 (Dec 1991).

034.193 **Characterization of frequency stability: frequency–domain estimation of stability measures.**
D. B. Percival.
Proc. IEEE, Vol. 79, No. 7, p. 961 – 972 (Jul 1991). Special issue: Time and frequency. J. Jespersen, D. W. Hanson (eds.).
The standard characterizations of frequency stability are, in the time domain, the Allan (or two sample) variance and, in the frequency domain, the spectral density function (SDF). The author describes SDF's that model sampled frequency stability data and that are related to the SDF's of the standard characterization. He reviews the fundamental role that SDF's play in the theory of stationary processes and processes with stationary backward differences.

High resolution imaging from Mauna Kea.
See Abstr. 009.001.

CAS Symposium No. 118: The 10th Meeting of Astronomical Instrumentation, Xinanjiang Zejiang (China), 3 – 6 Nov 1990.
See Abstr. 011.011.

Mini–workshop on large–size CCDs at ESO.
See Abstr. 011.023.

Electron image tubes and image intensifiers II. Proceedings.
See Abstr. 012.031.

Charge–coupled devices and solid state optical sensors.
See Abstr. 012.058.

Charge–coupled devices and solid state optical sensors II. Proceedings.
See Abstr. 012.059.

Gravitational astronomy: instrument design and astrophysical prospects. Proceedings.
See Abstr. 012.098.

Toward the global APT network. Proceedings. 3. New Zealand Conference on Photoelectric Photometry, Blenheim (New Zealand), 9 – 12 Mar 1989.
See Abstr. 012.120.

RISE telescope concept for CaK imaging.
See Abstr. 013.036.

Ensquared power for obscured circular pupils with off–center imaging.
See Abstr. 031.003.

Concentric systems for adaptation as spectrographs.
See Abstr. 031.042.

New, fast, high–performance single mirror spectrograph cameras.
See Abstr. 031.068.

Astrophysical functions of the reflector with diameter 1.05 m and its necessary auxiliary facilities.
See Abstr. 032.052.

InSb linear multiplexed FPAs for the CRAF/Cassini visible and infrared mapping spectrometer.
See Abstr. 035.089.

The Space Telescope imaging spectrograph 2048 CCD and its characteristics.
See Abstr. 035.164.

Observations with the scanner of the 6–m telescope in the Nasmyth–1 focus and automatized program package for spectrum reduction.
See Abstr. 036.083.

A CCD detector attached to the lower latitude meridian circle. (III) Several correction for the spherical astronomy.
See Abstr. 036.091.

Real–time image parameterization in high energy gamma–ray astronomy using transputers.
See Abstr. 036.106.

A new method to search for high energy gamma rays by track–calorimeter detectors at a high mountain.
See Abstr. 036.108.

The accuracy of infrared photometry with arrays.
See Abstr. 036.111.

Using image area to control CCD systematic errors in spaceborne photometric and astrometric time–series measurements.
See Abstr. 036.112.

Automatic plates measurement with images digitalization.
See Abstr. 036.121.

A system for line profile studies at the 150–foot tower on Mount Wilson.
See Abstr. 036.130.

Elimination of artifacts in interline charge–coupled device imagers.
See Abstr. 036.153.

Coordinate–system instability in the Askorekord measuring machine on temperature variations.
See Abstr. 036.158.

High–resolution imaging by optical aperture synthesis: first results from the MAPPIT project.
See Abstr. 036.226.

High spatial resolution techniques.
See Abstr. 036.276.

A CCD detector attached to the Lower Latitude Meridian Circle (V): Building up a high accurate celestial sphere frame using ground–based observations.
See Abstr. 036.303.

The importance of deadtime in photon–counting photoelectric photometry.
See Abstr. 036.363.

An introduction to frequency standards.
See Abstr. 044.016.

Synchronization and relativity.
See Abstr. 044.021.

Recent results and future prospects for the MACRO experiment.
See Abstr. 061.065.

Solar neutrinos and the history of the Sun.
See Abstr. 078.024.

Status and future of solar neutrino observations.
See Abstr. 080.085.

Solar neutrino experiments.
See Abstr. 080.122.

Interferometric seeing measurements at the La Palma Observatory.
See Abstr. 082.048.

OAUV programme for minor planets astrometry.
See Abstr. 098.041.

Multicolor observations of cataclysmic variables.
See Abstr. 113.021.

Status of searches for >TeV photons from SN 1987A.
See Abstr. 125.288.

Observations of TeV photons at the Whipple Observatory.
See Abstr. 143.021.

Gamma ray astrophysics at energies up to 10 GeV.
See Abstr. 143.037.

VHE and UHE gamma ray sources.
See Abstr. 143.038.

Gamma–ray astronomy above 100 GeV.
See Abstr. 143.055.

An upward–going muon background estimation.
See Abstr. 144.047.

CCD detectors applied to globular cluster research.
See Abstr. 154.037.

A feasibility study for detection of gravitational microlensing by baryonic dark matter via CCD photometry.
See Abstr. 155.023.

035 Space Instrumentation

035.001 **An eye on the violent universe.**
R. G. Nichols.
Astronomy, Vol. 19, No. 7, p. 44 – 49 (Jul 1991).
The Gamma Ray Observatory is now orbiting earth on a mission to probe the high–energy radiation from quasars, black holes, and exploding stars.

035.002 **A burst of gamma rays.**
R. Talcott.
Astronomy, Vol. 19, No. 10, p. 46 – 50 (Oct 1991).

035.003 **Anyone for a 94–inch telescope?**
S. Gonzaga.
Astronomy, Vol. 19, No. 12, p. 43 – 48 (Dec 1991).
Five talented amateur astronomers prepare to observe using the Hubble Space Telescope.

035.004 **Das System ASTRO 1 auf der Raumstation MIR.**
C. Elstner, G. Lichtenauer, W. Skarus.
Jenaer Rundsch., Jahrg. 36, Heft 3, p. 137 – 139 (1991).

035.005 **The ISO spacecraft.**
S. Ximénez de Ferrán.
ESA Bull., No. 67, p. 17 – 24 (Aug 1991).

035.006 **Performance of a balloon–borne magnet spectrometer for cosmic ray studies.**
R. L. Golden, C. Grimani, R. Hull, B. L. Kimbell, R. Park, S. A. Stephens, S. Stochaj, W. R. Webber, G. Basini, E. Bonaviri, F. M. Brancaccio, M. Ricci, J. F. Ormes, E. S. Seo, R. E. Streitmatter, F. Celletti, P. Spillantini, A. Codino, M. Menichelli, I. Salvatori, F. Bongiorno, V. Bidoli, A. Buccheri, M. P. de Pascale, A. Morselli, P. Picozza.
Nucl. Instrum. Methods Phys. Res., Sect. A, Vol. 306, No. 1/2, p. 366 – 377 (15 Aug 1991).
This paper summarizes the performance characteristics of the balloon–borne magnet spectrometer operated by New Mexico State University's Particle Astrophysics Laboratory. Particular emphasis has been placed on the rigidity resolution, including both random and systematic errors of the magnetic spectrometer system. Measurement of the performance characteristics has been greatly enhanced through the use of an imaging calorimeter as an independent aid in the identification of cosmic rays.

035.007 **Current measurements in space plasmas and the problem of separating between spatial and temporal variations in the field of a plane electromagnetic wave.**
V. V. Krasnosel'skikh, A. M. Natanzon, A. E. Reznikov, A. Yu. Schyokotov, S. I. Klimov, A. E. Kruglyi, L. J. C. Woolliscroft.
Adv. Space Res., Vol. 11, No. 9, p. 37 – 40 (1991). – See Abstr. 012.006 for the main entry.
A description of an instrument for measuring a current density vector is presented. This instrument uses 3 orthogonal Rogovskii coils. The need for such an instrument is discussed. Together with magnetic measurements it is possible to determine the wave vector of a plane electromagnetic wave. Test results for prototype coil are presented, together with a discussion of further developments.

035.008 **The AXAF low–energy transmission grating spectrometer LETGS: diagnostic capabilities for the study of stellar coronae.**
R. Mewe, J. R. Lemen, C. J. Schrijver.
Astrophys. Space Sci., Vol. 182, No. 1, p. 35 – 59 (Aug 1991).
The authors study the diagnostic capabilities of the high–resolution, Low–Energy Transmission Grating Spectrometer, LETGS, of NASA's planned Advanced X–ray Astrophysics Facility, AXAF, for optically thin stellar coronae. Spectra are simulated on the basis of isothermal and source loop models and are analyzed with particular emphasis on the extraction of the differential emission measure distribution. The AXAF–LETGS is shown to be particularly sensitive for plasma at temperatures between 0.5 and 15 MK. Emission from temperatures in excess of 20 MK can be observed, but the lack of strong spectral lines hampers accurate temperature determinations. The authors

simulate spectra of close binaries to demonstrate the observability of the Doppler effects associated with orbital motions. They present lists of spectral lines that can be used for density diagnostics, and simulate and compare various spectra at different electron densities.

035.009 **Die räumliche Verteilung von Ferninfrarotlinienstrahlung in externen Galaxien gemessen mit einem abbildenden Fabry–Pérot–Spektrometer.**
N. Geis.
Diss. (Dr.rer.nat.), Max–Planck–Institut für Physik und Astrophysik, Garching (Germany, F.R.). Inst. für Extraterrestrische Physik; München Univ. (Germany, F.R.). Fakultät für Physik. MPE—231, 157 p. (Dec 1991).
Contents: 1. Einleitung. 2. Astrophysikalische Grundlagen. 3. FIFI – Ein abbildendes Ferninfrarotspektrometer. 4. Beobachtung der räumlichen Verteilung von FIR–Linienstrahlung in externen Galaxien.

035.010 **The BATSE experiment on the Gamma Ray Observatory: early results.**
G. J. Fishman, C. A. Meegan, R. B. Wilson, W. S. Paciesas, G. N. Pendleton.
Bull. Am. Astron. Soc., Vol. 23, No. 2, p. 901 (1991). Abstract. – See Abstr. 012.008 for the main entry.

035.011 **SIGHT: a balloon borne hard X–ray telescope.**
A. Parsons, B. Sadoulet, T. K. Edberg, S. Weiss, J. Wilkerson, K. Hurley, R. P. Lin, G. Smith.
Bull. Am. Astron. Soc., Vol. 23, No. 2, p. 905 (1991). Abstract. – See Abstr. 012.008 for the main entry.

035.012 **The status of the MAMA detector for STIS.**
C. L. Joseph, J. G. Timothy, V. Argabright.
Bull. Am. Astron. Soc., Vol. 23, No. 2, p. 906 (1991). Abstract. – See Abstr. 012.008 for the main entry.

035.013 **Cryogenic performance of the SIRTF prototype secondary mirror assembly.**
M. T. Stier, J. C. Doyle, M. A. Duffy, K. S. Gottschalck, S. N. Gullapalli, J. W. Neuner.
Bull. Am. Astron. Soc., Vol. 23, No. 2, p. 906 (1991). Abstract. – See Abstr. 012.008 for the main entry.

035.014 **The Wisconsin Ultraviolet Photo–Polarimeter Experiment (WUPPE).**
K. H. Nordsieck, A. D. Code, M. R. Meade, B. L. Babler, C. M. Anderson, D. E. Michalski, R. H. Pfeifer, T. E. Jones.
Bull. Am. Astron. Soc., Vol. 23, No. 2, p. 906 (1991). Abstract. – See Abstr. 012.008 for the main entry.

035.015 **HST update: science amid setbacks.**
R. Tresch–Fienberg.
Sky Telesc., Vol. 82, No. 3, p. 242 – 246 (Sep 1991).
As the Hubble Space Telescope churns out discoveries, NASA settles on a plan to fix its flawed optics. But some astronomers worry that the price may be too high.

035.016 **Exploring the high–energy universe.**
B. Margon.
Sky Telesc., Vol. 82, No. 6, p. 607 – 612 (Dec 1991).
After a decade without any major space missions, American astronomers probing the most energetic phenomena in the cosmos hope to launch a new generation of orbiting X–ray and gamma–ray telescopes.

035.017 **Highlights from the Hopkins Ultraviolet Telescope.**
A. F. Davidsen.
Bull. Am. Astron. Soc., Vol. 23, No. 2, p. 958 (1991). Abstract. – See Abstr. 012.008 for the main entry.

035.018 **Observations with the Wisconsin Ultraviolet Photopolarimetry Experiment (WUPPE).**
A. D. Code.
Bull. Am. Astron. Soc., Vol. 23, No. 2, p. 971 (1991). Abstract. – See Abstr. 012.008 for the main entry.

035.019 **The universe as seen by the Ultraviolet Imaging Telescope (UIT).**
T. P. Stecher.
Bull. Am. Astron. Soc., Vol. 23, No. 2, p. 971 (1991). Abstract. – See Abstr. 012.008 for the main entry.

035.020 **The EUVE spectrometer.**
P. Jelinsky.
Bull. Am. Astron. Soc., Vol. 23, No. 2, p. 972 (1991). Abstract. – See Abstr. 012.008 for the main entry.

035.021 **Modelling the HST point spread function.**
C. J. Burrows, H. Hasan.
Bull. Am. Astron. Soc., Vol. 23, No. 2, p. 977 (1991). Abstract. – See Abstr. 012.008 for the main entry.

035.022 **Development of the Burst And Transient Source Experiment (BATSE).**
J. M. Horack.
National Aeronautics and Space Administration, Washington, DC (USA), NASA–RP—1268. 324 p. (Sep 1991).
BATSE, one of four instruments on the Gamma Ray Observatory, consists of eight identical detector modules mounted on the corners of the spacecraft. Developed at the Marshall Space Flight Center, BATSE is the most sensitive gamma ray burst detector flown to date. Details of the assembly and test phase of the flight hardware development are presented. Results and descriptions of calibrations documented extensively.

035.023 **Telescopio espacial Hubble (HST).**
L. de Juan, A. M. Romero, A. Sanchez, J. Zamorano.
Opt. Pura Apl., Vol. 23, No. 2, p. 73 – 94 (1990).
The authors describe the HST instrumentation and the research fields upon it will be performed.

035.024 **Accelerator beam experiments of a prototype cosmic ray heavy ion telescope.**
T. Kohno, K. Munakata, T. Imai, A. Yoneda, C. Kato, M. Matsuoka, K. Nishijima, T. Doke, J. Kikuchi, T. Kashiwagi, N. Hasebe, H. J. Crawford.
J. Phys. Soc. Jpn., Vol. 60, No. 11, p. 3967 – 3975 (Nov 1991).
A Heavy Ion cosmic ray Telescope (HIT) has been scheduled to be launched into the synchronous orbit in 1993 aiming to observe elemental and isotopic composition of solar energetic particles and galactic cosmic rays. In this paper, the authors report the results of accelerator beam experiments of a prototype model of HIT.

035.025 **The Solar Ultraviolet Network (SUN).**
L. Damé.
International Conference on Mechanisms of Chromospheric and Coronal Heating, p. 73 – 76 (1991). – See Abstr. 012.011 for the main entry.
SUN is a UV and Visible Space Interferometer dedicated to high resolution Solar Atmospheric Physics. It has been proposed for flight on the Space Station as part of the SIMURIS Mission which is currently undergoing an Assessment Study at ESA. The contribution briefly presents the aims and current design status of the proposed investigation.

035.026 **The OSL/HRTS VUV CCD development program.**
D. G. Socker, C. Korendyke, M. Marchywka.
Bull. Am. Astron. Soc., Vol. 23, No. 2, p. 1037 – 1038 (1991). Abstract. – See Abstr. 012.012 for the main entry.

035.027 **The Multi–Spectral Solar Telescope Array.**
R. B. Hoover, A. B. C. Walker Jr., J. Lindblom, T. W. Barbee Jr.
Bull. Am. Astron. Soc., Vol. 23, No. 2, p. 1038 (1991). Abstract. – See Abstr. 012.012 for the main entry.

035.028 **X–ray imaging telescope for small satellite payload.**
U. Desai.
Bull. Am. Astron. Soc., Vol. 23, No. 2, p. 1056 (1991). Abstract. – See Abstr. 012.012 for the main entry.

035.029 **A balloon–borne payload for imaging hard X–rays and gamma–rays from solar flares.**
C. J. Crannell, B. R. Dennis, J. P. Norris, L. E. Orwig, E. J. Schmahl, F. L. Lang, R. Starr, M. E. Greene, G. J. Hurford, W. N. Johnson, K. S. Wood.
Bull. Am. Astron. Soc., Vol. 23, No. 2, p. 1056 (1991). Abstract. – See Abstr. 012.012 for the main entry.

035.030 **Results from the MSFC solar hard X–ray imaging Fourier telescope. Numerical models.**
J. W. Campbell, J. M. Davis, A. G. Emslie.
Bull. Am. Astron. Soc., Vol. 23, No. 2, p. 1056 (1991). Abstract. – See Abstr. 012.012 for the main entry.

035.031 **Photometric calibration of the GSFC SERTS–3 rocket experiment.**
R. J. Thomas, W. M. Neupert, G. L. Epstein, C. E. Condor, W. T. Thompson.
Bull. Am. Astron. Soc., Vol. 23, No. 2, p. 1061 (1991). Abstract. – See Abstr. 012.012 for the main entry.

035.032 **Multi–spectral solar telescope array. V. Temperature diagnostic response to the optically thin solar plasma.**
C. E. DeForest, C. C. Krankelborg, M. J. Allen, E. S. Paris, T. D. Willis, J. F. Lindblom, R. H. O'Neal, A. B. C. Walker Jr., T. W. Barbee Jr., R. B. Hoover, T. W. Barbee III.
Opt. Eng., Vol. 30, No. 8, p. 1125 – 1133 (Aug 1991).
The authors have developed compact soft X–ray, extreme ultraviolet (EUV), and far–ultraviolet multilayer coated telescopes for the study of the solar chromosphere, corona, and corona/solar wind interface. In addition to permitting traditional normal incidence optical configurations to be used at soft X–ray/EUV wavelengths, multilayer coatings also allow a narrow wavelength band ($\lambda/\Delta\lambda \sim 15 - 100$) to be selected for imaging. The resulting telescopes provide a very powerful and flexible diagnostic instrument for the study of both the fine–scale structure of the chromosphere/corona interface and the large–scale structure of the corona and corona/solar wind interface. The authors report on the ability of the MSSTA payload to obtain temperature diagnostic information about the optically thin solar plasma. They also discuss applications of this information to studies of coronal structure.

035.033 **The solar stellar irradiance comparison experiment.**
T. N. Woods, G. J. Rottman.
Bull. Am. Astron. Soc., Vol. 23, No. 2, p. 1069 (1991). Abstract. – See Abstr. 012.012 for the main entry.

035.034 **Response of the OSSE detectors to 30 – 200 MeV neutrons.**
G. M. Frye Jr., T. L. Jenkins, S. E. Stansfield, R. L. Kinzer, J. D. Kurfess, W. N. Johnson, M. Strickman, R. Byrd, C. Foster.
Bull. Am. Astron. Soc., Vol. 23, No. 2, p. 1072 (1991). Abstract. – See Abstr. 012.012 for the main entry.

035.035 **Solar gamma–ray spectroscopy with BATSE.**
H. S. Hudson.
Bull. Am. Astron. Soc., Vol. 23, No. 2, p. 1074 (1991). Abstract. – See Abstr. 012.012 for the main entry.

035.036 **The second generation Wide–Field/Planetary Camera of the Hubble Space Telescope.**
K. Leschly, D. Allestad, L. Herrell.
J. Br. Interplanet. Soc., Vol. 44, No. 10, p. 477 – 486 (Oct 1991).

035.037 **Diagnosing the optical state of the Hubble Space Telescope.**
A. H. Vaughan.
J. Br. Interplanet. Soc., Vol. 44, No. 10, p. 487 – 494 (Oct 1991).

035.038 **Infrared imaging of extrasolar planets.**
D. J. Diner, E. F. Tubbs, S. L. Gaiser, R. P. Korechoff.
J. Br. Interplanet. Soc., Vol. 44, No. 10, p. 505 – 512 (Oct 1991).
This paper describes an optical system for direct detection, in the infrared, of planets orbiting other stars. The proposed system consists of a large aperture (~16m) space–based telescope to which is attached a specialized imaging instrument containing a set of optical signal processing elements to suppress diffracted light from the central star. Starlight suppression is accomplished using coronagraphic apodization combined with rotational shearing interferometry.

035.039 **The Voyager ultraviolet spectrometers – astrophysical observations from the outer solar system.**
S. H. Linick, J. B. Holberg.
J. Br. Interplanet. Soc., Vol. 44, No. 10, p. 513 – 520 (Oct 1991).

035.040 **Hubble Space Telescope optical performance.**
C. J. Burrows.
Workshop on the First Year of HST Observations, p. 96 – 105 (1991). – See Abstr. 012.017 for the main entry.
The author summarizes the present imaging performance of the observatory and compares it with the expected performance when corrective optics (COSTAR and WFPC 2) are installed on–orbit.

035.041 **Introduction to the Goddard High Resolution Spectrograph (GHRS).**
J. C. Brandt.
Workshop on the First Year of HST Observations, p. 106 – 109 (1991). – See Abstr. 012.017 for the main entry.
This paper presents the GHRS from the perspective of the Preliminary Desing Review (PDR), which began on December 12, 1978.

035.042 **Status of the Goddard High Resolution Spectrograph in May 1991.**
D. Ebbets, J. Brandt, S. Heap.
Workshop on the First Year of HST Observations, p. 110 – 122 (1991). – See Abstr. 012.017 for the main entry.
This presentation summarized the state of our knowledge about HST pointing accuracy, target acquisition procedures, sensitivity, spectral resolution, stray and scattered light, wavelength calibration, photometric precision and time resolution.

035.043 **Early operations with the High Speed Photometer.**
J. W. Pericival, R. C. Bless, M. J. Nelson.
Workshop on the First Year of HST Observations, p. 123 – 130 (1991). – See Abstr. 012.017 for the main entry.
The performance of the High Speed Photometer (HSP) during the Orbital Verification (OV) and the Science Verification (SV) programs of the Hubble Space Telescope (HST) is described. The HSP is operating as designed, and all hardware is fully operational. The HSP has been seriously affected by the degraded point spread function (PSF) of the telescope system, the telescope pointing calibration, and the jitter in the spacecraft pointing.

035.044 **Early commissioning astrometry performance of the Fine Guidance Sensors.**
G. F. Benedict, W. H. Jefferys, Q. Wang, A. Whipple, E. Nelan, D. Story, R. L. Duncombe, P. Hemenway, P. J. Shelus, B. McArthur, J. McCartney, O. G. Franz, L. Wasserman, T. Kreidl, W. F. van Altena, T. Girard, L. W. Fredrick.
Workshop on the First Year of HST Observations, p. 131 – 146 (1991). – See Abstr. 012.017 for the main entry.
The authors discuss astrometry with the Fine Guidance Sensors and explore various factors limiting their performance. These results were obtained before starting either the Orbital Verification or Science Verification programs.

035.045 **Faint Object Camera in–flight performance. Geometric distortion, stability and plate scale.**
D. Baxter.
Workshop on the First Year of HST Observations, p. 220 – 224 (1991). – See Abstr. 012.017 for the main entry.
The geometric distortion characteristics of the Faint Object Camera have been analysed in great depth and it has been found that the distortion pattern is remarkably stable.

035.046 **Future missions for gamma–ray astronomy – SIGMA and the Nuclear Astrophysics Explorer.**
P. Durouchoux, J. Matteson.
10. Santa Cruz Summer Workshop in Astronomy and Astrophysics: Supernovae, p. 291 – 304 (1991). – See Abstr. 012.052 for the main entry.

035.047 **In–flight performance of the FOC: early assessment of the absolute sensitivity.**
W. B. Sparks.
Workshop on the First Year of HST Observations, p. 225 – 228 (1991). – See Abstr. 012.017 for the main entry.
Observations with the Faint Object Camera on the Hubble Space Telescope in the f/96 imaging mode indicate an absolute sensitivity consistent with nominal (as given in the Instrument Handbook) for wavelengths longward of about 2500 to 3000 Å. Shortward of that, there is a smooth decline reaching approximately 60% relative to the baseline by 1200 Å. No secular changes have been identified at this stage.

035.048 **In–flight performance of the FOC: flat field response.**
P. Greenfield.
Workshop on the First Year of HST Observations, p. 229 – 233 (1991). – See Abstr. 012.017 for the main entry.
This paper addresses the spatial response of the FOC detectors.

035.049 **Background noise rejection in the Faint Object Spectrograph.**
E. I. Rosenblatt, E. A. Beaver, J. B. Linsky, R. W. Lyons.
Workshop on the First Year of HST Observations, p. 234 – 237 (1991). – See Abstr. 012.017 for the main entry.
The authors have modeled the background noise of the FOS "blue" MgF_2 detector to investigate the optimal strategies for reducing the noise level in spectra. Simulated background data has allowed to determine the rejection thresholds and frame times that yield the highest S/N ratio for a stellar source of any given flux level.

035.050 **Detection of binaries with the FGS: the transfer function mode data analysis.**
B. Bucciarelli, M. G. Lattanzi, L. G. Taff, O. G. Franz, L. H. Wasserman, E. Nelan.
Workshop on the First Year of HST Observations, p. 238 – 244 (1991). – See Abstr. 012.017 for the main entry.

035.051 **On orbit measurement of HST baffle rejection capability.**
P. Y. Bély, D. Daou, O. Lupie.
Workshop on the First Year of HST Observations, p. 267 – 268 (1991). – See Abstr. 012.017 for the main entry.

035.052 **Multi–spectral solar telescope array II: Soft X–ray/EUV reflectivity of the multilayer mirrors.**
T. W. Barbee Jr., J. W. Weed, R. B. Hoover, M. J. Allen, J. F. Lindblom, R. H. O'Neal, C. C. Kankelborg, C. E. DeForest, E. S. Paris, A. B. C. Walker Jr., T. D. Willis, E. Gluskin, P. Pianetta, P. C. Baker.
Opt. Eng., Vol. 30, No. 8, p. 1067 – 1075 (Aug 1991). Paper presented at SPIE Conference on Multilayer and Grazing Incidence X–Ray/EUV Optics, San Diego, CA (USA), July 1991.
The authors have developed seven compact soft X–ray/EUV (XUV) multilayer–coated and two compact FUV interference–film–coated Cassegrain and Ritchey–Chrétien telescopes for a rocket–borne observatory, the Multi–Spectral Solar Telescope Array. They report on extensive measurements of the efficiency and spectral bandpass of the XUV telescopes.

035.053 **Normal–incidence soft X–ray telescopes.**
E. Spiller, R. A. McCorkle, J. S. Wilczynski, L. Golub, G. Nystrom, P. Z. Takacs, C. Welch.
Opt. Eng., Vol. 30, No. 8, p. 1109 – 1115 (Aug 1991). Paper presented at SPIE Conference on X–Ray/EUV Optics for Astronomy, Microscopy, Polarimetry, and Projection Lithography as Part of SPIE's International Symposium on Optical and Optoelectronic Applied Science and Engineering, San Diego, CA (USA), 9 – 13 July 1990.
Photos obtained during 5 min of observation time from the flight of a 25 cm–diam normal–incidence soft X–ray (λ = 63.5Å) telescope are analyzed and the data are compared to the results expected from tests of the mirror surfaces. These tests cover a range of spatial periods from 25 cm to 1Å. The photos demonstrate a resolution close to the photon shot noise limit and a reduction in the scattering of the multilayer mirror compared to a single surface for scattering angles above 1′, corresponding to surface irregularities with spatial periods below 10 μm. The results are used to predict the possible performance of future telescopes. The authors conclude that sounding rocket observations might be able to reach a resolution around 0.1″.

035.054 **Multi–spectral solar telescope array IV: The soft X–ray and extreme ultraviolet filters.**
J. F. Lindblom, R. H. O'Neal, A. B. C. Walker Jr., F. R. Powell, T. W. Barbee Jr., R. B. Hoover, S. F. Powell.
Opt. Eng., Vol. 30, No. 8, p. 1134 – 1141 (Aug 1991). Paper presented at SPIE Conference on X–Ray/EUV Optics for Astronomy, Microscopy, Polarimetry, and Projection Lithography as Part of SPIE's International Symposium on Optical and Optoelectronic Applied Science and Engineering, San Diego, CA (USA), 9 – 13 July 1990.
The multilayer mirrors used in the normal–incidence optical systems of the Multi–Spectral Solar Telescope Array (MSSTA) are efficient reflectors for soft X–ray/extreme ultraviolet (EUV) radiation at wavelengths that satisfy the Bragg condition. However, these same mirrors are also excellent reflectors in the visible, ultraviolet, and far–ultraviolet part of the spectrum, where normal incidence reflectivities can exceed 50%. The MSSTA instrument uses various combinations of thin foil filters composed of aluminum, carbon, tellurium, potassium bromide, beryllium, molybdenum, rhodium, and phthalocyanine to achieve the desired radiation rejection characteristics. The authors discuss issues concerning the design, manufacture, and predicted performance of MSSTA filters.

035.055 **HiRES: a high–resolution stigmatic extreme ultraviolet spectroheliometer for sudies of the fine–scale structure of the solar chromosphere, transition region, and corona.**
J. G. Timothy, T. E. Berger, J. S. Morgan, A. B. C. Walker Jr., S. K. Jain, A. K. Saxena, J. C. Bhattacharyya, M. C. E. Huber, G. Tondello, G. Naletto.
Opt. Eng., Vol. 30, No. 8, p. 1142 – 1149 (Aug 1991). Peper presented at SPIE Conference on X–Ray/EUV Optics for Astronomy, Microscopy, Polarimetry, and Projection Lithography as Part of SPIE's International Symposium on Optical and Optoelectronic Applied Science and Engineering, San Diego, CA (USA), 9 – 13 July 1990.

The authors describe the design of a high-resolution stigmatic extreme-ultraviolet spectroheliometer, which consists of a 45 cm Gregory telescope coupled to a spectrometer employing a single toroidal diffraction grating in a normal-incidence Rowland circle mounting and an imaging pulse-counting multianode microchannel array (MAMA) detector system. The spectroheliometer will produce spatially resolved spectra of the chromosphere, transition region, and corona with an angular resolution of 0.4" or better, a spectral resolution $\lambda/\Delta\lambda$ of about 10^4 in first order, and a temporal resolution of the order of seconds. Because of the geometric fidelity of the MAMA detector system, the spectroheliometer will be able to determine Doppler shifts to a resolution of at least 2 mÅ at wavelengths near 600Å ($\sim$1.0 km s^{-1}), depending on the level of the accumulated signal.

035.056 **Imaging polarimeters for solar extreme ultraviolet astronomy.**
R. B. Hoover, S. Fineschi, J. M. Fontenla, A. B. C. Walker Jr.
Opt. Eng., Vol. 30, No. 8, p. 1169 – 1176 (Aug 1991). Paper presented at SPIE Conference on X-Ray/EUV Optics for Astronomy, Microscopy, Polarimetry, and Projection Lithography as Part of SPIE's International Symposium on Optical and Optoelectronic Applied Science and Engineering, San Diego, CA (USA), 9 – 13 July 1990.

The authors describe new EUV/FUV (100Å $\leqslant \lambda \leqslant$ 1500Å) polarimeter instrument concepts for solar research. These instruments are designed to observe linear polarization in EUV/FUV spectral lines originating in the outer solar atmosphere, specifically: (1) a new coronagraph/polarimeter operating at 1215.7Å (neutral hydrogen Lyman α), which could observe this line in the near solar corona and lead to the first direct measurements of both strength and direction of coronal magnetic fields and (2) a new multilayer EUV imaging polarimeter, operating at wavelengths of strong helium emission lines (e.g., 304Å, 584Å), which could observe impact polarization phenomena and provide information concerning the relative importance of thermal and nonthermal processes in solar flares. The authors discuss several instrument configurations and provide theoretical calculations and performance predictions.

035.057 **Polarimetry of extreme ultraviolet lines in solar astronomy.**
S. Fineschi, R. B. Hoover, J. M. Fontenla, A. B. C. Walker Jr.
Opt. Eng., Vol. 30, No. 8, p. 1161 – 1168 (Aug 1991). Paper presented at SPIE Conference on X-Ray/EUV Optics for Astronomy, Microscopy, Polarimetry, and Projection Lithography as Part of SPIE's International Symposium on Optical and Optoelectronic Applied Science and Engineering, San Diego, CA (USA), 9 – 13 July 1990.

Several mechanisms can induce a detectable amount of linear polarization in spectral lines emitted by the outer solar atmosphere at EUV/FUV wavelengths: (1) Polarization in FUV lines (up to 20%) can be originated by resonance scattering of radiation anisotropically illuminating the emitting atoms. (2) Impact line polarization can arise from anisotropic collisional excitation of the EUV-emitting atoms by particles (electrons, protons) with non-Maxwellian velocity distributions. The authors suggest how new technological developments in the production of ultrasmooth, low-scatter flow-polished mirror substrates and high-quality multilayer and interference film coatings can make possible some new optical instruments for the observation of these polarization effects. They describe several FUV/EUV polarimeter designs based on these polarization optics.

035.058 **SSCT for the observation of cosmic gamma rays in the GeV region.**
T. Doke, T. Kashiwagi, J. Kikuchi, H. Murakami, A. Nakamoto, K. Kasahara, T. Yuda, K. Masuda, S. Torii, Y. Muraki, V. V. Akimov, N. G. Leikov (*N. G. Lejkov*).
AIP Conf. Proc., No. 220, p. 300 – 304 (Apr 1991). Current Physics Microform No.: 9106B1434. – See Abstr. 012.020 for the main entry.

A scintillating fiber sampling calorimeter telescope (SSCT(SF)) is proposed which is to be operated on board a satellite for the observation of cosmic gamma rays ranging from 200 MeV to 200 GeV. From a preliminary test, the authors expect that the angular resolution of SSCT(SF) is, at worst, almost the same as SSCT(Si), and the energy resolution could be better than 30% at 1 GeV. Probably almost the same energy resolution is obtainable.

035.059 **Architectural design of a ground-based deep-space optical reception antenna.**
E. L. Kerr.
Opt. Eng., Vol. 30, No. 4, p. 446 – 451 (Apr 1991).

An architectural design of a ground-based antenna for receiving optical communications from deep space is presented. A channel capacity of 100 kbits/s from Saturn or 5 Mbits/s from Mars requires a 30 cm-diameter transmitter and a 10 m-diameter reception antenna. The f/0.5 primary mirror will be hexagonally segmented and will have a surface roughness tolerance of 2 μm rms to effect a substantial savings relative to the cost of an astronomical optical imaging telescope of the same diameter. A typical Saturn-to-Earth data link can reduce its source power requirement from 8.8 W to 2 W of laser output by employing a Fraunhofer filter instead of a conventional multilayer dielectric filter.

035.060 **Hybrid gamma cascade detector system.**
R. J. Wilkes.
AIP Conf. Proc., No. 220, p. 329 – 333 (Apr 1991). Current Physics Microform No.: 9106B1463. – See Abstr. 012.020 for the main entry.

Emulsion chambers provide a unique opportunity to observe gamma-ray sources in the 0.1 – 10 TeV energy range. A new hybrid detector system is proposed, employing a scintillator ribbon fiber counter system to provide particle arrival time and direction information as well as an energy estimate, supplementing the accurate particle identification and energy determination capabilities of the emulsion chamber. The counter system is intended to be self-contained and undemanding, and will work efficiently in long-duration balloon flights.

035.061 **Coronagraphs for use outside the atmosphere.**
A. V. Lenskij.
Opt. Spectrosc., Vol. 70, No. 4, p. 537 – 540 (Apr 1991). Current Physics Microform No.: 9203X0299.

A possible line of development of coronographs with external screening is considered. The diffraction properties of a diaphragm with two cusps specially arranged in an external screen are studied. On the basis of a theoretical estimate of the intensity of the light diffracted in specific directions by this type of high-efficiency external occulting system, the possibilities are assessed of doing without the Lyot system and using as the principal lens of the coronograph a fully adequate objective with the ordinary requirements for the choice of glasses, polishing, and coatings of the lenses or mirrors.

035.062 **First results from the Goddard High-Resolution Spectrograph: a demonstration of spectral resolution and experiments with deconvolution.**
G. M. Wahlgren, D. S. Leckrone, S. N. Shore, D. J. Lindler, R. L. Gilliland, D. C. Ebbets.
Astrophys. J., Lett., Vol. 377, No. 1, p. L41 – L44 (10 Aug 1991). The Second Special Hubble Space Telescope Issue.

High-quality spectra of the sharp-lined star χ Lupi were obtained with the Goddard High Resolution Spectrograph (GHRS). The authors display spectra for the various resolving powers achievable with the echelle and G160M gratings. The effect of spherical aberration upon the spectral resolution is found to be negligible when using the small science aperture. The resolution of the large science aperture spectra is degraded by less than a factor of 2. Efforts with spectral deconvolution using several techniques show that it is possible to regain much of the spectral resolution lost in the large science aperture, if high S/N ratio spectra are obtained.

035.063 **X-ray astronomy in its thirtieth year.**
K. A. Pounds.
Q. J. R. Astron. Soc., Vol. 32, No. 4, p. 355 – 366 (Dec 1991).
Both currently operational X-ray satellites, GINGA and ROSAT, have a significant British involvement, providing access to a wide variety of broad-band spectral, timing and imaging data, and building on the strong position developed in the mid-1980s. This review will illustrate the scientific returns from GINGA, with two particular examples of the relevant science capabilities of that mission, and then describe the current status of the ROSAT All Sky Survey, the first phase of which was completed a few weeks ago. Finally, the author will look forward to the prospects for JET-X, an exciting new instrument being developed for flight on the Soviet Spectrum-X mission in late 1994.

035.064 **Optical design considerations for next-generation space and lunar telescopes.**
D. Korsch.
Conference on Space Astronomical Telescopes and Instruments as Part of SPIE's Technical Symposia on Optical Engineering and Photonics in Aerospace Sensing (OE/Aerospace Sensing '91), p. 111 – 118 (1991). – See Abstr. 012.030 for the main entry.
The optical requirements for future large space or lunar telescopes and their influence upon the fundamental system parameters are discussed. Two telescope concepts, a four-reflection three-mirror telescope and a spherical-primary four-mirror telescope are introduced.

035.065 **Foundation, excavation and radiation shielding concepts for a 16-m large lunar telescope.**
K. M. Chua, S. W. Johnson.
Conference on Space Astronomical Telescopes and Instruments as Part of SPIE's Technical Symposia on Optical Engineering and Photonics in Aerospace Sensing (OE/Aerospace Sensing '91), p. 119 – 134 (1991). – See Abstr. 012.030 for the main entry.
Fundamental concepts of engineering activities on the Moon and how they can be applied to the establishment of a 16 m large lunar telescope [LLT] are discussed. The LLT system will include a regolith covered housing, the size of a small room, which will contain sensitive electronic equipment including charge coupled devices which need protection from cosmic radiation effects. A brief discussion is made on radiation, radiation transport and radiation effects on electronics and on humans. Radiation protection techniques and the different emplacement schemes for the LLT instrument housing for radiation protection are suggested. A structural concept of an early lunar based telescope is also presented.

035.066 **Construction of the 16 m large lunar telescope (LLT).**
H. A. Omar.
Conference on Space Astronomical Telescopes and Instruments as Part of SPIE's Technical Symposia on Optical Engineering and Photonics in Aerospace Sensing (OE/Aerospace Sensing '91), p. 135 – 146 (1991). – See Abstr. 012.030 for the main entry.
NASA is proposing building a 16 m telescope on the moon surface for extra terrestrial observations. This paper summarizes the findings of a study to identify different design options for a podestal to support the primary structure which supports the primary and secondary mirrors. At this early stage of the telescope's development, the main objectives of this study were to identify problems associated with constructing the telescope on the lunar surface, suggest solutions to those problems using technologies available today and finally to identify new technologies and recommend new studies that will enhance our understanding of the lunar environment and enable us to design more efficient systems. This paper addresses these objectives in detail.

035.067 **The 4 meter lunar engineering telescope.**
K. Peacock, J. A. Giannin, C. C. Kilgus, P. Y. Bely, B. S. May, S. A. Cooper, G. H. Schlimm, C. Sounder, K. Ormond, E. Cheek.
Conference on Space Astronomical Telescopes and Instruments as Part of SPIE's Technical Symposia on Optical Engineering and Photonics in Aerospace Sensing (OE/Aerospace Sensing '91), p. 147 – 159 (1991). – See Abstr. 012.030 for the main entry.
The 16 m diffraction limited lunar telescope incorporates a primary miror with 312 one-meter segments; 3 nanometer active optics surface control with laser metrology and hexapod positioners; a space frame structure with one-millimeter stability; and a hexapos mount for pointing. This paper presents a 4.33 m engineering telescope concept. The primary mirror, made up of 18 one-meter hexagonal segments, is sized to provide interesting science as well as engineering data. The optics are configured as a Ritchey-Chretien with a Coude relay to the focal plane beneath the surface. An active optics processor and piezoelectric actuators operate to maintain the end-to-end optical configuration established by wave front sensing using a guide star. The mirror segments, consisting of a one-centimeter thick faceplate on 30-cm deep ribs, maintain the surface figure to a few nanometers under lunar gravity and thermal environment.

035.068 **System design for lunar-based optical and submillimeter interferometers.**
R. Gershman, M. J. Mahoney, M. D. Rayman, M. Shao, G. C. Snyder.
Conference on Space Astronomical Telescopes and Instruments as Part of SPIE's Technical Symposia on Optical Engineering and Photonics in Aerospace Sensing (OE/Aerospace Sensing '91), p. 160 – 167 (1991). – See Abstr. 012.030 for the main entry.
The Space Exploration Initiative will provide opportunities to explore the further reaches of the universe as well as the moon and Mars. Astrophysical observatories located on the moon can provide combinations of resolutions, sensitives, and spectral coverage from the ultraviolet to the submillimeter that may never be available to Earth-based observatories and may never be practical in orbiting systems. The Jet Propulsion Laboratory has been studying the design and performance of lunar optical and submillimeter interferometer systems. Both interferometers as presently conceived, will consist of 12 collecting apertures arranged in a wye. The current status of the system aspect of these activities are described, including launch packaging and deployment strategies as well as design choices to contend with the lunar environment.

035.069 **Optical system design for a Lunar Optical Interferometer.**
M. M. Colavita, M. Shao, B. E. Hines, B. M. Levine, R. Gershman.
Conference on Space Astronomical Telescopes and Instruments as Part of SPIE's Technical Symposia on Optical Engineering and Photonics in Aerospace Sensing (OE/Aerospace Sensing '91), p. 168 – 181 (1991). – See Abstr. 012.030 for the main entry.
The moon offers particular advantages for interferometry, including a vacuum environment, a large stable base on which to assemble multi-kilometer baselines, and a cold nighttime temperature to allow for passive cooling of optics for high I.R sensitivity. A baseline design for a Lunar Optical Interferometer which exploits these features is presented. The instrument operates in the visible to mid-I.R region, and is designed for both astrometry and synthesis imaging. The design uses a Y-shaped array of 12 siderostats, with maximum arm lengths of ~ 1 km. With 1 m apertures, the instrument will have a point-source imaging sensitivity of ~ 29 mag; with the laser metrology system, astrometry at the microarcsecond level will be possible.

035.070 **In-flight performance of the Faint Object Camera of the Hubble Space Telescope.**
P. Greenfield, D. Baxter, P. Hodge, R. Jedrzejewski, W. B. Sparks, F. Macchetto, F. Paresce, R. Hook, P. Jakobsen, R. Laurance, A. Nota, N. Towers.
Conference on Space Astronomical Telescopes and Instruments as Part of SPIE's Technical Symposia on Optical Engineering and Photonics in Aerospace Sensing (OE/Aerospace Sensing '91), p. 16 – 39 (1991). – See Abstr. 012.030 for the main entry.
This paper describes the Faint Object Camera – launched with the Hubble Space Telescope nearly a year ago – and its performance to date. In particular, the detector efficiency, the spatial uniformity of response, distortion characteristics, detector and sky background, detector linearity, spectrograph, and operation are discussed. The effect of the severe spherical aberration of the telescope's primary mirror on the camera's point spread function is reviewed, as well as the impact it has on the general performance of the cameras. Finally, the scientific implications of the performance and the spherical aberration are outlined.

035.071 **SALSA: a Synthesis Array for Lunar Submillimeter Astronomy.**
M. J. Mahoney, K. A. Marsh.
Conference on Space Astronomical Telescopes and Instruments as Part of SPIE's Technical Symposia on Optical Engineering and Photonics in Aerospace Sensing (OE/Aerospace Sensing '91), p. 182 – 193 (1991). – See Abstr. 012.030 for the main entry.
A conceptual design is described for a lunar submillimeter wavelength interferometer called SALSA, a Synthesis Array for Lunar Submillimeter Astronomy. Its design is greatly simplified over conventional submillimeter wavelength arrays because it takes advantage of a beam waveguide to relay signals to a centrally-located receiver system. The array is optimized to synthesize images at wavelengths between 60 and 300 μm, and it has an angular resolution of 10 mas at 60 μm. The baseline design for SALSA consists for twelve, 3.5 m diameter antennas arranged in a Y-shaped configuration consisting of three 0.5 km long arms, each with four antennas distributed according to a power-law function of distance.

035.072 **Lunar dust: implications for astronomical observatories.**
S. W. Johnson, K. M. Chua, J. O. Burns, F. A. Slane.
Conference on Space Astronomical Telescopes and Instruments as Part of SPIE's Technical Symposia on Optical Engineering and Photonics in Aerospace Sensing (OE/Aerospace Sensing '91), p. 194 – 207 (1991). – See Abstr. 012.030 for the main entry.
The pervasiveness of the lunar fine-grained particulates, or dust, was alluded to many times on subsequent trips to the Moon. The dust poses one of the unanswered riddles for lunar observatory design, construction, and operations. This paper discusses the properties, origins, and previous operational experiences with the dust and emphasizes the implications for doing World-class astronomy on the Moon. The paper suggests what the mechanisms are that may govern the behavior of the fine particles; establishes working hypotheses for mitigating the dust hazard; and outlines future experiments, both on the Moon and in terrestrial simulations, that will assist in establishing effective and suitable means of limiting deleterious effects of dust on observatory operations.

035.073 **Effects of the lunar environment on optical telescopes and instruments.**
C. L. Johnson, K. L. Dietz.
Conference on Space Astronomical Telescopes and Instruments as Part of SPIE's Technical Symposia on Optical Engineering and Photonics in Aerospace Sensing (OE/Aerospace Sensing '91), p. 208 – 218 (1991). – See Abstr. 012.030 for the main entry.
Environmental aspects of a large lunar-based ultraviolet/visible/infrared telescope are discussed. The environmental concerns addressed include: galactic cosmic ray effects on telescope electronics, lunar dust obscuration of and damage to optical surfaces, and micrometer cratering of the optics and support structure.

035.074 **Lunar liquid mirror telescopes.**
E. F. Borra, R. Content.
Conference on Space Astronomical Telescopes and Instruments as Part of SPIE's Technical Symposia on Optical Engineering and Photonics in Aerospace Sensing (OE/Aerospace Sensing '91), p. 219 – 227 (1991). – See Abstr. 012.030 for the main entry.
The authors give a brief summary of the advantages of using a liquid mirror in a lunar telescope. They present interferometric measurements of a 1.5 m diameter liquid mirror which show that liquid mirrors can have the very high optical quality one needs in a lunar telescope. The authors discuss some of the liquid metals that could be used.

035.075 **New way of making a lunar telescope.**
P. C. Chen.
Conference on Space Astronomical Telescopes and Instruments as Part of SPIE's Technical Symposia on Optical Engineering and Photonics in Aerospace Sensing (OE/Aerospace Sensing '91), p. 228 – 233 (1991). – See Abstr. 012.030 for the main entry.
The author describes a novel concept for making a lunar telescope either in the form of an optical interferometer or as a large aperture segmented array. Individual mirror elements can be made in situ by the replication process. An arrangement of magnets or electromagnets and high temperature superconductors are used to move and point the mirror elements. The telescope can be fabricated and operated by preprogramming supplemented by remote control from earth. The advantages of this approach is that it employs proven technology, requires very little transportation of materials, little or no excavation, and is within the playload capability of current launch vehicles. A prototype is being made to demonstrate feasibility.

035.076 **Structural design considerations for the Space Infrared Telescope Facility.**
P. D. MacNeal, M. C. Lou.
Conference on Space Astronomical Telescopes and Instruments as Part of SPIE's Technical Symposia on Optical Engineering and Photonics in Aerospace Sensing (OE/Aerospace Sensing '91), p. 236 – 254 (1991). – See Abstr. 012.030 for the main entry.
To achieve the mission goals of the Space Infrared Telescope Facility (SIRTF), many technical challenges will need to be overcome. Structural challenges will focus on: stabilization of the primary mirror; minimizing heat leaks to the spacecraft while maintaining structural stiffness; developing a lightweight vapor cooled shield system; and trading off the relationship between launch capability, structural weight, and mission life. To assess design feasibility and to identify parameters that might impose constraints on performance, such as frequencies of vibration, structural concepts for both the telescope and spacecraft were developed and evaluated. Trade studies of key design features were accomplished utilizing finite-element modeling and analysis.

035.077 **The Infrared Space Observatory optical subsystem.**
C. Singer, J. A. Massoni, B. Moßbacher, C. Cinotti.
Conference on Space Astronomical Telescopes and Instruments as Part of SPIE's Technical Symposia on Optical Engineering and Photonics in Aerospace Sensing (OE/Aerospace Sensing '91), p. 255 – 264 (1991). – See Abstr. 012.030 for the main entry.
The Infrared Space Observatory (ISO), a project of the European space agency (ESA), is an astronomical satellite operating at wavelengths from 2.5 to 200 μm. It will be launched in 1993. The ISO optical subsystem is a cryogenically cooled telescope with its baffling system (main baffle and sunshade). The telescope, a 60 cm Ritchey-Chretien type, focuses the mean to the four scientific instruments located in its focal plane. The

extremely low temperature, 1.8K, is provided by a cryostat, filled with superfluid He.

035.078 **The manufacture of ISO mirrors.**
E. Ruch.
Conference on Space Astronomical Telescopes and Instruments as Part of SPIE's Technical Symposia on Optical Engineering and Photonics in Aerospace Sensing (OE/Aerospace Sensing '91), p. 265 – 278 (1991). – See Abstr. 012.030 for the main entry.

The Infrared Space Observatory (ISO) telescope is designed to operate at cryogenic temperatures of about 4K. The telescope consists of three mirrors made of fused silica substrate working in a Ritchey–Chretien configuration: – the primary mirror is aspherical, has an aperture diameter of 640 mm, and a fast F number of F/1.6. The manufacturing steps of these mirrors are described. The wavefront error measurements have shown diffraction limited performance for each mirror as well as for the whole telescope.

035.079 **Structural design of the Large Deployable Reflector (LDR).**
C. M. Satter, M. C. Lou.
Conference on Space Astronomical Telescopes and Instruments as Part of SPIE's Technical Symposia on Optical Engineering and Photonics in Aerospace Sensing (OE/Aerospace Sensing '91), p. 279 – 300 (1991). – See Abstr. 012.030 for the main entry.

The Large Deployable Reflector (LDR) is one of the major astronomical missions proposed by NASA for the early 2000's. This orbiting observatory will have a 20 m aperture segmented primary reflector for infrared/submillimeter observation in the 30 to 1000 μm wavelength range. Development of the LDR poses significant challenges to technologies in many areas, including those related to space telescope structural design. This paper describes the requirements for the LDR structural system, considerations in the design of major structural elements, and system configuration options. Details of the modeling and analysis effort, along with significant analysis results, are presented. Design modifications to enhance system disturbance rejection characteristics and reduce predicted responses to the anticipated disturbances are discussed.

035.080 **Fixing the Hubble Space Telescope.**
J. H. Crocker.
Conference on Space Astronomical Telescopes and Instruments as Part of SPIE's Technical Symposia on Optical Engineering and Photonics in Aerospace Sensing (OE/Aerospace Sensing '91), p. 2 – 8 (1991). – See Abstr. 012.030 for the main entry.

The flawed primary mirror of the Hubble Space Telescope was a bitter disappointment to the scientific community. Fortunately, methods to correct the problem have been determined. This paper discusses one such method now under intense study by NASA. The proposal is to install corrective optics in front of the existing science instruments using a device called COSTAR (Corrective Optics Space Telescope Axial Replacement). It is expected that COSTAR can essentially recover the expected capability of HST.

035.081 **Advanced X–ray Astrophysics Facility (AXAF) science instruments.**
C. E. Winkler, C. C. Dailey, N. P. Cumings.
Conference on Space Astronomical Telescopes and Instruments as Part of SPIE's Technical Symposia on Optical Engineering and Photonics in Aerospace Sensing (OE/Aerospace Sensing '91), p. 301 – 313 (1991). – See Abstr. 012.030 for the main entry.

The AXAF–science instruments (SI's) have undergone several changes since they were selected in 1985. AXAF will feature a grazing incidence telescope at one end observatory and accommodations for up to four SI's at the opposite end, each occupying a 90–degree sector at the telescope focus. The focal plane SI's selected for design and development include the High Resolution Camera (HRC), the AXAF CCD Imaging Spectrometer (ACIS), and the X–Ray Spectrometer (XRS). A possible fourth focal plane SI is the Bragg Crystal Spectrometer (BCS). This paper summarizes the overall AXAF program, giving particular emphasis to the SI's.

035.082 **High resolution imaging with multilayer soft X–ray, EUV and FUV Telescopes of modest aperture and cost.**
A. B. C. Walker Jr., J. F. Lindblom, J. G. Timothy,
R. B. Hoover, T. W. Barbee Jr., P. C. Baker, F. R. Powell.
Conference on Space Astronomical Telescopes and Instruments as Part of SPIE's Technical Symposia on Optical Engineering and Photonics in Aerospace Sensing (OE/Aerospace Sensing '91), p. 320 – 333 (1991). – See Abstr. 012.030 for the main entry.

The development of multilayer reflective coatings now permits soft X–ray, EUV and FUV radiation ($\lambda\lambda \sim 40$ Å – 2000 Å) to be efficiently imaged by conventional normal incidence optical configurations. Telescopes with quite modest apertures ($\sim$0.1 – 0.5 m), can, in principle, achieve images with resolutions ($\sim$0.1″ or better) which would require apertures of 1.25 m or more at visible wavelengths. The authors review the progress which has been made in developing compact telescopes for ultra–high resolution imaging of the sun at soft X–ray, EUV and FUV wavelengths, including laboratory test results and astronomical images obtained with rocket–borne multilayer telescopes. They discuss the factor which limit the resolution which has been achieved so far, and the problems which must be addressed to attain, and surpass the 0.1″ level. The authors also describe the application of these technologies to the development of solar telescopes for future space missions.

035.083 **Towards an ESA strategy for optical interferometry outside the atmosphere. Recommendations by the ESA "Space Interferometry Study Team" (SIST).**
J. E. Noordam.
Conference on Space Astronomical Telescopes and Instruments as Part of SPIE's Technical Symposia on Optical Engineering and Photonics in Aerospace Sensing (OE/Aerospace Sensing '91), p. 344 – 346 (1991). – See Abstr. 012.030 for the main entry.

A study group of the European Space Agency has recommended the launch of a 100 m class optical Space interferometer around the year 2005. The conclusions, recommendations and target design parameters for such an interferometer are summarised in this paper. ESA has now asked a new study group to consider the desirability of an interferometer on the Moon, covering the entire wavelength range from ultra–violet to sub–mm, with a Space interferometer as a possible intermediate step.

035.084 **Hubble Extra Solar Planet Interferometer.**
M. Shao.
Conference on Space Astronomical Telescopes and Instruments as Part of SPIE's Technical Symposia on Optical Engineering and Photonics in Aerospace Sensing (OE/Aerospace Sensing '91), p. 347 – 356 (1991). – See Abstr. 012.030 for the main entry.

This paper describes a proposed third–generation Hubble instrument for extra–solar planet detection, the Hubble Extra–Solar Planet Interferometer (HESPI). This instrument would be able to achieve starlight cancellation at the 10^6 to 10^8 level, given a stellar wavefront with phase errors comparable to the present Hubble telescope wavefront. At 10^6 starlight cancellation, HESPI would be able to detect a Jupiter–like planet next to a star at a distance of $\sim$10 pc, for which there are $\sim$400 candidate stars. This paper describes a novel approach for starlight suppression, using a combination of active control and single–

mode spatial filters, to achieve starlight suppression far below the classical limit set by scattering due to microsurface imperfections.

035.085 **A semi active telescope for the french PRONAOS submillimetric mission.**
M. Duran, P. Luquet, F. Buisson, B. Cousin.
Conference on Space Astronomical Telescopes and Instruments as Part of SPIE's Technical Symposia on Optical Engineering and Photonics in Aerospace Sensing (OE/Aerospace Sensing '91), p. 357 - 376 (1991). - See Abstr. 012.030 for the main entry.

The PRONAOS balloon borne gondola to be flown in 1992 will be dedicated to astronomical observation in the submillimetric wavelength domain (100 μm, 1 mm). A surface accuracy better than 10 μm (RMS) associated to an areal mass lower than 20 kg per m^2 have led to select for its 2 m primary mirror: a 6 actively controlled petals configuration. Each of them is supported by 3 linear actuators providing 3 degrees of freedom: the piston and the two tilts. A set of displacement sensors allows to derive through a control algorithm the actuators commands required to maintain the mirror geometry within its optimal configuration whatever the environmental conditions are. After a description of the basic design of the different units constituting the telescope, emphasis is be given to the alignment technique and to the results of the optical performance measurements.

035.086 **ORFEUS alignment concept.**
R. Graue, D. Kampf, H. Rippel, G. Witte.
Conference on Space Astronomical Telescopes and Instruments as Part of SPIE's Technical Symposia on Optical Engineering and Photonics in Aerospace Sensing (OE/Aerospace Sensing '91), p. 377 - 385 (1991). - See Abstr. 012.030 for the main entry.

ORFEUS (Orbital and Retrievable Far and Extreme Ultraviolet Spectrometer) is a short term scientific space payload, scheduled to be launched by the STS in Jan. 1993. ORFEUS comprises two alternatively operating spectrometers (Echelle and Rowland) implemented in a CFC telescope with a 4 m tube length and an aperture of 1000 mm. The spectrometers have to be integrated into the horizontal telescope aligned within a special antigravity device to reduce optical surface deformations and to ensure the optical performance of the primary mirror. The alignment of all optical components will be performed in the visible spectral range. The Echelle detector shall be finally adjusted with UV light sources.

035.087 **Stellar photometry with the PC on HST.**
P. Seitzer.
Meeting on Precision Photometry: Astrophysics of the Galaxy, p. 261 - 264 (1991). - See Abstr. 012.027 for the main entry.

035.088 **Satellite borne laser for adaptive optics reference.**
A. H. Greenaway.
Conference on Space Astronomical Telescopes and Instruments as Part of SPIE's Technical Symposia on Optical Engineering and Photonics in Aerospace Sensing (OE/Aerospace Sensing '91), p. 386 - 393 (1991). - See Abstr. 012.030 for the main entry.

Low power (2mW) lasers mounted on a small satellite in a highly eccentric orbit can provide a bright and spectrally well-defined reference source for calibration of ground-based adaptive optic systems. Because the reference is spectrally well-defined it can be efficiently filtered in broad-band imaging applications and yet can provide a very bright ($\leqslant +5$ mag) reference source for wavefront detectors when imaging faint sources. Dependent on the size of the atmospheric isoplanatic patch, the satellite reference may be useful for calibrating observations of selected objects for periods in excess of 1 hour, leading to limiting magnitudes for detection of up to +30. The area of sky for which the reference is valid is restricted (order 1 sq degree of sky per telescope per year). The reference is valid for phasing aperture synthesis telescope arrays of kilometric scale.

035.089 **InSb linear multiplexed FPAs for the CRAF/Cassini visible and infrared mapping spectrometer.**
C. Niblack, M. Blessinger, J. Forsthoefel, C. Staller, H. Sobel.
Conference on Space Astronomical Telescopes and Instruments as Part of SPIE's Technical Symposia on Optical Engineering and Photonics in Aerospace Sensing (OE/Aerospace Sensing '91), p. 403 - 413 (1991). - See Abstr. 012.030 for the main entry.

This paper provides a review of a custom multiplexer circuit designed for use with a 256 element InSb linear array in the CRAF/Cassini VIMS instruments. The requirements, operating noise model and test results from a prototype 1 × 64 array are discussed. The infrared focal plane array (FPA) preliminary design and the impact of the new multiplexer on the instruments' predicted performance are be discussed. Emphasis is placed on the multiplexer that was designed for the CRAF/Cassini missions.

035.090 **Initial performance of the High Speed Photometer.**
E. Richards, J. Percival, M. Nelson, E. Hatter, J. Fitch, R. White.
Conference on Space Astronomical Telescopes and Instruments as Part of SPIE's Technical Symposia on Optical Engineering and Photonics in Aerospace Sensing (OE/Aerospace Sensing '91), p. 40 - 48 (1991). - See Abstr. 012.030 for the main entry.

The Hubble Space Telescope High Speed Photometer has four image dissector tubes, two with UV sensitive photocathodes, two sensitive to the near UV and to visual light, and a single red sensitive photomultiplier tube. The HSP is capable of photometric measurements from 1200 Å to 7500 with time resolution of 11 microseconds and has no moving parts. An initial analysis of the on-orbit engineering performance of the HSP is presented with changes in operating procedures resulting from the primary mirror spherical aberration and experience gained during the verification period.

035.091 **Removal of adsorbed gases with CO_2 snow.**
R. R. Zito.
Conference on Space Astronomical Telescopes and Instruments as Part of SPIE's Technical Symposia on Optical Engineering and Photonics in Aerospace Sensing (OE/Aerospace Sensing '91), p. 427 - 433 (1991). - See Abstr. 012.030 for the main entry.

During the outgassing of orbiting astronomical observatories, the condensation of molecular species on optical surfaces can create difficulties for astronomers. The problem is particularly severe in ultraviolet astronomy where the adsorption of only a few atomic layers of some substances can be very damaging. In this paper the removal of adsorbed atomic layers using carbon dioxide snow is discussed.

035.092 **Station keeping strategy for multiple spacecraft interferometry.**
A. B. DeCou.
Conference on Space Astronomical Telescopes and Instruments as Part of SPIE's Technical Symposia on Optical Engineering and Photonics in Aerospace Sensing (OE/Aerospace Sensing '91), p. 440 - 451 (1991). - See Abstr. 012.030 for the main entry.

The feasibility of multiple spacecraft station keeping for submilimeter and optical interferometry is examined. A condition for interferometry is that two or more spacecraft must control their relative positions with better than 1 mm accuracy indefinitely in both radial and transverse directions although separated by as much as 1 Km in LEO and 100 Km in GEO. They must also maneuver through a useful area of the U-V plane of an arbitrary astronomical source. The problem is first outlined and a solution which utilizes gravity gradient to do most of the work and ion thrusters for additional maneuvering is proposed. An inertial position and attitude control strategy is suggested which utilizes existing or soon to be available sensors and actuators.

035.093 **Surface control techniques for the segmented primary mirror in the large lunar telescope.**
A. D. Gleckler, K. P. Pflibsen, B. L. Ulich, D. D. Smith.
Conference on Space Astronomical Telescopes and Instruments as Part of SPIE's Technical Symposia on Optical Engineering and Photonics in Aerospace Sensing (OE/Aerospace Sensing '91), p. 454 - 471 (1991). - See Abstr. 012.030 for the main entry.

The large lunar telescope is a proposed moon-based telescope which incorporates a sixteen-meter segmented primary mirror. An error budget is developed for the active control system of the primary mirror. A control methodology for the primary mirror is then described. A trade study is conducted in which some types of tilt sensors are examined. In addition, two independent methods of quantitatively verifying the performance of the telescope using a phase retrieval algorithm or an image sharpening technique, both of which are based on the quality of a stellar image, are presented.

035.094 **Faint Object Spectrograph (FOS) early performance.**
R. Harms, J. Fitch.
Conference on Space Astronomical Telescopes and Instruments as Part of SPIE's Technical Symposia on Optical Engineering and Photonics in Aerospace Sensing (OE/Aerospace Sensing '91), p. 49 - 65 (1991). - See Abstr. 012.030 for the main entry.

The Faint Object Spectrograph (FOS) instrument aboard the Hubble Space Telescope (HST) provides astronomers a moderate spectral resolution, low light level, analytical instrument sensitive throughout the wavelength region from below 120 nm to beyond 800 nm. The on-orbit performance of the HST + FOS instrument is described and illustrated with examples of initial scientific results. The effects of the spherical aberration upon isolated point sources and in complex fields such as the nuclei of galaxies are analyzed. It may be possible to install corrective optics into the HST which will eliminate the spherical aberration for the FOS (and some of the other instruments). A brief description of the more promising ideas and calculations of the expected improvements in performance are provided.

035.095 **Comparison of the angular resolution limit and SNR of the Space Hubble Telescope and the large ground-based telescopes.**
D. Souilhac, D. Billerey.
Conference on Space Astronomical Telescopes and Instruments as Part of SPIE's Technical Symposia on Optical Engineering and Photonics in Aerospace Sensing (OE/Aerospace Sensing '91), p. 503 - 526 (1991). - See Abstr. 012.030 for the main entry.

This paper compares and discusses the attempts to overcome the "seeing" resolution limit imposed by spherical aberration for the Hubble Space Telescope and by atmospheric turbulences for the case of ground-based telescopes. From recently reposed experimental data it is shown that wavefront reconstruction algorithms are capable of restoring nearly fully the Hubble's spherical aberration of the primary mirror. Also from previously reported results of ground-based telescopes which utilize the powerful spectral imaging technique at high and low light levels, the authors compare for different object magnitudes the two types of telescope resolutions "seeing" resolution, SNR and point source resolution.

035.096 **Hubble Space Telescope optics – problems and solutions.**
C. J. Burrows.
Conference on Space Astronomical Telescopes and Instruments as Part of SPIE's Technical Symposia on Optical Engineering and Photonics in Aerospace Sensing (OE/Aerospace Sensing '91), p. 528 - 533 (1991). - See Abstr. 012.030 for the main entry.

The Hubble Space Telescope suffers from spherical aberration caused by a spacing error in the null lens used to fabricate and test the primary mirror. The assumed conic constant of the primary mirror for all the corrective optics, -1.0139(5), is consistent with measurements by four major independent methods. Aligning the new optics will be very demanding, because of the large slope of the wavefront to be corrected. If the images are to be diffraction limited, the pupil at the corrective element must be aligned to better than one percent of its diameter. Some other residual effects of the spherical aberration will remain after installation of the corrective optics, primarily in the pointing and collimation of the telescope.

035.097 **Scientific results from the Hubble Space Telescope Fine Guidance Sensors.**
L. G. Taff.
Conference on Space Astronomical Telescopes and Instruments as Part of SPIE's Technical Symposia on Optical Engineering and Photonics in Aerospace Sensing (OE/Aerospace Sensing '91), p. 66 - 77 (1991). - See Abstr. 012.030 for the main entry.

The Fine Guidance Sensors of the Hubble Space Telescope have two roles to play. They are the ultimate pointing and control instruments onboard the spacecraft and they are the primary astrometric instruments of the observatory. The Fine Guidance Sensors have already serendipitiously discovered their first binary star from the Hubble Space Telescope Guide Star Catalog, have been exercised in most of their observing modes, and are able to fulfill their pointing and control functions up to their original specifications (in the absence of external influences). In addition, the imperfections in the primary mirror have minimally affected the performance of the Fine Guidance Sensors (especially when compared to the other scientific instruments). A up-to-date summary of FGS engineering and science is presented.

035.098 **Reconstruction of the Hubble Space Telescope mirror figure from out-of-focus stellar images.**
C. Roddier, F. Roddier.
Conference on Space Astronomical Telescopes and Instruments as Part of SPIE's Technical Symposia on Optical Engineering and Photonics in Aerospace Sensing (OE/Aerospace Sensing '91), p. 78 - 84 (1991). - See Abstr. 012.030 for the main entry.

From the illumination recorded in two defocused stellar images it is possible to reconstruct both the amplitude and the phase of the incoming wave front viewed from the focal plane, providing a powerful diagnostic tool for telescope optics. The authors present results obtained from defocused stellar images taken in flight by the Hubble Space Telescope.

035.099 **The Next Generation Space Telescope: a large UV-IR successor to HST.**
G. Illingworth.
Conference on Space Astronomical Telescopes and Instruments as Part of SPIE's Technical Symposia on Optical Engineering and Photonics in Aerospace Sensing (OE/Aerospace Sensing '91), p. 86 - 97 (1991). - See Abstr. 012.030 for the main entry.

The Next Generation Space Telescope (NGST) is a uniquely powerful scientific tool. Many fundamental astronomical problems can be tackled only with filled-aperture telescopes that combine high spatial resolution with large light-gathering capability. An 8 m class passively-cooled telescope in high earth orbit would have unprecedented power for problems as diverse as planet searches around nearby stars to the way in which galaxies formed in the young universe. It will build upon the discoveries and astronomical understanding of many decades of research with astronomical observatories, and is the natural successor to HST and the new generation of large 10 m class ground-based telescopes.

035.100 **System concepts for a large UV/optical/IR telescope on the Moon.**
M. E. Nein, B. Davis.
Conference on Space Astronomical Telescopes and Instruments as Part of SPIE's Technical Symposia on Optical Engineering and Photonics in Aerospace Sensing (OE/Aerospace Sensing '91), p. 98 - 110 (1991). - See Abstr. 012.030 for the main entry.

In the next century, return to the Moon for scientific exploration and as a waypoint for travel to Mars will become a reality. Because of the advantages of astronomical observations from the Moon, scientists and engineers are now developing concepts for lunar telescope which can be constructed in conjunction with the buildup of a lunar base. To assess the systems and technological requirements, the NASA has conducted concept studies of a 16 m aperture large lunar telescope (LLT) and a 4 m aperture precursor telescope, both operating in the ultraviolet/visible/infrared spectral region.

035.101 **In-flight performance of the Goddard High Resolution Spectrograph of the Hubble Space Telescope.**
J. Troeltzsch, D. Ebbets, H. Garner, A. Tuffli, R. Breyer, J. Kinsey, C. Peck, D. Lindler, J. Feggans.
Conference on Space Astronomical Telescopes and Instruments as Part of SPIE's Technical Symposia on Optical Engineering and Photonics in Aerospace Sensing (OE/Aerospace Sensing '91), p. 9 – 15 (1991). – See Abstr. 012.030 for the main entry.

The Goddard High Resolution Spectrograph (GHRS) has completed Orbital Verification and is well into the Science Verification phase of its mission. Tests using calibration lamps and standard UV stars have confirmed the instrument sensitivity and spectral resolving powers $\lambda/\Delta\lambda$ = 2000, 20000, and 90000. Observations have been successfully performed using both the 2.0 × 2.0″ aperture, and the smaller 0.25 × 0.25″ aperture. The extended point spread function caused by the spherical aberration of the HST primary mirror has been well measured, and observing methods to deal with it have been developed. Numerous science assessment observations of interesting astronomical targets have been completed, and indicate the extreme usefulness of the GHRS to the scientific community.

035.102 **The Hubble Space Telescope.**
A. Boggess.
IAU Colloquium No. 123: Observatories in Earth orbit and beyond, p. 3 – 7 (1990). – See Abstr. 012.038 for the main entry.

The Hubble Space Telescope was launched from the Kennedy Space Center on April 24, 1990. Its initial check-out indicates that all sub-systems of the satellite are working very well, with two key exceptions: The line-of-sight pointing is subject to occasional jitter apparently induced by thermal stresses in the solar arrays; this is expected to be overcome. The telescope mirrors are found to contain approximately 0.5 wave rms of spherical aberration which cannot be overcome by any controls on board the satellite. This defect will limit the scientific performance of the telescope in the short run. However, the aberration can be fully corrected in the optical designs of future replacement instruments, and the delivery schedules of these instruments are being accelerated.

035.103 **Early results from the Cosmic Background Explorer (COBE).**
J. C. Mather, M. G. Hauser, C. L. Bennett, N. W. Boggess, E. S. Cheng, R. E. Eplee Jr., H. T. Freudenreich, R. B. Isaacman, T. Kelsall, C. M. Lisse, S. H. Moseley Jr., R. A. Shafer, R. F. Silverberg, W. J. Spiesman, G. N. Toller, J. L. Weiland, S. Gulkis, M. Janssen, P. M. Lubin, S. S. Meyer, R. Weiss, T. L. Murdock, G. F. Smoot, D. T. Wilkinson, E. L. Wright.
IAU Colloquium No. 123: Observatories in Earth orbit and beyond, p. 9 – 18 (1990). Also published as report ESA-SP 314, p. 25 –31 (Dec 1990) and in Adv. Space Res. (1991) Vol. 11, No. 2, p. 181 –191 (1991). – See Abstr. 012.038 for the main entry.

The Cosmic Background Explorer, launched November 18, 1989, has nearly completed its first full mapping of the sky with all three of its instruments: a Far Infrared Absolute-Spectrophotometer (FIRAS) covering 0.1 to 10 mm, a set of Differential Microwave Radiometers (DMR) operating at 3.3, 5.7, and 9.6 mm, and a Diffuse Infrared Background Experiment (DIRBE) spanning 1 to 300 μm in ten bands. A preliminary map of the sky derived from DIRBE data is presented. There are no significant anisotropies in the microwave sky detected. At shorter wavelengths, the sky spectrum and anisotropies are dominated by emission from "local" sources of emission within our Galaxy and Solar System. The spacecraft, instrument designs, and data reduction methods are described.

035.104 **The status of the DIRBE instrument on the COBE.**
M. G. Hauser, J. C. Mather, T. Kelsall, H. Moseley, R. Silverberg, T. L. Murdock, G. Smoot, R. Weiss, E. L. Wright.
IAU Colloquium No. 123: Observatories in Earth orbit and beyond, p. 19 (1990). Abstract. – See Abstr. 012.038 for the main entry.

035.105 **The automatic spacecraft *GRANAT*.**
R. Sunyaev.
IAU Colloquium No. 123: Observatories in Earth orbit and beyond, p. 21 – 25 (1990). – See Abstr. 012.038 for the main entry.

The spacecraft *GRANAT* with its scientific instruments package is designed to conduct detailed studies of compact and diffuse space sources in X-Rays and soft gamma rays. The main scientific tasks to be solved by the scientific instruments package are: Producing images of the parts of the sky and localizing sources of space radiation in the studied parts of the sky with high accuracy; Spectral studies in time of point sources; Measuring linear polarization of X-ray sources; Patrol type monitoring of the sky in order to discover and study variable sources of radiation (bursts).

035.106 **The Hipparcos mission: will it be a scientific success?**
M. A. C. Perryman.
IAU Colloquium No. 123: Observatories in Earth orbit and beyond, p. 27 – 33 (1990). – See Abstr. 012.038 for the main entry.

The Hipparcos astrometry satellite was launched on 8 August 1989, and after spacecraft and payload commissioning, commenced the routine data acquisition phase on 26 November 1989. Having failed to reach its planned geostationary orbit, major revisions in the mission operations were made. The final astrometric accuracies attainable by the Hipparcos mission will be influenced by the spacecraft and payload performances on the one hand, and by fraction of useful data and mission lifetime on the other. A measurement duration of at least 18 months is mandatory if the astrometric parameters (parallaxes and proper motions) are to be decoupled through the data reductions procedures. It is shown that the present indications of the satellite performances, and the significant progress already made in the data reductions, indicate that the difficulties of the "revised" Hipparcos mission have been largely overcome, and that the target accuracies could still be achievable.

035.107 **The International Ultraviolet Explorer (IUE).**
Y. Kondo.
IAU Colloquium No. 123: Observatories in Earth orbit and beyond, p. 35 – 40 (1990). – See Abstr. 012.038 for the main entry.

The International Ultraviolet Explorer (IUE) was launched into a geosynchronous orbit on 26 January 1978. It is equipped with a 45 cm mirror and spectrographs operating in the far-ultraviolet (1150 – 2000 Å) and the mid-ultraviolet (1900 – 3200 Å) wavelength regions. In a low-dispersion mode, the spectral resolution is some 6 – 7 Å. In a high-dispersion echelle mode, the resolution is about 0.1 Å at the shortest wavelength and about 0.3 Å at the longest. By the end of 1989, 1870 papers, using IUE observations, have been published in referend journals. During the same period, over 1700 different astronomers from all over the world used the IUE for their research.

035.108 **X-ray astronomy satellite Ginga.**
F. Makino.
IAU Colloquium No. 123: Observatories in Earth orbit and beyond, p. 41 – 48 (1990). – See Abstr. 012.038 for the main entry.

The X-ray astronomy satellite Ginga carries three scientific instruments, the Large Area proportional Counters (LAC), All

Sky X-ray Monitor (ASM) and Gamma-ray Burst Detector (GBD). The LAC is the main instrument with an effective area of 4000 cm^2 giving it the highest sensitivity to hard X-rays so far achieved. Ginga observed about 250 targets up to the end of 1989.

035.109 **Extreme and far ultraviolet astronomy from Voyagers 1 and 2.**
J. B. Holberg.
IAU Colloquium No. 123: Observatories in Earth orbit and beyond, p. 49 - 57 (1990). - See Abstr. 012.038 for the main entry.
The instrumental characteristics, observational capabilities and scientific results of the Voyager 1 and 2 ultraviolet spectrometers are reviewed. These instruments provide current and ongoing access to low resolution spectra for a wide variety of astronomical sources in the 500 to 1700 Å band. Observations of the brightest OB stars and hot subluminous stars as faint as $V = 15$ mag, are possible. In the EUV, at wavelengths shortward of 900 Å, several new sources have been detected and a host of potential sources ruled out. In the Far UV, particularly at wavelengths between 900 and 1200 Å, Voyager is capable of observing a wide range of stellar and non-stellar sources.

035.110 **Röntgen satellite.**
J. Trümper.
IAU Colloquium No. 123: Observatories in Earth orbit and beyond, p. 61 - 62 (1990). Abstract. - See Abstr. 012.038 for the main entry.

035.111 **The Gamma-Ray Observatory.**
D. A. Kniffen.
IAU Colloquium No. 123: Observatories in Earth orbit and beyond, p. 63 - 70 (1990). - See Abstr. 012.038 for the main entry.
The Gamma-Ray Observatory (GRO), a Shuttle-launched free-flying Observatory is currently schedules for launch in November 1990. The mission provides nearly six orders of magnitude in spectral coverage, from about 50 keV to about 30 GeV, with a sensitivity over the entire range over an order of magnitude greater than that of previous observations. The 16,000 kg Observatory contains four instruments on a stabilized platform. The mission duration is expected to be from six to ten years.

035.112 **The Advanced X-ray Astrophysics Facility.**
M. C. Weisskopf.
IAU Colloquium No. 123: Observatories in Earth orbit and beyond, p. 71 - 79 (1990). - See Abstr. 012.038 for the main entry.
The Advanced X-Ray Astrophysics Facility (AXAF) is a remarkable and unique scientific endeavor. It will be a space-based earth orbiting X-ray observatory, designed to address fundamental questions in astronomy and astrophysics. It is remarkable in that it represents enormous advances in observational capability. It is unique in that it will be the best means of obtaining high resolution, spectrally resolved X-ray images of astronomical objects for the rest of this century and beyond.

035.113 **The Astro-D mission.**
Y. Tanaka.
IAU Colloquium No. 123: Observatories in Earth orbit and beyond, p. 81 - 87 (1990). - See Abstr. 012.038 for the main entry.
Astro-D, the fourth X-ray astronomy mission of ISAS following Ginga, is a high-throughput imaging and spectroscopic X-ray observatory, schedules for launch in early 1993. It utilizes multilayer conical X-ray mirrors which provide a large effective area over a wide range energy range up to 12 keV. The focal plane instruments comprize a set of CCD cameras and imaging gas scintillation proportional counters. Main features of Astro-D are described, and the important astrophysical objectives are discussed.

035.114 **The X-ray Timing Explorer.**
H. V. Bradt, A. M. Levine, E. H. Morgan, R. A. Remillard, J. H. Swank, B. L. Dingus, S. S. Holt, K. M. Jahoda, R. E. Rothschild, D. E. Gruber, P. L. Hink, R. M. Pelling, J. G. Jernigan.
IAU Colloquium No. 123: Observatories in Earth orbit and beyond, p. 89 - 110 (1990). Originally appeared in Advances in Space Research (Bradt, Swank and Rothschild 1990), Pergamon Press. - See Abstr. 012.038 for the main entry.
The capabilities of the X-ray Timing Explorer (XTE) are described with particular attention paid to current scientific problems it will address from galactic neutron star systems to active galactic nuclei. It features a low-background continuous 2 - 200 keV response with large apertures (a 0.63 m^2 proportional counter array and a 0.16 m^2 dual rocking Na I/Cs I scintillation array). Rapid response (in hours) to temporal phenomena, e.g. transients, is obtained by virtue of a scanning all-sky monitor and rapid maneuverability. XTE will carry out detailed energy-resolved studies of phenomena close to neutron stars.

035.115 **The Joint European Telescope for X-ray astronomy (JET-X).**
K. A. Pounds.
IAU Colloquium No. 123: Observatories in Earth orbit and beyond, p. 111 - 118 (1990). - See Abstr. 012.038 for the main entry.
JET-X is one of the main payload elements in the Soviet Spectrum-X Gamma mission scheduled for launch in 1993. It consists of two Wolter I X-ray telescopes feeding cooled arrays of CCD's. The combination of good angular resolution and long exposures during the four day orbit of Spectrum-X Gamma, will allow the JET-X telescopes to reach a sensitivity level ten times fainter than that of the Einstein Observatory deep survey. The CCD's will provide good spectral data for many thousands of brighter sources over an energy band ~0.3-10 keV. A brief description is given of the design and anticipated performance of JET-X together with illustrations of the anticipated scientific results.

035.116 **SODART telescope on SPECTRUM-RÖNTGEN-GAMMA and its instrumentation.**
H. W. Schnopper.
IAU Colloquium No. 123: Observatories in Earth orbit and beyond, p. 119 - 128 (1990). - See Abstr. 012.038 for the main entry.
SPECTRUM-RÖNTGEN-GAMMA (SRG) is one of a new series of large astronomical missions being planned by the Soviet Union and is scheduled for launch in mid-1993. Mission objectives include broad and narrow band imaging spectroscopy over a wide range of energies from EUV through gamma rays with particular emphasis on extragalactic objects. The design of the Soviet Danish Röntgen Telescope (SODART) consists of two thin foil, conical shell approximations to Wolter 1 geometry. Each telescope has an aperture of 60 cm, a focal length of 8 m, a field of view of 1° and is designed to have a halfpower width of $\leq 2'$. Images and spectra will be recorded with position sensitive proportional counters with spectral resolution as good as 13% at 6 keV. An objective Bragg crystal panel, placed in front of one of the telescopes, will be capable of high resolution spectroscopic studies.

035.117 **ESA's X-ray Astronomy Mission, XMM.**
B. G. Taylor, A. Peacock.
IAU Colloquium No. 123: Observatories in Earth orbit and beyond, p. 129 - 140 (1990). - See Abstr. 012.038 for the main entry.
ESA's X-ray Astronomy Mission, XMM, scheduled for launch in 1998, is the second of four cornerstones of ESA's long term science program Horizon 2000. Covering the range from about 0.1 to 10 keV, it will provide a high throughput of 5000 cm^2 at 7 keV with three independant telescopes, and have a spatial resolution better than 30″. Broadband spectrophotometry is provided by CCD cameras while reflection gratings provide medium resolution spectroscopy (resolving power of about 400)

in the range 0.3 – 3 keV. A 30 cm UV/optical telescope is boresighted with the X-ray telescopes to provide simultaneous optical counterparts to the numerous serendipitous X-ray sources which will be detected during every observation.

035.118 **The SAX mission for X-ray astronomy.**
R. C. Butler, L. Scarsi.
IAU Colloquium No. 123: Observatories in Earth orbit and beyond, p. 141 – 150 (1990). – See Abstr. 012.038 for the main entry.
The satellite for X-ray astronomy SAX, to be launched at the end of 1993, is devoted to systematic, integrated and comprehensive, studies of galactic and extra-galactic sources in the energy band 0.1 – 200 keV. The basic scientific objectives can be summarized as follows: Broad band spectroscopy ($E/\Delta E = 12$) from 0.1 – 10 keV with imaging resolution of 1′; Continuum and line spectroscopy ($E/\Delta E = 5-20$) in the energy range 3 – 200 keV; Variability studies of bright source energy spectra on timescales from milliseconds to days and months; Systematic long term variability studies over the entire sky down to a source intensity of 1 mCrab. The payload is described in detail.

035.119 **The Extreme Ultraviolet Explorer mission.**
S. Bowyer.
IAU Colloquium No. 123: Observatories in Earth orbit and beyond, p. 153 – 169 (1990). – See Abstr. 012.038 for the main entry.
The Extreme Ultraviolet Explorer mission is described. The satellite is now scheduled to be launched in September, 1991. For the first six months, an all sky survey will be carried out covering 90 to 820 Å, or essentially the entire EUV bandpass. This EUV survey will be made in four bands, or colors, $\lambda\lambda$90 – 150 Å; 170 – 250 Å, 400 – 600 Å and 550 – 750 Å. A portion of the sky which is free from the normally intense 304 Å geocoronal helium background will be surveyed at greater sensitivity; the wavelength coverage of this band is from 90 to 400 Å. Following the sky survey portion of the mission, spectroscopy of individual sources will be carried out. Three spectrometers employing novel variable line spaced gratings will provide spectra with ~1 Å resolution over the band from 90 to 800 Å.

035.120 **Lyman the Far Ultraviolet Spectroscopic Explorer.**
W. Moos.
IAU Colloquium No. 123: Observatories in Earth orbit and beyond, p. 171 – 176 (1990). – See Abstr. 012.038 for the main entry.
The Lyman Far Ultraviolet Spectroscopic Explorer mission will obtain high resolution spectra (R 30,000) in the 912 – 1200 Å region with sufficient sensitivity to study faint sources throughout the galaxy and at large extragalactic distances. This spectral region provides unique tools for solving problems is cosmology (e.g. deuterium), galactic structure and evolution (e.g. O VI and H_2), stellar evolution e.g. O VI) and planetary science (e.g. H_2). Recent advances in optical and detector technology, which make this goal possible with a moderate size experimental package, also enable the spectral coverage to extend down to 100 Å with good sensitivity and only a minimal increase in complexity. Thus a secondary goal is to cover the 100 – 912 Å region with moderate spectral resolution.

035.121 **Far and extreme ultraviolet astronomy with ORFEUS.**
G. Krämer, J. Barnstedt, N. Eberhard, M. Grewing, W. Gringel, C. Haas, A. Kaelble, N. Kappelmann, J. Petrik, I. Appenzeller, J. Krautter, H. Mandel, R. Östreicher, S. Bowyer, M. Hurwitz.
IAU Colloquium No. 123: Observatories in Earth orbit and beyond, p. 177 – 184 (1990). – See Abstr. 012.038 for the main entry.
ORFEUS (Orbiting and Retrievable Far and Extreme Ultraviolet Spectrometer) is a 1 m normal incidence telescope for spectroscopic investigations of cosmic sources in the far and extreme ultraviolet spectral range. The instrument will be integrated into the freeflyer platform ASTRO-SPAS. The authors describe the telescope with its two spectrometers and their capabilities i.e. spectral range, resolution and overall sensitivity. The main classes of objects to be observed with the instrument are discussed and two examples of simulated spectra for the white dwarf HZ43 and an O9-star in the LMC are shown.

035.122 **The SPECTRUM-UV project.**
IAU Colloquium No. 123: Observatories in Earth orbit and beyond, p. 185 – 190 (1990). – See Abstr. 012.038 for the main entry.
The Ultra-Violet Space Telescope "SPECTRUM-UV" consists of a 170 cm diffraction limited telescope for spectroscopy and imaging at $\lambda \geqslant 912$ Å and a co-aligned assembly of two 50 cm telescopes for imaging and spectroscopy in the 400 to 1200 Å range and of four 20 cm, multilayer coated telescopes for narrow band imaging in the 100 to 400 Å range. The observatory is an international facility to be launched in a highly eccentric orbit in the mid-1990s.

035.123 **The Space Infrared Telescope Facility (SIRTF).**
G. G. Fazio, P. Eisenhardt.
IAU Colloquium No. 123: Observatories in Earth orbit and beyond, p. 193 – 203 (1990). – See Abstr. 012.038 for the main entry.
The Space Infrared Telescope Facility (SIRTF) is a one-meter class observatory for infrared astronomy that will be launched into high earth orbit by NASA in the late 1990's. SIRTF's three focal plane instruments will permit imaging and spectroscopy over most of the infrared spectrum with sensitivities of 100 to 10,000 times their predecessors. This paper briefly reviews SIRTF's capabilities, science objectives, and current status.

035.124 **The Infrared Space Observatory.**
C. J. Cesarsky, M. F. Kessler.
IAU Colloquium No. 123: Observatories in Earth orbit and beyond, p. 205 – 214 (1990). – See Abstr. 012.038 for the main entry.
The Infrared Space Observatory (ISO) is an astronomical satellite, which will operate at wavelengths from 3 – 200 μm. The satellite essentially consists of a large cryostat containing at launch about 2300 litres of superfluid helium to maintain the Ritchey-Chrétien telescope, the scientific instruments and the optical baffles at temperatures between 2K and 8K. The telescope has a 60 cm diameter primary mirror and is diffraction-limited at a wavelength of 5 μm. ISO's instrument complement consists of four instruments, namely: a photo-polarimeter (3 – 200 μm), a camera (3 – 17 μm), a short wavelength spectrometer (3 – 45 μm) and a long wavelength spectrometer (45 – 180 μm).

035.125 **IRTS: Infrared Telescope in Space.**
T. Matsumoto.
IAU Colloquium No. 123: Observatories in Earth orbit and beyond, p. 215 – 222 (1990). – See Abstr. 012.038 for the main entry.
IRTS is a small cryogenically cooled telescope onboard the small space platform SFU (Space Flyer Unit). The IRTS telescope has an aperture of only 15 cm diameter, but is optimized to observe diffuse extended infrared sources. Four focal plane instruments are being developed. IRTS covers a wide wavelength range from near-infrared to submillimeter region, and has a capability for spectroscopic measurement.

035.126 **FIRST – Far Infrared and Submillimetre Space Telescope.**
U. O. Frisk.
IAU Colloquium No. 123: Observatories in Earth orbit and beyond, p. 223 – 230 (1990). Also published in reports ESA-SP 315, p. 13 – 18, (Nov 1990) and ESA-SP 314, p. 37 – 45 (Dec 1990). – See Abstr. 012.038 for the main entry.
FIRST is an element of the ESA long term science program. Currently in a detailed study phase it is foreseen as a large (4.5 – 8 m) diameter passively cooled telescope equipped with a combination of photometer/camera and very high resolution spectrometers. The spectrometers will utilize both direct detection and heterodyne techniques to cover the wavelength band

85 – 600 μm. The photometers will have both bolometer and photoconductor detector arrays.

035.127 **A submillimeter mission for the 1990s: SMMM.**
T. G. Phillips.
IAU Colloquium No. 123: Observatories in Earth orbit and beyond, p. 231 – 249 (1990). Also published as report ESA-SP 314, p. 221 – 228 (Dec 1990). – See Abstr. 012.038 for the main entry.

This report describes a 'Moderate Mission' spacecraft concept, designed to make physical and chemical studies of high red-shift galaxies, molecular clouds, star-forming regions and planets. SMMM will utilize a roughly 4 m ambient telescope with a liquid-helium cooled focal plane. It will achieve a complete sub-mm spectroscopic survey of sources betwenn 100 and 700 μm. In addition to the spcetroscopy, SMMM will have a far-infrared camera with a bolometer array.

035.128 **The Submillimeter Wave Astronomy Satellite.**
G. J. Melnick.
IAU Colloquium No. 123: Observatories in Earth orbit and beyond, p. 251 (1990). Abstract. Also published as report ESA-SP 314, p. 251 – 256 (Dec 1990). – See Abstr. 012.038 for the main entry.

035.129 **International VLBI Satellite (IVS).**
R. T. Schilizzi.
IAU Colloquium No. 123: Observatories in Earth orbit and beyond, p. 255 – 262 (1990). – See Abstr. 012.038 for the main entry.

IVS is under study in ESA as a second generation space VLBI observatory. The mission concept calls for a 25 m diameter radio telescope in space. Orbiting the Earth and observing in concert with the established ground-based VLBI arrays in Europe, USA, USSR and Australia, IVS will provide high quality images of galactic and extragalactic radio sources at wavelengths spanning the radio band from decimetres to millimetres with resolution as high as 10 μarcsec and sensitivity equal to those of ground-based images.

035.130 **VSOP, a space VLBI programme.**
H. Hirabayashi.
IAU Colloquium No. 123: Observatories in Earth orbit and beyond, p. 263 – 269 (1990). – See Abstr. 012.038 for the main entry.

VSOP, VLBI Space Observatory Programme, is an approved space VLBI programme for the study of very compact radio sources with the synthesized aperture of 30,000 km diameter, by connecting an orbiting radio observatory with ground-based radiotelescopes. The VSOP satellite carrying a 10 m antenna with 1.6, 5, and 22 GHz band receivers. Imaging of active galactic nuclei, star forming regions and stellar objects, and radioastrometry are main scientific targets.

035.131 **Soho – an observatory to study the solar interior and the solar atmosphere.**
A. I. Poland, V. Domingo.
IAU Colloquium No. 123: Observatories in Earth orbit and beyond, p. 277 – 284 (1990). – See Abstr. 012.038 for the main entry.

The Solar and Heliospheric Observatory, Soho, is a joint venture of ESA and NASA. The main objectives of Soho are: a) the study and understanding of solar coronal phenomena; and b) the study of the solar structure and interior dynamics from its core to the photosphere. The Soho spacecraft will be three-axis stabilized and located in a halo orbit around the L1 Lagrangian point (approximately 1% of the distance from the Earth to the Sun). It is currently scheduled for launch in July 1995.

035.132 **An overview of the Orbiting Solar Laboratory.**
D. S. Spicer.
IAU Colloquium No. 123: Observatories in Earth orbit and beyond, p. 285 (1990). Abstract. – See Abstr. 012.038 for the main entry.

035.133 **The Planetenteleskop mission.**
G. Neukum.
IAU Colloquium No. 123: Observatories in Earth orbit and beyond, p. 286 (1990). Abstract. – See Abstr. 012.038 for the main entry.

035.134 **The Astrometric Imaging Telescope: near-term discovery and study of other planetary systems.**
E. H. Levy, G. D. Gatewood, R. J. Terrile.
IAU Colloquium No. 123: Observatories in Earth orbit and beyond, p. 287 (1990). Abstract. – See Abstr. 012.038 for the main entry.

035.135 **Ultraviolet polarimetry.**
A. D. Code, K. H. Nordsieck, C. M. Anderson.
IAU Colloquium No. 123: Observatories in Earth orbit and beyond, p. 291 (1990). Abstract. – See Abstr. 012.038 for the main entry.

035.136 **The Hopkins Ultraviolet Telescope.**
A. Davidsen.
IAU Colloquium No. 123: Observatories in Earth orbit and beyond, p. 292 (1990). Abstract. – See Abstr. 012.038 for the main entry.

035.137 **The Ultraviolet Imaging Telescope for Astro 1.**
T. P. Stecher.
IAU Colloquium No. 123: Observatories in Earth orbit and beyond, p. 293 (1990). Abstract. – See Abstr. 012.038 for the main entry.

035.138 **The Broad Band X-Ray Telescope (BBXRT) on Astro-1.**
P. J. Serlemitsos.
IAU Colloquium No. 123: Observatories in Earth orbit and beyond, p. 294 (1990). Abstract. – See Abstr. 012.038 for the main entry.

035.139 **United States launch vehicle systems.**
R. B. Krause.
IAU Colloquium No. 123: Observatories in Earth orbit and beyond, p. 325 – 332 (1990). – See Abstr. 012.038 for the main entry.

This paper highlights the current capabilities of the mixed fleet and summarize the plans for new or modified United States launch vehicles through the first decade of the next century.

035.140 **ESA's Space Transportation Programme.**
J. Feustel-Buechel, W. Wamsteker.
IAU Colloquium No. 123: Observatories in Earth orbit and beyond, p. 333 – 338 (1990). – See Abstr. 012.038 for the main entry.

The ARIANE and HERMES programmes of ESA are shortly described.

035.141 **Launch vehicles of ISAS.**
Y. Tanaka.
IAU Colloquium No. 123: Observatories in Earth orbit and beyond, p. 339 – 342 (1990). – See Abstr. 012.038 for the main entry.

The Institue of Space and Astronautical Science (ISAS) has developed a series of launch vehicles for delivering modest scale scientific satellites into space. The present ISAS launch vehicle, M-3S II, carries a 770 kg payload into a low earth orbit (LEO).

035.142 **Lunar-based astronomy.**
H. J. Smith.
IAU Colloquium No. 123: Observatories in Earth orbit and beyond, p. 365 – 375 (1990). – See Abstr. 012.038 for the main entry.

The Moon offers for astronomy a truly impressive array of advantages, many of which are briefly reviewed in this paper.

035.143 **SIXA: the solid state spectrometer array onboard SPECTRUM-X-GAMMA.**
O. Vilhu, H. Sipilä, V. J. Kämäräinen, I. Taylor, E. Laegsgaard, G. Leppelmeier, H. W. Schnopper.
IAU Colloquium No. 123: Observatories in Earth orbit and beyond, p. 433 – 438 (1990). – See Abstr. 012.038 for the main entry.
The SPECTRUM-X-GAMMA mission is scheduled for launch in 1993. Mission objectives include broad and narrow band imaging spectroscopy over a wide range of energies from the EUV through gamma rays, with an emphasis on studying galactic and extragalactic X-ray sources. Moderate resolution spectroscopy will be done by the segmented solid state detector SIXA (Silicon X-ray Array). The detector will consist of 19 segments (Si(Li)), each with a diameter of about 8 mm. The spectral resolution of 160 eV (at 6 keV), combined with the large collecting area, provide good opportunities for time-resolved iron line spectroscopy (6 – 8 keV).

035.144 **The X-ray Large Array.**
K. S. Wood, P. Hertz, J. P. Norris, P. F. Michelson.
IAU Colloquium No. 123: Observatories in Earth orbit and beyond, p. 439 – 442 (1990). – See Abstr. 012.038 for the main entry.
XLA is an X-ray facility with an aperture substantially greater than 1 m^2, nominally 100 m^2. Most of this area is devoted to a large array of collimated proportional counters. There is also a ~ 1 m^2 coded aperture. It extends observational parameter space by several orders of magnitude in timing resolution, sensitivity to variability, and angular resolution. This will lead to a qualitatively new kind of X-ray astrophysics that can be applied to the study of a broad range of astrophysical objects. XLA is thus both an advanced timing mission and a general purpose facility whose principal uses are in areas that are not well covered in other aspects of the planned High Energy Astrophysics program.

035.145 **The Stellar X-Ray Polarimeter for the SPECTRUM-X-Gamma mission.**
P. Kaaret, R. Novick, C. Martin, P. Shaw, J. R. Fleischman, T. Hamilton, R. Sunyaev (*R. A. Syunyaev*), I. Lapshov, E. Silver, K. Ziock, M. Weisskopf, R. Elsner, B. Ramsey, G. Chanan, G. Manzo, S. Giarrusso, A. Santangelo, E. Costa, L. Piro, G. Fraser, J. F. Pearson, J. E. Lees, G. C. Perola, E. Massaro, G. Matt.
IAU Colloquium No. 123: Observatories in Earth orbit and beyond, p. 443 – 449 (1990). – See Abstr. 012.038 for the main entry.
The authors describe an X-ray polarimeter which will be flown on the SPECTRUM-X-Gamma mission. The instrument exploits three distinct physical processes to measure polarization: Bragg reflection from a graphite crystal, Thomson scattering from a metallic lithium target, and photoemission from a Cesium Iodide photocathode. These three processes allow polarization measurements over an energy band of 0.3 keV to 12 keV.

035.146 **A low energy gas scintillation proportional counter for the SAX-X-ray astronomy satellite.**
A. N. Parmar, A. Smith, M. Bavdaz.
IAU Colloquium No. 123: Observatories in Earth orbit and beyond, p. 457 – 461 (1990). – See Abstr. 012.038 for the main entry.
The payload of the Italian/Dutch satellite SAX will include a set of four concentrators each with a geometric area of 90 cm^2. Imaging GSPCs will be located at the focal planes of the concentrators. The Space Science Department of ESA will provide one of these GSPCs which will be sensitive to X-rays with energies between 0.1 – 10 keV. In order to achieve such a low-energy energy response, a driftless configuration and a thin plastic window have been adopted. At 6 keV the collecting area will be 50 cm^2 and the energy and angular resolutions 8% and 1.6′ FWHM, respectively.

035.147 **An X-ray all sky monitor for a Japanese Experimental Module on the space station.**
M. Matsuoka, N. Kawai, T. Imai, M. Yamauchi, A. Yoshida, T. Kohno, A. Yoneda, H. Tsunemi.
IAU Colloquium No. 123: Observatories in Earth orbit and beyond, p. 463 – 468 (1990). – See Abstr. 012.038 for the main entry.
The authors propose an X-ray all sky monitor for Japanese Experimental Module (JEM) on the space station. Position determination of gamma-ray bursts could be achieved with accuracy less than one degree observing the X-ray component of the burst. Weak X-ray sources such as active galactic nuclei could be also monitored with time resolution less than one day. The X-ray all sky monitor will work to discover X-ray novae and transient phenomena and give alarm for further detailed observations. The obtained data will be also used for archival study.

035.148 **The Astro Mission.**
T. R. Gull.
IAU Colloquium No. 123: Observatories in Earth orbit and beyond, p. 469 (1990). Abstract. – See Abstr. 012.038 for the main entry.

035.149 **The UV imager for the Israeli scientific satellite.**
N. Brosch.
IAU Colloquium No. 123: Observatories in Earth orbit and beyond, p. 471 – 474 (1990). – See Abstr. 012.038 for the main entry.
Israel will orbit a satellite dedicated to scientific research. One of the two experiments studied for deployment on this platform is a three-channel imager in the ultraviolet. The design provides very significant scientific returns in a small payload and for a moderate cost.

035.150 **ORFEUS-SPAS: the Berkeley EUV Spectrometer.**
S. Bowyer, M. Hurwitz.
IAU Colloquium No. 123: Observatories in Earth orbit and beyond, p. 475 – 480 (1990). – See Abstr. 012.038 for the main entry.
The Berkely spectrograph for ORFEUS SPAS will obtain high-resolution ($\lambda/\Delta\lambda > 7000$) spectra of point sources between 390 and 1200 Å. The Berkeley instrument incorporates a set of four novel spherically figured, varied line-space gratings used in a geometry that is similar to that of the classic Rowland mount to span this large wavelength interval. Two spectral detector units containing curved MCPs and delay-line anodes encode the arriving photons in digital format for telemetry. An additional optic directs the image of the source in the entrance aperture onto a sealed FUV detector which is used to track the source as it drifts during an observation, enabling a post-flight reconstruction of the spacecraft pointing vector. This in turn will allow to define the wavelength of each recorded photon with precision.

035.151 **An observatory for mapping the far UV diffuse galactic emission line background.**
F. L. Roesler, J. Harlander, R. J. Reynolds.
IAU Colloquium No. 123: Observatories in Earth orbit and beyond, p. 481 – 486 (1990). – See Abstr. 012.038 for the main entry.
A new instrumental concept for interference spectroscopy called Spatial Heterodyne Spectroscopy (SHS) is described. This instrument as currently demonstrated could provide important information on the structure, excitation, and dynamics of the $\cong 10^5$K component of the interstellar medium by providing velocity-resolved (20 km s^{-1}) maps of the faint FUV emission line background over a hemisphere of the sky within a 5 – 6 year observation period.

035.152 **SANTA MARIA: an orbiting multispectral observatory.**
C. Morales, L. Sabau, A. Giménez, A. L. Broadfoot, B. R. Sandel, R. Stalio, A. Talavera, A. Bucconi.
IAU Colloquium No. 123: Observatories in Earth orbit and beyond, p. 493 – 496 (1990). – See Abstr. 012.038 for the main entry.
SANTA MARIA is a small, low cost satellite which will perform imaging spectroscopy in the range from 200 to 7000 Å of

astrophysical and solar system sources. It will also perform continuous measurements of the Earth plasmasphere and the solar flux.

035.153 **Cryogenic testing of optics for ISOCAM.**
J. K. Davies.
IAU Colloquium No. 123: Observatories in Earth orbit and beyond, p. 497 – 499 (1990). – See Abstr. 012.038 for the main entry.
The ROE ISOCAM optical test facility is described. The results of testing of lenses shows good agreement with theory and remarkable consistency.

035.154 **EDISON: a second generation infrared space observatory.**
H. A. Thronson Jr., T. G. Hawarden, C. M. Mountain, J. K. Davies, T. J. Lee, M. Longair.
IAU Colloquium No. 123: Observatories in Earth orbit and beyond, p. 501 – 505 (1990). – See Abstr. 012.038 for the main entry.
EDISON is a large–aperture telescope under study for the second generation of infrared space observatories, whether in orbit or on the moon. The authors are presently evaluating a design for a 2.5 m observatory to be launched in collaboration with European astronomers as part of the NASA Explorer program. In this presentation, they describe possible spectroscopic and spectrophotometric studies of very faint infrared sources that will require large–aperture space telescopes working at the celestial background limit.

035.155 **High–resolution imaging spectroscopy at terahertz frequencies.**
R. L. Brown, A. R. Kerr, A. R. Thompson, F. R. Schwab.
IAU Colloquium No. 123: Observatories in Earth orbit and beyond, p. 509 – 515 (1990). – See Abstr. 012.038 for the main entry.
HISAT, a multi–element heterodyne interferometer attached to Space Station Freedom, will provide spectroscopic images with unprecedented detail of those submillimeter lines of C, O and C^+ which are critical diagnostics of UV excitation in the Galaxy. With the arsecond angular resolution achievable from the space station, HISAT will reveal (1) the distribution of sources of ultraviolet radiation in the Galaxy; (2) the effective temperature of the UV radiation as a function of galactocentric radius; (3) the chemical and isotopic enhancement of atomic carbon and oxygen with galactic radius; (4) the propagation of UV radiation in molecular clouds and its stimulative, or inhibitive, effect on star formation; (5) the density structure, clumpiness or fragmentation, of molecular clouds throughout the Galaxy.

035.156 **The Martian magnetic soils: proposal for a soil magnetometer.**
J. M. D. Coey, J. M. Knudsen, D. H. Ryan.
2. COSPAR Colloquium: The environmental model of Mars, p. 85 – 88 (1991). – See Abstr. 012.039 for the main entry.
The problem of the strong magnetism of the martian soils and suggestions regarding the iron minerals that may be responsible are reviewed. Various magnetic measurements could be carried out on a Mars lander or rover mission with a view to identifying the strongly–magnetic iron minerals. An instrument based on rotating permanent magnets is proposed which would measure hysteresis loops of soil samples.

035.157 **Transmission grating spectroscopy and the advanced X–ray astrophysics facility.**
M. L. Schattenburg, C. R. Canizares, D. Dewey, K. A. Flanagan, M. A. Hamnett, A. M. Levine, K. S. K. Lum, R. Manikkalingam, T. H. Markert, H. I. Smith.
Opt. Eng., Vol. 30, No. 10, p. 1590 – 1600 (Oct 1991). Revision of a paper presented at the "SPIE Seminar called X–ray Instrumentation in Astronomy 2, San Diego, CA (USA), Aug 1989".
The use of transmission gratings with grazing–incidence telescopes in celestial X–ray astronomy is reviewed. The basic properties of transmission grating spectrometers and the use of "phased" gratings to enhance the diffraction efficiency are outlined. Special attention is given to the AXAF high–energy transmission grating (HETG).

035.158 **The current ability of HST to reveal morphological structure in medium–redshift galaxies.**
I. R. King, S. A. Stanford, P. Seitzer, M. A. Bershady, W. C. Keel, D. C. Koo, N. Weir, S. Djorgovski, R. A. Windhorst.
Astron. J., Vol. 102, No. 4, p. 1553 – 1568 (Oct 1991). Current Physics Microform No.: 9110D2085.
A pair of FOC f/48 images was taken on a compact galaxy with $z \sim 0.3$; and a neighboring field, including a number of medium–redshift galaxies, was covered with the WFC. The conclusions are as follows: (1) The FOC48 (whose field is unlikely to include more than the single galaxy that is targeted) gives a good resolved image of a compact galaxy at blue magnitude $J = 20.5$ in a single–orbit exposure; longer exposures would (a) add detail to compact galaxies, or (b) allow morphological classification of objects of lower surface brightness. (2) WFC images have a survey capability that can include many galaxies per field, with sufficient resolving power to distinguish clearly between galaxies and stars down to the level of 0.″2 (FWHM), depending on signal–to–noise ratio, and a reasonable capacity for morphology. (3) Although some morphological detail can be discerned in even the aberrated images, deconvolutions very much enhance the ability to see structural detail. Even at the low S/N that is provided by single–orbit exposures the more sophisticated restoration methods offer some advantage over simple Fourier or Lucy techniques. (4) Where image restorations are involved, it is of great importance to have more than one image of each field, so as to compare results and see what is real, and also to facilitate the removal of cosmic rays.

035.159 **Measurement of the magnetic field vector from a rotating spacecraft.**
M. K. Trubetskov, E. G. Eroshenko, I. P. Lyannaya, A. A. Ruzmajkin, D. D. Sokolov, V. A. Styazhkin, A. M. Shukurov.
Kosm. Issled., Tom 29, Vyp. 4, p. 597 – 603 (Jul–Aug 1991). In Russian. English translation in Cosmic Res., Vol. 29, No. 4.
The authors propose a method of restoration of magnetic field components measured by a vector magnetometer onboard of a spacecraft which rotates at an angular velocity unknown in advance. The available magnetometer data are used for a precise estimation of the spacecraft angular velocity. The latter estimate, combined with knowledge of the initial orientation of the spacecraft, allows to restore the components of the external magnetic field in an inertial frame. The proposed method is applied to analysis of magnetic field measurements near Mars by "Phobos 2". The angular velocity of the spacecraft is estimated to an accuracy higher than 0.1 percent.

035.160 **The MPE/UCB far–infrared imaging Fabry–Perot interferometer (FIFI).**
A. Poglitsch, J. W. Beeman, N. Geis, R. Genzel, M. Haggerty, E. E. Haller, J. Jackson, M. Rumitz, G. J. Stacey, C. H. Townes.
3. Haystack Observatory Conference on the Interstellar Medium: Skylines – atoms, ions and molecules – new results in spectral line astrophysics, p. 451 – 455 (1991). Also published as report ESA–SP–314, Dec 1990, p. 149 – 152. – See Abstr. 012.051 for the main entry.
FIFI is an imaging spectrometer with two or three Fabry–Perot interferometers in series for airborne astronomical observations in the far–infrared range. It employs 5 × 5 arrays of photoconducting detectors and offers spectral resolutions as small as 2 km/s. Cryogenic optics minimizes thermal background radiation and provides for in–flight step tunable spatial resolution. The frequency–shopping mode allows for line detection in

extended objects. Absolute internal flux calibration ensures adequate "flat fielding" of the array elements.

035.161 **The sensitivity of EGRET to gamma-ray polarization.**
J. R. Mattox.
Exp. Astron., Vol. 2, No. 2, p. 75 - 84 (1991).
A Monte Carlo simulation shows that the EGRET gamma-ray telescope aboard the GRO satellite does not have sufficient sensitivity to detect linear polarization, even for 100% polarized gamma-ray sources. This is confirmed by analysis of calibration data. Several data selection techniques suggested to enhance polarization sensitivity have been evaluated and found to not significantly improve sensitivity.

035.162 **A new mask-antimask coded-aperture telescope for hard X-ray astronomy.**
J. Braga, T. Villela, U. B. Jayanthi, F. d'Amigo, J. A. Neri.
Exp. Astron., Vol. 2, No. 2, p. 101 - 113 (1991).
A new imaging balloon-borne telescope for hard X-rays in the energy range from 30 to 100 keV is described. The imaging capability is provided by the use of an extended URA-based coded-mask. With only one motor and suitable stop pins, one can rotate a carbon-fiber wheel with most of the mask elements attached to it by 180°, and a bar, which is also part of the mask pattern and is allowed to rotate freely over the wheel, by 90°; this combined rotation creates an antimask of the original mask, except for the central element. The authors show that the use of antimasks is a very effective method of eliminating systematic variations in the background map over the position-sensitive detector area. The expected sensitivity of the instrument for the 30 - 100 keV range is of the order of 7×10^{-5} photons $cm^{-2}s^{-1}keV^{-1}$, for an integration time of 10^4 seconds at a residual atmosphere of 3.5 g cm^{-2}. This telescope will provide imaging observations of bright galactic hard X-ray sources with an angular resolution of $\sim2°$ in a 10° by 10° FOV. Results of computer simulations of the imaging system are reported.

035.163 **Radiation concerns for the Solar-A Soft X-Ray Telescope.**
L. Acton, M. Morrison, J. Janesick, T. Elliott.
Conference on Charge-Coupled Devices and Solid State Optical Sensors II as Part of the SPIE/IS and T Symposium on Electronic Imaging Science and Technology, p. 123 - 139 (1991). - See Abstr. 012.059 for the main entry.
The CCD camera of the Soft X-ray Telescope for the Japanese Solar-A Mission utilizes at 1024 x 1024 virtual phase CCD. This sensor will be subject to radiation in the form of trapped protons from the earth's radiation belts and soft X-rays (0.2-4 keV) in the solar image. Proton damage produces "dark spikes" or pixels of enhanced dark current. During the preparation of this camera it has been discovered that exposure to soft X-rays creates "permanent" ionization damage in the gate insulator, resulting in flat-band shift, dark current increase, loss of charge transfer efficiency, and, ultimately, total unpinning of the sensor. The solution of this problem is discussed.

035.164 **The Space Telescope imaging spectrograph 2048 CCD and its characteristics.**
W. A. Delamere, T. Ebben, D. Murata-Seawalt, M. M. Blouke, R. Reed, B. Woodgate.
Conference on Charge-Coupled Devices and Solid State Optical Sensors II as Part of the SPIE/IS and T Symposium on Electronic Imaging Science and Technology, p. 288 - 297 (1991). - See Abstr. 012.059 for the main entry.
Development of 2 2048^2 CCD for a second generation Space Telescope instrument has produced some very encouraging devices. The first experimental lot of 10 devices have very few defects, dark currents of less than 12 electrons/pixel/hour at -80°, readout noise levels of less than 4 electrons rms and excellent charge transfer efficiency at signal levels of less than 10 electrons.

035.165 **100-μm array polarimetry from the Kuiper Airborne Observatory: instrumentation, techniques, and first results.**
S. R. Platt, R. H. Hildebrand, R. J. Pernic, J. A. Davidson, G. Novak.
Publ. Astron. Soc. Pac., Vol. 103, No. 669, p. 1193 - 1210 (Nov 1991).
The University of Chicago far-infrared array polarimeter, STOKES, is the first multiple-beam polarimeter for far-infrared astronomy. Observations are made from the NASA Kuiper Airborne Observatory. Two orthogonal components of linear polarization are detected simultaneously by corresponding pairs of bolometers in two 32-detector arrays. Removal of the effects induced by the polarimeter and telescope from the measured signals is complicated by the multiple optic axes defined by the 32 independent pixels of the instrument. Novel observing and data-analysis techniques are used to overcome the inherent difficulties of array polarimetry. Results from the first observing flights with the new instrument are reported for the molecular clouds W3 and W51. The measurements show that the magnetic-field structure in both clouds is nonuniform on the scale of 0.5 - 1.5 pc. Preliminary results from the second flight series have yielded approximately 40 new measurements in the Sgr A complex. These results indicate that modifications made since the first flights have significantly improved the performance of STOKES.

035.166 **An imager for gamma-ray astronomy - balloon prototype.**
G. di Cocco, C. Labanti, E. Rossi, F. Schiavone, A. Spizzichino, A. Traci, G. Malaguti, A. J. Bird, T. Carter, A. J. Dean, A. J. M. Gomm, K. J. Grant, M. Corba, E. Quadrini, M. Rossi, G. Villa, B. M. Swinyard.
SPIE Conference on EUV, X-Ray, and Gamma-Ray Instrumentation for Astronomy II as Part of SPIE International Symposium on Optical Applied Science and Engineering, p. 102 - 112 (1991). - See Abstr. 012.065 for the main entry.
A novel low energy astronomical gamma-ray detector is being developed for future satellite missions. A small prototype of a 3D detector has been realized for test in a balloon mission. The detector is based on a 12 x 8 array of position sensitive CsI (Tl) bars, typically 15 cm long with 1.3 x 1.3 cm cross section, viewed at each end by photodiodes. The detector includes four 1.3 x 1.3 x 2.5 cm CsI (Tl) scintillators located above the main array in order to evaluate the low energy response of the imager. The detector constitutes an active block of 2400 cm^3 of scintillator that can operate in the 0.2 - 10 MeV energy range. The overall experiment is briefly described in general and preliminary results of laboratory tests are presented.

035.167 **Imaging the Sun in hard X-rays: spatial and rotating modulation collimators.**
J. W. Campbell, J. M. Davis, A. G. Emslie.
SPIE Conference on EUV, X-Ray, and Gamma-Ray Instrumentation for Astronomy II as Part of SPIE International Symposium on Optical Applied Science and Engineering, p. 155 - 179 (1991). - See Abstr. 012.065 for the main entry.
Several approaches to imaging hard X-rays emitted from solar flares have been proposed or are planned for the nineties including the spatial modulation collimator (SMC) and the rotating modulation collimator (RMC). A basic photon counting, Monte Carlo "end-to-end" model telescope was employed using the Astronomical Image Processing System (AIPS) for imaging reconstruction. The resulting solar flare hard X-ray images compared against typical observations indicated that both telescopes show promise for the future.

035.168 **Goldhelox: a project to view the X-ray Sun.**
M. Fair.
SPIE Conference on EUV, X-Ray, and Gamma-Ray Instrumentation for Astronomy II as Part of SPIE International Symposium on Optical Applied Science and Engineering, p. 182 - 192 (1991). - See Abstr. 012.065 for the main entry.
The "Goldhelox" project ("GOLD" for the color of the sun and "HELOX" for heliocentric observations in X-rays) main

scientific objective is to construct a high resolution soft X-ray telescope to take rapid succession, full disk pictures of the sun, hopefully during Solar Max. The authors fell that this project will do much to increase the knowledge in the area of X-ray optics and space-based physics.

035.169 **An X-ray interferometric observatory.**
C. Martin.
SPIE Conference on EUV, X-Ray, and Gamma-Ray Instrumentation for Astronomy II as Part of SPIE International Symposium on Optical Applied Science and Engineering, p. 203 - 220 (1991). - See Abstr. 012.065 for the main entry.

X-ray interferometry has the potential for observations of unprecedented angular resolution and far-reaching astrophysical consequences. The author presents a new concept for an X-ray interferometric observatory, that obtains an angular resolution 10^5 times better than AXAF. Using a grazing incidence fringe spectrum design, the X-ray interferometer operates over a broad band (0.2 - 8 keV), with ~18 concentric Wolter-Schwarzschild mirror shells of diameter 0.6 - 2.4 meters. The interferometer is capable of obtaining detailed images of nearby stellar coronae, imaging the broad-line regions and jets of many AGNs, and resolving the accretion disks of a selection bright AGNs and QSOs.

035.170 **Characteristics of a high pressure gas proportional counter filled with xenon.**
H. Sakurai, B. D. Ramsey.
SPIE Conference on EUV, X-Ray, and Gamma-Ray Instrumentation for Astronomy II as Part of SPIE International Symposium on Optical Applied Science and Engineering, p. 20 - 27 (1991). - See Abstr. 012.065 for the main entry.

The characteristics of a conventional cylindrical geometry proportional counter filled with high pressure xenon gas up to 10 atm, were fundamentally investigated for use as a detector in hard X-ray astronomy. It is shown that proportional counters at high pressure operate at weaker reduced electric field than low pressure counters. The characteristics of a parallel grid proportional counter at low pressure showed similar pressure dependence. It is suggested that this is the fundamental reason for the degradation of resolution observed with increasing pressure.

035.171 **Performance characteristics of the imaging MAMA detector systems for SOHO, STIS, and FUSE/Lyman.**
J. G. Timothy.
SPIE Conference on EUV, X-Ray, and Gamma-Ray Instrumentation for Astronomy II as Part of SPIE International Symposium on Optical Applied Science and Engineering, p. 221 - 233 (1991). - See Abstr. 012.065 for the main entry.

The configurations and performance characteristics of the different detector systems are described and the plans for further development of the Advanced Technology MAMA system discussed.

035.172 **Silicon X-ray array detector on SPECTRUM-X-GAMMA satellite.**
H. Sipilä, P. Huttunen, V. J. Kämäräinen, O. Vilhu, J. Kurki, G. W. Leppelmeier, I. Taylor, A. Niemelä, E. Lægsgaard, R. Sunyaev.
SPIE Conference on EUV, X-Ray, and Gamma-Ray Instrumentation for Astronomy II as Part of SPIE International Symposium on Optical Applied Science and Engineering, p. 246 - 255 (1991). - See Abstr. 012.065 for the main entry.

The SPECTRUM X-GAMMA mission objectives include broad and narrow band imaging spectroscopy from EUV to gamma rays. The major telescopes on board are two SODART X-ray telescopes with 8-meter focal lenghts. One of them includes the Finnish silicon X-ray array detector (SIXA) experiment as one of its focal plane instruments. This paper gives an updated description of the SIXA instrument concept. The results from preamplifier development tests are summarized and different detector concepts compared. Thermal questions connected with the detector unit and passive cooler design are discussed.

035.173 **Recent developments using GaAs as an X-ray detector.**
T. J. Sumner, S. M. Grant, A. Bewick, J. P. Li, N. J. C. Spooner, K. Smith, S. P. Beaumont.
SPIE Conference on EUV, X-Ray, and Gamma-Ray Instrumentation for Astronomy II as Part of SPIE International Symposium on Optical Applied Science and Engineering, p. 256 - 264 (1991). - See Abstr. 012.065 for the main entry.

There have been recent developments in the manufacture of GaAs detectors for high energy physics applications and dark matter searches which have resulted in working devices made from LEC and HB starting material. This offers the promise of routine manufacture or reproducible devices at a modest cost. The most advanced of the techniques is that of Schottky diodes on LEC material. Results are presented demonstrating the performance of such devices (3.0 mm × 5.0 mm × 200 μm) as X-ray detectors, including their low temperature operation. An alternative technique using charge collection in bulk HB material at temperatures down to liquid helium is also briefly described.

035.174 **High resolution, two-dimensional imaging, microchannel plate detector for use on a sounding rocket experiment.**
B. C. Bush, D. M. Cotton, O. H. Siegmund, S. Chakrabarti, W. Harris, J. Clarke.
SPIE Conference on EUV, X-Ray, and Gamma-Ray Instrumentation for Astronomy II as Part of SPIE International Symposium on Optical Applied Science and Engineering, p. 290 - 301 (1991). - See Abstr. 012.065 for the main entry.

The authors discuss a high resolution michrochannel plate (MCP) imaging detector to be used in measurements of Doppler-shifted hydrogen Lyman α line emission from Jupiter and the interplanetary medium. Operating at nominal voltage, the four plate configuration provides a gain of 1.2×10^7 electrons per incident photon. The wedge-and-strip anode has two-dimensional imaging capabilities, with a resolution of 40 microns FWHM over a one centimeter diameter area. The detector has a high quantum efficiency while retaining a low background rate. A KBr photocathode is used to enhance the quantum efficiency of the bare MCPs to a value of 35% at Lyman α.

035.175 **One dimensional proportional counters for X-ray all-sky monitors.**
M. Matsuoka, M. Yamauchi, H. Nakamura, M. Kondo, N. Kawai, A. Yoshida, T. Imai.
SPIE Conference on EUV, X-Ray, and Gamma-Ray Instrumentation for Astronomy II as Part of SPIE International Symposium on Optical Applied Science and Engineering, p. 2 - 8 (1991). - See Abstr. 012.065 for the main entry.

The authors are preparing an X-ray observation system with a wide field of view for HETE (High Energy Transient Experiment) and also proposing MAXI (Monitor of All-sky X-ray Image) for Japanese Experimental Module on the Space Station. Both systems require one dimensional proportional counters. For this purpose the authors are developing one dimensional proportional counters. They have made basic tests on three kinds of proportional counters; (1) proportional counters with resistive wires, (2) proportional counters with backgammon cathodes and (3) microstrip proportional counters.

035.176 **V-groove diffraction grating for use in an FUV spatial heterodyne interferometer.**
D. M. Cotton, B. C. Bush, S. Chakrabarti, B. Bach.
SPIE Conference on EUV, X-Ray, and Gamma-Ray Instrumentation for Astronomy II as Part of SPIE International Symposium on Optical Applied Science and Engineering, p. 313 - 318 (1991). - See Abstr. 012.065 for the main entry.

A need has arisen for efficient, blazed, symmetric gratings for use as beam splitters in far and extreme ultraviolet interferometers. In particular, the development of an all-reflection, far ultraviolet spatial heterodyne interferometer can benefit tremendously from such a grating. To fulfill this need, the authors have manufactured a mechanically ruled grating with a V-groove profile blazed for H Lyman α at 1216 Å. The authors present the grating performance at Lyman α in the context of its application to the spatial heterodyne interferometer.

035.177 **Progress with pn–CCDs for the XMM Satellite Mission.**
H. Väuninger, D. Hauff, P. Lechner, G. Lutz, W. Kink, N. Meidinger, G. Metzner, P. Predehl, C. Reppin, L. Strüder, J. Trümper, E. Kendziorra, R. Staubert, V. Radcka, P. Rehak, S. Rescia, G. Bertuccio, E. Gatti, A. Longoni, M. Sampietro, N. Findeis, P. Holl, J. Kemmer, C. von Zanthier.
SPIE Conference on EUV, X–Ray, and Gamma–Ray Instrumentation for Astronomy II as Part of SPIE International Symposium on Optical Applied Science and Engineering, p. 330 – 339 (1991). – See Abstr. 012.065 for the main entry.
Pn–CCDs are being developed as focal plane detectors for ESA's X–ray Multi–Mirror satellite mission (XMM), to be launched at the end of this century. As a part of the European Photon Imaging Camera (EPIC) the pn–CCDs will convert the incoming X–ray radiation with high quantum efficiency, low readout noise, excellent background rejection, timing in the μsec regime, radiation tolerance up to several hundreds of krads and a position resolution tailored according to the angular resolution of the telescope. This article focuses on the position resolution capabilities of fully depleted pn–CCDs, some recent results in the noise analysis and preliminary results on 10 MeV proton damage.

035.178 **The CCD focal plane imaging detector for the JET–X instrument on Spectrum R–G.**
A. Wells, C. Castelli, A. Holland, K. J. McCarthy, J. E. Spragg, C. H. Whitford.
SPIE Conference on EUV, X–Ray, and Gamma–Ray Instrumentation for Astronomy II as Part of SPIE International Symposium on Optical Applied Science and Engineering, p. 357 – 373 (1991). – See Abstr. 012.065 for the main entry.
The two co–aligned X–ray imaging telescopes in the JET–X instrument on Spectrum–R–G use passively cooled X–ray sensitive CCDs in the focal plane imagers. Development of the CCD for JET–X, the design and performance of the focal plane assembly are reviewed and improvements to both the low energy and high energy quantum efficiency are described. Performance data on spatial and energy resolution, and the rejection efficiency to charged particle background, obtained with prototype CCDs, are presented.

035.179 **Instrument design and test results of the new all–reflection spatial heterodyne spectrometer.**
B. C. Bush, D. M. Cotton, J. S. Vickers, S. Chakrabarti.
SPIE Conference on EUV, X–Ray, and Gamma–Ray Instrumentation for Astronomy II as Part of SPIE International Symposium on Optical Applied Science and Engineering, p. 376 – 384 (1991). – See Abstr. 012.065 for the main entry.
An all–reflection spatial heterodyne spectrometer (SHS) has been recently developed. The advantages over conventional high–resolution grating spectrometers are that the SHS requires no mechanical scanning, a self–compensating optical design permits easy alignment, and it is much smaller than other spectrometers of comparable resolution. Since all beam–splits and recombinations occur by reflection off of a diffraction grating, the interferometer is capable of operating well into the extreme ultraviolet (EUV) and possibly into the soft X–ray region. A description of the design and the characteristics of the instrument is presented.

035.180 **Performance of micro–strip proportional counters for X–ray astronomy on spectrum–roentgen–gamma.**
C. Budtz–Jørgensen, A. Bahnsen, F. E. Christensen, M. Møhl Madsen, C. Olesen, H. W. Schnopper.
SPIE Conference on EUV, X–Ray, and Gamma–Ray Instrumentation for Astronomy II as Part of SPIE International Symposium on Optical Applied Science and Engineering, p. 429 – 437 (1991). – See Abstr. 012.065 for the main entry.
DSRI will provide a set of four imaging proportional counters for the Danish–Soviet X–ray telescopes XSPECT/SODART. The sensor principle is based on the novel micro–strip proportional counter (MSPC), where the strip electrodes are deposited by photolithography onto a rigid substrate. The MSPC offers many advantages: A uniform gas gain, an excellent energy resolution, the possibility to match the strip pitch to the desired positron resolution, a fast charge collection and low operating voltages. The MSPC is capable of operating at high counting rates (10^5ph s^{-1}) and the electronics is designed to match this capability.

035.181 **High resolution decoding of multi–anode microchannel array detector.**
D. B. Kasle, J. S. Morgan.
SPIE Conference on EUV, X–Ray, and Gamma–Ray Instrumentation for Astronomy II as Part of SPIE International Symposium on Optical Applied Science and Engineering, p. 52 – 58 (1991). – See Abstr. 012.065 for the main entry.
The Multi–Anode Microchannel Array (MAMA) is a photon counting detector with utilizes a photocathode for photon to electron conversion, a microchannel plate for signal amplification and a proximity focused anode array for position sensitivity. The detector electronics decode the position of an even through coincidence discrimination. The decoding algorithm which associates a given event with the appropriate pixel is determined by the geometry of the array. A new algorithm incorporated into a CMOS Application Specific Integrated Circuit decoder which improves the pixel spatial resolution is described.

035.182 **Dead–time effects in microchannel–plate imaging detectors.**
M. V. Zombeck, G. W. Fraser.
SPIE Conference on EUV, X–Ray, and Gamma–Ray Instrumentation for Astronomy II as Part of SPIE International Symposium on Optical Applied Science and Engineering, p. 90 – 100 (1991). – See Abstr. 012.065 for the main entry.
The authors report on laboratory measurements of the count rate characteristics of an MCP detector as a function of illuminated area. These measurements have been made by illuminating the imaging detector (the ROSAT flight back–up) with soft X–rays collimated by a series of four pinholes varying in diameter from 0.005 mm to 0.100 mm. A model for the observed experimental behavior and consequences of the experimental results for the interpretation of space observations and for the ground testing of EUV and X–ray telescopes are discussed.

035.183 **Imaging gas scintillation proportional counters for ASTRO–D.**
T. Ohashi, K. Makishima, M. Ishida, T. Tsuru, M. Tashiro, T. Mihara, Y. Kohmura, H. Inoue.
SPIE Conference on EUV, X–Ray, and Gamma–Ray Instrumentation for Astronomy II as Part of SPIE International Symposium on Optical Applied Science and Engineering, p. 9 – 19 (1991). – See Abstr. 012.065 for the main entry.
The design and performance of imaging gas scintillation proportional counters (IGSPCs) developed for ASTRO–D are described. The 10 μm beryllium window has enabled the X–ray sensitivity down to 0.7 keV for a sealed–off gas counter, and a position–sensitive phototube equipped with multi–wire anode gives a good imaging capability. The laboratory model of the IGSPC shows an energy resolution 8% FWHM at 6 keV and a position resolution 0.5 mm FWHM with an effective area of 50 mm diameter.

035.184 **The first year of observations with the Hubble Space Telescope.**
A. L. Kinney, S. P. Maran.
Publ. Astron. Soc. Pac., Vol. 103, No. 670, p. 1237 – 1249 (Dec 1991). Invited review paper.
The Hubble Space Telescope has gathered unique information that supports many very significant scientific investigations during its early orbital operations, despite the limitations imposed by an unexpected spherical aberration in the primary mirror. These early results, gathered primarily with the HST

cameras and spectrographs, are reviewed, with emphasis on extragalactic, stellar, and nebular astronomy.

035.185 **The Hard X-ray Telescope (HXT) for the SOLAR-A mission.**
T. Kosugi, K. Makishima, T. Murakami, T. Sakao, T. Dotani, M. Inda, K. Kai, S. Masuda, H. Nakajima, Y. Ogawara, M. Sawa, K. Shibasaki.
Sol. Phys., Vol. 136, No. 1, p. 17 – 36 (Nov 1991). Special issue: The YOHKOH (SOLAR-A) mission. For the hardcover edition see 003.013.

The Hard X-ray Telescope (HXT) is a Fourier-synthesis imager; a set of spatially-modulated photon count data are taken from 64 independent subcollimators and are Fourier-transformed into an image by using procedures such as the maximum entropy method (MEM) or CLEAN. The HXT takes images of solar flares simultaneously in four energy bands, nominally 15 (or 19) – 24, 24 – 35, 35 – 57, and 57 – 100 keV, with an ultimate angular resolution as fine as ~ 5 arcsec and a time resolution 0.5 s. Each subcollimator has a field of view wider than the solar disk. The total effective area of the collimator/detector system reaches $\sim 70\ \mathrm{cm}^2$, about one order of magnitude larger than that of the HINOTORI hard X-ray imager. Thanks to these improvements, HXT will for the first time enable us to take images of flares at photon energies above ~ 30 keV. These higher-energy images will be compared with lower-energy ones, giving clues to the understanding of nonthermal processes in solar flares, i.e., the acceleration and confinement of energetic electrons. It is of particular importance to specify the acceleration site with regard to the magnetic field figuration in a flaring region, which will be achieved by collaborative observations between HXT and the Soft X-ray Telescope on board the same mission.

035.186 **The Soft X-ray Telescope for the SOLAR-A mission.**
S. Tsuneta, L. Acton, M. Bruner, J. Lemen, W. Brown, R. Caravalho, R. Catura, S. Freeland, B. Jurcevich, M. Morrison, Y. Ogawara, T. Hirayama, J. Owens.
Sol. Phys., Vol. 136, No. 1, p. 37 – 67 (Nov 1991). Special issue: The YOHKOH (SOLAR-A) mission. For the hardcover edition see 003.013.

The Soft X-ray Telescope (SXT) of the SOLAR-A mission is designed to produce X-ray movies of flares with excellent angular and time resolution as well as full-disk X-ray images for general studies. A selection of thin metal filters provide a measure of temperature discrimination and aid in obtaining the wide dynamic range required for solar observing. The co-aligned SXT aspect telescope will yield optical images for aspect reference, white-light flare and sunspot studies, and, possibly, helioseismology. This paper describes the capabilities and characteristics of the SXT for scientific observing.

035.187 **The Wide Band Spectrometer on the SOLAR-A.**
M. Yoshimori, K. Okudaira, Y. Hirasima, T. Igarashi, M. Akasaka, Y. Takai, K. Morimoto, T. Watanabe, K. Ohki, J. Nishimura, T. Yamagami, Y. Ogawara, I. Kondo.
Sol. Phys., Vol. 136, No. 1, p. 69 – 88 (Nov 1991). Special issue: The YOHKOH (SOLAR-A) mission. For the hardcover edition see 003.013.

The SOLAR-A spacecraft has spectroscopic capabilities in a wide energy band from soft X-rays to gamma-rays. The Wide Band Spectrometer (WBS), consisting of three kinds of spectrometers, soft X-ray spectrometer (SXS), hard X-ray spectrometer (HXS) and gamma-ray spectrometer (GRS), is installed on SOLAR-A to investigate plasma heating, high-energy particle acceleration, and interaction processes. The SXS observations enable to study the thermal evolution of flare plasma by obtaining time series of electron temperature and emission measures of hot plasma; the HXS observations enable to study the electron acceleration and heating mechanisms by obtaining time series of the electron spectrum; and the GRS observations enable to study the high-energy electron and ion acceleration and interaction processes by obtaining time series of electron and ion spectra.

035.188 **The Bragg Crystal Spectrometer for SOLAR-A.**
J. L. Culhane, E. Hiei, G. A. Doschek, A. M. Cruise, Y. Ogawara, Y. Uchida, R. D. Bentley, C. M. Brown, J. Lang, T. Watanabe, J. A. Bowles, R. D. Deslattes, U. Feldman, A. Fludra, P. Guttridge, A. Henins, J. Lapington, J. Magraw, J. T. Mariska, J. Payne, K. J. H. Phillips, P. Sheather, K. Slater, K. Tanaka, E. Towndrow, M. W. Trow, A. Yamaguchi.
Sol. Phys., Vol. 136, No. 1, p. 89 – 104 (Nov 1991). Special issue: The YOHKOH (SOLAR-A) mission. For the hardcover edition see 003.013.

The Bragg Crystal Spectrometer (BCS) is one of the instruments which makes up the scientific payload of the SOLAR-A mission. The spectrometer employs four bent germanium crystals, views the whole Sun and observes the resonance line complexes of H-like Fe XXVI and He-like Fe XXV, Ca XIX, and S XV in four narrow wavelength ranges with a resolving power ($\lambda/\Delta\lambda$) of between 3000 and 6000. The spectrometer has approaching ten times better sensitivity than that of previous instruments thus permitting a time resolution of better than 1 s to be achieved. The principal aim is the measurement of the properties of the 10 to 50 million K plasma created in solar flares with special emphasis on the heating and dynamics of the plasma during the impulsive phase. This paper summarizes the scientific objectives of the BCS and describes the design, characteristics, and performance of the spectrometers.

035.189 **Recent advances in detectors for the EUV.**
M. L. Lampton.
1. Berkeley Colloquium on Extreme Ultraviolet Astronomy, p. 353 – 363 (1991). – See Abstr. 012.045 for the main entry.

The author reviews the principal techniques that are available to sense low-light-level EUV images, and presents recent results on the most promising of these detectors.

035.190 **In-orbit, EUV radiometric calibration of satellite instrumentation.**
P. L. Smith, W. H. Parkinson, M. Kock, M. C. E. Huber, M. Kühne.
1. Berkeley Colloquium on Extreme Ultraviolet Astronomy, p. 390 – 394 (1991). – See Abstr. 012.045 for the main entry.

Contents: 1. Introduction. 2. Calibration of spectroscopic telescopes. 3. EUV calibration lamp.

035.191 **The XUV wide field camera for ROSAT.**
J. P. Pye, K. A. Pounds, M. G. Watson, A. Wells.
1. Berkeley Colloquium on Extreme Ultraviolet Astronomy, p. 409 – 426 (1991). – See Abstr. 012.045 for the main entry.

Contents: 1. Introduction. 2. Design. 3. The observing program. 4. Background. 5. Scientific capability of WFC. 6. Predicted numbers of sources.

035.192 **The EUV capabilities of the Hopkins Ultraviolet Telescope.**
A. F. Davidsen, S. T. Durrance, K. S. Long, R. A. Kimble, C. W. Bowers.
1. Berkeley Colloquium on Extreme Ultraviolet Astronomy, p. 427 – 436 (1991). – See Abstr. 012.045 for the main entry.

Hopkins Ultraviolet Telescope (HUT) possesses unique capabilities for spectroscopic observations in the EUV, from 420 to 912 Å. The EUV observations of HUT during its mission aboard the shuttle will provide pioneering data that address a wide range of astrophysical problems.

035.193 **The *ORFEUS* mission.**
M. Grewing, I. Appenzeller, W. Gringel, N. Kappelmann, H. Mandel, J. Petrik, G. Krämer, N. Eberhard, A. Kälble, J. Krautter, R. Östreicher, J. Riegger.
1. Berkeley Colloquium on Extreme Ultraviolet Astronomy, p. 437 – 441 (1991). – See Abstr. 012.045 for the main entry.

The authors briefly describe the main characteristics of the ASTRO-SPAS platform, the *ORFEUS* FUV/EUV telescope, and one of the two spectrographs (the FUV echelle). The second spectrograph, which covers both the FUV and EUV ranges, is described by Hurwitz and Bowyer (this volume).

035.194 **The Berkeley EUV spectrometer for the *ORFEUS* mission.**
M. Hurwitz, S. Bowyer.
1. Berkeley Colloquium on Extreme Ultraviolet Astronomy, p. 442 – 447 (1991). – See Abstr. 012.045 for the main entry.
Contents: 1. Introduction and background. 2. The Berkeley spectrometer. 3. Spectral resolution error budget. 4. Instrument sensitivity and performance.

035.195 ***ALEXIS*: a narrow–band survey/monitor for the ultrasoft X–ray sky.**
W. C. Priedhorsky, F. A. Córdova, M. D. Ulibarri, H. L. Marshall, P. W. Vedder, J. J. Bloch, B. W. Smith, O. H. W. Siegmund, J. V. Vallerga.
1. Berkeley Colloquium on Extreme Ultraviolet Astronomy, p. 464 – 477 (1991). – See Abstr. 012.045 for the main entry.
Contents: 1. Introduction. 2. Optical design. 3. Microchannel plate detectors. 4. Background considerations. 5. Compatibility with minisatellite carrier: experiment envelope. 6. Ground support hardware and software. 7. Scientific objectives and sensitivity.

035.196 **Annealing of far infrared photoconductors with infrared diode lasers.**
U. Schießl, M. Tacke, J. Wolf, D. Lemke.
Exp. Astron., Vol. 2, No. 3, p. 139 – 148 (1991).
High sensitivity infrared detectors made from extrinsic semiconductors are used for astronomical applications on satellite missions in the earth orbit. High energy radiation reduces the sensitivity and changes the calibration of these photoconductors drastically. This process is reversible through on board annealing. Laser emission of mid infrared diode lasers was used for efficient annealing with low power dissipation.

035.197 **A very large telescope for neutrino and gamma astronomy.**
G. Iaselli, F. Nuzzo, F. Romano, A. Rossi, G. Susinno, A. Grillo, A. Marini, F. Ronga, V. Valente, P. Bernardini, P. Pistilli, M. Ambrosio, G. Barbarino, B. Bartoli, V. Silvestrini, G. Bressi, A. Lanza, M. Cambiaghi, S. Ratti, M. Bonori, M. Conversi, G. d'Agostini, M. de Vincenzi, P. Lipari, F. Massa, M. Mattioli, A. Nigro, R. Cardarelli, R. Santonico, L. de Cesare, G. Grella, M. Guida, G. Marini, G. Romano, G. Vitiello.
Vulcano Workshop 1988: Frontier objects in astrophysics and particle physics, p. 329 – 350 (1989). – See Abstr. 012.079 for the main entry.
The authors discuss the feasibility of a telescope consisting in a sampling array for extensive air showers measure combined with a muon tracking device. The sampling array will extend over a surface of $\sim 10^6 m^2$ while the muon tracking device will cover $\sim 10^4 m^2$. The telescope should be performed with a combination of streamer tubes and resistive plates counters and would became the most powerfull device to study high energy neutrinos and gamma rays astronomy as well as cosmic ray physics in the highest energy ($\sim 10^{18}$eV) region.

035.198 **A secondary wobbling mechanism for a balloon–borne infrared telescope.**
H. Matsuhara, N. Hiromoto, H. Shibai, T. Nakagawa, H. Okuda, T. Maihara.
Rev. Sci. Instrum., Vol. 62, No. 11, p. 2535 – 2539 (Nov 1991). Current Physics Microform No.: 9112A0013.
A wobbling mechanism for a secondary mirror has been developed for a balloon–borne infrared telescope. Friction of the wobbling mechanism is negligibly small, and hence the wobbling mechanism is very reliable for the use in a severe environment at balloon altitudes. Motion is controlled by servo electronics, whose transfer function includes the second–order differential term of the error signal in order to improve the waveform. Good performance of the drive mechanism has been confirmed in two balloon flights in 1988 at an altitude of 31 km.

035.199 **Sensitivity changes in the CNRS ultraviolet spectrometer aboard OSO–8.**
P. Lemaire.
ESA J., Vol. 15, No. 3/4, p. 237 – 242 (1991).
The CNRS solar ultraviolet multichannel experiment on the OSO–8 spacecraft operated from Jun 1975 to Sep 1978. At the beginning of orbital operations, a strong sensitivity change appeared in the ultraviolet channels. After a brief description of the instrumentation, the main optical parameters and the principle of the experiment, and the integration procedures and preflight calibration control, the in–orbit measurements of sensitivity are presented and the results are critically discussed.

035.200 **IVS – a second generation space VLBI mission.**
G. Pilbratt.
IAU Colloquium No. 131: Radio interferometry – theory, techniques, and applications, p. 102 – 106 (1991). – See Abstr. 012.089 for the main entry.
The International VLBI Satellite is under assessment study by the European Space Agency and the USSR Academy of Sciences. The IVS mission is conceived as a second generation space VLBI mission with the goal of providing higher sensitivity, greater maximum resolution and a wider spectral coverage compared to the precursor missions and, in addition, some single–dish capability.

035.201 **Simulations of space VLBI.**
D. W. Murphy.
IAU Colloquium No. 131: Radio interferometry – theory, techniques, and applications, p. 107 – 111 (1991). Also published in Adv. Space Res., Vol. 11, No. 2, p. 415 – 420 (1991). – See Abstr. 012.089 for the main entry.
The use of simulations to examine various trade–offs and estimating the scientific return from the space VLBI missions RADIOASTRON and VSOP is discussed. The impact of the on–board spacecraft constraints is determined.

035.202 **Space ground interferometer.**
A. I. Savin, M. B. Zaxon, L. I. Matveyenko.
IAU Colloquium No. 131: Radio interferometry – theory, techniques, and applications, p. 112 – 114 (1991). – See Abstr. 012.089 for the main entry.
A space project for studying ecological Earth problems is being carried out by means of radio techniques. A 30 m prototype antenna has already been deployed and tested. The radio telescope will be launched in 1994 into a circular orbit having an altitude 600 km and an inclination of 65°. The planned mission time is $\geqslant 1.5$ year; 25% of this will be available for VLBI observations at wavelengths 6 and 18 cm.

035.203 **Background in space–borne low–energy γ–ray telescopes.**
A. J. Dean, F. Lei, P. J. Knight.
Space Sci. Rev., Vol. 57, No. 1/2, p. 109 – 186 (Jul 1991).
In this paper the sources of background noise in space–borne gamma–ray telescopes are reviewed with the aim to obtain a quantitative analysis of individual contributions, as derived from the various classes of irradiative particle fluxes. The estimated contributions from the individual sources are combined in order to evaluate the total background level of a given telescope in a specific radiation environment, which for practical considerations generally relates to the orbit choice and detailed design of the telescope. The published background noise spectra of a number of past missions are compared to the computed values so as to provide an assessment of the validity of the overall calculations. The level of agreement achieved indicates that a good understanding of the sources of background noise exists. Finally some possibilities for the improvement of the sensitivity of future gamma–ray telescopes, in terms of the reduction of the background noise, are discussed.

035.204 **Quantitative assessment of the science return from an orbiting, imaging, optical interferometer.**
D. L. Meier, S. P. Synnott, P. J. Dumont, D. L. Jones.
IAU Colloquium No. 131: Radio interferometry – theory, techniques, and applications, p. 368 – 371 (1991). – See Abstr. 012.089 for the main entry.
The authors describe an investigation of the science potential of the filled Mills cross orbiting, imaging, optical interferometer proposed by Synnott et al. and the constraints that science places on the design of the instrument. Imaging simulations were performed with this design in order to determine its capabilities in the area of spatial resolution, dynamic range, and sensitivity. In addition, specific astronomical objects of high interest were chosen as candidate program sources to determine what science was desired for these objects, what spatial resolution, etc. was needed to perform that science, and whether or not this interferometer design could do the science.

035.205 **Aperture synthesis imaging from the Moon.**
J. O. Burns.
IAU Colloquium No. 131: Radio interferometry – theory, techniques, and applications, p. 420 – 427 (1991). – See Abstr. 012.089 for the main entry.
Four candidate imaging aperture synthesis concepts are described for possible emplacement on the Moon beginning in the next decade. These include an optical interferometer with 10 μarcsec resolution, a submillimeter array with 6 mas resolution, a Moon–Earth VLBI experiment, and a very low frequency interferometer in lunar orbit.

035.206 **JET–X – a Joint European X–ray Telescope for SPECTRUM–X.**
G. C. Stewart.
Workshop on Iron Line Diagnostics in X–Ray Sources, p. 267 – 273 (1991). – See Abstr. 012.088 for the main entry.
The Joint European X–ray Telescope, JET–X, is one of the core instruments of the scientific payload of the USSR SPECTRUM–X astrophysics mission due for launch in 1993. The JET–X instrument is described and its scientific performance and capability discussed particularly in the context of spectroscopy of the iron–line complex at 7 keV.

035.207 **X–ray spectroscopy with the SODART/XSPECT Telescope.**
H. W. Schnopper.
Workshop on Iron Line Diagnostics in X–Ray Sources, p. 274 – 283 (1991). – See Abstr. 012.088 for the main entry.
The SOviet–DAnish Röntgen Telescope (SODART) on board the Spectrum Röntgen Gamma (SRG) satellite will be equipped with three different instruments devoted to X–ray spectroscopy. Each of the two thin foil telescopes has an 8 m focal length, a 60 cm diameter, a 1 deg field–of–view (FOV), a <2 arcmin half–power width and ca. 1700 and 1200 cm^2 collecting area at 2 and 8 keV, respectively. Broad band (0.25–25 keV) imaging spectroscopy over the full FOV will be obtained from low– and high–energy imaging proportional counters with energy resolution as good as 28 and 14% (FWHM) at 1.5 and 6 keV, respectively, and with an efficiency $>70\%$ (2–25 keV). The central 18 arcmin of the FOV of one telescope will be viewed by an array of 19 cooled silicon detectors with energy resolution of 140 and 160 eV at 2 and 6 keV, respectively, and efficiency $>85\%$ (2–20 keV). Each detector element will subtend a 3.4 arcmin FOV. A full FOV objective crystal spectrometer in front of the other telescope will obtain high resolution imaged spectra of point and extended sources with an energy resolution of 2, 1 and 2 eV at 0.7, 2.5 and 6.7 keV, respectively.

035.208 **Iron line spectroscopy with the AXAF calorimeter.**
S. S. Holt.
Workshop on Iron Line Diagnostics in X–Ray Sources, p. 289 – 299 (1991). – See Abstr. 012.088 for the main entry.
The AXAF calorimeter is a non–dispersive X–ray spectrometer that will be part of the AXAF focal plane instrumentation.

035.209 **The Hubble Space Telescope. Useful in spite of the mirror.**
G. Meylan.
Europhys. News, Vol. 22, No. 11, p. 206 – 209 (Dec 1991).
Assessments of the Hubble Space Telescope and comparisons with ground–based observations of several well–studied astronomical objects demonstrate the instrument's unique observational capabilities.

035.210 **Hubble, Hubble, toil and trouble.**
J. J. Caldwell.
J. R. Astron. Soc. Can., Vol. 85, No. 4, p. 178 (Aug 1991). Abstract. – See Abstr. 012.091 for the main entry.

035.211 **Spectroscopic science with the Hubble Space Telescope.**
J. Hutchings.
J. R. Astron. Soc. Can., Vol. 85, No. 4, p. 178 (Aug 1991). Abstract. – See Abstr. 012.091 for the main entry.

035.212 **The Space Telescope and the problems of cosmology.**
P. Crane.
12. Autumn School of Physics. The physical universe – the interface between cosmology, astrophysics and particle physics, p. 102 – 109 (1991). – See Abstr. 012.097 for the main entry.
The status of the Hubble Space Telescope (HST) as of 1990 October is reviewed and several photographs of astrophysically interesting objects are presented to demonstrate current capabilities. Observations relevant to cosmology which may be pursued by HST are reviewed.

035.213 **X–ray astronomy: the unkindest cut.**
E. Marshall.
Science, Vol. 254, No. 5031, p. 508 – 510 (25 Oct 1991).
A 30–year compaign for a Hubble–sized X–ray telescope has broad support from astronomers and survived a bet with Congress, but the launch date just slipped again.

035.214 **Hubble Space Telescope.**
P. Tenjes.
Calendar of the Tartu Astronomical Observatory for the year, 1992.
p. 37 – 38 (1991). In Estonian. See Abstr. 046.016.

035.215 **Design of a digital filter for ghost removal for the camera of the Galileo spacecraft.**
B. Turgeon, C. Cunningham, C. Anger.
J. R. Astron. Soc. Can., Vol. 85, No. 4, p. 199 (Aug 1991). Abstract. – See Abstr. 012.091 for the main entry.

035.216 **Soft X–ray telescope (SXT).**
S. Tsuneta, L. Acton.
International SOLAR–A Science Meeting: Flare physics in solar activity maximum 22, p. 18 – 19 (1991). Abstract. The full paper appeared in Sol. Phys., Vol. 136, see Abstr. 035.186. – See Abstr. 012.104 for the main entry.

035.217 **The hard X–ray Telescope (HXT) on board SOLAR–A.**
T. Sakao, HXT group.
International SOLAR–A Science Meeting: Flare physics in solar activity maximum 22, p. 20 – 21 (1991). Abstract. The full paper appeared in Sol. Phys., Vol. 136, see Abstr. 035.185. – See Abstr. 012.104 for the main entry.

035.218 **The SOLAR–A Bragg crystal spectrometer.**
J. L. Culhane, E. Hiei, R. D. Bentley, C. M. Brown, G. A. Doschek, U. Feldman, J. Lang, T. Watanabe.
International SOLAR–A Science Meeting: Flare physics in solar activity maximum 22, p. 22 – 23 (1991). Abstract. The full paper appeared in Sol. Phys., Vol. 136, see Abstr. 035.188. – See Abstr. 012.104 for the main entry.

035.219 **The wide band spectrometer on the SOLAR–A.**
M. Yoshimori, K. Okudaira, Y. Hirasima, T. Igarashi, M. Akasaka, Y. Takai, T. Watanabe, K. Ohki, J. Nishimura, T. Yamagami, I. Kondo.
International SOLAR–A Science Meeting: Flare physics in solar activity maximum 22, p. 24 – 25 (1991). Abstract. The full paper appeared in Sol. Phys., Vol. 136, see Abstr. 035.187. – See Abstr. 012.104 for the main entry.

035.220 **First results from the ROSAT XRT.**
J. H. M. M. Schmitt.
Adv. Space Res., Vol. 11, No. 11, p. 115 – 123 (1991). – See Abstr. 012.107 for the main entry.
The author summarises the first results obtained from the ROSAT X–ray telescope after one month of operation. The background encountered by the ROSAT position sensitive proportional counters (PSPC) is lower than anticipated, thus increasing the point source detection sensitivity both in the all sky survey as well as in pointed observations. The gain of the ROSAT PSPC is highly uniform over the face of the counter and constant in time to better than one percent and thus allows the extraction of reliable pulse height spectra.

035.221 **The U.K. Wide Field Camera on ROSAT – first results.**
K. A. Pounds, A. A. Wells.
Adv. Space Res., Vol. 11, No. 11, p. 125 – 134 (1991). – See Abstr. 012.107 for the main entry.
In the first month since launch of ROSAT all essential spacecraft and payload subsystems have been checked out and shown to be working. First in–orbit data from the UK Wide Field Camera (WFC) are reviewed following a description of the WFC instrumentation.

035.222 **A 40–cm UV (2000 Å) balloon–borne imaging telescope: results and current work.**
B. Milliard, J. Donas, M. Laget.
Adv. Space Res., Vol. 11, No. 11, p. 135 – 138 (1991). – See Abstr. 012.107 for the main entry.
A 40–cm UV (2000 Å) telescope equipped with an image intensifier plus film was designed to be flown on a stabilized balloon–borne gondola. Five flights totaling ~10 hours exposure time have provided nominal quality observations in different astronomical fields as the light distributions and contents of galactic globular clusters, the star formation rates as traced by the intermediate mass stars in resolved galaxies or in clusters of galaxies, and counts in low extinction regions.

035.223 **Imaging detectors for FUV and EUV wavelengths.**
G. W. Fraser.
Adv. Space Res., Vol. 11, No. 11, p. 155 – 166 (1991). – See Abstr. 012.107 for the main entry.
Microchannel plate (MCP) detectors are currently the detectors of choice for astrophysical imaging and spectroscopy in the wavelength range 50 – 2000 Å. In this paper, the author firsts describe alternative detectors for the EUV and FUV bands. In particular, he considers the extent to which back–illuminated CCDs may, in theory or in practice, replace MCPs in future satellite missions.

035.224 **Fundamental and practical limitations of FUV/EUV photon–counting image detectors.**
M. Lampton.
Adv. Space Res., Vol. 11, No. 11, p. 167 – 170 (1991). – See Abstr. 012.107 for the main entry.
In EUV and FUV space astronomy application, the best contemporary detector designs are based on the use of MCPs, due to their ability to deliver photon counting performance with high efficiency. The author discusses how the centroid position encoder techniques overcome the four chief limitations of the discrete–wire readouts – their limited format size, their flat focal surface, their fundamental hex–channel vs. square–pixel moire pattern, and their image undersampling. With these limitations overcome, microchannel based image systems can deliver the performance demanded by the forthcoming generation of application in space astronomy.

035.225 **FUV and EUV spectrographs: development in the 1980s for astronomy in the 1990s.**
W. Cash.
Adv. Space Res., Vol. 11, No. 11, p. 171 – 179 (1991). – See Abstr. 012.107 for the main entry.
This is a review of the rapid developments of the technology for spectrographs to support the unique and difficult requirements of FUV and EUV astronomy. It summarizes the progress these spectrographs will allow, highlights the challenges that FUV and EUV astronomy will present for the following generation of observatory design, and discusses some of the possible routes to overcoming those challenges.

035.226 **A spectrometer for studies of diffuse emission in the 900 – 1200 Å band.**
J. Edelstein, S. Bowyer.
Adv. Space Res., Vol. 11, No. 11, p. 181 – 184 (1991). – See Abstr. 012.107 for the main entry.
A rich assortment of emission processes from the interstellar medium should generate complex spectra in the 900 – 1200 Å regime. However, the detection of these lines is a difficult instrumental task since geocoronal and interplanetary airglow lines form an unwanted background requiring rejection by at least six orders of magnitude. The authors describe here the design and performance of an imaging spectrometer. This spectrometer has been designed for 2.5 Å spectral resolution, low instrumental scattering, and a high sensitivity to diffuse emission.

035.227 **Optical coating technology for the EUV.**
J. F. Osantowski, R. A. M. Keski–Kuha, H. Herzig, A. R. Toft, J. S. Gum, C. M. Fleetwood.
Adv. Space Res., Vol. 11, No. 11, p. 185 – 201 (1991). – See Abstr. 012.107 for the main entry.
Advances in optical coating and materials technology have been one of the key motivators for the development of missions such as the Far Ultraviolet Spectroscopic Explorer recently selected by NASA for an Explorer class mission in the mid 1990's. The authors review the performance of a range of candidate coatings for normal incidence and glancing incidence applications, and discuss strengths and problem areas for their use in space. The importance of recent developments in multilayer films, chemical vapor deposited SiC mirrors, and SiC films are discussed in the context of EUV instrumentation design.

035.228 **The Extreme Ultraviolet Explorer mission.**
S. Bowyer, R. F. Malina.
Adv. Space Res., Vol. 11, No. 11, p. 205 – 215 (1991). – See Abstr. 012.107 for the main entry.
The Extreme Ultraviolet Explorer (EUVE) mission, currently scheduled for launch in Sep 1991, is described. The primary purpose of the mission is to survey the celestial sphere for astronomical sources of extreme ultraviolet (EUV) radiation. The survey will be accomplished with the use of three EUV telescopes, each sensitive to a different segment of the EUV band. A fourth telescope will perform a high sensitivity search of a limited sample of the sky in the shortest wavelength bands. The second phase of the mission will be devoted to spectroscopic observations of EUV sources. The performance of the instrument components is described. A newly invented, high efficiency grazing incidence spectrometer using variable line–space gratings will provide spectral data with ~1 Å resolution. An end–to–end model of the mission has been constructed.

035.229 **The Berkeley EUV spectrometer for ORFEUS.**
M. Hurwitz, S. Bowyer.
Adv. Space Res., Vol. 11, No. 11, p. 217 – 220 (1991). – See Abstr. 012.107 for the main entry.
The Space Astrophysics Group is currently fabricating a novel EUV spectrometer for the ORFEUS–SPAS mission. It uses a set of four varied line–space spherical diffraction gratings to obtain high resolution spectra of point sources at wavelengths between

390 and 1200 Å. The spectra are recorded with two detector units, each containing curved surface microchannel plates and a delay–line anode readout system. An independent optical system detects the image of the source in the entrance aperture and tracks the source as it drifts during an observation, enabling a reconstruction of the spectra postflight. The overall system performance is discussed and illustrated by synthetic spectra.

035.230 **The Lyman Far Ultraviolet Spectroscopic Explorer.**
W. Moos.
Adv. Space Res., Vol. 11, No. 11, p. 221 – 227 (1991). – See Abstr. 012.107 for the main entry.

The Lyman Far Ultraviolet Spectroscopic Explorer mission will obtain high resolution spectra in the 912 to 1200 Å region with sufficient sensitivity to study faint sources throughout the galaxy and at large extragalactic distances. This spectral region provides unique tools for solving problems in cosmology, galactic structure and evolution, stellar evolution, and planetary science. The transmission of the glancing incidence telescope will enable extension of the spectral coverage with moderate resolution down to 100 Å. The Phase A instrument design consists of a 70 cm Wolter type II telescope feeding either near normal incidence gratings on the Rowland circle for the longer wavelengths or glancing incidence EUV gratings for coverage down to 100 Å.

035.231 **The scintillation gamma–spectrometer for the determination of the element composition of Martin rocks on "Phobos" spacecraft.**
Yu. A. Surkov, L. P. Moskaleva, A. G. Mityugov, V. P. Kharyukova, S. E. Zajtseva, G. G. Smirnov, O. P. Shcheglov, V. L. Gimadov, V. N. Rasputnyj, L. N. Myasnikova, S. S. Bulychev.
Kosm. Issled., Tom 29, Vyp. 6, p. 933 – 943 (Nov – Dec 1991). In Russian. English translation in Cosmic Res., Vol. 29, No. 6.

The scientific loading of "Phobos" spacecraft included the gamma–spectrometer for determination of the element composition of Martian rocks. Ten background spectra on the mission route and 80 spectra over the Martian surface in some equatorial areas were measured by means of this instrument. In the process of preparation of the instrument for the flight the complex of the pre–flight tests was conducted.

035.232 **Hubble Space Telecope eller "historiens dyreste fejlkonstruktion".**
M. I. Andersen.
Astron. Tidsskr., Årg. 24, Nr. 3, p. 127 – 137 (1991).

035.233 **Das Kameraexperiment FREGAT an Bord der Phobos–Sonde.**
M. Danz, T. Mangoldt, U. Pihan, U. Weidlich, R. Berghänel, H. Lorenz, G. M. Richter.
Sterne, Band 67, Heft 5, p. 241 – 254 (1991).

The first year of HST observations. Proceedings.
See Abstr. 012.017.

Space astronomical telescopes and instruments. Proceedings.
See Abstr. 012.030.

Extreme ultraviolet astronomy. A selection of papers presented at the First Berkeley Colloquium on Extreme Ultraviolet Astronomy.
See Abstr. 012.045.

EUV, X–ray, and gamma–ray instrumentation for astronomy II. Proceedings.
See Abstr. 012.065.

Geoscience and a lunar base. A comprehensive plan for lunar exploration. Proceedings.
See Abstr. 012.117.

Infrared astronomy takes center stage.
See Abstr. 013.014.

Fabrication and characterization of replicated and lacquercoated grazing incidence optics for X–ray astronomy.
See Abstr. 031.014.

All–refractive telescope for next generation inter–satellite communication.
See Abstr. 031.039.

Metallic alternative to glass mirrors (active mirrors in aluminium). A review.
See Abstr. 031.050.

A method of making ultralight primary mirrors.
See Abstr. 031.051.

Project of a three reflection telescope for wide field ultraviolet observations.
See Abstr. 031.053.

Recent advances in EUV astronomy with multilayers.
See Abstr. 031.067.

A flat array large telescope concept for use on the moon, earth, and in space.
See Abstr. 031.071.

1024 x 1024 CCD cameras under development and operation at Lockheed PARL.
See Abstr. 034.032.

EUV/FUV response characteristics of photographic films for the multi–spectral solar telescope array.
See Abstr. 034.059.

Imaging pulse–counting detector systems for space ultraviolet astrophysics missions.
See Abstr. 034.079.

Further performance characteristics of a high sensitivity 64 x 64 element InSb hybrid focal plane array.
See Abstr. 034.080.

Si:Ga focal plane arrays for satellite and ground–based telescopes.
See Abstr. 034.081.

Microchannel–plate detectors for space–based astronomy.
See Abstr. 034.082.

The fabrication of toroidal and coma–corrected toroidal diffraction gratings from spherical master gratings using elastically–deformable substrates: a progress report.
See Abstr. 034.083.

High–dynamic–range MCP structures.
See Abstr. 034.114.

A high efficiency imaging extreme ultraviolet spectrometer.
See Abstr. 034.118.

Charge coupled devices (CCDs) in X–ray astronomy.
See Abstr. 034.128.

The HST spherical aberration and its effects on images.
See Abstr. 036.018.

Effect of PSF on HST spectrographs.
See Abstr. 036.019.

HST image processing: determination of best focus and effects of photon noise.
See Abstr. 036.020.

Phase retrieval using HST images.
See Abstr. 036.021.

MEM: new techniques, applications, and photometry.
See Abstr. 036.022.

On the problem of extremely narrow, color and position dependent, point spread functions.
See Abstr. 036.028.

Optical interferometers in space: what the program can do for HST; what HST can do for the program.
See Abstr. 036.034.

Maximum entropy deconvolution of a Wide Field Camera image of R 136.
See Abstr. 036.037.

Colors and isochrones for some HST Faint Object Camera bandpasses.
See Abstr. 036.043.

Scattered light in the echelle modes of the Goddard High Resolution Spectrograph aboard the HST: analysis of in–orbit science data.
See Abstr. 036.045.

Deconvolution and photometry on HST–FOC images.
See Abstr. 036.085.

Some algorithms and procedures useful to analyse HST–FOC images.
See Abstr. 036.096.

VLBI with TDRSS.
See Abstr. 036.134.

The EXOSAT results database.
See Abstr. 036.139.

Science observations with the IUE using the one gyro mode.
See Abstr. 036.142.

Knowledge based automated scheduling and planning tools for IUE.
See Abstr. 036.143.

The requirement to the baseline knowing of a space VLBI.
See Abstr. 036.240.

High spatial resolution techniques.
See Abstr. 036.276.

Lucy deconvolution of Hubble Space Telescope images.
See Abstr. 036.284.

A model for the Infrared Space Observatory photometer.
See Abstr. 036.341.

Attitude determination using star tracker data.
See Abstr. 036.343.

Artificial intelligence approaches to astronomical observation scheduling.
See Abstr. 036.347.

The ALEXIS project: current status and performance.
See Abstr. 051.002.

The Solar–A mission and its scientific aims.
See Abstr. 051.028.

First results from the X–ray astronomy mission ROSAT.
See Abstr. 051.030.

Future solar system missions.
See Abstr. 051.034.

The Ulysses mission in the high latitude heliosphere.
See Abstr. 051.035.

PRISMA: a space facility for studying rotation and activity.
See Abstr. 051.048.

The SPECTRUM–UV Project.
See Abstr. 051.049.

The space observatory Astro.
See Abstr. 051.050.

EXOSAT EUV observations.
See Abstr. 051.056.

Stellar corona and stellar flare observations with the Extreme Ultraviolet Explorer.
See Abstr. 051.057.

The SOLAR–A mission: an overview.
See Abstr. 051.058.

The *EUVE* mission.
See Abstr. 051.061.

The Lyman Far Ultraviolet Spectroscopic Explorer.
See Abstr. 051.063.

***Santa Maria*: an international orbiting observatory for multispectral (200–7000 Å) observations of astrophysical and solar system objects.**
See Abstr. 051.064.

Early results from the Far Infrared Absolute Spectrophotometer (FIRAS).
See Abstr. 051.065.

The Extreme Ultraviolet Explorer mission.
See Abstr. 051.068.

Capabilities of SODART/XSPECT for iron line detection.
See Abstr. 051.072.

Operational capabilities of AXAF for iron line diagnostics.
See Abstr. 051.073.

The proposed Israeli X–ray satellite.
See Abstr. 051.074.

The SOLAR–A mission.
See Abstr. 051.078.

Mass–spectrometric in situ analysis of solid–state extraterrestrial samples.
See Abstr. 051.083.

The SAC–B project: objective and instrumentation.
See Abstr. 051.087.

ASCHOT – astrophysical Schmidt orbital telescope.
See Abstr. 051.090.

High resolution gamma–ray and hard X–ray spectroscopy of solar flares.
See Abstr. 076.005.

The composition and plasma signature of a large dust impact on the Giotto spacecraft.
See Abstr. 106.029.

GHRS chromospheric emission line spectra of the red giant α Tau.
See Abstr. 112.031.

Theoretical colours and isochrones for some Hubble Space Telescope colour systems – II.
See Abstr. 113.043.

IUE far-ultraviolet spectra of Capella and γ Draconis for comparison to HST/GHRS GTO observations.
See Abstr. 114.033.

Very low mass stars and brown dwarfs.
See Abstr. 117.245.

Early results from the ROSAT Wide Field Camera.
See Abstr. 126.045.

Ultraviolet spectroscopic studies of the interstellar medium with the Hubble Space Telescope.
See Abstr. 131.067.

First results from BBXRT.
See Abstr. 142.010.

Balloon X-ray astronomical observations: a China-Japan collaborative transoceanic flight.
See Abstr. 142.051.

Gamma ray astrophysics at energies up to 10 GeV.
See Abstr. 143.037.

Space experiment in gamma-ray astronomy.
See Abstr. 143.048.

Un an d'astronomie γ à haute résolution avec le télescope SIGMA: du centre de la Galaxie à la nova de la Mouche.
See Abstr. 143.061.

Search for fast ultraviolet variability in NGC 5548 and non-linearities in IUE's SWP camera.
See Abstr. 158.074.

COBE's FIRAS: update on refining measurements of the cosmic microwave background radiation spectrum.
See Abstr. 161.082.

A millimeter-wave anisotropy experiment (MAX) to search for anisotropy in the cosmic background radiation on medium angular scales.
See Abstr. 161.356.

036 Methods of Observation and Reduction, Data Processing

036.001 **Direct imaging of nonsolar planets with infrared telescopes using apodized coronagraphs.**
S. M. Watson, J. P. Mills, S. L. Gaiser, D. J. Diner.
Appl. Opt., Vol. 30, No. 22, p. 3253 – 3262 (1 Aug 1991). Current Physics Microform No.: 9109A0181.

This research examines the use of modified Lyot coronagraphs with monolithic and segmented infrared telescopic systems for imaging nonsolar planets. These systems are investigated with the aim of reducing the effects of stellar diffracted energy on the planet image in the final image plane. A square telescope objective is best for this purpose. The associated coronagraph is composed of a cross-shaped apodizer in the first focal plane and either a square Lyot stop or circular corner Lyot stops in the corners of the pupil plane. The authors examine the consequences of segmenting the aperture and the effects of various segment spacings and random piston and tilt errors of the individual segments. A system to correct for the misalignments is proposed.

036.002 **Structures on galaxy counts from a wavelet analysis.**
E. Slezak, A. Bijaoui, G. Mars.
Bull. Inf. Cent. Données Stellaires, No. 38, p. 1 – 2 (Dec 1990). – See Abstr. 012.001 for the main entry.

036.003 **Upgrades to the "RPHOT" photometry package.**
R. Nolthenius.
I.A.P.P.P. Commun., No. 45, p. 18 – 23 (Sep – Nov 1991). – See Abstr. 012.002 for the main entry.

036.004 **Detached binaries and simple computer codes for fewer rocky roads.**
P. B. Etzel.
I.A.P.P.P. Commun., No. 45, p. 25 – 28 (Sep – Nov 1991). – See Abstr. 012.002 for the main entry.

036.005 **Analysis of rare signals at the Baksan underground scintillation telescope.**
E. N. Alekseev, L. N. Alekseeva, V. N. Zakidyshev, A. Yu. Reutov, A. E. Chudakov.
JETP Lett., Vol. 54, No. 8, p. 429 – 433 (25 Oct 1991). Current Physics Microform No.: 9201X2039.

Data accumulated since 1980 at the Baksan scintillation telescope in the course of a search for neutrinos from supernovae have been used to analyze the trajectories constructed from the individual counts of detectors over a time $\Delta t \leqslant 1$ s. An excess of signals amounting to $\sim 7\sigma$ has been found in the direction of the Sun. Difficulties in interpreting the effect are discussed.

036.006 **The use of statistics in astronomical observations.**
M. D. Joner.
I.A.P.P.P. Commun., No. 45, p. 41 – 44 (Sep – Nov 1991). – See Abstr. 012.002 for the main entry.

036.007 **Reduction, storage and distribution of helioseismic data for the GONG network.**
M. Trueblood.
I.A.P.P.P. Commun., No. 45, p. 45 – 51 (Sep – Nov 1991). – See Abstr. 012.002 for the main entry.

036.008 **A method of continuum estimation in spectral classification experiments.**
J. Pelt.
Bull. Inf. Cent. Données Stellaires, No. 38, p. 95 – 107 (Dec 1990).

A flexible method of continuum or pseudo-continuum estimation for digitally recorded spectra is presented. The properties of the algorithm are evaluated considering the requirements for spectral classification experiments.

036.009 **Minimization of the small powers of deviations as one of the robust methods.**
A. V. Loktin.
Bull. Inf. Cent. Données Stellaires, No. 38, p. 115 – 117 (Dec 1990).
On the basis of numerical experiments it is shown that in the case of asymmetric distribution of deviations the minimization of small powers of deviations can give the estimates near the mode of distribution.

036.010 **A new empirical approximation of the full characteristic curve of the negative.**
A. M. Sobolev, E. F. Kuznetsova (*Eh. F. Kuznetsova*), A. A. Kalinin.
Bull. Inf. Cent. Données Stellaires, No. 38, p. 119 – 123 (Dec 1990).
A new approximation function of the full characteristic curve (FCC) of the negatives is suggested. The function is flexible enough to represent intrinsic geometrical form of FCC, has only 2 (sometimes 3) free parameters and simple analytical expressions for the derivatives and inversion function.

036.011 **Propagation of errors in transformation of empirical distributions.**
A. E. Vasilevsky (*A. E. Vasilevskij*).
Bull. Inf. Cent. Données Stellaires, No. 38, p. 125 – 126 (Dec 1990).
It is shown that the use of regression lines in the procedure of transformation of empirical distributions is correct at least for smooth and monotonous functions connecting the arguments of distributions.

036.012 **Modelling and the interface between description and model: the role of the extreme terms of a sample on the behavior of density estimates.**
M. Broniatowski.
Bull. Inf. Cent. Données Stellaires, No. 38, p. 127 – 136 (Dec 1990).

036.013 **Die Entstehung eines CCD–Bildes.**
W. Bickel.
Sterne Weltraum, Jahrg. 30, Nr. 12, p. 760 – 764 (Dec 1991).

036.014 **ESA's on–line directory of space and earth–science data.**
M. James, L. Fusco, D. Lloyd.
ESA Bull., No. 67, p. 99 – 103 (Aug 1991).
Research and operational data needs in the 1990s require integrated data–management systems accessible to users at many sites. The ESA Prototype International Directory (ESA PID) is the European node of an international system of directories in order to support users of data from past and current missions.

036.015 **Adaptive Optik für extrem scharfe Sternbilder.**
R. Vaas.
Naturwiss. Rundsch., Vol. 44, No. 7, p. 263 – 268 (Jul 1991).

036.016 **Photometrie mit der ST–4 Kamera.**
S. Korth.
KPM, Jahrg. 6, No. 17, p. 9 – 11 (Aug 1991).
In this paper the hard– and software capabilities of the ST–4 CCD camera are described and reviewed. As test object an unfiltered frame of the open cluster M67 was used.

036.017 **Eine Methode zur Bestimmung genäherter Kleinplanetenörter.**
M. Kretlow.
KPM, Jahrg. 6, No. 17, p. 25 – 28 (Aug 1991).

036.018 **The HST spherical aberration and its effects on images.**
R. L. White, C. J. Burrows.
Workshop on the Restoration of HST Images and Spectra, p. 2 – 6 (1991). – See Abstr. 012.003 for the main entry.
The very first image taken a few weeks after the launch of the Hubble Space Telescope (HST) showed evidence for spherical aberration in the HST optics. The results of the next month of testing proved conclusively that the HST primary mirror has about 1/2 wave RMS of spherical aberration (λ = 5000 Å). The recently published paper by Burrows et al. (1991) gives a detailed description of the HST spherical aberration. In this paper the authors briefly summarize the problem and discuss its effects on HST imaging science.

036.019 **Effect of PSF on HST spectrographs.**
R. L. Gilliland.
Workshop on the Restoration of HST Images and Spectra, p. 7 – 12 (1991). – See Abstr. 012.003 for the main entry.
Basic capabilities of the two primary HST spectrographs are reviewed. The impact of a poorer than expected PSF enters through loss of throughput for small entrance apertures, loss of spectral resolution for observations through large apertures, and an unavoidable loss of spatial purity for observations in crowded fields. The extent to which deconvolution/restoration techniques can regain lost spectral resolution, and at what cost, for observations through large apertures is examined.

036.020 **HST image processing: determination of best focus and effects of photon noise.**
J. W. Beletic.
Workshop on the Restoration of HST Images and Spectra, p. 13 – 20 (1991). – See Abstr. 012.003 for the main entry.
The primary mirror of the Hubble Space Telescope (HST) suffers from spherical aberration that amounts to about 8 waves of wavefront error at the mirror edge. The deliterious effects of this aberration can be partially compensated by defocus. In this paper the author presents a methodology for optimizing the focus setting for image reconstruction.

036.021 **Phase retrieval using HST images.**
A. N. Lasenby.
Workshop on the Restoration of HST Images and Spectra, p. 21 – 30 (1991). – See Abstr. 012.003 for the main entry.
This talk discusses a new algorithm for achieving the phase reconstruction from out–of–focus images, and shows applications to both an existing millimetre telescope in Hawaii, and the HST itself.

036.022 **MEM: new techniques, applications, and photometry.**
N. Weir, S. Djorgovski.
Workshop on the Restoration of HST Images and Spectra, p. 31 – 38 (1991). – See Abstr. 012.003 for the main entry.
The authors examine a new implementation of the Maximum Entropy (MaxEnt) method of image restoration for direct astronomical images. Recent advances in the theory and application of MaxEnt have led to a significantly improved version of the algorithm, rendering virtually all previous MaxEnt codes obsolete. The authors suggest how the MaxEnt result may be used as an initial input to a standard least squares model fitting program to achieve unbiased photometry on objects detected in the deconvolved image. They demonstrate a first attempt at this hybrid approach on a simulated HST WFC image of a dense star cluster.

036.023 **Block iterative restoration of astronomical images from the Hubble Space Telescope.**
D. J. Lindler.
Workshop on the Restoration of HST Images and Spectra, p. 39 – 49 (1991). – See Abstr. 012.003 for the main entry.

036.024 **Telescope beam–profile diagnostics and the solar limb.**
C. A. Lindsey, T. L. Roellig.
Astrophys. J., Vol. 375, No. 1, p. 414 – 422 (1 Jul 1991).
The authors describe the basic method for determining the solar limb brightness profile properly corrected for spurious limb darkening caused by the far wings of the resolving beams encountered in large far–infrared and radio telescopes. When the far wings of the beam can be independently measured, this problem is usually amenable to standard deconvolution procedures. In practice, however, the far wings of the beam are best

determined using the solar limb itself. While one does not know a priori the solar brightness profile inside the limb, one may assume that the brightness profile outside the limb drops sharply to zero at submillimeter and millimeter wavelengths. One is thus interested in the problem of deconvolution, where one does not initially know the convolving kernel, but is partially informed of the source distribution. Under a broad range of well-defined cases, solutions to this deconvolution problem are unique to within the discrimination provided by the core of the beam profile. The authors apply the theory to solar limb scans made recently on the James Clerk Maxwell Telescope to show solar limb brightening in 850 μm radiation.

036.025 **Deconvolution of simulated HST frames by MIM.**
J. Pfleiderer.
Workshop on the Restoration of HST Images and Spectra, p. 50 – 55 (1991). – See Abstr. 012.003 for the main entry.

The minimum-information method (MIM) is based on a diophantic approximate solution of a linear system of equations. The data is convolved with the PSF and then deconvolved with the PSF convolved with itself. The quadratic smoothness constraint – minimization of structural information – is equivalent to an unconstrained deconvolution by a Prussian-helmet type modification of the deconvolving function. MIM compares favorably with other methods, as MEM, CLEAN, inverse or Wiener filter, or Lucy's algorithm. Advantages are a good resolution for point sources and extended sources, good recognition of weak extended features, and a linear response that makes photometry rather safe. Possible applications include images, visibilities, spectra, and time series. The author gives examples with a simulated HST image.

036.026 **Modifications of the Lucy–Richardson iteration for restoring Hubble Space-Telescope imagery.**
D. L. Snyder.
Workshop on the Restoration of HST Images and Spectra, p. 56 – 61 (1991). – See Abstr. 012.003 for the main entry.

The relationship of the Lucy–Richardson iteration method for restoring images to methods used for forming images of radionuclide distributions is identified. On the basis of this relationship, several modifications of the Lucy–Richardson iteration are suggested. These include flat-field corrections for detector nonuniformity, corrections for background radiation and read-out noise, and the inclusion of regularization to suppress noise and edge artifacts.

036.027 **Iterative Maximum Likelihood Estimator and Bayesian algorithms for image reconstruction in astronomy.**
J. Llacer, J. Nuñez.
Workshop on the Restoration of HST Images and Spectra, p. 62 – 70 (1991). – See Abstr. 012.003 for the main entry.

This paper discusses Maximum Likelihood Estimator and Bayesian methods of image reconstruction as applied to astronomy. The motivation for the algorithms is given in terms of the statistical distribution of the data to be reconstructed. Target functions and iterative formulas for Poisson distributed data and a new iterative method for Poisson data with additive Gaussian noise are described. The concept of feasible images is discussed in some detail and the expected benefits of using reconstruction methods yielding such images are illustrated by describing some results obtained in medical tomography.

036.028 **On the problem of extremely narrow, color and position dependent, point spread functions.**
A. Penny.
Workshop on the Restoration of HST Images and Spectra, p. 71 – 73 (1991). – See Abstr. 012.003 for the main entry.

This talk is concerned with the problems that the special nature of the HST point spread function will cause in the deconvolution of HST images. The discussion concentrates on the use of the Planetary Camera mode of the WF/PC.

036.029 **Remarks on hard modeling vs. image processing, circumstellar disks, and model validation.**
D. Geman.
Workshop on the Restoration of HST Images and Spectra, p. 74 – 79 (1991). – See Abstr. 012.003 for the main entry.

This contribution is a slightly revised version of the introductory section of "Statistical inference on the shape of circumstellar disks based on HST observations," co-authored with P. Dupuis, J. Horowitz, G. Reynolds, and S. Strom. The modifications are intended to emphasize the author's remarks about "off-the-shelf" solutions and performance evaluation.

036.030 **Restoration with increased sampling – images and spectra.**
L. B. Lucy.
Workshop on the Restoration of HST Images and Spectra, p. 80 – 87 (1991). – See Abstr. 012.003 for the main entry.

With respect to both images and spectra, the desire for wide coverage coupled with the limited array size of digitized detectors commonly results in a poor sampling of the instrumental response function. This led to work at ESO (Lucy and Baade 1989) in which a rebinning option was incorporated into a deconvolution package. A prime motivation for this development was in fact the HST since, according to specifications, it was expected to poorly sample the point spread function in the f/48 imaging modes. A further reason for rebinning is the displeasing appearance of modern astronomical images. This results partly from the coarse gridding imposed by the detector but most especially from pixel-to-pixel noise, since this destroys continuity and thus makes the eye acutely aware of the grid. In addition to this aesthetic consideration, one can well imagine that this prominence of the grid is deleterious for morphological and classification investigations of images.

036.031 **Deconvolution of Hubble Space Telescope data: computer simulations and laboratory experiments.**
T. Reinheimer, G. Weigelt.
Workshop on the Restoration of HST Images and Spectra, p. 88 – 94 (1991). – See Abstr. 012.003 for the main entry.

The authors have studied the deconvolution of simulated HST data by computer simulations and laboratory experiments. They discuss their results obtained with the roll deconvolution technique (Lohmann and Weigelt 1979, Müller and Weigelt 1987), inverse filtering, the van-Cittert method (van-Cittert 1931) and the Gerchberg method (Gerchberg 1974). The advantage of the computer and laboratory experiments is that it is possible to study the dependence of the SNR in the reconstructed image on aberrations and photon noise in the raw image.

036.032 **Image construction from the IRAS survey and data fusion.**
T. R. Bontekoe.
Workshop on the Restoration of HST Images and Spectra, p. 95 – 102 (1991). – See Abstr. 012.003 for the main entry.

The IRAS survey data can be used successfully to produce images of extended objects. The major difficulties, viz. non-uniform sampling, different response functions for each detector, and varying signal-to-noise levels for each detector for each scan, have been resolved. The results of three different image construction techniques are compared: co-addition, constrained least-squares, and maximum entropy. The maximum entropy result is superior. An image of the galaxy M51 with an average spatial resolution of 45 arcsec, using 60 micron survey data is presented. This exceeds the telescope diffraction limit of 1 minute of arc, at this wavelength. Data fusion is a proposed method for combining data from different instruments, with different spatial resolutions, at different wavelengths. Direct estimates of the physical parameters, temperature, density and composition, can be made from the data without prior image (re-)construction. An increase in the accuracy of these parameters is expected as the result of this more systematic approach.

036.033 **A Wiener filter version of Blind Iterative Deconvolution.**
P. Nisenson, C. Standley, D. Gay.
Workshop on the Restoration of HST Images and Spectra, p. 103 – 112 (1991). – See Abstr. 012.003 for the main entry.

Blind Iterative Deconvolution (BID) is a technique which was originally proposed for the correction of the effects of atmospheric turbulence on astronomical images. In this technique, both the restored image and the degrading point spread function (PSF) may be recovered from a single high–signal–to–noise ratio short exposure (speckle) image. At the Center for Astrophysics, the authors have implemented a modified version of BID using the general approach proposed by Ayers and Dainty (1988). The basic approach is to alternately deconvolve the original data by the PSF and then by the restored image. A set of physical constraints are applied after each iteration. The iterations are continued until an image and PSF are found which give the original data when convolved (with each other) and which adhere to the physical constraints.

036.034 **Optical interferometers in space: what the program can do for HST; what HST can do for the program.**
D. L. Meier.
Workshop on the Restoration of HST Images and Spectra, p. 113 – 120 (1991). – See Abstr. 012.003 for the main entry.

The Hubble Space Telescope is the first imaging optical interferometer to be launched into space. It is of the Fizeau class, forming its fringes in the image plane onto CCD detectors, thereby directly detecting the "dirty image". An unaberrated HST would have had little need for deconvolution of the dirty image with the point spread function. However, with the spherical aberration in the primary, deconvolution and other techniques commonly used in interferometric imaging are important. Some techniques which have been applied to the imaging simulations of optical interferometers at JPL are described, including image pre–processing which allows the CLEAN algorithm to work well on optical images. The author suggests a general restoration procedure for images obtained with HST, optical interferometers, and radio interferometers which is a hybrid of CLEAN and the Quantified Maximum Entropy Method.

036.035 **Restoration experiments at the ST–ECF.**
H.–M. Adorf, J. R. Walsh, R. N. Hook.
Workshop on the Restoration of HST Images and Spectra, p. 121 – 126 (1991). – See Abstr. 012.003 for the main entry.

Some experiments carried out with four different restoration methods on simulated HST Faint Object Camera f/96 images are described. The restored images were qualitatively and quantitatively evaluated. The two direct linear methods employed, inverse filtering and Wiener–filtering, generally delivered sharper images at the expense of more small–scale artifacts. The two iterative non–linear methods tested, maximum entropy and Lucy–restoration, delivered smoother extended structures with fewer artifacts. To the extent the methods could be tested, none of them restored both point sources and extended structures entirely satisfactorily.

036.036 **Linear restoration techniques: not all bad.**
R. J. Hanisch.
Workshop on the Restoration of HST Images and Spectra, p. 127 – 131 (1991). – See Abstr. 012.003 for the main entry.

Linear image deconvolution techniques such as the Fourier inverse and Wiener filter are often rejected for use in high fidelity image restoration work because of several types of artifacts that can be induced in the restored data. However, for certain scientific objectives the linear techniques can provide acceptable, and in some cases, preferable results. In this paper the Fourier inverse and Wiener filter techniques are reviewed, and sample restorations of HST images from both the Wide Field/Planetary Camera and the Faint Object Camera are shown.

036.037 **Maximum entropy deconvolution of a Wide Field Camera image of R 136.**
K. Horne.
Workshop on the Restoration of HST Images and Spectra, p. 132 – 138 (1991). – See Abstr. 012.003 for the main entry.

The author presents results of a test using the maximum entropy method (MEM) to deconvolve an image of R 136, the central star cluster of the 30 Doradus nebula, from data taken with HST's Wide–Field Camera. The deconvolved image reveals hundreds of stars within a radius of 2 arcseconds.

036.038 **Restoration of HST images using the Lucy method with oversampling.**
R. L. White.
Workshop on the Restoration of HST Images and Spectra, p. 139 – 143 (1991). – See Abstr. 012.003 for the main entry.

The author has been doing a number of experiments in the restoration of Hubble Space Telescope (HST) images using the Lucy method (Lucy 1974). Since this method is well described elsewhere, in this short paper he describes only the method used to implement oversampling and shows some results.

036.039 **Controlling the quality of photometeric data from unattended robotic telescopes.**
G. P. McCook.
Bull. Am. Astron. Soc., Vol. 23, No. 2, p. 873 – 874 (1991). Abstract. – See Abstr. 012.008 for the main entry.

036.040 **Initial results from the Four College APT: pulsations, outbursts, spots and eclipses from stars to quasars.**
E. F. Guinan, G. P. McCook.
Bull. Am. Astron. Soc., Vol. 23, No. 2, p. 874 (1991). Abstract. – See Abstr. 012.008 for the main entry.

036.041 **Photometric functions for photoclinometry and other applications.**
A. S. McEwen.
Icarus, Vol. 92, No. 2, p. 298 – 311 (Aug 1991).

Least–squared fits to the brightness profiles across a disk or "limb darkening" described by Hapke's photometric function are found for the simpler Minnaert and lunar–Lambert functions. The simpler functions are needed to reduce the number of unknown parameters in photoclinometry, especially to distinguish the brightness variations of the surface materials from that due to the resolved topography. The variation in limb darkening, pronounced for smooth surfaces, is reduced or eliminated when the Hapke parameters are in the range typical of most planetary surfaces; this result simplifies the problem of photoclinometry across terrains with variable surface materials.

036.042 **Recognition of γ–ray events in EGRET using neural networks.**
C. A. Meetre, J. P. Norris.
Bull. Am. Astron. Soc., Vol. 23, No. 2, p. 905 (1991). Abstract. – See Abstr. 012.008 for the main entry.

036.043 **Colors and isochrones for some HST Faint Object Camera bandpasses.**
R. A. Bell, G. Paltoglou.
Bull. Am. Astron. Soc., Vol. 23, No. 2, p. 906 (1991). Abstract. – See Abstr. 012.008 for the main entry.

036.044 **Crowded field stellar photometry of 30 Dor.**
S. R. Heap, E. M. Malumuth, J. D. Neill.
Bull. Am. Astron. Soc., Vol. 23, No. 2, p. 906 – 907 (1991). Abstract. – See Abstr. 012.008 for the main entry.

036.045 **Scattered light in the echelle modes of the Goddard High Resolution Spectrograph aboard the HST: analysis of in–orbit science data.**
J. A. Cardelli, B. D. Savage, D. C. Ebbets.
Bull. Am. Astron. Soc., Vol. 23, No. 2, p. 907 (1991). Abstract. – See Abstr. 012.008 for the main entry.

036.046 **A description of a proposed new algorithm for extracting spectral orders from IUE echelle high dispersion images.**
M. E. Van Steenberg, M. A. Smith.
Bull. Am. Astron. Soc., Vol. 23, No. 2, p. 907 (1991). Abstract. – See Abstr. 012.008 for the main entry.

036.047 **Hot DA white dwarfs as ultraviolet absolute flux standards.**
D. S. Finley, D. Koester.
Bull. Am. Astron. Soc., Vol. 23, No. 2, p. 907 (1991). Abstract. – See Abstr. 012.008 for the main entry.

036.048 **FITS formats for space data: ROSAT.**
R. A. White, R. Fink, R. Pisarski.
Bull. Am. Astron. Soc., Vol. 23, No. 2, p. 907 (1991). Abstract. – See Abstr. 012.008 for the main entry.

036.049 **High energy observations by the SMM satellite.**
W. T. Vestrand.
Bull. Am. Astron. Soc., Vol. 23, No. 2, p. 934 (1991). Abstract. – See Abstr. 012.008 for the main entry.

036.050 **FITSIO – a new Fortran–77 subroutine interface for reading–and writing FITS format files.**
W. D. Pence.
Bull. Am. Astron. Soc., Vol. 23, No. 2, p. 936 (1991). Abstract. – See Abstr. 012.008 for the main entry.

036.051 **Spectral data analysis on a personal computer.**
D. J. Lindler, J. K. Feggans, R. W. Thompson.
Bull. Am. Astron. Soc., Vol. 23, No. 2, p. 936 (1991). Abstract. – See Abstr. 012.008 for the main entry.

036.052 **ROSAT data processing and distribution system.**
R. L. Pisarski, P. E. Damon, E. R. Seufert, R. L. Fink, R. D. Price.
Bull. Am. Astron. Soc., Vol. 23, No. 2, p. 936 (1991). Abstract. – See Abstr. 012.008 for the main entry.

036.053 **Low–cost 3D visualisation techniques: an application to the 3–D distribution of nearby galaxies.**
D. Durand, M. J. Pierce.
Bull. Am. Astron. Soc., Vol. 23, No. 2, p. 954 – 955 (1991). Abstract. – See Abstr. 012.008 for the main entry.

036.054 **Application of an adaptive kernel technique for producing optical maps of clusters.**
K. Gebhardt, T. C. Beers, D. A. Ventimiglia, C. T. Vance.
Bull. Am. Astron. Soc., Vol. 23, No. 2, p. 955 (1991). Abstract. – See Abstr. 012.008 for the main entry.

036.055 **The EUVE software development.**
C. A. Dobson, C. A. Christian, E. Olson.
Bull. Am. Astron. Soc., Vol. 23, No. 2, p. 971 (1991). Abstract. – See Abstr. 012.008 for the main entry.

036.056 **Absolute spectrophotometry of the Sun. III. SEF–1 telescope automation system.**
K. A. Burlov–Vasil'ev, V. V. Kul'beda.
Kinematics Phys. Celest. Bodies, Vol. 7, No. 1, p. 67 – 70 (1991). English translation of Kinematika Fiz. Nebesn. Tel, Tom 7, No. 1, p. 72 – 77 (1991).
The hardware and software of a system that automates observations at the special–purpose SEF–1 telescope, which was designed for absolute spectrophotometry of the Sun, are described. The system can be used for measurements involving spectral analysis of the radiation, and its structure is open–ended for possible expansions. The automation system represented here makes it possible to observe under optimum conditions and acquire very high–quality information in large volumes.

036.057 **Proto–Jupiters.**
M. W. Pound, L. Blitz.
Bull. Am. Astron. Soc., Vol. 23, No. 2, p. 976 (1991). Abstract. – See Abstr. 012.008 for the main entry.

036.058 **A Bayesian period search method.**
P. C. Gregory, T. J. Loredo.
Bull. Am. Astron. Soc., Vol. 23, No. 2, p. 977 – 978 (1991). Abstract. – See Abstr. 012.008 for the main entry.

036.059 **Deprojection of emission in axially symmetric transparent systems.**
D. A. Leahy.
Astron. Astrophys., Vol. 247, No. 2, p. 584 – 589 (Jul 1991).
Methods for interpreting a two dimensional brightness distribution in terms of a three dimensional emissivity distribution are developed for the case that the emission is optically thin and has axial symmetry. Several simpler sub–cases are considered before the general case, with appropriate algorithms given for each of the cases. This work extends previously published work on this topic and presents iterative algorithms which are stable and give physically reasonable results.

036.060 **The transition to the true tangential point when using complete overlap plate method.**
Z. A. Galieva, N. N. Matveev, E. M. Loginova.
Kinematika Fiz. Nebesn. Tel, Tom 7, No. 4, p. 88 – 90 (Jul–Aug 1991). In Russian. English translation in Kinematics Phys. Celest. Bodies, Vol. 7, No. 4.
The transition to the true tangential point allows the model of six (instead of eight) linear constants to be used in astrometric reduction. That possibility was substantiated by A.A.Kiselev (1989).

036.061 **Correcting atmospheric–induced phase errors of a synthetic aperture antenna array by time series modeling and Kalman filtering.**
Zheng Yi, J. P. Basart, Y. S. Koh.
Radio Sci., Vol. 25, No. 6, p. 1145 – 1158 (Nov–Dec 1990).
A new technique is discussed and developed for correcting atmospherically induced errors in phase data collected by radio astronomy interferometers and synthetic aperture antenna arrays. The main features of this technique are modeling and filtering the information content of the complex visibility data in time sequences. In conjunction with other radio astronomy data processing algorithms, a time series modeling and parameter estimation technique is developed to obtain noise models and source models from observed phase data. These models can be in the form of stochastic difference equations, autocorrelation, power spectral density, and a state space format ready for further data processing. One useful application of this technique is for reducing atmospherically induced phase errors of small synthesis arrays that have too few antennas to apply self–calibration. Another application of this technique is for improving the performance of large synthesis arrays when the standard calibration methods are insufficient for correcting very noisy phase data. This technique has been tested using VLA and the Hat Creek millimeter interferometer.

036.062 **Restoration of turbulence–degraded images by the most–common method.**
Y. Glick, A. Baram, H. M. Loebenstein, Z. Azar.
Appl. Opt., Vol. 30, No. 27, p. 3924 – 3929 (20 Sep 1991). Current Physics Microform No.: 9111F1442.
A method for improving turbulence–degraded pictures is proposed and demonstrated. The algorithm is based on a study of the time distribution of the gray–level values of each pixel in a series of short–exposure video frames.

036.063 **Limited degree–of–freedom adaptive optics and image reconstruction.**
M. C. Roggemann.
Appl. Opt., Vol. 30, No. 29, p. 4227 – 4233 (10 Oct 1991). Current Physics Microform No.: 9201D0831.
The use of limited degree–of–freedom adaptive optics in conjunction with statistical averaging and a linear image reconstruction algorithm is addressed. Image reconstruction is traded for full predetection compensation. It is shown through analytic calculations that the average optical transfer function (OTF) is significant for high spatial frequencies in the case of imaging through atmospheric turbulence with an adaptive optics system composed of a Hartmann–type wave front sensor and a deformable mirror possessing far fewer actuators than one per atmospheric coherence diameter (r_0). Statistical averaging is used to overcome the effects of measurement noise and randomness in individual realizations of the OTF. The imaging concept and signal–to–noise considerations are presented.

036.064 **Feature enhancement of images using maximal contrast pixel to pixel.**
J. F. Sosa Urías.
Appl. Opt., Vol. 30, No. 32, p. 4598 – 4599 (10 Nov 1991). Current Physics Microform No.: 9201E1772.
A digital processing method is proposed to enhance image contrast with significant visibility of details. The technique is based on the local contrast associated with each pixel with a given surrounding point.

036.065 **Probability imaging: the statistics of speckle patterns of extended astronomical sources at high light levels.**
C. Aime, E. Aristidi.
J. Opt. Soc. Am. A, Vol. 8, No. 9, p. 1434 – 1441 (Sep 1991). Current Physics Microform No.: 9111F0352.
Assuming that the complex amplitude of the wave at the focus of a large telescope is Gaussian and making use of a technique of orthonormal expansion, the authors give a mathematical expression for the characteristic function of any order of the speckle pattern of an extended astronomical object as the inverse of the determinant of a matrix whose elements are defined by the spatial correlation function of the amplitude of the point–source speckle pattern and a diagonal matrix representing the astronomical object. This mathematical model is intended to be used in an inverse procedure to find the object most compatible with the observed spatial statistical properties. An example of application is given for a double star.

036.066 **Rechnergestützte Zeitregistrierungen und –vergleiche in astronomischen Beobachtungsgeräten.**
H. Potthoff.
Sterne, Band 67, Heft 4, p. 193 – 199 (1991).

036.067 **A new method for calibrating vector magnetograms.**
G. Cauzzi, D. M. Rust, J. W. O'Byrne.
Bull. Am. Astron. Soc., Vol. 23, No. 2, p. 1054 (1991). Abstract. – See Abstr. 012.012 for the main entry.

036.068 **The derivation of vector magnetic fields from Stokes profiles: derivative versus least squares fitting techniques.**
T. R. Metcalf, R. C. Canfield, D. L. Mickey.
Bull. Am. Astron. Soc., Vol. 23, No. 2, p. 1054 (1991). Abstract. – See Abstr. 012.012 for the main entry.

036.069 **Blending influence of Fe I 5250.6 Å spectral line on the Fe I 5250.2 Å spectral line and its implication for polarization measurements.**
K. S. Balasubramaniam.
Bull. Am. Astron. Soc., Vol. 23, No. 2, p. 1054 – 1055 (1991). Abstract. – See Abstr. 012.012 for the main entry.

036.070 **Problems associated with aligning MSFC transverse azimuth maps and H alpha images.**
E. A. West.
Bull. Am. Astron. Soc., Vol. 23, No. 2, p. 1055 (1991). Abstract. – See Abstr. 012.012 for the main entry.

036.071 **The practical application of the magnetic virial theorem: analytical results.**
J. E. Rhoads, J. A. Klimchuk, R. C. Canfield.
Bull. Am. Astron. Soc., Vol. 23, No. 2, p. 1055 (1991). Abstract. – See Abstr. 012.012 for the main entry.

036.072 **Further numerical tests for the progressive extension method (PEM) for extrapolation of solar force–free magnetic field.**
M. T. Sun, S. T. Wu, F. S. Weng, M. T. Song.
Bull. Am. Astron. Soc., Vol. 23, No. 2, p. 1055 (1991). Abstract. – See Abstr. 012.012 for the main entry.

036.073 **Considerations for flare related magnetic field measurements.**
J. M. Davis, J. M. Fontenla.
Bull. Am. Astron. Soc., Vol. 23, No. 2, p. 1055 (1991). Abstract. – See Abstr. 012.012 for the main entry.

036.074 **Solar tomography: deducing the three–dimensional structure of the corona from stereoscopic images of optically thin line emission.**
J. M. Davila.
Bull. Am. Astron. Soc., Vol. 23, No. 2, p. 1063 (1991). Abstract. – See Abstr. 012.012 for the main entry.

036.075 **Experimental demonstration of atmospheric compensation using multiple synthetic beacons.**
D. V. Murphy, C. A. Primmerman, B. G. Zollars, H. T. Barclay.
Opt. Lett., Vol. 16, No. 22, p. 1797 – 1799 (15 Nov 1991). Current Physics Microform No.: 9201D2161.
The authors present experimental results that demonstrate real–time, atmospheric–turbulence compensation of a bright star with the use of two synthetic beacons. Each beacon was used to measure the phase aberrations over only part of the telescope aperture, a configuration that is suitable for reducing focal–anisoplanatism error. This is the first demonstration of atmospheric compensation with the use of multiple synthetic beacons.

036.076 **Maximum likelihood image restoration. III. Algorithm. One–dimensional test problems.**
V. Yu. Terebizh, V. V. Biryukov.
Astrophysics, Vol. 33, No. 2, p. 475 – 482 (Sep – Oct 1990). English translation of Astrofizika, Tom 33, Vyp. 2, p. 305 – 315 (Oct 1990).
A method is proposed for numerical solution of the problem of maximizing a functional subject to probabilistic constraints in the method of maximal likelihood image restoration. Analysis of tests has revealed efficiency and stability of restoration by means of the method. A preliminary comparison is made of the efficiency of the maximal likelihood method and other methods of image restoration.

036.077 **Pulsars as instrumental polarization calibrators.**
K. M. Xilouris.
Astron. Astrophys., Vol. 248, No. 1, p. 323 – 327 (Aug 1991).
Radio–pulsars typically exhibit unique polarization properties, namely large linear polarization angle swing and significant amounts of both linear and circular polarization, in the short time of their duty cycle. A calibration method is presented which utilizes those properties in order to evaluate the instrumental polarization performance of a radio–polarimeter and monitor its time stability as well. Polarization observations of well studied pulsars made with the 18–cm receiving system of the 100–m Effelsberg radiotelescope were used to enable estimation of the Müller matrix which represents the instrumental distortion of the polarization state incident on a radiotelescope. The inverse of the

Müller matrix derived by this method, was then used to correct the systematic errors introduced to the polarization profiles observed similarly. The good agreement of the corrected profiles with well observed published profiles, supports the reliability of this method.

036.078 **Image construction from the IRAS survey.**
T. R. Bontekoe, D. J. M. Kester, S. D. Price, A. R. W. de Jonge, P. R. Wesselius.
Astron. Astrophys., Vol. 248, No. 1, p. 328 – 336 (Aug 1991).
IRAS survey data can be used successfully to produce images of extended objects. The major difficulties, viz. non–uniform sampling, different response functions for each detector, and varying signal–to–noise levels for each detector for each scan, have been resolved. The results of three different image construction techniques are compared: co–addition, constrained least–squares, and maximum entropy. The maximum entropy result is superior. The authors present an image of the galaxy M51 with an average spatial resolution of 45 arcsec, using 60 μm survey data. This exceeds the telescope diffraction limit of about 1 arcmin at this wavelength.

036.079 **Infrared computer solar spectrophotometer: principal operating characteristics and observational methods.**
A. G. Zuev, Eh. V. Kandrashov, A. V. Mikhajlov, V. G. Nikiforov, K. V. Rumyantsev, M. I. Rybakov.
Soln. Dannye, Byull., No. 7/90, p. 82 – 88 (Jul 1990). In Russian.
An observational method is described and principal characteristics (noise–to–signal ratio and nonlinearity) of an infrared computer solar spectrophotometer are given. An example of He I 10830 Å solar map and line profiles are presented.

036.080 **Interstellar scattering effects on the detection of narrow–band signals.**
J. M. Cordes, T. J. Lazio.
Astrophys. J., Vol. 376, No. 1, p. 123 – 134 (20 Jul 1991).
The authors have investigated the detection and decoding of narrow–band radio signals after propagation through the turbulent, ionized interstellar medium. For most lines of sight through the Galaxy, spectral broadening due to scattering is $\lesssim$0.1 Hz at 1 GHz. A few lines of sight may have as much as 5 Hz of broadening. Spectral broadening is therefore unimportant for the detection of hypothesized singals from extraterrestrial intelligence. Intensity scintillations, however, are of considerable importance. They both help and hinder detection: signals too weak to be detected without the scattering medium may be modulated above the detection threshold while, conversely, signals above threshold can be modulated below. For strong scattering (distances $\gtrsim$100 pc at 1 GHz), the authors demonstrate that multiple observations of a given target comprise a strategy that is superior to single observations even when the total time per target is held fixed. Decoding information carrying signals may encounter difficulties due to intensity scintillations (due to the ionosphere and solar wind as well as the interstellar medium) and, for distant objects, from temporal broadening like that seen from pulsars.

036.081 **ABCs of CCD's in astronomy II: image reduction.**
L. A. Wells.
I.A.P.P.P. Commun., No. 46, p. 1 – 8 (Dec 1991 – Feb 1992).

036.082 **Towards precision photometry.**
R. Miles.
I.A.P.P.P. Commun., No. 46, p. 17 – 26 (Dec 1991 – Feb 1992).
The author discusses some aspects of observational practice aimed at evaluating atmospheric transparency: one of the key steps towards achieving precision photometry.

036.083 **Observations with the scanner of the 6–m telescope in the Nasmyth–1 focus and automatized program package for spectrum reduction.**
V. L. Afanas'ev, V. A. Lipovetskij, V. P. Mikhajlov, E. A. Nazarov, A. I. Shapovalova.
Astrofiz. Issled. Izv. Spets. Astrofiz. Obs., Tom 31, p. 128 – 133 (1991). In Russian. English translation in Bull. Spec. Astrophys. Obs. – North Caucasus.
The hardware/software of the 1000–channels TV scanner of the 6–m telescope is described. The scanner is mounted on the spectrograph SP–124 in the Nasmyth focus. Program package for observations and automatic reduction of spectra is briefly reported.

036.084 **Search algorithm for weak periodicities in definite time intervals.**
M. Carbonell, J. L. Ballester.
Astron. Astrophys., Vol. 249, No. 1, p. 295 – 297 (Sep 1991).
An algorithm allowing the computation of the Lomb–Scargle periodogram in any predefined frequency interval with any desired resolution is presented. When the search of periodicities is restricted to definite time intervals, a significant saving of CPU time is obtained.

036.085 **Deconvolution and photometry on HST–FOC images.**
C. Barbieri, G. De Marchi, R. Ragazzoni.
Workshop on the First Year of HST Observations, p. 178 – 187 (1991). – See Abstr. 012.017 for the main entry.
Due to the peculiar characteristics of the PSF of HST, a careful analysis with many deconvolution experiments must be performed on HST–FOC images in order to understand their property. The authors briefly present some aspects of their work on this subject.

036.086 **Restoration of images degraded by telescope aberrations.**
T. Reinheimer, D. Schertl, G. Weigelt.
Workshop on the First Year of HST Observations, p. 245 – 248 (1991). – See Abstr. 012.017 for the main entry.
The spherical aberration of the Hubble Space Telescope (HST) causes a point spread function (psf) which consists of a central core of about 0.1 arcsec diameter and a halo of several arcsec diameter. The core contains only about 12% of the total psf intensity. The authors have performed laboratory simulations of images degraded by telescope aberrations and photon noise (10000 photon events per image). Spherical aberration was produced by suitable optics. The aberrated images were used to investigate the dependence of the reconstructed image on the applied image restoration method. The image reconstruction methods Wiener filtering, Clean, Gerchberg method, Lucy–Richardson method and MEM were compared.

036.087 **Coping with the Hubble Space Telescope's PSF: crowded field stellar photometry.**
E. M. Malumuth, J. D. Neill, D. J. Lindler, S. R. Heap.
Workshop on the First Year of HST Observations, p. 249 – 252 (1991). – See Abstr. 012.017 for the main entry.
In order to do stellar photometry in crowded fields with the HST, the authors have developed a simple approach that uses the known properties of the HST PSF. The following is a brief step by step description of this technique.

036.088 **A method for the automatic observation of solar active regions and the programming.**
Yang Xiaoning, Li Zhi.
Publ. Yunnan Obs., No. 1, p. 21 – 28 (Mar 1991). In Chinese.

036.089 **Test observation of the solar radio fast synchronous observation system at four wavelengths and its significance.**
Xie Ruixiang, Chen Guoqiang, Wang Congyun.
Publ. Yunnan Obs., No. 1, p. 29 – 39 (Mar 1991). In Chinese.

036.090 **A CCD detector attached to the lower latitude meridian circle. (I) The observational principle.**
Mao Wei, Li Zhiming, Hu Xiaochun, Zhang Jing, Peng Qingyu.
Publ. Yunnan Obs., No. 1, p. 61 – 67 (Mar 1991). In Chinese.
The observational principle, various instrument constants, the method for determining the errors and the accuracy of determining of the Lower Latitude Meridian Circle (LLMC) with a CCD detector are expounded, the absolute accuracy of location which can be attained with a CCD detector attached to the LLMC and the role of the construction of the highly accurate reference frame of the celestial sphere played by the LLMD with a CCD datector are discussed.

036.091 **A CCD detector attached to the lower latitude meridian circle. (III) Several correction for the spherical astronomy.**
Hu Xiaochun, Li Zhiming, Zhang Jing, Peng Qingyu, Mao Wei.
Publ. Yunnan Obs., No. 1, p. 77 – 86 (Mar 1991). In Chinese.
The curvature correction, which should be added to the determined value of the zenith distance in the meridian direction, and the corresponding corrections of the time of a star's passage across the prime vertical direction and of the determined value of the zenith distance are derived by means of the method of the projection ellipse. The influence of the variation in the angle between the projection of the right ascension circle of an object and the meridian plane on the determined value of the measured coordinate during the observation of the star carried out in the meridian direction is also derived.

036.092 **A CCD detector attached to the lower latitude meridian circle. (IV) The possibility of obtaining highly accurate location of a star with the LLMC.**
Li Zhiming, Hu Xiaochun, Peng Qingyu, Zhang Jing, Mao Wei.
Publ. Yunnan Obs., No. 1, p. 87 – 93 (Mar 1991). In Chinese.
The methods for the determination of the main errors of the LLMC are presented by comparing them with those of the traditional meridian circle, and the accuracies, especially the correction of the flexure of the tube of the circle, are estimated.

036.093 **Test for the method for correcting the atmospheric seeing and the diffraction limit of a telescope by means of the 2–dimensional DFT deconvolution. The application of DFT to astronomy (III).**
Yang Weimin, Zhang Bairong.
Publ. Yunnan Obs., No. 1, p. 94 – 100 (Mar 1991). In Chinese.
The method for carrying out the image reconstruction by means of the 2–dimensional discrete Fourier transform is introduced. A method for evaluating the optical transfer function by the spectrum fitting is also presented. The results obtained with the CCD system attached to the 1–m telescope show that the spatial resolution of the system increases by 275% and the binary stars that are mixed in the original image are obviously separated from each other after the processing.

036.094 **Several problems in preliminary processing of CCD raw images.**
Wu Guangjie.
Publ. Yunnan Obs., No. 2, p. 41 – 45 (Jun 1991). In Chinese.
It is discussed how to make and process a "bias", "flat", and a "sky background with fringe" in the preliminary processing of CCD observations.

036.095 **Optimum–processing for the noise in a low light level readout image from CCDs.**
Wu Junqin, Ye Binxun.
Publ. Yunnan Obs., No. 2, p. 46 – 51 (Jun 1991). In Chinese.
Based on analysing the statistical characteristic of the noise in CCD's read–out signal and on studying the method of pre–processing of CCD image at present, this paper brings up a new method for optimum–processing of noise in CCD's read–out signal. The new method has a promising result for depressing the CCD's read–out noise and other random noise.

036.096 **Some algorithms and procedures useful to analyse HST–FOC images.**
C. Barbieri, G. De Marchi, R. Ragazzoni.
Workshop on the First Year of HST Observations, p. 253 – 259 (1991). – See Abstr. 012.017 for the main entry.
Four procedures are briefly described. Emphasis is given to those algorithms the authors think particularly useful for this kind of space–based images.

036.097 **Deconvolution of an FOC image using a TIM–generated PSF.**
P. E. Hodge.
Workshop on the First Year of HST Observations, p. 260 - 263 (1991). – See Abstr. 012.017 for the main entry.
Science Data Analysts at the STScI have already computed a number of PSFs using the Telescope Image Modelling (TIM) software of Burrows and Hasan. These PSFs are the first of a catalog of PSFs that are to be prepared so that observers may deconvolve images taken with the Hubble Space Telescope. An image of a single star was deconvolved using the Lucy–Richardson algorithm with the appropriate TIM PSF.

036.098 **Rapid deconvolution of Hubble Space Telescope images on the NRL Connection Machine.**
P. Hertz, M. L. Cobb.
Workshop on the First Year of HST Observations, p. 264 – 266 (1991). – See Abstr. 012.017 for the main entry.
The authors have developed a rapid, highly parallel image space based convolution algorithm for use on the NRL 16k processor Connection Machine. HST images can be deconvolved in a few minutes. Images of Saturn and the LMC open cluster NGC 1850 were provided for testing the algorithm.

036.099 **Simulating the photoelectric recording of the circle and determining the positions of division lines.**
Zhu Zi, Liu Shaomin.
Publ. Shaanxi Astron. Obs., Vol. 14, No. 1, p. 50 – 57 (Aug 1991). In Chinese.
The method for simulating the photoelectric circle recording data with the computer is presented and the determination of center positions of division lines is discussed. For simulating the circle recording data, the authors assume that the function of the image profile of a circle division line is between the triangluar function and the Gaussian function, and the recording noise is of the normal distribution. The simulated output signals of the photoelectric circle recording are obtained by means of combining the function of the image profile with the noise. The method of determining the position of the division line via the linear least squares fit and the calculation formula are given.

036.100 **The methods for fringe detecting and tracking in optical long baseline interferometry.**
Wang Zhengming.
Publ. Shaanxi Astron. Obs., Vol. 14, No. 1, p. 65 – 70 (Aug 1991). In Chinese.
The two purposes of fringe detecting are to maintain a suitably stable fringe pattern and to estimate the complex degree of coherence of a source. Because of the atmospheric turbulence with high time frequencies and generally low light levels falling on the detectors, the detectors must operate in a digital photon counting mode with high time sensitivity. This paper describes some of the methods for detecting and tracking the fringe patterns, and explains that different methods are suitable to different observational objects and modes.

036.101 **Recent study on determination of total electron content from differential Doppler records.**
Fan Zhanyou.
Publ. Shaanxi Astron. Obs., Vol. 14, No. 1, p. 80 – 88 (Aug 1991). In Chinese.
A new method of determining the ionospheric total of electron content from differential Doppler records is introduced. Besides, a revised model of the centre altitude of the ionosphere is presented. Some results show that this new method and the revised model of the centre altitude of the ionosphere are successful and effective in calculating the total electron content from the differential Doppler records.

036.102 **Believability of signals from cosmic ray sources.**
M. Goodman.
AIP Conf. Proc., No. 220, p. 210 – 214 (Apr 1991). Current Physics Microform No.: 9106B1344. – See Abstr. 012.020 for the main entry.
This paper discusses some of the criteria by which an observer judges whether to believe a signal or limit that has been reported for a cosmic ray source. The importance of specifying the test before looking at the data is emphasized.

036.103 **Comparing coherent microwave emission from LPM and B–H showers.**
J. P. Ralston, D. W. McKay.
AIP Conf. Proc., No. 220, p. 295 – 299 (Apr 1991). Current Physics Microform No.: 9106B1429. – See Abstr. 012.020 for the main entry.
Coherent microwave emission offers a promising probe of cosmic ray events in the UHE region upwards of 10^{14}eV. It is possible that this means may allow detection of events which could not be detected in any other way, such as those initiated by UHE neutrinos from cosmic point sources.

036.104 **Pointwise automatic analysis of speckle photographs.**
Zhang Jun–jian, Sun Pei–mao, Chen Jing–nian, Ding Bo–ju.
Opt. Eng., Vol. 30, No. 4, p. 382 – 385 (Apr 1991). Also published in "International conference on optoelectronic science and engineering '90" edited by Wang Daheng, ISBN 0–8194–0273–7, Bellingham, WA (USA), SPIE, 1990, p. 629 – 631, SPIE Proc. Ser., Vol. 1230.
An automatic system for analyzing specklegrams point by point is described. A new method of evaluating Young's fringes based on making a TV camera scan both horizontally and vertically is proposed. Tracking, filtering, and edge locating techniques are developed to acquire fringe data. Experiments show that Young's fringes can be determined with an accuracy of 1% in spacing and 1° in orientation, and the displacement can be found with an accuracy of 1 μm in the range 20 to 100 μm.

036.105 **Detection of point sources with spark chamber gamma–ray telescopes.**
J. R. Mattox.
AIP Conf. Proc., No. 220, p. 316 – 320 (Apr 1991). Current Physics Microform No.: 9106B1450. – See Abstr. 012.020 for the main entry.
The two methods under consideration by the EGRET team for detecting point surces are cross correlation and maximum likelihood. The sensitivity of these methods have been analyzed numerically. Cross correlation is found to be $9 \pm 2\%$ more sensitive than maximum likelihood.

036.106 **Real–time image parameterization in high energy gamma–ray astronomy using transputers.**
M. Punch, D. J. Fegan.
AIP Conf. Proc., No. 220, p. 321 – 328 (Apr 1991). Current Physics Microform No.: 9106B1455. See also 53.036.071. – See Abstr. 012.020 for the main entry.
Recently, significant advances in very–high–energy gamma–ray astronomy have been made by parameterization of the Cherenkov images arising from gamma–ray initiated showers in the Earth's atmosphere. A prototype system to evaluate the use of transputers as a parallel–processing elements for real–time analysis of data from a Cherenkov imaging camera is described in this paper. The operation of and benefits resulting from such a system are described, and the viability of an applicaiton of the prototype system is discussed.

036.107 **Reconstruction of stellar images by phase retrieval based on the Fourier decomposition method.**
J. Ohtsubo, Y. Takahasi, N. Nakajima.
Opt. Eng., Vol. 30, No. 9, p. 1332 – 1336 (Sep 1991).
A new technique is proposed to reconstruct an astronomical image using phase retrieval from exponential filtered and unfiltered images based on the Fourier series expansion method. The amplitude of the Fourier transform of the filtered image is a mixture of the information of the amplitude and phase in the Fourier space corresponding to the unfiltered image. Therefore, the phase information can be derived from the Fourier modulus of the filtered image based on the Fourier decomposition method described. This method is applied to stellar speckle interferometry and the successful reconstruction of a double star image is obtained.

036.108 **A new method to search for high energy gamma rays by track–calorimeter detectors at a high mountain.**
F. Kajino, K. Hibino, M. Nishizawa, M. Ohnishi, M. Sakata, Y. Yamamoto.
AIP Conf. Proc., No. 220, p. 334 – 338 (Apr 1991). Current Physics Microform No.: 9106B1468. – See Abstr. 012.020 for the main entry.
The authors have carried out a Monte Carlo simulation to look for a possibility to observe high energy gamma rays at very high altitude of 5200 m such as Mt. Chacaltaya. Geomagnetic field effect is taken into account. Good advantage to set the detectors at such a mountain are that detection area can be easily taken about 10 to 100 times larger than the air borne experiment, an observation time can be taken about 100 to 1000 times, background flux can be reduced to 0.1 times. As a result the authors obtained a very good possibility to detect the gamma rays with energies around several hundred GeV although they depend on the energy spectrum of the sources.

036.109 **Data reduction of snapshot VLBI observations.**
Nan Rendong, Deng Jie, R. T. Schilizzi, C. R. Fanti.
Sci. China, Ser. A, Vol. 34, No. 5, p. 590 – 597 (May 1991). In Chinese.
Data reduction techniques have developed for generating images of radio sources from snapshot very long baseline interferometry observations with poor u–v coverage. The reliability of the radio source structures is discussed. Reference is made to 3C48, observed at 50 cm with a global array in a snapshot mode.

036.110 **Glare and celestial visibility.**
B. E. Schaefer.
Publ. Astron. Soc. Pac., Vol. 103, No. 665, p. 645 – 660 (Jul 1991).
Glare is when a bright source of light hides a nearby fainter source and is caused by scattering of light within the atmosphere, the telescope, and the eyeball. The author develops a model of the effects of glare on the visibility of astronomical sources as viewed with the human eye whether unaided or through a telescope. This model is tested and found to closely reproduce observations of lunar appulses, Galilean satellites, the Martian moons, well–known double stars, and lunar occultations. Glare calculations are then applied to a wide variety of situations of historical and astronomical interest.

036.111 **The accuracy of infrared photometry with arrays.**
M. D. Guarnieri, R. I. Dixon, A. J. Longmore.
Publ. Astron. Soc. Pac., Vol. 103, No. 665, p. 675 – 684 (Jul 1991).
The photometric accuracy and repeatability of an InSb array in the infrared camera IRCAM is investigated. Measurements of standard stars are described; these were made using the United

Kingdom Infrared 3.8-m Telescope (UKIRT) over five nights in May 1988. Observations were made at wavelengths of 1.25 μm (J) and 2.2 μm (K). The results show clearly that the camera–array combination is capable of a photometric accuracy of better than 1% when the standard–star images obtained with short on–chip exposure (0.5 sec) are flat fielded with long on–chip exposure (7 – 50 sec) sky flats. Linearity corrections determined in the laboratory were applied to the data and there was no evidence of residual nonlinearity of >2% over the magnitude range $6^m.5$ – $13^m.0$. The range of mean nightly zero points at K during the five nights was $0^m.03$, showing that both the site and the camera system were very stable over this period. At J the zero points were less well determined and had twice this range during three nights.

036.112 **Using image area to control CCD systematic errors in spaceborne photometric and astrometric time–series measurements.**
A. Buffington, C. H. Booth, H. S. Hudson.
Publ. Astron. Soc. Pac., Vol. 103, No. 665, p. 685 – 693 (Jul 1991).

The authors have investigated the effect of some systematic errors for high–precision time–series spaceborne photometry and astrometry, with a CCD as the detector. The "pixelization" of the images causes systematic error in astrometric measurements. They show that this pixelization noise scales as image radius $r^{-3/2}$. Subpixel response gradients, not correctable by the "flat field", and in conjunction with telescope pointing jitter, introduce further photometric and astrometric errors. They model subpixel gradients using observed properties of real flat fields. These errors can be controlled by having an image span enough pixels. Large images are also favored by CCD dynamic range considerations. However, magnified stellar images can overlap, thus introducing another source of systematic error. An optimum image size is therefore a compromise between these competing factors.

036.113 **Weighted slit extraction of low–dispersion IUE spectral data.**
A. L. Kinney, R. C. Bohlin, J. D. Neill.
Publ. Astron. Soc. Pac., Vol. 103, No. 665, p. 694 – 711 (Jul 1991).

The weighted slit extraction has been chosen by the IUE project as the standard extraction for low–dispersion spectra in the final IUE archives. In order to properly weight the data within the extraction slit and obtain the optimal evaluation of the true spectrum, a noise model is derived for IUE data. The exclusion of cosmic rays is demonstrated and the improvement in signal–to–noise ratio is quantified for spectra taken with a wide range of flux level. A "rogues' gallery" of problematic spectra, such as trailed spectra, multiple spectra, discontinuous spectra, and emission–line spectra, are extracted to demonstrate the robust nature of the method. Comparisons are made between the standard IUE Standard Image Processing System (IUESIPS) extraction, nonstandard narrow slit extraction, trailed IUESIPS extraction, and weighted extraction for these data. Other weighted slit extractions are briefly discussed. The advantages of the weighted slit extraction are discussed.

036.114 **A simple, photometrically accurate algorithm for deconvolution of optical images.**
W. C. Keel.
Publ. Astron. Soc. Pac., Vol. 103, No. 665, p. 723 – 729 (Jul 1991).

A modified version of the CLEAN deconvolution technique, incorporating a noise model, has been tested on simulated HST data. This straightforward algorithm displays many desirable features for astronomical image reconstruction, as shown by rigorous numerical tests: (1) The photometric scale of the data is preserved accurately. (2) Information on the local noise level due to Poisson statistics in the input data is preserved. (3) It can accommodate a spatially variable point–spread function. Some of these properties are not shared by any widely used deconvolution algorithm. Some considerations of computing speed, and application to ground–based images, are briefly discussed.

036.115 **Solar spectroscopy with a 100 × 100 diode array.**
A. D. Wittmann.
Reviews in Modern Astronomy 4, p. 61 – 68 (1991). – See Abstr. 003.008 for the main entry.

The 100 × 100 diode array was used to obtain high–resolution spectrograms of sunspots, pores, and the quiet photosphere. Extended time series of the solar granulation at the disc center were taken in two different spectral lines. The purpose of this investigation was to measure granular line shifts as a function of seeing on a statistical basis, and to try to extrapolate the measured quantities to ideal conditions. The present paper reports about the first results from 4 long time series that were taken on 1 May 1989 and on 6 – 8 July 1990.

036.116 **Error analysis for the removal of the interference fringes from a solar image obtained with a CCD system by means of the subtraction.**
Yang Xiaoning, Li Zhi.
Chin. J. Space Sci., Vol. 11, No. 4, p. 301 – 307 (Oct 1991). In Chinese.

This paper briefly reports the causes of the production of interference fringes of a CCD system. A method for removing the interference fringes, by means of the subtraction, from the solar Hα image obtained with the CCD system is also described. A method for estimating the error is proposed through tests and studies and the calculated results are analysed and discussed.

036.117 **The art of surface imaging.**
N. E. Piskunov.
IAU Colloquium No. 130: The Sun and cool stars – activity, magnetism, dynamos, p. 309 – 320 (1991). – See Abstr. 012.024 for the main entry.

The author intends to analyze the reliability of surface imaging of stars based on high resolution spectroscopy and the technique of inverse problem solution. Both astrophysical and mathematical aspects including different regularization methods are reviewed. The influence of the different factors on the resulting map is discussed and it is shown that the simultaneous use of different kinds of observational data (spectroscopy, photometry, polarimetry etc.) is very useful in providing additional constraints for the solution. The recent results in the surface imaging of Cp– and late–type stars show the way for further progress: the use of more adequate mathematical description of the stellar atmosphere and the simultaneous consideration of various surface inhomogeneities.

036.118 **Surface imaging of EI Eri.**
T. Hackman, N. E. Piskunov, M. Poutanen, K. G. Strassmeier, I. Tuominen.
IAU Colloquium No. 130: The Sun and cool stars – activity, magnetism, dynamos, p. 321 – 325 (1991). – See Abstr. 012.024 for the main entry.

The authors present maps of the temperature distribution on the spotted RS CVn–type binary EI Eri (HD 26337), obtained by means of the surface imaging technique (Piskunov, 1991). Images were calculated for two separate lines for two epochs of observations. For one epoch the authors also calculated an image using a blend of several lines. The reliability of the maps is confirmed by comparing the simultaneous photometric observations with the light curves calculated from the temperature maps.

036.119 **A new technique to measure magnetic field strength in active stars.**
J. Sánchez Almeida, R. J. García López.
IAU Colloquium No. 130: The Sun and cool stars – activity, magnetism, dynamos, p. 414 – 416 (1991). – See Abstr. 012.024 for the main entry.

The authors explore the use of the flux difference between the lines Fe I 5250 Å and Fe I 5247 Å as a technique to measure photospheric magnetic fields in late–type stars. The technique developed takes into account all the LTE physics of the problem, assuming a radial and homogeneous magnetic field distribution over the stellar surface. Some test calculations, in order to prove the feasibility of the method, are shown.

036.120 **Magnetic flux determination in late–type dwarfs.**
P. Rípodas, J. Sánchez Almeida, R. J. García López, M. Collados.
IAU Colloquium No. 130: The Sun and cool stars – activity, magnetism, dynamos, p. 417 – 419 (1991). – See Abstr. 012.024 for the main entry.
The authors present a very preliminary and simplified analysis designed to measure photospheric magnetic fields in late–type stars, using the Fe I 5247.06 Å and 5250.22 Å lines. They show how the use of the equivalent widths of the lines and differences in their depth can give a rapid estimation of the magnetic flux.

036.121 **Automatic plates measurement with images digitalization.**
A. López García, J. L. Valdés Navarro, G. Perez Añón.
1. Spain–USSR Workshop on Positional Astronomy and Celestial Mechanics, p. 67 – 70 (1991). – See Abstr. 012.025 for the main entry.
In the programme of selected minor planets observations of the Astronomical Observatory of Valencia the authors have developed an automated measurement process. They describe the mechanical characteristics of this instrument and some details of the image identification.

036.122 **V, (B–V) and DDO colors extracted from grism spectra.**
D. J. Bell, K. M. Yoss.
Meeting on Precision Photometry: Astrophysics of the Galaxy, p. 111 – 116 (1991). – See Abstr. 012.027 for the main entry.
Single–channel photometry becomes less accurate as fainter stars are observed, due to decreased signal–to–noise, increased errors due to sky brightness and increased interval of time between first and last observations, allowing sky brightness and extinction to change. The authors describe a multichannel procedure which permits simultaneous star and sky exposures, and which permits rapid acquisition of data over the entire sky. This procedure is much faster than the single–channel approach, allows observations past 16th magnitude, and produces accurate colors even on slightly subphotometric nights.

036.123 **Systematic color transformation effects in Strömgren photometry.**
J. Manfroid, C. Sterken.
Meeting on Precision Photometry: Astrophysics of the Galaxy, p. 139 – 146 (1991). – See Abstr. 012.027 for the main entry.
A major cause of systematic errors in photometry can be attributed to the color transformation. A thorough analysis of numerous observing runs performed with various photoelectric equipment in the Strömgren system shows how reliable the transformations in the different indices are. Conclusions are drawn about the choice of the standard stars defining the system.

036.124 **The promise and challenge of time–resolved CCD ensemble photometry.**
R. L. Gilliland.
Meeting on Precision Photometry: Astrophysics of the Galaxy, p. 265 – 274 (1991). – See Abstr. 012.027 for the main entry.
CCD photometry allows precise monitoring of stellar intensities for many stars simultaneously at good temporal resolution. The photometric precision attainable from CCDs is as good as that with photoelectric photometry, but through the available spatial information, CCD photometry is much more robust. The author reviews recent efforts to set empirical limits on the precision of time–differential CCD photometry of rich stellar ensembles. This observational approach holds the promise of allowing detection of stellar analogs to the solar 5–minute, p–mode oscillations.

036.125 **Photometric boxes in the $UBVB_1B_2V_1G$ system.**
M. Golay, B. Nicolet, N. Cramer.
Meeting on Precision Photometry: Astrophysics of the Galaxy, p. 303 – 307 (1991). – See Abstr. 012.027 for the main entry.

036.126 **A method for selecting Lambda Bootis candidates from Strömgren photometry and objective prism spectra of field stars.**
R. O. Gray.
Meeting on Precision Photometry: Astrophysics of the Galaxy, p. 309 – 312 (1991). – See Abstr. 012.027 for the main entry.
The λ Bootis stars are a rare class of chemically peculiar A–type stars. They are characterized by a weak Mg II > λ4481 I line, a K–line type of A0 or slightly later, and hydrogen–line types between A0 and F0. For their hydrogen–line types, the metallic–line spectrum is weak. The author has found that by combining photometric (uvby) data with objective–prism data it is possible to select a group of stars which have a much higher probability (about 25%) of being true λ Bootis stars. The procedure is described.

036.127 **Automatic photometric telescopes and photometric accuracy.**
C. Sterken, J. Manfroid.
Meeting on Precision Photometry: Astrophysics of the Galaxy, p. 335 – 339 (1991). – See Abstr. 012.027 for the main entry.
By means of the mean standard deviations of differential uvby magnitudes of constant stars, the authors discuss some effects which may negatively influence the precision of automatically performed measurements.

036.128 **A non–deterministic approach to Schmidt plate astrometry.**
M. G. Lattanzi, B. Bucciarelli.
Astron. Astrophys., Vol. 250, No. 2, p. 565 – 572 (Oct 1991).
Previous experience with the astrometric reduction of Schmidt plates suggests that the very local nature of the instrumentally induced deformations cannot be represented by a single function which maps measured coordinates into standard ones for an entire plate. The failure of such attempts is evident when comparing computed positions against a high precision astrometric catalog or when intercomparing partially overlapping plates; then clear patterns of systematic trends, especially concentrated toward the plate edges, emerge. The information present in such a pattern can be exploited to find an a posteriori point correction to the global plate solution. The authors have done so via the least squares collocation method, well known in photogrammetry and satellite geodesy, combined with a technique for the estimation of the correlations among the plate residuals. This method has shown to be capable of significantly improving the astrometric quality of a conventional Schmidt plate solution. The tests were performed on a sample of the material used for the construction of the Hubble Space Telescope Guide Star Catalog. The residuals showed a typical correlation length of $\sim 0\overset{\circ}{.}5$ with associated local distortions of $\sim 0\overset{''}{.}28$ (average value across the plate). In addition, the authors discuss the performance and applicability of this technique for absolute proper motions determination, and under different conditions encountered in practice.

036.129 **Detectors and detections of gravitational waves.**
M. J. Sarna.
Postepy Astron., Tom 39, Zesz. 4, p. 176 – 179 (1991). In Polish.

036.130 **A system for line profile studies at the 150–foot tower on Mount Wilson.**
R. K. Ulrich, L. Webster, J. E. Boyden, N. Magnone, R. S. Bogart.
Sol. Phys., Vol. 135, No. 2, p. 211 – 241 (Oct 1991).
The authors describe enhancements to the hardware and software for the 150–foot tower system on Mt. Wilson which make possible the acquisition of high precision line profile measurements. This system utilizes the 75–foot pit spectrograph with a photomultiplier detector system to scan line profiles repeatedly in order to study variations induced by the passage of waves vertically through the solar atmosphere. Oscillations of line profile parameters with an amplitude as low as 1.7 m s^{-1} have been detected with this system using integrated sunlight. Phase relations between oscillations of different parts of the line profile are appropriate to upward energy transport. Consistent

with the previous conclusion by Mein and Schmieder (1981), it is found that the magnitude of the energy transport is compatible with the 5-min oscillations making an important contribution to the heating of the low chromosphere.

036.131 **ERIC, a package for spectroscopic data reduction.**
J. Webb, D. Mills.
Gemini, No. 33, p. 6 – 10 (Sep 1991).

036.132 **On the derivation of selection functions from redshift survey data.**
M. A. Strauss, A. Yahil, M. Davis.
Publ. Astron. Soc. Pac., Vol. 103, No. 667, p. 1012 – 1019 (Sep 1991).
The authors describe a previously unrecognized effect in the derivation of luminosity functions and selection functions from existing redshift survey data, due to binning of quoted magnitudes and diameters. They correct for this effect in the Center for Astrophysics (CfA) and Southern Sky (SSRS) Redshift Surveys. This correction makes subtle but systematic changes in the derived density fields of the CfA survey, especially within 2000 km s^{-1} of the Local Group. The effect on the density field of the SSRS survey is negligible.

036.133 **Image quality, tip–tilt correction, and shift–and–add infrared imaging.**
J. C. Christou.
Publ. Astron. Soc. Pac., Vol. 103, No. 667, p. 1040 – 1048 (Sep 1991).
The image quality obtainable with shift–and–add (SAA) imaging for the recovery of diffraction–limited information is quantitatively investigated using data simulating an 8–m aperture at a site with near–infrared seeing conditions as found on Mauna Kea. This is compared to the image quality obtainable from image centroiding and wavefront tip–tilt correction. For good seeing conditions, image centroiding and SAA, which tracks the image peak, show similar performance containing $\sim$30% of the image power in a diffraction–limited component. However, as the seeing degrades, SAA consistently yields improved resolution maintaining significantly greater diffraction–limited information. This enhances the detection threshold by $\sim$2 magnitudes over the seeing–limited case for the seeing range studied. By comparison the gain due to image centroiding decreases to $<$1 magnitude at the same poor seeing limit.

036.134 **VLBI with TDRSS.**
R. P. Linfield.
IAU Colloquium No. 123: Observatories in Earth orbit and beyond, p. 271 – 274 (1990). – See Abstr. 012.038 for the main entry.
VLBI observations using a satellite in earth orbit and ground antennas in Japan and Australia were conducted in 1986, 1987, and 1988. Sources were detected on space–ground baselines at both observing frequencies: 2.3 and 15 GHz. The coherence on space–ground baselines for 340 s was 90% at 2.3 GHz and 76% at 15 GHz. Brightness temperatures in the range $1 - 4 \times 10^{12}$K were measured for 10 sources at 2.3 GHz and 6 sources at 15 GHz.

036.135 **Subarcsecond imaging at the WHT auxiliary port.**
P. Charles, T. Naylor, P. Hewett.
Gemini, No. 34, p. 24 – 25 (Dec 1991).

036.136 **Baseline errors on VLBI measurements. I. Analytic description.**
M. Massi, G. Tofani, G. Comoretto.
Astron. Astrophys., Vol. 251, No. 2, p. 732 – 736 (Nov 1991).
The VLBI response to an unpolarized point source is analyzed for the first time up to the first order. The aim is to determine analytically which are the error terms affecting a VLBI measurement that cannot be factorized into two antenna contributions and, therefore, cannot be corrected by the existing hybrid mapping procedures. In the final equation up to five of such errors, defined as baseline errors, can be distinguished. The physical origin of those errors can be traced in the polarization impurities of the feeds, in the differential Doppler shift at the two antennas and in frequency and time instabilities of parts of the acquisition systems.

036.137 **Science operations for future space astrophysics missions.**
G. R. Riegler.
IAU Colloquium No. 123: Observatories in Earth orbit and beyond, p. 317 – 321 (1990). – See Abstr. 012.038 for the main entry.
Plans for astrophysics science operations during the decade of the nineties are described from the point of view of a scientist who wishes to make a space–borne astronomical observation or to use archival astronomical data. In the process of preparing a proposal, making an observation, and carrying out data processing, analysis, and dissemination of results, the scientist will be able to use a variety of services and infrastructure, including the "Astrophysics Data System". The current status and plans for these science operations services are described.

036.138 **Occultation astronomy.**
F. J. Dyson.
IAU Colloquium No. 123: Observatories in Earth orbit and beyond, p. 413 – 415 (1990). – See Abstr. 012.038 for the main entry.
The author discusses the possibility to detect faint comets of the Kuiper Belt with an array of small Earth–based telescopes.

036.139 **The EXOSAT results database.**
A. N. Parmar, N. E. White.
IAU Colloquium No. 123: Observatories in Earth orbit and beyond, p. 462 (1990). Abstract. – See Abstr. 012.038 for the main entry.

036.140 **Merits of space VLBI missions for geodynamics.**
J. Adam, I. I. Mueller.
IAU Colloquium No. 123: Observatories in Earth orbit and beyond, p. 507 (1990). Abstract. – See Abstr. 012.038 for the main entry.

036.141 **Low frequency radio astronomy from earth orbit.**
K. W. Weiler, N. E. Kassim.
IAU Colloquium No. 123: Observatories in Earth orbit and beyond, p. 508 (1990). Abstract. – See Abstr. 012.038 for the main entry.

036.142 **Science observations with the IUE using the one gyro mode.**
C. Imhoff, R. Pitts, R. Arquilla, C. Shrader, M. Perez, J. Webb.
IAU Colloquium No. 123: Observatories in Earth orbit and beyond, p. 517 – 520 (1990). Also published as report ESA–SP 310, p. 595 – 596 (Aug 1990). – See Abstr. 012.038 for the main entry.
The science instrument onboard IUE consists of two spectrographs which span the wavelength range of 1150 to 3200 Å and offer two dispersions with resolutions of 6 Å and 0.2 Å. The science instrument includes a prime and redundant Field Error Sensor (FES) camera for target aquisition and offset guiding. Since launch, 4 of the 6 gyroscopes have failed. When the next gyro fails, a new attitude control system will be uplinked which will rely on the reamining gyro and the FSS for general 3–axis stabilization. In addition to the FSS, the FES cameras will be required to assist in maintaining fine attitude control during target aquisition. The results of this work are presented.

036.143 **Knowledge based automated scheduling and planning tools for IUE.**
C. R. Shrader.
IAU Colloquium No. 123: Observatories in Earth orbit and beyond, p. 525 – 530 (1990). – See Abstr. 012.038 for the main entry.
IUE has been successfully operated as a real time user–interactive space observatory for twelve years. It is expected to

continue operation for up to five additional years, but under increasing constraints. The option to operate IUE in a more automated, non user–interactive mode is under consideration. A sophisticated software system to support such an operation is a clear requirement. The conceptual framework of such a system is described. Results of a preliminary tests are presented for which a hypothetical four day schedule for space–craft activities at a time resolutions as low as ten minutes was generated.

036.144 **An eight month monitoring campaign of NGC 5548 with IUE: GEX data reduction and measurement of emission line and continuum fluxes.**
G. Reichert, G. Kriss, D. M. Crenshaw, D. Alloin, J. Clavel, J. H. Krolik, M. A. Malkan, H. Netzer, B. M. Peterson, W. Wamsteker.
Georgia State University Conference on Variability of Active Galactic Nuclei, p. 335 – 338 (1991). – See Abstr. 012.040 for the main entry.

From mid–Dec 1988 to mid–Aug 1989, NGC 5548 was observed using the IUE satellite once every four days. Here the authors present details concerning the Gaussian Extraction (GEX) techniques used to reduce the spectra and the multi–component fitting used to determine emission line and continuum fluxes. They specifically address the issues of how to determine uncertainties, both in the spectral fluxes and in the continuum and line flux measurements.

036.145 **Using variability data for meaningful time–series analysis.**
R. A. Edelson.
Georgia State University Conference on Variability of Active Galactic Nuclei, p. 358 (1991). – See Abstr. 012.040 for the main entry.

036.146 **UV variability of NGC 5548: geometry of the broad line region.**
J. H. Krolik.
Georgia State University Conference on Variability of Active Galactic Nuclei, p. 359 – 374 (1991). – See Abstr. 012.040 for the main entry.

036.147 **Algorithm for stereo pair matching.**
D. A. Usikov, M. I. Kolesnik.
2. COSPAR Colloquium: The environmental model of Mars, p. 47 – 50 (1991). – See Abstr. 012.039 for the main entry.

An algorithm to model the stereoscopic human perception is considered which can help the stereo pair matching and the construction of the 3–D digital model of terrain. This algorithm relies on the local correlation analysis at an image pyramid. The algorithm can be employed in mapping and technical vision systems.

036.148 **The reconstruction of 3–dimension scenes from stereo-pairs by the dynamic programming method.**
I. N. Sitenko, D. A. Usikov.
7. Jerusalem Winter School for Theoretical Physics: Quantum cosmology and baby universes, p. 137 – 141 (1991). – See Abstr. 012.036 for the main entry.

036.149 **Optimum determination of extinction and transformation in multicolor photometry.**
A. Schwarzenberg–Czerny.
Astron. Astrophys., Vol. 252, No. 1, p. 425 – 430 (Dec 1991).

The author describes the optimum method for reduction of the *UBV* observations. He uses the simulation for the quantitative comparison of the optimum reduction scheme and those described by Hardie (1962) and by Harris et al. (1981). The optimum method yields significantly smaller scatter of residuals and smaller errors of the extinction and transformation coefficients. In other words, employing the optimum method one can correct observations for effects of extinction and transformation with good accuracy while using reduced number of standard stars observations. Additionally, use of the method enables more sensitive tests for non–random errors in observations than hitherto.

036.150 **Computer processing and image processing of Hα solar flares.**
Sun Zhengwen, Cao Hongyan.
Publ. Yunnan Obs., No. 4, p. 232 – 233 (Dec 1990). – See Abstr. 012.047 for the main entry.

036.151 **Procedures for validating Hipparcos astrometric parameters.**
H. Lenhardt, R. Hering, H. G. Walter.
Manuscr. Geod., Vol. 16, No. 4, p. 254 – 265 (Aug 1991).

Astrometric parameter determination constitutes one of the final steps of the Hipparcos data processing and reflects directly the scientific aim of the mission. Therefore special emphasis is put on the validation of the results. The methods of validation consist of the refined adjustment of astrometric parameters, their verification by a series of statistical tests, and of the visualization of results by graphic displays. Using simulated data based on 4560 stars the data flow and computer software were checked by subjecting the astrometric parameters to the above methods followed by an assessment of the results. As a by–product a reference data set has been established for guiding purposes in conjunction with the real mission. The connection of the Hipparcos instrumental system to the extragalactic reference frame by means of radio stars is outlined and numerically practised by exploiting the currently available optical and radio positions. To enable external result verification several options have been devised which take advantage of stellar characteristics to be reproduced eventually by the Hipparcos results.

036.152 **Imaging performance analysis of adaptive optical telescopes using laser guide stars.**
B. M. Welsh.
Appl. Opt., Vol. 30, No. 34, p. 5021 – 5030 (1 Dec 1991). Current Physics Microform No.: 9203A1772.

The use of laser guide stars in conjunction with adaptive optical telescopes offers the possibility of nearly diffraction–limited imaging performance from large, ground–based telescopes. The author investigates the expected imaging performance of an adaptive telescope, using laser guide stars created in the mesospheric sodium (Na) layer. A 2 – 3 m class telescope is analyzed. The results of the investigation indicate that a 3 m adaptive telescope using a single Na guide star is capable of achieving a Strehl ratio of 0.57 and an angular resolution nearly matching that of diffraction–limited performance (0.05 arcsec).

036.153 **Elimination of artifacts in interline charge–coupled device imagers.**
B. T. Turko, G. J. Yates.
Opt. Eng., Vol. 30, No. 10, p. 1542 – 1546 (Oct 1991).

Charge–coupled devices (CCDs) of interline transfer design are especially useful in imaging at high frame rates. However, their sensitivity to ionization radiation and the reduced effective opacity for vertical charge transfer registers cause undesired image artifacts. Random white spots from the radiation and "ghost" images (or smear) generated in the registers may severely impair the image quality. An electronic method of eliminating these artificats is described. Special sequences of pulses clock the CCD, quickly dumping the unwanted charge. The fast readout of images, cleared of artifacts, follows immediately.

036.154 **Estimation of the accuracy of wavefront reconstruction by adaptive optical phase–conjugation systems.**
S. V. Butsev, V. Sh. Khismatulin.
Sov. J. Opt. Technol., Vol. 58, No. 10, p. 611 – 612 (Oct 1991). Current Physics Microform No.: 9204X0081.

The accuracy capabilities of reconstructing of the wavefront of astronomical objects by phase–conjugate adaptive optical systems using a Hartmann–type wavefront sensor for measuring distortions are analyzed.

036.155 **Video image – a new area of interests for astronomy amateurs.**
M. Borkowski, M. Zawilski.
Urania, Rok 62, Nr. 10, p. 269 – 274 (Oct 1991). In Polish.

036.156 ***JHKL'* photometry on the ESO system: systematic effects and absolute calibration.**
M. Bersanelli, P. Bouchet, R. Falomo.
Astron. Astrophys., Vol. 252, No. 2, p. 854 – 860 (Dec 1991).
The authors report on systematic effects affecting the accuracy of photometry in the near–IR. In particular, the influence of an inhomogeneous selection of standard stars, of the atmospheric transmission and of the conversion of magnitudes into monochromatic fluxes is described. The absolute calibration in the near–IR is discussed and a suitable calibration for the present ESO system is derived.

036.157 **Méthodes d'analyse des différents types de bruits perturbant le signal de sortie des oscillateurs.**
F. Vernotte, J. Groslambert, J. J. Gagnepain.
Journées 1991: Systèmes de référence spatio–temporels, p. 19 – 28 (1991). – See Abstr. 012.048 for the main entry.

036.158 **Coordinate–system instability in the Askorekord measuring machine on temperature variations.**
Yu. A. Shokin.
Kinematics Phys. Celest. Bodies, Vol. 7, No. 2, p. 79 – 85 (1991). English translation of Kinematika Fiz. Nebesn. Tel, Tom 7, No. 2, p. 87 – 95 (1991).
A procedure for study of an correction for the coordinate instability of the Askorekord instrument that results from temperature variations is proposed. The basic aspects of the behavior of the machine's coordinate system are considered. A procedure is given for calculation of the necessary corrections to the measured coordinates.

036.159 **Image detection and separation on Tautenburg Schmidt plates with the ASAG program package.**
M. Lange, G. M. Richter, R.–D. Scholz.
Astron. Nachr., Vol. 312, No. 4, p. 239 – 243 (1991).
The Babelsberg image processing system which will be used for the new automated measuring machine ASAG (Astronomisches Auswertegerät) includes as special feature an adaptive filtering technique. This technique allows a reduction of noise and the determination of the background in astronomical pictures and has already been used for the surface photometry of galaxies on Schmidt plates. For astrometric purposes a method of detecting and separating images in two–dimensional measuring rasters was developed within the frame of the ASAG software and algorithms.

036.160 **A set of control programs for automated data processing systems in a conversational mode.**
A. F. Nazarenko.
Bull. Spec. Astrophys. Obs. – North Caucasus, Vol. 32, p. 135 – 139 (1990). English translation of Astrofiz. Issled. Izv. Spets. Astrofiz. Obs., Tom 32, p. 166 – 172 (1990).
A set of control programs, on the basis of which a number of automated conversational astrophysical data processing systems are created, is described.

036.161 **A local complex for data collection and control of radiometers with maintenance of a parallel process mode.**
B. L. Erukhimov, O. P. Likhvan, A. V. Chepurnov, V. N. Chernenkov.
Bull. Spec. Astrophys. Obs. – North Caucasus, Vol. 32, p. 140 – 145 (1990). English translation of Astrofiz. Issled. Izv. Spets. Astrofiz. Obs., Tom 32, p. 173 – 180 (1990).
A system for data collection and control of radiometers using the mechanism of parallel processes is examined. The system is a development of the lower level of the earlier described multiple–user data collection and preliminary processing system based on the two–level multicomputer complex of the RATAN–600 radio telescope.

036.162 **Views from the OCA Schmidt Telescope.**
C. Pollas.
10. Santa Cruz Summer Workshop in Astronomy and Astrophysics: Supernovae, p. 724 – 726 (1991). – See Abstr. 012.052 for the main entry.

036.163 **Test d'une méthode de réduction globale des observations à l'astrolabe.**
F. Chollet, N–E. Najid.
Journées 1991: Systèmes de référence spatio–temporels, p. 153 – 158 (1991). – See Abstr. 012.048 for the main entry.
The authors give here the results of an evaluation of a new reduction method of astrolabe observations. An analysis of the conditional equation of observation shows that only five unknowns are necessary to have a correct parametrization of the results provided that the temporal variations of the apparent longitude and latitude are removed.

036.164 **The precision of radial–velocity determinations of solar–type stars by cross–correlation.**
K. Murdoch, J. B. Hearnshaw.
Astrophys. Space Sci., Vol. 186, No. 1, p. 137 – 149 (Dec 1991).
The theoretical limit to the precision (defined as the inverse of the square of the random error) of a radial–velocity determination of a solar–type star obtained from the technique of digital cross–correlation is constructed as a function of simple parameters of the recording of a stellar spectrum. The relationship is used to establish condition for maximizing the precision for a given object, exposure time, and observing instrument. Both fibre–fed and Cassegrain–mounted spectrographs are considered.

036.165 **Continuous calibration of radio observations.**
V. R. Amirkhanyan, A. A. Kapustkin.
Bull. Spec. Astrophys. Obs. – North Caucasus, Vol. 32, p. 163 – 165 (1990). English translation of Astrofiz. Issled. Izv. Spets. Astrofiz.. Obs., Tom 32, p. 204 – 206 (1990).
The proposed method of calibration enables continuous monitoring of the gain of the receiver without loss of the radiometer's sensitivity.

036.166 **On the astrometric observations, 1962.0 – 1991.0.**
Zheng–Xin Li, You–Fen Chen, Chang–Xia Qian.
Journées 1991: Systèmes de référence spatio–temporels, p. 204 – 212 (1991). – See Abstr. 012.048 for the main entry.
The astrometric observations, which have been obtained by the 155 optical instruments since the beginning of 1962, are stored in the database of the Shanghai Observatory. A brief presentation of the astrometric data during the period of 1962.0 – 1991.0 is given. The relations among the precision of the 5–day ERP results, the algorithm used in the reduction, the number or the weight of measurements made in a year, and the time of the measurement in a year, are discussed.

036.167 **First results of the photographic survey program of the northern sky.**
I. P. Vedenicheva, G. A. Ivanov, L. K. Pakulyak, T. P. Sergeeva, A. I. Yatsenko.
International Seminar on Problems of Astrometry and Space Geodynamics, p. 12 – 15 (1991). In Russian. – See Abstr. 012.049 for the main entry.
The method of the measuring input data file by using of the PPM and Fresneau catalogues, the measuring procedure with the automatic measuring machine PARSEK–K and the method of reduction are briefly described. The random mean square of the position and proper motion differences with PPM data appears to be $\pm 0''.35$ and $\pm 0''.005$ per year, accordingly.

036.168 **Colour and magnitude effects in optical astrometry observations; implications for re–reductions in Hipparcos reference frame.**
J. Hefty.
Journées 1991: Systèmes de référence spatio–temporels, p. 217 – 222 (1991). – See Abstr. 012.048 for the main entry.
The prepared project of re–reduction of past optical astrometry observations in the Hipparcos reference frame will be based on processing of individual star observations referred to unique celestial reference frame. The quite new approach for the calculation of the new series of Earth Rotation Parameters (ERP) involves the simultaneous derivation of ERP with other relevant parameters as nutation corrections, tidal effects, constant of refraction etc. The spectral class and magnitude dependence is analyzed in order to account for this effect in algorithms for calculations in Hipparcos reference frame.

036.169 **Creation of the star position catalogue by five CdC plate–block processing.**
V. S. Kislyuk, L. K. Pakulyak.
International Seminar on Problems of Astrometry and Space Geodynamics, p. 66 – 71 (1991). In Russian. – See Abstr. 012.049 for the main entry.
The method of rational processing of a great quantity of CdC plates for determination of the photographic sky survey input catalogue is given.

036.170 **Visible and infrared interferometry before and with the VLT.**
C. Perrier, R. Foy, A. Chalabaev.
Ann. Phys. (Paris) Colloq., Vol. 16, No. 4, p. 497 – 506 (Aug 1991).
After a brief recall of the basic principles of interferometric techniques both at visible and infrared wavelengths, the authors show from a few selected examples the contribution of these techniques in the field of the late stages of stellar evolution; indeed they provide unique constraints to model e.g.: late type evolved stars like Mira variables, protoplanetary nebulae and the supernova SN 1987A, which the authors discuss. As a conclusion, they emphasize the importance of the interferometric capabilities of the future ESO's Very Large Telescope for such astrophysical programs.

036.171 **Maximum likelihood image restoration. IV. Limiting resolving power under given alternative.**
V. Yu. Terebizh.
Astrofizika, Tom 33, Vyp. 3, p. 409 – 427 (Dec 1990). In Russian. English translation in Astrophysics, Vol. 33, No. 3.
The formal criterion is found for the choice of the blurred and noised images of an arbitrary object in the case when the alternative object is known. The criterion generalizes the analogous result by J. Harris (1964) to situations when a priori probabilities of the objects presence are not given and arbitrary probabilities of type–I and type–II mistakes (α and β, respectively) are allowed. The exact expression for the limiting resolving power under given α, β and signal to noise ratio Ψ is found. The limiting resolving power depends only on $t = (z_\alpha + z_\beta)/\Psi$, where z_ε is the quantile of the normal distribution. For bright objects, limiting resolution is proportional to $\Psi^{-1/2}$. A number of examples is considered that have self–dependent significance for practice.

036.172 **A sky–noise measurement and its implication for ground–based infrared astronomy in the 10 μm atmospheric window.**
H. U. Käufl, P. Bouchet, A. van Dijsseldonk, U. Weilenmann.
Exp. Astron., Vol. 2, No. 2, p. 115 – 122 (1991).
A spectral analysis of the fluctuations of the infrared sky radiance at $\lambda \approx 10\ \mu$m was made at ESO. The data are compared to literature. The consequences of the results on infrared observing for future large telescopes are discussed: the authors' data suggest that in order to achieve background noise limited performance in the 10 μm atmospheric window chopping with frequencies of ≈ 8 Hz and amplitudes of ≈ 10 arcsec is mandatory.

036.173 **Gain calibrating nonuniform image–array data using only the image data.**
J. R. Kuhn, H. Lin, D. Loranz.
Publ. Astron. Soc. Pac., Vol. 103, No. 668, p. 1097 – 1108 (Oct 1991).
An algorithm is developed for calibrating the spatial nonuniformity of image–array (CCD–type) detectors. Like other techniques this approach uses multiple, spatially displaced images. In circumstances where high–precision flat fields are not available by other means (i.e., sky flats) this technique is advantageous as it uses the data frames for gain calibration even when the array images extended, nonuniform, sources. Numerical experiments and direct observations with intrinsically uniform and quite nonuniform detectors show that this algorithm is useful when data frames are crowded with sources – circumstance where "median filtering" flatfielding techniques often fail. The algorithm described is robust and efficiently uses information from multiple data frames to determine pixel gain variations, as the authors demonstrate using visible and IR array observations of extended sources.

036.174 **An algorithm for the support of telescope microscheduling.**
J. D. Barrett, R. C. Thomas, J. E. F. Baruch.
Publ. Astron. Soc. Pac., Vol. 103, No. 669, p. 1218 – 1230 (Nov 1991).
The problem of optimizing the schedule for a night of observing is described. The problem is a dynamic one for the observer changing through the night as key observations are made and weather conditions vary. A software tool is developed as an advisor to give the observer an optimized schedule. The scheduler is designed with operation on a personal computer as an objective. The structure and basis of such a tool are discussed and various solutions evaluated using the artificial intelligence techniques of constraint satisfaction and heuristics. The solutions are evaluated on the archived schedules available over STARLINK from the INT group of telescopes on La Palma.

036.175 **The wavelet transform.**
A. Bijaoui.
3. ESO/ST–ECF Data Analysis Workshop, p. 3 – 15 (1991). – See Abstr. 012.061 for the main entry.
The Fourier transform is a tool widely used for many scientific purposes, but it is not adapted to the study of non–stationary signals. The Fourier analysis brings only a global information which is not sufficient to detect compact patterns. Gabor introduces a local Fourier analysis, taking into account a sliding window, leading to a time–frequency analysis. It is adapted only to situations where the coherence time is independant of the frequency. Morlet's wavelets corresponds to Gabor's wavelets with a coherence time proportional to the period. The use of the wavelet transform is essentially guided by the study of hierrarchically structured object. The wavelet transform is a very nice tool for studying fractal structures. The author starts this analysis on the structure of the universe from galaxy counts.

036.176 **Filtering and restoration with the wavelet transform.**
J.-L. Starck, A. Bijaoui.
3. ESO/ST–ECF Data Analysis Workshop, p. 17 – 22 (1991). – See Abstr. 012.061 for the main entry.
The Wavelet transform is a new tool which allows to get information both in time and frequency. The authors developed an algorithm based on the FFT with an isotropic wavelet. They get a very useful description of the pictures when the objects show hierarchical structures. The authors achieved some filtering methods which used the pyramidal set of transform coefficients: Wiener filtering based on multi–resolution, adaptative non linear filtering, hierarchical Wiener filtering, non linear hierarchical filtering. A deconvolution is performed from an application of classical iteratives techniques.

036.177 **Finding echelle orders by Hough transform.**
P. Ballester.
3. ESO/ST-ECF Data Analysis Workshop, p. 23 – 28 (1991). – See Abstr. 012.061 for the main entry.
An auto-adaptive, fast algorithm for the automatic detection of echelle orders is presented. It is based upon the Hough transform, which provides robustness against discontinuous, noisy data.

036.178 **Spatial demodulation of 2-D Fabry-Perot images.**
T. Bonev, K. Jockers.
3. ESO/ST-ECF Data Analysis Workshop, p. 29 – 35 (1991). – See Abstr. 012.061 for the main entry.
Piezoelectrically tuned Fabry-Perot interferometers provide an efficient new means for monochromatic imaging of faint extended astronomical sources. Optimum wavelength resolution is achieved when the Fabry-Perot fringe fills the image but the spatial modulation is still noticable. In this paper a method is described to remove this modulation and extract, in the presence of background continuum, a monochromatic image, the spectral width of which is small as compared with the instrumental profile of the FPI.

036.179 **Adaptive filtering and its application.**
G. M. Richter.
3. ESO/ST-ECF Data Analysis Workshop, p. 37 – 43 (1991). – See Abstr. 012.061 for the main entry.
The aim and the basic ideas of adaptive filtering are discussed. A particular realisation of this principles on the basis of an image transform is described and the use of its is demonstrated by some examples.

036.180 **An optimum background level for cosmic ray filtering?**
E. J. A. Meurs, V. H. Bonifácio, N. M. Lima.
3. ESO/ST-ECF Data Analysis Workshop, p. 45 – 50 (1991). – See Abstr. 012.061 for the main entry.
The condition for pixel correction by means of median filtering is examined for two important cases: a stellar profile and a background noise pixel. There appears to exist an optimum background level for which both object and background noise are treated in a similar way. The performance of a median filter applied to cosmic ray removal may then be optimized by adjusting the level of the background during the filtering process.

036.181 **Detecting cosmic ray hits on HST WF/PC images.**
F. D. Murtagh, H.-M. Adorf.
3. ESO/ST-ECF Data Analysis Workshop, p. 51 – 56 (1991). – See Abstr. 012.061 for the main entry.
The authors describe initial experiments in detecting cosmic ray hits on single Hubble Space Telescope Wide Field/Planetary Camera CCD-frames. Classifiers are trained on images with (what are considered to be) the cosmic ray hits marked. A range of classifiers are assessed. The IRAF cosmicrays task is found to be conservative, having a low detection rate combined with a low false alarm rate. The classical k-nearest neighbours method offers a versatile and broad-ranging alternative. With the rather homogeneous input data used, the various multilayer perceptron algorithms used did not appear to offer advantages over the variants of k-nearest neighbours.

036.182 **Non-linear filters based on ordered statistics.**
F. Pasian.
3. ESO/ST-ECF Data Analysis Workshop, p. 57 – 66 (1991). – See Abstr. 012.061 for the main entry.
In this paper, definitions, properties and applications of filters based on order statistics, often known as median-like filters, are briefly described. In particular, the properties of median filters are summarized, some of their possible extensions and generalizations are defined, and some applications are described. Attention is also dedicated to performance, and some sorting algorithms for the ordering step of the filters are compared.

036.183 **UBV CCD photometry with the ESO/MPG 2.2 m telescope.**
L. Labhardt, A. Spaenhauer, H. Schwengeler.
3. ESO/ST-ECF Data Analysis Workshop, p. 79 – 87 (1991). – See Abstr. 012.061 for the main entry.
The authors address some of the problems encountered in the context of deep UBV CCD photometry: limited performance of the available detector, optimum reproduction of the photometric system, extraction of instrumental magnitudes based on synthetic aperture photometry as well as model image profile fitting, and transformation to the standard system. The careful application of photometric measuring and reduction procedures yields photometry of very high internal consistency, the quality of which is limited by the considerable diversity of available UBV standard photometry.

036.184 **Image analysis context development in ESO-MIDAS environment at the IFCAI-CNR.**
M. C. Maccarone, M. Tripiciano.
3. ESO/ST-ECF Data Analysis Workshop, p. 89 – 93 (1991). – See Abstr. 012.061 for the main entry.
New commands related to image analysis and processing, together with procedure operating on tables, are presented as contextes in ESO-MIDAS environment, version 90NOV. The source programs have been written in FORTRAN-77 extended language.

036.185 **Contrast enhancement by galaxy modelling.**
P. Prugniel, M. R. Rosa.
3. ESO/ST-ECF Data Analysis Workshop, p. 95 – 100 (1991). – See Abstr. 012.061 for the main entry.
The subtraction of a smooth model of the surface brightness distribution from images of galaxies can be used to enhance structural and photometric information on embedded features. The authors demonstrate the application of a MIDAS compatible S/W package (PLEINPOT) on images of the southern Centaurus group galaxy NGC 5253 (E/S0 pec em). Numerous stellar and nonstellar objects are revealed and visible after subtraction of the galaxy main body light. As opposed to other methods, e.g. unsharp or median-filter masking, this linear method preserves the local structure and may therefore be usefully applied prior to a deconvolution procedure.

036.186 **How to remove easily the unpleasant column pattern.**
G. Pojmański.
3. ESO/ST-ECF Data Analysis Workshop, p. 101 – 105 (1991). – See Abstr. 012.061 for the main entry.
The author shows a very simple method of correcting the column pattern that is present sometimes in CCD images or sets of reticon spectra, using just the scientific frames. The correction function is constructed from the tails of the intensity distribution of the background (sky or continuum) which is supposed to be flat.

036.187 **Applications of maximum entropy techniques to HST data.**
N. Weir.
3. ESO/ST-ECF Data Analysis Workshop, p. 115 – 129 (1991). – See Abstr. 012.061 for the main entry.
The author describes an approach to Bayesian/Maximum Entropy based image restoration with applications to HST data. The problem of image reconstruction is addressed in terms of model fitting, rather than filtering or deconvolution. The author describes a new image model which, by explicitly incorporating a smoothness constraint, allows one to simultaneously restore extended and sharp structure in an image with little or no pre-processing of the data. The method also allows one to fit to multiple data sets simultaneously and to automatically detect and reject cosmic ray artifacts as a part of the restoration process. The author demonstrates the use of these techniques on both FOC and WFC HST images.

036.188 **HST image restoration activities at ST ScI.**
R. J. Hanisch.
3. ESO/ST–ECF Data Analysis Workshop, p. 131 – 135 (1991). – See Abstr. 012.061 for the main entry.
The Space Telescope Science Institute has been active in studying approaches to the restoration of HST images since the problem of the spherical aberration in the primary mirror was recognized. This work comprises several areas of research: image restoration techniques, evaluation of the effects on photometry, studies of the point-spread function for both the imaging and spectroscopic instruments, and an analysis of the effects of changes in telescope focus and spacecraft jitter on image restoration processing. These studies are summarized briefly in this report.

036.189 **Restoration of Space telescope wide field camera images.**
D. Schade.
3. ESO/ST–ECF Data Analysis Workshop, p. 137 – 142 (1991). – See Abstr. 012.061 for the main entry.
Image restoration experiments using the maximum entropy method have been performed on simulations of HST Wide–Field camera images in order to judge the potential effectiveness of image restoration algorithms as part of the data reduction procedure. The production of artifacts in the restored images and their correlations with actual objects in the images suggest that the interpretation of low level features near galaxies in terms of structure related to those galaxies should be done with great caution.

036.190 **The angular diameters of deconvolved stars.**
L. B. Lucy.
3. ESO/ST–ECF Data Analysis Workshop, p. 143 – 146 (1991). – See Abstr. 012.061 for the main entry.
Statistical experiments are described in order to determine the size limitations imposed on deconvolved stellar images by photon statistics.

036.191 **Functional and statistical approaches to the image restoration problem.**
V. Yu. Terebizh.
3. ESO/ST–ECF Data Analysis Workshop, p. 147 – 150 (1991). – See Abstr. 012.061 for the main entry.
The conventional approach to the image restoration problem, based on integral equations, contradicts the physical meaning of the smoothing process. The consistent way of consideration should proceed from a purely stochastic model.

036.192 **Restoring HST Faint Object Camera images.**
H.–M. Adorf.
3. ESO/ST–ECF Data Analysis Workshop, p. 151 – 160 (1991). – See Abstr. 012.061 for the main entry.
Hubble Space Telescope, despite its current spherical aberration, is a remarkable discovery instrument as is evidenced by the results obtained within the Science Assessment and Early Release observational programmes. However, its scientific potential can presently only be exploited by using sophisticated restoration techniques. This contribution reviews the status of the image restoration issue for HST's Faint Object Camera.

036.193 **Some experiments of HST FOC images.**
C. Barbieri, G. De Marchi, R. Ragazzoni.
3. ESO/ST–ECF Data Analysis Workshop, p. 161 – 165 (1991). – See Abstr. 012.061 for the main entry.
Due to the peculiar characteristics of the PSF of HST, a careful analysis with many deconvolution experiments must be performed on HST–FOC images in order to understand them. The authors briefly present a subset of their work on such a subject.

036.194 **Some algorithms and procedures useful to analyse HST FOC images.**
C. Barbieri, G. De Marchi, R. Ragazzoni.
3. ESO/ST–ECF Data Analysis Workshop, p. 167 – 172 (1991). – See Abstr. 012.061 for the main entry.
Four procedures chosen from the many the authors have developed in the framework of their work about reduction of HST–FOC frame are briefly described. Emphasis, during such a choice, was given to those algorithms the authors think particularly useful for such space–based images.

036.195 **Image sharpening – context–controlled enhancement and deconvolution.**
P. Linde, A. Johannesson.
3. ESO/ST–ECF Data Analysis Workshop, p. 173 – 178 (1991). – See Abstr. 012.061 for the main entry.
The authors present results from two different forms of image sharpening; context–controlled image enhancement and a fast implementation of the well known Richardson–Lucy deconvolution algorithm. Image enhancement is implemented using context images. The technique takes advantage of specialised hardware with parallel computing capability, making it possible to process typical CCD images in a couple of minutes. The Richardson–Lucy algorithm for image restoration, based on Bayes' equation of conditional probabilities, has been implemented to run on the same hardware. The speed is 54 seconds per iteration for a 512⊥ × 338 image and a PSF of 31 × 31 pixels. The authors show a sequence where a CCD image of a stellar field has been subjected to 3, 50 and 1000 interations of the algorithm.

036.196 **Deconvolution of HST images: simulations with observed and analytical PSF.**
F. Zavatti, O. Bendinelli, M. Gatti, G. Parmeggiani.
3. ESO/ST–ECF Data Analysis Workshop, p. 179 – 184 (1991). – See Abstr. 012.061 for the main entry.
A comparison between Richardson–Lucy and multigaussian deconvolutions is presented. It shows that, for single, round objects, multigaussian deconvolution may be preferable for speed, accuracy beeing essentially the same. A simulated object, similar to a globular cluster in M31, is used for both methods, while a fit of the FOC f/96 is the PSF for the multigaussian deconvolution.

036.197 **Initial experiments with DAOPHOT II and WFC images.**
P. B. Stetson.
3. ESO/ST–ECF Data Analysis Workshop, p. 187 – 199 (1991). – See Abstr. 012.061 for the main entry.
The author reports on recent modifications to the stellar photometry program DAOPHOT, and on the degree to which it can cope with the optical properties of the Wide–Field Camera of the Hubble Space Telescope.

036.198 **Photometric decomposition of galaxies.**
F. Simien.
3. ESO/ST–ECF Data Analysis Workshop, p. 201 – 211 (1991). – See Abstr. 012.061 for the main entry.
The separation of the main stellar components in a galaxy image consists in most cases in extracting the spheroid and the flat disk. Spiral galaxies have been studied extensively; recently, lenticulars and even many ellipticals have also provided an interesting problem of photometric decomposition. Several photometric models and decomposition algorithms are reviewed, and the role of the a priori information is emphasized. A few applications are presented.

036.199 **On the Theil's fitting method.**
A. Pegoraro.
3. ESO/ST–ECF Data Analysis Workshop, p. 213 – 217 (1991). – See Abstr. 012.061 for the main entry.
This paper present the first results of a revision of the Theil's fitting method. The overall method's scheme is retained, while a different formulation of the estimators are given and, in particular, a more general error estimator is proposed. Various

kind of functions are considered, not only the straight line, but also polynomials and some non–linear curves often employed in astronomy, such as the Gaussian and the Lorentz curve. The goodness of the method is tested through a number of Monte–Carlo simulations.

036.200 **The de–blooming technique of large field CCD images.** Jiang Zhaoji, Chen Jiansheng.
3. ESO/ST–ECF Data Analysis Workshop, p. 219 – 223 (1991). – See Abstr. 012.061 for the main entry.
With the development of large format CCD, especially when used to cover a large field, the blooming of the overload pixel due to the saturated bright stars in the field become a serious problem, blurring a significant part of the CCD frame. This paper presents a method based on the median filter to move out the influence of the blooming effects on both CCD image as well as on CCD photometry.

036.201 **Crowded field photometry with HST – a simulation.** P. Linde.
3. ESO/ST–ECF Data Analysis Workshop, p. 225 – 230 (1991). – See Abstr. 012.061 for the main entry.
To study various observational constraints, in particular the behaviour in crowded stellar fields, simulations of HST Faint Object Camera observations have been made. The simulated field is intended to represent the centre of the Large Magellanic Cloud. The photometric measurement accuracy in the simulated crowded field is affected by the large amounts of scattered light. This light gives a high and unequally distributed background level. However, by successively substracting the brighter stars, the author was able to simplify the structure of the background, and subsequently improve the accuracy of background estimations. To study alternative approaches, a deconvolution algorithm has been applied to the simulated field. The resulting images show a noticeable improvement in resolution and contrast.

036.202 **Comparison between different isophote–fitting programs.**
M. Stiavelli, W. W. Zeilinger, P. Prugniel.
3. ESO/ST–ECF Data Analysis Workshop, p. 231 – 236 (1991). – See Abstr. 012.061 for the main entry.
Three different packages to derive the isophotal properties of galaxies (PLEINPOT, ELLFIT, and a modified version of INMP) have been applied on test images to check the reliability, reproducibility, and robustness of their results. The main conclusion is that the properties of isophotes are well defined only at radii bigger than 2 pixels and good agreement between the various packages is only achieved when the input data have high S/N ratios.

036.203 **A proposed medium–term strategy for optical interferometry in space. Report to the ESA astronomy working group by the ESA Space Interferometry Study Team (SIST).**
J. E. Noordam, P. Y. Bely, M. Faucherre, A. H. Greenaway, F. Merkle, M. A. C. Perryman, P. H. Roussel, F. Vakili, S. Volonté, G. P. Weigelt.
European Space Agency, Paris (France). Space Science Dept., ESA–SP—1135. 65 p. (Aug 1990). ISBN 92–9092–116–1. Price Dfl. 30.00.
This report provides a summary of the scientific arguments leading to the requirements for space interferometry, both in terms of its astronomical objectives, and in view of the possibilities attainable from the ground. It is shown that, at optical wavelengths, space missions will be driven by the requirement to achieve diffraction limited images beyond the magnitude limits achievable from the ground. At ultraviolet and infrared wavelengths, similar considerations naturally apply but, in addition, other fundamental physical atmospheric properties enhance further the justification to go to space. It is shown that several issues, albeit important to the efficiency and applicability of both ground and space–based interferometry, are not critical to the scientific case and technological issues facing space interferometry. Amongst these topics are discussions of image reconstruction techniques (speckle, phase closure, etc.), and aperture distribution (redundant, non–redundant). Image restoration techniques and other technical issues have made rapid advances in the context of radio aperture synthesis and VLBI techniques, and, in this field, optical interferometry will be well placed to exploit these advances. There has also been considerable progress in methods, theory and results of speckle imaging from the ground.

036.204 **The astrophysical applications of high resolution imaging by optical interferometry.**
Wang Zhengming, Xu Jiayan.
Prog. Astron., Vol. 9, No. 3, p. 228 – 237 (Sep 1991). In Chinese.
The high resolution imaging technology based on optical interferometry will open up broad prospects for research in astrophysics. The authors describe its astrophysical applications, including studies on stellar physics, binary and multiple stars, quasars and nuclei of galaxies, protoplanetary, circumstellar shells, etc. The Mark III stellar interferometer and the Infrared Spatial Interferometer (ISI) have brought a few encouraging results for binary star measurements and circumstellar shell observations. The space astrometric interferometer, POINTS, will be of inestimably benefit for the survey of Cepheid parallaxes, the determination of masses of stars in binary systems, the study of the Galactic structure and the mass distribution in the Galaxy. The authors can only give a brief introduction for potential abilities of optical interferometry and aperture synthesis in the field of astrophysics.

036.205 **SOLAR–A reformatted data files and observing log.** M. D. Morrison, J. R. Lemen, L. W. Acton, R. D. Bentley, T. Kosugi, S. Tsuneta, Y. Ogawara, T. Watanabe.
Sol. Phys., Vol. 136, No. 1, p. 105 – 110 (Nov 1991). Special issue: The YOHKOH (SOLAR–A) mission. For the hardcover edition see 003.013.
All of the SOLAR–A telemetry data will be reformatted before distribution to the analysis computers and the various users. This paper gives an overview of the files which will be created and the format and organization which the files will use. The organization has been chosen to be efficient in space, to ease access to the data, and to allow for the data to be transportable to different machines. An observing log file will be created automatically using the reformatted data files as the input. It will be possible to perform searches with the observing log to list cases where instruments are in certain modes and/or seeing certain signal levels.

036.206 **A cross–correlation method for surface photometry.** S. Phillipps, J. Davies.
Mon. Not. R. Astron. Soc., Vol. 251, No. 1, p. 105 – 111 (1 Jul 1991).
The authors apply the cross–correlation technique familiar in spectral matching to the two–dimensional case of surface photometry of galaxies with low surface brightness, i.e. to a problem of surface photometry with low signal–to–noise ratio. Comparison with normal profiling methods shows a considerable improvement in the accuracy with which photometric parameters can be obtained. In particular, for simulated images representing objects like GP 1444a which are fairly large but have extremely low surface brightness, the authors find errors of only ~ 2 and 3% in the scalelength and central intensity, smaller by a factor ~ 6 than the errors obtained from intensity profiles.

036.207 **Luminosity functions from emission line selected samples.**
D. Schade.
Mon. Not. R. Astron. Soc., Vol. 251, No. 2, p. 310 – 315 (15 Jul 1991).
A method for computing luminosity functions from samples of objects selected by the presence of emission lines is presented. The calculation is an extension of existing methods for computing the accessible volume for magnitude limited samples. The detection probability for objects near a signal–to–noise threshold is incorporated and it is shown that the validity of an adopted

equivalent width distribution (required to compute the luminosity function) can be tested.

036.208 **On the validity of Davies' test for periodicity.**
A. F. L. Nemec, J. M. Nemec.
Mon. Not. R. Astron. Soc., Vol. 251, No. 2, p. 356 - 359 (15 Jul 1991).

Davies' test for periodicity, which is based on the phase dispersion minimization method, is shown to be of limited application. In his derivation of the null distribution of the test statistic (i.e. the distribution under the null hypothesis of no periodicity), Davies assumes that the period to be tested is known a priori. Consequently, although the test can be used to assess the statistical significance of predetermined periods, it cannot be used to test periods that are identified by conducting a period search of the data. In the latter case, Davies uses adjustment based on Bonferroni's Inequality to compensate for the fact that multiple trial periods have been examined. Until a better solution to this difficult problem has been found, the authors recommend the use of the non-parametric randomization test described by Nemec and Nemec.

036.209 **Davies' periodicity test revisited.**
S. R. Davies.
Mon. Not. R. Astron. Soc., Vol. 251, No. 3, p. 64P - 65P (1 Aug 1991).

Recently Davies (1990) proposed the use of Analysis of Variance (ANOVA), a well-known statistical technique, to test for periodicity. This proposal has been severely criticized by Nemec and Nemec (1991), but unfortunately their arguments are based on a serious misinterpretation of the work of Davies. The purpose of this paper is to reaffirm the usefulness of ANOVA, and to remove any ambiguities concerning its range of validity.

036.210 **Automated analysis of slitless spectra: stars.**
M. Beauchemin, E. F. Borra, S. Levesque.
Mon. Not. R. Astron. Soc., Vol. 252, No. 1, p. 163 - 170 (1 Sep 1991).

The authors report on the stellar spectral classification that can be achieved with very low-dispersion spectroscopy. They apply several methods to slitless spectra taken from the automated grens plates analysis project undertaken at Laval University. Broad-band photometry and template matching have been used to probe the global spectral information, while intermediate-band photometry has been applied to examine local features. Multivariate analysis (MVA) has also been carried out as an objective tool to complement the study. B and V photoelectric data and the Gunn and Stryker library of stellar spectra have been used as reference material to calibrate and/or compare the data. It appears that an accuracy in B–V of $\sim$0.1 mag at B$\sim$19 for stars bluer than B–V = 0.8 and $\sim$0.2 mag for redder stars can be reached by probing the large-scale spectral distribution. DDO-like intermediate-band results show that grens data behave similarly to the Gunn and Stryker library of standard stars for many indices and so the classification philosophy behind such a system can be applied to grens data. Results from MVA are coherent with the methods of classification mentioned above and give a powerful and objective means of cataloguing spectra of the same spectral shape. Rare objects can be extracted from the database by many of the methods explored.

036.211 **Radial velocities from objective-prism plates.**
J. Stock.
Objective-prism and other surveys - a meeting in memory of Nicholas Sanduleak, p. 75 - 76 (1991). Abstract. - See Abstr. 012.074 for the main entry.

036.212 **Laser beams untwinkle stars.**
F. Merkle.
Nature, Vol. 353, No. 6340, p. 115 - 117 (12 Sep 1991).

036.213 **Compensation of atmospheric optical distortion using a synthetic beacon.**
C. A. Primmerman, D. V. Murphy, D. A. Page, B. G. Zollars, H. T. Barclay.
Nature, Vol. 353, No. 6340, p. 141 - 143 (12 Sep 1991). Letter-to-the-editor.

Atmospheric turbulence limits the resolution of a ground-based astronomical telescope to much worse than the diffraction limit of the instrument. An adaptive-optics system, in which some part of the optical train of the telescope can be adjusted in real time, can compensate for atmospheric turbulence, but a beacon or guide-star is needed to allow measurement of the atmospheric optical distortion. It has been suggested that the measurement could be achieved by means of a synthetic beacon formed by atmospheric backscatter from a ground-based laser. The authors have performed an experiment to demonstrate atmospheric compensation using a synthetic beacon. With a pulsed dye laser to generate the beacon and a 241-channel adaptive-optics system to perform the phase correction, they obtained almost diffraction-limited resolution of star images in the visible part of the spectrum. The results indicate that there are no technical barriers to atmospheric compensation using synthetic beacons at ground-based observatories.

036.214 **Measurement of atmospheric wavefront distortion using scattered light from a laser guide-star.**
R. Q. Fugate, D. L. Fried, G. A. Ameer, B. R. Boeke, S. L. Browne, P. H. Roberts, R. E. Ruane, G. A. Tyler, L. M. Wopat.
Nature, Vol. 353, No. 6340, p. 144 - 146 (12 Sep 1991). Letter-to-the-editor.

The possibility of using an artificial "guide-star" to measure optical wavefront distortion caused by atmospheric turbulence has been discussed for some time, but few experimental data are available. The authors report experimental results demonstrating that atmospheric wavefront distortion can be measured by taking fast "snapshots" of a guide-star formed by light scattered from a laser beam focused in the upper atmosphere. These results agree with a theoretical prediction that the mean-square wavefront error created by using an artificial guide-star at a finite distance rather than a real (infinitely distant) star is proportional to the five-thirds power of the telescope aperture. Using this understanding of the physics, the authors have demonstrated continuous, real-time atmospheric compensation of a 1.5-m telescope using a high-pulse-rate laser, pulse-synchronized wavefront sensor and deformable mirror, and have been able to resolve the 1.3-arcsecond binary star 53 ξ Ursae Majoris in an exposure time of only one second.

036.215 **Analysis of lunar occultation data.**
W. P. Chen.
10. International Workshop on Maximum Entropy and Bayesian Methods, MaxEnt 90, p. 51 - 57 (1991). - See Abstr. 012.081 for the main entry.

Lunar occultation technique has been successfully applied in the near-infrared to obtain structural information of celestial objects with very high angular resolution. This paper reviews and evaluates the methods which have been used to analyze the data. Specifically, the least-squares modeling is adequate for data suggesting simple intensity structure, while inverse techniques are required for complicated objects. A technique incorporating Bayesian procedure (the Richardson-Lucy algorithm) is found particularly useful in restoration of lunar occultation data.

036.216 **The photometric accuracy of astronomical images restored with the MEMSYS3 code.**
J. G. Cohen.
10. International Workshop on Maximum Entropy and Bayesian Methods, MaxEnt 90, p. 59 - 65 (1991). - See Abstr. 012.081 for the main entry.

After reviewing the causes of blur in astronomical images obtained with optical telescopes, both ground based instruments and the Hubble Space Telescope, simulations are described which test the photometric accuracy of the maximum entropy image restoration algorithm used in the code MEMSYS3 version 2. It is found that at the faintest brightness levels, while the code can recover blurred out point sources, they are recovered systematically too faint. The size of the photometric error depends on the brightness of the source (relative to the noise of the background) and the crowding of the field. At present, the optimum technique to apply to blurred images of crowded fields where most of the sources are point sources seems to be to use a restored image to generate the list of objects, then feed this into a standard point spread function fitting code and use this on the original blurred frame. In that manner, the most crowded fields can be analyzed without losing photometric accuracy.

036.217 **Computational problems and signal processing in SETI.**
S. R. Deans, D. K. Cullers, R. Stauduhar.
10. International Workshop on Maximum Entropy and Bayesian Methods, MaxEnt 90, p. 67 – 75 (1991). – See Abstr. 012.081 for the main entry.

The Search for Extraterrestrial Intelligence (SETI), currently being planned at NASA, will require that an enormous amount of data be analyzed in real time by special–purpose hardware. Even though the SETI system design is not based on maximum entropy and Bayesian methods (partly due to the real–time processing constraint) it is expected that enough data will be saved to be able to apply these and other methods off line where computational complexity is not an overriding issue. Interesting computational problems that relate directly to the system design for processing such an enormous amount of data have emerged. Some of these problems are discussed, along with the current status on their solution.

036.218 **A subpixel deconvolution method for astronomical images.**
N. Weir, S. Djorgovski.
10. International Workshop on Maximum Entropy and Bayesian Methods, MaxEnt 90, p. 275 – 283 (1991). – See Abstr. 012.081 for the main entry.

The authors present results from the use of an image deconvolution method for achieving subpixel resolution in certain data regimes. Restorations are performed using a shell which utilizes the Gull–Skilling MEMSYS3 package. They have performed extensive tests using the method and have determined general limits of its applicability and resolving power. They present results from applying the technique to simulated and real images of sparse and closely crowded astronomical fields. Specific projects involved in this study include a search for close companions to quasars and the structural investigation of star cluster cores.

036.219 **Two new methods for retrieving an image from noisy, incomplete data and comparison with the Cambridge MaxEnt package.**
D. Ustundag, N. M. Queen, G. K. Skinner, J. E. Bowcock.
10. International Workshop on Maximum Entropy and Bayesian Methods, MaxEnt 90, p. 295 – 301 (1991). – See Abstr. 012.081 for the main entry.

The authors consider two new methods for retrieving an image from noisy, incomplete data, based on the principle of maximum entropy. The image is reconstructed by either placing grains one at a time into image pixels, or transferring them between randomly chosen pairs of pixels on the basis of certain deterministic rules. The performance of the methods is tested for an application in X–ray astronomy and compared with that of the Cambridge MaxEnt package. They also demonstrate the application of the Cambridge MaxEnt package to a real problem in X–ray astronomy.

036.220 **Maximum entropy image construction of the galaxy M51.**
A. Mohammad–Djafari, J. Idier.
10. International Workshop on Maximum Entropy and Bayesian Methods, MaxEnt 90, p. 313 – 318 (1991). – See Abstr. 012.081 for the main entry.

The authors explain the method they used to reconstruct the galaxy M51 image.

036.221 **The image reconstruction contest.**
T. R. Bontekoe.
10. International Workshop on Maximum Entropy and Bayesian Methods, MaxEnt 90, p. 319 – 324 (1991). – See Abstr. 012.081 for the main entry.

From the all–sky infra–red survey of the IRAS satellite, a dataset was prepared for an image reconstruction contest. The object for reconstruction was an image of the galaxy M51. The entries were judged by a panel of astronomers. The judging criterion was *which entry brings out the most astronomy*.

036.222 **Maximum entropy with Poisson statistics.**
D. R. T. Robinson.
10. International Workshop on Maximum Entropy and Bayesian Methods, MaxEnt 90, p. 337 – 341 (1991). – See Abstr. 012.081 for the main entry.

This paper presents the Bayesian MaxEnt solution to problems with data subject to Poisson statistics. Examples are shown of the deconvolution of Hubble Space Telescope images. The method is also applied to the reconstruction of a probability distribution given samples from the distribution.

036.223 **Accuracy of period determination.**
A. Schwarzenberg–Czerny.
Mon. Not. R. Astron. Soc., Vol. 253, No. 2, p. 198 – 206 (15 Nov 1991).

Periods of oscillation are frequently found using one of two methods: least–squares (LSQ) fit or power spectrum. Their errors are estimated using the LSQ correlation matrix or the Rayleigh resolution criterion $\delta\nu_R = 1/\Delta T$, respectively. In this paper the author demonstrates that both estimates are statistically incorrect. On the one hand the LSQ covariance matrix does not account for correlation of residuals from the fit. Neglect of the correlations may cause large underestimation of the variance. On the other hand the Rayleigh resolution criterion is insensitive to signal–to–noise ratio and thus does not reflect quality of observations. The author derives the correct variance estimates for the two methods. In the process he demonstrates that centre of the power spectrum line is a maximum likelihood estimate of frequency of the oscillation and demonstrates it is statistically equivalent to fitting of a sinusoid by LSQ, so the methods are statistically equivalent. The author's new and correct variance estimate is quite simple and practicable. It is using the autocorrelation function (ACF) of the residuals to determine their mean correlation length and is valid under certain assumptions. The author tested the extent to which the assumptions may be relaxed by numerical simulations.

036.224 **Improving the angular resolution of the Buckland Park air shower array.**
R. Meyhandan, R. W. Clay.
Proc. Astron. Soc. Aust., Vol. 9, No. 1, p. 113 – 114 (1991). – See Abstr. 012.078 for the main entry.

Air showers initiated by primary cosmic rays and gamma rays produce shower fronts which are curved. However, the arrival directions of air shower events have normally been fitted assuming a planar shower front. The authors present a technique which takes the average shower front shape into account to assign an improved shower direction after a first analysis assuming a plane front. They then examine the resulting angular resolution of the Buckland Park array.

036.225 **Maximum likelihood image restoration. IV. Limiting resolution for given alternative.**
V. Yu. Terebizh.
Astrophysics, Vol. 33, No. 3, p. 536 – 548 (Nov – Dec 1990). English translation of Astrofizika, Tom 33, Vyp. 3, p. 409 – 427 (Dec 1990).

A formal criterion is found for selection of blurred and noised images of an arbitrary object in the case when the alternative object is known. The criterion generalizes the analogous result of Harris to situations in which a priori probabilities for the presence of different objects are not known and arbitrary probabilities of errors of types I and II (respectively, α and β) are allowed. An explicit expression is obtained for the maximally attainable resolution for fixed α and β and signal-to-noise ratio ψ. It is shown that the limiting resolution depends only on the combination $t \equiv (a_\alpha + a_\beta)/\psi$, where z_ε is the quantile of the normal distribution. For bright objects, the limiting resolution is proportional to $\psi^{-1/2}$. Examples of independent practical interest are considered.

036.226 **High-resolution imaging by optical aperture synthesis: first results from the MAPPIT project.**
J. G. Robertson, T. R. Bedding, R. G. Marson,
P. R. Gillingham, R. H. Frater, J. D. O'Sullivan, R. P. Norris.
Proc. Astron. Soc. Aust., Vol. 9, No. 1, p. 162 – 163 (1991). – See Abstr. 012.078 for the main entry.

The authors have successfully demonstrated optical aperture synthesis at the 4-m Anglo-Australian Telescope. By using a multi-hole mask over the (re-imaged) primary mirror and recording the resulting fringe patterns with high time resolution, diffraction-limited images of sufficiently bright objects can be reconstructed. The data processing uses closure phases to overcome the effects of atmospheric turbulence. The authors show an image of the double star η Oph, with component separation 0.″45.

036.227 **The Mt John CCD system: status and first scientific results.**
W. Tobin.
Proc. Astron. Soc. Aust., Vol. 9, No. 1, p. 164 – 166 (1991). – See Abstr. 012.078 for the main entry.

A Photometrics CCD system containing a TH7882 chip has been in use at Mt John since 1989 October for imaging photometry. Extra control software has increased observing ease and image-header information. Photometrically the system appears to be performing satisfactorily. The blue helium star LSS 99 showed no variation exceeding $\sim 1\%$ over 40 min of observation. The blue eclipsing binary HV 1761 in the Small Magellanic Cloud is found to have approximately equal components.

036.228 **Analog and digital processing of the polarization and spectrozonal photos of the Moon.**
S. I. Ivanov, V. V. Novikov, Kh. G. Tadzhidinov.
Tr. Gos. Astron. Inst. im. Shternberga, Tom 62, p. 218 – 233 (1991). In Russian.

The results of computer processing of telescopic images of the Moon are given. The problems of digitizing, storing and superimposing of the images are considered. The main attention is given both to accounting of phase function in the computer programm of the two-parameter regioning of the surface of the Moon based on the light reflection and to the forming of differential images.

036.229 **A new technique for study of radial velocity changes.**
L. Ramsey, T. Brown, R. Gilliand, R. Noyes.
Symposium in Celebration of the 50. Anniversary of McDonald Observatory of the University of Texas at Austin: Frontiers of stellar evolution, p. 614 (1991). Abstract. – See Abstr. 012.085 for the main entry.

036.230 **Choice of filters for the detection of gravitational waves from coalescing binaries.**
B. S. Sathyaprakash, S. V. Dhurandhar.
Phys. Rev. D, Vol. 44, No. 12, p. 3819 – 3834 (15 Dec 1991). Current Physics Microform No.: 9101C0101.

Coalescing binaries are one of the most promising candidates for the detection of gravitational waves with laser interferometric gravitational-wave detectors. Signals from coalescing binaries will most probably not stand above the broadband noise of the detector. Their detection is possible by the use of special data analysis techniques such as matched filtering which takes advantage of the fact that the wave form can be fairly well predicted. The authors present an algorithm to choose a lattice set of filters by a criterion that every signal of a certain minimal strength is picked up by at least one filter of the set. The wave form is characterized by three parameters: the time of arrival, the mass parameter, and the phase of the signal. It is shown that it is enough to have just two filters corresponding to the phase of the signal. Determination of the lattice for various values of the mass parameter involves a knowledge of the cross correlation function of two chirp wave forms with different values of the parameters. It is shown that for a considerable range of the mass parameter, the peak value of the correlation function, in a certain approximation, does not depend on the absolute values of the parameters but only on their difference. An empirical relation for the distance, up to which a detector can see, is derived.

036.231 **Comets to quasars: surface photometry from standard stars, and the morphology of the galaxy-quasar interface.**
P. D. Usher.
Liller Symposium: Asteroids to quasars, p. 53 – 64 (1991). – See Abstr. 012.087 for the main entry.

Wide-angle phenomena in astronomy involve objects from the Solar System to the farthest reaches of the Cosmos. Pursuit of knowledge in these areas requires data from cameras of the Schmidt class, and hence the ability to quantify photographic information. Two fields of wide-angle research, viz. the photometry of large-scale structures in comets, and the uranography, taxonomy, and phenomenology, of quasars and related objects over large solid angles of the sky, are discussed in the context of Liller's seminal contributions to each. Recent advances in these areas are: (i) that standard stars can be used to derive and calibrate the surface brightness of extended structures like the Andromeda galaxy, and hold promise for the determination of a photometric history of comets past and present; and (ii) that the best data currently available suggest that M(B) = –24.0 is a safe and workable criterion by which to ensure that quasar samples to B = 18.5 mag are comprised wholly of objects that are unresolved by cameras of the 1.2 m Schmidt class.

036.232 **Mapping of overspread targets in radar astronomy.**
T. Hagfors, W. Kofman.
Radio Sci., Vol. 26, No. 2, p. 403 – 416 (Mar–Apr 1991).

The problem of delay-Doppler mapping of overspread targets, i.e., targets where the product of delay extent θ_t and Doppler width B_t is greater than unity is discussed. For a transmitter modulation waveform which is believed to be representative of continuous wave modulation, namely, a random Gaussian noise modulation, the effect of the "self-noise" and the additive noise on the output signal-to-noise ratio is evaluated, and expressions are derived which allows the signal-to-noise ratio to be predicted and the quality of the mapping assessed for various types of targets and modulation waveform properties. The particular case of the mapping of Mars with an existing facility is dealt with in detail, and it is shown that a limited amount of mapping with a resolution corresponding to 100 by 100 points on the planet is possible in certain regions.

036.233 **Stellar spectrum synthesis.**
J. Jugaku.
Liller Symposium: Asteroids to quasars, p. 141 – 145 (1991). – See Abstr. 012.087 for the main entry.

In the ultraviolet region of stellar spectra the absorption lines are numerous and are crowded so that the wings of these lines overlap. Abundance analysis using equivalent widths of stellar spectral lines require the features under investigation to be unblended. However, many of the elements whose abundances are key diagonstics of stellar structure and evolution only have lines in very crowded spectral regions. In such situations the method of spectrum synthesis enables one to obtain information from blended features.

036.234 **Self-noise in interferometers.**
M. Vivekanand, S. R. Kulkarni.
IAU Colloquium No. 131: Radio interferometry – theory, techniques, and applications, p. 1 – 5 (1991). – See Abstr. 012.089 for the main entry.

The authors study the distribution of noise in images synthesized by radio interferometers, focusing particularly on the noise due to source power fluctuations. For a total power interferometer, the r.m.s. fluctuation in the image consists of a term that is constant all over the map, which is the uncorrelated system noise divided by the number of telescopes in the interferometer, and a term that is the image itself. For a correlation interferometer the expression is not so simple but is qualitatively similar. The results are consistent with intuitive expectations in various observing situations. The authors find similarities in self–noise at optical and radio wavelengths.

036.235 **Source–noise in radio synthesis images.**
K. R. Anantharamaiah, A. A. Deshpande, V. Radhakrishnan, R. D. Ekers, T. J. Cornwell, W. M. Goss.
IAU Colloquium No. 131: Radio interferometry – theory, techniques, and applications, p. 6 – 9 (1991). – See Abstr. 012.089 for the main entry.
The authors discuss the distribution of source– and receiver–noise in radio synthesis images and show that the source–noise is maximum at the position of the source but also appears in the off–source region because of the sidelobes. Analytical expressions are derived for the rms noise at any location of both "total–power" and "correlation" images. The authors show that under certain conditions, deconvolution can remove the source noise from the off–source region in snap–shot images. Some of the results are verified experimentally.

036.236 **The effect of spatial coherence of sources on synthetic aperture mapping.**
D. F. V. James.
IAU Colloquium No. 131: Radio interferometry – theory, techniques, and applications, p. 10 – 14 (1991). – See Abstr. 012.089 for the main entry.
The interferometric mapping of astronomical objects relies on the van–Cittert Zernike theorem, one of the major results of the theory of partially coherent light. This theorem states that the degree of spatial coherence of the field from a distant spatially incoherent source is proportional to the Fourier transform of the intensity distribution across the source. Measurement of the degree of spatial coherence, by, for example, measuring the visibility of interference fringes, allows the object to be mapped by making an inverse Fourier transform. The author presents a summary of the results of an investigation into what happens when the distant source is not spatially coherent.

036.237 **Burst mode VLBI and pulsar applications.**
R. D. Wietfeldt, D. A. Frail.
IAU Colloquium No. 131: Radio interferometry – theory, techniques, and applications, p. 76 – 80 (1991). – See Abstr. 012.089 for the main entry.
The technique of Burst Mode (BM) VLBI is described and its particular application to the observations of pulsars is explored. The wideband pulsed radiation from these objects is exploited to acquire data at two ends of an interferometer baseline in high–speed bursts and only during the pulsar on–times. The result at the VLBI correlator is an improvement in sensitivity by as much as the inverse pulsar duty cycle, and is a consequence of the bandwidth expansion offered by BM and the intrinsic use of gating. As a result, precision astrometrical studies of pulsars will benefit through the greatly increased instantaneous bandwidths achieved by pulsar BM. The problems, implications, and limitations of acquiring and processing pulsar BM data are discussed.

036.238 **Using sub–aperture illumination to solve the short–spacing problem.**
D. P. Woody, N. Z. Scoville, L. G. Mundy.
IAU Colloquium No. 131: Radio interferometry – theory, techniques, and applications, p. 81 – 85 (1991). – See Abstr. 012.089 for the main entry.
A common problem in producing interferometer maps of objects with structure on the scale–size of the primary beam is the acquisition of UV data corresponding to telescope spacings of less than one telescope diameter. One technique for solving this "short–spacing" problem is to divide the aperture of one of the telescopes into smaller sub–apertures and measuring the visibilities between these sub–apertures. This technique is being tested using the Owens Valley Millimeter Interferometer.

036.239 **Phase control in the digital lens.**
K. Asuma, S. Iwase, K. Nishibori, J. Nakajima, E. Otobe, A. Tsuchiya, N. Watanabe, T. Daishido.
IAU Colloquium No. 131: Radio interferometry – theory, techniques, and applications, p. 90 – 93 (1991). – See Abstr. 012.089 for the main entry.
Phase and amplitude is controlled by manipulating the digitized radiation field. Application to the phase switching, and the calibration of phase and amplitude is discussed.

036.240 **The requirement to the baseline knowing of a space VLBI.**
L. R. Kogan, N. A. Eysmont.
IAU Colloquium No. 131: Radio interferometry – theory, techniques, and applications, p. 115 – 118 (1991). – See Abstr. 012.089 for the main entry.
It is shown that the influence of a baseline knowing error on the fringe rate change during coherent averaging time interval is extremely increasing for space VLBI due to more angle velocity of a space baseline rotation comparatively with ground VLBI. It is shown on the example of Radioastron orbit that the requirement to its orbit for such a short wavelength as 1.35 cm is determined by unknowing of the fringe rate change during coherent averaging time interval rather than by delay ambiguity and this requirement is equal several meters.

036.241 **Some highlights of interferometry in early radio astronomy.**
W. T. Sullivan III.
IAU Colloquium No. 131: Radio interferometry – theory, techniques, and applications, p. 132 – 149 (1991). – See Abstr. 012.089 for the main entry.
Two important episodes in the early development of interferometry in radio astronomy are traced in detail. The first is the use of the sea–cliff interferometer at the Radiophysics Laboratory in Sydney, first by Pawsey for solar observations and later by Bolton for radio star surveys. The second is the development of the Michelson interferometer and the phase switch by Ryle in Cambridge. This also was employed for important observations of the sun and radio stars.

036.242 **Multi–frequency synthesis.**
J. E. Conway.
IAU Colloquium No. 131: Radio interferometry – theory, techniques, and applications, p. 171 – 179 (1991). – See Abstr. 012.089 for the main entry.
A method of improving image fidelity by using observations at multiple frequencies is described. The author discusses the power and possible limitations of the technique. Results of narrow frequency range observations with MERLIN and global VLBI are presented. The author concludes with a consideration of the possible future impact of this technique.

036.243 **The use of beam–sets in the analysis of wide–field maps from the CLFST.**
E. M. Waldram.
IAU Colloquium No. 131: Radio interferometry – theory, techniques, and applications, p. 180 – 183 (1991). – See Abstr. 012.089 for the main entry.
The Cambridge Low Frequency Synthesis Telescope (CLFST) has baselines which are offset from the E–W direction by about 3°. The wide–field mapping problem has been overcome by the use of beam–sets, or artificial maps consisting of grids of synthesized beams, which incorporate the effects of both the geometric and chromatic aberration. By appropriate interpolation one can find a good approximation to the beam–shape at any point on a map.

036.244 **Imaging with non–coplanar data.**
R. A. Perley, T. J. Cornwell.
IAU Colloquium No. 131: Radio interferometry – theory, techniques, and applications, p. 184 – 187 (1991). – See Abstr. 012.089 for the main entry.
The general relation between interferometric data and sky brightness requires the use of three–dimensional Fourier transforms. Only in special cases does the standard two–dimensional Fourier relations result in images free of distortions. For the VLA, use of two–dimensional transforms introduces errors which are important at low frequencies and high resolutions. The authors review options and progress in solving the computational problems associated with the full solution.

036.245 **Adding short spacings to synthesis maps in the "sky"–domain.**
U. J. Schwarz, B. P. Wakker.
IAU Colloquium No. 131: Radio interferometry – theory, techniques, and applications, p. 188 – 191 (1991). – See Abstr. 012.089 for the main entry.
The authors discuss critically the methods used to complement synthesis maps with short–spacing information derived from single–dish observations. The conclusion is that from a single synthesis field the correct result cannot be derived, a large area has to be synthesized.

036.246 **The use of redundancy in interferometry – a comparison of redundancy and selfcal.**
M. Wieringa.
IAU Colloquium No. 131: Radio interferometry – theory, techniques, and applications, p. 192 – 196 (1991). – See Abstr. 012.089 for the main entry.
Many synthesis arrays currently in use have at least a few redundant baselines which can be used to derive model–independent telescope corrections. The author discusses the relative merits of redundancy–based versus standard selfcalibration methods with respect to sensitivity to noise and interference. He then derives an approach to optimally combine the two methods for partially redundant arrays.

036.247 **The bispectrum analysis technique in millimeter interferometry.**
K.–i. Morita.
IAU Colloquium No. 131: Radio interferometry – theory, techniques, and applications, p. 197 – 201 (1991). – See Abstr. 012.089 for the main entry.
The authors proposes an application of the bispectrum analysis technique in millimeter interferometry, which is used in the speckle masking technique. For the millimeter interferometry, in which atmospheric phase fluctuations are very serious, not only bispectrum phase but also bispectrum amplitude is good observable. The author developed an algorithm to reconstruct images directly from bispectrum. He made good maps with this algorithm in the case of both simple and compact sources.

036.248 **Self–calibration of M.O.S.T. data.**
A. D. Gray.
IAU Colloquium No. 131: Radio interferometry – theory, techniques, and applications, p. 202 – 205 (1991). – See Abstr. 012.089 for the main entry.
The Molonglo Observatory Synthesis Telescope (MOST), located near Canberra, Australia, is an east–west earth–rotation synthesis array which, unlike conventional synthesis arrays, generates multiple real–time fan–beams and forms images by the back–projection of the recorded intensities rather than by Fourier inversion of visibilities. The images produced in this fashion are often marred by the presence of radial artefacts emanating from strong sources due to residual calibration errors. An algorithm devised to self–calibrate these errors is discussed.

036.249 **High accuracy ($\ll 1\%$) polarisation measurements with the WSRT.**
J. E. Noordam.
IAU Colloquium No. 131: Radio interferometry – theory, techniques, and applications, p. 206 – 211 (1991). – See Abstr. 012.089 for the main entry.
The procedure for polarisation measurements with radio synthesis telescopes is fraught with uncertaincies and shaky assumptions. Therefore, their accuracy will be limited to a few percent at best. It is investigated here whether it will be possible to improve this accuracy by a factor 10 – 100 with the WSRT. Although the investigation is not yet finished, the tentative conclusion is that it may be possible, with crossed dipoles, to derive completely model–independent values for Q/I, U/I and V/I.

036.250 **Low frequency imaging and the non–isoplanatic atmosphere.**
C. R. Subrahmanya.
IAU Colloquium No. 131: Radio interferometry – theory, techniques, and applications, p. 218 – 222 (1991). – See Abstr. 012.089 for the main entry.
At low frequencies, the typical fields of view over which imaging is required is several degrees across, for which (a) the visibility and brightness distribution cannot be assumed to be related by a two–dimensional Fourier Transform, and (b) the propagation delays introduced by the atmosphere cannot be assumed constant within the field of view. These are adequately represented in a method discussed in this paper. Although compute–intensive, the method is suitable for implementation on parallel processors.

036.251 **VLA observations at 333 MHz: a search for pancakes.**
J. M. Uson, D. S. Bagri, T. J. Cornwell.
IAU Colloquium No. 131: Radio interferometry – theory, techniques, and applications, p. 223 – 227 (1991). – See Abstr. 012.089 for the main entry.
The P–band is the most recently completed observing band at the VLA. The aperture efficiency of the antennas is 0.4 and the system temperature is $\sim$130K. The primary beam is slightly off–axis and its half–power width is steeply frequency dependent, which induces abnormal spectral indices on off–axis ($\geqslant 1°$) sources. The authors have searched for Zel'dovich "pancakes" and reached a level of $\Delta S_{rms} \sim 3.5$ mJy for spectral channels of width $\Delta\nu \sim 100$ kHz. The authors find no pancake with $M_{HI} \gtrsim 5 \times 10^{14} M_{\odot}$.

036.252 **The three dimensional structure of astronomical sources through optimal inversion.**
E. Keto, W. Jeffrey.
IAU Colloquium No. 131: Radio interferometry – theory, techniques, and applications, p. 228 – 232 (1991). – See Abstr. 012.089 for the main entry.
The authors explore the application of optimal inversion techniques to astronomical data with a goal of developing a set of procedures for the determination of the three dimensional structure of astronomical sources. The authors show how constraints of symmetry and smoothness source allow to handle an inversion with an insufficiently sampled data base and mutually dependent solution parameters (mathematically ill–posed and ill–conditioned). The authors briefly discuss the background and motivation, and outline the procedure in general terms.

036.253 **Imaging for the Australia Telescope: present needs and future strategies.**
R. P. Norris.
IAU Colloquium No. 131: Radio interferometry – theory, techniques, and applications, p. 233 – 237 (1991). – See Abstr. 012.089 for the main entry.
The author outlines the strategy adopted to produce images from the Australia Telescope. The main production system is based on AIPS, but several departures from conventional strategy have been made.

036.254 **A modified algorithm for cleaning wide–field maps with extended structures.**
K. S. Dwarakanath, A. A. Deshpande, N. U. Shankar.
IAU Colloquium No. 131: Radio interferometry – theory, techniques, and applications, p. 242 (1991). Abstract. – See Abstr. 012.089 for the main entry.

036.255 **Application of an estimator–free information criterion (WIC) to aperture synthesis imaging.**
M. Ishiguro, K. I. Morita, M. Ishiguro.
IAU Colloquium No. 131: Radio interferometry – theory, techniques, and applications, p. 243 – 248 (1991). – See Abstr. 012.089 for the main entry.
A statistical criterion for stopping CLEAN procedure is proposed. The criterion is called WIC and an estimator of Kullback–Leibler information quantity which is a measure of the goodness of statistical models. A numerical example is given.

036.256 **Wide field mapping of 5C12.**
J. E. Okopi, L. B. Bååth.
IAU Colloquium No. 131: Radio interferometry – theory, techniques, and applications, p. 253 – 256 (1991). Abstract. – See Abstr. 012.089 for the main entry.

036.257 **Phase calibrators for MERLIN.**
A. R. Patnaik, I. W. A. Browne, P. N. Wilkinson, J. M. Wrobel.
IAU Colloquium No. 131: Radio interferometry – theory, techniques, and applications, p. 257 – 259 (1991). – See Abstr. 012.089 for the main entry.
The VLA was used to derive a catalogue of phase–reference sources for MERLIN. These sources can also be used for other interferometer arrays such as the VLBA. The positions of the sources in the catalogue are repeatable to about 5 mas rms.

036.258 **Observational test of the homogeneous array concept: an X–band image of the Crab Nebula.**
J. M. Uson, T. J. Cornwell.
IAU Colloquium No. 131: Radio interferometry – theory, techniques, and applications, p. 260 – 263 (1991). – See Abstr. 012.089 for the main entry.
The authors have performed an observational test of the "Homogeneous Array" concept utilized in the MMA design, whereby the array elements are used in total power mode to provide the shortest spacings. Using the VLA in interferometric mode and a VLBA element measuring total power, the authors have reconstructed a mosaic image of the Crab Nebula, which fits the data adequately and is of good visual appearance.

036.259 **Visibility based continuum subtraction.**
H. J. van Langevelde, B. Cotton.
IAU Colloquium No. 131: Radio interferometry – theory, techniques, and applications, p. 264 – 267 (1991). – See Abstr. 012.089 for the main entry.
The authors have been looking into the possibility to subtract out the continuum contribution in spectral line databases from the ungridded visibilities. It turns out that this simple method offers quite interesting possibilities in a lot of practical cases. The authors show how and why this method works. They have tried to compare the results of this method with the more conventional ways of subtracting the continuum.

036.260 **The multi–resolution CLEAN.**
B. P. Wakker, U. J. Schwarz.
IAU Colloquium No. 131: Radio interferometry – theory, techniques, and applications, p. 268 – 271 (1991). – See Abstr. 012.089 for the main entry.
The authors describe a modification to CLEAN which alleviates some problems for extended sources. This is accomplished by combining the results of a number of conventional CLEAN operations, each done at a different resolution. The algorithm is called "Multi–Resolution Clean" or "MRC". Experiments on model sources have shown that it works well even when the source is so extended that the usual CLEAN becomes impractical. For extended sources, MRC enhances the signal–to–noise ratio, resulting in an easier definition of the area of signal. Moreover, MRC is in principle faster than a standard CLEAN because less δ–functions are needed.

036.261 **Numerical experiments with low SNR data in radio interferometry.**
P. N. Wilkinson, P. Woodall.
IAU Colloquium No. 131: Radio interferometry – theory, techniques, and applications, p. 272 – 275 (1991). – See Abstr. 012.089 for the main entry.
The authors outline the results of numerical experiments on low SNR interferometer data. They confirm several theoretical analyses of the effect of averaging on visibility amplitude and closure phase data. However the triple product method for estimating the closure phase, which is useful in optical interferometry, does not provide a singificant improvement over existing practice in radio interferometry. The authors introduce a way of using the triple product to estimate visibility amplitudes from arrays dominated by a single large telescope.

036.262 **Hot spots in Cygnus A; a mosaiced map at 86 GHz.**
M. C. H. Wright, R. J. Sault.
IAU Colloquium No. 131: Radio interferometry – theory, techniques, and applications, p. 276 – 280 (1991). – See Abstr. 012.089 for the main entry.
The authors describe a new software package, MIRIAD, for making multichannel images of multiple–field aperture synthesis observations. As an example of the use of the new software they present a map of the radio galaxy Cygnus A. The authors used the BIMA array to map Cygnus A at 86 GHz with 2.28×3.06 arcsec resolution in 3 fields centered on the central radio soruce and the radio hot spots in each radio lobe. The 3 fields were combined and deconvolved to find the positions and flux densities of the radio source components.

036.263 **Linear polarization sensitive VLBI.**
D. H. Roberts, L. F. Brown, J. F. C. Wardle.
IAU Colloquium No. 131: Radio interferometry – theory, techniques, and applications, p. 281 – 288 (1991). – See Abstr. 012.089 for the main entry.
The authors describe the techniques currently used to produce linear polarization sensitive radio images at milliarcsecond resolution, and briefly discuss what will be possible in the next few years.

036.264 **VLBI phase–referencing for observations of weak radio sources.**
J. F. Lestrade.
IAU Colloquium No. 131: Radio interferometry – theory, techniques, and applications, p. 289 – 297 (1991). – See Abstr. 012.089 for the main entry.
Phase–referencing is a technique used in VLBI to extend the signal coherence time from a few minutes to a few hours in order to enhance significantly its sensitivity. With this technique, VLBI observations of milliJansky radio sources can be conducted for high–accuracy differential astrometry as well as imaging. The author describes the technique in some details and presents, as an example, a submilliarcsecond differential astrometric experiment design to identify the star responsible for the weak radio emission in the binary system Algol.

036.265 **EVN as a phase stable interferometer.**
L. B. Bååth, F. Mantovani.
IAU Colloquium No. 131: Radio interferometry – theory, techniques, and applications, p. 298 – 301 (1991). – See Abstr. 012.089 for the main entry.
A technique has been developed by which the data from the Mk3 correlator are read directly into the AIPS package. The data are Fourier transformed from delay lags into frequency channels and kept separately for each IF channel. The data are then corrected for multi– and single band delays. The authors have used this technique to solve for fringe solutions on a reference source and than apply the solution to correct the data on the

target source. In a recently analysed experiment they managed in this way to phase reference, using the European VLBI Network (EVN) at 18 cm, the quasar OQ208 and the radio quasar 1422 + 202. These sources are separated by more than 9° on the sky.

036.266 **Are the Band Width Synthesis delay and delay rate observables useful for VLBI imaging?**
P. Charlot.
IAU Colloquium No. 131: Radio interferometry – theory, techniques, and applications, p. 302 – 306 (1991). – See Abstr. 012.089 for the main entry.

The Band Width Synthesis (BWS) delay and delay rate observables measured by VLBI are used in astrometry to determine positions of extragalactic radio sources with a precision of 0.001″. At this level of precision, extragalactic radio sources exhibit extended structures that induce significant effects in the measured BWS delays and delay rates. The question whether closure BWS delays and delay rates could supplement the amplitudes and closure phases for VLBI imaging is studied. Preliminary results for the radio source NRAO140 observed in July 1983 are presented.

036.267 **Errors in high precision source position determination.**
P. Elosegui, J.-M. Marcaide, I. I. Shapiro.
IAU Colloquium No. 131: Radio interferometry – theory, techniques, and applications, p. 307 – 311 (1991). – See Abstr. 012.089 for the main entry.

The authors have made a detailed analysis of the systematic errors in the determination, from two sets of VLBI observations, of the position of the quasar 1038 + 528 A relative to the quasar 1038 + 528 B. This analysis confirms an apparent proper motion at $\lambda = 3.6$ cm of 26 ± 8 μas/yr of the core of the quasar 1038 + 528 A relative to the quasar 1038 + 528 B.

036.268 **A new method of analysis and reduction of VLBI observations.**
E. F. Arias, M. S. de Biasi.
IAU Colloquium No. 131: Radio interferometry – theory, techniques, and applications, p. 317 – 320 (1991). – See Abstr. 012.089 for the main entry.

The authors develope an inertial system of coordinates introducing a new observable: the arc between a pair of radio sources. The authors discuss how the conditional equation involving all the parameters is written in different systems when the observations are performed with a two and a three-element interferometers.

036.269 **Global fringe fitting applied to Mk3 VLBI data.**
L. B. Bååth.
IAU Colloquium No. 131: Radio interferometry – theory, techniques, and applications, p. 321 – 325 (1991). – See Abstr. 012.089 for the main entry.

A system has been developed where data from a Mk3 VLBI processor can be read into the AIPS-package. Multiband and singleband delays can then be fitted globally by station. The technique has been used to phaserefence a background quasar to two radiogalaxies in the Abell cluster A2634.

036.270 **mm VLBI – high resolution imaging.**
L. B. Bååth, R. S. Booth.
IAU Colloquium No. 131: Radio interferometry – theory, techniques, and applications, p. 326 – 329 (1991). – See Abstr. 012.089 for the main entry.

Development of receivers and the new technique to perform global fringe fitting on Mk3 data have now made it possible to produce hybrid maps at 100 GHz.

036.271 **Dual-frequency synthesis VLBI observations of 3C84 at 18 cm wavelength.**
J. A. Biretta, N. Bartel, J. Deng.
IAU Colloquium No. 131: Radio interferometry – theory, techniques, and applications, p. 334 – 337 (1991). – See Abstr. 012.089 for the main entry.

The authors present results of a multi-frequency synthesis VLBI experiment which employed observations at 1622 and 1707 MHz. Combining data from two frequencies gave a factor ~2 improvement in dynamic range over the single frequency images. The authors estimate map fidelity by comparing the images made separately from each frequency; while the dynamic range is ~1800, the fidelity index is only ~5.

036.272 **High-accuracy equation of equinoxes and VLBI astrometric modelling.**
A. M. Gontier, N. Capitaine.
IAU Colloquium No. 131: Radio interferometry – theory, techniques, and applications, p. 342 – 345 (1991). – See Abstr. 012.089 for the main entry.

In the course of implementing the use of nonrotating origin (Guinot 1979) in astrometric VLBI, for the transformation between celestial and terrestrial frames, the authors have compared the new approach with the classical one. They have shown that a difference exists between the two procedures at a few milliarcsecond level; this difference is due to the terms generally neglected when considering the equation of equinoxes in the classical representation of the Earth angle of rotation.

036.273 **MARK-III VLBI astrometry of pulsars.**
D. L. Jones, R. J. Dewey, C. R. Gwinn, M. M. Davis.
IAU Colloquium No. 131: Radio interferometry – theory, techniques, and applications, p. 346 – 349 (1991). – See Abstr. 012.089 for the main entry.

Differential 18-cm VLBI measurements of the position of PSR 1937 + 214 with respect to the quasar 1923 + 210 have been made to improve the knowledge of the offset between the planetary ephemeris reference frame and the extragalactic radio reference frame. The authors preliminary frame-tie result (VLBI – Timing position) is $+21 \pm 2$ mas in right ascension and $+14 \pm 9$ mas in declination. The authors describe the data analysis methods used and compare their results with other radio interferometric position measurements. The results illustrate the value of combining the large collecting area of the Arecibo telescope with the Mk-III VLBI recording system.

036.274 **Exploring the limits of phase referenced VLBI imaging; a progress report.**
R. C. Vermeulen, J. E. Conway.
IAU Colloquium No. 131: Radio interferometry – theory, techniques, and applications, p. 350 – 353 (1991). – See Abstr. 012.089 for the main entry.

The practical limits of phase referenced VLBI imaging are not well known. If it would be feasible to use target-reference separations of a few degrees, most of the sky would be accessible to the technique. Preliminary results at 6 cm using baselines of up to 1000 km are encouraging.

036.275 **State-of-the-art VLBI imaging: 3C345.**
A. E. Wehrle, S. C. Unwin.
IAU Colloquium No. 131: Radio interferometry – theory, techniques, and applications, p. 354 – 357 (1991). – See Abstr. 012.089 for the main entry.

Most VLBI images have low dynamic range because they are limited by instrumental effects such as calibration errors and poor u,v-coverage. The authors outline the method used to make a new image of the bright quasar 3C345 which has very high dynamic range (peak-to-noise of 5000:1) and which is limited by the thermal noise, not instrumental errors. Both the Caltech

VLBI package and the NRAO AIPS package were required to manipulate the data.

036.276 **High spatial resolution techniques.**
O. von der Lühe.
Solar interior and atmosphere, p. 688 – 726 (1991). – See Abstr. 003.028 for the main entry.
Observations of the solar surface from the ground are severely affected by turbulence of the Earth's atmosphere. Beside selecting observatory sites for excellent atmospheric quality, the careful design of the observing equipment, and the application of modern methods that undo the effects of the atmosphere or recover degraded data provide a wealth of information on small–scale processes. The effects of atmospheric turbulence on solar observations are reviewed, and methods that achieve high angular resolution on the Sun, such as adaptive optics and interferometry, are discussed. Some experiments that provide solar observations of high spatial resolution are presented.

036.277 **A test of radio interferometry by lunar reflection.**
L. Belkora, T. Hagfors, J. A. Phillips.
IAU Colloquium No. 131: Radio interferometry – theory, techniques, and applications, p. 433 – 436 (1991). – See Abstr. 012.089 for the main entry.
The authors discuss the use of the moon as a passive reflector for radio interferometry experiments with baselines on the order of the radius of the orbit of the moon. Because of the extreme loss in the lunar interferometer path, only exceptionally strong point sources are candidates for this kind of interferometry. As the moon is far from an ideal reflector due to the irregular surface, the data processing necessary to optimize the fringe detectability is rather complicated.

036.278 **On the possibility to detect NRP's by optical interferometry.**
F. Vakili, I. Percheron.
ESO Workshop on Rapid Variability of OB–Stars: Nature and Diagnostic Value, p. 77 – 79 (1991). – See Abstr. 012.092 for the main entry.

036.279 **Doppler imaging of spotted stars.**
M. Kürster.
ESO Workshop on Rapid Variability of OB–Stars: Nature and Diagnostic Value, p. 147 – 155 (1991). – See Abstr. 012.092 for the main entry.
After a brief review of the published literature some mathematical details of the Doppler imaging problem are given and solution strategies are discussed. A straightforward method for the localizing of star spots is presented for a simplified case. This method is the basis for a more elaborate (modified) CLEAN algorithm. Images obtained with the maximum entropy and CLEAN methods are compared.

036.280 **Simultaneous correlation analysis of combined gamma–ray data.**
K. N. Yu.
Astrophys. J., Vol. 382, No. 2, p. 647 – 651 (1 Dec 1991).
Expressions have been derived for the standard deviation of the total correlated flux, $\sigma(C_f^T)$, and the parent standard deviation of the underlying background $\sigma_0(C_f^T)$, of a γ–ray excess in a correlation analysis of raw γ–ray data combined from different sources, e.g., from different energy ranges of the same satellite, from the same energy range from different satellites, or from different energy ranges from different satellites. These expressions form the basis for the simultaneous correlation analysis method. The advantages of using simultaneous correlation analysis include the increase in the number of photons in the analysis, thus improving the statistics, and the enhancement of the signal received from a genuine source. This method can be utilized to analyze the SAS 2 data, the COS B data, the data from the GRO, and any combination of the above.

036.281 **Multivariate analysis and pattern recognition methods: a short review and some current directions.**
F. Murtagh.
12. Autumn School of Physics. The physical universe – the interface between cosmology, astrophysics and particle physics, p. 253 – 264 (1991). – See Abstr. 012.097 for the main entry.
In this article, the author aims to overview background aspects of a few important topics in the area of multivariate data analysis. In the case of graph–theoretic methods, these data analysis approaches have been widely used on astronomical data in recent years. Clustering by mode detection, partitioning and hierarchical clustering methods are briefly reviewed. Approaches based on graph theoretic methods and those based on variance minimization have different objectivities in many cases. The former may be more appropriate for detecting filamentary–type structures, whereas the latter may be found more effective for the general problem of condensing or summarizing data.

036.282 **A technique for precisely measuring stellar temperatures.**
D. F. Gray, H. L. Johanson.
J. R. Astron. Soc. Can., Vol. 85, No. 4, p. 183 (Aug 1991). Abstract. – See Abstr. 012.091 for the main entry.

036.283 **Optimum detection of an unknown periodic signal.**
P. C. Gregory.
J. R. Astron. Soc. Can., Vol. 85, No. 4, p. 189 (Aug 1991). Abstract. – See Abstr. 012.091 for the main entry.

036.284 **Lucy deconvolution of Hubble Space Telescope images.**
D. M. Anthony, C. C. Cunningham, J. J. Caldwell.
J. R. Astron. Soc. Can., Vol. 85, No. 4, p. 198 (Aug 1991). Abstract. – See Abstr. 012.091 for the main entry.

036.285 **Two–colour de–reddening – an automated statistical approach.**
E. J. Jylänne, M. P. Vatter FitzGerald, G. L. H. Harris.
J. R. Astron. Soc. Can., Vol. 85, No. 4, p. 198 – 199 (Aug 1991). Abstract. – See Abstr. 012.091 for the main entry.

036.286 **Photon–limited, high–resolution stellar imaging from triple–correlation and Knox–Thompson subplanes.**
J. Meng, G. J. M. Aitken.
J. R. Astron. Soc. Can., Vol. 85, No. 4, p. 205 – 206 (Aug 1991). Abstract. – See Abstr. 012.091 for the main entry.

036.287 **Maximum entropy imaging of contact binaries.**
P. D. Hendry, S. W. Mochnacki.
J. R. Astron. Soc. Can., Vol. 85, No. 4, p. 206 (Aug 1991). Abstract. – See Abstr. 012.091 for the main entry.

036.288 **The angular dependent reflected spectra by Compton scatterings.**
X.–M. Hua.
J. R. Astron. Soc. Can., Vol. 85, No. 4, p. 208 (Aug 1991). Abstract. – See Abstr. 012.091 for the main entry.

036.289 **An automated search for gravitational lenses performed on 25 grens plates.**
M. Beauchemin, E. F. Borra.
J. R. Astron. Soc. Can., Vol. 85, No. 4, p. 208 (Aug 1991). Abstract. – See Abstr. 012.091 for the main entry.

036.290 **On the analysis of physical wave trains.**
D. O. Gough, W. J. Merryfield, J. Toomre.
Conference on Challenges to Theories of the Structure of Moderate–Mass Stars, p. 265 – 270 (1991). – See Abstr. 012.096 for the main entry.
When a wave train whose constituent frequencies and wave numbers are unresolved by observation propagates through a inhomogeneous medium, beating between the components can contaminate deductions one might naively draw about the inhomogeneous background. This is a severe problem to anyone confronted with analysing helioseismic data with a view to

determining the structure of giant convective cells. The authors propose a procedure for analysing wave trains, based on approximating a packet at a single representative pure wave. They present some preliminary results of analysing artificial data. For simplicity, they have deliberately excluded some of the effects of wave interference, which must be faced by any means of analysis. Therefore they do not claim to have found a complete procedure for analysing real data.

036.291 **Untersuchungen zur Durchführung, Auswertung und astrometrischen Nutzung von photoelektrischen Sternbedeckungsbeobachtungen.**
G. Katerbaum.
Diss. (Dr.-Ing.), Technische Univ. Dresden (Germany). Fakultät fuer Bau-, Wasser- und Forstwesen. 104 p. (5 Oct 1990).

036.292 **Adaptive filtering in astronomical image processing.**
G. M. Richter, P. Böhm, H. Lorenz, A. Priebe, M. Capaccioli.
Astron. Nachr., Vol. 312, No. 6, p. 345 - 349 (1991).
An adaptive filter for processing of astronomical images is developed and described. The filter recognizes the local signal resolution (which usually varies strongly across the image) and adapts its own impulse response to this resolution.

036.293 **The SOLAR-A onboard data processor (DP).**
T. Kosugi.
International SOLAR-A Science Meeting: Flare physics in solar activity maximum 22, p. 26 - 27 (1991). Abstract. For more details see: Special issue in Sol. Phys., Vol. 136. - See Abstr. 012.104 for the main entry.

036.294 **Synthetische *UBVR*-Photometrien aus flußkalibrierten Spektren.**
J. Gochermann.
Astron. Nachr., Vol. 312, No. 6, p. 357 - 375 (1991). In German.
An accurate reconstruction of the *UBVRI* systems of Johnson and the Cousins systems used at the South African Astronomical Observatory, at Cerro Tololo Inter-American Observatory and at Mount Stromlo/Siding Springs Observatory has been attempted.

036.295 **Source of error in planetary photoclinometry.**
D. G. Jankowski, S. W. Squyres.
J. Geophys. Res., Vol. 96, No. E4, p. 20907 - 20922 (25 Nov 1991).
Photoclinometry has become a popular technique for the extraction of topography from digital images of planetary surfaces. The technique, however, is subjected to a number of error sources that can degrade results significantly. The authors present here a quantitative analysis of error sources in one-dimensional planetary photoclinometry. The technique is affected by at least seven error sources, which can be broken down into three categories: those arising from (1) the spacecraft image, (2) the planetary body, and (3) the scan line orientation. Spacecraft image error sources include noise, background offset, and quantization. Errors deriving from the planetary body include those induced by variations in photometric properties, such as albedo, by incorrectly compensated atmospheric effects, and by an incorrectly determined photometric function. Finally, errors will result for scan lines which do not lie perpendicular to topographic strike. The authors calculate slope errors for each of these sources, using the examples of Voyager imaging of Ganymede and Viking orbiter imaging of Mars.

036.296 **Analysis on the effects of spectral type error and magnitude difference of the 40 cm astrograph at Zô-Sè for determinations of radio star positions.**
Wang Shuhe, Xu Tongqi, Lu Peizhen.
Ann. Shanghai Obs., Acad. Sin., No. 12, p. 53 - 57 (1991). In Chinese.
The spectral type errors in difference zones of zenith distances for the 40 cm astrograph at Zô-Sè are estimated from theoretical calculations and statistical analysis respectively, and the relation between magnitude differences and observational residuals are estimated. It is shown from the above discussions that there are no significant effects of the spectral type error and the magnitude difference on optical determination of radio star positions, and they can be neglected in data reduction for the telescope.

036.297 **S-2 VLBI data correlation system.**
Liang Shiguang, Huang Xinyong.
Ann. Shanghai Obs., Acad. Sin., No. 12, p. 114 - 126 (1991). In Chinese.

036.298 **A software for radio source antenna temperature measurements.**
Kong Yi.
Ann. Shanghai Obs., Acad. Sin., No. 12, p. 134 - 139 (1991). In Chinese.

036.299 **The improvement of MK-3 field system at Shanghai Observatory.**
Xue Zhuhe.
Ann. Shanghai Obs., Acad. Sin., No. 12, p. 173 - 178 (1991). In Chinese.

036.300 **Towards a search for a taxonomy of celestial objects. I. Open clusters.**
H. L. Tignanelli, C. Mostaccio, S. Gordillo, A. Plastino, R. A. Vázquez, A. Feinstein.
Bol. Asoc. Argent. Astron., No. 36, p. 45 - 56 (1990). In Spanish. - See Abstr. 012.095 for the main entry.
In the present paper the authors introduce the method to develop a celestial objects taxonomy. The steps to follow are shown and its way of application explained. They are also given the preliminary results obtained when studying associations with more than 300 open clusters from our galaxy.

036.301 **Search for periodicity in the unequally-spaced data sequence.**
Jia Guibin, Xing Shiying, Ma Huaiyu, Qin Xuefen, Leng Shengquan.
Publ. Yunnan Obs., No. 3, p. 14 - 24 (1991). In Chinese.
The results both from simulations and practical applications show: based on Deeming's method, it is effective and preferable to introduce a few auxiliary ways into the power spectral analysis in improving the identification of spectral peak and the precise determination of period.

036.302 **Noises in observing stellar spectra using solid-state detectors and method to reduce them.**
Qin Songnian, Li Rufeng, Gao Cai.
Publ. Yunnan Obs., No. 3, p. 30 - 34 (1991). In Chinese.
Taking the No. 1 CCD system of Yunnan Observatory, attached to CCD-Coudé spectrographic system, as an example, the noises induced by area (CCDs) or linear solid-state detectors used for the spectra observations with large dispersion and the methods to reduce those noises, mainly the method for flatfielding, are discussed.

036.303 **A CCD detector attached to the Lower Latitude Meridian Circle (V): Building up a high accurate celestial sphere frame using ground-based observations.**
Mao Wei, Peng Qingyu, Zhang Jing, Hu Xiaochun, Li Zhiming.
Publ. Yunnan Obs., No. 3, p. 35 - 40 (1991). In Chinese.
The conditions required to construct a quasi-inertial celestial reference frame are discussed. The precessional constants and absolute proper motion of a star could be directly obtained by the observation of LLMC with a CCD. The possibility is analysed to establish a high accurate celestial sphere frame using the observational data of the instrument. The method is compared with those of space observation.

036.304 **A method for determining the absolute value of a photographic plate of Hα–SSHG.**
Li Kejun, Ding Yuji.
Publ. Yunnan Obs., No. 4, p. 55 – 60 (1991). In Chinese.

036.305 **Interférometrie des tavelures dans le visible.**
D. Bonneau.
J. Astron. Fr., No. 39, p. 6 – 7 (Jul 1991). – See Abstr. 012.099 for the main entry.

036.306 **Introduction à l'interferométrie et à la reconstruction d'images.**
C. Aime.
J. Astron. Fr., No. 39, p. 4 – 5 (Jul 1991). – See Abstr. 012.099 for the main entry.

036.307 **Interférométrie différentielle.**
R. G. Petrov.
J. Astron. Fr., No. 39, p. 20 – 21 (Jul 1991). – See Abstr. 012.099 for the main entry.

036.308 **CHARON/I3T. L'imagerie à haute résolution angulaire avec l'interférométrie à longue base.**
G. Schumacher, A. Boischot, P. Cruzalèbes, Y. Rabbia, J.-L. Starck.
J. Astron. Fr., No. 39, p. 24 – 25 (Jul 1991). – See Abstr. 012.099 for the main entry.

036.309 **The lunar selected areas program (SAP): how to get started.**
J. L. Benton Jr.
Strolling Astron., Vol. 35, No. 3, p. 100 – 103 (Sep 1991).

036.310 **Practical lunar photoelectric photometry.**
C. A. Kapral.
Strolling Astron., Vol. 35, No. 3, p. 104 – 110 (Sep 1991).
This report describes a lunar photoelectric photometry project that consisted of measuring the brightness of four lunar features through relative photometry with a solid-state photometer. It explains the selection criteria used and the observing procedure, with a worked example at the end of the paper.

036.311 **Image processing system applied to galaxies surface photometry.**
M. A. Bosio, C. J. Donzelli.
Bol. Asoc. Argent. Astron., No. 36, p. 311 – 315 (1990). In Spanish. – See Abstr. 012.095 for the main entry.
A sofware package for image processing in interactive form has been developed in the Digital Micro Vax II system. It was used to make a detailed photometric and morphological analysis of 9 early-type galaxies, 5 ellipticals an 4 S0. The processed images were obtained by a Hilger-Watts modified microdensitometer, which allows automatic digitalization of photographic plates.

036.312 **Digitalized reduction of interferograms.**
G. Goldes, G. Carranza.
Bol. Asoc. Argent. Astron., No. 36, p. 401 – 402 (1990). Abstract. In Spanish and English. – See Abstr. 012.095 for the main entry.

036.313 **A test of the optimization techniques for contact binaries photometric analysis.**
E. Lapasset, M. Gómez, R. Fariñas.
Bol. Asoc. Argent. Astron., No. 36, p. 423 – 424 (1990). Abstract. In Spanish and English. – See Abstr. 012.095 for the main entry.

036.314 **A new method of analysis and reduction of VLBI observations.**
E. F. Arias, M. S. De Biasi.
Bol. Asoc. Argent. Astron., No. 36, p. 446 – 449 (1990). Abstract. In Spanish and English. – See Abstr. 012.095 for the main entry.

036.315 **Astronomy and personal computers.**
J. B. Dunham.
Occultation Newsl., Vol. 5, No. 4, p. 95 – 97 (Aug 1991).

036.316 **Ultraviolet imaging of globular clusters and galaxies.**
R. W. O'Connell.
Adv. Space Res., Vol. 11, No. 11, p. 71 – 80 (1991). – See Abstr. 012.107 for the main entry.
Ultraviolet imaging of star clusters and galaxies is a subject still in its infancy. Here the authors review the techniques of UV imaging and its scientific potential for studies of advanced stellar evolution, the star formation history and interstellar medium of galaxies, active nuclei, and the evolutionary state of systems observed at high redshift.

036.317 **Astronomy and personal computers.**
J. B. Dunham.
Occultation Newsl., Vol. 5, No. 5, p. 114 – 116 (Nov 1991).

036.318 **A comprehensive system of occultation software.**
D. W. Dunham.
Occultation Newsl., Vol. 5, No. 5, p. 122 – 126 (Nov 1991). With a chart clarification in Vol. 5, No. 6, p. 145 (Dec 1991).

036.319 **A novel method for stabilizing singles rate in atmospheric Cerenkov telescopes.**
A. K. Tickoo, I. K. Kaul, S. K. Kaul, C. L. Bhat, R. K. Kaul, H. S. Rawat, V. K. Senecha, R. C. Rannot, M. L. Sapru, H. Razdan.
Bull. Astron. Soc. India, Vol. 19, Nos. 1 – 2, p. 90 (Mar – Jun 1991). Abstract. – See Abstr. 012.108 for the main entry.

036.320 **Spectroscopic reductions during RESPECT software.**
T. P. Prabhu, G. C. Anupama.
Bull. Astron. Soc. India, Vol. 19, Nos. 1 – 2, p. 97 – 104 (Mar – Jun 1991).

036.321 **Maximum likelihood image restoration. V. Incoherent fluxes.**
V. Yu. Terebizh, O. K. Cherbunina, Yu. G. Cherbunin, V. V. Biryukov.
Astrofizika, Tom 34, Vyp. 1, p. 91 – 102 (Feb 1991). In Russian. With a correction in Tom 34, Vyp. 1, p. 102 (1991). English translation in Astrophysics, Vol. 34, No. 1.
The general approach of maximum likelihood image restoration is adapted to the case, when a priori information concerning the Poisson distribution of stored events is available. The restoring algorithm has a simpler form, but the real gain due to such an information is appreciable only for very faint objects.

036.322 **Removal of subjectivity from solar astrolabe results by use of a C.C.D. camera. Application to solar diameter measurements.**
F. Laclare, G. Merlin.
C. R. Acad. Sci., Sér. II, Tome 313, No. 3, p. 323 – 330 (1 Aug 1991).
In order to avoid personal biases in measurements performed with the solar astrolabe and better to define the solar diameter, the authors have developed a system using a C.C.D. This acquisition can advantageously replace the visual observations on the existing instruments. The experimental results presented are in good agreement with those deduced from the visual program during the same period and they reveal a real improvement of measurement quality.

036.323 **Attractor reconstruction and dimensional analysis of chaotic signals.**
H. Atmanspacher, H. Scheingraber, W. Voges.
3. International Workshop on Data Analysis in Astronomy, p. 3 – 19 (1989). – See Abstr. 012.110 for the main entry.
Contents: Introduction. Attractors and their geometrical properties. Attractor reconstruction and scaling analysis. Numerical and experimental examples. Conclusions.

036.324 **On the possibility of determination of the equatorial speed and equatorial plane inclination of an astrophysical object in space.**
V. O. Gladyshev.
Astrofizika, Tom 34, Vyp. 2, p. 227 - 232 (Apr 1991). In Russian. English translation in Astrophysics, Vol. 34, No. 2.
The interdependence of the equatorial speed and equatorial plane inclination angle of an astrophysical object in space with spectral lines profile variations is are analyzed. The results of this analysis are discussed in terms of applicability of optical spectrometry methods.

036.325 **The Rayleigh statistic in the case of weak signals – applications and pitfalls.**
O. C. de Jager, B. C. Raubenheimer, J. W. H. Swanepoel.
3. International Workshop on Data Analysis in Astronomy, p. 21 - 30 (1989). - See Abstr. 012.110 for the main entry.
The distribution of the Rayleigh statistic for any kind of light curve is derived. It is shown that a two sigma (Gaussian) hole in an ON–source region with respect to an adjacent OFF–source region is still acceptable after a claimed periodic sinusoidal signal have been subtracted from the ON–source region. This is mainly due to the large effects of Rayleigh fluctuations. It is also shown that most of the present estimators of the signal strength of sinusoidal light curves are conservative in that they underestimate the signal strength.

036.326 **Stability in factor analyses with respect to small perturbations of the data and to departure from Gaussian law of the variables.**
M. L. Bougeard.
3. International Workshop on Data Analysis in Astronomy, p. 31 - 38 (1989). - See Abstr. 012.110 for the main entry.
The author considers the problem of approximations by sampling in the case of the principal component factor analysis. In a first part she recalls the asymptotic results of Anderson, Arconte, Davis, Dauxois, Pousse and Romain about the distribution of the proper values of the operators. In a second step, she recalls the principle of two non–parametric processes, namely jackknife and bootstrap. The author compares the robustness of the asymptotic results with the results obtained through non parametric processes by means of simulation and drawing of random samples in a given population.

036.327 **Maximum likelihood image restoration. VI. Rao–Kramér limit of restoration efficiency.**
V. Yu. Terebizh, V. V. Biryukov.
Astrofizika, Tom 34, Vyp. 2, p. 233 - 248 (Apr 1991). In Russian. English translation in Astrophysics, Vol. 34, No. 2.
The existence of the inner noise of radiation inevitably reduces the problem of restoration of a blurred and noised image to the search for the statistical estimation of unknown parameters (object intensities). The method–independent theoretical limit of efficency of image restoration exists which is given by the Rao–Kramér theorem. The general arguments as well as numerical and analytical examples indicate that maximum likelihood image restoration (MLIR) method reaches the limiting efficiency when light intensity at the image becames approximately equal to the mean level of external noise. The expressions for calculating the natural error corridor of restoration are given.

036.328 **Moving regression: a tool for the statistical and astrometrical analysis of solar observations.**
M. L. Bougeard, J. M. Grosbras, F. Laclare.
3. International Workshop on Data Analysis in Astronomy, p. 39 - 46 (1989). - See Abstr. 012.110 for the main entry.
The authors perform regressions on the solar observations made at the CERGA Observatory, through a moving window the size of which is chosen taking into account the physical constraints. After some statistical considerations, the results obtained with different sizes of window are analysed. Graphical comparisons are carried out on the estimators obtained by the usual method and by the new one for the 1978, 1979 and 1980 campaigns.

036.329 **Multivariate linear regression with self–estimation of errors.**
R. Bandiera, L. Hunt.
3. International Workshop on Data Analysis in Astronomy, p. 47 - 54 (1989). - See Abstr. 012.110 for the main entry.
A common approach to the analysis of a sample of data is to search for linear correlations between the variables that describe the sample. Most of the fitting techniques used in that context are derived from a statistically rigorous method of estimation, the maximum likelihood method. The authors have developed a specific technique for the case in which several measured mutually correlated quantities are available and all variables are affected by errors. The authors want to fit the data to a single correlation and determine the properties of the residual scatter. The method is based on the separation of all the correlations into two terms: one contains only the primary correlation (described by a single parameter) and the other represents residual scatter and is treated as a statistical dispersion.

036.330 **Maximum entropy image processing in gamma–ray astronomy.**
A. W. Strong, R. Diehl.
3. International Workshop on Data Analysis in Astronomy, p. 55 - 65 (1989). - See Abstr. 012.110 for the main entry.
Visual presentation is essential as a means for understanding data, comparing it with other wavelengths, and looking for the unexpected. Because of the particular problems associated with γ–ray observations, it is logical that considerable investment be made at this time in studying methods to obtain the best possible images for scientific analysis. The maximum entropy method is an obvious choice for imaging because of its characteristic treatment of the relation between image and data; it requires only that the detector response to any given image can be computed, and does not require any explicit relation between individual point in image and data space.

036.331 **Search for bursts of gamma–ray emission from the Crab pulsar in the COS–B data by adaptive cluster analysis.**
R. Buccheri, M. Busetta, C. M. Carollo.
3. International Workshop on Data Analysis in Astronomy, p. 67 - 74 (1989). - See Abstr. 012.110 for the main entry.
The photon arrival times detected by COS–B around the direction of the Crab pulsar have been analyzed by an adaptive cluster method in order to search for short bursts of gamma–ray emission from the source. As a result, two clusters of 3 and 6 photon arrival times within 264 s and 1813 s respectively were found with the corresponding residual phases having close correlation with the peaks of the Crab pulsed emission. 3σ upper limits on the existence of bursts of emission from the Crab are given.

036.332 **Statistical restoration of astronomical images.**
R. Molina, N. Perez de la Blanca, B. D. Ripley.
3. International Workshop on Data Analysis in Astronomy, p. 75 - 82 (1989). - See Abstr. 012.110 for the main entry.
It is well known that the theoretical intensity at a point from an astronomical image depends on S and h, were S represents the true underlying intensity and h is the point spread function (psf). The main problem in the restoration of astronomical images is to estimate S, knowledge of the psf and adequate assumptions about the noise process. This paper is divided in two parts. In the first the authors consider the problem of fitting the psf and in the second they propose methods to estimate S.

036.333 **Applications of decomposition algorithm in clustering methods.**
Z. A. Azmi.
3. International Workshop on Data Analysis in Astronomy, p. 91 - 102 (1989). - See Abstr. 012.110 for the main entry.
The aim of clustering is to partition a given collection of objects into a number of classes in such a way that objects within each class are strongly similar each other, while objects of different classes are appreciably less similar. MM (Minimax) Decomposition algorithm is of special importance to study

properties of non-fuzzy partitions derived from a number of algorithms providing convex decomposition of matrix U. The author suggests the possibility of the application of a decomposition algorithm in clustering methods in two approaches: (1) The application of non-metric similarity/dissimilarity measures which may be of value when the data do not admit the geometrical properties. (2) The application of fuzzy numbers instead of non-fuzzy numbers.

036.334 **Clustering analysis as a method of auto-classification and its application to the search for galaxy chains in two dimensions.**
Z. G. Deng, X. Y. Xia, H. Arp, G. Börner.
3. International Workshop on Data Analysis in Astronomy, p. 103 – 109 (1989). – See Abstr. 012.110 for the main entry.
Filaments and lines of galaxies on the sky have been discussed in recent years. The present paper outlines a method for quantitatively testing the connected nature of these features. Comparison with Monte Carlo simulations give probability values that the features are real. Two sky regions are tested with probably significant results.

036.335 **Object detection and analysis in digitised images: algorithmic components and software structure.**
F. Murtagh.
3. International Workshop on Data Analysis in Astronomy, p. 111 – 120 (1989). – See Abstr. 012.110 for the main entry.
The structures of major 2-dimensional image analysis software packages in optical astronomy are overviewed. Reference is made to the varied image analysis and pattern recognition algorithms in use.

036.336 **Automated classification of resolved galaxies.**
P. G. Ossorio, M. J. Kurtz.
3. International Workshop on Data Analysis in Astronomy, p. 121 – 128 (1989). Abstract. – See Abstr. 012.110 for the main entry.

036.337 **STSDAS: the Space Telescope Science Data Analysis System.**
R. J. Hanisch.
3. International Workshop on Data Analysis in Astronomy, p. 129 – 140 (1989). – See Abstr. 012.110 for the main entry.
A successful data analysis system must be able to accommodate alternate sources of data, and must have analysis routines both specific enough for use with the primary instrument and general enough for use with other astronomical data. Through its integration with the Image Reduction and Analysis Facility (IRAF) and through its full support for data transport in the FITS format STSDAS strives to meet this goal.

036.338 **IRAS calibration.**
D. J. M. Kester, T. J. R. Bontekoe, A. R. W. de Jonge, P. R. Wesselius.
3. International Workshop on Data Analysis in Astronomy, p. 141 – 147 (1989). – See Abstr. 012.110 for the main entry.
The Groningen Exportable Infrared Survey High-resolution Analysis (GEISHA) system consists of a collection of software tools which allow an astronomer to create infrared sky maps from raw IRAS data. There are 62 detectors in the focal plane in 4 wavelength bands. One of the central items in the project is the calibration. The procedure for the IRAS raw data calibration consists of flatfielding. For each detector scan a set of calibration constants is extracted by fitting the lower envelopes of the detectors of all scans to each other.

036.339 **Simulation of large-scale astronomical CCD surveys.**
E. E. Falco, M. J. Kurtz.
3. International Workshop on Data Analysis in Astronomy, p. 149 – 153 (1989). – See Abstr. 012.110 for the main entry.
The authors present the first stage of a project to develop a computer model of a CCD camera system dedicated to galaxy surveys. They plan to improve old and develop new algorithms for reduction and analysis of digital astronomical images. The authors intend to apply their model to testing theories and observations of the large-scale distribution of galaxies, and data reduction methods.

036.340 **Tests of reduction and analysis algorithms for astronomical images.**
M. J. Kurtz, E. E. Falco.
3. International Workshop on Data Analysis in Astronomy, p. 155 – 157 (1989). – See Abstr. 012.110 for the main entry.
The authors are beginning a long term program to create tools and methodologies for the automated reduction of data from large CCD surveys. It is important at this stage of the work to quantify the capabilities and limitations of the current state of the art procedures for dealing with data of this sort.

036.341 **A model for the Infrared Space Observatory photometer.**
A. Etemadi.
3. International Workshop on Data Analysis in Astronomy, p. 159 – 167 (1989). – See Abstr. 012.110 for the main entry.
The Infrared Space Observatory (ISO) consists of a cryogenically cooled telescope and instrumentation mounted on a 3-axis stabilised spacecraft. Photometer subexperiment (ISOPHOT) was designed to carry out photopolarimetry in the wavelength range 2 – 200 μm with a photometric accuracy of one percent. A software model of ISOPHOT has been recommended as a means of generating test data for the ISOPHOT scientific data analysis software. The author discusses the problems associated with this work, the overall design concept and the implementation.

036.342 **Algorithms for the analysis of data from the Extreme Ultraviolet Explorer.**
H. L. Marshall.
3. International Workshop on Data Analysis in Astronomy, p. 169 – 177 (1989). – See Abstr. 012.110 for the main entry.
An algorithm is described that is intended for use in the analysis of data from the Extreme Ultraviolet Explorer (EUVE) satellite. An improved technique for detecting sources has been developed. This method is shown to be computationally simpler than non-linear methods and more sensitive to point sources than previous linear techniques, because it incorporates more information from the instrument point spread function. It is shown how simulated data help test this algorithm and are useful in the design of the satellite hardware.

036.343 **Attitude determination using star tracker data.**
M. N. Boyarsky (*M. N. Boyarskij*).
3. International Workshop on Data Analysis in Astronomy, p. 187 – 190 (1989). – See Abstr. 012.110 for the main entry.
The paper describes the general structure and main algorithms of a software system developed in the Space Research Institute of the USSR Academy of Sciences for processing of star tracker data and spacecraft attitude determination using these data.

036.344 **Artificial intelligence: what can it do for astronomy?**
R. Albrecht.
3. International Workshop on Data Analysis in Astronomy, p. 191 – 204 (1989). – See Abstr. 012.110 for the main entry.
This paper summarizes the development and the applications of artificial intelligence techniques, and examines their application to astronomy. Areas of actual and possible use are identified. An attempt is made to extrapolate into future applications, emphasizing inter-disciplinary aspects and improvements of research methodology.

036.345 **Triangle interferometer and its optimization.**
Zhang Hui, Xu Jiayan, Xiao Jinhong, Wang Zhengming.
Publ. Shaanxi Astron. Obs., Vol. 14, No. 2, p. 95 – 101 (Dec 1991). In Chinese.
The optimal configuration of the triangle for determining the ERP is an equilateral triangle or an isosceles right triangle. Analysing the data observed by Connected Element Interferometer at Green Bank shows that the accuracy of the ERP solved

from three baselines' observations is much better than that of those solved from only two.

036.346 **Application of observations with two or three colours to interferometry.**
Xu Jiayan.
Publ. Shaanxi Astron. Obs., Vol. 14, No. 2, p. 102 – 105 (Dec 1991). In Chinese.
The observations with two bands in radio interferometry could eliminate the effect of ionosphere on the phase measurement. Also, the observations with two–colour in optical interferometry could eliminate the effect of turbulence on the star position measurement, and the remained effect of the fluctuation of the water vapour in the atmosphere could be eliminated by using three–colour observations.

036.347 **Artificial intelligence approaches to astronomical observation scheduling.**
M. D. Johnston, G. Miller.
3. International Workshop on Data Analysis in Astronomy, p. 205 – 214 (1989). – See Abstr. 012.110 for the main entry.
Automated scheduling will play an increasing role in future ground– and space–based observatory operations. Due to the complexity of the problem, artificial intelligence technology currently offers the greatest potential for the development of scheduling tools with sufficient power and flexibility to handle realistic scheduling situations. This paper summarizes the main features of the observatory scheduling problem, how AI techniques can be applied, and recent progress on AI scheduling for Hubble Space Telescope.

036.348 **OCAPI: an artificial intelligence tool for the automatic selection and control of image processing procedures.**
M. Thonnat, V. Clément.
3. International Workshop on Data Analysis in Astronomy, p. 215 – 224 (1989). – See Abstr. 012.110 for the main entry.
The authors present a general artificial intelligence tool, named OCAPI to manipulate the knowledge on image processing methods and on the use of libraries of procedures. The functionalities and the architecture of OCAPI are described. An example of utilization is developed for a pattern recognition problem in astronomy: the processing of an image of a galaxy or its description.

036.349 **An expert assistant supporting Hubble Space Telescope proposal preparation.**
H. M. Adorf, S. Di Serego Alighieri.
3. International Workshop on Data Analysis in Astronomy, p. 225 – 235 (1989). – See Abstr. 012.110 for the main entry.
The HST Expert Assistant is an advice giving expert system currently under development within the Artificial Intelligence Pilot Project of the Space Telescope European Coordinating Facility. The system aims at supporting astronomers who are planning observations with the Hubble Space Telescope. User requirements, knowledge elicitation and design issues, current status and future plans are described and discussed.

036.350 **Decision problems in the search for periodicities in gamma–ray astronomy.**
M. C. Maccarone, R. Buccheri.
3. International Workshop on Data Analysis in Astronomy, p. 237 – 244 (1989). – See Abstr. 012.110 for the main entry.
The basic steps of the analysis in the search for periodic signals present in a set of single arrival times are described, with particular attention to those nodes of the procedure for which external and subjective decisions are needed. Final aim is to study the possibility to connect the various aspects of the analysis in a support system able to help the decision making phase by using Artificial Intelligence tools and methods.

036.351 **Rule–based description and plausible classification of objects in digitized astronomical images.**
A. Accomazzi, G. Bordogna, P. Mussio, A. Rampini.
3. International Workshop on Data Analysis in Astronomy, p. 245 – 254 (1989). – See Abstr. 012.110 for the main entry.
A general approach to image interpretation founded on rule–based description and its plausible evaluation is tailored to astronomical image case by the study of the image model adopted by the astrophysicist and of his procedure of visual inspection of astronomical plates. The formal tools on which the approach is based are Bidimensional L–systems and Multivalued Logical Trees. They are briefly introduced and some examples of application are discussed.

036.352 **Parallelism in low–level computer vision – a review.**
V. Chaudhary, J. K. Aggarwal.
3. International Workshop on Data Analysis in Astronomy, p. 255 – 269 (1989). – See Abstr. 012.110 for the main entry.
The authors review various parallel algorithms and architectures used in Computer Vision. The problem of visual recognition is divided into three conceptual levels – low–level, intermediate–level and high–level. There are few conceptual difficulties in parallelizing low–level vision and most of them have been parallelized. However, not much work has been done in parallelizing intermediate and high–level vision. The authors present parallel algorithms for low–level vision.

036.353 **Performance of the mesh architecture in the analysis of sparse images.**
A. Machi, V. Di Gesu, F. Lupo.
3. International Workshop on Data Analysis in Astronomy, p. 271 – 280 (1989). – See Abstr. 012.110 for the main entry.
The aim of the paper is to show a parallel cluster algorithm on a mesh architecture useful in the analysis of sparse images. The algorithm computes the 1NN and an approximate version of the MST of a weighted undirected graph, which represents the image "on" pixels. Some indexes of performance are given on the basis of an experimental evaluation.

036.354 **Cosmic Background Explorer (COBE) satellite anisotropy experiment data analysis techniques.**
S. Torres, J. Aymon, C. Backus, C. Bennett, S. Gulkis, R. Hollenhorst, D. Hon, Q. H. Huang, M. Janssen, S. Keihm, L. Olson, G. Smoot, E. Wright.
3. International Workshop on Data Analysis in Astronomy, p. 319 – 334 (1989). – See Abstr. 012.110 for the main entry.
The COBE Differential Microwave Radiometer (DMR) experiment will measure the anisotropy of the cosmic microwave background radiation (CMBR). The initial phase of data analysis uses a "sparse matrix" algorithm to convert the differential temperature data into sky maps. The sky maps are then fitted to a "fast" multipole expansion in spherical harmonics. Since the CMBR anisotropy is very weak, powerful techniques are used to extract the angular unevenness of the sky from the low signal–to–noise data. Instrument signature and other systematic errors are subtracted by fitting models of these effects. Test results are presented.

036.355 **Some statistical problems encountered in the analysis of ultra high energy cosmic ray and ultra high energy gamma ray data.**
A. A. Watson.
3. International Workshop on Data Analysis in Astronomy, p. 335 – 349 (1989). – See Abstr. 012.110 for the main entry.
The subjects of ultra high energy cosmic rays and ultra high energy gamma rays are scientifically connected because, it is widely believed, ultra high energy gamma rays are produced through strong interactions of cosmic ray nuclei. The author outlines briefly the detection methods and problems associated with the two types of astronomy and draws attention to recent work which has been done to address the statistical questions which arise.

036.356 **Detecting gravitational waves by laser interferometers.**
A. Krolak.
3. International Workshop on Data Analysis in Astronomy, p. 393 – 404 (1989). Abstract. – See Abstr. 012.110 for the main entry.

036.357 **Data analysis in neutrino astronomy.**
M. Aglietta, G. Badino, G. Bologna, C. Castagnoli, A. Castellina, V. L. Dadykin, W. Fulgione, P. Galeotti, F. F. Kalchukov, V. B. Kortchaguin (*V. B. Korchagin*), P. V. Kortchaguin (*P. V. Korchagin*), A. S. Malguin (*A. S. Mal'gin*), V. G. Ryassny (*V. G. Ryasnyj*), O. G. Ryazskaya (*O. G. Ryazhskaya*), O. Saavedra, V. P. Talockin (*V. P. Talochkin*), G. Trinchero, S. Vernetto, G. T. Zatsepin, V. F. Yakushev.
3. International Workshop on Data Analysis in Astronomy, p. 405 – 411 (1989). – See Abstr. 012.110 for the main entry.
Contents: Introduction. The Mont Blanc LSD Neutrino Observatory. The data acquisition system. Search for bursts on real time and statistical analysis of data.

036.358 **Multi–access circle scanning system for DCMT.**
Li Huanxin.
Publ. Shaanxi Astron. Obs., Vol. 14, No. 2, p. 122 – 127 (Dec 1991). In Chinese.
A multi–access automatic circle scanning system for Chinese and Danish horizontal meridian circle has been developed in Shaanxi Astronomical Observatory. The system consists of a computer and related interface circuit and high quality drivers for fine step motors. The principles of all hardware and software are described.

036.359 **On board fault–tolerance real–time multitasking computer system.**
Zhang Xiuzhong.
Publ. Shaanxi Astron. Obs., Vol. 14, No. 2, p. 152 – 155 (Dec 1991). In Chinese.

036.360 **Possibility of adaptive image correction in an early visual–information processing stage.**
V. E. Semenov.
Opt. Spectrosc., Vol. 71, No. 2, p. 207 – 209 (Aug 1991). Current Physics Microform No.: 9206X1363.
This paper carries out a numerical experiment demonstrating the possibility of adaptive correction of a noisy image of an object generated by a three–layer neural net simulating a primary visual information processing neural net.

036.361 **Mount John photometric software.**
A. C. Gilmore.
South. Stars, Vol. 34, No. 3, p. 3 – 7 (18 Jun 1991). – See Abstr. 012.120 for the main entry.
A suite of programs has been developed by the author at Mt. John to operate a semi–automated photometer and to reduce the resulting photometric data. The programs enable quick production of standard photometry, usually based on mean extinction coefficients, with the minimum of observer input.

036.362 **Monte Carlo Simulationen der kosmischen Bestrahlung planetarer Oberflächen und Anwendung zur geochemischen Analyse Neutronen–induzierter Gammaspektren des Mars.**
G. Dagge.
Diss., Forschungszentrum Jülich GmbH (Germany). Inst. für Kernphysik; Technische Hochschule Aachen (Germany). Juel—2528, 173 p. (Sep 1991).
A Monte Carlo procedure for the simulation of the cosmic irradiation of planetary surfaces was developed by using the HERMES code system. The calculation procedure was successfully tested on experimental data from the lunar surface. The neutron and gamma–ray spectrum emitted from the martian surface was simulated for a variety of possible material compositions. These calculations are the basis for the analysis of future data from orbital neutron and gamma–ray spectroscopy with respect to the martian surface composition. A main aspect of these studies was the detection of water in the martian soil. The surface can be investigated to depths of 100 g/cm^2 by this method. The background signal at the detector site was calculated in order to show that the complete martian surface can be analyzed within a given time of one martian year.

036.363 **The importance of deadtime in photon–counting photoelectric photometry.**
W. Tobin, A. Wadsworth.
South. Stars, Vol. 34, No. 3, p. 49 – 55 (18 Jun 1991). – See Abstr. 012.120 for the main entry.
This paper discusses the method and necessary equipment for the determination of photometer deadtimes from measurements that can be made during the daytime.

036.364 **Precision in astronomical photometry.**
M. S. Bessell.
South. Stars, Vol. 34, No. 3, p. 95 – 100 (18 Jun 1991). – See Abstr. 012.120 for the main entry.
Astronomical photometry should be considered as low resolution spectrophotometry. In this way one better appreciates the interpretation of colors and the need for precise matching of passbands. The effects of atmospheric extinction are outlined and observing and reduction techniques are discussed.

036.365 **Astrometrie an Kometen.**
E. Meyer, H. Raab.
Sternenbote, Jahrg. 34, Nr. 12, p. 238 – 245 (1991).

036.366 **Die massenspektrometrische In–situ–Analyse von Festkörpern.**
J. Kissel.
Sterne, Band 67, Heft 6, p. 317 – 334 (1991).

036.367 **CCD photometry at Capilla Peak Observatory: a recipe for success.**
D. Beckert, M. Newberry, M. Zeilik.
South. Stars, Vol. 34, No. 3, p. 163 – 166 (18 Jun 1991). – See Abstr. 012.120 for the main entry.

036.368 **The use of CCD's for variable star differential photometry.**
G. W. Wolf.
South. Stars, Vol. 34, No. 3, p. 167 – 171 (18 Jun 1991). – See Abstr. 012.120 for the main entry.
High quality differential photometry of variable stars is possible under very "non–photometric" sky conditions using a CCD imaging stellar photometer. This ability greatly increases the number of usable observing nights at less–than–ideal observing sites.

036.369 **Coordinated multisite and multiwavelength observing programs.**
C. J. Butler.
South. Stars, Vol. 34, No. 3, p. 175 – 178 (18 Jun 1991). – See Abstr. 012.120 for the main entry.
A brief report is presented dealing with some recent recommendations on organization for multisite and multiwavelength observing programs. Some examples of such programs are given.

036.370 **Statistical and numerical investigations to optimize astrometrical reduction methods for Schmidt telescopes.**
S. Hirte.
Astron. Nachr., Vol. 312, No. 5, p. 335 (1991). Dissertation abstract.

036.371 **Come fotografare l'eclisse di sole.**
W. Ferreri.
Orione, Vol. 11, N. 3, p. 16 – 23 (May – Jun 1991).

The NSSDC Data Archiving and Distribution Service (NDADS).
See Abstr. 002.036.

The Second Palomar Sky Survey.
See Abstr. 002.063.

The activities of the NSSDC and ADC in the area of archiving photometric data.
See Abstr. 002.071.

Establishment and query of text database for the study of radio astronomy.
See Abstr. 002.124.

Planetary mapping.
See Abstr. 003.035.

Les astronomes et la naissance de la statistique.
See Abstr. 004.037.

High resolution imaging from Mauna Kea.
See Abstr. 009.001.

The restoration of HST images and spectra. Proceedings.
See Abstr. 012.003.

Precision photometry: astrophysics of the Galaxy. Proceedings.
See Abstr. 012.027.

Systèmes de référence spatio–temporels: métrologie et astrométrie.
See Abstr. 012.048.

3rd ESO/ST–ECF Data Analysis Workshop. Proceedings.
See Abstr. 012.061.

Maximum entropy and Bayesian methods. Proceedings.
See Abstr. 012.081.

Radio interferometry: theory, techniques, and applications. Proceedings.
See Abstr. 012.089.

Data analysis in astronomy. III. Proceedings.
See Abstr. 012.110.

High–resolution imaging in La Palma.
See Abstr. 013.054.

Status of the Perth Observatory automated supernova search program.
See Abstr. 013.091.

Interactive videotape with HyperCard.
See Abstr. 014.025.

Linear and circular digital spectral analysis of serial data.
See Abstr. 021.003.

Elements of a condition theory for the computational analysis of algorithms.
See Abstr. 021.029.

Ensquared power for obscured circular pupils with off–center imaging.
See Abstr. 031.003.

Evaluation of seeing on a 62–cm mirror.
See Abstr. 031.026.

Adaptive optics on a 3.6–m telescope: results and performance.
See Abstr. 031.046.

Efficiency of compensation for phase front distortions in adaptive optical systems when measuring the average phase–front tilt.
See Abstr. 031.054.

Differential adaptive optics for astronomical interferometry.
See Abstr. 031.055.

Of speckles, fringes, and an optical very large array.
See Abstr. 031.074.

Elementary software for optical design.
See Abstr. 031.081.

Pointing models for the Calar Alto 2.2 m and 3.5 m telescopes.
See Abstr. 032.018.

The Carlsberg Automatic Meridian Circle: La Palma.
See Abstr. 032.023.

The TNG – command architecture.
See Abstr. 032.024.

Primary mirror control system for the Galileo telescope.
See Abstr. 032.025.

Galileo project – workstation software system. Architecture design document.
See Abstr. 032.026.

Imaging optical interferometry.
See Abstr. 032.027.

Méthodes de haute résolution angulaire mono–pupille en infrarouge.
See Abstr. 032.044.

The express–control method of pointing the RATAN–600 antenna pattern in height direction.
See Abstr. 033.002.

Scaling posture of the antenna of a solar radio telescope and the design of the control program.
See Abstr. 033.005.

Hardware considerations for high–dynamic–range imaging.
See Abstr. 033.024.

A pulse calibration system for the VLBA.
See Abstr. 033.025.

Software systems for the Ryle Telescope.
See Abstr. 033.032.

GMRT mapping strategies.
See Abstr. 033.036.

The Smithsonian Astrophysical Observatory submillimeter wavelength array.
See Abstr. 033.043.

Progress of VLBI techniques and the role of the Chinese VLBI network.
See Abstr. 033.046.

Imaging with a homogeneous millimeter array.
See Abstr. 033.048.

Imaging characteristics of the proposed NRAO MMA, with and without a single large element, and a pointing correction algorithm.
See Abstr. 033.049.

The Canadian Geophysical LBI system: status and future plans.
See Abstr. 033.052.

Real–time acquisition, analysis, and display of photometric data at the Gettysburg College Observatory.
See Abstr. 034.011.

The drive system of the Spacewatch CCD–scanning telescope.
See Abstr. 034.015.

Solar 2D spectroscopy: a new MSDP instrument.
See Abstr. 034.039.

A thick CCD spectral observation system at the Yunnan Observatory.
See Abstr. 034.043.

Simultaneous observations with a tunable filter and the Echelle Spectrograph of the Vacuum Tower Telescope at Tenerife.
See Abstr. 034.068.

The RGO plate library and measuring machine facilities.
See Abstr. 034.087.

Computing facilities on La Palma.
See Abstr. 034.093.

Image recording complex with the CCD array of the 6–m telescope.
See Abstr. 034.099.

Position observations of geostationary artificial satellites at the Main Astronomical Observatory of the Ukrainian Academy of Sciences and estimate for their precision.
See Abstr. 034.106.

MONICA – the Wendelstein Observatory Monochromatic Imaging Camera.
See Abstr. 034.113.

IAU Working Group on Wide–Field Imaging.
See Abstr. 034.138.

Data analysis and algorithms for g.w. antennas.
See Abstr. 034.176.

The AXAF low–energy transmission grating spectrometer LETGS: diagnostic capabilities for the study of stellar coronae.
See Abstr. 035.008.

Observations with the Wisconsin Ultraviolet Photopolarimetry Experiment (WUPPE).
See Abstr. 035.018.

On orbit measurement of HST baffle rejection capability.
See Abstr. 035.051.

Architectural design of a ground–based deep–space optical reception antenna.
See Abstr. 035.059.

First results from the Goddard High–Resolution Spectrograph: a demonstration of spectral resolution and experiments with deconvolution.
See Abstr. 035.062.

Satellite borne laser for adaptive optics reference.
See Abstr. 035.088.

InSb linear multiplexed FPAs for the CRAF/Cassini visible and infrared mapping spectrometer.
See Abstr. 035.089.

Station keeping strategy for multiple spacecraft interferometry.
See Abstr. 035.092.

Surface control techniques for the segmented primary mirror in the large lunar telescope.
See Abstr. 035.093.

Reconstruction of the Hubble Space Telescope mirror figure from out–of–focus stellar images.
See Abstr. 035.098.

An observatory for mapping the far UV diffuse galactic emission line background.
See Abstr. 035.151.

High–resolution imaging spectroscopy at terahertz frequencies.
See Abstr. 035.155.

The current ability of HST to reveal morphological structure in medium–redshift galaxies.
See Abstr. 035.158.

The sensitivity of EGRET to gamma–ray polarization.
See Abstr. 035.161.

100–μm array polarimetry from the Kuiper Airborne Observatory: instrumentation, techniques, and first results.
See Abstr. 035.165.

Simulations of space VLBI.
See Abstr. 035.201.

Quantitative assessment of the science return from an orbiting, imaging, optical interferometer.
See Abstr. 035.204.

Aperture synthesis imaging from the Moon.
See Abstr. 035.205.

Determination of absolute proper motions for the connection of the HIPPARCOS system with a quasi–inertial system and the automation of astrometric measurements of Tautenburg Schmidt plates.
See Abstr. 041.013.

How to improve old astrometric observations by introducing more accurate corrections for astronomical refraction.
See Abstr. 041.019.

The link to extragalactic sources: observations of radio stars.
See Abstr. 041.024.

Determining the relative orientations of coordinate systems in the optical and radio bands from nonsynchronous observations of high–orbit satellites.
See Abstr. 043.001.

The reference frames in the data reduction of Hipparcos satellite.
See Abstr. 043.005.

Spectral composition of tiltmeter observations in Finland from 6 April 1977 to 28 November 1986.
See Abstr. 044.005.

Methods and techniques for precise time–interval measurements in laser ranging systems of 3rd generation.
See Abstr. 045.002.

Transformation: the link between the proposal and the Hubble Space Telescope database.
See Abstr. 051.023.

Proposal preparation by SPSS for scheduling on the Hubble Space Telescope.
See Abstr. 051.024.

The scheduling of science activities for the Hubble Space Telescope.
See Abstr. 051.025.

The scheduling efficiency for the Hubble Space Telescope during the first year of operation.
See Abstr. 051.026.

Routine science data processing of HST observations.
See Abstr. 051.027.

Computerized computation for first order non–spherical perturbations of satellites.
See Abstr. 052.027.

Optimal Cavendish–type experiment for the search of restrictions on light particles of quantum gravity.
See Abstr. 061.021.

A new way for testing of light deflection in earth orbit or beyond.
See Abstr. 066.119.

Dynamics of the solar granulation. I. Processing of MSDP spectra.
See Abstr. 071.023.

Time–resolved Stokes V polarimetry of small–scale magnetic structures on the Sun.
See Abstr. 075.034.

Tropospheric phenomena responsible for anomalous refraction at radio wavelengths.
See Abstr. 082.045.

Processing series of classical observations on a heliometer using a hindrancestable statistical analysis.
See Abstr. 094.019.

Present main sources of error in the reduction of LLR data.
See Abstr. 094.023.

Comparison of observations of the Martian satellites made in 1988 with ephemerides.
See Abstr. 097.037.

Synthetic curves of some observational properties of contact binary asteroids.
See Abstr. 098.052.

Structural analysis of comet Halley's CN coma by image processing techniques.
See Abstr. 103.102.

Analysis of visual plotting accuracy and sporadic pollution and consequences for shower association.
See Abstr. 104.027.

Méthode différentielle de détermination de mouvements propres absolus.
See Abstr. 111.010.

X–ray spectroscopy of stellar coronae.
See Abstr. 112.038.

Comparison of two semi–empirical calibrations of the uvbyβ system.
See Abstr. 113.019.

Photometric provision for the photographic survey of the north sky and MEGA programmes.
See Abstr. 113.037.

Measuring the stars.
See Abstr. 115.007.

Recent advances in the observation and analysis of stellar magnetic fields.
See Abstr. 116.029.

Surface imaging of rotating stars without Doppler information surface distribution of helium on He variable stars.
See Abstr. 116.090.

Determination of rotational velocities in B and Be stars: physical processes and methods.
See Abstr. 116.098.

Shot model parameters for Cygnus X–1 through phase portrait fitting.
See Abstr. 117.056.

Accretion disk image reconstructions in V4140 Sgr (NSV 12615).
See Abstr. 117.063.

Tomographic separation of composite spectra: the components of the O–star spectroscopic binary AO Cassiopeiae.
See Abstr. 120.011.

Collimation of stellar jets – constraints from the observed spatial structure. I. Data analysis methods.
See Abstr. 121.042.

Predicting white dwarf light curves.
See Abstr. 126.051.

On the observable signatures of a contracting protostellar fragment.
See Abstr. 131.237.

Speckles in interstellar radio–wave scattering.
See Abstr. 131.347.

Astrometry of SiO masers. First detections of circumstellar SiO maser emission at λ = 7 mm using three European VLBI stations.
See Abstr. 131.351.

The first VLA observations of radio sources at 73.8 MHz.
See Abstr. 141.012.

Locating very high energy gamma–ray sources with arcminute accuracy.
See Abstr. 143.027.

CCD photometry with a Schmidt telescope.
See Abstr. 153.062.

Image restoration of R136: the dense core of NGC 2070.
See Abstr. 153.068.

A new approach to aging globular clusters.
See Abstr. 154.150.

Search for fast ultraviolet variability in NGC 5548 and non–linearities in IUE's SWP camera.
See Abstr. 158.074.

Echo mapping of NGC 5548.
See Abstr. 158.145.

Super–shells of M82.
See Abstr. 158.194.

Outline of a Liquid Mirror Telescope QSO survey.
See Abstr. 159.132.

Analysis of absorption lines in quasar spectra.
See Abstr. 159.183.

High–resolution imaging of Virgo cluster galaxies. I. Observational techniques and first results.
See Abstr. 160.132.

Are the intermediate–scale cosmic background radiation anisotropies constrained by thermal emission of a very cold dust?
See Abstr. 161.156.

Positional Astronomy, Celestial Mechanics

041 Astrometry

041.001 **Unpublished measurements in the Potsdam zone of the Astrographic Catalogue.**
W. R. Dick.
Bull. Inf. Cent. Données Stellaires, No. 38, p. 19 – 20 (Dec 1990).
In the original Potsdam zone of the Astrographic Catalogue, measurements for only one third of the plates are published. The measurements carried out after publication had been stopped, are described. They are rather heterogeneous and of different importance. The measurements of 88 first-epoch plates are particularly valuable.

041.002 **A user's guide to astrometric data of radio stars.**
R. Hering, H. G. Walter.
Bull. Inf. Cent. Données Stellaires, No. 38, p. 209 – 213 (Dec 1990).
Astrometrical data in the optical and radio spectral ranges have been assembled for some 200 radio stars which are considered promising objects for linkage of optical and radio frames. Astrophysical data have been added to support observing programmes and data analysis. The arrangement of the catalogue data is briefly described.

041.003 **Umrechnung B1950.0 auf J2000.0.**
J. Jahn.
KPM, Jahrg. 6, No. 17, p. 4 – 8 (Aug 1991).
The author summarizes the IAU method for changing positions and orbital elements from the epoch B1950.0 to J2000.0.

041.004 **Optical positions of 221 radio stars obtained with the Bordeaux automatic meridian circle.**
Y. Réquième, J. M. Mazurier.
Astron. Astrophys., Suppl. Ser., Vol. 89, No. 2, p. 311 – 318 (Aug 1991).
From 1984 to 1990 the Bordeaux automatic meridian circle was used to determine accurate J2000 positions of 221 radio stars. The observations were strictly differential, a convenient number of FK5 stars being used each night to calculate the instrumental parameters. The mean positions were corrected for chromatic refraction and also for photocenter effect in the case of duplicity. The asymptotic accuracy is expected to be about 0.05″ for the Northern stars with a number of observations greater than 25. Mean position of 3C 273B is also given from 35 measurements obtained in the same conditions.

041.005 **Optical positions of radiostars. III.**
E. Costa, P. Loyola.
Astron. Astrophys., Suppl. Ser., Vol. 89, No. 2, p. 411 – 413 (Aug 1991).
Precise optical positions, relative to either the Perth 70 or the AGK3R catalogue, have been obtained for 44 radiostars proposed to the Hipparcos observing list. Precision levels between 0.″17 and 0.″28 were obtained.

041.006 **Astronomical computing: Precession of a star's position.**
Z. S. Krzeminski.
Sky Telesc., Vol. 82, No. 4, p. 408 – 410 (Oct 1991).

041.007 **VLBI astrometry of PSR 1937+21.**
R. J. Dewey, D. L. Jones, C. R. Gwinn.
Bull. Am. Astron. Soc., Vol. 23, No. 2, p. 946 (1991). Abstract. – See Abstr. 012.008 for the main entry.

041.008 **The catalogue of geodetical stars observed at Yunnan Observatory.**
Hu Hui, Cai Xin, Chen Cuixian, Wang Rui, Liao Yuanxiang, Zhang Zhiyuan.
Publ. Yunnan Obs., No. 2, p. 61 – 80 (Jun 1991). In Chinese.
The position corrections of 606 geodetical stars which passed across the circle of equal altitudes twice observed with the type II photoelectric astrolabe of Yunnan Observatory from 1987 to 1990 are given.

041.009 **J2000.0 positions and proper motions of 257 stars in the central part of the Praesepe astrometric standard region.**
Wang Jiaji, Jiang Peifang.
Acta Astron. Sin., Vol. 32, No. 2, p. 176 – 185 (Jun 1991). In Chinese. English translation in Chin. Astron. Astrophys., Vol. 15, No. 4, p. 397 – 407 (Dec 1991).
The authors present the J2000.0 positions and proper motions of 257 stars in the central 1.°5 × 1.°5 area of the Praesepe astrometric standard region with accuracies of 0.005 ∼ 0.10 arcsec for the positions in each direction and 0.0002 ∼ 0.006 arcsec per year for the proper motions in each direction.

041.010 **Influence of the reference catalogue errors on the determination of the parameters of fundamental systems using minor planet observation.**
D. D. Polojentsev (*D. D. Polozhentsev*).
1. Spain–USSR Workshop on Positional Astronomy and Celestial Mechanics, p. 19 – 22 (1991). – See Abstr. 012.025 for the main entry.
In the years 1931 – 1936 Prof. B. V. Numerov proposed a plan of minor planet observation for determination of the Fundamental Catalogue correction of equinox and equator. For realization of this plan it is very important to choose the appropriate reference catalogue. During the new cycle of minor planet observation it is very desirable at the same time to organise the re-observation of the reference catalogue in order to provide a better reference to the fundamental system.

041.011 **The results and prospects of use of minor planets observations for improvement of the systems of the fundamental catalogues.**
D. P. Duma.
1. Spain–USSR Workshop on Positional Astronomy and Celestial Mechanics, p. 139 – 142 (1991). – See Abstr. 012.025 for the main entry.
The main points of B. Numerov's and D. Brouwer's plans to use observations of selected minor planets for improvement of the zero-points and systematic errors of fundamental star catalogues are discussed. It is stated, that the realization of these proposals cannot be admitted to be successful owing to the lack of powerful meridian and photographic telescopes, the theories of minor planets motion accurate to ±0.″02, and the completed faint star catalogue. High-precision observations of minor planets are still needed for the zero-points improvement and for

perfection of the theories of motion of the solar system minor bodies.

041.012 **A model for reduction of solar observations to the system of a reference catalogue by means of a vertical circle under high–altitude conditions.**
K. G. Gnevysheva.
Astron. Tsirk., No. 1544, p. 29 – 30 (May – Jul 1990). In Russian.
"The point of reduction" at the time of the Sun's culmination was used for the reduction to a reference catalogue. The standard of one determination of the decliantion of the Sun is $\pm 0\farcs43$.

041.013 **Determination of absolute proper motions for the connection of the HIPPARCOS system with a quasi–inertial system and the automation of astrometric measurements of Tautenburg Schmidt plates.**
R.– D. Scholz.
Astron. Nachr., Vol. 312, No. 3, p. 224 (1991). Dissertation abstract.

041.014 **Tests statistiques de reconnaissance des étoiles non–simples observées par le satellite Hipparcos.**
M. Froeschlé, F. Mignard, J. L. Falin.
Journées 1991: Systèmes de référence spatio–temporels, p. 92 – 97 (1991). – See Abstr. 012.048 for the main entry.
The authors present the procedures used in the FAST consortium to recognize in the HIPPARCOS compressed data (method of five parameters) the multiplicity of the stellar objects. The principles applied to this recognition are described.

041.015 **Une estimation de la précision des mouvements propres et des parallaxes en astrométrie photographique.**
M. Rapaport, C. Ducourant, J. Colin, J. F. Lecampion.
Journées 1991: Systèmes de référence spatio–temporels, p. 107 – 109 (1991). – See Abstr. 012.048 for the main entry.
The authors present a preliminary program to estimate the accuracy that can be reached in parallaxes and proper motions determination using a Schmidt telescope and an automatic measuring machine.

041.016 **Erreurs systématiques des déterminations astronomiques de variation de coordonnées à l'astrolabe de Borowiec.**
M. Lehmann, J. Wiktorowski.
Journées 1991: Systèmes de référence spatio–temporels, p. 184 – 185 (1991). – See Abstr. 012.048 for the main entry.
This paper investigates the possible causes for the systematic differences in the zenith distances as derived between two successive groups of stars observed in the same night with Danjon's astrolabe OPL at Borowiec. Such systematic differences are shown to be strongly reduced when taking into account an empirical term proportional to the elapsed time from the sunset.

041.017 **Catalogues d'étoiles doubles et brillantes observées à Bucarest. Analyse de precision en ascension droite.**
M. Liculescu, M. Orzaru, P. Popescu.
Journées 1991: Systèmes de référence spatio–temporels, p. 188 – 190 (1991). – See Abstr. 012.048 for the main entry.
This paper presents an analysis of the differences $\Delta\alpha \cos\delta$ for 330 double stars (from the list of F. P. Scott) and for 599 bright stars (from the list of M. S. Zverev) as well as of the accuracy in right ascension as obtained from measurements which have been performed using Bucharest Meridian Circle.

041.018 **Meridian and astrophotographic observations of the reference stars in the fields with extragalactic radio and optical sources.**
V. V. Tel'nyuk–Adamchuk, I. I. Kumkova, S. V. Pasechnik, A. A. Molotaj, A. Ya. Gregul, N. D. Kanivets, N. A. Chernega, S. I. Kas'yan, Yu. A. Buzdugan.
International Seminar on Problems of Astrometry and Space Geodynamics, p. 41 – 47 (1991). In Russian. – See Abstr. 012.049 for the main entry.
The problems concerning mutual orientation of fundamental and radiointerferometric coordinate systems are discussed. The observation programs devoted to a solution of these problems are described.

041.019 **How to improve old astrometric observations by introducing more accurate corrections for astronomical refraction.**
K. Kurzyńska.
Journées 1991: Systèmes de référence spatio–temporels, p. 223 – 228 (1991). – See Abstr. 012.048 for the main entry.
The influence of both pure and anomalous refraction on the accuracy of observations made by various instruments is discussed. For astrolabe, it is shown that the use of a simplified refraction formula results in a systematic error. The adequate formula for the pure refraction calculation is proposed and the order of systematic error originating from the anomalous refraction is discussed on the basis of data collected for Poznań.

041.020 **The determination of selected extragalactic radio source positions by photographic observations.**
I. V. Ledovskaya.
International Seminar on Problems of Astrometry and Space Geodynamics, p. 47 – 50 (1991). In Russian. – See Abstr. 012.049 for the main entry.
Brief results of the works at the Central Astronomical Observatory connected with determination of the optical positions of extragalactic radio sources are given. The precision of extragalactic radio sources position is $0\farcs2$. A comparison of optical positions with their radio positions is made.

041.021 **Optical positions of extragalactic radio sources in the FK5 system.**
W. R. Dick, I. I. Kumkova.
International Seminar on Problems of Astrometry and Space Geodynamics, p. 51 – 54 (1991). – See Abstr. 012.049 for the main entry.
Optical positions of 11 objects from the IAU Commission 24 Working Group list of benchmark radio sources have been determined using Tautenburg Schmidt plates and the PPM as the reference catalogue. From comparison with the radio reference frame an r.m.s. position error of $0\farcs13$ in $\alpha \cos\delta$ and of $0\farcs18$ in δ has been derived.

041.022 **On the influence of atmospheric dispersion on the photographic positions of stars.**
V. P. Ryl'kov.
International Seminar on Problems of Astrometry and Space Geodynamics, p. 71 – 86 (1991). In Russian. – See Abstr. 012.049 for the main entry.
Due to change of the effective wavelength the centres of star images are shifted by $0.30 - 0\farcs37$ on the plate when observations are made at $z = 45°$ with respect to the position of their centres at $z = 0°$. In connection with that the role of the spectral distribution of reference stars is discussed. The mean systematic shift in declination $+0.298 \pm 0\farcs020$ for B, A reference stars relative to K ones was obtained for the Pleiades. This shift is decreased by 2 – 3 times when taking into account the chromatic correction for reference stars.

041.023 **J2000.0 positions and proper motions of 257 stars in the central part of the Praesepe astrometric standard region.**
Wang Jiaji, Jiang Peifang.
Chin. Astron. Astrophys., Vol. 15, No. 4, p. 397 – 407 (Dec 1991). English translation of Acta Astron. Sin., Vol. 32, No. 2, p. 176 – 185 (Jun 1991).
The authors give the J2000.0 positions and proper motions of 257 stars in the central $1\fdg5 \times 1\fdg5$ area of the Praesepe astrometric standard region with accuracies of $0.005 \sim 0.10$ arcsec for the position in each direction and $0.0002 \sim 0.006$ arcsec/yr for the proper motions in each direction.

041.024 **The link to extragalactic sources: observations of radio stars.**
A. Baudry, Y. Requieme.
IAU Colloquium No. 131: Radio interferometry – theory, techniques, and applications, p. 330 – 333 (1991). – See Abstr. 012.089 for the main entry.

Optical positions obtained with the Bordeaux automatic meridian circle (accuracy ~0.05″) are compared with VLA positions taken from the literature. Alignment of the FK5 and VLA frames are discussed, and other projects (Hipparcos, VLBI) are briefly presented.

041.025 **Beitrag der Schmidtspiegelastrometrie zur planetaren Geodäsie.**
K.-G. Steinert, C. Witschas.
Z. Vermessungswes., Jahrg. 116, Heft 6, p. 241 – 247 (Jun 1991). = Mitt. Lohrmann-Obs., Nr. 65.

A Conventional Inertial System (CIS) on the sky will be a radio reference system. From recent publications it is known that radio objects do not necessarily fulfill all conditions for a CIS. Therefore it seems to be optimal to combine a radio system with an optical one. The later one is only to be realized by means of large telescopes. The paper deals with investigations of the Tautenburg Schmidt Telescope performed for this purpose.

041.026 **Analytical description of the Fourth San Juan Meridian Circle Catalogue (CMSJ4).**
C. C. Mallamaci.
Bol. Asoc. Argent. Astron., No. 36, p. 248 – 263 (1990). In Spanish. – See Abstr. 012.095 for the main entry.

The fourth San Juan Meridian Circle Catalogue (CMSJ4) contains 364 FK4 stars, in the declination zone -30° to -90°, which formed the reference frame for the SRS program, observed with the Repsold meridian circle of the "Felix Aguilar" Astronomical Observatory. It is one of the three catalogues with which our research team has participated in the construction of the new international reference system FK5. The systematic components of the CMSJ4's (O – C) are described.

041.027 **Opsendelsen af HIPPARCOS lykkedes alligevel.**
E. Høg.
Astron. Tidsskr., Årg. 24, Nr. 3, p. 124 – 126 (1991).

Introduction to PPM-South. Preliminary version (May 1990).
See Abstr. 002.015.

PPM: der Nachfolger des SAO-Katalogs.
See Abstr. 002.019.

Positional reference stars in the Magellanic Clouds.
See Abstr. 002.028.

Santiago declination catalogue. II. A declination catalogue of 493 FK5 stars (equinox J2000.0).
See Abstr. 002.031.

Determination of positions and proper motions of stars from the plates of the northern sky photographic survey program.
See Abstr. 002.046.

Comparison of the Catalogues KCB and FK5.
See Abstr. 002.064.

On the "Ekat" catalogue and prospects of its application in photographic observations of minor planets.
See Abstr. 002.066.

Some results of determining the SAO catalogue systematic errors from observations of minor planets.
See Abstr. 002.068.

New reference catalogues for photographic observations of minor planets.
See Abstr. 002.069.

Analysis of observational catalogues and formation of the bright stars compiled catalogue system.
See Abstr. 002.089.

Thin prism survey for F and G stars at the south galactic pole.
See Abstr. 002.101.

Sydney Southern Star Catalogue (King and Lomb). Documentation for the machine-readable version.
See Abstr. 002.115.

The precise positions and proper motions of 1980 GC stars.
See Abstr. 002.123.

Tentative idea and practice of database of astrometry and geodynamics at Shanghai Observatory.
See Abstr. 002.125.

The Kenya expedition of Leiden Observatory.
See Abstr. 004.083.

CAS Symposium No. 106 and No. 107: Connections of reference systems and techniques of photographic astrometry, Yichang (China), 16 – 19 Nov 1990.
See Abstr. 011.009.

First Spain-USSR Workshop on Positional Astronomy and Celestial Mechanics. Proceedings.
See Abstr. 012.025.

Systèmes de référence spatio-temporels: métrologie et astrométrie.
See Abstr. 012.048.

Problems of astrometry and space geodynamics. Proceedings.
See Abstr. 012.049.

The programme FON-OBZOR-S.
See Abstr. 013.066.

Radio-stars and SAO-stars observations with the Repsold Meridian Circle of the "Felix Aguilar" Astronomical Observatory.
See Abstr. 013.107.

Improvement of type II photo-astrolabe and its preliminary results.
See Abstr. 032.028.

Position observations of geostationary artificial satellites at the Main Astronomical Observatory of the Ukrainian Academy of Sciences and estimate for their precision.
See Abstr. 034.106.

The micrometers of the Carlsberg Automatic Meridian Circle.
See Abstr. 034.107.

Das System ASTRO 1 auf der Raumstation MIR.
See Abstr. 035.004.

Early commissioning astrometry performance of the Fine Guidance Sensors.
See Abstr. 035.044.

The Hipparcos mission: will it be a scientific success?
See Abstr. 035.106.

Eine Methode zur Bestimmung genäherter Kleinplanetenörter.
See Abstr. 036.017.

The transition to the true tangential point when using complete overlap plate method.
See Abstr. 036.060.

Using image area to control CCD systematic errors in spaceborne photometric and astrometric time–series measurements.
See Abstr. 036.112.

A non–deterministic approach to Schmidt plate astrometry.
See Abstr. 036.128.

Procedures for validating Hipparcos astrometric parameters.
See Abstr. 036.151.

On the astrometric observations, 1962.0 – 1991.0.
See Abstr. 036.166.

First results of the photographic survey program of the northern sky.
See Abstr. 036.167.

Colour and magnitude effects in optical astrometry observations; implications for re–reductions in Hipparcos reference frame.
See Abstr. 036.168.

Creation of the star position catalogue by five CdC plate–block processing.
See Abstr. 036.169.

High–accuracy equation of equinoxes and VLBI astrometric modelling.
See Abstr. 036.272.

MARK–III VLBI astrometry of pulsars.
See Abstr. 036.273.

Untersuchungen zur Durchführung, Auswertung und astrometrischen Nutzung von photoelektrischen Sternbedeckungsbeobachtungen.
See Abstr. 036.291.

Analysis on the effects of spectral type error and magnitude difference of the 40 cm astrograph at Zô–Sè for determinations of radio star positions.
See Abstr. 036.296.

Astrometrie an Kometen.
See Abstr. 036.365.

Mutual orientation of the FK5 and VLBI reference frames using 251 radio sources and radio stars.
See Abstr. 043.002.

Determination of the angles of mutual orientation of radio and optical reference systems using radioastronomical and optical star positions.
See Abstr. 043.003.

Global and regional astrometric comparison between two dense VLBI extragalactic reference frames.
See Abstr. 043.004.

The reference frames in the data reduction of Hipparcos satellite.
See Abstr. 043.005.

Transformation from FK4 system to FK5 system.
See Abstr. 043.014.

Effects of systematic difference FK5–FK4 on the optical global solution of the Earth's Orientation Parameters.
See Abstr. 044.022.

Genauigkeitssteigerung in der geodätischen Astronomie durch den FK5 Sternkatalog. Neuauswertung von astronomisch bestimmten Längendifferenzen im Europäischen Längennetz (ELN).
See Abstr. 045.021.

TYCHO/HIPPARCOS a successful mission! First results and further improvements.
See Abstr. 051.031.

A method for improving the parameters of the solar system from ground–based and space observations of planetary satellites.
See Abstr. 091.042.

The accuracy of astrometric positions.
See Abstr. 098.001.

OAUV programme for minor planets astrometry.
See Abstr. 098.041.

New proper–motion stars south of declination –40° and right ascension between 04h 30m and 16h 00m.
See Abstr. 111.002.

Multichannel astrometric photometer–based parallaxes of evolved stars: χ Cygni, 51 Andromedae, and OP Andromedae.
See Abstr. 111.007.

Accuracy estimation for connection between proper motions of the Hipparcos catalogue and the optical catalogues by means of radio stars observations.
See Abstr. 111.016.

On realization of the photometric program for areas with galaxies in the Pulkovo plan of CFS (catalogue of faint stars).
See Abstr. 113.039.

Photometry of astrometric reference stars.
See Abstr. 113.040.

Mesures micrométriques d'étoiles doubles visuelles (5ème liste).
See Abstr. 118.008.

Binary star observations with the Hubble Space Telescope fine guidance sensors. I. ADS 11300.
See Abstr. 118.016.

Astrometry of SiO masers. First detections of circumstellar SiO maser emission at λ = 7 mm using three European VLBI stations.
See Abstr. 131.351.

Determination of proper motions in the Pleiades cluster.
See Abstr. 153.039.

Precise positions of reference stars in open clusters NGC 457 and NGC 581.
See Abstr. 153.077.

042 Celestial Mechanics, Figures of Celestial Bodies

042.001 **Comments on the symmetric difference quotient and its applications to the correction of orbits by R. A. Serafin and E. Wnuk.**
A. Brunini.
Celest. Mech. Dyn. Astron., Vol. 50, No. 4, p. 299 – 300 (1990/91).
A comment on the paper by R. A. Serafin and E. Wnuk published in Celest. Mech., Vol. 42, Nos. 1 – 4, p. 175 – 186 (1987/88). See Abstr. 46.042.052.

042.002 **The three–dimensional equilibria of magnetic–binary systems with oblate primaries.**
T. J. Kalvouridis.
Celest. Mech. Dyn. Astron., Vol. 50, No. 4, p. 301 – 311 (1990/91).
In this paper the three–dimensional equilibria in a magnetic–binary system with oblate primaries are studied. It is also examined how the primaries oblateness affect the equilibria configuration of the spherical case.

042.003 **Tame and chaotic behavior in the planar isosceles 3–body problem.**
F. N. Diacu.
Celest. Mech. Dyn. Astron., Vol. 50, No. 4, p. 313 – 324 (1990/91).
The author shows that every planar isosceles solution of the three–body problem encounters a collision of the symmetric particles, either forwards or backwards in time. Regularizing analytically this collision, the solution has at least a syzygy configuration and/or leads to a total collapse. Some further simple results support the intuitive image on the tame local behavior of the motion as long as it does not lead to a triple collision. As a main result the author proves that total collapse singularities can be regularized in a C^1–fashion with respect to time, for all values of the masses. Using symbolic dynamics, the chaotic character of the C^1–regularized solutions is pointed out.

042.004 **Periodic solutions around the collinear Langrangian points in the photogravitational restricted three–body problem: Sun–Jupiter case.**
O. Ragos, C. Zagouras.
Celest. Mech. Dyn. Astron., Vol. 50, No. 4, p. 325 – 347 (1990/91).
The evolution of the periodic orbits around the collinear equilibrium positions, belonging to the Strömgren families a, b and c, with the radiation pressure parameter of the more massive body is studied in the Sun–Jupiter system. These families are determined for a single value of the radiation pressure parameter and particularly when the radiation force of the more massive body is equal to one half of the gravitational attraction. Then the critical stability orbits of each family are transferred with the radiation parameter. The stability of each periodic solution is also studied.

042.005 **Hamiltonian dynamics of a rigid body in a central gravitational field.**
Wang Li–Sheng, P. S. Krishnaprasad, J. H. Maddocks.
Celest. Mech. Dyn. Astron., Vol. 50, No. 4, p. 349 – 386 (1990/91).
This paper is concerned with the dynamics of a rigid body of finite extent moving under the influence of a central gravitational field. A principal motivation behind this paper is to reveal the hamiltonian structure of the *n*–body problem for masses of finite extent and to understand the approximation inherent to modeling the system as the motion of point masses. To this end, explicit account is taken of effects arising because of the finite extent of the moving body. In the spirit of Arnold and Smale, exact models of spin–orbit coupling are formulated, with particular attention given to the underlying Lie group framework. Hamiltonian structures associated with such models are carefully constructed and shown to be non–canonical. Special motions, namely relative equilibria, are investigated in detail and the notion of a non–great circle relative equilibrium is introduced. The last part of the paper is concerned with rigorous derivations of nonlinear stability results for certain families of relative equilibria. Here Arnold's energy–Casimir method and Lagrange multiplier methods prove useful.

042.006 **Strong triple interactions in the general three–body problem.**
J. P. Anosova (*Zh. P. Anosova*).
Celest. Mech. Dyn. Astron., Vol. 51, No. 1, p. 1 – 15 (1991).
Strong three–body interactions play a decisive role in the course of the dynamical evolution of triple systems having positive as well as negative total energies. These interactions may produce qualitative changes in the relative motions of the components. In triple systems with positive or zero total energies the processes of formation, disruption or exchange of the components of binaries take place as the result of close approaches of the three single bodies or as the result of the passages of single bodies past wide or hard binaries. In the triple systems with negative energies, the strong triple interactions may result in an escape from the system as well as a formation of a hard final binary. This paper summarizes the general results of the studies of the strong three–body interactions in the triple systems with positive and negative energies. These studies were conducted at the Leningrad University Observatory by computer simulations during 1968–1989.

042.007 **Effect of perturbation on the location of libration point in the Robe restricted problem of three bodies.**
A. K. Shrivastava, D. Garain.
Celest. Mech. Dyn. Astron., Vol. 51, No. 1, p. 67 – 73 (1991).
The effect of small perturbation in the Coriolis and centrifugal forces on the location of libration point in the "Robe (1977) restricted problem of three bodies" has been studied. In this problem one body, m_1, is a rigid spherical shell filled with an homogeneous incompressible fluid of density ϱ_1. The second one, m_2, is a mass point outside the shell and m_3 is a small solid sphere of density ϱ_3 supposed to be moving inside the shell subject to the attraction of m_2 and buoyancy force due to fluid ϱ_1. Here the authors assume m_3 to be an infinitesimal mass and the orbit of the mass m_2 to be circular, and they also suppose the densities ϱ_1 and ϱ_3 to be equal. Then there exists an equilibrium point.

042.008 **Vertical stability of periodic solutions around the triangular equilibrium points.**
E. Perdios, C. G. Zagouras.
Celest. Mech. Dyn. Astron., Vol. 51, No. 1, p. 75 – 81 (1991).
The vertical stability character of the families of short and long period solutions around the triangular equilibrium points of the restricted three–body problem is examined. For three values of the mass parameter less or equal to the critical value of Routh (μ_R) i.e. for $\mu = 0.000953875$ (Sun–Jupiter), $\mu = 0.01215$ (Earth–Moon) and $\mu = \mu_R = 0.038521$, it is found that all such solutions are vertically stable. For $\mu > \mu_R$ vertical stability is studied for a number of "limiting" orbits extended to $\mu = 0.45$. The last limiting orbit computed by Deprit for $\mu = 0.044$ is continued to a family of periodic orbits into which the well known families of long and short period solutions merge. The stability characteristics of this family are also studied.

042.009 **Linear stability of a self–gravitating ring.**
D. J. Scheeres, N. X. Vinh.
Celest. Mech. Dyn. Astron., Vol. 51, No. 1, p. 83 – 103 (1991).
Stability of a self–gravitating ring about a central body is considered. The purpose is to derive a bound on the mass of the ring in order that the system will be linearly stable. The bound will, in some cases, be the best possible bound. The bound is also expanded as an asymptotic series. Comparisons of the authors'

result are made with respect to previous analyses performed by Tisserand, Pendse and Willerding.

042.010 **The generalized Hill problem. Case of an external field of force deriving from a central potential.**
B. Chauvineau.
Celest. Mech. Dyn. Astron., Vol. 51, No. 2, p. 119 - 129 (1991).
The purpose of this paper is to investigate the generalization of Hill's problem by using a central field of force deriving from a potential, not restricted to Newton's inverse square law. The author establishes the equations of motion, determines the equilibrium positions along with their linear stability.

042.011 **Secular resonances in the asteroid belt: theoretical perturbation approach and the problem of their location.**
A. Morbidelli, J. Henrard.
Celest. Mech. Dyn. Astron., Vol. 51, No. 2, p. 131 - 167 (1991).
In this paper a theoretical perturbation approach to the problem of the dynamics in secular resonance is exposed. This approach avoids any expansion of the main term of the Hamiltonian (linear term in the masses) with respect to the eccentricity or the inclination of the asteroid, in order to achieve results valid for any value of these variables. Moreover suitable action–angle variables are introduced to take properly into account the dynamics related to the motion of the argument of perihelion of the asteroid, which is relevant at high inclination. A class of secular resonances wider than that usually considered is found. An explicit computation of the location of the main secular resonances, estimating also the contribution of the quadratic term in the masses by means of classical series expansion, is reported in the last sections. The accuracy of computations obtained by series expansion is discussed in the paper.

042.012 **The main secular resonances ν_6, ν_5 and ν_{16} in the asteroid belt.**
A. Morbidelli, J. Henrard.
Celest. Mech. Dyn. Astron., Vol. 51, No. 2, p. 169 - 197 (1991).
In this paper an analytical model, suitable for a global description of the dynamics in a secular resonance of order 1, is derived from the general perturbation study developed in a previous paper (Morbidelli and Henrard (1991)). Such a model is then used to study the secular resonances ν_6, ν_5 and ν_{16}, and pictures illustrating the secular motion are obtained. The peculiarities of the ν_5 resonances are discussed in detail. The results are compared with those obtained by the theories of Yoshikawa and Nakai–Kinoshita. Some numerical simulations performed by Ch. Froeschlé and H. Scholl are discussed in the light of the new theoretical results. New numerical experiments on the ν_6 resonance are also presented.

042.013 **A general scheme of Newtonian interactions in systems of gravitating masses.**
J. Anosova (*Zh. P. Anosova*).
Astrophys. Space Sci., Vol. 181, No. 1, p. 147 - 152 (Jul 1991). Letter–to–the–editor.
This Letter reviews the results by computer simulations on the three–body problem carried out by the author at the Leningrad University Astronomical Observatory. The intensive systematic studies of triple systems with negative and positive total energies have yielded the general features of the evolution of these systems. The processes of formation of the wide and hard binaries have been studied in details. The scenario of the general class of the final motions of the triple systems with negative total energy is considered, the necessary conditions of disruption of these systems are formulated.

042.014 **Binary trees, symplectic matrices and the Jacobi coordinates of celestial mechanics.**
C. C. Lim.
Arch. Ration. Mech. Anal., Vol. 115, No. 2, p. 153 - 165 (5 Jul 1991).
A graph–theoretic algorithm for constructing the Jacobi coordinates in celestial mechanics is given. For every full binary tree with N leaves, there belongs a 6N × 6N symplectic matrix, which defines a Jacobi transformation. This correspondence yields a direct proof of the symplectic property for all the Jacobi coordinates, hitherto, only special examples of these transformations have been shown to be canonical.

042.015 **Orbital stability zones about asteroids.**
D. P. Hamilton, J. A. Burns.
Icarus, Vol. 92, No. 1, p. 118 - 131 (Jul 1991).
The authors have numerically investigated a three–body problem consisting of the Sun, an asteroid, and an infinitesimal particle initially placed about the asteroid. They assume that the asteroid has the following properties: a circular heliocentric orbit at R = 2.55 AU, an asteroid/Sun mass ratio of $\mu = 5 \times 10^{-12}$, and a spherical shape with radius R_A = 100 km; these values are close to those of the minor planet 29 Amphitrite. In order to describe the zone in which circum–asteroidal debris could be stably trapped, the authors pay particular attention to the orbits of particles that are on the verge of escape. They consider particles to be stable or trapped if they remain in the asteroid's vicinity for at least 5 asteroid orbits about the Sun, or about 20 y. Applying this criterion to particles started on circular orbits around the asteroid, one finds that, as the starting distance from the asteroid is varied, a fairly aburpt transition between trapped and untrapped objects occurs. The distance where the transition occurs is defined to be the critical distance.

042.016 **Preservation of the form of area integrals in averaging transformations.**
K. V. Kholshevnikov.
Sov. Astron., Vol. 35, No. 3, p. 325 - 327 (May 1991). Current Physics Microform No.: 9201X0607. English translation of Astron. Zh., Tom 68, Vyp. 3, p. 660 - 663 (1991).
Averaging transformations, performed to within an arbitrary power of a small parameter, are considered in a planetary problem. The invariance of the form of three area integrals is demonstrated. Hence it follows that one of the fundamental frequencies of secular motion of the nodes vanishes.

042.017 **Periodic motion around stable collinear equilibrium point in the photogravitational restricted problem of three bodies.**
O. Ragos, C. G. Zagouras, E. Perdios.
Astrophys. Space Sci., Vol. 182, No. 2, p. 313 - 336 (Aug 1991).
In a binary system with both bodies being luminous, the inner collinear equilibrium point L_1 becomes stable for values of the mass ratio and radiation pressure parameters in a certain region. The kind of periodic motions around L_1 is examined in this case. Second–order parameteric expansions are given and the families of periodic orbits generated from L_1 are numerically determined for several sets of values of the parameters. Short– and long–period solutions are identified showing a similarity in the character of periodicity with that around L_4. It is also found that the finite periodic solutions in the vicinity of L_1 are stable.

042.018 **New applications of Fatonu's problem.**
V. V. Radzievskij, V. P. Tomanov.
Astron. Vestn., Tom 25, No. 5, p. 593 - 598 (Sep - Oct 1991). In Russian. English translation in Solar Syst. Res., Vol. 25, No. 5.
The question of the motion of a small body m in the gravitational field of a massive body m_1 and a number of massive homogeneous parallel circles m_i (the ring of asteroids, planets with even distribution of their mass) is investigated. It is shown that all parabolic comets are isolated from Oort's cloud: they move inside of closed zero radial velocity surface with radius order 5000 AU. The influence of the asteroidal ring on the motion of Mercury's perihelion is investigated.

042.019 **Periodic solutions and their stability for the problem of gyrostat.**
F. M. El-Sabaa.
Astrophys. Space Sci., Vol. 183, No. 2, p. 199 - 213 (Sep 1991).
The Hamiltonian function of the problem of gyrostat is written in terms of Deprit's transform. The periodic solutions, using Lie's

transform, are given and the condition for their stability is carried out.

042.020 **The three components S, T, W of PPN perturbation for the rotating two–body problem.**
Han Tao.
Publ. Purple Mt. Obs., Vol. 10, No. 2, p. 128 – 135 (Jun 1991). In Chinese.
A PPN equation of motion for the rotating two–body problem is derived. This equation is then applied to obtain the three components S, T, W of rotating PPN perturbation for three conditions and the three components can be integrated. These components are suitable for conspicuous rotating effects with study of solar system or binary systems.

042.021 **Numerical research on the Sitnikov problem.**
Liu Jie, Zhou Jilin, Sun Yisui.
Chin. Astron. Astrophys., Vol. 15, No. 3, p. 339 – 344 (Sep 1991). English translation of Acta Astron. Sin., Vol. 32, No. 1, p. 107 – 112 (Mar 1991). See Abstr. 53.042.029.
Using the methods of Poincaré section and the largest Liapunov characteristic exponent in the Sitnikov problem the authors have derived a relation between the chaotic region and the eccentricity of the primary's orbit, which agrees very well with our previous theoretical result.

042.022 **Modern problems of computerization of celestial mechanics.**
A. G. Sokolsky (*A. G. Sokol'skij*).
1. Spain–USSR Workshop on Positional Astronomy and Celestial Mechanics, p. 5 – 8 (1991). – See Abstr. 012.025 for the main entry.

042.023 **The influence of orbital eccentricity of a disturbing body on chaotic region.**
Liu Jie.
Acta Astron. Sin., Vol. 32, No. 2, p. 222 – 224 (Jun 1991). In Chinese. English translation in Chin. Astron. Astrophys., Vol. 15, No. 4, p. 446 – 448 (Dec 1991).
The author applies the resonance overlap theory to the elliptic planar restricted three–body problem and obtains an estimation of the chaotic region: $q < c(e(p))\mu^{v(e(p))}$, where e(p) is the eccentricity of the perturbing body. When $e(p) = 0$, the result is equivalent to that of J. Wisdom. The author finds that the orbital eccentricity e(p) of primaries makes the size of the chaotic region larger, which proves M. Duncan's conjecture. It also provides a theoretical basis for the increase of size of the "feeding zone" of planetary embryos and may reduce or eliminate the following traditional problem. The numerical experiments attempting to form the terrestrial planets often produce an aggregate of planets that are too numerous, too small, and too closely spaced.

042.024 **Gaps of regular motion in Sitnikov's problem.**
J. M. Alfaro, C. Chiralt.
1. Spain–USSR Workshop on Positional Astronomy and Celestial Mechanics, p. 23 – 28 (1991). – See Abstr. 012.025 for the main entry.

042.025 **The isosceles three–body problem as a model for the three–body problem with two small masses.**
C. Chiralt Monleón, J. M. Alfaro.
1. Spain–USSR Workshop on Positional Astronomy and Celestial Mechanics, p. 29 – 41 (1991). – See Abstr. 012.025 for the main entry.
The Saturn's satellites Janus and Epimetheus are the first known bodies in the Solar System that have horseshoe orbits in a frame that rotates with uniform angular velocity. The authors model their movement by means of a planar three–body problem with two small similar masses that initially have nearly symmetric conditions. The aim of this paper is to find what kind of motion these bodies can present. The authors analyze the limit case, when both small masses are equal and in symmetric opposite positions, i.e. the isosceles three–body problem. Afterwards, by continuity they derive properties for the initial three–body problem.

042.026 **A method of numerical extending the periodic motions of Hamiltonian systems with respect to parameters.**
S. R. Karimov, A. G. Sokolsky (*A. G. Sokol'skij*).
1. Spain–USSR Workshop on Positional Astronomy and Celestial Mechanics, p. 43 – 48 (1991). – See Abstr. 012.025 for the main entry.
The method proposed is a generalization of the method described by Karimov and Sokol'skij (1989). It is represented in terms of the Hamiltonian formalism with all the advantages arising from it. A complete presentation of the method is given in the paper by Karimov and Sokol'skij (1990).

042.027 **On the determination of orbits of satellites of the distant planets and some requirements to their positional observations.**
Yu. V. Batrakov.
1. Spain–USSR Workshop on Positional Astronomy and Celestial Mechanics, p. 51 – 53 (1991). – See Abstr. 012.025 for the main entry.
The sources of errors of positional observations of the distant planetary satellites are discussed. Some recommendations aimed at increasing the accuracy of the orbits of such satellites are given.

042.028 **On effectiveness of using intermediate orbits for computing the perturbed motion.**
Yu. V. Batrakov, Eh. R. Mirmakhmudov.
1. Spain–USSR Workshop on Positional Astronomy and Celestial Mechanics, p. 71 – 73 (1991). – See Abstr. 012.025 for the main entry.
It is shown that an intermediate orbit having the second order contact with the perturbed trajectory at the initial point is expedient to use when computing the perturbed motion by Encke method.

042.029 **Precise ephemeris of minor planets.**
A. López García, J. López Ortí, R. López Machí, F. Marco.
1. Spain–USSR Workshop on Positional Astronomy and Celestial Mechanics, p. 75 – 81 (1991). – See Abstr. 012.025 for the main entry.
Related to the precise ephemeris of minor planets is the problem of improving the orbital elements by differential correction. In this paper the authors obtain a new set of normal equations for this correction.

042.030 **Present status of a new algorithm for orbital elements improvement.**
A. López Garcia, J. López Ortí, F. Marco.
1. Spain–USSR Workshop on Positional Astronomy and Celestial Mechanics, p. 83 – 93 (1991). – See Abstr. 012.025 for the main entry.
This paper contains the formulae to correct the orbital elements of minor planets and catalogue constants. The authors develop a set of matrix–based formulae for their use in a computer program.

042.031 **The "CELESTA" specialized system for analytical support of local methods of non–linear analysis in celestial mechanics.**
A. G. Sokolsky (*A. G. Sokol'skij*), S. R. Karimov, P. Yu. Salnikov, I. I. Shevchenko, N. N. Vasiliev (*N. N. Vasil'ev*).
1. Spain–USSR Workshop on Positional Astronomy and Celestial Mechanics, p. 143 – 146 (1991). – See Abstr. 012.025 for the main entry.

042.032 **Investigation of the observed field structure of objects from the point of view of integral geometry.**
G. A. Saiyan.
Astron. Zh., Tom 68, Vyp. 5, p. 1036 – 1045 (Sep–Oct 1991). In Russian. English translation in Sov. Astron., Vol. 35, No. 5.

042.033 **Final motions in non-hierarchical triple systems.**
T. A. Agekyan, Zh. P. Anosova.
Astron. Zh., Tom 68, Vyp. 5, p. 1099 – 1103 (Sep–Oct 1991). In Russian. English translation in Sov. Astron., Vol. 35, No. 5.
The scenario of the general class of final motions in triple systems with equal-mass bodies is considered. The computer simulations of the dynamics of triple systems have shown that the escape from the systems takes place, as a rule, after the state of "close triple approach" of the bodies; during this state, one component passes nearly the centre of inertia of the triple and escapes from it. The method of choice of coordinates and velocities of the bodies in this state of a system by means of certain parameters has been developed.

042.034 **On one kind of periodic solutions of the restricted problem of translatory-rotary motion of three bodies.**
L. Yu. Khorseva.
Astron. Zh., Tom 68, Vyp. 5, p. 1111 – 1114 (Sep–Oct 1991). In Russian. English translation in Sov. Astron., Vol. 35, No. 5.
Two families of periodic solutions of the problem of translatory-rotary motion of three rigid bodies are found near rectilinear points of libration by Lyapunov's method.

042.035 **On forming of binaries by close triple approaches with equal- and different-mass bodies.**
J. P. Anosova (*Zh. P. Anosova*), N. O. Kirsanov.
Comments Astrophys., Vol. 15, No. 5, p. 283 – 297 (Sep 1991).
Numerical simulations of 70,000 close triple approaches have been performed. A wide range of the radius r of a triple approach, and both Maxwell and normal distributions of the initial velocities of the bodies are considered. The probabilities P of formation of a binary with several approach parameters are determined. The strong correlation between P and the narrowness of an approach is confirmed. It is shown that P depends actually on the product $r\langle V^2\rangle$. In encounters of three bodies with different masses, mainly the most massive and the least massive bodies form the resulting binary; in this case the binary is considerably wider than if the bodies have equal masses. Eccentricities of resulting binaries as a rule are large: 50% of them are greater than 0.80.

042.036 **On secular perturbations of elements in the restricted parabolic three-body problem.**
A. G. Mamedov.
Astron. Zh., Tom 68, Vyp. 6, p. 1323 – 1327 (Nov–Dec 1991). In Russian. English translation in Sov. Astron., Vol. 35, No. 6.
Analytical formulae are obtained for secular perturbations of the eccentricity and of the pericentre argument in the restricted parabolic three-body problem. The perturbations obtained are compared with the results of numerical integration of the twice-averaged equations.

042.037 **The influence of oblateness on the motion of the apsid line in the three-body problem.**
B. R. Mushailov.
Astron. Zh., Tom 68, Vyp. 6, p. 1328 – 1331 (Nov–Dec 1991). In Russian. English translation in Sov. Astron., Vol. 35, No. 6.
In the framework of the restricted three-body problem with orbital commensurability of lower-order mean motions and with an account for the central body's oblateness, the evolution of the apsid line of a passively gravitating point mass is considered. Secular and periodical terms are selected in the obtained analytical expressions. The influence of the oblateness on the variation of the periodical component and of the angular velocity of the rotation of the apsid lines is analysed.

042.038 **Application of the stroboscopic method to the stellar three-body problem.**
J. F. Ling.
Astrophys. Space Sci., Vol. 185, No. 1, p. 51 – 61 (Nov 1991).
The stellar three-body problem has been approached by directly integrating the equations of motion in the orbital elements. The problem is set up in a barycentric chain with the orbits as perturbed ellipses. The integration was performed using the semi-analytical stroboscopic method, which is particularly useful for solving differential systems that depend on several slow variables and one fast, angular variable. Perturbation theory is applied and the solution for a particular order is obtained by means of successive approximations on the fundamental period of the fast variable.

042.039 **The Lissajous transformation. I. Basics.**
A. Deprit.
Celest. Mech. Dyn. Astron., Vol. 51, No. 3, p. 201 – 225 (1991).
A new canonical transformation is proposed to handle elliptic oscillators, that is, Hamiltonian systems made of two harmonic oscillators in a 1–1 resonance. The Lissajous transformation is defined in two ways: explicitly in terms of Cartesian variables, and implicitly by resolution of a partial differential equation separable in polar variables. Relations between the Lissajous variables, the common harmonic variables, and other sets of variables are discussed in detail.

042.040 **The Lissajous transformation. II. Normalization.**
A. Deprit, A. Elipe.
Celest. Mech. Dyn. Astron., Vol. 51, No. 3, p. 227 – 250 (1991).
Normalization of a perturbed elliptic oscillator, when executed in Lissajous variables, amounts to averaging over the elliptic anomaly. The reduced Lissajous variables constitute a system of cylindrical coordinates over the orbital spheres of constant energy, but the pole-like singularities are removed by reverting to the subjacent Hopf coordinates. The two-parameter coupling that is a polynomial of degree four admitting the symmetries of the square is studied in detail. It is shown that the normalized elliptic oscillator in that case behaves everywhere in the parameter plane like a rigid body in free rotation about a fixed point, and that it passes through butterfly bifurcations wherever its phase flow admits non isolated equilibria.

042.041 **The Lissajous transformation. III. Parametric bifurcations.**
B. R. Miller.
Celest. Mech. Dyn. Astron., Vol. 51, No. 3, p. 251 – 270 (1991).
The author examines a parametric family of cubic perturbed 1–1 resonant harmonic oscillators with an aim to understanding the phase flows of the reduced system. Variation of the parameters leads the system through five bifurcations of different types. Some bifurcations are due to passage through cases of discrete symmetry or integrability. A conjecture correlating degenerate equilibria in reduced systems with integrability is modified and reinforced.

042.042 **The Lissajous transformation. IV. Delaunay and Lissajous variables.**
A. Deprit, C. A. Williams.
Celest. Mech. Dyn. Astron., Vol. 51, No. 3, p. 271 – 280 (1991).
Relations are established between the Delaunay variables defined over a phase space E in four dimensions and the Lissajous variables defined over a four-dimensional phase space F when the latter is mapped onto E by a parabolic canonical transformation.

042.043 **Chaos im Kosmos?**
H.-E. Fröhlich.
Astron. Sch., Jahrg. 28, Heft 5, p. 4 – 6 (Oct 1991).

042.044 **The stability of the planetary triangular Lagrange points.**
S. Mikkola, K. A. Innanen.
IAU Colloquium No. 123: Observatories in Earth orbit and beyond, p. 533 – 536 (1990). – See Abstr. 012.038 for the main entry.
Numerical, self-consistent, n-body integrations of the solar system show significant indications of medium-term (i.e. several million-year) stability for the various planet-Sun L4, L5 configurations. A progress report of computations, emphasizing

the inner solar system, is given. There exist interesting possibilities for these locations (including the Earth) as the sites for longer term scientific applications, both pure and applied.

042.045 **Irregular behaviour and eccentricity increase in the trajectories of a dynamical system modelling asteroidal motions.**
H. Varvoglis.
Earth, Moon, Planets, Vol. 54, No. 3, p. 257 – 267 (Sep 1991).
Many trajectories of the third body are integrated numerically in a modified elliptical restricted three body problem (ERTBP) in which the eccentricity, e', of the orbit of the second primary varies sinusoidally with time. It is found that, in the case of the 2:1 resonance, the introduction of the time variability of e' modifies significantly the behaviour of the trajectories of the third body. In particular, its osculating eccentricity presents the following two notable features: (a) In all cases it shows a definite chaotic variation, which appears at significantly shorter time–scales than the one found by Wisdom in the e' = const. case. (b) In many cases it shows a significant increase, up and beyond the (critical) value $e_{crit} = 0.52$. As a result the third body approaches the first primary at distances smaller than 0.29 α' (where α' denotes the semi–major axis of the trajectory of the second primary around the first), which in the actual Sun–Jupiter–asteroid problem corresponds to the semi–major axis of Mars. The result might be of interest in the context of explaining the Kirkwood gaps at the resonances where the osculating eccentricity of asteroid trajectories calculated in the classical (e' = const.) ERTBP does not reach Mars crosser values.

042.046 **A search for planets in binary star systems: a new type of approach.**
V. Banfi.
Earth, Moon, Planets, Vol. 54, No. 3, p. 269 – 282 (Sep 1991).
A theoretical model, based on particular type of the restricted three–body problem, is here presented in order to demonstrate the existence of a possible planetary motion near the center of mass in a binary star system. The superposition principle is used, with the introduction of two fictitious negative masses in order to simulate the real two primary bodies system.

042.047 **A new light on the L_1, L_2 Lagrangian points.**
J. J. Rawal.
Earth, Moon, Planets, Vol. 54, No. 3, p. 283 – 288 (Sep 1991). Letter–to–the–editor.
The relation between the locations of L_1, L_2 Lagrangian points and the boundary to their respective satellite system is brought forth, in that, the Lagrangian points L_1, L_2 are seen to lie just on the boundary to their respective satellite system.

042.048 **On the representativity of initial conditions of triple systems.**
J. P. Anosova (*Zh. P. Anosova*), A. V. Nebukin.
Astron. Astrophys., Vol. 252, No. 1, p. 410 – 413 (Dec 1991).
The dependence of estimations of the basic characteristics of the dynamics of triple systems on the method of selecting representative samples of initial conditions has been studied by computer simulation. It is shown that the statistical estimation of the lifetime of the triples are virtually independent of the method of the choice of their initial configurations.

042.049 **Symplectic maps for the N–body problem.**
J. Wisdom, M. Holman.
Astron. J., Vol. 102, No. 4, p. 1528 – 1538 (Oct 1991). Current Physics Microform No.: 9110D2060.
The mapping method of Wisdom (1982) is generalized to encompass all gravitational N–body problems with a dominant central mass. The method is used to compute the evolution of the outer planets for a billion years. This calculation provides independent numerical confirmation of the result of Sussman & Wisdom (1988) that the motion of the planet Pluto is chaotic.

042.050 **The computation of orbits in indeterminate and uncertain cases.**
B. G. Marsden.
Astron. J., Vol. 102, No. 4, p. 1539 – 1552 (Oct 1991). Current Physics Microform No.: 9110D2071.
The GEM form of the Gaussian method of three–observation orbit determination is modified in order to obtain physically meaningful results in mathematically indeterminate cases. The modified procedure can enable the user to isolate situations where at least one observation is erroneous. A logical extension of the procedure then provides a mechanism for using only two observations of a minor planet to identify additional data. Though in some respects analogous to the "modified" Väisälä method, the new procedure is more convenient to apply in practice and has distinct advantages in the case of Apollo objects, which in the Väisälä method must unrealistically be presumed at aphelion. It is also shown that the three–observation procedure can readily be extended to incorporate additional observations. Further comments are made on the rather unfortunate manner in which the Gauss method is presented in the standard texts, including the one by Taff. There is a discussion of the relative performance of the GEM method and traditional orbit–determination methods in cases of discrete, nontrivial multiple solutions.

042.051 **Artificial satellites and the three–body problem: a general case.**
D. D. Dionysiou, A. K. Zois, K. G. Iakovidis.
Astrophys. Space Sci., Vol. 186, No. 1, p. 101 – 108 (Dec 1991).
The authors study a planar three–body problem, in which the satellites move along close orbits in a plane which forms an angle θ with the equatorial plane of the planet, the oblateness of which exercises a great effect. The practical importance of this problem arises from its applications. Differential equations of motion are given and particular solutions are shown to exist when the centers of masses are at the vertices of a nearly equilateral triangle or are collinear. If the first two terms of the Legendre series are taken with $\theta = 0$, the same results as by Aksenov (1988) are obtained.

042.052 **The restricted problem of a tri–axial rigid body and two spherical bodies with variable masses.**
S. M. Elshaboury.
Astrophys. Space Sci., Vol. 186, No. 2, p. 245 – 251 (Dec 1991).
The restricted problem of a tri–axial rigid body and two spherical bodies with variable masses is considered. The general solution of the equations of motion of the tri–axial body is obtained in which the motion of the spherical bodies is determined by the classic nonsteady Gyldén–Meshcherskii problem.

042.053 **On the optimal trajectories in gravitational fields allowing approximation by a central linear field.**
A. G. Azizov, N. A. Korshunova.
Kosm. Issled., Tom 29, Vyp. 4, p. 525 – 531 (Jul–Aug 1991). In Russian. English translation in Cosmic Res., Vol. 29, No. 4.
It is shown that for motions, allowing approximation of the gravitational field by a central linear field, the equations of Mayer's variational problem for a point with a limited mass flow rate permit an analytical solution. The case when the ellipse is the hodograph of the basis–vector corresponding to the gradation of zero and maximum thrust sections is selected and analyzed. Results are aaplied to the solution of the problem of change in velocity from local circular to elliptic in a central Newtonian field.

042.054 **Instabilities in first–order resonances.**
M. Šidlichovský.
Bull. Astron. Inst. Czech., Vol. 42, No. 3, p. 190 – 192 (Jun 1991).
The Laplace–Lagrange secular theory applied by Yokoyama et al. to the 3/1 and 2/1 asteroidal resonances is used to obtain formulae for the instability region of the first–order resonance $p = (p + 1)/p$. As this approach yields the instability region even for 3/2 resonance, where Hildas are observed, it is concluded that this calculation in itself cannot explain the origin of Kirkwood

gaps and that more refined methods allowing to calculate the maximum eccentricity have to be applied.

042.055 **A particular solution to the problem of two fixed centres with variable masses.**
P. Andrle.
Bull. Astron. Inst. Czech., Vol. 42, No. 5, p. 304 – 306 (Oct 1991).
It is shown that a particular Meshcherkij's type solution to the problem of two fixed centres can be found. This solution corresponds to the simultaneous escape of a test particle and the change of "oblateness" of the centres.

042.056 **Investigation of the structure of the observed field of an object from the standpoint of integral geometry.**
G. A. Saiyan.
Sov. Astron., Vol. 35, No. 5, p. 518 – 522 (Sep 1991). Current Physics Microform No.: 9203X1692. English translation of Astron. Zh., Tom 68, Vyp. 5, p. 1036 – 1045 (1991).

042.057 **Final motions in nonhierarchical triple systems.**
T. A. Agekyan, Zh. P. Anosova.
Sov. Astron., Vol. 35, No. 5, p. 551 – 553 (Sep 1991). Current Physics Microform No.: 9203X1725. English translation of Astron. Zh., Tom 68, Vyp. 5, p. 1099 – 1103 (1991).
Scenarios in the main class of final motions in triple systems with components of equal mass are considered. A study of the dynamics of triple systems has shown that the breakup of systems begins in most cases after a state of "close triple encounter" among the bodies, in which one of the components undergoes a close encounter with the center of inertia of the system and then leaves it. A method is developed for specifying the positions and velocities of the bodies for states of a system of this type using several parameters. The ranges of values of these parameters within which all points correspond to a situation leading to breakup of the triple system are found. The necessary conditions for the breakup of triple systems are formulated; 1) the triple encounter between the bodies must be close; 2) at the time t_0 of closest approach of the escaping component to the center of inertia of the system, the other two components must be moving apart; the acceleration imparted to the escaping component before this time will then be greater than its subsequent deceleration, and its resultant velocity can become higher than the critical velocity; 3) the directions of motion of the bodies at t_0 must be nearly orthogonal to the straight line connecting them at that time; 4) the closer a component passes to the center of inertia, the closer the approach of the three bodies, and the closer to orthogonal the directions of their motion, the more positive energy is carried off from the triple system in its breakup.

042.058 **One kind of periodic solutions of the restricted problem of translational–rotational motion of three bodies.**
L. Yu. Khorseva.
Sov. Astron., Vol. 35, No. 5, p. 558 – 559 (Sep 1991). Current Physics Microform No.: 9203X1732. English translation of Astron. Zh., Tom 68, Vyp. 5, p. 1111 – 1114 (1991).
Two families of periodic solutions of the problem of translational–rotational motion of three rigid bodies are found in the vicinity of collinear libration points by Lyapunov's method.

042.059 **The collinear central configuration of *n* equal masses.**
G. Buck.
Celest. Mech. Dyn. Astron., Vol. 51, No. 4, p. 305 – 317 (1991).
Moulton's Theorem says that given an ordering of masses $m_1,...,m_n$ there exists a unique collinear central configuration. The theorem allows one to ask the questions: What is the distribution of n equal masses in the collinear central configuration? What is the behavior of the distribution as $n \rightarrow \infty$? Central configurations are found to be attracting fixed points of a flow. This flow is studied in an effort to characterize the mass distribution. Specifically, for a collinear central configuration of n equal masses, a bound is found for the position of the masses furthest from the center of mass. Also some facts concerning the distribution of the inner masses are discovered.

042.060 **Solving Kepler's equation with high efficiency and accuracy.**
A. Nijenhuis.
Celest. Mech. Dyn. Astron., Vol. 51, No. 4, p. 319 – 330 (1991).
The author presents a method for solving Kepler's equation for elliptical orbits that represents a gain in efficiency and accuracy compared with those currently in use. The gain is obtained through a starter algorithm which uses Mikkola's ideas in a critical range, and less costly methods elsewhere. A higher–order Newton method is used thereafter. The author's method requires two trigonometric evaluations.

042.061 **Periodic motion around the triangular equilibrium points of the photogravitational restricted problem of three bodies.**
C. G. Zagouras.
Celest. Mech. Dyn. Astron., Vol. 51, No. 4, p. 331 – 348 (1991).
In this article the effect of radiation pressure on the periodic motion of small particles in the vicinity of the triangular equilibrium points of the restricted three body problem is examined. Second order parametric expansions are constructed and the families of periodic orbits are determined numerically for two sets of values of the mass and radiation parameters corresponding to the non–resonant and the resonant case. The stability of each orbit is also studied.

042.062 **Three–dimensional bifurcations of periodic solutions around the triangular equilibrium points of the restricted three–body problem.**
E. Perdios, C. G. Zagouras, O. Ragos.
Celest. Mech. Dyn. Astron., Vol. 51, No. 4, p. 349 – 362 (1991).
Vertically critical, planar periodic solutions around the triangular equilibrium points of the restricted three–body problem are found to exist for values of the mass parameter in the interval [0.03,0.5]. Four series of such solutions are completed. The families of three–dimensional periodic solutions that branch off these critical orbits are computed for $\mu = 0.3$ and are continued till their end. All orbits of these families are unstable.

042.063 **Generalized Hill's problem: some cases of complete integrability.**
B. Chauvineau.
Celest. Mech. Dyn. Astron., Vol. 51, No. 4, p. 363 – 377 (1991).
The author is investigating cases of integrability in the planar Hill's problem. The external potential U_{ext} is supposed to be time independent in a given uniformly rotating frame. Cases of integrability of the relative motion of two interacting particles in the vicinity of an equilibrium solution of U_{ext} are found. In all these cases, the form of the second integral is explicitly given, the first being the Jacobian one. Cases in which the interacting potential U between the two particles is of newtonian type are particularized.

042.064 **A numerical investigation of the one–dimensional newtonian three–body problem. III. Mass dependence in the stability of motion.**
S. Mikkola, J. Hietarinta.
Celest. Mech. Dyn. Astron., Vol. 51, No. 4, p. 379 – 394 (1991).
The one–dimensional Newtonian three–body problem is known to have stable (quasi–)periodic orbits when the masses are equal. The existence and size of the stable region is discussed here in the case where the three masses are arbitrary. The authors consider only the stability of the periodic (generalized) Schubart's (1956) orbit. If this orbit is linearly stable it is almost always surrounded by a region of stable quasi–periodic orbits and the size and shape of this stable region depends on the masses. The three–dimensional linear stability of the periodic orbits is also determined. Final results show that the region of stability has a complicated shape and some of the stable regions in the mass–plane are quite narrow. The non–linear three–dimensional stability is studied independently by extensive numerical integrations and the results are found to be in agreement with the linear stability analysis. The boundaries of stable region in the mass–

plane are given in terms of polynominal approximations. The results are compared with a similar work by Hénon (1977).

042.065 **The almost constant–speed two–body problem with resistance.**
A. G. Mavraganis.
Celest. Mech. Dyn. Astron., Vol. 51, No. 4, p. 395 – 405 (1991).
The almost constant–speed motion of a mass acted upon by a Newtonian attraction and a resisting force is treated. The equation of orbit is derived for a specific type of resistance which covers the familiar case of Danby's drag (αr^{-2}) whilst the vector invariants are obtained by direct operation on the vector form of the equation of motion.

042.066 **Mean values of particular functions in the elliptic motion.**
G. Metris.
Celest. Mech. Dyn. Astron., Vol. 52, No. 1, p. 79 – 84 (1991).

042.067 **Second–species solutions in the circular and elliptic restricted three–body problem. I. Existence and asymptotic approximation.**
G. Gomez, M. Olle.
Celest. Mech. Dyn. Astron., Vol. 52, No. 2, p. 107 – 146 (1991).
An analytical proof of the existence of some kinds of periodic orbits of second species of Poincaré, both in the circular and elliptic restricted three–body problem, is given for small values of the mass parameter. The proof uses the asymptotic approximations for the solutions and the matching theory developed by Breakwell and Perko. In the paper their results are extended to the elliptic problem and applied to prove the existence of second–species solutions generated by rectilinear ellipses in the circular problem and nearly–rectilinear ones in the elliptic case.

042.068 **Second–species solutions in the circular and elliptic restricted three–body problem. II. Numerical explorations.**
G. Gomez, M. Olle.
Celest. Mech. Dyn. Astron., Vol. 52, No. 2, p. 147 – 166 (1991).
The characteristic curves of symmetric periodic second–species solutions in the circular and elliptic RTBP are given for small $\mu > 0$. The behaviour in the neighbourhood of the bifurcation orbits, in the circular case, is described.

042.069 **A general timing conduction for consecutive collision orbits in the limiting case $\mu = 0$ of the elliptic restricted problem.**
K. C. Howell, S. M. Marsh.
Celest. Mech. Dyn. Astron., Vol. 52, No. 2, p. 167 – 194 (1991).
Consecutive collision orbits in the limiting case $\mu = 0$ of the elliptic restricted three–body problem are investigated, in particular those in which the infinitesimal mass collides twice with the smaller (massless) primary. A timing condition is presented that allows the extension of previous results to the case of arbitrary relative orientation of the orbits of the infinitesimal mass and the smaller primary. The timing condition is expressed in two general forms – in terms of orbit parameters and eccentric (or hyperbolic) anomalies at the times of collision – for the specific cases of elliptic, parabolic or hyperbolic orbits of the infinitesimal mass. Some families of solutions are presented.

042.070 **About stability of libration points in the restricted photogravitational three body problem.**
K. Goździewski, A. J. Maciejewski, Z. Niedzielska.
Celest. Mech. Dyn. Astron., Vol. 52, No. 2, p. 195 – 201 (1991).
Nonlinear stability of the triangular libration point in the photogravitational restricted three body problem was investigated in the whole range of the parameters. Some results obtained earlier are corrected. The method for proper determination of cases when stability cannot be determined by four order terms of the hamiltonian was proposed.

042.071 **Kepler's Equation and Newton's method.**
P. Colwell.
Celest. Mech. Dyn. Astron., Vol. 52, No. 2, p. 203 – 204 (1991).
Power series solutions of Kepler's Equation associated with Lagrange, Levi–Civita and Stumpff may be obtained by Newton's method as an algorithm for formal power series.

042.072 **On the origin of the planetary spin by accretion of planetesimals. II. Collisional orbits at the Hill surface.**
K. Tanikawa, N. Kikuchi, I. Sato.
Icarus, Vol. 94, No. 1, p. 112 – 125 (Nov 1991).
The structure of the totality of orbits starting at the Hill surface is investigated numerically and systematically in the framework of the planar–restricted three–body problem. The authors have obtained the following results. (1) The qualitative features of the totality of orbits at the Hill surface are clarified. (2) A general formula for the collisional probability as a function of the mass ratio μ, radius R, and Jacobi constant C is obtained for $\mu \leqslant 10^{-4}$, $2 \times 10^{-5} \leqslant R < 2 \times 10^{-4}$, and the range of C corresponding to $\Gamma \leqslant 2$, where Γ is the Jacobi constant for the Hill problem of the restricted three–body problem. The collisional probability is also given as a function of the velocity at the Hill surface. (3) The relative collisional probability as a function of (a, e) is obtained for $0.86 \leqslant a \leqslant 1.14$ and $0.0 \leqslant e \leqslant 0.1$.

042.073 **The influence of the orbital eccentricity of a primary on the chaotic region.**
Liu Jie.
Chin. Astron. Astrophys., Vol. 15, No. 4, p. 446 – 448 (Dec 1991). English translation of Acta Astron. Sin., Vol. 32, No.2, p. 222 – 224 (Jun 1991).
Applying the resonance overlap theory to the elliptic three–body model, it is shown that the effect of the eccentricity is to increase the size of the chaotic region.

042.074 **Some discussions about the ideal resonance.**
Zhao Changyin, Liu Lin.
Acta Astron. Sin., Vol. 32, No. 3, p. 268 – 276 (Sep 1991). In Chinese.
The analytical expression of the libration period of the ideal resonance in the neighborhood of an equilibrium point is given. It shows that the libration period is independent of the amplitude in the deep resonant region. The estimated expression of the libration width is also derived with a simple method. By numerical calculations with several resonant models, these two expressions have been verified.

042.075 **The chaotic region of an encounter–type orbit.**
Liu Jie, Sun Yisui.
Acta Astron. Sin., Vol. 32, No. 3, p. 277 – 287 (Sep 1991). In Chinese.
Under the assumption that in the restricted three–body problem the body 3 has small inclination and small eccentricity, the authors discuss the change of its orbital elements after an encounter with the body 2 which moves in an elliptical orbit. They also estimate the possible chaotic region and find that the eccentricity of the body 2 will enlarge the region.

042.076 **Erratum: "Capture probabilities for secondary resonances" [Icarus, Vol. 87, No. 2, p. 249 – 264 (Oct 1990)].**
R. Malhotra.
Icarus, Vol. 94, No. 2, p. 524 – 525 (Dec 1991). See Abstr. 52.042.063.

042.077 **Secular perturbations of elements in the restricted, parabolic three–body problem.**
A. G. Mamedov.
Sov. Astron., Vol. 35, No. 6, p. 664 – 666 (Nov 1991). Current Physics Microform No.: 9205X1474. English translation of Astron. Zh., Tom 68, Vyp. 6, p. 1323 – 1327 (1991).
Analytic equations are obtained for secular perturbations of the eccentricity and the argument of the pericenter in the restricted, parabolic three–body problem. Those perturbations

are compared with the results of numerical integration of twice-averaged equations.

042.078 **Influence of oblateness of the central body on motion of the line of apsides in the three-body problem.**
B. R. Mushailov.
Sov. Astron., Vol. 35, No. 6, p. 667 – 669 (Nov 1991). Current Physics Microform No.: 9205X1477. English translation of Astron. Zh., Tom 68, Vyp. 6, p. 1328 – 1331 (1991).
The evolution of the line of apsides of a passively gravitating particle is considered in the restricted three-body problem in the case of low-order orbital commensurability of mean motions and with allowance for the oblateness of the central body. The secular and periodic components are identified in the resulting analytic expressions. The influence of oblateness on the variation of the periodic component and the angular rotation rate of the line of apsides is analyzed.

042.079 **Relativistic effects resulting from the rotation of celestial bodies.**
Han Tao.
Publ. Purple Mt. Obs., Vol. 10, No. 4, p. 276 – 287 (Dec 1991). In Chinese.
In the PPN formalism, the first-order closed analytical solutions of rotating celestial bodies on the improvement of instantaneous orbital elements are obtained in the solar system. The semimajor axis and eccentricity are not affected by the secular relativistic effects, but are affected by the periodic relativistic effects. The inclination, longitude of ascending node, longitude of periastron and mean longitude of epoch are affected by both the secular and periodic relativistic effects. In two gravitational theories, the obtained results are applied to calculate the relativistic effects of the solar rotation on the motions of planets and minor planets, of the planetary rotation on the motions of natural satellites and of the Earth rotation on the motions of artificial satellites.

042.080 **Equations of motion of elliptic restricted problem of three bodies with variable mass.**
R. K. Das, A. K. Shrivastava, B. Ishwar.
Bull. Astron. Soc. India, Vol. 19, Nos. 1 – 2, p. 95 (Mar – Jun 1991). Abstract. – See Abstr. 012.108 for the main entry.

042.081 **Intermediary variable for expansions of functions in elliptic motion with great eccentricity.**
I. Stellmacher.
C. R. Acad. Sci., Sér. II, Tome 313, No. 1, p. 31 – 34 (4 Jul 1991).
On propose l'utilisation d'une variable intermédiaire à partir de laquelle les développements des fonctions des coordonnées du problème képlerien des deux corps, convergent uniformément pour les valeurs de l'excentricité supérieure à celle de Laplace.

042.082 **On the periodic solutions of a family of Hills equations.**
A. Makhlouf.
C. R. Acad. Sci., Sér. II, Tome 313, No. 2, p. 183 – 186 (18 Jul 1991).

042.083 **Expansions of functions of elliptic motion with large eccentricities.**
I. Stellmacher.
C. R. Acad. Sci., Sér. II, Tome 313, No. 12, p. 1403 – 1407 (5 Dec 1991).
The author presents a new simple algorithm under the form of Fourier series for the development of functions of coordinates in the Keplerian two-body problem, valid for any value of the eccentricity.

042.084 **Le problème des deux corps revisité.**
L. Duriez.
13. Ecole de Printemps d'Astrophysique de Goutelas: Les méthodes modernes de la mécanique céleste, p. 9 – 34 (1990). – See Abstr. 012.118 for the main entry.
Contents: 1. Introduction. 2. Formulation vectorielle du mouvement képlérien. 3. Eléments d'orbite. 4. Formulation hamiltonienne du mouvement képlérien. 5. Eléments osculateurs et équations du mouvement képlérien perturbé. 6. Conclusion.

042.085 **Le développement de la fonction perturbatrice.**
L. Duriez.
13. Ecole de Printemps d'Astrophysique de Goutelas: Les méthodes modernes de la mécanique céleste, p. 35 – 62 (1990). – See Abstr. 012.118 for the main entry.
Contents: 1. Introduction. 2. Développements du mouvement képlérien elliptique. 3. Développements de l'inverse de la distance. 4. Développement du potentiel de gravitation d'un corps non sphérique. 5. Remarques sur les propriétés de d'Alembert et de parité des séries. 6. Conclusion.

042.086 **Systèmes de variables et éléments.**
J. Laskar.
13. Ecole de Printemps d'Astrophysique de Goutelas: Les méthodes modernes de la mécanique céleste, p. 63 – 87 (1990). – See Abstr. 012.118 for the main entry.
Contents: 0. Introduction. 1. Repère barycentrique. 2. Repère héliocentrique. 3. Coordonnées de Jacobi. 4. Variables elliptiques de Poincaré. 5. Réduction du problème des N corps. 6. Conclusion.

042.087 **Perturbation methods in celestial mechanics.**
A. Milani.
13. Ecole de Printemps d'Astrophysique de Goutelas: Les méthodes modernes de la mécanique céleste, p. 109 – 150 (1990). – See Abstr. 012.118 for the main entry.
Contents: 1. Canonical transformations. 2. Small transformations, small perturbations. 3. Invariant tori. 4. Proper elements.

042.088 **Averaging Hamiltonian systems.**
S. Ferraz-Mello.
13. Ecole de Printemps d'Astrophysique de Goutelas: Les méthodes modernes de la mécanique céleste, p. 151 – 211 (1990). – See Abstr. 012.118 for the main entry.
Contents: 1. The theory of perturbations. 2. Hamiltonian mechanics. 3. The Hamilton-Jacobi theory. 4. Hori's averaging theory. 5. Geometry of resonances. 6. Averaging the elliptic asteroidal problem.

042.089 **1. Action-angle variables. 2. Semi-numerical perturbation method.**
J. Henrard.
13. Ecole de Printemps d'Astrophysique de Goutelas: Les méthodes modernes de la mécanique céleste, p. 213 – 247 (1990). – See Abstr. 012.118 for the main entry.
I. Action-angle variables.
1. One-degree of freedom. 2. The adiabatic invariant. 3. Saddle connection. 4. Two degrees of freedom separable systems.
II. Semi-numerical perturbation method.
1. First order perturbation theory. 2. Semi-numerical approach. 3. An application: resonance in the elliptic restricted problem. 4. Some numerical results.

042.090 **New insight on the stability problem from recent results in classical perturbation theory.**
A. Giorgilli.
13. Ecole de Printemps d'Astrophysique de Goutelas: Les méthodes modernes de la mécanique céleste, p. 249 – 284 (1990). – See Abstr. 012.118 for the main entry.
Contents: 1. Overview. 2. Integrable systems. 3. The theorem of Poincaré and the Arnold diffusion. 4. A simple proof of the theorem of Nekhoroshev.

042.091 **On a family of Hill's equations in the complex field.**
A. Makhlouf, F. Nahon.
Celest. Mech. Dyn. Astron., Vol. 52, No. 3, p. 277 – 291 (1991).
The variational equation of the periodic solution in an homogeneous potential of degree $2m$ leads to a family of Hill's equations. H. Yoshida solves these equations by a transformation to the Gauss equation of hypergeometric functions. The authors study directly this family for complex t, and in the case $m = 2$ they give a detailed description of critical solutions.

042.092 **Secular perturbations of fictitious satellites of Uranus.**
H. Kinoshita, H. Nakai.
Celest. Mech. Dyn. Astron., Vol. 52, No. 3, p. 293 – 303 (1991).
Secular perturbations of fictitious satellites that are initially circular and in the equatorial plane of Uranus are discussed. Satellites located in the region where the solar perturbation is dominant become highly eccentric and inclined with respect to the equator, and have a possibility to collide with Uranus. Satellites located in the region where the oblateness perturbation is dominant keep the original eccentricity and the inclination. A scenario of a possible extinction of outer satellites of Uranus is also discussed.

042.093 **Secular resonances from 2 to 50 AU.**
Z. Knežević, A. Milani, P. Farinella, C. Froeschlé, C. Froeschlé.
Icarus, Vol. 93, No. 2, p. 316 – 330 (Oct 1991).
By means of a new algorithm which generalizes the second–order and fourth–degree secular perturbation theory of Milani and Knežević, the authors find in the a–e–I proper elements space the location of secular resonances between the precession rates of the longitudes of perihelion and node of a minor body and the corresponding eigenfrequencies of the secular perturbations of the four outer planets.

042.094 **Expériences numériques et mécanique céleste.**
C. Froeschlé.
13. Ecole de Printemps d'Astrophysique de Goutelas: Les méthodes modernes de la mécanique céleste, p. 331 – 367 (1990). – See Abstr. 012.118 for the main entry.
Contents: (A) Surfaces de sections – applications de Poincaré. 1. Introduction. 2. La boite à outils des systèmes dynamiques. 3. Système à un degré de liberté. 4. Deux degrés de liberté. 5. Systèmes à plus de deux degrés de liberté.
(B) Les exposants caractéristiques de Lyapunov: une mesure du chaos.
1. Introduction. 2. Les exposants caractéristiques de Lyapunov. 3. Techniques numériques de calcul des ECLs. 4. Applications.

042.095 **Les résonances séculaires dans la Ceinture des Astéroids.**
H. Scholl, C. Froeschlé.
13. Ecole de Printemps d'Astrophysique de Goutelas: Les méthodes modernes de la mécanique céleste, p. 369 – 383 (1990). – See Abstr. 012.118 for the main entry.
Contents: 1. Introduction. 2. Les théories. 3. Etudes numériques.

042.096 **Intégrabilité du problème restreint elliptique.**
R. Dvorak.
13. Ecole de Printemps d'Astrophysique de Goutelas: Les méthodes modernes de la mécanique céleste, p. 385 – 403 (1990). – See Abstr. 012.118 for the main entry.
Contents: 1. Introduction. 2. Le problème restreint. 3. Une deuxième intégrale du problème restreint circulaire. 4. Le problème restreint elliptique. 5. Résultats qualitatifs et régions des mouvements. 6. Remarque finale.

Astronomical algorithms.
See Abstr. 003.020.

Methods of orbit determination for the microcomputer.
See Abstr. 003.021.

First Spain–USSR Workshop on Positional Astronomy and Celestial Mechanics. Proceedings.
See Abstr. 012.025.

Modern methods in celestial mechanics. Proceedings.
See Abstr. 012.118.

Soviet programs "Jupiter" and "Ceres" for improving orbits of bodies of solar system.
See Abstr. 013.051.

Tides in a space colony.
See Abstr. 015.021.

Problèmes de convergence d'algorithmes dans la détermination de trajectoires perturbées.
See Abstr. 021.034.

On the Lehmann–Filhés theorem as a consequence of one theorem of Ostrogradsky.
See Abstr. 021.038.

On the breakdown point of *M*–estimators with the convex object functions.
See Abstr. 021.044.

Manipulation des séries.
See Abstr. 021.053.

Symplectic integrators for long–term integrations in celestial mechanics.
See Abstr. 021.054.

On the analytical and numerical approximation of invariant manifolds.
See Abstr. 021.055.

The Connection Machine: principle and application in astronomy and astrophysics.
See Abstr. 021.056.

Mouvement du solide.
See Abstr. 022.172.

Umrechnung B1950.0 auf J2000.0.
See Abstr. 041.003.

The motion of a symmetric gyrostat around the center of mass in a circular orbit in the presence of viscoelastic bars and disc.
See Abstr. 052.001.

A complex analytic solution for the attitude motion of a near–symmetric rigid body under body–fixed torques.
See Abstr. 052.005.

The motion of artificial satellites in the set of Eulerian redundant parameters.
See Abstr. 052.008.

An analytic solution for Deprit's radial intermediary with drag in the equatorial case.
See Abstr. 052.011.

Lifetime of the ROHINI A satellite.
See Abstr. 052.013.

Perturbed motion of a satellite in the atmosphere based on the total density model.
See Abstr. 052.014.

Exact linearization of non–planar intermediary orbits in the satellite theory.
See Abstr. 052.015.

Analytical treatment of air drag and Earth oblateness effects upon an artificial satellite.
See Abstr. 052.016.

Motion of artificial satellites in the set of Eulerian redundant parameters. II.
See Abstr. 052.017.

A note on generalized lumped coefficients.
See Abstr. 052.022.

A note on classification of harmonic geopotential terms.
See Abstr. 052.023.

Drag perturbations of low eccentric orbits in the Martian atmosphere.
See Abstr. 052.024.

Towards a systematic definition of intermediaries in the theory of artificial satellites.
See Abstr. 052.025.

Automated design of gravity–assist trajectories to Mars and the outer planets.
See Abstr. 052.033.

Aerodynamic drag perturbations in artificial satellite nodal period.
See Abstr. 052.035.

Testing Einstein's theory of gravity by analyzing Lunar Laser Ranging data.
See Abstr. 066.208.

Formation of sharp ringlets by inelastic collisions.
See Abstr. 091.009.

On the existence of "stable and unstable" quantum levels in the gravitational field of the Sun.
See Abstr. 091.014.

Absence of "stable and unstable" quantum levels in the solar gravitational field.
See Abstr. 091.029.

Representation of numerical models of planetary motions by means of simple analytical theories.
See Abstr. 091.043.

Some properties of barycentric motion of the major planets and the Sun.
See Abstr. 091.044.

Significant high number commensurabilities in the main lunar problem. I. The Saros as a near–periodicity of the Moon's orbit.
See Abstr. 094.052.

On the ephemerides of occultations.
See Abstr. 096.002.

Orbits of Phobos and Deimos determined with consideration for "Phobos–2" on–board observations.
See Abstr. 097.011.

Motions of asteroids at the Kirkwood gaps. II. On the 5:2, 7:3, and 2:1 resonances with Jupiter.
See Abstr. 098.018.

Complex dynamical behaviour of the asteroid 2335 James associated with the secular resonances ν_5 and ν_{16}: numerical studies and theoretical interpretation.
See Abstr. 098.034.

The orbit of 203 Pompeja and the mass of Ceres.
See Abstr. 098.035.

On determination of the asteroid orbits and the systematic errors of fundamental star catalogues in presence of small perturbations.
See Abstr. 098.046.

The wavelet transform: a new tool for asteroid family determination.
See Abstr. 098.047.

Automated computation of general perturbations for the Flora asteroidal group.
See Abstr. 098.087.

The lifetime of binary asteroids vs gravitational encounters and collisions.
See Abstr. 098.098.

Saturn's position derived from meridian circle observations of Titan and Iapetus.
See Abstr. 100.004.

Angular momentum and tidal evolution of Saturn's system.
See Abstr. 100.013.

The orbits of Triton and Nereid from spacecraft and Earthbased observations.
See Abstr. 101.007.

Semi–analytical theory for the motion of Uranus' satellites.
See Abstr. 101.012.

Dynamics of comets in the outer planetary region. I. A numerical experiment.
See Abstr. 102.011.

Nongravitational effects in cometary motion.
See Abstr. 102.013.

On the possible cometary origin of Geminid type meteor streams.
See Abstr. 104.048.

The β Pictoris protoplanetary system. XII. Planetary perturbations in the disk and star–grazing bodies.
See Abstr. 112.003.

Effects of dynamic tides on the rate of apsidal motion in close binary systems.
See Abstr. 117.043.

On the nature of the anomalously slow apsidal motion of DI Herculis.
See Abstr. 119.005.

043 Astronomical Constants, Reference Systems

043.001 **Determining the relative orientations of coordinate systems in the optical and radio bands from nonsynchronous observations of high–orbit satellites.**
D. P. Duma, Yu. N. Ivashchenko.
Kinematics Phys. Celest. Bodies, Vol. 7, No. 1, p. 1 – 10 (1991). English translation of Kinematika Fiz. Nebesn. Tel, Tom 7, No. 1, p. 3 – 12 (1991).

Two methods of establishing a relationship between coordinate systems in the optical and radio bands on the basis of nonsynchronous position observations of high–orbit satellites are discussed. In the first procedure, a system of combinations of corrections to the elements of the satellite orbit is introduced and used for direct determination of the rotation angles of the axes of one system with respect to the other. In the second procedure, the relative–orientation elements of these coordinate systems are found from the differences between normal places obtained from radiointerferometric and optical observations. It is shown on the basis of model calculations and experience in the observation of geostationary satellites that the relative–orientation angles can be obtained at the level of accuracy with which the fundamental stellar coordinate system is constructed.

043.002 **Mutual orientation of the FK5 and VLBI reference frames using 251 radio sources and radio stars.**
V. V. Tel'nyuk–Adamchuk.
Astron. Tsirk., No. 1549, p. 27 – 28 (May – Aug 1991). In Russian.

Radio and optical positions of 203 compact extragalactic radio/optical sources and 65 radiostars have been compiled. The set of $\Delta\alpha$ and $\Delta\delta$ have been used to determine the mutual orientation angles between VLBI and FK5 reference frames. To fit the FK5 and radio frame together FK5 should be turned around X, Y, and Z axes clockwise by +0.00, +0.07, and 0.03 arcsec correspondingly. It is found that the FK5 equator is displaced by 0.03 arcsec to the South from that of VLBI.

043.003 **Determination of the angles of mutual orientation of radio and optical reference systems using radioastronomical and optical star positions.**
V. V. Tel'nyuk–Adamchuk, D. P. Duma, A. G. Svidunovich.
Kinematika Fiz. Nebesn. Tel, Tom 7, No. 4, p. 81 – 87 (Jul–Aug 1991). In Ukrainian. English translation in Kinematics Phys. Celest. Bodies, Vol. 7, No. 4.

Radiointerferometric and optical observations of radio stars have been collected. The number of stars with radiointerferometric coordinates is 60, but only 41 of them have been observed after 1980. There are large differences between radio and optical positions for three stars which are excluded from discussion. The positions and proper motions are obtained using collected meridian and photographic observations. Differences have been obtained in α and δ (in the sense "radio minus optical" positions) used for determination the angles of mutual orientation of these systems after reduction of the optical observations to the epoch of radio observations. The estimates of turning angles i_1, i_2, i_3 relative to the axes X, Y, Z are equal to –0.00, –0.07, –0.00, respectively, the m.s.e. of their calculation being equal to 0.02″. The angles have been also obtained using the results of observations (in the FK5 system) with Carlsberg automatic meridian circle and photoelectric instrument at Bordeaux. By accumulation of radiointerferometric observations for 200 radio stars it will be possible to study the mutual orientation of radio and optical systems and regional errors of the latter with error about 0.005″.

043.004 **Global and regional astrometric comparison between two dense VLBI extragalactic reference frames.**
E. F. Arias, J.-F. Lestrade.
Astron. Astrophys., Vol. 248, No. 1, p. 297 – 300 (Aug 1991).

An astrometric comparison of the two densest VLBI celestial reference frames produced by the Jet Propulsion Laboratory and by the Goddard Space Flight Centre is presented. The comparison indicates that there are global rotations of 0.″0015 between these two celestial frames which were produced with independent data and analysis. This comparison reveals also that there are regional deformations $\leqslant$0.″002 in the equatorial band of declination (–30°,30°) whose origins are presently difficult to identify.

043.005 **The reference frames in the data reduction of Hipparcos satellite.**
Ling Zhaofen.
Chin. J. Space Sci., Vol. 11, No. 4, p. 308 – 316 (Oct 1991). In Chinese.

The heliotropic, great–circle and Hipparcos telescope reference frames are explicitly defined in the light of the scanning law of the Hipparcos satellite. The differences between the reference frames adopted by the two data reduction consortia, FAST and NDAC, are presented. The abscissas and ordinates of 50 stars in the FAST and NDAC great–circle reference frame are simulated and the reason for using two data reduction consortia in ESA are discussed.

043.006 **HIPPARCOS–VLBI: the way towards an ideal primary celestial system.**
E. F. Arias.
Bol. Asoc. Argent. Astron., No. 36, p. 449 – 454 (1990). Abstract. In Spanish and English. – See Abstr. 012.095 for the main entry.

043.007 **Fundamental constants of astronomy of the year 1976.**
A. I. Nefed'eva, A. A. Krasikova.
Izv. Astron. Obs. Ehngel'gardt., Vyp. 55, p. 3 – 48 (1990). In Russian.

Contents: 1. The system of astronomical constants of the year 1896. 2. The system of fundamental astronomical constants of the year 1964. 3. The system of fundamental astronomical constants of the IAU of the year 1976. 4. The gravitational constant in astronomy. 5. Main constants. 6. Derived constants.

043.008 **Sur un modèle de référence terrestre commun à la géodésie et à la géophysique.**
C. Denis, A. I. Denis-Karafistan, C. Delire.
Journées 1991: Systèmes de référence spatio–temporels, p. 116 – 121 (1991). In French. – See Abstr. 012.048 for the main entry.

043.009 **Analytical algorithms of relativistic reduction of astronomical observations.**
V. A. Brumberg, P. Bretagnon, G. Francou.
Journées 1991: Systèmes de référence spatio–temporels, p. 141 – 148 (1991). – See Abstr. 012.048 for the main entry.

Using the analytical planetary theories VSOP87 (Bretagnon and Francou, 1988) and the relativistic theory of astronomical reference systems of Brumberg and Kopejkin (1989) the authors have derived the analytical expressions of the relativistic quantities enabling one to set the relationships between (1) TCB and TCG, (2) barycentric spatial coordinates and geocentric spatial coordinates and (3) observer's proper time and TCG.

043.010 **Une nouvelle méthode pour construire un repère céleste extragalactique plus exact.**
E. F. Arias, M. S. De Biasi.
Journées 1991: Systèmes de référence spatio–temporels, p. 177 – 179 (1991). – See Abstr. 012.048 for the main entry.

In the classical algorithm of analysis and resolution of VLBI observations, the astrometric and the geodetic parameters are simultaneously adjusted. Adopting a new strategy of observation which consists on ideally observing a pair of radio sources simultaneously with an interferometer, one can construct the celestial reference frame independently of the orientation of the Earth.

043.011 **The IERS extragalactic Celestial Reference Frame and its tie with HIPPARCOS.**
E. F. Arias, M. Feissel, J.-F. Lestrade.
IERS Tech. Note, No. 7, p. 1 – 73 (Dec 1991).
This Technical Note gives a presentation of analyses of VLBI extragalactic reference frames performed in the framework of IERS, in the course of the maintenance of its conventional celestial frame. In section 1 the definition of the IERS celestial reference system and its relationship with the 1991 IAU Resolution A4 on Reference Systems are described. Section 2 is devoted to the comparison of celestial frames obtained from independent observation programs by GSFC, JPL, NGS and USNO. Section 3 gives an estimation of the expected accuracy of the link of the HIPPARCOS frame to an extragalactic frame and its time evolution based on VLBI observations of radio stars.

043.012 **ITRF 90 and other realizations of the IERS Terrestrial Reference System for 1990.**
C. Boucher, Z. Altamimi.
IERS Tech. Note, No. 9, p. 1 – 143 (Dec 1991).
This report contains a detailed description of the IERS products related to the Terrestrial Reference System for the period of the IERS annual report for 1990. This description intends to be the continuation of the previous reports: the IERS Tech. Note, No. 4 (52.043.005) and No. 6 (53.043.053). In particular, methodological presentation is followed.

043.013 **Geocentric quasi–Fermi and geocentric harmonic coordinate systems.**
Han Chunhao, Huang Tianyi, Xu Bangxin.
Sci. China, Ser. A, Vol. 34, No. 7, p. 848 – 856 (Jul 1991).
This paper proves the transformation between the geocentric quasi–Fermi and the geocentric harmonic coordinate systems and points out that the difference between the two coordinate systems is negligible in practical applications.

043.014 **Transformation from FK4 system to FK5 system.**
Huang Tianyi, Xia Yifei.
Acta Astron. Sin., Vol. 32, No. 3, p. 258 – 267 (Sep 1991). In Chinese.
The different procedures of the transformation from the FK4 to the FK5 are reviewed, the application of the equinox correction and the selection of the transfer epoch are discussed, and some numerical examples for J2000.0 positions and proper motions are given.

043.015 **On the celestial reference frame independent of the dynamical planes of the Earth.**
Zhao Ming.
Acta Astron. Sin., Vol. 32, No. 3, p. 297 – 303 (Sep 1991). In Chinese. English translation in Chin. Astron. Astrophys., Vol. 16, No. 1, p. 103 – 108 (Jan – Mar 1992).
A celestial coordinate triad can be defined by an arbitrary group of objects (stars or radio sources). Referring to it, a Directly Defined Celestial Reference Frame (DCRF) can be constructed. In this paper, the author proves that the DCRF can be realized by means of ground–based techniques such as VLBI, meridian circle and so on, as well as the HIPPARCOS astrometry satellite. The DCRF can be applied to astrometry and geodesy. Such a frame is uniform and permanent. Therefore, he suggests that the traditional CCRF (Conventional Celestial Reference Frames) should be replaced by the DCRF.

043.016 **The celestial reference system of the International Earth Rotation Service (IERS).**
E. F. Arias.
Bol. Asoc. Argent. Astron., No. 36, p. 435 – 440 (1990). Abstract. In Spanish and English. – See Abstr. 012.095 for the main entry.

043.017 **Analysis of consistency between extragalactic celestial reference frames elaborated with technique of VLBI.**
E. F. Arias.
Bol. Asoc. Argent. Astron., No. 36, p. 440 – 446 (1990). Abstract. In Spanish and English. – See Abstr. 012.095 for the main entry.

The United States Naval astronomical expedition (1849 – 52) for the solar parallax.
See Abstr. 004.080.

CAS Symposium No. 106 and No. 107: Connections of reference systems and techniques of photographic astrometry, Yichang (China), 16 – 19 Nov 1990.
See Abstr. 011.009.

Systèmes de référence spatio–temporels: métrologie et astrométrie.
See Abstr. 012.048.

Problems of astrometry and space geodynamics. Proceedings.
See Abstr. 012.049.

Zur Veranschaulichung astronomischer Größen.
See Abstr. 014.046.

Procedures for validating Hipparcos astrometric parameters.
See Abstr. 036.151.

MARK–III VLBI astrometry of pulsars.
See Abstr. 036.273.

Influence of the reference catalogue errors on the determination of the parameters of fundamental systems using minor planet observation.
See Abstr. 041.010.

Determination of absolute proper motions for the connection of the HIPPARCOS system with a quasi–inertial system and the automation of astrometric measurements of Tautenburg Schmidt plates.
See Abstr. 041.013.

Meridian and astrophotographic observations of the reference stars in the fields with extragalactic radio and optical sources.
See Abstr. 041.018.

Optical positions of extragalactic radio sources in the FK5 system.
See Abstr. 041.021.

Beitrag der Schmidtspiegelastrometrie zur planetaren Geodäsie.
See Abstr. 041.025.

Effects of systematic difference FK5–FK4 on the optical global solution of the Earth's Orientation Parameters.
See Abstr. 044.022.

Preparing the new calculation of the Earth rotation parameters in the Hipparcos reference frame.
See Abstr. 044.036.

New reduction of Earth rotation parameters and Hipparcos reference frame.
See Abstr. 044.038.

Earth orientation and reference frame determinations, atmospheric excitation functions, up to 1990. (Annex to the IERS annual report for 1990).
See Abstr. 044.044.

The accurate post–Newtonian treatment of geodetic space measurements.
See Abstr. 066.161.

Accuracy estimation for connection between proper motions of the Hipparcos catalogue and the optical catalogues by means of radio stars observations.
See Abstr. 111.016.

Basic reference frame and variable stars.
See Abstr. 122.014.

Time variation of fundamental constants, preferred reference systems and CPT invariance.
See Abstr. 161.103.

On the variability of physical constants and conformal transformations.
See Abstr. 161.206.

044 Time and Latitude Determination, Earth Rotation, Polar Motion

044.001 **A Hamiltonian theory for an elastic Earth: first order analytical integration.**
J. Getino, J. M. Ferrándiz.
Celest. Mech. Dyn. Astron., Vol. 51, No. 1, p. 35 - 65 (1991).
In this article an approximate analytical integration is performed of the Hamiltonian corresponding to the rotational motion of an Earth whose elastic mantle is deformed by rotation and lunisolar attraction, using Deprit's perturbation method for the first order. Besides the usual terms, this Hamiltonian includes the perturbation of the kinetic energy and the elastic energy produced with the deformation, as well as their causes, the tidal and the centrifugal potential; these new terms have already been studied for the tidal deformation in previous articles (Getino and Ferrándiz, 1990). The effects of the deformation due to the centrifugal potential are studied in this article, following the same method as that used for the tidal deformation. Numerical tables of the periodic perturbations corresponding to the nutation in obliquity and longitude are obtained. As for the secular effects, a theoretical value of 457 days is obtained for Chandler's period.

044.002 **International Earth Rotation Service (IERS). Bulletin B.**
IERS Bull., B, No. 41, p. 1 - 6 (2 Jul 1991).
Contents: 1. Earth orientation parameters. 2. Daily interpolation of x, y, UT1, D, $d\psi$, $d\varepsilon$. 3. Normal values of the Earth orientation parameters at five-day intervals. 4. Duration of the day and angular velocity of the Earth. 5. Information on time scales. 6. Individual series of Earth orientation parameters.
Further data for the period May - Nov 1991 are given in Nos. 42 - 46 (Aug - Dec 1991).

044.003 **IRIS Earth Orientation Bulletin.**
IRIS Bull. A, No. 89, p. 1 - 7 (Jul 1991).
The Bulletin is the publication of record for the current determination of pole position, UT1, length of day, and nutation. Further informations are given in Nos. 90 - 94 (Aug - Dec 1991).

044.004 **Radiointerferometrische Beobachtungen zur Bestimmung der Polbewegung unter Benutzung langer Nord–Süd–Basislinien.**
A. Nothnagel.
Dtsch. Geod. Komm. Bayer. Akad. Wiss., Reihe C, Nr. 368.
Diss. (Dr.-Ing.), Bayerische Akademie der Wissenschaften, München (Germany, F.R.). Deutsche Geodätische Kommission; Bonn Univ. (Germany, F.R.). Landwirtschaftliche Fakultät. Verl. d. Inst. für Angewandte Geodäsie, Frankfurt (Germany, F.R.). 93 p. (1991). = Mitteilung Nr. 179 des Instituts für Angewandte Geodäsie.
This thesis describes a method of displaying both the observations and the horizon limitations of two stations forming a baseline in a unified system. This technique serves as a tool for improving the quality of observing schedules and it was used for the scheduling of two sets of measurements employing the Hartebeesthoek Radio Astronomy Observatory (HartRAO) in South Africa. In a series of 6 measurements separated by 5 to 10 days HartRAO observed together with two stations in the US and one station in Europe for Earth orientation determinations. In the second series, daily measurements of only 2 - 3 hours duration with a single north–south baseline were performed for polar motion monitoring filling the gaps between the network measurements. Both series were observed in 1986 and 1987. In the first year the observing schedules were prepared with a conventional approach while in 1987 the technique of displaying the observations relative to a baseline reference point was used. The analysis of the data as described in this thesis showed that the standard deviations of the Earth orientation parameters improved by up to 30% by the use of the modified scheduling strategy.

044.005 **Spectral composition of tiltmeter observations in Finland from 6 April 1977 to 28 November 1986.**
L. V. Rykhlova, G. S. Kurbasova, G. N. Shlikar'.
Sov. Astron., Vol. 35, No. 3, p. 313 - 319 (May 1991). Current Physics Microform No.: 9201X0595. English translation of Astron. Zh., Tom 68, Vyp. 3, p. 639 - 651 (1991).
Optimal methods of reconstruction and averaging of an experimental function and a zero-crossing spectral analysis are used to investigate an unevenly spaced run of almost ten years of tiltmeter measurements. The periods of all known short-period tidal waves are obtained. The 435-day period is detected, for the first time in tiltmeter measurements, with 99% probability in a analysis of long-period components. It is proposed to interpret this fact by expanding the description of the Laplace tidal force by introducing additional terms corresponding to perturbations of the orbital motion of the investigated body.

044.006 **Earth's variable rotation.**
R. Hide, J. O. Dickey.
Science, Vol. 253, No. 5020, p. 629 - 637 (9 Aug 1991).
Recent improvements in geodetic data and practical meteorology have advanced research on fluctuations in the Earth's rotation. The interpretations of the dynamics of the Earth-moon system and dynamical processes in the liquid metallic core of the Earth (where the geomagnetic field originates), other parts of the Earth's interior, and the hydrosphere and atmosphere. Fluctuations in the length of the day occuring on decadal time scales have implications for the topography of the core-mantle boundary and the electrical, magnetic, and other properties of the core and lower mantle.

044.007 **IERS Bulletin-A. International Earth Rotation Service.**
IERS Bull., A, Vol. 4, No. 27, p. 1 - 6 (3 Jul 1991).
Contents: 1. Observations - very long baseline interferometry, satellite laser ranging, University of Texas LLR. 2. Combined Earth orientation parameters - IERS rapid service, IERS final values. 3. Predictions - 90-day, long-term. 4. Celestial pole offset series - observations from VLBI, NEOS daily series, IERS final values, predictions.

Further informations covering the period Jul – Dec 1991 are published in Vol. 4, Nos. 28 – 52 (Jul – Dec 1991).

044.008 **Daily time differences.**
U.S. Nav. Obs., Ser. 4, No. 1274.
Naval Observatory, Washington, DC (USA). 7 p. (3 Jul 1991).
Further informations covering the period Jul – Dec 1991 are published in Nos. 1275 – 1299 (Jul – Dec 1991).

044.009 **Atomic time and the rotation of pulsars.**
B. Guinot, G. Petit.
Astron. Astrophys., Vol. 248, No. 1, p. 292 – 296 (Aug 1991).
Millisecond pulsars were quickly recognized as exciting objects enabling to re–examine the understanding in such various fields as cosmology, astrometry, solar system dynamics, and interstellar propagation. Also, the apparently unequalled stability of the period of pulsar rotation has sometimes led to consider them as the "clocks of the future". The authors review the relations between pulsar timing measurements and the current realization of time on Earth (i.e. Atomic Time), and ask in what respect the rotation of pulsars can help in the practical realization of a time scale, and what time scales should be used to analyse pulsar timing data.

044.010 **Quasi–biennial variation in the Earth's rotation and the solar neutrino flux.**
Feng Bo, Ding Hualing.
Publ. Shaanxi Astron. Obs., Vol. 14, No. 1, p. 71 – 75 (Aug 1991). In Chinese.
The earth's rotation shows a tendency to vary with a period of about 26 months. It was explained as the atomspheric oscillation. This study shows that above–mentioned tendency is correlative with the quasi–biennial variation in the solar neutrino flux.

044.011 **Investigations of Earth rotation variations.**
B. Kołaczek, A. Brzeziński, W. Kosek, J. Nastula, W. Hozakowski, B. Sołoducha.
Artif. Satell., Vol. 26, No. 1, p. 243 – 275 (1991).

044.012 **Astronomical observations of variations of Borowiec geographical coordinates in the years 1982 – 1990.**
M. Lehmann.
Artif. Satell., Vol. 26, No. 1, p. 277 – 288 (1991).
Determinations of time and geographical latitude carried out by means of Danjon's astrolabe OPL No. 31 in the years 1982 – 1988 have been reduced to the system FK4. After the introduction of the FK5 in 1989, the earlier observations were re–reduced. The geographical coordinate variations computed before and after introducing the FK5 catalogue, are given.

044.013 **Time generation and distribution.**
D. B. Sullivan, J. Levine.
Proc. IEEE, Vol. 79, No. 7, p. 906 – 914 (Jul 1991). Special issue: Time and frequency. J. Jespersen, D. W. Hanson (eds.).
This paper presents a broad overview of time and frequency technology, particularly those trends relating to the generation and distribution of time and frequency signals. The characterization of components and systems is also addressed.

044.014 **Astronomical time.**
D. D. McCarthy.
Proc. IEEE, Vol. 79, No. 7, p. 915 – 920 (Jul 1991). Special issue: Time and frequency. J. Jespersen, D. W. Hanson (eds.).
Astronomical time serves as the basis for civil time and is required to describe the orientation of the Earth with respect to an inertial reference frame. The definitions of astronomical time scales are reviewed as well as observational methods. The causes of the variations in the Earth's rotation are explained and a brief overview of international efforts to maintain astronomical time scales is presented.

044.015 **The past, present, and future of atomic time and frequency.**
N. F. Ramsey.
Proc. IEEE, Vol. 79, No. 7, p. 921 – 926 (Jul 1991). Special issue: Time and frequency. J. Jespersen, D. W. Hanson (eds.).
The early history of atomic time and frequency standards is reviewed. The most accurate and stable present standards are described. Prospective future improvements, particularly with trapped ions and atoms are discussed.

044.016 **An introduction to frequency standards.**
L. L. Lewis.
Proc. IEEE, Vol. 79, No. 7, p. 927 – 935 (Jul 1991). Special issue: Time and frequency. J. Jespersen, D. W. Hanson (eds.).
Quartz oscillators, and rubidium, cesium, and hydrogen atomic frequency standards are described in terms of the physics of operation. Sources of frequency instability and inaccuracy are given. This review is intended as an introduction to the practical aspects of these high performance frequency standards.

044.017 **Extratropical aspects of the 40–50 day oscillation in length–of–day and atmospheric angular momentum.**
J. O. Dickey, M. Ghil, S. L. Marcus.
J. Geophys. Res., Vol. 96, No. D12, p. 22643 – 22658 (20 Dec 1991).
Fluctuations in Earth rotation over time scales of 2 years or less are dominated by atmospheric effects; spectral analyses of length–of–day (LOD) and atmospheric angular momentum (AAM) data show significantly increased variability in the 40–50 day band. LOD and AAM fluctuations on the 40–50 day time scale have previously been linked to tropical, convectively driven waves of the type first described by Madden and Julian (1971) (referred to as MJ hereinafter). A significant spectral peak centered at 42 days has also been found, however, in the AAM of a 3–year (1120–day) perpetual–January simulation of the global atmosphere, performaned using a version of the University of California at Los Angles (UCLA) general circulation model (GCM) which does not give rise to MJ oscillations in the tropics. In the present work the 40–50 day oscillation is studied using the 12–year overlap between two records: (1) AAM data, compiled from the National Meteorological Center (NMC): and (2) LOD variation from the Jet Propulsion Laboratory Kalman–filtered Earth rotation series. The authors analyze the NMC records by latitude belts, in lighth of the UCLA GCM results, in order to identify possibly distinct sources of the AASM oscillation in the mid–latitudes and the tropics. Results suggest that two intraseasonal oscillations exist in the Earth–atmosphere system: a tropical, 50–day oscillation associated with the convectively driven MJ wave and a mid–latitude, 40–day oscillation associated with the interaction of nonzonal flow with topography.

044.018 **Satellite braodcasting of time and frequency signals.**
A. Sen Gupta, A. K. Hanjura, B. S. Mathur.
Proc. IEEE, Vol. 79, No. 7, p. 973 – 982 (Jul 1991). Special issue: Time and frequency. J. Jespersen, D. W. Hanson (eds.).
This paper discusses the advantages of broadcasting time from a satellite over the more traditional ground based methods such as short wave and low frequency standard time and frequency broadcasts. The paper describes several one way satellite time services currently operational and discusses, in particular, a broadcast service using the Indian domestic satellite INSAT. The signal format, transmitting and receiving setups and the methods for satellite position prediction are described in detail for the INSAT broadcast.

044.019 **Two–way time transfer via communication satellites.**
D. Kirchner.
Proc. IEEE, Vol. 79, No. 7, p. 983 – 990 (Jul 1991). Special issue: Time and frequency. J. Jespersen, D. W. Hanson (eds.).
Some information on the history of two–way satellite time transfer is provided and developments in satellite communications technology conducive to this time transfer technique are mentioned. The difference between one–way and two–way

satellite time transfer is explained, the advantages and disadvantages of the two-way method are pointed out, and the theory of the two-way technique using geostationary communication satellites is presented in detail showing the high accuracy potential of the method.

044.020 **GPS time transfer.**
W. Lewandowski, C. Thomas.
Proc. IEEE, Vol. 79, No. 7, p. 991 – 1000 (Jul 1991). Special issue: Time and frequency. J. Jespersen, D. W. Hanson (eds.).
GPS satellites have become the principal tool for national and international comparisons of atomic clocks. Using GPS, time transfer is now ten times more accurate than it used to be using LORAN-C. The present accuracy of GPS time transfer is of the order 10 – 20 ns for intercontinental distances and 2 – 3 ns within one continent. Although many improvements have been introduced during the past decade, the next should bring still further improvements leading to an ultimate accuracy at the subnanosecond level.

044.021 **Synchronization and relativity.**
G. M. R. Winkler.
Proc. IEEE, Vol. 79, No. 7, p. 1029 – 1039 (Jul 1991). Special issue: Time and frequency. J. Jespersen, D. W. Hanson (eds.).
The use of concepts and formulae of the Special and of the General Theory of Relativity has become a necessity in electronic systems that use high precision timing, especially space systems. The principles involved will be reviewed with emphasis on practical applications and operational standards used in timekeeping and remote synchronization.

044.022 **Effects of systematic difference FK5–FK4 on the optical global solution of the Earth's Orientation Parameters.**
Liao Dechun, Jin Wenjing, Xu Tongqi, Lu Peizhen.
Acta Astron. Sin., Vol. 32, No. 2, p. 165 – 175 (Jun 1991). In Chinese.
In accordance with the recommendation by the IAU, the IAU 1976 precession constant, the IAU 1980 nutation series, and the FK5 reference system have been applied to the data reduction of classical astrometry since 1984.0. Only the quasi-FK5 could be adopted during this period. This paper deals with the effects of the systematic difference FK5–FK4 on the Earth's Orientation Parameters (EOP) of 1987 – 1988 which were determined by the Analysis Centre of Optical Technique (ACOT) at Shanghai Observatory. By using the analytical method (Bien et al., 1978) the systematic differences FK5–FK4 (J2000.0) in positions and proper motions are obtained and listed. The comparison of EOP obtained in this paper with those obtained routinely by the ACOT, ΔX, ΔY and ΔU shows some systematic features clearly. By comparing EOP obtained in this paper and from the routine service of ACOT with those of IERS the mean deviations and RMS are obtained and listed. It is readily seen that the deviations of EOP obtained in this paper are smaller than those from the routine service of ACOT for all components but X component in 1988 and the RMS for both EOP are in the same level. The conclusion is that the effects of systematic differences FK5–FK4 on EOP derived by optical technique should be considered before the publication and application of FK5. The effects may reach a level of $0''.3$ for X and Y and $0^s.002$ for UT.

044.023 **Investigations of series of latitude observations at EAO for 1978 – 1984.**
V. V. Lapaeva, I. A. Urasina, A. P. Kisterskij.
Izv. Astron. Obs. Ehngel'gardt., Vyp. 55, p. 49 – 59 (1990). In Russian.
Results of recalculations of data of latitude observations made with the ZTL-180 at Engelhardt Astronomical Observatory (EAO) in the new astronomical constants IAU system are given. Recalculation is carried out in two star-catalogue systems: GC and the Compiled Catalogue of Latitude Stars of the Engelhardt Observatory/I. More exact values of the mean latitude and results of Z-term investigations are given.

044.024 **On possible causes of nonpolar latitude variations of EAO.**
A. P. Kisterskij, V. V. Lapaeva, I. A. Urasina.
Izv. Astron. Obs. Ehngel'gardt., Vyp. 55, p. 59 – 69 (1990). In Russian.
The influence of inaccurate Earth pole coordinates (x, y), of errors of the values of the micrometer screw revolution and of the variations of the level of the Kujbyshev water reservoir upon the latitude of the Engelhardt Observatory is considered.

044.025 **La nature des échelles de temps, définitions, notations.**
B. Guinot.
Journées 1991: Systèmes de référence spatio-temporels, p. 3 – 10 (1991). – See Abstr. 012.048 for the main entry.
The following concept is adopted: a time scale is a device which allows one to assign dates to events; dates are dimensionless; the relation between difference of dates and measurement of time duration requires the knowledge of a metric factor having the dimension of the unit of time. Although this concept is found, especially in relativistic theories, a number of contradictions are found in the literature, especially concerning the dimensioning of quantities used in astronomy.

044.026 **Maintenance préventive et stabilité des étalons commerciaux de frequence à jet de cesium.**
G. Fréon.
Journées 1991: Systèmes de référence spatio-temporels, p. 11 – 18 (1991). – See Abstr. 012.048 for the main entry.

044.027 **Transfert de temps par un satellite de télévision directe.**
F. Meyer, M. Brunet, M. Granveaud, F. Vernotte, M. Vincent.
Journées 1991: Systèmes de référence spatio-temporels, p. 38 – 43 (1991). – See Abstr. 012.048 for the main entry.
The TV signal transmitted by direct TV satellites such as TDF1 can be used to measure the time difference between two distant atomic clocks by measuring the instants of reception of a determined part of this signal on each time scale. This implies the knowledge of both the position of each clock and the satellite position. Actually, the imprecision of the satellite location generates errors of about 1 μs for baselines around 500 km. In order to reach 100 ns or less a triangulation can be carried out by using three reference stations. This paper presents the results obtained between three stations implanted in Paris, Toulouse and Besançon.

044.028 **Comparaison de deux techniques de transfert de temps par satellites: GPS et MITREX.**
P. Uhrich, R. Tourde, P. Grudler, F. Baumont.
Journées 1991: Systèmes de référence spatio-temporels, p. 44 – 50 (1991). – See Abstr. 012.048 for the main entry.
There are two well known techniques achieving nanosecond time transfer by satellite: Common-View GPS and MITREX. Both techniques are briefly described. Results are given of an international experiment on a link between Observatoire de la Côte d'Azur (France) and Graz (Austria), allowing a direct intercomparison between these two techniques connected to the same remote clocks.

044.029 **Accuracy of GPS time comparisons.**
W. Lewandowski, G. Petit, C. Thomas.
Journées 1991: Systèmes de référence spatio-temporels, p. 51 – 56 (1991). – See Abstr. 012.048 for the main entry.
Over intercontinental distances the accuracy of Global Positioning System (GPS) time transfer ranges from 10 to 20 ns. The principal error sources are the broadcast ionospheric model, the broadcast ephemerides and the local antenna coordinates. The authors realize for the first time the closure around the world obtained by the combination of three intercontinental time transfers after simultaneous reduction of the three major error sources. It gives the evidence of improvement in accuracy for GPS time transfer with using precise ephemerides, measured ionospheric delays and accurate antenna coordinates.

044.030 **On reduction of a infinite system of differential equations in the problem of the Earth's nutation.**
V. V. Bykova.
Kinematika Fiz. Nebesn. Tel, Tom 7, No. 6, p. 3 – 10 (Nov–Dec 1991). In Russian. English translation in Kinematics Phys. Celest. Bodies, Vol. 7, No. 6.
The Earth's nutation problem is reduced to the solution of the infinite system of linear differential equations using an expansion in spherical harmonics. The possibility of reducing this system for two models of the Earth is investigated: with and without the solid inner core. As numerical experiments show the consideration of the finite number of equations enables to obtain the true solution only for the model without inner core. The reduction is incorrect for the model with a solid inner core.

044.031 **Méthode d'analyse de séries par transformation en ondelettes. Application à la rotation de la Terre.**
D. Gambis.
Journées 1991: Systèmes de référence spatio–temporels, p. 98 – 102 (1991). – See Abstr. 012.048 for the main entry.
An analysis of short–term variations of the Earth's rotation is performed using the recently developed Wavelet Transform. This new approach is an alternative to the Fourier Transform. The Wavelet Transform is better adapted to studying the high frequency signals in the Earth's rotation characterized by non–periodic and transient oscillations (between 40 and 80 days).

044.032 **La nutation pour un modèle de Terre rigide. Resultats analytiques et integration numérique.**
J. Souchay.
Journées 1991: Systèmes de référence spatio–temporels, p. 149 – 152 (1991). – See Abstr. 012.048 for the main entry.
The theory of the nutation for a rigid Earth model serves as a basis for more sophisticated ones. Its determination must be as accurate as possible in order to deduce the coefficients of nutation for a model including the effects of the mantle elasticity and of the fluid core. The author presents here the different steps of improvements of this theory since it has been fully developed.

044.033 **Bruit géophysique et bruit de mesure du mouvement du pôle.**
N. Essaifi, M. Feissel.
Journées 1991: Systèmes de référence spatio–temporels, p. 159 – 164 (1991). – See Abstr. 012.048 for the main entry.
Time series of the pole coordinates are obtained at three to five day intervals by various analysis centres from VLBI and SLR observations. It is shown that despite the quasi commonality of the sets of observations analysed, the results are statistically independent at high frequency. This property is used to evaluate the effective error ellipses of the various solutions. The varialibity ellipse of the atmosphere driven oscillations under 100 d is derived.

044.034 **Observations of luni–solar and free core nutation.**
D. D. McCarthy, B. J. Luzum.
Astron. J., Vol. 102, No. 5, p. 1889 – 1895 (Nov 1991). Current Physics Microform No.: 9111D0895.
An analysis of the differences between observed nutation angles and the 1980 IAU Nutation Theory shows the existence of currently unmodeled effects. An empirical set of corrections to the 1980 IAU Nutation Theory is presented and compared with current geophysical models. A retrograde periodic variation (period $\approx$418 days), which may be related to the theoretical free core nutation, is seen.

044.035 **Déterminations de latitude et de temps rapportées à l'origine non–tournante.**
N. Capitaine, F. Chollet.
Journées 1991: Systèmes de référence spatio–temporels, p. 180 – 183 (1991). – See Abstr. 012.048 for the main entry.
A new procedure has been developed for deriving time and latitude from astrometric observations; this new procedure refers to the "non–rotating" origin (NRO) and uses the celestial pole coordinates for realizing the intermediate frame of date linked to the Celestial Ephemeris Pole (CEP). Comparisons between this new procedure and the classical procedure referring to the equinox and using the classical precession and nutation parameters have been performed. This paper shows comparisons for latitude, UT0–UTC and zenith distance. The new procedure is proposed to be used for the re–reduction of the existing latitude and time data referred to the Hipparcos catalogue.

044.036 **Preparing the new calculation of the Earth rotation parameters in the Hipparcos reference frame.**
J. Vondrák.
Journées 1991: Systèmes de référence spatio–temporels, p. 193 – 203 (1991). – See Abstr. 012.048 for the main entry.
The Commission 19 of the IAU (Rotation of the Earth) has set up a Working Group on Earth Rotation in the Hipparcos Reference Frame in order to prepare the algorithms to be used for the new reduction of the past astrometric observations since the beginning of the century. Instead of group results, star by star data will be used, all reduced in a unique system of astronomical constants and algorithms based on all known corrections.

044.037 **The scientific implications of the re–reductions in Hipparcos reference frame for Earth rotation studies.**
N. Capitaine.
Journées 1991: Systèmes de référence spatio–temporels, p. 213 – 216 (1991). – See Abstr. 012.048 for the main entry.
After recalling the present knowledge in Earth rotation and the precision which is presently achieved on the Earth Rotation parameters (ERP) by the new techniques such as VLBI or satellite and lunar laser ranging, this paper investigates what will be the scientific implications of the re–reduction of the existing astrometric data which will be performed by the IAU Working Group "Earth Rotation in the Hipparcos Reference Frame". One of the main advantages of such a re–reduction over previous reductions is to refer to a much better realization of the optical Celestial Reference System as well as to an improved realization of the Terrestrial Reference System using corrections for the deviation of the vertical and for plate motions. An other advantage of the planned re–reduction is to use very long series of observations as compared to similar series obtained by the new techniques.

044.038 **New reduction of Earth rotation parameters and Hipparcos reference frame.**
J. Vondrák.
International Seminar on Problems of Astrometry and Space Geodynamics, p. 106 – 112 (1991). – See Abstr. 012.049 for the main entry.
In order to prepare the collection of the relevant observations and to work out the new algorithms, the IAU Working Group on Earth rotation in the HIPPARCOS reference frame has been appointed.

044.039 **Analyses of variations of Chandler nutation and seasonal variation of polar motion in 1846 – 1988.**
A. Korsun', J. Nastula, B. Kolaczek, W. Kozek, W. Hozakowski.
International Seminar on Problems of Astrometry and Space Geodynamics, p. 112 – 114 (1991). – See Abstr. 012.049 for the main entry.
The long astrometric series of pole coordinate variations in the years 1846 – 1988 worked out by the Kiev Astronomical Observatory are at the present time the best data for investigation of variations of Chandler nutation.

044.040 **Calculation of the new series of the Earth orientation parameters in the HIPPARCOS reference frame.**
J. Vondrák.
Bull. Astron. Inst. Czech., Vol. 42, No. 5, p. 283 – 294 (Oct 1991).
The succesful lauchch of the HIPPARCOS astrometric satellite by the ESA in August 1989 enables the past observations, obtained by optical astrometry all over the world since the beginning of the century, to be re–reduced in order to get the

Earth orientation parameters in a homogeneous and unique celestial reference frame. Completely new algorithms, based on the information obtained from individual star observations rather than the traditional star group results, are proposed. In addition to the three fundamental Earth orientation parameters (coordinates of the instantaneous pole of rotation x, y and the integrated speed of the Earth's rotation UT1–UTC), the nutation angles (defining the position of the rotational axis in space) and some other relevent parameters will be obtained in a single global adjustment.

044.041 **Sub–milliarcsecond determination of pole position using Global Positioning System data.**
T. A. Herring, D. Dong, R. W. King.
Geophys. Res. Lett., Vol. 18, No. 10, p. 1893 – 1896 (Oct 1991).
The authors discuss the determination of pole–position variations from the analysis of 23 days of Global Positioning System (GPS) data continuously collected with a globally distributed, 21–station network. The comparision of the estimated pole positions from these data with results from 11 very long baseline interferometry (VLBI) experiments conducted during this same interval yields weighted root–mean–square differences about the mean differences of 0.5 mas for X– and Y–pole position estimates from an analysis using single–day arcs, and 1.0 and 0.8 mas from an analyis using three weekly multiday arcs. These results are of sufficient quality that GPS could provide an alternative technique to VLBI and satellite laser ranging (SLR) for regular monitoring of high–time–resolution variations in the position of the Earth's rotation axis.

044.042 **Comparison of new nutation series with numerical integration.**
J. Souchay, H. Kinoshita.
Celest. Mech. Dyn. Astron., Vol. 52, No. 1, p. 45 – 55 (1991).
To check the results given by the new tables of the nutation for the rigid Earth model in (Kinoshita and Souchay, 1990) the authors perform a calculation of this nutation by numerical integration, using the basic equations of Kinoshita (1977). Results show very small long period residuals, with a discrepancy of 0.06 mas sin Ω in longitude, –0.04 mas cos Ω in obliquity, without out–of–phase components. Moreover, the total residuals of short period do not exceed 0.1 mas in longitude, 0.06 mas in obliquity. All these results are considerably better than those found by previous numerical integrations, thus confirming the validity of the new tables.

044.043 **Sea tides and Earth's rotation.**
B. P. Pertsev.
Izv. Acad. Sci. USSR, Phys. Solid Earth, Vol. 26, No. 8, p. 628 – 632 (Mar 1990). Translated from Izv. Akad. Nauk SSSR, Fiz. Zemli.
Computations of the secular slowing of the rate of the Earth's rotation by sea tides were made. It is shown that the observed increase in length of day can be attributed to the braking effect of sea tides. The tidal forces of the Moon and Sun, acting on sea tides, also generate an appreciable secular moment relative to the axis lying in the plane of the Earth's equator.

044.044 **Earth orientation and reference frame determinations, atmospheric excitation functions, up to 1990. (Annex to the IERS annual report for 1990).**
IERS Tech. Note, No. 8, p. 1 – 165 (Oct 1991).
This publication presents results obtained by the IERS Analysis Centres for VLBI, LLR, SLR and AAM for Earth orientation parameters, polar motion, station coordinates and atmospheric angular momentum functions.

044.045 **BIPM. Circular T. Data for May 1991.**
BIPM Circ. T, No. 41.
Bureau International des Poids et Mesures, 92312 – Sèvres (France). 5 p. (1 Jul 1991).
Contents: 1. Coordinated Universal Time UTC. Computed values of UTC–UTC(k). 2. International Atomic Time TAI and local atomic time scales TA(k). 3. Notes on sections 1 and 2. 4. UTC–GPS time and TAI–GPS time. 5. UTC–GLONASS time. 6. Measurement of UTC(j)–UTC(k). 7. Duration of the TAI scale interval.
Further data for the period Jun – Oct 1991 are given in Nos. 42 – 46 (Aug – Dec 1991).

044.046 **Die Bestimmung der Ephemeridenzeitkorrektur aus Aufnahmen der totalen Sonnenfinsternis vom 30. Juni 1973.**
H. F. Haupt, M. G. Firneis, J. M. Fritzer.
Sitzungsber., Österr. Akad. Wiss., Math.–Naturwiss. Kl., Abt. II, Band 199, Heft 4 – 7, p. 131 – 142 (1990). = Mitt. Sonnenobs. Kanzelhöhe, Nr. 64.
Photographs of the partial phases of the 1973 June 30 total solar eclipse taken in Mauretania were used to derive the correction ΔT = ET–UT. The introduction describes the problems and history of ΔT determinations. After a brief description of the expedition, location, instruments, and the observations, the reduction is treated in some detail. Probably due to the improper exposure of the films the radius of the Sun turned out too small, while the lunar radius showed a peculiar increase after totality. Correspondingly, this increase resulted in a decrease of the Sun–Moon distances that had to be accounted for by the introduction of a correction function. After having applied this plausible correction several assumptions of ΔT were made in the computations and the one yielding the mean value of the residuals equal to zero, was taken as the ΔT of the date, namely $\Delta T_{1973.5} = +54^{s}.91$, which is in good agreement with other determinations.

044.047 **How large is the retrograde annual wobble?**
N. E. King, D. Carr Agnew.
Geophys. Res. Lett., Vol. 18, No. 9, p. 1735 – 1738 (Sep 1991).
The authors compare recent measurements of polar motion (wobble) made by space–geodetic methods with older optical measurements. Multitaper spectra of these data show that the retrograde (clockwise) annual wobble is much larger in the older data than the newer data, implying systematic errors in the older techniques. Two additional analyses show evidence for a small retrograde motion in the newer data, which appears to be consistent between data types. This implies that the excitation of the retrograde wobble must be about half of the prograde excitation. None of the available estimates of the excitation shows full agreement with these observations.

044.048 **Libration in the Earth's rotation.**
B. F. Chao, D. N. Dong, H. S. Liu, T. A. Herring.
Geophys. Res. Lett., Vol. 18, No. 11, p. 2007 – 2010 (Nov 1991).
External luni–solar torque exerted on the difference (B–A) of the Earth's two equatorial principal moments of inertia gives rise to two types of librational motions in the Earth's rotation: the semidiurnal libration in spin and the prograde diurnal libration in polar motion. Formulas for the librations considering a realistic Earth model and their tidal decompositions are derived and evaluated. The spin libration has a maximum peak–to–peak amplitude of 0.09 milliarcseconds, that of the polar libration is 0.06 milliarcseconds. Implications concerning their detectability and role in the tidal variation of Earth rotation are discussed.

044.049 **Latitude variations of the Potsdam Observatory.**
M. Meinig.
Astron. Nachr., Vol. 312, No. 6, p. 405 – 411 (1991). In German.
Latitude determinations with the photographic zenith tube PZT 2 of the geodetic–astronomical observatory Potsdam were analysed for the period 1981–90 in order to investigate the long term behaviour. The original data were corrected for polar motion. The series of corrected latitude values show besides a significant linear trend seasonal variations with variable parameters. Investigations for the Czeschoslovakian station Ondřejov lead to similar results. The cause for the obtained latitude variations could be systematic errors of the proper motions of the observed stars.

044.050 **The effects of solar gravitation field on the time delay for VLBI.**
Zhao Ming.
Ann. Shanghai Obs., Acad. Sin., No. 12, p. 1 – 4 (1991). In Chinese.
In the solar gravitational field, both the propagation direction and velocity of radio signal from a source could be changed. The former effects the apparent position of the source in celestial sphere, and the latter effects the "light time" of the signal. The evaluations of the extra–delay resulted from the gravitation effects were based on the change of apparent position by some authors or on the change of "light time" by others. The author concludes that these evaluations are not equivalent with each other and the former is wrong in the conception.

044.051 **Influence of the lunar libration models on UT0 determination from Lunar Laser Ranging.**
Xu Huaguan, Jin Wenjing.
Ann. Shanghai Obs., Acad. Sin., No. 12, p. 5 – 13 (1991). In Chinese.
The values of UT0–UTC were determined from global data of Lunar Laser Ranging during Jan. 1987 to Mar. 1989. The lunar libration models Eckhardt–500 and LLB200, were used in the data processing in order to analyse the effect of lunar libration models on UT0 determination. It is shown that there is a obvious systematic deviation between the two libration models.

044.052 **Anomalous residuals of optical astrometric observations at Shanghai Observatory and earthquake (Ms = 5) in town of Tai Cang.**
Liao Dechun, Zheng Dawei.
Ann. Shanghai Obs., Acad. Sin., No. 12, p. 14 – 18 (1991). In Chinese.
A statistical conclusion made by some Chinese scientists is that optical astrometric observations (universal time and latitude) will be affected ahead of 1 – 6 months by an earthquake happening in the region of 300 km far from the observatory with Ms $\geqslant$ 7 or 100 km with Ms $\geqslant$ 6.

044.053 **An analysis on the stability of the JYD series of polar coordinates.**
Qian Changxia, Li Zhengxin.
Ann. Shanghai Obs., Acad. Sin., No. 12, p. 26 – 30 (1991). In Chinese.
From the X, Y components of the JYD series of polar coordinates, the mean poles of epoch during the years 1949 – 1986 were obtained. The stability of the polar coordinates of JYD in long term was analysed. The JYD series has an obvious drift after the year of 1976.

044.054 **On the period and damping of polar motion.**
E. Groten, H. Lenhardt, S. M. Molodensky.
J. Geophys. Res., Vol. 96, No. B12, p. 20241 – 20256 (10 Nov 1991). Paper presented at the 18. Conference on Mathematical Geophysics, Jerusalem (Israel), June 1990.
The theory of the Chandler wobble is an old and rather complicated branch of mathematical geophysics which includes both the methods of the elastic and hydrodynamic equations solutions, describing the reaction of mantle, liquid core and ocean to the variable centripetal force. This theory has recently attracted increased attention in view of (1) the substantial improvement of geodetic measurements (such as very long baseline interferometry, satellite measurements, etc.) requires a corresponding refinement of the theory and (2) the last achievement in modeling the Earth's inner structure gave the possibility to calculate the period and damping of Chandler wobble with much higher accuracy. The comparison of these theoretical results with the results of measurements can isolate mantle anelasticity effects, thus permitting the determination of the Earth's anelastic properties at extremely low frequencies. Obviously, information of such kind is very important for modeling of many geodynamical processes. In this paper the influence of the mantle's anelasticity, the Earth's liquid core and oceanic pole tides on the Chandler wobble is considered.

044.055 **Mean latitude of Felix Aguilar Observatory: study of its variation.**
W. T. Manrique, J. F. Baldivieso.
Bol. Asoc. Argent. Astron., No. 36, p. 264 – 273 (1990). In Spanish. – See Abstr. 012.095 for the main entry.
The values of the latitude for each tenth of year are deduced from the curve which represents the variation of the instantaneous latitudes, as a result of systematic observations carried out during 15 consecutive years with the Danjon Astrolabe. A series of mean values are calculated. Thus, the necessary data for the curve which represents the variation of the mean latitude are obtained. These results are analysed.

044.056 **Variation of the instantaneous latitudes, pillar Danjon astrolabe–OAFA.**
W. Manrique, F. Baldivieso, A. Serafino, E. Actis, E. Alonso, R. Podestá.
Bol. Asoc. Argent. Astron., No. 36, p. 274 – 284 (1990). In Spanish. – See Abstr. 012.095 for the main entry.
Mean values are obtained from equal weights of instantaneous latitudes, derived from systematic observations of 4500 serie selected from the Fundamental Program, of higher weight than one, and which totalize approximately 135000 stellar passages. The observations correspond to a period of 15 consecutive years, realized with the Danjon Astrolabe.

044.057 **A discussion on the method "really visible time".**
Liu Ciyuan.
Publ. Shaanxi Astron. Obs., Vol. 14, No. 2, p. 111 – 116 (Dec 1991). In Chinese.
The author developed a method called "really visible time" to employ ancient Chinese occultation records for the study of long term variation of the Earth's rotation. It is proved that the result of "fitting method" is actually the same as "RVT".

044.058 **Changes in length of day: 500 BC to present.**
L. V. Morrison, F. R. Stephenson.
Geophys. Astrophys. Fluid Dyn., Vol. 60, No. 1–4, p. 419 (Nov 1991). Abstract. – See Abstr. 012.125 for the main entry.

044.059 **Small waves in UT1 caused by the inequality of the equatorial moments of inertia *A* and *B* of the Earth.**
J. Wünsch.
Astron. Nachr., Vol. 312, No. 5, p. 321 – 325 (1991).
A number of semidiurnal terms in UT1 arising from the inequality of the two equatorial moments of inertia *A* and *B* of the Earth are studied. Amplitudes are given using coefficients from Vondrák (1984). The most important waves correspond to the M2, S2, N2 and K2 tidal frequencies. At present, these are below the detection limit of geodetic VLBI experiments.

CAS Symposium No. 115: The 4th Symposium of Precise Time and Time Dissemination, Guilin (China), 15 – 19 Nov 1990.
See Abstr. 011.015.

Systèmes de référence spatio–temporels: métrologie et astrométrie.
See Abstr. 012.048.

Time Service of Borowiec Astronomical Latitude Observatory in the years 1985 – 1990.
See Abstr. 013.047.

The BIPM and the accurate measurement of time.
See Abstr. 013.050.

A process of extrapolation of polar coordinates with data in changeable length.
See Abstr. 021.002.

Time and frequency in fundamental metrology.
See Abstr. 022.171.

Atomic ion frequency standards.
See Abstr. 034.069.

Time synchronization using laser techniques.
See Abstr. 034.072.

Time Service and Astrolabe Station in Coimbra.
See Abstr. 034.139.

ITRF 90 and other realizations of the IERS Terrestrial Reference System for 1990.
See Abstr. 043.012.

The celestial reference system of the International Earth Rotation Service (IERS).
See Abstr. 043.016.

Zur Datumsfestlegung und Short–Arc Bahnverbesserung bei der Auswertung von GPS–Beobachtungen.
See Abstr. 045.019.

The correlation between the solar constant and the near–Earth space environment and Earth's rotation.
See Abstr. 080.054.

Relative motions between the inner core and the Earth's mantle due to attractions of the Sun and the Moon.
See Abstr. 081.040.

Core flow and length–of–day changes.
See Abstr. 081.063.

Core mantle coupling – length of day.
See Abstr. 081.064.

Variations common to the interplanetary magnetic field, the zonal atmospheric circulation and the earth's rotation.
See Abstr. 082.009.

The earth rotation responds to atmospheric and oceanic activity.
See Abstr. 082.029.

Trans–polar VGP shifts and Earth's rotation.
See Abstr. 084.075.

Solar flare and sudden change of Earth's rotation.
See Abstr. 085.027.

Non–linear equations for the rotation of a viscoelastic planet taking into account the influence of a liquid core.
See Abstr. 091.031.

Millisecond pulsars: nature's most stable clocks.
See Abstr. 126.022.

045 Astronomical Geodesy, Satellite Geodesy, Navigation

045.001 **A mobile satellite laser ranging system of 2–nd generation.**
H. Hurnik, R. Baranowski, W. Naskrecki, B. Furmann, S. Łoś.
Artif. Satell., Vol. 26, No. 1, p. 19 – 30 (1991).
The paper presents the design and actual status of the mobile laser system constructed and built in the Astronomical Observatory of the A. Mickiewicz University in 1986 – 1990.

045.002 **Methods and techniques for precise time–interval measurements in laser ranging systems of 3rd generation.**
J. Kalisz, R. Pełka, J. Łatka.
Artif. Satell., Vol. 26, No. 1, p. 31 – 49 (1991).
The report first describes the key techniques commonly used in laser ranging systems for precise measurements of time–intervals. Then follows a description of progress in this field obtained by the authors during the period 1987 – 1990. The current status of the Laser Ranging Station at the Borowiec Observatory is presented. After a general discussion of the modern time–interval digitizers for laser ranging applications the hardware and software details of the new time–interval digitizer T–1810B are described.

045.003 **Orbital problems in GPS interferometry.**
J. B. Zieliński.
Artif. Satell., Vol. 26, No. 1, p. 177 – 197 (1991).
Determination of the GPS orbits and the influence of the orbital errors on the geodetic determination was a topic of investigations during the last few years. In the paper the summary of some analyses is presented concerning the nature of the interferometric observations, the propagation of the orbital errors, correlations and covariances in geodetic GPS solutions.

045.004 **Method and program development for determining satellite orbits for geodetic purposes.**
M. Rutkowska, W. Pachelski, W. Popiński, W. Góral.
Artif. Satell., Vol. 26, No. 1, p. 199 – 211 (1991).

045.005 **Sequential orbit improvement using Kalman filtering technique in navigation system of aeronomic satellites.**
S. Oszczak, Z. Rzepecka.
Artif. Satell., Vol. 26, No. 1, p. 231 – 239 (1991).
The principles of the navigation system for aeronomic satellites are presented. For computation of satellite positions in the orbit the sequential estimation algorithm is proposed. Using the Kalman filtering technique the positions of the satellite as well as their accuracy can be calculated.

045.006 **An investigation of the three dimensional position dilution of precision.**
Zhuang Qixiang, Jian Shuguang, Fan Nenge.
Acta Astron. Sin., Vol. 32, No. 2, p. 113 – 120 (Jun 1991). In Chinese.
The definitions of variance and covariance matrices as well as the geometrical expression for the three dimensional position dilution of precision (PDOP) in the positon determination of a space target are described in this paper. The theoretical optimal solution for four station distribution is determined geometrically. The effects of observation model, station number, and order of stations to be arranged in PDOP are discussed. Combined with practical applications, some simulation computations are performed.

045.007 **The influence of relativistic effects on the results of satellite geodynamics, geodesy and navigation. Methods of investigation.**
N. V. Emel'yanov, A. V. Krivov.
Astron. Zh., Tom 68, Vyp. 4, p. 872 – 879 (Jul–Aug 1991). In Russian. English translation in Sov. Astron., Vol. 35, No. 4.
At the present time there are several methods of taking into account the relativistic effects in satellite problems. This paper aims at the investigation of the influence of the omission of relativistic effects on the results of practical solution of the problems in satellite geodynamics, geodesy, and navigation. The analysis is performed using modelization of motion of a satellite,

observational station, and light propagation due to the relativistic theory. This first part of the paper contains the formulation of the problem, the description of the general scheme of treatment, the accepted physical model, and computing program constructed to realize such an investigation. The results and conclusions are to be presented in the second part of the paper.

045.008 **The influence of relativistic effects on the results of satellite geodynamics, geodesy and navigation. Final results.**
N. V. Emel'yanov, A. V. Krivov.
Astron. Zh., Tom 68, Vyp. 5, p. 1093 – 1098 (Sep–Oct 1991). In Russian. English translation in Sov. Astron., Vol. 35, No. 5.
The influence of the omission of relativistic effects on the results of a practical solution of various satellite problems is investigated. The analysis is performed by modelling of the problems in their real formulation. The modelling method, formulae and description of computing program used for taking into account the relativistic effects are presented in the previous part of the paper. In this part, on the basis of the analysis, it is shown that the conclusion about the influence of the omission of relativistic effects on the final result depends essentially on the formulation of the problem. The quantities to be determined can be considerably distorted in some cases, whereas in other cases these distortions are not essential and can be neglected.

045.009 **Altimetry rank deficiency in crossover adjustment.**
R. Barzaghi, M. Brovelli, F. Sansò.
IAG Symposium: Determination of the geoid, p. 108 – 118 (1991). – See Abstr. 012.035 for the main entry.
Defining the problem of rank deficiency in a general manner for the analysis of altimetric data in the crossover mode supplies a nice framework in which both the empirical and the dynamical approaches can be seen in a unique formulation. The analysis of rank deficiency can easily be performed by using the concept of the continuous analog of a network. The rank deficiency analysis for the dynamical approach meets completely well known results in which it has been previously shown that the actual gravity field contains an invisible part.

045.010 **Spectral comparison of continuous Global Positioning System and strainmeter measurements of crustal deformation.**
D. M. Tralli.
Geophys. Res. Lett., Vol. 18, No. 7, p. 1285 – 1288 (Jul 1991).
Temporal power spectral density models of noise in continuous crustal deformation measurements obtained with the Global Positioning System (GPS) and high–quality strain–meters are compared. The crossover frequency at which GPS measurement noise is less than that of strainmeters is determined. Assuming GPS precision of 0.1 to 1 cm in horizontal components for baselines up to 100 km in length, local deformation monitoring with GPS may be preferable to strainmeters for observations of short–term deformation in under 6 months of continuous (at least daily) measurements. Short–period tropospheric path delays and multipath effects, which may obscure GPS–determined strain signals in local network measurements, also are discussed.

045.011 **Relativistic effects in satellite geodynamics, geodesy, and navigation. Methods.**
N. V. Emel'yanov, A. V. Krivov.
Sov. Astron., Vol. 35, No. 4, p. 432 – 436 (Jul 1991). Current Physics Microform No.: 9203X0618. English translation of Astron. Zh., Tom 68, Vyp. 4, p. 872 – 879 (1991).
A number of methods now exist for allowing for relativistic effects in problems involving Earth satellites. The influence of neglecting general relativity (GR) effects on the practical solution of problems of satellite geodynamics, geodesy, and navigation is studied in the present paper. The analysis is carried out by simulating the motion of the satellite and the station and light propagation in accordance with relativistic theory. The present first part of the paper contains a statement of the problem and a description of the general scheme for studying GR corrections, the chosen physical model, and the software developed for such a study. The results and conclusions will be given in the second part.

045.012 **Local geodetic survey for SLR Station in Kiev.**
W. Jaks, Ya. S. Yatskiv, N. T. Mironov, A. N. Samojlenko.
International Seminar on Problems of Astrometry and Space Geodynamics, p. 115 – 121 (1991). – See Abstr. 012.049 for the main entry.
The results of a three–dimensional geodetic survey in the vicinity of the SLR station in Kiev are presented. The survey was focused on the determination of approximate geocentric coordinates of SLR systems, the vectors between significant reference points and connection to the national geodetic network survey marker. Particular care was taken to ensure the factors connecting the reference points of the SLR systems, and survey marker were determined to uncertainties of a few centimeters (1 – 2 cm).

045.013 **Influence of relativistic effects on satellite geodynamics, geodesy and navigation. Research results.**
N. V. Emel'yanov, A. V. Krivov.
Sov. Astron., Vol. 35, No. 5, p. 547 – 550 (Sep 1991). Current Physics Microform No.: 9203X1721. English translation of Astron. Zh., Tom 68, Vyp. 5, p. 1093 – 1098 (1991).
The results of ignoring relativistic effects in the practical solution of various problems associated with the motion of Earth satellites are investigated. The analysis is carried out by simulating problems in their actual formulation. The modeling method, the equations, and a description of the software that allows for relativistic effects have been published in the preceding part of this paper. Here the authors show that the end result of ignoring relativistic effects depends sensitively on the formulation of the problem. The quantities being determined may be strongly distorted in some cases, and in others these distortions are insignificant.

045.014 **Application of the Global Positioning System to crustal deformation measurement. 1. Precision and accuracy.**
K. M. Larson, D. C. Agnew.
J. Geophys. Res., Vol. 96, No. B10, p. 16547 – 16566 (10 Sep 1991).
The authors assess the precision and accuracy of interstation vectors determined using the Global Positioning System (GPS) satellites. These vectors were between stations in California separated by 50–450 km. Using data from tracking the seven block I satellites in campaigns from 1986 through 1989, the authors examine the precision of GPS measurements over time scales of several days and a few years.

045.015 **Application of the global positioning system to crustal deformation measurement. 2. The influence of errors in orbit determination networks.**
K. M. Larson, F. H. Webb, D. C. Agnew.
J. Geophys. Res., Vol. 96, No. B10, p. 16567 – 16584 (10 Sep 1991).
Global Positioning System (GPS) measurements of a geodetic network in southern and central California have been used to investigate the errors introduced by adopting different sets of stations as fixed. Such fixed points, called fiducial stations, are necessary to eliminate the errors of imprecise satellites orbits, which otherwise would dominate the error budget for distances greater than tens of kilometers. These fiducial stations also define the reference frame of the crustal deformation network. Establishing the magnitude of the effect of changing the fiducial network is essential for crustal deformation studies, so that these artifacts of the differences between fiducial networks used for the data analyses are not interpreted as geophysical signals. Solutions for a crustal deformation network spanning distances up to 350 km were computed with a variety of fiducial networks.

045.016 **Error analysis of the GPS station keeping algorithm.**
Zhuang Qixiang.
Ann. Shanghai Obs., Acad. Sin., No. 12, p. 74 – 78 (1991). In Chinese.

A GPS station keeping algorithm is reviewed. To compare the elevation and azimuth predictions of GPS satellite with actual tracking data, an error analysis has been made. To predict one month, the prediction biases are about 0°.4 and 0°.6 for elevation and azimuth respectively, while, to predict one year, the prediction biases are about 2° and 5° for elevation and azimuth respectively.

045.017 **Satellite laser ranging observations at Shanghai Observatory in 1989.**
SLR Group.
Ann. Shanghai Obs., Acad. Sin., No. 12, p. 85 – 89 (1991). In Chinese.

045.018 **An introduction to the Global Positioning System and some geological applications.**
T. H. Dixon.
Rev. Geophys., Vol. 29, No. 2, p. 249 – 276 (May 1991).

045.019 **Zur Datumsfestlegung und Short–Arc Bahnverbesserung bei der Auswertung von GPS–Beobachtungen.**
D. Angermann.
Dtsch. Geod. Komm. Bayer. Akad. Wiss., Reihe C, Nr. 388. Diss. (Dr.–Ing.), Technische Univ. Berlin (Germany, F.R.). Fachbereich 7 – Bauingenieur– und Vermessungswesen. Verlag der Bayerischen Akademie der Wissenschaften, München (Germany, F.R.). 118 p. (1991). ISBN 3–7696–9434–1.

This dissertation deals with a detailed investigation of stability problems using short-arc methods for the evaluation of GPS phase data. The sensitivity analysis for the determination of orbit corrections is performed by means of redundancy decomposition of stochastic a priori information. With regard to the modelling of the GPS phase observations investigations of the datum definition in the satellite geodesy are carried out. A meaningful result of this investigation is that the z–component of the geocentre is defined only by the orbit disturbations mainly caused by the flattening of the Earth whereas the x– and y–component of the geocentre are defined additionally by the Earth rotation.

045.020 **GPO observations of a geostationary TV–satellite quartet.**
H. Boehnhardt.
Messenger, No. 65, p. 41 – 43 (Sep 1991).

045.021 **Genauigkeitssteigerung in der geodätischen Astronomie durch den FK5 Sternkatalog. Neuauswertung von astronomisch bestimmten Längendifferenzen im Europäischen Längennetz (ELN).**
W. Wende.
Das Projekt "Satelliten– und Inertialgeodäsie" der Bayerischen Kommission für die Internationale Erdmessung, p. 31 – 46 (1991). – See Abstr. 003.037 for the main entry.

For the establishment of a modern astronomical longitude reference system in Europe longitude differences between national reference stations of the Federal Republic of Germany, Italy, Spain, the Netherlands, France and Portugal have been measured by K. Kaniuth and the author with a Danjon–astrolabe in 1977 until 1980. The achieved internal accuracies are ±0,001 s by the use of an adjustment including parameters for catalogue errors of the Fourth Fundamental Catalogue FK4. Since the Fifth Fundamental Catalogue FK5 is available, a re–evaluation of the five longitude difference measurements, using the FK5 and an expanded model for the reduction of the observations, was carried out. It turns out, that the external accuracies of the longitude differences is to be expected at the same order as the internal by the use of the FK5.

Differential geodesy.
See Abstr. 003.007.

Das Projekt "Satelliten– und Inertialgeodäsie" der Bayerischen Kommission für die Internationale Erdmessung.
See Abstr. 003.037.

O. W. Struve's Levelling Project in European Russia.
See Abstr. 004.049.

Problems of astrometry and space geodynamics. Proceedings.
See Abstr. 012.049.

Borowiec laser station 1986 – 1990.
See Abstr. 013.046.

A new direct solution of the transformation problem of Cartesian into ellipsoidal coordinates.
See Abstr. 021.022.

Fast collocation.
See Abstr. 021.026.

Merits of space VLBI missions for geodynamics.
See Abstr. 036.140.

ITRF 90 and other realizations of the IERS Terrestrial Reference System for 1990.
See Abstr. 043.012.

Radiointerferometrische Beobachtungen zur Bestimmung der Polbewegung unter Benutzung langer Nord–Süd–Basislinien.
See Abstr. 044.004.

GPS time transfer.
See Abstr. 044.020.

Determination of the regional geoid from simulated GPS measurements by the ARISTOTELES solid–earth satellite.
See Abstr. 081.029.

Analyse von Lasermessungen zum Mond im Rahmen einer post–Newton'schen Theorie.
See Abstr. 094.013.

046 Ephemerides, Almanacs, Calendars, Chronology

046.001 **New evidence concerning the origin of the Julian period.**
R. L. Reese, E. D. Craun, M. Herrin.
Am. J. Phys., Vol. 59, No. 11, p. 1043 (Nov 1991). Abstract. Current Physics Microform No.: 9111A1605.

046.002 **Rocznik Astronomiczny na rok 1992.**
Instytut Geodezji i Kartografii, Warsaw (Poland), Tom 47. 145 p. (1992). ISSN 0209-0341.

046.003 **Length of the lunar crescent.**
B. E. Schaefer.
Q. J. R. Astron. Soc., Vol. 32, No. 3, p. 265 - 277 (Sep 1991).
The thin crescent Moon is observed not to extend the full 180° arc expected if the Moon had an ideal surface. The author reports on a collection of 65 crescent arc length observations made across North America on 1989 April 6. These observations show the crescent length to be independent of whether or not the observer used any optical aid. A theory to explain the crescent shortening is that the brightness falls off steeply to the cusps so that visual observers will not detect the cusps because they are fainter than the threshold for visual detection. Lunar surface brightnesses are calculated according to the accurate model by Hapke (1984), and it is found that all Danjon's (1932, 1936) collected observations and the authors' new data are well fitted by this model. One implication for lunar calendars is that the crescent cannot be detected if the Moon is within 7° of the Sun, although other effects will always cause the Moon to be invisible until the distance from the Sun is significantly greater.

046.004 **Newton's date for the crucifixion.**
J. P. Pratt.
Q. J. R. Astron. Soc., Vol. 32, No. 3, p. 301 - 304 (Sep 1991).

046.005 **Los calendarios.**
M. A. Di Paolo.
Rev. Astron., No. 246, p. 8 - 10 (Jun 1991).

046.006 **Chinese Astronomical Ephemeris 1992.**
Academia Sinica, Nanjing (China). Purple Mountain Observatory. 547 p. (1991). ISBN 7-03-002200-9. In Chinese.

046.007 **The Star of Bethlehem - a comet in 5 BC - and the date of the birth of Christ.**
C. J. Humphreys.
Q. J. R. Astron. Soc., Vol. 32, No. 4, p. 389 - 407 (Dec 1991).
Astronomical and historical evidence suggests that the Star of Bethlehem was a comet which was visible in 5 BC, and described in ancient Chinese records. A comet uniquely fits the description in Matthew of a star which newly appeared, travelled slowly through the sky against the star background and "stood over" Bethlehem. It is proposed that a remarkable sequence of three astronomical events stimulated the journey of the Magi.

046.008 **1993 Japanese Ephemeris.**
Maritime Safety Agency, Tokyo (Japan). Hydrographic Dept., Japan Oceanographic Data Center, Pub. No. 684. 515 p. (1992). ISSN 0373-3696. In Japanese and English.

046.009 **Apparent Places of Fundamental Stars 1993 for 1535 stars from the Fifth Fundamental Catalogue (FK5).**
R. Wielen, T. Lederle, H. Schwan (eds.).
Astronomisches Rechen-Institut, Heidelberg (Germany, F.R.), Braun, Karlsruhe (Germany, F.R.). 554 p. (1992). ISBN 3-7650-0093-0. ISSN 0174-254X. Price DM 66.00.

046.010 **The Star Almanac for Land Surveyors for the year 1993.**
HM Nautical Almanac Office, Cambridge (UK), Her Majesty's Stationery Office, London (UK). 96 p. (1992). ISBN 0-11-886945-0. Price £ 5.00.

046.011 **Astronomical ephemeris for the year 1992.**
N. Čabrić.
Vasiona, Année 39, No. 4, p. 57 - 68 (1991). In Croatian.

046.012 **Efemérides Astronómicas para o ano de 1992.**
Coimbra Univ. (Portugal). Observatório Astronómico. 268 p. (1992). ISSN 0870-1199.

046.013 **Anuario del Observatorio Astronómico de Madrid para 1992.**
Observatorio Astronómico, Madrid (Spain), Instituto Geográfico Nacional, Madrid (Spain). 470 p. (1991). ISBN 84-7819-033-3. ISSN 0373-5125. Price Pts. 800.00.

046.014 **Astronomical Phenomena for the Year 1993.**
Nautical Almanac Office, Washington, DC (USA); HM Nautical Almanac Office, Hailsham (UK), US Government Printing Office, Washington, DC (USA). 76 p. (1991). ISBN 0-11-886939-6. UK ed.: Her Majesty's Stationery Office, London (UK); ISBN 0-16-035922-8, Price £ 5.50.

046.015 **The selection of superior epoch in Chinese calendars during the period from East Han to Liu Song dynasty.**
Qu Anjing.
Acta Astron. Sin., Vol. 32, No. 4, p. 436 - 439 (Dec 1991). In Chinese.
This paper discusses and demonstrates the superior epoch selection of the extant 3 calendars from East Hun to Liu Song Dynasty. The results show that the names of day, tropical year and synodic month are the whole prerequisite for superior epoch selection in the Jing Chu Calendar. The superior epoch in the East Han Si Fen Calendar and Qian Xiang Calendar are determined by 4 prerequisites which include the lipse period.

046.016 **Tartu Tähetorni Kalender 1992. aastaks.** ***(Calendar of the Tartu Astronomical Observatory for the year 1992.)***
M. Jõeveer (comp.).
Eesti Teaduste Akadeemia, Tartu. Astrofüüsika ja Atmosfäärifüüsika Inst.; W. Struve nimeline Tartu Astrofüüsika Observatoorium (Estonia), Tallinn (Estonia), 68 aastakäik. 44 p. (1991).

046.017 **Supplément aux éphémérides de la Société Astronomique de France pour 1992.**
D. Savoie.
Obs. Trav., No. 28, p. 40 - 41 (1991).

046.018 **The Nautical Almanac for the year 1993.**
HM Nautical Almanac Office, London (UK); Nautical Almanac Office, Washington, DC (USA), Her Majesty's Stationery Office, London (UK). 357 p. (1992). ISBN 0-11-772618-4. Price £ 16.50. US ed.: Superintendent of Documents, US Government Printing Office, Washington, DC (USA).

046.019 **Philippine Astronomical Handbook 1992.**
Philippine Atmospheric, Geophysical and Astronomical Services Administration (PAGASA), Manila. 76 p. (1991). ISSN 0115-1207. Prepared by the Astronomical Publication Unit, Astronomy Research and Development Section, Atmospheric, Geophysical and Space Sciences Branch, Quezon City (Philippines).

046.020 **Ephémérides astronomiques et calendrier des événements célestes pour 1992.**
P. Cugnon, A. Koeckelenbergh, G. Evrard.
Ciel Terre, Vol. 107, No. 6, p. 1 - 76 (Nov - Dec 1991).

046.021 **Ephémérides Astronomiques 1993. Annuaire du Bureau des Longitudes. Calendriers - Soleil - Lune - planètes - astéroides - satellites - comètes - étoiles.**
Masson, Paris (France). 311 p. (1992). ISBN 2-225-82807-5.

046.022 **Calendar modification incorporating long–term variations in both the length of the tropical year and in the length of the day.**
F. Di Marzio.
Aust. J. Astron., Vol. 4, No. 2, p. 86 – 100 (Oct 1991).
The current Gregorian calendar has a quoted error of approximately one day in 19500 years. Ironically, however, small changes in the length of the tropical year produce errors of about one day in 5700 years and these errors occur with increasing regularity thereafter. Eventually, adjustments of one day will need to be made essentially every year. In order to account for such variations, a simple procedure is developed which enables calendar modifications to be made predictably as well as allowing an estimation of the length of the day in the distant past.

Lunar tables and programs from 4000 B.C. to A.D. 8000.
See Abstr. 002.098.

Astronomical algorithms.
See Abstr. 003.020.

Der "Zij Muhammad Shahi". Die Kalenderreform Jai Singhs II.
See Abstr. 004.005.

The system of formulating the calendar by observing the Beidou stars.
See Abstr. 004.007.

Inclusive counting as the source of the misunderstanding about the luni–solar calendar of Geminus prior to the octaeteris.
See Abstr. 004.016.

The use of astronomical tables by Albertus Magnus.
See Abstr. 004.081.

Soviet programs "Jupiter" and "Ceres" for improving orbits of bodies of solar system.
See Abstr. 013.051.

Precise ephemeris of minor planets.
See Abstr. 042.029.

On the ephemerides of occultations.
See Abstr. 096.002.

Comparison of observations of the Martian satellites made in 1988 with ephemerides.
See Abstr. 097.037.

The Comet Handbook 1992.
See Abstr. 103.006.

1993 Comet Handbook.
See Abstr. 103.016.

Space Research

051 Extraterrestrial Research Related to Astronomy and Astrophysics

051.001 **The ISO mission – a scientific overview.**
M. F. Kessler, A. Heske, L. Metcalfe, A. Salama.
ESA Bull., No. 67, p. 8 – 16 (Aug 1991).

051.002 **The ALEXIS project: current status and performance.**
J. J. Bloch, B. E. Edwards, W. C. Priedhorsky, D. Roussel–Dupré, B. W. Smith, O. H. Siegmund, T. Carone, J. Vallerga, S. Culley, J. Warren, H. Marshall, P. W. Vedder.
Bull. Am. Astron. Soc., Vol. 23, No. 2, p. 905 (1991). Abstract. – See Abstr. 012.008 for the main entry.

051.003 **Astro 1: an overview.**
T. R. Gull.
Bull. Am. Astron. Soc., Vol. 23, No. 2, p. 958 (1991). Abstract. – See Abstr. 012.008 for the main entry.

051.004 **The EUVE spectroscopic survey.**
J. Vallerga, P. Jelinsky, S. Bowyer.
Bull. Am. Astron. Soc., Vol. 23, No. 2, p. 969 – 970 (1991). Abstract. – See Abstr. 012.008 for the main entry.

051.005 **EUVE observations of the Earth and planets.**
G. R. Gladstone, S. Chakrabarti, W. K. Tobiska, J. Edelstein.
Bull. Am. Astron. Soc., Vol. 23, No. 2, p. 969 (1991). Abstract. – See Abstr. 012.008 for the main entry.

051.006 **EUVE observations of the interstellar medium.**
I. Hawkins, P. Jelinsky, J. Vallerga.
Bull. Am. Astron. Soc., Vol. 23, No. 2, p. 970 (1991). Abstract. – See Abstr. 012.008 for the main entry.

051.007 **EUVE observations of late–type stars.**
P. W. Vedder, P. Jelinsky, J. V. Vallerga, H. L. Marshall, K. Mukai, J. Warren, S. Culley, S. Bowyer.
Bull. Am. Astron. Soc., Vol. 23, No. 2, p. 970 (1991). Abstract. – See Abstr. 012.008 for the main entry.

051.008 **EUVE observations of cataclysmic variables.**
K. Mukai, J. V. Vallerga, T. E. Carone, J. Warren.
Bull. Am. Astron. Soc., Vol. 23, No. 2, p. 970 (1991). Abstract. – See Abstr. 012.008 for the main entry.

051.009 **EUVE observations of hot DA white dwarfs.**
G. Basri, D. S. Finley, P. Jelinsky.
Bull. Am. Astron. Soc., Vol. 23, No. 2, p. 970 – 971 (1991). Abstract. – See Abstr. 012.008 for the main entry.

051.010 **EUVE observations of active galactic nuclei.**
H. L. Marshall.
Bull. Am. Astron. Soc., Vol. 23, No. 2, p. 971 (1991). Abstract. – See Abstr. 012.008 for the main entry.

051.011 **The EUVE mission.**
S. Bowyer.
Bull. Am. Astron. Soc., Vol. 23, No. 2, p. 972 (1991). Abstract. – See Abstr. 012.008 for the main entry.

051.012 **The EUVE all–sky and deep surveys.**
R. F. Malina.
Bull. Am. Astron. Soc., Vol. 23, No. 2, p. 972 (1991). Abstract. – See Abstr. 012.008 for the main entry.

051.013 **SIMURIS: a high resolution solar physics interferometric mission in answer to the chromospheric and coronal heating problem.**
L. Damé.
International Conference on Mechanisms of Chromospheric and Coronal Heating, p. 66 – 72 (1991). – See Abstr. 012.011 for the main entry.

The SIMURIS mission, dedicated to ultrahigh resolution imaging and spectroscopy of the solar atmosphere from the convection zone to the high corona, has two major instruments: the Solar Ultraviolet Network (SUN) which is a 4 telescopes interferometer capable of 10 km spatial resolution on the Sun, and the Imaging Fourier Transform Spectrometer (IFTS) which provides multispectral instantaneous 2D velocity fields of the solar atmosphere. The main scientific objectives are outlined and the model payload is briefly described.

051.014 **The Solar Infrared Explorer (SIRE): a Small Explorer mission for solar physics.**
D. Deming, T. Kostiuk, D. Glenar.
Bull. Am. Astron. Soc., Vol. 23, No. 2, p. 1038 (1991). Abstract. – See Abstr. 012.012 for the main entry.

051.015 **Recent results of the Solar Disk Sextant experiment.**
S. Sofia, E. Maier, L. Twigg.
Bull. Am. Astron. Soc., Vol. 23, No. 2, p. 1050 (1991). Abstract. – See Abstr. 012.012 for the main entry.

051.016 **Space at JPL.**
W. I. McLaughlin.
Spaceflight, Vol. 33, No. 10, p. 350 – 355 (Oct 1991).

Contents: Mars Observer prepares for launch. Round and about the Solar System. Other planetary systems.

051.017 **SUSIM: a program to measure changes in the sun's ultraviolet (UV) spectral irradiance over solar cycle time scales.**
J. L. Lean, M. E. VanHoosier, G. E. Brueckner.
Bull. Am. Astron. Soc., Vol. 23, No. 2, p. 1068 (1991). Abstract. – See Abstr. 012.012 for the main entry.

051.018 **Probing solar particle acceleration during the 22nd solar maximum with GRO/OSSE.**
W. T. Vestrand, D. J. Forrest.
Bull. Am. Astron. Soc., Vol. 23, No. 2, p. 1072 (1991). Abstract. – See Abstr. 012.012 for the main entry.

051.019 **HST and dense stellar systems.**
I. R. King.
Workshop on the First Year of HST Observations, p. 1 – 5 (1991). – See Abstr. 012.017 for the main entry.
The author discusses the capabilities of HST in observing globular clusters, nearby galaxies, and distant galaxies.

051.020 **Hot stars and the HST.**
R. P. Kudritzki.
Workshop on the First Year of HST Observations, p. 68 – 69 (1991). – See Abstr. 012.017 for the main entry.
The author reports first HST observations obtained with the GHRS of the O3f star Melnick 42 in the 30 Doradus complex of the LMC.

051.021 **A review of planetary opportunities and observations with the Hubble Space Telescope.**
R. Beebe.
Workshop on the First Year of HST Observations, p. 147 – 160 (1991). – See Abstr. 012.017 for the main entry.
This review discusses the anticipated capabilities of the Hubble Space Telescope for observations of solar system bodies, reviews the results achieved from preliminary observations, suggests future observations and addresses improvements that will simplify and enhance planetary observations.

051.022 **Observations of Mars using Hubble Space Telescope Observatory.**
P. B. James, R. T. Clancy, S. W. Lee, R. Kahn, R. Zurek, L. Martin, R. Singer.
Workshop on the First Year of HST Observations, p. 161 – 177 (1991). – See Abstr. 012.017 for the main entry.
The authors have embarked on a three year program of Mars observations using the Hubble Space Telescope. During the initial phase of the project they have imaged Mars on five dates in a variety of spectral bands. Contents: Scientific objectives. Processing WFPC images of Mars. Preliminary results.

051.023 **Transformation: the link between the proposal and the Hubble Space Telescope database.**
M. L. McCollough, H. H. Lanning, K. E. Reinhard.
Workshop on the First Year of HST Observations, p. 270 – 275 (1991). – See Abstr. 012.017 for the main entry.

051.024 **Proposal preparation by SPSS for scheduling on the Hubble Space Telescope.**
K. E. Reinhard, H. H. Lanning, W. M. Workman III.
Workshop on the First Year of HST Observations, p. 276 – 280 (1991). – See Abstr. 012.017 for the main entry.
Preparation of a proposal for execution on board the Hubble Space Telescope encompasses a great deal of manually intensive work by Science Planning and Scheduling System (SPSS) personnel. The preparation task includes tracking of the work status, detailed analysis of the structure and contents of the proposal, modification of the database values as required for proper execution onboard, generation of scheduling windows, test scheduling, and incorporation of the commanding and proposal changes necessary for execution. This preparation process is shown.

051.025 **The scheduling of science activities for the Hubble Space Telescope.**
D. K. Taylor, K. E. Reinhard, H. H. Lanning, D. R. Chance, E. V. Bell II.
Workshop on the First Year of HST Observations, p. 281 – 287 (1991). – See Abstr. 012.017 for the main entry.
The Science Planning and Scheduling System (SPSS) is the operational software portion of the Science Operations Ground System (SOGS) responsible for scheduling science activities onboard the Hubble Space Telescope (HST). The authors show a chronological order of the activities and features of SPSS that take an observing proposal from Transformation to execution.

051.026 **The scheduling efficiency for the Hubble Space Telescope during the first year of operation.**
E. V. Bell II, K. E. Reinhard, H. H. Lanning.
Workshop on the First Year of HST Observations, p. 288 – 293 (1991). – See Abstr. 012.017 for the main entry.

051.027 **Routine science data processing of HST observations.**
D. A. Swade, S. B. Parsons, P. Van West, S. Baggett, M. Kochte, D. Macomb, A. Schultz, I. Wilson.
Workshop on the First Year of HST Observations, p. 294 – 297 (1991). – See Abstr. 012.017 for the main entry.
All science observations performed by the Hubble Space Telescope (HST) will be automatically processed by the Routine Science Data Processing (RSDP) pipeline at the Space Telescope Science Institute (STScI). Monitoring and maintenance of pipeline activity is the responsibility of the Post Observation Data Processing System (PODPS) branch.

051.028 **The Solar–A mission and its scientific aims.**
Y. Uchida, Y. Ogawara.
Royal Society Discussion Meeting: The physics of solar flares, p. 485 – 495 (1991). – See Abstr. 012.016 for the main entry.
Descriptions of the scientific instruments on board the Solar–A satellite are given. Relevant parameters and merits of the Solar–A experiments, consisting of co–aligned soft and hard X–ray telescopes with high sensitivities and spatial resolutions, together with Bragg and wide–band spectrometers, are presented. Some of the observational targets, both flare related and non–flare–related are discussed together with brief solar physical implications of these observations. The importance is stressed for collaboration with the ground–based optical and radio observatories to maximize the scientific outcome from this small but unique satellite for the investigation of flares during the solar activity maximum 22.

051.029 **New orbit determination system for INTERCOSMOS aeronomic satellite.**
P. Fraczyk, W. Jakś, J. A. Kaźmierski, S. Oszczak.
Artif. Satell., Vol. 26, No. 1, p. 213 – 220 (1991).

051.030 **First results from the X–ray astronomy mission ROSAT.**
B. Aschenbach.
Reviews in Modern Astronomy 4, p. 173 – 187 (1991). – See Abstr. 003.008 for the main entry.
The general summary of the past few months of ROSAT operations is that both the spacecraft and the scientific instruments perform as anticipated, sometimes even better. The scientific return, originally in the form of about 700 Mbit/day science data, is overwhelming, and it is clear that at present no

full account of the science can be given. Instead, the author restricts the presentation to a few highlights and tries to give something like a preview on what we expect ROSAT will contribute to astronomy and astrophysics.

051.031 **TYCHO/HIPPARCOS a successful mission! First results and further improvements.**
A. Wicenec.
Reviews in Modern Astronomy 4, p. 188 – 196 (1991). – See Abstr. 003.008 for the main entry.
The main part of this review is concerned with the TYCHO data, it's overall and special features and with data reduction improvements developed in the TYCHO Data Analysis Consortium (TDAC) during the implementation and verification of the revised mission.

051.032 **Overview of NASA's cosmic and heliospheric physics missions.**
W. V. Jones.
International School of Astroparticle Physics, p. 417 – 442 (1991). – See Abstr. 012.037 for the main entry.
The space missions supporting primarily the scientific goals of NASA's Cosmic and Heliospheric Physics Branch are briefly discussed. These include the long–standing Pioneers and Voyagers, the recently–launched (1990) Ulysses, the Small Explorer SAMPEX to be launched in 1992, the Explorer ACE to be launched in 1996, the Astromag investigations WiZard and LISA selected for Space Station Freedom, but now being considered for a free–flyer mission in the late 1990's, and the high priority future missions Solar Probe, Interstellar Probe, and UHCR experiment for the early 21st century.

051.033 **The science program of the European Space Agency.**
G. Cavallo.
International School of Astroparticle Physics, p. 443 – 449 (1991). – See Abstr. 012.037 for the main entry.
The science program of ESA is reviewed both in its historical development and in its scientific aspects, focusing in particular on the long term issues.

051.034 **Future solar system missions.**
G. A. Briggs.
IAU Colloquium No. 123: Observatories in Earth orbit and beyond, p. 297 – 306 (1990). – See Abstr. 012.038 for the main entry.
Some future NASA planetary exploration programs are reviewed.

051.035 **The Ulysses mission in the high latitude heliosphere.**
D. E. Page, R. G. Marsden, E. J. Smith, K.–P. Wenzel.
IAU Colloquium No. 123: Observatories in Earth orbit and beyond, p. 307 – 313 (1990). – See Abstr. 012.038 for the main entry.
Ulysses, a joint ESA/NASA mission launched in October 1990, will be the first to explore the high latitude heliosphere. Launch will be from the Shuttle and a Jupiter gravity assist will be used to send the spacecraft first over the southern solar pole approximately three and one half years after launch and then over the northern solar pole one year later. Instruments will be carried to study the solar wind, the heliospheric magnetic field, energetic solar particles, galactic cosmic rays, solar X–rays, cosmic gamma rays, cosmic dust and interstellar neutral helium. The radio signals used to track and transmit spacecraft data will be used also to sound the corona and to search for gravitational waves.

051.036 **Future deep space propulsion systems.**
E. Stuhlinger.
IAU Colloquium No. 123: Observatories in Earth orbit and beyond, p. 355 – 362 (1990). – See Abstr. 012.038 for the main entry.
Among several potential future deep space propulsion systems, the two which are closest to realization are selected for closer consideration: solar–electric, and nuclear–electric propulsion. In particular, the paper describes a manned Mars mission using a particle bed reactor and Brayton cycle converter as power source. Technical details of the design and the mission profile of a 4–astronaut expedition to Mars, and a proposed course of action for project implementation are presented.

051.037 **Relative merits of low–earth, eccentric, geosynchronous, and interplanetary orbits and sites in space.**
B. G. Taylor.
IAU Colloquium No. 123: Observatories in Earth orbit and beyond, p. 377 – 379 (1990). – See Abstr. 012.038 for the main entry.
Criteria for the best orbit for a space observatory are discussed.

051.038 **Humanity or robotics in space?**
M. Oda.
IAU Colloquium No. 123: Observatories in Earth orbit and beyond, p. 380 (1990). Abstract. – See Abstr. 012.038 for the main entry.

051.039 **Astrophysics from the Moon.**
M. J. Mumma.
IAU Colloquium No. 123: Observatories in Earth orbit and beyond, p. 381 – 390 (1990). – See Abstr. 012.038 for the main entry.
The moon offers mixed blessings as a site for astronomical observatories, and some of these are presented. Contents: Introduction; The principal themes; How do stars and planetary systems form? Are there Earth–like planets among the nearby stars? What is happening in our sun and in other stars? What happened in the early universe? Environmental and site considerations; Conclusion.

051.040 **Use of libration–point orbits for space observatories.**
R. W. Farquhar, D. W. Dunham.
IAU Colloquium No. 123: Observatories in Earth orbit and beyond, p. 391 – 395 (1990). – See Abstr. 012.038 for the main entry.
The Sun–Earth libration points, L_1 and L_2, are located 1.5 million km from the Earth towards and away from the Sun. Halo orbits about these points have significant advantages for space observatories in terms of viewing geometry, thermal and radiation environment, and delta–V expediture.

051.041 **Major observatories versus economy–class observatories in space.**
F. J. Dyson.
IAU Colloquium No. 123: Observatories in Earth orbit and beyond, p. 399 – 405 (1990). – See Abstr. 012.038 for the main entry.
Big space–instruments are essential for observing fine details of selected objects. Small space–instruments are essential for scanning large areas of sky, for discovering and cataloguing new objects, and for monitoring variable or transient sources of radiation. If astronomy is to flourish, both types of instrument must be provided. The problem is not to choose between big and small. The problem is to find the optimum mixture of big and small. Various examples from past history and from future plans illustrate the importance of maintaining a variety of different types of instrument.

051.042 **Environmental considerations of design alternatives for the exploration of Mars by balloon.**
T. F. Heinsheimer, L. D. Friedman.
2. COSPAR Colloquium: The environmental model of Mars, p. 11 - 24 (1991). - See Abstr. 012.039 for the main entry.
The Titan Mars Balloon Performance Model has been utilized to examine major environmental issues that will influence the choice of selection of balloon options both for the current mission and for quantitative evaluations of the various concepts for post-1994 missions.

051.043 **Mars sample return: site selection strategy and proposal in Aeolis region.**
N. Cabrol, E. Grin, A. Dollfus.
2. COSPAR Colloquium: The environmental model of Mars, p. 51 - 54 (1991). - See Abstr. 012.039 for the main entry.
A Rover Mission involves technical constraints which limit the planetary field of investigations to specific regions. This paper proposes one candidate-site considering these constraints and the geological interests. These two parameters lead to the evaluation of the scientific productivity of one site analysing the instrumentation in relationship with site criteria. A candidate-site is proposed in Aeolis region.

051.044 **Scientific rationale and technical elements of future Mars exploration in ESA.**
A. F. Chicarrro, G. E. N. Scoon.
2. COSPAR Colloquium: The environmental model of Mars, p. 69 - 83 (1991). - See Abstr. 012.039 for the main entry.
After a comprehensive analysis of the key relevant scientific issues in order to attain a much better understanding of the formation and evolution of Mars, ESA has identified three possible areas of European participation in the future international exploration of Mars. The need for a major and novel scientific return were considered as strong requirements during this study.

051.045 **Small mobile apparatus for Mars surface studies.**
A. L. Kemurjian (*A. L. Kemurdzhian*), P. N. Brodsky, R. K. Betkin, V. V. Gromov, G. N. Kozlov, A. E. Krasnov, B. E. Kutcherenko, M. I. Malenkov, B. N. Martinov, L. M. Muhin, D. F. Nenarokov, V. N. Petriga, K. M. Pichkhadze, G. N. Rogovsky (*G. N. Rogovskij*), P. S. Sologub.
2. COSPAR Colloquium: The environmental model of Mars, p. 113 - 120 (1991). - See Abstr. 012.039 for the main entry.
Designation, scientific purposes, development concepts of a small mobile apparatus (SMA) for Mars surface study, its performances, methods of SMA command and features of its delivering to Mars are described.

051.046 **The robotic exploration missions at Mars.**
G. E. Cunningham.
2. COSPAR Colloquium: The environmental model of Mars, p. 143 - 145 (1991). - See Abstr. 012.039 for the main entry.
A set of robotic missions to Mars are designed to provide information to enable the eventual human exploration of the planet. The strategy of the mission sequence including broad global assessments to specific site certification is described. The exploration and science information gained by each mission is discussed.

051.047 **Infrared universe - IRAS.**
B. T. Soifer.
Priroda, No. 8, p. 34 - 40 (Aug 1991). In Russian.
The astronomical satellite IRAS made jointly by the United States, the Netherlands and Great Britain allowed to see many celestial bodies (from galaxies to interstellar dust and gas) in a new range of the electromagnetic spectrum.

051.048 **PRISMA: a space facility for studying rotation and activity.**
P. Lemaire, T. Appourchaux, C. Catala, S. Catalano, S. Frandsen, A. Jones, W. Weiss.
NATO Advanced Research Workshop on Angular Momentum Evolution of Young Stars, p. 397 - 403 (1991). - See Abstr. 012.055 for the main entry.
The authors report on PRISMA, a space mission under assessment study at ESA, devoted to the study of the internal structure, rotation and activity of stars. Simultaneous observations of oscillation modes and activity for several stars well distributed over the HR diagram are expected to allow the mapping of the stellar structure from core to corona, as a function of mass and evolution. Instrument performancies, observation strategies and mission concepts are described.

051.049 **The SPECTRUM-UV Project.**
M. Rodonó, E. G. Tanzi, A. A. Boyarchuk, N. V. Steshenko.
NATO Advanced Research Workshop on Angular Momentum Evolution of Young Stars, p. 405 - 408 (1991). - See Abstr. 012.055 for the main entry.
A feasibility study of the Ultra-Violet Space Telescope SPECTRUM-UV, an international UV Observatory to be launched in a highly excentric (4 to 7 days) orbit in the mid-1990s, is being carried out by an international team. The main telescope features a 170-cm diffraction limited aperture for spectroscopy and imaging from the Lyman limit up to the visible domain. Two additional 50-cm telescopes for imaging and spectroscopy in the 400 to 1200 Å range and four 20-cm, multilayer coated telescopes for narrow band imaging in the 100 to 400 Å range are coaligned with the main telescope.

051.050 **The space observatory Astro.**
M. Gola.
Urania, Rok 62, Nr. 11, p. 295 - 297 (Nov 1991). In Polish.

051.051 **The submilimeter wave astronomy satellite.**
G. J. Melnick, A. Dalgarno, G. G. Fazio, J. Stauffer, P. Thaddeus, N. R. Erickson, P. F. Goldsmith, R. L. Snell, M. Harwit, D. J. Hollenbach, D. G. Koch, D. A. Neufeld, R. Schieder, G. F. Winnewisser.
3. Haystack Observatory Conference on the Interstellar Medium: Skylines - atoms, ions and molecules - new results in spectral line astrophysics, p. 439 - 450 (1991). - See Abstr. 012.051 for the main entry.
The Submillimeter Wave Astronomy Satellite (SWAS) was selected for flight as part of the Small Explorer Program. SWAS will consist of a 55 x 71-cm near-optical quality off-axis primary mirror, two heterodyne radiometers with Schottky barrier diode mixers, and a single broadband acousto-optical spectrometer. SWAS's primary objective will be to conduct a high spectral resolution survey of galactic molecular clouds in low-lying transitions of water, molecular oxygen, atomic carbon, and isotopic carbon monoxide. The AOS bandwidth is sufficient to permit simultaneous observation of all four lines, while the spectral resolution of the AOS, 1 MHz, will give SWAS a velocity resolution of approximately 0.6 km s^{-1}.

051.052 **Infrared observation in space - from IRAS to ISO.**
Qian Zhongyu.
Prog. Astron., Vol. 9, No. 3, p. 238 - 249 (Sep 1991). In Chinese.
This paper presents a brief history of space infrared observation for astronomy. Different means for infrared observation above the atmosphere - balloon, aircraft, rocket, and satellite are briefly described and emphasis is put on infrared satellite programs. For example, IRAS and its instruments, its target and the main results, the ISO program, its scientific goals and instruments are introduced in some details. At the end of this paper, the planned or possible programs for space infrared

activity towards the end of the century including SL2 IRT, COBE, IRTS, HST, SIRTF, FIRST, LDR are also mentioned.

051.053 **The Voyager encounter with Neptune.**
E. C. Stone, E. D. Miner.
J. Geophys. Res., Vol. 96, Suppl., p. 18903 – 18906 (30 Oct 1991). Special issue: Voyager 2 at Neptune.

051.054 **EUV astronomy on the Apollo–Soyuz mission: the first detection of stellar EUV sources and other astrophysical results.**
S. Bowyer.
1. Berkeley Colloquium on Extreme Ultraviolet Astronomy, p. 3 – 7 (1991). – See Abstr. 012.045 for the main entry.
A substantial set of results was derived from the data obtained with the EUV telescope flown on the Apollo–Soyuz mission. The author summarizes the major findings obtained on this mission.

051.055 **EUV results from Voyager.**
J. B. Holberg.
1. Berkeley Colloquium on Extreme Ultraviolet Astronomy, p. 8 – 14 (1991). – See Abstr. 012.045 for the main entry.
Contents: 1. Introduction. 2. Hot white dwarfs. 3. EUV limits for other objects (cataclysmic variables, nearby O and B stars, emission line sources). 4. Interstellar Medium. 5. The EUV sky background. 6. Very local interstellar medium.

051.056 **EXOSAT EUV observations.**
N. E. White.
1. Berkeley Colloquium on Extreme Ultraviolet Astronomy, p. 15 – 29 (1991). – See Abstr. 012.045 for the main entry.
The long wavelength/low energy response of the European X–ray Observatory, EXOSAT, extended further into the EUV than previous X–ray astronomy missions and provided the first detailed look at astronomical objects between 10 and 400 Å. This article reviews the key EUV results from the EXOSAT observatory.

051.057 **Stellar corona and stellar flare observations with the Extreme Ultraviolet Explorer.**
P. W. Vedder, P. Jelinsky, S. Bowyer, J. V. Vallerga, H. L. Marshall.
1. Berkeley Colloquium on Extreme Ultraviolet Astronomy, p. 120 – 129 (1991). – See Abstr. 012.045 for the main entry.
The authors describe the capabilities of *EUVE* for observing EUV emission from the coronae and flares of late–type stars. They estimate the number of sources to be detected in the six–month all–sky survey and discuss the use of fluxes measured in various bandpasses to estimate the temperatures of the EUV–emitting plasma. They also estimate the number of flares which will be detected during the survey and describe the capabilities of the *EUVE* spectrometers for making pointed spectroscopic observations of nearby late–type stars.

051.058 **The SOLAR–A mission: an overview.**
Y. Ogawara, T. Takano, T. Kato, T. Kosugi, S. Tsuneta, T. Watanabe, I. Kondo, Y. Uchida.
Sol. Phys., Vol. 136, No. 1, p. 1 – 16 (Nov 1991). Special issue: The YOHKOH (SOLAR–A) mission. For the hardcover edition see 003.013.
The SOLAR–A spacecraft is to be launched by the Institute of Space and Astronautical Science, Japan (ISAS) in August, 1991. As a successor of HINOTORI, this mission is dedicted principally to the study of solar flares, especially of high–energy phenomena observed in the X– and gamma–ray ranges. The SOLAR–A will be the unique space solar observatory during the current activity maximum period (1989 – 1992). With a coordinated set of instruments including hard X–ray and soft X–ray imaging telescopes as well as spectrometers with advanced capabilities, it will reveal many new aspects of flares and help better understand their physics, supporting international collaborations with ground–based observatories as well as theoretical investigations. An overview of this mission, including the satellite, its scientific instruments, and its operation, is given. Also the scientific objectives are briefly discussed.

051.059 **The *ALEXIS* project and the local interstellar medium.**
J. J. Bloch, B. W. Smith.
1. Berkeley Colloquium on Extreme Ultraviolet Astronomy, p. 298 – 303 (1991). – See Abstr. 012.045 for the main entry.
Contents: 1. Introduction. 2. *ALEXIS* as a tool for abundance and temperature diagnostics. 3. Spatial studies of the soft X–ray background with *ALEXIS*. 4. Conclusions.

051.060 **Spectroscopy of the ISM in emission using the *Extreme Ultraviolet Explorer* satellite.**
J. V. Vallerga, P. W. Vedder, P. Jelinsky, S. Bowyer.
1. Berkeley Colloquium on Extreme Ultraviolet Astronomy, p. 304 – 310 (1991). – See Abstr. 012.045 for the main entry.
With 25 independent wavelength bins, the *EUVE* spectrometer currently is the most sensitive diffuse EUV instrument longward of 200 Å planned to fly in orbit, and it will complement the broad– and narrow–band surveys at shorter wavelengths. The ~375 pointings of the *EUVE* spectrometer combined with the small angular scale ecliptic survey should greatly expand the knowledge of the spatial and thermal nature of the LISM.

051.061 **The *EUVE* mission.**
S. Bowyer, R. F. Malina.
1. Berkeley Colloquium on Extreme Ultraviolet Astronomy, p. 397 – 408 (1991). – See Abstr. 012.045 for the main entry.
Contents: 1. Introduction. 2. Overview of the mission. 3. Instrumentation. 4. End–to–end system modeling.

051.062 **A high–resolution, USSR–USA EUV telescope–spectrometer.**
S. Bowyer, R. Sunyaev (*R. A. Syunyaev*), J. C. Green.
1. Berkeley Colloquium on Extreme Ultraviolet Astronomy, p. 448 – 456 (1991). – See Abstr. 012.045 for the main entry.
The authors discuss a representative sample of the kinds of science possible with this instrument, they describe the instrument, and they detail its performance characteristics.

051.063 **The Lyman Far Ultraviolet Spectroscopic Explorer.**
W. Moos, S. D. Friedman.
1. Berkeley Colloquium on Extreme Ultraviolet Astronomy, p. 457 – 463 (1991). – See Abstr. 012.045 for the main entry.
Contents: 1. Introduction. 2. Scientific overview. 3. Instrumentation. 4. Mission operations. 5. Summary.

051.064 ***Santa Maria*: an international orbiting observatory for multispectral (200–7000 Å) observations of astrophysical and solar system objects.**
R. Stalio, C. Morales Duran, B. R. Sandel, T. E. Carone, F. Fumi, C. Morossi, A. Talavera, A. L. Broadfoot, R. S. Polidan, C. Batalli Cosmovici, A. Cassatella, J. B. Holberg, L. Sabau, G. Tondello.
1. Berkeley Colloquium on Extreme Ultraviolet Astronomy, p. 478 – 482 (1991). – See Abstr. 012.045 for the main entry.
Experiment concept and some aspects of the scientific program are described.

051.065 **Early results from the Far Infrared Absolute Spectrophotometer (FIRAS).**
J. C. Mather, E. S. Cheng, R. A. Shafer, R. E. Eplee, R. B. Isaacman, D. J. Fixsen, S. M. Read, S. S. Meyer, R. Weiss, E. L. Wright.
AIP Conf. Proc., No. 222, p. 43 – 52 (15 Apr 1991). Current Physics Microform No.: 9106E0043. – See Abstr. 012.069 for the main entry.

The Far Infrared Absolute Spectrophotometer (FIRAS) on the Cosmic Background Explorer (COBE) mapped 98% of the sky, 60% of it twice, before the liquid helium coolant was exhausted. The FIRAS covers the frequency region from 1 to 100 cm^{-1} with a 7° angular resolution. The spectal resolution is 0.2 cm^{-1} for frequencies less than 20 cm^{-1} and 0.8 cm^{-1} for higher frequencies. Preliminary results include: a limit on the deviations from a Planck curve of 1% of the peak brightness from 1 to 20 cm^{-1}, a temperature of 2.735 ± 0.06K, a limit on the Comptonization parameter y of 10^{-3}, on the chemical potential parameter μ of 10^{-2}, a strong limit on the existence of a hot smooth intergalactic medium, and a confirmation that the dipole anisotropy spectrum is that of a Doppler shifted blackbody. Although there are many unresolved issues about the data processing and analysis of systematic errors, some of which are described here, the instruments is expected to achieve its goal of an accuracy and rms sensitivity for νI_ν better than $10^{-9}W/m^2sr$, for each 7° beamwidth and each 5% spectral resolution element, for frequencies from 1 to 20 cm^{-1}.

051.066 **COBE attitude refinement.**
E. L. Wright, V. K. Kumar, I. Freedman, A. Rae, S. Basappa, R. Venkataraman, F. S. Patt.
AIP Conf. Proc., No. 222, p. 179 – 182 (15 Apr 1991). Current Physics Microform No.: 9106E0179. – See Abstr. 012.069 for the main entry.

Methods used to determine and validate the fine–aspect attitude (which contributes to the instrumental anisotropy limit) are presented for the Cosmic Background Explorer satellite. The focus is on the use of the Diffuse Infrared Background Experiment (DIRBE) star observations to refine the attitude and the methods used in the gyroscope calibration and attitude validation.

051.067 **The USSR Academy of Sciences and space research.**
G. A. Skuridin.
Istor.–Astron. Issled., Tom 21, p. 50 – 76 (1989). In Russian. – See Abstr. 003.017 for the main entry.

The author reports mainly on the first ten years of development of space research after 1957; on 4th of October 1957 the first artificial Earth satellite of the world was launched at Bajkonur (USSR).

051.068 **The Extreme Ultraviolet Explorer mission.**
B. Y. Welsh.
Proc. Astron. Soc. Aust., Vol. 9, No. 1, p. 156 – 157 (1991). – See Abstr. 012.078 for the main entry.

The Extreme Ultraviolet Explorer (EUVE) is a NASA funded astronomy mission which will operate in the 70–760 Å spectral band. The science payload has been designed and built by the Space Sciences Laboratory at the University of California, Berkeley and consists of three grazing incidence scanning telescopes and an EUV spectrometer/deep survey instrument. The author gives an overview of the planned mission profile and briefly discusses the instrumentation which comprises the science payload. The *EUVE* is scheduled for launch in late August 1991.

051.069 **Solar sailing: a new tool for solar system research.**
C. R. McInnes.
Vistas Astron., Vol. 34, Part 4, p. 369 – 408 (1991).

The concept of solar sailing is by no means a recent innovation, with its origins dating back to the Soviet pioneers of astronautics. However, it is only relatively recently that the versatility of solar sail spacecraft for solar system research has been investigated. Large scientific payloads can be carried on missions with multiple objectives, such as an asteroid tour. Other advanced missions are possible with, for example, the sail executing a circular heliocentric orbit, but displaced high above the ecliptic plane. The history of solar sailing, the engineering design of solar sail spacecraft and their many applications will also be discussed.

051.070 **Ulysses: a journey above the Sun's poles.**
E. J. Smith, D. E. Page, K.–P. Wenzel.
Earth Space, Vol. 4, No. 1, p. 10 – 14 (Sep 1991).

051.071 **The X–ray astronomy mission SAX.**
G. C. Perola.
Workshop on Iron Line Diagnostics in X–Ray Sources, p. 261 – 266 (1991). – See Abstr. 012.088 for the main entry.

SAX, acronym of "Satellite for Astronomy in the X–rays", is a project of the Italian Space Agency in collaboration with the Netherlands Agency for Aerospace Programs. The program entered the phase C/D (a continuous phase to completion) in April 1989, and the launch date is planned for the end of 1993.

051.072 **Capabilities of SODART/XSPECT for iron line detection.**
N. J. Westergaard, H. U. Nørgaard–Nielsen, F. E. Christensen, A. Hornstrup, C. Budtz–Jørgensen, B. Byrnak, H. W. Schnopper.
Workshop on Iron Line Diagnostics in X–Ray Sources, p. 284 – 288 (1991). – See Abstr. 012.088 for the main entry.

The Soviet mission Spectrum–X–Gamma is planned to be launched in 1993. The prime instrument is the SODART (Soviet Danish Röntgen Telescope) consisting of two telescope units, each with a thin foil nested mirror assembly and a number of focal plane instruments. These include gas detectors, a solid state detector and a polarimeter. One of the telescopes is equipped with an objective crystal spectrometer for high resolution spectroscopy of strong sources. The expected energy resolution and sensitivities for iron line detection are presented and compared with some current research goals.

051.073 **Operational capabilities of AXAF for iron line diagnostics.**
D. A. Schwartz.
Workshop on Iron Line Diagnostics in X–Ray Sources, p. 300 – 310 (1991). – See Abstr. 012.088 for the main entry.

The Advanced X–ray Astrophysics Facility (AXAF) will offer all great opportunities for iron line studies. With multiple gratings and focal plane instruments, and with uniquely high spatial and spectral resolution, it is not easy to give a simple summary of the capabilities. It will provide high resolution spectroscopy, $E/\Delta E \geqslant 1000$ for the K, L, and M line regions. The CCD imager provides simultaneous broad field of view and high spatial resolution (arcsec) observations of the K and L lines, with modest energy resolution. The microcalorimeter allows high spectral resolving power in 5×20 arcsec pixels, with high throughput. The throughput of both the CCD and calorimeter (net effective areas of 100 to 1000 cm^2) will be useful for detailed studies of thousands of sources, although other observatories will have comparable or greater effective areas for moderate resolution spectroscopy of K–shell iron lines.

051.074 **The proposed Israeli X–ray satellite.**
G. Shaviv, T. Weller, J. Felsteiner, B. Rosner, M. Livio, N. Lupu, I. Gertner, H. Abramovitz, N. Markevich, N. Niv.
Workshop on Iron Line Diagnostics in X–Ray Sources, p. 311 – 319 (1991). – See Abstr. 012.088 for the main entry.

About two years ago the Israel Space Agency issued a call for proposals for an experiment on board an Israeli satellite. The constraints were to use the Ofek 2 platform without any change in its design and specifications. The constraints were: an experimental volume of $0.5\ m^2$, total weight of payload 20 kg, total power consumption 20 W, about 0°.1 pointing accuracy and about 0°.1 drift in direction stability, survival in acceleration of 35 g during take off, and mission duration of two years. The authors propose to use the platform for a hard X–ray telescope.

051.075 **Ulysses. A voyage above the Sun's poles.**
K.–P. Wenzel, E. J. Smith.
Europhys. News, Vol. 22, No. 11, p. 203 – 205 (Dec 1991).

First results highlight the Ulysses spacecraft's potential to report on physical conditions and processes of which we have no direct knowledge in previously unexplored territory above the Sun's polar caps.

051.076 **ROSAT. A new look at the sky.**
J. E. Trümper.
Europhys. News, Vol. 22, No. 11, p. 210 – 212 (Dec 1991).

X–ray astronomy is receiving fresh stimulus from the ROSAT satellite observatory launched on 1 June 1990.

051.077 **SOHO and Cluster. Solar and geospheric plasmas and solar structure.**
V. Domingo, R. Schmidt.
Europhys. News, Vol. 22, No. 11, p. 213 – 215 (Dec 1991).

The SOHO/Cluster satellite missions will provide an unique opportunity to advance our understanding of the physics of the solar–terrestrial system.

051.078 **The SOLAR–A mission.**
Y. Ogawara.
International SOLAR–A Science Meeting: Flare physics in solar activity maximum 22, p. 15 – 17 (1991). Abstract. The full paper appeared in Sol. Phys., Vol. 136, see Abstr. 051.058. – See Abstr. 012.104 for the main entry.

051.079 **Capabilities and limitations of SOLAR–A.**
L. W. Acton, H. S. Hudson.
International SOLAR–A Science Meeting: Flare physics in solar activity maximum 22, p. 28 – 36 (1991). – See Abstr. 012.104 for the main entry.

The purpose of this paper is to present an overview useful to scientists, not familiar with Solar–A, who are interested in the capabilities of the mission for solar research. In order to keep the paper to a manageable size it will be assumed that the reader is generally familiar both with the experimental techniques of Solar–A and current research in solar high–energy physics. The authors do not provide detailed technical descriptions of the Solar–A instruments. They do, however, briefly discuss the capabilities of Solar–A in the context of solar activity research in the 1990's.

051.080 **Studies of high–energy flare phenomena from simultaneous observations of gamma–rays and energetic particles.**
M. Yoshimori.
International SOLAR–A Science Meeting: Flare physics in solar activity maximum 22, p. 110 – 113 (1991). – See Abstr. 012.104 for the main entry.

The HESP (High Energy Solar Physics) group in Japan has plans to advance the understanding of high–energy solar flare physics from gamma–ray, energetic charged particle and neutron observations in the 22nd solar maximum. The gamma–rays, energetic charged particles, and neutrons are observed with the SOLAR–A, GEOTAIL, and ground–based neutron monitor, respectively. The correlative studies of these high–energy photons and particles provide critically needed information for achieving a deep understanding of high–energy processes of particle acceleration, interaction, and transport in a complicated environment such as the turbulent and highly magnetized atmosphere of the flare region. The SOLAR–A gamma–ray spectrometer, GEOTAIL energetic particle instrument, and ground–based neutron momitor are outlined, and the scientific return expected from these correlative studies is described.

051.081 **Studies of X–ray flares by project CORONAS.**
I. A. Zhitnik, A. M. Urnov.
International SOLAR–A Science Meeting: Flare physics in solar activity maximum 22, p. 312 – 315 (1991). – See Abstr. 012.104 for the main entry.

A short description is given of the solar flare X–ray instrumentation and science program in the framework of the project CORONAS. The CORONAS project includes two successive missions, CORONAS–I and CORONAS–F on board the Soviet satellites AUOS, with launches planned in 1992 and 1993, respectively.

051.082 **Solar energetic particle observation by GEOTAIL satellite.**
T. Kohno.
International SOLAR–A Science Meeting: Flare physics in solar activity maximum 22, p. 316 – 319 (1991). – See Abstr. 012.104 for the main entry.

A program of solar energetic particle observation by GEOTAIL satellite is described. This satellite is scheduled to be launched in 1992. The High Energy Particle (HEP) sensors can observe electrons, protons and heavy particles with $Z < 45$. The high resolution isotopic observation from He to Ni will also be possible for the energy ranges of 4.5 – 110 MeV/nuc for C and 7.5 – 210 MeV/nuc for Fe.

051.083 **Mass–spectrometric in situ analysis of solid–state extraterrestrial samples.**
J. Kissel.
International School of Physics 'Enrico Fermi'. Course 111: Solid–state astrophysics, p. 169 – 195 (1991). – See Abstr. 012.106 for the main entry.

In this lecture the author discusses different methods of mass spectrometry that have been used successfully (or are about to be used) in space research.

051.084 **Space exploration of comets and primitive minor bodies.**
J. A. M. McDonnell.
International School of Physics 'Enrico Fermi'. Course 111: Solid–state astrophysics, p. 339 – 351 (1991). – See Abstr. 012.106 for the main entry.

Advances in the depth of knowledge of target objects chosen for specialized study in the solar system now leads, perhaps contrarily, to a need for a broader approach to future studies. Examine the fields of specialization such as meteors, cosmic dust, asteroids, comets, planetology, meteorology, solar physics, the solar wind and the interplanetary medium. Can they be the object of indepth isolated study? Can depth be achieved without the context of the particular specialization? Shown are a few of the major links between the fields – identified here only in terms of major cause and effect links. Further links are evident from the need for simultaneous comparison of data and observations in the field, namely synergistic studies where a common cause may have multiple effects. Linking perhaps more obviously similar fields such as the asteroids or comets, it should be noted thus the asteroid Chiron (diameter some 200 km) has recently sprouted a cometary tail. The author focuses in this seminar on cometary studies and in particular direct or in situ exploration.

051.085 **ROSAT – das weltweit begehrteste Observatorium.**
M. J. Schmidt.
Orion, Jahrg. 49, Nr. 245, p. 143 – 147 (Aug 1991).
Deutsch-Amerikanischer Röntgensatellit übertrifft alle Erwartungen – viele neue Ergebnisse.

051.086 **Gammastrahlenobservatorium durchmustert den Himmel nach Gammaquellen – Erste Erfolge.**
M. J. Schmidt.
Orion, Jahrg. 49, Nr. 246, p. 182 – 184 (Oct 1991).

051.087 **The SAC–B project: objective and instrumentation.**
H. S. Ghielmetti, A. M. Hernández, J. M. Gulich.
Bol. Asoc. Argent. Astron., No. 36, p. 419 – 421 (1990). Abstract. In Spanish and English. – See Abstr. 012.095 for the main entry.

051.088 **Les sciences de l'univers à l'Agence Spatiale Européenne.**
R.–M. Bonnet.
C. R. Acad. Sci., Sér. Gén., Vie Sci., Tome 8, No. 4, p. 259 – 281 (1991).

051.089 **A close encounter of the second kind – The Giotto Extended Mission (GEM).**
M. G. Grensemann, G. Schwehm.
ESA Bull., No. 68, p. 42 – 50 (Nov 1991).

051.090 **ASCHOT – astrophysical Schmidt orbital telescope.**
H. M. Tovmasyan (*G. M. Tovmasyan*), H. Lorenz, A. Priebe, G. M. Richter, K.–H. Schmidt, Y. M. Khodzhayants (*Yu. M. Khodzhayants*), M. N. Krmoyan, R. V. Isajanyan, A. Z. Zakharyan.
Astron. Nachr., Vol. 312, No. 5, p. 281 – 290 (1991).
The necessity for a wide–angle UV space telescope for astrophysical investigations is outlined. Some of the main scientific goals such as full sky survey and massive study of UV emission of extragalactic and galactic objects, which may be achieved by the use of the Astrophysical Schmidt Orbital Telescope ASCHOT, are presented. The concept of the telescope design, its optical system, and its system of guidance are described. Electronic devices which may be used as detectors of UV emission in the telescope are considered. The limiting sensitivity of the telescope is estimated: it is about 24^m for stellar objects at 1500 Å. The control system of the telescope is briefly described. At the end the organizations which take part in designing and manufacturing the telescope are listed.

The YOHKOH (SOLAR–A) mission. A series of invited contributions.
See Abstr. 003.013.

Transportation system for a lunar base.
See Abstr. 011.016.

The first year of HST observations. Proceedings.
See Abstr. 012.017.

Observatories in Earth orbit and beyond. Proceedings.
See Abstr. 012.038.

The environmental model of Mars. Proceedings.
See Abstr. 012.039.

Extreme ultraviolet astronomy. A selection of papers presented at the First Berkeley Colloquium on Extreme Ultraviolet Astronomy.
See Abstr. 012.045.

Iron line diagnostics in X–ray sources. Proceedings.
See Abstr. 012.088.

Astrophysics at FUV and EUV wavelengths. Proceedings. Topical Meeting of the COSPAR Interdisciplinary Scientific Commission E (Meeting E3) of the COSPAR 28. Plenary Meeting, The Hague (Netherlands), 25 Jun – 6 Jul 1990.
See Abstr. 012.107.

Geoscience and a lunar base. A comprehensive plan for lunar exploration. Proceedings.
See Abstr. 012.117.

Astronomy from space: the 1990's and beyond.
See Abstr. 013.010.

The EUVE guest observer program.
See Abstr. 013.018.

Space Telescope observations of white dwarf stars.
See Abstr. 013.070.

Space astrophysics in Japan.
See Abstr. 013.087.

CO_2 profiles in cometary analogs.
See Abstr. 022.163.

United States launch vehicle systems.
See Abstr. 035.139.

ESA's Space Transportation Programme.
See Abstr. 035.140.

Launch vehicles of ISAS.
See Abstr. 035.141.

The Hard X–ray Telescope (HXT) for the SOLAR–A mission.
See Abstr. 035.185.

The Soft X–ray Telescope for the SOLAR–A mission.
See Abstr. 035.186.

The Wide Band Spectrometer on the SOLAR–A.
See Abstr. 035.187.

The Bragg Crystal Spectrometer for SOLAR–A.
See Abstr. 035.188.

The XUV wide field camera for ROSAT.
See Abstr. 035.191.

The EUV capabilities of the Hopkins Ultraviolet Telescope.
See Abstr. 035.192.

The *ORFEUS* mission.
See Abstr. 035.193.

The Berkeley EUV spectrometer for the *ORFEUS* mission.
See Abstr. 035.194.

JET–X – a Joint European X–ray Telescope for SPECTRUM–X.
See Abstr. 035.206.

X–ray spectroscopy with the SODART/XSPECT Telescope.
See Abstr. 035.207.

Iron line spectroscopy with the AXAF calorimeter.
See Abstr. 035.208.

The EUVE software development.
See Abstr. 036.055.

Science operations for future space astrophysics missions.
See Abstr. 036.137.

SOLAR–A reformatted data files and observing log.
See Abstr. 036.205.

The SOLAR–A onboard data processor (DP).
See Abstr. 036.293.

Putting Einstein to the test – in space.
See Abstr. 066.106.

Magellan at Venus: new perspectives on the inner solar system.
See Abstr. 093.005.

The exploration of Venus: a Magellan progress report.
See Abstr. 093.074.

An engineering model of the atmosphere of Mars for the Mars–94 project (MA–90).
See Abstr. 097.014.

Landing site selection for the Mars 94 mission: a preliminary study.
See Abstr. 097.025.

Settlement of an engineering model of Martian atmosphere for Mars–94 project.
See Abstr. 097.029.

951 Gaspra: a pre–Galileo estimate of its surface evolution.
See Abstr. 098.048.

Fragmentation of dust grains by Giotto's impact–generated atmosphere during the encounter with comet Halley.
See Abstr. 103.097.

White dwarf stars.
See Abstr. 126.069.

The ROSAT Wide Field Camera identification programme.
See Abstr. 142.020.

Gamma ray astrophysics at energies up to 10 GeV.
See Abstr. 143.037.

Un an d'astronomie γ à haute résolution avec le télescope SIGMA: du centre de la Galaxie à la nova de la Mouche.
See Abstr. 143.061.

Cosmic Background Explorer (COBE) observations: new sky maps of the early Universe.
See Abstr. 161.240.

The diffuse infrared background: COBE and other observations.
See Abstr. 161.363.

052 Astrodynamics, Navigation of Space Vehicles

052.001 **The motion of a symmetric gyrostat around the center of mass in a circular orbit in the presence of viscoelastic bars and disc.**
S. A. El-Hafiez, M. E. Fares.
Celest. Mech. Dyn. Astron., Vol. 51, No. 2, p. 107 - 118 (1991).
The authors consider a symmetric gyrostat which has a homogeneous viscoelastic disc and two bars attached to it. Furthermore, the gyrostat has a rotor oriented inside it such that the rotor is statically and dynamically balanced. This system has a rotational motion around its center of mass in a circular orbit under a central gravitational field. Bending vibrations of the bars and the disc are accompanied by dissipation of energy, which is the cause of the evolution of the system's rotational motion. Using the method of separation of motion and averaging, the approximate equations describing the evolution of rotational motion in terms of Andoyer canonical variables are obtained. The stationary motions for the system are deduced, together with the conditions of its stability.

052.002 **On the determination of a preliminary orbit by the Laplace method.**
M. Yu. Klokacheva.
Astron. Zh., Tom 68, Vyp. 4, p. 863 - 871 (Jul-Aug 1991). In Russian. English translation in Sov. Astron., Vol. 35, No. 4.
Spherical coordinates, obtained from observations of a celestial body, are represented by polynomials. The required for Laplace's method approximate values of the first and second derivatives of spherical coordinates are determined by differentiation of those polynomials. An iteration process permitting the simultaneous improvement of both the orbital elements and the derivatives of spherical coordinates is described. Examples of geostationary satellites' orbit determination are given to elucidate the described method.

052.003 **Evolution of singular elliptic orbits of synchronous spacecraft.**
M. A. Vashkov'yak.
Cosmic Res., Vol. 29, No. 1, p. 123 - 133 (Jul 1991). English translation of Kosm. Issled., Tom 29, Vyp. 1, p. 133 -144 (1991).
A class of elliptic spacecraft orbits is studied whose period of revolution is close to a day, inclined at approximately 60°, of eccentricity 0.5 - 0.8 and with argument of perigee latitude $\pm 90°$. From a system of equations of motion, linearized around the singular solution of the Hill problem, taking into account the compression of the planet, the author obtains the dependences of the spacecraft's orbital elements on time, giving a satisfactory description of their variation on an interval of the order of five years.

052.004 **Influence of the geomagnetic field on periodic satellite motion relative to its center of mass.**
I. M. Aksenenkova.
Cosmic Res., Vol. 29, No. 1, p. 134 - 137 (Jul 1991). English translation of Kosm. Issled., Tom 29, Vyp. 1, p. 145 - 148 (1991).
An analogy is made between the perturbed Lagrange top problem in the case of small gravitational potential perturbations, and the problem of rotation of a satellite, with its center moving in a circular orbit in the equatorial plane, taking into account the Earth's magnetic field, modeled as a dipole with axis aligned with Earth's rotation axis. The role of perturbations in the latter problem is played by magnetization of the satellite's frame, and also the deflection of its intrinsic magnetic moment from the satellite's dynamic symmetry axis. A list is given of the resonance tori of the unperturbed problem, in the vicinity of which exist periodic solutions to the perturbed problem. The investigation is based on Poincare's theorem.

052.005 **A complex analytic solution for the attitude motion of a near-symmetric rigid body under body fixed torques.**
P. Tsiotras, J. M. Longuski.
Celest. Mech. Dyn. Astron., Vol. 51, No. 3, p. 281 - 301 (1991).
Although analytic solutions for the attitude motion of a rigid body are available for several special cases, a comprehensive theory does not exist in the literature for the more complicated problems found in spacecraft dynamics. In the present paper, analytic solutions in complex form are derived for the attitude motion of a near-symmetric rigid body under the influence of constant body-fixed torques. The solution is very compact, which enables efficient and rapid machine computation. Numerical simulations reveal that the solution is very accurate when applied to typical spinning spacecraft problems.

052.006 **Ephemeridenrechnung und Bahnbestimmung geostationärer Satelliten mit Hilfe der Taylorreihenintegration.**
O. Montenbruck.
Dtsch. Geod. Komm. Bayer. Akad. Wiss., Reihe C, Nr. 384. Diss. (Dr.rer.nat.), München Univ. (Germany, F.R.). Fakultät für Bauingenieur- und Vermessungswesen. Verlag der Bayerischen Akademie der Wissenschaften, München (Germany, F.R.). 98 p. (1991). ISBN 3-7696-9430-9.
Contents: 1. Potenzreihenintegration. 2. Implementierung. 3. Lösungseinschließung mit Intervallarithmetik. 4. Bahnmodell. 5. Beobachtungsgrößen. 6. Differentielle Bahnkorrektur. 7. Das Programm ORBIT.

052.007 **Determination of a preliminary orbit by the Laplace method.**
M. Yu. Klokacheva.
Sov. Astron., Vol. 35, No. 4, p. 428 - 432 (Jul 1991). Current Physics Microform No.: 9203X0614. English translation of Astron. Zh., Tom 68, Vyp. 4, p. 863 - 871 (1991).
The spherical coordinates of a celestial object obtained from observations are represented by polynomials. The approximate values of the first and second derivatives of the spherical coordinates, needed to determine the orbit by the Laplace method, are determined by differentiating these polynomials. An iterative process enabling one to improve simultaneously these derivatives and the orbital elements is described. Examples of the determination of the orbits of geostationary Earth satellites by the proposed method are given.

052.008 **The motion of artificial satellites in the set of Eulerian redundant parameters.**
M. A. Sharaf, M. E. Awad, S. A. A. Najmuldeen.
Earth, Moon, Planets, Vol. 55, No. 1, p. 21 - 44 (Oct 1991).
In this paper, the connections between orbit dynamics and rigid body dynamics are established throughout the Eulerian redundant parameters, the perturbation equations for any conic motion or artificial satellites are derived in terms of these parameters. A general recursive and stable computational algorithm is also established for the initial-value problem of the Eulerian parameters for satellites prediction in the Earth's gravitational field with axial symmetry. Applications of the algorithm are considered for cases of short and long term predictions. For the short-term prediction, the authors consider the problem of the final prediction of some typical ballistic missiles in the geopotential model with zonal harmonic terms up to J_{36}, while for the long-term prediction, the perturbed J_2 motion of Explorer 28 over 100 revolutions is considered.

052.009 **Correction for the influencene of the Moon's noncentral gravitational field on the motion of geodynamic artificial satellites.**
A. N. Marchenko.
Kinematics Phys. Celest. Bodies, Vol. 7, No. 2, p. 60 – 69 (1991). English translation of Kinematika Fiz. Nebesn. Tel, Tom 7, No. 2, p. 67 – 77 (1991).
The problem of correcting for the influence of the Moon's noncentral gravitational field on the motion of geodynamic artificial satellites is considered. It is solved by an approach based on introduction of instantaneous centers of attraction of the planets. The perturbed noncentral gravitational field function of the Moon is obtained in the Earth's noncentral field of attraction. A numerical algorithm is written to correct for the effect of the Moon's noncentral gravitational field on the motion of the satellite. It is tested in a calculation of the orbits of such geodynamic satellites as LAGEOS and Etalon. Numerical estimates are given for the contribution of the effect on arcs of various durations.

052.010 **On optimal launching of a spacecraft from the surface of the Moon to the circular orbit of its satellite.**
K. G. Grigor'ev, M. P. Zapletin, D. A. Silaev.
Kosm. Issled., Tom 29, Vyp. 5, p. 695 – 704 (Sep–Oct 1991). In Russian. English translation in Cosmic Res. Vol. 29, No. 5.

052.011 **An analytic solution for Deprit's radial intermediary with drag in the equatorial case.**
J. M. Franco.
Bull. Astron. Inst. Czech., Vol. 42, No. 4, p. 219 – 224 (Jul 1991).
An analytic solution to Deptrit's radial intermediary in the equatorial case with a specific drag model is obtained. The drag model considered is a force proportional to the vector velocity and inversely proportional to the square of the distance to the center of attraction. The integral of angular momentum for a central force field with drag is obtained by operating on the vector differential equation of motion. The solution is obtained in compact form in terms of Bessel functions of the first and second kind. Some numerical simulations to the solution are done.

052.012 **Perturbed motion of an orbiter in the Martian atmosphere.**
V. Mioc, C. Blaga, E. Radu.
Europhys. Lett., Vol. 16, No. 4, p. 327 – 330 (21 Sep 1991).
Using Sehnal's exponential law for density height–dependence $\varrho = \varrho_0 \cdot \exp[A_0 + A_1/h]$, the first–order drag perturbations in the motion of an orbiter evolving in the Martian atmosphere are determined over a nodal period by integrating the Newton–Euler equations. The results are more general than Sehnal's ones: the variations of all orbitals elements are obtained, Mars' oblateness and its atmospheric rotation are considered, the study of very low eccentric orbits is possible.

052.013 **Lifetime of the ROHINI A satellite.**
L. Sehnal, L. Pospíšilová.
Bull. Astron. Inst. Czech., Vol. 42, No. 5, p. 295 – 297 (Oct 1991).
The model of the thermosphere total density distribution TD88 is used to develop the second–order theory of the drag effects. Its usefulness is shown on the determination of the artificial satellites lifetime. The ROHINI A satellite was predicted by other authors using the numerical integration method with numerous integration steps. The authors needed just few adjustments of the elements successively determined to achieve even better accuracy of the lifetime determination.

052.014 **Perturbed motion of a satellite in the atmosphere based on the total density model.**
V. Mioc, E. Radu.
Bull. Astron. Inst. Czech., Vol. 42, No. 5, p. 298 – 304 (Oct 1991).
Using the TD model to describe the density distribution of the upper atmosphere, analytical expressions for the aerodynamic drag perturbations undergone by five independent orbital elements of an artificial satellite, over a nodal period, are established. All three components of the perturbing acceleration are taken into account. The variations of the orbital elements are obtained in a very concentrated form.

052.015 **Exact linearization of non–planar intermediary orbits in the satellite theory.**
J. M. Ferrándiz, A. Fernández–Ferreiros.
Celest. Mech. Dyn. Astron., Vol. 52, No. 1, p. 1 – 12 (1991).
The authors consider the reduction of the equations of motion for non–planar perturbed two body problems into linear form. It is seen that this can be easily accomplished for any element of the class of radial intermediaries to the satellite problem proposed by Deprit in 1981, since they have a functional dependence suitable for linearization. The transformation is worked out by using an adequate set of redundant variables. Four harmonic oscillators are obtained, of which two are coupled through gyroscopic terms. Their constant frequencies contain the secular contribution of the main problem of artificial satellite theory up to the order of the considered intermediary. Therefore, this result may well be interesting in relation to the study and prediction of accurate long–term solutions to satellite problems.

052.016 **Analytical treatment of air drag and Earth oblateness effects upon an artificial satellite.**
F. Delhaise.
Celest. Mech. Dyn. Astron., Vol. 52, No. 1, p. 85 – 103 (1991).
A development of an analytical solution for the motion of an artificial Earth satellite subject to the combined effects of Earth gravity and air drag is presented. The atmospheric model takes into account a linear variation of the density scale height with altitude, the rotation and the oblateness of the atmosphere. The perturbation theory is based upon Lie transforms. The secular and long–periodic terms as well as the short–periodic effects are included in the theory which is valid for small to moderate eccentricities and for all values of the inclination.

052.017 **Motion of artificial satellites in the set of Eulerian redundant parameters. II.**
M. A. Sharaf, M. E.–S. Awad, S. A.–S. A. Najmuldeen.
Earth, Moon, Planets, Vol. 55, No. 3, p. 223 – 231 (Dec 1991).
In this paper the full recurrent power series solution is established for J_2–gravity perturbed motion in terms of the Eulerian redundant parameters. Applications of the method for the problem of the final state prediction are illustrated by numerical examples of some typical ballistic missiles, a final state of very high accuracy is obtained for each case study.

052.018 **Orbit determination and analysis of Meteor 3 (1970–19 A) at 15th–order resonance.**
N. M. Harwood, G. G. Swinerd.
Planet. Space Sci., Vol. 39, No. 7, p. 961 – 974 (Jul 1991).
The orbit of Meteor 3 (1970–19 A) has been determined at 52 epochs between August 1978 and January 1980 during which the orbit was perturbed by 15th–order resonance with the geopotential. U.S. Navy, together with Hewitt camera, theodolite, kinetheodolite and visual observations were used in the orbit determination process. The variations in the inclination and eccentricity have been analysed to determine six lumped harmonics of order 15 and two of order 30. In the analysis of the eccentricity an air drag model incorporating an oblate atmosphere with a diurnal density variation was used. The variations

in the inclination and eccentricity have previously been analysed by King–Hele and Walker using orbits determined by the U.S. Naval Research Laboratory. The aim of this work is to find improved values of the lumped harmonics by using more accurately determined orbits.

052.019 **Application of a satellite aerodynamics model based on normal and tangential momentum accommodation coefficients.**
P. Moore, A. Sowter.
Planet. Space Sci., Vol. 39, No. 10, p. 1405 – 1419 (Oct 1991).
Tangential and normal momentum accommodation coefficients are reviewed as a parametrization of the gas–surface interaction for modelling aerodynamic forces on artificial Earth satellites. By reference to experimental data and other studies, a simple empirical relationship for the normal momentum coefficients, σ', is deduced whilst the tangential coefficient, σ, is taken to be invariant with the angle, ξ, between the incident stream and the surface normal. Parameters within these coefficients are estimated by least–squares reduction of orbital elements of ANS–1 (1974–70 A). Derived results are $\sigma = 0.93$ and $\sigma' = 1.11 - 0.17 \sec\xi$, yielding a drag coefficient for a sphere of 2.05 ± 0.03, which is less than 10% below the accepted values of 2.2 for heights near 300 km.

052.020 **Analysis of the resonance angle of Cosmos 1603 (1984–106A) near 14:1 resonance.**
P. Moore, M. S. Gilthorpe.
Planet. Space Sci., Vol. 39, No. 11, p. 1549 – 1558 (Nov 1991).
During 1987 the near–circular orbit of the U.S.S.R. satellite Cosmos 1603 (1984–106A) was strongly perturbed by 14th–order resonance terms in the Earth's gravity field. Due to near minimum solar activity, aerodynamic effects were subsidiary to resonance to the extent that, in a previous paper, lumped harmonics were recovered from the mean motion. The study is now taken a step further with lumped harmonics sought from the variation of the resonance parameter, $\Phi_{1,14}$. An analytical expression is developed for the variation of $\Phi_{1,14}$ with time due to the principal resonance term and then extended to include the second–order resonance effect. Air–drag is modelled by a small correction term. Lumped harmonics of order 14 and 28 are extracted from the resonance variable and compared against previous values from Cosmos 1603 and those from global gravity fields.

052.021 **The orbital resonances on the motions of artificial satellites.**
Liu Lin, Liao Xinhao, Zhao Zhangyin.
Prog. Astron., Vol. 9, No. 4, p. 298 – 308 (Dec 1991). In Chinese.
Based on the models of the perturbed ideal resonance, all kinds of orbital resonances on the motions of artificial satellites are analyzed, the characteristics of the corresponding orbital resonances and its evolutionary properties are given.

052.022 **A note on generalized lumped coefficients.**
J. Klokočník.
Bull. Astron. Inst. Czech., Vol. 42, No. 6, p. 389 – 392 (Nov 1991).
Generalized lumped coefficients, valid for resonant, m–daily and zonal perturbations with the relevant indices for $\bar{C}_{lm}$, $\bar{S}_{lm}$, $\bar{F}_{lmp}(I)$ and $G_{lpq}(e)$, are presented.

052.023 **A note on classification of harmonic geopotential terms.**
F. Kobrle.
Bull. Astron. Inst. Czech., Vol. 42, No. 6, p. 392 – 394 (Nov 1991).
The description of the motion of an artificial satellite in the Earth's gravity field is available using lumped coefficients. The limited number of such values then enables relatively high accuracy orbit determination. The geopotential harmonic terms have four indices and there is the selector equation to generate any combination of indices. The author found the relations between the parameters of the selector equations and the types of perturbation. Especially the case of orbital resonance was noticed.

052.024 **Drag perturbations of low eccentric orbits in the Martian atmosphere.**
V. Mioc, E. Radu.
Bull. Astron. Inst. Czech., Vol. 42, No. 6, p. 395 – 400 (Nov 1991).
The motion of an orbiter in low eccentric orbit in Mars' atmosphere is studied. The approximation of a spherically symmetrical, rotationless atmosphere is considered. The difference between the perturbed nodal and the corresponding Keplerian period is analytically estimated. The perturbations undergone by five independent orbital elements, over a nodal period, are established, too.

052.025 **Towards a systematic definition of intermediaries in the theory of artificial satellites.**
J. M. Ferrándiz, L. Floria.
Bull. Astron. Inst. Czech., Vol. 42, No. 6, p. 401 – 407 (Nov 1991).
This paper deals with obtaining intermediaries for the main problem in the theory of the Earth's artificial satellites by means of Lie canonical transformations. A suitable change of the independent variable renders the choice of the first–order terms in the new Hamiltonian natural, and the calculation of the respective periodic generating functions then becomes straightforward.

052.026 **An estimation of possible accuracies in determining low–altitude satellite by Chinese network.**
Sun Yiqin, Kang Ying, Zhu Wenyao.
Ann. Shanghai Obs., Acad. Sin., No. 12, p. 62 – 66 (1991). In Chinese.
Using the simulated observational data, the authors have estimated the possible accuracies in determining low–altitude geodetic satellite orbit and the geocentric coordinates of some tracking stations by Chinese network. They have also gained an insight into main errors affecting the orbit determination of the satellite, such as the shift of the tracking station coordinates, the unreality of atmospheric models as well as the gravity field models.

052.027 **Computerized computation for first order non–spherical perturbations of satellites.**
Zhou Ming.
Ann. Shanghai Obs., Acad. Sin., No. 12, p. 67 – 73 (1991). In Chinese.
The first order non–spherical perturbations of satellites including $J_2 J_3 J_4$ and $J_{2,2}$ terms have been automatically derived on IBM–PC computer by manipulating Broucke's symbolic Poisson series processory. The present report introduces the method and a set of relative subroutines.

052.028 **Motion of a satellite with flexible viscous–elastic booms in a non–central gravitational field.**
A. V. Shatina.
Kosm. Issled., Tom 29, Vyp. 6, p. 815 – 821 (Nov – Dec 1991). In Russian. English translation in Cosmic Res., Vol. 29, No. 6.
Motion of an artificial satellite (which is a plane disc with rigid and viscous–elastic booms) in the gravitational field of an asymmetrical planet in the equatorial plane is considered. Equations of the satellite's motion are derived by means of averaging in Delaunay's canonical elements.

052.029 **On motions of a spacecraft asymptotic to its regular precessions.**
B. S. Bardin.
Kosm. Issled., Tom 29, Vyp. 6, p. 822 – 827 (Nov – Dec 1991). In Russian. English translation in Cosmic Res., Vol. 29, No. 6.
The problem of solutions asymptotic to equilibrium position of an autonomous Hamiltonian system with two degrees of freedom in the case of second order resonance is investigated. Necessary and sufficient conditions and approximate analytical representation of asymptotic solutions are found. These results are used in the problem on motions of a dynamically symmetrical spacecraft asymptotic to its regular precessions in a circular orbit.

052.030 **Spatial rotations of a satellite in the circular problem of three bodies at fractional resonances.**
P. S. Krasil'nikov.
Kosm. Issled., Tom 29, Vyp. 6, p. 858 – 871 (Nov – Dec 1991). In Russian. English translation in Cosmic Res., Vol. 29, No. 6.
Spatial rotations of a satellite are considered around the center of masses in the circular problem of three bodies at fractional resonances $\Omega = \omega_0/2$, $\Omega = 2\omega_0$.

052.031 **Generation of multiple flyby orbits.**
M. Yu. Akhlebininskij, M. S. Konstantinov.
Kosm. Issled., Tom 29, Vyp. 6, p. 889 – 904 (Nov – Dec 1991). In Russian. English translation in Cosmic Res., Vol. 29, No. 6.
The problem posed in this paper can be defined as a new view to the generation of multiple flyby trajectories. Celestial bodies in Keplerian orbits are considered. An effective method for obtaining orbits for three bodies in elliptical and circular orbits are presented. The method has been applied to generation of asteroids flyby orbits. Some numerical examples are presented.

052.032 **On regularization of equatorial orbits.**
V. A. Kuz'minykh.
Kosm. Issled., Tom 29, Vyp. 6, p. 944 – 947 (Nov – Dec 1991). In Russian. English translation in Cosmic Res., Vol. 29, No. 6.

052.033 **Automated design of gravity–assist trajectories to Mars and the outer planets.**
J. M. Longuski, S. N. Williams.
Celest. Mech. Dyn. Astron., Vol. 52, No. 3, p. 207 – 220 (1991).
A new approach to planetary mission design is described which automates the search for gravity–assist trajectories. This method finds all conic solutions given a range of launch dates, a range of launch energies and a set of target planets. The new design tool is applied to the problems of finding multiple encounter trajectories to the outer planets and Venus gravity–assist trajectories to Mars. The last four–planet grand tour opportunity (until the year 2153) is identified. It requires an Earth launch in 1996 and encounters Jupiter, Uranus, Neptune, and Pluto. Venus gravity–assist trajectories to Mars for the 30 year period 1995–2024 are examined. It is shown that in many cases these trajectories require less launch energy to reach Mars than direct ballistic trajectories.

052.034 **Studying the trajectories of transfer to the halo–orbit near the libration point L_2 in the Earth–Sun system by using the Moon's gravity.**
M. L. Lidov, V. A. Lyakhova, N. M. Teslenko.
Pis'ma Astron. Zh., Tom 17, No. 12, p. 1124 – 1134 (Dec 1991). In Russian. English translation in Sov. Astron. Lett., Vol. 17, No. 6.
The mission scheme for transfer to the halo–orbit in the Earth–Sun system, proposed by R. W. Farquhar is studied in connection with the project "Relict–2". Using the Moon's gravity effect allows the spacecraft to achieve the halo–orbit with much smaller dimensions than at direct transfer. An economical computational technique has been developed to determine approximate trajectories in 1994. A preliminary analysis of these trajectories has been made.

052.035 **Aerodynamic drag perturbations in artificial satellite nodal period.**
V. Mioc, E. Radu.
Astron. Nachr., Vol. 312, No. 5, p. 327 – 334 (1991).
The difference caused by the atmospheric drag between the nodal period of artificial satellites and the corresponding Keplerian period is analytically studied for orbits with small eccentricities. Considering various effects in the distribution of the upper atmospheric density (described by the Total Density thermospheric model) and the atmospheric rotation as well, an analytic expression for the mentioned difference is established. Some particular cases are discussed.

052.036 **On the influence of the moon's gravitational field on the motion of the artificial satellites.**
A. N. Marchenko.
Manuscr. Geod., Vol. 16, No. 6, p. 360 – 366 (Dec 1991).
The problem of the computation of the influence of non–centricity of the lunar gravitational field on the motion of geodynamical satellites is considered. This problem is solved on the basis of the introduction of an instantaneous center of the planet's attraction. The analytical expression is constructed for the disturbing function of the non–central moon's gravitational field. The numerical algorithm is developed for computation of the considered effect. This algorithm is tested by means of the prediction of the "Lageos" and "Etalon–1" orbits. Quantitative estimations of perturbations from the moon's non–centricity are computed for different orbital arcs of satellites mentioned.

Prediction of solar activity with a neural network and its effect on orbit prediction.
See Abstr. 021.027.

Science observations with the IUE using the one gyro mode.
See Abstr. 036.142.

Periodic solutions and their stability for the problem of gyrostat.
See Abstr. 042.019.

Artificial satellites and the three–body problem: a general case.
See Abstr. 042.051.

The almost constant–speed two–body problem with resistance.
See Abstr. 042.065.

Relativistic effects resulting from the rotation of celestial bodies.
See Abstr. 042.079.

Sequential orbit improvement using Kalman filtering technique in navigation system of aeronomic satellites.
See Abstr. 045.005.

Use of libration–point orbits for space observatories.
See Abstr. 051.040.

Geodynamic parameters derived from 7 years of laser data on Lageos.
See Abstr. 081.051.

053 Artificial Satellites, Space Probes

053.001 **On variations of geostationary satellites' brightness at the boundary of the Earth's shadow.**
A. V. Didenko.
Astron. Tsirk., No. 1549, p. 29 – 30 (May – Aug 1991). In Russian.

053.002 **Satellite Digest – 241.**
Spaceflight, Vol. 33, No. 9, p. 298 – 299 (Sep 1991).
1991 September listing of satellite and spacecraft launches.

053.003 **Satellite Digest – 242.**
Spaceflight, Vol. 33, No. 10, p. 349 (Oct 1991).
1991 October listing of satellite and spacecraft launches.

053.004 **Vehicles moving on planets – "planetokhody".**
I. S. Bolkhovitinov, V. V. Gromov, A. A. Kemurdzhian, P. S. Sologub.
Zemlya Vselennaya, No. 5, p. 75 – 80 (Sep – Oct 1991). In Russian.

053.005 **Photographic positions of artificial geosynchronous satellites at Dresden Lohrmann Observatory.**
D. Böhme, E. Maase, H. Potthoff, K.-G. Steinert, S. Wächter, C. Witschas.
Mitt. Lohrmann-Obs., Tech. Univ. Dresden, Nr. 61.
Technische Univ. Dresden (Germany). Lohrmann Observatorium. 6 p. (1991). Paper contributed to the Fourth COGEOS Workshop, Pisa (Italy), 17 – 18 Jun 1991.

NSSDC Data Listing.
See Abstr. 002.116.

Spacecraft tables 1957 – 1990.
See Abstr. 002.134.

STEP. A Satellite Test of the Equivalence Principle.
See Abstr. 066.216.

Results from the LDEF Interplanetary Dust Experiment.
See Abstr. 106.015.

Theoretical Astrophysics

061 General Aspects (Nucleosynthesis, Elementary Particles, Neutrino Astronomy, etc.)

061.001 **Flavored cosmic strings.**
D. L. Chkareuli.
JETP Lett., Vol. 54, No. 4, p. 189 – 192 (25 Aug 1991). Current Physics Microform No.: 9111X1407.

String solutions related to the spontaneously broken gauge symmetry $SU(3)_H$ of the quark and lepton generations are analyzed. Solutions of a new class are found: binary Z_3–string configurations. The string cosmic networks which they form might make up most of the dark matter in the universe.

061.002 **Particle production by the formation of a global monopole.**
C. O. Lousto.
Int. J. Mod. Phys. A, Vol. 6, No. 20, p. 3613 – 3623 (20 Aug 1991).

The author studies pair production by the changing gravitational field of a global monopole during its formation in the very early universe after the breaking of a global symmetry. He obtains a result of the same order of magnitude as in the case of gauge strings $\varrho \sim (G\eta^2)^2/\tau^4$, where η is the mean value of the scalar field and τ is the time at which the phase transition occurs. He also discusses how a global monopole inside a mini–black–hole affects its final stages of evolution. He finds that neither the Hawking temperature nor the entropy–area relation is essentially modified by the presence of the monopole.

061.003 **Vacuum instability of spin–zero particles around a charged superconducting cosmic string.**
T. J. Allen, L. J. Tassie.
J. Phys. G, Vol. 17, No. 9, p. 1357 – 1362 (Sep 1991).

It is shown that the vacuum surrounding a line charge density may become unstable against the production of a charged scalar condensate. Such a situation could occur, for example, with a charged superconducting cosmic string.

061.004 **The reaction branching $^{31}P(p,\gamma)/^{31}P(p,\alpha)$ in the rp–process.**
C. Iliadis, U. Giesen, J. Görres, S. Graff, M. Wiescher, R. E. Azuma, J. King, M. Buckby, C. A. Barnes, T. R. Wang.
Nucl. Phys. A, Vol. 533, No. 1, p. 153 – 169 (14 Oct 1991).

The reactions $^{31}P(p,\gamma)^{32}S$ and $^{31}P(p,\alpha)^{28}Si$ have been studied in the energy range 280 – 620 keV to investigate the influence of low–energy resonances on the stellar reaction rates. Several new resonances have been observed and the resonance strengths for both reaction channels have been determined. The reaction rates have been calculated from the present results and are compared with the results of Hauser–Feshbach calculations.

061.005 **The $^{10}B(p,\alpha_0)^7Be$ reaction in the thermonuclear energy region.**
M. Youn, H. T. Chung, J. C. Kim, H. C. Bhang, K. H. Chung.
Nucl. Phys. A, Vol. 533, No. 2, p. 321 – 332 (21 Oct 1991).

Differential and total cross sections for the $^{10}B(p,\alpha_0)^7Be$ reaction at proton energies between 120 and 480 keV have been measured. The total cross section shows a weak structure at $E_p \sim 390$ keV. The S–factors agree with those of Szabo et al., but disagree with those of Burcham and Freeman by a factor of 1.3. The S–factors in this energy region are predominantly due to the tail of the 0.01 MeV resonance. The thermonuclear reaction rates show large discrepancies when compared with those of Caughlan et al. and Roughton et al.

061.006 **Inert neutrinos in supernovae.**
K. Kainulainen, J. Maalampi, J. T. Peltoniemi.
Nucl. Phys. B, Part. Phys., Vol. 358, No. 2, p. 435 – 446 (15 Jul 1991). Also published as report NORDITA—90/68–AP(prepr.), 17 p. (1990).

The authors study the constraints the observations of the supernova SN 1987A impose on the mixing of the electron neutrino with an inert neutrino. A part of the electron neutrinos produced in the core of the star convert themselves into inert neutrinos as a result of the mixing. They would then escape the core and cool the star very rapidly in contrast to the observations. It is shown that due to matter effects the mixing of relatively light states propagating in the core is strongly suppressed, so that essentially no bounds exist if the mass of the inert neutrino is less than 200 eV. The authors also give a wave packet treatment of the propagation and the mixing of a neutrino state in matter, and find a condition for the ceasing of the oscillations due to the separation of the wave packets.

061.007 **Electroweak baryogenesis in the two–doublet model.**
N. Turok, J. Zadrozny.
Nucl. Phys. B, Part. Phys., Vol. 358, No. 2, p. 471 – 493 (15 Jul 1991).

A new mechanism through which a net baryon asymmetry could be generated at the electroweak phase transition is discussed. It works efficiently in minimal extensions of the standard model such as occur in supersymmetric theories, where the Higgs sector involves more than one doublet.

061.008 **Massive neutrinos and the weak scale singlet majoron.**
G. Jungman, M. A. Luty.
Nucl. Phys. B, Part. Phys., Vol. 361, No. 1, p. 24 – 44 (26 Aug 1991).

The authors consider the phenomenology of the singlet majoron model of Chikashige, Mohapatra and Peccei in the case where lepton number is broken at the weak scale. This scenario naturally leads to a see–saw mechanism for the neutrino masses in which the light neutrinos are close to their experimental limits. A fourth generation can also be incorporated in a natural way. It is argued that this model is compatible with all experimental, astrophysical, and cosmological constraints at present, but can be stringently tested by experiments in the near future.

061.009 **Neutrino masses from galactic supernovae and large water Čerenkov detectors.**
D. Seckel, G. Steigman, T. P. Walker.
Nucl. Phys. B, Part. Phys., Vol. 366, No. 1, p. 233 – 251 (25 Nov 1991).

The range of μ and τ neutrino masses accessible to large water Čerenkov detectors via the observation of neutrinos from galactic supernovae is discussed. The authors show that the available neutrino mass window from detecting late ($t > 10$ s) arrivals is ~75 eV – 200 keV. A shortage of early ($t < 0.1$ s) events may signal a mass as small as 30 eV, but it seems unlikely that any technique using galactic supernovae may be used for masses as low as 10 eV.

061.010 **Nonlocal mass generation through spontaneous chiral–symmetry breaking for solar and stellar interiors.**
J. P. M. Lebrun.
Nuovo Cimento A, Vol. 104, No. 7, p. 1071 – 1073 (Jul 1991).
It is shown that the solution of integral equation for $\Sigma_2(x)$ in terms of the confluent hypergeometric function published recently leads to nonlocal mass generation. The effect occurs in the presence of neutrino and photon distributions, as in stellar interiors, and depends on finite–mass regulator for the photon.

061.011 **Indirect search for neutralinos at neutrino telescopes.**
A. Bottino, V. de Alfaro, N. Fornengo, G. Mignola, M. Pignone.
Phys. Lett. B, Vol. 265, No. 1/2, p. 57 – 63 (8 Aug 1991).
Under the assumption that cold dark matter is composed (at least partially) of neutralinos, capture and pair annihilation of these particles in the sun and the earth are evaluated in the framework of the minimal SUSY extension of the standard model (MSSM), taking into account the effects due to the radiative corrections to the masses of the Higgs bosons. The flux of the neutrinos produced in the neutralino–neutralino annihilation and the relevant rate of the muons generated by neutrino conversion in the earth and crossing a detector upwardly are calculated. The present experimental limits from underground detectors are employed to set constraints on the MSSM parameters, and the capabilities of future neutrino telescopes to further explore regions in parameter space are analyzed.

061.012 **Goldstone bosons versus domain walls bounded by cosmic strings.**
G. R. Dvali.
Phys. Lett. B, Vol. 265, No. 1/2, p. 64 – 68 (8 Aug 1991).
It is shown that two possible schemes of string bounded domain wall formation are closely related. Walls can be avoided if a certain type of couplings between scalars, responsible for string and wall formation, is excluded from the potential. This enlarges the continuous global symmetry of the model which, being broken, instead of domain walls gives rise to global strings and Goldstone bosons. A realistic example of spontaneously broken family symmetry is considered. It is shown that the existence of the axion in the model with local chiral flavour symmetry $SU(3)_H$ can solve the domain wall problem.

061.013 **Effects of the 17 keV neutrino in supernovae.**
J. Maalampi, J. T. Peltoniemi.
Phys. Lett. B, Vol. 269, No. 3/4, p. 357 – 360 (31 Oct 1991).
The authors discuss the astrophysical implications of the 17 keV neutrino, recently discovered in beta decay experiments. They show that due to their large mixing with the electron neutrinos enhanced by the matter effect, the 17 keV neutrinos are copiously produced in the core of supernovae. This may have substantial effects on the dynamics of the collapse. They consider alternative interpretations of the 17 keV neutrino and give a qualitative discussion on their consequences on the physics of supernovae.

061.014 **Stellar production cross section of $^{176}Lu^m$.**
W. R. Zhao, F. Käppeler.
Phys. Rev. C, Vol. 44, No. 1, p. 506 – 513 (Jul 1991). Current Physics Microform No.: 9108E1932.
The partial neutron capture cross section of ^{175}Lu to the isomer $^{176}Lu^m$ has been measured at $kT = 25$ keV via the activation technique. Systematic uncertainties were carefully evaluated by two independent methods for determination of the induced activities and by variation of the relevant experimental parameters. The final cross section of 1135 ± 30 mb was derived from six different activations. With this result, the production of the long–lived ground state $^{176}Lu^g$ ($t_{1/2} = 41$ Gyr) by s process neutron captures would not be sufficient to account even for the observed lutetium abundance. Consequently, other mechanisms, e.g., thermally induced transitions from the isomer to the ground state must have enhanced the yield of $^{176}Lu^g$ during the s process.

061.015 **Direct and valence neutron capture by ^7Li.**
J. E. Lynn, F. T. Jurney, S. Raman.
Phys. Rev. C, Vol. 44, No. 2, p. 764 – 773 (Aug 1991). Current Physics Microform No.: 9110D0194.
The authors have measured the cross section for neutron radiative capture by ^7Li at thermal neutron energy as 45.4 ± 3.0 mb. They have compared this value, and the available fast–neutron–capture cross–section data on ^7Li, with direct– and valence–capture theory. They recommend that the capture cross sections from the theoretical calculation, instead of those from a recent measurement, be used for nuclear astrophysics calculations.

061.016 **Projectile breakup by nuclear and Coulomb fields and application to astrophysically relevant radiative–capture processes.**
R. Shyam, G. Baur, P. Banerjee.
Phys. Rev. C, Vol. 44, No. 2, p. 915 – 918 (Aug 1991). Current Physics Microform No.: 9110D0345.
The interplay of nuclear and Coulomb contributions to breakup processes is studied theoretically. This is especially relevant for the extraction of astrophysical S factors for radiative–capture reactions. The authors study explicitly the breakup of ^6Li and ^7Li projectiles. Results of the calculations are in good agreement with recent experimental data. The different behavior of the nuclear and Coulomb amplitudes as a function of angle and energy helps to separate them from each other. This is vital for reliable extraction of astrophysical S factors of radiative–capture processes from breakup experiments.

061.017 **Direct reaction analysis of $^{19}F(p, \alpha)^{16}O$ below the Coulomb barrier.**
H. Herndl, H. Abele, G. Staudt, B. Bach, K. Grün, H. Scsribany, H. Oberhummer, G. Raimann.
Phys. Rev. C, Vol. 44, No. 3, p. R952 – R955 (Sep 1991). Current Physics Microform No.: 9111B1570.
The reaction $^{19}F(p, \alpha)^{16}O$ (g.s.) is analyzed in terms of the distorted–wave Born approximation at energies below the Coulomb barrier. The authors calculate differential cross sections at 250, 350, and 450 keV as well as excitation functions between 150 and 850 keV and compare with experimental data. The astrophysical S factor is determined. They conclude that a direct reaction mechanism alone can account for the observed sub–Coulomb cross section below about 500 keV.

061.018 **$^{35}Cl(n, p)^{35}S$ cross section from 25 meV to 100 keV.**
P. E. Koehler.
Phys. Rev. C, Vol. 44, No. 4, p. 1675 – 1678 (Oct 1991). Current Physics Microform No.: 9112A0777.
The author has measured the $^{35}Cl(n, p)^{35}S$ cross section from 25 meV to 100 keV. Strengths were determined for the resonances at $E_n = 398$, 4249, 16340, 27316, and 57700 eV. The astrophysical reaction rate, $N_A\langle\sigma v\rangle$, was calculated from the resonance strengths determined in the measurements, plus the thermal cross section. The role of this reaction in the nucleosynthesis of light elements in the s process as well as in explosive nucleosynthesis is discussed.

061.019 **Measurements of the Coulomb dissociation cross section of 156 MeV ^6Li projectiles at extremely low relative fragment energies of astrophysical interest.**
J. Kiener, H. J. Gils, H. Rebel, S. Zagromski, G. Gsottschneider, N. Heide, H. Jelitto, J. Wentz, G. Baur.
Phys. Rev. C, Vol. 44, No. 5, p. 2195 – 2208 (Nov 1991). Also publised as report KFK—4870, 47 p. (Apr 1991). Current Physics Microform No.: 9112C1549.
Coulomb dissociation of light nuclear projectiles in the electric field of heavy target nuclei has been experimentally investigated as an alternative access to radiative capture cross sections at low

relative energies of the fragments, which are of astrophysical interest.

061.020 **Three dimensional four–fermion (vector)⊗(vector) model is renormalizable and ultraviolet finite.**
N. V. Krasnikov, A. B. Kyatkin.
5. Seminar on Quantum Gravity, p. 188 – 194 (1991). – See Abstr. 012.004 for the main entry.
The authors show that the four–fermion (vector)⊗(vector) model is renormalizable in the framework of the 1/N expansion and moreover it is ultraviolet finite.

061.021 **Optimal Cavendish–type experiment for the search of restrictions on light particles of quantum gravity.**
V. M. Mostepanenko, I. Yu. Sokolov.
5. Seminar on Quantum Gravity, p. 213 – 232 (1991). – See Abstr. 012.004 for the main entry.
The new experiment of Cavendish–type is suggested on measuring deviations from the laws of gravitational and Casimir forces which is optimized for the search of hypothetical long–range forces and light elementary particles. The optimal configuration of test bodies for such experiment is found. It is shown that with the help of suggested experiment all known for today restrictions may be strengthened by several orders of magnitude.

061.022 **Inequalities which constrain negative energy in quantum field theory.**
L. H. Ford.
5. Seminar on Quantum Gravity, p. 330 – 341 (1991). – See Abstr. 012.004 for the main entry.
Negative energy densities and fluxes due to quantum coherence effects are discussed. The case of a massless scalar field in two and four dimensional flat spacetime and Reissner–Nordstrom spacetime is considered. It is argued that negative energy fluxes are constrained by inequalities that, in effect, state that on a timescale t, a (causally connected) system cannot absorb an energy that is more negative than about $-1/t$. It is further argued that these inequalities prevent macroscopic violations of the second law of thermodynamics and of cosmic censorship.

061.023 **The Dirac sea in accelerated frames or magnetic fields.**
S. Hacyan.
5. Seminar on Quantum Gravity, p. 385 – 390 (1991). – See Abstr. 012.004 for the main entry.

061.024 **Hamiltonian diagonalization in curved space.**
M. Castagnito, R. Ferraro, J. Sztrajman.
5. Seminar on Quantum Gravity, p. 391 – 403 (1991). – See Abstr. 012.004 for the main entry.
The authors review the well known Hamiltonian diagonalization criterion to define the ground state of a quantized field, in a curved space. The modes diagonalizing the Hamiltonian are found by solving an eigenvalue equation. This formalism is extended to the spin 1 case, and a non–trivial application in Schwarzschild geometry is shown.

061.025 **Experimental constraints on dark matter.**
J. Ellis.
5. Seminar on Quantum Gravity, p. 423 – 455 (1991). – See Abstr. 012.004 for the main entry.
The author discusses some of the particle candidates for dark matter than he personally finds the most plausible. For each one, he reviews the constraints on it that come from both accelerator and non–accelerator experiments. Among the new topics reviewed here are the constraints on dark matter candidates given by LEP, and a recent resurgence of interest in metastable particles from a hidden sector of string theory as possible candidates for dark matter.

061.026 **Die Geburt der Elemente. Das Innere der Sterne: Schmelztiegel für die Stoffe dieser Welt.**
N. Langer.
Bild Wiss., No. 9, p. 62 – 68 (Sep 1991).

061.027 **Cartan's torsion vector and spin–0 fields**
L. C. Garcia de Andrade.
7. Latin American Symposium on Relativity and Gravitation, p. 519 – 521 (1991). – See Abstr. 012.005 for the main entry.
Cartan's torsion vector is shown to be associated to a pair of massive particles of spin–0 (scalar fields) and massive photon (spin–1 particle).

061.028 **Soliton operators in two and three dimensions.**
J. Stephany.
7. Latin American Symposium on Relativity and Gravitation, p. 522 – 528 (1991). – See Abstr. 012.005 for the main entry.
The properties and applications of fermionic order and disorder variables in two and three dimensions, and their relation with bosonization are discussed.

061.029 **Quantization requires a Lagrangian.**
S. A. Hojman, L. C. Shepley.
7. Latin American Symposium on Relativity and Gravitation, p. 529 – 531 (1991). – See Abstr. 012.005 for the main entry.
It is shown that in order to quantize a classical system, a Lagrangian is required, if the condition that the coordinates commute is set.

061.030 **Second quantised Hamiltonian for light cone gauge heterotic superstrings.**
A. Restuccia.
7. Latin American Symposium on Relativity and Gravitation, p. 532 – 535 (1991). – See Abstr. 012.005 for the main entry.
The Hamiltonian of the second quantised field theory for the Heterotic Superstring is constructed from the complete closure of the second quantised field theory non–linear realisation of the ten–dimensional Super Poincare Algebra. The closure requires only quartic terms in the Hamiltonian.

061.031 **BRST effective actions on phase space.**
M. Caicedo, A. Restuccia.
7. Latin American Symposium on Relativity and Gravitation, p. 536 – 541 (1991). – See Abstr. 012.005 for the main entry.
By using the BRST approach of Kugo and Uehara on phase space a modified BFV approach to Quantum Field Theory is obtained. It is most appropriate to discuss the boundary conditions under which the BFV theorem is valid, a minimal set of boundary conditions in terms of ghost fields is found. The domains of integration for the BFV theorem for the relativistic particle and String Theory are briefly discussed.

061.032 **Quantum effects associated to a Dirac particle in an asymptotically uniformly accelerated frame.**
U. Percoco, V. M. Villalba.
7. Latin American Symposium on Relativity and Gravitation, p. 542 – 548 (1991). – See Abstr. 012.005 for the main entry.
The authors study the phenomena of creation of scalar and spin 1/2 massless particles in a system of coordinates associated to an observer moving with asymptotically uniform acceleration.

061.033 **The α–process and the r–process.**
R. D. Hoffman, S. E. Woosley.
Bull. Am. Astron. Soc., Vol. 23, No. 2, p. 975 (1991). Abstract. – See Abstr. 012.008 for the main entry.

061.034 **The minimal 17 keV neutrino model.**
E. D. Carlson, D. Land, L. Randall.
Phys. Lett. B, Vol. 264, No. 1/2, p. 132 – 136 (25 Jul 1991).
Recent results indicating the existence of a 17 keV neutrino can be explained within the context of the singlet majoron model. The authors consider the minimal such model, in which only one right–handed neutrino is included. Neutrinoless double beta decay limits are evaded naturally if this fourth neutrino is not too heavy. Cosmological problems are avoided by the annihilation of the 17 keV neutrinos into majorons. The absence of any astrophysical signal from the radiative decays of these neutrinos further constrains the model. Possible signatures of this model include a peak or shoulder in the background photons at 8.5 keV,

the appearance of either neutrinoless double beta decay or a kink in beta decay spectra, and a tau neutrino signal from future supernovae that is delayed by a few days.

061.035 **The r–process and nucleochronology.**
J. J. Cowan, F.–K. Thielemann, J. W. Truran.
Phys. Rep., Vol. 208, No. 4/5, p. 267 – 394 (Nov 1991).

The r–process of nucleosynthesis is the process which is responsible for the synthesis of approximately half of the nuclear species in nature that are more massive than iron. While the astrophysical site for r–process nucleosynthesis is not yet identified, its association with massive stars undergoing type II supernova events is strongly suggested. This can be deduced from the observation that r–process nuclei are already present in the oldest and most metal–deficient stars, which are the tracers of the chemical evolution of the Galaxy. The massive stars that become type II supernovae evolve fasted and contribute their ejecta to the interstellar medium at the earliest beginnings of the chemical evolution of the Galaxy. When utilizing the knowledge of the r–process production ratios of the long–lived chronometer nuclei, their observed ratios in primitive meteorites and our (limited) knowledge of star formation throughout galactic evolution, one can put limits on the duration of galactic nucleosynthesis, the age of the Galaxy and the universe. The latter has a large uncertainty but is comparable with ages estimated from globular clusters and cosmology. This paper provides a review of the basic physics underlying the r–process, the operation of the mechanisms for r–process nucleosynthesis, the possible, astrophysical sites, their time evolution in galactic evolution models, and the inferred ages of the Galaxy.

061.036 **Capture cross section measurements of krypton and xenon isotopes and the fundamental parameters of the s–process.**
H. Beer.
Astrophys. J., Vol. 375, No. 2, p. 823 – 832 (10 Jul 1991).

The capture cross sections of Kr and Xe isotopes have been determined by a fast cyclic activation technique. The data were used to perform s–process calculations with phenomenological models. The weak and the main s–process component were studied. Astrophysical parameters were determined in the frame of the model, i.e., iron seed abundance, the neutron exposure, average number of neutrons captured by the iron seed, the temperature dependence of the neutron exposure, neutron density, temperature, and electron density. The solar abundances of Kr and Xe were determined. From the ^{85}Kr branching the pulse width of a pulsed s–process was estimated. The isotopic anomaly Xe–S, s–process Xe, was investigated.

061.037 **Astrophysics and cosmology confront the 17 keV neutrino.**
E. W. Kolb, M. S. Turner.
Phys. Rev. Lett., Vol. 67, No. 1, p. 5 – 8 (1 Jul 1991). Current Physics Microform No.: 9108B2037.

Astrophysical and cosmological arguments constrain the properties of a 17 keV Dirac neutrino. It must have interactions beyond those of the standard electroweak theory to reduce its cosmic abundance by a factor of 200. A predicament arises because the additional helicity states of the neutrino necessary to construct a Dirac mass must have interactions strong enough to evade the astrophysical bound from SN 1987A, but weak enough to avoid violating the bound from primordial nucleosynthesis.

061.038 **Response of ^{127}I to solar neutrinos.**
J. Engel, S. Pittel, P. Vogel.
Phys. Rev. Lett., Vol. 67, No. 4, p. 426 – 429 (22 Jul 1991). Current Physics Microform No.: 9109B1101.

The authors use a configuration mixing quasiparticle Tamm–Dancoff approximation to calculate the expected event rate for an ^{127}I solar neutrino detector. The cross section for ^{8}B solar neutrinos is 2.2×10^{-42}cm^2, corresponding to about 13 solar neutrino units (SNU) in the standard solar model. This is a factor of more than 3 below an estimate by Haxton. This value may increase by about 30% when quadrupole correlations are included. The cross section for ^{7}Be neutrinos is 2.0×10^{-45}cm^2 (9.4 SNU). The total event rate per nucleus from all solar neutrinos is about 3 times that in ^{37}Cl.

061.039 **Strength function of ^{127}Xe and iodine–xenon neutrino detector.**
Yu. S. Lutostansky, N. B. Shul'gina.
Phys. Rev. Lett., Vol. 67, No. 4, p. 430 – 432 (22 Jul 1991). Current Physics Microform No.: 9109B1105.

The strength function $S_\beta(E)$ and the cross section of the reaction $\nu_e + {}^{127}I \rightarrow e^- + {}^{127}Xe$ are calculated in the framework of the theory of finite Fermi systems. The results are used to estimate the possibility of using an iodine–xenon detector in solar neutrino experiments. The advantages of the iodine–xenon detector over the current chlorine–argon one are discussed.

061.040 **Constraints from primordial nucleosynthesis on the mass of the τ neutrino.**
E. W. Kolb, M. S. Turner, A. Chakravorty, D. N. Schramm.
Phys. Rev. Lett., Vol. 67, No. 5, p. 533 – 536 (29 Jul 1991). Current Physics Microform No.: 9109C0555.

The familiar nucleosynthesis constraint on the number of neutrino species, $N_\nu \leqslant 3.4$, applies to massless neutrino species. An MeV–mass neutrino can have even greater impact, and the authors show that primordial nucleosynthesis excludes a τ neutrino mass from 0.3 to 25 MeV (Dirac) and 0.5 to 25 MeV (Majorana) provided that its lifetime $\tau_\nu \gtrsim 1$ sec; and from 0.3 to 30 MeV (Dirac) and 0.5 to 32 MeV (Majorana) for $\tau_\nu \gtrsim 10^3$sec. A modest improvement in the laboratory mass limit – from 35 to 25 MeV – would imply that the τ neutrino mass must be less than 0.5 MeV (provided $\tau_\nu \gtrsim 1$ sec).

061.041 **Three–neutrino decay of a 17–keV Dirac neutrino.**
K. S. Babu, R. N. Mohapatra, I. Z. Rothstein.
Phys. Rev. Lett., Vol. 67, No. 5, p. 545 – 547 (29 Jul 1991). Current Physics Microform No.: 9109C0567.

The authors construct a particle physics scenario which avoids the cosmological mass density constraint on a Dirac–type neutrino in the mass range 100 eV – 1 MeV by letting it decay primarily to three light Dirac neutrino final states. The nucleosynthesis constraints on the extra number of neutrinos (i.e., $\delta N_\nu \leqslant 0.3$) is satisfied in this case. Identifying the recently reported 17 keV neutrino with ν_τ, the authors predict its lifetime to be $10^8 \leqslant \tau_\nu \leqslant 10^{13}$sec. A hierarchy in the Dirac neutrino masses where $m_{\nu\tau} \approx 17$ keV and $m_{\nu e} \approx 10^{-3}$eV (so that the solar neutrino puzzle is resolved by the MSW mechanism) can naturally be realized in an extension of the standard model.

061.042 **Determination of the ^{13}N(p, γ)^{14}O reaction cross section using a ^{13}N radioactive ion beam.**
P. Decrock, T. Delbar, P. Duhamel, W. Galster, M. Huyse, P. Leleux, I. Licot, E. Liénard, P. Lipnik, M. Loiselet, C. Michotte, G. Ryckewaert, P. van Duppen, J. Vanhorenbeeck, J. Vervier.
Phys. Rev. Lett., Vol. 67, No. 7, p. 808 – 811 (12 Aug 1991). Current Physics Microform No.: 9109C2149.

The cross section for the astrophysically important ^{13}N(p, γ)^{14}O reaction has been measured directly with an intense and pure 8.2 MeV ^{13}N radioactive ion beam. The average value, for the 5.8 – 8.2 MeV ^{13}N energy range, is 106(30) μb. The partial γ width of the resonance which occurs in this reaction at a center–of–mass energy of 0.545 MeV has been deduced to be 3.8(1.2) eV. This value is compared with theoretical predictions and indirect determinations.

061.043 **Neutrino magnetic moment and the dicyclic group.**
D. Chang, W. Y. Keung, S. Lipovača, G. Senjanović.
Phys. Rev. Lett., Vol. 67, No. 8, p. 953 – 956 (19 Aug 1991). Current Physics Microform No.: 9110C0021.

The authors show a mechanism using a discrete symmetry, the dicyclic group Q_6, which naturally gives rise to a small mass but a large transitional magnetic moment for the neutrino, such that the anticorrelation between the solar neutrino intensity and the sunspot activity can be explained. The process of muonium –

antimuonium oscillation provides the most interesting constraint to the model besides neutrino physics.

061.044 **Geometrical neutrino mass hierarchy and a 17 keV ν_τ.**
K. S. Babu, R. N. Mohapatra.
Phys. Rev. Lett., Vol. 67, No. 12, p. 1498 – 1501 (16 Sep 1991). Current Physics Microform No.: 9111A1862.

The authors present an extension of the singlet majoron gauge model which realizes a novel scheme of geometrical neutrino mass hierarchy proposed recently by Glashow, wherein ν_e and ν_μ are Majorana particles with $m_{\nu e} \approx m_{\nu\mu} \approx 10^{-3}$eV while ν_τ is a Dirac particle with a mass of 17 keV. This model explains the solar neutrino deficit via the MSW mechanism and accounts for the recently reported anomaly in beta decay spectra in a natural manner without any undesirable fine tuning of parameters. An interesting consequence of the model is that ν_τ is short lived with a lifetime of $\approx 10^{-3} - 10^{-1}$sec.

061.045 **Economical neutrino oscillation.**
M. Kobayashi, C. S. Lim, M. M. Nojiri.
Phys. Rev. Lett., Vol. 67, No. 13, p. 1685 – 1687 (23 Sep 1991). Current Physics Microform No.: 9111C2059.

The authors consider pseudo–Dirac type neutrinos with Majorana masses which are smaller than their Dirac ones. It is shown that there are two possible kinds of resonant neutrino oscillations for each generation, whose interplay provides an economical way to solve the puzzles of solar and atmospheric neutrinos simultaneously.

061.046 **Can the Sun shed light on neutrino gravitational interactions?**
A. Halprin, C. N. Leung.
Phys. Rev. Lett., Vol. 67, No. 14, p. 1833 – 1835 (30 Sep 1991). Current Physics Microform No.: 9111D2055.

The authors have examined the effects of a large gravitational field on the phenomenon of neutrino oscillations as contemplated in the Mikheyev – Smirnov – Wolfenstein mechanism. They find that the Sun's gravitational field would amplify any small breakdown in the universality of the gravitational coupling by many orders of magnitude. A breakdown of only 1 part in 10^{14} would still make the gravitational effect comparable to the conventional weak interaction. The differing energy dependences of the two level–crossing mechanisms can therefore be used as a very sensitive tool to test the conventional universality hypothesis.

061.047 **Probing particle acceleration in close binary systems with the neutron capture line.**
W. T. Vestrand.
AIP Conf. Proc., No. 220, p. 222 – 226 (Apr 1991). Current Physics Microform No.: 9106B1356. – See Abstr. 012.020 for the main entry.

The author shows that if the reported detections of TeV and PeV gamma–ray signals from X–ray binaries are correct, then it is likely that sensitive low energy gamma–ray spectral measurements by NASA's Gamma–Ray Observatory will show evidence for a narrow neutron capture line at 2.22 MeV. It is argued that such a detection would provide an unambiguous signature of proton or heavier ion accleration within the binary system.

061.048 **Photoexcitation of 3.7 h $^{176}Lu^m$ and its astrophysical consequences.**
L. Lakosi, I. Pavlicsek, Á. Veres.
Acta Phys. Hung., Vol. 69, No. 3/4, p. 169 – 178 (16 Nov 1990).

Gamma–induced excitation of the 3.7 h half–life $^{176}Lu^m$ was studied by high activity ^{60}Co and ^{24}Na γ–sources as well as bremsstrahlung from a 4 MeV electron linear accelerator. Photoactivation exhibited a steep rise in the energy region concerned, from 1.17 to 4 MeV. Observed abundance of the 3.6×10^{10}y half-life ^{176}Lu was proposed to be used for determining the age of the neutron capture s–process in stellar nucleosynthesis. However, the situation is complicated by the β–radioactivity of the isomer. Photoactivation may reduce drastically the half–life in the stellar interior, so that it would reach the half–life of the isomer at 3.5×10^8K temperature, guaranteeing thermal equilibrium between ground and isomeric states. This makes ^{176}Lu unreliable as a cosmic chronometer.

061.049 **$^{17}O(n, \alpha)^{14}C$ cross section from 25 meV to approximately 1 MeV.**
P. E. Koehler, S. M. Graff.
Phys. Rev. C, Vol. 44, No. 6, p. 2788 – 2793 (Dec 1991). Current Physics Microform No.: 9203A1300.

The authors have measured the $^{17}O(n, \alpha)^{14}C$ cross section from thermal energy to approximately 1 MeV. The data are compared to measurements made via the inverse reaction. There are many differences between the two sets of data. The astrophysical reaction rate was calculated from the measured cross section. This reaction plays a role in the nucleosynthesis of heavy elements in nonstandard big–bang models. At big–bang temperatures, the experimental rate was found to be in fair agreement with the rate estimated from the previously known properties of states of ^{18}O in this region. Furthermore, using the available information from experiments, it was estimated that the $^{17}O(n, \alpha)^{14}C$ rate is approximately a factor of $10^3 - 10^4$ times larger than the $^{17}O(n, \gamma)^{18}O$ rate at big–bang temperatures. As a result, there may be significant cycling between ^{14}C and ^{17}O resulting in a reduction of heavy–element nucleosynthesis.

061.050 **Nuclear structure of ^{176}Lu and its astrophysical consequences. II. ^{176}Lu, a thermometer for stellar helium burning.**
N. Klay, F. Käppeler, H. Beer, G. Schatz.
Phys. Rev. C, Vol. 44, No. 6, p. 2839 – 2849 (Dec 1991). Current Physics Microform No.: 9203A1351.

On the basis of the improved level scheme of ^{176}Lu it is demonstrated that the stellar s–process production as well as the stellar beta decay rate of that nucleus depend strongly on temperature. This behavior results from the completely different half–lives of the ground state and the isomer; since these states are coupled by induced transitions in the hot stellar photon bath, the effective half–life of ^{176}Lu is drastically reduced. The 5^- state at 838.64 keV was identified as the most efficient mediating level. Consequently, ^{176}Lu can no longer be considered as a chronometer for the age of the s–process; instead, it can be interpreted as an s–process thermometer yielding a temperature range between 2.4×10^8 to 3.6×10^8K.

061.051 **^{176}Lu: An unreliable s–process chronometer.**
K. T. Lesko, E. B. Norman, R. M. Larimer, B. Sur, C. B. Beausang.
Phys. Rev. C, Vol. 44, No. 6, p. 2850 – 2864 (Dec 1991). Current Physics Microform No.: 9203A1362.

A level scheme of ^{176}Lu up to ~ 1400 keV excitation energy is deduced from a γ–γ coincidence experiment and previously published particle transfer data. 170 γ–ray transitions are placed between 85 levels. The authors identify 27 previously unknown levels and 131 previously unknown transitions in ^{176}Lu. With this γ–ray data one places the energy of the isomer at 122.9 keV. A level at 838.5 keV is found to decay with substantial strength to both the ground state and the 122.9 keV isomer. The resulting temperature sensitivity of its effective half–life rules out the use of ^{176}Lu as an s–process chronometer. The use of ^{176}Lu to determine s–process temperatures is discussed.

061.052 **Auf der Suche nach 'seltsamer' Materie.**
K. Lützenkirchen.
Naturwissenschaften, Vol. 78, No. 11, p. 489 – 496 (Nov 1991).

The author reports on a search for supermassive nuclei in nature with masses up to 10^7amu. Such exotic nuclei might consist, for example, of stable 'strange' matter, which comprises a mixture of up, down, and strange quarks. Supermassive nuclei are among the candidates of dark matter which is likely to constitute a major fraction of the mass of the Universe. In the experiments no clear indication of the presence of supermassive isotopes was observed.

061.053 **Erratum: "Pycnonuclear fusion rates" [Astrophys. J., Vol. 365, No. 1, p. 296 – 300 (10 Dec 1990)].**
S. Schramm, S. E. Koonin.
Astrophys. J., Vol. 377, No. 1, p. 343 (10 Aug 1991). See Abstr. 52.061.091.

061.054 **Non-standard neutrinos in the Sun and supernovae.**
J. T. Peltoniemi.
Comment. Phys.-Math. Chem.-Med., No. 132, p. 1 – 34 (1991). ISBN 951-653-233-0. = Diss., Univ. Helsinki.

Theoretical implications of the observations of neutrinos from the Sun and supernovae are considered. Special emphasis is laid on the effects of the neutrino mixing and of the magnetic moments of neutrinos. The magnetic moment of the electron neutrino is calculated for a general coupling to an arbitrary lepton. It is found that in models involving mirror leptons a magnetic moment large enough to solve the solar neutrino problem can be generated for the electron neutrino. It is argued that the best limit the supernova 1987A imposes on the magnetic moment of the electron neutrino is $\mu_{\nu e} < 2 \times 10^{-11} \mu_B$ if the right-handed neutrinos have no auxiliary interactions. In models involving mirror leptons the magnetic moment of an electron neutrino may be arbitrarily large without contradicting the observations of the neutrinos from SN 1987A. The mixing of the electron neutrino and a light ($m \lesssim 100$ eV) singlet neutrino is shown to be irrelevant for the dynamics of a supernova. For heavier neutrinos the ranges $2 \times 10^5 \mathrm{eV}^2 < \sin 2\theta_0 \, \delta m^2 < 10^8 \mathrm{eV}^2$ for $\delta m^2 \ll 10^9 \mathrm{eV}^2$ and $7 \times 10^{-9} < \sin^2 2\theta_0 < 10^{-2}$ for $\delta m^2 \gg 10^9 \mathrm{eV}^2$ can be ruled out irrespective of supernova models, but even stronger limits can be set in some specific cases.

061.055 **Cosmic neutrino radiation in the far ultraviolet and the E_6 model.**
J. Maalampi, M. Roos.
Phys. Lett. B, Vol. 263, No. 3/4, p. 437 – 440 (18 Jul 1991).

Observations of the cosmic ultraviolet background by the Berkeley UVX spectrometer sets an upper limit to the intensity of radiation due to possibly existing radiatively decaying primordial particles in the mass range 13.5 – 17.5 eV. If such particles exist with a number density equalling that of neutrinos today they must have a lifetime exceeding 4.6×10^{23}s. The authors explore the idea of Sciama that tau neutrinos ν_τ of mass 29 ± 1 eV with a lifetime of $(2 \pm 1) \times 10^{23}$s decay radiatively into ν_μ or ν_e, thereby being responsible for the ionization of interstellar and intergalactic hydrogen (the Gunn-Petersen effect). In the standard model this is impossible; however, in the context of an E_6 model the neutrinos may have radiative lifetimes of the order of 10^{23}s.

061.056 **Cosmology, astrophysics, and a 17 keV neutrino.**
L. M. Krauss.
Phys. Lett. B, Vol. 263, No. 3/4, p. 441 – 447 (18 Jul 1991).

Recent observations of structure below the endpoint in various beta decay spectra suggest that the cosmological consequences of a possible 17 keV neutrino admixture with the electron neutrino should be re-considered. The author derives constraints coming from the evolution of the universe, large scale structure, and big bang nucleosynthesis. When these are combined with constraints from astrophysical observables, including neutrinos from SN 1987A, the restrictions on models are quite severe. One model motivated by the recent data appears unacceptable, and the class of models which manages to satisfy all the constraints appears incompatible with the MSW solution to the solar neutrino problem.

061.057 **Direct detection of galactic axions with Rydberg atoms in an inhibited cavity regime.**
S. Matsuki, K. Yamamoto.
Phys. Lett. B, Vol. 263, No. 3/4, p. 523 – 528 (18 Jul 1991).

A new scheme to detect galactic axions directly with Rydberg atoms is proposed and discussed. Rydberg atoms are excited from the lower to the upper fine-structure states by absorbing the axions in an inhibited (non-resonant) cavity regime. The weak transition rate in axion absorption could be enormously enhanced due to the huge occupation number of the axions if the axions constitute our galactic halo, while the transition rate in photon absorption could be strongly suppressed in a cooled, inhibited cavity regime. Thus it would be possible to get a good signal-to-noise ratio for axion hunting.

061.058 **Multiple-scattering suppression of the bremsstrahlung emission of neutrinos and axions in supernovae.**
G. Raffelt, D. Seckel.
Phys. Rev. Lett., Vol. 67, No. 19, p. 2605 – 2608 (4 Nov 1991). Current Physics Microform No.: 9112F0583.

In a supernova core the nucleon collision rate Γ_{coll} far exceeds the oscillation frequency of neutrino or axion radiation originating from nucleon bremsstrahlung processes of the type $NN \to NN + \nu\bar{\nu}$. Therefore, including the imaginary part of the self-energy in the effective nucleon propagator, the rate for such processes is suppressed by a factor $[1 + (\Gamma_{\mathrm{coll}}/\xi \cdot T)^2]^{-1}$, where ξ is a numerical factor and T is the temperature. Previous SN 1987A bounds on axions, right-handed neutrinos, and other particles must be relaxed accordingly.

061.059 **Pb(^{6}Li, αd)Pb breakup experiment to test the feasibility of extracting the astrophysically relevant α + d capture cross section.**
J. Hesselbarth, K. T. Knöpfle.
Phys. Rev. Lett., Vol. 67, No. 20, p. 2773 – 2776 (11 Nov 1991). Current Physics Microform No.: 9201B1123.

The (α + d) breakup of 60 MeV ^{6}Li scattered from ^{208}Pb has been measured inside the grazing angle for c.m. energies of the fragments between 100 keV and 1.5 MeV. The integrated cross sections are found to be in agreement with Coulomb excitation theory, but the angular correlations of nonresonant breakup exhibit significant deviations from this theory. This shows that the measured breakup cross section cannot be related by first-order Coulomb excitation theory to the astrophysically relevant ^{4}He(d, γ)^{6}Li capture reaction.

061.060 **Dark matter and the equivalence principle.**
J. A. Frieman, B. A. Gradwohl.
Phys. Rev. Lett., Vol. 67, No. 21, p. 2926 – 2929 (18 Nov 1991). Current Physics Microform No.: 9201A2090.

If the dark matter in galaxies and clusters is nonbaryonic, it can interact with additional long-range fields that are invisible to experimental tests of the equivalence principle. The authors discuss the astrophysical and cosmological implications of a long range force coupled only to the dark matter and find rather tight constraints on its strength. If the force is repulsive (attractive), the masses of galaxy groups and clusters (and the mean density of the universe inferred from them) have been systematically underestimated (overestimated). Such an interaction also has unusual implications for the growth of large-scale structure.

061.061 **β^+ decays of ^{37}Ca: implications for the efficiency of the ^{37}Cl solar ν detector.**
A. García, E. G. Adelberger, P. V. Magnus, H. E. Swanson, O. Tengblad, D. M. Molz.
Phys. Rev. Lett., Vol. 67, No. 26, p. 3654 – 3657 (23 Dec 1991). Current Physics Microform No.: 9201E1866.

The authors studied ^{37}Ca β^+ decays using a mass separated ^{37}Ca beam and a proton telescope with 16 keV resolution. Delayed p's were detected in coincidence with γ's to identify p decays of ^{37}K daughter levels to excited states of ^{36}Ar. They observed 43 previously undetected proton groups, yielding an integrated B(GT) twice as large as that observed previously. Although the result increases the efficiency of the ^{37}Cl detector for counting solar ν's by 6%, its main effect is to reduce uncertainties in this efficiency arising from discrepancies between previous ^{37}Ca β decay and ^{37}Cl(p, n) results.

061.062 **Equation of state in metals and cold stars: evaluation of statistical models.**
D. Lai, A. M. Abrahams, S. L. Shapiro.
Astrophys. J., Vol. 377, No. 2, p. 612 – 628 (20 Aug 1991).

The Thomas – Fermi – Dirac (TFD) statistical model for electronic structure is refined by Weizsäcker gradient and correlation energy corrections. The resulting TFD–λWc model is then used to calculate the nonrelativistic, cold matter equation of state (EOS) in the Wigner – Seitz spherical cell approximation. The correction terms are important mainly at low densities near matter equilibrium (i.e., zero pressure). Inclusion of the gradient term removes many of the unphysical features of the TFD model. Results are summarized for several elements of astrophysical and theoretical importance including $_6C^{12}$, $_{12}Mg^{24}$, $_{13}Al^{27}$, $_{26}Fe^{56}$, and $_{92}U^{236}$. Nonspherical lattice structure effects are then incorporated and are found to be small except at very low density. An accurate EOS near matter equilibrium requires a full quantum–mechanical treatment of the electrons to handle shell–structure effects. The authors construct and critically evaluate a hybrid model which attempts to integrate shell–structure effects with a Thomas–Fermi statistical treatment of the valence electrons. Finally, the mass–radius relation of a low–mass white dwarf model is calculated using the TFD–λWc EOS. The maximum radius of a carbon white dwarf is $R/R_\odot = 3.9 \times 10^{-2}$ at a mass of $M/M_\odot = 2.4 \times 10^{-3}$.

061.063 **Introduction to the theory of particle physics.**
J. Ellis.
International School of Astroparticle Physics, p. 1 – 50 (1991). – See Abstr. 012.037 for the main entry.
Contents: 1. Constructing the standard model. 2. Precision tests of the standard model. 3. Beyond the standard model.

061.064 **A duality between strings and fivebranes.**
M. J. Duff, J. X. Lu.
International School of Astroparticle Physics, p. 155 – 182 (1991). – See Abstr. 012.037 for the main entry.
Not only does the heterotic string admit a heterotic fivebrane as a soliton but the heterotic fivebrane admits as a soliton a heterotic string. This provides further evidence that the two theories may be dual descriptions of the same physics, with the strong coupling regime of one theory described by the weak coupling regime of the other. To illustrate this, the authors show how the energy–momentum tensor of the quadratic Yang–Mills action associated with the string reduces to that of an elementary fivebrane, and how to the energy–momentum tensor of the quartic Yang–Mills action associated with the fivebrane reduces to that of an elementary string.

061.065 **Recent results and future prospects for the MACRO experiment.**
R. C. Webb.
International School of Astroparticle Physics, p. 353 – 374 (1991). – See Abstr. 012.037 for the main entry.
The MACRO experiment being staged at the Gran Sasso Laboratory in L'Aquila, Italy has been designed to study several astroparticle physics questions. The initial research program includes: a sensitive search for GUT monopoles, a search for astrophysical neutrino sources, a study of cosmic ray muons and a search for exotic particles in cosmic rays (WIMPs, etc.). This program is currently taking data with a single detector module. The author presents the latest results from this data taking and the collaboration's plans for the future running of the complete detector.

061.066 **Radioactive dating of the elements.**
J. J. Cowan, F.–K. Thielemann, J. W. Truran.
Annu. Rev. Astron. Astrophys., Vol. 29, p. 447 – 497 (1991). – See Abstr. 003.010 for the main entry.
This review emphasizes the determination of the age of the Galaxy from nuclear chronologies. Typically the most effective dating involves the use of the long–lived radioactive nuclei ^{187}Re, ^{232}Th, ^{238}U, and ^{235}U to set limits on the galactic age. The fact that all four of these cosmochronologically important radioactive isotopes are formed in a single astrophysical process, the rapid neutron capture process (r–process), allows reasonable estimates of the relative levels of production to be obtained. A brief discussion of the production ratios for these chronometers, as well as those formed in other nucleosynthesis processes, is presented. Effects on the integrated chemical and dynamical evolution of our Galaxy are discussed. Age determinations are compared to those from stellar sources (globular clusters and white dwarfs) and age constraints from cosmological considerations are summarized.

061.067 **Particle energization.**
G. Gisler.
Workshop on the Interpretation of Modern Synthesis Observations of Spiral Galaxies, p. 1 – 15 (1991). – See Abstr. 012.041 for the main entry.
A first–principles approach to the physics of particle energization is presented. The general physics of particle acceleration is then applied to a number of the classical astrophysical mechanisms for accelerating particles, with references to recent literature where these are used in specific circumstances. The solar flare is recommended as a microcosm for studying particle acceleration because many different processes seem to be occurring in close proximity, and there is abundant high time resolution data for diagnosing those processes. Finally, a list of possible sites and mechanisms for particle acceleration in spiral galaxies is presented.

061.068 **A critical look at cosmions.**
J. Kaplan, F. M. de Volnay, C. Tao, S. Turck–Chièze.
Astrophys. J., Vol. 378, No. 1, p. 315 – 325 (1 Sep 1991).
The authors have performed a thorough numerical investigation of the cosmion hypothesis as a solution to the solar neutrino puzzle, using the Saclay solar evolution code. The ensuing constraints on cosmion parameters appear as domains in the plane of cosmion – proton cross section versus cosmion mass. Various hypotheses on cosmion couplings are considered. Comparison with available experimental data on direct dark matter detection excludes most of the domain in the "vector" case. In the "axial" case the present results provide useful hints for experiments. The effect of cosmions on the sound speed near the center of the Sun, confronted with present knowledge and data on helioseismology, does not improve the comparison between standard predictions and helioseismological inversion.

061.069 **Neutrino astrophysics: new adjunct of astronomy and elementary particle physics.**
A. K. Mann.
10. Santa Cruz Summer Workshop in Astronomy and Astrophysics: Supernovae, p. 331 – 332 (1991). – See Abstr. 012.052 for the main entry.

061.070 **Neutrino inelastic scattering and prompt shocks in core bounce supernovae.**
A. Ray.
10. Santa Cruz Summer Workshop in Astronomy and Astrophysics: Supernovae, p. 362 – 364 (1991). – See Abstr. 012.052 for the main entry.

061.071 **Is there life after Fuller, Fowler and Newman?**
M. B. Aufderheide.
10. Santa Cruz Summer Workshop in Astronomy and Astrophysics: Supernovae, p. 372 – 374 (1991). – See Abstr. 012.052 for the main entry.

061.072 **Neutrino signal from rapidly rotating collapsed stellar iron cores.**
H.–T. Janka.
10. Santa Cruz Summer Workshop in Astronomy and Astrophysics: Supernovae, p. 408 (1991). – See Abstr. 012.052 for the main entry.

061.073 **Neutrino nucleosynthesis in massive stars.**
D. Hartmann.
10. Santa Cruz Summer Workshop in Astronomy and Astrophysics: Supernovae, p. 626 – 629 (1991). – See Abstr. 012.052 for the main entry.

061.074 **Possible gamma–ray signatures of an r–process event.**
B. S. Meyer, W. M. Howard.
10. Santa Cruz Summer Workshop in Astronomy and Astrophysics: Supernovae, p. 630 – 632 (1991). – See Abstr. 012.052 for the main entry.

061.075 **Mass splitting in the baryon octet.**
F. Hoyle.
Astrophys. Space Sci., Vol. 186, No. 2, p. 299 – 320 (Dec 1991).
The masses of the particles of the baryon octet have been calculated using the theory of the origin of mass described in a former paper (Hoyle, 1990).

061.076 **Particle masses in the baryon decouplet.**
F. Hoyle.
Astrophys. Space Sci., Vol. 186, No. 2, p. 321 – 330 (Dec 1991).
The masses of particles in the baryon decouplet have been obtained using the theory developed in two earlier papers (Hoyle, 1990 and 1991).

061.077 **Neutron capture in ^{122,123,124}Te: a critical test for s–process studies.**
K. Wisshak, F. Voß, F. Käppeler, G. Reffo.
Kernforschungszentrum Karlsruhe GmbH (Germany). Inst. für Kernphysik, KFK—4899. 98 p. (Nov 1991).
The neutron capture cross sections of ^{122,123,124,125,126}Te were measured in the energy range from 10 to 200 keV at the Karlsruhe Van de Graaff accelerator using gold as a standard. Neutrons were produced via the ^{7}Li(p, n)^{7}Be reaction by bombarding metallic Li targets with a pulsed proton beam. Capture events were registered with the Karlsruhe 4π Barium Fluoride Detector. Several runs have been performed under different experimental conditions to study the systematic uncertainties in detail. The cross section ratios were determined with an overall uncertainty of $\sim 1\%$. Maxwellian averaged neutron capture cross sections were calculated for thermal energies between kT = 10 and 100 keV by normalizing the cross section shape up to 600 keV neutron energy reported in literature to the present data. These stellar cross sections were used in an s–process analysis. With the classical approach the abundances of the three s–only isotopes ^{122,123,124}Te could be reproduced within the experimental uncertainties of $\sim 1\%$. The accuracy of the present data allowed also to derive constraints for the existing stellar models with respect to the effective neutron density. Furthermore, the p–process abundances for the tellurium isotopes are discussed.

061.078 **Generalized Tremaine – Gunn limits for bosons and fermions.**
J. Madsen.
Phys. Rev. D, Vol. 44, No. 4, p. 999 – 1006 (15 Aug 1991). Current Physics Microform No.: 9110D1053.
The Tremaine – Gunn phase–space constraint giving lower limits on the mass of dark matter particles that were once in kinetic equilibrium is generalized to arbitrary decoupling temperatures and chemical potentials for fermions as well as bosons. It is stressed that even in the cases where no exact limit exists (e.g., for bosons with a chemical potential approaching the mass) a statistical limit can be obtained. In most cases particle masses of several eV or more are necessary to explain the dark matter in galaxy halos.

061.079 **Perturbations on domain walls and strings: a covariant theory.**
J. Garriga, A. Vilenkin.
Phys. Rev. D, Vol. 44, No. 4, p. 1007 – 1014 (15 Aug 1991). Current Physics Microform No.: 9110D1061.
A covariant formalism is developed for describing perturbations on vacuum domain walls and strings. The treatment applies to arbitrary domain walls in (N + 1)–dimensional flat spacetime, including the case of bubbles of a true vacuum nucleating in a false vacuum. Straight strings and planar walls in de Sitter space, as well as closed strings and walls nucleating during inflation, are also considered. Perturbations are represented by a scalar field defined on the unperturbed wall or string world sheet. In a number of interesting cases, this field has a tachyonic mass and a nonminimal coupling to the world–sheet curvature.

061.080 **Sourceless Abelian gauge string in an expanding universe.**
J. R. Morris.
Phys. Rev. D, Vol. 44, No. 4, p. 1015 – 1019 (15 Aug 1991). Current Physics Microform No.: 9110D1069.
A model consisting of a sourceless Abelian gauge string in an expanding spacetime background is examined. Exact solutions are found which describe a radial inflow of gauge field energy that can be associated with the creation or destruction of a sourceless gauge string. The time rate of change of the string flux depends upon the expansion rate of the universe.

061.081 **Gamma–rays from electron, proton beam interactions with matter and/or radiation: application to Cyg X–1, Geminga and 3C 273.**
F. Giovannelli, L. Sabau Graziati, W. Bednarek, S. Karakula, W. Tkaczyk.
7. Course of the International School of Cosmic–Ray Astrophysics: Cosmic rays, supernovae, and the interstellar medium – A NATO Advanced Study Institute, p. 271 – 287 (1991). – See Abstr. 012.053 for the main entry.
The authors review the most recent studies on the electron–proton beam interactions with matter and/or radiation. The photon energy spectra from inverse compton scattering of an arbitrary background radiation by relativistic electron beams, the photon energy spectra from proton–proton interactions via π^0 decay, and the spectra of the secondary electrons from $\pi^\pm$ decay, for relativistic proton beams are discussed. The theoretical spectra have been succesfully used to fit the experimental data of Cyg X–1, Geminga, and 3C 273.

061.082 **Gamma–rays and neutrinos from accretion processes onto collapsed objects: application to 3C 273.**
F. Giovannelli, L. Sabau Graziati, S. Karakula, W. Tkaczyk.
7. Course of the International School of Cosmic–Ray Astrophysics: Cosmic rays, supernovae, and the interstellar medium – A NATO Advanced Study Institute, p. 289 – 306 (1991). – See Abstr. 012.053 for the main entry.
The authors review the methodology used by the authors in the past in order to develop the very important and difficult problem of the calculation of the energy production spectra of the γ–rays and neutrinos produced in the accretion processes onto collapsed objects. Because of the complexity of the problem they start in describing the physical processes in the case of spherical symmetric accretion. The results of these calculations will be used to evaluate the upper limit to the number of the black holes in the Galaxy and to fit the γ–ray COS B experimental data of the quasar 3C 273 and to predict the neutrino fluxes from the directions of 3C 273 and the Galactic Center.

061.083 **An analytical formulation of the double–pulse s–process model.**
H. Beer.
Astrophys. J., Vol. 379, No. 1, p. 409 – 419 (20 Sep 1991).
Recent studies of the astrophysical site of the s–process suggest a double–pulse s–process in the He–burning shell of low–mass asymptotic giant branch stars, where the ^{13}C(α, n) and the ^{22}Ne(α, n) neutron sources are active as well. Previous analyses have been performed in terms of a pulsed neutron s–process from intermediate–mass stars. The periodic neutron bursts were ascribed to the ^{22}Ne(α, n) reaction only. This paper reformulates the mathematics of the s–process in order to study the combined effect of repeated double pulses with different properties. Similar to the model of a single–pulse s–process, the effect of the pulses for the double–pulse model shows up at the branchings but, in contrast to the single–pulse model, also at the unbranched s–process path and at isomeric states populated by neutron capture. The new model contains the simpler models as special cases.

061.084 **A measurement of the $^{22}Ne(n, \gamma)^{23}Ne$ capture cross section at a stellar temperature of kT = 25 keV.**
H. Beer, G. Rupp, F. Voss, F. Käppeler.
Astrophys. J., Vol. 379, No. 1, p. 420 – 423 (20 Sep 1991).
The capture cross section of ^{22}Ne has been determined by a fast cyclic activation technique. The measurements were carried out at the Karlsruhe 3.75 MV pulsed Van de Graaff accelerator using the $^7Li(p, n)$ reaction close to the reaction threshold to generate neutrons with a distribution resembling a Maxwell spectrum of kT = 25 keV. The activation samples consisted of a mixture of enriched ^{22}Ne (99.9%) and natural Kr–gas contained in stainless steel spheres. The activity of the samples was counted with a high–resolution Ge(Li) detector via the characteristic 439 keV ^{23}Ne γ–ray line. The ^{22}Ne capture cross section at kT = 25 keV was found to be 66.0 ± 5.0 μbarn.

061.085 **Leptonic atoms (positronium and muonium) in astrophysical objects.**
V. V. Burdyuzha, V. L. Kauts, N. P. Yudin.
3. Haystack Observatory Conference on the Interstellar Medium: Skylines – atoms, ions and molecules – new results in spectral line astrophysics, p. 399 – 402 (1991). – See Abstr. 012.051 for the main entry.
Positrons and muons are produced in some astrophysical processes and interacting with electrons they produce the leptonic atoms – positronium (e^+e^-) and muonium (μ^+e^-). The production of μ^-e^+ and $\mu^+\mu^-$ is less probable because of its small density. Positronium is the lightest atom accompanying the process of the annihilation. In the laboratory, the annihilation line was observed by Curie and Joliet (1933). In astrophysical conditions, the 511 keV line was observed from the Sun and from the center of the Galaxy. Future missions with more sensitive spectrometers will lead to a new class of sources of annihilation lines. Astrophysical conditions for annihilation are the magnetosphere of pulsars, supernovae, the relativistic jets in compact radio sources, accreting disks around black holes, the evaporation of mini black holes, and the early Universe.

061.086 **High–resolution computations on inhomogeneous nucleosynthesis.**
H. Kurki–Suonio.
Workshop on Primordial Nucleosynthesis, p. 28 – 35 (1990). – See Abstr. 012.060 for the main entry.
To obtain accurate results of inhomogeneous nucleosynthesis with neutron diffusion it is necessary to have a good spatial resolution of the evolving density and abundance profiles. A closure baryon density seems to be ruled out because of helium overproduction.

061.087 **Astrophysical measurements: current and future.**
R. T. Rood.
Workshop on Primordial Nucleosynthesis, p. 36 – 48 (1990). – See Abstr. 012.060 for the main entry.
To constrain theories of primordial nucleosynthesis we need observational estimates of the abundances of ^2He, ^3He, ^4He, and ^7Li. To be more strictly correct we require the abundances of these isotopes 15 Gyr ago – i.e., we must correct for chemical evolution. Each of the abundances presents formidable obstacles, yet each is known to a precision which places significant constraints on cosmological parameters and particle physics. Still, with higher precision one could place tighter constraints and perhaps even explore the quark–hadron phase transition. The author summarizes the current status and near term future prospects with an emphasis on ^3He as an archetype.

061.088 **Measurement of cross sections: current experiments.**
H. P. Trautvetter.
Workshop on Primordial Nucleosynthesis, p. 63 – 76 (1990). – See Abstr. 012.060 for the main entry.
The experimental situation for light ion reactions is presented and the relations of the laboratory cross section to the quantities needed for stellar models are discussed. Two examples of nuclear reactions involving radioactive nuclei are given and compared with results obtained with indirect methods. In both cases the lifetime of the radioactive nucleus was long enough to use it as a target. For reactions on short lived isotopes, such as ^8Li, radioactive ion beam facilities are needed.

061.089 **Measurement of cross sections: new needs and possibilities.**
R. N. Boyd.
Workshop on Primordial Nucleosynthesis, p. 77 – 91 (1990). – See Abstr. 012.060 for the main entry.
Experimental studies of nuclear cross sections over the past several decades have done much to elucidate the processes of nucleosynthesis and energy generation in a variety of astrophysical sites. Recent studies have begun to focus on reactions which have not previously been studied either because they were experimentally difficult, or because they had been overlooked as being important in previous theoretical studies. This paper presents a proposed study of the $^{26}Al^m(p, \gamma)^{27}Si$ reaction as an example of an experiment which will soon be feasible. It then presents results of some new theoretical work on primordial nucleosynthesis which pinpoints some of the critical reactions for which no cross section measurements have been made. These primarily involve ^8Li, a key nucleus in one model of primordial nucleosynthesis. Finally, instrumentation needed for future measurements in nuclear astrophysics is discussed.

061.090 **Modified photon equation of motion as a test for the principle of equivalence.**
B. Bhawal, H. S. Mani, C. V. Vishveshwara.
Phys. Rev. D, Vol. 44, No. 4, p. 1323 – 1325 (15 Aug 1991). Current Physics Microform No.: 9110D1377.
The authors have considered a modified equation of motion for a photon, based on the principle of covariance. Some astronomical observations are used to place limits on the presence of the extra terms in the modified equation.

061.091 **Mikheyev – Smirnov – Wolfenstein effect with flavor–changing neutrino interactions.**
E. Roulet.
Phys. Rev. D, Vol. 44, No. 4, p. R935 – R938 (15 Aug 1991). Current Physics Microform No.: 9110D0989.
The author considers the effect that flavor – nondiagonal neutrino interactions with matter have on the resonant neutrino oscillations. It is shown that, even in the absence of ν mixing in a vacuum, an efficient conversion of the electron neutrinos from the Sun to another ν flavor can result, if the strength of this interaction is $\sim 10^{-2} G_F$. It is shown how this can be implemented in the minimal supersymmetric standard model with R–parity breaking. Here, the L–violating couplings induce neutrino masses, mixings, and the flavor – nondiagonal neutrino interactions that can provide a Mikheyev – Smirnov – Wolfenstein – like solution to the solar neutrino problem even for negligible vacuum mixings.

061.092 **Solar neutrino solutions with matter–enhanced flavor–changing neutral current scattering.**
V. Barger, R. J. N. Phillips, K. Whisnant.
Phys. Rev. D, Vol. 44, No. 6, p. 1629 – 1643 (15 Sep 1991). Current Physics Microform No.: 9111B1857.
The authors explore quantitatively recent suggestions that the solar neutrino puzzle may be explained by matter enhancement of nonstandard neutral current scattering of neutrinos on d quarks. These neutral currents introduce two new parameters, relating to flavor–diagonal and off–diagonal $\nu - d$ scattering, which affect the dynamical mixing of two neutrino flavors in the Sun. The authors consider this mixing in three alternative scenarios: (1) massless neutrinos, (2) massive neutrinos with off–diagonal but no new diagonal currents, and (3) massive neutrinos with both off–diagonal and new diagonal currents. The authors determine the regions in parameter space that give consistency between the standard solar model and existing solar neutrino data from the Homestake and Kamiokande II experiments, and give the corresponding predictions for future experiments with gallium and superfluid helium detectors. The authors also discuss the more general situation, with nonstandard neutral current

scattering from u quarks and electrons as well as from d quarks, and consider the effects of transmission through the Earth.

061.093 **Shapes of solar neutrino spectra: unconventional tests of the standard electroweak model.**
J. N. Bahcall.
Phys. Rev. D, Vol. 44, No. 6, p. 1644 – 1651 (15 Sep 1991).
Current Physics Microform No.: 9111B1872.
The shapes of the neutrino energy spectra produced by β decays or by nuclear reactions in the solar interior are compared to the corresponding spectral shapes that are produced in a terrestrial laboratory. The most important effects of the Sun, caused by thermal motion of the neutrino-emitting nuclei and by gravitational redshifts, are shown to be small. Any currently measurable difference between the observed shape of a solar neutrino energy spectrum and the spectrum shape for a terrestrial neutrino source must be due to a departure from the standard electroweak model.

061.094 **Rate of $^3H(^7Li, n_0)^9Be$ and big-bang nucleosynthesis.**
S. Barhoumi, G. Bogaert, A. Coc, P. Aguer,
J. Kiener, A. Lefebvre, J. P. Thibaud, F. M. Baumann,
H. Freiesleben, C. Rolfs, P. Delbourgo-Salvador.
Nucl. Phys. A, Vol. 535, No. 1, p. 107 – 119 (9 Dec 1991).
The differential cross sections for the $^3H(^7Li, n_0)^9Be$ reaction measured at 5 angles in the energy range E(c.m.) = 0.2 – 0.9 MeV using a pulsed ^7Li beam and time-of-flight technique. Absolute values of the cross section were obtained by comparison with the well-known cross section of $^3H(d, n)^4He$ at E_d = 1.0 MeV. The resulting reaction rates are obtained at temperatures relevant to big-bang nucleosynthesis, and consequences for primordial ^9Be abundances are discussed.

061.095 **Solar system abundances of the elements.**
Hu Zhongwei.
Prog. Astron., Vol. 9, No. 2, p. 83 – 95 (Jun 1991). In Chinese.
Recent results on element abundances of the solar photosphere, solar corona, solar wind, CI carbonaceous chondrites, etc. are reviewed. New tables on solar system element abundances and their isotopes (nuclides) are listed. Finally, some comparisons and discussions are made.

061.096 **Second angular moment closure on flux-limited neutrino diffusion theory.**
R. Dgani, J. Cernohorsky.
J. Quant. Spectrosc. Radiat. Transfer, Vol. 46, No. 1, p. 1 – 9 (Jul 1991).
The energy and momentum balance equations, as they arise in flux-limited neutrino diffusion theory (FNDT), are numerically solved as a coupled set. The closure explicitly includes the pressure. The results are compared with earlier calculations in which the standard FNDT approach was followed. In particular, the corrections to the fluxes and momentum transfer source terms are highlighted.

061.097 **Mass limits for dark matter particles derived from high-energy neutrinos from the Sun.**
N. Sato, K. S. Hirata, T. Kajita, T. Kifune, K. Kihara,
M. Nakahata, K. Nakamura, S. Ohara, Y. Suzuki, Y. Totsuka,
Y. Yaginuma, M. Mori, Y. Oyama, A. Suzuki, K. Takahashi,
H. Takei, T. Tanimori, M. Koshiba, T. Suda, T. Tajima,
K. Miyano, H. Miyata, M. Yamada, Y. Fukuda, K. Kaneyuki,
Y. Nagashima, M. Takita, E. W. Beier, L. R. Feldscher,
E. D. Frank, W. Frati, S. B. Kim, A. K. Mann,
F. M. Newcomer, R. van Berg, W. Zhang.
Phys. Rev. D, Vol. 44, No. 8, p. 2220 – 2240 (15 Oct 1991).
Current Physics Microform No.: 9112A1144.
High energy neutrinos could be emitted from the center of the Sun as annihilation products of heavy galactic dark matter particles. The authors analyze neutrino events observed by the Kamiokande detector to search for such neutrinos. The observed upper limits on solar high energy neutrinos from these data set limits on the mass and density of certain Galactic dark matter. The particles analyzed are heavy Dirac neutrinos (ν_D), heavy Majorana neutrinos (ν_M), Higgsinos (h), sneutrinos (nu) with three flavors, and photinos (gamma) as the lightest supersymmetric particles. Excluded mass regions are obtained for each dark matter candidate.

061.098 **Zee model on Majorana neutrino mass and magnetic moment.**
B. K. Pal.
Phys. Rev. D, Vol. 44, No. 8, p. 2261 – 2264 (15 Oct 1991).
Current Physics Microform No.: 9112A1185.
Some features of the Zee model of lepton number nonconservation are discussed. The solar neutrino puzzle can be solved within the framework of the Zee model which ensures a large transition magnetic moment, consistent with known experiments.

061.099 **Vacuum fluctuation inside a toroidal spacetime containing a cosmic string.**
G. E. A. Matsas.
Phys. Rev. D, Vol. 44, No. 8, p. 2593 – 2594 (15 Oct 1991).
Current Physics Microform No.: 9112A1517.
The author calculates the vacuum fluctuation $\langle\varphi^2\rangle$ of massless twisted and untwisted scalar fields outside a straight cosmic string in a flat Friedmann spacetime endowed with an openR × openS^1 × openR^2 topology.

061.100 **Solar neutrinos and the Mikheyev – Smirnov – Wolfenstein theory.**
H. A. Bethe, J. N. Bahcall.
Phys. Rev. D, Vol. 44, No. 10, p. 2962 – 2969 (15 Nov 1991).
Current Physics Microform No.: 9201B0690.
The observation of solar neutrinos by Kamiokande shows that the solar neutrino problem cannot be solved by changing the solar model. In combination with the observations with a chlorine detector, it makes the nonadiabatic form of the Mikheyev – Smirnov – Wolfenstein theory most likely, and determines $\Delta m^2 \sin^2\theta = 1.0 \times 10^{-8} eV^2$. Probably all neutrinos go through the resonance in the Sun, those from ^8B nonadiabatically, all others adiabatically. The latter emerge from the Sun in the higher-mass eigenstate ν_2 and have a probability $\sin^2\theta$ to be detected as ν_e. The gallium experiments, when done with sufficient accuracy, will be able to determine $\Delta m^2 = m^2(\nu_\mu) - m^2(\nu_e)$ within fairly close limits. If the day-night effect can be measured, it will further constrain these limits. The small value of $\Delta m^2 \sin^2\theta$ explains why the oscillation from ν_e to ν_μ has not been observed in the laboratory. From existing experiments, the temperature at the center of the Sun can be determined to be within about 6% of that derived from the standard solar model; future neutrino experiments may determine it to within 1%.

061.101 **Limits to the radiative decay of the axion.**
M. T. Ressell.
Phys. Rev. D, Vol. 44, No. 10, p. 3001 – 3020 (15 Nov 1991).
Current Physics Microform No.: 9201B0729.
An axion with a mass greater than 1 eV should be detectable through its decay into two photons. The author discusses the astrophysical and cosmological limits which define a small window of allowed axion mass above 3 eV. A firm upper bound to the axion's mass of $m_a \leqslant 8$ eV is derived by considering the effect of decaying axions upon the diffuse extragalactic background radiation and the brightness of the night sky due to axions in the halo of our Galaxy. The intergalactic light of clusters of galaxies is shown to be an ideal place to search for an emission line arising from the radiative decay of axions. An unsuccessful search for this emission line in three clusters of galaxies is then detailed. Limits to the presence of any intracluster line emission are derived with the result that axions with masses between 3 and 8 eV are excluded by the data, effectively closing this window of axion mass, unless a severe cancellation of axionic decay amplitudes occurs. The intracluster flux limits are then used to constrain the amplitude of any such model dependence.

061.102 **Energetic neutrinos from heavy–neutralino annihilation in the Sun.**
M. Kamionkowski.
Phys. Rev. D, Vol. 44, No. 10, p. 3021 – 3042 (15 Nov 1991). Current Physics Microform No.: 9201B0749.

Neutralinos may be captured in the Sun and annihilate therein producing high–energy neutrinos. Present limits on the flux of such neutrinos from underground detectors such as IMB and Kamiokande II may be used to rule out certain supersymmetric dark matter candidates, while in many other supersymmetric models the rates are large enough that if neutralinos do reside in the Galactic halo, observation of a neutrino signal may be possible in the near future. Neutralinos that are either nearly pure Higgsino or a Higgsino – gaugino combination are generally captured in the Sun by a scalar interaction with nuclei in which a virtual lightest Higgs boson is exchanged. If the squark mass is not much greater than the neutralino mass then capture of neutralinos that are primarily gaugino occurs predominantly by spin–dependent scattering off hydrogen in the Sun. Although only neutrinos from annihilation in the Sun are considered here, the neutrino signal from weakly interacting massive particle annihilation in the Earth should be of comparable strength.

061.103 **Semilocal cosmic strings.**
T. Vachaspati, A. Achúcarro.
Phys. Rev. D, Vol. 44, No. 10, p. 3067 – 3071 (15 Nov 1991). Current Physics Microform No.: 9201B0795.

The authors show that the presence of global symmetries in a Lagrangian can lead to U(1) flux–tube solutions even when the vacuum manifold is simply connected. The authors provide a model in which these flux tubes, called "semilocal" strings, occur and explicitly prove the existence of stable string solutions. The formation of semilocal strings in the early universe and the evolution of the network is discussed. It is also shown that semilocal magnetic monopoles cannot exist.

061.104 **Strange quark matter in physics and astrophysics.**
J. Madsen.
IUPAP Conference on Primordial Nucleosynthesis and Evolution of Early Universe, p. 69 – 78 (1991). – See Abstr. 012.071 for the main entry.

The physical properties of stable strange quark matter/quark nuggets are described. Laboratory searches are briefly discussed, but the main emphasis of the review is on the astrophysical consequences of strange matter. In particular on the possible formation and survival of quark nuggets in the cosmological quark–hadron phase transition, where new results regarding boiling of strange matter are mentioned. Also discussed are the rather stringent astrophysical limits, that can be derived on the galactic flux of quark nuggets. Finally it is pointed out, that much of the strange matter–discussion, including the flux–limits, can be generalized to other non–topological solitons carrying baryon number.

061.105 **Could cosmic QCD phase transition produce strange quark matter which survives until the present time?**
T. Kajino, K. Sumiyoshi.
IUPAP Conference on Primordial Nucleosynthesis and Evolution of Early Universe, p. 95 – 100 (1991). – See Abstr. 012.071 for the main entry.

Microscopic calculations using a chromoelectric flux tube model provide a low baryon penetrability through the phase boundary between the quark–gluon plasma and the hadron gas. The result is applied to the study of the evolution of baryon number density during the cosmic phase transition in quantum chromodynamics (QCD). Remarkable inhomogeneities of baryon number density result from a plausible range of QCD parameters constrained from lattice simulations. If the QCD phase transition is weakly first order, the strange quark matter could have been formed and survived evaporation and resolution in the hot early Universe.

061.106 **Reactions on carbon–14.**
L. H. Kawano.
IUPAP Conference on Primordial Nucleosynthesis and Evolution of Early Universe, p. 121 – 122 (1991). – See Abstr. 012.071 for the main entry.

In inhomogeneous nucleosynthesis, a series of neutron captures is thought to lead to the production of ^{14}C. The authors have studied the branching at ^{14}C and have deduced the importance of the $^{14}C(d,n)^{15}N$ reaction which may be the dominant channel for the production of heavier elements.

061.107 **Measurement of the cross section of the $^{12}C(n,\gamma)^{13}C$ reaction at stellar energy.**
Y. Nagai.
IUPAP Conference on Primordial Nucleosynthesis and Evolution of Early Universe, p. 125 – 126 (1991). – See Abstr. 012.071 for the main entry.

061.108 **Constraints on the coupling of weakly–interacting particles to matter from stellar evolution.**
N. Iwamoto.
IUPAP Conference on Primordial Nucleosynthesis and Evolution of Early Universe, p. 267 – 268 (1991). – See Abstr. 012.071 for the main entry.

061.109 **Galactic enrichment of s–elements: primary or secondary nature?**
R. Gallino, M. Busso, G. Picchio, C. M. Raiteri.
ESO/EIPC Workshop on Supernova 1987A and other Supernovae, p. 715 – 718 (Jun 1991). – See Abstr. 012.075 for the main entry.

061.110 **Big bang nucleosynthesis comes of age.**
G. Steigman.
Phys. Scr. T, Vol. T36 , p. 55 – 59 (1991). – See Abstr. 012.080 for the main entry.

Big Bang Nucleosynthesis (BBN), the synthesis of the light elements during the first thousand seconds in the evolution of the Universe, offers a laboratory for testing models of elementary particle physics under conditions inaccessible to even the largest terrestrial accelerators. The very recent SLC and LEP results on the width of the Z^0 now provide a validation of the BBN bound to the number of families of light neutrinos (N_ν) and a confirmation of the Cosmology/Particle Physics Connection. Since the BBN bound to N_ν is central to this connection, the history and evolution of this prediction is reviewed and its current status is explored along with a peek into the future.

061.111 **Primordial nucleosynthesis in an inhomogeneous universe.**
K. Sato, N. Terasawa.
Phys. Scr. T, Vol. T36 , p. 60 – 63 (1991). – See Abstr. 012.080 for the main entry.

From recent studies of nucleosynthesis in inhomogeneous universe, it has been pointed out that (1) $\Omega_B = 1$ model becomes consistent with light element observations, (2) heavy elements including r–process elements are abundantly synthesized, and (3) ^9Be plays a role of the indicator of inhomogeneity since ^9Be is also abundantly synthesized. These suggestions were, however, based on a simple two–zone model in which back diffusion of neutrons are neglected during the nucleosynthesis. In this short review, it is shown that (1) the upper limit of Ω_B which is consistent with light element observations is almost the same as that of homogeneous case, (2) the ^9Be abundance in the simple two–zone model is overestimated at least two orders of magnitude in plausible parameter space and (3) the abundances of CNO elements are very small in the proper range of baryon/photon ratio consistent with the abundances of light elements.

061.112 **Particle physics footprints of the invisible axion.**
R. D. Peccei.
Phys. Scr. T, Vol. T36, p. 218 – 221 (1991). – See Abstr. 012.080 for the main entry.

The authors discusses some particle physics expectations of a natural class of invisible axion models. These models require the presence of charged and neutral spin zero bosons of Fermi scale masses. Indeed, such excitations appear to be telltale signals for almost all viable cold dark matter scenarios.

061.113 **Time variation of fundamental constants. II. Quark masses as time–dependent parameters.**
P. Sisterna, H. Vucetich.
Phys. Rev. D, Vol. 44, No. 10, p. 3096 – 3108 (15 Nov 1991).
Current Physics Microform No.: 9201B0824.

Consistent bounds for the simultaneous variations of fundamental constants in the standard model of fundamental interactions are obtained from astronomical, astrophysical, geophysical, and laboratory data. Improving the analysis made in a previous work, the authors treat quark masses as time–dependent variables. QCD chiral invariance breaks down and the general scaling of strong energies has to be reanalyzed. The authors also discuss the existence of preferred reference systems in a world with time–varying fundamental parameters. As in the previous work, the bounds obtained exclude the Dirac large number hypothesis and, in general, any theory demanding a large variation of the fundamental constants.

061.114 **Cosmological bound on Dirac neutrino mass via $\gamma\gamma \to \pi^0 \to \nu\bar{\nu}$.**
W. P. Lam, K. W. Ng.
Phys. Rev. D, Vol. 44, No. 10, p. 3345 – 3347 (15 Nov 1991).
Current Physics Microform No.: 9201B1073.

The cosmological production of massive right–handed Dirac neutrinos via the pion–pole mechanism, $\gamma\gamma \to \pi^0 \to \nu\bar{\nu}$, is considered. Using the upper bound on the effective number of neutrino species during primordial nucleosynthesis, the authors obtain a limit on the branching ratio for the decay $\pi^0 \to \nu\bar{\nu}$, $R < 2.9 \times 10^{-13}$, and an upper bound on the mass of Dirac neutrinos, $m_\nu < 420$ keV.

061.115 **Capture rate of the ^{7}Li(n, γ)^{8}Li reaction by prompt gamma–ray detection.**
Y. Nagai, M. Igashira, N. Mukai, T. Ohsaki, F. Uesawa, K. Takeda, T. Ando, H. Kitazawa, S. Kubono, T. Fukuda.
Astrophys. J., Vol. 381, No. 2, p. 444 – 448 (10 Nov 1991).

The reaction ^{7}Li(n, γ)^{8}Li is a trigger reaction which bridges the instability gap at mass 8 while producing heavy elements during the rapid process of primordial nucleosynthesis in the inhomogeneous big–bang models. The reaction rate was remeasured at the most probable neutron energy of 30 keV by detecting prompt γ–rays from a captured state; 39.3 ± 6.0 μb was obtained. This value is 2 times larger than that recently measured and agrees with the estimated value from the thermal neutron capture cross section by using the 1/v law. The result is discussed in terms of both nucleosynthesis and the solar neutrino problem.

061.116 **Masses of neutrinos suggested from the anomalous atmospheric neutrino flux.**
S. Midorikawa, M. Honda, K. Kasahara.
Phys. Rev. D, Vol. 44, No. 11, p. R3379 – R3382 (1 Dec 1991).
Current Physics Microform No.: 9201E0973.

The atmospheric neutrino data taken by the Kamiokande and IMB–3 detectors are analyzed using various models for the fluxes. A translation formula is derived to compare different experiments and/or flux models. It is found that three–neutrino oscillations can explain both results consistently. If one neutrino–mass–squared difference dominantly determines the oscillations, it is greater than 1×10^{-4}eV^2 at the 1σ level almost independent of the specific model. From neutrino–induced down–going events at Kamiokande, a more stringent lower bound of 2×10^{-3}eV^2 is obtained at the 1σ level.

061.117 **Barrier penetration at astrophysical energies.**
N. Rowley, A. C. Merchant.
Astrophys. J., Vol. 381, No. 2, p. 591 – 596 (10 Nov 1991).

Barrier penetration coefficients, which take into account the exponential nature of the tail of the nuclear potential, are evaluated analytically for energies well below the Coulomb barrier. The formulae allow a simple discussion of theoretical uncertainties in the nonresonant part of astrophysical fusion cross sections arising from uncertainties in the tail of the potential, in particular the diffuseness. The systems p + ^{13}N and ^{12}C + ^{12}C are studied in detail. The derived expressions may also be useful in alpha decay.

061.118 **Conical bremsstrahlung in a cosmic–string spacetime.**
J. Audretsch, A. Economou.
Phys. Rev. D, Vol. 44, No. 12, p. 3774 – 3785 (15 Dec 1991).
Current Physics Microform No.: 9101C0056.

It is known that classical charged point particles moving on geodesics of the conical spacetime of a cosmic string lose energy via bremsstrahlung radiation. The authors present the corresponding quantum–field–theoretical treatment of the bremsstrahlung process in terms of cylindrical modes. Because of the localized absence of these modes, a local description of bremsstrahlung is possible. It is found that the bremsstrahlung is not well localized around the cosmic string: the corresponding radial distribution of the bremsstrahlung transition probability has an asymptotic r^{-1} radial falloff because of the significant soft–photon emission at arbitrarily large distances from the cosmic string. Correspondingly the total bremsstrahlung probability shows an infrared catastrophe. The authors give a physical interpretation of the catastrophe, based on the intuitive picture of quantum particles as probes with "extensions" approximately equal to their wavelengths, if they are massless, or to their Compton wavelengths if they are massive.

061.119 **Direct tests for solar neutrino mass, mixing, and Majorana magnetic moment.**
R. S. Raghavan, A. B. Balantekin, F. Loreti, A. J. Baltz, S. Pakvasa, J. Pantaleone.
Phys. Rev. D, Vol. 44, No. 12, p. 3786 – 3790 (15 Dec 1991).
Current Physics Microform No.: 9101C0068.

The present data on solar neutrinos may suggest matter effects resulting in (1) flavor conversion (Mikheyev – Smirnov – Wolfenstein effect) due to a nonzero neutrino mass and/or mixing; and (2) spin–flavor conversion due to a Majorana magnetic moment. Both possibilities can be explicitly tested in upcoming experiments (1) via the effect of Earth matter on ^{7}Be neutrinos and (2) by the detection of ^{8}B antineutrinos during solar activity.

061.120 **E_6 leptoquarks and the solar neutrino problem.**
E. Roulet.
Phys. Rev. D, Vol. 44, No. 12, p. 3971 – 3973 (15 Dec 1991).
Current Physics Microform No.: 9101C0253.

The possibility that nonconventional neutrino oscillations take place in the superstring – inspired E_6 models is considered. In this context, the influence of leptoquark – mediated interactions of the neutrinos with nucleons in the resonant flavor conversion is discussed. It is shown that this effect can be significant for $\nu_e - \nu_\tau$ oscillations if these neutrinos have the masses required in the ordinary Mikheyev – Smirnov – Wolfenstein effect, and may lead to a solution of the solar neutrino problem even in the absence of vacuum mixings. On the other hand, this model cannot lead to resonant behavior in the Sun if the neutrinos are massless.

061.121 **Molecules and chains in a strong magnetic field: statistical treatment.**
A. M. Abrahams, S. L. Shapiro.
Astrophys. J., Vol. 382, No. 1, p. 233 – 241 (20 Nov 1991).

The authors develop and employ a Thomas – Fermi – Dirac – Weizsäcker (TFD–λW) statistical model to investigate diatomic molecules and infinite molecular chains in strong magnetic fields. Such material may form on the surface of a neutron star where field strengths can reach $B \sim 10^{12}$G. The standard magnetic Thomas – Fermi – Dirac kinetic, potential, and exchange energy functionals are supplemented by a gradient correction to the kinetic energy. This correction leads to a model of electronic structure determined by a nonlinear system of two coupled partial differential equations with an eigenvalue. The numerical method used for solving this system in two spatial dimensions is detailed. In contrast to simpler treatments, molecules and chains can be bound in the TFD–λW model due to the gradient term. The authors present numerical solutions for a wide range of magnetic field strengths and elements. The calculations qualitatively reproduce many of the results of detailed quantum mechanical treatments. For example, the fractional binding energy is greatest for low atomic numbers and for strong magnetic fields.

061.122 **Nucleosynthesis in big–bang models.**
F. D. Santos.
12. Autumn School of Physics. The physical universe – the interface between cosmology, astrophysics and particle physics, p. 281 – 291 (1991). – See Abstr. 012.097 for the main entry.

In order to obtain more constraining bounds to cosmology and particle physics from primordial nucleosynthesis it is essential to take into account the most recent theoretical and observational data. The author gives an overview of the current status of big–bang nucleosynthesis, including a discussion of those aspects of the nuclear physics which are relevant in the quest for more accurate cosmological knowledge.

061.123 **Concocting a cosmic repice for matter.**
D. H. Freedman.
Science, Vol. 254, No. 5030, p. 382 – 383 (18 Oct 1991).

Taking their cues from accepted physics, two groups of researchers try to solve the mystery of the non–empty universe.

061.124 **Physicists close in on a weighty quarry.**
F. Flam.
Science, Vol. 254, No. 5036, p. 1298 – 1299 (29 Nov 1991).

After a year of sightings, the great (heavy) neutrino hunt seems about to capture the beast – or prove it a chimera.

061.125 **Search for ν_μ and ν_τ supernova neutrinos with massive liquid–scintillation detectors.**
P. Antonioli, W. Fulgione, P. Galeotti, L. Panaro.
Nuovo Cimento C, Vol. 14, N. 6, p. 631 – 637 (Nov–Dec 1991).

The authors have estimated the sensitivity of a large–mass liquid–scintillation detector to search for supernova neutrinos of different flavours. Events produced by ν_μ and ν_τ interactions can be identified by looking at the distorsion in the neutrino energy spectrum. They have shown that, overlapped to the main energy distribution produced by $\bar{\nu}_e$ interactions with protons, a peak at 15.11 MeV gives a possible signature of these neutrino flavours.

061.126 **Détermination de taux de réactions nucléaires conduisant à la nucléosynthèse stellaire du fluor.**
M. Kious.
Diss., Paris–11 Univ., 91 – Orsay (France). Centre de Spectrometrie Nucleaire et de Spectrometrie de Masse. CSNSM–T—9037, 221 p. (11 Oct 1990).

The fluorine stellar nucleosynthesis remains up to now unexplained. According to the current nuclear rates, the fluorine synthesized during the hydrogen or the helium burning is immediatly destroyed. The author started the study of resonant levels located close by the proton threshold in neon twenty, using a proton transfer reaction. The existence of a level close to the threshold appears to be of critical importance, and he determined its spectroscopic factor and its proton width. Through interferences with higher lying levels, the cross section of the fluorine plus proton reaction could be severely modified. Concerning the fluorine building up during helium burning phases, the author considered the consequences of some new measurements of fluorine alpha widths. The consequences for the fluorine stellar nucleosynthesis are investigated.

061.127 **Direct reaction mechanism in astrophysically relevant processes.**
H. Oberhummer, G. Staudt.
Nuclei in the cosmos, p. 29 – 59 (1991). – See Abstr. 003.038 for the main entry.

Contents: 1. Introduction. 2. Qualitative features of nuclear reaction mechanisms. 3. Nuclear reactions in astrophysical scenarios. 4. Optical potentials. 5. Theoretical descriptions of direct reactions. 6. Examples: transfer reactions with non–resonant behaviour; transfer reactions with resonant behaviour; direct capture. 7. Summary.

061.128 **Nuclear reaction rates – from laboratory experiments and in the stellar plasma.**
K. Langanke.
Nuclei in the cosmos, p. 61 – 88 (1991). – See Abstr. 003.038 for the main entry.

Contents: 1. Introduction. 2. Theoretical models for extrapolating measured data. 3. Example: the $^{12}C(\alpha, \gamma)^{16}O$ E1 and E2 capture reactions at astrophysical energies. 4. Electron screening in laboratory experiments. 5. Screening effects in the plasma.

061.129 **The s–process: branchings and chronometers.**
F. Käppeler.
Nuclei in the cosmos, p. 179 – 198 (1991). – See Abstr. 003.038 for the main entry.

Contents: 1. Introduction. 2. Chronometers – models and reality. 3. s–process concepts: the classical approach; stellar s–process models. 4. s–process time scales: helium shell burning; remixing to the interstellar medium and the age of the Galaxy; summary of potential chronometers. 5. Selected examples: ^{176}Lu – a defective cosmic clock; the branchings at ^{185}W and ^{186}Re – neutron density and r–process age. 6. Conclusions.

061.130 **Beta decay far from stability and double beta decay, and consequences for astrophysics.**
H. V. Klapdor–Kleingrothaus.
Nuclei in the cosmos, p. 199 – 232 (1991). – See Abstr. 003.038 for the main entry.

Contents: 1. Introduction. 2. Calculations of beta decay properties of nuclei far from stability. 3. Beta decay far from stability and the astrophysical r–process: synthesis of heavy elements in the universe. 4. The r–process and the age of the universe. 5. Age of the universe, cosmology and neutrino mass. 6. Double beta decay and neutrino mass. 7. Conclusion.

Elementary particles.
See Abstr. 003.011.

Neutrino physics.
See Abstr. 003.030.

Bausteine des Universums. Quarks und Leptonen.
See Abstr. 003.031.

The hidden universe.
See Abstr. 003.034.

Nuclei in the cosmos.
See Abstr. 003.038.

Quantum gravity. Proceedings.
See Abstr. 012.004.

Relativity and gravitation: classical and quantum. Proceedings.
See Abstr. 012.005.

Elementary particles and the universe. Essays in honor of Murray Gell–Mann.
See Abstr. 012.007.

Current problems in particle physics and cosmology. Proceedings. Workshop on Current Problems in Particle Physics and Cosmology, Ioannina (Greece), 25 – 29 Jun 1990.
See Abstr. 012.022.

Astroparticle physics. Proceedings.
See Abstr. 012.037.

Primordial nucleosynthesis. Proceedings.
See Abstr. 012.060.

Primordial nucleosynthesis and evolution of early universe. Proceedings.
See Abstr. 012.071.

The birth and early evolution of our universe. Proceedings. Nobel Symposium 79, Gräftåvallen (Sweden), 11 – 16 Jun 1990.
See Abstr. 012.080.

The physical universe: the interface between cosmology, astrophysics and particle physics. Proceedings.
See Abstr. 012.097.

Techniques and concepts of high–energy physics 6. Proceedings.
See Abstr. 012.102.

High energy physics and cosmology. Proceedings.
See Abstr. 012.105.

Particles, strings and supernovae. Volumes 1, 2. Proceedings.
See Abstr. 012.109.

Heavy ion physics and nuclear astrophysical problems. Proceedings.
See Abstr. 012.111.

Big bang, active galactic nuclei and supernovae. Proceedings.
See Abstr. 012.113.

Tests of fundamental laws in physics. Proceedings.
See Abstr. 012.116.

Correlation between the Maryland and Rome gravitational–wave detectors and the Mont Blanc, Kamioka and IMB particle detectors during SN 1987A.
See Abstr. 013.071.

Neutrino astronomy: Putting a project on ice.
See Abstr. 013.081.

Observation of muons using the polar ice cap as a Cerenkov detector.
See Abstr. 013.082.

The use of the reaction $^{9}Be(\alpha, \gamma n)^{12}C$ to determine the energy scale and the efficiency at low energies of stellar collapse neutrino detectors.
See Abstr. 022.011.

Experimental determination of stellar reaction rates.
See Abstr. 022.176.

Radioactive ion beams in nuclear astrophysics.
See Abstr. 022.177.

Neutrino and dark–matter detection at low temperature.
See Abstr. 034.022.

Techniques for detection of cosmic diffuse neutrinos and anti–neutrinos.
See Abstr. 034.121.

Data analysis in neutrino astronomy.
See Abstr. 036.357.

Overview of NASA's cosmic and heliospheric physics missions.
See Abstr. 051.032.

The Rosseland mean free–free Gaunt factor of the dense high–temperature stellar plasma.
See Abstr. 063.113.

On the possibility of nuclear processes in atmospheres of flare stars.
See Abstr. 064.049.

Neutron capture nucleosynthesis and the evolution of 15 and 30 $M_{\odot}$ mass–losing stars. II. The core carbon–burning phase.
See Abstr. 065.026.

Neutrino–nucleus interactions in core–collapse supernovae.
See Abstr. 065.030.

High–temperature neutrino–nucleus processes in stellar collapse.
See Abstr. 065.031.

Final evolutionary stages of massive stars.
See Abstr. 065.058.

On the production of ^{26}Al in AGB stars.
See Abstr. 065.076.

On the sodium excess in the atmospheres of yellow supergiants.
See Abstr. 065.101.

Effects of negatively charged massive particles on primordial stars.
See Abstr. 065.136.

Surface chemical composition, nuclear processing, and deep mixing: main sequence to first dredge–up.
See Abstr. 065.146.

On some problems of quantum and classical gravitation.
See Abstr. 066.009.

Lattice quantization of the $O(1,2)/O(2) \times Z$ sigma model in 4 dimensions.
See Abstr. 066.010.

No time and quantum gravity.
See Abstr. 066.012.

Quantum metric topology.
See Abstr. 066.013.

Wave functions constructed from an invariant sum–over–histories satisfy constraints: three derivations of the Wheeler–DeWitt equation.
See Abstr. 066.015.

Supersymmetric quantum wormhole states.
See Abstr. 066.016.

On the conformal factor problem in quantum gravity.
See Abstr. 066.017.

Uncertainty and equivalence principles in quantum general relativity.
See Abstr. 066.019.

Quantum field theory in space with discrete geometry.
See Abstr. 066.021.

Quantum theory of particles with Planckean masses.
See Abstr. 066.022.

Quantum string gravity near the Planck scale.
See Abstr. 066.023.

Vacuum polarization in a locally static multiply–connected space-time and a time machine problem.
See Abstr. 066.025.

A Clifford bundle approach to gravitational theory.
See Abstr. 066.048.

The geometrical functional integral for mini–superspace quantum gravity and the relativistic particle compared.
See Abstr. 066.066.

Exact solutions from one dimensional subspaces in Kaluza–Klein theory.
See Abstr. 066.072.

Physical models of white and grey holes and their astrophysical applications.
See Abstr. 066.092.

Fifth force from Kaluza–Klein unification.
See Abstr. 066.128.

The gravitational interaction of conical strings.
See Abstr. 066.130.

Some electromagnetic consequences of a geometric unified theory of gravitation and electromagnetism.
See Abstr. 066.132.

The mass formula of the spherically symmetric solution of the five–dimensional vacuum equations $R_{\mu\nu} = 0$.
See Abstr. 066.135.

Magnetic charge type equations from torsion.
See Abstr. 066.137.

Gauge–strings on Einstein – Rosen background.
See Abstr. 066.139.

Gauge invariant electromagnetic coupling with torsion potential.
See Abstr. 066.148.

Relativistic neutrino transport in stellar collapse.
See Abstr. 066.159.

Magnetic theory of gravitation.
See Abstr. 066.192.

Theory of an Earth–bound clock comparison experiment as test of the principle of equivalence.
See Abstr. 066.201.

Gravitational effects of light scalar particles.
See Abstr. 066.214.

More on the black hole S–matrix.
See Abstr. 067.018.

On a quantum mechanical model of a black hole.
See Abstr. 067.019.

Quantum–mechanical scattering of charged black holes.
See Abstr. 067.020.

Neutrino neutral current interactions in hot dense matter.
See Abstr. 067.033.

Proton acceleration in neutron star magnetospheres.
See Abstr. 067.047.

Reconciliation of neutron star masses and binding of the Λ in hypernuclei.
See Abstr. 067.068.

Astrophysical limits for binding energy of strange matter.
See Abstr. 067.096.

Superfluid hydrodynamics in rotating neutron stars. I. Nondissipative equations.
See Abstr. 067.164.

Cold equation of state in a strong magnetic field: effects of inverse β–decay.
See Abstr. 067.242.

Relativistic hadrons near accreting compact objects: dynamical and radiative signatures.
See Abstr. 067.251.

Nuclear interactions near accreting black holes.
See Abstr. 067.256.

Synchrotron emission of neutrino pairs in neutron stars.
See Abstr. 067.264.

Neutrino oscillations and solar models.
See Abstr. 080.074.

The role of low energy neutrinos in the solar neutrino problem.
See Abstr. 080.079.

On the matter–enhanced transitions of solar neutrinos in the absence of neutrino mixing in vacuum.
See Abstr. 080.080.

Real–time, directional measurement of ^{8}B solar neutrinos in the Kamiokande II detector.
See Abstr. 080.103.

Reconciling the time dependence of the chlorine and Kamiokande solar neutrino data.
See Abstr. 080.104.

A search for dark matters in the Kamiokande II.
See Abstr. 080.108.

Large solar flares: analysis of the events recorded by the Mont Blanc neutrino detector.
See Abstr. 080.119.

The Earth as an antineutrino star.
See Abstr. 081.028.

Neutrino tomography of the Earth.
See Abstr. 081.065.

Some graphic representations of solar–system abundances of nuclei.
See Abstr. 091.011.

Isotopically anomalous Ti in presolar SiC from the Murchison meteorite.
See Abstr. 105.012.

s–process Ba, Nd, and Sm in presolar SiC from the Murchison meteorite.
See Abstr. 105.047.

Calculations of neutrino heating supernovae.
See Abstr. 125.057.

Spectrophotometry of Cas A: implications for nucleosynthesis in massive stars.
See Abstr. 125.083.

The SN 1987A neutrino signal and the future.
See Abstr. 125.224.

Implications of the SN 1987A neutrinos for supernova theory and the mass of ν_e.
See Abstr. 125.225.

C I/CO ratio in interstellar dense clouds and decay photons of dark matter.
See Abstr. 131.325.

The lightest supersymmetric particle and the extragalactic γ–ray background.
See Abstr. 143.014.

Flux limits for relativistic cosmic photinos from the KAMIOKANDE neutrino experiment.
See Abstr. 144.013.

The ^{54}Mn clock and its implications for cosmic–ray propagation and Fe isotope studies.
See Abstr. 144.059.

Search for cosmic–ray magnetic monopoles using a three–loop superconductive detector.
See Abstr. 144.064.

Search for a flux of cosmic–ray magnetic monopoles with an eight–channel superconducting detector.
See Abstr. 144.065.

Experimental searches for strange quark matter in cosmic rays.
See Abstr. 144.082.

Strange quark matter in the cosmic radiation.
See Abstr. 144.083.

Cosmic ray search for strange quark matter with the MARCO detector.
See Abstr. 144.084.

Search for high–energy neutrinos from celestial sources in the Kolar Gold Fields experiment.
See Abstr. 144.090.

Nuclei in cosmic rays.
See Abstr. 144.118.

The galactic gradient of N/^{15}N.
See Abstr. 155.006.

Galactic evolution of lithium.
See Abstr. 155.029.

An alternative explanation of the evolution of element abundance ratios during the history of the Galaxy.
See Abstr. 155.101.

Reduced upper limits on the equivalent width of interstellar Li I 670.8 towards SN 1987A.
See Abstr. 156.019.

Heavy element abundances in the Magellanic Clouds.
See Abstr. 156.038.

Abundances in galaxies.
See Abstr. 157.352.

A non–baryonic galactic halo and the real galactic age.
See Abstr. 161.011.

Cosmological implications of axinos.
See Abstr. 161.030.

Cosmic abundances of stable particles: improved analysis.
See Abstr. 161.031.

Third quantization of gravity and the cosmological constant problem.
See Abstr. 161.044.

Neutrino degeneracy and cosmological nucleosynthesis, revisited.
See Abstr. 161.048.

One–loop quantum cosmology: the general semiclassical solution of the Wheeler–DeWitt equations and its applications.
See Abstr. 161.061.

Environment–induced superselection in cosmology.
See Abstr. 161.062.

The self–measurement of the Universe and the concept of time in quantum cosmology.
See Abstr. 161.063.

Quantum string theory in curved space–times.
See Abstr. 161.065.

Stably causal Gödel–type models.
See Abstr. 161.070.

Quantum processes in the gravitational field of a cosmic string.
See Abstr. 161.072.

Equilibrium configurations of superconducting cosmic strings in background fields.
See Abstr. 161.074.

Thermodynamics of the neutral string gas in electromagnetic background.
See Abstr. 161.075.

On multi–inflation cosmological models.
See Abstr. 161.076.

Kaluza–Klein models: from quantum state to observations.
See Abstr. 161.078.

Cosmological solutions from supersymmetric Kaluza–Klein theory.
See Abstr. 161.079.

Space–time dimension of the Universe.
See Abstr. 161.080.

On the origin of classical density perturbations in the inflationary Universe.
See Abstr. 161.081.

Theory of cosmological perturbations.
See Abstr. 161.089.

Resonant neutrino oscillations in the early Universe.
See Abstr. 161.098.

Time variation of fundamental constants, preferred reference systems and CPT invariance.
See Abstr. 161.103.

Creation of scalar and spin 1/2 particles in an expanding cosmological universe.
See Abstr. 161.107.

Anisotropy dissipation in the very early Universe.
See Abstr. 161.111.

Baryogenesis from collapsing topological defects.
See Abstr. 161.129.

Dark matter and the observability of the electroweak transition.
See Abstr. 161.130.

Large–scale structure from wiggly cosmic strings.
See Abstr. 161.137.

Primordial nucleosynthesis redux.
See Abstr. 161.147.

Cosmological models with generic particle decays.
See Abstr. 161.150.

The search for non–baryonic dark matter.
See Abstr. 161.155.

Modern cosmology and physics beyond the standard model.
See Abstr. 161.160.

Cosmology of three–generation Calabi–Yau superstring models.
See Abstr. 161.167.

Color breaking and X–, Y– and W–boson condensation on the cosmic string.
See Abstr. 161.187.

Entropy, inflation and the arrow of time.
See Abstr. 161.203.

Three lectures on cosmology.
See Abstr. 161.222.

Astroparticle physics and superstrings.
See Abstr. 161.226.

Measurements of the primordial abundance of helium.
See Abstr. 161.227.

Recent advances in big bang nucleosynthesis and its relation to particle physics.
See Abstr. 161.228.

Dark matter detection.
See Abstr. 161.243.

Dark matter: theory.
See Abstr. 161.244.

Quark– and gluon–jet emission from primordial black holes. II. The emission over the black hole lifetime.
See Abstr. 161.260.

Effect of neutrino heating in the early universe on neutrino decoupling temperatures and nucleosynthesis.
See Abstr. 161.261.

Strange matter lumps formed in the early universe.
See Abstr. 161.262.

First detection of beryllium in a very metal poor star: a test of the standard big bang model.
See Abstr. 161.264.

Cosmological constraints on the zero–point electromagnetic field.
See Abstr. 161.279.

Primordial nucleosynthesis without a computer.
See Abstr. 161.280.

Initial condition for the minimal isocurvature scenario.
See Abstr. 161.287.

Primordial nucleosynthesis: the next significant figure.
See Abstr. 161.303.

Inhomogeneous nucleosynthesis: extending the standard model.
See Abstr. 161.304.

Nuclear reaction uncertainties in standard and non–standard cosmologies.
See Abstr. 161.305.

Perspectives: nucleosynthesis and nuclear reactions in standard and inhomogeneous big bang models and beyond.
See Abstr. 161.306.

Motion of test particles and photons in the gravitational field of a cosmic string.
See Abstr. 161.307.

The cosmological quark–hadron transition.
See Abstr. 161.329.

Survival condition of primordial strange quark matter with low baryon penetrability.
See Abstr. 161.333.

Survival of quark nuggets and their present–day abundance.
See Abstr. 161.334.

The first three minutes: 1990 version.
See Abstr. 161.344.

Ultrarelativistic Bose–Einstein condensation in the Einstein universe and energy conditions.
See Abstr. 161.394.

Astronomical limits on the nature of the dark matter.
See Abstr. 161.404.

Dark matter theory: constraints from particle physics.
See Abstr. 161.405.

The cosmic neutrino background.
See Abstr. 161.406.

The new discussion of a neutrino mass and issues in the formation of large–scale structure.
See Abstr. 161.407.

The effect of some nonequilibrium processes on the primordial nucleosynthesis.
See Abstr. 161.426.

Baryogenesis, sphalerons, and the cogeneration of dark matter.
See Abstr. 161.432.

Sterile neutrinos in the early Universe.
See Abstr. 161.438.

Signatures of inhomogeneity in the early Universe.
See Abstr. 161.440.

Are galactic halos made of brown dwarfs or black holes?
See Abstr. 161.479.

LEP physics and the early Universe.
See Abstr. 161.487.

Reviving massive neutrinos for large–scale structure.
See Abstr. 161.503.

Nucleosynthesis: Beryllium and the Big Bang.
See Abstr. 161.521.

Big–bang nucleosynthesis – observational aspects.
See Abstr. 161.574.

Big bang nucleosynthesis: the standard model and alternatives.
See Abstr. 161.576.

Population III objects and lithium abundance.
See Abstr. 161.586.

Dark matter in the Universe.
See Abstr. 161.591.

CP–violation and electroweak baryogenesis.
See Abstr. 161.592.

Electroweak baryon number violation at high energies.
See Abstr. 161.593.

The alpha parameters of wormholes.
See Abstr. 161.595.

Can a man–made universe be created by quantum tunneling without an initial singularity?
See Abstr. 161.597.

Perspectives on particle physics and cosmology.
See Abstr. 161.601.

A hydrodynamic approach to cosmology: texture–seeded cold dark matter and hot dark matter cosmogonies.
See Abstr. 161.634.

Nonlinear matter clustering properties of a cold dark matter universe.
See Abstr. 161.635.

Forming cosmic structure with a 17 keV neutrino.
See Abstr. 161.650.

Matter and antimatter in the universe.
See Abstr. 161.658.

Introductory notes on particle physics and cosmology.
See Abstr. 161.659.

Lectures on particle physics and cosmology.
See Abstr. 161.664.

Cosmological and astrophysical aspects of the quark–hadron phase transition.
See Abstr. 161.665.

The role of boron in cosmology.
See Abstr. 161.674.

Primordial nucleosynthesis: beyond the standard model.
See Abstr. 161.677.

Production of heavy elements in inhomogeneous cosmologies.
See Abstr. 161.678.

062 Hydrodynamics, Magnetohydrodynamics, Plasma

062.001 **The force of ejection of a toroidal transient from an inhomogeneous magnetic field.**
A. F. Kharshiladze, K. G. Ivanov.
Geomagn. Aeron., Vol. 30, No. 1, p. 18 – 22 (Aug 1990). English translation of Geomagn. Aehron., Tom 30, No. 1, p. 13 – 16 (1990).

An analytic solution is obtained for the problem of the force of the ejection of an ideally conducting, axially symmetrical toroid of finite aspect ratio from an inhomogeneous field. The solution can be used in the calculation of the ejection of toroidal flare transients from the inhomogeneous field of the solar corona.

062.002 **How much energy can be stored in a three–dimensional force–free magnetic field?**
J. J. Aly.
Astrophys. J., Lett., Vol. 375, No. 2, p. L61 – L64 (10 Jul 1991).

The author develops simple physical arguments showing that any finite–energy force–free magnetic field occupying a half–space D (or the exterior of a "star–shaped" region) and having all its lines unknotted and tied to the boundary ∂D of D must have an energy which is not larger than that of the "open field" having the same flux distribution on ∂D. This result, which was first conjectured a few years ago by the author, has some important consequences for our understanding of the eruptive phenomena in the solar corona. In particular, it puts an absolute upper bound on the energy which can be released in a flare, and it precludes any spontaneous transition of the coronal field to an open state – thus allowing us to discard some popular models.

062.003 **Distribution functions in the statistical theory of convective MHD turbulence of an incompressible fluid.**
M. S. Alam Sarker, N. Kishore.
Astrophys. Space Sci., Vol. 181, No. 1, p. 29 – 42 (Jul 1991).

In this paper the authors define a hierarchy of distribution functions for simultaneous velocity, magnetic, and temperature fields. Some properties of the constructed distribution functions such as reduction, separation, and coincidence are discussed. Equations for the evolution of one– and two–point distribution functions have been derived. Finally, a comparison of the equation for the single–point distribution function in case of zero viscosity, negligible diffusivity, and infinite electrical conductivity is made with first equation of BBGKY hierarchy in the kinetic theory of gases.

062.004 **Energy degradation of fast electrons in hydrogen gas.**
Y. Xu, R. McCray.
Astrophys. J., Vol. 375, No. 1, p. 190 – 201 (1 Jul 1991).

The authors derive an equation for calculating the energy distribution of fast electrons in a partially ionized gas and provide a method to solve for the electron degradation spectrum and the energy deposition in different forms (ionization, excitation, or heating). Continuous degradation processes, such as Coulomb scattering, are included in the calculation. As an example, the authors calculate the energy degradation of fast electrons in a gas of pure hydrogen, considering excitations to the lowest 10 atomic levels. The Bethe approximation and the continuous slowing–down approximation are discussed and shown to be accurate to the order of 20% for electrons with initial energy of $\gtrsim$keV. The method and results can be used to determine heating, excitations, and ionizations by high–energy photoelectrons or cosmic–ray particles in various astrophysical circumstances, such as the interstellar medium, supernova envelopes, and QSO emission line clouds.

062.005 **Instability of partially–ionized superposed plasmas.**
A. Gupta, P. K. Bhatia.
Astrophys. Space Sci., Vol. 181, No. 1, p. 103 – 110 (Jul 1991).

The stability of the plane interface separating two viscous superposed partially–ionized plasmas of uniform densities has been studied. The whole system is assumed to be immersed in a uniform two-dimensional horizontal magnetic field and the stability analysis has been carried out through the normal mode technique. The dispersion relation has been derived for the case of two superposed plasmas of equal kinematic viscosities. The dispersion relation has been solved numerically for different values of the physical parameters involved. It is found that viscosity and collision frequency of ionized plasmas both have stabilizing influence on the growth rate of the unstable mode of disturbance.

062.006 **MHD free-convection and mass transfer flow through a porous medium with heat source.**
B. K. Jha, R. Prasad.
Astrophys. Space Sci., Vol. 181, No. 1, p. 117 – 123 (Jul 1991).
An analytical study is performed to examine the heat source characteristics on the free-convection and mass transfer flow past an impulsively started infinite non-conducting vertical plate of a viscous, incompressible eletrically conducting fluid under the action of a uniform magnetic field through porous medium. The effects of various parameters on the velocity field are extensively discussed.

062.007 **Mass transfer effects on the flow past an exponentially accelerated vertical plate with constant heat flux.**
B. K. Jha, R. Prasad, S. Rai.
Astrophys. Space Sci., Vol. 181, No. 1, p. 125 – 134 (Jul 1991).
An analytical study is performed to examine the mass-transfer characteristics on the free-convection flow of an incompressible viscous fluid past an exponentially accelerated infinite vertical plate under the action of constant heat flux. The effects of various parameters occurring in the problem, are extensively discussed.

062.008 **MHD free-convection flow of an elasto-viscous fluid past an infinite vertical plate.**
N. K. Samria, R. Prasad, M. U. S. Reddy.
Astrophys. Space Sci., Vol. 181, No. 1, p. 135 – 145 (Jul 1991).
The effect of a uniform transverse magnetic field on the free-convection flow of an electrically-conducting fluid (Walter's liquid *B*') past an infinite, vertical plate for both classes of impulsive as well as uniformly accelerated motion of the plate is discussed. The magnetic lines-of-force are assumed to be fixed relative to the plate. Expressions for the velocity field and shearing stress for both cases are obtained by the Laplace transform technique. The influence of the various parameters, entering into the problem, on the velocity field and shearing stress is extensively discussed.

062.009 **Expansion of a superbubble in a uniform magnetic field.**
K. M. Ferrière, M.-M. Mac Low, E. G. Zweibel.
Astrophys. J., Vol. 375, No. 1, p. 239 – 253 (1 Jul 1991).
The authors investigate the effects of interstellar magnetic fields on the evolution and structure of superbubbles. They first derive the governing equations for propagation of a radiative shock, and solve them analytically in the limit of high expansion velocity. A numerical code, able to solve these equations in the general case, is then presented and its results are discussed. The authors find that magnetic fields of the strength present in the Galactic disk do not significantly modify the overall shape and dimensions of a superbubble. These fields reduce the volume of the interior cavity by one-third on average during the expansion phase. The shell elongates slightly in the direction of the external field. Magnetic pressure dominates over gas pressure in most of the shell and forces it to thicken substantially. Finally, because of the transverse mass motions which take place from the magnetic poles to the equator, the column density at the poles is reduced by typically a factor of 10.

062.010 **Force-free magnetic fields: generating functions and footpoint displacements.**
R. Wolfson, R. Verma.
Astrophys. J., Vol. 375, No. 1, p. 254 – 263 (1 Jul 1991).
This paper presents analytic and numerical calculations that explore equilibrium sequences of bipolar force-free magnetic fields in relation to displacements of their magnetic footpoints. It is shown that the appearance of magnetic islands – sometimes interpreted as marking the loss of equilibrium in models of the solar atmosphere – is probably associated only with physically unrealistic footpoint displacements such as infinite separation or "tearing" of the model photosphere. This study suggests that the loss of equilibrium in bipolar configurations, sometimes proposed as a mechanism for eruptive solar events, probably requires either fully three-dimensional field configurations or nonzero plasma pressure. These results apply only to fields that are strictly bipolar, and do not rule out equilibrium loss in more complex structures such as quadrupolar fields.

062.011 **Unsteady free-convective flow and mass transfer in a rotating porous medium with Hall current. I.**
T. P. Singh, A. K. Agrawal, H. L. Agrawal, A. Raptis.
Astrophys. Space Sci., Vol. 181, No. 2, p. 171 – 181 (Jul 1991).
The Hall effect on the unsteady by hydromagnetic free-convection resulting from the combined effects of thermal and mass diffusion of an electrical-conducting liquid through a porous medium past an infinite vertical porous plate in a rotating system has been analysed. The expressions for the mean velocity, mean skin friction, and mean rate of heat transfer on the plate are derived. The effects of magnetic parameter M, Hall parameter m, Ekman number E, and permeability parameter K^* on the flow field are discussed with the help of graphs and tables.

062.012 **Nonlinear Rayleigh-Taylor instability in the presence of magnetic field and mass and heat transfer.**
M. H. Obied Allah, A. A. Yahia.
Astrophys. Space Sci., Vol. 181, No. 2, p. 183 – 194 (Jul 1991).
The nonlinear Rayleigh-Taylor instability in the presence of magnetic field and mass and heat transfer is studied using a simplified formulation. The method of multiple scale expansion is employed for the investigation. It is found that the nonlinear effects of magnetic field and mass and heat transfer stabilize the classically unstable system. A simple non-dimensional parameter is also found to characterize the stability of the system.

062.013 **Nonlinear break up of an infinite cylinder in the presence of a uniform axial magnetic fluid.**
R. K. Chhabra, S. K. Trehan.
Astrophys. Space Sci., Vol. 181, No. 2, p. 283 – 312 (Jul 1991).
A third-order theory has been developed to study the instability of an infinite cylinder in the presence of a uniform axial magnetic field. By use of strained coordinates, the growth of the surface waves is calculated for wave numbers less than the critical wave number. It is observed that the perturbed surface of an infinite cylinder assumes a non-sinusoidal shape resulting in secondary waves due to energy transfer between the harmonics of different orders.

062.014 **Magnetic field intensification and formation of field-aligned current in a non-uniform magnetic field.**
H. Murata.
J. Plasma Phys., Vol. 46, No. 1, p. 29 – 48 (Aug 1991).
A study is made of the induction of a direct field-aligned current due to Alfvén waves excited through an initial-boundary condition for an incompressible fluid with high conductivity in a non-uniform magnetic field. A simple but exact case is considered in which the direction of inhomogeneity of the magnetic field is perpendicular to that of the shear fluid motion at the boundary, and the fluid moves across the magnetic field. The boundary does not satisfy the frozen-flux condition, but has a large Lundquist number. The conductivity of the fluid held between the two boundaries is high, and the frozen-flux condition is approximately satisfied there. A preflare stage of the sun is investigated, and a reasonable magnetic field intensity and d.c. field-aligned current in the magnetic arcade are obtained for a given horizontal shear motion that is antisymmetric at both boundaries of the arcade

foot points, taking account of the observed horizontal fluid motion.

062.015 **Siphon flows in isolated magnetic flux tubes. IV. Critical flows with standing tube shocks.**
J. H. Thomas, B. Montesinos.
Astrophys. J., Vol. 375, No. 1, p. 404 – 413 (1 Jul 1991).
Critical siphon flows in arched, isolated magnetic flux tubes are studied within the thin flux tube approximation, with a view toward applications to intense magnetic flux concentrations in the solar photosphere. The authors present the results of calculations of the strength and position of the standing tube shock in the supercritical downstream branch of a critical siphon flow, as well as the flow variables all along the flux tube and the equilibrium path of the flux tube in the surrounding atmosphere. It is suggested that arched magnetic flux tubes, with magnetic field strength increased by a siphon flow, may be associated with some of the intense, discrete magnetic elements observed in the solar photosphere. The observational signature of such a structure at the solar surface would be a pair of magnetic elements of opposite polarity, separated by a distance of from one to several arcseconds, with slightly inclined magnetic fields and with an upflow in one element (the upstream footpoint of the arch) and a downflow and a somewhat greater magnetic field strength in the other element (the downstream footpoint).

062.016 **Cosmic–ray–driven galactic wind.**
H. Fichtner, H. J. Fahr, W. Neutsch, R. Schlickeiser, A. Crusius-Wätzel, H. Lesch.
Nuovo Cimento B, Vol. 106, No. 8, p. 909 – 925 (Aug 1991).
A simple one–dimensional Cartesian solution of the magnetohydrodynamic equations describing the global motion of interstellar matter including the effects of cosmic–ray and magnetic–field pressure is presented. First an adiabatic equation of state for the thermal background plasma and a loss–free cosmic–ray gas in the strong–scattering limit is considered. In this case only one type of flow is consistent with the physical constraints: a subsonic wind at the base which accelerates for all values of galactic height (z) but never becomes supersonic. Second, a nonadiabatic interstellar gas is discussed. Including additional heating causes an acceleration to higher velocities. The presented solutions can be used to model halo components of interstellar matter with large–scale heights.

062.017 **Theory of wave activity occurring in the AMPTE artificial comet.**
R. Bingham, V. D. Shapiro, V. N. Tsytovich, U. de Angelis, M. Gilman, V. I. Shevchenko.
Phys. Fluids, B, Vol. 3, No. 7, p. 1728 – 1738 (Jul 1991). Current Physics Microform No.: 9107F0366.
One of the main experiments of the AMPTE satellite mission was the release of neutral barium atoms in the solar wind. The barium atoms ionized by photoionization extremely rapidly forming a dense, expanding, plasma cloud that interrupted the solar wind flow creating diamagnetic cavities. On the upstream side of the cavity a region of compressed plasma and enhanced magnetic field was created as the result of being produced by the slowing down and deflection of the solar wind, and magnetic field line draping. Intense electrostatic and magnetic turbulence was observed by both the IRM and UKS satellites at the boundary of the diamagnetic cavity, with the most intense waves being detected near the outer boundary of the compressed region. This paper examines how the newly created expanding plasma couples to the solar wind by means of plasma–beam and current–driven instabilities. In particular, it is shown how lower–hybrid and lower–hybrid drift waves are generated by cross–field proton–barium streaming instabilities and cross–field electron currents. The saturation mechanism for these waves is considered to be the modulational instability, this instability can also lead to filamentation and coupling to magnetosonic modes, which are also observed. As the result of modulational instability the $k_{\|}$ component increases, which allows the heating and acceleration of electrons that is consistent with the observations.

062.018 **The resistive tearing instability for generalized resistivity models: applications.**
G. T. Birk, A. Otto.
Phys. Fluids, B, Vol. 3, No. 7, p. 1746 – 1754 (Jul 1991). Current Physics Microform No.: 9107F0384.
The dispersion relation for the resistive tearing mode is investigated with a general form of the resistivity that allows for evaluating the dependence of the mode on the anomalous dissipation caused by the lower–hybrid drift turbulence and the ion–acoustic turbulence. The coupling of the plasma dynamic and radiative processes due to a temperature–dependent Spitzer resistivity are also discussed. The dispersion relation is solved numerically for two–dimensional equilibrium configurations with applications to the Earth's magnetosphere and the solar corona. In the case of the parameter regimes of the magnetopause and the solar corona, growth rates that result in realistic time scales for the respective dynamic processes as flux transfer events and solar flares were found.

062.019 **Particle acceleration near X–type magnetic neutral lines.**
H. J. Deeg, J. E. Borovsky, N. Duric.
Phys. Fluids, B, Vol. 3, No. 9, p. 2660 – 2674 (Sep 1991). Current Physics Microform No.: 9109E1976.
The behavior of charged particles near X–type magnetic neutral lines is investigated. The magnetic field is taken to be hyperbolic and time stationary, with a uniform electric field perpendicular to the magnetic field. The general properties of the orbits of noninteracting particles in that field geometry are examined. Approximate analytic solutions to the orbit equations are derived. The kinetic–energy distribution function for accelerated particles is derived for particles flowing past the neutral line. These analytic derivations are supported by computer simulations, in which the relativistic equations of motion for single test particles are numerically solved and the orbits and kinetic–energy gains of the particles are obtained. A number of simulations were performed and one parameter at a time was varied. Applications of the results to magnetic–neutral–line regions in space physics and astrophysics are possible.

062.020 **Magnetohydrodynamic equilibria in the vicinity of an X–type neutral line specified by footpoint shear.**
J. M. Finn, Y. T. Lau.
Phys. Fluids, B, Vol. 3, No. 9, p. 2675 – 2683 (Sep 1991). Current Physics Microform No.: 9109E1991.
Results pertaining to two–dimensional magnetohydrodynamic equilibria in the presence of an X–type neutral line are presented. The solutions presented allow the specification of footpoint displacement (or safety factor q) that is finite as the separatrix is approached. These solutions are scale–invariant, or similarity, solutions. They are appropriate near the X line on length scales intermediate between the boundary layer width because of resistivity (or other nonideal effects) and the macroscopic length scale. Discussion is presented regarding the matching of these solutions to the physical conditions on boundaries where the field lines leave the system, e.g., the photospheric surface in the solar coronal context.

062.021 **Flux transfer events interpreted in terms of a generalized model for Petschek–type reconnection.**
V. S. Semenov, I. V. Kubyshkin, H. K. Biernat, M. F. Heyn, R. P. Rijnbeek, B. P. Besser, C. J. Farrugia.
Adv. Space Res., Vol. 11, No. 9, p. 25 – 28 (1991). – See Abstr. 012.006 for the main entry.
In Petschek's reconnection model MHD waves play the dominant role in energy conversion, and these waves are responsible for propagating the disturbances resulting from reconnection into the system at large. Petschek's model, originally developed to describe a steady–state configuration, may be generalised to describe unsteady phenomena by incorporating a time–varying reconnection rate. With this generalised model it is possible to unify the observations of high–speed flow signatures and flux transfer event signatures observed at the dayside magnetopause within a single description.

062.022 **Rôle of nonresonant Alfvén instabilities in solar system beam–plasmas.**
F. Verheest, G. S. Lakhina.
Adv. Space Res., Vol. 11, No. 9, p. 55 – 58 (1991). – See Abstr. 012.006 for the main entry.
Possible occurrence of nonresonant Alfvén instabilities in solar system beam–plasmas include hydromagnetic waves upstream of the Earth's bow shock, the parallel pick–up of interstellar neutrals by the solar wind and special type III radio bursts attributed to proton beams in solar coronal loops. For such two–beam models the authors use results from an earlier general multi–beam formalism, where Fowler's theorem for fluctuations in unstable plasmas is adapted in a novel way so as to obtain theoretical estimates for the magnetic field turbulence levels.

062.023 **Multiple–ion plasma boundaries.**
U. Motschmann, K. Sauer, T. Roatsch, J. F. McKenzie.
Adv. Space Res., Vol. 11, No. 9, p. 69 – 72 (1991). – See Abstr. 012.006 for the main entry.
Multiple–ion models of plasma boundaries are studied by analytical means and hybrid code simulations. Generalized Rankine–Hugoniot relations can be derived describing modified transition conditions at multiple–ion plasma boundaries. The results are applied to the plasma environments of Mars and comet Halley.

062.024 **Velocity distributions in the plasma sheet boundary layer.**
C. Y. Huang.
Adv. Space Res., Vol. 11, No. 9, p. 143 – 150 (1991). – See Abstr. 012.006 for the main entry.
It is agreed that the plasma sheet boundary layer is a site of highly dynamic plasma behavior, with fast flows, field–aligned currents and intense broadband electrostatic noise commonly detected in this region. Most of the studies in the literature focus on macroscopic plasma parameters, e.g. density, velocity and temperature. These studies, while useful, do not allow for detailed comparisons between mechanisms to produce the ion beams characteristic of the boundary layer. In this presentation three–dimensional ion velocity distributions at the high–latitude edges of the plasma sheet are shown, and the theories which have been suggested to account for the streaming ions are compared with the observations.

062.025 **Relaxation of non–equilibrium current sheets.**
U. Arendt.
Adv. Space Res., Vol. 11, No. 9, p. 167 – 170 (1991). – See Abstr. 012.006 for the main entry.
The dynamic evolution of a collisionless current sheet which is initially not in equilibrium is investigated by means of a one–dimensional electromagnetic implicit particle simulation code. In general a relaxation of a collisionless system into an exact equilibrium is not possible, although on macroscopic scales an equilibrium can be reached. The macroscopic view is inherently included in observations as well as in a particle simulation. Our simulation results show a filamentation of the distribution function in phase space.

062.026 **The formation of quasi–parallel shocks.**
P. J. Cargill.
Adv. Space Res., Vol. 11, No. 9, p. 209 – 218 (1991). – See Abstr. 012.006 for the main entry.
In a collisionless plasma, the coupling between a piston and the plasma must take place through either laminar or turbulent electromagnetic fields. Of the three types of coupling (laminar, Larmor and turbulent), shock formation in the parallel regime is dominated by the latter and in the quasi–parallel regime by a combination of all three, depending on the piston. In the quasi–perpendicular regime, there is usually a good separation between piston and shock. This is not true in the quasi–parallel and parallel regime. Hybrid numerical simulations for hot plasma pistons indicate that when the electrons are hot, a shock forms, but does not cleanly decouple from the piston. For hot ion pistons, no shock forms in the parallel limit.

062.027 **Magnetic field structures and related phenomena at quasi–parallel shocks.**
S. J. Schwartz.
Adv. Space Res., Vol. 11, No. 9, p. 231 – 240 (1991). – See Abstr. 012.006 for the main entry.
Under "quasi–parallel" conditions, when the upstream magnetic field is nearly aligned with the shock normal, collisionless shocks exhibit unsteady, complicated behaviour due primarily to the relative ease of particle motion from the downstream, shocked state to the upstream foreshock region. This paper summarizes and orders these field structures, from the low amplitude, nearly sinusoidal, low frequency waves observed far upstream to the large amplitude turbulent "pulsations" which are more traditionally associated with the shock itself.

062.028 **Collisions between quasi–parallel shocks.**
P. J. Cargill.
Adv. Space Res., Vol. 11, No. 9, p. 241 – 244 (1991). – See Abstr. 012.006 for the main entry.
The collision between pairs of quasi–parallel shocks is examined using hybrid numerical simulations. In the interaction, the two shocks are transmitted through each other leaving behind a hot plasma with a population of particles with energies in excess of 40 E_0, where E_0 is the kinetic energy of particles in the shock frame prior to the collision. The energization is more efficient for quasi–parallel shocks than parallel shocks. Collisions between shocks of equal strengths are more efficient than those that are unequal.

062.029 **On low frequency magnetic field turbulence upstream of Earth's bow shock.**
G. Mann.
Adv. Space Res., Vol. 11, No. 9, p. 249 – 252 (1991). – See Abstr. 012.006 for the main entry.
Low frequency magnetic field turbulence is described in terms of an ideal gas of circularly polarized Alfvén solitons. This approach leads to a magnetic field power spectrum with a broad local maximum and an exponential decay and is able to reproduce a special power spectrum obtained by the AMPTE/IRM spacecraft mission upstream of Earth's bow shock.

062.030 **Modeling Fermi acceleration at parallel shocks with Monte Carlo techniques.**
D. C. Ellison.
Adv. Space Res., Vol. 11, No. 9, p. 265 – 274 (1991). – See Abstr. 012.006 for the main entry.
The author describes a well–tested Monte Carlo simulation which combines a determination of parallel shock structure with first–order Fermi particle acceleration. By not calculating time ion scattering self–consistently, one is able to run the simulation long enough to model particle acceleration to energies seen in diffuse ions observed at Earth's bow shock.

062.031 **Magnetogravitodynamic instability of a cylindrical fluid jet acting upon a general, varying magnetic field.**
A. E. Radwan.
Astrophys. Space Sci., Vol. 182, No. 1, p. 85 – 96 (Aug 1991).
The instability of a fluid cylinder under the action of self–gravitating, inertia, and electromagnetic forces is developed. A general eigenvalue relation is derived analytically and the results are confirmed numerically. The jet is gravitationally unstable for small axisymmetric disturbances and stable for all other disturbances. The electromagnetic force has stabilizing and destabilizing influences not only in the axisymmetric mode but also in the non–axisymmetric disturbances. This is due to the fact that the pervading magnetic field is non–uniform. The vacuum field attains its maximum destabilizing influence if the perturbed and unperturbed vacuum fields are orthogonal. The Lorentz force for some states has the character of enlarging the gravitational unstable domain and decreasing the stable regions in all possible modes of perturbations. In other states the Lorentz force is

totally stabilizing but no magnetic field, however strong, can suppress the gravitational instability because the gravitational instability of sufficiently long waves will persist.

062.032 **Static charged dust distributions: sources of purely electromagnetic origin.**
R. N. Tiwari, S. Ray.
Astrophys. Space Sci., Vol. 182, No. 1, p. 105 – 110 (Aug 1991).
The present work is a continuation of the authors' earlier work on the charged dust sources of purely electromagnetic origin for static axisymmetric and static spherically-symmetric fields. Here their work is extended to the case of generalized static metric and have shown that a static charged dust distribution, irrespective of any symmetry conditions, can be only of purely electromagnetic origin. Incidentally, it follows from this result itself that the "Weyl–Majumdar–Papapetrou" class of static charged dust sources, which form an important class of astrophysical systems, are also of purely electromagnetic origin.

062.033 **A note on the mass–loaded MHD flow of the solar wind towards a cometary nucleus.**
K. R. Flammer, D. A. Mendis.
Astrophys. Space Sci., Vol. 182, No. 1, p. 155 – 162 (Aug 1991). Letter–to–the–editor.
In their pioneering work, Biermann et al. (1967) used the steady–state hydrodynamic equations to examine the dynamics of supersonic flow with mass addition. Assuming hypersonic flow at infinity, these authors showed that such a flow was possible only as long as the normalized flux $\hat{x}(\varrho u/\varrho_\infty u_\infty)$ was less than a critical value $\hat{x}_c$. At $\hat{x}_c$, the thermal Mach number became unity. In this brief note, the authors follow the treatment of Biermann et al., but investigate a 1–D steady state, mass accreting MHD flow in the more general case where the magnetic and thermal pressures at infinity are non–zero. Using the integrals of the single fluid MHD conservation equations and Faraday law, it is shown that a self–reversal of the flow occurs when the magnetosonic Mach number reduces to unity and $\hat{x}$ reaches a critical value which depends only on the ratio of specific heats of the plasma and the conditions at infinity.

062.034 **On the anomalous Doppler/inner Lindblad resonance.**
R. O. Dendy.
Plasma Phys. Controlled Fusion, Vol. 33, No. 9, p. 1069 – 1076 (Aug 1991).
There are strong formal similarities between the condition for anomalous Doppler resonance in plasmas (ω–$k_{\|}v_{\|} = -\Omega$, where ω is the frequency of the perturbing wave field, $k_{\|}v_{\|}$ measures the Doppler shift due to guiding–centre motion, and Ω is the cyclotron frequency) and the condition for inner Lindblad resonance in stellar discs (Ω_b–$\Omega_0 = -\varkappa$/m, where Ω_b is the frequency of the perturbing wave field, whose azimuthal harmonic number is m, Ω_0 is the natural frequency of equilibrium guiding–centre circulation, and $\varkappa$ is the epicyclic frequency). It is shown that this similarity reflects a close underlying correspondence between the two forms of resonance.

062.035 **Waves and instabilities.**
D. A. Gurnett.
Physics of the inner heliosphere 2. Particles, waves and turbulence, p. 135 – 157 (1991). – See Abstr. 003.002 for the main entry.
Contents: (1) Introduction. (2) Electrostatic waves: electron plasma oscillations, ion acoustic waves. (3) Electromagnetic waves: MHD waves and ion cyclotron waves, whistler–mode waves.

062.036 **Maxwell's equations in divergence form for general media with applications to MHD.**
M. H. P. M. van Putten.
Commun. Math. Phys., Vol. 141, No. 1, p. 63 – 77 (Oct 1991).
Maxwell's equations in media with general constitutive relations are reformulated in covariant form as a system of divergence equations without constraints. This reformulation enables one to express general electro–magneto–fluid problems as hyperbolic systems in divergence form. The author illustrates this method on the MHD problem. In the absence of constraints, a general representation is derived for the characteristic form for first–order systems of quasi–linear partial differential equations in vector fields and scalars. Using this covariant formulation of characteristics, one finds that the principle of covariance imposes a very rigid structure on the infinitesimally small amplitude waves in MHD. To demonstrate the power of the reformulation, the authors study numerically ultra–relativistic wave breaking using the divergence formulation of MHD.

062.037 **Relativistic kinetic shock waves in synchrotron sources: shock structure and non–thermal acceleration of positrons.**
M. Hoshino, J. Arons, Y. A. Gallant, A. B. Langdon, C. E. Max.
Bull. Am. Astron. Soc., Vol. 23, No. 2, p. 918 (1991). Abstract. – See Abstr. 012.008 for the main entry.

062.038 **Numerical simulation of a hypersonic jet.**
J. Burns, C. Loken, D. Clarke, M. Norman.
Bull. Am. Astron. Soc., Vol. 23, No. 2, p. 927 (1991). Abstract. – See Abstr. 012.008 for the main entry.

062.039 **Numerical observations of a jet–cloud collision in 3–D.**
D. A. Clarke, M. L. Norman, J. O. Burns, C. Loken.
Bull. Am. Astron. Soc., Vol. 23, No. 2, p. 951 (1991). Abstract. – See Abstr. 012.008 for the main entry.

062.040 **Asymmetric morphology of radio sources: numerical simulation of jets in atmospheres.**
P. E. Hardee, R. E. White III.
Bull. Am. Astron. Soc., Vol. 23, No. 2, p. 951 (1991). Abstract. – See Abstr. 012.008 for the main entry.

062.041 **Three–dimensional magnetostatic equilibrium.**
B. C. Low.
Bull. Am. Astron. Soc., Vol. 23, No. 2, p. 967 (1991). Abstract. – See Abstr. 012.008 for the main entry.

062.042 **Magnetoacoustic surface gravity waves.**
A. G. González, J. Gratton.
Sol. Phys., Vol. 134, No. 2, p. 211 – 232 (Aug 1991).
The effect of gravity on the surface magnetoacoustic waves may be important when considering applications to solar and laboratory plasmas. The linear magnetoacoustic waves, which may appear in a configuration with an interface between two plasmas or a plasma and an ordinary gas, are studied. Compressibility and gravity are taken into account. The different types of couplings between internal and surface modes are analyzed. Magnetoacoustic surface waves are studied in detail in a configuration consisting of an interface between an isothermal plasma and an ordinary gas. The possible regions where these modes may exist are discussed. A general way of grouping and classifying the complicated spectra of modes is presented. New groups of modes appear as a consequence of gravity and stratification, in addition to those already present in the absence of gravity. The results may be of help in studying more complicated cases.

062.043 **Shear–induced instability and arch filament eruption: a magnetohydrodynamic (MHD) numerical simulation.**
S. T. Wu, M. T. Song, P. C. H. Martens, M. Dryer.
Sol. Phys., Vol. 134, No. 2, p. 353 – 377 (Aug 1991).
The authors investigate, via a two–dimensional (nonplanar) MHD simulation, a situation wherein a bipolar magnetic field embedded in a stratified solar atmosphere (i.e., arch–filament–like structure) undergoes symmetrical shear motion at the footpoints. It was found that the vertical plasma flow velocities grow exponentially leading to a new type of global MHD–instability that could be characterized as a "Dynamic Shearing Instability", with a growth rate of about $\sqrt{8}\,\bar{V}_A a$, where $\bar{V}_A$ is the average Alfvén speed and a^{-1} is the characteristic length scale. The growth rate grows almost linearly until it reaches the same

order of magnitude as the Alfvén speed. Then a nonlinear MHD instability occurs beyond this point. This simulation indicates the following physical consequences: the central loops are pinched by opposing Lorentz foreces, and the outer closed loops stretch upward with the vertically-rising mass flow. This instability may apply to arch filament eruptions (AFE) and coronal mass ejections. To illustrate the nonlinear dynamical shearing instability, numerical examples are given for three different values of the plasma beta.

062.044 **Properties of mean field dynamos with non-axisymmetric α-effect.**
D. Moss, A. Brandenburg, I. Tuominen.
Astron. Astrophys., Vol. 247, No. 2, p. 576 – 579 (Jul 1991).
The authors investigate the influence of an azimuthally dependent α-effect on the properties of α^2 and $\alpha^2\Omega$-dynamos in spherical geometry, restricting themselves to odd parity solutions in linear theory. For all their linear models they find an exponentially growing mode, consisting of locked axisymmetric and nonaxisymmetric parts. A strong nonaxisymmetry in α substantially increases the linear growth rates at given dynamo number and can result in a marginal dynamo number that is significantly smaller than for an axisymmetric α with the same value of $\int \alpha dV$. The authors also report some exploratory nonlinear calculations and briefly discuss the relevance of their results to galactic dynamos and to stars.

062.045 **Registration of macro- and micro-inhomogeneities of plasma density in astrophysical objects.**
L. N. Kurochka.
Kinematika Fiz. Nebesn. Tel, Tom 7, No. 4, p. 31 – 37 (Jul–Aug 1991). In Russian. English translation in Kinematics Phys. Celest. Bodies, Vol. 7, No. 4.
Expedience of observations of extended astrophysical objects (like solar formations) using a broad-band filter behind the Balmer limit is indicated. These filtergrams may give information on macroscopic inhomogeneities of plasma density. It is shown that rather good knowledge about micro-inhomogeneities of plasma density (on the subtelescopic level) and extension of the volumes with different electron density can be received from intensity distribution in the Balmer or other hydrogen series precontinuum.

062.046 **Suppression of turbulent transport by a weak magnetic field.**
F. Cattaneo, S. I. Vainshtein.
Astrophys. J., Lett., Vol. 376, No. 1, p. L21 – L24 (20 Jul 1991). With plate L1.
Two-dimensional numerical simulations with high spatial resolution are used to study the effects of a large-scale magnetic field on its turbulent transport. It is commonly believed that the nonlinear back-reaction of the magnetic field on the turbulence becomes important when the field strength is close to equipartition. However, the authors show that turbulent diffusion is effectively reduced even when the large-scale field is much weaker than equipartition.

062.047 **Effects of Hall currents and Coriolis forces on the instability of a self-gravitating plasma of variable density.**
A. Gupta, P. K. Bhatia.
Astrophys. Space Sci., Vol. 182, No. 2, p. 171 – 187 (Aug 1991).
The hydromagnetic instability of a self-gravitating, incompressible rotating plasma of variable density has been examined in the presence of Hall currents. The system is assumed to be permeated by a variable horizontal magnetic field. The solution of the relevant linearized perturbation equations has been obtained by the normal mode technique through a variational principle which is shown to characterize the problem. Proper solutions have been obtained for a semi-infinite plasma having exponential density stratification along the vertical. The dispersion relation has been derived and solved numerically for different values of the physical parameters involved. It is found that Hall currents and Coriolis forces have both destabilizing influence as the growth rate of the unstable modes is found to increase with the increase of both Hall currents and Coriolis forces.

062.048 **Dynamic oscillatory reconnection at an X-type neutral point.**
A. N. McClymont, I. J. D. Craig, P. G. Watson.
Bull. Am. Astron. Soc., Vol. 23, No. 2, p. 1041 (1991). Abstract. – See Abstr. 012.012 for the main entry.

062.049 **Dynamic oscillatory reconnection in X-type neutral points.**
I. J. D. Craig, A. N. McClymont, P. G. Watson.
Bull. Am. Astron. Soc., Vol. 23, No. 2, p. 1041 (1991). Abstract. – See Abstr. 012.012 for the main entry.

062.050 **Nonlinear evolution of the resistive tearing mode instability with shear flow and viscosity.**
L. Ofman, P. J. Morrison, R. S. Steinolfson.
Bull. Am. Astron. Soc., Vol. 23, No. 2, p. 1042 (1991). Abstract. – See Abstr. 012.012 for the main entry.

062.051 **Numerical simulation of 3d MHD free shear layers.**
R. B. Dahlburg, T. A. Zang.
Bull. Am. Astron. Soc., Vol. 23, No. 2, p. 1042 (1991). Abstract. – See Abstr. 012.012 for the main entry.

062.052 **Stressed magnetic reconnection.**
A. B. Hassam.
Bull. Am. Astron. Soc., Vol. 23, No. 2, p. 1042 (1991). Abstract. – See Abstr. 012.012 for the main entry.

062.053 **The structure of radiative MHD shocks in the corona.**
P. Xu, T. G. Forbes.
Bull. Am. Astron. Soc., Vol. 23, No. 2, p. 1042 (1991). Abstract. – See Abstr. 012.012 for the main entry.

062.054 **Three-dimensional hydrodynamic stellar convection.**
M. Hossain, D. J. Mullan.
Bull. Am. Astron. Soc., Vol. 23, No. 2, p. 1047 (1991). Abstract. – See Abstr. 012.012 for the main entry.

062.055 **Comparison of relationships provided by numerical simulations of turbulent, compressible convection with mixing length theory.**
T. J. Lydon, S. Sofia.
Bull. Am. Astron. Soc., Vol. 23, No. 2, p. 1047 (1991). Abstract. – See Abstr. 012.012 for the main entry.

062.056 **On magnetic diffusion in a turbulent fluid.**
S. Vainshtein, F. Cattaneo, R. Rosner.
Bull. Am. Astron. Soc., Vol. 23, No. 2, p. 1049 (1991). Abstract. – See Abstr. 012.012 for the main entry.

062.057 **A sub-grid-scale magnetic resistivity formulation.**
M. L. Theobald, P. Fox, S. Sofia.
Bull. Am. Astron. Soc., Vol. 23, No. 2, p. 1049 (1991). Abstract. – See Abstr. 012.012 for the main entry.

062.058 **Modulation of ion acoustic waves.**
S. K. El-Labany.
Astrophys. Space Sci., Vol. 182, No. 2, p. 241 – 247 (Aug 1991).
The multiple scales perturbation theory is applied to a system of dispersive waves including plasma waves (ion-acoustic waves). Assuming the amplitude of waves are slowly varying function of space and time, the author finds that long-time slow modulation of the complex amplitude can be described by the non-linear Schrödinger equation. This result agrees with that obtained by Shimizu and Ichikawa using the reductive perturbation theory, and agrees exactly with the nonrelativistic limit of the nonlinear Schrödinger equation obtained by Nejoh using the stretching method.

062.059 **Effects of line–tying and non–uniformities on thermal instabilities and slow MHD modes.**
D. Hermans, A. W. Hood, L. Clifford, A. Milne.
International Conference on Mechanisms of Chromospheric and Coronal Heating, p. 405 – 407 (1991). – See Abstr. 012.011 for the main entry.

The thermal stability of a planar plasma is studied both analytically and numerically for two different models. For $\tau_{rad}/\tau_{\perp} \ll 1$, the growth rate for the thermal instabilities with small wavelengths perpendicular to the magnetic field lines is essentially the same as for the most unstable mode, i.e., a mode with large perpendicular wave length. This could indicate that the instability might be triggered by a mode with small scale structure across the magnetic field. When the magnetic field vanishes locally, localized modes exist.

062.060 **Heating in intense flux tubes.**
S. S. Hasan.
International Conference on Mechanisms of Chromospheric and Coronal Heating, p. 408 – 413 (1991). – See Abstr. 012.011 for the main entry.

Energy transport is considered in an intense flux tube on the Sun, with the aim of examining the role of various heating mechanisms and delineating their contributions to the energy budget. The temporal behaviour of an intense flux tube is studied, when its equilibrium is perturbed by a small downflow. It is found that overstable oscillations with a characteristic time of some 600 s are set up. The computed temperature structure appears to be compatible with observations in the upper layers of the tube. It appears that overstable oscillations are probably not important in heating the chromosphere and corona.

062.061 **Damping of shocks in magnetic flux tubes.**
A. Ferriz Mas, F. Moreno Insertis.
International Conference on Mechanisms of Chromospheric and Coronal Heating, p. 417 – 419 (1991). = Mitt. Kiepenheuer–Inst., Nr. 335. – See Abstr. 012.011 for the main entry.

The authors report on a study of the propagation and damping of slow–mode magnetohydrodynamic shock waves in a cylindrical magnetic flux tube embedded in a field–free homogeneous fluid. They carry out an analytical study of the nonlinear flow equations obtained in the thin flux tube approximation for magnetic flux concentrations. A direct motivation for this work in the context of the present conference is given by the fact that the conversion of kinetic and magnetic energy into thermal energy in flux tube shocks is possibly one of the main mechanisms responsible for the nonradiative heating of stellar chromospheres. Further examples of shock structures in magnetic tubes in the solar atmosphere are provided by the siphon flows.

062.062 **Alfvén waves.**
J. V. Hollweg.
International Conference on Mechanisms of Chromospheric and Coronal Heating, p. 423 – 434 (1991). – See Abstr. 012.011 for the main entry.

The authors discuss some of the basic questions which have been examined in terms of Alfvén waves. They will emphasize the question of whether enough wave energy can reach the corona, and point out a problem with powering short coronal loops. They will also discuss damping mechanisms, with emphasis on turbulence and propagation in stochastic magnetic fields.

062.063 **Heating in stochastic magnetic fields.**
R. N. Sudan.
International Conference on Mechanisms of Chromospheric and Coronal Heating, p. 448 – 463 (1991). – See Abstr. 012.011 for the main entry.

This review is concerned with the nature of micro–structure of the force–free magnetic field in coronal loops and how this micro–structure is created by random motions of the foot points of the field lines anchored in the convecting photosphere. This micro–structure enhances joule heating because the current is strongly filamented in fine scales. The effective Lundquist number is reduced by orders of magnitude from that based on the gross dimensions of the coronal loop. It examines whether the magnetic field lines are laminar or chaotic and if chaotic, over what fraction of the volume of the coronal loop. Finally, the propagation and dissipation of Alfvén waves in stochastic magnetic fields, where the field lines are chaotic in a significant fraction of the flux tube volume, is described.

062.064 **Resonance absorption heating.**
J. M. Davila.
International Conference on Mechanisms of Chromospheric and Coronal Heating, p. 464 – 479 (1991). – See Abstr. 012.011 for the main entry.

Perhaps the most striking feature of soft X–ray and ultraviolet observations of the solar corona is the presence of discrete, bright, loop–like structures. These pictures show the corona to be a highly inhomogeneous medium where the magnetic field actively mediates the heating process. The theory used to describe heating by waves in the presence of strong density gradients has become known as resonance absorption theory. Although the resonance absorption mechanism has been used in the laboratory to heat fusion plasmas for some time now, it was suggested as a mechanism for heating stellar coronae only about twelve years ago. In the intervening years, significant progress has been made toward a better understanding of the resonance absorption physics, and in the application of the process to solar and stellar plasmas.

062.065 **Magnetohydrodynamic surface waves.**
B. Roberts.
International Conference on Mechanisms of Chromospheric and Coronal Heating, p. 494 – 507 (1991). – See Abstr. 012.011 for the main entry.

The behaviour of magnetohydrodynamic surface waves propagating on a single magnetic interface is reviewed, including the effects of nonlinearity and stratification. Applications to the Sun, including sunspots and the global f–mode oscillation, are discussed.

062.066 **Magnetoacoustic–gravity surface waves.**
A. J. Miles, B. Roberts.
International Conference on Mechanisms of Chromospheric and Coronal Heating, p. 508 – 510 (1991). – See Abstr. 012.011 for the main entry.

The nature of magnetoacoustic–gravity surface waves at a single magnetic interface is examined for the case of parallel propagation. Above the interface is an isothermal medium permeated by a uniform horizontal magnetic field. The lower region is a field–free medium of different temperature to the magnetic atmosphere. In the absence of gravity the interface may support one or two surface modes determined by the temperatures of the two media. The effect of gravity on the modes is examined.

062.067 **Critical frequencies and reflection of fast mode waves in an isothermal atmosphere.**
B. A. Stark, Z. E. Musielak.
Bull. Am. Astron. Soc., Vol. 23, No. 2, p. 1060 (1991). Abstract. – See Abstr. 012.012 for the main entry.

062.068 **On the direction of the currents at magnetic neutral points.**
J. M. Fontenla, J. M. Davis.
Bull. Am. Astron. Soc., Vol. 23, No. 2, p. 1067 (1991). Abstract. – See Abstr. 012.012 for the main entry.

062.069 **Properties of non–parallel magnetoacoustic surface waves.**
R. Jain, B. Roberts.
International Conference on Mechanisms of Chromospheric and Coronal Heating, p. 511 – 513 (1991). – See Abstr. 012.011 for the main entry.

The properties of magnetoacoustic surface waves at a single magnetic interface, one side of which is field–free, are studied for the case of non–parallel propagation. In the absence of gravity the interface may support a slow surface wave or both slow and fast surface waves, depending upon the relative temperatures of the two media. The speed of propagation of the surface waves is reduced with increasing angle of propagation to the magnetic field. The effect of gravity on the modes is examined for an isothermal interface of a horizontal magnetic field above a field-free medium. The existence of a third surface wave, the f-mode, is discussed for the non-parallel propagation of waves. The dispersion relation for the f-mode is obtained analytically for small values of the Alfvén velocity.

062.070 **Viscous damping of magnetohydrodynamic surface waves.**
M. Ruderman.
International Conference on Mechanisms of Chromospheric and Coronal Heating, p. 514 - 516 (1991). - See Abstr. 012.011 for the main entry.

The magnetohydrodynamic surface wave propagation on a single magnetic interface in a cold plasma is considered. The ion viscosity is taken into account. The dispersion equation is obtained and investigated in the low-frequency approximation. The equation governing the evolution of small-amplitude nonlinear waves is derived. It is shown on the basis of a numerical solution of this equation, that at the limit of small viscosity the damping distance of nonlinear wave is independent of the viscosity. An asymptotic formula for the damping distance in the limit of small viscosity is derived. This formula shows that for typical coronal conditions the nonlinearity effect can decrease the wave damping distance as much as ten times in comparison with the results given by the linear theory.

062.071 **Coronal loop heating by the fast surface wave.**
A. S. de Assis, K. H. Tsui.
International Conference on Mechanisms of Chromospheric and Coronal Heating, p. 517 - 518 (1991). - See Abstr. 012.011 for the main entry.

Although coronal plasmas are considered to be collisional, it is well known that particle collisions do not provide effective energy dissipation. It is shown that the wave-particle resonant interactions can dissipate energy fast enough to sustain the coronal temperature.

062.072 **The formation of current sheets and coronal heating.**
E. R. Priest.
International Conference on Mechanisms of Chromospheric and Coronal Heating, p. 520 - 535 (1991). - See Abstr. 012.011 for the main entry.

The physical mechanisms for forming current sheets are reviewed. In a two-dimensional magnetic field, sheets may form near X-type neutral points or separatrices in response to footpoint motions (converging or shearing) or to magnetic instability. In three-dimensions they may develop near separators or separatric surfaces, but also more generally near any potential singular line in a sheared field. They may also be created in an MHD turbulent state as magnetic energy cascades to small scales. It is suggested that the closed magnetic regions of stellar coronae are filled with many small current sheets that are continually forming, filamenting and dissipating. Also a new self-consistent theory for heating by MHD turbulence is presented in which both the macroscopic state and the turbulent diffusivities depend on each other and the resulting heating depends only on the boundary motions and the geometry.

062.073 **Current sheet formation in force-free magnetic fields.**
G. Vekstein, E. R. Priest.
International Conference on Mechanisms of Chromospheric and Coronal Heating, p. 536 - 538 (1991). - See Abstr. 012.011 for the main entry.

The evolution of coronal magnetic fields in response to slow photospheric footpoint motions does not in general produce smooth a force-free equilibrium. The presence of separatrix lines results in current sheet formation even under an infinitesimal shearing deformation of a 2-D magnetic field. When a separatrix without neutral points is present a regular perturbation approach is valid and preserves the magnetic field geometry. In the case of a neutral X-point perturbation the scheme fails and a geometry with cusped triple points is formed. In general these current sheets could also trigger a solar flare.

062.074 **Two-dimensional magnetic neutral points.**
N. R. Strachan, E. R. Priest.
International Conference on Mechanisms of Chromospheric and Coronal Heating, p. 539 - 540 (1991). - See Abstr. 012.011 for the main entry.

Several types of magnetic neutral points are possible in two dimensions. X-points have four separatrix field lines meeting at the neutral point and are a particular kind of star point which have an even number of separatrices greater than two. Asymmetric star points have an odd number of separatrices, at least one of which is a current sheet. Those with three separatrices include: Y-points with the separatrices meeting at angles of $2\pi/3$; cusp-points with two of them touching at the neutral point; and T-points with two meeting the third at right angles.

062.075 **The significance of magnetic null points.**
K. Galsgaard, Å. Nordlund.
International Conference on Mechanisms of Chromospheric and Coronal Heating, p. 541 - 543 (1991). - See Abstr. 012.011 for the main entry.

An illustrative example is given how the field line connectivity changes when two magnetic null points merge and are transformed into a single magnetic minimum in space.

062.076 **Heating by field aligned DC Joule dissipation.**
D. S. Spicer.
International Conference on Mechanisms of Chromospheric and Coronal Heating, p. 547 - 561 (1991). - See Abstr. 012.011 for the main entry.

A short review of coronal heating by the Joule dissipation of DC magnetic field-aligned currents is given. The author summarizes what observational and theoretical constraints exist that constrain any model of coronal heating. He examines sources of field-aligned currents, the role of force-free current systems, and the significance of "voltage" driven and "current" driven current sources. The author then examines potential mechanisms for structuring field-aligned currents due to thermal instabilities and boundary effects and how they relate to "voltage" and "current" sources. Finally he examines the conditions required for field-aligned anomalous Joule heating.

062.077 **Current helicity and the generation of magnetic field aligned currents.**
N. Seehafer, K.-H. Rädler.
International Conference on Mechanisms of Chromospheric and Coronal Heating, p. 582 - 584 (1991). - See Abstr. 012.011 for the main entry.

The comparison of extrapolated force-free magnetic fields with chromospheric and coronal observations indicates that in active regions the current helicity, $H_C = \mathbf{B}\cdot\nabla\times\mathbf{B}$, is predominantly negative in the northern and positive in the southern hemisphere. There is a simple relation between current helicity and the α-effect. The magnetic fields and currents of active regions as a whole may be considered as fluctuations in the sense of mean-field theory of the global field of the Sun. Alternatively, properties of the magnetic fields and currents of individual active regions may be ascribed to mean fields. Both possibilities are discussed with reference to the relation between current helicity and the α-effect and the sign rule for the current helicity.

062.078 **Nonlinear evolution of a force-free arcade field driven by shear flow.**
N. Bekki, T. Tajima, J. W. Van Dam.
International Conference on Mechanisms of Chromospheric and Coronal Heating, p. 585 - 587 (1991). - See Abstr. 012.011 for the main entry.

The authors investigate a nonlinear dynamics of force-free arcade field driven by slow shear flow on a photospheric boundary analytically and numerically. The Woltjer-Taylor relaxation theory is generalized to obtain the time-dependent magnetohydrodynamics solution.

062.079 **Two-dimensional magnetic reconnection.**
M. Jardine.
International Conference on Mechanisms of Chromospheric and Coronal Heating, p. 588 – 600 (1991). – See Abstr. 012.011 for the main entry.
Magnetic reconnection is a process of fundamental importance in a magnetized plasma. It can not only change the topology of a magnetic field but it is also extremeley important as a mechanism for converting stored magnetic energy into kinetic and thermal energy. So it may be important for heating the corona of the Sun and perhaps of other stars too. The author reviews the basic physics of this process and the attempts that have been made, both analytical and numerical, to understand it.

062.080 **Magnetic field annihilation within a stagnation point flow.**
M. Jardine, E. R. Priest, H. R. Allen.
International Conference on Mechanisms of Chromospheric and Coronal Heating, p. 601 – 603 (1991). – See Abstr. 012.011 for the main entry.
The authors present a general family of models for magnetic field annihilation in a current sheet that has formed within a stagnation point flow. The different members of the family are characterised by the value of the vorticity imposed on the flow. The previously-known solution of Sonnerup and Priest (1975) is a special case obtained when the inflow vorticity is zero. The current sheet thickness is least for a positive vorticity and decreases as the resistivity or the viscosity is decreased. If the imposed vorticity is too large, the flow develops a cellular pattern and so may disrupt the process of magnetic field annihilation.

062.081 **Three-dimensional magnetic reconnection: basic concepts.**
M. Hesse, K. Schindler, J. Birn.
International Conference on Mechanisms of Chromospheric and Coronal Heating, p. 604 – 608 (1991). – See Abstr. 012.011 for the main entry.
A summary of the basic properties of magnetic reconnection in general three dimensional systems is presented. As an example, the authors discuss the evolution of a plasmoid in a three dimensional resistive MHD model of a magnetospheric substorm. Applications to the solar corona are suggested also.

062.082 **Nonlinear break-up of magnetohydrodynamic jet.**
R. K. Chhabra, S. K. Trehan.
Astrophys. Space Sci., Vol. 183, No. 1, p. 37 – 50 (Sep 1991).
A nonlinear analysis is presented to take into account the effect of a magnetic field on the capillary instability of a fluid jet, using the method of strained coordinates. The growth of the surface waves is calculated for wave number less than the critical wave number. A third-order solution shows that, as a result of the interaction among the lower harmonics, not only a higher harmonic appears; but there is also a feed back into the fundamental harmonic. It is important to note that the magnetic field has the effect of making the evolution curves of the disturbances symmetric about the axis as compared to the highly asymmetric pattern which prevails in its absence.

062.083 **The equilibria and evolutions of magnetized, rotating, isothermal clouds. V. The effect of the toroidal field.**
K. Tomisaka.
Astrophys. J., Vol. 376, No. 1, p. 190 – 198 (20 Jul 1991).
The structures of magnetized clouds are studied from the viewpoint of magnetohydrostatic equilibrium between self-gravity and magnetic and thermal pressure forces. A toroidal as well as a poloidal magnetic field is taken into account. A poloidal current flowing along the poloidal magnetic field forms a toroidal field which causes the cloud to be pinched towards the symmetry axis. In contrast to the poloidal field, which plays a role in supporting the cloud, the toroidal field has the effect of shrinking the cloud. A cloud that is pinched by the toroidal field, but has a low central density, often shows a prolate spheroidal shape. With increasing central density the cloud becomes a concave oblate spheroid. The maximum mass of the cloud supported by magnetic field and thermal pressure decreases with increasing toroidal field strength. Assuming that a cloud encounters a torsional Alfvén wave, and the magnetic field begins to be wound up, the cloud begins dynamical collapse if the cloud mass exceeds the marginally supported maximum mass. This indicates a new mode triggering star formation: toroidal field-induced collapse.

062.084 **A study of stellar point explosion in a self-gravitating radiative magneto-hydrodynamic medium.**
O. Nath, S. Ojha, H. S. Takhar.
Astrophys. Space Sci., Vol. 183, No. 1, p. 135 – 145 (Sep 1991).
A theoretical model of shock wave propagation in a self-gravitating radiative magneto-hydrodynamic medium has been studied. The effects of the magnetic field, radiation, and gravitation have been discussed separately. The results discussed depend upon the numerical variations of flow variables behind the shock.

062.085 **The interpretation of density sensitive line diagnostics from inhomogeneous plasmas. II. Non-isothermal plasmas.**
J. C. Brown, B. N. Dwivedi, Y. M. Almleaky, P. A. Sweet.
Astron. Astrophys., Vol. 249, No. 1, p. 277 – 283 (Sep 1991).
The problem of inferring density structure in inhomogeneous non-isothermal plasmas, from density sensitive line ratios, is discussed. It is shown how the concepts of emission measure differential in density $\zeta(n)$ and emission measure differential in temperature $\xi(T)$ can be generalized to analyse arbitrary plasmas. In the special case (pressure stratification) where surfaces of constant temperature S_T and of constant density S_n coincide, it is proven that it is possible in principle, if $\xi(T)$ is known from resonance lines, to derive information about the density (or pressure) distribution as a function of temperature, i.e. $n(T)$, using density sensitive lines, with an analytic solution in special cases. It is then investigated whether and when the predicted line strengths of a set of sources with different $n(T)$ but with the same $\xi(T)$ would be distinguishable using measurements of density sensitive lines, using the empirical model $n(T) = n_0(T/T_0)^{-\delta}$. It is found that the inferred value of δ and of the base density n_0 from two density sensitive line ratios is quite sensitive to the data, thus demonstrating the possibility of deducing $n(T)$. The approach was tested on hypothetical data using two lines ratios for ions of O IV and Fe IX in a temperature range 2×10^5–1.6×10^6K, values of δ and the electron pressure (base density) being satisfactorily recovered.

062.086 **Thermal instability in a two-component magnetic plasma.**
Yu. V. Vandakurov.
Pis'ma Astron. Zh., Tom 17, No. 11, p. 1031 – 1038 (Nov 1991). In Russian. English translation in Sov. Astron. Lett., Vol. 17, No. 6.
The author discusses the origin of discrepancies which arise under consideration either in the magnetohydrodynamic or in two-fluid approximations of the problem of thermal instability in a radiating magnetized homogeneous plasma for the case of an arbitrary angle between the wave vector and the equilibrium magnetic field. In the former approximation the displacement current is dropped, but this assumption is at variance with the full system of two-fluid plasma equations. Just for this reason there are discrepancies mentioned above. The author shows that in the case of of plasma motions approximately perpendicular to the magnetic field, the region of overstable perturbations may be even wider than in the case of strictly transversal motions.

062.087 **MHD discontinuities in a fine magnetic force tube.**
V. S. Semenov.
Geomagn. Aehron., Tom 31, No. 5, p. 769 – 774 (Sep – Oct 1991). In Russian. English translation in Geomagn. Aeron., Vol. 31, No. 5.

062.088 **MHD free and forced convection flow past an infinite vertical plate with heat source.**
B. K. Jha.
Astrophys. Space Sci., Vol. 183, No. 2, p. 169 – 175 (Sep 1991).
An analytical study is performed to examine the effects of magnetic field and temperature-dependent heat source on the free and forced convection flow past an infinite vertical plate. The expressions for the velocity field, penetration distance and skin-friction have been obtained by Laplace transform technique. The influence of the various parameters entering into the problem is extensively discussed.

062.089 **Role of magnetic field on transient forced and free-convection flow past an infinite vertical porous plate through porous medium with heat source.**
B. K. Jha.
Astrophys. Space Sci., Vol. 183, No. 2, p. 177 – 182 (Sep 1991).
Effects of magnetic field and permeability of the porous medium on unsteady forced and free-convection flow past an infinite vertical porous plate in presence of temperature-dependent heat source have been analysed. The Laplace transform method is used to obtain the expression for velocity field, skin friction, and leading edge effects. During the course of discussion, the effects of M (magnetic parameter), S (heat source parameter), λ (suction parameter), and K (permeability of porous medium) on velocity field, skin friction, and leading edge effect have been extensively discussed.

062.090 **Free convection in hydromagnetic flows of a viscous heat-generating fluid with wall temperature oscillation and Hall currents.**
H. S. Takhar, P. C. Ram.
Astrophys. Space Sci., Vol. 183, No. 2, p. 193 – 198 (Sep 1991).
A study is made of the free convection in hydromagnetic flows through a porous medium of a heat generating fluid past an infinite vertical porous plate. A strong magnetic field is imposed in a direction which is perpendicular to the free stream and makes an angle α to the vertical direction. The governing equations for the hydromagnetic fluid flow and the heat transfer are solved analytically. The influence of Hall currents, the permeability K and the inclination α upon the velocity field are shown in figures.

062.091 **Gravitational instability for some astrophysical systems.**
R. C. Sharma, T. Chand.
Astrophys. Space Sci., Vol. 183, No. 2, p. 215 – 224 (Sep 1991).
The gravitational instability of an infinite homogeneous self-gravitating mixture through porous medium in the presence of a variable horizontal magnetic field varying in vertical directions has been considered to include, separately, the effects due to suspended particles and collisions between ionized and neutral components. The dispersion relations in both cases have been obtained. It has been found that Jeans's criterion of instability holds good even if the effects due to suspended particles, collisions, porosity, and variable magnetic field are considered.

062.092 **Hall effects on oscillatory MHD boundary-layer flow under variable suction.**
M. U. S. Reddy, N. K. Samira, R. Prasad, H. N. Gupta.
Astrophys. Space Sci., Vol. 183, No. 2, p. 273 – 283 (Sep 1991).
The effects of Hall current on the oscillatory hydromagnetic boundary-layer flow under variable suction, past an infinite porous flat plate in the presence of a transverse magnetic field is discussed. The expressions for velocity and skin-friction are obtained and their variations for small and large frequency of oscillations are extensively discussed.

062.093 **Evolution to nonequilibrium in simple models of prominence filaments.**
E. G. Zweibel.
Astrophys. J., Vol. 376, No. 2, p. 761 – 770 (1 Aug 1991).
The author considers the support of cold plasma against gravity by a line-tied magnetic field. The normal component of the field at the boundaries, the mass per unit flux, and the magnetic shear are all specified. Fixing the first two of these quantities, and varying the third leads to evolutionary sequences. When the magnetic shear is less than a critical value, an infinite number of solutions exist. When the magnetic shear is sufficiently large, there are no solutions with the desired topology. These results support the idea that prominences erupt as a result of evolution to magnetic nonequilibrium.

062.094 **The interaction of collisionless shocks in astrophysical plasmas.**
P. J. Cargill.
Astrophys. J., Vol. 376, No. 2, p. 771 – 781 (1 Aug 1991).
The interaction between collisionless shocks, such as may arise in a number of situations in space plasmas, is examined by means of hybrid numerical simulations. The production of energetic particles by shock collisions is investigated. When the shocks are quasi-perpendicular, ions with energies greater than $10E_0$ are produced, where E_0 is the kinetic energy of the unshocked ions. The effect is optimized when the two colliding shocks have roughly equal strengths. As the shocks become closer to quasi-parallel, the number and energy of the accelerated ions increases. Energies in excess of $30E_0$ are obtained in the quasi-parallel regime, with a small fall-off when the shocks are exactly parallel. For quasi-parallel collisions, the ion energization can be attributed to the interaction of hot ions with both strong electromagnetic waves and the motional electric field located at the shocks. These results are applied to shock collisions in the distant solar wind, at planetary bow shocks as well as to the impulsive phase of solar flares.

062.095 **The effect of large Larmor radius on nonlinear Alfvén waves.**
N. Kumar, K. M. Srivastava.
Astrophys. Space Sci., Vol. 184, No. 1, p. 49 – 55 (Oct 1991).
The effect of large Larmor radius (LLR) on the nonlinear behaviour of Alfvén waves propagating parallel to a uniform magnetic field in a compressible fluid is investigated with the aid of LLR-MHD equations. It is shown that asymptotic evolution of these waves is governed by the modified nonlinear Schrödinger equation. The dispersion is provided by the large Larmor radius effect in the magnetic field equation. It is suggested that these calculations can have a bearing on the investigation of the structure of MHD waves in both laboratory and space plasmas.

062.096 **Magnetic flux tubes – nonlinear strings in relativistic magnetohydrodynamics.**
V. S. Semenov, L. V. Bernikov.
Astrophys. Space Sci., Vol. 184, No. 1, p. 157 – 166 (Oct 1991).
A special Lagrangian coordinate system is introduced, in which the magnetic flux tube is the main object of investigation. The flux tube is shown to behave as a nonlinear relativistic string in relation to tangential force. General and particular variational principles are also discussed. With the introduction of a Lagrangian coordinate system the relativistic magnetohydrodynamic equation of motion reduces to a set of nonlinear string equations. From these results it follows that a highly conductive plasma with a "frozen-in" magnetic field can be considered as a gas of nonlinear strings interacting through pressure forces. A method is developed which allows the multi-dimensional nonlinear problems of relativistic MHD with boundary layers to be reduced to a set of two-dimensional problems for flux tubes, i.e., nonlinear strings.

062.097 **Convection, spontaneous discontinuities, and stellar winds and X-ray emission.**
E. N. Parker.
Reviews in Modern Astronomy 4, p. 1 – 17 (1991). – See Abstr. 003.008 for the main entry.
The subject of this lecture is the origin of the solar corona, but the author begins with a brief recollection of the basic ideas on which present concepts of solar and stellar activity are based. Contents: 1. Introduction. 2. The active X-ray corona. 3. The magnetostatic theorem. 4. Energy considerations. 5. The coronal hole. 6. High speed solar wind. 7. Summary.

062.098 **Instabilities in astrophysical shear flows.**
W. Glatzel.
Reviews in Modern Astronomy 4, p. 104 – 116 (1991). – See Abstr. 003.008 for the main entry.
Shear flows are found in many astrophysical objects such as accretion disks, jets and rotating stars. The study of their stability is particularly important with respect to the occurrence of turbulence in accretion disks and the stability of stellar rotation laws. Properties and mechanism of shear instabilities relevant in various astrophysical situations are discussed.

062.099 **Shock waves in accretion disks.**
H. C. Spruit.
Reviews in Modern Astronomy 4, p. 197 – 207 (1991). – See Abstr. 003.008 for the main entry.
Contents: 1. Angular momentum transport processes:
Shear turbulence. Angular momentum loss due to a magnetic wind.
2. Angular momentum transport by spiral waves:
Selfgravitating spiral waves. Angular momentum transport by spiral shocks. Planet– generated shock waves. One armed spirals, eccentric instability.

062.100 **Dynamos in galaxies.**
F. Krause, R. Wielebinski.
Reviews in Modern Astronomy 4, p. 260 – 286 (1991). – See Abstr. 003.008 for the main entry.
Contents: 1. Introduction. 2. Observations of magnetic fields in galaxies. 3. The origin of cosmic magnetic fields. 4. The dynamo: a universal process in rotating systems. 5. Dynamo problem. 6. Disc dynamo models. 7. Conclusions.

062.101 **Diffusion of particles, heat and magnetic fields in compressible turbulent media.**
A. Z. Dolginov, N. A. Silant'ev.
IAU Colloquium No. 130: The Sun and cool stars – activity, magnetism, dynamos, p. 71 – 74 (1991). – See Abstr. 012.024 for the main entry.
A new method for the calculation of kinetic coefficients is presented. This method allows to obtain the distribution of scalar and vector fields (such as the temperature, the admixture particle number density and the magnetic field) in turbulent cosmic media with any value of $S = u_0\tau_0/R_0$. The explicit expression for the "turbulent" diffusivity D_T is obtained. In some cases D_T becomes negative, implying the clustering of the admixture particles in patches (a local increase of the temperature and magnetic fields). The magnetic α–effect is considered for the case $S \sim 1$.

062.102 **The spatial structure of homogeneous turbulence at high Reynolds number.**
A. Vincent, M. Meneguzzi.
IAU Colloquium No. 130: The Sun and cool stars – activity, magnetism, dynamos, p. 75 – 79 (1991). – See Abstr. 012.024 for the main entry.
A direct numerical simulation at resolution 240^3 is used to obtain a statistically stationary three–dimensional homogeneous and isotropic turbulent field at a Reynolds number around 1000 ($R_\lambda \approx 150$). The energy spectrum displays an inertial subrange extending over more than one decade. In the physical space, the strongest vorticity is organized in very elongated thin tubes. The typical length of a tube is around the integral scale of the flow and its thickness is a few dissipation scales.

062.103 **Compressible MHD in spherical geometry.**
L. Valdettaro, M. Meneguzzi.
IAU Colloquium No. 130: The Sun and cool stars – activity, magnetism, dynamos, p. 80 – 85 (1991). – See Abstr. 012.024 for the main entry.
The generation of a magnetic field by a conducting, compressible fluid inside a spherical shell is studied by direct numerical simulations. A pseudo–spectral method is used in order to resolve accurately all the scales present in the problem. The range of parameters considered is the following: a unit Prandtl number, Rayleigh numbers up to 100 times critical, Taylor number 625, an aspect ratio of 2, a Mach number slightly less than 1, and pressure and temperature scale heights of the order of the thickness of the shell. A dynamo effect is observed for magnetic Prandtl numbers larger than 1. The authors present the properties of the turbulent flow, the role of the helicity and of the differential rotation in the enhancement of the magnetic field, and the spectral properties of the flow fields.

062.104 **Vector potential magnetic null points.**
K. Galsgaard, Å. Nordlund.
IAU Colloquium No. 130: The Sun and cool stars – activity, magnetism, dynamos, p. 89 – 91 (1991). – See Abstr. 012.024 for the main entry.
Recent observations show that the surface distribution of magnetic fields on the solar surface is selfsimilar, with an approximately constant fractal dimension over a large range of horizontal scales. Also, recent ideas about flare energy release assumes a "fibrous" corona, with the energy release consisting of many small scale events. Prompted partly by these observations, the authors investigate the properties of self–similar three–dimensional magnetic fields.

062.105 **Magnetic tubes in overshooting compressible convection.**
R. L. Jennings, A. Brandenburg, Å. Nordlund,
R. F. Stein, I. Tuominen.
IAU Colloquium No. 130: The Sun and cool stars – activity, magnetism, dynamos, p. 92 – 94 (1991). – See Abstr. 012.024 for the main entry.
A magnetic tube is introduced into turbulent compressible penetrative convection. After being strongly advected, most of the magnetic flux is stored in the overshoot region. With rotation there are meridional travelling waves.

062.106 **Stable and unstable solutions of the nonlinear dynamo problem.**
R. Meinel.
IAU Colloquium No. 130: The Sun and cool stars – activity, magnetism, dynamos, p. 103 – 111 (1991). – See Abstr. 012.024 for the main entry.
A local potential approach to nonlinear dynamo models which allows the use of variational techniques to investigate the problem of stability is introduced. The method is applieds to models which include the back–reaction of the magnetic field on the fluid motion in a simplified way. A special application leads to a previously investigated one–dimensional dynamo model which shows a coexistence of a periodic solution (limit cycle) with two stable steady solutions of opposite polarities. The inclusion of some small–amplitude noise leads to interesting transition phenomena which may be of relevance to explain the behaviour of astrophysical dynamos. A simple dynamical system with a two–dimensional phase–space is used for illustration.

062.107 **Non–steady global magnetic fields in kinematic theory.**
J. H. G. M. van Geffen, P. Hoyng, C. Zwaan.
IAU Colloquium No. 130: The Sun and cool stars – activity, magnetism, dynamos, p. 129 – 134 (1991). – See Abstr. 012.024 for the main entry.
In order to find the damping time for the mean field $<\boldsymbol{B}>$, the authors analyse the mean magnetic energy $<\boldsymbol{BB}>$. They have developed a code to solve the equation for $<\boldsymbol{BB}>$ in a spherical shell (the convection zone), assuming axisymmetry. They report first results, which do not yet include differential rotation.

062.108 **Asymptotic methods in the nonlinear mean–field dynamo.**
D. D. Sokoloff (*D. D. Sokolov*), A. Shukurov, A. A. Ruzmaikin (*A. A. Ruzmajkin*).
IAU Colloquium No. 130: The Sun and cool stars – activity, magnetism, dynamos, p. 135 – 139 (1991). – See Abstr. 012.024 for the main entry.
The authors discuss the methods and results of analysis of nonlinear mean–field dynamo models based on α–quenching in two asymptotic regimes, namely for weakly and highly supercritical excitation. In the former case the spatial distribution of the

steady–state magnetic field is close to that given by the neutrally stable eigenfunction of the corresponding kinematic dynamo. In the latter case the magnetic field distribution within the main part of the dynamo volume is presumably determined by the balance between the Lorentz and Coriolis forces while near the boundaries boundary layers arise in which the field adjusts itself to the boundary conditions. The asymptotic behaviour of the highly supercritical $\alpha\omega$–dynamos is sensitive to the particular form of dependence of the mean helicity on magnetic field while α^2–dynamos are less sensitive to this dependence.

062.109 **$\alpha\Lambda$–dynamos.**
A. Brandenburg, D. Moss, M. Rieutord, G. Rüdiger, I. Tuominen.
IAU Colloquium No. 130: The Sun and cool stars – activity, magnetism, dynamos, p. 147 – 150 (1991). – See Abstr. 012.024 for the main entry.

In contrast to $\alpha\omega$–dynamos, where the angular velocity is arbitrarily prescribed, the authors consider here $\alpha\Lambda$–dynamos, for which the differential rotation and meridional circulation are solutions of the momentum equation. The non–diffusive parts of the Reynolds stress tensor are parameterized by the Λ–effect. In earlier investigations the authors have shown that the turbulent magnetic diffusivity has to be much smaller than the eddy viscosity, otherwise the dynamo is not oscillatory or else the contours of constant angular velocity are cylindrical, contrary to observations. In the present paper the effects of compressibility are investigated.

062.110 **Charged particle transport in a turbulent magnetized plasma: a reassessment and extension of quasilinear theory.**
U. Achatz, J. Steinacker, R. Schlickeiser.
Astron. Astrophys., Vol. 250, No. 1, p. 266 – 279 (Oct 1991).

Using a guiding center coordinate system the authors have rederived the Fokker–Planck equation for the movement of charged particles in a slab turbulence of transversal plasma fluctuations superimposed on a homogeneous magnetic field, including all phase–space variables. A comparison between the two approaches on this topic by Newman (1973) and Urch (1977) on the one hand and Hall and Sturrock (1967) on the other shows that the former has to be rejected for physical and mathematical reasons. In contrast to these previous papers the authors have included dispersive effects and considered a larger wavelength range than only Alfvén waves. Up to this time pitch–angle diffusion of electrons by Alfvén waves was presented with the problem that these particles see a "resonance gap" in pitch angle which they cannot cross if they are not highly relativistic. It is shown that by concerted action of Alfvén and ion cyclotron waves this can in principle be avoided. The situation becomes even better if whistler waves are also taken into account. The authors show that diffusion perpendicular to the magnetic field is only caused by waves with zero waves numbers which they doubt to exist. Finally they have derived from their equation a cosmic–ray transport equation assuming gyrophase and pitch–angle diffusion to be the fastest processes, thus making the use of nearly isotropic distributions realistic.

062.111 **On the structure of magnetized thin accretion disks of active galactic nuclei and their coronal radiations.**
Yang Lantian, Wu Xuebing, Yang Pibo.
Acta Astron. Sin., Vol. 32, No. 2, p. 154 – 164 (Jun 1991). In Chinese. English translation in Chin. Astron. Astrophys., Vol. 15, No. 4, p. 386 – 396 (Dec 1991).

By using magnetic stress in place of viscosity as the mechanism for angular momentum transfer, the effect of a frozen magnetic field on the structure of a geometrically thin accretion disk is examined. It is shown that the magnetized disk is quasi–Keplerian and its total luminosity is twice of that in the standard disk model. In the inner region, there exists a narrow cool region and the highly collimated jet is formed under the action of the azimuthal component of the magnetic field. The possibility that a magnetized corona may be formed near the surface of the accretion disk is discussed and the wide band radiation from the corona is studied. By using the magnetized disk–corona model the authors suggested that one can easily and reasonably explain the major AGN's properties such as the radiation, variation, the optical ultraviolet–soft X–ray "big bump", etc.

062.112 **Parametric study of the formation of magnetohydrodynamic shocks. I. Model.**
Y. C. Whang.
Astrophys. J., Vol. 377, No. 1, p. 250 – 254 (10 Aug 1991).

A model is developed to study the formation and propagation of planar MHD shocks. The model treats the shock discontinuity as a surface with zero thickness. The Rankine–Hugoniot relations are used to calculate the jumps in flow properties at all shock crossings. The shocks divide the domain of solution into several continuous flow regions. In each continuous flow region, the fluid is nondissipative, and the time–dependent MHD equations are studied using the method of characteristics. During the formation of a shock, because of the interaction of the shock with the ambient MHD flow, the shock speed and the shock strength change continuously until the shock reaches a fully developed state. The model will be used to obtain numerical solutions and to identify the controlling parameters for the formation of slow shocks. The overall objective is to determine the environmental conditions (in terms of dimensionless parameters) under which the MHD shocks produced are slow shocks, fast shocks, or a combination of slow and fast shocks.

062.113 **Parametric study of the formation of magnetohydrodynamic shocks. II. Slow shocks.**
Y. C. Whang.
Astrophys. J., Vol. 377, No. 1, p. 255 – 262 (10 Aug 1991).

As a coronal mass ejection pushes its way through preceding slower solar wind, the evolution of large disturbances leads to the formation of associated shocks. The formation process is studied using an MHD model in which the shock discontinuities are treated as surfaces with zero thickness and the exact Rankine–Hugoniot relations are used to calculate the jumps in flow properties at all shock crossings. A series of numerical solutions are carried out under the initial condition that the flow velocity is field–aligned in the rest frame of reference, and the magnetic field and the thermal pressure are constant throughout the flow field with $\beta = 0.1$ and the wave normal angle $\theta = 15°$. The solutions demonstrate the formation of planar slow shocks. The momentum impact compresses the plasma in the interaction region. The pressure fronts steepen to form a pair of forward and reverse slow shocks. The impact is equivalent to the collision of two streams. The dynamical impact of the ejecta on the ambient solar wind can be characterized by a reduced number density analogous to the reduced mass for the elastic collision between two particles.

062.114 **On the influence of turbulence of fragmentation of self–gravitating gas clouds.**
P. Battinelli, S. V. Vereshchagin, A. Di Fazio, R. Capuzzo–Dolcetta, V. A. Urpin.
Astron. Zh., Tom 68, Vyp. 4, p. 709 – 718 (Jul–Aug 1991). In Russian. English translation in Sov. Astron., Vol. 35, No. 4.

The processes of fragmentation of self–gravitating gas clouds may be accompanied by excitation of intense turbulent motions within the clouds. Turbulence may arise as a consequence of various causes, e. g., it may be generated by the motions of fragments formed during the collapse of a cloud. The authors study the influence of the turbulence parameters on the evolution of clouds with three different values of mass (800, 10^4 and $10^7 M_\odot$). An increase of the effective viscosity of gas due to turbulence results in a considerable heating of gas in the fragmentating cloud.

062.115 **Kinetics of energetic electrons, propagating through ion–sound plasma trubulence in solar flares.**
I. V. Kudryavtsev, Yu. E. Charikov.
Astron. Zh., Tom 68, Vyp. 4, p. 825 – 836 (Jul–Aug 1991). In Russian. English translation in Sov. Astron., Vol. 35, No. 4.

Propagation of energetic electrons through ion–sound plasma turbulence has been considered. The distribution function of these particles has been calculated, taking into account scattering on ion–sound waves. The authors have managed to demonstrate the possibility of particle–spectra formation with an inverse distribution over velocity. Those spectra may well turn out be unstable with regard to generation of oscillations. They have applied the obtained results to the problem of polarization calculations for hard X–ray emission.

062.116 **Excitation of ion cyclotron waves at the boundary of the magnetosphere.**
E. V. Belova, J. Blenski, M. Denis, L. M. Zelenyj, S. P. Savin.
Sov. J. Plasma Phys., Vol. 17, No. 8, p. 555 – 560 (Aug 1991).
Current Physics Microform No.: 9203X1573.

A model is proposed for the excitation of ion cyclotron waves in plasma jets close to the boundary of the magnetosphere. The source of these waves is taken to be nonuniform ion streams parallel to the magnetic field with transverse density and velocity gradients. It is shown that unlike the drift ion–cyclotron instability, which is one possible source of ion cyclotron waves in these regions, the instability caused by the variation in the ion velocity has no threshold. The growth rate and number of cyclotron harmonics that are excited are largest when both gradients are nonzero. Estimates of the saturation amplitude of the instability and also of the shape of the spectrum agree well with experimental data.

062.117 **The Helmholtz instability of magnetic flux sheets.**
Hu Wenrui.
Acta Astrophys. Sin., Vol. 11, No. 4, p. 378 – 382 (Oct 1991). In Chinese.

The Helmholtz instability is analyzed for the magnetic flux sheet, whose configuration consists of two discontinuous surfaces from a sheet magnetic flux region with finite thickness confined by plasma in both sides. The instability condition for the case of a sheet is obtained, and it agrees with that for the case of one discontinuous surface if the thickness of the sheet tends to infinity. The dispersion conditions at two discontinuous surfaces are also given. The results show that the influence of magnetic stabilization in the case of a sheet is relatively weaker than the case of one discontinuous surface, however, the influence of confined plasma is relatively stronger, although the growth rates of instability for both cases have the same order of magnitude.

062.118 **Boundary conditions in the problem of plasma ejection from the magnetosphere of an axisymmetric rotator.**
S. V. Bogovalov.
Astron. Zh., Tom 68, Vyp. 6, p. 1227 – 1239 (Nov–Dec 1991). In Russian. English translation in Sov. Astron., Vol. 35, No. 6.

The formulation of the problem of plasma ejection from the magnetosphere of an axisymmetric rotator is considered in dependence on physical conditions. The conditions defining the matter flux and the toroidal magnetic field are investigated. In the case of a hot plasma and weak magnetic field, these values are defined by the conditions of transition velocity of the plasma in terms of fast and slow magnetosound wave propagation velocity. The solar wind is an example of such a flow. In the case of a strong field and cold plasma, as in pulsars, the matter flux and toroidal magnetic field are defined by the processes which generate plasma. In this case, near the surface of pulsars and at the Alfven surface in the magnetosphere, the frozen–in condition is not fulfilled. The plasma acceleration to the velocity exceeding the fast magnetosound velocity occurs in the Alfven discontinuity. Conditions when the discontinuity appears are obtained. Electric current flowing from the polar cap proves to be equal to the Goldreich–Julian current. Energetic losses of the axisymmetric rotator in this case are comparable to those of a magnetic dipole in vacuum. The boundary conditions for a poloidal magnetic field are obtained, taking into account the discontinuity at the Alfven surface.

062.119 **Gravitational instability of a composite system.**
M. P. Bora, S. P. Talwar.
Astrophys. Space Sci., Vol. 184, No. 2, p. 193 – 203 (Oct 1991).

The problem of instability arising in a composite system consisting of an infinitely conducting hydromagnetic fluid interacting through gravitational forces with one or more than one neutral gas, is investigated, allowing for a possible relative streaming between the component fluids. Instability criteria are derived for special cases of a two–component (static or relatively streaming) system and for a three–component system consisting of two gases contra–streaming in the presence of a stationary background gas. It is found that for a static system only one unstable mode exists for wave numbers less than a critical value given by the square root of the sum of the squares of the Jeans's wave numbers for individual gases. However, for a configuration, where components are endured with characteristic streaming speeds, there are present simultaneously more than one unstable modes.

062.120 **Rayleigh–Taylor instability of a Hall plasma with arbitrary density gradient.**
P. D. Ariel.
Astrophys. Space Sci., Vol. 184, No. 2, p. 205 – 219 (Oct 1991).

The Rayleigh–Taylor instability of a plasma layer in the presence of a horizontal magnetic field is investigated, taking into account the effects of Hall currents and an arbitrarily large density gradient. It is shown for the general case that if the density decreases vertically upward, the system is thoroughly stable. For a plasma layer with exponentially varying density an approximate dispersion relation is obtained using the Galerkin's method. An analysis of the roots of the dispersion relation reveals that the Hall currents loosen the stabilizing influence of the magnetic field and impart instability to the system. For sufficiently large values of the density gradient and the Hall currents the system is thoroughly destabilized for all perturbations.

062.121 **Discrete phase changes within nonlinear steepened magnetosonic waves: comet Giacobini–Zinner.**
B. T. Tsurutani, E. J. Smith, B. Buti, H. Matsumoto, A. Brinca.
Geophys. Res. Lett., Vol. 17, No. 11, p. 1817 – 1820 (Oct 1990).

The authors illustrate the intricate complexities of nonlinear wave evolution and briefly discuss their consequences. They show examples of nonlinear steepened magnetosonic waves that have subintervals without phase rotation (linearly polarized) and other subintervals with discrete, rapid phase rotation. Some specific features are: (1) the trailing (~half) portion of the wave sometimes has no phase rotation and is simply compressive; (2) just behind the leading edge of the wave front where there is a sharp rotation (called a partial rotation), there is often a second region with no phase change; (3) between these two regions is an interval with a distinct phase change; (4) in the examples shown in this paper, the distinct phase changes have right–hand polarizations in the spacecraft frame; (5) the magnetosonic wave appears to split into two parts with time; the right–hand rotation is located where the splitting occurs. Possible explanations for the right–hand rotation are explored. One plausible explanation is that a right–hand wave is generated through the decay instability of a magnetosonic wave, i.e., a magnetosonic wave propagating towards the sun decays into a sound wave and a magnetosonic wave propagating away from the sun. If right–hand daughter waves are actually present, this will have important consequences. Even though these waves are present only ~10% (or less) of the time, this may allow significant wave–particle interactions to make the self–consistent second–order Fermi acceleration mechanism of energetic cometary ions feasible.

062.122 **Physics of MHD K–H instability in jets.**
Wu Ding, Wang Deyu.
3. Chinese Academy of Sciences and Max–Planck Society Workshop: High energy astrophysics: compact stars and active galaxies, p. 67 – 72 (1991). – See Abstr. 012.026 for the main entry.

The linear modal analysis of the cylindrical shear flow leads to an eigenvalue problem based on the differential equation. A detailed analysis of the solution is presented. From this, the following physical principles are extracted: (1) unstable modes only exist within a hemicycle in complex phase space. (2) Each mode has an associated "comoving point", i.e. there is always some fluid that comoves with the pattern. Near the comoving point, unstable modes are overreflected. (3) Short wavelengths, large mode number modes or long wavelengths pinch modes perturbations both for ordinary modes and reflecting modes can be stabilized due to the existence of shear layer. (4) Reflecting modes have a band structure in wave number vs growth rate space. For light jets, the growth rates are very small. (5) A shear layer suppresses pinching modes more effectively. (6) It is explained why a knotty morphology has been detected in only 2 or 3 objects, and why knots are usually formed near their parent nucleus.

062.123 **Dynamo magnetic–field generation in turbulent accretion disks.**
T. F. Stepinski.
Publ. Astron. Soc. Pac., Vol. 103, No. 666, p. 777 – 780 (Aug 1991). – See Abstr. 012.034 for the main entry.

Magnetic fields can play important roles in the dynamics and evolution of accretion disks. The presence of strong differential rotation and vertical density gradients in turbulent disks allows the $\alpha\omega$ dynamo mechanism to offset the turbulent dissipation and maintain strong magnetic fields. It is found that MHD dynamo magnetic–field normal modes in an accretion disk are highly localized to restricted regions of a disk. Implications for the character of real, dynamically constrained magnetic fields in accretion disks are discussed. The magnetic stress due to the mean magnetic field is found to be of the order of a viscous stress. The dominant stress, however, is likely to come from small–scale fluctuating magnetic fields. These fields may also give rise to energetic flares above the disk surface, providing a possible explanation for the highly variable hard X–ray emission from objects like Cyg X-1.

062.124 **MHD free–convection and mass transfer flow of an elasto–viscous fluid.**
B. K. Jha.
Astrophys. Space Sci., Vol. 185, No. 1, p. 129 – 135 (Nov 1991).

The effects of magnetic field and mass transfer on the flow of an elasto–viscous fluid past an infinite vertical plate, when the plate is moving with uniform velocity U, are discussed. The magnetic lines of force are assumed to be fixed relative to the plate. The Laplace transform method is used to obtain the expression for velocity. The effect of various parameters, occurring into the problem, is discussed with the help of tables.

062.125 **Nonresonant absorption of shear Alfvén waves.**
H. R. Strauss.
Geophys. Res. Lett., Vol. 18, No. 1, p. 77 – 80 (Jan 1991).

Resonant absorption of shear Alfvén waves is thought to be a likely candidate to explain heating of the solar corona and acceleration of the solar wind. A difficulty with the theory is that the absorption process is slow. It takes a long time $\sim S^{1/3}\tau_A$, where $S \gg 1$ is the magnetic Reynolds number, and τ_A a typical Alfvén wave propagation time, for wave fronts to steepen sufficiently for absorption to occur. Moreover, heating occurs in a very thin layer whose width scales as $S^{-1/3}$. A faster absorption mechanism is nonresonant absorption by compressional viscosity, in a curved magnetic field. Heating is nonresonant and is not localized to a narrow layer. The effect could be quite important where the solar coronal magnetic field is strongly curved, in the chromosphere. It could also be important on open field lines in the upper corona, where the compressional viscosity is large. It might imply that a significant part of outgoing Alfvén waves are absorbed in the corona.

062.126 **Nonlinear density waves in gaseous disks.**
Chi Yuan.
3. Chinese Academy of Sciences and Max–Planck Society Workshop: High energy astrophysics: compact stars and active galaxies, p. 277 – 285 (1991). – See Abstr. 012.026 for the main entry.

Nonlinear density waves can be resonantly excited in a gaseous disk by an external periodic potential. The potential may arise from a satellite in planetary rings, from an oval distortion or a bar in galactic disks, or from a binary in protostellar disks. Disks are very sensitive to the resonance excitation mechanism. Such waves are observed in Saturn's rings. They are expected to exist in other disk systems and probably play an important role in changing the structure and affecting the evolution of the disks. In this article, the author intends to present a comprehensive theory of nonlinear waves and its possible application to various problems of disk systems in astrophysics.

062.127 **Solitary dipole vortexes in rotating gravitating systems generation of turbulent spectrum.**
M. G. Abramian (*M. G. Abramyan*), E. E. Khachikian (*Eh. E. Khachikyan*), A. M. Fridman.
3. Chinese Academy of Sciences and Max–Planck Society Workshop: High energy astrophysics: compact stars and active galaxies, p. 314 – 318 (1991). – See Abstr. 012.026 for the main entry.

In this paper the nonlinear equation is obtained for describing the dynamics of perturbations with a small but finite amplitude in rotating gas system within the framework of the geostrophical approach. In the two special limiting cases of short–wave (with a wavelength much less than the Jeans wavelength $\lambda \ll \lambda_J$) and long–wave ($\lambda \gg \lambda_J$) perturbations the equations obtained are in the form of the Rossby's waves nonlinear equation with vectorial nonlinearity (Charney–Obukhov's or Hasegawa–Mima's equation). The Larichev–Reznik's stationary solution of this equation in the form of a solitary dipole vortex (modone) is obtained. The weak turbulence spectrum described by this equation leads to the known observable relations between the main parameters of the galactic gas population, in particular to the Solpiter's stellar mass spectrum.

062.128 **A chromospheric response to pulse beam heating.**
P. Heinzel.
Sol. Phys., Vol. 135, No. 1, p. 65 – 88 (Sep 1991).

Starting from the new flare models of Karlicky (1990) and Karlicky and Henoux (1991), the author presents the first time–dependent numerical simulations of hydrogen plasma excitation and ionization on time scales of less than one second. These time scales are consistent with the spiky behaviour of the kinetic temperature produced by non–thermal collisional processes. Such temperature spikes represent a chromospheric response to a series of short–duration electron beam pulses which are supposed to heat the flare atmosphere. Self–consistent numerical solution of a simplified, time–dependent, non–LTE problem for a three–level hydrogen atom model with continuum allows to predict theoretically a qualitative behaviour of the Hα line intensity variations on very short time intervals. The Hα temporal profiles, evaluated at the line center and for $\Delta\lambda = 1$ Å, can be qualitatively compared with some recent flare observations obtained with high temporal resolution.

062.129 **Parameterized rotating convection for core and planetary atmosphere dynamics.**
K. Zhang.
Geophys. Res. Lett., Vol. 18, No. 4, p. 685 – 688 (Apr 1991).

New types of convective instability and associated nonlinear phenomena in rapidly rotating spherical systems have been discovered through numerical simulations. The Prandtl number, defined as the ratio of the viscous and thermal diffusivities of a fluid, $P_r = \nu/\varkappa$, plays a crucial role in determining the fundamental features of both the instabilities and the corresponding nonlinear convection. The results shed new light on regimes of convection in the earth's core and the atmospheres of the major planets.

062.130 **Radiative transfer across a magnetic flux tube.**
S. Audic.
Sol. Phys., Vol. 135, No. 2, p. 275 – 297 (Oct 1991).

In the coming years, some solar telescopes will be able to yield the Stokes' parameters of polarized light with a resolution better than 1 arcsec. The author simulated the Stokes' parameters of a solar magnetic flux tube as seen with such a resolution. He showed that, observing with a line–of–sight not parallel to the axis of the flux tube (assumed vertical and axisymmetric), it is possible to see differences between different configurations of the magnetic field inside the flux tube (presence, and in what direction, of an azimuthal component of the field). Furthermore, along such a line-of-sight, the polarization profiles of any atomic line are strongly absorbed at the line center. The author then suggests a strategy to infer the structure of the magnetic field from observations at high spatial resolution.

062.131 **A numerical simulation of magnetic reconnection and radiative cooling in line-tied current sheets.**
T. G. Forbes, J. M. Malherbe.
Sol. Phys., Vol. 135, No. 2, p. 361 - 391 (Oct 1991).
The authors have used the radiative MHD equations for an optically thin plasma to carry out a numerical experiment related to the formation of "post"-flare loops. The numerical experiment starts with a current sheet that is in mechanical and thermal equilibrium, but which is unstable to both tearing-mode and thermal-condensation instabilities. The current sheet is line-tied at one end to a photospheric-like boundary and evolves asymmetrically. The effects of thermal conduction, resistivity variation, and gravity are ignored. In general, it is found that reconnection in the nonlinear stage of the tearing-mode instability can strongly affect the onset of condensations unless the radiative cooling time scale is much smaller than the tearing-mode time scale. When the ambient plasma β is less than 0.2, the reconnection enters a regime where the outflow from the reconnection region is supermagnetosonic with respect to the fast-mode wave speed. In the supermagnetosonic regime the most rapidly condensing regions occur downstream of a fast-mode shock that forms where the outflow impinges on closed loops attached to the photospheric-like boundary. A similar shock-induced condensation might occur during the formation of "post"-flare loops.

062.132 **Influence of turbulence on the fragmentation of self-gravitating gas clouds.**
P. Battinelli, S. V. Vereshchagin, A. Di Fazio, R. Capuzzo-Dolcetta, V. A. Urpin.
Sov. Astron., Vol. 35, No. 4, p. 349 - 353 (Jul 1991). Current Physics Microform No.: 9203X0535. English translation of Astron. Zh., Tom 68, Vyp. 4, p. 709 - 718 (1991).
The fragmentation of self-gravitating gas clouds may be accompanied by the excitation of intense turbulent motion. Turbulence may develop for various reasons, such as the motion of fragments formed in the gas. The influence of the turbulence characteristics on the evolution of clouds with three different masses (800, 10^4, and $10^7 M_\odot$) is investigated. The increase in the effective viscosity of the gas due to turbulence results in considerable heating of gas in a fragmenting cloud.

062.133 **Collective plasma radiation processes.**
D. B. Melrose.
Annu. Rev. Astron. Astrophys., Vol. 29, p. 31 - 57 (1991). - See Abstr. 003.010 for the main entry.
Emphasis is placed on models for coherent emission that involve saturation of a maser process. Theories of coherent emission are discussed from a general viewpoint. Theoretical developments in the specific contexts of plasma emission, ECME, and pulsar radio emission are reviewed.

062.134 **Kinetics of fast electrons in ion-acoustic waves in the turbulent plasma of solar flares.**
I. V. Kudryavtsev, Yu. E. Charikov.
Sov. Astron., Vol. 35, No. 4, p. 409 - 414 (Jul 1991). Current Physics Microform No.: 9203X0595. English translation of Astron. Zh., Tom 68, Vyp. 4, p. 825 - 836 (1991).
The kinetics of fast electrons propagating in the turbulent plasma of a solar flare are considered. The analytical form of the distribution function is obtained for the stationary case with allowance for scattering from ion-acoustic waves. The possibility of the formation of particle spectra with an inverted velocity distribuiton, which may be unstable against the generation of oscillations, is demonstrated. The results are used to calculate the degree of polarization of hard X-ray emission.

062.135 **Numerical study of compressible solar magnetoconvection with an open transitional boundary.**
H. Hanami, T. Tajima.
Astrophys. J., Vol. 377, No. 2, p. 694 - 706 (20 Aug 1991).
The authors study by computer simulation nonlinear evolution of magnetoconvection in a system with a dynamical open boundary between the convection region and corona of the Sun. In this model, the fluid is subject to the vertical gravitation, magnetohydrodynamics (MHD), and high stratification, through an MHD code with the MacCormack-Conner cell hybrid scheme. Initially the vertical fluid flux penetrates from the convectively unstable zone at the bottom into the upper diffuse atmosphere. As the instability develops, the magnetic fields are twisted by the convection motion and the folding of magnetic fields is observed. When the magnetic pressure is comparable to the thermal pressure in the upper layer of the convective zone, strong flux expulsion from the convective cell interior toward the cell boundary appears. The magnetic field patterns observed are those of concentrated magnetic flux tubes, accumulation of dynamo flux near the bottom boundary, pinched flux near the downdraft region, and the surface movement of magnetic flux toward the downdraft region.

062.136 **Dynamics of shock waves in a contracting cloud.**
M. Y. Amin, M. S. El-Nawawy, M. A. El Shalaby.
Astrophys. Space Sci., Vol. 185, No. 2, p. 195 - 209 (Nov 1991).
The contraction of an interstellar cloud is followed. The results indicate that there appear shock waves during contraction. In order to study the effect of shock waves, two models have been studied. The post-shock temperature for the two models are, respectively, 3006K and 2984K. The density increases by more than three orders of magnitude. The gas is generally cooled by atoms, molecules, and grains. The dominant cooling process changes according to the chemical composition and the temperature of the gas. The thermal equilibrium depends on the existing physical conditions.

062.137 **Unsteady forced and free-convection flow past an infinite vertical plate with constant heat flux.**
B. K. Jha.
Astrophys. Space Sci., Vol. 185, No. 2, p. 295 - 298 (Nov 1991).
An analytical study is performed to examine the unsteady forced and free-convection flow of a viscous, incompressible fluid when there is a constant heat flux between fluid and plate. The expressions for the velocity field, the penetration distance and the skin-friction have been obtained by the Laplace transform technique. The influence of the various parameters entering into the problem is extensively discussed.

062.138 **Numerical simulations of turbulent compressible flows.**
A. Pouquet, T. Passot, J. Léorat.
IAU Symposium No. 147: Fragmentation of molecular clouds and star formation, p. 101 - 118 (1991). - See Abstr. 012.043 for the main entry.
The authors give an overview of the use of numerical simulations in the modeling of turbulence in molecular clouds.

062.139 **Non-linear theory of particle dynamics in cosmic electromagnetic fields. I. Regular motion in a simple Alfvén wave.**
M. Carioli.
Astron. Astrophys., Vol. 251, No. 2, p. 700 - 712 (Nov 1991).
With this paper, the author begins a study of the non-linear dynamics of the non-thermal component of the cosmic environment. The problem of a test particle propagating in a sinusoidal Alfvén wave is of basic importance in the perturbation theory of cosmic ray dynamics. In fact, it is assumed that the character of the motion is diffusive because of resonant pitch angle scattering off fluctuations of the cosmic magnetic field. The non-linear theory of this problem shows a richer behaviour. The author finds an approximate invariant of the motion, which accurately describes the dynamics in the vicinity of resonances. This invariant shows that particles can be locked in a resonant configuration, instead of being scattered. Only one resonance is

admitted between a particle and an individual harmonic of the magnetic fluctuations, by the well known cyclotron resonance condition of quasi-linear theory. The author finds a non-linear resonance condition yielding an infinite number of resonances in the motion of a particle in a purely sinusoidal wave, parallel to the mean magnetic field. His results are not restricted to the case of small amplitude perturbations.

062.140 **Do random velocity fields inhibit or induce gravitational collapse and fragmentation?**
P. Chantry, R. Grappin, J. Léorat.
IAU Symposium No. 147: Fragmentation of molecular clouds and star formation, p. 396 – 397 (1991). – See Abstr. 012.043 for the main entry.

062.141 **Structure of radiatively cooled jets.**
M.-a. Kondo.
IAU Colloquium No. 123: Observatories in Earth orbit and beyond, p. 551 – 554 (1990). – See Abstr. 012.038 for the main entry.
Radiative cooling strongly affects the thermal structure of dense jets, such as in SS433, through free-free emission. From the dynamical aspect, the beam width of a cooled jet does not expand, unlike from an adiabatic jet. From the thermal aspect, cooling efficiency determines the ratio of X-ray region of high temperature to optical one of low temepratu re. However, this ratio is influenced by the heating due to contained high-energy particles, which produce synchrotron radiation in the tail of the jet. Extragalactic jets can also be considered in a similar way due to other energy loss mechanisms.

062.142 **Cosmic magnetism and the basic physics of the early stages of star formation.**
T. C. Mouschovias.
NATO Advanced Study Institute on The Physics of Star Formation and Early Stellar Evolution, p. 61 – 122 (1991). – See Abstr. 012.042 for the main entry.
Multifluid magnetohydrodynamics (nonideal MHD) is an indispensable tool for studying the physics of star formation. The physical origin of processes crucial to the formation of stars becomes clear, and the essence of their quantitative theoretical description becomes straightforward when a first-principles approach is taken. Progressively simpler forms of the equations are obtained and used to introduce concepts and describe processes such as ambipolar diffusion and the self-initiated formation and gravitational collapse of fragments (or cores); ohmic dissipation of magnetic flux; flux freezing in the electrons, ions, and neutral particles (through collisional coupling with ions and, possibly, charged grains); quasistatic and dynamical contraction of cloud cores; magnetic braking and the resolution of the thorny angular momentum problem; the magnetic Rayleigh-Taylor (or Parker) instability and formation of cloud complexes; equilibria of isothermal, magnetic clouds; the physical origin of the relation $B_c \propto (\varrho_c T_c)^{1/2}$ between the magnetic field strength and the gas density and temperature in cores of contracting, not necessarily isothermal, self-gravitating clouds; and critical masses for the collapse of interstellar clouds. Improvements to the Virial Theorem are also presented. Close contact with observations is emphasized and maintained throughout the paper.

062.143 **A determination of relativistic shock jump conditions using Monte Carlo techniques.**
D. C. Ellison, S. P. Reynolds.
Astrophys. J., Vol. 378, No. 1, p. 214 – 226 (1 Sep 1991).
In shocks with relativistic upstream flow velocities, the jump conditions cannot be determined in closed form because the ratio of specific heats of the gas may change in the shock. The authors use Monte Carlo techniques, assuming isotropic elastic scattering of all particles, to calculate jump conditions in parallel, relativistic, collisionless shocks in the absence of Fermi acceleration. The authors exhibit the shock velocity and compression ratios for arbitrary flow velocities and for any upstream temperature. Both single-component electron – positron plasmas and two-component proton – electron plasma are considered. For single-component plasmas, velocity ratios approach 3 smoothly as the upstream velocity approaches c. In single-component plasmas, elastic scattering guarantees that momentum and energy are conserved across the shock. In two-component plasmas, however, the elastic scattering of both components does not result in equal shocked electron and proton temperatures and will not, in general, lead to the continuity of both momentum and energy fluxes. It is shown that protons and electrons must share energy, directly or through the mediation of plasma waves, in order to satisfy the basic conservation conditions. For finite upstream temperatures, proton – electron plasma velocity ratios are always less than for single-component plasmas, but still approach 3 monotonically as the upstream velocity trends to c.

062.144 **Two-dimensional fluctuating flow and heat transfer to a relativistic radiating gas.**
A. R. Bestman.
Astrophys. Space Sci., Vol. 186, No. 1, p. 1 – 6 (Dec 1991).
The two-dimensional ideal flow of a neutrino gas past a vertical hot flat plate subjected to temperature fluctuation, is analysed within the framework of the continuum theory. If the difference between the wall temperature and the free-stream temperature is small, the first-order perturbed set of equations is reduced to a set of well-posed coupled linear boundary value problems for the temperture and the transverse velocity. The axial velocity and number density may then be determined.

062.145 **Prominence sheets supported by constant-current force-free fields. I. Imposition of normal magnetic field components at the current sheet and the photosphere.**
C. Ridgway, T. Amari, E. R. Priest.
Astrophys. J., Vol. 378, No. 2, p. 773 – 784 (10 Sep 1991).
The authors present an analytical model for the support of a prominence (represented by a sheet of mass and current) in a constant-current force-free field. The model produces both normal-type and inverse-type configurations which are in equilibrium everywhere along the sheet and for which the field is locally bounded.

062.146 **Stability of two superposed homogeneous fluids under uniform magnetic field.**
R. P. Singh, H. C. Khare.
Astrophys. Space Sci., Vol. 186, No. 1, p. 125 – 135 (Dec 1991).
The instability of two superposed homogeneous fluids is discussed under gravitational force and uniform magnetic field. The perturbation propagation is taken simultaneously along and perpendicular to streaming motion in the horizontal plane $z = 0$. The critical wave number k^* has been found and some special cases of interest are discussed.

062.147 **The structure of a tangential discontinuity: application to the cometary ionopause.**
K. R. Flammer, N. Omidi, K. B. Quest.
Geophys. Res. Lett., Vol. 18, No. 3, p. 369 – 372 (Mar 1991).
The kinetic structure of a tangential discontinuity interface separating an outflowing unmagnetized plasma from an inflowing, magnetized plasma is studied using 1-D hybrid simulations (particle ions and fluid electrons). By including the effects of an ion-neutral collisional force and plasma loss due to recombination, the authors conclude that two scale lengths control the structure of the cometary ionopause. The first is related to the ion Larmor radius of the outflowing ions which encounter the magnetic barrier, and the second is related to the balance between the ion-neutral collisional force and the electromagnetic JxB force.

062.148 **Acceleration of keV-electrons by a nearly perpendicular shock wave.**
M. Vandas.
Bull. Astron. Inst. Czech., Vol. 42, No. 3, p. 170 – 181 (Jun 1991).
The "fast Fermi process" – acceleration mechanism of electrons in a nearly perpendicular shock wave – uses in its formalism

the assumption of adiabatic behaviour of electrons. Via the direct electron trajectory calculations in model shock fields the author has proved that this formalism is valid also for keV–electrons despite the fact that these electrons do not behave adiabatically in the shock layer.

062.149 **Nonlinear Alfven waves in inhomogeneous plasmas.**
B. Buti.
Geophys. Res. Lett., Vol. 18, No. 5, p. 809 – 812 (May 1991).
Nonlinear Alfvén waves, in inhomogeneous plasmas, are shown to be governed by the MDNLS (Modified Derivative Nonlinear Schrödinger) equation. The inhomogeneity leads to acceleration (deceleration) of the Alfvén solitons depending upon plasma β (kinetic pressure/magnetic pressure) and on density gradients vis–a–vis direction of wave propagation. Potential applications, of the model, to space and astrophysical plasmas are pointed out.

062.150 **A hydrodynamical model for the fragmentation of the W49A star–forming region.**
E. Keto, J. Lattanzio, J. Monaghan.
3. Haystack Observatory Conference on the Interstellar Medium: Skylines – atoms, ions and molecules – new results in spectral line astrophysics, p. 183 – 184 (1991). – See Abstr. 012.051 for the main entry.
The authors present a 3–D hydrodynamical and radiative transfer simulation which suggests that the circular "necklace" of massive star formation in W49A may result from fragmentation via a ring instability during the collapse of a rotating cooling molecular cloud.

062.151 **Dynamo magnetic field modes in thin astrophysical disks: an adiabatic computational approximation.**
T. F. Stepinski, E. H. Levy.
Astrophys. J., Vol. 379, No. 1, p. 343 – 355 (20 Sep 1991).
The authors apply an adiabatic approximation to the calculation of turbulent MHD dynamo magnetic fields in thin disks. The approximation provides a unified approach to the problem of finding normal mode solutions for thin disk dynamos. The approach provides a general method for obtaining critical dynamo numbers and field structures for arbitrary forms of the relevant physical parameters in a disk. This paper investigates conditions under which magnetic fields generated by disk dynamos permeate the entire disk or are localized to restricted regions of a disk. Two specific cases of Keplerian disks are considered. In the first, magnetic field diffusion is assumed to be dominated by turbulent mixing, leading to a dynamo number independent of distance from the center of the disk. In the second, the dynamo number is allowed to vary with distance from the disk's center. Localization of dynamo magnetic field structures is found to be a general feature of disk dynamos, except in the special case of stationary modes in dynamos with constant dynamo number. Implications for the dynamical behavior of dynamo magnetized accretion disks are discussed. The results of this study are examined in the context of the protosolar nebula and accretion disks around compact objects.

062.152 **Simulated images of stellar jets.**
H. A. Feldman, A. C. Raga.
Astron. J., Vol. 102, No. 6, p. 2049 – 2053 (Dec 1991). Current Physics Microform No.: 9112C2025.
In many cases, the sources of jetlike Herbig–Haro objects are optically invisible, embedded objects which are usually detected at infrared wavelengths. The jets become visible only at distances of $\sim 10^{16}$–10^{17}cm from the source, where they emerge from the molecular cloud core that surrounds the central object. High spatial resolution images indicate that close to the position where the jets appear, their width is a decreasing function of distance from the source. The authors have simulated a string of linearly aligned emitting knots that emerges from a stratified medium. For this configuration, they compute the single scattering of photons in order to obtain intensity maps of the scattered + direct light that reaches the observer. The authors find that the observed width as a function of position close to the source in stellar jets can be reproduced from these calculations. Thus, they conclude that the effect of broadening toward the source observed in some stellar jets is the result of light scattered from the near environment of the jet, and not an actual hydrodynamic effect of the jet flow.

062.153 **Chaos driven by kinetic Alfven waves.**
W. W. Liu.
Geophys. Res. Lett., Vol. 18, No. 8, p. 1611 – 1614 (Aug 1991).
The author presents a simple model describing the effect of a kinetic Alfvén wave on the motion of a test electron along a closed magnetic field line. He shows that the electron motion can be strongly chaotic for a wide range of conditions. The chaos exhibits a broad–band power spectrum in the frequency domain, although only three principal frequencies are involved in the model. Phase space density of electrons exhibits complex but physically interesting evolution in response to the KAW. Typical of a chaotic situation, the scale of the electron distribution in phase space descends rapidly over the scale hierarchy to give rise to a growing set of irregularities, which may in turn generate strong electrostatic turbulences. This chain of events suggests a fundamental mechanism that couples physical processes of different scales.

062.154 **Diffuse alpha particles upstream of simulated quasi–parallel supercritical collisionless shocks.**
K. J. Trattner, M. Scholer.
Geophys. Res. Lett., Vol. 18, No. 10, p. 1817 – 1820 (Oct 1991).
One–dimensional hybrid simulations of quasi–parallel collisonless shocks have been performed. Alpha particles are included in the simulations self–consistently in such a way that sufficient simulation particles are present to give reasonable statistics. It is found that a fraction of the incoming solar wind protons and alpha particles is accelerated near the shock and escapes upstream. These particles constitute a diffuse upstream particle population. The density ratio bias factor, i.e., the density ratio of diffuse alpha particles to diffuse protons normalized to the solar wind He^{++} to H^+ density ratio varies between 0.5 and 4.0, and depends on the solar wind He^{++} to H^+ density ratio and on the shock Mach number. The density ratio bias factor for a low Mach number shock ($M_A = 4.6$) is $f \sim 0.5$ when the alpha particles are treated as test particles and increases to ~ 1 when the solar wind ratio of alpha to protons is 0.25. For a high Mach number shock of $M_A = 8.9$ with a very small alpha particle contribution to the solar wind the authors obtained a density ratio bias factor of $f \sim 4$.

062.155 **Role of highly–charged ions in astrophysical and laboratory plasmas.**
B. C. Fawcett.
5. Internationale Konferenz über die Physik hochgeladener Ionen, p. 1 – 5 (1991). – See Abstr. 012.064 for the main entry.
This presentation considers how the interaction between physical conditions in plasmas and atomic–structure properties affects the role played by highly–charged ions. It is emphasised that reliable and accurate atomic data are required to determine plasma conditions from the analysis of spectral line intensities. Atomic–structure problems which limit the quality of available atomic data are discussed. Spectroscopic observations quoted provide information on the behaviour of ions in plasmas.

062.156 **Magnetic X–lines in space plasmas.**
M. F. Heyn, V. S. Semenov.
2. International Workshop on Plasma Physics: Plasma Theoretical Problems in Astro– and Fusion Physics, p. 119 – 123 (1991). – See Abstr. 012.067 for the main entry.
Magnetic X–lines are sites of strong plasma activity and as such interesting objects for theoretical study. The intrinsic non–homogeneities of the associated current layers complicate the theoretical approach to the problem. In order to describe the large–scale evolution of reconnection, the classical surface wave analysis is modified to allow for a finite thickness of the "surface", the current layer. This analysis gives an insight into the distortion of the current layer around magnetic X–lines.

062.157 **Three–dimensional MHD–computations of magnetic reconnection.**
A. Otto.
2. International Workshop on Plasma Physics: Plasma Theoretical Problems in Astro– and Fusion Physics, p. 125 – 135 (1991). – See Abstr. 012.067 for the main entry.

In astrophyiscal plasmas the probably best investigated cases of magnetic reconnection occur in connection with geomagnetic activity. A potential application of a three–dimensional reconnection process occurs at the dayside magnetopause. Here magnetic "Flux Transfer Events" (FTE's) seem to play an important role for the transport of magnetic flux from the dayside magnetosphere to the magnetotail. In the framework of resistive MHD computations the author has analysed the dynamical evolution of three–dimensional magnetic reconnection. Considering a one–dimensional initial equilibrium state it is shown that a localized magnetic reconnection process develops if a spot of localized resistivity appears in the current layer (e.g. due to suitable fluctuations in the solar wind plasma). The three–dimensional reconnection process leads to the formation of structures which represent newly interconnected magnetic flux and show the typical bipolar B_N variation of FTE's. A close comparison with observations shows that many further properties of FTE's are recovered in the simulations.

062.158 **Multiple scattering of acoustic waves by a pair of uniformly magnetized flux tubes.**
T. J. Bodgan, D. C. Fox.
Astrophys. J., Vol. 379, No. 2, p. 758 – 775 (1 Oct 1991).

Motivated by the outstanding question of whether sunspot seismology can distinguish between the competing monolithic flux–tube and spaghetti sunspot models, the authors consider the interaction of an acoustic plane wave with a pair of uniformly magnetized flux tubes of circular cross section. It is shown that this interaction falls within one of three distinct categories depending upon the separation of the magnetic flux tubes and their individual scattering strengths. At the largest flux–tube separations the scattering is incoherent, with the net scattering cross section being just the sum of the scattering cross sections of the individual flux tubes. At intermediate separations phase coherence between the acoustic waves scattered by the individual flux tubes leads to net scattering cross sections that deviate considerably from the incoherent scattering cross section. Finally, a multiple scattering regime can exist at the smallest flux–tube separations, where the acoustic field scattered by a particular flux tube differs significantly from that which would be obtained in the absence of its companion magnetic flux tube. These estimates suggest that the interaction of the solar acoustic oscillations with a spaghetti sunspot will almost certainly show coherent phasing of the scattered waves and will most likely also exhibit multiple scattering effects. Thus the results of this work suggest that sunspot seismology will be able to distinguish between the monolithic flux–tube and the spaghetti sunspot models.

062.159 **Linear theory of rotating fluids using spherical harmonics. II. Time–periodic flows.**
M. Rieutord.
Geophys. Astrophys. Fluid Dyn., Vol. 59, No. 1–4, p. 185 – 208 (Nov 1991).

Expansion in spherical harmonics is used to solve linear equations of flows of homogeneous viscous fluids in a rotating frame. For a truncated series, analytical solutions are obtained for the radial functions. These solutions are used to investigate the modal properties of a viscous incompressible fluid in a spherical shell. The results are compared to the experimental data of Aldridge. The problem of identification of inertial modes in the Earth's outer core is also discussed.

062.160 **α–effect in generation of magnetic field by a flow of conducting fluid with internal scale in an axisymmetric volume.**
V. A. Zheligovsky.
Geophys. Astrophys. Fluid Dyn., Vol. 59, No. 1–4, p. 235 – 251 (Nov 1991).

A kinematic dynamic problem for the generation of magnetic field by a flow of electrically conducting fluid in an axisymmetric volume is considered in two cases. In the first case, the axisymmetric volume is of arbitrary shape and the flow velocity field is parameterized by a fast variable proportional to the azimuthal coordinate. In the second, the volume is a sphere and the velocity depends on three fast variables proportional to the spherical coordinates. A complete asymptotic expansion of the eigenvalues of the magnetic induction operator and of the corresponding modes of magnetic field is constructed. It is shown that the α–effect is recovered in the magnetohydrodynamic system under consideration.

062.161 **Dynamical buoyancy of hydrodynamic eddies.**
E. N. Parker.
Astrophys. J., Vol. 380, No. 1, p. 251 – 255 (10 Oct 1991).

The dynamical pressure reduction within a vortex tube produces both a tension along the tube and general buoyancy, analogous to magnetic flux tubes. The dynamical buoyancy causes convective cells to rise at speeds comparable to the rms fluid velocity within the cell. Consequently the convective cells in a stratified atmosphere are more active than indicated by the standard anelastic approximation. The coherent convective cells at each level actively crowd upward into the convective cells above, elbowing weaker cells out of the way and flattening themselves and others against the upper surface of the convective region. These effects can be seen in the recent SOUP observations of the solar granulation. Deeper in the convective zone the inhomogeneity of the buoyancy may explain the random character of the convective motions that turns up in recent numerical simulations.

062.162 **Propagation of magnetoacoustic waves in the solar atmosphere with random inhomogeneities of density and magnetic fields.**
M. Ryutova, M. Kaisig, T. Tajima.
Astrophys. J., Vol. 380, No. 1, p. 268 – 281 (10 Oct 1991).

The propagation of acoustic waves in a plasma containing an ensemble of tightly settled bundles of magnetic flux tubes is studied for the case where all the parameters of the medium – magnetic field, plasma density, temperature, etc. – change by an order of unity over a length small compared to the acoustic wavelength. Such a model of random and strong inhomogeneities can correspond, for example, to sunspots, the plage regions, and to other astrophysical objects, where domains of different size with different plasma parameters compose a dense conglomerate. The theoretical procedure is described for obtaining the average equations containing the nonlinearity of a wave, dispersion properties of a system, and dissipative effects. It is shown that the statistical properties of the medium determine different scenarios of wave propagation: in the predominance of dissipative effects the primary wave is damped away, and the efficiency of heating due to inhomogeneities is much greater than that in a homogeneous medium. This means that the regions containing closely packed inhomogeneities behave as a strong sink of incoming wave energy. The efficiency of the wave absorption crucially depends on the amplitude of the fluctuations of the background parameters rather than on their absolute values.

062.163 **Magnetic field – aligned electric potentials in nonideal plasma flows.**
K. Schindler, M. Hesse, J. Birn.
Astrophys. J., Vol. 380, No. 1, p. 293 – 301 (10 Oct 1991).

The authors discuss the electric field component parallel to the magnetic field arising from plasma flows which violate the frozen–in field condition of ideal magnetohydrodynamics. The quantity of interest is the potential $U = -\int E_{\parallel} ds$, where the integral is extended along field lines. It is shown that U can be directly related to magnetic field properties, expressed by Euler potentials, even when time–dependence is included. The results are applicable to the Earth magnetosphere, to solar flares, to aligned rotator models of compact objects, and to galactic rotation. These results also support the view that parallel electric fields associated with nonideal plasma flows might play an important role in cosmic particle acceleration.

062.164 **Ray–type spectra of plasma turbulence in coronal magnetic loops.**
Yu. M. Rosenraukh (*Yu. M. Rozenraukh*).
Sol. Phys., Vol. 136, No. 1, p. 117 – 132 (Nov 1991).
The problem of stationary spectra of Langmuir, *l*, and electromagnetic, *t*, waves excited in a magnetic trap (loop) by a group of suprathermal electrons, whose velocity distribution includes a "loss cone", is considered. Within the framework of weak turbulence theory, accurate spectra of *l*– and *t*–waves are found. These spectra have the form of thin rays in wavevector space. Forms of plasma emission radio lines of a homogeneous source near the plasma frequency and its second harmonic are determined.

062.165 **The EUV spectrum of thin plasmas.**
M. Landini, B. C. Monsignori–Fossi.
1. Berkeley Colloquium on Extreme Ultraviolet Astronomy, p. 292 – 297 (1991). – See Abstr. 012.045 for the main entry.
Each author uses different methods of calculating ionization balance and different sources of atomic data, making it rather difficult to merge the theoretical spectra and to link the regions observed by *Einstein* or *EXOSAT* to those observed by *IUE*. It therefore seems useful to perform the evaluation of the spectral distribution of the emission from thin plasmas up to 2000 Å using a homogeneous computation of ion abundances.

062.166 **Laboratory investigations of impact–generated plasma.**
D. A. Crawford, P. H. Schultz.
J. Geophys. Res., Vol. 96, No. E3, p. 18807 – 18818 (25 Oct 1991).
The production of magnetic fields by hypervelocity meteoroid impact has been proposed to explain the presence of paleomagnetic fields in some lunar samples as well as on the lunar surface. Impact–generated magnetic fields also may be significant for the paleomagnetic record on a variety of cratered surfaces in the solar system, such as the Moon, Mercury, Phobos, and asteroids. Previous experiments with the two–stage hydrogen light gas gun at the NASA Ames Vertical Gun Range demonstrated that hypervelocity impacts can produce impact–generated magnetic fields by the expansion of an impact–derived ionized vapor cloud (impact–generated plasma). Here the authors further characterize impact–generated magnetic fields and plasma with hypervelocity impact experiments at the NASA Ames Vertical Gun.

062.167 **On the nature of disk viscosity.**
J.–P. Zahn.
6. IAP Astrophysics Meeting – IAU Colloquium No. 129: Structure and emission properties of accretion disks, p. 87 – 97 (1991). – See Abstr. 012.046 for the main entry.

062.168 **Angular momentum transport in self–gravitating gaseous disks.**
J. C. Papaloizou, G. J. Savonije.
6. IAP Astrophysics Meeting – IAU Colloquium No. 129: Structure and emission properties of accretion disks, p. 99 – 108 (1991). – See Abstr. 012.046 for the main entry.
The authors discuss the evolution of thin self–gravitating gaseous disks that have unstable non–axisymmetric normal modes. The modes can have corotation situated either inside or just exterior to the disk outer boundary. The unstable modes are found to be important for redistributing the disk's mass and angular momentum on a dynamical timescale appropriate to the outer disk. Unstable non–axisymmetric modes may be crucial in driving mass accretion onto central galactic black holes or protostars embedded by self–gravitating gaseous disks.

062.169 **The role of magnetic fields in accretion disks.**
J. Heyvaerts.
6. IAP Astrophysics Meeting – IAU Colloquium No. 129: Structure and emission properties of accretion disks, p. 109 – 118 (1991). – See Abstr. 012.046 for the main entry.
The existence of magnetic fields in accreting objects is evidenced directly or indirectly in a number of them. These fields are sometimes surprisingly large. Also the association between accretion and collimated outflows, though it is not systematically observed, is at least suggested in a number of objects. The release of excess angular momentum is a condition for accretion to proceed. Though viscous transport mediated by an effective turbulen viscosity is often considered, this is by no means the only possible way in an MHD medium. Obviously the turbulence may be MHD in nature, but also stresses transmitted outside of the disk, by the linkage through a magnetized corona for example, could also play a role, and of course a rotating MHD wind blown from the accretion disk would carry angular momentum at infinity, more precisely it will communicate it to remote objects. In this paper the author reviews a few recent developments on these various aspects. Another question posed by the existence of the magnetic field in such objects is on its possible origin. Finally its role in heating a disk's corona, or in feeding flaring activity has been addressed.

062.170 **Spiral instability of magnetized accretion disks.**
M. Tagger, R. N. Henriksen, R. Pellat.
6. IAP Astrophysics Meeting – IAU Colloquium No. 129: Structure and emission properties of accretion disks, p. 119 – 122 (1991). – See Abstr. 012.046 for the main entry.

062.171 **The nonlinear development of gravitational instabilities in thick disks.**
J. E. Tohline.
6. IAP Astrophysics Meeting – IAU Colloquium No. 129: Structure and emission properties of accretion disks, p. 131 – 134 (1991). – See Abstr. 012.046 for the main entry.

062.172 **Magnetized accretion–ejection structures.**
G. Pelletier, R. E. Pudritz.
6. IAP Astrophysics Meeting – IAU Colloquium No. 129: Structure and emission properties of accretion disks, p. 161 – 164 (1991). – See Abstr. 012.046 for the main entry.

062.173 **Multigrid methods applied to an explosion at a plane density interface.**
S. J. Arthur, S. A. E. G. Falle.
Mon. Not. R. Astron. Soc., Vol. 251, No. 1, p. 93 – 104 (1 Jul 1991).
The astrophysical problem of a supernova explosion at the edge of a dense molecular cloud or a diffuse stellar wind bubble is simplified to that of a point explosion at a plane density interface separating two uniform media with density ratio α. This approximation gives rise to a self–similar flow which can be found by transforming the hydrodynamic equations to the similarity plane and looking for the steady solution there. A flux vector splitting scheme, together with the use of multigrid techniques provides the authors with a robust and efficient method of finding this steady solution. The high–resolution results so produced reveal the formation of a supersonic jet along the symmetry axis when the density ratio is large enough.

062.174 **Instability in oblique C–type shocks.**
M. Wardle.
Mon. Not. R. Astron. Soc., Vol. 251, No. 1, p. 119 – 127 (1 Jul 1991).
A sequence of steady, oblique, C–type shock waves differing in the angle between the ambient magnetic field and the direction of shock propagation, θ_s, are examined for stability to small perturbations in the plane containing the fluid velocities and the magnetic field. As θ_s is reduced the steady shocks become stronger and the growth rate and wavenumber of the fastest growing mode increase. The growing perturbations propagate across the face of the shock front at a speed of order the Alfvén speed in the pre–shock gas (typically $2\ \mathrm{km\ s^{-1}}$ in molecular clouds). The strength of the density perturbations in the neutral gas decreases relative to those in the ions as θ_s is reduced, and the perturbations in the neutral gas resemble ripples rather than the clumps found in the perpendicular case. It is noted that a similar

instability may arise in other situations in which ambipolar diffusion plays a major role.

062.175 **Maximum energy of semi-infinite magnetic field configurations.**
P. A. Sturrock.
Astrophys. J., Vol. 380, No. 2, p. 655 - 659 (20 Oct 1991).
This article addresses the conjecture by Aly that the maximum-energy state of magnetic field configurations that have a given flux distribution on a given surface is the open-field configuration. It is first shown that the existence of a maximum-energy configuration depends upon the topology of the source surface: for a multiply connected surface there is no maximum-energy state. However, the magnetic energy is known to be bounded above for a wide class of simply connected surfaces, and it is shown that in this case there must be a field configuration that has maximum energy. Furthermore, for this configuration the energy is stationary under arbitrary small footpoint displacements that leave the surface flux distribution unchanged. This condition leads to the result that either B_n, the normal component of magnetic field, or j_n, the normal component of current, must vanish at each point on the source surface. This condition is met by the open configuration, but not by any other configuration. The author discusses the implications of this result for our understanding of flares and of coronal mass ejections.

062.176 **Dynamical instabilities in self-gravitating annuli.**
D. M. Christodoulou, R. Narayan.
6. IAP Astrophysics Meeting - IAU Colloquium No. 129: Structure and emission properties of accretion disks, p. 413 - 414 (1991). - See Abstr. 012.046 for the main entry.

062.177 **Time-dependent diffusive acceleration of test particles at shocks.**
L. O'C. Drury.
Mon. Not. R. Astron. Soc., Vol. 251, No. 2, p. 340 - 350 (15 Jul 1991).
The acceleration of test particles at a steady plane non-relativistic shock is considered. Analytic expressions are found for the mean and the variance of the acceleration time distribution in the case where the diffusion coefficient has an arbitrary dependence on position and momentum. These expressions are used as the basis for an approximation scheme which is shown, by comparison with numerical solutions, to give an excellent representation of the time-dependent spectrum.

062.178 **Numerical simulations of thermal instabilities in stratified gases. II. Exploration of the parameter space.**
F. Reale, R. Rosner, A. Malagoli, G. Peres, S. Serio.
Mon. Not. R. Astron. Soc., Vol. 251, No. 3, p. 379 - 390 (1 Aug 1991).
The authors study the temporal evolution of density perturbations in an initially hydrostatic isothermal atmosphere, consisting of an optically thin, radiating, compressible plasma which is stratified by gravity and can be described as a perfect gas governed by standard hydrodynamics. They describe in detail the non-linear evolution of the perturbations using numerical techniques, and study the relative equilibrium between dynamic and thermal instabilities as governed by three independent control parameters: the initial density contrast of the perturbation, δ; the ratio of the local buoyancy oscillation (or Brunt-Väisälä) period to the local radiative cooling time, ζ; and the ratio of the perturbation radius to the local scaleheight, η. The evolution of each perturbation has been followed, whenever possible, on relatively long time scales (of the order of one buoyancy oscillation cycle) in order to allow us to sensibly understand the ultimate fate of the perturbation: collapse due to catastrophic cooling, or complete disruption due to hydrodynamic instabilities. The authors explored four orders of magnitude of initial density contrasts and ratios of buoyancy and cooling times, and one order of magnitude of the bubble dimensions. The results confirm and quantitatively define previous suggestions that thermal instabilities are unlikely to lead to inhomogeneities in cooling flows.

062.179 **First-order Fermi acceleration at oblique relativistic magnetohydrodynamic shocks.**
K. R. Ballard, A. F. Heavens.
Mon. Not. R. Astron. Soc., Vol. 251, No. 3, p. 438 - 448 (1 Aug 1991).
The authors investigate first-order Fermi acceleration at oblique relativistic magnetohydrodynamic (MHD) shock fronts. The jump conditions at the shock are solved using a full relativistic equation of state and allowing for various possible downstream conditions. Having obtained solutions to the jump conditions, the spectral indices of electrons accelerated by the shock front through the first-order Fermi mechanism are obtained by employing two methods: a Monte Carlo simulation and the semi-analytic theory of Kirk and Heavens. Good agreement is found for the case where irregularities in the magnetic field are assumed to be small. The authors find that synchrotron spectral indices in the range of 0 to 0.5 are expected for oblique relativistic MHD shocks, unless the Alfvén Mach number is not much greater than unity.

062.180 **Non-axisymmetric unstable modes in a thin differentially rotating gaseous disc.**
M. H. M. Heemskerk, G. J. Savonije.
6. IAP Astrophysics Meeting - IAU Colloquium No. 129: Structure and emission properties of accretion disks, p. 439 - 440 (1991). - See Abstr. 012.046 for the main entry.
The authors consider the linear stability of a thin differentially rotating gaseous disk with reflecting boundaries against non-axisymmetric perturbations and discuss the relevant conservation laws for the energy and angular momentum. They assume the unperturbed disc to be in centrifugal equilibrium with the gravitational attraction of a central mass M, and consider discs whose mass can be neglected relative to M.

062.181 **Globule formation in a disturbed converging shock.**
T. Kimura, M. Tosa.
Mon. Not. R. Astron. Soc., Vol. 251, No. 4, p. 664 - 669 (15 Aug 1991).
This paper presents results of hydrodynamical simulations of a cylindrical gas cloud strongly pushed by a surrounding high-pressure gas. The authors introduce a perturbation on the cloud surface and numerically follow its evolution. A distorted shell of high-density shocked gas is formed from the cloud surface. The distortion of the shocked shell increases as the shell converges toward the centre, and high-density gas clumps are formed behind the perturbed shock. The gas clumps gradually separate from the shell and eventually drift in the surrounding high-temperature region. For parent clouds with initial radius of 5–50 pc and density of 20–270 cm^{-3}, gas clumps are formed in the order of 10^6yr. Their size and density are 0.1–3.0 pc and 10^2–$10^4 cm^{-3}$, respectively, so their mass is several to several hundreds of solar masses, which is larger or close to the local Jeans mass. These physical properties of gas clumps are very similar to those of isolated dark globules, suggesting new mechanisms for the origin of isolated dark globules and stars.

062.182 **Radial oscillations of a gaseous torus with toroidal magnetic fields.**
R. U. Okada, J. Fukue, R. Matsumoto.
6. IAP Astrophysics Meeting - IAU Colloquium No. 129: Structure and emission properties of accretion disks, p. 495 - 496 (1991). - See Abstr. 012.046 for the main entry.
A steady axisymmetric model of MHD tori is constructed under the assumption that the distribution of the specific angular momentum is constant and that the magnetic field has only the toroidal component. Furthermore, the gas is isothermal and the Alfvén velocity is constant. It is found that there are several modes of the radial oscillations which are generated by the toroidal magnetic fields, and that the low frequency is the "magneto-gravity" mode. For the typical parameters of low

mass X–ray binaries, the period of the "magneto–gravity" mode is from a few to several tens of milliseconds.

062.183 **Mapping bow shocks in molecular hydrogen.**
M. D. Smith.
Mon. Not. R. Astron. Soc., Vol. 252, No. 3, p. 378 – 385 (1 Oct 1991).

Near–infrared imaging of Herbig–Haro objects in molecular clouds is now possible. Here, an attempt is made to simulate the H_2 emission–line maps with a model in which the H_2 emission is generated from a cool C–shock region in the oblique tail of a bow shock. It is shown that low–speed bows produce arcs and limb–brightened structures of conical appearance. High–speed bows yield more diffuse, tailed regions with blunter leading edges. The v = 2–1 S(1) distribution is more compact than that from v = 1–0 S(1). The location (relative to the optical line emission), the compactness and the identifying features also depend strongly on the orientation. In particular, for high–speed bows, the H_2 peak appears ahead (upstream) of the bow apex for motions close to the line–of–sight but is well behind the apex when the motion is in the plane of the sky. The magnetic field direction is especially important in low Alfvén number flows, corresponding to the high–field "shock–absorber" model. Then, a wider range of structures including distinct asymmetries are predicted. An initial comparison with observed bow structures is presented.

062.184 **Numerical simulations of axisymmetric accretion flows.**
H. Koide, T. Matsuda, E. Shima.
Mon. Not. R. Astron. Soc., Vol. 252, No. 4, p. 473 – 481 (15 Oct 1991).

The authors study properties of axisymmetric accretion flow on to a gravitating compact object from a uniform flow by performing pure hydrodynamic calculations. At first a comparison is made between a numerical solution and an analytic solution based on a ballistic orbit theory, and it is found that the analytic solution is an extremely good approximation even for Mach numbers as low as 1.4. Next, results are presented for Mach numbers of 0.6, 1.4, 2.4, 5 and 10, and for a ratio of specific heats of 5/3. The accretion rate, the stand–off distance of the bow shock and the stagnation point on the rear axis are computed based on various boundary conditions. An empirical formula for the accretion rate is proposed. It is found that the flow is not completely in steady state but a dome–like shock is formed quasi–periodically in front of the compact object for higher Mach number cases. It is argued that this shock perturbs the flow leading to the destruction of axisymmetry of the flow eventually.

062.185 **The Kelvin–Helmholtz instability for relativistic particle beams. II. Flows bounded by a simple shear layer.**
M. Birkinshaw.
Mon. Not. R. Astron. Soc., Vol. 252, No. 4, p. 505 – 527 (15 Oct 1991).

The growth of the Kelvin–Helmholtz instability in a sheared beam is investigated for beams bounded by linear gradients in the flow velocity and in the square of the sound speed. The properties of pressure perturbations on the beam are qualitatively similar to those on a beam bounded by a vortex layer except for modes which can interact strongly with waves in the shear layer. As a consequence of these interactions, some previously–unstable modes may be stabilized absolutely over a moderate range of wavelengths. At all wavelengths less than the stabilized range, a large number of modes are active, the eigenfunctions have extremely frequency–dependent forms, and the small–scale flow pattern will tend to become chaotic. If it is assumed that damping mechanisms restrict the amplitudes of modes with wavelengths less than the stabilized range, the sheared layer can prevent the Kelvin–Helmholtz instability from disrupting supersonic beams with astrophysically–interesting properties until after they have propagated many beam–radii. The distorted flows that result from the gradual growth of these long–wavelength modes resemble some radio jets.

062.186 **Three–dimensional simulations of the jets of extragalactic radio sources.**
C. I. Cox, S. F. Gull, P. A. G. Scheuer.
Mon. Not. R. Astron. Soc., Vol. 252, No. 4, p. 558 – 585 (15 Oct 1991).

The authors describe here the first fully three–dimensional simulations of the jets of extragalactic radio sources. Application of a steady precession to the jet leads to a variety of unsteady features in the resultant flow pattern, and gives structures which bear a striking resemblance to those observed. The simulations contain secondary hotspots which are formed in the manner of the dentist's drill model and are not properly described as splatter–spots, although material continues to flow into them after their creation. The authors propose explanations of the characteristics of multiple radio hotspots. They find that three–dimensional jets are more stable with respect to perturbations by an asymmetric backflow than are their counterparts in slab geometry.

062.187 **On the structure of thin magnetized accretion disks of active galactic nuclei and their coronal radiation.**
Yang Lantian, Wu Xuebin, Yang Pibo.
Chin. Astron. Astrophys., Vol. 15, No. 4, p. 386 – 396 (Dec 1991). English translation of Acta Astron. Sin., Vol. 32, No. 2, p. 154 – 164 (Jun 1991).

By taking magnetic stress in place of viscosity as the mechanism for angular momentum transfer, the effect of a frozen magnetic field on the structure of a geometrically thin accretion disk is examined. It is shown that the disk is quasi–Keplerian and its total luminosity is twice the luminosity in the standard disk model. In the inner region, there exists a narrow cool region and the highly collimated jet is formed under the action of the azimuthal component of the magnetic field. The authors also discuss the possibility that a magnetized corona be formed near the surface of the accretion disk and a wide band radiation issuing therefrom. The model suggested can easily and reasonably explain the major AGN properties such as the radiation variation, the "bumps" in the optical, ultraviolet and soft X–ray ranges, etc.

062.188 **Turbulence, fractals, and the solar granulation.**
P. N. Brandt, R. Greimel, E. Guenther, W. Mattig.
Applying fractals in astronomy, p. 77 – 96 (1991). – See Abstr. 003.023 for the main entry.

The authors give a brief, mostly qualitative introduction into the topics of convection and turbulence, and their description, mainly referring to laboratory experiments. Following Mandelbrot (1967), the concept of the fractal dimension is introduced and some earlier results of measurements of the fractal dimension in laboratory tubulence are discussed. The authors address the question whether hints of turbulence have been observed in the solar photosphere, and describe three independent methods of determining the fractal dimension of the solar granulation: the area–perimenter relation, the line and the plane intersection method. An analysis of a set of high resolution granulation photographs taken with the balloon–borne "Spectro–Stratoskop" telescope yields a fractal dimension of $d \approx 1.9$ for granules of diameters > 1.32 arcsec and of $d \approx 1.3$ for smaller granules, analysing the area–perimeter relation of approx. 40000 granules.

062.189 **Three–dimensional driven reconnection in an axially bounded magnetic field.**
H. R. Strauss.
Astrophys. J., Vol. 381, No. 2, p. 508 – 514 (10 Nov 1991).

A study of three–dimensional driven reconnection in nonvanishing magnetic fields is presented. The vanishing of the velocity components perpendicular to the magnetic field, along a magnetic line causes the formation of an intense localized current density. In turn, this causes large displacements of magnetic field lines from fluid elements moving with the fluid. Fluid elements, initially connected by magnetic field lines can become separated by a distance independent of the fluid resistivity. By projecting field lines onto planes transverse to the magnetic field, the topology of the reconnecting magnetic field can be seen to retain the essential features known from two-dimensional reconnection.

062.190 **Two-dimensional hydrodynamic calculations of the nonlinear development of the Goldreich – Schubert – Fricke instability in a rotating annulus.**
D. G. Korycansky.
Astrophys. J., Vol. 381, No. 2, p. 515 – 525 (10 Nov 1991).
The Goldreich – Schubert – Fricke (GSF) instability may be an important mechanism affecting the angular momentum distribution of rotating stars and gaseous planets. This paper presents two-dimensional nonlinear hydrodynamic calculations studying the effectiveness of the instability in transporting angular momentum in the equatorial zones of stars and planets which are stably stratified with respect to convection. The calculations were made by numerically integrating the two-dimensional axisymmetric Navier-Stokes equations, including viscosity and heat conduction. The initial state was that of a differentially rotating, stably stratified annulus in which angular momentum decreased with radius, and in which gravity was perpendicular to the rotation vector. The instability was followed into the nonlinear regime. The maximum rms velocity amplitude was found to correlate well with the product of the linear growth rate and radial length scale of the instability, consistent with the idea that the instability grows to an amplitude such that an eddy turnover time becomes equal to the growth time defined by the inverse of the growth rate. The time scale for angular momentum to be redistributed to a state of marginal stability was consistent with this picture. The results suggest that in physical situations a state of marginal stability will be maintained, since departures from such a state will be rapidly corrected.

062.191 **Introduction to solar system MHD.**
E. R. Priest.
Symposium on Geophysical and Astrophysical MHD as Part of the British Theoretical Mechanics Colloquium: Advances in solar system magnetohydrodynamics, p. 1 – 23 (1991). – See Abstr. 012.086 for the main entry.
Most of the solar systems are in the plasma state and its subtle non-linear interaction with the magnetic field is described for many purposes adequately by the equations of magnetohydrodynamics (MHD). These equations are also applicable to the electrically conducting liquid cores of planets, such as our Earth, where magnetic fields are being generated by dynamo action. Indeed, this realisation of the existence of a strong coupling between a magnetic field and a plasma has revolutionised our understanding of the solar system. Furthermore, the complementary remote sensing observations at high resolution of the Sun and in-situ spacecraft measurements of solar system plasmas have revealed a spectacular world of intriguing dynamic activity over a wide range of length- and time-scales.

062.192 **Dynamo models and Taylor's constraint.**
C. A. Jones.
Symposium on Geophysical and Astrophysical MHD as Part of the British Theoretical Mechanics Colloquium: Advances in solar system magnetohydrodynamics, p. 25 – 50 (1991). – See Abstr. 012.086 for the main entry.
Contents: 1. Introduction. 2. The rapid rotation limit. 3. Magnetoconvection. 4. Axisymmetric dynamo models.

062.193 **Problems of planetary dynamo theory.**
F. H. Busse.
Symposium on Geophysical and Astrophysical MHD as Part of the British Theoretical Mechanics Colloquium: Advances in solar system magnetohydrodynamics, p. 51 – 59 (1991). – See Abstr. 012.086 for the main entry.
An outline is given of a theory of the generation of magnetic fields by buoyancy-driven convection flows in rotating spherical shells subject to a spherically symmetric gravity force. The goal of this research is an understanding of the dynamo process and of the nonlinear equilibration of the magnetic field and of its dependence on the parameters of the problem rather than a realistic model of a planetary dynamo. After a brief outline of the spherical dynamo model the author discusses in more detail the problem of geomagnetic coremantle coupling. The paper closes with an outlook on future research.

062.194 **Convection near the Earth's core-mantle boundary?**
D. R. Fearn.
Symposium on Geophysical and Astrophysical MHD as Part of the British Theoretical Mechanics Colloquium: Advances in solar system magnetohydrodynamics, p. 61 – 75 (1991). – See Abstr. 012.086 for the main entry.
The Earth's core-mantle boundary (CMB) plays a vital role in controlling the evolution of both the core and the mantle. For example, the boundary layer at the base of the mantle regulates the rate at which the core is cooling. In turn, this has consequences for the rate of inner-core growth and therefore for the power available to drive the geodynamo. Contents: 1. Introduction. 2. Magnetic instabilities. 3. Convection with variable stratification. 4. Convection with variable field strength. 5. Summary and conclusions.

062.195 **Magnetic buoyancy.**
D. W. Hughes.
Symposium on Geophysical and Astrophysical MHD as Part of the British Theoretical Mechanics Colloquium: Advances in solar system magnetohydrodynamics, p. 77 – 104 (1991). – See Abstr. 012.086 for the main entry.
Magnetic buoyancy is of importance in a number of astrophysical contexts. The author discusses primarily the magnetic field of the solar interior. Contents: 1. Introduction. 2. The magnetic buoyancy of flux tubes. 3. Magnetic buoyancy as an instability mechanism: ideal MHD, the influence of diffusion and rotation, the non-linear evolution of magnetic buoyancy instabilities.

062.196 **Magnetohydrodynamic waves in the Sun.**
B. Roberts.
Symposium on Geophysical and Astrophysical MHD as Part of the British Theoretical Mechanics Colloquium: Advances in solar system magnetohydrodynamics, p. 105 – 136 (1991). – See Abstr. 012.086 for the main entry.
Taken overall, this brief outline of oscillatory phenomena observed in the Sun serves perhaps to show that the Sun is indeed a complex medium capable of supporting a large variety of wave phenomena. Contents: 1. Introduction. 2. Magnetohydrodynamic waves. 3. Surface waves. 4. Waves in magnetic flux tubes. 5. Stratification effects.

062.197 **Magnetohydrodynamic waves and wave heating in non-uniform plasmas.**
M. Goossens.
Symposium on Geophysical and Astrophysical MHD as Part of the British Theoretical Mechanics Colloquium: Advances in solar system magnetohydrodynamics, p. 137 – 172 (1991). – See Abstr. 012.086 for the main entry.
The present review will not try to give a complete description of the present status of MHD waves in the solar atmosphere. Instead it will attempt to outline basic properties of MHD waves in nonuniform plasmas and will focus on those aspects of MHD waves that are important for understanding wave heating theories of the solar corona.

062.198 **Weak solutions of magnetostatics: an aspect of current sheet theory.**
T. Amari.
Symposium on Geophysical and Astrophysical MHD as Part of the British Theoretical Mechanics Colloquium: Advances in solar system magnetohydrodynamics, p. 173 – 202 (1991). – See Abstr. 012.086 for the main entry.
This paper is only concerned with the processes of formation of a current sheet (CS) during the quasi-static evolution towards weak equilibrium of classical magnetostatic equilibria embedded in a perfectly conducting plasma. The author is mainly interested in the spontaneous process of classical magnetostatic equilibria embedded in a perfectly conducting plasma. Contents: 1. Introduction. 2. 2D quasi potential singular equilibria (QPSE):

spontaneous formation of a CS, asymptotic formation of QPSE. 3. 2D QFFSE: evolution towards a 2D force–free field, T– and X–like topology. 4. 3D force–free solutions.

062.199 **Magnetic reconnection in solar flares.**
M. Jardine.
Symposium on Geophysical and Astrophysical MHD as Part of the British Theoretical Mechanics Colloquium: Advances in solar system magnetohydrodynamics, p. 203 – 220 (1991). – See Abstr. 012.086 for the main entry.
Magnetic reconnection may be involved in creating the complex structures such as magnetic arcades within which flares occur; it may act as the trigger that initiates a flare and it is certainly responsible for the main energy release. Contents: 1. Introduction. 2. Reconnection and energy release. 3. Reconnection and flare initiation. 4. Reconnection theory. 5. Models for fast steady–state reconnection. 6. Numerical experiments.

062.200 **Recent developments in the theory of three–dimensional magnetic reconnection.**
M. Hesse.
Symposium on Geophysical and Astrophysical MHD as Part of the British Theoretical Mechanics Colloquium: Advances in solar system magnetohydrodynamics, p. 221 – 240 (1991). – See Abstr. 012.086 for the main entry.
Contents: 1. Introduction. 2. General magnetic reconnection for $\mathbf{B} \neq 0$. 3. The plasmoid and other flux ropes in the near–Earth environment. 4. Reconnection in the solar environment.

062.201 **Magnetic heating of the solar corona.**
M. A. Berger.
Symposium on Geophysical and Astrophysical MHD as Part of the British Theoretical Mechanics Colloquium: Advances in solar system magnetohydrodynamics, p. 241 – 256 (1991). – See Abstr. 012.086 for the main entry.
Contents: 1. Introduction. 2. Generation of coronal magnetic fields. 3. The Sturrock–Uchida model. 4. The Parker model. 5. The Heyvaerts–Priest model. 6. Dissipation of coronal magnetic fields.

062.202 **Hydromagnetic equilibrium.**
K. Tsinganos.
Symposium on Geophysical and Astrophysical MHD as Part of the British Theoretical Mechanics Colloquium: Advances in solar system magnetohydrodynamics, p. 257 – 279 (1991). – See Abstr. 012.086 for the main entry.
The methodology and state of the art of constructing analytical MHD equilibria is reviewed. The formalism is illustrated by two classes of analytical solutions. In the first, a typical example of MHD equilibrium in the uniform gravitational field of a polytropic atmosphere is constructed. The solution extends the familiar Kippenhahn–Schlüter model for a quiescent solar prominence by including flows along the magnetic field lines. In the second, a typical example of MHD equilibrium in the nonpolytropic stellar atmosphere surrounding the central gravitational field of a star is constructed. The solution extends the familiar Parker model of the solar/stellar wind by including a magnetic field without spherical symmetry and hydromagnetic critical points in the rotating outflow.

062.203 **MHD forces in astrophysical disks and jets.**
J. Heyvaerts.
Symposium on Geophysical and Astrophysical MHD as Part of the British Theoretical Mechanics Colloquium: Advances in solar system magnetohydrodynamics, p. 399 – 423 (1991). – See Abstr. 012.086 for the main entry.
Accretion disks and jets are found to be associated with several kinds of astrophysical objects which are briefly described. Observational evidence and theoretical arguments indicating that accretion disks are likely to be magnetized are presented, and the consequences for the structure and evolution of disks are discussed. It is shown that dynamo action in the disk results in a magnetized coronal halo around it. Being stressed by the disk's differential rotation, the halo is heated by a process similar to solar coronal heating. Conversely, Lorentz forces are exerted on the disk, which cause non–local interaction and redistribution of angular momentum. The disk, even at relatively low magnetization, may become non–keplerian, and eventually differential rotation may come to be frozen. A model incorporating these effects is outlined. If the disk is threaded by an open field configuration, the latter can support a rotating magnetized wind. It is shown that such a wind can remove angular momentum from the disk, thus allowing a secular contraction of it to take place. Finally, the shape of the wind in regions remote from its source is discussed. Recent results (Heyvaerts and Norman, 1989) show that the asymptotic form of the momentum equation perpendicular to flux surfaces implies that, for almost all such winds, the asymptotic structure consists of a family of nested cylindrical surfaces, potentially explaining the collimation of the observed "jets", in particular the famous jets from active galactic nuclei without resorting to any external confinement.

062.204 **Parametric study of the formation of magnetohydrodynamic shocks. III. Fast and slow shocks.**
Y. C. Whang.
Astrophys. J., Vol. 381, No. 2, p. 559 – 571 (10 Nov 1991).
The interaction of two colliding streams may evolve to form fast and slow shocks. The shock system and the interaction region are in a state of dynamical equilibrium with the colliding streams. The author uses dimensionless parameters to represent the two colliding streams and to study how various parameters control the transition of the nature of the shock system: from a pair of slow shocks, to a system of two slow shocks and two fast shocks, and finally to a pair of fast shocks. At decreasing β and θ (the angle between the wave normal and the field) the fast shocks become weaker and the slow shocks become stronger. At small β and θ, the shock system may consist of slow shocks only. At large momentum impact parameter Λ the formation of slow shocks becomes less likely and the strength of fast shocks is a monotonically increasing function of Λ. At increasing β and θ the fast shocks become stronger and the slow shocks diminish. The remnant of the slow shock may appear as a train of slow–mode wave structures. The author applies these results to interplanetary corotating shocks and shocks associated with solar coronal mass ejections (CME). It is explained why the corotating shocks are fast shocks and why shocks increase in strength between 1 and 10 AU. It is also shown that the CME–associated shocks are slow shocks and that the normals of these shocks make small angles with the magnetic field.

062.205 **Quasi–parallel collisionless shocks.**
J. R. Kan, M. E. Mandt, L. H. Lyu.
Space Sci. Rev., Vol. 57, No. 3/4, p. 201 – 236 (Aug 1991).
The magnetic field and plasma data from the ISEE 1, 2, and 3 spacecraft have greatly increased our knowledge of the quasi–parallel collisionless shock in space. Hybrid–code simulations have provided us with valuable insights into the physics of the quasi–parallel shock. Unfortunately, theoretical understanding of the nonlinear physics of the quasi–parallel shock is still in a qualitative stage of development. Generation of large–amplitude whistler waves and hydromagnetic waves observed in the quasi–parallel shock has been discussed either in terms of linear instabilities or qualitative nonlinear arguments. It appears that the ion reflection, ion heating, and leakage of the shock–heated downstream ions at the quasi–parallel shock can all be explained in terms of nonadiabatic scatterings of ions by the large–amplitude whistler–magnetosonic waves with frequencies near the ion gyrofrequency and wavelength near the ion inertial length. The nonadiabatic scattering is defined by the non–conservation of the magnetic moment. Future study of the quasi–parallel shock should focus on developing quantitative theoretical models for the nonlinear physical processes fundamental to the quasi–parallel shock.

062.206 **The plasma physics of shock acceleration.**
F. C. Jones, D. C. Ellison.
Space Sci. Rev., Vol. 58, No. 3/4, p. 259 – 346 (Nov 1991).
The notion that plasma shocks in astrophysical settings can and do accelerate charged particles to high energies is not a new

one. However, in recent years considerable progress has been achieved in understanding the role particle acceleration plays both in astrophysics and in the shock process itself. In this paper the authors briefly review the history and theory of shock acceleration, paying particular attention to theories of parallel shocks which include the backreaction of accelerated particles on the shock structure. The authors discuss in detail the work that computer simulations, both plasma and Monte Carlo, are playing in revealing how thermal ions interact with shocks and how particle acceleration appears to be an inevitable and necessary part of the basic plasma physics that governs collisionless shocks. The authors briefly describe some of the outstanding problems that sill confront theorists and observers in this field.

062.207 **Electron acceleration in a nonlinear shock model with applications to supernova remnants.**
D. C. Ellison, S. P. Reynolds.
Astrophys. J., Vol. 382, No. 1, p. 242 – 254 (20 Nov 1991).

The authors present Monte Carlo calculations of ion and electron spectra produced by Fermi acceleration in a steady state, plane–parallel, modified shock for Mach numbers of 170 and 43. The simulation assumes isotropic, elastic scattering in the local fluid frame, consistent with results from plasma simulations. The shock structure is calculated taking into account the back pressure of accelerated ions, and includes self–consistent pickup and acceleration of thermal ions. In a steady state, particle acceleration is balanced by particle escape; the authors assume that particles escape above some energy E_{max}, and show results for various values of E_{max} ranging up to 1 TeV. Results are described for two dependences of scattering mean free path on rigidity. The ion spectra include the entire continuous distribution function from thermal particles to E_{max}. The authors show that these shocks are extremely efficient, with up to 98% of the entering energy flux emerging as relativistic ions. Since the physics of electron injection is not understood, the authors inject electrons at various superthermal energies, producing the first electron spectra calculated in a modified shock profile. Electron spectra are steeper than ion spectra at virtually all energies relevant for radio emission from supernova remnants. The spectral results are compared with integrated radio spectra of young supernova remnants.

062.208 **Dynamical instability of a general relativistic rotating fluid.**
K. Nakayama, T. Hanawa.
Publ. Astron. Soc. Jpn., Vol. 43, No. 4, p. 649 – 656 (1991).

A necessary condition for instability is obtained for a steady, axisymmetric, general relativistic, rotating fluid. This represents a generalization of work by Nakayama (1991), who obtained a similar theorem for a Kerr metric. The authors have obtained the necessary condition for instability under the assumption that a perturbation is adiabatic and inviscid, and that the change in the metric due to the perturbation is negligibly small. In other words, the authors consider a dynamical instability but neglect any of thermal, viscous, self–gravitating, and gravitational radiation–induced instabilities. The necessary condition for instability is that the angular velocity ($\Omega \equiv U^{\Phi}/U^{t}$) is not constant in the fluid. This means that only rigid rotation is stable against a Kelvin–Helmholtz instability. The physical implication of the theorem is also discussed.

062.209 **Equation of state and opacity.**
W. Däppen, J. Keady, F. Rogers.
Solar interior and atmosphere, p. 112 – 139 (1991). – See Abstr. 003.028 for the main entry.

The three most important physical ingredients of stellar models are the nuclear–energy generation rates, the equation of state and the opacity. The authors begin this chapter with a discussion of the equation of state, not only because the equation of state plays an important role in stellar structure, but also because it is by itself a fundamental part of any opacity calculation. Furthermore, the computation of the equation of state is the simpler problem. There are two reasons for this. First, already at lower densities, where atoms exist (i.e. where many–body effects can be neglected), it suffices for the equation of state to know the energy levels of atoms (and their occupation). Second, at higher densities, where many–body effects become important (and where one cannot speak of atoms), there are at least roads to a correct treatment of the equation of state (the authors will show some of them in this chapter). Extending these techniques to opacity calculations faces the difficulty that again more detailed information about the quantum–mechanical many–body states is required.

062.210 **Mixing in the radiative zone.**
E. Schatzman.
Solar interior and atmosphere, p. 192 – 222 (1991). – See Abstr. 003.028 for the main entry.

This is a brief review of the present situation concerning the process of mixing in the radiative zone. After an introduction, the author presents the problem of penetration of the convective motion, and concludes that the penetration is likely to be small. He analyses the problem of turbulent mixing, with a discussion of the different mechanisms, which are usually supposed to be at the origin of the turbulence. Then he discusses the effect of rotation and time dependent mixing. The fact that these mechanisms are very unlikely to be active in the present Sun, and the difficulty of explaining the abundances of lithium as a function of mass and age suggest that some other mechanism should be present and efficient. The last section concerns the possible role of internal waves, which can as well carry angular momentum and create a random walk, equivalent to a diffusion effect, which can transport lithium to the burning level.

062.211 **Absorption of p–mode waves by magnetic fields.**
H. C. Spruit.
Conference on Challenges to Theories of the Structure of Moderate–Mass Stars, p. 121 – 134 (1991). – See Abstr. 012.096 for the main entry.

Vertical magnetic flux tubes can leak energy out or the p–mode wave guide in the form of transversal and longitudinal tube waves. A procedure is developed for calculating this energy loss for tubes whose diameter is much smaller than the wavelength of the exciting p–mode.

062.212 **Large eddy simulations of compressible convection.**
Xie Xin, J. Toomre.
Conference on Challenges to Theories of the Structure of Moderate–Mass Stars, p. 147 – 156 (1991). – See Abstr. 012.096 for the main entry.

Large–eddy simulation (LES) technique has been applied to two–dimensional compressible convection in a plane–parallel layer of ideal gas with intermediate density contrast. The general flow patterns consist of a series of rolls in the horizontal which are coherent throughout the entire vertical domain, much like findings by a number of investigators. The horizontal component of the subgrid–scale (SGS) force appears to enhance fluctuations of pressure and density, which in turn cause the resultant force, say f_{pg} arising from pressure gradient and gravity terms, to increase in magnitude in such a way that f_{pg}, besides balancing the inertial terms, can essentially offset of vertical component of the SGS force. Consequently, compared to a solution obtained omitting the SGS terms, the flow field of the LES solution only shows moderate differences. There is some indication that the overall SGS effect tends to reduce the asymmetry between the vertical flows in the interior of the domain while enhancing it in the base region.

062.213 **Vertically coherent compressible convective flows in a layer with a high density contrast.**
Xie Xin, J. Toomre.
Conference on Challenges to Theories of the Structure of Moderate–Mass Stars, p. 171 – 178 (1991). – See Abstr. 012.096 for the main entry.

Two–dimensional simulations are used to study compressible convection in a plane–parallel layer of ideal gas with a high density contrast across the layer. The convective flows are found to consist of rolls which are coherent throughout the entire vertical domain, similar to the results of a number of earlier investigations. Motions near the upper boundary are supersonic, though the Rayleigh number is moderate. The major effect of a high density stratification is to significantly rarefy the fluid in the upper part of the domain, hence increase its thermal diffusivity. The consequence appears to be twofold: Firstly, enhanced thermal diffusion makes non-adiabatic cooling more efficient, thus helps the flows to become supersonic more easily. Secondly, sufficient diffusion can broaden shocks to such an extent that it actually turns them into smooth transition regions.

062.214 **High wavenumber thermal convection enhanced in regions of partial ionization.**
M. P. Rast.
Conference on Challenges to Theories of the Structure of Moderate-Mass Stars, p. 179 – 186 (1991). – See Abstr. 012.096 for the main entry.

The linear stability of a compressible hydrogen fluid undergoing thermal ionization is examined. The ionziation formulation consistently includes particle number, latent heat and specific heat effects. It is found that the eigenfunctions of the convective modes are strongly peaked in the region of partial ionization. This tendency increases with increasing horizontal wavenumber. In addition, thermal diffusion is least effective in this region due to the increased specific heat of the fluid. Both these effects combine to yield elevated growth rates at high wavenumbers. This work implies the possible existence of shells of very high wavenumber convection embedded in stellar envelopes. Such shells may determine the depth of the peak source of acoustic emission and may mask large scale coherent flows below. Two-dimensional nonlinear simulations are in progress and are briefly discussed as well.

062.215 **The organization of turbulent convection.**
N. Brummell, F. Cattaneo, A. Malagoli, J. Toomre, N. E. Hurlburt.
Conference on Challenges to Theories of the Structure of Moderate-Mass Stars, p. 187 – 194 (1991). – See Abstr. 012.096 for the main entry.

Highly resolved numerical simulations are used to study three-dimensional, compressible convection. The viscous dissipation is sufficiently low that the flow divides itself in depth into two distinct regions: (i) an upper thermal boundary layer containing a smooth flow with a granular appearance, and (ii) a turbulent interior pierced by the strongest downflows from the surface layer. Such downflows span the whole depth of the unstable layer, are temporally coherent, and are thermodynamically well correlated. A remarkable property of such convection, once it becomes turbulent, is that the enthalpy and kinetic fluxes carried by the strong downflows nearly cancel, for they are of opposite sense and nearly equal in amplitude. Thus, although the downflows serve to organize the convection and are the striking feature that emerges from effects of compressibility, it is the small-scale, disorganized turbulent motions (between the coherent downflow structures) that serve as the principal carriers of net convected flux.

062.216 **Interaction of acoustic oscillations with time-dependent compressible convection.**
P. W. Jones, W. J. Merryfield, J. Toomre.
Conference on Challenges to Theories of the Structure of Moderate-Mass Stars, p. 213 – 220 (1991). – See Abstr. 012.096 for the main entry.

The authors present the results of numerical simulations in which acoustic waves are driven into a two-dimensional layer of compressible fluid which is undergoing convection. The energetics of the waves are analyzed by computing the work integrals associated with the modulation of the gas and turbulent pressures. The authors find that the relative importance of the turbulent pressure component of the work integral increases prominently with increasing wave frequency. Also, the time dependence of the convection leads to temporal irregularity in the component of wave driving and damping associated with the convection-pulsation coupling. The authors speculate that such irregularity may contribute to the aperiodicity observed in the light curves of Mira variables and other pulsating red giant stars.

062.217 **Towards a theory of wave transport.**
E. Knobloch.
Conference on Challenges to Theories of the Structure of Moderate-Mass Stars, p. 241 – 246 (1991). – See Abstr. 012.096 for the main entry.

A technique is described that should prove invaluable for the computation of effective diffusion coefficients due to wave motion.

062.218 **Transport in oscillatory flows.**
W. J. Merryfield, E. Knobloch.
Conference on Challenges to Theories of the Structure of Moderate-Mass Stars, p. 247 – 252 (1991). – See Abstr. 012.096 for the main entry.

The authors examine the degree to which the transport of a passive scalar contaminant is enhanced by an oscillatory cellular flow. The enhancement is found to depend on the Péclet number Pe and on the ratio η of the oscillation frequency to the shear rate. Analytical expressions are derived for the diffusivity enhancement in the limits of high and low η. Numerical solutions of the advection-diffusion equation confirm these results and connect these two régimes.

062.219 **Boundary conditions in the problem of plasma ejection from the magnetosphere of an axisymmetric rotator.**
S. V. Bogovalov.
Sov. Astron., Vol. 35, No. 6, p. 616 – 622 (Nov 1991). Current Physics Microform No.: 9205X1426. English translation of Astron. Zh., Tom 68, Vyp. 6, p. 1227 – 1239 (1991).

The author considers the problem of plasma ejection from the magnetosphere of an axisymmetric rotator as a function of its physical parameters. The conditions that govern the plasma stream and the toroidal magnetic field are determined. For a hot plasma and a weak magnetic field, those quantities are determined by the conditions of transition velocity of the plasma in terms of fast and slow magnetosonic waves propagation velocity. Such a situation occurs in the solar wind. For a strong field and a cold plasma, typical of pulsars, the mass flow and the toroidal magnetic field are determined by the mechanism that produces the plasma. Here the condition of freezing of the magnetic field into the plasma is violated near the pulsar surface and at the Alfvén surface in the magnetosphere. Acceleration of the plasma to velocities higher than the fast magnetosonic velocity occurs at the Alfvén discontinuity. The conditions under which the discontinuity develops are determined. The electric current flowing from the polar cap of the pulsar turns out to equal the Goldreich-Julian current. The energy loss from an axisymmetric rotator in this case is comparable to that from a magnetic dipole in vacuum. The boundary conditions on the poloidal magnetic field are obtained with allowance for a discontinuity in the flow of the cold plasma at the Alfvén surface.

062.220 **Compressible magnetic convection: formulation and two-dimensional models.**
P. A. Fox, M. L. Theobald, S. Sofia.
Astrophys. J., Vol. 383, No. 2, p. 860 – 881 (20 Dec 1991).

The authors present details on a formulation and numerical solution of the equations of time-dependent magnetized convection. The ADISM (alternating direction implicit on staggered mesh) method of Chan and Wolff is used to solve these equations in two spatial dimensions. This model is used to study the interaction between highly stratified compressible convection and magnetic fields under solar and stellar conditions. The authors present the results of the simulations in the context of previous work and highlight some qualitative properties of the convection – magnetic field interaction. For example, they find that the thermal relaxation time of a convective layer with magnetic fields is primarily influenced by the mean magnetic flux that permeates the layer. The simulations also demonstrate the presence of acoustic, magnetoacoustic, and Alfvén waves, in agreement with observations. Further, the authors present results on the transition between weak and strong magnetic fields, where the highly nonlinear behavior commences, and indicate the sensitivity of these results to the value of the magnetic resistivity, which can significantly influence the details of the interactions. Finally, the consequences of the present findings for modeling solar and stellar magnetic fields are discussed.

062.221 **Equation of state of partially ionized dense plasmas.**
F. J. Rogers.
International School of Physics 'Enrico Fermi'. Course 113: High-pressure equations of state: theory and applications, p. 77 - 99 (1991). - See Abstr. 012.100 for the main entry.

The present lecture describes methods for calculating the equation of state of partially ionized dense plasmas. The term dense plasma is used rather than strongly coupled plasma, since it is possible that at plasma conditions such that only a few levels can be observed spectroscopically the plasma coupling parameters are not large. Due mainly to their importance in theoretical astrophysics, the properties of partially ionized plasmas have been of interest for a long while. More recently, this interest has intensified due to the development of methods for producing partially ionized plasmas in the laboratory. This has opened up large programs of experimental investigation and of practical application.

062.222 **Statistical mechanics of high-pressure equations of state and freezing of classical fluids and dense plasmas.**
J. P. Hansen, G. Chabrier.
International School of Physics 'Enrico Fermi'. Course 113: High-pressure equations of state: theory and applications, p. 263 - 283 (1991). - See Abstr. 012.100 for the main entry.

This lecture is intended to give a brief overview of state-of-the-art statistical-mechanics methods for calculating the equation of state and the freezing line of dense classical fluids and of dense plasmas which are of interest in geophysics, astrophysics and inertial-confinement fusion. The recent theoretical advances are intimately linked to the considerable progress made over the last ten years by both static and dynamic experimental high-pressure techniques and to the continuous effort in developing sound interpretations of the various diagnostics of inertially confined plasmas. The emphasis is on approximation schemes based on a first-principles approach (i.e. starting from a given many-body Hamiltonian), which apply in general to simple fluids, rather than on more phenomenological calculations which are better adapted to the description of complex, multicomponent systems.

062.223 **Equation of state and related problems in astrophysics.**
E. Schatzman.
International School of Physics 'Enrico Fermi'. Course 113: High-pressure equations of state: theory and applications, p. 377 - 398 (1991). - See Abstr. 012.100 for the main entry.

The author considers the properties of astrophysical fluids in thermodynamical equilibrium or close to thermodynamical equilibrium. The solution of many astrophysical problems depends on the knowledge of these properties. To this effect, he presents briefly the domains of interest: they belong almost entirely to stellar physics, and especially to the domain of internal structure. To the exception of a very thin surface layer, the mean free path of ions and electrons and the mean free path of photons are very small compared to all macroscopic scales and the collision frequencies are extremely large compared to any characteristic time scales, evolution time scales or hydrodynamical time scales.

062.224 **Limits to pressure in outer layers of stars and velocity dispersions in outer regions of stellar systems.**
R. Mitalas.
J. R. Astron. Soc. Can., Vol. 85, No. 4, p. 215 (Aug 1991). Abstract. - See Abstr. 012.091 for the main entry.

062.225 **MHD wave coupling in inhomogeneous media.**
A. N. Wright.
Geophys. Res. Lett., Vol. 18, No. 11, p. 1951 - 1954 (Nov 1991).

The coupling of different MHD waves is important for an understanding of laboratory, space, and solar plasmas. In this paper the author investigates fast and Alfvén mode coupling in the cold plasma limit. The medium is assumed to have an invariant direction (e.g., slab or axisymmetric models) and carries an arbitrary solenoidal background magnetic field. Wave fields are Fourier analysed with wave number k_β in the invariant direction. The coupled linearized wave fields are then expanded in a power series (of k_β) to yield a decoupled hierarchy of inhomogeneous wave equations. The higher order terms in the series represent wave coupling phenomena such as the resonant excitation of Alfvén waves, and the damping of a magnetospheric cavity mode.

062.226 **A new approach to hot accretion disks.**
G. Björnsson, R. Svensson.
Workshop on Relativistic Hadrons in Cosmic Compact Objects, p. 53 - 56 (1991). - See Abstr. 012.103 for the main entry.

The authors present a new method for investigating hot two-temperature geometrically and effectively optically thin accretion disks including effects at mildly relativistic temperatures such as electron-positron pair production. They solve the disk structure equations analytically and show that the many studies of hot plasma clouds existing in the literature can be used directly to analyse the detailed disk structure. A number of new results were discovered using their approach. In particular they find equilibrium solutions at all radii and for all accretion rates, and they have discovered a critical value of the viscosity parameter, above which physically acceptable pair dominated disk solutions exist.

062.227 **Jets and hotspots in extragalactic radio sources.**
U. Achatz, H. Lesch, R. Schlickeiser.
Workshop on Relativistic Hadrons in Cosmic Compact Objects, p. 73 - 76 (1991). - See Abstr. 012.103 for the main entry.

Radio gaps, i.e. regions of strongly reduced radio (and also optical) intensity between the galactic nucleus (AGN) and the radio jets in active radio galaxies are a rather prominent feature of many of these sources. With the aim to explain this phenomenon the authors consider the evolution of the relativistic electron spectrum in such a source with the relevant plasma kinetic equations. They consider a possible acceleration process for relativistic electrons and positrons near the accreting central object which is based on the continuous interplay of stochastic Fermi acceleration processes and radiation losses of energetic particles. By the concerted action of synchrotron losses and adiabatic focussing in the ordered, spatially decaying magnetic field B_0 the isotropic relativistic electron distribution function rapidly develops into the oppositely directed beam distributions, explaining the formation of compact, superluminal radio jets near the AGN and the radio gap phenomenon. The stability of the electron/positron beam in the magnetized background plasma is studied by treating it as a cold beam. A lower limit for the mean free path for this isotropization process is calculated and compared with the length of the radio gaps. A discussion of the results is given.

062.228 **Particle acceleration at shock fronts in AGNs and jets.**
J. G. Kirk.
Workshop on Relativistic Hadrons in Cosmic Compact Objects, p. 91 - 100 (1991). - See Abstr. 012.103 for the main entry.

Several aspects of the theory of first order Fermi acceleration at shock fronts are discussed which may be relevant for active galactic nuclei (AGNs) and jets. In the test particle picture, recent work on acceleration at oblique shocks is reviewed which indicates that such shocks produce a harder spectrum than parallel shocks. They also tend to accumulate a substantial excess of energetic particles ahead of the front (snow plough effect). Next, models predicting the maximum energy to which electrons can be accelerated at hot spots in jets are discussed. These models are successful in explaining the observed synchrotron spectrum, but operate only if the Alfvén turbulence on each side of the shock front is of a very special nature. Finally, some recent work on the nonlinear problem of the back reaction of accelerated particles on a relativistic fluid flow is summarized, which indicates that higher acceleration efficiencies may be achieved in relativistic shocks than can be obtained in nonrelativistic shocks.

062.229 **Fermi particle acceleration in relativistic shocks: preliminary nonlinear results.**
D. C. Ellison.
Workshop on Relativistic Hadrons in Cosmic Compact Objects, p. 101 – 110 (1991). – See Abstr. 012.103 for the main entry.
The author has extended a previously developed and tested Monte Carlo simulation of parallel shocks to include relativistic flow velocities. This steady–state simulation treats thermal particle injection, determines the shock structure in the presence of accelerated particles, and simultaneously and self–consistently calculates the ratio of specific heats in the shocked plasma and the effects of particle escape. He presents preliminary results showing the absolute acceleration efficiencies and spectra for a range of parameters and shows that relativistic shocks can be extremely efficient particle accelerators.

062.230 **Cosmic ray transport in AGNs and jets.**
G. M. Webb.
Workshop on Relativistic Hadrons in Cosmic Compact Objects, p. 111 – 119 (1991). – See Abstr. 012.103 for the main entry.
The transport of cosmic rays in astrophysical fluid flows thought to be present in AGN's and jets is discussed in terms of diffusive particle transport equations; pitch angle evolution transport equations; and hydrodynamical equations governing the interaction of the energetic particles with the background flow. As an example of the use of these equations a discussion is given of the transport of energetic charged particles in an accretion flow onto a Schwarzschild black hole.

062.231 **Particle acceleration in relativistic shock waves with oblique magnetic fields in the presence of finite amplitude field perturbations.**
M. Ostrowski.
Workshop on Relativistic Hadrons in Cosmic Compact Objects, p. 121 – 124 (1991). – See Abstr. 012.103 for the main entry.
Non–monotonic changes of the accelerated particle spectral index and a decreasing particle density jump at the shocks are shown to occur in relativistic shock waves with oblique magnetic fields for increasing amplitude of magnetic field perturbations. Very steep particle energy spectra can be produced at such shocks.

062.232 **Influence of second–order Fermi acceleration and relativistic shock waves on nonthermal continuum emission in hot spots.**
W. M. Krülls.
Workshop on Relativistic Hadrons in Cosmic Compact Objects, p. 131 – 134 (1991). – See Abstr. 012.103 for the main entry.
The author has investigated modified descriptions of synchrotron spectra in hot spots beyond the standard model of diffusive shock acceleration. He has considered second order Fermi acceleration in combination with modified shock fronts. In addition the role of relativistic shock waves in hot spots is re–examined. In general it is always possible to get the same shape of synchrotron spectra as in the standard model of diffusive acceleration by pure second–order Fermi acceleration. Deviations of the shape of the distribution functions occur only near the injection momentum or for large Alfvén velocities. Hence it is possible to have combined spectra of first and second–order Fermi acceleration in hot spots. Due to the possibility of a power–law index $\alpha > -0.5$, flat spectra (3C 273) can be explained. In contrast, the relativistic shock waves are not able to generate cutoffs which are steep enough to fit the observed spectra. The author concludes that the shock wave velocities in hot spots seem to be smaller than 0.5c.

062.233 **Collimated relativistic winds driven by electromagnetic forces.**
T. Chiueh, Z.–Y. Li, M. C. Begelman.
Workshop on Relativistic Hadrons in Cosmic Compact Objects, p. 135 – 139 (1991). – See Abstr. 012.103 for the main entry.
The authors report their recent progress on two aspects of electromagnetically driven relativistic winds. First, they show that these winds, as the nonrelativistic counterpart, must always collimate to either current–free paraboloidal jets, or current–carrying cylindrical jets asymptotically. Second, they demonstrate the type of cylindrical jets by seeking a self–similar solution, which is a generalization of the self–similar nonrelativistic winds of Blandford and Payne, and the asymptotic structure of which can be shown to be stable against small–amplitude perturbations.

062.234 **Ultra–relativistic pulsar wind.**
Z.–Y. Li, M. C. Begelman, T. Chiueh.
Workshop on Relativistic Hadrons in Cosmic Compact Objects, p. 141 – 144 (1991). – See Abstr. 012.103 for the main entry.
The authors show that cold, ultra–relativistic hydromagnetic winds can have a kinetic energy flux much larger than the Poynting flux. Even slight deviation of streamlines from a perfect radial geometry can move the fast magnetosonic point to some finite radii comparable to that of the light cylinder. Once the flow becomes supermagnetosonic, divergence of the streamlines can accelerate it via the "magnetic nozzle" effect, thereby converting Poynting flux to kinetic energy. The authors discuss their results in the context of pulsar winds.

062.235 **Maximum energy of semi–infinite magnetic–field configurations.**
P. A. Sturrock.
International SOLAR–A Science Meeting: Flare physics in solar activity maximum 22, p. 224 – 227 (1991). Abstract. The full paper is published in Astrophys. J., Vol. 380, No. 2, p. 655 – 659 (20 Oct 1991). See Abstr. 062.175. – See Abstr. 012.104 for the main entry.

062.236 **Absorption of magnetoacoustic waves in the solar atmosphere with random inhomogeneities of density and magnetic fields.**
M. Kaisig, T. Tajima, M. Ryutova.
International SOLAR–A Science Meeting: Flare physics in solar activity maximum 22, p. 287 – 290 (1991). – See Abstr. 012.104 for the main entry.
Effects of a collection of strong and random inhomogeneities of the magnetic fields, plasma density, and temperature in the solar atmosphere on the propagation of magnetoacoustic waves of arbitrary amplitudes are numerically studied by using a one–dimensional code of ideal magnetohydrodynamics. The authors discuss the relevance to sunspots and magnetized regions of the solar atmosphere.

062.237 **Linear and nonlinear dust drift waves.**
P. K. Shukla, M. Y. Yu, R. Bharuthram.
J. Geophys. Res., Vol. 96, No. A12, p. 21343 – 21346 (1 Dec 1991).
The linear and nonlinear properties of low–frequency motion in inhomogeneous, magnetized, dusty plasmas are investigated. The dynamics of the dust grains are taken into account by means of a fluid model. A new type of low–frequency waves, the dust–drift waves, is shown to exist. The nonlinear coupling equations are derived, and the possibility of propagating vortex structures is discussed. The dust–drift waves may be relevant in astrophysical and cometary plasmas.

062.238 **Kinematic dynamos associated with large scale fluid motions.**
T. Nakajima, M. Kono.
Geophys. Astrophys. Fluid Dyn., Vol. 60, No. 1–4, p. 177 – 209 (Nov 1991). – See Abstr. 012.125 for the main entry.

A standard approach to the kinematic dynamo problem is that pioneered by Bullard and Gellman, which utilizes the toroidal–poloidal separation and spherical harmonic expansion of the magnetic and velocity fields. The authors re–examine the previous kinematic dynamo models, using the computer algebra approach initiated by Kono. They used the distribution of magnetic energies in various harmonics to infer the convergence of the results. They also examined the helicity distributions in these dynamo models, to see if there is any correlation between the helicity and efficiency of dynamo action. A successful dynamo can result both from the cases in which the helicity distributions are symmetric or antisymmetric with respect to the equator. In both cases, it appears that the dynamo action is efficient if the volume integral of helicity over a hemisphere is large.

062.239 **Nonlinear planetary dynamos in a rotating spherical shell. I. Numerical methods.**
C. F. Barenghi, C. A. Jones.
Geophys. Astrophys. Fluid Dyn., Vol. 60, No. 1–4, p. 211 – 243 (Nov 1991). – See Abstr. 012.125 for the main entry.

Results are presented of an investigation of nonlinear planetary dynamo models in the rapid rotation limit for the geometry of a spherical shell. The magnetic induction equation for axisymmetric fields is solved with a prescribed differential rotation and α–effect. The nonlinearity comes from the azimuthal geostrophic flow, whose magnitude is linked to Taylor's constraint as modified by core–mantle viscous coupling in the Ekman boundary layer. The objectives of this work are to determine whether Taylor's constraint is met when the magnetic field is strong and to solve the special numerical stability problems associated with the term which describes the modified Taylor's condition. A stable numerical method based on spectral approximations is presented to solve the general $\alpha^2\omega$–dynamo equations. Evidence of the existence of Taylor's states in the geometry of a spherical shell is demonstrated in both the α^2– and the $\alpha\omega$–limits.

062.240 **Parity coupling in α^2–dynamos.**
R. Hollerbach.
Geophys. Astrophys. Fluid Dyn., Vol. 60, No. 1–4, p. 245 – 260 (Nov 1991). – See Abstr. 012.125 for the main entry.

The modal α^2–dynamo of Hollerbach and Ierley is extended to include both parities, dipolar and quadrupolar. Two choices of α are considered. As before, the solutions approach the Taylor state, and the subsequent equilibration is independent of the viscosity in the asymptotic limit. The first choice of α yields stable steady–state pure–parity solutions of either type, and may yield some insight into the question of parity selection. The second choice of α yields rather complex time–dependent mixed–parity solutions, and demonstrates the extent to which parity coupling can affect the evolution of time–dependent solutions.

062.241 **Magnetic Reynolds number and model–Z geodynamo.**
A. P. Anufriev, I. Cupal.
Geophys. Astrophys. Fluid Dyn., Vol. 60, No. 1–4, p. 261 – 268 (Nov 1991). – See Abstr. 012.125 for the main entry.

The magnetic Reynolds number (MRN) is used in estimating the quantities appearing in the current theory of the nearly symmetric dynamo. In kinematic theory, all expansions can be developed in terms of the MRN, but this seems to be less advantageous in the case of hydromagnetic theory and in particular the model–Z. An attempt has been made to replace the role of the MRN in the nearly symmetric expansion by the amplitudes of the individual quantities, where the amplitudes of the non–axisymmetric quantities are considered to be small compared with those of the azimuthal quantities. The expansions of the theory are valid only when certain assumptions are made about the field amplitude, and the velocity and buoyancy forces, even though the amplitude of the resulting α–effect is unaffected. The α–effect can be altered only by rapid changes in the non–symmetric velocities in the generation region.

062.242 **Impact of large–scale meteoric bodies on a solid surface with splits and breaks.**
V. I. Bergelson, I. V. Nemchinov, T. I. Orlova, V. M. Khazins.
Sol. Syst. Res., Vol. 25, No. 3, p. 256 – 260 (Nov 1991). English translation of Astron. Vestn., Tom 25, No. 3, p. 344 – 349 (1991).

The propagation of a strong shock, caused by an impact of a large meteorite on the Earth is numerically analyzed. Discreteness of the geophysical structure is modelled by inserting thin low–density channels in front of the shock. The shock interaction with these channels leads to disintegration of the front, and formation of conic precursors with an intensive vortex motion inside of them. The applicability of the model for analysis of real shockwave processes in geophysical media is discussed.

062.243 **Nonlinear alpha–omega dynamos in a spherical shell.**
C. F. Barenghi, C. A. Jones, A. M. Soward, D. R. Fearn, M. R. E. Proctor.
Geophys. Astrophys. Fluid Dyn., Vol. 60, No. 1–4, p. 356 – 359 (Nov 1991). Abstract. – See Abstr. 012.125 for the main entry.

062.244 **Accretion disk magnetohydrodynamics and the origin of jets.**
R. D. Blandford.
NATO Advanced Research Workshop on Theory of Accretion Disks, p. 35 – 57 (1989). – See Abstr. 012.121 for the main entry.

Accretion disks are believed to be present in active galatic nuclei, binary X–ray sources, cataclysmic variables and protostars. Disks in all four types of source have been associated with collimated jets or bipolar outflows. It is argued that magnetic extraction of disk angular momentum provides a natural method for launching these jets and that poloidal field that passes through the disk is more likely to be effective than dynamo–generated field. Electromagnetic and hydromagnetic models for jet production are reviewed, emphasising their shortcomings and the dependence on physical conditions in the disk. Magnetic torques may also extract rotational energy from a central black hole. It is conjectured that, in practice, a black hole magnetosphere will be magnetically dominated and practically force–free and that no extra restrictions are imposed on its structure by the application of relativistic MHD theory. It is also argued that, despite recent concerns, the hole will be able to communicate its angular velocity to the magnetosphere and so drive a unipolar inductor.

062.245 **Convection–driven planetary dynamos.**
D. R. Fearn, C. C. Sellar, M. R. E. Proctor, C. F. Barenghi, C. A. Jones, A. M. Soward.
Geophys. Astrophys. Fluid Dyn., Vol. 60, No. 1–4, p. 387 – 388 (Nov 1991). Abstract. – See Abstr. 012.125 for the main entry.

062.246 **Formation of relativistic MHD jets in the magnetosphere of accretion disks.**
M. Camenzind.
NATO Advanced Research Workshop on Theory of Accretion Disks, p. 59 – 63 (1989). – See Abstr. 012.121 for the main entry.

Contents: 1. Magnetized jets in quasars. 2. Relativistic axisymmetric MHD. 3. Magnetized outflows from accretion disks.

062.247 **Thermal convection in differentially rotating systems.**
M. Kropp, F. H. Busse.
Geophys. Astrophys. Fluid Dyn., Vol. 61, No. 1–4, p. 127 – 148 (Dec 1991).

The onset of convection in a cylindrical fluid annulus is analyzed in the case when the cylindrical walls are rotating differentially, a temperature gradient in the radial direction is applied, and the centrifugal force dominates over gravity. The small gap approximation is used and no–slip conditions on the cylindrical walls are assumed. The relevance of the analysis for problems of convection in planetary and stellar atmospheres is briefly discussed.

062.248 **On thermonuclear convection: 1. Shellular instability.**
S. Ghosal, E. A. Spiegel.
Geophys. Astrophys. Fluid Dyn., Vol. 61, No. 1–4, p. 161 – 178 (Dec 1991).
As the sun evolves, a sharp compositional peak of He^3 builds up in the core. Nuclear reactions involving He^3 are very temperature sensitive, as a result, this He^3 layer is susceptible to thermal instability. The small horizontal wavenumber g–modes have large time scales, comparable to the thermal time scale. Using a two–layer model, the authors find that such "shellular modes" are the most unstable. As a result of nuclear heating, these modes may be excited in the solar core in a shallow layer confined to the He^3 zone. A possible effect of such shellular convection on the solar neutrino problem is discussed. The authors discuss the linear theory; the nonlinear effects will be treated in a subsequent paper.

062.249 **Nonsteady MHD jets from magnetized accretion disks – sweeping–magnetic–twist mechanism.**
K. Shibata, Y. Uchida.
NATO Advanced Research Workshop on Theory of Accretion Disks, p. 65 – 69 (1989). – See Abstr. 012.121 for the main entry.
By using two–dimensional axisymmetric MHD numerical simulations, the authors present an MHD mechanism (called sweeping–magnetic–twist mechanism) for the formation of astrophysical jets in which the directed flow are ejected along the rotation axis of an accretion disk. The acceleration of jet is due to the $\mathbf{J} \times \mathbf{B}$ force in the relaxing magnetic twist, which is produced by the interaction of the differentially rotating accretion disk with poloidal magnetic fields. It is shown that the accelerated gas is collimated by the $\mathbf{J} \times \mathbf{B}$ force and forms a supersonic bipolar jet which has a hollow cylindrical shell structure with helical motion in it. Interaction of the MHD jet with an interstellar condensation is also examined. Application to bipolar molecular flows in star forming regions is briefly discussed.

062.250 **Hydromagnetic $\alpha\Omega$–type dynamos with feedback from large scale motions.**
A. Brandenburg, D. Moss, G. Rüdiger, I. Tuominen.
Geophys. Astrophys. Fluid Dyn., Vol. 61, No. 1–4, p. 179 – 198 (Dec 1991).
Nonlinear axisymmetric mean–field $\alpha\Omega$–type dynamos in spherical shells of conducting incompressible fluid are computed, with differential rotation being generated by the Reynolds stress of anisotropic turbulence (Λ–effect). The correlation time of the turbulence is assumed to be short compared with the rotation period. Poloidal and toroidal fields can be in antiphase with equatorward field migration only when the angular velocity increases inwards and towards the poles. The energy of the mean magnetic field generated is usually less than the energy of the turbulent convective motions. The ratio between cycle period and rotational period can reach values of around fifty.

062.251 **The structure of magnetic neutral points in two dimensions.**
N. R. Strachan, E. R. Priest.
Geophys. Astrophys. Fluid Dyn., Vol. 61, No. 1–4, p. 199 – 224 (Dec 1991).
Magnetic neutral points are of central importance in many solar phenomena, such as flares, reconnection and coronal structures. Despite their importance, the mathematical nature of magnetic neutral points has not previously come under detailed mathematical scrutiny. The authors present a series of models for different types of neutral points in two dimensions.

062.252 **Steady flows in magnetic arcades – a class of exact MHD solutions.**
A. de Ville, E. R. Priest.
Geophys. Astrophys. Fluid Dyn., Vol. 61, No. 1–4, p. 225 – 234 (Dec 1991).
A class of exact solutions to the steady, two–dimensional magnetohydrodynamic equations in a cartesian geometry is presented. The method developed by Tsinganos (1981, 1982) is employed. These solutions may model siphon flow in magnetic arcades, the flow being driven by pressure differences between the footpoints. The basic result is that the presence of inertial forces causes the arcade to rise in comparison to the static case. Eventually, the arcade may be caused to erupt by continuing to increase the magnitude of inertial forces.

062.253 **Bounds on the instability of flows in accretion discs and tori.**
S. Kumar.
Geophys. Astrophys. Fluid Dyn., Vol. 61, No. 1–4, p. 235 – 276 (Dec 1991).
Bounds are placed on the growth rates of dynamical instability in shear flows in accretion discs and tori. It is shown that the critical layer lies in the flow even when the boundaries are free for flows in accretion discs, cylinders and isentropic, constant angular momentum tori (for which the relativistic effects are also considered). The central results are the bounds on growth rates in pseudo–barotropic tori, and the plausible destabilization of baroclinic stars and tori by self–gravity. No assumption is made about the length–scales of modes. An interesting but related result concerns the nature of trivial Lagrangian perturbations of toroidal flows and the implied conservation law for circulation.

062.254 **Numerical models of magnetic reconnection.**
M. Scholer.
Geophys. Astrophys. Fluid Dyn., Vol. 62, No. 1–4, p. 51 – 68 (Dec 1991). – See Abstr. 012.126 for the main entry.
This paper contains a review of numerical experiments concerning steady state magnetic reconnection. The Sweet/Parker current sheet model and Petschek's standing slow mode shock wave model for magnetic reconnection are reviewed briefly along with the recently developed unified theory of Priest and Forbes.

062.255 **Planetary magnetism re–visited.**
R. Hide.
Geophys. Astrophys. Fluid Dyn., Vol. 62, No. 1–4, p. 183 – 189 (Dec 1991). – See Abstr. 012.126 for the main entry.
This review paper outlines developments and outstanding questions in the study of self–exciting MHD dynamos, the Earth's core–mantle interface, sub–seismic oscillations of the Earth's core and the nature of the geomagnetic secular variations, and the origin of the magnetic fields of the outer planets.

062.256 **Magnetoconvection.**
N. O. Weiss.
Geophys. Astrophys. Fluid Dyn., Vol. 62, No. 1–4, p. 229 – 247 (Dec 1991). – See Abstr. 012.126 for the main entry.
Cowling investigated the effect of an imposed magnetic field on convection in order to explain the origin of sunspots. After summarizing the classical linear theory of Boussinesq magnetoconvection, this review proceeds to more recent nonlinear results. Thereafter, attention is focused on the astrophysically relevant problem of fully compressible magnetoconvection. Steady two–dimensional nonlinear solutions show two important effects: stratification introduces an asymmetry between rising and falling fluid, while compressibility leads to evacuated magnetic flux sheets. Time–dependent behaviour includes transitions between standing waves and travelling waves, as well as changes in horizontal scale, leading to the development of more complicated spatial structures. Work on three–dimensional models, which is now in progress, will lead to a better understanding of the structure of a sunspot.

062.257 **Parker–Jeans instability of gaseous disks.**
F. Nakamura, T. Hanawa, T. Nakano.
Publ. Astron. Soc. Jpn., Vol. 43, No. 5, p. 685 – 702 (1991).
The magnetohydrodynamical instability of self–gravitating magnetized gaseous disks is investigated by means of a normal–mode analysis. The growth rates and eigenfunctions of unstable perturbations are obtained numerically. The perturbed disks are shown graphically. Self–graviting magnetized gaseous disks, in general, suffer from two kinds of instabilities, the Jeans and Parker types. The former is driven by self–gravity and the latter

by a nonuniform magnetic field. These two kinds of instabilities coexist in the authors' model and have different characteristics. In the equilibrium model disk, the Jeans mode has a larger growth rate than does the Parker mode, when either the magnetic field is weak or the wavelength of the perturbation is long. The Parker instability dominates when the magnetic field is strong and the wavelength is short. The Jeans instability disturbs mainly the high–density region near the midplane; the Parker instability disturbs mainly the low–density region far from the midplane. The growth rate of the Jeans instability is maximum when the wave vector of the perturbation is parallel to the magnetic field. On the other hand, that of the Parker instability is maximum when the wavenumber in the direction perpendicular to the magnetic field is infinitely large.

062.258 **Triggering the collapse of self–gravitating clouds by torsional Alfvén waves.**
A. Habe, Y. Uchida, S. Ikeuchi, R. E. Pudritz.
Publ. Astron. Soc. Jpn., Vol. 43, No. 5, p. 703 – 718 (1991).

Magnetohydrodynamic braking mechanisms of rotation are related to the production of copious fluxes of nonlinear torsional Alfvén waves under a variety of conditions in an interstellar medium and within molecular clouds. The authors noted that these nonlinear torsional Alfvén waves may pinch ambient gas, sometimes leading to the gravitational collapse of gas clouds, and a spin up of the trapped ambient gas during propagation. They proposed this mechanism in order to interpret the observation that the infrared sources showing CO outflows in streamers tend to align along a spinning high–density helical streak–like structure in them, as seen in the case of L1641 in Orion. In the present paper, by adding a toroidal magnetic field the authors can simulate the magnetic pinching process of an isothermal cloud which is stabilized by a poroidal magnetic field, and then study the triggering of a gravitational instability of the cloud. Simulations are made by using a 3D smoothed particle magnetohydrodynamic code for the range of physical parameters and the mass of the clump appropriate for the situation of the present concern.

062.259 **Anisotropic relativistic magnetohydrodynamics and the problem of the sausage constriction in a pinch with a longitudinal magnetic field.**
B. A. Trubnikov, S. K. Zhdanov, V. P. Vlasov.
Sov. J. Plasma Phys., Vol. 17, No. 10, p. 690 – 694 (Oct 1991).
Current Physics Microform No.: 9206X0630.

The equations of ideal relativistic anisotropic magnetohydrodynamics are obtained from purely hydrodynamic considerations using the "magnetic analog of the Lorentz force", and the sausage instability in a plasma pinch with a longitudinal magnetic field and current is considered. The spectrum of the particles expelled from the constrictions is found. It is close to the spectrum of galactic cosmic rays (GCR). This suggests the hypothesis that the "gamma bursts" observed daily from artificial satellites are direct evidence of the presence in space of "broken–off" pinches which are responsible for the energy of the GCR.

062.260 **Hydrodynamic acceleration of particles in pinches with a longitudinal magnetic field.**
B. A. Trubnikov, S. K. Zhdanov, V. P. Vlasov.
Sov. J. Plasma Phys., Vol. 17, No. 10, p. 694 – 701 (Oct 1991).
Current Physics Microform No.: 9206X0634.

This paper is a continuation of the studies of Trubnikov et al. The authors consider a nonrelativistic pinch within a longitudinal magnetic field. See Abstr. 062.259.

062.261 **Non–axisymmetric shear instabilities in thick accretion disks.**
R. Narayan, J. Goodman.
NATO Advanced Research Workshop on Theory of Accretion Disks, p. 231 – 247 (1989). – See Abstr. 012.121 for the main entry.

Recent results on the Papaloizou–Pringle non–axisymmetric instability in thick accretion disks is reviewed. Considerable work has been done on "slender" tori and annuli; these are systems whose radial widths are small compared to their mean radii. The linear instabilities in slender configurations fall into two classes, (1) the principal branch, and (2) higher–order modes. The physical mechanism by which these modes become unstable has been identified. The idealized slender models have been extended in several ways.

062.262 **Images and line profiles from nonadiabatic jet models.**
A. C. Raga, S. Biro, J. Cantó, L. Binette.
Rev. Mex. Astron. Astrofis., Vol. 22, No. 2, p. 243 – 248 (Oct 1991).

The emission line spectrum observed in jet–like Herbig–Haro objects could be excited by internal "crossing shocks" in a nonadiabatic jet flow. The authors explore this possibility by presenting predictions of Hα and [S II] images and position–velocity diagrams (i.e., spatially resolved line profiles) from nonadiabatic, steady jet models. They find that while these models do predict some of the observed characteristics of stellar jets, they fail to reproduce in detail the observational results. This suggests that other excitation mechanisms might also be important in stellar jets.

062.263 **Instabilities of accretion flows caused by the interaction of internal gravity modes.**
W. Glatzel.
NATO Advanced Research Workshop on Theory of Accretion Disks, p. 249 – 258 (1989). – See Abstr. 012.121 for the main entry.

The modal structure of internal gravity modes in shearing systems is investigated using analytical methods. If the buoyancy frequency profile satisfies certain symmetry conditions, two types of instabilities are found, one of which can be explained by over–reflection, the second by mode resonances.

062.264 **Simulations of three–dimensional slender tori.**
J. F. Hawley.
NATO Advanced Research Workshop on Theory of Accretion Disks, p. 259 – 268 (1989). – See Abstr. 012.121 for the main entry.

Some considerations and conclusions that have been obtained from numerical simulations of slender, nonself–gravitating, fluid "tori" orbiting in a fixed central potential. The numerical work on the nonaxisymmetric "Papaloizou–Pringle" instability is summarized. The difficulties and pitfalls inherent in these time–dependent finite differencing calculations are reviewed. Preliminary three–dimensional simulations of slender torus modes are presented.

062.265 **Physics of accretion by spiral shock waves.**
H. C. Spruit.
NATO Advanced Research Workshop on Theory of Accretion Disks, p. 325 – 340 (1989). – See Abstr. 012.121 for the main entry.

The properties of disks accreting by global spiral shocks are reviewed. The discussion includes the formation of these waves, the way they transport angular momentum, a selfsimilar model for spiral shocked accretion and recent numerical simulations. These simulations suggest that at least in disks produced by mass overflow in close binaries the efficiency of accretion (in terms of an effective α–value) can be substantial.

062.266 **Spiral shocks in accretion disks: a preliminary numerical study.**
M. Różyczka, H. C. Spruit.
NATO Advanced Research Workshop on Theory of Accretion Disks, p. 341 – 354 (1989). – See Abstr. 012.121 for the main entry.

The response of accretion disks to tidal forces is studied numerically with the help of a 2–D, second order hydrodynamical code. The code is tested on analytical self–similar models of spiral shocks and proven capable of maintaining them stationary in the grid. The disks are assumed to reside in potential wells of primary components of binary systems. The models are not structured perpendicularly to the orbital plane, and only the flow in the orbital plane is studied. A uniform inflow condition applied to the outer boundary of the disk invariably leads to the formation of two-armed spiral waves resembling the analytical similarity solutions. Provided the disk approaches the Roche lobe, fairly strong shocks are observed even for mass ratios (secondary to primary) as small as 0.01. Stopping the inflow at the outer boundary modifies the spiral pattern and causes the disk to shrink rapidly.

062.267 **Mass transfer by tidally induced spiral shocks in an accretion disc.**
T. Matsuda, N. Sekino, E. Shima, K. Sawada, H. C. Spruit.
NATO Advanced Research Workshop on Theory of Accretion Disks, p. 355 – 371 (1989). – See Abstr. 012.121 for the main entry.
Numerical simulations of two–dimensional adiabatic inviscid flow in an accretion disc in a binary system or a proto–planetary system are performed using a second–order accurate implicit Roe upwind scheme. The purpose is to estimate quantitatively the amount of mass accretion rate due to spiral shocks produced by a less massive component. The mass ratio of the binary system, q, is varied. In the case of $q \geqslant 10^{-2}$, steady spiral shocks persist until the end of the calculations, i.e. about 15 – 20 rotation periods. If q is as small as 10^{-3}, i.e. the case of the Jupiter, the amplitude of the spiral waves is not very strong.

062.268 **Accretion disks and the link between an AGN and its host galaxy.**
M. C. Begelman, J. Frank, I. Shlosman.
NATO Advanced Research Workshop on Theory of Accretion Disks, p. 373 – 386 (1989). – See Abstr. 012.121 for the main entry.
The authors examine the possible role of accretion disks in bringing matter from the interstellar medium into the central regions of an active galactic nucleus (AGN). They outline a unified model for fuelling AGNs, in which the inflow is driven by global nonaxisymmetric gravitational instabilities on large scales, and on small scales forms a mildly self–gravitating disk of clouds in which angular momentum is transported by cloud–cloud collisions.

062.269 **Large–scale accretion flows in AGN.**
J. Frank, I. Shlosman, M. C. Begelman.
NATO Advanced Research Workshop on Theory of Accretion Disks, p. 387 – 395 (1989). – See Abstr. 012.121 for the main entry.
A two stage mechanism for fueling AGN is proposed which makes use of stellar dynamical and gas dynamical instabilities. The first stage adopts different forms depending on the type of host galaxy. In spiral hosts a stellar bar sweeps the interstellar material inward as a consequence of the gas losing angular momentum to the bar. In a second stage common to both types of host, the gaseous disk accumulated in the nuclear region of the galaxy goes bar unstable again and the material flows further in. Some observational and theoretical implications of this picture are discussed.

Numerical modeling in applied physics and astrophysics.
See Abstr. 003.022.

Planetary magnetospheric physics I. Proceedings. Symposium 6 and Topical Meeting of the COSPAR Interdisciplinary Scientific Commission D (Meetings D2, D4) of the COSPAR 28. Plenary Meeting, The Hague (Netherlands), 25 Jun – 6 Jul 1990.
See Abstr. 012.006.

Plasmas in astrophysics. Proceedings. 3. Texas–Mexico Conference on Astrophysics: Plasmas in astrophysics, Houston, TX (USA), 14 – 15 Feb 1991.
See Abstr. 012.034.

Structure and emission properties of accretion disks. Proceedings.
See Abstr. 012.046.

Theoretical problems in space and fusion plasmas. Proceedings.
See Abstr. 012.067.

Advances in solar system magnetohydrodynamics. Proceedings.
See Abstr. 012.086.

Challenges to theories of the structure of moderate–mass stars. Proceedings.
See Abstr. 012.096.

Theory of accretion disks. Proceedings.
See Abstr. 012.121.

Hybrid simulations of rotational discontinuities.
See Abstr. 022.014.

On the structure of rotational discontinuities with large phase angles.
See Abstr. 022.015.

Collisional rates and cooling within atomic hydrogen plasmas.
See Abstr. 022.108.

Theoretical He I line intensities in low–density plasmas.
See Abstr. 022.131.

Irradiation of solids: theory.
See Abstr. 022.161.

Experimental data on the nucleation and growth of solid particles from high–temperature gas mixtures.
See Abstr. 022.168.

Cartan's torsion vector and spin–0 fields
See Abstr. 061.027.

Quantization requires a Lagrangian.
See Abstr. 061.029.

On the transport of polarized radiation in a plasma with ion–acoustic turbulence.
See Abstr. 063.016.

Coherent radiation from energetic electron streams via collisionless bremsstrahlung in strong plasma turbulence.
See Abstr. 063.041.

Radiative transfer in cylindrical plasmas.
See Abstr. 063.084.

Emission at cyclotron harmonics due to coalescence of z–mode waves.
See Abstr. 063.091.

Relativistic electrons in photon fields: effects of triplet pair production on inverse Compton gamma–ray spectra.
See Abstr. 063.105.

On the influence of viscosity on the structure of the boundary layer of accretion disks.
See Abstr. 064.001.

Numerical hydrodynamics of counter flowing binary stellar winds.
See Abstr. 064.006.

Dynamos and angular momentum transport in accretion disks.
See Abstr. 064.009.

Dust formation in winds of long–period variables. I. Equations, method of solution, simple examples.
See Abstr. 064.021.

A powerful local shear instability in weakly magnetized disks. I. Linear analysis.
See Abstr. 064.050.

A powerful local shear instability in weakly magnetized disks. II. Nonlinear evolution.
See Abstr. 064.051.

Analytical studies of collimated winds. II. Topologies of 2–D helicoidal MHD solutions.
See Abstr. 064.053.

Thermosolutal–convective instability of a stellar atmosphere in the presence of suspended particles.
See Abstr. 064.056.

Nonlinear instability of the accretion line.
See Abstr. 064.058.

Thermosolutal–convective instability of a composite stellar atmosphere in the presence of a variable magnetic field.
See Abstr. 064.059.

Thermosolutal–convective instability of a composite plasma in a stellar atmosphere including the effects of variable magnetic field and rotation.
See Abstr. 064.073.

Asymptotic structure of hydromagnetically driven relativistic winds.
See Abstr. 064.075.

Shock waves in accretion disks.
See Abstr. 064.078.

Cosmic–ray–modified stellar winds. IV. The effect of self–generated Alfvén waves.
See Abstr. 064.087.

Turbulence, convection, and stability in accretion disks.
See Abstr. 064.093.

Thermosolutal–convective instability of a composite plasma in a stellar atmosphere including the effects of finite Larmor radius and Hall currents.
See Abstr. 064.118.

Magnetically driven jets and winds.
See Abstr. 064.128.

Dynamo action by internal waves in accretion disks.
See Abstr. 064.135.

Self–consistent description of convective domains in geometrically thin accretion disks.
See Abstr. 064.142.

Radiation hydrodynamics in accretion disks and the boundary layer.
See Abstr. 064.145.

Solving time dependent problems: the brute force method.
See Abstr. 064.153.

On turbulent transport in accretion discs.
See Abstr. 064.155.

Magnetohydrodynamical winds from an accretion disk: driving by the magnetic pressure and tension.
See Abstr. 064.158.

Time evolution of a magnetic seed–field in a turbulent accretion disk.
See Abstr. 064.159.

Two–temperature accretion disks with winds.
See Abstr. 064.161.

Magnetic accretion disks fall into two types.
See Abstr. 064.166.

Accretion driven by spiral waves – an example.
See Abstr. 064.168.

Convective viscosity in thin accretion disks.
See Abstr. 064.169.

A new luminosity limit for thin accretion disks.
See Abstr. 064.170.

Numerical simulations of flares on M dwarf stars. I. Hydrodynamics and coronal X–ray emission.
See Abstr. 064.174.

A model for tidally driven eccentric instabilities in fluid disks.
See Abstr. 064.175.

Simulations of tidally driven eccentric instabilities with application to superhumps.
See Abstr. 064.176.

On turbulent diffusion of magnetic fields and the loss of magnetic flux from stars.
See Abstr. 065.022.

Nonlinear stellar pulsations. I. Numerical methods, basic physics, initial models and first results.
See Abstr. 065.029.

Partially degenerate nonrelativistic isothermal gas spheres in the presence of a magnetic field.
See Abstr. 065.033.

Relativistic magnetic partially degenerate isothermal gas spheres.
See Abstr. 065.034.

Equations of state and bump Cepheids. I. Linear results.
See Abstr. 065.043.

A new model of stellar convection for compressible fluids.
See Abstr. 065.065.

Three–dimensional compressible hydrodynamic convection in the Sun and stars.
See Abstr. 065.105.

Symmetry breaking in stellar dynamos.
See Abstr. 065.113.

Stellar convection.
See Abstr. 065.115.

What is a stellar dynamo?
See Abstr. 065.120.

Convective penetration into stellar radiation zones.
See Abstr. 065.141.

Shock phenomena in a hydrodynamic model of a classical Cepheid.
See Abstr. 065.156.

Galilean limit of electrodynamics.
See Abstr. 066.057.

The Newtonian limit of electromagnetism.
See Abstr. 066.058.

The evolution of discontinuities in radiating spheres in the diffusion approximation.
See Abstr. 066.075.

Magnetostatic stiff–fluid model of cylindrical symmetry in general relativity.
See Abstr. 066.118.

Static spherically–symmetric perfect fluids with pressure equal to the energy density.
See Abstr. 066.163.

On the stability of wind accretion.
See Abstr. 067.039.

Oscillations of fluid disks rotating around compact objects.
See Abstr. 067.053.

Alfvén solitons in a dense ultrarelativistic electron–positron pulsar plasma.
See Abstr. 067.065.

Electrostatic drift vortices in a hot rotating strongly magnetized electron–positron pulsar plasma.
See Abstr. 067.066.

A model for variability of AGN.
See Abstr. 067.083.

Simple equations for general relativistic hydrodynamics in spherical symmetry applied to neutron star collapse.
See Abstr. 067.094.

A mechanism for gamma–ray bursts by Alfvén–wave acceleration in a nonuniform atmosphere.
See Abstr. 067.101.

Magnetohydrodynamic transonic flows around compact objects.
See Abstr. 067.175.

On the instability of spiral flows around compact objects.
See Abstr. 067.176.

Shocks in isothermal transonic flows in black hole geometry.
See Abstr. 067.177.

Inviscid hydromagnetic horizon boundary conditions.
See Abstr. 067.178.

Smoothed particle hydrodynamics and accretion processes around a massive black hole.
See Abstr. 067.192.

The collisional acceleration of relativistic electrons in the central regions of active galactic nuclei.
See Abstr. 067.211.

On the "flip–flop" instability of Bondi–Hoyle accretion flows.
See Abstr. 067.224.

On the possibility of excitating one–armed corrugation waves in relativistic accretion disks.
See Abstr. 067.226.

Cold super–fast magnetohydrodynamical winds from rapid rotators.
See Abstr. 067.227.

Spherical accretion onto black holes: a complete analysis of stationary solutions.
See Abstr. 067.234.

Formation of dense charged bunches in vacuum gaps.
See Abstr. 067.243.

The spectra of relativistic accretion disks.
See Abstr. 067.271.

Granulation: non–adiabatic patterns and shocks.
See Abstr. 071.060.

Loop ensemble models of solar active regions.
See Abstr. 072.060.

Magnetohydrodynamic turbulence dissipation and stochastic proton acceleration in solar flares.
See Abstr. 073.079.

Two–dimensional models of the solar chromosphere. I. The Ca II K line as a diagnostic: 1.5–D radiative transfer.
See Abstr. 073.102.

The 3–min oscillations of the solar chromosphere – a basic physical effect?
See Abstr. 073.104.

Magnetohydrodynamics of an impulsively heated, hard X–ray emitting filament.
See Abstr. 073.112.

The rotational motion in quiescent prominences.
See Abstr. 073.116.

Magnetoacoustic heating of the solar chromosphere.
See Abstr. 073.126.

EUV spectroscopy as a probe of astrophysical plasmas – learning from the Sun.
See Abstr. 073.151.

Subthreshold stochastic diffusion with application to selective acceleration of ^{3}He in solar flares.
See Abstr. 073.160.

Dynamical fine structure of a quiescent prominence.
See Abstr. 073.188.

Plasma flow in slowly diverging coronal structures.
See Abstr. 074.015.

Survival of discrete structures in the solar wind.
See Abstr. 074.065.

The phase mixing of Alfvén waves, coordinated modes, and coronal heating.
See Abstr. 074.071.

A model for the formation of solar prominences.
See Abstr. 074.091.

MHD parameters of the steady–state solar wind at 1 AU as functions of parameters at 18 Rs.
See Abstr. 074.096.

On the spontaneous formation of electric current sheets above a flexible solar photosphere.
See Abstr. 074.102.

Coronal current–sheet formation: the effect of asymmetric and symmetric shears.
See Abstr. 074.110.

Coronal evolution due to shear motion.
See Abstr. 074.116.

Dynamics of solar coronal magnetic fields.
See Abstr. 074.118.

Plasma waves caused by transient heat conduction in a coronal loop and anomalous resistivity.
See Abstr. 074.132.

Stationary siphon flows in thin magnetic flux tubes. II. Radiative heat exchange with the surroundings.
See Abstr. 075.026.

Magnetic flux transport and the Sun's dipole moment: new twists to the Babcock–Leighton model.
See Abstr. 080.014.

Theoretical spectrum of solar convection.
See Abstr. 080.154.

The effects of magnetic stretching in the convective overshoot region.
See Abstr. 080.157.

Preliminary investigations of magnetic field turbulence at a quasi–parallel shock.
See Abstr. 084.028.

MHD theory and the solar wind – magnetosphere interaction.
See Abstr. 084.059.

The Earth's dynamo.
See Abstr. 084.091.

Comparison of 2–dimensional MHD and semiempirical models of interplanetary shock wave propagation.
See Abstr. 106.033.

Shock interactions in the outer heliosphere.
See Abstr. 106.044.

X–ray lines in stellar coronae.
See Abstr. 112.111.

The structure of late–type contact binaries.
See Abstr. 117.135.

Herbig–Haro objects and stellar jets.
See Abstr. 121.009.

Models of spatially resolved recombination regions in H–H objects.
See Abstr. 121.110.

Supernova–remnant shock waves close up.
See Abstr. 125.032.

On the evolution of supernova remnants. II. Two–dimensional calculations of explosions inside pre–existing wind–driven bubbles.
See Abstr. 125.098.

The evolution of a supernova remnant in a strongly magnetized interstellar medium.
See Abstr. 125.111.

Barrel–like supernova remnants.
See Abstr. 125.141.

Non–equilibrium, non–LTE ionization in supernova remnants.
See Abstr. 125.146.

Magnetized supernova remnants with cosmic rays.
See Abstr. 125.149.

Supernova remnants in dense clouds. I. Blast–wave dynamics and X–ray irradiation.
See Abstr. 125.150.

Instability and clumping in SN 1987A.
See Abstr. 125.187.

A simulation of the collapse and fragmentation of cooling molecular clouds.
See Abstr. 131.010.

The dissipation of magnetohydrodynamic turbulence responsible for interstellar scintillation and the heating of the interstellar medium.
See Abstr. 131.068.

Fragmentation of elongated cylindrical clouds. III. Formation of binary and multiple systems.
See Abstr. 131.103.

A hydrodynamical study of fragmenting gas clouds.
See Abstr. 131.183.

Cloud formation by combined instabilities in galactic gas layers: evidence for a Q threshold in the fragmentation of shearing wavelets.
See Abstr. 131.217.

Partially dissociative jump shocks in molecular hydrogen.
See Abstr. 131.221.

Fragmentation of elongated cylindrical clouds. I. Isothermal clouds.
See Abstr. 131.222.

Shocks in dense molecular clouds: OMC–1 revisited.
See Abstr. 131.329.

Hydrodynamic cooling effects in expanding ionized nebulae.
See Abstr. 132.005.

On the dynamical evolution of planetary nebulae.
See Abstr. 134.020.

The hydrodynamics of aspherical planetary nebulae. I. Analytic evaluation of hydrodynamic difference schemes.
See Abstr. 134.041.

Hydrodynamical models of aspherical planetary nebulae.
See Abstr. 134.048.

Interaction of planetary nebulae with the interstellar medium: theory.
See Abstr. 134.051.

Iron L–shell line formation in diverse astrophysical environments.
See Abstr. 142.057.

Signatures of magnetic photon splitting in gamma–ray burst spectra.
See Abstr. 143.025.

A nonlinear density wave model of the spiral structure.
See Abstr. 151.065.

Gravothermal instability of anisotropic self–gravitating gas spheres: singular equilibrium solution.
See Abstr. 151.112.

Tumbling endstates for unstable rotating spheres: implications for rotation in elliptical galaxies.
See Abstr. 151.120.

Cooling of gas behind the explosive shocks and consequent star–formation in the central region of a disk galaxy.
See Abstr. 157.077.

Global magnetic structure in spiral galaxies.
See Abstr. 157.294.

Heavy, overpressured jets and the radio source Hercules A.
See Abstr. 158.186.

Outflows driven by cosmic–ray pressure in broad absorption line QSOs.
See Abstr. 159.173.

Cosmic–ray heating of cooling flows: a critical analysis.
See Abstr. 160.056.

Radio emission in a random magnetic field: radio haloes and the structure of the magnetic field in the Coma cluster.
See Abstr. 160.128.

Hydrodynamic simulations of the intracluster medium.
See Abstr. 160.181.

Equilibrium configurations of superconducting cosmic strings in background fields.
See Abstr. 161.074.

Thermodynamics of the neutral string gas in electromagnetic background.
See Abstr. 161.075.

A hydrodynamic approach to cosmology: texture–seeded cold dark matter and hot dark matter cosmogonies.
See Abstr. 161.634.

063 Radiative Transfer, Scattering

063.001 **Relativistic effects in supernova photospheres.**
P. H. Hauschildt, M. Best, R. Wehrse.
Astron. Astrophys., Vol. 247, No. 1, p. L21 – L24 (Jul 1991). Letter–to–the–editor.

The role of relativistic effects in the radiative transfer equation and the condition of radiative equilibrium is studied for supernova photospheres. It is found that these effects change significantly the equilibrium temperature stratification and the emergent spectra so that in abundance determinations errors up to about one order of magnitude may occur if they are neglected.

063.002 **Bremsstrahlung and energetic electrons in supernovae.**
D. D. Clayton, L.–S. The.
Astrophys. J., Vol. 375, No. 1, p. 221 – 238 (1 Jul 1991).

The authors present explicit Monte Carlo calculations of the rate of Compton scattering and of the spectrum of recoil electrons per gram at discrete depths within models of supernovae. The major objective is to calculate the bremsstrahlung emissivity within supernova interiors. Such bremsstrahlung power must exist in all supernovae that are bolometrically powered by radioactive decay. The total power in bremsstrahlung production is typically 1×10^{-3} of the total radioactive power emitted as ^{56}Co and ^{57}Co gamma rays that initiate these processes. The authors calculate the spectrum of the bremsstrahlung emissivity and show it to be $E^{-1.3}$ for a wide range of supernova models. They also calculate the radial and temporal dependence of that emissivity for two models of SN 1987A and for one model of a Type Ia and of a He core explosion model of a Type Ib. The spectrum emerging from the surface (the luminosity) is evaluated by Monte Carlo techniques. The spectral luminosity of the bremsstrahlung component increases slowly between 1 and 30 keV, even though the emissivity declines with energy within that energy band. The bremsstrahlung luminosity is found to dominate that of the primary scattered gammas for $E < 20$ keV.

063.003 **Compton scattering of polarized radiation in astrophysical objects.**
D. I. Nagirner, Yu. I. Poutanen.
Sov. Phys. – Dokl., Vol. 36, No. 6, p. 411 – 413 (Jun 1991). Current Physics Microform No.: 9202X0761.

063.004 **Stellar occultation of polarized light from circumstellar electrons. II. Flat envelopes viewed at arbitrary inclination.**
G. K. Fox, J. C. Brown.
Astrophys. J., Vol. 375, No. 1, p. 300 – 313 (1 Jul 1991).

The authors extend the treatment given in an earlier paper for the linear polarization produced from a flat circumstellar envelope with a finite–size light source viewed edge–on to the case of arbitrary inclination. They find that for an axisymmetric disk scattering region viewed at low inclination, stellar occultation enhances the net polarization, while for high inclination the polarization is reduced by occultation. The analysis is applied to the polarimetric variations expected from a rotating binary system in which the scattering material consists of a one–dimensional plume superposed on an axisymmetric disk. It is shown that the inclination of the system and the radial structure of the plume can be determined for sufficiently high inclinations ($i > 60°$). The extent to which this method can be generalized to an arbitrarily distributed two–dimensional scattering region is also analyzed via Fourier analysis. It is found that the inclination of the system is largely insensitive to the density distribution when the distribution function is of a simple power–law form.

063.005 **Exact solution of the equation of transfer in a finite exponential atmosphere by the method of Laplace transform and linear singular operators.**
S. Karanjai, T. K. Deb.
Astrophys. Space Sci., Vol. 181, No. 2, p. 267 – 275 (Jul 1991).

The equation which commonly appears in radiative transfer problem in a finite atmosphere having ground reflection according to Lambert's law is considered in this paper. The Planck's function $B_\nu(T)$ is taken in the form $B_\nu(T) = b_0 + b_1 e^{-\beta\tau}$. The exact solution of this equation is obtained for surface quantities in terms of the X–Y equations of Chandrasekhar by the method of Laplace transform and linear singular operators.

063.006 **On Geovenelli's approximate form of $\bar{J}(\tau)$.**
M. Karanjai, S. Karanjai.
Astrophys. Space Sci., Vol. 181, No. 2, p. 277 – 282 (Jul 1991).

The accuracy of the approximate replacement of the frequency mean $\bar{J}(\tau)$ of the total intensity $J_\nu(\tau)$ by $J_c + a(J_0(\tau) - J_c)$ by Geovenelli, where J_c is the total intensity at the continuum, $J_0(\tau)$

is that at the line center and a is independent of depth, is studied. Numerical calculations for strong solar lines show very high values of the residual intensities. A slight change in the approximation is then suggested to get a result quite consistent with observational and existing calculated data.

063.007 **The effects of electron scattering and wind clumping for early emission line stars.**
D. J. Hillier.
Astron. Astrophys., Vol. 247, No. 2, p. 455 – 468 (Jul 1991).
Using two independent codes the author has made a detailed study of the influence of electron scattering on line profiles arising from stellar winds around early emission line stars. These calculations indicate that the assumption of isotropic scattering gives acceptable answers when compared with more accurate calculations using a dipole phase function. On the other hand the use of angle–averaged redistribution functions can noticeably affect the strength of the predicted electron scattering wings. The author shows that the velocity (arising from thermal and turbulent motions) adopted for the Doppler line absorption profile can dramatically influence the strength of electron scattering wings for a line that is optically thick throughout most of its formation region. He has examined a crude Wolf–Rayet (W–R) stellar wind model with a density structure stratified in spherical shells in order to investigate the influence of density inhomogeneities on predicted line profiles, and hence on the inferred properties of W–R stars. Line profiles and line strengths in reasonable accord with observation can be reproduced, but for a lower mean mass loss rate. As a consequence profiles can be produced that have the correct line strength but with a significant reduction in the strength of the electron scattering wing. This may help to explain evidence that theoretical electron scattering wings are stronger than those observed. If correct, mass loss rates for W–R stars may have been overestimated.

063.008 **Scattering and polarization of light by rough and porous interstellar grains.**
J.–M. Perrin, J.–P. Sivan.
Astron. Astrophys., Vol. 247, No. 2, p. 497 – 504 (Jul 1991).
Interaction of light with rough or porous interstellar grains is studied. Scattering diagrams and polarization curves are presented for several roughness and porosity rates. Obsidian and glassy carbon particles are considered for size parameters either much smaller than or of the same order as the wavelength of the incident light. In the last case scattering diagrams and polarization curves are drastically affected by roughness and porosity. Implications on the structure of some reflection nebulae are discussed. Also, the results suggest the existence of interstellar grains larger than usually accepted.

063.009 **The ultraviolet extinction by hollow spherical particles of graphite.**
T. Inagaki, Y. Yamamoto, S. Yabushita.
Astrophys. Space Sci., Vol. 182, No. 1, p. 75 – 80 (Aug 1991).
Using the discrete dipole approximation, the authors have calculated the extinction efficiency of hollow spherical particles of graphite as a possible constituent of interstellar grains. The particles had a shell structure with the basal plane perpendicular to the radius. The calculations were made on the particles having the outer radius $R_0 = 10$ and 5 nm in the wave number region from 0.8 to 8.0 μm^{-1} using the anisotropic optical constants. It was found that the hollow particles with the inner radius $R_1 \cong 0.65\ R_0$ yield an extinction feature at 4.6 μm^{-1}, which fits fairly well to one observed in the interstellar extinction.

063.010 **Generation of gamma–rays in electron–photon cascades. I. Monoenergetic photon field.**
I. P. Ivanenko, V. V. Sizov.
Astrophys. Space Sci., Vol. 182, No. 1, p. 111 – 125 (Aug 1991).
The authors have studied an electron–photon cascade in an isotropic photon field as one of the possible gamma–ray generation mechanisms in compact sources. A cascade in a monoenergetic photon field and in a field of the power–law spectrum is considered. Two types of interaction of cascade particles with the photon field are taken into account: the inverse Compton scattering of electrons and the photoproduction of electron–positron pairs by gamma–quanta. The cascade characteristics have been obtained by numerical solution of the kinetic equations. A quantitative and qualitative analysis is made of the main regularities of the cascade. The authors have established simple relations between the spectra of cascade particles and the parameters of the phonon field and primary radiation. The magnetic field effect on the cascade characteristics is estimated. The results obtained are compared with the results of calculation of other investigators.

063.011 **Generation of gamma–rays in electron–photon cascades. II. Photon field of the power–law spectrum.**
I. P. Ivanenko, V. V. Sizov.
Astrophys. Space Sci., Vol. 182, No. 1, p. 127 – 140 (Aug 1991).
The authors discuss the results of calculation of electron–photon cascades in a photon field of the power–law spectrum, which are used to explain the feature of gamma–ray spectra of compact sources. The magnetic field effect on the cascade characteristics is discussed.

063.012 **On the long–wavelength absorption by interstellar dust.**
E. L. Wright.
Astrophys. J., Vol. 375, No. 2, p. 608 – 610 (10 Jul 1991).
Two proposed modifications to the standard theory for the long–wavelength absorption by interstellar dust grains are shown to be incorrect. The Mie theory for spheres or cylinders, or the discrete dipole approximation for irregular shapes, are based on Maxwell's equations for the electromagnetic field and should be used instead of the modified formula proposed by Duley, Jones, and Williams. The caluclated absorption cross sections should not be corrected for stimulated emission, since enhanced absorption from excited states exactly cancels the negative absorptions from stimulated emission.

063.013 **A non–LTE radiation transport model with complete zone–to–zone coupling.**
J. J. MacFarlane, M. Wolff.
Bull. Am. Astron. Soc., Vol. 23, No. 2, p. 968 (1991). Abstract. – See Abstr. 012.008 for the main entry.

063.014 **Near–infrared emission of neutral carbon from photon–dominated regions.**
V. Escalante, A. Sternberg, A. Dalgarno.
Astrophys. J., Vol. 375, No. 2, p. 630 – 634 (10 Jul 1991).
Detailed calculations are reported of the intensities of the near–infrared lines of neutral carbon atoms at $\lambda = 985.0$, 982.3, and 872.7 nm emitted from dense clouds subjected to intense radiation fields. There is a small contribution to the excitation of the metastable levels by electron impacts in the heated edge zones of the clouds, but it is insignificant compared to that from the recombination of the C^+ ions produced by photoionization. The lines observed in M42 and NGC 2024 can be interpreted as arising in gas with densities in excess of $10^5 cm^{-3}$ and radiation fields with intensities between 10^3 and 10^6 times the average interstellar field intensity. Radiative recombination of C^+ ions may also be an important source of the emission lines detected in the planetary nebulae NGC 6270 and NGC 7027.

063.015 **Line formation in low–mass dense core: Larson–Penston flow vs. inside–out collapse.**
Zhou Shu–dong.
Bull. Am. Astron. Soc., Vol. 23, No. 2, p. 976 (1991). Abstract. – See Abstr. 012.008 for the main entry.

063.016 **On the transport of polarized radiation in a plasma with ion–acoustic turbulence.**
I. M. Ojringel', V. V. Tirskij.
Astron. Tsirk., No. 1549, p. 1 – 2 (May–Aug 1991). In Russian.
The role of decay processes in the transport of polarized radiation is discussed as well as the influence of plasma effects upon its characteristics.

063.017 **Radiation of a spherical nebula due to a central star.**
A. K. Kolesov, V. V. Sobolev.
Astrophysics, Vol. 33, No. 2, p. 433 – 438 (Sep – Oct 1990). English translation of Astrofizika, Tom 33, Vyp. 2, p. 235 – 242 (Oct 1990).
The diffusion of radiation in a homogeneous spherical nebula of optical radius x under the influence of a central star is considered. An integro-differential equation that directly determines the intensity of the diffuse radiation emerging from the nebula is given. This equation is used to obtain asymptotic expressions, at $x \gg 1$, for the radiation intensity from the nebula and for the nebula's luminosity. Here, an asymptotic expression for the reflection coefficient of the nebula found earlier by the authors is used. The luminosities calculated in accordance with the asymptotic expression are compared with exact values. This comparison shows that the asymptotic expressions obtained in the present paper give good results already at not very large values of x.

063.018 **Partial redistribution modeling of the Ca II K line. Numerical method and solar applications.**
H. Uitenbroek.
Bull. Am. Astron. Soc., Vol. 23, No. 2, p. 1047 (1991). Abstract. – See Abstr. 012.012 for the main entry.

063.019 **A direct numerical approach to the Chandrasekhar's *H*-functions for arbitrary characteristic functions.**
K. Kawabata, T. Satoh, S. Ueno.
Astrophys. Space Sci., Vol. 182, No. 2, p. 249 – 260 (Aug 1991).
The purpose of this paper is to present a numerical technique to directly compute the Chandrasekhar's $H(\mu)$-function for anisotropic scattering in terms of the roots of the characteristic equation as well as the quadrature points of a certain degree n employed to approximate the definite integral involved in the basic equation. It is shown that the results obtained for some representative $H(\mu)$-functions using the present technique with relatively low-order formula (e.g., $n = 7$) are sufficiently accurate for all practical purposes.

063.020 **Stochastic acceleration of electrons in the presence of Coulomb collisions and synchrotron emissions.**
B. T. Park, V. Petrosian, R. J. Hamilton.
Bull. Am. Astron. Soc., Vol. 23, No. 2, p. 1064 (1991). Abstract. – See Abstr. 012.012 for the main entry.

063.021 **X-ray halos and the size distribution of the cosmic grains.**
S. Klose.
Astron. Astrophys., Vol. 248, No. 2, p. 624 – 632 (Aug 1991).
Arising as a result of scattering by interstellar particles detectable halos can surround celestial X-ray sources. Using van-de-Hulst's formulation of the theory of anomalous diffraction the author has constructed intensity profiles of X-ray halos for different size distributions of the grain population. The numerical calculations demonstrate that X-ray halos could be a tracer of the size distribution of the cosmic grains.

063.022 **X-ray halos as diagnostics of interstellar grains.**
J. S. Mathis, C.-W. Lee.
Astrophys. J., Vol. 376, No. 2, p. 490 – 499 (1 Aug 1991).
X-ray halos around point sources provide information regarding intestellar grains. The scattering is significantly affected by (1) the energy of the radiation; (2) the optical depth of the scattering, because of the effects of multiple scattering; (3) whether or not the grains are porous or fluffy; (4) the spatial distribution of dust along the line of sight. The scattering is not strongly affected by the amount of very small grains (or large molecules) present, by the differences among the various grain models which employ compact (nonporous) grains, or by the chemical composition of the grains if there are no absorption edges resolved by the observations. Multiple scattering for optical depths $\tau_{sca} \approx 1$ strongly affects the outer wings, dominating the total flux for $\tau_{sca} > 1.3$. The halo is also strongly affected by the distribution of dust along the line of sight, especially if the source lies within a zone with appreciably lower than average dust density. It is shown how the effects of such a distribution can be separated from the effects of multiple scattering and varying grain-size distributions. As an example, the Ginga observations of an occultation of the X-ray source GS 1741.2–2859 near the Galactic center by the Moon are analyzed.

063.023 **Emergent radiation from the atmosphere bounded by the two halves of the Lambert surface with different albedos.**
T. Takashima, K. Masuda.
Astrophys. Space Sci., Vol. 183, No. 2, p. 225 – 236 (Sep 1991).
For the evaluation of the effect of the nonuniform surface albedo to the emergent radiation from the atmosphere, the emergent radiation from the atmosphere bounded by the two-halves of the Lambert surface with different albedos is computed. The principal plane is assumed to be perpendicular to the boundary of surfaces. The atmosphere is assumed to be homogeneous, which is composed of aerosol, molecules, and adsorbent gases. Their optical thicknesses are 0.25, 0.23, and 0.02, respectively. The model aerosol is of the oceanic and water soluble types. In the computational procedure, the emergent radiation is approximated by the contributions due to the multiple scattering in the atmosphere, directly attenuated radiation, and radiation due to single scattering in the atmosphere which is reflected by the Lambert surface (up to 4 interactive radiative modes between atmosphere and surface). For quantitative analysis, results are compared with those of the atmosphere-uniform surface model, where the multiple scattering is considered.

063.024 **Cosmic-ray propagation in compressible turbulent media.**
M. E. Katz (*M. E. Kats*), M. Stehlík.
Astrophys. Space Sci., Vol. 183, No. 2, p. 259 – 271 (Sep 1991).
The influence of compressibility of media on both the statistical acceleration and the turbulent diffusion of cosmic-ray particles is investigated. The averaging over an ensemble of random velocity fields of the medium was performed in the kinetic equation. The kinetic coefficients, which are responsible for the particle acceleration, were obtained in the cases of weak and strong scattering due inhomogeneous magnetic fields.

063.025 **Solution of the equation of transfer for interlocked multiplets with Planck function as a nonlinear function of optical depth.**
S. Karanjai, T. K. Deb.
Astrophys. Space Sci., Vol. 184, No. 1, p. 57 – 63 (Oct 1991).
The equation of transfer for interlocked multiplets has been solved exactly by the method used by Busbridge and Stibbs (1954) for exponential form the Planck function $B_\nu(T) = b_0 + b_1 e^{-\beta\tau}$.

063.026 **Laser radiation in active amplifying media treated as a transport problem: transfer equation derived and exactly solved.**
S. Das Gupta, S. R. Das Gupta.
Astrophys. Space Sci., Vol. 184, No. 1, p. 77 – 142 (Oct 1991).
The flow of laser radiation in a plane-parallel cylindrical slab of active amplifying medium with axial symmetry is treated as a problem in radiative transfer. The appropriate one-dimensional transfer equation describing the transfer of laser radiation has been derived by an appeal to Einstein's *A*, *B* coefficients (describing the processes of stimulated line absorption, spontaneous line emission, and stimulated line emission sustained by population inversion in the medium) and considering the "rate equations" to completely establish the rational of the transfer equation obtained. The equation is then exactly solved and the angular distribution of the emergent laser beam intensity is obtained; its numerically computed values are given in tables and plotted in graphs showing the nature of peaks of the emerging laser beam intensity about the axis of the laser cylinder.

063.027 **Energy deposition by X–ray photoelectrons into interstellar molecular clouds.**
G. M. Voit.
Astrophys. J., Vol. 377, No. 1, p. 158 – 170 (10 Aug 1991).
In certain astrophysical systems, high–energy photoelectrons will deposit their kinetic energy into molecular gas, exciting numerous molecular and atomic levels. The author considers the deposition of energy by 100 – 10,000 eV electrons into molecular hydrogen and helium gas of cosmic abundances ($n_{He} = 0.2 \cdot n[H_2]$) and fractional ionization ranging from 10^{-12} to 10^{-1}. In order to account for energy deposition by both discrete and continuous degradation processes, the author introduces the loss spectrum, a form of the Spencer–Fano degradation spectrum particularly suited to describing electron energy deposition in a partially ionized gas. Rotational and vibrational excitations of H_2 are included, as well as electronic excitations and ionizations of H_2 and He^0. The calculated excitation and ionization yields drop as the fractional ionization increases. The dependences of these yields on fractional ionization are fitted to a form suitable for swift numerical computation.

063.028 **The polarization of radiation reflected diffusely by an inhomogeneous plane medium.**
M. T. Attia, E. M. Abulwafa.
Astrophys. Space Sci., Vol. 184, No. 2, p. 247 – 259 (Oct 1991).
The problem of determining the intensity and the degree of polarization of radiation emerging from an inhomogeneous finite plane medium for the case of Rayleigh scattering with internal energy source is considered. A system of coupled integral equations is obtained and solved by the Galerkin method. The degrees of polarization for homogeneous and inhomogeneous media are calculated for uniform and nonuniform sources.

063.029 **Polycyclic aromatic hydrocarbon optical properties and contribution to the acceleration of stellar outflows.**
I. Cherchneff, J. R. Barker, A. G. G. M. Tielens.
Astrophys. J., Vol. 377, No. 2, p. 541 – 552 (20 Aug 1991).
The optical constants of four polycyclic hydrocarbon (PAH) molecules (benzene, pyrene, pentacene, and coronene) are determined form their measured laboratory absorption spectra. The Planck mean of the radiation pressure cross section is computed for each molecule and for amorphous carbon (AC) grains, and semiempirically estimated for large PAH molecules up to 400 carbon atoms. Assuming that PAHs are present in carbon–rich stellar outflows, the radiation pressure forces acting on them are calculated and compared with the radiation forces on AC particles. The results show that PAHs possess very different optical properties from AC grains. Small PAHs may experience an "inverse greenhouse" effect in the inner part of the envelope, as they decouple from the gas close to the photosphere. The radiation pressure force on PAHs is always much less than the force at work on AC grains, and PAH molecules do not affect significantly the dynamics of the outflow.

063.030 **Continuum opacity produced by spectral lines in supernovae and similar expansions.**
R. V. Wagoner, C. A. Perez, M. Vasu.
Astrophys. J., Vol. 377, No. 2, p. 639 – 646 (20 Aug 1991).
Within rapidly expanding objects such as supernovae, the large velocity gradient Doppler – spreads the spectral lines to such an extent that even relatively weak transitions affect the propagation of radiation. It is shown here that under appropriate conditions (involving the separation between lines and the velocity gradient), the effects of many lines within a small frequency interval produce a contribution to the transfer equation which acts like an ordinary continuum opacity. This expansion opacity is a surprisingly simple function of the Sobolev optical depth of the lines. The authors evaluate the dependence of this opacity on frequency, temperature, density, velocity gradient, and heavy element abundance. The properties of 426,319 of the transitions compiled by Kurucz were used, and local thermodynamic equilibrium has been assumed. The results are compared with the earlier approximate analysis of Karp, Lasher, Chan, and Salpeter. An expression for the absorptive part of the opacity is also derived.

063.031 **Theoretical H II region models: the effects of stellar atmosphere models.**
I. N. Evans.
Astrophys. J., Suppl. Ser., Vol. 76, No. 3, p. 985 – 999 (Jul 1991).
Photoionization modeling has been used to compute several grids of theoretical H II region models in order to assess the degree to which the choice of the ionizing stellar atmosphere model affects the calibration of the emission–line diagnostic diagrams of Evans and Dopita and the semiempirical H II region abundance sequence calibration of Dopita and Evans. Emission–line diagnostic diagrams are presented and compared for model nebulae ionized by Hummer and Mihalas unblanketed LTE atmospheres, Kurucz line–blanketed LTE atmospheres, Mihalas unblanketed non–LTE atmospheres, and a truncated blackbody spectrum. The models demonstrate that for solar nebular and atmospheric abundances, there are only minor differences between H II region models ionized by the Hummer and Mihalas atmospheres, and models ionized by the Kurucz atmospheres. The unblanketed non–LTE stellar atmosphere models of Mihalas and truncated blackbody spectra are shown to be unsuitable for general H II region modeling. Changing the atmospheric metallicity relative to the nebular metallicity has a significant effect on predicted emission–line ratios for model nebulae ionized by the Hummer and Mihalas atmospheres, but a considerably smaller effect for models ionized by the Kurucz atmospheres.

063.032 **The symbiotic stellar lines at 6830 Å, 7088 Å; an origin in optically thick fluorescence.**
S. O. Kastner.
Astrophys. Space Sci., Vol. 185, No. 2, p. 265 – 275 (Nov 1991).
The question of the origin of the symbiotic stellar lines at 6830 and 7088 Å, recently proposed to arise from Raman scattering of ultraviolet O VI lines by hydrogen, is examined quantitatively by comparing line intensities expected from this process with the intensity of an Fe II line photoexcited by one of the same O VI lines. It appears that Raman scattering does not yield sufficient intensity to account for the symbiotic lines. An alternative mechanism for producing the lines exists, however, in off–line–center photoexcitation of optically thick hydrogen by the O VI lines. This gives intensities which are higher than Raman intensities, and which are more consistent with the Fe II fluorescent line intensity. Observed further redward shifts are also explained by the mechanism. It is therefore suggested that the proposed hypothesis of O VI–H I interaction may be valid, but that the process involved is probably not second–order Raman scattering but first–order fluorescent scattering.

063.033 **Total and effective radiative recombination coefficients.**
D. Péquignot, P. Petitjean, C. Boisson.
Astron. Astrophys., Vol. 251, No. 2, p. 680 – 688 (Nov 1991).
Total radiative recombination coefficient for all H, He, C, N, O, Ne ions and effective recombination coefficients for important optical and UV transitions of H, He, C, N, O ions are presented in the form of four parameter fits valid at low electron density and over a factor $\sim 10^3$ in $t = 10^{-4} T/z^2$, where T is the electron temperature and z the ionic charge.

063.034 **Stochastic simulation of fast particle diffusive transport.**
A. L. MacKinnon, I. J. D. Craig.
Astron. Astrophys., Vol. 251, No. 2, p. 693 – 699 (Nov 1991).
The authors address the solution of particle transport problems described by Fokker–Planck equations. It is pointed out that such equations are formally equivalent to coupled systems of stochastic ordinary differential equations which can be resolved very efficiently in computational form. Stochastic simulations are presented for the illustrative case of collision–dominated electron transport, and compared with analytical solutions obtained in the mean scattering approximation, and by spatial integration of the governing equation. As expected, pitch–angle dispersion smooths the spatial form of the electron energy deposition rate,

but more critically, the spatial form of the energy deposition is shown to be very sensitive to the initial conditions.

063.035 **A novel mechanism for nonthermal X–ray emission.**
K. Mannheim, W. M. Krülls, P. L. Biermann.
Astron. Astrophys., Vol. 251, No. 2, p. 723 – 731 (Nov 1991).

Some very interesting consequences from the theory of particle acceleration in extragalactic radio sources have been overlooked in the literature. Diffusive shock acceleration theory predicts both accelerated electrons and protons. Electrons cool via synchrotron radiation explaining the radio to optical continuum from low density hot spots in extragalactic radio jets. Relativistic protons immersed in this soft photon field produce extremely energetic photomesons (mainly pions) at a considerable rate. Moreover, the resulting photons from the pion–decay are degraded down to observable X– and γ–rays via an electromagnetic cascade. Introducing relativistic protons in equipartition with the magnetic field does not significantly alter the net–energetics of the jet. Here the authors present a case study of this natural mechanism for a sample of hot spots, viz. high surface–brightness regions at the working end of jets from Fanaroff–Riley type II galaxies or similar knots in jets, where diffusive shock acceleration has been shown to be consistent with detailed observations. It is shown that proton initiated cascade emission can dominate over synchrotron–self–Compton emission and bremsstrahlung even at X–ray frequencies. In the ROSAT–band the effective spectral index of the total spectrum is $\alpha_{keV} = 0.7(+0.2,-0.1)$. Contributions to the γ–ray background emission at a few MeV and close to 100 TeV could be important, if hot spots are representative of knots in extragalactic jets.

063.036 **Polarization of light and models of the circumstellar environment of young stellar objects.**
P. Bastien.
NATO Advanced Study Institute on The Physics of Star Formation and Early Stellar Evolution, p. 709 – 736 (1991). – See Abstr. 012.042 for the main entry.

This paper shows that polarization is the key to understanding important information about young stellar objects (YSOs) and their circumstellar disks. First, the properties of polarized of light are reviewed with particular concern about astronomical observations. Then two polarization mechanisms of interest for the study of star forming regions are discussed in some detail: dichroic extinction and scattering of light. Finally, these mechanisms are discussed in the context of models of YSOs. Parameters of disks around YSOs can now be deduced from linear polarization maps. The distribution of the circumstellar material around YSOs is discussed and the various theoretical density distributions which have been proposed are reviewed.

063.037 **Effects of triplet pair production on ultrarelativistic electrons in a soft photon field.**
C. D. Dermer, R. Schlickeiser.
Astron. Astrophys., Vol. 252, No. 1, p. 414 – 420 (Dec 1991).

An approximate formula is derived for the rate at which electrons lose energy through the triplet pair production process ($e\gamma \rightarrow ee^+e^-$). This expression is compared with the exact numerical results of Mastichiadis, and with energy–loss rates associated with Compton scattering, synchrotron emission, bremsstrahlung, and curvature radiation. This process is unlikely to be important in the formation of the spectrum of high energy cosmic–ray electrons, but could be important in models of ultra–high energy γ–ray sources, or in pair–cascade calculations in the magnetospheres of neutron stars.

063.038 **Rotation of the polarization plane near the inversion point for light scattered on rough surfaces.**
V. S. Degtyarev, L. O. Kolokolova, A. V. Morozhenko.
Astron. Tsirk., No. 1545, p. 35 – 36 (Aug – Sep 1990). In Russian.

The laboratory research of light scattered at different natural and man–made surfaces shows with confidence, that near the inversion point the rotation of the polarization plane occurs.

063.039 **The O VII soft X–ray spectrum and its application to hot plasmas in astrophysics.**
A. H. Gabriel, F. Bely–Dubau, P. Faucher, L. W. Acton.
Astrophys. J., Vol. 378, No. 1, p. 438 – 444 (1 Sep 1991).

The authors present a revised theory and atomic model for the line intensities emitted by O VII, taking into account all of the processes responsible for the emission. This is used to provide a revision of the density measurements made for solar active regions and during flares, as well as in an attempt to understand the spectrum of the Puppis A supernova remnant. In order to explain the strange intensity ratios observed from Puppis A, previous authors have proposed an interpretation based upon a high temperature thermal plasma ($> 5 \times 10^6$K), in a nonequilibrium ionization state. The authors present here an alternative model, based upon the assumed presence of a proportion of fast, nonthermal electrons imbedded in an otherwise thermal plasma at a temperature below 1×10^6K. This can adequately explain the observations without the necessity of invoking departures from the ionization balance.

063.040 **Pumping the interstellar (6, 3) ammonia maser.**
R. D. Brown, D. M. Cragg.
Astrophys. J., Vol. 378, No. 1, p. 445 – 454 (1 Sep 1991).

A numerical model of the excitation of ammonia in dense interstellar clouds predicts strong maser emission in various inversion doublet lines when infrared pumping to the lowest vibrationally excited state is included. Specifically the (6, 3) transition mases, while the (5, 3) and (4, 3) transitions are in normal emission, as observed in W51. Model conditions are a H_2 density between 10^8 and 10^{10}cm^{-3}, a NH_3 column density 10^{18}cm^{-2}, and infrared temperature greater than 200K, comparable with conditions required for OH and H_2O masing in regions of star formation. The intensity of the maser emission depends on the infrared field, suggesting that time variations may be due to fluctuations in an infrared source. Detailed modeling of the other ammonia maser lines awaits the calculation of more collision cross sections.

063.041 **Coherent radiation from energetic electron streams via collisionless bremsstrahlung in strong plasma turbulence.**
J. C. Weatherall, G. Benford.
Astrophys. J., Vol. 378, No. 2, p. 543 – 549 (10 Sep 1991).

Radiation produced by the scattering of energetic electron streams in regions of plasma turbulence can be more efficient than synchrotron emission. Plasma instability occurs naturally due to counterstreaming plasmas in the environment of galactic jets, and saturates in a turbulent state of intense localized regions of electrostatic field. Energetic electrons accelerate while traversing soliton–like electric field structures. Their radiation is extremely broad–band in frequency, a bremsstrahlung type spectrum in which the minimum impact parameter is the scale size of the soliton, assumed to be many Debye lengths. Radiation may extend many orders of magnitude higher in frequency than the plasma frequency, and is relativistically beamed. For a stream of electrons, the emission is partially coherent if the beam is bunched. A specific model for beam density fluctuation statistics yields a power–law radiation spectrum. Turbulent radiation in some parameter regimes can equal or exceed synchrotron emission. Such radiation may therefore significantly reduce the overall energy requirements for central engines of galactic radio sources.

063.042 **Recent advances in computational methods.**
G. B. Rybicki.
NATO Advanced Research Workshop on Stellar Atmospheres: Beyond Classical Models, p. 1 – 8 (1991). – See Abstr. 012.056 for the main entry.

This review focuses primarily on what is probably the most important development in computational methods in recent years, namely, the so–called Accelerated Lambda Iteration (ALI) methods. The roots and development of these methods are traced, and the major variants are discussed.

063.043 **Acceleration of convergence.**
L. Auer.
NATO Advanced Research Workshop on Stellar Atmospheres: Beyond Classical Models, p. 9 – 17 (1991). – See Abstr. 012.056 for the main entry.

Acceleration of the convergence of approximate operator iteration schemes is discussed. Algorithms based on both residual minimization and use of conjugate vector spaces are presented. It is shown that both give dramatic improvement, at very low computational cost, in the iterative solution of radiative transfer problems in the presence of scattering.

063.044 **Fast solution of radiative transfer problems with a multi–grid method.**
O. Steiner.
NATO Advanced Research Workshop on Stellar Atmospheres: Beyond Classical Models, p. 19 – 27 (1991). – See Abstr. 012.056 for the main entry.

A multi–grid method is used for the fast solution of two basic and simplistic radiative transfer problems: the line transfer of two–level atoms in a homogeneous slab, and a grey 2–D LTE model atmosphere in radiative equilibrium. The multi–grid method (W–cycle) is a combination of coarse grid corrections and smoothing steps, the latter being accelerated Λ–iterations. In contrast to classical acceleration schemes the present method makes use of the actual (approximate) solution of the problem, however only on a coarse grid on which computational costs are low. The new method preserves all the advantages of approximate operator techniques but considerably improves convergence. While the multi–grid line transfer calculations are faster by a factor of between 1 and more than 20 with respect to the CPU time needed by an accelerated Λ–iteration based on a non–local approximate operator, the 2–D slab problem shows speedup factors as high as 40. Grid refinement generally improves the convergence rate.

063.045 **Line blanketing without LTE: simple and complex spectra.**
L. Anderson.
NATO Advanced Research Workshop on Stellar Atmospheres: Beyond Classical Models, p. 29 – 38 (1991). – See Abstr. 012.056 for the main entry.

This paper reviews a number of numerical techniques for computing static stellar atmospheres and emergent flux distributions accounting for the blanketing and blending effects of all atomic transitions in full statistical equilibrium. The framework for calculation is the complete linearization of angle and frequency moments of the transfer equations; the so–called multi–frequency/multi–gray approach. The author discusses the algorithms, setting up the problem, and computational strategies.

063.046 **Global and local methods for 1–D problems; implementation aspects and CPU–time and memory scalings.**
M. Carlsson.
NATO Advanced Research Workshop on Stellar Atmospheres: Beyond Classical Models, p. 39 – 48 (1991). – See Abstr. 012.056 for the main entry.

The author describes the basic properties of the global Scharmer approximate operator applied to one–dimensional multi–level non–LTE radiative transfer problems. The approach takes fully into account the global nature of radiative transfer at the expense of relatively large requirements of computer storage. Special emphasis is given to implementation aspects. This global method is compared with an accelerated lambda iteration method using a local operator with Ng acceleration. A number of real applications are used to find typical convergence rates and the scaling of computer CPU–time and memory requirements. It is found that the global operator has advantages up to problems with 30 – 40 energy levels. For larger problems, the local operator is faster by up to a factor of two and the global operator also quickly becomes prohibitive in its memory requirement. With these methods it is possible to solve a 20–level problem in one minute and a 100–level problem in 30 minutes on a modern workstation.

063.047 **2–D axisymmetric line transport.**
J. I. Castor, P. G. Dykema, R. I. Klein.
NATO Advanced Research Workshop on Stellar Atmospheres: Beyond Classical Models, p. 49 – 59 (1991). – See Abstr. 012.056 for the main entry.

The methods used in the ALTAIR code for computing the transfer of spectral line radiation in two–dimensional axially–symmetric geometry are described. ALTAIR uses a variable–Eddington–tensor approach, in which the transfer equation of non–coherent line scattering is written in moment form, and the moments are closed with an assumed tensor relating the monochromatic pressure tensor and energy density; this Eddington tensor is obtained self–consistently using an accurate angle–dependent solution of the transfer equation. The finite element method for solving the moment system, and the discontinuous finite element method for solving the S_n equation of transfer are described. Two applications of the method are discussed: line formation in uniform cylinders with different length–diameter ratios, and monochromatic transfer on an irregular x–y mesh (The Mordant test problem).

063.048 **NLTE spectral line formation in three dimensions.**
Å. Nordlund.
NATO Advanced Research Workshop on Stellar Atmospheres: Beyond Classical Models, p. 61 – 68 (1991). – See Abstr. 012.056 for the main entry.

The solution of non–LTE line transfer problems in three spatial dimensions is discussed. Using a two–level atom model, the formation of Fe I lines in the solar photosphere is used as an illustration of a method which converges in few iterations, each of which requires a time which scales linearly with the number of degrees of freedom in the problem. The importance of Doppler shifts in providing increased opportunity for spatial couplings is stressed. The horizontally averaged line source function is increased significantly by the increased capture of continuum photons by the line.

063.049 **Iteration with approximate lambda operators, and its application to the expanding atmospheres of WR stars.**
W.–R. Hamann, L. Koesterke, U. Wessolowski.
NATO Advanced Research Workshop on Stellar Atmospheres: Beyond Classical Models, p. 69 – 80 (1991). – See Abstr. 012.056 for the main entry.

The multi–level non–LTE radiation transfer problem can be solved very efficiently by means of "iteration with approximate lambda operators" ("ALI"). A concise formulation of this technique is given, concentrating on those approximate operators which are "diagonal". The method requires the numerical solution of the equations of statistical equilibrium, which form a non–linear system here because they incorporate the approximate radiation tranfer. Its solution by the standard Newton–Raphson iteration becomes the most time–consuming part of an ALI computer code, if the number of atomic levels is large. The authors show how to apply a modern mathematical tool, known as "Broyden method", by which the computational effort is reduced drastically. No derivatives must be repeatedly calculated, nor matrices are to be inverted. The authors extensively applied the ALI technique to the spherically–expanding atmospheres of Wolf–Rayet stars, accounting for very complex model atoms of helium, hydrogen, nitrogen and carbon so far. These models allow, for the first time, the quantitative spectral analysis of these interesting objects. Some recent results are briefly summarized.

063.050 **Analytical methods of line formation theory: are they still alive?**
V. V. Ivanov.
NATO Advanced Research Workshop on Stellar Atmospheres: Beyond Classical Models, p. 81 – 104 (1991). – See Abstr. 012.056 for the main entry.

An overview of Analytical Radiative Transfer (ART) theory is presented. ART problems are classified in three logical trees and briefly discussed. The classifications are based on a) physics of interaction of radiation and matter; b) methods of solution; c) geometry of the medium. Several representative examples of nontrivial results from ART are discussed. The connection between the well-known $\sqrt{\varepsilon}$ law and scaling laws for finite slabs is elucidated. Three types of asymptotic results for the two-level frequency independent source function S are illustrated. ART problems of scattering polarization in spectral lines are discussed under the assumption of complete frequency redistribution. The $\sqrt{\varepsilon}$ law is generalized to incorporate polarization. The limiting limb polarization in the line core is expressed in terms of the components of the vector source function $S(\tau)$ for $\tau = 0$. For an isothermal atmosphere the largest possible core polarization at the limb for a Doppler broadened line is found.

063.051 **Iteration factors in the solution of the NLTE line transfer problem.**
E. Simonneau, O. Atanacković-Vukmanović.
NATO Advanced Research Workshop on Stellar Atmospheres: Beyond Classical Models, p. 105 – 110 (1991). – See Abstr. 012.056 for the main entry.
A simple and efficient mixed iteration procedure that uses iteration factors to solve the non-LTE line transfer problem is presented. By suitable choice of quasi-invariant iteration factors, the exact solution is reached within only a few iterations even under physical conditions very far from LTE.

063.052 **Analysis of ultraviolet P Cygni profiles in the spectra of O-type stars.**
F. Nemry, J. Surdej.
NATO Advanced Research Workshop on Stellar Atmospheres: Beyond Classical Models, p. 111 – 118 (1991). – See Abstr. 012.056 for the main entry.
In the framework of the analysis and interpretation of observed P Cygni line profiles, the authors describe and compare the nth-order moment method with the line profile fitting technique. They then point out that, to make a proper statistical study of the UV P Cygni profiles observed in the spectra of O-type stars, one has to take into account the severe pollution effects due to a forest of Fe IV and Fe V absorption lines. They discuss a method allowing for correcting the observed profiles for the presence of photospheric lines and give estimates of the error on $\dot{M}$ caused by this pollution.

063.053 **The $H+H^- \rightleftharpoons H_2(v,j)+e^-$ reaction: a consistent description of the associative detachment and dissociative attachment processes using the resonant scattering theory.**
J. M. Launay, M. Le Dourneuf, C. J. Zeippen.
Astron. Astrophys., Vol. 252, No. 2, p. 842 – 852 (Dec 1991).
The associative detachment process $H+H^- \rightarrow H_2(v,j) \rightarrow e^-$ and the inverse reaction $H_2(v,j)+e^-$, which leads to rovibrational excitation $[H_2(v',j')+e^-]$ and dissociative attachment $(H+H^-)$, have been calculated consistently using the resonant scattering theory and$_2$ detailed balance formulae. The evolution of the transient H^- molecular complex is governed by a local complex potential, obtained by combining the best short- and long-range potentials available in the literature. The corresponding state-to-state cross-sections for the two reverse reactions agree consistently better with experiment than previous calculations. Cross-sections and reaction rates for many energies in the temperature range of astrophysical interest are presented in tabular and graphical forms.

063.054 **Absorption line formation in an optically thick, weakly absorbing planetary atmosphere. I. Homogeneous slab.**
N. N. Fomin, Eh. G. Yanovitskij.
Kinematika Fiz. Nebesn. Tel, Tom 7, No. 5, p. 29 – 38 (Sep–Oct 1991). In Russian. English translation in Kinematics Phys. Celest. Bodies, Vol. 7, No. 5.
A rigorous asymptotic theory of absorption line formation in a homogeneous, weakly absorbing cloudly atmosphere is developed. The main formulae, that determine the profile and equivalent width of the absorption line are derived both for diffusely reflected and diffusely transmitted light. In the latter case, the profile and equivalent width do not depend on the zenith angle of the Sun and the direction to the receiver. Tables of the corresponding curves of growth are given. Formation of the "superweak" absorption line is discussed in detail, relevant to the case when the multiple scattering processes do not disturb the initial profile of the absorption coefficient of the line or molecular band.

063.055 **Absorption line formation in an optically thick, weakly absorbing planetary atmosphere. II. Vertically inhomogeneous medium.**
N. N. Fomin, Eh. G. Yanovitskij.
Kinematika Fiz. Nebesn. Tel, Tom 7, No. 5, p. 39 – 47 (Sep–Oct 1991). In Russian. English translation in Kinematics Phys. Celest. Bodies, Vol. 7, No. 5.
Absorption line formation in an optically thick, vertically inhomogeneous atmosphere is studied. It is assumed that the absorption is weak both in the line and continuum. The problem is rigorously solved for the following two cases: 1) isothermal slab with pressure varying in accordance with the barometric law; 2) polytropic model of the atmosphere (for the central line frequency). A method is described for computing the line profile in the general case of arbitrary dependence of the optical properties of the atmosphere on the optical depth. The method is based on reducing the corresponding boundary value problem to the Cauchy problem.

063.056 **Scattering cross section for randomly oriented particles of arbitrary shape.**
M. I. Mishchenko.
Kinematika Fiz. Nebesn. Tel, Tom 7, No. 5, p. 93 – 95 (Sep–Oct 1991). In Russian. English translation in Kinematics Phys. Celest. Bodies, Vol. 7, No. 5.
Light scattering by randomly oriented particles of arbitrary shape is considered. Waterman's T-matrix approach is used to derive a simple analytical formula for the total scattering cross section.

063.057 **Dynamics of and radiative transfer in inhomogeneous media.**
Å. Nordlund, R. F. Stein.
NATO Advanced Research Workshop on Stellar Atmospheres: Beyond Classical Models, p. 263 – 279 (1991). – See Abstr. 012.056 for the main entry.
The authors discuss mechanisms controlling the inhomogeneous structure of the upper parts of stellar convection zones, and the central role played by radiative energy transfer processes in the surface boundary layers. Radiative release of ionization energy occurs in a very narrow layer as hot, ascending plasma reaches the surface. Radiative reabsorption of energy maintains the temperature structure of stellar photospheres. The presence of spectral lines modifies the energy balance considerably, by increasing the effective mean opacity of layers which are optically thin in the continua.

063.058 **Fe II emission line diagnostics of the Sun and stars.**
P. Judge.
NATO Advanced Research Workshop on Stellar Atmospheres: Beyond Classical Models, p. 291 – 296 (1991). – See Abstr. 012.056 for the main entry.
Iron-group elements have a complex term structure owing to the incomplete 3d shell. The resulting line spectra are unique, although complex, spectral diagnostics since the lines span wide ranges of excitation conditions and optical depths. In this paper UV lines of Fe II, the most prominent ion of the iron group in astrophysical spectra, are examined in SKYLAB (SO82B) spectra of the Sun and IUE spectra of cool giant stars. The line excitation and formation are discussed and some implications for the structure of solar and stellar atmospheres are described.

063.059 **Chromospheric inhomogeneities in cool stars: possible effect on hydrogen line profiles.**
P. Heinzel.
NATO Advanced Research Workshop on Stellar Atmospheres: Beyond Classical Models, p. 297 – 302 (1991). – See Abstr. 012.056 for the main entry.
The authors propose to include chromospheric inhomogeneities like spicules and fibrils into the non-LTE modelling. They

introduce some kind of a hybrid chromospheric model which joins both 1D–semiempirical models and new models of the overlying inhomogeneous structures. As a result, averaging the hydrogen line profiles over the spectrograph slit length (which usually covers several fine–structure elements), they arrive a reasonable agreement between their theoretical Lyman profiles and OSO–8 observations. The same effect is to be expected when averaging over the whole stellar disk, which strongly suggests the importance of chromospheric inhomogeneities clearly observed on the Sun for UV–diagnostics of stellar atmospheres.

063.060 **On the influence of multi–dimensional radiative transfer on the energetic contribution of the Ca K line.**
P. Fabiani, J. Trujillo Bueno.
NATO Advanced Research Workshop on Stellar Atmospheres: Beyond Classical Models, p. 423 – 429 (1991). – See Abstr. 012.056 for the main entry.
By means of multi–dimensional radiative tranfer calculations in inhomogeneous configurations representative of the solar chromosphere the authors investigate how the radiative energy losses in the K line of Ca II deviate from the plane–parallel case as the horizontal structural length of the assumed inhomogeneous scenario diminishes.

063.061 **The opacity project and the practical utilization of atomic data.**
D. G. Hummer.
NATO Advanced Research Workshop on Stellar Atmospheres: Beyond Classical Models, p. 431 – 439 (1991). – See Abstr. 012.056 for the main entry.
The opacity project was organized to compute new opacities for stellar envelopes, but has produced as a secondary product a vast amount of atomic radiative data for almost all elements of astrophysical interest. The photoionization cross sections in many cases show extensive resonance structure, which introduces complications when they are used in stellar modeling. Possible ways of dealing with this problem are discussed. Finally, a method of presenting atomic data in standardized subroutines in order to eliminate the need for large, special purpose data sets is proposed.

063.062 **The extinction of starlight revisited.**
N. C. Wickramasinghe, B. Jazbi, F. Hoyle.
Astrophys. Space Sci., Vol. 186, No. 1, p. 67 – 80 (Dec 1991).
The rigorous Kerker–Matijević formulae for light scattering by co–axial double cylinders are used to calculate the extinction properties of hollow organic grains. A size distribution of such particles together with iron whiskers of radii 0.01 μm, silica spheres of radius 0.03 μm and free aromatic molecular clusters comprised of 50–100 atoms yield excellent agreement with the data on the extinction of starlight. The mass ratios of silica to organics and of iron to organics are in good accord with cosmic abundance constraints.

063.063 **New opacity calculations.**
R. L. Kurucz.
NATO Advanced Research Workshop on Stellar Atmospheres: Beyond Classical Models, p. 441 – 448 (1991). – See Abstr. 012.056 for the main entry.
The author has computed new opacities for model stellar atmospheres and envelopes using a large grant of Cray computer time at the San Diego Supercomputer Center. The opacities inlcude 58,000,000 atomic and diatomic molecular lines. They are available on magnetic tape. A solar photospheric model computed with the new opacities matches the observed energy distribution.

063.064 **Application of the superconvergence properties of the Galerkin approximation to radiative transfer.**
M. S. Abdel Krim, M. T. Attia, M. A. Madkour.
Astrophys. Space Sci., Vol. 186, No. 1, p. 89 – 99 (Dec 1991).
The superconvergence Galerkin approximation is introduced to calculate some functionals arising in radiative transfer problems. Forward and backward radiation heat fluxes are calculated for comparison.

063.065 **Extinction properties of infinitely long graphite cylinders.**
B. Jazbi, F. Hoyle, N. C. Wickramasinghe.
Astrophys. Space Sci., Vol. 186, No. 1, p. 151 – 155 (Dec 1991). Letter–to–the–editor.
The authors have calculated the extinction efficiencies of randomly oriented infintie graphite cylinders, including hollow cylinders, using the rigorous Kerker–Matijević formulae. The peak in the mid–ultraviolet extinction varies in wavelength with particle radius and cavity size in a way that makes such particles of limited interest as models of interstellar grains.

063.066 **The inability of the resonant Compton upscattering model of gamma–ray bursts to produce a third cyclotron harmonic.**
J. J. Brainerd.
Astrophys. J., Lett., Vol. 379, No. 2, p. L57 – L59 (1 Oct 1991).
The resonant Compton upscattering model for gamma–ray bursts cannot produce a third cyclotron harmonic under plausible physical conditions. If one observes a third harmonic in a burst spectrum, resonant Compton upscattering cannot be responsible for the burst emission. The second harmonic may be produced by this model for sufficiently large optical depths.

063.067 **Radiative energy transfer in extended photospheres of the components of close binary systems.**
Z. Kopal.
Astrophys. Space Sci., Vol. 186, No. 2, p. 253 – 276 (Dec 1991).
The aim of the present paper is to set up, and solve, the equations governing transfer of radiation in semi–transparent envelopes of the stars; and, in order to do so, to employ a system of curvilinear (non–orthogonal) three–dimensional coordinates in which the radial coordinate has been identified with equipotential surfaces. Such coordinates are particularly suitable to a treatment of the problems arising in close binary systems, which render the outcome more than any other amenable to observable tests, but which has so far received but very scant attention. The introductory section of this paper contains a statement of the problem; and its mathematical formulation in terms of Clairaut coordinates (cf. Kopal 1980, 1989) is outlined in Section 2; their methods in Section 3. Section 4 then contains an application to the problem of distribution of surface brightness ("limb–darkening") over the apparent discs of distorted components of close binary systems; while in Section 5 the same is done for radiative flux of distorted stars as a function of the phase ("gravity darkening"). The concluding Section 6 then contains an outline of additional problems arising in this connection, to which the author will turn in successive parts of this series.

063.068 **X–ray irradiation of interstellar grains in active galaxies: evaporation and infrared spectra.**
G. M. Voit.
Astrophys. J., Vol. 379, No. 1, p. 122 – 140 (20 Sep 1991).
X–ray irradiated dust grains are likely to be present in Seyfert galaxies, quasars, starburst galaxies, and perhaps near X–ray sources in our own Galaxy. Infrared observations have shown that small grains (<10 Å) subjected to ambient interstellar UV radiation will heat sporadically to more than 1000K and emit in the near infrared. X–rays will heat even larger grains in this way. A population of X–ray illuminated grains varying in size will radiate an infrared spectrum with a long–wavelength blackbody peak, at the equilibrium temperature of the large grains, plus some excess nonequilibrium continuum emission extending into the 1 – 3 μm range. X–rays will efficiently evaporate dust grains smaller than 10 Å, so X–ray illuminated dust should have no polycyclic aromatic hydrocarbon (PAH) features. The turnover shortward of 3 μm is robust and represents the onset of evaporative cooling in grains. UV emission stimulated by X–ray photoelectrons originating in the gas phase can substantially increase the strength of the near–IR excess.

063.069 **Radiation transport in dust in disk geometry. I. Application to externally heated interstellar clouds.**
G. F. Spagna Jr., C. M. Leung, M. P. Egan.
Astrophys. J., Vol. 379, No. 1, p. 232 – 244 (20 Sep 1991).
The problem of radiation transport is solved numerically for both one-dimensional spherical and two-dimensional disk geometries, to determine self-consistently the dust temperature distribution and the infrared energy spectrum of disk-shaped interstellar dust clouds which are heated externally by the ambient interstellar radiation field. The effects of source geometry (disk versus sphere, varying degree of disk flattening) and cloud opacity on the dust temperature distribution, radiation field anisotropy, and infrared energy spectrum are studied. It is found that the equilibrium grain temperature is not extremely sensitive to the cloud geometry or degree of disk flattening. Compared to a spherical model of the same opacity, a similar disk model shows a nearly identical temperature distribution and emergent flux spectrum. However, the internal radiation in the far-infrared is more anisotropic in disk models and is predominantly radially streaming. Estimates of dust mass based on observed flux may be uncertain by as much as a factor of 2 or 3, if spherical geometry is incorrectly assumed. The ratio of observed flux at 200 μm and 1.2 μm may be used as an indicator of disk geometry for unresolved clouds.

063.070 **The inverse problem of photometry of solar radiation reflected by an optically thick planetary atmosphere. Mathematical framework and weighting functions of the linearized inverse problem.**
E. A. Ustinov.
Kosm. Issled., Tom 29, Vyp. 4, p. 604 – 620 (Jul–Aug 1991). In Russian. English translation in Cosmic Res. Vol. 29, No. 4.
A general approach to the solution of the inverse problem of radiative transfer is considered, that is based on the solution of corresponding direct and adjoint problems of radiative transfer. This approach is applied to the case of an optically thick planetary atmosphere.

063.071 **On the effect of blue shift of supernovae emission lines due to the finiteness of light velocity.**
A. Eh. Grigoryan, T. G. Arshakyan.
Astrofizika, Tom 33, Vyp. 3, p. 371 – 378 (Dec 1990). In Russian. English translation in Astrophysics, Vol. 33, No. 3.
Possible causes of the emission line peak shift in an object with erupted envelope are considered. A new effect of blue shift of the emission line peak is revealed. It is shown that the latter is due to the fact that the simultaneously observed photons, arrived from the various points of the envelope, are emitted correspondingly in various moments of the envelope expansion. The considered effect is as large as the velocities of erupted envelope layers which are closer to the light velocity. The case of an isotropical expanding continuous envelope with $v \sim r$ kinematics is considered. A method of determining the blue shift and estimating the lower limit distribution of the envelope atoms participating in the line formation process is worked out. To make use of this method the radial velocity (–170 km/s) corresponding to the lower limit of H_α emission peak shift for SN 1987A for February 25, 1987 is estimated.

063.072 **Stellar occultation of polarized light from circumstellar electrons. III. General axisymmetric envelopes.**
G. K. Fox.
Astrophys. J., Vol. 379, No. 2, p. 663 – 675 (1 Oct 1991).
The effects of occultation and depolarization upon the theoretical polarization from axisymmetric (Thomson scattering) circumstellar envelopes are considered for a finite size, spherical light source. The resulting expression for the net polarization is found to have a complicated inclination dependence, such that the Brown and McLean result of $P \sim \sin^2 i$ no longer holds. The deviation of the variation of polarization with inclination from the Brown and McLean result is shown for three specific geometries. The results indicate that the McLean and Brown statistical analysis of the variation of polarization with V sin i is questionable (except for spheroidal envelopes). It is further found that for Be stars, which exhibit polarizations of up to 2%, the circumstellar envelope is required to be concentrated either towards the equatorial or polar regions of the star and the optical depth must be $\tau > 0.05$. It is alo shown that the ellipsoidal / spheroidal envelope model of Doazan and Thomas is unlikely to produce such a high degree of polarization.

063.073 **Radiative transfer in a spherical inhomogeneous medium with anisotropic scattering.**
S. A. El-Wakil, M. T. Attia, E. M. Abulwafa.
J. Quant. Spectrosc. Radiat. Transfer, Vol. 46, No. 1, p. 31 – 40 (Jul 1991).
The radiative heat flux at the boundary of a sphere containing an internal energy source and subject to general boundary conditions is obtained in terms of the albedo of the corresponding source-free problem with isotropic boundary condition. The relations obtained apply to the general case of anisotropic scattering in an inhomogeneous medium. The advantage of these relations is the result of the fact that there is no need to obtain a particular solution for specified internal sources. Therefore, calculations can be done easily for a non-uniform source distribution. The phase function is approximated by using a linear anisotropic relation. The linear coefficient is taken to be the sum of the coefficients of the Legendre expansion of the phase function. The resulting relations are used to calculate the partial heat flux and emissivity for a given internal energy source and inhomogeneous medium.

063.074 **On coupled conductive-radiative heat-transfer problems in a sphere.**
C. E. Siewert, J. R. Thomas Jr.
J. Quant. Spectrosc. Radiat. Transfer, Vol. 46, No. 2, p. 63 – 72 (Aug 1991).
A sphere-to-plane transformation technique and the spherical harmonics method are used, along with Hermite cubic splines, to define an iterative technique for solving a class of nonlinear radiative transfer problems in a sphere. Computational aspects of the technique are discussed, and the method is used to establish numerical results for several test problems.

063.075 **Solution of the special relativistic equation of radiative transfer in rapidly expanding spherical shells.**
P. H. Hauschildt, R. Wehrse.
J. Quant. Spectrosc. Radiat. Transfer, Vol. 46, No. 2, p. 81 – 98 (Aug 1991).
A new method for the solution of the radiative transfer equation in expanding spherical media based on the Discrete Ordinate Matrix Exponential (DOME) method is describe. The effects of advection and aberration terms on line profiles for scattering dominated shells as observed, e.g., in supernova photospheres are discussed. It is shown that these effects have a large influence on the radiation field in rapidly expanding shells and should therefore be included in model calculations.

063.076 **The effects of coherence and redistribution on radiative transfer in large velocity gradients.**
G. J. Pert.
J. Quant. Spectrosc. Radiat. Transfer, Vol. 46, No. 3, p. 165 – 169 (Sep 1991).
Radiative trasfer in an atmosphere with a velocity gradient is generally treated under the assumption of complete redistribution. Using Monte Carlo techniques the author explores the consequences of relaxing this condition within the Sobolev approximation that the velocity gradient is sufficiently large. Two cases of coherent scattering are considered, namely coherence in the atomic frame and coherence in the fluid frame. It is found by computer simulation that coherence in the atomic frame yields escape probabilities almost identical to complete redistribution, whereas coherence in the fluid frame doubles the rate at large optical depths. It is shown that the results are independent of the phase factor for scattering.

063.077 **Reflection of polarized light by plane–parallel slabs containing randomly–oriented, nonspherical particles.**
M. I. Mishchenko.
J. Quant. Spectrosc. Radiat. Transfer, Vol. 46, No. 3, p. 171 – 181 (Sep 1991).
The computation of the Stokes parameters of light reflected by plane–parallel, isotropic slabs composed of randomly–oriented, nonspherical particles of revolution is considered. Accurate numerical results are reported for homogeneous slabs containing monodisperse, randomly–oriented, homogeneous spheroids and Chebyshev particles.

063.078 **A novel collisional radiative model with a numerical bottom and an analytical top.**
D. A. Benoy, J. A. M. van der Mullen, B. van der Sijde, D. C. Schram.
J. Quant. Spectrosc. Radiat. Transfer, Vol. 46, No. 3, p. 195 – 210 (Sep 1991).
A cut–off procedure for numerical collisional radiative (CR) models is developed and is based on knowledge of the excitation flow in the collision–dominated part of an atomic system as revealed by analytical models. This new cut–off technique makes it possible to reduce the number of levels in a numerical CR–model drastically in comparison with the older, numerical cut–off technique used by Bates et al. in which the truncation level was demanded to be in Saha equilibrium. The model is applied to an ionizing and recombining argon plasma as an example of a more general technique which can also be applied to other atomic systems.

063.079 **A model for interface intensities in stochastic particle transport.**
G. C. Pomraning.
J. Quant. Spectrosc. Radiat. Transfer, Vol. 46, No. 4, p. 221 – 236 (Oct 1991).
The author considers particle transport and radiative transfer in a Markovian mixture of two immiscible fluids. An earlier reported balance equation for the ensemble–averaged intensity is rederived in a manner that generalizes to a derivation of a higher order balance–like equation for interface intensities. A closure is made in this higher order equation to obtain a relatively simple model for linear transport in a stochastic mixture, including the scattering interaction. This extends an earlier model that results from a lower order closure in the balance equation itself. Numerical results are given that compare the predictions of this model of stochastic transport with benchmark results available in the literature, as well as with predictions of the earlier model.

063.080 **Flux–limiting in stochastic transport.**
M. Sammartino, G. C. Pomraning.
J. Quant. Spectrosc. Radiat. Transfer, Vol. 46, No. 4, p. 237 – 249 (Oct 1991).
The authors consider generic particle transport, with radiative transfer as a special case, in a Markovian mixture of two immiscible fluids. A recently proposed model for the ensemble–averaged intensity consists of two coupled transport equations. In this paper, the authors develop the formalism which leads to a flux–limited diffusion approximation to this model in the form of two coupled diffusion equations. Such a flux–limiting diffusion description has proved extremely useful in the nonstochastic particle–transport setting. Numerical results are given for one–dimensional problems.

063.081 **Combined conduction and radiation in absorbing, emitting and anisotropically–scattering, concentric, spherical media.**
Chu Hsin–Sen, Weng Ling–Chia, Tseng Chung–Jen.
J. Quant. Spectrosc. Radiat. Transfer, Vol. 46, No. 4, p. 251 – 257 (Oct 1991).
The steady energy transfer by simultaneous conduction and radiation in an absorbing, emitting and anisotropically–scattering medium has been studied theoretically. The medium is confined between two concentric, gray, diffuse and/or specular spherical surfaces which are isothermal but at different temperatures. The radiative transfer equation is solved by using the P_3–approximation method which results in four coupled ordinary differential equations (ODEs). The set of the four ODEs and energy equation is solved as a two–point boundary–value problem. The effects of the ratio of the conduction–to–radiation parameter, single–scattering albedo, the scattering phase function, and boundary reflectivities on the temperatures and heat–flux distributions in the medium are examined in detail.

063.082 **Computational attributes of the integral form of the equation of transfer.**
J. I. Frankel.
J. Quant. Spectrosc. Radiat. Transfer, Vol. 46, No. 4, p. 329 – 342 (Oct 1991).
Difficulties can arise in radiative and neutron transport calculations when a highly anisotropic scattering phase function is present. In the presence of anisotropy, currently used numerical solutions are based on the integro–differential form of the linearized Boltzmann transport equation. In this paper, the author departs from classical thought and presents an alternative numerical approach based on application of the integral form of the transport equation. Use of the integral formalism facilitates the following steps: a reduction in dimensionality of the system prior to discretization, the use of symbolic manipulation to augment the computational procedure, and the direct determination of key physical quantities which are derivable through the various Legendre moments of the intensity. The approach is developed in the context of radiative heat transfer in a plane–parallel geometry, and results are presented and compared with existing benchmark solutions. Encouraging results are presented to illustrate the potential of the integral formalism for computation.

063.083 **The solution of radiative transfer problems in molecular bands without the LTE assumption by accelerated lambda iteration methods.**
A. A. Kutepov, D. Kunze, D. G. Hummer, G. B. Rybicki.
J. Quant. Spectrosc. Radiat. Transfer, Vol. 46, No. 5, p. 347 – 365 (Nov 1991).
An iterative method based on the use of approximate transfer (or Λ) operators, which was designed initially to solve multilevel NLTE line formation problems in stellar atmospheres, is adapted and applied to the solution of the NLTE molecular band radiative transfer in planetary atmospheres. The matrices to be constructed and inverted are much smaller than those used in the traditional Curtis matrix technique, which makes possible the treatment of more realistic problems (including rotational NLTE, overlapping of lines in the bands and overlapping of bands with continuua) using relatively small computers. This technique converges much more rapidly than straightforward iteration between the transfer equation and the equations of statistical equilibrium (Λ–iteration). A test application of this new technique to the solution of NLTE radiative transfer problems for optically–thick and thin bands (the 4.3 μm CO_2 band in the Venusian atmosphere and the 4.7 and 2.3 μm CO bands in the Earth's atmosphere) is described.

063.084 **Radiative transfer in cylindrical plasmas.**
G. J. Pert.
J. Quant. Spectrosc. Radiat. Transfer, Vol. 46, No. 5, p. 367 – 384 (Nov 1991).
The characteristic functions determining radiation transfer in cylindrical systems by both monochromatic radiation and spectral lines are derived. Functional forms for the scattering of monochromatic and Doppler–broadened lines with complete redistribution in a uniform medium are derived, and suitable approximations for rapid calculation identified.

063.085 **Particular solutions for the radiative transfer equation.**
N. J. McCormick, C. E. Siewert.
J. Quant. Spectrosc. Radiat. Transfer, Vol. 46, No. 6, p. 519 – 522 (Dec 1991).
Particular solutions for both the (formally exact) method of elementary solutions and the P_N method are derived for the case

of monochromatic radiative transfer in a homogeneous plane–parallel medium which contains a source that varies with position and direction.

063.086 **Polarized radiative transfer in an aerosol medium.**
S. A. El–Wakil, M. A. Madkour, E. M. Abulwafa.
J. Quant. Spectrosc. Radiat. Transfer, Vol. 46, No. 6, p. 523 – 529 (Dec 1991).

Polarized radiative transfer in a medium containing aerosols has been considered. The problem of determining the intensity and the degree of polarization of radiation emerging from a finite, plane medium, for the case of Rayleigh scattering with internal energy sources, is considered. A system of integral equations is obtained and solved by the Galerkin method. Parameters such as extinction, absorption and scattering coefficients are calculated using the Mie scattering theory for spherical particles. Average coefficients over different size distributions have been calculated. The degrees of polarization of the emerging radiation from homogeneous media with complex refractive indices are calculated for uniform and nonuniform internal sources.

063.087 **A new polarized atmospheric radiative transfer model.**
K. F. Evans, G. L. Stephens.
J. Quant. Spectrosc. Radiat. Transfer, Vol. 46, No. 5, p. 413 – 423 (Nov 1991).

A plane–parallel polarized radiative transfer model is described. The model is used to compute the radiance exiting a vertically inhomogeneous atmosphere containing randomly–oriented particles. Both solar and thermal sources of radiation are considered. A direct method of incorporating the polarized scattering information is combined with the doubling and adding method to produce a relatively simple formulation. Several numerical results are presented for verification and comparison.

063.088 **On a problem of multiple resonance scattering of gamma–ray quanta.**
N. B. Yengibarian (*N. B. Engibaryan*), A. K. Khachatrian.
J. Quant. Spectrosc. Radiat. Transfer, Vol. 46, No. 6, p. 565 – 575 (Dec 1991).

Multiple scattering of gamma–ray quanta in homogeneous media by Mössbauer isotopes is considered. The principal considerations relate to the case when the redistribution function corresponds to such physical processes as nucleus diffusion and electron–spin relaxation. An effective method for solving the transfer problem in semi–infinite media is suggested.

063.089 **Analytical solution of the linear problem of radiation transfer in a plane homogeneous layer.**
M. G. Mouradian.
J. Quant. Spectrosc. Radiat. Transfer, Vol. 46, No. 5, p. 477 – 481 (Nov 1991).

A linear, stationary problem of radiation transfer in a homogeneous plane layer is considered. An analytical expression is obtained for the general solution of the transfer equations. In the general solution of transfer problems for a layer of finite thickness, there occurs the reflection operator from a semi–infinite medium ϱ and a semi–group of transmission operators $X(\tau)$. The formulas obtained may be used for theoretical investigations, as well as for the approximate solution of transfer problems.

063.090 **Effects of aberration and advection on line formation.**
A. Peraiah.
Astrophys. J., Vol. 380, No. 1, p. 212 – 222 (10 Oct 1991).

The author studies the effects of inclusion of the aberration and advection terms in the equation of the line transfer. He considers spherical shells whose ratios of outer (B) to inner (A) radii (B/A) are 2, 5, and 10 and with line center optical depths 500, 1300, and 1350. The velocities of expansion of the medium are set at 1000, 2000, 3000, 4000, and 5000 km s^{-1}. The author calculates the line source functions and the corresponding profiles of lines. He considers the individual and combined effects of the transverse and radial velocity gradients, and those of aberration and advection in plane–parallel and spherically symmetric geometries on the line source functions and the emerging line profiles. Large effects are generated in the radiation field, when these physical mechanisms are taken into account in the transfer of line radiation. The author has tested the method for various velocity laws and found it to be stable.

063.091 **Emission at cyclotron harmonics due to coalescence of z–mode waves.**
D. B. Melrose.
Astrophys. J., Vol. 380, No. 1, p. 256 – 267 (10 Oct 1991).

It is proposed that solar spike radiobursts showing harmonic structure are due to emission resulting from the coalescence of z–mode waves. The required z–mode waves are assumed to be generated by electron – cyclotron maser emission and then should be concentrated in a small range of angles near perpendicular propagation, and in a small range of frequencies just above the cyclotron frequency Ω_e. Emission at the second harmonic $\omega \approx 2\Omega_e$ is due to coalescence of two z–mode waves, and is slightly above $2\Omega_e$, so that gyromagnetic absorption at $\omega = 2\Omega_e$ does not prevent its escape and is in a small range of angles about perpendicular propagation. Third harmonic emission is due to coalescence of second harmonic emission with another z–mode wave, fourth harmonic emission to coalescence of third harmonic emission with another z–mode wave, and so on. The condition for saturation of this sequence of emission processes is derived and shown to be satisfied under realistic conditions. The predicted maximum brightness temperature is determined by the effective temperature of the z–mode waves. The predicted polarization is nearly 100% in the sense of the x–mode, provided that the weaker o–mode component does not saturate, and zero if the o–mode saturates. The highest harmonic should have the highest polarization.

063.092 **Extreme ultraviolet extinction cross sections of fractal dust grains.**
I. Hawkins, E. L. Wright.
1. Berkeley Colloquium on Extreme Ultraviolet Astronomy, p. 333 – 340 (1991). – See Abstr. 012.045 for the main entry.

063.093 **EUV opacity with interstellar dust.**
P. G. Martin, F. Rouleau.
1. Berkeley Colloquium on Extreme Ultraviolet Astronomy, p. 341 – 349 (1991). – See Abstr. 012.045 for the main entry.

Contents: 1. Introduction. 2. Opacity of the gas phase. 3. Optical constants for generic dust carriers. 4. Electromagnetic scattering calculations. 5. Absorption and scattering by dust in the EUV.

063.094 **Meteorology of the broad line region.**
A. Robinson, E. Pérez, L. de la Fuente.
Workshop on Variability of Active Galaxies, p. 83 – 96 (1991). – See Abstr. 012.070 for the main entry.

The authors discuss the response of the broad emission lines to ionizing continuum variations in the context of the discrete cloud model. They consider cases in which the cloud system has a thick, spherical geometry and discuss several reverberation phenomena which may be relevant to the interpretation of observed light curves. They find that the lag obtained from cross–correlation analyses does not give a reliable indication of the effective "size" of the BLR; this can instead be obtained directly from the centroid of the transfer function. A notable property of this type of model is the relationship between the variability behaviour of an emission line and the shape of its profile. Computer simulations are presented which illustrate the evolution of the line profile in response to a "realistic" continuum light curve.

063.095 **An atlas of 2–dimensional transfer functions of the broad line region.**
E. Pérez, A. Robinson, L. de la Fuente.
Workshop on Variability of Active Galaxies, p. 97 – 104 (1991). – See Abstr. 012.070 for the main entry.

063.096 **Electron–ion coupling in Compton–heated plasmas.**
T. Schmutzler, H. Lesch.
Workshop on Variability of Active Galaxies, p. 221 – 224 (1991). – See Abstr. 012.070 for the main entry.
The authors present a physical mechanism of rapid electron–ion coupling in fully ionized thin plasmas which are illuminated by an anisotropic radiation source (BLR and NLR in AGN).

063.097 **Radiative widths and splitting of cyclotron lines in superstrong magnetic fields.**
G. G. Pavlov, V. G. Bezchastnov, P. Mészáros, S. G. Alexander.
Astrophys. J., Vol. 380, No. 2, p. 541 – 549 (20 Oct 1991).
The authors calculate the radiative line widths of the Landau levels in a superstrong magnetic field and give simple analytic expressions and fits for these, which are valid over a wide range of the principal quantum number and the magnetic field strength. If one does not allow for QED corrections, all levels but the first are doubly degenerate, corresponding to the two possible spin projections. However, the interaction with the QED vacuum removes this degeneracy, leading to an energy splitting of each level which exceeds the radiative line width for low Landau levels if the magnetic field is not too large ($B \lesssim 10^{13}$G). The authors present estimates of the splitting in various limits as a function of field strength and Landau number. The possibility of observing this splitting in accreting X–ray pulsars and in gamma–ray bursters is discussed.

063.098 **Towards better scattering properties for a mixture of realistic grains.**
P. Bastien, A.–C. Levasseur–Regourd.
6. IAP Astrophysics Meeting – IAU Colloquium No. 129: Structure and emission properties of accretion disks, p. 375 – 376 (1991). – See Abstr. 012.046 for the main entry.
A new method to obtain the elements of the scattering matrix for grains around young stellar objects is outlined. The authors take into account the scattering properties of protoplanetary grains as deduced from studies of cometary grains.

063.099 **X–ray spectra from dynamical tuning.**
D. J. Raine, M. D. O'Reilly.
6. IAP Astrophysics Meeting – IAU Colloquium No. 129: Structure and emission properties of accretion disks, p. 509 – 510 (1991). – See Abstr. 012.046 for the main entry.
By means of a simplified one–dimensional model the authors show that reprocessing by a hot corona of UV photons emitted in an accretion disc can lead to canonical X–ray spectra if the dynamics of the corona is taken into account.

063.100 **An efficient 3D radiative transfer method.**
L. G. Stenholm, H. Störzer, R. Wehrse.
6. IAP Astrophysics Meeting – IAU Colloquium No. 129: Structure and emission properties of accretion disks, p. 531 – 532 (1991). – See Abstr. 012.046 for the main entry.
The authors present a method for the solution of the transfer equation for unpolarized radiation in 3D media in which the corresponding systems of linear equations can be solved very effectively by recursion and iteration.

063.101 **The synchrotron and cyclo–synchrotron absorption cross–section.**
G. Ghisellini, R. Svensson.
Mon. Not. R. Astron. Soc., Vol. 252, No. 3, p. 313 – 318 (1 Oct 1991).
The authors emphasize the usefulness of the concept of a cross–section for the synchrotron and the cyclo–synchrotron absorption process. They show that the previously known relation between the cross–section and the single particle emissivity is most easily derived by applying the Einstein coefficients and their relations to a system with three energy levels. Using this relation and known emissivities the authors derive useful analytical expressions for the synchrotron and cyclotron absorption cross–section. As a side result they show that, although the angle averaged synchrotron cross–section is well known to be positive, the non–averaged cross–section is negative for photons whose direction of motion is outside the characteristic emission cone of angular width $1/\gamma$.

063.102 **Radiative transfer in axisymmetric dust clouds. II. Models of rotating protostars.**
A. Efstathiou, M. Rowan–Robinson.
Mon. Not. R. Astron. Soc., Vol. 252, No. 4, p. 528 – 534 (15 Oct 1991).
The authors present solutions of the axially symmetric radiative transfer problem in dust clouds, appropriate for modelling protostellar sources. In contrast to previous attempts to solve the axisymmetric problem, the method integrates the equation of radiative transfer exactly in the sense that it does not neglect any terms in the equation and treats multiple scattering from grains. The results of several computed models are presented showing the dependence of the emergent spectrum on model parameters and viewing angle.

063.103 **The dust opacity of spiral galaxies.**
Y. I. Byun, K. C. Freeman.
Proc. Astron. Soc. Aust., Vol. 9, No. 1, p. 86 – 88 (1991). – See Abstr. 012.078 for the main entry.
In order to find the best observable diagnostics for the amount of internal extinction within spiral galaxies, the authors have constructed models for disk galaxies with immersed dust layers. The radiative transfer problem, including both scattering and pure absorption, has been computed for a range of model galaxies. This reveals a set of superior diagnostics for the opacity. These include the behaviour of the radial colour and luminosity distributions, the amplitude of the asymmetry between the near and far sides of the major axis, and their dependence on the orientation of the galaxy with respect to the observer.

063.104 **Rotationally induced polarization in pure absorption spectral lines.**
G. W. Collins II, S. R. Cranmer.
Mon. Not. R. Astron. Soc., Vol. 253, No. 1, p. 167 – 174 (1 Nov 1991).
The authors show the existence of polarization in rotationally broadened pure absorption lines. The effect arises largely through the breaking of the axial symmetry about the line–of–sight by rotation so that the cancellation of polarized continuum radiation is incomplete. This symmetry breaking takes place as a result of the rotationally displaced local line profile covering part of the polarizing continuum on the apparent disc of the star. The result of this local shadowing of the disc by the line is a net polarization which varies across the rotationally broadened line. Since the effect is largely geometrical, it should be present in stars exhibiting even moderate rotation. While for simple models the average of the Q Stokes parameter across the line is zero, departures within the line may reach a substantial fraction of the local continuum polarization on the stellar surface. The effect is evaluated for the Struve–Unsöld model including limb darkening. In addition, results for more complicated models appropriate for early–type stars are considered.

063.105 **Relativistic electrons in photon fields: effects of triplet pair production on inverse Compton gamma–ray spectra.**
A. Mastichiadis.
Mon. Not. R. Astron. Soc., Vol. 253, No. 2, p. 235 – 244 (15 Nov 1991).
Pair production of a photon in the Coulomb field of an electron (often referred to as triplet pair production) despite the fact that it is a third–order QED process has a cross–section that exceeds the Compton cross–section for energies of collision $\geqslant 250\ m_e c^2$. In the present paper the author re–examines the problem of ultrarelativistic electrons entering a photon field by including triplet pair production as both an energy loss and re–injection mechanism for electrons. The author calculates the steady state electron distribution and photon spectrum for monoenergetic and power–law electron injections and compares them with the results one obtains in the pure Compton case, i.e. by ignoring triplet pair production.

063.106 **Blue shift of supernova emission lines due to finiteness of the speed of light.**
A. Eh. Grigoryan, T. G. Arshakyan.
Astrophysics, Vol. 33, No. 3, p. 513 – 517 (Nov – Dec 1990). English translation of Astrofizika, Tom 33, Vyp. 3, p. 371 – 378 (Dec 1990).

Possible reasons for displacement of the peak of an emission line of an object with an expanding shell are considered. A new effect, blue shift of an emission line peak, is established. It arises because simultaneously observed photons reaching the observer from points of the shell at different distances were emitted at different times of its expansion. The effect is more pronounced, the closer the speed of the ejected layers of the shell is to the speed of light. The case of an isotropically expanding continuous shell with kinematics $v \sim r$ is considered. A method is proposed for determining the displacement of the emission peak and estimating a lower limit on the basis of observational data under the assumption of a homogeneous distribution of the atoms of the shell that participate in forming the line. Application of this method to SN 1987A (February 25, 1987) leads to an estimate of the radial velocity (–170 km/sec) corresponding to the lower limit of the displacement of the Hα emission peak.

063.107 **An investigation of Brueckner's theory of line broadening with application to the sodium D lines.**
S. D. Anstee, B. J. O'Mara.
Mon. Not. R. Astron. Soc., Vol. 253, No. 3, p. 549 – 560 (1 Dec 1991).

Approximations in Brueckner's theory of spectral line broadening by collisions with neutral hydrogen atoms relevant to a solar-type atmosphere are discussed, and a modified theory for s–p transitions is presented. The theory utilizes explicit expressions for the interatomic interaction energy between a hydrogen atom in its ground state and general $m = 0, \pm 1$ p states, derived from second-order perturbation theory without exchange, allowing the removal of the Lindholm–Foley average over m states in the original Brueckner model. Approximate upper and lower bounds for the linewidth of the sodium D lines are derived, and these values are contrasted with available theoretical, experimental and solar empirical results. The removal of the Lindholm–Foley average is shown to reduce the D-state linewidths by about 30%, and an analysis of the interatomic separations important in the line broadening cross-section for the D lines has shown that there is little atomic overlap at the separations that are important.

063.108 **A disc-reflected component in the spectra of X-ray bursters.**
C. S. R. Day, C. Done.
Mon. Not. R. Astron. Soc., Vol. 253, No. 3, p. 35P – 38P (1 Dec 1991).

The authors show that a disc-reflection component, as seen in AGN, can be detected in the spectrum of X-ray bursts during the burst tail, and speculate on the use of the concomitant absorption edge as a diagnostic of the accretion disc.

063.109 **A model for the infrared emission from an OB star cluster environment.**
D. Leisawitz.
Astrophys. J., Suppl. Ser., Vol. 77, No. 4, p. 451 – 471 (Dec 1991).

A model for the infrared emission from the neighborhood of an OB star cluster is described. The distribution of gas and dust around the stars, properties of the dust, and the cluster and interstellar radiation fields are variable. The model can be applied to regions around clusters embedded to various degrees in their parental molecular clouds (i.e., compact H II regions, blister-type H II regions, and the tenuous H II regions ionized by naked O stars). The model is used to simulate IRAS observations of a typical blister H II region. Infrared surface brightness and spectral energy distributions are predicted and the impact of limited spatial resolution is illustrated. The model results are shown to be consistent with observations of the exemplary outer Galaxy OB cluster NGC 7380.

063.110 **Element separation by atomic diffusion.**
G. Michaud, S. Vauclair.
Solar interior and atmosphere, p. 304 – 325 (1991). – See Abstr. 003.028 for the main entry.

Atomic diffusion is a basic physical process and it should, in principle, be included in the standard solar model. In order to understand its possible importance in evolutionary calculations, the various physical processes that cause atomic diffusion are discussed in turn. Uncertainties in the atomic diffusion coefficients are considered. Atomic diffusion also appears to play a role in the outer layers of the Sun. The correlation between the first ionization potential and abundance anomalies observed in the solar wind and corona suggests that atomic diffusion is involved.

063.111 **A model of SVC radiation in magnetic dipole field and radiation varying with magnetic field. I. Active region model and SVC radiation source model.**
Zhao Renyang.
Sci. China, Ser. A, Vol. 34, No. 8, p. 969 – 976 (Aug 1991).

According to the combined mechanism including both the gyroresonance radiation and the bremsstrahlung, the author has calculated the radio radiation of the slowly varying component (SVC) and studied the radiative characteristics of the SVC radiation (such as the spectra of flux density, polarization degree, and brightness temperature, and the brightness distribution) and the geometrical features of the SVC radiation source (such as height and radius), and their variations with the magnetic field. Some very interesting results have been obtained and they are considerably in agreement with the observations.

063.112 **A model of SVC radiation in magnetic dipole field and radiation varying with magnetic field. II. SVC radiation mechanism and radiation variation with magnetic field.**
Zhao Renyang.
Sci. China, Ser. A, Vol. 34, No. 9, p. 1109 – 1118 (Sep 1991).

063.113 **The Rosseland mean free-free Gaunt factor of the dense high-temperature stellar plasma.**
N. Itoh, F. Kuwashima, K. Ichihashi, H. Mutoh.
Astrophys. J., Vol. 382, No. 2, p. 636 – 639 (1 Dec 1991).

The Rosseland mean free-free Gaunt factor of the dense high-temperature stellar plasma is calculated, based on both the accurate relativistic cross section and Sommerfeld's exact nonrelativistic cross section. A wide range of electron degeneracy is accurately taken into account. Comparison of the resulting free-free opacity with the electron conduction opacity is made.

063.114 **Discrete ordinate solution of the radiative transfer equation in the "polarization normal wave representation".**
A. Kylling.
Astrophys. J., Vol. 383, No. 1, p. 243 – 249 (10 Dec 1991).

The transfer equations for normal waves in finite, inhomogeneous and plane-parallel magnetoactive media are solved using the discrete ordinate method. The physical process of absorption, emission, and multiple scattering are accounted for, and the medium may be forced both at the top and bottom boundary by anisotropic radiation as well as by internal anisotropic sources. The computational procedure is numerically stable for arbitrarily large optical depths, and the computer time is independent of optical thickness.

063.115 **Partially dissociative jump shocks in molecular hydrogen.**
C. A. Chang, P. G. Martin.
J. R. Astron. Soc. Can., Vol. 85, No. 4, p. 207 (Aug 1991). Abstract. – See Abstr. 012.091 for the main entry.

063.116 **The theoretical variation of polarised light from stars exhibiting discrete UV absorption components.**
G. K. Fox.
ESO Workshop on Rapid Variability of OB-Stars: Nature and Diagnostic Value, p. 285 – 286 (1991). – See Abstr. 012.092 for the main entry.

063.117 **Forbidden components of He I and some related problems.**
A. E. Ringuelet, N. O. Rotstein, A. R. Diaz.
ESO Workshop on Rapid Variability of OB–Stars: Nature and Diagnostic Value, p. 289 (1991). Abstract. – See Abstr. 012.092 for the main entry.

063.118 **Stability of radiative shocks: nonplanar geometries.**
T. J. Gaetz.
J. R. Astron. Soc. Can., Vol. 85, No. 4, p. 207 – 208 (Aug 1991). Abstract. – See Abstr. 012.091 for the main entry.

063.119 **Radiation flux in a weakly absorbing atmosphere of arbitrary optical thickness.**
Eh. G. Yanovitskij.
Izv. Akad. Nauk SSSR, Fiz. Atmos. Okeana, Vol. 27, No. 11, p. 1241 – 1246 (Nov 1991). In Russian.

A strict asymptotic formula is derived for the radiation flux in a weakly absorbing homogeneous atmosphere of arbitrary optical thickness above a reflecting ground with an albedo A. The formula expresses the flux in some auxiliary quantities that are extensively tabulated. An integral of the radiative transfer equation is found, which is expressed in elementary functions.

063.120 **X–ray continuum emission caused by relativistic protons.**
K. Mannheim.
Workshop on Relativistic Hadrons in Cosmic Compact Objects, p. 57 – 66 (1991). – See Abstr. 012.103 for the main entry.

In this contribution the case is made that relativistic protons as predicted by diffusive shock acceleration can lead to X–ray emission exceeding synchrotron–self–Compton emission by electrons accelerated by the same process. Photomeson cooling of protons on ambient soft synchrotron photons of the primary electrons initiates a three–generation cascade of secondary particles terminating at X– to soft γ–ray energies. Pure nonthermal sources like hot spots in Fanarov–Riley type 2 galaxies or similar objects are proposed as optimum candidates to confirm predicted spectra.

063.121 **Effects of aberration and advection on line transfer in plane parallel geometry.**
A. Peraiah, B. A. Varghese.
Bull. Astron. Soc. India, Vol. 19, Nos. 1 – 2, p. 9 – 12 (Mar – Jun 1991).

The authors have investigated the effects of aberration and advection on line formation in a plane–parallel medium. They consider a two–level atom approximation and find that the line source function changes considerably depending on the optical depth.

063.122 **The inverse problem of photometry of solar radiation reflected by an optically thick planetary atmosphere. 2. Numerical aspects and requirements upon the geometry of observations.**
E. A. Ustinov.
Kosm. Issled., Tom 29, Vyp. 6, p. 917 – 932 (Nov – Dec 1991). In Russian. English translation in Cosmic Res., Vol. 29, No. 6.

The nonlinear inverse problem of radiative transfer formulated in a former paper by the author is considered. This inverse problem is destined for the retrieval of the angular and altitude dependence of the atmospheric scattering phase function from data of photometric observations over the planetary disc of the solar radiation reflected by a planetary atmosphere. Its iterative solution involves in its turn a solution of corresponding direct and adjoint problems of radiative transfer. The numerical algorithm is developed for solution of an adjoint problem based on spherical harmonics method. Angular oscillations appearing in computed weighting functions of the linearised inverse problem due to application of this method are eliminated using integration of the total source function of the equation of radiative transfer. The requirements placed upon the geometry of observations over the planetary disc are considered from the viewpoint of the information content of these observations.

063.123 **Formation of spectral lines in a medium of finite optical thickness with incoherent scattering of radiation.**
G. A. Arutyunyan.
Sov. Phys. – Dokl., Vol. 36, No. 10, p. 656 – 658 (Oct 1991). Abstract. Current Physics Microform No.: 9206X1904.

063.124 **Numerical approach to Zeeman line radiative transfer.**
Y. Takeda.
Publ. Astron. Soc. Jpn., Vol. 43, No. 5, p. 719 – 737 (1991).

An accelerated lambda iteration (ALI) method, a version of the operator perturbation technique, is formulated for applications to Zeeman line formation problems in the presence of magnetic fields. This approach has proven to be quite an effective and flexible numerical device, being applicable to extensive problems (e.g., LTE one–way integration problem, non–coherent scattering etc). In addition to its general formulation, a specialized practical version is also proposed which is limited to scattering (or multi–level) problems under the assumption of complete frequency redistribution (CRD), but requiring much less computing time. In order to examine the computational efficiency of this ALI method, numerical examples are presented concerning line formation in a magnetic field for several simple cases (LTE Milne–Eddington model, non–coherent CRD scattering angle–dependent coherent scattering), showing a reasonably rapid convergence with notable numerical stability. Comparisons with other recent numerical techniques confirm the distinguished superiority of the present method.

063.125 **The effect of far infrared radiation from dust on the departure coefficient from LTE.**
A. Qaiyum.
Bull. Astron. Soc. India, Vol. 19, Nos. 1 – 2, p. 91 (Mar – Jun 1991). Abstract. – See Abstr. 012.108 for the main entry.

063.126 **Be stars: solutions of the mass, momentum, and energy equations of the radiative transfer problem.**
L. S. Cidale, M. E. Iglesias, A. C. Vázquez, A. E. Ringuelet.
Rev. Mex. Astron. Astrofis., Vol. 22, No. 2, p. 306 (Oct 1991). Abstract. – See Abstr. 012.127 for the main entry.

063.127 **Line radiation from stationary accretion disks.**
J: Adam, D. E. Innes, G. Shaviv, H. Störzer, R. Wehrse.
NATO Advanced Research Workshop on Theory of Accretion Disks, p. 403 – 418 (1989). – See Abstr. 012.121 for the main entry.

The authors apply a finite difference method to the 3D radiative line transfer in accretion disks. A two–level–atom approximation with complete re–distribution is used. As a first step disks were assumed which are homogeneous in temperature and pressure. The authors give monochromatic images of the intensity distribution on the surface cataclysmic variables of disks for several assumptions on the scattering and disk thickness. In a parameter study for geometrically thick disks they show a variety of line profile shapes. The effect of a corona overlying a disk atmosphere on the Balmer line profiles is studied with a ray tracing technique. Using Stark profile functions and a parametrization of the coronal temperature, the authors show that the shape of the total profile strongly depends on the structure of the transition zone.

063.128 **Continuum spectra of accretion discs.**
G. Shaviv, R. Wehrse.
NATO Advanced Research Workshop on Theory of Accretion Disks, p. 419 – 444 (1989). – See Abstr. 012.121 for the main entry.

Contents: 1. Introduction. 2. The physical model and basis assumptions. 3. Method of solution. 4. Some general results. 5. Instabilities in the radiative transfer. 6. Comparison with observations.

063.129 **Influence of radiative transfer on the vertical structure of accretion disks.**
I. Hubeny.
NATO Advanced Research Workshop on Theory of Accretion Disks, p. 445 – 456 (1989). – See Abstr. 012.121 for the main entry.

A brief review of various theoretical approaches to model accretion disks is presented. Emphasis is given to models that determine self-consistently the structure of a disk together with the radiation field. It is argued that a proper treatment of the vertical structure is essential for calculating theoretical spectra to be compared with observations. It is demonstrated that neither the blackbody nor model stellar atmosphere fluxes provide a satisfactory approximation of the emergent radiation. Finally, several models calculated with various degree of sophistication in treatment of interaction of radiation with matter, and with different values of viscosity, are intercompared, and possible diagnostic consequences are briefly discussed.

063.130 **Relative strengths of multiplet lines as an indicator of stellar magnetic fields.**
Y. Takeda.
Publ. Astron. Soc. Jpn., Vol. 43, No. 6, p. 823 – 839 (1991).

The magnetic intensification of coupled Fe II lines belonging to a common multiplet is theoretically investigated, where the relative difference (δ) in the line–strength is a good indicator of the magnetic field (H). It is shown that δ is not only a function of H, but also depends upon various other parameters (e.g., Fe abundance, aspect–angle, line wavelength, micro–turbulence ξ); its behavior is controlled by the ratio of the Zeeman width to the Doppler width, as well as by the difference in component–patterns. Theoretical δ vs. H relations are applied to the hot Am star o Peg with the results of $H \sim 2-3$ kG and $\xi \sim 0$ km s^{-1}, suggesting that the conventionally derived ξ–value of ~ 2 km s^{-1} may simply reflect the magnetic effect (rather than some actual velocity field). As an alternative application, an empirical correlation between δ vs. H for cool Ap stars is examined and found to be consistent with the calculations.

Numerical modeling in applied physics and astrophysics.
See Abstr. 003.022.

Symposium on problems of the theory of radiative transfer, held in Leningrad, 22 – 24 October 1990.
See Abstr. 011.024.

Extreme ultraviolet astronomy. A selection of papers presented at the First Berkeley Colloquium on Extreme Ultraviolet Astronomy.
See Abstr. 012.045.

Stellar atmospheres: beyond classical models. Proceedings.
See Abstr. 012.056.

The Connection Machine: principle and application in astronomy and astrophysics.
See Abstr. 021.056.

Shape and clustering effects on the optical properties of amorphous carbon.
See Abstr. 022.068.

Forbidden lines of np^q ions. II. Line intensities.
See Abstr. 022.084.

Atomic data for resonance absorption lines. I. Wavelengths longward of the Lyman limit.
See Abstr. 022.085.

Regularities of Stark parameters along the periodic table.
See Abstr. 022.150.

Irradiation of solids: theory.
See Abstr. 022.161.

The angular dependent reflected spectra by Compton scatterings.
See Abstr. 036.288.

Second angular moment closure on flux–limited neutrino diffusion theory.
See Abstr. 061.096.

Relativistic kinetic shock waves in synchrotron sources: shock structure and non–thermal acceleration of positrons.
See Abstr. 062.037.

Non–linear theory of particle dynamics in cosmic electromagnetic fields. I. Regular motion in a simple Alfvén wave.
See Abstr. 062.139.

A hydrodynamical model for the fragmentation of the W49A star–forming region.
See Abstr. 062.150.

Simulated images of stellar jets.
See Abstr. 062.152.

Electron acceleration in a nonlinear shock model with applications to supernova remnants.
See Abstr. 062.207.

Absorption of p–mode waves by magnetic fields.
See Abstr. 062.211.

Particle acceleration at shock fronts in AGNs and jets.
See Abstr. 062.228.

Influence of second–order Fermi acceleration and relativistic shock waves on nonthermal continuum emission in hot spots.
See Abstr. 062.232.

Images and line profiles from nonadiabatic jet models.
See Abstr. 062.262.

On the possibility of nuclear processes in atmospheres of flare stars.
See Abstr. 064.049.

Approximate lambda–operators working at optimum convergence rate. II. Line transfer in expanding atmospheres.
See Abstr. 064.052.

Luminosity limit for alpha–viscosity accretion disks.
See Abstr. 064.057.

The contribution function for the depth of formation of lines in a stellar atmosphere.
See Abstr. 064.068.

Isothermal, Compton–heated coronae above accretion disks.
See Abstr. 064.080.

The interaction of two stellar winds with energy losses for emission and comptonization.
See Abstr. 064.089.

Computer codes for stellar atmospheric modeling.
See Abstr. 064.095.

Radiative transfer in expanding atmospheres – radiative acceleration of Wolf–Rayet envelopes?
See Abstr. 064.101.

Non–LTE analysis of hot stars including line blanketing.
See Abstr. 064.102.

A smooth source function method for including scattering in radiatively driven wind simulations.
See Abstr. 064.104.

Numerical simulation of photospheric convection: hydrodynamical test calculations.
See Abstr. 064.106.

Model atmospheres for SNe II during early stages.
See Abstr. 064.120.

New results on radiative forces on iron in hot white dwarfs.
See Abstr. 064.122.

Model atmospheres for accretion disks.
See Abstr. 064.136.

Irradiation of accretion disks around young objects. I. Near–infrared CO bands.
See Abstr. 064.144.

Line formation in accretion disks.
See Abstr. 064.148.

Irradiation of accretion disks around young objects: near infrared CO bands.
See Abstr. 064.152.

A non–LTE study of neutral calcium in late–type stars with special reference to Pollux.
See Abstr. 064.157.

Self consistent accretion disk corona.
See Abstr. 064.167.

Convective viscosity in thin accretion disks.
See Abstr. 064.169.

Model atmospheres for rotating B stars.
See Abstr. 064.178.

Simulation of the boundary layer in white dwarf with the SPH method.
See Abstr. 064.179.

On the nature of the excess 100 micron flux associated with carbon stars.
See Abstr. 064.182.

What can NLTE model atmosphere theory contribute to the understanding of line profile variability?
See Abstr. 064.186.

The vertical structure of T Tauri accretion disks. I. Heating by the central star.
See Abstr. 064.187.

Nonthermal excitation of helium in Type Ib supernovae.
See Abstr. 064.188.

Line formation in expanding extended atmospheres with spherical geometry.
See Abstr. 064.193.

Metal opacities and convective core overshooting in Population I stars.
See Abstr. 065.106.

Line profile variations caused by low frequency nonradial pulsations of rapidly rotating stars. II.
See Abstr. 065.129.

The effects of thermal radiation on some general relativistic stellar models.
See Abstr. 066.082.

Quantum–mechanical scattering of charged black holes.
See Abstr. 067.020.

Resonant Compton scattering in the magnetospheres of radio pulsars.
See Abstr. 067.027.

Line profiles from a disk around a rotating black hole.
See Abstr. 067.038.

Model atmospheres and X–ray spectra of bursting neutron stars.
See Abstr. 067.040.

Transient particle acceleration in strongly magnetized neutron stars.
See Abstr. 067.045.

Strahlungstransport in Atmosphären von Neutronensternen.
See Abstr. 067.048.

On spectra of X–ray bursters in the expansion phase.
See Abstr. 067.060.

High–polarization radiation from magnetized accretion disks.
See Abstr. 067.086.

Very high energy gamma–ray emission from accreting X–ray pulsars.
See Abstr. 067.119.

An empirical theory of pulsar emission.
See Abstr. 067.120.

Annihilation radiation in strong magnetic fields and gamma–ray burst spectra.
See Abstr. 067.127.

Hybrid accretion disks in active galactic nuclei. I. Structure and spectra.
See Abstr. 067.152.

The collisional acceleration of relativistic electrons in the central regions of active galactic nuclei.
See Abstr. 067.211.

Compton backscattered 511 keV annihilation line emission and the 170 keV line from the Galactic center direction.
See Abstr. 067.213.

Pair–density transitions in accretion disk coronae.
See Abstr. 067.215.

Existence of scattering atmospheres near luminous, magnetized compact objects.
See Abstr. 067.221.

Polarization properties of emission lines from relativistic accretion disks.
See Abstr. 067.225.

Gamma–ray bursts from sheared Alfvén waves.
See Abstr. 067.233.

Spherical accretion onto black holes: a complete analysis of stationary solutions.
See Abstr. 067.234.

Spectral signatures of the MeV–PeV radiation from accreting neutron stars.
See Abstr. 067.235.

Electron–positron pair production in compact sources with energetic protons.
See Abstr. 067.254.

Collisional acceleration of electrons in central regions of active galactic nuclei.
See Abstr. 067.259.

The formation mechanism of the lines of the C I solar infrared multiplet λ1069,5 nm.
See Abstr. 071.007.

Formation of the infrared emission lines of Mg I in the solar atmosphere.
See Abstr. 071.052.

Granulation: non–adiabatic patterns and shocks.
See Abstr. 071.060.

What is the difference between "emission" and "depression" contribution functions?
See Abstr. 071.065.

Effects of Mg II and Ca II ionization on ab–initio solar chromosphere models.
See Abstr. 073.101.

Two–dimensional models of the solar chromosphere. I. The Ca II K line as a diagnostic: 1.5–D radiative transfer.
See Abstr. 073.102.

Energy balance in the solar transition region. II. Effects of pressure and energy input on hydrostatic models.
See Abstr. 073.113.

Inferring chromospheric flare heating from hydrogen–line wings.
See Abstr. 073.118.

On solar emission lines.
See Abstr. 074.098.

The cyclotron maser instability: application to the planetary radio emission.
See Abstr. 091.033.

Optical properties of circumstellar silicate grains.
See Abstr. 112.004.

^{29}SiO $v = 1$ maser emission from evolved stars: pumping of ^{29}SiO by ^{28}SiO through infrared line overlaps.
See Abstr. 112.030.

CO ($J = 1$–0 and 2–1) mapping of IRC +10216: a hot core model for the gas kinetic temperature distribution and the mass–loss rate.
See Abstr. 112.032.

Fe II emission lines. I. Chromospheric spectra of red giants.
See Abstr. 112.057.

The investigations of "zodiacal light" of isolated AE–Herbig stars with non–periodic Algol–type minima.
See Abstr. 112.074.

The temperature of the circumstellar envelope of Alpha Orionis.
See Abstr. 112.133.

Multi–wavelength observations of infrared–bright carbon stars.
See Abstr. 112.145.

Zeeman–Doppler imaging of solar–type and Ap stars. IV. Maximum entropy reconstruction of 2D magnetic topologies.
See Abstr. 116.045.

Empirical velocity laws for cool giants. I. The symbiotic binary EG Andromedae.
See Abstr. 117.059.

Scattering of X–rays on the winds of supergiant X–ray binaries.
See Abstr. 117.138.

The optical emission from oscillating white dwarf radiative shock waves.
See Abstr. 117.146.

New polarimetric observations and a two pole model for the cyclotron emission from AM Herculis.
See Abstr. 117.191.

Emission lines from accretion disks.
See Abstr. 117.193.

Dust scattering in the high–state eclipse of Cen X–3.
See Abstr. 117.194.

GINGA observation of X1820–303 and its iron–line emission.
See Abstr. 117.267.

Polarization in binaries.
See Abstr. 117.298.

Reflected accretion disk emission lines in cataclysmic variables.
See Abstr. 117.313.

Dust continuum radiation from luminous young stellar objects.
See Abstr. 121.046.

Models for light scattering by dust around Herbig–Haro objects.
See Abstr. 121.063.

Molecular line emission models of Herbig–Haro objects. I. H_2 emission.
See Abstr. 121.099.

Type Ia supernovae: theoretical light curves for SN 1972e and SN 1981b.
See Abstr. 125.024.

Supernova remnants in dense clouds. I. Blast–wave dynamics and X–ray irradiation.
See Abstr. 125.150.

Analysis of SN 1987A polarimetry.
See Abstr. 125.172.

Modelling of the early light curve of SN 1987A with the multi–group time–dependent radiative transfer.
See Abstr. 125.212.

LP 790–29: preliminary model atmospheres for this strongly polarized carbon white dwarf.
See Abstr. 126.059.

Low–density photodissociation regions.
See Abstr. 131.091.

Photodissociation regions.
See Abstr. 131.098.

Synthetic linemaps for hierarchial clouds.
See Abstr. 131.192.

Partially dissociative jump shocks in molecular hydrogen.
See Abstr. 131.221.

The analysis of apparent optical depth profiles for interstellar absorption lines.
See Abstr. 131.257.

The position–velocity diagrams of protostellar discs.
See Abstr. 131.310.

A model for OH masers in star–forming regions.
See Abstr. 131.311.

Asymptotic theory for the spatial distribution of protostellar emission.
See Abstr. 131.356.

Model of interstellar grains.
See Abstr. 131.382.

Peculiar $^{15}NH_3$ toward NGC 7538–IRS 1.
See Abstr. 132.001.

H II region abundances: model oxygen line ratios.
See Abstr. 132.038.

Neutral oxygen in planetary nebulae: probing radiative transfer and nebular structure.
See Abstr. 134.031.

Photoionization modeling of Magellanic Cloud planetary nebulae. II.
See Abstr. 134.040.

N III line emission in planetary nebulae: not Bowen fluorescence.
See Abstr. 134.061.

Progress of research on the energy spectra of gamma–ray bursts.
See Abstr. 143.039.

The galactic diffuse component of soft X–rays and its source function.
See Abstr. 155.078.

The Galactic center radio source shines below the Compton limit.
See Abstr. 155.100.

Evolution of dust grains through a hot gaseous halo.
See Abstr. 157.270.

Scattering–induced optical polarization in thick accretion disks.
See Abstr. 158.006.

Formation of broad–line emission clouds in active galactic nuclei.
See Abstr. 158.041.

Variable soft X–ray excesses in active galactic nuclei from nonthermal electron–positron pair cascades.
See Abstr. 158.064.

A model for the spectral variability of BL Lacertae objects at high frequencies.
See Abstr. 158.083.

Steep radio spectra in high–redshift radio galaxies.
See Abstr. 158.181.

Echo images of broad–line regions in active galactic nuclei.
See Abstr. 158.195.

Modeling an accretion disk in NGC 985.
See Abstr. 158.196.

A two–phase model for the X–ray emission from Seyfert galaxies.
See Abstr. 158.204.

The high frequency properties of synchrotron radiation.
See Abstr. 158.223.

Accretion disk model for the broad line region of three active galactic nuclei.
See Abstr. 158.262.

NGC 985: new evidence for an accretion disk in a Seyfert 1 galaxy?
See Abstr. 158.263.

Wavelength dependent polarization in active galaxies.
See Abstr. 158.270.

The origin of the X–ray spectra of AGN.
See Abstr. 158.308.

Two phase accretion disks as a model for the X–ray emission from Seyfert galaxies.
See Abstr. 158.309.

Compton broadening of the iron line in NGC 3227.
See Abstr. 158.310.

The observations and studies of OH megamaser extragalaxies.
See Abstr. 158.316.

Relativistic enhancement of the Compton–reflected component in active galactic nuclei.
See Abstr. 158.322.

Multifrequency radio observations of Cygnus A: spectral aging in powerful radio galaxies.
See Abstr. 158.323.

The theory of formation of QSO spectra with broad absorption lines.
See Abstr. 159.030.

The visual wavelength continua of quasars: a reexamination.
See Abstr. 159.084.

Microlensing and the structure of active galactic nucleus accretion disks.
See Abstr. 159.108.

Theory of the production of quasar spectra with broad absorption lines.
See Abstr. 159.178.

A photoionization model for the optical line emission from cooling flows.
See Abstr. 160.134.

New photoionization models of intergalactic clouds.
See Abstr. 160.157.

Quantum string theory in curved space–times.
See Abstr. 161.065.

064 Stellar Atmospheres, Stellar Envelopes, Mass Loss, Accretion

064.001 **On the influence of viscosity on the structure of the boundary layer of accretion disks.**
W. Kley.
Astron. Astrophys., Vol. 247, No. 1, p. 95 – 107 (Jul 1991).

The author presents two-dimensional radiation hydrodynamical calculations of the structure of the boundary layer in cataclysmic variables using different values for the viscosity and varying also some of the system parameters (mass of the accreting star, mass accretion rate). The results depend strongly on the magnitude of the viscosity. For high and constant values of the kinematic viscosity ν ($\geqslant 10^{15}\mathrm{cm}^2\mathrm{s}^{-1}$) the flow in the BL is smooth and quasistationary. In this case the radial infall velocity of the gas near the star is supersonic. If a more realistic description of the turbulent viscosity is used, where ν is proportional to the local sound velocity times the pressure scale height (then on average $\nu \approx 10^{14}\mathrm{cm}^2\mathrm{s}^{-1}$) the flow is strongly time dependent and unstable. In the case when the flow becomes supersonic radial shocks are formed so that the flow remains mainly subsonic. The flow in the disk is very irregular consisting of eddies and even reversed (outwardly directed) flow. The luminosity of the boundary layer oscillates with an amplitude of about 1.5 mag on timescales of the Keplerian period at the stellar surface. This might explain some of the variability on short time scales (quasiperiodic oscillations, QPOs) observed in cataclysmic variables.

064.002 **A grid of synthetic spectra for the determination of effective temperature, gravity and metallicity of F, G, and K stars. I. Description of the method.**
R. Cayrel, M.-N. Perrin, B. Barbuy, R. Buser.
Astron. Astrophys., Vol. 247, No. 1, p. 108 – 121 (Jul 1991).

The computation of a grid of synthetic spectra in the wavelength region $\lambda\lambda$4780–5300 Å is described. With the use of this grid, a method is developed for deriving the stellar parameters temperature, gravity and metallicity from a low-resolution spectrum. The use of the grid is tested for 20 reference stars, in two ways. The stellar parameters for these 20 stars are derived (1) directly from the grid, (2) using the synthetic grid in conjunction with the observed spectrum of another reference star. A best fit is obtained by the second method, i.e., by using the grid in combination with a number of standard stars, in this way establishing corrections to the parameters obtained exclusively from the grid.

064.003 **Ambipolar diffusion in atmospheres of cool stars.**
A. E. Dudorov.
Pis'ma Astron. Zh., Tom 17, No. 6, p. 521 – 529 (Jun 1991). In Russian. English translation in Sov. Astron. Lett., Vol. 17, No. 3, p. 222 – 225 (1991).

The ambipolar diffusion of ions and electrons with magnetic field in the temperature minimum zone of the atmosphere of the Sun and cool stars is investigated. It is shown that in the local magnetostatic elements such magnetic ambipolar diffusion may be more efficient than the ambipolar diffusion under the pressure gradient by a factor of v_A/v_s where v_A and v_s are alfvenic and sound velocities. In the solar atmosphere the magnetic ambipolar diffusion may determine the dynamical and thermal behavior of magnetic flux tubes in active and ephemeral regions, small-scale spots elements and the chromospheric moustaches. In stars of late spectral classes the magnetic diffusion diminishes the buoyancy of magnetic flux tubes with cross section $R > H_p$ (H_p is the pressure height scale) under the photosphere and destroys the small-scale slender magnetic flux tubes with a magnetic field $B \geqslant (1 - 3)$ kGs.

064.004 **The UV-delay in dwarf nova outbursts as a consequence of details of the accretion disks' vertical stratification.**
W. J. Duschl.
Astron. Astrophys., Vol. 247, No. 2, p. 393 – 398 (Jul 1991).

The author presents results of time dependent model calculations for the evolution of dwarf nova accretion disks with self-consistent viscosity description. New features in the stationary vertical stratification, due to this modeling of viscosity, allow to explain quantitatively the halt in the visual wavelength band in the rise to maximum during a dwarf nova outburst. The same features are also responsible for the UV-delay during rise to maximum.

064.005 **Thermal and dissipative stability of convection-accretion disks.**
B. D. Scott.
Astron. Astrophys., Vol. 247, No. 2, p. 419 – 428 (Jul 1991).

Thermal and dissipative stability are examined for optically thick convection-accretion disks. The turbulent dissipation is calculated self-consistently within the context of a cascade model with a non-local normal-mode analysis. Model opacity functions with constant $\xi = \mathrm{dlog}\,\varkappa/\mathrm{dlog}\,T$ are considered. The variation of the Shakura-Sunyayev dissipation parameter (α) with the optical thickness is shown to exacerbate the well-known thermal-dissipative instability caused by large values of ξ, sharply reducing the instability threshold value from near 6 to near 2. In particular interest for convective protoplanetary disk models, the usual ice-grain opacity function ($\xi = 2$) leads to an unstable structure in the whole of the ice-grain region of a minimum-mass nebula. On the other hand, the Reynolds number tidal constraint on the formation of Jupiter can only be satisfied with a convective α for ξ near 2. One is thereby forced to look for additional sources for turbulent dissipation in the ice-grain region of the nebula.

064.006 **Numerical hydrodynamics of counter flowing binary stellar winds.**
J. Kallrath.
Astron. Astrophys., Vol. 247, No. 2, p. 434 – 446 (Jul 1991).

The dynamical and geometrical aspects of two quasi radial supersonic counter streaming gas flows are investigated numerically. Analytical approxmations of the boundary-layer as an envelope are derived from trajectorial models (two-fixed-centre problem, restricted three-body problem) and pressure equilibrium surfaces derived from an extended Newtonian approximation are used as initial data for the hyperbolic system of equations describing the conservation of mass, momentum and energy. The equations are solved by using a 2-D axial symmetric second order explicit Godunov scheme set up as a finite volume method on a 100 × 100 grid. By choice of a flux source term the author has modeled a free flow pattern, and the flow in the field of the (repulsive) two-fixed centre problem, which is used as a highly simplified model for the radiation field, for a set of parameters of a model (Wolf-Rayet/O) binary. In that test case a subsonic region with an extension of about six solar radii at the head-on collision point is established. The flow pattern visualized as contour plots of density, logarithmic pressure, and Mach number clearly shows the shock fronts and the contact discontinuity. Particularly, the physical parameters within the sub-sonic region are briefly discussed with respect to the possibility of emitting X-ray bremsstrahlung.

064.007 **A comparison of synthetic and observed Mg_2 indices for cool stars.**
R. K. Gulati, M. L. Malagnini, C. Morossi.
Astron. Astrophys., Vol. 247, No. 2, p. 447 – 454 (Jul 1991).

Synthetic spectra, based on models computed by using Kurucz's ATLAS8 code, have been derived in the wavelength range 4820–5380 Å in order to synthesize the Mg_2 index, which measures the MgH and Mg b at 5177 Å features. The atmospheric parameters are chosen so as to represent intermediate type stars of solar chemical composition. For a sample of dwarf and giant stars, comparison between synthetic and observed fluxes in the three wavelengths bands used to define the Mg_2 index shows systematic differences. The synthetic fluxes in the central band are generally lower than the observed ones, while in the blue and red bands they are systematically higher than the observed ones. Statistical tests reveal that the Mg_2 and the central bandpass flux discrepancies significantly correlate with both temperature and metallicity. The observed–to–synthetic spectra ratios are presented to illustrate the main features of the discrepancies. The grid of computed index values and their relationships with the basic atmospheric parameters effective temperature and surface gravity are also presented. Such a kind of results are a basic step towards the goal of providing a complete reference grid of spectral indices for the interpretation of the integrated light of stellar systems.

064.008 **Colors and stratifications of extended static model photospheres of M stars located on the FGB, AGB and supergiant branch.**
M. S. Bessell, J. M. Brett, M. Scholz, P. R. Wood.
Astron. Astrophys., Suppl. Ser., Vol. 89, No. 2, p. 335 – 366 (Aug 1991).

Colors and stratifications of a grid of spherically extended M giant and supergiant model photospheres, described in a previous paper, have been supplemented and interpolated in metallicity, gravity and photospheric extension so that they correspond to the parameter values of theoretical evolutionary (super–)giant branches for stars with masses of 1, 1.5, 2.25, 5, 10 and 15 $M_\odot$ and metallicities $\log Z/Z_\odot$ of +0.5, 0.0, –0.3, –0.6 and –1.0. These colors may be used, for example, for comparison with observed (super–)giant branches or for synthesizing integrated spectra of clusters and galaxies.

064.009 **Dynamos and angular momentum transport in accretion disks.**
E. T. Vishniac, Jin Li–ping, P. H. Diamond.
Phys. Fluids, B, Vol. 3, No. 8, p. 2374 – 2378 (Aug 1991). Current Physics Microform No.: 9109A2119.

The transport of angular momentum in astrophysical disks is one of the major issues in modern astrophysics. Here, recent work is reviewed that suggests that internal waves, analogous to deep ocean waves, play a critical role in transporting angular momentum in neutral disks and generating a magnetic dynamo in ionized disks. Previously, it was shown that low–frequency, slightly nonaxisymmetric ($|m| = 1$) waves in thin accretion disks could penetrate to small radii with a unique amplitude because of nonlinear saturation. Here, the ability of these waves to drive an α–Ω dynamo in a disk of thickness H and radius r and keplerian rotational frequency $\Omega(r) \sim r^{-3/2}$ is examined. The asymmetry in the wave distribution that creates a nonzero helicity follows from the fact that the fundamental waves all have a positive angular momentum flux. As a result, there will be a large–scale magnetic field driven by an α–Ω dynamo. It is also likely that small–scale fields, driven by higher–order wave modes, will contribute significantly to the local value of $B_r B_\varphi$. It is argued that the magnetic field saturates when its pressure is comparable to the thermal pressure and a crude model of the nonlinear transfer of power to small–scale turbulence is presented. The dynamo process creates a large–scale, axisymmetric toroidal field with $B_r \sim (H/r)^{3/2} B_\varphi$. Smaller–scale waves create small–scale fields with a maximum $b_r b_\varphi \sim (H/r)^{6/5} P$. In this model, viscous and thermal instabilities in radiation pressure dominated, and electron scattering regions in accretion disks appear to be substantially suppressed.

064.010 **The shape of dense structures in plasma envelopes of slowly rotating hot magnetic stars.**
P. A. Bespalov, S. S. Davydenko.
Pis'ma Astron. Zh., Tom 17, No. 9, p. 810 – 816 (Sep 1991). In Russian. English translation in Sov. Astron. Lett., Vol. 17, No. 5.

The geometrical structure of dense formations in plasma envelopes of rotating hot magnetic stars whose radiation pressure force on plasma exceeds the gravity force is found. The problems of stability of dense plasma structures are investigated. The influence of rotation effects on electromagnetic radiation of plasma star envelopes is noted.

064.011 **Spherically symmetric model atmospheres of late–type stars.**
P. D. Bennett.
Bull. Am. Astron. Soc., Vol. 23, No. 2, p. 902 (1991). Abstract. – See Abstr. 012.008 for the main entry.

064.012 **Numerical simulations of flares on M dwarf stars.**
C.–C. Cheng, R. Pallavicini.
Bull. Am. Astron. Soc., Vol. 23, No. 2, p. 902 (1991). Abstract. – See Abstr. 012.008 for the main entry.

064.013 **Polarization models for a two–component wind in early–type stars.**
J. P. Cassinelli, M. Taylor.
Bull. Am. Astron. Soc., Vol. 23, No. 2, p. 912 (1991). Abstract. – See Abstr. 012.008 for the main entry.

064.014 **Heating mechanism(s) for transition layers in giants.**
E. Böhm–Vitense, J. Mena–Werth.
Bull. Am. Astron. Soc., Vol. 23, No. 2, p. 942 (1991). Abstract. The full paper is published in Astrophys. J., Vol. 378, No. 2, p. 718 – 724 (10 Sep 1991). – See Abstr. 012.008 for the main entry.

064.015 **The SDSC grid from [–5] to [+1] at 2 km/s.**
R. L. Kurucz.
Bull. Am. Astron. Soc., Vol. 23, No. 2, p. 967 (1991). Abstract. – See Abstr. 012.008 for the main entry.

064.016 **Photon bubbles: overstability in a magnetized atmosphere.**
J. Arons.
Bull. Am. Astron. Soc., Vol. 23, No. 2, p. 967 (1991). Abstract. – See Abstr. 012.008 for the main entry.

064.017 **Hydrodynamic studies of accretion onto ONeMg white dwarfs using a large nuclear reaction network.**
S. Starrfield, M. Politano, J. W. Truran, W. M. Sparks, A. Weiss.
Bull. Am. Astron. Soc., Vol. 23, No. 2, p. 967 – 968 (1991). Abstract. – See Abstr. 012.008 for the main entry.

064.018 **The pressure ionization of metals in cool, DZ white dwarfs.**
E. R. Peterson.
Bull. Am. Astron. Soc., Vol. 23, No. 2, p. 968 (1991). Abstract. – See Abstr. 012.008 for the main entry.

064.019 **A model atmosphere interpretation of the line strengths and colors of Baade's window M giants.**
D. F. Carbon, G. C. Augason.
Bull. Am. Astron. Soc., Vol. 23, No. 2, p. 968 (1991). Abstract. – See Abstr. 012.008 for the main entry.

064.020 **Do we really know what the actual chromospheric heating requirements are?**
J. Trujillo Bueno.
International Conference on Mechanisms of Chromospheric and Coronal Heating, p. 60 – 62 (1991). – See Abstr. 012.011 for the main entry.

By means of multi–dimensional radiative transfer calculations it is shown that the plane–parallel approximation begins to fail to give proper account of the energetic role of spectral lines when the stellar atmospheric inhomogeneities have horizontal structural lengths smaller than 1000 km.

064.021 **Dust formation in winds of long–period variables. I. Equations, method of solution, simple examples.**
E. A. Dorfi, S. Höfner.
Astron. Astrophys., Vol. 248, No. 1, p. 105 – 114 (Aug 1991).

The authors present time–dependent numerical calculations of the formation and growth of dust particles in stellar winds. The production rate and the properties of the dust are obtained by solving the moment equations derived by Gail and Sedlmayr

(1988) and generalized by Gauger et al. (1990). The equations are reformulated to allow arbitrary motions of grid points. The numerical technique used is similar to an implicit formulation of the equations of radiation hydrodynamics in conservative form and will allow an incorporation into the existing radiation hydrodynamics code. In this first paper the authors calculate with a new numerical method the formation of carbon dust in stationary stellar winds around cool C-rich stars to compare these results with models available in the literature.

064.022 **What can solar and stellar ultraviolet observations tell about chromospheric and coronal heating mechanisms?**
J. L. Linsky.
International Conference on Mechanisms of Chromospheric and Coronal Heating, p. 166 – 178 (1991). – See Abstr. 012.011 for the main entry.

The determination of which heating mechanisms are responsible for the observed emission line fluxes is a difficult, nonunique process. The author discusses some of the spectral diagnostic problems encountered in deriving differential emission measures and methods for including all major emitters, especially Fe II, in determining the total power radiated from a plasma. The author proposes two scaling laws by which one may simply determine the total power radiated in the chromosphere (from the Mg II h and k lines) and the higher temperature layers (from the C IV 1549 Å feature). These scaling laws lead to estimates of the total power radiated and thus the mechanical heating rates for basal and saturated atmospheres. The heating rates for plages in RS CVn systems are consistent with saturated heating.

064.023 **He I 5876 Å line as an indicator of chromospheric heating in young F-type MS stars.**
R. J. García López, R. Rebolo, J. E. Beckman, C. D. McKeith.
International Conference on Mechanisms of Chromospheric and Coronal Heating, p. 179 – 181 (1991). – See Abstr. 012.011 for the main entry.

The authors have obtained observations of He I D_3 5876 Å absorptions in 31 F-type MS stars, from the Hyades and the Ursa Major Group, including 5 field stars. From these plus literature data, the authors show a clear separation between early to mid-F (0.3 < B–V < 0.42) and later type stars, in the plot of D_3 equivalent width W_λ D_3 vs Rossby number R_0. This constrasts with the published monotonic relation between L_x/L_{bol} and R_0 over the range early-F to M. It appears that a single, presumably magnetic, mechanism heats coronae throughout the range, whereas in early to mid-F stars the chromospheres are heated non-magnetically, as shown by the two regimes in D_3. Acoustic wave heating appears to satisfy many of the observational constraints.

064.024 **Chromospheric modelling of active regions on AU Mic.**
E. R. Houdebine.
International Conference on Mechanisms of Chromospheric and Coronal Heating, p. 182 – 184 (1991). – See Abstr. 012.011 for the main entry.

The author presents results of a successful attempt in modelling H_α and H_β high resolution profiles of a dMe star (the chromosphere of active regions of AU Mic). He shows that a very high transition region column density and a steep temperature gradient in the chromosphere are required to reproduce with accuracy the emission profiles of this very active star. The chromospheric structure is similar to those obtained when modelling solar flares, implying a very large and continuous chromospheric heating rate.

064.025 **New clues to atmospheric heating processes in luminous cool stars.**
A. K. Dupree.
International Conference on Mechanisms of Chromospheric and Coronal Heating, p. 185 – 199 (1991). – See Abstr. 012.011 for the main entry.

New spectroscopic and photometric observations of luminous cool stars enable to identify and map out the presence of various heating mechanisms in the color–magnitude diagram. Low mass stars on the main sequence show clear magnetic dynamo activity. As a star evolves off the main sequence and moves toward the supergiant region, or across and up the red giant branch, observations suggest that non-magnetic heating processes occur and eventually dominate in the atmospheres of giant and supergiant stars. The coolest and most luminous stars may have pure hydrodynamic chromospheres. Hybrid giant and supergiant stars appear to provide the link between the magnetically-dominated atmosphere and the hydrodynamic chromosphere.

064.026 **Chromospheres of cool non-Mira giant stars.**
H. R. Johnson, U. G. Jørgensen, D. G. Luttermoser.
International Conference on Mechanisms of Chromospheric and Coronal Heating, p. 200 – 205 (1991). – See Abstr. 012.011 for the main entry.

The authors review the density and velocity structure of the chromospheres and photospheres of the coolest non-mira M and carbon stars and subject them to the constraints imposed by known mass-loss rates. They note two severe problems. (1) In available chromospheric models, the density falls, within a short distance (< 1 R_*) above the photosphere, to less than the density needed to sustain the observed mass loss with the observed flow velocities. To keep the model chromospheric density above the mass-loss value requires large amounts of turbulent energy. (2) Even with reasonable amounts of turbulent energy, densities in all chromospheric models and some photospheric models become so low that mass flow may become significant. In addition, chromsopheres are likely inhomogeneous.

064.027 **What can observations of giants and supergiant stars tell us about chromospheric and coronal heating?**
M. Cuntz, R. E. Stencel.
International Conference on Mechanisms of Chromospheric and Coronal Heating, p. 206 – 221 (1991). – See Abstr. 012.011 for the main entry.

The authors discuss chromospheric and coronal activities in specific yellow and red giant and supergiant stars. Implications for heating mechanisms are described. Coronal yellow giant stars, which are rapid rotators, are most likely dominated by heating due to dynamo generated magnetic fields. The heating in noncoronal yellow giant stars, on the other hand, is probably dominated by the dissipation of acoustic waves. These stars have slow rotation rates and show no major chromospheric activities with the remarkable exception of the highly variable He I λ10830 line. However, short-period acoustic waves cannot explain the observed mass loss in these stars. Other types of outer atmospheric activity include pulsation and semiregular variabilities, present in Mira-type stars and red supergiants, respectively.

064.028 **Chromospheric activity in late-type giants and supergiants: constraints on heating theories.**
L. Pasquini, E. Brocato, R. Pallavicini.
International Conference on Mechanisms of Chromospheric and Coronal Heating, p. 222 – 224 (1991). – See Abstr. 012.011 for the main entry.

The authors analyse Ca II K line emission in a homogeneous sample of late-type giants and supergiants. Using the Wilson–Bappu relationship and colour-temperature scales, they construct an HR diagram and compare it with theoretical evolutionary tracks. They show that in spite of the errors involved in the determination of the fundamental stellar parameters, a clear relationship between chromospheric surface activity and stellar mass is present. The authors interpret the observed trends in terms of dynamo-driven activity and suggest that the chromospheres of yellow supergiants are most likely heated by a magnetic process. Less massive stars ascending along the Red Giant Branch, on the contrary, are probably heated by a non magnetic process.

064.029 **Coronal heating and the dividing line.**
B. Haisch.
International Conference on Mechanisms of Chromospheric and Coronal Heating, p. 225 – 227 (1991). – See Abstr. 012.011 for the main entry.

IUE and Einstein have shown that coronal and transition region emissions drop dramatically as stars evolve across the

Dividing Line. Diminishing dynamo activity due to decreasing rotation is likely, but a precise correlation has yet to be established. The angular momentum evolution of late-type stars needs further study.

064.030 **An infrared perspective on chromospheres.**
T. R. Ayres.
International Conference on Mechanisms of Chromospheric and Coronal Heating, p. 228 – 240 (1991). – See Abstr. 012.011 for the main entry.

The infrared spectrum beyond 1 μm is rich in diagnostics of temperature, velocity, and magnetic fields. It provides a view of chromospheric phenomena complementary to that of the visible, ultraviolet, and X-ray bands. Whereas the shorter wavelengths preferentially reveal the heated gas, the longer wavelengths map the relatively quiescent, cool material surrounding discrete chromospheric structures. A key example is carbon monoxide in the Sun: it presents a diagnostic dilemma that must be resolved before we can have any confidence in our understanding of the physical structure of the nearest, most scrutinized stellar chromosphere.

064.031 **Ambipolar diffusion of hydrogen in the atmosphere of Ap stars.**
J. Babel, G. Michaud.
Astron. Astrophys., Vol. 248, No. 1, p. 155 – 165 (Aug 1991).

High signal to noise observations of magnetic stars have allowed the precise determination of magnetic field geometry as well as of spots or rings of chemical abundance anomalies on the surface of several Ap stars. This has led to a surge of new constraints for the diffusion model. The failure of the simple diffusion model, has shown the need for a careful reinvestigation of the various processes competing in the atmosphere of Ap stars leading to the ring or patchy structures observed. The authors show that ambipolar diffusion of hydrogen could be one of the major processes affecting the chemical abundances supported in the atmosphere of Ap stars. It is important when magnetic field suppresses the hydrogen convection zone. Furthermore, its interaction with the large magnetic fields observed, could strongly modify the structure of the atmosphere in the line-forming region.

064.032 **Magnetic activity across the Hertzsprung–Russell diagram.**
C. Zwaan.
International Conference on Mechanisms of Chromospheric and Coronal Heating, p. 241 – 256 (1991). – See Abstr. 012.011 for the main entry.

Emissions from the outer atmospheres of cool stars are discussed using flux-color, flux-flux, and flux-rotation diagrams. Two distinct components have been identified: a basal emission and a magnetically controlled emission ("magnetic activity"). The tight power-law relations between the flux densities point to one single activity parameter determining the emissions from the outer atmospheres (this structure parameter complements the parameters describing stellar photospheres). This activity parameter is probably equivalent to the magnetic flux in the atmosphere, which in turn depends on the "dynamo efficiency". The empirical activity-rotation relation demonstrates that the dynamo efficiency depends on the rotation rate. The time dependence of the dynamo within evolving stars explains the distribution of activity of single stars across the HR-Diagram in a semi-quantitative fashion. Ways are pointed out how tidal effects may enhance the efficiency of dynamos in components of close binaries. Thus the distribution of magnetic activity across the HR-Diagram and the dependence on binary parameters are entirely attributed to the time-dependent dynamo action within the stellar interior. Clues to the magnetically controlled heating mechanisms are to be found in the flux-flux relations, the coronal temperature intervals, and particularly from observations of the processes in the intricate magnetic structure of the solar atmosphere. The basal emission is attributed to purely acoustic heating.

064.033 **Relations between activity and magnetic fields.**
C. J. Schrijver.
International Conference on Mechanisms of Chromospheric and Coronal Heating, p. 257 – 272 (1991). – See Abstr. 012.011 for the main entry.

Observational constraints on the relationship between radiative losses from the quiescent outer atmospheres of the Sun and cool stars and the underlying photospheric magnetic field are reviewed. The coronal radiative flux density appears to vary linearly with the magnetic flux density, at least when observing with sufficiently low spatial resolution. The chromospheric radiative losses increase almost like the square-root of the magnetic flux density. Transition-region flux densities show a dependence on the magnetic flux density which is intermediate to that of the chromospheric and coronal diagnostics; this may be the result of a combination of local heating like that which occurs in the chromosphere and heating by conduction of thermal energy from the corona. The possible causes for the non-linearity of flux-flux relationships between stellar fluxes and between spatially-resolved solar fluxes are discussed. Some observational constraints on the spatial and temporal variability of the heating mechanism(s) are given.

064.034 **Global electrodynamic coupling in stellar atmospheres.**
L. Cram.
International Conference on Mechanisms of Chromospheric and Coronal Heating, p. 282 – 286 (1991). – See Abstr. 012.011 for the main entry.

It is known that the dominant source of star-to-star variation in the level of chromospheric and coronal activity is star-to-star variation in the area covered by magnetic fields (i.e. the magnetic filling factor). Thus, detailed studies of the origin of chromospheric and coronal heating need to look beyond such variations if they are to explore the physical mechanisms that produce different heating rates in different parts of the atmosphere. One way to do this is to study flux-flux relations obtained either from spatially resolved solar observations or stellar data. The author offers a simple geometric interpretation of observed flux-flux relations that may shed some light on the nature of electrodynamic coupling in the atmospheres of the Sun and stars.

064.035 **Implications for coronal heating theories from stellar X-ray observations.**
R. Rosner.
International Conference on Mechanisms of Chromospheric and Coronal Heating, p. 287 – 299 (1991). – See Abstr. 012.011 for the main entry.

The author reviews the essential aspects of X-ray emission from late spectral type stars (principally the dwarf stars) from the point of view of theory: What can the variation of X-ray emission properties from such stars, and their correlations with other stellar properties, reveal about the processes which lead to coronal heating, and hence the observed X-ray emission? Specific topics discussed include the possible "saturation" of stellar surface X-ray fluxes at levels comparable to the brightest regions on the solar surface; and a variety of aspects of the stellar rotation-activity connection. In most of these cases, the underlying physics remains unclear. One key conclusion is that the observations reveal a dependence of X-ray emission levels on stellar rotation which is "unreasonably" well-correlated; that is, given the current state of coronal heating theories, it appears that the degree of correlation between stellar coronal X-ray luminosity and stellar rotation rate is far stronger than a naive application of available theories would predict.

064.036 **Empirical constraints on coronal heating processes.**
C. Jordan.
International Conference on Mechanisms of Chromospheric and Coronal Heating, p. 300 – 315 (1991). – See Abstr. 012.011 for the main entry.

Observations of the transition regions and coronae of main-sequence stars that constrain coronal heating theories are reviewed. The significance of the form of the emission measure distribution and of correlations between transition region and coronal fluxes is discussed. Existing measurements of emission line widths place limits on the available wave flux. Correlations can be found between coronal parameters and stellar rotation rates, without knowledge of the heating mechanism. Measurements of surface magnetic fields and their dependence on rotation are vital in understanding coronal heating. At present, heating via stochastic or twisted fields seems the most plausible mechanism.

064.037 **Minimum coronal energy requirements: constraints for heating mechanisms.**
R. Hammer.
International Conference on Mechanisms of Chromospheric and Coronal Heating, p. 316 – 318 (1991). – See Abstr. 012.011 for the main entry.
New calculations of the minimum energy requirements of magnetically open coronal regions on cool stars are presented. Several of the simplifying assumptions underlying previous calculations have been omitted; thus the present results are more accurate and can be extended to extreme regions of the parameter space. The models can be scaled from one star to another. For low gravity stars, the waviness of the emissivity as a function of temperature leads to multiple local minima of the coronal energy loss.

064.038 **Sun – hot star contrast in chromospheric/coronal Te(r). Nonradiative heating vs. outflow enhanced opacity.**
V. Doazan, R. N. Thomas.
International Conference on Mechanisms of Chromospheric and Coronal Heating, p. 319 – 323 (1991). – See Abstr. 012.011 for the main entry.
The authors present a condensed perspective on the contribution of studies of normal and peculiar hot stars to delineating facts/implications of chromosphere/coronae occurring throughout the HRD.

064.039 **Formation of chromospheres and coronae of accretion disks by viscous dissipation.**
R. Wehrse, H. Störzer, G. Shaviv.
International Conference on Mechanisms of Chromospheric and Coronal Heating, p. 324 – 326 (1991). – See Abstr. 012.011 for the main entry.
The formation of chromospheres and coronae of accretion disks around white dwarfs is studied by means of stationary non-grey models with parameterized opacities. It is shown that very high temperatures in the outer layers occur if the pressure dependence of the opacity is moderate or strong ($d\ln\varkappa/d\ln P > 0.25$) and the Shakura–Sunyaev parameter α is close to unity.

064.040 **Acoustic heating.**
P. Ulmschneider.
International Conference on Mechanisms of Chromospheric and Coronal Heating, p. 328 – 343 (1991). – See Abstr. 012.011 for the main entry.
Acoustic shock waves are a viable and prevalent heating mechanism both in early– and in late–type stars. Acoustic heating appears to be a dominant mechanism for situations where magnetic fields are weak or absent, as locally for certain stellar surface regions, and globally for very slowly rotating stars with intrinsically weak magnetic fields, like the basal flux stars, the giants and supergiants. It also seems to be an important mechanism in F–stars, where the acoustic flux production is at a maximum. For more rapidly rotating stars, acoustic heating is a weak background effect. The limiting shock strength behaviour and radiation damping of acoustic waves are discussed. Both the directly observed solar acoustic wave flux and the empirical solar chromospheric cooling flux agree with the theoretical limiting shock wave flux. The heating by pulsation–driven waves is by hydrodynamic shocks and thus is a type of acoustic heating. In strong magnetic fields slow–mode MHD waves or longitudinal tube waves are essentially acoustic waves which also dissipate by hydrodynamic shocks. These waves are discussed with the magnetic heating mechanisms.

064.041 **On the intrinsic difficulty of producing stellar coronae with acoustic waves.**
R. Hammer, P. Ulmschneider.
International Conference on Mechanisms of Chromospheric and Coronal Heating, p. 344 – 346 (1991). – See Abstr. 012.011 for the main entry.
The authors discuss theoretical limits on the potential of acoustic waves to produce coronae around inactive solar–like stars. For a star of solar mass and radius, short–period shock waves (which are produced most efficiently in the convection zone) could generate a corona only for base pressures and energy fluxes that are smaller than the observed values by several orders of magnitude. The energy budget of solar coronal holes could be supplied only by shock waves of extremely long periods ($\approx 2\ P_A$), which are difficult to produce in solar–like stars, and which suffer from propagation restrictions in the photosphere and chromosphere. This fundamental difficulty to produce coronae with acoustic waves is even aggravated towards stars of smaller gravity.

064.042 **The effect of waves on optically thin transition region lines.**
V. Hansteen.
International Conference on Mechanisms of Chromospheric and Coronal Heating, p. 347 – 349 (1991). – See Abstr. 012.011 for the main entry.
Numerical simulations of acoustic waves passing through plasma at transition region temperature and density are performed. The resultant effects on optically thin resonance line profiles are computed. The ionization balance of the emitting ion is solved consistently with the hydrodynamic equations. The results of the calculations show that acoustic waves can produce average line shifts on the order of the amplitude of the wave velocity.

064.043 **Nonlinear pulse propagation in a stratified atmosphere.**
G. Bodo, W. Kalkofen, S. Massaglia, P. Rossi.
International Conference on Mechanisms of Chromospheric and Coronal Heating, p. 353 – 355 (1991). – See Abstr. 012.011 for the main entry.
In a gravitationally stratified atmosphere, a pulse followed by eigenoscillations of the medium at the cutoff period may show motions that are better able than a model based on short–period waves to reproduce intensity enhancements and frequency shifts in the K line. In order to examine this question the authors study pulse propagation in an isothermal medium.

064.044 **The shock wave heating mechanism of pulsating star chromospheres.**
S. M. Andrievsky (*S. M. Andrievskij*), G. A. Garbunov.
International Conference on Mechanisms of Chromospheric and Coronal Heating, p. 356 – 358 (1991). – See Abstr. 012.011 for the main entry.
A mechanism is suggested which explains the origin and variability of chromospheric emission in the centres of the Mg II h and k lines in the spectra of pulsating stars. According to the model the emission originates from gas heated by a shock wave passing through the outer layers of the star's atmosphere.

064.045 **Ionization pumping.**
C. Lindsey.
International Conference on Mechanisms of Chromospheric and Coronal Heating, p. 359 – 365 (1991). – See Abstr. 012.011 for the main entry.
The authors consider the role ionization relaxation could play in the hydrodynamics of wave dissipation and heating in the chromosphere. While it is important to take proper dynamical account of ionization chemistry in chromospheric hydrodynamics, it is difficult to see at this point how ionization relaxation would contribute directly to chromospheric heating in the Sun and other stars.

064.046 **Recent developments in theories of wave generation.**
Z. E. Musielak.
International Conference on Mechanisms of Chromospheric and Coronal Heating, p. 369 – 379 (1991). – See Abstr. 012.011 for the main entry.

Theories of generation of acoustic and magnetohydrodynamic waves are briefly reviewed with a special emphasis on their applicability to stellar convection zones. Recent developments in these theories and possible directions for future studies are discussed. A new approach allowing to extend the Lighthill–Proudman theory of sound generation to a non–isothermal medium is also presented.

064.047 **Magnetic confinement, Alfvén wave reflection, and the origin of X–ray and mass loss "dividing lines".**
C.–H. An, R. Rosner, Z. E. Musielak, R. L. Moore, S. T. Suess.
International Conference on Mechanisms of Chromospheric and Coronal Heating, p. 445 – 447 (1991). – See Abstr. 012.011 for the main entry.

The authors discuss a simple model for the origin of the coronal and mass loss dividing lines which separate late–type giants and supergiants with and without X–ray emitting corona, and with and without significant mass loss. The basic physical effects are magnetic confinement for hot coronal material and large reflection efficiency of Alfvén waves in the cool exponential atmospheres.

064.048 **Coronal magnetic structure: the role of ideal MHD invariants.**
M. A. Berger.
International Conference on Mechanisms of Chromospheric and Coronal Heating, p. 570 – 581 (1991). – See Abstr. 012.011 for the main entry.

Magnetic helicity measures properties of field lines such as linking, knotting, and twisting. Severe bounds can be placed on its dissipation in a highly conducting plasma. Motions of field line endpoints at the photosphere inject helicity into the corona. An associated invariant is the net twist between two neighboring field lines. Random photospheric motions should twist neighboring lines through a few turns in one day. The current associated with this twist is small, suggesting that most coronal current resides in sheets. Other ideal invariants can be defined which measure the mutual braiding of three or more field lines. Coronal heating models which rely on the buildup of helicity and braiding will be reviewed.

064.049 **On the possibility of nuclear processes in atmospheres of flare stars.**
H. A. Bethe, G. A. Gurzadyan.
Astrophys. Space Sci., Vol. 183, No. 1, p. 1 – 10 (Sep 1991).

A resume of the main results obtained in the framework of fast electron hypothesis is presented. The fast electrons are possessed by strong constant properties, i.e., nearly monoenergetic in spectrum and with the energy of $\gamma \sim 3$. Fast electrons appear spontaneously in all kinds of cool dwarfs, and stimulate the generation of all types of flares in all ranges of electromagnetic waves – optical, ultraviolet, radio, X–rays. The general flare mechanisms are inverse Compton effect at the faint and moderate flares and nonthermal bremsstrahlung at the powerful ones. This hypothesis explains an enormous number of observational facts. In the second part, the problem of the spontaneous appearance of fast electrons above the stellar atmosphere is discussed. The nature of the fast electrons is universal and is independent of the physical conditions of the star or its atmosphere. Their origin has nothing in common also with the well known acceleration processes in stellar atmospheres. Nuclear decay processes seem to be most probable for the appearance of fast electrons in outer regions of stellar atmospheres. Some decay processes are examined with various probability of their realization. Creation of the radioactive nucleus ^{13}N as a result of capture of H by ^{12}C and its consequent decay with the escape of a positron seems most probable. The high effectivity of this process in binary systems is outlined.

064.050 **A powerful local shear instability in weakly magnetized disks. I. Linear analysis.**
S. A. Balbus, J. F. Hawley.
Astrophys. J., Vol. 376, No. 1, p. 214 – 222 (20 Jul 1991).

In this paper and a companion work, the authors show that a broad class of astrophysical accretion disks is dynamically unstable to axisymmetric disturbances in the presence of a weak magnetic field. Because of the ubiquity of magnetic fields, this result bears upon gaseous differentially rotating systems quite generally. This work presents a linear analysis of the instability. The instability is local and extremely powerful. The maximal growth rate is of order of the angular rotation velocity and is independent of the strength of the magnetic field, provided only that the energy density in the field is less than the thermal energy density. Unstable axisymmetric disturbances require the presence of a poloidal field component, and are indifferent to the presence of a toroidal component. The instability also requires that the angular velocity be decreasing outward. In the absence of a powerful dissipation process, there are no other requirements for instability. Fluid motions associated with the instability directly generate both poloidal and toroidal field components. The authors discuss the physical interpretation of the instability in detail. Conditions under which saturation occurs are suggested. The nonemergence of the classical Rayleigh criterion for shear instability in the limit of vanishing field strength is noted and explained.

064.051 **A powerful local shear instability in weakly magnetized disks. II. Nonlinear evolution.**
J. F. Hawley, S. A. Balbus.
Astrophys. J., Vol. 376, No. 1, p. 223 – 233 (20 Jul 1991).

The authors consider the dynamical evolution of an accretion disk undergoing Keplerian shear flow in the presence of a weak magnetic field. A linear perturbation analysis presented in a companion paper shows that such a flow is dynamically unstable; here, some nonlinear consequences of this instability are considered. The equations of compressible magnetohydrodynamics are solved using a two–dimensional finite–difference code. The Keplerian disk is threaded with a weak magnetic field that has a magnetic energy density much less than the thermal pressure. When perturbations are small, the numerical results are consistent with linear perturbation theory. Neither compressibility nor the presence of a toroidal field have a significant effect on the evolution of unstable modes. The most important dynamic effect is the redistribution of angular momentum leading to a strong interchange instability. The resulting radial motions produce field geometries that are conducive to reconnection, suggesting a possible mechanism for mode saturation. Even for very weak fields, whose most unstable wavelengths are small, nonlinear evolution results in the growth of structure on large scales. Total magnetic field energy increases by about one order of magnitude over the course of the simulations.

064.052 **Approximate lambda–operators working at optimum convergence rate. II. Line transfer in expanding atmospheres.**
J. Puls.
Astron. Astrophys., Vol. 248, No. 2, p. 581 – 589 (Aug 1991).

The author develops an approximate Λ–operator for the line transfer problem in the comoving frame (CMF) with the following properties: (i) it is parameter–free and optimized in analogy to the operator developed by Olson et al. (1986) for the monochromatic transfer in static atmospheres. (ii) It is purely local and intimately related to the formulation of the line transfer in the Sobolev approach. (iii) It can be calculated parallel to the formal solution in a small number of additional operations. With this newly derived operator it is possible to create a new generation of NLTE codes solving for the total range of expanding atmospheres of hot luminous stars (photosphere + wind = "unified atmosphere", see Gabler et al. 1989), applying the Sobolev approach and the CMF transfer in parallel, the latter in cases where a high accuracy is required or the Sobolev approximation is no longer well defined, i.e. in regions below the sonic point. Because of the structural and physical resemblence of the Sobolev and the local CMF operator, these two different approaches can be arbitrarily mixed in the system of statistical equilibrium equations. This mixture of the radiative transfer where the Sobolev solution costs almost no computing time will allow the treatment of a large number of metal ions in the "unified" approach.

064.053 **Analytical studies of collimated winds. II. Topologies of 2–D helicoidal MHD solutions.**
K. Tsinganos, E. Trussoni.
Astron. Astrophys., Vol. 249, No. 1, p. 156 – 172 (Sep 1991).
An analytic class of solutions of the full MHD equations for steady, rotating, nonspherically symmetric and helically collimated magnetized outflows from a gravitating central object is presented. By assuming a streamline geometry appropriate for helicoidally collimated outflows the required heating/cooling and temperature distribution to drive the outflow is calculated. The novel approach is followed of relaxing the restrictive assumption of a polytropic equation of state, which has been used in almost all available studies of rotating magnetized winds. It is found that the authors' solutions correspond to a polytropic relationship between pressure and density but with a variable index γ. The topology of the solutions is governed by a pair of hydromagnetic critical points that select a unique wind–type outflow solution. The parameter space of the solutions is fully explored by varying the strength of the magnetic field, rotation and collimation. Some general properties are briefly discussed in connection to astrophysical systems, such as coronal hole/helmet structures and helicoidal bipolar flows/surrounding denser disks.

064.054 **Dust–driven winds. II. Are the grains the real cause of the motion?**
N. Berruyer.
Astron. Astrophys., Vol. 249, No. 1, p. 181 – 191 (Sep 1991).
Different mechanisms have been suggested to explain the origin of the wind in the envelope of a red giant and, in particular, the mechanism of the radiation pressure acting on the dust. Conditions under which the presence of the dust grains may generate an instability in a static envelope and drive the motion of the gas component, are determined. It is shown that a minimum dust grain flux is necessary, for their presence to be at the origin of the wind and to impose motion.

064.055 **The outer layers of cool, non–Mira carbon stars.**
H. R. Johnson.
Astron. Astrophys., Vol. 249, No. 2, p. 455 – 464 (Sep 1991).
The author considers the outer layers and near circumstellar envelope (CSE) of a typical carbon star, drawing on available information regarding temperatures and densities in the photosphere (from theoretical models) and chromosphere (from empirical models based on ultraviolet observations) and densities and velocities in the distant CSE (from radio observations of CO lines). The outer atmosphere has generally been studied as two distinct regions: (a) a relatively thin high–temperature region (chromosphere) which is close to hydrostatic equilibrium and (b) a much more extended, cooler region of outflowing gas and dust. When the constraint of constant mass loss is applied to the chromospheric models from the literature, a clear discrepancy in density appears at the interface of the hydrostatic and hydrodynamic regimes. To extend the outer photosphere and chromosphere to match the mass loss density appears to require an injection of energy and momentum by some mechanism rather close to the stellar surface. Either significant grain formation must occur inside this point, or the flow must be initiated by some other mechanism, probably acoustically–hydrodynamically.

064.056 **Thermosolutal–convective instability of a stellar atmosphere in the presence of suspended particles.**
R. C. Sharma, A. Rajput.
Astrophys. Space Sci., Vol. 183, No. 2, p. 183 – 191 (Sep 1991).
Thermosolutal–convective instability of a stellar atmosphere is considered in the presence of suspended particles. The criteria for monotonic instability are derived which are found to hold good in the presence of uniform rotation and uniform magnetic field, separately, on the thermosolutal–convective instability in the presence of suspended particles.

064.057 **Luminosity limit for alpha–viscosity accretion disks.**
E. P. Liang, A. Wandel.
Astrophys. J., Vol. 376, No. 2, p. 746 – 749 (1 Aug 1991).
The authors report the existence of a luminosity limit for α–viscosity physically thin accretion disks around black holes, using a new formulation of the radiation equation bridging optically thick and thin regimes. For α close to unity this limit can be lower than the Eddington limit. Physically this limit is due to the combined effects of gas and radiation pressure which become too large to satisfy vertical hydrostatic balance at intermediate optical depths for sufficiently high luminosities. This effect was overlooked in previous treatments using only the optically thin or thick limits of the radiative equation.

064.058 **Nonlinear instability of the accretion line.**
N. Soker.
Astrophys. J., Vol. 376, No. 2, p. 750 – 756 (1 Aug 1991).
The author studies the instability of the flow along the accretion line in the two–dimensional Bondi – Hoyle – Lyttleton type accretion flow. By means of numerical simulations the author explores the nonlinear regime of the instability, thus extending a previous work that dealt with the linear regime using the WKB approximation. In contrast to the behavior in the linear regime, in the nonlinear regime the radial and tangential modes are coupled. The radial instability, which manifests itself as large density and velocity variations on short distances, grows far beyond the linear regime. The tangential instability, on the other hand, stays in the linear regime and is dominated by long–wavelength perturbations. This results from both the influence of the radial instability and the incoming material (the material accreted onto the accretion line).

064.059 **Thermosolutal–convective instability of a composite stellar atmosphere in the presence of a variable magnetic field.**
R. C. Sharma, V. Kumari.
Astrophys. Space Sci., Vol. 184, No. 1, p. 43 – 48 (Oct 1991).
The thermosolutal–convective instability of a composite stellar atmosphere is considered in the presence of variable horizontal magnetic field and collisional effects. The criteria for monotonic instability are obtained which generalize the criterion derived for thermal–convective instability in the absence of above effects.

064.060 **The Sun as a prototype in the study of stellar magnetic activity.**
C. J. Schrijver.
Reviews in Modern Astronomy 4, p. 18 – 42 (1991). – See Abstr. 003.008 for the main entry.
This paper presents a brief and basic summary of a selection of topics in the field of stellar magnetic activity. With magnetic activity the author means all phenomena, particularly those in the stellar atmospheres, which are directly or indirectly caused by the presence of a magnetic field in stars similar to the Sun, although large–scale flaring processes are excluded. This review concentrates on the peculiar circumstance that relationships between radiative flux densities from the outer atmospheres of a large variety of cool stars are often distinctly non–linear while exhibiting surprisingly little scatter despite the large range of fundamental properties of the stellar sample. This is all the more astounding in view of the complex and variable structure of the solar outer atmosphere. Why this is so, and how it may be explained by combining solar and stellar observations, constitutes the core of this review. The range of topics covered serves also to convey the complexity of the stellar outer atmosphere which nevertheless results in such simple flux–flux relationships.

064.061 **Stellar Zeeman analyses: effects of multi-component atmospheres.**
G. W. Marcy, G. Basri, J. A. Valenti.
IAU Colloquium No. 130: The Sun and cool stars – activity, magnetism, dynamos, p. 407 – 410 (1991). – See Abstr. 012.024 for the main entry.
The authors simulate the radiative transfer of Zeeman-broadened lines in an active star containing both quiet and flux-tube regions. A traditional one-atmosphere Zeeman analysis of these synthetic lines yields magnetic field measurements systematically inaccurate by up to 40%. However, two atmospheric components alone cannot reproduce the observed line strengths, luminosities or photometric constancy of active stars.

064.062 **Linear polarization and magnetic fields in cool stars.**
J. Huovelin, S. H. Saar.
IAU Colloquium No. 130: The Sun and cool stars – activity, magnetism, dynamos, p. 420 – 423 (1991). – See Abstr. 012.024 for the main entry.
Observations and theoretical models of braodband linear polarization (BLP) in cool stars are discussed. Two effects, (1) differential saturation in magnetically sensitive absorption lines and, (2) Rayleigh and Thomson scattering are considered as the possible causes of the BLP. Results of theoretical models, and comparisons with observations are summarized.

064.063 **Recent advances in our understanding of chromospheric and coronal heating mechanisms.**
J. L. Linsky.
IAU Colloquium No. 130: The Sun and cool stars – activity, magnetism, dynamos, p. 452 – 462 (1991). – See Abstr. 012.024 for the main entry.
The author summarizes some of the principal scientific results presented at the conference on "Mechanisms of chromospheric and coronal heating" (held 1990 in Heidelberg, Germany). The author concentrates on what is being learned from the empirical side, and summarizes the heating mechanisms discussed and for which types of stars they may be applicable.

064.064 **The statistical equilibrium and λλ670.776 and 670.791 nm absorption lines of Li I in red-giant spectra. Method.**
Ya. V. Pavlenko.
Astron. Zh., Tom 68, Vyp. 4, p. 776 – 781 (Jul–Aug 1991). In Russian. English translation in Sov. Astron., Vol. 35, No. 4.
The procedure of statistical equilibrium calculation for lithium in red-giant atmospheres without LTE is discussed. The method of calculation of opacity of cool stellar atmospheres (mean absorption) is investigated in detail.

064.065 **Limb darkening coefficients for stars with thin photospheres in WUBVR and UVBY systems derived from Kurucz's model atmospheres.**
A. A. Rubashevskij.
Astron. Zh., Tom 68, Vyp. 4, p. 799 – 807 (Jul–Aug 1991). In Russian. English translation in Sov. Astron., Vol. 35, No. 4.
Values of flux limb-darkening coefficients, determined from the condition of the flux conservation by replacing the model distribution of surface brightness with a linear law, were obtained in WUBVR and uvby systems from Kurucz's model atmospheres according to results of determination of their monochromatic values. A comparison with earlier results, derived from other non-gray models of stellar atmospheres, shows that in the interval of temperatures from 7000 to 8000K Kurucz's models appreciably underestimated the limb darkening. In this region the values of limb-darkening coefficients can be sometimes underestimated by more than 0.10 which is in discrepancy with results of light curves solutions for classical eclipsing systems. In the broader intervals of temperature, noticeable differences remain in the Balmer jump region. One of the causes of these discrepancies is discussed. The obtained results are compared with those derived from observations of the Sun and stars of the YZ Cas system.

064.066 **New lines, new models, new colors.**
R. L. Kurucz.
Meeting on Precision Photometry: Astrophysics of the Galaxy, p. 27 – 44 (1991). – See Abstr. 012.027 for the main entry.
The author has used his newly calculated iron group line list together with his earlier atomic and molecular line data to compute new opacities for the temperature range 2,000K to 200,000K. The author completed the grid of model atmospheres for [0.0], 2 km/s for the temperature range 3,500K to 50,000K, and for log g from 0.0 to 5.0. This grid will allow a consistent theoretical treatment of photometry from K stars to B stars. Preliminary results are reported for UBV and uvby photometry.

064.067 **A method for the determination of the extinction arising from circumstellar dust shells.**
Hu Jingyao.
Acta Astrophys. Sin., Vol. 11, No. 3, p. 234 – 237 (Jul 1991). In Chinese.
The extinction of stars with circumstellar dust shell is composed of two parts, i.e. the interstellar extinction A_{vi} and the circumstellar extinction A_{vc}. A method to derive A_{vc} using the spectral energy distribution (from UV to far infrared) is presented, on the assumptions: (1) The infrared radiation is reradiated from circumstellar dust grain, after they absorbed the UV and/or optical photons from the underlying stars. (2) The dust shell is spherically symmetric. The method is available for stars with temperatures hotter than 3500K, i.e. earlier than M0.

064.068 **The contribution function for the depth of formation of lines in a stellar atmosphere.**
L. Achmad, C. de Jager, H. Nieuwenhuijzen.
Astron. Astrophys., Vol. 250, No. 2, p. 445 – 450 (Oct 1991).
The authors have derived a function, describing the contribution of the various layers to the formation of spectral lines in a stellar atmosphere. The function is valid for absorption lines as well as for emission lines, and is based on the separation of the line and the continuum emergent radiation. In this respect, the function differs from the conventional contribution functions, which are based on the total radiation or on the line depression. The authors show that it does not appear possible to find a unique function using the concept of line depression, in contrast to the function developed in this paper.

064.069 **The statistical equilibrium and λλ670.776 and 670.791 nm absorption lines of Li I in red-giant spectra. Results.**
Ya. V. Pavlenko.
Astron. Zh., Tom 68, Vyp. 6, p. 1240 – 1246 (Nov–Dec 1991). In Russian. English translation in Sov. Astron., Vol. 35, No. 6.
The results of solution of the NLTE- problem for lithium in atmospheres of K and M giants (T_{eff} = 3600, 3800, 4000, 4400, 4800K, $\lg g = 1.7$) are presented. The opacity at the bound-bound and bound-free Li I transition frequencies is due to continuum sources, molecular bands and atomic lines. The LTE and NLTE curves of growth for the Li I doublet synthetic spectra in the region λ670.8 nm are obtained. The results have shown that NLTE effects are significant in red-giants' atmospheres ($\varepsilon_{NLTE}(Li)-\varepsilon_{LTE}(Li) \approx 0.40$ dex for lines of $W_\lambda = 5$ pm).

064.070 **On the possibility of determination of non-linear limb-darkening laws from models of stellar atmospheres and by an analysis of solutions of light curves of classical eclipsing systems.**
A. A. Rubashevskij.
Astron. Zh., Tom 68, Vyp. 6, p. 1247 – 1260 (Nov–Dec 1991). In Russian. English translation in Sov. Astron., Vol. 35, No. 6.
A method was proposed for the determination of parameters for non-linear limb-darkening laws derived from models of stellar atmospheres with the condition of both conservation of the flux of radiation at some spectral region and the best description of the distribution of surface brightness over their discs. On this basis an algorithm was proposed for obtaining non-linear limb-darkening laws in the process of solution of light curves of classical eclipsing variables for the simplest model of a

binary system. It was shown that, for the linear approximation to the real limb–darkening law, the solution of light curve leads to determination of u_R. The method was applied to obtaining parameters for the eclipsing variable YZ Cas and, in particular, limb–darkening coefficients of its stars and the non–linear limb–darkening law of an A–star in the red spectral region.

064.071 **Modelling viscosity in accretion disks.**
W. J. Duschl.
3. Chinese Academy of Sciences and Max–Planck Society Workshop: High energy astrophysics: compact stars and active galaxies, p. 25 – 31 (1991). – See Abstr. 012.026 for the main entry.
The author discusses the problem of identifying the source of viscosity in accretion disks, and describes a new approach that allows a self–consistent determination of the viscosity at least in convectively stratified domains. This model leads to reasonable values of the viscosity parameter α. Finally the author argues that one also can estimate α in radiative domains in a reliable way, and finds $O(\alpha) \approx 1$ there.

064.072 **The solitons solution on the nonlinear evolution of dynamical instability in a slender accretion torus.**
Wang Deyu, R. E. Taam, Wu Ding.
3. Chinese Academy of Sciences and Max–Planck Society Workshop: High energy astrophysics: compact stars and active galaxies, p. 32 – 38 (1991). – See Abstr. 012.026 for the main entry.
A dispersion relation for a two dimensional incompressible slender accretion torus has been obtained. This dispersion relation is similar to that of the linearized KdV equation. The dispersion relation results that the argument of the "planetlike" solution of non–linear dynamical instabilities in the numerical simulations is the solitons solution of KdV equation.

064.073 **Thermosolutal–convective instability of a composite plasma in a stellar atmosphere including the effects of variable magnetic field and rotation.**
K. C. Sharma, U. Gupta.
Astrophys. Space Sci., Vol. 184, No. 2, p. 261 – 271 (Oct 1991).
Thermosolutal–convective instability of a composite plasma in a stellar atmosphere is considered. The effect of a variable horizontal magnetic field and the simultaneous effect of a uniform rotation and a variable horizontal magnetic field have been considered on the thermosolutal–convective instability. The authors have derived the sufficient conditions for the existence of monotonic instability. It is found that the criteria for monotonic instability hold good in the presence of a variable horizontal magnetic field as well as in the presence of a uniform rotation and a variable horizontal magnetic field.

064.074 **Colliding stellar winds in WR + O binary systems.**
A. V. Myasnikov, S. A. Zhekov.
Astrophys. Space Sci., Vol. 184, No. 2, p. 287 – 295 (Oct 1991).
Two–dimensional calculations in the model of two colliding stellar winds are provided. The effects of energy losses on free–free emission and Comptonization are investigated. The expected X–ray emission of a typical WR + O binary system is calculated.

064.075 **Asymptotic structure of hydromagnetically driven relativistic winds.**
T. Chiueh, Z. Y. Li, M. C. Begelman.
Astrophys. J., Vol. 377, No. 2, p. 462 – 466 (20 Aug 1991).
The authors study the asymptotic properties of fully relativistic, steady, axisymmetric, hydromagnetic winds. They find that all flux surfaces generally converge to either cylinders or paraboloids that are nested around the rotation axis. First the authors exclude asymptotically horizontal flux surfaces solely from consideration of energy and magnetic flux conservation. Then, by taking into account the cross–field force balance, the authors show that asymptotically current–carrying paraboloidal and conical flux surfaces cannot occur arbitrarily close to the rotation axis. Asymptotically cylindrical flux surfaces, while having no difficulty in filling out the polar region, cannot extend to infinity transversely unless the terminal velocity on the last flux surface vanishes. Asymptotically cylindrical flux surfaces with finite transverse radius, either self–confined or externally confined, are possible. The authors show that the terminal Lorentz factor on any asymptotically cylindrical flux surface is directly proportional to the current enclosed within that surface, provided that the surface extends to a radius well beyond the light cylinder. The results presented here are potentially important for understanding the structures of relativistic jets in active galactic nuclei and of pulsar winds.

064.076 **Continuous energy distributions of accretion discs.**
G. Shaviv, R. Wehrse.
Astron. Astrophys., Vol. 251, No. 1, p. 117 – 132 (Nov 1991).
A model to calculate the continuum radiation emerging from accretion discs is presented. The method solves the disc equations in a self consistent way after the basic assumptions are made. The comparison with observations shows good agreement over a wide range of wavelengths. The reasons for the success of the model are analyzed and it is shown that the basic reason for this achievement is the fact that the solution is selfconsistent.

064.077 **Modelling stellar surface magnetic fields. I. Search strategies and uniqueness.**
M. J. Stift, M. Goossens.
Astron. Astrophys., Vol. 251, No. 1, p. 139 – 146 (Nov 1991).
Within the framework of the Rigid Rotator hypothesis, the authors propose an elaborate procedure for the modelling of stellar magnetic surface fields aimed at ensuring the uniqueness of the final models by means of appropriate search strategies. Two different field geometries are considered for this purpose, viz. Stift's (1975) eccentric tilted dipole and Goossens' (1979) centred dipole–qudrupole force–free magnetospheres. Using a sufficiently dense grid, the authors first explore the entire 6D or 4D parameter space for those models that best predict the observed integrated longitudinal field and integrated scalar field strength. In a second step, detailed calculations of Stokes profiles for a few selected transitions make it possible to pick out the most promising starting models for the final iterative line profile modelling procedure, often leading simultaneously to the resolution of the ambiguity between obliquity angle and inclination in axisymmetric models. Illustrating their method with the analysis of the magnetic variations of four well–observed Ap stars the authors show that the starting models derived are in reasonable agreement with the published spectra. Finally, they discuss the relation between their approach and previously proposed methods in the light of the constraints involved and point out ways towards improved physical consistency.

064.078 **Shock waves in accretion disks.**
H. C. Spruit.
3. Chinese Academy of Sciences and Max–Planck Society Workshop: High energy astrophysics: compact stars and active galaxies, p. 333 – 342 (1991). – See Abstr. 012.026 for the main entry.
After a brief review of the unsolved question whether the flow of gas in a Keplerian disk is unstable to ordinary shear instability, the author discusses the role played by shock waves, excited by the tidal force of a companion and by the mass transferring stream impacting on the disk. Recent numerical simulations of disks in binaries are also discussed.

064.079 **Alfvén waves in stellar winds.**
M. R. Barkhudarov.
Sol. Phys., Vol. 135, No. 1, p. 131 – 161 (Sep 1991).
A mathematical model for undamped, toroidal, small–amplitude Alfvén waves in a spherically–symmetric or equatorial stellar wind is developed in this paper. The equations are reduced to a very simple form by using real Fourier amplitudes and the ratio of the inward and outward propagating wave amplitudes, which is interpreted as a measure of the relative influence of wave reflection in the flow, on the solution at a given point. Special attention is paid to the applicability of the WKB approximation.

As an illustration of the analytical behaviour some numerical results are shown using a cool wind model. These are likely to express qualitatively the features of the Alfvén waves in any stellar wind, since the only assumptions about the flow used in the analytical study of the wave equations were that: the flow has small velocity at the base of the corona; it then passes through the critical point, and reaches its finite non–zero limit at infinity.

064.080 **Isothermal, Compton–heated coronae above accretion disks.**
E. C. Ostriker, C. F. McKee, R. I. Klein.
Astrophys. J., Vol. 377, No. 2, p. 593 – 611 (20 Aug 1991).
The authors examine the structure of Compton–heated coronae above accretion disks by the development of both approximate analytic solutions and self–consistent numerical solutions to determine the direct and scattered radiation reaching the base of the corona for a range of central source luminosities. The study generalizes previous work by treating the case of a central X–ray source at a finite distance above the disk midplane, while allowing for both flat and flaring disks. The corona is assumed vertically isothermal, with temperature set to half of the local Compton temperature. Analytic solutions for the optical depth from source to disk in both optically thin and optically thick cases show excellent agreement with detailed numerical results. The authors treat only singly scattered radiation and evaluate the range of radii for which multiple scattering is not important, for a given luminosity. It is found that the outer region of the corona is unaffected by multiple scattering in the interior, provided that the luminosity of the central source is sufficiently below the Eddington limit. For these luminosities, modification of the spectrum produced by the underlying disk due to the overlying corona is unimportant; for higher luminosities the effect may become significant.

064.081 **Statistical equilibrium and the 670.776 and 670.791 nm Li I absorption lines in the spectrum of red giants. Procedure.**
Ya. V. Pavlenko.
Sov. Astron., Vol. 35, No. 4, p. 384 – 386 (Jul 1991). Current Physics Microform No.: 9203X0570. English translation of Astron. Zh., Tom 68, Vyp. 4, p. 776 – 781 (1991).
The procedure for a non–LTE calculation of statistical equilibrium for lithium in the atmosphere of a red giant is analyzed. The method of calculating the opacity of matter in the atmosphere of a cool star (averaged over absorption) with allowance for absorption in atomic lines is investigated in detail.

064.082 **Accretion–disk thermal limit–cycle instability. I.**
S. Mineshige.
Astron. Her., Vol. 84, No. 8, p. 256 – 261 (1991). In Japanese.

064.083 **Accretion–disk thermal limit–cycle instability. II.**
S. Mineshige.
Astron. Her., Vol. 84, No. 9, p. 288 – 294 (1991). In Japanese.

064.084 **Darkening coefficients in the WUBVR and uvby systems for stars with a thin photosphere.**
A. A. Rubashevskij.
Sov. Astron., Vol. 35, No. 4, p. 396 – 400 (Jul 1991). Current Physics Microform No.: 9203X0582. English translation of Astron. Zh., Tom 68, Vyp. 4, p. 799 – 807 (1991).
Flux darkening coefficients determined by requiring flux conservation upon replacement of the model surface brightness distribution by a linear law are obtained in the WUBVR and uvby systems for stars with thin photospheres from Kurucz's model atmospheres, based on a determination of their monochromatic values. A comparison with earlier results obtained from other nongray model stellar atmospheres shows that Kurucz's models considerably underestimate limb darkening in the temperature range 7000 – 8000K. In that range, the darkening coefficient is sometimes underestimated by more than 0.10, at variance with the solution of light curves of classical eclipsing systems. Over a wider temperature range, pronounced differences remain in the vicinity of the Balmer jump. One reason for such discrepancies is briefly discussed. The results are compared with those obtained from observations of the Sun and the stars in the YZ Cas system.

064.085 **Kinematical models of maser shells formed by aspherical stellar outflows (U Orionis; OH 231.8+4.2; Orion–IRc2).**
P. F. Bowers.
Astrophys. J., Suppl. Ser., Vol. 76, No. 4, p. 1099 – 1125 (Aug 1991).
Kinematical models are presented of maser emission uniformly distributed throughout ellipsoidal shells with various orientations to the line of sight. The velocity fields consist of different combinations of outflow, rotation, or radial acceleration which may be produced by the central star. The models are intended to be a guide for analysis of complex, aspherical outflows. They do not depend on the evolutionary nature of the star, nor are they limited to interpretation of a specific molecule. The models are applied to three complex maser sources. The first case is the 1665 MHz OH emission associated with the classical Mira variable U Ori, which can be modeled by radial outflow in an ellipsoidal shell. The second case is the 1667 MHz OH emission associated with the bipolar nebula OH 231.8+4. The velocity field of the nebula is modeled as a prolate spheroid with the outflow velocity increasing from about 10 km s^{-1} in the equatorial plane to about 200 km s^{-1} along the polar axis, and the OH is distributed at latitudes within 75° of the plane. The third case consists of the low– and high–velocity outflows ostensibly associated with the luminous star IRc2 in the Orion molecular cloud. It is first shown that the vibrationally excited SiO emission, which occurs close to the star, can be modeled by an oblate spheroid with comparable components of rotation and outflow. It is then proposed that the kinematics of the low– and high–velocity outflows, which occur much farther from the star, can be roughly described by a prolate spheroid with the outflow velocity increasing with latitude from the equatorial plane.

064.086 **The radial–azimuthal stability of accretion disks: gas pressure contributions.**
M. R. McKee.
Astron. Astrophys., Vol. 251, No. 2, p. 689 – 692 (Nov 1991).
A radial–azimuthal stability analysis of a thin, alpha disk accretion flow is presented. The proportion of radiation pressure, P_r, of the unperturbed flow is allowed to vary according to the parameter $\beta = P_r/P$, where P is the total pressure. As is the case for a purely radial analysis, the disk is stable for $\beta \leqslant 0.6$. However, the coupling of radial and azimuthal perturbations eliminates the viscous instability for such non–radial modes for all values of β. The group velocity of the retrograde thermal mode is calculated as a function of β.

064.087 **Cosmic–ray–modified stellar winds. IV. The effect of self–generated Alfvén waves.**
Ko Chungming.
Astron. Astrophys., Vol. 251, No. 2, p. 713 – 722 (Nov 1991).
The modification of a stellar wind by galactic cosmic rays is investigated in a three–fluid hydrodynamical model. The model consists of a stellar wind plasma, cosmic rays and Alfvén waves. The cosmic rays are coupled to the wind plasma via hydromagnetic waves (say, Alfvén waves) which in turn are produced by the cosmic ray streaming instability. In this paper only self–generated Alfvén waves are considered. The equations governing the system are the overall mass, momentum, and energy of the three fluids; supplemented by the cosmic–ray energy equation and wave energy exchange equation. This constitutes a self–consistent picture. The positive pressure gradients of the cosmic rays and Alfvén waves assist in braking the stellar wind outflow. They also change the locations of the critical point and the termination shock to some extent. The transfer of energy from the thermal wind to the cosmic rays and waves results in outward going cosmic–ray and wave energy fluxes. The solar wind is taken as an example. When compared with the cosmic–ray–modified solar wind without waves, the presence of waves has a minor influence on the bulk flow. However, the cosmic ray pressure is larger in the

inner solar cavity due to the reduction of outward convection velocity. Moreover, if the diffusion coefficient of the cosmic rays is inversely proportional to the wave pressure, then significant modulation may come from a region outside the termination shock because of the very efficient wave generation by the shock.

064.088 **The reaction of accretion disks to strong input mass pulses.**
W. J. Duschl, M. Livio.
Georgia State University Conference on Variability of Active Galactic Nuclei, p. 29 – 31 (1991). – See Abstr. 012.040 for the main entry.
The authors present preliminary results of model calculations of the reaction of a viscous accretion disk to a strong input mass pulse. They show that under such conditions the evolution time scale of the disk is no longer dominated by the viscous time scale of the stationary pre–pulse disk but rather by the time scale corresponding to the mass flow rate of the mass pulse. Thus, the "reaction time" of a disk can be substantially shorter.

064.089 **The interaction of two stellar winds with energy losses for emission and comptonization.**
A. V. Myasnikov, S. A. Zhekov.
Astron. Tsirk., No. 1545, p. 1 – 2 (Aug – Sep 1990).
Two–dimensional calculations show: 1. X–ray luminosity decreases to $L_x \approx 10^{34}$ergs/s when the energy loss for comptonization is taken into account, but it is not enough to be in full consistency with X–ray observations. 2. X–ray emission absorption by WR wind having overabundant CNO elements reduced X–ray luminosity to the observational values, but the spectral shape differs greately from the observational one.

064.090 **The physics of disk winds.**
R. E. Pudritz, G. Pelletier, A. I. Gomez de Castro.
NATO Advanced Study Institute on The Physics of Star Formation and Early Stellar Evolution, p. 539 – 564 (1991). – See Abstr. 012.042 for the main entry.
The authors review the physics of disk winds and show that a hydromagnetic disk wind may be the simplest, self–consistent theory available for understanding a multitude of observed outflow properties. They briefly outline several of the key observational developments that constrain any theory of outflows from young stellar objects.

064.091 **Ionized winds from young stellar objects.**
N. Panagia.
NATO Advanced Study Institute on The Physics of Star Formation and Early Stellar Evolution, p. 565 – 593 (1991). – See Abstr. 012.042 for the main entry.
Observations and theory of the radio and infrared emission from ionized winds of young stellar objects are reviewed and discussed.

064.092 **The physics of neutral winds from low mass young stellar objects.**
A. Natta, C. Giovanardi.
NATO Advanced Study Institute on The Physics of Star Formation and Early Stellar Evolution, p. 595 – 621 (1991). – See Abstr. 012.042 for the main entry.
The physical structure of winds ejected by low luminosity, pre–main–sequence stars is investigated, using the results of numerical calculations as well as analytical approximations. The topics covered in this review include the ionization and excitation structure of hydrogen and of sodium, intensities and profiles of some interesting lines and the gas temperature. Line profile calculations can be compared to spectroscopic observations of suitable stars to derive wind parameters such as the rate of mass–loss and the gas temperature. An example of a diagnostic of this kind, using various infrared hydrogen recombination lines and the Na D doublet at 5900 Å in T Tauri stars, is briefly discussed. The temperatures derived from observations are compared to those produced by various heating mechanisms.

064.093 **Turbulence, convection, and stability in accretion disks.**
C. Meirelles Filho.
Astrophys. J., Vol. 378, No. 1, p. 266 – 270 (1 Sep 1991).
Allowance for turbulent pressure pulsations, convection, and dependence of the turbulence along z in accretion disks results in a turbulent scale length much smaller than the disk scale height, more effectiveness for the convective energy transport, and gas pressure dominance in the flow, even for sonic regimes. A solution for the convective disk exists only for $\beta_0 \geqslant 0.14$. Onset of instability depends on the turbulence level of the disk. For the adopted viscosity law, the disk will be stable up to $\alpha \approx 1$ (α is the turbulent Mach number). When applied to Galactic and extragalactic X–ray sources, these results suggest that the absence (presence) of fluctuations may indicate a subsonic (supersonic) turbulent regime as well as confirm (refute) the assumed viscosity law.

064.094 **On the mechanism of angular momentum transport in accretion disks.**
M. Livio, H. C. Spruit.
Astron. Astrophys., Vol. 252, No. 1, p. 189 – 192 (Dec 1991).
Recently, a new mechanism for angular momentum transport in accretion disks has been proposed. The mechanism is based on tidally induced perturbations which develop into coherent spiral shocks. Angular momentum transfer by the spiral shocks was shown to be quite effective for hot, two–dimensional disks. The authors use observational data on the mean recurrence time–scale of dwarf nova eruptions, in conjunction with the disk instability model, to obtain a relation between the viscosity parameter α and the mass ratio of the system. This relation can be compared to the effective "viscosity" obtained from the spiral–shocks scheme. The data show a weaker dependence on the mass ratio than predicted. Possible causes for the discrepancy are discussed.

064.095 **Computer codes for stellar atmospheric modeling.**
D. G. Hummer, I. Hubeny.
NATO Advanced Research Workshop on Stellar Atmospheres: Beyond Classical Models, p. 119 – 125 (1991). – See Abstr. 012.056 for the main entry.
The principal computer codes used for spectroscopic diagnostics of stellar atmospheres are summarized and briefly described.

064.096 **The quest for physical realism in stellar atmospheric modeling.**
D. Mihalas.
NATO Advanced Research Workshop on Stellar Atmospheres: Beyond Classical Models, p. 127 – 141 (1991). – See Abstr. 012.056 for the main entry.
The author attempts to summarize recent progress in, and future prospects of, the ongoing quest to improve the physical realism of models of stellar atmospheres.

064.097 **Unified NLTE model atmospheres including spherical extension and stellar winds: EUV–fluxes and the He II discrepancy in central stars of planetary nebulae.**
R. Gabler, R. P. Kudritzki, R. Méndez.
NATO Advanced Research Workshop on Stellar Atmospheres: Beyond Classical Models, p. 143 – 154 (1991). – See Abstr. 012.056 for the main entry.
The authors present NLTE model atmospheres for hot stars that avoid the artificial division between hydrostatic photosphere and supersonic stellar wind envelope. The models are of spherical geometry and extended over the entire sub– and supersonic atmosphere. They need only stellar effective temperature, gravity, radius and helium abundance as free parameters and are in principle self–consistent. The authors have calculated a large grid of models for central stars of planetary nebulae in the effective range from 40000K to 100000K. The calculated energy distributions are used to investigate the longstanding problem of the "Zanstra–discrepancy". It is shown that the increase of EUV flux in the unified models, which is caused by the presence of the wind velocity field in the layers of He II–groundstate continuum

formation leads to a much better agreement with the observations than was the case before, when plane–parallel hydrostatic models were used.

064.098 **Non–LTE models of Hα emission from accretion disks in nondegenerate eclipsing binaries.**
E. C. Olson.
Astron. J., Vol. 102, No. 4, p. 1423 – 1427 (Oct 1991). Current Physics Microform No.: 9110D1955.
Evidence is summarized for the existence of rotating accretion disks in some long–period nondegenerate eclipsing binary systems. These disks are optically thin in all continua except the Lyman. An accretion disk model which predicts the emission–line profile of Hα, including approximations for departures from local thermodynamic equilibrium, is described. Effects of pressure and kinetic temperature variations, departures from Keplerian rotation, disk partial eclipses by the cool star, and microturbulence on emitted profiles are briefly discussed.

064.099 **Non–LTE model atmosphere calculations with approximate lambda operators.**
S. Dreizler, K. Werner.
NATO Advanced Research Workshop on Stellar Atmospheres: Beyond Classical Models, p. 155 – 164 (1991). – See Abstr. 012.056 for the main entry.
The authors report on their state of the art ALI method for calculating static plane–parallel atmospheres. In particular the application of tridiagonal approximate lambda operators and a new numerical approach for the linearization procedure are presented. They used the ALI method to calculate a model grid for helium–rich sdO stars in the following parameter range: $40000K \leqslant T_{eff} \leqslant 80000K$, $4 \leqslant \log g/(cgs) \leqslant 8$, and $0.1 \leqslant n_{He}/n_H \leqslant 100$, including detailed helium, hydrogen, and carbon model atoms.

064.100 **NLTE model atmospheres for hot stars.**
T. Rauch, K. Werner.
NATO Advanced Research Workshop on Stellar Atmospheres: Beyond Classical Models, p. 165 – 174 (1991). – See Abstr. 012.056 for the main entry.
With the powerful Accelerated Lambda Iteration (ALI) method and todays enormous computer capacity we are able to treat model atoms in great detail. The authors present recent results of their NLTE model atmosphere calculations for hot sdO stars including very elaborate C IV, N IV and N V model ions simultaneously in order to determine the n_C/n_H and n_N/n_H abundance ratios.

064.101 **Radiative transfer in expanding atmospheres – radiative acceleration of Wolf–Rayet envelopes?**
J. Puls, A. W. A. Pauldrach.
NATO Advanced Research Workshop on Stellar Atmospheres: Beyond Classical Models, p. 175 – 189 (1991). – See Abstr. 012.056 for the main entry.
The authors present first results of investigations as to how far the winds of Wolf–Rayets are driven radiatively. This is done by comparing the acceleration required and obtained in the optically thin part of the wind. It turns out that the multi–line process can (in principle) provide an acceleration which may lift the stellar wind far beyond the so–called single–scattering limit, provided that the number of driving lines is large. For models based on present stellar and wind parameters deduced from atmospheric analyses, however, wind efficiencies of only 2 to 4 can be reached by radiative acceleration. In future calculations then, the uniqueness of these analyses has to be investigated critically in order to obtain a final answer.

064.102 **Non–LTE analysis of hot stars including line blanketing.**
W. Schmutz.
NATO Advanced Research Workshop on Stellar Atmospheres: Beyond Classical Models, p. 191 – 203 (1991). – See Abstr. 012.056 for the main entry.
A method is presented that permits the calculation of intensity–weighted effective opacity means which account for the blanketing effect of tens of thousands of spectral lines in expanding atmospheres. The coefficients are evaluated as a function of frequency and location in the atmosphere. By combining these effective opacities with an atmosphere code that solves the statistical equilibrium equations for non–LTE populations and the radiation transfer equations in the comoving frame, the author obtains a model atmosphere that accounts simultaneously and consistently for non–LTE, spherical extension, expansion, and line blanketing.

064.103 **The origin and development of instabilities in radiatively–driven stellar winds.**
J. I. Castor.
NATO Advanced Research Workshop on Stellar Atmospheres: Beyond Classical Models, p. 221 – 233 (1991). – See Abstr. 012.056 for the main entry.
The numerous observational indicators of instability in the radiatively–driven winds of hot stars are reviewed briefly, followed by a summary of the present theoretical understanding of the linear instability of such winds. This provides the motivation for the hydrodynamic simulations. A serious approximation that must be made in order to reduce the cost of the simulations to a reasonable level is discussed in some detail. The hydrodynamic methods are described briefly, and then the computational results for winds models computed in the absorption approximation are discussed. The most notable results pertain to the critical nature of the ratio v_{th}/a of the intrinsic line width to the sound speed. When this ratio is large, only a negligible wind results; when the ratio is small, the wind executes permanent self–excited oscillations; in an intermediate range the wind is globally stable, but acts as a powerful wave amplifier. The morphology of the oscillations is described and related to Abbott's linear theory, and the possible connection to observations is mentioned.

064.104 **A smooth source function method for including scattering in radiatively driven wind simulations.**
S. P. Owocki.
NATO Advanced Research Workshop on Stellar Atmospheres: Beyond Classical Models, p. 235 – 246 (1991). – See Abstr. 012.056 for the main entry.
The winds of the hot, luminous, OB–type stars are driven by line–scattering of the star's continuum radiation flux. Several kinds of observational evidence indicate that such winds are highly structured and variable, and it seems likely that a root cause of this variability is the known strong instability of the line–driving mechanism. This paper describes a new, smooth source function approximation method which enables computation of the diffuse force with only very modest (~20%) increase in timing. Initial results of simulations incorporating this approximation indicate the diffuse radiation can have dramatic effects in eliminating the intrinsic variability often seen in pure–absorption models, and in bringing the steady–state velocity laws into close agreement with previous results based on Sobolev theory and comoving frame transfer solutions. Nonetheless, such wind models remain advectively very unstable in that only quite small (~1%) base perturbations still lead to strong shocks and dense clumps in the wind within a stellar radius of the surface.

064.105 **Atmospheres of late–type giants.**
H. R. Johnson.
NATO Advanced Research Workshop on Stellar Atmospheres: Beyond Classical Models, p. 281 – 290 (1991). – See Abstr. 012.056 for the main entry.
Much of the research on atmospheres of cool giant stars is still found to be classical in nature. The author describes current developments in molecular opacities and discusses recently published model photospheres. Some of these go beyond classical models. Such distinctly non–classical topics as chromospheres, acoustic and hydrodynamic waves, and mass loss are then briefly discussed.

064.106 **Numerical simulation of photospheric convection: hydrodynamical test calculations.**
C. Reile, T. Gehren.
NATO Advanced Research Workshop on Stellar Atmospheres: Beyond Classical Models, p. 303 – 308 (1991). – See Abstr. 012.056 for the main entry.

As a first step in the simulation of photospheric granulation in cool stars the authors present a numerical treatment of the fully time-dependent inviscid hydrodynamical equations in conservative form and two dimensions assuming axisymmetric cylindrical geometry. The proposed explicit scheme is based on the control volume technique and uses the method of monotonic transport together with operator-splitting. Though for the moment one deals with an ideal gas and – for test purposes – adopts the adiabatic case, an arbitrary (e.g. LTE) equation of state and radiative energy transfer can be incorporated without difficulties. To demonstrate its versatility in the treatment of different hydrodynamical phenomena the authors apply their code to one- and two-dimensional test problems such as the planar shock tube, the (cylindrical) Sedov problem, and standing gravo-acoustic waves in an exponentially stratified isothermal medium.

064.107 **Stellar atmosphere theory as a spectroscopic tool. The example of hot stars.**
R. P. Kudritzki.
NATO Advanced Research Workshop on Stellar Atmospheres: Beyond Classical Models, p. 309 (1991). – See Abstr. 012.056 for the main entry.

064.108 **The winds of O-stars V: tests of the accuracy of the radiation driven wind models.**
H. J. G. L. M. Lamers.
NATO Advanced Research Workshop on Stellar Atmospheres: Beyond Classical Models, p. 311 – 316 (1991). – See Abstr. 012.056 for the main entry.

The observed line profiles of 26 O stars and one B supergiant taken from the IUE atlas were compared with theoretical line profiles.

064.109 **Diagnostics of Wolf-Rayet atmospheres.**
D. J. Hillier.
NATO Advanced Research Workshop on Stellar Atmospheres: Beyond Classical Models, p. 317 – 329 (1991). – See Abstr. 012.056 for the main entry.

To analyze Wolf-Rayet (W-R) spectra one must consider spherical geometry, non-LTE, and the transfer of radiation in an expanding atmosphere. A code incorporating all of these is briefly described. Limitations and problems with the currently available atomic data are discussed. Diagnostics available to allow a determination of the temperature structure, mass loss rate, and velocity law for W-R atmospheres are considered.

064.110 **Central stars of PN: spectral diagnostics based on model atmospheres vs. diagnostics based on the classical nebular approach.**
R. H. Mendez.
NATO Advanced Research Workshop on Stellar Atmospheres: Beyond Classical Models, p. 331 – 340 (1991). – See Abstr. 012.056 for the main entry.

Recent observational and theoretical progress in astronomical spectroscopy has made it possible to obtain useful information about the central stars of planetary nebulae directly from non-LTE spectroscopic studies. A comparison with information derived indirectly from nebular studies shows that the direct spectroscopic approach produces a much more consistent picture of the characteristics and evolution of central stars. The following data and asumptions should be avoided: hydrogen Zanstra temperatures, many uncertain nebular distances, and the assumption that many PN are completely optically thick in the H Lyman continuum.

064.111 **Spectral diagnostics of hot subdwarfs: successes and problems.**
K. Werner, U. Heber.
NATO Advanced Research Workshop on Stellar Atmospheres: Beyond Classical Models, p. 341 – 351 (1991). – See Abstr. 012.056 for the main entry.

The recent progress in spectral analysis of hot subdwarfs is closely related to the advent of new numerical techniques for the construction of sophisticated NLTE model atmospheres. The authors examine how computational details affect the analyses, and pay special attention to new model atmospheres for the hottest subdwarfs known, the PG 1159 stars.

064.112 **Line blanketing without LTE: the effect on diagnostics for B-type stars.**
L. Anderson, J. A. Grigsby.
NATO Advanced Research Workshop on Stellar Atmospheres: Beyond Classical Models, p. 365 – 373 (1991). – See Abstr. 012.056 for the main entry.

The authors show some general results from models of B-type stellar atmospheres. They compare the effect on typical diagnostics for the effective temperature and surface gravity of three stages of approximation: a) nonLTE, H, He, C, N, O only; b) LTE, H, He, C, N, O, Ne, Mg, Al, Si, Fe group; and c) nonLTE H, He, C, N, O, Ne, Mg, Al, Si, Fe group. Finally they show the present level of success in reproducing high-dispersion spectra obtained by IUE.

064.113 **Particle transport in magnetic stellar atmospheres.**
G. Michaud, J. Babel.
NATO Advanced Research Workshop on Stellar Atmospheres: Beyond Classical Models, p. 375 – 386 (1991). – See Abstr. 012.056 for the main entry.

The high signal to noise observations now available for the magnetic A star 53 Cam lead to a geographic distribution of chemical elements that is inconsistent with the simplest model based on diffusion for this object. The average abundances are explained by the diffusion model but not the horizontal inhomogeneities. More detailed particle transport calculations are needed. These must involve the use of polarized radiative transfer to calculate radiative accelerations. There then appears a horizontal component to the radiative acceleration that strongly modifies horizontal diffusion. When the magnetic field eliminates the H convection zone, the diffusion of trace species must be calculated in a region where hydrogen is partly ionized. The hydrogen ionization gradient can cause ionized trace elements to be dragged to the surface, even in the absence of radiative accelerations. Calculations have been carried out for a number of elements. It is finally argued that another physical process must play an important role and mass loss is suggested.

064.114 **Spectroscopic tests of late-type model atmospheres of dwarf stars.**
T. Gehren, C. Reile, W. Steenbock.
NATO Advanced Research Workshop on Stellar Atmospheres: Beyond Classical Models, p. 387 – 396 (1991). – See Abstr. 012.056 for the main entry.

The basic properties of horizontally stratified, homogeneous atmospheric models of cool dwarfs are compared with spectroscopic observations. Recent results of kinetic equilibrium calculations for stars of different metal abundances are discussed.

064.115 **Dust in the shells of cool giants and supergiants.**
E. Sedlmayr, J. M. Winters.
NATO Advanced Research Workshop on Stellar Atmospheres: Beyond Classical Models, p. 397 – 408 (1991). – See Abstr. 012.056 for the main entry.

The importance of dust for the atmospheric structure of late type giants and supergiants is discussed. Quantitative examples are given for stationary dust driven winds and for the generation of shocks by radiation pressure on dust in pulsating variables.

064.116 **NLTE analyses of massive OB stars.**
A. Herrero, J. M. Vilchez, R. P. Kudritzki, D. Kunze, K. Butler.
NATO Advanced Research Workshop on Stellar Atmospheres: Beyond Classical Models, p. 409 – 414 (1991). – See Abstr. 012.056 for the main entry.
The authors present the results of spectroscopic analyses of 12 massive OB stars using NLTE techniques and the theory of radiatively driven winds. They use both to derive the stellar masses and compare with evolutionary tracks. They find a discrepancy between the three methods and suggest possible directions for future work. The helium abundances are also compared with those predicted by the evolutionary tracks. The authors again find a discrepancy.

064.117 **40 years numerical stellar atmospheres. Concluding remarks and personal considerations.**
K. Hunger.
NATO Advanced Research Workshop on Stellar Atmospheres: Beyond Classical Models, p. 449 – 452 (1991). – See Abstr. 012.056 for the main entry.

064.118 **Thermosolutal–convective instability of a composite plasma in a stellar atmosphere including the effects of finite Larmor radius and Hall currents.**
K. C. Sharma, U. Gupta, R. C. Sharma.
Astrophys. Space Sci., Vol. 186, No. 2, p. 179 – 189 (Dec 1991).
Thermosolutal–convective instability of a composite plasma in a stellar atmosphere is considered to include the effects, separately, due to finite Larmor radius (FLR) and Hall currents in the presence of a uniform horizontal magnetic field. The sufficient conditions for the existence of monotonic instability are derived and are found to hold good both in the presence, separately, of FLR and Hall current effects.

064.119 **X–ray – illuminated stellar winds: optically thick wind models for massive X–ray binaries.**
I. R. Stevens.
Astrophys. J., Vol. 379, No. 1, p. 310 – 326 (20 Sep 1991).
In massive X–ray binary systems (MXRBs), the accretion – powered X–rays from the neutron star will alter the dynamics of the line–driven stellar wind of the early–type primary, by changing both the thermal and the ionization structure of the wind material. In a previous paper, Stevens and Kallman investigated this effect, neglecting optical depth effects in the transfer of X–rays through the stellar wind material. In this paper, the author extends this work and calculates radiative force multipliers $M(t, \xi, N_H)$, which now depend on the column of attenuating material N_H, as well as the ionization parameter ξ. Optical depth effects tend to suppress the effects of X–ray ionization on the force multipliers. The author has calculated a number of dynamical models for the winds of MXRBs with these force multipliers. In contrast to the optically thin models, self–consistent dynamical solutions for reasonable values of the X–ray luminosity L_x can be found. These solutions also reveal the presence of nonlinear mechanisms that affect the wind dynamics. The models find the existence of a self–consistent region of solutions at an X–ray luminosity of $L_x \sim 10^{36} \mathrm{erg\, s^{-1}}$, and suggest the possible existence of high–luminosity states.

064.120 **Model atmospheres for SNe II during early stages.**
M. Best, P. H. Hauschildt, R. Wehrse, G. Shaviv.
Ann. Phys. (Paris) Colloq., Vol. 16, No. 4, p. 397 – 411 (Aug 1991).
The photospheres of type II supernovae during the coasting phase are characterized by moderate temperatures, large radii, very low densities and high expansions velocities. These properties lead to large geometrical extensions and to electron scattering as the main source of opacity. Therefore these photospheres are very different from those of stars in hydrostatic equilibrium. In this review the authors first describe the input physics for models and essential numerical aspects of the model construction. In particular they stress the central role of an adequate treatment of the radiative transfer equation which should simultaneously take into account the sphericity, a high ratio of scattering to absorption, and relativistic effects. Subsequently they discuss details of the emergent spectra and their relevance for parameter determinations. Finally, model spectra are compared with observations of the brights supernovae SN 1979c and SN 1980k, which seem to be more average examples of type II supernovae than SN 1987A, and possibilities for using these objects for extragalactic distance determinations are investigated.

064.121 **Kinetic temperature structure in circumstellar envelopes.**
J. H. Kastner.
3. Haystack Observatory Conference on the Interstellar Medium: Skylines – atoms, ions and molecules – new results in spectral line astrophysics, p. 369 – 373 (1991). – See Abstr. 012.051 for the main entry.
The author describes a model of the kinetic temperature of circumstellar moleculare gas which accounts for the variation of mass loss rate from star to star. The model predicts that the gas temperature decreases with increasing mass loss due largely to less efficient heating from dust–gas collisions. In contrast to the "standard model" of CO emission, this formulation of kinetic temperature results in a sharp decrease in line intensity of very high values of the mass loss rate.

064.122 **New results on radiative forces on iron in hot white dwarfs.**
P. Chayer, G. Fontaine, F. Wesemael.
NATO Advanced Research Workshop – 7. European Workshop on White Dwarfs, p. 249 – 255 (1991). – See Abstr. 012.044 for the main entry.
Detailed computations of radiative forces on Fe are performed in H–rich and He–rich envelope models of hot white dwarfs, and equilibrium abundance profiles of Fe are obtained from those calculations. In the absence of other competing mechanisms, detectable amounts of Fe would be visible at the photosphere of a 0.6 $M_\odot$ white dwarf when $T_e \geqslant 50000$K for H–rich models, and when $T_e \geqslant 35000$K for He–rich models. In general, the abundance of Fe supported by the radiative forces increases monotonically with increasing effective temperature. Equilibrium abundances of Fe are larger in He–rich models than in H–rich ones at the same effective temperature.

064.123 **The modified hydrostatic equilibrium equations for stratified high gravity stellar atmospheres.**
K. Unglaub, I. Bues.
NATO Advanced Research Workshop – 7. European Workshop on White Dwarfs, p. 267 – 274 (1991). – See Abstr. 012.044 for the main entry.
A model atmosphere code for high gravity stars is in progress which takes into account the chemical stratification of hydrogen, helium and carbon. The normal hydrostatic equation is splitted into one equation for each species. These equations are coupled by the electric field and the ionization terms. The local electric field is obtained from the charge neutrality condition, the ionization terms are computed according to the method by Wildt (1936, 1937), solved analytically. Integration of the hydrostatic equilibrium equations from an inner layer outwards yields the run of partial pressures for each species with geometrical depth.

064.124 **Some effects of the UV radiation from white dwarfs on the accretion of interstellar hydrogen.**
C. P. Verdon, R. L. McCrory, R. Epstein, H. M. van Horn, M. P. Savedoff.
NATO Advanced Research Workshop – 7. European Workshop on White Dwarfs, p. 295 – 304 (1991). – See Abstr. 012.044 for the main entry.
The authors have carried out preliminary hydrodynamic calculations of the accretion of interstellar hydrogen by white dwarfs, and are beginning to study the effects of the ionizing UV radiation from the star upon the flow. Even for velocities as low as $\sim 2.82 \mathrm{~km\, s^{-1}}$, with a stellar temperature $T_{eff} = 10000$K, the parameter values they have selected for the trial calculations discussed here, the H II region around the star is highly

distorted. For this case, they find that the ionizing radiation does reduce the accretion rate of hydrogen, by about a factor of three.

064.125 **Convection in white dwarfs: application of CM theory to helium envelope WDS.**
I. Mazzitelli, F. D'Antona.
NATO Advanced Research Workshop – 7. European Workshop on White Dwarfs, p. 305 – 316 (1991). – See Abstr. 012.044 for the main entry.

The authors summarize the main physical properties of the new model for stellar turbolent convection by Canuto and Mazzitelli, which constitutes an improvement of description with respect to the classical Mixing Length Theory (MLT), and has been successfully applied to the solar model. They apply this theory to the study of convection in the envelope of a typical white dwarf of 0.55 $M_\odot$, and compare the results with those obtained by employing the MLT and leaving all the other physical and numerical inputs inaltered. The results are comforting in the description of the location of the blue edge of the DB instability strip, if this is at $T_{eff} \sim 25000$K, but a blue edge at $T_{eff} \sim 29000$K would be in strong contradiction with the new theory.

064.126 **Model atmospheres of massive hot stars.**
R. P. Kudritzki, R. Gabler, D. Kunze, A. W. A. Pauldrach, J. Puls.
Massive Stars in Starbursts Meeting, p. 59 – 96 (1991). – See Abstr. 012.062 for the main entry.

Contents: 1. Introduction. 2. Model atmospheres – the state of art. 3. Ionizing fluxes. 4. Winds and metallicity. 5. Detailed diagnostics.

064.127 **The collimation of a time–dependent stellar wind.**
S.–H. Kim, A. C. Raga.
Astrophys. J., Vol. 379, No. 2, p. 689 – 695 (1 Oct 1991).

It is well known that the interaction of an isotropic wind with a stratified environment can result in the formation of well–collimated jets. Here the authors study the problem of the collimation of a time–dependent stellar wind by nonadiabatic de Laval nozzle flows. It is found that both velocity and density variations of the wind from the source can cause large velocity variations in the jet that is produced by the nozzle flow. The largest velocity variations in the jet flow are obtained when the wind is collimated through a highly nonadiabatic nozzle. The authors also find that short–period time variations of the source (i.e., with periods < 30 yr, for typical stellar jet source parameters) are "filtered out" by the nozzle flow, so that they do not produce corresponding time variations in the stellar jet. This result might provide a natural explanation for the observed separation between knots of stellar jets. These indicate time scales (knot separation / jet velocity) of the order of 30 – 100 yr.

064.128 **Magnetically driven jets and winds.**
R. V. E. Lovelace, H. L. Berk, J. Contopoulos.
Astrophys. J., Vol. 379, No. 2, p. 696 – 705 (1 Oct 1991).

A simple but general theory for the formation and propagation of nonrelativistic jets and winds from magnetized accretion disks is derived from the equations of ideal magnetohydrodynamics (MHD). The theory characterizes a jet by its axial velocity, radius, angular rotation rate, and temperature as functions of axial distance z from the central object. Differential equations for these four variables are derived from the conservation laws of ideal MHD for the flow of mass, momentum, angular momentum, and energy in the jet. The jets are assumed to carry no net current because this is energetically favored. The magnetic field is essential in that the zz–component of the magnetic stress acts, in opposition to gravity, to "drive" matter through the slow magnetosonic critical point. Close to the disk, the gravitational field of the central object provides weak collimation of the flow. At much larger distances, good collimation can result from the focusing effect of the external medium. Further, the pressure of the external medium can act, in opposition to gravity, to "drive" the flow through the fast magnetosonic critical point. For a representative self–consistent disk/jet solution relevant to a protostellar system, the fraction of the accreted mass expelled in the jets is 0.1, the ratio of the power carried by the jets (kinetic and Poynting fluxes) to the disk luminosity is 0.66, and the ratio of the boundary layer to disk luminosities is $\lesssim 0.13$. The star's rotation rate is found to decrease with time even for rotation rates much less than the breakup rate.

064.129 **The shape of dense structures in the plasma envelopes of slowly rotating hot magnetic stars.**
P. A. Bespalov, S. S. Davydenko.
Sov. Astron. Lett., Vol. 17, No. 5, p. 342 – 345 (Sep 1991). Current Physics Microform No.: 9204X1289. English translation of Pis'ma Astron. Zh., Tom 17, No. 9, p. 810 – 816 (1991).

The authors have determined the geometrical shape of dense structures in the plasma envelopes of hot rotating magnetic stars where the radiation pressure force on the plasma exceeds gravity. They have investigated questions associated with the stability of the dense plasma structures. They comment on the effects of rotation on the electromagnetic radiation of these plasma envelopes.

064.130 **Hot stars – what can be learned from EUV spectroscopy?**
R.–P. Kudritzki, R. Gabler, J. Puls, J. H. M. M. Schmitt.
1. Berkeley Colloquium on Extreme Ultraviolet Astronomy, p. 130 – 152 (1991). – See Abstr. 012.045 for the main entry.

Contents: 1. Introduction. 2. The physics of hot star atmospheres. 2.1. Nonlocal thermodynamic equilibrium. 2.2. Stellar winds. 2.3. Unified model atmospheres: a solution of the EUV "Zanstra discrepancy". 3. What can be observed with *EUVE*? 3.1. Hot subdwarfs and central stars of planetary nebulae. 3.2. Massive OB stars. 3.3. The predicted EUV spectra of nearby B stars. 4. Conclusioons.

064.131 **EUV line fluxes from shocked wind models of B stars.**
J. Cassinelli, B. Y. Welsh, P. W. Vedder, J. MacFarlane, J. V. Vallerga.
1. Berkeley Colloquium on Extreme Ultraviolet Astronomy, p. 153 – 158 (1991). – See Abstr. 012.045 for the main entry.

The authors have investigated a driven–wave shock–model, and they have found it possible for a single shock–zone to explain the two–component X–ray model derived from *Einstein* observations by Swank (1985). The authors also use the empirically derived, differential emission measure to predict EUV line fluxes that could be measured with the spectrometers on the *EUVE* satellite.

064.132 **Observations and predictions of EUV emission from classical novae.**
S. Starrfield, W. M. Sparks, J. W. Truran, J. Krautter.
1. Berkeley Colloquium on Extreme Ultraviolet Astronomy, p. 168 – 176 (1991). – See Abstr. 012.045 for the main entry.

Contents: 1. Introduction. 2. Theoretical simulations of the outburst. 3. The early EUV emission of a nova. 4. The post–maximum EUV emission from novae in outburst. 5. Observations of novae. 6. Summary and conclusion.

064.133 **EUV constraints on boundary layer models for cataclysmic variables.**
R. A. Wade.
1. Berkeley Colloquium on Extreme Ultraviolet Astronomy, p. 186 – 192 (1991). – See Abstr. 012.045 for the main entry.

Contents: 1. Introduction. 2. Model computations. 3. Results. 3.1. General. 3.2. VW Hydri.

064.134 **Protostellar disks.**
S. Beckwith.
6. IAP Astrophysics Meeting – IAU Colloquium No. 129: Structure and emission properties of accretion disks, p. 33 – 42 (1991). – See Abstr. 012.046 for the main entry.

This review concentrates on three central issues: How are disks identified, are they actively accreting matter from their surroundings, and what clues do their properties give us about their subsequent evolution? There are several other in–depth reviews of these topics, some in this volume, to which the reader is referred for more detailed discussions. The author concentrates here on providing an overview for the recent results discussed in this conference.

064.135 **Dynamo action by internal waves in accretion disks.**
E. T. Vishniac, Jin Liping, P. H. Diamond.
6. IAP Astrophysics Meeting - IAU Colloquium No. 129: Structure and emission properties of accretion disks, p. 223 - 226 (1991). A summary of a work published in Astrophys. J., Vol. 365, No. 2, p. 648 - 659 (20 Dec 1990) - see Abstr. 52.064.075. - See Abstr. 012.046 for the main entry.

064.136 **Model atmospheres for accretion disks.**
I. Hubeny.
6. IAP Astrophysics Meeting - IAU Colloquium No. 129: Structure and emission properties of accretion disks, p. 227 - 235 (1991). - See Abstr. 012.046 for the main entry.

The present status of vertical structure models for accretion disks is briefly summarized. Particular emphasis is given to the problem of external irradiation of the disk by the central star. Numerical examples demonstrate that a more refined approach than usually used one based on a modification of the effective temperature, is needed in order to predcit accurately the temperature structure and the emergent radiation from the disk.

064.137 **Model atmospheres for accretion disks.**
R. Wehrse, G. Shaviv.
6. IAP Astrophysics Meeting - IAU Colloquium No. 129: Structure and emission properties of accretion disks, p. 237 - 246 (1991). - See Abstr. 012.046 for the main entry.

The authors describe model atmospheres for thin accretion disks which are calculated on the basis of the α model of Shakura and Sunyaev, in particular it is assumed here that the local energy generation rate is proportional to the (gas-) pressure.

064.138 **Post-outburst X-rays from accretion disk coronae.**
E. Meyer-Hofmeister, F. Meyer.
6. IAP Astrophysics Meeting - IAU Colloquium No. 129: Structure and emission properties of accretion disks, p. 257 - 260 (1991). - See Abstr. 012.046 for the main entry.

The model presented here describes a basic physical process, and except for the viscosity parameter, depends only on very elementary physical parameters.

064.139 **Predicted extreme-ultraviolet line emission for nearby main-sequence B stars.**
J. J. MacFarlane, J. P. Cassinelli, B. Y. Welsh, P. W. Vedder, J. V. Vallerga, W. L. Waldron.
Astrophys. J., Vol. 380, No. 2, p. 564 - 574 (20 Oct 1991).

The source of the X-ray emission from O and B stars is currently the subject of debate. Shocks propagating through the winds of O and B stars have previously been proposed to explain their observed X-ray luminosities. Strong shocks with velocity jumps of several hundred km s^{-1} can heat portions of the wind to 10^6- 10^7K, producing major sources of X-ray and extreme-ultraviolet (EUV) emission lines. Alternatively, a corona at the base of the wind might also be responsible for the X-ray emission. The authors investigate the characteristics of EUV lines produced in high-temperature X-ray emitting regions of early B stars. Estimates of EUV spectral line fluxes for single- and multi-temperature plasma models for three nearby main-sequence B stars (β CMa, α Vir, and β Cen A) are presented. The plasma models are constrained by X-ray spectral data, and are based on insights from previous calculations of shocks propagating through hot star winds. The results demonstrate the importance of EUV emission lines in determining the properties of the X-ray emitting regions of hot stars. The authors also show that EUV emission lines from B stars with low interstellar absorption may be detectable with the Extreme Ultraviolet Explorer (EUVE) satellite. The authors investigate the effects of Doppler spreading and the time-dependent attenuation of the overlying wind on line profiles, and discuss how spectral observations can be used to determine temperatures, velocities, and locations of the emitting plasma.

064.140 **Superoutburst phenomenon of the SU UMa-type dwarf novae.**
Y. Osaki.
6. IAP Astrophysics Meeting - IAU Colloquium No. 129: Structure and emission properties of accretion disks, p. 271 - 274 (1991). - See Abstr. 012.046 for the main entry.

The author proposes a new working model for the superoutburst phenomenon of the SU UMa stars based on instabilities in accretion disks. To explain this model, he first discusses intrinsic instabilities in accretion disks of catalysmic variable stars.

064.141 **The natural history of superhumps.**
R. Whitehurst.
6. IAP Astrophysics Meeting - IAU Colloquium No. 129: Structure and emission properties of accretion disks, p. 275 - 277 (1991). - See Abstr. 012.046 for the main entry.

The superhump phenomenon of SU UMa dwarf novae is qualitatively explained in terms of an accretion disk model.

064.142 **Self-consistent description of convective domains in geometrically thin accretion disks.**
W. J. Duschl.
6. IAP Astrophysics Meeting - IAU Colloquium No. 129: Structure and emission properties of accretion disks, p. 305 - 308 (1991). - See Abstr. 012.046 for the main entry.

The author discusses a newly proposed prescription for the viscosity parameter α in convective regions of geometrically thin Keplerian accretion disks that makes it no longer necessary to introduce α as a free parameter. If one interprets the viscosity in these accretion disks as turbulent viscosity, it is possible to determine it self-consistently together with convection turbulence. To describe both as manifestations of one single turbulent process, the mixing length model is used. It is found that anisotropy of the turbulence does not change the basic results. The details of the structure of accretion disks that follow from this model indicate that it is possible, additionally, to explain the UV-delay in dwarf novae.

064.143 **Modeling accretion disk boundary layers.**
O. Regev.
6. IAP Astrophysics Meeting - IAU Colloquium No. 129: Structure and emission properties of accretion disks, p. 311 - 321 (1991). - See Abstr. 012.046 for the main entry.

The outline of this article is as follows. First, the boundary layer (BL) problem is defined and several different approaches to its solution are briefly reviewed. Further, the method of matched asymptotic expansions (MAE) and its use in accretion disk BL problems are presented. Finally, recent results obtained for an FU Ori accretion disk BL using MAE are included.

064.144 **Irradiation of accretion disks around young objects. I. Near-infrared CO bands.**
N. Calvet, A. Patiño, G. Magris C., P. D'Alessio.
Astrophys. J., Vol. 380, No. 2, p. 617 - 630 (20 Oct 1991).

The authors calculate in an approximate way the effect of irradiation by the central star on the atmospheres of optically thick, physically thin, accretion disks around young objects, assuming radiative equilibrium. The net effect of irradiation is to increase the temperature in the atmosphere of a viscous accretion disk relative to the case when only the viscous flux goes through it. The temperature increase at a given depth depends on the adsorption and scattering coefficients characteristic of the atmosphere, and on the rate of irradiation. The effect on the near-infrared spectrum of the star - disk configuration is to decrease the strength of the absorption in the CO bands or to turn them into emission, depending on the irradiation rate and on the mass accretion rate. As the stellar effective temperature increases for a given mass accretion rate, the disk surface temperature increases and the CO bands turn into emission. On the other hand, as the mass accretion rate increases for a given star, the intrinsic atmospheric (viscous) temperature increases and the effect of irradiation becomes less important; the CO bands turn then into absorption. The stellar effective temperature – disk mass accretion rate diagram can be divided into regions where the near–infrared CO bands in the spectra appear either in absorption or in emission; observed spectra of young objects such as T Tauri stars, FU Ori objects, and Herbig Be/Ae stars are in qualitative agreement with those predicted using the corresponding stellar parameters and disk mass accretion rates.

064.145 **Radiation hydrodynamics in accretion disks and the boundary layer.**
W. Kley.
6. IAP Astrophysics Meeting – IAU Colloquium No. 129: Structure and emission properties of accretion disks, p. 323 – 326 (1991). – See Abstr. 012.046 for the main entry.

At the innermost part of the accretion disk where the disk grazes the surface of the star frictional interaction decelerates the disk material to stellar rotational velocity and a transition zone (boundary layer) is formed. Since in the most cases the star is slowly rotating, nearly half of the total accretion energy is liberated here. The inclusion of this boundary layer is crucial for a complete and selfconsistent solution of the accretion disk structure. The correct theoretical description of the boundary layer structure is necessarily a two–dimensional problem and has to be solved self–consistently. Therefore the author developed an algorithm which solves the coupled 2D radiation–hydrodynamic equations for axisymmetric flows. To investigate the influence of different physical parameters he ran different models varying distinct parameters individually.

064.146 **Protostellar and T Tauri winds.**
N. Calvet.
6. IAP Astrophysics Meeting – IAU Colloquium No. 129: Structure and emission properties of accretion disks, p. 343 – 351 (1991). – See Abstr. 012.046 for the main entry.

According to the present understanding of low–mass star formation the onset of a strong wind is required to: (1) stop the main phase of accretion and prevent further significant increase of the stellar mass, so that the object can start to be well described by the quasihydrostatic equilibrium evolutionary tracks of isolate models (although evolutionary tracks may be modified by disk accretion in $\approx 30\%$ of the star, (2) take away angular momentum, so that stars reaching their pre–main sequence phase of evolution have rotational velocities as low as observed, (3) clear up the environment so that the star appears in the optical range for the first time. It is then apparent that from very early times in its histroy the star is losing mass. This mass loss has implications on many different aspects, as the maintenance of clouds or the formation of planets. Here, however, the author refers only to those studies that pertain to the determination of the characteristics of the wind, namely loss rate, temperature, velocity fields, morphology, time–evolution, with special emphasis on the phases when the star is visible.

064.147 **Conference summary. Structure and emission properties of accretion disks.**
M. Livio.
6. IAP Astrophysics Meeting – IAU Colloquium No. 129: Structure and emission properties of accretion disks, p. 357 – 363 (1991). – See Abstr. 012.046 for the main entry.

Contents: 1. Accretion disk physics. (a) Transport of angular momentum. (b) Eruptions and variability. (c) Winds and jets. 2. Comments on the different classes of objects. (a) Cataclysmic variables. (b) "Protostellar" objects. (c) X–ray binaries. (d) Active galactic nuclei. 3. Conclusion.

064.148 **Line formation in accretion disks.**
J. Adam.
6. IAP Astrophysics Meeting – IAU Colloquium No. 129: Structure and emission properties of accretion disks, p. 369 – 370 (1991). – See Abstr. 012.046 for the main entry.

The author considers the spatial radiation transfer in an extended, differentially rotating cylindrical disk.

064.149 **A recipe for making hot accretion disks.**
G. Björnsson, R. Svensson.
6. IAP Astrophysics Meeting – IAU Colloquium No. 129: Structure and emission properties of accretion disks, p. 379 – 380 (1991). Also published in Astrophys. J., Lett., Vol. 371, No. 2, p. L69 – L72 (20 Apr 1991) – see Abstr. 53.064.069. – See Abstr. 012.046 for the main entry.

064.150 **Limb darkening for irradiated accretion disks and accretion disk parameters in Cyg X–1.**
N. G. Bochkarev, E. A. Karitskaya.
6. IAP Astrophysics Meeting – IAU Colloquium No. 129: Structure and emission properties of accretion disks, p. 383 – 384 (1991). – See Abstr. 012.046 for the main entry.

Analytical calculations in two–flux approximation have been made for optically thick disks with strong outer irradiation, the dissipation of gravitational energy in the disk being neglected. The authors pay their attention to optical and near–IR ranges mainly for X–ray binaries, i.e. to outer parts of the accretion disks.

064.151 **The influence of the self gravity in thick accretion discs.**
G. Bodo, A. Curir.
6. IAP Astrophysics Meeting – IAU Colloquium No. 129: Structure and emission properties of accretion disks, p. 387 – 388 (1991). – See Abstr. 012.046 for the main entry.

In this work the equilibrium structure of self gravitating disks is obtained by using the self–consistent field method (Ostriker and Mark, 1968).

064.152 **Irradiation of accretion disks around young objects: near infrared CO bands.**
N. Calvet, A. Patiño, G. Magris C., P. D'Alessio.
6. IAP Astrophysics Meeting – IAU Colloquium No. 129: Structure and emission properties of accretion disks, p. 399 – 400 (1991). – See Abstr. 012.046 for the main entry.

064.153 **Solving time dependent problems: the brute force method.**
S. K. Chakrabarti.
6. IAP Astrophysics Meeting – IAU Colloquium No. 129: Structure and emission properties of accretion disks, p. 403 – 404 (1991). – See Abstr. 012.046 for the main entry.

In this brief report, the author shows how it is possible to evolve an isothermal flow around a Newtonian star.

064.154 **Inverse–Compton gamma–ray emission from chaotic, early–type stellar winds and its detectability by Gamma Ray Observatory.**
W. Chen, R. L. White.
Astrophys. J., Lett., Vol. 381, No. 2, p. L63 – L66 (10 Nov 1991).

The authors give an estimate of the possible γ–ray luminosity from winds around hot stars and show that the predicted photon flux below 1 MeV from nearby stars can be greater than the detection threshold of the Oriented Scintillation Spectrometer Experiment on the Gamma Ray Observatory (GRO). Charged particles are first–order Fermi accelerated to relativistic energies by strong isothermal shocks generated by radiatively driven instabilities in the winds. Low–energy γ–rays of luminosity $\sim 10^{33}$erg s^{-1} are emitted by the relativistic electrons via inverse Compton scattering of the stellar UV photons and are predicted to be well correlated with the hard X–ray component observed by the Solid State Spectrometer of the Einstein Observatory from some nearby hot stars. Although these discrete γ–rays sources in general are only marginally observable, positive detection by GRO will provide important information on particle acceleration by shocks and will certainly help to complete the multiwavelength picture of the chaotic, early–type stellar winds.

064.155 **On turbulent transport in accretion discs.**
B. Dubrulle.
6. IAP Astrophysics Meeting – IAU Colloquium No. 129: Structure and emission properties of accretion disks, p. 419 – 420 (1991). – See Abstr. 012.046 for the main entry.

064.156 **Optically thin accretion discs in close binaries: non–LTE calculations.**
A. M. Dumont, S. Collin–Souffrin, A. R. King, J. P. Lasota.
6. IAP Astrophysics Meeting – IAU Colloquium No. 129: Structure and emission properties of accretion disks, p. 421 – 422 (1991). – See Abstr. 012.046 for the main entry.

064.157 **A non–LTE study of neutral calcium in late–type stars with special reference to Pollux.**
J. J. Drake.
Mon. Not. R. Astron. Soc., Vol. 251, No. 3, p. 369 – 378 (1 Aug 1991).
Detailed simultaneous radiative transfer–statistical equilibrium calculations have been undertaken for neutral calcium using model stellar atmospheres corresponding to a variety of late spectral types. The results are used to investigate non–LTE effects and trends with differing stellar parameters, and to estimate the likely influence of departures from LTE on model atmosphere analyses. The behaviour of individual Ca I atomic levels and lines are discussed in connection with calculations carried out for a model atmosphere corresponding to the K0 III giant Pollux (β Gem). Non–LTE effects in Ca I in solar–type dwarf stars are likely to be insignificant, but are expected to become increasingly significant toward lower gravity stars, with individual multiplets indicating widely different abundances in giant stars. Overionization effects are not expected to increase in severity in metal–poor stars due to a decrease in photospheric temperature gradient with decreasing metallicity, and in giant stars and hotter solar–type dwarfs because of enhancements of the Balmer continuous opacity in these spectral types.

064.158 **Magnetohydrodynamical winds from an accretion disk: driving by the magnetic pressure and tension.**
J. Fukue.
6. IAP Astrophysics Meeting – IAU Colloquium No. 129: Structure and emission properties of accretion disks, p. 429 – 430 (1991). – See Abstr. 012.046 for the main entry.
In order to describe MHD winds/jets from geometrically thin accretion disks with rotation, the MHD equations are written down in the orthogonal curvilinear coordinates whose axes are parallel and perpendicular to the streamlines, respectively. For the extreme case of the "steady" winds/jets, only the toroidal magnetic fields are taken into account. It is demonstrated that steady rotating wind–type MHD jets can be accelerated by the magnetic pressure and tension of toroidal magnetic fields, using a simple model.

064.159 **Time evolution of a magnetic seed–field in a turbulent accretion disk.**
G. Geertsema, A. Achterberg.
6. IAP Astrophysics Meeting – IAU Colloquium No. 129: Structure and emission properties of accretion disks, p. 431 – 432 (1991). – See Abstr. 012.046 for the main entry.
MHD turbulence can strongly enhance the transport of angular momentum in an accretion disk. To increase the understanding of turbulent activity in an accretion disk a computer program has been constructed which simulates MHD turbulence. Here the authors describe the evolution of a small magnetic seed–field. Starting from a magnetic energy many orders of magnitude smaller than the kinetic energy they find equipartion is reached well within one Kepler time.

064.160 **On the turbulent viscosity in accretion disks.**
I. Goldman.
6. IAP Astrophysics Meeting – IAU Colloquium No. 129: Structure and emission properties of accretion disks, p. 433 – 434 (1991). – See Abstr. 012.046 for the main entry.
The author considers a steady state accretion disk. He studies the turbulent viscosity in both the cases of a three–dimensional and a two–dimensional turbulence. Ii is concluded that in disk outbursts in cataclysmic variables there is a transition from three– to two–dimensional turbulence.

064.161 **Two–temperature accretion disks with winds.**
M. Kusunose.
6. IAP Astrophysics Meeting – IAU Colloquium No. 129: Structure and emission properties of accretion disks, p. 455 – 456 (1991). – See Abstr. 012.046 for the main entry.

064.162 **Thermal conditions in a protostellar disk.**
A. B. Makalkin, V. A. Dorofeyeva (*V. A. Dorofeeva*).
6. IAP Astrophysics Meeting – IAU Colloquium No. 129: Structure and emission properties of accretion disks, p. 475 – 476 (1991). – See Abstr. 012.046 for the main entry.
Vertical and radial temperature distributions in a protostellar (protoplanetary) disk are modeled. Two phases of disk evolution are considered: (1) The disk is deeply imbedded within an infalling envelope. (2) The disk surrounds a young star.

064.163 **Three–dimensional numerical simulation of accretion discs.**
T. Matsuda, K. Sawada.
6. IAP Astrophysics Meeting – IAU Colloquium No. 129: Structure and emission properties of accretion disks, p. 481 – 482 (1991). – See Abstr. 012.046 for the main entry.

064.164 **Stellar wind collision and dust formation in long–period, heavily interacting Wolf–Rayet binaries.**
V. V. Usov.
Mon. Not. R. Astron. Soc., Vol. 252, No. 1, p. 49 – 52 (1 Sep 1991).
The author estimates the rate of mass loss in the form of cold dense gas from a wide Wolf–Rayet binary like WR 140. At some distance from the binary, this gas can form a dust shell and reradiate the UV radiation in the IR region. In the case of WR 140 which has a high eccentricity the observed IR outbursts may be caused by the strong increase of the cold dense gas outflow from WR 140 at the periastron.

064.165 **Thermal and dissipative stability of convection–accretion disks.**
B. D. Scott.
6. IAP Astrophysics Meeting – IAU Colloquium No. 129: Structure and emission properties of accretion disks, p. 515 – 516 (1991). Cf. Astron. Astrophys., Vol. 247, No. 2, p. 419 – 428 (Jul 1991) – see Abstr. 54.064.005. – See Abstr. 012.046 for the main entry.

064.166 **Magnetic accretion disks fall into two types.**
K. Shibata, T. Tajima, R. Matsumoto, J. Fukue, R. Okada.
6. IAP Astrophysics Meeting – IAU Colloquium No. 129: Structure and emission properties of accretion disks, p. 517 – 518 (1991). Extended abstract. – See Abstr. 012.046 for the main entry.

064.167 **Self consistent accretion disk corona.**
N. Soker, J. C. Raymond.
6. IAP Astrophysics Meeting – IAU Colloquium No. 129: Structure and emission properties of accretion disks, p. 523 – 524 (1991). – See Abstr. 012.046 for the main entry.
The structure of accretion disk corona is calculated self consistently. X–ray radiative transfer is simulated by a Monte Carlo scheme. This work is in the progress, and only preliminary results are presented here. The method will allow the prediction of X–ray and UV characteristic emission as a function of inclination angle from accretion disks in X–ray binaries.

064.168 **Accretion driven by spiral waves – an example.**
E. Szuszkiewicz.
6. IAP Astrophysics Meeting – IAU Colloquium No. 129: Structure and emission properties of accretion disks, p. 541 – 542 (1991). – See Abstr. 012.046 for the main entry.

064.169 **Convective viscosity in thin accretion disks.**
A. Wandel, I. Goldman.
6. IAP Astrophysics Meeting – IAU Colloquium No. 129: Structure and emission properties of accretion disks, p. 549 – 550 (1991). – See Abstr. 012.046 for the main entry.

The authors analyze geometrically thin accretion disks where the primary mechanism for viscous dissipation of angular momentum is convective turbulence. Four cases are analyzed analytically: radiative or convective vertical energy transport, and gas or radiation pressure dominated disks.

064.170 **A new luminosity limit for thin accretion disks.**
A. Wandel, E. P. Liang.
6. IAP Astrophysics Meeting – IAU Colloquium No. 129: Structure and emission properties of accretion disks, p. 551 – 552 (1991). – See Abstr. 012.046 for the main entry.

The authors report a new luminosity limit for geometrically thin α disks, in the regime where the optically thin and thick solutions merge into a single solution of intermediate optical depth. For α close to unity this new limit can be lower than the Eddington limit.

064.171 **Emission lines from accretion disc atmospheres.**
J. H. Wood, S. Mineshige.
6. IAP Astrophysics Meeting – IAU Colloquium No. 129: Structure and emission properties of accretion disks, p. 559 – 560 (1991). – See Abstr. 012.046 for the main entry.

The authors calculate the vertical structure of optically thin atmospheres above a cool ($T_e \leqslant 5000$K) disc, which may be relevant to dwarf novae in quiescence.

064.172 **An analytical model for a Keplerian disc and a boundary layer around a rapidly rotating star.**
M. Colpi, M. Nannurelli, M. Calvani.
Mon. Not. R. Astron. Soc., Vol. 253, No. 1, p. 55 – 62 (1 Nov 1991).

The authors present an analytical model for a disc accreting on to a rapidly rotating star through a boundary layer. The disc is isothermal along the radial direction, and a polytropic equation of state is used to model the transport of energy along the vertical direction; the kinematic viscosity of the fluid is constant. If the star rotates near breakup, the disc angular velocity close to the stellar surface has no maximum, but it settles smoothly to the stellar value, the flow always remaining subsonic. The process is accompanied by a negligible transfer of angular momentum from the disc to the star; both features are general and do not depend on the model assumptions. The authors are able to determine the analytical relation between the specific angular momentum accreted by the star j_* and the stellar angular velocity Ω_*, for their isothermal model. The comparison with the recent calculations of Paczyński and Popham and Narayan has then shown that the magnitude of $j_*(\Omega_*)$ and its functional form depend rather sensitively on the way energy is transported in the disc along the radial direction, and on the viscosity law. The result might be relevant for models of neutron star formation from accretion induced collapse of massive rapidly rotating white dwarfs.

064.173 **Gravitational versus kinetic shifts in the radiation from accretion disks.**
J. Buitrago, F. Atrio, E. Mediavilla, C. García-Gómez.
Vulcano Workshop 1988: Frontier objects in astrophysics and particle physics, p. 321 – 324 (1989). – See Abstr. 012.079 for the main entry.

The radiation of light from matter rotating in the vicinitiy of compact objects (neutron stars or black holes) is expected to be affected from significant frecuency shifts which are two-fold in origin: (1) the velocity of rotation, in the inner parts of accretion disks, is of the order of the velocity of light, thereby producing Doppler shifts of pure kinematic origin. (2) the emitted radiation is affected from an additional redshift produced by the gravitational field of the central object. In this region, with a typical size of a few Schwarzschild radii, the local emitted radiation is shifted towards the blue or red while the action of the gravitational field always induce a shift towards lower frequencies. Due to the importance of these effects for an eventual confrontation with the observations, the authors shall briefly examine the essential aspects both in the kinematic situation as well as in the kinematic plus gravitational (general relativistic) case.

064.174 **Numerical simulations of flares on M dwarf stars. I. Hydrodynamics and coronal X-ray emission.**
C.-C. Cheng, R. Pallavicini.
Astrophys. J., Vol. 381, No. 1, p. 234 – 249 (1 Nov 1991).

In order to understand the large variety of physical parameters observed in stellar X-ray flares, and their time evolution, the authors have made a series of numerical simulations of flare loop models with different values of loop size, flare energy input, and initial loop conditions. The model solves the full set of mass, energy, and momentum equations in a magnetically confined loop structure which extends from the chromosphere to the corona. The computed results involve important physical processes such as the expansion of heated coronal gas, the evaporation of chromospheric material, the compression of the plasma at the loop footpoints, the gradual and/or catastrophic cooling of the gas during the decay phase. The expected X-ray emission is calculated and compared with X-ray light curves obtained by the EXOSAT satellite. The simulations can reproduce the general characteristics of the observed X-ray flares, including their intensities, time profiles, and average coronal temperatures. The authors find that it is possible to produce X-ray emission measures of up to $\sim 10^{52}\mathrm{cm}^{-3}$ with fairly small loops with a total length of only 2×10^9cm. Small- and medium-sized flares can thus be produced in short loops. Under physically realistic conditions, large loops with length of $\sim 8 \times 10^9$cm are required to produce the largest emission measures ($\sim 10^{53}\mathrm{cm}^{-3}$) observed in stellar flares.

064.175 **A model for tidally driven eccentric instabilities in fluid disks.**
S. H. Lubow.
Astrophys. J., Vol. 381, No. 1, p. 259 – 267 (1 Nov 1991).

Observations and simulations of superhump phenomena in close binary star systems and observations of eccentric planetary rings, such as the rings of Uranus, indicate that eccentric instabilities occur in disks and rings. The author investigates the eccentric stability of a fluid disk or ring in the presence of the tidal field of a companion object which is in circular orbit. The disk can have pressure, viscosity, and self-gravity. Eccentric Lindblad resonances that lie within a disk cause eccentricity growth by the following process. For a particular perturbing potential component $\varphi_m(r)\cdot\cos[m(\theta-\Omega_p t)]$ a simple tidal response is felt by the disk that scales with φ_m. This response couples with the imposed eccentricity to launch density waves at each eccentric Lindblad resonance, producing a response that scales with $e\varphi_m$. The coupling of these two responses produces a stress that increases eccentricity with growth rate that scales as ${\varphi_m}^2$. Considering an eccentric disk with slowly varying orbital elements, the author derives expressions for the growth rates of the eccentricity and the associated mechanical energy changes due to an eccentric Lindblad resonance.. For high m, the results apply to the eccentric planetary rings. For application to superhump phenomena, the $m = 3$ eccentric inner Lindblad resonance probably dominates over other eccentric Lindblad and eccentric corotational resonances.

064.176 **Simulations of tidally driven eccentric instabilities with application to superhumps.**
S. H. Lubow.
Astrophys. J., Vol. 381, No. 1, p. 268 – 277 (1 Nov 1991).

The author numerically analyzes the tidally induced eccentric evolution of a fluid disk having pressure and viscosity. He studies the effects of the 3:1 eccentric inner Lindblad resonance on disks in close binaries, as a model for superhumps, using a two-dimensional SPH code. First, orbit crossings near resonance are eliminated by applying only the $m = 3$ component of the perturbing potential. Eccentricity growth occurs as an instability through a two-stage process. In stage 1, initial eccentricity growth occurs exponentially in time while the disk dissipation is relatively low. In stage 2, eccentricity growth occurs much more slowly in time, due to elevated levels of dissipation and wave nonlinearities. Stage 1 reveals a very mild wave in the disk. Both the wave and the eccentricity grow exponentially in time. In stage 2, the wave develops further and sometimes dissipates through shocks. The wave is a two-armed, tightly wrapped, trailing, inwardly propagating disturbance, launched near the resonance. Apparent eccentricity changes occur as the disk becomes substantially lopsided. The author considers the effects of changing disk viscosity and the binary mass ratio. He also considers a model with the full potential of the companion and finds a somewhat reduced growth rate, but consistency with the overall picture.

064.177 **The origin of viscosity in thin accretion disks.**
E. T. Vishniac, L. Jin, P. Diamond.
Symposium in Celebration of the 50. Anniversary of McDonald Observatory of the University of Texas at Austin: Frontiers of stellar evolution, p. 625 (1991). Abstract. – See Abstr. 012.085 for the main entry.

064.178 **Model atmospheres for rotating B stars.**
G. W. Collins II, R. J. Truax, S. R. Cranmer.
Astrophys. J., Suppl. Ser., Vol. 77, No. 4, p. 541 – 606 (Dec 1991).
The authors present the results of extensive model atmosphere calculations applicable to rotating early-type stars. In particular, the nature of the "strong line pole-on" Be stars may be elucidated by these culculations. It appears that, next to rotational broadening, the dominant effect of extreme rotation on the formation of a spectral line is the "limb darkening" of the line which causes absorption lines to vanish at the limb. For lines that tend to weaken with decreasing temperature, rotation will cause the line to appear similar in strength to that of later type stars. The reverse is true for lines that increase in strength with decreasing temperature. Polarization of the ultraviolet Balmer continuum is shown to persist to much later spectral types than might be expected on the basis of ionization studies alone. In addition, it is shown that rotation behaves like a spatial filter for sharp absorption lines, resulting in the enhancement of the degree of polarization in the wings of these lines. It is suggested that the effect could be quite pronounced in the far-ultraviolet of the Balmer continuum, thus complicating the interpretation of the behavior of the polarization found in the spectra of Be stars.

064.179 **Simulation of the boundary layer in white dwarf with the SPH method.**
D. Molteni, G. Giannone.
Workshop on Iron Line Diagnostics in X-Ray Sources, p. 162 – 166 (1991). – See Abstr. 012.088 for the main entry.
The boundary layer between an accretion disk and the surface of a compact object is simulated using the Smoothed Particle Hydrodynamics (SPH) method. The equations include radiation, which is treated in the flux limited diffusion approximation. The integrated equations and some numerical simulations of the 3D motion of the gas are presented.

064.180 **Modelling the rotation in the wind of λ Cephei.**
P. A. Mazzali, C. A. Berger, H. F. Henrichs.
ESO Workshop on Rapid Variability of OB-Stars: Nature and Diagnostic Value, p. 223 – 226 (1991). – See Abstr. 012.092 for the main entry.
The observed variability of the N V, C IV and Si IV wind lines and of the He II 4686 Å near photospheric line of the O6 I(n)fp star λ Cephei is probably due to instabilities in the flow. Although spherically symmetric "unified" (atmosphere + wind) models can reproduce the deep photospheric line He I 4713 Å if photospheric rotational broadening is applied, they fail in their prediction of the He II 4686 Å line, whose formation region extends also into the lower part of the wind. This suggests that rotation has to be included in the wind model as well. In order to test the importance of rotation, the Munich Wind code and the "Wind-cum-Rotation" code were run for the stellar parameters of λ Cep and different values of v_{rot}. It is found that the broad emission peaks of the P-Cygni profiles can be explained by an amount of rotation similar to that inferred from the analysis of the photospheric line broadening. This is obviously of great encouragement for the continuation of this investigation with the inclusion in the model of a perturbed flow, which is probably responsible for the profile variations.

064.181 **Nonradial pulsation and line profile variability in early type stars.**
D. R. Gies.
ESO Workshop on Rapid Variability of OB-Stars: Nature and Diagnostic Value, p. 229 – 243 (1991). – See Abstr. 012.092 for the main entry.
The author first reviews the theoretical expectations for nonradial pulsation line profile variability, and then describes in detail how the model succeeds or fails in the case of ε Per, a well observed target.

064.182 **On the nature of the excess 100 micron flux associated with carbon stars.**
M. P. Egan, C. M. Leung.
Astrophys. J., Vol. 383, No. 1, p. 314 – 323 (10 Dec 1991).
IRAS data indicates that many carbon stars, especially those with optically thin dust shells, have large fluxes at 60 and 100 μm. It has been suggested that a remnant dust shell from an earlier mass-loss episode can explain the excess fluxes. Two hypotheses have been proposed to explain the origin of the remnant dust shell. The authors have constructed radiation transport models of dust around carbon stars: models with either a single C-rich dust shell or double shells. They find that single-shell models cannot explain the observed color distribution, while two-shell models with either a C-rich or an O-rich remnant shell can. To differentiate between the two scenarios, the authors examined the relation between mass-loss rate and the flux ratio at long wavelengths. They compared the model results with observations. They find that models with a single dust shell do not reproduce the observed relationship between mass-loss rate and flux ratio but the two-shell models do. The time scales required by the models and the necessity of a prior O-rich phase are consistent with the hypothesis that some O-rich stars evolve to form carbon stars. While one cannot rule out the existence of C-rich remnant shells, the authors conclude that they can only be formed after a phase of O-rich mass loss has occurred.

064.183 **Gravitational instabilities in a proto-planetary disk.**
H. Noh, E. T. Vishniac, W. D. Cochran.
Astrophys. J., Vol. 383, No. 1, p. 372 – 379 (10 Dec 1991).
The authors have studied gravitational instabilities in a thin, Keplerian proto-planetary disk with a uniform entropy. The disk is a two-fluid system, composed of dust and gas, coupled by gravity and friction. The growth rates and the eigenfunctions are computed by using the nonaxisymmetric linear perturbation method in two-dimensional cylindrical coordinates. Their study includes modes with high values of the azimuthal wavenumber (m). They find that for the usual values of input parameters their disk is unstable to nonaxisymmetric perturbations for some range of m in addition to m = 1. However, at very low masses, the disk becomes stable for all m except for m = 1. The growth rate has a well defined maximum in m and the number of radial nodes. The authors interpret this result. They also study the effect of different boundary conditions on the growth rate. The difference of the effect between these boundary conditions appears only for small m, and the growth rate using reflecting boundary conditions is higher than those using transmitting boundary conditions. The effects of various properties of dust on the instability, and the change of the growth rate due to the change of parameters are also shown.

064.184 **Mass loss of Be-stars.**
J. P. Koninx, A. G. Hearn.
ESO Workshop on Rapid Variability of OB-Stars: Nature and Diagnostic Value, p. 279 – 284 (1991). – See Abstr. 012.092 for the main entry.
Be stars show a fast, tenuous wind from the polar and meridional regions, and a slow, dense wind from the equatorial

region. Preliminary results obtained with the combination of the radiatively driven wind theory of Castor, Abbott and Klein (1975) and the sound wave driven wind theory of Pijpers and Hearn (1989) suggest that the observations can be explained by the addition of a sound–wave flux at the equatorial regions.

064.185 **Modeling utility of rapid variability diagnostics of OB stars.**
V. Doazan, R. N. Thomas.
ESO Workshop on Rapid Variability of OB–Stars: Nature and Diagnostic Value, p. 287 – 288 (1991). – See Abstr. 012.092 for the main entry.

064.186 **What can NLTE model atmosphere theory contribute to the understanding of line profile variability?**
K. P. Simon.
ESO Workshop on Rapid Variability of OB–Stars: Nature and Diagnostic Value, p. 335 – 343 (1991). – See Abstr. 012.092 for the main entry.
A brief description of non–LTE model photospheres and quantitative spectroscopy of early type stars is given. Specific problems of rapidly rotating stars are addressed and the applicability of these methods to the investigation of line profile variability due to non–radial pulsations is discussed.

064.187 **The vertical structure of T Tauri accretion disks. I. Heating by the central star.**
F. Malbet, C. Bertout.
Astrophys. J., Vol. 383, No. 2, p. 814 – 819 (20 Dec 1991).
The authors present the LTE formal solution for the vertical atmospheric temperature structure of an accretion disk illuminated by its central star. It is shown that there are two consequences of irradiation: (1) heating of the photospheric layers resulting in continuum radiation, and (2) heating of the optically thin layers resulting in a chromospheric–like temperature increase that may produce disk emission lines in these regions. The authors illustrate these results by computing the gray vertical temperature structure of a flat accretion disk. Because of the emergent intensity's anisotropy, the observed disk temperature distribution depends on the disk's viewing angle.

064.188 **Nonthermal excitation of helium in Type Ib supernovae.**
L. B. Lucy.
Astrophys. J., Vol. 383, No. 1, p. 308 – 313 (10 Dec 1991).
The presence of strong He I absorption troughs in the optical spectra of Type Ib SNs is attributed to nonthermal excitation. As for analogous effects in the premaximum spectra of SN 1987A, the early importance of nonthermal excitation in the Ib's is fundamentally due to a large–scale instability that mixes ^{56}Ni out into their envelopes shortly after explosion. Because of this mixing, the helium–rich envelopes of these SNs are, even at maximum light, not strongly shielded from the γ–rays emitted by ^{56}Co nuclei. In consequence, the helium atoms are subject to impact excitations and ionizations by the nonthermal electrons in the cascades created in the stopping of the MeV Compton electrons produced in the scattering of these γ–rays. A non–LTE treatment of this nonthermal excitation of helium is given using a recent model of a Type Ib SN due to Shigeyama et al. that invokes mixing of ^{56}Ni in order to reproduce observed light curves. The statistical equilibrium calculations predict gross overpopulations of the $n = 2$ levels and the resulting theoretical spectra show encouraging agreement with observation. The possible relevance of these calculations to the differences in spectra and light curves between the Ib's and the Ia's and Ic's is briefly discussed.

064.189 **Statistical equilibrium and the Li I 670.776 and 670.791 nm absorption lines in the spectrum of red giants. Results.**
Ya. V. Pavlenko.
Sov. Astron., Vol. 35, No. 6, p. 623 – 626 (Nov 1991). Current Physics Microform No.: 9205X1433. English translation of Astron. Zh., Tom 68, Vyp. 6, p. 1240 – 1246 (1991).
The results of a solution of the non–LTE problem for lithium in the atmospheres of K and M giants (T_{eff} = 3600, 3800, 4000, 4400, and 4800K, $\log g = 1.7$) are presented. Opacity at the frequencies of lithium bound–bound and bound–free transitions is determined by absorption in the contiuum, molecular bands, and atomic lines. The LTE and non–LTE curves of growth for resonance lines of the Li I doublet and synthetic spectra in the vicinity of 670.8 nm are obtained. The calculations show that non–LTE effects are significant in the atmospheres of red giants [$\varepsilon_{non-LTE}(Li) - \varepsilon_{LTE}(Li) \sim 0.4$ dex for lines with $W_\lambda = 5$ pm].

064.190 **Determination of nonlinear darkening laws from model stellar atmospheres and from the solution of light curves of classical eclipsing systems.**
A. A. Rubashevskij.
Sov. Astron., Vol. 35, No. 6, p. 626 – 634 (Nov 1991). Current Physics Microform No.: 9205X1436. English translation of Astron. Zh., Tom 68, Vyp. 6, p. 1247 – 1260 (1991).
A method is suggested for determining the parameters of nonlinear darkening laws obtained from model stellar atmospheres, under the condition of conservation of the radiative flux in a certain spectral range and the best description of the surface brightness distribution over the stellar disk. On the basis of that method, an algorithm is suggested for determining nonlinear darkening laws as part of the solution of light curves of classical eclipsing variables in the simplest model of a binary system. It is shown that in a linear approximation to the actual darkening law, the solution of the light curves leads to the determination of u_b. The method is used to find the parameters of the eclipsing variable YZ Cas, particularly the darkening coefficients of its stars and the nonlinear darkening law of the A star in the red.

064.191 **Main–sequence mass–loss: lithium dilution in the Hyades and the Sun.**
F. J. Swenson, J. Faulkner.
Conference on Challenges to Theories of the Structure of Moderate–Mass Stars, p. 345 – 352 (1991). – See Abstr. 012.096 for the main entry.
The authors investigate the consequences of possible main–sequence mass–loss on surface lithium abundances through the use of evolving stellar models. In particular, they attempt to model the lithium depletion observed in the Sun and the Hyades G–dwarfs. Mass–loss as an explanation for the lithium depletion has mixed success. In the case of the solar lithium problem the authors conclude that mass–loss can be a marginal explanation. In the case of the Hyades G–dwarfs they find, contrary to preliminary investigations, that mass–loss cannot be a significant contributor to the depletion pattern seen in these stars. Furthermore, it is shown that quasi–analytical means alone appear to be sufficiently accurate to model the consequences of mass–loss in these stars. With the results of their calculations the authors limit the amount of mass that may have been lost on the main–sequence by Hyades G–dwarfs.

064.192 **The collimation of a time–dependent stellar wind.**
S. H. Kim, A. C. Raga.
J. R. Astron. Soc. Can., Vol. 85, No. 4, p. 215 (Aug 1991). Abstract. – See Abstr. 012.091 for the main entry.

064.193 **Line formation in expanding extended atmospheres with spherical geometry.**
L. S. Cidale, A. E. Ringuelet.
Bol. Asoc. Argent. Astron., No. 36, p. 427 – 429 (1990). Abstract. In Spanish and English. – See Abstr. 012.095 for the main entry.

064.194 **MHD wave energy fluxes generated from convection zones of late type stars.**
Y.–J. Moon, H. S. Yun.
J. Korean Astron. Soc., Vol. 24, No. 2, p. 129 – 149 (Dec 1991).
An attempt has been made to examine the characteristics of acoustic and MHD waves generated in stellar convection zones (4000K $\leqslant T_{eff} \leqslant$ 7000K, $3 \leqslant \log g \leqslant 4.5$). The authors' study shows that acoustic wave fluxes are dominant in F and G stars, while the MHD waves are dominant in K and M stars, and that the MHD wave fluxes vary as $T_{eff}{}^4 \sim T_{eff}{}^7$ in contrast to the

acoustic fluxes, as T_{eff}^{10}. The gravity dependence, on the other hand, is found to be relatively weak; the acoustic wave fluxes $\propto g^{-0.5}$, the longitudinal tube wave fluxes $\propto g^{-0.3}$ and the transverse tube wave fluxes $\propto g^{0.3}$. In the case of the MHD body waves their gravity dependence is found to be nearly negligible. Finally the authors assessed the computed energy fluxes by comparing them with the observed fluxes F_{ob} of C IV (λ1549) lines and soft X-rays for selected main sequence stars. When they scaled the corrected wave fluxes down to F_{ob}, it is found that these slopes are almost in line with each other.

064.195 **Angular momentum transport in protostellar disks.**
D. N. C. Lin.
NATO Advanced Research Workshop on Theory of Accretion Disks, p. 89 – 104 (1989). – See Abstr. 012.121 for the main entry.

Analyses of angular momentum transport in protoplanetary disks are important for the investigation of star formation as well as the origin of the solar system. During the formation stages, the disk dynamics are regulated by mixing of infalling material and disk gas. After infall is switched off, convectively driven turbulence provides an effective angular momentum transfer which yields an evolutionary timescale of the order 10^{5-6}y. The evolution to "naked T Tauri stars" is shown. Protoplanetary formation modifies the structure and evolution of the disk.

064.196 **Protostellar discs.**
J. E. Pringle.
NATO Advanced Research Workshop on Theory of Accretion Disks, p. 105 – 111 (1989). – See Abstr. 012.121 for the main entry.

The author argues on both observational and theoretical grounds that protostellar discs are often large and massive. He proposes that for the short-lived stars there is evidence that such discs can exist through the mainsequence phase and become rejuvenated at the planetary nebula stage.

064.197 **Formation of viscous protostellar accretion disks.**
W. M. Tscharnuter.
NATO Advanced Research Workshop on Theory of Accretion Disks, p. 113 – 123 (1989). – See Abstr. 012.121 for the main entry.

The author discusses results and conclusions that have been derived from recent model calculations. Particular attention is drawn to the role of the initial conditions and symmetry assumptions for the stability behavior of protostellar cores. Furthermore, a sequence of collapse models referring to the formation of the solar nebula is briefly discussed. Finally, further arguments in favor of the physical reality and importance of core instabilities are presented on the basis of a simplified model.

064.198 **Formation and structure of protostellar accretion disks.**
R. E. Pudritz.
NATO Advanced Research Workshop on Theory of Accretion Disks, p. 125 – 133 (1989). – See Abstr. 012.121 for the main entry.

This paper summarizes two stages of work on disk structure, formation, and evolution.

064.199 **Nonlinear pulsation in the transonic region of geometrically thin accretion disks.**
R. Matsumoto, S. Kato, F. Honma.
NATO Advanced Research Workshop on Theory of Accretion Disks, p. 167 – 172 (1989). – See Abstr. 012.121 for the main entry.

Nonlinear hydrodynamical simulations are performed to study the evolution of axisymmetric instabilities in viscous accretion disks around a black hole or a nonmagnetized neutron star. When the viscosity parameter α, is sufficiently large ($\alpha > 0.05$), quasi-periodic pulsations are excited in the innermost transonic region of the disk. Numerical results also show that after the onset of the thermal instability in radiation pressure dominated region, a geometrically thin accretion disk evolves into a slim accretion disk.

064.200 **One-armed oscillations of disks and their applications.**
S. Kato.
NATO Advanced Research Workshop on Theory of Accretion Disks, p. 173 – 181 (1989). – See Abstr. 012.121 for the main entry.

The presence of low frequency oscillation modes in disks is demonstrated. The frequency is much smaller than the dynamical one of the disks. Such low-frequency modes are one-armed oscillations. In relativistic disks, in particular, they are one-armed corrugation waves fluttering in the vertical direction. Possible applications of these waves to some astronomical objects are discussed.

064.201 **Thermal and tidal instabilities in accretion disks of dwarf novae.**
Y. Osaki.
NATO Advanced Research Workshop on Theory of Accretion Disks, p. 183 – 205 (1989). = Contrib. Dep. Astron., Univ. Tokyo, No. 386. – See Abstr. 012.121 for the main entry.

The author discusses the thermal instability (the ordinary disk instability) and the tidal instability in accretion disks as causes of outbursts in dwarf novae. First he discusses the disk instability model for normal outbursts of dwarf novae; in particular he considers: the narrowness of transition fronts, the delay of UV flux to the optical at the rise of outburst, and the variation of disk radius in an outburst-quiescence cycle. He then considers the superoutburst and superhump phenomenon in SU UMa stars; particularly the tidal instability of accretion disks recently discovered by Whitehurst (1988), by which instability an accretion disk evolves into a non-axisymmetric form (an eccentric accretion disk). It is shown that the tidal instability in accretion disks is caused by a parametric resonance of particle orbits in accretion disks in binary systems. The author finally proposes a new working model of the superoutburst phenomenon based on the combined effects of the ordinary disk instability and the tidal instability. He concludes that the disk instability model can explain the basic features of both normal- and superoutbursts in dwarf novae.

064.202 **Hydrodynamic simulation of accretion disks in cataclysmic variables.**
M. Hirose, Y. Osaki.
NATO Advanced Research Workshop on Theory of Accretion Disks, p. 207 – 212 (1989). = Contrib. Dep. Astron., Univ. Tokyo, No. 387. – See Abstr. 012.121 for the main entry.

The non-axisymmetric configurations of accretion disks in cataclysmic variables are studied by two-dimensional hydrodynamical calculations. The time evolution of accretion disks under a constant mass supply rate from the secondary is followed until the system settles to a quasi-steady state. It is found that a binary system with comparable masses of component stars settles to a steady state of an elongated disk fixed in the rotating frame of the binary while in the case of low mass secondary it settles to a periodic oscillating state in which a non-axisymmetric (eccentric) structure is formed and it rotates slowly in the inertial frame of reference. The period of oscillation is a few percent longer than the orbital period of binary, and it offers a natural explanation to the "superhump" periodicity of SU UMa stars.

064.203 **Simulations of accretion flow in close binary stars.**
R. Whitehurst.
NATO Advanced Research Workshop on Theory of Accretion Disks, p. 213 – 220 (1989). – See Abstr. 012.121 for the main entry.

The dynamics of accretion flow in close binary stars is investigated via a numerical simulation of the accretion flow. The importance of the secondary's tidal torque is discussed with respect to the radius of the accretion disc. It is found that the steady-state disc has a fixed radius, which is within the primary's Roche lobe.

064.204 **Viscous evolution of accretion discs in the quiescence of dwarf novae.**
S. Mineshige, J. H. Wood.
NATO Advanced Research Workshop on Theory of Accretion Disks, p. 221 – 225 (1989). – See Abstr. 012.121 for the main entry.
The viscous evolution of accretion discs in the quiescence of dwarf novae is investigated semi–analytically. There are two key factors: the inward flow of the initial mass in the disc and the diffusion of the material added into the disc later. The results are compared with the brightness temperature distributions obtained by the eclipse mapping of OY Car and Z Cha.

064.205 **The structure of the boundary layer of accretion disks in cataclysmic binaries.**
W. Kley.
NATO Advanced Research Workshop on Theory of Accretion Disks, p. 289 – 295 (1989). – See Abstr. 012.121 for the main entry.
The author presents some calculations to study the structure of the interaction region of the inner accretion disk with the stellar surface layers in the case of a high mass flow rate through the disk.

064.206 **The dynamics of twisted accretion discs.**
S. Kumar.
NATO Advanced Research Workshop on Theory of Accretion Disks, p. 297 – 306 (1989). – See Abstr. 012.121 for the main entry.
The theory of twisted discs and its application is briefly reviewed. Such twisted discs are probably quite common, though their current observational relevance is limited to explaining long term periodicities in high mass X–ray binaries, the eclipse dip in ε Aur, and perhaps the S–shapes of jets of radio galaxies.

064.207 **A free accretion disk in SN 1987A?**
F. Meyer, E. Meyer–Hofmeister.
NATO Advanced Research Workshop on Theory of Accretion Disks, p. 307 – 311 (1989). – See Abstr. 012.121 for the main entry.
The observations of Middleditch et al. (1989) of a submillisecond pulsar in SN 1987A might indicate a remnant neutron star rotating near breakup. This suggests that sizeable angular momentum was present in the core of Sanduleak –69°202 before collapse. The authors speculate that not all angular momentum could be accreted and some matter was left after the collapse around the neutron star in critical rotation. They discuss here the development of such an accretion disk.

Mechanisms of chromospheric and coronal heating. Proceedings.
See Abstr. 012.011.

The Sun and cool stars: activity, magnetism, dynamos. Proceedings.
See Abstr. 012.024.

High energy astrophysics: compact stars and active galaxies. Proceedings.
See Abstr. 012.026.

White dwarfs. Proceedings.
See Abstr. 012.044.

Structure and emission properties of accretion disks. Proceedings.
See Abstr. 012.046.

Stellar atmospheres: beyond classical models. Proceedings.
See Abstr. 012.056.

Rapid variability of OB–stars: nature and diagnostic value.
See Abstr. 012.092.

Theory of accretion disks. Proceedings.
See Abstr. 012.121.

Force–free magnetic fields: generating functions and footpoint displacements.
See Abstr. 062.010.

Three–dimensional magnetostatic equilibrium.
See Abstr. 062.041.

Damping of shocks in magnetic flux tubes.
See Abstr. 062.061.

Viscous damping of magnetohydrodynamic surface waves.
See Abstr. 062.070.

The formation of current sheets and coronal heating.
See Abstr. 062.072.

Heating by field aligned DC Joule dissipation.
See Abstr. 062.076.

Two–dimensional magnetic reconnection.
See Abstr. 062.079.

Convection, spontaneous discontinuities, and stellar winds and X–ray emission.
See Abstr. 062.097.

Instabilities in astrophysical shear flows.
See Abstr. 062.098.

Shock waves in accretion disks.
See Abstr. 062.099.

Dynamo magnetic–field generation in turbulent accretion disks.
See Abstr. 062.123.

Dynamo magnetic field modes in thin astrophysical disks: an adiabatic computational approximation.
See Abstr. 062.151.

On the nature of disk viscosity.
See Abstr. 062.167.

Angular momentum transport in self–gravitating gaseous disks.
See Abstr. 062.168.

The role of magnetic fields in accretion disks.
See Abstr. 062.169.

Spiral instability of magnetized accretion disks.
See Abstr. 062.170.

The nonlinear development of gravitational instabilities in thick disks.
See Abstr. 062.171.

Magnetized accretion–ejection structures.
See Abstr. 062.172.

Radial oscillations of a gaseous torus with toroidal magnetic fields.
See Abstr. 062.182.

Numerical simulations of axisymmetric accretion flows.
See Abstr. 062.184.

High wavenumber thermal convection enhanced in regions of partial ionization.
See Abstr. 062.214.

A new approach to hot accretion disks.
See Abstr. 062.226.

Thermal convection in differentially rotating systems.
See Abstr. 062.247.

Relativistic effects in supernova photospheres.
See Abstr. 063.001.

Bremsstrahlung and energetic electrons in supernovae.
See Abstr. 063.002.

Stellar occultation of polarized light from circumstellar electrons. II. Flat envelopes viewed at arbitrary inclination.
See Abstr. 063.004.

Exact solution of the equation of transfer in a finite exponential atmosphere by the method of Laplace transform and linear singular operators.
See Abstr. 063.005.

The effects of electron scattering and wind clumping for early emission line stars.
See Abstr. 063.007.

Continuum opacity produced by spectral lines in supernovae and similar expansions.
See Abstr. 063.030.

Theoretical H II region models: the effects of stellar atmosphere models.
See Abstr. 063.031.

Polarization of light and models of the circumstellar environment of young stellar objects.
See Abstr. 063.036.

Line blanketing without LTE: simple and complex spectra.
See Abstr. 063.045.

NLTE spectral line formation in three dimensions.
See Abstr. 063.048.

Iteration with approximate lambda operators, and its application to the expanding atmospheres of WR stars.
See Abstr. 063.049.

Analysis of ultraviolet P Cygni profiles in the spectra of O–type stars.
See Abstr. 063.052.

Dynamics of and radiative transfer in inhomogeneous media.
See Abstr. 063.057.

The opacity project and the practical utilization of atomic data.
See Abstr. 063.061.

Radiative energy transfer in extended photospheres of the components of close binary systems.
See Abstr. 063.067.

Stellar occultation of polarized light from circumstellar electrons. III. General axisymmetric envelopes.
See Abstr. 063.072.

Effects of aberration and advection on line formation.
See Abstr. 063.090.

Towards better scattering properties for a mixture of realistic grains.
See Abstr. 063.098.

X–ray spectra from dynamical tuning.
See Abstr. 063.099.

Rotationally induced polarization in pure absorption spectral lines.
See Abstr. 063.104.

An investigation of Brueckner's theory of line broadening with application to the sodium D lines.
See Abstr. 063.107.

The theoretical variation of polarised light from stars exhibiting discrete UV absorption components.
See Abstr. 063.116.

Numerical approach to Zeeman line radiative transfer.
See Abstr. 063.124.

Influence of radiative transfer on the vertical structure of accretion disks.
See Abstr. 063.129.

Relative strengths of multiplet lines as an indicator of stellar magnetic fields.
See Abstr. 063.130.

Noise–driven radiative shocks: a new model for the optical quasi–periodic oscillations of the AM Herculis objects.
See Abstr. 065.002.

From wind to superwind: the evolution of mass–loss rates for Mira models.
See Abstr. 065.003.

Neutron capture nucleosynthesis and the evolution of 15 and 30 $M_\odot$ mass–losing stars. II. The core carbon–burning phase.
See Abstr. 065.026.

Model for R Coronae Borealis stars.
See Abstr. 065.064.

A new model of stellar convection for compressible fluids.
See Abstr. 065.065.

The influence of overshooting and stellar wind mass loss in the mass–luminosity relation.
See Abstr. 065.066.

The evolution of massive close binaries revised.
See Abstr. 065.067.

Helium inhomogeneities in main–sequence magnetic stars.
See Abstr. 065.077.

On the adiabatic pulsations of accretion disks and rotating stars.
See Abstr. 065.092.

Can neutron stars ablate their companions?
See Abstr. 065.093.

Three–dimensional compressible hydrodynamic convection in the Sun and stars.
See Abstr. 065.105.

Line profile variations caused by low frequency nonradial pulsations of rapidly rotating stars. II.
See Abstr. 065.129.

Excitation mechanism for nonradial pulsations in variable OB stars.
See Abstr. 065.130.

Nonradial pulsation, rotation, and episodic mass loss.
See Abstr. 065.131.

Theoretical aspects of asteroseismology of the rapidly oscillating Ap stars.
See Abstr. 065.152.

Images of accretion disks.
See Abstr. 067.026.

On the stability of wind accretion.
See Abstr. 067.039.

Model atmospheres and X–ray spectra of bursting neutron stars.
See Abstr. 067.040.

Steady spherical hypercritical accretion onto neutron stars.
See Abstr. 067.042.

The stability of slim accretion disks.
See Abstr. 067.043.

Variability of accretion discs around compact objects.
See Abstr. 067.046.

Radiative–hydrodynamical simulations of accretion disk coronae.
See Abstr. 067.061.

Calculations on the radiation spectrum of simplified stable accretion disk.
See Abstr. 067.078.

High–polarization radiation from magnetized accretion disks.
See Abstr. 067.086.

Diskoseismology: probing accretion disks. I. Trapped adiabatic oscillations.
See Abstr. 067.098.

On accretion by compact objects from a stellar wind.
See Abstr. 067.166.

Variability of accretion discs around compact objects.
See Abstr. 067.184.

Trapped oscillations of accretion disks.
See Abstr. 067.188.

Quasi–periodic oscillations in luminous accretion discs.
See Abstr. 067.189.

Variable line profiles from accretion disks and black hole mass.
See Abstr. 067.193.

Stability of slim accretion discs – effects of central mass and viscosity.
See Abstr. 067.203.

On the possibility of excitating one–armed corrugation waves in relativistic accretion disks.
See Abstr. 067.226.

Cold super–fast magnetohydrodynamical winds from rapid rotators.
See Abstr. 067.227.

Electron–positron pair production in compact sources with energetic protons.
See Abstr. 067.254.

On the double natured solution of two temperature accretion flows.
See Abstr. 067.266.

Magnetoacoustic heating of the solar chromosphere.
See Abstr. 073.126.

The confrontation of mean–field theories with numerical simulations.
See Abstr. 080.061.

Theoretical spectrum of solar convection.
See Abstr. 080.154.

Protostellar disks and planetary system formation.
See Abstr. 107.016.

Optical properties of circumstellar silicate grains.
See Abstr. 112.004.

CO (J = 1–0 and 2–1) mapping of IRC +10216: a hot core model for the gas kinetic temperature distribution and the mass–loss rate.
See Abstr. 112.032.

X–ray spectroscopy of stellar coronae.
See Abstr. 112.038.

Carbon stars associated with oxygen–rich circumstellar envelopes: EU Andromedae, BM Geminorum and V778 Cygni.
See Abstr. 112.049.

The extent of the molecular envelopes around evolved stars.
See Abstr. 112.055.

The light curve and changes in the circumstellar envelope around IRC+10216.
See Abstr. 112.059.

Dust disks around Vega–type stars.
See Abstr. 112.060.

The investigations of "zodiacal light" of isolated AE–Herbig stars with non–periodic Algol–type minima.
See Abstr. 112.074.

Molecules in evolved stars.
See Abstr. 112.079.

A 45 AU radius source around L1551–IRS 5: a possible accretion disk.
See Abstr. 112.091.

The light curves of periodic Be stars in NGC 3766.
See Abstr. 112.096.

X–ray lines in stellar coronae.
See Abstr. 112.111.

La química de las envolturas circunestelares de estrellas frias.
See Abstr. 112.113.

Structure of the extended atmosphere of Be stars and line profile variability.
See Abstr. 112.121.

Spectral line transients in λ Eridani and related Be stars.
See Abstr. 112.124.

The temperature of the circumstellar envelope of Alpha Orionis.
See Abstr. 112.133.

Are rapid variations of OB stars caused by circumstellar gaseous structures?
See Abstr. 112.144.

Modéliser les enveloppes circumstellaires?
See Abstr. 112.155.

Model of expanding atmosphere for HD 50138.
See Abstr. 112.158.

A grid of synthetic spectra for the determination of effective temperature, gravity and metallicity of F, G and K stars. II. Application to 41 stellar spectra taken in the Basel field of SA 141.
See Abstr. 114.001.

The formation of helium lines in the atmospheres of dMe stars.
See Abstr. 114.008.

Atmospheric model parameters and shock wave field for the supergiant alpha Carinae (F0 Ib).
See Abstr. 114.028.

Spectral analyses of Wolf–Rayet stars: hydrogen abundances in WN subtypes.
See Abstr. 114.034.

Modellatmosphärenanalyse von heißen Begleitern Roter Riesen und Überriesen.
See Abstr. 114.035.

A model–atmosphere analysis of Ap star HD 26385.
See Abstr. 114.065.

Temperatures, gravities and abundances of B stars: recent progress and remaining problems.
See Abstr. 114.075.

Chemical analysis of the atmosphere of the CP3 star π^1Boo. Revision of a previous analysis.
See Abstr. 114.080.

Atmospheric parameters of subluminous B stars.
See Abstr. 114.086.

A non–local thermodynamic equilibrium, line blanketed synthetic spectrum of Iota Herculis: C, Al, and Si lines.
See Abstr. 114.093.

New mechanism of light absorption in the atmospheres of Wolf–Rayet stars.
See Abstr. 114.101.

Oxygen abundances in halo stars.
See Abstr. 114.123.

The dependence of coronal temperatures on Rossby numbers.
See Abstr. 116.078.

On the wind geometry of the Wolf–Rayet star EZ Canis Majoris.
See Abstr. 116.085.

Rotational modulation and the line profile variables.
See Abstr. 116.093.

The formation and evolution of accretion disks in binary systems.
See Abstr. 117.014.

Binary X–ray sources: some aspects of wind accretion.
See Abstr. 117.086.

Accretion disks in binary stars.
See Abstr. 117.087.

The emission line of Her X–1 from the accretion column of magnetized neutron star.
See Abstr. 117.088.

The accretion column of DQ Herculis binaries.
See Abstr. 117.119.

Constraints on the physical properties of optical bullets in SS 433.
See Abstr. 117.141.

Can a disk model explain Beta Lyrae?
See Abstr. 117.143.

The optical emission from oscillating white dwarf radiative shock waves.
See Abstr. 117.146.

EUV emission from cool stars and stellar systems.
See Abstr. 117.177.

EUV emission from magnetic cataclysmic binaries.
See Abstr. 117.182.

Variability and structure of accretion disks in cataclysmic variables.
See Abstr. 117.185.

Accretion discs in low–mass X–ray binaries.
See Abstr. 117.186.

New polarimetric observations and a two pole model for the cyclotron emission from AM Herculis.
See Abstr. 117.191.

Is accretion via a disc in intermediate polars?
See Abstr. 117.192.

Emission lines from accretion disks.
See Abstr. 117.193.

Physical conditions in a gas around outer rim of Hercules X–1 accretion disk.
See Abstr. 117.199.

A theoretical model for the low–amplitude velocity variations in intermediate polars.
See Abstr. 117.204.

A search for evidence of accretion disk in two suspected Be–X transient binaries.
See Abstr. 117.205.

Evolution and outbursts properties of GK Per.
See Abstr. 117.208.

The Cygnus X–1 and SS433 accretion disks.
See Abstr. 117.221.

Search for effects of accretion driven optical flickering in the H Alpha emission line of Gamma Cassiopeiae.
See Abstr. 117.227.

The formation and evolution of accretion disks in binary systems.
See Abstr. 117.230.

The transition region structure of AR Lacertae.
See Abstr. 117.244.

A test of the starspot hypothesis for W–type W Ursae Majoris light curves.
See Abstr. 117.279.

Three–dimensional structure of accretion disks in close binary systems.
See Abstr. 117.317.

Possible properties of a VV Cephei type binary accretion disc.
See Abstr. 120.027.

Iron fluorescence lines in T Tauri stars.
See Abstr. 121.007.

Stochastic models of stellar winds in T Tauri–stars.
See Abstr. 121.021.

Ca II lines in T Tauri chromospheric models.
See Abstr. 121.031.

Polarization properties of YSO disks.
See Abstr. 121.066.

Disks around T Tauri stars.
See Abstr. 121.069.

FU Orionis disk models.
See Abstr. 121.070.

Effects of non–spherical mass loss and accretion on T Tauri line profiles.
See Abstr. 121.073.

On the near–infrared spectrum of FU Orionis.
See Abstr. 121.103.

Dust around young stars. The model of envelope of Ae–Herbig star WW Vulpeculae.
See Abstr. 121.108.

HD 160529: a new galactic luminous blue variable.
See Abstr. 122.005.

A double shock wave atmospheric model for the β Cephei star σ Scorpii?
See Abstr. 122.063.

UBVR colours of accretion disks and power–low sources of radiation with interstellar reddening.
See Abstr. 124.005.

Evolution of supernova remnants inside preexisting wind–driven cavities.
See Abstr. 125.035.

Circumstellar emission from SN 1987A.
See Abstr. 125.237.

The near–maximum–light spectrum of the Type Ic supernova 1987M.
See Abstr. 125.302.

On the evolutionary link between white dwarfs and central stars of planetary nebulae: NLTE analysis of PG 1144+005.
See Abstr. 126.003.

NLTE analysis of the unique pre–white dwarf H 1504+65.
See Abstr. 126.027.

PG1159 stars and the PNN–white dwarf connection.
See Abstr. 126.042.

The effective temperature of the DBV's, and the sensitivity of DB model atmospheres to input physics.
See Abstr. 126.057.

LP 790–29: preliminary model atmospheres for this strongly polarized carbon white dwarf.
See Abstr. 126.059.

Abundances in cool DZA and DAZ white dwarfs: new results using laboratory damping constants.
See Abstr. 126.060.

Evidence for fractionated accretion of metals on cool white dwarfs.
See Abstr. 126.061.

The Lyman alpha line wing in hydrogen–rich white dwarf atmospheres.
See Abstr. 126.063.

Atmospheric parameters for DA white dwarfs in the vicinity of the ZZ Ceti instability strip.
See Abstr. 126.064.

Outflow hydrodynamics in the eclipsing millisecond pulsar binaries PSR 1957+20 and PSR 1744–24A.
See Abstr. 126.076.

The observation and study of spectra of celestial SiO maser.
See Abstr. 131.288.

The time variation of celestial H_2O maser.
See Abstr. 131.289.

The position–velocity diagrams of protostellar discs.
See Abstr. 131.310.

Asymptotic theory for the spatial distribution of protostellar emission.
See Abstr. 131.356.

The effect of shock waves on the spectra of H II regions and planetary nebulae.
See Abstr. 132.016.

Optically thin dust models for IRAS point sources.
See Abstr. 133.004.

On the dynamical evolution of planetary nebulae.
See Abstr. 134.020.

The hydrodynamics of aspherical planetary nebulae. I. Analytic evaluation of hydrodynamic difference schemes.
See Abstr. 134.041.

Hydrodynamical models of aspherical planetary nebulae.
See Abstr. 134.048.

Analysis of the soft X–ray data from the central star of NGC 7293.
See Abstr. 134.055.

Neutrino–heated stars and broad–line emission from active galactic nuclei.
See Abstr. 158.154.

065 Stellar Structure and Evolution

065.001 **Diffusion and mixing in classical nova precursors.**
I. Iben Jr., M. Y. Fujimoto, J. MacDonald.
Astrophys. J., Lett., Vol. 375, No. 1, p. L27 – L29 (1 Jul 1991).
Numerical experiments have been conducted to determine the degree of enhancement of CNO elements in the envelope of an initially cold 1 $M_\odot$ carbon–oxygen white dwarf accreting hydrogen–rich material at rates expected in cataclysmic variables. Only ordinary particle diffusion and convective mixing have been considered, and the enhancements found are a consequence of the diffusive intermixing of hydrogen with carbon and oxygen; this mixing causes a hydrogen–burning thermonuclear runaway to begin below the surface of the original white dwarf. Convection mixes matter between the ignition point and the surface. Enhancements (relative to solar) vary from a factor of 10.5 at $\dot{M} = 10^{-8} M_\odot yr^{-1}$ to a factor of 26.6 at $\dot{M} = 10^{-10} M_\odot yr^{-1}$. Thus, if more mass is lost in a nova explosion than is gained between nova outbursts, no mixing mechanisms other than simple diffusion and convection need be invoked to account for the large enhancements of CNO elements observed in the spectra of many classical novae.

065.002 **Noise–driven radiative shocks: a new model for the optical quasi–periodic oscillations of the AM Herculis objects.**
M. T. Wolff, K. S. Wood, J. N. Imamura.
Astrophys. J., Lett., Vol. 375, No. 1, p. L31 – L34 (1 Jul 1991).
The authors have developed a model for the 0.3 – 1.2 Hz optical quasi–periodic oscillations (QPOs) observed in a number of AM Her–type binary systems. They suggest that the observed optical modulation is the result of shock oscillations induced by nonsteady accretion flows. It is shown that time–dependent models of radiative shock waves in nonsteady accretion flows onto magnetic white dwarfs with mass 0.6 $M_\odot$ and magnetic field strength of 30 MG can produce optical QPOs similar to those observed in the AM Her objects. Theoretical calculations have shown that shock oscillations cannot be sustained for these white dwarf parameters when the accretion rate is constant. The implications of this finding for our understanding of the temporal characteristics of the AM Her objects are discussed.

065.003 **From wind to superwind: the evolution of mass–loss rates for Mira models.**
G. H. Bowen, L. A. Willson.
Astrophys. J., Lett., Vol. 375, No. 2, p. L53 – L56 (10 Jul 1991).
Dynamical atmosphere models were calculated for a large grid of variables with Mira–like properties satisfying the Iben radius–luminosity–mass relationship for evolving AGB stars. Their masses ranged from 0.7 to 2.4 $M_\odot$, and their periods from 150 to 800 days. All were fundamental–mode pulsators, had solar metallicity, and included effects of dust. As a natural consequence of evolutionary changes in stellar parameters, the mass–loss rate increases as an approximately exponential function of time, reaching $\dot{M} \sim 10^{-5} - 10^{-4} M_\odot yr^{-1}$. Further evolution is dominated by the powerful wind, which strips the star's envelope from the core. This "superwind" is a remarkably robust effect. It occurs for all initial stellar masses and all modeling parameters that have been tested. Dust promotes the effect but is not essential for it. Models with very low metallicity also show the effect, but at higher luminosities. These results have intriguing implications for the number of supernovae in early low–metallicity populations and for the chemical evolution of galaxies.

065.004 **The faintest stars.**
J. B. Kaler.
Astronomy, Vol. 19, No. 8, p. 26 – 32 (Aug 1991).

065.005 **The evolution of intermediate–mass protostars. I. Basic results.**
F. Palla, S. W. Stahler.
Astrophys. J., Vol. 375, No. 1, p. 288 – 299 (1 Jul 1991).
The authors present a numerical calculation of the structure and evolution of protostars in the mass range from 1 to 8 $M_\odot$. They model the protostar as being built up from a spherically symmetric accretion flow, with a constant mass accretion rate of $10^{-5} M_\odot yr^{-1}$. The protostar is initially fully convective due to the steady state burning of deuterium near its center. At a mass of 2.4 $M_\odot$, there appears a radiative barrier, which prevents the convective transport of deuterium to the center. The region inside the barrier exhausts its deuterium and becomes radiatively stable, while the deuterium further out ignites in a subsurface shell. The shell burning swells the protostar, with a maximum radius of 10.5 $R_\odot$ being attained at 3.6 $M_\odot$. Further mass addition results in rapid gravitational contraction and the ignition of hydrogen via the CN cycle. The star joins the zero–age main sequence with a mass of 8 $M_\odot$. It is found that protostars over a significant mass range contain a substantial amount of deuterium. The ignition of this fuel during subsequent pre–main–sequence contraction could drive the surface convection that is apparently present in many Herbig Ae and Be stars.

065.006 **Collision cross section and effective energy generation rate for the reaction $H'(P\beta + \nu)H^2$ in stellar interiors.**
A. E. M. Khairozzaman.
Indian J. Phys., Part B, Vol. 65, No. 2, p. 143 – 149 (Apr 1991).
An attempt is made to relate the effective energy generation rate with the collision cross section for the reaction $H'(P\beta + \nu)H^2$ in stellar interiors. The energy generation rates are found to be greatly augmented due to the screening effect at the lower end of the hydrogen burning temperature. The method adopted may be equally useful for other steps of P–P cycle of reaction.

065.007 **The p–mode oscillation spectra of an evolving 1 $M_\odot$ Sun–like star.**
D. B. Guenther.
Astrophys. J., Vol. 375, No. 1, p. 352 – 365 (1 Jul 1991).
The p–mode frequencies, the first–order spacings $[\nu(n, l) - \nu(n-1, l)]$, and the second–order spacings $[\nu(n, l) - \nu(n-1, l+2)]$ of a 1 $M_\odot$ star, with solar composition, have been calculated for six models ranging from the zero–age main sequence to the base of the giant branch. The behavior of the frequency spacings, as a function of frequency and age, are analyzed in terms of the structure of the models. The calculated frequency spacings, when compared to observations of solar oscillations, reveal the Sun's deep interior to have the structure of a younger than 4.89 Gyr old stellar model (i.e., less concentrated than predicted by a standard stellar model), and the Sun's middle regions to have the structure of an older than 4.89 Gyr old stellar model (i.e., more concentrated than predicted by a standard stellar model). This structure is consistent with nonstandard modeling schemes proposed to reduce the predicted solar neutrino flux. For the most evolved models, near the base of the giant branch, mixed modes (g–mode–like oscillations in the interior and p–mode–like oscillations in the outer layers) occur and disrupt the near–constancy of the frequency spacings.

065.008 **Erratum: "A table of chemical yields of stellar populations" [Astron. Astrophys., Suppl. Ser., Vol. 87, No. 1, p. 109 – 118 (Jan 1991)].**
J. Köppen, N. Arimoto.
Astron. Astrophys., Suppl. Ser., Vol. 89, No. 2, p. 420 (Aug 1991). See Abstr. 53.065.029.

065.009 **Three–mode Cepheid pulsation. II. An example of the three–mode interacting hydrodynamic model of classical Cepheid.**
T. Ishida, M. Saitou.
Astrophys. Space Sci., Vol. 182, No. 1, p. 69 – 74 (Aug 1991).
The authors simulated two models of classical Cepheids with the same effective temperature and luminosity but different masses. Neither model has significant resonance among linear nonadiabatic periods of the fundamental, the first overtone, and the second overthone radial modes. The higher mass model is found to approach to the second overtone–only state. The lower mass model shows a complicated three–mode interaction and none of the amplitudes of three modes is negligible. The relation of the higher mass model to the shortest period Cepheid HR 7308, and that of the lower mass model to the double–mode Cepheids are discussed. It appears that three–mode nonresonant coupling should be considered to explain the behaviours of both models.

065.010 **The theory of supernova shocks: an analytical approach.**
P. M. Pizzochero.
Phys. Rep., Vol. 206, No. 3, p. 99 – 141 (Aug 1991).
The author presents an analytical study of the relations between the softness of the nuclear equation of state (EOS) which describes the dense matter in the core of a supernova, the structure of such a core under hydrostatic conditions and the initial strength of the shock wave which forms after bounce and eventually, if not stalled by energy losses, disrupts the whole star. He starts by developing a first–order perturbation theory for Newtonian polytropes around the $n = 1$ exact solution, and obtains explicit expressions for the various physical parameters of the polytope. He then points out that a dense pack of cold nuclear matter obeying a BCK–like EOS plus pressure from degenerate relativistic electrons and neutrinos can be described as an effective polytrope with index close to unity. Finally, the author outlines a phenomenological Newtonian model to describe the formation of the shock wave and its early propagation before it reaches the neutrinosphere.

065.011 **Der Lebenskreis der Sterne. Kugeln aus Staub und Gas.**
W. Knapp.
Bild Wiss., No. 9, p. 54 – 56 (Sep 1991).

065.012 **Riesen und Zwerge. Sterne sind Verwandlungskünstler.**
R. Kippenhahn.
Bild Wiss., No. 9, p. 58 – 61 (Sep 1991).

065.013 **Incompressibility of hot nuclear matter, general relativistic stellar collapse and shock propagation.**
J. A. Miralles, J. M. Ibáñez, J. M. Marti, A. Pérez.
Astron. Astrophys., Suppl. Ser., Vol. 90, No. 2, p. 283 – 299 (Oct 1991).
The authors have carried out research into the effect of considering two extreme kinds of Skyrme–type nuclear interactons on the outcome of stellar core collapsing models. Particular attention has been payed to dissociation effects generated by the shock. Neutrinos transport is included via a Multigroup Trapping Scheme.

065.014 **Approximating proto–stellar structure without a computer (The proto–sun at 50 $R_\odot$).**
J. F. Doorish.
Ir. Astron. J., Vol. 20, No. 2, p. 53 – 59 (Sep 1991).
This paper applies the basic method to the hydrostatic core of protostars (51.065.114) in order to model the "heart" of a forming star; the part that is unseen by any observational method except for a study of the infrared spectrum of the star forming region. A young, hot, forming star releases approximately 10^{49} ultraviolet photons, which are absorbed and reradiated by the surrounding accretion disk as infrared radiation. If one could see through this obscuring dust and look at the new forming star, one would basically see a spherical ball of gas in hydrostatic equilibrium since the accretion of matter takes place very gradually. This paper describes a method for approximating the structure of this hydrostatic core, and thereby allows to see beyond the obscuring dust.

065.015 **Additional evolutionary stellar models with core overshooting and mass loss: apsidal motion constants.**
A. Claret, A. Giménez.
Astron. Astrophys., Suppl. Ser., Vol. 91, No. 2, p. 217 – 227 (Dec 1991).
In order to extend the range of available chemical compositions and masses of the authors' previously published grid of stellar evolutionary models used to analyse apsidal motion observations, they have carried out new model calculations taking into account core overshooting and mass loss. Four additional chemical compositions were in fact adopted apart from the standard solar composition (X,Z) = (0.70,0.02). On the other hand, the authors have also extended the calculations for their standard model, as well as the non–standard one with solar composition, up to values of $\log M = 1.6$. The main reason to do this is to allow a straight comparison between observational data and theoretical models of massive binaries showing apsidal motion variations and also to permit the study of the effect of adopting different initial chemical compositions. Recent opacity tables from Los Alamos as well as revised rates of nuclear burning have been used and the internal stellar structure constants were computed for all the models presented.

065.016 **A simple dynamical model to study abrupt changes in stars.**
C. S. Zaidins, R. Tagg, R. Meglen.
Bull. Am. Astron. Soc., Vol. 23, No. 2, p. 968 – 969 (1991). Abstract. – See Abstr. 012.008 for the main entry.

065.017 **The evolution of supernova progenitors.**
T. A. Weaver, S. E. Woosley.
Bull. Am. Astron. Soc., Vol. 23, No. 2, p. 975 (1991). Abstract. – See Abstr. 012.008 for the main entry.

065.018 **The conductive propagation of nuclear flames.**
F. X. Timmes, S. E. Woosley.
Bull. Am. Astron. Soc., Vol. 23, No. 2, p. 975 (1991). Abstract. – See Abstr. 012.008 for the main entry.

065.019 **A potential asteroseismological test for convective overshooting theories.**
W. A. Dziembowski, A. A. Pamyatnykh.
Astron. Astrophys., Vol. 248, No. 1, p. L11 – L14 (Aug 1991). Letter–to–the–editor.
Among the modes that may be excited by opacity mechanism in δ Scuti stars there is one, at each $l > 0$, which is partially trapped in the region containing outer part of the convective core and the chemically inhomogenous zone left behind the shrinking core. The mode appears in the midst of the Main Sequence evolution as a transformation of the g_1 mode and evolves toward higher frequencies exhibiting the "avoided crossing" with consecutive p–modes. Determination of this mode frequency would provide a good measure of the extent of the overshooting from the convective core.

065.020 **Massive stars in galaxies.**
A. Maeder.
Q. J. R. Astron. Soc., Vol. 32, No. 3, p. 217 – 231 (Sep 1991).

065.021 **Structure of rapidly rotating axisymmetric stars. I. A numerical method for stellar structure and meridional circulation.**
Y. Eriguchi, E. Müller.
Astron. Astrophys., Vol. 248, No. 2, p. 435 – 447 (Aug 1991).
The authors present a numerical method to calculate the mechanical and thermal structure of stationary axisymmetric inviscid rotating stars in which the velocity of the meridional flow is very slow. They further assume that the centrifugal force can be derived from a rotational potential. With these assumptions surfaces of constant temperature, pressure, and density coincide, i.e. the structure and the flow can be treated separately. Thus the

authors first compute the structure of the rotating configuration and then calculate the flow pattern for the given configuration. In order to demonstrate how their method works, they have computed several sequences of stationary configurations for three different rotation laws. For the energy generation the point source model is assumed and two kinds of opacities (electron scattering opacity and Kramers opacity) are examined. The resulting stars consist of a convective core and a radiative envelope. In the uniformly rotating models the authors find a two-zone flow patter. Such a pattern was also found in previous investigations, which were based on a perturbation method for slowly rotating stars. The authors have further investigated models with two kinds of differential rotation laws. The flow patterns in these differentially rotating models are rather different from those of the uniformly rotating models. In particular, near the rotation axis the flow rising from the convective core does directly reach the surface of the star.

065.022 **On turbulent diffusion of magnetic fields and the loss of magnetic flux from stars.**
S. I. Vainshtein, R. Rosner.
Astrophys. J., Vol. 376, No. 1, p. 199 - 203 (20 Jul 1991).

The authors discuss the turbulent diffusion of magnetic fields in astrophysical objects, and the processes leading to magnetic field flux loss from such objects. They focus in particular on the suppression of turbulent diffusion by back-reaction of magnetic fields on small spatial scales, and on the constraint imposed on magnetic flux loss by flux-freezing within stars. It is shown that turbulent magnetic diffusion can be suppressed even for very weak large-scale magnetic fields, so that "standard" turbulent diffusion is incapable of significant magnetic flux destruction within a star. Finally, the authors show that magnetic flux loss via winds is generally ineffective, no matter what the value of the effective magnetic Reynolds number is.

065.023 **Rotational evolution of solar-type stars. I. Main-sequence evolution.**
K. B. MacGregor, M. Brenner.
Astrophys. J., Vol. 376, No. 1, p. 204 - 213 (20 Jul 1991).

The authors present a simple, parameterized model for the redistribution of angular momentum within the interiors of solar-type stars. By incorporating it with a description of angular momentum loss through the action of a magnetically coupled wind, the rotational histories of low-mass dwarf stars are traced for a variety of initial conditions and parameter specifications. The authors discuss the results of calculations for the rotational evolution of a 1 $M_\odot$ star, from the time of its arrival on the zero-age main sequence to the age of the present-day Sun. The best agreement with observational constraints is found for (1) a surface magnetic field strength which is largely independent of surface angular velocity Ω for rapid rotation, and approximately linearly dependent on Ω for $\Omega \approx \Omega_\odot$, and (2) a time scale for angular momentum transfer from the core to the convection zone which remains essentially constant throughout the evolution, with magnitude $\sim 10^7$yr. This value is approximately equal to the initial wind braking time.

065.024 **Wolf-Rayet stars of type WN/WC and mixing processes during core helium burning of massive stars.**
N. Langer.
Astron. Astrophys., Vol. 248, No. 2, p. 531 - 537 (Aug 1991).

Consequences of the recent finding that most WN/WC spectra probably originate from individual Wolf-Rayet stars for the internal structure of massive stars are discussed. Numerical models including the effect of slow-down or prevention of convective mixing due to molecular weight gradients are presented, in which a transition layer with a composition mixture of H- and He-burning ashes is formed above the convective He-burning core. These models are able to qualitatively account for the observed WN/WC frequency and agree quantitatively with the only WN/WC-composition determination so far. It is argued that the same transition layer may be responsible for the final blue loop which the SN 1987 A progenitor performed some 10^4yr before explosion. These results indicate that composition barrieres may be efficient in restricting convection during central helium burning, in contrast to computations relying on the Schwarzschild criterion for convection, with or without overshooting.

065.025 **On the orbital circularization of close binaries.**
I. Goldman, T. Mazeh.
Astrophys. J., Vol. 376, No. 1, p. 260 - 265 (20 Jul 1991).

The authors suggest a modified approach to the tidal circularization of short-period binaries with convective envelopes. The proposed model is based on Zahn's old theory and is motivated by his recent comment about the reduction of stellar viscosity due to the fact that the tidal variation time scale is comparable to or shorter than the typical convection turnover time scale. In the modified approximation the circularization time scale for close binaries is proportional to the binary period to the power of 10/3. This exponent is smaller than in previously suggested theories. The presently available data, though admittedly sparse, are compared with the different approximations. The new exponent seems to fit the data better than those of the previous theories.

065.026 **Neutron capture nucleosynthesis and the evolution of 15 and 30 $M_\odot$ mass-losing stars. II. The core carbon-burning phase.**
J.-P. Arcoragi, N. Langer, M. Arnould.
Astron. Astrophys., Vol. 249, No. 1, p. 134 - 140 (Sep 1991).

This paper presents the first detailed analysis of the neutron capture nucleosynthesis occurring in the carbon-burning core and in the helium-burning shell of massive (M_{ZAMS} = 15 and 30 $M_\odot$) mass-losing Population I stars during their central carbon-burning evolutionary phase. As such, this work is the natural continuation of a previous study (referred to as Paper I) of the s-process developing during central helium burning in the same stars. The adopted nucleosynthesis network is the same as in Paper I, and is fully coupled to the stellar evolution code in the way already described in that previous work. The authors show that the s-process taking place in the He-burning shell and in the C-burning core of the modeled stars is much less efficient than the one accompanying central He-burning, and in fact does not alter significantly the abundances derived in Paper I. This conclusion contradicts some previous studies which do not rely on the intimate coupling of nucleosynthesis and stellar evolution achieved here. In such conditions, a previous conclusion is confirmed that central He-burning in massive stars may be the main source of the galactic (solar-system) low-fluence s-process abundances, while shell He- and core C-burning could at most be responsible for the production of some specific nuclides of interest.

065.027 **Binary evolution with millisecond pulsar in a period gap.**
A. V. Fedorova, E. V. Ergma.
Pis'ma Astron. Zh., Tom 17, No. 11, p. 999 - 1007 (Nov 1991). In Russian. English translation in Sov. Astron. Lett., Vol. 17, No. 6.

The evolution of the binary with a millisecond pulsar in the period gap (P $\approx$ 1.8 ms, B = $5 \cdot 10^8$gauss) has been investigated in dependence on the fraction f of the impinging flux. It was shown that in the period gap the evolution is determined by four characteristic timescales τ_{gr}, τ_{evap}, τ_{nc}, τ_{sd} and small changes in f may lead to rather different evolutionary paths for the binary. For f = 0.015 - 0.017 the secondary fills its Roche lobe when the binary period is near 50 min. The results obtained in connection with the evolutionary stage of MXB 1916-05 are discussed.

065.028 **On the formation of WR + WR binaries.**
M. Vrancken, J. P. De Greve, L. Yungelson, A. Tutukov (*A. V. Tutukov*).
Astron. Astrophys., Vol. 249, No. 2, p. 411 - 416 (Sep 1991).

The formation and observability of WR + WR binaries is investigated, using Maeder's (1990) stellar evolution models, for primaries in the mass range 15 to 100 $M_\odot$. It is found that such binaries can only be formed for the largest masses in this range, and for a very narrow range of mass ratios close to 1. Due to severe constraints for the detection of such systems, they are not expected to be observed as spectroscopic or visual binaries. Moreover, those that might be observed are of the type WC+WC. Hence, all discovered WR stars of suspected type WN+WC are probably stars in a transition state.

065.029 **Nonlinear stellar pulsations. I. Numerical methods, basic physics, initial models and first results.**
E. A. Dorfi, M. U. Feuchtinger.
Astron. Astrophys., Vol. 249, No. 2, p. 417 – 427 (Sep 1991).
The authors introduce a new numerical method to compute nonlinear stellar pulsations. They describe some basic properties of this adaptive grid method and discuss several advantages and further improvements compared to the usual Lagrangian approach. Some first results provide additional justification for this new treatment. Starting from initial hydrostatic models the authors obtain periodic solutions in the case of Cepheids and an irregular behaviour in the case of extended cool objects. The nonlinear system of radiation hydrodynamics is solved implicitly in conservative form on a fully adaptive grid in order to provide a sufficient spatial resolution in regions of steep gradients. In their test calculations the authors adopt realistic constitutive relations according to Wuchterl (1990), Yorke (1979) and Baker and Kippenhahn (1962). In this first article they do not intend to relate the results directly to observations but want to demonstrate the ability of their method to model nonlinear stellar pulsation.

065.030 **Neutrino–nucleus interactions in core–collapse supernovae.**
S. W. Bruenn, W. C. Haxton.
Astrophys. J., Vol. 376, No. 2, p. 678 – 700 (1 Aug 1991).
Differential cross sections for charged–current and inelastic neutral–current neutrino–nucleus interactions have been computed. These are incorporated into a radiation hydrodynamics code, and a number of models have been evolved with the aim of determining the role of these rates in the evolution of the collapsing core and in the supernova mechanism. The authors focus, in particular, on the effect of these rates on the infall epoch, the prompt shock propagation phase, and the delayed–shock phase in the evolution of the core.

065.031 **High–temperature neutrino–nucleus processes in stellar collapse.**
G. M. Fuller, B. S. Meyer.
Astrophys. J., Vol. 376, No. 2, p. 701 – 716 (1 Aug 1991).
The authors investigate neutral current neutrino–nucleus interactions of interest in the supernova problem: endothermic and exothermic inelastic neutrino scattering processes; de–excitation of hot nuclei into neutrino pairs; and neutrino–antineutrino annihilation processes involving nuclei. A shell model based treatment including allowed and forbidden transitions is employed to compute the relevant weak strength distributions in nuclei. The authors discuss Fermi gas fitting formulae for the differential cross sections of all four processes for use in numerical stellar collapse calculations. Since target nuclei in stellar collapse are expected to be in highly excited states, thermal effects are explicitly incuded in the calculations. The authors find that low–energy neutrino–pair production from hot nuclei and exothermic neutrino–nucleus scattering may have an important role in determining the evolution of the lepton distribution functions in the infalling presupernova core.

065.032 **Bounds on KSVZ axions from high mass star evolution.**
Wang Jin.
Astrophys. Space Sci., Vol. 184, No. 1, p. 1 – 8 (Oct 1991).
The author studies the effects of axions on the theoretical stellar evolution of a 10 $M_\odot$ star from the Main–Sequence stage to a likely final stage. He has found that if the axion mass is less than 0.1 eV, an acceptable stellar evolution model which is consistent with observations is derived. If the axion mass is larger than 0.1 eV, such a star has a significantly different behaviour at the later stages of the evolution in the H–R diagram compared with a star without axions. The structure variables change greatly which is in contradiction with the observations.

065.033 **Partially degenerate nonrelativistic isothermal gas spheres in the presence of a magnetic field.**
H. P. Singh.
Astrophys. Space Sci., Vol. 184, No. 1, p. 65 – 76 (Oct 1991).
The effect of a weak poloidal magnetic field on the equilibrium structure of a partially degenerate nonrelativistic isothermal gas sphere has been studied. Models have been computed for varying degrees of central degeneracy and the effect of the magnetic field on the various parameters, e.g., mass, central condensation, moment of inertia, and oblateness has been looked into. It is found that as the central degeneracy parameter is increased, the central condensation in general decreases and, hence, the eccentricity of the configuration decreases.

065.034 **Relativistic magnetic partially degenerate isothermal gas spheres.**
H. P. Singh.
Astrophys. Space Sci., Vol. 184, No. 1, p. 143 – 146 (Oct 1991).
The effect of a poloidal magnetic field on the structure of isothermal gas spheres in hydrostatic equilibrium under the pressure of partially degenerate relativistic electrons and radiation has been considered. An equation of state involving Fermi–Dirac functions $F_2(\psi)$ and $F_3(\psi)$ has been used. Modifications to the values of various structural parameters have been tabulated for the cases $\psi_c = 0$, 2, 3, 5, and 10, ψ_c being the central degeneracy parameter.

065.035 **On numerical studies of solar/stellar convection.**
K. L. Chan, K. Serizawa.
IAU Colloquium No. 130: The Sun and cool stars – activity, magnetism, dynamos, p. 15 – 26 (1991). – See Abstr. 012.024 for the main entry.

065.036 **Turbulent convection: a new model.**
V. M. Canuto.
IAU Colloquium No. 130: The Sun and cool stars – activity, magnetism, dynamos, p. 27 – 32 (1991). – See Abstr. 012.024 for the main entry.
The author uses the latest models of turbulence to compute a new expression for the turbulent convective flux, F_c. The new values of F_c are up to ten times larger than those given by the mixing length theory, MLT. Astrophysical considerations indicate that the new model fares better with observational data than the MLT.

065.037 **Large scale convection in stars: towards a model for the action of coherent structures.**
M. Rieutord, J.–P. Zahn.
IAU Colloquium No. 130: The Sun and cool stars – activity, magnetism, dynamos, p. 33 – 36 (1991). – See Abstr. 012.024 for the main entry.
The authors show that, representing the descending fluid in a convection zone by a porous medium, the differential rotation of the (rising) fluid is very close to that in an axisymmetric model of the convection zone with anisotropic viscosity.

065.038 **Stellar convection as a low Prandtl number flow.**
J. M. Massaguer.
IAU Colloquium No. 130: The Sun and cool stars – activity, magnetism, dynamos, p. 57 – 61 (1991). – See Abstr. 012.024 for the main entry.
Thermal convection in the Sun and cool stars is often modeled with the assumption of an effective Prandtl number $\sigma \cong 1$. Such a parameterization results in masking of the presence of internal shear layers which, for small σ, might control the large scale dynamics. In this paper the author discusses the relevance of such layers in turbulent convection. Implications for heat transport – i.e. for the Nusselt number power law – are also discussed.

065.039 **Topological pumping in the lower overshoot layer.**
K. Petrovay.
IAU Colloquium No. 130: The Sun and cool stars – activity, magnetism, dynamos, p. 67 – 70 (1991). – See Abstr. 012.024 for the main entry.

Problems associated with topological pumping in the lower overshoot layer suggest a strongly turbulent and strongly differentially rotating upper radiative zone as the seat of the dynamo and as flux reservoir.

065.040 **Convective cores in stellar models.**
I. W. Roxburgh, M. Monteiro.
IAU Colloquium No. 130: The Sun and cool stars – activity, magnetism, dynamos, p. 95 – 97 (1991). – See Abstr. 012.024 for the main entry.

The condition for homogeneous radiative stellar models to be marginally stable to convection at the centre is investigated for the family of models where the opacity $\varkappa$ and energy generation ε are given by power laws in temperature and density $\varkappa = \varkappa_0 \varrho^\alpha T^{-\beta}$, $\varepsilon = \varepsilon_0 \varrho T^\eta$. The Naur–Osterbrock (1953) condition $6\eta > 6+10\beta-15\alpha$ is a necessary but not sufficient condition. A better estimate is obtained by taking the effective polytropic index $n = d \log P / d \log T - 1$ to be a linear function of temperature T throughout the model. This gives the condition $6\eta = 10\beta - 15\alpha + (12+4\beta)/(1+\alpha)$. The predictions of this condition agree well with results for a set of stellar models $0 \leqslant \alpha \leqslant 1$, $0 \leqslant \beta \leqslant 5$.

065.041 **Nonlinear nonaxisymmetric dynamos for active stars.**
D. Moss.
IAU Colloquium No. 130: The Sun and cool stars – activity, magnetism, dynamos, p. 112 – 116 (1991). – See Abstr. 012.024 for the main entry.

Recent observations seem to have detected large scale nonaxisymmetric structures on active giant stars. These structures are plausibly associated with underlying, dynamo generated, non-axisymmetric magnetic fields. Such developments have motivated the development of a computer code to solve the nonlinear mean field dynamo equation in spherical geometry with no imposed geometrical symmetries. The nonlinearity is a simple α–quenching. The nature of the stable solutions found depends quite sensitively on the assumed spatial distribution of both α–effect and differential rotation, and also on the degree of supercriticality of the dynamo. Such a dynamo model with stable, purely nonaxisymmetric solutions is described in this paper.

065.042 **Fossil magnetic fields and activity of young stars.**
A. E. Dudorov, E. E. Gorbenko.
IAU Colloquium No. 130: The Sun and cool stars – activity, magnetism, dynamos, p. 151 – 153 (1991). – See Abstr. 012.024 for the main entry.

The strength of fossil magnetic fields in young stars is estimated. The interaction of a fossil magnetic field with convection in stellar envelopes is investigated. The influence of such a magnetic field, transformed to a small scale, on the activity of young stars is discussed.

065.043 **Equations of state and bump Cepheids. I. Linear results.**
S. M. Kanbur.
Astron. Astrophys., Vol. 250, No. 2, p. 395 – 399 (Oct 1991).

The Hummer–Mihalas–Dappen (1988) and a simple Saha type equation of state are used to obtain linear pulsation characteristics of a grid of models spanning the Hertzsprung sequence (Cox 1974). The grid of models is taken from Simon and Davis (1983), who used the Los Alamos equation of state in their computations. The result is a sensitivity analysis of theoretical linear bump Cepheid models to the equation of state employed in the calculation. The results obtained with these equations of state are different, although not enough to resolve the Cepheid bump mass discrepancy (Stobie 1969; Simon and Lee 1976; Simon 1986).

065.044 **Magnetic fields and filling factors in late–type stars: predictions from dynamo theory.**
B. Montesinos, J. L. Fernández–Villacañas, C. Jordan.
IAU Colloquium No. 130: The Sun and cool stars – activity, magnetism, dynamos, p. 430 – 434 (1991). – See Abstr. 012.024 for the main entry.

The authors examine the link between observed magnetic fields and filling factors in main–sequence stars of types G and K, and the results obtained for these parameters from a simple dynamo model. They explore how the predicted magnetic fluxes for a selected sample of stars, vary with rotation, adopting two theoretical approaches to estimate the filling factor, and considering different expressions for the variation of the stellar angular velocity with depth.

065.045 **Activity of relatively close binaries.**
C. J. Schrijver, C. Zwaan.
IAU Colloquium No. 130: The Sun and cool stars – activity, magnetism, dynamos, p. 435 – 439 (1991). – See Abstr. 012.024 for the main entry.

Whereas the rotation–activity relationship for single cool stars can be described in terms of fundamental stellar properties, the activity in relatively close binaries is enhanced with respect to that of single stars by an amount which depends strongly on the properties of the companion. The authors consider mechanisms which could cause an excess in activity of binaries, and conclude that it is likely that the presence of a companion affects the interior structure of the stars in such a way that either the efficiency of the dynamo or of the atmospheric heating is enhanced.

065.046 **The Sun and cool stars – activity, magnetism, dynamos. Concluding summary: theoretical aspects.**
L. Mestel.
IAU Colloquium No. 130: The Sun and cool stars – activity, magnetism, dynamos, p. 515 – 521 (1991). – See Abstr. 012.024 for the main entry.

065.047 **Li depletion in F stars by internal gravity waves.**
R. J. García López, H. C. Spruit.
Astrophys. J., Vol. 377, No. 1, p. 268 – 277 (10 Aug 1991).

The authors argue that internal gravity waves generated by the convective envelope of a star may be effective in producing a weak mixing in its radiative interior. They estimate the spectrum of wavelengths and frequencies generated. Only the largest horizontal wavelengths and the lowest frequencies contribute significantly to the wave energy flux and the mixing at the Li–burning depth. At this layer, the combination of the dependence on spectral type, both for the wave energy flux and the shear rate induced by the waves, would produce a distribution of Li abundances similar to that of the Li gap found among F type stars of several open clusters and in the field. Quantitative agreement with the observed location and time scale of the gap is obtained, but only by increasing the intensity of the waves by a factor 15 above straightforward mixing–length estimates. The ratio of mixing length to scale height required to get the gap in the right spectral range is $\alpha = 1.6$. The blue edge of the gap is predicted to be sharp.

065.048 **Asymptotic theory of solar and stellar acoustic oscillations.**
S. V. Vorontsov.
Astron. Zh., Tom 68, Vyp. 4, p. 808 – 824 (Jul–Aug 1991). In Russian. English translation in Sov. Astron., Vol. 35, No. 4.

A second–order asymptotic theory is developed for high–frequency acoustic oscillations of a star. The theory is developed for a wide range of low– and intermediate–degree modes, taking into account the effects of gravity perturbation. The equation for the eigenfrequencies is obtained in a form convenient to be used in the helioseismological inverse problem.

065.049 **Formation of planetary systems during the evolution of close binary stars.**
A. V. Tutukov.
Astron. Zh., Tom 68, Vyp. 4, p. 837 – 842 (Jul–Aug 1991). In Russian. English translation in Sov. Astron., Vol. 35, No. 4.
A short description of modern scenarios of planetary system formation around single stars and products of merging of close binaries is presented. The frequency of occurrence of the scenarios in our Galaxy is estimated.

065.050 **Computer simulation of the formation of protostars.**
H. Pongracic, S. J. Chapman, J. R. Davies, M. J. Disney, A. H. Nelson, A. P. Whitworth.
Conference on Dynamics of Disc Galaxies, p. 405 – 408 (1991). – See Abstr. 012.028 for the main entry.
The authors report here computer simulations of protostars condensing out of interstellar gas. The simulations involve the supersonic collision of two clouds, which are intrinsically stable against contraction. These simulations are able to handle the necessary high dynamic ranges of density and linear size by combining the numerical techniques of Tree Code Gravity and a variant of Smoothed Particle Hydrodynamics.

065.051 **On lithium depletion: a possible discrimination of the physical processes.**
E. Schatzman.
Mem. Soc. Astron. Ital., Vol. 62, No. 1, p. 111 – 130 (1991). – See Abstr. 012.033 for the main entry.
A careful analysis in the case of the Hyades shows that lithium depletion is probably not due to a turbulent flow induced by rotation. It is suggested that internal waves can be responsible both for the transport of angular momentum and for the transport of lithium to the burning level. A first approach of the theory explains the internal distribution of angular velocities of the Sun and the lithium depletion of the Hyades.

065.052 **Lithium depletion by rotation induced diffusion.**
C. Charbonnel.
Mem. Soc. Astron. Ital., Vol. 62, No. 1, p. 131 – 138 (1991). – See Abstr. 012.033 for the main entry.
Relative effects of rotation induced mixing and microscopic diffusion on the lithium depletion in population I stars are computed and compared. An implicit numerical code is used to follow the evolution of the lithium concentration inside the star. The author takes into account a Skumanich law of rotational braking during the stellar life. Lithium depletion due to rotation induced mixing is computed in population I and population II stars and compared to the observations.

065.053 **A new turbulent convection theory and lithium depletion in pre–main sequence.**
I. Mazzitelli.
Mem. Soc. Astron. Ital., Vol. 62, No. 1, p. 139 – 149 (1991). – See Abstr. 012.033 for the main entry.
A new theory of overadiabatic convection, owing nothing to the traditional mixing length theory (MLT), has been recently developed by Canuto and Mazzitelli, 1991 (henceforth quoted as CM), and applied to the pre–main sequence (PMS) of pop I stars. Although the predictions from the CM theory for the temperature profiles in the subsurface regions of stars are quite different from what expected according to MLT, the physical stratifications at the bottom of the convective regions are not equally affected so that, in the end, the already established relations among star mass, chemical composition and lithium depletion in PMS are not modified. This is in a sense a welcome result, since it allows to definitely rule out the convection theory as a possible source of uncertainty in lithium depletion in PMS; on the other hand, it is also a strong warning for the interpretation of observational results, since the CM convection theory provides PMS evolutionary tracks, in the HR diagram, far different from those obtained according to the MLT.

065.054 **Hydrodynamical constraints from the lithium abundance in main sequence stars.**
G. Michaud, J. Richer.
Mem. Soc. Astron. Ital., Vol. 62, No. 1, p. 151 – 164 (1991). – See Abstr. 012.033 for the main entry.
Three processes appear to play a role for the Li abundance in main sequence stars. Gravitational settling explains the Li abundance gap in a very natural way in non rotating stars. In rotating stars, meridional circulation plays a role. The Li abundance for $T_{eff} < 6000K$ suggests the importance of some turbulence.

065.055 **Pre–MS lithium burning and the N(Li)–T_{eff} correlation in population I and II.**
F. D'Antona.
Mem. Soc. Astron. Ital., Vol. 62, No. 1, p. 165 – 170 (1991). – See Abstr. 012.033 for the main entry.
The author reviews the most simple–minded interpretation of the data on lithium abundances in the various stellar populations, namely that the only effective mechanisms for Li–depletion are pre–main sequence burning and dilution during the giant phase. In spite of the difficulties met by this scheme, several researchers seem to be confident that it can be maintained at least for population II stars. The author points out that also in this case we have to face a basic difficulty in the comparison with observations.

065.056 **Li in brown dwarfs and very low mass pre–MS stars.**
F. Pozio.
Mem. Soc. Astron. Ital., Vol. 62, No. 1, p. 171 – 179 (1991). – See Abstr. 012.033 for the main entry.
The author computed theoretical evolutionary histories of pre–main sequence contraction for low mass stars and brown dwarfs following in detail the burning of the light elements (D, Li, Be), in the range of mass $0.20 > M/M_\odot > 0.010$, for population I chemical composition ($Y = 0.25$, $Z = 0.02$). Particular attention was reserved to lithium and deuterium.The peculiar behavior of the brown dwarfs with mass of $0.015\ M_\odot$ was stressed that spend about 10^8years at luminosities $-3.5 > \log L/L_\odot > -4$. The evolutionary status was discussed and the lithium abundance of the objects of the pre–main sequence triple system UX Tau. For this purpose evolutionary tracks for masses up to $1.0\ M_\odot$ with the same chemical composition are also shown.

065.057 **The influence of internal gravity waves on the light elements depletion among F–type stars.**
R. J. García López, H. C. Spruit.
Mem. Soc. Astron. Ital., Vol. 62, No. 1, p. 183 – 185 (1991). – See Abstr. 012.033 for the main entry.

065.058 **Final evolutionary stages of massive stars.**
M. F. El Eid.
3. Chinese Academy of Sciences and Max–Planck Society Workshop: High energy astrophysics: compact stars and active galaxies, p. 297 – 313 (1991). – See Abstr. 012.026 for the main entry.
The final evolutionary stages of massive stars of $M \geqslant 10\ M_\odot$ are reviewed. Due to the complexity of this subject, this review is a personal selection of some important issues in this context. After presenting several observational facts about supernovae, the presupernova evolution of massive stars is outlined with special emphasis of the progenitor of SN 1987A. Some essential features of the type II supernova process are presented.

065.059 **Collisions of giant stars with compact objects: hydrodynamical calculations.**
F. A. Rasio, S. L. Shapiro.
Astrophys. J., Vol. 377, No. 2, p. 559 – 580 (20 Aug 1991).
Encounters between a $0.8\ M_\odot$ giant star and a $1.4\ M_\odot$ compact object on a parabolic orbit have been calculated using smooth particle hydrodynamics. Both the stellar core and the compact object are represented by point masses interacting with the gas through gravity only. It is found that all encounters with periastron distance $r_p \lesssim 2.5 \cdot R_G$, where R_G is the stellar radius,

lead to significant disruption of the stellar envelope, including some mass loss. When $r_p/R_G \lesssim 0.5$, the envelope is completely destroyed during the collision. About 40% of the gas escapes, while the rest forms a massive disk around the compact object, leaving behind the bare stellar core on a highly eccentric orbit. Subsequent close passages lead to continuous extraction of gas from the disk at a large rate, and the binary orbit may eventually stabilize while still retaining a large eccentricity. For $0.5 \lesssim r_p/R_G \lesssim 2.5$, complete destruction of the envelope is avoided during the first close passage. However, nonlinear effects tend to make the envelope expand by a large factor, so that subsequent close passages will likely destroy it. The authors confirm Bailyn's prediction that all captures of a neutron star by a red giant in globular clusters are likely to result in the eventual destruction of the red giant envelope. It is suggested that many millisecond pulsars in globular clusters may have been spun-up by accretion from a massive disk rather than from a binary companion.

065.060 **Irregular variability of stars and nonlinear dynamics.**
Y. Tanaka.
Astron. Her., Vol. 84, No. 12, p. 397 - 400 (1991). In Japanese.

065.061 **Asymptotic theory of acoustic oscillations of the Sun and stars.**
S. V. Vorontsov.
Sov. Astron., Vol. 35, No. 4, p. 400 - 408 (Jul 1991). Current Physics Microform No.: 9203X0586. English translation of Astron. Zh., Tom 68, Vyp. 4, p. 808 - 824 (1991).
A second-order asymptotic theory is constructed to describe high-frequency acoustic oscillations of a star. The theory is constructed for oscillations over a wide range of degrees of spherical harmonics with allowance for gravitational perturbations. The eigenfrequency equation is obtained in a convenient form for solving the inverse problem of solar seismology.

065.062 **Formation of planetary systems during the evolution of close binary stars.**
A. V. Tutukov.
Sov. Astron., Vol. 35, No. 4, p. 415 - 417 (Jul 1991). Current Physics Microform No.: 9203X0601. English translation of Astron. Zh., Tom 68, Vyp. 4, p. 837 - 842 (1991).
Present-day scenarios of the formation of planetary systems around single stars and the products of component mergers in close binaries are described briefly. The frequency of occurrence of the various scenarios in our Galaxy is estimated.

065.063 **The evolution of He-burning stars: horizontal and asymptotic branches in Galactic globulars.**
V. Castellani, A. Chieffi, L. Pulone.
Astrophys. J., Suppl. Ser., Vol. 76, No. 3, p. 911 - 977 (Jul 1991).
The authors present a grid of theoretical evolutionary models covering the horizontal-branch (HB) and the asymptotic giant branch (AGB) phases of globular cluster stars. The computations have been performed for a fixed amount of the original He ($Y = 0.23$) and for the metallicity values $Z = 10^{-4}$, 4×10^{-4}, 10^{-3}, 6×10^{-3} and 2×10^{-2}. For each value of the assumed metallicity, selected zero-age horizontal-branch models with evolutionary values for both M_c (mass of the He core) and dY (extra healium at the surface) have been evolved from the initial He-burning structure through the whole phase of central He burning and along the following He shell burning phase until the minimum in luminosity, marking the reignition of the H shell and the onset of the thermally pulsating phase. The authors provide extensive tabulations of the numerical results to be compared with globular cluster observations. They find that the luminosity of the AGB clump and the number ratio of AGB to HB stars should remain fairly constant with the cluster metallicity. However, the difference in bolometric luminosity between the AGB clump and the HB should be a rather sensitive function of the cluster metallicity, increasing when the cluster metallicity increases.

065.064 **Model for R Coronae Borealis stars.**
B. A. Whitney, N. Soker, G. C. Clayton.
Astron. J., Vol. 102, No. 1, p. 284 - 288 (Jul 1991). Current Physics Microform No.: 9107B1980.
The authors propose an evolutionary model in which R Coronae Borealis stars are in a second common-envelope phase, with a low mass companion orbiting well inside the envelope. The high mass loss rate in the first common-envelope phase turns the massive star into a hydrogen-deficient star. In the second common-envelope phase, as a result of shear induced by the inner envelope rotation and possibly by excited gravity-wave induced turbulence, carbon and oxygen are being dredged up from the core into the envelope. This evolutionary scenario smoothly connects R CrB stars to the double-mass-loss-episodes evolutionary scenario of hydrogen-deficient binaries suggested by Plavec in 1973. This model may have consequences for the nature of mass loss in R Coronae Borealis stars: The orbiting companion breaks spherical symmetry allowing for nonradial pulsations and asymmetric mass loss. The optical polarization and IR observations may point to axisymmetric mass loss with the same axis of symmetry for the different mass loss episodes.

065.065 **A new model of stellar convection for compressible fluids.**
M. Forestini, E. Lumer, M. Arnould.
Astron. Astrophys., Vol. 252, No. 1, p. 127 - 136 (Dec 1991).
A new model for stellar convection is presented that is able to reproduce the main features of the convective motions in compressible fluids. It is able, in contrast to the mixing length theory (MLT), to account for the asymmetry between the upward and downward motions in strongly stratified convective regions. The model has the major advantage of being readily workable in stellar evolution codes. It can be treated analytically in the limit of negligible kinetic fluxes, which is the only situation considered in this work. A hydrodynamical stellar evolution code is used to evaluate the effect of the model on the structure and evolution of pre-main sequence, main sequence, and asymptotic giant branch stars of low- and intermediate-mass. These results are compared with those derived from the same code in which the treatment of convection is replaced by the MLT. It is demonstrated that the model has no significant effect on the structure of pre-main sequence or main sequence stars (especially the Sun), as well as on their evolution tracks in the HR diagram, at least in the limit of negligible kinetic fluxes. The situation is quite different in the case of asymptotic giant branch stars, in view of the large extent of their convection envelopes. Asymptotic giant branch envelope and atmosphere models are evaluated in order to test the sensitivity of the results to the adopted values of the free parameters of the presented convection model.

065.066 **The influence of overshooting and stellar wind mass loss in the mass-luminosity relation.**
J. Figueiredo, J. P. De Greve, C. de Loore.
Astron. Astrophys., Vol. 252, No. 1, p. 155 - 158 (Dec 1991).
The effect of overshooting and stellar wind mass loss on the position of stellar models in the mass-luminosity diagram is examined. The average main sequence is well represented by $\log L/L_\odot = 2.47 \log M/M_\odot + 1.52$. The influence of variations in the amount of overshooting or the mass loss rate is rather small, and the exponent in the ML-relation varies around 2.5. The theoretical results are compared to corresponding relations derived from observations. The importance of the limiting mass values on the result is discussed.

065.067 **The evolution of massive close binaries revised.**
D. Vanbeveren.
Astron. Astrophys., Vol. 252, No. 1, p. 159 - 171 (Dec 1991).
Evolution computations of massive close binaries are presented using improved stellar wind mass loss rate formalisms during core hydrogen and core helium burning, improved physics and mass loss rates during the luminous blue variable phase of the most massive stars. The computations are compared with recent observations of WR stars.

065.068 **Convective penetration in stellar interiors.**
J.- P. Zahn.
Astron. Astrophys., Vol. 252, No. 1, p. 179 - 188 (Dec 1991).

The author recalls the evidence for convective penetration exhibited by geophysical fluids, laboratory experiments and computer simulations. A simple model is deduced from these observations, which serves to establish a relation between the subadiabatic extent L of a convective region and the velocity W of the penetrating motions: $L \sim W^{3/2}$. Assuming that this velocity obeys the usual scaling of thermal convection, it is found that the subadiabatic penetration at the bottom of a convective envelope is of the order of a pressure scale-height, and that above a convective core it amounts to a substantial fraction of the core radius. It is shown that the requirements for nearly adiabatic penetration are met deep enough in a stellar interior, and that the departures from adiabaticity are confined to a thin boundary layer. This allows the use of Roxburgh's integral constraint to predict the actual size of a convective core. However, that constraint is not applicable to a convective envelope, whose extent depends on a parameter of order unity which cannot be determined directly in this simple approach. Instead, this parameter ζ must be calibrated by comparing stellar models with the observations, as one proceeds for the parameter α of the mixing-length treatment.

065.069 **The status of stellar evolution models.**
D. A. VandenBerg.
Symposium on the Formation and Evolution of Star Clusters as Part of the 102. Summer Scientific Meeting of the Astronomical Society of the Pacific, p. 183 - 199 (1991). - See Abstr. 012.050 for the main entry.

Current uncertainties in opacity data, convection theory, and the treatment of the outer atmospheric layers in stellar models remain sufficiently large to have important ramifications for the T_{eff} scale of computed evolutionary tracks. As a result, the matching of predicted and observed colors should not be given high weight in the determination of star cluster ages from isochrone fits to C-M diagrams. Rather, the turnoff luminosity must continue to be regarded as the prime age indicator, which then demands accurate distances. But, to obtain precise cluster ages, it is also necesarry to know the chemical makeup of the member stars and to include the effects of rotation and diffusion, should these processes be important. Recent developments in these areas are recounted.

065.070 **Stellar models with envelope overshoot.**
M. Alongi, C. Chiosi, G. Bertelli, A. Bressan.
Symposium on the Formation and Evolution of Star Clusters as Part of the 102. Summer Scientific Meeting of the Astronomical Society of the Pacific, p. 223 - 226 (1991). - See Abstr. 012.050 for the main entry.

The authors show that a certain amount (≥ 0.7 of a pressure scale height) of non-local overshoot at the base of the outer convective envelope of low mass stars while climbing along the Red Giant Branch (RGB) toward central helium ignition, explains the shift of about 0.4 V mag required to bring the luminosity of the bump predicted in the theoretical RGB luminosity functions into agreement with the observational one. The results cast light on the role played by core and envelope overshoot. The former provides the correct hydrogen and helium-burning lifetimes for intermediate mass stars, whereas the latter reproduces the observed morphology of young LMC clusters and the correct location of the bump in the RGB luminosity function of Globular Clusters.

065.071 **RR Lyrae pulsations revisited.**
G. Kovács, J. R. Buchler, A. Marom.
Astron. Astrophys., Vol. 252, No. 2, p. L27 - L30 (Dec 1991). Letter-to-the-editor.

The new Livermore opacities are used to reexamine several outstanding problems in RR Lyrae pulsation. The inferred masses for the double-mode stars are found to depend sensitively on the metallicity. With the observed average [Fe/H] values and with a solar mixture the Oosterhoff II masses are found to be consistent with current (non CNO enhanced) stellar evolution computations; the Oosterhoff I masses, however, are higher by 0.05-0.10 $M_\odot$. The authors emphasize the need, on a star by star basis, not only for more reliable [Fe/H] values, but also for the other heavy element abundances. Sustained double-mode pulsation with the right periods is still not found possible with purely radiative models.

065.072 **Evolutionary properties of intermediate mass protostars.**
F. Palla.
NATO Advanced Research Workshop on Angular Momentum Evolution of Young Stars, p. 21 - 25 (1991). - See Abstr. 012.055 for the main entry.

The properties of young stars of intermediate-mass are discussed in the light of recent calculations of the protostellar accretion phase. In particular, the problem of the existence of surface convection in these stars will be addressed with reference to the strong activity manifested by the Herbig Ae/Be stars.

065.073 **Ekman pumping in a rotating protostar.**
D. Galli, F. H. Shu.
NATO Advanced Research Workshop on Angular Momentum Evolution of Young Stars, p. 33 - 40 (1991). - See Abstr. 012.055 for the main entry.

The authors consider the accretion phase during the formation of a (low mass) protostar. The spin-up resulting from accretion through the adjoining protostellar disk generates a meridional circulation (Ekman pumping) inside the protostar directed from the rotational poles to the equatorial plane. The adoption of appropriate analytical techniques allows to calculate the velocity field in the protostellar interior to the lowest order in a perturbative approach. The consequences for the evolution of an eventual protostellar magnetic field and of a magnetically coupled protostellar wind in the presence of this circulation are briefly discussed.

065.074 **Rotation in pre-main sequence stars: properties and evolution.**
J. Bouvier.
NATO Advanced Research Workshop on Angular Momentum Evolution of Young Stars, p. 41 - 62 (1991). - See Abstr. 012.055 for the main entry.

The rotational properties of pre-main sequence stars with a mass between 0.5 and 9 $M_\odot$ are reviewed. The impact of binarity and disk accretion upon early rotational evolution is discussed and unsolved problems related to the evolution of stellar angular momentum prior to the main sequence are outlined.

065.075 **Differential rotation of fully convective pre-main sequence stars.**
I. W. Roxburgh.
NATO Advanced Research Workshop on Angular Momentum Evolution of Young Stars, p. 83 - 88 (1991). - See Abstr. 012.055 for the main entry.

The internal rotation of fully convective pre-main sequence stars is investigated using anisotropic and inhomogeneous turbulent viscosity and heat transport coefficients. The models have a large scale circulation and differential rotation. Models with standard isotropic transport also have circulation and differential rotation; the circulation being driven by the latitudinal variation of convective and radiative heat flux in rotating stars in an analogous way to the driving of meridian circulation in radiative layers. Simple model calculations suggest that the differential rotation in pre-main sequence fully convective stars may be large and the stars may be heavily distorted.

065.076 **On the production of ^{26}Al in AGB stars.**
M. Forestini, G. Paulus, M. Arnould.
Astron. Astrophys., Vol. 252, No. 2, p. 597 - 604 (Dec 1991).

Low and intermediate mass stars in the range 0.8 to 6-8 $M_\odot$ on the main sequence evolve onto the asymptotic giant branch of the Hertzsprung-Russell diagram. In an AGB star, convective dredge-up lifts to the surface the ashes of H- and He-burning shells. Among those, ^{26}Al is particulary important because it plays a key role in γ-ray astronomy and cosmochemistry. The possible production of ^{26}Al in AGB stars has been examined in detail. The authors concentrate on the low end (1–3 $M_\odot$) of the mass spectrum. They have self-consistently calculated a 3 $M_\odot$ AGB model, from the T Tauri phase through the thermally pulsing phase. Special attention has been devoted to the nucleosynthesis, especially that of ^{26}Al. The contribution of the low mass AGB stars to the ^{26}Al enrichment of the galactic interstellar medium has also been estimated. The findings are that the $^{26}Al/^{27}Al$ abundance ratio in the envelope of the 3 $M_\odot$ model agrees with the mean values inferred for meteoritic grains, and that the AGB stars of 1–3 $M_\odot$ may have added ~ 0.1 $M_\odot$ of ^{26}Al to the interstellar medium over the last million years.

065.077 **Helium inhomogeneities in main-sequence magnetic stars.**
S. Vauclair, N. Dolez, D. O. Gough.
Astron. Astrophys., Vol. 252, No. 2, p. 618 – 624 (Dec 1991).

Helium abundance variations have been computed in models of main-sequence magnetic stars with masses in the range 1.6–5 $M_\odot$ (T_{eff} = 8000–19000K). To represent a predominantly dipolar magnetic field, regions in the stars have been schematically designated either polar, with vertical magnetic fields, or equatorial, with horizontal fields. A stellar wind is supposed to exist at the magnetic poles. The mass loss rate is varied as a free parameter in the computations. Under these conditions, helium is underabundant at the magnetic equators, while it may accumulate at the poles in the region of the first ionization of helium. This accumulation occurs if the mass loss rate is of the order of that caused by the solar wind. The helium peak is visible in the atmospheres of the hottest stars, which may account for the observed He rich stars. It lies below the photosphere and should not be visible in the coolest stars; but it can help to excite the pulsations of rapidly oscillating Ap stars by the $\varkappa$ mechanism.

065.078 **Lithium, rotation and age.**
E. Schatzman.
NATO Advanced Research Workshop on Angular Momentum Evolution of Young Stars, p. 223 – 242 (1991). – See Abstr. 012.055 for the main entry.

Lithium depletion as a function of time appears as a well established fact. After a review of the different physical processes which can be involved in this phenomenon, it appears that, very probably, stellar rotation does not play an important role. Stellar rotation is time dependant, lithium depletion can be used as a clock measuring stellar ages and can provide the relation between rotation and age, but this requires a calibration of the lithium-clock. This requests a well established model of lithium burning. Internal waves carry angular momentum and generate a mixing process which can be responsible for lithium depletion. The calibration of the lithium clock remains to be done.

065.079 **Lithium depletion induced by rotation in young stars.**
C. Charbonnel, S. Vauclair.
NATO Advanced Research Workshop on Angular Momentum Evolution of Young Stars, p. 259 – 264 (1991). – See Abstr. 012.055 for the main entry.

Lithium depletion in population I stars has been computed in the frame of Zahn's theory of rotation induced mixing. The implicit numerical code used to solve the turbulent and microscopic diffusion and nuclear destruction equations takes into account a Skumanich law of rotational braking during the stellar life. The results arc compared to the observations of lithium in galactic clusters. The depletion law follows the increase of the rotation velocities with T_{eff} in the F stars, and the lithium gap in the Hyades is well reproduced. The depletion curve for G stars can be accounted for if pre-main-sequence depletion is added.

065.080 **Rotational mixing and lithium in young stars.**
M. H. Pinsonneault.
NATO Advanced Research Workshop on Angular Momentum Evolution of Young Stars, p. 265 – 270 (1991). – See Abstr. 012.055 for the main entry.

The potential role of rotational mixing as an agent for lithium depletion is reviewed. For main sequence stars, mixing driven by angular momentum loss can cause lithium depletion for models with a wide range of masses. For young stars, however, the evolution of the structure inhibits lithium depletion in higher mass models. As a result, rotational mixing can cause a large dispersion in abundance for young low mass (late G and K) stars but not for young high mass (early G and F) stars. The properties of solar metallicity and metal-poor models are also compared.

065.081 **Theory of magnetic braking of late-type stars.**
A. Collier Cameron, Li Jianke, L. Mestel.
NATO Advanced Research Workshop on Angular Momentum Evolution of Young Stars, p. 297 – 314 (1991). – See Abstr. 012.055 for the main entry.

The authors review the basic theory of angular momentum loss via a magnetically channelled stellar wind. Various forms of power-law dependence of the instantaneous braking rate on the stellar rotation rate are derived. Observational constraints on the rotational evolution of young solar-type stars in open clusters and the field are used to calibrate and test the braking laws at rotation rates up to 100 times the solar rate.

065.082 **Pre-main sequence and main sequence rotational evolution: constraints on models derived from observations.**
K. B. MacGregor.
NATO Advanced Research Workshop on Angular Momentum Evolution of Young Stars, p. 315 – 331 (1991). – See Abstr. 012.055 for the main entry.

The author describes a simple, parametrized model for the redistribution of angular momentum within the interiors of solar-type stars. By combining the model with a computation of the rate of angular momentum removal by a magnetically-coupled wind, he can trace the rotational histories of low-mass dwarf stars for a variety of initial conditions and parameter specifications. He presents the results of calculations of the rotational evolution of a 1 $M_\odot$ star and compares these results with what is known from observations about the internal rotation of the Sun and the evolution of the surface rotation of sun-like stars.

065.083 **Evolutionary models of rotating stars.**
S. Sofia, M. Pinsonneault, C. P. Deliyannis.
NATO Advanced Research Workshop on Angular Momentum Evolution of Young Stars, p. 333 – 356 (1991). – See Abstr. 012.055 for the main entry.

The authors review the principal elements of the Yale code for evolving rotating stars (YREC). The various mechanisms to distribute angular momentum in the solar interior used in the code are discussed, both in regard to their properties and their consequences in the stellar models. In particular, two general types of mechanisms exist, one which depends on the velocity of rotation, and the other which depends on the radial gradient of this quantity. The main effects of rotation in stellar evolution are the surface velocity and the surface abundance of trace elements and CNO as a function of time.The authors computed stellar models including the effects of rotating for the Sun, open cluster stars, and halo stars. The solar models constructed in this fashion show that the present Sun rotates with increasing angular velocity in the deep interior.

065.084 **Evolution of massive stars in the Large Magellanic Cloud: models with semiconvection.**
N. Langer.
Astron. Astrophys., Vol. 252, No. 2, p. 669 – 688 (Dec 1991).

The author has computed two grids of models for the mass range $7\,M_\odot \leqslant M_{ZAMS} \leqslant 32\,M_\odot$ and with a metallicity of $Z = 0.005$ – corresponding roughly to the LMC. He invoked a convection model including semiconvection which was known to yield basic consistency with the observational constraints on the progenitor evolution of SN 1987A. He discusses the evolutionary tracks in the HR diagram and investigate the possible mass range of blue supernova progenitor stars. According to his models up to 40% of all supernovae, which are due to core collapse in single stars, may occur in blue supergiants in LMC-type galaxies, like SN 1987A; the models indicate that most blue supergiants should be nitrogen enriched. The author has compared the obtained evolutionary tracks with observations for the LMC by means of synthetic HR diagrams. He has furthermore computed several 15 and 20 $M_\odot$ sequences with various combinations of semiconvective and overshooting efficiency, in order to explore the consequences for internal and HRD evolution. He has found that already small amounts of overshooting – when used in addition to semiconvection – yield stellar tracks that predict almost no blue supergiant phase and have thus to be disregarded.

065.085 **Angular momentum transport, rotational instabilities, magnetic fields and mixing.**
I. W. Roxburgh.
NATO Advanced Research Workshop on Angular Momentum Evolution of Young Stars, p. 365 – 372 (1991). – See Abstr. 012.055 for the main entry.
The energetics of mixing material through a chemical composition gradient is considered. In regions where nuclear reactions are significant the rate of energy input required to maintain mixing against the growing composition gradient is of the order of the internal energy per nuclear time scale, this is greater than the rate of energy available from most sources except nuclear reactions themselves. Thus angular momentum transport by material motion is only likely to be effective in convective zones (and any associated overshoot regions), and possibly by instabilities themselves driven by nuclear reactions. Even though the angular velocity gradient may be unstable, transport of angular momentum by rotationally driven instabilities will be inhibited. Transport by magnetic fields does not require material motion through a composition gradient and is likely to be very effective in establishing almost uniform rotation. If the rotation is not constant then the differential rotation produces a toroidal field which reacts back on the differential rotation leading to torsional oscillations with a period of the local Alfven travel time. Neighbouring field lines rapidly get out of phase with each other creating a large gradient in magnetic field and substantial dissipation. This is likely to establish almost uniform rotation as this is the lowest energy state for a given angular momentum.

065.086 **The spin–down of main sequence stars based on observed magnetic field strength.**
C. Trigilio, G. Umana, S. Catalano, E. Marilli.
NATO Advanced Research Workshop on Angular Momentum Evolution of Young Stars, p. 373 – 378 (1991). – See Abstr. 012.055 for the main entry.
The authors try to reproduce the observed dependence of rotation decay of main sequence stars in the mass range 1.1 – 0.5 $M_\odot$. They adopt a radial magnetic field structure which yields the $t^{-1/2}$ relation and the observed magnetic field values as given by Saar (1990). They assume that the effective magnetic field at photospheric level is given by the product of the equipartition field B_{eq} by the filling factor f, and that the latter is proportional to the angular rotation ω. The results are consistent with the observed rotation rate decay as a function of the mass and do suggest that the whole star is slowed down.

065.087 **Effects of rotation on collapsing stellar iron cores.**
R. Mönchmeyer.
10. Santa Cruz Summer Workshop in Astronomy and Astrophysics: Supernovae, p. 358 – 361 (1991). – See Abstr. 012.052 for the main entry.

065.088 **A look at dissipation in stellar collapse.**
E. A. Baron.
10. Santa Cruz Summer Workshop in Astronomy and Astrophysics: Supernovae, p. 365 – 368 (1991). – See Abstr. 012.052 for the main entry.

065.089 **Explosion of massive Wolf–Rayet stars.**
M. F. El Eid.
10. Santa Cruz Summer Workshop in Astronomy and Astrophysics: Supernovae, p. 568 – 571 (1991). – See Abstr. 012.052 for the main entry.

065.090 **An iterative technique for computing rotating visco–polytropic models.**
V. S. Geroyannis.
Astrophys. Space Sci., Vol. 186, No. 1, p. 27 – 56 (Dec 1991).
The author develops and implements a "complex–plane iterative technique" for computing rotating visco polytropic models. The viscosity of the models is determined in view of the "relaxation condition" that differential rotation depends only on the distance from the rotation axis. To this purpose the author develops and implements a "viscosity–evaluated–upwards" technique which fits viscosity, within a few percent, with the required condition. He computes various models, and studies in detail some of them that satisfy, within a few percent, the above condition, and also the boundary condition of a vanishing surface stress vector.

065.091 **Radial oscillations of finite–temperature white dwarfs.**
H. P. Singh, M. K. Das.
Astrophys. Space Sci., Vol. 186, No. 1, p. 117 – 124 (Dec 1991).
Models of finite temperature completely degenerate stellar configurations are considered. The frequencies of fundamental radial modes of oscillations for these white dwarf models have been computed for different values of the central degeneracy parameter $1/y_0{}^2$ and for uniform temperatures of 20×10^6K and 10^8K. A variational formulation as well as a direct approach is employed to calculate the temperature induced modifications in the frequencies of oscillation of these white–dwarf models.

065.092 **On the adiabatic pulsations of accretion disks and rotating stars.**
J. R. Ipser, L. Lindblom.
Astrophys. J., Vol. 379, No. 1, p. 285 – 289 (20 Sep 1991).
The adiabatic pulsations of rotating Newtonian fluids, including nonbarotropic configurations, are represented completely in terms of two scalar potentials. The second–order equations that determine these potentials are more convenient to solve numerically than the standard Lagrangian displacement form of the equations. A variational principle is derived for these equations from which estimates of the frequencies of the modes may be obtained. This analysis generalizes previous work by describing the adiabatic pulsations of fluids having any (including nonbarotropic) equation of state, any (including differential) rotation law, and self–gravitation. It is applicable, therefore, to the pulsations of rotating stars or thick accretion disks.

065.093 **Can neutron stars ablate their companions?**
A. Levinson, D. Eichler.
Astrophys. J., Vol. 379, No. 1, p. 359 – 365 (20 Sep 1991).
Ablation of stars by energetic photons from a compact companion is considered for general parameters. It is concluded that pulsars with conventional parameters cannot ablate their companions with gamma rays at orbital separation of 3×10^{10}cm or more. It is estimated that in the absence of heat conductivity, line cooling is important when the incident radiation flux is less than 10^{15} ergs $s^{-1}cm^{-2}$. Millisecond pulsars may be able to ablate their companions more readily than in previous scenarios, if even a small fraction of the kilohertz radiation is absorbed by the atmosphere / wind of the companion and the heat is conducted to the surface. There are several scenarios in which low–mass companions can evolve from $m > 0.1\ M_\odot$ to $m < 0.05\ M_\odot$, provided the companion nearly fills its Roche lobe. If contact with the Roche lobe is not constantly maintained by angular momentum loss, and if self–sustained accretion cannot occur without this contact, then the lack of fading LMXBs can be attributed to accretion nulling, punctuated by periods of vigorous accretion.

065.094 **The initial stellar masses for the formation of white dwarfs, neutron stars and black holes.**
G. Meynet.
Ann. Phys. (Paris) Colloq., Vol. 16, No. 4, p. 459 – 479 (Aug 1991).
As is well known a star may end its nuclear lifetime as a white dwarf, neutron star, black hole or many, in certain circumstances, leave no remnant at all. The main question to be

addressed in this review is the following: what are the progenitors of these different final stages? After a brief review of the major physical principles governing stellar evolution, the author presents the different evolutionary scenarios resulting from numerical calculations. Particular attention will be paid to the effect of mass loss on theoretical determinations of the mass limits M_{WD} and M_{BH} which are respectively the maximum initial mass leading to the formation of a white dwarf and the minimum initial mass for the formation of a black hole. The author terminates this review by the presentation of some relevant observational results. The bulk of this paper is devoted to the discussion of the evolution of single Population I stars.

065.095 **Diffusion in late stages of stellar evolution.**
G. Michaud.
Ann. Phys. (Paris) Colloq., Vol. 16, No. 4, p. 481 – 495 (Aug 1991).

The abundances of the chemical elements observed at the surface of white dwarfs are the result of diffusion in the external regions of these stars. It is described how diffusion determines the abundances of such elements as carbon, that are now being observed in the far UV. Diffusion is shown to play a minor role in giants but its effects start to appear rapidly after the planetary nebula stage. While the basic the diffusion coefficients involved are uncertain in some cases. They pose interesting statistical physics problems specially for the thermal diffusion coefficients. Depending on the model used it can vary by orders of magnitude. A number of the possible sources of uncertainty of the diffusion coefficients are described as well as the astrophysical context where the coefficients are most uncertain.

065.096 **Why do stars become giants?**
A. Whitworth.
Ann. Phys. (Paris) Colloq., Vol. 16, No. 4, p. 515 – 526 (Aug 1991).

The author explains why stars become giants; and he reviews earlier work, indicating why there has been so little consensus on this question despite over fifty years of investigation. The mechanism causing giantness is not simple. It involves a complex natural conspiracy in which there are four critical roles. These roles are played by eight differential physical effects.

065.097 **New pre–white dwarf evolutionary tracks.**
T. Blöcker, D. Schönberner.
NATO Advanced Research Workshop – 7. European Workshop on White Dwarfs, p. 1 – 3 (1991). – See Abstr. 012.044 for the main entry.

The authors present new post–AGB evolutionary tracks that have been derived by the evolution of stellar models of 3 and 5 $M_\odot$ from the main sequence through the AGB with consideration of thermal pulses and of mass loss which increases rapidly with increasing luminosity.

065.098 **Influence of the phase diagram in the cooling of white dwarfs.**
J. Isern, E. Garcia–Berro, M. Hernanz, R. Mochkovitch.
NATO Advanced Research Workshop – 7. European Workshop on White Dwarfs, p. 5 – 16 (1991). – See Abstr. 012.044 for the main entry.

White dwarf interiors are made of a mixture of carbon, oxygen and other minor species, ^{22}Ne being the most abundant one. In this paper the authors show that the gravitational settling induced by the freezing can delay the cooling process during $\sim$1 Gyr depending on the adopted phase diagram. Furthermore, they show that ^{22}Ne can also settle down at the center as an outcome of solidification. In this case, the gravitational energy released keeps the white dwarf warm for 2 Gyrs. A similar effect can be expected from any other chemical component with higher atomic number present in the white dwarf interior, although its contribution would be probably minor because of its low abundance.

065.099 **An adiabatic survey for ZZ Ceti stars based on a finite element code.**
P. Brassard, G. Fontaine, C. Pelletier, F. Wesemael.
NATO Advanced Research Workshop – 7. European Workshop on White Dwarfs, p. 193 – 203 (1991). – See Abstr. 012.044 for the main entry.

The authors discuss briefly a new mathematical tool based on finite–element techniques to solve the adiabatic pulsation equations. They then present sample results of an extensive adiabatic survey which has been carried out for ZZ Ceti stars. They discuss the influence of cooling and of the thickness of the hydrogen layer on the profile of the Brunt–Väisälä frequency, on the period spectrum, and on the period changes in ZZ Ceti stars.

065.100 **Nonadiabatic nonradial pulsations for DAV white dwarf stars.**
A. N. Cox, D. E. Hollowell.
NATO Advanced Research Workshop – 7. European Workshop on White Dwarfs, p. 211 – 218 (1991). – See Abstr. 012.044 for the main entry.

New complete 0.6 $M_\odot$ models based on evolution studies including element diffusion and gravitational settling for white dwarfs have been studied for their nonadiabatic oscillations. The derived periods and growth rates confirm the earlier result that any thickness of the surface hydrogen can give the observed nonradial pulsations, providing it is at least thick enough to contain the surface convection zone. It is now found that a very unconventional and large ratio of the mixing length to pressure scale height is required to deepen the convection zone so that the pulsation driving mechanisms can operate for the low degree and order g–modes observed at the 13000K instability strip blue edge. These pulsation results were obtained from a Lagrangian–based eigensolution program.

065.101 **On the sodium excess in the atmospheres of yellow supergiants.**
N. Prantzos, A. Coc, J. P. Thibaud.
Astrophys. J., Vol. 379, No. 2, p. 729 – 733 (1 Oct 1991).

In order to explain quantitatively the reported Na excess at the surface of some yellow giant stars, the authors explore the possibility of an enhanced rate of the ^{20}Ne(p, γ) reaction. Detailed stellar nucleosynthesis calculations show that an enhancement factor f $\sim$ 10 at H–burning temperatures (T $\sim$ 2 – 4 $\times$ 10^7K) is required to explain the observations. A careful analysis of our current understanding of that reaction shows that such an enhancement is unlikely; in lack of reasonable alternatives, a reinvestigation of the Ne – Na cycle in the low–temperature regime seems worthwhile.

065.102 **Entropy of nonsingular self–gravitating polytropes and their TOV equation.**
Z. Zhang, X. Wang, H. Zhang, J. Shi.
Nuovo Cimento B, Vol. 106, No. 11, p. 1189 – 1194 (Nov 1991).

This paper has discussed the behaviour of total entropy of self–gravitating polytropes in the nonsingular region and proved that the self–gravitating polytropes in the nonsingular region satisfy the equation of general relativistic hydrostatic equilibrium (TOV equation). The authors obtain that the magnitude order of total entropy of the polytropes with mass $M_\odot$ is $(10^{40} - 10^{41})$erg K^{-1} by calculating the nonsingular entropy of self–gravitating polytropes with equation of state $p = K\varrho^{\Gamma}$, where $K \neq -1$, $-3-2\sqrt{2} < K < -3+2\sqrt{2}$, $K > 0$, Γ is a constant.

065.103 **High–mass stars with degenerate neutron cores.**
G. T. Biehle.
Astrophys. J., Vol. 380, No. 1, p. 167 – 184 (10 Oct 1991).

Thorne and Zytkow have shown that stars more massive than about 12 $M_\odot$ and containing a degenerate neutron core (1) must be red supergiants with luminosities $L \gtrsim 70{,}000\, L_\odot$, and (2) cannot produce their required luminosities by gradual, steady accretion of the massive envelope onto the neutron core. This paper investigates whether such a star can produce its required luminosity by nuclear burning in a nondegenerate region a few tens of kilometers thick, just above the neutron core. To do so, two models are constructed of rp burning (rapid, direct addition of protons onto nuclei) in such a star. These models show rather unequivocally that the rp–process can produce the required luminosity, and they make clear the basic physical and chemical structure of a star supported by the rp–process. The physical structure is unusual: convection extends continually from the photosphere down through the burning region and stops only at the burning region's base (just above the ~10 km radius degenerate core). This convection brings fuel for the rp–process into the burning region from throughout the envelope and deposits the products of rp burning back in the envelope, including the photosphere. It is proposed that such a star might be distinguishable from a red supergiant with a degenerate electron core by the presence of spectral lines of the hydrides of some of the rp reaction products, for example, ^{84}SrH. The lifetime of the star in its rp–burning phase, which is limited by the envelope's supply of light metals (^{12}C, ^{14}N, and ^{16}O), required as seeds for the rp–process, would be about the same as the lifetime of a normal red supergiant of the same mass with a degenerate electron core.

065.104 **Periodic stellar pulsations: stability analysis and amplitude equations.**
J. R. Buchler, P. Moskalik, G. Kovács.
Astrophys. J., Vol. 380, No. 1, p. 185 – 199 (10 Oct 1991).
The stability properties of nonlinear periodic stellar pulsations are studied within the amplitude equation formalism. Both nonresonant and resonant pulsations are considered. A comparison to a sequence of classical Cepheid models shows that the formalism provides a good description of the behavior of the Floquet coefficients and that it also captures the most important features of the Floquet eigenvectors. It thus helps shed new light on the behavior (bifurcations) of pulsating stars. In addition, the predictive powers of the analytical approach allow a systematic search for models with specific pulsational properties.

065.105 **Three–dimensional compressible hydrodynamic convection in the Sun and stars.**
M. Hossain, D. J. Mullan.
Astrophys. J., Vol. 380, No. 2, p. 631 – 654 (20 Oct 1991).
Using a pseudospectral code, the authors have obtained numerical solutions of three–dimensional compressible hydrodynamic convection in a stratified medium where the boundaries are open, and where radiative losses are strongest at the top. Results are presented for a series of three simulations in progressively deeper boxes: the boxes have depths of roughly H_p, $2H_p$, and $4H_p$ (where H_p is a pressure scale height). The code does not make the anelastic approximation: rather, it follows sound waves explicitly. The code remains stable for long time intervals: the integration times are such that sound waves have time to traverse the grid of order 100 times. By these times, a statistically steady state is well developed, with significant upward heat fluxes. The high stability of the code is due in part to the fact that the effective "grid Reynolds' number" is not greatly in excess of unity: as a result, the flows are organized spatially about time–dependent points of convergence and divergence, with convective flows superposed. Spatial organization is found in the form of vertical "stacking" of smaller structures on top of larger ones, with each structure being roughly H_p in vertical extent. In the context of coronal heating, fluxes of kinetic energy are of interest, both total fluxes and those in the horizontal direction. Also of interest from the point of view of resonant heating of coronal loops are the spectral distributions of power on a variety of length scales.

065.106 **Metal opacities and convective core overshooting in Population I stars.**
R. B. Stothers, C.–W. Chin.
Astrophys. J., Lett., Vol. 381, No. 2, p. L67 – L70 (10 Nov 1991).
New stellar opacities computed by Rogers and Iglesias are found to lead to revised evolutionary models for stars of intermediate to high mass that show remarkably close agreement with observed stars, under otherwise standard physical assumptions. Most or all of the earlier discrepancies, except for the problem of predicted apsidal motion constants in binaries that are perhaps still too large, appear to have been resoved by using these new opacities with their increased metal line contribution. Thus, the implied interior metallicity of massive stars in the solar neighborhood now becomes "normal". Furthermore, the effective distance of convective overshooting beyond the classical convective core boundary is confirmed to be small and probably less than 0.2 times the local pressure scale height.

065.107 **Masses of RRd variables using Livermore OPAL opacities.**
A. N. Cox.
Astrophys. J., Lett., Vol. 381, No. 2, p. L71 – L74 (10 Nov 1991).
Twenty–seven new stellar envelope models have been constructed using the new Livermore opacities to see if the discrepancy between the masses of the double–mode RR Lyrae variables using pulsation and evolution theories can be reconciled. For these extreme Population II variables ($Z = 0.0001$ to 0.0004), the large opacity increases found from the extensive line absorptions of iron around 250,000K seen in Population I compositions are not realized. If the Z is 0.001, period ratios decrease from the original Cox, Hodson, and Clancy values by more than 0.01, leading to double–mode masses more than 0.3 $M_\odot$ larger, but such a large Z is not reasonable for these stars. It is discovered that opacity decreases below a temperature of 10^5K, due to more narrow lines of hydrogen than used in the Los Alamos opacities, can also decrease theoretical period ratios. This opacity decrease produces steeper temperature increases and more shallow density gradients in the region where the pulsation eigenvectors can feel the stellar structure. This leads to longer periods for all modes, with the fundamental mode being increased in period the most. Thus it appears that period ratios should be smaller and theoretical double–mode masses about 0.1 $M_\odot$ larger than those using earlier opacities. Small opacity increases at about 250,000K can increase the mass of the RRd variables in the Oosterhoff I clusters by another 0.1 $M_\odot$. This essentially eliminates the inconsistency between masses from the two globular cluster types and between those determined from pulsation and evolution theories.

065.108 **Evolutionary sequences of stellar models of intermediate and high mass including convective core overshooting.**
C.–W. Chin, R. B. Stothers.
Astrophys. J., Suppl. Ser., Vol. 77, No. 2, p. 299 – 316 (Oct 1991).
Stellar evolution with parametrized overshooting from the convective core has been studied for the stellar mass range $M/M_\odot = 3 - 30$ and for the initial chemical composition range $X_e = 0.650 - 0.739$ (hydrogen) and $Z_e = 0.021 - 0.044$ (metals). Evolutionary sequences run from the zero–age main sequence to the end of core helium burning, but emphasis is placed here on the core helium–burning phase. Convective core overshooting during the previous main–sequence phase leads, in most cases, to a shortening of the blue loop on the H–R diagram that forms when helium is being depleted in the core. On the other hand, convective overshooting from the helium–burning core has the opposite effect. A larger initial metal abundance also tends to shorten the blue loop. In an extreme case the blue loop can be fully suppressed, but in a marginal case the normal trend can be reversed. The fraction of the core helium–burning lifetime that is spent in the blue phase is generally very sensitive to the physical input parameters. Detailed tables and plots of evolutionary tracks on the H–R diagram are provided.

065.109 **Evolution of stars with extended atmospheres.**
M. F. El Eid, P. Höflich.
ESO/EIPC Workshop on Supernova 1987A and other Supernovae, p. 29 – 36 (Jun 1991). – See Abstr. 012.075 for the main entry.
A simple method is presented for calculating stellar models with spherically–symmetric grey atmospheres in LTE–approximation. This is accomplished by an improvement of the surface boundary conditions and by introducing correction factors to the diffusion approximation obtained from the solution of the radiation transport equation. As a first application, the evolution

of the progenitor of SN 1987A (20 $M_\odot$ with LMC chemical composition) is calculated up to the onset of central carbon burning using the Schwarzschild criterion for convection without overshooting. The authors used new low-temperature ($T \leqslant 10^4$K) tables for the mean opacities. The modified treatment of the outer layers is found of minor influence in this case. For the same star, a new model describing mixing in semiconvective zones is also tested. A blue supergiant at the time of explosion has not been found in any of the present test computations.

065.110 **The pulsations of ZZ Ceti stars. III. The driving mechanism.**
A. J. Brickhill.
Mon. Not. R. Astron. Soc., Vol. 251, No. 4, p. 673 – 680 (15 Aug 1991).

The outer layers of the variable white dwarfs are in a state of partial ionization. During the pulsation cycle the base of the ionization zone is strongly heated by the radiative layers below, in phase with the pressure perturbation. If this excess heat is not quickly lost at the surface, then the driving effect is strong. The surface flux perturbation tends to be small and delayed in phase because the surface flux is remarkably insensitive to temperature changes in the deeper layers of the ionization zone. This insensitivity is closely associated with the well known inward divergence of the solutions for the equilibrium thermal structure in the convective layers. The mechanism which excites the oscillations could be called "convective driving".

065.111 **Effects of the equation of state in neutron stars and in stellar collapse.**
J. M. Lattimer, A. Burrows.
ESO/EIPC Workshop on Supernova 1987A and other Supernovae, p. 69 – 81 (Jun 1991). – See Abstr. 012.075 for the main entry.

Some effects of the equation of state of dense matter on the structure of neutron stars and on gravitational collapse supernovae are discussed. The authors discuss which nuclear parameters are most important for the equation of state, and constraints on them that have been obtained from laboratory experiments. It is then discussed how neutron stars and supernovae depend upon and, therefore, could constrain, these parameters. Some structural constraints on neutron stars are available from observations, but these cannot, at present, set limits on important nuclear parameters such as the bulk incompressibility parameter or the bulk symmetry energy. However, the mechanism of type II, or gravitational collapse, supernovae depends crucially on these nuclear parameters and upon other aspects of the equation of state. The authors examine the equation of state in the important regime just below nuclear density in which nuclei become deformed and undergo possible phase transitions to other shapes.

065.112 **The effects of radiative heat loss from the convective element, and of helium flux, on overshooting in stellar convective cores.**
M. Umezu.
Mon. Not. R. Astron. Soc., Vol. 252, No. 1, p. 56 – 62 (1 Sep 1991).

The non-local mixing-length theory of stellar convective cores by Maeder is extended to include the effects of radiative heat loss from the convective element, and of helium flux. The radiative heat loss tends to enlarge the overshooting region, although it barely changes the size of the overshooting core for the ordinarily accepted size of the convective element. The helium gradient seems negligible in the overshooting region.

065.113 **Symmetry breaking in stellar dynamos.**
R. L. Jennings, N. O. Weiss.
Mon. Not. R. Astron. Soc., Vol. 252, No. 2, p. 249 – 260 (15 Sep 1991).

The generation of magnetic fields in stars like the Sun can be described by an azimuthally averaged dynamo model. Solutions of the linear (kinematic) problem have pure dipole or quadrupole symmetry, i.e. toroidal fields that are antisymmetric or symmetric about the equator. These symmetries can only be broken at bifurcations in the non-linear regime, which lead to the appearance of spatially asymmetric mixed-mode solutions. The symmetries of dipole, quadrupole and mixed-mode solutions, whether steady or periodic, form the same group for any axisymmetric dynamo. To establish the bifurcation structure it is necessary to follow unstable as well as stable solutions. This is only feasible for simple systems and a minimal non-linear $\alpha\omega$ dynamo is studied in detail in order to illustrate the formation of mixed-mode periodic solutions and to distinguish between their symmetries. The results are applied to the Sun (where there are persistent deviations from dipole symmetry) and to other late-type stars.

065.114 **The pulsations of ZZ Ceti stars. IV. The instability strip.**
A. J. Brickhill.
Mon. Not. R. Astron. Soc., Vol. 252, No. 3, p. 334 – 341 (1 Oct 1991).

DA models become unstable to the observable g-mode oscillations when the hydrogen ionization zone is sufficient to excite modes with periods $\geqslant 100$ s. As the effective temperature decreases, modes of longer period are excited up to a maximum period of about 1000 s. Further cooling does not lead to the excitation of the high-order modes of still longer period (to a detectable amplitude) and the brightness changes associated with the excited modes decrease below the detection threshold. The position of the theoretical instability strip in the T_e–log g plane is thus determined. The width of this instability strip (at fixed g) is less than 1000K. It is argued that this value is consistent with the statistics of ZZ Ceti stars and that the much wider spread in temperature which has been found in several investigations may be due to observational errors. Calibration for the mixing length theory of convection on the basis of the mean temperature of the ZZ Ceti stars is in broad agreement with the solar calibration.

065.115 **Stellar convection.**
Xiong Darun.
Proc. Astron. Soc. Aust., Vol. 9, No. 1, p. 26 – 31 (1991). – See Abstr. 012.078 for the main entry.

Time-dependent convection theory and non-local convection theory are described. The difference between the convection theories results from the different treatment of the non-linear terms of the hydrodynamic equations. The author has obtained a better understanding of the thermodynamic coupling between convection and oscillations in comparison to the dynamic coupling.

065.116 **On equation of state interpolation errors in stellar interior calculations.**
B. Dorman, A. W. Irwin, B. B. Pedersen.
Astrophys. J., Vol. 381, No. 1, p. 228 – 233 (1 Nov 1991).

The authors present an investigation of the errors induced in stellar interior calculations by the adoption of equation of state (EOS) tables. They isolate the effects of interpolation by using a standard EOS formulation to create tables with various grid spacings, and then compare calculations utilizing them with the results of direct computation. The authors use EOS grid spacings from the literature, specifically those of Magni and Mazzitelli (MM), Fontaine, Graboske, and Van Horn (FGVH), and Mihalas, Däppen, and Hummer (MDH). Test calculations were performed for the low-mass zero-age main sequence (ZAMS) and for red giant evolution. It is found that the errors induced by too coarse a spacing produce a systematic shift of the models in the theoretical H–R diagram. The direct effect of the shift is probably smaller than other uncertainties in the giant branch models. However, the effect on the ZAMS is likely to produce serious errors in the calibration of solar models, and therefore of the mixing length parameter for giant branch evolutionary tracks. It is found that the "additive volume" method of composition interpolation can produce good results for gross stellar properties even with pure element tables. When the interpolation is between two tables separated by $\delta Y = 0.3$, the errors are reduced to negligible proportions. In sum, the authors demonstrate the fineness of the spacing of EOS tables required for acceptable errors in derived stellar properties.

065.117 **Pulsating stars and fractals.**
M.-J. Goupil, M. Auvergne, T. Serre.
Applying fractals in astronomy, p. 43 – 75 (1991). – See Abstr. 003.023 for the main entry.

Fractals are encountered in stellar pulsation theory as attractors, resulting from chaotic dynamics of pulsating stars. The authors review observational and theoretical investigations of chaotic stellar pulsations with particular emphasis of the physical information they provide about stellar interiors.

065.118 **Spectroscopic constraints on the evolution of massive stars.**
R. P. Kudritzki.
Symposium in Celebration of the 50. Anniversary of McDonald Observatory of the University of Texas at Austin: Frontiers of stellar evolution, p. 97 – 124 (1991). – See Abstr. 012.085 for the main entry.

After two decades of intensive work, the methods for quantitative investigation into the evolution of massive stars are at hand. The diagnostic techniques described in the paper, together with modern spectroscopic equipment attached to present–day large telescopes or forthcoming very large telescopes, appear to be very powerful and promising. Certainly, it will be possible to perform detailed quantitative spectroscopic studies of not only the massive stars in the Galaxy, but also of the luminous blue stars in all the Local Group galaxies. This will yield an alternative approach for determining the distances of these galaxies, plus the possibility of determining the metallicity of their younger population and the study of the evolution of massive stars in different galactic environments.

065.119 **Death and transfiguration – mass loss from cool evolved giant stars and evolution to planetary nebulae.**
G. R. Knapp.
Symposium in Celebration of the 50. Anniversary of McDonald Observatory of the University of Texas at Austin: Frontiers of stellar evolution, p. 229 – 263 (1991). – See Abstr. 012.085 for the main entry.

Evolved asymptotic giant branch stars are observed to be losing mass in the form of cool dusty molecular stellar winds at rates from about 10^{-8} to $10^{-4} M_{\odot}\ y^{-1}$. This review describes the observational methods used to measure these mass loss rates and uses the results to discuss the total rate of mass return to the interstellar medium from AGB stars, the chemical composition and the effect on the evolution of the central stars. New results on the evolution of stars between the AGB and planetary nebula phases are described; these suggest that for the massive stars at least the evolution may be extremely rapid, in agreement with theoretical models.

065.120 **What is a stellar dynamo?**
F. Cattaneo, D. W. Hughes, N. O. Weiss.
Mon. Not. R. Astron. Soc., Vol. 253, No. 3, p. 479 – 484 (1 Dec 1991).

Numerical simulations of turbulent stellar dynamos are now feasible. The characteristic time–scale for kinematic behaviour is related to the turnover time of the turbulent eddies. Results from idealized two–dimensional models show that the Lorentz force alters the velocity field, allowing transient magnetic activity to persist for intervals much longer than the expected turbulent decay time. For three–dimensional flows the authors define a characteristic time T_e for turbulent diffusion to act, based on the rate at which magnetic energy is dissipated, and assert that there is a dynamo only if the field survives for times much longer than T_e. This criterion is then applied to cyclic magnetic activity in late–type stars.

065.121 **CSPN and sdOs: observational hints on AGB and post–AGB stellar evolution.**
D. Husfeld, F. Eber.
Symposium in Celebration of the 50. Anniversary of McDonald Observatory of the University of Texas at Austin: Frontiers of stellar evolution, p. 603 (1991). Abstract. – See Abstr. 012.085 for the main entry.

065.122 **Photofission production of technetium in AGB stars: an alternative to the s–process.**
J. C. Lattanzio, R. A. Malaney.
Symposium in Celebration of the 50. Anniversary of McDonald Observatory of the University of Texas at Austin: Frontiers of stellar evolution, p. 603 (1991). Abstract. – See Abstr. 012.085 for the main entry.

065.123 **The secular and thermal spectrum of hydrogen–deficient carbon stars.**
A. Gautschy, W. Glatzel.
Mon. Not. R. Astron. Soc., Vol. 253, No. 3, p. 509 – 521 (1 Dec 1991).

Applying the Riccati method to the secular problem, secular and thermal spectra of a 0.9 $M_{\odot}$ HdC star are investigated in various evolutionary phases. The onset of shell burning is associated with a bewildering complexity of the thermal and secular spectrum and is interpreted using a simplified analytical model. Shell burning leads to the decoupling of different parts of the spectrum, one of which seems to be closely related to the strange mode II phenomenon.

065.124 **Implications of helium diffusion for globular cluster isochrones and luminosity functions.**
C. R. Proffitt, D. A. VandenBerg.
Astrophys. J., Suppl. Ser., Vol. 77, No. 4, p. 473 – 514 (Dec 1991).

Grids of evolutionary sequences for masses from 0.5 to 1.0 $M_{\odot}$ and metallicities [Fe/H] = –2.26, –1.26, and –0.65 (with [O/Fe] = 0.75, 0.55, and 0.30, respectively, and appropriate helium abundances) have been computed both with and without the inclusion of diffusion. These models have been interpolated to produce isochrones, for ages from 10 to 18 Gyr, and associated luminosity functions, for several power–law mass functions. The resultant age – luminosity relations indicate that the gravitational settling of helium can reduce the age at a given turnoff luminosity by about 10%, with considerably larger effects on turnoff temperatures and colors. Extensive tables of tracks, isochrones, and luminosity functions are provided for general use.

065.125 **Pre–main–sequence evolution.**
S. Sofia, S. Kawaler, R. Larson, M. Pinsonneault.
Solar interior and atmosphere, p. 140 – 161 (1991). – See Abstr. 003.028 for the main entry.

The pre–main–sequence (PMS) stage of a star's evolution begins when a subcondensation forms in a dense interstellar cloud, and it ends when the stellar core becomes sufficiently hot to begin nuclear processing of the hydrogen fuel. The PMS stage is very short lived in terms of evolution time scales. In the earlier stages, when the gravitational forces overcome internal pressure, the object is essentially in free fall. This dynamical stage comes to an end when hydrostatic equilibrium is re–established with the possible involvement of magnetic and rotational forces. At this point, evolution slows down, and it is regulated by the ability to shed internal energy and/or by the pace of dissipation of angular momentum (perhaps in an accretion disk). Meanwhile the structure of the object changes drastically from a radiative to a fully (or nearly fully) convective object. The details of the evolution process are still very sketchy, especially in the dynamic stage. The main uncertainty in the hydrodynamic stage involves the possible role of accretion. The authors describe the various possible scenarios envisioned in the evolution, state the uncertainties involved, and refer to existing or potential observational data which may help unravel the details of the PMS evolution. They follow recent evolution calculations that include rotation, and conclude that this important feature of the PMS can best be tested in terms of the chemical abundance of trace elements in the solar system, on the Sun, and on solar analogs of different ages.

065.126 **The stellar core collapse and explosion of SN II.**
Wang Yiren.
Prog. Astron., Vol. 9, No. 4, p. 263 – 270 (Dec 1991). In Chinese.

In this review, it is discussed in detail that the standard model for SN II involves stellar core collapse and bounce, mechanism of prompt explosion and of delayed explosion. Progress and problems in the theoretical research of SN II are presented briefly.

065.127 **Rayleigh–Taylor instability in the asymmetric supernova explosion.**
S. Yamada, K. Sato.
Astrophys. J., Vol. 382, No. 2, p. 594 – 602 (1 Dec 1991).
The authors studied matter mixing by Rayleigh–Taylor instability in a jetlike supernova explosion by numerical calculations of two–dimensional hydrodynamics. It is found that (1) the asymmetric explosion with 5% initial perturbation causes the same degree of matter mixing as the spherical one with 30% perturbation, and that (2) 30% initial perturbation in the jetlike explosion is sufficient to explain the mixing in SN 1987A suggested by many observations without any subsequent instability such as nickel bubble formation.

065.128 **Massive star evolution and SN 1987A.**
D. Arnett.
Astrophys. J., Vol. 383, No. 1, p. 295 – 307 (10 Dec 1991).
Our understanding of the evolution of massive stars has been challenged by SN 1987A, which may well have been the normal evolution of a single star of $\sim 20\ M_{\odot}$ of LMC abundances. A set of stellar evolutionary sequences, for mass $M/M_{\odot} = 15$, 20, and 25 and metallicity $z = 0.002$, 0.005, 0.007, 0.010, and 0.20 are presented; semiconvection is restricted to operating slower than the local thermal time scale. Using these sequences, simple models of the massive star content of the LMC agree moderately well with the new observational data of Fitzpatrick and Garmany. They imply that one has found few of the main–sequence supergiants in the LMC. Untangling the effects of various uncertainties will require more rigor in both observation and theory. Despite the variety of envelope behavior, the structure of the core at collapse is rather similar for the stars of a given mass. Variations due to different rates of mass loss are likely to be larger than those due to composition. There is an increasing tendency for dredge–up of helium core material with increasing heavy–element abundance.

065.129 **Line profile variations caused by low frequency nonradial pulsations of rapidly rotating stars. II.**
U. Lee, C. S. Jeffery, H. Saio.
ESO Workshop on Rapid Variability of OB–Stars: Nature and Diagnostic Value, p. 245 – 247 (1991). – See Abstr. 012.092 for the main entry.

065.130 **Excitation mechanism for nonradial pulsations in variable OB stars.**
H. Saio, U. Lee.
ESO Workshop on Rapid Variability of OB–Stars: Nature and Diagnostic Value, p. 293 – 302 (1991). – See Abstr. 012.092 for the main entry.
The authors discuss how low–frequency nonradial pulsations can be excited in rotating massive main–sequence stars which have convective cores. In this excitation mechanism, stellar rotation plays an essential role. The rotation can modify the property of low–frequency oscillations so strongly that negative energy oscillations are possible in a rotating convective region. The negative energy oscillations can be understood as the oscillations whose amplitudes grow when energy is extracted from the system. When a negative energy oscillation in the convective core is resonant with a positive energy oscillation, say g^{+}–mode oscillation, in the envelope, energy can flow from negative to positive oscillations to form a coupled overstable oscillation.

065.131 **Nonradial pulsation, rotation, and episodic mass loss.**
H. Ando.
ESO Workshop on Rapid Variability of OB–Stars: Nature and Diagnostic Value, p. 303 – 314 (1991). – See Abstr. 012.092 for the main entry.
The acceleration and/or deceleration of rotation by nonradial pulsation (NRP) is investigated in the realistic stellar models, and how the rotational modulation by NRPs depends on NRP characteristics is shown. Based on this, the basic mechanism of NRP and rotation interaction is investigated, and the possible stellar cycle driven by this interaction is proposed. Its application to the episodic mass loss in Be stars is discussed in detail.

065.132 **On the pulsation – evolution connection of early–type stars.**
A. Gautschy.
ESO Workshop on Rapid Variability of OB–Stars: Nature and Diagnostic Value, p. 315 – 329 (1991). – See Abstr. 012.092 for the main entry.
Observational evidence for nonradial pulsations in early type stars is reviewed. Attempts by theorists to explain the excitation of these pulsations are outlined. The influence of the internal structure on the frequency spectrum of nonradial oscillations is discussed for different sets of evolutionary stellar models. The author studies the possibility of excitation of nonradial pulsations by modal interaction.

065.133 **Mass determinations using the nonradial pulsational models with application to AB Cas star.**
M. D. Suran.
ESO Workshop on Rapid Variability of OB–Stars: Nature and Diagnostic Value, p. 331 – 332 (1991). – See Abstr. 012.092 for the main entry.

065.134 **From regularity to noise in variable stars.**
J. Perdang.
ESO Workshop on Rapid Variability of OB–Stars: Nature and Diagnostic Value, p. 349 – 361 (1991). – See Abstr. 012.092 for the main entry.
After a review of the types of oscillation theoretical stellar models can sustain, the author discusses the attempts at applying dynamical systems theory to the observational time series of Variable Stars. He concludes that there is a discrepancy between observations and theory, presumably related to an inadequate treatment of convection in the models. The author further associates with chaotic time series easily computed characteristic times (pseudo–periods) which may serve the purpose of diagnostic parameters for the underlying chaotic models.

065.135 **Possibilities to find observational evidence for chaos among early type stars.**
Z. Kollath, J. Nuspl.
ESO Workshop on Rapid Variability of OB–Stars: Nature and Diagnostic Value, p. 363 – 366 (1991). – See Abstr. 012.092 for the main entry.
The detection of the existence of chaos from observational data is very important, because the theoretical models of stellar pulsation have to predict this fact, i.e. it may be a check point of the theories. The analysis of observed chaotic processes is a new field in data processing. The methods of dynamic analysis of a one–dimensional observable appeared only in the early 80s. Since then, there were some attempts in the literature to find observational evidence for chaotic stellar pulsation among RV Tauri, Mira, semiregular and white dwarf stars. The authors review these recent analyses of light curves with criticism to describe the main characteristics of chaotic data processing. The main problems of the methods applied in these analyses are connected with the limited amount of data and the noise of the observations. Considering this fact the authors tried to find some general rules of observations according to a satisfactory analysis, performing different numerical tests on the methods of modern time series analyses of chaotic signals (phase space reconstruction and dimensional calculations) simulating the real variables. According to their results the authors make some suggestion about the necessary observations to detect chaos among the early type stars.

065.136 **Effects of negatively charged massive particles on primordial stars.**
R. N. Boyd, G. M. Fuller.
Astrophys. J., Vol. 383, No. 2, p. 615 - 620 (20 Dec 1991).
The effects of negatively charged non-strongly interacting massive particles, denoted as X^- particles, on stellar hot hydrogen burning are considered. Such particles would bind to nuclei and catalyze a very efficient hydrogen-burning cycle. This nuclear burning cycle would occur because the additional binding energy provided by binding the X^- particles to ^{8}Be nuclei creates a stable entity. Although such a cycle would not be likely to be of significance to present-day stars, it could significantly alter the range of masses, for which primordial Population III stars explode. The nucleosynthesis produced in these X^--induced supernovae can differ markedly from that of ordinary β-limited CNO cycle-induced explosions of very massive and supermassive stars. Thus the resulting abundances might provide constraints on the existence of intermediate lifetime X^- particles and/or the Population III initial mass function.

065.137 **Evolution of compact binary systems with X-ray heating.**
A. Harpaz, S. Rappaport.
Astrophys. J., Vol. 383, No. 2, p. 739 - 744 (20 Dec 1991).
The evolution of low-mass X-ray binary systems is calculated with the inclusion of heating of the secondary star by the X-radiation from the primary. The secondary is assumed to be nonsynchronously rotating with the orbit, so that the X-ray heating may be taken to be approximately uniform over its surface. The evolution of binary systems with secondary stars of initial mass 0.4 and 1 $M_\odot$ is computed. It is found that the expansion of the secondary due to the X-ray heating causes high rates of mass loss from the star for significant intervals of time, for both values of the secondary mass. This mass transfer is typically accompanied by an increase in the orbital period. The expansion of the secondary takes place in two distinct phases: (1) first, only the outermost layers of the star are heated and expand rapidly; (2) a long-term expansion phase causes the entire star to swell. The mass-loss rates during the first phase are very high, and they are decreasing (but are still super-Eddington) during the second stage of the expansion phases. When the mass transfer rate to the neutron star drops below the Eddington limit, the effects of the X-ray heating diminish sufficiently so that the secondary cannot be sustained in its fully bloated state. The secondary then shrinks within its Roche lobe, and mass transfer ceases until magnetic braking can bring the system back into contact. The authors discuss the relevance of these results to low-mass X-ray binaries and recycled radio pulsars.

065.138 **A conjecture regarding the evolution of dwarf stars into red giants.**
P. P. Eggleton, R. C. Cannon.
Astrophys. J., Vol. 383, No. 2, p. 757 - 760 (20 Dec 1991).
The authors investigate the conjecture that a barytrope with local nonconstant polytropic index n(r) is always less centrally condensed than the polytrope whose constant index n_m is max[n(r)], with $n_m < 5$. The conjecture is consistent with numerical results for 2×10^6 different barytropes. Although no exceptions are known, no *proof* of the conjecture does exist, except in a somewhat weaker form. If the conjecture is nevertheless true, a consequence is that a star can only have the high degree of central condensation generally found in computed models of red giants ($\varrho_c/\bar{\varrho} \gtrsim 10^8$), if somewhere within the star there is a region which has $n \gtrsim 5$. This means that in normal low-mass red giants, with degenerate helium cores, the cause of the high degree of central condensation is the molecular weight gradient in the hydrogen-burning shell, along with the thin isothermal nondegenerate helium shell sandwiched between the hydrogen-burning shell and the degenerate bulk of the helium core. Although the symptoms of a red giant manifest themselves in the envelope outside the burning shell, the underlying cause is to be found further in.

065.139 **Effects of drag luminosity in the common envelope phase of slow novae.**
M. Kato, I. Hachisu.
Astrophys. J., Vol. 383, No. 2, p. 761 - 765 (20 Dec 1991).
The effects of the drag luminosity during the decay phase of slow novae are examined under the assumption of spherical symmetry for two systems consisting of a 0.9 $M_\odot$ white dwarf and a 0.2 $M_\odot$ main sequence companion, and of a 0.6 $M_\odot$ white dwarf and a 0.2 $M_\odot$ main-sequence companion. The drag luminosity, L_{drag}, is 15% of the total diffusive luminosity, L_r, in the 0.9 + 0.2 $M_\odot$ system, while L_{drag} is 37% of L_r in the 0.6 + 0.2 $M_\odot$ system. The 0.6 $M_\odot$ white dwarf has a more massive envelope than the 0.9 $M_\odot$ white dwarf, and its density is much higher than that of the 0.9 $M_\odot$ white dwarf. Therefore, the drag luminosity is much larger in the 0.6 $M_\odot$ case than in the 0.9 $M_\odot$ case, because the frictional energy generation is proportional to the density of material at the companion orbits. It is also found that no optically thick wind occurs, mainly because the gravity of the companion suppresses the wind. As a result, the evolution time scale of the decay phase is not drastically changed for slow novae when the frictional process is included.

065.140 **Post-asymptotic giant branch nonradial instability strips.**
L. Stanghellini, A. N. Cox, S. Starrfield.
Astrophys. J., Vol. 383, No. 2, p. 766 - 778 (20 Dec 1991).
The authors have performed stability analyses, for nonradial g^+-mode pulsations of post-asymptotic giant branch (AGB) stellar models, to determine the location of their pulsational instability strips in the Hertzsprung-Russell diagram. It is assumed that the stars have undergone a major mass loss event either near the tip of the asymptotic giant branch or shortly thereafter and, therefore, consist of a 50% carbon - 50% oxygen core. All the models, which are analyzed for nonradial, nonadiabatic pulsational instabilities, are developed by using an evolutionary post-AGB track calculated by Iben and Hollowell. Their mass is 0.6 $M_\odot$, which is in good agreement with the peak of the observed distribution of white dwarf masses. Four different surface compositions are used: (1) 50% helium and 50% carbon, (2) 10% helium and 90% carbon, (3) 50% carbon and 50% oxygen, and (4) pure helium, all by mass. Using a Lagrangian, nonradial nonadiabatic pulsation stability program, the authors find pulsationally unstable models for all four compositions, but the instability strips narrow in effective temperature as the helium abundance increases. Pulsation driving occurs in the outer 10^{-9} or less of the star's mass. For the pure C + O surface composition, the authors obtain an instability strip that extends from the domain of the old planetary nebulae nuclei down the pre-white dwarf cooling track to the helium surface DB white dwarfs. The calculated results are compared both to the observed planetary nebula nuclei variables and the hot, pulsating DO variables.

065.141 **Convective penetration into stellar radiation zones.**
J.-P. Zahn.
Conference on Challenges to Theories of the Structure of Moderate-Mass Stars, p. 225 - 232 (1991). - See Abstr. 012.096 for the main entry.
The author recalls the evidence for convective penetration exhibited by geophysial fluids, laboratory experiments and computer simulations. A simple model is deduced from these observations, which serves to establish a relation between the subadiabatic extent of a convective region and the velocity of the penetrating motions. Assuming that this velocity obeys the usual scaling of thermal convection, the author finds that the subadiabatic penetration at the bottom of a convective envelope is of the order of a pressure scaleheight, and that above a convective core it amounts to a substantial fraction of the core radius. It is shown that the requirements for nearly adiabatic penetration are met deep enough in a stellar interior, and that the departures from adiabaticity are confined to a thin boundary layer. This allows the use of Roxburgh's integral constraint to predict the actual size of a convective core.

065.142 **Observational tests of convective core overshooting in stars of intermediate to high mass in the Galaxy.**
R. B. Stothers.
Astrophys. J., Vol. 383, No. 2, p. 820 – 836 (20 Dec 1991).

Fourteen tests for the presence of convective overshooting in large convecting stellar cores are discussed for stars with masses of 4 – 17 $M_\odot$, which are members of detached close binary systems and of open clusters in the Galaxy. Nine of these tests, including all the best tests, are either new or newly applied. A large amount of theoretical and observational data has been scrutinized and subjected to averaging in order to minimize accidental and systematic errors. Defining a ratio of the effective convective overshoot distance beyond the classical Schwarzschild core boundary to the local pressure scale height, the present investigation finds a conservative upper limit of $d/H_P < 0.4$ from at least four tests, and a tighter upper limit of $d/H_P < 0.2$ from one good test that is subject to only mild restrictions and is based on the maximum observed effective temperature of evolved blue supergiants. All the best test results, however, are formally consistent with the assumption $d/H_P = 0$ and are not sensitive to uncertainties about the initial chemical composition (or, probably, the opacities). Further analysis shows why the values of $d/H_P = 0.25 - 2$ that were derived in previous studies should be regarded as only upper limits.

065.143 **Hot bottom envelope burning in asymptotic giant branch stars.**
A. I. Boothroyd, I.–J. Sackmann.
J. R. Astron. Soc. Can., Vol. 85, No. 4, p. 204 – 205 (Aug 1991).
Abstract. – See Abstr. 012.091 for the main entry.

065.144 **Remarks on the evolution of low–mass stars: more or less opaque?**
J. Faulkner.
Conference on Challenges to Theories of the Structure of Moderate–Mass Stars, p. 293 – 308 (1991). – See Abstr. 012.096 for the main entry.

The author concentrates on a limited number of problems with understanding the structure or evolution of low–mass stars, where appropriate changes in opacity, or dominant mode of energy transport, may play a crucial role. The Hyades (classical lithium problem) and the Sun (high degree p–modes) support the need for enhanced Population I envelope opacities, as apparently confirmed by recent theoretical opacity calculations. The Sun (solar neutrino problem, perhaps the low–degree p–modes) and recent luminosity functions of globular clusters (a considerable excess of subgiant branch stars over conventional expectations) might require significantly enhanced central energy transport, as could be provided by accreted weakly interacting massive particles.

065.145 **Applications of stellar seismology to intermediate mass stars: on the main sequence and beyond.**
S. D. Kawaler.
Conference on Challenges to Theories of the Structure of Moderate–Mass Stars, p. 309 – 324 (1991). – See Abstr. 012.096 for the main entry.

The nonradial oscillations of stars in various regions of the HR diagram provide an alternate window into the processes of stellar evolution. In many cases, the non–radial pulsators play a crucial role in filling out the evolutionary picture. Two examples serve to illustrate this new probe: the rapidly oscillating Ap (ROAp) stars, and the pulsating white dwarfs. Because of the complex light curves that result from multiperiodic pulsations in these stars, global observing campaigns are required to fully decode their pulsation spectra. Such efforts can bear substantial scientific rewards. The observed frequency spacings in some of the ROAp stars allow identification of their evolutionary stage and, possibly, measurement of their rates of evolutionary expansion. For white dwarfs, seismological analysis using the observed *g*–mode pulsations allows to determine the compositional structure of their outer layers, measure their masses, and determine their rotation and cooling rates.

065.146 **Surface chemical composition, nuclear processing, and deep mixing: main sequence to first dredge–up.**
D. L. Lambert.
Conference on Challenges to Theories of the Structure of Moderate–Mass Stars, p. 325 – 344 (1991). – See Abstr. 012.096 for the main entry.

The standard theory of the structure of moderate mass stars predicts that, except for a few light nuclides, the composition of a main sequence star is identical to that of its parental interstellar cloud. Aspects of the depletion of Li in main sequence stars are discussed. Standard theory predicts that the composition of a red giant and its main sequence progenitor must differ because the giant's deep convective envelope brings to the surface material previously exposed to warm protons. Observational of Li, $^{12}C/^{13}C$ and $^{16}O/^{17}O/^{18}O$ in giants are related to the predictions for standard models.

065.147 **The equation of state of supernova and neutron star matter.**
W. Hillebrandt.
International School of Physics 'Enrico Fermi'. Course 113: High–pressure equations of state: theory and applications, p. 399 – 437 (1991). – See Abstr. 012.100 for the main entry.

Some of the basic ideas in connection with supernovae and neutron stars are reviewed. Recent approaches towards a realistic equation of state (EOS) of supernova matter, as well as their impact on supernova models are discussed. Neutron star matter EOS are presented together with a discussion of neutron star models and constraints on the EOS from observations. Finally, an outlook on future work is given.

065.148 **Comparing Victoria and Yale evolutionary tracks.**
B. B. Pedersen, D. A. VandenBerg, A. W. Irwin.
Conference on Challenges to Theories of the Structure of Moderate–Mass Stars, p. 353 – 358 (1991). – See Abstr. 012.096 for the main entry.

When the physics assumed in the published Yale evolutionary sequences is incorporated into the University of Victoria code, the latter produces tracks that differ from the former by only small amounts. This indicates that numerical methods are not a significant source of uncertainty.

065.149 **WIMP–like stellar models and the subgiant branch.**
J. Faulkner, F. J. Swenson.
Conference on Challenges to Theories of the Structure of Moderate–Mass Stars, p. 359 – 366 (1991). – See Abstr. 012.096 for the main entry.

WIMP–like isothermal cores in stellar models do not only result in reduced main–sequence lifetimes but also lead to such models spending a greater time than conventional models on the subgiant branch. Consequently, if WIMPs are effective in central energy transport, more stars should be seen than conventionally expected in this phase of evolution. While full luminosity functions have yet to be obtained, estimates may be made for two simple bases of comparison, by reference of conventional expectations normalized to red giant branch or main–sequence number densities. Two classical clusters, M3 and M92, appear to have anomalous excesses in their subgiant branches of an order that WIMP–like models can produce. Work needed to verify these preliminary results is indicated.

065.150 **Problems for the WIMP hypothesis?**
D. A. VandenBerg, P. B. Stetson.
Conference on Challenges to Theories of the Structure of Moderate–Mass Stars, p. 367 – 372 (1991). – See Abstr. 012.096 for the main entry.

If stars accrete WIMPs (Weakly Interacting Massive Particles) over their lifetimes, then they are predicted to develop and sustain central isothermal cores. This investigation examines whether models with small ($\leqslant 0.1\, M_\odot$) isothermal cores can be reconciled with a number of observations. Specifically, it is examined how well they can match the properties of the well–observed binary AI Phoenicis, whether they can explain the gap feature in the old open cluster M67, and whether they can provide a satisfactory fit to the observed composite luminosity function of M68/M92/NGC 6397. It is found that the hypothesis of an isothermal core is not easily rejected: of the three tests considered here, only the M67 comparison appears to pose some considerable difficulties for it. Interestingly, the luminosity function data seem to favor the isothermal–core models over their canonical counterparts.

065.151 **Nonadiabatic nonradial pulsations of R CrB stars.**
W. Glatzel.
Conference on Challenges to Theories of the Structure of Moderate–Mass Stars, p. 387 – 392 (1991). – See Abstr. 012.096 for the main entry.

Preliminary results of a theoretical study of nonadiabatic nonradial pulsations of R CrB stars are presented. Mode coupling instabilities of p–modes are found up to log $T_{eff} \approx 4.3$. The applicability of the Riccati method to nonadiabatic nonradial pulsations is demonstrated.

065.152 **Theoretical aspects of asteroseismology of the rapidly oscillating Ap stars.**
H. Shibahashi.
Conference on Challenges to Theories of the Structure of Moderate–Mass Stars, p. 393 – 410 (1991). – See Abstr. 012.096 for the main entry.

The author evaluates prospects of seismology of the rapidly oscillating Ap stars which are unique objects in the sense that their oscillations are influenced by strong magnetic fields. Seismic investigation of the rapidly oscillating Ap stars is expected to provide us with information about their magnetic fields. The author compares the observed power spectra with those theoretically expected where the magnetic axis is inclined to the rotation axis. An interesting characteristic of the rapidly oscillating Ap stars is that many of these stars are oscillating with frequencies higher than the theoretically reasonable critical frequency above which no standing modes are expected without energy leakage. The author discusses the character of these oscillations and their plausible physical explanations. Most of the rapid oscillations seem to be axially symmetric dipole modes with respect to the magnetic axis of the star. The author discusses an effect of the non–spherical outer layer as a possible selection mechanism for those modes.

065.153 **The seismology of stars.**
G. Vauclair, S. Vauclair.
Recherche, Vol. 22, No. 237, p. 1280 – 1287 (Nov 1991). In French.

065.154 **A convective model for solar and stellar flares.**
A. Grandpierre.
Mem. Soc. Astron. Ital., Vol. 62, No. 2, p. 401 – 407 (1991). Special issue: Stellar flares. B. R. Pettersen (ed.).

The author outlines a convective flare model. The basic problems of magnetic flare theories are pointed out. The magnetic flare theories as well as the convective flare theory are confronted with observations. It is suggested that a local runaway mechanism in solar and stellar cores is the ultimate cause of the flares. If this mechanism is driven by tides, it could provide an interpretation of the dependence of stellar activity cycle periods on rotation rates.

065.155 **Equilibrium structure of differentially rotating polytropic stars.**
A. K. Lal, C. Mohan, V. P. Singh.
Bull. Astron. Soc. India, Vol. 19, Nos. 1 – 2, p. 94 (Mar – Jun 1991). Abstract. – See Abstr. 012.108 for the main entry.

065.156 **Shock phenomena in a hydrodynamic model of a classical Cepheid.**
T. Ishida, M. Takeuti.
Publ. Astron. Soc. Jpn., Vol. 43, No. 6, p. 795 – 807 (1991).

The authors have compared a hydrodynamic model of a classical Cepheid containing 500 grid points with that containing 100 grid points. In linear models, the growth rates of the pulsation modes appear to grow as the number of grid points increases, and to saturate at about 200 meshes. In hydrodynamic models, some shock phenomena appear to be captured only in a 500–grid model. This indicates that a fine–grid model is very useful for studying the shock waves in a pulsating envelope. A newly constructed 500–grid model is applied to a study of the ionization/recombination fronts and shock waves in pulsating envelopes. The motion of discontinuities over an entire pulsation cycle is shown for the first time.

The Sun and cool stars: activity, magnetism, dynamos. Proceedings.
See Abstr. 012.024.

The problem of lithium. Proceedings. European Workshop on the Problem of Lithium in Honour of Evry Schatzman, Monte Porzio, Rome (Italy), 12 – 14 Sep 1990.
See Abstr. 012.033.

The physics of star formation and early stellar evolution. Proceedings.
See Abstr. 012.042.

White dwarfs. Proceedings.
See Abstr. 012.044.

Angular momentum evolution of young stars. Proceedings.
See Abstr. 012.055.

Frontiers of stellar evolution. Proceedings.
See Abstr. 012.085.

Rapid variability of OB–stars: nature and diagnostic value.
See Abstr. 012.092.

Challenges to theories of the structure of moderate–mass stars. Proceedings.
See Abstr. 012.096.

Stellar astrophysics at the IAC from 1979 to 1989.
See Abstr. 013.075.

Challenges to theories of the structure of moderate mass stars.
See Abstr. 013.102.

Experimental determination of stellar reaction rates.
See Abstr. 022.176.

Inert neutrinos in supernovae.
See Abstr. 061.006.

Nonlocal mass generation through spontaneous chiral–symmetry breaking for solar and stellar interiors.
See Abstr. 061.010.

The r–process and nucleochronology.
See Abstr. 061.035.

Capture cross section measurements of krypton and xenon isotopes and the fundamental parameters of the s–process.
See Abstr. 061.036.

Nuclear structure of ^{176}Lu and its astrophysical consequences. II. ^{176}Lu, a thermometer for stellar helium burning.
See Abstr. 061.050.

Multiple–scattering suppression of the bremsstrahlung emission of neutrinos and axions in supernovae.
See Abstr. 061.058.

Equation of state in metals and cold stars: evaluation of statistical models.
See Abstr. 061.062.

A critical look at cosmions.
See Abstr. 061.068.

Neutron capture in ^{122,123,124}Te: a critical test for s–process studies.
See Abstr. 061.077.

An analytical formulation of the double–pulse s–process model.
See Abstr. 061.083.

A measurement of the ^{22}Ne(n, γ)^{23}Ne capture cross section at a stellar temperature of kT = 25 keV.
See Abstr. 061.084.

Measurement of cross sections: current experiments.
See Abstr. 061.088.

Second angular moment closure on flux–limited neutrino diffusion theory.
See Abstr. 061.096.

Détermination de taux de réactions nucléaires conduisant à la nucléosynthèse stellaire du fluor.
See Abstr. 061.126.

The s–process: branchings and chronometers.
See Abstr. 061.129.

Instabilities in astrophysical shear flows.
See Abstr. 062.098.

Stable and unstable solutions of the nonlinear dynamo problem.
See Abstr. 062.106.

Two–dimensional hydrodynamic calculations of the nonlinear development of the Goldreich – Schubert – Fricke instability in a rotating annulus.
See Abstr. 062.190.

Mixing in the radiative zone.
See Abstr. 062.210.

Interaction of acoustic oscillations with time–dependent compressible convection.
See Abstr. 062.216.

Compressible magnetic convection: formulation and two–dimensional models.
See Abstr. 062.220.

Equation of state and related problems in astrophysics.
See Abstr. 062.223.

On thermonuclear convection: 1. Shellular instability.
See Abstr. 062.248.

Bremsstrahlung and energetic electrons in supernovae.
See Abstr. 063.002.

Line formation in low–mass dense core: Larson–Penston flow vs. inside–out collapse.
See Abstr. 063.015.

The Rosseland mean free–free Gaunt factor of the dense high–temperature stellar plasma.
See Abstr. 063.113.

Hydrodynamic studies of accretion onto ONeMg white dwarfs using a large nuclear reaction network.
See Abstr. 064.017.

The modified hydrostatic equilibrium equations for stratified high gravity stellar atmospheres.
See Abstr. 064.123.

Nonradial pulsation and line profile variability in early type stars.
See Abstr. 064.181.

Nonthermal excitation of helium in Type Ib supernovae.
See Abstr. 064.188.

Main–sequence mass–loss: lithium dilution in the Hyades and the Sun.
See Abstr. 064.191.

Self–similar gravitational collapse with volume energy loss.
See Abstr. 066.033.

Conformally symmetric spheres with surface pressure.
See Abstr. 066.059.

The effects of thermal radiation on some general relativistic stellar models.
See Abstr. 066.082.

Relation of gauge formalisms for pulsations of general relativistic stellar models.
See Abstr. 066.120.

Relativistic neutrino transport in stellar collapse.
See Abstr. 066.159.

Diffusion processes in the collapse of a radiating spherical body.
See Abstr. 066.215.

Supernovae: where and why do they break off?
See Abstr. 067.025.

The gravitational entropy of objects.
See Abstr. 067.084.

Normal–mode excitation from stellar collapse to a black hole: odd–parity perturbations.
See Abstr. 067.103.

Rotational properties of strange pulsar models.
See Abstr. 067.132.

On the origin of LMXRBS: the ONeMg case.
See Abstr. 067.133.

Self–similar scalar soliton stars in the thin–wall approximation.
See Abstr. 067.157.

Superfluid hydrodynamics in rotating neutron stars. I. Nondissipative equations.
See Abstr. 067.164.

Superfluid hydrodynamics in rotating neutron stars. II. Dissipative effects.
See Abstr. 067.165.

Diquark stars.
See Abstr. 067.218.

Quark–matter superconductivity and the cooling of compact stars.
See Abstr. 067.219.

On the Hawking radiation associated with an Oppenheimer–Snyder collapsing star.
See Abstr. 067.246.

Cosmic strings in the Sun?
See Abstr. 080.003.

On high degree *f* modes.
See Abstr. 080.049.

Our Sun. II. Early mass loss of 0.1 $M_\odot$ and the case of the missing lithium.
See Abstr. 080.050.

The effect of the Mihalas, Hummer, and Däppen equation of state and the molecular opacity on the standard solar model.
See Abstr. 080.092.

Gravitational settling in solar models.
See Abstr. 080.102.

Effects of opacity and equation of state on solar structure and oscillations.
See Abstr. 080.112.

Standard solar models.
See Abstr. 080.118.

Post–main–sequence solar evolution.
See Abstr. 080.136.

Helioseismic inversion.
See Abstr. 080.147.

Some aspects of the theory of solar oscillations.
See Abstr. 080.167.

s–process Ba, Nd, and Sm in presolar SiC from the Murchison meteorite.
See Abstr. 105.047.

The dynamical evolution of the protosolar nebula.
See Abstr. 107.004.

Atmospheric parameters of subluminous B stars.
See Abstr. 114.086.

Abundances of Tc and related elements in stars of type M, MS, and S.
See Abstr. 114.105.

Subgiant CH stars. II. Chemical compositions and the evolutionary connection with barium stars.
See Abstr. 114.115.

The lithium abundance in stars.
See Abstr. 114.116.

What has happened to Arcturus?
See Abstr. 114.130.

Brown dwarfs in young stellar clusters.
See Abstr. 115.001.

Tables of isochrones computed from stellar models with mass loss and overshooting.
See Abstr. 115.005.

Accurate masses and radii of normal stars.
See Abstr. 115.014.

The mass range of carbon stars.
See Abstr. 115.019.

About the stage of evolution of Wolf–Rayet stars.
See Abstr. 115.037.

Angular momentum evolution of young stars: summary of the workshop.
See Abstr. 116.072.

The dependence of coronal temperatures on Rossby numbers.
See Abstr. 116.078.

Rotational modulation and the line profile variables.
See Abstr. 116.093.

Effects of dynamic tides on the rate of apsidal motion in close binary systems.
See Abstr. 117.043.

The period gap and masses of cataclysmic variables.
See Abstr. 117.050.

The nature of the Be binary HR 2142.
See Abstr. 117.084.

Recent progress in understanding the secular evolution of cataclysmic binaries.
See Abstr. 117.091.

Cataclysmic variable evolution: clues from the underlying white dwarf.
See Abstr. 117.133.

The structure of late–type contact binaries.
See Abstr. 117.135.

Spin evolution and magnetic fields in cataclysmic variables.
See Abstr. 117.149.

Calibration of the α Centauri system: metallicity and age.
See Abstr. 118.001.

On the mass ratio of Capella.
See Abstr. 120.012.

Observational implications of protostellar collapse and accretion models.
See Abstr. 121.059.

Cepheids: theoretical models, instability strip and mass discrepancy.
See Abstr. 122.024.

Period changes and evolution in RV Tauri stars.
See Abstr. 122.026.

The RRd masses revisited.
See Abstr. 122.037.

NRP models of rapidly rotating δ Scuti stars.
See Abstr. 122.102.

The mode of pulsation in Mira variables determined by the observed acceleration data.
See Abstr. 122.105.

The white dwarf mass distribution in classical nova systems.
See Abstr. 124.001.

Type Ia supernovae: theoretical light curves for SN 1972e and SN 1981b.
See Abstr. 125.024.

Evolution of supernova remnants inside preexisting wind–driven cavities.
See Abstr. 125.035.

On the possible pulsation mass loss of a supernova precursor.
See Abstr. 125.043.

The progenitor of SN 1987A: evolutionary models with semiconvection.
See Abstr. 125.099.

A flux-limiter for neutrino transport in supernovae and protoneutron stars.
See Abstr. 125.103.

The entropy in supernova explosions.
See Abstr. 125.104.

Towards an understanding of the progenitor evolution of SN 1987A.
See Abstr. 125.207.

SN 1991T: Gamma-Ray Observatory's first supernova?
See Abstr. 125.335.

On the evolutionary link between white dwarfs and central stars of planetary nebulae: NLTE analysis of PG 1144+005.
See Abstr. 126.003.

NLTE analysis of the unique pre-white dwarf H 1504+65.
See Abstr. 126.027.

"Hidden" millisecond pulsars.
See Abstr. 126.041.

A measurement of the evolutionary timescale of the cool white dwarf G117-B15A with WET.
See Abstr. 126.047.

On the interpretation of the dP/dt measurement in G117-B15A.
See Abstr. 126.048.

The boundaries of the ZZ Ceti instability strip.
See Abstr. 126.049.

A study of period change rates in post-AGB stars. I. PG 1159-035.
See Abstr. 126.053.

The effective temperature of the DBV's, and the sensitivity of DB model atmospheres to input physics.
See Abstr. 126.057.

The rate of neutron star binary mergers in the universe: minimal predictions for gravity wave detectors.
See Abstr. 126.068.

The origin of the planet around PSR 1829-10.
See Abstr. 126.088.

A simulation of the collapse and fragmentation of cooling molecular clouds.
See Abstr. 131.010.

Fragmentation of elongated cylindrical clouds. III. Formation of binary and multiple systems.
See Abstr. 131.103.

A binary star formation mechanism through the fragmentation of prolate dense cores rotating end over end.
See Abstr. 131.203.

Newly formed massive stars.
See Abstr. 131.208.

The formation of low mass stars: theory.
See Abstr. 131.212.

Numerical studies of cloud collapse.
See Abstr. 131.213.

Fragmentation of elongated cylindrical clouds. II. Polytropic clouds.
See Abstr. 131.298.

Outflows in dark clouds: their role in protostellar evolution.
See Abstr. 131.324.

On the evolution of gravitationally unstable protostellar disks.
See Abstr. 131.355.

Core masses and abundances of low-excitation planetary nebulae in the Magellanic Clouds.
See Abstr. 134.075.

Hydrodynamics of collisions between binary stars.
See Abstr. 151.090.

Stellar encounters involving red giants in globular cluster cores.
See Abstr. 151.127.

On the physics of lithium depletion.
See Abstr. 153.015.

Radial velocities of turnoff stars in two intermediate-age open clusters.
See Abstr. 153.064.

NGC 1866 revisited: semiconvection versus overshoot.
See Abstr. 154.032.

On deriving globular-cluster luminosity functions from CCD observations, with a particular application to the main-sequence turnoff/subgiant branch in metal-poor clusters.
See Abstr. 154.046.

The HB width as a function of metals and helium.
See Abstr. 154.057.

Evolution versus pulsation on the horizontal branch.
See Abstr. 154.081.

Horizontal-branch evolution with time-dependent convective overshooting.
See Abstr. 154.082.

The role of binaries in cluster dynamical evolution.
See Abstr. 154.084.

Lithium in halo stars: constraining the effects of helium diffusion on globular cluster ages and cosmology.
See Abstr. 154.103.

Post-core collapse scenarios of globular clusters.
See Abstr. 154.104.

The Galactic globular cluster system: theoretical constraints for α-enhanced compositions.
See Abstr. 154.130.

Carbon isotopic abundances in giant stars in the CN-bimodal globular clusters NGC 6752 and M4.
See Abstr. 154.133.

Stellar evolution: Collisions and mergers.
See Abstr. 154.137.

Globular clusters as tests of stellar evolution.
See Abstr. 154.146.

Interpretation of the frequency distribution of isochrone ages of nearby F and G stars.
See Abstr. 155.010.

Galactic evolution of lithium.
See Abstr. 155.029.

On the formation of globular clusters. I. Dynamical limits on globular cluster metallicities.
See Abstr. 155.031.

Neutron star and black hole binaries in the Galaxy.
See Abstr. 155.077.

Wide binaries in the Woolley catalog.
See Abstr. 155.125.

The young Magellanic Cluster NGC 2004.
See Abstr. 156.023.

Massive star evolution in the LMC.
See Abstr. 156.026.

Detection of the helium flash gap in the integrated (U-B) versus (B-V) diagram for 624 Large Magellanic Cloud clusters.
See Abstr. 156.034.

Advanced evolutionary phases of large-metallicity low-mass stars and the problem of the ultraviolet excess of elliptical galaxies.
See Abstr. 157.301.

Neutrino-heated stars and broad-line emission from active galactic nuclei.
See Abstr. 158.154.

Perspectives: nucleosynthesis and nuclear reactions in standard and inhomogeneous big bang models and beyond.
See Abstr. 161.306.

Magnetohydrodynamic cosmologies.
See Abstr. 161.342.

066 Relativistic Astrophysics, Gravitation Theory

066.001 **General relativity and fifth force experiments.**
P. D. Mannheim.
Astrophys. Space Sci., Vol. 181, No. 1, p. 55 - 59 (Jul 1991).
The author calculates some general relativistic corrections to the Newtonian weak field limit of Einstein gravity, and discusses their possible relevance to considerations regarding the current set of fifth force experiments.

066.002 **An introduction to curved space-times.**
R. M. Williams.
European Organization for Nuclear Research, Geneva (Switzerland), CERN—91-06. 54 p. (9 Jul 1991). ISBN 92-9083-038-7. Lectures given in the Academic Training Programme of CERN 1989 - 1990. Price SFr 16.80.
These lectures focus on understanding relativity from a geometrical viewpoint, based on the use of space-time diagrams and without the tools of tensor calculus. After a brief discussion of flat space-times, curved space-times are introduced and it is shown how many of their properties may be deduced from their metric interval. The space-time around a spherically symmetric star and its possible collapse to form a black hole is described. Finally, some simple cosmological models are discussed, with emphasis on their causal properties and the existence of horizons. The titles of the lectures are: I. Flat space-times. II. Curved space-times. III. Spherical stars and stellar collapse. IV. Some simple cosmological models.

066.003 **Spectrum of gravitational waves in a double-inflation scenario.**
M. I. Zel'nikov, V. F. Mukhanov.
JETP Lett., Vol. 54, No. 4, p. 197 - 200 (25 Aug 1991). Current Physics Microform No.: 9111X1415.
The spectrum of relic gravitational waves formed in a double-inflation scenario might have obvious modulations.

066.004 **A practical approach to (nearly) elliptical gravitational lens models.**
P. Schneider, A. Weiss.
Astron. Astrophys., Vol. 247, No. 2, p. 269 - 275 (Jul 1991).
The authors consider the use of elliptical mass distributions in gravitational lens theory, since previous studies have shown that the deflection angle of a light ray traversing an elliptical lens can be obtained only by fairly costly numerical integration. A multipole expansion of an elliptical mass distribution is carried out, and the deflection potential ψ obtained as a series of trigonometric functions. The problem is thus reduced to computing (and interpolating) a finite number of one-dimensional functions. For a specific elliptical mass distribution $\varkappa$, this process is carried out explicitly. In this case, an approximate matter distribution $\hat{\varkappa}$ is constructed, which deviates from the original one as little as desired (determined by the order of truncation of the expansion), and for which the exact deflection angle can be easily calculated. The authors derive the corresponding equations and provide fairly explicit formulae, which they hope the potential user of their method will find convenient. They demonstrate the applicability of this method, and show that it is fast and accurate.

066.005 **Two aspects concerning the deflection of light.**
L. Hannibal.
Europhys. Lett., Vol. 16, No. 2, p. 129 - 132 (7 Sep 1991).
The question is posed whether the bending of sunlight in the Sun's own gravitational field has to be considered when the radius of the Sun is calculated from the observed angle. The general relativistic calculation yields a radius 3.0 km smaller than the standard value calculated by means of Euclidean geometry. The precise numerical calculation employed shows that the exact deflection of light rays observed under large angles differs substantially from the parameterized post-Newtonian (PPN) approximation used for an estimate of the PPN parameter γ.

066.006 **The source term of the field equation in the 5D STM theory of gravitation.**
Ma Guang-wen.
Astrophys. Space Sci., Vol. 181, No. 2, p. 331 – 334 (Jul 1991). Letter-to-the-editor.

There is a need of a source term $^{(5)}T_{ij}$ in the field equation of the Wesson 5D STM theory of gravity if one requires that the Wesson theory goes over to the Einstein theory of general relativity when the rest-mass m of a typical particle equals to a constant. A possible form of $^{(5)}T_{ij}$ is proposed.

066.007 **On the gravity theories and cosmology from strings.**
J. A. Casa, J. García-Bellido, M. Quirós.
Nucl. Phys. B, Part. Phys., Vol. 361, No. 3, p. 713 – 728 (9 Sep 1991).

The authors have studied the conformal properties and cosmological solutions of the four-dimensional gravitational action for different supersymmetric and non-supersymmetric string scenarios, assuming flat directions in the configuration space of moduli and dilaton fields. They have explicitly shown the equivalence between different conformal frames and chosen a preferred frame in which the relevant masses are constant. The theory of gravity for the radiation dominated era is general relativity. During the matter dominated era all analyzed examples are in conflict with the bounds on the time variation of gauge couplings and/or with the post-newtonian bounds of general relativity.

066.008 **Two-dimensional quantum cosmology.**
A. Cooper, L. Susskind, L. Thorlacius.
Nucl. Phys. B, Part. Phys., Vol. 363, No. 1, p. 132 – 162 (30 Sep 1991).

Two-dimensional quantum gravity coupled to conformally invariant matter with central charge $c > 25$ has been proposed as a toy model for quantum gravity in higher dimensions. The associated 'Wheeler-DeWitt equation' is non-linear and unstable to forming a condensate of baby universes. This will occur even in the classical $c \to \infty$ limit. Small fluctuations about this background describe the propagation of single universes and satisfy a more conventional linear Wheeler-DeWitt equation. The resulting two-dimensional cosmology depends on details of the non-linear dynamics. In particular the existence of a large-scale cosmological constant is determined by the behavior of a string theoretic tachyon potential near its minimum.

066.009 **On some problems of quantum and classical gravitation.**
M. A. Markov.
5. Seminar on Quantum Gravity, p. 3 – 17 (1991). – See Abstr. 012.004 for the main entry.

066.010 **Lattice quantization of the $O(1,2)/O(2) \times Z$ sigma model in 4 dimensions.**
B. S. DeWitt, E. Myers, R. Harrington, A. Kapulkin, J. de Lyra, S. K. Foong, T. Gallivan.
5. Seminar on Quantum Gravity, p. 18 – 26 (1991). – See Abstr. 012.004 for the main entry.

066.011 **Extended inflation and generalised scalar-tensor theories.**
A. Burd, A. Coley.
Phys. Lett. B, Vol. 267, No. 3, p. 330 – 336 (19 Sep 1991).

The authors study the class of Bergmann-Wagoner scalar-tensor theories from the viewpoint of dynamical systems theory. They relate this to the currently populator extended inflation models. They show that under a certain class of perturbation, these models are structurally stable (i.e., the global nature of the phase portrait of the dynamical system does not change); they also find a more general class of perturbations which make the system structurally unstable. The form of these perturbations is related to a possible potential term for the scalar field coupling to gravity.

066.012 **No time and quantum gravity.**
W. G. Unruh.
5. Seminar on Quantum Gravity, p. 44 – 59 (1991). – See Abstr. 012.004 for the main entry.

The role of time in the interpretation of Canonical Quantum Gravity is examined. In particular, some common claims about how time enters into the theory are examined and the key problems are listed. Three proposals are examined with respect to their treatment of time in General Relativity. In the third, a minor change in General Relativity is suggested which introduces an explicit external time, but difficulties with this theory are also pointed out.

066.013 **Quantum metric topology.**
C. J. Isham, Y. A. Kubyshin (*Yu. A. Kubyshin*), P. Renteln.
5. Seminar on Quantum Gravity, p. 60 – 81 (1991). – See Abstr. 012.004 for the main entry.

The authors construct a quantum theory on the set $M(X)$ of all metrics on a set X. For a finite X using an embedding of X into a vector space V they find a representation of $M(X)$ and quantize it with the aid of a Fock space construction. A simple model which can produce a change in metric topology is presented.

066.014 **Minisuperspace wavefunctions for cosmology and wormholes.**
D. N. Page.
5. Seminar on Quantum Gravity, p. 82 – 91 (1991). – See Abstr. 012.004 for the main entry.

Wormholes have generally been studied as real Euclidean classical solutions, but such solutions exist only for rather exotic forms of matter. One way around this difficulty, other than going to complex classical solutions, is to consider wormholes as solutions of the Wheeler-DeWitt equation. With boundary conditions requiring that these wavefunctions be regular everywhere and normalizable (square integrable), Hawking and the author have exhibited an infinite discrete set of explicit wormhole wavefunctions for gravity minimally coupled to a massless scalar field, and approximate solutions for the case of a massive scalar field.

066.015 **Wave functions constructed from an invariant sum-over-histories satisfy constraints: three derivations of the Wheeler-DeWitt equation.**
J. J. Halliwell.
5. Seminar on Quantum Gravity, p. 92 – 123 (1991). – See Abstr. 012.004 for the main entry.

The object of this paper has been to elucidate the connection between operator constraints on the wave functions in the Dirac quantization procedure and the symmetry properties of a sum-over-histories representation of such wave functions.

066.016 **Supersymmetric quantum wormhole states.**
P. D. D'Eath, H. F. Dowker, D. I. Hughes.
5. Seminar on Quantum Gravity, p. 124 – 143 (1991). – See Abstr. 012.004 for the main entry.

Quantum wormhole states have been defined by Hawking as physical states of the gravitational and matter fields in which the wave-function is exponentially damped for large compact 3-geometries, and is suitably regular as the 3-geometry collapses to zero. Here the existence of quantum wormhole states in a locally supersymmetric theory is investigated, since supersymmetry may eliminate the Planckian effective masses found in non-supersymmetric theories due to wormholes. The 4-dimensional theory of supergravity coupled to supermatter, consisting of a minimal massless complex scalar and spin-1/2 partner, leads by suitable reduction to a 1-dimensional mini-superspace theory which is invariant under local supersymmetry and coordinate transformations, and local rotations. The quantization of the 1-dimensional model is discussed.

066.017 **On the conformal factor problem in quantum gravity.**
Yu. V. Novozhilov, D. V. Vassilevich.
5. Seminar on Quantum Gravity, p. 144 – 153 (1991). – See Abstr. 012.004 for the main entry.
It is shown that kinetic energy of the quantized conformal field is positive in the case of the induced classical gravity as well as in the case of quantum gravity.

066.018 **w_∞–geometry and w_∞–gravity.**
K. S. Stelle.
5. Seminar on Quantum Gravity, p. 154 – 166 (1991). – See Abstr. 012.004 for the main entry.
The author describes the linear and non–linear realisations of w_∞ symmetry and some of the properties of its local field–theoretical realisation in two–dimensional field theories.

066.019 **Uncertainty and equivalence principles in quantum general relativity.**
H.-H. von Borzeszkowski.
5. Seminar on Quantum Gravity, p. 167 – 182 (1991). – See Abstr. 012.004 for the main entry.
Discussing gravitational blackbody radiation it is shown that the unification of general relativity and quantum mechanics, or what may be called quantum general relativity, does not generalize the conceptual framework of quantum field theory but imposes limitations on the validity of this conception. The results derived from quantum general relativity are compared to some aspects of quantum string theory.

066.020 **Ashtekar variables without hypersurfaces.**
G. Fodor, Z. Perjés.
5. Seminar on Quantum Gravity, p. 183 – 187 (1991). – See Abstr. 012.004 for the main entry.
A variant of Ashtekar's approach to the canonical treatment of the gravitational field is suggested. The essence of the authors' construction is an SU(2) spinor formalism referring to a congruence of observer's world lines.

066.021 **Quantum field theory in space with discrete geometry.**
V. E. Kur'yan.
5. Seminar on Quantum Gravity, p. 195 – 201 (1991). – See Abstr. 012.004 for the main entry.
This paper is devoted to develop consistent quantum gravity theory.

066.022 **Quantum theory of particles with Planckean masses.**
A. A. Grib, V. Yu. Dorofeev.
5. Seminar on Quantum Gravity, p. 202 – 212 (1991). – See Abstr. 012.004 for the main entry.

066.023 **Quantum string gravity near the Planck scale.**
G. Veneziano.
5. Seminar on Quantum Gravity, p. 235 – 250 (1991). – See Abstr. 012.004 for the main entry.
The structure of quantum string gravity, as it emerges from the study of gedanken experiments at Planckian energies, is outlined. According to the regime considered, Classical General Relativity phenomena dominate over (are dominated by) Quantum string–size effects. A consistent picture emerges whereby the fundamental string–length parameter shields the classical and quantum infinities of the point–particle limit.

066.024 **Physical effects in wormholes and time machine.**
V. P. Frolov, I. D. Novikov.
5. Seminar on Quantum Gravity, p. 266 – 284 (1991). – See Abstr. 012.004 for the main entry.
Physical effects in a spacetime with a traversable wormhole are considered. It is shown that the interaction of a wormhole with surrounding matter and with external gravitational field almost inevitably transforms it into a time machine.

066.025 **Vacuum polarization in a locally static multiply–connected spacetime and a time machine problem.**
V. P. Frolov.
5. Seminar on Quantum Gravity, p. 285 – 329 (1991). – See Abstr. 012.004 for the main entry.
The quantum field theory in a multiply connected locally static spacetime is considered. It is shown that the vacuum polarization effect in a spacetime with a non–potential gravitational field results in divergences of the renormalized stress–energy tensor for a quantum field. Possible applications of this result are discussed in connection with a time machine problem.

066.026 **Light like surface layers in general relativity.**
C. Barrabes, W. Israel.
5. Seminar on Quantum Gravity, p. 358 – 364 (1991). – See Abstr. 012.004 for the main entry.
Singular hypersurfaces in general relativity have been studied for a long time and the case of non lightlike surface layers is now well understood. Existing studies of lightlike surfaces either use adapted four–dimensional coordinates or are often confined to spherical symmetry. The authors' purpose in this work is to give a general coordinate–invariant prescription easily applicable to arbitrary surface layers.

066.027 **Einstein universe as a moving fluid cosmology.**
M. J. Rebouças, A. F. F. Teixeira.
J. Math. Phys., Vol. 32, No. 7, p. 1861 – 1865 (Jul 1991). Current Physics Microform No.: 9107E0205.
The combination of Synge's approach to Einstein's equations with energy conditions and the Segré classification of the energy–momentum tensor is used in the examination of Einstein's space–time. An alternative physical interpretation for the Einstein static model is discussed. An Einstein–type moving fluid universe is presented. The meaning of both static space–time and static cosmological models are discussed.

066.028 **Spherically symmetric density perturbation in relativistic cosmology.**
L. Hongya.
J. Math. Phys., Vol. 32, No. 8, p. 2279 – 2282 (Aug 1991). Current Physics Microform No.: 9108C1858.
Einstein's field equations for spherically symmetric perfect fluids are reduced to two equations plus three definitions, and from these the spherical density perturbation equation is derived. It is found that this perturbation equation can be separated into two equations by the method of separation of variables and can be solved explicitly in some special cases. An interesting result is that for matter–dominated era the perturbation equation is found to be of the Bonnor form of Newtonian theory.

066.029 **Curvature collineations in general relativity. I.**
G. S. Hall, J. da Costa.
J. Math. Phys., Vol. 32, No. 10, p. 2848 – 2853 (Oct 1991). Current Physics Microform No.: 9110E1040.
Although curvature collineations (curvature preserving transformations) have been studied within the context of general relativity for 20 years, there has been little attempt to study them systematically and there does not appear to have been a detailed mathematical investigation of their properties. This paper presents a discussion of the more general mathematical aspects of curvature collineations and suggests a program for studying them.

066.030 **Curvature collineations in general relativity. II.**
G. S. Hall, J. da Costa.
J. Math. Phys., Vol. 32, No. 10, p. 2854 – 2862 (Oct 1991). Current Physics Microform No.: 9110E1046.
This paper continues from the first one by investigating in detail many of the cases where curvature collineations can occur in space–time. It is based on a classification of the curvature tensor which is discussed in the first paper. This, together with the geometrical approach favored in this paper, leads to a rather general discussion of the problem which, it is hoped, does not

obscure those physical aspects of the situation that are important in Einstein's theory.

066.031 **Plane walls interacting with gravitational waves and matter fields.**
Wang Anzhong.
J. Math. Phys., Vol. 32, No. 10, p. 2863 – 2868 (Oct 1991). Current Physics Microform No.: 9110E1055.
Plane walls with reflection symmetry are studied. A method for generating such solutions to the Einstein equations is developed. It is found that plane wall solutions require very strong restrictions on the metric coefficients and on their derivatives. Using the generalized Bianchi identities, the interaction of plane walls with gravitational waves and matter fields is investigated. In particular, it is shown that a plane wall, in general, interacts with both gravitational plane waves and matter fields.

066.032 **O(3)–invariant tunneling at false vacuum decay in general relativity.**
V. A. Berezin, V. A. Kuzmin (*V. A. Kuz'min*), I. I. Tkachev.
5. Seminar on Quantum Gravity, p. 574 – 599 (1991). – See Abstr. 012.004 for the main entry.
The authors consider O(3)–invariant tunneling processes at false vacuum decay in general relativity. A general classification of bubble shell Euclidean trajectories is elaborated. Processes like spontaneous creation around a black hole of a double bubble with the new phase inside, tunneling of a black hole to the state of wormhole, etc, are discussed.

066.033 **Self–similar gravitational collapse with volume energy loss.**
M. V. Murzina, D. K. Nadezhin.
Sov. Astron., Vol. 35, No. 3, p. 282 – 289 (May 1991). Current Physics Microform No.: 9201X0564. English translation of Astron. Zh., Tom 68, Vyp. 3, p. 574 – 589 (1991).
A self–similar regime of gravitational collapse with volume energy loss is investigated. The first stage of collapse is cumulative: at a certain time, the density at the center becomes infinite (whereas the mass inside any sphere of finite radius is finite). In the second state of collapse, a point mass is formed at the center, which increases with time by the accretion of matter distributed in space. The limits of applicability of the self–similar solution are established: they are due to the onset of opacity of the collapsing matter to radiation that is carrying off energy. The self–similar solution is applied to the collapse of evolved stellar cores, as well as interstellar clouds of molecular hydrogen.

066.034 **Ein zartes Zittern. Die Suche nach den Gravitationswellen.**
K. Danzmann.
Bild Wiss., No. 8, p. 80 – 83 (Aug 1991).

066.035 **Smooth static solutions of the Einstein/Yang–Mills equations.**
J. A. Smoller, A. G. Wasserman, S. T. Yau, J. B. McLeod.
Commun. Math. Phys., Vol. 143, No. 1, p. 115 – 147 (Dec 1991).
The authors consider the Einstein/Yang–Mills equations in 3+1 space time dimensions with SU(2) gauge group and prove rigorously the existence of a globally defined smooth static solution. They show that the associated Einstein metric is asymptotically flat and the total mass is finite. Thus, for non-abelian gauge fields the Yang–Mills repulsive force can balance the gravitational attractive force and prevent the formation of singularities in spacetime.

066.036 **On two–dimensional supergravity and the quantum super–Möbius group.**
J. L. Gervais, B. Rostand.
Commun. Math. Phys., Vol. 143, No. 1, p. 175 – 199 (Dec 1991).
The authors describe the quantum–group structure of two–dimensional N = 1 supergravity in the conformal gauge. The operator–algebra of this (super–Liouville) theory is shown to correspond to a quantum deformation of the super–Möbius group, which provides a new solution of Yang and Baxter's equation. This structure is used to investigate the strong–coupling regime of the theory ($1 \leqslant d \leqslant 9$). For d = 5, a unitary truncation theorem – similar to the one of the bosonic case – is derived for the fundamental representation.

066.037 **Canonical quantum gravity.**
A. Ashtekar.
7. Latin American Symposium on Relativity and Gravitation, p. 2 – 32 (1991). – See Abstr. 012.005 for the main entry.
After a brief historical introduction to the subject of quantum gravity as a whole, a recently revived version of the canonical approach is outlined. The emphasis is on the conceptual framework and basic ideas involved in various constructions. The proofs and detailed arguments can be found in the original papers and monographs referred to in the text.

066.038 **Five lectures on planar gravity.**
R. Jackiw.
7. Latin American Symposium on Relativity and Gravitation, p. 74 – 97 (1991). – See Abstr. 012.005 for the main entry.
Contents: 1. Introduction. 2. Classical space–times. 3. Quantum dynamics. 4. Topological elaborations. 5. Relation to Chern–Simons gauge theory.

066.039 **Numerical relativity.**
R. A. Matzner.
7. Latin American Symposium on Relativity and Gravitation, p. 98 – 141 (1991). – See Abstr. 012.005 for the main entry.
Contents: 1. Introduction. 2. The Einstein equations. 3. Solution of the initial value problem. The York procedure. 4. Homogeneous and "simple" inhomogeneous data. 5. 1–dimensional initial data. 6. 1–d vacuum cosmology evolution. 7. 1–d results. 8. Numerical 3–d cosmology: inhomogeneous inflation. 9. Evolution of 3–dimensional data. 10. Asymptotically flat 3–dimensional gravity.

066.040 **The Yamabe constant.**
N. O'Murchadha.
7. Latin American Symposium on Relativity and Gravitation, p. 142 – 149 (1991). – See Abstr. 012.005 for the main entry.
The set of Riemannian three–metrics with positive Yamabe constant defines the space of independent data for the gravitational field. The boundary of this set is investigated, and it is shown that metrics close to the boundary satisfy the positive–energy theorem.

066.041 **On the classification of group–invariant connections.**
R. Bautista, J. Muciño, E. Nahmad-Achar, M. Rosenbaum.
7. Latin American Symposium on Relativity and Gravitation, p. 176 – 181 (1991). – See Abstr. 012.005 for the main entry.
Connection 1–forms on principal fiber bundles with arbitrary characteristic groups are considered, and a characterization of gauge–equivalent connections in terms of their associated holonomy groups is given. The authors then apply these results to group–invariant connections, which give an algebraic procedure for obtaining solutions to the gauge field equations and for classifying them into classes of equivalence.

066.042 **Non–local variables in general relativity.**
P. W. Lamberti, C. N. Kozameh.
7. Latin American Symposium on Relativity and Gravitation, p. 182 – 188 (1991). – See Abstr. 012.005 for the main entry.
A new formulation of the vacuum Einstein equations on asymptotically simple space–time is presented. This approach involves the introduction of two different non–local objects as the basic variables, rather than the usual (local) metric and connection, for general relativity. The first of these objects is the "light cone cut function", which analytically describes the intersection of the light cone from an arbitrary space–time point x^a, with null infinity. The second non–local variable is the holonomy operator, i.e., the parallel propagator around closed curves, associated with a subclass of loops that arises naturally on asymptotically simple space–times. Some kinematical properties of these variables as

well as the field equations satisfied by them are presented. Several implications of this formalism are discussed.

066.043 **Time and chaos in general relativity.**
J. Pullin.
7. Latin American Symposium on Relativity and Gravitation, p. 189 – 197 (1991). – See Abstr. 012.005 for the main entry.
Several recent computer studies of the Bianchi IX cosmology have given conflicting results, tending to suggest that the system is not chaotic. The author shows that the flaw underlying these studies is the use of an indicator of chaos which is not invariant under time reparametrizations. He shows an example of such a behaviour using the "unit disc approach", introduced by Misner and Chitre. He also suggests that great insight into the nature of the chaoticity of the model could be achieved using this latter approach.

066.044 **Lightlike contractions – singular limits of spacetimes.**
P. C. Aichelburg.
7. Latin American Symposium on Relativity and Gravitation, p. 198 – 203 (1991). – See Abstr. 012.005 for the main entry.
The technique of lightlike contractions in flat and curved space is described. The method consists in boosting a classical field configuration to the velocity of light by an appropriate generalized Lorentz transformation. The author gives examples of lightlike contractions of spacetimes belonging to the Lewis–Papapetrou class.

066.045 **Ghost fields in principal bundles.**
M. Socolovsky.
7. Latin American Symposium on Relativity and Gravitation, p. 204 – 210 (1991). – See Abstr. 012.005 for the main entry.
The author discusses the appearance of ghost and antighost fields in finite dimensional fiber bundles. As in the infinite dimensional case, the origin of the fields is the change of an integration domain from the base space to the total space of a smooth bundle. The use of coordinates however obscures the geometric meaning (if any) of these fields.

066.046 **Superalgebras of Duffin–Kemmer type.**
O. A. Sánchez–Valenzuela.
7. Latin American Symposium on Relativity and Gravitation, p. 211 – 215 (1991). – See Abstr. 012.005 for the main entry.
Duffin–Kemmer algebras are viewed as one of the homogeneous components of a superalgebra. In this setting, D–K algebras play a role analogous to that of Clifford algebras, and their infinite–dimensional counterpart, that of the Weyl algebras. Definitions are given and the basic constructions are reviewed.

066.047 **Superholomorphic structures, moment maps and super self dual Yang Mills connections over super Riemann surfaces.**
A. Mendoza.
7. Latin American Symposium on Relativity and Gravitation, p. 216 – 219 (1991). – See Abstr. 012.005 for the main entry.
The author shows, via moment maps, the existence of non trivial solutions of the global extension on super Riemann surfaces of the two dimensional reduction of the super self dual Yang Mills equations in four dimensions.

066.048 **A Clifford bundle approach to gravitational theory.**
W. A. Rodrigues Jr., Q. A. G. de Souza.
7. Latin American Symposium on Relativity and Gravitation, p. 220 – 226 (1991). – See Abstr. 012.005 for the main entry.
The authors present a Clifford bundle approach to gravitational theory, showing that Einstein's gravitational theory (GRT) admits a quite natural field interpretation, in the sense of Faraday, with the gravitational field living in a flat Minkowski space. Even more important is the fact that in the authors' theory the gravitational field, the generalized electromagnetic field (including monopoles) and the Dirac spinor field are particular sections of the Clifford bundle, being then objects of the same mathematical nature. This formalism suggests several generalizations of the physical theory discussed below.

066.049 **Canonical formalism, the initial value problem and the maximum principle.**
M. Lledo, A. Restuccia.
7. Latin American Symposium on Relativity and Gravitation, p. 227 – 232 (1991). – See Abstr. 012.005 for the main entry.
The authors present a different canonical formalism. It is an extension to singular Lagrangeans of the approach of Pontryagin. It always leads to a well posed initial value problem, solving the pathological cases of the canonical Dirac approach. Using this formalism, the authors can extend the Pontryagin Maximum Principle to singular Lagrangeans. Finally, they use the non canonical variables introduced in the formalism, to determine the correct domain of integration of the Lagrange multipliers in the functional integral.

066.050 **A class of interior solutions of the Kerr–Newman field.**
C. A. Lopez.
7. Latin American Symposium on Relativity and Gravitation, p. 234 – 240 (1991). – See Abstr. 012.005 for the main entry.
It is shown that a massive, charged, rotating shell can be maintained in equilibrium in a Kerr–Newman background. This fact allows the construction of stationary interior solutions of the Kerr–Newman field. They consist of a continuous superposition of thin shells with differential rotation. Although these solutions are made out of a charged exotic matter which does not satisfy the energy conditions, they are physically interesting as classical models of charged spinning particles in general relativity.

066.051 **Thermodynamics of spacetimes admitting a conformal symmetry parallel to the 4–velocity.**
R. A. Sussman.
7. Latin American Symposium on Relativity and Gravitation, p. 248 – 252 (1991). – See Abstr. 012.005 for the main entry.
Relativistic thermodynamics, singularities and global structure are examined for a simple class of perfect fluid spacetimes admitting a conformal symmetry parallel to the 4–velocity. It is shown that the compatibility between thermodynamical laws and field equations requires the physical material modelled by these fluids to have a vanishing coefficient of thermal conductivity. Relativistic Kinetic Theory models of collisionless gases is suggested as a possible application of these solutions.

066.052 **On the existence of affine collineations.**
S. A. Hojman, D. Nuñez.
7. Latin American Symposium on Relativity and Gravitation, p. 253 – 258 (1991). – See Abstr. 012.005 for the main entry.
Sufficient conditions on the Christoffel symbols and metric tensors to insure the existence of affine collineations in Riemann spaces are obtained. Several examples of physically relevant Riemannian spaces admitting affine collineations are presented and the constants of motion are constructed.

066.053 **Collineations in Gödel–type space–times.**
A. Melfo, L. Núñez, U. Percoco, V. M. Villalba.
7. Latin American Symposium on Relativity and Gravitation, p. 259 – 263 (1991). – See Abstr. 012.005 for the main entry.
Symmetries of Gödel–type space–times are studied. A curvature collineation is found for the general case. It is shown that homogeneous Gödel–type space–times do not posses proper Ricci collineations, with the only exception of the pure Gödel metric. Contracted Ricci collineations of physical interest are also presented.

066.054 **New solutions of the semiclassical Einstein equations.**
L. P. Chimento, N. A. Zuccala.
7. Latin American Symposium on Relativity and Gravitation, p. 271 – 277 (1991). – See Abstr. 012.005 for the main entry.
The authors present the back–reaction problem for conformally invariant free quantum fields in spatially flat Robertson–Walker spacetimes with nonzero cosmological constant containing classical sources. Exact zero–parameter families of solutions are found for a universe containing two perfect fluids. These solutions satisfy the classical Einstein equation for an effective perfect fluid. The time–symmetric bouncing solutions have an

initial and final de Sitter stage. Some of the singular solutions do not have particle horizons.

066.055 **Time shifts and non–linear interference in cylindrically symmetric Einstein–Maxwell solitons.**
A. D. Dagotto, R. J. Gleiser, C. O. Nicasio.
7. Latin American Symposium on Relativity and Gravitation, p. 280 – 286 (1991). – See Abstr. 012.005 for the main entry.
The authors present new results on the recently described "time shift" phenomena in the nonlinear (self) interaction of cylindrical solitonic gravitational waves, as they are reflected off the symmetry axis. Central in the analysis is the introduction of an appropriate descriptive framework in General Relativity, within which the authors can give an invariant meaning to concepts such as "localized perturbations" and their corresponding "time shifts".

066.056 **Cylindrically symmetric gravitational waves with rotational properties.**
H. Quevedo, B. Mashhoon.
7. Latin American Symposium on Relativity and Gravitation, p. 287 – 293 (1991). – See Abstr. 012.005 for the main entry.
The authors derive and investigate cylindrically symmetric solutions of Einstein's vacuum field equations which describe the propagation of rotating gravitational waves in empty space. It is shown that test particles that are initially at rest are "dragged" by the gravitational field of rotating waves. Rotating gravitational waves are characterized by curvature singularities which could be of significance for the discussion of the speed–of–light catastrophe.

066.057 **Galilean limit of electrodynamics.**
O. A. Reula, V. H. Hamity, S. Frittelli.
7. Latin American Symposium on Relativity and Gravitation, p. 301 – 307 (1991). – See Abstr. 012.005 for the main entry.
The final interest of the authors' work is to study the Newtonian limit as an approximation to General Relativity. In this paper they show, using the Galilean limit of electrodynamics with external sources as a test model, some of the problems that they will be confronted with, and the techniques that are introduced to attack them. The crucial physical issue, to define an asymptotic expansion of a class of solutions, is the selection of initial data which results of imposing regularity conditions in the nonrelativistic limit. The authors' model is an example of a more general class of systems which includes, hopefully, the gravitational field plus matter.

066.058 **The Newtonian limit of electromagnetism.**
V. H. Hamity, O. Reula.
7. Latin American Symposium on Relativity and Gravitation, p. 308 – 311 (1991). – See Abstr. 012.005 for the main entry.
The purpose of this note is to show some rigorous results about the Newtonian (or rather Galilean) limit of electromagnetism, namely the limit in which the speed of light is taken to be infinite. These results are directly available from the theory of hyperbolic systems with different time scales and are probably well known, but the authors treat them in detail here because their ultimate goal is to extend them to the general theory of relativity.

066.059 **Conformally symmetric spheres with surface pressure.**
M. Esculpi.
7. Latin American Symposium on Relativity and Gravitation, p. 314 – 319 (1991). – See Abstr. 012.005 for the main entry.
A known analytical solution for a metric which besides the spherical symmetry also accepts a one parameter group of conformal motions is generalized to the case when there exists a non–zero surface pressure. The behaviour of the isotropic solution under an adiabatic and slow contraction is studied in order to obtain some information about the stability of the model. A relation between the radial pressure and the energy density during the contraction is determined. As the sphere contracts the radial pressure decreases and the energy density increases. The sphere can not stand an adiabatic contraction.

066.060 **Numerical solution of the Boltzmann equation in general relativity: formalism.**
H. Harleston L.
7. Latin American Symposium on Relativity and Gravitation, p. 348 – 354 (1991). – See Abstr. 012.005 for the main entry.
Through the use of the Liouville operator in phase space, a form of the Boltzmann equation in spherical symmetry that makes it very amenable for numerical treatment is obtained. When coupled to the Einstein equations, the Boltzmann equation can then be used for the numerical study of either massless or massive particles interacting with ordinary matter.

066.061 **Consequences of one of I.S. Shklovsky's hypotheses for gravitating systems.**
T. B. Omarov.
Sov. Astron., Vol. 35, No. 3, p. 322 – 324 (May 1991). Current Physics Microform No.: 9201X0604. English translation of Astron. Zh., Tom 68, Vyp. 3, p. 656 – 660 (1991).
A gravitating system that is continuously dissipating due to asymmetrical explosions of bodies is considered. It is shown that the negative total energy of such a system increases in proportion to the number of its members. This dynamical effect is of interest in the light of Shklovsky's idea of the asymmetry of explosions of cosmic objects.

066.062 **2 + 1 gravity as an exactly solved system.**
H. Waelbroeck.
7. Latin American Symposium on Relativity and Gravitation, p. 418 – 426 (1991). – See Abstr. 012.005 for the main entry.
The author describes, in geometrical terms, the complete solution of Einstein's equations in three spacetime dimensions, with N pointlike sources, g wormholes and no cosmological constant. The spacetimes are represented as a cylinder carved out of Minkowski space by a one–parameter family of polygons, with Poincaré identifications of the edges. This is an explicit representation of Witten's solution of 2 + 1 gravity, the "moduli space of flat ISO (2,1) connections". The polygons (12g–12 parameters) are a generalization of Poincaré's fundamental polygons (6g–6 parameters). The doubling of the number of parameters allows for non–stationary solutions, i.e. universes with time–dependent Teichmüller parameters. The author examines the behavior and properties of these universes.

066.063 **A BRST approach to 2 + 1 gravity.**
G. Gonzáles, J. Pullin.
7. Latin American Symposium on Relativity and Gravitation, p. 427 – 431 (1991). – See Abstr. 012.005 for the main entry.
The authors analyze the BRST–Batalin–Fradkin–Vilkovisky structure of 2 + 1 vacuum gravity. They construct the BRST charge in the Hamiltonian approach and write a BRST–invariant action. They write the partition function and show the gauge in which it reduces to the covariant expression used by Witten.

066.064 **The algebra of traces in some Chern–Simons theories.**
F. Zertuche, L. F. Urrutia.
7. Latin American Symposium on Relativity and Gravitation, p. 432 – 436 (1991). – See Abstr. 012.005 for the main entry.
The authors report on two results in the algebra of observables of Chern–Simons theories with a group G in 2 + 1 dimensions: (1) they construct the algebra of observables for the Poincare case as a limiting process of the algebra corresponding to the De Sitter case and (2) they construct the algebra of observables for the conformal case.

066.065 **Einstein–Chern–Simons massive systems and self–dual spin–3.**
C. Aragone, P. J. Arias, A. Khoudeir.
7. Latin American Symposium on Relativity and Gravitation, p. 437 – 444 (1991). – See Abstr. 012.005 for the main entry.
The authors show that the massive spin–two system formed by the addition of second order linearized Einstein action and first order self dual gravity contains spin–1 ghosts, although it has the same structure as the similar Maxwell–Chern–Simons massive action arising from spontaneous symmetry breaking in abelian

vector systems. The addition of an auxiliary self–dual ghost-killing field restores the pure massive spin–two content of the system which has a non degenerate mass spectrum. The authors also present the massive pure spin–3 self–dual action which needs the presence of both an auxiliary self–dual vector field and a scalar–field.

066.066 **The geometrical functional integral for mini–superspace quantum gravity and the relativistic particle compared.**
J. Guven, J. D. Vergara.
7. Latin American Symposium on Relativity and Gravitation, p. 445 – 453 (1991). – See Abstr. 012.005 for the main entry.
The authors contrast the geometric functional integral quantization of a minisuperspace model for quantum gravity with that for a relativistic free particle. While formally the two models are very similar, there are differences assumed to have a genuine physical significance.

066.067 **One–dimensional subspaces of exact solutions of the n–dimensional Einstein's equations.**
T. Matos, G. Rodriguez.
7. Latin American Symposium on Relativity and Gravitation, p. 456 – 462 (1991). – See Abstr. 012.005 for the main entry.
The authors describe one method for finding exact solutions of the Einstein's equations in an n–dimensional Riemannian space, when the n–dimensional metric depends on two variables z and $\bar{z}$.

066.068 **Non perfect fluids in Kaluza–Klein cosmological models.**
A. Camacho, J. S. Garciá, O. Obregón.
7. Latin American Symposium on Relativity and Gravitation, p. 463 – 467 (1991). – See Abstr. 012.005 for the main entry.
The authors consider a flat Friedmann–Robertson–Walker model with D extra spacelike dimensions, the diagonal energy–momentum tensor is the one associated to a non perfect fluid. This means that in the $(3+D+1)$ space–time shear viscosity is present. Nevertheless, the energy–momentum tensor in the 4–dimensional world has the form of a perfect fluid, although the equation of state is not necessarily the one of a photon gas.

066.069 **The canonical structure of extrinsic gravity.**
V. Tapia.
7. Latin American Symposium on Relativity and Gravitation, p. 468 – 474 (1991). – See Abstr. 012.005 for the main entry.
In previous works the author has developed a model for the description of the gravitational interaction based on the use of extrinsic gravity. In this model the space–time is embedded into a higher–dimensional space and the embedding functions are considered as the dynamical variables. The model shows a number of advantages over the conventional theory of gravity. In this paper the author considers the canonical structure of extrinsic gravity. The calculation relies on the use of second-order field theory. The shift and lapse constraints are easily calculated and they satisfy the Dirac universal algebra for parametrisation invariant theories.

066.070 **Gravitational microlensing and the superluminal illusion.**
L. N. Mertz.
Bull. Am. Astron. Soc., Vol. 23, No. 2, p. 932 (1991). Abstract. – See Abstr. 012.008 for the main entry.

066.071 **Vanishing null energy solutions and Hamiltonian approach for membranes in the intrinsic light front gauge.**
C. Aragone, A. Restuccia, R. Torrealba.
7. Latin American Symposium on Relativity and Gravitation, p. 497 – 503 (1991). – See Abstr. 012.005 for the main entry.
The authors study the dynamics of the membrane using intrinsic light front coordinates for describing the 3–D world-volume. By means of a double Legendre transformation they obtain the null canonical formulation. This gauge can be completed with an extrinsic (in the ambient space time) light cone fixing. The authors find string like solutions, which are related to the recently pointed out instability of the classical relativistic membrane.

066.072 **Exact solutions from one dimensional subspaces in Kaluza–Klein theory.**
R. Becerril B., T. Matos Ch.
7. Latin American Symposium on Relativity and Gravitation, p. 504 – 509 (1991). – See Abstr. 012.005 for the main entry.
From the one dimensional subspace of the potential space two new stationary axisymmetric solutions to the vacuum five-dimensional Kaluza–Klein field equations are constructed. Both solutions have scalar potential and magnetic field, one describing a magnetic monopole and the other describing a magnetic dipole.

066.073 **Dynamic equations of test spinning bodies in the general theory of relativity.**
A. N. Aleksandrov.
Kinematics Phys. Celest. Bodies, Vol. 7, No. 1, p. 11 – 18 (1991). English translation of Kinematika Fiz. Nebesn. Tel, Tom 7, No. 1, p. 13 – 20 (1991).
Alternative descriptions of the rotational–translational motion of a test body in general relativity theory derived from the Mathisson–Papapetrou equations with Pirani's or Tulczyjew–Dixon's conditions are studied. Nonphysical solutions are excluded and the problem is formulated with initial data for the coordinates, velocity, and spin vector. Approximations in powers of the spin are considered and it is shown that the alternative equations agree in the first two approximations.

066.074 **A mass–radius relation for astrophysical objects and Weyl gravity.**
D. Kazanas, P. D. Mannheim.
Bull. Am. Astron. Soc., Vol. 23, No. 2, p. 974 (1991). Abstract. – See Abstr. 012.008 for the main entry.

066.075 **The evolution of discontinuities in radiating spheres in the diffusion approximation.**
W. Barreto, L. Herrera, L. Núñez.
Astrophys. J., Vol. 375, No. 2, p. 663 – 673 (10 Jul 1991).
The evolution of discontinuities in a general relativistic sphere free of singularities is studied. The energy transport mechanism through the fluid is diffusive. The distribution of matter is divided by a shock wave front into two regions. The equations of state at both sides of the shock are different, and the solutions are matched on it via the Rankine–Hugoniot conditions. The outer metric joins the Vaidya solution at the boundary surface of the sphere. Exploding models are obtained, and their dynamics are studied using a generalized compressibility coefficient for non-adiabatic systems.

066.076 **Nonlinear nature of gravitation and gravitational wave experiments.**
D. Christodoulou.
Phys. Rev. Lett., Vol. 67, No. 12, p. 1486 – 1489 (16 Sep 1991). Current Physics Microform No.: 9111A1850.
It is shown that gravitational waves from astronomical sources have a nonlinear effect on laser interferometer detectors on Earth, an effect which has hitherto been neglected, but which is of the same order of magnitude as the linear effects. The signature of the nonlinear effect is a permanent displacement of test mases after the passage of a wave train.

066.077 **Quasilocal mass constructions with positive energy.**
A. J. Dougan, L. J. Mason.
Phys. Rev. Lett., Vol. 67, No. 16, p. 2119 – 2122 (14 Oct 1991). Current Physics Microform No.: 9111E1673.
The authors propose a pair of new quasilocal constructions for the four–momentum and mass of the gravitational and matter fields within generic two–surfaces. It is shown that the momenta are future pointing when the dominant energy condition is satisfied on a spanning three–surface and the two–surface is suitably convex. The new definition gives zero in flat space and the correct results in linearized theory, at spacelike infinity and for round spheres. At null infinity at least one definition gives the Bondi mass. These definitions can be embedded in definitions of angular momentum.

066.078 **New test of nonsymmetric theories of gravity: observational limits on gravity-induced depolarization of solar spectral lines.**
M. D. Gabriel, M. P. Haugan, R. B. Mann, J. H. Palmer.
Phys. Rev. Lett., Vol. 67, No. 16, p. 2123 – 2126 (14 Oct 1991). Current Physics Microform No.: 9111E1677.
The authors show that theories which couple the antisymmetric part of a nonsymmetric tensor gravitational field to the electromagnetic field predict a depolarization of the Zeeman components of spectral lines emitted by extended, magnetically active regions near the limb of the Sun. This new effect is a consequence of the violation of the Einstein equivalence principle by these nonsymmetric alternatives to general relativity. The authors show that existing solar physics data limit the extent of such depolarization and imply that the Sun's antisymmetric charge $l_{\odot}^2$ must be less than $(535\ \mathrm{km})^2$ in Moffat's theory, the prototypical nonsymmetric theory of gravity.

066.079 **Erratum: "Conformally symmetric spheres in the thin wall approximation" [Astrophys. J., Vol. 353, No. 2, p. 579 – 584 (20 Apr 1990)].**
L. Herrera, J. Ibáñez, A. Di Prisco.
Astrophys. J., Vol. 376, No. 1, p. 367 (20 Jul 1991). See Abstr. 51.066.158.

066.080 **Gravitational field of rotating bodies.**
S. L. Parnovskij.
Zh. Ehksp. Teor. Fiz., Vol. 100, No. 5, p. 1423 – 1437 (Nov 1991). In Russian.
Two vacuum solutions for the Einstein equations, which can be regarded as challengers to the role of an external solution for the axial-symmetric steady rotating bodies prolate and oblate correspondingly, are obtained and analyzed. Both contain two infinite sets of arbitrary constants corresponding to the mass and angular momentum multipoles. In the absence of rotation and definite choice of the mass multipole moments the first solution goes over into the Zipoy–Voorhees metric and the second goes over into the known solution with the ring singularity. The connection between the first solution and van Stockum metric, which clarifies the interpretation of the latter, is analyzed.

066.081 **Gravitational effect of the celestial object with electrical charge or magnetic charge.**
Fan Junhui, Xie Guangzhong, Wang Yongjiu, Tang Zhiming.
Publ. Yunnan Obs., No. 2, p. 35 – 40 (Jun 1991). In Chinese.
The effect of the gravitational field of the celestial object with electrical charge or magnetic charge on gravitational redshift and the timelag of radar return wave is discussed.

066.082 **The effects of thermal radiation on some general relativistic stellar models.**
J. M. Aguirregabiria, J. Ibáñez, A. Di Prisco, L. Herrera.
Astrophys. J., Vol. 376, No. 2, p. 662 – 672 (1 Aug 1991).
The authors study the effects of heat flow on the evolution of two different families of self-gravitating, radiating spheres. The sensitivity of the system of differential equations, governing the evolution of the boundary surface, to the emission of radiation is brought out.

066.083 **Gravity from superstrings.**
M. Quiros.
Nucl. Phys., B Proc. Suppl., Vol. 22A, p. 104 – 114 (Jul 1991). – See Abstr. 012.022 for the main entry.
The author discusses the four-dimensional theory of gravity which arises from compactifications of the heterotic string. Its main feature is that the dilaton field is coupled to all non-gravitational matter in the universe. Its cosmological solutions correspont to general relativity of departures therefrom given by a JBD-like theory labelled by an $\tilde{\omega}$-parameter. Unlike in ordinary JBD gravity (where the scalar field is decoupled from the non- gravitational sector), bounds on the time variation of the gauge coupling constants provide very strong constraints on $\tilde{\omega}$.

066.084 **Mutually interacting quantum fields in curved space-times: the outcome of physical processes.**
J. Audretsch.
NATO Advanced Research Workshop on Quantum Mechanics in Curved Space-Time, p. 233 – 264 (1990). – See Abstr. 012.023 for the main entry.
During the last years many efforts have been made to study quantum field theory in given unquantized space-times. This external field approach for the influence of classical gravitational fields is generally regarded as some sort of semi-classical approximation to a full quantum theory of gravity. It is assumed that characteristic physical traits which show up in this approximation will be least heuristically important for the construction of the full theory. Furthermore the approximation has its own domain of application during the very early stages of the universe and outside black holes.

066.085 **Quantum strings in curved space times.**
N. Sanchez.
NATO Advanced Research Workshop on Quantum Mechanics in Curved Space-Time, p. 265 – 315 (1990). – See Abstr. 012.023 for the main entry.
The author discusses string theory in the context of quantum gravity, with specific treatment of string quantization in curved space times. The contents of these lectures is the following. In Lecture I he discusses thermal and ground state effects of quantum strings. In Lecture II he gives a new approach to string quantization in curved space time in order to take into account strong curvature effects of the geometry. In Lecture III he treats the case of boosted black-holes in the ultra relativistic limit.

066.086 **The probabilistic time and the semiclassical approximation of quantum gravity.**
M. A. Castagnino.
NATO Advanced Research Workshop on Quantum Mechanics in Curved Space-Time, p. 317 – 360 (1990). – See Abstr. 012.023 for the main entry.
The notion of "probabilistic time" is studied as a way to obtain quantum field theory in curved space-time from quantum gravity. It is shown how ideal and real clocks measure the probabilistic time and how the universe can be used as a real clock. The problem of the arrow of time is studied because it is related with the definition of probabilistic time.

066.087 **Anisotropic spheres in general relativity.**
H. Rago.
Astrophys. Space Sci., Vol. 183, No. 2, p. 333 – 338 (Sep 1991). Letter-to-the-editor.
A prescription originally conceived for perfect fluids is extended to the case of anisotropic pressures. The method is used to obtain exact analytical solutions of the Einstein equations for spherically-symmetric self-gravitating distribution of anisotropic matter. The solutions are matched to the Schwarzschild exterior metric.

066.088 **Astrophysical constraints on the gravitational constant.**
Wang Jin.
Astrophys. Space Sci., Vol. 184, No. 1, p. 31 – 36 (Oct 1991).
The author studies the astrophysical bounds on the change of the gravitational constant with time. He has found that $|\dot{G}/G| < 10^{-12}\mathrm{yr}^{-1}$ is the condition that has to be satisfied in order not to cause a conflict with the observations. This condition is in accord with the lower limits, the superstring theory predicts.

066.089 **Schwarzschild relativistic orbits.**
W. Robin.
Eur. J. Phys., Vol. 12, No. 5, p. 204 – 206 (Sep 1991).
The problem of predicting the orbit and precession of the perihelion of a planet, moving under the effects of a potential derived from the Schwarzschild metric in general relativity, is reviewed in the light of a recent paper by Weinstock. Following the work by McLoughlin and Smith the orbit and the precession are obtained directly from the conservation laws of energy and

angular momentum using elementary algebra and trigonometry and simple integration coupled with basic physical (observational) arguments.

066.090 **Measurements of a Machian transient mass fluctuation.**
J. F. Woodward.
Found. Phys. Lett., Vol. 4, No. 5, p. 407 – 423 (Oct 1991).
The prediction of a novel transient mass fluctuation expected in relativistic theories of gravitation that satisfy the "globally–local" Mach condition (that the cosmic gravitational potential is roughly equal to the square of the speed of light) is set forth. Cosmological consequences of the Mach condition are mentioned, and an experiment that tests the prediction is discussed. The apparatus, procedure, and results of this experiment are described. The predicted effect is seen with better than order of magnitude accuracy.

066.091 **A new description for a realistic inhomogeneous universe in general relativity.**
T. Futamase.
Prog. Theor. Phys., Vol. 86, No. 2, p. 389 – 399 (Aug 1991).
A new approximation scheme is proposed to describe a realistic inhomogeneous universe in general relativity. This scheme improves the former one and allows one to treat more general situations where the local effect of self–gravity of the matter dominates the effect of cosmic expansion. A new statistical way of averaging spacetime is also discussed.

066.092 **Physical models of white and grey holes and their astrophysical applications.**
A. P. Trofimenko.
Fizika, Vol. 23, No. 2, p. 175 – 191 (Jul 1991). In Russian.
On the basis of extended manifolds of general relativity astrophysical applications of white and grey holes are considered. Four types of anticollapsars are distinguished: canonical white and grey holes, light– and dark–grey holes. Bursts of these types of anticollapsars are similar, but relicts are essentially different. White and grey holes can be revealed in the form of bursts of gravitational and electromagnetic radiation. Quasars and cosmic voids can be associated with relicts of white and light–grey holes, and black holes do with relicts of grey and dark–grey holes. Possible manifestations of small mass black holes in cosmic bodies (millisecond pulsars, planets, etc.) are considered. The neutrino flow from microblack hole is estimated.

066.093 **Exact solutions for Einstein's equations for a spherical metric using Newman–Penrose formalism.**
C. Darierescu, A. Darierescu, M. Agop.
Stud. Cercet. Fiz., Vol. 42, No. 4, p. 329 – 333 (1990).
Using the Newman–Penrose formalism for a metric with spherical symmetry, the authors obtain the Schwarzschild and Reissner–Nordstrom solutions as exact solutions of Einstein's field equations.

066.094 **Metrica Plebanski ca solutie exacta a ecuatiilor Einstein–Maxwell.**
C. Dariescu, M. Agop, M. Dariescu.
Stud. Cercet. Fiz., Vol. 42, No. 6, p. 523 – 528 (1990).
Plebanski's metric in the Newman–Penrose formalism is obtained as an exact solution of Einstein–Maxwell's equations, using a nul tetradic frame starting from an orthonormal Cartan frame.

066.095 **Metrici statice cu simetrie cilindrica, solutie exacta a ecuatiilor de cimp Einstein–Maxwell.**
C. Dariescu, M. Dariescu, M. Agop.
Stud. Cercet. Fiz., Vol. 42, No. 6, p. 529 – 535 (1990).
Static metrics, with cylindrical symmetry, as an exact solution of Einstein–Maxwell equations using the Newman–Penrose formalism are obtained.

066.096 **Influence of the quadrupole field of bodies and their rotation on light propagation.**
S. A. Klioner.
Astron. Zh., Tom 68, Vyp. 5, p. 1046 – 1062 (Sep–Oct 1991). In Russian. English translation in Sov. Astron., Vol. 35, No. 5.
The differential equations of relativistic corrections in light propagation in the barycentric reference system due to the quadrupole field of bodies and their rotation have been obtained and solved. The influence of the oblateness and rotation of the body on the light deflection angle and on the time for electromagnetic radiation to propagate from one fixed point to another is investigated.

066.097 **Stability analysis of self–gravitating skyrmions.**
M. Heusler, S. Droz, N. Straumann.
Phys. Lett. B, Vol. 271, No. 1/2, p. 61 – 67 (14 Nov 1991).
The authors investigate the stability of their recently constructed self–gravitating skyrmions in the framework of small time–dependent perturbations. It is shown that the frequency spectrum for radial perturbations of the coupled Einstein–Skyrme equations can be reduced to the energy spectrum of a p–wave Schrödinger equation with a bounded effective potential. Bound states of this Schrödinger equation correspond to exponentially growing modes. It turns out that there are no such modes, as long as the relevant dimensionless coupling constant ($\varkappa$) is less than a critical value, $\varkappa_c$. For larger values of $\varkappa$ there is exactly one unstable mode. Therefore, the Skyrme 'stars' become unstable for $\varkappa > \varkappa_c$. The linear stability for $\varkappa < \varkappa_c$ is only a necessary condition for (non–linear) stability. Similar results are expected to hold for the new black hole solutions of the Einstein–Skyrme system.

066.098 **Three–dimensional gravity and string ghosts.**
S. Carlip, I. I. Kogan.
Phys. Rev. Lett., Vol. 67, No. 26, p. 3647 – 3649 (23 Dec 1991). Current Physics Microform No.: 9201E1859.
It is known that much of the structure of string theory can be derived from three–dimensional topological field theory and gravity. The authors show here that, at least for simple topologies, the string diffeomorphism ghosts can also be explained in terms of three–dimensional physics.

066.099 **Representations of quantum general relativity using Ashtekar's variables.**
J. Zegwaard.
Classical Quantum Gravity, Vol. 8, No. 7, p. 1327 – 1337 (Jul 1991).
On the basis of the link–class solutions for quantum general relativity found by Rovelli and Smolin in the loop representation, the author describes a possible procedure for assigning a Hilbert space structure of this set of solutions. On this Hilbert space, operators can be defined by going back to the connection representation; this is possible consistently since the "gauge choice" one makes by picking out specific loops in the link–classes only corresponds to general coordinate transformation on the observables connected to the operators. In this way the loop transform of Rovelli and Smolin is restricted to the physical states of the theory.

066.100 **On the nature of naked singularities in Vaidya spacetimes. II.**
I. H. Dwivedi, P. S. Joshi.
Classical Quantum Gravity, Vol. 8, No. 7, p. 1339 – 1348 (Jul 1991).
Continuing from earlier work, the structure of the naked singularity in Vaidya–Papapetrou models for the gravitational collapse of null dust is further analysed. The families of non–spacelike geodesics terminating at the singularity are classified and it is shown that the singularity exhibits a directional behaviour in terms of curvature growth along the singular geodesics.

066.101 **The numerical determination of apparent horizons.**
A. J. Kemball, N. T. Bishop.
Classical Quantum Gravity, Vol. 8, No. 7, p. 1361 – 1367 (Jul 1991).

The problem of determining apparent horizons in (3+1) numerical relativity is considered. The method proposed by Nakamura and co–workers is applied to initial data for several time–symmetric Schwarzschild black holes to determine the quantitative numerical properties of the algorithm. Several refinements are suggested to the method which lead to improved stability and convergence. Some circumstances in which it is not appropriate to use the method are found.

066.102 **Observer fields in Weylian spacetime models.**
V. Perlick.
Classical Quantum Gravity, Vol. 8, No. 7, p. 1369 – 1385 (Jul 1991).

Resting upon the Ehlers – Pirani – Schild axiomatic approach to spacetime theory, the author considers Weyl manifolds as reasonable spacetime models. He defines and discusses several concepts associated with a timelike vector field on such a spacetime mode. In particular, the author derives propagation equations for rotation, expansion and shear of V, proves some kinematical vorticity theorems and presents a new method how to measure length curvature operationally.

066.103 **Three–particle solutions in (2+1)–dimensional gravity.**
D. Lancaster, N. Sasakura.
Classical Quantum Gravity, Vol. 8, No. 8, p. 1481 – 1489 (Aug 1991).

The formalism for N–particle classical solutions in (2+1)–dimensional gravity is presented. Specific calculations are performed for the two– and three–particle cases, and scattering is discussed.

066.104 **On the uniqueness of boundaries at infinity of asymptotically flat spacetimes.**
B. Schmidt.
Classical Quantum Gravity, Vol. 8, No. 8, p. 1491 – 1504 (Aug 1991).

In the field of asymptotics various boundaries are attached to spacetime. The uniqueness of the point set and its differential structure added at infinity is shown for null infinity as well as for spacelike infinity of stationary solutions. A new definition of spacelike infinity for non–stationary spacetimes is given which involves the conformal and projective structures of spacetime.

066.105 **The semidegenerate configurations of a selfgravitating system of fermions.**
Gao Jiangong, M. Merafina, R. Ruffini.
Acta Astrophys. Sin., Vol. 11, No. 4, p. 297 – 305 (Oct 1991). In Chinese.

The semidegenerate equilibrium configurations of a selfgravitating system of fermions have usually been treated by an expansion of the thermodynamical quantitites around the fully degenerate values. It is shown how such an expansion is critical in the low density regimes where in general a distribution function far from the degenerate state applies. This point is explicitly illustrated by comparing and contrasting the results based on the first and second order expansion from the fully degenerate values with the exact ones. These exact values have been obtained by a direct computation of the Fermi integrals.

066.106 **Putting Einstein to the test – in space.**
A. Gibbons.
Science, Vol. 254, No. 5034, p. 939 – 941 (15 Nov 1991).

Gravity Probe B – an audacious experiment conceived 30 years ago to test general relativity – nears the launching pad.

066.107 **Principal null directions of the Curzon metric.**
R. Arianrhod, S. Fletcher, C. B. G. McIntosh.
Classical Quantum Gravity, Vol. 8, No. 8, p. 1519 – 1528 (Aug 1991).

The Curzon metric has an invariantly defined surface, given by the vanishing of the cubic invariant of the Weyl tensor. On a spacelike slice, this surface has the topology of a 2–sphere, and surrounds the singularity of the metric. In Weyl coordinates, this surface is defined by $R = m$ where $R = (\varrho^2+z^2)^{1/2}$. The two points where the z–axis, the axis of symmetry, cuts this surface, $z = m$ and $z = -m$, have the interesting property that, at both of them, the Riemann tensor vanishes. The Weyl tensor is of Petrov type D at all non–singular points of the z–axis, $\varrho = 0$, except at the two points where it is zero. Off the axis, the Weyl tensor is of type $I(M^+)$, and the metric asymptotically tends to flat spacetime away from the source. The principal null directions (PND) of the Weyl tensor are shown to be everywhere independent of the angular basis vector $\partial/\partial\Phi$, and their projections into a t = constant, Φ = constant plane are presented graphically.

066.108 **Lovelock–Cartan theory of gravity.**
A. Mardones, J. Zanelli.
Classical Quantum Gravity, Vol. 8, No. 8, p. 1545 – 1558 (Aug 1991).

The most general theory of gravity with torsion in D dimensions is discussed. The Lagrangian is required to be: (1) a D–form scalar under local Lorentz transformations; (2) a local polynomial of the vielbein, the spin connection, and their exterior derivatives; (3) constructed without the hodge dual (*–operation). Besides the purely metric Lovelock action, there is a series of torsion terms related to the Pontryagin classes in a way analogous to the relation between the Lovelock action and the Euler classes. Also, a family of global invariants of the differentiable structure of the manifold, constructed with the torsion, is identified.

066.109 **Charged null fluid and the weak energy condition.**
A. Ori.
Classical Quantum Gravity, Vol. 8, No. 8, p. 1559 – 1575 (Aug 1991).

The author uses the Einstein – Maxwell field equations to derive the fundamental equation of motion for charged null fluid. This equation of motion includes a Lorentz force term. Using this equation, it is shown that the charged null fluid always satisfies the weak energy condition. This result is in contrast to previous interpretations of the charged Vaidya solution (a special case of a charged null fluid), which produced violations of the weak energy condition. The errors in the previous interpretations are explained.

066.110 **Ashtekar formulation of general relativity and loop–space non – perturbative quantum gravity: a report.**
C. Rovelli.
Classical Quantum Gravity, Vol. 8, No. 9, p. 1613 – 1675 (Sep 1991).

The formulation of general relativity discovered by Ashtekar and the recent results obtained in non – perturbative quantum gravity using loop–space techniques are reviewed. The new formulation is based on the choice of a set of Lagrangian (and Hamiltonian) variables, instead of the spacetime metric. In terms of these new variables, the dynamical equations are remarkably simplified and a structural identity between general relativity and the Yang – Mills theories is revealed. In quantum gravity, the new formalism has overcome long–standing difficulties and led to unexpected results. A non – perturbative approach to quantum theory has been constructed in terms of the Wilson loops of the Ashtekar connection. This approach, denoted as loop–space representation, has led to the complete solution of the quantum diffeomorphism constraint in terms of knot states, to the discovery of an infinite – dimensional class of solutions to the quantum gravitational dynamics, and to certain surprising indications on the existence of a discrete structure of spacetime around the Planck length.

066.111 **Aspects of anomaloy structures in W gravity.**
K. Li, C. Pope.
Classical Quantum Gravity, Vol. 8, No. 9, p. 1677 – 1684 (Sep 1991).
The authors give a brief review of some of the recent developments in W gravity coupled to matter theories. They then extend the discussion to the case of minimal W conformal field theory, realized in terms of free bosons with background charges, and use this action to investigate some aspects of the anomaly structure and the covariantization of W gravity.

066.112 **Coleman – Weinberg effect in quantum gravity.**
R. Floreanini, E. Spallucci, R. Percacci.
Classical Quantum Gravity, Vol. 8, No. 9, p. L193 – L197 (Sep 1991). Letter–to–the–editor.
The authors describe an application of the method of the effective potential to a model theory of gravity with independent vierbein and O(1, 3) connection. Adopting a kind of mean field approximation, they find that a Coleman – Weinberg effect occurs: radiative corrections due to loops of the gauge field generate an effective potential for the vierbein, whose absolute minimum corresponds to Minkowski space.

066.113 **Exact two–soliton solutions to the Einstein gravitational field equations.**
F. T. To, P. C. W. Fung, C. Au.
Astrophys. Space Sci., Vol. 184, No. 2, p. 313 – 330 (Oct 1991).
Following their previous work on the existence of 1–soliton solution to the Einstein gravitational field equations in the presence of a spherically–symmetric static background field, the authors have found six sets of analytical 2–soliton solutions to the Einstein field equations under a certain ansatz in the absence of the stated background field. Numerical analysis shows that if the two solitons of the transverse nature are injected at space variable $z \rightarrow \pm\infty$, the longitudinal field component g_{33} will acquire non–zero values for a bounded spatial region at later time. The nature of the solitons becomes rather complex when they interact. The amplitude $g_{\mu\nu}$ of each soliton may change its magnitude resulting from the interaction. The authors have found that the evolution of one field component might be interpreted as the gravitational instanton in their solutions. They remark also that the total energy of the interacting solitons remains constant, as expected, at all time. These solutions correspond to the situation where the Riemann tensor is in general non–zero and there are truly non–trivial solutions.

066.114 **Relativistic dynamics of expanding sources.**
A. C. Sadun, L. A. Sadun.
Astrophys. Space Sci., Vol. 185, No. 1, p. 21 – 36 (Nov 1991).
Given a relativistically expanding source, the transformation of the emission spectrum from the comoving frame of the source to the observer's frame is calculated. In particular, the authors deal with two special cases, that of a uniformly expanding sphere and that of a cylinder. In each case they consider a power–law spectrum with two distinct regions – the optically thick and thin parts – and transform these regions separately. The results are discussed, and a generalized procedure for transforming a spectrum of unspecified form is outlined. This procedure takes into account any bulk motion by the source as well as expansion. Finally, applications of this procedure to astrophysical sources are briefly discussed.

066.115 **Hamiltonian formulation of general relativity.**
C. Teitelboim.
7. Jerusalem Winter School for Theoretical Physics: Quantum cosmology and baby universes, p. 1 – 63 (1991). – See Abstr. 012.036 for the main entry.
Contents: 1. Introduction. 2. Hamiltonian formulation of gauge theories (pre–BRST). 3. BRST Hamiltonian formulation of gauge theories. 4. Dynamics of the gravitational field. 5. Does the Hamiltonian vanish? General covariance as an "ordinary" gauge invariance. 6. Must time flow? 7. BRST charge of the gravitational field. 8. Elements of BRST theory.

066.116 **Wormholes and non–simply connected manifolds.**
S. W. Hawking.
7. Jerusalem Winter School for Theoretical Physics: Quantum cosmology and baby universes, p. 245 – 267 (1991). – See Abstr. 012.036 for the main entry.

066.117 **Probing the Higgs vacuum with general relativity.**
P. D. Mannheim, D. Kazanas.
Astrophys. Space Sci., Vol. 185, No. 2, p. 167 – 179 (Nov 1991).
The authors show that the structure of the Higgs vacuum could be revealed in gravitational experiments which probe the Schwarzschild geometry to only one order in MG/r beyond that needed for the classical tests of general relativity.

066.118 **Magnetostatic stiff–fluid model of cylindrical symmetry in general relativity.**
R. Bali, A. Tyagi.
Astrophys. Space Sci., Vol. 185, No. 2, p. 299 – 304 (Nov 1991).
The authors study the static stiff–fluid model for perfect fluid distributions in the presence of incident magnetic field. The magnetic field is surrounded by static stiff fluid of infinite electric conductivity and it is due to the electric current flowing along the z–axis. The various physical and geometrical properties together with the state of model in absence of magnetic field are also discussed.

066.119 **A new way for testing of light deflection in earth orbit or beyond.**
Qin Yiping.
IAU Colloquium No. 123: Observatories in Earth orbit and beyond, p. 555 – 558 (1990). – See Abstr. 012.038 for the main entry.
According to the principles of light deflection, different occulting times of an occultation respond to different basic theories of physics. An occulting way for testing of light deflection is earth orbit or beyond may be possible and is proposed. An application is studied and some suggestions are presented.

066.120 **Relation of gauge formalisms for pulsations of general relativistic stellar models.**
R. H. Price, J. R. Ipser.
Phys. Rev. D, Vol. 44, No. 2, p. 307 – 313 (15 Jul 1991). Current Physics Microform No.: 9108F0937.
There have been two recent reformulations of the equations for even–parity perturbations of general relativistic stellar models, in both of which fluid perturbation variables are absent in the final set of equations. The recent reformulation by Chandrasekhar and Ferrari uses the diagonal coordinate gauge and leads to a fifth order system of differential equations; the authors have recently presented a reformulation, based on the Regge – Wheeler coordinate gauge, which leads to a fourth order system. The authors show here that the Chandrasekhar – Ferrari equations for stellar perturbations can be reduced by one degree. They show that the extra degree of freedom, in all descriptions based on the diagonal gauge, is due to the fact that the diagonal gauge is an incomplete constraint on coordinates; a one degree of freedom set of gauge transformations can be made within the diagonal gauge. This previously unnoticed degree of freedom is responsible for the extra degree of freedom in the Chandrasekhar – Ferrari equations, and the related black–hole equations. It also provides an a priori solution with which those equations can be reduced.

066.121 **Radiation and false–vacuum bubble dynamics.**
K. M. Larsen, R. L. Mallett.
Phys. Rev. D, Vol. 44, No. 2, p. 333 – 339 (15 Jul 1991). Current Physics Microform No.: 9108F0963.
It is shown that child universes may be detected by an external observer via radiation emitted by a restricted class of black holes associated with a generalization of the model of Blau, Guendelman, and Guth. The equations of motion for false–vacuum bubbles in Vaidya–Mallett metrics are derived and numerically solved. It is demonstrated that the evolution of the bubble differs

from the standard Schwarzschild - de Sitter results. The luminosity as measured by an external observer is found to deviate from the normal Hawking luminosity, thus producing a signature for the existence of the child universe.

066.122 **Solutions to the Reissner - Nordström, Kerr, and Kerr - Newman problems in fourth-order conformal Weyl gravity.**
P. D. Mannheim, D. Kazanas.
Phys. Rev. D, Vol. 44, No. 2, p. 417 - 423 (15 Jul 1991). Current Physics Microform No.: 9108F1047.
The authors continue their study of the general structure of fourth-order conformal Weyl gravity. They present three new exact solutions to the theory. First, the authors present the complete and exact solution to the Reissner - Nordström problem associated with a static, spherically symmetric point electric and/or magnetic charge coupled to fourth-order conformal Weyl gravity. It is found that, unlike the familiar second-order Einstein case where the modification to the Schwarzschild metric is in the form of a term which behaves like $1/r^2$, in the fourth-order case the modification is found to behave like $1/r$, i.e., just like the Newtonian potential term. The authors present two further exact solutions to the theory, namely, those associated with the fourth-order Kerr and Kerr - Newman problems in which a stationary, axially symmetric rotating system with or without electric and/or magnetic charge is coupled to gravity.

066.123 **Gravitating topological matter in 2+1 dimensions.**
S. Carlip, J. Gegenberg.
Phys. Rev. D, Vol. 44, No. 2, p. 424 - 428 (15 Jul 1991). Current Physics Microform No.: 9108F1054.
The authors study topological matter minimally coupled to gravity in 2+1 dimensions. They show that the resulting system has a finite-dimensional physical phase space that can be exactly quantized. The model exhibits the mixing of "gravity" and "matter" degrees of freedom, and shows the impossibility of treating them independently.

066.124 **Traversible wormholes and the negative stress-energy problem in the nonsymmetric gravitational theory.**
J. W. Moffat, T. Svoboda.
Phys. Rev. D, Vol. 44, No. 2, p. 429 - 432 (15 Jul 1991). Current Physics Microform No.: 9108F1059.
The stress-energy tensor for a a general spherically symmetric matter distribution in the nonsymmetric gravitational theory (NGT) is determined using a heuristic argument. Using this tensor and the NGT field equations, it is shown that a wormhole threaded with matter must necessarily have a radial tension greater than the mass-energy density in the throat region. Hence, as in general relativity, a traversible wormhole in NGT must contain matter with a negative stress energy.

066.125 **Radiation in a massive Schwarzschild background.**
C. Kozameh, E. T. Newman, C. Rovelli.
Phys. Rev. D, Vol. 44, No. 2, p. 551 - 554 (15 Jul 1991). Current Physics Microform No.: 9108F1181.
The authors investigate the possibility of an important physical consequence arising from a recent theoretical observation by Kundu. This concerns the suppression of gravitational radiation from a weak source in the presence of a massive Schwarzschild background. The authors conclude that this result appears to be unlikely.

066.126 **A new approach to gravitational microlensing: time delays and the Galactic mass distribution.**
L. M. Krauss, T. A. Small.
Astrophys. J., Vol. 378, No. 1, p. 22 - 29 (1 Sep 1991).
Gravitational lensing of distant sources by massive compact objects such as stars produces multiple images which are too closely spaced to resolve. The combined image can be magnified however, resulting in an increase in apparent luminosity. This has been thought to be the sole signature for such "microlensing" events. The authors propose here a new signature which may be more sensitive and could give an unambiguous determination of the mass of the lensing object, independent of its distance. A time delay between the superposed images on the order of microseconds to milliseconds will occur for lensing by objects in our Galaxy. This might be measurable in sources such as radio pulsars, some stars, and perhaps even supernovae. The authors calculate in some detail microlensing probabilities (appropriate for any study of microlensing) from objects in the Galactic disk and bulge, and also halo objects such as massive stars or large black holes which may contribute significantly to the dark matter in the Galaxy. It is suggested that a study of such time delays could resolve many features of the mass distribution of luminous and dark matter in the Galaxy and might even provide new tests of general relativity.

066.127 **A geodesic study of the Erez-Rosen space-time.**
K. D. Krori, J. C. Sarmah.
Gen. Relativ. Gravitation, Vol. 23, No. 7, p. 801 - 810 (Jul 1991).
The authors present the nature of stable trajectories of neutral particles around an object with a quadrupole moment in both the equatorial ($\theta = \pi/2$) and the polar (Φ = constant) plane.

066.128 **Fifth force from Kaluza-Klein unification.**
Y. M. Cho, D. H. Park.
Gen. Relativ. Gravitation, Vol. 23, No. 7, p. 741 - 757 (Jul 1991).
The authors discuss the possible modification of Einstein's theory of gravitation due to the fifth force generated by the Kaluza-Klein dilaton. In particular, they discuss the gravitational redshift, the deflection of light, the precession of perihelia, and the time-delay of radar echo around a spherically symmetric black hole in a 7-dimensional Kaluza-Klein unification, which has E_2 as the internal isometry. It is shown that the long-range effect of the higher dimensional fifth force is characterized by the dilatonic charge carried by the black hole even when it is neutral.

066.129 **Five-dimensional Kaluza-Klein black holes coupled to massive scalar particles.**
Liu Hongya.
Gen. Relativ. Gravitation, Vol. 23, No. 7, p. 759 - 765 (Jul 1991).
Spherically symmetric solutions coupled to massive scalar particles in five-dimensional Kaluza-Klein theory are obtained. The solutions contain two event horizons. The inner horizon corresponds to the Schwarzschild black hole, and the outer one is a new horizon, which is produced by the massive scalar particles. It is found that the massive modes contribute an effective cosmological constant to the four-dimensional Einstein theory.

066.130 **The gravitational interaction of conical strings.**
C. Hellaby.
Gen. Relativ. Gravitation, Vol. 23, No. 7, p. 767 - 799 (Jul 1991).
Up to now calculations of the interaction of cosmic strings have neglected gravity. The author considers the purely gravitational interactions that occur at large distances, using the conical line singularity for the gravitational field of a string. He constructs spaces with multiple intersecting conical strings, which are exactly consistent with General Relativity, and which can be covered in a single Minkowski coordinate patch, using a Regge calculus type construction. It is shown that after two such strings pass through each other, they remain connected by another string. The branching rules governing the junction of three strings are derived. These rules apply to conical type strings in any smoothly curved background, whether they are straight or curved, moving or stationary, and the rules show that, at the junction, the three strings must be as coplanar as is possible in such a space. For these results to be matched to the short range results of field theory calculations, it is suggested that gravitational radiation must be considered.

066.131 **Simply-transitive homothety groups.**
J. D. Steele.
Gen. Relativ. Gravitation, Vol. 23, No. 7, p. 811 - 825 (Jul 1991).
In this paper the spin-coefficient formalism of Penrose and Rindler is applied to the problem of finding space-times that admit a simply transitive group of homotheties. The method is

illustrated by an application to algebraically special vacuum space-times. Vacuum metrics with a five-dimensional homothety group are also considered.

066.132 **Some electromagnetic consequences of a geometric unified theory of gravitation and electromagnetism.**
G. González Martín.
Gen. Relativ. Gravitation, Vol. 23, No. 7, p. 827 - 841 (Jul 1991).
The geometrical unified theory of gravitation and electromagnetism discussed in previous publications requires that electromagnetism be represented by an SU(2) subgroup. Electric matter should correspond to irreducible representations of the structure group, SL(4, R), induced from its compact subgroup, rotation SU(2) × electromagnetic SU(2). Results predict the quantization of electric charge, magnetic flux and angular momentum without requiring magnetic monopoles. Unexpectedly, the necessary quanta of charge and flux imply fractional quantization of transverse resistance, under certain conditions (fractional quantum Hall effect).

066.133 **Constructing physically valid fluid spheres in general relativity.**
H. Knutsen.
Gen. Relativ. Gravitation, Vol. 23, No. 7, p. 843 - 859 (Jul 1991).
The author extends a method by Goldman to include conditions which are both necessary and sufficient for the construction of physically acceptable relativistic fluid spheres. Conditions are given for having finite and non-negative pressure and finite and positive density inside the fluid sphere. Also given are conditions for the pressure gradient and the density gradient to be negative and for the speed of sound to be less than the speed of light in vacuum. Further, the condition are derived for the trace of the energy-momentum tensor to be positive and for the relativistic adiabatic index to be larger than 4/3. The author presents a class of new exact solutions, and shows that these models are physically valid. He calculates the total mass, the density and the physical radius for these fluid spheres. Some of these spheres turn out to be extremely compact with a radius of only several centimeters, but typically containing the mass of a planet. It is speculated whether these spheres could contain the hidden dark matter in the universe.

066.134 **Einstein vacuum field equations with a single non-null Killing vector.**
E. D. Fackerell, R. P. Kerr.
Gen. Relativ. Gravitation, Vol. 23, No. 8, p. 861 - 876 (Aug 1991).
It is shown that, in the case where there is a single non-null Killing vector, the vacuum Einstein field equations imply that there is a Ricci collineation in the quotient 3-space. Using coordinates adapted to the collineation vector, the authors derive a fourth order partial differential equation involving the metric of the quotient 3-space. It is shown that if this equation is satisfied, the Ernst potential may be obtained by integrating a total Riccati equation and a straightforward set of total differential equations. It is also shown that if the collineation vector is null, the metric of the quotient 3-space may be expressed in terms of two real Clebsch potentials.

066.135 **The mass formula of the spherically symmetric solution of the five-dimensional vacuum equations $R_{\mu\nu} = 0$.**
G. Lessner.
Gen. Relativ. Gravitation, Vol. 23, No. 8, p. 897 - 915 (Aug 1991).
Within the concept of extended elementary particles proposed by the author in some previous papers, the mass formula of the general stationary spherically symmetric and purely electrical solution of the 5-dimensional vacuum equations $R_{\mu\nu} = 0$ is calculated. The problem of how to establish a discrete mass spectrum is discussed in detail.

066.136 **The interplay between relativistic gravitational, centrifugal and electric forces: a simple example.**
M. A. Abramowicz, J. Bičák.
Gen. Relativ. Gravitation, Vol. 23, No. 8, p. 941 - 946 (Aug 1991).
The authors show that some striking properties of the circular motion of charged, ultrarelativistic particles close to the circular photon orbit around a charged black hole can be easily understood in terms of the centrifugal force vanishing at the circular photon orbit. The authors also comment briefly on a possible non-uniqueness of the recently proposed operational definition of the centrifugal force in static spacetimes.

066.137 **Magnetic charge type equations from torsion.**
R. T. Hammond.
Gen. Relativ. Gravitation, Vol. 23, No. 9, p. 973 - 980 (Sep 1991).
It is shown that torsion can be built from two independent vector fields, and that these vector fields obey, for the Lagrangian chosen, the equations of electromagnetism with magnetic charge from the two photon formalism. The equation of motion follows from the Bianchi identity of U_4 spacetime. The physical interpretation of these fields is discussed.

066.138 **Einstein - Maxwell theory in 2+1 dimensions.**
C. Nayak.
Gen. Relativ. Gravitation, Vol. 23, No. 9, p. 981 - 989 (Sep 1991).
Einstein - Maxwell theory in 2+1 dimensions is considered as a toy model for the 3+1 theory and as a means for understanding field theories without a background space-time geometry. Two possible approaches to the quantization of this theory are presented, one based on a Schwinger - Tomonaga time evolution equation and the other on a loop space construction, and the problems arising in both of these are discussed.

066.139 **Gauge-strings on Einstein - Rosen background.**
R. J. Slagter.
Gen. Relativ. Gravitation, Vol. 23, No. 9, p. 991 - 1006 (Sep 1991).
The full coupled Einstein scalar gauge field equations are obtained together with the high frequency perturbations on a cylindrical symmetric background. The vortex solution and the first order perturbations depend critically on the parameters of the model. In the static case, an exact solution of the vortex is found. In the nondiagonal case, an axial component of the gauge field is necessary in order to fulfil the energy equation.

066.140 **Solitonic solutions on a Bianchi II background generated by SHEEP algebraic manipulation.**
M. Bradley, A. Curir, M. Francaviglia.
Gen. Relativ. Gravitation, Vol. 23, No. 9, p. 1011 - 1021 (Sep 1991).
The exact solutions of Einstein's (vacuum) field equations corresponding to a (real-poles) three - solitonic perturbation of a Bianchi type II space-time are investigated by using computer algebra. Two-dimensional and three-dimensional plots of the relevant solitonic fields are obtained numerically.

066.141 **Quartic equations and classification of Riemann tensors in general relativity.**
J. E. Åman, R. A. d'Inverno, G. C. Joly, M. A. H. MacCallum.
Gen. Relativ. Gravitation, Vol. 23, No. 9, p. 1023 - 1055 (Sep 1991).
For space-times in general relativity, the Petrov classification of the Weyl conformal curvature and the Plebański or Serge classification of the Ricci tensor each depend on the properties of the roots of quartic equations. The coefficients in these quartic equations are in general complicated functions of the space-time coordinates. The authors review the general theory of quartic equations, and discuss algorithms for determining the existence and values of multiple roots. They consider the practical implementation of an algorithm and the consequent Petrov classification. Tests of programs embodying this algorithm, using

the computer algebra system CLASSI based on SHEEP, are reported.

066.142 **Perfect fluid spacetimes admitting curvature collineations.**
J. Carot, J. da Costa.
Gen. Relativ. Gravitation, Vol. 23, No. 9, p. 1057 – 1069 (Sep 1991).
Some basic concepts about curvature collineations are reviewed and the existing results on this topic are applied to the case of perfect fluids, giving a characterization of those amongst them which admit proper curvature collineations.

066.143 **Finsler spaces with Riemannian geodesics.**
I. W. Roxburgh.
Gen. Relativ. Gravitation, Vol. 23, No. 9, p. 1071 – 1080 (Sep 1991).
In Finsler spaces the interval $ds = F(x^i, dx^i)$ is an arbitrary function of the coordinates x_i and the coordinate increments dx^i with F homogeneous of degree one in the dx^i. It is shown that for Riemannian spaces with $ds_R{}^2 = g_{ij}dx^i dx^j$, and which admit a nontrivial covariantly constant tensor $H_{i..}(x^k)$, there is an associated Finsler space with the same geodesic structure. The subset of such Finsler spaces with $H_{i..}(x^k)$ a vector or second rank decomposable tensor is determined.

066.144 **Torsion vector and variable G.**
H. H. Soleng.
Gen. Relativ. Gravitation, Vol. 23, No. 10, p. 1089 – 1112 (Oct 1991).
In the Einstein – Cartan theory coupled to a classical spin fluid only the traceless part of the torsion field can be related to spin. Here the theory is generalized by allowing intrinsic dilation currents which couple algebraically to the torsion trace. The intrinsic dilation current is interpreted as the relative change of intrinsic length scales relative to the gravitational length scale. By a dilation current conservation law, one finds the scale function between the atomic length scales and the gravitational length scales. This is in effect a dynamic unification of the atomic and gravitational units of measurement. The theory is applied to the Friedmann– Robertson – Walker cosmology, and physical effects in the solar system are considered.

066.145 **Fluid spacetimes admitting covariantly constant vectors and tensors.**
A. A. Coley, B. O. J. Tupper.
Gen. Relativ. Gravitation, Vol. 23, No. 10, p. 1113 – 1142 (Oct 1991).
Spacetimes admitting a covariantly constant vector and satisfying the Einstein field equations for a perfect fluid, a viscous heat–conducting fluid, or an anisotropic fluid are studied. It is found that the only possible perfect fluid spacetimes are the Einstein static universe and "stiff–matter" spacetimes with an isolated spatial co–ordinate, while the possible viscous fluid and anisotropic fluid spacetimes must satisfy a number of strong restrictions. Examples illustrating most of the various possible situations are given. The paper conludes with a study of covariantly constant second–rank tensors in fluid spacetimes; the only possible solutions that do not also admit a covariantly constant vector are restricted to 2+2 spacetimes.

066.146 **The axially symmetric non–rotating vacuum solutions of Rosen's equations.**
Yu. Bozhkov.
Gen. Relativ. Gravitation, Vol. 23, No. 10, p. 1143 – 1150 (Oct 1991).
It is shown that all axially symmetric non–rotating solutions of Rosen's field equations can be expressed in terms of two harmonic functions. Furthermore, it is shown that the total energy of Rosen's metric is Mc^2.

066.147 **Anisotropic solutions in the 5D space – time – mass theory.**
W. L. Roque, W. M. Seiler.
Gen. Relativ. Gravitation, Vol. 23, No. 10, p. 1151 – 1168 (Oct 1991).
The authors present anisotropic families of cosmological solutions in the 5–dimensional space – time – mass theory of gravity. In particular, the 5D analogue of the Kasner solution of General Relativity is obtained. Some comments are given on the solutions.

066.148 **Gauge invariant electromagnetic coupling with torsion potential.**
R. T. Hammond.
Gen. Relativ. Gravitation, Vol. 23, No. 11, p. 1195 – 1203 (Nov 1991).
Electromagnetism is coupled to torsion in a gauge invariant manner by relaxing minimal coupling and introducing into the Lagrangian a term bilinear in the electromagnetic field tensor and the torsion potential. The resulting coupling between electromagnetism and torsion is examined and a solution corresponding to traveling coupled waves is given.

066.149 **Interaction between gravity and moving superconductors.**
Peng Huei, G. Lind, Y. S. Chin.
Gen. Relativ. Gravitation, Vol. 23, No. 11, p. 1231 – 1250 (Nov 1991).
The authors propose a unified phenomenological theory to investigate the interaction between arbitrarily moving superconductors and gravitational fields including the Newtonian gravity, gravitational waves, vector transverse gravitoelectric fields, and vector gravitomagnetic fields. In the limit of weak field and low velocity, the expressions for the induced electromagnetic and gravitational fields in the interior of a moving superconductor are obtained. The Meissner effect, London moment, DeWitt effect, effects of gravitational waves on a superconductor, and induced electric fields in the interior of a freely vibrating superconductor are recovered from these two expressions.

066.150 **Scalar fields in a seven – dimensional manifold behaving as Lorentz – covariant spinor fields in space–time.**
D. W. Ebner.
Gen. Relativ. Gravitation, Vol. 23, No. 11, p. 1279 – 1301 (Nov 1991).
A Lorentz – invariant model of vacuum is given in the form of a 7 – dimensional manifold endowed with a statistical metrical tensor. Certain scalar fields on this manifold behave then as spinor fields, when viewed from their space–time projection. This paper generalizes previous work from SO(3) – covariance to Lorentz – covariance.

066.151 **Cosmological solutions in the D = 5 Einstein – Maxwell theory coupled to matter.**
R. Balbinot, J. C. Fabris.
Gen. Relativ. Gravitation, Vol. 23, No. 12, p. 1307 – 1315 (Dec 1991).
The authors study the Einstein – Maxwell theory in five dimensions coupled to matter in two distinct ways. In the first, the Lagrangian is reduced to an effective four–dimensional one and is then coupled to matter; in the second, matter is introduced directly in the original five–dimensional theory. In both cases the authors use a nontrival configuration for the Maxwell potential. They find non–singular solutions which present a repulsive gravitational phase. When this phase is absent, the initial singularity is unavoidable.

066.152 **Propagation of self–gravitating density waves in the deDonder gauge: the physical importance of gauge solutions and the problem of initial conditions.**
B. Rose, M. Corona–Galindo.
Gen. Relativ. Gravitation, Vol. 23, No. 12, p. 1317 – 1335 (Dec 1991).
The propagation of perturbations on a spatially flat Robertson–Walker background is studied within linear perturbation

theory in deDonder gauge and for comparison in synchronous gauge. The metric perturbations should be determined uniquely by the density / pressure perturbations, therefore only two initial conditions, namely for the density contrast and its time derivative, should be needed. Since the number of fundamental solutions for the density perturbations is higher than 2 in both gauges (6 resp. 3), an additional reduction of possible initial conditions is needed. It is shown that the common treatment of excluding the so-called gauge solutions (solutions which can be gauged to zero in an already chosen gauge) leads to unphysical results. If gauge solutions are excluded, the density perturbation solutions are the same in both gauges. But the correct Newtonian limit – which is present in deDonder gauge but not in synchronous gauge – is bound to the differences in the two gauges for large spatial scales of perturbations. Furthermore, compressional wave solutions should vanish for infinite spatial scales of perturbations (isotropy), but this is guaranteed in deDonder gauge only by gauge solutions.

066.153 **A class of stationary charged dust solutions of Einstein's field equations.**
V. Perlick.
Gen. Relativ. Gravitation, Vol. 23, No. 12, p. 1337 – 1348 (Dec 1991).
The author considers a special class of stationary rotation charged dust solutions of Einstein's field equations without cosmological constant. In these space-times, the motion of freely falling particles and of light rays can be visualized by the motion of charged particles in an appropriate model magnetic field. Any curl-free magnetostatic field, given on an open subset of Euclidean 3-space, can serve as a model magnetic field for a charged dust solution in this sense. The simplest example, corresponding to a homogeneous model magnetic field, is given by a Som – Raychaudhuri space-time.

066.154 **Complex coordinates and Kähler potential for the Atiyah – Hitchin metric.**
D. Olivier.
Gen. Relativ. Gravitation, Vol. 23, No. 12, p. 1349 – 1362 (Dec 1991).
In the case of self dual metrics, the author details and puts together the Geroch three-dimensional formulation of the Einstein equations and the Boyer – Finley results on Killing vectors to obtain a set of complex coordinates and a Kähler potential for the Atiyah – Hitchin metric.

066.155 **Algebraic invariant curves in cosmological dynamical systems and exact solutions.**
C. G. Hewitt.
Gen. Relativ. Gravitation, Vol. 23, No. 12, p. 1363 – 1383 (Dec 1991).
The Einstein field equations for a number of classes of cosmological models have previously been written as polynominal systems of ordinary differential equations. The author shows that, for restricted parameter values, these equations admit algebraic invariant curves, which, in turn, lead to exact solutions of the field equations. This property explains the recent discovery of a number of exact solutions and is used to produce additional ones.

066.156 **Self-dual Maxwell field on a null cone.**
J. N. Goldberg.
Gen. Relativ. Gravitation, Vol. 23, No. 12, p. 1403 – 1413 (Dec 1991).
As an introduction to the use of self-dual variables on a null cone, the Hamiltonian for the Maxwell field is set up on a null cone in Minkowski space. In this treatment, the vector potential (the connection) and the self-dual components of the Maxwell field are treated as independent configuration space variables. Because the initial surface is a null cone, additional primary and secondary constraints arise. These constraints can be grouped into first class and second class. The elimination of the second class constraints, together with the reality conditions on the vector potential, reduce the independent phase space variables to two. In its final form the Hamiltonian can be expressed in terms of the product of the self-dual Maxwell field and its complex conjugate.

066.157 **Wormholes and the construction of compactified phases.**
E. I. Guendelman.
Gen. Relativ. Gravitation, Vol. 23, No. 12, p. 1415 – 1419 (Dec 1991).
In the context of classical general relativity, assuming the positive energy condition, the author studies the possibility of connecting a region where some dimensions are compact to another region where compactification does not exist. In some simple models, where compactification is achieved through the condensation of gauge fields, it is found that the matching of these two regions implies the existence of a wormhole. The author also discusses the possibility of spontaneous quantum nucleation of a compactified phase in the midst of an uncompactified space.

066.158 **The deconvolution of the quasar structure from microlensing light curves.**
B. Grieger, R. Kayser, T. Schramm.
Astron. Astrophys., Vol. 252, No. 2, p. 508 – 512 (Dec 1991).
Gravitational microlensing by compact objects (like stars) near the light path to a quasar can cause large brightness vriations, so called high amplification events (HAEs). In principle, the one-dimensional intrinsic luminosity profile – i.e. the surface luminosity integrated in one direction – of a quasar can be retrieved from the light curve of a HAE. The classical deconvolution leads, however, to the annoying problem of noise-amplification. In this paper the authors demonstrate how to use a non-classical inversion technique, the so-called regularization method, to solve this problem. The results of test calculations on simulated light curves are shown and compared with classical methods.

066.159 **Relativistic neutrino transport in stellar collapse.**
S. A. Bludman, P. J. Schinder.
10. Santa Cruz Summer Workshop in Astronomy and Astrophysics: Supernovae, p. 379 – 388 (1991). – See Abstr. 012.052 for the main entry.

066.160 **Gravitational radiation.**
T. Jarzebowski.
Urania, Rok 62, Nr. 9, p. 226 – 233 (Sep 1991). In Polish.

066.161 **The accurate post-Newtonian treatment of geodetic space measurements.**
M. Soffel.
Journées 1991: Systèmes de référence spatio-temporels, p. 134 – 140 (1991). – See Abstr. 012.048 for the main entry.
The modern, highly accurate post-Newtonian treatment of geodetic space measurements and related techniques is discussed. Special emphasis is given to the problems of reference frames, including time scales and astrometry.

066.162 **A measurement of planetary relativistic deflection.**
R. N. Treuhaft, S. T. Lowe.
Astron. J., Vol. 102, No. 5, p. 1879 – 1888 (Nov 1991). Current Physics Microform No.: 9111D0885.
The radio interferometric observations described here constitute the first measurement of the deflection of electromagnetic radiation by a planetary gravitational field. They also demonstrate a few hundred microarcsecond (μarcsec) angular astrometric technique, which can be applied over many-degree fields of sky. The ray path of the radio source P0201 + 113 passed within 200 arcsec of Jupiter ($\approx$10 Jovian radii) on 21 March 1988. Its angular position was measured 10 times over 4 hr on that date, with a similar measurement set on 2 April 1988 to detect the differential angular deflection of the ray path. According to general relativity, the expected gravitational bend of the ray path averaged over the March experiment duration was approximately 300 μarcsec projected onto the two California–Australia baselines over which it was measured. Measurement accuracies of the order of 160 μarcsec were obtained for each of the ten differential measurements. The χ^2 per degree of freedom of the

data for the hypothesis of general relativity was 0.6, suggesting that the modeled dominant errors due to system noise and tropospheric fluctuations fully accounted for the scatter in the measured angular deflections. The χ^2 per degree of freedom for the hypothesis of no gravitational deflection by Jupiter was 4.1, which rejects the no-deflection hypothesis with greater than 99.999% confidence. The system noise contributed about 70 μarcsec per combined-baseline differential measurement and tropospheric fluctuations contributed about 145 μarcsec. Unmodeled errors were assessed which could potentially increase the 160 μarcsec error by about 8%.

066.163 **Static spherically-symmetric perfect fluids with pressure equal to the energy density.**
R. B. S. Yadav, S. L. Saini.
Astrophys. Space Sci., Vol. 186, No. 2, p. 331 - 336 (Dec 1991). Letter-to-the-editor.

The present Letter represents an exact, static, and spherically-symmetric solution of Einstein's field equations for a homogeneous perfect fluid core surrounded by a field of Zeldovich's fluid which is asymptotically homaloidal.

066.164 **New exact solutions of the Einstein-Maxwell equations.**
V. Karas, D. Vokrouhlický.
Bull. Astron. Inst. Czech., Vol. 42, No. 4, p. 247 - 250 (Jul 1991).

Some interesting features of the solutions published by Gutsunaev and Manko are studied. The occurrence of a singularity is discussed and two new solutions of this series are given explicitly.

066.165 **First cooling below 0.1K of the new gravitational-wave antenna "Nautilus" of the Rome group.**
P. Astone, F. Bronzini, M. Bassan, E. Coccia, V. Fafone, I. Modena, P. Bonifazi, M. G. Castellano, M. Visco, C. Cosmelli, S. Frasca, G. V. Pallottino, G. Pizzella, P. Rapagnani, F. Ricci.
Europhys. Lett., Vol. 16, No. 3, p. 231 - 235 (14 Sep 1991).

The authors report on the first cryogenic test at ultralow temperature of the new gravitational-wave antenna "Nautilus" of the Rome group. The 2390 kg Al 5056 cylindrical-bar antenna has been cooled to a temperature of 95 mK.

066.166 **Decoherence and back reaction: the origin of the semiclassical Einstein equations.**
J. P. Paz, S. Sinha.
Phys. Rev. D, Vol. 44, No. 4, p. 1038 - 1049 (15 Aug 1991). Current Physics Microform No.: 9110D1092.

Two basic properties defining classical behavior are "decoherence" and "correlations between coordinates and momenta". The authors study how the correlations that define the semiclassical decohering histories of the relevant cosmological variables are affected by the interaction with an environment formed by unobserved ("irrelevant") degrees of freedom. For some quantum cosmological models, the authors analyze under what conditions the semiclassical coarse-grained histories obey the so-called semiclassical Einstein's equations (i.e., $G_{\mu\nu} = \varkappa\langle T_{\mu\nu}\rangle$). These equations are shown to be valid only as a description of adiabatic regions of histories, for which the interference effects have been suppressed. The authors also discuss the problem related to the existence of divergences in the decoherence factor of various quantum cosmological models.

066.167 **Geometric construction of the measure: minisuperspace quantum gravity and the relativistic particle compared.**
J. Guven, J. D. Vergara.
Phys. Rev. D, Vol. 44, No. 4, p. 1050 - 1058 (15 Aug 1991). Current Physics Microform No.: 9110D1104.

The authors examine the Polyakov functional - integral quantization of a minisuperspace model for quantum gravity using the relativistic free particle as a guideline. While formally the two models are very similar, there are differences which probably have a genuine physical significance.

066.168 **Quantum state of wormholes and path integral.**
L. J. Garay.
Phys. Rev. D, Vol. 44, No. 4, p. 1059 - 1066 (15 Aug 1991). Current Physics Microform No.: 9110D1113.

The quantum state of a wormhole can be represented by a path integral over all asymptotically Euclidean four-geometries and all matter fields which have prescribed values. The ground-state wave function is picked out by requiring that there be no matter excitations in the asymptotic region. Once the path integrals over the lapse and shift functions are evaluated, the requirement that the spacetime be asymptotically Euclidean can be accomplished by fixing the asymptotic gravitational momentum in the remaining path integral. It is claimed that no wave function exists which corresponds to asymptotic field configurations such that the effective gravitational constant is negative in the asymptotic region. The wormhole wave functions are worked out in minisuperspace models with massless minimal and conformal scalar fields.

066.169 **Quantum gravitational corrections to the functional Schrödinger equation.**
C. Kiefer, T. P. Singh.
Phys. Rev. D, Vol. 44, No. 4, p. 1067 - 1076 (15 Aug 1991). Current Physics Microform No.: 9110D1121.

The authors derive corrections to the Schrödinger equation which arise from the quantization of the gravitational field. This is achieved through an expansion of the full functional Wheeler - DeWitt equation with respect to powers of the gravitational constant. The correction terms fall into two classes: one describes the breakdown of the classical background picture, while the other corresponds to quantum gravitational corrections for the matter fields themselves. The corrections are numerically extremely tiny except near the big bang and the final stages of a black hole. They are also of principle significance for quantum field theories near the Planck scale.

066.170 **A new expression for the mass in the theory of gravitation.**
R. M. Avakyan.
Astrofizika, Tom 33, Vyp. 3, p. 429 - 436 (Dec 1990). In Russian. English translation in Astrophysics, Vol. 33, No. 3.

A new expression for the mass of a static spherically-symmetric configuration is found. It determinates the mass as a function of the pressure distribution in the configuration.

066.171 **The singularity structure of a soliton solution to the higher-dimensional Einstein equations.**
T. Azuma, M. Endo, T. Koikawa.
Prog. Theor. Phys., Vol. 86, No. 4, p. 833 - 840 (Oct 1991).

The authors study a stationary and axisymmetric solution to the higher-dimensional Einstein equations and investigate its singularity structure. The solution consists of two solitons in the four-dimensional part (i.e., the Kerr solution) and n solitons in the extra dimensions. Naked singularities appear on the symmetry axis (z-axis) and/or at the event horizons of the Kerr solution. In a certain choice of integration constants there are solutions with regular event horizons.

066.172 **Junction conditions in 2+1-dimensional gravity.**
J. Soda, Y. Yamanaka, H. Ooguchi.
Prog. Theor. Phys., Vol. 86, No. 4, p. 841 - 852 (Oct 1991).

Possible junctions among Minkowski, Cone and de Sitter space-times are exhausted. Using the positive energy condition for a string, the possible space-times are restricted. The motions of the string in various cases are solved.

066.173 **Spherically symmetric solutions in dimensionally reduced spacetimes with a higher-dimensional cosmological constant.**
D. L. Wiltshire.
Phys. Rev. D, Vol. 44, No. 4, p. 1100 - 1114 (15 Aug 1991). Current Physics Microform No.: 9110D1154.

Certain static solutions of D-dimensional gravity with a higher-dimensional cosmological constant Λ are studied. The solutions are taken to be spherically symmetric in the physical (m+2)–dimensional spacetime, where D = m+n+2 (or more generally the m–sphere is replaced by an arbitrary Einstein space), while the internal space is an arbitrary n–dimensional Einstein space. The global properties of all such solutions are derived by considering the equivalent dimensionally reduced system in m+2 dimensions, and by using techniques from the theory of dynamical systems after a judicious choice of variables. All solutions with a nonzero Λ are either found to contain naked singularities or not be asymptotically flat, as would be expected from the "no–hair" theorems.

066.174 **Exact solutions of Einstein gravity coupled to a scalar field with arbitrary potential.**
M. Cadoni.
Phys. Rev. D, Vol. 44, No. 4, p. 1115 – 1119 (15 Aug 1991). Current Physics Microform No.: 9110D1169.
The author presents a method to generate exact static solutions of the coupled Einstein – scalar–field equations with an arbitrary potential in D space–time dimensions. The spherically symmetric solutions corresponding to an exponential and a quadratic potential are explicitly derived and discussed. Some of these solutions indicate a way to circumvent the "no–hair" theorems.

066.175 **Gravitational Faraday rotation induced from interacting gravitational plane waves.**
Wang Anzhong.
Phys. Rev. D, Vol. 44, No. 4, p. 1120 – 1131 (15 Aug 1991). Current Physics Microform No.: 9110D1174.
The polarization of gravitational plane waves is studied. In particular, it is found that each interacting gravitational plane wave can be split into two parts: the shock part and the impulsive part. The plane of the polarization for the plane gravitational shock waves usually gets rotated due to two kinds of interaction: one is the nonlinear interaction between the oppositely moving plane gravitational shock waves, and the other is the interaction between the gravitational shock wave and the matter fields which are present. The change of polarization due to the nonlinear interaction is a gravitational analogue of Faraday rotation, but with the oppositely moving gravitational shock wave as the magnetic field and medium. The effect of the above two kinds of interaction on the impulsive gravitational plane waves is different from that on the shock waves. Only the interaction between the impulsive gravitational wave and the matter fields can change the polarization of the impulsive wave.

066.176 **Graviton–electron interactions.**
H. N. Saif.
Phys. Rev. D, Vol. 44, No. 4, p. 1140 – 1146 (15 Aug 1991). Current Physics Microform No.: 9110D1194.
First–order cross sections for the processes of graviton – Compton scattering, $ge^- \to \gamma e^-$, bremsstrahlung, and pair production by gravitons in the Coulomb field, are calculated. The calculations, which are linear in the gravitational coupling $\varkappa$, are obtained in the extreme relativistic limit.

066.177 **Gravitational scattering on a global monopole.**
P. O. Mazur, J. Papavassiliou.
Phys. Rev. D, Vol. 44, No. 4, p. 1317 – 1320 (15 Aug 1991).
The scattering amplitude and the total scattering cross section of massless particles propagating in the gravitational field of a global monopole are derived. It is found that the physical signature of such defects is a ringlike angular region where the scattering amplitude is very large. The size of this ringlike region is determined by the ratio of the global monopole mass to the Planck mass and its appearance stems from the fact that the metric of the global monopole is not asymptotically flat but rather displays the characteristic spherical angle defect. The situation is therefore very much reminiscent of scattering in the gravitational field of the cosmic string.

066.178 **Influence of the quadrupole field and rotation of objects on light propagation.**
S. A. Klioner.
Sov. Astron., Vol. 35, No. 5, p. 523 – 530 (Sep 1991). Current Physics Microform No.: 9203X1697. English translation of Astron. Zh., Tom 68, Vyp. 5, p. 1046 – 1062 (1991).
Differential equations describing the relativistic corrections to light propagation in a barycentric coordinate system that are induced by the quadrupole field of objects and their rotation are derived and solved. The influence of the flattening of an object and its spin on the total angle of light deviation and on the propagation time of electromagnetic radiation between two fixed points is investigated.

066.179 **Free parameters in the matching conditions for a cryogenic gravitational radiation detector.**
C. Cosmelli.
Europhys. Lett., Vol. 16, No. 1, p. 17 – 22 (1 Sep 1991).
The sensitivity of a resonant gravitational wave antenna coupled with a capacitive transducer to a current amplifier is written in terms of all the parameters of the system. The matching conditions that must be realized to obtain the maximum theoretical sensitivity are written as a function of the physical parameters of the transducer. It is shown that the energy coupling factor is independent of the mass as well as of the capacitance of the transducer. The only free parameter then, apart from a weak dependence on the material used, is shown to be the gap of the capacitor. This dependence suggests the strategy to be used to obtain the best sensitivity for the present systems.

066.180 **Spectral boundary conditions in one–loop quantum cosmology.**
P. D. d'Eath, G. V. M. Esposito.
Phys. Rev. D, Vol. 44, No. 6, p. 1713 – 1721 (15 Sep 1991). Current Physics Microform No.: 9111B1941.
For fermionic fields on a compact Riemannian manifold with a boundary, one has a choice between local and nonlocal (spectral) boundary conditions. The one–loop prefactor in the Hartle – Hawking amplitude in quantum cosmology can then be studied using the generalized Riemann ζ function formed from the squared eigenvalues of the four–dimensional fermionic operators. For a massless Majorana spin–1/2 field, the spectral conditions involve setting to zero half of the fermionic field on the boundary, corresponding to harmonics of the intrinsic three–dimensional Dirac operator on the boundary with positive eigenvalues. Remarkably, a detailed calculation for the case of a flat background bounded by a three–sphere yields the same value $\zeta(0) = 11/360$ as was found previously by the authors using local boundary conditions.

066.181 **Gravitons and loops.**
A. Ashtekar, C. Rovelli, L. Smolin.
Phys. Rev. D, Vol. 44, No. 6, p. 1740 – 1755 (15 Sep 1991). Current Physics Microform No.: 9111B1968.
The recently proposed loop representation is used to quantize linearized general relativity. The Fock space of graviton states and its associated algebra of observables are represented in terms of functionals of loops. The "reality conditions" are realized by an inner product that is chiral asymmetric, resulting in a chiral asymmetric ordering for the Hamiltonian, and, in an asymmetric description of the left– and right–handed gravitons. The formalism depends on an arbitrary "averaging" function that controls certain divergences, but does not appear in the final physical quantities. In spite of these somewhat unusual features, the loop quantization presented here is completely equivalent to the standard quantization of linearized gravity.

066.182 **Nucleation of a universe in (2+1)–dimensional gravity with a negative cosmological constant.**
Y. Fujiwara, S. Higuchi, A. Hosoya, T. Mishima, M. Siino.
Phys. Rev. D, Vol. 44, No. 6, p. 1756 – 1762 (15 Sep 1991). Current Physics Microform No.: 9111B1984.
The authors investigate the nucleation of a universe in a (2+1)–dimensional gravity model with a negative cosmological

constant. There are a variety of universes born from nothing by quantum tunneling. Utilizing the powerful technique of hyperbolic geometry, the authors explicitly construct three-manifolds which describe the nucleation of a higher-genus universe. They calculate the wave function of the universe in the WKB approximation.

066.183 **Inevitability of shell crossing in the gravitational collapse of weakly charged dust spheres.**
A. Ori.
Phys. Rev. D, Vol. 44, No. 8, p. 2278 - 2285 (15 Oct 1991).
Current Physics Microform No.: 9112A1202.

It is shown here that shell crossing is inevitable in the gravitational collapse of weakly charged dust spheres. That is, all the dust shells in the central part of the dust cloud are subject to shell crossing before they reach the stage of maximal contraction. This shell crossing, which is accompanied by a density singularity, indicates the breakdown of the dust model there. Most previous analyses of the gravitational collapse of charged dust spheres were based on the "free surface approach" - the assumption that the surface of the dust sphere moves like a test particle in Reissner - Nordström spacetime. Because of the occurrence of shell crossing, the free surface approach cannot be applied to the late stages of evolution. This makes it difficult to analyze, in general, what will be the final stage in the gravitational collapse of charged dust. However, for idealized models, in which the dust interior is considered to be self-similar and the shells are considered to collide inelastically, the shell crossing is shown to have a very significant effect on the causal structure: It completely prevents the gravitational bounce.

066.184 **Hoop conjecture for black-hole horizon formation.**
E. Flanagan.
Phys. Rev. D, Vol. 44, No. 8, p. 2409 - 2420 (15 Oct 1991).
Current Physics Microform No.: 9112A1333.

The hoop conjecture was proposed by Thorne in 1972 as a loosely formulated, necessary and sufficient condition for the formation of a horizon in nonspherical gravitational collapse. In this paper, the author discusses some issues that arise in formulating various precise interpretations of the conjecture, and reviews some of the small amount of literature that has appeared on this subject. The author also specializes the conjecture to static, vacuum, axisymmetric spacetimes and shows that it is satisfied in three particular cases of such spacetimes. He further proves a theorem for oblate geometries, giving a sufficient condition for the conjecture to be satisfied.

066.185 **Time as spacetime four-volume and the Ashtekar variables.**
L. Bombelli, W. E. Couch, R. J. Torrence.
Phys. Rev. D, Vol. 44, No. 8, p. 2589 - 2592 (15 Oct 1991).
Current Physics Microform No.: 9112A1513.

The authors consider a recently proposed theory of gravity, classically equivalent to Einstein's theory with the cosmological constant as an additional variable, in which spacetime volume plays the role of time. The authors develop a Hamiltonian formulation using Ashtekar's variables, set up the corresponding quantum theory, and show that the known loop state solutions of quantum general relativity are also solutions in the present theory.

066.186 **Observation of 1990 solar eclipse by a torsion pendulum.**
Luo Jun, Li Jianguo, Zhang Xuerong, V. Liakhovets, M. Lomonosov, A. Ragyn.
Phys. Rev. D, Vol. 44, No. 8, p. 2611 - 2613 (15 Oct 1991).
Current Physics Microform No.: 9112A1535.

During the solar eclipse of 22 July 1990 in the city of Bielomorsk (USSR), the authors repeated the torsion pendulum experiment of Saxl and Allen, who reported an anomalous period increase during the solar eclipse of 7 March 1970. The relative change in the pendulum's period associated with the eclipse was found to be less than 5.2×10^{-5} (90% confidence).

066.187 **Pulsar polarization measurements and the nonsymmetric gravitational theory.**
T. P. Krisher.
Phys. Rev. D, Vol. 44, No. 8, p. R2211 - R2213 (15 Oct 1991).
Current Physics Microform No.: 9112A1135.

Because of the breakdown of the Einstein equivalence principle in the nonsymmetric gravitational theory (NGT) of Moffat, orthogonally polarized electromagnetic waves can propagate at different velocities in a gravitational field. Moffat has proposed that galactic dark matter, in the form of cosmions, may act as a significant source of gravity in the NGT. The author discusses how observations of the highly polarized radiation from distant pulsars could provide significant limits on the strength of the coupling of cosmions in the NGT.

066.188 **Rapidly moving cosmic strings and chronology protection.**
A. Ori.
Phys. Rev. D, Vol. 44, No. 8, p. R2214 - R2215 (15 Oct 1991).
Current Physics Microform No.: 9112A1138.

Recently, Gott has provided a family of solutions of the Einstein equations describing pairs of parallel cosmic strings in motion. He has shown that if the strings' relative velocity is sufficiently high, there exist closed timelike curves (CTC's) in the spacetime. Here, the author shows that if there are CTC's in such a solution, then every $t = \text{const}$ hypersurface in the spacetime intersects CTC's. Therefore, these solutions do not contradict the chronology protection conjecture of Hawking.

066.189 **Repulsive gravity in tension stars.**
J. Katz, D. Lynden-Bell.
Mon. Not. R. Astron. Soc., Vol. 252, No. 1, p. 53 - 55 (1 Sep 1991).

In stars, pressure opposes the attractive force of gravity. In general relativity, if the pressure is negative (a tension) and $P < -1/3\, \varrho c^2$ then the resulting gravity is repulsive. Such material is invoked in cosmology to give the inflation of the Universe. In tension stars this repulsive gravity is balanced by the tension. The simplest tension stars only exist in esoteric situations beyond the neck of an Einstein-Rosen bridge in Schwarzschild space. Here the authors confine tension material within a massive shell of normal matter. The resulting object, while still repulsive inside, has an attractive exterior gravity and can, in principle, exist without horizons.

066.190 **Effect of gravity on false-vacuum decay rates for O(4)-symmetric bubble nucleation.**
D. A. Samuel, W. A. Hiscock.
Phys. Rev. D, Vol. 44, No. 10, p. 3052 - 3061 (15 Nov 1991).
Current Physics Microform No.: 9201B0780.

The self-gravity of quantum fields is often considered to be a negligible perturbation upon a background spacetime and not of much physical interest. Its importance is determined by the ratio of the mass of the field to the Planck mass. It is conceivable that either in the very early universe or even today a false-vacuum decay could occur associated with a field of appreciable mass. The effect of self-gravity upon false-vacuum decay was initially studied within the "thin-wall" approximation by Coleman and De Luccia. Their analysis involved the approximate solution of the coupled Euclideanized field and Einstein equations with the assumption of O(4)-symmetric bubble nucleation. In this paper the authors consider the range of validity of the "thin-wall" approximation by comparing the Coleman - De Luccia results with exact numerical results for a quartic polynomial potential. The authors also extend the analysis into regimes for which the "thin-wall" approximation is inapplicable. In the case of an initially de Sitter space decaying into Minkowski space, they find a smooth transition between the Coleman - De Luccia mode of bubble formation and the Hawking-Moss transition, wherein the entire spacetime tunnels "at once" to the maximum of the potential. In all cases considered, the inclusion of gravitational effects tends to thicken the wall of the bubble.

066.191 **Rotating frames and measurements of forces in general relativity.**
F. de Felice.
Mon. Not. R. Astron. Soc., Vol. 252, No. 2, p. 197 – 202 (15 Sep 1991).
It is argued that the phenomenon according to which an increase of the angular velocity causes more attraction than centrifugal repulsion, below the radius of the (spatially) circular photon orbit in the Schwarzschild space–time, has to be attributed to the strength of the gravitational field rather than to a centrifugal force becoming attractive. This is particularly evident in the Reissner–Nordström and in the internal Schwarzschild space–times.

066.192 **Magnetic theory of gravitation.**
Y. M. Cho.
IUPAP Conference on Primordial Nucleosynthesis and Evolution of Early Universe, p. 203 – 211 (1991). – See Abstr. 012.071 for the main entry.
Recently a generalization of Einstein's theory of gravitation has been proposed which can accommodate the gravitational monopoles. The generalization is made in such a way that the equations of motion become exactly symmetric under the dual transformation of the curvature tensor. The theory is characterized by a fundamental scale K which can be interpreted as a "magnetic" mass. In exact analogy with the Dirac's generalization of Maxwell's theory, the theory can be made sensible only if the energy E is quantized by the condition $\varkappa E = 2\pi n$, where n is an integer. The author reviews the recent development.

066.193 **Conformal gravity with torsion.**
Han Pin, W. F. Kao, Pen Ueli.
Chin. J. Phys., Vol. 29, No. 4, p. 283 – 288 (Aug 1991).
The authors study a conformal effective theory in the presence of the torsion field. A power–law inflationary solution in the constant gauge is obtained in $k = 0$ Friedmann–Robertson–Walker space. If $k \neq 0$, a formal result is also presented. Some technical details are discussed and remarked. It appears that the low energy relic from the torsion field deserves more study.

066.194 **A new expression for mass in the theory of gravitation.**
R. M. Avakyan.
Astrophysics, Vol. 33, No. 3, p. 548 – 553 (Nov – Dec 1990). English translation of Astrofizika, Tom 33, Vyp. 3, p. 429 – 436 (Dec 1990).
The gravitational field produced by a static spherically symmetric matter distribution is considered. In the framework of the theories of Newton and Einstein and Rosen's bimetric theory a new formula for the total mass of the spherical configuration is obtained; it relates the mass to the pressure distribution.

066.195 **Gravitational breather and topological properties of gravisolitons.**
V. Belinsky (*V. A. Belinskij*).
Phys. Rev. D, Vol. 44, No. 10, p. 3109 – 3115 (15 Nov 1991).
Current Physics Microform No.: 9201B0837.
It is shown that for a wide class of cosmological solutions to the vacuum Einstein equations the notions of gravisolitons and antigravisolitons with respect to some topological charge can be introduced. The presence of attractive forces between two gravitational solitons with charges of opposite signs and repulsive forces between solitons of the same charges is shown. The construction of the gravitational breather as a bound state of the gravisoliton and antigravisoliton is described.

066.196 **Consistency of finite–difference solutions of Einstein's equations.**
M. W. Choptuik.
Phys. Rev. D, Vol. 44, No. 10, p. 3124 – 3135 (15 Nov 1991).
Current Physics Microform No.: 9201B0852.
In the past, arguments have been advanced suggesting that certain finite–difference solutions of the 3+1 form of Einstein's equations suffer from a fundamental inconsistency. Specifically, it has been claimed that freely evolved solutions, where the constraint equations are not explicitly imposed after the initial time, will generally satisfy discrete versions of the constraints to a lower order in the basic mesh spacing h than the truncation order of the discretized evolution equations. This issue is reexamined here, and using the key observation, originally due to Richardson, that a numerical differentiation need not produce an $O(h^{p-1})$ quantity from an $O(h^p)$ one, it is argued that there should be no such inconsistency for convergent difference schemes. Numerical results from a study of spherically symmetric solutions of a massless scalar field minimally coupled to the gravitational field are presented in support of this claim.

066.197 **Scalar field, nonminimal coupling, and cosmology.**
M. Demiański, R. De Ritis, G. Marmo, G. Platania, C. Rubano, P. Scudellaro, C. Stornaiolo.
Phys. Rev. D, Vol. 44, No. 10, p. 3136 – 3146 (15 Nov 1991).
Current Physics Microform No.: 9201B0864.
The authors study the dynamics of a flat Friedmann – Robertson – Walker universe filled with a self–interacting scalar field nonminimally coupled to the gravitational field. Dynamical equations for the system can be derived from a pointlike Lagrangian. For this system an additional Nöther symmetry exists, provided that the coupling constant ξ is equal to 0 or 1/6. When $\xi = 1/6$, the scalar potential has to be constant. In this case one obtains an exact solution. The authors also analyze the behavior of the scalar field when $\xi \neq 0, 1/6$. Most of the considered solutions are unphysical, but there exists a very interesting case in which the effective cosmological constant is rapidly changing, which might lead to inflation.

066.198 **New exact solution for the exterior gravitational field of a charged spinning mass.**
A. Chamorro, V. S. Manko, T. E. Denisova.
Phys. Rev. D, Vol. 44, No. 10, p. 3147 – 3151 (15 Nov 1991).
Current Physics Microform No.: 9201B0875.
An exact asymptotically flat solution of the Einstein–Maxwell equations describing the exterior gravitational field of a charged rotating axisymmetric mass possessing an arbitrary set of multipole moments is presented explicitly.

066.199 **Gravitational field of a hedgehog and the evolution of vacuum bubbles.**
E. I. Guendelman, A. Rabinowitz.
Phys. Rev. D, Vol. 44, No. 10, p. 3152 – 3158 (15 Nov 1991).
Current Physics Microform No.: 9201B0880.
The gravitational field produced by a spherically symmetric "hedgehog" configuration in scalar field theories with global SO(3) symmetry is studied in the limit in which these models become nonlinear σ models. The same gravitational effect can be generated by a set of cosmic strings intersecting at a point, in the limit that one considers a continuous distribution of such intersecting strings in a spherically symmetric configuration. When the energy densities associated with the hedgehog are small, one obtains a static geometry, but for higher values, the resulting geometry is that of an anisotropic cosmology. The evolution of bubbles joining two phases, one of which contains a hedgehog is investigated. The role of such configurations in processes that lead to classical false vacuum destabilization and in the evolution of inflationary bubbles is discussed. The generalization of these results to the gauged case, i.e., to magnetic monopole hedgehogs, is discussed.

066.200 **Spherical shock waves in general relativity.**
Y. Nutku.
Phys. Rev. D, Vol. 44, No. 10, p. 3164 – 3168 (15 Nov 1991).
Current Physics Microform No.: 9201B0892.
The author presents the metric appropriate to a spherical shock wave in the framework of general relativity. This is a Petrov type N vacuum solution of the Einstein field equations, where the metric is continuous across the shock and the Riemann tensor suffers a step–function discontinuity. Spherical gravitational waves are described by type N Robinson–Trautman metrics. However, for shock waves the Robinson–Trautman

solutions are unacceptable because the metric becomes discontinuous in the Robinson–Trautman coordinate system. Other coordinate systems that have so far been introduced for describing Robinson–Trautman solutions also suffer from the same defect. The author presents the C^0–form of the metric appropriate to spherical shock waves using Penrose's approach of identification with warp.

066.201 **Theory of an Earth–bound clock comparison experiment as test of the principle of equivalence.**
G. I. Opat, W. G. Unruh.
Phys. Rev. D, Vol. 44, No. 10, p. 3342 – 3344 (15 Nov 1991).
Current Physics Microform No.: 9201B1070.

The comparison at sea level of the rates of two atomic clocks of differing geographical latitude is shown to provide a sensitive test of the principle of equivalence. Specifically, clocks of $1/10^{15}$ stability would effectively compare the rate of fall of the excitation energy of an atom with that of sea water to an accuracy of $1/10^{14}$ or better, a comparison which would be extremely difficult by normal techniques.

066.202 **Christodoulou's nonlinear gravitational–wave memory: evaluation in the quadrupole approximation.**
A. G. Wiseman, C. M. Will.
Phys. Rev. D, Vol. 44, No. 10, p. R2945 – R2949 (15 Nov 1991).
Current Physics Microform No.: 9201B0673.

Christodoulou has found a new nonlinear contribution to the net change in the wave form caused by the passage of a burst of gravity waves ("memory of the burst"). The authors argue that this effect is nothing but the gravitational wave form generated by the stress energy in the burst itself. They derive an explicit formula for this effect in terms of a retarded–time integral of products of time derivatives of wave zone gravitational wave forms. The resulting effect corresponds in size to a correction 2.5 post–Newtonian orders ($O(v/c)^5$) beyond the quadrupole approximation, and is therefore negligible for all but the most relativistic of systems. For gravitational bremsstrahlung from two stars moving at 300 km s^{-1}, the effect is much less than 10^{-10} of the usual linear quadrupole wave form, while for a system of coalescing binary compact objects the effect is estimated to be of order 10^{-1} for two neutron stars.

066.203 **The fractal structure of the quantum space–time.**
L. Nottale.
Applying fractals in astronomy, p. 181 – 200 (1991). Also published in Particle astrophysics – the early universe and cosmic structures, J. M. Alimi et al. (eds.), Editions Frontières (1990). See also 53.066.242. – See Abstr. 003.023 for the main entry.

The authors report on the present state of an attempt at understanding microphysics by basing on a "principle of scale relativity", according to which the laws of physics should apply to systems of reference whatever their scale. They suggest the achievement of such a principle by using the mathematical tool of fractals (Nottale 1989). Arguments are given for generally describing them as families of Riemannian space–times whose curvature tends to infinity when scale approaches zero. The authors develop the interpretation of the wave–particle duality as a property of families of geodesical lines in a fractal space–time and examplify it with the Young's hole and Einstein–Podolsky–Rosen experiments. Finally some possible consequences concerning gravitation are recalled, with the suggestion that Newton's law may break down for active gravitational masses smaller than the Planck mass.

066.204 **Wormholes in string theory.**
A. Lyons, S. W. Hawking.
Phys. Rev. D, Vol. 44, No. 12, p. 3802 – 3818 (15 Dec 1991).
Current Physics Microform No.: 9101C0084.

The authors discuss the wormhole effective interactions in string theory, thought of as a sum over two–dimensional field theories on different world sheets. The effective interactions are calculated in the "dilute wormhole approximation", initially by considering the Green's functions on higher–genus Riemann surfaces, and then by calculating the effect of a complete basis of wave functions on scattering amplitudes for a surface with a boundary. The sum over wormholes is equivalent to having a world sheet of trivial topology and summing over different space–time and matter–field backgrounds. To leading order these consist of the massless fluctuations, since the tachyon cancels out when a sum is done over different spin structures going through the wormhole. In this way the authors recover quantized general relativity as an effective theory, from a sum over field theories on higher–genus Riemann surfaces.

066.205 **Topological four–dimensional self–dual gravity.**
A. Nakamichi, A. Sugamoto, I. Oda.
Phys. Rev. D, Vol. 44, No. 12, p. 3835 – 3842 (15 Dec 1991).
Current Physics Microform No.: 9101C0117.

A theory of topological gravity with an independent vierbein and spin connection is formulated in four space–time dimensions. Topological symmetries, as well as local Lorentz transformations and diffeomorphisms. The gauge–fixing conditions of the topological symmetries are chosen in such a way that self–duality is preserved for both curvature and torsion two–forms. It is shown that these gauge–fixing conditions have nontrivial classical solutions. Nontrivial observables are also constructed.

066.206 **Generalized Friedmann – Robertson – Walker metric and redundancy in the generalized Einstein equations.**
W. F. Kao, Pen Ueli.
Phys. Rev. D, Vol. 44, No. 12, p. 3974 – 3977 (15 Dec 1991).
Current Physics Microform No.: 9101C0256.

A nontrivial redundancy relation, due to the differential structure of the gravitational Bianchi identity as well as the symmetry of the Friedmann – Robertson – Walker (FRW) metric, in the gravitational field equation is clarified. A generalized FRW metric is introduced in order to properly define a one–dimensional reduced problem which offers an alternative approach to obtain the gravitational field equations on FRW spaces.

066.207 **Legendre transforms in Ashtekar's theory of gravity.**
P. Peldán.
Classical Quantum Gravity, Vol. 8, No. 10, p. 1765 – 1783 (Oct 1991).

Legendre transforms from Ashtekar's Hamiltonian formulation of gravity to a metric–free second–order configuration space Lagrangian formulation, are performed for gravity with matter couplings and cosmological constant.

066.208 **Testing Einstein's theory of gravity by analyzing Lunar Laser Ranging data.**
J. Müller, M. Schneider, M. Soffel, H. Ruder.
Astrophys. J., Lett., Vol. 382, No. 2, p. L101 – L103 (1 Dec 1991).

Lunar Laser Ranging data of more than 20 years have been used to accurately determine parameters of the Earth–Moon system (e.g., the mass of the system, the lunar tidal acceleration etc.), the station – reflector geometry, and other physical parameters like the solar quadrupole moment J_2. In addition to these, parameters suitable for testing metric theories of gravity in the first post–Newtonian approximation have been determined with great accuracy. These are the PPN (parameterized post–Newtonian) parameters γ and β, the Nordtvedt parameter η, the geodetic precession of the lunar orbit Ω_{GP}, as well as a possible time variation of the gravitational constant G.

066.209 **Microscopic anisotropy in a Robertson–Walker cosmology.**
J. T. Lynch.
Classical Quantum Gravity, Vol. 8, No. 10, p. 1807 – 1813 (Oct 1991).

It is shown that in spinor–tetrad theory, with both left– and right–acting connection groups, there exists an exact Robertson–Walker cosmological solution with a matter field which possesses a polarized (as opposed to random) spin. The solution arises because the torsion vector, which appears anisotropically in the microscopic Dirac–like equation of the theory, enters the macroscopic Einstein–like gravitational equations in a purely scalar, isotropic form. The energy density of the cosmology is shown to be one quarter of that for the Einstein – de Sitter dust–filled universe. A pressure – density relation for the solution is also derived, although a conclusive physical interpretation for the solution remains unclear.

066.210 **Degenerate metrics and an empty black hole.**
I. Bengtsson.
Classical Quantum Gravity, Vol. 8, No. 10, p. 1847 – 1857 (Oct 1991).
Ashtekar's formulation of general relativity is polynomial, and admits degenerate "metrics". The author considers "metrics" which are everywhere degenerate on the initial data hypersurface. For a special class of them, the constraint and evolution equations are solved completely. The author then considers spherically symmetric spacetimes. A general form of the stationary solutions, in terms of three arbitrary functions, without gauge fixing, is given for both degenerate and non–degenerate metrics. It is shown that the Schwarzschild solution may be joined smoothly to a degenerate solution across the event horizon, yielding a "black hole" with a degenerate interior. Some comments on matter couplings are made.

066.211 **Energy–momentum tensor for minimally coupled charged perfect fluids with spin.**
L. L. Smalley, J. P. Krisch.
Classical Quantum Gravity, Vol. 8, No. 10, p. 1889 – 1894 (Oct 1991).
The authors minimally couple the electromagnetic field to gravity in Riemann–Cartan spacetimes using the self–consistent formalism for spinning fluids. The total Lagrangian is gauge invariant. The electromagnetic field equations, the spin transport equations and the symmetric stress–energy tensor are obtained.

066.212 **Gravitational Aharonov–Bohm effect in a locally flat spacetime.**
V. B. Bezerra.
Classical Quantum Gravity, Vol. 8, No. 10, p. 1939 – 1945 (Oct 1991).
A quantum particle is considered in the background gravitational field due to a tubular matter source with axial interior magnetic field and vanishing exterior magnetic field. It is shown that in the locally flat spacetime around this tubular matter source, the wavefunction, the energy spectrum and the probability current depend on the amount of matter and magnetic field. These dependences represent gravitational analogues of the Aharonov–Bohm effect in electrodynamics.

066.213 **Prolate collapse of string loops and domain walls.**
C. Barrabes.
Classical Quantum Gravity, Vol. 8, No. 10, p. L199 – L204 (Oct 1991). Letter–to–the–editor.
An exact derivation of the upper bound of the energy emitted as gravitational radiation is given for a prolate domain wall or string collapsing with the speed of light toward a black hole, under the assumption that cosmic censorship is valid. Estimates of the maximum deformation from spherical shape for a black hole to be formed are also derived.

066.214 **Gravitational effects of light scalar particles.**
T. Helbig.
Astrophys. J., Vol. 382, No. 1, p. 223 – 232 (20 Nov 1991).
The post–Newtonian metric of the general scalar–tensor theory with a massive scalar field is calculated. The result is used to recalculate several relativistic effects in the solar system. Experimental bounds for the massless scalar–tensor theory that have been obtained from these effects are used to obtain new bounds for the massive case. Furthermore, the author gives a counterexample to the conjecture that all theories with only one gravitational field obey the strong equivalence principle.

066.215 **Diffusion processes in the collapse of a radiating spherical body.**
R. Chan, L. Herrera, J. A. F. Pacheco, N. O. Santos.
Astrophys. J., Vol. 382, No. 1, p. 255 – 260 (20 Nov 1991).
The authors have further developed the model considered recently by Barreto et al. to study the heat conduction and photon diffusion processes in the collapse of a radiating spherical system. The spherical body is divided into two regions: the core, modeled with a nonadiabatic fluid that can produce radiation; and the envelope of dust and radiation that surrounds the core. The effective adiabatic index, the shear, and the temperature are analyzed for the heat conduction and the photon diffusion limits.

066.216 **STEP. A Satellite Test of the Equivalence Principle.**
R. Reinhard, P. W. Worden, C. W. F. Everitt.
Europhys. News, Vol. 22, No. 11, p. 216 – 218 (Dec 1991).
STEP aims to determine any difference in the rate of fall of test masses in an Earth–orbiting satellite to one part in 10^{17} of the total gravitational acceleration – a 10^6 factor improvement over existing measurements.

066.217 **Gravitation and hot big–bang cosmology.**
J. D. Barrow.
12. Autumn School of Physics. The physical universe – the interface between cosmology, astrophysics and particle physics, p. 1 – 20 (1991). – See Abstr. 012.097 for the main entry.
This lecture deals with mathematical and formal aspects of Newtonian and general relativistic gravitation as well as some higher–order lagrangian theories of gravity. It introduces the Friedman metrics and places them in the context of all solutions to Einstein's equations. The emphasis is upon basic principles and ideas that do not lean too heavily upon specific models.

066.218 **Structure of spacetime tangent bundle.**
H. E. Brandt.
Found. Phys. Lett., Vol. 4, No. 6, p. 523 – 536 (Dec 1991). Expanded version of a paper presented at "6. Marcel Grossmann Meeting on General Relativity, Kyoto (Japan), 23 – 29 Jun 1991".
The universal upper limit on attainable proper acceleration relative to the vacuum imposes restrictions on possible structures in the spacetime tangent bundle. Various features of the differential geometry of the spacetime tangent bundle are presented. Also, a modified Schwarzschild solution is obtained, and the associated gravitational redshift is calculated.

066.219 **Bianchi cosmologies: the role of spatial topology.**
A. Ashtekar, J. Samuel.
Classical Quantum Gravity, Vol. 8, No. 12, p. 2191 – 2215 (Dec 1991).
The role of spatial topology in the Hamiltonian description of Bianchi models is analysed. It turns out that, in general, the number of degrees of freedom of these models is not uniquely determined by the isometry group but depends in addition on the choice of topology. Consequently, the quantum theory is quite sensitive to this choice. Contrary to one's initial expectation, subtleties arise in the spatially open models – say with topology R^3 – rather than in closed models. Finally, it is shown that class B models cannot occur with closed topologies.

066.220 **Self–consistent spinning fluid theory in the cosmological context.**
M. A. P. Martins, E. P. Vasconcellos–Vaidya, M. M. Som.
Classical Quantum Gravity, Vol. 8, No. 12, p. 2225 – 2229 (Dec 1991).
It is shown that the assumed consistency equation of the self–consistent theory of a spinning fluid puts severe restrictions on the fluid configurations. If one drops the consistency equation, the spin kinetic energy gives rise to some novel effects similar to that of an imperfect fluid. Following the analogy of the imperfect fluid, the authors describe these effects as dissipative processes induced by spin, leading to entropy production.

066.221 **Tension shells and tension stars.**
J. Katz, D. Lynden–Bell.
Classical Quantum Gravity, Vol. 8, No. 12, p. 2231 – 2238 (Dec 1991).
In a tension star, the tension is so great that gravity becomes repulsive and is balanced by the cohesive effect of tension. Such repulsive gravity is the basis of the inflation of the universe. The simplest model of a tension star is Schwarzschild's internal solution but with a negative pressure. Such objects are static but inhibit the space on the far side of the Einstein–Rosen bridge. Static shells of matter in tension, including spherical domain walls, can be built in this region only.

066.222 **Gauge transformations in linearized theory of gravitational waves.**
P. Fortini, A. Ortolan.
5. Winter School on Hadronic Physics: Problems of fundamental modern physics II, p. 379 – 392 (1991). – See Abstr. 012.101 for the main entry.
The most important properties of gauge transformations in the linearized approximation of general relativity are reviewed. These properties are applied to solve the equations of motion of a light ray in the laboratory frame when a monochromatic plane gravitational wave impinges upon an interferometric device. From this theory a very interesting interpretation of the phase shift in the frame of the interferometer is derived.

066.223 **2D gravity coupled to discrete matter and random matrices.**
J. Zinn–Justin.
1990 Summer School in High Energy Physics and Cosmology, p. 202 – 221 (1991). – See Abstr. 012.105 for the main entry.
Recent progress in 2D gravity coupled to $d \leqslant 1$ matter, based on a representation of discrete gravity in terms of random matrices, is reported. Euclidean 2D gravity is also a theory of random surfaces. The matrices are $N \times N$ hermitian matrices because the surface are orientable. Continuum gravity is then recovered in a special large N double scaling limit. In the case of discrete matter the matrix problem can be in many cases solved by the introduction of suitable orthogonal polynominals. Alternatively in the continuum limit the orthogonal polynominal method can be shown to be equivalent to the construction of representation of the canonical commutation relations in terms of ordinary differential operators. The sum over topologies is then reduced to the solution of non–linear differential equations. Similar ideas lead to a solution of 2D gravity coupled to $d = 1$ matter. The problem can be formulated as quantum mechanics of hermitian matrices and solved by semiclassical methods.

066.224 **Aspects of anomaly structures in *W* gravity.**
K. Li, C. Pope.
1990 Summer School in High Energy Physics and Cosmology, p. 222 – 230 (1991). – See Abstr. 012.105 for the main entry.
The authors give a brief review of some of the recent developments in *W* gravity coupled to matter theories. They then extend the discussion to the case of minimal *W* conformal field theory, realised in terms of free bosons with background charges, and use this action to investigate some aspects of the anomaly structure and the covariantisation of *W* gravity.

066.225 **The Dirac equation in the Kerr metric with small angular momentum and the quasi bound state.**
Zhang Shiwei, Shen Yougen.
Ann. Shanghai Obs., Acad. Sin., No. 12, p. 90 – 94 (1991). In Chinese.
The Hamiltonian of the Dirac equation in the Kerr metric with small angular momentum is obtained by the Bergman–Brill–Wheeler prescription and the influence of a on the energy spectrum in the quasi bound state is derived.

066.226 **A test of Newton's inverse square law of gravitation using the 300–m tower of Erie, Colorado.**
J. Y. Cruz, J. C. Harrison, C. C. Speake, T. M. Niebauer, M. P. McHugh, P. T. Keyser, J. E. Faller, J. Mäkinen, R. B. Beruff.
J. Geophys. Res., Vol. 96, No. B12, p. 20073 – 20092 (10 Nov 1991).
Gravity was measured at eight different heights on a 300–m meteorological tower, using electrostatically nulled LaCoste and Romberg gravimeters. The observed values were corrected for tides, temperature, and gravimeter screw errors, and tested for systematic effects due to (wind–induced) tower motion (no such effects were found). These corrected results were compared with values predicted by means of Newton's inverse square law from surface gravity values. The differences exhibit no systematic trends, and their rms value is only 10 μGal, well within the errors of the experiment, as the estimated measurement errors increase from 9 μGal at the lowest platform to 14 μGal at the top and those of the predictions from 10 μGal to 23 μGal. These results set new constraints on the magnitude of any non–Newtonian gravitational force; if such a force is derived from a Yukawa potential, the absolute value of α must be less than 0.001 for $\lambda = 1000$ m.

066.227 **Some aspects of minimally relativistic Newtonian gravity.**
K. K. Nandi.
Z. Naturforsch., A, Vol. 46, No. 12, p. 1026 – 1032 (Dec 1991).
This paper aims to examine if the tests of general relativity can be predicted by a simpler approach based on minimal changes in the Newtonian gravity.

066.228 **Gravitational radiation connected with anticollapse and supernovae explosions.**
A. P. Trofimenko, V. S. Gurin.
Astrofizika, Tom 34, Vyp. 1, p. 83 – 89 (Feb 1991). In Russian. English translation in Astrophysics, Vol. 34, No. 1.
A model of D–bodies as anticollapsing objects (white holes) is considered in extended space–time manifold taking into account the field of gravitational radiation. The solution by the Robinson–Trautman type is used for the description of spherical gravitational wave space–time. It shows an oscillating character of space–time manifold global structure for the gravitationally radiating white hole. Such a model is applied to the explanation of energetics of the Supernova 1987A's gravitational burst. It has been pointed out that a creation of D–bodies is also accompanied by a burst of gravitational radiation.

066.229 **On the Tolman mass formula in the scalar–tensor theories of gravitation.**
M. R. Avakyan, L. Sh. Grigoryan, A. A. Saaryan.
Astrofizika, Tom 34, Vyp. 2, p. 265 – 270 (Apr 1991). In Russian. English translation in Astrophysics, Vol. 34, No. 2.
In the framework of scalar–tensor and scalar–tensor–bimetric theories of gravitation the total mass of matter and of the constant gravitational field is expressed in the form of an integral over the volume occupied by the matter.

066.230 **Lichtablenkung, Gravitationslinsen und Einsteinsche Ringe.**
K.–H. Lotze.
Sterne, Band 67, Heft 6, p. 297 – 316 (1991).

066.231 **Oscillating and inflating universe in SO(3)–gravitation theory.**
M. Mattes, M. Sorg.
Z. Naturforsch., A, Vol. 46, No. 11, p. 967 – 975 (Nov 1991).
The predictions of SO(3) gravitation theory are studied for a homogeneous, isotropic universe. Various types of oscillations are obtained for the pre–inflationary phase, where the geometry experiences violent fluctuations. These are ultimately terminating at either re–collaps or regular inflation.

Gauge field theories. An introduction with applications.
See Abstr. 003.003.

POSTPRINCIPIA. Gravitation for physicists and astronomers.
See Abstr. 003.006.

Search for gravitational radiation.
See Abstr. 004.010.

De l'influence de la gravitation sur la propagation de la lumière en théorie newtonienne. L'archéologie des trous noirs.
See Abstr. 004.025.

The happiest thought of Einstein's life.
See Abstr. 005.027.

Quantum gravity. Proceedings.
See Abstr. 012.004.

Relativity and gravitation: classical and quantum. Proceedings.
See Abstr. 012.005.

Current problems in particle physics and cosmology. Proceedings. Workshop on Current Problems in Particle Physics and Cosmology, Ioannina (Greece), 25 – 29 Jun 1990.
See Abstr. 012.022.

Quantum cosmology and baby universes. Proceedings.
See Abstr. 012.036.

Gravitational astronomy: instrument design and astrophysical prospects. Proceedings.
See Abstr. 012.098.

High energy physics and cosmology. Proceedings.
See Abstr. 012.105.

The Fifth Marcel Grossmann Meeting on Recent Developments in Theoretical and Experimental General Relativity, Gravitation and Relativistic Field Theories. Parts A, B. Proceedings.
See Abstr. 012.112.

General relativity and gravitational physics. Proceedings.
See Abstr. 012.115.

Observational project on gravitational microlensing.
See Abstr. 013.059.

Prospects of experimental gravitational wave research in Moscow State University.
See Abstr. 013.097.

Determining isometries of space–times using computer algebra.
See Abstr. 021.007.

Helmholtz conditions for differential equations of any order.
See Abstr. 021.008.

Interferometric gravitational wave detector using the transmission light on the Fabry–Perot cavity.
See Abstr. 022.010.

Heating by optical absorption and the performance of interferometric gravitational–wave detectors.
See Abstr. 022.019.

Search for anomalous spin–dependent forces using stored ion spectroscopy.
See Abstr. 022.037.

Submarine measurement of the Newtonian gravitational constant.
See Abstr. 022.066.

Modelling of the Virgo interferometer.
See Abstr. 034.142.

Ultralow temperature resonant–mass gravitational radiation detectors: current status and future prospects.
See Abstr. 034.143.

Development of 10 m laser interferometer gravitational wave antenna (TENKO–10) at ISAS.
See Abstr. 034.144.

Conceptual design of a Japanese interferometric gravitational wave detector.
See Abstr. 034.145.

Interferometric detection of gravitational waves: the VIRGO project.
See Abstr. 034.146.

Interferometer measurements using non–classical light.
See Abstr. 034.147.

A proposed large–scale laser interferometer gravitational wave detector in Australia.
See Abstr. 034.148.

The gravitational wave experiment of the Rome group. Status report for the resonant antennas Explorer and Nautilus.
See Abstr. 034.149.

Modern status and future plans of the Moscow University interferometer.
See Abstr. 034.150.

Low frequency noise in gravitational wave interferometers.
See Abstr. 034.151.

Some recent developments in interferometric gravitational–wave detection.
See Abstr. 034.152.

Coincidences between gravitational wave bar detectors.
See Abstr. 034.153.

Fabry–Perot resonators for interferometric GW antennas.
See Abstr. 034.154.

VIRGO: a ground based interferometric antenna for gravitational wave detection above 10 Hz.
See Abstr. 034.155.

Experimental gravitation at very low temperatures.
See Abstr. 034.156.

Detectors and detections of gravitational waves.
See Abstr. 036.129.

Choice of filters for the detection of gravitational waves from coalescing binaries.
See Abstr. 036.230.

Relativistic effects resulting from the rotation of celestial bodies.
See Abstr. 042.079.

Analytical algorithms of relativistic reduction of astronomical observations.
See Abstr. 043.009.

The effects of solar gravitation field on the time delay for VLBI.
See Abstr. 044.050.

Three dimensional four–fermion (vector)⊗(vector) model is renormalizable and ultraviolet finite.
See Abstr. 061.020.

Optimal Cavendish–type experiment for the search of restrictions on light particles of quantum gravity.
See Abstr. 061.021.

Inequalities which constrain negative energy in quantum field theory.
See Abstr. 061.022.

Hamiltonian diagonalization in curved space.
See Abstr. 061.024.

Cartan's torsion vector and spin–0 fields
See Abstr. 061.027.

Soliton operators in two and three dimensions.
See Abstr. 061.028.

Quantization requires a Lagrangian.
See Abstr. 061.029.

BRST effective actions on phase space.
See Abstr. 061.031.

Can the Sun shed light on neutrino gravitational interactions?
See Abstr. 061.046.

Modified photon equation of motion as a test for the principle of equivalence.
See Abstr. 061.090.

Vacuum fluctuation inside a toroidal spacetime containing a cosmic string.
See Abstr. 061.099.

Gravitational instability for some astrophysical systems.
See Abstr. 062.091.

Magnetic flux tubes – nonlinear strings in relativistic magnetohydrodynamics.
See Abstr. 062.096.

Two–dimensional fluctuating flow and heat transfer to a relativistic radiating gas.
See Abstr. 062.144.

Dynamical instability of a general relativistic rotating fluid.
See Abstr. 062.208.

Relativistic magnetic partially degenerate isothermal gas spheres.
See Abstr. 065.034.

On naked singularities in self–similar Tolman–Bondi spacetimes.
See Abstr. 067.022.

Oscillatory luminosity profiles and the contraction of radiating spheres in general relativity.
See Abstr. 067.024.

Emission of gravitational waves from roto–vibrational modes of a compressible and homogeneous star.
See Abstr. 067.035.

Non–radial pulsation of a boson star. I. Formulation.
See Abstr. 067.059.

Gravitational radiation from coalescing binary neutron stars. IV. Tidal disruption.
See Abstr. 067.071.

String theory and black holes.
See Abstr. 067.087.

Simple equations for general relativistic hydrodynamics in spherical symmetry applied to neutron star collapse.
See Abstr. 067.094.

General relativistic effects in astrophysical jets.
See Abstr. 067.100.

Normal–mode excitation from stellar collapse to a black hole: odd–parity perturbations.
See Abstr. 067.103.

The neutron star boundary layer and general relativity.
See Abstr. 067.170.

Charged black holes with scalar hair.
See Abstr. 067.209.

Gravitational radiation from dual neutron star elliptical binaries.
See Abstr. 067.214.

Trapped surfaces in the Schwarzschild geometry and cosmic censorship.
See Abstr. 067.220.

Self–gravitating magnetic monopoles, global monopoles and black holes.
See Abstr. 067.240.

The binding energy of a Schwarzschild sphere in *D* space–time dimensions.
See Abstr. 067.273.

Solar gravitational redshift from the infrared oxygen triplet.
See Abstr. 071.031.

General relativity and physics of the Earth.
See Abstr. 081.048.

Relativistic manifestations of gravitational fields in gravimetry and geodesy.
See Abstr. 081.054.

On the existence of "stable and unstable" quantum levels in the gravitational field of the Sun.
See Abstr. 091.014.

Absence of "stable and unstable" quantum levels in the solar gravitational field.
See Abstr. 091.029.

Analyse von Lasermessungen zum Mond im Rahmen einer post–Newton'schen Theorie.
See Abstr. 094.013.

Eclipsing binary stars as tests of gravity theories: new solutions for AS Camelopardalis.
See Abstr. 119.034.

On the registration of the events by Rome and Maryland gravitational–wave detectors during 23 February 1987.
See Abstr. 125.182.

On the detection of events of 23 February 1987 by the Rome and Maryland gravitational–wave detectors.
See Abstr. 125.186.

The rate of neutron star binary mergers in the universe: minimal predictions for gravity wave detectors.
See Abstr. 126.068.

A star–forming region in the gravitational lens 2237+030.
See Abstr. 131.379.

A scale invariant, general relativistic model for a galaxy in a dark matter background.
See Abstr. 151.009.

Probing Galactic halo dark matter with microlensing.
See Abstr. 155.004.

A feasibility study for detection of gravitational microlensing by baryonic dark matter via CCD photometry.
See Abstr. 155.023.

MG 1131+0456: discovery of the optical Einstein ring with the NTT.
See Abstr. 158.073.

Recent developments in gravitational lensing.
See Abstr. 158.098.

Enhanced probability of intensity outbursts via superluminal microlensing of quasars.
See Abstr. 159.028.

Gravitational micro–lensing for large sources.
See Abstr. 159.048.

Initial light curve of Q2237+0305.
See Abstr. 159.068.

Fifth image and photometric variability in 2237+0305 ("Einstein Cross").
See Abstr. 159.069.

Expected color variations of the gravitationally microlensed QSO 2237+0305.
See Abstr. 159.075.

Infrared K–band observations of the gravitational lens 2016+112.
See Abstr. 159.079.

Spectral differences in the QSO pairs Q1634+267A,B and Q2345+007A,B and the gravitational lens interpretation.
See Abstr. 159.082.

Interpreting the light curve of Q2237+0305.
See Abstr. 159.087.

The role of microlensing in estimates of the relative time delay for the gravitational images of Q 0957+561.
See Abstr. 159.113.

Thin and giant luminous arcs: a strong test of the lensing cluster mass distribution.
See Abstr. 160.151.

Robertson–Walker viscous fluid model.
See Abstr. 161.002.

Lie group for the Jordan and Brans–Dicke cosmology.
See Abstr. 161.003.

Inhomogeneous generalizations of Bianchi type VI_h models with perfect fluid.
See Abstr. 161.004.

Bianchi type III and Kantowski–Sachs cosmological models in Lyra geometry.
See Abstr. 161.005.

Relativistic fireworks.
See Abstr. 161.021.

Third quantization of gravity and the cosmological constant problem.
See Abstr. 161.044.

The arguments against "antigravity" and the gravitational acceleration of antimatter.
See Abstr. 161.059.

One–loop quantum cosmology: the general semiclassical solution of the Wheeler–DeWitt equations and its applications.
See Abstr. 161.061.

The self–measurement of the Universe and the concept of time in quantum cosmology.
See Abstr. 161.063.

Quantum string theory in curved space–times.
See Abstr. 161.065.

Quantum processes in the gravitational field of a cosmic string.
See Abstr. 161.072.

Equilibrium configurations of superconducting cosmic strings in background fields.
See Abstr. 161.074.

On multi–inflation cosmological models.
See Abstr. 161.076.

Kaluza–Klein models: from quantum state to observations.
See Abstr. 161.078.

Cosmological solutions from supersymmetric Kaluza–Klein theory.
See Abstr. 161.079.

Space–time dimension of the Universe.
See Abstr. 161.080.

On the origin of classical density perturbations in the inflationary Universe.
See Abstr. 161.081.

Theory of cosmological perturbations.
See Abstr. 161.089.

Spacetime and causal sets.
See Abstr. 161.092.

Polytropic gravitational wave production cosmological model.
See Abstr. 161.096.

Perturbations of the Robertson–Walker space: multicomponent sources and generalized gravity.
See Abstr. 161.099.

Perturbations in a singularity–free cosmological model.
See Abstr. 161.100.

Perturbations in a singularity–free cosmological model. II. The synchronous gauge.
See Abstr. 161.101.

Time variation of fundamental constants, preferred reference systems and CPT invariance.
See Abstr. 161.103.

Empty voids in the expanding universe.
See Abstr. 161.105.

Decoherence and backreaction in quantum cosmology.
See Abstr. 161.109.

The square root of the Wheeler–DeWitt equation in cosmology.
See Abstr. 161.112.

5–dimensional scalar–tensor cosmologies as 4–dimensional GR universes.
See Abstr. 161.115.

Spatially homogeneous vacuum space–time cosmological solutions for the Jordan Brans–Dicke theory in Bianchi–type universes.
See Abstr. 161.116.

Anisotropic degree measurement for Bianchi type models.
See Abstr. 161.117.

Cosmic strings of scalar fields: Brans–Dicke case.
See Abstr. 161.118.

Cosmological consequences of high–frequency oscillations of Newton's constant.
See Abstr. 161.131.

Supersymmetric Bianchi type IX cosmology.
See Abstr. 161.138.

On slowly–rotating cosmological viscous–fluid model universes in general relativity.
See Abstr. 161.140.

Can Brans–Dicke universes in five dimensions be interpreted as 4–dimensional general relativity cosmologies?
See Abstr. 161.141.

Some Bianchi–type cosmological models in a new scalar–tensor theory.
See Abstr. 161.143.

Gravitational lensing by large–scale structures.
See Abstr. 161.151.

Successful inflation in scalar–tensor theories of gravity.
See Abstr. 161.159.

On quantum gravity for homogeneous pure radiation universes.
See Abstr. 161.162.

Rotational perturbations of relativistic perfect–fluid model universes interacting with gravitational field.
See Abstr. 161.164.

A new gravitational interaction of cosmological importance.
See Abstr. 161.173.

The cosmological solution of a homogeneous, isotropic, ideal fluid at local rest.
See Abstr. 161.176.

The ground–state wavefunction of a radiation–dominated universe.
See Abstr. 161.193.

Existence of finitely perturbed Friedmann models via the Cauchy problem.
See Abstr. 161.194.

Singular self–adjoint quantum cosmological models in a slow–time gauge.
See Abstr. 161.197.

The gravitational effects of superconducting cosmic string.
See Abstr. 161.198.

Qualitative analysis of a class of inhomogeneous self–similar cosmological models. II.
See Abstr. 161.201.

On the variability of physical constants and conformal transformations.
See Abstr. 161.206.

The quantum mechanics of cosmology.
See Abstr. 161.208.

Introductory lectures on quantum cosmology.
See Abstr. 161.209.

Baby universes.
See Abstr. 161.210.

Critique of Coleman's theory of the vanishing cosmological constant.
See Abstr. 161.212.

Catastrophe and cosmology in Finsler spacetime.
See Abstr. 161.220.

Astroparticle physics and superstrings.
See Abstr. 161.226.

The double magnification bias by gravitational lensing.
See Abstr. 161.230.

Statistics of gravitational lensing: extended sources.
See Abstr. 161.233.

Viscous fluid universe filled with stiff fluid in general relativity.
See Abstr. 161.236.

Were all white holes in the early universe converted into black holes?
See Abstr. 161.256.

Quantum creation of topological defects during inflation.
See Abstr. 161.257.

Expressions for linearized perturbations in a massive scalar field dominated cosmological model.
See Abstr. 161.258.

Dependence of cosmological energy density irregularities on the shape of the scalar–field potential during inflation and "reheating".
See Abstr. 161.259.

Dynamics of the D–dimensional FRW – cosmological models within the superstring – generated gravity model.
See Abstr. 161.266.

Remarks about late stage homogeneous cosmological dynamics.
See Abstr. 161.267.

General vacuum solution for Brans – Dicke – Bianchi type V.
See Abstr. 161.268.

Inflation in the Einstein – Cartan cosmological model.
See Abstr. 161.269.

Dynamical reduction in multidimensional cosmological models.
See Abstr. 161.271.

Cosmology with G and Δ coupling scalars.
See Abstr. 161.273.

Comments on the computation of Liapunov exponents for the Mixmaster universe.
See Abstr. 161.274.

Models of voids in the expanding universe. II.
See Abstr. 161.278.

Cosmological constraints on the zero–point electromagnetic field.
See Abstr. 161.279.

Föppl's negative gravity masses and the repulsive cosmos.
See Abstr. 161.283.

Anisotropic and inhomogeneous solutions of the Einstein–Boltzmann equations for Friedmann–Robertson–Walker spacetimes.
See Abstr. 161.289.

Quantum cosmology and the creation of the universe.
See Abstr. 161.310.

Acceleration–free spherically symmetric inhomogeneous cosmological model with shear viscosity.
See Abstr. 161.328.

Dark matter from an oscillating Brans–Dicke scalar in the extended inflation model.
See Abstr. 161.367.

Gravity–wave insights to Bianchi type–IX universes.
See Abstr. 161.369.

Lorentz Chern–Simons terms in Bianchi cosmologies and the cosmic no–hair conjecture.
See Abstr. 161.372.

Gravitational lensing properties of curved cosmic strings.
See Abstr. 161.373.

Ultrarelativistic Bose–Einstein condensation in the Einstein universe and energy conditions.
See Abstr. 161.394.

Gravitational lensing frequencies: galaxy cross–sections and selection effects.
See Abstr. 161.554.

Unstable spacetime in the Jordan–Brans–Dicke gravity.
See Abstr. 161.600.

Physically permitted cosmological models with nonzero cosmological constant.
See Abstr. 161.628.

Hubble spheres and particle horizons.
See Abstr. 161.639.

The cosmological constant, third quantization and all that.
See Abstr. 161.648.

Quantum cosmology.
See Abstr. 161.666.

Generalized inflation from R^3 and $R\Box R$ terms.
See Abstr. 161.684.

067 Astrophysics of Compact Objects (Neutron Stars, Black Holes)

067.001 **Gravitational scattering of electromagnetic field by Schwarzschild black–hole.**
A. Bachelot.
Ann. Inst. Henri Poincare, Phys. Theor., Vol. 54, No. 3, p. 261 – 320 (Jul 1991).

The author studies the electromagnetic scattering by a spherical black–hole. Maxwell's equations are written in Schwarzschild coordinates: the electromagnetic tensor is replaced with electric and magnetic fields in a three dimensional absolute space. He introduces a set of wave operators, $W^{\pm}{}_0$, $W^{\pm}{}_1$, yielding an electromagnetic field given an asymptotic behavior far from the black–hole, $W^{\pm}{}_0$, and near the Schwarzschild radius, $W^{\pm}{}_1$, as universal time $t \to \pm\infty$. He interprets the results on the Kruskal universe: the existence of $W^{-}{}_0$, $W^{-}{}_1$ assures the characteristic Cauchy problem with data on the past horizons is well posed in the Schwarzschild submanifold, and the asymptotic completeness of $W^{+}{}_0$, $W^{+}{}_1$ allows to define the solution on the future horizons.

067.002 **Dynamics of charged particles near a black hole in a magnetic field.**
V. Karas, D. Vokrouhlický.
J. Phys., I, Vol. 1, No. 7, p. 1005 – 1012 (Jul 1991).

Flow lines of the plasma near a magnetised Kerr–Newman black hole are studied within the framework of the guiding centre approximation. Surfaces of constant magnetic and electric flux are constructed and the shape of the plasma horizon is discussed.

067.003 **The smallest stars in the Universe.**
J. B. Kaler.
Astronomy, Vol. 19, No. 11, p. 50 – 55 (Nov 1991).

067.004 **Pinning of proton vortex filaments in superfluid neutron star cores.**
D. M. Sedrakyan, A. D. Sedrakyan.
Sov. Phys. – JETP, Vol. 73, No. 2, p. 193 – 197 (Aug 1991). Current Physics Microform No.: 9112X0071. English translation of Zh. Ehksp. Teor. Fiz., Vol. 100, No. 2, p. 353 – 362 (Aug 1991).

The authors show that proton vortices, which are associated with remnant and generated pulsar magnetic fields, are pinned in superfluid neutron star cores. An asymmetric energy release at the surface of the superfluid core occurs as the result of the motion of the pinned vortices; this leads to the formation of two localized pulsar radio–emission sources. The authors study the evolution of the pulsar magnetic fields. They find that the magnetic field component at right angles to the rotational axis disappears exponentially with a characteristic time $\tau = 2p/\dot{p}$, where p is the pulsar period.

067.005 **Pinning of proton vortices in the superfluid cores of neutron stars.**
D. M. Sedrakyan, A. D. Sedrakyan.
Zh. Ehksp. Teor. Fiz., Vol. 100, No. 2, p. 353 – 362 (Aug 1991). In Russian. English translation in Sov. Phys. – JETP, Vol. 73, No. 2, p. 193 – 197 (Aug 1991).

067.006 **Strange–pulsar models.**
O. G. Benvenuto, J. E. Horvath, H. Vucetich.
Int. J. Mod. Phys. A, Vol. 6, No. 27, p. 4769 – 4830 (20 Nov 1991).
The authors review the theory and observational status of strange–pulsar models. After introduction of the subject, a summary of observational facts about pulsars is presented. The theory of quark matter and strange matter relevant to astrophysical applications is briefly discussed, and applied afterwards to type–II supernova theory and to pulsar models. A discussion of the comparison with observation shows the viability of strange–pulsar models.

067.007 **Limitations on the statistical description of black holes.**
J. Preskill, P. Schwarz, A. Shapere, S. Trivedi, F. Wilczek.
Mod. Phys. Lett. A, Vol. 6, No. 26, p. 2353 – 2361 (30 Aug 1991).
The authors argue that the description of a black hole as a statistical (thermal) object must break down as the extreme (zero–temperature) limit is approached, due to uncontrollable thermodynamic fluctuations. For the recently discovered charged dilaton black holes, the analysis is significantly different, but again indicates that a statistical description of the extreme hole is inappropriate. These holes invite a more normal elementary particle interpretation than is possible for Reissner–Nordström holes.

067.008 **Dual dilaton dyons.**
A. Shapere, S. Trivedi, F. Wilczek.
Mod. Phys. Lett. A, Vol. 6, No. 29, p. 2677 – 2686 (21 Sep 1991).
The authors find four–dimensional black hole solutions of the tree–level string effective action, that carry both electric and magnetic charge. The solutions involve non–trivial axion and dilaton fields, and are related to one another by a form of duality transformation. These black holes have drastically different global properties from the usual charged holes without axions and dilatons. The global causal structures of the effective metrics seen by different types of particles (for the same solution) can differ.

067.009 **Nonlinear perturbations of Einstein–Yang–Mills solitons and non–abelian black holes.**
Zhou Zhi–hong, N. Straumann.
Nucl. Phys. B, Part. Phys., Vol. 360, No. 1, p. 180 – 196 (12 Aug 1991).
The evolution of nonlinear perturbations of the regular Bartnik–McKinnon solution of the SU(2) Einstein–Yang–Mills equations is studied numerically. Depending on the initial perturbation, the authors find a collapse to a Schwarzschild black hole or an explosion of the gauge boson 'mini star'. This confirms their expectations from a previous linear stability analysis and provides also numerical evidence for the instability of the new 'colored' black hole solutions.

067.010 **Magnetic–dipole radiation reaction and constant regeneration of pulsar magnetic fields.**
N. Itoh.
Nuovo Cimento B, Vol. 106, No. 7, p. 819 – 822 (Jul 1991).
It is shown that radiation reaction forces due to magnetic–dipole radiation would cause a decay of the pulsar magnetic fields within a time scale much shorter than the lifetime of pulsars if it were not for the magnetic–field regeneration mechanism such as the dynamo mechanism. It is, therefore, concluded that the pulsar magnetic fields are constantly regenerated by some mechanism as in the case of the terrestrial magnetic field.

067.011 **Can young pulsars clarify the odd behaviour of pulsars?**
M. Akbar, A. Qadir.
Nuovo Cimento B, Vol. 106, No. 9, p. 1011 – 1016 (Sep 1991).
Though the basic model of the pulsar as a rotating neutron star is well established, its details may not be as clear as is often assumed. The pulsar drift, the value of the braking index and its relation to the age of the pulsar need to be explained. Suggestions for explaining these properties which are not in entire agreement with the standard model have been proposed. One such proposal includes a high initial pulsar frequency. With the millisecond pulsar expected in the supernova of 1987 it is possible that the proposal merits a more detailed investigation. Could young pulsars provide, more generally, tests for the theories of pulsar formation, evolution and radiation? The authors discuss here the possibility of testing it against the standard model.

067.012 **Stability of neutron and boson stars: a new approach based on catastrophe theory.**
F. V. Kusmartsev, E. W. Mielke, F. E. Schunck.
Phys. Lett. A, Vol. 157, No. 8/9, p. 465 – 468 (12 Aug 1991).
In the framework of general relativity, the authors investigate the stability of stars by classifying the singularities of smooth mappings. These maps arise from the functional dependence of the mass M and the particle number N of the star on its dynamical variables. The authors' new approach avoids the mathematical limitations of perturbation theory. In fact, it is capable of providing more information than perturbation theory and predicts novel oscillation and collapse regimes for a certain range of N. Moreover, the stability criteria are rather universal and apply to neutron stars as well as to boson stars.

067.013 **Entropy and Euclidean action bounds in black–hole thermodynamics.**
O. B. Zaslavskij.
Phys. Lett. A, Vol. 160, No. 4, p. 339 – 343 (25 Nov 1991).
It is shown that the generalized first and second laws of thermodynamics are sufficient conditions for Bekenstein's entropy upper bound to be valid for a black hole in equilibrium with a self–gravitating radiation. A lower bound for the Euclidean action of the system is obtained.

067.014 **Exact versus approximate solution of Einstein's equations for rotating neutron stars.**
F. Weber, N. K. Glendenning.
Phys. Lett. B, Vol. 265, No. 1/2, p. 1 – 5 (8 Aug 1991).
A redefined version of Hartle's perturbative method of solving Einstein's equations for rotating massive objects is applied for the investigation of the general relativistic Kepler frequency of a rotating neutron star. From the precise determination of rotating limiting mass neutron star models for seventeen representative neutron matter equations of state the authors find Kepler frequencies which are systematically larger by $\approx(10-15)\%$ than those obtained from the empirical formula of Haensel and Zdunik. The latter results however can be reproduced from Hartle's method as well by accounting for a small mass uncertainty of the limiting star model of $\lesssim 2\%$. In comparison with an exact numerical solution, Hartle's method is relatively easy to implement and should prove to be a practical tool for testing the compatibility of nuclear equations of state on pulsar periods.

067.015 **2D gravity and the black hole solution in 2D critical string theory.**
S. P. de Alwis, J. Lykken.
Phys. Lett. B, Vol. 269, No. 3/4, p. 264 – 270 (31 Oct 1991).
The authors discuss the relation between 2D gravity and critical string theory in two target space dimensions. In particular they consider the effect that the tachyon, and its back reaction on the metric, has on the interpretation of the critical theory as a non–critical theory coupled to world sheet gravity. They examine the generalizations of the black hole solution of the critical theory and argue that a generic feature of these solutions is that they have two horizons, similar to, but distinct from, the Reissner–Nordstrøm black hole. The authors also find some indication that the tachyon may destabilize a naked singularity. Finally they show that KPZ scaling is valid for the general solution of the critical theory, when it is interpreted as a theory of world sheet gravity.

067.016 **On a possible quadrupole component of the magnetic field near the neutron star surface in Hercules X–1.**
N. I. Shakura, K. A. Postnov, M. E. Prokhorov.
Pis'ma Astron. Zh., Tom 17, No. 9, p. 803 – 809 (Sep 1991). In Russian. English translation in Sov. Astron. Lett., Vol. 17, No. 5.
A complex variable pulse structure in Hercules X–1 can imply a configuration more complicated than a pure dipole configuration of the magnetic field near the neutron star surface. It is shown that in the framework of an ideal approximation of a pure dipole aligned with a quadrupole component, the X–ray pulse in this source can be formed both near the polar caps with different magnetic field strength and in a ring–like zone around the pole with weaker field.

067.017 **Asymmetric nuclear matter equation of state.**
I. Bombaci, U. Lombardo.
Phys. Rev. C, Vol. 44, No. 5, p. 1892 – 1900 (Nov 1991). Current Physics Microform No.: 9112C1246.
Systematic calculations of asymmetric nuclear matter have been performed in the framework of the Brueckner–Bethe–Goldstone approach in a wide range of both density and asymmetry parameter. The proton fraction in equilibrium with neutron matter has been determined from the beta–stability condition and its relevance to the superfluidity of neutron stars has been investigated.

067.018 **More on the black hole S–matrix.**
G. t' Hooft.
5. Seminar on Quantum Gravity, p. 251 – 265 (1991). – See Abstr. 012.004 for the main entry.
In previous work it was argued why it is practically inevitable to assume the existence of an S–matrix that describes particle absorption and production by a black hole, and the relationship between this S–matrix and string theory was derived. Here further properties are derived for the black hole horizon as a string world sheet. Knowledge of the standard model at the GeV scale leads to a precise formulation of the string degrees of freedom at this scale. Consequences of the Higgs mechanism and the confinement mechanism in the standard model for these string parameters at the corresponding scale are pointed out.

067.019 **On a quantum mechanical model of a black hole.**
V. A. Berezin.
5. Seminar on Quantum Gravity, p. 342 – 357 (1991). – See Abstr. 012.004 for the main entry.
A quantum mechanical model of a Schwarzschild black hole is constructed. This model is based on the simplest classical model, namely, a spherical thin dust shell as a source for a black hole mass. Such a mechanical system has only one degree of freedom which is quantized. The Klein–Gordon–type equation obtained exhibits some interesting features. This equation is an equation in finite differences and not a differential equation. There are nonlocalities near the singularities ($r = 0$) and at the horizon. For the large black holes this equation can be approximated by an ordinary differential equation, which has one solution in the interior regions of a black hole. This is an ingoing wave in the collapse region and an outgoing wave in the anticollapse regions. An approximate expression for the bound states is obtained.

067.020 **Quantum–mechanical scattering of charged black holes.**
J. Traschen, R. Ferrell.
5. Seminar on Quantum Gravity, p. 365 – 384 (1991). – See Abstr. 012.004 for the main entry.
The authors describe the quantum mechanical scattering of slowly moving maximally charged black holes. Their technique is to develop a canonical quantization procedure on the parameter space of possible static classical solutions. With this, they compute the capture cross sections for the scattering of two black holes. Finally, they discuss how quantization on this parameter space relates to quantization of the degrees of freedom of the gravitational field.

067.021 **The uniqueness of the Bekenstein black hole.**
B. C. Xanthopoulos, T. Zannias.
J. Math. Phys., Vol. 32, No. 7, p. 1875 – 1880 (Jul 1991). Current Physics Microform No.: 9107E0219.
The general static, spherically symmetric, asymptotically flat solution of the Einstein equations coupled to a conformal scalar field is determined; it depends on three free parameters. One of the parameters is eliminated by the requirement that the solution admits a smooth horizon and no naked curvature singularities, leading to the black hole solution discovered by Bekenstein.

067.022 **On naked singularities in self–similar Tolman–Bondi spacetimes.**
J. P. S. Lemos.
7. Latin American Symposium on Relativity and Gravitation, p. 241 – 247 (1991). – See Abstr. 012.005 for the main entry.
Bound, marginally–bound and unbound self–similar collapsing Tolman–Bondi spacetimes are analyzed. The degree of the inhomogeneity of the collapsing matter as a function of the binding energy necessary to form a naked singularity is given.

067.023 **Tidal disruption of stars in black hole clusters as a source of AGN luminosity.**
S. J. Stoeger, A. G. Pacholczyk, T. F. Stepinski.
Bull. Am. Astron. Soc., Vol. 23, No. 2, p. 921 (1991). Abstract. – See Abstr. 012.008 for the main entry.

067.024 **Oscillatory luminosity profiles and the contraction of radiating spheres in general relativity.**
R. Aquilano, W. Barreto, L. A. Núñez.
7. Latin American Symposium on Relativity and Gravitation, p. 320 – 329 (1991). – See Abstr. 012.005 for the main entry.
A seminumerical method by Herrera, Jiménez and Ruggeri is used to study the evolution of oscillatory general relativistic radiating spheres in the free streaming radiation approximation. Specific contraction schemes and particular oscillatory emission profiles are imposed on previously studied models. Firstly, the luminosity profile is obtained for radially oscillating contracting spheres. Secondly, the evolution of the boundary surface is analyzed when pulsating profiles are provided. It is found that, in both cases the oscillatory frequency of the surface coincides with the frequency of the pulsating radiation profile.

067.025 **Supernovae: where and why do they break off?**
L. Herrera, L. A. Núñez.
7. Latin American Symposium on Relativity and Gravitation, p. 337 – 347 (1991). – See Abstr. 012.005 for the main entry.
The authors discuss results from collapse calculations in which the shock is considered as an interface separating the quasi–static stiff core from the outer mantle. In the models considered the shock fades out and becomes a new boundary surface delimiting the compact homogeneous remnant. The resulting picture, becomes intelligible at the light of a recently proposed generalization of the concept of adiabatic index for systems where radiation flux are present.

067.026 **Images of accretion disks.**
C. A. Perez, R. V. Wagoner.
Bull. Am. Astron. Soc., Vol. 23, No. 2, p. 945 (1991). Abstract. – See Abstr. 012.008 for the main entry.

067.027 **Resonant Compton scattering in the magnetospheres of radio pulsars.**
S. J. Sturner, C. D. Dermer, F. C. Michel.
Bull. Am. Astron. Soc., Vol. 23, No. 2, p. 945 (1991). Abstract. – See Abstr. 012.008 for the main entry.

067.028 **No gamma–ray burst from white dwarf collapse.**
S. E. Woosley, E. Baron.
Bull. Am. Astron. Soc., Vol. 23, No. 2, p. 975 (1991). Abstract. – See Abstr. 012.008 for the main entry.

067.029 **The shear modulus of the neutron star crust and nonradial oscillations of neutron stars.**
T. Strohmayer, S. Ogata, H. Iyetomi, S. Ichimaru, H. M. Van Horn.
Astrophys. J., Vol. 375, No. 2, p. 679 – 686 (10 Jul 1991).

The authors have carried out new Monte Carlo calculations of the directionally averaged, effective shear moduli of both perfect and imperfect bcc crystalline Coulomb solids for conditions appropriate to the crusts of neutron stars. The differences between the perfect and imperfect crystalline states are comparable to the residual errors in the Monte Carlo calculations, except very near the freezing point of the crust. Both values differ substantially from the values previously used for the crusts, however. To illustrate one effect of the new shear modulus calculations, the authors have recomputed the nonradial oscillation modes of a particular model neutron star. The results can be summarized as follows: (1) the periods of the two interfacial (i_1 and i_2) modes are increased by about 10% compared to the previous results; (2) the s–mode periods are increased by about 30%; (3) the periods of the f– and p–modes are hardly affected at all, as expected, since these modes are not greatly influenced by the properties of the crust; (4) the periods of the surface g–modes also are not greatly affected; (5) the toroidal t–mode periods are increased by some 20% to 25%. These modes, together with the spheroidal s– and interfacial i–modes, are the most sensitive to the crustal properties.

067.030 **Lee–Wick model and soliton stars.**
R. Vinh Mau.
NATO Institut d'Etudes Scientifiques sur Physique des Particules, p. 335 – 342 (1990). – See Abstr. 012.010 for the main entry.

The properties of soliton stars, more specifically fermions soliton stars, are reviewed. In particular, it will be explained how these properties are strongly dependent on the choice of parameters in the Lee–Wick model. For example, it is shown that the features of the soliton stars described in reference were obtained with a very special choice of parameters in the Lee–Wick model, and a different choice can lead to soliton stars with very different properties, more like those of neutron stars. Then the possibility of a phase transition between these two types of soliton stars when a temperature dependence is introduced in the Lee–Wick model, is discussed. This gives some indications on a possible mechanism of evolution of Lee–Wick soliton stars from the early universe.

067.031 **Quantum black holes, strings and σ–models at $\theta = \pi$: Hawking quantum evaporation as world–sheet RG–flow.**
I. I. Kogan.
Mod. Phys. Lett. A, Vol. 6, No. 36, p. 3297 – 3311 (30 Nov 1991).

The author discusses the quantum black holes and even more generally the problem of quantum horizons in string theory. A toy model for quantum horizon is the two–dimensional O(3) σ–model. Using the interpretation of time as zero mode of the conformal factor of the world–sheet metric (Liouville field) the possible equivalence between two–dimensional renormalization group equations and the Hawking quantum evaporation formula is found. The infrared fixed points of the two–dimensional renormalization group corresponds to a final state of the quantum black hole. Using conjecture that such fixed points are described by σ–models with $\theta = \pi$ one suggests the axionic black holes as possible candidates to final (meta)stable states of black holes. The corresponding renormalization group picture is similar to the quantum Hall effect.

067.032 **Evidence against a strange ground state for baryons.**
R. R. Caldwell, J. L. Friedman.
Phys. Lett. B, Vol. 264, No. 1/2, p. 143 – 148 (25 Jul 1991).

The authors argue that if strange quark matter were the ground state of baryons at zero pressure and if some observed compact stars (or some cosmic rays) are strange, then the interstellar medium is so contaminated by strangelets that essentially all 'neutron' stars in the disk of the galaxy could have to be quark stars: Strangelet contamination would apparently vastly exceed the minimum abundance needed to seed all newly born neutron stars. Because observed glitches seem not to be compatible with quark stars, a strange quark ground state is unlikely.

067.033 **Neutrino neutral current interactions in hot dense matter.**
C. J. Horowitz, K. Wehrberger.
Phys. Lett. B, Vol. 266, No. 3/4, p. 236 – 242 (29 Aug 1991).

The response function for neutrino neutral current interactions are evaluated at finite temperature for a dense medium of electrons, protons and neutrons. The authors use real–time finite–temperature field theory. The ground state is described by the relativistic Hartree approximation to the $\sigma\omega$ model of quantum hydrodynamics. They compare the uncorrelated response with the response in random–phase approximation and discuss the influence of vacuum contributions. Results are presented for typical astrophysical scenarios in supernova collapse and neutron star cooling.

067.034 **On the impossibility of wave generation due to plasma motion in the weakly inhomogeneous magnetic field of pulsars.**
G. Z. Machabeli.
Plasma Phys. Controlled Fusion, Vol. 33, No. 11, p. 1227 – 1234 (Sep 1991).

It is shown that the waves cannot be generated in an infinitely large and weakly curved magnetic field (without the consideration of particle transversal motion).

067.035 **Emission of gravitational waves from roto–vibrational modes of a compressible and homogeneous star.**
P. Carini, R. Ruffini.
Astron. Astrophys., Vol. 248, No. 1, p. 115 – 123 (Aug 1991).

Based on Chandrasekhar's and Lebovitz's treatment of the roto–vibrational modes of a homogeneous compressible star, the energy losses and damping rates of each mode by gravitational radiation are computed as a function of the eccentricity of the equilibrium configuration. Numerical values are estimated for a neutron star. The limit of low eccentricity is considered, compared and contrasted to existing treatments.

067.036 **Inner structure of a charged black hole: an exact mass–inflation solution.**
A. Ori.
Phys. Rev. Lett., Vol. 67, No. 7, p. 789 – 792 (12 Aug 1991). Current Physics Microform No.: 9109C2130.

Recently, Poisson and Israel have shown how, when an electrically charged black hole is perturbed, its inner horizon becomes a singularity of infinite spacetime curvature – the mass–inflation singularity. In this paper the author constructs an exact mass–inflation solution of the Einstein–Maxwell equations, and uses it to analyze the mass–inflation singularity. It is found that this singularity is weak enough that its tidal gravitational forces do not necessarily destroy physical objects which attempt to cross it. The possible continuation of the spacetime through this weak singularity is discussed.

067.037 **Growing hair on black holes.**
S. Coleman, J. Preskill, F. Wilczek.
Phys. Rev. Lett., Vol. 67, No. 15, p. 1975 – 1978 (7 Oct 1991). Current Physics Microform No.: 9111E1529.

A black hole can carry quantum numbers that are not associated with massless gauge fields, contrary to the spirit of the "no–hair" theorems. In the Higgs phase of a gauge theory, electric charge on a black hole generates a nonzero electric field outside the event horizon. This field is nonperturbative in $\hbar$ and is exponentially screened far from the hole. It arises from the cloud of virtual cosmic strings that surround the black hole. In the confinement phase, a magnetic charge on a black hole generates a classical field that is screened at long range by nonperturbative effects. Despite the sharp difference in their formal descriptions, the electric and magnetic cases are closely similar in physical terms.

067.038 **Line profiles from a disk around a rotating black hole.**
A. Laor.
Astrophys. J., Vol. 376, No. 1, p. 90 – 94 (20 Jul 1991).
This paper presents calculations of profiles of lines emitted from an accretion disk around a rotating black hole. The calculations were done for a number of different line emissivity laws, with an emphasis on cases where most of the flux originates from the inner regions of the disk. The main results are as follows: (1) Most emissivity laws produce a broad single-peaked line. A double-peaked profile generally characterizes emission from the outer parts of the accretion disk. (2) The line is significantly asymmetric in most cases, having a steep blue wing and an extended red wing. (3) The line peak is blueshifted in most cases. In a few of these cases the blue wing of the line may extend up to an energy shift of 50%. A redshifted line is produced only by a nearly face-on disk, and the maximum redshift amounts to less than 5%. The strong relativistic effects, in particular a blueshift of more than 30%, are produced only in the inner-most regions of the accretion disk, and are therefore unique to the case of a rotating black hole. The results of this work are compared with observations of the $K\alpha$ iron line in Galactic X-ray sources and active galactic nuclei.

067.039 **On the stability of wind accretion.**
T. Matsuda, N. Sekino, K. Sawada, E. Shima, M. Livio, U. Anzer, G. Börner.
Astron. Astrophys., Vol. 248, No. 1, p. 301 – 314 (Aug 1991).
The authors present the results of numerical investigations of accretion flows past a compact object. The flows were taken uniform in density, temperature and velocity at large distances upstream. The accretion process was idealised by taking a completely absorbing inner boundary close to the compact object. The authors performed both 2D and 3D computations. They found that a very strong tendency towards non-steady flows exists both in the 2D cases and the 3D cases. Steady flows were only obtained if either the numerical resolution was low or if the central absorbing body was sufficiently large. The results indicate that accretion flow instabilities are not due to any perpendicular gradients at large upstream distances; they seem to be intrinsic to all accretion flows.

067.040 **Model atmospheres and X-ray spectra of bursting neutron stars.**
J. Madej.
Astrophys. J., Vol. 376, No. 1, p. 161 – 176 (20 Jul 1991).
This paper derives the model atmosphere equations corresponding to plane-parallel nongray atmospheres of very hot neutron stars in hydrostatic and radiative equilibrium. The equations take into account free-free opacity on hydrogen and helium ions and angle-averaged Compton scattering on electrons with fully relativistic thermal velocities. General-relativistic corrections in the equations correspond to a non-rotating central star. The nonlinear equation of transfer implements an exact photon redistribution function which traces precisely scattering events even with large photon-electron energy exchange. The equation of transfer is solved with the method of variable Eddington factors, and temperature corrections are given by the linearized equation of radiative equilibrium. Numerical results are presented as tables of surface fluxes for 20 model atmospheres ranging in effective temperature T_e^* from 8×10^6K to 3×10^7K and in surface gravity $\log g^* = 10^{15}$ cm s^{-2} or less. All the models consist of hydrogen and helium with the solar number abundance $N_{He}/N_H = 0.11$. It is found that the spectra of high-gravity models are practically identical with a blackbody spectrum shifted toward higher energies, whereas models approaching the Eddington limit develop a huge low-energy hump.

067.041 **Quantum self-gravitating systems of bosons and fermions.**
G. Ingrosso, D. Grasso, R. Ruffini.
Astron. Astrophys., Vol. 248, No. 2, p. 481 – 484 (Aug 1991).
The authors examine self-gravitating systems made of a mixture of condensed bosons and degenerate fermions in the Newtonian regime. They develop a numerical approach to determine the relevant quantities of the equilibrium configurations. An analytical treatment is also obtained in the limits in which the central density of one component is dominant. For the case in which the fermions are neutrons and the boson mass is greater than 10^{-7}eV c^{-2}, it is found: (1) For those configurations in which the central density of fermions dominates, the gravitational field confines the boson component to a region smaller than that occupied by the corresponding pure boson configuration. The size of such region ($R_B \sim (\hbar/m_B(G\varrho_F(0))^{1/2})^{1/2}$ is independent of the boson number and relates to the fermion density. (2) Both in the cases of a boson density smaller or larger than the fermion density, the bosons do not perturb significantly the fermions, which distribute according to the usual solution of Chandrasekhar.

067.042 **Steady spherical hypercritical accretion onto neutron stars.**
J. C. Houck, R. A. Chevalier.
Astrophys. J., Vol. 376, No. 1, p. 234 – 244 (20 Jul 1991).
The authors examine the structure of hypercritical accretion flows onto neutron stars, considering steady state, spherically symmetric flows with accretion rates in the range $10^{-4} \lesssim \dot{M} \leqslant 10^4 M_\odot yr^{-1}$. In this regime, the gravitational accretion energy is carried away by neutrinos. The models include pair production, radiation diffusion, and general relativistic effects. Pairs are the dominant pressure component near the neutron star, enhancing the pressure and density in that region by a factor of $\sim 11/4$. At accretion rates above $\sim 10^2 M_\odot yr^{-1}$, pair pressure dominates throughout the accretion envelope. Relativistic effects increase the pressure and density gradients in the envelope and decrease the envelope radii by as much as a factor of 2 compared to the Newtonian envelopes. Radiation diffusion becomes important when the accretion rate falls below $\sim 10^{-3} M_\odot yr^{-1}$ because the shock has moved out far enough to approach the outer edge of the photon trapping region. The luminosity exceeds 50% of the Eddington luminosity once $\dot{M} \lesssim 4 \times 10^{-4} M_\odot yr^{-1}$. The characteristic radiation temperature is relatively low, lying in the range 4×10^3K to 5×10^4K, because the effective photosphere is orders of magnitude larger than the neutron star radius. Applying these results to the case of SN 1987A, the authors find that the luminosity at infinity from accretion onto a central neutron star should reach 50% of the Eddington luminosity within ~ 6 months of the explosion, approaching the Eddington luminosity within a year.

067.043 **The stability of slim accretion disks.**
F. H. Wallinder.
Astron. Astrophys., Vol. 249, No. 1, p. 107 – 117 (Sep 1991).
Slim accretion disks are transonic, non-Keplerian, nonisothermal, and have an accretion rate $\dot{m} \sim 1$, where $\dot{m} = \dot{M}/\dot{M}_c$, $\dot{M}_c$ being a critical accretion rate, related to the Eddington one. A linear perturbation analysis of such disks is presented. The aim is twofold: firstly to find out what affects their stability in different parts of parameter space, and secondly to point out the possible connection between unstable modes and observed variability from various compact sources. It is shown that since the drift timescale decreases for smaller radii and/or higher accretion rates, slim disks become stabilized against the classical thermal and viscous instabilities. The onset of the stabilizing occurs as $\dot{m}$ is increased above ~ 0.2 and 1.0, in the inner and outer region, respectively. The standard Shakura-Sunyaev theory is found to be valid only for accretion rates lower than these, which gives rise to a partly unstable so-called middle branch in the slim disk relation between the surface density and the accretion rate. For a $10\ M_\odot$ central object and a viscosity parameter $\alpha = 10^{-3}$, the obtained thermal instability frequency ~ 0.1 Hz, and the corresponding acoustic one ~ 4 Hz, in the inner region. It is speculated

that these may be identified with the so-called normal and horizontal branch oscillations, respectively, observed in e.g. X-ray binaries. Some progress towards a qualitative, as well as quantitative, understanding of certain types of variability from compact sources may be the result.

067.044 **Superfluidity and magnetic fields of pulsars.**
D. M. Sedrakyan, K. M. Shakhabasyan.
Usp. Fiz. Nauk, Vol. 161, No. 7, p. 3 – 40 (Jul 1991). In Russian.
A review of the modern state of superfluidity theory in pulsars is presented. Superfluidity of hadronic matter in the neutron stars is considered. It is shown that a strong interaction between neutron and proton condensates leads to the drag current of superconducting protons and to the generation of the strong magnetic field $B = 10^{12}$Gs, which is parallel to the axis of rotation and constant in time. The value of this field depends on microscopic parameters of superfluid hadrons. The authors discuss the models explaining the origin of glitches and the postglitch relaxation. They calculate the coupling time between neutron superfluid and the right crust of the neutron star.

067.045 **Transient particle acceleration in strongly magnetized neutron stars.**
F. Melia, M. Fatuzzo.
Astrophys. J., Vol. 376, No. 2, p. 673 – 677 (1 Aug 1991).
Although several reasonable physical processes have been proposed to account for the initial energy release in transient events on strongly magnetized neutron stars, and although the manifestation of this activity as a burst of X- or γ-radiation evidently results from the Compton upscattering of soft photons by relativistic electrons, a self-consistent picture for the particle acceleration itself is lacking. Here, the authors develop a detailed model in which sheared Alfvén waves generated by the engergized stellar crust (e.g., due to a starquake) accelerate the charges to large Lorentz factors. It is shown that for canonical stellar parameters and a burst energy $\sim 10^{37}$ergs, the maximum attainable Lorentz factor is $\gamma_{max} \sim 50 \cdot m$, where m ($\sim 2 - 3\ cm^{-1}$) characterizes the shear of the magnetic perturbation. This value of γ_{max} is consistent with the particle energy required to produce the most energetic γ-rays (~ 100 MeV) observed in γ-ray bursts.

067.046 **Variability of accretion discs around compact objects.**
J. P. Lasota, D. Pelat.
Astron. Astrophys., Vol. 249, No. 2, p. 574 – 580 (Sep 1991).
The authors study the variability of inner regions of accretion discs around compact objects. This analysis is analogous to the approach of Taam and Lin (1984). Only the case in which viscous stresses are proportional to the total pressure is investigated. It is found that in the (locally) thermally unstable regime inner regions of the disc always become effectively optically thin ($\tau_{eff} < 1$) and that the evolution and disc structure depend critically on the treatment of the transitions between optically thick and optically thin configurations.

067.047 **Proton acceleration in neutron star magnetospheres.**
I. A. Smith, J. I. Katz, P. H. Diamond.
AIP Conf. Proc., No. 220, p. 232 – 236 (Apr 1991). Current Physics Microform No.: 9106B1366. – See Abstr. 012.020 for the main entry.
To explain the emission of TeV and PeV gamma-rays from accreting X-ray binary sources, protons must be accelerated to several times the gamma-ray energy. It was proposed that protons could be accelerated to these energies in the outer regions of a neutron star's closed magnetosphere. The authors show that at certain times the plasma in the accretion column of the neturon star may form a deep enough pool that the top portion becomes unstable to convective motions in spite of the strong magnetic field. The resulting turbulence produces fluctuations to the strength of the magnetic field that travel up the accretion column, taking energy out to the region of the energetic protons. The protons resonantly absorb this energy, and are accelerated to high energies. Including the synchrotron radiation losses of the protons, it is found that they can be accelerated to energies that are high enough to explain the gamma-ray observations.

067.048 **Strahlungstransport in Atmosphären von Neutronensternen.**
A. Piepenbrink.
Diss., München Univ. (Germany). Fakultät für Physik. 163 p. (23 Jul 1990).
Atmospheres of neutron stars have been modelled in dependence of chemical composition, gravity and effective temperature with an iterative method.

067.049 **Quantum mechanics of black holes in curved space-time.**
Zhang Zhenjiu, Huang Huanran, Bao Gong, He Changbai.
NATO Advanced Research Workshop on Quantum Mechanics in Curved Space-Time, p. 111 – 133 (1990). – See Abstr. 012.023 for the main entry.
The authors restrict their attention to the case of a real scalar field. The analysis of other linear fields of spin $s = 1$ or $s < 1$ is very similar, although important differences occur in the fermion case. Fields of $s > 1$ do not have a natural generalization to curved space-time. They assume the space-time ($M, g\mu\nu$) to be a c^{∞} 4-dimensional, globally hyperbolic, pseudo-Riemannian manifold M with metric $g\mu\nu$.

067.050 **Absorption cross section of a mini black hole.**
Fang Lizhi.
NATO Advanced Research Workshop on Quantum Mechanics in Curved Space-Time, p. 135 – 139 (1990). – See Abstr. 012.023 for the main entry.
Absorption of particles by a stellar formed black hole is mainly a classical problem, because the de Broglie wavelength λ of the particle is generally much smaller than the gravitational radius r_s of such a black hole. However, quantum effects should be considered for the absorption process of decoupled massive particles by the primordial black hole in the early universe, because the wavelength of such a non-relativistic particle can sometimes be larger than r_s of the mini black hole.

067.051 **Particle creation and vacuum polarization near black holes.**
V. P. Frolov.
NATO Advanced Research Workshop on Quantum Mechanics in Curved Space-Time, p. 141 – 201 (1990). – See Abstr. 012.023 for the main entry.
The author starts by considering the simplest possible case: the quantum theory of a scalar massless field in a flat spacetime. He briefly discusses the field quantization in an inertial frame of reference and the calculations of the local observables $\langle \hat{s}^2 \rangle$ and $\langle T_{\mu 7} \rangle$ which describe the quantum field fluctuations and its stress-energy tensor correspondingly. A simple geometrical model of a heated black body is also described and used for studying its thermal radiation.

067.052 **The final state of an evaporation black hole and the dimensionality of the space-time.**
V. de Sabbata, C. Sivaram.
NATO Advanced Research Workshop on Quantum Mechanics in Curved Space-Time, p. 485 – 501 (1990). – See Abstr. 012.023 for the main entry.

067.053 **Oscillations of fluid disks rotating around compact objects.**
D. K. Chakraborty, K. N. Mishra.
Astrophys. Space Sci., Vol. 183, No. 2, p. 323 – 332 (Sep 1991).
Frequency of global axisymmetric oscillations of a perfect fluid disk rotating around a compact object is investigated by trial function method. A formula for the frequency of local radial oscillation is presented.

067.054 **A new method calculating location and temperature of event horizon of an evaporating black hole.**
Zhao Zheng, Dai Xianxin.
J. Beijing Normal Univ. Nat. Sci., Vol. 26, No. 3, p. 198 – 199 (Jun 1990). In Chinese.

067.055 **The stability of non–singular spherical black hole.**
Yu Yunqiang, Fan Zuhui.
Acta Phys. Sin., Vol. 39, No. 7, p. 1179 – 1182 (Jul 1990). In Chinese.
The dynamics of a non–singular spherical black hole suggested by Gonzalez–Diaz is investigated. The authors show that it's matter configuration is in a peculiar equilibrium and unstable. It's formation is seldom and it is liable to transit to a normal black hole.

067.056 **The simplest black holes.**
R. B. Mann.
Found. Phys. Lett., Vol. 4, No. 5, p. 425 – 449 (Oct 1991).
A relativistic gravitational theory in (1+1) dimensions is presented which exhibits many of the qualitative features of (3+1)–dimensional general relativity. The field equations are simple enough for undergraduates to solve yet rich enough in structure to form a useful pedagogical example for exploring the qualitative features of relativistic gravitation. Black hole solutions to the field equations of the theory are derived and its relationship to Newtonian gravity is discussed in detail.

067.057 **Zur Elektrodynamik von schwarzen Löchern.**
N. Straumann.
Helv. Phys. Acta, Vol. 64, No. 8, p. 1175 – 1199 (Nov 1991).
After a short summary of the properties of isolated stationary black holes by the example of electrodynamics the physics in the neighbourhood of black holes is treated. In difference to the traditional four–dimensional space–time description the fundamental laws are written as equations of an "absolute space", over which the dynamical variables are developed as function of an "absolute time" similarly as in non–relativistic physics. Simultaneously the boundary conditions at the horizon are formulated as properties of a fictive conducting membrane. In this "membrane model", which was farly developed by different authors, the physics beyond black holes can intuitively much easier be understood. This is especially valid for the Blandford–Znajek process of the magnetic energy extraction from a black hole, which is possibly essential for an understanding of the quasars and other active galaxies.

067.058 **New black hole solutions with hair.**
S. Droz, M. Heusler, N. Straumann.
Phys. Lett. B, Vol. 268, No. 3/4, p. 371 – 376 (17 Oct 1991).
The authors present new black hole solutions which provide counter examples to the 'no hair' conjecture for non–linear matter fields. These have been found numerically for the non–linear sigma–model of Skyrme coupled to gravity. They are static and approach asymptotically the Schwarzschild solution, but have 'Skyrme hair'. They also describe self–gravitating generalizations of the well–known Skyrme solitons in flat spacetime. Both types of solutions are presumably stable.

067.059 **Non–radial pulsation of a boson star. I. Formulation.**
Y. Kojima, S. Yoshida, T. Futamase.
Prog. Theor. Phys., Vol. 86, No. 2, p. 401 – 410 (Aug 1991).
The authors investigate the non–radial pulsation of a boson star and the associated gravitational waves. As the first step they formulate the linearized perturbation around an equilibrium boson star in general relativity using the tensor harmonics and derive the perturbation equations for odd and even parity modes. It is found that the perturbations with odd parity do not couple with gravitational waves and the set of equations for even parity perturbations reduces to sixth–order system of differential equations.

067.060 **On spectra of X–ray bursters in the expansion phase.**
I. I. Lapidus.
Astrophys. J., Lett., Vol. 377, No. 2, p. L93 – L96 (20 Aug 1991).
The author presents two alternative approaches to the theoretical description of spectra of X–ray bursters showing radial expansion near the light curve maximum. The first one, using the LTE assumption, can explain the noticeable soft excess which is present in corresponding EXOSAT observations. The second one is based on numerical calculations of radiation transfer (taking into account spherical geometry, free–free processes, and Comptonization) and provides not only the qualitative scenario of the drastic softening of spectra as the expansion occurs, but also the ratio of maximum values (at the onset of contraction) and minimum values (during expansion) of Planckian fitting temperatures $T^{max}/T^{min} = 6.4 - 9$ which seems to be quite a good estimate for the value 7 – 7.5 observed in 4U/MXB 1820–30. The double–peaked structure of bursts with expansion is also a natural consequence of the scenario presented.

067.061 **Radiative–hydrodynamical simulations of accretion disk coronae.**
F. Melia, G. J. Zylstra, B. Fryxell.
Astrophys. J., Lett., Vol. 377, No. 2, p. L101 – L104 (20 Aug 1991). With plates L5 – L6.
The authors have developed an algorithm for performing detailed, two–dimensional, fully self–consistent radiative – hydrodynamical simulations of accretion disk coronae in X–ray–luminous compact sources. The calculation reported here, for an accreting neutron star radiating at 0.5 times the Eddington luminosity, reveals several striking features: (1) The corona is comprised of two main regions – an inner ($r \lesssim 1 \times 10^8$cm) highly dynamic portion whose vertical structure varies cyclically on a dynamical time scale (~ 0.2 s), and an outer, more stable zone in which the evaporated plasma rises to form a "sheath" that gradually merges into a wind at large radii. (2) The coronal structure shows a density inversion that contrasts sharply with the Gaussian profiles of earlier hydrostatic models. (3) Flow velocities as high as a few times 10^9cm s^{-1} are not uncommon in portions of the corona. The authors discuss the relevance of this point to the large measured width of the emission features seen in low–mass X–ray binaries and the Galactic black hole candidates.

067.062 **Thermal effects of non–equilibrium beta–processes in a strong magnetic field.**
G. A. Shul'man.
Astron. Zh., Tom 68, Vyp. 5, p. 984 – 991 (Sep–Oct 1991). In Russian. English translation in Sov. Astron., Vol. 35, No. 5.
The thermal energy is estimated which is released by the non–equilibrium beta–capture of ultrarelativistic electrons by nuclei in the partially degenerate matter of equilibrium stars in the quantum limit of the magnetic field frozen–in into their interiors. It is shown that, taking into consideration the influence of the magnetic field on the energy spectrum and density of the number of quantum states of the partially degenerate ultrarelativistic electron gas, the thermal energy release differs quantitatively from the one due to non–equilibrium beta–capture processes in the matter of degenerate magnetic stars. The limiting temperature is determined, to which can be heated the matter of a partially degenerate magnetic star in the quantum limit of the magnetic field frozen–in into its interiors.

067.063 **The age of pulsars and the phenomenological model of their evolution.**
Yu. I. Neshpor, A. A. Stepanyan.
Astron. Zh., Tom 68, Vyp. 5, p. 992 – 998 (Sep–Oct 1991). In Russian. English translation in Sov. Astron., Vol. 35, No. 5.
A model of evolution with an age parameter $\eta = L_R/-(dE/dt)$ where L_R is the radio luminosity and dE/dt is the total rotational energy loss, is proposed. Using the data on 292 pulsars, the authors have shown that variations of pulsars parameters, P, $\dot{P}$, β (P – period, $\dot{P}$ – time derivative of the period, β – angle between the magnetic moment and spin) fit into a self–consistent model, while the dependence of energy losses on the angle β corresponds to losses through magnetodipole radiation.

067.064 **Model photospheres for X–ray bursters near the Eddington limit.**
V. E. Zavlin, Yu. A. Shibanov.
Astron. Zh., Tom 68, Vyp. 5, p. 999 – 1008 (Sep–Oct 1991). In Russian. English translation in Sov. Astron., Vol. 35, No. 5.
Computations of neutron star photosphere models at the conditions typical for X–ray bursters with luminosities close to

the Eddington limit are presented. It is shown that spectra of radiation flux emitted by the photosphere at luminosities $L \geqslant 0.95\, L_{Edd}$ are similar to those emitted by semi-infinite homogeneous plasmas with saturated comptonization. Wien hump at high photon energies and a plateau at low energies are developed. These are associated with Compton heating of the surface layers by hot photons emerging from the deep regions, and with formation of an isothermal layer near the surface. Constraints are obtained on the photosphere parameters at which the layer is formed. An approach is outlined to improve the agreement between the semi-analytical approximations and the computer simulations. The results may be useful for an analysis of spectra of powerful X-ray bursters.

067.065 **Alfvén solitons in a dense ultrarelativistic electron–positron pulsar plasma.**
U. A. Mofiz.
International Centre for Theoretical Physics, Trieste (Italy), IC—91-178. 6 p. (Jul 1991).
The nonlinear propagation of Alfvén waves in a hot electron-positron plasma of a pulsar magnetosphere is investigated. Alfvén solitons are found to exist at ultrarelativistic ($T \gg mc^2$) temperature. Landau damping and three dimensional modulation perturbation of solitons are discussed.

067.066 **Electrostatic drift vortices in a hot rotating strongly magnetized electron–positron pulsar plasma.**
U. A. Mofiz.
International Centre for Theoretical Physics, Trieste (Italy), IC—91-177. 5 p. (Jul 1991).
Electrostatic drift wave in a hot rotating and strongly magnetized electron-positron pulsar plasma is considered. Using relativistic two fluid equations a pair of coupled nonlinear equations is derived. It is shown that the wave can propagate in the form of two-dimensional dipolar vortices at ultrarelativistic temperature ($T \gg mc^2$) of the plasma. The latter may affect the energy transport in the hot plasma, which can lead to a new turbulent state in the pulsar magnetosphere.

067.067 **Axion hair for dyon black holes.**
B. A. Campbell, N. Kaloper, K. A. Olive.
Phys. Lett. B, Vol. 263, No. 3/4, p. 364 – 370 (18 Jul 1991).
Gauge Chern-Simons couplings of pseudo-scalar bosons (axions) give $F\tilde{F}$ source terms for axions. Black holes carrying gauge charges of both electric and magnetic type then exhibit external axion field strength hair. The authors calculate the axion hair so produced in a simplified weak field case, and discuss its physical implications.

067.068 **Reconciliation of neutron star masses and binding of the Λ in hypernuclei.**
N. K. Glendenning, S. A. Moszkowski.
Phys. Rev. Lett., Vol. 67, No. 18, p. 2414 – 2417 (28 Oct 1991).
Current Physics Microform No.: 9112C0660.
By relating scalar and vector couplings of the hyperons to the inferred empirical binding of the Λ hyperon in saturated nuclear matter, the authors obtain compatibility of this binding energy with neutron star masses. Use of the observational constraint on the lower bound on the maximum neutron star mass and the upper bound of the couplings that are compatible with hypernuclear levels, places bounds on the reduction in neutron star mass that hyperons produce. For the best current estimate of nuclear-matter properties, the reduction in mass due to conversion of nucleons to hyperons is $(0.71 \pm 0.15) M_{\odot}$. Neutrons comprise a slight majority population in neutron stars with mass $\sim 1.5\, M_{\odot}$.

067.069 **The n = 1 colored black hole is unstable.**
P. Bizon, R. M. Wald.
Phys. Lett. B, Vol. 267, No. 2, p. 173 – 174 (12 Sep 1991).
The controversy about the stability of the $n = 1$ colored black hole is resolved. It is shown that although the unstable mode itself is badly behaved at the horizon, there exist perfectly regular initial perturbations which grow unboundedly in time.

067.070 **Phase transition of hot neutron matter to π^0-condensed state.**
M. Takahara, T. Takatsuka, K. Sato.
Prog. Theor. Phys., Vol. 86, No. 1, p. 57 – 71 (Jul 1991).
The authors investigate the equation of state (EOS) of hot neutron matter with π^0-condensed state on the basis of the Alternating-Layer-Spin model by solving the Hartree-Fock equation at finite temperature. They find that the phase transition to π^0-condensed state is suppressed considerably with temperature. They first calculate the entropy and specific heat of π^0-condensed state and find that they are much lower than those of normal state. They also find that the effective mass of nucleons in normal and π^0-condensed state is almost independent of temperature. The application to supernova explosions is discussed in detail, especially the evolution of the collapsing core along the iso-entropy line and its effects on the pressure.

067.071 **Gravitational radiation from coalescing binary neutron stars. IV. Tidal disruption.**
T. Nakamura, K.-i. Oohara.
Prog. Theor. Phys., Vol. 86, No. 1, p. 73 – 88 (Jul 1991).
The authors have performed three numerical simulations of coalescence of binary neutron stars using a Newtonian hydrodynamics code including the effect of radiation reaction due to emission of gravitational waves. In the first simulation, the authors start the calculation from a rotational equilibrium of two neutron stars of mass $m_1 = m_2 = 0.85\, M_{\odot}$ which touches with each other. This corresponds to a fission model of collapse of a rapidly rotating core. The radiated energy is 0.8% of the rest mass and the maximum amplitude of the gravitational waves is 2×10^{-21} if it occurs at a distance of 10 Mpc. The ratio of the rotational energy to the gravitational energy T/W is ~ 0.15 in the final stage of the numerical simulation. Since this value is greater than the secular instability limit, non-axisymmetric instability with the emission of gravitational waves occurs. In the other two simulations, two neutron stars have different masses; $m_1 = 1.70\, M_{\odot}$, $m_2 = 1.28\, M_{\odot}$ and $m_1 = 1.83\, M_{\odot}$, $m_2 = 0.97\, M_{\odot}$, respectively. The authors assume that the system is in a rotational equilibrium at first. Calculations are started when separation between neutron stars are 19 km and 21 km, respectively. In both cases, tidal disruption of the neutron star of smaller mass occurs. The radiated energy amounts to 2.9% and 2.3% of the total rest mass, respectively. The maximum amplitude of the gravitational waves at a distance of 10 Mpc is 4.4×10^{-21} and 3.4×10^{-21}.

067.072 **Two-dimensional charged black holes in string theory.**
N. Ishibashi, Li Miao, A. R. Steif.
Phys. Rev. Lett., Vol. 67, No. 24, p. 3336 – 3338 (9 Dec 1991).
Current Physics Microform No.: 9201E1324.
Exact string solutions corresponding to two-dimensional electrically charged black holes are constructed.

067.073 **The entropy change of imperfect fluid collapsing system in which the fluids' own gravitational field plays a role.**
Xiao Jiguang, Li Yuanjie.
Acta Astrophys. Sin., Vol. 11, No. 3, p. 213 – 219 (Jul 1991). In Chinese.
Making use of the comoving-frame, the authors discuss the energy-momentum tensor, the 4D-entropy current vector and the total entropy change of an imperfect fluid collapsing system, in which its own gravitational field plays a role, before the system's horizon appears. As a result, the total entropy of the collapsing system is unchanged or reduced, before the horizon appears. But according to original reference frames, the black hole entropy is higher by a factor of about 10^{20} than that of the corresponding collapsing matter. Therefore, when a black hole forms, the entropy value takes a sudden change, instead of increasing gradually to the order of magnitude of the black hole entropy in the course of collapsing.

067.074 **Line profile of radially flowing clouds with accretion disk.**
Cao Xinwu, Zhang Jialue.
Acta Astrophys. Sin., Vol. 11, No. 4, p. 321 – 326 (Oct 1991). In Chinese. English translation in Chin. Astron. Astrophys., Vol. 16, No. 1, p. 17 – 23 (Jan – Mar 1992).
According to the black hole model of active galactic nuclei, the light emitted from radially flowing clouds near a black hole should be affected by Doppler shift, gravitational redshift and deflection of light. Shielding of the light emitted from flowing clouds by the disk should be taken into account. In this paper, precise line profiles of radially flowing clouds near a Schwarzschild black hole for a distant observer are obtained by solving the kinetic equation of photons. The authors find that the results obtained may be useful in interpreting the profiles and variations of some emission lines in active galactic nuclei.

067.075 **An axionic Kerr black hole.**
Hsu Rueron, Lin Weifu.
Classical Quantum Gravity, Vol. 8, No. 8, p. L161 – L165 (Aug 1991). Letter–to–the–editor.
An exact axionic Kerr black hole solution, which has Kerr–like metric and non–vanishing axionic field strength, has been found by solving the stationary axisymmetric Einstein axion field equation. It indicates that there exists an axionic hair for the rotating black hole and the axionic hair can be detected by strings directly, without resorting to the Aharonov–Bohm process.

067.076 **Spin evolution of an accreting neutron star in low–mass X–ray binaries.**
A. V. Fedorova, E. V. Ergma.
Astron. Zh., Tom 68, Vyp. 6, p. 1274 – 1282 (Nov–Dec 1991). In Russian. English translation in Sov. Astron., Vol. 35, No. 6.
The results of an investigation of spin evolution of an accreting neutron star in low–mass X–ray binaries (LMXB) are presented. In the period gap, the evaporation of the neutron star companion is taken into account. Three possible ways of evolution of LMXB are discussed in relation to the neutron star spin evolution. It is found that the spin period of the neutron star is close to the local equilibrium period, which is determined by the local accretion rate and by the strength of the magnetic field of the neutron star.

067.077 **Line profile of radially flowing clouds with accretion disk.**
Cao Xinwu, Zhang Jialu.
3. Chinese Academy of Sciences and Max–Planck Society Workshop: High energy astrophysics: compact stars and active galaxies, p. 39 – 43 (1991). – See Abstr. 012.026 for the main entry.
In this paper precise line profile of radially flowing clouds near a Schwarzschild black hole seen from a distant observer is obtained by solving the kinetic equation of photons. Shielding of the light emitted from flowing clouds by the disk is considered in the calculation. The results obtained may be useful in interpreting the profiles and variations of some emission lines in active galactic nuclei.

067.078 **Calculations on the radiation spectrum of simplified stable accretion disk.**
Liu Renjie, You Junhan.
3. Chinese Academy of Sciences and Max–Planck Society Workshop: High energy astrophysics: compact stars and active galaxies, p. 52 – 56 (1991). – See Abstr. 012.026 for the main entry.
On the basis of a simplified theoretical model, the fundamental physical equations of accretion disk have been discussed and a theoretical deduction on the temperature distribution of disk has been made, in both cases of optical thinness and optical thickness. Numerical solutions of the temperature and thermal radiation spectrum of the accretion disk around a black hole of a binary system are given.

067.079 **On the limitation to luminosity of X–ray bursts.**
Li Qibin.
3. Chinese Academy of Sciences and Max–Planck Society Workshop: High energy astrophysics: compact stars and active galaxies, p. 94 – 99 (1991). – See Abstr. 012.026 for the main entry.
The effects which could reduce the disparity between the Eddington luminosity and observed super luminosity of X–ray bursts are reviewed. It is suggested that the Eddington luminosity should not be regarded as the limit of the X–ray burst because the X–ray bursts are eruptive processes so that the basic assumption for the Eddington luminosity as the luminosity limit for the stable stars is broken out. By driving the mass–radius relation from the observed luminosity and temperature and the dependence of the maximum stored energies of accretion matter upon the mass of neutron stars, the resulted mass–radial relation fits the theoretical configuration. It is concluded that the luminosity of X–ray burst should be limited by the stored fuel (He) before ignition of the accretion matter on the surface of neutron stars.

067.080 **Evolution of neutron–star magnetic fields.**
F. C. Michel.
Publ. Astron. Soc. Pac., Vol. 103, No. 666, p. 770 (Aug 1991). Abstract. – See Abstr. 012.034 for the main entry.

067.081 **Neutron stars and the distance to gamma–ray bursters.**
C. D. Dermer, K. C. Hurley.
Publ. Astron. Soc. Pac., Vol. 103, No. 666, p. 774 – 776 (Aug 1991). – See Abstr. 012.034 for the main entry.
Assuming that gamma–ray bursts originate from galactic neutron stars, the authors outline an analytic method for studying their statistical properties. If a significant fraction of all neutron stars are born with space velocities $\lesssim 100$ km s^{-1}, as suggested by studies of pulsar statistics, then the sampling distance to gamma–ray burst sources should be less than $\sim$several hundred pc. These results have important implications on theories of radio pulsar evolution and magnetic–field decay.

067.082 **Polarization of radiation from AGN accretion disks.**
H. H. Coleman, G. A. Shields.
Publ. Astron. Soc. Pac., Vol. 103, No. 666, p. 881 – 885 (Aug 1991). – See Abstr. 012.034 for the main entry.
A popular candidate for the Blue Bump often seen in AGN spectra is thermal radiation from an accretion disk. In models for disks surrounding supermassive black holes, such radiation would be emitted by a region of the disk dominated by electron–scattering opacity. Polarization predicted by standard, smooth, thin disk models disagrees with observations. The introduction of local surface roughness brings the predicted polarization more closely into line with observations. Further, the predicted polarization of radiation escaping from a thick radiation torus agrees well with observed values.

067.083 **A model for variability of AGN.**
E. T. Vishniac.
Publ. Astron. Soc. Pac., Vol. 103, No. 666, p. 886 – 887 (Aug 1991). – See Abstr. 012.034 for the main entry.

067.084 **The gravitational entropy of objects.**
Peng Qiuhe.
3. Chinese Academy of Sciences and Max–Planck Society Workshop: High energy astrophysics: compact stars and active galaxies, p. 289 – 296 (1991). – See Abstr. 012.026 for the main entry.
The Bekenstein's entropy of a black hole is 10^{18}times larger than the entropy of a star with the same mass. The purpose of this paper is to explain it by introducing the gravitational entropy of objects.

067.085 **1E 1740.7–2942: a black hole in a molecular cloud?**
I. F. Mirabel, M. Morris, J. Wink, J. Paul, B. Cordier.
Astron. Astrophys., Vol. 251, No. 2, p. L43 – L46 (Nov 1991). Letter–to–the–editor.
Observations of molecular transitions at millimeter wavelengths show that the most powerful compact source of radiation

in the Galaxy resulting from matter–antimatter annihilation may be associated with dense interstellar gas. Coincident with the position of the high energy source 1E 1740.7–2942, the authors find a molecular cloud. This cloud belongs to the inner molecular layer of the galactic center region, since it is located ~ 50 arcmin from the dynamical center of the Galaxy and has a Doppler shift of –140 km s^{-1}. At a distance of 8.5 kpc, it would have a size of ~ 3 pc and a mass of $\sim 5 \times 10^4 M_{\odot}$. The authors propose that this molecular cloud is the cold, high density interstellar environment needed for electron–positron annihilation, and that it could provide the fuel to directly feed the high energy source by a Bondi–Hoyle accretion mechanism. In this context, 1E 1740.7–2942 is viewed as compact object such as a stellar-mass black hole passing slowly through one of the many dense and massive clouds of the galactic center region. It has turned on as an X-ray/annihilation source because it is accreting $\sim 4 \times 10^{-8} M_{\odot}$ yr^{-1} from its cold, high density environment.

067.086 **High–polarization radiation from magnetized accretion disks.**
Yang Zhiliang, Yang Lantian, Yang Pibo.
Astron. Astrophys., Vol. 252, No. 1, p. 10 – 12 (Dec 1991).
The authors have calculated the polarization and intensity of radiation emergent from a geometrically thin and magnetized accretion disk, assuming that the radiation is synchrotron radiation scattered by thermal electrons in the disk. The results show that the degree of polarization and intensity depend sensitively on the orientation and on the total optical depth of the disk, and that the degree of polarization can exceed considerably the limit of 11.7% obtained with semi-infinite scattering atmospheres. The high degree of linear polarization and rapid variability of the disk radiation are consistent with observations of some QSOs and BL Lacertae objects.

067.087 **String theory and black holes.**
E. Witten.
Phys. Rev. D, Vol. 44, No. 2, p. 314 – 324 (15 Jul 1991). Current Physics Microform No.: 9108F0944.
An exact conformal field theory describing a black hole in two-dimensional space–time is found as a gauged Wess – Zumino – Witten model. For $k = 9/4$, the conformal field theory can be regarded as a classical solution of the same system that is probed in the $c = 1$ matrix model. The conformal field theory governing the space–time is regular at the Riemannian singularity, but it appears that generic perturbations blow up there. It is argued that the end point of the Hawking black hole evaporation is the standard space–time of the $c = 1$ matrix model, which should be regarded as an analog of the extreme Reissner – Nordström black hole of four-dimensional general relativity. The $c = 1$ model is thus a model of the quantum mechanics of matter interacting with a black hole.

067.088 **Rotational properties of strange pulsar models.**
O. G. Benvenuto, J. E. Horvath, H. Vucetich.
Phys. Rev. D, Vol. 44, No. 2, p. 555 – 556 (15 Jul 1991). Current Physics Microform No.: 9108F1185.
The authors present a study of the rotational properties of strange pulsars: strange matter stars capable of supporting glitches. It is shown that their differentiated internal structure implies a lower maximum rotational frequency than that of homogeneous strange stars. Nevertheless, they are able to fit the known pulsar properties.

067.089 **The contribution of light bending and redshift to the pulse characteristics of a pulsar in the case of smaller neutron stars.**
R. C. Kapoor.
Astrophys. J., Vol. 378, No. 1, p. 227 – 238 (1 Sep 1991).
An estimate has been made of the effect of light bending and redshift on the pulsar beam characteristics using the weak form of the Kerr metric, applicable to a slowly rotating neutron star. The calculations refer to the canonical narrow conical beam model of pulsar emission from a 1.4 $M_{\odot}$ neutron star with radius in the range 6 – 10 km and periods of 33 and 1.56 ms. The beam is found to diverge by a factor of 2 or less and to suffer an intensity reduction by an order of magnitude when emitted from close to the star's surface. This flattening of the pulse profile apparently is strongest for the smallest of the neutron stars, becoming insignificant for emission points located beyond $\cong 20$ km for all the neutron star radii considered. For a given emission location, the divergence is comparatively greater in the Schwarzschild background than in the rotational case, since, by virtue of the rotational terms in the metric, an aberration effect comes into play, which always tends to reverse the flattening effect of spacetime curvature by slightly squeezing the pulse in longitude. For instance, for a pulse emitted in the equatorial plane from beyond ~ 30 km in the case of the 1.56 ms period, aberration is so strong that it overcomes the divergence effect. This will result in an apparent reduction of the duty cycle.

067.090 **Magnetosphere of the rotation–powered pulsar: a DC circuit model.**
S. Shibata.
Astrophys. J., Vol. 378, No. 1, p. 239 – 254 (1 Sep 1991).
The global structure of the magnetosphere of rotation-powered pulsars is studied by analyzing a large-scale magnetospheric current system. It is assumed that this current system forms a DC circuit which includes the electromotive source of the rotating magnetic neutron star and three loads: the outer gap, the pair–plasma wind, and the inner gap. First the author constructs local models for the three parts of the magnetosphere (the two gaps and the wind) with adjustable free parameters, and then links them to a global model by imposing global link conditions. The global conditions are that (1) the sum of voltage gaps applied to the three loads is equal to the total electromotive force applied to the magnetospheric current circuit, (2) the total energy loss rate is the angular velocity of the star times the total angular momentum loss rate, and (3) the magnetic flux in the circuit is composed of an open flux in the pulsar wind and a closed flux in the outer gap. The procedure of the global modeling is analogous to obtaining operating points of an electric circuit when chracteristics of circuit elements are given. It is found that the outer gap is an indispensable element for the magnetospheric current circuit. The model for the Crab pulsar shows that about 99.4% of the total pulsar power is carried away by the pulsar wind, and that the remaining power is radiated from the inner and outer gaps.

067.091 **Limitations on the Penrose process.**
F. Fayos Valles, E. Llanta Salleras.
Gen. Relativ. Gravitation, Vol. 23, No. 8, p. 877 – 895 (Aug 1991).
It is well known that the Penrose process (PP) is a way for extracting energy from a black hole (BH). An analytical and partially numerical study of the PP is presented for a particular case: an incoming particle, at rest at infinity, decays into two photons inside the ergoregion of a Kerr BH, assuming that all particles follow equatorial orbits. It is shown that this process cannot exist, if the angular momentum of the BH is lower than a critical value. Considering the features of the non equatorial Kerr geodesics, it is conjectured that even when the incoming particle is not at rest at infinity and the trajectories are not equatorial, the critical value still holds. This lower limit on the angular momentum implies that not all the rotational energy of the BH could be extracted with the Penrose processes considered here.

067.092 **Surface structure of neutron stars and nuclear reactions with high magnetic fields.**
I. Fushiki.
10. Santa Cruz Summer Workshop in Astronomy and Astrophysics: Supernovae, p. 321 – 323 (1991). – See Abstr. 012.052 for the main entry.

067.093 **Multigroup simulation of protoneutron star cooling.**
H. Suzuki, K. Sato.
10. Santa Cruz Summer Workshop in Astronomy and Astrophysics: Supernovae, p. 324 – 327 (1991). – See Abstr. 012.052 for the main entry.

067.094 **Simple equations for general relativistic hydrodynamics in spherical symmetry applied to neutron star collapse.**
E. Gourgoulhon.
Astron. Astrophys., Vol. 252, No. 2, p. 651 – 663 (Dec 1991).
A complete system of simple equations for general relativity (GR) eulerian hydrodynamics in spherical symmetry is presented, with variables different from those commonly used. In particular, a "physical" fluid velocity field (instead of a momentum density) is introduced, for which the authors derive some GR generalization of the Euler equation. The other dynamical equations are found to be in a perfect newtonian form. As an application, the problem of stability and collapse of a neutron star near the maximum mass is investigated with a hydrodynamical code based on spectral methods. This latter integrates forward in time the whole system of equations, starting with an hydrostatic equilibrium model built upon a realistic equation of state. Depending whether the equilibrium is theoretically stable or unstable, small radial oscillations or catastrophic collapse are observed. As the central density increases, the development of the instability is shown to be due to the growth of the fundamental mode of the oscillations. When the star becomes unstable, the collapse is followed up to the formation of a black hole.

067.095 **A shot noise model for a two–component neutron star.**
A. Baykal, A. Alpar, Ü. Kiziloğlu.
Astron. Astrophys., Vol. 252, No. 2, p. 664 – 668 (Dec 1991).
The two component neutron star model and the response of the neutron star under the influence of external white torque noise are reviewed. A new simulation technique is developed for estimating the crust core coupling time and the moment of inertia of the superfluid in the neutron star from the irregularly sampled data. The model is applied to Vela X–1. In the analysis, it is found that moment of inertia of the superfluid cannot exceed 85% of the total moment inertia of the star if the crust core coupling time is longer than 1 d ($\tau > 1$ d). This is in agreement with the analysis of Boynton et al. (1984) on earlier data. This result indicates that either the core superfluid mixture is less than 85% of the total moment of inertia of the star or the crust core coupling time should be less than 1 d.

067.096 **Astrophysical limits for binding energy of strange matter.**
M. I. Krivoruchenko, B. V. Martemyanov.
Astrophys. J., Vol. 378, No. 2, p. 628 – 636 (10 Sep 1991).
Some of the observed pulsars can be identified quite reliably with ordinary neutron stars. If strange matter is stable, the central density in such neutron stars should be lower than the critical density of the quark–hadron phase transition occurring due to the strong interactions only with conservation of fractional concentration of strangeness. The quark matter being formed turns to more stable strange matter on a weak interaction time scale, converting neutron stars into strange stars or black holes. The existence of neutron stars therefore imposes constraints on the phenomenological parameters of the quark matter and gives limits for binding energy of strange matter. The authors consider stability requirements with respect to the quark–hadron phase transition in old and newly born hot neutron stars. They find that the existence of a neutron star of mass 1.3 $M_{\odot}$ rules out stable strange matter if neutron matter is described by the stiffest equations of state. The soft models, such as Reid ones, allow the existence of strange matter with binding energy of at most 50 MeV per baryon in absolute value.

067.097 **Neutrino transport in supernovae and protoneutron stars by Monte Carlo methods.**
H.–T. Janka.
10. Santa Cruz Summer Workshop in Astronomy and Astrophysics: Supernovae, p. 389 – 392 (1991). – See Abstr. 012.052 for the main entry.

067.098 **Diskoseismology: probing accretion disks. I. Trapped adiabatic oscillations.**
M. A. Nowak, R. V. Wagoner.
Astrophys. J., Vol. 378, No. 2, p. 656 – 664 (10 Sep 1991).
The authors analyze the normal modes of acoustic oscillations within thin accretion disks which are terminated by an innermost stable orbit around a slowly rotating black hole or weakly magnetized compact neutron star. The dominant relativistic effects, which allow modes to be trapped within the inner region of the disk, are approximated via a modified Newtonian potential. A general formalism is developed for investigating the adiabatic oscillations of arbitary unperturbed disk models. The authors explore the generic behavior via an expansion of the Lagrangian displacement about the plane of symmetry and by assuming separable solutions with the same radial wavelength for the horizontal and vertical perturbations. The lowest eigenfrequencies and eigenfunctions of a particular set of radial and quadrupole modes, which have minimum motion normal to the plane, are calculated. These modes correspond to the standard dispersion relation of disk theory.

067.099 **Time–dependent two–dimensional radiation hydrodynamics of accreting matter onto highly magnetized neutron stars: the evolution of photon bubbles.**
R. I. Klein, J. Arons.
NATO Advanced Research Workshop on Stellar Atmospheres: Beyond Classical Models, p. 205 – 220 (1991). – See Abstr. 012.056 for the main entry.
The authors present for the first time, the self–consistent solution of the two–dimensional, time–dependent equations of radiation–hydrodynamics governing the accretion of matter onto the highly magnetized polar caps of luminous X–ray pulsars. The calculations show a structure in the accretion column very different from previous one–zone uniform models. The authors have included all the relevant magnetic field corrections to both the hydrodynamics and the radiative transport. They include a new theory for the diffusion and advection of both radiation energy density and photon number density. For initially uniformly accreting models with super–Eddington flows, they have uncovered evidence of strong radiation–driven outflowing optically thin radiation filled regions of the accretion column embedded in optically–thick inflowing plasma. They follow the evolution of these photon bubbles for several dynamical timescales. The development of these photon "bubbles" indicates growth times on the order of a millisecond and show fluctuations on sub–millisecond timescales in agreement with a linear stability analysis. The photon bubbles are a consequence of the effect of radiative heat flux on the internal gravity waves in the strongly magnetized atmosphere and may result in observable fluctuations in the emitted luminosity leading to luminosity dependent changes in the pulse profile. This may provide important new diagnostics for conditions in accreting X–ray pulsars.

067.100 **General relativistic effects in astrophysical jets.**
D. Vokrouhlický, V. Karas.
Astron. Astrophys., Vol. 252, No. 2, p. 835 – 841 (Dec 1991).
General relativistic effects in the dynamics of optically thin jets are studied. Jet particles move along the rotation axis of the central compact object (Kerr black hole) in the radiation field of an accretion disk (relativistic α–viscosity thin disk). Attention is paid mainly to the effects in the close vicinity of the central object: what is the influence of rotation of the black hole, gravitational and Doppler redshift, and the bending of photon paths on the radiation field along particle's trajectory and how this may affect the terminal Lorentz factor of the jet. Within the model adopted, it is concluded that the most important general relativistic contribution comes only indirectly through the dependence of the local disk luminosity on the hole's rotation. These effects do not otherwise seem to affect the terminal speed of the jet significantly. Several typical particle trajectories outside of the rotation axis are constructed numerically.

067.101 **A mechanism for gamma–ray bursts by Alfvén–wave acceleration in a nonuniform atmosphere.**
K. A. Holcomb, T. Tajima.
Astrophys. J., Vol. 378, No. 2, p. 682 – 686 (10 Sep 1991).
It is widely believed that γ–ray bursts originate from neutron stars. A disturbance near the surface of a magnetized neutron star, whether due to a "starquake", an instability in an accretion disk, or some other phenomenon, can launch strong compressional Alfvén waves into the atmosphere. In this paper, it is

shown that such waves can accelerate particles by magnetic trapping; as the density of the atmosphere decreases away from the surface of the star, the condition for trapping can be satisfied for an extended time, resulting in acceleration of some particles to high energies. The mechanism of phase-locked acceleration naturally produces a fairly hard, power-law spectrum, when particle energy is converted to photons by bremsstrahlung. The photon spectra from this simple model can qualitatively reproduce several features of observed spectra. The mechanism can also account for the paucity of X-ray energy, relative to the energy in γ-rays, that is observed in γ-ray bursts.

067.102 **The maximum limit of the density variation parameter λ for a stable neutron star.**
R. K. Parui, H. N. K. Sarma.
Astrophys. Space Sci., Vol. 186, No. 1, p. 81 – 87 (Dec 1991).
A static spherically-symmetric model, based on an exact solution of Einstein's equation, gives the permissible matter density $\sim 2 \times 10^{14}$g cm^{-3}. The authors have estimated the upper limit of the density variation parameter (λ) and minimum mass limit of a superdense star like a neutron star. This limit gives an idea of the domain where the neutron abundance with negligible number of electrons and protons (may be treated as pure neutrons) and equilibrium in neutrons begins.

067.103 **Normal-mode excitation from stellar collapse to a black hole: odd-parity perturbations.**
E. Seidel.
Phys. Rev. D, Vol. 44, No. 4, p. 950 – 965 (15 Aug 1991). Current Physics Microform No.: 9110D1004.
Odd-parity perturbations of stellar models are studied numerically, with an emphasis on the gravitational radiation produced during collapse to a black hole. The stellar models are produced by a May-White stellar collapse code using a polytropic equation of state. A wide range of collapse models are studied, ranging from stars without pressure (dust collapse) to stars which collapse and then bounce at radii near the Schwarzschild radius. For all collapse models studied, the quasinormal modes (QNM's) of the black hole are excited to varying degrees after an initial, non-QNM transient wave form. Results are presented for some collapse models which show significantly damped and distorted wave forms. These models, which develop shocks that approach the surface of the star as it crosses the peak of the scattering potential, can produce an initial transient wave form unlike that which would be expected purely from quasinormal-mode excitation. Following this initial transient, the wave form settles into a more recognizable quasinormal wave form. This work indicates that while quasinormal ringing is a general feature of nonspherical collapse to a black hole, the details of the collapse can, in principle, affect both the amplitude and the shape of the emitted wave form.

067.104 **An introduction to matter at subnuclear densities.**
C. J. Pethick, D. G. Ravenhall.
NATO Advanced Study Institute on Neutron Stars: An Interdisciplinary Field, p. 3 – 20 (1991). Also published as Report NORDITA—91/49-AN(prepr.), 18 p. (1981). – See Abstr. 012.057 for the main entry.
An overview is given of important aspects of the physics of dense matter below nuclear matter density. Among topics discussed are electronic properties, the surface tension of neutron-rich and hot nuclei, the reduction of the Coulomb energy as a consequence of the close packing of nuclei, and the possibility of non-spherical nuclei. For the most part, the discussion of nuclear properties is based on generalizations of the liquid-drop model, but results of calculations based on more microscopic approaches are also given.

067.105 **The high density interiors of neutron stars.**
G. Baym.
NATO Advanced Study Institute on Neutron Stars: An Interdisciplinary Field, p. 21 – 36 (1991). – See Abstr. 012.057 for the main entry.
This talk reviews current understanding of the states of matter in the liquid interior of neutron stars – including nuclear matter, pion and kaon condensation, and normal and strange quark matter – and their effect on neutron star structure and dynamics. Learning about high density matter in the laboratory by means of ultrarelativistic heavy-ion collisions is also discussed.

067.106 **Pairing interactions in neutron stars.**
J. Wambach, T. L. Ainsworth, D. Pines.
NATO Advanced Study Institute on Neutron Stars: An Interdisciplinary Field, p. 37 – 48 (1991). – See Abstr. 012.057 for the main entry.
A microscopic theory for calculating $^1S_0$-pairing gaps for neutrons and protons in neutron star matter is presented. While short-range correlations are treated in a semi-empirical way, long-range fluctuations are incorporated consistently by solving coupled two-body equations for the particle-hole interaction and the scattering amplitude at the Fermi surface. The solutions determine the quasiparticle effective mass and the pairing matrix element from which the gaps are found in weak coupling BCS-theory. Implications for neutron star models are briefly discussed.

067.107 **Superfluid dynamics in neutron stars.**
M. A. Alpar.
NATO Advanced Study Institute on Neutron Stars: An Interdisciplinary Field, p. 49 – 56 (1991). – See Abstr. 012.057 for the main entry.
This lecture reviews some basic points of superfluid dynamics in neutron stars. Recent developments and open problems are pointed out.

067.108 **Neutron stars as cosmic hadron physics laboratories: what glitches teach us.**
D. Pines.
NATO Advanced Study Institute on Neutron Stars: An Interdisciplinary Field, p. 57 – 70 (1991). – See Abstr. 012.057 for the main entry.
Our present theoretical understanding of the crustal neutron superfluid is reviewed, with particular attention to the information concerning its existence and behavior which can be derived from observations of pulsar glitches and postglitch behavior.

067.109 **Neutron star plate tectonics.**
M. Ruderman.
NATO Advanced Study Institute on Neutron Stars: An Interdisciplinary Field, p. 71 – 86 (1991). – See Abstr. 012.057 for the main entry.
Spinning down (or up) neutron star crusts can be stressed beyond their yield strengths by neutron superfluid vortex line pinning or magnetic field stresses. The spin history of a fast pulsar determines radio emission features which can be compared to those observed in millisecond pulsars. In young warm spinning-down neutron stars such as the Crab pulsar, crust flow should be largely plastic. Crusts of cooler stars such as the Vela pulsar may crack and give period glitches. In more slowly rotating (cool) neutron stars the glitches should be associated with intense bursts of gamma-rays.

067.110 **Thermal emission of pulsars.**
H. Ögelman.
NATO Advanced Study Institute on Neutron Stars: An Interdisciplinary Field, p. 87 – 100 (1991). – See Abstr. 012.057 for the main entry.
Theoretical models of cooling and re-heating of single neutron stars is reviewed and compared with the observations of EINSTEIN and EXOSAT satellites. The expected contribution of the ROSAT satellite is discussed.

067.111 **The formation of neutron stars in the Galaxy.**
D. Bhattacharya.
NATO Advanced Study Institute on Neutron Stars: An Interdisciplinary Field, p. 103 – 123 (1991). – See Abstr. 012.057 for the main entry.
Two different ways for the formation of neutron stars have been suggested in the literature: gravitational collapse of the

cores of massive stars, and accretion-induced collapse of heavy white dwarfs. This article discusses constraints on both scenarios based on observations of supernovae, supernova remnants, binary radio pulsars and radio pulsars in globular clusters.

067.112 **Core collapse supernovae and neutron star formation.**
W. Hillebrandt.
NATO Advanced Study Institute on Neutron Stars: An Interdisciplinary Field, p. 125 – 141 (1991). – See Abstr. 012.057 for the main entry.

The author describes the micro- and macro-physics ingredients which enter into all numerical simulations of stellar collapse and supernova explosions. In particular, the nuclear equation of state and certain weak interaction rates seem to be of key importance. A second class of problems arises from the fact that certain approximative neutrino transport schemes are commonly used in numerical models which are far from being satisfactory. Finally, we may not even know the stellar properties at the onset of collapse with sufficient accuracy to make any firm statements concerning the final outcome of the collapse of the core of a massive star. The author discusses present day's favorite supernova models. It will be argued that given the obvious difficulties of prompt explosion models neutrino heating possibly supported by some convective energy transport seems to be the only promising scenario. Possibilities and difficulties to use supernova observations for the diagnostics of core collapse are discussed.

067.113 **Neutron star formation in close binary systems.**
K. Nomoto, H. Yamaoka, T. Shigeyama, S. Kumagai, T. Tsujimoto.
NATO Advanced Study Institute on Neutron Stars: An Interdisciplinary Field, p. 143 – 170 (1991). – See Abstr. 012.057 for the main entry.

Possible evolutionary scenarios for the formation of neutron stars in interacting binaries are summarized. (1) Firstly conditions for the occurrence of accretion-induced collapse of white dwarfs are examined in relation to the origin of low mass X-ray binaries and binary millisecond pulsars. The outcome of the evolution of accreting white dwarfs is summarized as functions of accretion rate and the initial mass of the white dwarf. (2) Secondly discussed are the neutron star formation in type Ib/Ic supernovae, which are likely to be the core collapse of helium stars in binary systems. The light curves of SNe Ib/Ic (maximum brightness and decline rate) are used to infer the progenitors' masses and the neutron star masses. (3) Thirdly the evolutionary origin of double neutron stars are discussed in relation to SNe Ib/Ic.

067.114 **Evolution of close binaries and the formation of millisecond radio pulsars.**
E. P. J. van den Heuvel.
NATO Advanced Study Institute on Neutron Stars: An Interdisciplinary Field, p. 171 – 196 (1991). – See Abstr. 012.057 for the main entry.

All millisecond pulsars share the special characteristics of the binary radio pulsars (on average much weaker magnetic fields and much faster spin that the general radio pulsar population), and a large fraction ($\sim$40%) of millisecond pulsars are found in binaries. All observational evidence indicates that the millisecond and binary pulsars are relatively old neutron stars that have been spun-up by accretion in binary systems (so-called "recycling"). The two classes of binary radio pulsars (PSR 1913+16 class and PSR 1953+29 class) appear to be the descendants of wide high-mass X-ray binaries and wide low-mass X-ray binaries, respectively. Low-mass X-ray binaries with short orbital periods (<0.5d) are expected to turn into millisecond binary radio pulsars which destroy (evaporate) their low-mass companions, presumably leaving a single millisecond pulsar in most cases. There is no longer clear evidence that magnetic fields of isolated neutron stars do decay spontaneously. Only neutron stars that have been recycled in binaries show clear evidence for magnetic field decay. These recycled pulsars are expected to highly contaminate (on the 50% level or more) the general population of single pulsars in the Galaxy – creating the (probably false) impression that the magnetic fields of isolated neutron stars decay on a relatively short timescale ($\sim 10^7$yrs).

067.115 **Vaporizing neutron stars in low-mass X-ray binaries and the statistics of millisecond pulsars.**
M. Tavani.
NATO Advanced Study Institute on Neutron Stars: An Interdisciplinary Field, p. 197 – 202 (1991). – See Abstr. 012.057 for the main entry.

Recent data on low-mass X-ray binaries (LMXBs) and msec pulsars (MSPs) pose a challenge to evolutionary theories which neglect the effects of disk and companion irradiation. The author discusses the main features of a radiation-driven evolutionary model that may be applicable to several LMXBs. According to this model, radiation from the accreting compact star in LMXBs "vaporizes" the accretion disk and the companion star by driving a self-sustained mass loss until a sudden accretion-turn off occurs.

067.116 **Companion winds excited by neutron star radiation in LMXBs and millisecond pulsars.**
J. Shaham.
NATO Advanced Study Institute on Neutron Stars: An Interdisciplinary Field, p. 203 – 218 (1991). – See Abstr. 012.057 for the main entry.

The lecture first describes the basic principles underlying formation of a wind from a stellar atmosphere by external heating. These ideas are then applied to the case of binary companions of neutron stars (NS) which are being heated by the NS radiation in low-mass X-ray binaries (LMXBs) and in binary millisecond pulsar systems. Among others, the possibilities of companion evaporation and of self-excited X-ray systems is discussed. Finally, a model for the newly discovered "windy" binary pulsar 1744-24A is presented. (orig.)

067.117 **The evolution of neutron star magnetic fields.**
D. Bhattacharya, G. Srinivasan.
NATO Advanced Study Institute on Neutron Stars: An Interdisciplinary Field, p. 219 – 233 (1991). – See Abstr. 012.057 for the main entry.

The authors review the arguments for and against the decay of the magnetic fields of neutron stars, in the light of most recent developments. They conclude that the field decay may only occur when the neutron star has been processed in an interacting binary. Some theoretical models for such an evolution are discussed.

067.118 **Electrical conductivity of neutron star cores and evolution of internal magnetic fields.**
D. G. Yakovlev.
NATO Advanced Study Institute on Neutron Stars: An Interdisciplinary Field, p. 235 – 242 (1991). – See Abstr. 012.057 for the main entry.

Physical input to the equations which govern the evolution of magnetic fields in non-superconducting parts of neutron star cores is discussed with emphasis to the conductivity of matter. Magnetic fields can be influenced by the effects associated with the balance of forces, by circulation (induced by the heat transport), by anisotropic resistivity of magnetized plasma (Ohmic decay of longitudinal and transverse currents and the Hall drift), etc. The timescales of these processes range from several minutes to the Universe age; the field lifetimes $\sim 10^4 - 10^7$yr might be expected.

067.119 **Very high energy gamma-ray emission from accreting X-ray pulsars.**
K. S. Cheng, K. N. Yu, T. Cheung, M. M. Lau.
Astrophys. J., Vol. 379, No. 1, p. 290 – 294 (20 Sep 1991).

Cheng and Ruderman have shown that electrostatic acceleration regions (accelerator gaps) above the inner part of a Keplerian accretion disk of an X-ray pulsar can accelerate positively charged particles (protons/ions) to extreme relativistic energies ($\sim 10^{15}$eV) toward the disk. The subsequent collisions between such ultrarelativistic protons (ions) and the disk material can produce very high energy (VHE) primary gamma rays via the decay of neutral pions. In this paper, the authors show that these "primary" gamma rays, with energies $E_\gamma > 3 \times 10^{12} B_6^{-1}$eV, will be converted to secondary $e^{\pm}$ pairs by the local magnetic field. Secondary synchrotron radiation by these secondary pairs combined with the surviving primary gamma rays produces the observed VHE gamma-ray spectrum. A comparison of the observed data with the model results is discussed.

067.120 **An empirical theory of pulsar emission.**
J. M. Rankin.
NATO Advanced Study Institute on Neutron Stars: An Interdisciplinary Field, p. 349 – 362 (1991). – See Abstr. 012.057 for the main entry.

A system of profile classification is reviewed which serves as a starting point for studying the emission characteristics of pulsars. Two types or mechanisms of pulsar radiation are identified which combine geometrically to produce five major species of profile. The core emission, which forms a pencil beam of radiation, is apparently produced close to the stellar surface within the entire polar-cap region by low γ particles. The conal emission, which consists of a hollow-conical beam, then seems to be emitted at heights of 10 to 20 stellar radii by currents of high γ particles travelling along some of the most peripheral of the "open" field lines.

067.121 **Plasma physics of accreting neutron stars.**
P. Ghosh, F. K. Lamb.
NATO Advanced Study Institute on Neutron Stars: An Interdisciplinary Field, p. 363 – 444 (1991). – See Abstr. 012.057 for the main entry.

Accretion of plasma by a neutron star is thought to be the fundamental process that powers many X- and γ-ray sources in the cosmos. Examples of such sources include the X-ray pulsars found in high- and low-mass binary systems and the luminous X-ray stars found in low-mass binary systems in the galactic bulge. These lectures provide an introduction to some of the plasma concepts and phenomena that are needed to understand such sources. The capture of material from the wind or from the atmosphere or envelope of a binary companion star is described and the resulting types of accretion flows discussed. The reasons for the formation of a magnetosphere around the neutron star are explained. The qualitative features of the magnetospheres of accreting neutron stars are then described and compared with the qualitative features of the geomagnetosphere. The conditions for stable flow and for angular and linear momentum conservation are explained in the context of accretion by magnetic neutron stars and applied to obtain rough estimates of the scale of the magnetosphere. Accretion from Keplerian disks is then considered in some detail. The radial structure of geometrically-thin disk flows, the interaction of disk flows with the neutron star magnetosphere, and models of steady accretion from Keplerian disks are described. Accretion torques and the resulting changes in the spin frequencies of rotating neutron stars are considered. The predicted behavior is then compared with observations of accretion-powered pulsars. Magnetospheric processes that may accelerate particles to very high energies, producing GeV and, perhaps, TeV γ-rays are discussed. Finally, the mechanisms that decelerate and eventually stop accreting plasma at the surfaces of strongly magnetic neutron stars are described.

067.122 **Accreting pulsars, gamma-ray bursters and LMXB as high energy polarized sources.**
P. Mészáros.
NATO Advanced Study Institute on Neutron Stars: An Interdisciplinary Field, p. 483 – 492 (1991). – See Abstr. 012.057 for the main entry.

The author discusses the significance of polarization producing mechanisms in determining the physical conditions in models of neutron star high energy sources, and the possible impact of polarization measurements at high energies in determining the geometry of the emission regions. Current experiments and their capabilities are reviewed. The author discusses calculations of the expected polarization from nonmagnetic neutron stars with accretion disks, and the pulse phase dependence of the degree and angle of polarization in magnetized accreting X-ray pulsars. Calculations of the polarization of gamma-ray burst sources and accreting pulsars are presented, with particular attention to the cyclotron line energy range, using relativistic two-photon processes.

067.123 **Accretion flows on to neutron stars.**
A. R. King.
NATO Advanced Study Institute on Neutron Stars: An Interdisciplinary Field, p. 493 – 500 (1991). – See Abstr. 012.057 for the main entry.

This article briefly reviews our understanding of accretion in both high- and low-mass X-ray binaries. In the first class the angular momentum of the captured material is rather uncertain, and the accretion flow may be very complex: the spin rates of the neutron stars in both supergiant and Be-star binaries suggest that the X-ray stage is very short. The nature of the accretion discs in the second group is a matter of debate, as observations imply vertical structure subtending large angles at the central neutron star.

067.124 **Slowly accreting neutron stars ("SANS").**
J. P. Lasota.
NATO Advanced Study Institute on Neutron Stars: An Interdisciplinary Field, p. 515 – 527 (1991). – See Abstr. 012.057 for the main entry.

The author discusses some physical processes occuring at the surface of a magnetized, slowly accreting neutron star (hereafter called SANS). He shows that thermonuclear runaway in the accreted matter may produce events very similar to γ-ray bursts.

067.125 **X-ray bursts.**
D. Q. Lamb.
NATO Advanced Study Institute on Neutron Stars: An Interdisciplinary Field, p. 531 – 544 (1991). – See Abstr. 012.057 for the main entry.

The author summarizes the observed features of X-ray bursts and the key elements of the thermonuclear flash model of this phenomenon. He then elucidates the full range of possible burst behavior using a global analysis of the stability of nuclear burning on accreting neutron stars.

067.126 **Isoperimetric inequalities in the physics of black holes.**
E. Malec.
Acta Phys. Pol., B, Vol. 22, No. 10, p. 829 – 858 (Oct 1991).

Results are reported on the following problems: (1) On the validity of the Penrose inequality. (2) On the hoop conjecture. (3) On the physical criteria for the formation of trapped surfaces. (4) On estimates of a total scalar curvature. A special emphasis is put on the last two points.

067.127 **Annihilation radiation in strong magnetic fields and gamma-ray burst spectra.**
G. G. Pavlov, A. D. Kaminker, P. G. Mamradze.
NATO Advanced Study Institute on Neutron Stars: An Interdisciplinary Field, p. 561 – 569 (1991). – See Abstr. 012.057 for the main entry.

The authors calculate the spectra of intensity and polarization of the annihilation radiation produced by one-dimensional thermal or power-law distributions of $e^{\pm}$ on the ground Landau level in strong magnetic fields $B \sim 10^{12}$G. The spectra depend strongly on the angle θ between the magnetic field and line-of-sight. The results are used for interpreting the annihilation features observed in some gamma-ray burst spectra. Narrowest features can be interpreted in terms of the one-dimensional thermal $e^{\pm}$ distributions. Broad features with power-law tails can be produced by power-law-distributed $e^{\pm}$ if the detected annihilation radiation is strongly beamed along the magnetic field or if it comes from an extended region with sufficiently inhomogeneous magnetic field.

067.128 **The search for TeV emission from neutron stars.**
O. C. De Jager.
NATO Advanced Study Institute on Neutron Stars: An Interdisciplinary Field, p. 571 – 577 (1991). – See Abstr. 012.057 for the main entry.

The author presents a review of the observation of very high energy γ-rays (above 100 GeV) from radio and X-ray pulsars. The best candidates for such emission are Crab, Her X-1, Vela X-1 and a few others discussed in this paper. However, the pulsed radiation is not always seen and none of the detections are of an exceedingly high statistical significance, except for the Crab nebula's unpulsed emission. The low signal to noise ratio usually encountered requires a careful approach to periodicity searches from X-ray pulsars where the spin periods change with accretion rate. The author therefore gives constraints and general hints for period searches from such objects.

067.129 **Thermal effects of nonequilibrium beta processes in a strong magnetic field.**
G. A. Shul'man.
Sov. Astron., Vol. 35, No. 5, p. 492 – 495 (Sep 1991). Current Physics Microform No.: 9203X1666. English translation of Astron. Zh., Tom 68, Vyp. 5, p. 984 – 991 (1991).

The thermal energy emitted in the nonequilibrium beta capture of ultrarelativistic electrons by nuclei in the partially degenerate matter of an equilibrium star in the quantum limit of the magnetic field frozen into its interior is determined. It is shown that with allowance for the influence of the magnetic field on the energy spectrum and number density of quantum states of the partially degenerate, ultrarelativistic electron gas, the release of thermal energy differs quantitatively from that resulting from nonequilibrium processes of beta capture in the matter of a degenerate magnetic star. The limiting temperature to which the matter of a partially degenerate magnetic star can be heated in the quantum limit of the magnetic field frozen into its interior is determined.

067.130 **Age of pulsars and a phenomenological model of their evolution.**
Yu. I. Neshpor, A. A. Stepanyan.
Sov. Astron., Vol. 35, No. 5, p. 495 – 498 (Sep 1991). Current Physics Microform No.: 9203X1669. English translation of Astron. Zh., Tom 68, Vyp. 5, p. 992 – 998 (1991).

A model of evolution with an age parameter $\eta = L_R/-(dE/dt)$ where L_R is the radio luminosity and dE/dt is the total rotational energy loss, is proposed. Using the data on 292 pulsars, the authors have shown that variations of pulsars parameters, P, $\dot{P}$, β (P – period, $\dot{P}$ – time derivative of the period, β – angle between the magnetic moment and spin) fit into a self-consistent model, while the dependence of energy losses on the angle β corresponds to losses through magnetodipole radiation.

067.131 **Model of photospheres for X-ray bursters near the Eddington limit.**
V. E. Zavlin, Yu. A. Shibanov.
Sov. Astron., Vol. 35, No. 5, p. 499 – 503 (Sep 1991). Current Physics Microform No.: 9203X1673. English translation of Astron. Zh., Tom 68, Vyp. 5, p. 999 – 1008 (1991).

The authors present calculations of model photospheres of neutron stars with a luminosity close to the Eddington limit, under conditions typical of the emitting regions of X-ray bursters. It is shown that at luminosity $L \geqslant 0.95\, L_{Edd}$, the photospheric emission spectrum becomes similar to the spectrum of a homogeneous semi-infinite medium with saturated Comptonization. A Wien hump develops in the high-energy range and a kind of plateau in the low-energy range. This is a consequence of Compton heating of surface layers of the photosphere by hotter photons escaping from the interior and the formation of a fairly deep layer of uniform temperature near the surface. The constraints on the photosphere parameters under which such a layer develops are obtained. The results are in qualitative agreement with the approximate analysis of Babul and Paczynski and of Titarchuk, although there is a significant quantitative difference in the spectral fluxes, particularly in the observed spectral range, which reaches an order of magnitude at $L \geqslant 0.99\, L_{Edd}$. It is shown that the reason for the difference is too crude a determination of the photospheric temperature in the approximate analysis, and a way to obtain a more accurate analytical approximation is indicated. The results can be used to analyze the spectra of strong bursts from X-ray bursters.

067.132 **Rotational properties of strange pulsar models.**
O. G. Benvenuto, J. E. Horvath, H. Vucetich.
Phys. Rev. D, Vol. 44, No. 4, p. 1321 – 1322 (15 Aug 1991). Current Physics Microform No.: 9110D1375.

The authors present a study of the rotational properties of strange pulsars: strange matter stars capable of supporting glitches. It is shown that their differentiated internal structure implies a lower maximum rotational frequency than that of homogeneous strange stars. Nevertheless, they are able to fit the known pulsar properties.

067.133 **On the origin of LMXRBS: the ONeMg case.**
J. Isern, R. Canal, J. Labay.
NATO Advanced Research Workshop – 7. European Workshop on White Dwarfs, p. 457 – 464 (1991). – See Abstr. 012.044 for the main entry.

The ignition of Ne-O in mass accreting white dwarfs, formed in close binary systems by mass loss from stars in the range of $8\, M_\odot \leqslant M \leqslant 12\, M_\odot$, is preceded by electron captures on ^{24}Mg and ^{24}Na. Electron captures on ^{20}Ne (or maybe on ^{24}Mg and ^{24}Na) are the triggering mechanism of explosive ignition at densities in the vicinity of 9.5×10^9g/cm^3. The outcome depends on the adopted propagation velocity of the thermonuclear burning front. Burning fronts propagating with hydrodynamical velocities ($\geqslant 0.02\, c_s$) lead to the total disruption of the star or, in some cases, to the formation of an "iron" white dwarf. Burning fronts propagating conductively lead to the formation of a neutron star.

067.134 **A model of unpulsed very high energy gamma rays from the Crab Nebula and pulsar.**
P. W. Kwok, K. S. Cheng, M. M. Lau.
Astrophys. J., Vol. 379, No. 2, p. 653 – 658 (1 Oct 1991).

The angular resolution of gamma-ray detectors does not allow one to separate the nebula from the pulsar in the Crab. It is generally assumed that the steady emission of gamma rays comes from the nebula. Using the "outer magnetospheric gap" model, the authors propose an alternative mechanism in which the steady emission of gamma rays could come from a compact region, a few light cylinder radii beyond the pulsar. In that region, the magnetic field has been transformed from a dipole field to a swirling field. In the model, two mono-energetic beams of primary e^+, if $\boldsymbol{\Omega}\cdot\mathbf{B}_s > 0$, or primary e^-, if $\boldsymbol{\Omega}\cdot\mathbf{B}_s < 0$, are streaming with a Lorentz factor $\sim 4 \times 10^7$ out from the outer gaps through the light cylinder. The gap primary $e^+(e^-)$ beam from the first outer gap can produce ~ 10 TeV primary very high energy (VHE) gamma rays through inverse Compton scattering with the IR photons from the second gap, and vice versa. The primary VHE gamma rays are sufficiently energetic to produce secondary $e^\pm$ pairs by collision with the same pulsed IR photons. Since the mean free path of these conversion processes is comparable to the local radius of curvature, the pitch angles of the secondary pairs are large enough to result in isotropic secondary TeV gamma rays. A comparison between the model spectra and the observed data is presented.

067.135 **Presence of a quadrupole component of the magnetic field near the surface of the neutron star in Hercules X-1.**
N. I. Shakura, K. A. Postnov, M. E. Prokhorov.
Sov. Astron. Lett., Vol. 17, No. 5, p. 339 – 342 (Sep 1991). Current Physics Microform No.: 9204X1286. English translation of Pis'ma Astron. Zh., Tom 17, No. 9, p. 803 – 809 (1991).

The complex variable pulse structure in the X-ray pulsar Her X-1 may be due to a magnetic-field configuration near the surface of the neutron star that is more complex than a dipole configuration. It is shown within the framework of the idealized approximation of a coaxial dipole and axial quadrupole that the pulse in this source may be formed in the region of polar caps with different magnetic–field strengths and in a ring–like zone around the magnetic pole with the weaker field.

067.136 **Structure of the one–way membrane regions of two superposed Reissner–Nordström black holes.**
Wang Yongcheng, Xia Ping.
Nuovo Cimento B, Vol. 106, No. 11, p. 1287 – 1298 (Nov 1991).
By use of the imaginary Weyl coordinates, the one–way membrane regions of two superposed Reissner–Nordström black holes are obtained. It has been shown that the coordinates of the horizon and one–way membrane region of one black hole are not affected by the other black hole if the two black holes are separated, and yet the structure of the horizons and one–way membrane regions are seriously distorted if the two black holes are intersected.

067.137 **On black holes in multidimensional theory with Ricci–flat internal spaces.**
S. B. Fadeev, V. D. Ivashchuk, V. N. Melnikov.
Phys. Lett. A, Vol. 161, No. 2, p. 98 – 100 (23 Dec 1991).
A generalization of the Tangherlini solution for the case of n internal Ricci–flat spaces is obtained. It is shown that in the $(2+d)$–dimensional section a horizon exists only in the trivial case when the internal–space factors are constant. The p–adic analog of the solution is also considered.

067.138 **How singular are black hole interiors?**
R. Balbinot, P. R. Brady, W. Israel, E. Poisson.
Phys. Lett. A, Vol. 161, No. 3, p. 223 – 226 (30 Dec 1991).
Ori has recently shown that an astronaut approaching the inner horizon of a black hole is not necessarily torn apart by tidal forces. This raises anew the possibility of astronavigation through black holes, perhaps to other universes. The authors re–examine this question in the light of hypotheses about probable conditions in the black hole core.

067.139 **Soliton stars at finite temperature.**
W. N. Cottingham, R. Vinh Mau.
Phys. Rev. D, Vol. 44, No. 6, p. 1652 – 1660 (15 Sep 1991). Current Physics Microform No.: 9111B1880.
The authors investigate the properties of soliton stars in the Lee–Wick model when a temperature dependence is introduced into this model. It is found that at some critical temperature, $T_c \sim 100-200$ MeV, a first–order phase transition occurs, leading to the formation of soliton stars with characteristics similar to those considered by Lee and Pang but with a much smaller mass and size. The authors study the evolution of these soliton stars with the temperature from the early universe to the present time.

067.140 **Strange stars.**
C. Alcock.
Nucl. Phys. B, Proc. Suppl., Vol. 24B, p. 93 – 102 (Dec 1991). – See Abstr. 012.068 for the main entry.
The general structure of strange stars is reviewed, paying particular attention to their similarities to and differences from neutron stars.

067.141 **Converting neutron stars into strange stars.**
A. Olinto.
Nucl. Phys. B, Proc. Suppl., Vol. 24B, p. 103 – 109 (Dec 1991). – See Abstr. 012.068 for the main entry.
If strange matter is formed in the interior of a neutron star, it will convert the entire neutron star into a strange star. The author reviews the proposed mechanisms for strange matter seeding and the possible strange matter contamination of neutron star progenitors. He discusses the conversion process that follows seeding and the recent calculations of the conversion timescale.

067.142 **Fast pulsars, variational bounds, other facets of compact stars.**
N. K. Glendenning.
Nucl. Phys. B, Proc. Suppl., Vol. 24B, p. 110 – 118 (Dec 1991). – See Abstr. 012.068 for the main entry.
The author derives a limit on the rotation period of a gravitationally bound star that is analogous to Ruffini's mass limit for a neutron star. He discusses the impact of gravitational radiation instabilities. The condition that self–bound stars can rotate faster is derived. The present status of searches for fast pulsars is reviewed. The composition of neutron stars is discussed in the context of nuclear and hypernuclear constraints. The phase transition from hadron to quark phase in the interior of dense neutron stars is discussed properly accounting for the fact that this first order phase transition involves two conserved charges. Hybrid stars are then described.

067.143 **Maximum rotation frequency of strange stars.**
J. L. Zdunik.
Nucl. Phys. B, Proc. Suppl., Vol. 24B, p. 119 – 124 (Dec 1991). – See Abstr. 012.068 for the main entry.
Two constraints on the rotation rate of strange stars are examined. The first one results from the observed mass of the binary pulsar, the second from the requirement that strange matter is the ground state of matter. Maximum Keplerian frequency is calculated in the bag model of strange matter. Damping of gravitational radiation reaction instabilities by shear and bulk viscosity is discussed.

067.144 **Strange matter and Type II supernovae.**
O. G. Benvenuto, J. E. Horvath, H. Vucetich.
Nucl. Phys. B, Proc. Suppl., Vol. 24B, p. 125 – 129 (Dec 1991). – See Abstr. 012.068 for the main entry.
It is shown that detonation of nuclear matter to strange matter may explain the origin of Type II supernovae.

067.145 **Towards a realistic strange–pulsar model.**
J. E. Horvath, O. G. Benvenuto, H. Vucetich.
Nucl. Phys. B, Proc. Suppl., Vol. 24B, p. 130 – 133 (Dec 1991). – See Abstr. 012.068 for the main entry.
The authors have started to study stellar models which can provide (in principle) some explanation to the observational facts connected with compact stars. One key point is contained in the low–density behavior of SM, which they address.

067.146 **Accreting strange stars.**
P. Haensel, J. L. Zdunik.
Nucl. Phys. B, Proc. Suppl., Vol. 24B, p. 139 – 143 (Dec 1991). – See Abstr. 012.068 for the main entry.
Spherical accretion of matter onto a strange star with a bare quark surface, and onto a strange star with a normal matter crust, is considered. Possibility of thermonuclear flashes for some accretion rates is discussed.

067.147 **Hybrid stars and the equation of state.**
A. Rosenhauer, E. F. Staubo.
Nucl. Phys. B, Proc. Suppl., Vol. 24B, p. 156 – 159 (Dec 1991). – See Abstr. 012.068 for the main entry.
Gross properties of hybrid stars consisting of a core of strange matter surrounded by ordinary neutron matter are investigated. The authors discuss star models based on phenomenological equations of state from nuclear reactions including a phase transition between the hadronic phase and the quark–gluon plasma. For certain parameters, such equations of state support the existence of hybrid stars: While the nuclear equation of state has a minor influence on the gross properties of hybrid stars, the identification of such objects could provide us with a rather stringent constraint on the vacuum energy density of the quark gluon plasma.

067.148 **Effects of quark matter superfluidity on compact stars cooling.**
O. G. Benvenuto, H. Vucetich, J. E. Horvath.
Nucl. Phys. B, Proc. Suppl., Vol. 24B, p. 160 – 161 (Dec 1991). Abstract. – See Abstr. 012.068 for the main entry.

067.149 **Gamma–ray bursts from coalescing strange stars.**
P. Haensel, P. Amsterdamski, B. Paczyński.
Nucl. Phys. B, Proc. Suppl., Vol. 24B, p. 162 – 165 (Dec 1991). – See Abstr. 012.068 for the main entry.

The rate of collisions between the neutron stars is $\sim 10^{-4}yr^{-1}$ in our Galaxy, and $\sim 10^{5}yr^{-1}$ within the Hubble distance. The collisions are the final phases of binary orbit decay driven by gravitational radiation. If strange stars exist then their collisions must release $\sim 10^{50}$erg in gamma–rays over ~ 0.2 s. Such events should be detectable out to ~ 1 Gpc with the current instruments.

067.150 **Rotational properties of strange stars.**
J. Miller, M. Colpi.
Nucl. Phys. B, Proc. Suppl., Vol. 24B, p. 166 – 169 (Dec 1991). – See Abstr. 012.068 for the main entry.

Results are presented from studies of the rotational properties of strange stars concerning, in particular, (1) the changes of mass and mean radius resulting from rotation; (2) the minimum rotation period consistent with stability to non–axisymmetric perturbations.

067.151 **Burning a neutron star into a strange star.**
M. L. Olesen, J. Madsen.
Nucl. Phys. B, Proc. Suppl., Vol. 24B, p. 170 – 174 (Dec 1991). – See Abstr. 012.068 for the main entry.

An approximate description of the conversion from neutrons to strange matter is used to investigate the conversion of neutron stars into strange stars through slow combustion using Bethe–Johnson's model I for neutron matter and the bag model for the quark phase. The burning time is calculated for various values of the bag constant and the initial mass of the star. Burning is found to take place on time scales from .1 seconds to a few minutes – for some parameter values a hybrid star is formed while for others the star seems to collapse into a black hole.

067.152 **Hybrid accretion disks in active galactic nuclei. I. Structure and spectra.**
A. Wandel, E. P. Liang.
Astrophys. J., Vol. 380, No. 1, p. 84 – 97 (10 Oct 1991).

The authors provide a unified treatment of the two distinct states of vertically thin AGN accretion disks: a cool ($T \sim 10^6$K) optically thick solution, and a hot ($T_e \sim 10^9$K), optically thin solution. The authors introduce a generalized formalism and a new radiative cooling equation valid in both regimes. They find a new luminosity limit $L < L_E$ at which the hot and cool α solutions merge into a single solution of intermediate optical depth. For the hot solutions, four cases are considered: different ion and electron temperatures or ions and electrons in thermal equilibrium, with Comptonized bremsstrahlung or unsaturated Compton cooling. Analytic solutions for the disk structure are given, and output spectra are computed numerically. This is used to demonstrate the prospect of fitting AGN broad–band spectra, containing both the UV bump as well as hard X–ray and gamma–ray tail, using a single accretion disk model. Such models make definite predictions on the observed spectrum, such as the relation between the hard X–ray spectral index, the UV–to–X–ray luminosity ratio, and a ~ 1 MeV feature.

067.153 **Black hole evaporation and ultrashort distances.**
T. Jacobson.
Phys. Rev. D, Vol. 44, No. 6, p. 1731 – 1739 (15 Sep 1991). Current Physics Microform No.: 9111B1959.

The role played by ultrahigh frequencies of ultrashort distances in the usual derivations of the Hawking effect is discussed and criticized. The question "would a black hole radiate if there were a Planck scale cutoff in the rest frame of the hole?" is posed. Guidance is sought from Unruh's fluid–flow analogue of black hole radiation, by taking into account the atomic nature of the fluid. Two arguments for black hole radiation are given which assume a Planck length cutoff. One involves the response of static accelerated detectors outside the horizon, and the other involves conservation of the expectation value of the stress tensor. Neither argument is conclusive, but both do strongly suggest that, in spite of reasonable doubt about the usual derivations of black hole radiation, a "safe" derivation avoiding our ignorance of ultrashort distance physics can probably be formulated.

067.154 **Stability of a shell around a black hole.**
P. R. Brady, J. Louko, E. Poisson.
Phys. Rev. D, Vol. 44, No. 6, p. 1891 – 1894 (15 Sep 1991). Current Physics Microform No.: 9111B2119.

The authors study the mechanical stability of a static, infinitely thin, spherically symmetric massive shell surrounding a classical Schwarzschild black hole. The shell is taken to have a non–negative surface energy density, and a speed of sound not greater than the speed of light. It is shown that the shell is stable against radial perturbations only outside a critical radius, which is always larger than the radius of the circular photon orbit. The surface energy density of a stable shell is always larger than twice the surface pressure, and thus satisfies the dominant energy condition by a wide margin. The authors briefly discuss the effects of Hawking radiation in view of a path–integral approach to black hole thermodynamics developed by York and collaborators. The results suggest that a macroscopic thermal equilibrium situation associated with the canonical ensemble in this approach may not be realizable with a thin matter shell in Lorentzian spacetime.

067.155 **EUV emission from gamma–ray–burst sources.**
D. Hartmann, J. Arons, S. E. Woosley.
1. Berkeley Colloquium on Extreme Ultraviolet Astronomy, p. 213 – 220 (1991). – See Abstr. 012.045 for the main entry.

Contents: 1. Introduction. 2. Galactic neutron stars. 3. Quiescent EUV emission from neutron stars. 4. Transient EUV emission from neutron stars.

067.156 **Prospects for observing EUV radiation from neutron stars.**
W. K. Rose.
1. Berkeley Colloquium on Extreme Ultraviolet Astronomy, p. 221 – 227 (1991). – See Abstr. 012.045 for the main entry.

Contents: 1. Thermal emission from neutron stars. 2. X–ray pulsars and low–mass X–ray binaries. 3. Accretion disks surrounding neutron stars. 4. Discussion of results. 5. Prospects for EUV observations.

067.157 **Self–similar scalar soliton stars in the thin–wall approximation.**
A. di Prisco, L. Herrera, M. Esculpi.
Phys. Rev. D, Vol. 44, No. 8, p. 2286 – 2294 (15 Oct 1991). Current Physics Microform No.: 9112A1210.

Two families of time dependent (analytical) solutions of Einstein equations, describing scalar soliton stars in the thin wall approximation, are found and discussed. The space–time within the star is self–similar, reinforcing thereby the hypothesis about the formation of these objects during a second order phase transition. Maximum and minimum values of the final surface potential, compatible with the asymptotic existence of the soliton star, are found.

067.158 **Effects of magnetohydrodynamic accretion flows on global structure of a Kerr black hole magnetosphere.**
S. Nitta, M. Takahashi, A. Tomimatsu.
Phys. Rev. D, Vol. 44, No. 8, p. 2295 – 2305 (15 Oct 1991). Current Physics Microform No.: 9112A1219.

In view of black hole models for central engines in active galactic nuclei, the authors discuss some effects of accreting matter on the global structure of the Kerr black hole magnetosphere. Based on the steady axisymmetric magnetohydrodynamic (MHD) equations in Kerr geometry, the authors analytically study the general relativistic Grad – Shafranov equation and the poloidal wind equation. The field line geometry is clearly given in the region where the plasma can be magnetically supported. Near the black hole, the magnetic support breaks down and the plasma begins to flow toward the event horizon. The authors consider the MHD conditions for the plasma inflows which should pass through the Alfvén critical point. It is found that the accumulation of magnetic field lines threading the horizon is suppressed for the rapidly rotating holes. This result means that the global shape of the field lines and the MHD energy extraction from the rotating black hole can drastically change as the spin-down of the black hole proceeds. The authors present an evolutionary model for the activity and the structure of the black hole magnetosphere.

067.159 **Variability of active galactic nuclei: a theorist's view.**
M. A. Abramowicz.
Workshop on Variability of Active Galaxies, p. 255 – 264 (1991). – See Abstr. 012.070 for the main entry.

067.160 **The stability of thick disks.**
W. Glatzel.
Workshop on Variability of Active Galaxies, p. 265 – 274 (1991). – See Abstr. 012.070 for the main entry.

The properties of instabilities occurring in thick disks are reviewed. Their origin and mechanism is discussed and a summary of the influence of various model parameters and stabilising effects is given.

067.161 **Fuelling accretion disks in AGN.**
J. Frank.
6. IAP Astrophysics Meeting – IAU Colloquium No. 129: Structure and emission properties of accretion disks, p. 135 – 141 (1991). – See Abstr. 012.046 for the main entry.

The mechanisms by which the engines powering active galactic nuclei (AGN) are supplied with gaseous fuel are reviewed. The working model adopted in this paper assumes disk accretion onto a massive black hole at the center of a galaxy, but the processes driving material toward the center of the host discussed here may well apply to a wider class of AGN models. The most promising mechanisms for fuelling activity at the center of galaxies seem to involve dynamical non-axisymmetric distortions of the constituent stars and gas in the host galaxy caused by gravitational instabilities, encounters with passing galaxies, and mergers. Furthermore, it is likely that different processes operate at large galactic scales depending on the environment and the morphology of the host, whereas the driving processes within ~ 1 kpc may be quite similar in both elliptical and spiral hosts.

067.162 **Accretion disk models of active galactic nuclei: a crucial role for reprocessing?**
M. C. Begelman, M. de Kool.
6. IAP Astrophysics Meeting – IAU Colloquium No. 129: Structure and emission properties of accretion disks, p. 143 – 152 (1991). – See Abstr. 012.046 for the main entry.

GINGA and EXOSAT data on the spectra and X-ray variability of AGN indicate that a large fraction of the primary X-ray emission is reprocessed by cold gas located close to the central black hole. However, it is not clear that this gas forms the inner part of a "standard" accretion disk. In particular, the observed behavior of the iron Kα fluorescence line is diffucult to explain in terms of the reprocessing of hard X-rays by a standard accretion disk, unless most of the X-ray luminosity is generated in local regions close to the disk, with a size smaller than R_g. The authors outline some alternatives to the standard disk model, both with and without a disk-like geometry. They briefly discuss the possible importance of reprocessing by a disk at large radii.

067.163 **Instabilities in thick disks.**
R. Narayan.
6. IAP Astrophysics Meeting – IAU Colloquium No. 129: Structure and emission properties of accretion disks, p. 153 – 160 (1991). – See Abstr. 012.046 for the main entry.

The properties of instabilities of thick accretion disks have been studied in fair detail by a number of investigators, and the results are reviewed briefly in this article.

067.164 **Superfluid hydrodynamics in rotating neutron stars. I. Nondissipative equations.**
G. Mendell.
Astrophys. J., Vol. 380, No. 2, p. 515 – 529 (20 Oct 1991).

Newtonian hydrodynamic equations are developed which describe the outer-core regions of neutron stars composed of superfluid neutrons, superconducting protons, and degenerate electrons and muons. The equations include couplings to the gravitational and electromagnetic fields. They also include the effects of rotation and forces due to the elastic properties of the neutron and proton vortices. The set of equations is closed by constructing a model of the total energy density and using it to express the dependent variables in terms of the independent variables. The low-frequency – long-wavelength limit of the equations is determined. The results are applicable to studies of the large-scale oscillations of rotating and magnetized neutron stars.

067.165 **Superfluid hydrodynamics in rotating neutron stars. II. Dissipative effects.**
G. Mendell.
Astrophys. J., Vol. 380, No. 2, p. 530 – 540 (20 Oct 1991).

The Newtonian superfluid hydrodynamic equations describing the outer-core regions of neutron stars are generalized to include dissipation. The effects of viscosity, thermal conductivity, and mutual friction (due to the scattering of electrons off the neutron and proton vortices) are included. The low-frequency – long-wavelength limit is taken to obtain a set of equations suitable for studies of p-modes in rapidly rotating neutron stars. An energy functional is constructed which determines the damping times due to the various forms of dissipation. The most important result is that mutual friction will be the dominant damping mechanism for the large-scale oscillations of rapidly rotating neutron stars for the temperature range $10^7\mathrm{K} \leqslant T < T_c$, where $T_c \sim 10^9\mathrm{K}$ is the superfluid transition temperature.

067.166 **On accretion by compact objects from a stellar wind.**
M. Livio.
6. IAP Astrophysics Meeting – IAU Colloquium No. 129: Structure and emission properties of accretion disks, p. 263 – 270 (1991). – See Abstr. 012.046 for the main entry.

The general problem of accretion from an infinite medium by a compact object is important in its astrophysical application for massive X-ray binaries, in which a neutron star or a black hole accretes from the wind of an OB companion and for white dwarfs (and main sequence stars) which accrete from the winds of cool giants.

067.167 **Thermal continuum from QSO disks.**
G. A. Shields.
6. IAP Astrophysics Meeting – IAU Colloquium No. 129: Structure and emission properties of accretion disks, p. 279 – 282 (1991). – See Abstr. 012.046 for the main entry.

Accretion disks are a promising source of the energy output of AGN and QSOs. Two outstanding questions are: (1) is the observed blue-UV continuum "bump" thermal emission from a disk, and (2) is the required accretion rate feasible? The author summarizes recent work supporting an affirmative answer to both questions.

067.168 **X-ray emission from accretion disks around massive black holes.**
L. Maraschi.
6. IAP Astrophysics Meeting – IAU Colloquium No. 129: Structure and emission properties of accretion disks, p. 283 – 287 (1991). – See Abstr. 012.046 for the main entry.

The author presents an improved computation of the temperature and spectrum of a standard disk, described by the equations of Shakura and Syunyaev (1973), with high viscosity and nearly critical accretion rate (Maraschi and Molendi 1990). Further, he refers to work in progress on a two phase disk model, in which an optically thick phase coexists with a hotter phase, heated by magnetic dissipation precesses and cooled via Comptonization of the soft photons provided by the optically thick component.

067.169 **Variability of active galactic nuclei.**
M. A. Abramowicz.
6. IAP Astrophysics Meeting – IAU Colloquium No. 129: Structure and emission properties of accretion disks, p. 299 – 303 (1991). – See Abstr. 012.046 for the main entry.

067.170 **The neutron star boundary layer and general relativity.**
W. Kluźniak.
6. IAP Astrophysics Meeting – IAU Colloquium No. 129: Structure and emission properties of accretion disks, p. 327 – 330 (1991). – See Abstr. 012.046 for the main entry.

067.171 **The candidate black hole A0620–00.**
C. Bartolini, A. Guarnieri, A. Piccioni, L. Solmi, M. Teodorani.
6. IAP Astrophysics Meeting – IAU Colloquium No. 129: Structure and emission properties of accretion disks, p. 373 – 374 (1991). – See Abstr. 012.046 for the main entry.

067.172 **Luminosity–function evolution of quasar accretion–disks.**
D. Caditz, A. Wandel.
6. IAP Astrophysics Meeting – IAU Colloquium No. 129: Structure and emission properties of accretion disks, p. 391 – 392 (1991). – See Abstr. 012.046 for the main entry.

067.173 **Line emission from accretion disks: observation vs. theory.**
M. Calvani, J. W. Sulentic, P. Marziani, W. Zheng.
6. IAP Astrophysics Meeting – IAU Colloquium No. 129: Structure and emission properties of accretion disks, p. 395 – 396 (1991). – See Abstr. 012.046 for the main entry.

067.174 **Soft X–ray pulses from neutron star glitches.**
K. A. Van Riper, R. I. Epstein, G. S. Miller.
Astrophys. J., Lett., Vol. 381, No. 2, p. L47 – L50 (10 Nov 1991).

Glitches release energy in the interiors of neutron stars. In some models of glitches, such as those involving the sudden transfer of angular momentum to the stellar crust from a more rapidly rotating neutron superfluid, the energy liberated is large enough ($\sim 10^{38}$– 10^{43}ergs) to noticeably raise the temperature of the stellar surface. These temperature enhancements last from hours to months, depending on the stellar equation of state and the depth at which the heating occurs. The authors describe how measurements of the amplitudes and time scales of the temperature changes can determine the depth and the magnitude of the internal heating and set bounds on the equation of state for dense neutron star matter. These variations may be observable with the ROSAT satellite or the upcoming LANL/Berkeley soft X–ray telescope ALEXIS.

067.175 **Magnetohydrodynamic transonic flows around compact objects.**
S. K. Chakrabarti.
6. IAP Astrophysics Meeting – IAU Colloquium No. 129: Structure and emission properties of accretion disks, p. 405 – 406 (1991). – See Abstr. 012.046 for the main entry.

067.176 **On the instability of spiral flows around compact objects.**
S. K. Chakrabarti.
6. IAP Astrophysics Meeting – IAU Colloquium No. 129: Structure and emission properties of accretion disks, p. 407 – 408 (1991). – See Abstr. 012.046 for the main entry.

067.177 **Shocks in isothermal transonic flows in black hole geometry.**
S. K. Chakrabarti.
6. IAP Astrophysics Meeting – IAU Colloquium No. 129: Structure and emission properties of accretion disks, p. 409 – 410 (1991). – See Abstr. 012.046 for the main entry.

067.178 **Inviscid hydromagnetic horizon boundary conditions.**
B. Punsly.
Phys. Rev. D, Vol. 44, No. 10, p. 2970 – 2982 (15 Nov 1991). Current Physics Microform No.: 9201B0698.

This article examines the state of a magnetized fluid or plasma arbitrarily close to the event horizon of a Kerr black hole. It is found that the bulk four–velocity attains a unique asymptotic form as viewed by all external observers and there are no stable equilibria arbitrarily close to the horizon. It is shown that if one attempts to suspend a plasma just outside the horizon with magnetic ropes, then the large–scale magnetic–field lines will be reconnected and the plasma will accrete toward the hole.

067.179 **Initial data for axisymmetric black–hole collisions.**
G. B. Cook.
Phys. Rev. D, Vol. 44, No. 10, p. 2983 – 3000 (15 Nov 1991). Current Physics Microform No.: 9201B0711.

The construction of initial data sets to be used in the simulation of black hole collisions is studied in the context of the conformal imaging formalism. An approach is described for evaluating a formal, infinite series solution of the momentum constraint equation. This solution allows for the construction of completely general three–dimensional configurations of two black holes with individually specifiable linear and angular momenta. Using this solution for the momentum constraint, the Hamiltonian constraint is solved numerically in the restricted case of axisymmetric configurations. Two codes for solving the Hamiltonian constraint are described, one based on bispherical coordinates and the other on Čadež coordinates. The physical context of data sets representing two equal sized holes with linear or angular momenta is explored. The results show several expected physical effects including the gravitational spin–spin interaction.

067.180 **Molecules in AGN's accretion discs.**
J. M. Huré, S. Collin–Souffrin, J. Le Bourlot, G. Pineau des Forêts.
6. IAP Astrophysics Meeting – IAU Colloquium No. 129: Structure and emission properties of accretion disks, p. 445 – 446 (1991). – See Abstr. 012.046 for the main entry.

067.181 **Flywheels; rapidly spinning, magnetized neutron stars in spherical accretion.**
S. Mineshige, M. J. Rees, A. C. Fabian.
Mon. Not. R. Astron. Soc., Vol. 251, No. 4, p. 555 – 563 (15 Aug 1991).

The behaviour of a rapidly spinning, magnetized neutron star embedded in plasma accreting spherically at a supercritical rate is investigated. Friction between the magnetosphere and the surrounding plasma is a source of power. In addition, when the magnetic field is asymmetric with respect to the rotation axis of the star, the rotating magnetosphere acts as a propeller, transferring kinetic energy to the material outside the magnetosphere. The authors quantify these processes and calculate the values of the Alfvén radius and of the density of the surrounding plasma. They find that, if the neutron star rotates rapidly and has a strong magnetic field, such objects manifest themselves as non–thermal X–ray sources. Otherwise the material outside the magnetosphere shrouds them in such high optical depth that any outgoing radiation is thermalized. Some observational implications are briefly discussed.

067.182 **Einstein observations and the internal dynamics of compact stars: further evidence against non–linear regimes of vortex creep.**
J. E. Horvath, O. G. Benvenuto.
Mon. Not. R. Astron. Soc., Vol. 251, No. 4, p. 670 – 672 (15 Aug 1991).

The authors present structural constraints on the Crab and Vela pulsars imposed by the simultaneous assumptions of (1) surface temperatures close to those observed by the Einstein Observatory satellite, and (2) validity of the vortex creep theory in the nonlinear regime for interpreting glitch observations and internal features predicted by it. The disagreement between both studies is quantified, thus pointing strongly to the need for linear regimes of creep, as recently suggested, or some alternative picture.

067.183 **Massive thin accretion discs. II. Polarization.**
A. Laor, H. Netzer, T. Piran.
6. IAP Astrophysics Meeting – IAU Colloquium No. 129: Structure and emission properties of accretion disks, p. 463 – 464 (1991). Cf. Mon. Not. R. Astron. Soc., Vol. 242, No. 4, p. 560 – 569 (15 Feb 1990) – see Abstr. 51.067.022. – See Abstr. 012.046 for the main entry.

067.184 **Variability of accretion discs around compact objects.**
J. P. Lasota, D. Pelat.
6. IAP Astrophysics Meeting – IAU Colloquium No. 129: Structure and emission properties of accretion disks, p. 467 – 468 (1991). Cf. Astron. Astrophys., Vol. 249, No. 2, p. 574 – 580 (Sep 1991) – see Abstr. 54.067.046. – See Abstr. 012.046 for the main entry.

067.185 **Inclination effects for Kerr disks.**
J. McDowell, O. Kuhn, M. Elvis, B. Wilkes.
6. IAP Astrophysics Meeting – IAU Colloquium No. 129: Structure and emission properties of accretion disks, p. 473 – 474 (1991). – See Abstr. 012.046 for the main entry.
The "Big Bump", the most striking feature in the continuum energy distributions of quasars, extends through the optical, UV and, probably, soft X–ray bands. The authors model the Bump with a blackbody accretion disk around a Kerr black hole, taking relativistic inclination effects into account.

067.186 **Relativistic electrons in pulsar magnetospheres: quantum corrections to the classical regime.**
A. A. da Costa, F. D. Kahn.
Mon. Not. R. Astron. Soc., Vol. 251, No. 4, p. 681 – 686 (15 Aug 1991).
A classical description of the motion of plasmas in pulsar magnetospheres has been developed following the single–particle approach. It is shown that this description needs to take into account quantum corrections due to radiative effects. Instead of deterministic trajectories the authors find stochastic ones and radiative effects are stronger than expected.

067.187 **Time–dependent X–ray emission from unstable black–hole accretion discs.**
S. Mineshige, S.–W. Kim, J. C. Wheeler.
6. IAP Astrophysics Meeting – IAU Colloquium No. 129: Structure and emission properties of accretion disks, p. 487 – 488 (1991). – See Abstr. 012.046 for the main entry.
Time–dependent X–ray emission from thermally unstable accretion discs around black holes are calculated based on the disc–instability model. The authors' results are in good agreement with the soft X–ray observations of X–ray transients, but the origin of the hard X–rays is still unknown. It is proposed that the observed hard X–rays originate in a hot ($kT > 10$ keV) accretion disc corona.

067.188 **Trapped oscillations of accretion disks.**
M. Nowak, R. V. Wagoner.
6. IAP Astrophysics Meeting – IAU Colloquium No. 129: Structure and emission properties of accretion disks, p. 493 – 494 (1991). – See Abstr. 012.046 for the main entry.

067.189 **Quasi–periodic oscillations in luminous accretion discs.**
T. Okuda, S. Mineshige.
6. IAP Astrophysics Meeting – IAU Colloquium No. 129: Structure and emission properties of accretion disks, p. 497 – 498 (1991). Cf. Mon. Not. R. Astron. Soc., Vol. 249, No. 4, p. 684 – 692 (15 Apr 1991) – see Abstr. 53.067.065. – See Abstr. 012.046 for the main entry.

067.190 **A model of pulsed gamma radiation from the X–ray binary Hercules X–1 / HZ Herculis.**
F. A. Aharonian (*F. A. Agaronyan*), A. M. Atoyan.
Astrophys. J., Vol. 381, No. 1, p. 220 – 227 (1 Nov 1991).
A model of pulsed very high energy and ultrahigh energy γ–radiation from X–ray binaries is proposed, which implies that the γ–rays are due to the bombardment of a cloud ejected from the companion normal star by the relativistic proton beam stationarily accelerated by the pulsar. In the framework of this model all the peculiarities of the γ–radiation observed from the X–ray binary Her X–1 are naturally explained, namely, (1) the γ–pulsation frequency shift with respect to the X–ray frequency; (2) the episodic nature of γ–ray events with typical burst duration $\Delta t \leqslant 1$ hr; (3) the absence of any correlation between the γ–ray events and the orbital phase of the binary; and (4) the observation of γ–ray events in the phase of the deep eclipse of the pulsar. The expected γ–ray spectra in a wide energy range of 100 MeV $< E <$ 1 PeV, as well as the possibility of experimental verification of the model proposed, are discussed.

067.191 **Time dependent accretion disk model in AGN.**
A. Siemiginowska.
6. IAP Astrophysics Meeting – IAU Colloquium No. 129: Structure and emission properties of accretion disks, p. 519 – 520 (1991). – See Abstr. 012.046 for the main entry.
An accretion disk around a supermassive black hole is a leading contender for the energy source in central regions of AGN. The results of the time evolution of the accretion disk with the central black hole equal to $10^8 M_\odot$ are presented.

067.192 **Smoothed particle hydrodynamics and accretion processes around a massive black hole.**
H. Sponholz.
6. IAP Astrophysics Meeting – IAU Colloquium No. 129: Structure and emission properties of accretion disks, p. 525 – 526 (1991). – See Abstr. 012.046 for the main entry.

067.193 **Variable line profiles from accretion disks and black hole mass.**
L. Stella.
6. IAP Astrophysics Meeting – IAU Colloquium No. 129: Structure and emission properties of accretion disks, p. 529 – 530 (1991). – See Abstr. 012.046 for the main entry.

067.194 **The physical structure of self–gravitating accretion disks in the inner parts of AGN's.**
H. Störzer, R. Wehrse.
6. IAP Astrophysics Meeting – IAU Colloquium No. 129: Structure and emission properties of accretion disks, p. 533 – 534 (1991). – See Abstr. 012.046 for the main entry.

067.195 **Magnetohydrodynamical energy extraction from a Kerr black hole.**
M. Yokosawa.
IUPAP Conference on Primordial Nucleosynthesis and Evolution of Early Universe, p. 419 – 420 (1991). – See Abstr. 012.071 for the main entry.
The author solves the freezing equation of the magnetic field and gives the exact nonstationary solution for the variation of magnetic field with a given axially symmetric accretion. He studies the dynamical evolution of the energy density distribution of the magnetic field and discusses the energy transfer through the surface of a stretched horizon.

067.196 **Inside a black hole.**
W. A. Hiscock.
Nature, Vol. 353, No. 6346, p. 702 – 703 (24 Oct 1991).

067.197 **Gamma rays and energetic particles from primordial black holes.**
F. Halzen, E. Zas, J. H. MacGibbon, T. C. Weekes.
Nature, Vol. 353, No. 6347, p. 807 – 815 (31 Oct 1991).
Black holes of almost arbitrarily small mass may have formed in the very early universe. Their presence today would be revealed by the energetic radiation they would produce by means of the quantum–gravitational Hawking mechanism, allowing observational limits to be set on their density today and on their past significance.

067.198 **Formation of a planet orbiting pulsar 1829–10 from the debris of a supernova explosion.**
D. N. C. Lin, S. E. Woosley, P. H. Bodenheimer.
Nature, Vol. 353, No. 6347, p. 827 – 829 (31 Oct 1991). Letter–to–the–editor.
The 10–Earth–mass planet in a nearly circular 0.7–AU orbit around PSR 1829–10 is unlikely to have survived the supernova, or especially the pre–supernova evolution of the star that became the pulsar. The authors describe how the planet might have been created inside the young supernova remnant. They propose that the planet formed from a rotationally supported disk of ~ 0.02 solar mass of heavy elements that fell back from the supernova explosion to an initial radius of about 1,000 km. Viscous evolution of the disk then concentrated most of its angular momentum into a small amount of material at the disk's outer extremity: 10 Earth mass at 10^{13}cm. Here, dust grains that had condensed and precipitated towards the midplane grew through cohesive collisions and gravitational instabilities into 100–km planetesimals, which coagulated into the planet on a million–year timescale. The authors find the presence of a second planet, more massive and more distant, unlikely, although residual planetesimals may provide the fuel for γ–ray bursts.

067.199 **Creation by stellar ablation of the low–mass companion to pulsar 1829–10.**
J. H. Krolik.
Nature, Vol. 353, No. 6347, p. 829 – 831 (31 Oct 1991). Letter–to–the–editor.
The pulsar 1829–10 has a remarkable companion of only ~ 10 Earth masses, which occupies a nearly circular orbit of 184–day period. Bailes et al. speculate that this companion is a planet which has either survived the earlier stellar evolution and supernova which created the pulsar, or else formed through some sort of coagulation process during the lifetime of the pulsar. The author argues here that another interpretation, which they excluded, may actually be more plausible: that the companion began its life as a star, and has been ablated down to its present mass by absorbing a portion of the pulsar's spindown energy. The similar phenomena have already been seen in two other binary pulsars, PSR 1957 + 20 and PSR 1744–24A, lends preliminary credence to this suggestion. The (surprisingly small) final mass of the remnant is determined by the interplay between decreasing spindown luminosity, recession of the companion from the pulsar as a result of its mass loss, and, most importantly, shrinkage of the companion due to convective cooling of its interior.

067.200 **A planet orbiting the neutron star PSR 1829–10.**
M. Bailes, A. G. Lyne, S. L. Shemar.
Nature, Vol. 352, No. 6333, p. 311 – 313 (25 Jul 1991). Letter–to–the–editor.
The authors report observations which imply the existence of a planet–sized companion orbiting a neutron star, the pulsar PSR 1829–10, whose motion can be seen by Doppler effects on the observed arrival times of the pulses from the rotating neutron star. The planet is about 10 times the mass of the Earth, and is in an almost circular six–month orbit. It is not clear whether it formed in the aftermath of the supernova that created the neutron star, or was pre–existing and somehow survived through the late phases of stellar evolution and neutron–star formation. In either case, the existence of the planet challenges conventional theories of the formation of neutron stars from supernovae and has important implications for the existence of planetary systems around other stars.

067.201 **The origin of the planet orbiting PSR 1829–10.**
P. Podsiadlowski, J. E. Pringle, M. J. Rees.
Nature, Vol. 352, No. 6338, p. 783 – 784 (29 Aug 1991). Letter–to–the–editor.
Bailes et al. have reported the discovery of a planet–mass object in a six–month orbit around the radio pulsar PSR 1829–10. The parameters of the orbit, in particular its circularity, make it unlikely that this planet existed before the supernova explosion that presumably created the pulsar, and somehow survived. The authors present two alternative explanations for the existence and orbital parameters of the claimed planet–mass object. In the first, the pulsar forms from the coalescence of two white dwarfs, and the planet condenses from a massive disk of material left behind. In the second, a neutron star collides with and cannibalizes the central star of a solar–type planetary system. The orbits of the inner planets are made elliptical by this collision, but then recircularized by drag forces due to the extended but short–lived envelope of the disrupted central star.

067.202 **Planet searches: It's all in the timing.**
D. Black.
Nature, Vol. 352, No. 6333, p. 278 – 279 (25 Jul 1991).
The author reports on planetary companions orbiting neutron stars, especially on the planet sized companion discovered by Bailes et al. in orbit around the pulsar PSR 1829–10. The presence of the companion is inferred from the timing of the pulsar's radio pulses.

067.203 **Stability of slim accretion discs – effects of central mass and viscosity.**
F. H. Wallinder.
Mon. Not. R. Astron. Soc., Vol. 253, No. 1, p. 184 – 192 (1 Nov 1991).
Slim accretion discs have a total luminosity of the order $L/L_E = \dot{m} \sim 1$, where L_E is the Eddington luminosity and $\dot{m} = \dot{M}/\dot{M}_c$, where $\dot{M}_c$ is a critical accretion rate, related to the Eddington one. The local stability properties of such discs are examined, in the three–dimensional parameter space spanned by the (α, μ, m) axes, where α and μ are two viscosity parameters, and $m = M/M_\odot$ the central mass. The authors suggest that various types of observed quasi–periodic behaviour may be connected with slim disc instabilities. If this turns out to be correct, the so–called normal and horizontal branch oscillations could be due to unstable thermal and acoustic modes, respectively. It is subsequently shown that some of the observed short–term (quasi–periodic) variability in active galactic nuclei may also originate from short–wavelength acoustic modes in the innermost region of the disc. Consequently, observational characteristics, in connection with stability theory, may yield estimates of basic accretion parameters. In the case of the Seyfert galaxy NGC 6814, this process seems to favour $(\alpha, \mu, m, \dot{m}) = (0.5, 0, 10^6, 10^{-2})$. The authors finally conclude that this line of work may provide additional evidence for both accretion discs and black holes, in various compact sources.

067.204 **Spectra of X–ray bursts at near–Eddington luminosities.**
G. G. Pavlov, Yu. A. Shibanov, V. E. Zavlin.
Mon. Not. R. Astron. Soc., Vol. 253, No. 2, p. 193 – 197 (15 Nov 1991).
The authors have developed numerical and analytical models of photospheres of X–ray bursters for the case when the burster luminosity is very close to the Eddington limit, $l = L/L_{Ed} = 0.9$–0.999. It is shown that the burst spectrum at observable energies can be described by a diluted Wien spectrum with the spectral temperature T coinciding with the electron temperature in an isothermal outer layer formed by Compton heating in the neutron star photosphere. The dilution factor is proportional to $(T_{eff}/T)^4$, where T_{eff} is the effective temperature. The authors obtained simple expressions for T/T_{eff}, and for the dilution factor, which are determined by the parameter $(1-l)/(3+5X)$,

where X is the hydrogen abundance. Fitting spectra observed around peaks of strong bursts may put additional constraints on masses and radii of neutron stars, and give new information on chemical composition of their photospheres and on distances to the bursters.

067.205 **Neutron superfluid spin–down and magnetic field decay in pulsars.**
P. B. Jones.
Mon. Not. R. Astron. Soc., Vol. 253, No. 2, p. 279 – 286 (15 Nov 1991).
The problem of magnetic flux expulsion from the superconducting interior of a neutron star is re–examined under the assumption that superfluid neutron vortices interact strongly with the two–dimensional lattice of proton vortices. The authors find the response of the proton–electron system to a Magnus force acting on the neutron vortices. An approximate upper limit is obtained for the Magnus force which would move a neutron vortex through the proton vortex lattice. Magnetic flux is expelled from the interior at a rate determined principally by the transport of field in the normally conducting solid crust rather than by the rotation spin–down time. It is shown that a large fraction of the original magnetic flux could remain in the superconducting interior of a binary neutron star, for example, Her X–1, after its initial rapid spin–down and would then be trapped by the inward radial movement of neutron vortices which follows. Both the strong surface fields of high–mass X–ray binaries and the field decay inferred from radio pulsar population surveys are consistent with the qualitative properties of the interior magnetic flux distribution expected for a standard form of neutron star internal structure.

067.206 **Quantum hair.**
J. Preskill.
Phys. Scr. T, Vol. T36 , p. 258 – 264 (1991). – See Abstr. 012.080 for the main entry.
A black hole can carry "quantum hair" that, while undetectable classically, can influence quantum physics outside the event horizon. The effect of quantum hair on black hole thermodynamics is computed here, in a particular field–theoretic model.

067.207 **O(3)–invariant tunnelling at false vacuum decay in general relativity.**
V. A. Berezin, V. A. Kuzmin, I. I. Tkachev.
Phys. Scr. T, Vol. T36 , p. 269 – 275 (1991). – See Abstr. 012.080 for the main entry.
The authors consider O(3)–invariant tunnelling processes at false vacuum decay in general relativity. A general classification of bubble shell Euclidean trajectories is elaborated. Processes like spontaneous creation around a black hole of a double bubble with the new phase inside and tunnelling of a black hole to the state of wormhole are discussed.

067.208 **The nature of pulsar subpulse drift.**
A. Z. Kazbegi, G. Z. Machabeli, G. I. Melikidze.
Aust. J. Phys., Vol. 44, No. 5, p. 573 – 584 (1991).
A possible mechanism for the explanation of pulsar subpulse drift is suggested. In the region of the open magnetic field lines the existence of an electron–positron plasma penetrated by a primary particle beam is assumed. There is a possibility of excitation of large–scale drift waves propagating transversely to the magnetic field lines. These waves can affect the fulfilment of the radio–wave generation conditions. If the pulsar angular velocity is near to the frequency of the drift waves one should observe regular drift phenomena.

067.209 **Charged black holes with scalar hair.**
K. Lee, E. J. Weinberg.
Phys. Rev. D, Vol. 44, No. 10, p. 3159 – 3163 (15 Nov 1991). Current Physics Microform No.: 9201B0887.
The authors describe a family of static black hole solutions arising in a theory with Einstein gravity coupled to electromagnetism and an axionlike scalar field. These solutions carry both electric and magnetic charge. In addition, the scalar field is spatially varying outside the horizon. Since this nontrivial scalar field is not required by any conservation law, these solutions shed further light on the black hole no–hair conjecture.

067.210 **On the circular polarization in pulsar emission.**
A. Z. Kazbegi, G. Z. Machabeli, G. I. Melikidze.
Mon. Not. R. Astron. Soc., Vol. 253, No. 3, p. 377 – 387 (1 Dec 1991).
A model explaining the existence and properties of the circular polarization in pulsar emission is offered. It is based on the properties of waves excited in pulsar magnetospheres by possible plasma instabilities. The cyclotron instability developing due to the anomalous Doppler effect is carefully investigated. The following characteristics of the emission polarization are explained: if circular polarization exists it should be confined to small angles and observed in the core–type emission; the maximum intensity of circular polarization in the profile window is near the intensity maximum of the core–type emission; the sense–reversals of circular polarization in some pulsars and their absence in others; the correlation between circular polarization and position–angle swings; the behaviour of circular polarization in PSR 1541 + 09, 1702–19 and others.

067.211 **The collisional acceleration of relativistic electrons in the central regions of active galactic nuclei.**
M. G. Baring.
Mon. Not. R. Astron. Soc., Vol. 253, No. 3, p. 388 – 400 (1 Dec 1991).
The acceleration of electrons to relativistic energies by collision with relativistic protons is examined in the environment of the compact central regions of active galactic nuclei. Coulomb collisions between the protons and cold electrons form a prominent scattering mechanism: only large–angle scatterings result in relativistic boosts to the electron energy. The cold electrons could be the end product of a pair cascade, such as is found in models of the X–ray and gamma–ray AGN continuum, while the protons are injected into the central regions of AGN by some acceleration process such as diffusive shock acceleration. The resulting electron excitation spectrum is determined, and is found to be generally of a power–law form. Coulomb scattering proves to be an efficient acceleration mechanism, and can provide a significant supply of non–thermal electrons for typical AGN conditions. If proton cooling in the collisions is included, very flat electron spectra can be obtained, otherwise the electron injection has an index of at least two. Bremsstrahlung collisions between the protons, and the cold electrons are also investigated and are found to contribute insignificantly to the accelerated electron population unless the proton Lorentz factor exceeds about 10^4. For a power–law distribution of protons, the Coulomb process is more efficient than bremsstrahlung at accelerating electrons to energy ε except when $\varepsilon \gtrsim 10^7 m_e c^2$. Other modes of proton cooling that are important for producing higher energy electrons are discussed briefly. It is concluded that the large–angle Coulomb collisional electron re–acceleration process may be quite important for pair cascade models of AGN spectra, as it can produce a significant fraction of the total electron injection luminosity.

067.212 **Neutron stars and black holes.**
F. K. Lamb.
Symposium in Celebration of the 50. Anniversary of McDonald Observatory of the University of Texas at Austin: Frontiers of stellar evolution, p. 299 – 388 (1991). – See Abstr. 012.085 for the main entry.
The radiation of neutron stars is powered by accretion, rotation, or internal heat. About 550 neutron stars have so far been detected in the Galaxy. Accreting black holes are thought to be the central engines of active galactic nuclei and of a handful of binary X–ray sources in the Galaxy. Although a variety of neutron matter equations of state have been proposed over the years, neutron star models based on recent many–body calculations of the equation of state are remarkably similar. Significant progress has been made in recent years in understanding radial and disk accretion by black holes. Radiation from pair plasmas

may make an important contribution to the X- and γ-ray spectra of active galactic nuclei and black holes in binary systems. At present, black holes in binary systems remain difficult to distinguish from neutron stars.

067.213 **Compton backscattered 511 keV annihilation line emission and the 170 keV line from the Galactic center direction.**
R. E. Lingenfelter, X.-M. Hua.
Astrophys. J., Vol. 381, No. 2, p. 426 - 438 (10 Nov 1991).
The authors show that Compton scattering of 511 keV electron-positron annihilation radiation produces a linelike reflection feature at ~170 keV from backscattered photons. Assuming simple models of clouds and accretion disks around a compact source, the authors explore the spectrum of Compton-scattered annihilation line emission for a range of geometries, opacities, and observing angles. They find that the linelike feature is produced under a wide variety of conditions. The authors further show that such Compton backscattering of slightly redshifted annihilation line emission from the inner edge of an accretion disk could account for the 170 keV line emission and higher energy continuum observed together with the 511 keV annihilation radiation from the direction of the Galactic center. Identification of the observed 170 keV line as a slightly redshifted annihilation line reflection feature provides strong new evidence that the source of this emission is a compact object surrounded by a disk of presumably accreting matter.

067.214 **Gravitational radiation from dual neutron star elliptical binaries.**
D. Hils.
Astrophys. J., Vol. 381, No. 2, p. 484 - 489 (10 Nov 1991).
The author derives general expressions for the gravitational radiation incident on Earth due to elliptical binary systems in the Galaxy. These results are applied to dual neutron star elliptical binaries. The calculations show that eccentric dual neutron star binaries lead to a moderate increase in gravitational flux density compared with circular systems for frequencies above approximately 10^{-4}Hz. Tables of various quantities such as average gravitational luminosity, number of sources per unit bandwidth, energy spectral flux density, and gravitational wave strain density are given.

067.215 **Pair-density transitions in accretion disk coronae.**
M. Kusunose, S. Mineshige.
Astrophys. J., Vol. 381, No. 2, p. 490 - 495 (10 Nov 1991).
The authors investigate the thermal and e^+e^- - pair equilibrium structure of two-temperature disk coronae above a cool ($T \sim 10^6$K) disk around a black hole of 10 $M_\odot$. Soft photons are assumed to be amply supplied from the cool disk. The authors then find two pair thermal equilibrium points for a given proton column density: the low state with very small pair density and the high state dominated by pairs. Two possible scenarios are discussed for the fate of a two-temperature corona. When the proton optical depth, τ_p, is relatively small and the temperature of input soft photons is low ($T_{soft} \lesssim 10^6$K), the corona will undergo a limit cycle between the high state and the low state on a time scale of milliseconds. As a consequence of Compton scattering of the soft photons, the emergent spectrum in the high state is rather flat with a big Wien bump at ~100 keV, whereas it is composed of a power-law component in the low state. When τ_p is relatively large (e.g., $\tau_p = 2$) or $T_{soft} \gtrsim 2 \times 10^6$K, in contrast, the plasma will cool to form a one-temperature corona. The emergent spectrum is then almost a blackbody with energy ~1 keV. Some observational consequences are briefly discussed in connection with the high - low spectral transition in Cyg X-1.

067.216 **Three-dimensional simulations of black hole tori.**
J. F. Hawley.
Astrophys. J., Vol. 381, No. 2, p. 496 - 507 (10 Nov 1991).
Two- and three-dimensional numerical simulations are employed to investigate the global hydrodynamic nonaxisymmetric instabilities in thick, constant angular momentum (l) accretion gas tori in orbit around a Schwarzschild black hole. A radially slender torus proves to be violently unstable, evolving into a highly nonaxisymmetric orbiting fluid blob, or "planet", a result consistent with previous simulations of slender tori. A radially-wide torus develops an m = 1 nonaxisymmetric density perturbation near the pressure maximum and a trailing spiral wave that extends through the outer part of the torus. This global wave transports angular momentum through the torus, causing the average angular momentum distribution to evolve slowly away from l = constant. The unstable mode drives an accretion flow from the torus into the black hole. The role of accretion is further investigated by modeling a wide torus with an inner boundary at the cusp of the Schwarzschild effective potential. In such a torus, accretion is present throughout the evolution; only modest unstable mode growth is observed, and saturation occurs at low amplitude. An accretion flow is effective at terminating or preventing the growth of the instability in three-dimensional tori, suggesting the possibility of an instability-regulated mechanism for accretion from a thick torus. Comparison simulations performed in the torus equatorial plane show qualitative similarities between the two- and three-dimensional system. The nonaxisymmetric instabilities do not appear to be sufficiently robust to rule out the thick torus as a viable accretion disk model.

067.217 **Spectral evolutionary synthesis models of metal-poor star-forming regions.**
K. Olofsson.
Symposium in Celebration of the 50. Anniversary of McDonald Observatory of the University of Texas at Austin: Frontiers of stellar evolution, p. 611 - 612 (1991). Abstract. - See Abstr. 012.085 for the main entry.

067.218 **Diquark stars.**
D. Kastor, J. Traschen.
Phys. Rev. D, Vol. 44, No. 12, p. 3791 - 3796 (15 Dec 1991).
Current Physics Microform No.: 9101C0073.
The authors study neutron stars with cores consisting of a mixture of constituent mass quarks and diquarks. Diquarks are colored, two-quark bound states which have been conjectured to exist in the density range above deconfinement. The authors compute an approximate equation of state for such a mixture. At relatively low densities this has the form of a polytrope with adiabatic index $\Gamma = 2$. It is found that the maximum-mass star with a quark-diquark core surrounded by a low-density envelope of nucleons has mass 1.79 $M_\odot$, radius 11.4 km, and central density 1.8×10^{15}g cm^{-3}.

067.219 **Quark-matter superconductivity and the cooling of compact stars.**
J. E. Horvath, O. G. Benvenuto, H. Vucetich.
Phys. Rev. D, Vol. 44, No. 12, p. 3797 - 3801 (15 Dec 1991).
Current Physics Microform No.: 9101C0079.
It has been suggested that quark matter is superfluid (superconducting) at the high densities and/or temperatures characteristic of compact star interiors. The authors present the first calculations of compact star cooling including this phase, showing that the cooling behavior could be dramatically changed with respect to former studies including a normal quark core. It is found that some of the presented scenarios tend to be closer to the observational data and open the possibility of subsequently modeling the histories of all known young and medium-aged (i.e., without significant reheating) pulsars, without having to postulate any fundamental difference between them. Moreover, improved observations might allow an astrophysical determination of the quark - matter critical temperature.

067.220 **Trapped surfaces in the Schwarzschild geometry and cosmic censorship.**
R. M. Wald, V. Iyer.
Phys. Rev. D, Vol. 44, No. 12, p. R3719 - R3722 (15 Dec 1991).
Current Physics Microform No.: 9101C0001.
The authors prove that in extended Schwarzschild spacetime there exists a family of Cauchy surfaces which come arbitrarily close to the black hole singularity at r = 0 but are such that there do not exist any outer trapped surfaces lying within the past of

any of these Cauchy surfaces. The authors argue that, in any spherically symmetric spacetime describing gravitational collapse to a Schwarzschild black hole, it should be possible to choose a spacelike slicing by Cauchy surfaces which terminates when a singularity (outside the matter) is reached, such that the region of the spacetime covered by the slicing contains no outer trapped surfaces. Thus, an important feature of the numerical collapse examples of Shapiro and Teukolsky, cited by them as evidence against cosmic censorship, does not appear to be qualitatively different from features which occur (for an appropriate choice of slicing) in the standard examples of collapse to a black hole.

067.221 **Existence of scattering atmospheres near luminous, magnetized compact objects.**
C. D. Dermer, S. J. Sturner.
Astrophys. J., Lett., Vol. 382, No. 1, p. L23 - L26 (20 Nov 1991).
The authors show that scattering atmospheres exist in the magnetospheres of luminous compact objects from considerations of resonance effects in magnetic Compton scattering. These regions of stable equilibria are formed by a balance between the forces of gravity and resonant Compton scattering, and could be important in spectral models of white dwarfs and neutron stars.

067.222 **Erratum: "Strangeness, condensation, nucleon superfluidity, and cooling of neutron stars" [Astrophys. J., Lett., Vol. 354, No. 1, p. L17 - L20 (1 May 1990)].**
D. Page, E. Baron.
Astrophys. J., Lett., Vol. 382, No. 2, p. L111 - L112 (1 Dec 1991). See Abstr. 51.067.126.

067.223 **On the pre-horizon regime in the Kerr metric.**
F. de Felice, S. Usseglio-Tomasset.
Classical Quantum Gravity, Vol. 8, No. 10, p. 1871 - 1880 (Oct 1991).
Sufficiently close to an event horizon an increase in the angular velocity induces more gravitational attraction than centrifugal repulsion on a circularly orbiting particle. At the same time the local compass of inertia tends to precess forward with respect to the sense of the orbital rotation, rather than backwards. A spacetime which gives rise to these effects is termed to be in a pre-horizon regime. Here the authors investigate the occurrence of this regime in the Kerr metric.

067.224 **On the "flip-flop" instability of Bondi-Hoyle accretion flows.**
M. Livio, N. Soker, T. Matsuda, U. Anzer.
Mon. Not. R. Astron. Soc., Vol. 253, No. 4, p. 633 - 636 (15 Dec 1991).
Using a highly simplified analytic treatment of the shock cone obtained in Bondi-Hoyle accretion flows past a compact object and the results of stability analyses of the accretion line, the authors demonstrate that the "flip-flop" instability discovered recently in numerical calculations, in which the wake oscillates from side to side, has a relatively simple physical interpretation. The results of the authors' analysis are in agreement with those obtained in multidimensional numerical hydrodynamic calculations.

067.225 **Polarization properties of emission lines from relativistic accretion disks.**
K. Chen, D. M. Eardley.
Astrophys. J., Vol. 382, No. 1, p. 125 - 133 (20 Nov 1991).
The authors investigate the polarization properties of line emission originating in a geometrically thin, relativistic, Keplerian disk. The detailed derivation of formulae for calculating the line profile as well as its polarization is presented. It is found that general relativistic effects on the line profile, especially on the polarization of the line, can be strong. Therefore, polarimetric observation will provide an invaluable diagnostic for the origin of the lines. The very broad X-ray Kα lines observed from many accretion powered sources are good candidates for the application of the present results. These lines provide an important clue to processes in the innermost region of the accretion disk.

067.226 **On the possibility of excitating one-armed corrugation waves in relativistic accretion disks.**
S. Kato.
Publ. Astron. Soc. Jpn., Vol. 43, No. 4, p. 557 - 568 (1991).
The possibility of excitating one-armed corrugation waves by viscous processes in geometrically thin, polytropic disks is examined for the case where the radial wavelength of the waves is shorter than the radial scale of the disk, but longer than the disk thickness. The effects of thermal processes on wave stability are neglected, since in the present case they are negligible in comparison with those of the viscous dynamical effects. The viscosity is assumed to have a given functional dependence on density and temperature. The results show that in the innermost region of relativistic disks one-armed corrugation waves are very close to neutral stability (slightly damped), even though they are highly damped in the outer region where the disks are non-relativistic. This result depends only slightly on parameters specifying the disks and the functional form of viscosity. The closeness to neutral stability in relativistic disks results from the fact that since the epicyclic frequency is low there, a density variation is essentially associated with the waves.

067.227 **Cold super-fast magnetohydrodynamical winds from rapid rotators.**
M. Takahashi.
Publ. Astron. Soc. Jpn., Vol. 43, No. 4, p. 569 - 587 (1991).
The author has studied relativistic ideal magnetohydrodynamical (MHD) winds in the stationary, axisymmetric magnetosphere of a rapidly rotating compact object without either gravitational or pressure effects. To obtain a highly accelerated wind, the author expects that a plasma injected from a rapid rotator should pass smoothly through both the Alfvén and fast magnetosonic points, where the poloidal velocity of the plasma equals either the Alfvén-wave speed or that of a fast magnetosonic wave; it would blow with a large asymptotic fast-Mach number. For a pressure-free limit it has been widely believed that the fast magnetosonic point is located only at infinity. However, the author shows that an X-type fast magnetosonic point can exist at a finite distance. A super-fast MHD wind is allowed when a dense plasma starts to blow near the light surface with a sufficiently large energy. The predicted asymptotic velocity of the super-fast MHD wind can exceed the velocity of the widely known "Michel's minimum energy solution", which has been interpreted as giving the upper limit for the asymptotic flow velocity of winds.

067.228 **Line profiles from a disk around a rotating black hole.**
A. Laor.
Workshop on Iron Line Diagnostics in X-Ray Sources, p. 205 - 208 (1991). Abstract published in Bull. Am. Astron. Soc., Vol. 22, No. 4, p. 1268 (1990) - see Abstr. 53.067.139; cf. also Astrophys. J., Vol. 376, No. 1, p. 90 - 94 (20 Jul 1991) - see Abstr. 54.067.038. - See Abstr. 012.088 for the main entry.

067.229 **Iron K-lines from accretion disks around Schwarzschild black holes.**
L. Stella, S. Campana.
Workshop on Iron Line Diagnostics in X-Ray Sources, p. 230 - 240 (1991). - See Abstr. 012.088 for the main entry.
Broad Fe K-shell emission lines at $\sim$6-7 keV are observed in a number of bright accreting galactic X-ray sources. In the case of the black hole candidates Cyg X-1 and 4U 1543-47 the line centroid energy is significantly lower than 6.4 keV, indicating a substantial redshift. Blending of different lines from different ionisation stages of iron and thermal broadening produce widths smaller than observed. The observed widths and redshifts are also unlikely to arise from Comptonisation. The iron K-line profile is most likely dominated by bulk plasma motions in the vicinity of the accreting collapsed object. A fully relativistic model is summarised for the emission-line profile produced by a Keplerian disk orbiting a Schwarzschild black hole. It is shown that the model predictions are in good agreement with the low resolution ($E/\Delta E \leqslant 10$) line profiles that have been obtained so far. The response of the line intensity to variations at the central source is

investigated. A detailed comparison with the recent Ginga results on the variability of the Fe–line intensity of the Seyfert galaxy NGC 6814 is carried out and constraints are derived on the disk geometry and mass of the central object. Finally a method to accurately measure the mass of disk accreting black holes in active galactic nuclei is outlined.

067.230 **Transient particle acceleration in strongly magnetized neutron stars. II. Effects due to a dipole field geometry.**
M. Fatuzzo, F. Melia.
Astrophys. J., Vol. 382, No. 2, p. 570 – 575 (1 Dec 1991).
Sheared Alfvén waves generated by nonradial crustal disturbances above the polar cap of a strongly magnetized neutron star ($B \lesssim 10^{12}$G) induce an electric field component parallel to **B**. Earlier calculations assuming a uniform magnetic field showed that this mechanism could successfully accelerate particles to Lorentz factors $\gamma \sim$ (50 cm)·m, where m ($\sim 2 - 3\ cm^{-1}$) characterizes the degree of shearing. However, magnetic field gradients in a realistic dipole field geometry cannot be ignored. Here, the authors determine the manner in which the strong radial dependence of **B** affects the propagation of these sheared Alfvén waves, and whether this magnetohydrodynamic process is still an effective particle accelerator. By comparison to the field gradients, the field–line curvature is small within a stellar radius of the polar cap, and the authors therefore use an approximate form of the dipole field that ignores the component perpendicular to the z–axis. Although the general field equation is quite complicated, a simple wavelike solution can still be obtained under the conditions of interest (i.e., $B \lesssim 10^{12}$G, and a particle number density $n_e \lesssim 10^{26} cm^{-3}$) for which the Alfvén phase velocity approaches c and n_e decouples from the wave equation. The results may be applicable to γ–ray burst sources.

067.231 **Neutron star crustal plate tectonics. II. Evolution of radio pulsar magnetic fields.**
M. Ruderman.
Astrophys. J., Vol. 382, No. 2, p. 576 – 586 (1 Dec 1991).
Spinning–down (or up) neutron star crusts can be stressed beyond their yield strengths by crust neutron superfluid vortex line pinning. Such stresses may then move crustal plates, and the magnetic field embedded in them, toward the equator in spinning–down radio pulsars. Consequences can include magnetic moment growth during initial spin–down of the youngest pulsars, followed by a much slower decrease twoard $B \sim 10^{11}$G in dead pulsars. Subsequent spin–up (e.g., by accretion from a companion) could lead to a variety of final spin periods and magnetic dipole moments depending upon the initial pulsar magnetic field configuration. Those stars in which essentially all of the flux from a polar spin hemisphere reenters the crust in that same hemisphere would be spun–up to millisecond periods as their magnetic dipoles are further diminished by spin–up, which pushes all magnetic flux into a small cap around the spin axis pole. The resulting magnetic dipole is then nearly orthogonal to the stellar spin. This contrasts with a spatially much more extended equatorial orthogonal moment expected in a spinning–down millisecond pulsar descended from a maximally spinning neutron star formed by the accretion–induced collapse of a white dwarf. Although both kinds of millisecond pulsar genesis lead to orthogonal rotators, radio pulse shape detail may allow discrimination between the two scenarios.

067.232 **Neutron star crustal plate tectonics. III. Cracking, glitches, and gamma–ray bursts.**
M. Ruderman.
Astrophys. J., Vol. 382, No. 2, p. 587 – 593 (1 Dec 1991).
When the growing vortex pinning stresses in a cool, spinning–down superfluid neutron star strain the crust beyond its elastic yield strength, large–scale crust cracking events may occur. These would result in pulsar timing glitches with magnitude and recurrence rates near those observed. In old or dead radio pulsars these sudden releases of stored elastic energy should give bursts of X–ray and gamma–rays whose number, energy, and rise time suggest those of the so far unidentified gamma–ray burst sources. The surface magnetic field of a spinning–down crust–cracking neutron star breaks up into large surface patches (platelets) which move apart from each other. Each platelet retains its original surface magnetic field, while the expanding new interplatelet crust regions have little or none. Pulsar spin–down torque observations would reflect the decrease in average surface dipole field, while the field strength inferred from a cyclotron resonance spectral feature above a platelet would remain high and independent of stellar age.

067.233 **Gamma–ray bursts from sheared Alfvén waves.**
F. Melia, M. Fatuzzo.
Astrophys. J., Lett., Vol. 383, No. 2, p. L57 – L60 (20 Dec 1991).
Sheared Alfvén waves in strongly magnetized neutron stars are efficient particle accelerators by virtue of the large electric field components E_z they generate parallel to the unperturbed magnetic field $\mathbf{B}_0$. For example, with an amplitude $B_{A0} = 10^{11}$G ($= 0.1 B_0$) and a shear wavenumber m $= 0.5$ (corresponding to a crustal perturbation length scale of ≈ 10 cm), these waves can accelerate electrons to a Lorentz factor $\gamma \sim 10^{4-5}$ within 5 km of the stellar surface, even under the influence of a Compton drag due to the thermal emission of a warm polar cap. Here, the authors apply this physical process to a study of gamma–ray bursts, taking into account both resonant and nonresonant scattering. Although the field is oscillatory, virtually all the charges are ejected from the system, resulting in very little backheating of the stellar surface. The particle number density (and hence the burst energy) is accounted for naturally in terms of B_{A0} and m, which in principle are known from the physical manifestation of the agent causing the crustal disturbance. The resulting γ–ray spectrum compares very favorably with the observations. This model restricts the geometry of the emission region, in the sense that only the Compton upscattering of soft photons from a warm polar cap will produce the correct spectral shape.

067.234 **Spherical accretion onto black holes: a complete analysis of stationary solutions.**
L. Nobili, R. Turolla, L. Zampieri.
Astrophys. J., Vol. 383, No. 1, p. 250 – 262 (10 Dec 1991).
The problem of stationary, spherical accretion onto a Schwarzschild hole is reinvestigated by the construction of a self–consistent model which incorporates all relevant physical processes taking place in an astrophysical plasma, apart from the presence of magnetic fields and dissipative processes. In particular, transfer of radiation through the accreting gas is treated in full generality using a completely relativistic formalism. A careful analysis of critical points and boundary conditions for radiation hydrodynamics equations is performed. The complete topology of solutions in the accretion rate – luminosity plane is obtained, showing the existence of two distinct branches of models with very different emission properties: stationary accretion reveals, therefore, a bimodal behavior. By means of a self–consistent study of the effects of Compton heating, both the upper and the lower bounds for the existence of high–luminosity solutions were derived. The stability of the two possible accretion regimes is also briefly discussed.

067.235 **Spectral signatures of the MeV–PeV radiation from accreting neutron stars.**
B. Rudak, P. Mészáros.
Astrophys. J., Vol. 383, No. 1, p. 269 – 279 (10 Dec 1991).
The authors discuss the formation of the high–energy gamma–ray and neutrino spectrum produced by the injection of a power–law proton distribution extending up to $E_p \sim 10^{16}$eV in the magnetosphere of accreting neutron stars. The resulting proton spectrum is calculated in the presence of advection and cooling by interactions with the X–ray spectrum of the neutron star, leading to several components of a secondary photon spectrum extending from the MeV to the PeV range, and a neutrino spectrum extending from about 0.1 TeV to 0.1 PeV. The shapes of the resulting high–energy secondary spectra depend sensitively on the accreting source's X–ray spectrum and energy density as well as the magnetic field strength involved. Detailed spectra are calculated for the specific examples of Her X–1 and Cyg X–3.

067.236 **Constraints for the binding energy of strange matter implied by existence of massive neutron stars.**
M. I. Krivoruchenko, B. V. Martemyanov.
Nucl. Phys. B, Proc. Suppl., Vol. 24B, p. 134 – 138 (Dec 1991). – See Abstr. 012.068 for the main entry.

Some of the observed pulsars can be identified with ordinary neutron stars. If strange matter is stable, the central density of the neutron stars should be lower than the critical density of the quark–hadron phase transition occurring due to the strong interaction only. The quark matter being formed turns to more stable strange matter on a weak interaction time scale, converting neutron stars into strange stars. The requirement of stability of old and newly born hot neutron stars with respect to the phase transition is used to constrain the binding energy of strange matter.

067.237 **Burning of strange quark matter and transport properties of QCD plasmas.**
H. Heiselberg, G. Baym, C. J. Pethick.
Nucl. Phys. B, Proc. Suppl., Vol. 24B, p. 144 – 147 (Dec 1991). – See Abstr. 012.068 for the main entry.

The authors derive the transport equations and the relevant transport coefficients for the burning of neutron matter into strange quark matter and solve the burning problem analytically for barely flammable neutron matter.

067.238 **Stability of bulk and quasi–bulk strange quark matter and strange stars.**
S. Chakrabarty.
Nucl. Phys. B, Proc. Suppl., Vol. 24B, p. 148 – 151 (Dec 1991). – See Abstr. 012.068 for the main entry.

Using a relativistic version of the Landau theory of Fermi liquids and a dynamical density dependent quark mass approach to confinement, the stability of bulk and quasi–bulk strange quark matter (SQM) has been studied. Using the same model an equation of state for SQM at $T = 0$ is derived. The stability and some global properties of strange stars have also been investigated using this equation of state.

067.239 **Light exotic particles in strange matter: energy flux and mean free path.**
R. N. Iha, A. Goyal, J. D. Anand.
Nucl. Phys. B, Proc. Suppl., Vol. 24B, p. 152 – 155 (Dec 1991). – See Abstr. 012.068 for the main entry.

The authors calculate the axion and photino energy flux from degenerate strange quark matter and compare it with the corresponding flux from nucleon matter. They also calculate the mean free path of axions in strange matter and examine the constraints on axion decay constant.

067.240 **Self–gravitating magnetic monopoles, global monopoles and black holes.**
G. W. Gibbons.
12. Autumn School of Physics. The physical universe – the interface between cosmology, astrophysics and particle physics, p. 110 – 133 (1991). – See Abstr. 012.097 for the main entry.

The properties of the simplest magnetic monopoles in Yang–Mills–Higgs theory in flat spacetime are by now quite well understood. Moreover the fact that almost any grand unified theory will admit monopole solutions has had a profound impact on modern cosmology since it provides a very strong argument in favour of inflation as a possible solution of the monopole problem. It is therefore rather surprising that comparatively little is known about the gravitational properties of magnetic monopoles and the relation between monopoles and black holes. The author reviews the current situation and makes some comments on it.

067.241 **Cooling of neutron stars.**
C. J. Pethick.
12. Autumn School of Physics. The physical universe – the interface between cosmology, astrophysics and particle physics, p. 265 – 274 (1991). – See Abstr. 012.097 for the main entry.

The subject of neutron stars is now a vast one, which it is difficult to survey briefly. In this article a discussion is given of various aspects of the cooling of neutron stars, a subject where there have been a number of recent developments. Emphasis is placed on the basic physical processes, rather than on detailed numerical estimates.

067.242 **Cold equation of state in a strong magnetic field: effects of inverse β–decay.**
D. Lai, S. L. Shapiro.
Astrophys. J., Vol. 383, No. 2, p. 745 – 751 (20 Dec 1991).

The authors study the influence of a strong magnetic field ($B > 10^{12}$G) on the degenerate matter equation of state appropriate to a neutron star. They focus on the regime dominated by relativistic electrons up to the neutron drip density, and determine the equilibrium matter composition and equation of state, allowing for inverse β–decay. Two different equilibrium models are adopted: an ideal neutron – proton – electron (npe) gas and the more realistic model of Baym, Pethick, and Sutherland, consisting of a Coulomb lattice of heavy nuclei embedded in an electron gas. For a sufficiently high field strength, the magnetic field has an appreciable effect, changing the adiabatic index of the matter and the nuclear transition densities. The authors also consider the influence of a strong field on some simple nonequilibrium processes, including β–decay and inverse β–decay (electron capture). The effects produced by the magnetic field are mainly due to the changes in the transverse electron quantum orbits and the allowed electron phase space induced by the field.

067.243 **Formation of dense charged bunches in vacuum gaps.**
F. C. Michel.
Astrophys. J., Vol. 383, No. 2, p. 808 – 813 (20 Dec 1991).

The author points out that the pair–production discharge mechanism originally applied to pulsars by Sturrock automatically produces dense bunches that can produce coherence at radio frequencies with sufficient intensity to simulate pulsar action, about 10^{-5} of the total spindown energy. Such action requires cascading in vacuum gaps as suggested by Cheng, Ruderman, and Sutherland. The natural action of such discharging would be to eliminate such gaps. Here the author suggests that it is the inclination of the neutron star magnetic moment producing large–amplitude radiation fields that drive away plasma otherwise trapped at large distances from the neutron star. Other questions, such as heating of the surface, gamma–ray luminosities etc., need to be addressed if such a mechanism is to be incorporated into a pulsar model.

067.244 **The evolution of neutron star surface layers.**
O. Blaes, R. Blandford, P. Madau.
J. R. Astron. Soc. Can., Vol. 85, No. 4, p. 204 (Aug 1991). Abstract. – See Abstr. 012.091 for the main entry.

067.245 **Evolution of the rotation of an accreting neutron star in a low–mass X–ray binary.**
A. V. Fedorova, E. V. Ergma.
Sov. Astron., Vol. 35, No. 6, p. 640 – 645 (Nov 1991). Current Physics Microform No.: 9205X1450. English translation of Astron. Zh., Tom 68, Vyp. 6, p. 1274 – 1282 (1991).

The evolution of the rotation of an accreting neutron star in a low–mass X–ray binary (LMXB) is calculated. Vaporization of the companion by the neutron star is taken into account in the gap in orbital periods. Three possible paths of LMXB evolution and their relationship to the rotational evolution of the neutron star are discussed. It is found that in the course of evolution, the rotation period of the neutron star is close to the local equilibrium period, determined by the current value of the accretion rate and the strength of the neutron star's magnetic field.

067.246 **On the Hawking radiation associated with an Oppenheimer–Snyder collapsing star.**
H. Salehi.
Classical Quantum Gravity, Vol. 8, No. 12, p. 2169 – 2190 (Dec 1991).

The author discusses on the basis of the geometrical optics approximation the development of the Hawking radiation in the spacetime of a black hole resulting from an Oppenheimer–Snyder collapsing star. To the extent that one can use this approximation, it is shown that the asymptotic flow of radiation to future null infinity does not depend on the specific parameter describing the initial condition for the stellar collapse. Further, for a large distance from the black hole a definition of the radial part of the radiation is given via the concept of normal ordering of the energy–momentum tensor operator with respect to a "local ground state" of the Schwarzschild metric. It is shown that the ambiguity in this definition, which results from the existence of many possible local ground states, does not affect the radial part, but enters only in other parts of the radiation.

067.247 **Normal modes of the Kerr black hole.**
K. D. Kokkotas.
Classical Quantum Gravity, Vol. 8, No. 12, p. 2217 – 2224 (Dec 1991).

A new way to derive the Schutz–Will formula for the calculation of the quasi normal mode (QNM) frequencies is presented. The method is based on the Bohr–Sommerfeld rule and it is applicable even for the frequencies with large damping. Using this method, the normal mode frequencies of the Kerr black hole have been calculated.

067.248 **Black hole radiation of Dirac particles in 1+1 dimensions.**
S. M. Morsink, R. B. Mann.
Classical Quantum Gravity, Vol. 8, No. 12, p. 2257 – 2268 (Dec 1991).

The authors investigate the Hawking radiation of massive spinor particles from a 1+1 dimensional black hole. Using the exact solution of the Dirac equation in the black–hole background, the vacuum expectation value of the spinor stress–energy tensor is calculated and shown to be equivalent to the energy density of black body radiation in 1+1 dimensions, yielding a temperature of $M/2\pi$ for the black hole. The results are the same as those found when a similar analysis is performed on massive scalar particles.

067.249 **Field quantization on incomplete manifolds and problem of particles creation in gravitational field.**
R. O. Grigor'ev.
Teor. Mat. Fiz., Vol. 89, No. 3, p. 473 – 480 (Dec 1991). In Russian.

A uniformly accelerated rod as a model of the black hole neighborhood is studied. It is shown that it is possible to build a quantization procedure in the accelerated system so that the field Hamiltonian remains hermitian as it must be in case of coordinate transformations. The absence of the heat background of the field excitation in the uniformly accelerated system is obtained as a result.

067.250 **Quantum–field approach to the theory of the relativistic jet activity of quasars and galactic nuclei.**
P. I. Fomin.
Ukr. Fiz. Zh., Vol. 36, No. 12, p. 1785 – 1795 (Dec 1991). In Russian.

A theory of the relativistic jet activity of quasars and galactic nuclei based on the idea that rotating black holes generate quantum vortices in QCD vacuum condensates is suggested.

067.251 **Relativistic hadrons near accreting compact objects: dynamical and radiative signatures.**
M. C. Begelman.
Workshop on Relativistic Hadrons in Cosmic Compact Objects, p. 1 – 14 (1991). – See Abstr. 012.103 for the main entry.

Ultrarelativistic protons in the vicinity of an accreting compact object lose energy by interacting with the ambient plasma and radiation field. In the process, they may create significant fluxes of gamma rays, electron–positron pairs, neutrinos and neutrons. The author outlines the processes by which these particles are created, and their ultimate fates. Pairs and gamma rays participate in an electromagnetic pair cascade, which ultimately contributes to (and may dominate) the nonthermal radiation output from the compact object. Virtually all of the neutrinos and a significant fraction of the neutrons escape, and may be responsible for observable radiative and dynamical effects originating far from the compact object.

067.252 **Can ultrarelativistic neutrons from the central engine drive the outflow in broad absorption line QSOs?**
M. de Kool, M. C. Begelman, M. Sikora.
Workshop on Relativistic Hadrons in Cosmic Compact Objects, p. 15 – 21 (1991). – See Abstr. 012.103 for the main entry.

The authors present a model that can explain many features of the very fast ($\sim 0.1c$) outflows that are observed in broad absorption line QSOs. In this model, the outflow is accelerated primarily by the pressure of ultrarelativistic protons. These protons are deposited in the wind, over a range of radii, by the decay of ultrarelativistic neutrons escaping from the central engine. Since the energy deposition is predicted to peak at distances of order 1 – 10 pc, the main acceleration will occur outside the broad emission line region, as is required by the observations. The absorbing gas occupies a very small fraction of the volume of the wind, and must be confined by it. The most likely confining mechanism is thermal pressure of the wind. The ionization parameter of the absorbing gas has to be sufficiently low for it to remain cool, which requires that the wind is very efficiently heated by the ultrarelativistic protons. This could be accomplished through the excitation and subsequent damping of Alfvén waves, or by a turbulent wind structure containing shocks that will drive the thermal and cosmic ray energy density to equipartition.

067.253 **Inflation of stars by TeV neutrinos.**
M. Czerny, M. Sikora, M. C. Begelman.
Workshop on Relativistic Hadrons in Cosmic Compact Objects, p. 23 – 25 (1991). – See Abstr. 012.103 for the main entry.

Irradiation by the mixture of TeV neutrinos and electromagnetic radiation has a strong impact on the structure of low–mass stars. The radii of such stars may be increased by a few orders of magnitude. This effect may have very important consequences for processes in the vicinity of supermassive black holes in active galactic nuclei and some X–ray binary systems.

067.254 **Electron–positron pair production in compact sources with energetic protons.**
A. A. Zdziarski.
Workshop on Relativistic Hadrons in Cosmic Compact Objects, p. 27 – 40 (1991). – See Abstr. 012.103 for the main entry.

Selected theoretical models of X–ray and γ–ray compact sources in which the available power is supplied to either thermal or nonthermal protons, and in which electron–positron pair production occurs are discussed. The pair production is due to electron and photon interactions in the ambient thermal electron plasma Coulomb–heated to a relativistic temperature and/or electromagnetic cascades initiated by decay of pions produced by energetic protons. The models assume either spherical or disk accretion onto a black hole or approximate the source as a one–zone region. The steady state pair abundance and instabilities related to pairs are discussed.

067.255 **Limit cycles in electromagnetic cascades in compact objects.**
B. Stern, R. Svensson.
Workshop on Relativistic Hadrons in Cosmic Compact Objects, p. 41 – 50 (1991). – See Abstr. 012.103 for the main entry.

Electromagnetic cascades possibly occurring near accreting compact objects have been discovered to show limit cycle behaviour. The power from accelerated protons gets converted by the cascade into soft radiation (X–rays and below) if the photon compactness is sufficiently large. Then the proton–photon system may develop limit cycles much like a prey–predator system with each component interchangeably dominating. This causes periodic large amplitude short time variability of the nonthermal luminosity from a compact object even if the acceleration or injection process is completely steady. Results both from detailed Monte Carlo simulations and from a simple phenomenological model are presented.

067.256 **Nuclear interactions near accreting black holes.**
V. I. Dokuchaev.
Workshop on Relativistic Hadrons in Cosmic Compact Objects, p. 51 – 52 (1991). Abstract. – See Abstr. 012.103 for the main entry.

067.257 **Pair cascade model constraints on energetic particle emission from active galactic nuclei.**
M. G. Baring.
Workshop on Relativistic Hadrons in Cosmic Compact Objects, p. 67 – 72 (1991). – See Abstr. 012.103 for the main entry.

The injection of extremely relativistic protons in the compact central region of an AGN creates electrons and neutrons by interaction with the ambient radiation and particles. The resulting electrons have very high Lorentz factors and participate in a Klein–Nishina pair cascade with UV photons. These high γ electrons cause a flattening of the gamma–ray spectra obtained from first–order pair cascade models, which is contrary to observations. This fact is used to place upper limits on the high γ electron injection luminosity in AGN central regions, and consequently deduce upper limits for luminosities of the relativistic neutrons that are produced and emitted from these compact regions.

067.258 **Pion production in strong magnetic fields: a model for gamma–ray emission from accreting X–ray pulsars.**
C. D. Dermer.
Workshop on Relativistic Hadrons in Cosmic Compact Objects, p. 77 – 89 (1991). – See Abstr. 012.103 for the main entry.

Hadrons accelerated to high energies in the accretion column of a binary X–ray pulsar can produce pions in collisions with accreting protons and ions. The ions decay promptly into charged particles and photons, giving rise to an electromagnetic cascade through pair and photon production processes in the strong magnetic field of the neutron star. The author gives a simplified analysis of this system for teragauss magnetic fields in the optically–thin limit, assuming that nonlinear interactions between the produced pairs and photons are small, and calculate photon spectra in the MeV range. Gamma–ray observations of rotating magnetized neutron stars in the MeV range can provide a new method for determining (1) the magnetic field strength at the gamma–ray production site, and (2) the neutron star inclination angle. These possibilities are primarily due to γ–B pair attenuation processes in the accretion column, and are not restricted to the specific pion production model for gamma–ray emission proposed here. The Gamma Ray Observatory, with unprecedented sensitivity at MeV energies, should be able to test these ideas.

067.259 **Collisional acceleration of electrons in central regions of active galactic nuclei.**
M. G. Baring.
Workshop on Relativistic Hadrons in Cosmic Compact Objects, p. 125 – 130 (1991). – See Abstr. 012.103 for the main entry.

Coulomb collisions between relativistic protons and cold electrons are investigated to determine the resulting electron excitation spectrum. The cold electrons could be the end products of a pair cascade, such as is used in models of the X–ray and gamma–ray AGN continuum, while the protons are injected into the central region of an AGN by some acceleration process. Large–angle Coulomb scattering is very efficient at producing relativistic electrons, and can be a significant supply of non–thermal electrons for typical AGN conditions. If proton cooling in the collisions is included, very flat electron spectra can be obtained, otherwise the electron injection has an index of at least two. This collisional acceleration may be important for AGN pair cascade models.

067.260 **A model of pulsed gamma–radiation from the X–ray binary Hercules X–1 / HZ Herculis.**
F. A. Aharonian (*F. A. Agaronyan*), A. M. Atoyan.
Workshop on Relativistic Hadrons in Cosmic Compact Objects, p. 145 (1991). Abstract. The full paper is published in Astrophys. J., Vol. 381, No. 1, p. 220 – 227 (1 Nov 1991). – See Abstr. 012.103 for the main entry.

067.261 **Interaction of a white hole with collapsing matter.**
T. K. Dey, S. Banerji.
Bull. Astron. Soc. India, Vol. 19, Nos. 1 – 2, p. 77 (Mar – Jun 1991). Abstract. – See Abstr. 012.108 for the main entry.

067.262 **On the physical nature of bursters.**
G. S. Saakyan, G. P. Alodzhants, A. V. Sarkisyan.
Astrofizika, Tom 34, Vyp. 1, p. 21 – 40 (Feb 1991). In Russian. English translation in Astrophysics, Vol. 34, No. 1.

The phenomenon of burster flashes is examined. The characteristic value for a compact binary system, consisting of a neutron and ordinary stars, is the rate of accretion $\dot{M}_E \approx 1.3 \times 10^{18}$g/s, which corresponds to Eddington limit of luminosity L_E. When the rate of accretion $\dot{M} < \dot{M}_E$, one has to deal with a source of soft X–ray emission, which becomes an X–ray pulsar in case of a sufficiently strong magnetic field. Bursters are objects with $\dot{M} > \dot{M}_E$ and apparently with a relatively weak magnetic field and, at last, objects with relativistic jets (e.g. SS433) $\dot{M} > \dot{M}_E$.

067.263 **A possible mechanism for the precession of extragalactic jets.**
Lu Jufu.
Chin. Sci. Bull., Vol. 36, No. 18, p. 1580 – 1581 (1991).

Many jets ejected from active galactic nuclei show an inversion symmetry in shape, indicating that the jet is precessing about an axis. If the jet is ejected along the spin axis of the black hole in the galactic nucleus, then it is the black hole that is precessing. This letter considers a possible mechanism that the precession of the hole is driven by the outer part of a surrounding, tilted accretion disk.

067.264 **Synchrotron emission of neutrino pairs in neutron stars.**
A. D. Kaminker, K. P. Levenfish, D. G. Yakovlev.
Pis'ma Astron. Zh., Tom 17, No. 12, p. 1090 – 1100 (Dec 1991). In Russian. English translation in Sov. Astron. Lett., Vol. 17, No. 6.

The neutrino energy loss rate Q is calculated and fitted by a simple expression for the synchrotron radiation of neutrino pairs by electrons in a degenerate relativistic electron gas with magnetic field **B** in the case when the electrons populate many Landau levels. In a wide parameter range the loss rate does not depend on the electron number density. In the envelopes of neutron stars the synchrotron losses are comparable or exceed other neutrino energy losses.

067.265 **Can the battery ($\nabla T \times \nabla n$) effect produce 10^{12}G magnetic fields in neutron stars?**
T. Yabe, N. Shibazaki, H. Hanami.
Publ. Astron. Soc. Jpn., Vol. 43, No. 5, p. L51 – L56 (1991).

The origin of strong magnetic fields in neutron stars has been a subject of dispute ever since the discovery of pulsars. Among a few important scenarios so far presented is one suggesting that magnetic fields in neutron stars may be generated by a thermally–driven battery process after the star is born. Here, the authors discuss the fluid motion induced in the liquid crust by this thermomagnetic process. They derive a generalized vorticity conservation formula and show that fluid vortices are also generated, together with the magnetic field, and growth with increasing field strength. The maximum field strength produced by this process is shown to be 10 G, unless the fluid velocity exceeds the velocity of sound. Even a dynamo action cannot enhance these fields beyond this critical value if the fluid motion and magnetic fields have the same origin. These results indicate that a thermomagnetic process may not be responsible for the origin of the neutron star magnetic fields.

067.266 **On the double natured solution of two temperature accretion flows.**
C. Meirelles Filho.
Rev. Mex. Astron. Astrofis., Vol. 22, No. 2, p. 191 – 196 (Oct 1991).
It is shown that non linear effects in the two temperature soft photon comptonized disc are exhibited when one allows for a variable coupling (ln Λ) for protons and electrons. Besides providing a new way to obtain the conventional solution to the disc equations, this procedure shows that the solution of the two temperature soft photon comptonized disc is double valued. For spherical accretion there will be solutions only for accretion rates above a critical minimum value, independently of cooling mechanisms. For soft comptonized accretion discs two temperature solutions only exist for accretion rates below a maximum value. Below this value, the solutions have quite different behaviour in the upper and lower branches. Solutions will be gas pressure dominated in the upper branch and radiation pressure dominated in the lower branch. For Cygnus X–1 the authors found that turbulence is, at least, sonic.

067.267 **Black hole accretion disc instability and soft X–ray transients.**
J. C. Wheeler, S. Mineshige.
NATO Advanced Research Workshop on Theory of Accretion Disks, p. 227 – 230 (1989). – See Abstr. 012.121 for the main entry.
Time dependent behaviour of the accretion discs in low–mass X–ray binaries are studied as a model for soft X–ray transients. For relevant accretion rates, the disc suffers a thermal instability leading to intermittent accretion onto the central compact object. It is shown that the effect of X–ray irradiation of the outer disc by the central disc is small in the case of black hole binaries, and so the propagation of the cooling waves is not prevented.

067.268 **Axially symmetric Schwarzschild interior solution for a very slowly rotating star.**
J. J. Rawal, V. D. Deshpande, V. B. Kelkar, R. P. Shenoy.
Bull. Astron. Soc. India, Vol. 19, Nos. 1 – 2, p. 76 (Mar – Jun 1991). Abstract. – See Abstr. 012.108 for the main entry.

067.269 **Generalized Schwarzschild static exterior solution.**
J. J. Rawal, V. D. Deshpande, R. P. Shenoy.
Bull. Astron. Soc. India, Vol. 19, Nos. 1 – 2, p. 77 (Mar – Jun 1991). Abstract. – See Abstr. 012.108 for the main entry.

067.270 **On the mass–radius relation for neutron stars.**
K. D. Abhyankar.
Bull. Astron. Soc. India, Vol. 19, Nos. 1 – 2, p. 105 – 107 (Mar – Jun 1991).

067.271 **The spectra of relativistic accretion disks.**
R. E. Taam, A. Fu.
NATO Advanced Research Workshop on Theory of Accretion Disks, p. 313 – 319 (1989). – See Abstr. 012.121 for the main entry.
The influence of general relativistic effects on the continuum spectra of geometrically thin accretion disks surrounding weakly magnetized neutron stars or black holes in compact X–ray sources is considered. Specific attention is focused on the effect of the gravitational redshift and relativistic Doppler shift on the emission spectrum and on the enhancement of the apparent accretion disk area due to gravitational light bending for a range of system inclination angles. It is shown that the effects of general relativity can be significant as compared to the Newtonian model especially for those cases where the inner edge of the accretion disk extends to the innermost stable orbit. As an application of these results the authors investigate the constraints on the mass transfer rate imposed by the lack of soft X–rays emitted during the quiescent state of the black hole candidate A 0620–00.

067.272 **The disk accretion of a tidally disrupted star onto a massive black hole.**
J. K. Cannizzo, H. M. Lee, J. Goodman.
NATO Advanced Research Workshop on Theory of Accretion Disks, p. 397 – 401 (1989). – See Abstr. 012.121 for the main entry.
The authors consider the tidal disruption of a star by a moderately massive black hole such as might be found in the nucleus of a galaxy like M31. The initial eccentric orbit of the stellar debris will circularize near the tidal radius after experiencing strong shocks. The authors study the evolution of the accretion disk produced by this torus using a time dependent α–disk model. They find that the light–to–mass ratio of the disk–plus–black–hole exceeds unity for several thousand years after disruption. Some fraction of galaxies should be extremely bright at far UV wavelengths if they contain black holes of mass $10^{6-8}M_{\odot}$.

067.273 **The binding energy of a Schwarzschild sphere in *D* space–time dimensions.**
C. Wolf.
Astron. Nachr., Vol. 312, No. 5, p. 299 – 306 (1991).
The author calculates the binding energy of a Schwarzschild gas sphere in *D* space–time dimensions. To the post–Newtonian approximation he demonstrates that the interior Schwarzschild solution does not exhibit an instability before the Schwarzschild radius is reached for $D = 4$. For $D = 5, 6$ he finds binding energy is negative to the post–Newtonian approximation. Extending the approximation to the post–post–Newtonian approximation he finds that the post–post–Newtonian contribution to the binding energy of a Schwarzschild gas sphere is positive for the cases $D = 4, 5, 6$. The effect of the post–post–Newtonian contribution to the binding energy on the stability of higher dimensional gas spheres is discussed.

The revised catalogue of DC–dwarfs (a variant of the year 1987).
See Abstr. 002.055.

Black holes. A selected bibliography.
See Abstr. 002.128.

De l'influence de la gravitation sur la propagation de la lumière en théorie newtonienne. L'archéologie des trous noirs.
See Abstr. 004.025.

High energy astrophysics: compact stars and active galaxies. Proceedings.
See Abstr. 012.026.

Plasmas in astrophysics. Proceedings. 3. Texas–Mexico Conference on Astrophysics: Plasmas in astrophysics, Houston, TX (USA), 14 – 15 Feb 1991.
See Abstr. 012.034.

Structure and emission properties of accretion disks. Proceedings.
See Abstr. 012.046.

Neutron stars: theory and observation. Proceedings.
See Abstr. 012.057.

Strange quark matter in physics and astrophysics. Proceedings. International Workshop on Strange Quark Matter in Physics and Astrophysics, Aarhus (Denmark), 20 – 24 May 1991.
See Abstr. 012.068.

Iron line diagnostics in X–ray sources. Proceedings.
See Abstr. 012.088.

Relativistic hadrons in cosmic compact objects. Proceedings.
See Abstr. 012.103.

Theory of accretion disks. Proceedings.
See Abstr. 012.121.

Spectroscopy in mega–Gauss fields: theory.
See Abstr. 022.044.

Atoms in very strong magnetic fields.
See Abstr. 022.139.

A burst of gamma rays.
See Abstr. 035.002.

The X–ray Timing Explorer.
See Abstr. 035.114.

Particle production by the formation of a global monopole.
See Abstr. 061.002.

Inequalities which constrain negative energy in quantum field theory.
See Abstr. 061.022.

The α–process and the r–process.
See Abstr. 061.033.

Equation of state in metals and cold stars: evaluation of statistical models.
See Abstr. 061.062.

Gamma–rays and neutrinos from accretion processes onto collapsed objects: application to 3C 273.
See Abstr. 061.082.

Constraints on the coupling of weakly–interacting particles to matter from stellar evolution.
See Abstr. 061.108.

Molecules and chains in a strong magnetic field: statistical treatment.
See Abstr. 061.121.

Multiple–ion plasma boundaries.
See Abstr. 062.023.

On the structure of magnetized thin accretion disks of active galactic nuclei and their coronal radiations.
See Abstr. 062.111.

Boundary conditions in the problem of plasma ejection from the magnetosphere of an axisymmetric rotator.
See Abstr. 062.118.

Dynamo magnetic field modes in thin astrophysical disks: an adiabatic computational approximation.
See Abstr. 062.151.

Angular momentum transport in self–gravitating gaseous disks.
See Abstr. 062.168.

The role of magnetic fields in accretion disks.
See Abstr. 062.169.

Spiral instability of magnetized accretion disks.
See Abstr. 062.170.

The nonlinear development of gravitational instabilities in thick disks.
See Abstr. 062.171.

Magnetized accretion–ejection structures.
See Abstr. 062.172.

Dynamical instabilities in self–gravitating annuli.
See Abstr. 062.176.

Numerical simulations of axisymmetric accretion flows.
See Abstr. 062.184.

On the structure of thin magnetized accretion disks of active galactic nuclei and their coronal radiation.
See Abstr. 062.187.

Boundary conditions in the problem of plasma ejection from the magnetosphere of an axisymmetric rotator.
See Abstr. 062.219.

A new approach to hot accretion disks.
See Abstr. 062.226.

Cosmic ray transport in AGNs and jets.
See Abstr. 062.230.

Ultra–relativistic pulsar wind.
See Abstr. 062.234.

Accretion disk magnetohydrodynamics and the origin of jets.
See Abstr. 062.244.

Generation of gamma–rays in electron–photon cascades. II. Photon field of the power–law spectrum.
See Abstr. 063.011.

Effects of triplet pair production on ultrarelativistic electrons in a soft photon field.
See Abstr. 063.037.

The inability of the resonant Compton upscattering model of gamma–ray bursts to produce a third cyclotron harmonic.
See Abstr. 063.066.

Radiative widths and splitting of cyclotron lines in superstrong magnetic fields.
See Abstr. 063.097.

X–ray spectra from dynamical tuning.
See Abstr. 063.099.

A disc–reflected component in the spectra of X–ray bursters.
See Abstr. 063.108.

Luminosity limit for alpha–viscosity accretion disks.
See Abstr. 064.057.

Asymptotic structure of hydromagnetically driven relativistic winds.
See Abstr. 064.075.

Isothermal, Compton–heated coronae above accretion disks.
See Abstr. 064.080.

The radial–azimuthal stability of accretion disks: gas pressure contributions.
See Abstr. 064.086.

Turbulence, convection, and stability in accretion disks.
See Abstr. 064.093.

X–ray – illuminated stellar winds: optically thick wind models for massive X–ray binaries.
See Abstr. 064.119.

Dynamo action by internal waves in accretion disks.
See Abstr. 064.135.

Conference summary. Structure and emission properties of accretion disks.
See Abstr. 064.147.

Two–temperature accretion disks with winds.
See Abstr. 064.161.

Three–dimensional numerical simulation of accretion discs.
See Abstr. 064.163.

An analytical model for a Keplerian disc and a boundary layer around a rapidly rotating star.
See Abstr. 064.172.

Gravitational versus kinetic shifts in the radiation from accretion disks.
See Abstr. 064.173.

Simulation of the boundary layer in white dwarf with the SPH method.
See Abstr. 064.179.

Nonlinear pulsation in the transonic region of geometrically thin accretion disks.
See Abstr. 064.199.

One–armed oscillations of disks and their applications.
See Abstr. 064.200.

Incompressibility of hot nuclear matter, general relativistic stellar collapse and shock propagation.
See Abstr. 065.013.

The conductive propagation of nuclear flames.
See Abstr. 065.018.

Binary evolution with millisecond pulsar in a period gap.
See Abstr. 065.027.

Neutrino–nucleus interactions in core–collapse supernovae.
See Abstr. 065.030.

High–temperature neutrino–nucleus processes in stellar collapse.
See Abstr. 065.031.

Collisions of giant stars with compact objects: hydrodynamical calculations.
See Abstr. 065.059.

Can neutron stars ablate their companions?
See Abstr. 065.093.

The initial stellar masses for the formation of white dwarfs, neutron stars and black holes.
See Abstr. 065.094.

Rayleigh–Taylor instability in the asymmetric supernova explosion.
See Abstr. 065.127.

Evolution of compact binary systems with X–ray heating.
See Abstr. 065.137.

The equation of state of supernova and neutron star matter.
See Abstr. 065.147.

An introduction to curved space–times.
See Abstr. 066.002.

Minisuperspace wavefunctions for cosmology and wormholes.
See Abstr. 066.014.

O(3)–invariant tunneling at false vacuum decay in general relativity.
See Abstr. 066.032.

Mutually interacting quantum fields in curved space–times: the outcome of physical processes.
See Abstr. 066.084.

Quantum strings in curved space times.
See Abstr. 066.085.

Physical models of white and grey holes and their astrophysical applications.
See Abstr. 066.092.

Stability analysis of self–gravitating skyrmions.
See Abstr. 066.097.

The numerical determination of apparent horizons.
See Abstr. 066.101.

Relation of gauge formalisms for pulsations of general relativistic stellar models.
See Abstr. 066.120.

Radiation and false–vacuum bubble dynamics.
See Abstr. 066.121.

Five–dimensional Kaluza–Klein black holes coupled to massive scalar particles.
See Abstr. 066.129.

The interplay between relativistic gravitational, centrifugal and electric forces: a simple example.
See Abstr. 066.136.

New exact solutions of the Einstein–Maxwell equations.
See Abstr. 066.164.

Inevitability of shell crossing in the gravitational collapse of weakly charged dust spheres.
See Abstr. 066.183.

Hoop conjecture for black–hole horizon formation.
See Abstr. 066.184.

Repulsive gravity in tension stars.
See Abstr. 066.189.

Christodoulou's nonlinear gravitational–wave memory: evaluation in the quadrupole approximation.
See Abstr. 066.202.

Prolate collapse of string loops and domain walls.
See Abstr. 066.213.

Diffusion processes in the collapse of a radiating spherical body.
See Abstr. 066.215.

Tension shells and tension stars.
See Abstr. 066.221.

Small black holes and volcanos.
See Abstr. 081.010.

General relativity and physics of the Earth.
See Abstr. 081.048.

SS 433 – again.
See Abstr. 117.083.

The synthesis of light curves of close binary systems with a relativistic object. Axisymmetric, geometrically thick disk around a relativistic object.
See Abstr. 117.085.

Binary X–ray sources: some aspects of wind accretion.
See Abstr. 117.086.

Accretion disks in binary stars.
See Abstr. 117.087.

The emission line of Her X–1 from the accretion column of magnetized neutron star.
See Abstr. 117.088.

Particle acceleration and main parameters of ultra–high energy gamma–ray binaries.
See Abstr. 117.089.

A fresh look at the low–mass X–ray binaries.
See Abstr. 117.092.

A possible TeV gamma–ray burst from Hercules X–1 with pulsations at X–ray period.
See Abstr. 117.124.

Scattering of X–rays on the winds of supergiant X–ray binaries.
See Abstr. 117.138.

A model for the 35 day variations in the pulse profile of Hercules X–1.
See Abstr. 117.150.

Neutron stars in X–ray binaries.
See Abstr. 117.158.

Optical light curves of X–ray binaries.
See Abstr. 117.159.

Observations of Z and atoll sources.
See Abstr. 117.161.

Unified model of X–ray spectra and QPOs in low mass neutron star binaries.
See Abstr. 117.162.

Soft X–ray transients.
See Abstr. 117.163.

Physical conditions in a gas around outer rim of Hercules X–1 accretion disk.
See Abstr. 117.199.

A search for evidence of accretion disk in two suspected Be–X transient binaries.
See Abstr. 117.205.

Evidence for twisted accretion disc from optical light curve of HZ Herculis/Hercules X–1.
See Abstr. 117.219.

The Cygnus X–1 and SS433 accretion disks.
See Abstr. 117.221.

Evidence that the compact object in SS433 is a neutron star and not a black hole.
See Abstr. 117.226.

Search for effects of accretion driven optical flickering in the H Alpha emission line of Gamma Cassiopeiae.
See Abstr. 117.227.

New information on the unique object SS433.
See Abstr. 117.234.

Black hole candidates in binaries.
See Abstr. 117.260.

Iron lines in X–ray spectra of high mass X–ray binaries.
See Abstr. 117.262.

Behaviour of the iron line in Cygnus X–1 during dips and persistent emission.
See Abstr. 117.264.

Narrow iron K line and warm iron K edge in pulse phase spectra of Hercules X–1.
See Abstr. 117.265.

GINGA observation of X1820–303 and its iron–line emission.
See Abstr. 117.267.

Low–flux hard X–ray observation of Cygnus X–1.
See Abstr. 117.276.

Synthesis of light curves of close binary systems containing a relativistic object. Axisymmetric, geometrically thick disk around the relativistic object.
See Abstr. 117.288.

Accretion disks in low mass X–ray binaries.
See Abstr. 117.312.

The compact object in SS433: neutron star or black hole?
See Abstr. 117.314.

The early optical evolution of X–ray Nova Muscae 1991.
See Abstr. 124.022.

Evidence for a black hole in the X–ray nova Muscae 1991.
See Abstr. 124.025.

The early evolution of Nova Muscae 1991: a candidate black hole?
See Abstr. 124.027.

Cosmic–ray transport and gamma–ray emission in supernova shells.
See Abstr. 125.047.

The central compact object in SN 1987A.
See Abstr. 125.270.

SN 1987A: the light curve.
See Abstr. 125.289.

Evolution of radio pulsars with allowance for some being formed in binary systems.
See Abstr. 126.007.

Zeeman and cyclotron lines from magnetic white dwarfs.
See Abstr. 126.019.

Evolution of luminosity and period of recycled pulsars.
See Abstr. 126.026.

On the evolution and birthrate of radio pulsars.
See Abstr. 126.031.

The rate of neutron star binary mergers in the universe: minimal predictions for gravity wave detectors.
See Abstr. 126.068.

The K parameter and evolution of the radio pulsars.
See Abstr. 126.074.

Outflow hydrodynamics in the eclipsing millisecond pulsar binaries PSR 1957+20 and PSR 1744–24A.
See Abstr. 126.076.

The origin of the planet around PSR 1829–10.
See Abstr. 126.088.

Two hard X–ray sources in 100 square degrees around the Galactic Center.
See Abstr. 142.001.

A common origin for the spectral differences between type II bursts and the spectral oscillations during QPO?
See Abstr. 142.012.

Broad–band X–ray spectra of black hole condidates, X–ray pulsars, and low–mass X–ray binaries. Results of "Kvant" observatory.
See Abstr. 142.014.

Detection of 0.8 Hz quasi–periodical oscillations from the black hole candidate GX 339–4.
See Abstr. 142.015.

EXOSAT observations of the X–ray burst source 4U 1702–42.
See Abstr. 142.017.

High–energy transients.
See Abstr. 142.019.

The spectrum and position of the hard galactic centre source 1E 1740.7–2942.
See Abstr. 142.025.

Modelling observed X–ray pulsar profiles.
See Abstr. 142.037.

Variability of active galactic nuclei and galactic QPO sources: a diagnosis.
See Abstr. 142.064.

Gamma–ray bursts from colliding strange stars.
See Abstr. 143.002.

The physics of gamma–ray bursts.
See Abstr. 143.007.

Passage of a comet through the magnetosphere of a neutron star: a possible cause of cosmic gamma–ray bursts.
See Abstr. 143.009.

New constraints on neutron star models of gamma–ray bursts. II. X–ray observations of three gamma–ray burst error boxes.
See Abstr. 143.015.

Gamma–ray emission from young neutron stars.
See Abstr. 143.019.

A new radiation mechanism for gamma–ray bursts.
See Abstr. 143.030.

Gamma–ray bursts.
See Abstr. 143.036.

High–energy background radiation from quiescent γ–ray burst sources.
See Abstr. 143.047.

Joint evolution of a galactic nucleus and central massive black hole.
See Abstr. 151.110.

Merging of two galaxies with central black holes.
See Abstr. 151.116.

The distribution of stars in a star cluster with a Kerr black hole in its center.
See Abstr. 151.146.

Compact objects in globular clusters.
See Abstr. 154.083.

Post–core collapse scenarios of globular clusters.
See Abstr. 154.104.

Evacuation of gas from globular clusters by winds from millisecond pulsars.
See Abstr. 154.136.

A feasibility study for detection of gravitational microlensing by baryonic dark matter via CCD photometry.
See Abstr. 155.023.

Gamma rays from the Galactic center region.
See Abstr. 155.049.

A search for massive compact halo objects in our Galaxy.
See Abstr. 155.102.

The statistics of slow interstellar accretion onto neutron stars.
See Abstr. 155.106.

A search for black holes in galaxy nuclei.
See Abstr. 157.049.

Galactic models with massive coronae. III. Giant elliptical galaxy M87.
See Abstr. 157.091.

Massive black holes in the nuclei of spiral galaxies: M31 and the problem of the initial conditions.
See Abstr. 157.266.

The iron line and high energy bump as X–ray signatures of cold matter in Seyfert 1 galaxies.
See Abstr. 158.004.

Formation of broad–line emission clouds in active galactic nuclei.
See Abstr. 158.041.

Physics, structure, and fueling of active galactic nuclei.
See Abstr. 158.085.

Studying the galactic central engine from space observatories.
See Abstr. 158.144.

A two–phase model for the X–ray emission from Seyfert galaxies.
See Abstr. 158.204.

The signatures of an accretion disk in the electromagnetic spectra of quasars and AGNs.
See Abstr. 158.221.

Accretion disk models for active galactic nuclei.
See Abstr. 158.239.

AGN disks: theory vs. observations.
See Abstr. 158.240.

Accretion models of some AGN and observational constraints.
See Abstr. 158.242.

Quasi–periodic continuum variability of NGC 6814.
See Abstr. 158.247.

Statistical properties of AGN emission line spectrum in relation with illuminated accretion disc model.
See Abstr. 158.251.

Soft X–ray variability and the accretion disk of NGC 5548.
See Abstr. 158.252.

Massive thin accretion discs. III. Comparison with the observations.
See Abstr. 158.253.

Accretion disk model for the broad line region of three active galactic nuclei.
See Abstr. 158.262.

Accretion disk modeling of the multi–wavelength spectra of Seyfert 1 galaxy Markarian 335.
See Abstr. 158.265.

The effect of anisotropic emission from thick accretion disks on the luminosity functions of AGN.
See Abstr. 158.267.

Accretion disk emission from a BL Lacertae object.
See Abstr. 158.269.

Iron lines in AGN and black hole candidates.
See Abstr. 158.300.

Spectral features in AGN: problems and questions.
See Abstr. 158.303.

The ionization structure of iron in AGN.
See Abstr. 158.304.

Iron K fluorescence and high energy bump from accretion disks in AGNs.
See Abstr. 158.305.

The structure and engines of AGNs.
See Abstr. 158.315.

Relativistic electron–positron beam formation in the framework of the two–flow model for active galactic nuclei.
See Abstr. 158.320.

Jet models and observed jets in a sample of superluminal active galactic nuclei.
See Abstr. 158.342.

The emerging picture of quasars.
See Abstr. 159.004.

Accretion–disk coronae in QSOs.
See Abstr. 159.051.

The ultraviolet spectrum of the bright QSO H1821+643 and its continuum energy distribution.
See Abstr. 159.111.

Thermal continuum emission of GQ Comae.
See Abstr. 159.115.

Outflows driven by cosmic–ray pressure in broad absorption line QSOs.
See Abstr. 159.173.

An attempt to explain the mechanism of creation of matter in an expanding nondecelerative universe.
See Abstr. 161.007.

Quantum string theory in curved space–times.
See Abstr. 161.065.

A "larger" universe.
See Abstr. 161.178.

Were all white holes in the early universe converted into black holes?
See Abstr. 161.256.

Quark– and gluon–jet emission from primordial black holes. II. The emission over the black hole lifetime.
See Abstr. 161.260.

Survival of quark nuggets and their present–day abundance.
See Abstr. 161.334.

Present and future limits on compact dark matter from gravitational lensing.
See Abstr. 161.401.

Are galactic halos made of brown dwarfs or black holes?
See Abstr. 161.479.

Population III objects and lithium abundance.
See Abstr. 161.586.

The evolution of black holes in an expanding universe.
See Abstr. 161.623.

Cosmological and astrophysical aspects of the quark–hadron phase transition.
See Abstr. 161.665.

Sun

071 Photosphere, Spectrum

071.001 **Die Feinstruktur der Sonnengranulation.**
A. Hanslmeier.
Sterne Weltraum, Jahrg. 30, Nr. 12, p. 724 – 728 (Dec 1991).

071.002 **Doppler width of CH lines.**
L. M. Punetha, G. C. Joshi.
Astrophys. Space Sci., Vol. 181, No. 2, p. 327 – 329 (Jul 1991). Letter–to–the–editor.
Twenty–one CH line profiles at the centre of the solar disk were Fourier–transformed to derive the Doppler widths. The Doppler width increases linearly with the equivalent width.

071.003 **Determining the solar–flare photosphere scale height from SMM gamma–ray measurements.**
R. E. Lingenfelter, X.–M. Hua.
Bull. Am. Astron. Soc., Vol. 23, No. 2, p. 924 (1991). Abstract. – See Abstr. 012.008 for the main entry.

071.004 **Ca II H_{2V} and K_{2V} cell grains.**
R. J. Rutten, H. Uitenbroek.
Sol. Phys., Vol. 134, No. 1, p. 15 – 71 (Jul 1991).
The bright Ca II H_{2V} and K_{2V} grains, which are intermittently present in the interiors of network cells in quiet–Sun areas, should provide important diagnostics of the dynamical interaction between the quiet photosphere and the chromosphere above it, but their nature has so far eluded identification. The authors review the extensive observational literature on these grains and on related phenomena. They resolve various contradictions, connect hitherto unconnected observations, distill new constraints and relate signatures in the measurement domain to signatures in the Fourier domain. They then review interpretations and simulation efforts to illustrate modeling options. The authors conclude that the grains are a hydrodynamical phenomenon in which magnetic fields do not play a major role. The grains are due to interference between a pervasive standing oscillation with about a 180 s periodicity and an 8 Mm horizontal wavelength in the chromosphere and the wave trains of the evanescent p–mode interference pattern in the upper photosphere. The roles of short–period waves, shock formation and granular piston excitation and the issue of long–lived patterning remain open.

071.005 **Continuum contrast and center to limb variation of solar magnetic elements observed in the photosphere.**
D. J. Swearingen, S. R. Walton, G. A. Chapman, J. K. Lawrence.
Bull. Am. Astron. Soc., Vol. 23, No. 2, p. 960 (1991). Abstract. – See Abstr. 012.008 for the main entry.

071.006 **On the structure of spectral line gap regions.**
F. Kneer, M. von Uexküll.
Astron. Astrophys., Vol. 247, No. 2, p. 556 – 564 (Jul 1991). = Mitt. Kiepenheuer–Inst. Nr. 341.
A photographic spectrogram with high spatial and spectral resolution, taken from a plage region close to the disc centre of the Sun with the Gregory–Coudé telescope at the Observatorio del Teide, Tenerife, is analysed. The spectral region around 6150 Å contains lines of neutral and singly ionised iron, vanadium and sodium which selected to extract information on the structure of line gap regions. Theoretical line profile calculations for comparison with the observations are based on the following model assumptions: a hot, slender tube with the magnetic field perpendicular to the solar surface is embedded in a surrounding gas that is cooler than the average quiet solar atmosphere. Hydrostatic equilibrium is assumed for the magnetic tube and the surrounding gas to calculate pressure and density. The cool gas moves downward with a speed of several km s^{-1}, while the gas in the tube is at rest. By suitable parameter choices and averaging it is possible to reproduce the observed intensity profiles. However, with this type of velocity field, the authors are unable to explain the reported asymmetries of the Stokes V profiles from network boundaries and plages.

071.007 **The formation mechanism of the lines of the C I solar infrared multiplet λ1069,5 nm.**
N. G. Shchukina.
Kinematika Fiz. Nebesn. Tel, Tom 7, No. 4, p. 38 – 47 (Jul–Aug 1991). In Russian. English translation in Kinematics Phys. Celest. Bodies, Vol. 7, No. 4.
The formation mechanisms of the C I infrared multiplet lines for the solar atmosphere are discussed. The NLTE level population of the multiplet is caused by two circumstances: the pumping of the lower level by ultraviolet photons of the C I lines $\lambda\lambda$165.72 and 258.29 nm in the upper photosphere and lower chromosphere; overrecombination and followed cascade, transitions throughout the atmosphere due to deficit of photons in the infrared bound–free continua and high–excitation lines. The latter increases the multiplet levels population compensating the depopulation through pumped near–ultraviolet lines that connect the lower level of the multiplet to levels close to the continuum. Ultraviolet 2.4 – 3.8 eV photons do not affect essentially the resulting distribution of departure coefficients.

071.008 **On the dynamics of granulation in active regions and the heating problem.**
A. Nesis, A. Hanslmeier, R. Hammer, R. Komm, W. Mattig.
International Conference on Mechanisms of Chromospheric and Coronal Heating, p. 36 – 38 (1991). = Mitt. Kiepenheuer–Inst., Nr. 333. – See Abstr. 012.011 for the main entry.
Using spectrograms of high spatial resolution the authors investigated the functional interdependence between the convective intensity fluctuations at $\tau_{5000} = 1$ and the line broadening of magnetically sensitive absorption lines in active and nonactive regions. They found an extra broadening in the intergranular regions. The variation of this broadening with height in the photosphere can be attributed to the variation of the magnetic filling factor as well as to enhanced nonthermal velocities.

071.009 **Generation of electric currents anad waves on magnetic flux tubes by horizontal velocities in the photosphere.**
T. D. Tarbell, G. L. Slater, Z. A. Frank, R. A. Shine, K. P. Topka.
International Conference on Mechanisms of Chromospheric and Coronal Heating, p. 39 – 41 (1991). – See Abstr. 012.011 for the main entry.
Horizontal velocities in the photosphere are an important "driver" in theories of coronal heating because they generate field–aligned currents by slow twisting and tangling of magnetic field lines and wave modes by higher frequency perturbations. As

a first attempt to provide observational constraints for theories of these processes, the authors present temporal power spectra and other statistics for measured horizontal velocities and for their curl and divergence.

071.010 **Observations of the birth and death of mesogranules.**
G. W. Simon, A. M. Title, R. A. Shine, Z. Frank, R. Muller, H. Auffret.
Bull. Am. Astron. Soc., Vol. 23, No. 2, p. 1034 (1991). Abstract. – See Abstr. 012.012 for the main entry.

071.011 **Modeling power spectra of the solar convection.**
G. P. Ginet, G. W. Simon.
Bull. Am. Astron. Soc., Vol. 23, No. 2, p. 1034 (1991). Abstract. – See Abstr. 012.012 for the main entry.

071.012 **A new theoretical model photosphere.**
R. L. Kurucz.
Bull. Am. Astron. Soc., Vol. 23, No. 2, p. 1047 (1991). Abstract. – See Abstr. 012.012 for the main entry.

071.013 **Frequency redistribution in solar resonance line wings.**
K. Gayley.
Bull. Am. Astron. Soc., Vol. 23, No. 2, p. 1047 (1991). Abstract. – See Abstr. 012.012 for the main entry.

071.014 **Solar granulation: simulations versus observations.**
N. E. Hurlburt.
Bull. Am. Astron. Soc., Vol. 23, No. 2, p. 1048 (1991). Abstract. – See Abstr. 012.012 for the main entry.

071.015 **On the fractal dimension of granulation.**
J. Newbury, S. L. Keil.
Bull. Am. Astron. Soc., Vol. 23, No. 2, p. 1048 (1991). Abstract. – See Abstr. 012.012 for the main entry.

071.016 **Granulation spectroscopy: first results from VTT–Tenerife.**
A. Nesis, A. Hanslmeier, R. Hammer, R. Komm, W. Mattig, J. Staiger.
Bull. Am. Astron. Soc., Vol. 23, No. 2, p. 1048 (1991). Abstract. – See Abstr. 012.012 for the main entry.

071.017 **Power spectra of flows and magnetic fields in the solar photosphere.**
T. D. Tarbell, G. L. Slater, Z. A. Frank, K. P. Topka, G. Scharmer, W. Schmidt.
Bull. Am. Astron. Soc., Vol. 23, No. 2, p. 1048 (1991). Abstract. – See Abstr. 012.012 for the main entry.

071.018 **Rotation rate of the supergranulation pattern.**
D. H. Hathaway, E. J. Rhodes Jr., S. Korzennik, A. Cacciani.
Bull. Am. Astron. Soc., Vol. 23, No. 2, p. 1051 (1991). Abstract. – See Abstr. 012.012 for the main entry.

071.019 **The effect of the Kelvin–Helmholtz instability on photospheric flows.**
J. T. Karpen, S. K. Antiochos, R. B. Dahlburg, D. S. Spicer.
Bull. Am. Astron. Soc., Vol. 23, No. 2, p. 1059 (1991). Abstract. – See Abstr. 012.012 for the main entry.

071.020 **Spatial correlation of spectrum line properties with magnetic flux.**
H. P. Jones.
Bull. Am. Astron. Soc., Vol. 23, No. 2, p. 1068 (1991). Abstract. – See Abstr. 012.012 for the main entry.

071.021 **Geometric line elevation in solar limb faculae.**
G. Stellmacher, E. Wiehr.
Astron. Astrophys., Vol. 248, No. 1, p. 227 – 231 (Aug 1991).
Spectra of markedly visible white light faculae near the solar limb are used to determine the geometric displacement, Δh, of line cores with respect to their adjacent continuum. The observed Δh are plotted as a function of the excitation potential and, respectively, equivalent width of the different lines. Comparison with calculated Δh from a photospheric LTE model shows that the observed facula values significantly exceed the photospheric ones.

071.022 **Granular and intergranular line profiles in solar active and quiet regions.**
A. Hanslmeier, W. Mattig, A. Nesis.
Astron. Astrophys., Vol. 248, No. 1, p. 232 – 236 (Aug 1991).
Two spectra obtained with high spatial resolution in active and non active regions of the sun have been analysed. The enhanced fluctuations at subgranular scales in active regions observed in the power spectra of intensity and velocity, provide evidence for the existence of small magnetic fluxtubes. Using the brightest and darkest continuum intensity as indicators for granulum–intergranulum, the authors have determined granular and intergranular line profiles. In Ca^+ active regions, the intergranular profiles are changed more than the granular profiles. Also the full width at half maximum of the intergranular line profiles is enhanced in the active region. These results have been obtained without any polarization equipment and suggest that the magnetic elements are predominantly located in the intergranular regions.

071.023 **Dynamics of the solar granulation. I. Processing of MSDP spectra.**
T. Roudier, P. Mein, R. Muller, J. Vigneau, J. M. Malherbe, C. Coutard, R. Hellier.
Astron. Astrophys., Vol. 248, No. 1, p. 237 – 244 (Aug 1991).
High resolution ($\approx 0\farcs5$) two–dimensional spectra have been obtained with the Multichannel Subtractive Double Pass (MSDP) spectrograph of the Pic du Midi Observatory in a quiet region at the solar disk centre. The authors describe in this paper the adapted image processing which provides two–dimensional intensity and velocity maps of the solar granulation, at two intensity levels of the line NaD_2. A quantitative evaluation of the spatial resolution is determined both for the intensity and velocity granulation fields.

071.024 **Dynamics of the solar granulation. II. Statistical analysis: power spectra, coherence, phase.**
T. Roudier, R. Muller, P. Mein, J. Vigneau, J. M. Malherbe, O. Espagnet.
Astron. Astrophys., Vol. 248, No. 1, p. 245 – 255 (Aug 1991).
Granulation velocity fields provided by the Multichannel Subtractive Double Pass of the Coupole Tourelle (Pic du Midi Observatory) are analysed. Two methods of filtering have been applied in order to eliminate the 5–min oscillations. An inspection of the velocity maps, filtered for 5–min oscillations, confirms that the bright granules at the continuum level are well correlated with rising material and that the velocity fluctuations penetrate much higher into the photosphere than the intensity fluctuations. Both the velocity power spectrum and the coherence phase spectra (between velocity and intensity) indicate that the large granules (of size close to 3 arcsec) carry the major part of the convective energy, in this layer. The one–dimensional power spectrum exhibits a linear decrease in the high wavenumber side, not so distant to the –5/3 value which is expected for a purely Kolmogoroff spectrum. However the influence of 5–min filtering and the seeing corrections are not accurate enough for a definite conclusion about the nature of the solar granules.

071.025 **Structure elements of the photosphere and their variations with the solar activity cycle.**
M. N. Stoyanova.
Soln. Dannye, Byull., No. 10/90, p. 85 – 90 (Oct 1990). In Russian.
Structure inhomogeneity of the photospheric surface and its variation with the cycle phase has been studied on the basis of long–term data of the Kislovodsk Solar Station. Characteristic sizes of small cells, coinciding in scale with mezogranulation have been found using a series of photoheliograms measured with a micrometer. The distributions of cell sizes for the maximum and minimum of the solar cycle show a definite tendency to larger

scales of cells at the minimum of the solar cycle. Larger cells were studied statistically. Power spectra were constructed and significant periods of brightness variations were detected. The derived sizes of quasistationary cells show that there is a tendency to their enlargement at the minimum of the solar cycle.

071.026 **Fitting of stratospheric frames of solar granulation.**
I. M. Kulagina, V. G. Nikiforov.
Soln. Dannye, Byull., No. 12/90, p. 87 – 91 (Dec 1990). In Russian.

From data obtained by a stratospheric station a series of images of solar granulation is discussed.

071.027 **The solar abundance of iron: a "final" word.**
E. Biémont, M. Baudoux, R. L. Kurucz, W. Ansbacher, E. H. Pinnington.
Astron. Astrophys., Vol. 249, No. 2, p. 539 – 544 (Sep 1991).

New accurate radiative lifetimes have been measured by a laser–induced fluorescence technique for 12 selected levels of Fe^+ of astrophysical interest. These new results combined with recent theoretical transition probabilities provide another contribution in order to try to solve the controversy raised during the last months concerning the content of iron in the Sun. The abundance value, $A_{Fe} = 7.54 \pm 0.03$, obtained from an extensive sample of Fe II lines, supports the "low" value of the iron abundance, in agreement with the meteoritic result.

071.028 **A redetermination of the solar iron abundance based on new Fe I oscillator strengths.**
H. Holweger, A. Bard, A. Kock, M. Kock.
Astron. Astrophys., Vol. 249, No. 2, p. 545 – 549 (Sep 1991).

The authors derive the solar abundance of iron from photospheric Fe I lines whose *f*–values have been determined recently in a hollow–cathode experiment in combination with accurate life–time measurements. Taking into account a small (+0.03 dex) abundance correction for photospheric non–LTE effects they infer a solar iron abundance of $\log N(\mathrm{Fe}) = 7.50 \pm 0.07$ on the scale $\log N(\mathrm{H}) = 12$. This result is in close agreement with that derived earlier from lines of ionized iron whose *f*–values have been obtained by the same technique, but it is 0.17 dex smaller than the currently accepted standard value based on absorption furnace *f*–values. The new result strengthens the evidence that the solar Fe/Si ratio agrees with the meteoritic value found in C1 chondrites.

071.029 **First identification of pure rotation lines of NH in the infrared solar spectrum.**
M. Geller, A. J. Sauval, N. Grevesse, C. B. Farmer, R. H. Norton.
Astron. Astrophys., Vol. 249, No. 2, p. 550 – 552 (Sep 1991).

High resolution solar spectra obtained from space (ATMOS Fourier Transform Spectrometer) have permitted the authors to detect for the first time pure rotation lines of NH of the $X^3\Sigma$ ground state near 600–900 cm^{-1}. All these very weak NH solar lines have never been observed before in the laboratory.

071.030 **Granular structure and its connection with the photospheric large scale structures.**
Kh. I. Abdusamatov.
Soln. Dannye, Byull., No. 1/91, p. 88 – 96 (Jan 1991). In Russian.

An analysis is made of the results of a two–dimensional photometric reduction of spectrograms ($\lambda\lambda$5040 – 5060 Å) and photoheliograms of granulation ($\lambda 4600 \pm 200$ Å) with the spatial resolution 0".6 and 0".25 respectively. The velocity of the rising and lowering of matter and its correlation with brightness fluctuations of details in the continuum increases with the growth of absolute values of the latter. A gathering of large granules into mesogranular cells with the average scale (diameter) 6 – 9" is noted, to which an enhanced rising of matter corresponds. Formation of large granules and mesogranules is considered as a result of interaction of penetrating convective outbursts with photospheric layers. All the four existing scales of photospheric structures, corresponding to granulation, mesogranulation, supergranulation and gigantic granules (cells) are considered as genetically interconnected convection formations and reflect processes, taking place in corresponding different inner layers of the convection zone of the Sun.

071.031 **Solar gravitational redshift from the infrared oxygen triplet.**
J. C. LoPresto, C. Schrader, A. K. Pierce.
Astrophys. J., Vol. 376, No. 2, p. 757 – 760 (1 Aug 1991).

The authors have observed the difference in wavelength between the infrared oxygen triplet $\lambda\lambda$7772 – 7775 in the Sun and in the laboratory. Converted to velocities, the disk values, extrapolated to the limb, give a redshift of 655 m s^{-1} compared to the value of 636 m s^{-1} predicted from the principle of equivalence. Observations of the chromospheric oxygen lines in emission just off the Sun's limb give a velocity of 627 m s^{-1} or 0.99 of the Einstein prediction.

071.032 **An analytical form of solar limb darkening in the 0.3 – 2.4 μm range.**
A. P. Sarychev, E. M. Roshchina, E. A. Makarova.
Soln. Dannye, Byull., No. 7/91, p. 84 – 87 (Jul 1991). In Russian.

An analytical function proposed by the authors is analysed. This function approximates reliably the experimental data of solar limb darkening in the quasi–continuum for a wide spectral range. Detailed observational data by A. K. Pierce et al. (1977) and averaged data by E. A. Makarova et al. (1991), that were obtained by generalisation of vast observational material have been used for the evaluation of quality of this approximation. It is shown that the modern data of solar limb darkening can be described by this approximation function. The parameters of this function describe not only the limb darkening but the source function with the optical depth.

071.033 **Is the spectral type of the Sun constant?**
P. C. Keenan.
Publ. Astron. Soc. Pac., Vol. 103, No. 665, p. 642 – 644 (Jul 1991).

The spectral type of the whole Sun was monitored from 1988 August 25 through 1990 October 2. No changes as large as half a spectral subdivision were detected. The spectral type remained at G2 V.

071.034 **On some features of the photosphere's brightness distribution at the extreme limb of the Sun from eclipse observations.**
L. A. Akimov, I. L. Belkina, N. P. Dyatel.
Astron. Tsirk., No. 1543, p. 23 – 24 (Mar – Apr 1990). In Russian.

The mean surface brightness distriutions of the quiet photosphere was obtained from the flash spectra of the total solar eclipses on 22 September, 1968, 10 July, 1972 and 31 July, 1981. It was found that the gradients of the mean darkening curves near the zero point of the height scale decrease with increasing of solar activity.

071.035 **Hydrodynamics of the solar photosphere: model calculations and spectroscopic observations.**
M. Steffen, B. Freytag.
Reviews in Modern Astronomy 4, p. 43 – 60 (1991). – See Abstr. 003.008 for the main entry.

The authors describe detailed hydrodynamical calculations for the solar photosphere. Based on the resulting physically self–consistent parameter–free models they calculate synthetic spectra in order to be able to make a direct comparison with various spectroscopic observations. They conclude that their time–dependent radiation–hydrodynamics models seem to be a reasonable representation of the real solar photosphere, which, on the other hand, is obviously quite different from the picture suggested by classical mixing length models.

071.036 **Center–to–limb variation of spectral lines and quiet sun atmospheric model.**
Ding Mingde, Fang Cheng.
Chin. J. Space Sci., Vol. 11, No. 4, p. 241 – 250 (Oct 1991). In Chinese.

Spectral lines of hydrogen and ionized calcium, as well as their center–to–limb variations, observed at Kitt Peak National Solar Observatory, are used to diagnose the VAL–C', quiet Sun atmospheric model. The PANDORA computer program is used to perform the detailed non–LTE model calculations, especially for hydrogen and calcium. The main discrepanicies between the computed spectral lines and the observed data are: (1) The residual intensity in Hβ and λ8498 line centers is evidently higher, while that in Ca II H and K lines are relatively lower than the observed values. (2) The center–to–limb enhancement coefficient of the residual intensities of Hβ, and all the Ca II lines (resonance and infrared lines) near $\Delta\lambda = 0.5$ Å is a little larger than indicated by the observation. However, all other spectral features observed can be reproduced fairly well. Finally, the possible causes which can make the computed values deviate from the oberved ones is discussed, and a comment on revising the VAL–C' model is given.

071.037 **The He I λ10830 Å line as an indicator of the chromospheric and coronal activity of the Sun.**
A. G. Shcherbakov, Z. A. Shcherbakova.
IAU Colloquium No. 130: The Sun and cool stars – activity, magnetism, dynamos, p. 252 – 265 (1991). – See Abstr. 012.024 for the main entry.

The behaviour of the He I λ10830 Å line profile in spectra of solar feature is reviewed. The connection of line characteristics with various phenomena on the solar disc, such as plages (faculae), sunspots, jets, prominences, coronal holes, flares, fine structure and emission above the solar limb, is discussed. The authors present a short summary of the mechanisms for helium excitation in the solar chromosphere. Time variations of the line profile, observed in quiet and active regions, and He I λ10830 Å absorption in the Sun as a star are discussed and compared with other indexes of solar activity. It is shown that the equivalent width of the helium line varies in an 11 year cycle and reflects the 27 day periodicity of the rotation of the Sun. It is also shown that the He I line is a more sensitive index of activity than the K–index.

071.038 **The rotational modulation of Ca II K in the Sun.**
I. Sattarov, A. Hojaev (*A. S. Khodzhaev*).
IAU Colloquium No. 130: The Sun and cool stars – activity, magnetism, dynamos, p. 266 – 267 (1991). – See Abstr. 012.024 for the main entry.

071.039 **Fe I line asymmetries and shifts caused by pressure broadening.**
V. Kršljanin, I. Vince, S. Erkapić.
IAU Colloquium No. 130: The Sun and cool stars – activity, magnetism, dynamos, p. 277 – 278 (1991). – See Abstr. 012.024 for the main entry.

In order to test the validity of the usual neglect of pressure line shifts in investigations of Fe I line asymmetries and shifts, the authors calculated synthetic line bisectors and shifts of 96 moderate and weak photospheric lines, taking into account both hydrogen– and electron–impact shifts and neglecting the velocity fields. Absolute pressure line shifts obtained are less than or appr. equal to 50 m/s and the largest relative shifts along the bisectors are about 40 m/s. The authors found pronounced dependences of both shifts and bisectors on the energy of the upper level in transitions and on the transition array.

071.040 **Average data on solar limb darkening in the quasi–continuum in the 300 – 2400 nm spectral range.**
E. A. Makarova, E. M. Roshchina, A. P. Sarychev.
Astron. Zh., Tom 68, Vyp. 4, p. 885 – 889 (Jul–Aug 1991). In Russian. English translation in Sov. Astron., Vol. 35, No. 4.

Average data for solar limb darkening were obtained by generalisation of measurements of many authors. These data can be replaced by an approximation function with 3 parameters depending on the wavelength at the same level of correctness. The analytical form of the average data is convenient for practical use.

071.041 **The solar Mg I spectrum from ATMOS. I. Identification and preliminary discussion.**
J. T. Jefferies.
Astrophys. J., Vol. 377, No. 1, p. 337 – 342 (10 Aug 1991).

As an initial step in a program to understand some puzzling aspects of the infrared solar Mg I spectrum, the author has examined the set of high–resolution, high S/N ratio lines of this spectrum obtained in the 1985 ATMOS experiment on Spacelab 3. From laboratory data, supplemented by theoretical expressions, wavenumbers were obtained for the Mg I transitions lying in or near the range covered by ATMOS, namely 650 to 4800 cm^{-1} (2.3 – 16 μm). Lines involving principal quantum numbers up to $n = 10$ are seen in the data; the author describes some striking features of the Mg I spectrum and indicates some directions of interest for future analytical studies.

071.042 **The solar–stellar connection – polarimetry.**
D. Clarke.
Vistas Astron., Vol. 34, Part 3, p. 303 – 324 (1991). – See Abstr. 012.032 for the main entry.

The status of the solar–stellar connections that can be made through polarimetry is presented by giving brief reviews of the work in the stellar field and of mechanisms that produce polarization in local areas on the Sun. Results are described from novel measurements of the whole solar disk made at the Glasgow Observatory Solar Polarimetric Experimental Laboratory (GOSPEL), these suggesting that the Sun can be considered as being a polarimetric variable star.

071.043 **The variation of the solar granulation structure in active and non–active regions.**
A. Hanslmeier, A. Nesis, W. Mattig.
Astron. Astrophys., Vol. 251, No. 1, p. 307 – 311 (Nov 1991).

With the aid of a coherence analysis between line center velocities, continuum brightness variations and residual intensities the authors studied the height dependent evolution of overshooting convective elements in the solar photosphere in magnetically active and non active regions. They found evidence of a structuring influence of the magnetic field on the height dependent evolution of temperature and velocity patterns. From a comparison of intensity and velocity frequency distributions in non active and active regions enhanced downward motions in active regions were found. The results are in agreement with small fluxtube concepts.

071.044 **Average data on solar limb darkening in the 300 – 2400 nm quasicontinuum.**
E. A. Makarova, E. M. Roshchina, A. P. Sarychev.
Sov. Astron., Vol. 35, No. 4, p. 441 – 443 (Jul 1991). Current Physics Microform No.: 9203X0627. English translation of Astron. Zh., Tom 68, Vyp. 4, p. 885 – 889 (1991).

The average solar limb darkening is obtained by generalizing measurements by many authors. Without increasing their error, these measurement can be replaced by a single approximating function containing three parameters that depend on wavelength. The proposed analytical form of the average data is convenient for practical use.

071.045 **Selected examples of bisector and line parameter variation over a granular–intergranular region.**
A. Hanslmeier, W. Mattig, A. Nesis.
Astron. Astrophys., Vol. 251, No. 2, p. 669 – 674 (Nov 1991).

Four examples of bisector and corresponding line parameter variations over a granular–intergranular region are discussed. These detailed case studies permit a direct comparison with hydrodynamical model calculations. Generally, the variation of the line parameters is found to be more similar than for data where granular and intergranular areas are averaged. However, the transition from granulum to intergranulum was found to be non monotonic and at the position where the velocity gradients reach a minimum, the slope of the line parameter curves also changes.

071.046 **Photospheric and chromospheric observations of AR 5395 by 26 cm solar telescope at Yunnan Observatory.**
Wu Mingchan, Li Zhikai, Luan Di, Li Zhi.
Publ. Yunnan Obs., No. 4, p. 31 – 33 (Dec 1990). – See Abstr. 012.047 for the main entry.

071.047 **The height dependence of velocity – intensity fluctuations and several non–dimensional parameters in the solar photosphere.**
R. Komm, W. Mattig, A. Nesis.
Astron. Astrophys., Vol. 252, No. 2, p. 812 – 820 (Dec 1991). = Mitt. Kiepenheuer-Inst. Nr. 346.

The authors have investigated the quiet photosphere with an analysis of autocorrelation functions (ac.f.s) of intensity- and velocity fluctuations and derived a correlation lengthscale (*L*). They have found that all ac.f.s are getting broader with height for the whole centre–to–limb variation. For the lengthscale *L*, there is no significant centre–to–limb variation but a general height dependence. In the lower photospheric layers (<150 km), *L* is of the order 5×10^2km, while in the higher layers *L* increases to 7×10^2km, which indicates that the "size" of the photospheric structures increases with height. Then the authors calculated the magnitude of the circulation ($L \times v$) which is almost constant of the order $5.5 \times 10^2 \mathrm{km}^2\mathrm{s}^{-1}$ in the whole photosphere. The vorticity is $2.5 \times 10^{-3}\mathrm{s}^{-1}$ in the lower photosphere and $10^{-3}\mathrm{s}^{-1}$ in the higher layers. In active regions, the circulations reduced remarkably and also the vorticity. The authors deduced the height dependence of several independent non–dimensional parameters to discuss the influence of dissipative processes on the granular motions. The Reynolds number decreases from 5×10^9 at the continuum layers to 10^8 at the temperature minimum, the Péclet number decreases from 70 to 1, the magnetic Reynolds number is of the order 5×10^5, and the Rossby number decreases from 10^3 to 2×10^2. Hence, it is concluded that granules are some kind of buoyantly rising turbulent structures which disintegrate due to turbulent mixing with the environment.

071.048 **Time variability of the central residual intensities of solar Fraunhofer lines.**
B. T. Babij.
Kinematics Phys. Celest. Bodies, Vol. 7, No. 2, p. 14 – 19 (1991). English translation of Kinematika Fiz. Nebesn. Tel, Tom 7, No. 2, p. 16 – 22 (1991).

For moderately strong Fraunhofer lines the observed scattering of central residual intensities decreases with an increase of excitation potentials; for weak lines such a dependence is absent. It is shown that up to some level below the surface the solar photosphere may be considered to be unstationary. Below that level the undisturbed solar photosphere may be viewed as a stationary formation.

071.049 **The decay of granular motions and the generation of gravity waves in the solar photosphere.**
R. Komm, W. Mattig, A. Nesis.
Astron. Astrophys., Vol. 252, No. 2, p. 827 – 834 (Dec 1991). = Mitt. Kiepenheuer-Inst. Nr. 338.

The authors have investigated the photosphere with a coherence analysis of rms–velocities, and confirmed that there is a distinction between the granular structures of the lower photosphere and the secondary structures of the higher photosphere. They also show that the conversion of motions occurs well below a height of 200 km. Further, they have focused their attention on the small-scale structures of the higher photosphere (above 170 km). In these layers they have found structures in the wavenumber range from $2.5 \mathrm{Mm}^{-1}$ to $7.0 \mathrm{Mm}^{-1}$ with a "dominant" scale of about $4.0 \mathrm{Mm}^{-1}$ ($\sim 2''\!.2$). Therefore, it is concluded that the secondary motions are generated by the decaying granular motions. In addition, the authors have analyzed the ratio of horizontal to vertical velocity power spectra as a function of the wavenumber. In the lower photospheric layers this ratio is the same for all wavenumbers, while in the higher photospheric layers this ratio shows a maximum at the dominant scale. Also, the authors have calculated the vertical mass flow in the photosphere and found a strong decrease with height. By calculating the Brunt–Väisälä period and comparing it with the time scale of the small–scale motions, they show that the gravity waves, observed by others, are generated in the mid–photospheric layers where a generation of the secondary motions has been found. Taking all these results into account, it is concluded that the secondary motions of the higher photospheric layers can at least partly be identified with internal gravity waves.

071.050 **Time variability of absorption lines in the solar spectrum.**
R. I. Kostyk.
Kinematika Fiz. Nebesn. Tel, Tom 7, No. 5, p. 48 – 50 (Sep–Oct 1991). In Russian. English translation in Kinematics Phys. Celest. Bodies, Vol. 7, No. 5.

The conclusion of Babij (1991) obtained from an analysis of variation of central depths of Fraunhofer lines observed in 1960 – 1985 seems to be doubtful. The conclusion is that up to some level (in the region where weak absorption lines are formed) the solar photosphere is in stationary state, and above that level the solar photosphere may be considered as unstationary.

071.051 **2–D radiation–hydrodynamics models of the solar photosphere.**
M. Steffen.
NATO Advanced Research Workshop on Stellar Atmospheres: Beyond Classical Models, p. 247 – 261 (1991). – See Abstr. 012.056 for the main entry.

The author reports on the results obtained from detailed numerical simulations of solar granular convection in 2 dimensions. Depending on the size of the cylindrical computational domain he found steady state or time–dependent solutions for the convective flow. Comparison with spectroscopic observations strongly suggests that the time–dependent models are more realistic. Moreover, these models are shown to be in basic agreement with 3–dimensional simulations of solar granulation.

071.052 **Formation of the infrared emission lines of Mg I in the solar atmosphere.**
E. S. Chang, E. H. Avrett, P. J. Mauas, R. W. Noyes, R. Loeser.
Astrophys. J., Lett., Vol. 379, No. 2, p. L79 – L82 (1 Oct 1991).

The authors study the origin of the Mg I emission lines at 7 and 12 μm in the solar atmosphere. They use a 41 level atomic model, including collisions with hydrogen atoms as well as with electrons, and including the charge exchange process $Mg^+ + H(n) \leftrightarrow Mg(nl) + H^+$. Using an average quiet Sun atmospheric model, the authors calculate emission line profiles that resemble the observed ones, that is, broad absorption troughs with narrow central emission, and significant limb brightening. The charge exchange reactions are significant, but the effects of high–n coupling between Mg and Mg^+ together with radiative losses between low–n transitions are of greater importance. The results confirm that the emission cores are formed no higher than the temperature minimum region, and that the emission is caused by non–LTE effects rather than by the chromospheric temperature rise. This photospheric origin explains why the brightness temperature of the trough, or minimum intensity in the line wings, is unrelated to the temperature minimum between the photosphere and the chromosphere. The calculations indicate that the line core is sensitive to magnetic fields located almost 400 km above those measured in ordinary magnetograms.

071.053 **Multicolor continuum analysis of the solar granulation in quiet and active regions.**
C. U. Keller, S. Koutchmy.
Astrophys. J., Vol. 379, No. 2, p. 751 – 757 (1 Oct 1991).

The authors analyze a set of narrow-band filtergrams of the solar photosphere recorded in three widely separated true continuum windows and in Mg I b_1 with a spatial resolution of about $0''\!.5$. The authors study the influence of small-scale magnetic fields on the granulation by comparing various statistical parameters in selected active and quiet regions. With the same data, the temperature stratification of faculae at the level of continuum formation is investigated. In active regions there exist more points with enhanced continuum intensity as compared with quiet regions, and there is more power at small spatial scales in active regions, whereas more power exists at large scales in quiet regions. Quiet regions near and far away from a large sunspot do not show any significant difference. It is concluded that changes in the granular pattern near sunspots are caused by the small-scale magnetic fields often found near sunspots. Faculae, identified by their brightness in the Mg I b_1 wing, show an enhanced continuum intensity and a reduced ratio between the blue and the red continuum, which is consistent with current flux tube models.

071.054 **The solar spectrum.**
R. L. Kurucz.
Solar interior and atmosphere, p. 663 – 669 (1991). With 7 tables. – See Abstr. 003.028 for the main entry.

This chapter contains a discussion of what is known about the solar spectrum as a whole from the ultraviolet to the infrared. The data that are available are described and information is provided on how the reader can obtain paper or magnetic tape copies for particular applications, be it in atomic or molecular spectroscopy, or in solar, atmospheric, planetary, cometary or stellar physics.

071.055 **The photosphere as a radiative boundary.**
L. S. Anderson, E. H. Avrett.
Solar interior and atmosphere, p. 670 – 687 (1991). – See Abstr. 003.028 for the main entry.

The authors review the role of the photosphere as a radiative boundary for the solar interior, concentrating on semi-empirical and a priori models for the structure and emergent radiation. Contents: 1. Introduction. 2. Plane-parallel models. 3. Departures from plane-parallel structure. 4. Dynamical convection models.

071.056 **High-resolution observations of the solar granulation.**
K. P. Topka, A. M. Title.
Solar interior and atmosphere, p. 727 – 747 (1991). – See Abstr. 003.028 for the main entry.

The opportunity to observe the solar surface from space and from excellent groundbased sites, combined with enhanced digital image recording and computer technology, has led to improved high-resolution observations of solar granulation in the last decade. Today, granulation movies over 1 hr long are available with spatial resolution of nearly 1/3″ (=240 km on the solar surface). These new data have modified our conception of solar granulation by showing it to be a more complicated combination of different phenomena than was previously appreciated. The smaller granules, for example, may be the result of the fragmentation of larger granules due to turbulent small-scale flows. The new data have been used to detect large-scale horizontal flows on the Sun's surface, by the direct measurement of the proper motion of granules during their lifetimes. These new results may also have important implications for the heating of the solar chromosphere and corona.

071.057 **Model for the relationship of granulation and supergranulation.**
V. Krishan.
Solar interior and atmosphere, p. 1029 – 1043 (1991). – See Abstr. 003.028 for the main entry.

A mechanism is proposed for producing the observed solar supergranulation from the photospheric granulation by a dissipative decay of two dimensional turbulence, which leads to concentration of the energy spectrum to the longest wavelengths. This concentration of convective eddies by selective dissipation to the scale with the maximum available spatial dimension and with a much longer time scale is verified by mode-mode coupling seen in computer simulations as well as in laboratory experiments. Theoretical predictions for these granulation scales and magnetic structures can be tested by high quality observations of the solar surface.

071.058 **Solar element abundances.**
N. Grevesse, E. Anders.
Solar interior and atmosphere, p. 1227 – 1234 (1991). Appendix A. – See Abstr. 003.028 for the main entry.

071.059 **Wide-band solar spectroscopy on the James Clerk Maxwell telescope with a polarizing Fourier transform spectrometer.**
T. A. Clark, D. A. Naylor, G. J. Tompkins, G. R. Davis, W. D. Duncan.
J. R. Astron. Soc. Can., Vol. 85, No. 4, p. 190 (Aug 1991). Abstract. – See Abstr. 012.091 for the main entry.

071.060 **Granulation: non-adiabatic patterns and shocks.**
Å. Nordlund, R. F. Stein.
Conference on Challenges to Theories of the Structure of Moderate-Mass Stars, p. 141 – 146 (1991). – See Abstr. 012.096 for the main entry.

The authors present, in graphical form, some results from numerical simulations of the solar granulation. They compare synthetic granulation images with observations of the solar granulation, and illustrate the corresponding pressure and velocity fields. In particular, the non-adiabatic part of the pressure fluctuation, which is a major source of stochastic excitation of p-modes, is shown.

071.061 **The supergranulation spectrum.**
D. H. Hathaway, E. J. Rhodes Jr., A. Cacciani, S. G. Korzennik.
Conference on Challenges to Theories of the Structure of Moderate-Mass Stars, p. 163 – 169 (1991). – See Abstr. 012.096 for the main entry.

Full-disk Dopplergram observations obtained at the 60-foot tower of the Mount Wilson Observatory with the Cacciani sodium magneto-optical filter were analyzed to determine the spectrum of the solar supergranulation. Individual Dopplergrams were averaged together using a weighted average over 20-minute intervals to remove the p-mode oscillations. The Doppler signals due to the motion of the observer, the solar rotation, differential rotation, and limb shift were then removed from the data to produce Dopplergrams dominated by the supergranular flows. These data were mapped to heliographic coordinates and projected onto the spherical harmonics. Synthetic data were constructed and passed through the same analysis procedures to determine the actual spectrum required to reproduce the results. No corrections for seeing were included in the analysis. Although the image was sampled at 8 arcsec resolution, the effects of seeing may alter the actual slope of the spectrum and make it somewhat flatter.

071.062 **Electric field strengths from the solar 12 micron lines.**
E. S. Chang, W. G. Schoenfeld.
Astrophys. J., Vol. 383, No. 1, p. 450 – 458 (10 Dec 1991).

The 12 μm and other similar lines in the ATMOS solar spectra are computer-fitted to obtain the emission and absorption components and their relative line center displacement. It is found that the emission peaks agree very well with laboratory wavenumbers and so the displacements are attributed to the absorption components. Interpreted as quadratic Stark shifts due to a quasi-static electric field, the displacements yield electric field strengths on the order of 100 V cm^{-1}, corresponding to an electron density of $10^{12} cm^{-3}$. Also discussed are properties of these lines such as the widths, the equivalent widths, and the heights of formation of both the absorption and the emission components.

071.063 **Is ionized CH present in the solar spectrum?**
K. Sinha, B. M. Tripathi.
Bull. Astron. Soc. India, Vol. 19, Nos. 1 – 2, p. 81 (Mar – Jun 1991). Abstract. – See Abstr. 012.108 for the main entry.

071.064 **A method describing profile asymmetry of spectral lines.**
Gu Xiaoma, Li Qiusha.
Chin. Sci. Bull., Vol. 36, No. 19, p. 1618 – 1621 (Oct 1991).

The authors propose a method of asymmetry factor for researching the asymmetric field of a spectral line profile and calculating the distribution of asymmetric profiles of the loop prominence system of February 18, 1984. An interesting result is obtained after comparing it with the line of-sight velocity field of the loop system.

071.065 **What is the difference between "emission" and "depression" contribution functions?**
E. A. Gurtovenko, V. A. Sheminova, A. P. Sarychev.
Sol. Phys., Vol. 136, No. 2, p. 239 – 250 (Dec 1991).

The right approach to the problem of the depth of Fraunhofer line formation enables us to distinguish the main processes involved in the formation of a Fraunhofer line and to locate the depths for these processes in the photosphere. Physical and analytical (with the help of contribution functions) comprehension of these processes facilitates considerably the understanding of the notion of the depth of formation of Fraunhofer lines. This makes it possible to account for the connection of these processes with the origin and depth of location of different solar phenomena which are being investigated by using Fraunhofer lines.

Accurate oscillator strengths of astrophysical interest for neutral oxygen.
See Abstr. 022.034.

A discussion of the formula in the calculation of the column density at second level of hydrogen atoms.
See Abstr. 022.050.

Temperature–sensitive line ratios of the lithium–like ion O VI.
See Abstr. 022.109.

Narrow passband filter solar observations.
See Abstr. 034.033.

Solar 2D spectroscopy: a new MSDP instrument.
See Abstr. 034.039.

The Mees CCD imaging spectrograph.
See Abstr. 034.084.

Absolute spectrophotometry of the Sun. III. SEF–1 telescope automation system.
See Abstr. 036.056.

Solar spectroscopy with a 100 × 100 diode array.
See Abstr. 036.115.

Error analysis for the removal of the interference fringes from a solar image obtained with a CCD system by means of the subtraction.
See Abstr. 036.116.

A system for line profile studies at the 150–foot tower on Mount Wilson.
See Abstr. 036.130.

Turbulence, fractals, and the solar granulation.
See Abstr. 062.188.

On Geovenelli's approximate form of $\mathfrak{J}(\tau)$.
See Abstr. 063.006.

Partial redistribution modeling of the Ca II K line. Numerical method and solar applications.
See Abstr. 063.018.

NLTE spectral line formation in three dimensions.
See Abstr. 063.048.

Dynamics of and radiative transfer in inhomogeneous media.
See Abstr. 063.057.

Fe II emission line diagnostics of the Sun and stars.
See Abstr. 063.058.

New opacity calculations.
See Abstr. 063.063.

Possibilities of spectropolarimetrically studying eruptive formations in the solar atmosphere.
See Abstr. 072.090.

AR5395 solar flare in relation to the longitudinal current density in the photosphere.
See Abstr. 073.127.

On the spontaneous formation of electric current sheets above a flexible solar photosphere.
See Abstr. 074.102.

Neutralized and unneutralized current patterns in the solar corona.
See Abstr. 074.103.

Dark points and microflares observed in the He I λ10830 line and their relation to the quiet sun magnetic fields.
See Abstr. 075.021.

The effect of non–linear oscillations in magnetic flux tubes on Stokes *V* asymmetry.
See Abstr. 075.027.

Study of small scale magnetic flux and the corresponding velocity pattern.
See Abstr. 075.029.

Direct evidence of magnetic reconnection in a photospheric layer on the Sun.
See Abstr. 075.046.

Fe XIV and other line identifications in the solar EUV spectrum.
See Abstr. 076.022.

On the phase shift between solar radio emission at the wavelength 10.7 cm and Lα–emission.
See Abstr. 077.025.

Modeling mesogranules and exploders on the solar surface.
See Abstr. 080.015.

Observations of waves and oscillations.
See Abstr. 080.017.

On the role of the photospheric magnetic network in the 11–year variation of total solar irradiance.
See Abstr. 080.039.

Contribution and response functions for Ca II lines in different atmospheric models.
See Abstr. 080.082.

Observations of the phase differences between irradiances and velocity for low–degree solar acoustic modes.
See Abstr. 080.087.

On the results of application of the nonadiabatic criterion of the stability of equilibrium.
See Abstr. 080.089.

Convective structures in the Sun.
See Abstr. 080.107.

Molecules in the Sun.
See Abstr. 080.109.

Large-scale velocity fields.
See Abstr. 080.134.

Sun-as-a-star spectrum variability.
See Abstr. 080.135.

Waves in the solar atmosphere.
See Abstr. 080.166.

072 Sunspots, Faculae, Activity Cycles, Solar Patrol

072.001 **Active longitudes on the Sun and their reflection in interplanetary space.**
V. Bumba, L. Heina.
Geomagn. Aeron., Vol. 30, No. 4, p. 465 - 467 (Feb 1990). Paper presented at the 5. Symposium on Solar Terrestrial Physics, Samarkand (USSR), 2 - 6 Oct 1989. English translation of Geomagn. Aehron., Tom 30, No. 4, p. 550 - 553 (1990).

Plots of the longitudinal distribution of various manifestations of solar activity according to the papers of a number of authors are analyzed. The slopes of the active longitudes found are calculated. A similar plot of these slopes reveals the existence of two main directions of active longitudes that rotate more rapidly than the Carrington rotation. The average value of the synodic rotation of the first direction is 26.77 days and of the second direction is 27.16 days. A combined plot of all active longitudes shows that the active longitudes of various activity indices, and according to data of various authors, overlap to a definite extent and that both directions cross in the course of several 11-year cycles without interruption - the first with a step of 45 - 55 rotations, and the second with a step of 200 rotations of the Sun. Both these active longitudes have, their own reflection in the two main slopes of the sector boundaries of the IMF shown by Svalgaard and Wilcox and rotating synoptically in 26.84 and 27.44 days.

072.002 **The 27-day recurrence of phenomena on the Sun.**
V. F. Chistyakov.
Geomagn. Aeron., Vol. 30, No. 4, p. 557 - 559 (Feb 1990). Paper presented at the 5. Symposium on Solar Terrestrial Physics, Samarkand (USSR), 2 - 6 Oct 1989. English translation of Geomagn. Aehron., Tom 30, No. 4, p. 663 - 665 (1990).

The 27-day rotation period of the Sun is unstable and related to the secular, 22-, and 11-year activity cycles as well as to global torsional oscillations of spots with a period of 1 year. Prediction of phenomena on the basis of a 27-day calendar must be made with these regularities taken into account.

072.003 **Effects of global solar waves from data on variations in the Wolf number, cosmic-ray intensity, and geomagnetic activity.**
V. G. Kavlashvili, B. D. Naskidashvili, L. Kh. Shatashvili.
Geomagn. Aeron., Vol. 30, No. 4, p. 562 - 564 (Feb 1990). Paper presented at the 5. Symposium on Solar Terrestrial Physics, Samarkand (USSR), 2 - 6 Oct 1989. English translation of Geomagn. Aehron., Tom 30, No. 4, p. 667 - 669 (1990).

A correlation diagram-periodogram analysis of the mean daily sunspot number, mean daily galactic cosmic-ray (GCR) intensity, and mean daily geomagnetic-activity index A_p is used to obtain cross-correlation coefficients as a function of frequency. Peaks are observed in the cross correlation functions at frequencies corresponding to periods T = 36, 27, 19, and 13 days. The heights of the peaks for T = 36 and 19 days are above the 95% confidence level. In contrast to mean power spectra, where frequencies corresponding to periods T = 27 and 19 days dominate, the frequency spectrum of the cross-correlation coefficient in the present case is dominated by frequencies T corresponding to T = 36 and 19 days. The authors suggest that these frequencies are caused by highly efficient transport of wave energy from below the photosphere into the solar chromosphere and corona by global Rossby waves.

072.004 **The distribution of sunspots over the Sun.**
W. Uddin, M. C. Pande, V. K. Verma.
Astrophys. Space Sci., Vol. 181, No. 1, p. 111 - 115 (Jul 1991).

The distribution of the sunspots for the period 1967–1987 (solar cycles 20 and 21) is presented here. The authors find that the ± 11-20° latitude belt is most prolific for the occurrence of various spot types irrespective of magnetic-field ranges. Furthermore, longitudinally sunspots occur most prolifically at six or more places on the Sun. Spatially 7–9 zones are present in each hemisphere (north or south) of the Sun where about 50% sunspots occur and occupy only about 4% area of the Sun. During the above cycles at least 5 flare zones were regularly present in each hemisphere. The existing models cannot explain these active zones on the Sun. Thus, the present analysis emphasizes the need for a new magnetic models of the Sun.

072.005 **The periodicity of AR 5395 evolution.**
Lin Hua-an, Wang Shi-jin.
Chin. J. Space Sci., Vol. 11, No. 3, p. 161 - 167 (Jul 1991). In Chinese.

There are periodical characteristics in the evolution of AR 5395 shown by analysing the observational data of X-ray flares and an area of a sunspot group. For AR 5395 the variation of peak flux F_x of soft X-rays exhibits a periodicity about 24.3 hours, the periodicity for the occurrence rate N_x of X-ray flares is about 12.2 hours, and the area S_x of sunspot groups varies with 24.4 hours period. It seems that AR 5395 is in a state of energy store - release-store cycle of large flares with a periodicity about 24 hours.

072.006 **The evolutionary features of the magnetic configuration in the active regions and the short-term prediction of the strong SWF.**
Zhang Gui-qing.
Chin. J. Space Sci., Vol. 11, No. 3, p. 176 - 186 (Jul 1991). In Chinese.

The magnetograms of the longitudinal field in 18 active regions for April - December, 1988 are analysed in a previous paper. Some magnetic precursors of SWF flares (flares causing strong Short Wave Fadeout) are distinguished from the magnetograms of the longitudinal field. There is a close relation between the SWF flares and the magnetic precursors.

072.007 **Sonnenfleckenrelativzahlen der SONNE-Gruppe und des S.I.D.C. für April 1991.**
Sterne Weltraum, Jahrg. 30, Nr. 7, p. 456 (Jul 1991).
Further listings of sunspot numbers concerning the period May 1991 – Sep 1991 are published in Nr. 8–9, p. 532 (Aug–Sep 1991), Nr. 10, p. 610 (Oct 1991), Nr. 11, p. 681 (Nov 1991), and Nr. 12, p. 754 (Dec 1991), respectively.

072.008 **Solar active region evolution and flare build-up.**
K. T. Strong, J. D. Hawley.
Bull. Am. Astron. Soc., Vol. 23, No. 2, p. 924 (1991). Abstract. – See Abstr. 012.008 for the main entry.

072.009 **Determination of the parameters of sunspot magnetic fields from the structure of the superpenumbra.**
M. M. Molodenskij, L. I. Starkova, B. P. Filippov.
Sov. Astron., Vol. 35, No. 3, p. 300 – 305 (May 1991). Current Physics Microform No.: 9201X0582. English translation of Astron. Zh., Tom 68, Vyp. 3, p. 612 – 623 (1991).
The direction field defined by fibrils in the chromosphere provides enough information, as shown earlier, to solve the boundary-value problem for the B field (the Dirichlet problem) if the directions of fibrils in a plane are known. In some cases, the direction field is known only in certain annular zones around sunspots. This is sufficient to determine the main integral characteristics of the sunspot magnetic field. The relative values of the dipole and quadrupole moments and the total vertical current are obtained for observed structures. The authors transform from the direction field to the envelope of a family of straight lines on a circle. The envelope turns out to be insensitive to the choice of the circle, but characterizes the field structure fairly completely. Envelopes of tangents to fibrils in the superpenumbra are constructed for a number of sunspots, and the corresponding parameters are found. The relative magnitude of the azimuthal component, characterizing the current in a sunspot, is 0.2 – 0.3.

072.010 **Area and magnetic field of a sunspot during slow dissipation.**
A. A. Solov'ev.
Sov. Astron., Vol. 35, No. 3, p. 306 – 309 (May 1991). Current Physics Microform No.: 9201X0588. English translation of Astron. Zh., Tom 68, Vyp. 3, p. 624 – 631 (1991).
In accordance with the method given in the first part of this paper, the author calculates a model of a dissipating sunspot that incorporates all the main features of this process that are presently known: the linear phase of decay, the slow dissipation in small sunspots, the round cell or moat around a sunspot, and the slow variation of the magnetic field throughout the process. Relevant numerical calculations are given.

072.011 **Sunspot numbers.**
Sky Telesc., Vol. 82, No. 2, p. 215 (Aug 1991).
Sunspot numbers for Mar and Apr 1991 are given. Further lists concerning the time span May 1991 – to Aug 1991 are given in No. 3, p. 326 (Sep 1991), No. 4, p. 437 (Oct 1991), No. 5, p. 565 (Nov 1991), and No. 6, p. 685 (Dec 1991), respectively.

072.012 **Infrared CCD observations of umbral dots.**
M. W. Ewell Jr.
Bull. Am. Astron. Soc., Vol. 23, No. 2, p. 938 (1991). Abstract. – See Abstr. 012.008 for the main entry.

072.013 **The magnetic fields of active regions. VII. East-west inclination of magnetic field lines.**
R. F. Howard.
Sol. Phys., Vol. 134, No. 2, p. 233 – 246 (Aug 1991).
Magnetogram data are analyzed to study east-west magnetic flux differences interpreted as the component of magnetic field line inclination at the photospheric level in a plane parallel to the solar equator. This component is determined by comparing average east-west pairs of flux values at equal distances from the central meridian. The average inclination of a whole region is such as to trail the rotation (incline toward the east) by about 1.9 deg. Leading and following polarities tilt toward each other by about 16 deg. Growing regions are strongly inclined to the west (to lead the rotation) with large differences between leading and following portions. Decaying regions are slightly inclined to the east with more normal differences between leading and following portions. These results concerning growing and decaying regions are seen with greater amplitude for reversed polarity regions. As the activity cycle progresses, the average inclination of the field lines of the following portions of regions varies from about 10 to about 3 deg (leading the rotation), and the average difference in inclination of the leading and following portions of regions decreases monotonically during the cycle from nearly 20 to about 11 deg. A slight difference is seen between the average east-west inclination angles of regions that are rotating faster than average and those that are rotating slower than average in the sense that slower regions are slightly inclined toward the east and faster regions toward the west. Some of these results may be related to the location or nature of the subsurface flux tubes to which the active regions fields are connected and also, perhaps, to the nature of this connection.

072.014 **Electric currents and Hα emission in two active regions on the Sun.**
V. I. Abramenko, S. I. Gopasyuk, M. B. Ogir'.
Sol. Phys., Vol. 134, No. 2, p. 287 – 297 (Aug 1991).
The authors investigated the structure of magnetic field and vertical electric currents in two active regions through a comparison of the observed transverse field with the potential field, which was computed according to Neumann boundary-value problem for the Laplace equation using the observed H_z-value. Electric currents were calculated from the observations of the transverse magnetic field. There exist two systems of vertical electric currents in active regions: a system of local currents and a global one. The global current is about 2×10^{12}A. In the leading part of the active regions it is directed upward and in the tail downward. Flare activity is closely connected with the value and direction of both local and global currents: the flares tend to apear in places with upward currents. The luminosity of Hα flocculi is also connected with vertical electric currents; the brighter the flocculi, the more frequently they appear in places of upward electric currents. The sensitivity of Hα emission to the sign of the current suggests that charged particles accelerated in the upper parts of magnetic loops may be responsible for these formations. Joule heating might be important for flocculi, if plasma conductivity is about 5×10^8c.g.s.e. A model of a flare is suggested based on current redistribution in a system of emerging loops due to changes of loop inductance.

072.015 **Weak magnetic fields and solar irradiance variations.**
J. K. Lawrence, G. A. Chapman, S. R. Walton.
Astrophys. J., Vol. 375, No. 2, p. 771 – 774 (10 Jul 1991). With plate 9.
NOAA active region 5643 was observed from 1989 August 17 to 21. Sets of video spectra and spectroheliograms including the Fe I line at 6302.5 Å were made at least daily with the San Fernando Observatory 28 cm vacuum telescope and vacuum spectroheliograph. These give simultaneous, co-registered digital images representing monochromatic continuum intensity, line core intensity and line-of-sight magnetic field. Three different critieria are used to define the pixels representing the quiet Sun and the facular portions of the images. These criteria are the magnetic field strength, the line core intensity, and the distribution of continuum intensities. Each of these definition schemes is used to estimate the irradiance change due to facular emission. The magnetic field and the continuum intensity distribution definitions give estimates which agree closely. The line core intensity definition leads to larger estimates of the facular irradiance contribution. Some model-dependent investigations of the contrasts and sizes of individual facular elements are also presented.

072.016 **Solar patrol.**
Soln. Dannye, Byull., No. 7/90, p. 1 – 76 (Jul 1990). In Russian.
Contents: Daily maps of the Sun and of magnetic fields of sunspots. Sunspot groups for 26 July 1990. Synoptic charts of

active formations on the Sun. Forecast of the form of the solar corona. One-dimensional scans of the Sun in the centimeter range.

072.017 **Solar patrol.**
Soln. Dannye, Byull., No. 8/90, p. 1 – 78 (Aug 1990). In Russian.
Contents: Daily maps of the Sun and of magnetic fields of sunspots. Synoptic charts of active formations on the Sun. Forecast of the form of the solar corona. One-dimensional scans of the Sun in the centimeter range.

072.018 **Solar patrol.**
Soln. Dannye, Byull., No. 9/90, p. 1 – 76 (Sep 1990). In Russian.
Contents: Daily maps of the Sun and of magnetic fields of sunspots. Sunspot groups for 29 July and 8 August 1990. Synoptic charts of active formations on the Sun. Forecast of the form of the solar corona. One-dimensional scans of the Sun in the centimeter range.

072.019 **Solar patrol.**
Soln. Dannye, Byull., No. 10/90, p. 1 – 70 (Oct 1990). In Russian.
Contents: Daily maps of the Sun and of magnetic fields of sunspots. Synoptic charts of active formations on the Sun. Forecast of the form of the solar corona. One-dimensional scans of the Sun in the centimeter range.

072.020 **Solar patrol.**
Soln. Dannye, Byull., No. 11/90, p. 1 – 75 (Nov 1990). In Russian.
Contents: Daily maps of the Sun and of magnetic fields of sunspots. Sunspot groups for 23 and 26 October 1990. Synoptic charts of active formations on the Sun. Forecast of the form of the solar corona. One-dimensional scans of the Sun in the centimeter range.

072.021 **Solar patrol.**
Soln. Dannye, Byull., No. 12/90, p. 1 – 76 (Dec 1990). In Russian.
Contents: Daily maps of the Sun and of magnetic fields of sunspots. Synoptic charts of active formations on the Sun. Forecast of the form of the solar corona. One-dimensional scans of the Sun in the centimeter range.

072.022 **American relative sunspot numbers for July 1991.**
Sol. Bull. (AAVSO), Vol. 47, No. 7, p. 1 – 2 (Jul 1991).
Sunspot numbers for Aug – Dec 1991 are given in Vol. 47, Nos. 8 – 12 (Aug – Dec 1991).

072.023 **On the magnetic field activity in solar active regions.**
J. Linke.
International Conference on Mechanisms of Chromospheric and Coronal Heating, p. 42 – 43 (1991). – See Abstr. 012.011 for the main entry.
A new method of analysis of the complexity and of the reactivity of ensembles of magnetic features in the solar active regions is proposed and briefly discussed.

072.024 **Electric currents and flares in solar active regions.**
R. C. Canfield.
Bull. Am. Astron. Soc., Vol. 23, No. 2, p. 1025 (1991). Abstract. – See Abstr. 012.012 for the main entry.

072.025 **Flaring active region structures in the corona, transition region, and lower atmosphere observed in the SAROC.**
D. Moses, G. E. Brueckner, J. W. Cook, K. P. Dere, C. M. Korendyke, D. G. Socker, J.-D. F. Bartoe.
Bull. Am. Astron. Soc., Vol. 23, No. 2, p. 1026 (1991). Abstract. – See Abstr. 012.012 for the main entry.

072.026 **The 154-day and related periodicities of solar activity are subharmonics of a 25.8-day clock.**
T. Bai, P. A. Sturrock.
Bull. Am. Astron. Soc., Vol. 23, No. 2, p. 1027 (1991). Abstract. – See Abstr. 012.012 for the main entry.

072.027 **The distribution of longitudinal currents in sunspots.**
L. K. Wilkinson.
Bull. Am. Astron. Soc., Vol. 23, No. 2, p. 1030 – 1031 (1991). Abstract. – See Abstr. 012.012 for the main entry.

072.028 **Current systems and particle precipitation in AR 6233.**
J.-F. de La Beaujardière, R. C. Canfield.
Bull. Am. Astron. Soc., Vol. 23, No. 2, p. 1031 (1991). Abstract. – See Abstr. 012.012 for the main entry.

072.029 **Three-component electric current density in a unipolar sunspot with twisted field.**
V. A. Osherovich, H. A. Garcia, M. J. Hagyard.
Bull. Am. Astron. Soc., Vol. 23, No. 2, p. 1031 (1991). Abstract. – See Abstr. 012.012 for the main entry.

072.030 **Sunspot umbral and penumbral oscillations in Hα.**
R. Shine, T. Tarbell, A. Title, K. Topka, Z. Frank, K. Smith.
Bull. Am. Astron. Soc., Vol. 23, No. 2, p. 1033 (1991). Abstract. – See Abstr. 012.012 for the main entry.

072.031 **Active regions as a percolation phenomenon.**
D. G. Wentzel, P. E. Seiden.
Bull. Am. Astron. Soc., Vol. 23, No. 2, p. 1034 (1991). Abstract. – See Abstr. 012.012 for the main entry.

072.032 **CoMStOC III: Measuring magnetic fields in active region coronal plasma.**
J. T. Schmelz, G. D. Holman, J. W. Brosius, R. F. Willson.
Bull. Am. Astron. Soc., Vol. 23, No. 2, p. 1045 (1991). Abstract. – See Abstr. 012.012 for the main entry.

072.033 **A statistical study of nonthermal mass motions in solar active regions.**
J. L. R. Saba, K. T. Strong.
Bull. Am. Astron. Soc., Vol. 23, No. 2, p. 1046 (1991). Abstract. – See Abstr. 012.012 for the main entry.

072.034 **Convective flows around sunspot-like objects.**
P. A. Fox, S. Sofia, K. L. Chan.
Bull. Am. Astron. Soc., Vol. 23, No. 2, p. 1048 (1991). Abstract. – See Abstr. 012.012 for the main entry.

072.035 **On the scaling of the axial magnetic field gradient for twisted sunspots.**
A. Skumanich.
Bull. Am. Astron. Soc., Vol. 23, No. 2, p. 1051 (1991). Abstract. – See Abstr. 012.012 for the main entry.

072.036 **Stokes polarimetry of a sunspot from the Swedish Solar Observatory at La Palma.**
K. P. Topka, Z. A. Frank, T. D. Tarbell, A. M. Title, G. Scharmer.
Bull. Am. Astron. Soc., Vol. 23, No. 2, p. 1052 (1991). Abstract. – See Abstr. 012.012 for the main entry.

072.037 **Fine-scale magnetic field in a sunspot penumbra and adjacent photosphere.**
Z. A. Frank, G. B. Scharmer, C. Keller, H. Lundstedt.
Bull. Am. Astron. Soc., Vol. 23, No. 2, p. 1052 (1991). Abstract. – See Abstr. 012.012 for the main entry.

072.038 **Field geometry of sunspots inferred from inclination effects.**
A. M. Title, Z. A. Frank, R. A. Shine, T. D. Tarbell.
Bull. Am. Astron. Soc., Vol. 23, No. 2, p. 1052 (1991). Abstract. – See Abstr. 012.012 for the main entry.

072.039 **Mapping of three-dimensional sunspot fields by inversion line and symmetry.**
H. Zirin.
Bull. Am. Astron. Soc., Vol. 23, No. 2, p. 1053 (1991). Abstract. – See Abstr. 012.012 for the main entry.

072.040 **Using the 'even-odd' sunspot cycle variation to predict maximum amplitude for cycle 23.**
R. M. Wilson.
Bull. Am. Astron. Soc., Vol. 23, No. 2, p. 1054 (1991). Abstract. – See Abstr. 012.012 for the main entry.

072.041 **The maximum of solar activity cycle 22.**
H. S. Ahluwalia.
Bull. Am. Astron. Soc., Vol. 23, No. 2, p. 1054 (1991). Abstract. – See Abstr. 012.012 for the main entry.

072.042 **Zeeman splitting and continuum measurements of sunspots at 1.56 μm.**
G. Kopp, D. Rabin, C. Lindsey.
Bull. Am. Astron. Soc., Vol. 23, No. 2, p. 1055 – 1056 (1991). Abstract. – See Abstr. 012.012 for the main entry.

072.043 **Mapping the thick-target energy loss regions in AR 5395 and AR 5569.**
R. A. Schwartz, S. A. Drake.
Bull. Am. Astron. Soc., Vol. 23, No. 2, p. 1065 (1991). Abstract. – See Abstr. 012.012 for the main entry.

072.044 **Hα imaging spectroscopy during the Max '91 Campaign.**
J. F. de La Beaujardière, K. A. Blais, R. C. Canfield.
Bull. Am. Astron. Soc., Vol. 23, No. 2, p. 1071 (1991). Abstract. – See Abstr. 012.012 for the main entry.

072.045 **A connection of the number of dark points observed on the solar disk in the He I 10830 Å line and the number of light knots in the K Ca II line with the solar cycle.**
L. D. Parfinenko, N. K. Parfinenko.
Soln. Dannye, Byull., No. 7/90, p. 92 – 97 (Jul 1990). In Russian.
Using TV spectroheliograms in the He I 10830 Å line it is shown that the number of dark helium points observed on the solar disc in 1986 – 1988 changed in the antiphase with the Wolf numbers. About 500 dark helium points were observed at the minimum of solar activity, their number should decrease to several dozens at the maximum. The number of light knots in the K Ca II line varies insignificantly in the phase with the Wolf numbers, the correlation coefficient being equal to 0.81.

072.046 **On some peculiarities of the spatial distribution of sunspot groups at the declining phase of the solar cycles.**
Yu. I. Vitinskij.
Soln. Dannye, Byull., No. 7/90, p. 97 – 103 (Jul 1990). In Russian.
Peculiarities of latitudinal and longitudinal distributions of sunspot groups at the declining phase of solar cycles NN 18 – 21 are given. It is shown that the latitudinal distribution of the number of sunspot groups at the declining phase of the cycle is more stable than that at the maximum and minimum of solar activity, when it is subject to a secular variation.

072.047 **On some peculiarities of the spatial distribution of sunspot groups in the ascending phase of a solar cycle.**
Yu. I. Vitinskij.
Soln. Dannye, Byull., No. 8/90, p. 88 – 93 (Aug 1990). In Russian.
Peculiarities of latitudinal and longitudinal distributions of sunspot groups in the ascending phase of solar cycles Nos. 19 – 22 have been found. It is shown that the latitudinal distribution of the number of sunspot groups in the ascending phase undergoes a considerable cycle-to-cycle variation, without manifesting, however, a definite secular variation of the N-S asymmetry of the sunspot group number with the phase of the solar cycle. It is found that the longitudinal distribution of summarized maximum sunspot areas in the ascending phase of the cycle is characterized with the least inhomogeneity and the least stability of active longitudinal intervals.

072.048 **Peculiarities of oscillatory processes in sunspot groups SD 135/1984 and SD 136/1984.**
Yu. A. Nagovitsyn, G. F. Vyal'shin.
Soln. Dannye, Byull., No. 9/90, p. 91 – 96 (Sep 1990). In Russian.
On the basis of coordinate measurements with the "Helicor" method and simultaneous observations of sunspot magnetic fields with two solar telescopes at Kislovodsk and Pulkovo oscillations in two selected sunspot groups were studied. The non-solid, radial-asynchronous character of torsional oscillations is shown; their period increased from 45 min to 120 min with the passage from the central part of the sunspot to its periphery. Meridional oscillations are a uniform process for a force tube of a bipolar sunspot group as a whole. The horizontal velocity of propagation of magnetic field disturbances over a sunspot was estimated as 1.8 km·sec^{-1}.

072.049 **A system of periods in the range from tens to hundreds days in the temporal series of the solar activity indices.**
R. N. Ikhsanov, E. V. Miletskij.
Soln. Dannye, Byull., No. 10/90, p. 75 – 85 (Oct 1990). In Russian.
Using daily values of different solar activity indices – Wolf numbers (solar cycles 14 – 21), sunspot areas (cycles 19 – 21), UV emission flux in the lines h and K Mg II (λ = 280 mkm) (cycle 21) and the total solar emission flux (cycle 21) – their power spectra were calculated with the FFT for each cycle. On the basis of an analysis of their spectra a conclusion is made that there is a system of periods in the series of these indices. The values of these periods vary insignificantly from cycle to cycle and are faintly dependent on the type of indices, unlike the amplitudes which can vary in a wide range. It is concluded that there might be a mechanism on the Sun, which generates the system of periodic oscillations of solar activity. A supposition is made that periodic energy impulses emerging as a result of coupling between different modes of solar proper oscillations can be such a mechanism.

072.050 **Average monthly and average yearly Wolf numbers from 1971 – 1988.**
Soln. Dannye, Byull., No. 10/90, p. 91 – 92 (Oct 1990). In Russian.

072.051 **Heliographic latitude and rotation velocity of sunspots in the Maunder minimum.**
V. F. Chistyakov.
Soln. Dannye, Byull., No. 11/90, p. 80 – 85 (Nov 1990). In Russian.
Distribution and rotation velocity of sunspots in the Maunder minimum are typical of low solar activity periods: latitudes $\leqslant 20°$; average latitude in the maximum year $\leqslant 12.^{\circ}5$. The rotation of the sunspot group is nearly rigid with a light poleward acceleration.

072.052 **On the variation of dispersion of the longitudinal distribution of sunspot groups in the 11-year solar cycle.**
B. M. Rubashev.
Soln. Dannye, Byull., No. 12/90, p. 77 – 79 (Dec 1990). In Russian.
Using the data for 1910 – 1963 it is shown that the standard deviation of the relative (with respect to the total annual sunspot groups number) number of sunspot groups in the 40° intervals of the heliographic longitude has an 11-year cycle, which is in antiphase with the corresponding cyclic run of the mean yearly Wolf numbers.

072.053 **Solar patrol.**
Soln. Dannye, Byull., No. 1/91, p. 1 – 87 (Jan 1991). In Russian.
Contents: Daily maps of the Sun and of magnetic fields of sunspots. Synoptic charts of active formations on the Sun.

Forecast of the form of the solar corona. One–dimensional scans of the Sun in the centimeter range.

072.054 **Solar patrol.**
Soln. Dannye, Byull., No. 2/91, p. 1 – 72 (Feb 1991). In Russian.
Contents: Daily maps of the Sun and of magnetic fields of sunspots. Synoptic charts of active formations on the Sun. Forecast of the form of the solar corona. One–dimensional scans of solar radio radiation in the centimeter range for January 1991. One–dimensional scans of solar radio radiation in the centimeter range.

072.055 **Solar patrol.**
Soln. Dannye, Byull., No. 3/91, p. 1 – 96 (Mar 1991). In Russian.
Contents: Daily maps of the Sun and of magnetic fields of sunspots. Sunspot groups for 31 January, 9 and 20 February 1991. Synoptic charts of active formations on the Sun. Forecast of the form of the solar corona. One–dimensional scans of solar radio radiation in the centimeter range.

072.056 **Solar patrol.**
Soln. Dannye, Byull., No. 4/91, p. 1 – 81 (Apr 1991). In Russian.
Contents: Daily maps of the Sun and of magnetic fields of sunspots. Synoptic charts of active formations on the Sun. Forecast of the form of the solar corona. One–dimensional scans of the Sun in the centimeter range.

072.057 **Solar patrol.**
Soln. Dannye, Byull., No. 5/91, p. 1 – 77 (May 1991). In Russian.
Contents: Daily maps of the Sun and of magnetic fields of sunspots. Sunspot groups for 21 April 1991. Synoptic charts of active formations on the Sun. Forecast of the form of the solar corona. One–dimensional scans of solar radio radiation in the centimeter range.

072.058 **Solar patrol.**
Soln. Dannye, Byull., No. 6/91, p. 1 – 94 (Jun 1991). In Russian.
Contents: Daily maps of the Sun and of magnetic fields of sunspots. Synoptic charts of active formations on the Sun. Forecast of the form of the solar corona. One–dimensional scans of solar radio radiation in the centimeter range.

072.059 **Solar patrol.**
Soln. Dannye, Byull., No. 7/91, p. 1 – 78 (Jul 1991). In Russian.
Contents: Daily maps of the Sun and of magnetic fields of sunspots. Synoptic charts of active formations on the Sun. One–dimensional scans of solar radio radiation in the centimeter range.

072.060 **Loop ensemble models of solar active regions.**
S. F. Brown.
Astron. Astrophys., Vol. 249, No. 1, p. 243 – 249 (Sep 1991).
The source of heating in the solar corona is still an unresolved problem despite four decades of study. Choosing between competing theories is made difficult by the lack of constraints placed on the heating mechanism by observations. Simple static models of coronal loops have proved extremely successful at describing the physical conditions in observed loops, but are completely insensitive to the details of the heating mechanism. The author here examines an active region model constructed from an ensemble of simple loops placed in a dipole magnetic structure. He demonstrates that the emission pattern produced by the model region depends sensitively on the assumed heating function. Emission patterns were calculated for six heating functions. All the models with realistic emission patterns had heating functions with an inverse dependence on loop volume. The two best studied heating theories, resonant electrodynamic heating and topological dissipation, both have an inverse volume dependence. Heating theories without such dependence can be rejected.

072.061 **Fractal properties of sunspots.**
L. M. Zelenyj, A. V. Milovanov.
Pis'ma Astron. Zh., Tom 17, No. 11, p. 1013 – 1019 (Nov 1991). In Russian. English translation in Sov. Astron. Lett., Vol. 17, No. 6.
A fractal model of sunspots is presented. The formation of sunspot fractal structure is regarded as a result of magnetic tubes aggregation to fractal cluster. The expressions for umbral and penumbral magnetic fields as well as analytical expressions for the sunspot fractal dimension corresponding to the minimum of the free energy of the cluster are found. The relation between fractal dimension and the plasma distribution function is discussed. The parameters of the distribution function corresponding to the observational value of fractal dimension are calculated.

072.062 **Evolution characteristics of the fine–structure of AR4811 and the associated flare activity.**
Li Zhikai, Li Zhi, Wu Mingchan, Luan Di.
Publ. Yunnan Obs., No. 1, p. 1 – 6 (Mar 1991). In Chinese.
The evolution characteristics of the fine–structure of AR4811 and the associated flare activity are analysed by using the high–space–resolution photographic observations of the sunspot group and H_α chromosphere obtained with the first high–resolution photospheric and chromospheric telescope in China. It is pointed out that (1) the newly emerging magnetic flux with the same polarities as those of the original magnetic fields plays an important role in the course of the evolution of the spot group but has no greater contributions to the flare activity, (2) the frequent activity of the related filaments is an indication of flares and (3) the simultaneous co–existence of the mutual extrusion, shearing and rotation of the magnetic fields with opposite polarities is a possible mechanism of the storage of energy.

072.063 **Some features of the solar activities during the rising phase of cycle 22.**
Hong Qinfang.
Publ. Yunnan Obs., No. 1, p. 15 – 20 (Mar 1991). In Chinese.
The activities of sunspots and proton flares observed during the rising phase of cycle 22 are statistically analysed. There are really some distinctive features of the solar activities during this period. Four strong active regions are also described.

072.064 **The meridional motions of stable recurrent sunspots.**
G. Lustig, H. Wöhl.
Astron. Astrophys., Vol. 249, No. 2, p. 528 – 532 (Sep 1991). = Mitt. Kiepenheuer–Inst. Nr. 349.
The data bases of the Greenwich Photoheliographic Results and the similar data from the Kanzelhöhe are used to determine meridional motions of a special sample of identical stable recurrent sunspots within the period from 1948 until 1976. The motions are determined by two different methods: from linear fits of the latitudes during the sunspot disk passage and from mean latitudes at about central meridian passages of recurrencies. The latter method is more reliable and allows also to determine an indication for a braking of the meridional motions with aging of the spots.

072.065 **Variations in the strength of recurrent geomagnetic activity in solar cycles 11 to 21.**
G. K. Rangarajan.
Proc. Indian Acad. Sci., Earth Planet. Sci., Vol. 100, No. 1, p. 49 – 54 (Mar 1991).
From maximum entropy spectral analysis of short lengths of Aa indices of geomagnetic activity, the characteristics of the strength of the 27–day and 13.5–day signals for each of the solar cycles 11 to 21 are highlighted. It is shown that the 13.5–day signal is a near permanent feature in geomagnetic activity. The Hale–cycle (22 yr) effect could be seen in the average magnitude of the 27–day signal with greater strengths in the even cycles. No

clear annual variation in the strength is noticed, contrary to some earlier known results.

072.066 **Complexes of activity and coronal holes.**
V. G. Banin, S. A. Yazev.
Soln. Dannye, Byull., No. 2/91, p. 78 – 83 (Feb 1991). In Russian.
Complexes of activity (CA), after their active phase, transform into extensive unipolar magnetic regions (UMR), which, in principle, create the necessary conditions for the production of coronal holes (CH). The intercomparison of CA and CH synoptic maps for the declining branch of cycle No. 21 shows that there does, indeed, exist a number of signature indicating a relationship between these features. It is concluded that there appears to be a causal association of CH with CA.

072.067 **Absolute horizontal oscillations of sunspots and global organization of their periods.**
Yu. A. Nagovitsyn, E. Yu. Nagovitsyna.
Soln. Dannye, Byull., No. 2/91, p. 83 – 88 (Feb 1991). In Russian.
The parameters of absolute horizontal oscillations of 29 sunspots have been calculated using the data of the new Debrecen photoheliographic catalog for 1977. It is shown that the measure of difference between the periods of longitudinal (T_λ) and meridian (T_φ) oscillation modes AST = $(T_\lambda - T_\varphi)/(T_\lambda + T_\varphi)$ is the function of the heliographic latitude.

072.068 **Quasi–biennial activity variations in the Sun's polar zones.**
V. I. Makarov, V. V. Makarova, A. G. Tlatov.
Soln. Dannye, Byull., No. 2/91, p. 89 – 90 (Feb 1991). Abstract. – See Abstr. 012.021 for the main entry.

072.069 **Quasi–biennial cycles and phases of the 11–year solar cycles.**
Yu. I. Vitinskij.
Soln. Dannye, Byull., No. 2/91, p. 90 – 91 (Feb 1991). Abstract. – See Abstr. 012.021 for the main entry.

072.070 **Two systems of large–scale organization of solar activity with characteristic semi–annual and biennial lifetimes.**
E. V. Ivanov.
Soln. Dannye, Byull., No. 2/91, p. 91 – 92 (Feb 1991). Abstract. – See Abstr. 012.021 for the main entry.

072.071 **Methodological aspects in the analysis of quasi–biennial oscillations (QBO) in solar activity.**
E. V. Ivanov.
Soln. Dannye, Byull., No. 2/91, p. 92 – 93 (Feb 1991). Abstract. – See Abstr. 012.021 for the main entry.

072.072 **Dynamics of quasi–biennial variations of sunspot activity indices.**
G. V. Kuklin, L. A. Plyusnina.
Soln. Dannye, Byull., No. 2/91, p. 95 (Feb 1991). Abstract. – See Abstr. 012.021 for the main entry.

072.073 **The phase of quasi–biennial variations in the solar and geomagnetic activity.**
G. S. Ivanov–Kholodnyj, V. E. Chertoprud.
Soln. Dannye, Byull., No. 2/91, p. 96 – 97 (Feb 1991). Abstract. – See Abstr. 012.021 for the main entry.

072.074 **Two types of quasi–biennial variations in the Sun and in interplanetary space.**
Yu. R. Rivin.
Soln. Dannye, Byull., No. 2/91, p. 97 – 98 (Feb 1991). Abstract. – See Abstr. 012.021 for the main entry.

072.075 **Two–year variations of solar activity, interplanetary space and cosmic rays parameters.**
V. P. Okhlopkov.
Soln. Dannye, Byull., No. 2/91, p. 88 – 100 (Feb 1991). Abstract. – See Abstr. 012.021 for the main entry.

072.076 **Two classes of the solar activity cycles.**
V. F. Chistyakov.
Soln. Dannye, Byull., No. 4/91, p. 91 – 95 (Apr 1991). In Russian.
A new classification of the solar activity cycles is offered: 1) the activity cycles of the solar globe with durations 11, 22, 100 and 200 years; 2) the activity cycles of the solar core with durations 2, 5 and 10 years.

072.077 **Quasi–biennial oscillations of sunspot activity zones during cycle 21.**
A. V. Mordvinov.
Soln. Dannye, Byull., No. 3/91, p. 97 (Mar 1991). Abstract. – See Abstr. 012.021 for the main entry.

072.078 **Quasi–biennal variations of solar activity asymmetry.**
G. A. Bazilevskaya.
Soln. Dannye, Byull., No. 3/91, p. 98 – 99 (Mar 1991). Abstract. – See Abstr. 012.021 for the main entry.

072.079 **Evolution of a strong pulsating sunspot center and its chromospheric activity.**
Yu. M. Slonim, K. F. Kuleshova.
Soln. Dannye, Byull., No. 5/91, p. 84 – 91 (May 1991). In Russian. With 1 plate.
The evolution of a pulsating sunspot center is studied, followed for 11 passages across the solar surface during April 1989 to January 1990 and achieving peak in September, when a large group of 4300 m.e.h. was formed in it. The real height of the stable loop system is determined that was due to the 3B flare formed near the W–limb on August 14. It is noted that the real height of the system R decreased from 80 × 10^3km on August 14 – 15 to 70 × 10^3km on August 16 – 17. The decrease of h was due to the reconnection of the loop legs.

072.080 **Statistical distribution of daily Wolf numbers in the epochs of extrema of 11–year solar cycles Nos. 19 – 20.**
B. M. Rubashev, E. V. Miletskij.
Soln. Dannye, Byull., No. 5/91, p. 92 – 95 (May 1991). In Russian.
It is shown that daily Wolf numbers are distributed statistically differently within certain gradations in each year of the epoch of extrema of solar cycles 19 – 20. It seems impossible to adjust a uniform continuous statistical distribution of any of the considered years. Therefore the authors can draw the conclusion that there is a composition of different distributions for each year.

072.081 **The Wilson effect of sunspots.**
Ye Shihui.
Publ. Purple Mt. Obs., Vol. 10, No. 2, p. 92 – 104 (Jun 1991). In Chinese.
This paper presents a short account on the history and significance of the study of the Wilson effect of sunspots, discusses the theoretical basis of the justification or establishement of sunspot models with the observational data of the Wilson effect, and also describes the main achievements in research of this effect and the existing questions. Besides, the author proposes a new and comprehensive definition of the Wilson effect and some suggestions for further work.

072.082 **A preliminary analysis on solar activity during the interval of international cooperative observation (July 19 – 25, 1990).**
Chen Xiangyang.
Publ. Purple Mt. Obs., Vol. 10, No. 2, p. 109 – 113 (Jun 1991). In Chinese.

072.083 **Moving periodogramanalysis of data on solar activity and solar–terrestrial relations over the past 400 years.**
G. E. Kocharov, A. N. Peristykh.
Soln. Dannye, Byull., No. 3/91, p. 107 – 115 (Mar 1991). In Russian.

The results of a comparative moving periodogramanalysis of data on solar activity and solar–terrestrial relations over the past 400 years are considered. The data used include the time series of the annual Wolf number, aa–index and recurrence of borealis aurorae, abundances of cosmogenic isotopes ^{10}Be in dated polar ice cores, cosmogenic ^{14}C and stable ^{13}C isotopes in tree rings and monthly air temperature in Central England. It is shown that a cycle with a period of about 22 years not only, manifests itself but moreover dominates in all the analysed series corresponding to the Maunder minimum (1645 – 1715 AD). The peculiarities of the Schwabe cycle manifestation over the past three centuries are considered.

072.084 **On the peculiarities of spatial–time distribution of sunspot groups at heliographic latitudes $\geqslant$40°.**
Yu. I. Vitinskij.
Soln. Dannye, Byull., No. 7/91, p. 79 – 84 (Jul 1991). In Russian.

A list of high latitude sunspot groups ($\varphi \geqslant \pm 40°$) compiled from the data of the Soviet Solar Service for 1955 – 1990 is given. It is shown that they are a special class of features of solar activity, and do not determine the main peculiarities of the 11–year solar cycle and the secular cycle. It seems that they are related to a longer cycle or to long intervals of very high activity.

072.085 **On the helical structure in a complex flare–active sunspot group.**
Z. B. Korobova, Eh. I. Mogilevskij.
Astron. Tsirk., No. 1543, p. 19 – 20 (Mar – Apr 1990).

Proper motions of sunspots and the development of a helical configuration in a large flare–active group of March 1989 are discussed. The helical structures persistence, rigid rotation in the topological center, concentration of flares in this group suggest the existence of a stationary auto–wave in this active region complex.

072.086 **Solar research at Potsdam: papers on the structure and dynamics of sunspots.**
J. Staude.
Reviews in Modern Astronomy 4, p. 69 – 89 (1991). – See Abstr. 003.008 for the main entry.

The investigation of sunspots at Potsdam has a tradition of many decades. After some historical remarks the present review considers efforts to determine the magnetic field structure in solar active regions, especially in sunspots, and the modelling of the thermodynamic structure and of the oscillatory regime of umbrae. The double vector magnetograph of the Potsdam Solar Observatory "Einsteinturm" enables to measure the full Stokes vector in the profiles of 2 independently chosen lines. The instrumental calibration and elimination of signal distortions provide the necessary conditions for the theoretical calibration in terms of a transfer theory of polarized radiation. By means of the latter the spatial distribution of the magnetic field vector and related quantities such as electric currents can be derived. Modelling efforts include the derivation of the height dependence of thermodynamic quantities in umbrae from the subphotosphere up to the corona, moreover, the interpretation of umbral oscillations by a model of resonant transmission of magnetoatmospheric waves through umbral models.

072.087 **Sun–as–a–star: its convective signature and the activity cycle.**
W. Livingston.
IAU Colloquium No. 130: The Sun and cool stars – activity, magnetism, dynamos, p. 246 – 251 (1991). – See Abstr. 012.024 for the main entry.

From May 1980 to April 1990 the author has observed several times a year the solar irradiance spectrum from 5000 Å to 6300 Å. During the ascending phase of the solar cycle, spectrum line asymmetry diminishes about 15% as the total magnetic field increases. At some epoch near the time of total magnetic field maximum, line asymmetry ceases to track surface magnetism and abruptly assumes a value appropriate to solar minimum.

072.088 **The prolonged minima and maxima of solar activity.**
V. N. Dermendjiev, G. T. Buyukliev, Y. Y. Shopov.
IAU Colloquium No. 130: The Sun and cool stars – activity, magnetism, dynamos, p. 268 – 270 (1991). – See Abstr. 012.024 for the main entry.

The authors study a new indirect index of solar activity – the intensity of luminescence of cave flowstone microzones. This index correlates directly with the solar activity. Using a time series of the index with resolution 4 – 5 pixels/year some of the statistical properties of the solar activity cycle were studied during 28 prolonged minima and 21 prolonged maxima.

072.089 **The main component analysis of the longitudinal distribution of solar activity.**
L. Hejna, H. Wöhl.
IAU Colloquium No. 130: The Sun and cool stars – activity, magnetism, dynamos, p. 275 – 276 (1991). – See Abstr. 012.024 for the main entry.

In this contribution, preliminary results of the main component analysis of Bartels diagram of time series of daily values of sunspot group numbers for solar cycles 18, 19 and 20 are presented. The results obtained suggest that the most significant feature in the longitudinal distribution of sunspot activity is the existence of preferred solar hemisphere alternating with a mean period of 2.5 Bartels rotations.

072.090 **Possibilities of spectropolarimetrically studying eruptive formations in the solar atmosphere.**
S. A. Kazantsev, N. M. Firstova, A. V. Gubin, N. A. Lankevich.
Opt. Spectrosc., Vol. 70, No. 5, p. 580 – 583 (May 1991). Current Physics Microform No.: 9203X0662.

Based on measurements of the linear polarization of the hydrogen H_α line radiated by emissive formations in active region No. 5669 (28 August – 12 September 1989) on the solar disk, the possibilities for a spectropolarimetric method of determining energy transport into the chromosphere by beams of superheated charged particles are analyzed.

072.091 **An explanation of penumbral dimensions.**
Zhou Daoqi.
Acta Astrophys. Sin., Vol. 11, No. 3, p. 275 – 282 (Jul 1991). In Chinese. English translation in Chin. Astron. Astrophys., Vol. 15, No. 4, p. 453 – 460 (Dec 1991).

Using short wave approximation, the author has solved magnetohydrodynamic disturbance equation including Joule heating and thermal conduction. It is possible that the penumbral filament structure results from the development of an instability in the magnetic field of the penumbral current. It is found that (1) there are two lower limits of penumbral filament width and length according to the instability condition respectively; (2) there is no overstable state mode propagating along the filaments and it is a very near non–overstable state mode propagating normal to the filaments; (3) the penumbral filament is related to the gravitation; (4) the magnetic field of sunspot is twisted when penumbral structure appears.

072.092 **The possible influence of the orbital motion of the major planets on sunspot activities.**
Yang Zhigen.
Acta Astrophys. Sin., Vol. 11, No. 4, p. 370 – 377 (Oct 1991). In Chinese.

The rate of change of the angular momentum of the Sun's center about the center of mass of the solar system caused by the overall planetary orbital motion denoted by $\dot{J}_\odot$ has been calculated in ecliptic coordinates of the barycenter of the solar system with reference to the epoch JD = 2451545.0. Various short periodicities were shown in the polynomial of $\dot{J}_\odot$. It has been indicated from the adoption of the multi–stage filter and the power spectrum analysis with Marple Algorithm that the results of the same period narrow band filtering of $\dot{J}_{\odot}$ series and relative sunspot numbers (SP) series are closely correlated with each other, and the coincidence of the spectral character between these two series is confirmed. The periodicities approaching 314.7, 149.9, 118.1, and 94.5 days coincide closely with the theoretical results. In conclusion, it is highly probable that the orbital motion of the planets is a very important dynamical factor affecting sunspot activities.

072.093 **Electric current in a unipolar sunspot with an untwisted field.**
V. A. Osherovich, H. A. Garcia.
Geophys. Res. Lett., Vol. 17, No. 12, p. 2273 – 2276 (Nov 1990).
The authors apply the Return Flux (RF) sunspot model to a round, unipolar sunspot observed by H. Kawakami (1983). Solving the magnetohydrostatic problem using the gas pressure deficit between the umbral and quiet–Sun atmospheres as a source function, the authors find a distribution of electric current density in an untwisted, unipolar sunspot as a function of height and radial distance from the sunspot center. Maximum electric current density is about 32 mA m^{-2} at the bottom of the sunspot.

072.094 **Convective flows around sunspot–like objects.**
P. A. Fox, S. Sofia, K. L. Chan.
Sol. Phys., Vol. 135, No. 1, p. 15 – 42 (Sep 1991).
Results are given for calculations of convective flows around objects in the outer layers of the Sun that have similar characteristics to small sunspots. These objects are allowed to radiatively (diffusively) exchange heat with their surroundings, but convective motions and exchange are absent. This assumption is based on the simple presumption that a sunspot magnetic field maintains pressure equilibrium with the surrounding medium and prevents convective exchange with that medium. The flow structure around the object, and the question of the overall balance or redistribution of the emerging heat flux as suggested by earlier empirical models, are studied and discussed. After a period of adjustment, shortly after the sunspot–like object is placed into the domain, the layer readjusts itself so that most of the heat flux actually reappears at the surface, although some fraction of the flux is carried horizontally far from the object. There is no indication of long term storage of the heat flux that would normally appear in the place where the object resides. Finally, when the object is removed, the surrounding medium responds very quickly and soon returns to the undisturbed state before the object was in place. The present numerical treatment includes restrictions that may influence aspects of the heat redistribution, convective flows and time scales. In particular, the shape of the object and its size (somewhat smaller than a sunspot) are important, as is the number of spatial dimensions and the treatment of some boundary conditions. Since all of these issues require further investigation, some discussion is presented regarding the applicability of our results to real sunspots.

072.095 **VLA observations of solar active regions at 6 and 20 cm.**
C. E. Alissandrakis, M. R. Kundu,
K. R. Shevgaonkar.
Astron. Astrophys., Vol. 251, No. 1, p. 276 – 284 (Nov 1991).
The authors present high resolution observations of two active regions at 6 and 20 cm over a period of 5 d, together with Hα and photospheric magnetic fields. The large scale emission at 20 cm is associated with the Hα plage. In one region the strongest source was over the neutral line, near the tip of an active region filament, which indicates that the emission probably originated in small scale coronal loops. In the second region the peak of the emission was near a well developed sunspot. Neither region showed evidence of large scale loops joining their preceding and following parts. Several other sources were observed at 20 cm; a source associated with an Hα plage region crossed by a filament and one associated with a small bipolar region are briefly discussed. The 6 cm emission from a well developed spot showed clearly the characteristics expected from gyroresonance model computations. The detailed shape of the source shows that the magnetic field of the spot was not potential. The sense of the required electric current is the same as that inferred from the analysis of chromospheric observations.

072.096 **Velocity measures in the sunspot Rome No. 7490 (July 5, 1979).**
M. G. Adam, A. D. Petford.
Sol. Phys., Vol. 135, No. 2, p. 319 – 326 (Oct 1991).
Material motions on the solar surface have been deduced from the wavelength shift of Fe I λ6302.5 Å, measured over the umbra and inner penumbra of a spot for which the magnetic field configuration has already been established with some confidence. The two vector fields are considered together in detail and the results support the "convective roll" sunspot model (Spruit, Galloway). For the magnetic field regions, both material flow along the field lines and field line motions are derived. A small upward motion only is deduced for the field free regions.

072.097 **Cycle latitude effects for sunspot groups.**
R. F. Howard.
Sol. Phys., Vol. 135, No. 2, p. 327 – 337 (Oct 1991).
Digitized Mount Wilson sunspot data from 1917 to 1985 are analyzed to examine meridional motion and rotation properties as a function of latitude and distance (ξ) from the average latitude of activity (ξ_0) in each hemisphere. Latitude dependence similar to previous results is found, but only for spot groups whose areas are decreasing from one day to the next. A previous study of active region magnetic fields, using this technique of motions as a function of the average latitude of activity, had shown meridional motions on average toward ξ_0. In this analysis of spot data some evidence is seen for motion away from ξ_0, with some slight evidence for faster rotation equatorward of ξ_0 and slower motion poleward of ξ_0, similar to the torsional oscillation phenomenon. For reasons that are not clear, both of these effects are significantly more pronounced for sunspot groups whose areas are decreasing.

072.098 **Sunspot group areas and the latitude distance from the average latitude of activity.**
R. F. Howard.
Sol. Phys., Vol. 135, No. 2, p. 339 – 342 (Oct 1991).
Digitized Mount Wilson sunspot data from 1917 to 1985 are analyzed to examine group areas as a function of latitude distance (ξ) from the central latitude of activity in each hemisphere. On average these group areas are larger for the smallest values of $|\xi|$. The effect is similar to that seen for the magnetic field of active regions. It is concluded that this is fundamentally a ξ dependence, and not a latitude dependence. The suggestion is made that the cause of this effect is the influence of large–scale convective motions on the rising flux tubes that make up the active regions. The smaller flux tubes (spot groups) are more easily displaced in latitude during their ascent to the surface by this velocity field than are the larger flux tubes.

072.099 **Activité solaire: décembre 1989.**
M. J. Martres, G. Zlicaric.
Astronomie, p. 13 (Jul – Sep 1991).

072.100 **On the dynamics of the large–scale magnetic fields of the Sun and the sunspot cycle.**
Z. Mouradian, I. Soru–Escaut.
Astron. Astrophys., Vol. 251, No. 2, p. 649 – 654 (Nov 1991).
The purpose of the present article is to analyse the solar activity cycle from the point of view of large scale magnetic fields (LSMFs), i.e. the "polarity background" (unipolar magnetic regions) and the "polar islands" (bipolar magnetic regions) with or without active regions. Using synoptic charts of the magnetic field (Stanford, McIntosh) and daily KPNO and Mount Wilson observations, a scenario has been developed that is based on the dynamic of the LSMFs. The activity cycle appears to be asymmetrical, with the north polar background predominating at the minimum and the south polar background at the maximum. The polarity islands in the background cover a large area and as they develop, they "push" the magnetic field background towards the poles and thereby establish an equator–to–pole

circulation at the solar surface. This circulation is also affected by the differential rotation of the Sun. A fractal dimension analysis shows that the polarity islands become "turbulent" as sunspot activity grows.

072.101 **Characteristics of sunspot group 5395.**
Zhao Aidi.
Publ. Yunnan Obs., No. 4, p. 25 – 27 (Dec 1990). – See Abstr. 012.047 for the main entry.

072.102 **Sunspot's motions and complex magnetic fields within the AR 5395.**
Li Weibao, Hong Qinfang, Wang Hongzhao.
Publ. Yunnan Obs., No. 4, p. 28 (Dec 1990). Abstract. – See Abstr. 012.047 for the main entry.

072.103 **AR5395: fine structures, magnetic shear and flares.**
Li Zhikai, Wu Mingchan, Luan Di, Li Zhi.
Publ. Yunnan Obs., No. 4, p. 53 – 57 (Dec 1990). – See Abstr. 012.047 for the main entry.

072.104 **Wolf's numbers predictions for the 22–nd cycle.**
O. V. Chumak, Z. N. Chumak.
Astron. Tsirk., No. 1545, p. 27 – 28 (Aug – Sep 1990). In Russian.
A new solar cycle predictions method is proposed. The cycle 22 predictions diagram is presented.

072.105 **Angular momentum transport by Reynolds stresses determined from the analysis of 100–year sunspot motions and its variations with solar cycle.**
L. Paternò, D. Spadaro, R. A. Zappalà, F. Zuccarello.
Astron. Astrophys., Vol. 252, No. 1, p. 337 – 342 (Dec 1991).
Some aspects of the latitudinal momentum transport by Reynolds stresses, as deduced from the analysis of the Greenwich sunspot data for the period 1874–1976 by using only very young sunspot groups (age $\leqslant$ 3 d), were investigated in order to study the equatorial acceleration maintenance and its possible correlations with the equatorial angular velocity time variations and the solar cycle. The results, which are consistent with those of previous authors, indicate the presence of a net angular momentum flux towards the equator which is more vigorous at higher latitudes, and sufficient on average to maintain the observed equatorial acceleration. The angular momentum transport time variations show a significant periodicity correlated with the solar cycle. Also the equatorial velocity time variations show significant periodicities, which in turn may depend on the angular momentum transport time variations. The scenario which emerges from these results is discussed in the framework of the theories of differential rotation and activity cycle.

072.106 **Solar flares and magnetic field configuration in a solar active region Boulder 5395.**
Liu Xinping, Li Wei.
Publ. Yunnan Obs., No. 4, p. 75 – 78 (Dec 1990). – See Abstr. 012.047 for the main entry.

072.107 **On the energy storage in a solar active region AR 5395 by rotation movement.**
Liu Xinping.
Publ. Yunnan Obs., No. 4, p. 79 (Dec 1990). Abstract. – See Abstr. 012.047 for the main entry.

072.108 **On the energy storage in a solar active region AR 5395 by shearing magnetic field.**
Liu Xinping, Li Jing, Liu Guilin.
Publ. Yunnan Obs., No. 4, p. 80 (Dec 1990). Abstract. – See Abstr. 012.047 for the main entry.

072.109 **The periodicity of flare occurrence in evolution of AR 5395.**
Lin Huaan, Wang Shijin.
Publ. Yunnan Obs., No. 4, p. 81 (Dec 1990). Abstract. – See Abstr. 012.047 for the main entry.

072.110 **Sunspots and their magnetic fields within AR5629.**
Li Weibao, Hong Qinfang, Wang Hongzhao, Zhang Bairong.
Publ. Yunnan Obs., No. 4, p. 171 – 173 (Dec 1990). – See Abstr. 012.047 for the main entry.

072.111 **Spatial variation of the magnetic field inclination in a sunspot penumbra.**
D. Degenhardt, E. Wiehr.
Astron. Astrophys., Vol. 252, No. 2, p. 821 – 826 (Dec 1991).
Stokes–V spectra from a sunspot penumbra at $\vartheta = 11\overset{\circ}{.}5$ show significant spatial fluctuations. Due to its deep formation layer, the Fe 6842.7 line is totally split (B = 2200 Gs) showing field fluctuations of the order of 10%. The spatial variations of the Stokes–V spectra are strongly affected by telescopic phase retardation thus allowing only a rough estimate of the inclination angle. Its fluctuation by $7° < \Delta\psi < 13\overset{\circ}{.}5$ to the solar surface is spatially related to the field strength fluctuation in the sense that steeper fields are stronger. Doppler shifts of the Stokes–V give a slight indication for upward motions at locations of steeper field inclination. A relation with the continuum intensity fluctuation is only found in the centre–side penumbra – possibly due to the steeper angle–of–view. The absence of a similar correlation in the limb–side spectra agrees with earlier findings that field strength and Evershed effect are not related to penumbral fine–structures.

072.112 **Time variations of the tangential component of velocity in the Evershed effect.**
S. A. Druzhinin, A. A. Pevtsov, V. I. Levkovskij, M. V. Nikonova.
Kinematika Fiz. Nebesn. Tel, Tom 7, No. 5, p. 51 – 60 (Sep–Oct 1991). In Russian. English translation in Kinematics Phys. Celest. Bodies, Vol. 7, No. 5.
Results of observations of the tangential component of velocity in the Evershed effect (torsional oscillations of sunspots) are presented. For six sunspots, the spectral composition of line–of–sight velocity signals from two areas of the penumbra located symmetrically relative to the sunspot umbra has been studied. The areas chosen are located perpendicular to the direction toward the solar image centre, i.e., gas rotation velocities in a sunspot were measured. The observations were made in lines of Fe I λ543.45 nm and H_β λ486.13 nm using a dissector–tube, whose electronic scanning control permits measuring (quasi–simultaneously in two channels) spectral line shifts on two sections of its photocathode. The results obtained confirm the presence of torsional oscillations of sunspots at the photospheric level with a period of about one hour. For the first time such oscillations have been measured at the chromospheric level (with a period of about 30 minutes). Based on observations of two sunspots during 3 and 4 days, respectively, it is concluded that these sunspots involve torsional oscillations with a period of several days.

072.113 **Formation of the K I 7699 Å line in sunspots.**
B. Caccin, M. Carlsson, M. T. Gomez, G. Severino.
NATO Advanced Research Workshop on Stellar Atmospheres: Beyond Classical Models, p. 415 – 422 (1991). – See Abstr. 012.056 for the main entry.
The authors use the version MUL20 of the NLTE code by M. Carlsson with improved opacities, and an K I atomic model to build synthetic profiles of the K I resonance line in spot umbra model atmospheres. They estimate several effects on line formation due to the spot atmospheric model, using spot models based on different approaches, and conclude that the core reversal should not be due to the residual NLTE sensitivity of the line source function to a chromospheric temperature rise. The authors also explore the effects due to the Zeeman splitting and to a possible insufficient correction for stray light, using simple

models to be included in the line formation. Finally they consider downflows as a possible cause of the line core asymmetry.

072.114 **Forecast of the maximum of the next 11–year cycle of sunspots No. 23.**
M. Kopecký.
Bull. Astron. Inst. Czech., Vol. 42, No. 3, p. 157 – 158 (Jun 1991).
Based on the dependence of the maximum relative R_M–number of the odd 11–year cycle on the R_M of the previous even cycle it is forecast that R_M should exceed 200 in the next 11–year sunspot cycle No. 23.

072.115 **When did the latest minimum of the 80–year sunspot period occur?**
M. Kopecký.
Bull. Astron. Inst. Czech., Vol. 42, No. 3, p. 158 – 160 (Jun 1991).
The pattern of the average mightiness of large sunspot groups, determined from Greenwich and Pulkovo catalogues of large sunspot groups, indicates that the minimum of the 80–year period of sunspot groups occurred in 11–year cycle No. 20, and that 11–year cycles Nos. 21 and 22 are already on the ascending branch of the 80–year period. The significance of this fact for prognosticating the height of the maxima in the future 11–year cycles is discussed.

072.116 **Active region and flare relationships.**
P. A. Boeder.
IPS Radio and Space Services, Chatswood (Australia); Department of Administrative Services, Darlinghurst (Australia). IPS Radio and Space Services, IPS–TR—91–01. 86 p. (May 1991). ISSN 0812–7395.
This report explores the properties of flares and of flare–producing regions on the sun, as based upon information contained within the IPS "Unify" Flare Database. The goal of the report is to examine the biases contained within the database and to provide information which is of use in forecasting or which would warrant further investigation.

072.117 **Photometry of the umbral structure of a fast–developing sunspot.**
M. Sobotka, S. Šimberová, V. Bumba.
Bull. Astron. Inst. Czech., Vol. 42, No. 4, p. 250 – 256 (Jul 1991). With plates 3, 4.
A high–quality white–light photograph of an umbra of a large complex spot was analysed. Digital image processing methods were used. The extensively inhomogeneous structure of the umbra, related to the fast dynamics of development, displayed many features (dark cores, bright dots, light bridge, etc.), intensities and temperatures of which were studied in detail. In the dark cores a relatively low temperature amounting to 3600K at $\tau_{500} = 1$ was obtained. A relation between the intensity of the individual bright umbral dots and their dark surroundings was found.

072.118 **The relation of the sunspot magnetic field and penumbra–umbra radius ratio.**
A. Antalová.
Bull. Astron. Inst. Czech., Vol. 42, No. 5, p. 316 – 320 (Oct 1991).
Indirect proportionality between the sunspot magnetic field B_L and penumbra–umbra radius ratio k was found. The observed functions $B_L = f(k)$ are given (74 regular sunspots observed from 1968 to 1976, and 15 regular sunspots observed in 1977). It is illustrated statistically that the smaller value B_L is observed in the regular sunspot with larger value k. This result is in qualitative coincidence with Saniga's alternative view of the sunspot phenomenon.

072.119 **The Sun in 1990.**
H. Barnes.
South. Stars, Vol. 34, No. 4, p. 288 – 292 (Sep 1991).
The activity of the Sun in 1990, as measured by sunspot number data, is presented, and comparison data for other recent years also provided.

072.120 **Motions, fields, and flares in the 1989 March active region.**
H. Wang, F. Tang, H. Zirin, Ai Guoxiang.
Astrophys. J., Vol. 380, No. 1, p. 282 – 286 (10 Oct 1991). With plates 1 – 11.
The active region of 1989 March, NOAA AR 5395, was the most prolific flare – producing region in recent history. The region produced some 200 flares, of which 50 were class M flares and 10 were class X. The authors present here the results of joint observations of this region at the Big Bear and Huairou Solar Observatories. While some cloudy periods occurred, the region was observed every day from limb to limb for significant periods, and nine of the ten class X flares were recorded. The salient characteristic of this region is that it was a great δ group, dominated by spots of following (*f*) polarity, which moved rapidly westward, producing large changes in magnetic structure which increased the shear and led to great flares. Aside from its great size, the region was unusual in that normally *p* spots dominate and move westward. In this case there was a 4:1 flux imbalance; 80% of the flux measured was of *f* polarity. The new phenomena observed were the following: (1) The major **f** spot in the region moved with a near constant acceleration, eventually reaching 0.25 km s^{-1}. While there is no obvious connection between the acceleration rate and flare occurrence, there is sufficient kinetic energy to power the flares. (2) Small *p* and *f* spots moved out from either side of the large **f** spot, and curled around it in curved trajectories. (3) The moving penumbral material coalesced into new umbrae.

072.121 **Studies on a very flare–active δ group: peculiar δ spot evolution and inferred subsurface magnetic rope structure.**
K. Tanaka.
Sol. Phys., Vol. 136, No. 1, p. 133 – 149 (Nov 1991).
The complex subsurface magnetic rope structure of a very flare–active isolated δ group (McMath 13043, July 1974) is studied by means of high–resolution evolutionary data from BBSO magnetic and velocity data. This group showed unusually fast evolution accompanied by a number of intense flares occurring on the neutral line of a δ spot, and provided an excellent opportunity to study the inherent relation of flare occurrence to changes of the magnetic configuration. The author first examines the abnormal evolution of this group started by formation of a large, compact, reversed δ spot by squeezing of multipoles. The δ configuration was deformed by penetration into the opposite polarity umbra and its subsequent disappearance, decaying by rapid shear motions. Strong transverse fields over 4000 G were detected in the penumbrae and some umbral components. Combining these data with the August 1972 region, the evolution of these isolated δ groups is shown to decompose into two flare–associated elementary modes: (A) shearing produced by spot growth and (B) reduction of shear as spots disappear. The author proposes a model of an emerging twisted magnetic knot to explain the two modes and applies realistically to the present evolution. The inferred magnetic topological structure of this region consists of tightly twisted (sheet–like) knots and a long–winding twisted rope with an internally reversed loop and a hooked bottom structure. Their consecutive emergences are suggested to explain the abnormal evolution of this δ group. This result indicates that the origin of the concentrated flare activity in these isolated δ groups may be traced to internal magnetic activity responsible for forming anomalous magnetic ropes.

072.122 **Diagnostic features of sunspot regions.**
R. N. Singh.
Sol. Phys., Vol. 136, No. 1, p. 191 – 193 (Nov 1991). Letter–to–the–editor.
Five–minutes p–mode oscillations are heavily attenuated in the active sunspot region. A comparative study of wave modes and luminosity variations outside and inside the sunspot region is found to depict certain diagnostic features of sunspot regions.

072.123 **Study of the morphological correlation between photospheric transversal magnetic field and penumbral filaments in AR5395.**
Hong Qinfang, Ding Youji.
Publ. Yunnan Obs., No. 4, p. 58 – 60 (Dec 1990). – See Abstr. 012.047 for the main entry.

072.124 **Nonfacular solar luminosity variations.**
J. R. Kuhn, K. G. Libbrecht.
Astrophys. J., Lett., Vol. 381, No. 1, p. L35 – L37 (1 Nov 1991).
Solar limb photometry data obtained in 1990 shows a decline in the active region facular contribution to the solar irradiance from the previous year – in accord with the maximum in the sunspot cycle occurring in late 1989. In contrast, the nonfacular and nonsunspot contribution to the solar flux increased compared to 1989. This third component of the irradiance variation is consistent with a significant increase in the total mean solar irradiance from the value observed near the time of sunspot maximum. Evidently the irradiance maximum, like other solar activity indices, exhibits a phase lag with respect to the solar activity maximum as measured by the sunspot number.

072.125 **A possible cause for umbral oscillations.**
H. S. Yun, J. W. Lee, Y.–J. Moon.
Proc. Astron. Soc. Aust., Vol. 9, No. 1, p. 153 – 154 (1991). – See Abstr. 012.078 for the main entry.
An attempt has been made to examine the nature of 3–minute umbral oscillations in order to identify their origin. For this purpose the authors have calculated, on the basis of Musielak and Rosner (1987), the fast–mode energy spectra generated in a typical sunspot convection zone (Yun 1968). The computed energy spectrum is fed into the lower boundary of the subphotospheric resonant cavity to examine the characteristics of the wave propagation through the SS umbral atmosphere (Avrett 1981). The upward velocity spectra computed at various heights are presented and their behaviour compared with observations.

072.126 **An explanation of the dimensions of penumbral filaments.**
Zhou Daoqi.
Chin. Astron. Astrophys., Vol. 15, No. 4, p. 453 – 460 (Dec 1991). English translation of Acta Astrophys. Sin., Vol. 11, No. 3, p. 275 – 282 (Jul 1991).
Unstable perturbation modes exist in the magnetic field of a penumbral electric current and the author suggests the penumbral filaments are formed from the development of such modes. Under the short wave approximation the non–adiabatic dispersion equation is solved in the radial and transverse directions of the sunspot. From the condition of instability the length and width of the penumbral filament can be evaluated and it is found that the filament mode is static in the direction of the length and is non–moving in the direction of the width, that the penumbral filaments are a feature of the sunspot magnetic flow under gravity and that the presence of the filaments implies the existence of a twisted magnetic field.

072.127 **Provisional sunspot–numbers for June 1991.**
Yamamoto Circ., No. 2169, p. 2 (13 Jul 1991). In Japanese.
Sunspot numbers for July – November 1991 are given in Nos. 2170 – 2174.

072.128 **Activity development in the area connected with the 23.08.88 proton flare.**
A. B. Delone, N. A. Myshinskikh.
Tr. Gos. Astron. Inst. im. Shternberga, Tom 62, p. 182 – 187 (1991). In Russian.
Chromospheric activity in groups 5115 + 5122 and 5117 from 19.8. to 26.8.88 was studied using Hα filtergrams. Interaction of areas involved in the big proton flares of 23.8.88 was studied. X–ray pictures of these areas were obtained on 26.8.1988 by the "Phobos" spacecraft.

072.129 **Sunspot number variations: stochastic or chaotic.**
G. E. Morfill, H. Scheingraber, W. Voges, C. P. Sonett.
The Sun in time, p. 30 – 58 (1991). – See Abstr. 003.027 for the main entry.
The authors analyze the sunspot record over time scales of weeks or months, which gives information about the magnetic field transport in the solar convection zone and, possibly, also information about mechanisms responsible for the production and annihilation of sunspots. The analysis used is new: the authors employ the standard Fourier or power spectral analysis and supplement this with a correlation integral analysis. Problems exist due to the superposition of stochastic processes and the 11–yr (dynamo associated) sunspot cycle. There is some "memory" in the sunspot cycle that correlates the observed short–term fluctuations over a characteristic time period. This time period has been found to be ~ 50 days. In conjunction with numerical experiments on Boussinesq convection, this suggests an association with giant convection cells as the main driving process for the "memory" as well as the number of Fourier modes. The observed phenomenon that occasionally new sunspots appear in the same sites where others have just dissipated, is presumably related to this association.

072.130 **An emerging electric current/magnetic system with delta structure in AR2372.**
Ding Youji, Hong Qinfang.
Acta Astron. Sin., Vol. 32, No. 3, p. 225 – 232 (Sep 1991). In Chinese. With one plate. English translation in Chin. Astron. Astrophys., Vol. 16, No. 1, p. 60 – 70 (Jan – Mar 1992).
The emerging characteristics of a current/magnetic flux system with delta structure near the neutral line of the active region AR2372 in early April 1980 are described and explained with a simple MHD theory.

072.131 **The solar activity cycle.**
D. M. Rabin, C. R. DeVorc, N. R. Sheeley Jr., K. L. Harvey, J. T. Hoeksema.
Solar interior and atmosphere, p. 781 – 843 (1991). – See Abstr. 003.028 for the main entry.
Study of the solar cycle is entering a new era dominated by objective, precise measurements of magnetic, velocity and radiation fields over the surface of the Sun. This review emphasizes observations of photospheric magnetic flux during cycle 21 (1976 – 1986) and how these measurements have been used to model the cyclic variability of the heliospheric magnetic field. Indices of solar activity are discussed in terms of their potential to figure in theoretical or empirical models. Other recent data, such as measurements of large–scale surface flows and information on the Sun's internal rotation from helioseismology, as well as the magnetic flux observations, are considered in the context of Babcock's phenomenological model of the solar cycle: can this model still serve? Is there anything better to replace it?

072.132 **Active regions, sunspots and their magnetic fields.**
M. Semel, Z. Mouradian, I. Soru–Escaut, P. Maltby, D. Rees, M. Makita, T. Sakurai.
Solar interior and atmosphere, p. 844 – 889 (1991). – See Abstr. 003.028 for the main entry.
Surface magnetism is the progenitor of active regions, sunspots, and all related phenomena. This cause and effect is reversible so that, using well–established empirical laws, the

presence and morphology of photospheric magnetic fields can be deduced from active–region light emission structure. In the (simplifying) case of sunspots, MHD and thermodynamic theory find some success in the interpretation of the interaction of magnetic fields and solar plasma. Coronal magnetic fields also appear to be predictable by extrapolation techniques starting from the photospheric conditions. Alternatively, surface magnetism can be observed "directly" by means of the spectroscopic Zeeman effect and Stokes polarimetry. Eventually these empirical, theoretical and direct–measurement techniques must converge to identical results as we better understand the physics of active regions.

072.133 **Monitoring the sunspots of cycle 22.**
L. Enright.
J. R. Astron. Soc. Can., Vol. 85, No. 4, p. 173 (Aug 1991). Abstract. – See Abstr. 012.090 for the main entry.

072.134 **Vector magnetic fields in sunspots. I. Stokes profile analysis using the Marshall Space Flight Center magnetograph.**
K. S. Balasubramaniam, E. A. West.
Astrophys. J., Vol. 382, No. 2, p. 699 – 705 (1 Dec 1991). With plates 18 – 23.

The Marshall Space Flight Center vector magnetograph is a tunable filter magnetograph with a bandpass of 125 mÅ. The authors present results of the inversion of Stokes polarization profiles observed with this instrument centered on a sunspot to recover the vector magnetic field parameters and thermodynamic parameters of the spectral line forming region using the Fe I λ5250.2 spectral line using a nonlinear least–squares fitting technique. As a preliminary investigation, the authors also show that the recovered thermodynamic parameters could be better understood if one breaks down the fitted parameters like Doppler width, opacity ratio, and damping constant into more basic quantities like temperature, microturbulent velocity, or density parameter.

072.135 **Le nombre de Wolf. Suivi de l'activité solaire de 1984 à 1991.**
C. Courdurie.
Obs. Trav., No. 28, p. 5 – 11 (1991).

072.136 **Plasma motion and electric currents in an active region.**
V. I. Abramenko, S. I. Gopasyuk, M. B. Ogir'.
Bull. Crimean Astrophys. Obs., Vol. 81, p. 1 – 5 (1990). English translation of Izv. Krym. Astrofiz. Obs., Tom 81, p. 3 – 8 (1990). See Abstr. 53.072.026.

Studies have been made on the correlation between sunspot motion and transverse magnetic–field vortex structure for electric currents for two active regions. The vertical current component is represented as a sum of terms, one of which is due to rotation of the potential transverse–field vector and the other to change in the length of that vector. Details of the current components and that vortex structure give a complete picture for the plasma motion. Motion at an angle to the magnetic field predominates in an active region, while much less often there is motion along the lines of the transverse field. There are two global plasma vortices, one of which rotates clockwise and covers the leading part of an active region, while the other is counterclockwise and covers the entire tail part. That rotation extends to a considerable depth in the solar atmosphere (from the photosphere to the base of the sunspots) and leads to field spiraling, i.e., a global current.

072.137 **He I 10830 Å observations of active regions.**
E. Hiei, K. Ichimoto, G. Fang.
International SOLAR–A Science Meeting: Flare physics in solar activity maximum 22, p. 67 – 70 (1991). – See Abstr. 012.104 for the main entry.

A number of He I 10830 Å spectroheliograms were obtained with a 25–cm aperture coronagraph at the Norikura Solar Observatory. It is found: 1) a close relationship between the steady features in He I 10830 Å intensity, velocity fields, and the magnetic fields, 2) darkenings of points in active regions with a time scale of 10 – 20 minutes, and 3) various manifestations of flares in He I 10830 Å. The He I 10830 Å line will give sensitive diagnostics of chromospheric and coronal active phenomena.

072.138 **Solar Active Longitude recurring every 28 – 29 days acting differently in low and higher latitudes.**
V. Bumba.
Bull. Astron. Inst. Czech., Vol. 42, No. 6, p. 381 – 385 (Nov 1991).

The author discusses briefly the existence of Active Longitudes recurring every 28 – 29 days. These A. Ls. develop at higher heliophysical latitudes ($\varphi = \pm(20° - 40°)$) in the same way as the 27–day A. Ls. in the equatorial belt ($\varphi = \pm 20°$), but in this equatorial belt the slowly rotating A.L. acts as a trigger mechanism of greater activity, especially when it crosses the 27–day A.L.

072.139 **Proton flare and terrestrial effects with aurorae in the minimum phase of solar activity (Jan 1985).**
L. Křivský, M. A. Soliman, S. Šimberová, J. Flusser.
Bull. Astron. Inst. Czech., Vol. 42, No. 6, p. 386 – 389 (Nov 1991).

In the minimum phase between cycles, on 19 Jan 1985, a new spot group started and developed very quickly. On 21 Jan. 23 52 UT, a proton flare with position S 8°, W 38° occurred. This flare was connected and followed by some terrestrial effects (Forbush–Decrease, geomagnetic storm and polar aurorae). The proton flare commenced after the preceding sheer trend of flare activity, when the magnetic type of sunspot group reached class δ and both polarities of spots were compensated in their intensities. This event confirms and increases the number of earlier established features of active regions with proton flares, which play an important role in predictions. Digital image processing has been developed to improve the quality of the faint aurora.

072.140 **A simultaneous solar observing project.**
G. Garcia.
Strolling Astron., Vol. 35, No. 3, p. 124 – 126 (Sep 1991).

072.141 **A.L.P.O. Solar Section observations for rotations 1829 – 1832 (1990 May 15 to Sep 01).**
R. E. Hill.
Strolling Astron., Vol. 35, No. 3, p. 127 – 133 (Sep 1991).

072.142 **A.L.P.O. Solar Section observations for rotations 1833 – 1836 (1990 Sep 01 to Dec 19).**
R. E. Hill.
Strolling Astron., Vol. 35, No. 4, p. 168 – 174 (Dec 1991).

072.143 **Evolution of sunspots seen in molecular lines. I.**
K. Sinha, B. M. Tripathi.
Bull. Astron. Soc. India, Vol. 19, Nos. 1 – 2, p. 13 – 22 (Mar – Jun 1991).

With a view to assess the role of changing magnetic field strengths on line intensities, the results of equivalent width calculations of some lines of C_2, MgH and TiO molecules are presented here for standard photospheric and sunspot model atmospheres.

072.144 **Evolution of sunspots seen in molecular lines. II.**
K. Sinha, B. M. Tripathi.
Bull. Astron. Soc. India, Vol. 19, Nos. 1 – 2, p. 23 – 36 (Mar – Jun 1991).

Utilizing the model atmospheres proposed by Sobotka for umbrae of different sizes and those proposed by Maltby et al. for well–developed umbrae, at different phases of solar activity, the authors examine the molecules C_2, MgH and TiO for a study of sunspot evolution.

072.145 **Ein dreidimensionales Klassifizierungssystem für Sonnenflecken: die McIntosh–Klassifikation.**
R. E. Hill.
Sonne, Jahrg. 15, Nr. 60, p. 121 – 123 (Dec 1991).

072.146 **Axial tilt angles of sunspot groups.**
R. F. Howard.
Sol. Phys., Vol. 136, No. 2, p. 251 - 262 (Dec 1991).

Digitized Mount Wilson sunspot data from 1917 to 1985 are analyzed to examine tilt angles determined from the area-weighted positions of leading and following sunspots. These spot group tilt angles are examined in relation to other group characteristics to give information which may relate to the formation and evolution of sunspot groups and the magnetic connection of groups to subsurface magnetic flux tubes. The average tilt angle of all 24816 (multiple-spot) group observations in this study is found to be $+4.2 \pm 0.2$ deg, where the positive sign signifies that the leading spots lie equatorward of the following spots. Average tilt angles are found to be larger at higher latitudes, confirming earlier results. There is a strong negative correlation between average daily latitudinal motion (plus to poles) and group tilt angle. These results are discussed.

072.147 **Sunspots.**
H. U. Schmidt.
Geophys. Astrophys. Fluid Dyn., Vol. 62, No. 1-4, p. 249 - 270 (Dec 1991). - See Abstr. 012.126 for the main entry.

Some examples of research on structure and formation of sunspots are briefly recollected in historical sequence. They relate to many facets of sunspots, first: magnetic inhibition of convection, the conjecture of a flat penumbra, the stratification beneath the umbra, the observable magnetic profile, the Evershed effect as syphon flow, the concept of a magnetopause; next: cooling by Alfvén waves, evolution and stability, the "bright ring", the observed change of umbra brightness with the phase of the sunspot cycle, the hypothetical cluster of separate flux strands underneath the umbra, the profile of the magnetopause, the structure of the penumbra and the inclination of its field and finally: the concept of a deep penumbra with volume currents, exchange convection and the concept of a second current sheet separating umbra and penumbra.

072.148 **Einige Besonderheiten der Sonnenaktivität im Jahre 1989.**
H. Aurass, H. W. Urbarz.
Sterne, Band 67, Heft 5, p. 255 - 263 (1991).

072.149 **Un antico incostante sole. I.**
G. Poletto.
Orione, Vol. 11, N. 3, p. 24 - 30 (May - Jun 1991).

072.150 **Un antico incostante sole. II.**
G. Poletto.
Orione, Vol. 11, N. 4, p. 22 - 28 (Jul - Aug 1991).

The UVSP data archive.
See Abstr. 002.042.

Observing the Sun.
See Abstr. 003.026.

Sunspots: the historical background.
See Abstr. 004.062.

L'activité solaire et l'activité aurorale au XVIIe siècle.
See Abstr. 004.078.

The National Symposium of the Dynamics of Solar Active Regions, Dali (China), 5 - 11 Nov 1990.
See Abstr. 011.012.

CAS Symposium No. 119: The MHD of solar active regions and redshift precursor of flares, Xiamen (China), 4 - 8 Jan 1991.
See Abstr. 011.014.

From the solar core to the Earth's magnetosphere. Extended abstracts of reports. All-Union Meeting on the Quasibiennial Solar Variations, Pushchino (USSR), 25 - 27 Feb 1991.
See Abstr. 012.021.

22 MAX. Global character research of solar-terrestrial system. Special issue III. Proceedings. Workshop on major solar-geophysical events in March and August 1989, Beijing (China), 23 - 26 Apr 1990.
See Abstr. 012.047.

Solar research and programs of the DFG at Observatorio del Teide.
See Abstr. 013.077.

Our nearest star.
See Abstr. 014.006.

Through rose coloured glasses.
See Abstr. 014.014.

Following the Sun.
See Abstr. 014.016.

Prediction of solar activity with a neural network and its effect on orbit prediction.
See Abstr. 021.027.

Spectrograph distortion in sunspot line profiles.
See Abstr. 034.025.

A method for the automatic observation of solar active regions and the programming.
See Abstr. 036.088.

Multiple scattering of acoustic waves by a pair of uniformly magnetized flux tubes.
See Abstr. 062.158.

Magnetoconvection.
See Abstr. 062.256.

A model of SVC radiation in magnetic dipole field and radiation varying with magnetic field. I. Active region model and SVC radiation source model.
See Abstr. 063.111.

On the dynamics of granulation in active regions and the heating problem.
See Abstr. 071.008.

Geometric line elevation in solar limb faculae.
See Abstr. 071.021.

Granular and intergranular line profiles in solar active and quiet regions.
See Abstr. 071.022.

Structure elements of the photosphere and their variations with the solar activity cycle.
See Abstr. 071.025.

The He I λ10830 Å line as an indicator of the chromospheric and coronal activity of the Sun.
See Abstr. 071.037.

The variation of the solar granulation structure in active and non-active regions.
See Abstr. 071.043.

The height dependence of velocity - intensity fluctuations and several non-dimensional parameters in the solar photosphere.
See Abstr. 071.047.

Multicolor continuum analysis of the solar granulation in quiet and active regions.
See Abstr. 071.053.

Chromospheric response across an active region to impulsive energy releases during a two–ribbon subflare.
See Abstr. 073.029.

Statistics on the evolution of solar active regions for flare forecasting.
See Abstr. 073.071.

Spatial distribution of polar faculae in their 22–year cycle.
See Abstr. 073.078.

Evidence for the interaction of large scale magnetic structures in solar flares.
See Abstr. 073.109.

Chromospheric observation and preliminary analysis of active region AR5395 at Urumqi Astronomical Station.
See Abstr. 073.119.

The two types of flares in the active region AR 5395.
See Abstr. 073.120.

AR 5395: observations on the asymmetries of the Hα and Ca II K line profiles of the flares on March 9 and 14, 1989.
See Abstr. 073.121.

The material ejections associated with flares in AR 5395.
See Abstr. 073.122.

AR 5395: a motion of the dark matter in 2N flare on 9 March 1989.
See Abstr. 073.123.

AR5395: 3B flare on March 15, 1989.
See Abstr. 073.125.

AR5395 solar flare in relation to the longitudinal current density in the photosphere.
See Abstr. 073.127.

A microflare–related activation of a filament observed in Hα and C IV lines.
See Abstr. 073.128.

Fine structure of solar filaments. II. Dynamics of threads and footpoints.
See Abstr. 073.129.

Flare activity and evolution within AR5629.
See Abstr. 073.130.

Chromospheric observation and preliminary analysis of active region AR5629 at Urumqi Astronomical Station.
See Abstr. 073.131.

AR5629: a large loop–shape flare on 17 Aug. 1989.
See Abstr. 073.136.

AR5629: the observations of the flare loop of Aug. 17, 1989.
See Abstr. 073.137.

LDE flares in the 20th solar cycle. I. Comparison of the time behaviour of H–alpha grouped and LDE flares.
See Abstr. 073.142.

LDE flares in the 21st solar cycle (1976–1986). I. Comparison of the time occurrences of H–alpha and LDE flares.
See Abstr. 073.143.

Investigation of fine structure in active formations on the Sun.
See Abstr. 073.156.

Relative abundances in the lower solar transition region.
See Abstr. 073.161.

Refining flare types from current patterns.
See Abstr. 073.174.

Coronal dynamics of a quiescent active region.
See Abstr. 074.026.

The green coronal line above an active region at the solar limb on 5 Aug. of 1989.
See Abstr. 074.076.

Acceleration of electrons outside flares: coronal manifestation and possible origin.
See Abstr. 074.083.

Coronal mass ejections and streamers associated with the new cycle active regions at solar minimum.
See Abstr. 074.092.

Avalanches and the distribution of solar flares.
See Abstr. 074.099.

Coronal structures and the sunspot cycle.
See Abstr. 074.122.

Activity associated with coronal mass ejections at solar minimum: SMM observations from 1984 – 1986.
See Abstr. 074.133.

Das Sonnenobservatorium Einsteinturm in Potsdam. Erforschung solarer Magnetfelder und der Physik von Sonnenflecken.
See Abstr. 075.001.

Sunspot umbrae: observed correlation between magnetic field and temperature.
See Abstr. 075.007.

Nonpotential models of active region magnetic fields.
See Abstr. 075.013.

Three–dimensional stability of a model sunspot field.
See Abstr. 075.016.

Elementary bipoles of active regions and ephemeral active regions.
See Abstr. 075.022.

Dynamics of the latitude structure of the solar large–scale magnetic field at separate longitude intervals during 1915 – 1965.
See Abstr. 075.025.

A model of quasi–open magnetic field line region.
See Abstr. 075.028.

The large–scale magnetic field in the global solar cycle: observational aspects.
See Abstr. 075.039.

Evolution of large and small scale magnetic fields in the Sun.
See Abstr. 075.040.

The longer term evolution of magnetic field and mass flow in a decaying active region.
See Abstr. 075.042.

The merging of magnetic fluxes of similar polarity in solar active regions and the energetic flares.
See Abstr. 075.045.

Intrinsic magnetic discontinuities and solar X–ray emission.
See Abstr. 075.048.

The magnetic fields of active regions. VIII. Cycle latitude variations.
See Abstr. 075.049.

Vector magnetic field measurements and penumbral structure.
See Abstr. 075.052.

Evolution of vector magnetic field and velocity field in active region 5395.
See Abstr. 075.054.

The evolution characters of magnetic field and velocity field in solar active region AR5629.
See Abstr. 075.056.

Results of magnetographic line–ratio measurements.
See Abstr. 075.058.

Fine structures of chromospheric magnetic field and material flow in a solar active region.
See Abstr. 075.071.

A comparison of SERTS EUV spectroheliograms with solar active region and quiet sun magnetic field structure.
See Abstr. 076.010.

A search for hard X–ray microflares near solar minimum.
See Abstr. 076.020.

A comparison of theoretical Fe XII emission line strengths with EUV observations of a solar active region.
See Abstr. 076.023.

Theoretical analysis of iron X–ray spectra observed in solar active region.
See Abstr. 076.025.

Search for a 152 – 158 days periodicity in the occurrence rate of solar flares inferred from spectral data of radio bursts.
See Abstr. 077.002.

On the polarization of microwave emission from active regions: results from CoMStOC.
See Abstr. 077.015.

VLA observations during the ESP MAX 91 Campaign.
See Abstr. 077.019.

On the emission of local radio sources at the wavelengths 9.1 and 21 cm during the declining phase of the solar cycle.
See Abstr. 077.029.

Spectral peculiarities of local radio sources at decimetric wavelengths during the declining phase of the solar cycle.
See Abstr. 077.030.

Some statistical features of solar radio SVC in the rising branch of the cycle 22.
See Abstr. 077.040.

Observation of solar radio emission at 3.2 cm wavelength at UAS (*Urumqi Astronomical Station*).
See Abstr. 077.049.

The characteristics of the optics and radio emission of the active region 5395 of March, 1989.
See Abstr. 077.050.

Characteristics of radio emission of AR 5395.
See Abstr. 077.051.

Radio events generated in especially large active region AR 5395.
See Abstr. 077.052.

Microwave observations in active region 5395.
See Abstr. 077.053.

Microwave observation highlights of AR5629.
See Abstr. 077.055.

Large–scale features of the Sun at 20 centimeter wavelength.
See Abstr. 077.060.

Some statistical features of solar radio SVC in the rising branch of cycle 22.
See Abstr. 077.067.

Radiation characteristics of solar proton events during the decline of solar cycle 21 and the rise of cycle 22.
See Abstr. 078.002.

Complexes of activity and flares.
See Abstr. 078.008.

Some studies of solar proton events in relation to active region characteristics.
See Abstr. 078.017.

Magnetic flux transport and the Sun's dipole moment: new twists to the Babcock–Leighton model.
See Abstr. 080.014.

How deeply might sunspots and supergranules be anchored within the Sun?
See Abstr. 080.024.

Analysis of p–modes in a sunspot umbra.
See Abstr. 080.032.

The role of slow mode waves in p–mode absorption by sunspots.
See Abstr. 080.033.

Total irradiance and the solar cycle.
See Abstr. 080.037.

Measurements of facular contrast in the near ultraviolet.
See Abstr. 080.042.

A possible mechanism of energy transport of biennial oscillations in the Sun.
See Abstr. 080.056.

Seismology for the fine structure in the Sun's oscillations varying with its activity cycle.
See Abstr. 080.057.

The confrontation of mean–field theories with numerical simulations.
See Abstr. 080.061.

Waves of solar activity.
See Abstr. 080.066.

Solar rotation over solar cycle 21.
See Abstr. 080.073.

Dynamics of spot groups and rotation of the external convective layers in the Sun and magnetically active stars.
See Abstr. 080.095.

Solar irradiance variability from modern measurements.
See Abstr. 080.115.

Large–scale velocity fields.
See Abstr. 080.134.

A torsional oscillation in the rotation of the solar magnetic field.
See Abstr. 080.140.

Do changes in the photospheric magnetic network cause the 11 year variation of total solar irradiance?
See Abstr. 080.141.

A new solar cycle model including meridional circulation.
See Abstr. 080.153.

Simulation of the effect of active regions on intermediate–degree solar oscillations.
See Abstr. 080.158.

Waves in the solar atmosphere.
See Abstr. 080.166.

Variations in the Sun's radiative output.
See Abstr. 080.172.

Geomagnetic and solar data. March 1991.
See Abstr. 084.047.

Is the recurrent enhanced geomagnetic activity preceding the sunspot minimum related to magnetic fields of the new or old activity cycle?
See Abstr. 084.053.

Geomagnetic and solar data. February 1991.
See Abstr. 084.063.

Geomagnetic and solar data. April 1991.
See Abstr. 084.069.

Geomagnetic and solar data. May 1991.
See Abstr. 084.070.

22. Sonnenfleckenmaximum in Beziehung zur Polarlichtverzögerung.
See Abstr. 084.072.

Geomagnetic and solar data. June – August 1991.
See Abstr. 084.076.

Reconnection in substorms.
See Abstr. 084.089.

Length of the solar cycle: an indicator of solar activity closely associated with climate.
See Abstr. 085.016.

Could the sun be warming the climate?
See Abstr. 085.017.

Hemispherical asymmetries in sunspot areas and auroral frequencies.
See Abstr. 085.029.

Solar flares and terrestrial impacts of active region 5395.
See Abstr. 085.033.

Evolution of AR5395 and time sequence of solar–terrestrial events associated with AR5395.
See Abstr. 085.034.

Solar–terrestrial effects of AR 5395.
See Abstr. 085.035.

The lower atmospheric response to solar activity of March 1989.
See Abstr. 085.037.

Comparative terrestrial planet thermospheres. 1. Solar cycle variation of global mean temperatures.
See Abstr. 093.045.

The effect of the solar cycle on the maintenance of the nightside ionosphere of Venus.
See Abstr. 093.047.

Superthermal >36–eV ions observed in the near–tail region of Venus by the Pioneer Venus orbiter neutral mass spectrometer.
See Abstr. 093.053.

Solar cycle effects on the structure of the electron density profiles in the dayside ionosphere of Venus.
See Abstr. 093.077.

Solar activity influence upon the brightness curves of comet Churyumov–Gerasimenko (1982 VIII) and comet Halley (1986 III).
See Abstr. 103.028.

Pre–and main–sequence evolution of solar activity.
See Abstr. 107.020.

The luminosity variability of solar–type stars.
See Abstr. 115.033.

The past, present and future of solar magnetism: stellar magnetic activity.
See Abstr. 116.086.

Cosmic ray modulation during solar cycle 22.
See Abstr. 144.017.

The role of cross helicity in solar modulation of cosmic rays and its charge sign dependence.
See Abstr. 144.085.

The physics of cosmic ray modulation: heliospheric propagation during the 1987 minimum.
See Abstr. 144.087.

The effect of the changing polarity and neutral sheet of the IMF on the cosmic ray diurnal anisotropy at neutron monitor energies.
See Abstr. 144.101.

Time variations of ^{10}Be and solar activity.
See Abstr. 144.105.

Solar and terrestrial components of the atmospheric ^{14}C variation spectrum.
See Abstr. 144.106.

Nature of solar–cycle and heliomagnetic–polarity dependence of cosmic rays, inferred from their correlation with heliomagnetic spherical surface harmonics in the period 1976–1985.
See Abstr. 144.107.

Bi–annual variations in solar activity and cosmic rays.
See Abstr. 144.119.

073 Chromosphere, Flares, Prominences

073.001 **Short-term predictions of solar flares.**
V. A. Burov.
Geomagn. Aeron., Vol. 30, No. 4, p. 455 - 460 (Feb 1990). Paper presented at the 5. Symposium on Solar Terrestrial Physics, Samarkand (USSR), 2 - 6 Oct 1989. English translation of Geomagn. Aehron., Tom 30, No. 4, p. 538 - 544 (1990).

A review of present-day theoretical investigations of the problem of the accumulation and release of energy in solar flares permits advancing the opinion that only individual flare events are described by a concrete model and that a single model alone does not describe the entire diversity of flares. Consideration of the observational data does not permit claiming the existence of a single universal mechanism known today of flare events. It appears possible to treat the problem of prediction in terms of the algebra of logic (Boolean logic) and to compare the truth table with the often-used contingency table. The introduction of a number of very general assumptions permits forming a general approach to the development of predictive schemes and selection of the individual elements of the models and informative criteria. Experimental results are given on the testing of some prediction procedures. The author's procedure of routine short-term prediction of flares on the basis of the methods of instruction on pattern recognition implemented in the form of a set of programs is outlined. The results of the application of this procedure in 1986 - 1988 are presented.

073.002 **The nature of velocity fields in quiescent prominences.**
K. R. Bondal, M. C. Pande.
Astrophys. Space Sci., Vol. 181, No. 1, p. 161 - 163 (Jul 1991). Letter-to-the-editor.

It is shown that for certain definite conditions of symmetry imposed on the permitting magnetic field geometry for an isothermal case in Kippenhahn and Schlüter's (1957) model of a quiescent prominence, any irrotational velocity field would quickly get converted to rotational.

073.003 **Stokes-V in an active prominence.**
E. Wiehr, G. Stellmacher.
Astron. Astrophys., Vol. 247, No. 2, p. 379 - 382 (Jul 1991).

Stokes-V spectra of the Ca^{+}8542 emission were taken with a CCD in an active prominence showing extremely high line-of-sight velocities up to 25 km/s. The spatial resolution allows detecting prominence fine-structures as small as 1.1 arcsec. Exact field strengths were determined from the measured steepness of the individual emission line flanks. Values between 75 and 180 Gs are obtained. No indication is found for a dependence of the field strength on the spatial resolution or on the line-of-sight velocity, or on the emission brightness.

073.004 **Interrelation of soft and hard X-ray emissions during solar flares. I. Observations.**
R. M. Winglee, A. L. Kiplinger, D. M. Zarro, G. A. Dulk, J. R. Lemen.
Astrophys. J., Vol. 375, No. 1, p. 366 - 381 (1 Jul 1991).

Simultaneous observations of hard X-rays and of soft X-ray line emissions at high time resolution ($\cong 6$ s) from the SMM are used to determine the interrelation between the acceleration and heating of electrons and ions during impulsive solar flares. The study involves five flares (three disk and two limb flares). For all the flares, the hard X-rays have a power-law spectrum which breaks down during the rise phase and beginning of the decay phase, after which the spectrum changes to either a single power law or a power law which breaks up at high energies. The characteristics of the soft X-rays depend on the flare position. For the three disk flares, the line emissions show a blueshifted component (arising from upward acceleration of ions into the corona) with the maximum inferred velocity reaching its peak value during the decay phase. For limb flares the soft X-ray line emissions show only a nonthermally broadened stationary component (arising from acceleration and heating perpendicular to the magnetic field); this broadening reaches a maximum at the hard X-ray peak rather than during the decay phase. These observations are not consistent with models where the chromospheric outflow is driven by explosive heating since both the blueshifts and broadening are predicted to have the same dependence on the hard X-ray flux. It is argued instead that small-scale quasi-static electric fields are important for determining the acceleration of the X-ray-producing electrons and the outflowing chromospheric ions.

073.005 **Interrelation of soft and hard X-ray emissions during solar flares. II. Simulation model.**
R. M. Winglee, G. A. Dulk, P. L. Bornmann, J. C. Brown.
Astrophys. J., Vol. 375, No. 1, p. 382 - 403 (1 Jul 1991).

Recent observations indicated that quasi-static electric fields may be important in the acceleration of the electrons responsible for hard X-ray emissions during impulsive solar flares. Quasi-static fields are also required to drive and to close the currents carried by the accelerated electrons. Two-dimensional electrostatic particle simulations are presented which incorporate the effects of these fields on the particle dynamics as well as those arising from wave - particle interactions induced by the accelerated particles. In the model, the particle acceleration is initiated by a cross-field current in the corona. Because of the limited cross-field mobility of particles in a collisionless plasma, this current can only be closed in the chromosphere where cross-field transport is aided by collisions. As a result, strong quasi-static electric fields are set up which produce strong downward acceleration of electrons in the primary current region. On adjacent field lines, electrons are accelerated up into the corona to provide a return current. The induced perpendicular currents are initially provided by the ions, but, as the chromosphere becomes heated by precipitation of energetic electrons, perpendicular electron currents become important, and these currents in turn modify the particle acceleration in the latter part of the flare. The calculated properties of the soft and hard X-ray and microwave emissions from such a system are compared in detail with the observations presented in a companion paper.

073.006 **A thermal/nonthermal approach to solar flares.**
S. G. Benka.
National Aeronautics and Space Administration, Washington, DC (United States). Scientific and Technical Information Div.; National Aeronautics and Space Administration, Greenbelt, MD (United States)., NASA-TM—4323. 108 p. (Dec 1991).

A new approach for modeling solar flare high energy emissions is developed in which both thermal and nonthermal particles coexist and contribute to the radiation. The thermal/nonthermal distribution function is interpreted physically by postulating the existence of DC current sheets in the flare region. These current sheets then provide both primary plasma heating through Joule dissipation, and runaway electron acceleration. The physics of runaway acceleration is discussed. Theoretical hard X-ray spectra are calculated. A simple model for hard X-ray images of two-ribbon flares is presented. Theoretical microwave gyrosynchrotron spectra are calculated and analyzed, uncovering important new effects caused by the interplay of thermal and nonthermal particles. The theoretical spectra are compared with observed high resolution spectra of solar flares, and excellent agreement is found, in both hard X-rays and microwaves.

073.007 **On electron acceleration in current sheets of solar flares.**
Yu. Eh. Litvinenko, B. V. Somov.
Pis'ma Astron. Zh., Tom 17, No. 9, p. 835 - 840 (Sep 1991). In Russian. English translation in Sov. Astron. Lett., Vol. 17, No. 5.

A model is presented for electron acceleration in high-temperature turbulent current sheets. In this model different processes of fast particles generation in solar flares have a common cause, namely, magnetic energy dissipation in a region of magnetic reconnection. The spectral index $\gamma = 2 - 5$ and the total number of electrons accelerated during the main phase of a flare are found to correspond to data of observations.

073.008 **Energetics and dynamics in an explosive flare of March 1989.**
D. M. Zarro, J.–P. Wülser, R. C. Canfield.
Bull. Am. Astron. Soc., Vol. 23, No. 2, p. 924 (1991). Abstract. – See Abstr. 012.008 for the main entry.

073.009 **Simultaneous UV and X–ray observations of solar microflares.**
J. G. Porter, J. M. Fontenla, R. L. Moore, G. M. Simnett.
Bull. Am. Astron. Soc., Vol. 23, No. 2, p. 935 (1991). Abstract. – See Abstr. 012.008 for the main entry.

073.010 **Solar chromospheric filling factors and contrast.**
R. A. Donahue, S. R. Restaino.
Bull. Am. Astron. Soc., Vol. 23, No. 2, p. 942 (1991). Abstract. – See Abstr. 012.008 for the main entry.

073.011 **A two–dimensional model for a solar prominence: effect of an external magnetic field.**
R. Oliver, J. L. Ballester, E. R. Priest.
Sol. Phys., Vol. 134, No. 1, p. 123 – 144 (Jul 1991).
Using analytical approximations the authors study the effects of different external magnetic configurations on the half–width, mass, and internal magnetic structure of a quiescent solar prominence, modelled as a thin vertical sheet of cool plasma. The authors build up a zeroth–order model and analyse the effects produced by a potential coronal field or a constant–α force–free field. This model allows to obtain the half–width and mass of the prominence for different values of the external field, pressure and shear angle. Then the effects of these external magnetic configurations on a two–dimensional model proposed by Ballester and Priest (1987) are studied. The main effects are a change of the half–width with height, an increase of the mass, a decrease of the magnetic field strength with height and a change in the shape of the magnetic field lines.

073.012 **Flaring arches. III. The subflare of June 27, 1980, and its related extended arch.**
J. M. Fontenla, Z. Švestka, F. Fárník, F. Y. Tang.
Sol. Phys., Vol. 134, No. 1, p. 145 – 169 (Jul 1991).
The authors show detailed observations in X–rays, UV lines, and Hα of an extended arch, about 300000 km long, which developed as a consequence of a compact subflare. This subflare occurred in an "included" magnetic polarity of relatively low magnetic field strength (compared to that of the sunspots). The apparition of this big arch was preceeded by that of a smaller arch, about 30000 km long, which masked the polarity inversion line filament in the early phase of the subflare. The big arch which developed later, around the time of the main X–ray and UV spike of the subflare, connected the included polarity and the main leading sunspot of the region, and became fully developed in a few minutes. The fact that both arches were simultaneously observed in all spectral domains as well as their fine structure in Hα can only be explained by considering the arch as composed of several unresolved portions of material having widely different temperatures. The Hα observations can be interpreted as showing the appearance of this cool material as a result of condensation, but a more appealing interpretation is that there was almost simultaneous ejection of superhot (10^7K), hot (10^6K), mild (10^5K), anc cool (10^4K) material from the subflare site along previously existing magnetic tubes of much lower density. The termination of the subflare was marked by a rather hard X–ray and UV spike which appeared to originate in a different structure than that of the main spike. The material in the arch gradually cooled and drained down after the end of the subflare.

073.013 **Nonthermal velocities of Na I, Ba II, Sr II atoms in the lower chromosphere from eclipse observations.**
I. L. Belkina, N. P. Dyatel, G. P. Marchenko.
Kinematics Phys. Celest. Bodies, Vol. 7, No. 1, p. 39 – 44 (1991). English translation of Kinematika Fiz. Nebesn. Tel, Tom 7, No. 1, p. 43 – 48 (1991).
Intensities obtained for the resonant doublet D_1, D_2 during the solar eclipse of 10 July 1972 are used to determine the nonthermal velocities of Na I atoms in active and quite chromospheres. The average nonthermal velocity v_t is about 5 km/sec at heights of 1200 – 1600 km in undisturbed regions. It decreases to 4 km/sec in active regions. It is found from eclipse data for the lines of the Ba II and Sr II resonant doublets that v_t = 3 km/sec at heights of 900 – 1200 km (in the Ba II lines) and v_t = 6 km/sec at heights of 1200 – 1700 km (in the Sr II lines). Introduction of a radial (flow) velocity $V_T \approx 1.5$ km/sec for heights above 1200 km results in similar values of the microturbulent velocities for all lines: v_t = 4 km/sec for Sr II and Na I and v_t = 3 km/sec for Ba II. Therefore the lines of the Na I, Ba II, and Sr II resonant doublets are formed at heights of 1200 – 1700 km in regions with low nonthermal velocities (v_t = 3 – 4 km/sec, $V_T \approx 1.5$ km/sec).

073.014 **Vertical velocities and oscillations in quiescent filaments.**
Yi Zhang, O. Engvold.
Sol. Phys., Vol. 134, No. 2, p. 275 – 286 (Aug 1991).
Analysis of He I λ10830 Å spectral observations of a large, quiescent filament reveals a pronounced oscillatory behaviour of the vertical mass motion. The filament is situated in a quiet region more than 15° away from the nearest active region. It is concluded that the magnetic field of the quiescent filament, which occurs in the form of long thin flux ropes, moves with the gas and that there is no net mass flow perpendicular to the most frequently observed horizontal field lines. The oscillatory motion is accompanied by phase dependent variation of the He I line intensity which could possibly imply wave induced compression of the plasma.

073.015 **High–energy flare explosions driven by 3–dimensional X–type current loop coalescence.**
J.–i. Sakai, C. de Jager.
Sol. Phys., Vol. 134, No. 2, p. 329 – 352 (Aug 1991).
The authors present a model for high–energy solar flare explosions driven by 3–dimensional X–type current loop coalescence. The 3–dimensional X–type current loop coalescence, where two crossed flux–tubes interact at one point, is a fundamentally new process as compared to the 1–D and 2–D cases studied earlier. This process is studied by a first–order approach of the relevant variables near the point of coalescence; it appears to yield reliable information in a sufficiently large area around this point. It is shown that, following a strong plasma collapse due to the pinch effect, a point–like plasma explosion can be driven while fast magnetosonic shock waves can also be excited. The authors found that the conditions in the area producing the remarkable flare bursts of 21 May, 1984 were indeed such that the many flare spikes could have been due to 3–D explosive X–type current loop coalescence. The authors also show, by studying the conditions of shock formation in a gamma ray flare, that the time delay of γ–rays from the impulsive phase could be the time needed for the shock formation in the flaring region. Some general conclusions are drawn on the question why certain flares do emit γ–rays in the MeV energy range, and why other, apparently important and energetic flares, do not.

073.016 **Three solar filament disappearances associated with interplanetary low–energy particle events.**
B. Sanahuja, A. M. Heras, V. Domingo, J. A. Joselyn.
Sol. Phys., Vol. 134, No. 2, p. 379 – 394 (Aug 1991).
Three low–energy particle events (35 – 1600 keV) associated with interplanetary shocks, detected at 1 AU by ISEE–3, have been identified as originating in solar disappearing filaments instead of large flares. The observational characteristics of these non–flare generated events are similar to the ones of the other eleven events already known (i.e., absence of type II or IV bursts, weak X–ray emission, Hα brightening in the surroundings of the

filament disappearance, frequent presence of a double–ribbon event, slow propagation of the generated interplanetary shock, lack of shock deceleration).

073.017 **The effects of mass flow on the temperature and abundance structure of the solar transition region.**
D. T. Woods, T. E. Holzer.
Astrophys. J., Vol. 375, No. 2, p. 800 – 817 (10 Jul 1991).
The authors have computed the density and temperature structure of a multicomponent plasma consisting of electrons, protons, ionized helium, and a trace of minor ion species (carbon). The equations of force and energy balance for this model are developed allowing for the fact that species may have different temperatures. The authors have solved these equations using an electron temperature profile which is taken from inversion of the observed emission measure curve. For the case of downflows, the minor ion temperature can be significantly hotter than the electron temperature, and significant abundance enhancements are possible due to the slow–down of the minor species from the effect of the thermal force. The minor ion temperature enhancements relative to the electron temperature can have significant implications for the computation of detailed line profiles. In addition, the effect of the thermal force may be of importance to studies of the emission from asymmetrically heated loops in which the minor ion abundance will be enhanced in the downflowing leg and diminished in the upward flowing leg. This would tend to make the downflowing side brighter in a given line, and therefore contribute to the production of a net redshift in a spatially unresolved loop.

073.018 **K_{2V} cell grains and chromospheric heating.**
R. J. Rutten, H. Uitenbroek.
International Conference on Mechanisms of Chromospheric and Coronal Heating, p. 48 – 53 (1991). – See Abstr. 012.011 for the main entry.
The appearance of bright Ca II K_{2V} grains in the interiors of network cells on the solar surface is an intriguing phenomenon, rich in observational characteristics and possibly important in the energy budget of the quiet chromosphere. Unfortunately, their nature has eluded identification so far. The authors discuss various scenarios and list observables which may verify or discard these, and present predictions.

073.019 **Wave heating in chromospheric bright points.**
W. Kalkofen.
International Conference on Mechanisms of Chromospheric and Coronal Heating, p. 54 – 59 (1991). – See Abstr. 012.011 for the main entry.
Wave heating in the interior of supergranulation cells is reviewed and it is suggested that the bright points may be heated by pulses and by oscillations in their wakes with periods near 3 min, the acoustic cutoff period in the upper photosphere. Such waves create propagation channels by the amplification of vertically travelling waves and by focussing for wave periods near the cutoff period. The middle chromosphere is heated by shock dissipation, but the low chromosphere may partly be heated also by radiative energy transport down from the higher layers.

073.020 **The effects of electron scattering on the Si II 1816 line in the solar chromosphere.**
K. E. Rangarajan, D. M. Rao.
International Conference on Mechanisms of Chromospheric and Coronal Heating, p. 63 – 65 (1991). – See Abstr. 012.011 for the main entry.
The authors have considered the formation of Si II 1816 line in the solar chromosphere. The VAL model has been adopted as the input for the electron density and temperature. These preliminary calculations show that there are measurable changes in the profile shape of this line due to electron scattering.

073.021 **HRTS time series observations: chromospheric and coronal heating.**
Cheng Chung–chieh.
International Conference on Mechanisms of Chromospheric and Coronal Heating, p. 77 – 82 (1991). – See Abstr. 012.011 for the main entry.
The authors have studied HRTS time series observations of solar network boundaries, cell interiors, sunspot and active regions. The results show that the intensities of C IV, Si II, Fe II and UV continuum in these features fluctuate constantly, sometimes with large amplitudes of a factor of 2 in 30 sec. Downflow velocities are most common in the transition region. The only place where there are noticeable upflows is in the cell interior. Often neighboring emission features show temporal changes independent of each other. The observations do not show evidence of regular periodic waves. However, there is evidence that temporal variations in the temperature minimum region, the chromosphere, and the transition region follow each other, indicating some "excitation" agent propagating at a speed > 500 km sec^{-1}. The heating occurs irregularly in time, perhaps to energy releases in locally stressed magnetic fields.

073.022 **High spatial resolution observations of the solar transition region: spicules and microflares.**
J. W. Cook.
International Conference on Mechanisms of Chromospheric and Coronal Heating, p. 83 – 96 (1991). – See Abstr. 012.011 for the main entry.
The author reviews recent observations from the solar transition region, obtained with high spatial resolution, of spicules and of high velocity explosive events which may be a candidate for microflares. Microflares are a proposed mechanism for heating the transition region and corona from a large number of small spatial scale events with a rapid energy release.

073.023 **New models of the chromosphere and transition region.**
E. H. Avrett.
International Conference on Mechanisms of Chromospheric and Coronal Heating, p. 97 – 99 (1991). – See Abstr. 012.011 for the main entry.
The semiempirical quiet–Sun models of Vernazza, Avrett, and Loeser included a temperature plateau near 20000K in the transition region to account for the observed emission in the hydrogen Lyman α line. New energy–balance calculations by Fontenla, Avrett, and Loeser of the transition region including the effects of particle diffusion account for Lyman α emission without such a temperature plateau. The author found that the upward diffusion of neutral hydrogen causes the Lyman α line to be formed near $T = 40000$K rather than near 20000K as implied by optically–thin calculations and by his earlier semiempirical modeling based on local statistical equilibrium. New versions of the VAL component models, A, C, and F are shown.

073.024 **The role of particle diffusion in the lower transition region: revised interpretation of emission measures.**
E. H. Avrett, J. M. Fontenla.
International Conference on Mechanisms of Chromospheric and Coronal Heating, p. 100 – 102 (1991). – See Abstr. 012.011 for the main entry.
The authors have calculated energy–balance models of the solar chromosphere – corona transition region for $T < 10^5$K including the effects of particle diffusion. They show how the diffusion of atoms and ions affects the neutral and ionized hydrogen and helium number densities vs. temperature in the transition region. The results suggest that the use of emission measure techniques that ignore the effects of diffusion may be unreliable.

073.025 **Why heating is not necessary in the transition region or upper chromosphere.**
P. S. Cally.
International Conference on Mechanisms of Chromospheric and Coronal Heating, p. 103 – 108 (1991). – See Abstr. 012.011 for the main entry.
EuV observations allow to determine the temperature gradient in the transition region. Assuming the usual Spitzer thermal

conductivity, the heat flux may then be estimated. This is found to be three orders of magnitude too small to supply the radiative losses from the upper chromosphere. However, there are theoretical and observational reasons for believing that the lower transition region is substantially turbulent. Introducing a turbulent conductivity into simple models produces excellent agreement with the observed TR emission measures. Furthermore, it enhances the flux to the level where it may now easily supply the upper chromospheric losses.

073.026 **Height-dependent solar plage temperature distribution.**
A. Kucera, Z. Scherbakova (*Z. A. Shcherbakova*), E. Baranovsky (*Eh. A. Baranovskij*).
International Conference on Mechanisms of Chromospheric and Coronal Heating, p. 109 – 112 (1991). – See Abstr. 012.011 for the main entry.

The authors present a temperature model of solar plages based on simultaneous observations of UV, visible and IR spectral lines, both in plages and the nearby quiet chromosphere. The calculations are carried out with an atomic model consisting of nine H, six Ca II and thirteen He I levels. The authors established temperature distributions with a sharp increase on the height 1200 km. Results are compared with the two-component models of Ayres et al. (1986).

073.027 **Diagnostics of energy transport during the impulsive phase of the solar flare.**
Li Peng.
Bull. Am. Astron. Soc., Vol. 23, No. 2, p. 1025 (1991). Abstract. – See Abstr. 012.012 for the main entry.

073.028 **Energy release and transport in long-duration events.**
K. T. Strong, J. Hawley.
Bull. Am. Astron. Soc., Vol. 23, No. 2, p. 1025 – 1026 (1991). Abstract. – See Abstr. 012.012 for the main entry.

073.029 **Chromospheric response across an active region to impulsive energy releases during a two-ribbon subflare.**
V. Gaizauskas, M. Proulx.
Bull. Am. Astron. Soc., Vol. 23, No. 2, p. 1026 (1991). Abstract. – See Abstr. 012.012 for the main entry.

073.030 **The prime energy release of a solar flare.**
G. E. Brueckner, D. Moses, J. W. Cook, K. P. Dere, C. M. Korendyke, D. G. Socker, J.-D. F. Bartoe.
Bull. Am. Astron. Soc., Vol. 23, No. 2, p. 1026 (1991). Abstract. – See Abstr. 012.012 for the main entry.

073.031 **Characteristics of white light flares as derived from their observed center-to-limb distribution.**
D. F. Neidig, P. H. Wiborg, L. B. Gilliam.
Bull. Am. Astron. Soc., Vol. 23, No. 2, p. 1026 – 1027 (1991). Abstract. – See Abstr. 012.012 for the main entry.

073.032 **High-temperature flares observed in broadband soft X-rays.**
H. A. Garcia, P. S. McIntosh.
Bull. Am. Astron. Soc., Vol. 23, No. 2, p. 1027 (1991). Abstract. – See Abstr. 012.012 for the main entry.

073.033 **Nonthermal velocities in the solar transition region.**
K. P. Dere, H. E. Mason.
Bull. Am. Astron. Soc., Vol. 23, No. 2, p. 1028 (1991). Abstract. – See Abstr. 012.012 for the main entry.

073.034 **Cell center bright points as dynamic phenomena: search for a chromospheric velocity signature.**
J. W. Cook.
Bull. Am. Astron. Soc., Vol. 23, No. 2, p. 1028 (1991). Abstract. – See Abstr. 012.012 for the main entry.

073.035 **Observational evidence for EUV loops within the network.**
J. F. Dowdy Jr.
Bull. Am. Astron. Soc., Vol. 23, No. 2, p. 1028 (1991). Abstract. – See Abstr. 012.012 for the main entry.

073.036 **A search for polarization in Ellerman bombs.**
D. M. Rust, S. L. Keil.
Bull. Am. Astron. Soc., Vol. 23, No. 2, p. 1029 (1991). Abstract. – See Abstr. 012.012 for the main entry.

073.037 **First observations of macrospicules at 4.8 GHz at the solar limb in polar coronal holes.**
S. R. Habbal, R. D. Gonzalez.
Astrophys. J., Lett., Vol. 376, No. 1, p. L25 – L27 (20 Jul 1991). With plates L2 – L4.

The authors report on the first radio observations of macrospicules in solar coronal holes. These were made with the VLA at 4.8, 8.5, and 15 GHz. The macrospicules appeared most prominently at 4.8 GHz. The brightness temperatures measured in the macrospicules at these different wavelengths yield an empirical model consisting of a cool 8×10^3K core surrounded by a thin hotter sheath at $1 - 2 \times 10^5$K, with the density varying from 10^{10}cm^{-3} at the core to 2×10^9cm^{-3} at the outer edge. These characteristics are almost identical to those derived from EUV observations more than a decade ago. A novel characteristic of macrospicules revealed by the 4.8 GHz emission is the pinching off and ballooning at the top of the magnetic structure of the macrospicules. There is also evidence in these radio observations that this pinching off leads to the escape of some plasma from the location of the macrospicules. These observations point to the importance of microwave emission for exploring the properties of the magnetic structure of the solar limb.

073.038 **Numerical studies of atmospheric dynamics driven by energy deposition in the chromosphere.**
A. C. Sterling, J. T. Mariska, K. Shibata.
Bull. Am. Astron. Soc., Vol. 23, No. 2, p. 1029 (1991). Abstract. – See Abstr. 012.012 for the main entry.

073.039 **Diffusion of helium in the solar transition region.**
E. H. Avrett, J. M. Fontenla.
Bull. Am. Astron. Soc., Vol. 23, No. 2, p. 1029 (1991). Abstract. – See Abstr. 012.012 for the main entry.

073.040 **The generation of the chromospheric 3 min oscillation.**
W. Kalkofen, P. Rossi, G. Bodo, S. Massaglia.
Bull. Am. Astron. Soc., Vol. 23, No. 2, p. 1033 (1991). Abstract. – See Abstr. 012.012 for the main entry.

073.041 **A search for solar oscillations in the solar chromosphere using millimeter interferometric observations.**
M. R. Kundu, S. M. White.
Bull. Am. Astron. Soc., Vol. 23, No. 2, p. 1033 – 1034 (1991). Abstract. – See Abstr. 012.012 for the main entry.

073.042 **A catastrophe model for eruptive flares and prominences.**
P. A. Isenberg, T. G. Forbes.
Bull. Am. Astron. Soc., Vol. 23, No. 2, p. 1036 (1991). Abstract. – See Abstr. 012.012 for the main entry.

073.043 **Martens-Kuin models of normal and inverse polarity filament eruptions and coronal mass ejections.**
E. Hildner, D. F. Smith, R. S. Steinolfson.
Bull. Am. Astron. Soc., Vol. 23, No. 2, p. 1036 (1991). Abstract. – See Abstr. 012.012 for the main entry.

073.044 **Alfvénically-driven slow shocks in the solar chromosphere and corona.**
J. V. Hollweg.
Bull. Am. Astron. Soc., Vol. 23, No. 2, p. 1037 (1991). Abstract. – See Abstr. 012.012 for the main entry.

073.045 **Solar chromospheric heating.**
S. D. Jordan.
Bull. Am. Astron. Soc., Vol. 23, No. 2, p. 1037 (1991). Abstract. – See Abstr. 012.012 for the main entry.

073.046 **Shock amplification by radiation.**
M. Carlsson, R. Stein.
International Conference on Mechanisms of Chromospheric and Coronal Heating, p. 366 – 368 (1991). – See Abstr. 012.011 for the main entry.
The authors present self-consistent radiative hydrodynamic calculations of acoustic waves in the solar chromosphere. In addition to well known effects like radiative damping of waves, there is an amplification of the shock strength in the upper chromosphere due to different time-scales for radiative ionization and recombination. The radiative recombination lags behind in the front part of the shock leading to radiative heating. The effect gives a sharper, larger amplitude, shock.

073.047 **Irrotational motion in quiescent prominences.**
M. C. Pande, K. R. Bondal.
Astrophys. Space Sci., Vol. 182, No. 2, p. 201 – 206 (Aug 1991).
On the basis of Kippenhahn and Schlüter's magnetohydrostatic model of a quiescent prominence, an attempt has been made to study the effect of irrotational motion existing in the prominences on the magnetic field pattern in it, introducing an irrotational velocity field. It is found that, under such a condition, the magnetic field geometry in the model does not change.

073.048 **Ion heating by nonlinear Landau damping of Alfvén waves in solar flares.**
J. A. Miller.
Bull. Am. Astron. Soc., Vol. 23, No. 2, p. 1043 (1991). Abstract. – See Abstr. 012.012 for the main entry.

073.049 **Stochastic acceleration of electrons in solar flares in the presence of Coulomb collisions.**
R. J. Hamilton, V. Petrosian.
Bull. Am. Astron. Soc., Vol. 23, No. 2, p. 1043 – 1044 (1991). Abstract. – See Abstr. 012.012 for the main entry.

073.050 **Models fit to the "electron dominated" flares of March 6, 1989.**
V. Petrosian, J. M. McTiernan.
Bull. Am. Astron. Soc., Vol. 23, No. 2, p. 1044 (1991). Abstract. – See Abstr. 012.012 for the main entry.

073.051 **Derivation of parent electron spetra from bremsstrahlung hard X-ray spectra.**
C. M. Johns, R. P. Lin.
Bull. Am. Astron. Soc., Vol. 23, No. 2, p. 1044 (1991). Abstract. – See Abstr. 012.012 for the main entry.

073.052 **Avalanches and the distribution of solar flares.**
E. T. Lu, R. J. Hamilton.
Bull. Am. Astron. Soc., Vol. 23, No. 2, p. 1044 (1991). Abstract. – See Abstr. 012.012 for the main entry.

073.053 **Return currents in a nonuniform flare loop.**
A. G. Emslie.
Bull. Am. Astron. Soc., Vol. 23, No. 2, p. 1044 (1991). Abstract. – See Abstr. 012.012 for the main entry.

073.054 **The thermal and plasma-physical evolution of current sheets formed in the solar atmosphere by emerging flux.**
T. N. LaRosa.
Bull. Am. Astron. Soc., Vol. 23, No. 2, p. 1044 (1991). Abstract. – See Abstr. 012.012 for the main entry.

073.055 **Long-period oscillations of the chromospheric network.**
B. W. Lites, W. Kalkofen, R. J. Rutten.
Bull. Am. Astron. Soc., Vol. 23, No. 2, p. 1050 (1991). Abstract. – See Abstr. 012.012 for the main entry.

073.056 **The K line of Ca II in chromospheric bright points.**
P. Rossi, W. Kalkofen, H. Uitenbroek, G. Bodo, S. Massaglia.
Bull. Am. Astron. Soc., Vol. 23, No. 2, p. 1050 (1991). Abstract. – See Abstr. 012.012 for the main entry.

073.057 **Magnetoacoustic heating of the solar chromosphere.**
S. M. Chitre, J. M. Davila.
International Conference on Mechanisms of Chromospheric and Coronal Heating, p. 402 – 404 (1991). – See Abstr. 012.011 for the main entry.
In the present work the authors wish to examine the viability of the acoustic waves generated in the solar convection zone propagating radially outward through the overlying atmosphere and getting resonantly absorbed by the magnetic arches of the low-lying chromospheric canopy. They demonstrate that the resultant Poynting and acoustic flux that enters the magnetic canopy in the network regions is adequate to explain the energy losses from the chromospheric layer of the Sun.

073.058 **Heating of the solar chromosphere by MHD-waves.**
R. Erdélyi, M. Marik.
International Conference on Mechanisms of Chromospheric and Coronal Heating, p. 420 – 422 (1991). – See Abstr. 012.011 for the main entry.
The authors calculate a semi-theoretical model of the lower part of the chromosphere with the aid of the Vernazza et al. (1981) emission coefficients. The equality of the radiative emission coefficient and the absorption coefficient of weak MHD shock waves is assumed. The results show an increase of temperature with height in close agreement with observations and other models. For the period of the shocks about 45 s is found.

073.059 **Solar "hot spots".**
G. Paris, T. Bai, P. S. McIntosh.
Bull. Am. Astron. Soc., Vol. 23, No. 2, p. 1053 (1991). Abstract. – See Abstr. 012.012 for the main entry.

073.060 **The formation and support of prominences in magnetic loops and arcades.**
G. Van Hoven, Y. Mok, J. F. Drake.
Bull. Am. Astron. Soc., Vol. 23, No. 2, p. 1059 (1991). Abstract. – See Abstr. 012.012 for the main entry.

073.061 **Eruption of solar prominences.**
E. Hameiri, A. A. Blank.
Bull. Am. Astron. Soc., Vol. 23, No. 2, p. 1059 (1991). Abstract. – See Abstr. 012.012 for the main entry.

073.062 **Dynamical evidence of fine structure in the solar transition region.**
P. Brekke, O. Kjeldseth-Moe, J.-D. F. Bartoe, G. E. Brueckner.
Bull. Am. Astron. Soc., Vol. 23, No. 2, p. 1060 (1991). Abstract. – See Abstr. 012.012 for the main entry.

073.063 **Solar Doppler shift measurements in the Ne VII 465 Å emission line.**
J. T. Mariska, J. F. Dowdy.
Bull. Am. Astron. Soc., Vol. 23, No. 2, p. 1060 (1991). Abstract. – See Abstr. 012.012 for the main entry.

073.064 **First observations of macrospicules at 4.8 GHz at the solar limb in polar coronal holes.**
S. R. Habbal, R. D. Gonzalez.
Bull. Am. Astron. Soc., Vol. 23, No. 2, p. 1062 (1991). Abstract. – See Abstr. 012.012 for the main entry.

073.065 **A comparison of hard X-ray and high-time-resolution H-alpha data for a flare on 1989 June 30 at 1737 UT.**
L. E. Orwig, A. L. Kiplinger, G. J. Labow.
Bull. Am. Astron. Soc., Vol. 23, No. 2, p. 1064 (1991). Abstract. – See Abstr. 012.012 for the main entry.

073.066 **Differential emission–measure variations and the "Neupert effect".**
H. S. Hudson.
Bull. Am. Astron. Soc., Vol. 23, No. 2, p. 1064 (1991). Abstract. – See Abstr. 012.012 for the main entry.

073.067 **An interesting pair of homologous flares from region 5395 in March 1989.**
J. N. Kile, U. D. Desai.
Bull. Am. Astron. Soc., Vol. 23, No. 2, p. 1065 (1991). Abstract. – See Abstr. 012.012 for the main entry.

073.068 **A method for determining which flares are consistent with the thick–target model.**
B. R. Dennis, D. A. Batchelor, B. M. Uberall, D. M. Zarro.
Bull. Am. Astron. Soc., Vol. 23, No. 2, p. 1065 (1991). Abstract. – See Abstr. 012.012 for the main entry.

073.069 **Observations of two large off limb solar flares in the radio, X–rays, and Hα.**
T. A. Kucera, G. A. Dulk, R. M. Winglee, A. L. Kiplinger, T. S. Bastian, D. E. Gary.
Bull. Am. Astron. Soc., Vol. 23, No. 2, p. 1065 (1991). Abstract. – See Abstr. 012.012 for the main entry.

073.070 **The magnetic morphology of high–pressure plasmas in three October 1989 (AR 5747) flares.**
K. D. Leka, R. C. Canfield.
Bull. Am. Astron. Soc., Vol. 23, No. 2, p. 1066 (1991). Abstract. – See Abstr. 012.012 for the main entry.

073.071 **Statistics on the evolution of solar active regions for flare forecasting.**
P. L. Bornmann.
Bull. Am. Astron. Soc., Vol. 23, No. 2, p. 1066 – 1067 (1991). Abstract. – See Abstr. 012.012 for the main entry.

073.072 **High time resolution imaging of an impulsive flare at two wavelengths.**
A. L. Kiplinger, B. R. Dennis, L. Orwig, G. Labow, B. Runyan.
Bull. Am. Astron. Soc., Vol. 23, No. 2, p. 1071 (1991). Abstract. – See Abstr. 012.012 for the main entry.

073.073 **Millimeter imaging of energetic electrons in solar flares in conjunction with GRO experiments.**
M. R. Kundu, S. M. White.
Bull. Am. Astron. Soc., Vol. 23, No. 2, p. 1073 (1991). Abstract. – See Abstr. 012.012 for the main entry.

073.074 **Observations of linear Hα polarization in flares.**
T. R. Metcalf, J. P. Wülser, R. C. Canfield.
Bull. Am. Astron. Soc., Vol. 23, No. 2, p. 1073 (1991). Abstract. – See Abstr. 012.012 for the main entry.

073.075 **Numerical simulation of microflare evolution in the solar transition region and corona.**
A. C. Sterling, J. T. Mariska.
International Conference on Mechanisms of Chromospheric and Coronal Heating, p. 630 – 633 (1991). – See Abstr. 012.011 for the main entry.

A series of numerical simulations indicates that quasi–impulsive thermal energy releases of $\sim 10^{26}$ergs in a narrow region of the upper chromosphere of a quiet region coronal loop can provide a transient heating source for the corona. Lower energy events radiate away their energy too quickly to generate substantial temperature enhancements. The microflares seen in UV images may generate such energy releases, and may also be the source for some of the dynamic events seen in UV spectra.

073.076 **On some peculiarities of the flare of 23 August 1988.**
S. A. Yazev.
Soln. Dannye, Byull., No. 7/90, p. 77 – 81 (Jul 1990). In Russian.

An analysis is made of the preflare situation and of the flare of 23 August 1988. The flare was preceded by high activity in the filament region, and by the appearance of a dark filament–like structure which was later replaced by one of flaring ribbons. A weak separation of flare ribbons, the absence of H_α–observable loops connecting ribbons, and a manifestation of supergranulation cells in the structure of ribbons were observed. The observational data are briefly discussed; two possible alternative explanations for the role of a dark filament–like structure in the flare are proposed.

073.077 **About the depth of formation of white–light emission in solar flares.**
Eh. A. Baranovskij.
Soln. Dannye, Byull., No. 8/90, p. 79 – 84 (Aug 1990). In Russian.

A set of models for white–light flares was calculated with a thin layer (6 – 12 km) in the low chromosphere. It is shown that the specific shape of Hα and Hβ line profiles, which white–light flares often have – the presence of very extended wings, is well explained by the deep position of the dense emitting layer ($N = 10^{22}cm^{-2}$). The white–light emission for the model with $n_H = 10^{16}cm^{-3}$, T = 7300 – 7600K, $N = 10^{22}$ equals to about 12%. The emission consists of two parts – emission from the dense chromospheric layer and emission from the photosphere. The possibility of heating the chromosphere emitting layer by energetic electrons and by X–ray is discussed.

073.078 **Spatial distribution of polar faculae in their 22–year cycle.**
V. V. Makarova.
Soln. Dannye, Byull., No. 8/90, p. 84 – 88 (Aug 1990). In Russian.

Latitudinal and longitudinal distribution of polar faculae in the southern and northern hemisphere of the Sun has been considered for 1981 – 1988. It has been found that during the period in question there were more polar faculae in the northern hemisphere then in the southern hemisphere and this N–S asymmetry was very definite. The study of the longitudinal distribution of polar faculae shows that there is a stable E–W asymmetry of their annual number. A variation of the N–S and E–W asymmetries depends on the phase of the polar activity cycle. The reliability of some values of the asymmetry was tested using the χ^2 criterion.

073.079 **Magnetohydrodynamic turbulence dissipation and stochastic proton acceleration in solar flares.**
J. A. Miller.
Astrophys. J., Vol. 376, No. 1, p. 342 – 354 (20 Jul 1991).

Two MHD modes, Alfvén and fast magnetosonic waves, are capable of stochastically accelerating protons from super–Alfvénic to ultrarelativistic energies in solar flares. Acceleration occurs through the Landau resonance for fast magnetosonic waves, and through the cyclotron resonance for Alfvén waves. The Landau resonance, however, is also satisfied by thermal electrons in a flare plasma, resulting in most of the magnetosonic wave energy being dissipated on electron heating rather than on stochastic proton acceleration. In contrast, thermal electrons cannot satisfy the cyclotron resonance, leaving stochastic gyroresonant proton acceleration as the dominant linear wave–particle interaction for Alfvén waves. Alfvén waves are also subject to a nonlinear wave–particle interaction. Nonlinear Landau damping efficiently and selectively heats the ambient protons and can thus essentially preaccelerate a significant number of them out of a low–temperature distribution to super–Alfvénic speeds, where they can be subsequently accelerated to ultrarelativistic energies by gyroresonant interactions. A spectrum of Alfvén waves, therefore, can energize protons from low–temperature thermal to ultrarelativistic energies through a combination of nonlinear and linear wave–particle interactions.

073.080 **Observations from the solar maximum mission.**
K. T. Strong.
Royal Society Discussion Meeting: The physics of solar flares, p. 327 – 337 (1991). – See Abstr. 012.016 for the main entry.
SMM has not only produced a comprehensive database on a variety of active–Sun phenomena but also taught us how to coordinate observations with ground–based observatories and other spacecraft to produce a more holistic view of the complex energy release and transport processes. SMM had a unique combination of four imaging instruments which emphasized the spectroscopy of the solar atmosphere: the Hard X–ray Imaging Spectrometer, the Bent Crystal Spectrometer, the Flat Crystal Spectrometer, and the Ultraviolet Spectrometer/Polarimeter. To illustrate the power of these instruments, this paper describes the results of various studies of energy release, transport, and deposition phenomena during the onset of flares.

073.081 **Observations from the Hinotori mission.**
T. Sakurai.
Royal Society Discussion Meeting: The physics of solar flares, p. 339 – 347 (1991). – See Abstr. 012.016 for the main entry.
The satellite Hinotori was launched in 1981 by the Institute of Space and Astronautical Science of Japan. Two major experiments on board the Hinotori satellite were a hard X–ray imaging telescope with modulation collimators, and a high dispersion soft X–ray crystal spectrometer utilizing the Bragg diffraction of X–rays on quartz crystals. These two instruments have revealed for the first time that solar flares show varying characteristics depending on the environment of flaring regions, and that flares produce plasmas as hot as $3 - 4 \times 10^7$K.

073.082 **High–energy flare observations from the Solar Maximum Mission.**
W. T. Vestrand.
Royal Society Discussion Meeting: The physics of solar flares, p. 349 – 362 (1991). – See Abstr. 012.016 for the main entry.
The author reviews high–energy observations of solar flares with emphasis on measurements by instruments on the Solar Maximum Mission satellite. The observations show that, contrary to prior expectation, high–energy emission is a common property of solar flares. Direct interpretation of the measurements shows that both energetic ion and relativistic electron acceleration are ubiquitous flare phenomena and that those particles can be accelerated to high energies very rapidly. The author presents evidence that the high–energy emission from flares is anisotropic. He also reviews how studies of the nuclear line spectra have allowed derivation of characteristics of both the energetic ions and the target material. Altogether, these results clearly refute many of the old ideas about energetic flare particles and have stimulated new thinking about how particles are accelerated and transported during flares.

073.083 **The magnetohydrodynamics of energy release in solar flares.**
E. R. Priest.
Royal Society Discussion Meeting: The physics of solar flares, p. 363 – 380 (1991). – See Abstr. 012.016 for the main entry.
A large solar flare is thought to occur when a sheared magnetic arcade loses equilibrium or goes unstable and erupts, and drives magnetic reconnection in the stretched–out magnetic field lines. These two key processes of magnetic eruption and magnetic energy conversion by reconnection are reviewed briefly, with an account of recent analytical and numerical models. When the height or length of a prominence in a sheared coronal arcade is too great it may erupt and drive the formation and reconnection of a current sheet below it.

073.084 **Flare energy release: observational consequences and signatures.**
R. C. Canfield, J.-F. de La Beaujardiere, K. D. Leka.
Royal Society Discussion Meeting: The physics of solar flares, p. 381 – 388 (1991). – See Abstr. 012.016 for the main entry.
At Mees Solar Observatory a programme is underway of simultaneous polarimetric and spectroscopic observations that allow to explore the spatial relation between active region currents, flare particle acceleration and flare heating. The authors discuss several days observations of two flare–productive active regions. They compared the locations of vertical currents to the sites of non–thermal electron precipitation and high coronal pressure inferred from Hα line profiles and spectroheliograms. Without exception they found that the sites of significant energetic electron precipitation into the chromosphere were at the edges of regions of vertical current, not within them. In contrast, they found that the footpoints of high–pressure flare plasmas during the main phase of the observed flares all coincided very well with such currents.

073.085 **The role of magnetic loops in solar flares.**
R. Pallavicini.
Royal Society Discussion Meeting: The physics of solar flares, p. 389 – 400 (1991). – See Abstr. 012.016 for the main entry.
The inferred magnetic topology of solar flares is reviewed with emphasis on the implications for processes of energy release and transfer. It is shown that the spatial resolution of the observations obtained so far is still inadequate for solving many basic questions of solar flare research.

073.086 **Energetic particles in solar flares: theory and diagnostics.**
J. C. Brown.
Royal Society Discussion Meeting: The physics of solar flares, p. 413 – 424 (1991). – See Abstr. 012.016 for the main entry.
Recent progress and future prospects in diagnostics of energetic electrons and ions in the flares are reviewed, together with the roles they play in the flare as a whole. Most of the discussion centres on hard X–ray and gamma–ray and thermal plasma emission data, rather than on radio sources.

073.087 **Asymmetrical field of the Hα profile of the post–flare loop system of 1984 February 18 and the total radiation.**
Li Qiusha, Gu Xiaoma.
Publ. Yunnan Obs., No. 1, p. 7 – 14 (Mar 1991). In Chinese.
From the calculations of the asymmetrical factors of the Hα profile and the Hα total radiation of the post–flare loop system of Feb 18, 1984 carried out by means of the spectroheliographic data of the system, the asymmetrical field of the profile and the distribution of the Hα radiation of the loop prominence system are quantatively given.

073.088 **Flaring arch in AR5533 on 15 June, 1989.**
Di Luan, Li Zhikai, Kong Yongjin.
Publ. Yunnan Obs., No. 2, p. 10 – 13 (Jun 1991). In Chinese.
An event of flaring arch which occurred on June 15, 1989 in AR5533 is described, and a series of H_α and H_α off–band images of the arch which were obtained by 26 cm solar telescope at Yunnan Observatory are also given. Conclusion is made as follows: (1) The secondary footpoint of the event is 35000 km away from the primary footpoint. (2) The interval between increasing brightness of the small intense flare and emission mass reaching to the secondary footpoint is about 7 min. (3) The H_α emission material propagates along the arch with a mean speed of 130 km s^{-1}.

073.089 **The continuous emission spectrum of the flare of 1989 July 5.**
Zhong Shuhua, Cheng Xuekun, Luo Zhi.
Publ. Yunnan Obs., No. 2, p. 14 – 18 (Jun 1991). In Chinese.
The continuous emission spectra of H_α, H_β and H_γ wave bands of the solar flare of July 5, 1989 are given and analysed. Studies show that characteristics of spectra, time and configuration of the flare are similar to those of a white light flare and therefore the flare is very likely a white light one.

073.090 **Flare energy transport by conduction and radiation.**
M. E. Machado.
Royal Society Discussion Meeting: The physics of solar flares, p. 425 – 437 (1991). – See Abstr. 012.016 for the main entry.

As energy is released in coronal flare loops, strong temperature gradients may appear and large-scale flows can be set up. In the absence of, or in addition to, energetic particle beams, thermal conduction and mass motions are major processes that distribute energy throughout the flare atmosphere. The author reviews basic physical concepts of the energy transport processes, and their observational evidence. He then turns to radiative energy transport, an often ignored processes that may constitute a major transfer mechanism, particularly into and throughout the dense, optically thick, chromospheric and photospheric layers. The basic physics of radiative energy transport is reviewed, and some specific examples discussed in detail, particularly in connection with upper photosphere heating and white light flare events.

073.091 **Energetic particle production in flares.**
G. M. Simnett.
Royal Society Discussion Meeting: The physics of solar flares, p. 439 – 450 (1991). – See Abstr. 012.016 for the main entry.

Bursts of energetic ions and electrons are frequently seen in interplanetary space. Many events are correlated with major solar flares. Possible acceleration mechanisms are presented. It is argued that the particle production occurs following magnetic reconnection relatively high in the corona; the bulk of the energy goes into the ionic, rather than the electronic, component. The principal characteristics of energetic solar particle events are reviewed briefly. A model is outlined whereby acceleration of the electrons responsible for the impulsive hard X-ray burst occurs in the chromosphere as a result of the interaction of the primary ions. Observations supporting this hypothesis are discussed.

073.092 **High-temperature plasma in solar flares.**
G. A. Doschek.
Royal Society Discussion Meeting: The physics of solar flares, p. 451 – 460 (1991). – See Abstr. 012.016 for the main entry.

The thermal soft X-ray flare plasma is at temperatures ranging from a few million degrees up to about 40×10^6K. The author discusses some current problems in our understanding of the physical conditions in this plasma, particularly the ionization balance and the possible detection of non-thermal electrons. He discusses X-ray spectroscopic diagnostics that might help in resolving some of the issues.

073.093 **Spectroscopy of high-temperature solar flare plasmas.**
K. J. H. Phillips.
Royal Society Discussion Meeting: The physics of solar flares, p. 461 – 470 (1991). – See Abstr. 012.016 for the main entry.

Several examples of line ratios are given to deduce temperature and density conditions as well as element abundances. Analysis shows that there is a wide range of electron temperatures – generally from 2×10^6K to 20×10^6K – though sometimes even higher. Electron densities of around $10^{17} - 10^{18}$m^{-3} have been derived, higher values occurring at the flare peak or just before, and then declining. The physical conditions of the hot plasma are now precisely enough known from X-ray spectroscopy that models of flares which have been constructed in the past can be constrained.

073.094 **Spectrophotometric study of chromospheric inhomogeneities in an active region.**
Kh. I. Abdusamatov, M. N. Stoyanova.
Soln. Dannye, Byull., No. 4/91, p. 86 – 90 (Apr 1991). In Russian. With 1 plate.

A spectrophotometric study of absorption cells of the calcium chromosphere has been made on the basis of the high resolution spectra lines of H and K Ca II reduction. These spectra have been taken with the Pamir solar telescope. The results obtained enable the authors to make a conclusion that the intensity decrease, observed in some cells is due to a real decrease of the chromosphere's temperature inside these cells.

073.095 **Preliminary studies on the solar flare on Nov. 15, 1989.**
Jiang Zhibo.
Publ. Purple Mt. Obs., Vol. 10, No. 2, p. 105 – 108 (Jun 1991). In Chinese.

The class 2B flare observed on Nov 15, 1989 had following characteristics: (1) no preceeding dark filament in the active region; (2) the delta-type sunspots group was covered during the flare occurrence; (3) successive brightening of the second ribbon; (4) all types of radio bursts causing class III ionosphere disturbance. But neither energetic particle flux nor X-ray bursts were recorded corresponding to this event.

073.096 **The Lα (λ = 121.6 nm) solar plage contrasts calculations.**
E. A. Bruevich.
Soln. Dannye, Byull., No. 6/91, p. 104 – 106 (Jun 1991). In Russian.

The results of calculations of Lα plage contrasts based on experimental data are presented. A three-component model ideology of Lα solar flux using "Prognoz-10" and SME daily smoothed values of Lα solar flux are applied. The values of contrast are discussed and compared with experimental values based on "Skylab" data.

073.097 **The acceleration of electrons and ions in solar flares.**
V. K. Verma.
Astrophys. Space Sci., Vol. 183, No. 2, p. 317 – 321 (Sep 1991).

The author finds that gamma-ray line (GRL) emissions start later than the hard X-ray (HXR) emissions during impulsive and extended solar flares. Starting delay is more in the case of extended solar flares suggesting a slow acceleration of electrons and ions, in comparison to impulsive solar flares which indicate different acceleration mechanism for impulsive and extended solar flares. The author further infers that during solar flares, electrons and ions are accelerated simultaneously and the delay between HXR and GRL emissions results mainly due to differences in acceleration times of electrons and ions to attain energies required for producing HXR emissions for electrons and GRL emissions for ions. Therefore, he is of view that a single step acceleration mechanism may work in solar flares.

073.098 **Numerical classification of solar flares with one maximum of intensity using five optical variables.**
I. Ferro Ramos, G. A. Sosa.
Soln. Dannye, Byull., No. 6/91, p. 107 – 112 (Jun 1991). In Russian.

On the basis of five optical variables a partial numerical classification of solar flares is made. Ward's method with the use of Wishort's algorithm is used. The percentage of agreement, when the visual scale for brightness is included, is 80%, the classification permits to know directly the different ways in which the variables are combined in each class.

073.099 **Flares observed by the normal incidence X-ray telescope on 1989 September 11.**
M. Herant, F. Pardo, E. Spiller, L. Golub.
Astrophys. J., Vol. 376, No. 2, p. 797 – 802 (1 Aug 1991). With plates 8 – 10.

The authors report observations of two solar flares seen in soft X-rays during a sounding rocket flight of the Normal Incidence X-ray Telescope (NIXT) payload on 1989 September 11. The rocket was launched at the onset of a two-ribbon event. The flare in X-rays consists primarily of a single bright loop crossing the neutral line and having its footpoints at the southern ends of the ribbons; this loop accounts for more than two-thirds of the emission and is shown to have an electron density of $\sim 6 \times 10^{10}$cm^{-3} at a temperature of $\sim 6 \times 10^6$K. Within the remainder of each of the flare ribbons, a complicated coronal structure is observed, which is clearly interacting with the main flare loop. A second event in an active region at the limb began during the flight: its most striking feature is a strong correlation with Hα images taken at the same time, which indicates the coexistence of chromospheric and coronal temperature material in close proximity. The authors give possible interpretations of this phenomenon as either a sheath of X–ray emitting material surrounding the Hα ejection or as an admixture of hot and cool magnetic flux tubes, each ejecting the material which they contain.

073.100 **On solar prominences distribution according to the value of the angle between the prominence long axis and the line of sight.**
A. Sh. Akmamedova, M. G. Goshdzhanov, I.S. Kim.
Astron. Tsirk., No. 1543, p. 21 – 22 (Mar – Apr 1990). In Russian.

073.101 **Effects of Mg II and Ca II ionization on ab–initio solar chromosphere models.**
W. Rammacher, M. Cuntz.
Astron. Astrophys., Vol. 250, No. 1, p. 212 – 219 (Oct 1991).
The authors compute acoustically heated solar chromosphere models considering radiation damping by (non–LTE) emission from H^- and by Mg II and Ca II emission lines. The radiative transfer equations for the Mg II k and Ca II K emission lines are solved using the core–saturation method with complete redistribution. The Mg II k and Ca II K cooling rates are compared with the VAL model C. The authors include several substantial improvements over the work of Ulmschneider et al. (1987). First, they remove the dependency of the results from the core–wing separation points by improving the frequency integration method. Second, they treat the ionization equilibria of Mg II–III and Ca II–III in a non–LTE approximation using a two–level atom (instead of LTE). It is found that the rapid temperature rises caused by the ionization of Mg II are not formed in the middle chromosphere, but occur at larger atmospheric heights. The authors' models represent the temperature structure of the "real" solar chromosphere much better. This result is a major precondition for the study of ab–initio models for solar flux tubes based on MHD wave propagation and also for ab–initio models for the solar transition layer. The authors have checked the dependence of their results on the radiative transfer method by comparing their computations with those using the operator perturbation method of Kalkofen (1984) and those using the method of Carlsson (1986), which combines complete linearization techniques with approximate lambda operators.

073.102 **Two–dimensional models of the solar chromosphere. I. The Ca II K line as a diagnostic: 1.5–D radiative transfer.**
S. K. Solanki, O. Steiner, H. Uitenbroek.
Astron. Astrophys., Vol. 250, No. 1, p. 220 – 234 (Oct 1991).
Profiles of the Ca II K line are calculated for a 5 level atom and partial frequency redistribution along multiple rays (1.5–D) passing through two–dimensional (2–D) MHD models of magnetic flux tubes in the solar atmosphere. Temperature stratifications corresponding to various empirical chromospheric models are incorporated into the models. The influence of a number of model parameters on individual and spatially averaged line profiles is considered in detail. The profiles are also compared with observations. It is found that the present models can, at least qualitatively, reproduce a wide variety of observations if the temperature within the magnetic component lies between that of model F of Vernazza et al. (1981) and of model VALP of Ayres et al. (1986) and if the temperature in the non–magnetic part of the atmosphere corresponds to that of the COOLC model of Ayres et al.

073.103 **The quantitative research on the velocity field of a post–flare loop prominence system.**
Gu Xiaoma, Lin Jun, Li Qiusha.
Chin. J. Space Sci., Vol. 11, No. 4, p. 291 – 300 (Oct 1991). In Chinese.
The downward motion of the matter within the post–flare loop system of 1984 February 18 is studied under the joint action of solar gravity and the both gradients of dipole magnetic field pressure and atmospheric pressure. The two–dimensional line–of–sight velocity field is calculated by means of a numerical method under the assumption that the temperature keeps unchanged within the loop system and the loops are quasi–closed. In the system, some parameters as well as their distribution are derived from the curve fit of both velocity fields between theory and observation. The calculations indicate that the effects of density and magnetic field intensity within the loop system on the downward motion of matter is comparatively large while the effect of the temperature is much small.

073.104 **The 3–min oscillations of the solar chromosphere – a basic physical effect?**
B. Fleck, F. Schmitz.
Astron. Astrophys., Vol. 250, No. 1, p. 235 – 244 (Oct 1991).
The authors show that a fundamental physical effect which occurs in a plane atmosphere with a finite acoustic cut–off frequency provides a simple explanation of the 3–min oscillations of the solar chromosphere. Since the results are obtained on the assumption of an isothermal atmosphere it is concluded that there is no need for a chromospheric cavity as commonly assumed. The authors surmise that the 3–min oscillations are excited already at the base of the photosphere by long period disturbances like the 5–min oscillations.

073.105 **The effects of magnetic geometry on electron confinement in solar flares.**
K. G. McClements.
Vistas Astron., Vol. 34, Part 3, p. 325 – 330 (1991). – See Abstr. 012.032 for the main entry.
A Fokker–Planck code which includes the effects of a variable magnetic field is used to compute the time–averaged distribution of X–ray emitting electrons in a solar flare. Results are obtained for three different representations of the magnetic induction, each defining a loop structure in the solar corona. From these results are computed the rate at which electrons are lost from the system (via the footpoints of the loop), and the corresponding fraction of the total X–ray flux which will be produced by the loop footpoints. Generally speaking, the results are insensitive to the shape of the magnetic field, and correspond fairly closely with crude analytical estimates.

073.106 **Solar flare proton acceleration by Alfven turbulence.**
A. L. MacKinnon.
Vistas Astron., Vol. 34, Part 3, p. 331 – 342 (1991). – See Abstr. 012.032 for the main entry.
The acceleration of solar flare γ–ray producing protons by Alfven mode turbulence is addressed. The author improves on previous modelling of this process by allowing the turbulent energy density to decline as the proton energy density increases. Thus some discussion of the time–dependent behaviour of the accelerator is possible. The γ–ray line spectra, timing and absolute intensities place constraints on parameters of the accelerator. In particular, it is found that the acceleration region must comprise most of the flaring corona.

073.107 **Solar flares: a review.**
K. J. H. Phillips.
Vistas Astron., Vol. 34, Part 3, p. 353 – 365 (1991). – See Abstr. 012.032 for the main entry.
The past twenty years or so have seen an enormous increase in our understanding of solar flares. There has been much progress in their observation, both from ground–based telescopes and from spacecraft, and in theoretical interpretation. However, some very important questions remain unresolved, such as the nature of the primary energy release. The author reviews both observational and theoretical evidence and how this is being put together to form an acceptable picture of the phenomenon.

073.108 **Are the spicules a source of the coronal matter?**
R. A. Gulyaev.
Astron. Zh., Tom 68, Vyp. 5, p. 1108 – 1110 (Sep–Oct 1991). In Russian. English translation in Sov. Astron., Vol. 35, No. 5.
Measurements of the Doppler shifts of the H_α line and the EUV C I lines at the solar disk centre contradict the widely accepted picture of the ascending motion of the spicular matter with a velocity of about 25 km/s. The downward motions with a

speed of several km/s are rather predominant in the spicules, as well as in the structural elements of the chromosphere – corona transition region. Then, the spicules cannot be considered as a source of matter inflow into the corona. A certain role in the matter transfer into the corona may play chromospheric jets observed inside the cells of the chromospheric network.

073.109 **Evidcncc for the interaction of large scale magnetic structures in solar flares.**
C. H. Mandrini, P. Démoulin, J. C. Hénoux, M. E. Machado.
Astron. Astrophys., Vol. 250, No. 2, p. 541 – 547 (Oct 1991).

By modelling the observed vertical magnetic field of an active region AR 2372 by the potential field of an ensemble of magnetic dipoles, the authors derived the likely location of the separatrices, surfaces that separates cells of different field line connectivities, and of the separator which is the intersection of the separatrices. Four of the five off-band Hα kernels of a flare, that occurred less than twenty minutes before obtaining the magnetogram, are shown to have taken place near or at the separatrices. These Hα kernels are connected by field lines that pass near the separator. This indicates that the flare may have resulted from the interaction in the separator region of large scale magnetic structures.

073.110 **Electron acceleration by magnetosonic waves in solar flares.**
E. G. Zweibel, J.-F. de La Beaujardière.
Geophys. Res. Lett., Vol. 17, No. 11, p. 2051 – 2054 (Oct 1990). Paper presented at the Yosemite Conference on Transition Regions in Solar System Plasmas, Yosemite National Park, CA (USA), 6 – 9 Feb 1990.

Results of a simulation of electron acceleration by magnetosonic waves in a solar coronal loop are presented. The dependence of the energization rate on the strength and shape of the wave spectrum is discussed. Comparisons are made with the predictions of quasilinear theory and of a model describing stochastic acceleration by waves as isotropic diffusion in momentum space.

073.111 **Ca II K line asymmetries in two well-observed solar flares of October 18, 1990.**
C. Fang, E. Hiei, T. Okamoto.
Sol. Phys., Vol. 135, No. 1, p. 89 – 97 (Sep 1991).

Two-dimensional evolutions of two flares have been well observed in the Ca II K line with a CCD camera. There are two common characteristics for the flares: ~3 – 5 min before the impulsive phase, the heating already begins at the footpoints of the flares, but no asymmetry in line emission has been detected. After the onset of the impulsive phase, Ca II K line emission at the footpoints shows strong red asymmetry, with the maximum asymmetry occurring at the same time as the peak of the radio bursts. The maximum downward velocity is about 30 – 50 km s^{-1}. For flare 1, blue and red asymmetries were observed in two sides of the footpoint area. They developed and attained a maximum nearly at the same time and the inferred Doppler velocities are comparable. This implies that two mass jets started from a small region and ejected along a loop but in opposite directions with roughly equivalent momentum. A possible mechanism is discussed.

073.112 **Magnetohydrodynamics of an impulsively heated, hard X-ray emitting filament.**
G. Roumeliotis, A. G. Emslie.
Astrophys. J., Vol. 377, No. 2, p. 685 – 693 (20 Aug 1991).

The authors study the evolution of a twisted magnetic filament following the partial unwinding of its field. It is assumed that the magnetic energy released in the unwinding process causes impulsive heating of the plasma within the filament. The heated material expands, cooling adiabatically and distorting the surrounding magnetic field lines. The temperature and density of the heated material are sufficient to produce observable hard X-ray emission. The hard X-ray signature of such an impulsively heated magnetic filament is a brief ($\approx$100 ms) pulse of radiation with a simple time structure and a spectrum typical of solar bursts ($\gamma \approx 3$). Such fast bursts have been observed by the SMM, and in the present model their decay time structure is governed not by the energy release rate, buth rather by the physics of the atmospheric response to the energy release. More complicated hard X-ray bursts may be produced by the sequential and/or overlapping firing of many adjacent twisted magnetic filaments.

073.113 **Energy balance in the solar transition region. II. Effects of pressure and energy input on hydrostatic models.**
J. M. Fontenla, E. H. Avrett, R. Loeser.
Astrophys. J., Vol. 377, No. 2, p. 712 – 725 (20 Aug 1991).

The authors address the question of how energy is radiated by hydrogen lines and continua in hydrostatic energy-balance models of the transition region between the solar chromosphere and corona. The models assume that mechanical or magnetic energy is dissipated in the hot corona and is then transported toward the chromosphere down the steep temperature gradient of the transition region. This model can be applied to the lower portions of magnetic flux tubes in which the temperature varies from chromospheric to coronal values, and to either open field or loop models. The model properly takes into account the diffusion of hydrogen atoms and ions. The diffusion effects are used in both the statistical equilibrium and energy balance equations and, as shown earlier, give order of magnitude agreement between calculations and the observations of the hydrogen Lyman continuum and Lyman lines in the average quiet Sun. The present paper describes various models constructed for the transition region that correspond to different features of the solar atmosphere. The calculations give the magnitude and distribution of energy radiated by the lower transition region after being transported down from the corona. The relations between the downward energy flux, the pressure of the transition region, and the different hydrogen emissions are also derived.

073.114 **Flares on the Sun and other stars.**
B. Haisch, K. T. Strong, M. Rodonò.
Annu. Rev. Astron. Astrophys., Vol. 29, p. 275 – 324 (1991). – See Abstr. 003.010 for the main entry.

The authors concentrate on observations that show the solar-stellar connection or point to probable future stellar observations. Contents: 1. Introduction. 2. History of solar and stellar flare observations. 3. Cases of flare stars. 4. The flare in various spectral regions. 5. The physics of flares. 6. The analysis of flares. 7. Microflaring and coronal heating. 8. Miscellaneous topics. 9. For the future.

073.115 **Modelling the moustache phenomenon in network regions.**
G. Stellmacher, E. Wiehr.
Astron. Astrophys., Vol. 251, No. 2, p. 675 – 679 (Nov 1991).

The intensity enhancement in the inner line wings (moustache) of Mgb_1 5183 and Fe 5434.5 Å observed in Ca^+K bright points outside active regions is found to result from a superposition of a hot component (representing the magnetic structure) surrounded by a cool (intergranular) region. The I + V and I–V profiles of the magnetically sensitive line Fe 6302.5, observed striktly simultaneously with Mgb_1, are equally well modelled by the same procedure (which in an earlier paper had successfully been used to explain the gap effect of Zeeman insensitive lines in plages). The Mgb_1 moustaches are cospatial with the simultaneously observed Fe 6302.5 gaps. Their equal spatial extent indicates equal width of the temperature- and the magnetic structures of the Ca^+K region.

073.116 **The rotational motion in quiescent prominences.**
M. C. Pande, K. R. Bondal.
Earth, Moon, Planets, Vol. 54, No. 2, p. 119 – 124 (Aug 1991).

On the basis of Kippenhahn and Schlüter's magnetohydrostatic model of a quiescent prominence the authors have attempted to study the effect of a rotational velocity field in it. They find that a physically plausible solution is not possible in the vertical plane. A possibility, however, is shown in the horizontal plane, with certain assumptions to get equal velocity contours.

073.117 **^3He in the solar flares.**
G. F. Sitnik, L. M. Kozlova.
Astron. Tsirk., No. 1545, p. 21 – 22 (Aug – Sep 1990). In Russian.
From spectral observations near λ 10830 Å, a considerable enhancement of the helium isotope abundance was found in solar flares, as compared to that in quiescent regions and facula fields, which, on the contrary, do not differ in isotope abundance. Temporal variations of ^3He abundance are found in different stages of a flare, as well as its locality, limited by the flare region boundary. The importance of the problem discussed for nuclear heliophysics is stressed.

073.118 **Inferring chromospheric flare heating from hydrogen-line wings.**
K. G. Gayley, R. C. Canfield.
Astrophys. J., Vol. 380, No. 2, p. 660 – 676 (20 Oct 1991).
The authors consider strong flare heating that penetrates deeply into the solar chromosphere and apply a simple model, supported by a numerical radiative transfer simulation, to study how elevated free electron densities produce markedly increased Stark emission in the wings of hydrogen Lyman and Balmer lines. This impulsively generated emission is produced in a slab roughly a scale height in thickness where hydrogen is partially ionized. When flare heating persists for about a second or longer, energy balance and ionization equilibrium are attained, and both the Stark emission far in the wings and the total cooling in the slab scale with the square of the density. Thus a rough proportionality exists between observed Stark wing intensity and the total energy deposition in the partially ionized layer. The authors calculate this proportionality for various Lyman and Balmer lines, and show how one may infer from the observed line wings the magnitude of the nonthermal heating that penetrates the chromosphere. The results depend sensitively on the low-temperature radiative cooling by effectively optically thin metals.

073.119 **Chromospheric observation and preliminary analysis of active region AR5395 at Urumqi Astronomical Station.**
Zheng Ruimin, Huang Zhen, Bao Xingming.
Publ. Yunnan Obs., No. 4, p. 29 – 30 (Dec 1990). – See Abstr. 012.047 for the main entry.

073.120 **The two types of flares in the active region AR 5395.**
Zhang Bairong, Luan Di, Li Qiongying.
Publ. Yunnan Obs., No. 4, p. 34 – 37 (Dec 1990). – See Abstr. 012.047 for the main entry.

073.121 **AR 5395: observations on the asymmetries of the Hα and Ca II K line profiles of the flares on March 9 and 14, 1989.**
Fang Cheng, Hu Ju, Ji Guoping, Yin Suying, Zhang Qizhou.
Publ. Yunnan Obs., No. 4, p. 38 – 41 (Dec 1990). – See Abstr. 012.047 for the main entry.

073.122 **The material ejections associated with flares in AR 5395.**
Luan Di, Wu Mingchan, Li Zhi.
Publ. Yunnan Obs., No. 4, p. 42 – 43 (Dec 1990). – See Abstr. 012.047 for the main entry.

073.123 **AR 5395: a motion of the dark matter in 2N flare on 9 March 1989.**
Li Zhi, Luan Di, Li Zhikai.
Publ. Yunnan Obs., No. 4, p. 44 – 47 (Dec 1990). – See Abstr. 012.047 for the main entry.

073.124 **AR5395: the flare trigger and the evolution of magnetic fields.**
Chen Xiezhen.
Publ. Yunnan Obs., No. 4, p. 48 – 50 (Dec 1990). – See Abstr. 012.047 for the main entry.

073.125 **AR5395: 3B flare on March 15, 1989.**
Ma Kaiquan, Cao Ying.
Publ. Yunnan Obs., No. 4, p. 51 – 52 (Dec 1990). – See Abstr. 012.047 for the main entry.

073.126 **Magnetoacoustic heating of the solar chromosphere.**
J. M. Davila, S. M. Chitre.
Astrophys. J., Lett., Vol. 381, No. 1, p. L31 – L33 (1 Nov 1991).
Long-period acoustic waves generated in the solar convection zone can propagate radially outward through the overlying atmosphere and get resonantly absorbed in the magnetic arches of the low-lying chromospheric canopy. The resulting Poynting and acoustic flux that enters the magnetic canopy in the network regions is demonstrated to be adequate to account for the observed chromospheric emission.

073.127 **AR5395 solar flare in relation to the longitudinal current density in the photosphere.**
Chen Jimin, Li Wei, Liu Jianqiang, Liu Guilin.
Publ. Yunnan Obs., No. 4, p. 61 – 63 (Dec 1990). – See Abstr. 012.047 for the main entry.

073.128 **A microflare-related activation of a filament observed in Hα and C IV lines.**
B. Schmieder, J. Fontenla, E. Tandberg-Hanssen.
Astron. Astrophys., Vol. 252, No. 1, p. 343 – 352 (Dec 1991).
A filament in active region AR 2717 was observed in two lines formed at different temperatures (Hα at 10^4K and C IV at 10^5K) with the multi-channel-double-pass spectrograph and the ultraviolet spectrometer and polarimeter. The partial disparition brusque of the filament as observed in Hα was due to the heating of a filament section in the vicinity of a bright point. A propagating disturbance followed this event. A detailed analysis of C IV rasters shows that this disturbance was not a passive perturbation, but was itself triggering active phenomena at various locations along its path, resulting in energy releases. The authors suggest that this propagation of brightness was due to fast successive reconnections between fine loop-like structures of the filament.

073.129 **Fine structure of solar filaments. II. Dynamics of threads and footpoints.**
B. Schmieder, M. A. Raadu, J. E. Wilk.
Astron. Astrophys., Vol. 252, No. 1, p. 353 – 365 (Dec 1991).
A filament observed in an active region in June 1986 underwent a slow disparition brusque which lasted 4 d. Here the authors focus their study on the dynamical behaviour of the fine structures (scale ~350 km) in this filament. The fine structures were observed at the Pic du Midi with the Multi-Channel Subtractive Double Pass spectrograph during a period of 15 min. No substantial changes in intensity during this period were observed. However, the authors observed widespread continual changes in the velocity field and, for the region taken as a whole, there was virtually no correlation from one minute to the next. High velocities (up to ± 15 km s^{-1}) were detected using the first order differential cloud model method at the locations where the filament was anchored in the chromosphere. The authors argue that these footpoints are essentially the same kind of structures as the chromospheric bushes seen around supergranules, showing spicule-like activity modified by the local magnetic field geometry, and which may supply matter to the filament by direct injection. Elongated Doppler structures with coherent motions were observed parallel to the filament axis. These structures may be interpreted as magnetic flux tubes identifiable during their short rise time, or they may indicate the presence of coherent wave motions.

073.130 **Flare activity and evolution within AR5629.**
Hou Shuming.
Publ. Yunnan Obs., No. 4, p. 160 – 168 (Dec 1990). – See Abstr. 012.047 for the main entry.

073.131 **Chromospheric observation and preliminary analysis of active region AR5629 at Urumqi Astronomical Station.**
Zheng Ruimin, Huang Zhen, Bao Xingming.
Publ. Yunnan Obs., No. 4, p. 169 – 170 (Dec 1990). – See Abstr. 012.047 for the main entry.

073.132 **Big loop prominence system at the west limb of solar disk on the days 15 to 17, Aug. in 1989.**
Li Jiukang, Cao Ying, Li Qiongying.
Publ. Yunnan Obs., No. 4, p. 174 – 175 (Dec 1990). – See Abstr. 012.047 for the main entry.

073.133 **The characteristics and interpretation of Hα limb flare and radio burst on Aug. 16, 1989 in AR5629.**
Zhu Zuyan, Yao Jinxing.
Publ. Yunnan Obs., No. 4, p. 176 – 178 (Dec 1990). – See Abstr. 012.047 for the main entry.

073.134 **The morphological properties of two limb flares occurring in AR5629.**
Fan Zhongyu.
Publ. Yunnan Obs., No. 4, p. 179 (Dec 1990). Abstract. – See Abstr. 012.047 for the main entry.

073.135 **AR 5629: a super flare on 16 Aug. 1989.**
Luo Zhi, Ye Huilian, Xuan Jiayu, Lin Jun, Zhong Shuhua, Chen Xuekun.
Publ. Yunnan Obs., No. 4, p. 180 (Dec 1990). Abstract. – See Abstr. 012.047 for the main entry.

073.136 **AR5629: a large loop–shape flare on 17 Aug. 1989.**
Xuan Jiayu, Luo Zhi, Ye Huilian, Zhong Shuhua, Chen Xuekun.
Publ. Yunnan Obs., No. 4, p. 181 – 184 (Dec 1990). – See Abstr. 012.047 for the main entry.

073.137 **AR5629: the observations of the flare loop of Aug. 17, 1989.**
Li Wei.
Publ. Yunnan Obs., No. 4, p. 185 (Dec 1990). Abstract. – See Abstr. 012.047 for the main entry.

073.138 **AR5629: a loop prominence system and radio spike radiation 17 Aug., 1989.**
Gong Yuanfang, Shang Qiongzhen, Lu Songquan, Hu Hanming, Yang Rongbang, Li Jiukang.
Publ. Yunnan Obs., No. 4, p. 193 – 196 (Dec 1990). – See Abstr. 012.047 for the main entry.

073.139 **Multi–wave band SMM – VLA observations of an M2 flare and an associated coronal mass ejection.**
R. F. Willson, J. T. Schmelz, R. D. Gonzalez, K. R. Lang, K. L. Smith.
Astrophys. J., Vol. 378, No. 1, p. 360 – 371 (1 Sep 1991).

An M2 flare and an associated coronal mass ejection (CME) were observed by instruments on SMM as well as by the VLA and other ground–based observatories on 1988 September 30 as part of the International Solar Month campaign. The multi–wave band data show a gradual, slowly changing event which lasted several hours. The microwave burst emission originates in compact (10″– 30″), moderately circularly polarized ($\varrho_c \leqslant 60\%$) sources ($T_b = 0.5 - 20 \times 10^6$K) located near the sites of bright Hα and soft X–ray emission. These data were combined with estimates of electron temperature and emission measure obtained from Ca XIX and Fe XXV spectra to show that the microwave emission can be attributed to thermal gyrosynchrotron radiation in regions where the magnetic field strength is H = 425 – 650 G. The CME was observed by both ground– and space–based coronagraphs and a good estimate of the acceleration at low altitudes could be measured: 0.1240 ± 0.0466 km s^{-2}. From these observations, the flare and CME start times appear to be simultaneous. The event can be described by a "cartoon" where a global magnetic disturbance both triggers the flare and opens the field lines allowing the filament cavity to rise.

073.140 **Radio signature and Hα limb features of the 1979 March 9 and 1982 July 9 flares.**
H. Aurass, A. Krüger, B. Rompolt, I. Garczynska, V. V. Fomichev, I. M. Chertok, V. N. Ishkov, H. W. Urbarz.
Astron. Nachr., Vol. 312, No. 4, p. 245 – 256 (1991).

The authors present a comprehensive compilation of data for both events. Some typical Hα limb and type IV radio spectrogram features are found to be associated. The escape of matter out of loops visible in Hα via blob ejection or loop eruption is associated with fast drift burst signatures resp. early flare continuum emission in radio spectra and pulses in HXR. The comparison of both events reveals that the metric radio–HXR association is essentially a matter of particle escape conditions into the corona.

073.141 **Strong magnetic fields in solar flares: observation data and theoretical model.**
N. I. Lozitskaya, V. G. Lozitskij, A. A. Solov'ev.
Kinematika Fiz. Nebesn. Tel, Tom 7, No. 6, p. 40 – 47 (Nov–Dec 1991). In Russian. English translation in Kinematics Phys. Celest. Bodies, Vol. 7, No. 6.

Spectral observations with a circular polarization analyzer for five solar flares in 1981 and 1989 are studied. The observations are performed with the echelle spectrograph of the Kiev University Observatory. Direct measurements of the Zeeman splitting of the emission in the Fe I, Fe II, Na I, He I and H I lines show that the strength $H^e_{\parallel}$ of the longitudinal magnetic field is in the range between 10 – 20 and 350 mT. At the He I–emission formation level, magnetic strength is 120 – 180 mT. The strength $H^{ph}_{\parallel}$ at the photospheric level measured by the splitting of the Fraunhofer line profiles is systematically less than the strength $H^e_{\parallel}$ measured by splitting of the emission profiles. It might indicate the local amplification of the magnetic field in flares. A theoretical model for interpretation of the observation data is proposed. According to this model, strong magnetic fields in flares arise owing to the magnetic tube twists in the upper layer of the solar atmosphere.

073.142 **LDE flares in the 20th solar cycle. I. Comparison of the time behaviour of H–alpha grouped and LDE flares.**
A. Antalová, B. Viktorínová.
Bull. Astron. Inst. Czech., Vol. 42, No. 3, p. 133 – 144 (Jun 1991).

Numerical series of H–alpha and LDE flares have been tested and compared for the years 1969–1976, using methods of statistical analysis. The results have shown that the occurrence of H–alpha flares can be expressed by an exponential function. The analytical expression of the LDE (M–X) flare occurrence (their SXR duration exceeds 2 h) is a parabola or a set of spline functions. Time intervals with increased flare activities were observed in the years 1970, 1972 and 1974, whereas decreased flare activity was found in 1971 and 1975.

073.143 **LDE flares in the 21st solar cycle (1976–1986). I. Comparison of the time occurrences of H–alpha and LDE flares.**
B. Viktorínová, A. Antalová.
Bull. Astron. Inst. Czech., Vol. 42, No. 3, p. 144 – 157 (Jun 1991).

The similarities and differences of the temporal profiles of H–alpha grouped solar flares and LDE (M–X) flares during the 21st cycle are examined. The basic distribution function of the occurrence of LDE (M–X) flares in the 21st cycle is a parabola, its most active part being in the years 1978–1982. As opposed to the monotonic post–maximum (1980) decrease of the number of H–alpha grouped flares, the occurrence of LDE (M–X) flares has two comparable maxima: the first in 1980 and the second in 1982. The largest number LDE (M–X) flares originated in 1980 ($N_{1980} = 168$, $FI_{1980} = 2849$); lower than in 20th cycle ($N_{1970} = 225$, $FI_{1970} = 5677$). Important episodes of this first LDE flare pulse are November 1980 ($N = 38$) and April 1981 ($N = 36$). In 1982, the number of LDE (M–X) flares was smaller ($N_{1982} = 135$), but they were mightier and, consequently, the LDE flare index reached its maximum in 1982 ($FI_{1982} = 3547$).

The fact that two comparably intensive (1980, 1982) sources of interplanetary protons (accelerated during LDE flares) existed is independently supported by the solar modulation of cosmic ray protons, measured on ISEE 3 (Evenson and Mayer 1984). North–south LDE flare asymmetry during cycle 21 is also considered.

073.144 **Center–to–limb variations of characteristics of solar flare hard X–ray and gamma–ray emission.**
J. M. McTiernan, V. Petrosian.
Astrophys. J., Vol. 379, No. 1, p. 381 – 391 (20 Sep 1991).
The observed distribution of fluxes and spectral indices of flares at X–ray and gamma–ray range are analyzed with particular attention to variation with heliocentric longitude of these distributions and their moments. It is shown how these distributions can be used to set constraints on model parameters describing the characteristics of the accelerated electrons and flare plasma. The models assumed are the nonthermal thick target models, whereby accelerated electrons with power–law spectra and Gaussian pitch angle distributions are injected at the top of a coronal loop with two kinds of converging field geometries. The basic distributions are derived from data on all flares, irrespective of their position on the Sun, at low energies (hard X–rays) where the angular dependence of the parameters is negligible. A comparison of X–ray and gamma–ray spectral indices shows that the spectrum of the accelerated electrons must flatten at higher energies. The range of the Gaussian width of the pitch angle distribution of electrons beamed along the field lines and the convergence rate of the magnetic field are delimited. Isotropic pitch angle distributions or distributions peaking perpendicular to the field lines, and field convergence rates with more than a fivefold increase in the magnetic field from corona to the photosphere can be ruled out.

073.145 **The iron–to–calcium abundance ratio in the 20×10^6K plasma of medium and large solar flares.**
K. J. H. Phillips, U. Feldman.
Astrophys. J., Vol. 379, No. 1, p. 401 – 405 (20 Sep 1991).
The iron–to–calcium abundance ratio during several flares observed by instruments on the P78–1 and SMM spacecraft has been determined using the intensity ratio of a dielectronic satellite line of Fe XXIV and the resonance line of Ca XIX. Values of A(Fe)/A(Ca) equal to 6.8 ± 1.0 and 6.0 ± 1.5 were derived from these data sets, respectively; the small difference is attributable to errors in the calculated instrument sensitivities. These values fall between two other determinations from X–ray spectra, but are less than the photospheric value. They are fairly constant from flare to flare, contrasting with other recent work which indicates strong flare–to–flare variations.

073.146 **The $3s^23p\ ^2P - 3s3p^2\ ^4P$ intercombination lines of Fe XIV in the Sun.**
F. P. Keenan, P. L. Dufton, E. S. Conlon, M. B. Boylan, A. E. Kingston, K. G. Widing.
Astrophys. J., Vol. 379, No. 1, p. 406 – 408 (20 Sep 1991).
The authors have examined EUV spectra of solar flares obtained with the NRL S082A slitless spectrograph on board Skylab and have provisionally identified the five components of the Fe XIV $3s^23p\ ^2P - 3s3p^2\ ^4P$ intercombination multiplet. A comparison of the observed Fe XIV line intensities with theoretical predictions revals good agreement between theory and experiment, which tends to confirm these identifications and those made previously by Träbert, Hutton, and Martinson. The potential usefulness of the lines as electron density diagnostics is briefly discussed.

073.147 **Are spicules a source of coronal matter?**
R. A. Gulyaev.
Sov. Astron., Vol. 35, No. 5, p. 556 – 557 (Sep 1991). Current Physics Microform No.: 9203X1730. English translation of Astron. Zh., Tom 68, Vyp. 5, p. 1108 – 1110 (1991).
Measurements of the Doppler shift of the Hα line and EUV C I lines at the center of the solar disk contradict the widely held concept of ascending motion of spicule matter at a velocity ~25 km/sec. More likely, descending motion at a velocity of several km/sec predominates in spicules, as in formations in the transition zone from chromosphere to corona. In such a case, spicules cannot be considered to be a source of supply of matter to the corona. The so–called chromospheric jets, observed inside cells of the chromospheric network, may play a decisive role in the transport of matter into the corona.

073.148 **Solar gamma–ray–line flares, type II radio bursts, and coronal mass ejections.**
E. W. Cliver, H. V. Cane, D. J. Forrest, M. J. Koomen, R. A. Howard, C. S. Wright.
Astrophys. J., Vol. 379, No. 2, p. 741 – 750 (1 Oct 1991).
The authors use a Big Flare Syndrome (BFS) test to substantiate earlier reports of a statistically significant association between nuclear γ–ray line (GRL) flares (observed by the Gamma Ray Spectrometer on the SMM) and metric type II bursts from coronal shocks. The type II onset characteristically follows the onset of γ–ray emission with a median delay of 2 min. The authors find that 70% – 90% of GRL flares, for which coronagraph data were available, were associated with coronal mass ejections (CMEs). Gradual and impulsive GRL flares were equally well associated with CMEs. The CMEs were typically fast, with a median speed greater than 800 km s^{-1}. The authors discuss the following possible "non–BFS" explanations for the GRL – type II association: (1) the "type II shock" accelerates GRL protons in a "second step" process as it traverses low–lying coronal loops; (2) the observed shocks are blast waves resulting from the sudden deposition of energy in the low corona signaled by the GRL event; (3) a fast CME is the underlying independent cause of both GRL and type II emission from GRL flares.

073.149 **Electron acceleration in current sheets in solar flares.**
Yu. Eh. Litvinenko, B. V. Somov.
Sov. Astron. Lett., Vol. 17, No. 5, p. 353 – 356 (Sep 1991). Current Physics Microform No.: 9204X1300. English translation of Pis'ma Astron. Zh., Tom 17, No. 9, p. 835 – 840 (1991).
The authors present a model for electron acceleration in high–temperature turbulent current sheets (HTCS). In this model, the various processes that generate fast particles in solar flares have a common origin, namely, dissipation of magnetic energy in a region of magnetic reconnection. The spectral indices γ which are found (in the range 2 – 5), as well as the number of electrons accelerated during the main phase of a flare, are in agreement with observational data.

073.150 **Chromospheric and transition region diagnostics using emission–line intensities.**
R. G. Athay, K. P. Dere.
Astrophys. J., Vol. 379, No. 2, p. 776 – 787 (1 Oct 1991). With plate 5.
EUV emission–line intensities from the HRTS experiment on Spacelab 2 are analyzed to determine the causes of intensity variations from point to point on the solar surface. Chromospheric lines of O I and transition region lines of C II, Si IV and C IV exhibit saturation effects that limit the line intensities in the brighter regions. No such effect is found in lines of C I, Fe II, or Si III. The chromospheric lines saturate because the optical thickness exceeds the thermalization depth, whereas the transition region lines saturate because of the geometry of the unresolved fine structure. The latter is assumed to consist of elongated cylinder–like features oriented along magnetic lines of force that are mainly vertical. Saturation occurs when the optical thickness along the axis of the cylinders exceeds unity. Estimates based on the observed saturation effects give filling factors for the fine structure of 5×10^{-3} and 0.4 at 10^5K and 3×10^4K, respectively. The corresponding length of the cylinders are estimated at 1000 km and 40 km at the two temperatures. It is argued that in both chromospheric and transition region lines changes in density, temperature, and the volume of the emitting material each play important roles in the intensity changes from one point to another on the solar disk. The evidence suggests that changes in the volume of the emitting material overshadow the effects of density and temperature changes. Intensity ratios of line

pairs are shown to provide a reliable basis for density diagnostics if both lines are free from saturation effects and if they are formed at similar temperatures.

073.151 **EUV spectroscopy as a probe of astrophysical plasmas – learning from the Sun.**
G. A. Doschek.
1 Berkeley Colloquium on Extreme Ultraviolet Astronomy, p. 94 – 112 (1991). – See Abstr. 012.045 for the main entry.
The author discusses the new developments in plasma diagnostics and illustrates how spectroscopy has led to a better understanding of the solar atmosphere with particular emphasis on the transition region, corona and flares.

073.152 **Stability of prominences exposing helical–like patterns.**
B. Vršnak, V. Ruždjak, B. Rompolt.
Sol. Phys., Vol. 136, No. 1, p. 151 – 167 (Nov 1991).
The internal structure of prominences appearing as twisted tubes was studied. The sample embraced 15 stable and 13 eruptive prominences, exposing patterns which possibly reflect a helical configuration. The equivalent pitch angles (ϑ) of twisted fine structure features were measured. In some cases the evolution of the internal structure was followed and 49 independent measurements of the parameter ϑ were performed in total. The results are presented in the plane relating the parameter ϑ and the normalized prominence height. The eruptive prominences occupy the region characterized by $\vartheta > 50°$ and $h > 0.8\,d$, where h and d are the prominence height and the footpoint half–separation, respectively. All prominences characterized by $h < 0.6\,d$ or by $\vartheta < 35°$ were stable. Such a result is in good agreement with an order of magnitude treatment of the forces acting in a curved magnetic tube, anchored at both ends in the photosphere.

073.153 **The high Ne/C and Ne/O abundance ratios in the footpoints of solar flares.**
A. Shemi.
Mon. Not. R. Astron. Soc., Vol. 251, No. 2, p. 221 – 226 (15 Jul 1991).
High chromospheric abundance ratios of Ne/C and Ne/O were deduced from the gamma–ray spectroscopy of the 1981 April 27 flare. A similar high Ne/O ratio was observed in energetic solar particles from ^{3}He–rich flares. These high ratios are consistent neither with local galactic abundances, nor with the first ionization–potential (FIP) scenario (Meyer, 1985). The inconsistency is explained by temporal deviations from the quiet–chromosphere abundances in the nuclear (gamma–ray) line production region (the interaction region). During the flare soft X–ray radiation penetrates through the chromosphere and creates a slab–like region of non–thermal ionization ratios, i.e. ionized neon mixed with mostly neutral carbon and neutral oxygen below the interaction region. Charge–dependent mass transport upward from this slab causes an enhancement of Ne/C and Ne/O ratios at the interaction region. These considerations constrain the use of "gamma–ray" abundances as indicator of quiet–chromosphere abundances.

073.154 **Numerical simulations of microflare evolution in the solar transition region and corona.**
A. C. Sterling, J. T. Mariska, K. Shibata, Y. Suematsu.
Astrophys. J., Vol. 381, No. 1, p. 313 – 322 (1 Nov 1991).
Several observers report transient ultraviolet brightenings, often referred to as microflares, in the solar atmosphere. Here, the authors present the results of a series of one–dimensional numerical simulations examining possible relationships between microflares and the generation of dynamical chromospheric and transition region features. It is assumed that the microflare energy deposition is localized just below the transition region on one side of a 22,000 km long quiet region coronal loop. Low–energy (10^{24}ergs; ~ 1 ergs cm^{-3}s^{-1}) and medium–energy (10^{25}ergs; ~ 10 ergs cm^{-3}s^{-1}) microflares eject long–lived cool, dense gas plugs into the corona, with the gas plug traversing the loop apex in the medium energy case. In the case of high–energy (10^{26}ergs) microflares, the gas plug is rapidly (~ 10 s) heated to the temperature of the surrounding corona, and the results resemble the dynamics occurring in standard solar flare thick–target electron beam models. The gas flows and ejections which develop in these simulations do not resemble spicules, but they may represent some of the dynamic features seen in ultraviolet spectral lines. The high–energy microflare produces sustained coronal temperature increases of 2 – 4 times the initial (7.7×10^5K) value, and density increases of 3 – 6 times the initial value at the loop apex.

073.155 **Velocity gradients in the chromosphere–corona transition region.**
R. G. Athay, K. P. Dere.
Astrophys. J., Vol. 381, No. 1, p. 323 – 332 (1 Nov 1991). With plate 4.
Velocity data obtained from Doppler shifts of EUV spectral lines observed with the High Resolution Telescope and Spectrograph flown on Spacelab II are analyzed to determine the properties of the mean velocities $\bar{V}$ and their spatial variances V_σ, for chromospheric and transition region lines. The lines studied include the ions O I, C I, Fe II, C II, Si III, Si IV, and C IV. Observations were made using 11 radial slit positions covering a rectangular area $\sim 850''$ in length and $30''$ wide with a spatial resolution of $1''$ and $3''$ spacing between slits. The constancy of V_σ from center to limb together with a systematic maximum redshift in $\bar{V}$ in central disk regions is interpreted to mean that the plasma flow is divergent from the tops of closed field lines and that the cool plasma more or less fills the magnetic loops from which most of the radiation originates. The same is true of the hotter plasma. It is suggested that the crossfield component of ∇T is large and that the alternation from cool loops to hot loops occurs on spatial scales much less than $1''$.

073.156 **Investigation of fine structure in active formations on the Sun.**
V. S. Prokudina.
Tr. Gos. Astron. Inst. im. Shternberga, Tom 62, p. 163 – 181 (1991). In Russian.
On the basis of studies of Hα–filtergrams, obtained for a number of active phenomena, and of analysis of data of observations in other wavelength ranges, the characteristic features of fine structure of flare formations are revealed. The fine–filament and helical structure of eruptive filaments, caused by the development of various kinds of instabilities in the moving plasma, is analysed. The fine–structure features in loop systems and flare emission ribbons are investigated, and their characteristic timescales and dimensions are estimated; different versions, suggested for the interpretation of the observed phenomena, are discussed.

073.157 **Solar prominences.**
P. Démoulin.
Symposium on Geophysical and Astrophysical MHD as Part of the British Theoretical Mechanics Colloquium: Advances in solar system magnetohydrodynamics, p. 281 – 305 (1991). – See Abstr. 012.086 for the main entry.
A rapid overview of observations is given, and then it is tried to understand thermal and magnetic processes that lead to prominence formation. One section is devoted to what we can learn about the Sun, from prominences. The support and eruption of prominences are totally omitted.

073.158 **MHD of solar flares.**
A. W. Hood.
Symposium on Geophysical and Astrophysical MHD as Part of the British Theoretical Mechanics Colloquium: Advances in solar system magnetohydrodynamics, p. 307 – 326 (1991). – See Abstr. 012.086 for the main entry.
The author considers MHD models for the initial stages of solar flares.

073.159 **Soft X-ray emission from electron-beam – heated solar flares.**
J. T. Mariska, D. M. Zarro.
Astrophys. J., Vol. 381, No. 2, p. 572 – 577 (10 Nov 1991).
Using time-dependent numerical simulations and SMM observations of a solar flare on 1985 January 23, the authors examine the ability of an electron-beam – heating model to reproduce the rise phase of a flare as observed in soft X-ray lines of Ca XIX. The electron beam is parameterized by a peak flux, a low-energy cutoff, and a spectral index, and has a time dependence similar to the observed hard X-ray burst. For a spectral index of 6, only models with a low-energy cutoff of 20 keV reproduce the observed peak emission in the Ca XIX line complex. All models with a low-energy cutoff of 15 keV produce too much emission, while all models with a 25 keV cutoff give too little emission. None of the models reproduces the temporal behavior of the soft X-ray emission. The authors conclude that the electron-beam – heated component may represent only a small fraction of the energy released in the impulsive phase of this flare.

073.160 **Subthreshold stochastic diffusion with application to selective acceleration of ^{3}He in solar flares.**
S. Riyopoulos.
Astrophys. J., Vol. 381, No. 2, p. 578 – 582 (10 Nov 1991).
A mechanism of resonant acceleration driven by waves below the threshold for global chaos is discussed. It affects only resonant ions and proceeds through a stochastic network in phase space (Arnold web) characterizing this particular type of wave-particle interaction. A theory for selective acceleration of ^{3}He and ^{3}He-like ions through resonant subthreshold diffusion is proposed. It relies on the coincidence of two natural resonances, the two-ion (^{4}He, H) hybrid plasma resonance with the ion cyclotron resonance of ^{3}He. This occurs over a wide range of plasma parameters for relative abundances ^{4}He/H $\cong$ 0.25. It is argued that existing low-frequency fluctuations or high-frequency noise tend to amplify the selective diffusion of ^{3}He.

073.161 **Relative abundances in the lower solar transition region.**
G. A. Doschek, K. P. Dere, P. A. Lund.
Astrophys. J., Vol. 381, No. 2, p. 583 – 590 (10 Nov 1991). With plate 5.
The authors investigate properties of the lower solar transition region as determined from UV spectral lines recorded by the NRL High Resolution Telescope and Spectrograph (HRTS). The spectra were obtained from a rocket flight in 1975 July. The authors discuss variations of intensity ratios of strong lines of C IV, Si IV, C II, Si III, and O IV over the entire field of view of the HRTS instrument, which extends from disk center to the solar limb. The authors find evidence for abundance variations in the lower transition region, although it is not possible to state with absolute certainty that the line intensity variations are not due to emission measure effects. The largest apparent abundance variations appear to be in the active regions surrounding a sunspot. The C/Si ratio is lower in the active regions than anywhere else in the instrument field of view, implying a lower C/Si abundance ratio in the active regions. There is a possible correlation of line intensity variation in the active region with Doppler shift of the lines. The authors discuss the magnitude and reality of intensity ratio variations of all observed solar features, and measure monotonic center-to-limb intensity ratio variations due to the opacity of some of the spectral lines.

073.162 **Improved line ratio calculations involving $\Delta n = 1$ (2 – 3) transitions in O V and a reanalysis of Skylab observations of solar flares.**
F. P. Keenan, P. L. Dufton, L. K. Harra, E. S. Conlon, K. A. Berrington, A. E. Kingston, K. G. Widing.
Astrophys. J., Vol. 382, No. 1, p. 349 – 352 (20 Nov 1991).
New **R**-matrix calculations of electron impact excitation rates in O V are used to rederive theoretical electron density diagnostic emission-line ratios involving transitions between the n = 2 and 3 levels, which includes lines at 192.80, 192.90, 215.10, 215.25, 220.35, and 248.46 Å. A comparison of these diagnostics with observational data for two solar flares obtained with the NRL S-082A spectrograph on board Skylab reveals better agreement between theory and observation than was found previously. This provides experimental support for the improved accuracy of the atomic data employed in the present analysis.

073.163 **The observational evidence of the solar flares occurring in the redshift area of Hβ Dopplergrams.**
Ai Guoxiang, Zhang Hongqi, Li Jing, Li Wei, Chen Jimin.
Chin. Sci. Bull., Vol. 36, No. 15, p. 1275 – 1278 (Aug 1991).
The authors present a new phenomenon that the flares occurred on the redshift side of the inversion lines of the Hβ Doppler velocity fields observed half to two hours before. The features are described.

073.164 **The heating of the solar chromosphere.**
W. Kalkofen.
Solar interior and atmosphere, p. 911 – 932 (1991). – See Abstr. 003.028 for the main entry.
The chromospheric layers in the quiet Sun show three distinct regions. The chromosphere can be largely identified with the atmosphere inside magnetic elements (flux tubes). This chapter discusses the heating mainly of the bright points that are associated with the intranetwork magnetic fields. It reviews the relevant observations in line and continuum radiation and concludes that the energy dissipated by the 3-min waves is probably sufficient to heat the low and middle chromosphere in the bright points to the observed temperatures.

073.165 **Solar flares.**
H. Zirin, A. MacKinnon, S. M. P. McKenna-Lawlor.
Solar interior and atmosphere, p. 964 – 995 (1991). – See Abstr. 003.028 for the main entry.
Solar flares are discussed from three complementary viewpoints. The first is the radiative aspect which includes optical and radio observations, in addition to the magnetic circumstances of their origin. It is seen that flares are produced in regions of magnetic shear, a condition that arises mainly from sunspot motions. In this regard recipes for the prediction of flares are given. The second view point considers that flares produce great streams of energetic particles which carry away much of the flare energy. Gamma and X-ray studies of the spectra of these particles provide information on the flare's transient high-temperature plasma. A third consideration, satellite samples of flare-particle streams, provide diagnostic information on acceleration mechanisms pertinent to different kinds of flare event, while shedding complementary light on solar atmospheric composition at particular flare sites and on physical processes operating in the corona.

073.166 **Fine structure of the solar transition region: observations and interpretation.**
J. W. Cook, G. E. Brueckner.
Solar interior and atmosphere, p. 996 – 1028 (1991). – See Abstr. 003.028 for the main entry.
The authors review recent observations of the solar transition region and the temperature minimum obtained with high spatial resolution, primarily ultraviolet spectra and spectroheliograms from sounding rocket and Spacelab 2 flights of the High Resolution Telescope and Spectrograph. Information on intensities, velocities, emission measures and electron densities can be determined. The authors discuss observational results and the implications of observed fine structure for our ideas of the structure and physics of the solar atmosphere. Inclusion of fine structure in our ideas about the transition region affects the plausibility of one-dimensional average models of the atmosphere, the determination of temperature gradients, possible mechanisms for nonradiative heating, and the comparison of

transition region structures with corresponding high-resolution observations from the photosphere and corona.

073.167 **Observed evidences of the flare model related to filament current.**
Xu Aoao.
Acta Astron. Sin., Vol. 32, No. 4, p. 343 – 352 (Dec 1991). In Chinese.
The results show the intrinsic relationship between flare eruption and filament, i.e. the explosion of filaments causes the appearance of flares, and the energy of flares is transported by filament currents.

073.168 **Observation and analysis of Ca II K line asymmetry of solar flares.**
Yin Suying, Fang Cheng, Gan Weiqun.
Acta Astron. Sin., Vol. 32, No. 4, p. 353 – 361 (Dec 1991). In Chinese.
The Ca II K line spectra for eight flares are analysed in this paper. It is found that the red asymmetry is common for all flares. Such red asymmetry becomes most obvious around the maximum of the Ca II K line intensity. However the Ca II K line center does not show obvious Doppler shift for all flares studied here. It is shown that the asymmetry observed may be explained by the motion of material around the temperature minimum region.

073.169 **Exploration on the structure of the 1N/M7.1 flare on 30 Dec. 1982.**
Huang Youran, Yin Suying, Fang Cheng.
Acta Astron. Sin., Vol. 32, No. 4, p. 421 – 428 (Dec 1991). In Chinese.
The spectral microwave burst and HXRBS data of a 1N/M7.1 flare which occurred in the active region No. 4033, have been analysed. This flare seems to be a flaring arch. The intensity of the magnetic field and the structure of the flare loop have been figured out. It was found that there were 3 microwave bursts which occurred in a flaring loop system, and that the delay of the high frequency burst relative to the low frequency one is shortened successively. This can be explained by the decrease of the twist of magnetic lines of force in the flaring loop after each burst.

073.170 **The solar chromospheric supergranular network in 850 micron radiation.**
C. A. Lindsey, J. T. Jefferies.
Astrophys. J., Vol. 383, No. 1, p. 443 – 449 (10 Dec 1991).
The authors examine the first submillimeter solar observations of the chromospheric supergranular network, made on the 15 m James Clerk Maxwell Telescope on Mauna Kea in 850 μm radiation. These observations are useful for thermal diagnostics of the low and middle chromospheres of the quiet-Sun and magnetic regions, where mechanical heating of the atmospheric medium first becomes manifest. The models of Vernazza, Avrett and Loeser appear to be consistent with these observations.

073.171 **The sources of solar flares.**
H. Zirin.
International SOLAR-A Science Meeting: Flare physics in solar activity maximum 22, p. 9 – 12 (1991). – See Abstr. 012.104 for the main entry.
The author discusses some of the ideas in Tanaka's last paper. While for a long time it has been known that most, if not all, major flares are caused by shear associated with flux eruption and spot motion, the source of this behavior is unknown. Tanaka tried to explain some of the shear by the changing photospheric intersection with the three-dimensional twisted erupting flux loops. In this paper the author addresses the general problem of the origin of such features, and their possible connection with the ultimate source deep in the Sun.

073.172 **Flares appear on the red shift side near the inversion line of the H_β Dopplergram.**
Ai Guoxiang.
International SOLAR-A Science Meeting: Flare physics in solar activity maximum 22, p. 51 – 57 (1991). – See Abstr. 012.104 for the main entry.
It is discoved that flares appear on the red shift side near the inversion line of the H_β Dopplegram. The main features, measuring method and some explanations have been described.

073.173 **Microflares observed in He I 10830 and their relation to the quiet Sun magnetic fields.**
K. L. Harvey.
International SOLAR-A Science Meeting: Flare physics in solar activity maximum 22, p. 62 – 66 (1991). – See Abstr. 012.104 for the main entry.
This paper presents the preliminary results of a detailed study of three He I λ10830 microflares, with emphasis on the evolution of the underlying photospheric magnetic fields.

073.174 **Refining flare types from current patterns.**
V. I. Abramenko, S. I. Gopasyuk, M. B. Ogir'.
Bull. Crimean Astrophys. Obs., Vol. 81, p. 6 – 10 (1990). English translation of Izv. Krym. Astrofiz. Obs., Tom 81, p. 8 – 13 (1990). See Abstr. 53.073.023.
Measurements have been made on how the H_α emission from an active region is affected by magnetic-field processes that result in currents. Two types of flare occur. The first is the more common and arises in stable H_α flocculi. Here the current is due to rotation of the potential transverse field vector $H_\perp^0$, and the plasma moves at an angle to the field. These flares have considerable areas and show smooth development, while they occur mainly in regions where the current is upwards. The second type consists of a few flares that arise away from the flocculi, which occur in regions where the currents are due to length change in $H_\perp^0$ and the plasma moves along the transverse-field lines. These flares are usually compact and pulsed, and their radio emission spectra extend to shorter wavelengths.

073.175 **Magnetic morphology of nonthermal electron precipitation during three flares in a highly nonpotential active region.**
R. C. Canfield, K. D. Leka, J.-P. Wülser.
International SOLAR-A Science Meeting: Flare physics in solar activity maximum 22, p. 96 – 99 (1991). – See Abstr. 012.104 for the main entry.
NOAA active region 5747, during its October 1989 transit across the solar disk, showed highly nonpotential photospheric vector magnetic field structure and produced many solar flares, three of which were observed by the authors. They determined the photospheric distribution of the vertical current density and compared the locations of the major current systems to sites of nonthermal electron precipitation inferred from Hα profiles. The authors found that the sites of energetic electron precipitation are at the edges of these currents, not at their peaks.

073.176 **Numerical simulations of electron-beam-heated solar flares.**
J. T. Mariska.
International SOLAR-A Science Meeting: Flare physics in solar activity maximum 22, p. 153 – 156 (1991). – See Abstr. 012.104 for the main entry.
Using time-dependent numerical simulations and Solar Maximum Mission (SMM) observations of a solar flare on 1985 January 23, the author examines the ability of an electron-beam-heated model to reproduce the rise phase of a flare as observed in soft X-ray lines of Ca XIX. For a spectral index of 6, only models with a low-energy cutoff of 20 keV reproduce the observed peak emission in the Ca XIX line complex. None of the models reproduces the temporal behavior of the soft X-ray emission.

073.177 **A hydrodynamic thermal model of the impulsive phase of solar flares.**
C. Fang, W. Q. Gan, H. Q. Zhang.
International SOLAR-A Science Meeting: Flare physics in solar activity maximum 22, p. 157 - 160 (1991). - See Abstr. 012.104 for the main entry.

The authors have tried to make some improvements, together with introducing the chromospheric heating by coronal soft X-ray emission, and obtained the hydrodynamic model of a solar flare loop responding to the thermal energy deposition around the top of the loop.

073.178 **Material flow of a surge flare.**
Y. Nakagawa.
International SOLAR-A Science Meeting: Flare physics in solar activity maximum 22, p. 161 - 164 (1991). - See Abstr. 012.104 for the main entry.

The material flows observed in the surge flare of May 22, 1970 by Zirin (1971) are closely reproduced by a numerical simulation as the response of the atmosphere to a shearing motion of the foot points of evolving magnetic arcade of a linear dipole field. The computations are performed on the basis of the method of projected characteristics so that the time-dependent boundary conditions are treated in a physically self-consistent manner. It is found that no artificial viscosity is needed in the computation.

073.179 **Evolution of hot plasma in flares.**
Z. F. Svestka.
International SOLAR-A Science Meeting: Flare physics in solar activity maximum 22, p. 165 - 178 (1991). - See Abstr. 012.104 for the main entry.

This review discusses, in particular, X-ray manifestations of the development of hot flare plasma during the cooling phase of flares. In a preexisting coronal configuration, a small flare source can excite much larger coronal loops and arches, of which only the footpoints become visible if their density is low, or their whole structures can be seen if density is high enough. The main part of this review is devoted to the development of hot plasma in flaring structures. It is emphasized that X-ray observations of growing systems of loops in eruptive flares, and giant arches following these flares high in the corona, require large fields of view pointed for several hours at the active region in which the flare occurred. This makes their observations with SOLAR-A rather difficult.

073.180 **Thermodynamic evolution of flares.**
B. Sylwester, J. Sylwester, J. Jakimiec, S. Serio, F. Reale.
International SOLAR-A Science Meeting: Flare physics in solar activity maximum 22, p. 188 - 192 (1991). - See Abstr. 012.104 for the main entry.

The aim of this paper is to investigate the evolution of the basic thermodynamic parameters of a single flaring loop. A grid of simple models has been calculated using the PALERMO-HARVARD hydrodynamic code. It was assumed in the calculations that the extra heat responsible for flaring is deposited in the coronal portion of the loop, with the maximum energy deposition rate at the loop top. A constant cross-section was assumed. In all cases, the initial state of the loop corresponded to the steady state conditions related to the fixed background heating rate.

073.181 **Magnetic morphologies of solar flares.**
T. Hirayama.
International SOLAR-A Science Meeting: Flare physics in solar activity maximum 22, p. 197 - 201 (1991). - See Abstr. 012.104 for the main entry.

The author suggests magnetic morphologies for several flares using Hα filtergrams and a flare model of reconnection under an erupting magnetic flux rope. He selects those flares which, at first sight, seem difficult to understand in terms of the model. Some restricting conditions on the Petschek type magnetic reconnection model are empirically derived.

073.182 **A new kind of solar X-ray flares.**
Xu Yongxuan.
Publ. Purple Mt. Obs., Vol. 10, No. 4, p. 263 - 275 (Dec 1991). In Chinese.

A new kind of solar X-ray flares which has the following characteristics is discussed: a flare X-ray burst involves three phases. In every phase both the properties of X-ray emissions and the locations of the X-ray source are evidently different. The X-ray emissions show the impulsive properties in initial phase, in final phase the impulsive properties of the emissions are overwhelmed by the gradual properties. The X-ray source rises slowly in the development of a solar X-ray flare. The energy continues to be released from the initial phase of an event to its final phase. The author presents a view on the dynamical processes producing this kind of X-ray flares.

073.183 **Theoretical models of solar flares.**
K. Shibata.
International SOLAR-A Science Meeting: Flare physics in solar activity maximum 22, p. 205 - 218 (1991). - See Abstr. 012.104 for the main entry.

Recent progress in the understanding of the basic physical picture of solar flares is discussed from a theoretical point of view, with emphasis on magnetohydrodynamic processes, such as magnetic reonnection. Several models of CME (Coronal Mass Ejection) related flare and compact flare models are critically reviewed. The role of the successive emergence of twisted flux tubes is stressed, not only for modeling compact flares, but also for understanding CME related flares.

073.184 **On mechanisms of solar flares – some observational tests by using SOLAR-A.**
Y. Uchida, K. Shibata.
International SOLAR-A Science Meeting: Flare physics in solar activity maximum 22, p. 230 - 236 (1991). - See Abstr. 012.104 for the main entry.

The possibilities for using SOLAR-A observations to discriminate between the thus-far proposed flare mechanisms are disucssed. In this context, the authors point out some thus-far unnoticed restrictions of the "standard" models, and suggest that by relaxing these restrictions one can avoid the difficulties these conventional models encountered. These theoretical points indicate what SOLAR-A observations would be particularly useful for clarifying this topic.

073.185 **Plasma waves caused by transient heat conduction in a coronal loop as a trigger for impulsive solar flares.**
T. Takakura.
International SOLAR-A Science Meeting: Flare physics in solar activity maximum 22, p. 237 - 241 (1991). - See Abstr. 012.104 for the main entry.

Numerical simulation is made of the transient heat conduction during local heating in a model coronal magnetic loop with an applied axial electric current. The generation of the plasma waves caused by the local heating of the coronal loop may be a trigger for impulsive loop flares.

073.186 **The MSFC vector magnetograph, eruptive flares, and the SOLAR-A X-ray images.**
R. L. Moore, M. J. Hagyard, J. M. Davis, J. G. Porter.
International SOLAR-A Science Meeting: Flare physics in solar activity maximum 22, p. 324 - 329 (1991). - See Abstr. 012.104 for the main entry.

The authors first give an example of observations from the Marshall Space Flight Center vector magnetograph of critical magnetic field shear for large two-ribbon flares. They then present a "magnetic explosion" picture for these flares. Finally, they show an X-ray image of the onset of a large two-ribbon flare. The spatial resolution of this image is marginal but sufficient to suggest that, by having three times better resolution, the SOLAR-A soft X-ray images of large flares will reveal whether the eruptive model is a good or poor representation of the flare process.

073.187 **Magnetic reconnection in solar flares.**
T. G. Forbes.
Geophys. Astrophys. Fluid Dyn., Vol. 62, No. 1–4, p. 15 – 36 (Dec 1991). – See Abstr. 012.126 for the main entry.

The magnetic energy stored in the corona is the only plausible source for the energy released during large solar flares. Three flare models which have been extensively investigated are the emerging–flux model, the sheared–arcade model, and the magnetic–flux–rope model. All of these models can store and release magnetic energy efficiently provided that rapid magnetic reconnection occurs. However, only the magnetic–flux–rope model appears to provide a plausible mechanism for ejecting magnetic flux into interplanetary space.

073.188 **Dynamical fine structure of a quiescent prominence.**
P. Mein, N. Mein.
Sol. Phys., Vol. 136, No. 2, p. 317 – 333 (Dec 1991).

A quiescent prominence has been observed with the MSDP spectrograph at the Pic du Midi Observatory. Hα profiles are obtained simultaneously in a 2D field, allowing a statistical analysis. The standard deviations of Doppler shifts and line widths are investigated as functions of the line intensity. The observations are compared with numerical simulations assuming that the prominence is made of identical "threads", the velocity of which is distributed according to gaussian functions. The processing of simulations is very close to the processing of observations. The mixing by seeing effects and the transfer of radiation across several threads along the line of sight are considered. The results are improved by taking into account a slight scatter of source functions throughout the prominence. In the central parts of the prominence, the fit is obtained by assuming that the line–of–sight velocities of the threads have a gaussian probability function. In the edges, larger scatter of velocities, and two combined dispersions are suggested.

073.189 **Ion acceleration and plasma heating in solar flares.**
G. M. Simnett.
Mem. Soc. Astron. Ital., Vol. 62, No. 2, p. 359 – 388 (1991). Special issue: Stellar flares. B. R. Pettersen (ed.).

It has been known for half a century that the Sun is a transient source of energetic ions. Interactions of these ions with the solar atmosphere provide important signatures relevant to the study and interpretation of solar flares. Ion acceleration processes which have been proposed in recent years are reviewed. Observations which provide constraints on both the acceleration process and the importance of ions, energetically, to the flare energy budget, are presented. Chromospheric evaporation and plasma heating are discussed. It is concluded that from the primary energy release in flares, accelerated ions are more likely to transfer the energy than accelerated electrons. Direct electric field acceleration is a strong candidate for the accelerating mechanism.

The complete Hard X–Ray Burst Spectrometer event list, 1980 – 1989.
See Abstr. 002.023.

Observing the Sun.
See Abstr. 003.026.

Mechanisms of chromospheric and coronal heating. Proceedings.
See Abstr. 012.011.

The physics of solar flares. Proceedings.
See Abstr. 012.016.

Flare physics in solar activity maximum 22. Proceedings.
See Abstr. 012.104.

The progress in the research on radio rapid fluctuation in solar flares for the last year.
See Abstr. 013.063.

Two feeds with single common antenna operating at 15 and 21 cm wavelengths.
See Ábstr. 033.006.

An electrograph for measurement of macroscopic electric fields in prominences and flares.
See Abstr. 034.085.

An overview of the Orbiting Solar Laboratory.
See Abstr. 035.132.

Considerations for flare related magnetic field measurements.
See Abstr. 036.073.

Computer processing and image processing of Hα solar flares.
See Abstr. 036.150.

SIMURIS: a high resolution solar physics interferometric mission in answer to the chromospheric and coronal heating problem.
See Abstr. 051.013.

Probing solar particle acceleration during the 22nd solar maximum with GRO/OSSE.
See Abstr. 051.018.

Particle energization.
See Abstr. 061.067.

How much energy can be stored in a three–dimensional force–free magnetic field?
See Abstr. 062.002.

Registration of macro– and micro–inhomogeneities of plasma density in astrophysical objects.
See Abstr. 062.045.

On the direction of the currents at magnetic neutral points.
See Abstr. 062.068.

Evolution to nonequilibrium in simple models of prominence filaments.
See Abstr. 062.093.

The interaction of collisionless shocks in astrophysical plasmas.
See Abstr. 062.094.

Kinetics of energetic electrons, propagating through ion–sound plasma trubulence in solar flares.
See Abstr. 062.115.

A chromospheric response to pulse beam heating.
See Abstr. 062.128.

A numerical simulation of magnetic reconnection and radiative cooling in line–tied current sheets.
See Abstr. 062.131.

Kinetics of fast electrons in ion–acoustic waves in the turbulent plasma of solar flares.
See Abstr. 062.134.

Prominence sheets supported by constant–current force–free fields. I. Imposition of normal magnetic field components at the current sheet and the photosphere.
See Abstr. 062.145.

Maximum energy of semi–infinite magnetic field configurations.
See Abstr. 062.175.

Magnetic reconnection in solar flares.
See Abstr. 062.199.

Chromospheric inhomogeneities in cool stars: possible effect on hydrogen line profiles.
See Abstr. 063.059.

On the influence of multi–dimensional radiative transfer on the energetic contribution of the Ca K line.
See Abstr. 063.060.

Do we really know what the actual chromospheric heating requirements are?
See Abstr. 064.020.

What can solar and stellar ultraviolet observations tell about chromospheric and coronal heating mechanisms?
See Abstr. 064.022.

An infrared perspective on chromospheres.
See Abstr. 064.030.

Relations between activity and magnetic fields.
See Abstr. 064.033.

Ionization pumping.
See Abstr. 064.045.

A convective model for solar and stellar flares.
See Abstr. 065.154.

Determining the solar–flare photosphere scale height from SMM gamma–ray measurements.
See Abstr. 071.003.

The He I λ10830 Å line as an indicator of the chromospheric and coronal activity of the Sun.
See Abstr. 071.037.

Photospheric and chromospheric observations of AR 5395 by 26 cm solar telescope at Yunnan Observatory.
See Abstr. 071.046.

A method describing profile asymmetry of spectral lines.
See Abstr. 071.064.

The periodicity of AR 5395 evolution.
See Abstr. 072.005.

The evolutionary features of the magnetic configuration in the active regions and the short–term prediction of the strong SWF.
See Abstr. 072.006.

Solar active region evolution and flare build–up.
See Abstr. 072.008.

Electric currents and Hα emission in two active regions on the Sun.
See Abstr. 072.014.

Electric currents and flares in solar active regions.
See Abstr. 072.024.

Flaring active region structures in the corona, transition region, and lower atmosphere observed in the SAROC.
See Abstr. 072.025.

Evolution characteristics of the fine–structure of AR4811 and the associated flare activity.
See Abstr. 072.062.

Evolution of a strong pulsating sunspot center and its chromospheric activity.
See Abstr. 072.079.

An explanation of penumbral dimensions.
See Abstr. 072.091.

AR5395: fine structures, magnetic shear and flares.
See Abstr. 072.103.

The periodicity of flare occurrence in evolution of AR 5395.
See Abstr. 072.109.

Active region and flare relationships.
See Abstr. 072.116.

Motions, fields, and flares in the 1989 March active region.
See Abstr. 072.120.

Study of the morphological correlation between photospheric transversal magnetic field and penumbral filaments in AR5395.
See Abstr. 072.123.

An explanation of the dimensions of penumbral filaments.
See Abstr. 072.126.

Activity development in the area connected with the 23.08.88 proton flare.
See Abstr. 072.128.

He I 10830 Å observations of active regions.
See Abstr. 072.137.

Distribution and dynamics of a high temperature coronal plasma associated with chromospheric activity.
See Abstr. 074.017.

The thermal continuum in coronal loops: instability criteria and the influence of perpendicular thermal conductions.
See Abstr. 074.025.

Acceleration of electrons outside flares: coronal manifestation and possible origin.
See Abstr. 074.083.

Alfvén wave trapping, network microflaring, and heating in solar coronal holes.
See Abstr. 074.090.

A model for the formation of solar prominences.
See Abstr. 074.091.

Joule heating as an explanation for the differential emission measure structure and systematic redshifts in the Sun's lower transition region.
See Abstr. 074.095.

Avalanches and the distribution of solar flares.
See Abstr. 074.099.

Neutralized and unneutralized current patterns in the solar corona.
See Abstr. 074.103.

Concerning sources of eruptive flows on the Sun.
See Abstr. 074.104.

Coronal activity.
See Abstr. 074.113.

The bright points in the Ca II K–line and their relation to the inner network magnetic structures.
See Abstr. 075.005.

Magnetic field changes associated with a subflare and surge.
See Abstr. 075.008.

Evolution of vector magnetic field and the August 27, 1990 X–3 flare.
See Abstr. 075.009.

Dark points and microflares observed in the He I λ10830 line and their relation to the quiet sun magnetic fields.
See Abstr. 075.021.

The incorporation of the magnetic fluxes of similar polarity in active regions for Sept. – Oct. 1988 and relevant strong X–ray flares.
See Abstr. 075.044.

The merging of magnetic fluxes of similar polarity in solar active regions and the energetic flares.
See Abstr. 075.045.

On the structure of active–region magnetic fields. The homogeneous inverse problem.
See Abstr. 075.047.

Magnetic field structure of active regions. The homogeneous inverse problem.
See Abstr. 075.065.

Optical observations of flare–productive flux emergence.
See Abstr. 075.066.

The practical application of the magnetic virial theorem.
See Abstr. 075.068.

Fine structures of chromospheric magnetic field and material flow in a solar active region.
See Abstr. 075.071.

HXIS observations of weak X–ray sources.
See Abstr. 076.002.

The X–ray counterparts of UV microflares.
See Abstr. 076.003.

High resolution gamma–ray and hard X–ray spectroscopy of solar flares.
See Abstr. 076.005.

Solar flare X–ray observations with Ulysses.
See Abstr. 076.007.

Gamma ray observation from a behind–the–limb flare.
See Abstr. 076.008.

Stereoscopic observations of solar flare X–ray and gamma–ray sources using GRO instruments.
See Abstr. 076.015.

Automated search for solar microflares on the Burst And Transient Source Experiment.
See Abstr. 076.016.

A statistical approach to solar flare spectral analysis with BATSE.
See Abstr. 076.017.

Statistical properties of solar gamma–ray flares.
See Abstr. 076.019.

X–ray observations of filament eruption in the 1980 May 21 flare.
See Abstr. 076.021.

Some new statistical properties of solar hard X–ray, gamma ray bursts and their astrophysical implication for charged particle acceleration in solar flare.
See Abstr. 076.027.

Heating and flows in flares.
See Abstr. 076.033.

Studies of helium–like ion spectra with the SMM Flat Crystal Spectrometer.
See Abstr. 076.034.

Quality improvement of solar flare temperature and density diagnostics derived from BCS spectra.
See Abstr. 076.036.

The beam driven chromospheric evaporation model of large solar flares: a model getting "no respect" from the Sun.
See Abstr. 076.037.

The "onset" phase in the two ribbon flares.
See Abstr. 076.046.

Microwave characteristics of a behind–the–limb proton flare.
See Abstr. 077.004.

The attractor dimension of solar decimetric radio pulsations.
See Abstr. 077.022.

On a relationship between the flare activity index and radio emission of active regions.
See Abstr. 077.033.

The effects of plasma ambient parameters on spike emissions.
See Abstr. 077.042.

Some features of solar microwave emission of millisecond time scale.
See Abstr. 077.043.

The negative frequency drift of microwave bursts due to collisional and radiative losses of superthermal electrons.
See Abstr. 077.062.

Solar radio spike radiation coexisting with a loop prominence system.
See Abstr. 077.064.

Submillimeter solar limb profiles determined from observations of the total solar eclipse of 1988 March 18.
See Abstr. 077.068.

Quasi–periodic structures superimposed upon a large radio burst at 3.2 cm wavelength on May 23, 1990.
See Abstr. 077.071.

The observations and analyses of solar radio decimetric spike radiation.
See Abstr. 077.072.

Stochastic dynamics of protons in solar magnetic loops.
See Abstr. 078.005.

Particle injection following solar flares on 1980 May 28 and June 8: evidence for different injection time histories in impulsive and gradual events?
See Abstr. 078.009.

Charge states of solar energetic particles and correlations between particle intensities.
See Abstr. 078.010.

The features of the ground level enhancement event of cosmic rays on 29 September 1989.
See Abstr. 078.011.

The characters of cosmic ray events in Mar. & Aug. 1989.
See Abstr. 078.015.

Some studies of solar proton events in relation to active region characteristics.
See Abstr. 078.017.

Possible ground–level measurements of solar neutron decay protons during the 19 October 1989 solar cosmic ray event.
See Abstr. 078.018.

Multispacecraft observations of solar ^{3}He–rich events.
See Abstr. 078.019.

Accelerated particles in solar flares.
See Abstr. 078.022.

Interplanetary transport of decay protons from solar flare neutrons.
See Abstr. 078.027.

Acceleration of high–energy nuclei as viewed from their chemical composition.
See Abstr. 078.029.

Solar flare neutrinos.
See Abstr. 080.004.

Energy and equilibrium of electric currents in the atmosphere of the Sun.
See Abstr. 080.013.

Observations of waves and oscillations.
See Abstr. 080.017.

Line simulation of solar structures permeated by acoustic and MHD–waves.
See Abstr. 080.035.

Plasma heating by current sheets in solar active regions.
See Abstr. 080.045.

Large solar flares: analysis of the events recorded by the Mont Blanc neutrino detector.
See Abstr. 080.119.

On the SID durations.
See Abstr. 083.006.

A new technique for estimating D–region effective recombination coefficients under different solar flare conditions.
See Abstr. 083.022.

Solar flare and sudden change of Earth's rotation.
See Abstr. 085.027.

The distribution of solar flares and probable relations to biological effects.
See Abstr. 085.032.

Solar flares and terrestrial impacts of active region 5395.
See Abstr. 085.033.

Composition of solar flare noble gases preserved in meteorite parent body regolith.
See Abstr. 105.034.

Filament disappearances and associated shocks of May 1979.
See Abstr. 106.037.

What do stellar flares tell us about solar white light flares?
See Abstr. 116.016.

Comparison of solar and stellar flare ultraviolet continua.
See Abstr. 116.043.

On the role of accelerated particles in stellar flares.
See Abstr. 116.106.

Cosmic ray generation in space and laboratory plasmas.
See Abstr. 144.094.

074 Corona, Solar Wind

074.001 **Study of the structure of coronal inhomogeneities by the Faraday rotation of a radio signal.**
V. I. Bekker, M. V. Tokarov.
Geomagn. Aeron., Vol. 30, No. 1, p. 138 – 139 (Aug 1990). English translation of Geomagn. Aehron., Tom 30, No. 1, p. 163 – 165 (1990).
According to Helios–2 data, in the context of a model of the solar corona with a "bubble"–type transient, and considering the sector structure of the magnetic field of the Sun, an analysis is made of the Faraday rotation of a radio signal Λ traversing the solar corona. The electron distribution in the model considered is presented. Calculations are made of the dependence of Λ on a sighting parameter p and compared with the experimental data. The calculations of Λ(p) are in good agreement with the experimental data. This result is indicative of a realistic choice of the transient model.

074.002 **Vortex flows in the solar wind.**
I. S. Veselovskij.
Geomagn. Aeron., Vol. 30, No. 2, p. 248 – 249 (Oct 1990). English translation of Geomagn. Aehron., Tom 30, No. 2, p. 303 – 305 (1990).
The qualitative structure of vortex flows of the von Karman trail type near the heliomagnetic equator, which is a chain of "cyclones" or "anticyclones" traveling in a solar–wind stream, is discussed. Linearized equations of motion of a polytropic gas against the background of an axially symmetric supersonic expansion with a velocity minimum in the equatorial plane are used.

074.003 **Two sources of recurrent high–speed solar wind streams.**
M. I. Pudovkin, S. A. Zajtseva, N. A. Zakorina.
Geomagn. Aeron., Vol. 30, No. 2, p. 256 – 257 (Oct 1990). English translation of Geomagn. Aehron., Tom 30, No. 2, p. 310 – 312 (1990).
Variations of the mean diurnal values of the solar–wind velocity at the Earth's orbit during January – September 1979 are discussed. It has been shown that the value of the velocity in solar–wind streams formed near sector boundaries depends on the solar–radiation flux at a wavelength $\lambda = 10.7$ cm ($r = 0.59$); in streams generated within sectors, the velocity depends on the magnetic field intensity ($r = 0.38$) and does not depend on the radio emission power in the region of the source.

074.004 **Solar-wind parameters and magnetic activity during periods of uncompressed increases in the solar-wind density.**
N. M. Rudneva, P. M. Svidskij, G. I. Ol'.
Geomagn. Aeron., Vol. 30, No. 3, p. 315 - 318 (Nov 1990). English translation of Geomagn. Aehron., Tom 30, No. 3, p. 372 - 375 (1990).
The results of a statistical analysis of data on the solar-wind parameters and the ΣK_p indices of magnetic activity under conditions of extended increases of the solar-wind density against the background of a constant or decreasing velocity are given. It is shown that predominantly a low level of magnetic activity corresponds to such situations in the solar wind.

074.005 **Eruptive-active features on the Sun and the solar wind velocity in 1973 - 1976.**
N. M. Rudneva, P. M. Svidskij, G. I. Ol'.
Geomagn. Aeron., Vol. 30, No. 3, p. 319 - 322 (Nov 1990). English translation of Geomagn. Aehron., Tom 30, No. 3, p. 376 - 380 (1990).
It has been shown the the central parts of high-speed streams in the solar wind are projected in the overwhelming majority of cases onto quiet sections of the solar surface, and the periods of increasing and decreasing velocity can be projected onto the neighborhoods of eruptive-active features. The probability of attaining a velocity exceeding 700 km/s is appreciably increased if the outflow region of a high-speed stream borders on an active region. The question of the relation between the solar-wind velocity and processes in active regions is discussed.

074.006 **Radio range measurements of coronal electron densities at 13 and 3.6 centimeter wavelengths during the 1988 solar conjunction of Voyager 2.**
T. P. Krisher, J. D. Anderson, D. D. Morabito, S. W. Asmar, S. E. Borutzki, M. L. Delitsky, A. C. Densmore, P. M. Eshe, G. D. Lewis, M. J. Maurer, D. C. Roth, Y. H. Son, T. R. Spilker, D. N. Sweetnam, A. H. Taylor, G. L. Tyler, D. L. Gresh, P. A. Rosen.
Astrophys. J., Lett., Vol. 375, No. 2, p. L57 - L60 (10 Jul 1991).
The authors report on radio range measurements of total solar plasma delay obtained during the solar conjunction of the Voyager 2 spacecraft in 1988 December, which occurred near solar maximum activity in the 11 yr cycle. The radio range measurements were generated by the Deep Space Network at two wavelengths on the downlink from the spacecraft: 3.6 and 13 cm. A direct measurement of the integrated electron density along the ray path between the Earth stations and the spacecraft was obtained by differencing the range at the two wavelengths. Coronal electron density profiles have been derived during ingress and egress of the ray path, which approached the Sun to within $5\,R_\odot$. At $10\,R_\odot$, the derived density profiles yield $34079 \pm 611\ cm^{-3}$ on ingress and $49688 \pm 983\ cm^{-3}$ on egress. These density levels are significantly higher than observed near previous solar maxima.

074.007 **The structure of the transition region of the solar wind.**
N. A. Lotova, Ya. V. Pisarenko.
Geomagn. Aeron., Vol. 30, No. 4, p. 461 - 464 (Feb 1990). Paper presented at the 5. Symposium on Solar Terrestrial Physics, Samarkand (USSR), 2 - 6 Oct 1989. English translation of Geomagn. Aehron., Tom 30, No. 4, p. 545 - 549 (1990).
The dependence of the scattering angle of radio waves in the near-solar plasma on the radial distance R is investigated. A comparatively narrow region in which the scattering is sharply reduced precedes the extended transition region of enhanced scattering discovered earlier. This finding indicates a local reduction of the level of spatial fluctuations of the electron concentration. A similar structural peculiarity has also been discovered in the results of measurements of the average electron concentration N_e. The region of reduced scattering coincides with the zone of largest positive gradients of the solar-wind velocity and with the region in which the main acceleration of the near-solar plasma occurs.

074.008 **"Alfvénic" versus "standard" turbulence in the solar wind.**
R. Grappin, M. Velli, A. Mangeney.
Ann. Geophys., Vol. 9, No. 6, p. 416 - 426 (Jun 1991).
The authors study the variation of the properties of turbulence with stream structure, on time scales of hours and minutes, in the inner heliosphere at solar minimum. Between fast hot streams, this turbulence is found to show many properties typical of "standard" weakly compressible magnetohydrodynamic (MHD) turbulence such as excess of turbulent magnetic energy and a relative level of density fluctuation approximately equal to the turbulent Mach number squared. They discuss whether or not the more peculiar properties of Alfvénic turbulence, found within fast streams, represent some genuinely different state of MHD turbulence which might not relax towards standard turbulence at large distances from the sun and the ecliptic plane. The Ulysses spacecraft data should allow these possibilities to be distinguished.

074.009 **The structure of the meridian solar wind and its effect on the propagation of coronal mass ejections.**
Guo Wei-ping, Wang Jing-fang, Yao Yong-qiao, Liang Bai-xian.
Chin. J. Space Sci., Vol. 11, No. 3, p. 168 - 175 (Jul 1991). In Chinese.
A combined scheme is developed to solve the ideal MHD equations. Keeping certain precision, this model has ensured stability and small computing capacity. A two-dimensional helmet-streamer solution of the meridian solar wind near the sun is obtained. Using this solution as initial state, the effect of the helmet-streamer configuration on the propagation of coronal mass ejections is investigated. It is found that the configuration deflects the ejected mass toward the equator. This effect can explain some observational results satisfactorily.

074.010 **First scientific results from the Ulysses mission.**
R. G. Marsden, K.-P. Wenzel.
ESA Bull., No. 67, p. 78 - 83 (Aug 1991).
The successful launch of the Ulysses spacecraft on 6 October 1990 from the Space Shuttle "Discovery" marked the start of a five-year exploratory journey to study the Sun and its environment from a unique vantage point, a solar polar orbit. Although currently still in the ecliptic plane, the scientific payload carried by Ulysses is already returning new and valuable data.

074.011 **High-velocity fluxes of the solar wind and the fluxes of energetic electrons in the Earth's outer magnetosphere.**
K. G. Afanasiev (*K. G. Afanas'ev*), Ye. V. Gorchakov (*E. V. Gorchakov*), V. A. Iozenas, M. V. Ternovskaya.
Adv. Space Res., Vol. 11, No. 9, p. 283 - 286 (1991). - See Abstr. 012.006 for the main entry.

074.012 **Kinetic physics of the solar wind plasma.**
E. Marsch.
Physics of the inner heliosphere 2. Particles, waves and turbulence, p. 45 - 133 (1991). - See Abstr. 003.002 for the main entry.
Contents: (1) Introduction. (2) Particle velocity distributions. (3) Coulomb collisions and nonclassical transport. (4) Wave-particle interactions and anomalous transport: electrostatic waves and plasma instability, electromagnetic waves and plasma instability, electron heat flux regulation, regulation of the ion temperature anisotropy and heat flux, regulation of the electromagnetic ion-beam instability, some nonlinear aspects of wave-particle interactions. (5) Nonequilibrium thermodynamics. (6) Dissipation of MHD turbulence and particle heating. (7) Energy, momentum, and mass input into the solar wind: coronal boundary conditions, a brief account of coronal structures and associated interplanetary phenomena, interplanetary and coronal energy fluxes and solar wind constants of motion.

074.013 **MHD turbulence in the solar wind.**
E. Marsch.
Physics of the inner heliosphere 2. Particles, waves and turbulence, p. 159 – 241 (1991). – See Abstr. 003.002 for the main entry.

Contents: (1) Introduction. (2) Structures and fluctuations in the inner heliosphere. (3) MHD theory and turbulence diagnostics: basic equations, Alfvén waves and magnetosonic waves, analysis of random MHD fields, integral invariants of incompressible MHD, turbulence spectra, analysis techniques in the time domain. (4) Nature and morphology of MHD fluctuations: power spectra in terms of Elsässer variables, stream–structure dependence of Alfvénic fluctuations, compressive fluctuations. (5) Origin and evolution of solar wind MHD turbulence: sources of interplanetary fluctuations, radial evolution of Alfvénicity, radial evolution of spectra, self–similarity and origin of turbulence, evidence for the dissipation of turbulence. (6) Theoretical concepts of turbulence generation and evolution: numerical simulation, models of turbulence for inhomogeneous media, nonlinearities and cascading, compressibility and pseudosound. (7) Summary and conclusions.

074.014 **Solar Maximum Mission observations of activity associated with the coronal mass ejection of October 13, 1988.**
O. C. St. Cyr, J. T. Burkepile, J. L. R. Saba, R. Schwartz, E. Einfalt, S. A. Drake, W. T. Vestrand.
Bull. Am. Astron. Soc., Vol. 23, No. 2, p. 925 (1991). Abstract. – See Abstr. 012.008 for the main entry.

074.015 **Plasma flow in slowly diverging coronal structures.**
O. G. Badalyan.
Sov. Astron., Vol. 35, No. 3, p. 295 – 300 (May 1991). Current Physics Microform No.: 9201X0577. English translation of Astron. Zh., Tom 68, Vyp. 3, p. 602 – 611 (1991).

The outflow of hot isothermal plasma in a gravitational field in structures with less than radial divergence is investigated theoretically. A solution is found that passes through the critical point; its dependence of the parameters is investigated. It is found that the velocity is lower in such structures than in spherically symmetrical flow under the same physical conditions. The author has considered the case of large solar streamers earlier, and shown, in particular, that streamers are probably sources of low–velocity wind. Application of the theoretical results obtained here shows that large masses of gas can be transported outward in streamers.

074.016 **SMM's long, hard look at the solar corona.**
A. J. Hundhausen.
Bull. Am. Astron. Soc., Vol. 23, No. 2, p. 934 (1991). Abstract. – See Abstr. 012.008 for the main entry.

074.017 **Distribution and dynamics of a high temperature coronal plasma associated with chromospheric activity.**
W. M. Neupert, H. Zirin.
Bull. Am. Astron. Soc., Vol. 23, No. 2, p. 938 – 939 (1991). Abstract. – See Abstr. 012.008 for the main entry.

074.018 **VLBI angular broadening studies of plasma turbulence in the solar wind.**
T. Sakurai, S. R. Spangler, R. L. Mutel, J. Armstrong.
Bull. Am. Astron. Soc., Vol. 23, No. 2, p. 939 (1991). Abstract. – See Abstr. 012.008 for the main entry.

074.019 **Thermal equilibria of isobaric coronal magnetic arcades.**
C. D. C. Steele, E. R. Priest.
Sol. Phys., Vol. 134, No. 1, p. 73 – 97 (Jul 1991).

A coronal magnetic arcade can be thought of as consisting of an assembly of coronal loops. By solving equations of isobaric thermal equilibrium along each loop and assuming a base temperature of 2×10^4K, the thermal structure of the arcade can be found. The possible thermal equilibria can be shown to depend on two parameters L^*p^* and h^*/p^* representing the ratios of cooling (radiation) to conduction, and heating to cooling, respectively. Arcades can contain four types of loop: hot loops with summits hotter than 400000K; cool loops at temperatures less than 80000K along their lengths, hot–cool loops with cool summits and cool footpoints but hotter intermediate portions; and warm loops, cooler than 80000K along most of their lengths but with summits as hot as 400000K. Two possibilities for coronal heating are considered, namely a heating that is independent of magnetic field and a heating that is proportional to the square of the local magnetic field. When the arcade is sheared the thermal structure of the arcade may change, leading in some cases to non–equilibrium or in other cases to the formation of a cool core.

074.020 **Two–dimensional steady–state pressure structure in a coronal loop. I. Superposition of two Chandrasekhar–Kendall functions (n = 0, m = 0; n = 1, m = 0).**
S. Chandra, L. Prasad.
Sol. Phys., Vol. 134, No. 1, p. 99 – 107 (Jul 1991).

By expressing the magnetic field and fluid velocity in terms of two Chandrasekhar–Kendall functions (n = 0, m = 0; n = 1, m = 0) the authors investigated the steady–state pressure profile inside a solar coronal loop. For constant density loops, they found a two–dimensional (radial and axial) structure of pressure. At the base of the loop, the pressure is found to increase steeply outwards along the radius, whereas at the apex it decreases slowly. The radial variation of pressure is found to be minimum around L/5, where L is the length of the loop measured from one foot to another one. For axial variation, the authors found that along the loop axis the pressure increases from the base up to z = 3L/8 and then decreases up to the apex, whereas at the surface, the pressure decreases from the base up to the apex. This work is a modified version of the work of Krishan (1985).

074.021 **Comment on the preceding paper by Chandra and Prasad.**
V. Krishan.
Sol. Phys., Vol. 134, No. 1, p. 109 (Jul 1991). See 074.020.

074.022 **Kelvin–Helmholtz instability in an Alfvén resonant layer of a solar coronal loop.**
E. Uchimoto, H. R. Strauss, W. S. Lawson.
Sol. Phys., Vol. 134, No. 1, p. 111 – 121 (Jul 1991).

A Kelvin–Helmholtz instability has been identified numerically on an azimuthally symmetric Alfvén resonant layer in an axially bounded, straight cylindrical coronal loop. The physical model employed is an incompressible, reduced magnetohydrodynamic model including resistivity, viscosity, and density variation. The set of equations is solved numerically as an initial value problem. The linear growth rate of this instability is shown to be approximately proportional to the Alfvén driving amplitude and inversely proportional to the width of the Alfvén resonant layer. It is also shown that the linear growth rate increases linearly with m–1 up to a certain m, reaches its maximum value for the mode whose half wavelength is comparable to the Alfvén resonant layer width, and decreases at higher m's (m is the azimuthal mode number).

074.023 **A plasmoid model for the solar wind.**
T. Tamano.
Sol. Phys., Vol. 134, No. 1, p. 187 – 201 (Jul 1991).

A model describing magnetized plasmoids as a possible origin of the solar wind is discussed. The magnetized plasmoids are assumed to be created and accelerated to a very high speed through reconnection processes from small–scale magnetic loops. Afterward, the plasmoids are considered to be nearly in a relaxed state under magnetic helicity conservation and to expand freely and linearly. Characteristics of such plasmoids with finite β are examined. The results show remarkable agreement between the model predictions and spacecraft observations including temperature characteristics such as the dependence on the heliocentric distance and ion mass. The validity of the assumptions and the applicability of the model are also discussed.

074.024 **Evolution of magnetic clouds in the solar wind.**
S. J. Goldstein Jr.
Bull. Am. Astron. Soc., Vol. 23, No. 2, p. 960 (1991). Abstract. – See Abstr. 012.008 for the main entry.

074.025 **The thermal continuum in coronal loops: instability criteria and the influence of perpendicular thermal conductions.**
R. A. M. van der Linden, M. Goossens.
Sol. Phys., Vol. 134, No. 2, p. 247 – 273 (Aug 1991).
The linear MHD spectrum is investigated for cylindrical equilibrium models under typical coronal conditions. Non-ideal effects are included and attention is focussed on the thermal instability and the influence of perpendicular thermal conduction. It is shown that, when thermal conduction across magnetic field lines is neglected, the "classic" Alfvén and slow continua are supplemented by a new "thermal" continuum. Surprisingly, the existence of this non-ideal continuous spectrum appears to have been overlooked for a very long time. Unlike the Alfvén continuum modes and the slow continuum modes, the thermal continuum modes are exponentially growing or decaying in time. As with the Alfvén and slow conitnua, discrete modes may be present above or below the thermal continuum, depending upon the choice of equilibrium parameters. These modes are localized using a simple WKB approach. The knowledge of the thermal subspectrum is then exploited to find necessary and sufficient conditions for instability. The inclusion of perpendicular thermal conduction eliminates this thermal contiuum, but replaces it by a dense set of eigenmodes located at the same values of the grwoth rate. It is shown that perpendicular thermal conduction has no significant influence on the thermal stability of the equilibrium, though individual modes may be strongly influenced. The corresponding eigenfunctions still have a nearly-singular behaviour, but in addition they may contain a rapid spatial oscillation on a scale proportional to some power of the perpendicular thermal conduction coefficient. This rapid oscillation is confined to the most unstable part of the equilibrium and may be of relevance for the formation of fine-scale structure (threads) in prominences.

074.026 **Coronal dynamics of a quiescent active region.**
J. L. R. Saba, K. T. Strong.
Astrophys. J., Vol. 375, No. 2, p. 789 – 799 (10 Jul 1991).
The Flat Crystal Spectrometer (FCS) onboard SMM was used to investigate coronal X-ray line broadening in a nonflaring active region. The resulting map of FCS line profiles across the central X-ray bright portion of a large active region shows a clear variation in line width for the Mg XI resonance line at 9.17 Å, but no evidence of line shifts. Line ratios yield electron temperatures $T_e \sim 3\times10^6$K within 10% for most of the map. The brightest, broadest profiles are consistent with emission from the tops of loops overlying the magnetic neutral line. For the spatially averaged Mg XI resonance line profile, the "excess velocity" v_{ex}, corresponding to the excess width above the thermal value, is equal to the thermal velocity for Mg ions, 45 km s^{-1}; for some individual bright pixels, $v_{ex} \geqslant 60$ km s^{-1} with high significance ($\geqslant 5\sigma$). An instrumental source for the broadening is ruled out by the presence of narrow (i.e., only thermally broadened) line profiles in some areas of the map; contributions from higher temperature plasma are inconsistent with the observed lack of flux in the high-temperature FCS channels. The observed coronal line broadening is interpreted as a signature of the mechanism(s) responsible for coronal heating in active regions.

074.027 **Measurements of 1-Hz coronal oscillations at total eclipses and their implications for coronal heating.**
J. M. Pasachoff.
International Conference on Mechanisms of Chromospheric and Coronal Heating, p. 25 – 29 (1991). – See Abstr. 012.011 for the main entry.
As a test of predictions of solar coronal heating via surface Alfvén waves, the Author observed small coronal loops during the total solar eclipses of 1980 and 1983. The results indicate excess power in the 0.5 – 2 Hz range at the level of 1% in the coronal green line relative to continuum oscillation power. The results indicate that the surface Alfvén-wave models of Ionson and the subsequent Alfvénic models of Hollweg can supply adequate energy to heat the corona.

074.028 **Short-term oscillations in green and red coronal lines.**
V. Rusin, M. Minarovjech.
International Conference on Mechanisms of Chromospheric and Coronal Heating, p. 30 – 32 (1991). – See Abstr. 012.011 for the main entry.
Analysis of the 530.3 and 637.4 nm coronal line intensities is carried out on the basis of data sets obtained with a photoelectric photometer at Lomnicky Stit coronal station. The most significant oscillations (waves) have periods of 312 s in the green corona, and 296 s in the red one. Some other, shorter periods, i.e. of 222, 90,..., 30 s (important) etc. were detected too with less significance.

074.029 **Height-dependent short-period oscillations in the Fe XIV (530.3 nm) solar corona above a sunspot group crossing the limb.**
V. N. Dermendjiev.
International Conference on Mechanisms of Chromospheric and Coronal Heating, p. 33 – 35 (1991). – See Abstr. 012.011 for the main entry.
A sequence of 86 green coronal line spectra is analyzed in order to find short-period oscillations in the corona. It is shown that a tendency exists of the revealed periods to change with the height above the limb.

074.030 **On the temperature inhomogeneity of the lower solar corona.**
K. I. Nikol'skaya.
International Conference on Mechanisms of Chromospheric and Coronal Heating, p. 113 – 114 (1991). – See Abstr. 012.011 for the main entry.
Temperature inhomogeneities of the plasma in the active and undisturbed solar corona are investigated on the basis of ground eclipse observations of coronal spectra and monochromatic images. It is shown that the lower solar corona is a mixture of near isothermal fine structures of different temperatures. Fine multi-temperature structuring of the coronal plasma has to be taken into a account in theoretical developments of the heating mechanisms.

074.031 **Spectra of MHD turbulence in coronal active regions.**
D. Gomez, P. Martens, M. Herant, F. Pardo, L. Golub.
International Conference on Mechanisms of Chromospheric and Coronal Heating, p. 124 – 126 (1991). – See Abstr. 012.011 for the main entry.
The authors present results of a search for observational evidence for MHD turbulence in coronal active regions. They have chosen three active regions observed by NIXT in its rocket flight of September 11, 1989. For each of these regions the authors obtained a broadband omnidirectional power spectrum for the soft X-ray intensity, indicating the presence of a fully developed turbulent state.

074.032 **Observations of coronal bright points and implications for coronal heating mechanisms.**
S. R. Habbal.
International Conference on Mechanisms of Chromospheric and Coronal Heating, p. 127 – 131 (1991). – See Abstr. 012.011 for the main entry.
A study of the observational characteristics of coronal bright points shows that they are the preferential site for the variable emission from the quiet Sun. They seem to define the basic small scale closed magnetic structure in the quiet Sun. To account for the selective heating of loops in bright points to different temperatures, and for the characteristic radiative losses from the steady and variable components of the emission, viable coronal heating mechanisms should consist of a spectrum of energies and time scales.

074.033 **Modelling coronal active region emission patterns.**
S. F. Brown, C. J. Durrant.
International Conference on Mechanisms of Chromospheric and Coronal Heating, p. 132 – 134 (1991). – See Abstr. 012.011 for the main entry.
The authors examine an active region model constructed from an ensemble of simple loops placed in a dipole magnetic structure. The emission pattern in EUV and X–rays produced by the model depends sensitively on the model heating function. Of the six models tested, the most realistic emission patterns arose from heating functions with an inverse dependence on the loop volume. Both resonant heating and nanoflare heating theories predict such a dependence and therefore pass the test.

074.034 **The Fe ionization equilibrium in the solar corona with a non–Maxwellian distribution function.**
E. Dzifcakova.
International Conference on Mechanisms of Chromospheric and Coronal Heating, p. 135 – 136 (1991). – See Abstr. 012.011 for the main entry.
It is shown that for ionization equilibria in the solar transition layer and corona, non–Maxwellian distribution functions are important.

074.035 **Energetics of solar coronal holes.**
Lou Yu–qing.
International Conference on Mechanisms of Chromospheric and Coronal Heating, p. 137 – 139 (1991). – See Abstr. 012.011 for the main entry.
The author examines the energetics of a solar coronal hole in the framework of magnetohydrodynamics. Given typical values of the observed parameters, it is estimatet that the Poynting flux maintained by the solar rotation and a trailing magnetic field is potentially capable of driving and heating a wind stream in a coronal hole region.

074.036 **Properties of impulsive events in a polar coronal hole.**
S. Koutchmy, M. L. Loucif.
International Conference on Mechanisms of Chromospheric and Coronal Heating, p. 152 – 158 (1991). – See Abstr. 012.011 for the main entry.
Observations made at NSO/Sacramento Peak Observatory with the 16″ coronagraph allow the analysis of fast impulsive events. Broad–band Hα filtergrams and CCD–spectra taken at 1 to 45 arcsec above the solar chromospheric–limb are used. A deep well–developed coronal hole was chosen above a polar region, during years of minimum of sunspot activity. The authors present and analyze a whole range of impulsive events observed there in 1988. The results strongly suggest they could be the source of the fast wind. The recurrence of impulsive events is reported for the first time.

074.037 **Explosive instability in solar coronal loops.**
M. Ryutova.
International Conference on Mechanisms of Chromospheric and Coronal Heating, p. 159 – 161 (1991). – See Abstr. 012.011 for the main entry.
The presence of shear flows along the stretched magnetic field structures such as solar coronal loops change their dynamics essentially, and, particularly, can lead to the explosive instabilty which results in fast and efficient energy release.

074.038 **Dynamics of erupting 3D toroidal current loops.**
J. Chen.
Bull. Am. Astron. Soc., Vol. 23, No. 2, p. 1035 – 1036 (1991). Abstract. – See Abstr. 012.012 for the main entry.

074.039 **Signatures of coronal structures and turbulence in the solar wind.**
E. Marsch.
International Conference on Mechanisms of Chromospheric and Coronal Heating, p. 162 – 164 (1991). – See Abstr. 012.011 for the main entry.
Prominent structures and fluctuations in the solar wind appear to be closely associated with coronal magnetic structures and turbulence. Coronal hole related high–speed flows act as wave guides conducting Alfvénic fluctuations of solar origin into interplanetary space. Medium–sized plasma flow tubes seem to originate from coronal fine–structure presumed to be identical with the plumes seen in coronagraph pictures.

074.040 **The three–dimensional structure of coronal mass ejections.**
R. S. Steinolfson.
Bull. Am. Astron. Soc., Vol. 23, No. 2, p. 1036 (1991). Abstract. – See Abstr. 012.012 for the main entry.

074.041 **A novel way to convert Alfvén waves to heat in coronal holes: reflective damping.**
R. L. Moore.
Bull. Am. Astron. Soc., Vol. 23, No. 2, p. 1037 (1991). Abstract. – See Abstr. 012.012 for the main entry.

074.042 **Dark gaps in emission corona images.**
R. N. Smartt, Z. Zhang, D. A. Landman.
Bull. Am. Astron. Soc., Vol. 23, No. 2, p. 1045 (1991). Abstract. – See Abstr. 012.012 for the main entry.

074.043 **Element abundances in a coronal polar plume.**
K. G. Widing, U. Feldman.
Bull. Am. Astron. Soc., Vol. 23, No. 2, p. 1046 (1991). Abstract. – See Abstr. 012.012 for the main entry.

074.044 **A model for the anomalous elemental abundances of the solar corona and wind.**
S. K. Antiochos.
Bull. Am. Astron. Soc., Vol. 23, No. 2, p. 1046 (1991). Abstract. – See Abstr. 012.012 for the main entry.

074.045 **Coronal temperature estimate from intensity obtained during the total solar eclipse of 1988 March 18.**
M. Guhathakurta, G. J. Rottman, R. R. Fisher, F. Q. Orrall.
Bull. Am. Astron. Soc., Vol. 23, No. 2, p. 1046 (1991). Abstract. – See Abstr. 012.012 for the main entry.

074.046 **Reflection of Alfvén waves and heating in solar coronal holes.**
R. L. Moore, Z. E. Musielak, S. T. Suess, C.–H. An.
International Conference on Mechanisms of Chromospheric and Coronal Heating, p. 435 – 437 (1991). – See Abstr. 012.011 for the main entry.
Using a simple model coronal hole, the authors estimate the short–period period limit for the reflection of Alfvén waves in solar coronal holes. They find that the value of this limiting period and its dependence on temperature and field spreading encourage the view of Hollweg (1981, 1986) and Parker (1987, 1988) that coronal holes are heated by Alfvén waves.

074.047 **On propagation and absorption of Alfvén waves in coronal loops.**
Yu. D. Zhugzhda.
International Conference on Mechanisms of Chromospheric and Coronal Heating, p. 442 – 444 (1991). – See Abstr. 012.011 for the main entry.
The theories of loop resonances are considered. The possibility of phase–mixing and resonant absorption in the solar atmosphere is discussed.

074.048 **Resonant absorption of MHD waves in magnetic loops in the solar corona.**
M. Goossens.
International Conference on Mechanisms of Chromospheric and Coronal Heating, p. 480 – 485 (1991). – See Abstr. 012.011 for the main entry.
Results of recent large scale numerical computations on the stationary state of resonant absorption of MHD waves in coronal loops are discussed. Analytic theory is introduced as a helpfull tool for the interpretation of some of these numerical results. The need for further large scale numerical projects for

determining the temporal and/or non-linear behaviour of resonant absorption is indicated.

074.049 **On the time scales and the efficiency of solar coronal loop heating by resonant absorption.**
S. Poedts.
International Conference on Mechanisms of Chromospheric and Coronal Heating, p. 486 – 488 (1991). – See Abstr. 012.011 for the main entry.
The heating of solar coronal loops by the resonant absorption of Alfvén waves is studied in the framework of linearized compressible and resistive magnetohydrodynamics. The loops are approximated by straight cylindrical, axisymmetric plasma columns excited by an external source. The temporal evolution of this driven system is simulated by means of a very accurate numerical code. The time-scales and the efficiency of resonant absorption in solar coronal loops are investigated for different driving sources.

074.050 **Line-tying effects on stability and heating of solar coronal loops.**
G. Halberstadt, J. P. Goedbloed, S. M. Poedts,
R. A. M. Van der Linden.
International Conference on Mechanisms of Chromospheric and Coronal Heating, p. 489 – 491 (1991). – See Abstr. 012.011 for the main entry.
The effect of line-tying of solar coronal loops on both stability and heating by means or resonant absorption of Alfvén waves is investigated numerically. It is shown that line-tying alone is not stabilizing, whereas the joint effect of line-tying and neighbouring flux tubes may stabilize a coronal loop.

074.051 **Coronal loop heating by discrete Alfvén waves.**
C. A. Azevedo, A. S. de Assis, H. Shigueoka,
P. H. Sakanaka.
International Conference on Mechanisms of Chromospheric and Coronal Heating, p. 492 – 493 (1991). – See Abstr. 012.011 for the main entry.
The authors have modeled the solar coronal loop heating by discrete Alfvén waves. The authors have modeled the coronal loop as a semi-toroidal plasma. They have shown that the absorptions of the discrete Alfvén wave by the plasma through the viscosity can account for at least 20% of the necessary coronal power density of 10^{-9}erg cm^{-3}s^{-1}.

074.052 **Rapid plasmoid formation of double arcades.**
J. M. Finn, P. N. Guzdar, J. Chen.
Bull. Am. Astron. Soc., Vol. 23, No. 2, p. 1058 (1991). Abstract. – See Abstr. 012.012 for the main entry.

074.053 **Three dimensional reconnection for coronal plasmoid structures with and without field nulls.**
Y.-T. Lau, J. M. Finn.
Bull. Am. Astron. Soc., Vol. 23, No. 2, p. 1058 (1991). Abstract. – See Abstr. 012.012 for the main entry.

074.054 **The effect of strong shear on solar coronal magnetic fields.**
C. R. DeVore, S. K. Antiochos.
Bull. Am. Astron. Soc., Vol. 23, No. 2, p. 1058 (1991). Abstract. – See Abstr. 012.012 for the main entry.

074.055 **Modelling the large-scale structure of the solar corona.**
F. Bagenal, S. Gibson.
Bull. Am. Astron. Soc., Vol. 23, No. 2, p. 1058 (1991). Abstract. – See Abstr. 012.012 for the main entry.

074.056 **Three-dimensional reconnection in coronal loops.**
H. R. Strauss.
Bull. Am. Astron. Soc., Vol. 23, No. 2, p. 1059 (1991). Abstract. – See Abstr. 012.012 for the main entry.

074.057 **A dynamical model for heating of coronal loops.**
P. Beaufumê, B. Coppi, L. Golub.
Bull. Am. Astron. Soc., Vol. 23, No. 2, p. 1059 – 1060 (1991). Abstract. – See Abstr. 012.012 for the main entry.

074.058 **The effect of relative elemental abundances on broad-band X-ray diagnostics of solar coronal plasmas.**
K. Waljeski, D. Moses.
Bull. Am. Astron. Soc., Vol. 23, No. 2, p. 1061 (1991). Abstract. – See Abstr. 012.012 for the main entry.

074.059 **Coronal mass flows and turbulence from high resolution extreme ultraviolet observations.**
W. M. Neupert, J. W. Brosius, W. T. Thompson, R. J. Thomas.
Bull. Am. Astron. Soc., Vol. 23, No. 2, p. 1062 (1991). Abstract. – See Abstr. 012.012 for the main entry.

074.060 **Spectra of MHD turbulence in the solar corona.**
D. Gómez, P. C. H. Martens.
Bull. Am. Astron. Soc., Vol. 23, No. 2, p. 1062 (1991). Abstract. – See Abstr. 012.012 for the main entry.

074.061 **The scale sizes of CMEs and associated surface activity.**
D. F. Webb, K. L. Harvey, E. W. Cliver, S. W. Kahler.
Bull. Am. Astron. Soc., Vol. 23, No. 2, p. 1062 (1991). Abstract. – See Abstr. 012.012 for the main entry.

074.062 **Electric fields in post flare loops.**
S. Hinata.
Bull. Am. Astron. Soc., Vol. 23, No. 2, p. 1064 (1991). Abstract. – See Abstr. 012.012 for the main entry.

074.063 **Fe XIV and Fe X limb flux as proxies for space- and ground-based global solar parameters.**
R. C. Altrock, T. W. Henry.
Bull. Am. Astron. Soc., Vol. 23, No. 2, p. 1068 (1991). Abstract. – See Abstr. 012.012 for the main entry.

074.064 **Effect of coronal heating on coronal arcades.**
C. D. C. Steele, E. R. Priest.
International Conference on Mechanisms of Chromospheric and Coronal Heating, p. 544 – 546 (1991). – See Abstr. 012.011 for the main entry.
A coronal magneric arcade can be thought of as consisting of an assembly of coronal loops. Equations of thermal equilibrium are solved along each loop, assuming a base temperature of 20000K. Arcades can contain four types of loops: hot loops with summits hotter than about 400000K, cool loops completely at temperatures of less than 80000K, hot-cool loops with cool summits and footpoints but hotter intermediate portions and warm loops cooler than 80000K for most of their lengths but with summits as hot as 400000K. By adding the loop solutions the thermal structure of the whole arcade can be found.

074.065 **Survival of discrete structures in the solar wind.**
D. J. Mullan.
Astron. Astrophys., Vol. 248, No. 1, p. 256 – 262 (Aug 1991).
Structures ejected by the Sun into the solar wind are subject to decay processes during their transit from Sun to Earth. Here, the author estimates the survival time-scales for these structures against magnetic reconnection into the ambient field. Structures with initial dimensions less than a few tens of Mm probably cannot survive to the Earth's orbit.

074.066 **Micro/nanoflare coronal heating.**
E. N. Parker.
International Conference on Mechanisms of Chromospheric and Coronal Heating, p. 615 – 629 (1991). – See Abstr. 012.011 for the main entry.
Almost all strong continuous deformations of an initially uniform magnetic field in an infinitely conducting fluid cause the field to develop internal tangential discontinuities as the field relaxes to equilibrium. It is suggested that the dissipation at the tangential discontinuities in the bipolar magnetic fields of active

regions is the principal heat source for the solar X-ray corona. The dissipation is expected to be sporadic, so that the magnetic energy is converted to heat in small transient bursts which are called nanoflares. This theoretical conclusion accounts for the highly variable transition zone radiation and for the observed emission spectrum.

074.067 **Coronal heating by nanoflares: plasma dynamics of elementary events.**
R. A. Kopp, G. Poletto.
International Conference on Mechanisms of Chromospheric and Coronal Heating, p. 634 – 636 (1991). – See Abstr. 012.011 for the main entry.

The corona may be stochastically heated by a continuous succession of many small flarelike events commonly referred to as nanoflares. The authors extend a semianalytical "point" model developed previously for solar compact flares, to include gravitational plasma downfall during the late decay phase. Applying the model to conditions representative of nanoflares allows to predict the temporal variation of average plasma properties in such events and should ultimately facilitate a calculation of the spectral characteristics of a nanoflare-heated corona.

074.068 **Coronal heating by nanoflares: possible evidence of plasmoids in radio occultation data.**
D. J. Mullan.
International Conference on Mechanisms of Chromospheric and Coronal Heating, p. 637 – 639 (1991). – See Abstr. 012.011 for the main entry.

Coronal heating at multiple reconnection sites is expected to supply plasmoids to the corona with length scales of the order of a few hundred kilometers. Evidence for the existence of structures on these scales may have already been detected in radio occulation studies of the inner wind. Such plasmoids will contribute to accelerating the solar wind.

074.069 **Relative distribution of "hot" and "cold" coronal matter.**
V. L. Merzlyakov.
Soln. Dannye, Byull., No. 11/90, p. 92 – 95 (Nov 1990). In Russian.

During October–December, 1983, the spectrograms of the coronal lines $\lambda 5303$ Å and $\lambda 6374$ Å have been obtained at the 20–cm coronograph. The instrumental profile of the spectrograph was ≈ 0.4 Å. Influence of the height and the latitude on the intensity ratio of these lines was investigated. Using this investigation and some assumptions it was established that hot matter–cold matter ratio increased with height. The value of this ratio was 2.5 – 4.0 with the height intervals 33 – 97″ and the latitude intervals 5°N – 20°S.

074.070 **On the height variation of the intensities and the profile widths of the lines $\lambda 5303$ Å and $\lambda 6374$ Å in the solar corona.**
V. L. Merzlyakov.
Soln. Dannye, Byull., No. 12/90, p. 79 – 83 (Dec 1990). In Russian.

Spectrograms of the coronal lines $\lambda 5303$ Å and $\lambda 6374$ Å have been received at the 20–cm coronograph at the latitudes 5°N – 20°S during October – December, 1983. The study of the spectrograms shows that the value of the intensity gradient decreased with the increase of the profile width of the lines. Also, the value of the intensity gradient of the line $\lambda 5303$ Å decreased and the value of the intensity gradient of the line $\lambda 6374$ Å increased with the increase of the region brightness. The widths of the line profiles.

074.071 **The phase mixing of Alfvén waves, coordinated modes, and coronal heating.**
E. N. Parker.
Astrophys. J., Vol. 376, No. 1, p. 355 – 363 (20 Jul 1991).

Phase mixing has been a popular theoretical mechanism invoked to dissipate the photospheric plane shear Alfvén waves hypothesized to penetrate up into the coronal hole and solar wind, thereby maintaining the expanding coronal gas at 1.5×10^6K and serving as the prinicpal energy source for the solar wind. However, phase mixing requires an ignorable coordinate, which is not available in the presumably filamentary coronal hole. The filamentary density and temperature of the coronal hole couple the waves into a coordinated mode with a unique phase velocity ω/k, which provides the other popular theoretical mechanism – resonant absorption where the local Alfvén speed C is equal to ω/k. This provides coronal heating at radial distances $O(10\ R_\odot)$, where it serves to accelerate the solar wind to high velocity. But it does not provide the principal heat input in the first 1 – 2 $R_\odot$, required by Withbroe's analysis of the observed structure of the coronal hole.

074.072 **Coronal mass ejection.**
R. A. Harrison.
Royal Society Discussion Meeting: The physics of solar flares, p. 401 – 412 (1991). – See Abstr. 012.016 for the main entry.

The author summarizes the observational aspects of the transient solar coronal features known as coronal mass ejections. Recognizing the importance of understanding this form of solar activity, particularly in the light of relations to flare and prominence activity, and geomagnetic effects, he considers the spectrum of models which have been used to describe these events and assess their viability. The author finds most models to be unphysical and all represent a gross over simplification of solar conditions. In conclusion he sets up a cartoon model which best fits the observations and which should be further developed.

074.073 **Shape of the magnetic cloud in the solar wind.**
E. P. Romashets, K. G. Ivanov.
Geomagn. Aehron., Tom 31, No. 4, p. 583 – 586 (Jul – Aug 1991). In Russian. English translation in Geomagn. Aeron., Vol. 31, No. 4.

074.074 **Aspects of the present heliospheric research.**
H. J. Fahr.
Reviews in Modern Astronomy 4, p. 126 – 144 (1991). – See Abstr. 003.008 for the main entry.

First of all the question is raised why the Sun at all is surrouned by a circumsolar heliosphere, instead by a Strömgren sphere configuration. The fact that counts most in this context is the evident peculiar motion of the Sun relative to the ambient interstellar medium. An interesting puzzle thus is how the inner heliospheric plasma properties reflect this fact. The author studies the penetration of neutral interstellar atoms through the perturbed plasma region ahead of the solar system into the inner regions of the heliosphere where their presence has been proven by their typical resonance glows. First he gives the main outlines of the theory connected with this problem and then addresses in detail the question of the neutral hydrogen and oxygen presence in the heliosphere. The conclusion is reached that for interstellar ion densities of 10^{-2}cm^{-3} or greater neutral LISM hydrogen or oxygen will be depleted at least by factors of 0.5 or 0.1, respectively. The problem still stands whether or not the anomalous cosmic ray components showing up as He^+, O^+, Ne^+, N^+–ions of about 10 MeV/nucleon in the inner heliosphere can be ascribed to LISM neutrals entering the heliosphere and becoming ionized and reenergized there.

074.075 **Turbulence in the solar wind.**
E. Marsch.
Reviews in Modern Astronomy 4, p. 145 – 156 (1991). – See Abstr. 003.008 for the main entry.

The solar wind is, apparently, a turbulent medium permeated by fluctuations on a broad range of scales characterizing the plasma as a fluid and kinetic entity. The author briefly reviews some recent developments and observations on MHD fluctuations in the frequency range between some 10^{-6} and 10^{-2}Hz. Spectral analysis reveals that solar wind fluctuations represent ongoing turbulence, often with high Alfvénic correlations of coronal origin, which undergoes considerable spatial and spectral evolution in the radially expanding wind and is coupled to large–

scale plasma structures and magnetic field inhomogeneities. Theoretical attempts to model the transport of magnetohydrodynamic fluctuations are discussed. Finally, the dissipation of turbulence and concurrent heating of the wind are addressed.

074.076 **The green coronal line above an active region at the solar limb on 5 Aug. of 1989.**
Du Jinsheng, N. F. Tyagun, G. M. Khmyrov.
Chin. J. Space Sci., Vol. 11, No. 4, p. 286 – 290 (Oct 1991). In Chinese.
The green coronal line above the sunspot group No. 5629 of 5 Aug. of 1989 with excited continuum and bright knots is observed. According to the intensities of the continuum, the following parameters are estimated: the electron densities are $3 \cdot 10^9 - 10^{11}$/cm, half width is up to 1.2 Å, temperature is $4 - 5 \cdot 10^6$K without considering turbulent velocity, and the sizes of strong radiation regions along the direction of line of sight are $0.5 \cdot 10^9$cm.

074.077 **The solar corona over the recent saros.**
J. M. Pasachoff.
IAU Colloquium No. 130: The Sun and cool stars – activity, magnetism, dynamos, p. 285 – 287 (1991). – See Abstr. 012.024 for the main entry.
We are now at both the maximum of the solar activity cycle and at the most populated part of the saros. The author discusses the solar corona over the recent saros and its changes with the solar activity cycle. He considers the scientific value of eclipse studies and how they relate to other ongoing coronal studies on the Sun and other stars.

074.078 **Numerical simulation of shock waves in the heliosphere.**
A. G. Kosovichev, T. V. Stepanova.
Astron. Zh., Tom 68, Vyp. 6, p. 1283 – 1298 (Nov–Dec 1991). In Russian. English translation in Sov. Astron., Vol. 35, No. 6.
Results of modelling shock waves, generated by coronal transients and expanding flare loops, are discussed. Loss of hydrostatic equilibrium of a magnetic loop (as a result of an increase of magnetic field under the loop, or as a result of flare heating of plasma inside the loop) is the basis of modelling. The espanding magnetic loops act as a piston on the coronal plasma and produce shock waves propagating in the solar wind. The shock waves are computed in a one–dimensional, two–fluid approximation, taking into account processes of turbulent dissipation. The authors obtain shock wave parameters as functions of the initial perturbations of the magnetic loops.

074.079 **Short–term variability of solar wind number density, speed and dynamic pressure as a function of the interplanetary magnetic field components: a survey over two solar cycles.**
G. A. Bowe, M. A. Hapgood, M. Lockwood, D. M. Willis.
Geophys. Res. Lett., Vol. 17, No. 11, p. 1825 – 1828 (Oct 1990).
The variability of hourly values of solar wind number density, number density variation, speed, speed variation and dynamic pressure with IMF B_z and magnitude |B| has been examined for the period 1965–1986. The authors wish to draw attention to a strong correlation in number density and number density fluctuation with IMF B_z characterized by a symmetric increasing trend in these quantities away from $B_z = 0$ nT. The fluctuation level in solar wind speed is found to be relatively independent of B_z. The authors infer that number density and number density variability dominate in controlling solar wind dynamic pressure and dynamic pressure variability. It is also found that dynamic pressure is correlated with each component of IMF and that there is evidence of morphological differences between the variation with each component. Finally, the authors examine the variation of number density, speed, dynamic pressure and fluctuation level in number density and speed with IMF magnitude |B|. Again they find that number density variation dominates over solar wind speed in controlling dynamic pressure.

074.080 **A three–dimensional simulation of a coronal streamer.**
J. A. Linker, G. van Hoven, D. D. Schnack.
Geophys. Res. Lett., Vol. 17, No. 13, p. 2281 – 2284 (Dec 1990).
The authors have used a time–dependent magnetohydrodynamic (MHD) simulation to investigate the magnetic topology of an idealized streamerlike configuration in three dimensions. Starting from an initially current–free multipole field and a transonic flow, a quasi–steady, streamerlike configuration forms. The simulated streamer exhibits a closed magnetic field region bounded by a current layer and surrounded by open fields. This helmet region is surrounded by a current layer, with an enhanced density in the closed field and current sheet regions. The simulated streamer has a finite longitudinal extent, with the closed field region primarily confined to the center of the structure.

074.081 **Observations of disconnection of open coronal magnetic structures.**
D. J. McComas, J. L. Phillips, A. J. Hundhausen, J. T. Burkepile.
Geophys. Res. Lett., Vol. 18, No. 1, p. 73 – 76 (Jan 1991).
The authors have surveyed the Solar Maximum Mission coronagraph/polarimeter observations for evidence of magnetic disconnection of previously open magnetic structures and have identified several sequences of images consistent with this interpretation. Such disconnection occurs when open field lines above helmet streamers reconnect, in contrast to previously suggested disconnections of CMEs into closed plasmoids. In this paper a clear example of open field disconnection is shown in detail.

074.082 **Solar–wind krypton and solid/gas fractionation in the early solar nebula.**
R. C. Wiens, D. S. Burnett, M. Neugebauer, R. O. Pepin.
Geophys. Res. Lett., Vol. 18, No. 2, p. 207 – 210 (Feb 1991).
Krypton is the best candidate for determining limits on solid/gas fractionation in the early sun because of the smoothness of the odd–mass abundance curve in its mass region, which permits relatively precise interpolations of its abundance assuming no fractionation. Here the authors calculate the solar–system Kr abundance from solar–wind noble–gas ratios, determined previously by low–temperature oxidations of lunar ilmenite grains, normalized to Si by spacecraft solar–wind measurements. The estimated ^{83}Kr abundance of 4.1 ± 1.5 per 10^6 Si atoms is within uncertainty of estimates assuming no fractionation, determined from CI–chondrite abundances of surrounding elements. This is significant because it is the first such constraint on solid/gas fractionation, though the large uncertainty only confines it to somewhat less than a factor of two.

074.083 **Acceleration of electrons outside flares: coronal manifestation and possible origin.**
J. P. Raulin, R. F. Willson, A. Kerdraon, K.–L. Klein, K. R. Lang, G. Trottet.
Astron. Astrophys., Vol. 251, No. 1, p. 298 – 306 (Nov 1991).
Imaging observations at decimetric and metric wavelengths of the solar corona are used to investigate a short–lasting noise storm and the associated changes of the underlying active region plasma. It is shown that a new source appears in the active region in close temporal and spatial coincidence with the onset of the noise storm in the middle corona. The onset of the noise storm is delayed at longer wavelengths. At a given wavelength, the noise storm source undergoes a systematic slow movement with a significant component perpendicular to the magnetic field lines above the active region. The observations are tentatively attributed to the emission of nonthermal electrons in a system of expanding coronal loops, the expansion being initiated by the appearance of the new 21 cm source in the low atmosphere. The derived velocity of expansion is about 80 to 150 km s^{-1}. It is shown that the electrons emitting the noise storm cannot be provided by the high–energy tail of the Maxwellian in the new active region source, but originate either from a nonthermal population in this source or from acceleration at higher altitudes, in the structures which give rise to the noise storm. The new

source in the active region underneath the noise storm is then considered as triggering the expansion of the overlying loop system in which the noise storm is emitted.

074.084 **A giant coronal arch observed by Skylab.**
Z. Švestka.
Sol. Phys., Vol. 135, No. 2, p. 419 – 422 (Oct 1991). Letter–to–the–editor.

The limb event of 13 August 1973, observed by Skylab in soft X–rays, has all the typical characteristics of the giant post–flare arches observed by HXIS and FCS on board SMM in the 1980s. Skylab images provide us with 4 – 5 times better angular resolution than the SMM experiments and thus, for the first time, make it possible to distinguish the real fine structure of a giant post–flare arch. The image processing of the 13/14 August 1973 event is now in progress.

074.085 **Low–frequency waves in the solar wind near Neptune.**
M. Zhang, J. W. Belcher, J. D. Richardson, V. M. Vasyliunas, R. P. Lepping, N. F. Ness, C. W. Smith.
Geophys. Res. Lett., Vol. 18, No. 6, p. 1071 – 1074 (Jun 1991).

Plasma and magnetic field observations from the Voyager 2 spacecraft when it was outbound from Neptune reveal low–frequency waves in the solar wind which are clearly associated with the planet. The waves have frequencies below the proton cyclotron frequency f_{cp}, which is about 10^{-3} Hz during the periods waves are observed. The waves are present when the interplanetary magnetic field is oriented such that the spacecraft is connected to the bow shock by the magnetic field lines. The authors have identified the waves to be Alfvénic waves propagating at $\sim 140°$ to the ambient magnetic field and away from the bow shock. As at the other planets, these downstream waves are thought to be generated in the upstream region, where energetic protons created near the nose of the bow shock excite waves as they stream along solar wind magnetic field lines.

074.086 **Comparison of solar wind velocity measurements with a theoretical acceleration model.**
W. A. Coles, R. Esser, U.–P. Løvhaug, J. Markkanen.
J. Geophys. Res., Vol. 96, No. A8, p. 13849 – 13860 (1 Aug 1991).

Interplanetary radio scintillation (IPS) measurements of the solar wind velocity were made using the receiving antennas of the European Incoherent Scatter Facility (EISCAT) radar system in northern Scandinavia from June through October 1990. The observations, which cover the distance range from 11 to 90 R_S From Sun center, were taken with sufficient density to measure the same stream at two (or more) different distances. The deduced velocities are in the range $100 \lesssim u \lesssim 540$ km s^{-1}. The authors selected from 192 observations, 16 examples of streams observed with good radial alignment, of which 12 were observed unchanged for several days. The measured velocities are compared with calculations based on a two–fluid solar wind model with Alfvén waves. In eight cases the measurements are in good agreement with the model when a moderate amount of wave energy is added to the flow. In four cases the observed streams show low or moderate velocities below, say, 20 R_S but then accelerate fast at larger distances from the Sun. This delayed acceleration is much steeper than the acceleration in the model at these distances. In the remaining four cases the streams seem to reach their final velocities much closer to the base than in other cases, and they are not observed to accelerate much between 10 and 90 R_S. At these distances all related solar wind models give the same results; they all fit half the data, and none can fit the other half. The authors do not see an obvious modification to these models which will allow them to fit either the delayed acceleration or the early acceleration cases.

074.087 **Out–of–ecliptic tests of the inverse correlation between solar wind speed and coronal expansion factor.**
N. R. Sheeley Jr., E. T. Swanson, Y.–M. Wang.
J. Geophys. Res., Vol. 96, No. A8, p. 13861 – 13868 (1 Aug 1991).

The authors address the question of whether out–of–ecliptic measurements satisfy the inverse correlation between wind speed at 1 AU and flux tube divergence in the corona, already found from measurements in the ecliptic. Using the in–ecliptic calibration, the authors derive out–of–ecliptic speeds from coronal expansion factors determined from global observations of photospheric field and their current–free coronal extension. These derived speeds are compared with speeds inferred from interplanetary scintillation measurements during 1972–1988 and with in situ speeds measured by the Pioneer 11 spacecraft at 16°N latitude during 1984–1988. These three sets of wind speed show the same overall variation with latitude and time during the sunspot cycle, with higher latitudes having more years of fast wind than lower latitudes and all latitudes having slow wind at sunspot maximum. Although some detailed discrepancies are also present, the overall agreement is comparable to that achieved in the ecliptic plane.

074.088 **Estimates of solar wind velocity gradients between 0.3 and 1 AU based on velocity probability distributions from Helios 1 at perihelion and aphelion.**
S. Arya, J. W. Freeman.
J. Geophys. Res., Vol. 96, No. A8, p. 14183 – 14188 (1 Aug 1991).

The authors have estimated the average velocity gradients for the solar wind protons by comparing the velocity probability distributions at 0.3 and 1 AU. Assuming a power law radial dependence, they find, for a full 6–year data set, that the gradients for the lowest velocity ranges are about $R^{0.1}$ to $R^{0.14}$ and that the power index decrease steadily with increasing velocity until the slope is near zero for the high–speed solar wind. However, upon examining the solar cycle dependence the authors find that this trend for the velocity gradient to decrease with increasing velocity is a characteristic primarily of the increasing solar activity and solar maximum period and is almost absent in the solar minimum data. The solar wind above 500 km/s during solar minimum shows an average acceleration similar to the slow wind, about 55 to 85 km/s/AU. On the other hand, winds above 350 km/s from the period of increasing solar activity and solar maximum show essentially no average accleration beyond 0.3 AU.

074.089 **On the lifetime of the metastable state of hydrogen $2S_{1/2}$ in a turbulent plasma of the solar corona.**
E. B. Klejman, I. M. Ojringel', E. G. Stupko.
Astron. Tsirk., No. 1544, p. 25 – 26 (May – Jul 1990). In Russian.

The authors discuss a new possibility of diagnosing plasma turbulence in a nonequilibrium plasma of the solar corona from the lifetime of the metastable state of hydrogen $2S_{1/2}$.

074.090 **Alfvén wave trapping, network microflaring, and heating in solar coronal holes.**
R. L. Moore, Z. E. Musielak, S. T. Suess, C.–H. An.
Astrophys. J., Vol. 378, No. 1, p. 347 – 359 (1 Sep 1991).

The authors present new evidence that much of the heating in coronal holes is provided by Alfvén waves. This evidence comes from examining the reflection of Alfvén waves in an isothermal hydrostatic model coronal hole with open magnetic field. Reflection occurs if the wavelength is of the order of the scale height of the Alfvén velocity. For Alfvén waves with periods of about 5 min, and for realistic density, magnetic field strength, and magnetic field spreading in the model, the waves are reflected back down (trapped) within the model hole if the coronal temperature is only slightly less than 1.0×10^6K, but are not reflected and escape out the top of the model, if the coronal temperature is only slightly greater than 1.0×10^6K. Because the spectrum of Alfvén waves in real coronal holes is expected to peak around 5 min and the temperature is observed to be close to 1.0×10^6K, the sensitive temperature dependence of the trapping suggests that the temperature in coronal holes is regulated by heating by the trapped Alfvén waves. The authors discuss the heating of the quiet corona in general and the possible role of network microflares in particular. It is pointed out that the network microflares might be the source of the Alfvén waves that heat coronal holes.

074.091 **A model for the formation of solar prominences.**
S. K. Antiochos, J. A. Klimchuk.
Astrophys. J., Vol. 378, No. 1, p. 372 – 377 (1 Sep 1991).
The authors propose a model for the formation of prominence condensations in hot coronal loops. Previous studies have concentrated on cooling the hot plasma by decreasing the coronal heating rate. The difficulty with such models is that when the heating decreases, most of the loop mass is lost by draining onto the chromosphere. The authors argue that a prominence condensation is likely to be due to an increase in the heating. The key idea is that the heating increase is localized nearer to the chromospheric footpoints than to the loop midpoint. The authors present results of numerical simulations of hot loops that are initially heated uniformly, and then undergo heating increases that are concentrated away from the loop midpoint. The temperature at the midpoint first increases, but eventually it collapses to chromospheric values as a result of chromospheric evaporation. Hence, one obtains the curious result that increasing the heating causes cooling. The resulting densities and time scales agree well with observations. The implications of this model for coronal heating and prominence structure are discussed.

074.092 **Coronal mass ejections and streamers associated with the new cycle active regions at solar minimum.**
S. Kahler.
Astrophys. J., Vol. 378, No. 1, p. 398 – 406 (1 Sep 1991).
The author uses data from the coronagraph / polarimeter on the SMM spacecraft to examine the coronal streamers and coronal mass ejections (CMEs) associated with the appearance of the high latitude, new–cycle active regions at the beginning of solar cycle 22 in 1986 and 1987. Short-lived active regions, those lasting less than a disk passage, were not associated with streamers or CMEs. CMEs were generally associated with a high latitude region only when another strong active region lay within $\sim 30°$ of the longitude of the high latitude region. The CMEs appeared to arise from magnetic connections between the high latitude regions and low latitude, old–cycle regions or between the high latitude regions of opposite hemisphere. The development of streamers over the active regions appeared to be independent of the occurrence of CMEs. In some cases streamers formed over active regions with which no CMEs were observed, and in other cases CMEs were associated with regions with no overlying streamers. Streamers were also observed to survive CMEs associated with the underlying active regions.

074.093 **Erratum: "Variations of the hydrodynamic parameters of solar wind protons and α–particles according to measurements aboard the Prognoz–7 satellite" [Kosm. Issled., Tom 28, Vyp. 2, p. 218 – 225 (Mar–Apr 1990)].**
Yu. I. Ermolaev, V. V. Stupin, G. N. Zastenker.
Kosm. Issled., Tom 29, Vyp. 5, p. 800 (Sep–Oct 1991). In Russian. See Abstr. 51.074.021

074.094 **Erratm: "A new method to study the large–scale structure of the solar corona on the basis of solar wind measurments" [Kosm. Issled., Tom 28, Vyp. 6, p. 890 – 902 (Nov–Dec 1990)].**
Yu. I. Ermolaev.
Kosm. Issled., Tom 29, Vyp. 5, p. 800 (Sep–Oct 1991). In Russian. See Abstr. 52.074.092.

074.095 **Joule heating as an explanation for the differential emission measure structure and systematic redshifts in the Sun's lower transition region.**
G. Roumeliotis.
Astrophys. J., Vol. 379, No. 1, p. 392 – 400 (20 Sep 1991).
It is proposed that most of the Sun's lower transition region – consisting of plasma with temperature in the range 3×10^4K $< T < 3 \times 10^5$K – is contained in sheared magnetic boundary layers separating adjacent flux loops, where it is heated by the Joule dissipation of intense, field–aligned electric currents. To explore this idea, a one–dimensional hydrodynamic model is used to study the atmospheric structure along a Joule–heated field line embedded in a sheared magnetic boundary layer. When the electric current density is less than a certain critical value, the atmosphere along the field line evolves towards a "cool" equilibrium state, with $T_{max} < 8 \times 10^4$K. However, when the electric current density exceeds the critical value, the atmosphere evolves towards a "hot" equilibrium state, with 10^6K $> T_{max} \gg 8 \times 10^4$K. A transient downflow of the order of 10 km s^{-1} develops, when the electric current density is increased past the critical value. The proposed new model for the lower transition region explains the deep minimum in the differential emission measure curve near 10^5K, as well as the systematic redshifts observed in lower transition region lines.

074.096 **MHD parameters of the steady–state solar wind at 1 AU as functions of parameters at 18 Rs.**
D. Odstrčil.
Bull. Astron. Inst. Czech., Vol. 42, No. 4, p. 263 – 268 (Jul 1991).
Numerical simulation of interplanetary disturbances requires to know, as initial background, the steady–state solar wind. Its parameters are unambiguously determined by values of the MHD variables at the inner boundary. Usually, it is requested to specify here such values which determine the solar wind with values close to observations at 1 AU. This task is performed by trial computations and the main purpose of the paper is to provide some help. Position of the inner boundary is chosen at 18 Rs which is frequently used and lies in the supersonic flow region. Because the magnetic field has a little effect on the steady–state, the MHD task is split into solution of hydrodynamic equations and using simple relations to determine supplementary variables. Large set of such computations is performed and their results processed to graphical form. The presented procedure is compared with complete 1D HD and 2D MHD computations.

074.097 **Multifractal structure of speed fluctuations in recurrent streams at 1 AU and near 6 AU.**
L. F. Burlaga.
Geophys. Res. Lett., Vol. 18, No. 8, p. 1651 – 1656 (Aug 1991).
This paper demonstrates the existence of multifractal structure in the velocity fluctuations associated with recurrent streams at 1 AU and near 6 AU. There is a significant evolution of the multifractal spectrum from 1 AU to 6 AU. At 1 AU the fluctuations on all scales in the indicated range are related to the local speed gradients associated with the stream structure. Near 6 AU the multifractal structure of the speed fluctuations approaches the universal form for intermittent turbulence for moments $q > 5$, and it has approximately the value expected for a set of discontinuities for $q = 2$.

074.098 **On solar emission lines.**
G. P. Malik, L. K. Pande, V. S. Varma.
Astrophys. J., Vol. 379, No. 2, p. 788 – 795 (1 Oct 1991).
The authors draw attention to the large number of unidentified solar emission lines and the various ingenious ideas that have been proposed from time to time to resolve the problem. An attempt is made to tackle the problem on the basis of one fundamental system consisting of an electron and a proton, the Coulomb potential between which is regulated by temperature. Use is made of an equation, which has recently been given for the description of such a system.

074.099 **Avalanches and the distribution of solar flares.**
E. T. Lu, R. J. Hamilton.
Astrophys. J., Lett., Vol. 380, No. 2, p. L89 – L92 (20 Oct 1991).
The authors propose that the solar coronal magnetic field is in a self–organized state, thus explaining the observed power–law dependence of solar flare occurrence rate on flare size which extends over more than five orders of magnitude in peak flux. The physical picture that arises is that solar flares are avalanches of many small reconnection events, analogous to avalanches of sand in the models published by Bak and colleagues in 1987 and 1988. Flares of all sizes are manifestations of the same physical processes, where the size of a given flare is determined by the number of elementary reconnection events. The relation between

small-scale processes and the statistics of global flare properties, which follows from the self-organized magnetic field configuration, provides a way to learn about the physics of the unobservable small-scale reconnection processes. The authors present a simple lattice reconnection model which is consistent with the observed flare statistics. They discuss the implications for coronal heating and suggest some observational tests of this picture.

074.100 **The interaction effect of fast and slow solar wind streams in interplanetary space on wind characteristics at the Earth's orbit.**
V. G. Fainshtein.
Sol. Phys., Vol. 136, No. 1, p. 169 – 189 (Nov 1991).
By analyzing observational data, it has been possible to determine quantitative relationships that represent the role of the interaction of fast and slow solar wind (SW) streams in the formation of characteristic SW properties at the Earth's orbit. It is shown that maximum values of magnetic field B^M and density n^M peaks in the neighbourhood of the sector boundary (SB) at the base of the high-speed stream front are associated with solar wind characteristics such as the SW minimum velocity near the SB, the maximum velocity in the central part of the fast stream, and the slope β of the magnetic field neutral line to the solar equatorial plane at $R = 2.5\ R_{\odot}$. It is concluded that enhancements of absolute values of the z-component of the magnetic field, $|B_z|$, recorded at the Earth's orbit, are largely attributable, at sufficiently large values of β, to the interaction of different-velocity SW streams.

074.101 **Two-dimensional MHD equilibria in the solar corona.**
Wang Shui, S. T. Wu, A. H. Wang.
Chin. Astron. Astrophys., Vol. 15, No. 4, p. 470 – 479 (Dec 1991). English translation of Chin. J. Space Sci., Vol. 11, No. 2, p. 81 – 90 (Apr 1991).
Using a combined analytical and numerical method the authors have treated the question of two-dimensional MHD equilibrium in an inviscid compressible, perfect conducting plasma with an embedded magnetic field in the spherically symmetric gravitational field of the sun. Two solutions are obtained. (1) A steady, self-consistent plasma flow in a magnetic field with both a closed and an open region. In the open region, beyond a few solar radii, the plasma velocity exceeds the local sound and Alfvén velocities. (2) The plasma velocity is everywhere smaller than the local sound and Alfvén velocities and tends to zero at large radial distances.

074.102 **On the spontaneous formation of electric current sheets above a flexible solar photosphere.**
B. C. Low.
Astrophys. J., Vol. 381, No. 1, p. 295 – 305 (1 Nov 1991).
This paper treats the general tendency for a magnetic field in an electrically perfectly conducting plasma to seek an equilibrium state with embedded electric current sheets or magnetic tangential discontinuities, a hydromagnetic property first pointed out by Parker. This property is simplest to demonstrate theoretically with two-dimensional force-free magnetic fields in the infinite half-space above a plane boundary, where magnetic footpoints are rigidly anchored. The present paper attempts to answer to the criticism of this model by Karpen, Antiochos, and DeVores in 1990. Based on a numerical simulation, these authors have concluded that the formation of current sheets in this model depends on the use of rigid boundary conditions to the extent that current sheets would not form if the boundary is not rigid but is replaced by a gravitationally stratified photosphere of finite thickness. It is shown that the conclusion of these authors is not valid. This conclusion is based on a physical analysis and an explicit magnetostatic construction of a possible equilibrium state in which a current sheet has formed in the presence of a flexible photosphere. Some suggestions for future numerical studies are given.

074.103 **Neutralized and unneutralized current patterns in the solar corona.**
D. B. Melrose.
Astrophys. J., Vol. 381, No. 1, p. 306 – 312 (1 Nov 1991).
Observations of the vector magnetic field at the photosphere suggest that there is a net current flowing through the corona along coronal magnetic loops from one footpoint to the other. There is also evidence that flare kernels correlate with regions of high current. The absence of any apparent return current through the corona is inconsistent with this current being generated by stresses in the photosphere or in subphotospheric layers after the loop has emerged from below the photosphere. The possibility that a return current is present and has escaped detection is discussed. If there is no return current, then a rethinking of several widely accepted ideas on solar magnetic fields and their role in solar flares is required.

074.104 **Concerning sources of eruptive flows on the Sun.**
V. G. Eselevich.
Planet. Space Sci., Vol. 39, No. 7, p. 1031 – 1038 (Jul 1991).
By analysing and generalizing published experimental data, it is shown that: (1) the locations of sources of eruptive flows on the Sun and of related flares are displaced with respect to each other by an angle $\Delta\varphi \cong 20$–$40°$, on average; (2) there is not sufficient justification to believe that there exist eruptive flows unassociated with either flares or active filaments; and (3) eruptive flows have a fundamentally different nature as compared to quasi-stationary fast-speed flows originating from coronal holes (CH). The hypothesis is suggested, according to which eruptive flows are produced at locations of jump-like large-scale increases in portions of CH boundaries.

074.105 **The structure and length of tail-associated phenomena in the solar wind downstream from the Earth.**
S. W. H. Cowley.
Planet. Space Sci., Vol. 39, No. 7, p. 1039 – 1043 (Jul 1991).
The length of the geomagnetic tail, estimated on the basis of the open model of the magnetosphere, is generally taken to be $\sim 1000\ R_E$. Here the author shows that a tail-like structure consisting of field lines disconnected from the Earth and lying downstream from the tail reconnection region will typically remain present up to distances of $\sim 5000\ R_E$. At even larger distances, the most prominent terrestrial effect is expected to be substorm-associated plasmoids propagating in the solar wind.

074.106 **MHD turbulence in the solar wind.**
A. Mangeney, R. Grappin, M. Velli.
Symposium on Geophysical and Astrophysical MHD as Part of the British Theoretical Mechanics Colloquium: Advances in solar system magnetohydrodynamics, p. 327 – 356 (1991). – See Abstr. 012.086 for the main entry.
The basic characteristics of the wind necessary to understand the in situ measurements are described. The authors discuss how and to what extent the data in a medium frequency range support the idea of evolution of MHD turbulence and discuss the specificity of Alfvénic turbulence versus "standard" MHD turbulence in the context of the solar wind.

074.107 **Magnetohydrodynamic simulation of the radial evolution and stream structure of solar-wind turbulence.**
D. A. Roberts, S. Ghosh, M. L. Goldstein, W. H. Matthaeus.
Phys. Rev. Lett., Vol. 67, No. 27, p. 3741 – 3744 (30 Dec 1991).
Current Physics Microform No.: 9201E1952.
The authors present a unified interpretation of observations of interplanetary fluctuations in terms of nearly incompressible magnetohyrodynamics. Incompressive effects explain the rapid evolution of turbulence in slow wind containing the heliospheric current sheet. The relative constancy of the spectrum of "inward propagating" fluctuations compared to the rapid decline in "outward" fluctuations results from incompressive spectral transfer combined with strong dissipation of the outward fluctuations. Secondary compressive effects account for nearly pressure-balanced structures and the density fluctuation levels.

074.108 **Large-scale structure of solar wind and its relationship with solar corona: Prognoz 7 observations.**
Yu. I. Yermolaev (*Yu. I. Ermolaev*).
Planet. Space Sci., Vol. 39, No. 10, p. 1351 - 1361 (Oct 1991).
On the basis of selective measurements of proton and α-particle bulk parameters and magnetic field on board the Prognoz 7 satellite, the structure of solar wind is studied and compared with the known structure of solar corona. For that purpose the magnetic field value, B, proton temperature, T_p, ratio of thermal and magnetic pressures, $\beta = nkT_p/(B^2/8\pi)$, and relative helium abundance, n_α/n_p, are represented as distributions on the n-V_p (solar wind ion density-proton velocity) plane. The values and different behaviour of these parameters on the n-V_p plane allow the authors to identify five regions (types of solar wind streams), which may be connected with well known solar corona structure and events: (1) heliospheric current sheet; (2) streams from closed coronal magnetic field regions (streamers); (3) streams from open magnetic field regions (coronal holes); (4) disturbed solar wind streams at large angles relative to the solar meridians of non-stationary events in the corona; and (5) disturbed streams at small angles relative to the meridians of non-stationary events, including the matter of coronal mass ejections. Distributions of velocity difference, V_α-V_p, and temperature ratio, T_α/T_p, of α-particles and protons on the n-V plane show that preferential acceleration ($V_\alpha > V_p$) and heating ($T_\alpha/T_p \geqslant 4$) of α-particles are generally observed in streams from coronal holes.

074.109 **Long time variations in solar wind properties: possible causes versus observations.**
J. Geiss, P. Bochsler.
The Sun in time, p. 98 - 117 (1991). - See Abstr. 003.027 for the main entry.
Three decades of observation have given us an overview over averages and variations of the physical properties and the chemical composition in the solar wind. Significant time variability in the elemental composition is observed, and also systematic differences between average solar wind and solar composition have been established. Reasonable theoretical interpretations have been developed for the anomalies in solar wind abundances. The theoretical models that have been established to interpret elemental abundances and their variations indicate that with the exception of $^3He/^4He$ isotopic ratios should remain relatively unaffected. The few direct observations of isotopic abundances in the solar wind seem to confirm this prediction. Lunar surface material is examined.

074.110 **Coronal current-sheet formation: the effect of asymmetric and symmetric shears.**
J. T. Karpen, S. K. Antiochos, C. R. DeVore.
Astrophys. J., Vol. 382, No. 1, p. 327 - 337 (20 Nov 1991).
A major issue in coronal-heating research is whether or not current sheets can occur without a null point being present in the initial potential magnetic field. Several analytic studies contend that current sheets form along separatrices between magnetic flux systems whenever the footpoints of the configuration are moved continuously. Both symmetric and asymmetric systems have been investigated, all with the assumption that the magnetic field is line tied at the discontinuous photospheric boundary. The present work extends earlier studies of the antisymmetric case to systems with asymmetric and symmetric shears. Using a 2.5-dimensional numerical code, the authors studied the effects of an asymmetric shear imposed on a potential, quadrupolar magnetic field under two sets of atmospheric and boundary conditions: (1) a low-β plasma with line tying at the base; and (2) a hydrostatic equilibrium atmosphere with solar gravity, typical of the observed photosphere - chromosphere interface. The low-β simulation confirms the crucial role of the line-tying assumption in producing current sheets. The authors also examine the effects of a symmetric shear on the same hydrostatic equilibrium atmosphere. Again it is found that true current sheets do not form in the corona when a more realistic atmospheric model is considered. The amount of Ohmic dissipation in the thick currents is estimated to be two to four orders of magnitude below that required to heat the corona.

074.111 **Nonequilibrium ionization effects in asymmetrically heated loops.**
D. Spadaro, S. K. Antiochos, J. T. Mariska.
Astrophys. J., Vol. 382, No. 1, p. 338 - 343 (20 Nov 1991).
The authors have investigated the effects of nonequilibrium ionization on magnetic loop models with a steady siphon flow that is driven by a nonuniform heating rate. In particular, the authors have studied the model developed by Mariska to explain the observed redshifts of transition region emission lines, and they have computed the number densities of the ions of carbon and oxygen along the loop, with and without the approximation of ionization equilibrium. Considerable deviations from equilibrium have been found. In order to determine the consequences of these nonequilibrium effects on the EUV emission from the loop plasma, the authors calculated the profiles and wavelength positions of all the important emission lines due to carbon and oxygen. The results are in broad agreement with the conclusions reached by Mariska, although they show a significant diminution of the Doppler shifts, as well as modifications to the line widths. Apparently, the inclusion of nonequilibrium effects complicates the interpretation of the observed characteristics of the solar transition region by means of asymmetric heating models.

074.112 **Interaction of coronal loops.**
Zhang Zhenda, R. N. Smartt.
Acta Astron. Sin., Vol. 32, No. 3, p. 233 - 238 (Sep 1991). In Chinese. With 2 plates.
The occasional transient enhancement at the interaction regions of some loops as recorded in coronal images of Fe XIV (5303 Å) and Fe X (6374 Å), and the associated images in Hα have been analyzed. Estimates of the electron density are derived from the cooling time, and the values of dissipated energy are obtained by two separate methods.

074.113 **Coronal activity.**
S. Koutchmy, J. B. Zirker, R. S. Steinolfson, J. D. Zhugzda (*Yu, D. Zhugzhda*).
Solar interior and atmosphere, p. 1044-1086 (1991). - See Abstr. 003.028 for the main entry.
This chapter considers the rather extended, very inhomogeneous and dynamical part of the solar atmosphere called the corona. Coronal structures are first described from the point of view of quasi-static structure in order to consider their thermodynamical gross properties, including heating and magnetic fields. The solar-cycle-related variations of coronal structures are reviewed as well. Solar prominences are a well-observed ingredient of the corona; they form its cool component. Their properties, including magnetic field, thermal and velocity field structures are reviewed before considering the problem of their formation. Coronal mass ejections are the most dramatic phenomena occurring in a short time in the corona and are almost inevitably phenomena occurring in a short time in the corona and are almost inevitably related to a prominence ejection. Their observations are described with reference to several related solar phenomena in order to understand the drinving mechanism. Models of coronal mass ejections are described based on numerical simulations including consideration of waves. Finally, the very important and still open question of the waves in the solar corona is briefly discussed from a theoretical point of view, emphasizing the possible role of hydromagnetic waves.

074.114 **The solar wind and its coronal origins.**
G. L. Withbroe, W. C. Feldman, H. S. Ahluwalia.
Solar interior and atmosphere, p. 1087 - 1106 (1991). - See Abstr. 003.028 for the main entry.
Coronal holes are the most well-established coronal source of steady-state solar wind, the high-speed solar wind streams which typically have asymptotic flow speeds of 700 km s^{-1}. Coronal mass ejections associated with flares and/or eruptive prominences are another clearly identified source of solar wind, a transient

component that accounts for approximately 5% of the total solar-wind mass loss. The role of other coronal structures in the generation of the solar wind is less clear. Streamers and the interfaces between streamers and other coronal regions are likely sources of low-speed wind. Small-scale dynamical phenomena observed at the base of the corona (spicules, macrospicules and high-speed jets) and small-scale structures observed in polar coronal holes (polar plumes) may or may not play a significant role in supplying mass, momentum and energy to the solar wind. Improved measurements are required to determine the role, if any, of these small-scale structures in the generation of the solar wind.

074.115 **The connection between coronal bright points and the variability of the quiet-Sun extreme-ultraviolet emission.**
S. R. Habbal, E. Grace.
Astrophys. J., Vol. 382, No. 2, p. 667 - 676 (1 Dec 1991). With plates 16 - 17.

The authors determine the characteristics of the variable EUV emission from the quiet Sun by analyzing the EUV/Skylab data recorded at six wavelengths simultaneously scanning the chromosphere to the corona, from a coronal hole and a quiet region. Changes in the intensity of emission in consecutive spectroheliograms, 5.5 minute apart, are called significant if larger than σ (where σ is the root mean square deviation from the mean in a source free region). In both regions studied, the significant variable emission is preferentially localized in EUV bright points which form the enhanced EUV emission from the quiet Sun. The spatial density of the enhanced and variable emission has a marked temperature dependence. The authors interpret the minimum at 3×10^5K as evidence for two favored temperature distributions of closed magnetic field structures in the quiet Sun, one below 3×10^5K and one at coronal temperatures. The enhanced emission occupies between 10% and 25% of the solar surface, depending on the temperature, while the variable emission covers a smaller fraction, between 5% and 15% of the solar surface. These estimates are likely to vary with the solar cycle, since the number of bright points is known to be solar cycle dependent. The spatial density of the variable emission derived in this study represents the spatial density of random fluctuations and is an upper limit if the actual spatial scale of the variability is less than the 5" spatial resolution of the data.

074.116 **Coronal evolution due to shear motion.**
R. S. Steinolfson.
Astrophys. J., Vol. 382, No. 2, p. 677 - 687 (1 Dec 1991).

The evolution of the solar corona in response to photospheric motion of the magnetic field footpoints is simulated using numerical solutions of the compressible and dissipative MHD equations in the meridional plane of a spherical coordinate system. An initial dipole magnetic configuration in a static, exponential atmosphere is sheared by azimuthal motions in opposite directions on each side of the equatorial plane. A continuous iterative feedback between the instantaneous coronal magnetic field and the field-line footpoints forces the field to move with the surface motion so that there is no "resistive slippage". Simulations have been completed for values of plasma beta from 0.1 to 0.5, maximum shear velocities from 0.5 to 10.3 km s^{-1}, and with various amounts of resistive and viscous dissipation. In all cases the evolution proceeds in two qualitatively different stages. In the early stage, the field evolves gradually with the field lines expanding outward at a velocity not unlike the shear velocity. During the second stage the field begins to expand much more rapidly until it reaches velocities exceeding a characteristic Alfvén velocity. For a representative case with $\beta = 0.2$, maximum shear velocity 2.6 km s^{-1}, and no explicit dissipation, the eruptive behavior occurs after the field has been sheared by a maximum total distance of 0.6 $R_\odot$ from footpoint to footpoint over a period of 21 hr. Inclusion of the thermodynamics, gravity, and compressibility is shown to have only a quantitative effect on the onset of the eruptive phase, thereby illustrating that the primary interactions are between the dynamics and the magnetic field evolution.

074.117 **Energy balance in coronal funnels.**
D. Rabin.
Astrophys. J., Vol. 383, No. 1, p. 407 - 419 (10 Dec 1991).

Semi-analytic models are used to explore the energy balance in magnetic flux tubes over wide ranges of areal constriction, shape, length, and maximum temperature. Models in which thermal conduction balances radiation are emphasized, but enthalpy transport is also considered. The main results for conduction-dominated models are (1) magnetic constriction does not greatly alter the overall energy budget of the solar corona. Except for some funnels, the downward heat flux at the 10^6K level is typically within a factor of 2 of the value for an unconstricted flux tube of the same length, even for areal constrictions of 100 or more. (2) The empirical differential emission measure over the range $5.3 \lesssim \log T \lesssim 6.0$ is closely matched by a bowl-shaped funnel with a constriction factor of 4. (3) "Loop" scaling relationships of the form $T_h \propto (pL)^{1/3}$, where p is the gas pressure, L is the length of the tube and T_h is the temperature at the hot end, are derived for the full range of models. For a simple flow-dominated model, constriction can reduce the total energy requirement of the funnel by a factor of 5 in extreme cases but does not influence the differential emission measure.

074.118 **Dynamics of solar coronal magnetic fields.**
R. B. Dahlburg, S. K. Antiochos, T. A. Zang.
Astrophys. J., Vol. 383, No. 1, p. 420 - 430 (10 Dec 1991).

The authors present the results of a numerical simulation which is a three-dimensional time-dependent simulation of footpoint stressing in a geometry appropriate for the corona. They consider an arcade that is initially current-free and impose a smooth footpoint motion that produces a twist in the field of approximately 2π. They then stop the photospheric motion and follow the evolution until the field relaxes, primarily by resistive diffusion, to another current-free state. The most striking feature of the evolution is that, in response to photospheric motions, the field expands rapidly upward to minimize the stress. The authors conclude that the expansion has two important effects. First, it suppresses the development of "dips" in the field that would be capable of supporting dense, cool material. Second, the expansion inhibits ideal instabilities such as kinking. The results indicate that simple shearing of a single arcade is unlikely to lead to any significant manifestation of solar activity such as flares or prominences. The authors discuss effects that have not been included in the simulation and may possibly lead to such activity.

074.119 **Modeling the large-scale structure of the solar corona.**
F. Bagenal, S. Gibson.
J. Geophys. Res., Vol. 96, No. A10, p. 17663 - 17674 (1 Oct 1991).

The aim of this study is to find a quantitative description of both the magnetic field and the distribution of plasma in the lower corona that matches the white light images of the K-corona. The authors use the magnetostatic model of Bogdan and Low (1986) and data obtained by the High Altitude Observatory Mark III K-Coronameter stationed at Mauna Loa, Hawaii. To start with, the authors take the simplest, solar minimum case when the corona is approximately longitudinally symmetric. By varying parameters in the Bogdan and Low model they are able to quantitatively match the general characteristics of the lower corona at solar minimum: power law radial profiles of coronal brightness; enhanced brightness at the equator; uniform density depletion at the pole. Further, the authors find the set of parameters that best fit the data and investigate how well the model parameters are resolved by the data.

074.120 **Numerical simulation of shocks in the heliosphere.**
A. G. Kosovichev, T. V. Stepanova.
Sov. Astron., Vol. 35, No. 6, p. 646 - 652 (Nov 1991). Current Physics Microform No.: 9205X1456. English translation of Astron. Zh., Tom 68, Vyp. 6, p. 1283 - 1298 (1991).

Calculations of shocks produced by coronal transients and expanding flare loops are presented. The processes under consideration are based on loss of hydrostatic equilibrium in magnetic arches as the result of strengthening of the background magnetic field or heating of plasma in the arch. Rising magnetic arches act as a piston on the surrounding plasma, producing shocks in the solar wind. The shocks are calculated in the one-dimensional, two-temperature approximation with allowance for turbulent dissipative processes. The dependence of the shock parameters on the initial perturbation of the magnetic configuration is found.

074.121 **Magnetic structures in the corona.**
T. Sakurai.
International SOLAR-A Science Meeting: Flare physics in solar activity maximum 22, p. 245 - 258 (1991). - See Abstr. 012.104 for the main entry.
The research into the physics of the solar corona by using the SOLAR-A satellite is reviewed. Discussions are given on coronal heating mechanisms, the physics of coronal loops, coronal mass ejections, and the observational approach to these topics to be taken by the SOLAR-A satellite.

074.122 **Coronal structures and the sunspot cycle.**
J. M. Pasachoff.
International SOLAR-A Science Meeting: Flare physics in solar activity maximum 22, p. 283 - 286 (1991). - See Abstr. 012.104 for the main entry.
Several solar eclipses must be observed to study the corona at different phases of the solar-activity cycle. The author describes a series of coronal observations at various eclipses over the most recent saros. The most recent experiment studies a mechanism of coronal heating by measuring high-frequency intensity oscillations in coronal loops in the green line. The author also describes an application of the monthly variation in the daily sunspot numbers to the solar-neutrino problem.

074.123 **Coronal restructuring and coronal mass ejections as inferred from interplanetary magnetic flux ropes.**
K. Marubashi.
International SOLAR-A Science Meeting: Flare physics in solar activity maximum 22, p. 344 - 348 (1991). - See Abstr. 012.104 for the main entry.
The purpose of this paper is to identify problems concerned with coronal mass ejections (CMEs) which should be addressed by the soft X-ray imaging observations from the SOLAR-A spacecraft. For this purpose, a model is proposed for the origins of interplanetary magnetic flux ropes. This model describes the geometry of a magnetic bottle originating from restructuring of coronal magnetic fields. The model is then compared with observed features of CME-associated phenomena, and some outstanding questions about the evolution of CMEs are identified.

074.124 **A proposal for a study of the solar wind near the Sun by simultaneous observations with SOLAR-A and Kashima 34 m antenna.**
M. Tokumaru, H. Mori, T. Tanaka.
International SOLAR-A Science Meeting: Flare physics in solar activity maximum 22, p. 349 - 352 (1991). - See Abstr. 012.104 for the main entry.
The microwave interplanetary scintillation observations using Kashima 34 m antenna have disclosed that plasma fluctuations are enhanced around 10 - 30 solar radii where the solar wind is accelerated. The observed enhancement of plasma fluctuations can be interpreted as an evidence of Alfvén wave driving the solar wind. To clarify the relationship of enhanced fluctuations with coronal structures, the authors propose the collaborative observation of SOLAR-A satellite with Kashima 34 m antenna.

074.125 **Search for transient coronal holes by SOLAR-A soft X-ray telescope.**
T. Watanabe.
International SOLAR-A Science Meeting: Flare physics in solar activity maximum 22, p. 353 - 356 (1991). - See Abstr. 012.104 for the main entry.
The author proposes to perform a systematic search for new coronal holes which are expected to be formed in association with solar-filament disappearances, because, some disappearing-filament associated solar wind shock waves are seemed to be driven by transient high-speed streams. For an event associated with a solar filament disappearance on 22 April 1979, the relevant shock wave propagated with a constant speed for 14 hours, during which a transient coronal hole was observed.

074.126 **Application of SOLAR-A SXT data to flare and geomagnetic storm forecasting research.**
S.-I. Watari.
International SOLAR-A Science Meeting: Flare physics in solar activity maximum 22, p. 357 - 360 (1991). - See Abstr. 012.104 for the main entry.
In the view of forecasting solar flares and geomagnetic storms, it is important to examine features of active regions associated with flares, and solar phenomena (coronal mass ejections (CMEs), and coronal holes etc.) associated with geomagnetic storms.

074.127 **Two dimensional steady state pressure structure in a coronal loop.**
S. Chandra.
Bull. Astron. Soc. India, Vol. 19, Nos. 1 - 2, p. 81 (Mar - Jun 1991). Abstract. - See Abstr. 012.108 for the main entry.

074.128 **Kinematics of discrete components in solar coronal plasma in relation to coronal magnetic fields.**
K. P. Raju, J. N. Desai, T. Chandrasekhar, N. M. Ashok.
Bull. Astron. Soc. India, Vol. 19, Nos. 1 - 2, p. 82 (Mar - Jun 1991). Abstract. - See Abstr. 012.108 for the main entry.

074.129 **Possible wave amplitudes in shocks in the solar corona: predictions for Solar Probe.**
S. L. Moses, F. V. Coroniti, E. W. Greenstadt, B. T. Tsurutani.
J. Geophys. Res., Vol. 96, No. A12, p. 21397 - 21401 (1 Dec 1991).
Shocks and Alfven turbulence are frequently invoked as mechanisms to heat the solar corona and accelerate the solar wind. A primary objective of the Solar Probe Mission is to detect and characterize these nonthermal dissipation and energization processes. Although the solar wind in the corona is usually assumed to be sub-Alfvenic, temporal and spatial variability in the plasma parameters could lead to conditions under which weak, fast mode shocks might form. ISEE 3 data from the flanks of the Earth's bow shock show that even low Mach number, quasi-parallel shocks generate large-amplitude Alfven turbulence with $\delta B/B$ of the order of 1, and intense high-frequency electrostatic plasma waves. Using a model of the corona and scaling parameters to those expected in the regions to be traversed by Solar Probe ($r \geqslant 4R_S$), the authors suggest it is possible that such shocks might produce Alfven with magnetic field ampltiudes up to the order of 0.1 G and electric field amplitudes up to the order of 1 V/m; electrostatic waves near 100 kHz may have amplitudes of 0.1 V/m. Since the shock-generated Alfven waves would be Doppler shifted to frequencies of a few kilohertz because of the high spacecraft velocity at perihelion, detection of these waves imposes severe requirements on the designs of plasma wave and magnetic field sensors on Solar Probe.

074.130 **Waves from the Sun?**
M. Velli, R. Grappin, A. Mangeney.
Geophys. Astrophys. Fluid Dyn., Vol. 62, No. 1-4, p. 101 - 121 (Dec 1991). - See Abstr. 012.126 for the main entry.
Satellite observations of solar wind turbulence in the low frequency MHD domain show highly variable properties with respect to time and distance from the Sun: one of the markers of this variability is the degree of "Alfvénicity", which characterizes the relative level of quasi-incompressible waves propagating away from the Sun. To answer the question of the origin and the evolution of this wave spectrum one must investigate the propagation of MHD fluctuations through the highly inhomogeneous and spherically expanding solar wind. The authors discuss some aspects both of the linear propagation before the

critical point and of recent models for the evolution of the turbulence in the supersonic regions of the wind.

074.131 **Coronal loop heating by discrete Alfvén waves.**
C. A. De Azevedo, A. S. De Assis, H. Shigueoka, P. H. Sakanaka.
Sol. Phys., Vol. 136, No. 2, p. 295 - 301 (Dec 1991).
The authors have modeled the solar coronal active loop heating by discrete Alfvén waves. Discrete Alfvén waves (DAW) are a new class of Alfvén waves which can be described by the two-fluid model with finite ion-cyclotron frequency, or the MHD model with plasma current along the magnetic field line as shown by Appert, Vaclavik, and Villar (1984). The authors have modeled the coronal loop as a semi-toroidal plasma with the major toroidal radius much larger than the plasma radius. They have shown that the absorption of discrete Alfvén waves by the plasma through viscosity can account for at least 30% of the coronal heating rate density of 10^{-4}J m^{-3}s^{-1}.

074.132 **Plasma waves caused by transient heat conduction in a coronal loop and anomalous resistivity.**
T. Takakura.
Sol. Phys., Vol. 136, No. 2, p. 303 - 316 (Dec 1991).
Numerical simulation is made of the transient heat conduction during local heating in a model coronal magnetic loop with an axial electric current. It is assumed that a segment near the top of the normal coronal loop is heated to above 10^7K by a sufficiently small heat input as compared with the total flare energy. A hump appears in the velocity distribution of electrons moving down the temperature gradient with speeds slightly below the thermal one. Consequently, electron plasma waves are excited. The high intensity of the waves persists in the upper region of the loop for more than a second until the termination of the simulation. The energy density of the plasma waves normalized with respect to thermal density is $10^{-3.5}$ at maximum. A theoretical estimate gives an anomalous resistivity 5 orders of magnitude greater than an initial value. Based on the above result, the author proposes a model for impulsive loop flares.

074.133 **Activity associated with coronal mass ejections at solar minimum: SMM observations from 1984 - 1986.**
O. C. St.Cyr, D. F. Webb.
Sol. Phys., Vol. 136, No. 2, p. 379 - 394 (Dec 1991).
The authors have examined 73 coronal mass ejections observed by the coronagraph aboard NASA's Solar Maximum Mission between 1984 - 1986. The goal of this study was to determine the distribution of various forms of solar activity that were spatially and temporally associated with mass ejections during solar minimum phase. The authors present the statistical results of the analysis of these 73 mass ejections, and found that slightly less than half of the mass ejections had associations. This was attributed to the large fraction of slow mass ejections detected. Taken as a group, the slow mass ejections were infrequently associated with other forms of solar activity. The results indicate that the distribution of the various forms of activity related to mass ejections does not appear to change at different phases of the solar cycle. For those CMEs with associations eruptive prominences and soft X-ray events (especially long-decay events) were the most likely forms of activity to accompany the appearance of mass ejections. The result strengthens the interpretation that most coronal mass ejections are the result of the reconfiguration of a magnetic field structure surrounding a prominence, leading to the destabilization and eruption of the prominence and its overlying coronal structure. This phenomenon occurs in both quiescent prominences and in prominences found in active regions.

074.134 **The excitation mechanism of [Fe XIV] 5303 Å line in the inner regions of solar corona.**
K. P. Raju, J. N. Desai, T. Chandrasekhar, N. M. Ashok.
J. Astrophys. Astron., Vol. 12, No. 4, p. 311 - 317 (Dec 1991).
The line intensity of the green coronal line and the continuum intensity are derived from the filter and white light photographs of the solar corona obtained during the 1980 total solar eclipse. Ratio of the line to continuum intensity is plotted against the radial distance in various position angles. A simple model assuming an electron density dependence of the line and continuum intensities suggests a dominant collisional mechanism for the excitation of the line in the innermost regions.

Mechanisms of chromospheric and coronal heating. Proceedings.
See Abstr. 012.011.

Temperature-sensitive line ratios of the lithium-like ion O VI.
See Abstr. 022.109.

Suprathermal chemistry in space.
See Abstr. 022.162.

Solar Flare Telescope and 10-cm New Coronagraph.
See Abstr. 032.035.

The Mees CCD imaging spectrograph.
See Abstr. 034.084.

Multi-spectral solar telescope array. V. Temperature diagnostic response to the optically thin solar plasma.
See Abstr. 035.032.

Soho - an observatory to study the solar interior and the solar atmosphere.
See Abstr. 035.131.

Solar tomography: deducing the three-dimensional structure of the corona from stereoscopic images of optically thin line emission.
See Abstr. 036.074.

SIMURIS: a high resolution solar physics interferometric mission in answer to the chromospheric and coronal heating problem.
See Abstr. 051.013.

Ulysses: a journey above the Sun's poles.
See Abstr. 051.070.

The force of ejection of a toroidal transient from an inhomogeneous magnetic field.
See Abstr. 062.001.

Theory of wave activity occurring in the AMPTE artificial comet.
See Abstr. 062.017.

The resistive tearing instability for generalized resistivity models: applications.
See Abstr. 062.018.

Magnetohydrodynamic equilibria in the vicinity of an X-type neutral line specified by footpoint shear.
See Abstr. 062.020.

A note on the mass-loaded MHD flow of the solar wind towards a cometary nucleus.
See Abstr. 062.033.

Shear-induced instability and arch filament eruption: a magnetohydrodynamic (MHD) numerical simulation.
See Abstr. 062.043.

The structure of radiative MHD shocks in the corona.
See Abstr. 062.053.

Three-dimensional hydrodynamic stellar convection.
See Abstr. 062.054.

Alfvén waves.
See Abstr. 062.062.

Heating in stochastic magnetic fields.
See Abstr. 062.063.

Resonance absorption heating.
See Abstr. 062.064.

Viscous damping of magnetohydrodynamic surface waves.
See Abstr. 062.070.

Coronal loop heating by the fast surface wave.
See Abstr. 062.071.

Current sheet formation in force–free magnetic fields.
See Abstr. 062.073.

Heating by field aligned DC Joule dissipation.
See Abstr. 062.076.

Nonlinear evolution of a force–free arcade field driven by shear flow.
See Abstr. 062.078.

Two–dimensional magnetic reconnection.
See Abstr. 062.079.

Three–dimensional magnetic reconnection: basic concepts.
See Abstr. 062.081.

The interaction of collisionless shocks in astrophysical plasmas.
See Abstr. 062.094.

Convection, spontaneous discontinuities, and stellar winds and X–ray emission.
See Abstr. 062.097.

Parametric study of the formation of magnetohydrodynamic shocks. II. Slow shocks.
See Abstr. 062.113.

Boundary conditions in the problem of plasma ejection from the magnetosphere of an axisymmetric rotator.
See Abstr. 062.118.

Nonresonant absorption of shear Alfvén waves.
See Abstr. 062.125.

Numerical study of compressible solar magnetoconvection with an open transitional boundary.
See Abstr. 062.135.

Prominence sheets supported by constant–current force–free fields. I. Imposition of normal magnetic field components at the current sheet and the photosphere.
See Abstr. 062.145.

Acceleration of keV–electrons by a nearly perpendicular shock wave.
See Abstr. 062.148.

Diffuse alpha particles upstream of simulated quasi–parallel supercritical collisionless shocks.
See Abstr. 062.154.

Ray–type spectra of plasma turbulence in coronal magnetic loops.
See Abstr. 062.164.

Maximum energy of semi–infinite magnetic field configurations.
See Abstr. 062.175.

Magnetohydrodynamic waves and wave heating in non–uniform plasmas.
See Abstr. 062.197.

Magnetic heating of the solar corona.
See Abstr. 062.201.

Parametric study of the formation of magnetohydrodynamic shocks. III. Fast and slow shocks.
See Abstr. 062.204.

Boundary conditions in the problem of plasma ejection from the magnetosphere of an axisymmetric rotator.
See Abstr. 062.219.

Maximum energy of semi–infinite magnetic–field configurations.
See Abstr. 062.235.

On the intrinsic difficulty of producing stellar coronae with acoustic waves.
See Abstr. 064.041.

Coronal magnetic structure: the role of ideal MHD invariants.
See Abstr. 064.048.

Analytical studies of collimated winds. II. Topologies of 2–D helicoidal MHD solutions.
See Abstr. 064.053.

Alfvén waves in stellar winds.
See Abstr. 064.079.

Cosmic–ray–modified stellar winds. IV. The effect of self–generated Alfvén waves.
See Abstr. 064.087.

Flaring active region structures in the corona, transition region, and lower atmosphere observed in the SAROC.
See Abstr. 072.025.

CoMStOC III: Measuring magnetic fields in active region coronal plasma.
See Abstr. 072.032.

Loop ensemble models of solar active regions.
See Abstr. 072.060.

Complexes of activity and coronal holes.
See Abstr. 072.066.

Flaring arches. III. The subflare of June 27, 1980, and its related extended arch.
See Abstr. 073.012.

The effects of mass flow on the temperature and abundance structure of the solar transition region.
See Abstr. 073.017.

HRTS time series observations: chromospheric and coronal heating.
See Abstr. 073.021.

High spatial resolution observations of the solar transition region: spicules and microflares.
See Abstr. 073.022.

New models of the chromosphere and transition region.
See Abstr. 073.023.

The role of particle diffusion in the lower transition region: revised interpretation of emission measures.
See Abstr. 073.024.

Why heating is not necessary in the transition region or upper chromosphere.
See Abstr. 073.025.

Nonthermal velocities in the solar transition region.
See Abstr. 073.033.

Observational evidence for EUV loops within the network.
See Abstr. 073.035.

First observations of macrospicules at 4.8 GHz at the solar limb in polar coronal holes.
See Abstr. 073.037.

Numerical studies of atmospheric dynamics driven by energy deposition in the chromosphere.
See Abstr. 073.038.

Diffusion of helium in the solar transition region.
See Abstr. 073.039.

Martens–Kuin models of normal and inverse polarity filament eruptions and coronal mass ejections.
See Abstr. 073.043.

Alfvénically–driven slow shocks in the solar chromosphere and corona.
See Abstr. 073.044.

The formation and support of prominences in magnetic loops and arcades.
See Abstr. 073.060.

Dynamical evidence of fine structure in the solar transition region.
See Abstr. 073.062.

Solar Doppler shift measurements in the Ne VII 465 Å emission line.
See Abstr. 073.063.

First observations of macrospicules at 4.8 GHz at the solar limb in polar coronal holes.
See Abstr. 073.064.

Numerical simulation of microflare evolution in the solar transition region and corona.
See Abstr. 073.075.

Flare energy transport by conduction and radiation.
See Abstr. 073.090.

The quantitative research on the velocity field of a post–flare loop prominence system.
See Abstr. 073.103.

The effects of magnetic geometry on electron confinement in solar flares.
See Abstr. 073.105.

Are the spicules a source of the coronal matter?
See Abstr. 073.108.

Electron acceleration by magnetosonic waves in solar flares.
See Abstr. 073.110.

Energy balance in the solar transition region. II. Effects of pressure and energy input on hydrostatic models.
See Abstr. 073.113.

Multi–wave band SMM – VLA observations of an M2 flare and an associated coronal mass ejection.
See Abstr. 073.139.

The $3s^23p\ ^2P - 3s3p^2\ ^4P$ intercombination lines of Fe XIV in the Sun.
See Abstr. 073.146.

Are spicules a source of coronal matter?
See Abstr. 073.147.

Solar gamma–ray–line flares, type II radio bursts, and coronal mass ejections.
See Abstr. 073.148.

Chromospheric and transition region diagnostics using emission–line intensities.
See Abstr. 073.150.

EUV spectroscopy as a probe of astrophysical plasmas – learning from the Sun.
See Abstr. 073.151.

Numerical simulations of microflare evolution in the solar transition region and corona.
See Abstr. 073.154.

Velocity gradients in the chromosphere–corona transition region.
See Abstr. 073.155.

MHD of solar flares.
See Abstr. 073.158.

Relative abundances in the lower solar transition region.
See Abstr. 073.161.

Fine structure of the solar transition region: observations and interpretation.
See Abstr. 073.166.

Thermodynamic evolution of flares.
See Abstr. 073.180.

On mechanisms of solar flares – some observational tests by using SOLAR–A.
See Abstr. 073.184.

Plasma waves caused by transient heat conduction in a coronal loop as a trigger for impulsive solar flares.
See Abstr. 073.185.

Calculation of force–free coronal magnetic fields from vector magnetograms.
See Abstr. 075.012.

Nonpotential models of active region magnetic fields.
See Abstr. 075.013.

Energy buildup in sheared coronal magnetic fields.
See Abstr. 075.014.

A numerical experiment on the origin of MHD non–equilibrium.
See Abstr. 075.015.

Alfvén wave propagation in a solar magnetic structure.
See Abstr. 075.020.

Intrinsic magnetic discontinuities and solar X–ray emission.
See Abstr. 075.048.

Generation of coronal magnetic fields by random surface motions. I. Mean square twist and current density.
See Abstr. 075.055.

Influence of longitudinal electric fields on plasma instabilities in solar magnetic traps.
See Abstr. 075.057.

Steady flows in solar magnetic structures – a class of exact MHD solutions.
See Abstr. 075.061.

Very high resolution solar X–ray imaging.
See Abstr. 076.004.

Simulation of SXT response to XSST soft X–ray spectrum.
See Abstr. 076.040.

X–ray observations of global solar activity.
See Abstr. 076.042.

SXT observations of MHD turbulence in active regions.
See Abstr. 076.043.

Simulated SXT observations of coronal loops.
See Abstr. 076.044.

Coronal radio emission.
See Abstr. 077.009.

Implications of microwaves for heating and particle acceleration on the Sun.
See Abstr. 077.010.

Global streamer evolution.
See Abstr. 077.014.

Large–scale features of the Sun at 20 centimeter wavelength.
See Abstr. 077.060.

VLA supporting observations for SOLAR–A.
See Abstr. 077.077.

Effects of turbulence on the electron cyclotron–maser mechanism for solar microwave spike bursts.
See Abstr. 077.080.

Electron groups traced from the Sun to 1 AU.
See Abstr. 077.081.

Multispacecraft observations of solar ^{3}He–rich events.
See Abstr. 078.019.

On the differences in element abundances of energetic ions from corotating events and from large solar events.
See Abstr. 078.020.

History of the Sun during the past 4.5 Gyr as revealed by studies of energetic solar particles recorded in extraterrestrial and terrestrial samples.
See Abstr. 078.021.

Solar atmospheric responses to magnetic flux eruptions. I. A mechanism for coronal mass ejections.
See Abstr. 080.059.

Solar atmospheric responses to magnetic flux eruptions. II. Slow shocks near the sun.
See Abstr. 080.060.

Waves in the solar atmosphere.
See Abstr. 080.166.

Variation in the energy spectra of energetic ions at the Earth's bow shock.
See Abstr. 084.029.

Morphology category of geomagnetic storms and characteristics of interplanetary magnetic clouds.
See Abstr. 084.037.

Is the recurrent enhanced geomagnetic activity preceding the sunspot minimum related to magnetic fields of the new or old activity cycle?
See Abstr. 084.053.

Relationship between field line resonance at low geomagnetic latitudes and solar wind structures.
See Abstr. 084.056.

Examination of time–variable input effects in a nonlinear analogue magnetosphere model.
See Abstr. 084.057.

Observational aspects of solar wind – magnetosphere interactions from multipoint measurements made on 1 November, 1984.
See Abstr. 084.058.

MHD theory and the solar wind – magnetosphere interaction.
See Abstr. 084.059.

The solar wind flow around the dayside magnetosphere.
See Abstr. 084.060.

Energy coupling quantities in the solar wind – magnetosphere system.
See Abstr. 084.061.

Coupling processes at the magnetopause.
See Abstr. 084.062.

Non–steady–state solar wind – magnetosphere interaction.
See Abstr. 084.065.

Quasi–biennial variations of solar wind velocity, cosmic rays, solar and geomagnetic activity.
See Abstr. 085.025.

Global hybrid simulation of unmagnetized planets: comparison of Venus and Mars.
See Abstr. 093.055.

On the momentum transfer of the solar wind to the Martian topside ionosphere.
See Abstr. 097.018.

Solar wind interaction with Mars. "Phobos–2" results.
See Abstr. 097.044.

Momentum transport at the Mars magnetopause.
See Abstr. 097.051.

The solar wind interaction with Mars: consideration of Phobos 2 mission observations of an ion composition boundary on the dayside.
See Abstr. 097.052.

The solar wind interaction with Mars: Mariner 4, Mars 2, Mars 3, Mars 5, and Phobos 2 observations of bow shock position and shape.
See Abstr. 097.056.

Plasma wave system measurements of the Martian bow shock from the Phobos 2 spacecraft.
See Abstr. 097.058.

Interplanetary magnetic field control of the Mars bow shock: evidence for Venuslike interaction.
See Abstr. 097.059.

Emission cones in Jupiter's hectometric radiation.
See Abstr. 099.016.

The role of solar wind reconnection in driving the Neptune radio emission.
See Abstr. 101.072.

Discussion of plasma instability between solar wind and type–I comet tail.
See Abstr. 102.001.

Quasilinear diffusion rates of cometary ions.
See Abstr. 102.003.

Pitch angle scattering of cometary ions into monospherical and bispherical distributions.
See Abstr. 102.015.

Pitch–angle scattering of cometary ions: computer simulations.
See Abstr. 102.016.

Singularities in mass–loaded MHD flow: the cometary bow shock.
See Abstr. 102.023.

A self–consistent model for the particles and fields upstream of an outgassing comet. 1. Gyrotropic and isotropic ion distributions.
See Abstr. 102.025.

Quasi–linear evolution of ULF waves excited by cometary ion pickup.
See Abstr. 102.034.

Observations of solar wind ion charge exchange in the comet Halley coma.
See Abstr. 103.103.

A kick from the solar wind as the cause of comet Halley's February 1991 flare.
See Abstr. 103.107.

Quasi–linear pitch angle and energy diffusion of pickup ions near comet Halley.
See Abstr. 103.115.

Cosmic rays and magneto–acoustic instability of the solar–wind stream near the heliosphere shock wave.
See Abstr. 106.002.

A case study of very low cross–helicity fluctuations in the solar wind.
See Abstr. 106.008.

The influence of the interface between the heliosphere and the local interstellar medium on the penetration of the H atoms to the solar system.
See Abstr. 106.010.

Formation of the heliospheric neutral layer in the solar corona and interplanetary space.
See Abstr. 106.021.

Plasma wave generation near the inner heliospheric shock.
See Abstr. 106.031.

Numerical simulation of the April 24, 1981 interplanetary shock wave.
See Abstr. 106.032.

Statistical properties of shock–accelerated ions in the outer heliosphere.
See Abstr. 106.038.

The heliospheric current sheet (HCS) and high–speed solar wind: interaction effects.
See Abstr. 106.041.

The Sun and interplanetary magnetic field.
See Abstr. 106.042.

Shock interactions in the outer heliosphere.
See Abstr. 106.044.

Physical reasons and consequences of a three–dimensionally structured heliosphere.
See Abstr. 106.045.

Protoplanetary thermal metamorphism: the hypothesis of electromagnetic induction in the protosolar wind.
See Abstr. 107.021.

X–ray spectroscopy of stellar coronae.
See Abstr. 112.038.

Cosmic ray modulation.
See Abstr. 144.022.

Relation of cosmic ray scintillations to the solar wind parameters.
See Abstr. 144.045.

The role of corotating interaction regions in cosmic–ray modulation.
See Abstr. 144.079.

075 Magnetic Fields

075.001 **Das Sonnenobservatorium Einsteinturm in Potsdam. Erforschung solarer Magnetfelder und der Physik von Sonnenflecken.**
J. Staude.
Sterne Weltraum, Jahrg. 30, Nr. 8–9, p. 505 – 509 (Aug–Sep 1991).

075.002 **A diffusion current sheet with a big transverse component of the magnetic field on the Sun.**
V. M. Bardakov.
Pis'ma Astron. Zh., Tom 17, No. 9, p. 841 – 849 (Sep 1991). In Russian. English translation in Sov. Astron. Lett., Vol. 17, No. 5.
The question is put about a maximal possible velocity of reconnection in the regime of the diffusion current sheet (DCS). It is shown that for DCS, in which the relation of plasma concentrations inside and outside the sheet (N parameter) is much more than one, the maximal possible reconnection velocity exceeds as many as $N^{1/4}$ times the reconnection velocity determined in the neutral DCS approximation (Sweet–Parker current). When having the maximal reconnection velocity, the DCS has a big transverse component of the magnetic field (the ratio between the transverse and longitudinal components on the border of the sheet exceeds as many as N times the ratio between the sheet thickness and its width); along the sheet plasma is mainly accelerated by Lorentz force at the expense of the transverse component of the magnetic field.

075.003 **On the properties of toroidal flux tubes in the solar dynamo.**
K. Petrovay.
Sol. Phys., Vol. 134, No. 2, p. 407 – 410 (Aug 1991). Letter–to–the–editor.
Prompted by a recently revived debate concerning the structure of the toroidal magnetic field of the Sun, this letter points out that the observational evidence now suggests that the toroidal flux tubes have fluxes of 10^{14}Wb (10^{22} Mx) and flux densities of about 10 T (10^5G). It is proposed that such high flux densities may be produced by the work done on the flux tubes by the drag force due to differential rotation.

075.004 **Observations of waves and oscillations in solar magnetic fluxtube concentrations.**
B. Fleck, F.–L. Deubner.
International Conference on Mechanisms of Chromospheric and Coronal Heating, p. 19 – 21 (1991). – See Abstr. 012.011 for the main entry.
The authors investigate the dynamical behaviour of magnetic fluxtube concentrations by analyzing a high spatial resolution time series of spectra taken simultaneously in left–hand and right–hand circularly polarized light. The spectral range covered the lines of Fe 6302.5 Å and Fe 6301.5 Å. The dominant velocity signal derived from the Stokes I central wavelength and Stokes V zero–crossing wavelength are the 5 minute oscillations.

075.005 **The bright points in the Ca II K–line and their relation to the inner network magnetic structures.**
K. R. Sivaraman.
International Conference on Mechanisms of Chromospheric and Coronal Heating, p. 44 – 47 (1991). – See Abstr. 012.011 for the main entry.
The author presents observational evidence to show that the K_{2V} points bear a one to one correspondence with the inner network magnetic elements when the quiet Sun is viewed under good spatial resolution. This would have implications on the theories of chromospheric heating over quiet regions on the Sun.

075.006 **A true–field magnetogram.**
D. Rabin, L. Cole, D. Jaksha, K. Iwata.
Bull. Am. Astron. Soc., Vol. 23, No. 2, p. 1030 (1991). Abstract. – See Abstr. 012.012 for the main entry.

075.007 **Sunspot umbrae: observed correlation between magnetic field and temperature.**
W. Livingston.
Bull. Am. Astron. Soc., Vol. 23, No. 2, p. 1030 (1991). Abstract. – See Abstr. 012.012 for the main entry.

075.008 **Magnetic field changes associated with a subflare and surge.**
M. J. Hagyard, E. A. West, G. A. Gary, J. E. Smith.
Bull. Am. Astron. Soc., Vol. 23, No. 2, p. 1030 (1991). Abstract. – See Abstr. 012.012 for the main entry.

075.009 **Evolution of vector magnetic field and the August 27, 1990 X–3 flare.**
H. Wang, H. Zirin.
Bull. Am. Astron. Soc., Vol. 23, No. 2, p. 1030 (1991). Abstract. – See Abstr. 012.012 for the main entry.

075.010 **The practical application of the magnetic virial theorm: simulated magnetograph observations.**
J. A. Klimchuk, R. C. Canfield, J. E. Rhoads.
Bull. Am. Astron. Soc., Vol. 23, No. 2, p. 1031 (1991). Abstract. – See Abstr. 012.012 for the main entry.

075.011 **A regularization method for the extrapolation of solar magnetic fields.**
G. A. Gary, Z. Musielak.
Bull. Am. Astron. Soc., Vol. 23, No. 2, p. 1031 (1991). Abstract. – See Abstr. 012.012 for the main entry.

075.012 **Calculation of force–free coronal magnetic fields from vector magnetograms.**
Z. Mikić.
Bull. Am. Astron. Soc., Vol. 23, No. 2, p. 1032 (1991). Abstract. – See Abstr. 012.012 for the main entry.

075.013 **Nonpotential models of active region magnetic fields.**
A. A. van Ballegooijen.
Bull. Am. Astron. Soc., Vol. 23, No. 2, p. 1035 (1991). Abstract. – See Abstr. 012.012 for the main entry.

075.014 **Energy buildup in sheared coronal magnetic fields.**
R. Wolfson.
Bull. Am. Astron. Soc., Vol. 23, No. 2, p. 1035 (1991). Abstract. – See Abstr. 012.012 for the main entry.

075.015 **A numerical experiment on the origin of MHD non–equilibrium.**
P. C. H. Martens, M. T. Sun, S. T. Wu.
Bull. Am. Astron. Soc., Vol. 23, No. 2, p. 1035 (1991). Abstract. – See Abstr. 012.012 for the main entry.

075.016 **Three–dimensional stability of a model sunspot field.**
D. D. Schnack, Z. Mikić.
Bull. Am. Astron. Soc., Vol. 23, No. 2, p. 1036 – 1037 (1991). Abstract. – See Abstr. 012.012 for the main entry.

075.017 **A numerical simulation of two–dimensional radiative equilibrium in magnetostatic flux tubes.**
V. J. Pizzo, K. B. MacGregor, P. Kunasz.
Bull. Am. Astron. Soc., Vol. 23, No. 2, p. 1051 (1991). Abstract. – See Abstr. 012.012 for the main entry.

075.018 **A new numerical model of thin flux tubes in three dimensions.**
B. M. Patten, G. H. Fisher, E. E. DeLuca.
Bull. Am. Astron. Soc., Vol. 23, No. 2, p. 1051 (1991). Abstract. – See Abstr. 012.012 for the main entry.

075.019 **Formation of moving magnetic features.**
J. W. Lee.
Bull. Am. Astron. Soc., Vol. 23, No. 2, p. 1052 (1991). Abstract. – See Abstr. 012.012 for the main entry.

075.020 **Alfvén wave propagation in a solar magnetic structure.**
P. L. Similon, S. Zargham.
International Conference on Mechanisms of Chromospheric and Coronal Heating, p. 438 – 441 (1991). – See Abstr. 012.011 for the main entry.
The effect of magnetic field expansion and density stratification is studied for shear Alfvén wave propagation in an open magnetic structure emerging from the Sun. It is shown that because of magnetic expansion, the coefficient of transmission for wave periods of a minute is about 10%, one order of magnitude larger than the one obtained by assuming constant magnetic field intensity.

075.021 **Dark points and microflares observed in the He I λ10830 line and their relation to the quiet sun magnetic fields.**
K. L. Harvey.
Bull. Am. Astron. Soc., Vol. 23, No. 2, p. 1053 (1991). Abstract. – See Abstr. 012.012 for the main entry.

075.022 **Elementary bipoles of active regions and ephemeral active regions.**
S. F. Martin.
Bull. Am. Astron. Soc., Vol. 23, No. 2, p. 1053 (1991). Abstract. The full paper is published in Mem. Soc. Astron. Ital., Vol. 61, No. 2, p. 293 – 315 (1990). – See Abstr. 012.012 for the main entry.

075.023 **Evolution of the zonal solar magnetic field.**
J. T. Hoeksema.
Bull. Am. Astron. Soc., Vol. 23, No. 2, p. 1053 (1991). Abstract. – See Abstr. 012.012 for the main entry.

075.024 **Dynamics of modes of the global axisymmetrical magnetic field of the Sun.**
D. V. Erofeev.
Soln. Dannye, Byull., No. 9/90, p. 84 – 91 (Sep 1990). In Russian.
The observed pattern of the global axisymmetrical (zonal) magnetic field was investigated by the adaptive processing method. The zonal magnetic field consists of two components which essentially differ from each other. These components may have been regarded as two wave modes of the Sun's magnetic field generator. The first mode involves a greater part of the energy of the magnetic field and manifests a regular cyclic evolution of the field. The second mode represents nonregular poleward propagating waves and is excited as a succession of separate impulses which are accompanied by synchronous short-time disturbances of the first mode. The duration of these impulses is about 2 – 3 years. This property of the zonal magnetic field evolution suggests a non–linear interaction between two modes.

075.025 **Dynamics of the latitude structure of the solar large–scale magnetic field at separate longitude intervals during 1915 – 1965.**
V. I. Makarov, M. V. Kushnir.
Soln. Dannye, Byull., No. 11/90, p. 76 – 79 (Nov 1990). In Russian.
Dynamics of background magnetic fields has been studied on the basis of H–alpha synoptic charts for 1915 – 1965 at separate longitude intervals. Peculiarities of the polarity reversal of the magnetic field latitude migration of the zonal structures in dependence on the phase of the cycle is analyzed.

075.026 **Stationary siphon flows in thin magnetic flux tubes. II. Radiative heat exchange with the surroundings.**
D. Degenhardt.
Astron. Astrophys., Vol. 248, No. 2, p. 637 – 646 (Aug 1991).
Stationary siphon flows in arched magnetic flux tubes are studied. A pressure difference between the two footpoints of the tubes, which are anchored below the solar surface, drives the flow along the tubes. The radiative heat exchange between the tube and its surroundings is taken into account. The magnetic tubes are described by the thin flux tube approximation. The flow and the large scale mechanical equilibrium of the tube, which determines the path of the tube through the solar atmosphere, are calculated self–consistently. The energy exchange between the tube and its surroundings is approximated by the relaxation time approach. The flux tubes are embedded in a realistic solar model atmosphere. In the upstream part of the arch the flow is accelerated and reaches a velocity maximum of a few km/s near the top of the arch. In the downstream part the flow is decelerated. The material flow is nearly adiabatic in the deeper and upper atmospheric layers whereas the flow behavior is isothermal near the solar surface. This leads to a strong temperature depression in the upper parts of the flux tube arch. The shape of the arch can be asymmetric and it can span several thousands of kilometers above the surface. The simulation of the Evershed effect in terms of the calculated siphon flows is discussed.

075.027 **The effect of non–linear oscillations in magnetic flux tubes on Stokes *V* asymmetry.**
U. Grossmann–Doerth, M. Schüssler, S. K. Solanki.
Astron. Astrophys., Vol. 249, No. 1, p. 239 – 242 (Sep 1991). = Mitt. Kiepenheuer Inst. Nr. 351.
The present knowledge on, and interpretation of, the asymmetry of Stokes *V* of spectral lines thought to be formed in and around magnetic flux tubes is reviewed. Crude models of non–linear oscillations in the flux tube were investigated in order to explain the observed values of Stokes *V* amplitude and area asymmetry. It was found that flux tube models with quasi–oscillatory motions, consisting of a slow upflow and a rapid downflow in the magnetic region, as well as a downflow in the non–magnetic surroundings, may reproduce the observations, both asymmetries and zero–crossing shift of Stokes *V*.

075.028 **A model of quasi–open magnetic field line region.**
Ji Shuchen.
Publ. Yunnan Obs., No. 2, p. 81 – 83 (Jun 1991). In Chinese.
Over the bipolar sunspot in the active region, a qualitative model of quasiopen mangetic field line region is proposed to explain the relationship between the energetic electron beams in the corona and the rapid processes in solar flares.

075.029 **Study of small scale magnetic flux and the corresponding velocity pattern.**
C. E. Alissandrakis, H. C. Dara, S. Koutchmy.
Astron. Astrophys., Vol. 249, No. 2, p. 533 – 538 (Sep 1991).
The authors calculated two dimensional maps of the longitudinal magnetic field and the line of sight velocity of two small scale magnetic regions in the quiet sun, using a one–hour time sequence of filtergrams in the magnetically sensitive λ6103 Å Ca I line (with circular polarization measurements) obtained with the Sacramento Peak Vacuum Tower Telescope and the universal filter. After the elimination of the effect of the 5–min photospheric oscillations downflows smaller than $300\ m\ s^{-1}$ in both magnetic regions were found. The magnetic field intensity and the velocity are weakly correlated: high values of the magnetic field tend to occur in regions of small negative velocities. The magnetic field intensity increased during the observations, while some magnetic peaks moved with respect to others. The difference of the magnetic field intensity measured in the blue and the red wing indicates a line asymmetry; this asymmetry was found to be independent of the line of sight velocity.

075.030 **Quasi–biennial variation and latitude solar magnetic field reversal.**
E. E. Benevolenskaya, V. I. Makarov.
Soln. Dannye, Byull., No. 2/91, p. 89 (Feb 1991). Abstract. – See Abstr. 012.021 for the main entry.

075.031 **Biennial cycle in dynamics of large–scale magnetic fields.**
V. I. Makarov, K. S. Tavastsherna, N. S. Petrova.
Soln. Dannye, Byull., No. 3/91, p. 98 (Mar 1991). Abstract. – See Abstr. 012.021 for the main entry.

075.032 **Dynamo and two–years variation.**
A. A. Ruzmajkin.
Soln. Dannye, Byull., No. 3/91, p. 100 – 101 (Mar 1991). Abstract. – See Abstr. 012.021 for the main entry.

075.033 **Behaviour of even and odd *l*–modes of the solar magnetic field with the phase of the global solar cycle on the basis of Hα charts. Interval: 1915 – 1985.**
V. I. Makarov, K. S. Tavastsherna, M. H. Gokhale, K. R. Sivaraman.
Soln. Dannye, Byull., No. 6/91, p. 95 – 103 (Jun 1991). In Russian.
The properties of the latitude–time distribution of the area of predominant polarity of a large–scale magnetic field on the Sun have been studied in 10° latitude zones on Hα charts for 1915 – 1985. The power spectrum of the rotating symmetric harmonic, *l*–mode, when $m = 0$ has been calculated for the entire period in question and separated for the epoch of maxima and minima of the cycle. It is shown that alongside with the dipole component $l = 1$, the mode $l = 3$ (octapole) is dominant at some intervals of the minimum activity. The mode $l = 5$ is dominant at the maximum of activity. The intensity of even modes is 3 – 5 times smaller than that of odd modes. Two dominant periods are detected: 22 year and 7 year; they have the maximum intensity of the power spectrum at $l = 3$ and $l = 5$, respectively. The dependence of the phase $\varphi(l)/2\pi$ on *l*–mode is determined. The phase displacement between the activity of the large–scale magnetic field at the latitude larger 40° and the zone of sunspot activity is derived.

075.034 **Time–resolved Stokes V polarimetry of small–scale magnetic structures on the Sun.**
B. Fleck.
Reviews in Modern Astronomy 4, p. 90 – 103 (1991). – See Abstr. 003.008 for the main entry.
The author addresses problems of the dynamics of small–scale magnetic fluxtube concentrations by analyzing high spatial resolution time series of spectra of selected lines taken simultaneously in left–hand and right–hand circularly polarized light. The observations were obtained with the Vacuum Tower Telescope and the echelle spectrograph at Izaña, Tenerife. The spectral range covered the lines of Fe 6302.5 Å and Fe 6301.5 Å. All parameters extracted from the Stokes V and Stokes I profiles reveal significant spatio–temporal variations. The dominant oscillatory signal in the velocity data is due to the 5–min oscillations. Of special interest are the strong spatially and temporally varying differences between the (I + V) and (I–V) spectra in the magnetic structures. They cause large differences between the velocities deduced from the Stokes I profile and the Stokes V zero–crossing wavelength (up to 1 km/s), whereas their spatio–temporal averages are almost identical. The author observes also an anticorrelation between the amplitude and asymmetry of the Stokes V profile.

075.035 **Fractal flux tubes of the solar magnetic field.**
A. Ruzmaikin (*A. A. Ruzmajkin*), D. Sokoloff (*D. D. Sokolov*), T. Tarbell.
IAU Colloquium No. 130: The Sun and cool stars – activity, magnetism, dynamos, p. 140 – 146 (1991). – See Abstr. 012.024 for the main entry.
The small–scale solar magnetic field exceeding a given threshold forms a fractal set. A dimension of this fractal is found from magnetograms with varying linear resolution. The dimension depends on the value of the threshold magnetic field (multifractality). A simple dynamo model explaining the origin of the fractal magnetic structure is considered. The dynamo produces a magnetic field in the form of flux tubes with a fractal distribution of magnetic field across the tube. The observed dimension gives a possibility of estimating a degree of structuredness of the solar velocity field.

075.036 **The toroidal magnetic field inside the Sun.**
V. N. Krivodubskij, A. E. Dudorov, A. A. Ruzmaikin (*A. A. Ruzmajkin*), T. V. Ruzmaikina (*T. V. Ruzmajkina*).
IAU Colloquium No. 130: The Sun and cool stars – activity, magnetism, dynamos, p. 187 – 189 (1991). – See Abstr. 012.024 for the main entry.

075.037 **The transfer of large–scale magnetic field by radial inhomogeneity of the material density in the rotating convection zone.**
V. N. Krivodubskij, L. L. Kichatinov.
IAU Colloquium No. 130: The Sun and cool stars – activity, magnetism, dynamos, p. 190 – 192 (1991). – See Abstr. 012.024 for the main entry.
The influence of rotation on the transfer of the mean magnetic field of the Sun, caused by the radial inhomogeneity of the solar turbulent plasma density, is investigated. It turns out that the transfer directions of the poloidal and toroidal magnetic fields do not coincide.

075.038 **Diagnostics of the solar dynamo using the observed pattern of surface magnetic fields.**
J. O. Stenflo.
IAU Colloquium No. 130: The Sun and cool stars – activity, magnetism, dynamos, p. 193 – 212 (1991). – See Abstr. 012.024 for the main entry.
The solar surface represents for the solar dynamo an outer boundary that is directly accessible to observations. The evolutionary and rotational properties of the magnetic fields at this boundary can be empirically determined using existing synoptic magnetograph data. The derived properties provide detailed constraints on the underlying theory, such that an inversion approach to the dynamo problem becomes feasible. Ambiguities in the interpretation may be removed using the independent constraints from helioseismology.

075.039 **The large–scale magnetic field in the global solar cycle: observational aspects.**
V. I. Makarov, K. R. Sivaraman.
IAU Colloquium No. 130: The Sun and cool stars – activity, magnetism, dynamos, p. 213 – 217 (1991). – See Abstr. 012.024 for the main entry.
The global solar cycle is considered as an interaction of 3 types of activity: at low–latitude (sunspots), at high–latitude (polar–faculae) and the weak magnetic field. The properties of single and 3–fold reversals of the polar magnetic field are considered. The variation spectrum of the large–scale magnetic field of the Sun is analyzed in the range of 1 – 30 nHz. A dependence between the rate of a poleward meriodional flow and phase of the global cycle is discussed.

075.040 **Evolution of large and small scale magnetic fields in the Sun.**
P. A. Fox, M. L. Theobald, S. Sofia.
IAU Colloquium No. 130: The Sun and cool stars – activity, magnetism, dynamos, p. 218 – 222 (1991). – See Abstr. 012.024 for the main entry.
This paper discusses issues relating to the detailed numerical simulation of solar magnetic fields, those on the small scale which are directly observable on the surface, and those on larger scales whose properties must be deduced indirectly from phenomena such as the sunspot cylce. Results of simulations using the ADISM technique is presented to demonstrate the importance of the treatment of Alfvén waves, the boundary conditions, and the statistical evolution of small scale convection with magnetic

fields. To study the large scale fields and their time dependence, the magnetic resistivity plays an important role.

075.041 **A topological model of the solar magnetic field reversals.**
E. E. Benevolenskaya.
IAU Colloquium No. 130: The Sun and cool stars – activity, magnetism, dynamos, p. 234 – 236 (1991). – See Abstr. 012.024 for the main entry.

075.042 **The longer term evolution of magnetic field and mass flow in a decaying active region.**
Zhang Hongqi, Ai Guoxiang, Wang Haimin, H. Zirin, A. Patterson.
IAU Colloquium No. 130: The Sun and cool stars – activity, magnetism, dynamos, p. 271 – 274 (1991). – See Abstr. 012.024 for the main entry.

075.043 **Rotation of large scale patterns on the solar surface as determined from filament and millimeter data.**
S. Pohjolainen, B. Vršnak, H. Teräsranta, S. Urpo, R. Brajša, V. Ruždjak, S. Jurač, A. Schroll.
IAU Colloquium No. 130: The Sun and cool stars – activity, magnetism, dynamos, p. 279 – 281 (1991). – See Abstr. 012.024 for the main entry.
The rotation of large scale solar magnetic field patterns was studied using quiescent filaments and low temperature regions observed at 37 GHz as tracers.

075.044 **The incorporation of the magnetic fluxes of similar polarity in active regions for Sept. – Oct. 1988 and relevant strong X–ray flares.**
Zhang Guiqing.
Acta Astrophys. Sin., Vol. 11, No. 3, p. 289 – 291 (Jul 1991). In Chinese.
The incorporations of the magnetic fluxes of similar polarity in every active region for Sept. – Oct. 1988 are identified and the relevant class M, X flares are found. The results show that the incorporations of the magnetic fluxes of similar polarity only exist in a few active regions (4% of all) and its productivity for class M, X flares is very high.

075.045 **The merging of magnetic fluxes of similar polarity in solar active regions and the energetic flares.**
Zhang Guiqing.
Acta Astrophys. Sin., Vol. 11, No. 4, p. 362 – 369 (Oct 1991). In Chinese.
The merging of magnetic fluxes of similar polarity in 6 active regions is analysed. Almost all the merging processes in the 6 active regions were associated with energetic flares. Therefore, such merging may be used to predict the energetic flares. In order to make such predictions, two criteria are suggested as follows: (1) Among the magnetic fluxes of similar polarity participating in the merging, the strongest magnetic intensity is $\geqslant$320 G, the weakest is $\geqslant$80 G. (2) The maximum magnetic gradient near the merging area after the merging is $\geqslant$0.15 G/km.

075.046 **Direct evidence of magnetic reconnection in a photospheric layer on the Sun.**
Wang Jingxiu, Shi Zhongxian.
Acta Astrophys. Sin., Vol. 11, No. 4, p. 389 – 393 (Oct 1991). In Chinese. English translation in Chin. Astron. Astrophys., Vol. 16, No. 1, p. 71 – 77 (Jan – Mar 1992).
A direct evidence of magnetic reconnection in a photospheric layer on the Sun is presented for the first time with the aid of the time sequence of vector magnetograms on the photosphere. The main observational characterisitcs of this reconnection are given. The authors predict that reconnection in the photosphere should be commonly observed with vector magnetograms of high spatial and temporal resolution.

075.047 **On the structure of active–region magnetic fields. The homogeneous inverse problem.**
B. P. Filippov.
Astron. Zh., Tom 68, Vyp. 6, p. 1299 – 1308 (Nov–Dec 1991). In Russian. English translation in Sov. Astron., Vol. 35, No. 6.
Information on the tangential magnetic field directions in the chromosphere carried by fibril orientation on H_α filtergrams makes it possible to find the structure of the field above the chromosphere. In the potential approximation the problem is reduced to the one–dimensional Fredholm integral equation of the first kind. The known methods of regularization allow to solve this ill–posed problem. Results of numerical calculations of the model are presented.

075.048 **Intrinsic magnetic discontinuities and solar X–ray emission.**
E. N. Parker.
Geophys. Res. Lett., Vol. 17, No. 11, p. 2055 – 2058 (Oct 1990). Paper presented at the Yosemite Conference on Transition Regions in Solar System Plasmas, Yosemite National Park, CA (USA), 6 – 9 Feb 1990.
The fundamental theorem of magnetostatic equilibrium predicts that the bipolar magnetic fields of active regions on the Sun contain tangential discontinuities as an intrinsic part of their equilibrium. It appears that rapid reconnection at these many tangential discontinuities (nanoflares) is the primary source of heat that produces the X–ray emitting corona. The X–ray corona is then a cloud of nanoflares.

075.049 **The magnetic fields of active regions. VIII. Cycle latitude variations.**
R. F. Howard.
Sol. Phys., Vol. 135, No. 1, p. 43 – 55 (Sep 1991).
The Mount Wilson coarse array magnetograph data set is analyzed to determine characteristics of magnetic regions as a function of distance from the average latitude, ξ_0, of regions in each hemisphere, a quantity which varies during the activity cycle. Regions with normal polarity axis orientations are distributed asymmetrically about ξ_0 with the median latitude about 1 deg equatorward of ξ_0. Reversed polarity orientation regions show a somewhat broader and more symmetric distribution. Average sizes for regions at $\xi = 0$ (ξ_0) are nearly twice as large as those located at 10 deg latitude in either direction. Regions poleward of ξ_0 tend to show a net magnetic field biased toward the following polarity, and regions equatorward of ξ_0 are biased toward the leading polarity, both by around 10%. Neither region growth rates nor decay rates are related to ξ. The average polarity axis tilt angles of regions are lower for regions near the equator than for those nearer the poles. Meridional motions of young regions are shown to be toward ξ_0. Older regions do not show this behavior. This may be a magnetic effect. The various results are discussed in terms of a hypothetical subsurface magnetic flux tube which gives rise to the surface activity.

075.050 **Diffusion of magnetic flux elements on a fractal geometry.**
J. K. Lawrence.
Sol. Phys., Vol. 135, No. 2, p. 249 – 259 (Oct 1991).
Recent observations have indicated that magnetic field elements are distributed on the Sun in fractal patterns with dimension <2. The author suggests that the transport of magnetic field elements across the solar surface should be treated as diffusion on a fractal geometry. He reviews a semi–analytical, theoretical treatment of fractal diffusion. Comparison with observations of small–scale motions of solar magnetic flux concentrations indicates that fractal diffusion may be taking place with dimension in the range 1.3 to 1.8. It is shown that, compared to the predictions that would be made for two–dimensional diffusion, fractal diffusion in this range would lead to an increased level of in situ flux cancellation in decaying active regions by 7% to 35%. Other work in specialities outside of solar physics may be useful in explaining solar magnetic phenomena.

075.051 **Dynamical behavior of strong magnetic fields in the solar convection zone.**
S. Vainshtein, E. H. Levy.
Sol. Phys., Vol. 135, No. 2, p. 261 – 274 (Oct 1991).
Magnetic buoyancy is thought to play an important role in the dynamical behavior of the Sun's magnetic field in the convection zone. Magnetic buoyancy is commonly thought to cause inescapable rapid loss of toroidal flux from much of the convection zone, thereby suppressing effective operation of a solar dynamo. This paper re-examines the detailed character of magnetic buoyancy, especially as it is influenced by the magnetic field's effect on heat transport and temperature gradients in the convection zone. It is suggested that suppression of convective heat transport across strong magnetic flux tubes can alter the temperature within the tubes and can subdue, or even reverse, the effect of magnetic buoyancy.

075.052 **Vector magnetic field measurements and penumbral structure.**
B. Kálmán.
Sol. Phys., Vol. 135, No. 2, p. 299 – 317 (Oct 1991).
The alignment of penumbral fibrils along the direction of the transverse magnetic field is good in the center of the solar disk, but deteriorates near the limb. This effect was studied on the basis of 15 vector magnetograms from various observatories in the period 1966 – 1984 for 5 sunspot groups. The results can be described with a simple geometrical model, where the magnetic field vectors and the penumbral fibrils lie in the same vertical plane, the inclination of the penumbra to the solar surface is 0 – 5°, and the elevation angle of the magnetic field vector is 40 – 50°. An adequate fit to observations was achieved only when an assumed uncertainty with 25° r.m.s. standard error was introduced in the angle measurements. The results are similar to earlier measurements for the chromosphere, although in the chromosphere the alignment of structures along the transverse magnetic field is better.

075.053 **Fractal dimensions and the correlation "area–flux" for local magnetic fields on the Sun.**
O. V. Chumak.
Astron. Tsirk., No. 1545, p. 25 – 26 (Aug – Sep 1990). In Russian.
It is shown that in the limits of a self-similar fractal model the dependence "area magnetic structure – magnetic flux" is a power function. The power index is equal to 2/C, where C is the fractal dimension.

075.054 **Evolution of vector magnetic field and velocity field in active region 5395.**
Zhang Hongqi, Ai Guoxiang, Li Jing, Li Wei, Chen Jimin.
Publ. Yunnan Obs., No. 4, p. 64 – 74 (Dec 1990). – See Abstr. 012.047 for the main entry.

075.055 **Generation of coronal magnetic fields by random surface motions. I. Mean square twist and current density.**
M. A. Berger.
Astron. Astrophys., Vol. 252, No. 1, p. 369 – 376 (Dec 1991).
Random motions in the photosphere or convection zone of a star can generate currents in the coronal plasma. These currents are associated with twisted, braided or sheared magnetic structures. Dissipation of these currents and structures may be largely responsible for coronal heating. In the Sturrock–Uchida model, individual magnetic flux tubes in the corona acquire a mean-square twist owing to the random motions of their footpoints in the photosphere. The author investigates this model in detail, and derives expressions for the mean-square twist in terms of the statistical properties of the photospheric flow. The results presented here for the expected twist on a flux tube place strong constraints on where high current densities reside. In a current filament, neighboring field lines twist about each other by an amount proportional to the current. Neighboring field lines acquire a typical twist of a few turns after 24 h of random photospheric motions; this twist corresponds to an extremely small current density. For this reason current filaments probably carry only a small fraction of the total coronal current. Similarly, small-scale structures induced by a simple turbulent cascade should carry relatively little current. Only in singular current layers such as those produced by the collapse of an X-point can the current density be strong enough to produce significant heating or magnetic energy dissipation.

075.056 **The evolution characters of magnetic field and velocity field in solar active region AR5629.**
Liu Yang, Zhang Hongqi.
Publ. Yunnan Obs., No. 4, p. 186 – 188 (Dec 1990). – See Abstr. 012.047 for the main entry.

075.057 **Influence of longitudinal electric fields on plasma instabilities in solar magnetic traps.**
A. N. Krishtal', A. K. Yukhimuk.
Kinematics Phys. Celest. Bodies, Vol. 7, No. 2, p. 20 – 28 (1991). English translation of Kinematika Fiz. Nebesn. Tel, Tom 7, No. 2, p. 23 – 32 (1991).
A possible mechanism by which a longitudinal electric field may influence the stability of Langmuir and ion–acoustic oscillations in solar magnetic traps is examined, with post-flare loops as an example. It is shwon that the presence of a longitudinal electric field may result in the development of plasma instability. Expressions for the increments and estimates of development time are obtained for the linear stages of the instabilities, as well as the ranges of variation of the critical external electric field values as functions of perturbation wavelength. It is shown that the longitudinal electric fields observed in post-flare loops strongly influence the formation of ion–acoustic and Langmuir turbulence in the loops.

075.058 **Results of magnetographic line–ratio measurements.**
G. Bachmann.
Astron. Nachr., Vol. 312, No. 4, p. 257 – 263 (1991).
The author adapts the line-ratio method to magnetographic measurements. The method yields information on inhomogeneous magnetic fields without resolving their fine structure. Supplementing the magnetic field measurements, each day several polarization calibrations were made to eliminate the influence of the instrument. Modified theoretical calibration curves served for the transformation into inclination angles and magnetic field strength. The circular polarization along some scans across two sunspots practically shows the same distributions in both wings. From the amount of the circular polarization in faculae comparing it with results of model calculations the filling factors were determined. The results arranged between 3 and 11% nearly agree with the values by Solanki and Stenflo (1984). The line-ratio method in the version adapted to the author's magnetograph yielded from the V ratios in sunspots an average of 1880 G, but from the Q_0 ratios an average of 2240 G. In faculae both values are practically identical and run to 1980 G. These results show that the line-ratio method is superior to the old reduction procedure of magnetographic measurements.

075.059 **Effect of rotation of the Sun on turbulent transfer of a large-scale magnetic field into the convective zone.**
L. L. Kichatinov, V. N. Krivodubskij.
Kinematika Fiz. Nebesn. Tel, Tom 7, No. 6, p. 30 – 39 (Nov–Dec 1991). In Ukrainian. English translation in Kinematics Phys. Celest. Bodies, Vol. 7, No. 6.
The turbulent transfer of a large-scale magnetic field is investigated in the density-stratified convective zone with due regard for turbulence anisotropy excited by the rotation of the Sun. It is found that the transfer directions of poloidal and toroidal components of the magnetic field do not coincide. The values of different components of transfer velocity of the magnetic fields for the Spruit convective zone model are calculated. The possible role of the transfer effects in the dynamics of large-scale fields over a solar cycle is discussed.

075.060 **Diffusion current sheets with large transverse magnetic field component on the Sun.**
V. M. Bardakov.
Sov. Astron. Lett., Vol. 17, No. 5, p. 356 – 359 (Sep 1991). Current Physics Microform No.: 9204X1303. English translation of Pis'ma Astron. Zh., Tom 17, No. 9, p. 841 – 849 (1991).

The author examines the question of the maximum possible reconnection speed in the diffusion current sheet (DCS) regime. He shows that for a DCS with plasma concentration ratio inside to outside the sheet (N) much greater than unity, the maximum reconnection speed is $N^{1/4}$ times the reconnection speed determined in the neutral DCS approximation (the Sweet–Parker current). In the state with maximum reconnection speed, the DCS has a large transverse magnetic field component (the ratio of the transverse to longitudinal components at the sheet boundary is N times the sheet's thickness to width ratio), and the plasma is accelerated along the sheet mainly by the Lorentz force of the transverse magnetic field component. The conditions of thermal instability formation for a cold preflare DCS with a large transverse magnetic field component differ substantially from those studied using the neutral DCS approximation in the well–known work of Syrovatskij (1976).

075.061 **Steady flows in solar magnetic structures – a class of exact MHD solutions.**
A. de Ville, E. R. Priest.
Geophys. Astrophys. Fluid Dyn., Vol. 59, No. 1–4, p. 253 – 265 (Nov 1991).

A class of exact solutions to the steady, two–dimensional magnetohydrodynamic equations in a cylindrical geometry is presented. These may model both closed and open magnetic structures found in the solar atmosphere. For closed structures, it is found that increasing the flow speed causes the summit of the arcade of closed magnetic fieldlines to rise. Parameter ranges also exist where the solution has regions of open and closed field, and so the solutions may be relevant for modelling flows in solar magnetic structures such as coronal streamers, X–ray bright points, coronal plumes and coronal holes.

075.062 **A solar magnetic field model and its 3–D boundary element method solution.**
Yan Yihua, Yu Qing, Kang Feng.
Sol. Phys., Vol. 136, No. 1, p. 195 – 198 (Nov 1991). Letter–to–the–editor.

A force–free magnetic field model with constant α is established, and a boundary element method is proposed to solve the problem. The procedure ensures a unique solution as well as a finite magnetic energy content. The proposed formulation is effective in solving magnetic fields above the solar surface, and the validity of this procedure is demonstrated by satisfactory agreement between calculated and observed magnetograms.

075.063 **Filigree and flux tube physics.**
H. C. Spruit, M. Schüssler, S. K. Solanki.
Solar interior and atmosphere, p. 890 – 910 (1991). – See Abstr. 003.028 for the main entry.

This review covers the properties of the small–scale (outside sunspots) magnetic field from observational and theoretical points of view. Special emphasis is put on the physics of small isolated magnetic concentrations (flux tubes). Topics discussed include the basic observational properties, the origin and disappearance of the small–scale field, the properties of individual magnetic elements, their influence on the solar irradiance and their interaction with solar oscillations.

075.064 **Adaptive potential drops as energy dissipators in solar magnetic structures.**
K. F. Tapping.
J. R. Astron. Soc. Can., Vol. 85, No. 4, p. 178 – 179 (Aug 1991). Abstract. – See Abstr. 012.091 for the main entry.

075.065 **Magnetic field structure of active regions. The homogeneous inverse problem.**
B. P. Filippov.
Sov. Astron., Vol. 35, No. 6, p. 652 – 656 (Nov 1991). Current Physics Microform No.: 9205X1462. English translation of Astron. Zh., Tom 68, Vyp. 6, p. 1299 – 1308 (1991).

The information about the direction of the tangential magnetic field in the chromosphere incorporated into the orientation of fibrils in Hα filtergrams enables one to determine the structure of the field above the chromosphere under certain conditions. In the potential approximation, the problem reduces to a one–dimensional Fredholm integral equation of the first kind. Well–known regularization methods enable one to solve this ill–posed problem. The results of modeling calculations are given.

075.066 **Optical observations of flare–productive flux emergence.**
H. Kurokawa.
International SOLAR–A Science Meeting: Flare physics in solar activity maximum 22, p. 39 – 50 (1991). – See Abstr. 012.104 for the main entry.

The optical observations of the relationship between emerging magnetic fluxes and flares are reviewed and the characteristics essential to the flare–productive emerging flux regions are studied. Examining several essential parameters, i.e. location of emergence, orientation of bipolar axis, growth rate and magnetic shear development, the author concludes that the emergence of twisted magnetic flux ropes is the most important factor for the major flare activities. Some observational evidence of the emergence of the twisted magnetic flux ropes is demonstrated with high resolution Hα pictures.

075.067 **Small scale active phenomena observed with DST and Huairou magnetogram.**
M. Akioka.
International SOLAR–A Science Meeting: Flare physics in solar activity maximum 22, p. 58 – 61 (1991). – See Abstr. 012.104 for the main entry.

The author examines in detail the evolutionary change of fibrils in a small and short–lived active region. The video movie of high resolution Hα filtergram and high resolution magnetograms are used to reveal magnetic field change in the photosphere and chromosphere. The interaction of fibrils, called fibril merging, was found near the site of cancellation of magnetic field. Some small scale activities, subflares and a surge, were found around this site. Fibril merging may be due to magnetic reconnection of the neighbouring magnetic loops.

075.068 **The practical application of the magnetic virial theorem.**
J. A. Klimchuk, R. C. Canfield, J. E. Rhoads.
International SOLAR–A Science Meeting: Flare physics in solar activity maximum 22, p. 219 – 223 (1991). The full paper is published in Astrophys. J., Vol. 358, No. 1, p. 327 – 343 (20 Jan 1992). – See Abstr. 012.104 for the main entry.

The authors have performed simulated vector magnetograph observations to study the effects of random and systematic magnetic field measurement errors on the magnetic energies that may be estimated using the virial theorem.

075.069 **The reversal of the solar polar magnetic fields. II. Simulations of the large–scale fields.**
P. R. Wilson, P. S. McIntosh.
Sol. Phys., Vol. 136, No. 2, p. 221 – 237 (Dec 1991).

Observations of the first large–scale patterns of magnetic fields near the sunspot minimum of 1986 (the start of cycle 22) are presented using synoptic magnetic data provided by the National Solar Observatory and contour maps constructed from data provided by the Mount Wilson Solar Observatory. The latter are compared with simulated contour maps derived from numerical solutions of the flux transport equation using data from particular Carrington rotations as initial conditions. The simulated evolutions of the large–scale magnetic fields are qualitatively consistent with observed evolutions, but differ in several significant respects. Some of the differences can be removed by varying the diffusivity and the parameters of the large–scale velocity fields. It is concluded that the patterns of weak magnetic fields which comprise the large-scale features cannot be formed entirely by the diffusive decay of active regions. There must be a significant contribution to these patterns by non-random flux eruptions within the network structure, independent of active regions.

075.070 **Solar magnetic elements.**
M. Schüssler.
Geophys. Astrophys. Fluid Dyn., Vol. 62, No. 1-4, p. 271 - 281 (Dec 1991). - See Abstr. 012.126 for the main entry.

The interaction of magnetic fields with convective flows and radiative energy transport leads to a remarkable spatial organization of the magnetic flux permeating the solar photosphere: most of the flux outside sunspots is concentrated in small-scale structures of large fields strength. The properties of these flux concentrations are largely uniform suggesting the concept of a prototype structure, the magnetic element. This contribution gives a brief overview of the observational results and the theoretical concepts used to describe the physics of magnetic elements.

075.071 **Fine structures of chromospheric magnetic field and material flow in a solar active region.**
Zhang Hongqi, Ai Guoxiang, T. Sakurai, H. Kurokawa.
Sol. Phys., Vol. 136, No. 2, p. 269 - 293 (Dec 1991).

The authors analyze the relations between photospheric vector magnetic fields, chromospheric longitudinal magnetic fields and velocity fields in a solar active region. Agreements between the photospheric and chromospheric magnetograms can be found in large-scale structures or in the stronger magnetic structures, but differences also can be found in the fine structures or in other places, which reflect the variation of the magnetic force lines from the photosphere to the chromosphere. The chromospheric superpenumbral magnetic field, measured by the Hβ line, presents a spoke-like structure. It consists of thick magnetic fibrils which are different from photospheric penumbral magnetic fibrils. The outer superpenumbral magnetic field is almost horizontal. The direction of the chromospheric magnetic fibrils is generally parallel to the transverse components of the photospheric vector magnetic fields. The chromospheric material flow is coupled with the magnetic field structure. The structures of the Hβ chromospheric magnetic fibrils in the network are similar to Hβ dark fibrils, and the feet of the magnetic fibrils are located at the photospheric magnetic elements.

The Sun and cool stars: activity, magnetism, dynamos. Proceedings.
See Abstr. 012.024.

Simultaneous observations with a tunable filter and the Echelle Spectrograph of the Vacuum Tower Telescope at Tenerife.
See Abstr. 034.068.

An electrograph for measurement of macroscopic electric fields in prominences and flares.
See Abstr. 034.085.

A new method for calibrating vector magnetograms.
See Abstr. 036.067.

The derivation of vector magnetic fields from Stokes profiles: derivative versus least squares fitting techniques.
See Abstr. 036.068.

Blending influence of Fe I 5250.6 Å spectral line on the Fe I 5250.2 Å spectral line and its implication for polarization measurements.
See Abstr. 036.069.

Problems associated with aligning MSFC transverse azimuth maps and H alpha images.
See Abstr. 036.070.

The practical application of the magnetic virial theorem: analytical results.
See Abstr. 036.071.

Further numerical tests for the progressive extension method (PEM) for extrapolation of solar force-free magnetic field.
See Abstr. 036.072.

Considerations for flare related magnetic field measurements.
See Abstr. 036.073.

Force-free magnetic fields: generating functions and footpoint displacements.
See Abstr. 062.010.

Magnetic field intensification and formation of field-aligned current in a non-uniform magnetic field.
See Abstr. 062.014.

Siphon flows in isolated magnetic flux tubes. IV. Critical flows with standing tube shocks.
See Abstr. 062.015.

Heating in intense flux tubes.
See Abstr. 062.060.

Current helicity and the generation of magnetic field aligned currents.
See Abstr. 062.077.

Radiative transfer across a magnetic flux tube.
See Abstr. 062.130.

Multiple scattering of acoustic waves by a pair of uniformly magnetized flux tubes.
See Abstr. 062.158.

Maximum energy of semi-infinite magnetic field configurations.
See Abstr. 062.175.

Maximum energy of semi-infinite magnetic-field configurations.
See Abstr. 062.235.

The Sun as a prototype in the study of stellar magnetic activity.
See Abstr. 064.060.

On the structure of spectral line gap regions.
See Abstr. 071.006.

Spatial correlation of spectrum line properties with magnetic flux.
See Abstr. 071.020.

Granular and intergranular line profiles in solar active and quiet regions.
See Abstr. 071.022.

The rotational modulation of Ca II K in the Sun.
See Abstr. 071.038.

The variation of the solar granulation structure in active and non-active regions.
See Abstr. 071.043.

Model for the relationship of granulation and supergranulation.
See Abstr. 071.057.

Determination of the parameters of sunspot magnetic fields from the structure of the superpenumbra.
See Abstr. 072.009.

Area and magnetic field of a sunspot during slow dissipation.
See Abstr. 072.010.

The magnetic fields of active regions. VII. East–west inclination of magnetic field lines.
See Abstr. 072.013.

Electric currents and Hα emission in two active regions on the Sun.
See Abstr. 072.014.

Weak magnetic fields and solar irradiance variations.
See Abstr. 072.015.

Solar patrol.
See Abstr. 072.016.

Solar patrol.
See Abstr. 072.017.

Solar patrol.
See Abstr. 072.018.

Solar patrol.
See Abstr. 072.019.

Solar patrol.
See Abstr. 072.020.

Solar patrol.
See Abstr. 072.021.

On the magnetic field activity in solar active regions.
See Abstr. 072.023.

On the scaling of the axial magnetic field gradient for twisted sunspots.
See Abstr. 072.035.

Stokes polarimetry of a sunspot from the Swedish Solar Observatory at La Palma.
See Abstr. 072.036.

Fine–scale magnetic field in a sunspot penumbra and adjacent photosphere.
See Abstr. 072.037.

Field geometry of sunspots inferred from inclination effects.
See Abstr. 072.038.

Mapping of three–dimensional sunspot fields by inversion line and symmetry.
See Abstr. 072.039.

Solar patrol.
See Abstr. 072.053.

Solar patrol.
See Abstr. 072.054.

Solar patrol.
See Abstr. 072.055.

Solar patrol.
See Abstr. 072.056.

Solar patrol.
See Abstr. 072.057.

Solar patrol.
See Abstr. 072.058.

Solar patrol.
See Abstr. 072.059.

Two–year variations of solar activity, interplanetary space and cosmic rays parameters.
See Abstr. 072.075.

Solar research at Potsdam: papers on the structure and dynamics of sunspots.
See Abstr. 072.086.

Sun–as–a–star: its convective signature and the activity cycle.
See Abstr. 072.087.

VLA observations of solar active regions at 6 and 20 cm.
See Abstr. 072.095.

Velocity measures in the sunspot Rome No. 7490 (July 5, 1979).
See Abstr. 072.096.

On the dynamics of the large–scale magnetic fields of the Sun and the sunspot cycle.
See Abstr. 072.100.

Solar flares and magnetic field configuration in a solar active region Boulder 5395.
See Abstr. 072.106.

Spatial variation of the magnetic field inclination in a sunspot penumbra.
See Abstr. 072.111.

The relation of the sunspot magnetic field and penumbra–umbra radius ratio.
See Abstr. 072.118.

Motions, fields, and flares in the 1989 March active region.
See Abstr. 072.120.

Study of the morphological correlation between photospheric transversal magnetic field and penumbral filaments in AR5395.
See Abstr. 072.123.

An emerging electric current/magnetic system with delta structure in AR2372.
See Abstr. 072.130.

The solar activity cycle.
See Abstr. 072.131.

Active regions, sunspots and their magnetic fields.
See Abstr. 072.132.

Vector magnetic fields in sunspots. I. Stokes profile analysis using the Marshall Space Flight Center magnetograph.
See Abstr. 072.134.

Plasma motion and electric currents in an active region.
See Abstr. 072.136.

Solar Active Longitude recurring every 28 – 29 days acting differently in low and higher latitudes.
See Abstr. 072.138.

Stokes–V in an active prominence.
See Abstr. 073.003.

K_{2V} cell grains and chromospheric heating.
See Abstr. 073.018.

Irrotational motion in quiescent prominences.
See Abstr. 073.047.

The formation and support of prominences in magnetic loops and arcades.
See Abstr. 073.060.

Two–dimensional models of the solar chromosphere. I. The Ca II K line as a diagnostic: 1.5–D radiative transfer.
See Abstr. 073.102.

Evidence for the interaction of large scale magnetic structures in solar flares.
See Abstr. 073.109.

Magnetohydrodynamics of an impulsively heated, hard X–ray emitting filament.
See Abstr. 073.112.

Modelling the moustache phenomenon in network regions.
See Abstr. 073.115.

AR5395: the flare trigger and the evolution of magnetic fields.
See Abstr. 073.124.

Strong magnetic fields in solar flares: observation data and theoretical model.
See Abstr. 073.141.

Flares appear on the red shift side near the inversion line of the H_β Dopplergram.
See Abstr. 073.172.

Microflares observed in He I 10830 and their relation to the quiet Sun magnetic fields.
See Abstr. 073.173.

Magnetic morphology of nonthermal electron precipitation during three flares in a highly nonpotential active region.
See Abstr. 073.175.

Material flow of a surge flare.
See Abstr. 073.178.

First scientific results from the Ulysses mission.
See Abstr. 074.010.

The effect of strong shear on solar coronal magnetic fields.
See Abstr. 074.054.

Three–dimensional reconnection in coronal loops.
See Abstr. 074.056.

Survival of discrete structures in the solar wind.
See Abstr. 074.065.

Observations of disconnection of open coronal magnetic structures.
See Abstr. 074.081.

Acceleration of electrons outside flares: coronal manifestation and possible origin.
See Abstr. 074.083.

Avalanches and the distribution of solar flares.
See Abstr. 074.099.

On the spontaneous formation of electric current sheets above a flexible solar photosphere.
See Abstr. 074.102.

Neutralized and unneutralized current patterns in the solar corona.
See Abstr. 074.103.

Energy balance in coronal funnels.
See Abstr. 074.117.

Dynamics of solar coronal magnetic fields.
See Abstr. 074.118.

Magnetic structures in the corona.
See Abstr. 074.121.

Kinematics of discrete components in solar coronal plasma in relation to coronal magnetic fields.
See Abstr. 074.128.

Large scale patterns on the solar surface indicated by microwave observations.
See Abstr. 077.036.

Stochastic dynamics of protons in solar magnetic loops.
See Abstr. 078.005.

Some studies of solar proton events in relation to active region characteristics.
See Abstr. 078.017.

Magnetic flux transport and the Sun's dipole moment: new twists to the Babcock–Leighton model.
See Abstr. 080.014.

Chromosphere generation in magnetic flux–tubes.
See Abstr. 080.046.

Atmospheric heating in emerging flux regions.
See Abstr. 080.047.

On generation of low–frequency pulsations in magnetic waveguides in the solar atmosphere.
See Abstr. 080.052.

The confrontation of mean–field theories with numerical simulations.
See Abstr. 080.061.

The role of overshoot in solar activity: a direct simulation of the dynamo.
See Abstr. 080.064.

Waves of solar activity.
See Abstr. 080.066.

The solar dynamo.
See Abstr. 080.071.

Solar internal rotation, the boundary layer dynamo and latitude distribution of activity belts.
See Abstr. 080.072.

Concentrated magnetic fields in the solar atmosphere.
See Abstr. 080.101.

The solar dynamo.
See Abstr. 080.126.

The internal rotation and magnetism of the Sun from its oscillations.
See Abstr. 080.130.

A torsional oscillation in the rotation of the solar magnetic field.
See Abstr. 080.140.

Do changes in the photospheric magnetic network cause the 11 year variation of total solar irradiance?
See Abstr. 080.141.

A new solar cycle model including meridional circulation.
See Abstr. 080.153.

The effects of magnetic stretching in the convective overshoot region.
See Abstr. 080.157.

Is the recurrent enhanced geomagnetic activity preceding the sunspot minimum related to magnetic fields of the new or old activity cycle?
See Abstr. 084.053.

Magnetic history of the Sun.
See Abstr. 107.019.

Magnetic field measurements on cool stars.
See Abstr. 116.022.

The past, present and future of solar magnetism: stellar magnetic activity.
See Abstr. 116.086.

Nature of solar–cycle and heliomagnetic–polarity dependence of cosmic rays, inferred from their correlation with heliomagnetic spherical surface harmonics in the period 1976–1985.
See Abstr. 144.107.

076 UV, X, Gamma Radiation

076.001 **SMM and solar gamma–ray astrophysics.**
R. Ramaty.
Bull. Am. Astron. Soc., Vol. 23, No. 2, p. 934 (1991). Abstract. – See Abstr. 012.008 for the main entry.

076.002 **HXIS observations of weak X–ray sources.**
M. E. Machado.
Bull. Am. Astron. Soc., Vol. 23, No. 2, p. 1027 (1991). Abstract. – See Abstr. 012.012 for the main entry.

076.003 **The X–ray counterparts of UV microflares.**
J. G. Porter, J. M. Fontenla, R. L. Moore, G. M. Simnett.
Bull. Am. Astron. Soc., Vol. 23, No. 2, p. 1027 (1991). Abstract. – See Abstr. 012.012 for the main entry.

076.004 **Very high resolution solar X–ray imaging.**
L. Golub.
International Conference on Mechanisms of Chromospheric and Coronal Heating, p. 115 – 123 (1991). – See Abstr. 012.011 for the main entry.

The author reports results from a recent flight of the Normal Incidence X–ray Telescope (NIXT), a sounding rocket payload which has obtained sub–arcsecond images of the Solar X–ray corona. He found that the loops, which were observed by earlier experiments on Skylab and on sounding rockets, are resolved into numerous long, thin loops. The lengths of coronal loops vary from several arcseconds to several arcminutes, and in nearly all cases the transverse sizes of the loops are only a few arcseconds; the aspect ratios (length:width) range from ~ 2 to > 20. The author also observed that the coronal loops terminate very precisely in areas of enhanced chromospheric emission as seen, for instance, in Hα. These enhancements lie directly below the tarpered ends of coronal loops and are seen in network, plage, and in the penumbrae of sunspots; there are no hot loops terminating in sunspot umbrae.

076.005 **High resolution gamma–ray and hard X–ray spectroscopy of solar flares.**
P. T. Feffer, K. Hurley, D. Smith, R. P. Lin, M. Pelling, R. H. Pehl, P. N. Luke, N. W. Madden, G. Vedrenne.
Bull. Am. Astron. Soc., Vol. 23, No. 2, p. 1039 (1991). Abstract. – See Abstr. 012.012 for the main entry.

076.006 **Future prospects for solar gamma–ray astrophysics.**
R. Ramaty.
Bull. Am. Astron. Soc., Vol. 23, No. 2, p. 1040 (1991). Abstract. – See Abstr. 012.012 for the main entry.

076.007 **Solar flare X–ray observations with Ulysses.**
S. R. Kane, K. Hurley, F. Cotin, M. Niel, G. Vedrenne, J. C. Henoux, M. Sommer.
Bull. Am. Astron. Soc., Vol. 23, No. 2, p. 1042 (1991). Abstract. – See Abstr. 012.012 for the main entry.

076.008 **Gamma ray observation from a behind–the–limb flare.**
D. J. Forrest, T. Vestrand.
Bull. Am. Astron. Soc., Vol. 23, No. 2, p. 1043 (1991). Abstract. – See Abstr. 012.012 for the main entry.

076.009 **The solar UV continuum 1440 – 1680 Å.**
O. Kjeldseth–Moe, P. Brekke, J.–D. F. Bartoe, G. E. Brueckner.
Bull. Am. Astron. Soc., Vol. 23, No. 2, p. 1060 – 1061 (1991). Abstract. – See Abstr. 012.012 for the main entry.

076.010 **A comparison of SERTS EUV spectroheliograms with solar active region and quiet sun magnetic field structure.**
J. W. Brosius, W. T. Thompson, W. M. Neupert.
Bull. Am. Astron. Soc., Vol. 23, No. 2, p. 1061 (1991). Abstract. – See Abstr. 012.012 for the main entry.

076.011 **Comparison of SERTS spectroheliograms with ground–based observations.**
W. T. Thompson, W. M. Neupert, J. W. Brosius, H. P. Jones, B. Schmieder.
Bull. Am. Astron. Soc., Vol. 23, No. 2, p. 1061 – 1062 (1991). Abstract. – See Abstr. 012.012 for the main entry.

076.012 **Spectral hardening and softening in hard X–ray bursts.**
E. J. Schmahl, B. R. Dennis, A. Kiplinger.
Bull. Am. Astron. Soc., Vol. 23, No. 2, p. 1064 (1991). Abstract. – See Abstr. 012.012 for the main entry.

076.013 **Statistical studies of ISEE–3/ICE data of impulsive hard X–ray bursts.**
J. M. McTiernan, S. R. Kane.
Bull. Am. Astron. Soc., Vol. 23, No. 2, p. 1066 (1991). Abstract. – See Abstr. 012.012 for the main entry.

076.014 **High resolution solar gamma–ray and hard X–ray observations on a long–duration balloon flight from Antarctica.**
D. M. Smith, K. A. Anderson, P. T. Feffer, K. Hurley, R. P. Lin.
Bull. Am. Astron. Soc., Vol. 23, No. 2, p. 1071 (1991). Abstract. – See Abstr. 012.012 for the main entry.

076.015 **Stereoscopic observations of solar flare X–ray and gamma–ray sources using GRO instruments.**
S. R. Kane, K. Hurley, J. Laros, M. Sommer.
Bull. Am. Astron. Soc., Vol. 23, No. 2, p. 1073 (1991). Abstract. – See Abstr. 012.012 for the main entry.

076.016 **Automated search for solar microflares on the Burst And Transient Source Experiment.**
D. A. Biesecker.
Bull. Am. Astron. Soc., Vol. 23, No. 2, p. 1073 (1991). Abstract. – See Abstr. 012.012 for the main entry.

076.017 **A statistical approach to solar flare spectral analysis with BATSE.**
D. J. Forrest, W. T. Vestrand.
Bull. Am. Astron. Soc., Vol. 23, No. 2, p. 1074 (1991). Abstract. – See Abstr. 012.012 for the main entry.

076.018 **Lagging behind of the fluxes of ultraviolet emission relative to the index $F_{10.7}$.**
M. V. Krasinets, A. A. Nusinov.
Soln. Dannye, Byull., No. 7/91, p. 88 – 91 (Jul 1991). In Russian.
On the basis of the data on the UV emission flux in the line Hα ($I_{L\alpha}$) and on the radio emission flux at the wavelength 10.7 cm ($F_{10.7}$) for 1982 – 1988 it is shown that the values $F_{10.7}$ lag on the average one day behind the values $I_{L\alpha}$. It is likely to be caused by a decrease in the emission brightness and the difference of the law of the limb darkening.

076.019 **Statistical properties of solar gamma–ray flares.**
Xu Aoao, Yin Chunlin, Zhang Heqi, S. T. Wu.
Chin. Astron. Astrophys., Vol. 15, No. 3, p. 318 – 328 (Sep 1991). English translation of Acta Astron. Sin., Vol. 32, No. 1, p. 36 – 45 (Mar 1991). See Abstr. 53.076.010.
The authors studied the 120 solar gamma–ray events observed by the gamma–ray spectrometer on board the SMM satellite between 1980 March and 1983 June with particular emphasis on the basic properties of the active regions producing such events. Combined with observed data of Type–II and Type–IV metre wave radio bursts, they further discussed the physical mechanisms in solar flares. Their results show 1) that gamma–ray flares generally occur in morphologically and magnetically complicated regions; 2) that flares with gamma–ray lines are closely related to Type–II and Type–IV bursts; 3) that there are regions with multiple gamma–ray flares; 4) that a deep study of the electric current in the active region and plasma instabilities involving the current is important for understanding the basic physical process of flares; and 6) that different flares may have different particle acceleration mechanisms at work.

076.020 **A search for hard X–ray microflares near solar minimum.**
R. P. Lin, K. C. Hurley, D. M. Smith, R. M. Pelling.
Sol. Phys., Vol. 135, No. 1, p. 57 – 64 (Sep 1991).
Near solar maximum, hard X–ray microflares with peak 20 keV fluxes of $\gtrsim 10^{-2}(\mathrm{cm}^2\mathrm{s\ keV})^{-1}$, more than ten times smaller than for typical flares and subflares, can occur at the rate of about once every five minutes. The authors report on a search for hard X–ray microflares made on a long duration balloon flight in February 1987 near solar minimum, at a time when no active regions were on the Sun. No microflares were observed over a total observing time of 16.5 hours spread over three days, implying a statistical upper limit to their rate of occurrence about a factor of ten lower than observed near solar maximum. Thus hard X–ray microflaring appears to be an active region phenomenon, and apparently not associated with flaring of soft X–ray bright points.

076.021 **X–ray observations of filament eruption in the 1980 May 21 flare.**
D. A. Batchelor, K. P. Hindsley.
Sol. Phys., Vol. 135, No. 1, p. 99 – 105 (Sep 1991).
X–ray and Hα observations of an erupting filament, discussed herein, and other observations of the associated flare on 1980 May 21, suggest that an erupting filament played a major role in the X–ray flare. The authors show that pre–heated plasma existed and may have contributed part of the blue–shifted X–ray emission observed in the Ca XIX line, which was formerly attributed solely to chromospheric evaporation.

076.022 **Fe XIV and other line identifications in the solar EUV spectrum.**
S. O. Kastner.
Sol. Phys., Vol. 135, No. 2, p. 343 – 351 (Oct 1991).
A number of solar extreme ultraviolet lines, previously unidentified or given assignments at the time of observation, are found to be coincident in wavelength with transitions recently identified as aluminum–like or magnesium–like in the laboratory. Three tables summarize the results of this comparison, which does not imply positive identifications but suggests that several assignments may be valid. Additional assignments are proposed in other ions and a revised table of predicted and observed forbidden line wavelengths in the Ne I sequence is included.

076.023 **A comparison of theoretical Fe XII emission line strengths with EUV observations of a solar active region.**
F. P. Keenan, J. G. Doyle, S. S. Tayal, R. J. W. Henry.
Sol. Phys., Vol. 135, No. 2, p. 353 – 360 (Oct 1991).
New theoretical electron–density–sensitive Fe XII emission line ratios are derived using R–matrix electron impact excitation rate calculations. The authors have identified the Fe XII transitions in an active region spectrum obtained with the Harvard S–055 spectrometer on board Skylab at wavelengths of 364.0, 382.8, 1241.7, and 1349.4 Å. Electron densities are significantly larger than the typical active region measurements, but are similar to those derived from some active region spectra observed with the Skylab 2082A instrument.

076.024 **Density measurement from Al IX λ385.03/λ392.42 line ratio.**
B. N. Dwivedi, A. K. Gupta.
Sol. Phys., Vol. 135, No. 2, p. 415 – 417 (Oct 1991). Letter–to–the–editor.
Al IX EUV line ratio λ385.03/λ392.42 has been found to be density sensitive. Using this, an electron density of $3.7 \times 10^8 \mathrm{cm}^{-3}$ is found in the quiet Sun off–limb from Skylab observation reported by Vernazza and Reeves (1978).

076.025 **Theoretical analysis of iron X–ray spectra observed in solar active region.**
M. Cornille, J. Dubau, M. Loulergue, F. Bely–Dubau, P. Faucher.
5. Internationale Konferenz über die Physik hochgeladener Ionen, p. 159 – 160 (1991). – See Abstr. 012.064 for the main entry.
The theoretical analysis of X–ray spectra observed by the Flat Crystal Spectrometer of the Solar Maximum Mission in the 13 Å – 19 Å wavelength range is presented. The emission lines correspond to Fe ions (Fe XVI, Fe XVII, Fe XVIII) and some O–ions (O VII, O VIII). The fitting of a synthetic spectrum to the observed one shows that it is difficult to represent the emissive region with one temperature model.

076.026 **High–resolution solar spectrum between 2000 and 3100 Å.**
L. A. Hall, G. P. Anderson.
J. Geophys. Res., Vol. 96, No. D7, p. 12927 – 12932 (20 Jul 1991).
Solar spectra in the wavelength range 2000–3100 Å, measured near 40 km in the stratosphere in 1978 and 1983, have been combined and extrapolated to zero optical depth to provide a reference spectrum in 0.1–Å resolution. The wavelength scale has been carefully compared with wavelength standards, and a cyclic instrumental effect has been removed, resulting in a wavelength accuracy to within 0.04 Å. The absolute intensities have been normalized to the previously reported 1983 spectrum in 1.0–Å resolution.

076.027 **Some new statistical properties of solar hard X–ray, gamma ray bursts and their astrophysical implication for charged particle acceleration in solar flare.**
Zhang Heqi, Yin Chunlin, Xu Aoao.
Publ. Yunnan Obs., No. 4, p. 9 – 13 (Dec 1990). – See Abstr. 012.047 for the main entry.

076.028 **Numerical simulation of ultraviolet and X–ray microflares.**
A. C. Sterling, K. Shibata, J. T. Mariska.
International SOLAR–A Science Meeting: Flare physics in solar activity maximum 22, p. 71 – 74 (1991). – See Abstr. 012.104 for the main entry.

A series of numerical simulations indicates that thermal energy releases of $\sim 10^{25} - 10^{27}$ergs in the middle chromosphere can produce ejections into the corona in the form of pressure gradient generated jets, jets with pressure gradient and shock generated components, or high speed gas plugs. Heating of the chromosphere to X–ray emitting temperatures occurs in association with gas plugs, perhaps generating X–ray microflares observable by Solar–A. Chromospheric UV–microflares can occur in association with some jets, but do not generally occur with spicules.

076.029 **Particle acceleration and the locations of hard X–ray sources.**
H. Nakajima.
International SOLAR–A Science Meeting: Flare physics in solar activity maximum 22, p. 77 – 88 (1991). – See Abstr. 012.104 for the main entry.

This article reviews hard X–ray imaging observations of solar flares with emphasis on the results obtained with instruments on both the Hinotori and Solar Maximum Mission satellites. The major findings during the last solar maximum are as follows: (1) the hard X–ray flares show a wide variety but can be grouped into three categories: impulsive, gradual–hard, and hot–thermal; (2) the hard X–ray imaging observations seem to support the nonthermal electron beam model rather than the thermal model; (3) the energy release in the impulsive phase occurs in sheared magnetic loops; (4) the individual spikes of a multiple impulsive burst are emitted from rather stable (slightly different) sources in most cases, but from distinctly different loops in other cases. In the latter cases, the growth of the loop is not a continuous process, but jumps from one loop to a nearby (lateral or higher) loop. Finally (5) high–energy electrons with energies from a few tens of keV to several MeV are trapped in the tops of loops in the gradual–hard flares.

076.030 **The controversial relationship between hard X–ray and soft X–ray flares: causal or non–causal?**
B. R. Dennis.
International SOLAR–A Science Meeting: Flare physics in solar activity maximum 22, p. 89 – 95 (1991). – See Abstr. 012.104 for the main entry.

076.031 **The role of accelerated ions during the impulsive phase of flares and the production of hard X–rays.**
G. M. Simnett.
International SOLAR–A Science Meeting: Flare physics in solar activity maximum 22, p. 100 – 105 (1991). – See Abstr. 012.104 for the main entry.

It is becoming clear that ions play a more significant role in the energy transfer at the onset of solar flares than had previously been thought. The author compared the electron and ion beam models for flares and reviewed a proposal for generating the hard X–ray burst from an ion beam via secondary electron acceleration in the chromosphere.

076.032 **Physical implications of X–ray imaging observations.**
K. Ohki.
International SOLAR–A Science Meeting: Flare physics in solar activity maximum 22, p. 106 – 109 (1991). – See Abstr. 012.104 for the main entry.

076.033 **Heating and flows in flares.**
G. A. Doschek.
International SOLAR–A Science Meeting: Flare physics in solar activity maximum 22, p. 121 – 132 (1991). – See Abstr. 012.104 for the main entry.

A review is presented of some unresolved problems in high resolution soft X–ray spectroscopy of solar flares. The main emphasis is on the dynamics of the multi–million degree thermal plasma, as revealed by line broadening and blueshifted components of spectral lines. The author discusses some of the most recent results obtained from analysis of X–ray spectra. He also discusses some problems related to the temperature distribution and ionization balance in the flare plasma. The data from SOLAR–A should provide substantial new insight into all of these problems.

076.034 **Studies of helium–like ion spectra with the SMM Flat Crystal Spectrometer.**
K. J. H. Phillips.
International SOLAR–A Science Meeting: Flare physics in solar activity maximum 22, p. 133 – 136 (1991). – See Abstr. 012.104 for the main entry.

Analysis of X–ray spectra from the FCS instrument on SMM has given valuable information about the physical conditions in active regions and flares, notably temperatures and densities. Element abundances can also be derived from some line ratios. This discussion reviews some of the progress made over the past few years.

076.035 **X–ray spectra from Hinotori satellite and suprathermal electrons.**
T. Kato, K. Masai.
International SOLAR–A Science Meeting: Flare physics in solar activity maximum 22, p. 137 – 140 (1991). – See Abstr. 012.104 for the main entry.

The authors analysed the X–ray spectra of H–like and He–like iron ions from a solar flare observed by Hinotori satellite, taking into account a possible contribution of high energy electrons. The time behaviour of the relative densities of iron ions was calculated for a plasma of nonequilibrium ionization in the presence of suprathermal electrons of a few percent fraction. This model can reproduce the time evolution of the H–like and He–like spectra consistently.

076.036 **Quality improvement of solar flare temperature and density diagnostics derived from BCS spectra.**
J. Dubau, M. Cornille, F. Bely–Dubau.
International SOLAR–A Science Meeting: Flare physics in solar activity maximum 22, p. 141 – 145 (1991). – See Abstr. 012.104 for the main entry.

Many atomic data have been specially calculated to interpret the Bent Cristal Spectrometer data observed by SMM: H–like and He–like $n = 2$ levels excitation rates including radiative cascades, collisional resonance contributions, electron recombination and ionization, as well as, $K\alpha$ satellite line atomic parameters for ions with 2, 3, 4, 5,... electrons. Nevertheless few line intensities are still difficult to interpret and may indicate that some atomic or plasma process has been disregarded. It is expected that SOLAR A will greatly help to clarify this riddle.

076.037 **The beam driven chromospheric evaporation model of large solar flares: a model getting "no respect" from the Sun.**
U. Feldman.
International SOLAR–A Science Meeting: Flare physics in solar activity maximum 22, p. 146 – 152 (1991). – See Abstr. 012.104 for the main entry.

Since before the last solar maximum, theoretical and to some extent observational, flare physics were dominated by beam driven chromospheric evaporation models. Comparisons between theoretical predictions and observations from soft and hard X–ray radiations point to major model inconsistencies. An account of some of the inconsistencies as well as the requirements for a better soft X–ray model are presented.

076.038 **The electron distribution and SXT images of a coronal soft X–ray source.**
J. M. McTiernan, S. R. Kane, J. M. Loran.
International SOLAR–A Science Meeting: Flare physics in solar activity maximum 22, p. 179 – 183 (1991). – See Abstr. 012.104 for the main entry.

076.039 **Interpretation of multi–channel X–ray intensities from solar flares.**
A. H. Gabriel, F. Bely–Dubau, F. Millier.
International SOLAR–A Science Meeting: Flare physics in solar activity maximum 22, p. 184 – 187 (1991). – See Abstr. 012.104 for the main entry.

The authors investigate the analysis of multi–channel X–ray spectral intensities in terms of the electron energy distribution in a flare. This is related to the identification of thermal and non–thermal, as well as possible "super–hot" components. It is shown that the observations can be fitted by a wide range of different plasma conditions, making it necessary to constrain the range of models considered.

076.040 **Simulation of SXT response to XSST soft X–ray spectrum.**
W. A. Brown.
International SOLAR–A Science Meeting: Flare physics in solar activity maximum 22, p. 193 – 196 (1991). – See Abstr. 012.104 for the main entry.

In July 1982 the Lockheed Space Astronomy group performed a rocket experiment and obtained a unique solar X–ray spectrum from a flare. This work reports the result of simulating the response of the Solar–A Soft X–ray Telescope (SXT) to the soft X–ray spectrum actually observed. A set of hypothetical exposures was computed for comparison with the SXT response to theoretically generated X–ray spectra.

076.041 **High–energy particle acceleration during the implosion driven by 3–dimensional X–type current loop coalescence in solar flares.**
J.–i. Sakai.
International SOLAR–A Science Meeting: Flare physics in solar activity maximum 22, p. 228 – 229 (1991). Abstract. – See Abstr. 012.104 for the main entry.

076.042 **X–ray observations of global solar activity.**
L. Golub.
International SOLAR–A Science Meeting: Flare physics in solar activity maximum 22, p. 271 – 282 (1991). – See Abstr. 012.104 for the main entry.

High resolution observations of the Solar corona, mainly from the Skylab missions in 1973 – 74, have shown that the X–ray corona is variable on all observable temporal and spatial scales. Variability which has been observed ranges from X–ray bright point flares, with sizes of a few arcseconds and rise times of tens of seconds, to large scale structure evolution on time scales of months to years and size scales of a Solar radius. The author presents a sampling of the data, up to and including the most recent high resolution results from the NIXT rocket, in order to discuss the important new contributions which will be made by the SOLAR–A instruments.

076.043 **SXT observations of MHD turbulence in active regions.**
P. C. H. Martens, D. O. Gómez, G. Slater, L. Golub.
International SOLAR–A Science Meeting: Flare physics in solar activity maximum 22, p. 291 – 296 (1991). – See Abstr. 012.104 for the main entry.

The recent discovery from NIXT images that the Fourier transform of the X–ray intensity in active regions is a power–law, is consistent with 2D MHD turbulence theory. The authors briefly discuss this theory and its application to the heating of the solar corona. Then they demonstrate that SXT will be capable of observing similar spectra, even in compressed data–transfer mode between flare observations. Finally the authors discuss observing plans for verification of the hypothesis of coronal heating through turbulent MHD cascades.

076.044 **Simulated SXT observations of coronal loops.**
W. W. Dixon, J. A. Klimchuk, P. A. Sturrock, J. R. Lemen.
International SOLAR–A Science Meeting: Flare physics in solar activity maximum 22, p. 297 – 301 (1991). – See Abstr. 012.104 for the main entry.

The authors have simulated the appearance of two static coronal loops as they might be observed by the Soft X–ray Telescope onboard the SOLAR–A spacecraft. One loop corresponds to a non–flaring active region loop, and the other corresponds to a post–flare loop. The authors find that the loops have fundamentally different appearances: the quiescent loop is brightest at its apex, while the hotter post–flare loop is brightest at, or near, its base (depending on the particular X–ray filter assumed for the observations).

076.045 **Comparative analysis of X–ray flares.**
A. M. Hernández, M. G. Rovira, C. H. Mandrini, M. E. Machado.
Bol. Asoc. Argent. Astron., No. 36, p. 421 – 422 (1990). Abstract. In Spanish and English. – See Abstr. 012.095 for the main entry.

076.046 **The "onset" phase in the two ribbon flares.**
M. G. Rovira, C. H. Mandrini, A. M. Hernández, M. E. Machado.
Bol. Asoc. Argent. Astron., No. 36, p. 426 – 427 (1990). Abstract. In Spanish and English. – See Abstr. 012.095 for the main entry.

The complete Hard X–Ray Burst Spectrometer event list, 1980 – 1989.
See Abstr. 002.023.

The archive data set for the gamma–ray spectrometer on SMM.
See Abstr. 002.043.

The archive data set for the Gamma–Ray Spectrometer on SMM.
See Abstr. 002.053.

The YOHKOH (SOLAR–A) mission. A series of invited contributions.
See Abstr. 003.013.

Flare physics in solar activity maximum 22. Proceedings.
See Abstr. 012.104.

The MSFC solar GRO guest investigation.
See Abstr. 013.042.

The distribution of BATSE quick–look flare data through the SMM Data Analysis Center.
See Abstr. 013.043.

Atomic physics calculations relevant to solar flare spectra.
See Abstr. 022.053.

A balloon–borne payload for imaging hard X–rays and gamma–rays from solar flares.
See Abstr. 035.029.

Results from the MSFC solar hard X–ray imaging Fourier telescope. Numerical models.
See Abstr. 035.030.

Photometric calibration of the GSFC SERTS–3 rocket experiment.
See Abstr. 035.031.

The solar stellar irradiance comparison experiment.
See Abstr. 035.033.

Response of the OSSE detectors to 30 – 200 MeV neutrons.
See Abstr. 035.034.

Imaging the Sun in hard X–rays: spatial and rotating modulation collimators.
See Abstr. 035.167.

The Hard X–ray Telescope (HXT) for the SOLAR–A mission.
See Abstr. 035.185.

The Soft X–ray Telescope for the SOLAR–A mission.
See Abstr. 035.186.

The Wide Band Spectrometer on the SOLAR–A.
See Abstr. 035.187.

Soft X–ray telescope (SXT).
See Abstr. 035.216.

The hard X–ray Telescope (HXT) on board SOLAR–A.
See Abstr. 035.217.

SUSIM: a program to measure changes in the sun's ultraviolet (UV) spectral irradiance over solar cycle time scales.
See Abstr. 051.017.

The Solar–A mission and its scientific aims.
See Abstr. 051.028.

Capabilities and limitations of SOLAR–A.
See Abstr. 051.079.

Studies of high–energy flare phenomena from simultaneous observations of gamma–rays and energetic particles.
See Abstr. 051.080.

Studies of X–ray flares by project CORONAS.
See Abstr. 051.081.

Kinetics of energetic electrons, propagating through ion–sound plasma trubulence in solar flares.
See Abstr. 062.115.

Kinetics of fast electrons in ion–acoustic waves in the turbulent plasma of solar flares.
See Abstr. 062.134.

Stochastic acceleration of electrons in the presence of Coulomb collisions and synchrotron emissions.
See Abstr. 063.020.

The O VII soft X–ray spectrum and its application to hot plasmas in astrophysics.
See Abstr. 063.039.

The solar spectrum.
See Abstr. 071.054.

Mapping the thick–target energy loss regions in AR 5395 and AR 5569.
See Abstr. 072.043.

Einige Besonderheiten dcr Sonnenaktivität im Jahre 1989.
See Abstr. 072.148.

Interrelation of soft and hard X–ray emissions during solar flares. I. Observations.
See Abstr. 073.004.

Interrelation of soft and hard X–ray emissions during solar flares. II. Simulation model.
See Abstr. 073.005.

A thermal/nonthermal approach to solar flares.
See Abstr. 073.006.

Simultaneous UV and X–ray observations of solar microflares.
See Abstr. 073.009.

Flaring arches. III. The subflare of June 27, 1980, and its related extended arch.
See Abstr. 073.012.

High–energy flare explosions driven by 3–dimensional X–type current loop coalescence.
See Abstr. 073.015.

Derivation of parent electron spetra from bremsstrahlung hard X–ray spectra.
See Abstr. 073.051.

A comparison of hard X–ray and high–time–resolution H–alpha data for a flare on 1989 June 30 at 1737 UT.
See Abstr. 073.065.

Differential emission–measure variations and the "Neupert effect".
See Abstr. 073.066.

An interesting pair of homologous flares from region 5395 in March 1989.
See Abstr. 073.067.

A method for determining which flares are consistent with the thick–target model.
See Abstr. 073.068.

Millimeter imaging of energetic electrons in solar flares in conjunction with GRO experiments.
See Abstr. 073.073.

Numerical simulation of microflare evolution in the solar transition region and corona.
See Abstr. 073.075.

Observations from the solar maximum mission.
See Abstr. 073.080.

Observations from the Hinotori mission.
See Abstr. 073.081.

High–energy flare observations from the Solar Maximum Mission.
See Abstr. 073.082.

The role of magnetic loops in solar flares.
See Abstr. 073.085.

Energetic particles in solar flares: theory and diagnostics.
See Abstr. 073.086.

Energetic particle production in flares.
See Abstr. 073.091.

High–temperature plasma in solar flares.
See Abstr. 073.092.

Spectroscopy of high–temperature solar flare plasmas.
See Abstr. 073.093.

The Lα (λ = 121.6 nm) solar plage contrasts calculations.
See Abstr. 073.096.

The acceleration of electrons and ions in solar flares.
See Abstr. 073.097.

Flares observed by the normal incidence X–ray telescope on 1989 September 11.
See Abstr. 073.099.

The effects of magnetic geometry on electron confinement in solar flares.
See Abstr. 073.105.

Solar flare proton acceleration by Alfven turbulence.
See Abstr. 073.106.

Solar flares: a review.
See Abstr. 073.107.

Magnetohydrodynamics of an impulsively heated, hard X–ray emitting filament.
See Abstr. 073.112.

Multi–wave band SMM – VLA observations of an M2 flare and an associated coronal mass ejection.
See Abstr. 073.139.

Radio signature and Hα limb features of the 1979 March 9 and 1982 July 9 flares.
See Abstr. 073.140.

Center–to–limb variations of characteristics of solar flare hard X–ray and gamma–ray emission.
See Abstr. 073.144.

The iron–to–calcium abundance ratio in the 20×10^6K plasma of medium and large solar flares.
See Abstr. 073.145.

The $3s^2 3p\ ^2P - 3s3p^2\ ^4P$ intercombination lines of Fe XIV in the Sun.
See Abstr. 073.146.

Solar gamma–ray–line flares, type II radio bursts, and coronal mass ejections.
See Abstr. 073.148.

Chromospheric and transition region diagnostics using emission–line intensities.
See Abstr. 073.150.

Soft X–ray emission from electron–beam – heated solar flares.
See Abstr. 073.159.

Improved line ratio calculations involving $\Delta n = 1\ (2-3)$ transitions in O V and a reanalysis of Skylab observations of solar flares.
See Abstr. 073.162.

Numerical simulations of electron–beam–heated solar flares.
See Abstr. 073.176.

A hydrodynamic thermal model of the impulsive phase of solar flares.
See Abstr. 073.177.

Evolution of hot plasma in flares.
See Abstr. 073.179.

A new kind of solar X–ray flares.
See Abstr. 073.182.

On mechanisms of solar flares – some observational tests by using SOLAR–A.
See Abstr. 073.184.

The MSFC vector magnetograph, eruptive flares, and the SOLAR–A X–ray images.
See Abstr. 073.186.

Coronal dynamics of a quiescent active region.
See Abstr. 074.026.

Spectra of MHD turbulence in coronal active regions.
See Abstr. 074.031.

The effect of relative elemental abundances on broadband X–ray diagnostics of solar coronal plasmas.
See Abstr. 074.058.

Coronal mass flows and turbulence from high resolution extreme ultraviolet observations.
See Abstr. 074.059.

Micro/nanoflare coronal heating.
See Abstr. 074.066.

A giant coronal arch observed by Skylab.
See Abstr. 074.084.

Nonequilibrium ionization effects in asymmetrically heated loops.
See Abstr. 074.111.

A proposal for a study of the solar wind near the Sun by simultaneous observations with SOLAR–A and Kashima 34 m antenna.
See Abstr. 074.124.

Search for transient coronal holes by SOLAR–A soft X–ray telescope.
See Abstr. 074.125.

Application of SOLAR–A SXT data to flare and geomagnetic storm forecasting research.
See Abstr. 074.126.

The incorporation of the magnetic fluxes of similar polarity in active regions for Sept. – Oct. 1988 and relevant strong X–ray flares.
See Abstr. 075.044.

Intrinsic magnetic discontinuities and solar X–ray emission.
See Abstr. 075.048.

Joint microwave/hard X–ray observations of solar bursts during March 1989.
See Abstr. 077.017.

The association of solar millisecond radio spikes with hard X–ray emission.
See Abstr. 077.045.

VLA supporting observations for SOLAR–A.
See Abstr. 077.077.

Millimeter interferometric observations of solar flares during the SOLAR–A mission.
See Abstr. 077.078.

Albedo gamma–rays from cosmic ray interactions on the solar surface.
See Abstr. 080.055.

Sun–as–a–star spectrum variability.
See Abstr. 080.135.

On the SID durations.
See Abstr. 083.006.

A new technique for estimating D–region effective recombination coefficients under different solar flare conditions.
See Abstr. 083.022.

Study on the response of the D–region of the ionosphere to the solar bursts.
See Abstr. 083.035.

The scattering rate of solar 834 Å radiation by magnetospheric O^+ and O^{++}.
See Abstr. 084.041.

A simple algorithm for simulating the solar ultraviolet radiation at the Earth's surface: an application in determining the minimum erythema dose.
See Abstr. 085.031.

Cosmic ray positron production by gamma ray interactions on starlight.
See Abstr. 144.092.

Signatures of cosmic–ray interactions on the solar surface.
See Abstr. 144.109.

077 Radio, Infrared Radiation

077.001 **Some properties of the narrow bandwidth millisecond pulsations in solar radio bursts at frequenciens 2.5 and 2.85 GHz.**
Yu. F. Yurovskij.
Pis'ma Astron. Zh., Tom 17, No. 7, p. 629 – 636 (Jul 1991). In Russian. English translation in Sov. Astron. Lett., Vol. 17, No. 4.
It is shown that millisecond pulsations of the bursts intensity with 100% modulation have different time profiles at two neighbouring frequencies without any periodicity. The negative and positive pulses have comparable durations. It is supposed that pulsations arise due to refraction and dispersion of the radiowaves in the turbulent solar atmosphere.

077.002 **Search for a 152 – 158 days periodicity in the occurrence rate of solar flares inferred from spectral data of radio bursts.**
V. K. Verma, G. C. Joshi, W. Uddin, D. C. Paliwal.
Astron. Astrophys., Suppl. Ser., Vol. 90, No. 1, p. 83 – 87 (Oct 1991).
The authors have carried out a power spectrum analysis of the daily number of solar flares producing type I, II, III, IV, and V radio bursts for the period 1980 – 1984. The analysis shows that type II and IV radio bursts confirm the existence of a 152 – 158 days periodicity in the rate of occurrence, while type I, III and V radio bursts do not show this 152 – 158 days periodicity. The authors conclude that only type II and IV radio bursts should be used as a parameter which indicates the solar activity.

077.003 **Some properties of narrow–band millisecond pulsations in solar radio bursts at 2.5 and 2.85 GHz.**
Yu. F. Yurovskij.
Sov. Astron. Lett., Vol. 17, No. 4, p. 268 – 270 (Jul 1991). Current Physics Microform No.: 9202X0444. English translation of Pis'ma Astron. Zh., Tom 17, No. 7, p. 629 – 636 (Jul 1991).
Millisecond pulsations in burst intensity with 100% modulation depth in a narrow band do not have a definite periodicity, and their temporal profiles are considerably different even for nearby frequencies. The intensity peaks alternate with intensity minima of comparable duration. The author proposes that these properties may be a consequence of the usual refraction and dispersion of radio waves in the Sun's turbulent atmosphere.

077.004 **Microwave characteristics of a behind–the–limb proton flare.**
A. Krüger, J. Hildebrandt, B. Kliem, H. Aurass, J. Kurths, V. V. Fomichev, I. M. Chertok, L. Křivský.
Sol. Phys., Vol. 134, No. 1, p. 171 – 186 (Jul 1991). With a correction in Vol. 135, No. 2, p. 423 (Oct 1991).
Behind–the–limb flares provide a unique opportunity for the study of vertical source structures of microwave bursts and dynamic flare porcesses. Based on complex observational data related to the outstanding solar proton event on 16 February, 1984, the development of burst emission at a height $z \geqslant 200000$ km above the photosphere has been investigated. A comparison with the associated X–ray emission measured aboard various spacecraft yields a time lag of about $\leqslant 1$ min between the onset of the unocculted impulsive HXR–emission and the onsets of the X–ray and microwave emissions occulted by the solar limb. The lag corresponds to a range of speeds of the propagation of the flare volume of about 3000 – 5000 km s^{-1}. Considering competing transport agents that could account for such expansion of the source volume, a qualitative model of shock–wave activation of loops successively reaching into larger coronal heights is proposed. From a discussion of the possible emission processes involved, conclusions about the magnetic field, electron density, and particle energies have been obtained.

077.005 **Electron–cyclotron maser emission in solar microwave spike bursts.**
P. A. Robinson.
Sol. Phys., Vol. 134, No. 2, p. 299 – 314 (Aug 1991).
A new model is developed for electron–cyclotron maser emission from flaring loops, which incorporates competition between driving of the instability and maser–induced relaxation, together with interactions between small neighboring regions of unstable plasma. This results in a picture in which radiation is emitted in bursts from regions whose length scale is determined self–consistently by previous bursts, while the unstable plasma fluctuates about the point, close to marginal stability, at which driving of the instability is balanced by relaxation due to maser–induced electron diffusion. Under the conditions applicable to flaring loops, time scales of fundamental x–mode ($x1$) driving and saturation are approximately equal at ~ 1 ms, resolving a (10^4 – 10^6)–fold discrepancy in previous models and agreeing with the observed time scales of microwave spike bursts. Saturation effects are found to be especially effective in suppressing amplification of the most strongly growing modes. This suppression enables fundamental o–mode ($o1$) and second–harmonic x–mode ($x2$) emission to compete more effectively against $x1$ emission for the available free energy than has previously been estimated. Consideration of mode competition, burst time scales, suppression of growth due to overlap between amplification and adsorption bands, and escape of radiation through absorption layers to the observer, implies that the observed radiation probably escapes from the corona principally in the o–mode, either emitted directly as $o1$ radiation or mode converted from $x1$ emission.

077.006 **VLA observations of interacting flaring loops.**
M. R. Kundu, S. M. White, D. M. McConnell.
Sol. Phys., Vol. 134, No. 2, p. 315 – 327 (Aug 1991).
The authors present 4.9 GHz observations of an impulsive radio burst observed at the Very Large Array on 1981 May 16. The flare occurred in a complex active region containing several

spots. The radio burst lay at the edge of an active–region microwave source, close to a neutral line. The compact burst showed morphological evidence for the presence of two loops in the rise phase, with the subsequent burst peak lying between these loops. This suggests that interaction between the loops played some role in the initiation of the flare. The flare spectrum is consistent with thermal gyrosynchrotron emission. The main microwave peak was displaced from the nearest Hα kernels by about 10″, but there is strong evidence for post–flare loops coincident with the Hα kernels during the later stages of the event.

077.007 **Spectral structures of "line" type in microwave bursts emission on the Sun.**
A. P. Klassen.
Astron. Tsirk., No. 1549, p. 21 – 22 (May – Aug 1991). In Russian.
Observational results of radio bursts spectra in the range of 2.0 – 4.0 GHz are given. Spectral structures of "line" type with linewidth of $\Delta f/f < 0.1$ have been found.

077.008 **Formation of the infrared emission lines of Mg I in the solar atmosphere.**
E. S. Chang, E. H. Avrett, P. J. Mauas, R. Loeser, R. W. Noyes.
Bull. Am. Astron. Soc., Vol. 23, No. 2, p. 1029 – 1030 (1991). Abstract. – See Abstr. 012.012 for the main entry.

077.009 **Coronal radio emission.**
A. O. Benz.
International Conference on Mechanisms of Chromospheric and Coronal Heating, p. 140 – 149 (1991). – See Abstr. 012.011 for the main entry.
This review is focussed on (mostly solar) radio emissions outside flares and aims at phenomena possibly related to coronal heating.

077.010 **Implications of microwaves for heating and particle acceleration on the Sun.**
A. Krüger, J. Hildebrandt, S. Urpo.
International Conference on Mechanisms of Chromospheric and Coronal Heating, p. 150 – 151 (1991). – See Abstr. 012.011 for the main entry.
The capacity of microwave signatures for the diagnostics of hot plasma in the solar atmosphere is demonstrated by the use of emission models for the S–component and burst radiations. Observations with high spatial resolution permit a classification of characteristic temperature domains in active regions. Nonthermal energy dissipation is clearly indicated for the hottest parts of the S–component radiation. The relation between bulk heating and and nonthermal energy release is also considered for flare processes. Exciting evidence is provided for the existence of sporadic coronal mm–wave sources in giant arch structures which are also displayed by X–rays. Implications of these phenomena for the energization process and a possible relation between microbursts and gradual heating are discussed.

077.011 **Multifrequency observations of a remarkable solar radio burst.**
S. M. White, M. R. Kundu, T. S. Bastian, D. E. Gary, G. J. Hurford, T. Kucera, J. H. Bieging.
Bull. Am. Astron. Soc., Vol. 23, No. 2, p. 1043 (1991). Abstract. – See Abstr. 012.012 for the main entry.

077.012 **A thermal/non–thermal (TNT) model for solar microwave bursts.**
S. G. Benka, G. D. Holman.
Bull. Am. Astron. Soc., Vol. 23, No. 2, p. 1043 (1991). Abstract. – See Abstr. 012.012 for the main entry.

077.013 **VLA observations of radio filaments.**
N. Gopalswamy, S. M. White, M. R. Kundu.
Bull. Am. Astron. Soc., Vol. 23, No. 2, p. 1045 (1991). Abstract. – See Abstr. 012.012 for the main entry.

077.014 **Global streamer evolution.**
G. Thejappa, M. R. Kundu, N. Gopalswamy.
Bull. Am. Astron. Soc., Vol. 23, No. 2, p. 1045 (1991). Abstract. – See Abstr. 012.012 for the main entry.

077.015 **On the polarization of microwave emission from active regions: results from CoMStOC.**
G. D. Holman, J. W. Brosius, J. T. Schmelz, R. F. Willson.
Bull. Am. Astron. Soc., Vol. 23, No. 2, p. 1045 (1991). Abstract. – See Abstr. 012.012 for the main entry.

077.016 **Coordinated VLA–PHOENIX–SMM observations: microwave type U–bursts.**
M. J. Aschwanden, T. S. Bastian, A. O. Benz, B. R. Dennis.
Bull. Am. Astron. Soc., Vol. 23, No. 2, p. 1065 – 1066 (1991). Abstract. – See Abstr. 012.012 for the main entry.

077.017 **Joint microwave/hard X–ray observations of solar bursts during March 1989.**
D. E. Gary, G. J. Hurford.
Bull. Am. Astron. Soc., Vol. 23, No. 2, p. 1066 (1991). Abstract. – See Abstr. 012.012 for the main entry.

077.018 **The effects of source inhomogeneity on solar microwave bursts.**
G. J. Hurford, D. E. Gary.
Bull. Am. Astron. Soc., Vol. 23, No. 2, p. 1066 (1991). Abstract. – See Abstr. 012.012 for the main entry.

077.019 **VLA observations during the ESP MAX 91 Campaign.**
R. F. Willson, K. R. Lang.
Bull. Am. Astron. Soc., Vol. 23, No. 2, p. 1071 (1991). Abstract. – See Abstr. 012.012 for the main entry.

077.020 **Temporally, spectrally, and spatially resolved observations of a pulsating solar radio burst.**
T. S. Bastian, M. J. Aschwanden, D. E. Gary.
Bull. Am. Astron. Soc., Vol. 23, No. 2, p. 1072 (1991). Abstract. – See Abstr. 012.012 for the main entry.

077.021 **Summary of observations with the Owens Valley Solar Array during Max '91 Campaign 2.**
D. E. Gary, G. J. Hurford.
Bull. Am. Astron. Soc., Vol. 23, No. 2, p. 1072 (1991). Abstract. – See Abstr. 012.012 for the main entry.

077.022 **The attractor dimension of solar decimetric radio pulsations.**
J. Kurths, A. O. Benz, M. J. Aschwanden.
Astron. Astrophys., Vol. 248, No. 1, p. 270 – 276 (Aug 1991).
The temporal characteristics of decimetric pulsations and related radio emissions during solar flares are analyzed. Since the flux variations in general are not periodic, the authors make use of the statistical methods recently developed for nonlinear, dynamic systems. The quantitative characterization of 20 pulsation events gives good evidence that they typically have their origin in deterministic chaos corresponding to motion on a strange attractor of low dimensionality. The correlation dimension in general turned out to be a useful way to quantify the regularity and hidden order of the irregular or quasi–periodic pulsations. To judge the reliability of its estimate even from rather small data samples, the authors have studied an optimum choice of the alorithmic parameters as well as experimented with known models. The correlation dimension has generally been found to be in the range 2.5–3.5. A metric periodic oscillation and an extraordinarily regular event of "sudden reductions" have yielded a dimension of 2, and modulations on two type IV continua exhibited dimension around 4. These findings suggest that the source of pulsating emissions in the simplest case can be described by the orbit of a point in a three–dimensional phase space. In more complicated events, coinciding in general with more powerful energy release, four up to eight free parameters characterize the source.

077.023 **On fluctuations of radio emission of the quiet atmosphere of the Sun.**
S. I. Avdyushin, E. A. Borisova, Sh. Sh. Gusejnov,
S. P. Leonenko, B. A. Poperechenko, L. V. Yasnov.
Soln. Dannye, Byull., No. 7/90, p. 88 – 92 (Jul 1990). In Russian.
Fluctuations of radio emission of solar quiet regions are studied with the aim of detection of a number of parameters of wave processes, occurring in these regions. The observations were made at $\lambda = 5.2$ and 8.1 cm with the radio telescope THA–1500 in July – August 1963. Using the obtained data the crosscorrelation function and its spectrum have been calculated. Time spectra of power density are approximated by the function $f^{-\alpha}$, where α is in the interval 1.6 – 2.1 at $\lambda = 5.2$ cm and 1 – 1.7 at $\lambda = 8.1$ cm. Time delays of fluctuations have not been found at different wavelengths. A supposition is made that fluctuations of radio emission in the quiet atmosphere of the Sun are due to density disturbances in optical regions.

077.024 **Proton flare diagnostics using integral parameters of microwave bursts.**
E. Del Pozo.
Soln. Dannye, Byull., No. 9/90, p. 77 – 84 (Sep 1990). In Russian.
The dependences between the > 10 MeV proton flux intensity near the Earth and integral parameters (increasing phase flux and total flux) of solar microwave bursts at 6.7, 9 and 15 GHz are analyzed using data for 1966 – 1986. The heliolongitude attenuation of proton fluxes, conditions for proton escaping from the flare region (metric radio component), and the energy spectrum of particles (peak flux density radio spectrum) are taken into account.

077.025 **On the phase shift between solar radio emission at the wavelength 10.7 cm and Lα–emission.**
G. S. Ivanov–Kholodnyj, V. V. Katyushina, A. A. Nusinov.
Soln. Dannye, Byull., No. 10/90, p. 71 – 75 (Oct 1990). In Russian.
A comparison of daily intensities of the solar Lα–emission and solar radio flux at the wavelength 10.7 cm ($F_{10.7}$) for the period January 1982 – June 1988 has been made. It is shown that the lag of 27–day maxima of Lα–emission relative to those of $F_{10.7}$ is equal to 24 hours on the average and can be explained as due to a different velocity of the decrease of the active region of intensity at different intervals of the solar spectrum.

077.026 **On the difference of characteristics in radio spectra of proton and nonproton flares.**
A. G. Stupishin.
Geomagn. Aehron., Tom 31, No. 4, p. 712 – 715 (Jul – Aug 1991). In Russian. English translation in Geomagn. Aeron., Vol. 31, No. 4.

077.027 **Solar radio data. 1991 Jul – Sep.**
Sol. Radio Data, Part III/91.
Observatorium für solare Radioastronomie, Potsdam (Germany, F.R.). 2 p. (1991). In microfiche form.

077.028 **Solar radio data. 1991 Oct – Dec.**
Sol. Radio Data, Part IV/91.
Observatorium für solare Radioastronomie, Potsdam (Germany, F.R.). 1 p. (1991). In microfiche form.
This is the final issue of this series.

077.029 **On the emission of local radio sources at the wavelengths 9.1 and 21 cm during the declining phase of the solar cycle.**
L. P. Ipatova, L. V. Yasnov.
Soln. Dannye, Byull., No. 4/91, p. 82 – 85 (Apr 1991). In Russian.
Spatial peculiarities at the wavelengths 9.1 and 21 cm obtained from radio maps for 1970 – 1973 are investigated. The main results of this work are as follows: the distances between regions of radio emission generation increase at 9.1 and 21 cm during the declining phase of the solar cycle; there are radio sources only on radio maps at the wavelength 21 cm ($T_{21} > 10^5$K) that do not correlate with spot locations and even plage locations; the number of low frequency bursts increases and the number of wide range bursts decreases during the declining phase of the solar cycle. These results indicate an active region re–arrangement during the declining phase that is accompanied by other burst activity.

077.030 **Spectral peculiarities of local radio sources at decimetric wavelengths during the declining phase of the solar cycle.**
L. P. Ipatova, L. V. Yasnov.
Soln. Dannye, Byull., No. 5/91, p. 78 – 84 (May 1991). In Russian.
Spectral peculiarities of radio sources are investigated. These radio sources were observed at the wavelengths 9.1 and 21 cm from radio maps for 1970 – 1973.

077.031 **Space–time features in solar radio emission.**
L. P. Ipatova, L. V. Yasnov.
Soln. Dannye, Byull., No. 3/91, p. 99 – 100 (Mar 1991). Abstract. – See Abstr. 012.021 for the main entry.

077.032 **The characteristics of the solar microwave bursts of June 30, 1989.**
Yu Xingfeng.
Publ. Purple Mt. Obs., Vol. 10, No. 2, p. 114 – 121 (Jun 1991). In Chinese.
The two–peak strucutre, spectral characteristics, and time–delay of the microwave burst event of June 30, 1989 are described and analysed. (1) There were time delays of the two peaks from the high frequencies to low frequencies (1 – 15 GHz). The high frequency peak time preceded the low frequency peak time for 84 s. It seems that the efficiency of particle acceleration at the two peaks is different each other. (2) This event accompanied the proton event. (3) The spectral indices of two peaks at 9400 MHz and 3750 MHz changing with time were consistent with the time profile of the burst 9375 MHz.

077.033 **On a relationship between the flare activity index and radio emission of active regions.**
L. P. Ipatova, L. V. Yasnov.
Soln. Dannye, Byull., No. 7/91, p. 92 – 96 (Jul 1991). In Russian.
There is a relationship between the behaviour of S–component radiation spectra and the flare index. The maximum correlation coefficient of the average daily S–component flux density and the flare index corresponds to the 2800 MHz frequency during the declining phase and the minimum of 20th solar cycle.

077.034 **Spectral analysis of solar radio Type III burst groups.**
M. Pick, A. Mangeney, Zhao Renyang.
Chin. Astron. Astrophys., Vol. 15, No. 3, p. 298 – 305 (Sep 1991). English translation of Acta Astron. Sin., Vol. 32, No. 1, p. 28 – 35 (Mar 1991). See Abstr. 53.077.015.
Using the high time resolution and high sensitivity data on the intensity, polarization, position and size of Type III burst groups obtained at 5 frequencies with the Nançay radio heliograph, the authors have made a careful Fourier transform spectral analysis. Their study further confirms the existence of periodicity in Type III burst groups, most of the events have periods between 1 and 6 seconds and show multiple periods at 90% confidence level or better. Especially the authors found that the Type III burst groups can be divided into two classes, one with weak polarization and with periods over a wide range up to 6 seconds, the other with strong polarization and short periods less than 2 seconds and smaller size compact sources.

077.035 **Fast fine structure in solar radio bursts at 3 cm waves.**
Qin Zhihai, Jiang Shuying, Wei Shuanglin,
Chen Yunxia.
Chin. Astron. Astrophys., Vol. 15, No. 3, p. 362 – 363 (Sep 1991). English translation of Acta Astrophys. Sin., Vol. 11, No. 2, p. 194 – 196 (Apr 1991). See Abstr. 53.077.026.
Between 1989 March and 1990 April the authors carried out fast sampling observations of solar radio emission and found that

at 3 cm waves, the emission also has fast fine structures with time scales between 0.1 and 2 seconds.

077.036 **Large scale patterns on the solar surface indicated by microwave observations.**
B. Vršnak, S. Pohjolainen, H. Teräsranta, S. Urpo, R. Brajša, V. Ruždjak, A. Schroll, S. Jurač.
IAU Colloquium No. 130: The Sun and cool stars – activity, magnetism, dynamos, p. 282 – 284 (1991). – See Abstr. 012.024 for the main entry.

A large set of observations of the Sun at 37 GHz is analysed. An association of 99% is found between the regions of brightness temperature depression and the magnetic field inversion lines. Observations indicate a possible existence of "giant cells" with duration of 1 – 2 years and a longitudinal extension up to 90°.

077.037 **Close loop model of stochastic electron acceleration in solar gradual bursts of April 27, 1981.**
Yao Jinxing.
Acta Astron. Sin., Vol. 32, No. 2, p. 186 – 193 (Jun 1991). In Chinese. English translation in Chin. Astron. Astrophys., Vol. 15, No. 4, p. 408 – 415 (Dec 1991).

A model of stochastic particle acceleration by whistlers in a turbulent plasma in a closed magnetic loop is used to explain the observational features of the April 27, 1981 bursts during the gradual phase.

077.038 **Nonlinear saturation of an electron–cyclotron maser.**
G. D. Flejshman, Yu. E. Charikov.
Astron. Zh., Tom 68, Vyp. 4, p. 719 – 731 (Jul–Aug 1991). In Russian. English translation in Sov. Astron., Vol. 35, No. 4.

A modern data set is described regarding the observations of nonstationary microwave emission generated at the solar–flare site. It is pointed out that, as one turns to magnetic bremsstrahlung maser radiation in order to give an explanation to microwave spikes, a development of instabilities in plasma proves to be of primary importance in the process of saturation of the above–mentioned radiation. Quasilinear relaxation of an unstable distribution of fast non–relativistic electrons with account of the feedback influence of Langmuir turbulence generated by the electrons is discussed. It is shown that the characteristic time–scale of stabilization of the instability can reach into the range of several microseconds – tens of microseconds. It means that the spikes which have been observed in the flares apparently have a fine temporal structure. Other processes are discussed which could effect the stabilization of magnetic bremsstrahlung instability.

077.039 **MHD oscillations in solar radio spike radiation.**
Zhao Renyang, Jin Shengzhen, Fu Qijun, Li Xiaocong.
Acta Astrophys. Sin., Vol. 11, No. 3, p. 269 – 274 (Jul 1991). In Chinese. English translation in Chin. Astron. Astrophys., Vol. 15, No. 4, p. 439 – 445 (Dec 1991).

In the solar spike radiation of the microwave outburst observed on May 16, 1981, the authors found the presence of 1.4 – 1.6 s quasi–periodic oscillations. In the present paper, the observations are analysed in terms of MHD waves. The authors consider that the fast magnetoacoustic waves ("sausage" mode) propagating inside and outside a loop can modulate the magnetic field and the pitch angle distribution of the electron beams in the source region. These affect the growth rates of the electron–cyclotron–maser instability and give rise to the quasi–periodic oscillations in the spike radiation. In addition, the authors quantitatively estimated some relevant physical parameters.

077.040 **Some statistical features of solar radio SVC in the rising branch of the cycle 22.**
Zhang Bairong, Ma Yuan.
Acta Astrophys. Sin., Vol. 11, No. 3, p. 283 – 286 (Jul 1991). In Chinese. English translation in Chin. Astron. Astrophys., Vol. 15, No. 4, p. 449 – 452 (Dec 1991).

Based on the relations between the apparent area of major sunspots on the solar disk and the daily total flux at the wavelengths 2 cm, 3.4 cm, 6 cm, 10.7 cm and 21.2 cm for the period of Jan. 1987 to Nov. 1989, the statistical features of the solar radio slow variation component (SVC) for some major active regions during the rising stage of the cycle 22 are analysed. It is found that several large active regions were located below the average curve in the S – Ay diagram, and the increase in the radio flux density was insignificant as compared with the corresponding increase in the sunspot area.

077.041 **Temporal fine structure of solar radio burst at 2 cm waveband.**
Gao Zhengmin, Li Ziying, Huo Caiping, Fan Ying.
Acta Astrophys. Sin., Vol. 11, No. 3, p. 287 – 288 (Jul 1991). In Chinese.

The temporal fine structure (FS) superposed on the solar microwave burst was observed at the 2 cm band during April – May, 1990. It has the characteristic of low intensity and short duration similar with millimeter wave FS. The intensity of the FS and its duration are given. The temporal profile of FS event on 23 May has been presented.

077.042 **The effects of plasma ambient parameters on spike emissions.**
Huang Guangli, Wang Deyu.
Acta Astrophys. Sin., Vol. 11, No. 4, p. 352 – 361 (Oct 1991). In Chinese. English translation in Chin. Astron. Astrophys., Vol. 16, No. 1, p. 91 – 102 (Jan – Mar 1992).

On the basis of the theory of electron–cyclotron maser instabilities, the effects of the plasma ambient parameter ω_{pe}/Ω_e on the properties of the spike emission are discussed, including the physical meaning of this parameter, its magnitude and the polarization and harmonic structure of the radio spikes which are more sensitive to the parameter.

077.043 **Some features of solar microwave emission of millisecond time scale.**
Jin Shengzhen, Li Xiaocong, Wang Shulan, Ji Huirong, Yan Yuanyi, Zhao Bing.
Acta Astrophys. Sin., Vol. 11, No. 4, p. 394 – 396 (Oct 1991).

In June 1990, a new microwave polarimeter was put into operation at Beijing Astronomical Observatory. The polarimeter receives right and left hand circular polarizations at 2545 and 2645 MHz frequencies, signal from each of the four channels can be sampled in every 0.2 ms or 2 ms. Since July 1990, over one hundred "spike" and fine structure events have been recorded. Some interesting features are demonstrated in this paper.

077.044 **Spatial, spectral, and polarization properties of radio emission of the 3 February, 1983 proton flare.**
L. M. Bakunin, V. G. Ledenev, V. P. Nefedyev (*V. P. Nefed'ev*), N. N. Potapov, G. Ya. Smol'kov, A. M. Uralov, G. P. Chernov, V. V. Fomichev, V. P. Melnikov, T. S. Podstrigach, O. V. Epifanova, A. V. Khokhlov, S. Enome.
Sol. Phys., Vol. 135, No. 1, p. 107 – 129 (Sep 1991).

A complex analysis in different radio ranges of the evolutionary features of the 3 February, 1983 flare has shown that the flare is a prolonged ($\sim$15 hr) process of energy release and particle acceleration that gradually extends to still greater zones of the active region magnetosphere in both area and altitude. Observations from the Siberian Solar Radio Telescope at $\lambda = 5.2$ cm indicate that the flare was preceded by quasi–periodical brightness enhancements with a period of 6 – 7 min of two sources of size 20″ and with a brightness temperature of $\sim 10^7$K. During the flare maximum phase, a type II burst with harmonic structure and the subsequent type FC II continuum with fine structure were both observed in the meter band. It has been found that zebra–structure appearances correlate with the Hα–flare kernel brightenings at loop tops. The observed characteristics of the type II burst and of the type FC II continuum treated in this paper are interpreted in terms of the complex flare flow structure, involving forward and backward shock waves.

077.045 **The association of solar millisecond radio spikes with hard X–ray emission.**
M. Güdel, M. J. Aschwanden, A. O. Benz.
Astron. Astrophys., Vol. 251, No. 1, p. 285 – 297 (Nov 1991).
A large set of decimetric and microwave observations of solar millisecond radio spikes is compared and correlated with simultaneous hard X–ray observations in the 25–438 keV range, recorded by the Hard X–Ray Burst Spectrometer on board the Solar Maximum Mission satellite. Radio spikes have time scales of 20–200 ms, while hard X–ray bursts rarely show fine structures shorter than 1 s. The main conclusions of a statistical analysis of (a) the association rate, (b) the correlation degree, and (c) relative time delays between hard X–ray and radio spike emissions, are presented.

077.046 **Nonlinear parametric instability and millisecond solar radio spikes.**
Wang Deyu, Li Dingyi.
Sol. Phys., Vol. 135, No. 2, p. 393 – 405 (Oct 1991).
A model of nonlinear parametric instability (NLPI) modulation is suggested to explain the millisecond spikes in solar decimetric radio emission. The radio emission energy will periodically transfer to two other waves, and the intensity of the radio emission will be modified by NLPI processes in the corona, when the matching conditions of three–wave coupling are satisfied. This model can simultaneously explain the duration of spikes, why the millisecond spikes have not been observed in other solar radio emission, and the relation between the duration of the spikes and the intensity and the frequency of the radio emission as well.

077.047 **Nonlinear saturation of a cyclotron maser.**
G. D. Flejshman, Yu. E. Charikov.
Sov. Astron., Vol. 35, No. 4, p. 354 – 360 (Jul 1991). Current Physics Microform No.: 9203X0540. English translation of Astron. Zh., Tom 68, Vyp. 4, p. 719 – 731 (1991).
Recent observations of nonsteady emission from solar flares are presented. The primary role of the development of plasma inatabilities in the saturation of magnetic bremsstrahlung maser emission, used to explain the radio observations, is pointed out. Quasilinear relaxation of an unstable distribution of fast electrons is investigated with allowance for feedback from the Langmuir turbulence generated by them. It is shown that the characteristic stabilization time can range from several to tens of microseconds, and thereby determine the temporal fine structure of the microwave emission. Other possible processes affecting the regime of stabilization of this instability are discussed.

077.048 **Dependence of the power spectra of solar radio–emission at $\lambda = 9.1$ cm on the heliolatitude.**
L. P. Ipatova, L. V. Yasnov.
Astron. Tsirk., No. 1545, p. 23 – 24 (Aug – Sep 1990). In Russian.
There are some periods in the power spectra of radio emission at $\lambda = 9.1$ cm, which values are independent to heliolatitude. Such periods point to the existence of the long–living spatial patterns of radio emission with solid rotation.

077.049 **Observation of solar radio emission at 3.2 cm wavelength at UAS (*Urumqi Astronomical Station*).**
Sun Qinghua.
Publ. Yunnan Obs., No. 4, p. 82 – 85 (Dec 1990). – See Abstr. 012.047 for the main entry.

077.050 **The characteristics of the optics and radio emission of the active region 5395 of March, 1989.**
Zhou Shurong.
Publ. Yunnan Obs., No. 4, p. 86 – 90 (Dec 1990). – See Abstr. 012.047 for the main entry.

077.051 **Characteristics of radio emission of AR 5395.**
Wu Hongao.
Publ. Yunnan Obs., No. 4, p. 91 – 95 (Dec 1990). – See Abstr. 012.047 for the main entry.

077.052 **Radio events generated in especially large active region AR 5395.**
Jiang Shuying, Qin Zhihai, Chen Yunxia.
Publ. Yunnan Obs., No. 4, p. 96 – 100 (Dec 1990). – See Abstr. 012.047 for the main entry.

077.053 **Microwave observations in active region 5395.**
Ji Shuchen, Xie Ruixiang, Huang Zhankui.
Publ. Yunnan Obs., No. 4, p. 101 – 105 (Dec 1990). – See Abstr. 012.047 for the main entry.

077.054 **AR 5395: preliminary analysis of the double peak structures of microwave bursts.**
Gao Zhengmin, Li Weidong, Li Ziying, Huo Caiping, Fan Ying, Fu Qijun.
Publ. Yunnan Obs., No. 4, p. 106 – 109 (Dec 1990). – See Abstr. 012.047 for the main entry.

077.055 **Microwave observation highlights of AR5629.**
Zhou Aihua, Xu Fuying.
Publ. Yunnan Obs., No. 4, p. 189 – 192 (Dec 1990). – See Abstr. 012.047 for the main entry.

077.056 **Solar radio observations during solar–terrestrial joint–observations in March and August 1989.**
Fu Qijun, Liu Yuying, Li Xiaocong, Wang Shulan, Chen Zhijun, Zhao Bing, Zhai Pujiang.
Publ. Yunnan Obs., No. 4, p. 197 – 200 (Dec 1990). – See Abstr. 012.047 for the main entry.

077.057 **AR5629: a great solar radio outburst on August 15, 1989.**
Ji Shuchen, Xie Ruixiang, Zhao Risheng, Huang Zhankui.
Publ. Yunnan Obs., No. 4, p. 201 – 204 (Dec 1990). – See Abstr. 012.047 for the main entry.

077.058 **The 152–day periodicity of the occurrence rate of solar microwave bursts.**
Liu Yuying.
Publ. Yunnan Obs., No. 4, p. 243 – 248 (Dec 1990). – See Abstr. 012.047 for the main entry.

077.059 **Homologous radio bursts and their exciting functions.**
Li Xiaocong, Wang Shulan.
Publ. Yunnan Obs., No. 4, p. 249 – 252 (Dec 1990). – See Abstr. 012.047 for the main entry.

077.060 **Large–scale features of the Sun at 20 centimeter wavelength.**
N. Gopalswamy, S. M. White, M. R. Kundu.
Astrophys. J., Vol. 379, No. 1, p. 366 – 380 (20 Sep 1991).
The authors report the results of an investigation of the characteristics of large–scale coronal structures such as active regions, plages, filaments, and coronal holes using data obtained with the VLA at 1.5 GHz during the period 1988 September 11 – 17. The radio data were supplemented with He $\lambda 10830$, Hα, and Ca K spectroheliograms. The authors perform a statistical analysis of some of the characteristics of the active regions. Most of the active region sources were $\sim 100''$ in size, with bridges between regions common; lower brightness temperature regions showed a higher degree of polarization in general. The maximum polarization was found at the edge of active regions but well within the associated plages. The degree of polarization from bright active regions was small ($\leqslant 20\%$). A newly emerging region was polarized to a high degree ($\sim 50\%$) implying a very high magnetic field (180 G) in that region. The 20 cm radio emission is interpreted in terms of thermal bremsstrahlung emission, and the magnetic field and the density in the 20 cm sources have been estimated. A large number of quiescent as well as plage filaments were present on the disk. Associated with these filaments there were brightness depressions at 20 cm below the quiet Sun emission. The brightness temperature observed over a filament is consistent with the contribution from coronal cavities.

077.061 **Spectral observations of solar radio bursts in the range of 2.0–4.5 GHz by a new digital radio spectrometer.**
A. Tlamicha.
Bull. Astron. Inst. Czech., Vol. 42, No. 4, p. 257 – 260 (Jul 1991). Wit plates 1, 2.
Solar radio spectra in the range of 2.0–4.5 GHz observed by using a digital radio spectrometer at the Astronomical Observatory Ondřejov are presented. A brief description of the observational equipment and examples of solar radio spectra observed in the period April 10 – December 29, 1989 are given. The correlation of fast drift bursts and type III bursts is presented.

077.062 **The negative frequency drift of microwave bursts due to collisional and radiative losses of superthermal electrons.**
M. Karlický.
Bull. Astron. Inst. Czech., Vol. 42, No. 4, p. 260 – 263 (Jul 1991).
In this paper the effect of collisional and radiative losses of superthermal electrons in a compact radio source on resulting microwave radio spectrum is estimated. The expected negative frequency drift of microwave bursts is compared with observations.

077.063 **Spectrum of microwave bursts and a possible method to obtain.**
Fu Qijun.
Publ. Yunnan Obs., No. 4, p. 241 – 242 (Dec 1990). – See Abstr. 012.047 for the main entry.

077.064 **Solar radio spike radiation coexisting with a loop prominence system.**
Gong Yuangfang, Shang Qiongzhen, Hu Hanming, Lü Songquan, Li Jiukang, Yang Rongbang.
Chin. Sci. Bull., Vol. 36, No. 22, p. 1890 – 1893 (Nov 1991).

077.065 **Closed loop model of turbulent acceleration for the gradual phase of the radio burst of 1981 April 27.**
Yao Jinxing.
Chin. Astron. Astrophys., Vol. 15, No. 4, p. 408 – 415 (Dec 1991). English translation of Acta Astron. Sin., Vol. 32, No. 2, p. 186 – 193 (Jun 1991).
The author proposes a closed loop model of turbulent stochastic acceleration for the gradual phase of the radio burst of 1981 April 27. It can explain the various observed features in the spectrum, the rather long time of rise, the absence of X–ray bursts below 56 keV and of any detectable proton events.

077.066 **MHD oscillations in the solar spike radiation.**
Zhao Renyang, Jin Shengzhen, Fu Qijun, Li Xiaocong.
Chin. Astron. Astrophys., Vol. 15, No. 4, p. 439 – 445 (Dec 1991). English translation of Acta Astrophys. Sin., Vol. 11, No. 3, p. 269 – 274 (Jul 1991).
In this paper the observed 1.4 – 1.6 s quasi–periodic oscillations in the spike radiation of the microwave outburst of 1981 May 16 are analysed in terms of MHD waves. The authors point out that the fast magnetoacoustic waves ("sausage" mode) propagating inside and outside a loop can modulate the magnetic field and the pitch angle distribution of the electron beams in the source region. The growth rate of the electron–cyclotron–maser instability is then affected to give rise to the quasi–periodic oscillations. Quantitative estimates of relevant physical parameters are given.

077.067 **Some statistical features of solar radio SVC in the rising branch of cycle 22.**
Zhang Bairong, Ma Yuan.
Chin. Astron. Astrophys., Vol. 15, No. 4, p. 449 – 452 (Dec 1991). English translation of Acta Astrophys. Sin., Vol. 11, No. 3, p. 283 – 286 (Jul 1991).
For the rising branch of cycle 22, 1987 January – 1989 November, the authors plot the daily total flux S at each of the five wavelengths 2.0, 3.4, 6.0, 10.7 and 21.2 cm against the apparent sunspot area of the dominant sunspot A_y and find that, for several large active areas, the locus is located below the average regression line, and the increase in the radio flux is insignificant compared to the increase in the sunspot area.

077.068 **Submillimeter solar limb profiles determined from observations of the total solar eclipse of 1988 March 18.**
T. L. Roellig, E. E. Becklin, J. T. Jefferies, G. A. Kopp, C. A. Lindsey, F. Q. Orrall, M. W. Werner.
Astrophys. J., Vol. 381, No. 1, p. 288 – 294 (1 Nov 1991).
Observations were made of the extreme solar limb in six far–infrared wavelength bands ranging from 30 to 670 μm using the Kuiper Airborne Observatory during the total eclipse of the Sun on 1988 March 18. By observations of the occultation of the solar limb by the Moon, it was possible to obtain a spatial resolution of 0".5 normal to the limb. The results at wavelengths of 100 μm and less were found to be essentially equivalent to those obtained in 1986 by Lindsey et al., and the authors discuss here in detail only the results of the 200, 360, and 670 μm observations. The solar limb was found to be extended with respect to the visible limb at all of these wavelengths, with the extension increasing with wavelength. Limb brightening was observed to increase slightly with increasing wavelength, and no sign of a sharp emission spike at the extreme limb was found at any of these wavelengths. The observations can be well fitted by a chromospheric model incorporating cool dense spicules in the lower chromosphere.

077.069 **Spectral observations of solar radio bursts at 2.0 – 4.5 GHz. January – August 1991.**
A. Tlamicha, Z. Ambrožová, T. Sigmund.
Publ. Astron. Inst. Czech. Acad. Sci., No. 78.
Československá akademie věd, Ondřejov. Astronomický ústav. 139 p. (1991).
Spectral observations of solar radio bursts at 2.0 – 4.5 GHz taken at the Astronomical Observatory Ondřejov between January and August 1991 are presented. Tables and computer outputs in equidensities and in time intensity profiles of the bursts are also given.

077.070 **Absolute calibration of solar radio flux density at 9375 MHz.**
Zhou Shurong, Xu Fuying, Yu Xingfeng.
Acta Astron. Sin., Vol. 32, No. 3, p. 239 – 246 (Sep 1991). In Chinese.
The authors have developed a large pyramidal horn and two black–body boxes to measure solar radio fluxes and to correct the gain of a paraboloid antenna at 9375 MHz. During the absolute calibration, they obtained solar radio fluxes with an optimal pyramidal horn at 9375 MHz. The results indicate that using an optimal pyramidal horn to observe the Sun and two black bodies with different temperatures as the temperature standards at the aperture plane of the horn is the best and most reliable method for absolute calibration of both solar radio emissive fluxes and an antenna gain.

077.071 **Quasi–periodic structures superimposed upon a large radio burst at 3.2 cm wavelength on May 23, 1990.**
Qin Zhihai, Li Chunsheng, Wei Shuangling, Cheng Yunxia.
Acta Astron. Sin., Vol. 32, No. 3, p. 247 – 251 (Sep 1991). In Chinese. English translation in Chin. Astron. Astrophys., Vol. 16, No. 1, p. 78 – 83 (Jan – Mar 1992).
This paper describes a lot of quasi–periodic structures superimposed upon a large solar radio burst which was observed at 3.2 cm wavelength on May 23, 1990. This large burst corresponded to a M8.7/1B flare. The average period of these quasi–periodic structures is about 1.5 s and their relative amplitudes range from 2% to 5%. A preliminary analysis shows that there exists a linear relation between the flux and repetition of the quasi–periodic structures. The authors suggest that these quasi–periodic structures may be the modulated gyro–synchrotron emissions due to a MHD wave oscillating along the corresponding magnetic field lines.

077.072 **The observations and analyses of solar radio decimetric spike radiation.**
Wang Min, Xie Ruixing, Wang Congyun.
Acta Astron. Sin., Vol. 32, No. 3, p. 252 – 257 (Sep 1991). In Chinese. English translation in Chin. Astron. Astrophys., Vol. 16, No. 1, p. 84 – 90 (Jan – Mar 1992).

Observational results at 1420 MHz during January 1988 to June 1989 of solar radio radiation are analysed. The decimetric millisecond spikes are found in about 20% of all radio bursts. The relation between the decimetric millisecond spikes and radio bursts is discussed, and it is found that spikes always appeared at the rise and maximum phase of impulsive radio bursts. The association of decimetric millisecond spikes with Hα flares is found to be high (90%). The decimetric millisecond spikes are also frequently accompanied by type III bursts and soft X-ray bursts.

077.073 **Physics of the infrared spectrum.**
D. Deming, D. E. Jennings, J. Jefferies, C. Lindsey.
Solar interior and atmosphere, p. 933 – 963 (1991). – See Abstr. 003.028 for the main entry.

The authors describe the diagnostic value and principal results derived from solar studies at wavelengths exceeding 1.6 μm. The infrared is a favorable region to conduct studies of the solar magnetic field. The high-n emission lines in the 12-μm spectrum are of special interest. However, the LTE or NLTE nature of the lines, and the mechanism of their excitation, remain poorly understood. The far-infrared continuum is an excellent thermometer for the upper photosphere and chromosphere, allowing study of the average thermal state and the compressional effects of wave motions. Observations of limb brightening at far-infrared wavelengths have shown that the structure of the chromosphere is spatially inhomogeneous, even at the lowest chromospheric altitudes. Time-series observations in the far-infrared show that the chromosphere exhibits a substantial thermal response to the 5-min oscillations. Further progress in far-infrared studies will result from the new generation of large-aperture submillimeter telescopes, and from the development of the theory of radiative transfer in inhomogeneous media.

077.074 **A non-linear modulation model for the fine structures in the solar short centimetre wave bursts.**
Wang Deyu.
Acta Astron. Sin., Vol. 32, No. 4, p. 362 – 368 (Dec 1991). In Chinese.

A theoretical model has been suggested which can satisfactorily interpret the inversely proportional relation between the observed flux and the repetition rates of fine structures in the solar short centimeter wave bursts. Moreover, this theoretical model anticipates that the modulation period is related to the radio frequency, while the modulated amplitude of the fine structures is independent of the radio flux.

077.075 **Solar microwave radiation maps measured at Metsähovi Radio Research Station in 1990.**
S. Urpo, S. Pohjolainen, H. Teräsranta, K. Karlamaa.
Helsinki Univ. Technol., Metsähovi Radio Res. Stn., Rep. Ser. A, No. 7.
Helsinki University of Technology, Espoo (Finland). Metsähovi Radio Research Station. 123 p. (1991). ISBN 951-22-0755-9.

Annually the 14 m radio telescope at Metsähovi has been reserved for solar measurements for 4 – 10 weeks. Measurements include maps of the whole or part of the Sun, tracking of active regions and solar oscillation monitoring. Most of the solar maps have been measured using receivers operating at 22.2 GHz or at 36.8 GHz. This report contains selected maps measured in 1990. A list of all measured maps during this period is included.

077.076 **Polarized radio-burst spectra from local sources on the Sun.**
L. I. Tsvetkov.
Bull. Crimean Astrophys. Obs., Vol. 81, p. 11 – 18 (1990). English translation of Izv. Krym. Astrofiz. Obs., Tom 81, p. 14 – 22 (1990). See Abstr. 53.077.013.

Observations made in 1980-2 are reported on the fluctuations in circularly-polarized radio emission from eight local sources. The observations were made with the RT-22 radio telescope with a polarimeter operating at 2.25 cm. Spectral analyses for the range 1.3 – 8.3 mHz are presented. The amplitude ratios indicate amplification of the 3 – 5 minute fluctuations in the emission from active regions. The spectra are related to unipolar, bipolar, and compound spot goups and are characterized by distinct lines in the three, five, and ten minute period ranges. Almost simultaneous radio and optical observations have been combined on some pulsating sources on the Sun. It is suggested that there is partial excitation in the 2.6 – 4.0 mHz range for distinct modes in the atmospheric resonator. The mean fluctuation amplitude is only slightly dependent on the source position because of the wide angle in the solar microwave modulation zone.

077.077 **VLA supporting observations for SOLAR-A.**
K. R. Lang.
International SOLAR-A Science Meeting: Flare physics in solar activity maximum 22, p. 264 – 270 (1991). – See Abstr. 012.104 for the main entry.

High-resolution VLA maps at 20-cm wavelength can be used to resolve coronal loops within individual active regions and thereby reveal the coronal magnetic interaction that can trigger solar flares. These VLA maps can complement simultaneous SXT SOLAR-A observations.

077.078 **Millimeter interferometric observations of solar flares during the SOLAR-A mission.**
M. R. Kundu, S. M. White.
International SOLAR-A Science Meeting: Flare physics in solar activity maximum 22, p. 338 – 343 (1991). – See Abstr. 012.104 for the main entry.

The authors present the results of the first high-spatial-resolution interferometric observations of solar flares at millimeter wavelengths, carried out with the Berkeley-Illinois-Maryland Array (BIMA). The observations represent an improvement of an order of magnitude in both sensitivity and spatial resolution compared with previous solar observations at these wavelengths. Most of the flares occurring within the field of view during the observations have been detected by BIMA, including both very impulsive and longer-duration events. It appears that millimeter burst sources are not much smaller than microwave sources. During the SOLAR-A mission the authors plan to obtain both imaging data as well as dedicated patrol observations of flare time profiles at millimeter wavelengths.

077.079 **The observation of a "negative burst" with high spatial resolution.**
V. P. Maksimov, V. P. Nefedyev (*V. P. Nefed'ev*).
Sol. Phys., Vol. 136, No. 2, p. 335 – 342 (Dec 1991).

The case of a short-time "disappearance" of the S-component source over the active region is described and its association with events in Hα is discussed. Under the assumption that in this case the radio source was observed to be covered by coronal mass ejection caused by filament eruption, estimates are made of the velocity of the ejection, its linear size, and of the optical thickness.

077.080 **Effects of turbulence on the electron cyclotron-maser mechanism for solar microwave spike bursts.**
P. A. Robinson.
Sol. Phys., Vol. 136, No. 2, p. 343 – 360 (Dec 1991).

It is shown that small magnetic perturbations can significantly alter the rates of cyclotron growth, absorption, mode conversion, and refraction because of the sensitive dependence of these processes on the field strength in narrow layers. In particular, growth lengths are increased, absorption depths decreased, mode conversion becomes more effective, and turbulent refraction leads to isotropization of the emission. The criteria for significant effects to occur are derived and it is shown that they can be met by the few-percent field perturbations observed in coronal loops.

077.081 **Electron groups traced from the Sun to 1 AU.**
B. V. Jackson, Y. Leblanc.
Sol. Phys., Vol. 136, No. 2, p. 361 – 377 (Dec 1991).

The authors trace electrons from the Sun by a variety of proxy methods – solar flare positions, and metric and kilometric type III radio bursts from the Sun until they can be observed in situ as electrons at the ISEE–3 spacecraft. The study extends over the period of operation of the electron experiment on ISEE 3 from August 1978 to November 1979. By carefully restricting timing within the data sets involved, a peak in the number of flares is found associated with in situ electrons near 60° west solar longitude. This peak shows that type III bursts can be fairly limited in spatial extent, and that the best connection with the solar surface to the flare is along the Archimedean magnetic field spiral. The authors use this spatial determination to define an "average" beam shape for an event which they assume to be representative of the distribution in space of each electron group. The electron numbers at 2 and 29 – 45 keV energies combined with this average beam shape are used to approximate the total numbers of electrons and energy per burst for individual events. The authors find that the total number of electrons and total energy for events varies significantly with flare type; that on the average brighter flares are associated with more electrons.

Scaling posture of the antenna of a solar radio telescope and the design of the control program.
See Abstr. 033.005.

Telescope beam–profile diagnostics and the solar limb.
See Abstr. 036.024.

Infrared computer solar spectrophotometer: principal operating characteristics and observational methods.
See Abstr. 036.079.

Test observation of the solar radio fast synchronous observation system at four wavelengths and its significance.
See Abstr. 036.089.

The Solar Infrared Explorer (SIRE): a Small Explorer mission for solar physics.
See Abstr. 051.014.

Collective plasma radiation processes.
See Abstr. 062.133.

Emission at cyclotron harmonics due to coalescence of z–mode waves.
See Abstr. 063.091.

A model of SVC radiation in magnetic dipole field and radiation varying with magnetic field. I. Active region model and SVC radiation source model.
See Abstr. 063.111.

A model of SVC radiation in magnetic dipole field and radiation varying with magnetic field. II. SVC radiation mechanism and radiation variation with magnetic field.
See Abstr. 063.112.

An infrared perspective on chromospheres.
See Abstr. 064.030.

The formation mechanism of the lines of the C I solar infrared multiplet λ1069,5 nm.
See Abstr. 071.007.

The solar spectrum.
See Abstr. 071.054.

Electric field strengths from the solar 12 micron lines.
See Abstr. 071.062.

A connection of the number of dark points observed on the solar disk in the He I 10830 Å line and the number of light knots in the K Ca II line with the solar cycle.
See Abstr. 072.045.

VLA observations of solar active regions at 6 and 20 cm.
See Abstr. 072.095.

Einige Besonderheiten der Sonnenaktivität im Jahre 1989.
See Abstr. 072.148.

First observations of macrospicules at 4.8 GHz at the solar limb in polar coronal holes.
See Abstr. 073.037.

A search for solar oscillations in the solar chromosphere using millimeter interferometric observations.
See Abstr. 073.041.

Millimeter imaging of energetic electrons in solar flares in conjunction with GRO experiments.
See Abstr. 073.073.

The characteristics and interpretation of Hα limb flare and radio burst on Aug. 16, 1989 in AR5629.
See Abstr. 073.133.

AR5629: a loop prominence system and radio spike radiation 17 Aug., 1989.
See Abstr. 073.138.

Multi–wave band SMM – VLA observations of an M2 flare and an associated coronal mass ejection.
See Abstr. 073.139.

Radio signature and Hα limb features of the 1979 March 9 and 1982 July 9 flares.
See Abstr. 073.140.

Solar gamma–ray–line flares, type II radio bursts, and coronal mass ejections.
See Abstr. 073.148.

Exploration on the structure of the 1N/M7.1 flare on 30 Dec. 1982.
See Abstr. 073.169.

The solar chromospheric supergranular network in 850 micron radiation.
See Abstr. 073.170.

Acceleration of electrons outside flares: coronal manifestation and possible origin.
See Abstr. 074.083.

A proposal for a study of the solar wind near the Sun by simultaneous observations with SOLAR–A and Kashima 34 m antenna.
See Abstr. 074.124.

Rotation of large scale patterns on the solar surface as determined from filament and millimeter data.
See Abstr. 075.043.

Lagging behind of the fluxes of ultraviolet emission relative to the index $F_{10.7}$.
See Abstr. 076.018.

Nature of the relationship between the fluxes of solar cosmic ray electrons and protons and the parameters of microwave bursts.
See Abstr. 078.012.

Some studies of solar proton events in relation to active region characteristics.
See Abstr. 078.017.

078 Cosmic Radiation

078.001 **Solar flares accompanied by the generation of high-energy neutrons according to the supermonitor data.**
A. T. Filippov, M. L. Basalaev, Yu. F. Rudnev,
V. L. Yanchukovskij, V. K. Korotkov, N. G. Serebrenitskij.
Geomagn. Aeron., Vol. 30, No. 3, p. 323 – 326 (Nov 1990). English translation of Geomagn. Aehron., Tom 30, No. 3, p. 381 – 385 (1990).
A combined analysis of bursts of solar cosmic rays with the generation of bursts of γ rays of various energies and the 5-minute data of supermonitors of Siberia and the Far East for 24 events in the period 1979 – 1984 is given. It is shown that during some solar flares accompanied by the generation of γ rays with energies of tens of MeV a statistically significant increase of the flux of particles recorded by the neutron supermonitors is noted.

078.002 **Radiation characteristics of solar proton events during the decline of solar cycle 21 and the rise of cycle 22.**
N. K. Pereyaslova, M. N. Nazarova, I. E. Petrenko.
Geomagn. Aeron., Vol. 30, No. 4, p. 478 – 481 (Feb 1990). Paper presented at the 5. Symposium on Solar Terrestrial Physics, Samarkand (USSR), 2 – 6 Oct 1989. English translation of Geomagn. Aehron., Tom 30, No. 4, p. 566 – 570 (1990).
The authors present some data on variations in the radiation characteristics of solar proton events as a function of phase in the solar cycle obtained by comparison of the radiation characteristics of solar proton events during solar cycle 22 with similar periods during cycles 20 and 21. They construct probability distributions for the radiation characteristics during cycle 22, and carry out a comparison with model distributions for the rising phases of cycles 20 and 21.

078.003 **The ionization of solar cosmic ray in the ionospheric D-region.**
Zong Qiu-gang, Ye Zong-hai.
Chin. J. Space Sci., Vol. 11, No. 3, p. 201 – 208 (Jul 1991). In Chinese.
Solar cosmic ray which propagates to the ionospheric D-region causes the PCA phenomena. The results show that the electron production rate of solar cosmic rays varies with different importances of solar cosmic rays and its energy spectral index. When the importance of a solar cosmic ray event is the same (the integration flux is the same), the larger the energy spectral index is, the larger the electron production rate at high altitudes (>55 km) is; the smaller the energy spectral index is, the larger the electron production rate at low height (>55 km) is. The distribution of the electron production rate with height have obvious two peaks, one is at 60 km that is mainly caused by protons, the other is at 85 km that is mainly caused by particles of $z \geqslant 2$.

078.004 **Energetic particles in the inner solar system.**
H. Kunow, G. Wibberenz, G. Green,
R. Müller-Mellin, M.-B. Kallenrode.
Physics of the inner heliosphere 2. Particles, waves and turbulence, p. 243 – 342 (1991). – See Abstr. 003.002 for the main entry.
Contents: (1) Introduction. (2) Modulation of galactic cosmic radiation. (3) Interplanetary propagation of solar energetic particles. (4) Coronal propagation. (5) Particle acceleration in solar flares. (6) Particle acceleration in corotating interplanetary regions: features of corotating events and their relation to the solar wind, corotating events throughout the solar system and solar cycle dependence, models of corotating event generation and transport.

078.005 **Stochastic dynamics of protons in solar magnetic loops.**
V. N. Senatorov, V. M. Tomozov.
Kinematics Phys. Celest. Bodies, Vol. 7, No. 1, p. 27 – 30 (1991). English translation of Kinematika Fiz. Nebesn. Tel, Tom 7, No. 1, p. 30 – 33 (1991).
It is shown that stochastization of particles by Alfvén waves is accompanied by effective acceleration in loop structures of solar magnetic fields.

078.006 **Solar cosmic rays.**
E. O. Flückiger.
Nucl. Phys., B Proc. Suppl., Vol. 22B, p. 1 – 20 (Jul 1991). – See Abstr. 012.019 for the main entry.
The paper reviews a selection of typical solar cosmic ray data, including the unprecedented sequence of relativistic solar particle events that occurred in 1989, to present some of the main results obtained in recent years from the analysis of solar cosmic ray observations, and to make a synopsis of open questions and still unexplained phenomena.

078.007 **Two-year variations of solar neutrino fluxes, galactic cosmic rays, solar and geomagnetic activities.**
V. P. Okhlopkov.
Soln. Dannye, Byull., No. 2/91, p. 100 – 101 (Feb 1991). Abstract. – See Abstr. 012.021 for the main entry.

078.008 **Complexes of activity and flares.**
V. G. Banin, S. A. Yazev, G. M. Khmyrov.
Soln. Dannye, Byull., No. 3/91, p. 116 – 119 (Mar 1991). In Russian.
The existence of a wide class of proton LDE-flares is associated with their belonging to complexes of activity (CA). A distinctive feature of these flares is the long decay time of X-ray emission which can be brought about by excitation of an extensive system of CA coronal loops. This hypothesis receives some support by results of an intercomparison of variations in time sequences of the power of CA cores and of the number of LDE-events on the descending branch of cycle No. 21.

078.009 **Particle injection following solar flares on 1980 May 28 and June 8: evidence for different injection time histories in impulsive and gradual events?**
M.-B. Kallenrode, G. Wibberenz.
Astrophys. J., Vol. 376, No. 2, p. 787 – 796 (1 Aug 1991).
The authors discuss the injection time histories of one solar particle event on 1980 June 8 and a group of four homologous events on 1980 May 28 and compare them with the electromagnetic radiation emitted by the parent flares. Comparison with electromagnetic radiation in all events shows electron injection to be nearly simultaneous with the particle acceleration as indicated by the increase in electromagnetic radiation. In contrast, the proton injection differs from one event to the other. In the June 8 event, it is delayed by about 12 min with respect to the electron injection, occurring simultaneouly with a second increase in electromagnetic radiation. In the May 28 events not only the electrons, but also the nuclei, are injected simultaneously with the onset of the electromagnetic radiation. These observations are in agreement with a rapid acceleration of ions, as inferred from the observations of γ-flares. The authors suggest an acceleration model in which both classes of events start with an initial energy release, called impulsive phase, and develop then in a different manner. In the long-duration events, a gradual phase accelerating relativistic particles by a coronal shock follows about 10 min after the initial energy release. In the impulsive events, an intermediate phase or a second step in the impulsive phase occurs immediately after the impulsive phase, accelerating electrons to relativistic energies and nuclei up to energies of some tens of MeV per nucleon on time scales of 1 s.

078.010 **Charge states of solar energetic particles and correlations between particle intensities.**
Huang Yongnian.
Chin. J. Space Sci., Vol. 11, No. 4, p. 251 – 257 (Oct 1991). In Chinese.

A possible ionic charge state distribution of He, C, N, O, Ne, Mg, Si, S and Fe in solar energetic particle events is presented. The number of different ionic orders in the distributions of elements increases with the elemental mass except for He. The correlations between the mean particle intensities for the elements as a function of mean intensities of C in the energy range 2 – 3 MeV per nucleon can be explained in terms of the increase of ionic order number with mass for the elements. The ionic charge states cannot be explained in a model in which the coronal temperature determines a charge equilibrium. It is concluded that ionic charges of solar energetic particles can be produced in a solar flaring region rather than in an interplanetary shock region.

078.011 **The features of the ground level enhancement event of cosmic rays on 29 September 1989.**
Ye Zonghai.
Chin. J. Space Sci., Vol. 11, No. 4, p. 258 – 266 (Oct 1991). In Chinese.

The ground level enhancement (GLE) from the solar proton flare of 29 Sept. 1989 is recorded by the world wide network of the neutron monitors and μ–meson telescopes. This GLE is the greatest event since the event on 23 Feb. 1956. The features of the GLE including the intensity time profiles, the spectrum, the anisotropy and the geographical distribution etc. are studied. The influences of the GLE on the Earth's environment are described briefly.

078.012 **Nature of the relationship between the fluxes of solar cosmic ray electrons and protons and the parameters of microwave bursts.**
V. F. Mel'nikov, T. S. Podstrigach, E. I. Dajbog, V. G. Stolpovskij.
Cosmic Res., Vol. 29, No. 1, p. 87 – 94 (Jul 1991). English translation of Kosm. Issled., Tom 29, Vyp. 1, p. 95 – 103 (1991).

The interrelation between flare solar cosmic ray fluxes and the parameters of microwave bursts is analyzed. The correlation between the protons and electron fluxes and the radio emission is significantly enhanced if the frequency of the spectral maximum of the burst and its effective duration are included in the radio index along with the maximum intensity. It is shown that the relationship between the solar cosmic ray proton fluxes with the effective burst duration is stronger then for the electron fluxes. This is explained by the difference in the dynamics of accelerated electrons and protons in flare arches of different sizes. The conditions for the escape of accelerated protons and electrons into interplanetary space are examined. The combination of results obtained may be used for diagnostics of the total number of high–energy protons in the flare arch and the conditions for their escape into interplanetary space from the flare radio emission in the miscrowave and meter–decameter ranges.

078.013 **Detection of 0.5 – 15 GeV solar protons on 29 September 1989 at Australian stations.**
J. E. Humble, M. L. Duldig, D. F. Smart, M. A. Shea.
Geophys. Res. Lett., Vol. 18, No. 4, p. 737 – 740 (Apr 1991).

A major cosmic ray ground–level enhancment, the largest in 33 years, occurred on 29 September 1989 during which intensity enhancements at Australian observatories range up to a maximum of 344% at Mt. Wellington. The Darwin neutron monitor (cutoff rigidity 14.1 GV) recorded an ~13% increase in the five–minute counting rate indicating that solar particles up to at least 15 GeV must have been present. Surface muon detectors at Hobart and Mawson recorded increased fluxes, but the event was not recorded by underground muon detectors at either station. Preliminary analysis of the solar particle flux during the initial phase of the event shows a hard spectrum approximated equally well by an exponential spectrum with a P_0 of 2.0 GV or by a modified power law spectrum of exponent ~–2.9. Particles arriving at the detectors from non–vertical directions make a significant contribution to the total increase recorded at mid and low latitude stations.

078.014 **The characteristics of great cosmic ray storm events in March 1989.**
Ye Zonghai, Xiao Shaoyu, Xue Shunsheng.
Publ. Yunnan Obs., No. 4, p. 115 – 119 (Dec 1990). – See Abstr. 012.047 for the main entry.

078.015 **The characters of cosmic ray events in Mar. & Aug. 1989.**
Xiao Shaoyu, Yu Shaohua, Liu Bin, Li Runshen, Wu Cuifang.
Publ. Yunnan Obs., No. 4, p. 219 – 222 (Dec 1990). – See Abstr. 012.047 for the main entry.

078.016 **The characteristics of cosmic ray intensity on 4 – 17 August 1989.**
Xue Shunsheng, Ye Zonghai.
Publ. Yunnan Obs., No. 4, p. 223 – 226 (Dec 1990). – See Abstr. 012.047 for the main entry.

078.017 **Some studies of solar proton events in relation to active region characteristics.**
T. B. Chakravorti, T. K. Das, A. K. Sen, M. K. Das Gupta.
Bull. Astron. Inst. Czech., Vol. 42, No. 3, p. 165 – 170 (Jun 1991).

A study has been made of the proton–associated active regions that occurred during the period from 1966 to 1984. Important results obtained are as follows: Proton events are mostly found to develop in regions of sunspots α_p, β_p, $\beta\nu$ and δ configurations. 64% of the proton associated sunspots have field strengths in the range of 1500–2500 G. A considerable number of proton–associated active regions have flare index values greater than 700, whereas non–proton–associated active regions are entirely absent in these values. Proton and non–proton–associated active regions have almost the same rate of H–alpha flare productivity. The number of prompt metric bursts (mostly of type III) increases with the proton flux. In 70% cases, the Ca–plage index, radio emission flux and maximum intensity of 9.1 cm radio measurement attain their peak value on the day of occurrence of proton events, the values of these quantities decreasing with days both before and after the day of onset. Most of the proton events are associated with H–alpha flares displaying F– and U–type visual features.

078.018 **Possible ground–level measurements of solar neutron decay protons during the 19 October 1989 solar cosmic ray event.**
M. A. Shea, D. F. Smart, M. D. Wilson, E. O. Flückiger.
Geophys. Res. Lett., Vol. 18, No. 5, p. 829 – 832 (May 1991).

The high energy solar proton event of 19 October 1989 had an atypical particle anisotropy during the initial phase with the Deep River, Ottawa and Goose Bay neutron monitors in the Eastern Canadian region recording the intensity increase about 25 minutes before comparable detectors in Europe and Western Canada. Using the asymptotic cones of acceptance for these neutron monitors, the authors show that this early onset in the particle intensity increase may be the detection of relativistic protons that are the decay product of solar flare generated relativistic neutrons.

078.019 **Multispacecraft observations of solar ^{3}He–rich events.**
D. V. Reames, M.–B. Kallenrode, R. G. Stone.
Astrophys. J., Vol. 380, No. 1, p. 287 – 292 (10 Oct 1991).

The authors report on a systematic search of He isotope observations on the Helios 1 and 2 spacecraft during the occurrence of 66 ^{3}He–rich time periods observed on ISEE 3 over a 4 yr interval. Seven time periods show possible event associations but only three show confirmed events with similar ^{3}He/^{4}He ratios of Helios and ISEE 3. The authors use kilometric radio observations of type III bursts produced by electrons associated with the ^{3}He–rich events to map the interplanetary field lines outward from the events. The observations are best explained in terms of a narrow cone of emission of the particles from these events.

078.020 **On the differences in element abundances of energetic ions from corotating events and from large solar events.**
D. V. Reames, I. G. Richardson, L. M. Barbier.
Astrophys. J., Lett., Vol. 382, No. 1, p. L43 – L46 (20 Nov 1991).
The authors have examined the abundances of energetic ions accelerated from high-speed solar wind streams by shock waves formed at corotating interaction regions (CIRs) where high-speed streams overtake the lower speed solar wind. The observed element abundances appear to represent those of the high-speed solar wind, unmodified by the shock acceleration. These abundances, relative to those in the solar photosphere, are organized by the first ionization potential (FIP) of the ions in a way that is different from the FIP effect commonly used to describe differences between abundances in the solar photosphere and those in the solar corona, solar energetic particles (SEPs), and the low-speed solar wind. In contrast, the FIP effect of the ion abundances in the CIR events is characterized by a smaller amplitude of the difference between high-FIP and low-FIP ions and by elevated abundances of He, C, and S. It is likely that these differences in abundances can be understood in terms of higher temperatures and shorter time scales in the chromospheric fractionation region of coronal holes.

078.021 **History of the Sun during the past 4.5 Gyr as revealed by studies of energetic solar particles recorded in extraterrestrial and terrestrial samples.**
D. Lal, R. E. Lingenfelter.
The Sun in time, p. 221 – 231 (1991). – See Abstr. 003.027 for the main entry.
Recent observations of solar particles and radiation have greatly improved our understanding of a variety of physical processes operating in both the near-surface region and the interior of the Sun. These observations include studies of solar irradiance, solar wind, and a great variety of solar energetic particles and radiation: heavier nuclei, protons, neutrons, gamma rays, X-rays and neutrinos. Observations and studies of the effects of the interactions of energetic solar particles in terrestrial and extraterrestrial material enabled to deduce important information of the prehistory of such solar particles all the way back to 4.5 Gyr. The authors give a brief overview of what we have learned and can learn about the history of energetic particles emitted by the Sun.

078.022 **Accelerated particles in solar flares.**
R. Ramaty, G. M. Simnett.
The Sun in time, p. 232 – 259 (1991). – See Abstr. 003.027 for the main entry.
There is considerable evidence that the Sun has been a source of energetic particles over much of its life. It is of current interest to understand the ways in which this might be achieved in the present Sun. The authors first review the modern evidence from space-borne instruments for energetic solar-particle emission. They then relate this to the fluences at the Sun by detailed examination of the gamma-ray, hard X-ray and radio emissions produced at the time of energetic solar flares. Possible acceleration mechanisms are reviewed, together with interaction models of both relativistic and nonrelativistic ions and electrons in the solar atmosphere. The correlation between the particle populations escaping from the Sun and those actually accelerated in the solar atmosphere is dependent on the trapping, propagation and escape from the corona. Different acceleration phases and their relationship to other transient phenomena such as coronal mass ejections are discussed. The authors also compare the current average rate of accelerated ion production in solar flares with the ion production rates at earlier phases in the evolution of the Sun suggested by flare activity on T Tauri stars and isotopic anomalies in meteorites.

078.023 **Solar-cosmic ray fluxes during the last ten million years.**
R. C. Reedy, K. Marti.
The Sun in time, p. 260 – 287 (1991). – See Abstr. 003.027 for the main entry.
The fluxes of energetic solar particles in the vicinity of the Earth in the past can be determined from nuclides that they produced in the top centimeter of lunar rocks. Activity-vs-depth profiles of short-lived radioactivities measured in the top centimeter of lunar rocks agree with profiles calculated with directly measured solar-proton fluxes since about 1965 and were used with indirect observations to get solar-proton fluxes back to 1956. Lunar-rock profiles for long-lived radionuclides have been used to infer solar-proton fluxes averaged over several time periods in the past. New results are reported for solar-proton-produced ^{81}Kr measured in lunar rock 68815. Activites of 76000-yr ^{59}Ni can be used to get fluxes of solar α particles averaged over the last $\sim 10^5$yr. The average solar-proton fluxes in the past are not greatly different from those observed during the last three 11-yr solar cycles. The authors discuss the work that needs to be done to determine more and better fluxes of energetic particles from the Sun in the past.

078.024 **Solar neutrinos and the history of the Sun.**
K. Wolfsberg, G. E. Kocharov.
The Sun in time, p. 288 – 313 (1991). With 2 plates. – See Abstr. 003.027 for the main entry.
Neutrinos may be the only particles created in the thermonuclear processes of the Sun that escape directly. Therefore, their measurement provides important tests of solar models and of fundamental properties of neutrinos. There is a serious discrepancy between the results of the Homestake and Kamiokande experiments that measure the high-energy neutrino flux at the present time and the values predicted by the Standard Solar Model. Interpretations of this famous puzzle fall into two classes. New detectors have been proposed to measure other features of the current neutrino fluxes. Geochemical experiments can measure long-lived nuclides in the Earth's crust that were produced by solar neutrinos and can offer information on the neutrino fluence over millions of years. The molybdenum-technetium experiment, also currently underway, is one such experiment.

078.025 **Long-term changes in composition of solar particles implanted in extraterrestrial materials.**
J. F. Kerridge, P. Signer, R. Wieler, R. H. Becker, R. O. Pepin.
The Sun in time, p. 389 – 412 (1991). – See Abstr. 003.027 for the main entry.
Analysis of lunar surface samples for elements implanted therein by solar corpuscular radiation reveals evidence for the following compositional changes over a time period between 1.5 and 3 Gyr: 50% decreases in the ratios ^{4}He/^{36}Ar and Xe/^{36}Ar; a 20% increase in the ratio ^{3}He/^{4}He; a 3% increase in the ratio ^{20}Ne/^{22}Ne; and a 50% increase in the ratio ^{15}N/^{14}N. The causes of these changes are not resolved at this time but may include (a) a change in acceleration conditions of the solar wind, (b) a change in flux of solar energetic particles relative to that of the solar wind, and (c) a change in composition of the solar convective zone. There is good evidence for a long-term decrease in the solar-wind flux.

078.026 **Solar flare heavy-ion tracks in extraterrestrial objects.**
J. N. Goswami.
The Sun in time, p. 426 – 444 (1991). – See Abstr. 003.027 for the main entry.
Lunar samples, meteorites and interplanetary dust particles contain fossil tracks produced by energetic solar flare heavy ions. Solar flare track records in certain primitive meteorites provide us with information on the activity of the Sun more than 4.4 Gyr ago. Similar records in lunar samples allow us to probe the level of solar activity at different epochs during the last 1 Gyr. Solar flare track records in interplanetary dust particles serve as a potential source for information about solar activity over the past thousand years. Recent investigations of primitive meteorites that combine solar flare track and noble gas studies suggest the possibility that the Sun had gone through an active early (T Tauri) phase.

078.027 **Interplanetary transport of decay protons from solar flare neutrons.**
D. Ruffolo.
Astrophys. J., Vol. 382, No. 2, p. 688 – 698 (1 Dec 1991).
The author has developed models for the propagation of protons produced by the decay of solar flare neutrons. These

models are applied to the observation of such protons on 1982 June 3 and 1984 April 25. The decay protons are produced with the same direction and velocity as the parent neutrons, and pitch–angle scattering of the energetic protons transforms a monodirectional beam into an isotropic distribution. The author has performed numerical simulations treating the deposition, pitch–angle scattering, and adiabatic focusing of neutron–decay protons. Fitting these simulations to the measurements of the time evolution of the decay proton flux strongly constrains the interplanetary scattering mean free path, λ, to be 0.26 ± 0.05 AU on 1982 June 3 and 0.37 ± 0.06 AU on 1984 April 25. Power–law exponents for the energy spectra of parent neutrons in the energy range 26.5 MeV to 147 MeV have values from −1.7 to −1.9 for the 1982 flare, and from −1.2 to −1.4 for the 1984 flare. In addition, the evolution of the neutron–decay proton flux seems to imply that the neutron emission was either nonisotropic or occurred over a time of several minutes.

078.028 **Direct solar neutrons detected by neutron monitors on 24 May 1990.**
M. A. Shea, D. F. Smart, K. R. Pyle.
Geophys. Res. Lett., Vol. 18, No. 9, p. 1655 – 1658 (Sep 1991).

The ground–level energetic solar particle event on 24 May 1990 has an extremely impulsive onset in the particle intensity that the authors attribute to a significant flux of direct solar neutrons impacting the top of the atmosphere. This is the largest solar neutron event reported to data with a 25.3% increase recorded by the Climax neutron monitor in the time interval 2050–2055 UT. At least 7 neutron monitors in North America recorded this increase simultaneously, the magnitude of which is ordered by the air mass along the line of sight to the sun at the time of the event. The initial solar neutrons arrive within the same minute as the maximum in both the soft X–ray and the H–alpha emission observed at the earth, whereas the onset of the world–wide increase of solar protons occurred approximately 15 minutes later.

078.029 **Acceleration of high–energy nuclei as viewed from their chemical composition.**
K. Sakurai.
International SOLAR–A Science Meeting: Flare physics in solar activity maximum 22, p. 114 – 117 (1991). – See Abstr. 012.104 for the main entry.

High energy electromagnetic emissions associated with solar flares are causally related to the generation of energetic particles such as so–called solar cosmic rays. These particles seem to be accelerated by some transient mechanism in the flare regions. Taking into account the observed properties as their energy spectra and chemical composition, it becomes possible to find the acceleration mechanism of high–energy particles in solar flares.

078.030 **A system of loop interplanetary traps of solar cosmic rays on June 1974.**
G. P. Lyubimov, E. A. Chuchkov.
Kosm. Issled., Tom 29, Vyp. 6, p. 910 – 916 (Nov – Dec 1991). In Russian. English translation in Cosmic Res., Vol. 29, No. 6.

An increasing of solar protons with energy 1 – 15 MeV during June 8 – 12, 1974 is studied. The source of this increasing was the solar flare on July 6, 1974 which carried particles of solar cosmic rays into a system of quasisteady interplanetary loop traps. The solar and interplanetary data are analysed in detail. The solar cosmic ray data are approximated by a model.

Solar and galactic cosmic rays. Proceedings. 12. European Cosmic Ray Symposium, Nottingham (UK), 15 – 21 Jul 1990.
See Abstr. 012.019.

Accelerator beam experiments of a prototype cosmic ray heavy ion telescope.
See Abstr. 035.024.

Studies of high–energy flare phenomena from simultaneous observations of gamma–rays and energetic particles.
See Abstr. 051.080.

Solar energetic particle observation by GEOTAIL satellite.
See Abstr. 051.082.

Activity development in the area connected with the 23.08.88 proton flare.
See Abstr. 072.128.

Proton flare and terrestrial effects with aurorae in the minimum phase of solar activity (Jan 1985).
See Abstr. 072.139.

Three solar filament disappearances associated with interplanetary low–energy particle events.
See Abstr. 073.016.

Energetic particle production in flares.
See Abstr. 073.091.

Subthreshold stochastic diffusion with application to selective acceleration of ^{3}He in solar flares.
See Abstr. 073.160.

Solar flares.
See Abstr. 073.165.

Ion acceleration and plasma heating in solar flares.
See Abstr. 073.189.

Coronal mass ejections and streamers associated with the new cycle active regions at solar minimum.
See Abstr. 074.092.

High–energy particle acceleration during the implosion driven by 3–dimensional X–type current loop coalescence in solar flares.
See Abstr. 076.041.

Proton flare diagnostics using integral parameters of microwave bursts.
See Abstr. 077.024.

Large solar flares: analysis of the events recorded by the Mont Blanc neutrino detector.
See Abstr. 080.119.

Near–earth sporadic radio emission caused by cosmic rays of isolated solar flares.
See Abstr. 085.001.

Do meteorites contain irradiation records from exposure to an enhanced–activity Sun?
See Abstr. 105.048.

Statistical properties of shock–accelerated ions in the outer heliosphere.
See Abstr. 106.038.

Cosmic ray positron production by gamma ray interactions on starlight.
See Abstr. 144.092.

Signatures of cosmic–ray interactions on the solar surface.
See Abstr. 144.109.

079 Solar Eclipses

079.001 **The makings of a prize eclipse.**
O. Gingerich.
Sky Telesc., Vol. 82, No. 1, p. 15 – 17 (Jul 1991).
Why the July 11th eclipse is the best until 2132.

079.002 **Observations of total solar eclipses in the stratosphere.**
S. Koutchmy.
Zemlya Vselennaya, No. 5, p. 33 – 37 (Sep – Oct 1991). In Russian.

079.003 **Observing solar eclipses.**
J. M. Pasachoff.
Liller Symposium: Asteroids to quasars, p. 67 – 82 (1991). – See Abstr. 012.087 for the main entry.

079.004 **Halbschatten – etwas genauer betrachtet.**
P. Richter.
Nachr. Olbers-Ges. Bremen, Nr. 155, p. 2 – 5 (Oct 1991).

Solar eclipse 1991 July 11

079.005 **Die totale Sonnenfinsternis am 11. Juli 1991.**
K. Löchel.
Sterne Weltraum, Jahrg. 30, Nr. 7, p. 468 – 470 (Jul 1991).

079.006 **The great eclipse of 1991.**
D. di Cicco.
Sky Telesc., Vol. 82, No. 6, p. 589 – 595 (Dec 1991).

079.007 **Eclipse science from Hawaii.**
K. Krisciunas.
Sky Telesc., Vol. 82, No. 6, p. 597 – 600 (Dec 1991).

079.008 **Eclipse plans for NSO/SP.**
J. B. Zirker, S. Koutchmy, R. L. Coulter,
L. J. November, R. N. Smartt.
Bull. Am. Astron. Soc., Vol. 23, No. 2, p. 1063 (1991). Abstract. – See Abstr. 012.012 for the main entry.

079.009 **Expedition to South California.**
Eh. V. Kononovich.
Zemlya Vselennaya, No. 6, p. 61 – 64 (Nov – Dec 1991). In Russian.

079.010 **Expedition to Brazil.**
N. I. Dzyubenko, I. S. Kim, O. G. Mazuura.
Zemlya Vselennaya, No. 6, p. 65 – 67 (Nov – Dec 1991). In Russian.

079.011 **Photographing the 11 July 1991 solar eclipse.**
M. Watson.
J. R. Astron. Soc. Can., Vol. 85, No. 4, p. 173 (Aug 1991). Abstract. – See Abstr. 012.090 for the main entry.

079.012 **L'eclisse totale di sole dell'11 luglio 1991.**
S. De Meis.
Orione, Vol. 11, N. 3, p. 8 – 15 (May – Jun 1991).

Precision of medieval Islamic eclipse measurements.
See Abstr. 004.079.

Accuracy of eclipse observations recorded in medieval Arabic chronicles.
See Abstr. 004.084.

Astronomical computing: North America's solar eclipse.
See Abstr. 014.027.

Come fotografare l'eclisse di sole.
See Abstr. 036.371.

Die Bestimmung der Ephemeridenzeitkorrektur aus Aufnahmen der totalen Sonnenfinsternis vom 30. Juni 1973.
See Abstr. 044.046.

Observation of 1990 solar eclipse by a torsion pendulum.
See Abstr. 066.186.

On some features of the photosphere's brightness distribution at the extreme limb of the Sun from eclipse observations.
See Abstr. 071.034.

Nonthermal velocities of Na I, Ba II, Sr II atoms in the lower chromosphere from eclipse observations.
See Abstr. 073.013.

Coronal temperature estimate from intensity obtained during the total solar eclipse of 1988 March 18.
See Abstr. 074.045.

The solar corona over the recent saros.
See Abstr. 074.077.

Submillimeter solar limb profiles determined from observations of the total solar eclipse of 1988 March 18.
See Abstr. 077.068.

080 Atmosphere, Figure, Internal Constitution, Neutrinos, Rotation, etc.

080.001 **The location of the source of high-frequency solar acoustic oscillations.**
P. Kumar, E. Lu.
Astrophys. J., Lett., Vol. 375, No. 1, p. L35 - L39 (1 Jul 1991).
Recently Libbrecht and Jefferies et al. have reported regular peaks in the solar oscillation power spectrum extending well above 5.3 mHz, the maximum frequency of trapped acoustic modes. Kumar et al. argued that these peaks are primarily due to the interference of traveling waves which are excited due to acoustic emission from turbulent convection. In contrast with the standing wave P-mode frequencies below 5.3 mHz, the positions of the high-frequency interference peaks (HIPs) are dependent on the location of the source of the acoustic oscillations. Here, the authors make use of this dependence to determine the acoustic source strength as a function of depth. They find that the acoustic source profile, and thus the convective velocity, is peaked ~200 km deeper than what is expected from standard mixing length theory.

080.002 **Time variations in Kamiokande solar neutrino data.**
P. Raychaudhuri.
Mod. Phys. Lett. A, Vol. 6, No. 22, p. 2003 - 2007 (20 Jul 1991).
Solar neutrino flux ($E_\nu \geqslant 7.5$ MeV) data from 1st January to April 1990 as measured in the Kamiokande solar neutrino experiment have been analyzed statistically and have found that the solar neutrino data varies with the solar activity cycle with very high statistical significance (>98% confidence level). Average solar neutrino flux data in the sunspot minimum range cannot be equal to twice the average solar neutrino flux data in the sunspot maximum range, which suggests that the neutrino flip through the magnetic field of the convection zone of the sun is not responsible for the solar neutrino flux variation. Thus the variation of solar neutrino flux with the solar activity cycle suggests that the solar activity cycle is due to the pulsating character of the nuclear energy generation inside the core of the sun.

080.003 **Cosmic strings in the Sun?**
Wang Jin.
Mod. Phys. Lett. A, Vol. 6, No. 23, p. 2101 - 2106 (30 Jul 1991).
The author applies Chudnovsky and Vilenkin's (1988) idea of the possible existence of cosmic strings in the sun. He made stellar evolution calculation for a one solar mass star with solar age. Stellar evolution with cosmic strings at solar age gives radius and luminosity of the star which are in contradiction with observation.

080.004 **Solar flare neutrinos.**
G. E. Kocharov, G. A. Koval'tsov, I. G. Usoskin.
Nuovo Cimento C, Vol. 14, N. 4, p. 417 - 422 (Jul-Aug 1991).
The theory of high-energy neutrinos generation in the solar atmosphere during flares is developed. A possible effect in different neutrino detectors is discussed.

080.005 **Differential rotation and isorotation surfaces in the convective, slowly rotating envelope of the Sun.**
N. A. Plieva.
Pis'ma Astron. Zh., Tom 17, No. 8, p. 741 - 753 (Aug 1991). In Russian. English translation in Sov. Astron. Lett., Vol. 17, No. 4.
It is shown that turbulent convection in a density stratified medium redistributes the angular momentum in such a way that there arises differential rotation. Angular velocity decreases at high latitudes and increases at low ones. The inversion region of the sign of the angular velocity radial gradient belongs to the middle latitudes. It is shown that the solution obtained does not contradict to observations and helioseismology data.

080.006 **Detecting neutrinos from the Sun.**
S. R. Elliott, R. G. H. Robertson.
Contemp. Phys., Vol. 32, No. 4, p. 251 - 263 (Jul-Aug 1991).
The Sun shines brightly to our eyes, but is three times dimmer in neutrinos than it should be for the amount of energy it is producing. How that can happen has been a mystery for twenty years. It was long thought to be an error in our understanding of how the Sun works, but new experiments and new ideas about particle physics are lending support to the possibility that it is the neutrino itself that was not understood. If so, a revolution in particle physics and our understanding of the universe may be at hand. The authors describe the remarkable experiments that have brought us to this point, and new ones to come.

080.007 **Advances in helioseismology.**
K. G. Libbrecht.
Bull. Am. Astron. Soc., Vol. 23, No. 2, p. 930 - 931 (1991). Abstract. The full paper is published in Science, Vol. 253, No. 5016, p. 152 - 157 (12 Jul 1991). - See Abstr. 012.008 for the main entry.

080.008 **The impact of SMM data on SMM (simple-minded modeling).**
A. G. Emslie.
Bull. Am. Astron. Soc., Vol. 23, No. 2, p. 934 (1991). Abstract. - See Abstr. 012.008 for the main entry.

080.009 **5 minute oscillations and large scale spatial structures.**
S. R. Restaino, L. Bertello.
Bull. Am. Astron. Soc., Vol. 23, No. 2, p. 938 (1991). Abstract. - See Abstr. 012.008 for the main entry.

080.010 **Differential rotation and isorotation surfaces in a slowly rotating convective envelope of the Sun.**
N. A. Plieva.
Sov. Astron. Lett., Vol. 17, No. 4, p. 314 - 319 (Jul 1991). Current Physics Microform No.: 9202X0490. English translation of Pis'ma Astron. Zh., Tom 17, No. 8, p. 741 - 753 (Aug 1991).
The author shows that in a rotating density-stratified medium turbulent convection redistributes the flux of angular momentum in such a way as to cause differential rotation. The angular velocity increases (decreases) with depth at thigh (low) latitudes. The region where the radial gradient of angular velocity changes sign is at midlatitudes. The author shows that his solution is consistent with the observations and helioseismological data.

080.011 **Advances in helioseismology.**
K. G. Libbrecht, M. F. Woodard.
Science, Vol. 253, No. 5016, p. 152 - 157 (12 Jul 1991).
Globally coherent oscillation modes were discovered in the sun about a decade ago, providing a unique seismological probe of the solar interior. Current observations detect modes that are phase-coherent for up to 1 y, with surface velocity amplitudes as low as 2 mm s^{-1}, and thousands of mode frequencies have been measured to accuracies as high as 1 part in 10^5. This article discusses the properties of these oscillation modes and the ways in which they are adding to our understanding of the structure and dynamics of the sun.

080.012 **Ground-based modelling of solar irradiance variations.**
J. K. Lawrence, G. A. Chapman, A. D. Herzog, S. R. Walton, H. S. Hudson, B. M. Fisher.
Bull. Am. Astron. Soc., Vol. 23, No. 2, p. 960 (1991). Abstract. - See Abstr. 012.008 for the main entry.

080.013 **Energy and equilibrium of electric currents in the atmosphere of the Sun.**
Yu. Eh. Litvinenko, B. V. Somov.
Kinematics Phys. Celest. Bodies, Vol. 7, No. 1, p. 19 – 26 (1991). English translation of Kinematika Fiz. Nebesn. Tel, Tom 7, No. 1, p. 21 – 29 (1991).
An expression derived from the virial theorem for the magnetic energy of a large–scale electric current, and current sheet in particular, above the photosphere of an active region is discussed. It is shown that the erroneous conclusion that the magnetic energy of the current is equal to its gravitational energy is based on unjustified application of a "mirror–current" model to the Sun's atmosphere. The wave processes initiated by generation of electric currents in the corona are discussed. It is shown that proper application of the virial theorem admits the presence of equilibrium currents in the solar atmosphere. The energy of the currents may be large enough to explain flare activity.

080.014 **Magnetic flux transport and the Sun's dipole moment: new twists to the Babcock–Leighton model.**
Y.–M. Wang, N. R. Sheeley Jr.
Astrophys. J., Vol. 375, No. 2, p. 761 – 770 (10 Jul 1991).
The authors discuss the mechanism for the generation of the Sun's poloidal magnetic field in the general framework of the Babcock–Leighton model. There are two distinct contributions to this process, the first being associated with the axial tilts of bipolar magnetic regions (BMRs) as they erupt at the solar surface, the second arising from the subsequent transport of the erupted flux over the surface. The authors suggest that the action of Coriolis forces on rising, expanding flux loops can account for both the magnitude and the latitudinal variation of the average BMR tilt angles. After a BMR has erupted, supergranular diffusion and meridional flow acting in combination can cause its axial dipole moment to grow or decay, depending mainly on the BMR's latitude of eruption. For realistic rates of diffusion and flow, the net effect of flux transport over the sunspot cycle is to reduce the Sun's axial dipole below the strength that it would have attained in the absence of any transport. In this sense, the "α–effect" in the Babcock–Leighton dynamo is inherent in the axial tilts of the BMRs; the surface transport acts mainly to redistribute the poloidally oriented component of the BMR flux, producing highly concentrated polar fields.

080.015 **Modeling mesogranules and exploders on the solar surface.**
G. W. Simon, A. M. Title, N. O. Weiss.
Astrophys. J., Vol. 375, No. 2, p. 775 – 788 (10 Jul 1991).
Correlation tracking of high–resolution observations of solar granulation reveals a horizontal flow field containing both the supergranulation pattern and a much smaller pattern which corresponds to mesogranulation. Exploding granules (exploders) occur preferentially near the centers of mesogranules. This raises the question: are mesogranules just collections of exploders? Radial outflows in exploders and mesogranules can be modeled by superposing Gaussian source functions. This model is used to explore the relationship between mesogranules and exploders. Although the authors demonstrate that there is a mathematical equivalence between mesogranules and exploders distributed normally about the mesogranule centers, the results indicate that the observed mesogranular velocity pattern is not consistent with a flow pattern generated by exploders dropped randomly on the solar surface. Detailed comparisons with observations suggest that the averaged mesogranular velocity is produced by a combination of a persistent outflow from a source together with exploders distributed randomly about its center. A similar analysis shows that supergranules are not the result of random occurrences of mesogranules.

080.016 **The Sun as the source of VHE neutrinos.**
I. V. Moskalenko, S. Karakula, W. Tkaczyk.
Astron. Astrophys., Vol. 248, No. 1, p. L5 – L6 (Aug 1991). Letter–to–the–editor.
During the last decade proposals to build large underground/underwater neutrino detectors have been discussed actively. The authors argue here that the Sun is the source of very high energy (VHE) neutrinos which are the result of cosmic ray interactions with the solar matter. The ratio of the flux of the solar $\tilde{\nu}_\mu$ and ν_μ neutrinos to the Earth's atmospheric ones for the $\approx 0\overset{\circ}{.}5$ angular resolution detector is ≈ 1 for the energy range $\leqslant 200$ GeV and is much greater than 1 for the energy > 200 GeV. The possible applications of such measurements are discussed.

080.017 **Observations of waves and oscillations.**
F.–L. Deubner.
International Conference on Mechanisms of Chromospheric and Coronal Heating, p. 6 – 18 (1991). – See Abstr. 012.011 for the main entry.
Phase relations between velocity and brightness fluctuations observed in spectral lines in the visible are utilized both to diagnose the dynamical state of the solar atmosphere, and to examine the various oscillatory motions as potential carriers of mechanical energy dissipated in the upper layers. After a presentation of recent improvements of the diagnostic techniques, acoustic waves in the upper photosphere and resonant oscillations in the chromosphere are reviewed in the heating context. Modifications of the phase relations in the presence of localized magnetic fields are discussed, and a heating process operating at the footpoints of magnetic canopies is suggested.

080.018 **The generation of long period acoustic waves by solar global oscillations.**
B. Fleck, F. Schmitz.
International Conference on Mechanisms of Chromospheric and Coronal Heating, p. 22 – 24 (1991). – See Abstr. 012.011 for the main entry.
The authors study the response of a stratified isothermal atmosphere to long period disturbances with frequencies below the acoustic cut–off frequency ω_{co}. Simulating the 5–min and other long period oscillations by a sinusoidally moving piston, they find that a few scale heights above the piston the atmosphere does not oscillate anymore with the excitation period but with the cut–off period (≈ 180 s). They therefore suggest that the observed chromospheric 3–min oscillations are not due to a cavity but to a very simple – and in principle well known – basic effect of the dynamical behaviour of atmospheres, namely the excitation of the "cut–off frequency mode".

080.019 **Search for evidence for an oblique rotator within the Sun.**
P. A. Sturrock, T. Bai.
Bull. Am. Astron. Soc., Vol. 23, No. 2, p. 1028 (1991). Abstract. – See Abstr. 012.012 for the main entry.

080.020 **The formation, structure and dissipation of direct electric current systems in the solar atmosphere.**
G. Roumeliotis.
Bull. Am. Astron. Soc., Vol. 23, No. 2, p. 1029 (1991). Abstract. – See Abstr. 012.012 for the main entry.

080.021 **Seismic analysis of the Sun from intermediate and high degree p–modes.**
S. G. Korzennik.
Bull. Am. Astron. Soc., Vol. 23, No. 2, p. 1032 (1991). Abstract. – See Abstr. 012.012 for the main entry.

080.022 **Characteristics of intermediate–degree solar p–mode line widths.**
T. L. Duvall Jr., S. M. Jefferies, J. W. Harvey, Y. Osaki, M. A. Pomerantz.
Bull. Am. Astron. Soc., Vol. 23, No. 2, p. 1032 (1991). Abstract. – See Abstr. 012.012 for the main entry.

080.023 **On the detection of g–mode signatures in p–mode signals.**
J. Kennedy, S. M. Jefferies, F. Hill.
Bull. Am. Astron. Soc., Vol. 23, No. 2, p. 1032 (1991). Abstract. – See Abstr. 012.012 for the main entry.

080.024 **How deeply might sunspots and supergranules be anchored within the Sun?**
E. J. Rhodes Jr., S. G. Korzennik, D. H. Hathaway, A. Cacciani.
Bull. Am. Astron. Soc., Vol. 23, No. 2, p. 1033 (1991). Abstract. – See Abstr. 012.012 for the main entry.

080.025 **On the generation of magnetic tube waves in the solar convection zone.**
Z. E. Musielak, R. Rosner, P. Ulmschneider, P. Gail.
Bull. Am. Astron. Soc., Vol. 23, No. 2, p. 1037 (1991). Abstract. – See Abstr. 012.012 for the main entry.

080.026 **Heating of the solar atmosphere by spicules.**
Cheng Qing–qi, P. Ulmschneider, P. Korevaar.
International Conference on Mechanisms of Chromospheric and Coronal Heating, p. 350 – 352 (1991). – See Abstr. 012.011 for the main entry.

The authors simulate a solar spicule numerically with a single quasi–impulsive wave, which is generated in the photosphere and propagates upward along a flux tube. In the calculations radiation, ionization, conduction and mechanical heating are taken into account. From the rise and fall of the spicular material it is shown that the observed downflow at the temperature of 10^5K in the chromosphere–corona transition region is a direct result of the spicular upflow, and that the spicules can play a significant role in the energy balance and the heating of the upper solar atmosphere.

080.027 **Generation of acoustic flux derived from numerical simulations of the solar granular convection.**
M. Steffen, A. Krüss, H. Holweger.
International Conference on Mechanisms of Chromospheric and Coronal Heating, p. 380 – 385 (1991). – See Abstr. 012.011 for the main entry.

The authors have performed hydrodynamical calculations of convective flows in photospheric and subphotospheric layers of the Sun accounting for the basic physics of compressible convection in a stratified atmosphere, including ionization of hydrogen and helium as well as non–local multi–dimensional radiative transfer (grey approximation in LTE). Based on these numerical simulations of granular convection they estimate the acoustic flux generated at the top of the solar convection zone. Using different methods to "measure" the the amount of energy propagating in these models into the higher atmospheric layers as acoustic–gravity waves, the authors conclude that convective motions in the subphotospheric layers supply ample acoustic energy to account for the observed chromospheric radiation losses.

080.028 **Magnetoacoustic waves and their generation by convection.**
R. F. Stein, Å. Nordlund.
International Conference on Mechanisms of Chromospheric and Coronal Heating, p. 386 – 401 (1991). – See Abstr. 012.011 for the main entry.

Solar convection consists of a nearly uniform, warm, diverging upflow in which is embedded cool, converging, filamentary downdrafts. This convective flow generates acoustic waves (primarily resonant modes of the computational box), and where magnetic fields are present, it generates torsional, kink and sausage magnetic flux tube waves. As acoustic waves propagate upward, they reach a height (in the mid chromosphere) where the photospheric magnetic flux tubes have spread to fill the entire surface and $P_{gas} = P_{magnetic}$. Where this occurs, about 1/3 of the acoustic flux is converted to a fast magnetoacoustic wave flux and the rest is reflected. The transmitted fast magnetoacoustic waves are severely refracted and may be totally internally reflected because of the increase in Alfvén speed with height. The difficulty in getting energy to the upper chromosphere and corona may be overcome by waves ducted along the magnetic flux tubes. Such waves are produced directly by the convective flow and may also be produced by coupling to the acoustic wave flux incident on the chromospheric magnetic canopy. Wave energy that reaches the upper chromosphere and corona is dissipated there by phase mixing, resonant absorption and non–linear coupling to compressive magnetoacoustic waves that shock.

080.029 **On the dissipation of magnetic fields in the solar convection zone.**
E. E. DeLuca, G. H. Fisher.
Bull. Am. Astron. Soc., Vol. 23, No. 2, p. 1048 – 1049 (1991). Abstract. – See Abstr. 012.012 for the main entry.

080.030 **Properties of high–frequency and high–wavenumber solar oscillations.**
D. Fernandes.
Bull. Am. Astron. Soc., Vol. 23, No. 2, p. 1049 (1991). Abstract. – See Abstr. 012.012 for the main entry.

080.031 **An acoustic Poynting vector for solar p–mode oscillations.**
D. C. Braun, C. Lindsey.
Bull. Am. Astron. Soc., Vol. 23, No. 2, p. 1049 (1991). Abstract. – See Abstr. 012.012 for the main entry.

080.032 **Analysis of p–modes in a sunspot umbra.**
M. Penn, B. LaBonte.
Bull. Am. Astron. Soc., Vol. 23, No. 2, p. 1049 (1991). Abstract. – See Abstr. 012.012 for the main entry.

080.033 **The role of slow mode waves in p–mode absorption by sunspots.**
Y. Fan, G. H. Fisher, A. N. McClymont.
Bull. Am. Astron. Soc., Vol. 23, No. 2, p. 1049 – 1050 (1991). Abstract. – See Abstr. 012.012 for the main entry.

080.034 **The solar equatorial internal rotation rate estimated from combined South Pole and NSO/Sac Peak helioseismic data sets.**
F. Hill, S. M. Jefferies, M. A. Pomerantz, T. L. Duvall, J. W. Harvey.
Bull. Am. Astron. Soc., Vol. 23, No. 2, p. 1050 (1991). Abstract. – See Abstr. 012.012 for the main entry.

080.035 **Line simulation of solar structures permeated by acoustic and MHD–waves.**
W. Rammacher.
International Conference on Mechanisms of Chromospheric and Coronal Heating, p. 414 – 416 (1991). – See Abstr. 012.011 for the main entry.

Mg II–k and Ca II–K line profiles generated by acoustic and MHD–waves of different periods and fixed energy flux are computed.

080.036 **A numerical simulation of atmospheric responses due to emerging flux from sub–photospheric layers.**
S. T. Wu, M. T. Song, E. Tandberg–Hanssen.
Bull. Am. Astron. Soc., Vol. 23, No. 2, p. 1059 (1991). Abstract. – See Abstr. 012.012 for the main entry.

080.037 **Total irradiance and the solar cycle.**
H. S. Hudson, R. C. Willson.
Bull. Am. Astron. Soc., Vol. 23, No. 2, p. 1067 (1991). Abstract. – See Abstr. 012.012 for the main entry.

080.038 **Total solar irradiance variations compared with ground–based photometry at the SFO.**
G. A. Chapman, J. K. Lawrence, H. S. Hudson.
Bull. Am. Astron. Soc., Vol. 23, No. 2, p. 1067 (1991). Abstract. – See Abstr. 012.012 for the main entry.

080.039 **On the role of the photospheric magnetic network in the 11–year variation of total solar irradiance.**
K. L. Harvey, P. V. Foukal.
Bull. Am. Astron. Soc., Vol. 23, No. 2, p. 1068 (1991). Abstract. – See Abstr. 012.012 for the main entry.

080.040 **Multi–variate spectral analysis of solar irradiance variations.**
J. M. Pap.
Bull. Am. Astron. Soc., Vol. 23, No. 2, p. 1069 (1991). Abstract. – See Abstr. 012.012 for the main entry.

080.041 **Modelling solar UV irradiance.**
J. M. Pap.
Bull. Am. Astron. Soc., Vol. 23, No. 2, p. 1069 (1991). Abstract. – See Abstr. 012.012 for the main entry.

080.042 **Measurements of facular contrast in the near ultraviolet.**
T. Moran.
Bull. Am. Astron. Soc., Vol. 23, No. 2, p. 1069 (1991). Abstract. – See Abstr. 012.012 for the main entry.

080.043 **Alfvén wave propagation in the solar atmosphere.**
P. L. Similon, S. Zargham.
Bull. Am. Astron. Soc., Vol. 23, No. 2, p. 1070 (1991). Abstract. – See Abstr. 012.012 for the main entry.

080.044 **Joule heating in the Sun's lower transition region.**
G. Roumeliotis.
International Conference on Mechanisms of Chromospheric and Coronal Heating, p. 562 – 563 (1991). – See Abstr. 012.011 for the main entry.

The author proposes that most of the plasma in the Sun's lower transition region (3×10^4K $<$ T $<$ 3×10^5K) is contained in sheared magnetic boundary layers separating adjacent magnetic flux loops, and that this plasma is heated by the Joule dissipation of intense, field–aligned electric currents.

080.045 **Plasma heating by current sheets in solar active regions.**
B. Kliem, N. Seehafer.
International Conference on Mechanisms of Chromospheric and Coronal Heating, p. 564 – 566 (1991). – See Abstr. 012.011 for the main entry.

A comparison of thresholds and time scales of tearing, coalescence, and kinetic current–driven instabilities shows that current sheets which are continuously driven, e.g. by photospheric plasma motions causing continual energy input into the superphotospheric magnetic field, develop internal dynamics also for stationary (but not static) boundary conditions at high magnetic Reynolds numbers. This may explain the fragmentation of solar flare energy release (frequently observed at radio wavelengths and in hard X–rays). It is expected that the main part of released magnetic energy goes into turbulent motions and finally into heat in the vicinity of the current sheet.

080.046 **Chromosphere generation in magnetic flux–tubes.**
J. C. Hénoux, B. V. Somov.
International Conference on Mechanisms of Chromospheric and Coronal Heating, p. 567 – 569 (1991). – See Abstr. 012.011 for the main entry.

The localization of the magnetic field in concentrated flux–tubes leads to thermal structures in the upper solar atmosphere. There the difference in the perpendicular to magnetic field velocities of ions, electrons and neutrals, produces a privileged trapping of ions inside flux–tubes. The electric current required to confine these ions can be generated in the photosphere. This DC current, together with a 1 km s^{-1} plasma upflow in the upper photosphere, would generate a chromosphere inside the flux–tubes.

080.047 **Atmospheric heating in emerging flux regions.**
K. Shibata, S. Nozawa, R. Matsumoto, T. Tajima, A. C. Sterling.
International Conference on Mechanisms of Chromospheric and Coronal Heating, p. 609 – 614 (1991). – See Abstr. 012.011 for the main entry.

Using a two–dimensional MHD code, the authors study the nonlinear dynamics and associated heating of a magnetic loop emerging from the convection zone into the photosphere, chromosphere, and corona.

080.048 **A calibration of solar envelope models using the frequencies of intermediate–degree solar acoustic oscillations.**
A. A. Pamyatnykh, S. V. Vorontsov, W. Däppen.
Astron. Astrophys., Vol. 248, No. 1, p. 263 – 269 (Aug 1991).

The frequency dependence of the phase shift, which arises from the reflection of internal acoustic waves (p–modes) at the outer solar layers, can be determined from the accurately measured frequencies of intermediate–degree p–modes. This dependence is sensitive to the complicated structure of the outer solar layers, down to the depth of the second ionization zone of helium. It can be used to calibrate solar envelope models. The authors report the results of the direct tests of a variety of solar envelope models using the phase shift inferred from the solar oscillation frequencies.

080.049 **On high degree *f* modes.**
M. Gabriel.
Astron. Astrophys., Vol. 248, No. 1, p. 277 – 280 (Aug 1991).

The author shows that for spherical stars the equations of oscillation have terms which prevent the ratio of $\delta p/p$ to $\delta x/x$ to go to zero for high degree f modes. As a result there remains always a difference $\Delta\omega^2$ between the dimensionless eigenvalue of a spherical star and that of a plane atmosphere. $\Delta\omega^2$ is evaluated using two different approximations (Picard's and the perturbation methods). They both lead to values of $\Delta\omega^2$ of the order of a few units (while $\omega^2 \sim l$). The magnitude of $\Delta\omega^2$ can be used to probe the structure of the outer layers of the Sun as f modes contains at least as much information as low order p modes of high degree.

080.050 **Our Sun. II. Early mass loss of 0.1 $M_\odot$ and the case of the missing lithium.**
A. I. Boothroyd, I.–J. Sackmann, W. A. Fowler.
Astrophys. J., Vol. 377, No. 1, p. 318 – 329 (10 Aug 1991).

The authors have computed detailed solar models, taking early main–sequence mass loss into account. The authors have followed the nuclear destruction (and production) of ^{7}Li throughout the interior structure of the star during the entire evolution, keeping careful account of the changes in the surface lithium abundance due to mass loss and convection. Initial masses of M_i = 1.8, 1.5, 1.2, and 1.1 $M_\odot$ were considered, each with mass–loss time scales of τ_m = 0.16, 0.5, and 0.985 Gyr. A solar model of initial mass $M_i \approx 1.1\ M_\odot$ can solve the case of the missing solar lithium, producing the observed surface lithium depletion of a factor of 100. Furthermore, this lithium depletion is almost independent of the mass–loss time scale. With this mass loss of only $\Delta M \approx 0.1\ M_\odot$, there are only minor changes in the presolar helium abundance, the solar metallicity, the mixing–length parameter, and the predicted solar neutrino catpure rates. The authors estimate that this mass loss would cause the convective envelope to just reach the layers where ^{9}Be has been burned, resulting in little surface ^{9}Be depletion, in agreement with solar observations. Main–sequence mass loss reduces the required presolar helium abundance (which must always exceed the primordial abundance); this constrains the mass loss. The luminosity history of the M_1 = 1.1 $M_\odot$ case is different from that of the standard solar model, with an initial decline during the mass–loss period (rather than a steady brightening) but with smaller total variation.

080.051 **The quantities $D_{n,l}$ as measures of small frequency separations in the Sun and their origin.**
T. Van Hoolst, P. Smeyers.
Astron. Astrophys., Vol. 248, No. 2, p. 647 – 655 (Aug 1991).

The origin of the small frequency separations between 5–min solar oscillations of low degree l and radial order n and of low degree $L+2$ and radial order n–1 is examined.

080.052 **On generation of low-frequency pulsations in magnetic waveguides in the solar atmosphere.**
B. M. Nakaryakov, N. S. Petrukhin, S. M. Fajnshtejn.
Pis'ma Astron. Zh., Tom 17, No. 11, p. 1008 – 1012 (Nov 1991). In Russian. English translation in Sov. Astron. Lett., Vol. 17, No. 6.

An analytical conclusion of Schrödinger's non-linear equation describing the non-linear dynamics of the envelope of eigenmodes of a tube in the Sun's atmosphere within the ideal magnetic layer approximation is given. The generation of low-frequency pulsations arising as a result of instability of the packets of captured modes is investigated; estimates of the period of low-frequency pulsations for photospheric magnetic tubes are made.

080.053 **The primary results of research on solar oscillations by using magnetic field telescope.**
Zhang Heng, Cao Ying.
Publ. Yunnan Obs., No. 2, p. 1 – 9 (Jun 1991). In Chinese.

The time series data of λ5324.19 Å velocity field from Magnetic Field Telescope at Beijing Observatory have been analysed. Two (l–ν) spectra for zonal modes and sectorial modes were obtained respectively. Five-minute oscillation and the oscillation with period longer than ten minutes exist in both power spectra. From the power spectrum, the authors have recognized f mode, p_1 modes and p_2 modes with the degree between 10 and 60. Also both power spectra indicate the oscillations with period of 48.3 min and degree l equal to 12 and 20, they can be classfied into g modes with order n probably equal to 7 and 10 respectively.

080.054 **The correlation between the solar constant and the near-Earth space environment and Earth's rotation.**
Feng Bo.
Publ. Shaanxi Astron. Obs., Vol. 14, No. 1, p. 76 – 79 (Aug 1991). In Chinese.

The correlation analyses are made between the yearly mean values of the solar constant and the critical frequency of the F2 layers, the stratosphere temperature, the ground temperature, and the variation of the length of a day for the period of 1979 – 1987. The results show that the correlation coefficients are more than +0.8, except the correlation coefficient between the solar constant and the ground temperature is −0.7.

080.055 **Albedo gamma-rays from cosmic ray interactions on the solar surface.**
D. Seckel, T. Stanev, T. K. Gaisser.
AIP Conf. Proc., No. 220, p. 227 – 231 (Apr 1991). Current Physics Microform No.: 9106B1361. – See Abstr. 012.020 for the main entry.

The authors estimate the flux of gamma-rays that result from collisions of high energy galactic cosmic rays with the solar atmosphere. The high energy photon flux should be observable by the GRO, although details of the results are sensitive to assumptions about cosmic ray propagation through the magnetic fields of the inner solar system. The authors also estimate the neutrino and neutron fluxes resulting from the same process.

080.056 **A possible mechanism of energy transport of biennial oscillations in the Sun.**
E. I. Mogilevskij.
Soln. Dannye, Byull., No. 2/91, p. 93 – 94 (Feb 1991). Abstract. – See Abstr. 012.021 for the main entry.

080.057 **Seismology for the fine structure in the Sun's oscillations varying with its activity cycle.**
W. A. Dziembowski, P. R. Goode.
Astrophys. J., Vol. 376, No. 2, p. 782 – 786 (1 Aug 1991).

The authors inverted the symmetric part of the fine structure in the 1986 and 1988 solar oscillation data of Libbrecht and Woodard to find statistically significant evidence for a steady megagauss toroidal field at the bottom of the convective envelope. The analysis confirms the sizable amplitude of a cycle-dependent near-surface perturbation which appears to have its origin in the fibril field.

080.058 **Spin-(flavor) precession and short wavelength vacuum oscillation as a solution for the solar neutrino puzzle.**
P. K. Mohapatra.
Mod. Phys. Lett. A, Vol. 6, No. 37, p. 3467 – 3472 (7 Dec 1991).

The author investigates the possibility of spin-(flavor) precession combined with short wavelength vacuum oscillation as a solution for the solar neutrino puzzle. A large frozen-in magnetic field inside the sun with a neutrino magnetic moment of the order of 10^{-10} Bohr magneton can completely depolarize the ν_{eL} resulting in a factor of half of the emitted number. With a short wavelength vacuum oscillation and maximal mixing, the number of ν_{eL}'s reaching the earth is reduced by another factor of half; this explains the Homestake chlorine experiment. The difference between the Homestake and the Kamiokande-II experiments can be attributed to the contribution to the Cherenkov radiation in the latter through the neutral current and electromagnetic interactions of the components which are inert in the former.

080.059 **Solar atmospheric responses to magnetic flux eruptions. I. A mechanism for coronal mass ejections.**
Hu Youqiu, Zhang Huinan.
Chin. Astron. Astrophys., Vol. 15, No. 3, p. 284 – 291 (Sep 1991). English translation of Acta Astrophys. Sin., Vol. 11, No. 2, p. 181 – 186 (Apr 1991). See Abstr. 53.080.096.

The authors carried out a numerical simulation of atmospheric responses to magnetic flux eruptions from the base of a helmet streamer. The results show that the material originally contained in the streamer is compressed and lifted upward by the erupting magnetic field, forming a high density bright loop. At the same time, fast magnetosonic waves appear in front of the loop and a low density dark region, below it. These results suggest that magnetic flux eruption is probably an important triggering mechanism of loop-like coronal mass ejection.

080.060 **Solar atmospheric responses to magnetic flux eruptions. II. Slow shocks near the sun.**
Hu Youqiu, Zhang Huinan.
Chin. Astron. Astrophys., Vol. 15, No. 3, p. 292 – 297 (Sep 1991). English translation of Acta Astrophys. Sin., Vol. 11, No. 2, p. 188 – 193, (Apr 1991). See Abstr. 53.080.097.

The observed speeds of coronal mass ejections are often below the estimated Alfvén speed but above the sound speed for the background solar corona. This circumstance suggests that coronal mass ejections may form slow shocks in the corona. This paper presents a numerical simulation of the slow shock triggered by a coronal mass ejection, taking a coronal streamer with an equatorial current sheet as the background and a magnetic flux eruption at the base of the streamer as the driving mechanism. The results show that the current sheet has a significant influence on the structure of the shock wave. The slow shock occurs only on the outside of the current sheet and is gradually being transformed into a fast shock within the sheet. A fast magnetosonic wave exists in front of the slow shock, and it deflects the background magnetic field substantially and reduces the background density by more than 10% over there. The slow shock has a limited extent in latitude from the edge of the current sheet to ±20° and has a concave upward shape.

080.061 **The confrontation of mean-field theories with numerical simulations.**
F. Krause.
IAU Colloquium No. 130: The Sun and cool stars – activity, magnetism, dynamos, p. 3 – 14 (1991). – See Abstr. 012.024 for the main entry.

The investigation of magnetic phenomena on the Sun is confronted with the problem of turbulently moving electrically conducting media in rotating objects. One way of attacking this complicated problem, which was at first successful, was the development of mean-field magnetohydrodynamics, where the cooperative action of the small-scale turbulence was taken into account by certain average effects. The development of more and more powerful computers now offers possibilities of calculating the small-scale phemomena in a direct way. This paper is an attempt to compare these approaches, at least as far as there are comparable results.

080.062 **Magnetoconvection patterns in rotating convection zones.**
P. H. Roberts.
IAU Colloquium No. 130: The Sun and cool stars – activity, magnetism, dynamos, p. 37 – 56 (1991). – See Abstr. 012.024 for the main entry.

In addition to the well-known granulation and supergranulation of the solar convection zone (the "SCZ"), the presence of so-called "giant cells" has been postulated. These are supposed to span the entire thickness of the SCZ and to stretch from pole to pole in a sequence of elongated cells like a "cartridge belt" or a bunch of "bananas" strung uniformly round the Sun. Conclusive evidence for the existence of such giant cells is still lacking, despite strenuous observational efforts to find them. After analyses of sunspot motion, Ribes and others believe that convective motions near the solar surface occurs in a pattern that is the antithesis of the cartridge belt: a system of "toroidal" or "doughnut" cells, girdling the Sun in a sequence that extends from one pole to the other. Galloway, Jones and Roberts have recently tried to meet the resulting theoretical challenge, with the mixed success reported in this paper.

080.063 **Convective overshoot as a source of helicity.**
R. L. Jennings.
IAU Colloquium No. 130: The Sun and cool stars – activity, magnetism, dynamos, p. 62 – 66 (1991). – See Abstr. 012.024 for the main entry.

In order to understand what happens in the Sun when convection overshoots into the radiation zone an idealized model of penetrative convection with rotation is studied. The author highlights two properties of the model which occur with parameters relevant to the Sun. Firstly rotation allows motions to persist far beneath the convection zone, and secondly the profile of helicity with depth is dominated by a local maxima just beneath the convection zone. This second result has consequences in dynamo theory.

080.064 **The role of overshoot in solar activity: a direct simulation of the dynamo.**
A. Brandenburg, R. L. Jennings, Å. Nordlund, R. F. Stein, I. Tuominen.
IAU Colloquium No. 130: The Sun and cool stars – activity, magnetism, dynamos, p. 86 – 88 (1991). – See Abstr. 012.024 for the main entry.

The authors investigate convective overshoot in a layer of electrically conducting fluid. The radiative conductivity is assumed to be larger in the lower part of the layer which makes it stable to convective motions, yet penetrative convection from the upper layer can occur. The numerical resolution is 63^3 gridpoints. The authors observe a dynamo effect for magnetic Reynolds numbers around one thousand when a magnetic seed field is rapidly concentrated to form flux tubes. Later the average magnetic field is expelled from the convectively unstable regions, but it accumulates in the interface between the convection zone and the radiative interior.

080.065 **Rotational effects on Reynolds stresses in the solar convection zone.**
P. Pulkkinen, I. Tuominen, A. Brandenburg, Å. Nordlund, R. F. Stein.
IAU Colloquium No. 130: The Sun and cool stars – activity, magnetism, dynamos, p. 98 – 100 (1991). – See Abstr. 012.024 for the main entry.

Three-dimensional hydrodynamic simulations are carried out in a rectangular box. The angle between gravity and rotation axis is kept as an external parameter in order to study the latitude-dependence of convection. Special attention is given to the horizontal Reynolds stress and the Λ-effect (Rüdiger, 1989). The results of the simulations are compared with observations and theory and a good agreement is found.

080.066 **Waves of solar activity.**
M. R. E. Proctor, E. A. Spiegel.
IAU Colloquium No. 130: The Sun and cool stars – activity, magnetism, dynamos, p. 117 – 128 (1991). – See Abstr. 012.024 for the main entry.

The authors develop a theory of the sunspot cycle predicated on the assumption that the observed bands of activity are packets of dynamo waves. An approximate equation is proposed to describe the dynamics of these packets, using standard ideas from bifurcation theory. The authors show that in a certain limit the system can be described in terms of a slowly-evolving solitary wave, and that periodic behavior, like that of the observed butterfly diagram, can easily be found. Generalizations of the theory are discussed.

080.067 **The Sun's internal differential rotation from helioseismology.**
P. R. Goode.
IAU Colloquium No. 130: The Sun and cool stars – activity, magnetism, dynamos, p. 157 – 171 (1991). – See Abstr. 012.024 for the main entry.

Well-confirmed helioseismic data from several groups using various observational techniques at different sites have allowed to determine the differential rotation in the outer half of the Sun's interior. The resulting rotation law is simple – the surface differential rotation persists through much of the convection zone with a transition toward solid body rotation beneath. There is no appealing evidence for a rapidly rotating core. There is however, weak evidence for a solar cycle dependence of the Sun's internal rotation.

080.068 **Λ-effect, meridional flow and the differential solar rotation.**
G. Rüdiger, I. Tuominen.
IAU Colloquium No. 130: The Sun and cool stars – activity, magnetism, dynamos, p. 172 – 177 (1991). – See Abstr. 012.024 for the main entry.

Rotation laws for the solar convection zone are produced by the Λ-effect in rotating anisotropic turbulence fields. The authors use the structure of the turbulence parameters introduced by Rüdiger and Kichatinov (1990), based on a simplified turbulence model. Neglecting the meridional circulation, for small inverse Rossby numbers the angular velocity isolines are spherical, while for increasing inverse Rossby number they approach more and more the helioseismologically derived shape. This simple picture becomes more complicated if the meridional circulation is allowed to act as an angular momentum transporter.

080.069 **Differential rotation as an axisymmetric resonant mode of convection.**
K. L. Chan, H. G. Mayr.
IAU Colloquium No. 130: The Sun and cool stars – activity, magnetism, dynamos, p. 178 – 181 (1991). – See Abstr. 012.024 for the main entry.

080.070 **The solar internal rotation and its implications.**
L. Paternò.
IAU Colloquium No. 130: The Sun and cool stars – activity, magnetism, dynamos, p. 182 – 186 (1991). – See Abstr. 012.024 for the main entry.

The present internal rotation of the Sun as deduced from helioseismological data has many implications which concern the location of the solar cylce dynamo mechanism, core magnetism and history of the angular momentum. These aspects and their mutual connections are briefly discussed.

080.071 **The solar dynamo.**
A. Brandenburg, I. Tuominen.
IAU Colloquium No. 130: The Sun and cool stars – activity, magnetism, dynamos, p. 223 – 233 (1991). Also published as report NORDITA—91/7A (prepr.), 12. p. (1991). – See Abstr. 012.024 for the main entry.

The traditional $\alpha\Omega$-dynamo as a model for the solar cycle has been successful in explaining the butterfly diagram, phase relations between poloidal and toroidal field, and polar branch migration features. Observational and theoretical achievements in recent years have however shaken this picture. The current trend is towards dynamos operating in the overshoot region of the convection zone. Nevertheless, there are many open questions and a consistent picture has not been established. The authors compare recent approaches and discuss remaining problems.

080.072 **Solar internal rotation, the boundary layer dynamo and latitude distribution of activity belts.**
G. Belvedere, M. R. E. Proctor, G. Lanzafame.
IAU Colloquium No. 130: The Sun and cool stars – activity, magnetism, dynamos, p. 237 – 240 (1991). – See Abstr. 012.024 for the main entry.
The authors suggest that the latitude distribution of solar activity belts and the related equatorward or poleward migration of different tracers of the solar cycle are a natural consequence of the internal radial profile of angular velocity via the working of a dynamo in the boundary layer beneath the convection zone. This has been confirmed by the results of a non–linear dynamo model in a very thin spherical shell which show that dynamo action may reasonably take place in the boundary layer and reproduce the observed surface phenomenology.

080.073 **Solar rotation over solar cycle 21.**
E. Ribes, I. Vince, P. Mein, E. N. Ferreira.
IAU Colloquium No. 130: The Sun and cool stars – activity, magnetism, dynamos, p. 241 – 245 (1991). – See Abstr. 012.024 for the main entry.
Having measured the rotation rate of sunspots through solar cycle 21 the authors have found that the mean differential rotation averaged over a seven year record is similar to the grand average differential rotation determined by Howard et al. (1984) over the period 1921 – 1982. However, the rotation rate does change from year to year. These changes are evidenced by a steepening or a flattening of the mean differential rotation profile, as well as significant changes in the equatorial rate. The presence of a time–dependent pattern of azimuthal rolls inferred from the meridional circulation pattern of the sunspots offers a qualitative explanation of the observed rotation rates.

080.074 **Neutrino oscillations and solar models.**
C.–X. Chen, M. L. Cherry.
Astrophys. J., Lett., Vol. 377, No. 2, p. L105 – L108 (20 Aug 1991).
If neutrino oscillations in the solar interior are responsible for the suppressed ^{8}B flux measured by the Homestake ^{37}Cl experiment and the Kamiokande II water Cerenkov detector, then a comparison of the gallium, chlorine, and water results may make possible a determination of the neutrino mass difference and mixing angle. Present experimental results disfavor a neutrino oscillation solution with large neutrino mass differences, but cannot yet be used to choose among a variety of lower mass oscillation solutions and oscillations coupled with nonstandard solar models. In particular, although nonstandard solar models with elevated core temperatures have received little attention, there are theoretical justifications for such models, and the expected counting rates from such models coupled with neutrino oscillations can give results entirely consistent with the observations.

080.075 **The history of activity of the Sun and its rotational period on the ZAMS.**
K. Stępień.
IAU Colloquium No. 130: The Sun and cool stars – activity, magnetism, dynamos, p. 288 – 293 (1991). – See Abstr. 012.024 for the main entry.
The rotation period of the Sun after it reached ZAMS is estimated from the present rotation rate, average X–ray emission flux and average calcium emission flux. Taking into account all existing uncertainties it is concluded that this initial period was within the range 1 – 9 days, with the most probable value 2 – 3 days. Possible influence of the solar activity on evolution of life on the Earth is briefly discussed.

080.076 **Characteristics of intermediate–degree solar p–mode line widths.**
S. M. Jefferies, T. L. Duvall Jr., J. W. Harvey, Y. Osaki, M. A. Pomerantz.
Astrophys. J., Vol. 377, No. 1, p. 330 – 336 (10 Aug 1991).
The authors present measurements of the solar p–mode line widths over the frequency range $1.87 < \nu < 4.93$ mHz) and degree range $1 \leqslant l \leqslant 150$). These results are based on high quality spectra obtained from intensity observations made at the South Pole at a time of low solar activity (1987.9). The line width is observed to vary with mode frequency and degree. The frequency variation has been explained in terms of radiative and convective damping of the modes. The observed variation with degree (at constant ν) is shown to exceed the S^{-1} variation that is expected on theoretical grounds. Here S is the travel time of a mode from its lower turning point in the solar interior, to its reflection at the solar surface. These deviations may indicate the presence of two possible damping mechanisms, in addition to radiative and convective damping. The authors suggest that (1) the extra damping for high l–modes with $l \geqslant 25$ may be due to the absorption of mode power by active regions, and (2) the extra damping for low l–modes with $l \lesssim 40$ may be caused by an increased leakage of wave energy into the corona by modes that tunnel through the temperature minimum region.

080.077 **Toward a seismic model of the Sun.**
A. G. Kosovichev, A. V. Fedorova.
Astron. Zh., Tom 68, Vyp. 5, p. 1015 – 1029 (Sep–Oct 1991). In Russian. English translation in Sov. Astron., Vol. 35, No. 5.
New observational data of p–mode frequencies of intermediate degrees $l = 4 - 140$ are used to determine the radial structure of the Sun, assuming spherical symmetry and hydrostatic equilibrium. A solar model based on the solutions of the helioseismological inverse problem is constructed. The frequencies of the seismic solar model are closer to the observed ones as compared with those of a standard evolutionary model. The remaining frequency discrepancies are primarily related to the poorly determined structure of the subphotospheric zone of turbulent convection. The deviations of the seismic model from the standard model are suspected to be due to mixing in the solar core. The authors calculate a set of theoretical solar models assuming partial mixing in the core. By comparing the calculated models with the seismic model of the Sun the authors estimate parameters of mixing. A more accurate estimation of the solar core structure can be made using low–degree ($l = 0 - 3$) p–mode freuqencies. However, the observed frequencies of the modes are not consistent with the seismic model. In particular, the inconsistency is in a higher central density estimated from the data. It may be caused by systematic frequency shifts which were not taken into account in the observations.

080.078 **The solar–neutrino problem.**
V. de Sabbata, C. Sivaram.
Nuovo Cimento A, Vol. 104, No. 10, p. 1577 – 1579 (Oct 1991).
The problem of solar neutrinos is discussed in the case of a large magnetic field in the innermost region of the core ($<0.05\,R_\odot$) and of neutrinos with magnetic moment. Some predictions are made as regards experiments with ^{37}Cl and ^{71}Ga and of Kamiokande type.

080.079 **The role of low energy neutrinos in the solar neutrino problem.**
Y. Ono, D. Suematsu.
Phys. Lett. B, Vol. 271, No. 1/2, p. 165 – 171 (14 Nov 1991).
The authors study the time variation of a solar neutrino in the light of spin–flavour precession. It seems to exist parameter regions that can explain the different time variation features of the solar neutrino flux in the Homestake and Kamiokande–II experiments. It is emphasized that the intermediate–energy neutrinos in these parameter regions play an important role in the results of the Homestake experiment.

080.080 **On the matter–enhanced transitions of solar neutrinos in the absence of neutrino mixing in vacuum.**
M. M. Guzzo, S. T. Petcov.
Phys. Lett. B, Vol. 271, No. 1/2, p. 172 – 178 (14 Nov 1991).
An analytic description of the matter–enhanced two–neutrino transitions of the solar neutrinos in the case when the neutrino mixing is induced effectively in matter by neutrino flavour–changing neutral current interactions with the particles of matter is derived. A proof that the indicated mechanism of generation of neutrino mixing provides a possible solution of the solar neutrino problem is given.

080.081 **Search for neutrinos from the Sun using the reaction $^{71}Ga(\nu_e, e^-)^{71}Ge$.**
A. I. Abazov, O. L. Anosov, E. L. Faizov, V. N. Gavrin, A. V. Kalikhov, T. V. Knodel, I. I. Knyshenko, V. N. Kornoukhov (*V. N. Kornokhov*), S. A. Mezentseva, I. N. Mirmov, A. V. Ostrinsky, A. M. Pshukov, N. E. Revzin, A. A. Shikhin, P. V. Timofeyev (*P. V. Timofeev*), E. P. Veretenkin, V. M. Vermul, G. T. Zatsepin, T. J. Bowles, B. T. Cleveland, S. R. Elliott, H. A. O'Brien, D. L. Wark, J. F. Wilkerson, R. Davis Jr., K. Lande, M. L. Cherry, R. T. Kouzes.
Phys. Rev. Lett., Vol. 67, No. 24, p. 3332 – 3335 (9 Dec 1991).
Current Physics Microform No.: 9201E1320.
A radiochemical ^{71}Ga – ^{71}Ge experiment using 30 tons of gallium to determine the primary flux of neutrinos from the Sun has begun operation at the Baksan Neutrino Observatory in the USSR. Assuming that the extraction efficiency for ^{71}Ge atoms produced by solar neutrinos is the same as from natural Ge carrier, the authors observed the capture rate to be 20 [+15, –20](stat) ± 32(syst) solar neutrino units (SNU), resulting in a limit of less than 79 SNU (90% C.L.). This is to be compared with 132 SNU predicted by the standard solar model.

080.082 **Contribution and response functions for Ca II lines in different atmospheric models.**
S. A. Grigoryeva (*S. A. Grigor'eva*), I. P. Turova, R. B. Teplitskaya.
Sol. Phys., Vol. 135, No. 1, p. 1 – 13 (Sep 1991).
The authors have calculated profiles of five Ca II lines using three models of the solar atmosphere with different mean temperature gradients in the chromosphere. Contribution functions and response functions to temperature and velocity variations for the line centers and for the K_2 and H_2 emission peaks were found. Disturbances were specified locally at each point of the chromosphere, but for temperature disturbances the other thermodynamical parameter distributions were redetermined. Effective heights of the contribution and response functions, and also the width of the layer reacting to disturbances, depend substantially on the line strength and on the temperature gradient. Contribution curves and response functions to velocity variations differ drastically in form but, in spite of this, their centers of gravity and widths have a similar behaviour. The response function to temperature variations differs from the former two in all its characteristics.

080.083 **Recent results on the solar diameter.**
A. D. Wittmann, E. Alge, M. Bianda.
Sol. Phys., Vol. 135, No. 2, p. 243 – 248 (Oct 1991).
Using optically identical telescopes at different sites, the authors have measured the solar diameter with a drift–scan technique. The results are discussed. Although a small residual increase of the solar diameter during the last ten years seems to be indicated, it is concluded that most – if not all – of the observed variations are due to variable seeing conditions, and that there is still no conclusive evidence for a genuine solar variation with amplitudes in excess of about $\pm 0.3''$.

080.084 **Does the solar interior rotate rigidly?**
A. Gould.
Astrophys. J., Vol. 377, No. 2, p. 707 – 711 (20 Aug 1991).
If heat transport by weakly interacting massive particles (WIMPs) is the solution to the solar neutrino problem, these WIMPs would also be extremely efficient at transporting momentum. They would cause the region of the Sun interior to a critical radius, R_{crit}, to rotate in a distinctive fashion, usually rigidly. The critical radius may be as great as $\sim 0.5\,R_\odot$ for the lightest WIMPs, where $R_\odot$ is the solar radius. The boundary between the affected and unaffected regions would be extremely sharp, $\delta R_{crit} \lesssim 0.02\,R_\odot$. This characteristic rotation profile could be observed or ruled out, given sufficient improvement in the measurement of the azimuthal splitting of solar p–modes.

080.085 **Status and future of solar neutrino observations.**
K. Lande, B. Cleveland, T. Daily, R. Davis, J. Distel, C. K. Lee, A. Weinberger, P. Wildenhain, J. Ullman.
International School of Astroparticle Physics, p. 333 – 352 (1991). – See Abstr. 012.037 for the main entry.
The continuous twenty year record of these measurements of the solar neutrino flux with the Homestake chlorine detector, indicates that the average solar neutrino flux is 2.2 ± 0.3 SNU and that this ν_e flux appears to vary with the 11 year solar activity cycle. Higher ν_e fluxes are observed during solar quiet periods and lower ν_e fluxes during solar active periods. When the Homestake data is combined with the Kamiokande results, the region of overlap between the two experiments is for an observed to predicted ^8B neutrino flux ratio of about 0.4 and very little low energy neutrino flux. If the Kamiokande results are corrected for MSW effects, neutral current scatterings by non–electron neutrinos, then both the ^8B and the ^7Be fluxes are about 1/3 of the Standard Solar Model predictions.

080.086 **Seismic observations of the solar interior.**
D. Gough, J. Toomre.
Annu. Rev. Astron. Astrophys., Vol. 29, p. 627 – 685 (1991). – See Abstr. 003.010 for the main entry.
Contents: 1. Introduction. 2. Solar evolution. 3. Properties of modes. 4. Observational principles. 5. Inversion of data. 6. Inference of hydrostatic structure. 7. The neutrino problem. 8. Rotation and other subsurface flows. 9. Mode excitation and decay. 10. Solar cycle variations. 11. Asteroseismology.

080.087 **Observations of the phase differences between irradiances and velocity for low–degree solar acoustic modes.**
C. J. Schrijver, A. Jiménez, W. Däppen.
Astron. Astrophys., Vol. 251, No. 2, p. 655 – 668 (Nov 1991).
The authors derive phase differences and gains between broadband relative irradiance signals and full–disk velocity variations for $l = 0$, 1, and 2 modes with order ranging from 12 to 32 (1.8–4.6 mHz). They use irradiance measurements from the IPHIR instrument flown on the PHOBOS 2 mission to Mars. The data were obtained during the second half of 1988. The authors compare these data with contemporaneous measurements of photospheric velocities obtained at Tenerife during 4 intervals within the 155 d observation period of PHOBOS 2. They find that the phase difference between irradiance at 500 nm (5 nm FWHM) and velocity is roughly constant with an average value of $-119° \pm 3°$ from 2.5 up to about 4.3 mHz, with a slight increase suggested by the data running up to 4.6 mHz. The large effects of noise make the results below 2.5 mHz uncertain. It is also shown that the phase differences between the green (500 nm) and red (865 nm) channels, derived from a continuous data set covering 155 d, do not differ significantly from zero below 2.8 mHz, while a small difference of about $13° \pm 5°$ exists at higher frequencies, with the red signal ahead of the green. This jump appears to coincide with a jump in the separation between the $l = 0$ and $l = 2$ eigenfrequencies. The ratio of amplitudes between irradiance at 500 nm and velocity is $(0.228 \pm 0.015) \times 10^{-6}$ s cm^{-1}, and between the red and green irradiances 0.556 ± 0.011, both with no significant change as a function of frequency.

080.088 **Application of the continuous orthonormalization and adjoint methods to the computation of solar eigenfrequencies and eigenfrequency sensitivities.**
R. D. Rosenwald, G. F. Rabaey.
Astrophys. J., Suppl. Ser., Vol. 77, No. 1, p. 97 – 117 (Sep 1991).
Two new methods of analysis for solar seismology, called the continuous orthonormalization (CON) and adjoint methods, are

developed. Their use enables both solar eigenfrequencies and eigenfrequency sensitivities (partial derivatives with respect to solar model parameters) to be computed more accurately and efficiently than with existing methods. The CON method integrates an eighth-order nonlinear system of ordinary differential equations (ODEs) which defines the linear adiabatic nonradial oscillation modes of the Sun. All normal modes of oscillation are treated identically, regardless of their type (pressure, gravity, or fundamental) or their predominant location inside the Sun. The adjoint method integrates a related eighth-order linear inhomogeneous system of ODEs. From the resultant solution, an eigenfrequency's partial derivatives with respect to an extensive set of solar model parameters may be computed simultaneously. Numerical tests confirm the validity of the two new methods. Eigenfrequencies obtained via the CON method agree (at the 1% level) with the corresponding eigenfrequencies obtained via finite-difference or mesh approaches. Eigenfrequency sensitivities for the three different types of modes (pressure, gravity, and fundamental) obtained via the adjoint method are presented. The results clearly show the distinctive features of these three classes of modes.

080.089 **On the results of application of the nonadiabatic criterion of the stability of equilibrium.**
G. F. Sitnik.
Astron. Tsirk., No. 1544, p. 27 – 28 (May – Jul 1990). In Russian.
Using the nonadiabatic criterion of the stability of equilibrium, the possibility of development of convection and of its basis oscillations with periods of 2 – 3 minutes, in the stable (according to Schwarzschild) layers of the photosphere is shown. For certain initial conditions, outflows of matter into the chromosphere and corona are also possible.

080.090 **The Sun as seen by a resonant scattering spectrometer.**
H. B. van der Raay.
Astron. Astrophys., Vol. 252, No. 1, p. 366 – 368 (Dec 1991).
When observing the rotating solar disk surface with a high-resolution spectrometer, a wavelength-positional correlation exists which results in a weighting of the contributions to the observed signal over the projected solar image dependent on the relative line-of-sight velocity between the observer and the solar surface. Hence when determining solar oscillations from measurements based on two differing line-of-sight velocities the phase difference of the observed mode at the two positions of observation can lead to confusing results especially for the higher order modes.

080.091 **Observational constraints on theories of the solar differential rotation.**
B. R. Durney.
Astrophys. J., Vol. 378, No. 1, p. 378 – 397 (1 Sep 1991).
Helioseismic observations are providing increasingly reliable information on the dependence of the inner solar angular velocity, Ω, on the spatial coordinates. The azimuthal momentum equation in the solar convection zone (SCZ) depends on Ω, $<u_r u_\varphi>$, $<u_\theta u_\varphi>$ and on the meridional motions. Here **u** designates the turbulent convective velocities, and the brackets denote an appropriate average. The meridional motions can be decomposed into flows with one, two, three, ... cells per hemisphere characterized by their stream functions $\psi_2(r)$, $\psi_4(r)$, $\psi_6(r)$, ... The velocity correlations, $<u_i u_j>$, can be evaluated with the help of a formalism for differential rotation developed previously, if the length $l = l_r = \tau <u_r{}^2>^{1/2}$ and the ratios l_r/l_φ and l_θ/l_φ (designated by **RR**) are known. Here τ and l_i stand for the eddy lifetime and eddy dimensions, respectively. The formalism, without determining these ratios, provides strong clues about their dependence on $\alpha = 4\Omega^2 T/g(\nabla \Delta T)$, where T, g, $\nabla \Delta T$ are the temperature, gravity, and superadiabatic gradient. The dimensionless parameter α estimates the effect of rotation on convection. For all realistic values of **RR**, $\psi_2(r)$ is negative, i.e., the flow rises at the equator and sinks at the poles. It is shown that such flows together with the fact that $<u_r u_\varphi>$ is negative allow for a natural explanation of the helioseismic observations near the solar surface (with increasing depth, the angular velocity first increases and then decreases).

080.092 **The effect of the Mihalas, Hummer, and Däppen equation of state and the molecular opacity on the standard solar model.**
Y.-C. Kim, P. Demarque, D. B. Guenther.
Astrophys. J., Vol. 378, No. 1, p. 407 – 412 (1 Sep 1991).
Improvements to the Yale Rotating Stellar Evolution Code by incorporating the Mihalas – Hummer – Däppen equation of state, an improved opacity interpolation routine, and the effects of molecular opacities, calculated at Los Alamos, have been made. The effect of each of the improvements on the standard solar model has been tested independently by computing the corresponding solar nonradial oscillation frequencies. According to these tests, the Mihalas – Hummer – Däppen equation of state has very little effect on the model's low l p-mode oscillation spectrum compared to the model using the existing analytical equation of state implemented in the code. On the other hand, the molecular opacity does improve the model's oscillation spectrum. The effect of molecular opacity on the computed solar oscillation frequencies is much larger than that of the Mihalas – Hummer – Däppen equation of state. Together, the two improvements to the physics reduce the discrepancy with observations by 10 μHz for the low l modes.

080.093 **The depth of the solar convection zone.**
J. Christensen-Dalsgaard, D. O. Gough, M. J. Thompson.
Astrophys. J., Vol. 378, No. 1, p. 413 – 437 (1 Sep 1991).
The transition of the temperature gradient between being subadiabatic and adiabatic at the base of the solar convection zone gives rise to a clear signature in the sound speed. Helioseismic measurements of the sound speed therefore permit a determination of the location of the base of the convection zone. The authors have tested two techniques by applying them to artificial data, obtained by adding simulated noise to frequencies computed from two different solar models. The determinations appear to be relatively insensitive to uncertainties of the physics of the solar interior, although, if present, a large-scale megagauss magnetic field or some other phenomenon not normally incorporated into theoretical models at the base of the convection zone might have a significant effect on the results. On the assumption that this is not the case, the authors conclude, from an analysis of observed frequencies of solar oscillation, that the depth of the solar convection zone is (0.287 ± 0.003) solar radii.

080.094 **Internal solar rotation.**
D. O. Gough.
NATO Advanced Research Workshop on Angular Momentum Evolution of Young Stars, p. 271 – 279 (1991). – See Abstr. 012.055 for the main entry.
The helioseismic determination of the angular velocity distribution within the Sun is discussed. In particular, the bulk of the radiative interior of the Sun is rotating at more-or-less the same rate as the photosphere. The latitudinal variation of angular velocity appears to change quite abruptly immediately beneath the base of the convection zone. Associated with that change is no doubt a meridional circulation which must be confined to only a thin layer; the photospheric lithium and beryllium abundances set constraints on the possible extent of substantial flow. There is slight evidence that the angular velocity changes with the solar cycle.

080.095 **Dynamics of spot groups and rotation of the external convective layers in the Sun and magnetically active stars.**
A. F. Lanza, M. Rodonò, R. A. Zappalà.
NATO Advanced Research Workshop on Angular Momentum Evolution of Young Stars, p. 289 – 296 (1991). – See Abstr. 012.055 for the main entry.
The observed kinematics of sunspot groups and a simplified model, which describes the coupling between magnetized and unmagnetized plasmas, are used to derive some information on

the rotation of the subphotospheric layers of the Sun. A comparison with present theoretical models and an extrapolation to some aspects of the phenomenology observed in RS CVn binaries are presented.

080.096 **The internal rotation of the Sun: implications on the history of its angular momentum.**
L. Paternò.
NATO Advanced Research Workshop on Angular Momentum Evolution of Young Stars, p. 357 – 364 (1991). – See Abstr. 012.055 for the main entry.

Recent helioseismological data have revealed that the internal rotation of the Sun may not differ very much from that observed on its surface. This implies that the angular momentum loss must be compatible with a slowing–down of the entire Sun. If we look at the rotation of solar type stars we cannot avoid considering the difficulty of reconciling the initial sharp spin–down in a very short time scale with the subsequent soft one leading to the present angular velocity of the Sun, unless some very effective internal instability mechanism takes place. This may involve the presence of magnetic fields in the Sun's core and is related to the location of the activity cycle dynamo mechanism.

080.097 **Magnetic fields and motions in the solar atmosphere.**
S. I. Gopasyuk.
Kinematika Fiz. Nebesn. Tel, Tom 7, No. 5, p. 3 – 10 (Sep–Oct 1991). In Russian. English translation in Kinematics Phys. Celest. Bodies, Vol. 7, No. 5.

The results of observations of magnetic field structure and motions in the active region of the Sun are analysed in the review. Typical classes of motions are derived. A connection of motions with magnetic field structures and electric currents is established. The importance of the magnetic field structure in formation of some typical plasma motions is indicated. Similarity of the magnetic field structure and relations of plasma motion with the magnetic field in active and undisturbed regions is noted.

080.098 **Models for temperature distribution in the solar atmosphere.**
B. Alam.
Indian J. Radio Space Phys., Vol. 20, No. 2, p. 135 – 139 (Apr 1991).

A temperature model for the transition region – inner corona is deduced by assuming a constant conductive flux from higher temperature region towards the lower temperature region. Moreover, using the observed temperatures, a general model is also fitted, which appears to represent the whole temperature regime of the solar atmosphere. These models are then compared with earlier models. Besides, a comparative study of all of these models (earlier and present ones) with various sets of parameters is also presented. The temperature distribution given by all these models do not differ much from each other. Consequently, one from each region, i.e. TR2 for the transition region – inner corona and OC2 for the outer corona, may be chosen as representative models. The combination of these two models then represents the whole solar atmosphere.

080.099 **Construction of a seismic model of the Sun.**
A. G. Kosovichev, A. V. Fedorova.
Sov. Astron., Vol. 35, No. 5, p. 507 – 514 (Sep 1991). Current Physics Microform No.: 9203X1681. English translation of Astron. Zh., Tom 68, Vyp. 5, p. 1015 – 1029 (1991).

New observational data on the frequencies of p modes of natural solar oscillations of intermediate degrees $l = 4 - 140$ are used to determine the radial structure of the Sun, assuming spherical symmetry and hydrostatic equilibrium. On the basis of the solutions of the inverse problem of solar seismology obtained, an improved solar model is constructed, the oscillation frequencies of which agree considerably better with observations than the standard evolutionary model. The remaining discrepancy is associated mainly with the uncertain structure of the subphotospheric turbulent convection zone. A likely reason for this departure of the internal solar structure from the standard model may be mixing of matter in the central core. The authors calculate a class of theoretical models with partial mixing in the core. The degree of mixing is estimated by comparing them with the seismic model of the Sun. A more accurate determination of the parameters of the solar core is possible from the frequencies of modes of low degree, $l = 0 - 3$. The available observations are not consistent with the authors' seismic model, however, as manifested particularly in a higher density at the center of the Sun reconstructed from these data. This may be related to systematic frequency shifts that were neglected in the observations.

080.100 **Early solar mass loss: a potential solution of the weak sun paradox.**
T. E. Graedel, I.–J. Sackmann, A. I. Boothroyd.
Geophys. Res. Lett., Vol. 18, No. 10, p. 1881 – 1884 (Oct 1991).

The weak sun paradox, in which early planetary temperatures are known to have remained above the freezing point of water despite presumably lower early solar luminosity, has customarily been resolved by imposing high early CO_2 atmospheric abundances. Speculative new solar models incorporating early solar mass loss provide a possible solution to the case of "missing" solar lithium; at the same time, they predict higher early (4.5–3.8 Gyr BP) solar luminosities. These higher luminosities have the potential to produce planetary temperatures within the liquid water range indicated by observations without requiring extremely high CO_2 concentrations.

080.101 **Concentrated magnetic fields in the solar atmosphere.**
M. Schüssler.
2. International Workshop on Plasma Physics: Plasma Theoretical Problems in Astro– and Fusion Physics, p. 1 – 8 (1991). – See Abstr. 012.067 for the main entry.

The solar atmosphere provides the opportunity to study magnetohydrodynamic and plasma processes on scales and under circumstances inaccessible to experimental investigation on Earth. In this contribution the author gives a summary of the present knowledge about magnetic elements, the small–scale structures into which most of the magnetic flux in the solar photosphere is concentrated. They represent an example for the formation of dissipative structures in systems far from thermal equilibrium. He discusses in particular the observationally determined properties of magnetic elements as well as theoretical concepts and results concerning their formation, stationary state, instabilities and decay.

080.102 **Gravitational settling in solar models.**
C. R. Proffitt, G. Michaud.
Astrophys. J., Vol. 380, No. 1, p. 238 – 250 (10 Oct 1991).

The authors present calculations of the effects of gravitational settling and atomic diffusion in solar models, with the goal of constraining the actual amount of gravitational settling that has occurred in the Sun. Most of the discrepancies between the results of previous workers are explained. The authors use the observed solar Li abundance to constrain the amount of turbulent mixing below the solar convection zone and thereby set a lower limit to the amount of gravitational settling in the Sun. Of the total increase in the central He mass fraction during evolution to the solar age, only about 0.013 out of the total increase of ≈ 0.37 is due to gravitational settling. The gravitational settling of metals can, by changing the opacity, have structural effects that are almost as large as the changes resulting from the gravitational settling of He alone. If diffusion is added to evolutionary models without changing any of the physical parameters (such as α and the initial He abundance), the total change in the central He abundance, as compared with a model without diffusion, is much larger than when both models are constrained to reproduce the solar luminosity and radius at a given age. Gravitational settling reduces the surface He abundance by roughly 10% below its initial value. The predicted reduction in the heavy element abundance is somewhat smaller but is also subject to greater uncertainties. If the turbulent mixing needed to account for the current solar Li abundance is included, the change in the surface abundances are moderated, and the abundance profiles below the convection zone are modified. However, turbulent mixing below the surface convection zone cannot reduce the amount of surface settling by more than a factor of 2. Larger amounts of turbulence would cause the destruction of too much Li and/or Be.

080.103 **Real-time, directional measurement of ^{8}B solar neutrinos in the Kamiokande II detector.**
K. S. Hirata, K. Inoue, T. Ishida, T. Kajita, K. Kihara, M. Nakahata, K. Nakamura, S. Ohara, N. Sato, Y. Suzuki, Y. Totsuka, Y. Yaginuma, M. Mori, Y. Oyama, A. Suzuki, K. Takahashi, M. Yamada, M. Koshiba, K. Nishijima, T. Suda, T. Tajima, K. Miyano, H. Miyata, H. Takei, Y. Fukuda, E. Kodera, Y. Nagashima, M. Takita, K. Kaneyuki, T. Tanimori, E. W. Beier, L. R. Feldscher, E. D. Frank, W. Frati, S. B. Kim, A. K. Mann, F. M. Newcomer, R. van Berg, W. Zhang.
Phys. Rev. D, Vol. 44, No. 8, p. 2241 – 2260 (15 Oct 1991). Current Physics Microform No.: 9112A1165.
The method of ^{8}B solar neutrino measurement by means of the reaction $\nu_e e \rightarrow \nu_e e$ in the Kamiokande II detector is described in detail. A data sample of 1040 live detector days in the time period January 1987 through April 1990 yields a clear directional correlation of the solar neutrino – induced electron events with respect to the Sun and a measurement of the differential electron energy distribution. The measured flux of ^{8}B solar neutrinos from the subsamples of 450 days at electron energy threshold $E_e \geqslant 9.3$ MeV, and 590 days at $E_e \geqslant 7.5$ MeV, relative to calculations of the ^{8}B flux based on the standard solar model, are 0.46 ± 0.05(stat)± 0.06(syst) times the prediction of Bahcall and Ulrich, and 0.70 ± 0.08(stat)± 0.09(syst) times the prediction of Turck-Chièze et al. The shape of the recoil electron energy distribution is consistent with that expected from the product of the known shape of the neutrino flux from ^{8}B β decay and the cross section for $\nu_e e \rightarrow \nu_e e$ scattering. Within the statistical error, there is no evidence in the solar neutrino signal for a significant time variation.

080.104 **Reconciling the time dependence of the chlorine and Kamiokande solar neutrino data.**
K. S. Babu, R. N. Mohapatra, I. Z. Rothstein.
Phys. Rev. D, Vol. 44, No. 8, p. 2265 – 2277 (15 Oct 1991). Current Physics Microform No.: 9112A1189.
The authors present a detailed analysis of the effect of a large transition magnetic moment of the electron neutrino (i.e., $\mu_{\nu e} \approx 10^{-11}\mu_B$) on the solar neutrino flux measured in the chlorine and Kamiokande experiments. For $\Delta m_{\nu\mu} - \nu_e{}^2 \approx 10^{-8}$eV^2 and $\mu_\nu \approx 10^{-11}\mu_B$, the combination of matter effects and spin precession can explain the apparent anticorrelation of the neutrino flux with the sunspot activity in the chlorine experiment, as well as the lack of time variation in the Kamiokande experiment. For this choice of parameters, the authors predict that the gallium experiment should show strong anticorrelation with the solar activity, with a maximum flux of 80 SNU during the sunspot minimum

080.105 **Low-l p-mode solar eigenfrequency measurements from the Birmingham Network.**
Y. Elsworth, R. Howe, G. R. Isaak, C. P. McLeod, R. New.
Mon. Not. R. Astron. Soc., Vol. 251, No. 1, p. 7P – 9P (1 Jul 1991).
The authors present the results of observations of oscillations of the Sun over a period of eight years from 1981.

080.106 **Oscillations in the solar atmosphere: the 5-min peak as a consequence of wave reflection at the photosphere.**
G. Worrall.
Mon. Not. R. Astron. Soc., Vol. 251, No. 3, p. 427 – 437 (1 Aug 1991).
The vertical propagation of compression waves from beneath the solar atmosphere is considered. It is shown that at the step-up in critical frequency which occurs at the photosphere, the phase difference between reflected and incident waves is strongly dependent on frequency, and is zero at about 3 mHz. This can explain why velocity power spectra deduced from photospheric spectral lines, have maxima at that frequency, period 5 min, if all modes are excited to equal velocity amplitudes, that is, to equal energy density, in the region just beneath the photosphere. Higher in the atmosphere, there is a step-down in critical frequency into the corona, the whole constituting a quasi-rectangular barrier in critical frequency. Standing-wave components can be set up, by partial reflection, at frequencies higher than the top of the barrier which is at about 4.5 mHz. The first of these has a period of about 200 s, and may be responsible for the peak observed at that period in power spectra from chromospheric spectral lines. For waves of frequency higher than 4.5 mHz, the relatively cool layer between the photosphere and corona acts as an acoustic interference filter which is most transparent at those frequencies at which standing waves are set up within it.

080.107 **Convective structures in the Sun.**
G. W. Simon, N. O. Weiss.
Mon. Not. R. Astron. Soc., Vol. 252, No. 1, p. 1P – 5P (1 Sep 1991).
Observed patterns of motion at the solar surface reflect the structure of subphotospheric convection, which controls the distribution of angular momentum and magnetic fields. The authors interpret mesogranules as secondary features associated with supergranular circulation, and suggest that coupling between mesogranules and granulation triggers the spasmodic formation of exploding granules. Supergranules are expected to generate isolated sinking plumes. The authors argue that these plumes can penetrate to the base of the convective zone, and that there is no organized structure on larger scales.

080.108 **A search for dark matters in the Kamiokande II.**
Y. Suzuki.
IUPAP Conference on Primordial Nucleosynthesis and Evolution of Early Universe, p. 363 – 369 (1991). – See Abstr. 012.071 for the main entry.
A search has been made for high energy neutrino events from the Sun in the Kamiokande II detector as annihilation products of galactic cold dark matters. The author found no excess neutrino events from the Sun and sets limits on the mass and the density of the dark matter candidates.

080.109 **Molecules in the Sun.**
K. Sinha.
Proc. Astron. Soc. Aust., Vol. 9, No. 1, p. 32 – 36 (1991). – See Abstr. 012.078 for the main entry.
In view of the already available reviews on the subject, the progress after 1972 is summarised. Newly identified species and molecular transitions are listed. How the studies of solar molecules can be helpful in understanding the solar atmosphere, in structuring better models and in deriving molecular parameters with some degree of confidence, is stated.

080.110 **s-process yields from type II supernovae: contributions to the solar composition from iron to zirconium.**
C. M. Raiteri, M. Busso, R. Gallino, G. Picchio.
ESO/EIPC Workshop on Supernova 1987A and other Supernovae, p. 709 – 713 (Jun 1991). – See Abstr. 012.075 for the main entry.

080.111 **Neutrino properties from SN 1987A and from future observations of GSC.**
A. Dar.
Vulcano Workshop 1988: Frontier objects in astrophysics and particle physics, p. 67 – 101 (1989). – See Abstr. 012.079 for the main entry.
The early neutrino signal reported by the Mont Blanc group on Feb 23, 1987 was not produced by a neutrino burst from SN 1987A. The second neutrino burst that was reported by the KAMIOKANDE II, the IMB and the Baksan groups is consistent with the general theoretical picture of supernovae explosions, but it also has unexpected features. Analysis of the neutrino burst yields new limits on the lifetime, mass, mixing, and electric charge of the ν_e, and new limits on the magnetic moment and the radiative lifetime of light neutrinos. These limits imply that neither decay, nor mixing, nor electric charge, nor magnetic

moment of the ν_e are responsible for the solar neutrino problem. Similar limits on the properties of the ν_μ and the ν_τ require the observation of a gravitational stellar collapse in the Galaxy with more advanced neutrino telescopes which are under construction. Such limits, when obtained, will resolve the question whether relic neutrinos from the big bang can close the universe.

080.112 **Effects of opacity and equation of state on solar structure and oscillations.**
J. A. Guzik, A. N. Cox.
Astrophys. J., Vol. 381, No. 1, p. 333 – 340 (1 Nov 1991).
The evolution and p–mode oscillation frequencies of several solar models are presented, highlighting the effects of uncertainties in the opacity and equation of state. The authors have incorporated the Däppen, Mihalas, Hummer, and Mihalas ("MHD") equation of state tables into their evolution and oscillation calculations, and have compared the results with those of models calculated using the analytical Iben equation of state. The nonadiabatic low–degree p–mode oscillation frequencies derived from models using either of these equations of state agree with observations to within a few microhertz. The MHD equation of state, with more detailed treatment of envelope ionization regions, improves agreement with observations for p–modes of degree l = 40 – 300. For higher degree p–modes, observational uncertainties make if difficult to determine how much improvement the MHD equation of state provides over the analytical Iben procedure. The authors incorporated small opacity increases (10% – 15%) in the $2 - 5 \times 10^6$K temperature region below the bottom of the convection zone, as indicated by the new Iglesias and Rogers Livermore opacities, to adjust the convection zone depth and fine–tune agreement between calculated and observed p–mode frequencies.

080.113 **The Sun and asteroseismology.**
T. M. Brown.
Symposium in Celebration of the 50. Anniversary of McDonald Observatory of the University of Texas at Austin: Frontiers of stellar evolution, p. 139 – 183 (1991). – See Abstr. 012.085 for the main entry.
The study of p–mode oscillations on the Sun has become a powerful means of probing the interior of the nearest star. A natural extension of these studies would be to examine similar oscillations on more distant stars, a process that would be part of the somewhat broader field of asteroseismology. The author reports on the current status and future prospects for such studies on stars of roughly solar type. Although there is not yet general agreement that p–modes have been detected on any such star other than the Sun, there is good evidence that such detections will occur in the near future. Moreover, recent theoretical work has begun to show what could be learned if precision studies of oscillations could be made on other stars. Taken together, these recent advances suggest that this area of asteroseismology may add in significant ways to our knowledge of stellar structure and evolution.

080.114 **Results of a study of the spectral composition of direct solar radiation at several sites in Mongolia.**
G. Bastukh, G. F. Sitnik, D. Khaltar.
Tr. Gos. Astron. Inst. im. Shternberga, Tom 62, p. 253 – 267 (1991). In Russian.
Some results of direct solar radiation spectral measurements are given for several sites in Mongolia. Statistical study of direct solar radiation in 4 spectral regions was made for different solar altitudes and spectral transmission. Coefficients allowing to determine the fraction of direct solar radiation in those spectral regions in the integral flux are estimated.

080.115 **Solar irradiance variability from modern measurements.**
C. Fröhlich, P. V. Foukal, J. R. Hickey,
H. S. Hudson, R. C. Willson.
The Sun in time, p. 11 – 29 (1991). – See Abstr. 003.027 for the main entry.
Direct measurements from satellites of the solar "constant" – the total irradiance at mean Sun–Earth distance – during more than ten years show variations over time scales from minutes to years and decades. At high frequencies solar oscillations contribute to the variance. The most important influences are related to solar activity: during the passage of active regions on the solar disk (sunspots and faculae) changes of a few 0.1% lasting for several days are observed. The effect of spots can be well reproduced by the projected sunspot index, whereas the influence of faculae have to be modeled from proxy data like the Ca–K plage index or the He I index. Long–term trends have also been detected which are connected to the 11–yr solar activity cycle showing a difference between the maximum and minimum of nearly 0.1% with the irradiance being higher during solar maximum. This modulation can be modeled by proxy data as the He I index or the 10.7–cm flux indicating the importance of the magnetic network. Moreover, it can also be explained by observed temperature variations with latitude over the solar cycle. Whether the latter is physically linked to the explanation by a varying bright network is not yet clear.

080.116 **The variability of the solar diameter.**
E. Ribes, B. Beardsley, T. M. Brown, P. Delache,
F. Laclare, J. R. Kuhn, N. V. Leister.
The Sun in time, p. 59 – 97 (1991). – See Abstr. 003.027 for the main entry.
The authors discuss possible variations of the solar radius, which are of interest both because of their astrophysical significance, and because of their potential relation to the terrestrial climate. Observations of the solar radius span more than three centuries, and display considerable variation. The authors argue that most of this dispersion arises because the various observation methods do not directly measure a true radius, e.g., the radius of a surface of given optical depth. Rather, they measure properties of the Sun's limb darkening function, and are affected by many sources of degradation of the solar image. Those uncertainties make interpretation of such observations difficult. For these reasons, the authors discuss mainly long series of visual observations made by a single observer, or recent photoelectric observations. They find evidence for periodicities in the apparent radius that occur in both modern and historical records. The magnitude of the observed variations is quite different in visual and photoelectric observations, suggesting that the process responsible for the periodicities is either one that modifies the solar limb darkening function, or one that causes systematic variations in image blurring by the Earth's atmosphere. A connection between solar magnetic activity and apparent radius seems likely, with evidence for such a relation dating back as far as the Maunder minimum. Interesting questions that cannot yet be answered concerning the apparent radius variations include whether they have any connection with a deep–seated unknown phenomenon or with the recently observed time–varying asphericity of the Sun, or with the quasi–biennial oscillation seen in the Earth's stratospheric circulation.

080.117 **Theory of luminosity and radius variations.**
H. C. Spruit.
The Sun in time, p. 118 – 158 (1991). – See Abstr. 003.027 for the main entry.
Theoretical calculations of luminosity and radius variations are reviewed. As sources for these variations, two kinds of effects are considered: (i) temporary conversion of thermal energy inside the convection zone into a different energy form (such as magnetic fields); and (ii) the effect of surface magnetic fields on the energy emissivity of the solar surface.

080.118 **Standard solar models.**
D. S. P. Dearborn.
The Sun in time, p. 159 – 174 (1991). – See Abstr. 003.027 for the main entry.
Solar/stellar comparisons provide improved understanding of stellar astronomy, cosmology and various problems in physics. These and the physical processes included in stellar evolution calculations are presented, along with those of nonstandard phenomena. The bases of physical models are discussed in

conjunction with observational data by which solar models and their uncertainties may be judged.

080.119 **Large solar flares: analysis of the events recorded by the Mont Blanc neutrino detector.**
M. Aglietta, G. Badino, G. Bologna, C. Castagnoli, A. Castellina, V. L. Dadykin, W. Fulgione, P. Galeotti, F. F. Kalchukov (*F. F. Khalchukov*), I. V. Korolkova (*I. V. Korol'kova*), P. V. Kortchaguin (*P. V. Korchagin*), V. A. Kudryavtsev, A. S. Malguin (*A. S. Mal'gin*), L. Periale, V. G. Ryassny (*V. G. Ryasnyj*), O. G. Ryazhskaya, O. Saavedra, G. Trinchero, S. Vernetto, V. F. Yakushev, G. T. Zatsepin.
Astrophys. J., Vol. 382, No. 1, p. 344 – 348 (20 Nov 1991).
The results of the analysis of events recorded by the Mont Blanc neutrino detector during the occurrence of 19 large solar flares, in the period 1988 August – 1990 March (including the two powerful flares on 1989 September 29 and October 19) are reported and discussed. No significant neutrino signal was found in time coincidence with solar flares. Upper limits (90% C.L.) obtained on the integral neutrino and antineutrino flux of different flavors are $F(\bar{\nu}_e) < 1.3 \times 10^{10} cm^{-2}$ [$10 < E_\nu < 100$ MeV], $F(\nu_e) < 2.3 \times 10^{10} cm^{-2}$ ($25 < E_\nu < 100$ MeV), $F(\nu_\mu) < 3.8 \times 10^{11} cm^{-2}$ ($20 < E_\nu < 100$ MeV), and $F(\bar{\nu}_\mu) < 4.0 \times 10^{11} cm^{-2}$ ($20 < E_\nu < 100$ MeV). In particular, the ν_e flux is considerably lower than the flux required to explain the excess of ν_e captures observed by the Homestake ^{37}Cl solar neutrino experiment during runs 19, 27, and 71, when large solar flares occurred.

080.120 **The global Sun.**
J.–C. Pecker.
Solar interior and atmosphere, p. 1 – 30 (1991). – See Abstr. 003.028 for the main entry.
After a definition of the various terms used to identify the solar layers, from the center to the exterior, and a physical description of these layers, the author shows that various couplings are controlling the physics in the core and the outer layers, and even the planets. One of these couplings is between convection, rotation and magnetism (the dynamo), and another coupling is between solar activity and planetary physics. These couplings allow to use observed data (oscillations, neutrinos, emergence of active regions, and of course their evolution) to infer properties of the solar interior. The theoretical knowledge of the Sun must take into account the existence of these couplings, as well as the existence of another type of coupling, the one that links the past and the present states of our Sun.

080.121 **Nuclear energy generation in the solar interior.**
P. D. MacD. Parker, C. E. Rolfs.
Solar interior and atmosphere, p. 31 – 50 (1991). – See Abstr. 003.028 for the main entry.
The present status of our knowledge of nuclear energy generation in the solar interior is reviewed, together with a discussion of the role of the solar neutrino measurements as tests of our understanding of the solar interior. The theory of nuclear reaction rates is reviewed, including the effects of screening. Recent energy production rates are given for the many reactions of the p–p chains as well as the CNO cycles. The current disagreement between observations and predictions for the solar neutrino output does not seem to be any fault of nuclear reaction uncertainties.

080.122 **Solar neutrino experiments.**
R. Davis Jr., A. N. Cox.
Solar interior and atmosphere, p. 51 – 85 (1991). – See Abstr. 003.028 for the main entry.
Observational methods for detecting solar neutrinos are reviewed, and the current results presented. Four operating systems, the Homestake Mine ^{37}Cl radiochemical method, the real–time neutrino scattering detector at Kamiokande II, in Japan, and two radiochemical gallium–germanium experiments are discussed in some detail. The results from these operating detectors are compared to solar model calculations and various neutrino oscillation processes that could alter the primary electron neutrino flux from the Sun. It is of great interest to study the solar neutrino particle spectrum with a number of detectors that have a potential for exploring the solar energy processes and the interactions of neutrinos. A few old and new solar neutrino detectors are reviewed that have promise for revealing new information about the neutrinos from the Sun.

080.123 **Theory of neutrino oscillations and the Sun.**
S. P. Rosen.
Solar interior and atmosphere, p. 86 – 111 (1991). – See Abstr. 003.028 for the main entry.
This chapter describes the enhancement by matter of neutrino oscillations with small mixing angles and indicates how this phenomenon can be used to solve the solar neutrino problem. There are three possible solutions to the ^{37}Cl experiment and they yield different predictions for the Kamiokande II solar neutrino–electron scattering experiment and for the ^{71}Ga experiment. Present data tend to favor the nonadiabatic solution but do not exclude the large–angle one. They do, however, definitely disfavor the adiabatic solution.

080.124 **The computation of standard solar models.**
R. K. Ulrich, A. N. Cox.
Solar interior and atmosphere, p. 162 – 191 (1991). – See Abstr. 003.028 for the main entry.
Procedures for calculating standard solar models with the usual simplifying approximations of spherical symmetry, no mixing except in the surface convection zone, no mass loss or gain during the solar lifetime, and no separation of elements by diffusion are described. The standard network of nuclear reactions among the light elements is discussed including rates, energy production and abundance changes. Several of the equation of state and opacity formulations required for the basic equations of mass, momentum and energy conservation are presented. The usual mixing–length convection theory is used for these results. Numerical procedures for calculating the solar evolution, and current evolution and oscillation frequency results for the present Sun by some recent authors are given.

080.125 **Recent development in solar convection theory.**
K. L. Chan, Å. Nordlund, M. Steffen, R. F. Stein.
Solar interior and atmosphere, p. 223 – 274 (1991). – See Abstr. 003.028 for the main entry.
In recent years, the theory of solar (and stellar) convection has made fundamental advances due to the increasing cost effectiveness of supercomputers and the constant improvement of numerical techniques. It is expected that the numerical approach will become a dominant trend for the future. The authors report on these new advances. References to theoretical studies on phenomena related to solar convection are compiled. The authors then discuss three numerical studies of solar convection in greater detail, so as to provide the readers with some general understanding of the numerical techniques being used and the results obtained: The discussion starts with a two–dimensional study of the spectroscopic properties of solar granules. While the two–dimensional limitation is severely detrimental to some important hydrodynamical processes, it is both economical and able to provide some initial understanding of the gross features of solar convection. Next they discuss the testing of the well–known mixing–length theory with three–dimensional numerical experiments. An example is also given of applying the numerically gained knowledge to analytical study, in this case the behavior of compressible convection as a heat engine. The third case describes a realistic, three–dimensional simulation of solar granulation; many observational features of solar granules are faithfully reproduced. It is the most sophisticated numerical calculation of this sort today.

080.126 **The solar dynamo.**
E. E. DeLuca, P. A. Gilman.
Solar interior and atmosphere, p. 275 – 303 (1991). – See Abstr. 003.028 for the main entry.
The authors discuss the present state of our understanding of the origin of the Sun's magnetic field. They begin with an

introduction to the theory of magnetic field generation in rotating, conducting fluids. Next they consider a dilemma that has persisted for some 15 yr, namely the inconsistency of the kinematic and dynamic convection zone dynamo models. A resolution of this dilemma has been suggested by the recent helioseismology observations on the rotation rate as a function of latitude and depth. These observations, together with other observational and theoretical constraints, suggest that the solar dynamo operates not in the convection zone, but rather in a thin layer between the convection zone and the radiative interior. A model of such a dynamo is presented and discussed. The authors conclude by discussing the problems posed by the placement of the dynamo below the convection zone.

080.127 **Oscillation observations.**
F. Hill, F.-L. Deubner, G. Isaak.
Solar interior and atmosphere, p. 329 - 400 (1991). - See Abstr. 003.028 for the main entry.

This chapter reviews recent observations of solar oscillations. The oscillations discussed are global and local 5-min p-modes, the 160-min oscillation and oscillations in the solar atmosphere. Experimental and data reduction methods are described. Summaries of recent results are provided, including measurements of frequencies, amplitudes, line widths and splittings. Other topics include active-region tomography, solar cycle changes, the chromospheric cavity and diameter measurements.

080.128 **Theory of solar oscillations.**
J. Christensen-Dalsgaard, G. Berthomieu.
Solar interior and atmosphere, p. 401 - 478 (1991). - See Abstr. 003.028 for the main entry.

This chapter provides theoretical background for understanding the properties of solar oscillations and their relation to the structure of the Sun. Although accurate numerical computations are required to match the precision of the observed frequencies, a great deal of insight can be obtained from considering simple asymptotic approximations. In particular, the wide range of modes observed allows a separation of the effects of different parts of the Sun; this allows inversion of the observed frequencies to determine localized properties of the solar interior. Also, changes in the assumed physics of solar models, within the range of uncertainty in the physics, lead to frequency changes that are substantial compared with the intrinsic precision of the observations. Thus as probes of the solar interior, the frequencies are both sensitive and straightforward to interpret.

080.129 **The solar rotation.**
K. G. Libbrecht, C. A. Morrow.
Solar interior and atmosphere, p. 479 - 500 (1991). - See Abstr. 003.028 for the main entry.

The authors review the rotation of the Sun, with emphasis on the observations. While it has been known for centuries that at the solar surface the equator rotates with a higher angular velocity than the poles, different surface measurements suggest that the near-surface rotation is not as simple as one might expect. Recently we have been able to infer the interior rotation rate from helioseismology. These new observations indicate that surface-like rotation extends through the convection zone, changing to solid-body rotation in the upper part of the radiative interior; observations are inconclusive regarding the rotation of the deep core of the Sun. The dynamics that produces the observed solar rotation profile remains a mystery.

080.130 **The internal rotation and magnetism of the Sun from its oscillations.**
W. A. Dziembowski, P. R. Goode.
Solar interior and atmosphere, p. 501 - 518 (1991). - See Abstr. 003.028 for the main entry.

The study of solar oscillations has revealed knowledge of the internal rotation of the Sun and something of its internal magnetic field. The authors present the formalism needed to determine the internal rotation from oscillation data. Equations are developed that describe centrifugal distortion and results are given. They sketch the formalism required to treat poloidal and toroidal magnetic fields inside the Sun. Results are presented for a toroidal field concentrated near the base of the convection zone and for assumed relic poloidal and toroidal fields in the deep interior.

080.131 **The inversion problem.**
D. O. Gough, M. J. Thompson.
Solar interior and atmosphere, p. 519 - 561 (1991). - See Abstr. 003.028 for the main entry.

The authors discuss some fundamental problems associated with inverting helioseismic data to infer properties of the solar interior, and review various methods used to invert frequencies of normal modes of oscillation. After discussing linear methods, the authors present and compare different asymptotic methods that have been used to infer the internal solar sound speed. They also discuss numerical inversions for solar structure, and address the issue of inverting for more than one function. Methods of inverting for the nonspherically symmetric structure of the Sun are presented.

080.132 **Solar gravity modes.**
H. Hill, C. Fröhlich, M. Gabriel, V. A. Kotov.
Solar interior and atmosphere, p. 562 - 617 (1991). - See Abstr. 003.028 for the main entry.

The internal gravity modes, or g-modes, of the Sun may be important both in affecting the internal structure of the Sun and in furnishing a diagnostic probe of the solar interior. The internal structure could be altered by core mixing due to unstable g-modes or by nonlinear effects due to large-amplitude g-modes located in the core. On the other hand, small-amplitude g-modes operating in the linear regime offer the possibility of studying in detail the present state of internal solar structure such as the internal rotation rate, the Brunt-Väisälä frequency, the speed of sound and the mean molecular weight. In all of these roles, the solar g-modes may be important. The g-modes may be a contributor to the solar neutrino paradox and/or they may be a source of information leading to a more complete understanding of the physics responsible for the paradox. The observational work on solar g-modes does not exclude any of these possibilities at this time.

080.133 **Oscillation mode excitation.**
A. N. Cox, S. M. Chitre, S. Frandsen, P. Kumar.
Solar interior and atmosphere, p. 618 - 660 (1991). - See Abstr. 003.028 for the main entry.

The excitation of the oscillation modes in the Sun is very different from that for the previously known variable stars. A review of the normal pulsation mechanisms seen in many classes of variable stars shows that they actually are operating in the Sun. However, most, but not all, studies of the solar mode excitation predict that radiative damping, and damping by convective processes also, overwhelm the driving to stabilize all radial and nonradial modes. This is in accord with the observations that frequently show measurable widths of the lines in the p-mode oscillation spectrum. These line widths indicate mode lifetimes of days for the p-modes. Most calculations predict that solar g-modes are stable, leading to the question of how they then can ever be observed. However, there is a possibility that low-degree and low-order g-modes could be just slightly unstable. Improvements to the predictions of lowest-order p-mode excitation by the inclusion of better radiative intensity formulations at the top of the convection zone and in the photosphere indicate even more mode stability. Calculations that show how convection may drive solar p-modes are presented. Arguments about three-mode couplings that are not strongly damping lead to the conclusion that the solar p-modes are probably stochastically driven by coupling with convection. Current disagreements among authors are discussed.

080.134 **Large–scale velocity fields.**
R. F. Howard, L. L. Kichatinov, R. S. Bogart, E. Ribes.
Solar interior and atmosphere, p. 748 – 778 (1991). – See Abstr. 003.028 for the main entry.

Results in the area of solar rotation over the past few years are reviewed. Considerable effort has gone into rotation determinations from tracers – especially sunspots in recent years. The present status of our knowledge of solar supergranulation, mesogranulation and giant–scale convection is briefly reviewed. Theoretical suggestions are discussed which may serve to reconcile theory and observations of giant–scale solar convection. Global meridional circulation in the photosphere is suggested by models of the solar rotation. The solar rotation shows a latitude and cycle dependence. The effect is clearly seen in the magnetic tracers (sunspot data) as well as the spectroscopic observations, although the latter are affected by systematic errors which cannot easily be removed from a real solar signal. This variability can be understood in terms of a time–dependent convective toroidal roll pattern. The connection of the rolls with rigidly rotating layers favors a solar dynamo located below the convective zone, possibly in the radiative interior which rotates like a solid body. Such a new picture of the rotation and convection has to be confirmed. Undoubtedly helioseismology will further add to our knowledge of the solar rotation in the near future.

080.135 **Sun–as–a–star spectrum variability.**
W. Livingston, R. F. Donnelly, V. Grigoryev (*V. Grigor'ev*), M. L. Demidov, J. Lean, M. Steffen, O. R. White, R. L. Willson.
Solar interior and atmosphere, p. 1109 – 1160 (1991). – See Abstr. 003.028 for the main entry.

The Sun is observed as a star in order to determine luminosity change, detect minute variability in average granulation and facular signals, and to use as a standard against which other stars may be compared. In this regard, topics discussed include: total irradiance variability as measured from space by the Activity Cavity Radiometer Irradiance Monitor and Earth Radiation Budget radiometers; Fraunhofer line heights of formation and examples of their variability in visible wavelengths; ultraviolet and extreme ultraviolet irradiance variability as observed from space; the magnetic origin of irradiance change; and the observed mean magnetic field of the Sun.

080.136 **Post–main–sequence solar evolution.**
P. Demarque, D. B. Guenther.
Solar interior and atmosphere, p. 1186 – 1223 (1991). – See Abstr. 003.028 for the main entry.

The authors' objective is to discuss some current research frontiers in solar–stellar physics. First, they describe and update the standard evolution of a 1–$M_\odot$ star from the zero–age main sequence to the helium flash. Then, they focus on two areas of solar–stellar research which are presently in a state of rapid development. These two areas are: (1) the theory of solar and stellar nonradial acoustic (p–mode) oscillations, which can be used to test the pressure stratification in the stellar interior; and (2) the theory of the evolution of rotating stars, which predicts the evolution of surface rotational velocities and the mixing of chemical elements to the surface. The authors note in addition that oscillations and rotation are closely interrelated, because each individual p–mode is also split in frequency by internal rotation in the region through which it propagates. Observations of p–mode splittings thus can provide information on the state of rotation of the solar and stellar interiors.

080.137 **Solar interior models.**
J. A. Guzik, Y. Lebreton.
Solar interior and atmosphere, p. 1235 – 1248 (1991). Appendix B. – See Abstr. 003.028 for the main entry.

The authors present the interior properties of four solar models from the recent astrophysical literature. The first three are standard solar models, each with somewhat different input physics. The fourth is an unconventional model designed to address the solar neutrino problem, with reduced central opacity to simulate the effects of weakly interacting massive particles.

080.138 **Solar p–mode frequencies.**
P. L. Pallé.
Solar interior and atmosphere, p. 1249 – 1255 (1991). Appendix C. – See Abstr. 003.028 for the main entry.

The author presents some mathematical formulae which have been derived from observed frequencies and which allow the reader, not only to reproduce the observed frequencies accurately, but also to extrapolate them to limits beyond present observations.

080.139 **The surface lithium and beryllium depletions of the Sun.**
Xiong Darun, M. Kondo, W. Unno.
Acta Astron. Sin., Vol. 32, No. 4, p. 333 – 342 (Dec 1991). In Chinese.

The surface lithium and beryllium depletions have been calculated for three models of the solar convective envelope. The results of the numerical calculations show that the surface beryllium has no significant depletion, but the surface lithium is depleted too quickly for the model consistent with the helioseismological data. This means that the depth of the convective overshooting zone is overestimated. A reasonable explanation may be that there is a magnetic field of about 10^3 – 10^4 gauss in the lower convection zone. The magnetic field will brake extensive convective overshooting. Therefore, the surface lithium depletion will slow down. Other evidences of a magnetic field in the lower convection zone of the Sun are discussed.

080.140 **A torsional oscillation in the rotation of the solar magnetic field.**
H. B. Snodgrass.
Astrophys. J., Lett., Vol. 383, No. 2, p. L85 – L87 (20 Dec 1991). With plate L3.

The pattern rotation rate for the Sun's magnetic field, determined by cross–correlating Mount Wilson full disk λ5250.2 (Fe I) magnetograms spaced a full solar rotation apart, closely parallels at all latitudes the photospheric plasma rotation profile determined from the Doppler shifts of the same spectral line. When an 11 yr running mean is subtracted, a torsional oscillation is revealed, in the form of an equatorward – migrating pattern of fast and slow zones. Although the magnetic rotation torsional pattern is similar enough to its much–studied Doppler counterpart to provide confirmation, there are significant differences between the two – the magnetic pattern is strongest (~ 20 m s^{-1}) at high latitudes, weakens at sunspot latitudes where the Doppler pattern is strongest, and is offset at all latitudes by $\sim 10°$ toward the equator, so that its slow zones approximately coincide with the maximal shear zones of the Doppler pattern. These zones appear to be forerunners to the wings of the magnetic flux (sunspot) butterflies of the activity cycle.

080.141 **Do changes in the photospheric magnetic network cause the 11 year variation of total solar irradiance?**
P. Foukal, K. Harvey, F. Hill.
Astrophys. J., Lett., Vol. 383, No. 2, p. L89 – L92 (20 Dec 1991). With plate L4.

Changes in the area of the photospheric magnetic network over the sunspot cycle have been put forward as the "missing component" required to explain the 11 yr variation of total solar irradiance observed by spaceborne radiometers. The authors show that this explanation is consistent with recent measurements of the photometric contrast of magnetic faculae and with their measurement of the network area change during cycle 21.

080.142 **Solar oscillations and the physics of the solar interior.**
J. Christensen–Dalsgaard.
Conference on Challenges to Theories of the Structure of Moderate–Mass Stars, p. 11 – 36 (1991). – See Abstr. 012.096 for the main entry.

The variety and precision of the observed frequencies of solar oscillations allow a detailed investigation of the foundation of computations of stellar models. This is of obvious importance to the understanding of the structure of other stars of moderate mass. Furthermore, it may be hoped to use the solar data to obtain information about the basic properties of matter under solar conditions. To exploit the solar data it is crucial to understand and limit the uncertainties in the computation of solar models and frequencies. In addition, it is necessary to explore the sensitivity of the models to the properties of the physics of the solar interior and to the simplifying assumption made in the calculations. The author discusses numerical tests of the calculations, including comparisons of independently computed models. Furthermore, he presents results of modifying the assumed physics, in particular the equation of state and the opacity. At the level of the observational precision the frequencies are highly sensitive to even quite subtle details of the physics.

080.143 **Helium gravitational settling below the solar convection zone.**
S. Vauclair, C. Charbonnel.
Conference on Challenges to Theories of the Structure of Moderate-Mass Stars, p. 37 – 41 (1991). – See Abstr. 012.096 for the main entry.

Helium gravitational settling time scales have been computed below the solar convection zone in evolutionary models. Helium could be depleted in the solar outer layers by as much as 20% in the standard models. However the necessity of explaining also the lithium destruction in the Sun and solar type stars reduces the expected helium depletion to a few percent.

080.144 **Influence of the upper layers of the sun on the *p*-mode frequencies.**
J. Christensen-Dalsgaard, F. Pérez Hernández.
Conference on Challenges to Theories of the Structure of Moderate-Mass Stars, p. 43 – 50 (1991). – See Abstr. 012.096 for the main entry.

The authors consider the effect of uncertainties in solar models of the *p*-mode frequencies of moderate degree ($40 \leqslant l \leqslant 100$). This is done by means of an asymptotic relation that provides a simple way of estimating differences in *p*-mode frequencies of two solar models. Also, the authors have investigated the possibility for separating different contributions to the difference between the models, on the basis of the frequencies of the modes. Analysis of observed frequencies suggests that the solar envelope helium abundance Y_s is close to 0.25; however, this result may be sensitive to other uncertainties in the solar models.

080.145 **Accuracy tests for the computation of solar models.**
M. Gabriel.
Conference on Challenges to Theories of the Structure of Moderate-Mass Stars, p. 51 – 55 (1991). – See Abstr. 012.096 for the main entry.

The author discusses the choice of the difference scheme and of the variables. He also shows that a slight modification to the usual difference method allows a significant increase of the accuracy and the computation of a solar model with a reasonable number of meshpoints.

080.146 **Comparison of solar models with Los Alamos and Livermore opacities.**
I. W. Roxburgh.
Conference on Challenges to Theories of the Structure of Moderate-Mass Stars, p. 57 – 59 (1991). – See Abstr. 012.096 for the main entry.

Models of the present Sun were computed using the Los Alamos and the new Livermore opacities. The relative abundance of heavy elements was the Grevesse mixture and the ratio Z/X was taken as 0.02765. The models were iterated to have $R = R_{sun}$ and $L = L_{sun}$ at an age of $t_o = 4.6 \times 10^9$years. The differences between the two models are presented.

080.147 **Helioseismic inversion.**
M. J. Thompson.
Conference on Challenges to Theories of the Structure of Moderate-Mass Stars, p. 61 – 80 (1991). – See Abstr. 012.096 for the main entry.

The application of inverse theory enables to use the measured frequencies of normal modes of oscillation of the Sun to make inferences about the structure and dynamics of the solar interior. Some basic concepts and a few inversion techniques are discussed. Results on solar rotation and radial stratification are reviewed, and finally the potential for investigating other moderate-mass stars seismologically is considered.

080.148 **An inversion for the rotation rate in the solar interior.**
J. Schou.
Conference on Challenges to Theories of the Structure of Moderate-Mass Stars, p. 81 – 86 (1991). – See Abstr. 012.096 for the main entry.

The results of an inversion of a set of preliminary rotational splittings from 1986 and 1988 from Big Bear Solar Observatory are presented. The splittings and the results of the inversions for the 1986 case are essentially identical to those in Christensen-Dalsgaard and Schou (1986), and there appear to be no significant differences between the two years. The results from both years seem to confirm the earlier result that the rotation rate is surface-like and constant on radii in the part of the convection zone that is resolved by this mode set and independent of both radius and latitude in the radiative interior. Since the observations used here only span degrees from 5–60 it has not been possible to perform an inversion for the rotation rate inside approximately 0.4 $R_\odot$ and in the outer part of the convection zone, and it is therefore not possible to see whether there is a bump in the rotation rate near the surface, as indicated by Hill et al. (1988). There are, however, indications that the rotation rate is not constant throughout the convection zone.

080.149 **The role of *f* modes in the inversion of high-*l* rotational splittings.**
D. A. Haber, F. Hill, J. Toomre.
Conference on Challenges to Theories of the Structure of Moderate-Mass Stars, p. 87 – 92 (1991). – See Abstr. 012.096 for the main entry.

The contribution of the solar *f* modes of oscillation to the inversion of high-degree rotational splitting data is examined. The authors find that the *f* modes play an important role in such inversions as revealed by the magnitude of their weighting coefficients c_i. This may be attributed to the single-peaked structure of the *f*-mode kernels, which is similar to that of the desired averaging kernels and in contrast to the many-peaked shape of the higher-order kernels. The high weight placed on the *f* modes in the inversions raises issues for observational techniques since the *f* modes possess modest power levels and their detection is influenced by the choice of spatial filtering.

080.150 **Implication of the *l* dependence and the frequency dependence of long-term frequency variation.**
H. Shibahashi.
Conference on Challenges to Theories of the Structure of Moderate-Mass Stars, p. 101 – 110 (1991). – See Abstr. 012.096 for the main entry.

The author shows that the *l* dependence of the frequency variation between 1986 and 1988 reported by Libbrecht and Woodard (1990) is useful to infer how deeply the outer layer of the sun has been changed in this period. Based on an asymptotic formula, the *l* dependence of the frequency change is derived in the case that the structural change is limited only in the region where the acoustic cut-off frequency dominates over the Lamb frequency. By comparing the formula with the observed *l* dependence of frequency variation, the author sets constraints on the extent of the structural change in the outer layer of the sun in this period. He derives furthermore a formula to invert the frequency dependence of the frequency variation. By applying this formula to the observed frequency variation between 1986 and 1988, he concludes that the structural change in the outer layer of the sun in this period is confined within the hydrogen ionization zone and the layer 200 km below the photosphere.

080.151 **A new inversion for the hydrostatic stratification of the Sun.**
W. Däppen, D. O. Gough, A. G. Kosovichev, M. J. Thompson.
Conference on Challenges to Theories of the Structure of Moderate–Mass Stars, p. 111 – 120 (1991). – See Abstr. 012.096 for the main entry.
Inversions for the spherically symmetric component of the hydrostatic stratification of the Sun are presented. These employ the Backus–Gilbert optimally localized averaging procedure applied to oscillation multiplet frequencies in the range 1.5–3 mHz of modes with $4 \leqslant l \leqslant 140$ determined by Libbrecht et al. (1990) from observations carried out in 1986. The authors also obtain an estimate of the helium abundance in the solar convective envelope.

080.152 **Prospects of solar quantum chaos.**
J. Perdang.
Conference on Challenges to Theories of the Structure of Moderate–Mass Stars, p. 135 – 140 (1991). – See Abstr. 012.096 for the main entry.
The authors have carried out a preliminary statistical analysis of the spacings of the observed solar p modes (Libbrecht data). They find some evidence that the distribution of the spacings violates the Poisson distribution, which should hold under conditions of spherical symmetry of the equilibrium state of the Sun. The relevance of this results in connection with inversion problems is briefly emphasised.

080.153 **A new solar cycle model including meridional circulation.**
Y.–M. Wang, N. R. Sheeley Jr., A. G. Nash.
Astrophys. J., Vol. 383, No. 1, p. 431 – 442 (10 Dec 1991).
The authors present a kinematic model for the solar cycle which includes not only the transport of magnetic flux by supergranular diffusion and a poleward bulk flow at the Sun's surface, but also the effects of turbulent diffusion and an equatorward "return flow" beneath the surface. The rotational shearing of a subsurface poloidal field generates toroidal flux that erupts at the surface in the form of bipolar magnetic regions. The large–scale toroidal field is destroyed both by "unwinding" as the local poloidal field reverses its polarity, and by diffusion as the toroidal flux is transported equatorward by the subsurface flow and merged with its opposite hemisphere counterpart. The inclusion of meridional circulation allows stable oscillations of the magnetic field, accompanied by the equatorward progression of flux eruptions, to be achieved even in the absence of a radial gradient in the angular velocity. The authors describe an illustrative case in which a subsurface flow speed of order 1 m s^{-1} and subsurface diffusion rate of order 10 $km^2 s^{-1}$ yield 22 yr oscillations in qualitative agreement with observations.

080.154 **Theoretical spectrum of solar convection.**
H. M. Antia, S. M. Chitre.
Conference on Challenges to Theories of the Structure of Moderate–Mass Stars, p. 157 – 162 (1991). – See Abstr. 012.096 for the main entry.
The theoretical power spectrum of solar velocity field and flux variation is obtained with amplitudes of the unstable convective eigenmodes determined by linearly superposing these modes to yield the model convective flux profile. The numerical results are compared with the observed spectrum of solar convection and the technique is applied to other stars.

080.155 **Convection and its influence on oscillations.**
R. F. Stein, Å. Nordlund.
Conference on Challenges to Theories of the Structure of Moderate–Mass Stars, p. 195 – 212 (1991). – See Abstr. 012.096 for the main entry.
The authors investigate the interaction between p–mode oscillations and convection using a realistic, three–dimensional simulation of the upper solar convection zone. These oscillations are excited at the eigenfrequencies of the simulation volume. Their frequency is different than that found from one–dimensional mixing length models. Their resonant cavity becomes larger when overshooting into the photosphere is possible, which lowers the mode frequencies, while interaction with the inhomogeneities in the sound speed and the motions generated by the convection tends to raise the mode frequencies. The modes are excited stochastically by non–adiabatic fluctuations in the gas pressure caused by the switch from convective to radiative energy transport at the solar surface.

080.156 **The vibrational stability of the Sun.**
N. J. Balmforth, D. O. Gough.
Conference on Challenges to Theories of the Structure of Moderate–Mass Stars, p. 221 – 224 (1991). – See Abstr. 012.096 for the main entry.
The authors studied the pulsational stability of solar acoustic oscillations in an earlier investigation (Balmforth and Gough, 1990). To take some account of the interaction between convection and pulsation they used a time–dependent formulation of local mixing–length theory developed by Gough (1977). Unfortunately, in the linear stability analysis, this theory leads to unrealistic spatial oscillations of the linear eigenfunctions in the lower layers of the convection zone where the thermal stratification is nearly adiabatic. The theory also predicts excessively sharp resonances in the interaction between the pulsation and the convection. This artificially increases the sensitivity of the growth rate to variations in frequency, and degrades any comparison with the observations of solar oscillation line widths. To alleviate these problems the authors have incorporated a nonlocal formulation of the mixing–length theory into the time–dependent formalism, using the Eddington approximation in the manner of Spiegel (1963).

080.157 **The effects of magnetic stretching in the convective overshoot region.**
G. H. Fisher, A. N. McClymont, Chou Deanyi.
Conference on Challenges to Theories of the Structure of Moderate–Mass Stars, p. 233 – 240 (1991). – See Abstr. 012.096 for the main entry.
The authors consider the fate of a magnetic flux tube lying initially near the bottom of the solar convective overshoot region, which is the weakly stable boundary larger at the base of the convection zone perturbed by convective motions. Stretching of the flux tube, for instance by differential rotation, decreases its density, causing it to rise quasi–statically until it reaches the top of the overshoot region and enters the buoyantly unstable convection zone. The authors refer to this as "destabilization" of the flux tube.

080.158 **Simulation of the effect of active regions on intermediate–degree solar oscillations.**
D. A. Haber, F. Hill.
Conference on Challenges to Theories of the Structure of Moderate–Mass Stars, p. 253 – 258 (1991). – See Abstr. 012.096 for the main entry.
The effect of active regions on the solar oscillations is simulated by artifically suppressing the velocities in full–disk quiet–sun Doppler images. An image of the active Sun is used as a template to mask the regions within which the velocities are reduced by as much as 35%. Two–dimensional l–ν power spectra from both the quiet Sun and the masked data are compared. The authors find that the addition of one large active region does not significantly alter the power in the oscillations, but the presence of several simulated regions decreases the power in the global modes by up to 25%, in agreement with observations. The shapes of the oscillation spectral lines do not substantially change, suggesting that frequencies measured from these spectra will not be significantly altered.

080.159 **Effects of spatial filtering on high–l power spectra and rotational splitting inversions.**
D. A. Haber, F. Hill, J. Toomre.
Conference on Challenges to Theories of the Structure of Moderate–Mass Stars, p. 259 – 264 (1991). – See Abstr. 012.096 for the main entry.
The effects of the spatial filtering algorithm on the results of an inversion of high–degree solar oscillation data are examined. In

attempting to isolate sectoral modes, the Fourier transforms in longitude are accompanied by three different spatial filterings in the orthogonal direction. The filters are: simple averaging in the plane of the sky, averaging along lines of constant heliographic longitude with uniform weighting, and another with weighting based on appropriate Legendre functions. The choice of spatial filter changes the distribution of power in the two-dimensional spectrum of the oscillations, particularly in the f and p ridges. These variations in the power distribution affect the determination of the rotational splittings, thereby influencing the results of inversions which emphasize information contained in the low-order ridges.

080.160 **Solar oscillation ring diagrams from Mt. Wilson full-disk magneto-optical Dopplergrams.**
F. Hill, E. J. Rhodes Jr., S. G. Korzennik, A. Cacciani, T. M. Brown.
Conference on Challenges to Theories of the Structure of Moderate-Mass Stars, p. 271 - 275 (1991). - See Abstr. 012.096 for the main entry.

Three-dimensional power spectra of solar oscillations have been computed from moderate-resolution full disk Doppler images obtained with the Magneto-Optical Filter at Mt. Wilson. Slices of the spectra at constant frequency reveal the ring structures that are analogous to the ridges in two-dimensional spectra. Ring diagrams obtained at different heliographic positions show large differences in the structure of the rings. These variations can be attributed to the changing effective spatial resolution of the observations across the disk. After correction for this effect, and for terrestrial seeing, the rings will be used to map the horizontal flows in the convection zone as a function of position and depth.

080.161 **Measurement of intermediate- and high-degree ($20 < l < 600$) p-mode solar oscillation power and energy.**
E. J. Rhodes Jr., T. M. Brown, A. Cacciani, S. G. Korzennik, R. K. Ulrich.
Conference on Challenges to Theories of the Structure of Moderate-Mass Stars, p. 277 - 284 (1991). - See Abstr. 012.096 for the main entry.

The authors present measurements of the total model power and energy of both intermediate- and high-degree ($20 < l < 600$) solar p-mode oscillations which have been corrected to first order for the combined effects of atmospheric seeing, image motion due to imperfect tracking, and the point spread function of the optics. These power and energy estimates have been obtained from an average of 20 separate zonal l-ν power spectra, which were obtained from observations obtained at the 60-Foot Solar Tower of the Mt. Wilson Observatory between July 1 and 20, 1988.

080.162 **Further evidence for radial variations in the solar equatorial angular velocity profile.**
E. J. Rhodes Jr., A. Cacciani, S. G. Korzennik, R. K. Ulrich.
Conference on Challenges to Theories of the Structure of Moderate-Mass Stars, p. 285 - 292 (1991). - See Abstr. 012.096 for the main entry.

One year ago the authors presented evidence that the rotationally-induced frequency splitting of intermediate- and high-degree sectoral p-mode oscillations vary systematically as a function of the degree of the modes. In this paper they present the results of several additional analyses of their 20-day long time series of Dopplergrams which they have carried out since the meeting held in December 1989 in Hakono.

080.163 **Chromospheric and coronal activities in the quiet Sun originating from photospheric 5-minute oscillations.**
Y. Suematsu, A. Takeuchi.
International SOLAR-A Science Meeting: Flare physics in solar activity maximum 22, p. 259 - 263 (1991). - See Abstr. 012.104 for the main entry.

The authors numerically investigated the non-linear response of the solar atmosphere to 5-min oscillations, assuming that atmospheric motions are restricted to flow along magnetic field lines. It was found that 5-min oscillations can generate shock waves along magnetic field lines with inclination angles of about 50 degrees from vertical in the chromosphere. It is suggested that coronal activities driven by the shock waves might be observable with the Soft X-ray Telescope on board SOLAR-A.

080.164 **The effect of Kelvin-Helmholtz instability on rising flux tubes in the convection zone.**
S. Z. D'Silva, A. R. Choudhuri.
Sol. Phys., Vol. 136, No. 2, p. 201 - 219 (Dec 1991).

If the solar dynamo operates at the bottom of the convection zone, then the magnetic flux created there has to rise to the surface. When the convection zone is regarded as passive, the rising flux is deflected by the Coriolis force to emerge at rather high latitudes, poleward of typical sunspot zones. Choudhuri and D'Silva (1990) included the effects of convective turbulence on the rising flux through (a) giant cell drag and (b) momentum exchange by small-scale turbulence. The momentum exchange mechanism could enable flux tubes of radii not more than a few hundred km to emerge radially at low latitudes, but the giant cell drag mechanism required unrealistically small flux tube radii (a few meters for a reasonable giant cell upflow) to counteract the Coriolis force. The authors now include the additional effect of Kelvin-Helmholtz instability in a symmetrical flux ring caused by the azimuthal flow induced during its rise. The azimuthal flow crosses the threshold for the instability only if there is a giant cell upflow to drag the flux tubes appreciably. In the absence of such a drag, as in the case of a passive convection zone or in the case of momentum exchange by small-scale turbulence, the azimuthal velocity never becomes large enough to cause the instability, leaving the results of the previous calculations unaltered. The giant cell drag, aided by Kelvin-Helmholtz instability, however, becomes now a viable mechanism for curbing the Coriolis force - 10^4G flux tubes with radii of a few hundred km being dragged radially by upflows of 70 m s^{-1}.

080.165 **A gas dynamic model of the rotating solar atmosphere.**
O. Nath.
Sol. Phys., Vol. 136, No. 2, p. 263 - 268 (Dec 1991).

A model is presented of a solar atmosphere which is heated by the periodic passage of shock waves. The outer atmosphere rotates and is assumed not to affect the strength of the shock waves. This constant shock strength hypothesis is used as the basis of the model of the outer solar atmosphere. From the model it is concluded that the chromospheric temperature rise and flow Mach number are slightly affected by the rotation of the atmosphere.

080.166 **Waves in the solar atmosphere.**
B. Roberts.
Geophys. Astrophys. Fluid Dyn., Vol. 62, No. 1-4, p. 83 - 100 (Dec 1991). - See Abstr. 012.126 for the main entry.

It is pointed out that there are a great variety of wave phenomena in the solar atmosphere, an understanding of which rests upon a full appreciation of the dual roles of magnetic structuring and gravitational stratification in determining the modes. The basic waves of magnetic flux tubes are described in some detail and applied to photospheric magnetic flux tubes, sunspots, and coronal loops. Oscillations in quiescent prominences and the influence of the chromospheric magnetic canopy on p-modes are also discussed.

080.167 **Some aspects of the theory of solar oscillations.**
J. Christensen-Dalsgaard.
Geophys. Astrophys. Fluid Dyn., Vol. 62, No. 1-4, p. 123 - 152 (Dec 1991). - See Abstr. 012.126 for the main entry.

The study of nonradial oscillations of stars, based on the seminal work by Cowling half a century ago, is increasingly providing information about the interiors of stars. Observations of solar oscillations have led to detailed inferences about the structure and rotation of the solar interior. Nonradial oscillations have also been studied in white dwarfs and in rapidly oscillating Ap stars, and indications of solar-like oscillations have been detected in a few stars. The author gives an overview of some properties of nonradial oscillations, and discusses the diagnostic potential of low-degree modes for probing the core of the Sun and other stars.

080.168 **Observations of solar oscillations.**
F. L. Deubner.
Geophys. Astrophys. Fluid Dyn., Vol. 62, No. 1–4, p. 153 – 172 (Dec 1991). – See Abstr. 012.126 for the main entry.
The most recent observational activities and results in the field of solar oscillations are reviewed. The author addresses the dynamics and structure of the solar interior and of the atmosphere up to the transition layer, while global, local and atmospheric oscillations are being discussed. The topics of this review include: line width and amplitude limits of the p-modes; cycle variations of frequencies and amplitudes; internal solar rotation rate; p-mode absorption in spots and antipodal focussing of p-modes; diagnostics in the UV and in the IR; phase diagnostics in the atmosphere; chromospheric bright points.

080.169 **Solar neutrino rate and solar activity.**
L. I. Dorman, A. W. Wolfendale.
Geophys. Astrophys. Fluid Dyn., Vol. 62, No. 1–4, p. 173 – 182 (Dec 1991). – See Abstr. 012.126 for the main entry.
A study is made of the correlation of the rates from the solar neutrino experiment of Davis et al. (1989) with sunspot numbers for various periods of time and as a function of solar latitude. There is a measure of confirmation for the idea of a modulation of the detected rate by the magnetic field in the solar convection zone. If true it follows that the true rate, corrected for loss, is to inconsistent with the predictions of the standard model.

080.170 **The solar dynamo.**
M. Stix.
Geophys. Astrophys. Fluid Dyn., Vol. 62, No. 1–4, p. 211 – 228 (Dec 1991). – See Abstr. 012.126 for the main entry.
This review begins with remarks on contributions by T.G. Cowling to the problem of the solar magnetism. Then the author discusses problems connected to first-order smoothing in mean-field dynamo theory. Next, the reasons for a solar dynamo operating near the base of the outer convection zone are recalled, and models are reviewed. Finally, the author addresses the variety of possible solutions to the dynamo equations: stationary and oscillatory, symmetric and antisymmetric with respect to the equator, axisymmetric and longitude-dependent. Limit cycles and chaotic fields which appear in the non-linear regime are mentioned in this context.

080.171 **Asymptotic inversion of the solar oscillation frequencies: sound speed in the solar interior.**
S. V. Vorontsov, H. Shibahashi.
Publ. Astron. Soc. Jpn., Vol. 43, No. 5, p. 739 – 753 (1991).
The authors report on the results of a nonlinear inversion of the solar p-mode frequencies to infer the distribution of the sound speed inside the Sun. An inversion technique which is based on second-order asymptotic theory of acoustic oscillations is described. This technique is applied to three data sets: (1) the eigenfrequencies of the evolutionary solar model of Shibahashi et al. (1983); (2) the eigenfrequencies of solar model 1 of Christensen-Dalsgaard (1982); and (3) the observational solar oscillation frequencies of Libbrecht et al. (1990) and Jiménez et al. (1988). The authors mutually compared the results in order to test the possibilities of the inversion technique and to compare both models with observational data.

080.172 **Variations in the Sun's radiative output.**
J. Lean.
Rev. Geophys., Vol. 29, No. 4, p. 505 – 535 (Nov 1991).
By how much does the Sun's radiation vary? Although the Sun has long been an object of immense fascination, fundamental information about the magnitude and variability of its radiative output, sought for over a century, is only now approaching a level satisfactory for geophysical applications. During the past decade, satellite instruments have measured, simultaneously, both the Sun's spectrally integrated radiative output and its ultraviolet spectrum. These data have been analyzed in terms of their relationships to ground-based observations that characterize different aspects of the Sun's 11-year activity cycle, allowing estimates of solar radiative output variations over times scales from days to decades and interpretations of these variations in the broader context of the variable Sun. Uncertainties still remain to be answered by the next generation of solar radiometers, which commenced observations with the launch of the Upper Atmosphere Research Satellite in September 1991, near the maximum of solar cycle 22.

Solar variability data on CD-ROM NGDC-05.
See Abstr. 002.051.

The Solar Maximum Mission data analysis center.
See Abstr. 002.052.

The Sun in time.
See Abstr. 003.027.

Solar interior and atmosphere.
See Abstr. 003.028.

The first solar neutrino conference: Irvine 1972.
See Abstr. 004.009.

21. Meeting of the Solar Physics Division of the American Astronomical Society (AAS), Huntsville, AL (USA), 9 – 11 Apr 1991. Abstracts of presented papers.
See Abstr. 012.012.

The Sun and cool stars: activity, magnetism, dynamos. Proceedings.
See Abstr. 012.024.

Advances in solar system magnetohydrodynamics. Proceedings.
See Abstr. 012.086.

Challenges to theories of the structure of moderate-mass stars. Proceedings.
See Abstr. 012.096.

T. G. Cowling Memorial Symposium on Solar System Plasmas as Part of the 16. General Assembly of the European Geophysical Society (EGS), Wiesbaden (Germany), 22 – 26 Apr 1991.
See Abstr. 012.126.

The Solar Radio Astronomy Programme at the National Research Council of Canada.
See Abstr. 013.034.

Solar research at the IAC.
See Abstr. 013.074.

Program at ITP: helioseismology – probing the interior of a star.
See Abstr. 013.096.

Challenges to theories of the structure of moderate mass stars.
See Abstr. 013.102.

On the 2-dimensional rotational inversion problem.
See Abstr. 021.045.

(^{6}Li,^{6}He) measurements as an alternative calibration for solar neutrino detectors.
See Abstr. 022.059.

The Bartol/NASA/NSO high-degree helioseismometer.
See Abstr. 034.029.

On the analysis of physical wave trains.
See Abstr. 036.290.

Removal of subjectivity from solar astrolabe results by use of a C.C.D. camera. Application to solar diameter measurements.
See Abstr. 036.322.

Quasi–biennial variation in the Earth's rotation and the solar neutrino flux.
See Abstr. 044.010.

Recent results of the Solar Disk Sextant experiment.
See Abstr. 051.015.

Ulysses. A voyage above the Sun's poles.
See Abstr. 051.075.

SOHO and Cluster. Solar and geospheric plasmas and solar structure.
See Abstr. 051.077.

Nonlocal mass generation through spontaneous chiral–symmetry breaking for solar and stellar interiors.
See Abstr. 061.010.

Response of ^{127}I to solar neutrinos.
See Abstr. 061.038.

Can the Sun shed light on neutrino gravitational interactions?
See Abstr. 061.046.

Non–standard neutrinos in the Sun and supernovae.
See Abstr. 061.054.

β^+ decays of ^{37}Ca: implications for the efficiency of the ^{37}Cl solar ν detector.
See Abstr. 061.061.

A critical look at cosmions.
See Abstr. 061.068.

Mikheyev – Smirnov – Wolfenstein effect with flavor–changing neutrino interactions.
See Abstr. 061.091.

Solar neutrino solutions with matter–enhanced flavor–changing neutral current scattering.
See Abstr. 061.092.

Shapes of solar neutrino spectra: unconventional tests of the standard electroweak model.
See Abstr. 061.093.

Solar system abundances of the elements.
See Abstr. 061.095.

Mass limits for dark matter particles derived from high–energy neutrinos from the Sun.
See Abstr. 061.097.

Zee model on Majorana neutrino mass and magnetic moment.
See Abstr. 061.098.

Solar neutrinos and the Mikheyev – Smirnov – Wolfenstein theory.
See Abstr. 061.100.

Energetic neutrinos from heavy–neutralino annihilation in the Sun.
See Abstr. 061.102.

Direct tests for solar neutrino mass, mixing, and Majorana magnetic moment.
See Abstr. 061.119.

E_6 leptoquarks and the solar neutrino problem.
See Abstr. 061.120.

Siphon flows in isolated magnetic flux tubes. IV. Critical flows with standing tube shocks.
See Abstr. 062.015.

Magnetoacoustic surface gravity waves.
See Abstr. 062.042.

On magnetic diffusion in a turbulent fluid.
See Abstr. 062.056.

A sub–grid–scale magnetic resistivity formulation.
See Abstr. 062.057.

Heating in intense flux tubes.
See Abstr. 062.060.

Magnetohydrodynamic surface waves.
See Abstr. 062.065.

Critical frequencies and reflection of fast mode waves in an isothermal atmosphere.
See Abstr. 062.067.

Vector potential magnetic null points.
See Abstr. 062.104.

Magnetic tubes in overshooting compressible convection.
See Abstr. 062.105.

The Helmholtz instability of magnetic flux sheets.
See Abstr. 062.117.

Numerical study of compressible solar magnetoconvection with an open transitional boundary.
See Abstr. 062.135.

Multiple scattering of acoustic waves by a pair of uniformly magnetized flux tubes.
See Abstr. 062.158.

Dynamical buoyancy of hydrodynamic eddies.
See Abstr. 062.161.

Propagation of magnetoacoustic waves in the solar atmosphere with random inhomogeneities of density and magnetic fields.
See Abstr. 062.162.

Turbulence, fractals, and the solar granulation.
See Abstr. 062.188.

Magnetic buoyancy.
See Abstr. 062.195.

Magnetohydrodynamic waves in the Sun.
See Abstr. 062.196.

Magnetohydrodynamic waves and wave heating in non–uniform plasmas.
See Abstr. 062.197.

Equation of state and opacity.
See Abstr. 062.209.

Mixing in the radiative zone.
See Abstr. 062.210.

Compressible magnetic convection: formulation and two–dimensional models.
See Abstr. 062.220.

Absorption of magnetoacoustic waves in the solar atmosphere with random inhomogeneities of density and magnetic fields.
See Abstr. 062.236.

On thermonuclear convection: 1. Shellular instability.
See Abstr. 062.248.

The structure of magnetic neutral points in two dimensions.
See Abstr. 062.251.

Steady flows in magnetic arcades – a class of exact MHD solutions.
See Abstr. 062.252.

Emission at cyclotron harmonics due to coalescence of z–mode waves.
See Abstr. 063.091.

An investigation of Brueckner's theory of line broadening with application to the sodium D lines.
See Abstr. 063.107.

Element separation by atomic diffusion.
See Abstr. 063.110.

Ambipolar diffusion in atmospheres of cool stars.
See Abstr. 064.003.

The Sun as a prototype in the study of stellar magnetic activity.
See Abstr. 064.060.

Modeling utility of rapid variability diagnostics of OB stars.
See Abstr. 064.185.

Main–sequence mass–loss: lithium dilution in the Hyades and the Sun.
See Abstr. 064.191.

The p–mode oscillation spectra of an evolving 1 $M_\odot$ Sun–like star.
See Abstr. 065.007.

Rotational evolution of solar–type stars. I. Main–sequence evolution.
See Abstr. 065.023.

On numerical studies of solar/stellar convection.
See Abstr. 065.035.

Stellar convection as a low Prandtl number flow.
See Abstr. 065.038.

Topological pumping in the lower overshoot layer.
See Abstr. 065.039.

The Sun and cool stars – activity, magnetism, dynamos. Concluding summary: theoretical aspects.
See Abstr. 065.046.

Asymptotic theory of solar and stellar acoustic oscillations.
See Abstr. 065.048.

Asymptotic theory of acoustic oscillations of the Sun and stars.
See Abstr. 065.061.

Evolutionary models of rotating stars.
See Abstr. 065.083.

Three–dimensional compressible hydrodynamic convection in the Sun and stars.
See Abstr. 065.105.

Symmetry breaking in stellar dynamos.
See Abstr. 065.113.

Pre–main–sequence evolution.
See Abstr. 065.125.

Remarks on the evolution of low–mass stars: more or less opaque?
See Abstr. 065.144.

On the structure of spectral line gap regions.
See Abstr. 071.006.

Solar gravitational redshift from the infrared oxygen triplet.
See Abstr. 071.031.

Center–to–limb variation of spectral lines and quiet sun atmospheric model.
See Abstr. 071.036.

Granulation: non–adiabatic patterns and shocks.
See Abstr. 071.060.

Weak magnetic fields and solar irradiance variations.
See Abstr. 072.015.

Angular momentum transport by Reynolds stresses determined from the analysis of 100–year sunspot motions and its variations with solar cycle.
See Abstr. 072.105.

Nonfacular solar luminosity variations.
See Abstr. 072.124.

A possible cause for umbral oscillations.
See Abstr. 072.125.

The 3–min oscillations of the solar chromosphere – a basic physical effect?
See Abstr. 073.104.

Magnetoacoustic heating of the solar chromosphere.
See Abstr. 073.126.

EUV spectroscopy as a probe of astrophysical plasmas – learning from the Sun.
See Abstr. 073.151.

On solar emission lines.
See Abstr. 074.098.

Stationary siphon flows in thin magnetic flux tubes. II. Radiative heat exchange with the surroundings.
See Abstr. 075.026.

The toroidal magnetic field inside the Sun.
See Abstr. 075.036.

The transfer of large–scale magnetic field by radial inhomogeneity of the material density in the rotating convection zone.
See Abstr. 075.037.

Diagnostics of the solar dynamo using the observed pattern of surface magnetic fields.
See Abstr. 075.038.

Dynamical behavior of strong magnetic fields in the solar convection zone.
See Abstr. 075.051.

Effect of rotation of the Sun on turbulent transfer of a large–scale magnetic field into the convective zone.
See Abstr. 075.059.

Adaptive potential drops as energy dissipators in solar magnetic structures.
See Abstr. 075.064.

Density measurement from Al IX λ385.03/λ392.42 line ratio.
See Abstr. 076.024.

Submillimeter solar limb profiles determined from observations of the total solar eclipse of 1988 March 18.
See Abstr. 077.068.

Solar neutrinos and the history of the Sun.
See Abstr. 078.024.

The atmosphere's response to solar irradiation.
See Abstr. 085.052.

Magnetic history of the Sun.
See Abstr. 107.019.

The luminosity variability of solar–type stars.
See Abstr. 115.033.

The evolution of angular momentum in solar mass stars.
See Abstr. 115.034.

What can other stars tell us about the Sun?
See Abstr. 115.036.

Observations relating to stellar activity, magnetism and dynamos.
See Abstr. 116.041.

Magnetic activity in pre–main–sequence stars.
See Abstr. 121.093.

Asteroseismology.
See Abstr. 122.115.

Signatures of cosmic–ray interactions on the solar surface.
See Abstr. 144.109.

Earth

081 Structure, Figure, Gravity, Orbit, etc.

081.001 **The flow within the Earth created by the relative rotation of the mantle and solid core.**
L. V. Nikitina, A. A. Ruzmajkin.
Geomagn. Aeron., Vol. 30, No. 1, p. 99 – 102 (Aug 1990). English translation of Geomagn. Aehron., Tom 30, No. 1, p. 127 – 131 (1990).
A large–scale flow, including differential rotation and meridional circulation, arises in the liquid density–stratified outer core as a result of the difference in angular velocities between the mantle and solid core. The steady–state difference in angular velocities is maintained by a balance between the moments of the external (tidal) forces and the internal viscous and magnetic friction.

081.002 **A Hamiltonian theory for an elastic Earth: elastic energy of deformation.**
J. Getino, J. M. Farrándiz.
Celest. Mech. Dyn. Astron., Vol. 51, No. 1, p. 17 – 34 (1991).
The authors study the perturbation due to the deformation of the elastic mantle by a tidal body force. In a previous publication (Getino and Ferrándiz, 1989) they defined two canonical systems of variables – they gave them the names of elastic variables of Euler and Andoyer respectively. Next, using them, they obtained the canonical expression of rotational kinetic energy. In the present paper, using the same variables, the authors build up the elastic energy which is produced by the deformation of the elastic mantle. They show that the three terms $m = 0, 1, 2$ corresponding to the second order of the development in spherical harmonics of the perturbing potential, a tidal potential, are of the same order of magnitude. In addition, the numerical integration for a particular Earth Model (Takeuchi's Model 2) is performed, with the aim of obtaining a numerical estimate of the coefficients which intervene in both this energy and the previously mentioned kinetic energy.

081.003 **Gravitational field strength inside the Earth.**
L. Hodges.
Am. J. Phys., Vol. 59, No. 10, p. 954 – 956 (Oct 1991). Current Physics Microform No.: 9110D1766.

081.004 **Physical properties at the top of the core and core surface motions.**
D. Jault, J. L. le Mouël.
Phys. Earth Planet. Inter., Vol. 68, No. 1/2,special issue, p. 76 – 84 (Aug 1991). Paper presented at the 15. General Assembly of the European Geophysical Society (EGS), Copenhagen (Denmark), 23 – 27 Apr 1990.
Changes of the Earth's magnetic field, with time constants of 5 – 100 years, have been used to constrain the fluid motions at the top of the core. The authors recall arguments to support the hypothesis that the horizontal components of the Navier–Stokes equation reduce, at the top of the core, to a geostrophic equilibrium.

081.005 **On the excitation, detection and damping of core modes.**
D. J. Crossley, J. Hinderer, H. Legros.
Phys. Earth Planet. Inter., Vol. 68, No. 1/2,special issue, p. 97 – 116 (Aug 1991). Paper presented at the 15. General Assembly of the European Geophysical Society (EGS), Copenhagen (Denmark), 23 – 27 Apr 1990.
The authors continue work on the possibility that internal oscillations in the Earth's core might be detectable by a gravity meter at the surface. As a possible excitation mechanism they use earthquakes and for the damping they consider the contribution of anelasticity in both solid and liquid parts of the Earth. The calculation of the eigenfunctions is restricted in this paper to a non–rotating Earth model.

081.006 **Terrestrial mantle siderophiles and the lunar impact record.**
C. F. Chyba.
Icarus, Vol. 92, No. 2, p. 217 – 233 (Aug 1991).
A simple analytical fit to the lunar cratering record, scaled from final to transient crater diameters, then to impactor masses, implies a total mass $\sim 1.0 \times 10^{20}$kg incident on the Moon subsequent to the solidification of the lunar crust $\sim$4.4 Gyr ago. About half this mass would be retained, and a comparable lunar mass would be eroded. These results are in good agreement with geochemical estimates of the meteoritic component mixed into the lunar crust, which give $(0.4 - 1.5) \times 10^{20}$kg. Gravitationally scaling to Earth, and taking account of the statistical probability that the largest impactors incident on Earth were more massive than the largest incident on the Moon, gives an estimate of 1.5×10^{22}kg of material accumulated by Earth subsequent to 4.4 Gyr ago. This result is in excellent accord with geochemical estimates of post–core formation meteoritic input.

081.007 **Killer crater in the Yucatán?**
J. K. Beatty.
Sky Telesc., Vol. 82, No. 1, p. 38 – 40 (Jul 1991).
Geologists now guardedly point to the Yucatán peninsula as the long–sought site of the impact that triggered the widespread extinction of life on Earth 65 million years ago.

081.008 **Early differentiation of the earth and the problem of mantle siderophile elements: a new approach.**
V. R. Murthy.
Science, Vol. 253, No. 5017, p. 303 – 306 (19 Jul 1991).
The long–standing problem of the excess abundances of siderophile elements in the mantle can be resolved by considering an equilibrium core–mantle differentiation in the earth at 3000 to 3500K. The high–temperature differentiation results in mantle siderophile element abundances that closely match the observed values. Some lithophile (light) elements could enter the core in this process as is necessary to account for its low density. The abundances of siderophile elements in the mantle are consistent with the conclusion derived from the recent physical models that the earth was molten during accretion.

081.009 **Profiles of elastic properties and density for a chondritic model of a nondifferentiated earth.**
O. L. Kuskov, V. A. Miklishanskij.
Izv. Acad. Sci. USSR, Phys. Solid Earth, Vol. 25, No. 9, p. 739 – 742 (Apr 1990). Translated from Izv. Akad. Nauk SSSR, Fiz. Zemli.

081.010 **Small black holes and volcanos.**
A. P. Trofimenko.
Fizika, Vol. 22, No. 3, p. 545 – 548 (Jul–Sep 1990). Letter–to–the–editor.
The volcan model with radiation of micro–black hole in magma chamber as the source of energy is considered. At the earth surface near a volcan the neutrino flow from a micro–black hole is estimated. It has the specific spectrum and appears to be enough big in order to be perceptable by contemporary devices.

081.011 **Informatii asupra structurii fizice a litosferei din România dedusa din studiul exploziilor industriale.**
V. Smalbergher, F. Radulescu, M. Anghel.
Stud. Cercet. Fiz., Vol. 42, No. 3, p. 245 – 252 (1990).
The paper presents seismic velocities of the P and S–waves, and the Poisson ratio in the median crust and upper mantle of some important geological units of Romania which were obtained from the Romanian seismological network records of the large industrial explosions produced in the period 1971–1985.

081.012 **Consideratii asupra discontinuitatii seismice crusta–manta din Romania.**
A. Pompilian, F. Radulescu.
Stud. Cercet. Fiz., Vol. 42, No. 3, p. 287 – 293 (1990).
The paper presents the characteristics of seismic waves generated by the lower crust and the crust–mantle transition recorded in deep seismic sounding and in reflection studies. Results of seismic works show some features of the crust–mantle transition in the Carpathian fore–deep, the Moesian Platform and the Pannonian Depression.

081.013 **Relative estimation of the mineral ages using uranium migration.**
A. Danis.
Stud. Cercet. Fiz., Vol. 42, No. 6, p. 519 – 522 (1990).
Using the uranium fission track micromapping technique the correlation between the age and uranium migration from inclusions was studied. It is shown that during geological time, as function of the mineral, its age and its uranium migration speed, the pattern of the track, clusters corresponding to the uranium inclusions get a typical feature. Thus for a bulk polished geological sample it is possible to establish an age succession of the constituent minerals as a function of the track cluster patterns. Also, it is shown that knowing the migration speed of the uranium in a mineral it is possible to estimate the age of this mineral by measuring the migration distance on the micromapping.

081.014 **A parametrized solid earth tide model and ocean tide loading effects for global geodetic baseline measurments.**
H. G. Scherneck.
Geophys. J. Int., Vol. 106, No. 3, p. 677 – 694 (Sep 1991).
An observation model for earth tide displacements in application to Very Long Baseline Interferometry and similar precise geodetic techniques is developed. It incorporates effects from anelastically, ellipsoidal figure, and fluid core resonance. Based on a harmonic development of the external potential, the model follows the familiar Love number concept. The transfer function of the earth to each harmonic is formulated in terms of coupled harmonics in the space domain and fulfils the causality condition in the time domain. Solve–for parameters can be chosen flexibly. The guideline, however, has been to provide a minimum set of well–defined and well–resolvable tide response parameters for analysis of observations. The aim of accuracy for tide displacements prediction is below 1 mm. Being the major perturbation of the solid earth tide, ocean tide loading effects are computed, and the accuracy of the models involved is discussed.

081.015 **The Earth's equatorial principal axes and moments of inertia.**
H. S. Liu, B. F. Chao.
Geophys. J. Int., Vol. 106, No. 3, p. 699 – 702 (Sep 1991).
Let the Earth's equatorial principal moments of inertia be A and B, where A < B, and the corresponding principal axes be a and b. Explicit formulae are here derived for determining the orientation of a and b axes and the difference B–A using C_{22} and S_{22}, the two gravitational harmonic coefficients of degree 2 and order 2. For the Earth, the a axis lies along the (14.93°W, 165.07°E) diameter, and the b axis lies perpendicular to it along the (75.07°E, 104.93°W) diameter. The difference B–A is $7.260 \times 10^{-6} MR^2$. These quantities for other planets are contrasted, and geophysical implications are discussed.

081.016 **The precision of density estimation deep in the Earth.**
B. A. Bolt.
Q. J. R. Astron. Soc., Vol. 32, No. 4, p. 367 – 388 (Dec 1991).
A critical outline, spanning the years since the first edition (1924) of Jeffreys' "The Earth", is given of the methods of estimating density inside the Earth. In particular, the nonlinear nature of this inverse problem and the role in parameter resolution of the stochastic uncertainties of seismic velocity distributions and terrestrial eigen–spectra are stressed. At present, few well–resolved density models have been independently constructed for sensitivity comparisons. In the next decade, there are bright prospects for improved precision and resolution of both radially–symmetric and laterally–varying density models.

081.017 **On the consequences of strong stable stratification at the top of Earth's outer core.**
J. Bloxham.
Geophys. Res. Lett., Vol. 17, No. 12, p. 2081 – 2084 (Nov 1990).
The author considers the consequence of strong stable stratification at the top of the Earth's fluid outer core, concentrating on the generation of the geomagnetic secular variation. He assumes that the core near the core–mantle boundary is both strongly stably stratified and free of Lorentz forces: he finds that this set of assumptions severely limits the class of possible motions, none of which is compatible with the geomagnetic secular variation. Relaxing either assumption is adequate: tangentially geostrophic flows are consistent with the secular variation if one relaxes the assumption that the core is strongly stably stratified (but retain the assumption that Lorentz forces are negligible); purely toroidal flows (consisting in this case of zonal flows together with horizontally polarized MAC–waves) may explain the secular variation if one includes Lorentz forces. The author is unable to discriminate between these two cases.

081.018 **OSU89A/B potential coefficient models.**
R. H. Rapp, N. K. Pavlis.
IAG Symposium: Determination of the geoid, p. 7 – 9 (1991). – See Abstr. 012.035 for the main entry.
The OSU89A and B models reflect an improvement over earlier OSU models. This improvement comes about because of three main factors: improved combination modeling; the improved satellite model, and the improved terrestrial and altimeter derived data base. Tests document the quality of the models. The model can be improved using the direct incorporation of satellite altimeter data into the solution, an improved satellite potential field, and improved terrestrial data.

081.019 **The aliasing effect in coefficients estimation.**
A. Albertella, F. Migliaccio, F. Sansò.
IAG Symposium: Determination of the geoid, p. 10 – 15 (1991). – See Abstr. 012.035 for the main entry.
The authors' scientific goal is a better knowledge of a global model of the gravity field, in terms of the coefficients of the expansion of the gravity potential into spherical harmonics.

081.020 **GRIM-4 – a new global earth gravity model (status report).**
P. Schwintzer, C. Reigber, G. Balmino, R. Biancale.
IAG Symposium: Determination of the geoid, p. 16 – 17 (1991). Abstract. – See Abstr. 012.035 for the main entry.

081.021 **Geoid recovery at 0.5 degree resolution from global satellite gradiometry data sets.**
M. Vermeer.
IAG Symposium: Determination of the geoid, p. 18 – 29 (1991). – See Abstr. 012.035 for the main entry.

An efficient technique for global geoid recovery was developed operating in the frequency domain and intended for use with gridded global sets of satellite gravity gradiometry (and similar) data. A global buried masses grid is obtained as an intermediate solution, from which geoid undulations, gravity disturbance components and disturbing gravity gradient components can be regenerated by direct use of Newton-type formulae. Test computations using OSU86F are described showing the functionality, accuracy and resource requirements of the algorithms implemented and their use for geoid determination.

081.022 **Compact approximation of the geoid height on the base of modified high resolution global geopotential models.**
M. V. Belikov, M. S. Petrovskaya.
IAG Symposium: Determination of the geoid, p. 30 – 39 (1991). – See Abstr. 012.035 for the main entry.

081.023 **An improved method to the Earth's potential coefficients estimation.**
Z. Martinec.
IAG Symposium: Determination of the geoid, p. 40 – 48 (1991). – See Abstr. 012.035 for the main entry.

The scalar boundary value problem is treated numerically in order to obtain a new method for computing the potential coefficients of the Earth's gravity field. Only the gravity modulus, and not its direction, is assumed to be given on the topographic surface in addition to the value of the potential. On the other hand, only the geocentric radius of the points on the topographic surface is unknown, while their angular position is known. The test of the method on the synthetic data sets showed that the iterative procedure converges rapid to the actual fields. An immediate data processing of gravity data measured on the physical surface without any continuation to a reference surface belongs to the main advantages of the presented method.

081.024 **Altimetry based geoid determination at the German Processing and Archiving Facility within the ERS-1 project.**
W. Bosch, T. Gruber.
IAG Symposium: Determination of the geoid, p. 75 – 85 (1991). – See Abstr. 012.035 for the main entry.

This paper presents the concept for combined geoid determination by a high degree least squares harmonic analysis with an as rigorous as possible mathematical approach. The approach will be realized at the German Processing and Archiving Facility (D-PAF), operating as one element of the off-line ground segment for the first remote sensing satellite ERS-1 of the European Space Agency (ESA). The paper first introduces the embedding of geoid determination within the ERS-1 project. This is followed by a foundation of high degree least squares harmonic analysis (LSHA) and a discussion of numerical aspects. Then, the LSHA is applied to the Earth gravity field (or its geoid). Observation equation for surface data are derived, the use of satellite altimetry is discussed, and the inclusion of a priori information from most recent results of satellite-only gravity field modeling is shown.

081.025 **The determination of oceanic geoid using modified Hotine integral.**
G. Zelin, L. Yecai.
IAG Symposium: Determination of the geoid, p. 86 – 94 (1991). – See Abstr. 012.035 for the main entry.

By means of theoretical analysis and practical computation, the paper points out that Hotine integral is superior to Stokes integral in the determination of oceanic geoid.

081.026 **A strategy for gross-error detection in satellite altimeter data applied in the Baltic-Sea area for enhanced geoid and gravity determination.**
C. C. Tscherning.
IAG Symposium: Determination of the geoid, p. 95 – 107 (1991). – See Abstr. 012.035 for the main entry.

Using least-squares collocation (LSC) with altimeter track biases and tilts as additional unknowns it is possible to compute adjusted altimeter heights. A comparison of the difference between the original and the adjusted value with the error estimate obtained from LSC may then be used to find suspected gross-errors. The procedure was used in two areas in the Baltic Sea with and without taking advantage of high quality gravity data in addition to a spherical harmonic expansion (IFE88E2) complete to degree 360. Gross-errors were found in 1% (12) of the 1232 values used. The additional use of gravity values reveiled only a few errors (4) in addition to these already detected using a 3-sigma level.

081.027 **Estimation of the disturbing potential components and comparison with GPS and astrogeodetic data.**
I. N. Tziavos, D. Arabelos.
IAG Symposium: Determination of the geoid, p. 141 – 151 (1991). – See Abstr. 012.035 for the main entry.

The optimal combination of surface gravity data with the OSU86F geopotential model (GM) complete to degree and order 360 is studied in a North-American test area. Experiments using gravity data and different truncations of the geopotential expansion are carried out in order to obtain for absolute geoidal heights an accuracy close to 0.40 m and for relative geoidal heights 2 – 3 ppm.

081.028 **The Earth as an antineutrino star.**
M. Kobayashi, Y. Fukao.
Geophys. Res. Lett., Vol. 18, No. 4, p. 633 – 636 (Apr 1991).

Antineutrinos are emitted in the Earth upon β decays of radioactive elements and reach the surface with little interaction with terrestrial materials. These antineutrinos are anticipated to carry unique information about the chemistry and energetics of the Earth. The authors make a systematic search for the candidate target materials that undergo nuclear reactions with incoming antineutrinos and therefore that could be used as an antineutrino detector. The target nuclides most preferable for the measurement are ^{3}He and ^{1}H. For ^{3}He the authors suggest a liquid ^{3}He scintillation counter, of which practicability depends on the feasibility of amassing at least 10 tons of ^{3}He. For ^{1}H one may design a H_2O or liquid H_2 scintillation counter with an appropriate activator, which should be capable of detecting scintillation light in a huge tank containing at least 10^4tons of H_2O or 10^3tons of H_2.

081.029 **Determination of the regional geoid from simulated GPS measurements by the ARISTOTELES solid-earth satellite.**
P. N. A. M. Visser, K. F. Wakker, B. A. C. Ambrosius.
IAG Symposium: Determination of the geoid, p. 173 – 182 (1991). – See Abstr. 012.035 for the main entry.

081.030 **A global geoid computation by a solution of the bipotential equation.**
J. Engels.
IAG Symposium: Determination of the geoid, p. 311 – 320 (1991). – See Abstr. 012.035 for the main entry.

The boundary value problem with given disturbing potential values and given radial derivatives on a known topography is considered; it is solved by means of an expansion of the disturbing potential in global base functions (spherical harmonics).

081.031 **The geoid and its computation from the gravimetric boundary value problem.**
E. Grafarend, B. Heck, J. Engels.
IAG Symposium: Determination of the geoid, p. 321 – 332 (1991). – See Abstr. 012.035 for the main entry.

Since the geoid is situated partially in the exterior, partially in the interior of the earth's body, the gravity potential of the earth has to be determined in both domains. The authors focus on the potential in the interior domain, which was calculated by solving an oblique boundary value problem of Poisson's equation.

081.032 **Some integral formulas for a non-spherical Earth.**
L. E. Sjöberg.
IAG Symposium: Determination of the geoid, p. 333 – 338 (1991). – See Abstr. 012.035 for the main entry.

Solutions to the free and the fixed geodetic boundary problems are presented for a non-spherical Earth. The solutions become the classical integral formulas for a spherical boundary.

081.033 **The boundary equation approach to the geoid.**
S. L. Svensson.
IAG Symposium: Determination of the geoid, p. 357 – 364 (1991). – See Abstr. 012.035 for the main entry.

The main problem of physical geodesy is to find the shape of the earth from observations of the gravitational field such as geopotential numbers, gravity anomalies, gravity gradients, and deflections of the vertical. The author concentrates upon the Molodensky problem.

081.034 **The geoid, Green's functions and the gravimetric boundary value problem for the spheroid earth.**
N. C. Thông.
IAG Symposium: Determination of the geoid, p. 365 – 373 (1991). – See Abstr. 012.035 for the main entry.

The method to determine the Geoid from the gravimetric data on the known physical surface of the earth and from the mass density model leads to the fixed gravimetric boundary value problem for both the exterior and interior domain of the earth. A geometrically as well as physically spheroidal approximation is outlined. The corresponding Green's functions are derived. They are, in contrast to the spherical case, also dependent on the azimuth. The criterion for the continuous differentiability of the solutions is also derived.

081.035 **Correlation of strain representations of the potential anomalies with the geoid.**
E. Livieratos, I. N. Tziavos.
IAG Symposium: Determination of the geoid, p. 374 – 381 (1991). – See Abstr. 012.035 for the main entry.

In this paper, the strain field decomposed into dilatational (isotropic strain) and maximum shear (anisotropic strain) part is compared with the geoidal representation of the potential anomalies. It is shown the high degree of correlation between the dilatational isotropic strain field and the geoid and the poor correlation between the maximum shear anisotropic field and the geoid in the short wavelengths which means that the geoidal representation of the anomalous potential is almost exclusively of dilatational isotropic character.

081.036 **Eine approximative Lösung der fixen gravimetrischen Randwertaufgabe im Innen- und Außenraum der Erde.**
J. Engels.
Dtsch. Geod. Komm. Bayer. Akad. Wiss., Reihe C, Nr. 379.
Diss. (Dr.-Ing.), Stuttgart Univ. (Germany, F.R.). Fakultät 2 – Bauingenieur- und Vermessungswesen. Verlag der Bayerischen Akademie der Wissenschaften, München (Germany, F.R.). 115 p. (1991). ISBN 3-7696-9425-2.

This thesis constructs the geoid from the solution of an exterior as well as an interior boundary value problem, i.e. it does not employ reductions. In the first chapter the boundary value problems (especially that of the interior domain) are mathematically formulated. In the second chapter the author considers the boundary value problem in the exterior space; he quotes uniqueness and existence theorems and discusses several solution methods. He presents in detail the method of series expansions with spherical harmonics, which is well suited for global calculations; he derives a linear equation system, which has to be solved for the unknown series coefficients. Accordingly the third chapter deals with the interior boundary value problem. In the fourth chapter a reference potential for both domains is constructed. In the last chapter numerical simulation computations are presented, whereby a good agreement with the reduction methods is found.

081.037 **A comparison between two internal structure models of the Earth.**
C. Z. Zhang.
Earth, Moon, Planets, Vol. 54, No. 2, p. 129 – 143 (Aug 1991).

In view of the elastic deformation of the Earth the author performed the comparative study for the Earth's models 1066 A and PREM, calculated the static Love numbers from degree 2 to 30, and discussed the relative variations of the second degree Love numbers and their combinations due to the variation of the position of the core-mantle boundary, due to the redistribution of V_p, V_s, and ϱ in the lower mantle, and due to the possible rigidity in the outer core. From the discussions mentioned above it is recommended that the Standard Earth Model (SEM) should include two kinds of models – one is oceanless, and another has an oceanic surface. Finally, the author calculated the astronomic-geodetic parameters, which are consistent with the primary constants in the IERS Standards, of the SEM.

081.038 **A note on the thermal component of the equation of state in solids.**
V. Čelebonović.
Earth, Moon, Planets, Vol. 54, No. 2, p. 145 – 149 (Aug 1991). Letter-to-the-editor.

A simple method for determining the thermal component of the equation of state of solids under high pressure is proposed. Application to the interior of the Earth gives results in agreement with recent geophysical data.

081.039 **Long wave radiation of the Earth.**
R. Schreiber.
Urania, Rok 62, Nr. 7 – 8, p. 194 – 198 (Jul – Aug 1991). In Polish.

081.040 **Relative motions between the inner core and the Earth's mantle due to attractions of the Sun and the Moon.**
H. Jochmann.
Astron. Nachr., Vol. 312, No. 4, p. 265 – 273 (1991). = Mitt. Zentralinst. Phys. Erde, Nr. 1867.

The response of the inner core motion with respect to the mantle due to the attraction of the Sun and the Moon is studied. It could be shown that some periodic constituents of polar motion and the dipole part of the geomagnetic westward drift, which correspond with astronomical nutation periods, are in accordance with the applied hypothesis of a relative precession and nutation of the inner core.

081.041 **On the influence of underground water level variations on the gravitational field of the Earth.**
P. D. Dvulit, A. A. Korsun'.
International Seminar on Problems of Astrometry and Space Geodynamics, p. 128 – 132 (1991). In Russian. – See Abstr. 012.049 for the main entry.

It is established that the underground water level variations (about 2 m) lead to a free-fall acceleration variation with an amplitude of 30×10^{-8} m/s^2, that in turn causes variation of deflection of the vertical ($\sim 0.001''$) as well as the height variation of the quasi-geoid (up to ~ 2 mm). Therefore the underground water level variations should be taken into consideration during precise gravimetric, astrometric and geodetic works.

081.042 **Toroidal oscillations of a transradially isotropic elastic sphere.**
A. Gupta, S. J. Singh.
Proc. Indian Acad. Sci., Earth Planet. Sci., Vol. 99, No. 3, p. 383 – 391 (Sep 1990).

In order to consider the effect of anisotropy on the periods of the oscillations of the Earth, the problem of toroidal oscillations of a transradially isotropic elastic sphere is considered. At each point, the medium is assumed to be transversely isotropic about the radius through the point. The roots of the frequency equation are obtained for different values of the anisotropy parameter.

081.043 **Thermal convection in a spherical shell with a variable radius ratio.**
S. J. Drew.
Geophys. Astrophys. Fluid Dyn., Vol. 59, No. 1–4, p. 165 – 183 (Nov 1991).
The author considers, as a model for the Earth's core, an electrically conducting fluid sphere with a solid concentric inner core. He investigates the effect of the growth of an inner core on convection due to a uniform distribution of heat sources, and finds that increasing the inner boundary radius is initially beneficial to convection. Further, increasing the magnetic field strength increases the range of this effect.

081.044 **A geodynamo powered by luni–solar precession.**
J. P. Vanyo.
Geophys. Astrophys. Fluid Dyn., Vol. 59, No. 1–4, p. 209 – 234 (Nov 1991).
Controlled spin and precession of liquid filled oblate spheroidal cavities in laboratory experiments that simulate luni–solar forced precession of the mantle and a simplified analytical liquid flow model, indicate Earth core flows approximated as spin about a core axis that lags behind the precessing spin axis of the mantle by a small angle ($\sim 0.01°$). Core spin is slower than mantle spin by an amount consistent with westward drift of the magnetic field ($\sim 0.2°$/yr). Theoretical and experimental evidence argue against inertia waves being effective in preventing such motion. This precessional flow model is consistent with geodynamo requirements indicating a kinematic over a magnetohydrodynamic dynamo in that viscous core–mantle coupling may dominate magnetic coupling, with only negligible topographic coupling. While the model appears to provide geodynamo requirements, it does not preclude simultaneous density gradient convective mechanisms.

081.045 **Some improvements of the solid earth tide model.**
S. Y. Zhu, C. Reigber, F.-H. Massmann.
Manuscr. Geod., Vol. 16, No. 4, p. 215 – 220 (Aug 1991).
The solid earth tide model adopted by the International Earth Rotation Service (IERS) Standards (McCarthy ed., 1989) is improved in three ways: (1) Three different nominal Love numbers for long period, diurnal and semi–diurnal tides are introduced instead of one nominal value. (2) In some places the mean radius of the earth is used instead of the equatorial one. (3) In the computation of the frequency dependent Love numbers of the diurnal tides a modification according to the recent improved estimate of the Free Core Nutation (FCN) period is introduced. Simulations and real data analysis have shown that the effect of these improvements is non–negligible. Taking the satellite Starlette as an example, their effect on the orbit is about ± 1.5 mm/day in radial direction, ± 3 cm/day in cross–track direction and 9 cm/day in along track direction. In the differential orbit correction process of a 7–day Starlette arc an average improvement of the residual rms is 6 mm. The maximum residuum is reduced by 2.3 cm, the effect on the satellite position vector is a few cm (up to 8 cm).

081.046 **Nonlinear inverse potential problems and construction of two–layer models of Earth and Moon.**
I. L. Prutkin.
Izv. Acad. Sci. USSR, Phys. Solid Earth, Vol. 25, No. 11, p. 913 – 918 (Jun 1990). Translated from Izv. Akad. Nauk SSSR, Fiz. Zemli, Vol. 25, No. 11 (1989).
An improved method is proposed for solving nonlinear three–dimensional inverse potential problems. Gravitational and magnetic models of relief of the core–mantle boundary are constructed on its basis. They are compared with one another and with a model constructed on the basis of seismological data. It is demonstrated in the example of construction of gravitational two–layer lunar models that the proposed method can be used in the localization of inhomogeneities.

081.047 **S_1: a small but perplexing component of the tide.**
G. Godin.
Dtsch. Hydrogr. Z., Band 43, Heft 2, p. 73 – 83 (1990).
S_1, of exact diurnal periodicity, is detectable in most water level records. Its unexpected and persistent presence in almost all records probably is due to a diurnal inequality in the tidal forces actually exerted on the Earth, caused by a slight rotational asymmetry in the shape of the geoid and being felt during the course of each rotation.

081.048 **General relativity and physics of the Earth.**
A. P. Trofimenko.
Fizika, Vol. 23, No. 3, p. 247 – 272 (Sep 1991).
Geophysical manifestations of otons (objects of general relativity) are discussed. Values of energy and flow of neutrino from a microblack hole which can supply energetics of volcanos are specified. The energy of catastrophic volcano explosions is pointed out to coincide in the order of value with the energy of exploding black holes. The question is put about registration of high energy neutrino from exploding black holes by means of deep–water detectors. Short–time variations of gravitational potential derivatives (otonic graviimpulses) are described which are produced by fastmoving otons in the Earth. Expressions for oton mass are obtained through measurable physical quantities. The question of otonic graviimpulses registration is analysed.

081.049 **The "Terrestrial Maria" hypothesis: earth in the context of the inner solar system.**
M. Connors.
J. R. Astron. Soc. Can., Vol. 85, No. 4, p. 175 (Aug 1991). Abstract. – See Abstr. 012.090 for the main entry.

081.050 **Thermal equilibration of the Earth following a giant impact.**
T. Spohn, G. Schubert.
Geophys. J. Int., Vol. 107, No. 1, p. 163 – 170 (Oct 1991).
The Earth is found to equilibrate thermally after a giant impact melts a significant portion of the planet on a time–scale of a few million years. A simple thermal evolution model for the cooling of a terrestrial planet, a prescribed fraction of which is melted during a giant impact, is presented.

081.051 **Geodynamic parameters derived from 7 years of laser data on Lageos.**
P. Gegout, A. Cazenave.
Geophys. Res. Lett., Vol. 18, No. 9, p. 1739 – 1742 (Sep 1991).
Seven years (1984 through 1990) of Laser data on Lageos have been analysed to recover absolute station velocities and j_2. Relative motions between stations located in stable tectonic regions agree well with the NUVEL–1 model predictions within 5 mm/yr rms. Relative motion across the San Andreas Fault is found to be $-33. \pm 1.$ mm/yr. Solution for the secular variation of j_2 is $(-2.8 \pm 0.04) \times 10^{-11}$/yr.

081.052 **Tektites in Cretaceous–Tertiary boundary rocks on Haiti and their bearing on the Alvarez impact extinction hypothesis.**
G. A. Izett.
J. Geophys. Res., Vol. 96, No. E4, p. 20879 – 20905 (25 Nov 1991).
Relic tektites are associated with a Pt–group metal abundance anomaly and shocked minerals in a thin marl bed that marks the K–T boundary on Haiti. The presence of these three impact–produced materials at the precise K–T boundary enormously strengthens the Alvarez impact extinction hypothesis.

081.053 **The Sudbury structure: controversial or misunderstood?**
R. A. F. Grieve, D. Stöffler, A. Deutsch.
J. Geophys. Res., Vol. 96, No. E5, p. 22753 – 22764 (25 Dec 1991).

081.054 **Relativistic manifestations of gravitational fields in gravimetry and geodesy.**
S. M. Kopejkin.
Manuscr. Geod., Vol. 16, No. 5, p. 301 – 312 (Oct 1991).
The theoretical principles and methods of the relativistic gravimetry and geodesy are established in the framework of General Relativity. A gravitational field and properties of geocentric and topocentric coordinate systems are described by the specific forms of metric tensor, which is obtained from the Einstein equations with the help of the post–Newtonian iterations. By matching of asymptotic, post–Newtonian expansions of the metric tensor in the geocentric and topocentric coordinates the author has derived the relationship between these reference systems as well as relativistic corrections to the Earth's gravity and its gradient. Besides, a new term in the Newtonian definition of gravity was discovered.

081.055 **The statistics of the Earth's gravity field, revisited.**
C. Jekeli.
Manuscr. Geod., Vol. 16, No. 5, p. 313 – 325 (Oct 1991).
A common interpretation of the Earth's residual gravity field is that of a stationary, stochastic process. With only one realization of the process at hand, ergodicity is invoked to perform spectral and covariance analyses, as well as statistical prediction and estimation. To conduct statistical hypothesis testing, however, the probability distribution function of the residual gravity field may need to be known. The underlying probability law of a stationary, ergodic gravity field can be Gaussian (normal) on the infinite plane, but not on the sphere. In fact, it is argued that for the local gravity field to be a Gaussian process, it must be interpreted as a subset of a process strictly on the plane (not on the sphere in planar approximation). With this interpretation, tests may be applied to determine whether a given local gravity field is normally distributed. One particular test is applicable to the residual anomaly field after a long–wavelength reference field has been removed. Applied to a section of the residual North Carolina gravity field, the test suggests that it is normally distributed.

081.056 **Error variances–covariances of GEM–T1: their characteristics and implications in geoid computation.**
R. H. N. Haagmans, M. van Gelderen.
J. Geophys. Res., Vol. 96, No. B12, p. 20011 – 20022 (10 Nov 1991).
GEM–T1 is the first Earth potential model for which a full error variance–covariance matrix of its coefficients was made available. This matrix offers the opportunity to look into the properties of the error covariance functions for geoid heights from GEM–T1. They appear to be rather anisotropic and inhomogeneous. Geoid error covariance functions are used in regional geoid computations, where usually a set of potential coefficients is combined with gravity anomalies. In order to see the effect of simplification of the error variance–covariance propagation in geoid computations, three cases were considered: the full matrix, the diagonal elements only, and the limitation to the use of error degree variances. The resulting variations in the estimated geoid errors are shown in two test areas for different cap sizes. They can reach 50 cm for a zero cap size. Finally, test computations are presented of geoid determination by least squares collocation. Since the error covariance of the observations is included in the collocation operator, the geoid heights depend upon the choice of the type of error covariance propagation of the Earth potential coefficient set, as it is a part of the observation. For the three considered cases, variations in the geoid heights amount up to 45 cm.

081.057 **An improved error assessment for the GEM–T1 gravitational model.**
F. J. Lerch, J. G. Marsh, S. M. Klosko, G. B. Patel, D. S. Chinn, E. C. Pavlis, C. A. Wagner.
J. Geophys. Res., Vol. 96, No. B12, p. 20023 – 20040 (10 Nov 1991).
Several tests have been designed to estimate the correct error variances for the GEM–T1 gravitational solution that was derived exclusively from satellite tracking data. The basic method uses both independent an dependent subset data solutions and produces a coefficient by coefficient estimate of the model uncertainties. The GEM–T1 errors have been further analyzed using a method based on eigenvalue–eigenvector analysis, which calibrates the entire covariance matrix. Dependent satellite data sets and independent altimetric, resonant satellite, and surface gravity data sets all confirm essentially the same error assessment. The calibration test results yield very stable calibration factors, which vary only by approximately 10% over the range of tests performed. Based on these calibrated error estimates, GEM–T1 is a significantly improved solution, which to degree and order 8 is twice as accurate as earlier satellite derived models like GEM–L2. Also, by being complete to degree and order 36, GEM–T1 is more complete and has significantly reduced aliasing effects that were present in previous models.

081.058 **Gravity field error analysis: applications of Global Positioning System receivers and gradiometers on low orbiting platforms.**
E. J. O. Schrama.
J. Geophys. Res., Vol. 96, No. B12, p. 20041 – 20052 (10 Nov 1991).
The concept of a Global Positioning System (GPS) receiver as a tracking facility and a gradiometer as a separate instrument on a low orbiting platform offers a unique tool to map the Earth's gravitational field with unprecedented accuracies. The former technique allows determination of the spacecraft's ephemeris at any epoch to within 3–10 cm; the latter permits the measurement of the tensor of second–order derivatives of the gravity field to within 0.01–0.0001 Eötvös units depending on the type of gradiometer. An analytical technique is described and applied for any error analysis of gravity field parameters from gradiometer and GPS observation types. Results are discussed for the configuration proposed on Aristoteles, indicating that a "GPS only" solution may be computed up to degree and order 85, whereas a combined GPS/gradiometer experiment may result in an acceptable solution up to degree and order 240.

081.059 **Cold–nuclear fusion within the core of the Earth?**
L. R. McHargue, P. E. Damon, T. D. Swindle.
AIP Conf. Proc., No. 228, p. 611 – 615 (10 May 1991). Current Physics Microform No.: 9108D1757. – See Abstr. 012.122 for the main entry.
It is proposed that the core of the Earth where metal hydrides are subject to high pressure and temperature conditions is a suitable site for "cold–nuclear" fusion to proceed. This is compatible with recent experimental evidence, that shows that low levels of cold–nuclear fusion may take place in hydrides, that hydrogen is soluble in iron at high pressures, and in some models up to 0.4% by weight of the core of the Earth may be hydrogen. If it is assumed that ^{3}He is produced within the Earth by cold–nuclear fusion, and production and escape to the atmosphere has attained steady–state, the terrestrial fusion rate λ_f is between 0.3×10^{-25} to 4.4×10^{-25} fusions per second, per deuteron. In contrast to terrestrial radiogenic heat production, cold–nuclear fusion can produce only 10^{-5} of the observed terrestrial heat flow.

081.060 **Cold nuclear fusion in the Earth.**
E. P. Palmer.
AIP Conf. Proc., No. 228, p. 616 – 645 (10 May 1991). Current Physics Microform No.: 9108D1762. – See Abstr. 012.122 for the main entry.
Evidence that p–d or d–d fusion could be occurring in the Earth stimulated the original laboratory search for cold fusion. That evidence is reviewed. It is found that the geologic ratio of heat to ^{3}He is too high to be explained by the usually accepted fusion processes. Laboratory evidence indicates that fusion can be made to occur in processes of fracture and high strain rate similar to Earth processes. An extension of the Oppenheimer–Phillips theory of neutron tunneling advanced to illustrate alternate fusion paths which could explain the high heat/^{3}He ratio. The search for fusion as a source for additional heat and

non-primordial ^3He is further stimulated by recent data and analysis indicating that radioactivity can supply less than five percent of the Earth's heat budget.

081.061 **Towards a realistic theory of the geodynamo.**
S. I. Braginsky.
Geophys. Astrophys. Fluid Dyn., Vol. 60, No. 1–4, p. 89 – 134 (Nov 1991). This article was presented in Session 1.06 of the "Scientific Assembly of IAGA held in Exeter (UK), 24 Jul – 4 Aug 1989". – See Abstr. 012.125 for the main entry.

The physics of the geodynamo is discussed. The main processes relevant for the buoyancy driven geodynamo are isolated. The successive stages of development of geodynamo theory are briefly described. The mechanism of local turbulence in the Earth's core is explained, and an estimate is presented of the turbulent transport of density inhomogeneities in the Earth's core. The significance of this turbulent transport to the geodynamo mechanism is stressed. The general scheme of the complete geodynamo theory of the future is outlined.

081.062 **On unique determination of toroidal or geostrophic flow in the Earth's core.**
D. Gubbins.
Geophys. Astrophys. Fluid Dyn., Vol. 60, No. 1–4, p. 165 – 176 (Nov 1991). – See Abstr. 012.125 for the main entry.

Fluid flow at the surface of the Earth's core can be determined from changes in the radial component of magnetic field at the core surface by using the frozen–flux hypothesis, which treats the core as a perfect conductor and the mantle as an insulator. The flow determination is non–unique; the ambiguity may be reduced by placing restrictions on the allowed flows and by using horizontal components of field. The determination is unique if the flows are toroidal and overdetermined if they are tangentially geostrophic.

081.063 **Core flow and length–of–day changes.**
A. Jackson, J. Bloxham.
Geophys. Astrophys. Fluid Dyn., Vol. 60, No. 1–4, p. 399 – 401 (Nov 1991). Abstract. – See Abstr. 012.125 for the main entry.

081.064 **Core mantle coupling – length of day.**
J. L. le Mouël.
Geophys. Astrophys. Fluid Dyn., Vol. 60, No. 1–4, p. 409 – 410 (Nov 1991). Abstract. – See Abstr. 012.125 for the main entry.

081.065 **Neutrino tomography of the Earth.**
A. Nicolaidis, M. Jannane, A. Tarantola.
J. Geophys. Res., Vol. 96, No. B13, p. 21811 – 21818 (10 Dec 1991).

Neutrino oscillations inside matter are very sensitive to the density of the medium traversed by the neutrinos. If the neutrino mixing angle (so far unknown) is not very small, then the neutrino oscillation length is of the size of the Earth, and one can use the phenomenon of neutrino matter oscillations to study the Earth's density distribution. Although neutrino attenuation data have already been proposed as a means to infer the density structure of the Earth, the authors show that neutrino oscillation data could also be used to make a density tomograph of the Earth. A numerical experiment in which neutrino beams from an accelerator are sent across the Earth, and are detected at six different locations on the Earth's surface, shows the feasibility of using neutrino oscillation data, with the advantage of using only currently available neutrino energies.

Tentative idea and practice of database of astrometry and geodynamics at Shanghai Observatory.
See Abstr. 002.125.

150 Jahre Bessel–Ellipsoid, 1841 – 1991.
See Abstr. 004.022.

The role of impacts in the history of the early Earth.
See Abstr. 011.020.

Determination of the geoid – present and future. Proceedings.
See Abstr. 012.035.

Tidal research in geodynamics at Warsaw Space Research Centre (1985 – 1990).
See Abstr. 013.049.

Nonspherical boundary value problem by pseudodifferential operators.
See Abstr. 021.019.

Coerciveness and geometry in the linear gravimetric boundary value problem – reduced version.
See Abstr. 021.020.

Holes in boundary and out–of–boundary data.
See Abstr. 021.021.

Stokes formula using Fast Fourier Techniques.
See Abstr. 021.023.

Simulation of regional gravity field recovery from satellite gravity gradiometer data using collocation and FFT.
See Abstr. 021.024.

The GEMT1 variance–covariance matrix, its characteristics and applications in geoid computations.
See Abstr. 021.025.

Partitioning of platinum group elements in the Fe–O–S system to 11 GPa and their fractionation in the mantle and meteorites.
See Abstr. 022.175.

Fundamental constants of astronomy of the year 1976.
See Abstr. 043.007.

Sur un modèle de référence terrestre commun à la géodésie et à la géophysique.
See Abstr. 043.008.

A Hamiltonian theory for an elastic Earth: first order analytical integration.
See Abstr. 044.001.

Application of the Global Positioning System to crustal deformation measurement. 1. Precision and accuracy.
See Abstr. 045.014.

Application of the global positioning system to crustal deformation measurement. 2. The influence of errors in orbit determination networks.
See Abstr. 045.015.

An introduction to the Global Positioning System and some geological applications.
See Abstr. 045.018.

A note on generalized lumped coefficients.
See Abstr. 052.022.

A note on classification of harmonic geopotential terms.
See Abstr. 052.023.

Parameterized rotating convection for core and planetary atmosphere dynamics.
See Abstr. 062.129.

Linear theory of rotating fluids using spherical harmonics. II. Time–periodic flows.
See Abstr. 062.159.

α–effect in generation of magnetic field by a flow of conducting fluid with internal scale in an axisymmetric volume.
See Abstr. 062.160.

Convection near the Earth's core–mantle boundary?
See Abstr. 062.194.

Kinematic dynamos associated with large scale fluid motions.
See Abstr. 062.238.

Nonlinear planetary dynamos in a rotating spherical shell. I. Numerical methods.
See Abstr. 062.239.

Magnetic Reynolds number and model–Z geodynamo.
See Abstr. 062.241.

Nonlinear alpha–omega dynamos in a spherical shell.
See Abstr. 062.243.

The impacts of giant asteroids and geomagnetic inversions.
See Abstr. 084.006.

Theory of the geodynamo and the problem of core–mantle coupling.
See Abstr. 084.079.

The Earth's dynamo.
See Abstr. 084.091.

The faint young Sun problem.
See Abstr. 085.051.

Equations of state in geophysics and in planetary physics.
See Abstr. 091.052.

Physical properties of terrestrial planets and their satellites: an introduction.
See Abstr. 091.054.

Magma generation on Mars: amounts, rates, and comparisons with Earth, Moon, and Venus.
See Abstr. 097.013.

Origin of the Archean Earth impactors.
See Abstr. 098.050.

Spacewatch: Our asteroid–pelted planet.
See Abstr. 098.084.

Near miss of the Earth by a small asteroid.
See Abstr. 098.085.

Isotopic modelling of the early evolution of the Earth.
See Abstr. 107.013.

082 Atmosphere (Refraction, Scintillation, Extinction, Airglow, Site Testing)

082.001 **Urban light pollution: The effect of atmospheric aerosols on astronomical observations at night.**
J. H. Joseph, Y. J. Kaufman, Y. Mekler.
Appl. Opt., Vol. 30, No. 21, p. 3047 – 3058 (20 Jul 1991). Current Physics Microform No.: 9109D0391.
The transfer of diffuse city light from a localized source through a dust–laden atmosphere with optical depth <0.5 has been analyzed in the source–observer plane on the basis of an approximate treatment. The effect on several types of astronomical observation at night has been studied, considering different size distributions and amounts as well as particle shapes of the aerosols. The model is applied to conditions at the Wise Astronomical Observatory in the Negev desert and limiting backgrounds for spectroscopy, photometry, and photography of stars and extended objects have been calculated for a variety of signal–to–noise ratios. Applications to observations with different equipment at various distances from an urban area of any size are possible. Due to the use of signal–to–noise ratios, the conclusions are different for the different experimental techniques used in astronomy.

082.002 **Polarimetry of a 22° halo.**
G. P. Können, J. Tinbergen.
Appl. Opt., Vol. 30, No. 24, p. 3382 – 3400 (20 Aug 1991). Current Physics Microform No.: 9109A0310. Topical Meeting on Light and Color in the Open Air, Washington, DC (USA), 11 – 13 Jul 1990.
The linear polarization and intensity of a 22° halo has been measured simultaneously at seven wavelengths as a function of scattering angle. The polarization pattern is found to be dominated by a narrow peak centered at the halo angle. The amount of polarization in this peak is much higher than expected from Fresnel refraction alone. The observations are explained with a birefringence–diffraction halo polarization model. The data demonstrate the power of halo polarimetry as a tool for detection and identification of birefringent crystals in terrestrial or extraterrestrial atmospheres.

082.003 **Mie theory model of the corona.**
J. A. Lock, Y. Leiming.
Appl. Opt., Vol. 30, No. 24, p. 3408 – 3414 (20 Aug 1991). Current Physics Microform No.: 9109A0336. Topical Meeting on Light and Color in the Open Air, Washington, DC (USA), 11 – 13 Jul 1990.
The authors performed a calculation of the corona colors that employed Mie theory to obtain the scattered light intensity. The scattered intensity was integrated over the visible spectrum for a number of different cloud droplet size distributions. The results were converted to chromaticity coordinates, convolved with the angular size of the sun, and plotted on the 1931 CIE chromaticity diagram. The results were compared to observations of multiple–ring coronas.

082.004 **Luminescence of the upper atmosphere at λ630 nm according to Vertikal–6 and –7 rocket measurements.**
A. V. Pavlov.
Geomagn. Aeron., Vol. 30, No. 2, p. 214 – 219 (Oct 1990). English translation of Geomagn. Aehron., Tom 30, No. 2, p. 261 – 267 (1990).
The concentrations of $O(^1D)$, $N(^2D)$ and the volumetric rate V of the luminescence of the atmosphere at λ630 nm are computed for the conditions of the Vertikal–6 and –7 experiments during periods of low and high solar activity. At F_1 and F_2 heights, the largest contribution to the formation of $O(^1D)$ is introduced by the process of excitation of atomic oxygen by photo–electrons; with a growth in solar activity, the relative role of this process is

reduced. Using measurements of the cross section for the formation of $O(^1D)$ in atomic oxygen–electron collisions in the low–energy region, an analytic expression is derived for the rate of formation of $O(^1D)$ owing to thermal electrons, and the generally accepted expression for the rate of cooling of thermal electrons in this process is reviewed.

082.005 **Study of the spatial correlations of the Earth's twilight horizon brightness field by analyzing the photographs from space.**
G. M. Grechko, S. V. Kireev, V. P. Savinykh.
Izv. Akad. Nauk SSSR, Fiz. Atmos. Okeana, Vol. 27, No. 7, p. 776 – 782 (Jul 1991). In Russian.
This paper describes the investigation of the fluctuations of the Earth's twilight horizon brightness field by the correlation methods. Realizations of the brightness field of light scattered by aerosols were obtained from the photographs of the Earth's limb during sunset. Images were received on the orbital space station "Salyut–6" in May, 1981 with an MKF–6 camera.

082.006 **Checking of the stability of the atmospheric optical properties.**
T. Z. Muldashev, V. E. Pavlov, Ya. A. Tejfel'.
Izv. Akad. Nauk SSSR, Fiz. Atmos. Okeana, Vol. 27, No. 8, p. 831 – 841 (Aug 1991). In Russian.
The absolute brightness indicatrices for the clear atmosphere in the visible region of spectrum are computed using the spherical harmonics technique. The effects of three aerosol fractions (the Aitken cores, submicron and hard–dispersed particles) on the brightness of the sky are studied. Some limitations of the technique used for checking the stability of the earth atmosphere optical properties are detected. The approximation relations are derived and presented for the determination of the aerosol optical depth at any time moment from the observations of the brightness indicatrices.

082.007 **The response of the middle atmosphere to the effects of galactic cosmic rays.**
P. I. Velinov, L. N. Mateev.
Geomagn. Aeron., Vol. 30, No. 4, p. 500 – 504 (Feb 1990). Paper presented at the 5. Symposium on Solar Terrestrial Physics, Samarkand (USSR), 2 – 6 Oct 1989. English translation of Geomagn. Aehron., Tom 30, No. 4, p. 593 – 598 (1990).
An improved model is devised for the interaction of cosmic rays with the middle atmosphere. Profiles of the ionization q(h) are calculated as a function of the composition (protons, α particles, L, M, H, and VH groups of the nucleus) and energy spectrum of the penetrating high–energy particles. A computer program is set up to realize the Gauss algorithm for the solution of multiple integrals. The q(h) profiles are plotted in the 25 – 80 km range, together with the N(h) profiles of the 50 – 80 km layer for periods of maximum and minimum solar activity.

082.008 **A global numerical model of the thermosphere, ionosphere, and protonosphere of the Earth.**
A. A. Namgaladze, Yu. N. Koren'kov, V. V. Klimenko, I. V. Karpov, F. S. Bessarab, V. A. Surotkin, T. A. Glushchenko, N. M. Naumova.
Geomagn. Aeron., Vol. 30, No. 4, p. 515 – 521 (Feb 1990). Paper presented at the 5. Symposium on Solar Terrestrial Physics, Samarkand (USSR), 2 – 6 Oct 1989. English translation of Geomagn. Aehron., Tom 30, No. 4, p. 612 – 619 (1990).
A numerical hydrodynamic model is presented of the thermosphere, ionosphere, and protonosphere of the Earth, which describes the global distribution of the concentrations, velocities, and temperatures of neutral and charged particles as well as the electric fields of magnetospheric and thermospheric origin at altitudes from 80 km to 15 Earth radii with account taken of the noncoincidence of the geographic and geomagnetic axes of the Earth. The results of calculations for quiet equinoctial conditions at a minimum of the solar activity that are presented illustrate the appreciable influence even of comparatively weak electric fields on the global structure of thermospheric circulation and the distribution of electron concentration in the F_2 region and in the protonosphere.

082.009 **Variations common to the interplanetary magnetic field, the zonal atmospheric circulation and the earth's rotation.**
D. Djurovic, P. Pâquet.
Q. J. R. Meteorol. Soc., Vol. 117, No. 499, p. 571 – 585 (Apr 1991).
A study of the cyclic fluctuations in the earth's rotation, the geomagnetic field, the global atmospheric circulation and solar activity reveals the existence of 50–day and 120–day oscillations common to all. The authors' results as well as the results obtained by radiometric measurements from the satellites Nimbus–7 and the Solar Maximum Mission suggest that their origin lies in the physical processes in the sun. The mechanism remains unexplained but from this work it follows that the role of the interplanetary magnetic field could very well be important.

082.010 **Atmospheric fluctuations – deviations from simple Kolmogorov approximations and their effect on high resolution astronomy.**
M. Bester, W. C. Danchi, C. G. Degiacomi, L. J. Greenhill, C. H. Townes.
Bull. Am. Astron. Soc., Vol. 23, No. 2, p. 894 (1991). Abstract. – See Abstr. 012.008 for the main entry.

082.011 **Atmospheric phase fluctuations measured with the Mark III stellar interferometer.**
D. F. Buscher, J. T. Armstrong, D. Mozurkewich, C. S. Denison, M. Colavita, M. Shao.
Bull. Am. Astron. Soc., Vol. 23, No. 2, p. 894 – 895 (1991). Abstract. – See Abstr. 012.008 for the main entry.

082.012 **Atmospheric contributions to image degradation.**
D. L. Walters.
Bull. Am. Astron. Soc., Vol. 23, No. 2, p. 895 (1991). Abstract. – See Abstr. 012.008 for the main entry.

082.013 **Lunisolar atmospheric tides. III.**
R. Brahde.
Aust. J. Phys., Vol. 44, No. 1, p. 87 – 100 (1991).
In two earlier papers (46.082.074, 50.082.072) the atmospheric tide in Oslo (Norway) was studied using pressure data for 23 continuous years. In the present paper a similar study based on pressure data from Batavia (Indonesia) is presented. The result is that the tidal wave caused by the lunisolar tide is represented by a one–day and a half–day oscillation with mean amplitudes of 0.11 and 0.33 mb respectively. The amplitude spectrum reveals amplitudes of up to 1 mb of dynamic origin.

082.014 **Astronomical seeing at several sites in the USSR.**
P. V. Shcheglov, A. Eh. Gur'yanov.
Sov. Astron., Vol. 35, No. 3, p. 310 – 313 (May 1991). Current Physics Microform No.: 9201X0592. English translation of Astron. Zh., Tom 68, Vyp. 3, p. 632 – 638 (1991).
Histograms of estimates of the diameter β''_{atm}(FWHM) $= 10.1/r_0$ (r_0 is Fried's parameter in cm) of purely atmospheric blurring of a stellar image at the zenith for $\lambda = 0.5\ \mu m$ in a long exposure are compared for 10 sites in the USSR (including observatories in the Crimea, the Caucasus, and Central Asia) as a result of photoelectric observations of scintillation of Polaris between 1977 to 1987. The best seeing statistics were those measured at the observatory on Mt. Sanglok in Tadzhikistan (median $\beta_{atm} = 0''.7$ or $r_0 = 15$ cm, 39% of β_{atm} estimates $\leqslant 0''.6$). In this respect, Mt. Sanglok exceeds Mt. Maidanak in Uzbekistan (median $\beta_{atm} = 1''.1$ or $r_0 = 9$ cm, 6% of β_{atm} estimates $\leqslant 0''.6$), and in a comparison with the best foreign sites, it is inferior only to Mauna Kea Observatory in Hawaii.

082.015 **Possible good observatory sites in India.**
J. C. Bhattacharyya.
National Large Telescope Workshop: Scientific requirements and requirement of site, p. 99 – 105 (1991). – See Abstr. 012.009 for the main entry.

082.016 **Measurements for selecting a site.**
S. N. Tandon.
National Large Telescope Workshop: Scientific requirements and requirement of site, p. 106 – 109 (1991). – See Abstr. 012.009 for the main entry.

082.017 **On interferometric evaluation of a telescope site.**
J. N. Desai.
National Large Telescope Workshop: Scientific requirements and requirement of site, p. 110 – 111 (1991). – See Abstr. 012.009 for the main entry.

082.018 **Characterization of seeing using interferometric method.**
A. K. Saxena.
National Large Telescope Workshop: Scientific requirements and requirement of site, p. 112 – 123 (1991). – See Abstr. 012.009 for the main entry.

The paper describes different methods for the measurement of atmospheric seeing. Interferometric method having advantages over other methods and providing modulation transfer functions directly has been suggested for more reliable data. The method can be adopted for site survey programme for the Indian large telescope project.

082.019 **Requirements of a site for large optical/IR telescopes and techniques of measuring "seeing" and "precipitable" water vapour.**
K. V. K. Iyengar.
National Large Telescope Workshop: Scientific requirements and requirement of site, p. 124 – 128 (1991). – See Abstr. 012.009 for the main entry.

082.020 **The physics of "seeing".**
P. Venkatakrishnan.
National Large Telescope Workshop: Scientific requirements and requirement of site, p. 129 – 131 (1991). – See Abstr. 012.009 for the main entry.

The properties of images formed through turbulence are reviewed. The causes of turbulence are also listed and some suggestions are made for site selection based on this physical reasoning.

082.021 **Solar site testing: techniques and evaluation methods.**
K. R. Sivaraman.
National Large Telescope Workshop: Scientific requirements and requirement of site, p. 132 – 139 (1991). – See Abstr. 012.009 for the main entry.

The parameters required to be measured in site survey work for selecting an excellent site for establishing major solar telescope facilities are described. The techniques to be adopted for these measurements as well as the criteria for evaluation of the site from the survey data to judge the worthiness of the site for astronomical work are also discussed.

082.022 **Site survey for high spatial resolution solar observations.**
R. Jain.
National Large Telescope Workshop: Scientific requirements and requirement of site, p. 140 – 147 (1991). – See Abstr. 012.009 for the main entry.

The requirements and their measurements for selecting a solar astronomy site to obtain high spatial resolution solar observations for long uninterrupted periods are discussed. A new approach for optical testing and hence to evaluate such a site through the photography of granulation and determining sunshine hours is also proposed. Possibilities are explored for location of a suitable site.

082.023 **Site testing for high resolution solar observations.**
V. P. Gaur.
National Large Telescope Workshop: Scientific requirements and requirement of site, p. 148 – 153 (1991). – See Abstr. 012.009 for the main entry.

082.024 **Effect of a finite outer scale on the Zernike decomposition of atmospheric optical turbulence.**
D. M. Winker.
J. Opt. Soc. Am. A, Vol. 8, No. 10, p. 1568 – 1573 (Oct 1991). Current Physics Microform No.: 9111F0486.

Phase distortions in optical systems induced by atmospheric turbulence are investigated with the use of Zernike polynomial decompositions. An analytic solution for the variances of the Zernike coefficients is found for the case of Kolmogorov turbulence with a finite outer–scale length. It is shown that the effect of finite outer scale is to attenuate low–order Zernike components, even when the outer–scale length is much larger than the optical aperture. Effects are investigated for constant outer–scale size and for height–dependent outer scales. It is found that seeing effects on large telescopes are dependent more on the magnitude of the outer scale than on the shape of the outer–scale vertical profile.

082.025 **Atmospheric–turbulence measurements using a synthetic beacon in the mesospheric sodium layer.**
R. A. Humphreys, C. A. Primmerman, L. C. Bradley, J. Herrmann.
Opt. Lett., Vol. 16, No. 18, p. 1367 – 1369 (15 Sep 1991). Current Physics Microform No.: 9111C2185.

The authors have performed the first measurements of wave fronts propagated through atmospheric turbulence with the use of a synthetic beacon in the mesospheric sodium layer. The synthetic–beacon wave fronts showed reasonable agreement with those from reference stars. The sodium–layer beacon performance was also compared with that of beacons produced by Rayleigh backscatter in the 6 – 20 km altitude range.

082.026 **Optical phenomena in summer 1908.**
V. A. Romejko.
Astron. Vestn., Tom 25, No. 4, p. 482 – 489 (Jul – Aug 1991). In Russian. English translation in Sol. Syst. Res., Vol. 25.

The analysis of optical phenomena in the Earth atmosphere, which took place in summer 1908, shows that the origin of the noted phenomena may be explained both by the presence of powerful noctilucent clouds, initiated by the explosion of the Tungus object, and the volcanic activity in 1907 – 1908.

082.027 **Nature of the anomalous illumination of the sky related to the Tunguska event.**
V. A. Bronshtehn.
Astron. Vestn., Tom 25, No. 4, p. 490 – 504 (Jul – Aug 1991). In Russian. English translation in Sol. Syst. Res., Vol. 25.

A quantitative interpretation is presented of the anomalous illumination of the sky observed on the night June 30 – July 1, 1908 (and on the subsequent nights) over the vast region from Western Siberia to the British Isles.

082.028 **Low–intensive emissions of the night sky.**
V. I. Krasovskij.
Geomagn. Aehron., Tom 31, No. 6, p. 1105 – 1108 (Nov – Dec 1991). In Russian. English translation in Geomagn. Aeron., Vol. 31, No. 6.

082.029 **The earth rotation responds to atmospheric and oceanic activity.**
Li Zhian, Lin Qiao, Zheng Dawei.
J. Beijing Normal Univ. Nat. Sci., Vol. 26, No. 3, p. 192 – 197 (Jun 1990). In Chinese.

The interannual and irregular variations of earth rotation are analyzed. The results show that there are quite identity of the interannual variations of earth and atmosphere and ocean of tropical Pacific–El Niño events.

082.030 **The absorption and scattering properties of Saharan and volcanic dust.**
R. H. Garstang.
Observatory, Vol. 111, No. 1104, p. 239 – 243 (Oct 1991).

082.031 **The effect of telluric absorption on the intensity of He I λ10830.**
J. Kingdon, G. J. Ferland.
Publ. Astron. Soc. Pac., Vol. 103, No. 666, p. 752 – 758 (Aug 1991).
The observed intensity of He I λ10830 in planetary nebulae is roughly half of that expected by standard level population calculations. Several recent studies have suggested that the physical processes populating excited states of helium are not now understood. In this paper, absorption of λ10830 by water vapor in the Earth's atmosphere is examined as a means of explaining this discrepancy. The authors find that this effect can reduce the intensity of He I λ10830 by as much as 20% – 25% in the well-studied planetary nebula NGC 7027, depending on the time of year. This effect accounts for ~50% of the discrepancy discussed by previous workers. The ramifications of these results are discussed, and radial-velocity "danger zones", in which the line is most affected, are given to aid in the assessment of this effect in other objects. The authors also provide tables to determine the observing seasons when the line is least affected by telluric absorption.

082.032 **Status of the OH contribution to the surface glow detected from the Dynamics Explorer satellite.**
R. A. Copeland, T. G. Slanger.
Geophys. Res. Lett., Vol. 17, No. 13, p. 2341 – 2344 (Dec 1990).
A Fabry–Perot interferometer on the DE–2 satellite measured a surface glow near 732 nm at an altitude of 250 km. By comparison with the mesospheric nightglow, the hydroxyl radical (OH) was implicated as the fluorescing species. Unfortunately, the identification was not clear cut, since the spectrum showed an additional feature not observed in the nightglow. Calculations indicated that this feature could originate from ground state OH vibrational levels well above the energetic limit of $\upsilon = 9$ for the reaction of $H + O_3$. Until recently, existing spectroscopic data were inadequate for a positive identification of all the features in the Fabry–Perot spectrum. New laboratory data coupled with accurate spectroscopic fitting indicate that the position of the additional feature is consistent with its identification as the $P_2(5)$ doublet of the (11,5) vibrational band of OH.

082.033 **A model of the brightness of moonlight.**
K. Krisciunas, B. E. Schaefer.
Publ. Astron. Soc. Pac., Vol. 103, No. 667, p. 1033 – 1039 (Sep 1991).
A knowledge of the brightness of moonlight is needed for detailed calculations of the limiting magnitude of astronomical detectors, whether they be visual, photographic, or electronic. The previous literature contains no method for making even crude estimates and has few actual measurements of moonlight brightness. In this paper the authors report new measurements of the sky brightness from the 2800–m level of Mauna Kea. In addition, they present a model for predicting the moonlight as a function of the Moon's phase, the zenith distance of the Moon, the zenith distance of the sky position, the angular separation of the Moon and sky position, and the local extinction coefficient. The model equations can be quickly calculated on a pocket calculator. A comparison of the model with the lunar data and with some Russian solar data shows the accuracy of the predictions to range from 8% to 23%.

082.034 **Near infrared imaging of hydroxyl wave structure over an ocean site at low latitudes.**
M. J. Taylor, M. J. Hill.
Geophys. Res. Lett., Vol. 18, No. 7, p. 1333 – 1336 (Jul 1991).
Coordinated observations of wave structure in the near infrared hydroxyl (OH) nightglow emission have been made from Maui, Hawaii using a suite of narrow angle and all–sky TV cameras. Two sets of data were obtained, the first in conjunction with the ALOHA–90 campaign and the second during the subsequent new moon period. Well formed, short period (<20 min) wave patterns of comparable morphology, dynamics and abundance to those regularly imaged from mid–latitude mountain sites were detected on several high volcanic peaks, the patterns were not consistent with gravity waves generated by the interaction of strong winds with the local island topography. This suggests that other mid–latitude wave patterns may also not be of mountain origin. The wave patterns imaged during ALOHA–90 were of significantly lower constrast than those detected later. This effect may be related to changes in the characteristics of the middle atmosphere that occur shortly after the spring equinox.

082.035 **Observations of the OH Meinel (6,2) and O_2 atomspheric (0,1) nightglow emissions from Maui during the ALOHA–90 campaign.**
J. H. Hecht, R. L. Walterscheid.
Geophys. Res. Lett., Vol. 18, No. 7, p. 1341 – 1344 (Jul 1991).
Observations are reported from a CCD night–glow camera deployed on Maui during ALOHA–90. These data show that there were periods of both strong and weak gravity wave activity, that background winds in the direction of wave propagation can be derived using air–glow data, and that there may be a connection between sporadic Na and large–scale wave activity. These data are in fair to good agreement with models of wave induced airglow fluctuations.

082.036 **Temporal variations in the hydroxyl nightglow observed during ALOHA–90.**
D. N. Turnbull, R. P. Lowe.
Geophys. Res. Lett., Vol. 18, No. 7, p. 1345 – 1348 (Jul 1991).
Ground–based observations of the hydroxyl nightglow were made with a Fourier transform spectrometer from Haleakala, Maui during the ALOHA–90 campaign. Vibrational level populations and rotational temperature were obtained with a temporal resolution of 1 minute. Temperature precision was ± 1K. Highly correlated in–phase variations in temperature and population were seen on all nights. However, a long term trend in the vibrational population was not reflected in the temperature.

082.037 **Coincident imaging and spectrometric observations of zenith OH nightglow structure.**
M. J. Taylor, D. N. Turnbull, R. P. Lowe.
Geophys. Res. Lett., Vol. 18, No. 7, p. 1349 – 1352 (Jul 1991).
During the ALOHA–90 campaign a novel comparative study was made between near infrared wave structure imaged in the zenith using a CCD camera and that detected at infrared wavelengths by a Fourier Transform Spectrometer. Coincident measurements were made briefly on several occasions and for an extended period on 31 March. The temporal variations imaged in the near infrared structure during this night almost completely matched those detected in the OH (3,1) band spectrometer data when similar viewing fields were compared. However, the image data also displayed small scale wave forms that were not resolved by the larger field instrument. These structures exhibited significant changes in brightness and position on a time scale much shorter than the local Brunt–Väisälä period indicating that very high resolution measurements are necessary to investigate short period (<20 min) upper atmsopheric wave motions.

082.038 **Airglow rotational temperature measurements during the ALOHA campaign.**
R. J. Niciejewski, J. H. Yee.
Geophys. Res. Lett., Vol. 18, No. 7, p. 1353 – 1356 (Jul 1991).
Two instruments sensitive to nocturnal airglow emissions were flown aboard the NCAR Electra aircraft during the 1990 ALOHA campaign – a Fourier transform spectrometer and a 1/4 meter grating spectrometer. The two instruments were chosen to provide complementary data in the near infrared spectral region, the former between 6000 and 10,000 cm^{-1} and the latter between ~11,300 and ~11,600 cm^{-1}. Coordinated measurements were acquired by both instruments of several ro–vibrational transitions within the ground state of the Meinel band of hydroxyl and one electronic transition of the O_2 At system. Rotational

temperatures were deduced from both species, including several transitions within the Meinel system.

082.039 **Observations of O_2 ($^1\Sigma$) and OH nightglow during the ALOHA–90 campaign.**
J. H. Yee, R. Niciejewski, M. Z. Luo.
Geophys. Res. Lett., Vol. 18, No. 7, p. 1357 – 1360 (Jul 1991).
Two spectroscopic instruments, a 1/4 meter Ebert–Fastie spectrometer and a Michelson interferometer, were flown aboard the NCAR Electra during the ALOHA–90 campaign. These instruments were designed to carry out measurements of the brightnesses and rotatinal temperatures of the O_2 Atmospheric (0–1) band and several OH Meinel bands simulataneously with sodium lidar observations. The spectrometer results taken during the nights of March 25 and March 31 are reported. Wave–like oscillations in both O_2 and OH airglow brightness were clearly present during both flights and a strong positive correlation of OH band brightness with sodium column density was found. A slightly weaker correlation of O_2 band brightness with sodium column density was also detected. The measured brightness ratio between the OH (7–4) and (7–3) bands is found to be 11.2 ± 0.3, which is in good agreement with the calculated ratios of 12.8 by Mies (1974) and 10.2 by Langhoff et al. (1986).

082.040 **Simultaneous lidar and airglow temperature measurements in the mesopause region.**
J. R. Yu, H. Latifi, C. Y. She, J. H. Yee, R. J. Niciejewski.
Geophys. Res. Lett., Vol. 18, No. 7, p. 1361 – 1363 (Jul 1991).
A scientific flight over Fort Collins, CO (40.6°N, 105°W) was conducted on the night of April 15–16, 1990, during the ALOHA–90 campaign. In this Colorado mission, a direct comparison between measured mesospheric O_2 and OH rotational temperatures with the airglow instruments on board the Electra and simultaneously measured Na temperatures by a ground–based narrowband Na lidar has been made. This first comparison resulted in general agreement in measured temperatures between different instruments.

082.041 **Horizontal and vertical structure of the major sporadic sodium layer events observed during ALOHA–90.**
T. J. Kane, C. A. Hostetler, C. S. Gardner.
Geophys. Res. Lett., Vol. 18, No. 7, p. 1365 – 1368 (Jul 1991).
Sporadic Na (Na_S) layers have been observed frequently at low– and high–altitude sites primarily in the Northern Hemisphere. Na_S layers are very dense and very thin ($\sim$1 km FWHM) and can form over periods ranging from a few minutes to several hours. Radar and lidar observations by the authors' groups at Arecibo have shown conclusively that Na_S layers are linked to the formation of sporadic E (E_S) layers. Large Na_S events were observed on several flights during ALOHA–90 to extend over horizontal distances of approximately 1800 km. Atmospheric tidees and gravity waves were observed to have an effect on the vertical motion of these layers. The authors describe the 22 March event which was observed from Maui (21°N) to near the equator (5°N) and the 27 March event which was observed from Maui (156°W) to almost the International Data Line ($\sim$172°W).

082.042 **Formation characteristics of sporadic Na layers observed simultaneously by lidar and airglow instruments during ALOHA–90.**
C. S. Gardner, T. J. Kane, J. H. Hecht, R. L. Walterscheid, J. H. Yee, R. J. Niciejewski, R. P. Lowe, D. N. Turnbull.
Geophys. Res. Lett., Vol. 18, No. 7, p. 1369 – 1372 (Jul 1991).
Sporadic Na (Na_S) layers were observed by the airborne lidar during ALOHA–90 on the 22, 25 and 27 March flight missions. Perturbations in the O_2 and OH nighttime airglow emission intensities and temperature were also observed by instruments on the aircraft and Haleakala Crater (20.8°N, 156.2°W) during these events. The most striking correlation between the airglow and lidar measurements occurred during the northbound flight leg of the 25 March mission. When the Na_S layer formed at 90.7 km, while the Electra aircraft was between 750 and 500 km south of Haleakala, the O_2 temperatures near 95 km above the Electra and Haleakala increased by approximately 45K. The data for this night suggest a connection between Na_S and a large–scale wave, and suggest that the wave is tidal in nature. The data also suggest that some Na_S layers can form very quickly over large geographic areas. Fast chemical processes are required to generate the large amounts of atomic Na involved in some of these events.

082.043 **Gravity wave–driven fluctuations in OH nightglow from an extended, dissipative emission region.**
G. Schubert, R. L. Walterscheid, M. P. Hickey.
J. Geophys. Res., Vol. 96, No. A8, p. 13869 – 13880 (1 Aug 1991).
The theory of gravity wave–driven fluctuations in the OH nightglow from an extended source region is generalized to account for effects of eddy kinematic viscosity ν and eddy thermal diffusivity $\varkappa$.

082.044 **On the nonlinear response of airglow to linear gravity waves.**
J. R. Isler, T. F. Tuan, R. H. Picard, U. Makhlouf.
J. Geophys. Res., Vol. 96, No. A8, p. 14141 – 14152 (1 Aug 1991).
The question of whether a linear gravity wave will give rise to nonlinear effects in ground–based airglow observations is important for the proper interpretation of gravity wave dynamics. In this paper the authors obtain a closed form solution for the integrated airglow response to a linear gravity wave, containing all the higher–order nonlinear response terms. A comparison is made to the linear response, and the higher orders are seen to be significant. In addition, the wave–induced airglow intensity fluctuations are shown to be much greater than the corresponding major species density fluctuations.

082.045 **Tropospheric phenomena responsible for anomalous refraction at radio wavelengths.**
C. E. Coulman.
Astron. Astrophys., Vol. 251, No. 2, p. 743 – 750 (Nov 1991).
Anomalous refraction events cause apparent displacements of radio sources by several seconds or even tens of seconds of arc. Measurements of humidity fluctuations during certain meteorological processes which frequently occur in the troposphere are used to predict the likely phase variations and angular displacements of radio sources. Phenomena studied include the dispersal of the nocturnal inversion, mixing through the inversion which caps the planetary boundary layer, convective structure in that layer and in the sub–cloud layer and the passage of cold fronts. The anomalous refraction effects estimated from these observational data are generally in agreement with published radioastronomical observations.

082.046 **On the short–term variations of the atmospheric transparency.**
N. K. Bektasova, L. K. Karimova, A. V. Kharitonov.
Astron. Tsirk., No. 1545, p. 39 – 40 (Aug – Sep 1990). In Russian.
It is shown that the observed short–term variations may be described by a dynamic process with 2 – 3 nonlinear modes.

082.047 **Why we can see better from space.**
A. I. Lazarev.
Priroda, No. 8, p. 86 – 87 (Aug 1991). In Russian.

082.048 **Interferometric seeing measurements at the La Palma Observatory.**
N. S. Nightingale, D. F. Buscher.
Mon. Not. R. Astron. Soc., Vol. 251, No. 1, p. 155 – 166 (1 Jul 1991).
The authors describe an interferometric seeing monitor which measures the wavefront perturbations due to the atmosphere directly, rather than depending on an assumed model for these perturbations. It is able to measure simultaneously both the spatial and temporal parameters of the seeing (r_0 and t_0) and to test theoretical models. The authors report results from the commissioning run on the William Herschel Telescope. They find

that the seeing at optical wavelengths is well described by a model where the perturbations are due to turbulence with Kolmogorov statistics, and show evidence that the outer scale of this turbulence is about 2 m at the La Palma Observatory.

082.049 **Atmospheric transmission at submillimetre wavelengths from Mauna Kea.**
D. A. Naylor, T. A. Clark, A. A. Schultz, G. R. Davis.
Mon. Not. R. Astron. Soc., Vol. 251, No. 2, p. 199 – 202 (15 Jul 1991).
Measurements of the submillimetre solar spectrum through the 350– and 450–μm windows of the atmosphere using a Michelson interferometer on the James Clerk Maxwell Telescope have been used to determine the transmission spectrum of the atmosphere above Mauna Kea, Hawaii. This spectrum is shown to be in close agreement with that generated by the FASCOD synthesis program using the HITRAN database. The majority of absorption features can be attributed to H_2O (which dominates the overall window shape), O_2 and O_3, although several significant but so far unassigned absorption features are noted. These results demonstrate the feasibility, under drier and more stable atmospheric conditions, of obtaining broad–band, intermediate resolution spectra from the JCMT with a Fourier spectrometer.

082.050 **On the accuracy of atmospheric extinction determination during spectrophotometric observations of stars.**
G. A. Terez, Eh. I. Terez.
Kinematika Fiz. Nebesn. Tel, Tom 7, No. 5, p. 68 – 74 (Sep–Oct 1991). In Russian. English translation in Kinematics Phys. Celest. Bodies, Vol. 7, No. 5.
An analysis of the possible accuracy of determining atmospheric extinction for spectrophotometric observations of stars by the numerical modeling method is made. A dependence of final values of the root–means–square errors of extraterrestrial magnitudes on the number of stars observed during the night and their relative positions is obtained.

082.051 **Special features of the astronomical climate for the observation site in Bolivia.**
G. A. Alekseeva, A. A. Arkharov, V. D. Galkin, V. V. Novikov.
Kinematika Fiz. Nebesn. Tel, Tom 7, No. 6, p. 71 – 73 (Nov–Dec 1991). In Russian. English translation in Kinematics Phys. Celest. Bodies, Vol. 7, No. 6.
An analysis of spectrophotometric data shows that the observation site in Tarija (Bolivia) is characterized by moderate astronomical climate. The clear nights prevail in autumn and winter (from May to July). The amount of water vapour in the atmosphere varies from 5 to 30 mm of precipitated water and even in winter it reaches 15 to 20 mm. The extinction coefficient sharply increases with the increase of the atmospheric moisture.

082.052 **Frequency of clear sky over the Ukrainian central part.**
Ya. S. Yatskiv, N. T. Mironov.
International Seminar on Problems of Astrometry and Space Geodynamics, p. 121 – 128 (1991). In Russian. – See Abstr. 012.049 for the main entry.
General information and regularities on frequency of clear sky (stability, diurnal and seasonal periodicities) over the Ukrainian central part are given on the basis of meteorological observations at 0, 3, 6, – 21 hour of mean solar time for each months of 1979 – 1983. Sites have been selected, which can be characterized as having the maximum frequency of clear sky in daytime and nighttime, for the location of sattelite laser ranging systems.

082.053 **The NASA/MSFC Global Reference Atmospheric Model – 1990 version (GRAM–90). Part 1. Technical/users manual.**
C. G. Justus, F. N. Alyea, D. M. Cunnold, W. R. Jeffries III, D. L. Johnson.
National Aeronautics and Space Administration, Washington, DC (USA), NASA–TM—4268(Part 1). 88 p. (Apr 1991).
A new (1990) version of the NASA/MSFC Global Reference Atmospheric Model (GRAM–90) is presented and discussed. GRAM–90 incorporates extensive new data, mostly collected under the Middle Atmosphere Program (MAP), to produce a completely revised middle atmosphere model (20 to 120 km). The NASA Marshall Engineering Thermosphere (MET) model has also been incorporated into the GRAM–90 program, at all orbital altitudes greater than 120 km. Part I of this report serves as a technical description of the GRAM–90 program, with extensive documentation on the program operation and usage. Sample results are presented, in the form of height–latitude and latitude–longitude plots. Comparisons are made between the empirically based GRAM–90 model and results from a "first principles" stratospheric circulation model (SCM) in the 20– to 80–km height region. Specific areas of improvement of GRAM–90 over the earlier GRAM model results are also noted. A test case example is also included.
For Part II see Abstr. 53.002.070.

082.054 **Tropospheric extinction brightness and excess path length in relation to communication radar and radio–astronomical studies.**
A. K. Sen, P. K. Karmakar, M. Rahaman, K. S. Zand, M. K. Sengupta.
Indian J. Radio Space Phys., Vol. 20, No. 2, p. 79 – 81 (Apr 1991).
Tropospheric radio wave propagation in millimetre wave band is mainly influenced by water vapour content. The analysis of the radiosonde data reveals the variation of communication parameters such as brightness temperature, extinction etc. with water vapour content at 94 GHz and the height up to which the contribution of water vapour towards brightness temperature, extinction and excess path lengths exists. The results show that water vapour content at the height around 2 ± 0.5 km influences more towards the parameters.

082.055 **Dust and light pollution.**
R. H. Garstang.
Publ. Astron. Soc. Pac., Vol. 103, No. 668, p. 1109 – 1116 (Oct 1991).
The author has refined his model for the prediction of the brightness of the night sky due to man–made light pollution by the addition of an ozone layer, by the use of a more accurate representation of the atmospheric molecular density variation as a function of height, and by using a better mathematical representation of the scattering angular function of aerosols. Each of these modifications leads to a small reduction in the predicted brightness of the night sky. He has also added to the model a thin layer of dust of arbitrary optical thickness and height above sea level. Most of the calculations have been performed for Kitt Peak National Observatory. Most calculations have used scattering and absorption coefficients appropriate for volcanic clouds; a few calculations refer to desert dust. Light pollution is reduced by a dust cloud of moderate density whose altitude is below about 10 km and increased for dust clouds at greater altitudes. Observations from good sites are not likely to be greatly affected by the increases in light pollution caused by volcanic clouds at altitudes of order 20 km.

082.056 **On the dynamics and composition of the high latitude thermosphere.**
T. L. Killeen, F. G. McCormac, A. G. Burns, J. P. Thayer, R. M. Johnson, R. J. Niciejewski.
J. Atmos. Terr. Phys., Vol. 53, No. 9, p. 797 – 815 (Sep 1991). Paper presented at the Symposium on Middle Atmospheric Electrodynamics held at IAGA Meeting, Exeter (UK), 1989.
Recent experimental measurements of the dynamics of the neutral upper thermosphere have demonstrated the important roles of ion–drag and Joule heating processes in establishing the basic neutral wind morphology and controlling neutral composition, particularly in the high–latitude region. Instruments on the Dynamics Explorer–2 spacecraft (DE 2), for example, were capable of measuring the three–dimensional vector neutral wind and ion drift in the thermosphere along the orbital track, together with constituent densities and temperatures. Ground–based optical and radar measurements of winds and temperatures from

observatories in Greenland have contributed additional measurements of thermospheric neutral wind velocities and ionospheric parameters. The comprehensive nature of these various data sets has enabled more stringent experimental constraints to be placed on the numerical models of the region (thermosphere–ionosphere general circulation models, TIGCMs), leading to an improved theoretical understanding of the important physical processes that control thermospheric circulation and variability. In addition, the measurements have enabled the development of semi-empirical models of thermosphere dynamics which can be used in various theoretical studies.

082.057 **Night–sky brightness from the visibility of stars near the horizon.**
A. R. Upgren.
Publ. Astron. Soc. Pac., Vol. 103, No. 670, p. 1292 – 1295 (Dec 1991).

Hundreds of naked–eye observations of bright stars near the southern horizon have been made on clear moonless nights over the interval 1977 – 91. They reveal a consistency in their visibility from one season or year to another, but a deterioration over the entire time interval reveals a possible increase in light pollution and possibly air pollution as well. The method is constrained to the sky near the northern and southern horizons where stars remain at near constant altitude over a suitable time interval, but the repeatability assures that it is sound. The limiting magnitude is given as a function of zenith distance, and color and seasonal effects do not appear to influence the results.

082.058 **Mesospheric temperature inferred from daytime observation of the O_2 atmospheric (0,0) band system.**
J. W. Heller, A. B. Christensen, J. H. Yee, W. E. Sharp.
J. Geophys. Res., Vol. 96, No. A11, p. 19499 – 19506 (1 Nov 1991).

A rocket–borne Ebert–Fastie spcectrometer was used to measure the dayglow $O_2(b^1\Sigma^+_g)$ atmospheric (0,0) band centered at 761.9 nm. The tangent altitude for the observations ranged from 63 to 101 km. Analysis of the (0,0) band observations is presented. The peak intensity of the (0,0) band was 3.7 ± 0.2 megarayleigh at 87 km. The results were compared to a multiple–scattering radiative transfer model that includes a detailed photochemical calculation of the sources and sinks of the O_2 excited state. The kinetic temperature was determined by analysis of the rotational development of the O_2 (0.0) band. The average temperature between 94 and 101 km was 182 ± 7K.

082.059 **Atmospheric constituent density profiles from full disk solar occultation experiments.**
J. D. Lumpe, C. S. Chang, D. J. Strickland.
J. Quant. Spectrosc. Radiat. Transfer, Vol. 46, No. 6, p. 483 – 506 (Dec 1991).

An algorithm is presented for the retrieval of atmospheric number density profiles by inversion of full disk solar occultation data. A general derivation of the normalized absorption profile obtained from such measurements is first given, assuming an arbitrary solar intensity distribution. Using Fourier transform techniques, the resulting integral equation is then solved exactly to yield an expression for the number density in terms of the measured absorption profile. This exact mathematical solution of the integral equation eliminates the need for an iterative solution based on an assumed model density profile. The formal solution is implemented numerically by introducing a suitable representation of the data in terms of Laguerre polynomials, and model calculations are presented for the special case of a uniform solar intensity distribution. The results of these calculations are used to demonstrate the convergence properties of the algorithm and its suitability for accurate reproduction of several representative model density profiles with both constant and altitude–dependent scale heights. Finally, a simple numerical calculation is performed to estimate the effect of nonuniform solar intensity distributions on the retrieved density.

082.060 **Organised observations of night–sky brightness in Japan during 1987–1989.**
H. Kosai, S. Isobe.
Proc. Astron. Soc. Aust., Vol. 9, No. 1, p. 180 – 183 (1991). – See Abstr. 012.078 for the main entry.

This paper describes the Star Watch Program in Japan to monitor night–sky brightness.

082.061 **The altitude dependence of the $OH(X^2\Pi)$ vibrational distribution in the nightglow: some model expectations.**
I. C. McDade.
Planet. Space Sci., Vol. 39, No. 7, p. 1049 – 1057 (Jul 1991).

A number of different OH Meinel band excitation models are used to investigate the extent to which quenching and vibrational deactivation processes could cause the nightglow OH vibrational distributions to vary with altitude. The various models, which are based upon those described by McDade and Llewellyn (1988), are used to calculate the steady state OH vibrational distributions throughout the 80–100 km region. The results of the calculation show that, irrespective of the basic model assumptions, the OH vibrational distributions should not be strongly altitude dependent and that the emission profiles of the Meinel bands from the v' levels 1 through 9 should not differ in altitude by more than one or two kilometers. The results of some recent laboratory studies pertaining to the Meinel band excitation mechanism are also discussed and it is shown that stepwise collisional vibrational deactivation is unlikely to play a major role in controlling the observed nightglow OH vibrational distribution.

082.062 **Simultaneous intensity, temperature and imaging measurements of short period wave structure in the OH nightglow emission.**
M. J. Taylor, P. J. Espy, D. J. Baker, W. R. Pendleton Jr., R. J. Sica, P. C. Neal.
Planet. Space Sci., Vol. 39, No. 8, p. 1171 – 1188 (Aug 1991).

A low light TV camera co–aligned with a narrow field Michelson interferometer have been used to investigate how variations in the intensity and temperature of the near infrared hydroxyl nightglow emission are related to the passage of short period mesospheric gravity waves. The observations were made from Sacramanto Peak, NM (32.8°N, 105.8°W) on the night of 14/15 June 1983, during a spectacular nightglow display that also exhibited wave structure visible to the dark adapted eye. A coherent wave–like pattern of horizontal wavelength 23 ± 1 km moving uniformly with a velocity of 28 ± 2 m s^{-1} (apparent period 13.7 ± 1.2 min), was measured at several azimuths over an interval of 2.5 h. The induced intensity and temperature perturbations were wave–like in nature and exhibited a high degree of correlation. The intensity variations were large, up to 30%, yet the brightest wave forms were ≤ 10K hotter than the adjacent dark forms. The amplitude of the intensity fluctuation was found to be ~ 8 times larger than the temperature variation. In all cases the intensity wave led the thermal wave, but only by a small amount. High contrast wave patterns of this type are uncommon and are a principal feature of "bright night" displays.

082.063 **High latitude summer observations of the hydroxyl airglow.**
R. P. Lowe, K. L. Gilbert, D. N. Turnbull.
Planet. Space Sci., Vol. 39, No. 9, p. 1263 – 1270 (Sep 1991).

Spectra of the hydroxyl night airglow in the near infrared region were obtained under high latitude summer conditions at Poker Flat, Alaska, during August 1986. Rotational temperatures and relative populations have been deduced from the spectra for vibrational levels from 2 to 8. During the observing period the nightly average rotational temperature increased from 155 to 173K, but the nightly variation of temperature generally did not exceed 10K. On one night, a short period nearly sinusoidal variation of both intensity and temperature was observed, with small phase shifts which varied with vibrational level. Correlation of intensity and temperature variation on two nights gave a value of ~ 8 for Krassovsky's parameter, η. Differences in the average rotational temperature as measured from bands arising from different vibrational levels were noted, but in general they did not exceed the possible calibration uncertainty.

082.064 **The temperature dependence of airglow emissions from the upper mesosphere and lower thermosphere.**
B. R. Clemesha, H. Takahashi, P. P. Batista, Y. Sahai, D. M. Simonich.
Planet. Space Sci., Vol. 39, No. 10, p. 1397 – 1404 (Oct 1991).
An analysis of a long time-series of measurements of the airglow emissions, OH(9,4) 775.0 nm, Na D 589 nm, O I 557.7 nm and $O_2(0,1)$ 864.5 nm atmosphere band, shows that the nocturnal mean intensities of all these emissions are highly correlated with the nocturnal temperature means, as determined from the rotational structure of the OH(9,4) emission. After removing seasonal and long-term variations, the closest correlation is observed between the OH(9,4) emission intensity and the corresponding rotational temperature, followed by the NaD, and the O I 557.7 nm and O_2 A emissions. The regression coefficients, (dI/I)/(dT/T), where I is intensity and T is temperature, are between 8 and 9 for the Na D and oxygen emissions, and about 5.5 for the OH band. It is clear that these regression coefficients do not reflect the temperature dependences of the various reactions involved. It seems probable that the observed correlations are the result of correlated temperature and density changes in the upper atmosphere.

082.065 **The NRAO Millimeter Array site survey.**
F. N. Owen.
IAU Colloquium No. 131: Radio interferometry – theory, techniques, and applications, p. 124 – 127 (1991). – See Abstr. 012.089 for the main entry.
The measurements being made to pick a site for the NRAO Millimeter Array are described. The instruments, techniques and preliminary results are discussed, along with the strategy for picking the final site for the array.

082.066 **Ultraviolet spectroscopy and remote sensing of the upper atmosphere.**
R. R. Meier.
Space Sci. Rev., Vol. 58, No. 1/2, p. 1 – 185 (Oct 1991).
The Earth's ultraviolet airglow contains fundamental diagnostic information about the state of its upper atmosphere and ionosphere. Our understanding of the excitation and emission processes which are responsible for the airglow has undergone dramatic evolution from the earliest days of space research through the past several years during which a wealth of new information has been published from high-resolution spectroscopy and imaging experiments. This review of the field begins with an overview of the phenomenology: how the Earth looks in the ultraviolet. Next the basic processes leading to excitation of atomic and molecular energy states are discussed. These concepts are developed from first principles and applied to selected examples of day and night airglow; a detailed review of radiation transport theory is included. This is followed by a comprehensive examination of the current status of knowledge of individual emission features seen in the airglow, in which atomic physics issues as well as relevant atmospheric observations of major and minor neutral and ionic constituents are addressed. The use of airglow features as remote sensing observables is then examined for the purpose of selecting those species most useful as diagnostics of the state of the thermosphere and ionosphere. Imaging of the plasmasphere and magnetosphere is also briefly considered. A summary of upcoming UV remote sensing missions is provided.

082.067 **Untersuchung der Flugzeugkondensstreifen und ihrer klimatischen Auswirkungen in Kärnten.**
H. F. Haupt, M. Rotter.
Wetter und Leben, 43. Jahrg., p. 173 – 188 (1991). = Mitt. Sonnenobs. Kanzelhöhe, Nr. 77.
It has been suspected that the increasing sky pollution by condensation trails had deteriorated the observing conditions at the Austrian Solar Observatory Kanzelhöhe as well as the general climatic conditions in Carinthia. An investigation on site was made from November 1985 to October 1986 by visual recording and by the use of photographs with an all sky camera to check the frequency, extension and persistence of the contrails. Mean sizes and lifetimes for different types of contrails were derived. About 15% of the observing days at the solar station were hampered by contrails. There seemed to be no significant influence to the mean temperature and cloudiness in Carinthia.

082.068 **Die Nachtbewölkung von Innsbruck.**
J. Pesch.
Wetter und Leben, 43. Jahrg., p. 143 – 150 (1991).
To assess the possibilities of observation of the starry sky it is shown that, analogous to the classification customary in climatology of bright and dull days, cloudiness at night can be defined as absolutely clear, bright, short time bright intervals, and unsuitable for star observation. It is demonstrated that synoptical observations of flight weather stations can be used. Based on the overall night cloudiness records of the Innsbruck Flight Weather Bureau over a period of 20 years (1947 – 1966) mean values are developed and discussed. As is to be expected for the Alps favourable night observation weather prevailed in autumn during the months of September and October, followed by the winter months January to March.

082.069 **Photographisch–photometrische Bestimmung der atmosphärischen Extinktion in Innsbruck.**
J. Pesch.
Sitzungsber., Österr. Akad. Wiss., Math.-Naturwiss. Kl., Abt. II, Band 200, Heft 1 – 10, p. 1 – 12 (1991).
Observations of the atmospheric extinction in Innsbruck were made using a photographic method of extrafocal images in a small well defined wavelength range (λ_{eff} = 407.5 nm). The instrument and the method of observation and reduction are described in some detail. For the observing period 1953/54 a mean extinction coefficient $\bar{k} = 0\overset{m}{.}607 \pm 0\overset{m}{.}025$ was found from 78 nights. For a few very good nights it could be shown that not the Alpine Föhn was the main cause for the fine transparency but rather downstreaming warm and dry airmasses from ridges and wedges of large and high reaching anticyclons.

082.070 **Molecular absorption of radiation in the 8 – 13 μm atmospheric window (review).**
V. N. Aref'ev.
Izv. Akad. Nauk SSSR, Fiz. Atmos. Okeana, Vol. 27, No. 11, p. 1187 – 1225 (Nov 1991). In Russian.
Laboratory and field measurements results of molecular atmospheric gases and water vapor continuum absorptions are reviewed.

082.071 **The search for an astronomical site at some Argentinian bases in the Antarctica.**
J. H. Calderón, P. G. Recabarren, M. J. Lombardo, H. A. Ochoa.
Bol. Asoc. Argent. Astron., No. 36, p. 316 – 325 (1990). In Spanish. – See Abstr. 012.095 for the main entry.
Córdoba Astronomical Observatory has begun a site testing program in Argentine scientific stations at Antarctica. Measurements in order to estimate sky quality were made during the 1989 polarnight at the Belgrano II station, which offerred the best conditions according to its geographical position and meteorological conditions.

082.072 **Sky conditions for infrared observations at Kavalur.**
K. K. Scaria, R. M. Nair, K. R. Sivaraman.
Bull. Astron. Soc. India, Vol. 19, Nos. 1 – 2, p. 65 – 73 (Mar – Jun 1991).
The authors present the extinction coefficients in the J, H, K near infrared bands derived for Kavalur during the years 1987, 1988 and 1989. Humidity and temperature observations made at Kavalur during these years are used to ascertain the most suitable months for infrared observations.

082.073 **Optically permitted emissions excited in the atmosphere by an impulse source of ultraviolet radiation.**
N. V. Eliseev, V. A. Kiselev, S. I. Kozlov.
Kosm. Issled., Tom 29, Vyp. 6, p. 950 – 951 (Nov – Dec 1991). In Russian. English translation in Cosmic Res., Vol. 29, No. 6.

082.074 **Unusual solar halos over La Silla.**
D. Hutsemekers.
Messenger, No. 66, p. 18 – 20 (Dec 1991).

082.075 **The Earth's dust cloud and atmospheric oxygen.**
V. N. Lebedinets.
Sol. Syst. Res., Vol. 25, No. 3, p. 261 – 272 (Nov 1991). English translation of Astron. Vestn., Tom 25, No. 3, p. 350 – 363 (1991).
Hypotheses of "minicomets" and the content, in their snow nuclei, of high-molecular organic compounds in amounts of a few percentage points provide, for the first time, a physical mechanism explaining the formation of the Earth's dust cloud with a dust concentration higher than the interplanetary background by two or three orders of magnitude. The easily evaporating organic particles and loose chondrite dust particles are registered very poorly by punch sensors of micrometeorites but quite well by acoustic and optical sensors. The influx of minicomets is the main source of oxygen in the entire stratum of the atmosphere and water in the thermosphere, mesosphere, and the upper stratosphere. Swarms of minicomets may exist on the orbits of certain large comets. At an encounter with a swarm, a brief rise of the dust content in the upper atmosphere and in the near-Earth space may be observed.

082.076 **Mesospheric clouds and the physics of the mesopause region.**
G. E. Thomas.
Rev. Geophys., Vol. 29, No. 4, p. 553 – 575 (Nov 1991).
During the last decade, new satellite observations of polar mesospheric clouds and ground-based radar and lidar sounding of the high-latitude summertime mesosphere have stimulated new interest in this highly unusual atmospheric region. Computer simulations indicate that the ice condensation model for mesospheric cloud formations is correct in broad outline. This review contains a brief historical account of early mesospheric cloud studies with an emphasis on the two main theories: cosmic dust and ice condensation models. Current topics are described, including wave-cloud interactions, the dynamics and momentum budget of the high-latitude polar mesosphere, new satellite observations, polar mesospheric summer radar echoes, heterogeneous chemistry, and implications of apparent secular changes in cloud brightness for global change of the atmosphere as a whole. There are major uncertainties in the theoretical understanding of the ice formation process, particularly in how the ice particles are nucleated. Future progress will require new research efforts in three separate directions: observations (ground based, rocket, and satellite), laboratory experiments, and theoretical modeling.

082.077 **The small-comet hypothesis.**
A. J. Dessler.
Rev. Geophys., Vol. 29, No. 3, p. 355 – 382 (Aug 1991). With a correction in Vol. 29, No. 4, p. 609 – 610 (Nov 1991).
According to the small-comet hypothesis, small comets strike the Earth approximately 20 times per minute, each small comet nominally containing 100 tons of water-ice. The primary observations interpreted as evidence for these small comets are dark spots in the Earth's atomic-oxygen UV dayglow seen by the UV imager on the Dynamics Explorer satellite. These small comets must disintegrate near Earth and then sublimate within a few seconds, the water vapor expanding to form clouds of water vapor, nominally 50 km in diameter, that temporarily block the spacecraft's view of the dayglow, thus producing the dark spots. In this review the author examines problems in basic mechanisms underlying the small-comet hypothesis. These include inconsistencies with known geophysical phenomena, conflicting results from independent searches for evidence of the presence of small comets, and inconsistencies within the small-comet hypothesis itself.

082.078 **Spectroscopic measurement of HO_2, H_2O_2, and OH in the stratosphere.**
J. H. Park, B. Carli.
J. Geophys. Res., Vol. 96, No. D12, p. 22535 – 22542 (20 Dec 1991).
Stratospheric concentrations of HO_2, H_2O_2, and OH have been retrieved simultaneously from the far infrared emission spectra obtained with a balloon-borne Fourier transform spectrometer in June 1983 at 32°N latitude. Retrieved concentrations of HO_2 and H_2O_2 are reported in this paper, along with vertical distributions of OH which were reported in an earlier paper for the afternoon, sunset, and nighttime periods for altitudes from 26 to 38 km. HO_2 distributions are obtained with uncertainties that are about the same as OH for the same vertical range and for the afternoon and sunset periods. H_2O_2 concentration is obtained at an altitude of 30 km for the period that covers afternoon and sunset hours. The retrieved concentrations of these HO_x species agree well with other individually measured results and the steady state photochemical predictions. The ratio HO_2/OH at around 32 km seems to increase from the afternoon period to the sunset period.

082.079 **Laboratory studies of large protonated water clusters under the conditions of formation of noctilucent clouds in the summer mesopause.**
X. Yang, A. W. Castleman.
J. Geophys. Res., Vol. 96, No. D12, p. 22573 – 22578 (20 Dec 1991).
Studies of the formation and growth of large protonated water clusters in a fast flow tube reactor at temperatures equivalent to those under which NLC are known to form offer, for the first time, the possibility to experimentally investigate some of the mechanisms of NLC formation at the molecular level. Importantly, bimodal cluster distributions obtained in the laboratory experiments are similar to those found in situ and also are in accordance with predictions based on ion-induced nucleation theories. The experiments show a strong dependence of temperature and water partial pressure on the production of large protonated water clusters of sizes similar to those observed during NLC displays. A consideration of the thermodynamics of the clustering reactions and their unimolecular dissociation kinetics provides a further understanding of the conditions necessary for the observation of the protonated water clusters both in the laboratory and in situ. Investigations of chemical reactions of the protonated water clusters with a variety of neutral reactants were also made, and the findings enable an assessment to be made of the chemical environment existing during NLC formation.

082.080 **An unusual display of noctilucent clouds.**
M. S. Zalcik.
J. Geophys. Res., Vol. 96, No. D12, p. 22709 – 22715 (20 Dec 1991).
On the morning of July 1, 1990, an intense display of noctilucent clouds as seen from Namao, Alberta, completely disappeared in 30 min. Observations of the clouds from Slave Lake, Alberta, indicate that the display lasted only slightly longer from that site. The noctilucent clouds' rapid degeneration may have been due to induced heating of their formation zone, perhaps by a coincident aurora.

The aurora. Sun-Earth interactions.
See Abstr. 003.015.

ISPP-7. Controlled active global experiments (CAGE). Proceedings.
See Abstr. 012.083.

Infrared spectroscopy as a tool to study earth's atmosphere.
See Abstr. 013.045.

High-resolution imaging in La Palma.
See Abstr. 013.054.

Progress on light pollution control at the David Dunlap Observatory.
See Abstr. 013.100.

Simularea numerica a proceselor microfizice din ceata.
See Abstr. 022.061.

Determination of absolute strengths of N_2 quadrupole lines from high–resolution ground–based IR solar observations.
See Abstr. 022.098.

Adaptive optics experiments using synthetic beacons.
See Abstr. 031.007.

Evaluation of seeing on a 62–cm mirror.
See Abstr. 031.026.

A new type of telescope flow control – the unclosure.
See Abstr. 032.004.

Telescope enclosure flow visualization.
See Abstr. 032.005.

Mirror, dome, and natural seeing at CFHT.
See Abstr. 032.017.

Wind–tunnel study of an observatory dome with a circular aperture.
See Abstr. 032.022.

Experimental demonstration of atmospheric compensation using multiple synthetic beacons.
See Abstr. 036.075.

Towards precision photometry.
See Abstr. 036.082.

Test for the method for correcting the atmospheric seeing and the diffraction limit of a telescope by means of the 2–dimensional DFT deconvolution. The application of DFT to astronomy (III).
See Abstr. 036.093.

Glare and celestial visibility.
See Abstr. 036.110.

Compensation of atmospheric optical distortion using a synthetic beacon.
See Abstr. 036.213.

Measurement of atmospheric wavefront distortion using scattered light from a laser guide–star.
See Abstr. 036.214.

High spatial resolution techniques.
See Abstr. 036.276.

How to improve old astrometric observations by introducing more accurate corrections for astronomical refraction.
See Abstr. 041.019.

On the influence of atmospheric dispersion on the photographic positions of stars.
See Abstr. 041.022.

Extratropical aspects of the 40–50 day oscillation in length–of–day and atmospheric angular momentum.
See Abstr. 044.017.

Earth orientation and reference frame determinations, atmospheric excitation functions, up to 1990. (Annex to the IERS annual report for 1990).
See Abstr. 044.044.

EUVE observations of the Earth and planets.
See Abstr. 051.005.

Analytical treatment of air drag and Earth oblateness effects upon an artificial satellite.
See Abstr. 052.016.

Emergent radiation from the atmosphere bounded by the two halves of the Lambert surface with different albedos.
See Abstr. 063.023.

The solution of radiative transfer problems in molecular bands without the LTE assumption by accelerated lambda iteration methods.
See Abstr. 063.083.

Radiation flux in a weakly absorbing atmosphere of arbitrary optical thickness.
See Abstr. 063.119.

A simple algorithm for simulating the solar ultraviolet radiation at the Earth's surface: an application in determining the minimum erythema dose.
See Abstr. 085.031.

The lower atmospheric response to solar activity of March 1989.
See Abstr. 085.037.

A possible effect of the solar activity on atmospheric ozone content.
See Abstr. 085.038.

Studies of the correlation between the global radiation, ultraviolet radiation, cosmic radiation, sunspot number and meteorological parameters.
See Abstr. 085.040.

The research advance and problems about 11–year solar cycle and quasi–biennial oscillation of atmosphere.
See Abstr. 085.048.

The atmosphere's response to solar irradiation.
See Abstr. 085.052.

Role of carbon dioxide in cooling planetary thermospheres.
See Abstr. 091.017.

Atmospheric variability on a zonally symmetric land planet.
See Abstr. 091.056.

Comparative terrestrial planet thermospheres. 1. Solar cycle variation of global mean temperatures.
See Abstr. 093.045.

Nitrogen airglow sources: comparison of Triton, Titan, and Earth.
See Abstr. 101.050.

Terrestrial cometary tail and lunar corona induced by small comets: predictions for Galileo.
See Abstr. 102.014.

The small–comet hypothesis.
See Abstr. 102.029.

The micrometeoroids in the thermosphere and mesosphere.
See Abstr. 105.028.

The production of cosmogenic isotopes in the Earth's atmosphere and their inventories.
See Abstr. 144.104.

Solar and terrestrial components of the atmospheric ^{14}C variation spectrum.
See Abstr. 144.106.

The temperature of the cosmic microwave background radiation at 3.8 GHz: results of a measurement from the South Pole site.
See Abstr. 161.605.

083 Ionosphere

083.001 **Readjustment of the large–scale structure of the high–latitude ionosphere upon a change in the UV radiation of the Sun and the Earth's thermosphere.**
A. G. Kolesnik, V. I. Chernyshev, V. N. Platonov.
Geomagn. Aeron., Vol. 30, No. 1, p. 122 – 125 (Aug 1990). English translation of Geomagn. Aehron., Tom 30, No. 1, p. 149 – 152 (1990).
Numerical experiments are conducted on the basis of a three–dimensional time–dependent model of the ionosphere to investigate the readjustment of the large–scale structure of the high–latitude ionosphere upon a simultaneous change in the UV radiation of the Sun and the Earth's thermosphere. It is shown that in contrast to the mid–latitude ionosphere, a breakdown in the steady state in the high–latitude ionosphere is accompanied by a more extended recovery time, which is related to the dominant process of plasma convection under the action of an electric field of magnetospheric origin. The upshot of this is that the high–latitude ionosphere is located in a nonsteady state practically constantly.

083.002 **The role of super–thermal electrons in the formation of luminous regions in the vicinity of a cosmic body.**
A. Yu. Ol'khovatov.
Geomagn. Aeron., Vol. 30, No. 1, p. 136 – 137 (Aug 1990). English translation of Geomagn. Aehron., Tom 30, No. 1, p. 161 – 163 (1990).
Questions relating to the formation of large regions of luminescence around a cosmic body are considered. On the basis of a comparison of the data from meteor observations and the results of rocket sounding of the ionosphere, it is concluded that these phenomena are caused by heating of the ambient electrons with subsequent disturbance of the atmospheric components by electron impact. The assumption is advanced that the heating mechanism of the ambient electrons is associated with the formative mechanism of sporadic ionospheric layers (ionosphere inhomogeneities).

083.003 **Calculation of the electron temperature in the ionosphere–plasmasphere system.**
O. A. Gorbachev, Yu. V. Konikov, I. M. Siderov, G. V. Khazanov.
Geomagn. Aeron., Vol. 30, No. 3, p. 415 – 417 (Nov 1990). English translation of Geomagn. Aehron., Tom 30, No. 3, p. 495 – 497 (1990).
In the context of coordinated models of ionosphere–plasmasphere interactions, calculations are made of the electron temperature using the standard approach and taking into account the transfer equation for heat flows. It is shown that the main differences in the calculated temperature distributions are observed during the periods of sunrise and sunset and grow with increasing height and number of the geomagnetic force tube. These deviations result from the effects of the nonsteady nature of the heat flow, which, if taken into account, also reduces to an acceleration of the heating and cooling processes of the electrons in the morning and evening hours, respectively.

083.004 **The high–latitude ionosphere. Model description and analysis of observational results applied to problems of prognosis.**
G. I. Ostrovskij.
Geomagn. Aeron., Vol. 30, No. 3, p. 418 – 420 (Nov 1990). English translation of Geomagn. Aehron., Tom 30, No. 3, p. 498 – 500 (1990).
An analysis is made of the experimental data obtained at vertical–sounding stations and by the satellite Kosmos–1809 during the period February – June 1987. It is shown that in most cases perturbations in the D and E regions of the ionosphere are not accompanied by disturbances in the F_2 region.

083.005 **Dynamics of nonlinear internal gravity waves at ionosphere F–region heights.**
V. Yu. Belashov.
Geomagn. Aeron., Vol. 30, No. 4, p. 536 – 538 (Feb 1990). Paper presented at the 5. Symposium on Solar Terrestrial Physics, Samarkand (USSR), 2 – 6 Oct 1989. English translation of Geomagn. Aehron., Tom 30, No. 4, p. 637 – 641 (1990).
A two–dimensional evolutionary equation describing the dynamics of nonlinear wave disturbances in the neutral component at F–region heights of the ionosphere is derived. A broad range of solutions is investigated numerically, and the excitation of traveling disturbances of the F–layer electron concentration by nonlinear internal gravity waves is considered. As an example, the model problem is solved for moderate– and large–scale wave effects of a solar eclipse.

083.006 **On the SID durations.**
T. Ataç.
Astrophys. Space Sci., Vol. 181, No. 1, p. 157 – 159 (Jul 1991). Letter–to–the–editor.
This Letter is based on analysis of data on Sudden Ionospheric Disturbances (SID) listed in Solar Geophysical Data (SGD). The mean durations of 1272 SIDs taken from the SGD lists were calculated with respect to their associated X–ray flares. In general, all of the calculated mean durations showed a large scatter in the mean values. The comparison of the values by the estimated Solar Geophysical Data Center SID durations indicated an apparent discrepancy in the mean duration relative to that estimated by SGD for the case of low levels of SID importance. In the author's opinion this apparent discrepancy suggests a need for a correction in SGD's assumptions in the importance range from –1 to +1.

083.007 **Electron density structures in the polar F region.**
J. Buchau, B. W. Reinisch.
Adv. Space Res., Vol. 11, No. 10, p. 29 – 37 (1991). – See Abstr. 012.013 for the main entry.
Arctic campaigns conducted over the last decade, using digital ionospheric sounders and all–sky imaging photometers have allowed an ordering of the typically very irregular behavior of the polar cap ionosphere, and a preliminary definition of the solar cycle effects. Under quiet magnetic conditions the polar cap is populated with sun–aligned F layer arcs. Under magnetically active conditions large patches of enhanced plasma densities move rapidly from the cusp region across the polar cap. Both arcs and patches are highly structured. Summer and winter ionospheric soundings in the central polar cap under various solar cycle conditions, establish the observed pattern as a start for the systematic modeling of this most turbulent region.

083.008 **Model and observation comparison of the universal time and IMF by dependence of the ionospheric polar hole.**
J. J. Sojka, R. W. Schunk, W. R. Hoegy, J. M. Grebowsky.
Adv. Space Res., Vol. 11, No. 10, p. 39 – 42 (1991). – See Abstr. 012.013 for the main entry.
The polar ionospheric F–region often exhibits regions of marked density depletion. These depletions have been observed by a variety of polar orbiting ionospheric satellites over a full range of solar cycle, season, magnetic activity, and universal time (UT). An empirical model of these observations has recently been developed to describe the polar depletion dependence on these parameters.

083.009 **IMF influence upon the electron density of the midlatitude F2 layer.**
L. Třísková.
Adv. Space Res., Vol. 11, No. 10, p. 43 – 46 (1991). – See Abstr. 012.013 for the main entry.
The IMF polarity influences the nighttime midlatitude electron density causing uncertainty in the prediction of foF2.

083.010 **An influence of IMF–By on the ion density planetary distribution during disturbed period 20 – 23 October 1981: satellite and ground–based results.**
N. A. Kilifarska, Ts. P. Dachev, A. S. Besprozvannaya, T. I. Schuka.
Adv. Space Res., Vol. 11, No. 10, p. 47 – 50 (1991). – See Abstr. 012.013 for the main entry.
The planetary distribution of the disturbed ion density (Ni) at fixed magnetic latitudes in both hemispheres, and in the evening and morning sectors, is examined using DE–B satellite and ground–based vertical sounding data. It is found that (1) the amplitude or (in some cases) the shape changes of the disturbed Ni longitude variations are asymmetrical in both hemispheres as well as in the morning and evening sectors; (2) the amplitude of "storm" variations is different at height of F2–layer peak and in the upper ionosphere (800 – 900 km). These irregularities in the ionospheric storm reaction are interpreted in terms of IMF influence on the magnetospheric convection pattern and its asymmetry – not only between evening and morning sectors in a given hemisphere, but also between northern and southern hemispheres.

083.011 **Data needed for an analytical description of the electron density profile.**
K. Rawer.
Adv. Space Res., Vol. 11, No. 10, p. 57 – 65 (1991). – See Abstr. 012.013 for the main entry.
Constraints are needed when applying the analytic profile description using LAY–functions. The reliability of these is tested experimentally. Some have a well–defined geophysical meaning. Others have been suggested which can be used to study the morphology of the terrestrial ionosphere. The data needed to provide these are indicated, gaps are identified.

083.012 **Statistical studies of the N(h) profile of the ionosphere using automatically scaled digital ionograms.**
B. W. Reinisch, L. F. McNamara, T. Bullett, R. R. Gamache.
Adv. Space Res., Vol. 11, No. 10, p. 81 – 87 (1991). – See Abstr. 012.013 for the main entry.
The advent of digital vertical incidence ionosondes and automatic scaling has brought with it the capability of performing large–scale analyses of the detailed structure of the ionosphere. In this paper, the authors use about 6,000 Digisonde 256 ionograms recorded at Millstone Hill, Massachusetts during 1988 and scaled automatically by ARTIST, to investigate relationships which exist between various profile parameters and to investigate the accuracy of estimates of these parameters based on simple models of the ionosphere.

083.013 **Ionosonde observations of the E–F valley and comparison with incoherent scatter radar profiles.**
C.–F. Chen, B. D. Ward, B. W. Reinisch, M. J. Buonsanto, R. R. Gamache.
Adv. Space Res., Vol. 11, No. 10, p. 89 – 92 (1991). – See Abstr. 012.013 for the main entry.
Modern digital ionosondes represent the most practical method for routine measurement of electron density profiles in the ionosphere. However, inversion of ionogram data to produce an N(h) profile suffers from the well known valley problem. Regions of decreased ionization between the ionospheric layers do not reflect the radio wave and the resulting discontinuity leads to uncertainty in the true height above the valley region. Incoherent scatter radar is capable in principle of providing direct information on the electron density variation within the E–F valley. Colocation of a Digisonde 256 sounder with the Millstone Hill incoherent scatter radar affords the opportunity to carry out detailed comparisons.

083.014 **A D–region electron density model based on rocket borne wave propagation data.**
M. Friedrich, K. M. Torkar.
Adv. Space Res., Vol. 11, No. 10, p. 101 – 104 (1991). w – See Abstr. 012.013 for the main entry.
The temporal and spatial coverage by rocket borne radio wave propagation experiments is rather poor. These data, however, are considered the most reliable in the lower ionosphere. Published electron density profiles were fitted to simple functions of solar zenith angle at different latitudes. Where applicable comparisons are made to the corresponding results from IRI.

083.015 **Electron and ion temperature data for ionospheric modelling.**
D. Bilitza.
Adv. Space Res., Vol. 11, No. 10, p. 139 – 148 (1991). – See Abstr. 012.013 for the main entry.
The major sources for ionospheric temperature data are the incoherent scatter radars and the insitu instruments (Langmuir probe, retarding potential analyzer) flown on several long–lasting satellite missions. The radar data are best suited for investigations of the short and long–term variations at the radar location, whereas the satellite data provide the means to understand the global characteristics of electron and ion temperatures. Both techniques have been considerably improved over the last decade and comparisons between simultaneous measurements show excellent agreement. This paper provides an overview over the different data sets and discusses their volume, their temporal and spatial resolution, their accuracy, and their availability. The state of the art of empirical modelling of ionospheric plasma temperatures is examined and present shortcomings and future data needs are pointed out.

083.016 **Empirical models of the latitudinal variations of Te and Ne in the ionosphere at solar maximum.**
L. H. Brace, R. F. Theis.
Adv. Space Res., Vol. 11, No. 10, p. 159 – 166 (1991). – See Abstr. 012.013 for the main entry.
Global spectral models of electron temperature (T_e) and density (N_e) were produced earlier using Dynamics Explorer–2 (DE–2) Langmuir probe measurements made between 1981 and 1983. The models were derived for the fixed altitudes of 300, 500, and 850 km. Unfortunately, the DE–2 data base is not adequate to fully define all of these variables because of the eccentricity of the orbit, the slow evolution of the orbit, and the brevity of the mission. In the present work the authors take a different approach by modelling only the latitudinal variations observed in narrow altitude and local time bands and using measurements obtained during intervals short compared to a season. This approach allows the data to accurately define latitudinal terms up to 17 th order. To illustrate this improved latitudinal resolution, selected models of T_e and N_e at 500 km are compared with latitude profiles from the earlier global models and with the DE–2 measurements for the same local times and altitudes.

083.017 **The relationship between electron density and electron temperature under quiet conditions in the auroral zone.**
A. R. Breen, P. J. S. Williams.
Adv. Space Res., Vol. 11, No. 10, p. 167 – 170 (1991). – See Abstr. 012.013 for the main entry.
A theoretical model accurately predicts the electron temperature in the daytime F–region given the observed values of electron density, ion temperature, the daily average value of 10.7–cm solar radiation and the MSIS–86 model of the neutral atmosphere. The model was originally developed using data from Malvern, St. Santin and Arecibo and has now been confirmed using data from EISCAT on geomagnetically quiet days. The quality of EISCAT data is good enough to allow the comparison of predicted and observed values of T_e on an hourly basis, including estimates of cooling by heat conduction. The agreement is very good for all data from quiet conditions.

083.018 **Ionospheric ion composition from satellite measurements made during 1970 – 1980: altitude profiles.**
W. R. Hoegy, J. M. Grebowsky, L. H. Brace.
Adv. Space Res., Vol. 11, No. 10, p. 173 – 182 (1991). – See Abstr. 012.013 for the main entry.
Ion mass spectrometers were carried by a number of satellites in the 1970s. The ion composition measurements from two of these missions, the Orbiting Geophysical Observatory–6 and the Atmosphere Explorer–C, have been collected into an ion

composition data base to evaluate several widely used empirical and theoretical models for the species H^+, He^+, N^+, O^+, NO^+, N_2^+, and O_2^+. The data base covers all latitudes and local times, and the altitude range from 150 km to 1200 km, but here the authors present altitude plots of the ion densities at noon and at dip latitudes of 20 – 40°N. The satellite data are compared with an early ion density profile by Johnson, with the Koehnlein and IRI–90 empirical models, and with the Utah State University theoretical ionosphere model. These comparisons serve to verify some aspects of the models, but they also reveal some outstanding differences.

083.019 **Origin of the ionospheric current loops which give rise to the solar–daily variations of the geomagnetic field.**
V. I. Ermakov.
Geomagn. Aehron., Tom 31, No. 6, p. 969 – 973 (Nov – Dec 1991). In Russian. English translation in Geomagn. Aeron., Vol. 31, No. 6.

083.020 **Ionospheric science and geomagnetism.**
H. Rishbeth.
Q. J. R. Astron. Soc., Vol. 32, No. 4, p. 409 – 421 (Dec 1991).

083.021 **Derivations of coupled equations appropriate to the lower regions of the ionosphere.**
S. S. De, B. K. Sarkar, B. Ghosh, M. Mal, B. Ghosh, S. Bandyopadhyay, S. K. Adhikari, A. C. Sen.
Astrophys. Space Sci., Vol. 184, No. 2, p. 281 – 285 (Oct 1991).
Coupled equations are derived for different ionospheric parameters through susceptibility dyadic appropriate to the medium. The nature of dispersion within the medium is studied.

083.022 **A new technique for estimating D–region effective recombination coefficients under different solar flare conditions.**
A. C. Balachandra Swamy.
Astrophys. Space Sci., Vol. 185, No. 1, p. 153 – 164 (Nov 1991).
Under solar flare conditions, the intensity of the solar X–rays below 10 Å increases by several orders of magnitude, while the increase in intensity of H–Lα is small. Photo–ionization rates in the various wavelength bands in the XUV spectrum have been presented graphically as a function of altitude under quiet, M3, X4, and outstanding flare conditions to show the relative importance of solar X–rays below 10 Å in the height range between 50 and 90 km. Presuming the total time constant for recombination of the ions with electrons remains constant at each altitude under different flare conditions, one can obtain the effective recombination coefficient ψ_{eff} under these conditions with a knowledge of the quiet time recombination coefficient, the production rate profiles and λ profiles. The importance of the ratio λ of negative ions to electrons below 70 km in lowering the effective electron loss rates has been highlighted.

083.023 **Sudden ionospheric oscillations accompanyed with sudden commencement of geomagnetic storm of March 8 – 20, 1989.**
Li Jun, Li Libin, Wu Zhenhua, Ning Baiqi, Wan Weixing.
Publ. Yunnan Obs., No. 4, p. 133 – 136 (Dec 1990). – See Abstr. 012.047 for the main entry.

083.024 **Magnetic disturbances and equivalent ionospheric currents during the great magnetic storm of March 13 – 14, 1989.**
Xu Wenyao, Lin Yunfang, Zeng Xiaoping.
Publ. Yunnan Obs., No. 4, p. 137 – 140 (Dec 1990). – See Abstr. 012.047 for the main entry.

083.025 **Nocturnal enhancement of TEC induced by magnetic storm of March 13, 1989.**
Jiang Herong, Yang Meihua, Liu Yuling, Wu Shunzhi.
Publ. Yunnan Obs., No. 4, p. 141 – 144 (Dec 1990). – See Abstr. 012.047 for the main entry.

083.026 **TEC fluctuation due to the severe storms in March 1989.**
Li Jun, Li Libin, Wu Zhenhua, Ning Baiqi, Wan Weixing.
Publ. Yunnan Obs., No. 4, p. 145 – 147 (Dec 1990). – See Abstr. 012.047 for the main entry.

083.027 **Behavior of the lower ionosphere at middle and lower latitudes during the solar events of March, 1989.**
Huang Zerong, Zhang Xunjie, Ruan Xueqin, Li Wentao.
Publ. Yunnan Obs., No. 4, p. 148 – 151 (Dec 1990). – See Abstr. 012.047 for the main entry.

083.028 **Vortex structures in the ionosphere and the magnetosphere of the Earth.**
V. M. Chmyrev, V. A. Marchenko, O. A. Pokhotelov, L. Stenflo, A. V. Streltsov, Å. Steen.
Planet. Space Sci., Vol. 39, No. 7, p. 1025 – 1030 (Jul 1991).
Model nonlinear equations, which have stationary solutions in the form of monopole and dipole vortices and vortex chains, are presented. The authors show that there is a qualitative agreement with satellite observations in the magnetosphere, as well as with the all–sky camera ionospheric data.

083.029 **Numerical studies of dynamo effect driven by planetary waves with 2–day period.**
Chen Peiren, Zhou Liqun.
Acta Geophys. Sin., Vol. 34, No. 6, p. 672 – 679 (Nov 1991). In Chinese.
A global distribution and the time variation of the dynamio electric field driven by the 2–day wave traveling westwards with longitudinal wave number 3 have been computed. When this electric field transfers to the F region along the geomagnetic field line, it will play an important role in electrodynamic processes in the F region. It may be the cause of the 2–day oscillation of equatorial ionization anomaly.

083.030 **Incoherent scatter observations of the high latitude ionosphere and implications for solar wind – magnetosphere – ionosphere coupling.**
H. Todd.
2. International Workshop on Plasma Physics: Plasma Theoretical Problems in Astro– and Fusion Physics, p. 157 – 175 (1991). – See Abstr. 012.067 for the main entry.
The ionospheric plasma characteristics and the technique of "incoherent" scatter are briefly reviewed. Attention is then focussed on the ionospheric bulk ExB flow in the F region above 200 km as deduced from measurements made using the European Incoherent Scatter (EISCAT) facility in northern Scandinavia. The EISCAT "UK–Polar" experiment was designed to measure flows in the auroral zone with a temporal resolution sufficient to resolve the the expected ionospheric effects of the coupling processes at the magnetopause. The latter are often observed to produce variations on timescales of a few minutes. Some ideas and results are discussed which relate to our understanding of solar wind – magnetosphere – ionosphere coupling.

083.031 **Digisonde observation of night time spread–F at Hainan Island.**
Zhang Qiwei, Guo Jianshan, Xu Chufu, Wang Zuogui.
Publ. Yunnan Obs., No. 4, p. 258 (Dec 1990). Abstract. – See Abstr. 012.047 for the main entry.

083.032 **QBO effect in the ionosphere.**
Chen Peiren.
Publ. Yunnan Obs., No. 4, p. 259 (Dec 1990). Abstract. – See Abstr. 012.047 for the main entry.

083.033 **Numerical studies of ionospheric currents and electric fields driven by planetary waves with a two–day period.**
Chen Peiren, Zhou Liqun.
Publ. Yunnan Obs., No. 4, p. 260 – 261 (Dec 1990). Abstract. – See Abstr. 012.047 for the main entry.

083.034 **Nonlinear analysis of the nonlocal gradient drift instability in the daytime equatorial electrojet.**
Cai Hengjin, Chen Peiren.
Publ. Yunnan Obs., No. 4, p. 262 (Dec 1990). Abstract. – See Abstr. 012.047 for the main entry.

083.035 **Study on the response of the D–region of the ionosphere to the solar bursts.**
Pan Liande, Wang Suqin, Ma Lanxiang.
Publ. Yunnan Obs., No. 4, p. 263 – 266 (Dec 1990). – See Abstr. 012.047 for the main entry.

Enlarged space and ground data base for ionospheric modelling. Proceedings. Topical Meeting of the COSPAR Interdisciplinary Scientific Commission C (Meeting C5) of the COSPAR 28. Plenary Meeting, The Hague (Netherlands), 25 Jun – 6 Jul 1990.
See Abstr. 012.013.

Electron transport in atmospheric gases. II. Energy spectra.
See Abstr. 022.052.

Dissociative recombination of the ground state of $N^+{}_2$.
See Abstr. 022.081.

Recent study on determination of total electron content from differential Doppler records.
See Abstr. 036.101.

Long wave radiation of the Earth.
See Abstr. 081.039.

A global numerical model of the thermosphere, ionosphere, and protonosphere of the Earth.
See Abstr. 082.008.

Ultraviolet spectroscopy and remote sensing of the upper atmosphere.
See Abstr. 082.066.

Effect of solar flares on low–frequency wave propagation (0.5 – 10 kHz) in the Earth–ionosphere waveguide.
See Abstr. 085.019.

084 Aurorae, Geomagnetic Field, Magnetosphere

084.001 **Structure and properties of the Earth's plasmasphere. Experimental data and problems of their interpretation (review).**
K. I. Gringauz, V. S. Bassolo.
Geomagn. Aeron., Vol. 30, No. 1, p. 1 – 12 (Aug 1990). English translation of Geomagn. Aehron., Tom 30, No. 1, p. 1 – 17 (1990).

The results of direct measurements of the concentration and temperature of the plasma are discussed, together with the mass and charge makeup of the plasmasphere ions. For the most part, the data were obtained during the last decade. It is shown that the data confirm the existence of two zones in the plasmasphere: an inner zone, which is stable, quasi–stationary, and cold; and an outer zone, which is usually warm and essentially nonstationary, with large longitudinal gradients of the parameters. The limit of the outer zone, the plasmapause, is clearly expressed during the evening–nighttime hours and is usually washed out in the day sector. It has a complex and nonstationary structure, changes strongly as a function of geomagnetic conditions, and is asymmetric in local time, with the evening and midday projections of the plasmapause being reliably distinguished. A comparison of these data with the theoretical constants leads to the conclusion that at present there is no generally accepted model that adequately describes all the basic features of the observed structure and dynamics of the plasmapause.

084.002 **The properties of the primary dynamo–generator in the open magnetosphere.**
V. V. Shelomentsev.
Geomagn. Aeron., Vol. 30, No. 3, p. 332 – 338 (Nov 1990). English translation of Geomagn. Aehron., Tom 30, No. 3, p. 393 – 401 (1990).

The possible properties of a generator of electromagnetic energy on the open magnetopause of the tail as an element of the equivalent circuit of the magnetosphere are investigated. It is concluded that the primary generator is a voltage source which has the usual electrical coupling through the current with the magnetospheric–ionospheric load. The variations of the source are controlled by the electromagnetic conditions in the solar wind, leading the time of the current variation in view of the presence of inductance in the circuit. The power of the generator is determined both by the external (the energy function ε) and internal (the effective conductivity of the load) parameters. However, the latter does not indicate the presence of strong positive feedback between the generator and the load.

084.003 **Current instability and the generation of kinetic Alfvén waves in the Earth's magnetosphere.**
Yu. M. Vojtenko, A. N. Krishtal', P. P. Malovichko, A. K. Yukhimuk.
Geomagn. Aeron., Vol. 30, No. 3, p. 339 – 342 (Nov 1990). English translation of Geomagn. Aehron., Tom 30, No. 3, p. 402 – 406 (1990).

The linear and quasi–linear stages of development of a current instability of kinetic Alfvén waves in the Earth's magnetosphere at altitudes $h \gtrsim 3R_E$ are discussed. The results obtained are used to interpret the low–frequency radiation observed with the help of the DE1 satellite. It is shown that the low–frequency turbulence observed above auroral zones may be caused by the development of a current instability of kinetic Alfvén waves.

084.004 **Structure of the magnetotail plasma sheet.**
I. I. Alekseev, Kh. V. Malova.
Geomagn. Aeron., Vol. 30, No. 3, p. 343 – 346 (Nov 1990). English translation of Geomagn. Aehron., Tom 30, No. 3, p. 407 – 411 (1990).

The authors discuss a model for the magnetotail current in the defined–field approximation. The current transverse to the sheet is due to free particles. They show that phase–space restrictions on particle motion determine the structure of the current sheet. If the particles incident on the sheet have small pitch angles, this leads to an increase in particle density near the neutral sheet, which in turn leads to a paramagnetic current and a decrease in the flux of large–pitchangle particles near the neutral sheet, as well as a diamagnetic current in the vicinity of the neutral sheet.

084.005 **Explanation for seasonal variations of the geomagnetic field in the polar cap using the polarity of interplanetary sectors.**
A. A. Danilov, A. K. Gnutova, G. A. Makarov.
Geomagn. Aeron., Vol. 30, No. 4, p. 485 – 487 (Feb 1990). Paper presented at the 5. Symposium on Solar Terrestrial Physics, Samarkand (USSR), 2 – 6 Oct 1989. English translation of Geomagn. Aehron., Tom 30, No. 4, p. 575 – 578 (1990).

Data obtained at high-latitude stations are used to analyze the daily and seasonal variations in the geomagnetic field under various polarities of the sectors in the interplanetary magnetic field. The authors obtain a qualitative estimate of the contribution from known current systems to these variations. They show that the geomagnetic effect of the interplanetary magnetic field sectors on the polar cap involves changes in the return current for the auroral electrojets and development of a polar electrojet.

084.006 **The impacts of giant asteroids and geomagnetic inversions.**
Yu. D. Kalinin.
Geomagn. Aeron., Vol. 30, No. 4, p. 555 – 556 (Feb 1990). Paper presented at the 5. Symposium on Solar Terrestrial Physics, Samarkand (USSR), 2 – 6 Oct 1989. English translation of Geomagn. Aehron., Tom 30, No. 4, p. 662 – 663 (1990).

Geomagnetic inversions are one of the consequences of the impacts on the Earth of giant asteroids. The energy of these impacts propagates within the Earth in the form of seismic waves. Reaching the lower mantle, they cause disruptions of its integrity. Blocks of material of the lower mantle separate, and if their density is greater than the density of the material of the liquid part of the Earth's core, they sink into it. As a result, relief with amplitudes of the order of 12 km, detectable from seismic data, arises on the boundary of the core with the mantle. Part of the blocks of magnetic material reaches the surface of the solid subcore of the Earth and creates relief of this surface with amplitudes of up to 23 km, also revealed by seismic data. Blocks in large number simultaneously sinking disrupt the meridional convection in the liquid part of the core, which creates conditions for the onset of a geomagnetic inversion.

084.007 **Variations in the elements of the geomagnetic field during the late Permian.**
D. K. Nurgaliev.
Geomagn. Aeron., Vol. 30, No. 4, p. 589 – 591 (Feb 1990). Paper presented at the 5. Symposium on Solar Terrestrial Physics, Samarkand (USSR), 2 – 6 Oct 1989. English translation of Geomagn. Aehron., Tom 30, No. 4, p. 690 – 692 (1990).

The author studies a "lingulian clay" core (age $\sim$255 million years) that shows seasonal laminations and determine the length of the time interval covered ($\sim$3250 yr). The natural remanent paleomagnetism was determined by thermomagnetic relaxation. Variations with periods of $\sim$75, $\sim$275, $\sim$550, $\sim$1150, and $\sim$3600 yr were detected. The author discusses a model for the paleomagnetic secular variations based on the paleomagnetic data.

084.008 **Reaction of $N(^2D)$ with O_2: not a major source of $O(^1D)$ atoms in aurorae.**
V. Singh.
Indian J. Radio Space Phys., Vol. 20, No. 1, p. 76 – 77 (Feb 1991).

The reaction of $N(^2D)$ with O_2 as a source of $O(^1D)$ proposed by Rusch et al. (1978) is restudied in the light of recent measured reaction rate coefficients. The present results indicate that this reaction is not a major source of $O(^1D)$ atoms in aurorae.

084.009 **On the nature of low frequency turbulence in the foot of strong quasi-perpendicular shocks.**
V. V. Krasnosel'skikh, M. A. Balikhin, H. S. C. Alleyne, S. I. Klimov, W. A. C. Mier-Jedrzejowicz, A. K. Pardaens, A. Petrukovich, D. J. Southwood, T. Vinogradova, L. J. C. Woolliscroft.
Adv. Space Res., Vol. 11, No. 9, p. 15 – 18 (1991). – See Abstr. 012.006 for the main entry.

Low frequency fluctuations of the magnetic field in the vicinity of the Earth's bow shock are presented. Observations from the Prognoz-10 and AMPTE UK satellites are analysed using cospectral techniques. For high Mach number quasi-perpendicular shocks the main low frequency fluctuations seem to be due to whistler waves which are generated by non-linear evolution of the shock front. This is in contradiction to previous proposals that these waves are generated by an instability driven by the reflected ions.

084.010 **Experimental derivation of ELF waves dispersion relations and evidence of wave coupling in the Earth's bow shock foot region from the results of the "Prognoz 10" satellite.**
S. A. Romanov, S. I. Klimov, P. A. Mironenko.
Adv. Space Res., Vol. 11, No. 9, p. 19 – 24 (1991). – See Abstr. 012.006 for the main entry.

The ELF plasma turbulence in the foot region of the Earth's bow shock is investigated. The electric current density and magnetic field fluctuation data are used to derive dispersion relations for the ELF waves in the foot region of the quasiperpendicular collisionless shock. The evidence of nonlinear coupling processes in the shock foot region is found and a mechanism for the initial stage of energy transformation at the bow shock is discussed.

084.011 **Observations of waves on the dayside by the AMPTE UKS satellite and their role in processes.**
S. N. Walker, A. P. Ameen, H. S. C. Alleyne, L. J. C. Woolliscroft.
Adv. Space Res., Vol. 11, No. 9, p. 29 – 32 (1991). – See Abstr. 012.006 for the main entry.

Wave observations from the AMPTE UKS satellite are analysed. Wave data are presented for magnetopause crossings and other dayside regions and phenomena in order to address the problem of the role of plasma waves in processes such as reconnection. Wave measurements for all known UKS flux transfer events have been analysed to show morphological trends. The observed wave power is never sufficient to account for the dissipation. Indications that the wave power may occur at other frequencies, or in very short bursts are discussed.

084.012 **Observations and theory of whistler-mode waves in the vicinity of the Earth's magnetopause.**
S. S. Sazhin, S. N. Walker, L. J. C. Woolliscroft.
Adv. Space Res., Vol. 11, No. 9, p. 33 – 36 (1991). – See Abstr. 012.006 for the main entry.

Wave power measurements based on the AMPTE UKS satellite data in the ELF/VLF frequency range can show a spin modulation effect. Analysis of this spin modulation for data collected near the Earth's magnetopause indicates that these waves are probably whistler-mode waves propagating parallel or almost parallel to the ambient magnetic field. The energy source of these waves is identified with the anisotropy of the electron distribution function. The intensification of these waves near the earthward boundary of the Earth's magnetopause is explained in terms of wave trapping in this region.

084.013 **Impulsive plasma waves observed by DE-1 in the magnetosphere.**
T. Ondoh, Y. Nakamura, S. Watanabe.
Adv. Space Res., Vol. 11, No. 9, p. 41 – 44 (1991). – See Abstr. 012.006 for the main entry.

Impulsive plasma waves with durations less than 100 msec have been found in DE-1 wide-band electric field data received at Kashima, Japan. The waves are associated with a strong narrow-band ELF hiss, and were observed at geocentric distances from 3.1 to 4.9 R_δ in the low-latitude nightside magnetosphere. The waves are discussed in terms of Landau resonant trapping of magnetospheric electrons by the associated whistler-mode ELF hiss.

084.014 **On the distribution of polar electrostatic emissions and their relationships to auroral hiss and direction finding measurements of saucer emission in polar regions observed by Isis, Kyokko and Akebono satellites.**
T. Yoshino.
Adv. Space Res., Vol. 11, No. 9, p. 45 – 50 (1991). – See Abstr. 012.006 for the main entry.
Since 1976 a total of 3500 orbits of data from the Isis-1 and 2, Kyokko and Akebono satellites have been analysed to produce some new observational results about polar VLF and ELF emissions at Arctica and Antarctica. By mapping the distribution of the occurrence of the emissions in the polar regions the source of Electrostatic Ion Cyclotron (EIC) emissions has been located.

084.015 **Role of electromagnetic noise in the interhemispheric plasma exchange along closed field lines.**
N. Singh.
Adv. Space Res., Vol. 11, No. 9, p. 51 – 54 (1991). – See Abstr. 012.006 for the main entry.
Satellite observations have demonstrated that the equatorially trapped superthermal hydrogen ions are a common occurrence in the outer plasmaspheric flux tubes. Such ions occur in conjunction with broadband waves over a frequency spectrum from the proton cyclotron frequency to the lower-hybrid frequency. Trapped ions are produced by the transverse heating of the ions by the waves. The effect of such ion heating on the interhemispheric plasma flow is studied here.

084.016 **The acceleration of charged particles in magnetic current sheets.**
S. W. H. Cowley.
Adv. Space Res., Vol. 11, No. 9, p. 99 – 106 (1991). – See Abstr. 012.006 for the main entry.
The basic concepts of charged particle acceleration in magnetic current sheets are reviewed, and related to conditions at the dayside magnetopause and in the geomagnetic tail. Particular topics discussed include the effect of magnetosheath flow on plasma acceleration at the dayside magnetopause, and the acceleration of ions in the tail during field "dipolarization" at substorm onset.

084.017 **A two-dimensional, time-dependent, near-Earth magnetotail.**
E. Whipple, R. Puetter, M. Rosenberg.
Adv. Space Res., Vol. 11, No. 9, p. 133 – 142 (1991). – See Abstr. 012.006 for the main entry.
The plasma particle distributions in the near-earth region of the Earth's magnetosphere appear to have a complicated structure both in velocity space and real space. Recent observations and theoretical work have emphasized the role of non-isotropic particle distributions in this region. A kinetic approach may be necessary to describe field/plasma configurations in the near-earth magnetotail. The authors present such an approach for obtaining slowly time-dependent, self-consistent plasma and magnetic field configurations for this region, using anisotropic particle distributions.

084.018 **Particle motion in the tail current sheet.**
T. W. Speiser.
Adv. Space Res., Vol. 11, No. 9, p. 151 – 159 (1991). – See Abstr. 012.006 for the main entry.
Theory of particle motion in current sheets is reviewed.

084.019 **Stability of the Earth's neutral sheet.**
A. T. Y. Lui.
Adv. Space Res., Vol. 11, No. 9, p. 161 – 165 (1991). – See Abstr. 012.006 for the main entry.
The stability of the Earth's neutral sheet is analyzed. It is found that for conditions just prior to substorm expansion onset, the cross-field current instability (CFCI) can be triggered. Numerical solution of the dispersion equation shows that the CFCI has a growth time comparable to the substorm onset time. The excited waves have a mixed polarization in the lower hybrid frequency range and propagate obliquely to the local magnetic field. The ensuing turbulence of this instability can lead to a reduction of the cross-tail current and the formation of a substorm current wedge as observed.

084.020 **A quantitative study of ion transport during substorms.**
D. C. Delcourt, J. A. Sauvaud.
Adv. Space Res., Vol. 11, No. 9, p. 171 – 175 (1991). – See Abstr. 012.006 for the main entry.
The transport of magnetospheric plasma during dynamical reconfiguration of the geomagnetic field is investigated by means of three-dimensional single particle codes. It is shown that the "tail-like" to "dipole-like" magnetic transitions which accompany the development of substorms, yield rapid injections of mid-tail ($\sim 10 - 15$ R_E) ions into close L-shells. It is demonstrated that, during this transport, the guiding center is at stake depending upon particle mass and/or charge state.

084.021 **Regular and chaotic particle motion in sheared magnetic field reversals.**
J. Büchner, L. M. Zelenyi (*L. M. Zelenyj*).
Adv. Space Res., Vol. 11, No. 9, p. 177 – 182 (1991). – See Abstr. 012.006 for the main entry.
Space plasma current sheets, like those in the plasmasheet of the Earth's magnetotail or in the magnetopause, create strongly curved magnetic fields. The adiabaticity of charged particles motion in them may be destroyed and deterministic chaos may appear. In this paper the authors investigate the influence of a magnetic shear, caused by a finite guide field component, on the chaotic particle motion. They derive the adiabaticity parameter in such fields and describe the routes to chaos.

084.022 **Ion distribution in the plasma sheet of the Earth's magnetotail.**
J. Büchner, M. Ashour-Abdalla, L. M. Zelenyi (*L. M. Zelenyj*).
Adv. Space Res., Vol. 11, No. 9, p. 183 – 187 (1991). – See Abstr. 012.006 for the main entry.
The influence of the chaotic scattering on the formation of the stationary ion distribution in the Earth's magnetotail is studied using a model, which includes variations of the magnetotail field in the sun-Earth direction. It is shown, how the chaotic scattering permanently redistributes ions between the central plasma sheet and the plasma sheet boundary layer of the tail.

084.023 **Particle distributions in collisionless magnetic reconnection: an implicit PIC description.**
D. W. Hewett, G. E. Francis, C. E. Max.
Adv. Space Res., Vol. 11, No. 9, p. 193 – 201 (1991). – See Abstr. 012.006 for the main entry.
Evidence from magnetospheric and solar flare research supports the belief that collisionless magnetic reconnection can proceed on the Alfven-wave crossing timescale. Reconnection behavior that occurs this rapidly in collisionless plasmas is not well understood because underlying mechanisms depend on the details of the ion and electron distributions in the vicinity of the emerging X-points. The authors use the direct implicit Particle-In-Cell (PIC) code AVANTI to study the details of these distributions as they evolve in the self-consistent E and B fields of magnetic reconnection.

084.024 **The energetic ion signature of an O-type neutral line in the geomagnetic tail.**
R. F. Martin Jr., D. F. Johnson, T. W. Speiser.
Adv. Space Res., Vol. 11, No. 9, p. 203 – 206 (1991). – See Abstr. 012.006 for the main entry.

084.025 **Periodic oscillations of the quasi-parallel bow shock as observed with the AMPTE/CCE spacecraft.**
R. J. Strangeway, L. J. Zanetti.
Adv. Space Res., Vol. 11, No. 9, p. 219 – 222 (1991). – See Abstr. 012.006 for the main entry.
The AMPTE/CCE spacecraft has, on occasion, encountered the Earth's bow shock near apogee. During one encounter, on 1 Nov 1984, the bow shock was observed to pass back and forth over the spacecraft several times in a periodic fashion. The period

of the oscillation was about 20 sec, and an upstream wave with a similar period was observed in the solar wind just prior to the multiple shock crossings. These data may provide support for the wave-like structure of quasi-parallel shocks predicted in simulations.

084.026 **Some comments on the ion distribution function evolution in the quasiparticle shock.**
J. Šafránková, Z. Němeček, O. Santolík.
Adv. Space Res., Vol. 11, No. 9, p. 223 - 226 (1991). - See Abstr. 012.006 for the main entry.
The paper is devoted to a detailed study of the ion distribution function evolutiuon in the quasiparallel bow shock based on the data from the Prognoz 10 satellite. As follows from our analysis, in the foreshock region the plasma flux is divided into many partial fluxes with differing drift velocities. The authors study the evolution and gradual relaxation of these beams in various time intervals through the foreshock region. The relaxation processes are studied with respect to the different behaviour of protons and alpha particles. The results are compared with a similar analysis carried out for quasiperpendicular shocks.

084.027 **Ion distributions associated with magnetic field fluctuations in the vicinity of quasi-parallel shocks.**
R. L. Kessel.
Adv. Space Res., Vol. 11, No. 9, p. 245 - 248 (1991). - See Abstr. 012.006 for the main entry.
The author examines 3D ion distributions from the AMPTE-UKS ion instrument associated with magnetic field structures near the Earth's bow shock. He looks at an embedded and isolated region on day 304 of 1984 which are typical of these structures and searches for origins of the observed ions from four source populations.

084.028 **Preliminary investigations of magnetic field turbulence at a quasi-parallel shock.**
M. A. Balikhin, H. S. C. Alleyne, V. V. Krasnosel'skikh, M. Dunlop, A. K. Pardaens, D. J. Southwood, A. E. Sumner, L. J. C. Woolliscroft.
Adv. Space Res., Vol. 11, No. 9, p. 253 - 256 (1991). - See Abstr. 012.006 for the main entry.
AMPTE UKS data from the magnetometer and wave experiments are presented along with some electron and ion parameters taken from the AMPTE UKS database for a crossing of a quasi-parallel bow shock. Different regions in the shock crossing are identified on the basis of the magnetic field turbulence characteristics and the wave spectra.

084.029 **Variation in the energy spectra of energetic ions at the Earth's bow shock.**
K. J. Trattner, E. Möbius, B. Klecker, M. Scholer.
Adv. Space Res., Vol. 11, No. 9, p. 261 - 264 (1991). - See Abstr. 012.006 for the main entry.
It has been recognized that collisionless shocks are efficient particle accelerators. The model of Lee (1982) describes particle acceleration and wave generation at the earth's bow shock in a self-consistent way. The acceleration energy is provided by the solar wind and is a function of solar wind density and velocity. The authors have performed a statistical analysis of the correlation of solar wind parameters with variation of the ion energy spectra by making use of energetic ion measurements with the time-of-flight spectrometer SULEICA on AMPTE/IRM. It is shown that the differential particle flux in the energy range from 10 - 30 keV correlates with the solar wind density thus supporting a solar wind source and that the spectral slope correlates with the solar wind velocity.

084.030 **Energetic protons upstream of the Earth's bow shock: statistical study.**
K. Kudela, M. Slivka, V. Stolpovski (*V. Stolpovskij*), V. Lutsenko, S. Fischer.
Adv. Space Res., Vol. 11, No. 9, p. 279 - 282 (1991). - See Abstr. 012.006 for the main entry.
Many upstream events were analyzed previously at energies above 30 keV. Here the authors report the results of a statistical study of protons $E_p > 15$ keV and electrons $E_e > 30$ keV in the upstream region based on Prognoz-10 measurements.

084.031 **Long-period geomagnetic field fluctuations and the structure of the turbopause.**
Yu. Karadzhaev, T. A. Gorbunova.
Geomagn. Aehron., Tom 31, No. 4, p. 725 - 727 (Jul - Aug 1991). In Russian. English translation in Geomagn. Aeron., Vol. 31, No. 4.

084.032 **Structural variances in the geomagnetic field.**
R. T. Vasil'ev, M. A. Vasil'eva, S. V. Pastushenko, N. P. Yanovskaya.
Geomagn. Aehron., Tom 31, No. 4, p. 760 - 762 (Jul - Aug 1991). In Russian. English translation in Geomagn. Aeron., Vol. 31, No. 4.

084.033 **Location and dynamical properties of auroral arcs during activations of different intensity.**
T. V. Kozelova, M. I. Pudovkin, T. A. Kornilova.
Geomagn. Aehron., Tom 31, No. 5, p. 786 - 791 (Sep - Oct 1991). In Russian. English translation in Geomagn. Aeron., Vol. 31, No. 5.

084.034 **The latitudinal profile of the auroral electron precipitation.**
V. S. Smirnov, B. B. Gvozdevskij.
Geomagn. Aehron., Tom 31, No. 5, p. 936 - 937 (Sep - Oct 1991). In Russian. English translation in Geomagn. Aeron., Vol. 31, No. 5.

084.035 **Influence of spatial distribution and data errors on the secular variation at sea stations when constructing global models of secular variations.**
T. I. Zvereva, L. G. Kas'yanenko.
Geomagn. Aehron., Tom 31, No. 6, p. 1084 - 1093 (Nov - Dec 1991). In Russian. English translation in Geomagn. Aeron., Vol. 31, No. 6.

084.036 **Quasi-biennial oscillations in geomagnetic aa-index.**
V. P. Bobova.
Soln. Dannye, Byull., No. 3/91, p. 103 - 104 (Mar 1991). Abstract. - See Abstr. 012.021 for the main entry.

084.037 **Morphology category of geomagnetic storms and characteristics of interplanetary magnetic clouds.**
Zhang Gongliang.
Sci. China, Ser. A, Vol. 34, No. 6, p. 717 - 731 (Jun 1991).
Six typical geomagnetic storms are analyzed from the aspect of varieties of morphology categories of magnetic storms with sudden commencement caused by interplanetary magnetic clouds. The high-density structure in front of magnetic cloud is an important factor in determining the property of the initial phase of geomagnetic storm. Under similar conditions for the magnetic field of mangetic cloud, the velocities of both the background solar wind and the magnetic cloud affect the development and intensity of magnetic storm.

084.038 **Modelo de la variación secular del campo geomagnético en España. Período 1970 - 1988.**
A. Garcia, J. M. Torta, J. J. Curto, L. F. Alberca.
Publ. Obs. Ebro, Misc., No. 39.
Observatorio del Ebro, Roquetes (Spain). 14 p. (1991).

084.039 **The Matachewan dyke swarm, Canada: an early Proterozoic magnetic field reversal.**
H. C. Halls.
Earth Planet. Sci. Lett., Vol. 105, No. 1-3, p. 279 - 292 (Jul 1991).
Consistent age relationships between oppositely magnetized dykes of the 2.45 Ga Matachewan dyke swarm suggest that only a single magnetic field reversal occurred during the period of igneous activity. The magnetic field throughout most of this time was characterized by a SSW declination and shallow negative

inclination but reversed toward the waning stages of magmatism. The new paleomagnetic data provide the oldest known magnetic reversal for which the relative reversal sense is known.

084.040 **A 65 Ma palaeomagnetic pole for the Pacific plate from the skewness of magnetic anomalies 27r–31.**
G. D. Acton, R. G. Gordon.
Geophys. J. Int., Vol. 106, No. 2, p. 407 – 420 (Aug 1991).
The authors use the skewness of seafloor-spreading magnetic anomalies 27r–31 to determine a revised Late Maastrichtian – early Palaeocene palaeomagnetic pole for the Pacific plate. They used numerical experiments to estimate the potential information in magnetic profiles that had not been previously analysed for skeweness information.

084.041 **The scattering rate of solar 834 Å radiation by magnetospheric O^+ and O^{++}.**
R. R. Meier.
Geophys. Res. Lett., Vol. 17, No. 10, p. 1613 – 1616 (Sep 1990).
Several recent proposals to image sunlit magnetospheric O^+ have been difficult to assess because of the lack of detailed knowledge of the solar spectrum at 834 Å. An NRL solar spectrograph flown on a sounding rocket in 1962 actually had sufficient resolving power to allow the calculation of scattering rates. Using these solar data, examples of scattering rates for a range of thermal and bulk velocities are presented for both O^+ and O^{++}, and the effects of ionospheric albedo are estimated. The scattering rates are significantly lower than previously estimated for hot or moving plasma and exhibit substantial structure.

084.042 **Changes in the intensity of the geomagnetic field.**
D. Walton.
Geophys. Res. Lett., Vol. 17, No. 12, p. 2085 – 2088 (Nov 1990).
A new, more reliable technique has been used to determine the ancient intensity of the earth's magnetic field using samples of baked clay from eight sites in the Eastern Mediterranean. The resultant power spectrum reveals a single strong component with a period of 1430 years. The reproducibility of the results obtained with the new method also makes limited archaeomagnetic dating with a precision of between about one and two hundred years between 900 BC and AD 400 possible.

084.043 **A southern hemisphere record of the Matuyama–Brunhes polarity reversal.**
B. M. Clement, D. V. Kent.
Geophys. Res. Lett., Vol. 18, No. 1, p. 81 – 84 (Jan 1991).
The authors present a record of the Matuyama–Brunhes (0.73 Ma) polarity transition from a southern hemisphere deep-sea sediment core (V16–58; 46°S, 30°E). The transition is recorded across at least 30 cm of section and is defined by a nearly 180° change in directions which occurs during a low in the relative intensities. An increase in the $^{10}Be/^{9}Be$ ratios associated with the intensity low suggests that the relative intensities may document a decrease in the strength of the geomagnetic field. The virtual geomagnetic pole (VGP) path is neither clearly near- or far-sided but tracks approximately 120° west of the site longitude. Considered with other Matuyama–Brunhes transition records, this southern hemisphere record supports previous interpretations that the Matuyama–Brunhes transitional field was not dipolar. Instead, the V16–58 transition together with other records of this reversal recently obtained from this longitudinal sector suggests that nonzonal terms which are symmetric about the equator were important during this transitional field.

084.044 **Indications of low dimensionality in magnetospheric dynamics.**
D. A. Roberts, D. N. Baker, A. J. Klimas, L. F. Bargatze.
Geophys. Res. Lett., Vol. 18, No. 2, p. 151 – 154 (Feb 1991).
Using three separate but related approaches, the authors examine the question of whether the dynamic response of the magnetosphere to the solar wind input may be described by a low-order system of equations.

084.045 **Combined EISCAT and All–Sky camera observations of auroral structures.**
L. Girard, C. Senior.
Geophys. Res. Lett., Vol. 18, No. 7, p. 1185 – 1188 (Jul 1991).
Discrete arc structures were identified from EISCAT data on March 11, 1984 which were simultaneously recorded on All–Sky camera pictures from the Kilpisjärvi station in Finland. From the electric field measurements supplied by the tri-static UHF facility, some privileged orientation has been found through the requirement of a uniform tangential electric field across the arc structure, which corresponds to the preferential arc alignment as deduced from camera pictures. Additional fold structures, resolved by the All–Sky camera, develop along the preexisting arcs which still organize the electric field fluctuations in their vicinity.

084.046 **Transpolar auroras, their particle precipitation, and IMF B_y component.**
K. Makita, C.-I. Meng, S.-I. Akasofu.
J. Geophys. Res., Vol. 96, No. A8, p. 14085 – 14096 (1 Aug 1991).
Transpolar auroras, their associated particle precipitation, and their occurrence with respect to the IMF B_y polarity are examined on the basis of DMSP F6 auroral images and the corresponding particle data.

084.047 **Geomagnetic and solar data. March 1991.**
H. E. Coffey.
J. Geophys. Res., Vol. 96, No. A8, p. 14195 (1 Aug 1991).

084.048 **Periodicity of geomagnetic field reversal and conclusions concerning its nature.**
A. V. Dolitskij.
Izv. Acad. Sci. USSR, Phys. Solid Earth, Vol. 26, No. 8, p. 718 – 720 (Mar 1990). Translated from Izv. Akad. Nauk SSSR, Fiz. Zemli.

084.049 **The Earth's magnetosphere.**
M. Saunders.
Symposium on Geophysical and Astrophysical MHD as Part of the British Theoretical Mechanics Colloquium: Advances in solar system magnetohydrodynamics, p. 357 – 397 (1991). – See Abstr. 012.086 for the main entry.
The author begins by placing the magnetosphere together with the reasons and techniques for studying it in context. Then he reviews the overall structure and characteristics of the magnetosphere, emphasising those regions and processes where MHD is fundamental. He focusses on the key advances (experimental and theoretical) in our exploration of magnetospheric MHD during the past few years, and concludes with a "look into the future".

084.050 **Imaging the Earth's magnetosphere: effects of plasma flow and temperature.**
D. E. Garrido, R. W. Smith, D. S. Swift, S.-I. Akasofu.
Planet. Space Sci., Vol. 39, No. 11, p. 1559 – 1571 (Nov 1991).
The authors simulate EUV images of the magnetosphere at 83.4 nm for cases where the plasma has a non-zero drift velocity with respect to the sun and variable plasma temperature.

084.051 **The prediction of geomagnetic activities during the 1989 joint observation.**
Shi Enqi, Zheng Tirong, Xu Wenyao.
Publ. Yunnan Obs., No. 4, p. 273 – 276 (Dec 1990). – See Abstr. 012.047 for the main entry.

084.052 **On the generation of ULF and VLF emission in an inhomogeneous plasma in the plasmapause region.**
N. I. Izhovkina, V. I. Larkina.
Kosm. Issled., Tom 29, Vyp. 4, p. 593 – 596 (Jul–Aug 1991). In Russian. English translation in Cosmic Res., Vol. 29, No. 4.
Experimental results and calculations of generation of ULF and VLF waves in an inhomogeneous plasma in the plasmapause region are described. It is assumed that the inhomogeneous

plasma density in this region may be presented by tubes aligned along geomagnetic field lines.

084.053 **Is the recurrent enhanced geomagnetic activity preceding the sunspot minimum related to magnetic fields of the new or old activity cycle?**
V. Bumba.
Bull. Astron. Inst. Czech., Vol. 42, No. 3, p. 161 – 165 (Jun 1991).
The author compares the new index of recurrence of enhanced geomagnetic activity with the latitudinal and longitudinal distribution of solar magnetic fields and prominences for the recent two eleven–year cycles of solar activity. It is demonstrated that the streamers of the solar wind causing the considerable enhancement and regularity in recurrences of large geomagnetic disturbances during the descending phase and before the end of an eleven–year cycle, are anchored into the system of magnetic fields formed by the fields of the old eleven–year cycle of solar activity.

084.054 **Dynamics of high–energy trapped radiation in the inner radiation belt of the Earth.**
N. N. Volodichev, A. A. Gusev, Yu. V. Mineev, G. I. Pugacheva, P. I. Shavrin.
Kosm. Issled., Tom 29, Vyp. 5, p. 790 – 794 (Sep–Oct 1991). In Russian. English translation in Cosmic Res. Vol. 29, No. 5.

084.055 **An assessment of a magnetospheric model by tracing the energetic particle trapping boundary.**
B. Popielawska, D. Zwolakowska.
Geophys. Res. Lett., Vol. 18, No. 8, p. 1489 – 1492 (Aug 1991).
The authors use the two most recent empirical magnetosphere models of Tsyganenko (1987; 1989) to find the energy dependent boundary of isotropic precipitation due to the nonadiabatic particle motion in the nightside current sheet. Model predictions are compared with the earlier published satellite observations of this boundary at low altitudes. A dramatic difference has been found between the predicted position of the isotropy boundary based on T87 and T89 models. The most recent magnetosphere model of Tsyganenko featuring the warped current sheet with thickness dependent both on XGSM and YGSM predicts that the isotropy boundary is situated at much lower latitudes than inferred from the previous models. This is in agreement with observations. However, even this model with very stretched tail magnetic field lines has difficulties in explaining the frequently observed very steep profile of critical rigidity vs. L.

084.056 **Relationship between field line resonance at low geomagnetic latitudes and solar wind structures.**
M. Vellante, U. Villante.
Geophys. Res. Lett., Vol. 18, No. 8, p. 1501 – 1504 (Aug 1991).
The link between the pulsation activity in the local resonant band at a low latitude station and the macroscale structure of the interplanetary medium has been investigated. The experimental results show a clear dependence of the amplitude of the local resonant events on the interplanetary magnetic field magnitude and a preferable occurrence in correspondence of the compression regions preceding the high velocity solar wind streams. These features appear consistent with an upstream wave source for these events.

084.057 **Examination of time–variable input effects in a nonlinear analogue magnetosphere model.**
D. N. Baker, A. J. Klimas, D. A. Roberts.
Geophys. Res. Lett., Vol. 18, No. 8, p. 1631 – 1634 (Aug 1991).
The authors' previously developed nonlinear dynamical models of the magnetosphere exhibit deterministic chaotic behavior as the solar wind input which drives the systems increases from low to high levels. In this paper the authors employ the plasma physical analogue model (an extention of the damped, harmonic–oscillatory dripping faucet model) to consider explicitly the effect of time–varying the inputs.

084.058 **Observational aspects of solar wind – magnetosphere interactions from multipoint measurements made on 1 November, 1984.**
C. J. Farrugia, S. Elliott, M. P. Freeman.
2. International Workshop on Plasma Physics: Plasma Theoretical Problems in Astro– and Fusion Physics, p. 9 – 29 (1991). – See Abstr. 012.067 for the main entry.
The authors present and discuss observations on magnetic fields and plasmas made by various spacecraft in the solar wind, in the magnetosheath, and in the outer magnetosphere during a strong compression of the terrestrial magnetosphere. The day in question is November 1, 1984. They supplement the spacecraft observations by simultaneous low latitude ground magnetograms. By intercomparisons of data sets the authors are able to illustrate a number of features related to the effect of the solar wind on the magnetosphere.

084.059 **MHD theory and the solar wind – magnetosphere interaction.**
R. P. Rijnbeek.
2. International Workshop on Plasma Physics: Plasma Theoretical Problems in Astro– and Fusion Physics, p. 63 – 75 (1991). – See Abstr. 012.067 for the main entry.
In this article a theoretical basis for understanding the solar wind – magnetosphere interaction and interpreting the associated phenomena is discussed. Magnetohydrodynamic (MHD) analysis provides the main foundation of present–day knowledge. Although MHD allows only a restricted study of the plasma behaviour, it offers an entry point into many problems of space plasma physics in general, and magnetospheric physics in particular. Such problems include the flow of the solar wind around the magnetosphere, and coupling between the solar wind and the magnetosphere, which gives rise to such phenomena as the auroral lights and geomagnetic field disturbances.

084.060 **The solar wind flow around the dayside magnetosphere.**
B. P. Besser, R. P. Rijnbeek.
2. International Workshop on Plasma Physics: Plasma Theoretical Problems in Astro– and Fusion Physics, p. 77 – 90 (1991). – See Abstr. 012.067 for the main entry.
The interaction between the solar wind and the terrestrial magnetosphere is a large topic of investigation in the subject of space plasma physics. An important problem in this respect is the description of the solar wind flow around the magnetosphere. In this article the basic features of this problem are considered, and some of the progress in theoretical investigations and data interpretation are reviewed. In doing so, emphasis is placed on the results obtained from magnetohydrodynamic (MHD) theory.

084.061 **Energy coupling quantities in the solar wind – magnetosphere system.**
G. K. F. Rabl.
2. International Workshop on Plasma Physics: Plasma Theoretical Problems in Astro– and Fusion Physics, p. 91 – 103 (1991). – See Abstr. 012.067 for the main entry.
The interaction between the solar wind and the magnetic field of the Earth generates a vast magnetic cavity, called magnetosphere. This dynamic region shows rapid changes in response to both solar wind plasma/interplanetary magnetic field conditions and internal instabilities. The main goal of the topic termed "solar wind – magnetosphere coupling" is to investigate how the solar wind transfers energy into the magnetosphere, and which solar wind quantities most efficiently drive magnetospheric activities. The solar wind – magnetosphere coupling problem is reviewed based on observations of various solar wind parameters and quantities representing auroral activities such as geomagnetic indices as well as the non–thermal terrestrial kilometric radiation.

084.062 **Coupling processes at the magnetopause.**
H. K. Biernat.
2. International Workshop on Plasma Physics: Plasma Theoretical Problems in Astro– and Fusion Physics, p. 105 – 117 (1991). – See Abstr. 012.067 for the main entry.
The problem of the solar wind – magnetosphere interaction is discussed with particular reference to the Earth's dayside

magnetopause. An overview is given of the coupling processes at the magnetopause. The most widely discussed processes are "viscous-like" interaction and "magnetic reconnection". From the global point of view it has been concluded that reconnection of magnetic field lines is the dominant of the two. The most widely applied model of reconnection is that, suggested by Petschek, in which most of the conversion of magnetic energy and plasma energy occurs via magnetohydrodynamic waves. From spacecraft data we know that there exist transient changes of the magnetic field and other plasma parameters, so-called "flux transfer events". For the description of such phenomena the Petschek model has been generalized for unsteady conditions and a finite length of the reconnection line.

084.063 **Geomagnetic and solar data. February 1991.**
H. E. Coffey.
J. Geophys. Res., Vol. 96, No. A7, p. 11647 (1 Jul 1991).

084.064 **The ARIES auroral modelling campaign: characterization and modelling of an evening auroral arc observed from a rocket and a ground-based line of meridian scanners.**
A. Vallance Jones, R. L. Gattinger, F. Creutzberg, F. R. Harris, A. G. McNamara, A. W. Yau, E. J. Llewellyn, D. Lummerzheim, M. H. Rees, I. C. McDade, J. Margot.
Planet. Space Sci., Vol. 39, No. 12, p. 1677 - 1705 (Dec 1991).
An auroral arc system excited by soft electrons was studied with a combination of in situ rocket measurements and optical tomographic techniques, using data from a photometer on a horizontal, spinning rocket and a line of three meridian scanning photometers.

084.065 **Non-steady-state solar wind - magnetosphere interaction.**
J. Lemaire, M. Roth.
Space Sci. Rev., Vol. 57, No. 1/2, p. 59 - 108 (Jul 1991).
Most of the theories proposed to explain the interaction between the solar wind and the geomagnetic field are stationary descriptions based on ideal MHD. In this review an alternative, non-stationary description is discussed.

084.066 **Giotto's mission to planet Earth.**
K.-H. Glassmeier, F. M. Neubauer, G. Brach, H. Marschall, M. H. Acuña, L. F. Burlaga, F. Mariani, G. Musmann, N. F. Ness, M. K. Wallis, E. Ungstrup, H. U. Schmidt.
Geophys. Res. Lett., Vol. 18, No. 9, p. 1663 - 1666 (Sep 1991).
After its successful encounter with comet P/Halley and a four-year hibernation period ESA's Giotto spacecraft has been reactivated in February 1990 and performed the first-ever Earth gravity-assisted maneuver on July 2, 1990 to be retargeted for comet P/Grigg-Skjellerup. This swing-by is of unique scientific interest due to Giotto's hyperbolic, high-inclination orbit. Here, the authors report on scientific results of the Giotto magnetic field experiment. Due to the high fly-by velocity and the relative quietness of the magnetosphere during the swing-by period these measurements present a snapshot view of the Earth magnetosphere with clearly identified inbound and outbound bow shock and magnetopause crossings. The outbound crossings are of particular interest as surface waves at the polar magnetopause at a distance of 28 R_E as well as a strong quasi-perpendicular bow shock at a distance of about 64 R_E are observed.

084.067 **Magnetic reconnection in the Earth's distant magnetotail. I. Streaming tearing instability.**
Wang Shui, Zheng Huinan.
Acta Geophys. Sin., Vol. 34, No. 6, p. 663 - 671 (Nov 1991). In Chinese.
A magnetic reconnection process caused by the streaming instability in the Earth's distant magnetotail has been studied numerically. The results show that after a brief phase of exponential growth, the growth rate of the streaming tearing instability is greatly reduced when the magnetic island width becomes comparable to the plasma sheet thickness and finally the magnetic island width saturates.

084.068 **Simultaneous optical observations of transpolar arcs in the two polar caps.**
J. D. Craven, J. S. Murphree, L. A. Frank, L. L. Cogger.
Geophys. Res. Lett., Vol. 18, No. 12, p. 2297 - 2300 (Dec 1991).
Optical observations with DE 1 and Viking demonstrate conclusively that transpolar arcs can be present simultaneously at polar latiudes in both hemispheres. The arcs observed here are aligned parallel to one another on opposite sides of the magnetic pole, and relative to the noon-midnight meridian are rotated slightly in the counterclockwise direction when viewed from above the North Pole.

084.069 **Geomagnetic and solar data. April 1991.**
H. E. Coffey.
J. Geophys. Res., Vol. 96, No. A10, p. 17913 (1 Oct 1991).

084.070 **Geomagnetic and solar data. May 1991.**
H. E. Coffey.
J. Geophys. Res., Vol. 96, No. A10, p. 17915 - 17917 (1 Oct 1991).

084.071 **Identification of the aurorals $O(^1S)$ precursor in photometric time sequences of pulsating aurora.**
G. G. Shepherd, K. V. Paulson, S. Brown, W. A. Gault, A. Moise, B. H. Solheim.
Geophys. Res. Lett., Vol. 18, No. 11, p. 1939 - 1942 (Nov 1991).
Previous studies of photometric time sequences from pulsating aurora have established that the $O(^1S)$ metastable leading to the emission of the auroral green line is excited by two processes, one direct, the other indirect, with the indirect precursor having a lifetime of about 0.1 s. In this paper the authors report on new high time-resolution measurements which extend the observations of precursor lifetime down to about 0.01 s. Simultaneous observations of pulsations in the Vegard-Kaplan band system were also made in order to obtain direct measurements of the lifetimes of the $N_2(A^3\Sigma^+{}_u)$ metastable; these were found to be essentially equal to those of the precursor over the lifetime range from 0.01 s to 0.1 s, and less well-correlated for longer lifetimes. This result identifies the $O(^1S)$ precursors as $N_2(A^3\Sigma^+{}_u)$, at least over the altitude range 97 to 125 km.

084.072 **22. Sonnenfleckenmaximum in Beziehung zur Polarlichtverzögerung.**
L. Schlamminger.
Sonne, Jahrg. 15, Nr. 60, p. 123 - 125 (Dec 1991). In German.

084.073 **Palaeointensity and reversal frequency - are they related?**
J. Shaw, G. Sherwood.
Geophys. Astrophys. Fluid Dyn., Vol. 60, No. 1-4, p. 135 - 140 (Nov 1991). - See Abstr. 012.125 for the main entry.
It has been predicted that during long periods when the Earth's magnetic field is of a single polarity the field should be stronger. The authors have carried out palaeointensity experiments on rock samples from periods of high, medium and low reversal frequency and can find no evidence to support the above prediction.

084.074 **Sequential reversals from the southern hemisphere.**
M. Ruocco.
Geophys. Astrophys. Fluid Dyn., Vol. 60, No. 1-4, p. 141 - 148 (Nov 1991). - See Abstr. 012.125 for the main entry.
The results of a study of five successive geomagnetic reversal records are presented. The polarity transitions have been recorded in Mar del Plata, Argentina. This sequence of reversals includes the Gauss-Matuyama, Lower and Upper Olduvai, and Lower and Upper Jaramillo polarity transitions as recorded in sedimentary sequences.

084.075 **Trans–polar VGP shifts and Earth's rotation.**
N. A. Mörner.
Geophys. Astrophys. Fluid Dyn., Vol. 60, No. 1–4, p. 149 – 155 (Nov 1991). – See Abstr. 012.125 for the main entry.

At about 13,200 BP and 1000 BP, the geomagnetic field experienced trans–polar VGP shifts between main centra in Arctic Siberia and Arctic Canada, respectively. Both shifts are associated with major changes in the Earth's rate of rotation indicating a causal connection. The 1000 BP VGP shift and change in rotation is also associated with a decrease in the geomagnetic dipole field strength. The trans–polar VGP shifts seem to be of dipolar nature. They seem to be consistent with the displacement of the symmetry axis of two rotating bodies.

084.076 **Geomagnetic and solar data. June – August 1991.**
H. E. Coffey.
J. Geophys. Res., Vol. 96, No. A12, p. 21415 – 21417 (1 Dec 1991).

084.077 **The secular variation of the Earth's magnetic field.**
J. Bloxham, A. Jackson.
Geophys. Astrophys. Fluid Dyn., Vol. 60, No. 1–4, p. 365 – 366 (Nov 1991). Abstract. – See Abstr. 012.125 for the main entry.

084.078 **Bubble dynamics and the geodynamo.**
J. W. M. Bush, J. Bloxham, H. A. Stone.
Geophys. Astrophys. Fluid Dyn., Vol. 60, No. 1–4, p. 373 – 375 (Nov 1991). – See Abstr. 012.125 for the main entry.

084.079 **Theory of the geodynamo and the problem of core–mantle coupling.**
F. H. Busse.
Geophys. Astrophys. Fluid Dyn., Vol. 60, No. 1–4, p. 375 (Nov 1991). Abstract. – See Abstr. 012.125 for the main entry.

084.080 **Paleointensities of the geomagnetic field: methods, accuracy, and variation during reversals and excursions.**
R. S. Coe.
Geophys. Astrophys. Fluid Dyn., Vol. 60, No. 1–4, p. 379 – 382 (Nov 1991). – See Abstr. 012.125 for the main entry.

084.081 **Recent progress in estimating uncertainty in geomagnetic field modeling.**
R. A. Langel.
Geophys. Astrophys. Fluid Dyn., Vol. 60, No. 1–4, p. 37 – 88 (Nov 1991). – See Abstr. 012.125 for the main entry.

Recent work pertaining to estimating error and accuracies in geomagnetic field modeling is reviewed from a unified viewpoint and illustrated with examples. The formulation of a finite dimensional approximation to the underlying infinite dimensional problem is developed.

084.082 **Secular variation of the geomagnetic field for the past 1600 years.**
R. L. Dubois.
Geophys. Astrophys. Fluid Dyn., Vol. 60, No. 1–4, p. 384 – 387 (Nov 1991). – See Abstr. 012.125 for the main entry.

084.083 **A numerical method for simulating the geodynamo.**
G. A. Glatzmaier, P. H. Roberts.
Geophys. Astrophys. Fluid Dyn., Vol. 60, No. 1–4, p. 389 (Nov 1991). Abstract. – See Abstr. 012.125 for the main entry.

084.084 **Geomagnetic field acceleration.**
V. Golovkov.
Geophys. Astrophys. Fluid Dyn., Vol. 60, No. 1–4, p. 392 (Nov 1991). Abstract. – See Abstr. 012.125 for the main entry.

084.085 **Palaeomagnetic observations of the reversing geodynamo.**
K. A. Hoffman.
Geophys. Astrophys. Fluid Dyn., Vol. 60, No. 1–4, p. 397 – 398 (Nov 1991). Abstract. – See Abstr. 012.125 for the main entry.

084.086 **Modelling the geomagnetic main field and recent secular variation.**
D. R. Barraclough.
Geophys. Astrophys. Fluid Dyn., Vol. 60, No. 1–4, p. 3 – 35 (Nov 1991). – See Abstr. 012.125 for the main entry.

This paper describes the data available for global modelling in terms of their accuracy and distribution. It then reviews methods for modelling the geomagnetic field and its secular variation, including techniques such as harmonic spline expansions and stochastic inversion that have been specifically developed to produce field descriptions that are acceptably smooth at the core–mantle boundary. Finally, two sets of models are intercompared, one set having been produced using these specialised techniques and the other using more traditional methods. It is found that the former techniques have most effect for the earliest epoch (1945) considered, when the global data distribution was rather poor.

084.087 **Long–term geomagnetic field secular variation during periods of stable polarity.**
S. P. Lund.
Geophys. Astrophys. Fluid Dyn., Vol. 60, No. 1–4, p. 413 – 414 (Nov 1991). Abstract. – See Abstr. 012.125 for the main entry.

084.088 **Turbulent dynamo as source of geomagnetic variations and reversals.**
A. A. Ruzmajkin.
Geophys. Astrophys. Fluid Dyn., Vol. 60, No. 1–4, p. 424 – 425 (Nov 1991). Abstract. – See Abstr. 012.125 for the main entry.

084.089 **Reconnection in substorms.**
K. Schindler.
Geophys. Astrophys. Fluid Dyn., Vol. 62, No. 1–4, p. 37 – 49 (Dec 1991). – See Abstr. 012.126 for the main entry.

This paper attempts to deal with magnetospheric activity from the kinematic point of view, using the homogeneous subset of Maxwell's equations and Ohm's law. This approach gives insight into three major substorm processes: Plasma entry through the magnetopause (including impulsive penetration), plasmoid formation in the magnetotail and near–Earth magnetic reconfiguration. The central quantity is a generalized electric potential arising from electric field components parallel to the magnetic field. The present discussion is also extended to solar activity and to deviations from isorotation with possible consequences for the origin of cosmic rays.

084.090 **Three–dimensional magnetohydrodynamic simulations of processes at the Earth's magnetopause.**
A. Otto.
Geophys. Astrophys. Fluid Dyn., Vol. 62, No. 1–4, p. 69 – 82 (Dec 1991). – See Abstr. 012.126 for the main entry.

At the Earth's magnetopause magnetic reconnection is believed to be the cause of magnetic flux transfer events or structures which are interpreted in terms of rotational discontinuities. The author reviews main results of two kinds of three–dimensional MHD simulations. Using the local approximation he also presents some new results which might be relevant to the formation of boundary layers at the magnetopause.

084.091 **The Earth's dynamo.**
A. M. Soward.
Geophys. Astrophys. Fluid Dyn., Vol. 62, No. 1–4, p. 191 – 209 (Dec 1991). – See Abstr. 012.126 for the main entry.

Recent developments in geodynamo theory have advanced along two distinct tracks. On the one hand, a mean field approach is adopted. With reasonable assumptions about the nature of the α– and ω–effects of mean field theory, axisymmetric dynamo solutions are constructed which include the nonlinear feed back of the Lorentz force on the motion. On the other hand, the nature of the asymmetric waves, convection and turbulence, which occur in the presence of a prescribed magnetic field are considered. The Earth's rapid rotation Ω causes these asymmetric disturbances to be helical and so give rise to the α–effect. Recently numerical solutions of the complete set of equations governing a convectively driven dynamo have been obtained. The Earth is an $\alpha\omega$–dynamo with a magnetic field predominantly azimuthal in the fluid core.

The global paleomagnetic database. Design, installation and use with ORACLE.
See Abstr. 002.118.

Paläomagnetismus und Archäomagnetismus.
See Abstr. 003.012.

The aurora. Sun–Earth interactions.
See Abstr. 003.015.

Humboldt and the Earth's magnetic field.
See Abstr. 004.020.

L'activité solaire et l'activité aurorale au XVIIe siècle.
See Abstr. 004.078.

Planetary magnetospheric physics I. Proceedings. Symposium 6 and Topical Meeting of the COSPAR Interdisciplinary Scientific Commission D (Meetings D2, D4) of the COSPAR 28. Plenary Meeting, The Hague (Netherlands), 25 Jun – 6 Jul 1990.
See Abstr. 012.006.

Theoretical problems in space and fusion plasmas. Proceedings.
See Abstr. 012.067.

2. Symposium on Reversals, Secular Variation and Dynamo Theory, Santa Fe, NM (USA), 6 – 10 Aug 1990.
See Abstr. 012.125.

The visibility of the aurora borealis.
See Abstr. 014.039.

Particle simulations of collisionless magnetic reconnection in magnetospheric current sheets.
See Abstr. 022.016.

Numerical simulation of flux transfer events at the magnetopause: electron dynamics.
See Abstr. 022.017.

Ion acceleration in a current sheet with magnetic islands.
See Abstr. 022.018.

Electron transport in atmospheric gases I. Integral characteristics.
See Abstr. 022.051.

Electron transport in atmospheric gases. III. Angular distributions.
See Abstr. 022.054.

SOHO and Cluster. Solar and geospheric plasmas and solar structure.
See Abstr. 051.077.

The resistive tearing instability for generalized resistivity models: applications.
See Abstr. 062.018.

Velocity distributions in the plasma sheet boundary layer.
See Abstr. 062.024.

On low frequency magnetic field turbulence upstream of Earth's bow shock.
See Abstr. 062.029.

Modeling Fermi acceleration at parallel shocks with Monte Carlo techniques.
See Abstr. 062.030.

Three–dimensional magnetic reconnection: basic concepts.
See Abstr. 062.081.

Acceleration of keV–electrons by a nearly perpendicular shock wave.
See Abstr. 062.148.

Three–dimensional MHD–computations of magnetic reconnection.
See Abstr. 062.157.

Recent developments in the theory of three–dimensional magnetic reconnection.
See Abstr. 062.200.

Quasi–parallel collisionless shocks.
See Abstr. 062.205.

MHD wave coupling in inhomogeneous media.
See Abstr. 062.225.

Planetary magnetism re–visited.
See Abstr. 062.255.

Variations in the strength of recurrent geomagnetic activity in solar cycles 11 to 21.
See Abstr. 072.065.

Proton flare and terrestrial effects with aurorae in the minimum phase of solar activity (Jan 1985).
See Abstr. 072.139.

High–velocity fluxes of the solar wind and the fluxes of energetic electrons in the Earth's outer magnetosphere.
See Abstr. 074.011.

The structure and length of tail–associated phenomena in the solar wind downstream from the Earth.
See Abstr. 074.105.

On the consequences of strong stable stratification at the top of Earth's outer core.
See Abstr. 081.017.

Relative motions between the inner core and the Earth's mantle due to attractions of the Sun and the Moon.
See Abstr. 081.040.

Ionospheric science and geomagnetism.
See Abstr. 083.020.

Sudden ionospheric oscillations accompanyed with sudden commencement of geomagnetic storm of March 8 – 20, 1989.
See Abstr. 083.023.

Magnetic disturbances and equivalent ionospheric currents during the great magnetic storm of March 13 – 14, 1989.
See Abstr. 083.024.

Vortex structures in the ionosphere and the magnetosphere of the Earth.
See Abstr. 083.028.

Incoherent scatter observations of the high latitude ionosphere and implications for solar wind – magnetosphere – ionosphere coupling.
See Abstr. 083.030.

The plasma environment of the Earth.
See Abstr. 085.012.

Origin of quasi–biennial cycle.
See Abstr. 085.024.

Hemispherical asymmetries in sunspot areas and auroral frequencies.
See Abstr. 085.029.

Why radiation belts exist.
See Abstr. 091.046.

Upstream waves at Mercury, Venus and Earth: comparison of the properties of one Hertz wave.
See Abstr. 092.001.

Plasma wave system measurements of the Martian bow shock from the Phobos 2 spacecraft.
See Abstr. 097.058.

Comparison of plasma wave measurements in the bow shocks at Earth, Jupiter, Saturn, Uranus and Neptune.
See Abstr. 101.016.

Theory of diffusion of meteor trains in the geomagnetic field.
See Abstr. 104.044.

Diffusion of the IMF into the far magnetosphere tail.
See Abstr. 106.001.

An estimate of the penetration rate of the IMF into the magnetosphere due to the development of a trough instability.
See Abstr. 106.003.

Acceleration of particles at the front of a bow shock.
See Abstr. 144.068.

Cosmic ray generation in space and laboratory plasmas.
See Abstr. 144.094.

Evidence for trapped anomalous cosmic ray oxygen ions in the inner magnetosphere.
See Abstr. 144.113.

085 Solar-terrestrial Relations

085.001 **Near–earth sporadic radio emission caused by cosmic rays of isolated solar flares.**
A. V. Dubnik, V. M. Kartashev, A. V. Lazarev, E. S. Shmatko.
Geomagn. Aeron., Vol. 30, No. 1, p. 103 – 105 (Aug 1990). English translation of Geomagn. Aehron., Tom 30, No. 1, p. 132 – 134 (1990).
An analysis is given of the sporadic radio emission of near–earth outer space at the frequencies 38 and 325 MHz in the vicinity of the starting times of isolated flares on the Sun. Groups of sporadic radio flares delayed by 0 – 6 h, which are probably caused by charged particles accelerated on the Sun and arriving in the vicinity of the earth, have been detected. The suggestion is advanced that groups of short ($\sim$0.2 sec) flares are generated by bunches of electrons precipitating from the capture zones under the influence of the action of streams of solar cosmic rays on the earth's magnetosphere.

085.002 **"Effective" solar activity indices for the prediction of the state of the ionosphere.**
A. A. Nusinov, G. I. Ostrovskij.
Geomagn. Aeron., Vol. 30, No. 1, p. 119 – 121 (Aug 1990). English translation of Geomagn. Aehron., Tom 30, No. 1, p. 146 – 149 (1990).
"Effective" solar activity indices based on the laws of variations of the short–wavelength ultraviolet solar radiation fluxes are proposed. It is shown that the use of such indices leads to more accurate results in connection with calculations of the state of the ionosphere over a wide range of activity levels.

085.003 **Common mechanisms in the solar–terrestrial system – part of the STEP international program (a review).**
M. I. Pudovkin, V. S. Semenov.
Geomagn. Aeron., Vol. 30, No. 3, p. 301 – 311 (Nov 1990). English translation of Geomagn. Aehron., Tom 30, No. 3, p. 353 – 367 (1990).
This review is devoted to one of the themes of the STEP program "Common mechanisms in the solar–terrestrial system." Common structural features are isolated: boundary layers, current systems, flare loops and ejections, double layers, magnetospheres, and universal mechanisms responsible for their formation are discussed. The role of self–consistent electrodynamic couplings in transport processes in a plasma is illustrated by the example of diffusion of plasma in a magnetic field. In conclusion the trends in the development of the physics of cosmic plasma are discussed.

085.004 **Variations of the equatorial boundary of polar aurorae as a function of solar sources.**
L. S. Evlashin, T. A. Khviyuzova.
Geomagn. Aeron., Vol. 30, No. 3, p. 421 – 423 (Nov 1990). English translation of Geomagn. Aehron., Tom 30, No. 3, p. 500 – 502 (1990).
The equatorial boundary of the auroral oval is considered as a function of the type and parameters of solar sources such as flares, coronal holes and winking filaments. It is established that the increase in intensity and the position of a solar source relative to the central meridian leads to an increase in auroral activity, expansion of the auroral oval and a decrease in the geomagnetic latitude of the equatorial boundary of polar aurorae.

085.005 **Electron density fluctuations in the supercorona and the terrestrial effects of solar activity.**
V. P. Yakubov, E. Yu. Chuzhkov.
Geomagn. Aeron., Vol. 30, No. 4, p. 475 – 477 (Feb 1990). Paper presented at the 5. Symposium on Solar Terrestrial Physics, Samarkand (USSR), 2 – 6 Oct 1989. English translation of Geomagn. Aehron., Tom 30, No. 4, p. 563 – 565 (1990).
Circumsolar radio occultation data and vertical ionospheric sounding data are combined to carry out a cross–correlation analysis of the relationship between the mean daily variation in solar activity, the fluctuations in solar wind electron density, and the critical frequency of the ionosphere. The authors show that variations in solar activity as reflected by the variations in electron density within the solar supercorona propagate towards the Earth at the solar wind velocity. They conclude that radio occultation observations of the supercorona can be used to predict solar activity and the condition of the ionosphere several days in advance.

085.006 **Quasi–periodic two–year variations in cosmic rays, solar activity, solar wind, and geomagnetic activity.**
O. G. Rogava, L. Kh. Shatashvili.
Geomagn. Aeron., Vol. 30, No. 4, p. 482 – 484 (Feb 1990). Paper presented at the 5. Symposium on Solar Terrestrial Physics, Samarkand (USSR), 2 – 6 Oct 1989. English translation of Geomagn. Aehron., Tom 30, No. 4, p. 571 – 574 (1990).
The authors study the characteristics of the quasi–periodic two–year variations in the intensity of primary cosmic rays over a wide energy range from a combination of observational material obtained by a worldwide network of ground–based and stratospheric observatories. They identify one characteristic fo these quasi–periodic two–year variations in cosmic ray intensity associated with the anomalous changes in the solar magnetic moment during 1979 and 1982. These data on the quasi–periodic two–year variations in the cosmic rays are compared with data on

the quasi-periodic two-year variations in solar activity, geomagnetic activity, and the solar wind. They then provide an appropriate interpretation of the results.

085.007 **Longitudinal variations of the ion composition at F_1-region heights associated with sudden changes in solar-wind parameters January 25, 1974.**
A. V. Mikhajlov, Yu. L. Terekhin, V. V. Mikhajlov, C.- U. Wagner.
Geomagn. Aeron., Vol. 30, No. 4, p. 496 - 499 (Feb 1990). Paper presented at the 5. Symposium on Solar Terrestrial Physics, Samarkand (USSR), 2 - 6 Oct 1989. English translation of Geomagn. Aehron., Tom 30, No. 4, p. 588 - 592 (1990).

The longitudinal picture of the development of an ionosphere disturbance at F_1-layer heights, associated with sudden changes in the solar-wind parameters on January 25 - 26, 1974, is examined in the middle-latitude region (50°N) in conjunction with AE-C satellite observations of the neutral and ionic components and the temperatures of the neutral atmosphere and plasma. A sharp decrease in $[O^+]$ and an increase in $[NO^+]$ and $[O_2{}^+]$ were observed at night-sector longitudes during the period of growth of the disturbance. It is shown that the observed middle-latitude distribution of the disturbance of the nighttime composition is linked with features of the diurnal variations of the thermosphere circulation.

085.008 **Does saturation of the Earth's magnetosphere exist at high solar activity?**
M. Kopecky.
Geomagn. Aeron., Vol. 30, No. 4, p. 560 - 561 (Feb 1990). Paper presented at the 5. Symposium on Solar Terrestrial Physics, Samarkand (USSR), 2 - 6 Oct 1989. English translation of Geomagn. Aehron., Tom 30, No. 4, p. 665 - 666 (1990).

Indices characterizing the level of solar activity in the 11-year cycles on the whole are defined on the basis of the Wolf numbers and the total area of the sunspots, and their interrelation with indices of geomagnetic activity is analyzed. It is shown that the response of geomagnetic activity to a change in solar activity is of a nonlinear nature and that some saturation of the Earth's magnetosphere occurs at a high level of solar activity.

085.009 **A new index for the recurrence of high geomagnetic activity and the relationship between this index and the solar activity cycle.**
V. Bumba, L. Heina.
Geomagn. Aeron., Vol. 30, No. 4, p. 565 - 567 (Feb 1990). Paper presented at the 5. Symposium on Solar Terrestrial Physics, Samarkand (USSR), 2 - 6 Oct 1989. English translation of Geomagn. Aehron., Tom 30, No. 4, p. 669 - 671 (1990).

The authors propose and calculate a new index for the 27-day recurrence of large geomagnetic disturbances for the past 100 years. The time variation in this index is compared against the latitude distribution of solar-magnetic fields and prominences for cycles 20 and 21. They show that the solar-wind flows which lead to significant enhancement of large geomagnetic disturbances and the regular recurrence period of large geomagnetic disturbances during the decline in solar activity at the end of each 11-year solar cycle are correlated with the magnetic field in the preceding 11-year cycle.

085.010 **Solar cycles and climatic oscillations.**
V. F. Chistyakov.
Geomagn. Aeron., Vol. 30, No. 4, p. 598 - 600 (Feb 1990). Paper presented at the 5. Symposium on Solar Terrestrial Physics, Samarkand (USSR), 2 - 6 Oct 1989. English translation of Geomagn. Aehron., Tom 30, No. 4, p. 698 - 700 (1990).

Climatic oscillations with periods of 100, 200, and 2600 yr are positively correlated with similar oscillations in solar activity: higher solar activity is correlated with warmer climate. Geological data from varved clays indicate that the length of the solar-climate cycle has decreased from 23.4 yr to 11 yr over the past 2.5 thousand million years. During the great Ice Ages, these cycles were 12 yr long; during the interglacial periods, they were 10 yr long. It is suggested that these oscillations may be associated with variations in solar activity and solar luminosity.

085.011 **Interplanetary perturbation-induced effects on polar ozone level.**
M. de Petris, M. Gervasi, S. Masi, G. Moreno, M. Parisi, B. Melchiorri Olivio, M. Storini.
Ann. Geophys., Vol. 9, No. 6, p. 381 - 386 (Jun 1991). Part of this paper was presented at "A.G.U.-Spring Meeting and SCOSTER-S.T.P. Symposium".

The authors study the possible linkage between the ozone depletion of the Antarctic atmosphere and the interplanetary travelling perturbations coming from solar wind sources. It turns out that, during solar cycle 21, a significant change has occurred in the solar wind regime, which may have affected the terrestrial environment and hence the ozone equilibrium.

085.012 **The plasma environment of the Earth.**
S. W. H. Cowley.
Contemp. Phys., Vol. 32, No. 4, p. 235 - 250 (Jul-Aug 1991).

The author discusses the structure and dynamics of the Earth's plasma environment and how it is influenced by its interaction with the outer atmosphere of the Sun, as revealed by both space-based and ground-based observations over the past thirty years. They begin with a discussion of the basic physical principles involved, and then apply these principles to the solar wind and interplanetary magnetic field, and their interaction with the Earth's magnetic field. The structure and flow of magnetospheric plasma populations then follows, together with consequent magnetosphere-ionosphere interactions.

085.013 **Possible coincidence of solar-oscillation periods with periods of the natural vibration of the Earth.**
V. P. Bobova, B. M. Vladimirskij, S. A. Zajtseva, N. G. Savina, M. I. Pudovkin.
Kinematics Phys. Celest. Bodies, Vol. 7, No. 1, p. 31 - 38 (1991). English translation of Kinematika Fiz. Nebesn. Tel, Tom 7, No. 1, p. 34 - 42 (1991).

Stable periods that have been found in the geomagnetic AE-index from seismic measurements and radio and optical observations of the solar oscillations in the range 10 - 200 min are compared with one another. The variations of the AE-index include periods of the Earth's natural torsional vibrations and periods (about 90%) attributed to unidentified natural vibrations of the Earth. On the other hand, nonaccidental agreement is noted between the periods in the AE-index and periods found from radio and optical observations of the Sun. It is established that many of the periods discerned from geophysical measurements agree accurate to ± 0.5 min with solar oscillation periods. Possible variant explanations of this agreement are formulated. It is not yet possible to make choices among them.

085.014 **Solar UV variability during solar cycle 22 as measured by the NOAA-9 SBUV/2 Mg II index.**
M. T. DeLand, R. P. Cebula, B. M. Schlesinger.
Bull. Am. Astron. Soc., Vol. 23, No. 2, p. 1068 (1991). Abstract. - See Abstr. 012.012 for the main entry.

085.015 **Short-period oscillations of the Caspean sea level and solar-geophysical data.**
M. A. Nuzhdina.
Soln. Dannye, Byull., No. 11/90, p. 85 - 91 (Nov 1990). In Russian.

Short-period oscillations ($P \leqslant 12$ years) of the Caspean sea level are analysed in connection with the similar oscillations in the solar and geomagnetic activity by the spectral and cross-spectral methods.

085.016 **Length of the solar cycle: an indicator of solar activity closely associated with climate.**
E. Friis-Christensen, K. Lassen.
Science, Vol. 254, No. 5032, p. 698 - 700 (1 Nov 1991).

It has recently been suggested that the solar irradiance has varied in phase with the 80- to 90-year period represented by the envelope of the 11-year sunspot cycle and that this variation is causing a significant part of the changes in the global temperature. This interpretation has been criticized for statistical reasons and because there are no observations that indicate significant changes in the solar irradiance. A set of data that supports the suggestion of a direct influence of solar activity on global climate is the variation of the solar cycle length. This record closely matches the long-term variations of the Northern Hemisphere land air temperature during the past 130 years.

085.017 **Could the sun be warming the climate?**
R. A. Kerr.
Science, Vol. 254, No. 5032, p. 652 – 653 (1 Nov 1991).
A new correlation between solar variations and climate change hints, yet again, at a sun-climate connection.

085.018 **f_0F2 as a function of the solar activity indices.**
M. V. Krasinets, A. A. Nusinov.
Geomagn. Aehron., Tom 31, No. 4, p. 741 – 743 (Jul – Aug 1991). In Russian. English translation in Geomagn. Aeron., Vol. 31, No. 4.

085.019 **Effect of solar flares on low-frequency wave propagation (0.5 – 10 kHz) in the Earth-ionosphere waveguide.**
N. D. Borisov, A. P. Dubovoj, D. S. Fligel', A. A. Yaroslavtsev.
Geomagn. Aehron., Tom 31, No. 5, p. 792 – 800 (Sep – Oct 1991). In Russian. English translation in Geomagn. Aeron., Vol. 31, No. 5.

085.020 **Effects of solar and geomagnetic activity in f_0E routine data.**
G. S. Ivanov-Kholodnyj, A. K. Stankevich, V. E. Chertoprud.
Geomagn. Aehron., Tom 31, No. 5, p. 848 – 858 (Sep – Oct 1991). In Russian. English translation in Geomagn. Aeron., Vol. 31, No. 5.

085.021 **An indirect index of solar activity.**
V. Dermendjiev, Ya. Shopov, G. Buyukliev, I. Panajotova.
Geomagn. Aehron., Tom 31, No. 5, p. 923 – 925 (Sep – Oct 1991). In Russian. English translation in Geomagn. Aeron., Vol. 31, No. 5.

085.022 **Connection of generation of the high-latitude pulsations *vlp* with solar wind irregularities.**
O. V. Bol'shakova, B. I. Klajn, N. A. Kurazhovskaya.
Geomagn. Aehron., Tom 31, No. 6, p. 989 – 996 (Nov – Dec 1991). In Russian. English translation in Geomagn. Aeron., Vol. 31, No. 6.

085.023 **Quasi-biennial variation of the recurrent geomagnetic activity.**
D. I. Ponyavin.
Soln. Dannye, Byull., No. 2/91, p. 99 (Feb 1991). Abstract. – See Abstr. 012.021 for the main entry.

085.024 **Origin of quasi-biennial cycle.**
M. A. Nuzhdina.
Soln. Dannye, Byull., No. 3/91, p. 101 – 102 (Mar 1991). Abstract. – See Abstr. 012.021 for the main entry.

085.025 **Quasi-biennial variations of solar wind velocity, cosmic rays, solar and geomagnetic activity.**
L. Kh. Shatashvili, O. G. Rogava.
Soln. Dannye, Byull., No. 3/91, p. 102 – 103 (Mar 1991). Abstract. – See Abstr. 012.021 for the main entry.

085.026 **Filter for isolation of quasi-biennial variations in helio-geophysical parameters.**
Yu. R. Rivin.
Soln. Dannye, Byull., No. 3/91, p. 105 – 106 (Mar 1991). Abstract. – See Abstr. 012.021 for the main entry.

085.027 **Solar flare and sudden change of Earth's rotation.**
Feng Bo.
Chin. Astron. Astrophys., Vol. 15, No. 3, p. 329 – 335 (Sep 1991). English translation of Acta Astron. Sin., Vol. 32, No. 1, p. 73 – 78 (Mar 1991). See Abstr. 53.085.012.
It is shown that it is only when, in a shock pair produced by a large solar flare, the velocity of the following shock is clearly greater than that of the leading shock that a sudden change in the Earth's rotation can occur.

085.028 **Response of the middle atmosphere to the solar proton events of August – December, 1989.**
G. C. Reid, S. Solomon, R. R. Garcia.
Geophys. Res. Lett., Vol. 18, No. 6, p. 1019 – 1022 (Jun 1991).
Intense solar activity during the period August – December, 1989 gave rise to several major energetic-particle events, which were accompanied by greatly enhanced ionization rates and NOy production in the polar regions of both hemispheres. A two-dimensional model of the chemistry and dynamics of the middle atmosphere has been used to calculate the production and subsequent fate of the NOy and its effect on ozone concentrations and temperatures. In the sunlit southern polar cap, NO increases as large as a factor of 20 are estimated near 60 km altitude, with column density enhancements of 55%. Corresponding peak ozone depletions of about 20% are calculated near 40 km in late October 1989, with predicted temperature decreases of about 3–3.5K. Effects in the northern polar regions are considerably smaller, due to the lack of sunlight during the peak phase of the events.

085.029 **Hemispherical asymmetries in sunspot areas and auroral frequencies.**
L. Schlamminger.
Sol. Phys., Vol. 135, No. 2, p. 407 – 413 (Oct 1991).
1052 aurorae boreales and 554 aurorae australes recorded during the nineteenth century at medium latitudes $\leqslant 55°$N or $\leqslant 55°$S are compared statistically with the known hemispherical asymmetry of the sums of the areas of sunspots. According to the present study, the solar hemispherical asymmetry may be accompanied by an analogous pattern of the hemispherical frequency of auroral days. For the number of auroral days in each hemisphere beyond the two auroral ovals, a remarkable degree of phase equality with the sunspot areas during the second Gleissberg cycle can be demonstrated.

085.030 **On the efficiency of solar heating in the middle atmosphere.**
M. G. Mlynczak, S. Solomon.
Geophys. Res. Lett., Vol. 18, No. 7, p. 1201 – 1204 (Jul 1991).
The authors calculate the efficiency at which solar ultraviolet radiation absorbed in the Hartley band of ozone is directly converted to heat in the terrestrial mesosphere and lower thermosphere (50–110 km).

085.031 **A simple algorithm for simulating the solar ultraviolet radiation at the Earth's surface: an application in determining the minimum erythema dose.**
A. Katsambas, C. Andoniou, J. Stratigos, I. Arvanitis, F. Zolota, C. Varotsos, C. Cartalis, D. N. Asimakopoulos.
Earth, Moon, Planets, Vol. 53, No. 3, p. 191 – 204 (Jun 1991).
The purpose of this study is to present a new simple algorithm for simulating the solar UV radiation reaching the surface of the Earth. The algorithm is capable of providing both the direct and diffuse solar UV radiation for different atmospheric conditions, times and geographic locations. An application of the model relates to the calculation of the minimum energy input of UV radiation to be received by a human to allow an erythema development.

085.032 **The distribution of solar flares and probable relations to biological effects.**
B. Petropoulos, C. Poulakos.
Earth, Moon, Planets, Vol. 55, No. 1, p. 1 – 19 (Oct 1991).
The authors compared the monthly distribution of the total number of flares to the monthly mortality from cardiovascular

and myocardial infarctions for the time period 1974–1983, and compared correlation coefficients. Periodicities of 3 and 5 months (150 d) have been found in both distributions.

085.033 **Solar flares and terrestrial impacts of active region 5395.**
Zhang Qin.
Publ. Yunnan Obs., No. 4, p. 16 – 19 (Dec 1990). – See Abstr. 012.047 for the main entry.

085.034 **Evolution of AR5395 and time sequence of solar–terrestrial events associated with AR5395.**
Wang Jialong.
Publ. Yunnan Obs., No. 4, p. 20 – 24 (Dec 1990). – See Abstr. 012.047 for the main entry.

085.035 **Solar–terrestrial effects of AR 5395.**
Ye Zonghai, Jiang Herong, Zhang Zhongmou.
Publ. Yunnan Obs., No. 4, p. 110 – 114 (Dec 1990). – See Abstr. 012.047 for the main entry.

085.036 **Low latitude pulsations during SC storm in March 13, 1989.**
Yang Shaofeng.
Publ. Yunnan Obs., No. 4, p. 128 – 132 (Dec 1990). – See Abstr. 012.047 for the main entry.

085.037 **The lower atmospheric response to solar activity of March 1989.**
Qin Guotai, He Aiqing.
Publ. Yunnan Obs., No. 4, p. 155 (Dec 1990). Abstract. – See Abstr. 012.047 for the main entry.

085.038 **A possible effect of the solar activity on atmospheric ozone content.**
Zhang Zhongmou, Song Zhi, Wang Yingjian.
Publ. Yunnan Obs., No. 4, p. 156 – 159 (Dec 1990). – See Abstr. 012.047 for the main entry.

085.039 **Ionospheric effects of solar events in August 1989.**
Liang Shangqi, Zhang Shunrong, Chen Jingwu.
Publ. Yunnan Obs., No. 4, p. 205 – 208 (Dec 1990). – See Abstr. 012.047 for the main entry.

085.040 **Studies of the correlation between the global radiation, ultraviolet radiation, cosmic radiation, sunspot number and meteorological parameters.**
W. E. Alnaser, M. J. Alothman.
Astrophys. Space Sci., Vol. 186, No. 2, p. 191 – 203 (Dec 1991).
The correlation between the ratio of the global irradiation to the extraterrestrial solar radiation (H/H_0), and the ratio of the ultraviolet solar irradiation to the extraterrestrial solar radiation (H_u/H_0) on a horizontal surface at Bahrain ($\Phi = 26°$), and some terrestrial and solar parameters (the monthly average relative humidity, temperature, relative sunshine duration, cosmic radiation intensity, and sunspot number) have been studied. Moreover, the role of the solar effects and the terrestrial effects on the global and the solar ultraviolet radiation has been studied. A detailed investigation has been carried between the level of the cosmic radiation received at Bahrain and the sunspot number. It was concluded that as the solar activity increases, cosmic radiation and sunspot number play a predominant effect on the correlation of (H/H_0) and (H_u/H_0). Furthermore, the correlation between cosmic radiation and sunspot number also increases.

085.041 **The solar UV related changes in total ozone from a solar rotation to a solar cycle.**
S. Chandra.
Geophys. Res. Lett., Vol. 18, No. 5, p. 837 – 840 (May 1991).
The Nimbus–7 TOMS version 6 data, corrected for the instrument degradation, are analyzed to delineate the solar UV related changes in total ozone (TOZ) against background signals of dynamical origin. It is shown that the solar UV related change in TOZ over a solar cycle is about 1.5 percent that may be attributed to about 6 percent change in the solar UV flux near 200 nm. This estimate is also consistent with the solar UV related changes in TOZ over a time scale of a solar rotation. In the solar rotation case, ozone lags the solar UV by 3–4 days and its sensitivity to solar UV change is a factor of 2–3 less than for the solar cycle case. Both these effects are attributed to chemical time constants in the lower stratosphere that are comparable to the period of a solar rotation.

085.042 **The severe geomagnetic storm group generated by the solar activities in March 1989.**
Zhang Gongliang, Gao Yufen, Nie Huashan, Li Zhuohua, Zhang Qinghua.
Publ. Yunnan Obs., No. 4, p. 120 – 121 (Dec 1990). Abstract. – See Abstr. 012.047 for the main entry.

085.043 **The general effects and model analysis of major magnetic storms in March, 1989.**
Kong Nan, Xiang Jingtian, Gao Meiqing, Wen Xiaolei.
Publ. Yunnan Obs., No. 4, p. 122 – 127 (Dec 1990). Abstract. – See Abstr. 012.047 for the main entry.

085.044 **The effects of solar event on whistler activities in March 1989.**
Dong Aiying.
Publ. Yunnan Obs., No. 4, p. 152 – 154 (Dec 1990). – See Abstr. 012.047 for the main entry.

085.045 **Analyses and study on the ionospheric disturbances associated with the solar bursts for the periods of March and August, 1989.**
Pan Liande, Wang Suqin, Ma Lanxiang.
Publ. Yunnan Obs., No. 4, p. 209 – 213 (Dec 1990). – See Abstr. 012.047 for the main entry.

085.046 **Ionospheric effects over Beijing of solar flare events in mid–August of 1989.**
Xu Chufu.
Publ. Yunnan Obs., No. 4, p. 214 – 218 (Dec 1990). – See Abstr. 012.047 for the main entry.

085.047 **Synthetic observation and analyses of the strong flare events occurred in August, 1989.**
Liu Wantong, Huang Zhen, Zheng Ruimin.
Publ. Yunnan Obs., No. 4, p. 227 – 231 (Dec 1990). – See Abstr. 012.047 for the main entry.

085.048 **The research advance and problems about 11–year solar cycle and quasi–biennial oscillation of atmosphere.**
Huang Zhen, Lu Daren.
Publ. Yunnan Obs., No. 4, p. 267 – 271 (Dec 1990). – See Abstr. 012.047 for the main entry.

085.049 **Effects of solar UV radiation on the minor constituents in the middle atmosphere.**
Wang Yingjian, Zhang Zhongmou.
Publ. Yunnan Obs., No. 4, p. 272 (Dec 1990). Abstract. – See Abstr. 012.047 for the main entry.

085.050 **Sun–Earth connections: The great solar storms of 1989.**
W. S. Kurth.
Nature, Vol. 353, No. 6346, p. 705 – 706 (24 Oct 1991).

085.051 **The faint young Sun problem.**
J. F. Kasting, D. H. Grinspoon.
The Sun in time, p. 447 – 462 (1991). – See Abstr. 003.027 for the main entry.
The Sun was almost certainly much less bright during the early stages of its main–sequence lifetime. In the absence of some compensating factor, the Earth's mean surface temperature would therefore have been below the freezing point of water prior to ~2 Gyr ago. Geologic evidence for liquid water as early as 3.8 Gyr ago implies that the Earth was never this cold. This discrepancy can be resolved if the greenhouse effect of the early atmosphere was much larger than today, caused by an increase in

atmospheric carbon dioxide (CO_2) concentrations by a factor of ~1000 or more compared to today. Such an increase could have resulted from feedbacks inherent in the carbonate–silicate geochemical cycle that controls the atmospheric CO_2 level over long time scales. Indeed, if the continents were originally much smaller, the rate of silicate weathering would have been slow, and the CO_2 partial pressure could have been as high as 10 bar. In this case, the early Earth could have been quite warm (~85°C), despite the low solar luminosity.

085.052 **The atmosphere's response to solar irradiation.**
D. M. Hunten, J.-C. Gérard, L. M. François.
The Sun in time, p. 463 – 497 (1991). – See Abstr. 003.027 for the main entry.

The variability of solar ultraviolet and corpuscular emissions is described, along with their known effects on the atmosphere. The discussion excludes variations in the total output and their possible climatic effects. Subtle variations are seen at and above 30 km, growing to factor–of–2 changes above 200 km. All claims of correlation of the solar cycle with weather and climate must be viewed with great suspicion, if only because any plausible driving forces are exceedingly small. The primitive atmosphere was deficient in the oxygen and ozone that keep the varying ultraviolet away from the surface, and there are some possibilities for effects that might have survived in the geologic record. Even further back, the abundances and isotopic patterns of surviving noble gases suggest an epoch during which a greatly enhanced ultraviolet flux drove a hydrodynamic escape flow of hydrogen.

085.053 **Long–term history of climate ice ages and Milankovitch periodicity.**
A. Berger.
The Sun in time, p. 498 – 510 (1991). – See Abstr. 003.027 for the main entry.

During the first half of the 20th century, Köppen, Spitaler and Milankovitch considered high winter and low summer insolation as favoring glaciation. After Köppen and Wegener related the Milankovitch new radiation curve to Penck and Brückner's subdivision of the Quaternary, there was a long–lasting debate as to whether or not such changes in the insolation can explain the Quaternary glacial–interglacial cycles. In the 1970s, with improvements in radiometric dating and in acquiring and interpreting the long continuous geologic records, with the advent of computers and with the development of more sophisticated astronomical and climate models, the astronomical theory and in particular its Milankovitch version were revived. Now, most geologic, astronomical and climatological obstacles have been overcome and the causal role of the orbital variations in long–term variation of climate over the last Myr is taken almost universally for granted.

085.054 **Quaternary glaciations: theory and observations.**
M. Ghil.
The Sun in time, p. 511 – 542 (1991). – See Abstr. 003.027 for the main entry.

This chapter reviews the geochemical and micropaleontological records of Quaternary glaciations, and the history of glacial theories. It discucsses a very simple model of self–sustained climatic oscillations, coupling the radiation balance of the atmosphere and oceans with the mass balance of ice sheets and with the isostatic adjustment of bedrock. The model's spectral response to small, quasi–periodic insolation variations, due to secular changes in the Earth's orbital parameters is described. This response exhibits the known peaks at 100 kyr, 40 kyr and 20 kyr and predicts additional peaks at 15 kyr, 13 kyr and 10 kyr. Observational evidence for these latter peaks is presented. Other models of Quaternary glaciations are reviewed and confronted with the data. Open questions are mentioned. In particular, the net, globally and annually averaged change in solar radiation at the top of the atmosphere is very small (0.1%) for orbital changes with very long periods, as well as for solar irradiance changes with periods of one or two decades. This leaves a substantial role to internal climate variability on either time scale.

085.055 **Solar variability captured in climatic and high–resolution paleoclimatic records: a geologic perspective.**
R. Y. Anderson.
The Sun in time, p. 543 – 561 (1991). – See Abstr. 003.027 for the main entry.

Natural fiters in the ocean–atmosphere system and in geologic processes mask expression of the 11–yr cycle of solar variability in climatic and paleoclimatic records. However, paleoclimatic records from ~170 Myr and 250 Myr old varved evaporite sediments, believed to have been regulated by thermal climatic forcing, are cyclic at Gleissberg–, Maunder– and Sporer–cycle frequencies. Also long–term changes in the El Nino/Southern Oscillation and/or Quasi–biennial Oscillation climate system may be modulated by solar variability. If the cyclicity recorded in these climatic and paleoclimatic records can be linked to solar forcing through a physical model, some rare ancient sediments will provide information about solar variability over a broad range of frequencies.

085.056 **Solar–terrestrial relationships in recent sea sediments.**
G. C. Castagnoli, G. Bonino, A. Provenzale.
The Sun in time, p. 562 – 586 (1991). – See Abstr. 003.027 for the main entry.

The calcium carbonate ($CaCO_3$) and the bulk thermoluminescence (TL) profiles of two recent Ionian Sea cores that span the last four millennia are analyzed. The cores have been dated with a precision of 1% over the last 20 centuries and the constancy of the sedimentation rate has been verfied. The study of the $CaCO_3$ and the TL time series obtained in this way reveals the existence of several outstanding features and well–defined cyclic variations. The high precision of the time scale allows for carefully determining the periodicities present in the signal and for exploring their possible solar origin. The analysis of the carbonate profile reveals the presence of a century–scale wave with period of about 206 yr, modulated by a long component with period of ~2000 yr. In addition, the presence of some other spectral lines are revealed.

085.057 **Seasonal and solar cycle variations in high–latitude thermospheric winds.**
A. L. Aruliah, D. Rees, A. Steen.
Geophys. Res. Lett., Vol. 18, No. 11, p. 1983 – 1986 (Nov 1991).

Thermospheric wind measurements have been collected systematically every winter for over nine years from a high–latitude site at Kiruna, Sweden (67.8°N, 20.4°E). The database contains 1242 nights of data collected with a Fabry–Perot Interferometer (FPI), perhaps the largest single – site database of thermospheric winds. This analysis shows a marked seasonal and solar cycle variation. Particularly at high solar activity, sunward winds of the evening period (16–20 UT) are more than 50% stronger at Spring than at Autumn equinox. This large asymmetry in the behavior of high–latitude thermospheric winds at spring and autumn equinox has not yet been predicted by model simulations.

085.058 **Correlations between solar activity and the atmosphere: an unphysical explanation.**
M. L. Salby, D. J. Shea.
J. Geophys. Res., Vol. 96, No. D12, p. 22579 – 22596 (20 Dec 1991).

The aurora. Sun–Earth interactions.
See Abstr. 003.015.

The Sun in time.
See Abstr. 003.027.

A. L. Chizhevskij's views on solar–terrestrial relations.
See Abstr. 004.047.

From the solar core to the Earth's magnetosphere. Extended abstracts of reports. All–Union Meeting on the Quasibiennial Solar Variations, Pushchino (USSR), 25 – 27 Feb 1991.
See Abstr. 012.021.

22 MAX. Global character research of solar–terrestrial system. Special issue III. Proceedings. Workshop on major solar–geophysical events in March and August 1989, Beijing (China), 23 – 26 Apr 1990.
See Abstr. 012.047.

Objectives and scope of RISE.
Scc Abstr. 013.035.

Some theoretical topics for RISE.
See Abstr. 013.037.

SunRISE at the NSF.
See Abstr. 013.038.

RISE–related programs at NASA.
See Abstr. 013.039.

NOAA's program of research on solar influences in global change.
See Abstr. 013.040.

RISE and the NRC's Committee on Global Change. Working Group on Solar Influences on Global Change.
See Abstr. 013.041.

The solar–terrestrial sciences in ninety's.
See Abstr. 013.061.

The progress of solar–terrestrial sciences in China.
See Abstr. 013.099.

Search for periodicity in time series through prediction.
See Abstr. 021.046.

SOHO and Cluster. Solar and geospheric plasmas and solar structure.
See Abstr. 051.077.

Variations in the strength of recurrent geomagnetic activity in solar cycles 11 to 21.
See Abstr. 072.065.

The phase of quasi–biennial variations in the solar and geomagnetic activity.
See Abstr. 072.073.

Moving periodogramanalysis of data on solar activity and solar–terrestrial relations over the past 400 years.
See Abstr. 072.083.

Two–year variations of solar neutrino fluxes, galactic cosmic rays, solar and geomagnetic activities.
See Abstr. 078.007.

The history of activity of the Sun and its rotational period on the ZAMS.
See Abstr. 080.075.

Origin of the ionospheric current loops which give rise to the solar–daily variations of the geomagnetic field.
See Abstr. 083.019.

Quasi–biennial oscillations in geomagnetic aa–index.
See Abstr. 084.036.

Is the recurrent enhanced geomagnetic activity preceding the sunspot minimum related to magnetic fields of the new or old activity cycle?
See Abstr. 084.053.

Solar and terrestrial components of the atmospheric ^{14}C variation spectrum.
See Abstr. 144.106.

Planetary System

091 Physics and Dynamics of the Planetary System

091.001 **Models of giant planets with a variable ratio of ice to rock.**
V. N. Zharkov, T. V. Gudkova.
Ann. Geophys., Vol. 9, No. 5, p. 357 - 366 (May 1991).
Numerical techniques are used to construct new models of the internal structure of Jupiter, Saturn, Uranus and Neptune. New, significantly increased estimates of the abundance of helium give a new insight into the internal structures of Jupiter and Saturn and their cores. It is confirmed that the mass of Jupiter's core is several times smaller than Saturn's. Jupiter-like models of Uranus and Neptune with two-layer molecular envelopes are constructed. Models of the giant planets do not support Mizuno's hypothesis of approximately equal masses of embryos of the planets, on which the accretion of gas takes place. The assumption about the great depletion of carbon in Jupiter is discussed.

091.002 **Die Staubringe der äußeren Planeten.**
E. Grün.
Sterne Weltraum, Jahrg. 30, Nr. 7, p. 429 - 434 (Jul 1991).

091.003 **Der Strahlungshaushalt der terrestrischen Planeten Venus, Erde und Mars.**
D. Spänkuch.
Naturwissenschaften, Vol. 78, No. 8, p. 347 - 354 (Aug 1991).
The radiation budget of planets substantially determines their climates. For the terrestrial planets having their own imperceptible heat source compared to the solar radiation, the radiation budget supplies energy that can be transformed and used in their atmospheres and at their surfaces. The intercomparison and analysis of the radiation budget of the terrestrial planets, due to quite different atmospheric compositions and external conditions, provide a deeper insight into the relations of the different components of the radiation budget.

091.004 **On the origin and early evolution of terrestrial planet atmospheres and meteoritic volatiles.**
R. O. Pepin.
Icarus, Vol. 92, No. 1, p. 2 - 79 (Jul 1991).
A simple analytic model of mass fractionation in hydrodynamic escape is applied to the problem of the origin and evolution of terrestrial planet atmospheres and meteoritic noble gases from primordial compositions. It is shown that isotopic distributions of xenon, krypton, argon, and neon on Earth can be generated by two stages of hydrodynamic loss, first of a hydrogen-rich, isotopically solar primitive atmosphere, and then of Kr, Ar, and Ne subsequently outgassed from deep within the planet. A similar model is constructed for derivation of meteoritic noble gases from solar-like and nucleogenetic components carried into the accretion disk in the mantles and interiors of extrasolar grains.

091.005 **Observations of Mars, Uranus, Neptune, Io, Europa, Ganymede, and Callisto at a wavelength of 2.66 mm.**
D. O. Muhleman, G. L. Berge.
Icarus, Vol. 92, No. 2, p. 263 - 272 (Aug 1991). = Contrib. No. 4699 Div. Geol. Planet. Sci., Calif. Inst. Technol., Pasadena, CA (USA).
Brightness temperature observations at a wavelength of 2.66 mm of Uranus, Neptune, and the Galilean Satellites calibrated against Mars are presented. A detailed procedure is reported for using an accurate thermal model of Mars as a calibrator which yields brightness temperatures for Uranus and Neptune of 122 ± 5K and 114 ± 7K in excellent agreement with published "absolutely calibrated" measurements of these bodies. The measurements for Io, Europa, and Callisto are consistent with thermal models of the lunar type but Ganymede is strongly anomalous with its 2.7-mm brightness temperature remaining very low and equal to those at centimeter wavelengths.

091.006 **Erratum: "Numerical simulations of dense collisional systems" [Icarus, Vol. 90, No. 2, p. 254 - 270 (Apr 1991)].**
H. Salo.
Icarus, Vol. 92, No. 2, p. 367 - 368 (Aug 1991). See Abstr. 53.091.017.

091.007 **The origin of the systematic component of planetary rotation for a planet in a circular orbit in a 2-D disk.**
D. M. Kary, J. J. Lissauer.
Bull. Am. Astron. Soc., Vol. 23, No. 2, p. 961 (1991). Abstract. - See Abstr. 012.008 for the main entry.

091.008 **Ammonium hydrosulfide and clouds in the atmospheres of the giant planets.**
K. Yu. Ibragimov, A. A. Solodovnik.
Kinematics Phys. Celest. Bodies, Vol. 7, No. 1, p. 53 - 58 (1991). English translation of Kinematika Fiz. Nebesn. Tel, Tom 7, No. 1, p. 58 - 63 (1991).
The physicochemical properties of two possible compounds - ammonium hydrosulfide (NH_4SH) and ammonium sulfide $(NH_4)_2S$ - that may be formed in a reaction of ammonia NH_3 with hydrogen sulfide H_2S are discussed, and the probability of their formation is analyzed on the basis of the Le Chatelier principle. It is shown that the conditions of their formation on the basis of available data on the concentration ratio of the reagents (NH_3 and H_2S) in the atmospheres of giant planets make the appearance of enough NH_4SH for cloud formation highly problematic. Accordingly, the authors propose as an alternative candidate for a cloud-forming role ammonium sulfide $(NH_4)_2S$, for whose formation the conditions in the atmospheres of the giant planets are more favorable. The possible spatial localization of $(NH_4)_2S$ clouds is estimated, and the result is used in an attempt to identify this compound as one of the chromophores.

091.009 **Formation of sharp ringlets by inelastic collisions.**
K. Shida, K. Suzuki, T. Kawai.
J. Phys. Soc. Jpn., Vol. 60, No. 11, p. 3953 - 3966 (Nov 1991).
The authors investigate the possibility that the sharp ringlets observed around giant planets of the solar system are generated through inelastic collisions. First, a model is defined, and then ringlet formation is shown to occur under certain conditions on the rebound coefficients. A three-dimensional case with 2000-body was then simulated. If the condition is satisfied, particles under the gravitational influence of a planet are shown to evolve from a three-dimensional torus to a two-dimensional disc, and finally to one-dimensional ringlets. The form is thus simplified through inelastic collisions. This result is contrary to the common

notion that collisions among the constituents of planetary rings would cause the rings to lose their sharp features.

091.010 **Modelling of the stratiform clouds in the atmospheres of giant planets with consideration of the components intersolubility. II.**
M. V. Bujkov, K. Yu. Ibragimov, G. A. Kirienko, A. M. Pirnach.
Kinematika Fiz. Nebesn. Tel, Tom 7, No. 4, p. 23 – 30 (Jul–Aug 1991). In Russian. English translation in Kinematics Phys. Celest. Bodies, Vol. 7, No. 4.
The authors have examined the conditions of condensation of the aqueous–ammonia clouds for the case of noticeable deviation from the Raul–Henry law. Experimental phase diagrams of the aqueous–ammonia solution and Antuan's law were used in the calculations of the equilibrium saturation vapour above the aqueous ammonia solution drop for variable concentration. The authors believe these results are more accurate when compared with Clausius' law. They have applied them to modelling of the Jovian clouds. An analysis of 49 models has shown significant changes both in the macro– and microstructure of clouds. As illustration, two variants of the model are given: for strong and weak interpenetration of water and ammonia clouds.

091.011 **Some graphic representations of solar–system abundances of nuclei.**
I. Török.
Acta Phys. Hung., Vol. 69, No. 3/4, p. 191 – 209 (16 Nov 1990).
In the last years several hundred new isotopes were synthesized in different laboratories all over the world. A new version of our half–life systematics is under progress. The task is divided into four parts: the even–even, the even–odd, the odd–even, and the odd–odd nuclei are displayed in different diagrams. Some three–dimensional views of the data are incorporated. The logarithm of the abundances is shown as the Z–axis, the X–axis is the neutron number, the Y–axis is the proton number. All the four parity groups of nuclides are shown in four different representations.

091.012 **Dust rings around planets.**
E. Grün.
Reviews in Modern Astronomy 4, p. 157 – 164 (1991). – See Abstr. 003.008 for the main entry.
Rings have been observed around all giant planets. Micron sized dust particles are constituents of most rings. Because of various dynamic effects acting on small particles dust rings are short–lived phenomena or have to be continuously replenished. Descriptions of the ring systems are given and the significance of dust is discussed.

091.013 **The formation of terrains antipodal to major impacts.**
A. W. Watts, R. Greeley, H. J. Melosh.
Icarus, Vol. 93, No. 1, p. 159 – 168 (Sep 1991).
In 1975, Schultz and Gault suggested that seismic energy from a major impact could cause extensive crustal fracturing and surface disruption on the part of the planet directly opposite the impact. Disrupted terrains of this type have been identified on Mercury antipodal to the Caloris impact basin and on the Moon antipodal to Imbrium. Large impact basins are found on many icy satellites. The Simplified Arbitrary Lagrangian Eulerian code was used to calculate the antipodal pressures generated by large impacts into stylized icy satellites and planets that have different compositions and core sizes. It was found that in a planet with a high–density core surrounded by a lower density mantle the incident pressure wave is refracted away from the antipode by the core and the generated antipodal pressure will be low. In a planet with a low velocity core surrounded by a mantle with a higher wave speed the core will focus the wavefront and the antipodal pressure will be high; in both cases (for equal impacts) the magnitude is dependent upon the size of the core relative to the planet.

091.014 **On the existence of "stable and unstable" quantum levels in the gravitational field of the Sun.**
E. N. Kramer, Yu. M. Gorbanev.
Astron. Zh., Tom 68, Vyp. 5, p. 1030 – 1035 (Sep–Oct 1991). In Russian. English translation in Sov. Astron., Vol. 35, No. 5.
A statistical analysis of average sizes of photographic and radar meteor orbits does not give a basis for a statement concerning the existence of "stable and unstable" quantum levels in the solar gravitational field. Elementary numerical examples illustrate the invalidity of ideas on mega– or macro–quantization of orbits in the solar system.

091.015 **Complexes of minor bodies in the solar system.**
Yu. V. Obrubov.
Astron. Zh., Tom 68, Vyp. 5, p. 1063 – 1073 (Sep–Oct 1991). In Russian. English translation in Sov. Astron., Vol. 35, No. 5.
A search for parent bodies for 22 short–period meteroid streams with account of long–period planetary perturbations was carried out. Five minor body complexes are found among short–period comets, Earth–crossing asteroids and meteroid streams. There are ten members in the major complex: two comets – P/Schwassmann–Wachmann 3 and P/Pons–Winnecke; four asteroids – 1984 KD (3671), 4788 PL, 1987 SJ3, 1987 PA; and four meteor showers – Sigma Leonids, Alpha Bootids, Tau Herculids and June Bootids. The results a of numerical simulation of orbital evolution for three short–period comets are presented.

091.016 **Improvement of parameters of the unified relativistic theory of motion of the planets of the solar system.**
T. I. Afanas'eva, M. D. Kislik, Yu. F. Kolyuka, O. K. Margorin, V. F. Tikhonov.
Astron. Zh., Tom 68, Vyp. 6, p. 1309 – 1314 (Nov–Dec 1991). In Russian. English translation in Sov. Astron., Vol. 35, No. 6.
Results are given of an improvement of parameters of the unified relativistic theory of motion of the inner solar–system planets as determined from observations of 1960 – 1989.

091.017 **Role of carbon dioxide in cooling planetary thermospheres.**
R. D. Sharma, P. P. Wintersteiner.
Geophys. Res. Lett., Vol. 17, No. 12, p. 2201 – 2204 (Nov 1990).
A new value of the rate coefficient for the deactivation of the bending mode of carbon dioxide by atomic oxygen at low temperatures is derived from the observation of 15 μm emission from the atmosphere of the Earth. This new value gives a cooling rate for the lower thermosphere that is two to three times the rate previously calculated, and it may resolve a long–standing problem in the Mars–Venus aeronomy.

091.018 **A mechanism for bursty radio emission in planetary magnetospheres.**
H. K. Wong, M. L. Goldstein.
Geophys. Res. Lett., Vol. 17, No. 12, p. 2229 – 2232 (Nov 1990).
Bursty radio emissions are often observed from the polar magnetospheres of the Earth, Jupiter, Saturn, and Uranus in addition to the smooth radio emissions commonly detected. The authors show that in plasma regimes in which the electron plasma frequency is less than the electron cyclotron frequency, anisotropic electron beams or gyrating electron beams can excite directly broadband electromagnetic radiation. The largest growth is for right–hand X–mode radiation with frequencies above the electron cyclotron frequency. This instability can produce bursty, broadband emission, consistent with some of the properties of the radiation observed from the magnetized planets.

091.019 **Interior structure of the giant planets.**
V. N. Zharkov.
Astron. Vestn., Tom 25, No. 6, p. 627 – 649 (Nov – Dec 1991). In Russian. English translation in Sol. Syst. Res., Vol. 25, No. 6.
An account of the principal ideas and data related to the problem of construction of models of the interior structure of Jupiter, Saturn, Uranus and Neptune is given. The following problems are considered: the establishment of a concept of

Jupiter and Saturn as hydrogenic planets; the theory of figures of hydrostatic equilibrium rotating planets; the gas–liquid, adiabatic models of the giant planets; analysis of observational data; element abundances and groups of cosmochemical substances; equation of state; models of interior structure of Jupiter, Saturn, Uranus and Neptune; the time of formation of Jupiter; the two–components model of formation of terrestrial planets and the role of Jupiter in the formation of the Earth; the role of Jupiter in the formation of the giant planets.

091.020 **Stochastic behaviour of orbits of the planets in the process of their accumulation.**
I. N. Ziglina.
Astron. Vestn., Tom 25, No. 6, p. 703 – 722 (Nov – Dec 1991). In Russian. English translation in Sol. Syst. Res., Vol. 25, No. 6.

The evolution of the orbital elements of a growing planet during the accumulation process is considered. The planetary orbit undergoes perturbations as a result of random encounters and collisions with bodies of its accretion zone and also under gravitational perturbations from an already formed massive planet (Jupiter).The Fokker–Planck equation describing the behaviour of the distribution function of orbital elements of the growing planet is worked out and solved. The present mean values of eccentricities and inclinations of orbits of the terrestrial planets can be explained in the case of their accumulation from a swarm of bodies with mean meass $\sim 10^{-2} M_{\delta}$ and with mean orbital eccentricities and inclinations ~ 0.2.

091.021 **Electroseismic effect on the terrestrial planets.**
A. V. Kozenko, M. A. Yaroslavskij.
Astron. Tsirk., No. 1544, p. 33 – 34 (May – Jul 1990). In Russian.

It is shown that the electroseismic effect can exist not only on the Earth, but on the other terrestrial planets. In contrast with the Earth, on Mars and Venus this effect can provide some weathering.

091.022 **Computer simulation of regolith vertical structure and exposure age of small satellites of planets.**
N. N. Evsyukov.
Earth, Moon, Planets, Vol. 54, No. 1, p. 1 – 12 (Jul 1991).

The simulation has shown that the regolith layers several hundred meters thick can be formed on the small planetary satellites of Phobos and Deimos type providing that almost all the material lost at meteorite impacts return to the satellite. The existence of a global layer on Deimos that filled the craters for the depth of about 5 m can be explained by the crater presence of diameter of about 4 km and the age of less than 350 Myr on its hemisphere almost unstudied. Dust belts with dust concentrations by 3 orders greater than around the Earth can exist in the region of satellite orbits.

091.023 **Sources of error in the photoclinometric determination of planetary topography: a reappraisal.**
N. D. Efford.
Earth, Moon, Planets, Vol. 54, No. 1, p. 19 – 58 (Jul 1991).

The technique of photoclinometry has frequently been used to determine planetary topography without proper consideration of possible sources of error. Previous studies of error sources have been limited in extent and have overlooked the importance of factors such as atmospheric scattering and the choice of a surface photometric function. This paper adopts a thorough and more direct approach to error analysis, whereby known topography is compared with photoclinometric profiles derived from synthetic quantised reflectance scans. Instrumental and geometric sources of error are found to exert a minimal influence on profiles in practice, provided that sufficient care is taken in the selection of images and the extraction of scans from those images. Environmental factors – relating to the scattering properties of the surface and, if present, atmosphere – are far more important. It is found that Hapke's equation is a far superior model of surface optical properties. In the case of an atmosphere, the need to correct for scattering by aerosols or suspended dust becomes more acute as atmospheric opacity increases and as particle scattering becomes more forward–biased. To perform this correction, a model for the combined reflectance of surface and atmosphere must be used when deriving profiles. Two case studies – of a small impact crater on Triton and a dust–mantled basaltic lava flow on Mars – are presented here.

091.024 **The long–term dynamical behavior of small bodies in the Kuiper belt.**
H. F. Levison.
Astron. J., Vol. 102, No. 2, p. 787 – 794 (Aug 1991). Current Physics Microform No.: 9108A1723.

The results are presented from a new numerical technique to determine the long–term dynamical behavior of objects in the Kuiper belt. According to this model, it is possible for some objects that formed in a well–defined region of the Kuiper belt to have lifetimes of approximately $5x10^9$yr, after which they leave the region between 30 and 100 AU. Therefore, the Kuiper belt can be the source of short–period comets. Objects tend to diffuse through a region 70 AU in extent on timescales that are on the order of the age of the solar system. Objects that form close to the orbit of Neptune have a significant chance of evolving to orbits with $a > 100$ AU. These objects can be effectively stored in these orbits for a very long time.

091.025 **Modeling stratiform clouds in the atmospheres of giant planets with consideration of the mutual solubility of condensed components. I.**
M. V. Bujkov, K. Yu. Ibragimov, G. A. Kirienko,
A. M. Pirnach.
Kinematics Phys. Celest. Bodies, Vol. 7, No. 2, p. 1 – 8 (1991). English translation of Kinematika Fiz. Nebesn. Tel, Tom 7, No. 2, p. 3 – 10 (1991).

The formation mechanism of two–component clouds is examined with consideration of the mutual solubility of the components. Equations are derived to describe condensational droplet growth in the NH_3–H_2O system. It is shown that the time to establishment of the equilibrium concentration is very short (no more than fractions of a second), whence follows the need to consider mutual solubility. General aspects of the formation of mixed clouds are studied for the simplest case of a linear concentration dependence of the saturating specific contents. The equation system is solved first for the case of an isolated volume, followed by the transition to the real atmosphere. A sample calculation of a Jovian water–ammonia cloud is included.

091.026 **Influence des effets de surface sur la réduction astrométrique des satellites des planètes.**
P. Descamps.
Journées 1991: Systèmes de référence spatio–temporels, p. 165 – 170 (1991). – See Abstr. 012.048 for the main entry.

Until now, the surface effects have never been taken into account in the astrometric reduction of the satellites observations. Generally, the satellites are considered in the theories as single points. However, we know that the methods of reduction must take into account the influence of the law of the reflexion of the light by the surfaces of the satellites, and the phase effect. The improvement of the methods of observations and analysis could so permit to benefit of the full accuracy of these observations. For this purpose, we have recently developed a modelisation of the lightcurves of the mutual phenomena. The aim of this work is mainly to improve the interpretation of the observations in order to get better astrometric corrections of the theoretical model of the motions.

091.027 **Tests of random density models of terrestrial planets.**
W. M. Kaula, P. D. Asimow.
Geophys. Res. Lett., Vol. 18, No. 5, p. 909 – 912 (May 1991).

Random density models are analyzed to determine the low degree harmonics of the gravity field of a planet, and therefrom two properties: an axiality P_l, the percent of the degree variance in the zonal term referred to an axis through the maximum for degree l; and an angularity E_{ln}, the angle between the maxima for two degrees l, n. The random density distributions give solutions reasonably consistent with the axialities and angularities for the low degrees, $l < 5$, of Earth, Venus, and Moon, but not for Mars, which has improbably large axialities and small angularities. Hence the random density model is an unreliable predictor for the non–hydrostatic second–degree gravity of Mars, and thus for the moment–of–inertia, which is more plausibly close to 0.365 MR^2.

091.028 **Some remarks on the dynamics of the Jovian atmospheres.**
R. S. Lindzen.
Geophys. Astrophys. Fluid Dyn., Vol. 58, No. 1–4, p. 123 – 131 (Jul 1991).
This paper examines the nature and role of lunar tides on the outer planets. It is shown that lunar tidal forcing on the outer planets is much stronger than it is on Earth. It is also shown that the dynamics of large, rapidly rotating planets (i.e., Jupiter, Saturn, Uranus and Neptune) favors higher order meridional modes. The author discusses how this might contribute to the banded structure of the outer planets.

091.029 **Absence of "stable and unstable" quantum levels in the solar gravitational field.**
E. N. Kramer, Yu. M. Gorbanev.
Sov. Astron., Vol. 35, No. 5, p. 515 – 517 (Sep 1991). Current Physics Microform No.: 9203X1689. English translation of Astron. Zh., Tom 68, Vyp. 5, p. 1030 – 1035 (1991).
A statistical analysis of the mean sizes of photographic and radar meteor orbits provides no basis for stating that "stable and unstable" quantum levels exist in the solar gravitational field. Elementary numerical examples illustrate the invalidity of the ideas of mega– or macro–quantization of orbits in the solar system.

091.030 **Complexes of minor solar system bodies.**
Yu. V. Obrubov.
Sov. Astron., Vol. 35, No. 5, p. 531 – 537 (Sep 1991). Current Physics Microform No.: 9203X1705. English translation of Astron. Zh., Tom 68, Vyp. 5, p. 1063 – 1073 (1991).
The parent bodies for 22 short–period meteoroid swarms are sought on the basis of qualitative allowance for long–period planetary perturbations. Five complexes of minor bodies are found, which include short–period comets, asteroids of the Apollo, Amor, and Aten groups, and meteor showers. The largest complex includes comets Schwassmann–Wachmann 3 and Pons–Winnecke, asteroids 1984 KD (3671), 4788 PL, 1987 SJ3, and 1987 PA, and the σ Leonid, α Bootid, τ Herculid, and June Bootid meteor showers. The results of a numerical investigation of the evolution of the orbits of three short–period comets are given.

091.031 **Non–linear equations for the rotation of a viscoelastic planet taking into account the influence of a liquid core.**
M. Lefftz, H. Legros, J. Hinderer.
Celest. Mech. Dyn. Astron., Vol. 52, No. 1, p. 13 – 43 (1991).
Non–linear equations governing the temporal evolution of the vector of instantaneous rotation are developed for an Earth with a homogeneous mantle having a viscoelastic Maxwell rheology and with a homogeneous inviscid fluid core. This general theory is investigated using the angular momentum theorem applied to the coupled core–mantle system. It allows to study the influence upon the planetary rotation of a quasi–rigid rotational motion in the liquid core. It also enables to investigate the consequences of excitation sources (e.g. pressure), located at the core–mantle interface. Especially, the influence of viscoelastic variations in the inertia tensors resulting from the rotation itself or from various excitation sources are detailed with the help of a Love number formalism. The equations of the linear theory for an elastic Earth with a liquid core, and the non–linear theory for a viscous planet with a quasi–fluid behavior are shown to be particular cases of the authors' generalized system of equations. Some planetological applications may be derived from the quasi–fluid approximation.

091.032 **Atmospheric mass loss from non–magnetic planets via plasma processes.**
S. J. Bauer.
2. International Workshop on Plasma Physics: Plasma Theoretical Problems in Astro– and Fusion Physics, p. 31 – 38 (1991). – See Abstr. 012.067 for the main entry.
Following a brief discussion of physical conditions leading to the escape of atmosphere constituents from a planetary gravitational field, i.e. availability of particles having outwards velocities greater than the planetary escape velocity and absence of collisions, occurring in a planetary exosphere, specific "non–thermal" escape processes involving the plasma component of planetary atmospheres such as charge transfer, ion–molecule and dissociative recombination reactions are discussed. Especially the latter reaction leads to the formation of a hot atom corona around Venus, Mars and Titan. Interaction of the solar wind with these non magnetic bodies can remove atmospheric constituents from their coronae leading to mass loss rates of up to about 1 kg/s.

091.033 **The cyclotron maser instability: application to the planetary radio emission.**
H. O. Rucker.
2. International Workshop on Plasma Physics: Plasma Theoretical Problems in Astro– and Fusion Physics, p. 53 – 61 (1991). – See Abstr. 012.067 for the main entry.
Five planets are known to be emitters of non–thermal radiation: Earth, Jupiter, Saturn, Uranus, and Neptune. Several of the properties of the radio planets are quite similar, suggesting that for particular radio components the same emission processes take place. A theory has been known which can explain the main features and processes of the generation mechanism of planetary radio emission, the so–called cyclotron maser instability (CMI–) theory. An important aspect of the CMI–theory is the necessary inclusion of the relativistic effect on the usual cyclotron resonance condition. With great success the cyclotron maser theory can explain the terrestrial kilometric radiation, certain features of the Jovian decametric radio emission, and the Saturnian, Uranian, and Neptunian kilometric radiation. The source location of the respective radio components can be attributed to these auroral regions where the electron cyclotron frequency approximately corresponds to the frequency of the generated emission.

091.034 **Plasma waves in planetary magnetospheres.**
W. S. Kurth, D. A. Gurnett.
J. Geophys. Res., Vol. 96, Suppl., p. 18977 – 18992 (30 Oct 1991). Special issue: Voyager 2 at Neptune.
With the completion of the Voyager 2 encounter with Neptune we have now surveyed the plasma wave spectra of five planetary magnetospheres: Earth, Jupiter, Saturn, Uranus, and Neptune. Here the authors provide a first general comparison of the various plasma wave modes at each of the planets with the use of a common formal for displaying the spectra.

091.035 **Charged dust dynamics: orbital resonance due to planetary shadows.**
M. Horányi, J. A. Burns.
J. Geophys. Res., Vol. 96, No. A11, p. 19283 – 19290 (1 Nov 1991).
First, the authors discuss the dynamics of a weakly charged dust grain orbiting in the equatorial plane of a planet surrounded by a rigidly corotating magnetosphere, where the magnetic field can be represented by a centered aligned dipole field. If the grain has a constant charge, one can introduce an effective one–dimensional potential, from which one finds that the perturbation due to electrostatic forces induces a motion of the pericenter, similar to the effect of the planetary oblateness. However, the electric charge on an orbiting dust grain is generally not constant. Accordingly, the authors explore the case where the charge varies periodically due to the modulation of the photoelectron current that occurs as the grain enters and leaves the planetary shadow. Thus the electromagnetic perturbation resonates with the orbital period and can modify the orbit's size and eccentricity. The

authors demonstrate this effect both numerically and analytically for small grains comprising the Jovian ring. They show that the resulting changes are periodic, and their amplitude is much bigger than the amplitude of the periodic changes due to light-pressure perturbation or the secular changes due to resonant charge variations that develop over a comparable time span.

091.036 **EUV planetary astronomy.**
P. D. Feldman, F. Bagenal.
1. Berkeley Colloquium on Extreme Ultraviolet Astronomy, p. 252 – 260 (1991). – See Abstr. 012.045 for the main entry.
Contents: 1. Introduction. 2. Jovian planets. 2.1. Magnetospheric plasmas: the Io plasma torus. 2.2. Aurora. 2.3. Upper atmospheres. 3. Satellites. 4. Terrestrial planets. 5. Comets. 6. Conclusion.

091.037 **Streamline formalism and ring orbit determination.**
P.-Y. Longaretti, N. Borderies.
Icarus, Vol. 94, No. 1, p. 165 – 170 (Nov 1991).
The orbital parameters of the Uranian rings are commonly obtained by fitting to ring occultation data precessing inclined ellipses with constant elliptic elements and constant precession rates; these elements are referred to as "geometric" elements. On the theoretical side, many results were derived by considering the streamlines of the flow of particles and studying their dynamical evolution with the help of Gauss perturbation equations. This approach involves some kind of "mean" elements. The purpose of this paper is to give a theoretical basis of these approaches.

091.038 **Coherent backscatter model for the unusual radar reflectivity of icy satellites.**
B. Hapke, D. Blewett.
Nature, Vol. 352, No. 6330, p. 46 – 47 (4 Jul 1991). Letter-to-the-editor.
Radar is a powerful technique for probing the surfaces and subsurfaces of solar system bodies. Inner solar system bodies reflect radar in an almost specular fashion, with low reflectivity and little polarization. The icy satellites of Jupiter, by contrast, show high reflectivities, diffuse scattering laws and unusual polarization properties. According to the coherent backscatter model, most of the received power from icy satellites is multiply reflected by particles about a wavelength in size located randomly under the surface of the regolith. The authors have constructed a laboratory analogue of this model by reflecting laser light off polystyrene beads suspended in water, and find that this model reproduces the unusual polarization ratios observed in the radar data. This implies that the regoliths of icy satellites are weakly absorbing matrices of small refractive index containing imbedded scatterers separated by distances of the order of a wavelength. No structures of special shape or other geometrical or optical properties are required.

091.039 **Solar system: A plurality of worlds.**
S. J. Weidenschilling.
Nature, Vol. 352, No. 6332, p. 190 – 192 (18 Jul 1991).

091.040 **Planetary rings: Ever decreasing circles.**
L. W. Esposito.
Nature, Vol. 354, No. 6349, p. 107 (14 Nov 1991).
The author reports on explanations for the rapid evolution and possibly short lifetimes of planetary rings.

091.041 **What is it that supports the rings of planets?**
N. N. Gor'kavyj, A. M. Fridman.
Zemlya Vselennaya, No. 6, p. 72 – 74 (Nov – Dec 1991). In Russian.

091.042 **A method for improving the parameters of the solar system from ground-based and space observations of planetary satellites.**
S. N. Vashkov'yak, N. V. Emel'yanov.
Tr. Gos. Astron. Inst. im. Shternberga, Tom 62, p. 3 – 28 (1991). In Russian.
The method of joint improving the numerous dynamic and kinematic parameters of solar system bodies motion is described. The formulae for the construction of conditional equations are given. The theories of satellites and planets motion are assumed to be known. Among the refining parameters are the orbital elements of satellites and planets, planet pole coordinates, the corrections of the difference between dynamic and stellar reference frames. Refinement of the parameters is carried out using the ground-based photographic, Doppler and spacecraft observations. In the paper a brief description of the resolving algorithm is given.

091.043 **Representation of numerical models of planetary motions by means of simple analytical theories.**
N. V. Emel'yanov, A. A. Kanter.
Tr. Gos. Astron. Inst. im. Shternberga, Tom 62, p. 29 – 39 (1991). In Russian.
A method for determining the parameters of the analytical theory in a fixed time interval so that the coordinate deviations from the numerical solution would be minimal is proposed. As a numerical model the DE200/LE200 solution may be taken. Some remarks concerning the software constructed by the authors as well as computation examples are given.

091.044 **Some properties of barycentric motion of the major planets and the Sun.**
V. P. Dolgachev, L. M. Domozhilova, A. I. Khlystov.
Tr. Gos. Astron. Inst. im. Shternberga, Tom 62, p. 111 – 118 (1991). In Russian.
From results of a numerical integration of equations of motion of the major planets and the Sun the trajectory of motion of the solar center relative to the solar system has been obtained for 1973 – 2008 years. A formula connecting the heliocentric and barycentric distances of a planet with the solar distance from the barycenter has been derived. It is shown that for evaluation of the Sun's influence on geophysical processes the geocentric motion of the Earth is more important in comparison with the terrestrial barycentric motion.

091.045 **Reports of planetary astronomy – 1991.**
National Aeronautics and Space Administration, Washington, DC (USA). NASA Office of Space Science and Applications, NASA-TM—4329. 202 p. (Oct 1991).
This publication is a compilation of summaries of reports written by principal investigators funded through the Planetary Astronomy Program of NASA's Solar System Exploration Division, Office of Space Science and Applications. The summaries are designed to provide information about scientific research projects conducted in the Planetary Astronomy Program in 1990 – 1991 and to facilitate communication and coordination. Highlights of recent accomplishments in planetary astronomy are included.

091.046 **Why radiation belts exist.**
J. A. Van Allen.
Earth Space, Vol. 4, No. 2, p. 5 – 7 (Oct 1991).
The author describes some of the reasons for the existence of radiation belts around the Earth and other planets and some of their basic properties.

091.047 **Love numbers of the terrestrial planets.**
Zhang Chengzhi.
Acta Astron. Sin., Vol. 32, No. 3, p. 288 – 296 (Sep 1991). In Chinese.
Using the latest geodetic parameters of the terrestrial planets, the author constructed the parametric models of the internal structure of the terrestrial planets. By means of these models he computed their static Love numbers from order 2 to 30 and listed

the results. Then he discussed the influence of the outer core rigidity on the second-order Love numbers.

091.048 **Wide-band planetary spectroscopy on the James Clerk Maxwell telescope with a polarizing Fourier transform spectrometer.**
G. R. Davis, D. A. Naylor, G. J. Tompkins, T. A. Clark, W. D. Duncan.
J. R. Astron. Soc. Can., Vol. 85, No. 4, p. 190 (Aug 1991). Abstract. – See Abstr. 012.091 for the main entry.

091.049 **Solid C≡N bearing material on outer solar system bodies.**
D. P. Cruikshank, L. J. Allamandola, W. K. Hartmann, D. J. Tholen, R. H. Brown, C. N. Matthews, J. F. Bell.
Icarus, Vol. 94, No. 2, p. 345 – 353 (Dec 1991).
Using telescopic observations by the authors and other observers, they have identified cyano-group containing molecules in the very dark solids on the surfaces of a few D-class asteroids, the dust of some comets, and low-albedo hemisphere of Iapetus, and the rings of Uranus, through spectroscopic detection of the 2.2-μm overtone of the C≡N stretching fundamental mode. The occurrence of this band on all four classes of small solar system bodies may be diagnostic of the duration of exposure and degree of modification of surface materials, and may also establish a link between outer solar system and interstellar materials.

091.050 **Brine volcanism and the interior structures of asteroids and icy satellites.**
J. S. Kargel.
Icarus, Vol. 94, No. 2, p. 368 – 390 (Dec 1991).
Cryovolcanism is among the foremost processes responsible for modifying the surfaces of icy satellites. Volcanic brine petrogenesis in ammonia-deficient satellites should mainly involve eutectic melting in relevant salt-water systems. Carbonaceous chondrites provide useful insights into the compositions of salts and aqueously altered rock in icy satellites and asteroids. Many aspects of the petrogenesis and physical volcanology of icy satellite brines should be well described in the system H_2O–$MgSO_4$–Na_2SO_4. Minor components include sulfates of K, Ni, Mn, and Ca. Soluble salts may have profound effects on the geology and structure of icy satellites and asteroids. In some models late episodes of water volcanism are facilitated by high buoyant forces due to the relatively high densities of sulfate-rich mantle and crustal layers. In other models early hypersaline brine volcanism quickly yields to plutonic magmatism due to low crustal densities.

091.051 **Nonlinear satellite wakes in planetary rings. II. Phase-space kinematics.**
G. R. Stewart.
Icarus, Vol. 94, No. 2, p. 436 – 450 (Dec 1991).
An explicit expression is derived for the phase-space density of a planetary ring perturbed by a nearby satellite. The derivation is facilitated by working in guiding center variables instead of local position and velocity variables and by neglecting collisions between ring particles. The usual equations for perturbed streamlines are recovered by taking first-order moments of the phase-space density. The local surface density and the local mean velocity in a nonlinear satellite wake are obtained in the form of convergent infinite series. Unlike previous estimates based on streamline crowding, this surface density is positive definite because the finite velocity dispersion of ring particles is taken into account. The finite width of streamlines limits the maximum value of the local surface density that can result from streamline crowding. These results suggest that collisional stresses between neighboring streamlines should depend on the second-order derivatives of the streamline parameters because the local mean velocity is itself a strong function of the first-order derivatives of the stream-line parameters.

091.052 **Equations of state in geophysics and in planetary physics.**
J. Leliwa-Kopystyński.
International School of Physics 'Enrico Fermi'. Course 113: High-pressure equations of state: theory and applications, p. 439 – 464 (1991). – See Abstr. 012.100 for the main entry.

091.053 **Improvement of the parameters of the unified relativistic theory of motion of solar system planets.**
T. I. Afanas'eva, M. D. Kislik, Yu. F. Kolyuka, O. K. Margorin, V. F. Tikhonov.
Sov. Astron., Vol. 35, No. 6, p. 657 – 660 (Nov 1991). Current Physics Microform No.: 9205X1467. English translation of Astron. Zh., Tom 68, Vyp. 6, p. 1309 – 1314 (1991).
Improved parameters for the unified relativistic theory of motion of the inner solar system planets based on 1960 – 1989 observations are presented.

091.054 **Physical properties of terrestrial planets and their satellites: an introduction.**
D. Moehlmann.
International School of Physics 'Enrico Fermi'. Course 111: Solid-state astrophysics, p. 353 – 370 (1991). – See Abstr. 012.106 for the main entry.
Planetary science has advanced significantly due to new results of space research. Close-up spacecraft observations have been made for all of the terrestrial planets and their satellites. Landers on Moon, Venus and Mars gave first results of contact measurements. In situ measurements on the lunar surface have also been carried out by astronauts. Because of this new quality of data of planetary exploration, planetary science has been transformed into a field of interdisciplinary interest, increasingly including disciplines of physical Earth sciences. These lecture notes are aimed at giving an introduction to some aspects of planetary physics as a basis for understanding the scientific background of actual planetary exploration. Following the "definition" of what is a planet and a summary of the mechanical properties of terrestrial planets, their gravitational fields, surface figures, internal structures and magnetic fields, theories of the origin of the planetary system and its satellite systems are described in some detail. This discussion is centred on the "terrestrial planets", which are partially molten "stony" solid bodies with masses of the order of 10^{24}kg (or less). They failed to incorporate or to capture a (dominating) amount of gaseous matter from the preplanetary disk, as the "giant planets" with their more massive cores (and their dominating atmospheres) did. Consequently, "giant planets" and planetary atmospheres are not discussed in this context of solid-state astrophysics.

091.055 **The random component of planetary rotation.**
J. J. Lissauer, V. S. Safronov.
Icarus, Vol. 93, No. 2, p. 288 – 297 (Oct 1991).
The authors compute the root mean square rotational angular momentum accumulated by a planet as a result of stochastic off-center impacts of large planetesimals during the accretionary epoch. Planets which accrete from large planetesimals have, on the average, shorter spin periods and greater obliquities than those which accrete exclusively from small bodies. The authors use the observed spin periods and obliquities of the planets in order to estimate the masses of the largest bodies to impact each planet during its history. Most, if not all, planets accreted one or more individual bodies having at least 1% of the final planetary mass. Several planets probably suffered at least one collision with a body having $\geqslant 5\%$ of the planet's final mass. These results support giant impact theories for the origin of the Moon and Charon, and models which explain the excess iron in Mercury by the partial collisional stripping of that planet's mantle.

091.056 **Atmospheric variability on a zonally symmetric land planet.**
Leung Laiyung, G. R. North.
J. Climate, Vol. 4, No. 8, p. 753 – 765 (Aug 1991).
Atmospheric variability on a zonally symmetric planet in the absence of external forcing anomalies is studied. With idealized

boundary conditions and by using perpetual equinox solar forcing, a 15-year long stationary time series of the atmosphere is simulated with the NCAR Community Climate Model (CCM0). This provides sufficient time samples for realistic study of the properties of the atmosphere. Zonally averaged and space-time statistics for the surface air temperature field on this planet are presented. In search of a simple statistical model for atmospheric variability, the space-time spectra obtained from the CCM simulation are fitted statistically with a stochastic energy balance model. The space-time spectra for three zonal wavenumbers are found to be fitted satisfactorily by the stochastic model with only five parameters. The ability of the stochastic model for describing atmospheric variability on a more realistic planet can be tested.

091.057 **Small solar system bodies.**
P. Farinella.
13. Ecole de Printemps d'Astrophysique de Goutelas: Les méthodes modernes de la mécanique céleste, p. 405 – 431 (1990). – See Abstr. 012.118 for the main entry.

Contents: 1. Introduction. 2. Asteroids. 3. Comets. 4. Interplanetary material and meteorites. 5. Planetary rings and the Roche limit. 6. Satellites. 7. Tidal evolution of orbits. 8. From stones to planets.

091.058 **Giant planets: theoretical and observational aspects.**
A. P. Vid'machenko.
Sol. Syst. Res., Vol. 25, No. 3, p. 203 – 214 (Nov 1991). English translation of Astron. Vestn., Tom 25, No. 3, p. 277 – 292 (1991).

This article discusses theoretical aspects and results of ground based and space observations of the atmospheres, ionospheres, and magnetospheres of the giant planets of the solar system. Consideration is given to the photochemistry of the main constituents and the energy balance in the upper atmospheres, to the magnetic fields, and to the dynamics of and models for the general and local circulation in the atmospheres of the giant planets.

Bibliography (April – June 1989).
See Abstr. 002.079.

Bibliography (April – June 1989).
See Abstr. 002.080.

Bibliography (April – June 1989).
See Abstr. 002.081.

Bibliography (July – September 1989).
See Abstr. 002.082.

Bibliography (July – September 1989).
See Abstr. 002.092.

A bibliography of planetary geology and geophysics principal investigators and their associates, 1990 – 1991.
See Abstr. 002.117.

The new solar system.
See Abstr. 003.016.

Seeing the solar system. Telescopic projects, activities, and explorations in astronomy.
See Abstr. 003.024.

Planetary mapping.
See Abstr. 003.035.

Planetary geosciences – 1988.
See Abstr. 003.036.

William H. Wright's early planetary photography in the near infrared and near ultraviolet.
See Abstr. 005.004.

Laboratory research for planetary atmospheres. A selection of papers. 2. International Conference on Laboratory Research for Planetary Atmospheres (ICLRPA), Charlottesville, VA (USA), 21 Oct 1990.
See Abstr. 012.073.

Advances in solar system magnetohydrodynamics. Proceedings.
See Abstr. 012.086.

Solid-state astrophysics. Proceedings.
See Abstr. 012.106.

Soviet programs "Jupiter" and "Ceres" for improving orbits of bodies of solar system.
See Abstr. 013.051.

Planetesimals.
See Abstr. 014.009.

The orbit of Mercury.
See Abstr. 014.020.

A technique for measuring the electric properties of planetary grounds.
See Abstr. 022.088.

Collisional cooling as an environment for planetary research.
See Abstr. 022.110.

Optical constants of solid methane and ethane from 10,000 to 450 cm^{-1}.
See Abstr. 022.112.

Methane spectral line widths and shifts, and dependences on physical parameters.
See Abstr. 022.113.

Herman-Wallis factors in the C_2H_2 ν_5 infrared fundamental near 14 μm.
See Abstr. 022.115.

Pressure-induced absorption in the infrared: a data base for the modelling of planetary atmospheres.
See Abstr. 022.117.

New measurements of the 6190-Å band of methane.
See Abstr. 022.118.

A laboratory facility for mid-UV absorption spectroscopy of molecular compounds for planetary atmospheres.
See Abstr. 022.119.

High-resolution, VUV (147-201 nm) photoabsorption cross sections for C_2H_2 at 195 and 295K.
See Abstr. 022.121.

Electron impact excitation and dissociation of N_2 via the $b^1\Pi_u$ state.
See Abstr. 022.126.

Cross sections for direct and dissociative ionization of NH_3 and CS_2 by electron impact.
See Abstr. 022.127.

Laboratory simulations of large scale fragmentation events.
See Abstr. 022.132.

Laboratory reflectance measurements of analogues to "dirty" ice surfaces on atmosphereless solar system bodies.
See Abstr. 022.134.

New emission bands in the high resolution spectrum of molecular nitrogen between 107.7 and 124.2 nm.
See Abstr. 022.146.

Absolute state–selected total cross sections for the $O^+(^4S)+CO_2$ reactions.
See Abstr. 022.173.

Abosolute state–selected total cross sections for the $O^+(^2D,^2P)+N_2$ reactions.
See Abstr. 022.174.

Photometric functions for photoclinometry and other applications.
See Abstr. 036.041.

Glare and celestial visibility.
See Abstr. 036.110.

Source of error in planetary photoclinometry.
See Abstr. 036.295.

On the determination of orbits of satellites of the distant planets and some requirements to their positional observations.
See Abstr. 042.027.

Chaos im Kosmos?
See Abstr. 042.043.

The stability of the planetary triangular Lagrange points.
See Abstr. 042.044.

Symplectic maps for the N–body problem.
See Abstr. 042.049.

Secular resonances from 2 to 50 AU.
See Abstr. 042.093.

Fundamental constants of astronomy of the year 1976.
See Abstr. 043.007.

EUVE observations of the Earth and planets.
See Abstr. 051.005.

A review of planetary opportunities and observations with the Hubble Space Telescope.
See Abstr. 051.021.

Nonlinear density waves in gaseous disks.
See Abstr. 062.126.

Parameterized rotating convection for core and planetary atmosphere dynamics.
See Abstr. 062.129.

Laboratory investigations of impact–generated plasma.
See Abstr. 062.166.

Introduction to solar system MHD.
See Abstr. 062.191.

Problems of planetary dynamo theory.
See Abstr. 062.193.

Recent developments in the theory of three–dimensional magnetic reconnection.
See Abstr. 062.200.

Convection–driven planetary dynamos.
See Abstr. 062.245.

Thermal convection in differentially rotating systems.
See Abstr. 062.247.

Planetary magnetism re–visited.
See Abstr. 062.255.

Absorption line formation in an optically thick, weakly absorbing planetary atmosphere. I. Homogeneous slab.
See Abstr. 063.054.

Absorption line formation in an optically thick, weakly absorbing planetary atmosphere. II. Vertically inhomogeneous medium.
See Abstr. 063.055.

The inverse problem of photometry of solar radiation reflected by an optically thick planetary atmosphere. Mathematical framework and weighting functions of the linearized inverse problem.
See Abstr. 063.070.

Radiation flux in a weakly absorbing atmosphere of arbitrary optical thickness.
See Abstr. 063.119.

The inverse problem of photometry of solar radiation reflected by an optically thick planetary atmosphere. 2. Numerical aspects and requirements upon the geometry of observations.
See Abstr. 063.122.

A model for tidally driven eccentric instabilities in fluid disks.
See Abstr. 064.175.

Schwarzschild relativistic orbits.
See Abstr. 066.089.

The possible influence of the orbital motion of the major planets on sunspot activities.
See Abstr. 072.092.

Multiple–impact mechanisms on Venus and other solid planets and satellites.
See Abstr. 093.038.

Sixteen years of stellar occultations.
See Abstr. 096.003.

Characteristics of spectra from the Martian bow shock and comparison with Venus, Earth, AMPTE, Jupiter, and Saturn.
See Abstr. 097.054.

The theory of Neptunian arcs. I. Stability of single epiton.
See Abstr. 101.126.

A hypothesis on the Oort clouds of planets.
See Abstr. 107.008.

Temperatures in the protoplanetary disk and their influence on the formation of planets.
See Abstr. 107.011.

The origin of the systematic component of planetary rotation. I. Planet on a circular orbit.
See Abstr. 107.018.

092 Mercury

092.001 **Upstream waves at Mercury, Venus and Earth: comparison of the properties of one Hertz wave.**
D. S. Orlowski, G. K. Crawford, C. T. Russell.
Geophys. Res. Lett., Vol. 17, No. 13, p. 2293 – 2296 (Dec 1990).
Previous studies have shown that the Venus foreshock region contains low–frequency upstream waves similar to those in the terrestrial foreshock, but perhaps with different amplitudes than at Earth. Herein, the authors compare the properties of a second class of upstream waves, analogous to the so–called 1 Hz waves at Earth. The waves observed at Mercury, Venus, and Earth have very similar properties, i.e., propagation angles less than 55 degrees to the magnetic field and less than 35 degrees to the solar wind flow direction. The waves occur exclusively on the field lines connected to the bow shock. They are most commonly left–hand elliptical polarized with similar fractional amplitudes, approximately 0.1 of the background field strength. Their amplitudes decrease with increasing distance from the shock. The observed frequencies are similar for Mercury, Venus and Earth when scaled by the interplanetary magnetic field. If, as generally assumed at Earth, these waves arise in regions of backstreaming electrons, these results imply that similar electron foreshocks occur at Earth, Venus and Mercury despite differences in bow shock size and the nature of the obstacle to the solar wind.

092.002 **Ice right under the Sun.**
C. R. Chapman.
Nature, Vol. 354, No. 6354, p. 504 – 505 (19–26 Dec 1991).
The author reports on the discovery of a polar ice cap on Mercury, announced at the 23rd Annual Meeting of the Division for Planetary Sciences of the American Astronomical Society, Palo Alto, CA (USA), 4 – 8 Nov 1991.

092.003 **Mercury.**
M. S. Dimitrijević.
Vasiona, Année 39, No. 5, p. 70 – 73 (1991). In Serbo–Croatian.
A short review of our knowlege about Mercury is presented with the emphasis on possibilities for amateur observations.

092.004 **Source processes for the alkali metals in the atmosphere of Mercury.**
D. E. Shemansky, T. H. Morgan.
Geophys. Res. Lett., Vol. 18, No. 9, p. 1659 – 1662 (Sep 1991).
A large (fivefold) increase in Mercury's potassium (K) column abundance on 14 October 1987, has been reported by Sprague et al. (1990) [SKH], who attributed the enhancement to diffusion through the surface in the Caloris Basin, from depths of order 10 km. The postulated source rate is much larger than any previously estimated diffusion rate, and if true certainly affects consideration of the origin of other atmospheric species. However, Killen et al. (1991) have pointed out that the claim is not supported by the published observations of K or sodium (Na) as a whole. Sprague et al. (1991) have responded by further hypothesizing the existence of several other sources of gas diffusing out of the regolith, all of which are time variable. In any case the SKH data indicate large variations in abundance, and it is important to understand the cause. With this issue in mind the authors have examined the available abundance estimates for correlation with possible controlling physical parameters. The authors have found a significant correlation between the average zenith K column abundance and indices of solar activity.

092.005 **Terrace width variations in complex Mercurian craters and the transient strength of cratered Mercurian and lunar crust.**
A. C. Leith, W. B. McKinnon.
J. Geophys. Res., Vol. 96, No. E4, p. 20923 – 20932 (25 Nov 1991).
Widths of slump terraces of complex craters can be used to determine the effective cohesion of the cratered region during crater collapse. The authors have measured terrace widths for complex craters on Mercury: these generally increase outward toward the rim for a given crater, and the width of the outermost major terrace is generally an increasing function of crater diameter. Similar observations apply to lunar complex craters, but the widths of the outermost slump terraces of Mercurian complex craters are less than those of similarly sized lunar complex craters. Using the terrace widths on Mercury and a gravity–driven slump model, the authors estimate the strength of the cratered region immediately after impact to be ~1–2 MPa. Comparison with the previous study of lunar complex craters by Pearce and Melosh indicates that the transient strength of cratered Mercurian crust is not greater than that of the Moon. The strength estimates only vary slightly with the geometric model used to restore the outermost major terrace to its precollapse configuration and are consistent with independent strength estimates from the simple–to–complex crater depth/diameter transition, in particular, the most recent depth/diameter study of Mercurian craters by Pike. Thus, contrary to previous work, the difference in "target properties" between Mercury and the Moon may be small, and systematic morphological differences between craters on the two worlds may be largely caused by the factor of 2 difference in surface gravity.

092.006 **Is the mass of Mercury 1/6000000?**
R. L. Branham Jr.
Bol. Asoc. Argent. Astron., No. 36, p. 409 – 411 (1990). Abstract. In Spanish and English. – See Abstr. 012.095 for the main entry.

New applications of Fatonu's problem.
See Abstr. 042.018.

093 Venus

093.001 **Venus: planet of fire.**
R. Burnham.
Astronomy, Vol. 19, No. 9, p. 32 – 41 (Sep 1991).

093.002 **Magellans erste Venus–Runde.**
D. Fischer.
Sterne Weltraum, Jahrg. 30, Nr. 11, p. 664 – 666, 668 – 670 (Nov 1991).

093.003 **Venus revealed.**
P. Cattermole.
Astron. Now, Vol. 5, No. 12, p. 33 – 38 (Dec 1991).

093.004 **Variations in plasma wave intensity with distance along the electron foreshock boundary at Venus.**
G. K. Crawford, R. J. Strangeway, C. T. Russell.
Adv. Space Res., Vol. 11, No. 9, p. 93 – 96 (1991). – See Abstr. 012.006 for the main entry.

Plasma waves are observed in the solar wind upstream of the Venus bow shock by the Pioneer Venus Orbiter. The authors find that the peak wave intensity may increase along the foreshock boundary from the tangent point to a maximum value at several Venus radii, then decrease in intensity with subsequent increase in distance. These observations could be associated with the instability process: the instability of the distribution function increasing with distance from the tangent point to saturation at the peak.

093.005 **Magellan at Venus: new perspectives on the inner solar system.**
S. C. Solomon.
Bull. Am. Astron. Soc., Vol. 23, No. 2, p. 873 (1991). Abstract. – See Abstr. 012.008 for the main entry.

093.006 **Die nackte Venus. Nahaufnahmen vom Abendstern.**
H. M. Hahn.
Bild Wiss., No. 8, p. 16 – 20 (Aug 1991).

093.007 **Venus in the radar spotlight.**
J. K. Beatty.
Sky Telesc., Vol. 82, No. 1, p. 24 – 30 (Jul 1991).

Come along on a wild ride across the pitted, fractured, and lava–flooded landscape of Earth's neighbor world, as seen through the radar eyes of NASA's Magellan spacecraft.

093.008 **The dark side of Venus: near–infrared images and spectra from the Anglo–Australian Observatory.**
D. Crisp, D. A. Allen, D. H. Grinspoon, J. B. Pollack.
Science, Vol. 253, No. 5025, p. 1263 – 1266 (13 Sep 1991).

Near–infrared images and spectra of the night side of Venus taken at the Anglo–Australian Telescope during Feb 1990 reveal four new thermal emission windows at 1.10, 1.18, 1.27, and 1.31 μm, in addition to the previously discovered windows at 1.74 and 2.3 μm. Images of the Venus night side show similar bright and dark markings in all windows, but their contrast is much lower at short wavelengths. The 1.27–μm window includes a bright, high–altitude O_2 airglow feature in addition to a thermal contribution from the deep atmosphere. Simulations of the 1.27– and 2.3–μm spectra indicate water vapor mixing ratios near 40 ± 20 ppm by volume between the surface and the cloud base.

093.009 **The Galileo Venus encounter.**
T. V. Johnson, C. M. Yeates, R. Young, J. Dunne.
Science, Vol. 253, No. 5027, p. 1516 – 1518 (27 Sep 1991).

The Galileo spacecraft passed Venus on its way to Jupiter on 10 Feb 1990, less than 4 months after launch from Earth. Because Galileo's instruments were selected for broad–based planetary exploration, the spacecraft was able to obtain a wide range of measurements during the Venus encounter. Together with ground–based observations conducted during the encounter, these observations have yielded more accurate information about the planet's plasma environment, cloud patterns, and the possible existence of lightning.

093.010 **Magnetic field studies of the solar wind interaction with Venus from the Galileo flyby.**
M. G. Kivelson, C. F. Kennel, R. L. McPherron, C. T. Russell, D. J. Southwood, R. J. Walker, C. M. Hammond, K. K. Khurana, R. J. Strangeway, P. J. Coleman.
Science, Vol. 253, No. 5027, p. 1518 – 1522 (27 Sep 1991).

During the 10 Feb 1990 flyby of Venus, the Galileo spacecraft skimmed the downstream flank of the planetary bow shock. This provided an opportunity to examine both the global and the local structure of the shock in an interval during which conditions in the solar wind plasma were quite steady. The data show that the cross section of the shock in planes transverse to the flow is smaller in directions aligned with the projection of the interplanetary magnetic field than in directions not so aligned. Ultralow–frequency waves were presented in the unshocked solar wind, and their amplitude peaked when the spacecraft was downstream of the foreshock. At large distances down the tail, the Mach number of the flow normal to the shock is low, thus providing the opportunity of study repeated crossings of the collisionless shock in an interesting parameter regime.

093.011 **Lightning and plasma wave observations from the Galileo flyby of Venus.**
D. A. Gurnett, W. S. Kurth, A. Roux, R. Gendrin, C. F. Kennel, S. J. Bolton.
Science, Vol. 253, No. 5027, p. 1522 – 1525 (27 Sep 1991).

During the Galileo flyby of Venus the plasma wave instrument was used to search for impulsive radio signals from lightning and to investigate locally generated plasma waves. A total of nine events were detected in the frequency range from 100 kHz to 5.6 MHz. Although the signal are weak, lightning is the only known source of the signals. Near the bow shock two types of locally generated plasma waves were observed, low–frequency electromagnetic waves from about 5 to 50 Hz and electron plasma oscillation at about 45 kHz. The plasma oscillations have considerable fine structure, possibly because of the formation of soliton–like wave packets.

093.012 **Energetic particles at Venus: Galileo results.**
D. J. Williams, R. W. McEntire, S. M. Krimigis, E. C. Roelof, S. Jaskulek, B. Tossman, B. Wilken, W. Stüdemann, T. P. Armstrong, T. A. Fritz, L. J. Lanzerotti, J. G. Roederer.
Science, Vol. 253, No. 5027, p. 1525 – 1528 (27 Sep 1991).

At Venus the Energetic Particles Detector (EPD) on the Galileo spacecraft measured the differential energy spectra and angular distributions of ions $\gtrsim 22$ keV and electrons $\gtrsim 15$ keV in energy. The only time particles were observed by EPD was in a series of episodic events near closest approach. Angular distributions were highly anisotropic, ordered by the magnetic field, and showed ions arriving from the hemisphere containing Venus and its bow shock. The spectra showed a power law form with intensities observed into the 120– to 280–keV range. Comparisons with model bow shock calculations show that these energetic ions are associated with the venusian foreshock–bow shock region.

093.013 **Plasma observations at Venus with Galileo.**
L. A. Frank, W. R. Paterson, K. L. Ackerson, F. V. Coroniti, V. M. Vasyliunas.
Science, Vol. 253, No. 5027, p. 1528 – 1531 (27 Sep 1991).

Plasma measurements were obtained with the Galileo spacecraft during an approximately 3.5–hour interval in the vicinity of Venus on 10 Feb 1990. Several crossings of the bow shock in the local dawn sector were recorded before the spacecraft passed into the solar wind upstream from this planet. Although observations of ions of the solar wind and the postshock magnetosheath plasmas were not possible, solar wind densities and bulk speeds were determined from the electron velocity distributions. Ions streaming into the solar wind from the bow shock were detected. A search for pickup ions from the hot hydrogen and oxygen planetary coronas yielded an upper limit for these densities in the range of 10^{-3} ion/cm^{-1}, which is consistent with densities expected from current models of neutral gas densities.

093.014 **Images from Galileo of the Venus cloud deck.**
M. J. S. Belton, P. J. Gierasch, M. D. Smith, P. Helfenstein, P. J. Schinder, J. B. Pollack, K. A. Rages, A. P. Ingersoll, K. P. Klaasen, J. Veverka, C. D. Anger, M. H. Carr, C. R. Chapman, M. E. Davies, F. P. Fanale, R. Greeley, R. Greenberg, J. W. Head III, D. Morrison, G. Neukum, C. B. Pilcher.
Science, Vol. 253, No. 5027, p. 1531 – 1536 (27 Sep 1991).

Images of Venus taken at near–infrared (NIR) show that the morphology and motions of large–scale features change with depth in the cloud deck. Poleward meridional velocities, seen in both spectral regions, are much reduced in the NIR. The images follow the changing state of the upper cloud layer downwind of the subsolar point, and the zonal flow field shows a longitudinal periodicity that may be coupled to the formation of large–scale planetary waves.

093.015 **Middle infrared thermal maps of Venus at the time of the Galileo encounter.**
G. S. Orton, J. Caldwell, A. J. Friedson, T. Z. Martin.
Science, Vol. 253, No. 5027, p. 1536 – 1538 (27 Sep 1991).

Images of the disk of Venus, taken at wavelengths between 8 and 22 μm, were obtained a few days after the Galileo spacecraft's closest approach on 8 Feb 1990; these images show variations in the thickness of the main H_2SO_4 cloud deck and the overlying temperature structure. Several features are qualitatively similar to those of earlier observations. Equatorial limb darkening is quantitatively similar to that of previous observations; fairly constant at wavelengths up to 20 μm, where limb darkening increases substantially. In contrast to what was found in previous observations, polar and equatorial limb darkening are nearly the same at most wavelengths.

093.016 **Ground–based near–infrared imaging observations of Venus during the Galileo encounter.**
D. Crisp, S. McMuldroch, S. K. Stephens, W. M. Sinton, B. Ragent, K.–W. Hodapp, R. G. Probst, L. R. Doyle, D. A. Allen, J. Elias.
Science, Vol. 253, No. 5027, p. 1538 – 1541 (27 Sep 1991).

Near–infrared images of Venus, obtained from a global network of ground–based observatories during Jan and Feb 1990, document the morphology and motions of the night–side near–infrared markings before, during, and after the Galileo Venus encounter. A dark cloud extended halfway around the planet at low latitudes ($< \pm 40°$) and persisted throughout the observing program. It had a rotation period of 5.5 ± 0.15 days. The remainder of this latitude band was characterized by small–scale dark and bright markings with rotation periods of 7.4 ± 1 days. The different rotation periods for the large dark cloud and the smaller markings suggests that they are produced at different altitudes. Maps of the water vapor distribution show no evidence for large horizontal gradients in the lower atmosphere of Venus.

093.017 **Galileo infrared imaging spectroscopy measurements at Venus.**
R. W. Carlson, K. H. Baines, T. Encrenaz, F. W. Taylor, P. Drossart, L. W. Kamp, J. B. Pollack, E. Lellouch, A. D. Collard, S. B. Calcutt, D. Grinspoon, P. R. Weissman, W. D. Smythe, A. C. Ocampo, G. E. Danielson, F. P. Fanale, T. V. Johnson, H. H. Kieffer, D. L. Matson, T. B. McCord, L. A. Soderblom.
Science, Vol. 253, No. 5027, p. 1541 – 1548 (27 Sep 1991).

During the 1990 Galileo Venus flyby, the Near Infrared Mapping Spectrometer investigated the night–side atmosphere of Venus in the spectral range 0.7 to 5.2 μm. Multispectral images at high spatial resolution indicate substantial cloud opacity variations in the lower cloud levels, centered at 50 km altitude. Zonal and meridional winds were derived for this level and are consistent with motion of the upper branch of a Hadley cell. Northern and southern hemisphere clouds appear to be markedly different. Spectral profiles were used to derive lower atmosphere abundances of water vapor and other species.

093.018 **Galileo Ultraviolet Spectrometer experiment: initial Venus and interplanetary cruise results.**
C. W. Hord, C. A. Barth, L. W. Esposito, W. E. McClintock, W. R. Pryor, K. E. Simmons, A. I. F. Stewart, G. E. Thomas, J. M. Ajello, A. L. Lane, R. W. West, B. R. Sandel, A. L. Broadfoot, D. M. Hunten, D. E. Shemansky.
Science, Vol. 253, No. 5027, p. 1548 – 1550 (27 Sep 1991).

The Galileo Extreme Ultraviolet Spectrometer obtained a spectrum of Venus atmospheric emissions in the 55.0– to 125.0–nm wavelength region. Emissions of helium (58.4 nm), ionized atomic oxygen (83.4 nm), and atomic hydrogen (121.6 nm), as well as a blended spectral feature of atomic hydrogen (Lyman–β) and atomic oxygen (102.5 nm), were observed at 3.5–nm resolution. During the Galileo spacecraft cruise from Venus to Earth, Lyman–α emission from solar system atomic hydrogen was measured. The dominant source of the Lyman–α emission is atomic hydrogen from the interstellar medium. A model of Galileo observations at solar maximum indicates a decrease in the solar Lyman–α flux near the solar poles.

093.019 **Venus: eastern and western elongations 1987 – 89.**
J. McCue, J. Nichol.
J. Br. Astron. Assoc., Vol. 101, No. 5, p. 285 – 288 (Oct 1991).

093.020 **First results of radar survey of Venus by Magellan spacecraft.**
A. T. Bazilevskij.
Astron. Vestn., Tom 25, No. 5, p. 548 – 568 (Sep – Oct 1991). In Russian. English translation in Solar Syst. Res., Vol. 25, No. 5.

A review of the high–resolution radar survey till May 1991 is given. High–quality imagery, altimetric and radiophysical properties maps are obtained. Analysis of these data led to the discovery of a number of new phenomena. In particular are found: 1) pancake features with a diameter of 20 – 25 km, evidently formed by eruptions of viscous non–basaltic lava; 2) long "channels" probably formed by flow of low–viscous liquids being possibly komatite, carbonatite or sulfur lavas; 3) areas of polygonal, sometimes very regular grid–like, fissured state on volcanic plains; 4) enigmatic flows originating within ejecta blankets of impact craters; 5) traces of wind erosion/accumulation of loose material. Western part of Aphrodite Terra, which was considered in some works as possible analog of midoceanic spreading zones of the Earth, is observed to be tessera and does not look like a spreading zone. Reliable evidences of plate tectonics on Venus have not been found. The survey is going on.

093.021 **Venus: search for possible four–day brightness variations in the near IR–spectrum.**
V. M. Klimenko.
Astron. Tsirk., No. 1543, p. 27 – 28 (Mar – Apr 1990). In Russian.

The results of photometric observations of Venus in two narrow bands made in August 1988 are presented. It is found that Venus magnitude in $\lambda = 0.79$ μm varies from day to day in the range of $0^m.04$ and correlates significantly with changes that occur in $\lambda = 0.38$ μm (amplitude are $\sim 0^m.1$ and periodicity of 4 – 5 days). Further observations are needed to help in understanding of the superrotation of Venus' atmosphere.

093.022 **Calculation of the phase correction in observations of Venus with the Danjon astrolabe.**
N. M. Gaftonyuk, S. G. Shubin, V. N. Tryapitsyn.
Astron. Zh., Tom 68, Vyp. 4, p. 843 – 852 (Jul–Aug 1991). In Russian. English translation in Sov. Astron., Vol. 35, No. 4.

A new formula is proposed to take into account the phase influence in Venus observations with the Danjon astrolabe. The formula is applied to data processing of Venus observations obtained in Simeiz in 1981 – 1988.

093.023 **Measurement of winds in Venus' upper mesosphere based on Doppler shifts of the 2.6–mm ^{12}CO line.**
K. P. Shah, D. O. Muhleman, G. L. Berge.
Icarus, Vol. 93, No. 1, p. 96 – 121 (Sep 1991).
The authors present interferometric observations of the CO 2.6–mm line which fulfill the goal of measuring winds in Venus' upper mesosphere. The authors describe the actual analysis of the CO spectra, the models of both the zonal and the subsolar–to–antisolar winds employed and the necessary noise reduction. The radiative transfer model is presented which calculated the mesospheric altitude associated with the measurements. The results and their implications are discussed.

093.024 **Venus: a total mass estimate.**
W. L. Sjogren, G. B. Trager, G. R. Roldan.
Geophys. Res. Lett., Vol. 17, No. 10, p. 1485 – 1488 (Sep 1990).
Reductions of four independent blocks of Pioneer Venus Orbiter Doppler radio tracking data have produced very consistent determinations of the GM of Venus (the product of the universal gravitational constant and total mass Venus). These eatimates have uncertainties that are significantly smaller than any values published to date. The value of GM is also consistent with previously published results in that it falls within their one–sigma uncertainties. The value of 324858.60 ± 0.05 km^3/sec^2 is the authors' best estimate.

093.025 **A signature of auroral precipitation in the nightside ionosphere of Venus.**
J. L. Fox, H. A. Taylor Jr.
Geophys. Res. Lett., Vol. 17, No. 10, p. 1625 – 1628 (Sep 1990).
The authors show here that the densities of mass 28–ions measured by the Pioneer Venus Orbiter ion mass spectrometer (OIMS) on the nightside of Venus are highly variable and show little correlation with the values of the O^+ densities. The authors have determined the total production rates of mass–28 ions in the chemical equilibrium region and find that this production rate cannot be explained by known chemical production reactions. The authors propose that the "excess" production is due to precipitation of electrons into the nightside thermosphere.

093.026 **Electron plasma oscillations in the Venus foreshock.**
G. K. Crawford, R. J. Strangeway, C. T. Russell.
Geophys. Res. Lett., Vol. 17, No. 11, p. 1805 – 1808 (Oct 1990).
Plasma waves are observed in the solar wind upstream of the Venus bow shock by the Pioneer Venus Orbiter. These wave signatures occur during periods when the interplanetary magnetic field through the spacecraft position intersects the bow shock, thereby placing the spacecraft in the foreshock region. The electron foreshock boundary is clearly evident in the data as a sharp onset in wave activity and a peak in intensity. Wave intensity is seen to drop rapidly with increasing penetration into the foreshock. The peak wave electric field strength at the electron foreshock boundary is found to be similar to terrestrial observations. A normalized wave spectrum was constructed using measurements of the electron plasma frequency and the spectrum was found to be centered about this value. These results, along with polarization studies showing the wave electric field to be field aligned, are consistent with the authors' interpretation of the waves as electron plasma oscillations.

093.027 **Blob tectonics: a prediction for Western Aphrodite Terra, Venus.**
R. R. Herrick, R. J. Phillips.
Geophys. Res. Lett., Vol. 17, No. 12, p. 2129 – 2132 (Nov 1990).
It is proposed that the Western Aphrodite Terra region of Venus was formed by four diapirs of heated mantle material, or blobs, rising to the surface in a region of convective upwelling. The four blobs are in different stages of evolution, producing different topography, gravity anomalies, and morphology. The authors show that this model is consistent with available data and make specific predictions that can be tested with Magellan radar data.

093.028 **Dynamic properties of the upper dayside ionosphere of Venus based upon radio–occultation data.**
L. N. Samoznaev.
Cosmic Res., Vol. 29, No. 1, p. 95 – 104 (Jul 1991). English translation of Kosm. Issled., Tom 29, Vyp. 1, p. 104 – 114 (1991).
Electron density profiles obtained using radio–occultation from the "Venera" spacecraft and direct measurements onboard the "Pioneer–Venus" space vehicle are compared. A systematic difference between profiles dating back to different years is noted and attributed to changing conditions of interaction of the upper ionosphere with the solar wind. Data on solar wind and ionospheric pressures are analyzed for periods of high and low solar activity. In order to explain the observed difference in the location of the ionopause for the data from the "Venera" and "Pioneer–Venus" spacecraft, the gas–dynamical model of interaction of the ionosphere with the solar wind is discussed and calculations are performed on the dependence of the ionopause altitude on solar zenith angle at different solar wind pressures and different levels of solar activity. It is necessary to take into account induced magnetic field effects in the ionization balance equations in order to explain the electron density profiles in the upper ionosphere.

093.029 **A two–dimensional kinematic dynamo model of the ionospheric magnetic field at Venus.**
T. E. Cravens, D. Wu, H. Shinagawa.
Geophys. Res. Lett., Vol. 17, No. 12, p. 2261 – 2264 (Nov 1990).
The ionosphere of Venus was observed by the magnetometer on the Pioneer Venus Orbiter (PVO) to be permeated by a large–scale magnetic field for conditions of high solar wind dynamic pressure. The ionospheric conductivity is very high above 170 km and magnetic flux is carried by the plasma flow both vertically downward and in an antisunward direction. Ohmic dissipation of the electric current allows the magnetic field to diffuse for altitudes below 150–170 km. A number of one–dimensional kinmatic dynamo and MHD models have been developed to explain the observed vertical magnetic field profiles, but did not include the effects of horizontal transport. The authors have constructed a two–dimensional kinematic dynamo model of the dayside ionospheric magnetic field at Venus in which the 2–D magnetic induction equation (i.e., magnetic diffusion–convection equation) is numerically solved using specificed plasma velocities. The authors demonstrate that the basic features of the vertical magnetic field profile, such as the presence of a magnetic layer at 170 km, remain unaltered by horizontal flow effects. However, they also show that horizontal plasma flow can strongly affect the magnetic field for altitudes above 300 km.

093.030 **Evidence for unusually high densities of plasma in the Venusian ionosheath.**
D. S. Intriligator, L. H. Brace, S. H. Brecht, W. C. Knudsen, F. L. Scarf, R. J. Strangeway, H. A. Taylor Jr.
Geophys. Res. Lett., Vol. 18, No. 1, p. 61 – 64 (Jan 1991).
An unusual occurrence was recorded by Pioneer Venus Orbiter (PVO) in the Venusian ionosheath on November 10–11, 1981 (orbit 1071): more than an order of magnitude higher densities of ions and electrons were observed than have ever been measured by PVO before then or since. This region of enhanced densities covered an extended region in the dayside ionosheath. The purpose of this paper is to briefly present the authors' analysis of these unique observations.

093.031 **Asymmetries in the location of the Venus and Mars bow shock.**
T.–L. Zhang, K. Schwingenschuh, C. T. Russell, J. G. Luhmann.
Geophys. Res. Lett., Vol. 18, No. 2, p. 127 – 129 (Feb 1991).
An examination of observations of the position of the terminator bow shock at Venus and Mars shows that the terminator bow shock varies varies with the angle between the local bow shock normal and the upstream magnetic field, θ_{BN}.

The part of the shock on the quasi-parallel side is closer to the planet than the part on the of the quasi-perpendicular side, a result which had been suggested by an earlier computer simulation by Thomas and Winske (1990). This bow shock asymmetry is observed to be larger at Mars than at Venus.

093.032 **Steady-state plasma transition in the Venus ionosheath.**
H. Pérez-de-Tejada, D. S. Intriligator, R. J. Strangeway.
Geophys. Res. Lett., Vol. 18, No. 2, p. 131 – 134 (Feb 1991).

The results of an extended analysis of the plasma and electric field data on the Pioneer Venus Orbiter (PVO) are presented. The authors report the persistent presence of a plasma transition embedded in the flanks of the Venus ionosheath between the bow shock and the ionopause. This transition is identified by the repeated presence of characteristic bursts in the 30 kHz channel of the electric field detector of the PVO. The observed electric field signals coincide with the onset of different plasma conditions in the inner ionosheath where more rarified plasma fluxes are measured. The repeated identification of this intermediate ionosheath transition in the PVO data indicates that it is present as a steady state feature of the Venus plasma environment. The distribution of PVO orbits in which the transition is observed suggests that it is more favourably detected in the vicinity of and downstream from the terminator.

093.033 **Geology and structure of Beta Regio, Venus: results from Arecibo radar imaging.**
D. A. Senske, J. W. Head, E. R. Stofan, D. B. Campbell.
Geophys. Res. Lett., Vol. 18, No. 6, p. 1159 – 1162 (Jun 1991).

Arecibo radar images of a portion of the equatorial region of Venus provide the first high resolution (1.5– to 2.0–km) synoptic coverage of Beta Regio. Within this area, tessera, a complex deformed terrain, is identified as a major geologic unit with the largest region corresponding to a plateau on the east flank of the highland. Three models are proposed to explain the origin and evolution of Beta Regio and are identified as Mantle Plume/Passive Crust, Mantle Plume/Active Crust, and Mantle Plume/Crustal Spreading. The mantle Plume/Passive Crust model appears to be the most consistent with the geology in this region and suggests that a plume disrupts a preexisting region of tessera.

093.034 **Calculation of the phase correction in observations of Venus with the Danjon astrolabe.**
N. M. Gaftonyuk, S. G. Shubin, V. N. Tryapitsyn.
Sov. Astron., Vol. 35, No. 4, p. 418 – 422 (Jul 1991). Current Physics Microform No.: 9203X0604. English translation of Astron. Zh., Tom 68, Vyp. 4, p. 843 – 852 (1991).

A formula is proposed to allow for the phase effect in observations of Venus with the Danjon astrolabe. The formula is used to process observations of Venus at Simeiz in 1981 – 1988.

093.035 **Galileo flies by Venus, the Earth and the Moon.**
A. T. Bazilevskij.
Astron. Vestn., Tom 25, No. 6, p. 677 – 685 (Nov – Dec 1991). In Russian. English translation in Sol. Syst. Res., Vol. 25, No. 6.

A short summary of scientific observations of Venus, the Earth and the Moon by Galileo spacecraft is given. TV and IR observations of Venus give new information on structure and dynamics of the atmosphere of this planet. Observations of the atmosphere and surface of the Earth will be used for calibration of Galileo scientific instruments. During the fly-by of the Moon images of its western hemisphere have been obtained. This area is very promising for the study because it combines both well-studied parts of the nearside and poorly-known parts of the farside. Multicolor imagery obtained permits to map the areal distribution of major types of lunar rocks.

093.036 **Cometary water on Venus: impact modelling and possible evidence.**
G. Kolovos, H. Varvoglis, L. Pylarinou.
Earth, Moon, Planets, Vol. 54, No. 2, p. 103 – 117 (Aug 1991). With a correction in Vol. 55, No. 3, p. 301 (Dec 1991).

The present small amount of water in the atmosphere of Venus, in connection with the estimated short time scale of water loss from this planet, has led to the hypothesis that the water concentration in the Venusian atmosphere is in a dynamical equilibrium, where the losses are counter-balanced by a suitable water source. The main candidate water sources are: (a) outgassing from the Venusian surface, due to volcanic activity and (b) cometary impacts. The lack of observational evidence of cometary impacts is usually attributed either to the rarity of such events or to the fact that presently their observational signature is not well understood. The authors report on a photographic evidence of a short duration dark feature on Venus observed on 18 May 1988. After eliminating the possibilities of a film defect and an interference from an artificial Earth satellite or an interplanetary object, it is concluded that this feature was the signature of an event that took place on the upper haze layer of the Venusian atmosphere. It is proposed that this event was the impact of a small ($\sim 10^{10}$g) comet-like object consisting mainly of water. This impact caused a temporary evaporation of the sol H_2SO_4 particles of the upper haze layer and, consequently, a decrease of the albedo of the region around the point of entrance of the comet in the Venusian cloud layers. This region of lower albedo appears as a dark feature in the reported photograph. The authors' model accounts for the short duration of the feature as well as for its shape.

093.037 **The thermal state of the Venusian mantle.**
J. Arkani-Hamed.
Earth, Moon, Planets, Vol. 55, No. 1, p. 51 – 68 (Oct 1991).

Assuming that the surface topography of the Ovda and Thetis plateaus, and that of the saddle region between Thetis and Atla plateaus, is due to surface subsidence of oceanic-type thermal boundary layers, the author calculated the temperature at 200 km beneath the plateaus to be about 1600°C, and that beneath the saddle region to be about 1400–1500°C. The total subsidence of Ovda plateau in the last 3/4 of its existence, i.e. between 200 km and 800 km off the postulated ridge axis, suggests that the plateau is probably a young feature, less than 40 m.y. old. The spreading plate models imply a half-spreading rate of 2.3–5.3 cm/yr for the plateaus and 2–2.8 cm/yr for the saddle region.

093.038 **Multiple-impact mechanisms on Venus and other solid planets and satellites.**
K. D. Trego.
Earth, Moon, Planets, Vol. 55, No. 1, p. 69 – 72 (Oct 1991). Letter-to-the-editor.

While atmospheric pressure can be considered a major multiple impact mechanism of Venus, probable mechanisms for multiple impacts on other solid planetary bodies cannot be isolated with certainty.

093.039 **The influence of mantle water and heterogeneity on the viscous relaxation of Venus impact craters.**
M. R. Rovetta.
Geophys. Res. Lett., Vol. 18, No. 3, p. 471 – 474 (Mar 1991).

Empirical flow laws for periodotite and diabase were used to model the viscous relaxation of large impact craters on Venus. The models were then compared to the depths of the three largest impact craters ($\geqslant$60 km diameter) measured by Venera 15/16 radar altimetry. The maximum crater retention times predicted by models that assumed $\geqslant$0.03% wt. water in the upper mantle were inconsistent with the observed depths of these craters and a mean minimum age of 0.5 Ga. Models that assumed an upper mantle water content of $\leqslant$0.003% wt. were consistent with the observations and a maximum crater retention time of 1.0 Ga. This result was insensitive to up to 60% by volume diabase mixed into the upper mantle as inclusions at the several kilometer scale and to the presence of a basalt-eclogite or melting transition at a depth of 100 km.

093.040 **A two dimensional shock capturing, hydrodynamic model of the Venus ionosphere.**
A. F. Nagy, A. Körösmezey, J. Kim, T. I. Gombosi.
Geophys. Res. Lett., Vol. 18, No. 5, p. 801 – 804 (May 1991).

A two-dimensional, time-dependent, shock capturing single-species (O^+) hydrodynamic model of the Venus ionosphere, which solves the coupled continuity, momentum and energy equations for the altitude range of 150–500 km and the solar zenith angle range of 0°–180° has been developed and is presented in this paper. It was again demonstrated that the introduction of topside heat inflows leads to calculated dayside electron and ion temperatures, which are consistent with the measured values. In order to reproduce the measured electron temperatures, which are roughly constant over all SZA's, the heat inflows had to be reduced significantly over the nightside compared to the dayside values. The calculated transterminator ion flows are supersonic and relatively close to the observed average values. The model predicts a deceleration shock at a SZA of about 135°, consistent with the ion temperature and velocity observations.

093.041 **Wrinkle ridges in Lunae Planum Mars: implications for shortening and strain.**
J. B. Plescia.
Geophys. Res. Lett., Vol. 18, No. 5, p. 913 - 916 (May 1991).
Photoclinometric data for Lunae Planum wrinkle ridges indicate average relief of ~130 m and distinct elevation discontinuities across the ridge of 55 m. Modelling ridges as the result of thrust faulting and associated upper–plate folding indicates shortening across individual ridges of ~131 m (90% faulting). Total shortening at 20°N across Lunae Planum is ~1840 m corresponding to a regional compressive strain of 0.29%. Strain appears uniform across Lunae Planum, although it is accommodated by larger and fewer structures in the west than in the east.

093.042 **Studies on the dynamics of Venus ionosphere: effects of varying the ionopause height and the onset of turbulence.**
R. P. Singhal, R. C. Whitten.
Indian J. Radio Space Phys., Vol. 20, No. 2, p. 150 - 156 (Apr 1991).
The conservation equations of plasma dynamics in the upper ionosphere of Venus have been solved by using a spectral method in the horizontal and finite differencing in the vertical direction. The effect of varying the ionopause height on the computed nightside ion densities is investigated. These ion densities show a sharp decrease as the ionopause altitude is reduced to 300 km. The effect of viscous forces on the horizontal plasma flow is investigated for a wide range of values of the coefficient of viscosity. The Reynolds numbers characteristics of the flow are calculated and the conditions for the onset of turbulence discussed. The influence of magnetic fields on viscosity is also discussed.

093.043 **Waves in Venus' middle and upper atmosphere; implication of Pioneer Venus probe data above the clouds.**
A. Seiff, D. B. Kirk.
J. Geophys. Res., Vol. 96, No. A7, p. 11021 - 11032 (1 Jul 1991). - See Abstr. 012.063 for the main entry.
Probe data showing the presence of waves in Venus's middle and upper atmosphere (Seiff et al., 1989; Seiff and Kirk, 1982) are critically reevaluated and extended to 138 km, near the level of in situ data taken by the Pioneer Venus orbiter. The observations lead to the suggestion that the thermospheric waves are solar–fixed, induced either by the major subsidence across the terminators or as continuations upward of waves in the middle atmosphere. The wave structure in the large probe sounding below 100 km is similar to, but does not quantitatively support the solar–tidal model of Pechmann and Ingersoll, which gives much larger amplitudes and different wave phases.

093.044 **Pioneer Venus orbiter search for Venusian lightning.**
W. J. Borucki, J. W. Dyer, J. R. Phillips, P. Pham.
J. Geophys. Res., Vol. 96, No. A7, p. 11033 - 11044 (1 Jul 1991). - See Abstr. 012.063 for the main entry.
During 1988 and 1990, the star sensor aboard the Pioneer Venus orbiter (PVO) was used to search for optical pulses from lightning on the nightside of Venus. In a previous effort to use the star sensor for this purpose (Borucki et al., 1981) the authors reported an upper limit to the lightning activity, but they were unable to show that lightning was being detected because the signal rate was indistinguishable from the false alarm rate. Because the periapsis altitude has increased by nearly a factor of ten since 1979, the star scanner views a much larger area of Venus than it did previously. This increased viewing area should provide an increased signal rate because the amplitude of optical pulses should still be above the detection threshold of the sensor if the flashes are as bright as terrestrial flashes. Because the false alarm rate did not increase, the increased viewing area allows a more sensitive search for lightning activity. Useful data were obtained for 53 orbits in 1988 and 55 orbits in 1990. During this period, approximately 83 s of search time plus 7749 s of control data were obtained. The authors again find no optical evidence for lightning activity.

093.045 **Comparative terrestrial planet thermospheres. 1. Solar cycle variation of global mean temperatures.**
S. W. Bougher, R. G. Roble.
J. Geophys. Res., Vol. 96, No. A7, p. 11045 - 11056 (1 Jul 1991). - See Abstr. 012.063 for the main entry.
Global average models for the thermospheres of Venus, Earth, and Mars are used to calculate the solar cycle variations of the global mean temperatures using the compositional profiles for each planet.

093.046 **Plasma motion in the Venus ionosphere: transition to supersonic flow.**
R. C. Whitten, A. Barnes, P. T. McCormick.
J. Geophys. Res., Vol. 96, No. A7, p. 11057 - 11064 (1 Jul 1991). - See Abstr. 012.063 for the main entry.
A remarkable feature of the ionosphere of Venus is the presence of nightward supersonic flows at high altitude near the terminator. In general the steady flow of an ideal gas admits a subsonic–supersonic transition only in the presence of special conditions, such as a convergence of the flow followed by divergence, or external forces. Here, the authors show that the relatively high pressure dayside plasma wells up slowly, and at high altitude it is accelerated horizontally through a relatively constricted region near the terminator toward the low–density nightside. In effect, the plasma flows through a "nozzle" that is first converging, then diverging, permitting the transition to supersonic flow. Analysis of results from previously published models of the plasma flow in the upper ionosphere of Venus shows how such a "nozzle" is formed. The model plasma does indeed accelerate to supersonic speeds, reaching sonic speed just behind the terminator. The computed speeds prove to be close to those observed by the Pioneer Venus orbiter, and the ion transport rates are sufficient to produce and maintain the nightside ionosphere.

093.047 **The effect of the solar cycle on the maintenance of the nightside ionosphere of Venus.**
A. J. Kliore, J. G. Luhmann, M. H. G. Zhang.
J. Geophys. Res., Vol. 96, No. A7, p. 11065 - 11072 (1 Jul 1991). - See Abstr. 012.063 for the main entry.
Pioneer Venus radio occultation measurements of the nightside ionosphere of Venus collected from 1979 to 1986 have made it possible to study its behavior at times of both solar maximum and solar minimum. Although some solar maximum measurements are similar in nature to those observed at solar minimum, which have an average peak density of about $7 \times 10^3 cm^{-3}$, others show much higher peak densities, reaching values of about $4 \times 10^4 cm^{-3}$. These elevated peak densities also occur at higher altitudes. The integrated electron column density for these measurements are also much higher, indicating the presence of substantial ionization above the main peak. The magnitudes of both the peak density and the integrated content above the peak are anticorrelated with solar wind dynamic pressure, leading to the interpretation that these enhancements during solar maximum are due to transterminator transport of O^+ ions from the dayside when the solar wind dynamic pressure is low enough to permit a sufficiently high dayside ionopause. The resulting ionization peak can be many times the concentration produced by energetic electron fluxes impacting the neutral atmosphere on the nightside, which apparently form the remaining source of the nightside peak at such times during solar maximum, when

transterminator flow is cut off by high solar wind pressure depressing the dayside ionopause, and during solar minimum, when the ionopause is always depressed.

093.048 **Magnetization of the ionospheres of Venus and Mars: results from radio occultation measurements.**
R. Woo, A. J. Kliore.
J. Geophys. Res., Vol. 96, No. A7, p. 11073 - 11082 (1 Jul 1991).
- See Abstr. 012.063 for the main entry.

In situ measurements by the Pioneer Venus orbiting spacecraft, conducted during solar maximum only, have shown that magnetization (permeation of large-scale magnetic fields) of the ionosphere of Venus occurs under high solar wind dynamic pressure and that this takes place most frequently near the subsolar region. Magnetization is also revealed in electron density profiles by the presence of disturbed topside plasma and/or a ledge near the altitude of 200 km. In this paper the authors use remote sensing radio occultation measurements to study magnetization of the ionospheres of Venus and Mars based on these characteristics.

093.049 **A comprehensive magnetohydrodynamic model of the Venus ionosphere.**
H. Shinagawa, J. Kim, A. F. Nagy, T. E. Cravens.
J. Geophys. Res., Vol. 96, No. A7, p. 11083 - 11096 (1 Jul 1991).
- See Abstr. 012.063 for the main entry.

The MHD Venus ionospheric model developed by Shinagawa and Cravens (1988) has been improved by including the energy equations for ions and electrons in a self-consistent manner. This new model reproduces observed electron density and magnetic field profiles very well, and the basic MHD processes of the Venus ionosphere, as described by Shinagawa and Cravens (1988), remain virtually unchanged.

093.050 **The ionopause current layer at Venus.**
T. E. Cravens, H. Shinagawa.
J. Geophys. Res., Vol. 96, No. A7, p. 11119 - 11132 (1 Jul 1991).
- See Abstr. 012.063 for the main entry.

Magnetic fields are induced by the solar wind in the ionospheres of planets such as Venus and Mars, which possess only weak, or nonexistent, intrinsic magnetic fields. The magnetometer on the Pioneer Venus Orbiter (PVO) observed large-scale magnetic fields in the ionosphere when the solar wind dynamic pressure was high. During high solar wind dynamic pressure conditions the ionopause was observed to be broad and at low altitudes, whereas during low solar wind dynamic pressure conditions the ionopause was located at high altitudes ($z > 300$ km) and was narrow ($\Delta z \approx 20$ km). The authors review the results of recent theoretical models of the magnetic field in the ionosphere of Venus, and, in particular, the one-dimensional MHD model of Shinagawa and Cravens (1988) and the two-dimensional kinematic dynamo model of Cravens et al. (1990). The results of these models are used as a starting point for an explanation of the behavior of the ionopause.

093.051 **Venus ionospheric "clouds": relationship to the magnetosheath field geometry.**
M. Ong, J. G. Luhmann, C. T. Russell, R. J. Strangeway, L. H. Brace.
J. Geophys. Res., Vol. 96, No. A7, p. 11133 - 11144 (1 Jul 1991).
- See Abstr. 012.063 for the main entry.

Venus plasma "clouds", apparently detached regions of ionospheric plasma above the main ionosphere of Venus observed by the Langmuir probe on the Pioneer Venus orbiter, are thought to be evidence for bulk removal of ionospheric ions by the solar wind interaction. In addition to the primary feature of an electron density enhancement above the ionopause, the characteristic signature of a cloud may include a change in the sign of the B_x (sunward/antisunward) component of the surrounding magnetosheath magnetic field. The plasma wave instrument also sometimes observed enhanced electric field noise in the clouds. In an effort to learn more about clouds and their cause(s), the authors analyzed the magnetic field control of cloud positions relative to the planet by rotating their observed locations into a coordinate system in which all of the transverse upstream (interplanetary) magnetic fields were aligned. The results indicate that clouds are scattered around the periphery of the planet in the terminator plane. There is no evidence of a concentration of clouds where the magnetosheath magnetic field is most strongly "draped", as suggested by some earlier studies of selected occurrences.

093.052 **The magnetic barrier at Venus.**
T. L. Zhang, J. G. Luhmann, C. T. Russell.
J. Geophys. Res., Vol. 96, No. A7, p. 11145 - 11154 (1 Jul 1991).
- See Abstr. 012.063 for the main entry.

The magnetic barrier at Venus is a region within which the magnetic pressure dominates all other pressure contributions. The barrier is formed in the inner region of the dayside magnetosheath to transfer solar wind momentum flux to the ionoshpere. Passes through the dayside magnetosheath and ionopause with Pioneer Venus have allowed the authors to probe the magnetic barrier directly. These passes have been used to construct altitude profiles of the barrier. The magnetic barrier is strongest at the subsolar point and weakens as expected with increasing solar zenith angle. The existence of a north-south asymmetry in the barrier strength is also demonstrated. The magnetic barrier is about 200 km thick at the subsolar point and 800 km thick at the terminator which is comparable with the so-called "mantle". The authors find that the magnetic barrier transfers most of the solar wind dynamic pressure to the ionsophere via the enhanced magnetic pressure. The convected field gasdynamic model is found to predict the correct bow shock location if the magnetic barrier is treated as the obstacle.

093.053 **Superthermal >36-eV ions observed in the near-tail region of Venus by the Pioneer Venus orbiter neutral mass spectrometer.**
W. T. Kasprzak, J. M. Grebowsky, H. B. Niemann, L. H. Brace.
J. Geophys. Res., Vol. 96, No. A7, p. 11175 - 11188 (1 Jul 1991).
- See Abstr. 012.063 for the main entry.

The Pioneer Venus orbiter neutral mass spectrometer (ONMS) has made measurements of >36-eV ions in the altitude range 1300-3700 km for solar zenith angles greater than 120°. The observations cover a time period from late 1982 to 1989. The superthermal ions form part of the near-tail region of Venus termed the ionotail. Statistics describing the details of the ion composition, direction, solar activity effects, and correlation with the prevailing magnetic field are discussed in this paper.

093.054 **A comparison of induced magnetotails of planetary bodies: Venus, Mars, and Titan.**
J. G. Luhmann, C. T. Russell, K. Schwingenschuh, Ye. Yeroshenko (*E. G. Eroshenko*).
J. Geophys. Res., Vol. 96, No. A7, p. 11199 - 11208 (1 Jul 1991).
- See Abstr. 012.063 for the main entry.

The Pioneer Venus orbiter (PVO), PHOBOS 2, and Voyager 1 spacecraft have together provided observations of three planetary bodies with induced magnetotails: Venus, Mars, and Titan. During the extended mission of PVO, the tail of Venus was probed at an altitude of ~1.3 planetary radii, which provided a more appropriate basis for comparison with the Mars data (at ~2.7 planetary radii), and Titan data (~2.5 planetary radii downstream), than the previously analyzed Venus tail data obtained near PVO apoapsis (~12 planetary radii). A parallel examination of the magnetic properties of these tails at downstream distances within 3 planetary radii is presented and discussed.

093.055 **Global hybrid simulation of unmagnetized planets: comparison of Venus and Mars.**
S. H. Brecht, J. R. Ferrante.
J. Geophys. Res., Vol. 96, No. A7, p. 11209 - 11220 (1 Jul 1991).
- See Abstr. 012.063 for the main entry.

Results from three-dimensional hybrid particle simulations of the solar wind interaction with the planets Mars and Venus are presented. The simulations produce shocks and magnetic barriers which are asymmetric. These results are qualitatively in agreement with data. In the absence of an ionosphere the subsolar shock standoff distance was found to agree with the observations if the Hall current is limited. It was also found that the solar wind interaction with Mars and Venus was substantially different. The interaction with Venus can be generally viewed as a magnetized interaction. The Mars interaction is very kinetic in nature and appears not to have a shock in the classic sense.

093.056 **ULF waves upstream of the Venus bow shock: properties of one–Hertz waves.**
D. S. Orlowski, C. T. Russell.
J. Geophys. Res., Vol. 96, No. A7, p. 11271 – 11282 (1 Jul 1991). – See Abstr. 012.063 for the main entry.

Pioneer Venus orbiter data are used to examine the properties of a class of ULF upstream waves with relatively high observed frequencies (1.1–1.3 Hz). In the spacecraft frame these waves are most often left–hand elliptically polarized. They have amplitudes up to 3 nT (near the bow shock) and propagate obliquely to the magnetic field at angles from 10° to 50°. The waves appear almost immediately after the spacecraft crosses the magnetic field tangent line to the bow shock surface into the region of connected field lines. The amplitude of these waves decreases with distance from the shock measured along the magnetic field line. The authors have measured along the magnetic field line. They have used the cold plasma dispersion relation in order to study the propagation of these waves and to explain their observed polarization. Calculated group velocities indicate that those waves have sufficient upstream velocities to propagate from the shock into the solar wind. The totality of observations, including the observed wave damping and the observation of right–handed waves, seems best explained by a source of right–handed whistler mode waves at the bow shock.

093.057 **Geology and tectonics of the Themis Regio–Lavinia Planitia–Alpha Regio–Lada Terra area, Venus: results from Arecibo image data.**
D. A. Senske, D. B. Campbell, J. W. Head, P. C. Fisher, A. A. Hine, A. de Charon, S. L. Frank, S. T. Keddie, K. M. Roberts, E. R. Stofan, J. C. Aubele, L. S. Crumpler, N. Stacy.
Earth, Moon, Planets, Vol. 55, No. 2, p. 97 – 161 (Nov 1991).

New radar images obtained from the Arecibo Observatory (resolution 1.5–4.0 km) for portions of the southern hemisphere of Venus are discussed.

093.058 **Geology and tectonics of Beta Regio, Guinevere Planitia, Sedna Planitia and Western Eistla Regio, Venus: results from Arecibo image data.**
D. A. Senske, D. B. Campbell, E. R. Stofan, P. C. Fisher, J. W. Head, N. Stacy, J. C. Aubele, A. A. Hine, J. K. Harmon.
Earth, Moon, Planets, Vol. 55, No. 2, p. 163 – 214 (Nov 1991).

New radar images (resolution 1.5–2.0 km) obtained from the Arecibo Observatory are used to assess the geology of a portion of the equatorial region of Venus (1°S to 45°N and from 270° easterward to 30°). Nine geologic units are mapped on the basis of their radar characteristics and their distribution and correspondences with topography are examined.

093.059 **Compressible convection in a viscous Venusian mantle.**
A. M. Leitch, D. A. Yuen.
J. Geophys. Res., Vol. 96, No. E1, p. 15551 – 15562 (25 Aug 1991).

Finite element simulations of axisymmetric spherical shell compressible convection were carried out to investigate the effect of various surface boundary conditions in a Venusian mantle. The authors employed a thermal expansivity α which decreased with depth, a uniform viscosity an order of magnitude greater than the Earth's, and zero and chondritic quantities of internal heating. As long as hot plumes from the core–mantle boundary were strong, the convection pattern was typical of that for variable α flow; that is, it was characterized by steady upflowing regions, unsteady collections of downflowing plumes, and large aspect ratio cells. Increases in the internal heating or the temperature T_0 at the top of the convecting layer weakened the hot plumes and therefore decreased the width of the cells. A rigid surface increased the internal temperature and also decreased the width of convection cells. Extensive regions of subadiabaticity were found in the mantle. The authors compare their results with those for fully three–dimensional convection under similar conditions (Schubert et al., 1990).

093.060 **Induced magnetic fields at the surface of Venus inferred from Pioneer Venus Orbiter near–periapsis measurements.**
J. G. Luhmann.
J. Geophys. Res., Vol. 96, No. E3, p. 18831 – 18840 (25 Oct 1991).

Although magnetic field measurements have not been made on the surface of Venus or, for that matter, below the Pioneer Venus Orbiter (PVO) periapsis altitude of ~150 km, the theory of the large–scale ionospheric magnetization that results from the solar wind interaction can be used in concert with the available data to infer the existence of nonzero fields down to the surface. Here it is argued that magnetometer measurements in the upper part of the "diffusion region" of the ionosphere can be extrapolated to lower altitudes because of the particularly simple physics that determines the altitude profile of the large–scale magnetic field in this region. The magnetic field at the bottom of the ionosphere inferred by this method using the PVO data is up to tens of nanoteslas when large–scale ionospheric fields are present, suggesting the likely presence of fields of similar strength below. Given the negligible intrinsic field of Venus inferred from the PVO measurements, these fields must result from the solar wind interaction. The implications for Mars, which may have an analogous solar wind interaction, is that magnetometers on the surface could regularly observe varying fields of similar magnitude and origin.

093.061 **Stress analysis of Tellus Regio, Venus, based on gravity and topography: comparison with Venera 15/16 radar images.**
D. R. Williams, L. Gaddis.
J. Geophys. Res., Vol. 96, No. E3, p. 18841 – 18859 (25 Oct 1991).

The tectonics of the Tellus Regio highland of Venus have been examined using Pioneer Venus altimetry and gravity data on Venera 15/16 radar images. A thin elastic shell model has incorporated the Pioneer Venus data to calculate both global, long–wavelength and regional, shorter–wavelength stresses for various assumed values of crust, lithosphere, and mantle thickness and modes of compensation. The effects of these parameters on the calculated stress magnitudes have been examined. The model also allows calculation of crustal thickness and mantle density anomalies. The resulting calculated stress fields have been compared to the surface morphology observed in the Venera 15/16 radar images and interpreted in terms of the stress history.

093.062 **Absolute wind velocities in the lower thermosphere of Venus using infrared heterodyne spectroscopy.**
J. J. Goldstein, M. J. Mumma, T. Kostiuk, D. Deming, F. Espenak, D. Zipoy.
Icarus, Vol. 94, No. 1, p. 45 – 63 (Nov 1991).

Absolute wind velocities in the thermosphere of Venus were retrieved using NASA/Goddard Space Flight Center infrared heterodyne spectrometers at the NASA IRTF and the McMath Solar Telescope, Dec 1985 to Mar 1987. Measurement of beam–integrated Doppler shifts in the nonthermal emission core of the $^{12}C^{16}O_2$ line provided sampling of the lower thermospheric wind field projected along the line of sight. The spectrometer's diffraction–limited beam yielded the spatial resolution necessary for circulation model discrimination. Qualitative analysis of beam–integrated winds indicated a dominant subsolar–to–antisolar circulation in the lower thermosphere. Beam–integrated winds were modeled with a 100 × 100 element grid over the beam. Horizontal wind velocity was derived from a two–parameter model wind field comprised of subsolar–antisolar and zonal components. Acceptable model fits to the individual observing periods indicated a dominant subsolar–antisolar flow. No temporal variation in the wind field could be discerned at the sensitivity of the fits. A single, acceptable model fit was found common to all observing periods indicating the wind field may have been constant Dec 1985 to Mar 1987.

093.063 **Venus.**
IAU Circ., No. 5365, p. 1 (14 Oct 1991).

093.064 **The tectonic evolution of western Ishtar Terra, Venus.**
A. Lenardic, W. M. Kaula, D. L. Bindschadler.
Geophys. Res. Lett., Vol. 18, No. 12, p. 2209 – 2212 (Dec 1991).
The origin and evolution of Ishtar Terra is modeled using numerical simulations of crust/mantle interaction on Venus. Based on their modeling, the authors favor a two phase evolutionary sequence for Ishtar. The first phase is driven by bulk mantle flow due to thermal convection and results in the formation of a thickened crustal block over a mantle downwelling. The second phase, a result of the thermal perturbation exerted on the mantle by the region of thickened crust, is driven by the sinking of a lithospheric slab. The slab–driven tectonic pulse leads to low angle subduction and orogenesis at the peripheries of the crustal block. Throughout its evolution Ishtar remains coupled to flow within the deep mantle, due to the lack of an asthenosphere, and ultimately bulk mantle flow comes to dominate as the effects of the slab–driven pulse wane.

093.065 **Physical processes in the plasma mantle on Venus.**
K. Szegö, V. S. Shapiro, V. I. Shevchenko,
R. Z. Sagdeev, W. T. Kasprzak, A. F. Nagy.
Geophys. Res. Lett., Vol. 18, No. 12, p. 2305 – 2308 (Dec 1991).
The Phobos–2 observations established the presence of a turbulent region on the dayside, called the mantle, just outside the magnetic barrier at Mars. This mantle is the interaction region of cold Martian ions and the shocked solar wind. This work presents the results of a study, which analyzed data from ten Pioneer Venus orbits, in order to see whether similar wave particle interaction processes also exist in the corresponding region around Venus. The first conclusion is that the apparent physical processes in the mantle are indeed similar around Venus and Mars. The planetary thermal O^+ ions outside the ionopause interact with the shocked solar wind and excited electrostatic waves close to the lower hybrid frequency. These waves propagate inwards, heating first the electron and deeper down in the ionosphere the thermal ion population. The observed superthermal ions are believed to be the product of this wave particle interaction process. The authors' second conclusion is that the wave energy transferred to the thermal electrons is of the right magnitude ($\sim 4 \times 10^9 eV\,cm^{-2}s^{-1}$) to provide the supplemental heat source necessary to reconcile observed and calculated electron temperatures in the ionosphere.

093.066 **Venus ionospheric tail rays: spatial distributions and interplanetary magnetic field control.**
M. Ong, J. G. Luhmann, C. T. Russell, R. J. Strangeway,
L. H. Brace.
J. Geophys. Res., Vol. 96, No. A10, p. 17751 – 17762 (1 Oct 1991).
Venus tail rays were first observed in the Pioneer Venus Langmuir probe data. These rays appear to be filaments of ionospheric plasma extending far downstream from the nightside ionosphere. The characteristic signatures of tail rays are in many ways similar to those of plasma "clouds" observed above the terminator ionopause. These signatures include a localized electron density enhancement, and occasional magnetic field reversal in the B_x component and/or enhanced electric field noise. Here, selected Pioneer Venus orbits from the interval between orbits 700 and 3500 during the period November 1980 to May 1988 are surveyed for tail ray structures as seen in electron density time series from the Orbiter Langmuir probe. The results of the present study suggest that tail rays are a normal feature of the steady solar wind interaction with Venus but are not generally associated with a central tail plasma sheet. Instead, a terminator source, or multiple sources, is suggested.

093.067 **Testing radio bursts observed on the nightside of Venus for evidence of whistler mode propagation from lightning.**
V. S. Sonwalkar, D. L. Carpenter, R. J. Strangeway.
J. Geophys. Res., Vol. 96, No. A10, p. 17763 – 17778 (1 Oct 1991).
Radio burst events recorded on the nightside of Venus by the orbiting electric field detector (OEFD) on Pioneer Venus Orbiter (PVO) have been interpreted as originating in subionospheric lightning. This lightning source interpretation has been subjected to repeated challenges. In the present work, methods have been developed for testing the hypothesis that particular burst events were the result of whistler mode propagation of signals from subionospheric lightning sources. The tests allow prediction of the resonance cone angle, wave normal direction, refractive index, wave dispersion, and wave polarization and are believed to represent an improved way of categorizing OEFD burst data for purposes of investigating source/propagation mechanisms. The tests, which are capable of refinement, were applied to observations from 11 periods along seven orbits. The results obtained from the limited data sample indicate that there are at least two main categories of burst events, one for which the assumed vertical wave normal angle was within the allowed cone of angles for whistler mode propagation and one for which this was not the case. Lightning is thus considered to be a candidate source for at least some of the OEFD bursts.

093.068 **Venera–15 and –16 middle atmosphere profiles from radio occultations: polar and near–polar atmosphere of Venus.**
O. I. Yakovlev, S. S. Matyugov, V. N. Gubenko.
Icarus, Vol. 94, No. 2, p. 493 – 510 (Dec 1991).
Results of 42 radio occulation measurements of the thermal structure of the Venus atmosphere at altitudes of 42 to 90 km, carried out with the help of Venera–15 and Venera–16 satellites during three occultation cycles from Oct 1983 to Sep 1984, are presented. The measurements were made in polar and near–polar regions of the southern and northern hemispheres. The results point out the similarity of the thermal structures of both hemispheres. In particular, the southern hemisphere has the same cold collar region as the northern. The thermal structure of the atmosphere in both hemispheres depends mainly on the latitude and only weakly on solar illumination conditions. Comparison of the results with data of ratio occultations obtained by the Pioneer Venus orbiter leads to the conclusion that the thermal structure of the atmosphere persists over a time interval on the order of 5 years.

093.069 **Kinetic simulation of the Kelvin–Helmholtz instability at the Venus ionopause.**
V. A. Thomas, D. Winske.
Geophys. Res. Lett., Vol. 18, No. 11, p. 1943 – 1946 (Nov 1991).
Two dimensional hybrid simulations with particle ions and fluid electrons are used to calculate the kinetic evolution of the Kelvin–Helmholtz instability at a boundary similar to the ionopause at the planet Venus. The magnetic field has been chosen to be essentially transverse to the plasma flow velocity, which is the most unstable case for the Kelvin–Helmholtz instability. Effects analogous to those found in MHD simulations are recovered, including a mode cascade to longer wavelengths. In addition, isolated structures on the order of the ion gyroradius are formed which can cross the boundary in either direction. These structures may be related to flux ropes commonly observed near the ionopause of Venus. The time evolution of these structures may represent transport across the boundary layers and the smearing of the shear layer may give rise to internal

structure in the boundary. Similar processes occur in magnetized layers akin to the Earth's magnetopause.

093.070 **Corona structures on Venus: models of origin.**
E. R. Stofan, D. L. Bindschadler, J. W. Head, E. M. Parmentier.
J. Geophys. Res., Vol. 96, No. E4, p. 20933 - 20946 (25 Nov 1991).

Coronae on Venus are circular to elongate structures with maximum widths of 150–1000 km characterized by annuli of concentric ridges surrounding complex interiors. The features have raised topography relative to the surroundings, they are associated with volcanic activity, and most are partially surrounded by a peripheral trough. Variations in morphology between individual coronae are due to differences in their stage of evolution and/or differences in the relative significance of the geologic processes that occur in each stage. The authors examine models for three processes that may be involved in corona origin and evolution: (1) a hotspot or rising mantle diapir model, (2) a sinking mantle diapir model, and (3) gravitational relaxation of topography.

093.071 **A mantle plume model for the Equatorial Highlands of Venus.**
W. S. Kiefer, B. H. Hager.
J. Geophys. Res., Vol. 96, No. E4, p. 20947 - 20966 (25 Nov 1991).

The Equatorial Highlands of Venus consist of four main structures, Atla, Beta, Ovda, and Thetis regiones. Each has a circular to oval–shaped planform and rises 4–5 km above the mean planetary radius. These highlands are associated with long–wavelength geoid highs, with amplitudes ranging from 35 m at Ovda to 120 m at Atla. They also contain topographic valleys, interpreted as extensional rift zones, and Beta is known to contain shield volcanoes. These characteristics are all consistent with the Equatorial Highlands being formed by mantle plumes. Some smaller highlands, such as Bell Regio, Eistla Regio, and the Hathor/Innini/Ushas region, may also be plume related, but most coronae are unlikely to be the direct result of plume activity. The authors have modeled plumes using a cylindrical, axisymmetric finite element code and a depth–dependent, Newtonian rheology. They compare their model results with profiles of geoid and topography across Ata, Beta, Ovda, and Thetis; the best model fits are for Beta and Atla. Assuming whole mantle convection and that Earth and Venus have similar mantle heat flows, Venus must lack an Earth–like low–viscosity zone in its upper mantle in order to satisfy the observed geoid and topography for these features. This conclusion is consistent with the long–wavelength admittance spectrum of Venus and with the observed differences in the slopes of the geoid spectra for the two planets. One explanation for the different viscosity structures of the two planets could be that the mantle of Venus is drier than Earth's mantle.

093.072 **Mantle downwelling and crustal convergence: a model for Ishtar Terra, Venus.**
W. S. Kiefer, B. H. Hager.
J. Geophys. Res., Vol. 96, No. E4, p. 20967 - 20980 (25 Nov 1991).

The Ishtar Terra region contains the highest topography known on Venus, over 10 km above mean planetary radius, as well as abundant tectonic features, many of apparently compressional origin. These characteristics suggest that Ishtar is a crustal convergence zone overlying downwelling mantle. In order to explore quantitatively the implications of this hypothesis for Ishtar's origin, the authors present and discuss models of viscous crustal flow driven by gradients in topography. They use the results of their model to constrain the thickness of the crust in the Ishtar region. Although the authors treat Ishtar as a crustal convergence zone, their crustal flow model shows that under some circumstances near–surface material may actually flow away from Ishtar, providing a possible explanation for graben-like structures in Fortuna Tessera.

093.073 **Polarization of the impulsive signals observed in the nightside ionosphere of Venus.**
R. J. Strangeway.
J. Geophys. Res., Vol. 96, No. E5, p. 22741 - 22752 (25 Dec 1991).

The impulsive plasma wave bursts detected by the Pioneer Venus Orbiter electric field detector in the nightside ionosphere of Venus have been attributed to atmospheric lightning. However, it has also been argued that the wave bursts are generated locally by plasma instabilities. The waves associated with local instabilities are most probably electrostatic in nature, while lightning–generated waves should be whistler mode waves, at least at the lowest frequencies. It may be possible to identify the wave modes through analysis of the wave polarization. The authors show that for typical ionospheric parameters the whistler mode wave electric field should be polarized predominantly perpendicular to the ambient magnetic field. However, the signals are often impulsive in nature, with durations much less than the spacecraft spin period, and an individual wave event is likely to be aliased since the event may occur at arbitrary spin phase and arbitrary intensity. Statistical analysis of the wave polarization using data acquired in the third nightside periapsis season shows that the 100–Hz data suffer from interference, which may be due to the interaction of the spacecraft with the ambient plasma. The interference is removed through visual inspection of the data, and the authors show that the 100–Hz waves are polarized perpendicular to the ambient magnetic field provided they restrict the data to those intervals in which the magnetic field is sufficiently far from horizontal to allow vertical propagation within the whistler mode resonance cone. The 100–Hz waves detected outside of the resonance cone are polarized parallel to the magnetic field, as are the waves at higher frequency. The waves consequently fall into two classes: whistler mode waves which are most likely due to atmospheric lightning, since the low phase speed of the whistler mode argues against an in situ instability, and a mode that is polarized parallel to the ambient field. This latter mode may be analogous to the anomalous parallel polarized wave fields detected in the terrestrial ionosphere above thunderstorms.

093.074 **The exploration of Venus: a Magellan progress report.**
S. Saunders.
Mercury, Vol. 20, No. 5, p. 130 - 144 (Sep - Oct 1991).

093.075 **The 1988 – 89 western (morning) apparition of the planet Venus: visual and photographic observations.**
J. L. Benton Jr.
Strolling Astron., Vol. 35, No. 3, p. 116 - 123 (Sep 1991).

093.076 **The 1989 – 90 eastern (evening) apparition of the planet Venus: visual and photographic observations.**
J. L. Benton Jr.
Strolling Astron., Vol. 35, No. 4, p. 157 - 167 (Dec 1991).

093.077 **Solar cycle effects on the structure of the electron density profiles in the dayside ionosphere of Venus.**
A. J. Kliore, J. G. Luhmann.
J. Geophys. Res., Vol. 96, No. A12, p. 21281 - 21289 (1 Dec 1991).

Since 1975, United States and Soviet Venus–orbiting spacecraft have carried out more than 190 radio occulation measurements of the vertical electron density strucure of the dayside ionosphere of Venus: 148 of these have been produced by the Pioneer Venus orbiter between 1979 and 1989; Venera 9 and 10 provided 13 measurements in 1975; and Venera 15 and 16 made 31 measurements in 1983. These measurements were made at various times of the solar cycle, including solar maximum and solar minimum. It has therefore been possible to observe the changes in the electron density structure of the dayside ionosphere of Venus that were brought about by changing solar activity. Some of the resulting observations are presented and discussed in detail.

093.078 **Small-scale plasma irregularities in the nightside Venus ionosphere.**
J. M. Grebowsky, S. A. Curtis, L. H. Brace.
J. Geophys. Res., Vol. 96, No. A12, p. 21347 - 21359 (1 Dec 1991).
The individual volt-ampere curves from the Pioneer Venus orbiter electron temperature probe showed evidence for small-scale density irregularities, or short-period plasma waves, in regions of the nightside ionosphere where the orbiter electric field detector observed waves in its 100-Hz channel. A survey of the nightside volt-ampere curves has revealed several hundred examples of such irregularities (approximately 2% of all the current-voltage (I-V) sweeps from below 2000 km and 7% of those below 200 km). The I-V structures correspond to plasma density structures with spatial scale sizes in the range of about 100-2000 m, or alternatively they could be viewed as waves having frequencies extending toward 100 Hz. They are often seen as isolated events, with spatial extent along the orbit frequently less than 80 km. The density irregularities or waves occur in or near prominent gradients in the ambient plasma concentrations both at low altitudes where molecular ions are dominant and at higher altitudes in regions of reduced plasma density where O^+ is the major ion. Electric field 100-Hz bursts occur simultaneously, with the majority of the structured I-V curves providing demonstrative evidence that at least some of the E field signals are produced within the ionosphere. Potential sources for the irregularities are extraionospheric particle precipitation and spatial structures and turbulence resulting from the day to night ion transport within the ionosphere.

093.079 **Occurrence characteristics of VLF bursts in the nightside ionosphere of Venus.**
C.-M. Ho, R. J. Strangeway, C. T. Russell.
J. Geophys. Res., Vol. 96, No. A12, p. 21361 - 21369 (1 Dec 1991).
Many of the impulsive VLF signals observed by the Pioneer Venus orbiter electric field detector (OEFD) in the nightside ionosphere have been interpreted as arising from lightning in the Venus atmosphere. In order to determine the characteristics of the source, and to compare with terrestrial lightning, the authors determine the normalized occurrence rate for individual bursts at Venus, as opposed to previous studies of activity in 30-s intervals. They establish burst identification criteria which take into account the decay constant of the instrument to calculate the burst rate. Under the assumption that all OEFD observations in the nightside ionosphere are due to impulsive signals, the authors find that the 100-Hz channel has the highest burst rate, about 0.20/s in the altitude range 150-180 km and in the postmidnight local time sector. The burst occurrence rates for all four frequency channels decrease with increasing altitude. Burst rates at frequencies above 100 Hz are greater in the premidnight hours while 100-Hz signals peak after midnight; however, these burst rates vary from orbit to orbit. The 100-Hz low-frequency burst rate has a stronger dependence on the magnetic field strength than the higher-frequency rates. Most 100-Hz bursts are closely spaced and occur independently of signals at higher frequencies. These dependences suggest that the low- and high-frequency signals have different propagation mehcanisms. The properties of the bursts are generally consistent with a lightning source, and the planet-wide burst rate at Venus may be comparable to or larger than the terrestrial lightning rate.

The Venus table of the Dresden codex and the movements of the planet Venus.
See Abstr. 004.070.

Chapman Conference on Venus and Mars: Atmospheres, Ionospheres, and Solar Wind Interactions, Balatonfüred (Hungary), 4 - 8 Jun 1990.
See Abstr. 012.063.

Laboratory measurement of the millimeter wave properties of liquid sulfuric acid (H_2SO_4).
See Abstr. 022.111.

Measurements of pure CO_2 absorption at high densities near 2.3 μm.
See Abstr. 022.116.

The solution of radiative transfer problems in molecular bands without the LTE assumption by accelerated lambda iteration methods.
See Abstr. 063.083.

Role of carbon dioxide in cooling planetary thermospheres.
See Abstr. 091.017.

Physical properties of terrestrial planets and their satellites: an introduction.
See Abstr. 091.054.

Upstream waves at Mercury, Venus and Earth: comparison of the properties of one Hertz wave.
See Abstr. 092.001.

Photo-ionization balance and ion composition model on the ionospheres of Mars and Venus.
See Abstr. 097.002.

Magma generation on Mars: amounts, rates, and comparisons with Earth, Moon, and Venus.
See Abstr. 097.013.

On possible discrete dusty annular formations around Mars and Venus.
See Abstr. 097.039.

Comparison of observed plasma and magnetic field structure in the wakes of Mars and Venus.
See Abstr. 097.053.

The solar wind interaction with Mars: Mariner 4, Mars 2, Mars 3, Mars 5, and Phobos 2 observations of bow shock position and shape.
See Abstr. 097.056.

Plasma wave system measurements of the Martian bow shock from the Phobos 2 spacecraft.
See Abstr. 097.058.

Interplanetary magnetic field control of the Mars bow shock: evidence for Venuslike interaction.
See Abstr. 097.059.

094 Moon

094.001 **A strontium and neodymium isotopic study of Apollo 17 high–Ti mare basalts: resolution of ages, evolution of magmas, and origins of source heterogeneities.**
J. B. Paces, S. Nakai, C. R. Neal, L. A. Taylor, A. N. Halliday, D.-C. Lee.
Geochim. Cosmochim. Acta, Vol. 55, No. 7, p. 2025 – 2043 (Jul 1991).
A combined Sr and Nd isotopic study of 15 Apollo 17 high–Ti mare basalts was undertaken to investigate geochronological and compositional differences between previously identified magma types.

094.002 **The case for a younger Imbrium basin: new ^{40}Ar–^{39}Ar ages of Apollo 14 rocks.**
F. J. Stadermann, E. Heusser, E. K. Jessberger, S. Lingner, D. Stöffler.
Geochim. Cosmochim. Acta, Vol. 55, No. 8, p. 2339 – 2349 (Aug 1991).
Petrographically and chemically well–characterized rocks of the Apollo 14 landing site have been dated by the ^{40}Ar–^{39}Ar technique. The samples represent two different formations at the Apollo 14 site. The first group of samples has exposure ages close to 26 Ma and ^{40}Ar–^{39}Ar ages ranging from 3.86 to 4.09 Ga, which indicates that breccia 14063 was assembled 3.86 Ga ago. The second group of samples displays lower ^{40}Ar–^{39}Ar ages ranging from 3.73 to 3.85 Ga. The authors interpret breccia 14063 as part of a Nectarian megabreccia unit which was incorporated into the Fra Mauro Formation as a megablock by secondary mass wasting of the ejecta blanket of a local pre–Imbrian crater. The remaining samples are thought to represent clasts of the Apollo 14 subregolith basement megabreccia defined as the Fra Mauro Formation. Consequently, the reported ages of these clasts confirm previous suggestion that the Fra Mauro Formation and therefore the Imbrium basin is only 3.75 Ga old and not 3.85 Ga as commonly assumed.

094.003 **Mysteries of the Moon.**
D. Benningfield.
Astronomy, Vol. 19, No. 12, p. 50 – 55 (Dec 1991).

094.004 **Modeling the non–grey body thermal emission from the full moon.**
K. J. Vogler, P. E. Johnson, R. W. Shorthill.
Icarus, Vol. 92, No. 1, p. 80 – 93 (Jul 1991).
In an attempt to understand the non–grey–body brightness temperature spectra of the moon and Galilean satellites the authors have developed a series of thermophysical computer models of solid–surfaced planetary bodies. Surface roughness is modeled as paraboidal holes (craters) of a specified depth–to–diameter ratio. The model explains, to first order, the behavior of the published limbscans and disk–integrated infrared brightness temperature spectra for the full moon.

094.005 **A suspected dome on the southern floor of the lunar crater Schickard.**
K. W. Abineri.
J. Br. Astron. Assoc., Vol. 101, No. 4, p. 230 – 232 (Aug 1991).
Work continues on the floor and walls of Schickard, based on the microscopic study of Orbiter IV microfilm, using low power photomicrography to construct photographic outlines for the preparation of accurate drawings at the microscope. This project is intended to include the examination of all the various types of feature present on the floor and in the walls, with a view to searching for likely endogenic and exogenic modifications to the original structure of Schickard. Since the Orbiter IV microfilm frame is an exceptionally good example in terms of the detail shown, most of the areas photographed on to transparencies are worthy of close inspection. Clearly the limitations of the Orbiter IV image data must be considered.

094.006 **Models of structure and origin of the Moon.**
E. L. Ruskol.
Astron. Vestn., Tom 25, No. 4, p. 408 – 421 (Jul – Aug 1991). In Russian. English translation in Sol. Syst. Res., Vol. 25.
The fundamental data about the inner structure of the Moon obtained from space and Earth–based investigations are briefly summarized. The early thermal history and the orgin of the Moon remain still not clarified. The attempt of comparison of two extreme variants of the origin of the Moon from a circumterrestrial disk is made: for an instant formation of the disk due to the megaimpact of a large protoplanetary body into the mantle of growing Earth and for a continuously replenished disk from the feeding zone of the Earth (co–accretion). Both models of the origin of the Moon need a careful quantitative development with the account of early thermal history of the Earth, particularly for the case of megaimpact.

094.007 **Geological structure of the Moon.**
A. L. Sukhanov.
Astron. Vestn., Tom 25, No. 4, p. 422 – 430 (Jul – Aug 1991). In Russian. English translation in Sol. Syst. Res., Vol. 25.
A short summary is given of present–day knowledge of the main structures of the lunar surface: highlands, maria, basins, impact craters and volcanical structures, the data of their internal structure and of covering regolith. The morphological features are described which give the possibility to establish their genesis and time of formation and to construct the stratigraphic succession from the crust formation to the present time.

094.008 **Selenodesy and cartography of the Moon.**
V. S. Kislyuk.
Astron. Vestn., Tom 25, No. 4, p. 431 – 438 (Jul – Aug 1991). In Russian. English translation in Sol. Syst. Res., Vol. 25.
This paper reviews a series of recent studies on the gravity field as well as the geometrical and dynamical figures of the Moon. The selenodetic reference frames and catalogues are considered. The generalized selenodetic and selenodynamic parameters as well as a short information on mapping of the Moon are given.

094.009 **Evidence for metasomatism of the lunar highlands and the origin of whitlockite.**
C. R. Neal, L. A. Taylor.
Geochim. Cosmochim. Acta, Vol. 55, No. 10, p. 2965 – 2980 (Oct 1991).
The presence of rare earth element–rich (REE–rich) whitlockite in the relatively primitive, REE–poor lunar highlands presents a paragenetic problem to the petrologist. Comparison with occurrences in mare basalt mesostasis assemblages demonstrates that whitlockite in the lunar highlands has not resulted from the crystallization of the residual interstitial melt. The authors envisage that whitlockite found in the lunar highlands is a post–cumulus metasomatic phase precipitated from a melt formed when urKREEP underwent silicate liquid immiscibility. The basic immiscible melt thus produced was able to percolate through the lunar highlands due to its low viscosity and act as a metasomatizing agent.

094.010 **Investigation of the lunar surface by means of remote sensing methods.**
V. V. Shevchenko, Yu. G. Shkuratov, N. V. Opanasenko.
Astron. Vestn., Tom 25, No. 5, p. 569 – 577 (Sep – Oct 1991). In Russian. English translation in Solar Syst. Res., Vol. 25, No. 5.
The review includes the main results of remote sensing investigation of the lunar surface on the basis of terrestrial telescopes and spacecraft data. Remote sensing is the process of investigating space distribution of static fields, flux of particles, electromagnetic radiation. An analysis of these lunar surface characteristics is given. The magnetic field is studied by direct and

indirect methods. Chemical composition of the surface rocks is determined by registration of alpha, gamma and X-rays. Data about mineral composition are based on studying the visual, ultraviolet and infrared albedo using methods of colorimetry and spectroscopy. The structure of the surface layer is studied by the radar method and by means of photometric and polarimetric investigations. Data about the characteristic space distribution were used for thematic lunar mapping.

094.011 **Dependence between polarization inversion angle and albedo for the Moon and laboratory samples.**
N. V. Opanasenko, Yu. G. Shkuratov.
Astron. Tsirk., No. 1543, p. 25 – 26 (Mar – Apr 1990). In Russian.
Correlations between polarization inversion angles (α_{inv}) and albedo (A) are plotted. Approximately linear dependence was found for α_{inv} and P_{max}.

094.012 **Observations of potassium in the tenuous lunar atmosphere.**
R. W. H. Kozlowski, A. L. Sprague, D. M. Hunten.
Geophys. Res. Lett., Vol. 17, No. 12, p. 2253 – 2256 (Nov 1990).
Spectra showing the potassium D1 emission line above the Moon's bright limb (subsolar point) were obtained at first quarter the night of 29 April 1989. The number density at the surface is 9.5 ± 1 atoms cm^{-3}. Linear regression analyses find equally good fits to a one term model with a scale height of 80 km and corresponding temperature of 600 K, or a two term model with a thermal component accommodated to the subsolar surface temperature (T = 395K, scale height = 52 km; n_t = 9.6K atoms $cm^{-2}sec^{-1}$) and with a nonthermal component (represented as T = 2500K, scale height = 329 km; n_{nt} = 1.1 K atoms $cm^{-2}sec^{-1}$). In either case there is a large nonthermal component and a deficiency of atoms equilibrated to the surface temperature. The calculated thermalization rate of the nonthermal component through encounters with the lunar surface gives a source strength for the thermal component a factor of 7 greater than loss by photoionization. A possible explanation for the low thermalized population is the existence of a fast and efficient way of depleting the thermal component and supplying the nonthermal by momentary adsorption on the surface and subsequent photo–desorption by the considerable visible solar irradiance. Also important, this plausible mechanism for the redistribution of thermal and nonthermal components may completely mask the original velocity distribution of the source population.

094.013 **Analyse von Lasermessungen zum Mond im Rahmen einer post–Newton'schen Theorie.**
J. Müller.
Dtsch. Geod. Komm. Bayer. Akad. Wiss., Reihe C, Nr. 383.
Diss. (Dr.–Ing.), München Univ. (Germany, F.R.). Fakultät für Bauingenieur– und Vermessungswesen. Verlag der Bayerischen Akademie der Wissenschaften, München (Germany, F.R.). 132 p. (1991). ISBN 3-7696-9429-5.
The analysis of Lunar Laser Ranging (LLR) data provides an excellent tool for the determination of parameters concerning the Earth–Moon–System and to test Einstein's theory of gravity in the first post–NEWTONian (PN) approximation. All equations of the LLR model will be formulated at the first PN level and their influences in the analysis of the LLR data will be estimated. Based upon these results, the existing programs will be changed and new determinations of the parameters will be performed. Finally the equations for the analysis of Very Long Baseline Interferometry (VLBI) observations will be derived consistently to the LLR equations to be used for a later combination of both observation data.

094.014 **$^{40}Ar/^{39}Ar$ ages of six Apollo 15 impact melt rocks by laser step heating.**
G. B. Dalrymple, G. Ryder.
Geophys. Res. Lett., Vol. 18, No. 6, p. 1163 – 1166 (Jun 1991).
The authors have obtained 15 high resolution (21–51 step) $^{40}Ar/^{39}Ar$ age spectra on six Apollo 15 impact melt rocks of different compositions using a continuous laser system on submilligram subsamples and on single crystal plagioclase clasts. Four of the six samples gave reproducible age with well–defined intermediate temperature plateaus over 48% or more of the ^{39}Ar released; the plateaus are interpreted as crystallization ages. Samples 15304,7,69, 15294,6,21, and 15314,26,156 gave virtually identical plateau ages whose weighted mean is $3{,}870 \pm 6$ Ma. These three melt rocks differ in composition and likely formed in three separate impact events. Sample 15356,9 gave replicate plateau ages that average $3{,}836 \pm 12$ Ma and date a fourth and younger impact event. The age spectra for samples 15308,9 and 15414,3,36 increase with increasing increment temperature and may have been formed in or affected by impacts at about 2,700 Ma and 3,870 Ma, respectively. So far there continues to be no convincing evidence in the lunar record for impact melts older than about 3.9 Ga.

094.015 **Petrology of lunar rocks.**
M. I. Korina.
Astron. Vestn., Tom 25, No. 6, p. 686 – 702 (Nov – Dec 1991). In Russian. English translation in Sol. Syst. Res., Vol. 25, No. 6.
Multiple available data on lunar rocks are compiled. The highlands and maria rocks are reviewed in a short form in terms of mineralogy, chemistry and petrology. A reasonable classification of these rocks was constructed. Principal features of rocks geochemistry are discussed. Petrological hypotheses of the lunar rocks origin are summarised and their mutual contradictions are outlined. The important role of accretional impact processes for the highland material formation is pointed out.

094.016 **Astrometric methods of lunar observations.**
N. G. Rizvanov.
Izv. Astron. Obs. Ehngel'gardt., Vyp. 54, p. 85 – 97 (1989). In Russian.
A brief historical survey of astrometric methods of lunar observations is given.

094.017 **The analysis of selenodetic systems.**
L. I. Rakhimov, N. G. Rizvanov.
Izv. Astron. Obs. Ehngel'gardt., Vyp. 54, p. 97 – 102 (1989). In Russian.
The authors discuss a method for analyzing the accuracy of selenodetic reference systems by comparing catalogue and measured crater positions on large–scale lunar photographs with star positions.

094.018 **Maps of the marginal zone of the Moon, constructed with taking into consideration the macrorelief of the Moon.**
A. A. Nefed'ev, Yu. A. Nefed'ev, V. S. Borovskikh.
Izv. Astron. Obs. Ehngel'gardt., Vyp. 55, p. 69 – 109 (1990). In Russian.
For the construction of maps of the marginal zone of the Moon the authors used classical heliometric observations of the physical libration of the Moon. The observational material consists of nine heliometric series and includes 10484 measured heights. The mutual position of the center of the lunar limb and the mass center of the Moon was determined by Yakovkin's second model of the figure of the Moon. Readings of heights are taken from a common zero level.

094.019 **Processing series of classical observations on a heliometer using a hindrancestable statistical analysis.**
Yu. A. Nefed'ev, V. M. Bezmenov.
Izv. Astron. Obs. Ehngel'gardt., Vyp. 55, p. 110 – 117 (1990). In Russian.
The possibility of applying a special statistical analysis to processing of long–time selenodetical observation series is practically realized. The authors present results of a reduction of these series with different variations of statistical parameters.

094.020 **Update on transient lunar phenomena.**
P. Moore.
South. Stars, Vol. 34, No. 4, p. 285 – 287 (Sep 1991).
A new account of Transient Lunar Phenomena is presented, together with an appeal for more observers and data.

094.021 **Remote sensing of potential lunar resources. 1. Near–side compositional properties.**
J. R. Johnson, S. M. Larson, R. B. Singer.
J. Geophys. Res., Vol. 96, No. E3, p. 18861 – 18882 (25 Oct 1991).
New telescopic CCD multispectral imaging of the lunar near side and 330–870 nm spectroscopy of selected regions are presented as aids in estimating compositional differences relevant to locating potential lunar resources such as ilmenite ($FeTiO_3$) and solar wind–implanted ^3He and H. Conversion of 400/560 nm CCD ratio images to weight percent TiO_2 using a version of the Charette et al. (1974) empirical relation allows construction of a new TiO_2 abundance map for the lunar maria accurate to ± 2 wt % TiO_2. This map provides a consistent view of TiO_2 distribution among mature mare soils and can be used to estimate areas potentially rich in ilmenite.

094.022 **Revived interest in the lunar atmosphere.**
T. H. Morgan, S. A. Stern.
Earth Space, Vol. 4, No. 2, p. 10 – 11, 14 (Oct 1991).

094.023 **Present main sources of error in the reduction of LLR data.**
Nie Zhaoming, Li Jinling, Jin Wenjing.
Acta Astron. Sin., Vol. 32, No. 4, p. 402 – 408 (Dec 1991). In Chinese.
The post–residuals, obtained from the adjusted LLR data, are usually much larger than the errors of measurements. For this sake, the influence of both the uncertainty of the orientation of the reference frame of the ephemeris and the inaccuracy of the moon's Love numbers on the calculated distance, from an observing station to the reflectors, is analysed in this article. By means of the uncertainties of the obliquity of the ecliptic and of the equinox contained by DE200/LE200 and the relative errors of the moon's Love numbers, the values of the distance errors are estimated. Finally, the results of comparison of these values with the accuracy of the reduction of the LLR data obtained from 1970 to 1987 show that the uncertainties of the orientation of the reference frame of the ephemeris and the inaccuracy of the moon's Love numbers are the main sources of error in the reduction of the LLR data at present.

094.024 **Lunar ferroan anorthosites and mare basalt sources: the mixed connection.**
G. Ryder.
Geophys. Res. Lett., Vol. 18, No. 11, p. 2065 – 2068 (Nov 1991).
Global overturn of a hot, gravitationally unstable lunar mantle immediately following the solidification of a magma ocean (and essentially complete by 4.4 Ga) explains several characteristics of lunar petrology. Lunar mare basalt sources are inferred to be depleted in europium and alumina. These depletions are consensually attributed to complementary plagioclase floating from a magma ocean. However, the connection cannot be so simple and direct: in contrast to the mare basalt source parent magma, the ferroan anorthosite parent magma was more evolved by virtue of its lower Mg/Fe ratio and Ni abundances, although less evolved in its poverty of clinopyroxene constituents, flat rare earth pattern, and lower incompatible element abundances. The europium anomaly in mare sources is inferred to be present at 400 km depth, too deep to have been directly influenced by plagioclase crystallization. Massive overturning of the post–magma ocean mantle would have carried down clinopyroxene, ilmenite, and phases containing fractionated rare earths, europium anomalies, and some heat–producing radionuclides. These phases contributed to deep mare basalt sources. Upward–moving phases would have been magnesian mafic minerals; their immediate melting on pressure release contributed the magnesian suite of plutonic norites and troctolites that post–date the anorthosites in the highlands crust.

094.025 **Diapirism and the origin of high TiO_2 mare glasses.**
P. C. Hess.
Geophys. Res. Lett., Vol. 18, No. 11, p. 2069 – 2072 (Nov 1991).
High TiO_2 mare picritic glasses are derived from cumulate source regions that are only modestly endowed with ilmenite–enriched crystallization products. These sources are mobilized by the heat derived from the primitive interior and evolve into diapirs which rise adiabatically from depths in excess of 700 km. As these diapirs undergo pressure–release melting, they also stir in significant portions of the surrounding mantle.

094.026 **Thermodynamic models of the chemistry of lunar volcanic gases.**
B. Fegley Jr.
Geophys. Res. Lett., Vol. 18, No. 11, p. 2073 – 2076 (Nov 1991).
Thermodynamic models and mass balance arguments are used to constrain the chemistry of lunar volcanic gases. The results predict that lunar gases were dominated by reduced C and S gases such as CO, COS, CS_2, S_2. The more oxidized gases CO_2 and SO_2 were also important, but only in limited temperature ranges. Gases such as Cl_2, CCl_4, and CF_4 were more abundant than HF and HCl, which were the two major H compounds in the lunar gases. Chlorides and fluorides were important species for transporting many volatile and ore–forming metals, and the implications for fractionating and concentrating metals into lunar "ore–deposits" merit further study.

094.027 **Partitioning behavior of Ni, Co, Mo, and W between basaltic liquid and Ni–rich metal: implications for the origin of the moon and lunar core formation.**
V. J. Hillgren.
Geophys. Res. Lett., Vol. 18, No. 11, p. 2077 – 2080 (Nov 1991).
A recent model of the origin of the Moon in a giant impact put forth by O'Neill (1991) states that the bulk abundances of siderophile elements in the Moon would be provided by the impactor plus a late reduced veneer. A small Ni–rich core would then separate setting the siderophile element abundance pattern in the lunar mantle. The author tests this hypothesis using simple mass balance and the results of an experimental study that he has conducted on the partitioning of moderately siderophile elements between basalt and Ni–rich metal. The author finds this hypothesis to be viable under a limited set of conditions.

094.028 **Additional complexity in the lunar crust: petrology of sodic anorthosites and sulfur–rich, ferroan noritic anorthosites.**
M. D. Norman, G. J. Taylor, K. Keil.
Geophys. Res. Lett., Vol. 18, No. 11, p. 2081 – 2084 (Nov 1991).
Lunar breccia 67016 contains two suites of unusual highlands rocks: sulfur–rich, noritic anorthosites, and sodic anorthosites. Mineral compositions of the sulfur–rich noritic anorthosites closely match those of ferroan anorthosites. Unusually large amounts of sulfides, and sulfidation reactions involving olivine in these clasts probably reflect endogeneous volatile transfer. The sodic anorthosites are mafic–poor, with mineral compositions intermediate between those of pristine ferroan and alkali anorthosites. Rare examples of pristine rocks with similar mineral compositions are known, and may represent a distinct class of lunar crustal rocks.

094.029 **Mineral compositions in pristine lunar highland rocks and the diversity of highland magmatism.**
M. G. Bersch, G. J. Taylor, K. Keil, M. D. Norman.
Geophys. Res. Lett., Vol. 18, No. 11, p. 2085 – 2088 (Nov 1991).
High precision electron microprobe analyses of olivine and pyroxene in pristine lunar highland rocks confirm the dichotomy between ferroan anorthosites and the Mg–suite. Ferroan–anorthosites plot as coherent trends, consistent with formation in a complex global magma system. Lack of coherent compositional trends in the Mg–suite rocks indicates derivation from numerous magmas.

094.030 **Observations of the lunar sodium exosphere.**
A. E. Potter Jr., T. H. Morgan.
Geophys. Res. Lett., Vol. 18, No. 11, p. 2089 – 2092 (Nov 1991).
Measurements of sodium resonance emission showed that the sodium exosphere of the Moon extends up to 1500 km altitude above the equatorial limb. The line-of-sight sodium atom densities at various altitudes were derived from the measured sodium emission intensities, using a calibration based on brightness of the lunar limb. The data were fit to a simple two-parameter model of the exosphere. Above the equatorial limb, this yielded apparent temperatures of 950–1150K, and surface densities of sodium atoms of 29–46 atoms/cm^3. Above the north pole, adjacent to the polar terminator, the temperature was much less, 450K, and the surface density rose to 65 atoms/cm^3. Sodium emission could not be detected above the equatorial terminator, indicating a decrease to levels below about 200 Rayleighs, difficult to distinguish from the sodium nightglow. Solar wind impact on the surface does not seem to be an important factor, because the sodium density changed very little when the Moon was shielded from the direct solar wind by the Earth's magnetotail. This suggests that solar photon radiation plays a major role in maintenance of high kinetic temperatures in the sodium exosphere.

094.031 **The atomic sodium exosphere/coma of the Moon.**
W.-H. Ip.
Geophys. Res. Lett., Vol. 18, No. 11, p. 2093 – 2096 (Nov 1991).
A Monte Carlo method is used to study the random walk process of sodium atoms created by meteoroid impacts on the lunar surface. The collisional surface interaction is parameterized by a thermal accomodation coefficient of 0.5. Sodium atoms launched with initial velocities well below the surface escape velocity of 2.38 km s^{-1} will be recycled and contribute to the formation of an exosphere with a temperature of about 500K on the sunlit side. For large launch velocities (>2 km s^{-1}) there will be systematic transport of the escaping sodium atoms to the antisolar direction as a result of the acceleration effect of the solar radiation pressure. An extended coma with a significant elongation in the antisolar direction can form this way. Recent spectroscopic and imaging observations are consistent with this picture. The present model calculations suggest that meteoroid impact effect (and solar wind sputtering) should be the major supplier of the lunar sodium exosphere; the possible existence of an efficient loss mechanism of the sodium atoms during surface interaction is also indicated.

094.032 **Imaging observations of the extended sodium atmosphere of the moon.**
M. Mendillo, J. Baumgardner, B. Flynn.
Geophys. Res. Lett., Vol. 18, No. 11, p. 2097 – 2100 (Nov 1991).
Pilot ground-based optical imaging observations of the extended lunar environment were conducted at McDonald Observatory in February 1991. One set of images taken on February 20 revealed emission at the combined D1 (5896 Å) and D2 (5890 Å) lines of neutral sodium extending out to about 5 lunar radii (R_M) from the center of the Moon on the sunward side and much fainter emission extending out to about 15–20 R_M on the antisunward side. Peak emission (D1 + D2) above the Moon's limb at a distance of 1.8 R_M along the luni-solar axis was measured to be 900 Rayleighs (R), decreasing with distance as r^{-4}. The morphology of the emitting region is qualitatively that of a comet, i.e., a bright coma centered on the Moon and an extended tail trailing away in the antisunward direction. The tail exhibits a decrease in intensity coincident with the shadow cast by the moon. The sunward extent of the sodium is used to calculate a surface ejection speed of about 2 km s^{-1}, consistent with recent estimates of a meteor impact surface vaporization source. The data agree well with previous observations made within 1 lunar radius above the surface. Modeling yields a surface density (52 atoms cm^{-3}) decreases with distance as r^{-5} in the solar direction.

094.033 **A possible meteor shower on the Moon.**
D. M. Hunten, R. W. H. Kozlowski, A. L. Sprague.
Geophys. Res. Lett., Vol. 18, No. 11, p. 2101 – 2104 (Nov 1991).
Over the 3-day period from 12 to 14 October, 1990, the sodium abundance in the lunar atmosphere at 80° South increased by 60%, while interspersed measurements at the equator showed no substantial change. The source is suggested to be an unknown meteor shower with a radiant near the south ecliptic pole. A low relative velocity of $\sim$20 km/sec, combined with small particle masses, would keep the shower below the detectability threshold of radar. The stream could evolve from a reasonable asteroidal or cometary orbit with perihelion somewhat greater than 1 astronomical unit (AU) and a major axis of a few AU. The short residence time of lunar sodium makes it more favorable than the terrestrial sodium layer for detection of such an event.

094.034 **The Apollo 17 ultraviolet spectrometer: lunar atmosphere measurements revisited.**
P. D. Feldman, D. Morrison.
Geophys. Res. Lett., Vol. 18, No. 11, p. 2105 – 2108 (Nov 1991).
Ultraviolet resonance fluorescence of solar radiation provides the most sensitive means of detecting the expected major constituents of the tenuous daytime lunar atmosphere. Such an experiment was carried out with the Apollo 17 Ultraviolet Spectrometer in December 1972 (Fastie et al, 1973) and produced only upper limits of the number density of H, H_2, O, C, N, CO and two noble gases near the surface of the moon. The complete data set of 47 terminator crossing observations, which were not utilized in the earlier analysis, has been re-examined and more stringent upper limits to the column emission rates for several species have been derived. These results, together with most recent values for the atomic and molecular fluorescence efficiencies, lead to more definitive limits on the content of the lunar atmosphere. The revised upper limit on atomic oxygen density allows for the possibility of comparable O/Na ratios in the atmosphere of both the Moon and Mercury.

094.035 **The Apollo lunar surface water vapor event revisited.**
J. W. Freeman Jr., H. K. Hills.
Geophys. Res. Lett., Vol. 18, No. 11, p. 2109 – 2112 (Nov 1991).
On March 7, 1971, the first sunrise following the Apollo 14 mission, the Suprathermal Ion Detector Experiment (SIDE) deployed at the Apollo 14 site reported an intense flux of ions whose mass per charge was consistent with water vapor. The authors examine the amount of water and discuss the various acceleration processes responsible for accelerating ions into SIDE. They conclude that during most of the event the observed water vapor ions were accelerated by the negative lunar surface electric potential and secondly that this event was probably the result of mission associated water vapor, either from the LM ascent and descent stage rockets or from residual water in the descent stage tanks.

094.036 **Exospheric transport restrictions on water ice in lunar polar traps.**
R. R. Hodges Jr.
Geophys. Res. Lett., Vol. 18, No. 11, p. 2113 – 2116 (Nov 1991).
There is little doubt that at least 10^{17}g of water has accreted on the moon as a result of the reduction of ferric iron at the regolith surface by solar wind protons, the vaporization of chondrites, and perhaps comet impacts. Lacking an efficient escape mechanism, most of this water (or its progeny) is probably on the moon now. If the water were to have migrated to permanently shaded cold traps near the lunar poles, ice deposits with densities greater than 1000 g cm^{-2} would cover the traps, providing accessible resources. However, exospheric transport considerations suggest that the actual amount of water ice in the cold traps is probably too small to be of practical interest. The alternative is global assimilation of most of the water into the regolith, a process that must account for about 30 micromoles of water per gram of soil.

094.037 **Large scale lunar horizon glow and a high altitude lunar dust exosphere.**
H. A. Zook, J. E. McCoy.
Geophys. Res. Lett., Vol. 18, No. 11, p. 2117 – 2120 (Nov 1991).

All three Apollo 17 astronauts sketched a "lunar horizon glow", seen from orbit above the Moon. It is shown that the shape of the glow is compatible with scattering of sunlight off a gas or dust at high altitudes above the Moon. The authors' mathematical modeling best simulates the glow with submicron dust grains whose spatial density varies with altitude above the Moon as $\exp(-H/H_s)$, where H_s is in the range of 5 to 20 km. These dust grains are probably electrically charged and ejected above the lunar surface by local electric fields.

094.038 **Absence of large shield volcanoes and calderas on the Moon: consequence of magma transport phenomena?**
J. W. Head, L. Wilson.
Geophys. Res. Lett., Vol. 18, No. 11, p. 2121 – 2124 (Nov 1991).
Large shield volcanoes and calderas are not observed on the Moon. Theoretical assessments show the melt ascending from the mantle will stall at rheological traps or at the shallower density trap at the base of the low–density lunar crust. Magma will then only reach the surface along dikes propagated from these deep reservoirs, and such events will be infrequent and characterized by very large–volume eruptions. These conditions favor the formation of sinuous rilles, single long flows, and widespread flow units, rather than abundant small eruptions from a central edifice. The absence of large shield volcanoes and calderas on the Moon is interpreted to be related to the fact that shallow neutral buoyancy zones and magma reservoirs are not favored on the Moon.

094.039 **Evolution of lunar topography by impact processes.**
S. Sugita, T. Matsui.
Geophys. Res. Lett., Vol. 18, No. 11, p. 2125 – 2128 (Nov 1991).
Using a simple model of lunar topographic evolution resulting from the formation of impact craters, the authors try to constrain the total number of impactors striking the Moon after the solidification of the lunar crust. The numerical results indicate that the evolution of the Moon's hyposometric curve is strongly influenced by the total number of craters and is less affected by their size distribution. It is suggested that the total number of craters necessary to explain the standard deviation of the present hypsometeric curve is roughly equivalent to the number density of craters on the oldest lunar crust, corresponding to about 10^{-4} lunar mass of impactors. This may be the total mass of impactors striking after solidification of the lunar crust, more specifically, after about 4.4 to 4.5 Ga ago when the viscous degradation of craters became ineffective.

094.040 **Bullialdus: strengthening the case for lunar plutons.**
C. M. Pieters.
Geophys. Res. Lett., Vol. 18, No. 11, p. 2129 – 2132 (Nov 1991).
Although many craters expose material of a composition different from that of the local surroundings, Bullialdus has excavated material representing three distinct stratigraphic zones that occur in the upper 6 km of crust, the top two of which are gabbroic and the deepest of which is noritic. This three–component stratigraphy at Bullialdus provides strong evidence that the lunar crust includes pockets of compositionally layered material reminiscent of mafic layered plutons. When combined with previous information on the compositional diversity at other large craters, these remote analyses obtained in a geologic context substantially strengthen the hypothesis suggested from lunar samples that plutons play an integral role in lunar crustal evolution.

094.041 **The distribution of olivine in the crater Copernicus.**
P. G. Lucey, B. R. Hawke, K. Horton.
Geophys. Res. Lett., Vol. 18, No. 11, p. 2133 – 2136 (Nov 1991).
The authors employed multispectral imaging in the visble and near–IR at four wavelengths (0.73, 0.96, 1.45, and 1.99 μm) of Copernicus crater in order to map the distribution of olivine–rich, pyroxene–poor material known previously to occur in the central peak complex. The authors identified three additional portions of the crater to exhibit spectral characteristics similar to those of the central peaks strongly suggesting the presence of material similar to that exposed in the central peaks. These areas are a scarp forming a portion of the northern rim of Copernicus, and two slump blocks in the north wall which may have been derived from the same portion of the ejecta now exposed in the rim scarp. These occurrences decrease the minimum allowable depth for this unusual material in the Copernicus target site though still represent some of the deepest material exposed by Copernicus.

094.042 **Alphonsus crater: floor fracture and dark–mantle deposit distribution from new 3.0–cm radar images.**
S. H. Zisk, B. C. Campbell, G. H. Pettengill, R. Brockelman.
Geophys. Res. Lett., Vol. 18, No. 11, p. 2137 – 2140 (Nov 1991).
The lunar crater Alphonsus is characterized by numerous fractures or graben, and by endogenic dark–halo craters. Existing maps of fractures from analysis of lunar photography may be biased by the east–west solar illumination. This paper presents new high–resolution, dual–polarization 3.0–cm wavelength radar images of Alphonsus with radar illumination from northernly directions, and uses these data to better map the locations of both the graben and a variety of dark–mantle deposits. The distribution of fractures, and several graben which cut the crater floor and central ridge, are cited as possible evidence for simultaneous, post–Imbrium uplift of both structures. Some of the endogenic dark halo deposits are more extensive in depolarized radar images than in photographs; these extensions are attributed in some cases to more distant emplacement of pyroclastic material, and in others to fortuitous connections with smoother, less cratered portions of the Alphonsus floor.

094.043 **Remote sensing studies of the Orientale region of the moon: a pre–Galileo view.**
B. R. Hawke, P. G. Lucey, G. J. Taylor, J. F. Bell,
C. A. Peterson, D. T. Blewett, K. Horton, G. A. Smith,
P. D. Spudis.
Geophys. Res. Lett., Vol. 18, No. 11, p. 2141 – 2144 (Nov 1991).
The authors have acquired both an extensive data base of visible and near–infrared spectra and multispectral images (ultraviolet and visible) of the Orientale region of the Moon. The results show that the eastern Inner Rook Mountains are composed of anorthosite. Portions of the main ring of the Humorum basin and the inner ring of Grimaldi are also composed of anorthosite. Other deposits within the Orientale basin and the major ejecta unit outside the basin are dominated by noritic anorthosites; mature surfaces have spectra nearly identical to those taken of areas in the vicinity of the Apollo 16 site. Thus, it appears that Orientale ejecta are more mafic than materials that are thought to have originated at depth (i.e., the anorthosites in the Inner Rook Mountains). There are large areas that have mare basalt signatures mixed with highland rock types, indicating the existence of ancient, pre–Oriental mare volcanism. Present mare surfaces in the Orientale region contain basalts of intermediate to very low TiO_2 content.

094.044 **Infrared reflectance spectra (4–12 μm) of typical lunar samples.**
D. B. Nash.
Geophys. Res. Lett., Vol. 18, No. 11, p. 2145 – 2148 (Nov 1991).
A laboratory study of the mid–infrared spectral reflectance properties of typical Apollo lunar rock and soil samples was conducted using a modern interferometric spectrometer. The results show that characteristic spectral features exist in the 4–12 micrometer range that are diagnostic of rock composition and mineralogy, especially for the plagioclase–rich rocks and soils. This is the first phase of a comprehensive infrared reflectance study underway to better understand the use of infrared emission spectroscopy for mapping the surface composition of the Moon and other silicate bodies.

094.045 **Evaluation of the sensitivity of reflectance ratios to mafic minerals in the lunar regolith.**
S. M. Larson, J. R. Johnson, R. B. Singer.
Geophys. Res. Lett., Vol. 18, No. 11, p. 2149 – 2152 (Nov 1991).

The 950/560 nm spectral ratio has been traditionally used as a qualitative indicator or lunar soil maturity because the 950 nm band falls within the major Fe^{2+} mineral absorption bands near 1 μm characteristic of relatively immature, crystalline surfaces. Since there are several other processes unrelated to maturation that may affect this ratio value, the authors have evaluated the 950/730 nm ratio as a potentially better measure of the relative 1–μm band strength. They find that the 950/730 nm ratio appears to be much more sensitive to the 1 μm band depth, but feel that other unknown variables preclude using these single ratios to provide meaningful quantitative results at this time. Because of problems with near–IR scattering in vidicon detectors used in previous studies, the authors have measured the scattering properties of their two CCDs and conclude that any scattering is negligible.

094.046 **A reevaluation of spectral ratios for lunar mare TiO_2 mapping.**
J. R. Johnson, S. M. Larson, R. B. Singer.
Geophys. Res. Lett., Vol. 18, No. 11, p. 2153 – 2156 (Nov 1991).
The empirical relation established by Charette et al. (1974) between the 400/560 nm spectral ratio of mature mare soils and weight percent TiO_2 has been used extensively to map titanium content in the lunar maria (e.g., Johnson et al., 1977; Johnson et al., 1991). Relative reflectance spectra of mare regions show that a reference wavelength further into the near–IR, e.g., > 700 nm, could be used in place of the 560 nm band to provide greater contrast (a greater range of ratio values) and hence a more sensitive indicator of titanium content. An analysis of 400/730 nm ratio values derived from both laboratory and telescopic relative reflectance spectra suggests that this ratio provides greater sensitivity to TiO_2 content than the 400/560 nm ratio. The increased range of ratio value is manifested in higher contrast 400/730 nm ratio images compared to 400/560 nm ratio images. This potential improvement in sensitivity encourages a reevaluation of the original Charette et al. (1974) relation using the 400/730 nm ratio.

094.047 **Lunar neutron leakage fluxes as a function of composition and hydrogen content.**
W. C. Feldman, R. C. Reedy, D. S. McKay.
Geophys. Res. Lett., Vol. 18, No. 11, p. 2157 – 2160 (Nov 1991).
Areas of the Moon that might have enhanced concentrations of hydrogen are discussed. The fluxes of thermal, epithermal, and fast neutrons escaping from the lunar surface were calculated for eight different compositions and several hydrogen contents. These flux ratios depend on compositon and are very sensitive to the hydrogen contents. Their measurement using flight–proven techniques from lunar orbit should allow detection and identification of hydrogen residing in either permanently shaded cold traps near the lunar poles or mature regoliths. Knowledge of composition, especially of Fe and Ti, improves the thresholds for hydrogen detection and also could allow identification of Gd and Sm.

094.048 **Expected extreme ultraviolet spectrum of the lunar surface.**
B. C. Edwards, W. C. Priedhorsky, B. W. Smith.
Geophys. Res. Lett., Vol. 18, No. 11, p. 2161 – 2164 (Nov 1991).
The moon was recently observed to be a source of very soft X–ray emission. The emission was most intense at wavelengths longer than 62 Å and was attributed to Thomson scattering of solar X–rays. This observation prompted the authors to study the emissions expected from the lunar surface in the wavelength range between 90 and 500 Å. Photons in this wavelength range scatter inefficiently. Instead, the solar X–rays are absorbed in the first several microns of lunar regolith. The absorbed X–rays can excite the surface elements and result in fluorescent emission. The authors find that much of the L– and M–shell extreme ultraviolet fluorescence, in the wavelength range betwen 90 and 500 Å, have higher peak intensities than the scattered solar spectrum. The total integrated fluorescent emission is also higher than the total scattered solar radiation. The L–shell fluorescent radiation can be an indicator of the surface abundances of Si, Al, Mg and other major lunar elements.

094.049 **Lunar surface composition and solar wind–induced secondary ion mass spectrometry.**
R. C. Elphic, H. O. Funsten III, B. L. Barraclough, D. J. McComas, M. T. Paffett, D. T. Vaniman, G. Heiken.
Geophys. Res. Lett., Vol. 18, No. 11, p. 2165 – 2168 (Nov 1991).
The Moon has no strong global magnetic field and only a tenuous atmosphere, so solar wind ions ($\sim$95% H^+, 5% He^{++}) directly bombard the lunar surface, sputtering atoms and secondary ions from the exposed grains of the regolith. The secondary ions potentially provide surface composition information through secondary ion mass spectrometry (SIMS), a standard laboratory surface composition analysis technique. In this paper the authors report the results of laboratory SIMS experiments on lunar soil simulants using solar wind–like ions. They find that H^+ and He^{++}, while not efficient sputterers, nevertheless produce significant fluxes of secondary lunar ions, including Na^+, Mg^+, Al^+, Si^+, K^+, Ca^+, Ti^+, Mn^+ and Fe^+. The authors predict that lunar surface secondary–ion fluxes range between $\sim$10 and 10^4ions $cm^{-2}s^{-1}$, depending on the species.

094.050 **Lunar surface: sputtering and secondary ion mass spectrometry.**
R. E. Johnson, R. Baragiola.
Geophys. Res. Lett., Vol. 18, No. 11, p. 2169 – 2172 (Nov 1991).
The authors combine laboratory and Apollo observations to describe the sputtering of the lunar surface and the composition of the ejecta with special reference to O. The atmospheric inventory appears to be dominated by micrometeorite vaporization of lunar grains. Sputtering effects are observable in the local plasma due to ion ejection, in the extended atmospheric through energetic neutral ejection, and grain surfaces through the chemical fractionation of the redeposited sputter–ejecta. Ionization of the micrometeorite–vapor also contributes to the local plasma.

094.051 **Spectral–chemical analysis of lunar surface materials.**
R. Jaumann.
J. Geophys. Res., Vol. 96, No. E5, p. 22793 – 22807 (25 Dec 1991).
Samples of lunar rocks and soils that have been examined in the laboratory both spectroscopically and chemically were analyzed with the goal of identifying and quantifying correlations between reflectance spectra and compositional properties of lunar materials. The method used to quantify spectral variations within the set of spectral samples is principal component analysis. The correlations between the chemical analysis of lunar samples and their spectral quantities are determined by using multivariate analysis. The combination of both methods yields functional relationships between the principal axes of variation and the chemical constituents, in particular, FeO, TiO_2, SiO_2, Al_2O_3, MgO, CaO, and MnO, and with the content of amorphous materials (glass) in lunar samples. To first order, the nature of these relationships is found to be linear. The results define a spectral–chemical model that is applicable to derive abundances of chemical constituents and amorphous components from individual spectra. The accuracy of the model is better than $\pm$2.4 wt % and $\pm$4% for chemical constituents and amorphous components, respectively, making the model also sensitive to remotely sensed spectra of the lunar surface.

094.052 **Significant high number commensurabilities in the main lunar problem. I. The Saros as a near–periodicity of the Moon's orbit.**
E. Perozzi, A. E. Roy, B. A. Steves, G. B. Valsecchi.
Celest. Mech. Dyn. Astron., Vol. 52, No. 3, p. 241 – 261 (1991).
A description is given of certain historically known cycles associated with high–number near commensurabilities among the synodic, anomalistic and nodical lunar months and the

anomalistic year. Using eclipse records, the JPL ephemeris and results from three-body numerical integrations, any dynamical configuration of the Earth-Moon-Sun system (within the framework of the main lunar problem) is shown to repeat itself closely after a period of time equal in length to the classical Saros cycle of 18 years and 10 or 11 days. The role played by mirror configurations in reversing solar perturbations on the lunar orbit is examined and it is shown that the Earth-Moon-Sun system moves in a nearly periodic orbit of period equivalent to the Saros. The Saros cycle is therefore the natural averaging period of time by which solar perturbations can be most effectively removed in any search into the long term evolution of the lunar orbit.

094.053 **Geochemical comparison of impact glasses from lunar meteorites ALHA81005 and MAC88105 and Apollo 16 regolith 64001.**
J. W. Delano.
Geochim. Cosmochim. Acta, Vol. 55, No. 11, p. 3019 – 3029 (Nov 1991).

Most glasses that occur in lunar highland regolith are quenched droplets of impact melt. The chemical compositions of these glasses are equivalent, in the absence of volatile losses, to the original target materials. The compositional range of impact glasses in a regolith reflects the chemical diversity that existed throughout the region up to the time of system closure (e.g., breccia formation). Since these glasses are product of widespread and random sampling, both in terms of space and time, they can be used for geochemical exploration of the Moon.

094.054 **Impact melts in the MAC88105 lunar meteorite: interferences for the lunar magma ocean hypothesis and the diversity of basaltic impact melts.**
G. J. Taylor.
Geochim. Cosmochim. Acta, Vol. 55, No. 11, p. 3031 – 3036 (Nov 1991).

The MAC88105 lunar meteorite contains three major types of impact melt breccias. The most abundant type is clast-laden, fine-grained, and rich in Al_2O_3; these clasts constitute most of the meteorite. Their abundance and aluminous nature indicate that the MAC88105 source area was very aluminous. The second type of impact melt has a basaltic bulk composition similar to many other lunar impact melts. The third impact-melt type resembles a prominent melt group at Apollo 16. These data show that basaltic impact melts are compositionally diverse.

094.055 **Paired lunar meteorites MAC88104 and MAC88105: a new "FAN" of lunar petrology.**
C. R. Neal, L. A. Taylor, Lui Yungang, R. A. Schmitt.
Geochim. Cosmochim. Acta, Vol. 55, No. 11, p. 3037 – 3049 (Nov 1991).

New lunar meteorite MAC88104/5 represents an exciting new opportunity to study a potentially unsampled region of the Moon. The authors have analyzed six thin sections by electron microprobe and three bulk samples in order to determine the chemical characteristics of this new lunar sample. While the exact site of origin for the lunar meteorites cannot be pinpointed, it is evident that they were derived from a relatively KREEP-free ferroan anorthosite terrain.

094.056 **A ferroan region of the lunar highlands as recorded in meteorites MAC88104 and MAC88105.**
B. L. Jolliff, R. L. Korotev, L. A. Haskin.
Geochim. Cosmochim. Acta, Vol. 55, No. 11, p. 3051 – 3071 (Nov 1991).

Lunar meteorites MAC88104 and MAC88105 were recovered on 13 Jan 1989 on a blue ice field near the MacAlpine Hills, Victoria Land, Antarctica. The authors have studied a number of samples from both meteorites and report the results of the petrographic and compositional characterization of the bulk meteorites and selected clasts. The data presented here confirm that MAC88104 and MAC88105 are paired. The authors show that MAC88104/5 exhibits some unique features compared to other lunar material.

094.057 **MAC88105 – a regolith breccia from the lunar highlands: mineralogical, petrological, and geochemical studies.**
C. Koeberl, G. Kurat, F. Brandstätter.
Geochim. Cosmochim. Acta, Vol. 55, No. 11, p. 3073 – 3087 (Nov 1991).

The new large lunar meteorite MAC88105 is a dense breccia, with lithic and mineral clasts and fragments set into a welded matrix. It is a regolith breccia which shows some recrystallization and evidence for a late shock event. The chemistry and mineralogy of MAC88105 is different from that of the other lunar meteorites and suggests a different source, which is supported by cosmic-ray and noble gas data. At this time it seems likely that about four individual impact events have been responsible for delivering the seven highland meteorites. MAC88105 and other lunar meteorites are important as they are probably random samples from the lunar surface and may thus be more representative than the Apollo and Luna rocks which come from only a small area of the moon.

094.058 **Geochemistry and petrography of the MacAlpine Hills lunar meteorites.**
M. M. Lindstrom, S. J. Wentworth, R. R. Martinez, D. W. Mittlefehldt, D. S. McKay, Wang Mingsheng, M. E. Lipschutz.
Geochim. Cosmochim. Acta, Vol. 55, No. 11, p. 3089 – 3103 (Nov 1991).

Mac-Alpine Hills 88104 and 88105 are anorthositic lunar meteorites recovered from the same area in Antarctica which are demonstrated to be paired samples of the same fall. Petrographic studies of matrix and clasts show that MAC88104/5 is a polymict breccia dominated by impact melt clasts.

094.059 **Lunar highland meteorites and the composition of the lunar crust.**
H. Palme, B. Spettel, K. P. Jochum, G. Dreibus, H. Weber, G. Weckwerth, H. Wänke, A. Bischoff, D. Stöffler.
Geochim. Cosmochim. Acta, Vol. 55, No. 11, p. 3105 – 3122 (Nov 1991).

Major, minor, and trace element data obtained by neutron activation techniques and by spark source mass spectrometry on two lunar meteorites MAC88104 and MAC88105 are reported. Both MAC samples were also analysed for their contents and isotopic compositions of rare gases. Additional SSMS-data were obtained on four lunar highland meteorites previously found in Antarctica: ALHA81005, Y791197, Y82192, and Y86032. MAC88104 and MAC88105 are very similar in chemistry, suggesting that they are pieces of a single fall event. The bulk chemical composition of MAC88104/5 is not very different from the other lunar highland meteorites. However, mafic element concentrations (Mg, Cr, Mn, etc.) are slightly lower in MAC88104/5 than in the other lunar highland meteorites.

094.060 **The MacAlpine Hills lunar meteorite and implications of the lunar meteorites collectively for the compositions and origin of the Moon.**
P. H. Warren, G. W. Kallemeyn.
Geochim. Cosmochim. Acta, Vol. 55, No. 11, p. 3123 – 3138 (Nov 1991).

The authors report on MacAlpine Hills 88104 and 88105, a paired sample that is the most recently discovered of four meteorites derived from the highlands portion of the lunar surface and is also, at 724 g mass, the largest lunar meteorite in captivity. The authors characterize the rock as a whole as well as a number of its component clasts. They interpret the new data, and the new data for lunar meteorites collectively, as constraints on the nature, origin, and evolution of the ancient nonmare crust, and on the bulk composition and origin of the Moon.

094.061 **History of the paired lunar meteorites MAC88104 and MAC88105 derived from noble gas isotopes, radionuclides, and some chemical abundances.**
O. Eugster, J. Beer, M. Burger, R. C. Finkel, H. J. Hofmann, U. Krähenbühl, T. Michel, H. A. Synal, W. Wölfli.
Geochim. Cosmochim. Acta, Vol. 55, No. 11, p. 3139 – 3148 (Nov 1991).
Noble gas isotopes, radionuclides, and chemical abundances were studied in the lunar meteorites MAC88104 and MAC88105 collected in the MacAlpine Hills area of Antarctica. The concentrations of the noble gas isotopes and the radionuclide activities in the two meteorites are essentially identical, proving that the two meteorites are paired.

094.062 **Exposure histories of lunar meteorites: ALHA81005, MAC88104, MAC88105, and Y791197.**
K. Nishiizumi, J. R. Arnold, J. Klein, D. Fink, R. Middleton, P. W. Kubik, P. Sharma, D. Elmore, R. C. Reedy.
Geochim. Cosmochim. Acta, Vol. 55, No. 11, p. 3149 – 3155 (Nov 1991).
The cosmogenic radionuclides ^{41}Ca, ^{36}Cl, ^{26}Al, and ^{10}Be in the Allan Hills 81005, MacAlpine Hills 88104, MacAlpine Hills 88105, and Yamato 791197 meteorites were measured by accelerator mass spectrometry. These four lunar meteorites experienced similar histories. A comparison of the cosmogenic nuclide concentrations in MacAlpine Hills 88104 and MacAlpine Hills 88105 clearly indicates that they are a pair from the same fall.

094.063 **Exposure histories of the lunar meteorites: MAC88104, MAC88105, Y791197, and Y86032.**
S. Vogt, D. Fink, J. Klein, R. Middleton, B. Dockhorn, G. Korschinek, E. Nolte, G. F. Herzog.
Geochim. Cosmochim. Acta, Vol. 55, No. 11, p. 3157 – 3165 (Nov 1991).
Four lunar meteorites, MacAlpine Hills 88104, MacAlpine Hills 88105, Yamato 791197, and Yamato 86032 were analyzed for the cosmogenic radionuclides ^{10}Be, ^{26}Al, ^{36}Cl, and ^{41}Ca. From these and published data, histories of exposure to cosmic rays were modelled in terms of two-stage irradiations each with a long first stage on the Moon. From the number and exposure histories of lunar meteorites the authors infer a production rate on the order of 5 Ma^{-1} and an arrival rate worldwide of about 3×10^6meteorites Ma^{-1}. These results suggest that each impact event large enough to produce lunar meteorites sends a large number of them to the Earth.

094.064 **The natural thermoluminescence of meteorites: III. Lunar and basaltic meteorites.**
D. W. G. Sears, P. H. Benoit, H. S. J. D. Batchelor, S. Symes.
Geochim. Cosmochim. Acta, Vol. 55, No. 11, p. 3167 – 3180 (Nov 1991).
Natural thermoluminescence (TL) data have been obtained for the lunar meteorite MacAlpine Hills 88104/5 and for 65 eucrites, howardites, diogenites, and mesosiderites in order to investigate their recent thermal and radition histories. All these meteorites have low levels of natural TL compared to chondrites. Although they are clearly not paired, the possibility that MAC88104/5 and ALHA81005 were ejected from the moon by the same event should be considered, since diverse rock types are found in close proximity on the lunar surface. The natural TL data confirm most previous published pairings among basaltic meteorites and suggest others.

Automated information system for the Lunar Nomenclature data base.
See Abstr. 002.062.

Lunar tables and programs from 4000 B.C. to A.D. 8000.
See Abstr. 002.098.

Infrared and visible brightness characteristics of the lunar surface.
See Abstr. 003.018.

Planetary geosciences – 1988.
See Abstr. 003.036.

Transportation system for a lunar base.
See Abstr. 011.016.

Geoscience and a lunar base. A comprehensive plan for lunar exploration. Proceedings.
See Abstr. 012.117.

Langzeitbeobachtungen des Mondes.
See Abstr. 014.047.

The origin of the Moon and the single impact hypothesis. IV.
See Abstr. 021.006.

The connection of optical characteristics with the chemical composition of the Earth's and Moon's magnetic rock.
See Abstr. 022.145.

Lunar automatic photometric telescopes (LAPT's).
See Abstr. 032.055.

Lunar dust: implications for astronomical observatories.
See Abstr. 035.072.

Analog and digital processing of the polarization and spectrozonal photos of the Moon.
See Abstr. 036.228.

The lunar selected areas program (SAP): how to get started.
See Abstr. 036.309.

Practical lunar photoelectric photometry.
See Abstr. 036.310.

Influence of the lunar libration models on UT0 determination from Lunar Laser Ranging.
See Abstr. 044.051.

Length of the lunar crescent.
See Abstr. 046.003.

Correction for the influencene of the Moon's noncentral gravitational field on the motion of geodynamic artificial satellites.
See Abstr. 052.009.

On the influence of the moon's gravitational field on the motion of the artificial satellites.
See Abstr. 052.036.

Testing Einstein's theory of gravity by analyzing Lunar Laser Ranging data.
See Abstr. 066.208.

Long–term changes in composition of solar particles implanted in extraterrestrial materials.
See Abstr. 078.025.

Solar flare heavy–ion tracks in extraterrestrial objects.
See Abstr. 078.026.

Terrestrial mantle siderophiles and the lunar impact record.
See Abstr. 081.006.

Nonlinear inverse potential problems and construction of two–layer models of Earth and Moon.
See Abstr. 081.046.

Thermal equilibration of the Earth following a giant impact.
See Abstr. 081.050.

Physical properties of terrestrial planets and their satellites: an introduction.
See Abstr. 091.054.

Small solar system bodies.
See Abstr. 091.057.

Terrace width variations in complex Mercurian craters and the transient strength of cratered Mercurian and lunar crust.
See Abstr. 092.005.

Galileo flies by Venus, the Earth and the Moon.
See Abstr. 093.035.

Multiple–impact mechanisms on Venus and other solid planets and satellites.
See Abstr. 093.038.

Magma generation on Mars: amounts, rates, and comparisons with Earth, Moon, and Venus.
See Abstr. 097.013.

Terrestrial cometary tail and lunar corona induced by small comets: predictions for Galileo.
See Abstr. 102.014.

A lunar meteorite found outside the Antarctic.
See Abstr. 105.044.

MacAlpine Hills 88104 and 88105 lunar highland meteorites: general description and consortium overview.
See Abstr. 105.057.

Mineral–chemical comparisons of MAC88105 with Yamato lunar meteorites.
See Abstr. 105.058.

095 Lunar Eclipses

095.001 **Partial lunar eclipse 1991 December 21.**
B. W. Soulsby.
Aust. J. Astron., Vol. 4, No. 2, p. 71 – 76 (Oct 1991).

Precision of medieval Islamic eclipse measurements.
See Abstr. 004.079.

Accuracy of eclipse observations recorded in medieval Arabic chronicles.
See Abstr. 004.084.

096 Lunar and Planetary Occultations

096.001 **Photoelectric observations of lunar occultations of six binaries in Pleiades.**
Qian Bochen, Fan Qingyuan.
Chin. Astron. Astrophys., Vol. 15, No. 3, p. 336 – 338 (Sep 1991). English translation of Acta Astron. Sin., Vol. 32, No. 1, p. 103 – 106 (Mar 1991). See Abstr. 53.096.002.

On 1987 December 31, the authors observed lunar occultations in Pleiades and obtained a series of diffraction curves of stars. After reduction, six stars were found to be double stars, they are SAO 076131, 076140, 076155, 076183, 076236 and 076259. This is the first run of photoelectric occultation observation with the 1.56 metre telescope.

096.002 **On the ephemerides of occultations.**
V. N. L'vov, R. I. Smekhacheva.
1. Spain–USSR Workshop on Positional Astronomy and Celestial Mechanics, p. 99 – 100 (1991). – See Abstr. 012.025 for the main entry.

At the Institute of Theoretical Astronomy a system of software packages for a personal computer has been elaborated for predictions of occultations of stars by solar system objects.

096.003 **Sixteen years of stellar occultations.**
J. Elliot.
Liller Symposium: Asteroids to quasars, p. 27 – 51 (1991). – See Abstr. 012.087 for the main entry.

Stellar occultations by solar system bodies have emerged from an infrequent "astronomical happening" in the 1950's and 60's, to a routine tool used by a growing number of planetary astronomers in the 1970's and 80's. The increasing popularity of the technique is due to (i) the 10000–fold improvement in spatial resolution that it affords over ordinary groundbased methods for probing the outer solar system and (ii) several technical advancements that have allowed occulations to be observed more frequently and with continually improving signal–to–noise ratios. In addition to large telescopes at fixed locations, the observing platforms have included portable telescopes on the ground, the Voyager spacecraft, and NASA's Kuiper Airborne Observatory. Major results from the occultation technique include: the discovery of thermal waves in the Martian atmosphere: the discovery and comprehensive study of the Uranian ring system; the accurate measurement of the sizes and shapes of several asteroids; the discovery of waves and other features in the Saturnian rings; the probing of the upper atmospheres of the gas giant planets; the discovery of the Neptune "arcs"; and most recently, the detection of an atmosphere surrounding Pluto. In the future, further development of the occultation technique can be expected with observations from the Hubble Space Telescope and the use of CCD's for occultation predictions and observations.

096.004 **Eastern hemisphere grazing occultation supplement for 1992.**
D. W. Dunham.
International Occultation Timing Association. 8 p. (1991). Supplement to Occultation Newsl.

096.005 **European grazing occultation supplement for 1992.**
D. W. Dunham.
International Occultation Timing Association. 13 p. (1991).
The paths for grazing occultations of stars as faint as magnitude 7.5 (to mag. 9.0 during total lunar eclipses) that are predicted to occur in Europe and nearby areas during 1992 are shown on six maps.

096.006 **A precise predicting method for stellar occultation by planets.**
Pa Dasa.
Ann. Shanghai Obs., Acad. Sin., No. 12, p. 79 – 84 (1991). In Chinese.

096.007 **Grazing occultations.**
D. Stockbauer.
Occultation Newsl., Vol. 5, No. 4, p. 89 – 91 (Aug 1991).

096.008 **Solar system occultations during 1991.**
D. W. Dunham.
Occultation Newsl., Vol. 5, No. 4, p. 92, 97 – 105 (Aug 1991).

096.009 **New double stars.**
T. Murray.
Occultation Newsl., Vol. 5, No. 5, p. 116 – 119 (Nov 1991).

096.010 **Solar system occultations during 1992.**
D. W. Dunham.
Occultation Newsl., Vol. 5, No. 6, p. 130 – 145, 147 – 150 (Dec 1991).

096.011 **Archenhold occultation summary. I. First results.**
K. Guhl, B. Stecklum.
Astron. Nachr., Vol. 312, No. 5, p. 315 – 319 (1991).
First results of lunar occultation observation and reduction at the Archenhold Observatory are presented. Using the 50 cm Cassegrain telescope under urban observing conditions acceptable results for 16 events have been obtained including one suspected occultation binary.

Analysis of lunar occultation data.
See Abstr. 036.215.

Untersuchungen zur Durchführung, Auswertung und astrometrischen Nutzung von photoelektrischen Sternbedeckungsbeobachtungen.
See Abstr. 036.291.

A comprehensive system of occultation software.
See Abstr. 036.318.

The diameter of (9) Metis from the occultation of SAO 190531.
See Abstr. 098.088.

Occultations by Uranus and Neptune: 1991–1999.
See Abstr. 101.053.

Pluto–Charon stellar occultation candidates: 1990–1995.
See Abstr. 101.055.

Lunar occultation and near–infrared speckle observations of DG Tauri, FV Tauri, FW Tauri and GG Tauri.
See Abstr. 121.018.

The angular diameter of R Leonis at 2.16 μm.
See Abstr. 122.035.

Estimations of brightness distributions across stellar disks from an analysis of IR observations of lunar occulatations. Mira variables.
See Abstr. 122.041.

Brightness distributions over stellar disks estimated from IR observations of lunar occultations. Mira variables.
See Abstr. 122.058.

097 Mars, Mars Satellites

097.001 **Effective diffusion in bounded random–inhomogeneous media and threshold formation of global duststorms on Mars.**
S. G. Chefranov.
Izv. Akad. Nauk SSSR, Fiz. Atmos. Okeana, Vol. 27, No. 7, p. 746 – 754 (Jul 1991). In Russian.

The propagation of admixture or simple waves through turbulent finite media is considered. It is shown that the spatial localization of admixture or waves which is characterized by the negative value of the effective eddy diffusivity D_e is possible in such case. Unlike D_e the threshold criteria for the exponential growth of the mass of admixture (i.e. for the medium "settlement") are found to be independent of both the type of boundary conditions and the functional type of local formations of admixture production. The orographic inhomogeneity effect on both the value of D_e and the frequency of formation of local duststorms that result in the global dusting of the Mars atmosphere under the supercritical conditions is estimated.

097.002 **Photo–ionization balance and ion composition model on the ionospheres of Mars and Venus.**
A. C. Balachandra Swamy.
Indian J. Radio Space Phys., Vol. 20, No. 1, p. 68 – 75 (Feb 1991).

Photo–ionization rates on the ionospheres of Mars and Venus have been computed using the latest available data such as Hinteregger solar fluxes, photo–ionization and photo–absorption cross–sections and the neutral atmospheric models. Photochemical equilibrium has been assumed in the height range 100 – 300 km. The derived $N_e(h)$ profiles are compared with the observed $N_e(h)$ profiles. There is thermal equilibrium ($T_e = T_g$) in the case of the Martian ionosphere. However, there is thermal imbalance ($T_e > T_g$) on the Venusian ionosphere above 160 km. Ion density profiles have been derived by the application of appropriate photochemistry. There is an excellent agreement between the observed and computed ion density profiles below 200 km in both the ionospheres.

097.003 **Aerosol phenomena in the Martian atmosphere from KRFM experiment data.**
V. I. Moroz, E. V. Petrova, L. V. Ksanfomaliti, L. Esposito, J.-P. Bibring, M. Combes, A. Soufflot, O. F. Ganpantserova, N. V. Goroshkova, A. V. Zharkov, G. E. Nikitin.
Pis'ma Astron. Zh., Tom 17, No. 6, p. 548 – 559 (Jun 1991). In Russian. English translation in Sov. Astron. Lett., Vol. 17, No. 3, p. 235 – 240 (1991).

Photometric limb to limb profiles of Mars were obtained in 8 narrow bands between 315 and 550 nm from the "Phobos-2" spacecraft. Tentative results of an analysis are presented in terms of optical properties of the atmosphere and surface. Imaginary part of the refraction index χ from 0.0075 to 0.02 for 315 nm and from 0.005 to 0.01 for 550 nm was estimated for the "constant" dust haze, using the Mie theory for spherical particles. These values of χ are few times higher than those obtained by laboratory tests of terrestrial analogous materials including basalt, andesit and montmorillonite. Two explanations are possible: influence of the irregular shape of particles and/or presence of some more absorbing substances (as goethit). Particles sizes a few tenths micron having refraction index 1.55 are compatible with the discussed model of the "constant" haze. The shape of the near equatorial photometric profile on 550 nm can be explained by a slightly absorbed atmosphere (with the optical depth 0.4 and imaginary part of the refraction index 0.0075). Icy particles with the same average sizes as in the haze (r_m = 0.4 microns) but with more narrow size distribution can explain the bright spot above the Arsia Mons. Optical depth of $\tau = 0.1$ and column mass density of $7 \cdot 10^{-5} g \cdot cm^{-2}$ of the icy clouds were evaluated.

097.004 **The Martian dynamo.**
A. Ruzmaikin (*A. A. Ruzmajkin*).
Phys. Earth Planet. Inter., Vol. 67, No. 3/4, p. 268 – 274 (Jul 1991).

Mars has a well–conducting core composed of a mixture of iron and sulphur. It is probably liquid and differentially rotating. The main problem is whether the core is stable stratified or convective. About 10 – 15% of sulphur is enough to drive a compositional convection over a great part of Martian evolution if an inner solid solid core was formed. Convection in a liquid rotating stratified core results in a mean helicity, which, together with differential rotation and turbulent magnetic diffusivity, is a source for mean field dynamo. The magnitude of the magnetic field generated can explain the origin of magnetization of the shergottie, nakhlite and chassignite meteorites. However, the dynamo probably does not work at present, in accordance with a weak large–scale magnetic field on the Martian surface estimated from the Mars–2, –3, –5, and Phobos mission results. This field is interpreted as an external product of the magnetized mantle having a slightly elliptical form. Some other possible explanations for the outer magnetic field are discussed.

097.005 **Waves and cold plasma observations near Mars.**
C. M. C. Naim, R. Grard, A. Skalsky (*A. Skal'skij*), J. G. Trotignon.
Adv. Space Res., Vol. 11, No. 9, p. 87 – 91 (1991). – See Abstr. 012.006 for the main entry.

This paper summarizes the plasma and wave observations performed in the night sector of Mars with the Plasma Wave System. Embedded in the magnetic field structure of the Martian tail, cold plasmas with densities in the range 10 – 65 cm^{-3} are observed in association with broadband wave activity extending from a few Hz up to several kHz; these plasmas have characteristics analogous to the plasma clouds observed at Venus.

097.006 **Ground–based infrared measurements of the global distribution of ozone in the atmosphere of Mars.**
F. Espenak, M. J. Mumma, T. Kostiuk, D. Zipoy.
Icarus, Vol. 92, No. 2, p. 252 – 262 (Aug 1991).

The authors report measurements of the global distribution of ozone in the atmosphere of Mars, based on Doppler–limited infrared spectroscopy during the period 3 – 7 June 1988. Ozone measurements were obtained at eight beam positions over a range of Martian latitudes (80°S to 20°N) and local solar hour angles (-0.5^h to $+5.5^h$). The total O_3 column abundance at each position was retrieved by fitting the lines with synthetic spectra generated by a radiative transfer program. The retrieved O_3 column burdens of this investigation are less than 2.2 μm–atm for all latitudes sampled, consistent with seasonal abundances predicted by the models of Liu and Donahue, and Shimazaki and Shimizu.

097.007 **Radar images of Mars.**
D. O. Muhleman, B. J. Butler, A. W. Grossman, M. A. Slade.
Science, Vol. 253, No. 5027, p. 1508 – 1513 (27 Sep 1991).

Full disk images of Mars have been obtained with the use of the VLA to map the radar reflected flux density. The surface of Mars was illuminated with continuous wave radiation at a wavelength of 3.5 cm. The reflected energy was mapped in individual 12–minute snapshots with the VLA in its largest configuration; fringe spacings as small as 67 km were obtained. The images reveal near–surface features including a region in the Tharsis volcano area, over 2000 km in eaast–west extent, that displayed no echo to the very low level of the radar system noise. The feature is interpreted as a deposit of dust or ash with a density less than about 0.5 g cm^{-3} and free of rocks larger than 1 cm across. The deposit must be several meters thick and may be much deeper. The strongest reflecting geological feature was the south polar ice cap. The cap image is interpreted as arising from nearly pure CO_2 or H_2O ice with a small amount of martian dust and a depth greater than 2 to 5 m.

097.008 **The opposition of Mars, 1988.**
R. McKim.
J. Br. Astron. Assoc., Vol. 101, No. 5, p. 265 – 283 (Oct 1991).

097.009 **Methods and results of investigations of aerosols of the Martian atmosphere in the periods of its high transparency.**
A. V. Morozhenko.
Kinematika Fiz. Nebesn. Tel, Tom 7, No. 4, p. 3 – 22 (Jul–Aug 1991). In Russian. English translation in Kinematics Phys. Celest. Bodies, Vol. 7, No. 4.

A critical analysis of the existing methods and results of determination of the optical properties of the Martian atmosphere in the periods of its high transparency was carried out. The estimations obtained with direct (Viking's measurements from the surface of the plane) and distance methods, turned out not to agree. The analysis of the landing sites of these spacecraft lead to the conclusion that the optical properties of the Martian atmosphere change largely during carrying out the photometric experiments. The origin of these variations can be dissipation (during morning measurements) and condensation (during evening experiments) of the haze. It is shown that, if this fact is not taken into account, one can obtain overestimated, and in some cases even fictitious values of the optical thickness. It is concluded that the problem of the determination of the optical properties of the Martian atmosphere is not solved and this circumstance must be taken into account when planning new space missions.

097.010 **Composition, structure and gravitational field of Mars.**
V. N. Zharkov, E. M. Koshlyakov, K. I. Marchenkov.
Astron. Vestn., Tom 25, No. 5, p. 515 – 547 (Sep – Oct 1991). In Russian. English translation in Solar Syst. Res., Vol. 25, No. 5.

This review considers the problems, related to the interior structure of the planet, including the questions of chemical composition and petrologic models. Mars is divided into three basic regions: the crust, the mantle, and the core. The thickness of the crust may be very high, approaching $\sim 150 - 200$ km. The depth of the core–mantle boundary is equal to $\sim 1600 \pm 200$ km. The mantle is divided into three basic regions: the olivine zone, the phase transition zone and the spinel zone. The basic impurity in the iron core are sulfur and probably hydrogen. The Martian core may be in liquid state. The correlation of temperature distribution with effective viscosity distribution leads to a "low" temperature distribution in the interior of Mars. The thickness of the Martian lithosphere is equal to $\sim 500 \pm 200$ km. The interpretation of the gravitational field with help of the Green function method is considered. For these purposes the authors used for the gravitational field and topography the spherical harmonic expansion up to 18 degree and order. The stress level in the Martian lithosphere is equal to several hundred bars. The order of magnitude of stresses in the mantle is equal to ~ 30 bars.

097.011 **Orbits of Phobos and Deimos determined with consideration for "Phobos–2" on–board observations.**
V. A. Shor.
1. Spain–USSR Workshop on Positional Astronomy and Celestial Mechanics, p. 55 – 61 (1991). – See Abstr. 012.025 for the main entry.

The improvement of orbital parameters of Phobos and Deimos taking into account 37 observations of Phobos and 8 observations of Deimos made from "Phobos–2" spacecraft as well as all available observations of the satellites are given. Values of the secular accelerations of the satellites found by different authors are compared.

097.012 **A new look at dust and clouds in the Mars atmosphere: analysis of emission–phase–function sequences from global Viking IRTM observations.**
R. T. Clancy, S. W. Lee.
Icarus, Vol. 93, No. 1, p. 135 – 158 (Sep 1991).

An analysis of emission–phase–angle sequences taken by tghe IRTM instrument on board the Viking orbiters yields a wide variety of results regarding dust and cloud opacities, dust and cloud optical scattering properties, and surface albedos. The results for over 75 regions spanning the entire Viking period of observations and latitudes from 75°S to 75°N are presented.

097.013 **Magma generation on Mars: amounts, rates, and comparisons with Earth, Moon, and Venus.**
R. Greeley, B. D. Schneid.
Science, Vol. 254, No. 5034, p. 996 – 998 (15 Nov 1991).

Total extrusive and intrusive magma generated on Mars over the last ~ 3.8 billion years is estimated at $654 \times 10^6 km^3$ or 0.17 km^3/yr, substantially less than rates for Earth (26 to 34 km^3/yr) and Venus (< 20 km^3/yr) but much more than for the Moon (0.025 km^3/yr). When scaled to Earth's mass the martian rate is much smaller than that for Earth or Venus and slightly smaller than for the Moon.

097.014 **An engineering model of the atmosphere of Mars for the Mars–94 project (MA–90).**
V. I. Moroz, V. V. Kerzhanovich, V. A. Krasnopol'skij.
Cosmic Res., Vol. 29, No. 1, p. 1 – 79 (Jul 1991). English translation of Kosm. Issled., Tom 29, Vyp. 1, p. 3 – 84 (1991).

The Soviet mission to Mars that is planned for the 1994 launch window (Mars–94) consists of a Mars orbiter and a whole series of descent elements: a balloon, small stations, penetrators, and even a small Mars rover. The landings of these elements, the balloon flight, and the evolution of the orbiter's orbit depend on the characteristics of the atmosphere. The paper presented is a summary of the data on these characteristics which takes into account the characteristics of the Mars–94 project. The balloon is most sensitive to them; therefore, the most attention is paid to the layer of the atmosphere below 10 km. Both data from the published literature and also results from additional research carried out especially for the Mars–94 project are used. The MA–90 model has been recommended by the Space Research Institute of The Academy of Sciences of the Ukrainian SSR as the initial data for the Mars–94 project.

097.015 **The fast atomic oxygen corona extent of Mars.**
W.–H. Ip.
Geophys. Res. Lett., Vol. 17, No. 13, p. 2289 – 2292 (Dec 1990).

The production of fast oxygen atoms in the Martian exosphere as a result of the electron dissociative recombination of $O^+{}_2$ ions is re–evaluated for the time period appropriate to the Phobos 2 mission. Because of the prevalent conditions of solar maximum activities in 1989, both the upper atmosphere and ionsphere should have more extended structures than those encountered during the Viking Orbiter (VO) mission in 1976 which was near solar minimum. The author compares the distributions of the hot oxygen atoms for these different levels of solar activities and finds that the number density of the hot oxygen atoms during the Phobos 2 mission should be a factor of two larger than estimated for the time period of the VO mission. The non–thermal escape rate of the exospheric oxygen atoms is thus comparable to the ionospheric escape rate measured in the Martian magnetotail. Furthermore, the flux of the oxygen ions picked up in the hot atomic corona should be more intense than previously thought.

097.016 **An estimate of the amount of material removed from the Martian fretted terrain.**
A. Dimitriou.
Geophys. Res. Lett., Vol. 17, No. 13, p. 2461 – 2464 (Dec 1990).

Three models are proposed for the former extent of a continuous heavily cratered plateau within the present area of fretted terrain between 270°W and 360°W. To the south and west, each model area limit is well constrained by the northern boundary of the unfretted heavily cratered plateau. The northern limit of each model is defined by the gradational northern limit of fretted terrain. The eastern limits are distinct for each model. Model 1 assumes a once continuous plateau was dropped to the north along NW–trending normal boundary faults in the Ismenius Lacus SE subquadrangle. Model 2 includes model area 1 and places the former limit of the plateau 400 km farther east along a trace subparallel to the present plateau margin in this subquadrangle. Model 3 includes model area 2 and uses the rim of the proposed Borealis basin in Casius SW subquadrangle as the eastern limit of the former plateau. The estimates for the amount of material removed vary from $1.04 \pm 0.60 \times 10^6 km^3$ for model 1 through $1.53 \pm 0.80 \times 10^6 km^3$ for model 2 to $1.99 \pm 1.01 \times 10^6 km^3$ for model 3. This implies that this area of fretted terrain is not a significant source of sediment for the northern lowlands of Mars.

097.017 **Photochemical regulation of CO on Mars.**
E. Chassefière.
Geophys. Res. Lett., Vol. 18, No. 6, p. 1055 – 1058 (Jun 1991).
The balance between the production and loss rates of CO in the Martian atmosphere is examined in the photochemical frame defined by Parkinson and Hunten (1972). The contribution of the present ideas to the understanding of the recent Earth–based and space observations of the Martian CO is evaluated. Some new directions of research are proposed.

097.018 **On the momentum transfer of the solar wind to the Martian topside ionosphere.**
R. Lundin, E. M. Dubinin (*Eh. M. Dubinin*), H. Koskinen, O. Norberg, N. Pissarenko (*N. Pisarenko*), S. W. Barabash.
Geophys. Res. Lett., Vol. 18, No. 6, p. 1059 – 1062 (Jun 1991).
Hot plasma measurements from the Soviet Phobos–2 spacecraft in the Martian magnetosphere suggests that the solar wind interaction with Mars is cometary–like, with mass loading of the solar wind and ion pick–up occurring also outside the subsolar bow–shock. The interaction is characterized by a pronounced decrease of the solar wind speed inside what has been termed the mass–loading boundary (MLB). Well outside the MLB, the ion pick–up process acts in a "normal" sense. There ions pick up approximately the solar wind velocity – independent of mass. Inside the MLB, the momentum loss of solar wind ions is more pronounced – heavy ions of Martian origin taking up most of the solar wind ion momentum flux. The heavy mass–loading of solar wind ions in the innermost part of the Martian boundary layer (near the "magnetopause") leads to a "loaded" ion pick–up. The process can be understood as internal loading of an MHD–dynamo, propelled by a driver plasma – the solar wind. The ASPERA ion composition and momentum data is consistent with such a pick–up process. Inside the "magnetopause" ions of Martian origin are accelerated up to energies close to those of the solar wind protons. The authors propose two types of acceleration processes, one similar to that acting within the Earth's auroral acceleration region (acting in the presence of an ambient magnetic field), another a pick–up process acting within a limited spatial region.

097.019 **Photometric properties of Phobos' regolith based on the Phobos mission data.**
L. V. Ksanfomaliti.
Astron. Vestn., Tom 25, No. 6, p. 650 – 676 (Nov – Dec 1991). In Russian. English translation in Sol. Syst. Res., Vol. 25, No. 6.
The paper reports the results of a Phobos spectrophotometry in the range 300 – 600 nm. It is based on data obtained by means of the spectrophotometer KRFM on board of the Phobos spacecraft in 1989. Phobos' reflectivity has little in common with that in previously published papers. It does not agree with reflectivity features of carbonaceous chondrites and it does not permit unique identification with any other type of meteoritic material. In the 300 – 600 nm spectral range, a wide variety of reflectivity features was observed to be connected with details of the relief.

097.020 **Gravity studies of the Tharsis area on Mars.**
P. Janle, E. Erkul.
Earth, Moon, Planets, Vol. 53, No. 3, p. 217 – 232 (Jun 1991). = Contrib. Nr. 421, Inst. Geophys. Univ. Kiel.
The Tharsis rise on Mars with a diameter of about 8000 km and an elevation up to 10 km shows extensive volcanism and an extensional fracture system. Gravity models of four profiles are in accordance with a total Airy isostatic compensation of the whole rise with mean crustal thicknesses of 50 and 100 km. But two regions exhibit significant mass deficits: (i) the area between Olympus Mons and the three large Tharsis volcanoes and (ii) central Tharsis. This can be explained by (1) a heated upper mantle, (2) a chemically modified upper mantle, (3) a crustal thickening, or (4) a combination of these three processes. Gravity modelling results in a different isostatic state of the three Tharsis volcanoes. Pavonis Mons is not compensated, Ascraeus Mons is highly or totally compensated, and Arsia Mons is medium or not compensated. The large, flat volcanic structure Alba Patera has been explained by a hot spot with an evolution of a mantle diaper. The results have shown that the Tharsis rise is a very complex structure. The central and eastern part of the rise is characterized by extensional features and a mass deficit (Extensional Province). The western part is dominated by many volcanic features and a central elongated mass deficit (Volcanic Province). The northern part consists of Alba Patera. It seems unlikely that the whole rise has been generated by one stationary large axisymmetric plume or hot spot. There could have been one or more active hot spots with an evolution in space and time.

097.021 **Surface locations of potential aquifer ecosystem on Mars.**
K. D. Trego.
Earth, Moon, Planets, Vol. 53, No. 3, p. 237 – 238 (Jun 1991). Letter–to–the–editor.
The origins of outflow channels and sites of suspected karst terrain on Mars possibly indicate where aquifers once existed. Aquifers on Mars are potential hosts to ecosystem development because of the sustained presence of liquid water and subsurface warm temperature.

097.022 **Mars boundary layer simulations: comparison with Viking lander and entry observations.**
R. M. Haberle, H. C. Houben.
2. COSPAR Colloquium: The environmental model of Mars, p. 3 – 6 (1991). – See Abstr. 012.039 for the main entry.
Diurnal variations of wind and temperature in the lower Martian atmosphere are simulated with a boundary layer model that includes radiative heating in a dusty CO_2 atmosphere, turbulence generated by convection and/or shear stresses, a surface heat budget, and time varying pressure forces due to sloping terrain. Model results for early northern summer are compared with Viking lander observations to determine the model's strengths and weaknesses, and suitability as an engineering model.

097.023 **The NASA environmental models of Mars.**
D. I. Kaplan.
2. COSPAR Colloquium: The environmental model of Mars, p. 25 – 34 (1991). – See Abstr. 012.039 for the main entry.
The document "Environment of Mars, 1988" is a compilation of the latest understanding of the Martian environment. Scientific models are presented of the Martian atmosphere and of the Martian surface. Astrodynamic information about Mars is also provided. Two engineering models of Mars are under development. The first engineering model is the Mars Global Reference Atmospheric Model. This model will account for seasonal, diurnal, surface topography, and dust storm effects upon the atmosphere. It can also simulate appropriate random density perturbations along any trajectory path through the atmosphere. The second engineering model is the Mars Terrain Simulator. This software program builds pseudo–Martian terrains by layering the effects of geological processes upon one another. Output pictures of the constructed surfaces can be "viewed" from any vantage point under any illumination conditions.

097.024 **Exploration of the atmosphere and climate system of Mars.**
C. Leovy.
2. COSPAR Colloquium: The environmental model of Mars, p. 35 – 45 (1991). – See Abstr. 012.039 for the main entry.
The present understanding of the components of the martian general circulation and the role of the general circulation in controlling the long term balances of carbon dioxide, dust, and

water are briefly reviewed. An observational strategy for improving the understanding of the general circulation and its role in regulating the climate system is proposed. The importance of coordinated in situ and remote sensing measurements over an extended time period is emphasized.

097.025 **Landing site selection for the Mars 94 mission: a preliminary study.**
F. M. Costard, F. X. Chaffaut, P. H. Masson, D. Mège, J. P. Peulvast.
2. COSPAR Colloquium: The environmental model of Mars, p. 55 – 58 (1991). – See Abstr. 012.039 for the main entry.
According to topographic and engineering constraints, a selection of potential landing sites for the Mars 94 mission is proposed. Three sites have been chosen in Acidalia, Utopia and Arcadia Planitiae. These plains exhibit various geological units such as volcanic rocks or eolian and fluvial sediments and contain subsurface ground–ice.

097.026 **Are Phobos and Deimos carbonaceous chondrites? Mössbauer spectroscopy on carbonaceous chondrites and their relation to the Martian moons.**
M. B. Madsen, J. M. Knudsen, L. Vistisen, H. G. Jensen.
2. COSPAR Colloquium: The environmental model of Mars, p. 59 – 62 (1991). – See Abstr. 012.039 for the main entry.
The Martian moons Phobos and Deimos are relatively small irregularly shaped objects that represent a large class of small Solar System bodies. Based on spectral reflectance measurements and comparison with meteorites the Martian moons are believed to be of carbonaceous chondrite composition. The carbonaceous chondrites are characterized by a very high average oxidation state of the iron in them compared to that of all other types of meteorites. By use of Mössbauer spectroscopy carbonaceous chondrites of the classes CI, CM, CV and CO are easily distinguished from each other and from other classes of meteorites. The possibility of identifying the Martian moons as distinct types of carbonaceous chondrites based on Mössbauer spectra of the surface is investigated.

097.027 **Thermal environment of Mars.**
T. Z. Martin.
2. COSPAR Colloquium: The environmental model of Mars, p. 89 – 100 (1991). – See Abstr. 012.039 for the main entry.
Thermal radiation from Mars arises principally from the surface layer. The atmosphere emits weakly under clear conditions, but when dust is present, the atmosphere absorbs sunlight directly and also radiates more effectively. The effect of increasing dust in the atmosphere is to reduce diurnal temperature variation at the surface, while enhancing atmospheric thermal variation. Even in regions of polar night, large atmospheric thermal effects can be induced by global dust storms. Maximum temperatures occur near the equator in summer or in local sites with sunward facing slopes, where low surface thermal inertias permit values up to about 310K. Atmospheric temperatures near the surface are driven by radiation and convection in the daytime, and are limited by radiation at night. There is considerable evidence that Mars' thermal state and climate can be significantly different from year to year, due to the variation in dust activity.

097.028 **Sensing the physical properties at the Martian surface.**
A. Dollfus.
2. COSPAR Colloquium: The environmental model of Mars, p. 101 – 104 (1991). – See Abstr. 012.039 for the main entry.
Photopolarimetry enables a characterization of grain size for the dusty layer which covers the surface of Mars and was implemented with the instrument VPM on board the soviet orbiter MARS–5. Radiometry is sensing the compaction of the ground immediately below the exposed surface analysed by polarimetry and was operated with the instrument IRTM on board spacecraft Viking. Combination of these data produced characterizations of soil surface physical properties over several representative terrains of the Martian provinces Mare Erythraeum and Thaumasia.

097.029 **Settlement of an engineering model of Martian atmosphere for Mars–94 project.**
P. Mauroy, A. Vargas.
2. COSPAR Colloquium: The environmental model of Mars, p. 105 – 108 (1991). – See Abstr. 012.039 for the main entry.
The Soviet–French project MARS–94 is based on the exploration of Mars by an aerostat, with daily take–off and landing. The sizing of the aerostat requires the use of engineering model of the Martian atmosphere in the accessible regions of the planet to calculate the extreme performances of the vehicle. The objective of the present work is to evaluate the possibility to settle extreme atmospheric models leading to extreme sizing conditions.

097.030 **Regolith formation on Phobos and Deimos.**
M. Banaszkiewicz, W.–H. Ip.
2. COSPAR Colloquium: The environmental model of Mars, p. 109 – 112 (1991). – See Abstr. 012.039 for the main entry.
Regolith formation on the Martian satellites is considered as a stochastic process. Each cratering event is further divided into five elementary processes: impact, cratering, launching of ejecta, blanket formation and recapturing of ejecta from circum -Mars orbits. Peculiarities of the blanket shape for Phobos and Deimos ejecta, found from calculations, are indicated. The possible scenario of the regolith development on the satellites is discussed.

097.031 **Accretion of meteoritic material onto Mars: implications for the surface, atmosphere, and moons.**
G. J. Flynn.
2. COSPAR Colloquium: The environmental model of Mars, p. 121 – 124 (1991). – See Abstr. 012.039 for the main entry.
Micrometeorites provide a continuous, planet–wide flux of chondritic material which is accreted onto the surface of Mars and its moons. The micrometerorite contribution will perturb the chemical and mineralogical composition of the soil on Mars as well as on the Martian moons. Since the soil of Mars appears to be of uniform composition, direct measurement of the flux of micrometeorites at Mars by a future Mars orbiting spacecraft coupled with a determination of the meteoritic fraction of the soil from a Mars Sample Return Mission should allow inference of the global soil production rate on the planet. The micrometerorite contribution to the regoliths of Phobos and Deimos may be sufficient to coat these moons with a layer of chondritic meterotic material sufficient to obscure the underlying bedrock, making inference of the bulk mineralogy by reflection spectroscopy or other surface analysis techniques difficult or even impossible.

097.032 **Exploration of the Dao Vallis – Hadriaca Patera region on Mars by Mars balloon exploration vehicles.**
F. R. West.
2. COSPAR Colloquium: The environmental model of Mars, p. 125 – 128 (1991). – See Abstr. 012.039 for the main entry.
Reasons are listed for interest in exploring the Dao Vallis – Hadriaca Patera region of Mars. A tentative plan is outlined for exploring this region by balloon and possibly a surface rover.

097.033 **Martian Mössbauer spectrometer.**
J. Galazka–Friedman, A. Kotlicki, A. Ślawska–Waniewska, A. Witek.
2. COSPAR Colloquium: The environmental model of Mars, p. 129 – 132 (1991). – See Abstr. 012.039 for the main entry.
Investigation of soil on the surface of Mars by means of the Mössbauer spectrometer is proposed. Two main goals that can be achieved by Mössbauer spectroscopy are discussed: identification of iron containing coumpounds and determination of the Fe^{2+}/Fe^{3+} ratio. The parameters of the spectrometer to be sent on the surface of Mars are given. Various objects in the martian regolith that can be investigated by Mössbauer spectroscopy are indicated.

097.034 **How to lay a safety path for rover in 3–dimension scene?**
I. N. Sitenko, D. A. Usikov.
2. COSPAR Colloquium: The environmental model of Mars, p. 133 – 136 (1991). – See Abstr. 012.039 for the main entry.
The problem of automatic laying of safety route for rover in 3–dimensional scene is regarded. The map of distances, obtained

by algorithm of scene reconstruction based on stereopairs is used as the source map for the described algorithm. The landscape of martian surface type with slopes stawn with stones, rocks with possible clefts and craters is considered. The task of algorithm is to put route from the spot of rover to a distance point of the scene.

097.035 **Surface knowledge and risks to landing and roving: the scale problem.**
R. D. Bourke.
2. COSPAR Colloquium: The environmental model of Mars, p. 147 – 153 (1991). – See Abstr. 012.039 for the main entry.
The Martian surface in the vicinity of the two Viking landers proved to be rougher than expected when the landing sites were chosen. Retrospectively, hazards at the scale of the lander were not detectable nor predictable with the information available to the Viking team. Future surface missions, both landers and rovers, will want higher confidence that surface conditions are consistent with their capabilities prior to landing commitment. Analysis done by the NASA Mars Rover Sample Return study team found that rover performance was significantly increased and lander risk was significantly decreased if one meter objects and slopes >15 degrees (over baselines comparable to the lander and rover size) could be detected in the landing and roving zones. This implies that accurate surface models, obtained either by direct measurement or inference, are an important component in mission risk management. This suggests that prior to major surface missions, additional validated surface models will be required.

097.036 **The impact of Mars surface characteristics on rover design.**
D. L. S. Pivirotto.
2. COSPAR Colloquium: The environmental model of Mars, p. 155 – 160 (1991). – See Abstr. 012.039 for the main entry.
The characteristics of the surface of Mars will have a profound affect on the design and operation of roving vehicles. The design impacts of surface conditions, and the implications of knowing these conditions well or poorly, are described. It is a clear benefit to have orbital imaging, and radar subsurface sounding data at a scale of about one meter before designing the rover. In addition, knowledge of the terrain to one meter scale is important for efficient traverse and operations planning.

097.037 **Comparison of observations of the Martian satellites made in 1988 with ephemerides.**
F. Colas, J. E. Arlot.
Astron. Astrophys., Vol. 252, No. 1, p. 402 – 409 (Dec 1991).
This paper gives an analysis of 813 intersatellite observations of the satellites of Mars, Phobos and Deimos, made in 1988 during the opposition of Mars with the Sun. These observations have been made with an anti–blooming CCD target allowing the presence on a frame of the satellites and the planet simultaneously. Details are given on the procedure used for the measurement of the images and on the reduction performed: the small size of the field did not allow the use of reference stars. The accuracy of the observations has been tested by comparison with the new ephemerides ESAPHO and ESADEI. The r.m.s. of these observations is about 0.15 arcsec and a study of the defects of the method is made.

097.038 **Old and new mysteries of Mars.**
S. R. Brzostkiewicz, D. Górnicza.
Urania, Rok 62, Nr. 10, p. 258 – 269 (Oct 1991). In Polish.

097.039 **On possible discrete dusty annular formations around Mars and Venus.**
Yu. K. Gulak, I. A. Dychko.
Kinematika Fiz. Nebesn. Tel, Tom 7, No. 5, p. 27 – 28 (Sep–Oct 1991). In Russian. English translation in Kinematics Phys. Celest. Bodies, Vol. 7, No. 5.
The positions of the possible annular dusty belts around Mars and Venus are calculated on the basis of the statistical theory of quasi–isoenergetic complexes in the form of radial standing waves.

097.040 **The motion of Mars' pole. I. Rigid body precession and nutation.**
J. L. Hilton.
Astron. J., Vol. 102, No. 4, p. 1510 – 1527 (Oct 1991). Current Physics Microform No.: 9110D2042.
The total precession in longitude for a solid–core, rigid Mars model is found to be $\Delta\Psi = -7''.296 \pm 0''.021$ yr^{-1}. This precession includes the motion resulting from the extremely long–period (>10,000 yr) nutation components that result from changes in Mars' orbit with time. These same very long–period nutation components also contribute a "precession" in latitude of $\Delta\varepsilon = 0''.4255 \pm 0''.0012$ yr^{-1}. The nutation components in longitude with amplitudes larger than 0''.001 consist of eight solar nutation terms and one nutation each contributed by the motions of the nodes of Phobos and Deimos and a single nutation from the direct gravitational torque of Jupiter. The nutation components in obliquity consist of six solar terms and one nutation term each from the motion of the nodes of Phobos and Deimos. The amplitudes and periods of the solar nutation terms are in agreement with the nutation found by Reasenberg & King (1979). The solar precession rate also agrees with Reasenberg and King's value with the two small corrections resulting from the change in the orientation of Mars' orbit with time added to it. A single nutation in longitude driven by Jupiter is the only significant planetary contribution at the milliarcsecond level.

097.041 **Observations astrométriques CCD à tres long foyer des satellites proches des planètes: les problèmes de mesure et de réduction.**
F. Colas.
Journées 1991: Systèmes de référence spatio–temporels, p. 171 – 176 (1991). – See Abstr. 012.048 for the main entry.
During the 1988 opposition of Mars, the author used a CCD camera to measure the positions of Phobos and Deimos. He obtained more than 800 positions of the satellites. He has obtained high precision satellite–satellite positions. The problems of the reduction were: the non uniform background and of the astrometric calibration of the target. To estimate the quality of the positions, the author used a high precision ephemeris and then computed the O–C. The mean square error is about 0.15 arcsec. The reason of this calibration error is analysed. It is shown that it is possible to increase this precision to about 0.05 arcsec.

097.042 **Location of the Martian bow shock measurements by the plasma wave system on Phobos 2.**
J. G. Trotignon, R. Grard, S. Klimov.
Geophys. Res. Lett., Vol. 18, No. 3, p. 365 – 368 (Mar 1991).
During the early months of 1989, the Plasma Wave System, PWS, onboard the Soviet spacecraft Phobos–2 made plasma and electric–field spectrum measurements near the Martian bow shock. A total of 26 shock boundaries were unambiguously identified in the PWS data coming from the high–resolution mode. The location of these boundaries was modeled and the results compared with other studies.

097.043 **Physical properties of Phobos' regolith.**
L. Ksanfomaliti, V. Moroz, S. Murchie, D. Britt, N. Goroshkova, T. Duxbury, B. Shukov, E. Kührt, B. Murray, G. Nikitin, E. Petrova, C. Pieters, A. Soufflot, P. Fisher, J. W. Head.
Kosm. Issled., Tom 29, Vyp. 4, p. 621 – 640 (Jul–Aug 1991). In Russian. English translation in Cosmic Res. Vol. 29, No. 4.
The paper presents results of two experiments on board of the PHOBOS spacecraft: spectrophotometry in the range 300 – 600 nm and thermal radiometry (intrinsic thermal emission in the range 6 – 50 μm). Thermophysics of the regolith is similar to that of the Moon. The reflectivity of the Phobos regolith is heterogeneous along the studied traces and depends mostly on

features of the relief, first of all on craters and their age. In all cases the albedo which is low by its absolute value rises from 350 to 600 nm. Among the closest analogs of the surface there are optically modified mafic rocks. The awaited similarity of the reflected spectra with carbonaceous chondrites is not confirmed.

097.044 **Solar wind interaction with Mars. "Phobos-2" results.**
T. K. Breus, A. M. Krymskij, Eh. M. Dubinin, E. G. Eroshenko, V. Ya. Mitnitskij, N. F. Pisarenko, V. A. Styazhkin, S. V. Barabash.
Kosm. Issled., Tom 29, Vyp. 5, p. 741 - 753 (Sep-Oct 1991). In Russian. English translation in Cosmic Res. Vol. 29, No. 5.

The authors investigate the characteristics of the boundary which separates the region of solar wind (SW) dominance and plasma of planetary origin - the planetopause - discovered in experiments at "Phobos-2". It is shown that the characteristics of this boundary are close to the characteristics of analogous chemical boundaries - the cometopause near comet Halley and the chemical boundary existing on Venus in periods of high SW dynamical pressure. A numerical simulation of solar wind interaction with Mars has shown that at maximum solar activity interaction of the dense neutral atmosphere of Mars with SW can lead to a charge exchange of SW protons with the neutral atmosphere and to an essential decrease of SW proton concentration in the transition region near the planetopause.

097.045 **The magnetic field and magnetosphere of Mars.**
Sh. Sh. Dolginov.
Kosm. Issled., Tom 29, Vyp. 5, p. 754 - 789 (Sep-Oct 1991). In Russian. English translation in Cosmic Res. Vol. 29, No. 5.

A combined analysis of magnetograms measured at both day- and night-side of Mars and the solar wind fields and plasma measurements by "Mars-2,3,5" and "Phobos-2" indicates that Mars possesses an intrinsic dipole field controlled by the solar wind. Topological differences between the Martian and Venusian magnetospheric tails are shown. In the framework of the precession dynamo model one can reveal a similarity between the Martian and the Earth's magnetic field, and the derived Martian core site agrees with that obtained from the modern, "working" model of the internal structure of Mars under a natural presumption that the conductivities of the liquid cores of the two planets obey the inequality $\sigma_M \leqslant \sigma_E$. The above results and small inclination of the dipole axis to the rotation axis imply that the observed dipole field is associated with a present-day dynamo-activity in the Martian liquid core.

097.046 **Penetration of chondrites into Mars' atmosphere.**
V. Padevět.
Bull. Astron. Inst. Czech., Vol. 42, No. 4, p. 206 - 215 (Jul 1991).

The physical theory of meteors is capable of forecasting the depth of penetration of chondrites of types H, L, C2, C1 with weights in excess of 0.1 kg into Mars' atmosphere. As compared to the Earth's atmosphere, corresponding bolids end 20 to 30 km lower on Mars, whereas the altitudes of their beginnings are lower by about 10 km. The calculated impact energies of chondrites on Mars' surface enable to determine the lower limit of the dimensions of the impact craters.

097.047 **Theory of rotation of Phobos.**
I. Pešek.
Bull. Astron. Inst. Czech., Vol. 42, No. 5, p. 271 - 282 (Oct 1991).

The perturbations in Phobos' rotation due to the attraction of aspheric Mars and the Sun were derived semi-analytically under the assumption of presence of free librations. Particular care was given to resonances and free and forced term coupling. Influence of dissipation is briefly discussed. The forced parts of libration components are compared with the solutions by Chapront-Touzé, and Borderies and Yoder.

097.048 **Picked-up protons near Mars: PHOBOS observations.**
S. Barabash, Eh. Dubinin, N. Pissarenko (*N. F. Pisarenko*), R. Lundin, C. T. Russell.
Geophys. Res. Lett., Vol. 18, No. 10, p. 1805 - 1808 (Oct 1991).

The measurements carried out by the plasma spectrometer, ASPERA, onboard the PHOBOS-2 spacecraft show that protons, originating in the extended hydrogen corona of Mars, were observed at altitudes $\leqslant 7500$ km. The cyclotron instability of these pickup ions appears to generate Alfven waves observed by the MAGMA magnetometer. Analysis of the plasma data shows that weak pitch-angle diffusion of a ring-distribution of pickup protons occurs. The altitude profiles of pickup proton fluxes and number densities of the parent hydrogen atoms are derived.

097.049 **Viking 1 electron observations at Mars.**
F. S. Johnson, W. B. Hanson.
J. Geophys. Res., Vol. 96, No. A7, p. 11097 - 11118 (1 Jul 1991). - See Abstr. 012.063 for the main entry.

The Viking retarding potential analyzer (RPA) flown to Mars in 1976 had negative retarding potential sweeps in addition to the positive ion sweeps. An analysis of the electron mode sweeps made in Viking 1 above the ionosphere is presented here.

097.050 **Light elements in Mars meteorites.**
C. T. Pillinger.
Priroda, No. 12, p. 59 - 61 (Dec 1991). In Russian.

Investigations of SNC-meteorites are aimed at a better understanding of the peculiarities of Mars. This information is necessary for planning of the space missions, for example Mars-94, Mars Observer and Marsenet.

097.051 **Momentum transport at the Mars magnetopause.**
H. Perez-de-Tejada.
J. Geophys. Res., Vol. 96, No. A7, p. 11155 - 11164 (1 Jul 1991). - See Abstr. 012.063 for the main entry.

The conditions leading to the transport of momentum of the shocked solar wind to the Mars magnetosphere are examined. It is argued that planetary pickup ions born in the magnetosheath and scattered across the magnetopause by local turbulent waves carry that momentum and deliver it to the magnetospheric plasma. It is further suggested that as the pickup ions experience momentum scattering interactions with the wave field in the velocity shear adjacent to the magnetosphere they are subjected to a gradual internment within that region of space. The end effect of this phenomenon is that the pickup ions deliver a larger amount of momentum to the local flow than what they can subtract from it. Calculations of the efficiency of the process lead to values of the effective mean free path of the pickup ions of the order of a few hundred kilometers.

097.052 **The solar wind interaction with Mars: consideration of Phobos 2 mission observations of an ion composition boundary on the dayside.**
T. K. Breus, A. M. Krymskii (*A. M. Krymskij*), R. Lundin, Eh. M. Dubinin, J. G. Luhmann, Ye. G. Yeroshenko (*E. G. Eroshenko*), S. V. Barabash, V. Ya. Mitnitskii (*V. Ya. Mitnitskij*), N. F. Pissarenko (*N. F. Pisarenko*), V. A. Styashkin (*V. A. Styazhkin*).
J. Geophys. Res., Vol. 96, No. A7, p. 11165 - 11174 (1 Jul 1991). - See Abstr. 012.063 for the main entry.

This paper describes the features of the boundary in the plasma ion composition near Mars which separates the region dominated by the solar wind protons from the plasma of planetary origin. This boundary was detected by the ASPERA experiment on Phobos 2. It is argued that the features of this boundary seem to be similar to those of other composition boundaries detected elsewhere: the cometopause near comet Halley, and a boundary in the ion composition which appears near Venus during periods of high solar wind dynamic pressure. Numerical modeling of the solar wind interaction with Mars supports the idea that during solar maximum the interaction of the Martian neutral atmosphere with the solar wind can result in a composition transition from solar wind to planetary ions in the low–altitude magnetosheath. This transition occurs because of charge exchange of solar wind protons with the neutral atmosphere and photoionization.

097.053 **Comparison of observed plasma and magnetic field structure in the wakes of Mars and Venus.**
E. Dubinin (*Eh. M. Dubinin*), R. Lundin, W. Riedler,
K. Schwingenschuh, J. G. Luhmann, C. T. Russell, L. H. Brace.
J. Geophys. Res., Vol. 96, No. A7, p. 11189 – 11198 (1 Jul 1991).
– See Abstr. 012.063 for the main entry.

Plasma and magnetic field observations from the Phobos 2 spacecraft at Mars and the Pioneer Venus orbiter (PVO) at Venus show that there are some notable similiarities in the structure of the low–altitude magnetotails at both of these weakly magnetized planets. In particular, it is found that when conditions in the interplanetary medium are steady and the orbit sampling geometry is appropriate, two magnetic tail lobes, with an intervening "plasma sheet" or "central tail ray" in the approximate location of the dividing current sheet, are present. This behavior is seen in both the Phobos 2 ASPERA plasma analyzer data and in the PVO Langmuir probe data. In the Phobos 2 data, the tail ray is found to be composed primarily of antisunward streaming oxygen (O^+) plasma which has a bulk velocity consisent with an energy close to that of the upstream solar wind plasma. The PVO Langmuir probe experiment also detected two (or more) additional cold plasma structure flanking the central feature; Phobos 2 data, on the other hand, show a proton plasma "boundary layer" flanking the central (mostly O^+) tail ray or plasma sheet, with sporadic fluxes or rays of O^+ ions. If the latter considered is to be the magnetosheath (solar wind plasma) at the tail boundary, it is mainly the common central tail O^+ features that suggest that there are common planetary ion acceleration and magnetotail formation processes at work in the low–altitude wakes of Mars and Venus. On the other hand, an important contribution from picked–up exospheric hydrogen in the wake at Mars cannot be ruled out.

097.054 **Characteristics of spectra from the Martian bow shock and comparison with Venus, Earth, AMPTE, Jupiter, and Saturn.**
S. L. Moses, C. F. Kennel, R. J. Strangeway, R. Grard,
C. Nairn.
J. Geophys. Res., Vol. 96, No. A7, p. 11221 – 11226 (1 Jul 1991).
– See Abstr. 012.063 for the main entry.

The authors present plasma wave spectra from crossings of the Martian bow shock by the Phobos 2 spacecraft and compare them with spectra from bow shocks at other planets and the AMPTE lithium release. They find that the shock spectra of the inner planets are very similar in shape, and their wave energy densities, when normalized to the upstream electron thermal energy density, are comparable. However, the Martian shock spectrum most resembles the spectrum from AMPTE, which may indicate either the presence of pickup particles or processes involving ions with gyroradii that are large compared with the size of the interaction region. The normalized wave amplitudes at Mars are midway between the high normalized amplitudes found at Jupiter and Saturn and the lower, normalized amplitudes at Earth. This is consistent with a trend that has recently been shown to occur out to Neptune's bow shock.

097.055 **Plasma and wave observations in the night sector of Mars.**
C. M. C. Nairn, R. Grard, A. Skalsky, J. G. Trotignon.
J. Geophys. Res., Vol. 96, No. A7, p. 11227 – 11234 (1 Jul 1991).
– See Abstr. 012.063 for the main entry.

The Phobos 2 spacecraft, initially injected into an elliptical orbit around Mars on January 29, 1989, was subsequently transferred, on February 18, 1989, to a nearly circular orbit, close to that of the Phobos moon, with an areocentric radius of the order of 9600 km. The spacecraft remained in this orbit until the end of the mission, on March 27, 1989. This paper summarizes the plasma and wave observations carried out in the night sector of Mars with the plasma wave system (PWS). Embedded in the magnetic field structure of the Martian tail, cold electron enhancements ("tail rays") with densities in the range 10–65 cm^{-3} are observed in association with broadband wave activity extending from a few hertz up to several kilohertz; these enhancements appear to have characteristics analogous to enhancements observed at Venus. The ion outflow through the Martian eclipse region is estimated from Langmuir probe measurements to be of the order of 10^{25}ions/s.

097.056 **The solar wind interaction with Mars: Mariner 4, Mars 2, Mars 3, Mars 5, and Phobos 2 observations of bow shock position and shape.**
J. A. Slavin, K. Schwingenschuh, W. Riedler, Ye. Yeroshenko
(*E. G. Eroshenko*).
J. Geophys. Res., Vol. 96, No. A7, p. 11235 – 11243 (1 Jul 1991).
– See Abstr. 012.063 for the main entry.

Observations taken by Mariner 4, Mars 2, Mars 3, Mars 5, and Phobos 2 are used to model the shape, position, and variability of the Martian bow shock for the purpose of better understanding the interaction of this planet with the solar wind. Emphasis is placed upon comparisons with the results of similar analyses at Venus, the only planet known to have no significant intrinsic magnetic field.

097.057 **Derivation of electron densities from differential potential measurements upstream and downstream of the bow shock and in the magnetosphere of Mars.**
A. Pedersen, C. Nairn, R. Grard, K. Schwingenschuh.
J. Geophys. Res., Vol. 96, No. A7, p. 11243 – 11252 (1 Jul 1991).
– See Abstr. 012.063 for the main entry.

A new technique is developed where the potential difference between a conductive spacecraft at floating potential and an electric field probe, negatively biased in relation to the spacecraft, with a current of 50 nA, is used to determine electron densities in the range 1–70 cm^{-3}. The density can be obtained with an acceptable accuracy without any knowledge of the electron energy, provided the latter does not exceed 100 eV. This technique has been applied to the Phobos 2 spacecraft data to determine characteristic densities in the solar wind near Mars, in the foot of the bow shock, in the planetosheath, and at the top of the ionosphere. In the latter region, measurements have also been made using a Langmuir probe on board the spacecraft. A magnetic barrier, with a low plasma density, is observed between the planetosheath and the top of the ionosphere. Pronounced electron density increases are observed on the flanks and in the tail, some correlated and some anticorrelated with variations in the magnetic field strength.

097.058 **Plasma wave system measurements of the Martian bow shock from the Phobos 2 spacecraft.**
J. G. Trotignon, R. Grard, S. Savin.
J. Geophys. Res., Vol. 96, No. A7, p. 11253 – 11264 (1 Jul 1991).
– See Abstr. 012.063 for the main entry.

The high–resolution data of the electric field observations performed by the plasma wave system (PWS) during some of the Martin bow shock intersections by Phobos 2 were analyzed. Plasma and wave detectors are very useful instruments for locating the shock transition region and studying structures in the upstream region, such as the foot or the electron foreshock. The electron plasma oscillations that develop in the latter give access to the plasma density of the solar wind. Shock surface models derived from the PWS data are compared to those obtained by other authors, and attention is paid to similarities and differences between the electric field measurements obtained for Mars, Venus, and Earth.

097.059 **Interplanetary magnetic field control of the Mars bow shock: evidence for Venuslike interaction.**
T. L. Zhang, K. Schwingenschuh, H. Lichtenegger, W. Riedler,
C. T. Russell, J. G. Luhmann.
J. Geophys. Res., Vol. 96, No. A7, p. 11265 – 11270 (1 Jul 1991).
– See Abstr. 012.063 for the main entry.

The Mars bow shock location and shape have been determined by examining the PHOBOS spacecraft magnetometer data. Observations show that the position of the terminator bow shock varies with interplanetary magnetic field orientation in the same way as at Venus. The shock is farthest from Mars in the direction of the interplanetary electric field, consistent with the idea that mass loading plays an important role in the solar wind interaction with Mars. The authors also find that the shock cross section at the terminator plane is asymmetric and is controlled by the interplanetary magnetic field as expected from the asymmetric propagation velocity of the fast magnetosonic wave. Comparing with earlier mission data, the authors show that the Mars shock location varies with solar activity. The shock is farther from Mars during solar maximum. Thus the solar wind interaction with Mars appears to be Venuslike, with a magnetic moment too small to affect significantly the solar wind interaction.

097.060 **The dynamics of submicron-sized dust particles lost from Phobos.**
M. Horányi, M. Tátrallyay, A. Juhász, J. G. Luhmann.
J. Geophys. Res., Vol. 96, No. A7, p. 11283 - 11290 (1 Jul 1991).
- See Abstr. 012.063 for the main entry.
The dynamics of submicron-sized dielectric particles lost from the Martian moon Phobos are studied in connection with the possible detection of dust by the Phobos 2 spacecraft. The motion of these small dust grains is influenced not only by gravity but also by solar radiation pressure and electromagnetic forces. The plasma environment of Mars is described by applying a hybrid gasdynamic-cometary model. Some of the submicron-sized grains ejected at speeds on the order of a few tens meters per second can stay in orbit around Mars for several months forming a nonuniform and time-dependent dust halo. The lifetime of the particles depends on their size, on the actual interplanetary parameters (constant or varying with a periodicity of 28 days) and also on the orbital position of Mars at the time of ejection since there is a 24° obliquity between the orbit of Phobos and that of Mars.

097.061 **On the possible source of the ionization in the nighttime Martian ionosphere. 1. Phobos 2 HARP electron spectrometer measurements.**
M. I. Verigin, K. I. Gringauz, N. M. Shutte, S. A. Haider, K. Szego, P. Kiraly, A. F. Nagy, T. I. Gombosi.
J. Geophys. Res., Vol. 96, No. A11, p. 19307 - 19314 (1 Nov 1991).
The measurements of electron spectra in the Martian magnetosphere by the HARP instrument on board the Phobos 2 orbiter are presented. The energy of the electrons (a few tens of electron volts) is sufficient for the impact ionization of the planetary neutral gas, and the characteristic flux of electrons ($\sim 10^8 cm^{-2} s^{-1}$) could produce the nightside ionospheric layer with a peak density of a few thousands of electrons per cubic centimeter, which corresponds to densities observed earlier during radio occultations of the Mars 4 and 5 and Viking 1 and 2 spacecraft. The possibility of magnetospheric electron precipitation into the nightside atmosphere of Mars is in agreement with the mainly induced nature of the magnetic field in the planetary magnetotail (as at Venus), while the variability of the Martian nightside ionosphere may be explained by the partial screening of the atmosphere by a weak intrinsic magnetic field of the planet.

097.062 **On the problem of the Martian atmosphere dissipation: Phobos 2 TAUS spectrometer results.**
M. I. Verigin, K. I. Gringauz, G. A. Kotova, N. M. Shutte, H. Rosenbauer, S. Livi, A. K. Richter, W. Riedler, K. Schwingenschuh, K. Szego.
J. Geophys. Res., Vol. 96, No. A11, p. 19315 - 19320 (1 Nov 1991).
The measurements of proton spectra obtained by the TAUS spectrometer on board the Phobos 2 spacecraft in elliptical orbits near Mars are presented. A strong deceleration of the solar wind upstream of the Martian bow shock was revealed. It can be caused by the mass loading of the plasma flow by ions originating from the hot oxygen/hydrogen corona of Mars and/or by protons specularly reflected from the bow shock. In the first case the deceleration of the solar wind by about 100 km/s implies that the hot oxygen corona of Mars could be several times denser than it was anticipated to be (at least during the observation period that was close to solar cycle maximum). Furthermore, the loss of planetary oxygen through the corona appears to be the main process of oxygen loss from Mars. The upper limit of loss rate for such a process is determined to be 10^{26} oxygen atoms or 2.5 kg of oxygen per second.

097.063 **Detectability of Martian carbonates from orbit using thermal neutrons.**
W. C. Feldman, B. M. Jakosky.
J. Geophys. Res., Vol. 96, No. E1, p. 15589 - 15598 (25 Aug 1991).
Numerical simulations of Martian neutron leakage flux spectra are made in order to explore the detectability of magnesium and calcium carbonate deposits using neutron spectroscopic techniques from orbit. The primary signature of such deposites is found to be an enhanced thermal amplitude. Although this enhancement is weakened by (1) partial burial beneath an aeolian regolith blanket, (2) admixture with regolith on a microscopic (centimeter) or macroscopic (tens of centimeters) scale, and (3) reduction in the areal size of the deposit, near-surface stratigraphies of carbonates hypothesized by some authors as possible on Mars are still detectable by simple neutron sensors from orbit. However, the large variations in the magnitude of the thermal neutron enhancements caused by different carbonate deposit configurations found in this study require a combined gamma ray and neutron analysis for their unique specification.

097.064 **Origin of periodically spaced wrinkle ridges on the Tharsis Plateau of Mars.**
T. R. Watters.
J. Geophys. Res., Vol. 96, No. E1, p. 15599 - 15616 (25 Aug 1991). With a correction in Vol. 96, No. E5, p. 22829 - 22832 (25 Dec 1991).
Ridged plains are a major geologic unit on Mars and are a predominant unit on the Tharsis Plateau, a region that has undergone uplift, extension, and extensive shield and flood volcanism. These units, probably flood volcanic in origin, are characterized by landforms classed as wrinkle ridges. Wrinkle ridges are interpreted to be folds, resulting from buckling followed by reverse to thrust faulting (flexure-fracture). A prominent characteristic of many of the wrinkle ridges on Mars, particularly those in the Tharsis ridge system, is the periodic nature of their spacing. The periodic spacing has been evaluated in six major provinces on the Tharsis Plateau with regions further divided into domains based largely on variation in ridge orientation. The average spacing of the Tharsis ridges, based on 2934 measurements, is 30 km. In an effort to account for the periodic nature of the wrinkle ridges, the ridged plains material is modeled as both a single member and a multilayer with frictionless contacts that has buckled at a critical wavelength of folding.

097.065 **Reconciliation of stress and structural histories of the Tharsis region of Mars.**
K. L. Tanaka, M. P. Golombek, W. B. Banerdt.
J. Geophys. Res., Vol. 96, No. E1, p. 15617 - 15634 (25 Aug 1991).
The authors present a new compilation of the structural and stratigraphic evolution of the Tharsis region of Mars that incorporates recent advances in understanding its stratigraphy, and they introduce a lithospheric deformation model that can account for the observations.

097.066 **Graben and extension in northern Tharsis, Mars.**
J. B. Plescia.
J. Geophys. Res., Vol. 96, No. E3, p. 18883 - 18895 (25 Oct 1991).
A morphometric analysis of north-south trending graben at 35°N in northern Tharsis is used to provide the first quantitative measurements of crustal extension in and strain for areas of the Martian crust that have undergone both extreme and mild

deformation. These estimates also demonstrate the viability of the Viking data for making such quantitative assessment.

097.067 **CO_2 condensation and the climate of early Mars.**
J. F. Kasting.
Icarus, Vol. 94, No. 1, p. 1 – 13 (Nov 1991).
A one-dimensional, radiative-convective climate model was used to reexamine the question of whether early Mars could have been kept warm by the greenhouse effect of a dense, CO_2 atmosphere. The new model differs from previous models by considering the influence of CO_2 clouds on the convective lapse rate and on the planetary radiation budget. Condensation of CO_2 decreases the lapse rate and, hence, reduces the magnitude of the greenhouse effect. This phenomenon becomes increasingly important at low solar luminosities and may preclude warm (0°C), globally averaged surface temperatures prior to ~ 2 billion years ago.

097.068 **Mars volatile evolution: evidence from stable isotopes.**
B. M. Jakosky.
Icarus, Vol. 94, No. 1, p. 14 – 31 (Nov 1991).
Constraints on the history of the Martian volatile system can be determined by using observations of the ratios $^{18}O/^{16}O$, $^{17}O/^{16}O$, $^{13}C/^{12}C$, and D/H in atmospheric species and in various components of the SNC meteorites which are thought to be pieces of the Martian surface. These observations are compared with the variations expected from processes which can fractionate the different isotopes. All of the available observations and models are synthesized for the first time in order to provide a coherent view of the evolution of both CO_2 and H_2O over geologic time. While no scenarios for atmospheric evolution can be absolutely and uniquely ruled out, a strongly favored scenario does exist. In this favored model, fractionation of oxygen by escape to space is diluted by the exchange with a nonatmospheric reservoir; it is unlikely that there is sufficient CO_2 adsorbed in the regolith to serve this function, so the polar-water-ice deposits must be exchanging with the atmospheric water over geologic time. Exchange of oxygen between atmospheric CO_2 and H_2O probably provides the observed fractionation between those species.

097.069 **Infrared solar occultation sounding of the Martian atmosphere by the Phobos spacecraft.**
V. A. Krasnopolsky (*V. A. Krasnopol'skij*), O. I. Korablev, V. I. Moroz, A. A. Krysko (*A. A. Krys'ko*), J. E. Blamont, E. Chassefiere.
Icarus, Vol. 94, No. 1, p. 32 – 44 (Nov 1991).
A spectrometer designed at the Space Research Institute, Moscow, for solar occultation measurements of the Martian atmosphere from the Phobos spacecraft has the spectral range 2706 – 2740 and 5293 – 5372 cm^{-1} in the first and second diffraction orders, respectively, with resolving power $\lambda/\Delta\lambda = 1200$. Data on the water vapor distribution in the Martian atmosphere were derived from nine occultation profiles at low altitudes and give a mean mixing ratio of 150 ppm below 12 km decreasing to 3 ppm at 40 km. The total amount of the water vapor is 10 μm of precipitable water. Mean scale height of water vapor is 3.8 km between 12 and 45 km. Small-scale variations of the water vapor density, though close to the experimental errors, may be caused by tidal waves with an amplitude of 2K near 20 km and 5K at 40 km with a further increase to 15K at 100 km.

097.070 **Resonant obliquity of Mars?**
W. R. Ward, D. J. Rudy.
Icarus, Vol. 94, No. 1, p. 160 – 164 (Nov 1991).
The obliquity of Mars executes large-scale oscillations as a result of spin axis and orbit plane precessions. This motion is an important driver of the system and exhibits two dominant periodicities: (1) oscillations driven by the differential spin axis and orbit precession rates during which the obliquity may change by $\gtrsim 20°$ over $\sim 10^5$y and (2) amplitude modulations over $\sim 10^6$y due to changes in the orbital inclination. This paper discusses a third time scale for change involving a possible spin-orbit resonance.

097.071 **Making Mars habitable.**
C. P. McKay, O. B. Toon, J. F. Kasting.
Nature, Vol. 352, No. 6335, p. 489 – 496 (8 Aug 1991).
Mars is believed to be lifeless, but it may be possible to transform it into a planet suitable for habitation by plants, and conceivably humans. The success of such an enterprise would depend on the abundance, distribution and form of materials on the planet that could provide carbon dioxide, water and nitrogen.

097.072 **Ancient oceans, ice sheets and the hydrological cycle on Mars.**
V. R. Baker, R. G. Strom, V. C. Gulick, J. S. Kargel, G. Komatsu, V. S. Kale.
Nature, Vol. 352, No. 6336, p. 589 – 594 (15 Aug 1991). With a correction in Vol. 354, No. 6348, p. 86 – 87 (7 Nov 1991).
A variety of anomalous geomorphological features on Mars can be explained by a conceptual scheme involving episodic ocean and ice-sheet formation. The formation of valley networks early in Mars' history is evidence for a long-term hydrological cycle, which may have been associated with the existence of a persistent ocean. Cataclysmic flooding, triggered by extensive Tharsis volcanism, subsequently led to repeated ocean formation and then dissipation on the northern plains, and associated glaciation in the southern highlands until relatively late in martian history.

097.073 **Early outgassing of Mars supported by differential water solubility of iodine and xenon.**
D. S. Musselwhite, M. J. Drake, T. D. Swindle.
Nature, Vol. 352, No. 6337, p. 697 – 699 (22 Aug 1991). Letter-to-the-editor.
The martian atmosphere has a high $^{129}Xe/^{132}Xe$ ratio compared with any on Earth and most meteorites. The $^{129}Xe/^{132}Xe$ ratio in the martian atmosphere is also high relative to the martian mantle. As ^{129}Xe is the daughter product of the extinct nuclide ^{129}I, a means of fractionating iodine from xenon early in martian history appears necessary to account for the $^{129}Xe/^{132}Xe$ ratios of its known reservoirs. Crystal/melt partitioning will fractionate iodine from xenon in the right sense, but the fractionation is probably inadequate in magnitude; differences in the silicate melt solubilities of iodine and xenon would cause fractionation in the wrong direction. The authors present a model to account for the martian xenon data which relies on the very different solubilities of the two elements in water to fractionate them after outgassing. Atmospheric xenon is lost by impact erosion during heavy bombardment, followed by release of ^{129}Xe produced from ^{129}I decay in the crust.

097.074 **The magnetic field and the magnetosphere of the planet Mars.**
Sh. Sh. Dolginov, L. N. Zhuzgov.
Planet. Space Sci., Vol. 39, No. 11, p. 1493 – 1510 (Nov 1991).
The Phobos-2 plasmagrams and magnetograms measured on the dayside of Mars are used to identify the shock front, the transition region and magnetopause crosses in the same way as these boundaries are identified near the Earth's magnetosphere.

097.075 **On the Schumann resonances on Mars.**
A. I. Sukhorukov.
Planet. Space Sci., Vol. 39, No. 12, p. 1673 – 1676 (Dec 1991).
A conductive ionosphere and a totally non-conductive layer of the atmosphere close to the surface of the planet form a quasispherical concentric resonator. This provides in principle for the possibility of the existence of global resonances of an electromagnetic field generated by thunderstorm activity or by hydromagnetic waves excited in an upper ionosphere and transformed into ordinary electromagnetic waves while penetrating the resonator. The author has obtained an estimate of resonance frequencies of a Martian resonator: $f_1 \sim 13$–14 Hz, $f_2 \sim 24$–26 Hz, $f_3 \sim 35$–38 Hz, etc. for two essentially different models of electron density distribution in the low ionosphere of

Mars. The corresponding estimated quality values are low: $Q_n \sim 2$–4. A relatively wide range of the quality variation depending on a model of averaged altitudinal electron density distribution in the low ionosphere of Mars yields the criterion for an adequate model.

097.076 **First absolute wind measurements in the middle atmosphere of Mars.**
E. Lellouch, J. J. Goldstein, S. W. Bougher, G. Paubert, J. Rosenqvist.
Astrophys. J., Vol. 383, No. 1, p. 401 – 406 (10 Dec 1991).

The first absolute wind measurements in the middle atmosphere of Mars (40 – 70 km) were obtained from Doppler shifts in the J = 2–1 CO transition at 230.538 GHz. During the 1988 opposition, this line was observed at 100 kHz resolution with the IRAM 30 m telescope. The high S/N of the data allowed measurement of winds with a 1 σ absolute line–of–sight accuracy of 20 m s^{-1}. These wind measurements, together with the temperature measurements of Deming et al., provide the first experimental rough picture of the middle atmosphere circulation of Mars, in general agreement with the Jaquin axisymmetric middle atmosphere model and the current Mars GCM model of Pollack et al.

097.077 **Un comportement anormal de la région polaire sud de Mars durant l'opposition de 1990.**
S. Ebisawa.
Obs. Trav., No. 28, p. 31 – 36 (1991).

097.078 **The role of thermal vapor diffusion in the subsurface hydrologic evolution of Mars.**
S. M. Clifford.
Geophys. Res. Lett., Vol. 18, No. 11, p. 2055 – 2058 (Nov 1991).

Various lines of geomorphic evidence suggest that large bodies of groundwater were present on Mars throughout much of its early geologic history. In this paper the hydrologic response of groundwater to the thermal evolution of the early martian crust is considered.

097.079 **The particle size of Martian aeolian dunes.**
K. S. Edgett, P. R. Christensen.
J. Geophys. Res., Vol. 96, No. E5, p. 22765 – 22776 (25 Dec 1991).

The effective particle size of unconsolidated materials on the Martian surface can be determined from thermal inertia, due to a pore size dependence of thermal conductivity at Martian atmospheric pressures. Because dunes consist of a narrow range of well–sorted, unconsolidated particles, they provide for a test of the relationship between particle size and thermal inertia calculated from midinfrared emission data for the Martian surface. The authors use two independent approaches. First, thermal inertia data indicate that Martian dunes have an average particle size of about 500 ± 100 μm, or medium to coarse sand. Second, the authors determine expected dune particle sizes from grain trajectory calculations and the particle size transition from suspension to saltation. They find a Martian transition grain size of about 210 μm. Both approaches indicate that Martian dune sand should be coarser than terrestrial dune sand. Thus, while terrestrial dune grains are in the fine to medium sand range, the average Martian dune sediments are probably medium to coarse sands.

097.080 **Structural development of Coprates Chasma and western Ophir Planum, Valles Marineris rift, Mars.**
R. A. Schultz.
J. Geophys. Res., Vol. 96, No. E5, p. 22777 – 22792 (25 Dec 1991).

A portion of Valles Marineris was mapped in detail in order to clarify the dominant processes responsible for the formation of Coprates Chasma.

097.081 **Near–Mars space.**
J. G. Luhmann, L. H. Brace.
Rev. Geophys., Vol. 29, No. 2, p. 121 – 140 (May 1991). With a correction in Vol. 29, No. 4, p. 607 – 608 (Nov 1991).

The space environment of Mars is largely determined by its lack of a substantial planetary magnetic field. In contrast to Earth, Mars does not have radiation belts or an auroral "oval", and its perturbation of the interplanetary medium or solar wind is confined to a relatively small volume. Moreover, the direct interaction of the solar wind plasma with the Martian atmosphere produces a distinctive local energetic particle population of planetary origin, which provides evidence of the possibly significant scavenging of certain elements from Mars over the life of the solar system. In this paper the authors describe the general characteristics of near–Mars space, as it is known today, including both the ambient interplanetary environment and the ionosphere, upper atmosphere, and regions above that are affected by the presence of Mars.

097.082 **Existence of a swarm of particles in the vicinity of the orbit of Phobos.**
A. V. Krivov, L. L. Sokolov, K. V. Kholshevnikov, V. A. Shor.
Sol. Syst. Res., Vol. 25, No. 3, p. 233 – 242 (Nov 1991). English translation of Astron. Vestn., Tom 25, No. 3, p. 317 – 326 (1991).

The existence of a swarm of particles in the vicinity of the orbit of Phobos is demonstrated. The dimensions and density of the swarm material have been determined; it is five to six orders of magnitude denser than the interplanetary density. The mechanism for forming the dust torus is the following. The fall of meteorites onto Phobos leads to the ejection into outer space of a mass that is many times larger than the mass of the impactor. The meteorids which are scattered inside the torus are gradually swept out by Phobos. The balance between ejection and sweeping out leads to an equilibrium state after less than 100 years.

097.083 **Ozone heating in the Martian atmosphere.**
B. L. Lindner.
Icarus, Vol. 93, No. 2, p. 354 – 361 (Oct 1991).

Ozone is shown to be a minor, although nonnegligible, heat source in the Martian atmosphere, in contradiction to earlier reports which had suggested a larger contribution to the atmospheric heat budget by ozone. It is further shown that the observed hemispherical asymmetry in ozone abundance is unlikely to be a significant factor in the observed hemispherical asymmetry in the permanent polar caps. Ozone heating may still be an important consideration in the formation and decay of clouds and snowfall, although not as important as previously suggested.

097.084 **Origin of the Martian crustal dichotomy: evaluating hypotheses.**
G. E. McGill, S. W. Squyres.
Icarus, Vol. 93, No. 2, p. 386 – 393 (Oct 1991).

Any hypothesis for the origin of the Martian global dichotomy should survive three elementary tests: (1) it must account for the observed plan shape and apparent depth of the Martian northern lowland, (2) it must be physically consistent, and (3) it must be compatible with available geological and geophysical data. At present, there are three contending types of hypotheses for the origin of the dichotomy: creation by some endogenic process or processes, creation by a single mega–impact, or creation by several overlapping large impacts. None of these hypotheses can survive all three tests without the incorporation of additional processes. While the authors believe that aspects of the multiple–impact hypothesis are required to account for some topographic details of the boundary and of the lowland, other processes appear better able to account for the dichotomy as a whole. These could include an early mega–impact, endogenic processes, or some combination of these.

097.085 **Martian paleolakes and waterways: exobiological implications.**
D. H. Scott, J. W. Rice Jr., J. M. Dohm.
Origins Life Evol. Biosphere, Vol. 21, No. 3, p. 189 – 198 (1991).

Mars and the development of life.
See Abstr. 003.033.

The environmental model of Mars. Proceedings.
See Abstr. 012.039.

Chapman Conference on Venus and Mars: Atmospheres, Ionospheres, and Solar Wind Interactions, Balatonfüred (Hungary), 4 – 8 Jun 1990.
See Abstr. 012.063.

Dissociative recombination of the ground state of $N^+{}_2$.
See Abstr. 022.081.

Spectral signature of oxidized pyroxenes: implications to remote sensing of terrestrial planets.
See Abstr. 022.129.

The Martian magnetic soils: proposal for a soil magnetometer.
See Abstr. 035.156.

Measurement of the magnetic field vector from a rotating spacecraft.
See Abstr. 035.159.

The scintillation gamma–spectrometer for the determination of the element composition of Martin rocks on "Phobos" spacecraft.
See Abstr. 035.231.

Das Kameraexperiment FREGAT an Bord der Phobos–Sonde.
See Abstr. 035.233.

Algorithm for stereo pair matching.
See Abstr. 036.147.

The reconstruction of 3–dimension scenes from stereopairs by the dynamic programming method.
See Abstr. 036.148.

Mapping of overspread targets in radar astronomy.
See Abstr. 036.232.

Monte Carlo Simulationen der kosmischen Bestrahlung planetarer Oberflächen und Anwendung zur geochemischen Analyse Neutronen–induzierter Gammaspektren des Mars.
See Abstr. 036.362.

Observations of Mars using Hubble Space Telescope Observatory.
See Abstr. 051.022.

Environmental considerations of design alternatives for the exploration of Mars by balloon.
See Abstr. 051.042.

Mars sample return: site selection strategy and proposal in Aeolis region.
See Abstr. 051.043.

Scientific rationale and technical elements of future Mars exploration in ESA.
See Abstr. 051.044.

Small mobile apparatus for Mars surface studies.
See Abstr. 051.045.

The robotic exploration missions at Mars.
See Abstr. 051.046.

Perturbed motion of an orbiter in the Martian atmosphere.
See Abstr. 052.012.

Drag perturbations of low eccentric orbits in the Martian atmosphere.
See Abstr. 052.024.

Multiple–ion plasma boundaries.
See Abstr. 062.023.

Role of carbon dioxide in cooling planetary thermospheres.
See Abstr. 091.017.

Computer simulation of regolith vertical structure and exposure age of small satellites of planets.
See Abstr. 091.022.

Sources of error in the photoclinometric determination of planetary topography: a reappraisal.
See Abstr. 091.023.

Tests of random density models of terrestrial planets.
See Abstr. 091.027.

Physical properties of terrestrial planets and their satellites: an introduction.
See Abstr. 091.054.

Asymmetries in the location of the Venus and Mars bow shock.
See Abstr. 093.031.

Multiple–impact mechanisms on Venus and other solid planets and satellites.
See Abstr. 093.038.

Comparative terrestrial planet thermospheres. 1. Solar cycle variation of global mean temperatures.
See Abstr. 093.045.

Magnetization of the ionospheres of Venus and Mars: results from radio occultation measurements.
See Abstr. 093.048.

A comparison of induced magnetotails of planetary bodies: Venus, Mars, and Titan.
See Abstr. 093.054.

Global hybrid simulation of unmagnetized planets: comparison of Venus and Mars.
See Abstr. 093.055.

Mössbauer spectroscopy of SNC meteorites.
See Abstr. 105.032.

Interplanetary variability in particle fluxes recorded by the low energy charged particle detector SLED ($\sim$30 keV – $>$30 MeV) during the Cruise Phase of the Phobos Mission to Mars and its moons.
See Abstr. 106.009.

098 Minor Planets

098.001 **The accuracy of astrometric positions.**
B. Koch.
Minor Planet Bull., Vol. 18, No. 3, p. 23 - 25 (Jul - Sep 1991).
The accuracy of precise astrometric positions is discussed for the example of the asteroids 1989 PB and 1990 KA. If certain photographic and measuring requirements are met, positions can be obtained to an accuracy of better than 2 arcsec with focal lengths of at least 1000 mm.

098.002 **Photoelectric photometry of 4 Vesta.**
C.-I. Lagerkvist, T. Oja.
Minor Planet Bull., Vol. 18, No. 3, p. 25 - 27 (Jul - Sep 1991).
UBV photometry was performed of 4 Vesta during 3 nights in Nov. and Dec. 1990. Composite lightcurves and color curves are presented and a variation of the colors with rotational phase was found in the sense that Vesta is more red just before lightcurve maximum.

098.003 **General report of position observations by the ALPO Minor Planets Section for the year 1990.**
F. Pilcher.
Minor Planet Bull., Vol. 18, No. 3, p. 28 - 31 (Jul - Sep 1991).
Observations of positions of minor planets by members of the Minor Planets Section in calendar year 1990 are summarized, along with significant discrepancies from the predicted magnitudes noted for a few planets.

098.004 **Precise astrometric positions of 1991 AQ and 1990 PA.**
B. Koch.
Minor Planet Bull., Vol. 18, No. 3, p. 34 (Jul - Sep 1991).

098.005 **Astrometric positions of minor planets.**
J. Sudol, G. Garber.
Minor Planet Bull., Vol. 18, No. 3, p. 34 (Jul - Sep 1991).

098.006 **Astrometric positions of minor planets.**
R. Chanal.
Minor Planet Bull., Vol. 18, No. 3, p. 34 (Jul - Sep 1991).

098.007 **Astrometric positions of minor planets.**
J. Pryal.
Minor Planet Bull., Vol. 18, No. 3, p. 36 (Jul - Sep 1991).

098.008 **Rotation period and lightcurve of asteroid 338 Budrosa.**
R. G. Hutton, J. Licandro.
Minor Planet Bull., Vol. 18, No. 4, p. 37 - 38 (Oct - Dec 1991).
Photoelectric observations in V and B filters of the asteroid 338 Budrosa were made from Estación de Altura "Dr. Carlos U. Cesco" of Félix Aguilar Observatory during the 1990 apparition. The synodic rotational period, lightcurve amplitude and average B–V color found for 338 Budrosa are given.

098.009 **Photoelectric photometry of asteroid 125 Liberatrix.**
R. G. Hutton.
Minor Planet Bull., Vol. 18, No. 4, p. 39 - 40 (Oct - Dec 1991).
Photoelectric observations in V and B filters of the asteroid 125 Liberatrix were made from Estación de Altura "Dr. Carlos U. Cesco" of Félix Aguilar Observatory during the 1990 apparition. The 3.97 ± 0.1 hour synodic rotational period, and 0.695 ± 0.014 average B–V color agree with previous determinations.

098.010 **The 4 January 1991 U.T. occultation of star SAO93228 by asteroid Vesta: event timing using the SSP-3 photometer.**
G. J. Frey.
I.A.P.P.P. Commun., No. 45, p. 86 - 88 (Sep - Nov 1991).

098.011 **The end of civilization.**
G. L. Verschuur.
Astronomy, Vol. 19, No. 9, p. 50 - 54 (Sep 1991).
Will a minor planet cause a catastrophe?

098.012 **Bedeckung von SAO 147658 durch (521) Brixia.**
H.-J. Bode, R. Büchner, W. Zimmermann.
Sterne Weltraum, Jahrg. 30, Nr. 8–9, p. 548 - 549 (Aug–Sep 1991).

098.013 **Neue Namen von Kleinplaneten.**
J. Jahn.
KPM, Jahrg. 6, No. 17, p. 14 - 24 (Aug 1991).

098.014 **Entdeckungen von Erdkreuzern 1985 bis 1991.**
G. Monz.
KPM, Jahrg. 6, No. 18, p. 31 - 33 (Dec 1991).

098.015 **Neue Namen von Kleinplaneten.**
J. Jahn.
KPM, Jahrg. 6, No. 18, p. 34 - 35 (Dec 1991).

098.016 **Positionsbeobachtungen von Kleinen Planeten und Kometen.**
J. Jahn.
KPM, Jahrg. 6, No. 18, p. 38 - 39 (Dec 1991).

098.017 **The interpretation of light curves of minor planets.**
A. Yu. Kogan, K. Yu. Pimenov.
Pis'ma Astron. Zh., Tom 17, No. 9, p. 850 - 856 (Sep 1991). In Russian. English translation in Sov. Astron. Lett., Vol. 17, No. 5.
The telescopic images of minor planets look like twinkling points. Their brightness varies due to their irregular shape and rotation. The brightness records may be processed in such a way that it makes it possible (under some a priori assumptions) to determine uniquely both the rotation parameters of an asteroid and its shape. Mathematically, the problem deals with the inversion of an integral transformation relative to the Gaussian spherical mapping. A numerical inversion algorithm is presented as well as some results of the shape reconstruction from the computer-generated light curves.

098.018 **Motions of asteroids at the Kirkwood gaps. II. On the 5:2, 7:3, and 2:1 resonances with Jupiter.**
M. Yoshikawa.
Icarus, Vol. 92, No. 1, p. 94 - 117 (Jul 1991).
The behavior of asteroids at the 5:2, 7:3, and 2:1 resonances with Jupiter is studied in the scheme of the elliptical restricted three-body problem. The origin of the Kirkwood gaps at the 3:1 and 5:2 resonances can be explained by the scenario of collision with major planets. However, within the restriction of this study, the results obtained here do not support this scenario for the gaps at the 7:3 and 2:1 resonances. In order to reveal the formation mechanism of these gaps, one should study more general cases, that is, the motions of asteroids in three-dimensional space under the perturbation of several planets.

098.019 **Shape and size of asteroid 4 Vesta: speckle interferometry and polarimetry.**
V. S. Tsvetkova, V. N. Dudinov, S. B. Novikov, Ye. A. Pluzhnik, Yu. G. Shkuratov, V. G. Vakulik, A. P. Zheleznyak.
Icarus, Vol. 92, No. 2, p. 342 - 349 (Aug 1991).
Speckle interferometric data for Vesta and diffraction-limited images of Vesta's disk are presented for three measurements near the opposition of 1988. It is argued that polarimetric data can be utilized to estimate a contribution of albedo features to Vesta's lightcurve, and thus to derive the geometric component of the lightcurve. The analysis of speckle data and polarimetric data together with the results of simulation of speckle interferometry

for Vesta enables one to conclude that Vesta must have an almost equilibrium shape.

098.020 **The spin vector of asteroid 10 Hygiea.**
T. Michałowski, F. P. Velichko, M. Lindgren, T. Oja, C.-I. Lagerkvist, P. Magnusson.
Astron. Astrophys., Suppl. Ser., Vol. 91, No. 1, p. 53 - 59 (Nov 1991).

Photoelectric lightcurves of asteroid 10 Hygiea from the 1985, 1988 and 1989 apparitions are presented. A sidereal rotation period of $27^{h}.65909$ was obtained. This is much longer than previously published values. The spin vector has ecliptic coordinates close to one of the two solutions ($\lambda_0 = 299°$, $\beta_0 = -39°$) and ($\lambda_0 = 112°$, $\beta_0 = -41°$). An ellipsoidal model of Hygiea has axis ratios of about $a/b = 1.36$ and $b/c = 1.04$.

098.021 **Ursprung und Entwicklung der Asteroiden.**
R. P. Binzel, M. A. Barucci, M. Fulchignoni.
Spektrum Wiss., No. 12, p. 110 - 117 (Dec 1991).

098.022 **Physical studies of asteroids. XXIII: Photometric observations of the asteroids 6, 32, 196, 243, 416, 532 and 1580.**
A. Erikson, G. Cutispoto, H. Debehogne, G. Hahn, C.-I. Lagerkvist, M. Lindgren, P. Magnusson.
Astron. Astrophys., Suppl. Ser., Vol. 91, No. 2, p. 259 - 264 (Dec 1991).

Photometric studies of seven asteroids observed at the European Southern Observatory (La Silla, Chile) with the 50 cm ESO telescope are presented. For these asteroids (6 Hebe, 32 Pomona, 196 Philomela, 243 Ida, 416 Vaticana, 532 Herculina and 1580 Betulia) composite lightcurves are presented. Mean colour indexes, calculated for some of the nights, are also given.

098.023 **Phase trajectories of asteroids as a function of inclination in the case of first-order resonances.**
I. A. Gerasimov.
Sov. Astron., Vol. 35, No. 3, p. 327 - 328 (May 1991). Current Physics Microform No.: 9201X0609. English translation of Astron. Zh., Tom 68, Vyp. 3, p. 664 - 666 (1991).

The influence of the orbital inclination of an asteroid on its phase trajectory in the case of first-order resonance is considered. It is shown that allowance for the inclination is necessary in calculating the characteristics of the motion of resonance asteroids.

098.024 **Photometric properties of the minor planets: observations of 6 Hebe in 1987 and 1989.**
A. J. Hollis.
J. Br. Astron. Assoc., Vol. 101, No. 5, p. 289 - 290 (Oct 1991).

098.025 **Die Entdeckung des kleinen Planeten (3910) Liszt.**
E. W. Elst, F. Börngen.
Sterne, Band 67, Heft 4, p. 223 - 224 (1991).

098.026 **Keeping track of comets and minor planets.**
G. V. Williams.
Spaceflight, Vol. 33, No. 8, p. 276 - 280 (Aug 1991).

098.027 **Asteroids: package of surprise.**
C. Peebles.
Spaceflight, Vol. 33, No. 10, p. 342 - 343 (Oct 1991).

098.028 **Orbital elements of numbered minor planets.**
E. Bowell, G. V. Williams, B. G. Marsden, C. M. Bardwell, S. Nakano, H. Kaneda, T. Kobayashi, L. D. Schmadel, H. Oishi.
Minor Planet Circ., Nos. 18485 - 19346 (Jul - Dec 1991).

The numbered minor planets are listed according to their definitive number. Newly numbered objects are indicated by an asterisk. The names of the orbit computers are given behind the respective M.P.C. numbers:
(193), (271) 18999, (287) 18600 Bowell; (348) 19274 Williams; (412) 18999, (736) 18775, (793) 18600, (842) 19274, (857), (1032), (1046) 18601, (1164) 18999, (1610), (1656), (1672), (1785) 18601, (1815) 18999, (1869) 18601 Bowell; (1932) 18999 Williams; (1969), (1991) 18999, (2113) 18775, (2147), (2148) 18601, (2169) 18999, (2175) 18601, (2197) 18775, (2217) 18601, (2255) 18602, (2265) 18776, (2290) 18602, (2300) 18776, (2306) 18602, (2328) 18776, (2329), (2334) 18602, (2338) 18999, (2339), (2349), (2363) 18602, (2381) 18776, (2401), (2403) 18602, (2405) 18999, (2409), (2413) 18602, (2484), (2493) 18603, (2504) 18776, (2535) 19000, (2544) 18603, (2545) 18776, (2563), (2571) 19000, (2606), (2614) 18776, (2653) 18603, (2692), (2702) 18776, (2710) 19000, (2721), (2727) 18776, (2738) 18777, (2739) 18603, (2752) 18777, (2755) 19000, (2761), (2800) 18603, (2802) 18777, (2807) 19000, (2853) 18603, (2854) 18777, (2862) 19000, (2884) 18603, (2889) 18777, (2907) 19000, (2916) 18603, (2918) 18777, (2927), (2932) 18603, (2933), (2980), (2981), (2982) 18604, (2993), (3019), (3027), (3032) 18777, (3033) 18604, (3110) 18777, (3116) 18604, (3134) 18777, (3135) 18604, (3149) 18778, (3160), (3170), (3229), (3231) 18604, (3247) 18778, (3252) 18604, (3255), (3265) 18778, (3297), (3345), (3370) 18605, (3371), (3374), (3394), (3410), (3435) 18778, (3464) 19000, (3524) 18778, (3552) 18605, (3569) 19000, (3601) 18778, (3612) 19000, (3620) 18778, (3623) 18605, (3696), (3759) 18779, (3801) 18605, (3807) 18779, (3812) 18605, (3843), (3845) 18779, (3898) 19000 Bowell; (3903) 19274 Williams; (3952) 18605, (3954) 18779, (3959) 19001 Bowell; (3966) 19275 Williams; (3969) 18605, (3972), (4014) 18779, (4025) 19001, (4026) 18779, (4027) 19001, (4040) 18779, (4087) 18605, (4088), (4095) 19001, (4097), (4114) 18605 Bowell; (4117) 18606 Marsden; (4126) 18606, (4145) 19001, (4146) 18606 Bowell; (4158) 19001 Williams; (4163) 18606, (4173), (4344) 18779, (4371) 18606 Bowell; (4413) 19275 Williams; (4529) 18780, (4755) 19001 Bowell; (4878)*, (4879)* 18606 Bardwell; (4880)* 18607 Nakano; (4881)* 18607 Bowell; (4882)* 18607–18608 Nakano; (4883)* 18608 Kaneda; (4884)* 18608 Nakano; (4885)* 18608–18609 Kobayashi; (4886)* 18609, (4887)* 18609–18610 Bowell; (4888)* 18610 Nakano; (4889)* 18610 Schmadel; (4890)* 18610–18611 Bardwell; (4891)* 18611 Nakano; (4892)*18611 Kaneda; (4893)* 18611–18612 Williams; (4894)*, (4895)* 18612 Bardwell; (4896)* 18612–18613 Oishi; (4897)* 18613 Nakano; (4898)*18613–18614, (4899)*, (4900)* 18614 Bardwell.

For a continuation of this list see Abstr. 098.029.

098.029 **Orbital elements of numbered minor planets.**
S. Nakano, C. M. Bardwell, K. Ichikawa, H. Kaneda, E. Bowell, T. Kobayashi, B. G. Marsden, G. V. Williams, T. Urata, D. W. E. Green, K. Kawanishi.
Minor Planet Circ., Nos. 18485 - 19346 (Jul - Dec 1991).

This list is a continuation of Abstr. 098.028. The numbered minor planets are listed according to their definitive number. Newly numbered objects are indicated by an asterisk. The names of the orbit computers are given behind the respective M.P.C. numbers:
(4901)* 18614–18615 Nakano; (4902)* 18615 Bardwell; (4903)* 18615 Ichikawa; (4904)* 18615–18616 Nakano; (4905)* 18616 Kaneda; (4906)* 18616 Bowell; (4907)* 18616–18617 Kobayashi; (4908)* 18780 Marsden; (4909)* 18780, (4910)* 18780–18781 Nakano; (4911)* 18781 Bardwell; (4912)* 18781 Kobayashi; (4913)* 18782 Bardwell; (4914)* 18782 Nakano; (4915)* 18782, (4916)* 18783 Marsden; (4917)* 18783, (4918)* 18783–18784 Bardwell; (4919)* 18784 Nakano; (4920)* 18784 Marsden; (4921)* 18784–18785 Nakano; (4922)* 18785, (4923)* 18785–18786 Williams; (4924)* 18786 Bowell; (4925)* 18786 Bardwell; (4926)* 18786–18787 Bowell; (4927)* 18787 Kobayashi; (4928)* 18787–18788 Bardwell; (4929)* 18788 Nakano; (4930)* 18788 Marsden; (4931)* 18788–18789 Bardwell; (4932)* 18789, (4933)* 18789–18790, (4934)* 18790 Nakano; (4935)* 18790 Marsden; (4936)* 18790–18791, (4937)* 18791 Bardwell; (4938)* 18791–18792, (4939)* 18792, (4940)* 18792–18793 Nakano; (4941)* 18793 Urata; (4942)* 18793–18794, (4943)* 18794 Nakano; (4944)* 18794 Marsden; (4945)* 18794–18795 Nakano; (4946)* 18795 Bardwell; (4947)* 18795 Marsden; (4948)* 18796 Nakano; (4949)* 18796 Kaneda; (4950)* 18796–18797 Williams; (4951)*, (4952)* 18797 Nakano; (4953)* 18797–18798 Marsden; (4954)* 18798–18799 Williams; (4955)* 18799 Kaneda; (4956)* 18799, (4957)* 18799–18800, (4958)* 18800 Williams; (4959)* 18800–

18801 Urata; (4960)* 18801 Bowell; (4961)* 19001–19002 Kaneda; (4962)* 19002 Green; (4963)* 19002 Marsden; (4964)* 19002–19003 Nakano; (4965)* 19003, (4966)* 19003–19004 Marsden; (4967)*, (4968)* 19004 Williams; (4969)* 19004–19005 Marsden; (4970)* 19005 Williams; (4971)* 19005–19006 Marsden; (4972)* 19006 Williams; (4973)* 19006 Kaneda; (4974)* 19006–19007 Williams; (4975)*, (4976)* 19007 Kaneda; (4977)* 19007–19008, (4978)* 19008 Williams; (4979)* 19275 Nakano; (4980)* 19275, (4981)* 19275–19276, (4982)* 19276, (4983)* 19276–19277 Bardwell; (4984)* 19277 Kaneda; (4985)* 19277 Marsden; (4986)* 19277–19278, (4987)* 19278 Nakano; (4988)* 19278 Marsden; (4989)* 19278–19279 Bardwell; (4990)* 19279 Williams; (4991)* 19279–19280 Marsden; (4992)*, (4993)* 19280, (4994)* 19281, (4995)* 19281–19282, (4996)* 19282 Nakano; (4997)* 19282 Bardwell; (4998)* 19282–19283 Urata; (4999)* 19283 Bardwell; (5000)* 19283–19284, (5001)* 19284 Nakano; (5002)* 19284 Kawanishi; (5003)* 19285 Williams; (5004)* 19285 Nakano; (5005)* 19285–19286 Kawanishi; (5006)*, (5007)* 19286, (5008)* 19286–19287 Nakano; (5009)* 19287 Bardwell; (5010)* 19287 Nakano; (5011)* 19287–19288 Williams; (5012)* 19288 Nakano.

098.030 **Orbital elements of unnumbered minor planets.**
E. Bowell, H. Kaneda, T. Kobayashi, S. Nakano, G. V. Williams, A. Lowe, H. Oishi, D. W. E. Green, B. G. Marsden, C. M. Bardwell, R. Nagata.
Minor Planet Circ., Nos. 18485 - 19346 (Jul - Dec 1991).

The unnumbered minor planets are sorted according to their provisional designations. The names of the orbit computers are given behind the respective M.P.C. numbers:
[1929 CY] 18617 Bowell; [1929 VS] 18617, [1930 UX] 19008 Kaneda; [1931 VS] 19008–19009 Kobayashi; [1937 NN] 19009 Kaneda; [1938 HA] 18617 Nakano; [1952 SW_1] 19288 Williams; [1953 GN], [1953 TA_1] 19009, [1964 BF] 19010 Kaneda; [1965 DC] 18618 Lowe; [1971 UD_1] 18618 Bardwell; [1972 RF] 18801 Green; [1972 RX_1] 19010 Kaneda; [1973 SQ_3] 19288–19289 Bowell; [1973 SS_4] 19289 Williams; [1973 UR_5] 19010 Kaneda; [1974 PC] 18802 Marsden; [1975 LR] 19010 Kaneda; [1975 TR_2] 19010–19011 Bowell; [1975 VR_5] 18802 Marsden; [1976 GN_2] 19011 Williams; [1976 QP] 19289 Kaneda; [1976 SK_3] 18802 Bowell; [1976 UD_4] 19011, [1976 YB_2] 19289, [1977 EL] 19011 Williams; [1977 EC_2] 18618 Oishi; [1977 EW_5] 19290, [1977 EA_6] 19012 Kaneda; [1977 FN] 19012 Kobayashi; [1977 NN] 18618 Nakano; [1977 OL_1] 19290 Kaneda; [1977 QN_2] 18619 Kobayashi; [1977 RG] 18802 Green; [1977 UO_5] 19290 Oishi; [1978 NY_7] 18619 Nakano; [1978 RX_7] 19290, [1978 RH_9] 18802–18803 Marsden; [1978 RD_{10}] 18803 Green; [1978 SL_6] 19012 Kobayashi; [1978 SC_7] 19291 Kaneda; [1978 VW_2] 19291 Nakano; [1978 VB_6] 18619 Marsden; [1978 VT_8] 18619–18620 Kaneda; [1978 WB] 18620, [1978 WU] 19012, [1979 HH_3] 19013 Williams; [1979 MB_4] 18803 Bowell; [1979 MS_4] 19291 Williams; [1979 MX_5] 18803–18804, [1979 MK_6] 18804 Bowell; [1979 MU_8] 18620 Williams; [1979 QX_3] 19291 Kaneda; [1979 QB_{10}] 18620 Williams; [1979 SW_2] 19013 Kaneda; [1979 SU_{11}] 18804 Green; [1979 UC_4] 18804, [1980 DD_1] 19013, [1980 EF] 18804–18805 Bowell; [1980 FR_1] 19291–19292 Williams; [1980 FT_3] 18805 Nakano; [1980 GG] 18620 Bowell; [1980 PB_1] 19013 Kaneda; [1980 RP] 18805, [1980 RL_2] 19014 Bowell; [1980 RB_8], [1980 TW_5] 19014 Williams; [1980 TK_6] 18621 Bowell; [1980 TE_7] 19292 Nakano; [1980 TQ_{14}] 18805 Williams; [1980 UC] 19014 Bowell; [1980 UL] 19015, [1981 DZ] 18805–18806, [1981 DT_2] 18806 Williams; [1981 DB_3] 19015 Kaneda; [1981 ES_4] 18806 Bowell; [1981 EA_{12}] 18806 Green; [1981 EW_{20}] 18621, [1981 EY_{27}], [1981 ET_{47}] 18807 Bowell; [1981 GM_1] 19015, [1981 QQ_2] 19015–19016, [1981 QP_3] 19016, [1981 SE] 18621 Williams; [1981 SY_1] 18807 Nakano; [1981 SQ_2] 18808, [1981 SZ_6] 18621 Williams; [1981 TT] 18808 Bowell; [1981 TM_3] 18808 Nakano; [1981 YO_1] 18622 Marsden; [1982 FF_2] 19292 Nakano; [1982 FP_3] 18622 Bowell; [1982 JR_1] 18622, [1982 QG] 19292–19293 Kaneda; [1982 QM] 19016 Williams; [1982 SJ_1] 19293 Kaneda; [1982 SG_4] 18622–18623 Marsden; [1982 SL_6] 19017 Williams; [1982 UQ_6] 18808 Marsden; [1982 UB_7], [1983 AB] 18623 Williams; [1983 AA_3] 18808–18809 Bardwell; [1983 EU] 19293 Bowell; [1983 GQ] 19293–19294 Nakano; [1983 HB_1] 18623–18624 Marsden; [1983 JQ] 18809 Green; [1983 RL] 19017 Williams; [1983 RB_2] 19294 Kaneda; [1983 RM_2] 18624 Williams; [1983 RC_4] 18624 Bardwell; [1984 DZ] 19294 Williams; [1984 EG] 18624 Marsden; [1984 HL_1] 18624–18625 Bowell; [1984 HP_1] 19017, [1984 MR] 18809 Kaneda; [1984 SA] 19294 Nagata; [1984 SW_5], [1984 SH_6] 18625, [1984 UD_3] 19017 Williams.
For a continuation of this list see Abstr. 098.031.

098.031 **Orbital elements of unnumbered minor planets.**
H. Kaneda, G. V. Williams, E. Bowell, B. G. Marsden, R. Nagata, T. Kobayashi, D. W. E. Green, C. M. Bardwell, S. Nakano, K. Ichikawa.
Minor Planet Circ., Nos. 18485 - 19346 (Jul - Dec 1991).

This list is a continuation of Abstr. 098.030. The unnumbered minor planets are sorted according to their provisional designations. The names of the orbit computers are given behind the respective M.P.C. numbers:
[1985 BH] 19018 Kaneda; [1985 CV_1] 19295, [1985 CC_2] 18809–19810, [1985 FD], [1985 JY] 19295, [1985 QP] 19295–19296 Williams; [1985 QL_4] 19018 Bowell; [1985 SC_3] 19018, [1985 TG_3] 18626, [1985 TW_3] 19296 Williams; [1985 UQ] 19018 Bowell; [1985 VE] 19019 Kaneda; [1986 GV] 18626, [1986 PC_1] 18810 Williams; [1986 PS_4] 18810 Marsden; [1986 PX_4] 19296–19297 Williams; [1986 QS] 19297 Nagata; [1986 RD] 19297, [1986 RY_5] 18810 Kaneda; [1986 SD_2] 18810–18811 Williams; [1986 TD_2] 19296 Bowell; [1986 TB_5] 18811 Williams; [1986 TB_7] 19298 Kaneda; [1986 TZ_{11}] 19019 Williams; [1986 VT] 19019 Bowell; [1986 VF_5] 19019 Williams; [1986 WN_7] 19298 Bowell; [1986 WO_7] 19298 Kaneda; [1986 WP_8] 19019–19020 Williams; [1987 BB] 18626 Kaneda; [1987 DC_6] 18626–18627 Marsden; [1987 DD_6] 18811 Kaneda; [1987 EQ] 18811–18812 Marsden; [1987 EV] 18627 Williams; [1987 JG] 18812, [1987 QF_3] 19020 Bowell; [1987 QW_7] 19020 Kobayashi; [1987 RC_1] 19298–19299 Williams; [1987 RT_3] 18812 Marsden; [1987 SC_1] 18812 Bowell; [1987 SL_1] 19299 Kaneda; [1987 SQ_1] 19020, [1987 SM_4] 18812–18813 Williams; [1987 ST_{11}] 18813 Green; [1987 SV_{11}] 19299 Kaneda; [1987 SN_{12}] 18813, [1987 UN] 19299 Williams; [1987 UW] 19021 Bardwell; [1987 UW_1], [1987 WT_1] 19021 Kaneda; [1987 WS_3] 18627 Bardwell; [1987 YL_1] 18627–18628, [1988 AF], [1988 BX_1], [1988 BY_1] 18628, [1988 CA] 18629 Williams; [1988 CA_1] 18813–18814 Kaneda; [1988 CL_2] 19300 Nakano; [1988 EL] 19021–19022, [1988 EA_2] 19300, [1988 ER_2] 18629, [1988 FK] 19022 Williams; [1988 HE] 18814 Marsden; [1988 PL] 19300, [1988 PV] 18629, [1988 PL_1] 19300, [1988 RO] 19022, [1988 RF_1] 18629–18630, [1988 RS_4] 19301 Williams; [1988 RS_6] 19022 Kobayashi; [1988 RF_9] 18630 Nakano; [1988 RO_{11}] 19022–19023 Bowell; [1988 RJ_{13}] 18630 Williams; [1988 UT] 19023 Bowell; [1988 VK] 18814 Marsden; [1988 VK], [1988 VN_3] 19023 Williams; [1988 VP_3] 18814 Bowell; [1988 VN_4] 18814–18815 Williams; [1988 VR_5] 18815, [1988 VE_7] 19024, [1988 VM_9] 18815 Kaneda; [1988 XK] 19024, [1988 XO] 18630 Williams; [1988 XT] 19301 Kobayashi; [1988 XD_1] 19301 Kaneda; [1988 XE_1] 18631 Marsden; [1988 XU_1] 18815, [1988 XY_1] 19302 Kaneda; [1989 AU_1] 18816 Williams; [1989 AN_6] 18816 Ichikawa; [1989 BB_1] 18631 Williams; [1989 BR_1] 19024 Kaneda; [1989 CA] 19024–19025 Kobayashi; [1989 CD] 19302 Bowell; [1989 CL_1] 18816, [1989 CV_1] 19302 Nakano; [1989 CK_2] 19025 Williams; [1989 CW_2] 18631 Bowell; [1989 CY_2] 19302–19303, [1989 CJ_5] 19025 Kaneda; [1989 ES] 18816–18817 Williams; [1989 GJ] 18817, [1989 SA_3] 19025 Bowell; [1989 SN_5] 18631 Kaneda; [1989 TN] 19025–19026 Bowell; [1989 TT_1] 19026 Williams; [1989 TL_{15}] 19026 Kaneda; [1989 UL_1] 19303 Williams; [1989 UU_3] 18632 Kaneda; [1989 UY_3] 18817 Bowell; [1989 WD_3] 18632, [1989 YB_6] 18817, [1989 YS_6] 19026 Kaneda.
For a continuation of this list see Abstr. 098.032.

098.032 **Orbital elements of unnumbered minor planets.**
C. M. Bardwell, T. Kobayashi, G. V. Williams, B. G. Marsden, E. Bowell, H. Kaneda, S. Nakano, D. W. E. Green, K. Ichikawa.
Minor Planet Circ., Nos. 18485 - 19346 (Jul - Dec 1991).

This list is a continuation of Abstr. 098.031. The unnumbered minor planets are sorted according to their provisional designations. The names of the orbit computers are given behind the respective M.P.C. numbers:

[1990 BG] 18817–18818 Bardwell; [1990 BU] 18632 Kobayashi; [1990 BS_1] 18632 Williams; [1990 CH] 18818 Marsden; [1990 DA] 18818–18819 Williams; [1990 DL] 18819 Bowell; [1990 DK_3] 18819 Marsden; [1990 DU_3] 19027 Kaneda; [1990 EO] 18819 Bowell; [1990 EU] 19027 Kaneda; [1990 ES_1] 18819–18820 Williams; [1990 ES_3] 19303 Nakano; [1990 EL_7] 19303, [1990 FD_1] 18820, [1990 FR_1] 19303–19304, [1990 FW_1] 18820 Williams; [1990 HA] 18820 Bardwell; [1990 HC_1] 19304 Williams; [1990 HF_1] 18633 Marsden; [1990 KG_2] 18820 Williams; [1990 MJ] 18821 Bardwell; [1990 OD] 19027 Bowell; [1990 OB_2], [1990 OK_2] 18633, [1990 QG] 18633–18634 Williams; [1990 QD_2], [1990 QG_2] 18821, [1990 QB_4] 18822 Green; [1990 QF_5] 19304, 1990 RT_2] 19304–19305, [1990 RC_3] 19305 Williams; [1990 RE_7], [1990 RC_8] 19305 Kaneda; [1990 RS_{17}] 19305, [1990 SB] 18822, [1990 SW] 19305–19306, [1990 SG_4] 18822–18823, [1990 ST_8], [1990 SJ_{16}] 19306 Williams; [1990 SQ_{16}] 19027–19028, [1990 SX_{16}] 19028 Kaneda; [1990 SM_{28}] 19306, [1990 TR] 18634 Williams; [1990 TQ_1), [1990 TE_7] 18823 Bowell; [1990 TU_8] 19307 Williams; [1990 UF_2] 18823 Bowell; [1990 UG_3] 18634 Kaneda; [1990 VA] 18823 Bardwell; [1990 VA] 19307 Marsden; [1990 VU_1] 18634, [1990 VD_2] 19307, [1990 VX_2] 18823–18834, [1990 WA] 18824 Williams; [1990 WW_2] 18824, [1990 WZ_2] 18634 Bardwell; [1990 WZ_2] 18825, [1990 WU_5] 19307 Williams; [1990 XZ] 18635 Bardwell; [1991 AR_1] 18635 Marsden; [1991 AX_1] 18635 Kaneda; [1991 BB] 18635 Bardwell; [1991 BB] 18825, [1991 CF] 19308 Williams; [1991 CQ] 18825 Bardwell; [1991 CA_2] 18635–18636 Kaneda; [1991 DG] 18825 Bardwell; [1991 DS] 19028 Bowell; [1991 DT] 19308 Kaneda; [1991 EE] 18825 Marsden; [1991 EA_1] 18636 Kobayashi; [1991 FH] 18826, [1991 GB_1] 18636 Williams; [1991 GP_1] 18826 Bardwell; [1991 GE_2] 18636 Marsden; [1991 GQ_2] 18636; [1991 GC_6], [1991 GC_7], [1991 GV_8] 18637 Kaneda; [1991 GE_9] 19308 Nakano; [1991 GY_9] 18637, [1991 GZ_9] 18637–18638 Williams; [1991 GG_{10}], [1991 GP_{10}], [1991 GQ_{10}] 18826, [1991 JG] 18638 Kaneda; [1991 JW] 18827 Williams; [1991 JX] 18638 Marsden; [1991 JX] 18827 Williams; [1991 JY] 18638 Bardwell; [1991 JA_1] 18638 Green; [1991 JD_1], [1991 JG_1] 18639 Marsden; [1991 JN_1] 18639 Williams; [1991 JS_1] 18639 Kaneda; [1991 JY_1], [1991 JZ_1] 18827 Bardwell; [1991 JR_2], [1991 KA] 18640 Nakano; [1991 LD] 18640 Kaneda; [1991 LH] 18640 Marsden; [1991 LW], [1991 LC_1] 18641 Williams; [1991 LE_1] 19308 Bardwell; [1991 MA] 18641 Williams; [1991 NL], [1991 NP] 19309 Bardwell; [1991 NQ] 18827 Marsden; [1991 NY] 18641, [1991 NG_1], [1991 NH_1] 18828 Williams; [1991 NK_1] 19028 Kaneda; [1991 NL_1] 18828 Williams; [1991 NS_1] 19028–19029 Kaneda; [1991 NX_1] 18828 Ichikawa; [1991 NA_2] 18829 Williams; [1991 NP_2] 19029 Kaneda; [1991 NR_2] 19309 Bardwell; [1991 NS_2] 19029 Williams; [1991 OA] 18642 Marsden; [1991 OA] 18829, [1991 OW], [1991 OZ] 19029 Williams; [1991 PE] 18829 Marsden; [1991 PJ] 19029–19030 Kaneda; [1991 PK] 19030 Williams; [1991 PE_1] 19309 Marsden; [1991 PM_1] 19309 Williams; [1991 PQ_1] 19030 Kaneda; [1991 PT_1] 18829 Williams.
For a continuation of this list see Abstr. 098.033.

098.033 **Orbital elements of unnumbered minor planets.**
H. Kaneda, B. G. Marsden, G. V. Williams, K. Ichikawa, D. W. E. Green, S. Nakano, T. Urata, E. Bowell.
Minor Planet Circ., Nos. 18485 - 19346 (Jul - Dec 1991).

This list is a continuation of Abstr. 098.032. The unnumbered minor planets are sorted acording to their provisional designations. The names of the orbit computers are given behind the respective M.P.C. numbers:
[1991 PO_2] 19310 Kaneda; [1991 PE_3] 19310 Marsden; [1991 PE_5] 18829–18830, [1991 PE_5] 19310 Williams; [1991 PJ_5] 19310–19311, [1991 PS_6] 19311 Marsden; [1991 PM_8] 19030 Ichikawa; [1991 PR_{10}] 19031 Williams; [1991 PR_{10}] 19311 Green; [1991 PA_{11}] 19031 Williams; [1991 PH_{11}] 19311 Kaneda; [1991 PM_{11}] 19031 Nakano; [1991 PW_{12}] 19311–19312 Kaneda; [1991 PN_{13}] 19031 Nakano; [1991 PY_{14}], [1991 PF_{15}] 19032 Williams; [1991 PF_{15}] 19312 Kaneda; [1991 PW_{16}] 19312 Marsden; [1991 QB], [1991 QC] 19032 Kaneda; [1991 QD] 19033 Nakano; [1991 QE] 19033, [1991 RB] 18830, [1991 RB] 19033, [1991 RC] 18830, [1991 RC] 19033 Marsden; [1991 RG] 19312 Kaneda; [1991 RJ] 19033 Williams; [1991 RN] 19033–19034, [1991 RA_1] 19034 Nakano; [1991 RS_1] 19034 Kaneda; [1991 RJ_2] 19034, [1991 RJ_2] 19313 Marsden; [1991 RX_2] 19313 Kaneda; [1991 RZ_2] 19313 Marsden; [1991 RT_3] 19313 Kaneda; [1991 RC_4] 19313–19314 Urata; [1991 RW_4] 19314 Marsden; [1991 RB_5] 19314 Urata; [1991 RJ_5] 19034–19035 Williams; [1991 RQ_7] 19314 Kaneda; [1991 RX_{10}] 19315 Nakano; [1991 SY] 19315 Williams; [1991 SF_1] 19315 Nakano; [1991 SL_1] 19315, [1991 SM_1] 19316 Kaneda; [1991 TM] 19035 Bowell; [1991 TT], [1991 TU], [1991 TB_1] [1991 TB_2] 19316 Marsden; [1991 TE_2] 19316–19317 Nakano; [1991 TK_2] 19317 Williams; [1991 TF_3], [1991 VA] 19317 Marsden; [2023 P–L] 18642 Nakano; [2197 P–L] 18830 Williams; [2557 P–L] 19317 Kaneda; [2702 P–L] 19035 Bowell; [4026 P–L] 18830, [4100 P–L] 19317–19318 Williams; [4581 P–L] 18642 Nakano; [4606 P–L] 18830–18831 Bowell; [4614 P–L], [4882 P–L], [6030 P–L] 19318, [6114 P–L] 19035, [6207 P–L] 19318–19319, [7581 P–L] 18831 Williams; [7637 P–L] 18642 Nakano; [7643 P–L] 19319 Williams; [9540 P–L] 18831 Green; [1039 T–3] 18834 Bowell; [1053 T–2] 19328, [1054 T–3] 19330 Williams; [1070 T–2] 18643 Marsden; [1079 T–2] 19328 Williams; [1089 T–1], [1104 T–1] 19319 Green; [1114 T–1] 19319–19320, [1170 T–2] 19329 Williams; [1171 T–1], [1217 T–1], [1220 T–1] 19320, [1232 T–1] 19320–19321 Green; [1266 T–2] 19035 Williams; [1269 T–2] 18831, [1280 T–1] 19321, [1281 T–2] 18832, [1287 T–1] 19321 Green; [1289 T–2] 18832 Bowell; [1324 T–2] 19036, [1607 T–2] 18832, [2041 T–3] 19036 Williams; [2146 T–1] 19321 Green; [2192 T–3] 18834–18835 Marsden; [2224 T–2] 18832–18833 Green; [2232 T–2] 18833 Marsden; [2246 T–] 19322 Green; [2252 T–2] 19328 Williams; [2259 T–1] 19322 Green; [2272 T–3] 19330 Williams; [2281 T–2] 18833 Marsden; [2288 T–3] 19331 Bowell; [2312 T–1] 19322, [2908 T–2] 19329 Williams; [3019 T–3] 19331 Bowell; [3058 T–1] 19322–19323, [3070 T–2] 19036 Williams; [3105 T–1], [3128 T–1], [3174 T–1] 19323 Green; [3178 T–2] 19329–19330, [3181 T–2] 19330 Williams; [3196 T–1] 19323–19324 Green; [3201 T–2] 18834 Marsden; [3227 T–1], [3233 T–1], [3252 T–1] 19324, [3271 T–1] 19324–19325, [3289 T–1], [3300 T–1] 19325 Green; [3395 T–3] 19331, [3424 T–3] 19331–19332, [3854 T–3] 19332 Williams; [4050 T–1] 19325 Green; [4050 T–3] 19332 Williams; [4098 T–1] 19325–19326 Green; [4121 T–1] 19326 Williams; [4179 T–3] 18835 Green; [4181 T–3] 18835 Williams; [4192 T–1] 19326 Green; [4206 T–1] 19326, [4255 T–1], [4283 T–1], [4298 T–1], [4321 T–1] 19327, [4408 T–1] 19328, [5016 T–3] 18835, [5016 T–3] 19036–19037 Williams; [5192 T–3] 18836, [5490 T–2] 18834 Kaneda.

098.034 **Complex dynamical behaviour of the asteroid 2335 James associated with the secular resonances ν_5 and ν_{16}: numerical studies and theoretical interpretation.**
C. Froeschlé, A. Morbidelli, H. Scholl.
Astron. Astrophys., Vol. 249, No. 2, p. 553 - 562 (Sep 1991).

A forward integration in the frame of the Sun–Jupiter–Saturn model over 1 Myr shows that the asteroid 2335 James is located in the secular resonance ν_{16}. James is a ω–librator during the whole period. A complex orbital evolution appears in a backwards integration over 5 Myr: James is temporarily located also in the ν_5 secular resonance. The argument of perihelion ω alternates between circulation and libration. A theoretical explanation for this complex behaviour is provided. In a Sun–Mars–Jupiter–Saturn model, close encounters of James with Mars occur without ejecting James from the ν_{16} resonance region. The integration of a cloud of test objects surrounding the orbit of James confirms the chaotic nature of this resonance regions as indicated by theory.

098.035 **The orbit of 203 Pompeja and the mass of Ceres.**
E. Goffin.
Astron. Astrophys., Vol. 249, No. 2, p. 563 - 568 (Sep 1991).

From an analysis of the motion of the minor planet 203 Pompeja a value for the mass of Ceres is determined. The result, $(4.7 \pm 0.3) \times 10^{-10} M_{\odot}$, is the lowest value obtained up to now. The importance of adding Ceres as a perturbing body in the determination of minor-planet orbits is emphasized, and the value of re-reducing older observations is discussed.

098.036 **Asteroid 4 Vesta: relation between V brightness variations and absorption in the pyroxene band 0.92 μm.**
V. D. Vdovichenko, F. P. Velichko, S. M. Gajsin, S. A. Mosina.
Astron. Tsirk., No. 1543, p. 29 – 30 (Mar – Apr 1990). In Russian.

Detailed spectrophotometric observations of the first pyroxene band profile of asteroid 4 Vesta in the 1988 opposition showed that the band varies with asteroid rotation mainly in the depth. The amplitude of variations reaches 6 – 8% of the band's intensity. The moments of maximum bandpower, beginning and end of depression coincide with small peculiarities of the V–lightcurve.

098.037 **Asteroid–asteroid interactions – dynamically irrelevant?**
M. Hoffmann.
Reviews in Modern Astronomy 4, p. 165 – 172 (1991). – See Abstr. 003.008 for the main entry.

Structural investigations of the distribution of the asteroids have so far concentrated on the interactions with the major planets, especially Jupiter, and the resulting effects in element space. Here some less known structures of their distribution in actual space are shown; the radial number density and the orthogonal orbit density. A closer look at the orbit densities at different length scales reveals structures that are difficult to explain by gravitational influences from the major planets. This is particularly conspicious if special subgroups like the original Hirayama families are considered. A further limit for inter–asteroid–interactions comes from the distribution of locations of actual close enounters in space. A comparison with the gravitational cross section densities of the terrestrial planets and their influences on the population of planet crossing asteroids indicates that some interactions with the asteroids should lead to observable effects but the involved timescales are difficult to estimate because of the inhomogeneity of the asteroid distribution.

098.038 **On the program of observations of selected minor planets for 1991 – 2000.**
Yu. V. Batrakov, A. S. Vashkevich, V. A. Shor.
1. Spain–USSR Workshop on Positional Astronomy and Celestial Mechanics, p. 9 – 13 (1991). – See Abstr. 012.025 for the main entry.

Some results of observations of twenty selected minor planets according to the Orelskaya's program in 1976 – 1990 are summarized. It is concluded that the continuation of the selected planet observations is necessary. The fifteen brightest minor planets have been chosen for such a program and some measures to improve the accuracy of positional observations of these planets are outlined.

098.039 **Orbital elements of one–opposition minor planets.**
Minor Planet Circ., Nos. 18485 – 19346 (Aug – Nov 1991).
The orbits are published in Nos. 18598 – 18600 (25 Aug 1991), 18773 – 18775 (23 Sep 1991), 18995 – 18999 (23 Oct 1991), and 19259 – 19274 (21 Nov 1991), respectively.

098.040 **New names of minor planets.**
Minor Planet Circ., Nos. 18485 – 19346 (Aug – Nov 1991).
Citations of new minor planet names are published in Nos. 18643 – 18648 (25 Aug 1991), and 19332 – 19342 (21 Nov 1991), respectively.

098.041 **OAUV programme for minor planets astrometry.**
A. López García, J. L. Valdés Navarro,
J. A. López Ortí, R. López Machí.
1. Spain–USSR Workshop on Positional Astronomy and Celestial Mechanics, p. 15 – 18 (1991). – See Abstr. 012.025 for the main entry.

The authors present the status of their program for minor planets astrometry that begun in 1984. It includes systematic observations of the selected minor planets of the ITA accessible to their instrument and conditions. They are now extending the observations to other bright asteroids. To be able to start such a program the authors had to develop the photographic equipment and the mechanical and electronic devices for plate measuring as well as the software support to whole process. In this work they describe briefly the equipments and programs involved.

098.042 **Ephemerides of minor planets and comets.**
Minor Planet Circ., Nos. 18485 – 19346 (Aug – Nov 1991).
The ephemerides are published in Nos. 18648 – 18664 (25 Aug 1991), 18836 – 18846 (23 Sep 1991), 19037 – 19054 (23 Oct 1991), and 19342 – 19346 (21 Nov 1991), respectively.

098.043 **Identifications and identification changes, corrected and deleted observations of minor planets.**
Minor Planet Circ., Nos. 18485 – 19346 (Aug – Nov 1991).
The data are published in Nos. 18487 – 18489 (25 Aug 1991), 18665 – 18666 (23 Sep 1991), 18850 (23 Oct 1991), and 19056 – 19057 (21 Nov 1991), respectively.

098.044 **Index to orbital elements of comets and minor planets.**
Minor Planet Circ., Nos. 19057 – 19065 (21 Nov 1991).
The index refers to both comet and minor planet orbits published in MPC 18011 – 19057. The list is divided into the parts (1) comets, (2) numbered minor planets, (3) unnumbered minor planets, and (4) planets from the P–L, T–2, and T–3 surveys.

098.045 **Observations of minor planets.**
Minor Planet Circ., Nos. 18485 – 19346 (Aug – Nov 1991).
Positional data are published in Nos. 18497 – 18596 (25 Aug 1991), 18669 – 18772 (23 Sep 1991), 18854 – 18994 (23 Oct 1991), and 19071 – 19256 (21 Nov 1991), respectively. Observations made at the following observatories are published:
Akou, Bassano Bresciano, Boyden Obs., Burlington remote site, Calar Alto, Caussols, Cerro Tololo, Chorzów, Cockfield, Colleverde di Guidonia, Crimean Astrophys. Obs., Dynic Astron. Obs., Eldagsen, ESO, Farra d'Isonzo, Fujieda, Geisei, Goethe Link Obs., Haute Provence, JCPM Kagoshima Stn., JCPM Sapporo Stn., JCPM Yakiimo Stn., Kani, Karasuyama, Kiryuu Obs., Kitami, Kitt Peak, Kitt Peak (Steward Obs.), Kiyosato and Otomo, Kleť, Kushiro, Lick Obs., Linz, Lowell Obs., Lowell Obs. Anderson Mesa Stn., Mauna Kea Obs., McDonald Obs., Middlebury, Minami–Oda, Miyasaki Obs., Mount John Obs., Nihondaira Obs. Oohira Stn., Nyukasa, Oak Ridge Obs., Oizumi, Ojima, Okutama, Otomo, Palomar, Perth, Pic–du–Midi, Prague–Petrin, Resse Obs., San Marcello Pistoiese, San Vittore, Santa Lucia Stroncone, Seewalchen, Siding Spring, Sollies–Pont, Sormano, Susono, Tautenburg, Toyota, Uenohara, Victoria (Climenhaga Obs.), Yatsugatake South Base Obs., Yatsuka, YGCO Chiyoda Stn., Yorii, Zimmerwald.

098.046 **On determination of the asteroid orbits and the systematic errors of fundamental star catalogues in presence of small perturbations.**
D. P. Duma, P. M. Fedij.
1. Spain–USSR Workshop on Positional Astronomy and Celestial Mechanics, p. 95 – 98 (1991). – See Abstr. 012.025 for the main entry.

The results for the estimates of errors in calculated geocentric positions of asteroids caused by neglecting their mutual perturbations and probable error of the adopted value of the Jupiter's mass as well as the difference of geocentric coordiantes of planets based on DE200 and DE202 are discussed.

098.047 **The wavelet transform: a new tool for asteroid family determination.**
P. Bendjoya, E. Slézak, C. Froeschlé.
Astron. Astrophys., Vol. 251, No. 1, p. 312 – 330 (Nov 1991).

An analysis based on the wavelet transform has been performed in order to identify families in the proper element space of asteroids. This multi–scale analysis has been made on a set of 4100 proper elements computed by a new second order fourth–degree secular perturbation theory (Milani and Knežević 1990).

A metric has been defined in the proper element space in order to give a dynamical definition of families. This metric is related to the incremental velocity needed for orbital change after the ejection from a fragmented parent body. A degree of significance has been established from the same analysis of a quasi-random distribution, matching the large scale structure of the real distribution. The influence of the metric has been tested as well as the sensitivity of families to small variations of the proper elements within the accuracy of the theory. A cross comparison has been made between these results and the families obtained from the same data by another method (Zappalà et al. 1990). A good agreement between these two different methods has been established not only for the well known populous families, but also for the little ones the existence of which seemed before to depend on the method used.

098.048 **951 Gaspra: a pre-Galileo estimate of its surface evolution.**
N. Namiki, R. P. Binzel.
Geophys. Res. Lett., Vol. 18, No. 6, p. 1155 – 1158 (Jun 1991).

The authors attempt to quantitatively model the collisional environment and surface evolution of asteroid 951 Gaspra prior to the flyby of the Galileo spacecraft later this year. They compute Gaspra's average intrinsic collisional probability to be $5.4\times10^{-8}km^{-2}yr^{-1}$ with a root mean square impact velocity of 5.0 km sec^{-1}. The projectile population and impact strength for Gaspra are highly uncertain and the authors consider a range of values. Based on these models, Gaspra is unlikely to be a primordial object surviving from the beginning of the solar system. It is most likely a second or multiple generation fragment forming $\sim 10^8$yr ago. For a low impact flux, the crater size distribution directly reveals the size distribution of the impacting population. For a non-equilibrium surface, the authors present model-dependent isochrons for estimating the age of the surface. The model sets lower limits for Gaspra's age ($\sim 10^9$yr) and flux for the case of a high impact flux producing an equilibrium surface.

098.049 **Determination of positions of celestial bodies from photographic observations.**
S. S. Tokhtas'ev, S. K. Fomin, I. A. Dubyago, M. I. Kibardina.
Izv. Astron. Obs. Ehngel'gardt., Vyp. 55, p. 118 – 125 (1990). In Russian.

Positions of 3 asteroids and 7 comets observed in 1982 – 1988 are given.

098.050 **Origin of the Archean Earth impactors.**
K. D. Trego.
Earth, Moon, Planets, Vol. 53, No. 3, p. 233 – 235 (Jun 1991). Letter-to-the-editor.

A survey of current near-Earth asteroids and inner solar system comets suggests C class asteroids may have been the impactors during the Archean period resulting in the formation of spherule beds because of their carbonaceous composition and large diameters.

098.051 **Asteroidal jet-stream Flora A.**
J. Klačka.
Earth, Moon, Planets, Vol. 55, No. 1, p. 45 – 50 (Oct 1991).

Problem of the existence of the most significant jet-stream Flora A is analysed. On the basis of the data analysis it is shown that Flora A consists mostly of very faint asteroids with (1) oppositions of perihelia during autumn months, (2) very low inclinations. Due to the favourable observing conditions just during September, the concentration of proper longitudes of perihelia is then explained as a consequence of the observational selection effects. The low inclinations explain the concentration in proper longitudes of nodes just in the area corresponding to that to the jet-stream Flora A. Thus, the most important asteroidal jet-stream Flora A is fully explained as a consequence of observational selection effects.

098.052 **Synthetic curves of some observational properties of contact binary asteroids.**
M. Hoffmann.
Astron. Nachr., Vol. 312, No. 3, p. 201 – 207 (1991).

The shape of some asteroids may be deduced by other ground based means than rotation light curves or advanced techniques like speckle imaging or radar observations: the analysis of high precision astrometry and lunar occultation light curves. Calculations of the phenomena among the cosmogonically interesting case of a double asteroid configuration, when the objects are in contact with each other, are presented with special emphasis on the involved shadow phenomena.

098.053 **The rotation of (51) Nemausa.**
L. K. Kristensen.
Astron. Nachr., Vol. 312, No. 3, p. 209 – 220 (1991).

The problem of interpretation of lightcurves and pole determination is illustrated by a discussion of all observations of (51) Nemausa from 1887 to 1990. A new concept is the introduction of a fictitious convex body which can in principle be directly determined from the observations. The importance of the first and third harmonics in the lightcurves is demonstrated. Different oppositions are connected by the synodic period $7^h.78293$ and the proposed pole is: $19^h45^m + 41°$ (1950). The rotation is probably retrograde and there are indications of a red spot.

098.054 **Initial refinement of minor planet orbits for pointing and observation with the Hubble Space Telescope.**
A. L. Whipple, P. D. Hemenway, D. Ingram.
Astron. J., Vol. 102, No. 2, p. 816 – 822 (Aug 1991). Current Physics Microform No.: 9108A1752.

Refinements of the orbital elements for 24 minor planets that will be observed with the Fine Guidance Sensors (FGSs) on the Hubble Space Telescope are presented. The accuracy of these orbits is discussed in the context of the ephemeris requirements for target acquisition with the FGSs. Comparisons with standard catalog orbits are made to evaluate the suitability of the use of the catalog orbits for HST pointing. The authors find that the orbits published in the most recent Minor Planet Circulars are capable of positional predictions that are accurate at the 1 arcsec level. However, there are still many orbits that have not been revised in the last 10 yr and these should not be used for critical pointing ephemerides. In many cases the orbital elements will have to be refined using recent ground-based observations before a minor planet can be reliably observed by HST.

098.055 **Statistics of asteroid axial rotation.**
Yu. V. Aleksandrov.
Kinematics Phys. Celest. Bodies, Vol. 7, No. 2, p. 9 – 13 (1991). English translation of Kinematika Fiz. Nebesn. Tel, Tom 7, No. 2, p. 11 – 15 (1991).

It is shown that asteroids can be classified into three groups in accordance with the varying degrees of influence of collisional evolution on the parameters of their axial rotation on the basis of the variations of axial rotation velocity with diameter and of specific angular momentum with mass. The population of rotation velocities in each group can be described by a continuous Bose-Einstein distribution as a system with a variable number of quasi-particles.

098.056 **Photometry and polarimetry of asteroid 47 Aglaja.**
G. P. Chernova, D. F. Lupishko, V. G. Shevchenko, N. N. Kiselev, R. Salles.
Kinematika Fiz. Nebesn. Tel, Tom 7, No. 5, p. 20 – 26 (Sep-Oct 1991). In Russian. English translation in Kinematics Phys. Celest. Bodies, Vol. 7, No. 5.

Light-curves and phase dependences of magnitude, colour and polarization of C-type asteroid 47 Aglaja were observed for 33 nights in September – November 1989. The magnitude-phase dependence obtained for phase angles $\alpha = 0.1 - 12.9°$ shows the lack of a sharp opposition increase of Aglaja brightness at $\alpha = 0 - 2°$ found for high-albedo E-type asteroids. The difference of opposition effect mechanisms for low and high albedo asteroids and a probable reason of the difference is indicated. A polarimetry of Aglaja was carried out for the first time and parameters of polarization phase curve were obtained. The improved value of the rotation period (P = 13.178±0.005 hr), estimations of albedo (p_v = 0.073) and diameter (D = 122.0 km) are given.

098.057 **Asteroid radar astrometry.**
S. J. Ostro, D. B. Campbell, J. F. Chandler, I. I. Shapiro, A. A. Hine, R. Velez, R. F. Jurgens, K. D. Rosema, R. Winkler, D. K. Yeomans.
Astron. J., Vol. 102, No. 4, p. 1490 - 1502 (Oct 1991). Current Physics Microform No.: 9110D2022.

The authors report all useful measurements of time delay and Doppler frequency for asteroid radar echoes obtained at Arecibo and Goldstone during 1980–1990. New radar astrometry is presented for 23 near-Earth asteroids and three mainbelt asteroids. These measurements, which are orthogonal to optical, angular-position measurements, and typically have a fractional precision between 10^{-5} and 10^{-8}, permit significant improvement in estimates of orbits and hence in the accuracy of prediction ephemerides. The authors also report estimates of radar cross section and circular polarization ratio for all asteroids observed astrometrically during 1980–1990.

098.058 **Galileo support observations of asteroid 951 Gaspra.**
J. D. Goldader, D. J. Tholen, D. P. Cruikshank, W. K. Hartmann.
Astron. J., Vol. 102, No. 4, p. 1503 - 1509 (Oct 1991). Current Physics Microform No.: 9110D2035.

The authors report observations of 951 Gaspra in support of the Galileo spacecraft encounter. Photometric observations yield a synodic rotational period of 7.04246±0.00006 hr (0.293436±0.000003 days), a mean absolute visual magnitude of 11.8026±0.0025, and a slope parameter of 0.285±0.005. The authors review another determination of the slope parameter and suggest that the sub-Earth latitude was changing significantly during the opposition, and so neither the authors nor other investigators were able to sample uniquely the asteroid's phase function during the 1990 opposition. The authors suggest adopting a slope parameter of 0.25 from combining available observations, and they also place limits on a possible pole orientation. The infrared reflectance spectrum (0.8–2.5 μm) shows an absorption band near 1 μm attributed to olivine and pyroxene, and a weak second pyroxene band centered near 2.0 μm. This mineral signature plus the overall redness of the spectrum suggest that Gaspra originated in the lower mantle of a differentiated parent body. Astrometric positions obtained during the course of these observations are also reported.

098.059 **Polarimetry of CMEU asteroids. II. Peculiarity of the M-type asteroids.**
I. N. Bel'skaya, N. N. Kiselev, D. F. Lupishko, G. P. Chernova.
Kinematika Fiz. Nebesn. Tel, Tom 7, No. 6, p. 11 - 14 (Nov–Dec 1991). In Russian. English translation in Kinematics Phys. Celest. Bodies, Vol. 7, No. 6.

Results of measurements of the depth P_{min} of the negative branch on the phase curve of polarization are presented for asteroids 147, 217, 325, 347 and 796. According to the values of P_{min} and U–V, four of them have been assigned to the M-type and only 147 Protogenea is assigned to C or other low-albedo type. It is shown that M-type asteroids are distinguished from other ones by the value range of parameter $P_{min} \cdot p_V$ (p_V is the albedo), forming a certain sequence. It is supposed that it may be connected with different content of metal on the M-type asteroid surfaces.

098.060 **The origins of the asteroids.**
R. P. Binzel, M. A. Barucci, M. Fulchignoni.
Sci. Am., Vol. 256, No. 4, p. 66 - 71 (Oct 1991). For the German translation see Abstr. 098.021.

Once regarded as "vermin of the skies", asteroids are now known to offer important clues to the birth of the solar system. They are the remnants of a planet that failed to form.

098.061 **Interpreting the light curves of asteroids.**
A. Yu. Kogan, K. Yu. Pimenov.
Sov. Astron. Lett., Vol. 17, No. 5, p. 360 - 362 (Sep 1991). Current Physics Microform No.: 9204X1307. English translation of Pis'ma Astron. Zh., Tom 17, No. 9, p. 850 - 856 (1991).

Asteroids are twinkling point sources, their brightness being modulated by rotation. Having recorded the visible flux from an asteroid as a function of time, one can – with a few a priori assumptions – uniquely determine the figure of the asteroid and its rotation parameters. From a mathematical standpoint, the problem involves the inversion of an integral transform associated with the Gaussian spherical mapping. The authors propose a numerical approach to solving that problem, and present the reconstructed shape of a computer-modeled asteroid.

098.062 **1991 JX.**
IAU Circ., No. 5308, p. 1 (17 Jul 1991).

098.063 **1991 OA.**
IAU Circ., No. 5314, p. 1 (30 Jul 1991).
Further information is given in Nos. 5322, 5323, 5324.

098.064 **1991 EE.**
IAU Circ., No. 5326, p. 1 (15 Aug 1991). Further information is given in No. 5349.

098.065 **1991 RB.**
IAU Circ., No. 5337, p. 1 (7 Sep 1991).
Further information is given in No. 5344.

098.066 **1991 RC.**
IAU Circ., No. 5338, p. 1 (7 Sep 1991).
Further information is given in No. 5350.

098.067 **Doublet craters and the tidal disruption of binary asteroids.**
H. J. Melosh, J. A. Stansberry.
Icarus, Vol. 94, No. 1, p. 171 - 179 (Nov 1991).

Out of approximately 28 known terrestrial impact craters with diameters greater than 20 km, at least 3 are doublets created by the nearly simultaneous impact of objects of comparable size. The large size and separation of these doublets rule out atmospheric breakup as the agent of dispersion, while previous investigations of fragmentation within the Roche Limit are similarly unable to account for the dispersion of the doublets. The authors investigate the possibility that tidal disruption of a population of contact binary asteroids might account for doublet craters. After detailed orbital integrations one finds that although such asteroids are often disrupted by tidal forces well outside the Roche limit, the magnitude of the separations, averaged over a random population of impactors, is too small to account for the observed doublet crates. The only rational explanation one can find for the occurrence of doublet craters is an initial population of kilometer-size Earth-crossing objects containing between 10 and 20% of binaries separated by four times or more of their mean diameter.

098.068 **Asteroid 4 Vesta: simultaneous visible and ultraviolet IUE observations.**
M. C. Festou, S. A. Stern, G. P. Tozzi.
Icarus, Vol. 94, No. 1, p. 218 - 231 (Nov 1991).

On 6 – 7 Oct 1990, 16 long-wavelength primary camera (LWP) spectra were recorded with the long-wavelength primary spectrograph of the IUE to obtain the resolved UV lightcurve of asteroid 4 Vesta. Simultaneously, the fine error sensor (FES) of the telescope was used to obtain Vesta's near-B lightcurve. Deep LWP and short-wavelength primary camera (SWP) spectra were also collected to establish the reflectivity curve of Vesta in the 1700 – 3200 Å region. The FES data show that the rotation period is of the order of 5 h; combined with an old observation obtained in similar geometrical conditions, a new, accurate value of the rotation period is derived. The UV and optical lightcurves have one single minimum and one sigle maximum. The UV spectrum is redder than that of the Sun. The asteroid albedo increases linearly with wavelength from 2300 to 3000 Å, and this characteristic probably holds over the entire 2000 – 3500 Å range. The intrinsic amplitude of Vesta's lightcurve is of the order of 10% and does not vary with wavelength from UV to near–IR. The present observations are compared with existing data to develop a self–consistent picture of Vesta's surface properties.

098.069 **1991 RJ_2.**
IAU Circ., No. 5357, p. 1 (3 Oct 1991).

098.070 **1991 TT and 1991 TU.**
IAU Circ., No. 5367, p. 1 (15 Oct 1991).

098.071 **1991 TB_1.**
IAU Circ., No. 5368, p. 1 (15 Oct 1991).

098.072 **1991 TB_2.**
IAU Circ., No. 5371, p. 1 (19 Oct 1991).

098.073 **1991 TF_3.**
IAU Circ., No. 5373, p. 1 (24 Oct 1991).

098.074 **1991 VA.**
IAU Circ., No. 5383, p. 1 (12 Nov 1991).

098.075 **1991 VB.**
IAU Circ., No. 5384, p. 1 (12 Nov 1991).
Further information is given in No. 5417.

098.076 **1991 VE.**
IAU Circ., No. 5385, p. 1 (12 Nov 1991).

098.077 **1991 VG.**
IAU Circ., No. 5387, p. 1 (13 Nov 1991).
Further information is given in Nos. 5388, 5401, 5402.

098.078 **1991 VH.**
IAU Circ., No. 5390, p. 1 (18 Nov 1991).
Further information is given in Nos. 5409, 5410.

098.079 **1991 VK.**
IAU Circ., No. 5392, p. 1 (21 Nov 1991).

098.080 **(3288) Seleucus.**
IAU Circ., No. 5392, p. 1 (21 Nov 1991).

098.081 **1991 VL.**
IAU Circ., No. 5393, p. 1 (22 Nov 1991).
Further information is given in No. 5411.

098.082 **1991 WA.**
IAU Circ., No. 5396, p. 1 (4 Dec 1991).

098.083 **1991 XB.**
IAU Circ., No. 5399, p. 1 (9 Dec 1991).
Further information is given in No. 5418.

098.084 **Spacewatch: Our asteroid–pelted planet.**
D. Steel.
Nature, Vol. 354, No. 6351, p. 265 – 267 (28 Nov 1991).

098.085 **Near miss of the Earth by a small asteroid.**
J. V. Scotti, D. L. Rabinowitz, B. G. Marsden.
Nature, Vol. 354, No. 6351, p. 287 – 289 (28 Nov 1991). Letter–to–the–editor.

The 0.91–m Spacewatch Telescope, on Kitt Peak in Arizona, is being used to search for new Earth–approaching asteroids. On the night of 18 January 1991 a streaked image extending over more than 161 arcseconds was recorded. Further observations over the next 4.6 hours showed that the images were of an object travelling in an independent orbit around the Sun; it was given the asteroidal designation 1991 BA. Orbital computations, presented, indicate that 1991 BA was only 0.0053 AU from the Earth at the time of discovery, closing to 0.0033 AU at the last detection. Extrapolation of the calculated orbit shows that the object passed only 0.0011 AU (170,000 km) from the Earth 12 hours after it was found. The observed brightness translates into an absolute visual magnitude 28.9, corresponding to a diameter of only 5 – 10 m. 1991 BA is the closest and smallest asteroid yet observed outside the Earth's atmosphere.

098.086 **Minor planet 3900 Knežević.**
M. Jeličić.
Vasiona, Année 39, No. 5, p. 69 – 70 (1991). In Croatian.

The basic data concerning the minor planet 3900 Knežević and the Belgrade astronomer Zoran Knežević are given.

098.087 **Automated computation of general perturbations for the Flora asteroidal group.**
M. F. He, Zhangjie.
Proc. Astron. Soc. Aust., Vol. 9, No. 1, p. 149 (1991). – See Abstr. 012.078 for the main entry.

The general perturbations of the asteroids of the Flora group ($975'' < n < 1150''$, n is the mean daily motion) due to Jupiter have been automatically computed on an IBM 4341 computer with the Hansen method by manipulating Broucke's Symbolic Poisson Series Processor, and the results have been compared with that computed previously with the Bohlin group method.

098.088 **The diameter of (9) Metis from the occultation of SAO 190531.**
W. M. Kissling, G. L. Blow, W. H. Allen, J. Priestley, P. Riley, P. Daalder, M. George.
Proc. Astron. Soc. Aust., Vol. 9, No. 1, p. 150 – 152 (1991). – See Abstr. 012.078 for the main entry.

The observations in this paper were carried out as part of a program co–ordinated by the Occultation Section of the Royal Astronomical Society of New Zealand. On 6 August 1989 minor planet (9) Metis occulted the magnitude 8.7 star SAO 190531. The occultation was observed by two visual observers in Tasmania, and at one photoelectric and two visual sites in New Zealand. From the five chords obtained the authors have deduced an average cross–sectional diameter for this minor planet of 173.5 km at the time of the occulation. This is in good agreement with previous radiometric and polarimetric results.

098.089 **Theory of motion of asteroid 108 Hecuba. Simonin's work.**
I. A. Gerasimov, B. R. Mushailov.
Tr. Gos. Astron. Inst. im. Shternberga, Tom 62, p. 91 – 110 (1991). In Russian.

Hecuba's theory of motion being worked out by Simonin under limiting conditions of the three–body model is presented, periodical solution here is regarded as a basis. Osculating orbital elements are calculated by integration of a system of linear and uniform equations.

098.090 **1991 VG.**
Yamamoto Circ., No. 2173, p. 1 (16 Nov 1991). In Japanese.

Further information is given in No. 2175.

098.091 **1991 VH, 1991 VK, 1991 VL, 1991 WA, 1991 XB.**
Yamamoto Circ., No. 2174, p. 2 (15 Dec 1991). In Japanese.

098.092 **Asteroids.**
J. Veverka.
Liller Symposium: Asteroids to quasars, p. 3 – 24 (1991). – See Abstr. 012.087 for the main entry.

098.093 **The size distribution of main-belt asteroids from IRAS data.**
A. Cellino, V. Zappalà, P. Farinella.
Mon. Not. R. Astron. Soc., Vol. 253, No. 4, p. 561 – 574 (15 Dec 1991).
The IRAS data base on asteroid albedos and diameters has been used to derive size distributions for a set of about 4000 numbered main-belt asteroids and for various subsets defined according to the semimajor axis range and family membership. Different procedures have been used to obtain albedo estimates for the (mostly small) members of the sample not observed by IRAS, allowing the authors to assess the sensitivity of the derived size distributions to the assumptions used for this purpose.

098.094 **The prediction and discovery of a Martian Trojan asteroid.**
K. A. Innanen.
J. R. Astron. Soc. Can., Vol. 85, No. 4, p. 151 – 157 (Aug 1991).

098.095 **Astronomers capture an asteroid.**
F. J. Howell.
J. R. Astron. Soc. Can., Vol. 85, No. 4, p. 174 (Aug 1991). Abstract. – See Abstr. 012.090 for the main entry.

098.096 **Galileo's frustrating asteroid pursuit.**
R. A. Kerr.
Science, Vol. 254, No. 5030, p. 381 – 382 (18 Oct 1991).
Astronomer's first closeup of an asteroid probably won't answer the most burning question – what is it made of?

098.097 **Space weathering of a non-spherical, rotating body.**
R. O. Redman.
J. R. Astron. Soc. Can., Vol. 85, No. 4, p. 193 (Aug 1991). Abstract. – See Abstr. 012.091 for the main entry.

098.098 **The lifetime of binary asteroids vs gravitational encounters and collisions.**
B. Chauvineau, F. Mignard, P. Farinella.
Icarus, Vol. 94, No. 2, p. 299 – 310 (Dec 1991).
The authors quantitatively assess the long-term stability of binary asteroids with respect to (1) gravitational encounters with passing asteroids and (2) physical collisions. The effects of gravitational encounters are estimated by a simple physical model, which is based on the distinction between penetraring, close, and far encounters, and on the assumption of a random-walk accumulation of the corresponding energy changes. Explicit computations of lifetimes vs disruption by the above-mentioned processes are then made as a function of the main parameters of the systems (separation, size of the components), and estimates are given for a few specific asteroids for which a binary structure has been inferred from different types of observations. The main conclusion is that, especially when bodies of comparatively small sizes are concerned, the lifetimes of asteroids binaries are limited by the occurrence of disruptive collisions, while gravitational encounters are generally less important.

098.099 **Observations of minor planets. XI.**
F. Börngen, L. D. Schmadel.
Astron. Nachr., Vol. 312, No. 6, p. 393 – 403 (1991). In German.
A summary is given about the Minor Planet survey performed in 1990 on Tautenburg Schmidt plates. 561 asteroids were observed and 2104 positions are calculated for them. These are 430 asteroids with provisional designations (364 of them have new designations) and 131 numbered objects. One-opposition orbits have been computed for 160 Tautenburg asteroids. Tautenburg observations could give a tribute to 49 planets numbered in the period of this report. Among them are six asteroids discovered in Tautenburg.

098.100 **Characteristics of new asteroids discovered at the Crimean Astrophysical Observatory.**
N. S. Chernykh.
Bull. Crimean Astrophys. Obs., Vol. 81, p. 19 – 24 (1990). English translation of Izv. Krym. Astrofiz. Obs., Tom 81, p. 23 – 28 (1990). See Abstr. 53.098.039.
Features are described for 450 new minor planets discovered in the last 20 years at the observatory during an observational program in collaboration with the Institute of Theoretical Astronomy. The orbit-element distributions for the new asteroids are compared with those for ones known up to 1970. There are certain differences associated with the observational selection.

098.101 **Asteroids, meteorites, interplanetary and cometary dust.**
E. K. Jessberger.
International School of Physics 'Enrico Fermi'. Course 111: Solid-state astrophysics, p. 107 – 127 (1991). – See Abstr. 012.106 for the main entry.
The aim of this lecture is to put together relevant facts about minor bodies in the solar system. Some of these are known since long and can be found in textbooks, some have been very recently revealed by laboratory and space experiments. Certainly the lecture cannot fully cover all four research areas. The author also does not attempt to concisely synthesize the results, but tries to draw connecting lines from one area to the other.

098.102 **Il pianetino (4062) Schiaparelli Scoperto all'Osservatorio "S. Vittore".**
E. Colombini.
Astronomia UAI, N. 10, p. 25 (Nov – Dec 1991).

098.103 **Rotation period and light curve of asteroid 423 Diotima.**
R. G. Hutton.
Bol. Asoc. Argent. Astron., No. 36, p. 193 – 200 (1990). In Spanish. – See Abstr. 012.095 for the main entry.
The asteroid 423 Diotima was observed photoelectrically in V filter during Oct 1987 at Observatorio Municipal de Mercedes. The synodic rotational period and amplitude are obtained. The values confirm the solution proposed by Zappala et al. (1985) to the ambiguity presented in previous determinations of the rotational period and convert the asteroid in a peculiar object.

098.104 **Asteroiden (51) Nemausa.**
L. K. Kristensen.
Astron. Tidsskr., Årg. 24, Nr. 3, p. 115 – 123 (1991).

098.105 **Rotation of asteroids.**
F. P. Velichko, D. F. Lupishko.
Sol. Syst. Res., Vol. 25, No. 3, p. 189 – 202 (Nov 1991). English translation of Astron. Vestn., Tom 25, No. 3, p. 259 – 276 (1991).
This survey is devoted to a generalization and analysis of available data on the rotational parameters of asteroids from the point of view of the cosmogonic problems of the asteroid belt. It includes discussion of rotation rates, sense of rotation, orientation of rotation axes, and precession of rotation axes. It concludes with remarks about the main problems for further research on asteroid rotation.

098.106 **Zehn neue Tautenburger Planetoidenentdeckungen.**
F. Börngen.
Sterne, Band 67, Heft 5, p. 265 – 269 (1991).

098.107 **The Anglo-Australian near-Earth asteroid survey.**
D. Steel, R. H. McNaught.
Aust. J. Astron., Vol. 4, No. 2, p. 42 – 48 (Oct 1991).
The Anglo-Australian near-Earth asteroid survey began in 1990 May and has so far resulted in the discovery of eight Apollo-type and three Amor-type asteroids, four comets, one object of a new type (1991 DA, which appears asteroidal in nature but has a comet-like orbit stretching from Mars out past Uranus), many recoveries and precoveries, and twenty-one supernovae as a by-product.

On the "Ekat" catalogue and prospects of its application in photographic observations of minor planets.
See Abstr. 002.066.

Some results of determining the SAO catalogue systematic errors from observations of minor planets.
See Abstr. 002.068.

New reference catalogues for photographic observations of minor planets.
See Abstr. 002.069.

Comparison of catalogues through asteroid observations.
See Abstr. 002.127.

First Spain–USSR Workshop on Positional Astronomy and Celestial Mechanics. Proceedings.
See Abstr. 012.025.

Soviet programs "Jupiter" and "Ceres" for improving orbits of bodies of solar system.
See Abstr. 013.051.

Scanning with charge–coupled devices.
See Abstr. 013.094.

Iron and chromium absorption bands in the spectra of terrestrial pyroxenes: application to mineralogic remote sensing of asteroid surfaces.
See Abstr. 022.039.

Laboratory simulations of large scale fragmentation events.
See Abstr. 022.132.

Angular momentum transfer in low velocity oblique impacts: implications for asteroids.
See Abstr. 022.155.

Impact experiments. 3: Catastrophic fragmentation of aggregate targets and relation to asteroids.
See Abstr. 022.156.

Suprathermal chemistry in space.
See Abstr. 022.162.

Eine Methode zur Bestimmung genäherter Kleinplanetenörter.
See Abstr. 036.017.

Influence of the reference catalogue errors on the determination of the parameters of fundamental systems using minor planet observation.
See Abstr. 041.010.

The results and prospects of use of minor planets observations for improvement of the systems of the fundamental catalogues.
See Abstr. 041.011.

Secular resonances in the asteroid belt: theoretical perturbation approach and the problem of their location.
See Abstr. 042.011.

The main secular resonances ν_6, ν_5 and ν_{16} in the asteroid belt.
See Abstr. 042.012.

Orbital stability zones about asteroids.
See Abstr. 042.015.

New applications of Fatonu's problem.
See Abstr. 042.018.

Precise ephemeris of minor planets.
See Abstr. 042.029.

Present status of a new algorithm for orbital elements improvement.
See Abstr. 042.030.

Irregular behaviour and eccentricity increase in the trajectories of a dynamical system modelling asteroidal motions.
See Abstr. 042.045.

The computation of orbits in indeterminate and uncertain cases.
See Abstr. 042.050.

Instabilities in first–order resonances.
See Abstr. 042.054.

Some discussions about the ideal resonance.
See Abstr. 042.074.

Averaging Hamiltonian systems.
See Abstr. 042.088.

Secular resonances from 2 to 50 AU.
See Abstr. 042.093.

Les résonances séculaires dans la Ceinture des Astéroids.
See Abstr. 042.095.

Space exploration of comets and primitive minor bodies.
See Abstr. 051.084.

Generation of multiple flyby orbits.
See Abstr. 052.031.

Tektites in Cretaceous–Tertiary boundary rocks on Haiti and their bearing on the Alvarez impact extinction hypothesis.
See Abstr. 081.052.

The impacts of giant asteroids and geomagnetic inversions.
See Abstr. 084.006.

Complexes of minor bodies in the solar system.
See Abstr. 091.015.

Complexes of minor solar system bodies.
See Abstr. 091.030.

Brine volcanism and the interior structures of asteroids and icy satellites.
See Abstr. 091.050.

Physical properties of terrestrial planets and their satellites: an introduction.
See Abstr. 091.054.

Small solar system bodies.
See Abstr. 091.057.

Solar system occultations during 1991.
See Abstr. 096.008.

Solar system occultations during 1992.
See Abstr. 096.010.

Comets, asteroids and interstellar grains.
See Abstr. 102.008.

Are the Geminids and 3200 Phaethon related genetically?
See Abstr. 104.007.

Meteoroid streams: formation, evolution, associations with comets and asteroids.
See Abstr. 104.008.

The structure and evolution of the Taurid Complex.
See Abstr. 104.040.

Midinfrared (2.5 – 13.5 μm) reflectance spectra of powdered stony meteorites.
See Abstr. 105.007.

Phosphates in pallasite meteorites as probes of mantle processes in small planetary bodies.
See Abstr. 105.045.

A hypothesis on the Oort clouds of planets.
See Abstr. 107.008.

Evolution of the protoplanetary disk and the problem of mass deficiency in the asteroid belt.
See Abstr. 107.009.

Secular resonances in the primitive solar nebula.
See Abstr. 107.015.

Protoplanetary thermal metamorphism: the hypothesis of electromagnetic induction in the protosolar wind.
See Abstr. 107.021.

099 Jupiter, Jupiter Satellites

099.001 **Faraday rotation of Jupiter's decametric radiation.**
M. Y. Boudjada, A. Lecacheux.
Astron. Astrophys., Vol. 247, No. 1, p. 235 – 246 (Jul 1991).
Jovian decametric emission (DAM) is known to exhibit, contrary to other low-frequency planetary radio emission, a high degree of linear polarization. As it travels through the Io plasma torus and the terrestrial ionosphere, DAM is subject to Faraday rotation that may exceed $\sim 10^2$ radius at 30 MHz. The authors report recent observations of DAM's Faraday rotation obtained with the array of 144 log-periodic antennas at Nançay (France) coupled to a wide band (10–40 MHz) sweep frequency polarimeter. The study of the Faraday effect on Jovian emission shows that the usual simple approximation where the rotation of the polarization plane varies as the inverse of the frequency squared (ν^{-2}) cannot explain the observations. A second order term proportional to (ν^{-4}), resulting from the difference between the *X*- and *O*-mode ray paths and the high density gradients of the ionosphere, must be included in order to account for the observations. For the three most accurate measurements, done on three Io–B emissions, the major axis of the polarization ellipse at Jupiter is found, on average, to be at position angle +15 from the jovian northern pole; unlike, five Io–A events exhibit really a random distribution.

099.002 **Io's volcanic activity during the 1989 – 90 apparition.**
F. J. Melillo.
I.A.P.P.P. Commun., No. 45, p. 89 – 91 (Sep – Nov 1991).

099.003 **Ganymede and Callisto: surface textural dichotomies and photometric analysis.**
B. J. Buratti.
Icarus, Vol. 92, No. 2, p. 312 – 323 (Aug 1991).
Voyager imaging observations of Ganymede and Callisto have been reduced and combined with ground-based telescopic data to produce complete solar phase curves of the satellites' leading and trailing hemispheres. The curves, along with disk-resolved measurements, were fit to scattering models to derive hemispherical values of the physical parameters describing the surface.

099.004 **Light curve model for the Galilean satellites during Jovian eclipse.**
A. Mallama.
Icarus, Vol. 92, No. 2, p. 324 – 331 (Aug 1991).
A model for the light curve of a Galilean satellite eclipsed by Jupiter's shadow is developed. This model allows center-of-figure information on the satellite's position to be derived from photometric eclipse data.

099.005 **HUT observations of neutral sulfur emissions from Io.**
S. T. Durrance, P. D. Feldman, H. W. Moos, A. F. Davidsen, W. P. Blair, C. W. Bowers, W. V. Dixon, R. C. Henry, G. A. Kriss, J. W. Kruk, O. Vancura, H. C. Ferguson, R. A. Kimble, K. S. Long.
Bull. Am. Astron. Soc., Vol. 23, No. 2, p. 940 (1991). Abstract. – See Abstr. 012.008 for the main entry.

099.006 **HUT spectra of the Io torus.**
H. W. Moos, P. D. Feldman, S. T. Durrance, A. F. Davidsen, W. P. Blair, C. W. Bowers, W. V. Dixon, R. C. Henry, G. A. Kriss, J. W. Kruk, O. Vancura, H. C. Ferguson, R. A. Kimble, K. S. Long.
Bull. Am. Astron. Soc., Vol. 23, No. 2, p. 940 (1991). Abstract. – See Abstr. 012.008 for the main entry.

099.007 **Molecular origin of Io's fast sodium.**
N. M. Schneider, J. T. Trauger, J. K. Wilson, D. I. Brown, R. W. Evans, D. E. Shemansky.
Science, Vol. 253, No. 5026, p. 1394 – 1397 (20 Sep 1991).
Neutral sodium emissions encircling Jupiter exhibit an intricate and variable structure that is well matched by a simple loss process from Io's atmosphere. These observations imply that fast neutral sodium is created locally in the Io plasma torus. Sodium-bearing molecules may be present in Io's upper atmosphere, where they are ionized by the plasma torus and swept downstream. The molecular ions dissociate and dissociatively recombine on a short time scale, releasing neutral fragments into escape trajectories from Jupiter. This theory explains a diverse set of sodium observations, and it implies that molecular reactions (particularly electron impact ionization and dissociation) are important at the top of Io's atmosphere.

099.008 **Periodic hot-spot distribution on Io.**
A. Yamaji.
Science, Vol. 254, No. 5028, p. 89 – 91 (4 Oct 1991).
The Jovian satellite Io has active volcanic hot spots. The point-to-point correlation of the hot-spot distribution indicates that the hot spots preferentially make chains. The arrangement of the chains is periodic with a typical spacing of 120 km. The chains exhibit concordant trends with stresses imposed by the tidal deflection of the lithosphere, suggesting that the hot spots are formed along fissures in the lithosphere. The typical spacing may be controlled by lithosphere thickness.

099.009 **Time-series images of Jupiter at 10 microns.**
B. Fisher, H. Hudson, B. Jones, R. Piña, R. Puetter.
Bull. Am. Astron. Soc., Vol. 23, No. 2, p. 961 (1991). Abstract. – See Abstr. 012.008 for the main entry.

099.010 **Relative spectral reflectivity features and altitude differences of Jupiter's cloud belts.**
V. G. Tejfel', G. A. Kharitonova.
Kinematics Phys. Celest. Bodies, Vol. 7, No. 1, p. 45 – 52 (1991). English translation of Kinematika Fiz. Nebesn. Tel, Tom 7, No. 1, p. 49 – 57 (1991).

Analysis of latitudinal variations of the relative spectrophotometric gradient on spectrograms of Jupiter's central meridian show that this gradient is much smaller in the $\lambda\lambda$ 0.40 – 0.50 μm band than at $\lambda\lambda$ 0.50 – 0.68 μm in dark SEB and NEB cloud belts, in contrast to the situation in bright zones. This property is also confirmed by photoelectric spectrophotometry. One possible explanation is that the effective upper boundary of the clouds in the dark equatorial belts is approximately 15 – 20 km lower than the cloud limit in the bright zones.

099.011 **Erkundung der Radioplaneten durch Voyager.**
H. O. Rucker.
Phys. Unserer Zeit, Jahrg. 22, Nr. 4, p. 149 – 155 (Jul 1991).

099.012 **Possible detection of Jovian global oscillations.**
F.-X. Schmider, B. Mosser, E. Fossat.
Astron. Astrophys., Vol. 248, No. 1, p. 281 – 291 (Aug 1991).

The detection of global acoustic modes is a powerful test of the internal structure of astronomical objects. Such oscillations can take place on non–variable stars as well as on giant planets. After the success of helioseismology, a lot of teams tried to detect such oscillations on solar–type stars. Several specific instruments were therefore designed, aiming to measures periodical radial velocity variations with amplitude as small as a few meters per second. The stellar seismometer designed in Nice and based on an optical resonant sodium cell has been successful in the detection of oscillations on Procyon and α Cen in 1984. A new version of this instrument has been built in 1987, providing a sensitivity better than 50 cm s^{-1} on Procyon with a 3.6 m telescope. This instrument has been first tested on Jupiter in November 1987. The observations proved the existence of strong velocity variations of the planet in the 10–20 min periods range. This paper describes the principle of the measurement and the performance reached by the instrument. The result of the authors' observations of Jupiter is displayed and the Jovian origin of the measurements is discussed. It is concluded that the observed variations can only be explained as global acoustic modes. In addition, a first interpretation of the frequencies of the modes and their consequences for the knowledge of the internal structure of the planet is given.

099.013 **Eclipses of Jupiter's satellites in 1992 for slow–scan CCD observers.**
A. Mallama.
I.A.P.P.P. Commun., No. 46, p. 46 – 47 (Dec 1991 – Feb 1992).

099.014 **On the elliptical polarization of Jupiter's decametric radio emission.**
D. B. Melrose, G. A. Dulk.
Astron. Astrophys., Vol. 249, No. 1, p. 250 – 257 (Sep 1991).

The authors consider the origin of the nearly–unqiue, 100% elliptical polarization of Jupiter's decametric radio emission. After investigating the transfer of polarized radiation when coupling of the Stokes parameters is important, they concur with earlier authors that the density in and near the source region must be so low that the polarization remains fixed along the ray path; this requires $n_e \lesssim 1$ to 5 cm^{-3} for $f = 30$ MHz and proportionally lower at lower frequencies. They determine (i) the polarization of cyclotron maser radiation in these circumstances, (ii) that the dispersion relation of the rarified plasma composed of energetic, anisotropic electrons is like that in vacuo, (iii) that the growth rate is sufficient to saturate the maser and accounts for the observed brightness temperature. Possible sources of plasma in and near the source region in Jupiter's inner, polar magnetosphere are considered. The principal source, the ionosphere, has been observed by the Pioneer and Voyager spacecraft to be tightly confined by gravity, with the electron density decreasing exponentially with scale height $H \lesssim 1000$ km.

099.015 **Refraction effects on the Galileo probe telemetry carrier frequency.**
D. H. Atkinson, T. R. Spilker.
Radio Sci., Vol. 26, No. 1, p. 31 – 43 (Jan–Feb 1991).

As the Galileo probe Relay Radio Link (RRL) signal propagates outward through the Jovian atmosphere, the atmosphere will manifest itself in two ways. First, the geometric path length of the signal is increased, resulting in a small change of the RRL signal departure angle from the probe (transmitter). Secondly, the velocity of the signal is decreased. For a spherical, static atmosphere with a known profile of refractivity versus altitude the effects of refraction on the RRL frequency can be found using a variation of standard ray–tracing techniques, whereby the ray departure angle is found by an iterative process. From the dispersive characteristics of a mixture of hydrogen and helium with trace amounts of methane and ammonia a simple model of the Jovian atmosphere is constructed assuming spherical symmetry and uniform mixing. The contribution to the RRL Doppler frequency arising from refraction is calculated, and its effect on the Doppler wind measurements is discussed.

099.016 **Emission cones in Jupiter's hectometric radiation.**
C. H. Barrow.
Astron. Astrophys., Vol. 250, No. 1, p. 245 – 252 (Oct 1991).

The jovian hectometric radio emission (HOM) is generally believed to be generated at frequencies close to the gyrofrequency along high magnetic latitude field lines. The emission is beamed into a radiation pattern having the form of a hollow cone with apex at the point of emission and axis parallel to the magnetic field direction. Propagation is in the R–X mode. A qualitative model is proposed which invokes the observed HOM polarization characteristics and continuously radiating sources on field lines having footprints at all longitudes, along a narrow band of high magnetic latitudes close to, or perhaps within, the northern and southern auroral zones. HOM is received by an observer when the emission cone orientation is correct. Solar wind control of the HOM is re–examined for indications that the HOM may be associated with the jovian aurora by some process analogous to the generation of auroral kilometric radiation at the Earth but the results are inconclusive. This is not critical to the model, however.

099.017 **The orbital resonance among the Galilean satellites I, II, III of Jupiter.**
Zhao Zhangyin, Liu Lin.
Acta Astron. Sin., Vol. 32, No. 2, p. 121 – 128 (Jun 1991). In Chinese. English translation in Chin. Astron. Astrophys., Vol. 15, No. 4, p. 366 – 374 (Dec 1991).

The orbital resonance among the three inner Galilean satellites (Io, Europa, Ganymede) of Jupiter is discussed by using the perturbed ideal resonance model. The numerical results show that the characteristics of resonance among the three inner Galilean satellites are consistent with observational facts.

099.018 **On the theory of decameter S–emission of Jupiter.**
A. G. Boev, M. Yu. Luk'yanov.
Astron. Zh., Tom 68, Vyp. 4, p. 853 – 862 (Jul–Aug 1991). In Russian. English translation in Sov. Astron., Vol. 35, No. 4.

The paper proposes a novel mechanism for generation of Jovian decameter S–radioemission; the mechanism is based on Cherenkov instability of an electron beam, able to arise inside the Io–Jupiter flux tube. This instability results in anomalous–dispersion plasma waves excited near the local upper–hybrid frequency; these waves are then converted to R–X modes on a fluent inhomogeneity at a diffusion–tube edge. It is shown that the proposed mechanism permits to explain fairly well a number of experimentally observed facts: cone–like radiation pattern, the anomalous angle – frequency dependence, the dynamic drift of S–bursts over all frequency ranges. The estimations, made for Jupiter intrinsic parameters, demonstrate this mechanisms to be able to provide the observed values of emission energy fluxes.

099.019 **The Io sulfur torus in 1981.**
R. J. Oliversen, F. Scherb, F. L. Roesler.
Icarus, Vol. 93, No. 1, p. 53 – 62 (Sep 1991).

A Fabry–Perot/CCD spectrometer was used to obtain images of the Io torus in emission lines of [S II] λ6716, λ6731 and [S III] λ9531 in 1981 Feb and Mar on the 2.1–m telescope at KPNO. The [S II] and [S III] images showed a large variation in brightness and radial extent. There is an indication that [S II] and [S III] emissions in the warm torus are correlated. The [S II] and [S III] emissions in the warm torus also have similar scale heights along the magnetic field lines of $\sim 0.6\ R_J$– $0.72\ R_J$. The east–west asymmetry in the [S II] images taken at similar magnetic longitudes, but 2.5 Jovian rotations apart, supports the theory of convective motions suggested by others.

099.020 **Io's volcanic and sublimation atmospheres.**
M. A. Moreno, G. Schubert, J. Baumgardner,
M. G. Kivelson, D. A. Paige.
Icarus, Vol. 93, No. 1, p. 63 – 81 (Sep 1991).

Computer simulations of SO_2 and H_2S frost sublimation and SO_2 day and nightside volcanic atmospheres on Io are presented. The models include a surface temperature which is a function of distance from the volcanic vent or subsolar point, sublimation and condensation with a sticking coefficient of unity, and a crude radiative transfer approximation. The full three–dimensional, axisymmetric gas dynamic equations are solved numerically using a time explicit, finite volume formulation. Neither a SO_2 frost sublimation atmosphere nor a SO_2 Pele–type eruption centered on the dayside extends into the nightside.

099.021 **Hydrogen–helium mixtures at megabar pressures: implications for Jupiter and Saturn.**
J. E. Klepeis, K. J. Schafer, T. W. Barbee III, M. Ross.
Science, Vol. 254, No. 5034, p. 986 – 989 (15 Nov 1991).

Models of Jupiter and Saturn postulate a central rock core surrounded by a fluid mixture of hydrogen and helium. These models suggest that the mixture is undergoing phase separation in Saturn but not Jupiter. State–of–the–art total energy calculations of the enthalpy of mixing for ordered alloys of hydrogen and helium confirm that at least partial phase separation has occurred in Saturn and predict that this process has also begun in Jupiter.

099.022 **The left–hand polarization sense of the Jovian decameter radiation.**
M. Y. Boudjada, F. Genova.
Astron. Astrophys., Suppl. Ser., Vol. 91, No. 3, p. 453 – 467 (Dec 1991).

Decameter radiation from the Jovian southern hemisphere, which is left–hand polarized, is studied statistically from 95 events. The authors used the data of the Nançay spectropolarimeter that has operated since 1986. From the variation of the intensity with time and frequency they analyse the occurrence in the Io–CML plane and the spectral characteristics of the left–hand emissions. The main Jovian bursts consist of two families: vertex–early arcs (VEA) and vertex–late arcs (VLA). The authors show that the Io–controlled emissions IoA and IoD are associated with the first family (VEA) and IoC with the second (VLA). Non–Io controlled emissions are observed only in the non–Io A–C region which have the same dynamic spectra (VLA) as Io–controlled emissions. The high frequency limit of left–hand polarized radiation is usually between 18 and 25 MHz, for both Io and non–Io events. Left–hand radiation has never been observed at frequencies higher than 25 MHz, impling that the maximum field strength in Jupiter's southern hemisphere is approximately 9 Gauss.

099.023 **Jupiter's north and south polar aurorae with IUE data.**
T. A. Livengood, H. W. Moos.
Geophys. Res. Lett., Vol. 17, No. 12, p. 2265 – 2268 (Nov 1990).

Observations of Jupiter's auroral H_2 emissions with the IUE spacecraft, from 1978 through 1990, show that aurorae at the two poles are similar in several respects: (1) emission intensity is modulated with magnetic longitude, north peak emission at $\lambda_{III}(1965) \sim 200°$, south peak at $\lambda_{III}(1965) \sim 20°$; (2) emission intensity and degree of modulation are comparable for both poles; (3) attenuation by hydrocarbons varies in phase with the intensity, peak attenuation coinciding approximately with peak emission. Increases in the hydrocarbon column are an indicator of either enhanced penetration depth and thus energy of the auroral primaries, or of increases in the high–altitude hydrocarbon content caused by modification of the local atmosphere.

099.024 **Plasma transport in the Io torus: the importance of microscopic diffusion.**
Y. Mei, R. M. Thorne.
Geophys. Res. Lett., Vol. 18, No. 2, p. 119 – 122 (Feb 1991).

The authors consider the question of whether the distribution of mass in the Io plasma torus is consistent with the concept of interchange eddy transport. Specifically, the flux tube content exhibits a gradual decrease with increasing radial distance from the source near Io without any evidence for substantial density irregularity associated with the plasma source or loss. Using a simple one dimensional numerical model to simulate macroscopic interchange eddy transport, the authors demonstrate that this smooth equilibrium distribution of mass can occur but only with the inclusion of a minimal level of small scale microscopic mixing at a rate approaching Bohm diffusion. Otherwise the system exhibits a chaotic appearance which never approaches an equilibrium distribution. The authors discuss various physical mechanisms for the microscopic diffusion process which is required to provide a sufficiently rapid mixing of material between the macroscopic eddies.

099.025 **The Jovian ionospheric E region.**
Y. H. Kim, J. L. Fox.
Geophys. Res. Lett., Vol. 18, No. 2, p. 123 – 126 (Feb 1991).

The authors have constructed a model of the Jovian ionosphere that includes direct photoionization of hydrocarbon molecules. A high resolution solar spectrum was synthesized from Hinteregger's solar maximum spectrum (F79050N) and high resolution cross sections for photoabsorption by H_2 bands in the range 842 to 1116 Å were constructed. Two strong solar lines and about 30% of the continuum flux between 912 to 1116 Å penetrate below the methane homopause despite strong absorption by CH_4 and H_2. The authors find that hydrocarbons (mainly C_2H_2) are ionized at a maximum rate of $55\ cm^{-3}sec^{-1}$ at 320 km above the ammonia cloud tops. The hydrocarbon ions produced are quickly converted to more complex hydrocarbon ions through reactions with CH_4, C_2H_2, C_2H_6, and C_2H_4. The authors find that a hydrocarbon ion layer is formed near 320 km that is about 50 km wide with a peak density in excess of $1x10^4 cm^{-3}$.

099.026 **Collisionless reconnection in Jupiter's magnetotail.**
G. Zimbardo.
Geophys. Res. Lett., Vol. 18, No. 4, p. 741 – 744 (Apr 1991).

The author presents the first quantitative study of collisionless reconnection in Jupiter's magnetotail. Recently it has been shown that collisionless reconnection can occur in the Earth's magnetotail quasi–neutral sheet when the electrons are subject to chaotic pitch angle diffusion. This happens when the magnetic field is very taillike, i.e., when the curvature parameters $\varkappa_e = B_z/B_0(L_z/\varrho_{e0})^{1/2}$ decreases to ≈ 1.6. For Jupiter's magnetotail, a self–consistent axisymmetric equilibrium model is used to compute the magnetic field and $\varkappa_e$, and it is found that the condition for electron chaotization is satisfied in a region large enough to allow the growth of tearing instability. The favorite site for reconnection is situated at about 60 R_J, and the growth time is estimated to be about 370 s, i.e., much shorter than the plasma transit time through the chaotization region. The inductive e.m.f along a 20 R_J X line is estimated to be of order 100 kV – 1 MV, and should accelerate the ions dawnward and the electrons duskward.

099.027 **Spectra and complete polarization state of two, Io–related, radio storms from Jupiter.**
A. Lecacheux, A. Boischot, M. Y. Boudjada, G. A. Dulk.
Astron. Astrophys., Vol. 251, No. 1, p. 339 – 348 (Nov 1991).
The authors report on the complete Stokes polarimetry, as a function of both frequency and time, of two, Io–related radio bursts from Jupiter, one from source Io–A and the other from source Io–B. The observations were made with the spectropolarimeter at Nancay, France, fed by two oppositely–polarized antenna arrays of 72 elements each.

099.028 **A tentative identification of Jovian global oscillations.**
B. Mosser, F. X. Schmider, P. Delache, D. Gautier.
Astron. Astrophys., Vol. 251, No. 1, p. 356 – 364 (Nov 1991).
The authors propose here the identification of the Jovian frequency spectrum obtained by Schmider et al. (1990, 1991), from Doppler shift observations of the solar sodium D line reflected by the planet, as a signature of Jovian global pressure modes. This spectrum, derived from 4 nights of observations, presents high–amplitude peaks whose origin is undoubtedly not noise, but it suffers from the daily interruption in the observations. Echelle diagram analysis has allowed the determination of a regular pattern, corresponding to the regularity indicated by the asymptotic theory (Tassoul 1980) revised for Jovian seismology (Provost et al., in preparation). The observed equidistance is not a direct measurement of the characteristic frequency ν_0, but is closely related to it. As expected two frequencies dominate the spectrum: the rotational frequencies of the Earth (11.6 μHz) and of Jupiter (28.0 μHz). However, it has been possible to construct an échelle diagram based on a frequency which satisfies the three following criteria: regular échelle diagram, commensurability neither with 28 μHz, nor with multiples of this frequency, satisfactory agreement with the theoretically expected asymptotic pattern. From the identification based on the échelle diagram, the authors derive first observational conclusions about Jupiter's interior. The physical implications of the low value of the characteristic frequency ν_0 and of the large amplitudes of detected oscillations are discussed.

099.029 **Jovian UV aurorae, IR aurorae, and particle precipitations: a common origin?**
R. Prangé.
Astron. Astrophys., Vol. 251, No. 2, p. L15 – L18 (Nov 1991). Letter–to–the–editor.
The morphological features of the mid–IR (hydrocarbons) north auroral "hot spot" are compared with those of the UV aurora as derived from the Voyager observations, and with various possible loci of particle precipitations. This suggests that all three processes must occur in the same region, delimited by the "UVS auroral oval". The longitudinal profile of the IR emission is then shown to be similar to model profiles of the electron flux precipitated in a diffuse aurora, and the best fit model parameters are those which also fit the UVS brightness distribution. The author concludes that either the precipitations are the cause (direct or indirect) of both the UV and mid–IR aurorae, or all three phenomena derive from a common initial process in the middle magnetosphere.

099.030 **Theory of the decameter S–radio emission of Jupiter.**
A. G. Boev, M. Yu. Luk'yanov.
Sov. Astron., Vol. 35, No. 4, p. 422 – 427 (Jul 1991). Current Physics Microform No.: 9203X0608. English translation of Astron. Zh., Tom 68, Vyp. 4, p. 853 – 862 (1991).
A new mechanism for generating the decameter S–radio emission of Jupiter is suggested in this paper, based on Cherenkov instability of an electron beam that can develop in the Io–Jupiter flux tube. As a result of this instability, plasma waves with anomalous dispersion are generated in the tube at the local upper hybrid resonance frequency, and are then transformed into fast extraordinary waves at the smooth inhomogeneity of the tube boundary, which comes about through diffusion. It is shown that the proposed mechanism explains rather well a number of experimentally observed facts: the conical beam pattern, the anomalous angular frequency sweep, the frequency drift over a wide frequency range. Estimates under conditions typical of Jupiter show that this mechanism is also capable of providing the observed radiative flux.

099.031 **Drift wave instability in the Io plasma torus.**
T. S. Huang, T. W. Hill.
J. Geophys. Res., Vol. 96, No. A8, p. 14075 – 14084 (1 Aug 1991).
The authors present a linear normal–mode analysis of low–frequency electrostatic drift waves in the Io plasma torus, providing a kinetic description of the centrifugal interchange instability of the torus regulated by coupling to Jupiter's ionosphere.

099.032 **Charged–coupled device spectral images of spatially resolved regions of Jupiter in the 6190– and 8900–Å ammonia bands during the 1989 opposition.**
F. Moreno, A. Molina, L. M. Lara.
J. Geophys. Res., Vol. 96, No. A8, p. 14119 – 14128 (1 Aug 1991).
High–quality charge–coupled device spectroscopic observations at different regions of Jupiter in the spectral regions $\lambda\lambda$ 6000–6600 and 8700–9200 Å at a resolution of 1.2 Å have been obtained. These spectra allowed the authors to obtain the center–to–limb curves of the equivalent widths of the CH_4 6190–Å band, the NH_3 6450–Å band, and the band depth of the 8900–Å band, as well as their latitudinal distribution. Some asymmetries in the center–to–limb distribution of the equivalent widths of the CH_4 6190–Å and NH_3 6450–Å bands were detected, as well as a very prominent decrease toward the planetary limbs for the ammonia band. The latitudinal variations of the CH_4 6190–Å and NH_3 6450–Å equivalent widths were strongly correlated for fractional radii less than 0.4, with a maximum of absorption at the North Equatorial Belt. The South Equatorial Belt, undetectable at visible and near–infrared continuum wavelengths, was discernable at $\lambda = 8900$ Å. Model computations for the Equatorial Region assuming three aerosol layer models gave some constraints on tropospheric atmospheric parameters during the 1989 Jupiter opposition data.

099.033 **Jupiter en 1989 – 1990: opposition du 27 décembre 1989.**
M. Jacquesson, G. Farroni.
Astronomie, p. 15 – 19 (Dec 1991).

099.034 **Great disturbances in Jupiter's atmosphere.**
V. A. Zinov'ev.
Astron. Tsirk., No. 1544, p. 31 – 32 (May – Jul 1990). In Russian.
A disparition of the South Equatorial Belt (SEB) in Jupiter's atmosphere was observed on August 24, 1989. Such events were observed also in 1936 – 1937 and in the 17–th century (Hooke, Cassini).

099.035 **Radio images of Jupiter's synchrotron radiation at 6, 20, and 90 cm.**
I. de Pater.
Astron. J., Vol. 102, No. 2, p. 795 – 805 (Aug 1991). Current Physics Microform No.: 9108A1731.
VLA observations of Jupiter's nonthermal radiation at wavelengths of 6, 20, and 90 cm are presented and compared. The spatial resolution in all images is 0.25 R_J (R_J = Jovian radius). The brightness distribution of Jupiter is very similar at each of the three wavelengths, although the radiation peaks at 6 cm are at a slightly larger distance from Jupiter than those at the longer wavelengths. This is due to the higher rate of synchrotron radiation losses at the shorter wavelengths. Radial profiles through the images clearly show the presence of a "shoulder" or flattening in the intensity at $\sim$2.5 R_J due to absorption effects by the satellite Amalthea. In addition, the so–called "hot region" in Jupiter's radiation belts is clearly present at all three wavelengths: images of the planet at a cml (central meridian longitude) of $\sim$30°–40° show the radiation peak to the right of Jupiter to be much brighter than the peak at the left, while images at a cml of $\sim$210°–220° show the opposite. The difference in brightness

between the two radiation peaks is similar at all three wavelengths. Measurements of the total intensity confirm earlier results which showed a flattening of the radio spectrum at longer wavelengths, or a decrease in spectral index with increasing wavelength: $\alpha = 0.35 \pm 0.05$ between 6 and 20 cm, and $\alpha = 0.13 \pm 0.04$ between 20 and 90 cm.

099.036 **An analysis of the 1985 observations of mutual phenomena of the Galilean satellites.**
F. A. Franklin, J. Africano, W. Allen, K. Aksnes, W. Weller, P. Birch, C. Blanco, I. Coulson, F. Franklin, J. Goguen, W. Sinton, R. Marcialis, R. Millis, L. Wasserman, T. Nakamura, M. Soma, J. Walters, J. Westfall, R. Jones, D. Rettig, T. Langhans, C. Rowe, J. E. Solheim, D. Skillman.
Astron. J., Vol. 102, No. 2, p. 806 – 815 (Aug 1991). Current Physics Microform No.: 9108A1742.

This paper derives midtimes and relative satellite positions from nearly 200 light curves of mutual eclipses and occultations of the Galilean satellites that occurred in 1985. The authors show that, at least for observations of the highest quality, the standard errors on the midtimes are smaller than 1 s and close to 0.01 arcsec on the positional offsets in right ascension and declination.

099.037 **CCD photometry and spectroscopy of the outer Jovian satellites.**
J. Luu.
Astron. J., Vol. 102, No. 3, p. 1213 – 1225 (Sep 1991). Current Physics Microform No.: 9109D1617.

The author presents new charge–coupled device (CCD) photometry and spectroscopy of the eight outer Jovian satellites obtained from November 1987 to February 1989. The photometry consists of light curves for J7 – J12, plus BVRI colors for J6, J8 – J13. Spectroscopic data include reflectance spectra for J6, J8 – J11. The satellites are spectrally diverse, with reflectivity gradients ranging from neutral ($0.4\% \pm 0.1\%/10^3$Å) to red ($12\% \pm 1\%/10^3$Å). The satellites thus resemble a mixture of C– and D–type asteroids. Similarities between the satellites and the Trojans (diverse colors, low albedos) are consistent with the hypothesis that these two groups of objects share a common origin (Kuiper, 1956). However, the satellites seem to have less extreme shapes than the Trojans, perhaps due to the longer amount of time they spent in the proto–Jupiter nebula. Physical properties of the satellites are generally consistent with, but do not prove, the capture origin theory (Pollack et al., 1979).

099.038 **What is new in the world of giant planets?**
S. R. Brzostkiewicz, D. Górnicza.
Urania, Rok 62, Nr. 7 – 8, p. 198 – 208 (Jul – Aug 1991). In Polish.

099.039 **Position observations of Jovian satellite VI at the Main Astronomical Observatory AS UkrSSR in 1987 – 1988.**
E. M. Izhakevich, S. V. Kaltygina, S. P. Major, S. V. Shatokhina.
Kinematics Phys. Celest. Bodies, Vol. 7, No. 2, p. 88 – 89 (1991). Translated from Kinematika Fiz. Nebesnykh Tel (1991) v. 7(2) p. 98–99, 104

The results of 12 observations of Himalia made at Goloseevo and on Mt. Maidanak in 1987 – 1988 are presented and compared with the ephemeris.

099.040 **Global high–resolution imaging of hotspots on Io.**
B. A. McLeod, D. W. McCarthy Jr., J. D. Freeman.
Astron. J., Vol. 102, No. 4, p. 1485 – 1489 (Oct 1991). Current Physics Microform No.: 9110D2017.

The authors present 3.8 μm images of the Jovian satellite Io at an unprecedented ground–based resolution of 0."18. The images were obtained through two–dimensional speckle interferometry using the 6.86 m aperture of the Multiple Mirror Telescope in November 1989 and March 1990. To process the images the authors developed new speckle reduction techniques appropriate for objects with a size that is a significant fraction of the field of view. The images show three volcanic hotspots, including Pele, detected here for the first time since the Voyager 1 flyby in 1979.

099.041 **Refraction of Jovian decametric radio emission. II. Distortion of the L–radiation pattern.**
A. V. Arkhipov.
Kinematika Fiz. Nebesn. Tel, Tom 7, No. 5, p. 11 – 19 (Sep–Oct 1991). In Russian. English translation in Kinematics Phys. Celest. Bodies, Vol. 7, No. 5.

It is shown that when a realistic plasma model is used, refraction can account for the observed radiation pattern of L–emission as well as the asymmetry of a peak flux density there. An electrical current influence on the Jovian decametric radio emission refraction is considered for the first time.

099.042 **The three white ovals and adjacent belts in the South Temperate region of Jupiter, 1940 – 1990.**
J. Rogers, D. Herbert.
J. Br. Astron. Assoc., Vol. 101, No. 6, p. 351 – 360 (Dec 1991).

099.043 **Proton energy deposition and EUV emission in the Jovian and the Saturnian atmosphere.**
V. Singh.
Indian J. Radio Space Phys., Vol. 20, No. 2, p. 140 – 149 (Apr 1991).

A model of the interaction of energetic protons with the atmospheres of Jupiter and Saturn is developed to examine the possibility that the high latitude ultraviolet aurorae observed on both planets are at least partly excited by proton precipitation. The range of proton energies compatible with the spectral composition of the ultraviolet auroral spectra of Jupiter and Saturn is found to be $\sim 1 - 1000$ keV. It is suggested that ~ 1 MeV protons are viable candidates to explain the secondary peak observed in the longitudinal distribution of the Jovian auroral intensity. It is also shown that the H_2 bands/H_β ratio calculated in the model is compatible with the intensity of the visible arcs observed at Jupiter's high latitude limb with the Voyager camera.

099.044 **Ray–tracing of electromagnetic waves in planetary magnetospheres – application to the low–frequency radio emission of Jupiter.**
H. P. Ladreiter, Y. Leblanc.
2. International Workshop on Plasma Physics: Plasma Theoretical Problems in Astro– and Fusion Physics, p. 39 – 52 (1991). – See Abstr. 012.067 for the main entry.

The magnetosphere of Jupiter is one of the most complex in our solar system. The most powerful radio components are the Jovian hectometric radiation and the decametric component. At the orbit of the satellite Io, Jupiter is encircled by the so–called Io plasma torus which is fed by particles from volcanic origin at the surface of Io. The peak electron number density exceeds 3000/ccm, therefore the Io torus acts like an obstacle for the radiation at hectometric wavelengths. In order to determine the sources of the hectometric emission, the authors have performed a ray–tracing study including all effects of refraction that the wave suffers during the propagation from the source to the point of observation. Throughout a three–dimensional investigation they have modeled the hectometer observations as recorded by the Planetary Radio Astronomy experiment aboard Voyager 1 and 2. Hectometric source locations at high magnetic altitudes at both hemispheres fit best the emission pattern as observed by the Voyager spacecraft and are in agreement with the maser cyclotron mechanism.

099.045 **Polarization of Jovian hectometric emission.**
A. Ortega–Molina, A. Lecacheux.
J. Geophys. Res., Vol. 96, No. A7, p. 11441 – 11454 (1 Jul 1991).

The authors have studied the polarization properties of the Jovian hectometric (HOM) emission observed in the low–frequency band 1.2–1326 kHz by the planetary radio astronomy (PRA) experiment aboard the Voyager spacecraft. They have shown that the complex morphology of HOM can be explained by simultaneous radiations from two independent 100% oppositely polarized sources. Analysis of data recorded under different antenna–source configurations indicates that the existence of a linear component in HOM is very unlikely. The authors show that the apparent degree of circular polarization measured by the PRA instrument, which practically ranges from -1 to +1, is the result of fluctuations in the relative intensities of the two HOM circular components. The observed distribution of the degree of circular polarization is accurately reporduced by assuming that the intensity distribution of each individual HOM circular component follows an exponentially decreasing probability law. The emission mode, source location, and beaming of the HOM components are inferred from the polarization properties of the emission. A global model for the hectometric and decametric (DAM) Jovian emissions in which the left and right HOM components are identified with the left and right non–Io DAM components, respectively, is outlined.

099.046 **Comment on "Bremsstrahlung X–rays from Jovian auroral electrons".**
J. H. Waite Jr.
J. Geophys. Res., Vol. 96, No. A11, p. 19529 - 19532 (1 Nov 1991).
A comment on the paper by D. D. Barbosa published in J. Geophys. Res., Vol. 95, No. A9, p. 14969 - 14976 (1 Sep 1990). See Abstr. 52.099.046. A reply by D. D. Barbosa is given in J. Geophys. Res., Vol. 96, No. A11, p. 19533 - 19534 (1 Nov 1991).

099.047 **Ganymede and Callisto: complex crater formation and planetary crusts.**
P. M. Schenk.
J. Geophys. Res., Vol. 96, No. E1, p. 15635 - 15664 (25 Aug 1991).
New depth/diameter, rim, and peak height and width measurements of fresh craters on Ganymede and Callisto are utilized to explore not only crater mechanics on icy satellites but intersatellite crater scaling and crustal properties and composition. Significant systematic differences in crater morphology on icy and rocky planets are confirmed.

099.048 **Modeled EUV spectrum of Jupiter's aurora.**
J. Clarke, G. R. Gladstone, S. Chakrabarti, W. Harris, H. Waite.
1. Berkeley Colloquium on Extreme Ultraviolet Astronomy, p. 261 - 267 (1991). - See Abstr. 012.045 for the main entry.
Contents: 1. Introduction. 2. Nature of the exciting charged particles. 3. The EUV emission spectrum of precipitating S and O ions. 4. Conclusions.

099.049 **Modeling of Jupiter's millimeter wave emission utilizing laboratory measurements of ammonia (NH_3) opacity.**
J. Joiner, P. G. Steffes.
J. Geophys. Res., Vol. 96, No. E2, p. 17463 - 17470 (25 Sep 1991). With a correction in Vol. 96, No. E5, p. 22827 (25 Dec 1991). - See Abstr. 012.073 for the main entry.
Radiative transfer models which have been used to compute Jupiter's millimeter wave emission do not agree well with the existing radio astronomical observations. This apparent discrepancy has gone large unexplained due to lack of laboratory absorption data at these wavelengths coupled with uncertainties in the calibration of millimeter wave observations. Gaseous ammonia (NH_3) is the largest source of millimeter wave opacity on Jupiter. Previous laboratory measurements at 7.5– to 9.38–mm were inconclusive as to which theoretical line shape most accurately describes the millimeter absorptivity of NH_3. The authors have made additional laboratory absorption measurements of gaseous ammonia at a shorter wavelength (3.2 mm) where the theoretical line shapes can be better evaluated. They have conducted measurements at a temperature of 210K, at pressures ranging from 1 to 2 atm., and in a mixture consisting of 85.56% hydrogen (H_2), 9.37% helium (He), and 5.07% ammonia (NH_3). The authors give a revised formalism for computing NH_3 absorption in an H_2/He atmosphere using the Ben–Reuven line shape. They investigate several other potential millimeter wavelength absorbers and give revised formalisms for computing their absorption. The authors have compiled a list of Jupiter's reliable millimeter wavelength observations. They compare their list of observations with synthetic emission spectra utilizing their revised expressions for computing absorption.

099.050 **Jupiter.**
IAU Circ., No. 5345, p. 1 (14 Sep 1991).

099.051 **A disturbance in Jupiter's high–speed north temperate jet during 1990.**
A. Sanchez–Lavega, I. Miyazaki, D. Parker, P. Laques, J. Lecacheux.
Icarus, Vol. 94, No. 1, p. 92 - 97 (Nov 1991).
During Feb 1990 a remarkable disturbance developed in the strongest Jovian atmospheric jet located at $\approx 24°N$ planetographic latitude. Telescopic images at visible wavelengths revealed the outbreak of two bright features with an initial length scale ≈ 6000 km, the first discovered on Feb 10 and the second a week later. Both resided on the peak of the eastward jet, drifting rapidly with zonal velocities of 165 and 170 m sec^{-1}, respectively. To the west of the bright features a pattern of alternating dark and light spots developed over the next few days. These features moved with a lower mean elocity of 108 m sec^{-1} and had a mean east–west length of 10,000 km. Their separation in longitude showed a bimodal distribution. The disturbance expanded zonally, encircling the whole latitudinal band at 24°N 3 months after its outburst, generating a new temperate belt.

099.052 **Elusive H_3^+ ion tracked down.**
A. Dalgarno.
Nature, Vol. 353, No. 6344, p. 502 - 503 (10 Oct 1991).
The author reports on the molecular ion H_3^+ which has been discovered in the atmosphere of Jupiter and on its relevance for interstellar clouds.

099.053 **Images of aurorae on Jupiter from H_3^+ emission at 4 μm.**
S. J. Kim, P. Drossart, J. Caldwell, J.–P. Maillard, T. Herbst, M. Shure.
Nature, Vol. 353, No. 6344, p. 536 - 539 (10 Oct 1991). Letter–to–the–editor.
Since their discovery by Voyager, aurorae on Jupiter have been regularly observed over the past ten years, from space at ultraviolet wavelengths and from the Earth in the thermal infrared band (8 - 13 μm). The authors present images of near–infrared H_3^+ emission, recorded simultaneously with high–resolution spectra. Having both imaging and spectral data provides them with a much improved opportunity to study the spatial and spectral characteristics of the H_3^+ emission. They find that northern and southern emissions show strong spatial variation at the wavelengths of H_3^+ emission. They suggest that the appearance of the jovian auroral regions at these wavelengths may be explained by a combination of localized H_3^+ emission, reflection of sunlight by polar haze and stratospheric CH_4 emission in the 3.3 - 3.6 μm range.

099.054 **Imaging Jupiter's aurorae from H_3^+ emissions in the 3 - 4 μm band.**
R. Baron, R. D. Joseph, T. Owen, J. Tennyson, S. Miller, G. E. Ballester.
Nature, Vol. 353, No. 6344, p. 539 - 542 (10 Oct 1991). Letter–to–the–editor.
Since H_3^+ was first spectroscopically detected on Jupiter, there has been considerable interest in using this simple molecular ion to probe conditions existing in the planet's auroral regions. The authors present a series of images of Jupiter recorded at wavelengths sensitive to emission by H_3^+, which reveal the spatial distribution of excited H_3^+ molecular ions in the jovian ionosphere, as seen from Earth. They believe that they provide high–spatial–resolution images of polar aurorae on Jupiter. They suggest that the intensity of the auroral emission can vary on a timescale of an hour, a shorter period than had previously been

noted. They also find that the spatial distribution of H_3^+ emissions correlates only partially with the loci of auroral activity inferred from ultraviolet and longer-wavelength infrared observations. The H_3^+ emission may therefore be controlled by auroral processes that are different from those responsible for the ultraviolet and infrared emissions.

099.055 **The quasiquadrennial oscillation of Jupiter's equatorial stratosphere.**
C. B. Leovy, A. J. Friedson, G. S. Orton.
Nature, Vol. 354, No. 6352, p. 380 – 382 (5 Dec 1991). Letter–to–the–editor.

A new 11–year record of Jupiter's stratospheric temperature shows an equatorial temperature oscillation with an apparent period of 4 – 5 years. The authors compare this oscillation to two long–period oscillations of zonal winds in the Earth's equatorial stratosphere, and propose that the same mechanism – forcing by the stresses of vertically propagating waves – is responsible for the oscillations on both planets. Jupiter's temperature oscillation has been observed for slightly more than two cycles and closely resembles the temperature signatures of the Earth's semiannual oscillation and quasibiennial oscillation. If the mechanisms responsible for these oscillations are indeed similar, Jupiter's oscillation indicates that there is significant vertical momentum transport due to equatorially trapped atmospheric waves, and it may provide a means for probing the poorly understood process of the generation of these waves by convection.

099.056 **The orbital resonance among the Galilean satellites I, II and III of Jupiter.**
Zhao Changyin, Liu Lin.
Chin. Astron. Astrophys., Vol. 15, No. 4, p. 366 – 374 (Dec 1991). English translation of Acta Astron. Sin., Vol. 32, No. 2, p. 121 – 128 (Jun 1991).

It is shown that an ideal resonance model for the first three Galilean satellites will reproduce the relevant observed facts.

099.057 **On the formation of Valhalla basin (Callisto).**
G. A. Lejkin, A. N. Sanovich.
Tr. Gos. Astron. Inst. im. Shternberga, Tom 62, p. 202 – 217 (1991). In Russian.

The morphology of multi–ring Valhalla basin is defined more precisely. A method of concentricity estimation of the basin rims and ringes is proposed. It is shown that the structure is not concentric. The hypotheses of multi–ring structure formation on ice moons are discussed. A new hypothesis of the Valhalla basin formation is suggested.

099.058 **Determination of ionic abundances in the Io torus using the Hopkins Ultraviolet Telescope.**
H. W. Moos, P. D. Feldman, S. T. Durrance, W. P. Blair, C. W. Bowers, A. F. Davidsen, W. V. Dixon, H. C. Ferguson, R. C. Henry, R. A. Kimble, G. A. Kriss, J. W. Kruk, K. S. Long, O. Vancura.
Astrophys. J., Lett., Vol. 382, No. 2, p. L105 – L108 (1 Dec 1991).

During the Astro–1 mission in 1990 December, the Hopkins Ultraviolet Telescope (HUT) was used to obtain a 415 – 1864 Å spectrum of the emissions from the Io torus with ~3 Å resolution. The ~1274 s observation of the east ansa took place during orbital night when spectral contamination by terrestrial airglow was minimal. A list of the 31 strongest features shows that, apart from five features that are totally or partially terrestrial, all are due to O^+, S^+, S^{++} and S^{+++}. There is no evidence for elements other than oxygen and sulfur. However, the spectrum shows many additional weaker emission features so that the existence of minor species at present cannot be ruled out. Because the HUT spectrum contains well resolved lines for all four of the major species, the mixing ratio (N_i/n_e) for each ion can be determined without using models of the spectral emissions or a detailed geometric model of the torus plasma. The relative concentration of O^+ determined from the HUT observation is 38% less than that obtained from the Voyager UVS data.

099.059 **Vibrationally excited H_2 in the outer planets thermosphere: fluorescence in the Lyman and Werner bands.**
T. Majeed, J. C. McConnell, R. V. Yelle.
Planet. Space Sci., Vol. 39, No. 11, p. 1591 – 1606 (Nov 1991).

The authors have considered the impact of fluorescence of ground state H_2 on the distribution of the vibrational levels of H_2 in the upper atmospheres of Jupiter and Saturn for non–auroral latitudes. For $v \geq 3$, for the conditions studied, this is the most important source of vibrationally excited H_2 compared with other sources, such as photoelectron induced fluorescence, dissociative recombination of H_3^+ ions, and direct vibrational excitation of H_2 by photoelectron impact. Combining the Voyager limb observations of H_2 band emissions on Saturn, theoretical calculations of the H_2 fluorescence distribution, and column constraints of Jovian H_2 airglow, the authors estimate that some of the higher vibrational levels may have effective temperatures >3000K on both Jupiter and Saturn. In turn, the vibrational population of $v \geq 4$ levels are sufficiently increased by the fluorescence source that the chemical sink for the ionization is enhanced. As a result, ionospheric densities may be greatly affected. The authors also show that the vertical ion flows induced by horizontal neutral winds or dynamo electric fields must play some role in maintaining the plasma peaks at higher altitudes.

099.060 **The upper ionospheres of Jupiter and Saturn.**
T. Majeed, J. C. McConnell.
Planet. Space Sci., Vol. 39, No. 12, p. 1715 – 1732 (Dec 1991).

The authors use a 1–D chemical diffusive model, in conjunction with the measured neutral atmospheric structure, to analyze the Voyager RSS electron density profiles for the ionospheres of Jupiter and Saturn.

099.061 **Detection of ^{13}C–ethane in Jupiter's atmosphere.**
G. Wiedemann, G. L. Bjoraker, D. E. Jennings.
Astrophys. J., Lett., Vol. 383, No. 1, p. L29 – L32 (10 Dec 1991).

High-resolution ^{12}C– and ^{13}C–ethane spectra of Jupiter were acquired with the Kitt Peak 4 m Fourier spectrometer and the Goddard postdisperser in 1987 June. A relative abundance ratio $^{12}C/^{13}C$ of 94 ± 12 was derived from the measurements. This nearly terrestrial value indicates little or no fractionation of carbon isotopes, when ethane is produced in the photolysis of methane in Jupiter's atmosphere.

099.062 **Tidal origin of the Laplace resonance and the resurfacing of Ganymede.**
R. Malhotra.
Icarus, Vol. 94, No. 2, p. 399 – 412 (Dec 1991). = Contrib. No. 4994 Div. Geol. Planet. Sci.

The author presents a new scenario for the tidal origin of the Laplace resonance. The inner three Galilean satellites are supposed to have formed in orbits well away from the 2/1 pair-wise mean motion commensurabilities and the three body Laplace resonance. The dynamics is controlled by the evolution of the 2/1 critical frequencies. It is shown that the Laplace relation may have been established from within a previous three body resonance. The attendant enhanced tidal heating in Ganymede during that high eccentricity period provides a possible explanation for the geologically younger surface of Ganymede compared to that of its "twin satellite", Callisto.

099.063 **An unusual change in the Jovian Lyman–alpha bulge.**
M. A. McGrath.
Geophys. Res. Lett., Vol. 18, No. 11, p. 1931 – 1934 (Nov 1991).

A very unusual change is reported in the morphology of the Jovian Ly–α bulge, a region of enhanced Ly–α emission near the Jovian spin equator at a fixed system III magnetic longitude (λ_{III}) that has persisted since its discovery in the late 1970s. Observations from 1979 – March 1989 consistently showed Ly–α emission that peaked near $\lambda_{III} \sim 100°$ then dropped and remained constant in the nonbulge region from $\lambda_{III} \sim 200$–$360°$. During this time the bulge had a full width half maximum (FWHM) of ~100–130°. In September 1989 and March and May 1991 this nomimal morphology was substantially different: the Ly–α brightness showed a much broader peak (FWHM ~ 200°) centered near 40° meaning that the bulge encompassed nearly the entire equatorial region of the planet and the previously flat nonbulge region was virtually nonexistent. These observations show a fundamental change in the morphology of the Ly-α bulge which was not observed in the previous 11 years. Present theories for the bulge invoke asymmetric processes or densities in the Jovian ionosphere. The change in bulge morphology may imply a significant change in the pattern of large-scale magnetospheric convection, in turn implying a possible change in longitudinal density structure of the Io plasma torus between March and Sept. 1989. The possibility that the observed change in bulge morphology is related to long-term solar activity also cannot be ruled out.

099.064 **Jupiter's dayglow H Lyα emission line profile.**
J. T. Clarke, G. R. Gladstone, L. Ben Jaffel.
Geophys. Res. Lett., Vol. 18, No. 11, p. 1935 – 1938 (Nov 1991).

IUE spectra of Jupiter's low latitude H Lyα emission line profile have been obtained on and off the "Lyα bulge" at the longitude of the central meridian (CML), near the limb of the planet, and at midlatitudes on the CML. All equatorial locations show line broadening beyond the IUE resolution of 0.14 Å. The equatorial lines are significantly broader than at midlatitudes on the CML, and near the limb the lines are broader than on the CML, both on and off the bulge. The observed line shapes are not consistent with pure resonant scattering of solar Lyα by upper atmospheric H atoms with a variable H abundance: calculated profiles near the limb have much less intense cores. The broadening of the bulge line profiles on the CML and limb compared with off-bulge regions demonstrates that the bulge effect is a broadening of the line, not an increase in the line core flux. The authors propose that there is either a substantial broadened internal source and/or the upper atmosphere has considerable Doppler broadening (winds or turbulence) to increase the resonant scattering line width.

099.065 **Two-dimensioinal model of slow-mode expansion fan at Io.**
P. H. Krisko, T. W. Hill.
Geophys. Res. Lett., Vol. 18, No. 11, p. 1947 – 1950 (Nov 1991).

The authors develop a 2-D model for the standing slow-mode expansion fan that is expected to exist downstream of the Jovian moon Io. The leading edge of the expansion fan makes an angle of 45° with the upstream magnetic field direction, and the fan width is about 114°. The plasma flow returns to its upstream direction by way of a slow-mode shock behind Io where the MHD parameters return asymptotically to their upstream conditions. The magnetic field perturbation within the fan is much smaller than that associated with the Alfvén wing, which lies farther upstream.

099.066 **Jupiter en 1989 – 1990: opposition du 27 décembre 1989.**
G. Farroni, M. Jacquesson.
Obs. Trav., No. 27, p. 3 – 27 (1991).

099.067 **The thermal gradient of Callisto constrained by Asgard basin: rheological and chemical implications.**
P. Allemand, P. G. Thomas.
J. Geophys. Res., Vol. 96, No. E4, p. 20981 – 20988 (25 Nov 1991).

Using the width of concentric grabens formed during the relaxation of the Asgard basin on Callisto, in combination with rheological flow laws of ice and laws concerning the variations in strain rate around a relaxing crater, the surface thermal gradient of Callisto at -4 b.y. is estimated to be near 0.5K km^{-1}. It is assumed that the graben width depends on the depth of the brittle/ductile transition zone. Around the Asgard basin, the depth of this brittle/ductile transition zone varies from 23 km (at 200 km from the crater rim) to 4 km (at 800 km from the crater rim). It is assumed that the thermal characteristics were constant over the extended zone and that the difference in the depth of the brittle/ductile transition zone is only due to differences in the strain rate around the basin during the relaxation of the initial cavity. Calculations show that Newtonian flow law for ice gives a strain rate of external relaxation at the crater rim of about $10^{-14}s^{-1}$, while a power law gives a strain rate of about $10^{-6}s^{-1}$, assuming a surface temperature between 100 and 150K. The order of magnitude of the calculated thermal gradient is independent of the chosen flow law. This calculated gradient is, however, one order of magnitude lower than those calculated theoretically with a chondritic abundance of radioactive elements in the silicate part of Callisto.

099.068 **The sodium zenocorona.**
W. H. Smyth, M. R. Combi.
J. Geophys. Res., Vol. 96, No. E5, p. 22711 – 22727 (25 Dec 1991).

A recent narrow-band-filtered CCD image by Mendillo et al. (1990) has shown that a sodium corona, produced near Io, extends at least 400 Jupiter radii in the planet's equatorial plane. Isophotes indicate that the polar to equatorial extents are in ~1 to 3 proportions. The purpose of this paper is to explore the nature of this sodium zenocorona and to understand its implications for Io, the planetary magnetosphere, and for the other (i.e., K, O, S) zenocoronae that are too faint to be currently observed.

099.069 **Thermally driven diffusion of SO_2 within the surface of Io.**
P. E. Meade, B. M. Jakosky.
J. Geophys. Res., Vol. 96, No. E5, p. 22729 – 22740 (25 Dec 1991).

The presence of sulfur dioxide (SO_2) on Io, together with the fact that the surface layer of Io has extremely high porosity, suggests the possibility of diffusion of this volatile within the surface, as well as exchange between the surface and an atmosphere. The authors investigate the former possibility through the development of a surface layer thermal model and subsequent calculations of the thermally driven diffusion flux of SO_2 within the layer. The major factors affecting the diffusion process are the temperature and temperature gradient in the surface layer throughout the day (which are results of the thermal model), and the porosity and grain size in the surface layer. The results indicate that the net transport of SO_2 in the near-surface region is downward into the subsurface, causing near-surface depletion of SO_2. Near-surface depletion would result in a layer of reduced thermal inertia overlying the bulk of the surface, consistent with thermal eclipse observations of Io. For the authors' nominal model with 10-μm grains and a porosity of 85%, the peak net diurnal downward flux reaches nearly $8x10^{-3}g\ cm^{-2}period^{-1}$.

099.070 **Observations of Jupiter during the 1989 – 90 apparition.**
R. W. Schmude Jr.
Strolling Astron., Vol. 35, No. 3, p. 111 – 115 (Sep 1991).

099.071 **Le volcanism de Io observé depuis la Terre.**
O. Hainaut.
Ciel, Vol. 53, p. 267 – 271 (Oct 1991). In French.

099.072 **A three-dimensional MHD simulation of plasma flow past Io.**
J. A. Linker, M. G. Kivelson, R. J. Walker.
J. Geophys. Res., Vol. 96, No. A12, p. 21037 – 21054 (1 Dec 1991).

The authors describe a time-dependent three-dimensional magnetohydrodynamic (MHD) simulation of plasma flow past Jupiter's satellite Io. The simulation starts with a uniform, magnetized flow interacting with a finite conducting sphere (representing Io), and the integration is carried forward in time until a quasi-steady state develops. Previously the authors have reported on the presence of standing slow mode perturbations in the flow, attached to Io. In this paper they present details of the simulation model and describe the overall features of the interaction.

099.073 **Pressure anisotropy and radial stress balance in the Jovian neutral sheet.**
C. P. Paranicas, B. H. Mauk, S. M. Krimigis.
J. Geophys. Res., Vol. 96, No. A12, p. 21135 - 21140 (1 Dec 1991).
By examining particle and magnetic field data from the Voyager 1 and 2 spacecraft, the authors have found signatures indicating that the ($\gtrsim$28 keV) particle pressure parallel to the magnetic field is greater than the pressure perpendicular to the field within the nightside neutral sheet (three nightside neutral sheet crossings, with favorable experimental conditions, were used). By incorporating the pressure anisotropy into the calculation of radial forces within the nightside neutral sheet, the authors find that (1) force balance is approximately achieved and (2) the anisotropy force term provides the largest contribution of the other particle forces considered (pressure gradients and the corotation centrifugal force). With regard to the problem of understanding the balance of radial forces within the dayside neutral sheet (McNutt, 1984; Mauk and Krimigis, 1987), the nightside pressure anisotropy force is larger than the dayside pressure gradient forces at equivalent radial distances; however, a full accounting of the dayside regions remains to be achieved.

099.074 **Prediction of the Ulysses Jovian hectometric observations.**
H. P. Ladreiter, Y. Leblanc.
J. Geophys. Res., Vol. 96, No. A12, p. 21207 - 21212 (1 Dec 1991).
The Ulysses unified radio and plasma wave (URAP) experiment is well suited to observe the Jovian hectometric radio emission (HOM) and so complement the observations made by the low-frequency band receiver on Voyager. When the spacecraft encounters Jupiter, in February 1992, observations will be possible from considerably higher magnetic latitudes, in both hemispheres, than those accessible to Voyager. It should also be possible to determine the source location, as URAP can measure the direction of the incoming waves and all four Stokes parameters. Based upon the HOM beaming and source model of Ladreiter and Leblanc (1990), the following predictions can be made: (1) During the inbound trajectory, Ulysses will, at first, be close to the Jovian equatorial plane when equal amounts of right-handed (RH) and left-handed (LH) circularly polarized HOM should be observed from the northern hemisphere and the southern hemisphere, respectively. During the three days immediately before closest approach, the spacecraft will move to a higher northerly magnetic latitude where it will be outside the main HOM emission beam so that a relatively weak RH polarized signal should be observed. (2) At closest approach, Ulysses will be very close to the low-frequency HOM sources and will cross the HOM shadow zone, which extends out to about 20 Jovian radii (R_J) close to the magnetic equatorial plane. (3) During the outbound trajectory, the spacecraft will be at a fairly high southerly magnetic latitude and outside the main HOM beam, so that a relatively weak LH polarized signal should be observed.

099.075 **The effect of electron screening and velocity distribution on proton–deuterium fusion rates in Jupiter.**
G. S. Chulick, R. A. Rice, Y. E. Kim.
AIP Conf. Proc., No. 228, p. 584 - 590 (10 May 1991). - See Abstr. 012.122 for the main entry.
Coulomb screening, in conjunction with a particle velocity distribution, enhances the rate of p–D fusion at extremely low temperature ($\approx$1 eV). As a particular case, the authors demonstrate, given an appropriately strong screening potential, that p–D fusion may provide a significant portion of the excess heat radiated by Jupiter.

099.076 **Nuclear fusion in Jupiter.**
M. Gajda, J. Rafelski.
AIP Conf. Proc., No. 228, p. 591 - 610 (10 May 1991). Current Physics Microform No.: 9108D1737. - See Abstr. 012.122 for the main entry.
The authors study nuclear fusion occurring according to conventional wisdom in the planet Jupiter. In particular, they consider if in a standard evolutionary model of Jupiter a significant part of Jupiter's luminosity has been due to nuclear fusion at any time during its evolution. Nuclear rate equations in dense matter allowing for screening and pressure effects have been integrated in time.

099.077 **A color map of Jupiter from spectrophotometric observations in 1987.**
V. G. Tejfel', A. N. Aksenov, N. V. Priboeva, O. M. Starodubtseva, G. A. Kharitonova.
Sol. Syst. Res., Vol. 25, No. 3, p. 215 - 222 (Nov 1991). English translation of Astron. Vestn., Tom 25, No. 3, p. 293 - 301 (1991).
Based on the photometric reduction of 338 spectrograms of Jupiter's central meridian obtained during three nights in October 1987, a map of its color distribution has been compiled with different modifications, and the limb darkening coefficients of the north and south polar regions have been determined for different longitudes. Results are shown of a preliminary analysis of the features of the latitude and longitude distrubutions of the color differences on Jupiter during this period.

099.078 **Modeling the precipitation flux in the Jovian auroral zones. 1. The model and its application to the UV auroral emissions.**
R. Prangé, M. Elkhamsi.
J. Geophys. Res., Vol. 96, No. A12, p. 21371 - 21389 (1 Dec 1991).
The IUE observatory and the UVS instrument on board the Voyager spacecraft have measured EUV Jovian auroral emission and given evidence for its strong longitudal dependence. However, limited spatial resolution has not allowed the derivation of any information concerning the latitudinal structure, nor even the exact location of the excitation process. This study is an attempt to relate the atmospheric auroral features to the magnetospheric mechanisms which generate the auroral precipitation of particles and, if possible, to allow the identification of the particle species, energy, L shell of injection, and equatorial anisotropy.

099.079 **Near-infrared observations of the Jovian ring and small satellites.**
P. D. Nicholson, K. Matthews.
Icarus, Vol. 93, No. 2, p. 331 - 346 (Oct 1991).
On 3 December 1988 the authors obtained 119 images of the Jovian ring, using the Caltech Cassegrain infrared camera at a wavelength of 2.2 μm. The measurements indicate a ring radius at maximum brightness of 42$''$.9, or 126,100 $\pm$ 500 km, compatible with Voyager data at high phase angles. Many of the images also show either Metis or Adrastea, the two small satellites associated with the ring and previously seen only in Voyager images. From a set of 98 images spanning a period of 5.7 hr, the authors have determined circular orbital elements for both satellites. The derived orbits are consistent with those determined from Voyager images, but the greatly extended temporal baseline provided by combining the measurements with those of Synnott permits to substantially improve the mean motions.

Nineteenth century investigations of periodicities in the Jovian atmosphere.
See Abstr. 004.053.

Simulation of the spectral content of the 7.8-μm emission from Jupiter at high resolution.
See Abstr. 022.065.

Dielectronic recombination rate coefficients for S^{q+} (q = 1 - 5) ions.
See Abstr. 022.094.

Laboratory spectra of ^{13}C ethane.
See Abstr. 022.114.

Alteration of the UV–visible reflectance spectra of H_2O ice by ion bombardment.
See Abstr. 022.122.

$CH_4/NH_3/H_2O$ spark tholin: chemical analysis and interaction with Jovian aqueous clouds.
See Abstr. 022.158.

The use of CCD for observations of large planets.
See Abstr. 034.026.

Thermal and dissipative stability of convection–accretion disks.
See Abstr. 064.005.

A measurement of planetary relativistic deflection.
See Abstr. 066.162.

Modelling of the stratiform clouds in the atmospheres of giant planets with consideration of the components intersolubility. II.
See Abstr. 091.010.

Interior structure of the giant planets.
See Abstr. 091.019.

Modeling stratiform clouds in the atmospheres of giant planets with consideration of the mutual solubility of condensed components. I.
See Abstr. 091.025.

Charged dust dynamics: orbital resonance due to planetary shadows.
See Abstr. 091.035.

Modeling the non–grey body thermal emission from the full moon.
See Abstr. 094.004.

Comparison of plasma wave measurements in the bow shocks at Earth, Jupiter, Saturn, Uranus and Neptune.
See Abstr. 101.016.

100 Saturn, Saturn Satellites

100.001 **A recent cometary origin for Saturn's rings?**
L. Dones.
Icarus, Vol. 92, No. 2, p. 194 – 203 (Aug 1991).

Gravitational torques and erosion by micrometeroids may cause rapid evolution of Saturn's rings, possibly suggesting formation of the rings within the least 10^8y. Optically thin and/or narrow rings at Jupiter, Uranus, and Neptune may form repeatedly by disruption of inner satellites. However, present-day creation of Saturn's more massive rings in this way is unlikely. An alternative scenario for ring origin is tidal distruption of a large comet, such as Chiron, passing within Saturn's Roche radius. Subsequent collisions among the fragments lead to a flat, equatorial ring. Assuming a nominal flux of Saturn-crossing bodies, 10 – 100 close passage occur in 4.5 Gy. Most close passages leave no bound debris, so that approximately one ring capture event occurs in the age of the solar system. This rate is comparable to the expected frequency of disruption of ring parent satellites by cometary bombardment. These results imply that the rings result from a recent unlikely tidal or collisional event, or that the arguments for rapid evolution of Saturn's rings are erroneous and the rings are primordial.

100.002 **Fine scale structure in the Saturn equatorial storm.**
L. F. Huber, R. F. Beebe, C. D. Barnet,
J. A. Westphal, E. J. Groth, J. A. Holtzman, R. White.
Bull. Am. Astron. Soc., Vol. 23, No. 2, p. 940 (1991). Abstract. – See Abstr. 012.008 for the main entry.

100.003 **The greenhouse and antigreenhouse effects on Titan.**
C. P. McKay, J. B. Pollack, R. Courtin.
Science, Vol. 253, No. 5024, p. 1118 – 1121 (6 Sep 1991).

There are many parallels between the atmospheric thermal structure of the Saturnian satellite Titan and the terrestrial greenhouse effect; these parallels provide a comparison for theories of the heat balance of Earth. Titan's atmosphere has a greenhouse effect caused primarily by pressure–induced opacity of N_2, CH_4, and H_2. H_2 is a key absorber because it is primarily responsible for the absorption in the wave number 400 to 600 cm^{-1} "window" region of Titan's infrared spectrum. The concentration of CH_4, also an important absorber, is set by the saturation vapor pressure and hence is dependent on temperature. In this respect there is a similarity between the role of H_2 and CH_4 on Titan and that of CO_2 and H_2O on Earth. Titan also has an antigreenhouse effect. This effect reduces the surface temperature by 9K whereas the greenhouse effect increases it by 21K. The net effect is that the surface temperature (94K) is 12K warmer than the effective temperature. If the haze layer were removed, the antigreenhouse effect would be greatly reduced, the greenhouse effect would become even stronger, and the surface temperature would rise by over 20K.

100.004 **Saturn's position derived from meridian circle observations of Titan and Iapetus.**
D. B. Taylor, L. V. Morrison, M. Rapaport.
Astron. Astrophys., Vol. 249, No. 2, p. 569 – 573 (Sep 1991).

Carlsberg and Bordeaux meridian circle observations of Titan and Iapetus made in the years 1987 to 1989 are analysed for corrections to their orbital elements and to the JPL Development Ephemeris, DE200, position of Saturn. The only significant correction to the elements is to the mean longitude of Iapetus. The corrections to the position of Saturn agree well with the latest ephemeris, DE202.

100.005 **Resonance structures in Saturn's rings probed by radio occultation. I. Methods and examples.**
P. A. Rosen, G. L. Tyler, E. A. Marouf.
Icarus, Vol. 93, No. 1, p. 3 – 24 (Sep 1991).

Optical depth profiles of Saturn's rings derived from Voyager 1 radio occultation reveal a remarkable array of wave–like features. Many are the signatures of spiral density waves and bending waves excited by gravitational resonances with Saturn's satellites. Twenty–nine wave features – 18 in Ring A, 1 in Ring B, and 10 in Ring C – are analyzed, individually and comparatively. Of the wave features originally seen in the data, only 20 could be identified readily with known resonances; all 8 still unexplained wave features are located in Ring C. For 14 weak density waves in Ring A, modeling of the waveform near resonance with linear density wave theory gives independent estimates of surface mass density, as well as reliable estimates of resonance location. The methods are described, and applied to the Mimas 5:3, Prometheus 12:11, and Mimas 4:1 density waves as illustration.

100.006 **Resonance structures in Saturn's rings probed by radio occultation. II. Results and interpretation.**
P. A. Rosen, G. L. Tyler, E. A. Marouf, J. J. Lissauer.
Icarus, Vol. 93, No. 1, p. 25 – 44 (Sep 1991).

Twenty–nine wave features in the Voyager 1 radio occultation profiles of Saturn's rings are examined, individually and comparatively, by the wavelength dispersion and waveform modeling methods described in Abstr. 100.005. Surface density estimates from waveform modeling are 20 to 40% larger than dispersion-derived values, suggesting accumulation of mass near resonance. Model waveforms of overlapping waves excited by the satellites Janus and Epimetheus agree well with observed morphologies in the linear region near resonance, and imply new mass estimates for Janus and Epimetheus.

100.007 **Micrometeorite erosion of the main rings as a source of plasma in the inner Saturnian plasma torus.**
M. K. Pospieszalska, R. E. Johnson.
Icarus, Vol. 93, No. 1, p. 45 – 52 (Sep 1991).
Micrometeorite bombardment is thought to erode the main rings of Saturn and to produce a plasma locally, accounting for transient ring phenomena and the supply of water molecules to Saturn's atmosphere. This process is shown here to be also a significant source of water molecules and molecular ions in the region between the outer edge of the main rings and Enceladus, adding to those neutrals and plasma produced by sputtering of the icy satellites. However, the magnitude and distribution of the ring source depends on the micrometeorite flux and the energy distribution of the neutral ejecta, both of which are uncertain. Limiting cases are examined.

100.008 **CCD spectroscopic observations of Saturn, Uranus, Neptune, and Titan during the 1990 apparitions.**
F. Moreno, A. Molina, J. L. Ortiz.
Icarus, Vol. 93, No. 1, p. 88 – 95 (Sep 1991).
High signal-to-noise CCD long-slit spectra of Saturn, Uranus, Neptune, and Titan have been recorded at the coude focus of the 1.52-m telescope at Calar Alto Observatory (Spain) on Jun 1990. The spectra were obtained in different spectral regions in the wavelength range 4900 – 9300 Å. The spatial resolution was 1″2, which allowed to obtain spatially resolved spectra on different regions across Saturn's disk. The model atmospheres for Saturn are consistent with an enrichment of carbon of a factor of two with respect to the solar abundance. The derived haze optical thickness is 13 ± 2, in accordance with previous measurements carried out during the 1986 apparition.

100.009 **Erratum: "Radio observations of Saturn as a probe of its atmosphere and cloud structure" [Icarus, Vol. 80, No. 1, p. 77 – 103 (Jul 1989)].**
F. H. Briggs, P. D. Sackett.
Icarus, Vol. 93, No. 1, p. 179 (Sep 1991). See Abstr. 50.100.010.

100.010 **Improvement of the Iapetus' orbital elements on the basis of the analytical theory of motion.**
S. A. Gasanov.
Astron. Zh., Tom 68, Vyp. 5, p. 1086 – 1092 (Sep–Oct 1991). In Russian. English translation in Sov. Astron., Vol. 35, No. 5.
The orbital elements of Iapetus are improved. For determination of the elements observations at the time interval of 1974 – 1984 and the analytical theory of motion, constructed by means of the Laplace method, were used.

100.011 **Sputter-produced plasma as a measure of satellite surface composition: the Cassini mission.**
R. E. Johnson, E. C. Sittler Jr.
Geophys. Res. Lett., Vol. 17, No. 10, p. 1629 – 1632 (Sep 1990).
Measurements of the co-orbiting neutral cloud or the plasma produced by sputtering of the icy Saturnian satellites can be used to determine the relative abundance of a minority surface species which would be difficult to determine from reflectance spectra. This is due to the fact that the sputter source rates, hence the plasma supply rates, are directly proportional to the bulk concentrations of mixed solids or clathrates, although the surface grains may be depleted in the most volatile species.

100.012 **Evidence of Saturn's magnetic field anomaly from Saturnian kilometric radiation high-frequency limit.**
P. Galopeau, A. Ortega-Molina, P. Zarka.
J. Geophys. Res., Vol. 96, No. A8, p. 14129 – 14140 (1 Aug 1991).
The detailed analysis of Voyager observations of Saturn kilometric radiation (SKR), taking into account the polarization response of the planetary radio astronomy experiment, has allowed the authors to study separately the high-frequency limit of SKR emitted from the northern and southern auroral regions of the planet. Interpretation of the variations of these high-frequency limits, in the frame of a model inspired by the cyclotron maser mechanism has allowed them to demonstrate unambiguously the existence of a magnetic "anomaly" postulated for years, in the near-surface Saturnian field. Its location and amplitude are estimated, and it is shown that, although it accounts for the observed SKR modulation, it could not be detected by the spacecraft-borne magnetometers which passed to far from the planetary surface. This demonstrates that the current degree of comprehension of planetary radio emission makes them a unique and very pertinent tool for the fine study of planetary magnetic fields.

100.013 **Angular momentum and tidal evolution of Saturn's system.**
M. Burša.
Earth, Moon, Planets, Vol. 54, No. 3, p. 241 – 255 (Sep 1991).
It has been demonstrated that dynamically the Saturnian system is analogous to the Jovian system; however, it is not an analogue of the Solar system as a whole. The departures in the figure parameters of the tri-axial Saturnian satellites orbiting in 1:1 resonance, from equilibrium figure parameters are not large in general, and the tidal and centrifugal distorting forces can be supposed to be responsible for the actual figures. The estimates for different dynamical parameters of the system support the hypothesis that the tri-axial satellites in 1:1 resonance were formed from the same protoplanetary nebula that gave rise to Saturn.

100.014 **On the stability of Hyperion's orbit.**
B. R. Mushailov.
Astron. Tsirk., No. 1545, p. 29 – 30 (Aug – Sep 1990). In Russian.
The motion of Hyperion within the framework of the restricted three-body problem and with regard to Saturn's oblateness shows Lyapunov's stability. The author comes to the conclusion that Hyperion's orbit is stable in the case of small perturbations.

100.015 **Stratospheric depletion of CO on Titan.**
E. Chassefière, M. Cabane.
Geophys. Res. Lett., Vol. 18, No. 3, p. 467 – 470 (Mar 1991).
A deep stratrospheric depletion of CO is found in Titan's atmosphere comparing millimetric and infrared observations (Marten et al., 1988; Lutz et al., 1983). The authors suggest that adsorption of CO on aerosols in the region of Titan's atmosphere where methane is expected to condense may change the vertical distribution of CO below ≈ 40 km. Aerosols slowly pervading through the methane condensation region could transport enough CO to generate the observed tropospheric enrichment provided the sticking coefficient ε (fraction of collisions yielding adsorption) is 6×10^{-4}, assuming the same eddy diffusion coefficient K as in the Earth troposphere. In the case of a lower mixing rate, generally favored, smaller values of ε are allowed (down to $\approx 10^{-6}$). Besides, the final radius of methane particles, just above the region where methane evaporates (below ≈ 10 km), must be 100 μm within a factor of 10. The hypothesis presented in this paper, if it is confirmed by laboratory measurements, could be an indirect evidence for large amounts of CH_4 ice in the Titanian atmosphere.

100.016 **Upstream waves in Saturn's foreshock.**
M. B. Bavassano-Cattaneo, P. Cattaneo, G. Moreno, R. P. Lepping.
Geophys. Res. Lett., Vol. 18, No. 5, p. 797 – 800 (May 1991).
Waves in the foreshock of Saturn have been studied using data from the plasma and magnetic field experiments onboard

Voyager 1. The appearance of the waves, in the plasma as well as in the magnetic field data, is related to magnetic connection of the spacecraft to the bow shock. Their relative amplitude $\Delta B/B$ lies between 0.3 and 0.8 and their period in the spacecraft frame is ~ 500 s. The waves are elliptically polarized and propagate at $\sim 30°$ to the average magnetic field direction.

100.017 **Improvement of the orbital elements of Iapetus: analytical theory of motion.**
S. A. Gasanov.
Sov. Astron., Vol. 35, No. 5, p. 544 - 547 (Sep 1991). Current Physics Microform No.: 9203X1718. English translation of Astron. Zh., Tom 68, Vyp. 5, p. 1086 - 1092 (1991).
The orbital elements of Iapetus have been improved. Observations over the interval 1974 - 1984 and an analytical theory of motion constructed by the Laplace method have been used to determine the elements.

100.018 **Thermal stress tectonics on the satellites of Saturn and Uranus.**
J. Hillier, S. W. Squyres.
J. Geophys. Res., Vol. 96, No. E1, p. 15665 - 15674 (25 Aug 1991).
Most of the major Saturnian and Uranian satellites have undergone some form of extensional tectonism and/or resurfacing early in their histories. The authors examine the stresses created by the early thermal evolution of these satellites and consider the geologic implications of these stresses.

100.019 **Saturn.**
IAU Circ., No. 5328, p. 1 (21 Aug 1991).
Further information is given in No. 5341.

100.020 **Satellites of Saturn and Neptune.**
IAU Circ., No. 5347, p. 1 (16 Sep 1991).

100.021 **Convection and lithospheric strength in Dione, an icy satellite of Saturn.**
O. Forni, A. Coradini, C. Federico.
Icarus, Vol. 94, No. 1, p. 232 - 245 (Nov 1991).
Finite element numerical simulations for the structure of convection cells within a self-gravitating, Boussinesq fluid sphere with variable viscosity are used to determine temperature distribution, stream functions, velocities, and intensities of the stresses within Dione, and icy satellite of Saturn. A non-Newtonian, i.e., stress-dependent, viscosity is considered. The initial temperature profile has been determined taking into account the amount of energy trapped as heat during the accumulation process and the other energy source is the radioactivity of the long-lived isotopes. The obtained results seem to show a long duration endogenic activity in agreement with the geological obserations.

100.022 **The Great White Spot and disturbances in Saturn's equatorial atmosphere during 1990.**
A. Sanchez-Lavega, F. Colas, J. Lecacheux, P. Laques, I. Miyazaki, D. Parker.
Nature, Vol. 353, No. 6343, p. 397 - 401 (3 Oct 1991).
A giant storm, the Great White Spot, erupted at the end of September 1990 as a localized, bright cloud system close to Saturn's equator. Its evolution produced a complex planetary disturbance which affected the whole equatorial region a month later. Similar spots have appeared approximately once every saturnian year (about 30 Earth years), implying that a seasonal change - in solar heating, for example - may be responsible for their occurrence.

100.023 **Saturn.**
Yamamoto Circ., No. 2171, p. 2 (14 Sep 1991). In Japanese.

100.024 **Saturn.**
Br. Astron. Assoc. Circ., No. 711, p. 4 (4 Oct 1991).

100.025 **Satellites of Saturn and Neptune.**
Br. Astron. Assoc. Circ., No. 711, p. 8 (4 Oct 1991).

100.026 **Non-LTE models of Titan's upper atmosphere.**
R. V. Yelle.
Astrophys. J., Vol. 383, No. 1, p. 380 - 400 (10 Dec 1991).
Models for the thermal structure of Titan's upper atmosphere, between 0.1 mbar and 10^{-2}nbar are presented. The calculations include non-LTE heating/cooling in the rotation-vibration bands of CH_4, C_2H_2, and C_2H_6, absorption of solar IR radiation in the near-IR bands of CH_4 and subsequent cascading to the ν_4 band of CH_4, absorption of solar EUV and UV radiation, thermal conduction and cooling by HCN rotational lines. The calculations predict a well-developed mesopause with a temperature of 135 - 140K at an altitude of approximately 600 km and pressure of ~ 0.1 μbar. The mesopause is at a higher pressure than predicted by earlier calculations. The accuracy of the LTE approximation for source functions and heating rates is discussed. It found that C_2H_6 acts as a heat source near the mesopause by absorbing radiation from the warm stratosphere; consequently, the temperature at the mesopause depends sensitively on the C_2H_6 abundance. Because of the strong coupling between photochemistry and thermal structure the agreement between calculated and observed temperatures lend support to the photochemical model of Yung et al.

100.027 **Evidence for seasonal change on Titan.**
C. C. Cunningham, H. P. White, J. J. Caldwell, D. M. Anthony.
J. R. Astron. Soc. Can., Vol. 85, No. 4, p. 186 (Aug 1991).
Abstract. - See Abstr. 012.091 for the main entry.

100.028 **Near ultraviolet spectroscopy of Saturn with the International Ultraviolet Explorer satellite.**
H. P. White, J. Caldwell, R. Wagener.
J. R. Astron. Soc. Can., Vol. 85, No. 4, p. 193 (Aug 1991).
Abstract. - See Abstr. 012.091 for the main entry.

100.029 **The shape of Tethys.**
P. C. Thomas, S. F. Dermott.
Icarus, Vol. 94, No. 2, p. 391 - 398 (Dec 1991).
The shape of Tethys has been determined from Voyager limb data to be a triaxial ellipsoid of approximately equilibrium shape. The mean radius, after allowance for limb topography, is 529.9 ± 1.5 km. Using the satellite mass determined by Kozai from observations of the libration period and the libration amplitude of the Mimas-Tethys resonance, the authors deduce that the mean density of the satellite is 1.00 ± 0.02 g/cm^3. The principal axes of the satellite are 535.6, 528.2, and 525.8 km with relative errors of about 1.5 km.

100.030 **Nonradial oscillations of Saturn.**
M. S. Marley.
Icarus, Vol. 94, No. 2, p. 420 - 435 (Dec 1991).
The periods of the natural oscillation modes of the Jovian planets are sensitive to the interior states of these planets. Since the f-mode oscillations of Saturn may be detected through their gravitational signature in the rings of that planet, this paper examines the sensitivity of these modes to variations in the assumed state of the planetary interior. Mode periods are calculated for a variety of interior models, envelope compositions, and hydrogen phase transition pressures spanning the range of the current uncertainty in these quantities. It is found that the low degree f-modes are most sensitive to the size of the planetary core while the higher degree modes are most sensitive to the rotation state of the outer layers of the planet. The physical causes of these period variations are discussed.

100.031 **Structure and particle properties of Saturn's E ring.**
M. R. Showalter, J. N. Cuzzi, S. M. Larson.
Icarus, Vol. 94, No. 2, p. 451 - 473 (Dec 1991).
The authors have systematically reanalyzed a large subset of the available photometric data on Saturn's E ring. In order to compare data sets with vastly different viewing geometries on this

broad, vertically extended ring, they have calibrated every observation based on a common three-dimensional model. The ring shows a density peak at a location indistinguishable from the Enceladus orbit, with a radial offset of <3000 km. It is found that a very simple power-law model describes the ring's normal optical depth profile with orbital radius. The ring shows a general increase in vertical thickness with distance from the planet. However, this trend is violated near the density peak, where a localized decrease in thickness by ~30% is observed. No power-law size distribution is compatible with the available photometry. A narrow distribution of slightly nonspherical particles of radius 1.0 ± 0.3 μm provides the best fit to the data. This highly peculiar size distribution clearly indicates that the ring does not originate from collisional or disruptive processes, and is therefore unlike any other known ring.

100.032 **Multifrequency radio observations of Saturn at ring inclination angles between 5 and 26 degrees.**
I. de Pater, J. R. Dickel.
Icarus, Vol. 94, No. 2, p. 474 – 492 (Dec 1991).
The authors observed Saturn with the VLA at 1.3 – 20 cm for ring inclination angles between 5° and 26°. An image at a wavelength of 3.4 mm was obtained using the three-element interferometer at the Hat Creek Radio Observatory, while the rings were fully open. The Saturnian disk is featureless at 1.3 and 2 cm, but shows a bright band at midnorthern latitudes at 6 and 20 cm. In addition, at 6 cm wavelength, the planet is less limb darkened toward the north pole than expected from model calculations. It is suggest the latitudinal brightness variation to be due to variations in the location of the NH_4SH cloud. The data are consistent with nearly pure ice particles of sizes between a few millimeters and a few meters, with the number of particles having a power law distribution approximately proportional to the radius cubed.

100.033 **Determination of Saturn ring particle density: the first result of a new method.**
N. N. Gor'kavyj, T. A. Tajdakova.
International School of Physics 'Enrico Fermi'. Course 111: Solid-state astrophysics, p. 423 – 429 (1991). – See Abstr. 012.106 for the main entry.

100.034 **Saturno nel 1990: gli ovali chiari equatoriali.**
P. Tanga.
Astronomia UAI, N. 10, p. 16 – 23 (Nov – Dec 1991).

100.035 **Transformation method of the initial values in calculating ephemeris of the satellites of Saturn.**
Qiao Rongchuan, Shen Kaixian.
Publ. Shaanxi Astron. Obs., Vol. 14, No. 2, p. 128 – 135 (Dec 1991). In Chinese.
A set of transformation formula of the initial values in improving the orbit of the satellites of Saturn via numerical integration are described, which also can be applied to other calculation of ephemerides in celestial mechanics. The program is given as appendix.

100.036 **Titan's surface and troposphere, investigated with ground-based, near-infrared observations.**
C. A. Griffith, T. Owen, R. Wagener.
Icarus, Vol. 93, No. 2, p. 362 – 378 (Oct 1991).
New observations of Titan's near-infrared spectrum combined with points taken from Fink and Larson's (1979) spectrum provide information on Titan's haze, possible clouds, surface albedo, and atmospheric abundance of H_2. The observations suggest the existence of an optically thin CH_4 cloud layer. The optical depths that are derived for Titan's haze and clouds are small enough to allow one to sense the surface of Titan at 4900, 6250, and 7700 cm^{-1}. The most plausible interpretation of the albedos determined at these wavenumbers suggests a surface dominated by "dirty" water ice.

100.037 **New observations of Saturn's co-orbital satellites.**
D. P. Hamilton, P. D. Nicholson, K. Mathews, S. S. Leroy.
News Lett. Astron. Soc. N.Y., Vol. 3, No. 9, p. 19 – 20 (Feb 1991). – See Abstr. 012.119 for the main entry.

A laboratory facility for mid-UV absorption spectroscopy of molecular compounds for planetary atmospheres.
See Abstr. 022.119.

High-resolution, low-temperature photoabsorption cross sections of C_2H_2, PH_3, AsH_3, and GeH_4, with application to Saturn's atmosphere.
See Abstr. 022.120.

Block iterative restoration of astronomical images from the Hubble Space Telescope.
See Abstr. 036.023.

Linear restoration techniques: not all bad.
See Abstr. 036.036.

The isosceles three-body problem as a model for the three-body problem with two small masses.
See Abstr. 042.025.

Nonlinear density waves in gaseous disks.
See Abstr. 062.126.

A comparison of induced magnetotails of planetary bodies: Venus, Mars, and Titan.
See Abstr. 093.054.

Hydrogen-helium mixtures at megabar pressures: implications for Jupiter and Saturn.
See Abstr. 099.021.

What is new in the world of giant planets?
See Abstr. 099.038.

Proton energy deposition and EUV emission in the Jovian and the Saturnian atmosphere.
See Abstr. 099.043.

Vibrationally excited H_2 in the outer planets thermosphere: fluorescence in the Lyman and Werner bands.
See Abstr. 099.059.

The upper ionospheres of Jupiter and Saturn.
See Abstr. 099.060.

Comparison of plasma wave measurements in the bow shocks at Earth, Jupiter, Saturn, Uranus and Neptune.
See Abstr. 101.016.

Nitrogen airglow sources: comparison of Triton, Titan, and Earth.
See Abstr. 101.050.

101 Uranus, Neptune, Pluto, Transplutonian Planets

101.001 **Neptune's weather forecast: cloudy, windy and cold.**
S. S. Limaye.
Astronomy, Vol. 19, No. 8, p. 38 - 43 (Aug 1991).

101.002 **Rotationally resolved midultraviolet studies of Triton and the Pluto/Charon system I: IUE results.**
S. A. Stern, N. Brosch, E. S. Barker, G. R. Gladstone.
Icarus, Vol. 92, No. 2, p. 332 - 341 (Aug 1991).
The authors present a uniform analysis of the full set of usable IUE long wavelength spectra of Pluto+Charon and Triton. Their objective is to study the variation in UV photometric properties of these objects with rotational phase. Comparing Pluto to Triton, they found that Pluto's UV albedo is much lower than Triton's, but that the amplitude of Pluto's UV lightcurve is greater than Triton's. The lower UV albedo of Pluto's surface indicates that fresh frost may have been more recently deposited on Triton than on Pluto.

101.003 **Triton and Nereid astrographic observations from Voyager 2.**
R. A. Jacobson.
Astron. Astrophys., Suppl. Ser., Vol. 90, No. 3, p. 541 - 563 (Nov 1991).
This article describes the reduced astrographic observations of Triton and Nereid derived from Voyager 2 imaging data. The data set contains 496 sets of spacecraft–centered right ascension and declination observations and includes all of the observations used in Voyager encounter operations. The details of the conversion process from imaging to astrographic observations are given. The effect of using the astrographic rather than imaging form in ephemeris improvement is evaluated.

101.004 **Interior structure of Neptune: comparison with Uranus.**
W. B. Hubbard, W. J. Nellis, A. C. Mitchell,
N. C. Holmes, S. S. Limaye, P. C. McCandless.
Science, Vol. 253, No. 5020, p. 648 - 651 (9 Aug 1991).
Measurements of rotation rates and gravitational harmonics of Neptune made with the Voyagert 2 spacecraft allow tighter constraints on models of the planet's interior. Shock measurements of material that may match the composition of Neptune, the so–called "ice", have been carried out to pressures exceeding 200 gigapascals. Comparison of shock data with inferred pressure–density profiles for both Uranus and Neptune shows substantial similarity through most of the mass of both planets.

101.005 **An explanation for Neptune's ring arcs.**
C. C. Porco.
Science, Vol. 253, No. 5023, p. 995 - 1001 (30 Aug 1991).
The Voyager mission revealed a complex system of rings and ring arcs around Neptune and uncovered six new satellites, four of which occupy orbits well inside the ring region. Analysis of Voyager data shows that a radial distortion is traveling through the ring arcs, a perturbation attributable to the nearby satellite Galatea. Moreover, the arcs appear to be azimuthally confined by a resonant interaction with the same satellite, yielding a maximum spread in ring particle semimajor axes of 0.6 km and a spread in forced eccentricities large enough to explain the arcs' 15–km radial widths. Additional ring arcs discovered in the course of this study give further support to this model.

101.006 **Pluto–Charon: a double planet on the ragged edge of the solar system.**
S. A. Stern.
Bull. Am. Astron. Soc., Vol. 23, No. 2, p. 962 (1991). Abstract. – See Abstr. 012.008 for the main entry.

101.007 **The orbits of Triton and Nereid from spacecraft and Earthbased observations.**
R. A. Jacobson, J. E. Riedel, A. H. Taylor.
Astron. Astrophys., Vol. 247, No. 2, p. 565 - 575 (Jul 1991).
The authors discuss the determination of improved orbits for the Neptunian satellites Triton and Nereid. Their primary results are the final set of model parameters which generate orbits that best fit both the Earthbased satellite observations and data acquired by the Voyager spacecraft during the Neptune encounter. The authors assess the accuracy of the improved orbits and compare them with the orbits generated prior to the encounter. They also provide mean elements as a geometrical representation for the orbits.

101.008 **The photometry of Pluto in two colors.**
R. I. Kiladze, V. D. Kukhianidze.
Astron. Vestn., Tom 25, No. 4, p. 439 - 441 (Jul - Aug 1991). In Russian. English translation in Sol. Syst. Res., Vol. 25.
The results of two–color photoelectric observations of Pluto, carried out at the Abastumani Astrophysical Observatory during 1986 - 1990, are described. The decrease of color–index of this planet is found out in 1990. Pluto's albedo (at least in blue light) reached minimum in 1988. This result contradicts to the point of view of Buie et al. (1987), Cruikshank et al. (1980) and Stern et al. (1988), explaining the mean brightness decrease of Pluto by the change of physical conditions on its surface, namely, by ice sublimation, which may occur as a result of the change of distance between Pluto and the Sun.

101.009 **The Neptunian arcs as chains of epitons in the complete transparent ring.**
N. N. Gor'kavyj.
Pis'ma Astron. Zh., Tom 17, No. 11, p. 1020 - 1030 (Nov 1991). In Russian. English translation in Sov. Astron. Lett., Vol. 17, No. 6.
Gor'kavyj (1989) suggested that arc confinement is possible if an arc is a set of epicyclic vortices (epitons) situated in the dust ring. The particles of such elliptical vortices are rotating around their mass center in epicycles. A strict equality of the big semiaxes of the particles ensures identical periods of orbital rotation, therefore the vortex is stable and does not disperse along the orbit. The continuous narrow ring making equal the big semiaxes of epitons is a pledge for stability of many tens of vortices. "Voyager–2" photographed a regular chain of compact clumps situated in the arcs of Neptune. The existence of such a structure is a direct confirmation of the epiton model; no other hypotheses good as they were envisage a possibility for the existence of the inner arc structure.

101.010 **Latitudinal and longitudinal oscillations of cloud features on Neptune.**
L. A. Sromovsky.
Science, Vol. 254, No. 5032, p. 684 - 686 (1 Nov 1991).
Voyager observations suggest that three of Neptune's major cloud features oscillate in latitude by 2° to 4° and that two of them simultaneously oscillate in longitude by 7.8° and 98° about their mean drift longitudes. The observations define most convincingly the two orthogonal oscillations of the second dark spot (near 53° south). These oscillatons have similar periods near 800 hours and approximately satisfy a simple advective model in which a latitudinal oscillation produces a phase–shifted longitudinal oscillation proportional to the local wind shear.

101.011 **Source location of the Uranus smooth low–frequency emission.**
G. K. F. Rabl, H. P. Ladreiter, H. O. Rucker, Y. Leblanc.
Ann. Geophys., Vol. 9, No. 7, p. 487 - 494 (Jul 1991).
Two very different source locations, polar and equatorial, have been suggested in the literature for the smooth low frequency

emission (SLF) from Uranus, detected by the Voyager 2 Planetary Radio Astronomy (PRA) experiment. These possibilities are considered by studying the occultation of possible source regions by the planet with respect to the spacecraft. It is shown that a polar source region over a limited range of magnetic longitude can be excluded, while a polar source distributed along all magnetic longitudes is compatible with visibility studies. However, comparison of observations with computed intensity contours, for polar and equatorial sources, definitely favors the equatorial source model.

101.012 **Semi-analytical theory for the motion of Uranus' satellites.**
D. Lazzaro.
Astron. Astrophys., Vol. 250, No. 1, p. 253 – 265 (Oct 1991).

A complete semi-analytical theory for the motion of the five greatest Uranian satellites is presented. The classical Laplace-Souillart theory was adopted considering the perturbations due to the mutual interaction between the satellites, due to the gravitational attraction of the Sun and due to Uranus' potential. Analytical solutions were obtained for the eccentricities, the inclinations, the arguments of the pericenters and the longitudes of the nodes. Semi-analytical solutions were developed for the semi-major axis and the mean longitudes. Exhaustive comparisons show that the present theory is complete from the theoretical point of view giving results which are perfectly compatible with those obtained by other authors.

101.013 **On the origin of Uranus and Neptune magnetic fields.**
A. A. Ruzmaikin (*A. A. Ruzmajkin*), S. V. Starchenko.
Icarus, Vol. 93, No. 1, p. 82 – 87 (Sep 1991).

The Uranus and Neptune magnetic fields discovered by Voyager 2 can be explained by a dynamo acting in a thin conductive convective shell existing at the bottom of the icy oceans of the planets. The main helicity and differential rotation are the source for the dynamo which effectively excites non-axisymmetric modes of the mean magnetic field. Estimates of the magnetic field amplitude in the nonlinear regime and of the inclination between the magnetic moment and the rotation axis are given.

101.014 **Smooth radio emission and a new emission at Neptune.**
C. Sawyer, J. W. Warwick, J. H. Romig.
Geophys. Res. Lett., Vol. 17, No. 10, p. 1645 – 1648 (Sep 1990).

The Planetary Radio Astronomy (PRA) experiment (Warwick et al., 1977) on the spacecraft Voyager 2 observed three types of smooth emission: (1) Numerous recurrent episodes are modeled by filled emission cones from midlatitude conjugate sources. (2) An "equatorial" feature seen soon after closest approach includes electron cyclotron harmonic emission above the upper hybrid resonance, as well as smooth recurrent emission; its strange appearance is a result of rapid change in Voyager's magnetic latitude. (3) Unique "high-latitude" emission is seen near closest approach during Voyager's single excursion to high north (+) magnetic latitude when f_c, the electron cyclotron frequency at the spacecraft, lies in the observable range. The stronger component covers a broad band of frequencies above $2f_c$; its sensitivity to magnetic field identifies it as extraordinary (X) mode. The weaker component extends smoothly through $f = f_c$ and is identified as ordinary (O) mode. At each frequency f the observed sense of polarization reverses when $f = 2f_c$.

101.015 **Low-frequency radio emissions at Neptune.**
W. S. Kurth, D. D. Barbosa, D. A. Gurnett, R. L. Poynter, I. H. Cairns.
Geophys. Res. Lett., Vol. 17, No. 10, p. 1649 – 1652 (Sep 1990).

The Voyager 2 plasma wave receiver detected weak radio emissions from Neptune's magnetosphere in the frequency range of 3–60 kHz. The emissions occurred in bursts lasting for typically 1.5 hours, often occurring twice per planetary rotation. Most of these radio bursts were detected within several degrees of the magnetic equatorial plane. During the passage through the magnetosphere, electrostatic upper hybrid resonance bands were observed close to the magnetic equator in conjunction with intensifications of the radio emissions at frequencies close to and above the upper hybrid bands. Further, near closest approach, the radio emissions were observed to cross the right-hand cutoff frequency with no apparent attenuation. The authors conclude that the Neptunian radio emissions below about 60 kHz are produced by mode conversion from the upper hybrid waves and propagate in the ordinary mode into beams within about 12° of the magnetic equator. There is also evidence of an extraordinary mode emission at about 60 kHz which is apparently generated by an entirely different source from the escaping continuum radiation.

101.016 **Comparison of plasma wave measurements in the bow shocks at Earth, Jupiter, Saturn, Uranus and Neptune.**
S. L. Moses, F. V. Coroniti, C. F. Kennel, W. S. Kurth, D. A. Gurnett.
Geophys. Res. Lett., Vol. 17, No. 10, p. 1653 – 1656 (Sep 1990).

The authors present plasma wave measurements from the Voyager 2 crossing of Neptune's bow shock and compare them with measurements from the bow shocks of Earth, Jupiter, Saturn and Uranus. The wave amplitudes above 0.01 f_p, when normalized to the solar wind ion thermal energy density at each planet, are significantly higher at the outer planets than at Earth. Despite the differences in amplitude the shock spectra of all the planets can be fitted to curves of similar form in this frequency range. The total normalized electric field energy densities exhibit an exponential dependence on ion thermal Mach number. Magnetosheath wave energies are comparable at all of the planets when normalized to the downstream plasma pressure.

101.017 **Electrostatic electron and ion cyclotron harmonic waves in Neptune's magnetosphere.**
D. D. Barbosa, W. S. Kurth, I. H. Cairns, D. A. Gurnett, R. L. Poynter.
Geophys. Res. Lett., Vol. 17, No. 10, p. 1657 – 1660 (Sep 1990).

Voyager 2 observations of electrostatic electron and ion cyclotron waves detected in Neptune's magnetosphere are presented. Both types of emission appear in a frequency band above the electron and ion (proton) cyclotron frequencies, respectively, and are tightly confined to the magnetic equator occurring within a few degrees of it. The electron cyclotron waves are interpreted as electron Bernstein modes including an intense upper hybrid resonance emission excited by an unstable loss cone distribution of low-density superthermal electrons. The ion cyclotron waves are interpreted as hydrogen Bernstein modes including an intense lower hybrid resonance emission excited by an unstable ring distribution of low-density pickup N^+ ions deriving from the satellite Triton.

101.018 **Magnetospheric interaction with Triton's ionosphere.**
D. F. Strobel, A. F. Cheng, M. E. Summers, D. J. Strickland.
Geophys. Res. Lett., Vol. 17, No. 10, p. 1661 – 1664 (Sep 1990).

The large electron densities measured by the Voyager radio occultation experiment are attributed to the precipitation of magnetospheric electrons with energy >10 keV. Because the ionospheric electric Pedersen conductivity of Triton is $\sim(1\text{–}2)\times10^4$ mho and the Alfvén conductance is ~ 3.5 mho, direct convective flow of plasma into the essentially infinitely conducting ionosphere is negligible. Magnetospheric electrons are transported to Triton's ionopause by curvature drift as a result of weak magnetic field line draping in a sub-Alfvenic plasma interaction with Triton. At the ionopause energetic electrons have a high probability of elastic and inelastic scattering and precipitate into the upper atmosphere. The average power dissipation is estimated to be $\sim(2\text{–}3)\times10^8$ W.

101.019 **Satellite sweeping of electrons at Neptune and Uranus.**
J. F. Cooper.
Geophys. Res. Lett., Vol. 17, No. 10, p. 1665 – 1668 (Sep 1990).

Knowledge of satellite sweeping parameters at Neptune and Uranus, and of their functional dependences on particle energy and pitch angle, can be critical in the proper identification of parent absorbers for observed absorption signatures in regions where OTD (Offset, Tilted Dipole) models are valid representations of the measured magnetic fields. In this paper, the author calculates critical electron energies for longitudinal drift resonance, snowplow (i.e., strong) absorption, leapfrog, and corkscrew effects, using a reduced version of OTD that neglects nonaxial dipole offsets. Earlier analytic work on sweeping rates is extended to give the radial dependence of these rates within the minimum-L region and to set limits on diffusion of electrons with the simplifying approximation that leapfrog effects are ignored.

101.020 **Triton torus and Neptune aurora.**
A. F. Cheng.
Geophys. Res. Lett., Vol. 17, No. 10, p. 1669 – 1672 (Sep 1990).
Triton is shown to be the dominant source of plasma for $L \gtrsim 7$ in the magnetosphere of Neptune. Triton maintains a neutral hydrogen torus of average density comparable to or greater than that of the Titan torus at Saturn. The Triton torus may be detectable in H Lyman-α emissions. However, the energy source from plasma outward transport and mass loading in the Triton torus is insufficient to explain the Neptune aurora. It is proposed that Neptune's aurora is driven mainly by a solar wind interaction.

101.021 **The Triton torus revisited.**
J. D. Richardson, A. Eviatar, M. L. Delitsky.
Geophys. Res. Lett., Vol. 17, No. 10, p. 1673 – 1676 (Sep 1990).
Prior to the Voyager encounter with Neptune, Delitsky et al. (1989) predicted that a torus of ions emanating from Triton would be discovered. The authors revisit those predictions in light of the Voyager results. Sputtering of Triton's atmosphere can produce the heavy ion densities inferred at Triton's orbit by the Voyager plasma experiment if the ion residence time is about 30 days. The torus is found to be longitudinally asymmetric near Triton, with peak densities at longitudes of 170° and 350°. The total nitrogen flux due to sputtering is about $2 \times 10^{21} s^{-1}$. The consequences of larger escape fluxes of both N_2 and H_2 (suggested by Broadfoot et al. (1989)) are investigated; it is difficult to reconcile large escape fluxes with the plasma and ultraviolet spectrometer observations.

101.022 **Convection in Neptune's magnetosphere.**
T. W. Hill, A. J. Dessler.
Geophys. Res. Lett., Vol. 17, No. 10, p. 1677 – 1680 (Sep 1990).
The authors assume that nonthermal escape from Triton's atmosphere produces a co-orbiting torus of un-ionized gas (presumably nitrogen and hydrogen) that subsequently becomes ionized by electron impact to populate a partial Triton plasma torus analogous to the Io plasma torus in Jupiter's magnetosphere. Centrifugal and magnetic-mirror forces confine the ions to a plasma sheet located between the magnetic and centrifugal equators. The ionization rate, and hence the torus ion concentration, is strongly peaked at the two points (approximately 180° apart in longitude) at which Triton's orbit intersects the plasma equator. During the course of Neptune's rotation these intersection points trace out two arcs roughly 75° in longitudinal extent, which the authors take to be the configuration of the resulting (partial) plasma torus. The implied partial ring currents produce a quadrupolar (four-cell) convection system that provides rapid outward transport of plasma from the arcs. Ring current shielding, however, prevents this convection system from penetrating very far inside the plasma-arc distance. The authors suggest that this convection/shielding process accounts for the radial confinement of $\gtrsim$150 keB trapped particles within $L \approx 14.3$ as observed by the Voyager LECP instrument.

101.023 **Plasma convection in Neptune's magnetosphere.**
R. S. Selesnick.
Geophys. Res. Lett., Vol. 17, No. 10, p. 1681 – 1684 (Sep 1990).
The magnetosphere of Neptune changes its magnetic configuration continuously as the planet rotates, leading to a strong modulation of the convection electric field. Even though the corotation speed is considerably larger, the modulation causes the small convection speed to have a cumulative effect, much like the acceleration of particles in a cyclotron. A model calculation shows that plasma on one side off the planet convects out of the magnetosphere in a few planetary rotations, while on the other side it convects slowly planetward. The observation of nitrogen ions from a Triton plasma torus may provide a critical test of the model.

101.024 **Hot plasma parameters in Neptune's magnetosphere.**
S. M. Krimigis, B. H. Mauk, A. F. Cheng, E. P. Keath, M. Kane, T. P. Armstrong, G. Gloeckler, L. J. Lanzerotti.
Geophys. Res. Lett., Vol. 17, No. 10, p. 1685 – 1688 (Sep 1990).
Energy spectra of energetic protons and electrons ($E_p \gtrsim 28$ keV, $E_e \gtrsim 22$ keV, respectively) obtained with the Low Energy Charged Particle (LECP) instrument during the Voyager 2 encounter with Neptune on August 24–25, 1989 are presented. The proton spectral form was a power law ($dj/dE = KE^{-\gamma}$), outside the orbit of Triton ($\sim 14.3\ R_N$); inside that distance, it was found to be a hot ($kT \approx 60$ keV) Maxwellian distribution. Such distributions, observed in other planets as well, have yet to be explained theoretically. Similarly, the electron spectral form changed from a simple power law outside Triton to a two-slope power law with a high energy tail inside. Intensity and spectral features in both proton and electron fluxes were identified in association with the crossings of the Triton and 1989 N1 L-shells, but these features do not occur simultaneously in both species. Such signatures were manifested by relative peaks in both kT and γ spectral indices. Peak proton pressures of $\sim 2 \times 10^{-9}$ dynes cm^{-2}, and $\beta \sim 0.2$ were measured at successive magnetic equatorial crossings, both inbound and outbound. These parameters show Neptune's magnetosphere to be relatively undistorted by hot plasma loading, similar to that of Uranus and unlike those of Saturn and Jupiter. Trapped electron fluxes at Neptune, as at Uranus, exceed the whistler mode stably trapped flux limit. Whistler-induced pitch angle scattering of energetic electrons in the radiation belts can yield a precipitating energy flux sufficient to drive Neptune's aurora.

101.025 **Low-energy plasma in Neptune's magnetosphere.**
J. D. Richardson, R. L. McNutt Jr.
Geophys. Res. Lett., Vol. 17, No. 10, p. 1689 – 1692 (Sep 1990).
Plasma data from the region of Neptune's magnetosphere between $L = 6.5$ and $L = 13.5$ are analyzed. Selecting only spectra where both light and heavy ions are present, the authors derive density and temperature profiles under the assumption that the ions are H^+ and N^+. These results are used to calculate the total flux shell content (for a shell of unit width in L), N, of each ion in this region. Values of N derived from inbound and outbound data are similar, indicating the plasma distribution functions are isotropic. Average values of NL^2 are 3×10^{32} for H^+ and 1×10^{32} for N^+. Profiles of NL^2 for both H^+ and N^+ increase outwards, suggesting that Triton is the probable source of both these plasma components. Evidence is presented for significant inbound-outbound asymmetries in the plasma morphology.

101.026 **Aurora and airglow on the night side of Neptune.**
B. R. Sandel, F. Herbert, A. J. Dessler, T. W. Hill.
Geophys. Res. Lett., Vol. 17, No. 10, p. 1693 – 1696 (Sep 1990).
The authors have examined the latitude-longitude distribution of emissions detected by the Voyager ultraviolet spectrometer on the dark hemisphere of Neptune. The emissions have two significant geographic features: (1) a broad peak near longitude 60°W that extends rather uniformly over the entire range of observed latitudes (55°S to 50°N); and (2) a brighter, narrower peak near the south pole and 240°W. The authors interpret the first peak as due to excitation of the night side atmosphere by photoelectrons from the magnetically conjugate, sunlit atmosphere. The second peak can plausibly be attributed to a southern aurora; the field geometry would then seem to require a conjugate (and probably brighter) northern aurora that escaped detection poleward of the latitude range sampled by the UVS data. The power for such an aurora could be extracted from Neptune's rotation by the injection of plasma at Triton's orbit at a rate $dm/dt \sim 1$ kg/s.

101.027 **Global magnetic anomaly and aurora of Neptune.**
A. F. Cheng.
Geophys. Res. Lett., Vol. 17, No. 10, p. 1697 – 1700 (Sep 1990).
The large offset and tilt of Neptune's dipole magnetic field combine to create a global magnetic anomaly, analogous to but much more important than Earth's South Atlantic Anomaly. Energetic particle precipitation loss within the Neptune anomaly creates "Atmospheric drift shadows" within which particle fluxes are greatly reduced. The energetic particle dropout observed by Voyager near closest approach occurred near the predicted times when Voyager passed within the atmospheric drift shadows. Extremely soft, structured burst of ions and electrons within the drift shadow may result from plasma wave–induced pitch angle scattering of trapped particles confined near the magnetic equator. The dropout does not necessarily imply that Voyager passed through an Earth–like discrete auroral zone, as earlier reported. The ion and electron fluxes observed within the dropout period correspond to particles that must precipitate to Neptune's atomsophere within the anomaly region. This anomaly precipitation can account for a major portion of the ultraviolet emissions previously identified as Neptune aurora.

101.028 **Energetic charged particle angular distributions near ($r \leqslant 2\ R_N$) and over the pole of Neptune.**
B. H. Mauk, M. Kane, E. P. Keath, A. F. Cheng,
S. M. Krimigis, T. P. Armstrong, N. F. Ness.
Geophys. Res. Lett., Vol. 17, No. 10, p. 1701 – 1704 (Sep 1990).
Energetic ion (>28 keV) and electron (>22 keV) pitch angle distributions very close to ($r \leqslant 2\ R_N$) and over the north planetary pole of Neptune are presented using data from the Low Energy Charged Particle (LECP) and Magnetometer (MAG) Experiments on the Voyager 2 spacecraft. The particle data are temporally structured and spectrally soft; a similarity with Earth–like auroral signatures has previously been noted. However, the pitch angle distributions (showing trapped distributions at high magnetic latitudes) do not support an Earth–like auroral interpretation, and alternative explanations for the temporal dynamics must be sought. Between $r \sim 1.6$ and $2.0\ R_N$ and in the vicinity of the magnetic equator, the higher energy ion and electron pitch angle distributions ($E \geqslant 80$ keV) display dramatic "bite–outs" at ~90°. This bite–out feature could be caused by interactions with the newly discovered ring 1989N3R.

101.029 **The magnetosphere of Neptune: its response to daily rotation.**
G.–H. Voigt, N. F. Ness.
Geophys. Res. Lett., Vol. 17, No. 10, p. 1705 – 1708 (Sep 1990).
The Neptunian magnetospehre periodically changes every eight hours between a "pole–on" magnetosphere with only one polar cusp and an "Earth–type" magnetosphere with two polar cusps. In the "pole–on" configuration, the tail current sheet has an almost circular shape with plasma currents closing entirely within the magnetosphere. Eight hours later the tail current sheet assumes an almost flat shape with plasma currents touching the magnetotail boundary and closing over the tail magnetopause. The authors have calculated magnetic field and tail current sheet configurations in a three–dimensional model, but have investigated the plasma– and thermodynamic conditions in a simplified two–dimensional MHD equilibrium magnetosphere. By matching the asymptotic plasma sheet $\beta = 2\mu_0 P/B^2$ values in the two models, they found to their surprise that the free energy F in the tail region of the two–dimensional model becomes independent of the dipole tilt angle. The authors conjecture the Neptunian magnetotail might assume quasi–static equilibrium states that make the free energy of the system independent of its daily rotation.

101.030 **He 584 Å dayglow at Neptune.**
C. D. Parkinson, J. C. McConnell, B. R. Sandel,
R. V. Yelle, A. L. Broadfoot.
Geophys. Res. Lett., Vol. 17, No. 10, p. 1709 – 1712 (Sep 1990).
The Voyager 2 Ultraviolet Spectrometer measured the emission intensity of the He resonance line at 584 Å to be 0.34(+0.2/–0.15) R on the day side of Neptune. The authors' calculations of the He 584 Å intensity at Neptune using a partial frequency redistribution and inhomogeneous atmsopheric models show that the product of the volume mixing ratio (mole fraction), f_{He}, and the eddy diffusion coefficient at the homopause, K_h, is $f_{He}K_h = 10^7 cm^2 s^{-1}$ with upper and lower bounds to the uncertainty of about a factor of 3 and 9, respectively. If f_{He} is taken as the current Infrared Interferometer Spectrometer working value, $f_{He} = 0.19$, then $K_h = 5x10^7 cm^2 s^{-1}$ with a similar uncertainty. This range of K overlaps that obtained from analysis of the hydrocarbon distributions on Neptune (Broadfoot et al., 1969).

101.031 **On the ionosphere of Triton: an evaluation of the magnetospheric electron precipitation and photoionization effects.**
W.–H. Ip.
Geophys. Res. Lett., Vol. 17, No. 10, p. 1713 – 1716 (Sep 1990).
In order to explore the origin and physical nature of the ionosphere of Triton, a simple 2–dimensional electron drift model is developed to approximate the ionization effect of magnetospheric electrons. The derived profile of ion production rate is then applied to a steady–state photochemical model calculation. It is shown that, as a result of the very low CH_4 mixing ratio in Triton's atmosphere, the most dominant ion in the upper ionosphere could be N^+. There are, however, difficulities in explaining the electron number density profile below 500 km as inferred from the radio science experiment on Voyager 2. One major factor is the increasing importance of the ion chemistry involving N^+ and H_2 from CH_4 photolysis in this region. This problem may be partially resolved if one invokes a more efficient electron transport in the atomsphere or the possible presence of the metallic ions like Si^+, Ne^+ and Na^+ which do not react with N_2 and H_2 in the lower ionosphere.

101.032 **Triton: topside ionosphere and nitrogen escape.**
Y. L. Yung, J. R. Lyons.
Geophys. Res. Lett., Vol. 17, No. 10, p. 1717 – 1720 (Sep 1990).
The principal ion in the ionosphere of Triton is N^+. Energetic electrons of magnetospheric origin are the primary source of ionization, with a smaller contribution due to photoionization. To explain the topside plasma scale height, the authors postulate that N^+ ions escape from Triton. The loss rate is $3.4x10^7 cm^{-2}s^{-1}$ or $7.9x10^{24}$ions s^{-1}. Dissociative recombination of ${N^+}_2$ produces neutral exothermic fragments that can escape from Triton. The rate is estimated to be $8.6x10^6$N $cm^{-2}s^{-1}$ or $2.0x10^{24}$atoms s^{-1}. Implications for the magnetosphere of Neptune and Triton's evolution are discussed.

101.033 **The ionosphere of Triton.**
T. Majeed, J. C. McConnell, D. F. Strobel,
M. E. Summers.
Geophys. Res. Lett., Vol. 17, No. 10, p. 1721 – 1724 (Sep 1990).
The authors have used a model of the atomospheric temperature structure and composition inferred from the Voyager UVS solar occultations together with a 1–D chemical diffusive model to interpret the Voyager RSS ingress measurements of Triton's electron density. Although ${N^+}_2$ is the major ion created, N^+ produced by dissociative ionization is the dominant ion. Reaction of thermospheric H_2, produced by Lyman–α–dissociation of CH_4 in the lower atmosphere, is the major loss for N^+ ions and maintains these ions in PCSS below 600 km. Solar EUV ionization cannot generate electron densities at the magnitude measured by the RSS experiment and an additional ionization source $\sim 3x10^8$ions $cm^{-2}s^{-1}$ is required. The ionosphere may undergo a transition from PCSS to diffusive control if the N^+ ion production rates rates were greater than the H_2 flux derived from CH_4. In this case the upward flowing H_2 is totally converted to H by reaction with N^+ and the remaining N^+ ions recombine radiatively to create an ionosphere under diffusive control above the peak.

101.034 **Nitrogen ion cluster in Triton's atomsphere.**
M. L. Delitsky, R. P. Turco, M. Z. Jacobson.
Geophys. Res. Lett., Vol. 17, No. 10, p. 1725 – 1728 (Sep 1990).
The formation of nitrogen cluster ions $[{N^+}_2(N_2)_n, N^+(N_2)_n]$ should occur easily in Triton's thin cold nitrogen atmosphere.

These ions are formed stepwise. At low altitudes, thermodynamics (temperature and pressure) favor cluster sizes up to and including $N^+{}_2(N_2)_{40}$. Such ions are close to the threshold (critical size) to nucleate, forming a stable nitrogen ice aerosol seen as extended hazes in Voyager images of Triton's limb. The critical ion radius for nucleation is about 8.3 Å. As temperature increases and pressure decrease with altitude, cluster ion formation becomes less favorable and small cluster sizes predominate. At altitudes >200 km, clusters are thermodynamically unstable and only $N^+{}_2$ and N^+ are present. Because electron recombination with atomic ions (N^+) is very slow, a dense ionosphere should form where N^+ recombination via N^+ clustering with N_2 is unfavorable. Model calculations yield peak electron concentrations above 200 km, as detected by Voyager radio occultation measurements. The sharp ledge at the bottom of the ionosphere is caused by the onset of rapid electron–ion recombination associated with cluster ion formation.

101.035 **The photochemistry of methane in the atmosphere of Triton.**
D. F. Strobel, M. E. Summers, F. Herbert, B. R. Sandel.
Geophys. Res. Lett., Vol. 17, No. 10, p. 1729 – 1732 (Sep 1990).
The photochemistry of methane in Triton's atmosphere is driven by approximately equal contributions of solar and local interstellar medium (LISM) Lyα and in combination with weak vertical mixing, $K \sim 4x10^3$ in the winter hemisphere and $8x10^3 cm^2 s^{-1}$ in the summer hemisphere, yields the observed CH_4 scale heights of ~9 km below 70 km altitude (Broadfoot et al., 1989). The hemispheric difference in K is consistent with a seasonal meridional circulation with upwelling in the summer hemisphere and downwelling in the winter hemisphere. Methane photolysis produces H_2 molecules which are removed from the photolysis region by an upward flux ($\sim 1.5x10^8 cm^{-2} s^{-1}$, normalized to the surface) to the ionosphere where ion reactions convert H_2 to H. This results in a hydrogen escape flux of $3x10^8 cm^{-2} s^{-1}$. The H column density is $\sim(5–10) \times 10^{14} cm^{-2}$ and resonantly scatters ~26 R of solar Lyα, consistent with the day–night asymmetry observed by the Voyager UVS. The inferred surface mixing ratios of CH_4 indicate subsaturated conditions. The CH_2 photolysis rate is sufficient to generate a smog of condensed C_2H_2, C_2H_4, C_2H_6, and C_4H_2 particles in the lowest 30 km of Triton's atmosphere with an optical depth consistent with Voyager imaging results.

101.036 **The atmosphere of Neptune: results of radio occultation measurements with the Voyager 2 spacecraft.**
G. F. Lindal, J. R. Lyons, D. N. Sweetnam, V. R. Eshleman, D. P. Hinson, G. L. Tyler.
Geophys. Res. Lett., Vol. 17, No. 10, p. 1733 – 1736 (Sep 1990).
Radio tracking data acquired during Voyager 2's occultation by Neptune have been used to study the thermal structure and composition of Neptune's troposphere and stratosphere.

101.037 **Post Voyager comparisons of the interiors of Uranus and Neptune.**
M. Podolak, R. T. Reynolds, R. Young.
Geophys. Res. Lett., Vol. 17, No. 10, p. 1737 – 1740 (Sep 1990).
The recent Voyager flyby of Uranus and Neptune has provided refined values for the gravitational moments and rotation periods of those planets. Using these new paramters, models of the interiors of these planets show that their density distributions are very similar. This lends support to the conjecture that their compositions are similar as well. The models are indeed consistent with such a conjecture. The difference in the internal heat sources of these two planets may be due to the fact that heat transport from the interior of Uranus is inhibited by a statically stable interior.

101.038 **A model of dust production in the Neptune ring system.**
J. E. Colwell, L. W. Esposito.
Geophys. Res. Lett., Vol. 17, No. 10, p. 1741 – 1744 (Sep 1990).
Voyager 2 images of the Neptunian ring system show it to be relatively dusty with the fraction of micro-sized dust particles in the rings between about 0.2 and 0.7 (Smith et al. 1989). The authors apply a numerical model developed for the Uranus dust bands to the Neptune system. The results show that (1) the suggested bimodal distribution of dust fractions in the rings is not related to ring optical depth and may be due to different ring particle size distributions. (2) Collision velocities between the macroscopic ring particles must be ~1 m/sec if they are to produce dust fractions above 1/2. (3) The small moons of Neptune must have regoliths for meteoroid excavation to produce the dust population at Voyager ring plane crossing. (4) Dust optical depths interior to 1989N3R are negligible unless there are unseen macroscopic source bodies in that region.

101.039 **Observations of Neptunian rings by Voyager photopolarimeter experiment.**
L. J. Horn, J. Hui, A. L. Lane, J. E. Colwell.
Geophys. Res. Lett., Vol. 17, No. 10, p. 1745 – 1748 (Sep 1990).
The Voyager photopolarimeter experiment detected the Neptunian rings 1989N1R and possibly 1989N2R. These rings were also photographed by the Voyager imaging cameras in August 1989. The authors' firm detection of 1989N1R has an equivalent depth (the product of the radial width and mean normal optical depth) of 0.77 ± 0.13 km, while a less certain detection of 1989N2R has an equivalent depth of 0.7 ± 0.2 km. Several statistical techniques were used to search for additional material in the Neptune ring system and none was identified at a high confidence level.

101.040 **The coupled orbital and thermal evolution of Triton.**
M. N. Ross, G. Schubert.
Geophys. Res. Lett., Vol. 17, No. 10, p. 1749 – 1752 (Sep 1990).
The coupled orbital and thermal evolution of Triton is simulated with a model of the satellite having temperature dependent k/Q (k is the second degree Love number and Q is the tidal dissipation factor). Large changes in orbital and interior properties occur in a short time span of less than 10^8 years. The peak tidal heating rate exceeds the radioactive heating rate by a factor of at least 10^3. The ability of tidal heating to overwhelm solid state convection and cause global melting is unequivocally demonstrated. Triton's initial thermal state and composition are shown to control the length of time that passes between capture from heliocentric orbit and the occurrence of swift and dramatic changes in orbital internal properties. This time interval could have been several billon years.

101.041 **Rheology of solid methane and nitrogen: applications to Triton.**
J. Eluszkiewicz, D. J. Stevenson.
Geophys. Res. Lett., Vol. 17, No. 10, p. 1753 – 1756 (Sep 1990).
The existing information about the rheological properties of solid methane and nitrogen is reviewed. The main limitation is the absence of creep data for solid nitrogen, but estimated properties are derived on the basis of simple scaling arguments for van der Waals bonded materials. On Triton solid methane and most likely nitrogen are too soft to support any topography over billions of years. Topographic support by water ice–rich materials seems almost certain. Nitrogen and methane do not intermix by solid–state diffusion. The weakness (absence?) of the opposition effect on Triton can be explained by annealing of nitrogen grains, provided the latter have sizes characteristic of freshly deposited frost.

101.042 **Koyaanismuuyaw: the hypothesis of a perennially dichotomous Triton.**
J. M. Moore, J. R. Spencer.
Geophys. Res. Lett., Vol. 17, No. 10, p. 1757 – 1760 (Sep 1990).
The Voyager 2 observation that, at the height of a "major" southern summer, the southern "cap" of Triton extends nearly to the equator, and the northern temperature regions (where they can be seen) are relatively dark even though nitrogen frost should be accumulating at those latitudes, has prompted the authors to propose a hypothesis that there has been a net non-reversable flux of surficial volatiles on Triton from the northern hemisphere to the southern hemisphere over much of the satellite history and up to the present. The north–south hemispherical albedo

difference is permanent and overwhelms the seasonal effect on volatile transport. This hypothesis assumes that volatile erosion, transport, and deposition has been in the form of sublimation, advection, and precipitation.

101.043 **Spectral geometric albedo and bolometric Bond albedo of Neptune's satellite Triton from Voyager observations.**
R. M. Nelson, B. J. Buratti, B. D. Wallis, W. D. Smythe, L. J. Horn, A. L. Lane, M. J. Mayo, K. E. Simmons.
Geophys. Res. Lett., Vol. 17, No. 10, p. 1761 – 1764 (Sep 1990).

The authors have calculated the spectral geometric albedo and bolometric Bond albdeo of Triton by combining the published data from the Voyager spacecraft Photopolarimeter (PPS) experiment and the Voyager imaging science (ISS) experiment. The authors have used the PPS ultraviolet (0.25 μm) and infrared (0.75 μm) filter data and the ISS violet (0.41 μm) and green (0.56 μm) filter data to produce geometric albedos and phase integrals at each of the four wavelengths. The resulting spectral geometric albedo is not inconsistent with the presence of weak absorption feature in Triton's spectrum near 0.75 μm. The phase integrals were determined by Russell's approximation and by 2–point Gaussian quadrature. The geometric albedo at the uv, v, g and ir filters is 0.59, 0.68 0.81, and 0.75 respectively. The resulting bolometric Bond albedo ($\sim$0.65) is consistent with the 38K, unit emissivity, daytime surface temperature for Triton reported as a best fit to the data of the Voyager Infrared Spectrometer and Radiometer (IRIS). The authors' results are also in agreement with the 37.5K temperature of nitrogen at an inferred basal pressure of 14 μbar as reported by the Voyager Ultraviolet Spectrometer investigation.

101.044 **Global color and albedo variations on Triton.**
A. S. McEwen.
Geophys. Res. Lett., Vol. 17, No. 10, p. 1765 – 1768 (Sep 1990).

Global multispectral mosaics of Triton have been produced from Voyager approach images; six spectral units are defined and mapped. The margin of the south polar cap (SPC) is scalloped and ranges in latitude from $+10°$ to $-30°$. A bright fringe is closely associated with the cap's margin; from it, diffuse bright rays extend north–northeast for hundreds of kilometers. Thus, the rays may consist of fringe materials that were redistributed by northward–going Coriolis–deflected winds. From 1977 to 1989, Triton's full–disk spectrum changed from markedly red and uv–dark to nearly neutral white and uv–bright. This spectral change can be explained by new deposition of nitrogen frost over both the northern hemisphere and parts of a formerly redder SPC. Frost deposition in the southern hemisphere during southern summer is possible over relatively high albedo areas of the cap (Stansberry et al., 1990), which helps to explain the apparent stability of the unexpectedly large SPC and the presence of the bright fringe.

101.045 **Nitrogen frost migration on Triton: a historical model.**
J. R. Spencer.
Geophys. Res. Lett., Vol. 17, No. 10, p. 1769 – 1772 (Sep 1990).

The author presents the result of numerical simulations of the seasonal migration of nitrogen frost on Triton, constrained by Voyager observations of atmospheric pressure, temperature, and albedo distribution. Most of the exposed nitrogen is probably seasonal frost, whose migration can produce major variations in atmosheric pressure. For instance, models explored here predict a tenfold pressure drop in the coming decade. The observed albedo patterns can be understood if fresh nitrogen frost is relatively dark but brightens with increasing insolation in a manner analogous to the Martian southern CO_2 cap.

101.046 **Zonally averaged thermal balance and stability models for nitrogen polar caps on Triton.**
J. A. Stansberry, J. I. Lunine, C. C. Porco, A. S. McEwen.
Geophys. Res. Lett., Vol. 17, No. 10, p. 1773 – 1776 (Sep 1990).

The authors have analyzed Voyager four color imaging data of Triton to calculate the bolometric hemispheric albedo as a function of latitude and longitude. Zonal averages of these data have been incorporated into a thermal balance model involving insolation, reradiation and latent heat of sublimation of N_2 ice for the surface. The current average bolometric albedo of Triton's polar caps is 0.8, implying an effective temperature of 34.2K and a surface pressure of N_2 of 1.6 μb for unit emissivity. This pressure is an order of magnitude lower than the surface pressure of 18 μb inferred from Voyager data, a discrepancy that can be reconciled if the emissivity of the N_2 on Triton's surface is 0.66. The model predicts that Triton's surface north of 15°N latitude is experiencing deposition of N_2 frosts, and are the bright portions of the south polar cap near the equator. This result explains why the south cap covers nearly the entire southern hemisphere of Triton. Initial results from a sasonal model are also decribed briefly, although they do not agree with the Voyager observations.

101.047 **The dipole tilt angle effect on the magnetosphere of Neptune – a laboratory simulation.**
S. Minami, K. Hashimoto, Y. Takeya.
Geophys. Res. Lett., Vol. 17, No. 11, p. 1885 – 1888 (Oct 1990).

Recent observations of Neptune by Voyager 2 reveal that the dipole tilt angle is almost 50 degrees – close to that of Uranus. When comparing the magnetospheres of planets, it is therefore important to understand configurations of magnetic field lines with large dipole tilt angles. This paper deals with the laboratory characteristics of Neptune's magnetosphere whose dipole tilt angle dominates the configuration of its magnetic field lines and particle intrusions from interplanetary space. The dipole tilt angle has a considerable effect on the formation of the asymmetrical structure of the field lines and the cylindrical tail current sheet in the magnetosphere. The experiment also shows the effect of the x–component of the interplanetary magnetic field to form an asymmetrical magnetospheric structure.

101.048 **A simple inertial model for Neptune's zonal circulation.**
M. Allison, J. T. Lumetta.
Geophys. Res. Lett., Vol. 17, No. 12, p. 2269 – 2272 (Nov 1990).

Voyager imaging observations of zonal cloud–tracked winds on Neptune revealed a strongly subrotational equatorial jet with a speed approaching 500 m s^{-1} and generally decreasing retrograde motion toward the poles. The wind data are interpreted with a speculative but revealingly simple model based on steady gradient flow balance and an assumed global homogenization of potential vorticity for shallow layer motion. The prescribed model flow profile relates the equatorial velocity to the mid–latitude shear, in reasonable agreement with the available data, and implies a global horizontal deformation scale $L_D \approx 3000$ km.

101.049 **Evidence for a diurnally rocking plasma mantle at Neptune.**
M. Zhang, V. M. Vasyliunas, G. L. Siscoe, R. P. Lepping, N. F. Ness.
Geophys. Res. Lett., Vol. 17, No. 13, p. 2285 – 2288 (Dec 1990).

Voyager's post–encounter trajectory at Neptune allows the directions of the magnetic field in the magnetosheath produced by the draping of the typical interplanetary field and by alignment with the magnetotail to be distinguished. Changes of the field from the draped to the magnetotail–aligned direction accompanied by decreases of plasma velocity, density, and temperature, observed at intervals of approximately a Neptunian day, are consistent with the assumption that Voyager repeatedly encountered a plasma mantle region extending well into the magnetosheath and modulated by the rotation of the planet's magnetic dipole. Other interpretations are either implausible or inconsistent with the observations. Previously reported periodic velocity decreases in the magnetosheath of Uranus can be interpreted in the same way. Extended plasma mantles may thus be a general property of planetary magnetospheres interacting with the magnetized solar wind. Analogous effects at Earth cannot yet be observed because of lack of suitable spacecraft missions, but their existence is suggested by theoretical arguments.

101.050 **Nitrogen airglow sources: comparison of Triton, Titan, and Earth.**
D. F. Strobel, R. R. Meier, M. E. Summers, D. J. Strickland.
Geophys. Res. Lett., Vol. 18, No. 4, p. 689 - 692 (Apr 1991).
Detailed calculations of the emission rates of nitrogen airglow in the upper atmospheres of Triton and Titan by direct solar excitation, photoelectrons, and magnetospheric electrons are compared with the Voyager Ultraviolet Spectrometer (UVS) observations. The principal spectral features of Triton's airglow are shown to be consistent with precipitation of magnetospheric electrons with power dissipation $\sim 5\times10^8$W. Solar excitation rates of the dominant N_2 and N^+ emission features are factors of 2–7 weaker than magnetospheric electron excitation. On Titan the calculated disk center and bright limb N^+ (1085 Å) intensities by solar excitation agree with observed values, whereas if the 970 Å feature is mostly N_2 c_4 band emission, the calculated photoelectron excitation intensity is only 40% of the observed value. The difference could be due to magnetospheric electron precipitation. But the calculated LBH intensity by photoelectrons is in agreement with observations and suggests that magnetospheric electrons play a minor role in Titan's UV airglow. On the Earth solar/photoelectron excitation explains the observed N^+ (1085 Å) and LBH intensities, accounts for only 40% of the N^+ (916 Å) intensity, and overpredicts substantially the N_2 c_4 (0–0) band intensity.

101.051 **Gesichter Plutos.**
M. Reichstein.
Astron. Sch., Jahrg. 28, Heft 5, p. 7 - 10 (Oct 1991).

101.052 **Mass and orbit estimation of planet X via a family of comets.**
R. Neuhäuser, J. V. Feitzinger.
Earth, Moon, Planets, Vol. 54, No. 3, p. 193 - 202 (Sep 1991).
The crucial assumption of this paper is that the observed clustering of aphelion distances of intermediate-period comets in the 70–90 AU range is due to the influence of a tenth planet, called Planet X. The authors contribute to the search for Planet X by means of a new and extended evaluation of a family of comets assumed to be Planet X's family of comets. Averaging the aphelion distances of comets that belong to a transplutonian family of comets gives Planet X's semi-major axis $a_X = (83.0 \pm 5.3)$ AU. The comets' orbits also yield the upper limit of the planet's orbital eccentricity $e_X \cong 0.019$. If this planet played an important part in "sending" quasi-periodic comet showers to the inner solar system, it is possible to calculate its orbital inclination $i_X = 46\overset{\circ}{.}1 \pm 3\overset{\circ}{.}6$. By distributing all planetary masses into the heliocentric, torus-like zones, in which the planets were formed, the density distribution of the primordial solar nebula is obtained. Extrapolating this distribution gives the mass of Planet X, $M_X = 5.1(+3.6,-2.4)\ M_{Earth}$. A few plausible assumptions (e.g. Uranus and Neptune perturbations being caused by Planet X) lead to Planet X's actual location with declination and ecliptic longitude being $\delta = 57° \pm 17°$ and $\lambda = 54° \pm 34°$, respectively (1989.5 position). In addition, the authors give Planet X's apparent brightness dependent on its unknown albedo. All those properties and predictions are more or less in agreement with earlier work on Planet X.

101.053 **Occultations by Uranus and Neptune: 1991–1999.**
A. R. Klemola, D. J. Mink.
Astron. J., Vol. 102, No. 1, p. 389 - 394 (Jul 1991). Current Physics Microform No.: 9107B2085.
The results of a photographic plate search are presented for stars as faint as $m_v = 14$ which may be occulted by Uranus or Neptune or their rings between 1 January 1991 and 31 December 1999. Circumstances for the closest approach of Uranus to 76 stars and Neptune to 18 stars are presented. Occultations by Neptune's ring "arcs" are predicted in 1992, 1997, and 1999.

101.054 **Pecularities of seasons of a year on Pluto.**
T. Z. Dworak.
Urania, Rok 62, Nr. 11, p. 297 - 303 (Nov 1991). In Polish.

101.055 **Pluto–Charon stellar occultation candidates: 1990–1995.**
E. W. Dunham, S. W. McDonald, J. L. Elliot.
Astron. J., Vol. 102, No. 4, p. 1464 - 1484 (Oct 1991). Current Physics Microform No.: 9110D1996.
The authors have carried out a search to identify stars that might be occulted by Pluto or Charon during the period 1990–1995 and part of 1996. This search was made with an unfiltered CCD camera operated in the strip scanning mode, and it reaches an R magnitude of approximately 17.5 - about 1.5 mag fainter than previous searches. Circumstances for each of the 162 potential occultations are given, including an approximate R magnitude of the star, which allows estimation of the signal-to-noise ratio (S/N) for observation of each occultation. The faintest stars in the authors' list would yield an S/N of about 20 for a 1 s integration when observed with a CCD detector on an 8 m telescope under a dark sky. The authors' astrometric precision (± 0.2 arcsec, with larger systematic errors possible for individual cases) is insufficient to serve as a final prediction for these potential occultations, but is sufficient to identify stars deserving of further, more accurate, astrometric observations. Statistically, the authors expect about 32 of these events to be observable somewhere on Earth. The number of events actually observed will be substantially smaller because of clouds and the sparse distribution of large telescopes. Finder charts for each of the 91 stars involved are presented.

101.056 **The story of Pluto is continued.**
S. R. Brzostkiewicz, D. Górnicza.
Urania, Rok 62, Nr. 12, p. 328 - 334 (Dec 1991). In Polish.

101.057 **Neptune's cosmic ray cutoff.**
R. S. Selesnick, E. C. Stone.
Geophys. Res. Lett., Vol. 18, No. 3, p. 361 - 364 (Mar 1991).
The Voyager 2 cosmic ray experiment observed the signature of a magnetic cutoff of cosmic ray protons immediately after the closest approach to Neptune. The cutoff signature shows that the simultaneous observation of a trapped particle dropout could not have occurred at high magnetic latitudes, but is likely to be a drift shadow of the planet caused by the large offset of the particle drift shells. The OTD2 dipole model of Neptune's magnetic field predicts a drift shadow at approximately the correct times. In addition, the cutoff signature is similar to that predicted by the OTD2 model, although the model is not accurate throughout the observation period. The similarity is consistent with small estimated corrections to the predicted signature based on the locally observed magnetic field values, indicating that the L shell values derived from the model are approximately valid after the closest approach time.

101.058 **Structure and dynamics of the Uranian magnetotail: results from hot plasma and magnetic field observations.**
M. Kane, B. H. Mauk, E. P. Keath, S. M. Krimigis.
J. Geophys. Res., Vol. 96, No. A7, p. 11485 - 11500 (1 Jul 1991).
Analysis of the Voyager 2 low-energy charged particle (LECP) and magnetic field (MAG) data in the Uranian magnetotail has shown this system to have many features similar to those seen at Earth. Isotropization of core magnetosphere 28–43 keV ions have been observed at the trapping boundary upon entry into the magnetotail near L = 17, an effect seen in ion populations at Earth and attributed to neutral sheet scattering. An extensive and persistent plasma sheet boundary layer was found to contain field-aligned ion streams with a variety of angular and energetically dispersive effects. Streaming distributions in the magnetotail horn region map to the auroral emission region observed by the Voyager ultraviolet spectrometer (UVS). A quiet-time near-tail region is distinguished from a distant disturbed region, where evidence of a substorm is observed. There is evidence suggesting that the distant plasma sheet is not corotating with the planet. Proton density gradients in this disturbed region indicate the presence of detached plasma structures. During what appears to be an occuring substorm, energy dispersionless field-aligned streaming, seen within the neutral sheet region, indicates close proximity to a substormlike acceleration. Extension observations

of the Earth's magnetotail provide a model for ordering a variety of particle and magnetic phenomena observed in the Uranian magnetotail.

101.059 **The helium abundance of Neptune from Voyager measurements.**
B. J. Conrath, D. Gautier, G. F. Lindal, R. E. Samuelson, W. A. Shaffer.
J. Geophys. Res., Vol. 96, Suppl., p. 18907 – 18920 (30 Oct 1991). Special issue: Voyager 2 at Neptune.
Voyager radio occultation and infrared spectrometer measurements are used to obtain an estimate of the helium abundance in the atmosphere of Neptune. It is found that the shape of the measured spectrum cannot be well matched by spectra calculated from atmospheric models that include only gaseous opacity. The most plausible explanation of the observed spectral shape appears to require the existence of an additional opacity source associated with clouds or hazes. The data can be fit with either a tropospheric or a stratospheric cloud model.

101.060 **The albedo, effective temperature, and energy balance of Neptune, as determined from Voyager data.**
J. C. Pearl, B. J. Conrath.
J. Geophys. Res., Vol. 96, Suppl., p. 18921 – 18930 (30 Oct 1991). Special issue: Voyager 2 at Neptune.
Data from the Voyager infrared spectrometer and radiometer (IRIS) investigation are used in determining the albedo, effective temperature, and energy balance of Neptune. From broadband radiometric observations made at phase angles of 14° and 134°, together with measurements at intermediate phase angles from the literature, an orbital mean value for the bolometric Bond albedo, $\bar{A} = 0.290 \pm 0.067$, is obtained; this yields an equilibrium temperature $T_{eq} = 46.6 \pm 1.1$K. From thermal spectra obtained over latitudes from pole to pole an effective temperature $T_{eff} = 59.3 \pm 0.8$K is derived. This represents a substantial improvement over previously determined values. The energy balance of Neptune is therefore $E = 2.61 \pm 0.28$, which is in agreement with previous results. The reduced uncertainty in this value is due to the improved determination of the effective temperature.

101.061 **Thermal structure and dynamics of Neptune's atmosphere from Voyager measurements.**
B. J. Conrath, F. M. Flasar, P. J. Gierasch.
J. Geophys. Res., Vol. 96, Suppl., p. 18931 – 18940 (30 Oct 1991). Special issue: Voyager 2 at Neptune.
Spatially resolved Voyager infrared spectra from a global mapping sequence are used to characterize the thermal structure of the atmosphere of Neptune between approximately 50 and 100 mbar. A zonal mean meridional temperature cross section is obtained, which shows a minimum in the upper tropospheric and lower stratospheric temperatures at southern mid latitudes wtih maxima at the equator and high southern latitudes. This is qualitatively similar to the structure observed on Uranus, even though the obliquities and internal heat fluxes of the two planets are quite different. Thermal wind calculations show zonal wind speeds decreasing with altitude at all latitudes. Within the framework of a linear radiative–dynamical model, the thermal structure and cloud top winds together imply a frictional damping time comparable in magnitude to the radiative relaxation time; similar results have previously been obtained for the other three giant planets. Periodogram techniques were used to search the observed longitude variations of temperature for possible planetary–scale wave featues. Data from two latitude bins are analyzed; one of the bins includes the Great Dark Spot, and the other is centered on the southern mid–latitude minimum in the zonal mean temperature.

101.062 **Winds of Neptune: Voyager observations of cloud motions.**
S. S. Limaye, L. A. Sromovsky.
J. Geophys. Res., Vol. 96, Suppl., p. 18941 – 18960 (30 Oct 1991). Special issue: Voyager 2 at Neptune.
High temporal and spatial resolution images acquired from Voyager cameras have been used to measure cloud motions to improve the meridional profile of the zonal mean circulation on Neptune. A wide range of atmospheric periods between 12 and 21 hours is revealed by the average cloud motions, consistent with the previous observations. New observations have expanded latitudinal coverage, improved the determination of streak motions, especially near 30°N, and added statistical weight to latitudes already covered by previous measurements. Both new and earlier observations have been subjected to quality control procedures to reduce dispersion caused by erroneous observations. The resulting data set emphasizes short time intervals to maximize target recognition. Most reliable cloud targets last less than one Neptune rotation, many only a fraction of it. The wind observations have a high degree of variability, some of which is due to variability of motions; but at many latitudes it reflects observation errors arising from rapid evolution of Neptune's clouds. Zonal motions and morphology suggest the global circulation to be symmetric about the equator.

101.063 **Hydrocarbons in Neptune's stratosphere from Voyager infrared observations.**
B. Bézard, P. N. Romani, B. J. Conrath, W. C. Maguire.
J. Geophys. Res., Vol. 96, Suppl., p. 18961 – 18976 (30 Oct 1991). Special issue: Voyager 2 at Neptune.
Emission from the acetylene and ethane bands at 729 and 822 cm^{-1} detected in the Voyager infrared spectra of Neptune has been analyzed. A large selection of low–spatial resolution spectra was used to derive the disk–averaged abundances of C_2H_2 and C_2H_6.

101.064 **Low–energy ions near Neptune.**
J. D. Richardson, J. W. Belcher, M. Zhang, R. L. McNutt Jr.
J. Geophys. Res., Vol. 96, Suppl., p. 18993 – 19012 (30 Oct 1991). Special issue: Voyager 2 at Neptune.
This paper provides a synopsis of the ion observations near Neptune, from the upstream solar wind through the magnetosheath, magnetosphere, and back out through the magnetosheath to the solar wind. The main emphasis is on the magnetosphere; complete density and temperature profiles are shown for this region. Observations are consistent with the presence of two ion species, H^+ and N^+, which increase in density and temperature with decreasing distance from Neptune. Calculations of the flux tube content and energy invariant imply that N^+ escapes directly from Triton's ionosphere, whereas H^+ is created by ionization of a large neutral H cloud emanating from Triton and extending inward to 8 R_N. Inside $L = 8$, substantial plasma losses occur, precipitation into Neptune's atmosphere is one loss mechanism, and charge exchange with a hypothesized Neptune–centered H cloud may also be important. Given expected source rates of about $1x10^{25}$ions/s, transport times must be very fast, of the order of 10^5s at $L = 10$.

101.065 **A mysterious plasma wave emission and the determination of plasma densities in Neptune's inner magnetosphere.**
S. L. Moses, F. V. Coroniti.
J. Geophys. Res., Vol. 96, Suppl., p. 19013 – 19022 (30 Oct 1991). Special issue: Voyager 2 at Neptune.
One of the strongest plasma signals observed during the Voyager 2 encounter with Neptune is a narrowband emission between 3.0 and 4.3 kHz that was detected over a period of roughly 2 hours around closest approach. The emission occurs below the electron cyclotron frequency and the low–frequency cutoff of the radio continuum radiation. Of the naturally occurring signals in the Earth's auroral zone and in Jupiter's magnetosphere this emission most resembles trapped Z mode waves found near the left–hand cutoff frequency. Using this identification, the authors obtain a plasma density profile that is independent of the plasma temperature. These densities greatly exceed those measured by the plasma science instrument on Voyager but are lower than estimates based on other models of Neptunian plasma wave phenomenology. If this wave mode is not a natural emission, it might arise from an unusual interaction of the spacecraft with the cold, dense ambient plasma.

101.066 **The magnetic field of Neptune.**
J. E. P. Connerney, M. H. Acuña, N. F. Ness.
J. Geophys. Res., Vol. 96, Suppl., p. 19023 – 19042 (30 Oct 1991).
Special issue: Voyager 2 at Neptune.

A spherical harmonic model of the planetary magnetic field of Neptune is obtained from the Voyager 2 encounter observations. The model is obtained by partial solution to the underdetermined inverse problem (eighth–order and degree spherical harmonic model) using generalized inverse techniques. Dipole and quadrupole Schmidt–normalized spherical harmonic coefficients are determined, or resolved, independently of higher–order terms. Much additional information is also obtained regarding octupole and higher–order components of the field. Model parameter resolution and uniqueness are addressed using new visual representations of the "resolution matrix". The model yields a dipole of magnitude 0.14 G R^3_N, tilted by 47° toward 72° west longitude. Neptune's quadrupole is relatively large, equal to or exceeding in magnitude the (surface) dipole field. The octupole is likewise very large but less well constrained. The magnetic fields of Neptune and Uranus are very much alike and equally unlike those of the other known magnetized planets. Characteristics of this complex magnetic field are illustrated, using contour maps of the field on the planet's surface, and discussed in the context of dynamo generation in the relatively poorly conducting "ice mantle".

101.067 **Restrictions on the characteristics of Neptunian lightning.**
M. L. Kaiser, P. Zarka, M. D. Desch, W. M. Farrell.
J. Geophys. Res., Vol. 96, Suppl., p. 19043 – 19048 (30 Oct 1991).
Special issue: Voyager 2 at Neptune.

Although lightning–triggered whistlers were detected by the Voyager 2 plasma wave experiment at Neptune, only four possible lightning sferics were detected by the planetary radio astronomy instrument. This low level of detection places strict limits on the characteristics of Neptunian lightning, ruling out lightning with power levels as high as those observed at Saturn or Uranus, but leaving the intriguing possibility that lightning slightly stronger than normal terrestrial lightning flashes but with much lower occurrence rate could exist. Possible scenarios to explain the observations are examined, including reduced NH_3 concentration in the planet's atmosphere and an unusually slow risetime discharge process.

101.068 **Evidence of auroral plasma cavities at Uranus and Neptune from radio burst observations.**
W. M. Farrell, M. D. Desch, M. L. Kaiser, W. Calvert.
J. Geophys. Res., Vol. 96, Suppl., p. 19049 – 19060 (30 Oct 1991).
Special issue: Voyager 2 at Neptune.

Radio bursts originating from the stronger magnetic polar regions of both Uranus and Neptune were detected by the planetary radio astronomy experiment during the Voyager 2 encounters with the planets. It has previously been demonstrated that these bursts are beamed into a broad, hollow emission pattern from their auroral sources. The authors now show that the bursts at both planets also manifest similar detailed patterns, with the waves beamed into two separate and distinct radiation cones at intermediate wave frequencies. This double–cone emission pattern is predicted by relativistic cyclotron resonance theory, and application of this theory to the observed emission pattern yields the plasma density structure within the radio source region. Calculations indicate that at both Uranus and Neptune the plasma–to–cyclotron frequency ratio can drop well below 0.01 within the active region. Such low values indicate that the southern auroral zones at both planets contain an auroral plasma cavity that is similar to that found in Earth's nightside auroral zone.

101.069 **The magnetosphere of Neptune: hot plasmas and energetic particles.**
B. H. Mauk, E. P. Keath, M. Kane, S. M. Krimigis,
A. F. Cheng, M. H. Acuña, T. P. Armstrong, N. F. Ness.
J. Geophys. Res., Vol. 96, Suppl., p. 19061 – 19084 (30 Oct 1991).
Special issue: Voyager 2 at Neptune.

A comprehensive overview is provided of the hot plasmas and energetic particles ($\gtrsim$20 keV) observed in the vicinity of Neptune by the low energy charged particle (LECP) experiment on the Voyager 2 spacecraft.

101.070 **Voyager 2 electron observations in the magnetosphere of Neptune.**
M. Zhang, J. D. Richardson, E. C. Sittler Jr.
J. Geophys. Res., Vol. 96, Suppl., p. 19085 – 19100 (30 Oct 1991).
Special issue: Voyager 2 at Neptune.

This paper describes electron data obtained during the Neptune encounter by the Voyager 2 plasma science experiment. The authors derive the densities and temperatures of low–energy (10–5950 eV) electrons and the electrostatic potential of the spacecraft near Neptune. The data indicate that the escape of charged particles from Triton and the local ionization of atoms in the neutral torus originating from Triton are the major plasma sources. The authors infer that this neutral torus of hydrogen atoms has a density of about 300 cm^{-3} and an inner boundary at 8 R_N. The data near Neptune exhibit signatures suggesting that both precipitation into Neptune's atmosphere and ring absorption are important plasma loss mechanisms. Plasma transport in the magnetosphere appears to be very fast; the diffusion coefficient is $1x10^{-7}L^3R^2{}_Ns^{-1}$.

101.071 **Emission characteristics and source location of the smooth Neptunian kilometric radiation.**
H. P. Ladreiter, Y. Leblanc, G. K. F. Rabl, H. O. Rucker.
J. Geophys. Res., Vol. 96, Suppl., p. 19101 – 19110 (30 Oct 1991).
Special issue: Voyager 2 at Neptune.

The observations of the planetary radio astronomy (PRA) experiment aboard Voyager 2 reveal the existence of smooth and bursty radio emission. Both recur in a rather regular pattern with a 16.1–hour period (the Neptunian spin period). The authors describe the phenomenology of the smooth component in terms of frequency, polarization, and occurrence in magnetic longitude and latitude.

101.072 **The role of solar wind reconnection in driving the Neptune radio emission.**
M. D. Desch, W. M. Farrell, M. L. Kaiser, R. P. Lepping,
J. T. Steinberg, L. A. Villanueva.
J. Geophys. Res., Vol. 96, Suppl., p. 19111 – 19116 (30 Oct 1991).
Special issue: Voyager 2 at Neptune.

The only remote diagnostic of conditions within the outer planets' magnetospheres is the highly variable flux of low–frequency radio waves. As at the other radio planets, Neptune radio emission also manifests, on a time scale of days, major intensity fluctuations that are indicative of a solar wind energy–coupling process of some kind. The authors find that the merging of interplanetary magnetic field lines with Neptune's magnetosphere is the best predictor of emitted radio energy. By contrast, viscouslike energy coupling processes, such as might be caused by solar wind density of bulk speed fluctuations, are apparently ineffective in driving the radio emission.

101.073 **Modeling of whistler ray paths in the magnetosphere of Neptune.**
J. D. Menietti, D. Tsintikidis, D. A. Gurnett, D. B. Curran.
J. Geophys. Res., Vol. 96, Suppl., p. 19117 – 19122 (30 Oct 1991).
Special issue: Voyager 2 at Neptune.

A three–dimensional ray tracing program is used to investigate the propagation and dispersion of lightning–generated whistlers in the magnetosphere of Neptune. A simple empirical magnetospheric plasma model based on observational constraints is used for the plasma density distribution, and an offset tilted dipole model is used for the magnetic field. Source positions for the whistlers are chosen along magnetic field lines with $L < 3$ in agreement with observations. Whistlers with frequencies of about 10 kHz generally propagate very near the resonance cone angle for most of the ray path, which produces dispersions considerably larger than those predicted for quasiparallel propagation. Dispersions computed using the authors' "standard" plasma density model are over 30,000 s $\sqrt{Hz}$ for "one–hop" whistlers which is comparable to observed values. In general, the whistlers were found to propagate about 10° in magnetic longitude during one hop and to cross L shells. These results strongly support the existence of whistlers at Neptune.

101.074 **Theory of ring sweeping of energetic particles.**
C. P. Paranicas, A. F. Cheng.
J. Geophys. Res., Vol. 96, Suppl., p. 19123 – 19130 (30 Oct 1991).
Special issue: Voyager 2 at Neptune.

Because the effective "area" of the Neptunian rings is larger than that of the inner moons, the sweeping of energetic particles by the rings is perhaps the dominant process for particle loss in the magnetosphere within 5 R_N. In this paper the authors describe a theory for calculating the absorption probability of energetic charged particles by the rings. The authors include the effects of a large tilt and an offset between the planet and dipole centers. They find that the probability of absorption for protons is so high that the sweeping lifetime is only a few times the gradient–curvature drift period. For electrons the sweeping lifetime is even less. The pitch angle dependence for sweeping manifests itself strongly only at large equatorial pitch angles. Lower–energy particles have higher absorption rates by the rings.

101.075 **Satellite sweeping of energetic particles at Neptune.**
C. P. Paranicas, A. F. Cheng.
J. Geophys. Res., Vol. 96, Suppl., p. 19131 – 19136 (30 Oct 1991).
Special issue: Voyager 2 at Neptune.

The calculation of the absorption rate of charaged particles by planetary satellites introduced by Paonessa and Cheng (1987) is generalized to include an arbitrary offset of the dipole center from the planet center, appropriate for Neptune. The absorption rates calculated for particles of fixed L shell, energy, and pitch angle reflect the features of the complicated geometry of the dipole and the moons. This absorption probability is found to be insignificant compared with that of the rings at L shells to which both sets of absorbers map. However, at larger radii the sweeping rate is controlled by the moons, and the corresponding absorption features provide a starting point for understanding the Voyager energetic particle observations.

101.076 **The electron absorption signature of 1989N1.**
R. S. Selesnick, E. C. Stone.
J. Geophys. Res., Vol. 96, Suppl., p. 19137 – 19148 (30 Oct 1991).
Special issue: Voyager 2 at Neptune.

A deep minimum in the energetic ($\gtrsim 1$ MeV) electron intensity at Neptune is associated with absorption by the satellite 1989N1. The absorption signature is modeled by assuming that the electron losses are replenished by diffusion due to violation of the third adiabatic invariant. It is predominantly a steady state feature, or macrosignature, for which approximate solutions of the diffusion equation lead to an estimate of the radial diffusion coefficient, $D_{LL} \approx 2\mathrm{x}10^{-8}\mathrm{s}^{-1}$ at $L = 5$. The model suggests a strong L dependence for D_{LL} ($\sim L^7$), but an outward displacement of the macrosignature by ~ 0.3 unit in L from the location predicted by the model, and differences in the observed and model energy spectra, may indicate that processes such as pitch angle and energy diffusion need to be included for an accurate assessment of the transport coefficients. An approximate solution is also found for the time–dependent, two–dimensional diffusion equation including particle drifts and satellite absorption.

101.077 **Magnetopause and cusp observtions at Neptune.**
A. Szabo, G. L. Siscoe, A. J. Lazarus,
R. L. McNutt Jr., R. P. Lepping, N. F. Ness.
J. Geophys. Res., Vol. 96, Suppl., p. 19149 – 19152 (30 Oct 1991).
Special issue: Voyager 2 at Neptune.

The fortuitous timing of the Voyager 2 flyby of Neptune provided a unique opportunity to carry out plasma and magnetic field measurements in a subsolar cusp region of a planet. The measured plasma parameters near the inbound magnetopause strongly suggest that a rotational discontinuity separates the magnetosheath and the high–altitude cusp region. After crossing the magnetopause, a thick layer of magnetosheathlike plasma was observed which can be accounted for by two models: (1) either a cusplike layer of stagnant plasma separated from the magnetosheath by a tangential discontinuity, or (2) or a mantle-like layer of dynamic plasma separated from the sheath by a rotational discontinuity. The plasma observations are more consistent with the mantle interpretation. Distinguishing the mantle and the stagnant cusp models is not possible on geometrical grounds because Voyager 2 crossed the magnetic polar region on the subsolar side. Inside the cusp region, there is an overall decrease in the ion density and a concurrent increase in the magnetic field strength which is consistent with a slow mode expansion fan such as is encountered in Earth's polar regions.

101.078 **Remote sensing of Neptune's bow shock: evidence for large–scale shock motions.**
I. H. Cairns, C. W. Smith, W. S. Kurth, D. A. Gurnett,
S. Moses.
J. Geophys. Res., Vol. 96, Suppl., p. 19153 – 19170 (30 Oct 1991).
Special issue: Voyager 2 at Neptune.

The Voyager 2 spacecraft observed high levels of Langmuir waves before the inbound crossing of Neptune's bow shock, thereby signifying magnetic connection to the bow shock. The Langmuir waves occurred in multiple bursts throughout two distinct periods separated by an 85–min absence of wave activity. The authors use the times of onsets, stable maxima, and disappearances of the waves, together with the magnetic field direction and spacecraft position, to perform a "remote–sensing" analysis of the shape and location of Neptune's bow shock prior to the inbound bow shock crossing.

101.079 **On the lack of a magnetic signature of Triton's magnetospheric interaction on the Voyager 2 flyby trajectory.**
F. M. Neubauer, A. Lüttgen, N. F. Ness.
J. Geophys. Res., Vol. 96, Suppl., p. 19171 – 19176 (30 Oct 1991).
Special issue: Voyager 2 at Neptune.

There is strong, albeit indirect, evidence for a sub–Alfvénic and transsonic interaction between Triton and the Neptunian magnetosphere. A new inspection of magnetic field data around Triton encounter shows no evidence for the interaction. It is shown that this is due to the geometry of Triton's Alfvénic wings, which were probably too far from the spacecraft during the encounter. The situation is complicated by finite ion gyroradius effects, which tend to smear out the simple Alfvén wings such as encountered at Io.

101.080 **Micron–size particles detected near Neptune by the Voyager 2 plasma wave instrument.**
D. A. Gurnett, W. S. Kurth, L. J. Granroth, S. C. Allendorf,
R. L. Poynter.
J. Geophys. Res., Vol. 96, Suppl., p. 19177 – 19186 (30 Oct 1991).
Special issue: Voyager 2 at Neptune.

This paper presents a detailed analysis of the dust impacts observed by the Voyager 2 plasma wave instrument in the vicinity of Neptune. The objectives of the paper are (1) to give a detailed description of the signals produced by the dust impacts, (2) to interpret the impact noise in terms of the number density and mass of the particles, and (3) to discuss the origin of the particles.

101.081 **Dust distribution around Neptune: grain impacts near the ring plane measured by the Voyager planetary radio astronomy experiment.**
B. M. Pedersen, N. Meyer–Vernet, M. G. Aubier, P. Zarka.
J. Geophys. Res., Vol. 96, Suppl., p. 19187 – 19196 (30 Oct 1991).
Special issue: Voyager 2 at Neptune.

During the Voyager 2 flyby of Neptune the planetary radio astronomy (PRA) experiment recorded an intense noise near the equatorial plane around 3.4 and 4.2 R_N, as already observed during previous Voyager ring plane crossings at Saturn and Uranus. This noise is interpreted as being due to impact ionization of dust grains striking the spacecraft. The authors deduce a power law index of the grain mass distribution of about 2. The PRA system is sensitive to particle with radii larger than ≈ 1.6 μm, and the largest particles, detected near the ring plane, are evaluated to have a radius of ~ 10 μm. The spatial dust distribution along the spacecraft trajectory around the two equatorial crossings is found not be be symmetrical with respect to the ring plane and spread over wide regions: over $\sim 2\ R_N$ perpendicularly to the equatorial plane with the densest part concentrated within ~ 700 km. The vertical optical depth τ of this dense region is found to be 10^{-6}–10^{-8}.

101.082 **Triton's surface properties: a preliminary analysis from ground–based, Voyager photopolarimeter subsystem, and laboratory measurements.**
B. J. Buratti, A. L. Lane, J. Gibson, H. Burrows, R. M. Nelson, D. Bliss, W. Smythe, V. Garkanian, B. Wallis.
J. Geophys. Res., Vol. 96, Suppl., p. 19197 – 19202 (30 Oct 1991). Special issue: Voyager 2 at Neptune.
The surface properties of Triton have been determined from a threefold approach involving analysis of ground–based and Voyager photopolarimeter subsystem observations of its phase curve, theoretical modeling, and laboratory measurements of the phase functions of a wide variety of materials. The physical parameters derived include the single–scattering albedo, the single–particle phase function, the compaction state of the optically active portion of the regolith, and the degree of macroscopic roughness. The authors find that Triton has a high single–scattering albedo (0.96 ± 0.01 at $0.75\ \mu m$) and an unusually compacted surface possibly similar to that of Europa. Preliminary results suggest that Triton's single–particle phase function and the macroscopically rough character of its surface are similar to those of most other icy satellites. Laboratory measurements of the phase curve of a sample of a high–albedo, compact volatile–contaminant analogue are in reasonable agreement with Triton's phase curve and provide evidence that the model parameters derived for Triton's surface are physically realistic.

101.083 **Voyager photometry of Triton: haze and surface photometric properties.**
J. Hillier, P. Helfenstein, A. Verbiscer, J. Veverka.
J. Geophys. Res., Vol. 96, Suppl., p. 19203 – 19210 (30 Oct 1991). Special issue: Voyager 2 at Neptune.
The authors have analyzed Voyager whole–disk observations of Triton at violet ($0.41\ \mu m$), blue ($0.48\ \mu m$), and green ($0.56\ \mu m$) filter wavelengths, using a model which combines an improved version of Hapke's photometric equation with a thin atmospheric haze layer in the appropriate spherical geometry. The model accurately describes the phase curves over a wide range of phase angles ($10° \leqslant \alpha \leqslant 159°$) and agrees well with disk–resolved brightness scans along the photometric equator and mirror meridian. An upturn in the phase curves seen at the highest phase angles observed (Smith et al., 1989) can be explained by including scattering in a thin atmospheric haze layer with optical depths ranging from 0.066 in the violet to 0.036 for the green filter date. The haze is forward scattering with $g \sim +0.6$, requiring particle sizes of $\gtrsim 0.2\ \mu m$. The haze single–particle scattering albedo is near unity at all three wavelengths, suggesting the haze particles are relatively neutral ice condensates. The geometric albedo atmospheric parameters are derived. The bolometric Bond albedo of Triton calculated from the authors' results is 0.85 ± 0.05. If the 16–μbar N_2 atmosphere detected by the Voyager radio occultation experiment is in vapor equilibrium with the surface (therefore implying a surface temperature of 37.7K), then the Bond albedo implies a surface emissivity of 0.46 ± 0.16.

101.084 **The wavelength dependence of Triton's light curve.**
J. Hillier, J. Veverka, P. Helfenstein, A. McEwen.
J. Geophys. Res., Vol. 96, Suppl., p. 19211 – 19216 (30 Oct 1991). Special issue: Voyager 2 at Neptune.
Using Voyager observations, the authors demonstrate that Triton's orbital light curve is strongly wavelength dependent, a characteristic which readily explains some of the apparent discrepancies among pre–Voyager telescopic measurements. Specifically, the authors find a light curve amplitude (peak to peak) that decreases systematically with increasing wavelength from about 0.08 magnitude (peak to peak) near 200 nm to less than 0.02 magnitude near 1000 nm. Peak brightness occurs near 90° orbital longitude (leading hemisphere). The brightness variation across this hemisphere is close to sinusoidal; the variation across the darker hemisphere is more complex. The decrease in light curve amplitude with increasing wavelength appears to be due to a decrease in contrast among surface markings, rather than to atmospheric obscuration. The authors' model also explains the observed decrease in the ampltiude of Triton's light curve at visible wavelengths over the past decade, a decrease related to the current migration of the subsolar latitude toward the south pole: the authors predict that this trend will continue into the 1990s.

101.085 **On the microphysical state of the surface of Triton.**
J. Eluszkiewicz.
J. Geophys. Res., Vol. 96, Suppl., p. 19217 – 19230 (30 Oct 1991). Special issue: Voyager 2 at Neptune.
On the basis of an analysis of the microphysical processes involved in pressureless sintering of particulate materials and an examination of the physical conditions likely to prevail on Triton, it is argued that a well–annealed transport nitrogen layer can form on Triton at 37K on a seasonal time scale (~ 100 Earth years), provided the initial grain diameter is less than $1\ \mu m$. Slightly larger grains (in the 1–10 μm range) are allowed if grain growth does not hinder densification, but grains larger than $100\ \mu m$ will certainly not achieve seasonal densification. If densification is to occur simultaneously with deposition, an initial grain diameter of $< 0.3\ \mu m$ is required. The proposed layer is polycrystalline, the final grain diameters being 10–30 μm. In the north this layer is currently forming, whereas in the south it is the remnant of the layer formed during the last winter. The presence of such a layer offers a natural explanation for most of what we see on the surface of Triton: the 2.16–μm absorption, the apparent smoothness at visible and infrared wavelengths, the albedo contrast between north and south, and the plumes.

101.086 **Search for glazed surfaces on Triton.**
P. Lee, P. Helfenstein, J. Veverka.
J. Geophys. Res., Vol. 96, Suppl., p. 19231 – 19240 (30 Oct 1991). Special issue: Voyager 2 at Neptune.
Near–infrared reflectance spectra of Triton obtained from earth since 1978 shows several absorption bands between 0.8 and 4 μm, features attributed to CH_4 and N_2 and inferred to respect materials in the condensed state. Clark et al (1983) had previously proposed a frost grain metamorphism model for icy satellite surfaces, according to which freshly condensed CH_4 and N_2 crystals would rapidly reconfigure themselves into a glaze or grow to unusually large sizes under the pressure and temperature conditions prevailing on Triton. In either case, the long path lengths apparently needed to explain the prominence of the absorptions could be provided. More recently, the Voyager 2 flyby of Triton revealed the existence of topographically smooth frozen "lava lake" and active geyserlike plumes. The presence of the former, and the solid–state greenhouse mechanism proposed by Smith et al. (1989) and Brown et al. (1990) to power the latter, both suggest the likely existence of surficial clear ice. Thus smooth, even glazed areas of N_2 and CH_4 ice might occur on Triton. The authors report the results of a search, at the pixel scale, for such glazed areas in Voyager 2 high–resolution images of Triton.

101.087 **CH_4 and haze in Triton's lower atmosphere.**
F. Herbert, B. R. Sandel.
J. Geophys. Res., Vol. 96, Suppl., p. 19241 – 19252 (30 Oct 1991). Special issue: Voyager 2 at Neptune.
Voyager 2 ultraviolet spectrometer occultation observations at Triton have revealed two constituents of the troposphere: CH_4 and another absorber visible between 1400 and 1600 Å below about 20 km altitude. The CH_4 appears to be saturated at the surface at both entrance and exit occultation sites. The density scale height and wavelength dependence of the long–wavelength absorber are consistent with Rayleigh scattering in N_2. However, the inferred N_2 column abundance is inconsistent with the Voyager radio science (RSS) occultation measurement, and so the N_2 hypothesis is discarded. The most attractive alternative hypothesis for the identity of the long–wavelength opacity is

Rayleigh scattering by an aerosol haze. EUV Rayleigh scattering requires that the haze particles be very small ($\leqslant 0.03$ μm) and abundant. These characteristics suggest formation as photochemical smog or condensing N_2 with a great abundance of nucleation centers. Condensation onto ions allows rapid formation of very small particles, so possibly ions created by penetrating magnetospheric charged particles (0.1 to several MeV) are nucleating the haze. Small particle size also promotes long residence times against settling, essential for accumulatin to significant abundance and suggesting the eddy transport dominates settling. If this is the case, the haze scale height likely approximates the gas scale height, and the implied temperature is around 38K. This is lower than the 50K estimated by the RSS and is also too low for the dust devil hypothesis for the visible atmospheric plumes.

101.088 **Voyager observations of Nereid.**
P. Thomas, J. Veverka, P. Helfenstein.
J. Geophys. Res., Vol. 96, Suppl., p. 19253 – 19260 (30 Oct 1991). Special issue: Voyager 2 at Neptune.

The authors have analyzed observations of Nereid obtained by Voyager 2 over a 12–day interval at solar phase angles of 25°–96° to determine the satellite's size and basic phtometric properties. Nereid's radius is found to be 170 ± 25 km: its geometric albedo ($\lambda = 0.48$ μm) exceeds 0.16 and is probably close to 0.2. There is no evidence of rotation light curve in the Voyager observations, any light curve amplitude being definitely less than 0.15 mag (total amplitude). Hapke's photometric model yields a geometric albedo of about 0.2, a phase intergral of 0.5, and a Bond albedo of 0.1. In terms of albedo, color, and photometric properties, Nereid is similar to some of the icy satellites of Uranus but is quite distinct from the group of six small inner satellites of Neptune discovered by Voyager. Although these satellites share Nereid's neutral color, they are much darker and have much higher phase coefficients. The authors conclude that Nereid is unlike these "carbonaceous" objects and is probably a "dirty" icy body. The satellite's spin state (period and spin axis direction) cannot be determined from the Voyager data.

101.089 **Neptune's small inner satellites.**
P. Thomas, J. Veverka.
J. Geophys. Res., Vol. 96, Suppl., p. 19261 – 19268 (30 Oct 1991). Special issue: Voyager 2 at Neptune.

Voyager 2 images revealed six small satellites orbiting within five planetary radii of Neptune. The largest, Proteus (1989N1), has a mean radius of 208 ± 8 km; the smallest, Naiad (1989N6), is 29 ± 6 km in radius. The four satellites for which resonable radius measurements can be made all have geometric albedos of 0.06 at 0.48 μm. The albedos and phase coefficients of these satellites are very smilar to those measured for the inner satellites of Uranus but are quite distinct from the values of Nereid ($p \approx 0.16$–0.20, $\beta = 0.028$ magnitude/degree). Color measurements (0.4 to 0.6 μm) of Proteus show a neutral color. The two largest satellites are resolved as irregular objects and have limb roughness comparable to those of similar–sized satellites of other planets. One image of Proteus shows a 150–km crater with some overlapping small craters; this feature allows measurements of a rotation rate consistent with a synchronous period. The irregular shape and heavily cratered, photometrically bland surface of Proteus are consistent with characteristics of a primitive, undifferentiated object. Evidently, Proteus has escaped the tidal heating believed to have been responsible for the remarkable geologic processes on the similar–sized bodies Miranda and Enceladus and perhaps Mimas.

101.090 **A control network of Triton.**
M. E. Davies, P. G. Rogers, T. R. Colvin.
J. Geophys. Res., Vol. 96, No. E1, p. 15675 – 15682 (25 Aug 1991).

A control network for Triton has been computed using a bundle–type analytical triangulation program. The network contains 105 points that were measured on 57 Voyager 2 pictures. The adjustment contained 1010 observation equations and 382 normal equations and resulted in a standard measurement error of 13.36 μm. The authors determined coordinates of the control points, the camera orientation angles at the times when the pictures were taken, and Triton's mean radius. A separate statistical analysis confirmed Triton's radius to be 1352.6 ± 2.4 km. Attempts to tie the control network around the satellite were unsuccessful because discontinuities exist in high–resolution coverage between 66° and 289° longitude, north of 38° latitude, and south of 78° latitude.

101.091 **Whistler waves associated with the Uranian bow shock: outbound observations.**
C. W. Smith, H. K. Wong, M. L. Goldstein.
J. Geophys. Res., Vol. 96, No. A9, p. 15841 – 15852 (1 Sep 1991).

The encounter of the Voyager 2 spacecraft with the Uranian planetary system led to a series of outbound crossings of the Uranian bow shock between January 27 and January 30 of 1986. Examination of magnetic field data recorded in close proximity to the shock reveals a series of whistler wave events that appear to result from processes associated with the shock. Two wave events display two separate and simultaneous wave enhancements each, one at approximately 0.2 Hz and the second more nearly at 1 Hz in the spacecraft reference frame. The actual frequencies of the waves vary with the two events. Both wave enhancements of the two events are right–hand polarized in the spacecraft frame of reference. The authors have examined these wave events using high–resolution magnetic field data and conclude that they are analogous to those whistler waves upstream of the Earth's bow shock that are driven by beams of electrons.

101.092 **Monte Carlo simulation of photoelectron energization in parallel electric fields: electroglow on Uranus.**
R. P. Singhal, A. Bhardwaj.
J. Geophys. Res., Vol. 96, No. A9, p. 15963 – 15972 (1 Sep 1991).

A Monte Carlo simulation of photoelectron energization and energy degradation in H_2 gas in the presence of parallel electric fields has been carried out. Numerical yield spectra which contain information about the electron energy degradation process and can be used to calculate the yield for any inelastic event are obtained. The variation of yield spectra with incident electron energy, electric field, pitch angle, and cutoff limit has been studied. The yield function is employed to determine the photoelectron fluxes. H_2 Lyman and Werner band excitation rates and integrated column intensity are computed for three different electric field profiles taking various low–energy cutoff limits. It is found that an electric field profile with peak value of 4 mV/m at neutral number density of $3 \times 10^{10} cm^{-3}$ produces enhanced volume emission rates of H_2 bands ($\lambda < 1100$ Å) explaining about 20% of the observed electroglow emission on Uranus. The effect of solar zenith angle and solar cycle variation on peak excitation rate is discussed.

101.093 **Neptune.**
IAU Circ., No. 5331, p. 1 (31 Aug 1991).

101.094 **Equatorial waves in the stratosphere of Uranus.**
D. P. Hinson, J. A. Magalhães.
Icarus, Vol. 94, No. 1, p. 64 – 91 (Nov 1991).

The authors have identified and characterized an equatorial wave in the stratosphere of Uranus through analysis of radio occultation data from Voyager 2. The methodology relies on two physical phenomena: (1) atmospheric waves modulate the ambient background structure of a planetary atmosphere, producing a distinctive pattern of spatial variations in temperature and density; and (2) this type of spatial modulation in turn causes fluctuations in the phase and amplitude of radio signals received from an occulted spacecraft. The authors compared predictions of the linear theory for equatorial waves with observations of the stratosphere at both immersion and emersion. The observed quasi–periodic vertical variations in atmospheric density agree closely with theoretical predictions for a wave propagating vertically through the observed background structure of the stratosphere, providing compelling evidence for the presence of an atmospheric wave. Quantitative comparisons between the measurements at immersion and emersion yielded constraints on the meridional and zonal structure of the wave. The ratio of amplitudes at the two locations is consistent with a wave confined in latitude near the equator. Moreover, the observations at immersion and emersion are correlated, suggesting a wave of planetary scale. These constraints allowed one to identify two equatorial wave modes as equally likely alternatives for explaining the observations.

101.095 **A search for distant satellites of Pluto.**
S. A. Stern, R. A. Fesen, E. S. Barker, J. W. Parker, L. M. Trafton.
Icarus, Vol. 94, No. 1, p. 246 - 249 (Nov 1991).

The stability domain for satellite orbits around Pluto extends ≈100 times further than the radius of Charon's orbit. The authors report here the results of a deep CCD imaging search for faint, distant satellites of Pluto. The search was made in the region extending from 6 arcsec from Pluto to beyond the edge of Pluto's stability domain, during oppositions in 1990 and 1991. No detections were made, and the authors conservatively set 90%-confidence upper limits on the absence of potential satellites brighter than 20.6 ± 0.5 m in the region 6 to 10 arcsec from Pluto and brighter than 22.6 ± 0.5 m in the region from 10 arcsec to the edge of Pluto's stability domain. At the 90% confidence level this allows us to rule out the presence of any object with Charon-like albedo with a radius $> 58 \pm 19$ km in the region 6 - 10 arcsec from Pluto.

101.096 **The Uranian system: an overview.**
J. T. Bergstralh, E. D. Miner.
Uranus Colloquium, p. 3 - 25 (1991). - See Abstr. 012.077 for the main entry.

The Uranian system is a unique and important element in our understanding of the origin and evolution of the solar system. Its atmospheric composition offers additional cosmogonic clues. The atmospheric structure reveals family resemblances with the other giant planets, most notably in the form of unexpectedly high exospheric temperatures. The discovery of 9 narrow, dark rings around Uranus in 1977 was a great surprise, and provided our first inkling that ring systems are the rule, not the exception, for the giant planets. The satellites of Uranus reveal a surprisingly diverse range of surface characteristics for such small icy bodies. The Voyager encounter returned the first reliable measurements of their masses and radii, and therefore their bulk densities. Uranus possesses a strong (~1 gauss) magnetic field with a bizarre geometry. This implies that at least part of the interior is a convecting fluid, further implying an adiabatic temperature profile and a warm interior.

101.097 **Models of Uranus' interior and magnetic field.**
M. Podolak, W. B. Hubbard, D. J. Stevenson.
Uranus Colloquium, p. 29 - 61 (1991). - See Abstr. 012.077 for the main entry.

Our understanding of Uranus' interior structure comes from the results of theoretical models. The authors review the methods for computing such models. They discuss the various physical assumptions employed, and assess their strengths and weaknesses. The accretion theory of the formation of the planets is summarized and the implications for the structure and composition of Uranus are presented. The authors then discuss the results of the most recent modeling efforts. Finally, the chemistry of Uranus' atmosphere and deep interior is reviewed in the context of these models, and the implications of these models for theories of the formation of Uranus are discussed.

101.098 **The upper atmosphere of Uranus.**
D. F. Strobel, R. V. Yelle, D. E. Shemansky, S. K. Atreya.
Uranus Colloquium, p. 65 - 109 (1991). - See Abstr. 012.077 for the main entry.

The upper atmosphere of Uranus is predominantly H_2. The helium abundance in the upper atmosphere is not well known, but is probably less than 10% by volume. Atomic hydrogen is the dominant constituent in the exosphere. The thermosphere is warm, reaching an asymptotic temperature of ~800K. The dynamo hypothesis can provide in situ particle acceleration and excitation at the pressure levels where limb observations indicate that ultraviolet emission originates. An extended ionosphere was detected.

101.099 **Photochemistry and vertical mixing.**
S. K. Atreya, B. R. Sandel, P. N. Romani.
Uranus Colloquium, p. 110 - 146 (1991). - See Abstr. 012.077 for the main entry.

The only known photochemically active volatile in the atmosphere of Uranus is methane. The photolysis of methane produces primarily ethylene, ethane and acetylene. The latter two species condense near the tropopause. In addition, formation and subsequent condensation of polyacetylenes is likely near the 0.1 mbar level. Transport of the hydrocarbon ices to the deeper troposphere, followed by evaporation or pyrolysis and high-pressure chemistry and subsequent convection appears to be responsible for the stability of methane in the upper atmosphere. The vertical mixing on Uranus, as characterized by the eddy diffusion coefficient, is sluggish. This chapter discusses theory, observations, and their interpretation in terms of the relevant photochemistry and atmospheric mixing on Uranus.

101.100 **Spectroscopy and chemistry of the atmosphere of Uranus.**
B. Fegley Jr., D. Gautier, T. Owen, R. G. Prinn.
Uranus Colloquium, p. 147 - 203 (1991). - See Abstr. 012.077 for the main entry.

The authors give a comprehensive review of the chemistry and spectroscopy of the atmosphere of Uranus. Earth-based, Earth-orbital, and Voyager 2 observations covering the ultraviolet, visible, infrared and radio wavelength regions are discussed. Chemical models imply that both equilibrium and non-equilibrium (e.g., photochemistry and vertical mixing) processes are important on Uranus. In particular, vertical mixing is predicted to be an important source of N_i in the troposphere of Uranus.

101.101 **Thermal structure and energy balance of Uranus.**
B. J. Conrath, J. C. Pearl, J. F. Appleby, G. F. Lindal, G. S. Orton, B. Bézard.
Uranus Colloquium, p. 204 - 252 (1991). - See Abstr. 012.077 for the main entry.

The basic properties of the atmospheric temperature field of Uranus have been established through a combination of Earth-based and Voyager measurements. Stellar occultation observations indicate both spatial and temporal variability at μbar pressure levels. Voyager radio occultation and infrared measurements as well as Earth-based full-disk infrared observations have established the tropospheric and stratospheric vertical structure. The latitude dependence of the upper tropospheric temperatures has been determined from Voyager infrared measurements; remarkably little contrast is found. Voyager measurements of reflected solar energy and planetary thermal emission have established new constraints on the global energy balance of Uranus. Analyses of the data carried out thus far have established an upper limit on the internal heat flux of 14% of the absorbed solar flux.

101.102 **Uranus atmospheric dynamics and circulation.**
M. Allison, R. F. Beebe, B. J. Conrath, D. P. Hinson, A. P. Ingersoll.
Uranus Colloquium, p. 253 - 295 (1991). - See Abstr. 012.077 for the main entry.

Uranus, with its 98° obliquity, apparently negligible internal heat source and methane-enriched, hydrogen-helium atmosphere, is a unique study in planetary meteorology. The available data indicate atmospheric rotation at midlatitudes nearly 200 m s^{-1} faster than that of the planetary magnetic field (presumably tied to the deep interior). While the speed of the differential motions is comparable to that on Jupiter and Saturn, the inferred shape of the zonal wind profile bears a surprising resemblance to the terrestrial circulation. Despite the current uncertainties, the planet's unique external forcing and at least superficially simple flow structure (compared with the other

Jovian planets) make the Uranian meteorology an important target for further observational and theoretical investigation.

101.103 **Clouds and aerosols in the Uranian atmosphere.**
R. A. West, K. H. Baines, J. B. Pollack.
Uranus Colloquium, p. 296 – 324 (1991). – See Abstr. 012.077 for the main entry.

Properties of at least three types of cloud and haze layers in the Uranian atmosphere are becoming evident as a result of data accumulted from ground-based instruments, the IUE satellite, and Voyager experiments. Studies of groundbased and IUE spectrophotometry data suggested an optically thin stratospheric haze, an optically thicker ($0.4 \leqslant \tau \leqslant 1$) but surprisingly thin (considering the CH_4 abundance) methane cloud near or above the 2-bar level, and a deeper cloud top of unknown composition near the 3-bar level. Photometric observations of Uranus obtained by the imaging and photopolarimeter instruments on the Voyager 2 spacecraft have significantly refined this picture of the atmospheric structure.

101.104 **Dynamics and structure of the Uranian rings.**
R. G. French, P. D. Nicholson, C. C. Porco, E. A. Marouf.
Uranus Colloquium, p. 327 – 409 (1991). – See Abstr. 012.077 for the main entry.

Most of the ten known narrow rings are eccentric and inclined, and the γ and δ rings have self-excited normal modes. Several of the rings have a close connection with two small satellites, Cordelia and Ophelia, which straddle the ε ring and provide strong evidence for ring confinement by shepherds. The rings are remarkably diverse in their radial structure and azimuthal variations, and a detailed comparison is presented of high-resolution ring profiles obtained from the Voyager 2 radio and stellar occultation observations. The central elements of the dynamics of ring confinement by shepherding are presented.

101.105 **Particle properties and processes in Uranus' rings.**
L. W. Esposito, A. Brahic, J. A. Burns, E. A. Marouf.
Uranus Colloquium, p. 410 – 465 (1991). – See Abstr. 012.077 for the main entry.

The Uranian rings provide some confirmations but also many new challenges for ring physics. Current analytic and numerical models can explain many details, but must be extended to be fully realistic. Occultation observations of the ε ring are interpreted to yield an effective size of the ring particles that exceeds 70 cm, a surface mass density that exceeds 80 g cm^{-2} and a ring vertical thickness greater than tens of meters for solid ice particles. As proposed for regions of the other known ring systems, new ring material can be continually created by the destruction of small moons near the rings. This phenomenon may explain the youthful appearance of the Uranian rings. Otherwise, the confinement of the rings over the age of the solar systems requires some strong mechanism which is yet to be understood.

101.106 **Origin of the Uranian satellites.**
J. B. Pollack, J. I. Lunine, W. C. Tittemore.
Uranus Colloquium, p. 469 – 512 (1991). – See Abstr. 012.077 for the main entry.

The satellites of Uranus formed within a disk of material that surrounded this planet during its early history, just as the regular satellites of the other giant planets formed from analogous circumplanetary disks. Sources for these disks include the primordial solar nebula and the envelopes of the outer planets. The authors describe four alternative models for the genesis of the satellite-forming disk of Uranus and compare their predictions with relevant physical, dynamical and chemical properties of the Uranian system, as well as those of the Jovian and Saturnian systems, to discriminate among them. After the Uranian satellites formed, tidal interactions between the larger ones and Uranus caused their orbits to change significantly over the age of the solar system, resulting, in some cases, in significant tidal heating over restricted periods of time.

101.107 **Physical properties of the Uranian satellites.**
R. H. Brown, T. V. Johnson, S. Synnott, J. D. Anderson, R. A. Jacobson, S. F. Dermott, P. C. Thomas.
Uranus Colloquium, p. 513 – 527 (1991). – See Abstr. 012.077 for the main entry.

The Voyager encounter has provided a wealth of information regarding the satellites' radii, masses, mean densities and consequently their internal structures. The authors review the latest results concerning the physical properties of the satellites of Uranus. Topics covered are the sizes, shapes, topography, masses, densities and models of internal structures of the five major satellites. For the 10 small satellites discovered by Voyager 2, the authors discuss their sizes and shapes. They then compare the physical properties of the large satellites of Uranus to those other satellites in the outer solar system, particularly those of Jupiter and Saturn, and discuss the implications that these comparisons have for understanding the origin and evolution of the satellites of Uranus.

101.108 **Uranus satellites: surface properties.**
J. Veverka, R. H. Brown, J. F. Bell.
Uranus Colloquium, p. 528 – 560 (1991). – See Abstr. 012.077 for the main entry.

The authors review the post-Voyager knowledge of the photometric, calorimetric, spectral and thermal properties of the satellites of Uranus. While albedo variations of at least a factor of 2 exist, color differences (within the limited spectral range of the Voyager cameras) are almost absent (Miranda) or subdued (Oberon). In the case of Titania, the strong opposition effect reported by groundbased observers was confirmed. Voyager did not observe the opposition parts of the phase curves of the other satellites. Voyager thermal observations of Ariel and Miranda suggest that both have highly porous regoliths, thermophysically similar to those of Jupiter's icy satellites. The Voyager data do provide some useful constraints: there is strong evidence that albedo patterns on the satellites are related to geology, and little if any that they are correlated with charged particle dose.

101.109 **Geology of the Uranian satellites.**
S. K. Croft, L. A. Soderblom.
Uranus Colloquium, p. 561 – 628 (1991). – See Abstr. 012.077 for the main entry.

Six of the satellites of Uranus were seen in sufficient detail by Voyager 2 to allow geologic analysis. All of the satellites except the smallest, Puck, show evidence of cryovolcanic resurfacing: global on the largest four satellites, local in the spectacular coronae on Miranda. The only evidence of compressional deformation on the Uranian satellites is a set of ridges on Titania. As a group, the Uranian satellites are somewhat more active geologically than similarly sized Saturnian satellites. This may be due to the presence of volatile ices such as NH_3, CH_4, N_2, etc., which would then by inference be largely absent in the Saturnian satellites.

101.110 **Cratering of the Uranian satellites.**
W. B. McKinnon, C. R. Chapman, K. R. Housen.
Uranus Colloquium, p. 629 – 692 (1991). – See Abstr. 012.077 for the main entry.

The satellites of Uranus exhibit various crater populations that provide evidence concerning the history of the impactor flux in the Uranus system over time. Relative absences of craters suggest that resurfacing or other crater removal processes operated on three of the satellites during more recent epochs. The authors summarize available crater counts and their interpretations. The authors find that the published crater counts on the Uranian satellites, despite mutual inconsistencies, can be interpreted as compatible with cratering by the heliocentric population of cometary bodies that was responsible for much of the cratering of the satellites of Jupiter and Saturn. There may also be a component due to secondary debris from cratered or disrupted Uranian moons.

101.111 **Miranda.**
R. Greenberg, S. K. Croft, D. M. Janes, J. S. Kargel, L. A. Lebofsky, J. I. Lunine, R. L. Marcialis, H. J. Melosh, G. W. Ojakangas, R. G. Strom.
Uranus Colloquium, p. 693 – 735 (1991). – See Abstr. 012.077 for the main entry.

Miranda is the innermost and smallest of the major Uranian satellites, so it was long expected to be simply a cratered snowball, devoid of endogenically derived surface features. Drawing on the observed geology, photometry and geophysical data, the authors examine various processes and properties that may have contributed to Miranda's evolution. The coronae, unique to Miranda, may have formed by relaxation of topographic highs, by lithospheric stress driven by density anomalies in the asthenosphere, or by diapirs either breeching the surface or feeding large–scale volcanic flooding through pre–existing crack structure. Current data are insufficient to choose among the competing scenarios, and there may have been significant differences in the process for each corona even if all have fundamentally similar origins. Miranda remains tantalizingly enigmatic, but comparison with other icy moons defines the variety of physical and chemical processes involved in their formation and evolution.

101.112 **The magnetic field and magnetospheric configuration of Uranus.**
N. F. Ness, J. E. P. Connerney, R. P. Lepping, M. Schulz, G.–H. Voigt.
Uranus Colloquium, p. 739 – 779 (1991). – See Abstr. 012.077 for the main entry.

The Voyager 2 measurements at Uranus have revealed a remarkable planetary magnetic field and magnetospheric configuration. The magnetic center was found to be displaced by 0.3 R_U from the planetary center, this being a significantly larger fraction of the planetary radius than are the offsets for Earth, Jupiter and Saturn. The stagnation–point distance to the solar–wind magnetopause boundary was found to be 18 R_U. This means that all five major satellites of Uranus lie deep within the trapped radiation belts. The large angular separation between the dipole and rotation axes, in combination with the equatorial orbits of the satellites, results in the complex and dynamic radiation–belt structure observed by Voyager 2. The large inclination of the Uranian rotation axis leads to a special magnetic–tail configuration at this epoch.

101.113 **The plasma environment of Uranus.**
J. W. Belcher, R. L. McNutt Jr., J. D. Richardson, R. S. Selesnick, E. C. Sittler Jr., F. Bagenal.
Uranus Colloquium, p. 780 – 830 (1991). – See Abstr. 012.077 for the main entry.

In the course of the Voyager 2 encounter with Uranus, an extended magnetosphere filled with a tenuous plasma was detected. This low–energy plasma consists of protons and electrons, with no significant heavy ion contribution, and with a density in the regions sampled by the spacecraft of at most three electrons per cubic centimeter. The plasma electrons and ions exhibit both a thermal component (with temperatures of tens of eV) and a hot component (with temperatures of a few keV). The source of the thermal component of the plasma is either the planetary ionosphere or the neutral hydrogen corona surrounding Uranus, whereas the hot component is convected in from the magnetotail, with probably an ionospheric source. Uranus also possesses a well–developed magnetotail and plasma sheet similar in many respects to those of the Earth. The bow shock observed inbound is a high Mach number quasi–perpendicular shock and shows detailed structure in the transition region similar to that seen at the Earth.

101.114 **Energetic particles at Uranus.**
A. F. Cheng, S. M. Krimigis, L. J. Lanzerotti.
Uranus Colloquium, p. 831 – 893 (1991). – See Abstr. 012.077 for the main entry.

This chapter reviews the energetic particle measurements by the Low Energy Charged Particles and Cosmic Ray instruments on the Voyager 2 spacecraft in the magnetosphere of Uranus. Upstream ion events were observed outside the Uranian bow shock, probably produced by ion escape from the magnetosphere. However, no energetic neutral particle fluxes were observed. The magnetospheric composition is totally dominated by protons, with only a trace abundance of H_2^+ and no evidence for He or heavy ions; the Uranian atmosphere must be the principal plasma source. These results and plasma wave data suggest the existence of structures analogous to Earth's plasmasphere and plasmapause. The Uranian magnetosphere is the most Earth–like of the giant planet magnetospheres so far studied.

101.115 **Uranus as a radio source.**
M. D. Desch, M. L. Kaiser, P. Zarka, A. Lecacheux, Y. LeBlanc, M. Aubier, A. Ortega–Molina.
Uranus Colloquium, p. 894 – 925 (1991). – See Abstr. 012.077 for the main entry.

The morphology of the Uranus radio emission is complex, with at least 7 distinct components, each manifesting different characteristics in the frequency and time domains. The active radio regions of the planet are located above the magnetic poles, in the magnetic equator region near the planet, and in the density cavity between the magnetopause and the dense inner magnetosphere. Analysis of the radio emissions had led to (1) an accurate measure of the internal rotation period of the planet; (2) information on local plasma conditions and on the structure of the downstream magnetopause; (3) additional evidence for a substorm–like response of the magnetosphere to the solar wind; and (4) information on dayside and nightside ionospheric electron densities. Uranus is the fourth planet known to emit low–frequency radio waves and the fifth thought to have lightning.

101.116 **Wave–particle interactions in the magnetosphere of Uranus.**
W. S. Kurth, D. A. Gurnett, F. L. Scarf, F. V. Coroniti.
Uranus Colloquium, p. 926 – 958 (1991). – See Abstr. 012.077 for the main entry.

The Voyager 2 encounter of Uranus provided observations of plasma waves in and near the magnetosphere. The observations include electrostatic waves upstream of the bow shock, turbulence in the shock, Bernstein emissions and whistler–mode waves in the magnetosphere, broadband electrostatic noise in the magnetotail, and many other types of plasma waves which have yet to be clearly identified. Wave–particle interactions provide the channels through which the waves can accelerate, scatter or thermalize the plasmas. The most spectacular example in the case of Uranus is the extremely intense whistler–mode activity in the inner magnetosphere which is the source of strong pitch–angle diffusion. The resulting electron precipitation is sufficient to produce the auroral emissions observed by Voyager.

101.117 **Icy clues to Triton's origins.**
W. B. McKinnon.
Nature, Vol. 354, No. 6353, p. 431 (12 Dec 1991).

The author reports on results from a meeting of the Division for Planetary Sciences of the American Astronomical Society, Palo Alto, CA (USA), 4 – 8 Nov 1991.

101.118 **An explanation for both the large inclination and eccentricity of the dipole–like field of Uranus and Neptune.**
S.–I. Akasofu, L.–H. Lee, T. Saito.
Planet. Space Sci., Vol. 39, No. 9, p. 1259 – 1262 (Sep 1991).

It is shown that the offset tilted dipole model of Uranus and Neptune, deduced from the spherical harmonics analysis of the Voyager magnetic field observation, can be represented fairly well by the combined field of an axial (parallel or anti–parallel to the rotation axis) and an auxiliary dipole; the latter is roughly oriented in the East–West direction and is located near the surface of the core in low latitude. The present dynamo theories of planetary magnetism consider an axial dipolar field as an essential element, since the planetary rotation plays a vital role in the dynamo process. On the other hand, the auxiliary dipoles may be a result of leakage of the toroidal field, like a pair of sunspots on the photosphere, which is also an essential part of the dynamo process. Thus, the authors' suggestion has a sound physical basis and may throw some light on the great puzzle of the large inclination and eccentricity of the dipole-like field of Uranus and Neptune. This idea may also be applicable to some pulsars.

101.119 **Triton's atmosphere: a source of N and H for Neptune's Magnetosphere.**
M. E. Summers, D. F. Strobel.
Geophys. Res. Lett., Vol. 18, No. 12, p. 2309 - 3312 (Dec 1991).
Mass loading of the Neptunian magnetosphere occurs primarily by thermal escape of H, H_2, and N from Triton's upper atmosphere. The global escape rate of hydrogen is $\sim 7x10^{25}s^{-1}$, determined by the global average methane photolysis rate, whereas the escape rate of nitrogen for the authors' preferred model is $\sim 3.4x10^{25}s^{-1}$, and is controlled by the global and orbital average energy deposition rate due to precipitating magnetospheric electrons. The escape rate of H^+ and N^+ is $<4\%$ of the neutral escape rate and implies that mass loading of the Neptunian magnetosphere is not localized to Triton's corona. The ratio of hydrogen to nitrogen escape rates for the authors' preferred model is $\sim 2:1$, comparable to the $[H^+]/[N^+]$ abundance ratio inferred for Neptune's magnetosphere.

101.120 **Planet X: going, going... but not quite gone.**
R. Matthews.
Science, Vol. 254, No. 5037, p. 1454 - 1455 (6 Dec 1991).
The case for a tenth planet comes under fire at an astronomical conclave, but a handful observers aren't giving up just yet.

101.121 **The mass of Neptune.**
D. A. Pierce.
Icarus, Vol. 94, No. 2, p. 413 - 419 (Dec 1991).
Estimation of the mass of Neptune have ranged from 6037 to 20,039. This paper is a discussion of all the attempts (discovered by the author) to determine the mass of the planet Neptune, from the prediscovery estimations of Adams and Le Verrier in 1845 and 1846 to the recent accurate determination made possible by the 1989 Neptune flyby of Voyager 2. It is shown that the two fundamental mass determination methods - from planetary perturbations and from satellite motions - yield results of nearly equal accuracy for the mass of Neptune.

101.122 **The remote planets: 1987 - 1990 report.**
R. W. Schmude Jr.
Strolling Astron., Vol. 35, No. 4, p. 153 - 156 (Dec 1991).

101.123 **New aspects of the binary planet Pluto–Charon.**
V. Burwitz, K. Reinsch, M. W. Pakull, P. Bouchet.
Messenger, No. 66, p. 23 - 26 (Dec 1991).

101.124 **The angular momentum of the Pluto–Charon system: considerations about its origin.**
G. Tancredi, J. A. Fernández.
Icarus, Vol. 93, No. 2, p. 298 - 315 (Oct 1991).
The formation conditions of the binary system Pluto–Charon in the protoplanetary disk are analyzed. Particular attention is paid to its high content of angular momentum that would make the system rotationally unstable in the ideal case that Pluto and Charon coalesced into a single body. Among several possible formation scenarios, the acquisition of such a large amount of angular momentum is explained as the result of the accumulation process of planetesimals in heliocentric orbits with masses up to a few tenths of the mass of the proto–Pluto itself. Numerical simulations of accreting planetesimals show that a significant fraction (5 - 20%) of the Pluto–sized bodies formed reach rotational instability. Under these circumstances, the Pluto–Charon system may have already formed before it acquired its highly eccentric and inclined orbit.

101.125 **Influence of hypothetical Transneptunian planets on certain solar system bodies.**
M. V. Bazhkova, E. N. Makarova, Yu. D. Medvedev.
Sol. Syst. Res., Vol. 25, No. 3, p. 228 - 230 (Nov 1991). English translation of Astron. Vestn., Tom 25, No. 3, p. 309 - 311 (1991).
The perturbations of the hypothetical planets proposed by Brady, Guliev, and Radzievskij on Comet Halley and the outer planets Uranus and Neptune are calculated by numerical integration. The results indicate that the probability of existence of the planets of Brady and Guliev is small, but the existence of Radzievskij's planet is not ruled out by the available observations of Uranus, Neptune, and Comet Halley.

101.126 **The theory of Neptunian arcs. I. Stability of single epiton.**
N. N. Gor'kavyj, T. A. Tajdakova, N. M. Gaftonyuk.
Pis'ma Astron. Zh., Tom 17, No. 12, p. 1105 - 1115 (Dec 1991). In Russian. English translation in Sov. Astron. Lett., Vol. 17, No. 6.
The fundamental problem of the arc dynamics is: why do they not disperse along the orbit? Gor'kavyj (1989) suggested that arc is a chain of epicyclic vortices in the complete ring. The authors test this model with a direct numerical simulation, which takes into account inelastic collision between identical spherical particles. It is shown that the vortex may be regarded as a stable formation if there exists a natural mechanism which arranges the particles of the vortex in a collisionless order.

101.127 **The theory of Neptunian arcs. II. Dynamics of epitons in the ring.**
N. N. Gor'kavyj, T. A. Tajdakova.
Pis'ma Astron. Zh., Tom 17, No. 12, p. 1116 - 1123 (Dec 1991). In Russian. English translation in Sov. Astron. Lett., Vol. 17, No. 6.
The authors study the dynamics of epicyclic vortices (epitons) plunged into a dust ring. The numerical calculation which takes into account the vortex selfgravitation has shown that the epiton is compressed in the dust flow and drifts along the radius towards a most dense part of the ring. This means the stability of the vortex position in the dust ring. It is shown that the formation of the epiton chain is due to the interaction of the selfgravitating vortices produced by dust particles of the ring.

101.128 **Particle sizes of the Uranus delta ring's inner diffuse companion through comparison of RSS and PPS Voyager occultation data.**
J. Hui, L. J. Horn, A. L. Lane.
Icarus, Vol. 93, No. 2, p. 347 - 353 (Oct 1991).
The Voyager 2 spacecraft encountered the Uranian system in January 1986. Several occultations of the Uranian rings were observed including a radio science (RSS) earth occultation and two stellar occultations. The photopolarimeter (PPS) and ultraviolet spectrometer (UVS) observed the stars σ Sagittarii and β Persei as they were occulted by the δ ring. An inner diffuse companion of the δ ring was detected in the RSS data taken at $\lambda = 3.6$ cm and in both the PPS and the UVS σ Sagittarii occultation data taken at $\lambda = 0.26$ and $\lambda = 0.11$ μm, respectively. The companion has also been observed in Earth–based stellar occultation observations. Using the PPS σ Sagittarii and RSS data, the authors found the inner diffuse companion of the δ ring to have an average width of 12 km with an average RSS equivalent depth of 0.60 ± 0.080 km and an average PPS equivalent depth of 0.36 ± 0.054 km. From comparing the width and equivalent depths between the two sets of data, the authors found that the particles that contribute the most to the integrated opacities of the companion are particles which have sizes in the several centimeter or greater regime.

101.129 **A new spectrum of Triton near the time of the Voyager encounter.**
W. M. Grundy, U. Fink.
Icarus, Vol. 93, No. 2, p. 379 - 385 (Oct 1991).
The authors present a new spectrum of Triton from 5200 to 10,000 Å in which the 8900–Å methane ice absorption band can

be clearly seen and accurately measured. The data were obtained during the summer of 1989, just before the Voyager II spacecraft encountered the Neptune system. By combining data from the spacecraft with the telescopic data, the authors have shown that the absorption is entirely caused by CH_4 ice on Triton's surface. They have modeled Triton's spectrum with a simple Hapke-type model which allowed them to set a 20-μm lower limit on the mean grain size of methane ice on Triton.

Uranus. Invited reviews.
See Abstr. 012.077.

Off-center point explosion in a radially stratified medium – Kompaneets approximation.
See Abstr. 022.029.

The absorption coefficient of the liquid N_2 2.15-μm band and application to Triton.
See Abstr. 022.064.

Symplectic maps for the N-body problem.
See Abstr. 042.049.

Secular perturbations of fictitious satellites of Uranus.
See Abstr. 042.092.

The Voyager encounter with Neptune.
See Abstr. 051.053.

Low-frequency waves in the solar wind near Neptune.
See Abstr. 074.085.

Sources of error in the photoclinometric determination of planetary topography: a reappraisal.
See Abstr. 091.023.

Streamline formalism and ring orbit determination.
See Abstr. 091.037.

What is new in the world of giant planets?
See Abstr. 099.038.

CCD spectroscopic observations of Saturn, Uranus, Neptune, and Titan during the 1990 apparitions.
See Abstr. 100.008.

Thermal stress tectonics on the satellites of Saturn and Uranus.
See Abstr. 100.018.

Satellites of Saturn and Neptune.
See Abstr. 100.020.

Satellites of Saturn and Neptune.
See Abstr. 100.025.

102 Comets (Origin, Structure, Atmospheres, Dynamics)

102.001 **Discussion of plasma instability between solar wind and type-I comet tail.**
Guo She-yu, Li Zhong-yuan.
Chin. J. Space Sci., Vol. 11, No. 3, p. 187 – 192 (Jul 1991). In Chinese.

The "two-stream" plasma instability between the solar wind and a comet tail in a force-free field is discussed. The data come from the exploration for comet Giacobini-Zinner by the International Cometary Explorer Spacecraft. The conclusion shows that a "two stream" instability exists in the inner part of a comet if a resonable plasma distribution is considered.

102.002 **Two possible acceleration mechanisms in comet tail.**
Li Zhong-yuan, Guo She-yu, Tong Yi.
Kexue Tongbao, Vol. 36, No. 17, p. 1453 – 1456 (Sep 1991).

102.003 **Quasilinear diffusion rates of cometary ions.**
P. H. Yoon, L. F. Ziebell.
Phys. Fluids, B, Vol. 3, No. 8, p. 2124 – 2132 (Aug 1991). Current Physics Microform No.: 9109A1869.

The freshly created ions in the upstream region of the cometary bow shock usually form a ring-beam velocity distribution in the solar wind frame of reference. These newborn ions subsequently excite a variety of instabilities which in turn cause the ions to diffuse in pitch angle space to form a quasispherical shell velocity distribution. Such a process can be described by quasilinear kinetic theory. In the present paper, the detailed properties of the initial diffusion rates associated with the cometary newborn ions are studied for various physical parameters including the so-called injection angle α (that is, the average initial pitch angle associated with the ions) ranging from 0° (pure beam distribution) to 90° (pure ring distribution). It is shown that initial diffusion rates are strongly dependent upon the angle α, such that for $0° \leqslant \alpha \leqslant 60°$ (quasiparallel regime) the pitch angle diffusion rate remains relatively small, while for $60° \leqslant \alpha \leqslant 90°$ (quasiperpendicular regime) the diffusion rate increases substantially.

102.004 **The directional dependence of cometary magnetic energy density in the quasi-parallel and quasi-perpendicular regimes.**
R. H. Miller, T. I. Gombosi, S. P. Gary, D. Winske.
Adv. Space Res., Vol. 11, No. 9, p. 79 – 82 (1991). – See Abstr. 012.006 for the main entry.

The direction of propagation of low frequency magnetic fluctuations generated by cometary ion pick-up is examined by means of one-dimensional electromagnetic hybrid simulations. The newborn ions are injected at a constant rate, and the helicity and direction of propagation of magnetic fluctuations are explored for cometary ion injection angles of 0° and 90° relative to the solar wind magnetic field.

102.005 **An expansion of the secular and resonant parts of the disturbing function in the theory of motion of long-periodic comets.**
V. V. Emel'yanenko.
Pis'ma Astron. Zh., Tom 17, No. 9, p. 857 – 864 (Sep 1991). In Russian. English translation in Sov. Astron. Lett., Vol. 17, No. 5.

A method is proposed for the expansion of the disturbing function in the restricted circular three-body problem for cometary orbits. The formula is obtained for the development of functions in the elliptic motion as the power series of $\varepsilon = (1-e)/2$ (e is the eccentricity of the cometary orbit). For long-period comets the convergence region of these series reaches such large distances where planetary perturbations can be neglected. The formulae and the recurrence relations for the calculations of the secular and resonant parts of the disturbing function are found.

102.006 **Steady–state injection of short–period comets from the trans–Neptunian cometary belt.**
W.-H. Ip, J. A. Fernández.
Icarus, Vol. 92, No. 2, p. 185 – 193 (Aug 1991).
To examine the scenario of the existence of a trans–Neptunian comet belt as an important source region of the short–period comets, the dynamical perturbation on a system of such icy planetesimals by a few Mars– to Earth–sized planetoids in highly eccentric orbits is considered. In the context of the authors' planetoid–scattering model with three or four large planetoids orbiting in highly eccentric orbits, the total mass of the trans–Neptunian comet belt between the orbit of Neptune and 200 AU could be on the order of 0.002 $M_{\odot}$, at the present time.

102.007 **Modeling of the thermal behavior and of the chemical differentiation of cometary nuclei.**
S. Espinasse, J. Klinger, C. Ritz, B. Schmitt.
Icarus, Vol. 92, No. 2, p. 350 – 365 (Aug 1991).
Taking into account the low values of average density reported for comet P/Halley, a model of porous comet nucleus has been developed. The model includes heat conduction through the solid matrix and heat and mass transfer due to the vapor phase circulating in the pore system. The model is based on the resolution of two symmetric diffusion equations through the whole nucleus, one describing the transport of matter and the other the transport of heat. These equations are linked by the source term. Several phenomena are evidenced such as the depletion of volatiles in the subsurface layers and their enrichment in deeper layers. The important role of CO and CO_2 in controlling the propagation of the phase transition is emphasized. The relation between the composition of the nucleus and gas production rates is discussed.

102.008 **Comets, asteroids and interstellar grains.**
D. E. Brownlee.
Bull. Am. Astron. Soc., Vol. 23, No. 2, p. 905 (1991). Abstract. – See Abstr. 012.008 for the main entry.

102.009 **Growth and disruption of composite dust grains in the environment of a comet.**
J. S. Neff.
Bull. Am. Astron. Soc., Vol. 23, No. 2, p. 941 (1991). Abstract. – See Abstr. 012.008 for the main entry.

102.010 **On the production of metallic atoms and ions in cometary comae at large heliocentric distances.**
S. Ibadov.
Astron. Tsirk., No. 1549, p. 23 – 24 (May – Aug 1991). In Russian.
Processes producing metallic atoms and ions in cometary atmospheres are connected with the problem of ionization of the inner coma of comets. It is shown that the irradiation of cometary dust by energetic hydrogen atoms, generated within the cometopause (i.e., at cometocentric distances $r \leqslant 10^5$km) due to the charge–exchange of the solar wind protons on cometary molecules, may be one of the basic processes producing metallic atoms and ions in the cometary coma at large heliocentric distances.

102.011 **Dynamics of comets in the outer planetary region. I. A numerical experiment.**
A. Manara, G. B. Valsecchi.
Astron. Astrophys., Vol. 249, No. 1, p. 269 – 276 (Sep 1991).
In order to study the dynamical evolution of comets in the region of the outer planets, 100 fictitious cometary orbits are generated with random orbital parameters, and their motions are integrated over 1000 revolutions each. The individual orbital evolutions are examined and compared to the results of previous studies concerning both fictitious and real comets. Attention is given to the effects of close planetary encounters, and in particular to those in which the initial orbit of the comet does not cross that of the planet that it encounters; the abundance and effectiveness of this type of encounters and its relevance to the dynamics of comets in general is discussed.

102.012 **The regularity of the change of matter concentration in a cometary head as a function of the heliocentric distance.**
R. S. Osherov, K. I. Churyumov.
Astron. Vestn., Tom 25, No. 5, p. 603 – 608 (Sep – Oct 1991). In Russian. English translation in Solar Syst. Res., Vol. 25, No. 5.
The earlier detected dependence of matter concentration in a cometary head on the heliocentric distance is confirmed on the example of visual observations of comet Halley.

102.013 **Nongravitational effects in cometary motion.**
Yu. D. Medvedev.
1. Spain–USSR Workshop on Positional Astronomy and Celestial Mechanics, p. 101 – 113 (1991). – See Abstr. 012.025 for the main entry.
The procedure of computing the nongravitational forces which act on a cometary nucleus has been described. The surface of the nucleus is represented by a polynomial in terms of spherical functions. The nongravitational effects in the cometary motion are supposed to result from the sublimation of the water ice molecules from the nucleus surface. The sublimation of the matter gives rise to two effects: (1) reactive pressure on the nucleus and, as a consequence, change of the orbital and rotational motions of the latter, (2) the displacement of the photocentre. The latter has been calculated taking into account nucleus surface figure, the perturbed rotation of the nucleus and the comet heliocentric distances. Position observations of P/Halley of the 1986 apparition have been used to determine the parameters of the rotation motion of the nucleus along with the coefficients of the spherical polynomial representing the nucleus surface.

102.014 **Terrestrial cometary tail and lunar corona induced by small comets: predictions for Galileo.**
A. J. Dessler, B. R. Sandel, V. M. Vasyliunas.
Geophys. Res. Lett., Vol. 17, No. 12, p. 2257 – 2260 (Nov 1990).
A search for small comets near 1 AU is an objective of the Galileo mission. If small comets are as numerous and behave as has been proposed, two near–Earth signatures of small comets should be observable by the UVS experiment on the Earth flybys of Galileo: (1) a comet–like tail of Earth created by small comets that come close to Earth, breakup and vaporize, but just miss the atmosphere and proceed back into interplanetary space, and (2) a corona surrounding the Moon induced by lunar impact of small comets.

102.015 **Pitch angle scattering of cometary ions into monospherical and bispherical distributions.**
R. H. Miller, S. P. Gary, D. Winske, T. I. Gombosi.
Geophys. Res. Lett., Vol. 18, No. 6, p. 1063 – 1066 (Jun 1991).
Low frequency magnetic fluctuations generated by solar wind/cometary ion distributions cause pitch–angle scattering of cometary ions. The nature of this pitch angle scattering is examined by means of one–dimensional hybrid simulations, in which newborn ions are created at a constant rate, and for various injection angles, α, relative to the magnetic field. Pitch angle scattering in the quasi–perpendicular regime ($\alpha > 60°$) results in a bispherical velocity distribution while in the quasi–parallel regime ($0 \leqslant \alpha \leqslant 60°$), pitch angle scattering can result in either a mono or bispherical distribution depending on the injection angle and initial beam velocity.

102.016 **Pitch–angle scattering of cometary ions: computer simulations.**
S. P. Gary, R. H. Miller, D. Winske.
Geophys. Res. Lett., Vol. 18, No. 6, p. 1067 – 1070 (Jun 1991).
Pitch–angle evolution of newborn cometary ions is studied by means of one–dimensional electromagnetic hybrid computer simulations of homogeneous plasmas. Newborn ions are injected into the simulations at a constant rate, with a velocity relative to the solar wind which makes an angle α with respect to the ambient magnetic field. The simulations are done with relatively weak ion injection rates commensurate with those in the distant environment of comet Halley. In response to the linear temporal growth of the fluctuating magnetic field energy, the injected ions

pitch-angle scatter toward isotropy in both the quasiparallel ($0° \leqslant \alpha \lesssim 60°$) and quasiperpendicular ($60° < \alpha \leqslant 90°$) regimes. For the injection of cometary oxygen ions the simulations show pitch angle scattering rates that are essentially independent of the solar wind ion/cometary ion relative speed, and increase as the square root of the injection rate. Furthermore, the oxygen ion pitch angle scattering rate at perpendicular injection is approximately twice that in the quasiparallel regime, in qualitative agreement with observations at comet Halley.

102.017 **Numerical analysis of structures in a cometary dust tail.**
K. Beisser.
Astron. Nachr., Vol. 312, No. 3, p. 223 (1991). Dissertation abstract.

102.018 **Nongravitational effects and the aging of periodic comets.**
H. Rickman, L. Kamél, C. Froeschlé, M. C. Festou.
Astron. J., Vol. 102, No. 4, p. 1446 – 1463 (Oct 1991). Current Physics Microform No.: 9110D1978.

The authors use statistics of nongravitational effects on short-period comets, paying some attention also to absolute magnitudes, in order to deduce observational constraints on the physical process of cometary aging and to quantify the relevant time scales.

102.019 **On the problem of origin of nearly parabolic comets.**
V. V. Radzievskij.
Kinematika Fiz. Nebesn. Tel, Tom 7, No. 6, p. 21 – 29 (Nov–Dec 1991). In Russian. English translation in Kinematics Phys. Celest. Bodies, Vol. 7, No. 6.

The correlation $E(p|a^{-1})$ between parameters p and semi-major axes a^{-1} of nearly parabolic comets is investigated. It is established that this correlation is negative for elliptical orbits with $p > 2.7$ AU (correlation coefficient $K = -0.94$) and it is positive for orbits with $p < 1.7$ AU ($K = +0.99$). It is shown that all nuances of that correlation my be equally explained in the frame of two hypotheses of origin of nearly parabolic comets: 1) comet eruption by massive trans–Plutonian planets. 2) formation of comets in situ.

102.020 **Evolution of the Oort cloud angular momentum: numerical simulation.**
A. Lopatnikov, L. Marochnik (*L. S. Marochnik*), L. Mukhin (*L. M. Mukhin*), R. Sagdeev (*R. Z. Sagdeev*), D. Usikov (*D. A. Usikov*), G. Zaslavsky (*G. M. Zaslavskij*).
Astrophys. Space Sci., Vol. 186, No. 2, p. 225 – 243 (Dec 1991).

This paper considers the evolution of a flat svarm of cometary bodies (under the effect of the passage of stars), initially moving in one direction along the circular orbits with radii $1.4 \times 10^4 < r < 2 \times 10^4$AU and along elliptic orbits with semi-major axes $5 \times 10^3 < a < 1 \times 10^4$AU and with perihelia within $50 < q < 100$ AU. Numerical simulation shows that the original flat belt of comets is thermalizing. Its root-mean-square z-coordinate grows with r. A cometary cloud forms with a dense flattened inner core and a rarefield halo (the proper Oort cloud). The value $\beta = N_{core}/N_{halo}$ varies within a wide range (up to the order of magnitude) depending on the model used (N_{core} and N_{halo} are the numbers of comets in the core and the halo, respectively). The hypothesis of a massive Oort cloud (Marochnik et al. 1988) implies that the Oort cloud should have a large angular momentum. This paper employs numerical simulation to calculate Oort cloud models to which the initially flat located at the periphery of the solar nebular rotating cometary swarms is evolving in time. The loss of the initial angular momentum over the time of the Oort cloud evolution is not large.

102.021 **Randomization of dust-ejecta motions and the observed morphology of cometary heads.**
Z. Sekanina.
Astron. J., Vol. 102, No. 5, p. 1870 – 1878 (Nov 1991). Current Physics Microform No.: 9111D0876.

The computer code that has been used over years to study the dust-coma morphology of comets and to generate their synthetic images is upgraded by introducing two categories of random perturbations into the motions of dust particles ejected from discrete sources of activity on the nucleus surface. The new technique is applied to three types of morphological features: multiple concentric halos, a system of spiral jets, and a sunward fan-shaped coma. The results indicate that inclusion of random perturbations substantially enhances the capability for generating synthetic images that faithfully simulate the observed appearance of dust comets. By increasing the magnitude of perturbations in the synthetic images beyond a certain limit, it is possible to "erase" morphological features characteristic of the ejection process. This implies that the lack of distinct morphology in the head of a comet cannot be interpreted as the absence of discrete sources of activity on the nucleus surface. The effects of varying random perturbations on the appearance of the computer-generated images are investigated in detail and the suspected major role that imperfect flow collimation (or flow divergence) in dust jets plays with regard to the observed coma morphology is discussed.

102.022 **Motion of cometary ice grain in the central field of forces.**
P. Žáček, M. Wolf.
Bull. Astron. Inst. Czech., Vol. 42, No. 4, p. 241 – 247 (Jul 1991).

The terms for the reactive force effect of the ice grains, and for absorption and emission of radiation by those grains are derived in this paper. These results are used to solve the problem of motion of these grains in a cometary tail under central field of forces (gravity, pressure of radiation and reactive force).

102.023 **Singularities in mass-loaded MHD flow: the cometary bow shock.**
I. Kh. Khabibrakhmanov, A. J. Coates, V. L. Galinsky.
Geophys. Res. Lett., Vol. 18, No. 8, p. 1509 – 1512 (Aug 1991).

The authors present a one-dimensional model of the mass-loading of the solar wind by cometary ions which predicts a singularity in the mass-loaded flow at $M = 2$. Further, a subshock occurs when the flow speed reaches $M \approx 1.15$. The shape of the cometary bow shock in two dimensions is predicted, by requiring that the flow Mach number of the shock is 2 taking the velocity component normal to the shock surface. The Mach number results compared favourably with observations at comet Halley.

102.024 **Expansion of the secular and resonance parts of the perturbing function in the theory of motion of long-period comets.**
V. V. Emel'yanenko.
Sov. Astron. Lett., Vol. 17, No. 5, p. 363 – 366 (Sep 1991). Current Physics Microform No.: 9204X1310. English translation of Pis'ma Astron. Zh., Tom 17, No. 9, p. 857 – 864 (1991).

A method is proposed for expansion of the perturbing function in the restricted circular three-body problem for cometary orbits. A formula is derived for representing functions of elliptical motion as power series in $\varepsilon = (1-e)/2$ (where e is the eccentricity of the cometary orbit). It is found that for long-period comets the domain of convergence of these series extends to distances such that planetary perturbations may be ignored. Formulae and recursion relations are found for computing the secular and resonance parts of the perturbing function.

102.025 **A self-consistent model for the particles and fields upstream of an outgassing comet. 1. Gyrotropic and isotropic ion distributions.**
K. R. Flammer, D. A. Mendis, T. G. Northrop, E. C. Whipple.
J. Geophys. Res., Vol. 96, No. A9, p. 15907 – 15916 (1 Sep 1991).

The magnetic field and solar wind flow parametes in the vicinity of a weakly outgassing comet are determined using a self-consistent model which treats the cometary ions kinetically. Two different assumptions are made concerning the cometary ion distribution function; the pickup cometary ions form either a velocity space gyrotropic ring distribution or a velocity space isotropic shell distribution in the solar wind frame of reference. The individual currents due to the solar wind and cometary ions, which determine the magnetic field, are calculated in each case. Using theoretically determined parameters which reflect the conditions expected at comet Kopff (a typical short-period comet and one possible target for the future CRAF/Cassini mission) for various heliocentric distances, the authors are able to determine how the global field and flow characteristics are influenced by the nature of the cometary ion distribution function.

102.026 **Near-parabolic cometary flux in the outer Solar System. II.**
S. Yabushita, T. Tsujii.
Mon. Not. R. Astron. Soc., Vol. 252, No. 1, p. 151 – 155 (1 Sep 1991).
The flux of comets with semimajor axes ~20000 AU in the outer planetary region is estimated on the presumption that the major planets and the galactic tide control the dynamics of comets. The flux is found to be a rapidly increasing function of q, the perihelion distance. The supply rate is given by $\dot{n}_{SP} = (0.0015 \sim 0.003)\ \dot{n}_{LP}$, where $\dot{n}_{LP}$ is the flux of comets which have semimajor axes between 10000 and 20000 AU before entering the planetary region. If the lifetime of short-period (SP) comets is typically 2×10^4yr, it would be possible to regard SP comets as originating from the Oort cloud. If, however, it is as short as 2×10^3yr there is a discrepancy of a factor 10 between the observed and predicted supply rates. The effects of factors such as the discovery probability and magnitude difference between long-period (LP) and SP comets are discussed.

102.027 **Gravitationally unbound comets move in predominantly retrograde orbits.**
J. J. Matese, P. G. Whitman, D. P. Whitmire.
Nature, Vol. 352, No. 6335, p. 506 – 508 (8 Aug 1991). Letter-to-the-editor.
Comets are presumed to enter the inner solar system from the Oort cloud. Provided the perturbing effects of planetary encounters are taken into account, the original orbital energy of a comet can be calculated, and is negative or positive according to whether the comet's orbit is bound or unbound. The Oort effect is the tendency for the original energies of long-period (>200-yr) comets to fall within a narrow range: about 25% of such comets have original energies in the upper 0.2% of the total energy range. In addition 10% of long-period comets are unbound, and it has been found that positive original energy correlates with distance of closest approach to the Sun. The authors report a further correlation: unbound comets are more likely to move in retrograde orbits. They suggest that this anomaly comes about because of the omission from the orbital energy determination of non-gravitational effects arising from enhanced volatility.

102.028 **Alfven wave amplifications and generation at cometary bow shocks.**
W.-H. Ip, J. F. McKenzie.
Planet. Space Sci., Vol. 39, No. 7, p. 1045 – 1048 (Jul 1991).
The possible effect of hydromagnetic wave amplification and generation at cometary bow shocks is investigated by means of a planar hydromagnetic shock model. For the inbound crossing of comet Halley's bow shock, it is found that the transmission of the sunward-streaming hydromagnetic waves through the bow shock could lead to the enhancement of its wave amplitude by a factor of 2–3 and the generation of Alfvén waves with the opposite helicity. The plasma observations during inbound crossing of comet Halley's bow shock provide some evidence for such effect.

102.029 **The small-comet hypothesis.**
A. J. Dessler.
Earth Space, Vol. 4, No. 4, p. 10 – 15 (Dec 1991).
After more than 5 years of intense scientific debate, the jury may be reaching a verdict on the idea that Earth is being showered by tiny comets.

102.030 **Two possible acceleration mechanisms in comet tails.**
Li Zhongyuan, Tong Yi, Guo Sheyu.
Chin. Sci. Bull., Vol. 36, No. 17, p. 1453 – 1456 (Sep 1991).

102.031 **The formation and role of water dimers in a cometary atmosphere, with application to comet P/Halley.**
J. F. Crifo, Z. Slanina.
Astrophys. J., Vol. 383, No. 1, p. 351 – 355 (10 Dec 1991).
The formation of water dimers in a cometary coma is comprehensively modeled, using recent thermochemical data and plausible chemical schemes. For P/Halley at the time of the 1986 flyby missions, the computed dimer mole fraction is everywhere smaller than 10^{-5}. The model is also used to confirm the earlier results of Crifo concerning the partial recondensation of the coma water into large clusters, which unavoidably follows the formation of the dimers.

102.032 **Analysis of the aphelia of 179 long period comets considering the effect of the vertical galactic tidal force.**
A. Brunini, D. M. Canosa.
Bol. Asoc. Argent. Astron., No. 36, p. 225 – 241 (1990). In Spanish. – See Abstr. 012.095 for the main entry.
The authors analyze the incoming direction of 179 long period comets. They have taken into account the gravitational influence of the planets as well as the Galaxy as a whole. The authors have made an improvement of some previous works on the matter, including comets observed up to 1989. An asymmetry of aphelia is detected that cannot be explained as due to observational selection. It is concluded that the vertical galactic tidal force gives account of the observed distribution.

102.033 **Monte Carlo simulation for the distribution of perihelia of long period comets**
R. Gulati, M. Khanna, S. D. Sharma, P. S. Goraya.
Bull. Astron. Soc. India, Vol. 19, Nos. 1 – 2, p. 94 (Mar – Jun 1991). Abstract. – See Abstr. 012.108 for the main entry.

102.034 **Quasi-linear evolution of ULF waves excited by cometary ion pickup.**
M. A. Lee, S. P. Gary.
J. Geophys. Res., Vol. 96, No. A12, p. 21319 – 21327 (1 Dec 1991).
An analytical model is presented which describes the quasi-linear evolution of both ultralow-frequency (ULF) magnetic fluctuations and cometary ion flux in the solar wind due to continuous ion pickup. The derived solution yields both the initial exponential growth of the total wave intensity and the time-asymptotic growth proportional to time obtained in numerical simulations using a one-dimensional hybrid code with a constant rate of cometary ion injection (Gary et al., 1988, 1989). For pickup rates appropriate to cometary environment, the derived fluctuation intensity and pickup ion flux are in general accord with ULF wave observations and water group ion pitch angle distributions upstream of the bow waves at both comet Giacobini-Zinner and comet Halley.

102.035 **Statistical confirmation of the tidal concept of the origin of nearly parabolic comets.**
V. V. Radzievskij.
Sol. Syst. Res., Vol. 25, No. 3, p. 222 – 227 (Nov 1991). English translation of Astron. Vestn., Tom 25, No. 3, p. 302 – 308 (1991).
Predicted effects of comet statistics following from the preceding article are confirmed by statistical data to a high degree of accuracy.

102.036 **Statistical test of the hypothesis of comet ejection.**
V. P. Tomanov.
Sol. Syst. Res., Vol. 25, No. 3, p. 230 – 233 (Nov 1991). English translation of Astron. Vestn., Tom 25, No. 3, p. 312 – 316 (1991).
Some parameters of the system of nearly parabolic comets are analyzed statistically. No factors are found to confirm the hypothesis of comet ejection.

Astrophysics' contribution to the study of comets in the XIX century.
See Abstr. 004.060.

Optical reflection spectroscopy of mineral and ice surfaces in comet simulation experiments.
See Abstr. 022.003.

Clathrate hydrate formation in amorphous cometary ice analogs in vacuo.
See Abstr. 022.038.

Laboratory simulation, a tool for comet research.
See Abstr. 022.069.

KOSI?
See Abstr. 022.070.

Visible and near IR albedo measurements of ice/dust mixtures.
See Abstr. 022.071.

Energy balance of the KOSI 4 experiment.
See Abstr. 022.072.

Thermal histories of the KOSI samples.
See Abstr. 022.073.

The effect of non–volatile porous layers on temperature and vapor pressure of underlying ice.
See Abstr. 022.074.

Gas release from ice/dust mixtures.
See Abstr. 022.075.

KOSI: gas drag derived from ice/dust–particle trajectories.
See Abstr. 022.076.

Measurement of the volatile component in particles emitted from an ice/dust mixture.
See Abstr. 022.077.

Dust emission of mineral/ice mixtures: residue structure and dynamical parameters.
See Abstr. 022.078.

Cometary analogue material: preparation, composition, and thin section petrography.
See Abstr. 022.079.

Comet simulation experiments as a prelude to the CRAF and ROSETTA missions.
See Abstr. 022.080.

Studies of proton irradiated H_2O+CO_2 and H_2O+CO ices and analysis of synthesized molecules.
See Abstr. 022.123.

Ion irradiation experiments relevant to cometary physics.
See Abstr. 022.124.

Irradiation effects in a comet's outer layers.
See Abstr. 022.125.

Cross sections for direct and dissociative ionization of NH_3 and CS_2 by electron impact.
See Abstr. 022.127.

Implications of C_2H photochemistry on the modeling of C_2 distributions in comets.
See Abstr. 022.128.

Thermal budget of multicomponent porous ices.
See Abstr. 022.130.

An experimental study of the sublimation of water ice and the release of trapped gases.
See Abstr. 022.157.

Molecule formation by fast particles in astronomical objects.
See Abstr. 022.166.

Occultation astronomy.
See Abstr. 036.138.

New applications of Fatou's problem.
See Abstr. 042.018.

Theory of wave activity occurring in the AMPTE artificial comet.
See Abstr. 062.017.

A note on the mass–loaded MHD flow of the solar wind towards a cometary nucleus.
See Abstr. 062.033.

The structure of a tangential discontinuity: application to the cometary ionopause.
See Abstr. 062.147.

Tektites in Cretaceous–Tertiary boundary rocks on Haiti and their bearing on the Alvarez impact extinction hypothesis.
See Abstr. 081.052.

The Earth's dust cloud and atmospheric oxygen.
See Abstr. 082.075.

The small–comet hypothesis.
See Abstr. 082.077.

The long–term dynamical behavior of small bodies in the Kuiper belt.
See Abstr. 091.024.

Mass and orbit estimation of planet X via a family of comets.
See Abstr. 101.052.

Influence of hypothetical Transneptunian planets on certain solar system bodies.
See Abstr. 101.125.

Analysis of the Swings effect and Greenstein effect in Comet P/Halley.
See Abstr. 103.093.

Sulfur–bearing ions in the ionosphere of comet Halley.
See Abstr. 103.099.

Are the Geminids and 3200 Phaethon related genetically?
See Abstr. 104.007.

Meteoroid streams: formation, evolution, associations with comets and asteroids.
See Abstr. 104.008.

The structure and evolution of the Taurid Complex.
See Abstr. 104.040.

A hypothesis on the Oort clouds of planets.
See Abstr. 107.008.

103 Comets (Individual Objects)

103.001 **Bahnstatistik von Kometen.**
J. Jahn.
KPM, Jahrg. 6, No. 17, p. 29 – 32 (Aug 1991).

103.002 **VdS–Archiv Kometen.**
A. Kammerer, J. Jahn.
KPM, Jahrg. 6, No. 17, p. 33 – 43 (Aug 1991).
Observations of the comets Levy (1990c), P/Metcalf–Brewington (1991a) and Arai (1991b) are listed.

103.003 **Die schwächeren Kometen des Jahres 1990.**
A. Kammerer.
KPM, Jahrg. 6, No. 18, p. 12 – 14 (Dec 1991).

103.004 **Die periodischen Kometen des Jahres 1992.**
J. Jahn.
KPM, Jahrg. 6, No. 18, p. 20 – 29 (Dec 1991).

103.005 **Interessante Kometen bis ins Jahr 2020.**
J. Jahn, M. Kempf, M. Kretlow.
KPM, Jahrg. 6, No. 18, p. 40 – 45 (Dec 1991).

103.006 **The Comet Handbook 1992.**
S. Nakano (comp.).
Oriental Astronomical Association. Comet Section, PO Box 32, Sumoto, Hyogo–ken, 656–91 (Japan). 144 p. (1992).

103.007 **Ephemerides of comets.**
Minor Planet Circ., Nos. 18485 – 19346 (Aug – Nov 1991).
Ephemerides for the following comets are given in Nos. 18648 – 18652 (25 Aug 1991), 18836 – 18840 (23 Sep 1991), 19037 – 19039 (23 Oct 1991), and 19343 – 19344 (21 Nov 1991), respectively:
1986 VIII P/Machholz, 1989u P/Kearns–Kwee, 1990c Levy, 1990e P/Wolf–Harrington, 1990g McNaught–Russell, 1990i Tsuchiya–Kiuchi, 1990m P/Harrington–Abell, 1991a P/Metcalf–Brewington, 1991d Shoemaker–Levy, 1991i P/Kowal 1, 1991l Helin–Lawrence, 1991o P/Chernykh, 1991p P/Shoemaker 1, 1991q P/Levy, 1991r Helin–Alu, 1991t P/Hartley 2, 1991v McNaught–Russell, 1991w McNaught–Russell, 1991y P/McNaught–Hughes, 1991z P/Shoemaker–Levy 5, 1991a_1 Shoemaker–Levy, P/Grigg–Skjellerup, P/Tsuchinshan 1.

103.008 **Orbital elements of comets.**
S. Nakano, B. G. Marsden.
Minor Planet Circ., Nos. 18485 – 19346 (Aug – Nov 1991).
The comets are listed according to their Roman numeral designation or their preliminary designation. The names of the orbit computers are given behind the respective M.P.C. numbers: 1986 VIII P/Machholz 18598 (Marsden); 1990d P/Peters–Hartley 19257 (Nakano); 1990j P/Mueller 2 19258 (Nakano); 1990k P/Holt–Olmstead 19257 (Nakano); 1990l P/Mueller 3 19257 (Nakano); 1990o P/Shoemaker–Levy 1 19257 (Nakano); 1991a P/Metcalf–Brewington 18597 (Nakano); 1991o P/Chernykh 18598 (Marsden); 1991p P/Shoemaker 1 18598 (Marsden); 1991q P/Levy 18597, 18773, 19258 (Marsden); 1991r Helin–Alu 18598, 18773 (Marsden); 1991t P/Hartley 2 18598 (Marsden); 1991v McNaught–Russell 18773, 18995 (Marsden); 1991w McNaught–Russell 18773, 18994, 19257 (Marsden); 1991x P/Spacewatch 18995, 19258 (Marsden); 1991y P/McNaught–Hughes 19258 (Marsden); 1991z P/Shoemaker–Levy 5 19258 (Marsden); 1991a_1 Shoemaker–Levy 19258 (Marsden).

103.009 **Observations of comets.**
Minor Planet Circ., Nos. 18485 – 19346 (Aug – Nov 1991).
Observations made at the following stations are published in Nos. 18489 – 18497 (25 Aug 1991), 18666 – 18669 (23 Sep 1991), 18851 – 18854 (23 Oct 1991), and 19065 – 19071 (21 Nov 1991), respectively:
Calar Alto, Cambridge, Chiyoda, Conder Brow, Crimean Astrophys. Obs., Dynic Astron. Obs., ESO, Farra d'Isonzo, Geisei, JCPM Sapporo Stn., Kharkow, Kitt Peak, Kitt Peak (Steward Obs.), Klet, Linz, Mauna Kea Obs., Mount John Obs., Nikolaev, Oak Ridge Obs., Ohtsu, Oizumi, Otomo, Palomar, Pian dei Termini, Prague, Pulkovo, Santa Lucia Stroncone, Sendai Obs. Ayashi Stn., Siding Spring, Skalnaté Pleso, Sormano, Stakenbridge, Sternberg Astron. Inst. Middle–Asia Mountain Obs., Victoria, Yatsugatake South Base Obs., Zelenchukskaya, Zimmerwald.
Observations of the following comets – sorted according to their provisional designation or their Roman numeral designation – are published:
1983 XI P/Tempel 1, 1983 XIII P/Kopff, 1984 XXI P/Arend–Rigaux, 1984 XXII P/Schaumasse, 1984 XXIII Levy–Rudenko, 1985 II Shoemaker, 1985 XVIII P/Shoemaker 3, 1986 VIII P/Machholz, 1987 VI P/Howell, 1987 X P/Grigg–Skjellerup, 1988 XI P/Shoemaker–Holt 2, 1988 XIV P/Tempel 2, 1988 XX Yanaka, 1989 VI P/Churyumov–Gerasimenko, 1989 XI P/Gunn, 1989 XV P/Schwassmann–Wachmann 1, 1989 XXI Helin–Roman–Alu, 1989t P/Wild 2, 1989u P/Kearns–Kwee, 1989h_1 P/van Biesbroeck, 1990c Levy, 1990e P/Wolf–Harrington, 1990g McNaught–Hughes, 1990h P/Johnson, 1990j P/Mueller 2, 1990k P/Holt–Olmstead, 1990n P/Taylor, 1991a P/Metcalf–Brewington, 1991b Arai, 1991d Shoemaker–Levy, 1991h P/Takamizawa, 1991j P/Hartley 1, 1991k P/Mrkos, 1991l Helin–Lawrence, 1991n P/Faye, 1991o P/Chernykh, 1991p P/Shoemaker 1, 1991q P/Levy, 1991r Helin–Alu, 1991s P/Wirtanen, 1991t P/Hartley 2, 1991u P/Arend, 1991v McNaught–Russell, 1991w McNaught–Russell, 1991x P/Spacewatch, 1991y P/McNaught–Hughes, 1991z P/Shoemaker–Levy 5, 1991a_1 Shoemaker–Levy.

103.010 **A tentative identification of methanol as the progenitor of the 3.52–μm emission feature in several comets.**
S. Hoban, M. Mumma, D. C. Reuter, M. DiSanti, R. R. Joyce, A. Storrs.
Icarus, Vol. 93, No. 1, p. 122 – 134 (Sep 1991).
The authors observed comets P/Brorsen–Metcalf (1989 X), Okazaki–Levy–Rudenko (1989 XIX), Austin (1989c_1), and Levy (1990c) in the 3.44 – 3.64 μm spectral region. Continuum emission was detected in all four comets. The organic feature at 3.4 μm and an emission feature at 3.52 μm were detected in all comets except Okazaki–Levy–Rudenko. The spatial brightness profile of the 3.52–μm emission feature was measured for the first time. The observed spatial profile is consistent with that expected for uniform outflow from a central unresolved source and exhibits an optocenter which is displaced from that of the dust emission by 630 ± 90 km in the antisunward direction. The authors conclude that the molecule which gives rise to the emission is either directly sublimated from the cometary surface or is the dissociation product of a very short–lived parent species.

103.011 **Early and unidentified apparitions of short–period comets.**
L. Kresák, M. Kresáková.
Astron. Astrophys., Vol. 251, No. 1, p. 331 – 338 (Nov 1991).
A set of criteria was developed for finding short–period comets among those for which only parabolic orbits are available. These criteria indicate that the basic Catalogue of Cometary Orbits by Marsden (1989) includes about 18 such cases with revolution periods less than 20 years. About 15 of them were observed before the discovery of the first known object of this type. One of them, comet 1678, was already definitely identified with P/d'Arrest (Carusi et al., 1991). Another 14 of them – comets 1230, 1345, 1351, 1457 I, 1491 II, 1499, 1539, 1577 II, 1585, 1702, 1743 I, 1833, 1949 III, and 1963 IX – were almost

certainly of short period (p_s = 95 to 99%). For comets 568, 1080 and 1457 II a short revolution period appears probable (p_s = 80%), for comets 1245, 1293, 1618 I and 1618 III possible (p_s = 50%), and for comets 390, 1757 and 1860 IV not impossible (p_s = 20%). There are some cases of orbital similarity with the currently known short-period comets (P/Finlay, P/Denning-Fujikawa, P/Biela, P/d'Arrest, P/Barnard 3 and P/Denning), which do not exclude the possibility that they may be earlier apparitions of these comets. Any unambiguous identification, however, will be very difficult without historical records of additional returns. At the same time, it would be very important for the understanding of the physical lifetimes of comets, their longterm dynamical evolution, and of the variable nongravitational forces acting on them.

103.012 **An analysis of the spectrophotometric and photometric observations of comets 1967 II, 1968 I, 1968 V, and 1968 VI.**
A. A. de Almeida.
Earth, Moon, Planets, Vol. 53, No. 3, p. 205 - 216 (Jun 1991).
Emission fluxes of CN, C_2 and C_3 carbon-bearing molecular species observed in the coma of comets Rudnicki (1967 II), Ikeya-Seki (1968 I), Whitaker-Thomas (1968 V), and Honda (1968 VI) are analysed in the framework of Haser model. CN, C_2 and C_3 production rates are determined using recently derived fluorescence efficiencies. The dependence of CN, C_2 and C_3 production rates on the heliocentric distance and the possible correlations among these radicals is studied. It is shown that comets Ikeya-Seki (1968 I) and Honda (1968 VI) have the same mean color indices (*B-V*) and (*U-B*).

103.013 **Systematic displacements between gas and dust in the sunward fans of two low-activity comets.**
H. Spinrad.
Astron. J., Vol. 102, No. 3, p. 1207 - 1212 (Sep 1991). Current Physics Microform No.: 9109D1611.
IRAS-Araki-Alcock (1983 VII) and P/Tempel 2 (1988 XIV), two low-production comets with sunward ejection fans, showed unusual displacements between their radical emission-line gas and the scattered-light dust continuum on series of long-slit spectra. For IRAS-Araki-Alcock the typical Δx peak offset was only ~200 km (gas leading toward the Sun), while for P/Tempel 2 the corresponding offset was nearly 2000 km. The author illustrates the phenomenon, but does not present an adequate quantitative explanation.

103.014 **Bright comets in the year 1990.**
M. Ilić.
Vasiona, Année 39, No. 5, p. 80 - 82 (1991). In Serbo-Croatian.
Observations of three bright comets: comet Austin (1989c_1), Comet Levy (1990c), and periodic comet Encke with the 12-cm reflector in the year 1990 are presented.

103.015 **Orbits for HOAA 1992.**
OAA Comput. Sect. Circ., No. NK 557, p. 1 - 2 (15 Dec 1991).
Concerning comets: P/Kearns-Kwee, P/Wild 2, McNaught-Hughes (1990g), P/Hartley 1, P/Levy, P/Machholz, P/Takamizawa, Helin-Lawrence (1991*l*), P/Spitaler, P/Schaumasse.

103.016 **1993 Comet Handbook.**
S. Nakano, D. W. E. Green.
Int. Comet Q., Vol. 14, No. 2a, p. H1 - H52 (May 1992).
This special issue is the 7th edition of the ICQ Comet Handbook, a collection of up-to-date (as of April 1992) orbital elements and ephemerides of comets for the year 1993.

103.017 **Peculiarities of gas and dust production rates in comets P/Halley (1986 III), P/Giacobini-Zinner (1985 XIII), P/Hartley-Good (1985 XVII) and P/Thiele (1985 XIX).**
K. I. Churyumov, V. K. Rosenbush (*V. K. Rozenbush*).
Astron. Nachr., Vol. 312, No. 6, p. 385 - 391 (1991).
An IHW narrow-band photometry of comets P/Halley (1986 III), P/Giacobini-Zinner (1985 XIII), P/Hartley-Good (1985 XVII) and P/Thiele (1985 XIX) is presented. Molecular column densities and production rates of CN, C_2, and C_3 species observed in the coma are calculated in the framework of the Haser model. A comparative analysis of gas and dust production rates in the atmospheres of these comets is given. The effective comet nuclear radii have been estimated by different methods: for comet P/Hartley-Good R_{eff} ~ 1.7 km, for P/Thiele R_{eff} ~ 2 km and for P/Giacobini-Zinner R_{eff} ~ 1.3 km.

103.018 **Periodic comets for the visual observer in 1992.**
A. Hale.
Int. Comet Q., Vol. 13, No. 4, p. 145 - 147 (Oct 1991).

103.019 **Recent news and research concerning comets.**
D. W. E. Green.
Int. Comet Q., Vol. 13, No. 4, p. 147 - 148 (Oct 1991).

103.020 **Tabulation of comet observations.**
Int. Comet Q., Vol. 13, No. 4, p. 149 - 184 (Oct 1991).
Concerning comets: 1988 V Liller, 1989 XV Schwassmann-Wachmann 1, 1989 XIX Okazaki-Levy-Rudenko, 1989 XXII Aarseth-Brewington, 1989t P/Wild 2, 1989c_1 Austin, 1989d_1 P/Schwassmann-Wachmann 3, 1989e_1 Skovichenko-George, 1989h_1 P/van Biesbroeck, 1990a P/Wild 4, 1990b Černis-Kiuchi-Nakamura, 1990c Levy, 1990e P/Wolf-Harrington, 1990r P/Honda-Mrkos-Pajdušáková, 1990g McNaught-Hughes, 1990i Tsuchiya-Kiuchi, 1991a P/Metcalf-Brewington, 1991b Arai, 1991c P/Swift-Gehrels, 1991d Shoemaker-Levy, 1991h P/Takamizawa, 1991j P/Hartley 1, 1991k P/Mrkos, 1991*l* Helin-Lawrence, 1991n P/Faye, 1991*o* P/Chernykh, 1991p P/Shoemaker 1, 1991q P/Levy, 1991v Helin-Alu, 1991s P/Wirtanen, 1991t P/Hartley 2, 1991y P/McNaught-Hughes, 1991z P/Shoemaker-Levy 5, 1991a_1 Shoemaker-Levy, 1991b_1 P/Shoemaker-Levy 6, P/Encke, P/Machholz, P/Arend-Rigaux.

103.021 **Post-perihelion survival of comets with small *q*.**
J. E. Bortle.
Int. Comet Q., Vol. 13, No. 3, p. 89 - 91 (Jul 1991).
The absolute magnitudes of 85 comets with perihelion distances (*q*) less than 0.51 AU were examined, along with their observational histories. From these data, a formula is derived for the prediction of post-perihelion survival/non-survival of intrinsically faint comets with small values of *q*.

103.022 **Recent news and research concerning comets.**
D. W. E. Green.
Int. Comet Q., Vol. 13, No. 3, p. 91 - 94 (Jul 1991).

103.023 **Tabulation of comet observations.**
Int. Comet Q., Vol. 13, No. 3, p. 94 - 140 (Jul 1991).
Concerning comets: 1957 III Arend-Roland, 1957 V Mrkos, 1958 III Burnham, 1959 III Bester-Hoffmeister, 1959 VI Alcock, 1959 VIII P/Giacobini-Zinner, 1960 I P/Wild 1, 1960 II Burnham, 1960 III P/Schaumasse, 1960 VIII P/Finlay, 1961 V Wilson-Hubbard, 1961 VIII Seki, 1962 III Seki-Lines, 1962 VIII Humason, 1963 I Ikeya, 1963 III Alcock, 1963 V Pereyra, 1964 VI Tomita-Gerber-Honda, 1964 VIII Ikeya, 1964 IX Everhart, 1979 X Bradfield, 1980 X P/Stephan-Oterma, 1980 XII Meier, 1980 XIII P/Tuttle, 1981 II Panther, 1982 I Bowell, 1982 IV P/Grigg-Skjellerup, 1982 VII P/d'Arrest, 1986 III P/Halley, 1987 IV Shoemaker, 1987 XII P/Hartley 3, 1987 XIX P/Schwassmann-Wachmann 2, 1987 XXIX Bradfield, 1987 XXX Levy, 1987 XXXII McNaught, 1987 XXXIII P/Borrelly, 1987 XXXVI P/Parker-Hartley, 1987 XXXVII P/Helin-Roman-Alu 1, 1988 III Shoemaker-Holt, 1988 V Liller, 1988 VIII P/Ge-Wang, 1988 XI P/Shoemaker-Holt 2, 1988 XIV P/

Tempel 2, 1988 XV Machholz, 1988 XX Yanaka, 1988 XXI Shoemaker, 1988 XXIV Yanaka, 1989 III Shoemaker, 1989 V Shoemaker–Holt–Rodriquez, 1989 VIII P/Pons–Winnecke, 1989 X P/Brorsen–Metcalf, 1989 XI P/Gunn, 1989 XIII P/Lovas 1, 1989 XVIII McKenzie–Russell, 1989 XIX Okazaki–Levy–Rudenko, 1989 XX P/Clark, 1989 XXI Helin–Ronan–Alu, 1989 XXII Aarseth–Brewington, 1989h P/Van Biesbroeck, 1989t P/Wild 2, 1989u P/Kearns–Kwee, 1989c_1 Austin, 1989e_1 Skorichenko–George, 1990a P/Wild 4, 1990b Černis–Kiuchi–Nakamura, 1990c Levy, 1990e P/Wolf–Harrington, 1990f P/Honda–Mrkos–Pajdušáková, 1990g McNaught–Hughes, 1990i Tsuchiya–Kiuchi, 1990n P/Taylor, 1990*o* P/Shoemaker–Levy 1, 1991a P/Metcalf–Brewington, 1991b Arai, 1991c P/Swift–Gehrels, 1991d Shoemaker–Levy, 1991e P/Shoemaker–Levy 3, 1991h P/Takamizawa, 1991j P/Hartley 1, 1991k P/Mrkos, 1991*l* Helin–Lawrence, 1991n P/Faye, 1991q P/Levy, 1991t P/Hartley 2, P/Encke, P/Schwassmann–Wachmann 1.

103.024 **Comet corner.**
D. E. Machholz.
Strolling Astron., Vol. 35, No. 3, p. 135 – 137 (Sep 1991).

103.025 **The first 25 years of the A.L.P.O. Comets Section.**
D. W. E. Green, D. D. Meisel, D. Milon.
Strolling Astron., Vol. 35, No. 4, p. 145 – 152 (Dec 1991).

103.026 **Comet corner.**
D. E. Machholz.
Strolling Astron., Vol. 35, No. 4, p. 175 – 176 (Dec 1991).

103.027 **Microwave observations of hydrogen sulfide and searches for other sulfur compounds in comets Austin (1989c_1) and Levy (1990c).**
J. Crovisier, D. Despois, D. Bockelée-Morvan, P. Colom, G. Paubert.
Icarus, Vol. 93, No. 2, p. 246 – 258 (Oct 1991).

The authors detected hydrogen sulfide through observations of its 1_{10}–1_{01} rotational line at 2 mm in comets Austin (1989c_1) and Levy (1990c) using the IRAM 30–m radiotelescope. In comet Levy, the 2_{20}–2_{11} line of H_2S was also observed; the relative intensities of the two lines suggest a cold rotational distribution for H_2S. The authors derive molecular production rates using an out-of-equilibrium excitation model which takes into account collisions and vibrational radiative excitation.

103.028 **Solar activity influence upon the brightness curves of comet Churyumov–Gerasimenko (1982 VIII) and comet Halley (1986 III).**
K. I. Churyumov, V. S. Filonenko.
Pis'ma Astron. Zh., Tom 17, No. 12, p. 1135 – 1142 (Dec 1991). In Russian. English translation in Sov. Astron. Lett., Vol. 17, No. 6.

A statistically reliable correlation between the outbursts of brightness and its variations of comet Churyumov-Gerasimenko and the level of solar activity has been found out. It is shown that despite the results obtained by S. V. Orlov, P/Halley's integral brightness can correlate with changes in the indices of solar activity and solar wind velocity.

Comet 1975 IX Kobayashi–Berger–Milon

103.029 **Cometary ray closing rates: comet Kobayashi–Berger–Milon.**
E. P. Moore.
Astron. Astrophys., Vol. 247, No. 1, p. 247 – 251 (Jul 1991).

Cometary ray closing rates are measured for a sequence of a dozen plates of Comet Kobayashi-Berger-Milon (1975 IX) on July 31, 1975. The angular velocity is a factor of three to six higher than previously published in the literature for other comets. A deceleration of one to two deg h^{-2}, with no secular trend, is measured. Comparisons are made between these observations and some existing models.

Comet 1984 XIII Austin

103.030 **Polarimetric properties of comet Austin.**
A. K. Sen, U. C. Joshi, M. R. Deshpande.
Mon. Not. R. Astron. Soc., Vol. 253, No. 4, p. 738 – 742 (15 Dec 1991).

Model calculations have been made to explain the polarization observed during the passage of comet Austin. The observations were spread over four nights during the pre– and post–perihelion periods and were carried out at the IHW interference filters. The Austin data are compared with that of comet P/Halley reported in an earlier work (Sen et al., 1991). The comparison, with the help of Mie theory, indicates that the polarizations in the two comets are probably caused by scattering due to two different types of grains, characterized by a change either in complex refractive index or size distribution or both. There are indications that comet Austin may contain finer grains as compared to P/Halley. Also a possible interpretation of these data with the help of scattering by large fluffy dark grains (Dollfus) is discussed.

Comet 1989 XIX Okazaki–Levy–Rudenko

103.031 **Molecular emission from the head of Comet Okazaki–Levy–Rudenko (1989r).**
B. B. Sanwal, B. S. Rautela.
Earth, Moon, Planets, Vol. 54, No. 2, p. 125 – 128 (Aug 1991).

Spectrophotometric scans of the head of Comet Okazaki–Levy–Rudenko during three nights in October 1989 are presented. An estimate of the CN and C_2 production rates and column densities have been made.

103.032 **Identification of H_2O emissions in comet Okazaki–Levy–Rudenko (1989r).**
O. M. Mamadov.
Astron. Tsirk., No. 1545, p. 43 – 44 (Aug – Sep 1990). In Russian.

In November 1989, at Sanglock observatory, the spectra of comet Okazaki–Levy–Rudenko have been taken with dispersion 28 A/mm, in the spectral range 7700–8260 Å. A detailed identification showed rotational structure of CN red system, Phillips System of C_2 and numerous emissions of H_2O molecules. This is the first detection of H_2O emissions in a comet by groud-based observations in optical wavelength region.

Comet 1990 V Austin

103.033 **Near–ultraviolet spectroscopic features in Comet Austin (1989c_1).**
J. H. Valk, C. R. O'Dell, A. L. Cochran, W. D. Cochran, C. B. Opal.
Bull. Am. Astron. Soc., Vol. 23, No. 2, p. 939 (1991). Abstract. – See Abstr. 012.008 for the main entry.

103.034 **The spectrum of comet Austin.**
T. G. Slanger.
Science, Vol. 253, No. 5018, p. 452 – 453 (26 Jul 1991). With a response by J. Green, W. Cash, T. Cook, S. A. Stern.

103.035 **Constraints on noble gas abundances in comet Austin (1989c_1).**
S. A. Stern, J. C. Green, W. Cash, T. A. Cook.
Bull. Am. Astron. Soc., Vol. 23, No. 2, p. 961 (1991). Abstract. – See Abstr. 012.008 for the main entry.

103.036 **Enhanced radio source scintillation due to comet Austin (1989c_1).**
P. Janardhan, S. K. Alurkar, A. D. Bobra, O. B. Slee.
Aust. J. Phys., Vol. 44, No. 5, p. 565 – 571 (1991).

Enhanced scintillations in the direction of the quasar 2204 + 29 (3C441) were observed on 13 May 1990 when the tail of comet

Austin passed in front of it. Comparison with previous observations at 103, 327 and 408 MHz of comet Halley and at 408 MHz of comet Wilson show that proper occultation geometry is essential for observing enhanced scintillations. It has been shown that the solar elongation during such observations should be large, typically greater than 60° and in no case less than 30° at 103 MHz. At the time of the occultation the scintillation index was greater than that expected for this source by a factor of 3. The r.m.s. electron density variation, at a distance of 0.9 A.U. from the sun and 7.3° downstream of the nucleus, was found to be 6 cm^{-3} as compared with 1 cm^{-3} for the normal solar wind at 1 A.U. The corresponding scale sizes of the turbulence were found to be much finer than normally found in interplanetary scintillation caused by the solar wind.

Comet 1990 XIX McNaught–Russell

103.037 **Comet McNaught–Russell (1991g).**
OAA Comput. Sect. Circ., No. NK 555, p. 1 (4 Sep 1991).

Comet 1990 XX Levy

103.038 **HUT observations of Comet Levy (1990c).**
P. D. Feldman, A. F. Davidsen, W. P. Blair, C. W. Bowers, W. V. Dixon, S. T. Durrance, R. C. Henry, G. A. Kriss, J. W. Kruk, H. W. Moos, O. Vancura, H. C. Ferguson, R. A. Kimble, K. S. Long, T. R. Gull.
Bull. Am. Astron. Soc., Vol. 23, No. 2, p. 939 (1991). Abstract. – See Abstr. 012.008 for the main entry.

103.039 **Observations of comet Levy (1990c) with the Hopkins Ultraviolet Telescope.**
P. D. Feldman, A. F. Davidsen, W. P. Blair, C. W. Bowers, W. V. Dixon, S. T. Durrance, H. C. Ferguson, R. C. Henry, R. A. Kimble, G. A. Kriss, J. W. Kruk, K. S. Long, H. W. Moos, O. Vancura, T. R. Gull.
Astrophys. J., Lett., Vol. 379, No. 1, p. L37 – L40 (20 Sep 1991).
Observations of comet Levy (1990c) were made with the Hopkins Ultraviolet Telescope during the Astro-1 Space Shuttle mission on 1990 December 10. The spectrum, covering the wavelength range 415 – 1850 Å at a spectral resolution of 3 Å, shows the presence of carbon monoxide and atomic hydrogen, carbon, and sulfur in the coma. Aside from H I Lyman-β, no cometary features are detected below 1200 Å, although cometary O I and O II would be masked by the same emissions present in the day airglow spectrum. The 9.4×116 arcsecond aperture corresponds to $12{,}000 \times 148{,}000$ km at the comet. The derived production rate of CO relative to water is 0.11 ± 0.02, compared with 0.04 ± 0.01 derived from IUE observations (made in 1990 September) which sample a much smaller region of the coma. This suggests the presence of an extended source of CO, as was found in comet Halley. Upper limits on Ne and Ar abundance are within one order of magnitude of solar abundances.

103.040 **The detection of a strong 3.28-μm emission feature in Comet Levy.**
J. K. Davies, S. F. Green, T. R. Geballe.
Mon. Not. R. Astron. Soc., Vol. 251, No. 1, p. 148 – 151 (1 Jul 1991).
An infrared spectrum of Comet Levy was obtained on 1990 August 27 when the comet was 1.4 AU from the Sun. In addition to the broad emission feature at 3.4 μm seen in a number of comets, the spectrum clearly shows emission at 3.28 μm which is stronger, relative to the 3.4-μm feature, than in any other comet yet observed. The ratio of emission in the 3.4-μm band to the rate of water production is the lowest yet observed in a comet and is consistent with the tendency for the ratio to decrease with increasing heliocentric distance.

103.041 **Comet Levy (1990c).**
IAU Circ., No. 5353, p. 1 (25 Sep 1991).

103.042 **Comet Levy (1990c): groundbased photometric results.**
D. G. Schleicher, R. L. Millis, D. J. Osip, P. V. Birch.
Icarus, Vol. 94, No. 2, p. 511 – 523 (Dec 1991).
Narrowband filter photometry of Comet Levy (1990c) was obtained on 51 nights in the interval from 3 Jun 1990 to 19 Apr 1991. The emission bands of OH, NH, CN, C_3, and C_2 were isolated, in addition to continuum points at 3650 and 4845 Å. Relative abundances among the species were typical of most comets previously observed. All species showed pronounced asymmetry about perihelion, with production rates for most species being about twice as large before perihelion as after over the heliocentric distance range of 1.0 – 3.0 AU. The OH production rate's asymmetry was nearly double that observed for the other species, with maximum production occurring more than a month before other species.

Comet 1990 XXII McNaught–Russell

103.043 **Comet McNaught–Russell (1991w).**
IAU Circ., No. 5335, p. 1 (5 Sep 1991).
Further information is given in No. 5339.

103.044 **Comet McNaught–Russell (1991w).**
OAA Comput. Sect. Circ., No. NK 567, p. 1 (24 Dec 1991).

103.045 **Comet McNaught–Russell (1991w).**
Yamamoto Circ., No. 2171, p. 1 – 2 (14 Sep 1991). In Japanese.

Comet 1990 XXVI Arai

103.046 **Comet Arai (1991b).**
OAA Comput. Sect. Circ., No. NK 554, p. 1 – 3 (4 Sep 1991).

Comet 1990g McNaught–Hughes

103.047 **Comet McNaught–Hughes (1990g).**
IAU Circ., No. 5335, p. 1 (5 Sep 1991).

Comet 1991a_1 Shoemaker–Levy

103.048 **Comet Shoemaker–Levy (1991a_1).**
IAU Circ., No. 5363, p. 1 (10 Oct 1991).
Further information is given in Nos. 5380, 5414.

103.049 **Comet Shoemaker–Levy (1991a_1).**
Yamamoto Circ., No. 2172, p. 1 (13 Oct 1991). In Japanese.

Comet 1991d Shoemaker–Levy

103.050 **Comet Shoemaker–Levy (1991d).**
IAU Circ., No. 5379, p. 1 (6 Nov 1991).
Further information is given in Nos. 5398, 5415.

103.051 **Comet Shoemaker–Levy (1991d).**
Yamamoto Circ., No. 2174, p. 1 (15 Dec 1991). In Japanese.

103.052 **Comet Shoemaker–Levy (1991d).**
OAA Comput. Sect. Circ., No. NK 558, p. 1 – 3 (24 Dec 1991).

Comet 1991g_1 Zanotta–Brewington

103.053 **Comet Zanotta–Brewington (1991g_1).**
IAU Circ., No. 5412, p. 1 (24 Dec 1991).
Further information is given in Nos. 5413, 5419.

103.054 **Comet Zanotta–Brewington (1991g_1).**
Yamamoto Circ., No. 2175, p. 1 (25 Dec 1991). In Japanese.

Comet 1991*l* Helin–Lawrence

103.055 **Comet Helin–Lawrence (1991*l*).**
IAU Circ., No. 5380, p. 1 (6 Nov 1991).
Further information is given in No. 5410.

Comet 1991r Helin–Alu

103.056 **Comet Helin–Alu (1991r).**
IAU Circ., No. 5305, p. 1 (13 Jul 1991).
Further information is given in No. 5374.

103.057 **Comet Helin–Alu (1991r).**
Yamamoto Circ., No. 2169, p. 2 (13 Jul 1991). In Japanese.

103.058 **Comet Helin–Alu (1991r).**
OAA Comput. Sect. Circ., No. NK 564, p. 1 - 2 (24 Dec 1991).

Comet 1991v McNaught–Russell

103.059 **Comet McNaught–Russell (1991v).**
IAU Circ., No. 5333, p. 1 (2 Sep 1991).
Further information is given in Nos. 5369, 5413.

103.060 **Comet McNaught–Russell (1991v).**
Yamamoto Circ., No. 2171, p. 1 (14 Sep 1991). In Japanese.

103.061 **Comet McNaught–Russell (1991v).**
OAA Comput. Sect. Circ., No. NK 566, p. 1 (24 Dec 1991).

Periodic comet Arend

103.062 **Periodic comet Arend (1991u).**
IAU Circ., No. 5322, p. 1 (9 Aug 1991).

103.063 **Periodic comet Arend (1991u).**
Yamamoto Circ., No. 2170, p. 1 (17 Aug 1991). In Japanese.

Periodic comet Brorsen–Metcalf

103.064 **P/Brorsen–Metcalf: photometry, grain color and organic crust.**
G. A. Baratta, F. A. Catalano, F. Leone, G. Strazzulla.
Astron. Astrophys., Vol. 251, No. 1, p. 349 - 355 (Nov 1991).

Photometric observations of comet P/Brorsen–Metcalf obtained before its passage at perihelion are presented. The standard filters of the International Halley Watch, C_2, BC, C_3, CN, UC have been used. Gas production rates have been derived. The authors find these vary as r^{-n} (where r is the comet–Sun distance) with $n = 3\,(C_2)$, $1.1\,(C_3)$ and $3.9\,(CN)$. Production rates of solids follow the same law as gas with $n = 2.3$. The blue color exhibited by cometary grains may be interpreted as due to small grains (radius $< 0.1\ \mu$m) dominating (by number) over larger ones. The observations are discussed in the light of a model predicting the presence of an external organic cometary crust and radiation processed organic materials coming out from fissures and patches in the crust.

Periodic comet Chernykh

103.065 **Periodic comet Chernykh (1991o).**
IAU Circ., No. 5341, p. 1 (10 Sep 1991).
Further information is given in Nos. 5347, 5391.

103.066 **Periodic comet Chernykh.**
OAA Comput. Sect. Circ., No. NK 562, p. 1 - 6 (24 Dec 1991).

Periodic comet d'Arrest

103.067 **Observations of periodic comet d'Arrest in 1678 and implications for its evolutionary history.**
A. Carusi, L'. Kresák, M. Kresáková, G. B. Valsecchi.
Astron. Astrophys., Vol. 252, No. 1, p. 377 - 384 (Dec 1991).

The identity of Comet 1678 La Hire with P/d'Arrest is demonstrated, which makes this the oldest known observation of a short-period comet of the Jupiter family. Different modifications of linkages of the strongly perturbed orbit over 25 unobserved perihelion passages, and four close encounters with Jupiter, are applied and evaluated. The best one, starting with the 1851 osculating orbit and assuming constant nongravitational forces in the past, represents the 1678 observations to $\pm 0\overset{\circ}{.}2$, well within their positional uncertainty. At the same time, the 1989 perihelion passage is represented to $0\overset{d}{.}01$. Since the fitted A_2 parameter is about the average of the mean value from the period 1851–1989 (4% lower) and the value obtained with a statistically significant term for the varying perihelion distance (5% higher), the nongravitational accelerations appear to be extraordinarily stable. This peculiarity is tied to a uniquely strong recurrent perihelion asymmetry of the light curve. The observations of P/d'Arrest from 1678 imply that at that time, 47 revolutions ago, both its absolute brightness and the shape of its light curve were the same as at present, within the range of observational accuracy.

Periodic comet Encke

103.068 **Comet Encke motion and photocentre shift.**
Yu. A. Chernetenko.
1. Spain–USSR Workshop on Positional Astronomy and Celestial Mechanics, p. 115 - 123 (1991). - See Abstr. 012.025 for the main entry.

When constructing the numerical theory of Comet Encke it was supposed that the brightness centre lays on the line of the vector of the nongravitational acceleration and the photocentre shift does not depend on the heliocentric distance. Such a reduction takes down the mean residual and explains partly the fluctuation of the nongravitational acceleration in individual apparitions. In systematical respect this influence is practically absent.

103.069 **Encke, the comet.**
Z. Sekanina.
J. R. Astron. Soc. Can., Vol. 85, No. 6, p. 324 - 376 (Dec 1991).

The history of observations, orbital investigations, and physical studies of Encke's comet is reviewed and its significance in the quest for understanding the nature and evolution of comets is discussed.

103.070 **The evolution of P/Encke's light curve: no secular fading, a vanishing perihelion asymmetry.**
L. Kamél.
Icarus, Vol. 93, No. 2, p. 226 – 245 (Oct 1991).
An investigation of the evolution of the light curve of P/Encke in the period 1832 – 1987 has been made. Contrary to other analyses, no evidence for a secular fading was found. What has been reported by others as a secular fading is instead a result of a shift in the time of maximum brightness, from 3 weeks before perihelion in the 1840s to a few days after perihelion in the latest apparitions. This shift has also caused the perihelion asymmetry of the light curve to decrease.

Periodic comet Faye

103.071 **Periodic comet Faye (1991n).**
IAU Circ., No. 5325, p. 1 (15 Aug 1991).
Further information is given in Nos. 5328, 5348, 5356, 5362, 5366, 5382, 5407.

103.072 **P/comet Faye (1991n).**
Br. Astron. Assoc. Circ., No. 711, p. 5 – 6 (4 Oct 1991).

Periodic comet Gale

103.073 **Periodic comet Gale (1927 VI = 1938).**
OAA Comput. Sect. Circ., No. NK 549, p. 1 – 3 (4 Apr 1991).

Periodic comet Giacobini–Zinner

103.074 **Comparison of magnetosonic wave and water group ion energy densities at comet Giacobini–Zinner.**
K. Staines, A. Balogh, S. W. H. Cowley, P. M. de F. Forster, R. J. Hynds, T. S. Yates, T. R. Sanderson, K.-P. Wenzel, B. T. Tsurutani.
Adv. Space Res., Vol. 11, No. 9, p. 83 – 86 (1991). – See Abstr. 012.006 for the main entry.
Magnetosonic waves are generated by the instability of the new-born cometary water group ion distribution. The authors investigate the energy transfer between components of the water group ion velocity distributions measured at comet Giacobini–Zinner, caused by wave–particle scattering, and compare the thermal energy density of the ions with the energy density of the waves.

Periodic comet Grigg–Skjellerup

103.075 **Linkage of all the apparitions of the periodic comet Grigg–Skjellerup during 1808–1988.**
G. Sitarski.
Acta Astron., Vol. 41, No. 3, p. 237 – 253 (1991).
As an observational material of the comet, 440 positional observations made in the period 1808–1988 were collected. The 433 observations from 1922–1988 were selected and weighted yielding 820 residuals for the orbit improvement. The observations before 1922, six in 1902 and one in 1808, of a low accuracy were also used but with a very small weight. Nongravitational effects in the comet's motion have been investigated in two ways: either as a secular change da/dt of the semi-major axis of the comet's orbit or as the Marsden's nongravitational parameters A_1, A_2, A_3. A new procedure was applied to the Marsden's method for nongravitational effects. It allowed the author to find the angular parameters η, I, Φ for the rotating nucleus of the comet and to express by them the traditional parameters A_1, A_2, A_3, being now complicated functions of time. Assuming the existence of the precessional motion of the spin-axis of the cometary nucleus, the author took into account this phenomenon by the simplest way on accepting $d\Phi/dt$ and dI/dt as constant quantities. The six parameter model for the nongravitational effects, based on the rotating nucleus, appeared to be very good to link successfully all the apparitions of the comet. The evolution of the comet's orbit as well as of the angular parameters of the rotating nucleus are presented in a table of orbital elements.

103.076 **Prediction for the return of Comet P/Grigg–Skjellerup in 1992.**
G. Sitarski.
Acta Astron., Vol. 41, No. 4, p. 297 – 308 (1991).
A close encounter of the spacecraft GIOTTO with Comet P/Grigg–Skjellerup, expected on July 10, 1992, requires an exact ephemeris of the comet for July 1992. Prediction for the next return of the comet is based on the model of nongravitational comet's motion (elaborated earlier) using all the observations of the comet from the interval 1808–1988. Nongravitational effects varying with time were taken into account applying the Marsden's method, but using the angular parameters η, I, Φ for the rotating nucleus of the comet and including the precessional motion of the spin-axis of the nucleus. A displacement of the photometric and gravitational centers along the radius-vector r, expressed by $\Delta r = D\,r^{-3}\exp(-r^2/2)$ was considered when using the observations for the orbit improvement. An exactness of the prediction was estimated by means of testing the prediction for the comet's return in 1987. Ephemeris of the comet for 1992 and the space ephemeris of the comet for July 1992 are computed.

103.077 **Periodic comet Grigg–Skjellerup.**
IAU Circ., No. 5315, p. 1 (31 Jul 1991).
Further information is given in No. 5344.

103.078 **Giotto to visit comet P/Grigg–Skjellerup in 1992.**
G. Schwehm, T. Morley, H. Boehnhardt.
Messenger, No. 65, p. 37 – 39 (Sep 1991).

Periodic comet Halley

103.079 **Interpretation of the ion mass spectra in the mass per charge range 25–35 amu/e obtained in the inner coma of Halley's comet by the HIS-sensor of the Giotto IMS experiment.**
J. Geiss, K. Altwegg, E. Anders, H. Balsiger, W.-H. Ip, A. Meier, M. Neugebauer, H. Rosenbauer, E. G. Shelley.
Astron. Astrophys., Vol. 247, No. 1, p. 226 – 234 (Jul 1991).
The IMS–HIS double-focusing mass spectrometer that flew on the Giotto spacecraft covered the mass per charge rage from 12 to 56 (amu/e). By comparing flight data, calibration data and results of model calculations of the ion population in the inner coma, the absolute mass scale is established, and ions in the mass-range 25 to 35 are, identified. Ions resulting from protonation of molecules with high proton affinity are relatively abundant, enabling to estimate relative source strengths for H_2CO, CH_3OH, HCN, and H_2S, providing for the first time a positive in situ measurement of methanol. Also upper limits for NO and some hydrocarbons are derived.

103.080 **On the energetic ion species detected by EPONA at comet Halley.**
G. Erdös, P. W. Daly, E. Kirsch, S. McKenna-Lawlor.
Ann. Geophys., Vol. 9, No. 4, p. 291 – 297 (Apr 1991).
The anisotropy of energetic ions due to solar wind convection is investigated at comet Halley on the basis of data obtained by EPONA aboard Giotto. A simple method is introduced to determine the ion velocity from the counting rates detected in two sectors with different viewing angles relative to the Sun. The calculated ion velocities imply that the cometary ions observed up to 8 million km from the nucleus are unlikely to be protons and most probably comprised oxygen (or water group) ions. Consequences accruing to the appearance of heavy ions at large distances from the nucleus are discussed.

103.081 **The spin state and homogeneity of comet Halley's nucleus.**
M. J. S. Belton, W. H. Julian, A. J. Anderson, B. E. A. Mueller.
Icarus, Vol. 93, No. 2, p. 183 – 193 (Oct 1991).
The authors determine a model for the spin state of comet Halley's nucleus that simultaneously satisfies imaging data from the Vega and Giotto encounters, and a wide range of ground-based data including observations of CN-jets, CN-shells, C_2 production rates, and photometric variability. The model is that of an excited, axially symmetric, rotator whose shape is taken to be that of a prolate spheroid. The model is characterized by five, localized, areas on the nucleus which dominate the observed activity. One of these, located near the "waist" of the nucleus, appears to be solely responsible for the initiation of CN-shell boundaries seen propagating through the coma and may also be responsible for photometric activity seen at large heliocentric distances on approach to the sun. This behavior may indicate the presence of either a large-scale chemical inhomogeneity in the nucleus or, possibly, an extensive region on the surface with unusual physical structure.

103.082 **Estimation of the composition of the comet Halley dust grains.**
L. M. Mukhin, G. G. Dol'nikov, E. N. Evlanov,
O. F. Prilutskij, M. N. Fomenkova.
Pis'ma Astron. Zh., Tom 17, No. 7, p. 637 – 644 (Jul 1991). In Russian. English translation in Sov. Astron. Lett., Vol. 17, No. 4.
A new approach to the determination of the composition of comet Halley's dust component is suggested in this paper with dust particles mass taken into account. Dust particles mass is estimated by using signals measured at the moment of dust grain impact on target. It is shown that using weighting factors makes mean ionic abundances rather close to solar abundances. Mean ionic abundances of the main rock-forming elements of particles of different mass or different types are also considered.

103.083 **Jets physics in comet P/Halley.**
V. Formisano, G. Bellucci, F. Tripodi, E. Amata.
Nuovo Cimento C, Vol. 14, N. 4, p. 319 – 334 (Jul–Aug 1991).
CCD images of comet P/Halley taken in April 1986 are analysed. These images show the presence of jets. The jet maximum intensity is at <1000 km from the nucleus for dust, at $(20-30)\cdot10^3$km for H_2O^+ and at $(60-90)\cdot10^3$km for C_2, implying $2\cdot10^4$s or 10^5s formation time for the last two molecules. Comparison of various images shows that H_2O^+ and C_2 jets are in very different directions, the first one coinciding with the main dust jet, the second one appearing as a continuation in space of a very short dust jet. It is speculated that dust grains have very different sizes and chemical composition from the moment they leave the cometary nucleus, and that grains in the small dust jet undergo fragmentation and chemical reactions to produce C_2 jets.

103.084 **Ultra low frequency waves at the magnetic pile-up boundary of comet P/Halley.**
C. Mazelle, G. Belmont, K.-H. Glassmeier, D. Le Quéau,
H. Rème.
Adv. Space Res., Vol. 11, No. 9, p. 73 – 77 (1991). – See Abstr. 012.006 for the main entry.
Around the magnetic pile-up boundary observed by Giotto at 135,000 km from the nucleus of comet Halley, plasma and magnetic features display evidence of ultra-low frequency modes. Outside the pile-up region, a large region is characterized by an anti-correlation between the electron plasma density and the magnetic field strength. The study of the electron pressure tensor and the anisotropy of the distribution function lead to the identification of these modes as mirror modes. Inside the boundary, quasi-monochromatic large amplitude waves are observed with a very close correlation between the magnetic field magnitude and the electron density and have been identified as fast mode waves.

103.085 **Plasma parameters near the comet Halley bow shock.**
A. J. Coates, A. D. Johnstone, R. L. Kessel,
D. E. Huddleston, B. Wilken, K. Jockers, F. M. Neubauer,
H. Rème.
Adv. Space Res., Vol. 11, No. 9, p. 227 – 230 (1991). – See Abstr. 012.006 for the main entry.
The authors analyse the available plasma data from the instruments on Giotto for the inbound bow shock crossing at comet Halley. Proton, cometary heavy ion, electron and magnetic field data are combined on the inbound pass to yield the pressures of all the plasma components, and hence the characteristic plasma speeds and the Mach number of the flow.

103.086 **Cometary ion distributions near the pickup energy outside comet Halley's bow shock.**
T. I. Gombosi, M. Neugebauer, A. D. Johnstone, A. J. Coates,
D. E. Huddleston.
Adv. Space Res., Vol. 11, No. 9, p. 275 – 278 (1991). – See Abstr. 012.006 for the main entry.
This paper compares calculated and measured energy spectra of implanted H^+ and O^+ ions on the assumption that the pick-up geometry is quasi-parallel and about 1% of the waves generated by the cometary pickup process propagates backward (towards the comet). The model provides a good description of the implanted O^+ and H^+ energy distribution near the pickup energies.

103.087 **Chemical composition of the dust component of Comet Halley.**
L. M. Mukhin, G. G. Dol'nikov, E. N. Evlanov,
O. F. Prilutskij, M. N. Fomenkova.
Sov. Astron. Lett., Vol. 17, No. 4, p. 271 – 274 (Jul 1991). Current Physics Microform No.: 9202X0447. English translation of Pis'ma Astron. Zh., Tom 17, No. 7, p. 637 – 644 (Jul 1991).
The authors propose a new way to determine the composition of the dust component of Comet Halley that takes dust particle masses into consideration. These masses are estimated using the signals measured upon particle impact on a target. They show that using weighting factors makes the mean ionic composition of cometary dust similar to that of the Sun. Finally, they compare the mean ion abundances of the principal rock-forming elements among particles of differing mass and composition.

103.088 **The 16.0 March 1986 disconnection event in Comet Halley.**
J. C. Brandt, Y. Yi, C. E. Randall.
Bull. Am. Astron. Soc., Vol. 23, No. 2, p. 939 (1991). Abstract. – See Abstr. 012.008 for the main entry.

103.089 **Analysis of the Pioneer-Venus Lyman-α image of the hydrogen coma of comet P/Halley.**
W. H. Smyth, M. R. Combi, A. I. F. Stewart.
Science, Vol. 253, No. 5023, p. 1008 – 1010 (30 Aug 1991).
Comet Halley passed within 0.27 AU of Venus on 4 Feb 1986, 5 days before perihelion. This provided a unique opportunity to observe the comet's coma with the ultraviolet spectrometer orbiting the planet aboard the Pioneer Venus Orbiter spacecraft when the coma was otherwise obscured from Earth's view by the sun's glare. More than 9000 data points acquired systematically over the 5-day period from 2 to 6 Feb were combined to construct an excellent Lyman-α image of the hydrogen coma. The Lyman-α image was successfully reproduced with a comprehensive physical model, thereby verifying and documenting the underlying chemical kinetics and dynamics of the hydrogen coma.

103.090 **Morphology of the nucleus of comet P/Halley.**
P. J. Stooke, A. Abergel.
Astron. Astrophys., Vol. 248, No. 2, p. 656 – 668 (Aug 1991).
The nucleus of comet P/Halley has been modelled using techniques developed for mapping non-spherical bodies. The rotation assumptions of Sagdeev et al. (1986) were used to estimate sub-spacecraft and sub-solar latitudes and longitudes for Vega and Giotto images. An initial triaxial ellipsoid model of

the nucleus was altered by changing radii locally until the limbs and terminators of most images could be reproduced. The shape thus defined is illustrated cartographically, demonstrating techniques which may be applied to many other small bodies in future. The inability of the model to successfully reproduce all features of the images indicates that the rotation state is still not fully understood.

103.091 **Comet P/Halley's dust production rate at Giotto encounter derived from Halley Multicolour Camera observations.**
N. Thomas, H. U. Keller.
Astron. Astrophys., Vol. 249, No. 1, p. 258 – 268 (Sep 1991).
Halley Multicolour Camera observations of the inner coma of comet Halley on the night of March 13th 1986 from on board ESA's spacecraft Giotto showed that the intensity seen along the line of sight through the point of closest approach of Giotto to the cometary nucleus was 2.4 times greater than the average intensity around a cricle at the same impact parameter. The result implies that in situ measurements of dust, when combined with assumptions of isotropy, will over–estimate the cometary mass loss. Optical constants for the dust have been determined by application of the dust size distribution and the total mass impacting the spacecraft. Observational evidence suggests that the in situ measurements may have studied points in the coma where larger particles were more numerous and where particle terminal velocities were high. Lower velocities and smaller particle sizes may be present (for example above the nightside of the nucleus) leading to lower values of the total dust production rate.

103.092 **Spectrophotometry of Halley's comet in near IR region.**
V. F. Esipov, P. P. Korsun, O. Mamadov, V. G. Parusimov.
Astron. Vestn., Tom 25, No. 5, p. 599 – 602 (Sep – Oct 1991). In Russian. English translation in Solar Syst. Res., Vol. 25, No. 5.
Five negatives of Halley's comet were selected for photometrical treatment from the series of spectrograms near IR region of 6000 – 9000 Å. The molecular bands CO (3–0), (4–1), (10–2), (11–3); NH_2(0,6,0), (0,5,0); CN (2–0), (3–1), (4–2), (5–3) in the red system and C_2(2–0) in Phillips system were identified. The asymmetrical distribution has been detected from cross sections of CN and it was explained by near nuclear region jets influence.

103.093 **Analysis of the Swings effect and Greenstein effect in Comet P/Halley.**
W. A. Jaworski, J. B. Tatum.
Astrophys. J., Vol. 377, No. 1, p. 306 – 317 (10 Aug 1991).
Spectra have been obtained of Comet P/Halley using the Image Photon Counting System on the AAT in 1986 April. The CN (0, 0) violet system is analyzed to demonstrate the Swings resonance – fluoresence mechanism and multiple spectral images are used in an attempt to create a velocity map of the CN gas throughout the coma by application of the Greenstein effect. The theoretical calculations of the Swings effect are found to be in good agreement with the observed intensities of the R–branch lines of Comet P/Halley. This validates the assertion of an optically thin coma. In addition, 13 lines interspersed throughout the R–branch are identified as lines belonging to the ^{12}CN (1, 1) band and the ^{13}CN (0, 0) band. The ^{13}CN (0, 0) isotope R(7) and R(8) lines are used to obtain the $^{12}C/^{13}C$ ratio in the comet. A value of 89 ± 17 is found in agreement with the solar value. The Greenstein effect is seen observationally and velocity profiles through the coma are obtained by modeling the R(9)/R(12) ratio of the CN (0, 0) band. Changes in radial velocity as large as 1.57 km s^{-1} are found to exist within spatial distances of 4500 km.

103.094 **Velocity of individual large dust particles ejected from comet P/Halley.**
K. Richter, W. Curdt, H. U. Keller.
Astron. Astrophys., Vol. 250, No. 2, p. 548 – 555 (Oct 1991).
During the final approach phase of ESA's spacecraft Giotto to comet P/Halley several attitude excursions were detected by distinct spin rate changes of the spacecraft and corresponding changes of the viewing direction of the on board Halley Multicolour Camera. These attitude excursions were caused by individual large dust particles with lower limits for the mass between 0.5 and 10 mg derived by elementary mechanics. For 5 discrete impact events Keplerian trajectories of the particles and associated parameters of their emissions from the nucleus have been calculated. Ejection velocities of about 40 m s^{-1} have been found for two particles in the 1 mg mass range and of $\leqslant 10$ m s^{-1} for particles in the 10 mg range for the most probable scenario, where the recorded heavy particles start their trajectories in sunward directed jets during a period around perihelion. The outflow velocities of the 10 mg particles are significantly lower than predicted by gas dynamic calculations. The discrepancy could lead to an alternative interpretation of the observation assuming that the particles found near the nucleus had been released from the nucleus long time ago on a time scale of orbital periods. Similar size particles were found in long dust trails of short period comets. This alternative would influence the derivation of the particle size distribution and accumulated mass flux from the observations by the Giotto instruments.

103.095 **Hypervelocity dust particle impacts observed by the Giotto magnetometer and plasma experiments.**
F. M. Neubauer, K.–H. Glassmeier, A. J. Coates, R. Goldstein, M. H. Acuña, G. Musmann.
Geophys. Res. Lett., Vol. 17, No. 11, p. 1809 – 1812 (Oct 1990).
The authors report thirteen very short events in the magnetic field of the inner magnetic pile–up region of comet Halley observed by the Giotto magnetometer experiment together with simultaneous plasma data obtained by the Johnstone plasma analyzer and the ion mass spectrometer experiments. The events are due to dust impacts in the milligram range on the spacecraft at the relative velocity between the cometary dust and the spacecraft of 68 km/sec. They are generally consistent with dust impact events derived from spacecraft attitude perturbations by the Giotto camera. Their characteristic shape generally involves a sudden decrease in magnetic field magnitude, a subsequent overshoot beyond initial field values and an asymptotic approach to the initial field somewhat reminiscent of the magnetic field signature after the AMPTE releases in the solar wind. These observations give a new way of analyzing ultra–fast dust particles incident on a spacecraft.

103.096 **Distributed coma sources and the CH_4/CO ratio in comet Halley.**
D. C. Boice, W. F. Huebner, M. J. Sablik, I. Konno.
Geophys. Res. Lett., Vol. 17, No. 11, p. 1813 – 1816 (Oct 1990).
Early analyses of the Ion Mass Spectrometer (IMS) data from the Giotto flyby of Comet P/Halley indicated significant abundances of CH^+_n (n = 1 to 4). The source of these ions was assumed to be frozen CH_4 in the nucleus. An abundance of about 2% CH_4 was consistent with this interpretation, resulting in a ratio of CH_4/CO that is greater than the predicted limits for interstellar clouds or the solar nebula. However, subsequent analyses of data from the Giotto Neutral Mass Spectrometer (NMS) indicate distributed sources of CO and H_2CO in the coma that are most likely associated with organic (CHON) particles, rich in CH–bearing compounds that decay and produce CH^+_n species. The authors present a model that qualitatively accounts for the measured spatial distribution of CO and H_2CO and indicates that most of the CH^+_n deduced from the Giotto IMS data may originate from organic compounds in the coma dust and not from volatiles released directly from the nucleus. As a consequence, the absence of CH_4 in the icy phase of the nucleus is consistent with the observations with an upper limit of about 0.5% to the ratio CH_4/CO. This is an important criterion for theories of comet formation.

103.097 **Fragmentation of dust grains by Giotto's impact–generated atmosphere during the encounter with comet Halley.**
H. Porsche, J. Hoell.
Astron. Astrophys., Vol. 251, No. 2, p. L19 – L22 (Nov 1991). Letter–to–the–editor.
During its flight through comet Halley's atmosphere the European spacecraft Giotto had to overcome a great number of high–energy impacts of gas molecules and dust grains. To protect

the spacecraft it was equipped with a dual sheet bumper shield. Only heavy dust grains with masses $\geqslant 10^{-9}$kg could have penetrated the outer shield. Within the innermost region of the cometary coma the impacting dust generated an unstable atmosphere of secondary particles (mainly Al^+ ions) in front of the spacecraft. When the arriving particles entered this artificial atmosphere numerous collisions reduced their mass below the critical value for penetrating the shield. Only far from encounter, namely at much more than 1000 km distance to the cometary nucleus, the atmosphere around the spacecraft was so tenuous that large dust grains could perforate the outer shield.

103.098 **Spectrophotometric study of P/Halley during post perihelion passage.**
P. S. Goraya, B. S. Rautela, B. B. Sanwal, H. K. Duggal, N. K. Chandel.
Earth, Moon, Planets, Vol. 54, No. 1, p. 13 – 17 (Jul 1991).

Spectrophotometric observations of P/Halley are presented for four nights in the optical region (λλ3200–7000 Å) during post–perihelion period. Emission features due to CN, CH, C_3 and C_2 molecules are identified and the total apparent fluxes in each emission band are measured. It is found that the comet display large variations in the emission bands as well as continuum.

103.099 **Sulfur–bearing ions in the ionosphere of comet Halley.**
M. L. Marconi, D. A. Mendis, D. L. Mitchell, R. P. Lin, A. Korth, H. Réme.
Astrophys. J., Vol. 378, No. 2, p. 756 - 762 (10 Sep 1991).

The radial density profiles of the ions corresponding to 35, 45, 47, and 48 amu, inside the ionopause of comet Halley, obtained by the heavy ion analyzer (PICCA) on board the Giotto spacecraft are presented. A photochemical model of the outflowing atmosphere / ionosphere, which emphasizes the sulfur chemistry in an H_2O–dominated atmosphere, is used to support the identification of the four ions as H_3S^+, HCS^+, H_3CS^+, and SO^+, respectively. Both the observed values as well as the radial profiles of the ion densities are in fair agreement with the observations in the inner portion of the ionosphere. While the observed increases in the ion densities just inside the ionopause, following the initial radial decreases, cannot be explained by this simple spherically symmetric flow model, it can nevertheless be shown that they are not a consequence of the photochemistry. The authors speculate that the increased ion densities may be associated with the existence of a thin "inner" shock layer just inside the ionopause, where both lateral flow and dissociative recombination play important roles in determining the ion densities.

103.100 **Swan bands in the spectrum of comet Halley.**
A. A. Atai, D. I. Shestopalov.
Kinematika Fiz. Nebesn. Tel, Tom 7, No. 6, p. 15 – 20 (Nov–Dec 1991). In Russian. English translation in Kinematics Phys. Celest. Bodies, Vol. 7, No. 6.

Spectrophotometric observations of comet Halley made at the Shemakha Astrophysical Observatory (December, 1985 – January, 1986) are analysed. The emission bands of Swan C_2 (2,0) λ438.2 nm and $CN(B^2\Sigma^+–X^2\Sigma^+)$ near λ420.0 nm have been identified. It is shown that flux ratios of the Swan bands in the spectra of comet Halley, lg [F(0,1)/F(0,0)] and lg [F(1,0)/F(0,0)], depend on the heliocentric distance: for the pre–perihelion as $r^{-0.43\pm0.06}$ and $r^{-0.26\pm0.30}$; for the post–perihelion as $r^{0.33\pm0.40}$ and $r^{1.43\pm0.38}$, respectively. The excitation temperature T_{exc} of C_2 molecules in comet Halley's atmosphere (for thermodynamical equilbrium) is 5900 – 6500K for the C_2 (1,0) λ473.7 nm and 5900 – 6700K for the C_2 (0,1) λ563.5 nm. It is established that formation of C_2 molecules in comet Halley's atmosphere before and after perihelion occurs with different velocity.

103.101 **Estimate of the mass of comet Halley's dust particles based on data of PUMA experiment.**
E. N. Evlanov, O. F. Prilutskij, M. N. Fomenkova.
Kosm. Issled., Tom 29, Vyp. 4, p. 641 – 646 (Jul–Aug 1991). In Russian. English translation in Cosmic Res. Vol. 29, No. 4.

The mass of cometary dust particles the spectra of which were measured during VEGA mission in PUMA experiment is estimated. All the of spectra were subdivided into groups in accordance with the values of analogue signals measured at the moment of dust particle impact on target. Absolute mass values for each group are determined from a comparison of obtained distribution and data of SP–2 dust counter. It is shown that dust particle mass is in the range $5{\cdot}10^{-17}$ to $5{\cdot}10^{-12}$g for PUMA–1 instrument and in the range $2{\cdot}10^{-16}$ to $5{\cdot}10^{-12}$g for PUMA–2 instrument. The dependence of dust particles properties on their mass is also considered.

103.102 **Structural analysis of comet Halley's CN coma by image processing techniques.**
R. Schulz.
3. ESO/ST–ECF Data Analysis Workshop, p. 73 – 78 (1991). – See Abstr. 012.061 for the main entry.

The evolution of jet and shell structures in comet Halley's neutral CN coma was investigated by using several special image processing techniques. Four different methods are presented to enhance the two structures as well as their temporal changes. These are the construction of radial derivatives and the division of original images by time averaged profiles, mean radial profiles or consecutive images, respectively. Both structures are likewise existent in each image and a transition between the CN jets and CN shells can be pointed out by the first and the last named method.

103.103 **Observations of solar wind ion charge exchange in the comet Halley coma.**
S. A. Fuselier, E. G. Shelley, B. E. Goldstein, R. Goldstein, M. Neugebauer, W.–H. Ip, H. Balsiger, H. Rème.
Astrophys. J., Vol. 379, No. 2, p. 734 – 740 (1 Oct 1991).

Giotto Ion Mass Spectrometer / High Energy Range Spectrometer observations of solar wind ions show charge exchange effects and solar wind compositional changes in the coma of comet Halley. As the comet was approached, the He^{2+} to proton density ratio increased from 2.5% in the solar wind to ~4% about 1 hr before closest approach, after which time it decreased to ~1%. Abrupt increases in this ratio from 2.5% to 4.5% were also observed in the beginning and near the end of the so–called mystery region (8.6 – 5.5×10^5km from the comet along the spacecraft trajectory). These abrupt increases in the density ratio were well correlated with enhanced fluxes of keV electrons. The general increase and then decrease of the He^{2+} to proton density ratio is quantitatively consistent with a combination of the addition of protons of cometary origin to the plasma and loss of plasma through charge exchange of protons and He^{2+}. In agreement with the solar wind proton and He^{2+} observations, solar wind oxygen and carbon ions were observed to charge exchange from higher to lower charge states with decreasing distance to the comet. The more abrupt increases in the He^{2+} to proton and the He^{2+} to O^{6+} density ratios in the mystery region require a change in the solar wind ion composition in this region, while the correlation with energetic electrons and properties of the exit from this region indicate processes associated with the comet.

103.104 **Comment on "Ion acceleration at the contact surface of comet P/Halley".**
R. Goldstein.
J. Geophys. Res., Vol. 96, No. A7, p. 11641 (1 Jul 1991).

A comment on the paper by C. Gurgiolo and J. D. Winningham published in J. Geophys. Res., Vol. 95, No. A10, p. 17051 – 17073 (1 Oct 1990). See Abstr. 52.103.158. A reply by C. Gurgiolo and J. D. Winningham is given in J. Geophys. Res., Vol. 96, No. A7, p. 11643 – 11646 (1 Jul 1991).

103.105 **Comet Halley's outburst.**
D. W. Hughes.
Mon. Not. R. Astron. Soc., Vol. 251, No. 1, p. 26P – 29P (1 Jul 1991).

It is suggested that the outburst suffered by Comet Halley (1986 III) in early 1991 February, was caused by the ejection from the nucleus of 1.4×10^{13}g of dust, this being about 0.02% of the cometary mass. The energy required was in the range 2×10^{19}

to $2x10^{23}$erg. The most likely source of this energy was impact with a member of an outer asteroid belt, the impacting object having a diameter of between about 2.6 and 60 m.

103.106 **Carbon suboxide in comet Halley?**
W. T. Huntress Jr., M. Allen, M. Delitsky.
Nature, Vol. 352, No. 6333, p. 316 – 318 (25 Jul 1991). Letter–to–the–editor.
Observational data acquired during the recent appearance of comet Halley pose a puzzle about the nature and distribution of elemental carbon and carbonaceous material in its nucleus and coma. The authors propose that an additional contribution from carbon suboxide (C_3O_2) in the coma dust and the nucleus material is consistent with the observational data. A production rate in the coma for C_3O_2 of about 0.03 – 0.04 times that of water would provide the distributed source of elemental carbon and CO within 10^4km of the nucleus that is required to explain the data from the Giotto spacecraft and from ground–based observations.

103.107 **A kick from the solar wind as the cause of comet Halley's February 1991 flare.**
D. S. Intriligator, M. Dryer.
Nature, Vol. 353, No. 6343, p. 407 – 409 (3 Oct 1991). Letter–to–the–editor.
Knowledge of the interaction of comets with the solar wind was greatly advanced during the 1986 passage of comet Halley, which was studied in situ and by remote sensing more thoroughly than any previous comet flyby. Of particular interest and novelty was the large flare that Halley was seen to produce on 12 February 1991, when it was 14.3 AU from the Sun and 18° below the ecliptic plane. There have been further outbursts, most recently on 17 March. The authors note that the Sun has been unusually active in recent months, and suggest that a shock wave, generated by the solar flare and propagating through the interplanetary medium, could have caused the comet to flare. For example, they show that a solar flare on 31 January could plausibly have produced a shock wave that would have reached Halley on 12 February, and would have been sufficiently strong to crack the comet's crust of fluffy ice.

103.108 **Why did Halley hiccup?**
P. Weissman.
Nature, Vol. 353, No. 6347, p. 793 – 794 (31 Oct 1991).
The author reports on the behaviour of comet Halley, the flaring in brightness again in February 1991 14 AU from the Sun and on possible explanations of this phenomenon.

103.109 **Comet P/Halley: spatial distributions and scale lengths for C_2, CN, NH_2, and H_2O.**
U. Fink, M. R. Combi, M. A. DiSanti.
Astrophys. J., Vol. 383, No. 1, p. 356 – 371 (10 Dec 1991).
Spatial profiles were obtained for emissions by C_2, CN, NH_2, and O I(^{1}D) from comet P/Halley long–slit spectroscopic exposures on 12 dates, extending from 1985 October to 1986 May. Haser model scale lengths were fitted to these data. The extended time coverage allowed to check for consistency between the various dates. Not unexpectedly, the time varying production rate of P/Halley severely affected the profiles after perihelion. This is demonstrated in two profile sequences on adjacent dates, March 1/2 and April 14/15. Because of the time–varying production rate, it was difficult to obtain reliable Haser model scale lenghts after perihelion. The authors' preperihelion analysis yielded Haser model scale lengths of sufficient consistency that they can be used for production rate determinations.

103.110 **Eruption on Comet Halley.**
M. Jõeveer.
Calendar of the Tartu Astronomical Observatory for the year, 1992.
p. 31 (1991). In Estonian. See Abstr. 046.016.

103.111 **Photometry of comet P/Halley on 1986 March 19.**
A. K. Sen, M. R. Deshpande, U. C. Joshi.
Bull. Astron. Soc. India, Vol. 19, Nos. 1 – 2, p. 47 – 51 (Mar – Jun 1991).
Comet P/Halley was photometrically observed in the coma and tail regions using IHW filters, at a heliocentric distance of 0.99 AU, on 1986 March 19. It was found that the maximum flux in the coma occurs for the C_2 molecule, whereas the fluxes in C_3 and CN are almost equal. At a distance of about 5.5×10^4km from the nucleus across the dust tail no flux was observed in CO^+ and H_2O^+, whereas the fluxes in CN, C_3 and C_2 were found lower as compared to the coma.

103.112 **Imaging polarimetry of comet P/Halley.**
A. K. Sen, U. C. Joshi, M. R. Deshpande, C. D. Prasad.
Bull. Astron. Soc. India, Vol. 19, Nos. 1 – 2, p. 82 (Mar – Jun 1991). Abstract. – See Abstr. 012.108 for the main entry.

103.113 **Estimation of electron density in the ion–tail of comet Halley using 103 MHz IPS observations.**
A. K. Sharma, S. K. Alurkar, P. Janardhan.
Bull. Astron. Soc. India, Vol. 19, Nos. 1 – 2, p. 82 (Mar – Jun 1991). Abstract. – See Abstr. 012.108 for the main entry.

103.114 **Observational and dynamical constraints on the rotation of comet P/Halley.**
N. H. Samarasinha, M. F. A'Hearn.
Icarus, Vol. 93, No. 2, p. 194 – 225 (Oct 1991).
The authors have carried out an extensive search to determine all rotational states of the nucleus of comet P/Halley which are capable of explaining a wide variety of observational data. They find that the orientation of the large end of the long axis during the Vega–1 encounter must be reversed from that suggested by Sagdeev et al. (1989) in order to satisfy some of the constraints. Neither pure rotations nor short axis modes (SAMs) are capable of satisfying the constraints. The most likely solutions are long axis modes.

103.115 **Quasi–linear pitch angle and energy diffusion of pickup ions near comet Halley.**
D. E. Huddleston, A. D. Johnstone, A. J. Coates, F. M. Neubauer.
J. Geophys. Res., Vol. 96, No. A12, p. 21329 – 21341 (1 Dec 1991).
The process of ion pickup in the environment of Halley's comet is studied in order to see if velocity diffusion driven by the observed level of turbulence can explain the observed development of the implanted ion distribution. The theoretical description used is based on a quasi–linear approach and considers the implantation and transport of cometary ions along solar wind flow lines. To make such a study requires some way of extrapolating the measurements on the Giotto trajectory into the upstream region; models for mass loading and turbulence are used. A simplified kinetic equation describing the source, convection and quasi–linear velocity diffusion of the heavy cometary ions is solved numerically along flow lines parallel to the Sun–comet line. Full two–dimensional (pitch angle and velocity) distributions are obtained at positions along the Giotto trajectory, which are compared with measurements. This study finds that quasi–linear theory, with the authors empirical model for the observed turbulence level, produces the right order of pitch angle diffusion.

103.116 **Comet Halley observations: An amateur participation in the International Halley Watch programme.**
G. Wolf.
South. Stars, Vol. 34, No. 5, p. 309 – 313 (Dec 1991).
Summary data are presented for ~800 individual visual observations of comet Halley, made over the period 1985 Sep 12 to 1987 May 1, as part of the International Halley Watch. The data are reduced to 152 records.

Periodic comet Hartley 2

103.117 **Periodic comet Hartley 2 (1991t).**
IAU Circ., No. 5304, p. 1 (12 Jul 1991).
Further information is given in Nos. 5312, 5324, 5340, 5359, 5372, 5391.

103.118 **Periodic comet Hartley 2 (1991t).**
Yamamoto Circ., No. 2170, p. 1 (17 Aug 1991). In Japanese.
Further information is given in No. 2172.

103.119 **P/comet Hartley 2 (1991t).**
Br. Astron. Assoc. Circ., No. 710, p. 3 – 4 (2 Aug 1991).
Further information is given in No. 711.

103.120 **Periodic comet Hartley 2.**
OAA Comput. Sect. Circ., No. NK 565, p. 1 – 4 (24 Dec 1991).

Periodic comet Kowal 1

103.121 **Periodic comet Kowal 1.**
OAA Comput. Sect. Circ., No. NK 556, p. 1 (4 Sep 1991).

Periodic comet Kowal 2

103.122 **Periodic comet Kowal 2 (1991f_1).**
IAU Circ., No. 5406, p. 1 (17 Dec 1991).
Further information is given in Nos. 5407, 5415.

103.123 **Periodic comet Kowal 2 (1991f_1).**
Yamamoto Circ., No. 2175, p. 1 (25 Dec 1991). In Japanese.

Periodic comet Levy

103.124 **Periodic comet Levy (1991q).**
IAU Circ., No. 5296, p. 1 (2 Jul 1991).
Further information is given in Nos. 5306, 5313, 5325, 5352, 5375, 5413.

103.125 **Periodic comet Levy (1991q).**
Yamamoto Circ., No. 2169, p. 1 – 2 (13 Jul 1991). In Japanese.
Further information is given in No. 2170.

103.126 **Periodic comet Levy (1991q).**
Br. Astron. Assoc. Circ., No. 710, p. 5 – 6 (2 Aug 1991).
Further information is given in No. 711.

Periodic comet Machholz

103.127 **Periodic comet Machholz (1986 VIII).**
IAU Circ., No. 5301, p. 1 (9 Jul 1991).
Further information is given in Nos. 5302, 5318, 5320.

103.128 **Periodic comet Machholz (1986 VIII).**
Yamamoto Circ., No. 2169, p. 1 (13 Jul 1991). In Japanese.
Further information is given in No. 2170.

103.129 **P/comet Machholz (1986 VIII).**
Br. Astron. Assoc. Circ., No. 710, p. 1 – 2 (2 Aug 1991).

Periodic comet McNaught–Hughes

103.130 **Periodic comet McNaught–Hughes (1991y).**
IAU Circ., No. 5354, p. 1 (1 Oct 1991).
Further information is given in Nos. 5360, 5374.

103.131 **Periodic comet McNaught–Hughes (1991y).**
Yamamoto Circ., No. 2172, p. 1 – 2 (13 Oct 1991). In Japanese.

Periodic comet Metcalf–Brewington

103.132 **Periodic comet Metcalf–Brewington.**
OAA Comput. Sect. Circ., No. NK 551, p. 1 – 5 (8 Aug 1991).

Periodic comet Mrkos

103.133 **Periodic comet Mrkos (1991k).**
OAA Comput. Sect. Circ., No. NK 561, p. 1 – 2 (24 Dec 1991).

Periodic comet Mueller 3

103.134 **Periodic comet Mueller 3 (1990*l*).**
OAA Comput. Sect. Circ., No. NK 553, p. 1 (4 Sep 1991).

Periodic comet Peters–Hartley

103.135 **Periodic comet Peters–Hartley.**
OAA Comput. Sect. Circ., No. NK 550, p. 1 – 2 (4 Aug 1991).

Periodic comet Schwassmann–Wachmann 1

103.136 **The development of the CO^+ coma of comet P/Schwassmann–Wachmann 1.**
A. L. Cochran, W. D. Cochran, E. S. Barker, A. D. Storrs.
Icarus, Vol. 92, No. 1, p. 179 – 183 (Jul 1991).
The authors present spectroscopic observations of comet P/Schwassmann–Wachmann 1 in which they observed the development of the CO^+ coma for the first time. With these data it is shown that the maximum rise time of the CO^+ gas is 1.2 days. This yields a maximum creation time scale for CO^+ gas of 1×10^5sec at 5.8 AU. Evidence is also presented that the dust events and the gas events in this comet are not necessarily connected.

103.137 **Periodic comet Schwassmann–Wachmann 1**
IAU Circ., No. 5321, p. 1 (9 Aug 1991).
Further information is given in Nos. 5349, 5396, 5404.

103.138 **Occultation by P/Schwassmann–Wachmann 1.**
IAU Circ., No. 5417, p. 1 (27 Dec 1991).

103.139 **Periodic comet Schwassmann–Wachmann 1.**
Yamamoto Circ., No. 2170, p. 2 (17 Aug 1991). In Japanese.
Further information is given in Nos. 2174, 2175.

103.140 **Periodic comet Schwassmann–Wachmann 1.**
OAA Comput. Sect. Circ., No. NK 548, p. 1 - 18 (6 Apr 1991).

Periodic comet Shoemaker 1

103.141 **Periodic comet Shoemaker 1 (1991p).**
IAU Circ., No. 5336, p. 1 (5 Sep 1991).
Further information is given in No. 5361.

103.142 **Periodic comet Shoemaker 1.**
OAA Comput. Sect. Circ., No. NK 563, p. 1 - 3 (24 Dec 1991).

Periodic comet Shoemaker–Levy 1

103.143 **Periodic comet Shoemaker–Levy 1 (1991o).**
OAA Comput. Sect. Circ., No. NK 552, p. 1 - 2 (27 Aug 1991).

Periodic comet Shoemaker–Levy 2

103.144 **Periodic comet Shoemaker–Levy 2 (1990p).**
OAA Comput. Sect. Circ., No. NK 547, p. 1 (2 Feb 1991).

Periodic comet Shoemaker–Levy 3

103.145 **Periodic comet Shoemaker–Levy 3 (1991e).**
OAA Comput. Sect. Circ., No. NK 559, p. 1 - 2 (23 Dec 1991).

Periodic comet Shoemaker–Levy 4

103.146 **Periodic comet Shoemaker–Levy 4 (1991f).**
OAA Comput. Sect. Circ., No. NK 560, p. 1 - 2 (24 Dec 1991).

Periodic comet Shoemaker–Levy 5

103.147 **Periodic comet Shoemaker–Levy 5 (1991z).**
IAU Circ., No. 5359, p. 1 (4 Oct 1991).
Further information is given in Nos. 5361, 5376, 5416.

103.148 **Periodic comet Shoemaker–Levy 5 (1991z).**
Yamamoto Circ., No. 2172, p. 2 (13 Oct 1991). In Japanese.

Periodic comet Shoemaker–Levy 6

103.149 **Periodic comet Shoemaker–Levy 6 (1991b_1).**
IAU Circ., No. 5382, p. 1 (10 Nov 1991).
Further information is given in Nos. 5383, 5386, 5397, 5411, 5416.

103.150 **Periodic comet Shoemaker–Levy 6 (1991b_1).**
Yamamoto Circ., No. 2173, p. 1 - 2 (16 Nov 1991). In Japanese.

Periodic comet Shoemaker–Levy 7

103.151 **Periodic comet Shoemaker–Levy 7 (1991d_1).**
IAU Circ., No. 5389, p. 1 (18 Nov 1991).
Further information is given in Nos. 5394, 5397, 5412.

103.152 **P/Shoemaker–Levy 7 (1991d_1).**
Yamamoto Circ., No. 2174, p. 1 (15 Dec 1991). In Japanese.

Periodic comet Spacewatch

103.153 **Periodic comet Spacewatch (1991x).**
IAU Circ., No. 5341, p. 1 (10 Sep 1991).
Further information is given in Nos. 5343, 5351.

103.154 **Periodic comet Spacewatch (1991x).**
Yamamoto Circ., No. 2171, p. 1 (14 Sep 1991). In Japanese.
Further information is given in No. 2172.

Periodic comet Swift–Gehrels

103.155 **Nongravitational motion of Comet P/Swift–Gehrels.**
M. Bielicki, G. Sitarski.
Acta Astron., Vol. 41, No. 4, p. 309 - 323 (1991).

The authors collected 143 observations of the comet made during its four apparitions in: 1889/90, 1973, 1981/82, and 1991. The observations were elaborated for each apparition separately by the Bielicki's method which is presented in the paper. To link all the apparitions it was necessary to take into account a secular deceleration in the comet's motion. The authors applied Marsden's method for nongravitational effects in two ways: either with constant nongravitational parameters A_i, $i = 1, 2, 3$, or with $A_i(t)$ depending on time as connected with the rotating and precessing nucleus of the comet. The latter solution was a basis for the ephemeris of the comet during its expected return in 2000.

103.156 **Periodic comet Swift–Gehrels.**
OAA Comput. Sect. Circ., No. NK 546, p. 1 - 4 (15 Jan 1991).

Periodic comet Takamizawa

103.157 **Periodic comet Takamizawa (1991h).**
IAU Circ., No. 5303, p. 1 (11 Jul 1991).
Further information is given in No. 5318.

Periodic comet Tempel 2

103.158 **Comprehensive model for the nucleus of Periodic Comet Tempel 2 and its activity.**
Z. Sekanina.
Astron. J., Vol. 102, No. 1, p. 350 - 388 (Jul 1991). Current Physics Microform No.: 9107B2046.

This paper presents a comprehensive analysis and synthesis of an existing database on Period Comet Tempel 2, aimed at interpreting information on the physical properties of the comet's nucleus, its environment, and its activity. The result of this effort is a synergistic nucleus model, which entails the determinations of the object's dimensions, its rotation characteristics, its gross surface morphology, and the mode of its activity, including diurnal and seasonal variations in the production of water and dust.

Periodic comet Tsuchinshan 1

103.159 **Periodic comet Tsuchinshan 1 (1991c_1).**
IAU Circ., No. 5383, p. 1 (12 Nov 1991).

103.160 **Periodic comet Tsuchinshan 1 (1991c_1).**
Yamamoto Circ., No. 2173, p. 2 (16 Nov 1991). In Japanese.

Periodic comet Tsuchinshan 2

103.161 **Periodic comet Tsuchinshan 2 (1991e_1).**
IAU Circ., No. 5403, p. 1 (13 Dec 1991).

103.162 **Periodic comet Tsuchinshan 2 (1991e_1).**
Yamamoto Circ., No. 2175, p. 2 (25 Dec 1991). In Japanese.

Periodic comet van Biesbroeck

103.163 **Periodic comet van Biesbroeck (1989h_1).**
IAU Circ., No. 5314, p. 1 (30 Jul 1991).
Further information is given in Nos. 5345, 5352.

Periodic comet Wild 2

103.164 **Periodic comet Wild 2 (1989t).**
IAU Circ., No. 5299, p. 1 (7 Jul 1991).

Periodic comet Wirtanen

103.165 **Periodic comet Wirtanen (1991s).**
IAU Circ., No. 5303, p. 1 (11 Jul 1991).
Further information is given in Nos. 5323, 5346, 5367, 5389, 5400.

103.166 **Periodic comet Wirtanen (1991s).**
Yamamoto Circ., No. 2169, p. 2 (13 Jul 1991). In Japanese.

103.167 **P/comet Wirtanen (1991s).**
Br. Astron. Assoc. Circ., No. 711, p. 7 – 8 (4 Oct 1991).

Astrophysics' contribution to the study of comets in the XIX century.
See Abstr. 004.060.

Medieval observations of comets in Latvia.
See Abstr. 004.061.

Scanning with charge–coupled devices.
See Abstr. 013.094.

Suprathermal chemistry in space.
See Abstr. 022.162.

CO_2 profiles in cometary analogs.
See Abstr. 022.163.

Comets to quasars: surface photometry from standard stars, and the morphology of the galaxy–quasar interface.
See Abstr. 036.231.

Astrometrie an Kometen.
See Abstr. 036.365.

Space exploration of comets and primitive minor bodies.
See Abstr. 051.084.

A close encounter of the second kind – The Giotto Extended Mission (GEM).
See Abstr. 051.089.

Multiple–ion plasma boundaries.
See Abstr. 062.023.

Discrete phase changes within nonlinear steepened magnetosonic waves: comet Giacobini–Zinner.
See Abstr. 062.121.

Towards better scattering properties for a mixture of realistic grains.
See Abstr. 063.098.

Complexes of minor bodies in the solar system.
See Abstr. 091.015.

Complexes of minor solar system bodies.
See Abstr. 091.030.

Small solar system bodies.
See Abstr. 091.057.

Cometary water on Venus: impact modelling and possible evidence.
See Abstr. 093.036.

Positionsbeobachtungen von Kleinen Planeten und Kometen.
See Abstr. 098.016.

Keeping track of comets and minor planets.
See Abstr. 098.026.

Ephemerides of minor planets and comets.
See Abstr. 098.042.

Determination of positions of celestial bodies from photographic observations.
See Abstr. 098.049.

Asteroids, meteorites, interplanetary and cometary dust.
See Abstr. 098.101.

The regularity of the change of matter concentration in a cometary head as a function of the heliocentric distance.
See Abstr. 102.012.

Nongravitational effects in cometary motion.
See Abstr. 102.013.

Nongravitational effects and the aging of periodic comets.
See Abstr. 102.018.

Randomization of dust–ejecta motions and the observed morphology of cometary heads.
See Abstr. 102.021.

Singularities in mass–loaded MHD flow: the cometary bow shock.
See Abstr. 102.023.

A self–consistent model for the particles and fields upstream of an outgassing comet. 1. Gyrotropic and isotropic ion distributions.
See Abstr. 102.025.

Alfven wave amplifications and generation at cometary bow shocks.
See Abstr. 102.028.

The formation and role of water dimers in a cometary atmosphere, with application to comet P/Halley.
See Abstr. 102.031.

Perseids and comet P/1862 III.
See Abstr. 104.033.

Visual meteor results from the international Halley watch.
See Abstr. 104.034.

Perseid meteors and periodic comet Swift–Tuttle.
See Abstr. 104.039.

The Leonids, Comet Biela and Biela's associated meteoroid stream.
See Abstr. 104.041.

Prediction of meteor shower associated with P/Hartley 2 (1991t).
See Abstr. 104.043.

The orbit of the Lyrid meteor stream.
See Abstr. 104.046.

The composition and plasma signature of a large dust impact on the Giotto spacecraft.
See Abstr. 106.029.

A hypothesis on the Oort clouds of planets.
See Abstr. 107.008.

Observations of B stars and comet Austin from 900 – 1180 Å.
See Abstr. 114.133.

104 Meteors, Meteor Streams

104.001 **The energy spectra of meteor trails in the radio range.**
V. I. Bojkov, T. V. Kazakova, A. V. Karpov, M. M. Katsevman.
Geomagn. Aeron., Vol. 30, No. 2, p. 239 – 242 (Oct 1990). English translation of Geomagn. Aehron., Tom 30, No. 2, p. 292 – 296 (1990).
The spectral transformations and method of calculation of the energy spectra of radio waves scattered by a set of meteor trails formed in a given region of space in a specified time interval are given. Astronomical and atmospheric factors as well as the parameters of the measuring equipment are taken into account. Experiments confirm the calculations both in absolute value and also in shape: the calculated and experimental spectra at the points of the maxima agree, at the other points the differences do not exceed 7 dB, and the average deviations are no larger than tenths of a dB.

104.002 **Will the lion roar again?**
K. Croswell.
Astronomy, Vol. 19, No. 11, p. 44 – 49 (Nov 1991).
Concerning the Leonid meteor shower bursts.

104.003 **Lyriden, Eta Aquariden en Alfa Scorpiiden.**
J. van't Leven.
Radiant, Jaarg. 13, Nr. 4, p. 92 – 94 (Aug 1991).

104.004 **De vuurbol van 24 maart 1933.**
P. Jenniskens.
Radiant, Jaarg. 13, Nr. 5, p. 118 – 119 (Oct 1991).

104.005 **Winter 1990: Geminiden, Monocerotiden en σ–Hydrusiden.**
P. Jenniskens.
Radiant, Jaarg. 13, Nr. 6, p. 126 – 133 (Dec 1991).

104.006 **Major fireball seen from Ireland on 1991 February 20.**
H. Miles.
J. Br. Astron. Assoc., Vol. 101, No. 5, p. 291 – 292 (Oct 1991).

104.007 **Are the Geminids and 3200 Phaethon related genetically?**
E. N. Kramer, I. S. Shestaka.
Astron. Tsirk., No. 1549, p. 25 – 26 (May – Aug 1991).
The authors can state that even assuming the comet nature of Phaethon (to say nothing of an asteroidal one) one can not explain the ejection of Geminid swarm from this object.

104.008 **Meteoroid streams: formation, evolution, associations with comets and asteroids.**
P. B. Babadzhanov, Yu. V. Obrubov.
Astron. Vestn., Tom 25, No. 4, p. 387 – 407 (Jul – Aug 1991). In Russian. English translation in Sol. Syst. Res., Vol. 25.
The review of modern notions on the formation and evolution of meteoroid streams and their associations with comets and Apollo–Aten asteroids is given. The results of investigations of the Quadrantid, Geminid, Taurid, Eta–Aquarid and Orionid meteoroid streams dynamics show the possibility of producing up to eight related meteor showers by one meteoroid stream. Five new small body complexes in the Solar system were revealed.

104.009 **Influx of meteoritic matter to the Earth.**
Yu. I. Voloshchuk, B. L. Kashcheev.
Astron. Vestn., Tom 25, No. 4, p. 453 – 465 (Jul – Aug 1991). In Russian. English translation in Sol. Syst. Res., Vol. 25.
Estimations of the influx of meteoritic matter to the Earth, analysis and comparison of orbital element distributions of meteor bodies in the wide range of masses from 10^{-6} to 10 g, calculated by photographic and radar observations of meteor trails in the Earth's atmosphere are given.

104.010 **Velocity distribution and flux density of sporadic meteor bodies.**
P. B. Babadzhanov, V. M. Kolmakov.
Astron. Vestn., Tom 25, No. 4, p. 476 – 481 (Jul – Aug 1991). In Russian. English translation in Sol. Syst. Res., Vol. 25.
A new method for determining the distribution function of meteor bodies according to velocities from observations is described. This method gives the possibility of calculating the flux density of meteor bodies at any interval of meteor velocities.

104.011 **The front echo of meteors.**
A. Yu. Ol'khovatov.
Geomagn. Aehron., Tom 31, No. 4, p. 750 – 751 (Jul – Aug 1991). In Russian. English translation in Geomagn. Aeron., Vol. 31, No. 4.

104.012 **Determination of meteoric velocity from photographic observations.**
A. F. Zausaev, A. N. Pushkarev.
Astron. Vestn., Tom 25, No. 5, p. 609 – 612 (Sep – Oct 1991). In Russian. English translation in Solar Syst. Res., Vol. 25, No. 5.
An algorithm for computing the preatmospheric velocity by means of the corresponding meteor photographs (one of them

tracking through the rotating shutter) is proposed. The preatmospheric velocity is determined by solution of a system of equations which expresses the meteor path as a function of time.

104.013 **The complex of short–periodic orbits of Geminid–type.**
I. S. Shestaka.
Astron. Tsirk., No. 1543, p. 33 – 34 (Mar – Apr 1990). In Russian.
An analysis of meteor orbits made it possible to single out the meteor swarms observed from January up to December. They originated from bodies with orbits similar to those of the Geminids.

104.014 **The statistical confirmation of meteor swarms' evolution.**
I. S. Shestaka.
Astron. Tsirk., No. 1543, p. 35 – 36 (Mar – Apr 1990).
The author made an analysis of about 3000 meteor showers which had been observed by different methods from the thirties of the last century to the eighties of the present century. There is a possibility of some swarms' dissipation and formation of others, such processes however take much more time than that for regular meteor observations covering only a century and a half.

104.015 **The 1990 Orionids.**
R. Koschack, P. Roggemans.
WGN, Vol. 19, Nr. 4, p. 115 – 130 (Aug 1991).

104.016 **Computer simulation of Earth–grazing fireballs.**
D. W. Olson, R. L. Doescher, K. M. Watson.
WGN, Vol. 19, Nr. 4, p. 130 – 131 (Aug 1991).
A computer simulation is presented for the trajectory of the Earth–grazing fireball observed over Czechoslovakia and Poland on Oct. 13, 1990.

104.017 **Brightness distribution of meteors and fireballs.**
V. V. Martynenko, A. S. Levina.
WGN, Vol. 19, Nr. 4, p. 132 – 136 (Aug 1991).
About 16000 meteor data gathered in the USSR during the period 1957 – 1975 were used to determine meteoroid flux as a function of magnitude. The luminosity function thus obtained shows a distinctive bend between magnitudes –6 and –8. In general, the results agree well with those obtained in other studies.

104.018 **Telescopic observations of Geminid persistent trains.**
M. Vints.
WGN, Vol. 19, Nr. 4, p. 138 – 140 (Aug 1991).

104.019 **Large–scale structure of the Perseid meteor shower from long–basis observations.**
A. I. Grishchenyuk.
WGN, Vol. 19, Nr. 4, p. 142 – 147 (Aug 1991).
The existence is examined of large–scale structures in the Perseid stream. For this, meteor counts in subsequent short–length intervals from several stations were correlated. In addition, a joint cross–spectral analysis was performed using FFT techniques on the data of August 12 – 13, 1986. The results indicate that large–scale structures were only evident in the early '80s, which may have been caused by the parent body's encounter.

104.020 **One–hour outburst of the 1991 Perseids surprises Japanese observers!**
P. Roggemans, M. Gyssens, J. Rendtel.
WGN, Vol. 19, Nr. 5, p. 181 – 184 (Oct 1991).
While overall Perseid activity was rather normal in 1991, a short one–hour outburst was witnessed in Japan, yielding ZHRs of over 400. The position of this outburst in the Perseid activity profile coincides with the first peak of the double maximum found in 1988 and 1989 and points towards the presence of a new filament of particles connected with the return of the Perseids' parent comet P/Swift–Tuttle.

104.021 **A photographic Geminid campaign in Southern France.**
C. ter Kuile, P. Jenniskens, M. de Lignie.
WGN, Vol. 19, Nr. 4, p. 150 – 154 (Aug 1991).

104.022 **The 1990 Geminids.**
P. Roggemans, R. Koschack.
WGN, Vol. 19, Nr. 5, p. 184 – 193 (Oct 1991).
A global analysis of the 1990 Geminids is presented, based on 11255 meteors seen by 83 observers from 13 countries. The results confirm the general picture of the stream's activity profile.

104.023 **Bulletin 1 of the International Leonid Watch.**
P. Brown.
WGN, Vol. 19, Nr. 5, p. 193 – 197 (Oct 1991).
An overview is given of primary and secondary aims of the International Leonid Watch (ILW). An inventory of useful observing techniques is made.

104.024 **Shower meteor colors.**
A. McBeath.
WGN, Vol. 19, Nr. 5, p. 198 – 205 (Oct 1991).
An analysis of visual colors for the δ–Aquarid, Perseid, Taurid and Geminid meteor showers as recorded by the JAS Meteor Section observers in recent years is presented and discussed. A general confirmation of the trends shown by sporadic meteors is found, though Geminid meteors appear somewhat richer in their variety of colored meteors than other types.

104.025 **Some results from five years of Danish radio observations.**
G. M. Kristensen.
WGN, Vol. 19, Nr. 5, p. 206 – 209 (Oct 1991).

104.026 **Visual double station observations of Taurids and Leonids in 1990.**
L. R. Bellot, F. R. Andrés.
WGN, Vol. 19, Nr. 5, p. 210 – 211 (Oct 1991).
A report is given on the trajectory calculations for two visual double–station meteors observed in Spain in the fall of 1990, the first being a Southern Taurid and the second a Leonid.

104.027 **Analysis of visual plotting accuracy and sporadic pollution and consequences for shower association.**
R. Koschack.
WGN, Vol. 19, Nr. 6, p. 225 – 241 (Dec 1991).
An analysis of the plotting accuracy and of the sporadic pollution for visual meteor observations is given. It is found that both factors limit the observability of minor showers to ZHR $\geqslant 3$. Based on the results of the analysis, rules are developed for minor shower observations.

104.028 **The 1991 Perseids from the United Kingdom.**
A. McBeath.
WGN, Vol. 19, Nr. 6, p. 242 – 243 (Dec 1991).

104.029 **The 1991 Perseids from Crimea and Siberia.**
A. I. Grishchenyuk.
WGN, Vol. 19, Nr. 6, p. 243 – 244 (Dec 1991).

104.030 **Observations of Quadrantids meteor shower in 1990.**
Feng Zhanliang.
Acta Astron. Sin., Vol. 32, No. 2, p. 129 – 133 (Jun 1991). In Chinese.
In Jan. 1990, the Quadrantid meteor shower could be seen in Asia and was observed successfully in China. The analysis of observations shows: the maximum of activity of the Quadrantid meteor shower took place between 19^h15^m and 20^h40^m on Jan. 3(UT), peak at 20^h15^mUT, or 282°.64 mean solar longitude.

104.031 **Dynamics and dimensions of short–period meteorid streams. Interrelations between meteor showers.**
P. B. Babadzhanov, Yu. V. Obrubov.
Astron. Zh., Tom 68, Vyp. 5, p. 1074 – 1085 (Sep–Oct 1991). In Russian. English translation in Sov. Astron., Vol. 35, No. 5.
At the early stage of evolution, meteroid streams may be considered as elliptical rings of relatively small thickness. The influence of planetary perturbations can essentially increase the stream width and its thickness. As a result, one stream may produce several couples of meteor showers active in different seasons of the year. The 22 short–period meteroid streams under review may theoretically produce 102 meteor showers. The existence of 72 showers is confirmed by photographic and radar observations.

104.032 **The height distribution of radio meteors: comparison of observations at different frequencies on the basis of standard echo theory.**
D. I. Steel, W. G. Elford.
J. Atmos. Terr. Phys., Vol. 53, No. 5, p. 409 – 417 (May 1991).
A comparison is made between the observed height distributions of underdense radio meteor echoes measured at frequencies of 2, 6, 26 and 54 MHz and a simple model based on standard theory. This theory takes account of the initial radius of the train, the finite formation time due to the meteor velocity, and diffusion in the time between radar sweeps. The main features of the measured VHF (26 and 54 MHz) height distributions are predicted by this model, with peaks below 100 km and few meteors detected above 105 km. The main features of the measured HF (2 and 6 MHz) distributions are also successfully predicted, with peaks at $\sim$105 km. The model indicates few 6 MHz echoes from above 115 km, in line with the observational data, although for the data used there is an instrumental cut–off just above this height. It is suggested that even at 2 MHz perhaps 50% of all meteor trains above 105 km remain undetected. A comparison of the model and the measured 2 MHz distribution, which displays many meteors to at least 130 km, reveals that the true height distribution continues to rise above 105 km and would peak above 110 km were it not for the limited detectability of such high altitude meteors, even at such low frequencies.

104.033 **Perseids and comet P/1862 III.**
E. N. Kramer, I. S. Shestaka, A. K. Markina, V. I. Musij.
Astron. Tsirk., No. 1545, p. 31 – 32 (Aug – Sep 1990).
The authors considered the mean orbit of the Perseid swarm and found the point of mutual intersection of this orbit and comet P/1862 III's orbit. The authors are of the opinion that the formation of the Perseid swarm was the result of subsequent destruction of products of the initial nucleus.

104.034 **Visual meteor results from the international Halley watch.**
V. Porubčan, A. Hajduk, B. A. McIntosh.
Bull. Astron. Inst. Czech., Vol. 42, No. 3, p. 199 – 204 (Jun 1991).
Almost 300 observers from all over the world took part in visual observations of the meteor showers associated with Comet P/Halley under the direction of the International Halley Watch. The authors present here a preliminary analysis of the accumulated data on the Eta Aquarid (1984–1987) and Orionid (1982–1987) meteor showers. Due to a large spread in the reliability of the meteor counts resulting from the varying skilfullness of the observers, selection criteria were established which eliminated one quarter of the data. The activity of the Eta Aquarid shower appears to be much higher than that of the Orionid shower. The former has a sharp but typical double maximum at solar longitude 44°.5 and 46°.5 whereas the latter has multipeaked activity with a flat maximum between 206°.5–210°.5. The total activity period of each of the showers exceeds one month and the half–maximum period is about 7–8 d. The results are consistent with the ribbon model of the Halley particle stream.

104.035 **On the radiant determination of meteors.**
P. Pecina.
Bull. Astron. Inst. Czech., Vol. 42, No. 5, p. 307 – 316 (Oct 1991).
The article deals with the problem of determination of meteor radiants. It is shown that the classical method utilizing known formula for the zenith attraction correction and for the velocity magnitude correction is not quite correct when observed quantities are employed directly as its input. This is caused since these formulae have been derived under the assumption of gravitational action only, while the input data are usually obtained by solving the equation of motion only taking into account the drag force. To overcome this shortcoming, the author proposes to integrate numerically the equation of motion taking into account simultaneously both the drag and the gravitational force from the observed heights to the standard reference level of 150 km, where the gravitational effect certainly outweighs the drag and where the classical formulae can be used without any doubt. It is also demonstrated, how the final results can be affected by the use of a particular method of fitting the data. As a byproduct, the parameter p of the relation $\tau(v) \sim v^p$ and the product $\mu\sigma$ were obtained from the light–curve data. All findings are visualized by the results of processing selected fireball data observed in the European network. The author has found that the differences between the results applying the classical and his method are not enormous, but are outside the standard deviation boxes in many cases and, therefore, the new approach should be employed when processing precise data.

104.036 **Diurnal, seasonal and shower period variations in radio meteor rates over Waltair.**
M. Srirama Rao, K. K. Jyothi, K. Niranjan.
Indian J. Radio Space Phys., Vol. 20, No. 2, p. 102 – 106 (Apr 1991).
The diurnal and seasonal variations in the sporadic radio meteor rates as observed by the VHF back scatter radar at 33.85 MHz are reported. The diurnal meteor rates exhibit a sinusoidal trend with an early morning peak and evening trough during each month. The diurnal variations of the radio meteor rates during some of the major meteor shower periods are discussed.

104.037 **Dynamics and sizes of short–period meteoroid swarms. Related meteor showers.**
P. B. Babadzhanov, Yu. V. Obrubov.
Sov. Astron., Vol. 35, No. 5, p. 538 – 543 (Sep 1991). Current Physics Microform No.: 9203X1712. English translation of Astron. Zh., Tom 68, Vyp. 5, p. 1074 – 1085 (1991).
In the early stage of evolution, meteoroid swarms consist of elliptical rings of relatively small thickness. Planetary perturbations can significantly change both the width of a meteoroid swarm and its thickness. As a result, one meteoroid swarm may produce several pairs of related meteor showers, which become active at different times of the year. Thus, the 22 short–period swarms (with periods of revolution less than six years) that the authors have analyzed may theoretically produce 102 meteor showers. The existence of 72 meteor showers has been confirmed by photographic and radar observations.

104.038 **Energy deposition in antimatter meteors.**
P. M. Papaelias.
Earth, Moon, Planets, Vol. 55, No. 3, p. 215 – 222 (Dec 1991).
Antimatter meteors, like ordinary ones, can be heated during their infall flight. However, this could happen by a completely different process than in the case of koinomatter meteors, since in the latter case the annihilation interactions mechanism is absent. In case of antimatter meteors, the temperature may be increased mainly due to the energy deposition effect, caused by the passage of the annihilation secondaries penetrating throughout the meteor. The energy deposition of the secondary particles produced in matter antimatter annihilation interactions as a function of the dimensions of an antimatter meteor is described in this paper.

104.039 **Perseid meteors and periodic comet Swift–Tuttle.**
IAU Circ., No. 5330, p. 1 (28 Aug 1991).
Further information is given in Nos. 5340, 5342, 5345.

104.040 **The structure and evolution of the Taurid Complex.**
D. I. Steel, D. J. Asher, S. V. M. Clube.
Mon. Not. R. Astron. Soc., Vol. 251, No. 4, p. 632 – 648 (15 Aug 1991).
The structure of the Taurid meteoroid complex is investigated using orbital element measurements from the IAU Meteor Data Center. The complex is found to have been formed during the last ~10 kyr, this time-scale corresponding to a probable late stage in the evolution of the parent object, a giant comet which was apparently captured into a small-perihelion, short-period orbit ~20 kyr ago and which, in an initial highly active phase, gave rise to the material that now broadly constitutes the zodiacal cloud. Models of the evolution of the complex under gravitational perturbations suggest that meteoroids must have originally left the parent object near perihelion, but also allow the possibility that fragmentations have occurred when large disintegration products collided with objects in the asteroid belt, more or less as described by Whipple and Hamid (1952). Such a model may be entertained, for example, if the core of the evolved giant comet has previously undergone devolatilization, producing a high degree of fragility in the constituent debris.

104.041 **The Leonids, Comet Biela and Biela's associated meteoroid stream.**
P. B. Babadzhanov, Z. Wu, I. P. Williams, D. W. Hughes.
Mon. Not. R. Astron. Soc., Vol. 253, No. 1, p. 69 – 74 (1 Nov 1991).
Of all the meteor showers regularly seen from Earth, the Leonids produces by far the most spectacular display, through only at intervals of about 33 years. Comet Biela is the best documented example of a cometary nucleus dividing into two parts, a break-up that considerably enhanced the comet's associated meteroid stream. An intriguing suggestion was made by Bosler and Roure, namely that the division of comet Biela came about because of its passage through the Leonid stream in 1846. The authors have investigated the dynamics of the two orbits and confirm that the minimum distance between the orbits of Biela and the Leonid stream is very small. It was further discovered that in 1832, Comet Biela was close to the main bulk of the Leonid stream. The future of the enhanced stream caused by the break up of comet Biela is also investigated and it is predicted that this stream will start to intersect the Earth's orbit in 2120, producing a new meteor shower. The possibility that the passage of the nucleus of Biela through the densest part of the Leonid stream could cause the break-up of Biela is also discussed.

104.042 **Perseids 1991.**
Yamamoto Circ., No. 2170, p. 1 (17 Aug 1991). In Japanese.
Further information is given in No. 2171.

104.043 **Prediction of meteor shower associated with P/Hartley 2 (1991t).**
K. Ohtsuka.
Tokyo Meteor Network Rep., No. 11, p. 65 – 67 (Sep 1991). In Japanese.
The meteor activity associated with P/Hartley 2 (1991t) was expected on November 9.6 UT.

104.044 **Theory of diffusion of meteor trains in the geomagnetic field.**
W. Jones.
Planet. Space Sci., Vol. 39, No. 9, p. 1283 – 1288 (Sep 1991).
The author presents a solution, in closed form, of the problem of the diffusion of a meteor train in the geomagnetic field from an initial line density. The solution, in terms of effective diffusion coefficients depending on direction, is essentially exact within the quasi-neutrality approximation, and enables detailed calculations to be performed across the entire range of angles of train to field and relevant heights. The author generalises the solution to take approximate account of the fact that the train is formed at a finite radius, rather than as a line density, the general nature of the conclusions being unchanged. The results show that for heights of 95 km and above, the diffusion is severely inhibited by the field if θ, the angle between the train axis and the field lines, is close to zero. This effect diminishes very rapidly as θ is increased to about 2°. However, while the effective diffusion coefficient in the plane of train and field then remains close to the zero field ambipolar value right up to 90°, the effective coefficient in the direction of the normal to the plane of train and field drops steadily to its $\theta = 0$ value at $\theta = 90°$. Even for a height as low as 95 km, the magnetic field can reduce the diffusion in the normal direction by a factor of two below the zero field value. This corresponds to a change of almost 5 km in "diffusion height", that is, the height of an underdense meteor estimated on the basis of the exponential decay of its radar echo.

104.045 **Oblique-scatter of radio waves from meteor trains: full-wave calculations.**
J. Jones, W. Jones.
Planet. Space Sci., Vol. 39, No. 9, p. 1289 – 1296 (Sep 1991).
The authors describe the results of full-wave calculations, based upon the oblique-scattering theory of radio waves from meteor trains of Jones and Jones. They discuss the implications of the results of these calculations for forward-scatter meteor radar systems. For underdense trains, the present work confirms the findings of the authors' earlier approximate treatment. For overdense trains, the duration of the phase during which the train reflects like a metallic cylinder is governed by the oblique, rather than the forward-scatter angle. When the plasma in an overdense train becomes subcritical the reflection coefficient decays exponentially with time, but at a rate which is considerably less than that given by previous treatments. A number of approximate formulae are given, which may be useful in the design of forward-scatter meteor communications systems, and in the interpretation of astronomical observations obtained in this way.

104.046 **The orbit of the Lyrid meteor stream.**
B. A. Lindblad, V. Porubčan.
Bull. Astron. Inst. Czech., Vol. 42, No. 6, p. 354 – 359 (Nov 1991).
The radiant motion and orbit of the Lyrid meteor stream is determined from 14 precisely reduced photographic meteors observed in the period 1941 – 1985. The scatter in the orbital elements is extremely small. The mean Lyrid orbit is practically identical with that of P/Comet 1861 I. The mean photographic orbit is compared with that determined from radar measurements.

104.047 **On the preablation heating of rotating meteoroids. I. Theory.**
P. Pecina.
Bull. Astron. Inst. Czech., Vol. 42, No. 6, p. 359 – 376 (Nov 1991).
The paper deals with the problem of mathematical description of meteoroid heating during pre-ablation of its atmospheric motion. Only the terms linear in temperature are considered, i.e. the radiation is not taken into account.

104.048 **On the possible cometary origin of Geminid type meteor streams.**
A. K. Terent'eva, O. A. Bayuk.
Bull. Astron. Inst. Czech., Vol. 42, No. 6, p. 377 – 378 (Nov 1991).
Gravitational perturbations during close encounters of minor bodies with the planets of the Earth's group (Mercury – Mars) are considered as a possible mechanism of formation of short-period orbits of meteor streams. One of the versions of the known Laplace scheme of multiple application of the unperturbed two-body problem is used. It is shown that the transformation of cometary orbits ($e = 0.9$, $q = 0.14$ to 0.24 AU, $q' = 4 - 5$ AU) into Geminid-type orbits is possible. Single encounters of the

comet with Earth and Venus play a dominant role in this transformation.

104.049 **Meteors Section news.**
R. D. Lunsford.
Strolling Astron., Vol. 35, No. 3, p. 134 – 135 (Sep 1991).

104.050 **Meteors Section news.**
R. D. Lunsford.
Strolling Astron., Vol. 35, No. 4, p. 177 – 180 (Dec 1991).

104.051 **Combined simulation of the flight and explosion of a meteoroid in the atmosphere.**
V. P. Korobejnikov, P. I. Chushkin, L. V. Shurshalov.
Sol. Syst. Res., Vol. 25, No. 3, p. 242 – 255 (Nov 1991). English translation of Astron. Vestn., Tom 25, No. 3, p. 327 – 343 (1991).
A set of gas–dynamic models is proposed for studying the complex process of interaction of a meteoroid with the Earth's atmosphere when its flight ends with explosive breakup in the air. This set includes a model of an equivalent explosion of a semi–infinite cylindrical charge, a model of the radiative–thermal action, a model of the flight, and a model of the explosion of the flying body. An analysis is carried out on the example of the Tunguska cosmic body, for which improved estimates of its main parameters are obtained. The hypothesis that the Tunguska body is genetically related to comet Encke seems very likely.

104.052 **Number and spatial density of Orionid meteors in 1987.**
G. N. Sizonov, R. L. Khotinok.
Sol. Syst. Res., Vol. 25, No. 3, p. 286 – 287 (Nov 1991). English translation of Astron. Vestn., Tom 25, No. 3, p. 381 – 383 (1991).
Visual observations of the Orionid meteor shower were conducted in Armavir ($\lambda = 41°09'$, $\varphi = 45°$). The hourly number, influx, spatial density, relative activity each night, and also the luminosity function exponent and mass parameter were determined. The maximum influx occurred 21 – 22 October. The Sun's longitude at maximum influx was 208.8°. The hourly number during the night of 21 – 22 October, accounting for the group observability coefficient, and for continuously changing radiant height, was 28.0, and the influx was $I = 1.2 \cdot 10^{-13} m^{-2} sec^{-1}$.

104.053 **Die Feuerkugel vom 3. November 1988.**
J. Rendtel, D. Heinlein.
Sterne, Band 67, Heft 5, p. 289 – 294 (1991).

The Geminid meteor shower: a possible pre–discovery observation in 1834.
See Abstr. 004.018.

The stationary radiant debate revisited.
See Abstr. 004.019.

Elaboration of an apparatus on the basis of CCD for meteor observations.
See Abstr. 034.066.

The Earth's dust cloud and atmospheric oxygen.
See Abstr. 082.075.

The role of super–thermal electrons in the formation of luminous regions in the vicinity of a cosmic body.
See Abstr. 083.002.

Complexes of minor bodies in the solar system.
See Abstr. 091.015.

Complexes of minor solar system bodies.
See Abstr. 091.030.

A possible meteor shower on the Moon.
See Abstr. 094.033.

Penetration of chondrites into Mars' atmosphere.
See Abstr. 097.046.

The micrometeoroids in the thermosphere and mesosphere.
See Abstr. 105.028.

Possible hazards of meteorite falls.
See Abstr. 105.055.

105 Meteorites, Meteorite Craters

105.001 **Noble gas studies in vapor–growth diamonds: comparison with shock–produced diamonds and the origin of diamonds in ureilites.**
J. Matsuda, K. Fukunaga, K. Ito.
Geochim. Cosmochim. Acta, Vol. 55, No. 7, p. 2011 – 2023 (Jul 1991).

105.002 **An ion microprobe study of corundum in the Murchison meteorite: implications for ^{26}Al and ^{16}O in the early solar system.**
A. Virag, E. Zinner, S. Amari, E. Anders.
Geochim. Cosmochim. Acta, Vol. 55, No. 7, p. 2045 – 2062 (Jul 1991).

105.003 **Oxygen isotope studies of ordinary chondrites.**
R. N. Clayton, T. K. Mayeda, J. N. Goswami, E. J. Olsen.
Geochim. Cosmochim. Acta, Vol. 55, No. 8, p. 2317 – 2337 (Aug 1991).
Several stages in the evolution of ordinary chondritic meteorites are recorded in the oxygen isotopic composition of the meteorites and their separable components (chondrules, fragments, clasts, and matrix). The oxygen isotopic compositions of chondrites serve as "fingerprints" for identification of genetic association with other meteorite types (achondrites and iron) and for recognition of source materials in meteoritic breccias.

105.004 **Comment on "Actinide abundances in ordinary chondrites" by Hagee et al. [Geochim. Cosmochim. Acta, Vol. 54, No. 10, p. 2847 – 2858 (Oct 1990)].**
W. A. Myers, P. K. Kuroda.
Geochim. Cosmochim. Acta, Vol. 55, No. 8, p. 2355 – 2358 (Aug 1991). See Abstr. 52.105.188.

105.005 **Response to comment by W. A. Myers and P. K. Kuroda on "Actinide abundances in ordinary chondrites".**
B. Hagee, T. J. Bernatowicz, F. A. Podosek, M. L. Johnson, D. S. Burnett, M. Tatsumoto.
Geochim. Cosmochim. Acta, Vol. 55, No. 8, p. 2359 – 2363 (Aug 1991). See Abstr. 105.004.

105.006 **Major and trace element concentrations for chondrules of the Boxian (LL4) chondrite and their implications.**
Chen Jiang-feng, Jiang Shan, Chai Zhi-fang, Ma Shu-lan, Ma Jian-guo.
Chin. J. Space Sci., Vol. 11, No. 3, p. 217 – 225 (Jul 1991). In Chinese.

Al, Mg, Mn, Na, V, La, Sm, Eu, Sc, Cr, F, Ni, Co, Au and Ir concentrations for 14 chondrules from the Boxian (LL4) chondrite are analyzed by INAA. The average bulk chondrule compositions normalized to Mg and CI chondrite for the Boxian chondrite are identical to the ones for other unequilibrated ordinary chondrites (UOC's). Relations among the element concentrations for chondrules from the Boxian chondrite are discussed using the clustering and factor analysis. A few precursor chemical components are implied, siderophile elements, refractory and moderately refractory lithophile elements and moderately volatile elements etc. This conclusion is compatible to the results of other UOC's.

105.007 **Midinfrared (2.5 – 13.5 μm) reflectance spectra of powdered stony meteorites.**
J. W. Salisbury, D. M. D'Aria, E. Jarosewich.
Icarus, Vol. 92, No. 2, p. 280 – 297 (Aug 1991).

Reflectance spectra (2.5 to 13.5 μm) of 60 powdered meteorite samples representing 50 different meteorites are presented for comparison with spectral measurements of asteroids. Powdered samples were used as analogous of asteroidal regoliths. These spectra show that most powdered meteorite samples have undergone alteration, even if only exposed to water vapor in the air, and many have been contaminated by volatile hydrocarbons characterized by absorption bands near 3.45 μm. In contrast, primary macromolecular hydrocarbons do not display the 3.45-μm bands, or, in fact, any other detectable spectral features in these reflectance spectra.

105.008 **De Glatton meteoriet.**
R. Hutchison.
Radiant, Jaarg. 13, Nr. 4, p. 89 – 90 (Aug 1991).

105.009 **De l'Aigle.**
P. Jenniskens.
Radiant, Jaarg. 13, Nr. 5, p. 119 (Oct 1991).

105.010 **Seeing stars in a handful of dust.**
F. Flam.
Science, Vol. 253, No. 5018, p. 380 – 381 (26 Jul 1991).

Interstellar material brought to Earth in meteorites is yielding clues to the nuclear furnaces that forged it 5 billion years ago and more, in the interiors of giant stars.

105.011 **A search for superheavy elements in meteorites using a neutron multiplicity detector.**
J. C. Barton, C. J. Hatton, J. E. McMillan.
J. Phys. G, Vol. 17, No. 12, p. 1901 – 1907 (Dec 1991).

A search for superheavy element fission events in meteoritic samples was made using a novel neutron multiplicity detector which was operated underground. No differences could be detected between samples of meteoritic and terrestrial materials. All multiple neutron events could be attributed either to very small amounts of fissile material contaminating the detector materials or to the effects of cosmic rays. An upper limit of $1.3\,kg^{-1}y^{-1}$ for the fission rate in meteorites is derived. Assuming the half life to be 10^9y, this translates to a concentration of $<1.0\times10^{-15}kg\,kg^{-1}$ which is below all previous measurements. Further analysis of the data suggests that Runcorn's lunar core hypothesis remains just barely tenable.

105.012 **Isotopically anomalous Ti in presolar SiC from the Murchison meteorite.**
T. R. Ireland, E. K. Zinner, S. Amari.
Astrophys. J., Lett., Vol. 376, No. 2, p. L53 – L56 (1 Aug 1991).

The Ti-isotopic compositions of 14 large (8 – 23 μm) interstellar SiC grains from the Murchison meteorite have been measured. Three SiC grains with normal isotopic compositions of C, N, and Si also have normal Ti, but the other 11 grains with anomalous C-, N-, and Si-isotopic compositions also have anomalous Ti. The anomalous Ti qualitatively bears the signature of neutron-capture nucleosynthesis with excesses in the minor isotopes ^{46}Ti, ^{47}Ti, ^{49}Ti, and ^{50}Ti, relative to the major isotope ^{48}Ti. The anomalies range in magnitude up to 15% for ^{46}Ti and ^{50}Ti with smaller anomalies at ^{47}Ti and ^{49}Ti relative to ^{48}Ti when normalized to the solar abundances. The Si- and Ti-isotopic compositions are correlated in the sense that the grains with the largest ^{50}Ti excesses have the largest ^{29}Si excesses. While the anomalies are qualitatively consistent with the patterns predicted by n-capture nucleosynthesis, the correlations between anomalies in different isotopes differ substantially from predictions for the He-burning shell of AGB stars and core He-burning in massive stars. However, uncertainties in n-capture cross sections are large and require refinement before a full evaluation of the predictions for nucleosynthesis in proposed stellar sources can be made.

105.013 **The flight's moment and the trajectory of the 1908 June 30 Tunguska fireball according to observations of witnesses.**
A. A. Yavnel'.
Astron. Vestn., Tom 25, No. 4, p. 505 – 511 (Jul – Aug 1991). In Russian. English translation in Sol. Syst. Res., Vol. 25.

105.014 **Xenon and neon from acid-resistant residues of Inman and Tieschitz.**
R. H. Nichols Jr., C. M. Hohenberg, C. M. O'D. Alexander, C. T. Olinger, J. W. Arden.
Geochim. Cosmochim. Acta, Vol. 55, No. 10, p. 2921 – 2936 (Oct 1991).

105.015 **Geochromatography on the parent body of the carbonaceous chondrite Ivuna.**
M. R. Wing, J. L. Bada.
Geochim. Cosmochim. Acta, Vol. 55, No. 10, p. 2937 – 2942 (Oct 1991).

105.016 **The influence of bulk composition and dynamic melting conditions on olivine chondrule textures.**
H. C. Connolly Jr., R. H. Hewins.
Geochim. Cosmochim. Acta, Vol. 55, No. 10, p. 2943 – 2950 (Oct 1991).

105.017 **^{244}Pu-Xe formation and gas retention age, exposure history, and terrestrial age of angrites LEW86010 and LEW87051: comparison with Angra dos Reis.**
O. Eugster, T. Michel, S. Niedermann.
Geochim. Cosmochim. Acta, Vol. 55, No. 10, p. 2957 – 2964 (Oct 1991).

The authors have studied noble gas isotopic abundances in the two angrites LEW86010 and LEW87051. Many characteristics of LEW86010 are similar to those observed for the basaltic achondrite Angra dos Reis, until recently the only known angrite. The authors calculate an atomic ratio ^{244}Pu/^{150}Nd $= 1.36\times10^{-3}$ that corresponds to a fission Xe retention age 19 ± 40 Ma younger than that of Angra dos Reis. Thus, LEW86010 formed contemporaneously with Angra dos Reis for which previous analyses yielded a formation age of 4550 Ma.

105.018 **Terrestrial impact: the record in the rocks.**
R. A. F. Grieve.
Meteoritics, Vol. 26, No. 3, p. 175 – 194 (Sep 1991). = Contrib. Geol. Surv. Canada, No. 16491.

Approximately 130 terrestrial craters are currently known. They range up to 140 km, and perhaps as much as 200 km, in diameter and from recent to ~2 billion years in age. The known sample, however, is highly biased to geologically young craters on the better known cratonic areas. The sample is also deficient in small (D < 20 km) craters compared to other planetary bodies. These biases are largely the result of active terrestrial geologic processes and their effects have to be considered when interpreting the record. Unlike their planetary counterparts, terrestrial impact craters are mostly recognized not by morphology but by the occurrence of characteristic shock metamorphic effects. Their study has led to models of shock wave attenuation and an understanding of the character and formation of various impact–lithologies, including impact melt rocks. They, in turn, aid in interpreting the nature of extraterrestrial samples, particularly samples from the lunar highlands.

105.019 **Mineralogy of interstitial rim materials of the Y74123 and Y790981 ureilites and their origin.**
H. Ogata, H. Mori, H. Takeda.
Meteoritics, Vol. 26, No. 3, p. 195 – 201 (Sep 1991).
Y74123 is an olivine–rich, relatively unshocked ureilite and contains more interstitial pigeonitic materials than do ureilites which have been reported previously. Thus, Y74123 is especially suited for detailed study of the interstitial materials. The authors have studied these materials by optical microscope, electron microprobe, scanning electron microscope, high resolution transmission electron microscope (TEM) and analytical TEM to gain a better understanding of their nature and origin. These observations are consistent with the idea that the interstitial materials in this ureilite are a mixture of residual liquids of high Ca melts and shock–produced partial melts of olivine and pigeonite.

105.020 **Cosmic spherules in the geologic record.**
S. Taylor, D. E. Brownlee.
Meteoritics, Vol. 26, No. 3, p. 203 – 211 (Sep 1991).
Cosmic spherules obtained from the Greenland ice cap, deep–sea sediments, and ancient oceanic deposits of Eocene and Jurassic age were chemically and texturally compared. The proportions of spherule types were found to change as a function of time, with the number of iron spherules increasing as the age of the sample increased. It is not yet possible, however, to determine if the variation in spherule types represents a real change in the meteoroid complex or is a result of differential weathering of the spherules in the Earth's environment. The age of these spherules ranges from less than 3000 years to about 190 million years. Although only a few spherules older than 500000 years are chemically unaltered, textural information is available even from the oldest spherules and, by comparison with contemporary spherules, gives some insight into the original compositions of the particles.

105.021 **Further finds of the Derrick Peak meteorite, transantarctic mountains, and implications for terrestrial age.**
M. P. Kirkbride, M. A. Bradshaw, F. J. Harmsen.
Meteoritics, Vol. 26, No. 3, p. 213 – 216 (Sep 1991).
Nine additional iron meteorite fragments weighing a total of 72 kg were recovered from the Derrick Peak area in late 1988. One iron was located in the Onnum Valley, 6 km south of the previous finds. Geochemical analysis indicates that all irons belong to a single meteorite shower, greatly increasing the known extent of the fall zone. Kamp and Lowe have previously estimated the terrestrial age of the meteorite from glacial geological evidence. The location of the 1988 finds supports Kamp and Lowe's interpretation that the meteorites lie in situ, but recent revisions of the chronology of Cenozoic glacial history of the region reduce the interpreted terrestrial age.

105.022 **Anomalous Ne enrichments in tektites.**
K. Matsubara, J. Matsuda.
Meteoritics, Vol. 26, No. 3, p. 217 – 220 (Sep 1991).
The authors measured noble gases and Ne isotopic compositions of five tektites collected from three different strewn fields. The elemental abundance patterns of noble gases in all samples show anomalous Ne enrichments relative to air. Ne isotopic compositions in tektites are in good agreement with that of atmospheric Ne, suggesting that Ne has diffused in from the atmosphere.

105.023 **On the production of cosmogenic nuclides in meteoroids by galactic protons.**
R. Michel, P. Dragovitsch, P. Cloth, G. Dagge, D. Filges.
Meteoritics, Vol. 26, No. 3, p. 221 – 242 (Sep 1991).
A purely physical model is presented describing the depth– and size–dependence of the production of cosmogenic nuclides in meteoroids with radii up to 85 cm and in planetary surfaces by galactic cosmic ray protons. The model is based on Monte Carlo calculations of the intra– and internuclear cascades, by which depth– and size–dependent spectra of primary and secondary protons and of secondary neutrons are derived, and on experimental and theoretical thin–target cross sections of the underlying nuclear reactions. Model calculations are presented and compared with experimental data. An analysis of experimental depth profiles in four H– and L–chondrites demonstrates, that the new model is well capable of describing depth– and size–dependences of production rates of cosmogenic nuclides. Moreover, it is possible to determine exposure ages for these meteorites on the basis of the theoretical ^{21}Ne production rates.

105.024 **The frequency of meteorite falls: comments on two conflicting solutions to the problem.**
I. Halliday, A. T. Blackwell, A. A. Griffin.
Meteoritics, Vol. 26, No. 3, p. 243 – 249 (Sep 1991).
Recent estimates of the current rate of meteorite falls have been derived from camera network observations and from the statistics of the recovery of small fragments in Roosevelt County, New Mexico. The results are discordant. The integrated sky coverage for the camera data is an order of magnitude greater than the long exposure over small areas in New Mexico, yet the number of inferred events is comparable. The authors examine potential problems and find no effects other than random ones to bias the camera data. New data on the total number of suitable fireballs indicate that the flux from the camera network is already close to the upper limit imposed by the count of over 700 fireballs. By contrast, the calibration of the decay time for small fragments in New Mexico appears insecure and could account for a factor–of–three discrepancy. The large area of a typical strewn field relative to the small search areas is also a serious problem that remains uncalibrated.

105.025 **Classification of Streaky Bay, Mangalo, Ethiudna and Crockers Well: stony meteorites from South Australia.**
M. Wallace, A. Pring.
Meteoritics, Vol. 26, No. 3, p. 250 (Sep 1991).
Four meteorites from South Australia are described and classified. Ethiudna and Crockers Well were previously thought to be paired and were previously described only as chondrites.

105.026 **The Meteoritical Bulletin, No. 71.**
F. Wlotzka.
Meteoritics, Vol. 26, No. 3, p. 255 – 262 (Sep 1991).
Place of fall/find, class and type of individual specimens, total weight and circumstances of fall/find of some meteorites are given.

105.027 **The radiant of the Tunguska meteorite from visual observations.**
I. T. Zotkin, A. N. Chigorin.
Astron. Vestn., Tom 25, No. 5, p. 613 – 620 (Sep – Oct 1991). In Russian. English translation in Solar Syst. Res., Vol. 25, No. 5.
Data of about 100 eyewitnesses are analysed. The radiant location is determined: azimuth 126°, height 20°, error ±12°. This point is related to beta–Taurids meteor stream and comet Encke.

105.028 **The micrometeoroids in the thermosphere and mesosphere.**
F. Kamijo.
Astrophys. Space Sci., Vol. 184, No. 1, p. 147 – 152 (Oct 1991).
The variation of temperature and radius of typical micrometeoroids in the low thermosphere and mesosphere is calculated theoretically, and the formation of the spherule is explained consistently with this model. The fluffy particle floating in the

stratosphere is also consistent with the author's calculation, since the particle of radius 10 μ and initial velocity 20 km s^{-1} cannot be heated up to the sublimation temperature, even if it were not fluffy. The calculation coincides exactly with heights and velocities of meteor streams. The formation of the secondary particles from the evaporated vapour is also calculated, and it is concluded that they are very few and small, compared with the primary particles.

105.029 **Iron–magnesium order–disorder in an orthopyroxene crystal from the Johnstown meteorite.**
G. M. Molin, S. K. Saxena, E. Brizi.
Earth Planet. Sci. Lett., Vol. 105, No. 1–3, p. 260 – 265 (Jul 1991).

Equilibrium reversals of Fe^{2+}–Mg distribution between the M1 and M2 sites of an orthopyroxene from the Johnstown meteorite were achieved at several temperatures between 700 and 1000°C. One single crystal was used for the whole thermal treatment and for collecting all the X–ray data after quenching. The intracrystalline ion exchange for the bulk chemical composition is given. A transmission electron microscopy (TEM) study of part of the crystal showed the presence of very thin augite lamellae and Guinier–Preston zones indicating a relatively rapid cooling of the host rock at temperatures close to 1000°C. The new temperature scale yields a relatively high quenching temperature of 379($\pm$8)°C for the pyroxene which appears consistent with a rapid cooling (estimated few degrees per hundred years) of a magmatic cumulate excavated by an impact on its parental body.

105.030 **High temperature rims around chondrules in primitive chondrites: evidence for fluctuating conditions in the solar nebula.**
D. A. Kring.
Earth Planet. Sci. Lett., Vol. 105, No. 1–3, p. 65 – 80 (Jul 1991).

Rims or rim sequences surrounding chondrules have been identified in carbonaceous and unequilibrated ordinary chondrites. These chondrule rims include three chemical subtypes: Fe,Ca–rich and Fe,Ni–metal–rich rims, which occur predominantly in Kainsaz (CO3), and ferromagnesian rims which occur in Kainsaz (CO3), Allende (CV3), Renazzo (CR2), Chainpur (LL3), Semarkona (LL3), Krymka (L3)8 and Tieschitz (H3). The compositions of minerals in these rims are often drastically different from those in the underlying chondrule cores, indicating that the solar nebula was chemically heterogeneous. In many cases the compositions of the rims require an environment that was much more oxidizing than a solar composition gas. Particularly interesting is that some of the Fe,Ca–rich chondrule rims are remarkably similar to some of the rims around refractory inclusions, suggesting that chondrules and refractory inclusions experienced late, coeval processing. The textures of the chondrule rims suggest they formed at high temperatures and that they accreted onto chondrules that had already solidified. The lengthscale of the thermal heterogeneities necessary to make available hot material that could accrete to cold chondrules has been calculated to be less than 10 km, implying there were localized heat sources in the solar nebula.

105.031 **Evidence for multiple sources of diamond from primitive chondrites.**
S. S. Russell, J. W. Arden, C. T. Pillinger.
Science, Vol. 254, No. 5035, p. 1188 – 1191 (22 Nov 1991).

Fine–grained diamonds, the most abundant form of circumstellar dust isolated from primitive meteorites, have elemental and isotopic characteristics that are dependent on the host meteorite type. Although the ratio of carbon to nitrogen evolves in a distinctive way during combustion of diamond separates, metamorphic degassing of nitrogen is not the main cause of the differences in nitrogen content. The data suggest that intrinsic differences must have been inherited by the diamonds at the time of their formation and that the diamonds were distributed heterogeneously in the solar nebula during condensation. However, the hypothesis that a distinct nitrogen carrier remains hidden within the diamond cannot be ruled out.

105.032 **Mössbauer spectroscopy of SNC meteorites.**
J. M. Knudsen, M. B. Madsen, L. Vistisen.
2. COSPAR Colloquium: The environmental model of Mars, p. 7 – 10 (1991). – See Abstr. 012.039 for the main entry.

The achondrite Nakhla, belonging to the group of SNC meteorites, is more oxidized than the ordinary achondrites. Besides the Fe^{3+} in the titanomagnetite, Nakhla includes other compounds carrying Fe^{3+}. By means of Mössbauer spectroscopy the authors demonstrate that in some of the pyroxenes separated from Nakhla the Fe^{3+} is distrubuted throughout the crystals, i.e. the oxidation of the pyroxenes in Nakhla is not only a surface phenomenon.

105.033 **Application of the Mössbauer effect to the study of meteorites – a review.**
R. B. Scorzelli.
Hyperfine Interact., Vol. 66, No. 1–4, p. 249 – 257 (Nov 1991). Paper presented at the Latin American Conference on the Applications of the Mössbauer Effect (LACAME '90), Havanna (Cuba), 29 Oct – 2 Nov 1990.

Mössbauer and X–ray diffraction studies revealed that segregation and ordering of Fe–Ni alloys occurs in meteorites in close resemblance with irradiation effects in these alloys. In Santa Catharina (35% Ni) and Twin City (30% Ni) ataxites, the ordering mechanism leads to the formation of rather large amounts of the 50–50 Fe–Ni ordered phase with tetragonal symmetry (tetrataenite). Furthermore this ordered phase has also been detected in metal particles of chondrites.

105.034 **Composition of solar flare noble gases preserved in meteorite parent body regolith.**
M. N. Rao, D. H. Garrison, D. D. Bogard, G. Badhwar, A. V. Murali.
J. Geophys. Res., Vol. 96, No. A11, p. 19321 – 19330 (1 Nov 1991).

The isotopic composition (long–term average) of solar flare (SF) Ne has been determined by three isotope correlation techniques applied to data measured on chemically etched pyroxene separates prepared from the Kapoeta meteorite, which is known to contain implanted solar gases. The SF $^{20}Ne/^{22}Ne$ ratio obtained is 11.6$\pm$0.2 and confirms previous determinations of this SF ratio in lunar and meteoritic samples. The same SF Ne composition is also obtained by applying an ordinate intercept technique to the same data set. The ordinate intercept technique was also applied to the Ar and He data, on which the three–isotope correlation technique cannot be applied.

105.035 **The subsurface character of meteor crater, Arizona, as determined by ground–probing radar.**
J. A. Pilon, R. A. F. Grieve, V. L. Sharpton.
J. Geophys. Res., Vol. 96, No. E1, p. 15563 – 15576 (25 Aug 1991).

Ground–probing radar has provided new two–dimensional stratigraphic and structural information on Meteor Crater, Arizona. Four 50–MHz transects inside the crater and one 100–MHz transect on a northeast radial across the exterior ejecta blanket were carried out. The frequencies used were in relation to the desired depth of investigation: greater than 50 m at 50 MHz and ~30 m for 100 MHz. The results are presented and discussed.

105.036 **Sub–ice topography and meteorite finds near the Allan Hills and the near Western Ice Field, Victoria Land, Antarctica.**
G. Delisle, J. Sievers.
J. Geophys. Res., Vol. 96, No. E1, p. 15577 – 15588 (25 Aug 1991).

The authors have investigated the glaciological causes of large meteorite concentrations on blue ice fields west and southwest of the Allan Hills in Antarctica.

105.037 **Fassaite composition trends during crystallization of Allende type B refractory inclusion melts.**
S. B. Simon, L. Grossman, A. M. Davis.
Geochim. Cosmochim. Acta, Vol. 55, No. 9, p. 2635 - 2655 (Sep 1991).

105.038 **The Carlisle Lakes-type chondrites: a new grouplet with high $\Delta^{17}O$ and evidence for nebular oxidation.**
M. K. Weisberg, M. Prinz, H. Kojima, K. Yanai,
R. N. Clayton, T. K. Mayeda.
Geochim. Cosmochim. Acta, Vol. 55, No. 9, p. 2657 - 2669 (Sep 1991).
Carlisle Lakes, ALH85151, and Y75302 are similar ungrouped chondrites which have petrologic and bulk compositional similarities to the ordinary chondrites, but are more oxidized; and their oxygen isotopic compositions differ. They represent a new grouplet which the authors call the Carlisle Lakes-type chondrites. They have the highest $\Delta^{17}O$ values (up to 2.91) measured to date. The whole chondrites and most of their chondrules plot on the same mass fractionation line on an oxygen 3-isotope diagram. The Carlisle Lakes-type chondrites may have originally been derived from an ordinary chondrite-like precursor which was later oxidized, prior to its final lithification.

105.039 **Systematic study of sulfur isotopic composition in iron meteorites and the occurrence of excess ^{33}S and ^{36}S.**
X. Gao, M. H. Thiemens.
Geochim. Cosmochim. Acta, Vol. 55, No. 9, p. 2671 - 2679 (Sep 1991).

105.040 **Carbon-14 content of the Antarctic meteorite, MacAlpine Hills 88105.**
A. J. T. Jull, D. J. Donahue.
Geochim. Cosmochim. Acta, Vol. 55, No. 9, p. 2681 - 2682 (Sep 1991).

105.041 **Age and geomorphic history of Meteor Crater, Arizona, from cosmogenic ^{36}Cl and ^{14}C in rock varnish.**
F. M. Phillips, M. G. Zreda, S. S. Smith, D. Elmore,
P. W. Kubik, R. I. Dorn, D. J. Roddy.
Geochim. Cosmochim. Acta, Vol. 55, No. 9, p. 2695 - 2698 (Sep 1991).
Using cosmogenic ^{36}Cl buildup and rock varnish radiocarbon, the authors have measured the exposure age of rock surfaces at Meteor Crater, Arizona. The ^{36}Cl measurements on four dolomite boulders ejected from the crater by the impact yield a mean age of 49.7 ± 0.85 ka, which is in excellent agreement with an average age of 49 ± 3 ka obtained from thermoluminescence studies on shock-metamorphosed dolomite and quartz.

105.042 **In situ ^{10}Be-^{26}Al exposure ages at Meteor Crater, Arizona.**
K. Nishiizumi, C. P. Kohl, E. M. Shoemaker, J. R. Arnold,
J. Klein, D. Fink, R. Middleton.
Geochim. Cosmochim. Acta, Vol. 55, No. 9, p. 2699 - 2703 (Sep 1991). Letter-to-the-editor
A new method of dating the surface exposure of rocks from in situ production of ^{10}Be and ^{26}Al has been applied to determine the age of Meteor Crater, Arizona. A lower bound on the crater age of 49200 ± 1700 y has been obtained by this method.

105.043 **Does debris from the formation of other planetary systems impact Earth?**
T. G. Brophy.
Icarus, Vol. 94, No. 1, p. 250 - 254 (Nov 1991).
The author calculates the expected flux of small and large particles impacting the Earth using model size distributions for the ejected debris from extrasolar planet formation. The flux of extrasolar impactors of large meteorite size and smaller at Earth should be detectable given likely size distributions, if the standard Solar System formation model is applicable to other planetary systems. The author proposes that the report of impactors with extrasolar velocities at the Long Duration Exposure Facility and by the Kharkov radar may be due to these particles, and that a number of recovered meteorites may be extrasolar. Conversely, if it can be determined that no meteorites or small meteoroids are extrasolar planetary debris, then the standard Solar System formation model will be constrained.

105.044 **A lunar meteorite found outside the Antarctic.**
D. H. Hill, W. V. Boynton, R. A. Haag.
Nature, Vol. 352, No. 6336, p. 614 - 617 (15 Aug 1991). Letter-to-the-editor.
Our knowledge of the Moon's surface composition has come from samples returned by the Apollo and Luna missions, and from eleven lunar meteorites, all of which were discovered in Antarctica. The authors report the discovery of a new lunar meteorite, Calcalong Creek, in a desert region of Australia which is analogous to Antarctica in its ability to preserve meteorites of different types. On the basis of a diagnostic Fe/Mn ratio of 73-78, and other element abundances, they conclude that Calcalong Creek is a lunar breccia, containing both highland and mare materials. Calcalong Creek will add to our understanding of lunar petrology, and as the first non-Antarctic lunar meteorite, may also shed new light on the transfer of impact ejecta from the Moon to the Earth.

105.045 **Phosphates in pallasite meteorites as probes of mantle processes in small planetary bodies.**
A. M. Davis, E. J. Olsen.
Nature, Vol. 353, No. 6345, p. 637 - 640 (17 Oct 1991). Letter-to-the-editor.
The stony-iron pallasite meteorites, which are generally believed to come from the core-mantle boundaries of asteroids, contain small quantities of phosphate minerals. The authors use trace element analyses of these phosphates to investigate the magnetic history of the silicate portions of pallasites. It is suggested that some pallasites may come from regions of their parent bodies much nearer the surface than the core-mantle boundary.

105.046 **Space messengers.**
R. L. Khotinok.
Zemlya Vselennaya, No. 4, p. 54 - 57 (Jul - Aug 1991). In Russian.
Concerning the meteorites Akmolla, Dengli, Omolon, Sukhoj Liman, Sychevka, Uchkuduk, Dar'inskoe, Ojogos-Yar, Sterlitamak.

105.047 **s-process Ba, Nd, and Sm in presolar SiC from the Murchison meteorite.**
E. Zinner, S. Amari, R. S. Lewis.
Astrophys. J., Lett., Vol. 382, No. 1, p. L47 - L50 (20 Nov 1991).
Six grain size fractions in the size range $0.2 - 3.0\ \mu m$ of interstellar SiC from the Murchison carbonaceous meteorite have been analyzed by ion microprobe mass spectrometry for Ba isotopes (^{134}Ba through ^{138}Ba), one fraction also for Nd (^{142}Nd through ^{146}Nd) and Sm (^{147}Sm through ^{149}Sm) isotopes. All elements show isotopic abundance patterns characteristic for the s-process. The $^{134}Ba/^{136}Ba$ ratios are distinct from solar, indicating that s-Ba in SiC is different from that in the solar system. $^{138}Ba/^{136}Ba$ ratios decrease with grain size. Although the other isotopic ratios for Ba and those for Nd and Sm differ in detail from theoretical predictions, the general agreement of the data with models of s-process nucleosynthesis support an AGB star origin for the relatively fine-grained SiC of this study. ^{136}Ba in SiC is up to 85% pure s-process. If the (isotopically normal) remainder is from the AGB envelope, Ba/Si was $\approx 170 - 300$ times solar in the He-burning shell where the s-Ba was synthesized.

105.048 **Do meteorites contain irradiation records from exposure to an enhanced-activity Sun?**
M. W. Caffee, C. M. Hohenberg, R. H. Nichols Jr.,
C. T. Olinger, R. Wieler, A. Pedroni, P. Signer, T. D. Swindle,
J. N. Goswami.
The Sun in time, p. 413 - 425 (1991). - See Abstr. 003.027 for the main entry.
Discovery of solar-flare tracks in some of the grains of gas-rich meteorites indicates that ancient irradiation records can be

preserved and the scarcity of such grains suggests that only about 1% of solar-system material has been involved in pre-compaction irradiation, thus removing some of the early objections. Recently, an active early Sun has been proposed to explain the quantity of spallogenic ^{21}Ne observed in meteoritic grains containing solar-flare tracks (reflecting pre-compaction irradiation). This hypothesis has been challenged by other investigators who would explain such effects by conventional galactic cosmic-ray irradiation of near-surface material on the meteorite's parent body. The authors review the evidence both for and against the active early solar origin of these effects in gas-rich meteorites.

105.049 **Thermal histories of CO3 chondrites: application of olivine diffusion modelling to parent body metamorphism.**
R. H. Jones, D. C. Rubie.
Earth Planet. Sci. Lett., Vol. 106, No. 1-4, p. 73 - 86 (Sep 1991).
The petrologic sequence observed in the CO3 chondrite group has been suggested to be the result of thermal metamorphism on a parent body. The authors have developed a model to examine the possibility that chondrule and matrix olivines equilibrated in situ, during parent body metamorphism. The model considers Fe-Mg interdiffusion between chondrule and matrix olivines. Zoning profiles comparable to those observed in chondrule olivines from partially equilibrated members of the series are reproduced successfully. Metamorphism of CO3 chondrites on a parent body is therefore a viable model for the observed equilibration. Results indicate that peak metamorphic temperatures experienced 500°C, and that the range of peak temperatures between unequilibrated and equilibrated subtypes was relatively narrow, around 100°C.

105.050 **^{41}Ca in iron falls, Grant and Estherville: production rates and related exposure age calculations.**
D. Fink, J. Klein, R. Middleton, S. Vogt, G. F. Herzog.
Earth Planet. Sci. Lett., Vol. 107, No. 1, p. 115 - 128 (Oct 1991).
Recent technical developments in accelerator mass spectrometry have enabled the authors to measure routinely and with a precision of 5-7% the ^{41}Ca ($T_{1/2}$ = 104 ky) contents of extraterrestrial samples weighing approximately 100 mg. In essence, these advances have elevated ^{41}Ca to the role of a new and potent cosmogenic radioisotope with wide-ranging applications. The authors present here the results from the first phase of their ^{41}Ca cosmogenic studies program, aimed at establishing baseline concentrations and trends in selected meteorites and the use of ^{41}Ca in estimating exposure ages and pre-atmospheric meteorite radii. The average ^{41}Ca saturation activity recorded in four small iron falls is 24 ± 1 dpm/kg. This result, together with measurements at the center and surface of the large iron Grant, indicates that production of ^{41}Ca from spallation on iron is weakly dependent on shielding to depths as large as 250 g/cm^2. The authors estimate the ^{41}K-^{41}Ca exposure age of Grant to be 330 ± 50 My, and an upper limit to its terrestrial age of 43 ± 15 ky. A comparison of the ^{41}Ca contents of stony and metallic material separated from the mesosiderite Estherville identifies low-energy neutron capture on native Ca as a second important channel of production. The authors find that the ^{41}Ca signal in the stone phase from three meteorites correlates with their size, and that the inferred low-energy neutron fluxes vary by a factor of at least 20.

105.051 **Ce anomalies in the LEW85300 eucrite: evidence for REE mobilization during Antarctic weathering.**
C. Floss, G. Crozaz.
Earth Planet. Sci. Lett., Vol. 107, No. 1, p. 13 - 24 (Oct 1991).
There is abundant evidence for widespread REE mobilization in Antarctic eucrites, resulting in Ce and Eu anomalies and lower whole-rock REE abundances relative to typical eucrites. Using secondary ion mass spectrometry (SIMS), the authors investigated the microdistribution of Ce anomalies in the Antarctic polymict eucrite LEW85300. Ce anomalies (positive and/or negative) are present in all three of the minerals (plagioclase, pyroxene and silica) analysed. The anomalies occur in grains from all four clasts as well as in mineral and lithic fragments from the surrounding matrix; thus there appears to be no lithologic association. Silica, which has been heavily shocked and altered, exhibits a LREE-enriched pattern with negative Ce and Eu anomalies, consistent with a derivation of the REE from Ca-phosphate dissolution: trivalent REE are leached from the phosphates, with preferential retention of tetravalent Ce, and redeposited on silica. Pyroxene is also widely affected by REE mobilization, due to an extensive, shock-induced microcrack network along cleavage planes. A tendency for the largest Ce anomalies to occur at low LREE concentrations suggests that pyroxene itself experienced REE leaching with concomitant Ce retention. Ce anomalies are rare in plagioclase because this mineral lacks an extensive network along which REE can be mobilized. The widespread REE mobilization and evidence for other forms of weathering requires that trace element data be used cautiously, especially for Antarctic eucrites.

105.052 **Remanent magnetic properties of unbrecciated eucrites.**
S. M. Cisowski.
Earth Planet. Sci. Lett., Vol. 107, No. 1, p. 173 - 181 (Oct 1991).
The remanent magnetic properties of five unbrecciated eucrites, ranging from the coarse-grained cumulate Moore County, to the quenched melt rock ALH 81001, have been investigated in order to assess the strength of the magnetic field associated with their parent body during their formation.

105.053 **The Vaca Muerta meteorite.**
M. Jõeveer.
Calendar of the Tartu Astronomical Observatory for the year, 1992.
p. 33 - 34 (1991). In Estonian. See Abstr. 046.016.

105.054 **The natural thermoluminescence of meteorites. II. Meteorite orbits and orbital evolution.**
P. H. Benoit, D. W. G. Sears, S. W. S. McKeever.
Icarus, Vol. 94, No. 2, p. 311 - 325 (Dec 1991).
Natural thermoluminescence (TL) data for 26 meteorites for which orbital elements have been estimated are reported. Calculations of equilibrium natural TL level in ordinary chondrites suggest that TL should be a sensitive indicator of perihelion. Meteorites with perihelia <0.85 AU should have very low levels of natural TL, while meteorites with perihelia >0.85 AU should show a range of natural TL with significant scatter as a result of slight variations in dose rate (shielding) and albedo. The data presented here generally agree with the theory and suggest an effective dose rate of ~ 3 rad/y for most meteorites. Comparison with cosmic ray exposure ages indicates that natural TL level is also partly related to exposure age. Meteorites with high (>20 Ma) exposure ages generally have a lower and more constrained range of natural TL than meteorites with ages <20 Ma ages. The authors suggest that this reflects orbital evolution, since mature meteorite orbits evolve to lower perihelia.

105.055 **Possible hazards of meteorite falls.**
C. E. Spratt.
J. R. Astron. Soc. Can., Vol. 85, No. 5, p. 263 - 280 (Oct 1991).
A survey of those meteorites that have struck buildings and people is presented. Included is a tabulation of some well-known meteorite showers which have occurred over the past two and a half centuries. A statistical break-down of the probability that a building or structure might actually be struck by a meteorite is also included.

105.056 **The Vaca Muerta mesosiderite.**
H. Pedersen, H. Lindgren, C. De Bon.
Messenger, No. 65, p. 39 - 41 (Sep 1991).

105.057 **MacAlpine Hills 88104 and 88105 lunar highland meteorites: general description and consortium overview.**
M. M. Lindstrom, C. Schwarz, R. Score, B. Mason.
Geochim. Cosmochim. Acta, Vol. 55, No. 11, p. 2999 – 3007 (Nov 1991).

MacAlpine Hills 88104 and 88105 are new lunar meteorites returned from Antarctica by the 1988 – 1989 US meteorite collection team. MAC88105 is the largest lunar meteorite yet found (662 g). An origin on the Moon is supported by the mineral, bulk, and oxygen isotope compositions.

105.058 **Mineral–chemical comparisons of MAC88105 with Yamato lunar meteorites.**
H. Takeda, H. Mori, J. Saito, M. Miyamoto.
Geochim. Cosmochim. Acta, Vol. 55, No. 11, p. 3009 – 3017 (Nov 1991).

The new lunar meteorite MAC88105 has been studied by mineral–chemical techniques and was compared with the Yamato lunar meteorites. Clast types and pyroxene compositions of MAC88105 indicate that evidence for a pairing with any known lunar meteorites is unlikely. Clast–laden vitric breccia and comminuted mineral fragments in glassy matrix are common components as in other lunar meteorites, but granulitic clasts are not as common as in other lunar meteorites. The preservation of glassy materials indicates that MAC88105 had a metamorphic annealing history different from other lunar meteorites.

105.059 **Chemical studies of H chondrites: III. Regolith evolution of the Fayetteville chondrite parent.**
Xiao Xiaoyue, M. E. Lipschutz.
Geochim. Cosmochim. Acta, Vol. 55, No. 11, p. 3407 – 3415 (Nov 1991).

Data for Ag, Au, Bi, Cd, Co, Cs, Ga, In, Rb, Sb, Se, Te, Tl, and Zn in consortium samples of solar–gas– and track–rich dark matrix, and dark and light clasts in the Fayetteville H chondrite regolith breccia were determined by radiochemical neutron activation analysis. The matrix proved chemically homogeneous at a level never before observed in an ordinary chondrite.

105.060 **Atmospheric fragmentation of meteorites according to the strength theory.**
V. I. Tsvetkov, A. Ya. Skripnik.
Sol. Syst. Res., Vol. 25, No. 3, p. 273 – 279 (Nov 1991). English translation of Astron. Vestn., Tom 25, No. 3, p. 364 – 371 (1991).

The fragmentation of large meteoric bodies in the Earth's atmosphere is caused by mechanical stresses. The stresses at a susceptible body point are calculated from equations of elasticity theory. An important factor is the inhomogeneity of the material. Transition from experimental strength properties of a material to evaluation of fracturing loads should be done on the basis of the strength theory. Since breaking loads are associated with the aerodynamic head and, therefore, of the velocity, it is possible to estimate the geocentric velocity and to pass to astronomic parameters of the bodies generating meteorite showers.

105.061 **Peculiarities of distribution of the long–lived cosmogenic radionuclides ^{53}Mn, ^{10}Be, and ^{21}Al in meteorites.**
V. A. Alekseev.
Sol. Syst. Res., Vol. 25, No. 3, p. 279 – 286 (Nov 1991). English translation of Astron. Vestn., Tom 25, No. 3, p. 372 – 380 (1991).

The distributions of the saturated (corrected for cosmic ray exposure age of meteorite) contents of the long–lived cosmogenic radionuclides ^{53}Mn ($T_{1/2}$ = 3.7 My), ^{10}Be (1.51 My), and ^{26}Al (0.705 My) in 300 non–Antarctic and Antarctic ordinary chondrites (H, L, and LL) are studied. The average contents of the radionuclides in non–Antarctic H chondrites of low exposure ages ($t \leqslant 8$ My) were found to be higher than those in non–Antarctic H chondrites of high exposure ages ($t > 8$ My). This difference is directly proportional to the radionuclide half–life and equal to $(29 \pm 8)\%$ for ^{53}Mn. A similar trend was obtained in Antarctic chondrites. No such differences were found in the average contents of radionuclides in the different age groups of L and LL chondrites. The effect can be explained in the framework of the following assumptions: (1) a large number of H chondrites with exposures ages in the region of the main peak (maximum at $t \sim 6$ My) belong to a compact group of bodies existing on the resonant orbit; (2) the inclination of this orbit to the ecliptic is changed periodically (with a period of about 1 – 2 My). The changes in inclination stimulate the changes in degree of cosmic ray modulation, and consequently the rate of production of cosmogenic radionuclides in meteorites.

105.062 **Trace element analysis of ureilites: new constraints on their petrogenesis.**
A. H. Spitz, W. V. Boynton.
Geochim. Cosmochim. Acta, Vol. 55, No. 11, p. 3417 – 3430 (Nov 1991).

Six ureilites were analyzed using neutron activation analysis for Ca, Sc, Cr, Mn, Fe, Co, Ni, Zn, Ga, REE, W, Re, Os, Ir, and Au. The authors examined bulk samples as well as acid–treated samples. Rare earth elements (REEs) in ureilites are quite depleted and display light and heavy rare earth enrichments. In terms of the new trace–element results, ureilities do not fall into the coherent groups that other workers have defined by chemical or petrographic characteristics. Trace elements do provide additional constraints on the models for the petrogenesis of ureilities.

105.063 **Iodine–xenon, chemical, and petrographic studies of Semarkona chondrules: evidence for the timing of aqueous alteration.**
T. D. Swindle, J. N. Grossman, C. T. Olinger, D. H. Garrison.
Geochim. Cosmochim. Acta, Vol. 55, No. 12, p. 3723 – 3734 (Dec 1991).

The authors have performed INAA, petrographic, and noble gas analyses on seventeen chondrules from the Semarkona meteorite (LL3.0) primarily to study the relationship of the I–Xe system to other measured properties. They observe a range of $\geqslant 10$ Ma in apparent I–Xe ages. The three latest apparent ages fall in a cluster, suggesting the possibility of a common event. The initial $^{129}I/^{127}I$ (R_0) ratio is apparently related to chondrule type and/or mineralogy. The authors conclude that the variations in R_0 represent variations in ages, and that later event(s), possibly aqueous alteration, preferentially affected chondrules with non-porphyritic textures and/or sulfide–rich exteriors about 10 Ma after the formation of the chondrules.

105.064 **Thermoluminescence constraints on the metamorphic, shock, and brecciation history of basaltic meteorites.**
J. D. Batchelor, D. W. G. Sears.
Geochim. Cosmochim. Acta, Vol. 55, No. 12, p. 3831 – 3844 (Dec 1991).

Induced thermoluminescence measurements have been performed on eighteen eucrites, thirteen howardites, six diogenites, and fifteen mesosiderites in order to help clarify their metamorphic, shock, and brecciation history. The TL sensitivities of the basaltic meteorites reflect their bulk composition, and the effects of (1) igneous processes (diogenites and mesosiderites) and (2) the mixing of materials of different compositions (e.g., diogenitic and eucritic) during brecciation (howardites) can be observed. The temperature of the dominant TL peak observed for basaltic meteorites, and experiments in which four eucrites with diverse petrographic properties were annealed at various temperatures in the laboratory, suggests the metamorphic equilibrium temperatures for most basaltic meteorites were $\leqslant 800$°C. This estimate is generally consistent with palaeotemperatures determined from two–pyroxene, olivine–pyroxene, and oxygen isotope equilibria.

105.065 **Shock metamorphism of ordinary chondrites.**
D. Stöffler, K. Keil, E. R. D. Scott.
Geochim. Cosmochim. Acta, Vol. 55, No. 12, p. 3845 – 3867 (Dec 1991).

A revised petrographic classification of progressive stages of shock metamorphism of ordinary chondrites is proposed. Six stages of shock (S1 to S6) are defined, based on shock effects in olivine and plagioclase as recognized by thin section microscopy.

The meteorite and tektite collector's handbook. A practical guide to their acquisition, preservation and display.
See Abstr. 003.019.

Planetary geosciences – 1988.
See Abstr. 003.036.

Determination of major, minor and trace constituents in meteorites by inductively coupled plasma–atomic emission spectroscopy.
See Abstr. 022.056.

"Polarization shadows" on the negative branch of the polarization curve.
See Abstr. 022.091.

Partitioning of platinum group elements in the Fe–O–S system to 11 GPa and their fractionation in the mantle and meteorites.
See Abstr. 022.175.

Solar system abundances of the elements.
See Abstr. 061.095.

Impact of large–scale meteoric bodies on a solid surface with splits and breaks.
See Abstr. 062.242.

Solar flare heavy–ion tracks in extraterrestrial objects.
See Abstr. 078.026.

Tektites in Cretaceous–Tertiary boundary rocks on Haiti and their bearing on the Alvarez impact extinction hypothesis.
See Abstr. 081.052.

The Sudbury structure: controversial or misunderstood?
See Abstr. 081.053.

Optical phenomena in summer 1908.
See Abstr. 082.026.

Nature of the anomalous illumination of the sky related to the Tunguska event.
See Abstr. 082.027.

On the origin and early evolution of terrestrial planet atmospheres and meteoritic volatiles.
See Abstr. 091.004.

Geochemical comparison of impact glasses from lunar meteorites ALHA81005 and MAC88105 and Apollo 16 regolith 64001.
See Abstr. 094.053.

Impact melts in the MAC88105 lunar meteorite: interferences for the lunar magma ocean hypothesis and the diversity of basaltic impact melts.
See Abstr. 094.054.

Paired lunar meteorites MAC88104 and MAC88105: a new "FAN" of lunar petrology.
See Abstr. 094.055.

A ferroan region of the lunar highlands as recorded in meteorites MAC88104 and MAC88105.
See Abstr. 094.056.

MAC88105 – a regolith breccia from the lunar highlands: mineralogical, petrological, and geochemical studies.
See Abstr. 094.057.

Geochemistry and petrography of the MacAlpine Hills lunar meteorites.
See Abstr. 094.058.

Lunar highland meteorites and the composition of the lunar crust.
See Abstr. 094.059.

The MacAlpine Hills lunar meteorite and implications of the lunar meteorites collectively for the compositions and origin of the Moon.
See Abstr. 094.060.

History of the paired lunar meteorites MAC88104 and MAC88105 derived from noble gas isotopes, radionuclides, and some chemical abundances.
See Abstr. 094.061.

Exposure histories of lunar meteorites: ALHA81005, MAC88104, MAC88105, and Y791197.
See Abstr. 094.062.

Exposure histories of the lunar meteorites: MAC88104, MAC88105, Y791197, and Y86032.
See Abstr. 094.063.

The natural thermoluminescence of meteorites: III. Lunar and basaltic meteorites.
See Abstr. 094.064.

Are Phobos and Deimos carbonaceous chondrites? Mössbauer spectroscopy on carbonaceous chondrites and their relation to the Martian moons.
See Abstr. 097.026.

Penetration of chondrites into Mars' atmosphere.
See Abstr. 097.046.

Light elements in Mars meteorites.
See Abstr. 097.050.

Doublet craters and the tidal disruption of binary asteroids.
See Abstr. 098.067.

Asteroids, meteorites, interplanetary and cometary dust.
See Abstr. 098.101.

Combined simulation of the flight and explosion of a meteoroid in the atmosphere.
See Abstr. 104.051.

What heated the parent meteorite planets?
See Abstr. 107.022.

The relict magnetism of meteorites.
See Abstr. 107.023.

Comparison of interstellar and meteoritic organic matter at 3.4 μm.
See Abstr. 131.239.

106 Interplanetary Matter, Interplanetary Magnetic Field, Zodiacal Light

106.001 **Diffusion of the IMF into the far magnetosphere tail.**
A. V. Kitaev.
Geomagn. Aeron., Vol. 30, No. 1, p. 17 – 19 (Aug 1990). English translation of Geomagn. Aehron., Tom 30, No. 1, p. 23 – 26 (1990).

A magnetohydrodynamic model of the diffusion penetration of the IMF into the tail of the magnetosphere is considered. The magnetosphere is approximated by an axisymmetric cavity that changes into a circular cylinder at large distances from the Earth. An asymptotic solution is found for the problem of the diffusion of the transverse component of the magnetic field from the flux inside the cylindrical cavity.

106.002 **Cosmic rays and magneto–acoustic instability of the solar–wind stream near the heliosphere shock wave.**
S. V. Chalov.
Geomagn. Aeron., Vol. 30, No. 1, p. 24 – 27 (Aug 1990). English translation of Geomagn. Aehron., Tom 30, No. 1, p. 32 – 36 (1990).

The instability of fast magneto–acoustic waves propagating near the equatorial plane ahead of a heliospheric shock front where the solar wind is retarded by the cosmic–ray pressure gradient is investigated. The unstable waves with the highest increment are propagated normal to the equatorial plane. The direction of the propagation of these waves depends on the polarity of the interplanetary magnetic field.

106.003 **An estimate of the penetration rate of the IMF into the magnetosphere due to the development of a trough instability.**
I. I. Alekseev, Yu. P. Mal'tsev.
Geomagn. Aeron., Vol. 30, No. 1, p. 106 – 108 (Aug 1990). English translation of Geomagn. Aehron., Tom 30, No. 1, p. 134 – 137 (1990).

The penetration of the interplanetary magnetic field (IMF) into the magnetosphere due to the development of a trough instability of a thin layer of heated plasma near the magnetopause is described. Breakdown of the layer leads to the fact that magnetic field diffusion related to dissipation occurs on a scale of the order of the Larmor radius of the particles. This scale is determined by the sizes of the individual filaments formed in the nonlinear stage of development of the instability. The time of penetration of the IMF into the magnetosphere is specified by the "mixing" time of filaments of the heated plasma $\approx 2R/V_A \approx 0.5$ h for the terrestrial magnetosphere.

106.004 **Anisotropic motion of charged particles in interplanetary space. Some analytic solutions.**
I. G. Kurganov, V. M. Ostryakov.
Geomagn. Aeron., Vol. 30, No. 2, p. 155 – 159 (Oct 1990). English translation of Geomagn. Aehron., Tom 30, No. 2, p. 182 – 189 (1990).

The anisotropic stage of the propagation of charged particles in the interplanetary plasma is considered on the basis of the kinetic equation. New analytic solutions are obtained for the cases of weak pitch–angle diffusion with small and arbitrary angles of injection in a nonuniform magnetic field. The spatial and temporal spreading of the angular distribution of particles is here characterized by a Gaussian curve. The steady–state problem of strong pitch–angle diffusion of particles for a particular form of flat turbulence spectrum existing at large heliocentric distances is also solved.

106.005 **The neutral surface of the interplanetary magnetic field and the variations of galactic cosmic rays.**
D. Z. Borisov, I. S. Samsonov, Z. N. Samsonova.
Geomagn. Aeron., Vol. 30, No. 2, p. 250 – 252 (Oct 1990). English translation of Geomagn. Aehron., Tom 30, No. 2, p. 305 – 307 (1990).

Proceeding from the concept of a region flattened towards the neutral surface of the IMF which modulates cosmic rays, the currents (first harmonic) of galactic cosmic rays in the heliosphere are analyzed. During periods of greatest flattening of the modulating region at maximum separation from the neutral surface, cases of the disappearance of the currents of galactic cosmic rays (GCR) are observed. These cases depend on the width of high–speed streams of the solar wind. Such behavior of GCR currents as well as the absence of the arrival of solar protons are a consequence of the separation of the observer from the neutral surface.

106.006 **The probable existence of two subclasses of interplanetary magnetic clouds.**
M. S. Bobrov.
Geomagn. Aeron., Vol. 30, No. 3, p. 312 – 314 (Nov 1990). English translation of Geomagn. Aehron., Tom 30, No. 3, p. 368 – 371 (1990).

Two samples of interplanetary magnetic clouds are compared with each other: 1) sporadic solar–wind streams. They were selected by the criterion of their genetic relation to solar flares or the disappearance of filaments; 2) Klein–Burlaga objects. They were selected by the criterion of a looplike configuration of the magnetic field. For both types of objects there were hodographs of the magnetic field vector, whose main peculiarities turned out to be practically identical. It has been shown that approximately 15 Klein–Burlaga objects should be genetically related either to flares or to the disappearance of filaments, i.e., they belong to the subclass of sporadic streams. The other 31 objects are evidently related to some other kind of features on the Sun. It has been concluded that interplanetary magnetic clouds are subdivided into two subclasses: sporadic streams and the Klein–Burlaga objects.

106.007 **The correlation vector of cosmic rays and its relation to the spirality of the interplanetary magnetic field.**
M. E. Kats, K. Kudela, M. Steglik, M. Slivka.
Geomagn. Aeron., Vol. 30, No. 3, p. 327 – 331 (Nov 1990). English translation of Geomagn. Aehron., Tom 30, No. 3, p. 386 – 392 (1990).

A relation has been established between the correlation vector of cosmic rays and the spectral function of the spirality of the interplanetary magnetic field on the basis of the kinetic equation. In the case of isotropic and two–dimensional turbulence a relation is given between the spectral indices of the correlation vector p and the spirality $\nu = p+1$. An analysis is made of experimental data on measurements of the IMF by the ISEE–3 spacecraft and measurements of the cosmic–ray intensity at the Vostok station. The calculated correlation vector is directed perpendicularly to the ecliptic plane and gives a relation between the spectral indices $p = \nu-1$.

106.008 **A case study of very low cross–helicity fluctuations in the solar wind.**
C. Y. Tu, E. Marsch.
Ann. Geophys., Vol. 9, No. 5, p. 319 – 332 (May 1991).

Two periods of very low cross–helicity fluctuations observed by Helios 2 near 1 AU in 1976 have been studied in detail. It is shown that in both cases a nearly zero cross–helicity occurs together with an extremely low Alfvén ratio and a –5/3 power–

law for the turbulent energy spectra. The simultaneous occurrence of these features is apparently not influenced by compressible fluctuations. Explaining the nature of the zero cross-helicity fluctuations poses a new theoretical problem, for which no satisfactory solution exists at the present time. However, a possible scenario is suggested in which the observed fluctuations mainly result from convecting static structures, not involving any velocity variations but merely "Magnetic Field Directional Turnings" (MFDTs). The associated evolutions of both the cross-helicity and the Alfvén ratio in the inner solar system may qualitatively be explained by including these structures in extended solar wind turbulence models. The difficulties remaining with the scenario of MFDTs have also been pointed out.

106.009 **Interplanetary variability in particle fluxes recorded by the low energy charged particle detector SLED (~30 keV – >30 MeV) during the Cruise Phase of the Phobos Mission to Mars and its moons.**
S. M. P. McKenna-Lawlor, V. V. Afonin, K. I. Gringauz, E. Keppler, E. Kirsch, A. K. Richter, M. Witte, D. O'Sullivan, A. Thompson, K. Kecskemety, A. Somogyi, L. Szabo, A. Varga, R. Marsden.
Ann. Geophys., Vol. 9, No. 5, p. 348 – 356 (May 1991).

Two lightweight telescope detector systems with the capability to monitor electron and ion fluxes within an energy range spanning 34 keV to a few tens of MeV, were launched on the twin spacecraft of the Soviet Phobos Mission to Mars and its moons in July 1988. Solar-related particle enhancements recorded during the Cruise Phase, and also in the near Martian environment, over the interval 19 July 1988 – 27 March 1989 while the interplanetary medium was in course of changing over from solar-minimum to solar-maximum dominated conditions, are presented.

106.010 **The influence of the interface between the heliosphere and the local interstellar medium on the penetration of the H atoms to the solar system.**
V. B. Baranov, M. G. Lebedev, Yu. G. Malama.
Astrophys. J., Vol. 375, No. 1, p. 347 – 351 (1 Jul 1991).

A self-consistent gasdynamic model of the solar wind interaction with the local interstellar medium (LISM), which takes into account the mutual influence of the plasma component (electrons and protons) of LISM and its H atoms that penetrate into the heliosphere, was constructed by Baranov et al. This model, however, has a number of defects. In particular, the motion of the H atoms was described by hydrodynamical equations, although the mean free path of the H atoms and the characteristic length of the problem are comparable. In this paper, the first step of an iterative method using Monte Carlo simulation is realized. To illustrate the effect of the plasma component of the LISM on the penetration of H atoms into the heliosphere, the authors calculate the trajectories of H atoms by the Monte Carlo method. The bulk velocity and the number density of H atoms as functions of the distance from the Sun for different values of polar angle are analyzed. The spatial distribution of the scattered solar Lyα radiation for different values of LISM parameters is calculated and compared with the observations of Prognoz 6.

106.011 **Voyager plasma wave observations near the outer planets.**
W. S. Kurth.
Adv. Space Res., Vol. 11, No. 9, p. 59 – 68 (1991). – See Abstr. 012.006 for the main entry.

With the completion of the reconnaissance of the magnetospheres of the outer planets by Voyager one has a first order view of the existence of plasma waves in a wide variety of settings with which to begin the task of intercomparison. Voyager carried the first plasma wave receivers to Jupiter, Saturn, Uranus, and Neptune and we now can make some very strong statements about the likelihood of several types of plasma wave modes and where in a planetary magnetosphere these emissions will occur.

106.012 **Dependence of amplitude and spectrum of ELF turbulence on Θ_{Bn}.**
N. L. Borodkova, O. L. Vaisberg (*O. L. Vajsberg*), M. N. Nozdrachev, V. N. Smirnov, G. N. Zastenker.
Adv. Space Res., Vol. 11, No. 9, p. 257 – 259 (1991). – See Abstr. 012.006 for the main entry.

Analysis of several low Mach number interplanetary shocks have shown that the steepness of the amplitude spectrum of electric field fluctuations in the frequency range from several Hz to about 20 Hz is different at quasi-parallel shock front compared to the front of quasi-perpendicular shock. The amplitude of electric field fluctuations in the ion sound frequency range is also varying with the angle between the shock normal and magnetic field direction, Θ_{Bn}. The authors extend their analysis to sample of bow shock crossings using the data of Prognoz-8 in order to verify the Θ_{Bn} dependence of ELF noise for supercritical shocks.

106.013 **Magnetic clouds.**
L. F. E. Burlaga.
Physics of the inner heliosphere 2. Particles, waves and turbulence, p. 1 – 22 (1991). – See Abstr. 003.002 for the main entry.

Contents: Magnetic clouds: definition and properties, origin and composition, model of the local structure, large-scale structure, motion and stability, nonthermal particles. Effects of magnetic clouds on geomagnetic activity and on galactic cosmic rays. Magnetic clouds and compound streams. Questions concerning magnetic clouds.

106.014 **Interplanetary slow shocks.**
A. K. Richter.
Physics of the inner heliosphere 2. Particles, waves and turbulence, p. 23 – 44 (1991). – See Abstr. 003.002 for the main entry.

Contents: (1) Introduction. (2) Generation of interplanetary slow shocks: steepening of slow-mode waves, magnetic field line merging, interaction of interplanetary discontinuities. (3) Solar wind constraints on the development and evolution of interplanetary slow shocks. (4) Constraints on the unique identification of interplanetary slow shocks from space observations: the classical way, additional conditions, implications of checking against tangential discontinuities, examples.

106.015 **Results from the LDEF Interplanetary Dust Experiment.**
J. P. Oliver, J. D. Mulholland, S. F. Singer, W. J. Cooke, N. L. Montague, J. L. Weinberg, P. C. Kassel, J. J. Wortman.
Bull. Am. Astron. Soc., Vol. 23, No. 2, p. 940 – 941 (1991). Abstract. – See Abstr. 012.008 for the main entry.

106.016 **Types of interplanetary shocks and Forbush decreases.**
Badruddin, B. Venkatesan, A. G. Ananth.
Sol. Phys., Vol. 134, No. 2, p. 395 – 401 (Aug 1991).

Two types of interplanetary shocks have been identified and classified into two groups, those associated with a helium-enhancement and those not associated with any helium-enhancement. The cosmic-ray intensity decreases at Calgary neutron monitor are studied with respect to the arrival time of the two groups of shocks. The observations show that large Forbush decreases are caused by shocks associated with the helium-enhancement; and those not associated with He shocks show comparatively a small decrease in cosmic-ray intensity.

106.017 **The asteroidal and meteoric components of the zodiacal light.**
N. B. Divari.
Astron. Vestn., Tom 25, No. 4, p. 466 – 475 (Jul – Aug 1991). In Russian. English translation in Sol. Syst. Res., Vol. 25.

A review of the existing hypotheses of the origin of the interplanetary dust is presented. Two components seem to exist in the zodiacal light: asteroidal (or ecliptic) and meteoric (or interplanetary). The first component resulting from the dust in the asteroidal ring is seen in the visible region as a zodiacal band. The meteoric component responsible for the light diffused throughout the sky is provided by the dust in the inner part of the solar system. These particles have the same origin as the meteoroids. The necessity to reconsider the time of dust particles spiralling to the Sun under the Poynting–Robertson effect with regard of perturbations from planets is emphasized.

106.018 **Large–scale structure of the interplanetary magnetic field as derived from geomagnetic data for 1984 1989.**
A. N. Zajtsev, V. A. Popov.
Soln. Dannye, Byull., No. 12/90, p. 83 – 86 (Dec 1990). In Russian.

Large–scale magnetic variations of the sector structure of the interplanetary magnetic field are considered in connection with the general character of the development of solar activity and dynamics of the heliospheric current layer from the minimum to the maximum of solar activity. The presence of a two–sector structure is noted. The data on the determination of the sector structur of the interplanetary magnetic field are given on the basis of 1984 – 1989 ground–based observations at the Vostok and Resolute Bay stations.

106.019 **Nonstationary transport of flare–generated charged particles in the interplanetary space.**
M. F. Bakhareva.
Geomagn. Aehron., Tom 31, No. 4, p. 577 – 582 (Jul – Aug 1991). In Russian. English translation in Geomagn. Aeron., Vol. 31, No. 4.

106.020 **Neutral surface of the interplanetary magnetic field and intensity of galactic cosmic rays.**
I. S. Samsonov, D. Z. Borisov, Yu. F. Rudnev, Z. N. Samsonova.
Geomagn. Aehron., Tom 31, No. 4, p. 717 – 719 (Jul – Aug 1991). In Russian. English translation in Geomagn. Aeron., Vol. 31, No. 4.

106.021 **Formation of the heliospheric neutral layer in the solar corona and interplanetary space.**
V. P. Mikhajlutsa.
Geomagn. Aehron., Tom 31, No. 6, p. 961 – 968 (Nov – Dec 1991). In Russian. English translation in Geomagn. Aeron., Vol. 31, No. 6.

106.022 **Restoration of polarity of the interplanetary magnetic field in the past.**
D. I. Ponyavin.
Geomagn. Aehron., Tom 31, No. 6, p. 1094 – 1096 (Nov – Dec 1991). In Russian. English translation in Geomagn. Aeron., Vol. 31, No. 6.

106.023 **Abundance analysis for low–Z elements in interplanetary dust particles.**
Wu Bei, Xu Yinlin.
Publ. Purple Mt. Obs., Vol. 10, No. 2, p. 122 – 127 (Jun 1991). In Chinese.

The authors analyse the abundances of low–Z elements in two IDPs, from their mass spectra with secondary ion spectrometry, and EDS (X–ray Energy Dispersive Spectra). The results show that the abundances are similar to those of the solar system. So the two IDPs are chondritic, and they may have more information of the early solar system.

106.024 **Interplanetary intermediate shock.**
Hu Youqiu.
Acta Geophys. Sin., Vol. 34, No. 4, p. 397 – 403 (Jul 1991). In Chinese.

A hybrid shock formed by linking up different types of shocks may exist in the interplanetary space. As a necessary component of a hybrid shock, the intermediate shock occurs in the vicinity of the heliospheric electric current sheet with its front concave towards the sun. This conclusion is confirmed by observations to a certain extent, and has important relations with the 3–dimensional properties of interplanetary shocks.

106.025 **Circumsolar forbidden spheres for micrometeoroids.**
V. Mioc, A. V. Pop, I. Giurgiu.
Stud. Cercet. Fiz., Vol. 42, No. 7, p. 601 – 606 (1990).

The motion of a micrometeoroid in the neighbourhood of a star, under the only influences of gravitation and radiation pressure, is studied. A critical value for the radius of the grain (for a given density), under which this one cannot approach indefinitely the star (whatever be its initial speed), is pointed out. Such forbidden spheres around the Sun are determined for micrometeoroids of different dimensions, densities and velocities.

106.026 **Heliospheric shocks and catastrophe theory.**
L. F. Burlaga.
Geophys. Res. Lett., Vol. 17, No. 10, p. 1633 – 1636 (Sep 1990).

Various configurations of forward and reverse shocks that occur in the outer heliosphere can be classified using catastrophe theory. The existence of a forward shock is associated with a local maximum of a polynomial, and the existence of a reverse shock is associated with a local minimum of a polynomial. A configuration with N forward shocks and N reverse shocks corresponds to a polynomial with N maxima and N minima. The formation of forward and reverse shocks corresponds to the creation of maxima and minima of a polynomial, which is described by the separatrices of the catastrophes A_k. The coalescence of two forward (reverse) shocks corresponds to the situation when two maxima (minima) of a polynomial have equal values, and the interaction of a forward shock with a reverse shock corresponds to a polynomial with a local maximum equal to a local minimum; these situations are described by the Maxwell sets of the appropriate catastrophes.

106.027 **Flow downstream of the heliospheric terminal shock: magnetic field kinematics.**
S. Nerney, S. T. Suess, E. J. Schmahl.
Astron. Astrophys., Vol. 250, No. 2, p. 556 – 564 (Oct 1991).

The authors have developed a kinematic model of the interplanetary magnetic field in the heliosheath beyond the solar wind terminal shock. The purpose of the calculation is to evaluate the possible importance of MHD effects in that region of space. The need for this evaluation arises because the interplanetary magnetic field is compressed across the terminal shock and further amplified by the decreasing flow speed beyond the shock. This effect is enhanced because the shocked subsonic wind is turned back by the streaming local interstellar medium. Streamlines which approach the stagnation point before turning in the downstream direction lead to the strongest effects due to the extreme slowing of the solar wind and consequent compression of the embedded magnetic field. The magnetic volume force therefore cannot be neglected on streamlines that approach the heliopause in the upstream direction, where the volume containing them is a large fraction of the overall volume of the heliosheath in the upstream direction. The increase in the magnetic pressure may act to bring the upstream terminal shock significantly closer to the Sun, potentially reconciling a conflict between models and observations.

106.028 **Multifractal structure of the interplanetary magnetic field: Voyager 2 observations near 25 AU, 1987 – 1988.**
L. F. Burlaga.
Geophys. Res. Lett., Vol. 18, No. 1, p. 69 – 72 (Jan 1991).

This paper analyzes the large–scale fluctuations of the interplanetary magnetic field strength observed by Voyager 2 from 23.3 AU to 27.8 AU during the period from day 190, 1987 to day 345, 1988.

106.029 **The composition and plasma signature of a large dust impact on the Giotto spacecraft.**
R. Goldstein, B. E. Goldstein, H. Balsiger, A. J. Coates, W. Curdt, H. U. Keller, F. M. Neubauer, C. Perry, J. Zarnecki.
J. Geophys. Res., Vol. 96, No. A8, p. 13739 – 13748 (1 Aug 1991).

At about 14,800 km from the comet Halley nucleus, on the inbound leg, at least six of the sensors onboard the Giotto spacecraft observed an unusual, brief ($\sim$ 30 to 500 ms) event: The ion mass spectrometer data show a brief flow of energetic (up to several hundred electron volts) plasma consisting of protons, water group, and heavier ions. The Johnstone plasma analyzer data show a short burst of plasma, while the dust impact detector system data show an impact event in four of its detectors. The magnetometer signature of the event shows two brief dips in the field. The sudden change in the spacecraft attitude and spin rate observed by the camera at that same time has been interpreted as the result of a large ($\geqslant 5$ mg) dust particle impact on the front bumper shield of the spacecraft. In addition, at about the same time the spacecraft star–tracker suffered damage. Here the authors give the first report combining direct measurements of the composition and dynamics of a dust impact plasma cloud, the dust particle mass and the location of the impact on the spacecraft. Analysis of the data indicate that the impacting particle was water or ice–bearing, possibly loosely compacted, and was composed of one or more of: carbon, nitrogen, and silicon.

106.030 **Interplanetary Lyman–α observations with UVS on Voyager: data, first analysis, implications for the ionization lifetime.**
R. Lallement, J. L. Bertaux, E. Chassefière, B. Sandel.
Astron. Astrophys., Vol. 252, No. 1, p. 385 – 401 (Dec 1991).

Measurements of the interplanetary Lyα background were collected by the UV Spectrometer (UVS) instrument during the whole cruise of Voyager 1 and 2. A fraction of these measurements taken during the 1977–1983 period (heliocentric distance 1–18 AU) is presented and compared with results from current models of the interaction between the Sun and the neutral interstellar gas. There are two different sets of data: measurements along fixed lines of sight (FL), and sky maps collected during special roll maneuvers (RM) of the spacecraft. For FL data line–of–sight geometry and time are coupled, while RM maps show the whole H distribution on some particular dates. The analysis of both datasets leads to the conclusion that the same H atom lifetime cannot fit all the data. The actual ionization rate is inferred from the intensity gradient in the maximum emissivity region observed from a sidewind Voyager position at 6 AU, yielding a lifetime of $\cong 10^6$s at 1 AU, whereas the upwind/downwind intensity ratio in the inner solar system favors 2×10^6s, as measured by Voyager, Prognoz, and Pioneer Venus instruments. It is suggested that there is an excess of Lyα emission in the downwind region which forces the model toward too high values of the lifetime. Possible explanations are discussed, like incorrect modelling, or additional source of H atoms. It is argued that this cannot be easily accounted for by a variation with time of the solar parameters. Finally, the H density at infinity is revised upward as a result of the reduced lifetime.

106.031 **Plasma wave generation near the inner heliospheric shock.**
W. M. Macek, I. H. Cairns, W. S. Kurth, D. A. Gurnett.
Geophys. Res. Lett., Vol. 18, No. 3, p. 357 – 360 (Mar 1991).

There is mounting evidence that the Voyager 1 and 2 and Pioneer 11 spacecraft may approach the inner (termination) heliospheric shock near the end of this century. It is argued here, by analogy with planetary bow shocks, that energetic electrons backstreaming from the heliospheric shock along the magnetic field should be unstable to the generation of Langmuir waves by the electron beam instability. Analytic expressions for the cutoff velocity, corresponding to the beam speed of the electrons backstreaming from the shock, are derived for a standard solar wind model.

106.032 **Numerical simulation of the April 24, 1981 interplanetary shock wave.**
D. Odstrčil.
Bull. Astron. Inst. Czech., Vol. 42, No. 3, p. 181 – 189 (Jun 1991).

The paper deals with the application of the 1 D HD computer code to the simulation of the interplanetary shock wave generated on April 24, 1981. This event is simulated, in terms of density, velocity and temperature, by a pulse introduced at 18 Rs into a steady–state solar wind. The observed data were used to specify all significant parameters of the steady–state solar wind and the introduced shock wave. The short duration of the input pulse caused the shock wave to be initially highly decelerated. Special attention is given to the type II radio emission associated with this shock and measured by the Czechoslovak–Soviet experiment aboard the Prognoz–8 satellite. From the given analysis it follows that the emission is generated in front of the shock front at the blended fundamental and harmonic plasma frequency.

106.033 **Comparison of 2–dimensional MHD and semiempirical models of interplanetary shock wave propagation.**
M. Karlický, Z. Smith, M. Dryer.
Bull. Astron. Inst. Czech., Vol. 42, No. 5, p. 320 – 328 (Oct 1991).

Shock wave velocities at 126 points in the solar equatorial plane, within 1 AU, are obtained from a 2–D magnetohydrodynamic model (Wu et al. 1983). These simulations are compared with those from a semiempirical model of shock wave propagation by Vörös and Karlický (1989). Shocks with the three different initial velocities, $v_s = 1000, 1500$ and 2000 km s^{-1}, were considered. Although agreement in the exponential dependence ($N \cong 0.5$) of shock wave velocities ($v_s \sim 1/R^N$) was found, the two models differ: (a) in the early stages of shock wave propagation, and (b) in the shock shapes. A small E–W asymmetry of shock wave propagation was found in the MHD simulations.

106.034 **The structure of helical interplanetary magnetic fields.**
M. L. Goldstein, D. A. Roberts, C. A. Fitch.
Geophys. Res. Lett., Vol. 18, No. 8, p. 1505 – 1508 (Aug 1991).

The interplanetary magnetic field is known to be highly helical. Although the detailed spatial structure of the fields has yet to be elucidated, the helicity spectrum has been conjectured to result from a random walk in direction of a constant magnitude magnetic field vector. The authors demonstrate that a model using three–dimensional fluctuations with variations in |B| gives a good fit to the helicity spectrum as well as to other properties of the interplanetary magnetic field.

106.035 **New observations of the low frequency interplanetary radio emissions.**
W. S. Kurth, D. A. Gurnett.
Geophys. Res. Lett., Vol. 18, No. 10, p. 1801 – 1804 (Oct 1991).

Recent Voyager 1 observations reveal reoccurrences of the low frequency interplanetary radio emissions. Three of the new events are weak transient events which rise in frequency from the range of 2–2.5 kHz to about 3 kHz with drift rates of ~1.5 kHz/year. The first of the transient events begins in mid–1989 and the more recent pair of events both were first detected in late 1990. In addition, there is an apparent onset of a 2–kHz component of the emission beginning near day 70 of 1991. The new transient emissions are barely detectable on Voyager 1 and are below the threshold of detectability on Voyager 2, which is less sensitive than Voyager 1. The new activity provides new opportunities to test various theories of the triggering, generation, and propagation of the outer heliospheric radio emissions and may signal a response of the source of the radio emissions to the increased solar activity associated with the recent peak in the solar cycle.

106.036 **Helios spacecraft photometer observation of elongated corotating structures in the interplanetary medium.**
B. V. Jackson.
J. Geophys. Res., Vol. 96, No. A7, p. 11307 – 11318 (1 Jul 1991).

The Helios spacecraft zodiacal light photometers can be used to measure persistent elongated structures at interplanetary distances from the Sun. These data give the heliospheric poistions of these structures from their east to west motion with time. An outward motion of the material within each structure can be determined from its curvature. This technique has been used to measure over 40 of these structures from solar minimum to solar maximum (1976 through 1979) using both Helios A and Helios B observations. The positions of the structures measured range from low latitude to as high as 50°. The speeds determined have a mean value of 301 ± 47 km s^{-1}. The modeling, which determines densities for these structures from 0.15 to over 1.0 AU, indicates that these heliospheric structures can have densities many times above the ambient and that the density in excess of an r^{-2} ambient can differ greatly along their curved extents at any given time. Most of the structures measured in this sample either remain approximately the same density or decrease in density with height above the Sun. This implies that these structure are generally remnants of features maintained by the physical processes which formed them near the surface of the Sun (as streamers) and that they are not caused by interplanetary interactions.

106.037 **Filament disappearances and associated shocks of May 1979.**
H. V. Cane, I. G. Richardson, K. Harvey.
J. Geophys. Res., Vol. 96, No. A11, p. 19525 – 19528 (1 Nov 1991).

In late May 1979 a shock was seen at each of two locations separated by about 90° in heliolongitude and at about the same time. A shock was detected near the Earth, and a shock was detected at Helios 1 when it was off the Sun's west limb. Some authors have considered that these shocks were part of the same event, which originated with a filament dispperance near 65°W. The authors discuss the observations which show that a large complex filament, extending from 20°W to 70°W, disappeared in several places, probably generated more than one shock. The shock detected near Earth was probably associated with a filament section which disappeared near the central meridian. If this is the case, then the interplanetary scintillation observations of this time period can be explained without invoking highly anisotropic shock propagation.

106.038 **Statistical properties of shock-accelerated ions in the outer heliosphere.**
L. J. Lanzerotti, R. E. Gold, D. J. Thomson, R. E. Decker, C. G. Maclennan, S. M. Krimigis.
Astrophys. J., Lett., Vol. 380, No. 2, p. L93 – L96 (20 Oct 1991). With plate L6.

Reported here are statistical studies of interplanetary protons ($\sim$0.5 – 1.5 MeV) as measured by the Voyager 2 spacecraft in the outer heliosphere between $\sim$1 and $\sim$22 AU. The acceleration processes which produce the fluctuations in intensity of the protons throughout the interval studied are caused by solar flare – generated shock waves and corotating shock structures in the solar wind. It is found that the proton count rates can be represented by a lognormal distribution, independent of the phase of the solar cycle in which the measurements were made. Power spectra of the logarithm of the counts show that for frequencies in the range $\sim$0.02 – 0.5 cycles per day, the spectra can be characterized as $\sim f^{-2}$. At lower frequencies, the spectra turn over and flatten. Over the entire frequency range studied, $\sim$0.002 – 0.5 cycles per day, the spectra are therefore non-fractal. The rollover in the spectra at $\sim$100 days probably represents the reorganization time for solar wind streams within the coronal source at the Sun. The lognormal distributions and the $\sim f^{-2}$ spectra are both consistent with the statistical net acceleration occurrences for interplanetary ions being a product of random probabilities occurring in an approximately steplike fashion in association with interplanetary shocks.

106.039 **Reconciling ZIP and IRAS observations of the zodiacal background.**
L. J. Rickard, S. W. Stemwedel, S. D. Price.
AIP Conf. Proc., No. 222, p. 183 – 186 (15 Apr 1991). Current Physics Microform No.: 9106E0183. – See Abstr. 012.069 for the main entry.

106.040 **Puzzles of the zodiacal light.**
N. B. Divari.
Zemlya Vselennaya, No. 6, p. 3 – 9 (Nov – Dec 1991). In Russian.

106.041 **The heliospheric current sheet (HCS) and high-speed solar wind: interaction effects.**
V. G. Eselevich, V. G. Fainshtein.
Planet. Space Sci., Vol. 39, No. 8, p. 1123 – 1131 (Aug 1991).

By analysing 75 sections of the heliospheric current sheet (HCS) at the Earth's orbit, it is shown that, for a stream-associated HCS, an additional enhancement of plasma density n_{max} and magnetic field B_{max} results from collisions with a high-speed solar wind (SW) inside the sheet as compared with their values for a horizontal HCS, corresponding to the type of stream-free HCS. In this case the effect of additional increase of n_{max} and B_{max} is the stronger, the larger is the slope θ of the HCS section considered to the solar equatorial plane and the higher is the value of SW velocity gradient $\Delta V/\Delta\varphi$ at points n_{max} and B_{max}. Quantitative characteristics of these relations are obtained.

106.042 **The Sun and interplanetary magnetic field.**
E. J. Smith.
The Sun in time, p. 175 – 201 (1991). – See Abstr. 003.027 for the main entry.

The interplanetary magnetic field (IMF) serves as a link between the Sun, the response of the Earth to solar activity and variations in galactic cosmic radiation. The IMF originates as a solar-coronal magnetic field that is transported into space by the solar wind. The close connection between solar magnetic fields and the origin and structure of the solar wind is described. The solar wind forms the heliosphere, a cavity containing the magnetized solar plasma from which the interstellar plasma and field are excluded. The entry of galactic cosmic rays into the heliosphere and their strong interaction with the IMF are discussed, this topic being of primary importance to the production and temporal variations of radiogenic elements. The profound influence of the IMF on geomagnetic activity and the aurora is discussed within the context of merging or reconnection with the planetary field. The physical connection is thus established between solar magnetic fields, magnetic storms and aurora. The state of the solar wind and IMF during the Maunder minimum is considered and an explanation for the (relative) absence of sunspots and aurora is proposed. The mechanism is an interruption of the oscillatory solar dynamo, a consequent reduction in the heating of the corona, a cessation of the supersonic solar wind and a weakening or absence of southward-directed magnetic fields in the vicinity of the Earth.

106.043 **Cosmoids: solution to the Pioneer 10 and 11 meteoroid measurement enigma.**
M. Dubin, R. K. Soberman.
Planet. Space Sci., Vol. 39, No. 11, p. 1573 – 1590 (Nov 1991).

A theory characterizing interplanetary matter and the zodiacal light has been derived from the dust experiments on the Pioneer 10 and 11 spacecraft to explain the extremely disparate results of the distribution of particles and the zodiacal light measurements. The theory based on these measurements requires a population in the inner solar system of cold meteoroids, cosmoids, composed mainly of volatile molecules, moving in highly eccentric orbits and seldom surviving a single perihelion passage. Impulsive or bursty disruptions from solar heating appears necessary to explain the measured rapid orders of magnitude increases in optical cross section. The integrated scattering from the dispersed material appears to account for nearly all the zodiacal light, with polarization from scattering by micron sized particles. The sublimation time in sunlight opposes the gravitational flux increase expected in approaching the sun. The three Pioneer w 10/11 dust experiments were: the Imaging Photopolarimeter (IPP) for the zodiacal light (ZL), the Meteoroid Detection Experiment (MDE) that measured penetrations of 25 (Pioneer 10) and 50 (Pioneer 11) μm thick walls of pressurized gas cells, and the Asteroid/Meteoroid Experiment or "Sisyphus" that remotely measured meteoroids from reflected sunlight and ZL between meteoroid "events".

106.044 **Shock interactions in the outer heliosphere.**
Y. C. Whang.
Space Sci. Rev., Vol. 57, No. 3/4, p. 339 – 388 (Aug 1991).

Observations of plasma and magnetic fields by Pioneer 10 and 11 and Voyager 1 and 2 reveal that MHD shocks are an important component of the large-scale wind structures in the outer heliosphere. This review discusses recent progress in simulation studies of the nonlinear evolution of the solar wind structures, and in particular concentrates on the theoretical development and applications of the shock interaction model.

106.045 **Physical reasons and consequences of a three-dimensionally structured heliosphere.**
H. J. Fahr, H. Fichtner.
Space Sci. Rev., Vol. 58, No. 3/4, p. 193 - 258 (Nov 1991).

The authors discuss reasons both of solar and of interstellar origin giving rise to a pronounced three-dimensional structure of the expanding solar wind and thus of the global configuration of the heliosphere. The present observational knowledge of these structurings is reviewed, and all attempts to theoretically model these solar wind structures are critically analysed with respect to their virtues and flaws. It is especially studied by what mechanisms interstellar imprints on the actual type of solar wind expansion can be envisaged.

106.046 **On the importance of impact erosion.**
J. Klačka.
Bull. Astron. Inst. Czech., Vol. 42, No. 6, p. 379 - 381 (Nov 1991). With plates 3, 4.

The problem of the importance of the impact erosion on the orbital evolution of interplanetary dust particles is discussed. The impact erosion can be expressed in terms of reduction in radius of particles $\mathrm{d}s/\mathrm{d}t = -\varepsilon^{D} s^{1.5}/r^{2.3}$, where $\varepsilon^{D} \cong 10^{-6}\mathrm{cm}^{-0.5}\mathrm{yr}^{-1}$, if the falloff of particle density as $r^{-1.3}$ is assumed. The effect of the impact erosion can be neglected in comparison to the corpuscular sputtering for particles of radii $s < 10^{-3}$cm.

106.047 **Structure and evolution of compound streams at $\leqslant 1$ AU.**
K. W. Behannon, L. F. Burlaga, A. Hewish.
J. Geophys. Res., Vol. 96, No. A12, p. 21213 - 21225 (1 Dec 1991).

Two cases of compound stream formation and evolution at heliocentric distances of $\leqslant 1$ AU have been studied using simultaneous data from both multipoint in situ and global sources. The sequences of events analyzed occurred around December 16, 1978, and February 19, 1979, respectively. The sources of the multipoint observations were Helios A, Helios B, IMP 8, and ISEE 3. The global observations were obtained by interplanetary scintillation analysis of radio sources. It was found that the radio and in situ data complemented each other and that both were needed for a unique interpretation of complex, time-dependent flows. The cases studied show that a single corotating stream can interact with different transient flows at different times, and a single transient flow can interact with different corotating streams at different times. In each instance, a compound stream is formed. This work further shows that the cosmic ray intensity at a point depends on the global compound stream configuration in the vicinity of the point.

106.048 **Comparison of Doppler scintillation and in situ spacecraft plasma measurements of interplanetary disturbances.**
R. Woo, R. Schwenn.
J. Geophys. Res., Vol. 96, No. A12, p. 21227 - 21244 (1 Dec 1991).

Since 1979 the radio path of the Pioneer Venus orbiter (PVO) spacecraft has spent considerable time probing the near-Sun solar wind off the limbs of the Sun. On occasion, Helios 1 has also been in position above the Sun's limb at the same solar longitude to simultaneously observe in situ the solar wind plasma. These fortuitious circumstances, along with the availability of near-continuous measurements, have made it possible for the first time to carry out detailed comparisons between Doppler scintillation and in situ plasma measurements and to improve our understanding of Doppler scintillation transients.

Physics of the inner heliosphere 2. Particles, waves and turbulence.
See Abstr. 003.002.

The interplanetary plasma.
See Abstr. 013.060.

A brief review, recent development in interplanetary physics and magnetospheric physics.
See Abstr. 013.062.

Spectroscopy of solid materials simulating cosmic dust.
See Abstr. 022.164.

Rôle of nonresonant Alfvén instabilities in solar system beam-plasmas.
See Abstr. 062.022.

Acceleration of keV-electrons by a nearly perpendicular shock wave.
See Abstr. 062.148.

Parametric study of the formation of magnetohydrodynamic shocks. III. Fast and slow shocks.
See Abstr. 062.204.

Active longitudes on the Sun and their reflection in interplanetary space.
See Abstr. 072.001.

Two types of quasi-biennial variations in the Sun and in interplanetary space.
See Abstr. 072.074.

Two-year variations of solar activity, interplanetary space and cosmic rays parameters.
See Abstr. 072.075.

Three solar filament disappearances associated with interplanetary low-energy particle events.
See Abstr. 073.016.

Solar-wind parameters and magnetic activity during periods of uncompressed increases in the solar-wind density.
See Abstr. 074.004.

Radio range measurements of coronal electron densities at 13 and 3.6 centimeter wavelengths during the 1988 solar conjunction of Voyager 2.
See Abstr. 074.006.

The structure of the transition region of the solar wind.
See Abstr. 074.007.

Kinetic physics of the solar wind plasma.
See Abstr. 074.012.

Survival of discrete structures in the solar wind.
See Abstr. 074.065.

Aspects of the present heliospheric research.
See Abstr. 074.074.

Short-term variability of solar wind number density, speed and dynamic pressure as a function of the interplanetary magnetic field components: a survey over two solar cycles.
See Abstr. 074.079.

The interaction effect of fast and slow solar wind streams in interplanetary space on wind characteristics at the Earth's orbit.
See Abstr. 074.100.

Magnetohydrodynamic simulation of the radial evolution and stream structure of solar-wind turbulence.
See Abstr. 074.107.

Coronal restructuring and coronal mass ejections as inferred from interplanetary magnetic flux ropes.
See Abstr. 074.123.

Electron groups traced from the Sun to 1 AU.
See Abstr. 077.081.

Interplanetary transport of decay protons from solar flare neutrons.
See Abstr. 078.027.

A system of loop interplanetary traps of solar cosmic rays on June 1974.
See Abstr. 078.030.

Variations common to the interplanetary magnetic field, the zonal atmospheric circulation and the earth's rotation.
See Abstr. 082.009.

Explanation for seasonal variations of the geomagnetic field in the polar cap using the polarity of interplanetary sectors.
See Abstr. 084.005.

Transpolar auroras, their particle precipitation, and IMF B_y component.
See Abstr. 084.046.

Electron density fluctuations in the supercorona and the terrestrial effects of solar activity.
See Abstr. 085.005.

Small solar system bodies.
See Abstr. 091.057.

Interplanetary magnetic field control of the Mars bow shock: evidence for Venuslike interaction.
See Abstr. 097.059.

Asteroids, meteorites, interplanetary and cometary dust.
See Abstr. 098.101.

Growth and disruption of composite dust grains in the environment of a comet.
See Abstr. 102.009.

The structure and evolution of the Taurid Complex.
See Abstr. 104.040.

Magnetic history of the Sun.
See Abstr. 107.019.

Analytic solution of the steady–state isotropic boundary value problem of the propagation of galactic cosmic rays in interplanetary space.
See Abstr. 144.001.

Study and effect of magnetic clouds on the transient modulation of cosmic–ray intensity.
See Abstr. 144.016.

Cosmic ray modulation.
See Abstr. 144.022.

Transport of energetic particles in the interplanetary medium.
See Abstr. 144.023.

The role of corotating interaction regions in cosmic–ray modulation.
See Abstr. 144.079.

Cosmic ray unidirectional latitude gradient: evidence for north–south asymmetric solar modulation.
See Abstr. 144.080.

Forbush decreases and interplanetary magnetic field disturbances: association with magnetic clouds.
See Abstr. 144.081.

The role of cross helicity in solar modulation of cosmic rays and its charge sign dependence.
See Abstr. 144.085.

The low–frequency spectral behavior of cosmic ray intensity.
See Abstr. 144.086.

The physics of cosmic ray modulation: heliospheric propagation during the 1987 minimum.
See Abstr. 144.087.

An observation of a heliospheric magnetic cycle dependence for the integral radial gradient of E > 60 MeV cosmic rays.
See Abstr. 144.088.

The effect of the changing polarity and neutral sheet of the IMF on the cosmic ray diurnal anisotropy at neutron monitor energies.
See Abstr. 144.101.

Variations of the cosmic–ray flux with time.
See Abstr. 144.103.

Hubble diagram for QSOs with middle and weak interplanetary scintillation.
See Abstr. 159.056.

On the cosmological significance of the apparent deficit of small interplanetary scintillation sources.
See Abstr. 161.416.

107 Cosmogony

107.001 **Collapse of a dusty medium and Titius–Bode law in natural units.**
A. Eh. Filippov.
JETP Lett., Vol. 54, No. 7, p. 351 – 354 (10 Oct 1991). Current Physics Microform No.: 9201X0437.

If a protoplanetary cloud forms as a pancake of collapsing dust, the system acquires a special length scale: the orbit of the most massive planet. This scale determines the sequence of the orbits of other planets. In a zeroth approximation, these other orbits satisfy a simple rule involving resonances of orbital periods.

107.002 **Numerical models of giant planet formation with rotation.**
D. G. Korycansky, P. Bodenheimer, J. B. Pollack.
Icarus, Vol. 92, No. 2, p. 234 – 251 (Aug 1991).

The authors present models of giant planet formation including the accretion and transport of angular momentum. The calculations are carried out in a one–dimensional quasispherical approximation, assuming hydrostatic equilibrium. The approximation of quasi–sphericity limits the calculations to relatively slowly rotating models. The calculations are carried up to masses approximately equal to that of Saturn. Angular momentum transport is highly effective, producing models with a strong concentration of angular momentum in the outer convective zones that develop at envelope masses above 30 $M_♁$. Near the end of the calculation, gas accretion was turned off, and the models allowed to contact. As a result of the loss of angular momentum, however, the final planetary models have only one-third to one-half the present-day value of angular momentum of Saturn, indicating that yet larger amounts of angular momentum may have been present during formation.

107.003 **Occurrence of earth-like bodies in planetary systems.**
G. W. Wetherill.
Science, Vol. 253, No. 5019, p. 535 - 538 (2 Aug 1991).

Present theories of terrestrial planet formation predict the rapid "runaway formation" of a planetary embroyos. The size of the embryos increase with heliocentric distance. These embryos then merge to form planets. In earlier Monte Carlo simulations of the merger of these embryos it was assumed that embryos did not form in the asteroid belt, but this assumption may not be valid. Simulations in which runaways were allowed to form in the asteroid belt show that, although the initial distributions of mass, energy, and angular momentum are different from those observed today, during the growth of the planets these distributions spontaneously evolve toward those observed. Even when a larger planet analogous to "Jupiter" does not form, an Earth-sized planet is almost always found near Earth's heliocentric distance. These results suggest that occurrence of Earth-like planets may be a common feature of planetary systems.

107.004 **The dynamical evolution of the protosolar nebula.**
S. P. Ruden, J. B. Pollack.
Astrophys. J., Vol. 375, No. 2, p. 740 - 760 (10 Jul 1991).

The authors calculate the global evolution of protostellar nebulae for a variety of initial disk masses and angular momenta. The constant α of the viscous stress is calibrated in terms of the properties of turbulent thermal convection. As the evolution proceeds, the nebula viscously spreads and is partially accreted by the protostar. The consequent decrease of the surface density renders portions of the disk optically thin in the vertical direction. When the Rosseland mean optical depth decreases below a critical value of order unity, it is assumed that the disk is unable locally to sustain thermal convection and that the turbulent viscosity and energy dissipation cease. It is also assumed that once convection subsides, no other sources of turbulent stress take its place. The resulting disk evolutionary properties are calculated and compared with the solar system. The authors find that (1) moderately viscous disks with $\alpha \approx 10^{-2}$ evolve to marginally optically thin conditions in $\sim 6 \times 10^5$yr; (2) the resulting disk radii are primarily sensitive to the initial nebular angular momentum and insensitive to the initial disk mass; (3) significant amounts of mass ($\sim 0.1\ M_\odot$) can remain behind in a nonturbulent disk if the initial disk angular momentum is high; (4) the heating of the disk surface inside r $\lesssim$ 1 AU by protosolar illumination can halt thermal convection after $\sim 5 \times 10^5$yr. It is speculated that the termination of convective turbulent viscosity in the outer solar nebula marks the inception of the giant planet building epoch.

107.005 **Einige neuere Ergebnisse zur Entstehung des Sonnensystems.**
P. Bochsler.
Orion, Jahrg. 49, Nr. 247, p. 224 - 228 (Dec 1991).

107.006 **Magnetic fields in the solar nebula and the angular momentum transfer.**
M. Kiguchi, S. Narita, T. Terasawa, C. Hayashi.
IAU Symposium No. 147: Fragmentation of molecular clouds and star formation, p. 441 - 442 (1991). - See Abstr. 012.043 for the main entry.

The authors are carrying out a project to investigate how the primodial solar disk is formed and evolves. One of the central problems of this project is to study the angular momentum transfer in the nebula caused by the magnetic braking or the turbulent mixing in the boundary layer between the surface of the disk and its high temperature corona envelope. Here, they summarize the result of analysis for the magnetic braking in 1-dimensional approximation.

107.007 **The evolution of a global magma ocean with olivine flotation.**
S. Franck, M. Riedel.
Earth, Moon, Planets, Vol. 54, No. 1, p. 59 - 65 (Jul 1991). Letter-to-the-editor.

The authors investigate the thermal evolution of a global magma ocean divided by a dunite septum into a shallow and a lower part with help of simple stability estimations and cooling time calculations. It comes out that the septum is not stable against thermal convection so that this mechanism for a separated chemical evolution of the two mantle regions does not work.

107.008 **A hypothesis on the Oort clouds of planets.**
J. J. Rawal.
Earth, Moon, Planets, Vol. 54, No. 2, p. 89 - 102 (Aug 1991).

However, there are no quantitative arguments to prove, but there is a good reasoning to believe that the Oort clouds existed surrounding the planets. The short period comets, Apollo-Amor-Toro-asteroids and such other objects may be the relics of the Oort clouds that existed surrounding the planets long ago.

107.009 **Evolution of the protoplanetary disk and the problem of mass deficiency in the asteroid belt.**
T. V. Ruzmajkina, K. I. Usovich.
Astron. Tsirk., No. 1545, p. 41 - 42 (Aug - Sep 1990). In Russian.

Abnormal viscosity in the asteroidal zone of the protoplanetary disk is supposed to be the cause of mass deficiency in the asteroid belt. Effects of inhomogeneous viscosity and accretion on the surface density of the disk are treated with numerical calculations.

107.010 **At the "sources" of the contemporary theory of the origin of the Earth.**
E. L. Ruskol, V. S. Safronov.
Priroda, No. 9, p. 72 - 79 (Sep 1991). In Russian.

Contents: 1) Origin of the protoplanetary disk. 2) Formation of planets. 3) Initial state and evolution of the Earth.

107.011 **Temperatures in the protoplanetary disk and their influence on the formation of planets.**
A. B. Makalkin, V. A. Dorofeeva.
Priroda, No. 9, p. 79 - 87 (Sep 1991). In Russian.

Contents: 1) The role of temperatures in planetary cosmogony. 2) Astrophysical data on protoplanetary disks. 3) Cosmochemical data on the temperatures in the protoplanetary disk. 4) Models of the protoplanetary disk. 5) Consequences for the planets.

107.012 **Formation of the planetary system of the Sun.**
A. V. Vityazev.
Priroda, No. 9, p. 87 - 93 (Sep 1991). In Russian.

Contents: 1) From interstellar dust to protoplanetary bodies. 2) From planetesimals to planets. 3) Cosmochemical problems of disk evolution. 4) Early evolution of the Earth.

107.013 **Isotopic modelling of the early evolution of the Earth.**
I. A. Azbel', I. N. Tolstikhin.
Priroda, No. 9, p. 93 - 99 (Sep 1991). In Russian.

107.014 **The origin of the angular momentum distribution in the solar nebula.**
A. D. Chernin.
Astrophys. Space Sci., Vol. 186, No. 1, p. 159 - 160 (Dec 1991). Letter-to-the-editor.

Supersonic non-central collision and coalescence of interstellar matter clouds is suggested as a physical process that could lead to the formation of a solar nebula with an appropriate distribution of the spin angular momentum.

107.015 **Secular resonances in the primitive solar nebula.**
A. Lemaitre, P. Dubru.
Celest. Mech. Dyn. Astron., Vol. 52, No. 1, p. 57 – 78 (1991).
The authors present here a very simple model that could explain the relatively high eccentricities and inclinations observed in the minor planet belt. This model is based upon the sweeping of the secular resonances ν_6 and ν_{16} through the belt due to the gravitational effect of the dissipation of a primitive solar nebula. The sweeping of the ν_{16} secular resonance (responsible for the high inclinations) is very sensitive to the density profile of the nebula. For the model to work a density profile proportional to ϱ^{-k} with k between 1.0 and 1.5 is necessary.

107.016 **Protostellar disks and planetary system formation.**
D. N. C. Lin.
6. IAP Astrophysics Meeting – IAU Colloquium No. 129: Structure and emission properties of accretion disks, p. 123 – 127 (1991). – See Abstr. 012.046 for the main entry.

107.017 **Planetary accretion rates: analytical derivation.**
R. Greenberg, W. F. Bottke, A. Carusi,
G. B. Valsecchi.
Icarus, Vol. 94, No. 1, p. 98 – 111 (Nov 1991).
Analytic formulae for estimating planetary accretion rates have long been available for conditions where random motion of planetesimals controls velocities relative to a growing planet. Recent Monte Carlo studies have explored a wider range of conditions, but gave impact rates only for the specific conditions under which those numerical experiments were performed. Now analytical formulae, which are valid even when keplerian shear controls the approach velocity, and which accurately reproduce the Monte Carlo results, give accretion rates over a wide range of conditions relevant to the problem of planet formation.

107.018 **The origin of the systematic component of planetary rotation. I. Planet on a circular orbit.**
J. J. Lissauer, D. M. Kary.
Icarus, Vol. 94, No. 1, p. 126 – 159 (Nov 1991).
The rotation rate of a planet which accretes from small, solid planetesimals is computed as a function of the eccentricities of the planetesimals' orbits. Planets acquire spin angular momentum from the relative motion of planetesimals at impact. The calculations presented herein assume that the planet is spherical and that its orbit is circular. Unless otherwise stated, the planetesimal distribution is also assumed to be uniform in semimajor axis. The spin angular momentum produced by a particular collision can be either positive or negative; due to symmetries these contributions nearly cancel. Thus, in order to obtain meaningful nonzero results, perturbation expansions must be carried through to second order in the ratio of the planet's radius to the amplitude of the radial excursions of the planetesimals, and numerical simulations require the calculation of a very large number of impacts. One finds that nongravitating planets accreted from the disk of planetesimals rotate very slowly in the prograde direction; their spin period is longer than their orbital period. In simulations which start with a Rayleigh distribution of planetesimal eccentricities, one finds that the maximum prograde rotation rates are much slower than those observed in the Solar System, suggesting that other factors are important.

107.019 **Magnetic history of the Sun.**
E. H. Levy, A. A. Ruzmaikin (*A. A. Ruzmajkin*),
T. V. Ruzmaikina (*T. V. Ruzmajkina*).
The Sun in time, p. 589 – 632 (1991). – See Abstr. 003.027 for the main entry.
This review gives a brief overview of the major magnetic effects and behavior that characterized and influenced aspects of the gross history of the Sun. The chapter surveys the behavior of magnetic fields in the Sun from the early effect of the interstellar field on protosolar collapse through to the modern solar magnetic cycle. Emphasis is given to the evolution of the Sun's magnetic field, its memory of the past magnetic states, and its loss of memory, as these depend on physical episodes, conditions and phenomena during the Sun's development.

107.020 **Pre–and main–sequence evolution of solar activity.**
F. M. Walter, D. C. Barry.
The Sun in time, p. 633 – 657 (1991). – See Abstr. 003.027 for the main entry.
The magnetic activity on single solar–like stars declines with stellar age. This has important consequences for the influence of the Sun on the early solar system. The authors review what is meant by stellar activity, and how it is measured. They discuss stellar activity in the pre–main–sequence phase of evolution, and conclude that the classical T Tauri stars do not exhibit solar–like activity, while the naked T Tauri stars do. The emission surface fluxes of the naked T Tauri stars are similar to those of the youngest main–sequence G stars. The authors discuss the activity–decay relations, and conclude that the best representation for solar–like stars is a decay proportional to $\exp(A \times t^{0.5})$, where A is a function of line excitation temperature. From these decay laws one can determine the interdependences of the activity, age and rotation periods. The authors discuss the fluxes of ionizing photons at the Earth early in its history, and show that there was sufficient fluence to account for the observed isotopic ratios of the noble gases. Finally, they discuss some of the observational tasks remaining.

107.021 **Protoplanetary thermal metamorphism: the hypothesis of electromagnetic induction in the protosolar wind.**
F. Herbert, C. P. Sonett, M. J. Gaffey.
The Sun in time, p. 710 – 739 (1991). – See Abstr. 003.027 for the main entry.
Observed taxonomic types of asteroids are identified with thermal histories deduced from asteroidal reflection spectra and meteoritic compositional trends. These thermal histories are interpreted assuming electromagnetic–induction heating arising from an active pre–main–sequence Sun. The immediate cause of the hypothetical induction is a strong electric field present in a high–density solar wind. The existence of an asteroidal heat source dependent on a particular evolutionary phenomenon of the Sun such as a dense plasma outflow can be used to deduce or constrain aspects of solar history if the fossil evidence is strong enough. The authors argue that the fossil evidence is compelling enough to attempt to deduce from it some basic features of the plasma outflow such as its duration (possibly 1 Myr) and time–integrated mass loss (of order 10^{-2} to $10^{-1} M_\odot$). One possible model that yields the dense equatorial ionized solar wind needed by the induction model is the "X–celerator" mechanism of Shu et al.

107.022 **What heated the parent meteorite planets?**
J. A. Wood, P. Pellas.
The Sun in time, p. 740 – 760 (1991). – See Abstr. 003.027 for the main entry.
The authors examine the plausibility of the two most widely discussed mechanisms for heating the small planetesimals in which the meteorites formed: the decay of short–lived ^{26}Al in the planetesimals, and joule heating by electric currents in the planetesimals induced by the motion past them of an early intense solar wind plasma. Both hypotheses are seen to have problems. Noncumulate eucrites (meteoritic fragments of lava erupted on the surface(s) of early planetesimals) did not contain enough ^{26}Al to leave detectable excess ^{26}Mg signatures; if ^{26}Al melted the lava, the latter had to wait underground 1.2 Myr or more while its ^{26}Al decayed before erupting, a requirement that may be difficult to satisfy mechanically. In the case of inductive heating, it appears increasingly unlikely that there was an epoch of solar system history in which an intense (i.e., T Tauri) solar wind streamed past young planetesimals. The wind appears to have blown (mostly in poleward directions) only while the solar nebula (accretion disk) was present and shielding any planetesimals that had formed. Solar wind inductive heating cannot be modeled accurately because the parameters most important to modeling, such as the early solar wind flux in the ecliptic plane and the electrical conductivity function of the material in planetesimals, are very poorly known. No model has been published that makes conservative assumptions across the board and achieves melting in planetesimals.

107.023 **The relict magnetism of meteorites.**
S. M. Cisowski, L. L. Hood.
The Sun in time, p. 761 – 784 (1991). – See Abstr. 003.027 for the main entry.

The remanent magnetism recorded in chondritic and achondritic meteorites can potentially constrain the existence and properties of large-scale magnetic fields in the early solar system. Paleointensity experiments involving repeated heatings of selected samples suggest that carbonaceous chondrites experienced magnetic fields of the order of one tenth to several gauss during low-temperature ($\sim 300°C$) metamorphic events on their parent bodies occurring within a few tens of Myr of protosolar collapse. A possible source of the magnetism recorded in carbonaceous chondrites and chondrules is a large-scale magnetic field in the protoplanetary nebula. Limited directional evidence from oriented chondrules suggests that the field was oriented mainly in the ecliptic plane. An alternative possibility is a solar field embedded in a protosolar wind after dispersal of the nebular gas. Intrinsic parentbody fields and fields due to local surface processes such as impacts may have contributed to the magnetization of all meteorite classes but particularly to the achondrites and ordinary chondrites.

107.024 **Dynamical limits on dark mass in the outer solar system.**
G. D. Quinlan, D. Hogg, S. Tremaine.
J. R. Astron. Soc. Can., Vol. 85, No. 4, p. 187 (Aug 1991). Abstract. – See Abstr. 012.091 for the main entry.

107.025 **Kuiper prize lecture: Some problems in the formation of the planets.**
V. S. Safronov.
Icarus, Vol. 94, No. 2, p. 260 – 271 (Dec 1991).

The author considers several problems of the evolution of the solar nebula, the precipitation of dust particles toward its central plane, and the possibility that a gravitational instability developed in the dust subdisk. A solution of the dispersion equation is given for the development of an instability in a two-component (gas-dust) rotating disk under the assumption that there is no instability in the gaseous component. This solution is compared to one found earlier for one-component disks. A runaway scenario for the accumulation of preplanetary bodies is discussed. I find that a transition to higher relative velocities of bodies and a slower growth of planet embryos began already before the sum of masses of the embryos reached one-tenth the mass of all the other bodies. Mechanisms for removal of almost all the initial mass of solids from the asteroidal zone are considered.

107.026 **Sublimation and reformation of icy grains in the primitive solar nebula.**
J. I. Lunine, S. Engel, B. Rizk, M. Horanyi.
Icarus, Vol. 94, No. 2, p. 333 – 344 (Dec 1991). = Contrib. 90-37 Theor. Astrophys. Program, Univ. Arizona.

The authors examine the frictional heating, sublimation, and recondensation of grains free-falling into the solar nebula from a surrounding interstellar cloud. The amount of water ice sublimated varies over a wide range – from over 90% of the grain mass at 30 AU from the nebular center to less than 10% beyond 100 AU. They conclude that essentially all of the water sublimated eventually recondenses, because the cold nebular gas beyond 10 AU is able to hold only a small fraction as vapor. The expansion of the sublimating gas from the grain surface and abundance of cold grains implies that most of the returns to the solid phase near nebular ambient temperatures ($\sim 50K$). Such a process could lead to at least two populations of grains: (1) essentially unaltered interstellar grains which did not sublimate due to drag or accretion shock heating and (2) a component comprised of water ice cocondensed, after heating, with more volatile gases at nebular ambient temperatures, yielding volatile-rich amorphous phases.

107.027 **Gas dynamic heating of chondrule precursor grains in the solar nebula.**
L. L. Hood, M. Horanyi.
Icarus, Vol. 93, No. 2, p. 259 – 269 (Oct 1991).

Conditions under which gas dynamic processes operating within the solar nebula could have been responsible for melting of chondrule precursor grains are investigated. Gas-grain energy and momentum transfer are treated in the free molecular flow approximation, including both drag heating due to grain relative motion and heating due to collisions with gas molecules in thermal motion. The effect of thermally emitted radiation on grain heating and cooling is also considered. It is found that melting of chondrule-sized grains is most easily achieved in optically thick, dust-rich zones where radiative cooling of a grain is balanced by thermal radiation from surrounding grains.

107.028 **Mass spectrum and velocity dispersions during planetesimal accumulation. I. Accretion.**
P. Barge, R. Pellat.
Icarus, Vol. 93, No. 2, p. 270 – 287 (Oct 1991).

The accumulation of the planetesimals has been investigated taking account of the simultaneous evolution of mass distribution and velocity dispersions. Under the assumption that collision frequencies are much smaller than the orbital frequency the velocity dispersions are determined from a previously developed formalism which works as long as the velocity dispersions remain approximately greater than the "Hill velocity"; this formalism is extended to the lower velocities through a model based on the numerical computations of Ida (1990). On the other hand the evolution of the mass distribution is derived from the simplified Oort-Van de Hulst formulation of the coagulation equation, which is sufficient to investigate the main trends of the accumulation of the planetesimals.

107.029 **Interpretazione di H. Poincaré della ipotesi cosmogonica di P. S. Laplace sull'origine del sistema solare.**
V. Banfi.
G. Astron., Vol. 17, N. 1 – 2, p. 59 – 64 (Mar – Jun 1991). Paper presented at "La cosmologia nella cultura dell '900". Conferenza Internazionale nel centenario della nascita di Hubble, Torino (Italy), 21 – 23 Nov 1990.

The Sun in time.
See Abstr. 003.027.

The fight without defeat: the 1951 conference on planetary cosmogony.
See Abstr. 015.014.

Accretional evolution of a planetesimal swarm. 1. A new simulation.
See Abstr. 021.004.

Tidal disruption of inviscid planetesimals.
See Abstr. 021.005.

Velocity distribution of fragments formed in a simulated collisional disruption.
See Abstr. 022.020.

The influence of orbital eccentricity of a disturbing body on chaotic region.
See Abstr. 042.023.

On the origin of the planetary spin by accretion of planetesimals. II. Collisional orbits at the Hill surface.
See Abstr. 042.072.

The influence of the orbital eccentricity of a primary on the chaotic region.
See Abstr. 042.073.

Shock waves in accretion disks.
See Abstr. 062.099.

Thermal and dissipative stability of convection-accretion disks.
See Abstr. 064.005.

Thermal conditions in a protostellar disk.
See Abstr. 064.162.

Thermal and dissipative stability of convection–accretion disks.
See Abstr. 064.165.

Gravitational instabilities in a proto–planetary disk.
See Abstr. 064.183.

Solar–wind krypton and solid/gas fractionation in the early solar nebula.
See Abstr. 074.082.

Early solar mass loss: a potential solution of the weak sun paradox.
See Abstr. 080.100.

Thermal equilibration of the Earth following a giant impact.
See Abstr. 081.050.

Models of giant planets with a variable ratio of ice to rock.
See Abstr. 091.001.

The origin of the systematic component of planetary rotation for a planet in a circular orbit in a 2–D disk.
See Abstr. 091.007.

Stochastic behaviour of orbits of the planets in the process of their accumulation.
See Abstr. 091.020.

The long–term dynamical behavior of small bodies in the Kuiper belt.
See Abstr. 091.024.

Solar system: A plurality of worlds.
See Abstr. 091.039.

The random component of planetary rotation.
See Abstr. 091.055.

Small solar system bodies.
See Abstr. 091.057.

CCD photometry and spectroscopy of the outer Jovian satellites.
See Abstr. 099.037.

Angular momentum and tidal evolution of Saturn's system.
See Abstr. 100.013.

The angular momentum of the Pluto–Charon system: considerations about its origin.
See Abstr. 101.124.

Comets, asteroids and interstellar grains.
See Abstr. 102.008.

An ion microprobe study of corundum in the Murchison meteorite: implications for ^{26}Al and ^{16}O in the early solar system.
See Abstr. 105.002.

High temperature rims around chondrules in primitive chondrites: evidence for fluctuating conditions in the solar nebula.
See Abstr. 105.030.

Does debris from the formation of other planetary systems impact Earth?
See Abstr. 105.043.

Formation and evolution of the solar nebula.
See Abstr. 131.401.

9783662123782VOL01

ASTRONOMY AND ASTROPHYSICS ABSTRACTS

A Publication of the Astronomisches Rechen-Institut Heidelberg
Produced in Cooperation
with the Fachinformationszentrum Karlsruhe

Astronomy and Astrophysics Abstracts is Prepared
Under the Auspices of the International Astronomical Union

Volume 54 B
Literature 1991, Part 2

Edited by
G. Burkhardt U. Esser H. Hefele I. Heinrich W. Hofmann
D. Krahn V. R. Matas L. D. Schmadel R. Wielen G. Zech

Springer-Verlag Berlin Heidelberg GmbH

Astronomisches Rechen-Institut, Mönchhofstraße 12–14,
W-6900 Heidelberg 1, Fed. Rep. of Germany
Telex: 461 336 ARIHD D
Telefax: 0 62 21-40 52 97
Telephone: 0 62 21-40 50
Director: Prof. Dr. Roland Wielen

Astronomy and Astrophysics Abstracts
Department Head: Dr. Lutz D. Schmadel
Editors-in-Chief: Inge Heinrich, Dr. Lutz D. Schmadel

ISBN 978-3-662-12378-2 ISBN 978-3-662-12376-8 (eBook)
DOI 10.1007/978-3-662-12376-8

Originally published by Springer-Verlag Berlin Heidelberg New York in 1992
Softcover reprint of the hardcover 1st edition 1992

Library of Congress Catalog Card Number 72-104650.

Media conversion: Daten- und Lichtsatz-Service, Würzburg.

55/3140-5 4 3 2 1 0 – Printed on acid-free paper

Contents

PART A

Stellar Systems, Galaxy, Extragalactic Objects, Cosmology

151 Stellar Systems (Kinematics, Dynamics)

151.001 **Higher order resonant orbits.**
J. Greiner.
Celest. Mech. Dyn. Astron., Vol. 50, No. 4, p. 387 – 394 (1990/91).
The author presents higher–order resonances of the tube and boomerang orbits in the co–rotating meridional plane of an axisymmetric galactic potential. Additionally, mirror forms of these orbits are given.

151.002 ***N*–body simulation of two–dimensional stationary stellar distributions: dissipation of energy and development of the field of velocity.**
K. O. Thielheim, Liu Bulin.
Astrophys. Space Sci., Vol. 181, No. 1, p. 43 – 53 (Jul 1991).
The procedure for the simulation of two–dimensional *N*–body configurations described in this paper has been devised to investigate the dissipation of energy and the development of velocity in given sub–regions of stellar systems, especially in the central region of a galaxy. In these simulations massive particles are injected into the given spatial region at a pre–determined constant time rate and at a given distribution of velocity vectors. The mass points are allowed to interact gravitationally and special attention is given to the influence of near–by encounters. After a certain period of time a stationary (equilibrium) situation settles down with a certain distribution of mass points in position and velocity space. Preliminary results obtained so far demonstrate the reliability of this procedure for *N*–body simulation and for the investigation of the dissipation of energy and of the development of velocity fields.

151.003 **Theory of the radial–orbit instability as a universal cause of structure formation in stellar systems.**
V. L. Polyachenko.
JETP Lett., Vol. 54, No. 10, p. 555 – 558 (25 Nov 1991). Current Physics Microform No.: 9202X0373.
An analytic theory is derived for a basic instability of collisionless gravitating systems: the radial–orbit instability. Radial orbits arise in a natural way upon (for example) the formation of galaxies during the collapse of a stellar cloud from an initial state with an extremely low density. The theory shows that the instability occurs if the gravitational potential of the system leads to a precession of orbits with small angular momenta in the direction in which the stars are revolving in these orbits.

151.004 **Dynamics of "clouds" of protogalactic medium in the gravitational potential of dark matter.**
A. D. Chernin, M. B. Shakenov.
Pis'ma Astron. Zh., Tom 17, No. 7, p. 563 – 568 (Jul 1991). In Russian. English translation in Sov. Astron. Lett., Vol. 17, No. 4.
Simple numerical models of protogalactic clouds are developed taking into account their mutual gravity and gravitational field of dark matter. The clouds are shown to gain considerable relative velocities and angular momentum which is higher than the spin momentum of giant spiral galaxies. This momentum can be transformed almost completely into spin momentum of condensation which is formed as a result of a binary non–central collision and subsequent coalescence of two massive gaseous clouds.

151.005 **The signs of stochasticity in the dynamics of model triplets of galaxies.**
A. V. Ivanov, A. D. Chernin.
Pis'ma Astron. Zh., Tom 17, No. 7, p. 569 – 574 (Jul 1991). In Russian. English translation in Sov. Astron. Lett., Vol. 17, No. 4.
Dynamical instability is studied for an ensemble of model triplets of galaxies. The maximum Lyapunov exponents and the characteristic decay times of autocorrelation functions for dynamical variables demonstrate that the timescale of stochastization of the system is considerably less than its lifetime. Dark matter, if distributed in the volume of the system, stimulates dynamical instability of the system and speeds up its stochastization.

151.006 **Erratum: "Extended stellar hydrodynamics for galactic disks" [Astrophys. J., Vol. 368, No. 1, p. 79 – 104 (10 Feb 1991)].**
P. Amendt, P. Cuddeford.
Astrophys. J., Vol. 375, No. 1, p. 427 (1 Jul 1991). See Abstr. 53.151.049.

151.007 **Nature of the universal structure of elliptical galaxies.**
V. L. Polyachenko.
Pis'ma Astron. Zh., Tom 17, No. 8, p. 691 – 697 (Aug 1991). In Russian. English translation in Sov. Astron. Lett., Vol. 17, No. 4.
From the marginal stability condition for a system created under a collapse of a collisionless spherical cloud, the universal energy distribution for stars with low angular momenta is derived. The distribution coincides qualitatively with that resulting from the recent N–body simulations. It should take the place of the well–known Lynden–Bell's distribution which cannot be reproduced in the simulations mentioned.

151.008 **Collective instabilities in stellar environments of galactic nuclei as source of their activity and cause of starbursts.**
V. L. Polyachenko.
Pis'ma Astron. Zh., Tom 17, No. 10, p. 877 – 883 (Oct 1991). In Russian. English translation in Sov. Astron. Lett., Vol. 17, No. 5.
The mechanism of starbursts and supporting the nucleus activity connected with the bar–mode instability is re–examined. It turns out that only the bar–mode which is due to the alignment of star orbits may be effective under conditions of central galactic regions. It is true also for excitement of inner bar by a galaxy–companion (Noguchi's scheme). The conventional bar–mode connected with fast rotation is suppressed by a great gravitational potential exterior relative to the active disk. For gravitational potentials in which the oscillatory mode occurs instead of the instability of purely–radial orbits, different instability is possible which is a far analog of the cone instability in plasma. This instability is studied in linear approximation within the framework of the simplest analytical model.

151.009 **A scale invariant, general relativistic model for a galaxy in a dark matter background.**
R. Hojman, L. Peña, N. Zamorano.
7. Latin American Symposium on Relativity and Gravitation, p. 330 – 336 (1991). – See Abstr. 012.005 for the main entry.
A general relativistic model for a galaxy is studied. It considers four different spherically symmetric regions, concentric, properly

matched and with a specific equation of state in each one of them. In the case of a spiral galaxy, it reproduces the orbital velocity observed for stellar objects. The dependence of the total mass (dark and luminous) on the radius in cluster and field spiral galaxies (displayed as a log(mass) vs. log(radius)), published recently by a group of astronomers, sets limits on the sizes of the different shells introduced here. The authors have also studied the Newtonian isothermal sphere as a model for a galaxy dominated by dark matter, to compare the mass distribution along the radius with the observational results. Finally, if the assumptions made here, reflect in some degree the real world, a characteristic jump in the slope of the total mass of the galaxy along the radial direction, may have a counterpart in the jump observed in the mass of some of the spiral galaxies.

151.010 **Colliding galaxies.**
J. Barnes, L. Hernquist, F. Schweizer.
Sci. Am., Vol. 265, No. 2, p. 26 – 33 (Aug 1991).

Violent encounters between galaxies appear surprisingly common. Computer models and observations indicate that these cosmic collisions may form elliptical galaxies and activate quasars.

151.011 **Global stability of self–similar Newtonian gaseous disks against axisymmetric perturbations.**
J. P. S. Lemos, A. J. Kalnajs, D. Lynden–Bell.
Astrophys. J., Vol. 375, No. 2, p. 484 – 491 (10 Jul 1991).

Solutions for self–gravitating disks in equilibrium with a velocity law given by $v \sim R^{-\beta}$ are found. These disks are self–similar, if they are of infinite extent. So, the intrinsic properties of these disks at one radius are scaled versions of the properties at any other radius. Global and local stability criteria are given and compared. The breathing mode is also analyzed.

151.012 **Dynamics of "clouds" of the protogalactic medium in the gravitational potential of dark matter.**
A. D. Chernin, M. B. Shakenov.
Sov. Astron. Lett., Vol. 17, No. 4, p. 241 – 243 (Jul 1991). Current Physics Microform No.: 9202X0417. English translation of Pis'ma Astron. Zh., Tom 17, No. 7, p. 563 – 568 (Jul 1991).

Simple numerical models of the dynamics of protogalactic clouds in their intrinsic gravitational field and the external gravitational field of dark matter are constructed. It is shown that these fragments can acquire considerable relative velocities and orbital momentum greater than the spin momentum of giant spiral galaxies. This orbital momentum can be transformed almost entirely into spin momentum of the condensation formed from the noncentral collision and subsequent coalescence of two massive gas clouds.

151.013 **Stochastic dynamics of model triples of galaxies.**
A. V. Ivanov, A. D. Chernin.
Sov. Astron. Lett., Vol. 17, No. 4, p. 243 – 245 (Jul 1991). Current Physics Microform No.: 9202X0419. English translation of Pis'ma Astron. Zh., Tom 17, No. 7, p. 569 – 574 (Jul 1991).

The authors study the development of dynamical instability in an ensemble of model triples of galaxies. The largest Lyapunov exponent and the typical autocorrelation decay times of dynamical quantities indicate that the system randomization time is much less than the system lifetime. If dark matter is present in the system and is distributed throughout, it will enhance dynamical instability and accelerate the rate at which the system becomes stochastic.

151.014 **Dynamics of elliptical galaxies.**
D. Merritt.
Bull. Am. Astron. Soc., Vol. 23, No. 2, p. 937 (1991). Abstract. – See Abstr. 012.008 for the main entry.

151.015 **Orbital decay in aspherical galaxies. II. Triaxial systems.**
T. S. Statler.
Astrophys. J., Vol. 375, No. 2, p. 544 – 564 (10 Jul 1991).

The orbital evolution of small satellite galaxies or globular clusters sinking into triaxial parent systems is calculated for parent galaxies with Stäckel (separable) potentials. Analytic expressions for the instantaneous decay rates of change of the orbital turning points are derived for arbitrary anisotropic drag forces. Orbit averaging is then used to compute evolutionary paths in turning–point space. While such a path strictly represents the average evolution of an ensemble of satellites distributed at random phase within the same initial orbit, it is demonstrated that it is often a good approximation for a single satellite if the satellite's mass is sufficiently small. Under the assumption that the drag forces can be approximated by the anisotropic version of Chandrasekhar's local dynamical friction formula, the effects of various sorts of velocity anisotropy and ordered motion (internal streaming) in the parent galaxy are estimated. In particular, the circumstances that favor orbital circularization, planarization, or margin crossing are identified. A set of simulations using a restricted (not self–gravitating) N–body code is presented, the results of which are consistent with the predictions of orbit–averaged local drag. Finally, some of the consequences of the sinking process, including the further evolution of stars and gas tidally stripped from satellites and the formation of "X–structures", boxy ellipticals, and gaseous disks or rings, are discussed.

151.016 **Nature of the universal structure of elliptical galaxies.**
V. L. Polyachenko.
Sov. Astron. Lett., Vol. 17, No. 4, p. 292 – 295 (Jul 1991). Current Physics Microform No.: 9202X0468. English translation of Pis'ma Astron. Zh., Tom 17, No. 8, p. 691 – 697 (Aug 1991).

The universal energy distribution for stars with low angular momentum is derived from the marginal stability condition for a system obtained from the collapse of a collisionless spherical cloud. This distribution is in qualitative agreement with the distributions obtained in recent N–body simulations. It should supplant the well–known Lynden–Bell distribution (which in the simplest cases coincides with the Fermi distribution), which cannot be reproduced in the simulations.

151.017 **Ring formation in off–axis collisions of galaxies.**
R. A. Gerber, S. A. Lamb, D. S. Balsara.
Bull. Am. Astron. Soc., Vol. 23, No. 2, p. 953 (1991). Abstract. – See Abstr. 012.008 for the main entry.

151.018 **N–body simulations of post–collisional ring galaxies.**
T. von Sadovszky, T. Colegrove, L. Spight.
Bull. Am. Astron. Soc., Vol. 23, No. 2, p. 953 (1991). Abstract. – See Abstr. 012.008 for the main entry.

151.019 **What damps small perturbations in stellar systems?**
E. Maoz.
Astrophys. J., Vol. 375, No. 2, p. 687 – 690 (10 Jul 1991).

A relaxation mechanism analogous to Landau damping in plasma is commonly believed to play a significant role in damping perturbations. However, a detailed calculation for the gravitational case shows that the significance of this process has been greatly overestimated. Landau's mechanism is found to have a negligible contribution for damping perturbations on the entire range of stable waves, even on scales as large as the Jeans length. Damping of all stable modes in stellar systems is due virtually entirely to phase mixing.

151.020 **Galactic wind and metal enrichment in spheroidal star systems.**
L. Angeletti, P. Giannone.
Astron. Astrophys., Vol. 248, No. 1, p. 45 – 54 (Aug 1991).

The closed one–zone model was applied to the evolution of spheroidal star systems (from globular clusters to elliptical galaxies) in order to determine the occurrence of global winds driven by supernovae and the stellar metal enrichment. The initial conditions to be used in the computations were selected so that the final binding energies of the system models agreed with those given, on the average, by observations. An iterative procedure was devised in order to fullfil this requirement. The results show that, when the bulk of stars was formed, protogalaxies should have been more tightly bound than assumed in previous one–zone models. Moreover, the impulsive global wind setting the end of the star formation was rather unconspicuous, whereas the subsequent gradual mass loss might have been much more effective. The authors discuss some observational constraints on the stellar metallicity, and the mass and metal content of the ejected gas. They suggest that major observational properties can be met very satisfactorily with a rapid star formation rate and an initial mass–power function with exponent between 0.95 and 1.35.

151.021 **Kinematics of stars in Kovalsky–Kapteyn figures. I.**
R. B. Shatsova, G. B. Anisimova.
Astrophysics, Vol. 33, No. 2, p. 466 – 474 (Sep – Oct 1990). English translation of Astrofizika, Tom 33, Vyp. 2, p. 291 – 304 (Oct 1990).

For a small area of the sky, the polar diagram of position angles of the proper motions of stars (Kovalsky–Kapteyn (KK) figure) is approximated by an ellipse around the centroid of the figure. This is possible not only in the case of an exponential ellipsoidal velocity distribution. The motion of the Sun and the rotation of the Galaxy, taken to be the same for all stars of the area, are determined from the centroid coordinates. The vertex longitude and ratio of dispersions along the principal axes of the ellipsoid of the star velocities are determined from the elements of the ellipse.

151.022 **Resonance excitation of spiral density waves in a gaseous disk. II. A nonlinear theory and application to the 3 kiloparsec arm.**
C. Yuan, Y. Cheng.
Astrophys. J., Vol. 376, No. 1, p. 104 – 114 (20 Jul 1991).

A nonlinear theory of spiral density waves in a gaseous disk is developed. The disk is thin, viscous, and self–gravitating. The gas in the disk is assumed to obey the isothermal law. The waves are generated near the Lindblad resonance by a periodic disturbance through an excitation mechanism. The method is formulated in general terms and may have applications to any disk systems in astrophysics. Here, it is used to analyze the 3 kpc arm problem in the Galaxy. The earlier suggestion of Yuan is confirmed that either a minor oval distortion or an uneven distribution of mass in the center can excite a spiral density wave which has a radial velocity and mass concentration in excellent agreement with the observations of the 3 kpc arm. The correct treatment of the nonlinearity and viscosity provides reliable results for the nonlinear density waves either in a gaseous disk or in the gas component of a galactic disk.

151.023 **Stellar systems following the $R^{1/m}$ luminosity law.**
L. Ciotti.
Astron. Astrophys., Vol. 249, No. 1, p. 99 – 106 (Sep 1991).

The author studies the structural and dynamical properties of isotropic spherical galaxies with surface luminosity profile following the $R^{1/m}$–law. For these models the luminosity density, the potential, the velocity dispersion and other related deprojected properties are presented in detail, together with analytical expressions for their asymptotic behavior at small and large radii. The parameter m for the computed models covers the range $<2,10>$. Moreover, the author constructs and studies the dimensionless self–consistently generated distribution function $\tilde{D}\tilde{F}_m(\tilde{E}_m)$, and the normalized differential energy distribution $N_m(\tilde{E}_m)$, where $\tilde{E}_m = E/\psi_m(0)$ is the relative binding energy E, normalized to the central value of the potential of the model, $\psi_m(0)$. The two functions $\tilde{D}\tilde{F}_m(\tilde{E}_m)$ and $N_m(\tilde{E}_m)$ are calculated under the assumptions that the velocity dispersion tensor is isotropic and that the mass–to–light ratio (M/L) is constant.

151.024 **Stability of nonrotating stellar systems. II. Prolate shell–orbit models.**
D. Merritt, L. Hernquist.
Astrophys. J., Vol. 376, No. 2, p. 439 – 457 (1 Aug 1991).

The authors investigate the dynamical stability of nonrotating prolate galaxy models constructed from thin long–axis tube orbits ("shell" orbits). Models more elongated than $\sim$E6 (axis ratio of $\sim$2:5) are unstable to bending modes that rapidly increase the velocity dispersion perpendicular to the long axis and decrease the model's elongation. The authors obtain approximate representations of the spatial forms of the fastest growing modes and their growth rates. Most of the evolution is due to two modes: a symmetric (banana–shaped) bending and an antisymmetric (S–shaped) bending. The instability is similar to the "firehose" instability of a thin self–gravitating slab, except that it persists in models with velocity anisotropies that are much less extreme than the critical value for instability of the slab. The authors present a simple model that reproduces the basic features of the instability in the prolate geometry. These results provide support for the hypothesis of Fridman and Polyachenko that the absence of elliptical galaxies flatter than about E6 is due to dynamical instability.

151.025 **Galactic disk warps.**
K. Kuijken.
Astrophys. J., Vol. 376, No. 2, p. 467 – 479 (1 Aug 1991).

Warped equilibrium configurations of galactic disks embedded in flattened haloes are considered. The relevant torques are calculated in their full nonlinearity, extending the results of linear–mode calculations. It is shown that for two specific examples, NGC 2841 and M33, it is possible to reproduce closely the observed warp shapes with these models. It is further demonstrated that the observed shapes of a galaxy's warp and rotation curve do not suffice to determine the disk mass uniquely, but can only provide upper limits on the disk–to–halo mass ratio. In the case of M33, this limit is about one–third of a maximal disk decomposition of the rotation curve. Qualitative differences between the linear theory and the present work are discussed.

151.026 **Interpreting asymmetric line profiles in merger remnants.**
M. Balcells.
Astron. Astrophys., Vol. 249, No. 2, p. L9 – L12 (Sep 1991). Letter–to–the–editor.

Absorption line profile asymmetries in ellipticals with peculiar core kinematics, and their role as diagnostics of mergers, are analyzed using self–consistent numerical merger models. The remnants of mergers involving a denser secondary component show asymmetric line–of–sight velocity distributions at their cores. Skewness is negative where the rotation curve is positive and positive where the rotation curve is negative, in close agreement with observations. In the models, the asymmetry is not due to a disk of accreted secondary material; it is intrinsic to the distribution of primary particles, and has been added during the merger. It is argued that the asymmetries are a consequence of the transfer of the orbital energy and angular momentum to the primary particles during the merger; for the mergers studied here, profile asymmetries are a relic of the formation dynamics rather than the signature of superimposed components.

151.027 **On the stability of gaseous disks of galaxies which have a local minimum on the rotation curve.**
M. A. Buzina, A. G. Morozov, Yu. M. Torgashin, A. V. Zasov, A. A. Strigachev.
Astron. Tsirk., No. 1543, p. 1 – 2 (Mar – Apr 1990). In Russian.

Hydrodynamical stability of gaseous disks of galaxies is evaluated for a two–parametric model of rotation curves with local minimum. Out of 20 of galaxies considered about half lay near the margin of stability if velocity dispersion of gas is $c_s \approx 10$ km/s.

151.028 **Simulations of the effect of spiral arms on the cloud–ensemble velocity dispersion.**
M. Thomasson, K. J. Donner, B. G. Elmegreen.
Astron. Astrophys., Vol. 250, No. 2, p. 316 – 323 (Oct 1991).
The authors use *N*–body simulations to study the effect of spiral heating, viscous heating, and cooling by inelastic collisions on the molecular cloud ensemble velocity dispersion. The model galaxies contain collisionless stars and inelastically colliding gas clouds. The potential from the flocculent or multiple–arm spiral patterns that develop in the simulations is an important heating mechanism for the cloud population. The spiral heating and the viscous heating are balanced by the cooling from inelastic collisions between the clouds, leading to a steady–state velocity dispersion for the cloud population that is approximately constant with the galactocentric radius. This situation is very different from the stellar velocity dispersion, which shows a strong radial dependence, corresponding to an approximately radius–independent value of Toomre's stability parameter Q. For a typical simulation showing a multiple–arm spiral structure, the cloud dispersion corresponds to about 3 km s^{-1}. The spiral arms not only heat the cloud population, they also cause increased cooling due to the large number of dissipative collisions in the arms. The heating dominates, however, in that a more pronounced spiral pattern leads to a higher cloud–ensemble velocity dispersion.

151.029 **Modes of oscillation of a uniformly rotating, homogeneous spheroid of stars.**
P. O. Vandervoort.
Astrophys. J., Vol. 377, No. 1, p. 49 – 71 (10 Aug 1991).
This paper describes an investigation of the oscillations and the stability of a family of stellar systems which are uniformly rotating, homogeneous spheroids. The members of this family are distinguished by the shapes of their figures, their rotation rates, and the shapes of their velocity ellipsoids. The characteristic value problem governing the normal modes of oscillation is formulated in a Lagrangian representation. Solutions of the characteristic value problem are obtained explicitly for seven distinct families of modes. These include counterparts of the second–harmonic modes of oscillation in the classical Maclaurin spheroids, and modes that can exhibit so–called "firehose" instabilities. The Eulerian perturbations of the gravitational potentials of these modes are polynomials of degrees 2 and 3, respectively, in the Cartesian components of the position. For each family of modes examples are presented of the dependence of the characteristic frequencies on the structures and kinematics of the equilibrium configurations, and conditions for instability are delineated. The results of this work are of interest as a guide in the investigation of the stability of more "realistic" models of stellar systems, as a source of exact models with which to test N–body codes, and as a guide in the Lagrangian formulation of variational and matrix methods for more general investigations of the oscillations and stability of stellar systems.

151.030 **Instability of a nonlinearly pulsating model of a stellar system. Volume perturbations: Camm's model.**
S. N. Nuritdinov.
Astron. Zh., Tom 68, Vyp. 4, p. 763 – 775 (Jul–Aug 1991). In Russian. English translation in Sov. Astron., Vol. 35, No. 4.
The stability of a nonlinearly pulsating model constructed by the author is studied; the basis of the model is the well–known Camm's sphere. The total dispersion equation in integral and differential forms for a general case is derived. It was found that the dispersion equation does not depend on the azimuth wave number. A study of physics of instability shows that "egg–shaped" (N = 3) and "ring–shaped" (N = 4) perturbations can be instable with overstability and exponential nature. Critical amplitudes of pulsations and increments of the perturbations are computed. Among the studied perturbations the "egg–shaped" instability is the most dangerous (even more than for N = m = 2). The results show that in an invesgtiation of theoretical problems of formation and evolution of corresponding large–scale structures, the inclusion of small nonstationarity into the equilibrium model is necessary.

151.031 **Stochastic model of fragmented structures in galaxies.**
I. K. Rozgacheva.
Astron. Zh., Tom 68, Vyp. 5, p. 897 – 908 (Sep–Oct 1991). In Russian. English translation in Sov. Astron., Vol. 35, No. 5.
An inhomogeneous, stationary rotating disk has been considered. The medium was assumed to have a zero pressure. The author shows that the ensemble of the density inhomogeneities can be a mixing system. The space structure of a collective gravitational field of density inhomogeneities is analysed. This field proves to have a large–scale spiral structure. The density inhomogeneities are distributed on a disk in accordance with this structure, its fractal dimension depending on the number of density inhomogeneities.

151.032 **Spiral instabilities and bars in N–body simulations.**
J. A. Sellwood.
Conference on Dynamics of Disc Galaxies, p. 123 – 136 (1991). – See Abstr. 012.028 for the main entry.
The main topic of this review is the evidence from N–body simulations for a recurrent cycle of groove instabilities as the source of spiral structure in galaxies. The author gives a very brief outline of the physical mechanism of groove modes and then presents some new evidence for the recurrent cycle in a realistic galaxy model. He reviews the recently discovered buckling modes of stellar bars as a case study of how apparently innocent and widely used approximations can turn out to be invalid.

151.033 **Shadow orbits and the gravitational N–body problem.**
G. D. Quinlan, S. Tremaine.
Conference on Dynamics of Disc Galaxies, p. 143 – 148 (1991). – See Abstr. 012.028 for the main entry.
The authors briefly describe a novel measure of the accuracy of N–body integrations. In spite of limitations, the existence of shadow orbits shows that numerical integrations of N–body systems can yield orbits that closely resemble true orbits, and puts the validity of these integrations on a sounder footing.

151.034 **Bars: orbits, gas flows and observational input.**
L. Athanassoula.
Conference on Dynamics of Disc Galaxies, p. 149 – 166 (1991). – See Abstr. 012.028 for the main entry.
The author reviews briefly the morphological characteristics of the main families of periodic orbits in a barred galaxy. Simulations show that the flow of gas is intimately linked to the properties of the periodic orbits. The gas density in and around the bar is low, except for the center and two narrow lanes, which are the loci of shocks. The analysis of a sample of early type strongly barred galaxies shows that real bars have two properties not included in the models. Namely their isophotal contours are more rectangular like than elliptical like, and they have a non constant axial ratio. A discussion of some possible pitfalls in the analysis of observations, as well as a comparison of theoretical and observational results is also given.

151.035 **The 3D dynamics of barred galaxies with central mass concentrations.**
H. Hasan, D. Pfenniger, C. Norman.
Conference on Dynamics of Disc Galaxies, p. 167 – 170 (1991). – See Abstr. 012.028 for the main entry.
The stellar dynamics at the transition between a strongly barred rotating potential and a dense spherical central mass concentration is investigated. This situation is generic for all barred galaxies accreting or accumulating mass at the centre. The main families of 3D periodic orbits and their bifurcations are calculated for three different concentrations of the central mass. The main effect occurs around the inner Lindblad resonance (ILR) where a large region of instability develops. Owing to the central spherical potential, the radial ILR is always near a corresponding vertical ILR. Therefore the pronounced instability is radial as well as vertical, helping the z–heating of disk stars and the dissolution of the bar into a bulge.

151.036 **Non–stationary wave and the motion of stars in plane systems.**
A. S. Baranov.
Astron. Zh., Tom 68, Vyp. 6, p. 1160 – 1173 (Nov–Dec 1991). In Russian. English translation in Sov. Astron., Vol. 35, No. 6.
Effects of non–stationary action on the dynamical evolution of non–interacting particles, which are in a certain regular field, are studied. Various forms of the disturbance of the spiral–wave type with slowly changing parameters have been considered. Qualitative features of the action of the non–stationary disturbance have been revealed. There is a category of captured stars which move together with the resonant zone and fill the phase space in the domains, not contiguous to each other. The noncaptured stars pass in the intervals between these domains and are not subject essentially to the action of the disturbance. It is shown that in the vicinity of separatrices irreversible changes are accumulated, due to both redistribution of the phase domains of different behaviour and to chaotization of motion in the resonant zones. It is concluded that during the time interval when a certain zone is a resonant one, a temporary local decrease of density must occur.

151.037 **The potential of early–type barred galaxies: towards a self–consistent model for NGC 936.**
H. Wozniak.
Conference on Dynamics of Disc Galaxies, p. 175 – 179 (1991). – See Abstr. 012.028 for the main entry.
In order to get a realistic expression for the potential, the author has computed it directly from CCD frames. He applied this method to a sample of 17 early–type barred galaxies and, for NGC 936, he has computed orbits in the calculated potential.

151.038 **Bar–driven gas fueling of galactic nuclei.**
D. Friedli, W. Benz, L. Martinet.
Conference on Dynamics of Disc Galaxies, p. 181 – 184 (1991). – See Abstr. 012.028 for the main entry.
Non–interacting barred starburst galaxies exist. By means of numerical simulations the authors study the influence of various gaseous and stellar parameters, and derive some quantitative estimations of this phenomenon.

151.039 **Unexpected behaviour in the rotation of perturbed barred galaxies.**
M. Sundin, B. Sundelius.
Conference on Dynamics of Disc Galaxies, p. 185 – 190 (1991). – See Abstr. 012.028 for the main entry.
The rotational velocity of a perturbed, barred galaxy has been found to increase or decrease depending on a combination of the mass of the perturber and the angle between the major axis of the bar and the perturber.

151.040 **Dissolution of a bar by merging.**
D. Pfenniger.
Conference on Dynamics of Disc Galaxies, p. 191 – 194 (1991). – See Abstr. 012.028 for the main entry.

151.041 **Interacting galaxies.**
B. Sundelius, K. J. Donner.
Conference on Dynamics of Disc Galaxies, p. 195 – 207 (1991). – See Abstr. 012.028 for the main entry.
The authors review specifically Swedish contributions to the field.

151.042 **Direct measurement of dynamical friction on circular orbits.**
K. J. Donner, B. Sundelius.
Conference on Dynamics of Disc Galaxies, p. 209 – 214 (1991). – See Abstr. 012.028 for the main entry.

151.043 **Modeling the NGC 5194/95 interaction.**
S. Engström, L. Athanassoula.
Conference on Dynamics of Disc Galaxies, p. 215 – 218 (1991). – See Abstr. 012.028 for the main entry.
This paper presents the first steps in modeling the interacting pair of galaxies NGC 5194/95. The authors found two groups of solutions in the parameter space surveyed.

151.044 **Molecules in disks.**
F. Combes.
Conference on Dynamics of Disc Galaxies, p. 219 – 233 (1991). – See Abstr. 012.028 for the main entry.
Considerable progress has been made in our knowledge of the distribution of molecular gas in disks. The author summarizes recent results obtained with high spatial resolution. The molecular gas is one of the best tracer of the spiral structure and its peculiar streaming motions. This is illustrated by gas–cloud simulations of the M51 disk, that allow to determine the location of resonances, and the gas orbits in the rotating frame.

151.045 **Parabolic–type orbits for a model of galactic potentials.**
E. I. Timoshkova.
Astron. Zh., Tom 68, Vyp. 6, p. 1315 – 1322 (Nov–Dec 1991). In Russian. English translation in Sov. Astron., Vol. 35, No. 6.
The two–parameter family of parabolic–type orbits in the meridional galactic plane is found for the problem of perturbed stellar motion in a rotationally–symmetric field of a galaxy. The connection of this family with the stationary solutions is determined. The analytical formulae for star coordinates and velocities are derived in a closed form for the cases of periodic and asymptotic motions on the trajectories.

151.046 **A dynamical study of the frequency of merging galaxies. III. Mergers involving disk galaxies in a single crossing time or orbital period.**
T. K. Chatterjee.
Astrophys. Space Sci., Vol. 184, No. 2, p. 273 – 280 (Oct 1991).
A study of the expected frequency of merging galaxies, involving disk systems, is conducted on the basis of the collision theory using the impulsive approximation. The results indicate that the expected frequency of such mergers is several orders of magnitude smaller than the observational values if mergers taking place in a single crossing time are considered. The expected frequency of elliptical pairs, spiral–elliptical pairs and spiral pairs in merger is of the same order of magnitude; though spiral pairs seem to have marginally more abundance as compared to the other two types. The frequency of galaxy mergers in several orbital periods will be further investigated in the continuing series of papers.

151.047 **Flattening of tilted protogalactic disks.**
Lu Jufu, D. N. Lin.
3. Chinese Academy of Sciences and Max–Planck Society Workshop: High energy astrophysics: compact stars and active galaxies, p. 48 – 51 (1991). – See Abstr. 012.026 for the main entry.
The authors study the viscous evolution of, and the star formation in, a twisted gaseous protogalactic disk. They show that flattening of such a disk occurs on a viscous evolution timescale, and that the gas in the disk must form stars on the same timescale. The result could be used to explain the observational characters of the thin disk component of S and S0 galaxies.

151.048 **Dissipational galaxy formation. I. Effects of gasdynamics.**
N. Katz, J. E. Gunn.
Astrophys. J., Vol. 377, No. 2, p. 365 – 381 (20 Aug 1991).
The authors present numerical simulations of hierarchical galaxy formation including gasdynamics. These simulations are conducted using a new, general – purpose program for evolving self–gravitating systems in three dimensions. The gravitational forces are calculated with a hierarchical tree algorithm, while the gasdynamic properties are determined by an approach known as smoothed particle hydrodynamics. In this method the complete

thermodynamic state of the gas is known everywhere, so dissipational effects can be included by allowing the gas to cool radiatively. These simulations model the collapse of isolated constant density perturbations, initially in solid–body rotation and in Hubble flow. The perturbations consist of dark and baryonic matter in a 10 to 1 ratio. Small–scale power is added using the Zel'dovich approximation assuming a power law slope of either −2.5 or 0. The simulations produce two component systems that resemble spiral galaxies: a thin disk made of gas and a dark matter halo. The disk transfers more than 50% of its original angular momentum to the dark halo and forms at an angle of $\sim 30°$ to the rotation axis of the dark halo. The disks form with a warp that is also a consequence of the angular momentum transport that acts during the collapse. Beyond ~ 5 kpc the gaseous disk has a flat rotation curve and an exponential surface density profile.

151.049 **The disk heating problem.**
C. Lacey.
Conference on Dynamics of Disc Galaxies, p. 257 – 278 (1991). – See Abstr. 012.028 for the main entry.

Observations show that in the solar neighbourhood, the velocity dispersions of disk stars increase with age roughly as $\sigma \propto \tau^p$, with $p \approx 0.2 - 0.5$, while the dependence on galactocentric radius is approximately $\sigma \propto \exp(-R/2h_R)$, h_R being the radial scalelength of the disk. The most likely explanation of this is disk heating by a combination of transient spiral density waves increasing random motions in the plane, and giant molecular clouds scattering stars and transferring some of this energy into vertical random motions. Many spiral disks have probably also been heated by accretion of satellite galaxies.

151.050 **Spiral arms and the velocity dispersions of stars and gas.**
M. Thomasson, K. J. Donner, B. G. Elmegreen.
Conference on Dynamics of Disc Galaxies, p. 279 – 283 (1991). – See Abstr. 012.028 for the main entry.

The authors use N–body simulations to study the effect of spiral heating, viscous heating, and cooling by inelastic collisions on the molecular cloud ensemble velocity dispersion. Heating by transient spiral arms is shown to be an important effect. The velocity dispersion for the cloud population is independent of galactocentric radius in the simulations, in agreement with observations, and it is of the correct order of magnitude. In contrast, the stellar velocity dispersion corresponds to a constant value of Q_{star}.

151.051 **The cold interstellar gas and the stability of galactic discs: vertical structure at equilibrium.**
A. B. Romeo.
Conference on Dynamics of Disc Galaxies, p. 285 – 288 (1991). – See Abstr. 012.028 for the main entry.

151.052 **The cold interstellar gas and the stability of galactic discs: local finite–thickness effect.**
A. B. Romeo.
Conference on Dynamics of Disc Galaxies, p. 289 – 292 (1991). – See Abstr. 012.028 for the main entry.

151.053 **Warps.**
J. Binney.
Conference on Dynamics of Disc Galaxies, p. 297 – 309 (1991). – See Abstr. 012.028 for the main entry.

At least half of galactic disks are warped. It seems unlikely that the intergalactic magnetic field could be large enough to generate warps as far in as they are observed. Probably warps are a manifestation of heavy halos, and arise because the principal axes of halos at large radii do not coincide with the principal axes of the embedded galaxies. It is not yet clear why this should be so, but there is a good prospect that *n*–body simulations currently under way will solve the riddle.

151.054 **Galactic disk warps.**
K. Kuijken.
Conference on Dynamics of Disc Galaxies, p. 311 – 315 (1991). – See Abstr. 012.028 for the main entry.

Many galactic disks are warped. A coherent dynamical explanation involves the interaction between the disk and a surrounding nonspherical halo whose plane of symmetry does not coincide with the disk's. In this paper, the detailed predictions of the warp shapes are calculated for two well–studied warped disk galaxies (M33 and NGC 2841), and compared to the observations. It is possible to obtain very good agreement between the predictions from this model and the observations.

151.055 **Dynamics and evolution of acquired gaseous discs.**
D. Bettoni, G. Galletta.
Conference on Dynamics of Disc Galaxies, p. 317 – 322 (1991). – See Abstr. 012.028 for the main entry.

151.056 **Pattern speeds of density waves.**
A. J. Kalnajs.
Conference on Dynamics of Disc Galaxies, p. 323 – 332 (1991). – See Abstr. 012.028 for the main entry.

Numerical simulations of disk galaxies without massive haloes show that the pattern speeds of density waves are high enough to place co–rotation in the middle of the disk and the outer Lindblad resonance near the optical edge. The observed ring structures in external galaxies catalogued by Buta can be identified as manifestations of the Lindblad type resonances. The outer rings are very close to the locations of the outer Lindblad resonance. With the help of a rotation curve one can convert that location into a pattern speed. The key feature in ring formation is the intersection of particle orbits at the Lindblad resonances. This causes large scale collisions and shocks in the gaseous component, and a two streamed flow in the stellar component. The gaseous component evolves into an outer ring. Kapteyn's star streaming in the solar neighbourhood fits in very nicely in the above scheme. It requires to be located just outside the outer Lindblad resonance radius. If the density wave force field in our Galaxy comes from a bar, then its major axis must be oriented along a line which is 56° ahead of us in the direction of galactic rotation.

151.057 **Feature velocities in numerical simulations.**
S. Engström.
Conference on Dynamics of Disc Galaxies, p. 333 – 336 (1991). – See Abstr. 012.028 for the main entry.

The problem of interpreting observations of the kinematics of a galactic disk is discussed. The author examines the pattern speed of spiral arms, or, when no constant value can be determined, the winding rate of the features. Feature velocities thus obtained are compared to an observationally feasible method for measuring pattern speeds – the Tremaine–Weinberg (1984) method. General agreement is found.

151.058 **Spiral chaos in an orbiting patch.**
A. Toomre, A. J. Kalnajs.
Conference on Dynamics of Disc Galaxies, p. 341 – 358 (1991). – See Abstr. 012.028 for the main entry.

This paper illustrates how gravity polarizes various lumpy shearing disks – such as the ones containing giant molecular clouds or meter–sized ice balls or simple mass points – into kaleidoscopes of streaky features which come and go individually but which together manage to look very much like the spiral arms of Sc galaxies, or like the inferred 100–meter microstructure in Saturn's rings, or also like many of the ragged "instabilities" met in computer simulations of thin disk galaxies. This paper offers an almost–final commentary on a long series of numerical experiments with gravitating particles in idealized shearing sheets. All this effort has produced some beautifully accurate test cases or illustrations for swing amplification, spiral wakes, rates of heating and even some dynamical friction.

151.059 **Recurrent swing amplification induced by nonlinear amplitude effects.**
B. Fuchs.
Conference on Dynamics of Disc Galaxies, p. 359 – 363 (1991). = Astron. Rechen–Inst. Heidelberg, Mitt. Ser. A, Nr. 230. – See Abstr. 012.028 for the main entry.

First results of a study of the role of nonlinear amplitude effects of spiral density waves in the dynamics of disk galaxies are reported. It is shown that the nonlinearities may provide by two–wave interactions a continuous "seed" for the linear swing amplification mechanism.

151.060 **Merger of binary globular clusters: case of unequal masses.**
J. Makino, K. Akiyama, D. Sugimoto.
Astrophys. Space Sci., Vol. 185, No. 1, p. 63 – 78 (Nov 1991).

Merger process of binary globular clusters is discussed for a pair of unequal–mass components. The authors calculated the case of mass ratio 1:0.5 by means of an *N*–body code with 6144 particles in total. They have found the following: The mass exchange between the components takes place through the Roche–lobe overflow. In the early stages, however, the dynamical evolution is mainly governed by escape of particles from the system. As the particles escape carrying angular momentum with them, the separation between the component clusters shrinks. The time–scale of this shrinkage depends upon the size of the clusters. When a critical separation is reached, the orbital angular momentum is transferred unstably to the spins of the component clusters. This is the process of the synchronization instability which was found in a previous study on binary cluster of equal masses. As a result the component clusters merge into a single cluster. The structures of the mergers are quite similar among different cases except for the central cores which retain their initial central concentrations. In particular, the ellipticity and the rotation curve are quite close each other among models of different initial radii and of different mass ratios.

151.061 **Potentials for stationary point–axial stellar system models.**
J. M. Juan–Zornoza, J. Sanz i Subirana.
Astrophys. Space Sci., Vol. 185, No. 1, p. 95 – 100 (Nov 1991). Presented at the 12th European Astronomical Meeting of the IAU on "European Astronomers Look to the Future", Davos (Switzerland), 8 – 11 Oct 1990.

Some potentials for a stationary stellar system model with point–axial symmetry and equatorial plane of symmetry that verifies Chandrasekhar's postulates have been obtained.

151.062 **Stability of oblate galaxy models.**
F. Robijn, T. de Zeeuw.
Conference on Dynamics of Disc Galaxies, p. 373 – 376 (1991). – See Abstr. 012.028 for the main entry.

This is a progress report on ongoing investigation of the stability of self–consistent galaxy models by means of a normal–mode analysis. Currently, the authors are examining oblate models.

151.063 **Formation of disk galaxies.**
S. D. M. White.
Conference on Dynamics of Disc Galaxies, p. 377 – 387 (1991). – See Abstr. 012.028 for the main entry.

The disk components of galaxies contain a large fraction of all stars, and are the sites of almost all current star formation. They are thin, centrifugally supported, dynamically cold, and relatively young. They occur predominantly in regions of low galaxy density, and are apparently surrounded by halos of dark matter containing at least ten times the observed stellar mass. They show systematic relations between their characteristic properties (size, luminosity, rotation speed) with relatively little scatter, suggesting that they can be considered roughly as a one–parameter family of structures. These properties can all be understood if disks form slowly by the cooling and settling of gas initially associated with the halo dark matter. These processes could have occurred soon after the initial equilibration of the present halo, or they could be more protracted, continuing today in a small scale analogy of galaxy cluster cooling flows. The latter model can be worked out in some detail and predicts that isolated bright spirals should still have substantial gaseous halos which should be detectable in soft X–rays.

151.064 **Numerical simulation of the formation of galaxy discs.**
J. R. Davies, A. H. Nelson.
Conference on Dynamics of Disc Galaxies, p. 401 – 404 (1991). – See Abstr. 012.028 for the main entry.

The authors present computer simulation of the collapse of a proto–galactic cloud. The main control parameter for the final morphology is the initial velocity dispersion in the dark matter component.

151.065 **A nonlinear density wave model of the spiral structure.**
M. G. Abramian (*M. G. Abramyan*).
3. Chinese Academy of Sciences and Max–Planck Society Workshop: High energy astrophysics: compact stars and active galaxies, p. 286 – 288 (1991). – See Abstr. 012.026 for the main entry.

Apart from the theoretical interest, the development of the nonlinear theory of perturbations of gaseous gravitating discs also has practical interest since it can help explain the spiral structure of galaxies.

151.066 **Instability of a nonlinearly pulsating model of a star system. Volume perturbations: the Camm model.**
S. N. Nuritdinov.
Sov. Astron., Vol. 35, No. 4, p. 377 – 383 (Jul 1991). Current Physics Microform No.: 9203X0563. English translation of Astron. Zh., Tom 68, Vyp. 4, p. 763 – 775 (1991).

The stability of volume perturbations imposed on a nonlinearly pulsating model, constructed by the author on the basis of the well–known Camm equilibrium sphere, is investigated. The total dispersion relation is derived in integral and differential form. It turns out not to depend on the azimuthal wave number. An analysis of the physics of the instability of "ovoid" ($N = 3$) and "annular" ($N = 4$) perturbations shows the existence of oscillatory and periodic branches of instability. The corresponding critical values of the pulsation amplitude and the characteristic rise times of perturbation amplitude are calculated. Among the perturbations studied, the ovoid instability is the most dangerous (even more than for $N = m = 2$). The model that is isotropic with respect to velocity turns out to be unstable with respect to oscillations. The results show that in an analysis of theoretical problems on the origin and evolution of the corresponding large–scale structures, at least slight nonstationarity must be incorporated into the initial model.

151.067 **The few–body problem in astrophysics.**
M. Valtonen, S. Mikkola.
Annu. Rev. Astron. Astrophys., Vol. 29, p. 9 – 29 (1991). – See Abstr. 003.010 for the main entry.

The authors draw attention to areas of astrophysics where the few–body problem has obtained a central position and also discuss some new areas where its importance has still to be proven.

151.068 **Structure and dynamics of elliptical galaxies.**
T. de Zeeuw, M. Franx.
Annu. Rev. Astron. Astrophys., Vol. 29, p. 239 – 274 (1991). – See Abstr. 003.010 for the main entry.

The authors focus on the equilibrium dynamics of elliptical galaxies. They summarize recent work on dynamical models, with emphasis on triaxial systems. Observational efforts are reviewed and implications for the formation processes are discussed. Possible future work is outlined.

151.069 **Structure and dynamics of 3D N–body barred galaxies.**
D. Pfenniger, D. Friedli.
Astron. Astrophys., Vol. 252, No. 1, p. 75 – 93 (Dec 1991).

A detailed study of 3D stable, self–gravitating and rotating N–body barred galaxies is presented. Static and evolutionary descriptions are considered. In order to gain insight into the intricate dynamics taking place in bars, the models are examined at four levels: (1) global shape and kinematics, (2) potential and resonances, (3) periodic orbits and their stability, and (4) phase space occupation by the particles.

151.070 **The exact form of autonomous oscillations of Camm's phase model of a spherical stellar cluster.**
V. A. Antonov, S. N. Nuritdinov.
Astron. Tsirk., No. 1545, p. 3 – 4 (Aug – Sep 1990). In Russian.
Two interesting properties of Camm's equilibrium model are given. The analytical exact form of its autonomous oscillations associated with volume perturbations is found by Jacobi's polynomials. The total stability of the model with respect to arbitrary asymmetric oscillations is proved.

151.071 **Computer simulations of encounters between massive black holes and binaries.**
J. G. Hills.
Astron. J., Vol. 102, No. 2, p. 704 – 715 (Aug 1991). Current Physics Microform No.: 9108A1640.
The authors simulated over $8x10^5$ encounters of binaries with black holes having masses that range from 10 to 10^4 times that of the binary components. The authors made the calculations at many collision velocities and impact parameters, which allowed them to find the velocity–dependent cross sections for exchange collisions, for binary dissociation, and for changing the orbital energy and eccentricity. The authors also found the distribution of closest approaches of the stars to the black hole. They determined the probability of the black hole tidally disrupting a star in the close encounters that produce exchange collisions. Exchange collisions dominate over tidal dissociation if the sum of the radii of the stars is less than the binary semimajor axis. Even encounters of black holes with soft binaries can lead to exchange collisions in which the black hole captures one binary component and ejects the other at high speed.

151.072 **Velocity fields of disks in triaxial potentials.**
P. J. Teuben.
Workshop on Warped Disks and Inclined Rings Around Galaxies, p. 40 – 49 (1991). – See Abstr. 012.054 for the main entry.
Velocity fields of disks in triaxial halos are modeled and discussed. Residual velocities from a tilted–ring model fit along a galactic circle show a unique sin 3θ signature, and provide a sensitive measure of the triaxiality of a potential.

151.073 **Dynamics of polar rings.**
L. S. Sparke.
Workshop on Warped Disks and Inclined Rings Around Galaxies, p. 85 – 95 (1991). – See Abstr. 012.054 for the main entry.

151.074 **Mergers and the structure of disk galaxies.**
L. Hernquist.
Workshop on Warped Disks and Inclined Rings Around Galaxies, p. 96 – 111 (1991). – See Abstr. 012.054 for the main entry.
Simple estimates suggest that mergers with less–massive companions can have a profound influence on the structure of disks in spiral galaxies. This review focuses on such events where the mass ratio between the companion and victim disk is quite small, say $\lesssim 10\%$.

151.075 **Formation of polar rings.**
H.–W. Rix, N. Katz.
Workshop on Warped Disks and Inclined Rings Around Galaxies, p. 112 – 121 (1991). – See Abstr. 012.054 for the main entry.
This paper reports on preliminary steps in investigating the formation of polar rings or gas rings and disk in early–type galaxies. The authors briefly discuss the numerical method, outline in general the initial conditions for the simulations, and describe the simplest possible scenario, the disruption of a purely gaseous satellite on a circular orbit in a spherical potential.

151.076 **On the velocity fields of elliptical galaxies.**
T. S. Statler.
Astron. J., Vol. 102, No. 3, p. 882 – 892 (Sep 1991). Current Physics Microform No.: 9109D1286. With plate 26.
A family of self–consistent, maximum entropy dynamical models is presented for the triaxial "perfect ellipsoids". These models are projected in different viewing geometries, to explore the possible morphologies of the two–dimensional radial–velocity fields of elliptical galaxies. It is found that, typically, about half the time one should expect to see an "ordinary" velocity field, with a mixture of major– and minor–axis rotation. For other lines of sight, the velocity fields are more complicated, showing multiple peaks, steep gradients, strongly twisted contours, or kinematically distinct or counter–rotating cores. The origin of each of these velocity field morphologies is explained in terms of competing contributions from the major families of tube orbits.

151.077 **Gas–dynamical models of settling disks.**
D. M. Christodoulou, J. E. Tohline.
Workshop on Warped Disks and Inclined Rings Around Galaxies, p. 122 – 130 (1991). – See Abstr. 012.054 for the main entry.

151.078 **Evolutionary processes affecting galactic accretion disks.**
T. Steiman–Cameron.
Workshop on Warped Disks and Inclined Rings Around Galaxies, p. 131 – 142 (1991). – See Abstr. 012.054 for the main entry.
The author tries to interpret the structure of nonsteady–state disks in the context of the geometry of the host galaxy and the history of the disk.

151.079 **Particle simulations of polar rings.**
T. Quinn.
Workshop on Warped Disks and Inclined Rings Around Galaxies, p. 143 – 151 (1991). – See Abstr. 012.054 for the main entry.

151.080 **A bending instability in prolate stellar systems.**
D. Merritt.
Workshop on Warped Disks and Inclined Rings Around Galaxies, p. 152 – 158 (1991). – See Abstr. 012.054 for the main entry.
Prolate stellar systems are unstable to bending modes when sufficiently elongated. The purpose of this contribution is to report the existence of a bending instability in a family of nonrotating prolate models. The instability appears to persist in models as round as 1:3, i.e. with "Hubble types" of E6 or E7.

151.081 **Merger origin of starburst galaxies.**
L. Hernquist.
Workshop on Warped Disks and Inclined Rings Around Galaxies, p. 170 – 180 (1991). – See Abstr. 012.054 for the main entry.
Observational evidence supports the view that some starbursts are triggered by collisions between galaxies. The evolution of galaxies merging to form starbursts entails both collisionless and gas–dynamics. These two components are coupled since they interact with one another gravitationally. It also apears that the response that ultimately drives gas into galactic nuclei involves non–linear processes, such as shock waves, and dissipation through radiative cooling.

151.082 **Warps and bulges.**
J. Pitesky.
Workshop on Warped Disks and Inclined Rings Around Galaxies, p. 215 – 224 (1991). – See Abstr. 012.054 for the main entry.

A review of the observational literature shows that stellar disk warps occur more frequently in galaxies that have smaller nuclear bulges. The author has modeled the effect of a freely-moving bulge potential on the shape of normal warping modes in self-gravitating disks, and has examined the effect of changing the bulge mass, scale length and rotation rate on these modes. In these modes, the bulge is tipped relative to the central disk. The warping mode remains discrete, but flattens out as the bulge rotation rate increases. The warp amplitude is also dependent on the length of the bulge effective radius.

151.083 **Time evolution of galactic warps.**
P. Hofner, L. S. Sparke.
Workshop on Warped Disks and Inclined Rings Around Galaxies, p. 225 – 236 (1991). – See Abstr. 012.054 for the main entry.

The authors have modeled the time evolution of a galactic disk initially tilted with respect to an oblate dark halo. The disk is modeled as a set of concentric rings free to tilt about the center, embedded in a fixed halo potential. When a disk that has a discrete mode of bending is initially inclined with a constant angle to the symmetry plane of an oblate dark halo, one finds that the inner part of the galactic disk settles towards the mode shape. Even if the disk as a whole has no discrete bending mode, the inner part still settles into a coherent warp with no sense of spirality. The authors ran models with parameters corresponding to the observed systems NGC 4013, NGC 4565 and NGC 2903. Comparing the models with the observed shape of the warp allows to place constraints on the core radius and to find a lower limit for the oblateness of the dark halo.

151.084 **Are warps normal modes?**
S. Casertano.
Workshop on Warped Disks and Inclined Rings Around Galaxies, p. 237 – 246 (1991). – See Abstr. 012.054 for the main entry.

151.085 **Disk warping in a slewing potential.**
E. C. Ostriker.
Workshop on Warped Disks and Inclined Rings Around Galaxies, p. 247 – 255 (1991). – See Abstr. 012.054 for the main entry.

Due to gradual infall of cosmic matter with skew angular momentum, a flattened galactic halo is expected to reorient. As it follows this slewing external potential, the embedded galactic disk warps such that torques from local self-gravity balance those from the external potential, in the slewing frame. Such a warp does not precess, and persists only while the potential slews.

151.086 **Transient spiral patterns in close galaxy encounters.**
K. J. Donner, S. Engström, B. Sundelius.
Astron. Astrophys., Vol. 252, No. 2, p. 571 – 582 (Dec 1991).

The authors consider the spiral patterns generated by close interactions of galaxies with the aim of obtaining a more systematic understanding of the types of patterns that are formed in the early stages of galaxy interactions. On the basis of a simulated survey of direct perturbations they identify the most characteristic forms of transient tidal patterns during about one rotation period of the disc at the radius of closest approach of the disturber. For a representative set of parameters these patterns occur in a universal sequence. The authors show that the duration of a typical encounter is short enough so that the impulse approximation is valid over a large fraction of the disc. Using the impulse approximation one can analytically calculate the positions of caustics in the disc and thus reproduce the characteristic patterns found in the simulations. In particular, the tidal tail is produced by two distinct mechanisms, leading in some circumstances to the formation of a double arm. The spiral pattern generated under these circumstances cannot be described as a density wave with a determinate pattern velocity. Since the qualitative patterns produced in different interactions are so insensitive to modifications of the models, the use of observed patterns to impose constraints on the parameters of interactions has to be based on carefully defined criteria.

151.087 **The velocity dispersion of giant molecular clouds. II. Mathematical and numerical refinements.**
C. F. Gammie, J. P. Ostriker, C. J. Jog.
Astrophys. J., Vol. 378, No. 2, p. 565 – 575 (10 Sep 1991).

It has been proposed by Fukunaga and Jog and Ostriker that gravitational scattering of clouds off one another in encounters caused by differential rotation can account for the velocity dispersion of the largest molecular clouds in our Galaxy. Angular momentum transfer in cloud–cloud scatterings increases the eccentricity, or epicyclic amplitude, of the clouds. This input of random energy is ultimately balanced by dissipative cloud collisions, leading to equilibrium. The authors recalculate the energy input to the clouds using the proper linearized equations of motion, including the Coriolis force and allowing for changes in the guiding center. Perturbation theory gives a result in the limit of distant encounters and small initial epicyclic amplitudes. Direct integration of the equations of motion allows the study of the strong encounter regime. The result of the numerical integrations for the two–dimensional (planar) velocity dispersion is $\sigma = 5.1$ km s^{-1} for a fiducial cloud of mass $5 \times 10^5 M_\odot$ at $R_0/2$, slightly smaller than the value obtained by Jog and Ostriker. In an Appendix the authors calculate the accretion rate for a molecular cloud in the Galactic disk.

151.088 **Modeling edge–on elliptical galaxies: are two integrals enough?**
M. R. Merrifield.
Astron. J., Vol. 102, No. 4, p. 1335 – 1342 (Oct 1991). Current Physics Microform No.: 9110D1867.

The author tests the hypothesis that highly flattened elliptical galaxies are edge–on axisymmetric systems that depend on only the two classical integrals of motion. By deprojecting the observed kinematics of such a system, the author shows that simple global parameterizations of the degree of orbital anisotropy may not be appropriate. He therefore presents a method that tests the fit to a two integral model without requiring such a parameterization. This test uses the integrated properties of the galaxy from a number of kinematic cuts to search for the systematic effects characteristic of a third integral. If a third integral is detected, one can also obtain a measure of its effect on the shape of the velocity ellipsoid. Application of this test to existing kinematic data from the galaxies NGC 4697 and NGC 1052 reveals the presence of a third integral in both systems. The case is particularly compelling for NGC 4697, where the observed kinematics are most naturally explained by a third integral that causes a modest degree of anisotropy in the velocity ellipsoid.

151.089 **Chaos in a galaxy model with nucleus and bulge components.**
N. D. Caranicolas, K. A. Innanen.
Astron. J., Vol. 102, No. 4, p. 1343 – 1347 (Oct 1991). Current Physics Microform No.: 9110D1875.

A galaxy model, with a disk halo and a spherical nucleus, is used to study the transition from regular to chaotic motion for stars moving in the r–z plane. Stars with values of angular momentum L_z less than or equal to a critical value L_{zc}, moving near the galactic plane, are scattered to the halo upon encountering the nuclear region, and subsequently display chaotic motion. A linear relationship exists between the critical angular momentum and the mass of the nucleus M_n. Some theoretical arguments are presented to explain this behavior. Similar results are obtained if one uses the same model with a spherical bulge instead of a nucleus. A comparison is made between L_{zc} and the circular angular momentum L_{z0} at different distances from the center of the galaxy. The results suggest that, for disk galaxies with prominent nuclei, the stars in an associated bulge may be in predominantly chaotic orbits.

151.090 **Hydrodynamics of collisions between binary stars.**
J. Goodman, L. Hernquist.
Astrophys. J., Vol. 378, No. 2, p. 637 – 655 (10 Sep 1991).

Hydrodynamical processes in collisions between two binary stars, presumed to have formed by tidal capture, are investigated using a smoothed particle hydrodynamics code. Stellar mergers occur frequently in such events, when the binaries approach one another sufficiently closely that they could, in principle, convert substantial internal binary binding energy to external translational energy. Relative to binary–binary collisions involving point masses, hydrodynamical effects reduce the average translational energy produced by a factor ~ 3. This may be sufficient to eliminate tidal capture binaries as a viable direct energy source in star clusters, although they may still contribute indirectly through formation and stellar evolution of mergers. Multiple mergers are common, giving rise to remnants consisting of three or even all four stars. The relevance of these results to the dynamics of star clusters is discussed, along with possible implications for the hypothesis that blue stragglers form by stellar mergers.

151.091 **Dissolution of tidally limited star clusters in galaxies by passing giant molecular clouds or other massive objects.**
R. Wielen.
Symposium on the Formation and Evolution of Star Clusters as Part of the 102. Summer Scientific Meeting of the Astronomical Society of the Pacific, p. 343 – 349 (1991). – See Abstr. 012.050 for the main entry.

The author presents and applies improved formulae for the dissolution time of star clusters by passing massive objects (e.g. giant molecular clouds) in which he takes into account the effect of the stationary tidal field of the parent galaxy.

151.092 **On one type of models of spherical stellar clusters.**
V. M. Danilov.
Kinematika Fiz. Nebesn. Tel, Tom 7, No. 5, p. 75 – 86 (Sep–Oct 1991). In Russian. English translation in Kinematics Phys. Celest. Bodies, Vol. 7, No. 5.

The estimates of the time of star transitions between halo and core in spherical clusters as a result of stellar encounters are given. Numerical integration of gross–evolution equations for quasi–stationary and nonstationary (in a regular field) stellar clusters has been performed. There are three possible types of "halo–core" structure evolution in the quasi–stationary stellar clusters depending upon the initial conditions. All the open stellar clusters evolve near a certain sequence in the space of gross–parameters ξ and μ during the most of their lifetime. The parameters ξ and μ are the ratio of core radius to halo radius and the ratio of core mass to halo mass, respectively. The majority of open clusters must be concentrated near the sequence $\xi = \xi(\mu)$. This sequence corresponds to the balance of rates of transitions between core and halo during the cluster relaxation. The results of comparison of the author's cluster model calculations with observations and numerical N–body experiments are discussed.

151.093 **The effect of dynamical friction on globular cluster motion in triaxial galaxies.**
R. Capuzzo–Dolcetta, E. Pesce, M. Vietri.
Symposium on the Formation and Evolution of Star Clusters as Part of the 102. Summer Scientific Meeting of the Astronomical Society of the Pacific, p. 370 – 372 (1991). – See Abstr. 012.050 for the main entry.

The possible relevance of dynamical friction in triaxial potential is investigated, and some preliminary results of a more comprehensive work are given. In order to check quantitatively the role of dynamical friction in such environments, the authors have numerically integrated a set of orbits for a $10^6 M_\odot$ globular cluster moving in the field of a self–consistent model of triaxial galaxy. The authors find that the dynamical friction deceleration is particularly efficient on globular clusters moving on box orbits.

151.094 **Effect of tidal fields on star clusters.**
D. Chernoff, M. Weinberg.
Symposium on the Formation and Evolution of Star Clusters as Part of the 102. Summer Scientific Meeting of the Astronomical Society of the Pacific, p. 373 – 376 (1991). – See Abstr. 012.050 for the main entry.

The authors follow the dynamical evolution of a star cluster in a galactic tidal field using a restricted N–body code. They find large asymmetric distortions in the outer profile of the cluster in the first 10 or so crossing times as material is lost. Prograde stars escape preferentially and establish a potentially observable retrograde rotation in the halo. The authors present the rate of particle loss and compare with the prescription proposed by Lee and Ostriker (1987).

151.095 **Hydrodynamic simulations of close triple encounters.**
S. L. W. McMillan, S. R. Cranmer, S. A. Shorter, L. Hernquist.
Symposium on the Formation and Evolution of Star Clusters as Part of the 102. Summer Scientific Meeting of the Astronomical Society of the Pacific, p. 418 – 420 (1991). – See Abstr. 012.050 for the main entry.

Tidal binary systems play an important dynamical and observational role in our understanding of globular clusters. However, their very nature means that their interactions with other stars cannot be easily modelled by simple technqies, and hydrodynamical effects are likely to be critical in determining the outcome of those events. The authors have performed a series of simulations of collisions between close binaries and other cluster members, covering a fairly representative range of initial orbital configurations. They find that most "interesting" three–body encounters actually lead to mergers of two or all three of the stars involved, with around 5 – 10% of the total mass ejected from the triple system (and probably also from the cluster).

151.096 **Star cluster evolution with primordial binaries.**
S. L. W. McMillan, P. Hut, J. Makino.
Symposium on the Formation and Evolution of Star Clusters as Part of the 102. Summer Scientific Meeting of the Astronomical Society of the Pacific, p. 421 – 423 (1991). – See Abstr. 012.050 for the main entry.

Recent observations indicate that there may be a significant population of primordial binary systems in several globular clusters. In order to investigate this possibility, the authors have followed the evolution of a number of star clusters initially containing of order 20% binaries by mass, using direct numerical integration, for periods of up to 30 core collapse times. The authors discuss the implications of some of their simulations for the structure of dense globular clusters.

151.097 **Binaries in a medium of fast low–mass objects.**
A. Gould.
Astrophys. J., Vol. 379, No. 1, p. 280 – 284 (20 Sep 1991).

For a binary in a medium of fast low–mass objects, Heggie's Law can be made precise. Heggie's Law states broadly that hard binaries ($E_b \gg m\sigma^2$) get harder and soft binaries ($E_b \ll m\sigma^2$) get softer, where E_b is the binding energy of the binary and m and σ are the mass and velocity dispersion of the ambient objects. In the limit of fast low–mass objects, the transition between hard and soft binaries is $E_b = (3/2)m\sigma^2$. Objects traveling faster than V_b, the binary orbital speed, contribute to the viscous (hardening) term by encounters with individual members of the binary. Slower objects contribute by being captured and then expelled. For roughly equal mass binaries, fast objects dominate viscosity. This is just the opposite of the case of dynamical friction on single bodies, where slow objects dominate viscosity. Heggie's Law applies only when the binary center of mass is roughly in equilibrium with the medium. Far from equilibrium, new effects come into play. A binary traveling at supersonic speeds through even a very cold medium will tend to soften unless its orbital axis is aligned with its velocity to within 55°.

151.098 **Approximate third integrals in axisymmetric galaxies and the tilt of the velocity ellipsoid in the solar neighborhood.**
S. M. Kent, T. de Zeeuw.
Astron. J., Vol. 102, No. 6, p. 1994 – 2000 (Dec 1991). Current Physics Microform No.: 9112C1970.
The authors present three different methods for calculating approximate third integrals of the motion for orbits in axisymmetric potentials. All three methods are based on approximating the orbits by motion in a separable potential. The results are applied to the Carlberg–Innanen potential for the Galaxy and allow a determination of the tilt of the velocity ellipsoid in the solar neighborhood.

151.099 **Stellar kinematics in Kovalsky–Kapteyn figures. II.**
R. B. Shatsova, G. B. Anisimova.
Astrofizika, Tom 33, Vyp. 3, p. 379 – 393 (Dec 1990). In Russian. English translation in Astrophysics, Vol. 33, No. 3.
The possibilities of a new method for studying Kovalsky–Kapteyn figures, described in the first part of the work, are checked with SAO catalogue. The obtained Solar apex latitude and longitude, the angular velocity of the Galaxy rotation, the vertex longitude, the elongation of the velocity ellipsoid are not far from standard. The larger value of Oort's constant A is accounted to the adoption of pure Galaxy rotation scheme. The positive results of examination mean that the suggested method can be a fine basis for further perfection and application.

151.100 **Stochastic model of fragmented structures in galaxies.**
I. K. Rozgacheva.
Sov. Astron., Vol. 35, No. 5, p. 447 – 452 (Sep 1991). Current Physics Microform No.: 9203X1621. English translation of Astron. Zh., Tom 68, Vyp. 5, p. 897 – 908 (1991).
An inhomogeneous, steadily rotating disk is considered. The pressure of the medium is assumed to be negligible. It is shown that the ensemble of small inhomogeneities in the medium is a mixing system. The spatial structure of the collective gravitational field of the density inhomogeneities is analyzed. This field is found to have large-scale spiral structure. Density inhomogeneities are distributed over the disk in accordance with this structure and form a fragmented spiral, the fractal dimension of which depends on the number of inhomogeneities.

151.101 **Collective instabilities in the stellar environment of a galactic nucleus: a source of activity and an initiator of starbursts.**
V. L. Polyachenko.
Sov. Astron. Lett., Vol. 17, No. 5, p. 371 – 374 (Sep 1991). Current Physics Microform No.: 9204X1318. English translation of Pis'ma Astron. Zh., Tom 17, No. 10, p. 877 – 883 (1991).
The author re-examines the starburst mechanism and the support for active galactic nuclei derived from bar-mode instability. It turns out that only a bar mode that results from the alignment of stellar orbits provides an effective mechanism under the conditions encountered near a galactic center. This is true even if an internal bar is excited by a companion galaxy (the Noguchi scheme). Ordinary bar modes associated with rapid rotation are suppressed by the large gravitational potential external to an active disk. The author shows that if the gravitational potential is such that an oscillatory mode occurs instead of purely radial orbits, then the system may be subject to a distant relative of the plasma cone instability. The simplest analytical model is employed to study this instability to first order.

151.102 **The secular instability of axisymmetric collisionless star clusters.**
H. E. Kandrup.
Astrophys. J., Vol. 380, No. 2, p. 511 – 514 (20 Oct 1991).
Recently developed Hamiltonian techniques are used to investigate the stability of an axisymmetric, collisionless equilibrium, characterized by a distribution function $f_0(E, J_z, m)$ depending on both the particle energy E and azimuthal angular momentum J_z, as well as the particle mass m. The conserved total energy ε associated with a phase-preserving linear perturbation δf is constructed, and it is shown that ε is not positive definite. Specifically, it is demonstrated explicitly that, at least provided the function $f_0(E, J_z, m)$ is analytic in J_z and its derivative $F_J \equiv \partial f_0/\partial J_z$ is nonvanishing when $J_z = 0$, there exist nonaxisymmetric perturbations corresponding to a nontrivial change in the density $\delta\varrho$, which decrease ε. This indeterminacy in the sign of ε does not necessarily imply a linear instability. However, it does imply a secular instability in the presence of any dissipation, such as that provided by gravitational radiation reaction, which forces the time derivative $d\varepsilon/dt$ to be negative. It thus follows that, in the same sense as is true for rotating, axisymmetric, perfect fluid stars, "generic" rotating, axisymmetric equilibria $f_0(E, J_z, m)$ are unstable or at least marginally unstable in the presence of dissipation.

151.103 **The shape of the invisible halo: N-body simulations on parallel supercomputers.**
M. S. Warren, W. H. Zurek, P. J. Quinn, J. K. Salmon.
AIP Conf. Proc., No. 222, p. 216 – 223 (15 Apr 1991). Current Physics Microform No.: 9106E0216. – See Abstr. 012.069 for the main entry.
The authors study the shapes of halos and the relationship to their angular momentum content by means of N-body ($N \sim 10^6$) simulations. Results indicate that in relaxed halos with no apparent substructure: i) The shape and orientation of the isodensity contours tends to persist throughout the virialized portion of the halo. ii) Most ($\approx 70\%$) of the halos are prolate. iii) The approximate direction of the angular momentum vector tends to persist throughout the halo. iv) For spherical shells centered on the core of the halo the magnitude of the specific angular momentum is approximately proportional to their radius. v) The shortest axis of the ellipsoid which approximates the shape of the halo tends to align with the rotation axis of the halo. This tendency is strongest in the fastest rotating halos.

151.104 **N-body simulations: from galaxies to cosmology.**
E. Bertschinger.
AIP Conf. Proc., No. 222, p. 297 – 311 (15 Apr 1991). – See Abstr. 012.069 for the main entry.
Gravitational N-body simulations are increasingly used to complement analytical and observational studies in cosmology. This article reviews some of the key results obtained by and the numerical methods used for N-body simulations in both galaxy dynamics and cosmology. New results are presented from large particle-mesh and particle-particle/particle-mesh simulations.

151.105 **Spheroidal galaxy models with anisotropic stresses.**
N. W. Evans, D. Lynden-Bell.
Mon. Not. R. Astron. Soc., Vol. 251, No. 2, p. 213 – 220 (15 Jul 1991).
This paper applies a method for finding the kinematical properties of galaxy models without the explicit construction of distribution functions. For any axisymmetric mass density generated by a Stäckel potential, an ansatz is introduced for the stress tensor, which enables a unique solution of the equations of stellar hydrodynamics to be deduced. The methods are used to find velocity ellipsoids for oblate and prolate Kuzmin–Kutuzov models. The authors find that the velocity ellipsoids can be rounder for flatter systems, which illustrates the limitations of the virial equations.

151.106 **Bar-disc angular momentum exchange.**
B. Little, R. G. Carlberg.
Mon. Not. R. Astron. Soc., Vol. 251, No. 2, p. 227 – 242 (15 Jul 1991).
The authors use a model galaxy with an artificial bar to explore the angular momentum exchange induced by bar-like ($m = 2$) perturbations to a disc. They provide an analytic formula for this exchange which predicts to within ~ 1 per cent the interaction between a weak bar and a non-self-gravitating disc. The authors also present the results of N-body experiments which suggest

that the neglect of disc self–gravity in analytic theory leads to: (1) an underestimate of the slowdown rate of a bar which is dominated by corotation and outer Lindblad resonances; (2) an overestimate of the spin–up rate of a bar which is dominated by inner Lindblad resonances. Additional simulations suggest that the importance of disc self–gravity is generally enhanced by increasing the age of the bar, but is lessened by increasing the temperature of the disc. All these effects have natural explanations, and estimates of their magnitude are provided for several particular cases.

151.107 **Triggering of repetitive starbursts in merging galaxies.**
M. Noguchi.
Mon. Not. R. Astron. Soc., Vol. 251, No. 2, p. 360 – 368 (15 Jul 1991).

Merging galaxies are often observed to be the site of active star formation. In order to investigate the effect of mergers on the gas dynamics and the star formation process, the authors have carried out numerical simulations for the merger of two disc galaxies which contain gas clouds as well as stars.

151.108 **Anisotropic gaseous models for the evolution of star clusters.**
P. D. Louis, R. Spurzem.
Mon. Not. R. Astron. Soc., Vol. 251, No. 3, p. 408 – 426 (1 Aug 1991).

The authors present two anisotropic gaseous models for star clusters. Both are extensions of the standard isotropic gaseous model where the heat flux is proportional to the local temperature gradient and the coefficient of thermal conductivity is a function of the local Jeans length and the local relaxation time, but they differ in the relation between the third–order moments of the velocity distribution. The authors derive self–similar precollapse solutions for both models. They find that, in both solutions, the power–law index is $\alpha = 2.23$ and the ratio of the radial and tangential mean–square velocity dispersion grows to $\sigma_r^2/\sigma_t^2 \cong 1.30$ in the outer parts of the halo which is in excellent agreement with higher order fluid–dynamical models including anisotropy. The authors demonstrate that these results do not depend sensitively on the assumptions in the models. They then show that, for the boundary conditions of a growing singular isothermal sphere with a central energy source, only one of the models has a self–similar post–collapse solution. The implications of this fact for future time–dependent calculations through core collapse are discussed.

151.109 **Recovering galactic orbits by perturbation theory.**
O. E. Gerhard, P. Saha.
Mon. Not. R. Astron. Soc., Vol. 251, No. 3, p. 449 – 467 (1 Aug 1991).

With the aim of finding useful representations for the orbit structure in galaxy potentials, the authors have applied three Hamiltonian perturbations methods to a perturbed isochrone sphere. They find that axisymmetric galaxy potentials can be treated successfully as perturbed spherical systems, for axial ratios at least as flat as 0.6 in density. In these potentials the orbital tori can be constructed by perturbation theory and a corresponding approximate invariant ("third integral") can be found. The authors test these tori by comparing with numerically integrated orbits, and by calculating the component of the Hamiltonian flow normal to them. They show that they give an excellent description of the orbits in the meridional plane of the system, but the methods are also applicable to orbits with azimuthal motion. The methods the authors investigate are: (1) a first–order averaging method, which is less successful in this case, for reasons they explain; (2) a resonant method, which gives a good description of both loops and boxes in the meridional plane; (3) a higher order method based on the Kolmogorov–Arnold–Moser theory. This is restricted to orbit families already present in the unperturbed system – in this case loops – but improves accuracy at higher orders.

151.110 **Joint evolution of a galactic nucleus and central massive black hole.**
V. I. Dokuchaev.
Mon. Not. R. Astron. Soc., Vol. 251, No. 4, p. 564 – 574 (15 Aug 1991).

The general scheme of the joint evolution of a stellar galactic nucleus and central massive black hole (BH) is deduced in the homology approximation. Physically distinctive stages of this evolution crucially depend on the distribution and motion of stars near the BH which determine the rate of their penetrating into the "loss–cone" and successive tidal disruption or swallowing by the hole. Joint evolution leads either to contraction or expansion of the stellar cluster around the BH, depending on the parameters of the total system. Respectively, a massive BH in the centre of the host galaxy exists either in the excited state as powerful central source or in the ground state as a "dead" quasar.

151.111 **Formation and evolution of the disc in spiral galaxies: viscous models without infall.**
S. S. Olivier, G. R. Blumenthal, J. R. Primack.
Mon. Not. R. Astron. Soc., Vol. 252, No. 1, p. 102 – 115 (1 Sep 1991).

The authors study models for the evolution of the disc of spiral galaxies that include the effects of viscosity but assume no continuous infall of gas on to the disc. The time–scale for star formation is assumed to be of the same order as the time–scale for viscous redistribution of the mass and angular momentum in the disc. The authors confirm that these models are able to reproduce the observed radial distribution of gas and stars for present day spirals for a variety of model parameters and initial conditions. Furthermore, the authors generally find that the evolution of both the stellar and gas surface density profiles is approximately self–similar over at least 90% of the history of the disc, and that the size of the initial gas disc is approximately equal to the present Holmberg radius for the galaxy. Implications of these results for the evolution of spiral structure and for the interpretation of damped Lyα lines in the spectra of high–redshift QSOs are considered. A discussion of the possible physical mechanisms for the viscosity and the associated time–scales is also presented.

151.112 **Gravothermal instability of anisotropic self–gravitating gas spheres: singular equilibrium solution.**
R. Spurzem.
Mon. Not. R. Astron. Soc., Vol. 252, No. 2, p. 177 – 189 (15 Sep 1991).

A re–investigation of linear perturbation theory is presented, which examines the hydrostatic re–adjustment of an isolated self–gravitating gas sphere to a redistribution of energy. Such a model describes a stellar system by the common equations of gas in hydrostatic equilibrium but with the effect of the anisotropic velocity distribution on the pressure gradient. The author takes as equilibrium models the singular isothermal solution with and without anisotropy. The total variation of the Boltzmann entropy resulting from a perturbation of the system caused by a redistribution of heat (i.e. rms kinetic energy of the stars) is calculated for anisotropic solutions to first order as well as to second order for the isotropic equilibrium. The extremized eigenfunctions which represent the entropy and anisotropy perturbation functions, are determined analytically. They exhibit gravothermal behaviour in the central region where heat is removed. It is also found that the anisotropy re–adjusts non–thermally in the sense that the system departs from isotropy although the total entropy increases.

151.113 **A dynamical instability of bars in disk galaxies.**
N. Raha, J. A. Sellwood, R. A. James, F. D. Kahn.
Nature, Vol. 352, No. 6334, p. 411 – 412 (1 Aug 1991). Letter–to–the–editor.

Strong, rapidly rotating, persistent bars readily form in numerical simulations of initially axisymmetric disk galaxies. The authors present the results of three–dimensional simulations which show that the bar is dynamically unstable to buckling out of the galactic plane (the majority of previous disk galaxy simulations have been strictly two-dimensional). Stars acquire large motions normal to the plane, giving the bar a peanut shape through a mechanism that one suggests to be the fire-hose instability. Bars may be weakened or even destroyed by this instability; thus the fraction of disk galaxies containing strong bars today could be lower than the fraction in which they have formed in the past. The instability may also account for the peanut morphology of many galaxies, and because it leads to a less flattened stellar system with increased central density, it may play a part in the formation of galactic bulges.

151.114 **Inner secrets of ellipticals.**
J. Binney.
Nature, Vol. 354, No. 6350, p. 186 - 187 (21 Nov 1991).

151.115 **Origin of kinematic subsystems in elliptical galaxies.**
L. Hernquist, J. E. Barnes.
Nature, Vol. 354, No. 6350, p. 210 - 212 (21 Nov 1991). Letter-to-the-editor.

Elliptical galaxies were once thought to be smooth, featureless stellar systems with little or no substructure, but increasingly sophisticated observations are challenging this point of view. Some ellipticals contain small central disks which may be counter-rotating or otherwise kinematically decoupled from the rest of the galaxy. It seems unlikely that these disks could form during the monolithic collapse of a slowly rotating proto-galaxy, as the most plausible outcome of such evolution is a system with a simple rotation pattern. Like other kinds of fine structure in elliptical galaxies, these subsystems have been widely interpreted as evidence for multiple formation events or episodes. The authors demonstrate the formation of a counter-rotating central gas disk in a merger of two gas-rich disk galaxies of equal mass. Such a structure may well account for the unusual gas kinematics found in the merger remnant NGC 7252. Continued star formation in such gaseous disks may produce central components with decoupled kinematics, resembling the cores of some elliptical galaxies.

151.116 **Merging of two galaxies with central black holes.**
T. Ebisuzaki, J. Makino, S. K. Okumura.
Nature, Vol. 354, No. 6350, p. 212 - 214 (21 Nov 1991). Letter-to-the-editor.

Observations of elliptical galaxies show a positive correlation between total luminosity and core radius, defined as the radius at which surface luminosity becomes half of the central value. This has been taken as evidence against the idea that ellipticals are the result of galaxy mergers, because in simulated mergers the core radius remained almost constant. In those simulations, the galaxies contained no central black holes. The authors present simulations of the merging of identical galaxies with and without central black holes, and find that when black holes are included, the merged galaxy acquires an isothermal core comparable in mass to the sum of the two initial black holes. Furthermore, the ratio of the core radius to the half-mass radius is approximately the same as the ratio of the black hole mass to the total galaxy mass, a result also consistent with observational evidence. These results, which can be understood by means of simple analytical arguments, suggest that most elliptical galaxies contain central black holes with masses comparable to the mass of their cores.

151.117 **The hydrodynamical interaction of galaxies.**
M. Umemura, S. Yoshioka.
IUPAP Conference on Primordial Nucleosynthesis and Evolution of Early Universe, p. 565 - 566 (1991). - See Abstr. 012.071 for the main entry.

The authors present 3D hydrodynamical calculations on interacting/merging galaxies. They find that nearly head-on collisions of galaxies lead to new galaxy formation. The dissipative effect of gaseous component significantly affects on merging process of interacting galaxies.

151.118 **Dynamical evolution of globular clusters with mass spectrum.**
Lee Hyungmok.
Proc. Astron. Soc. Aust., Vol. 9, No. 1, p. 41 - 44 (1991). - See Abstr. 012.078 for the main entry.

The author presents a series of numerical models describing the dynamical evolution of globular clusters with a mass spectrum, based on integration of the Fokker-Planck equations. He includes three-body binary heating and a steady galactic tidal field. A wide range of initial mass functions is adopted and the evolution of the mass function is examined. The mass function begins to change appreciably during the post-collapse expansion phase due to the selective evaporation of low mass stars through the tidal boundary. One signature of highly evolved clusters is thus the significant flattening of the mass function. The age (in units of the half-mass relaxation time) increases very rapidly beyond about 100 signifying the final stage of cluster disruption. This appears to be consistent with the sharp cut-off of half-mass relaxation times at near 10^8years for the Galactic globular clusters.

151.119 **Abel models for triaxial systems.**
H. Dejonghe, D. Laurent.
Mon. Not. R. Astron. Soc., Vol. 252, No. 4, p. 606 - 636 (15 Oct 1991).

The authors present the general theory of equilibrium components of collisionless stellar systems in a Stäckel potential, for which the relation between mass density and distribution function can be written as an Abel integral. These Abel-components are generalizations of the well-known spherical models by Eddington and Osipkov-Merritt. Their most important properties include analytical simplicity, which is convenient, and the freedom of an arbitrary function in their definition, which makes it conceivable to combine them in order to model galaxies more realistically. The authors further show that the Abel-components can be divided into three classes with different dynamical properties, and they discuss their orbital contents.

151.120 **Tumbling endstates for unstable rotating spheres: implications for rotation in elliptical galaxies.**
J. C. B. Papaloizou, P. L. Palmer, A. J. Allen.
Mon. Not. R. Astron. Soc., Vol. 253, No. 1, p. 129 - 146 (1 Nov 1991).

The authors present 10^5 particle numerical simulations for the evolution of unstable, radially anisotropic spherical systems which have a modest net rotation about the z axis. Reliability of these experiments is demonstrated by continuity of the results with small changes in the parameters as well as qualitative agreement with the analytic theory presented. The results are directly applicable to the shapes of rotating elliptical galaxies. In simulations of weakly unstable systems isophotal twists associated with interacting subsystems developed. The authors constructed systems with counter rotating cores which had a similar ratio of streaming to random velocity to that observed in some elliptical galaxies. This feature was long lived, thus demonstrating that these systems do not need to have formed recently.

151.121 **Collapse of a protogalactic cloud.**
S. Yoshioka.
NATO Advanced Research Workshop on Observational Tests of Inflation, p. 477 - 478 (1991). - See Abstr. 012.076 for the main entry.

The author investigates the process of galaxy formation through collapse of a protogalactic cloud consisting from gas and dark matter using a three-dimensional hydrodynamical code.

151.122 **Kinematics of stars in Kovalsky-Kapteyn figures. II.**
R. B. Shatsova, G. B. Anisimova.
Astrophysics, Vol. 33, No. 3, p. 518 - 527 (Nov - Dec 1990). English translation of Astrofizika, Tom 33, Vyp. 3, p. 379 - 393 (Dec 1990).

The possibilities of the new method of studying Kovalsky-Kapteyn figures described in part I of the study are tested using the SAO catalog. Values close to the standard ones are obtained for the longitude and latitude of the sun's apex, the angular velocity of the rotation of the Galaxy, the longitude of the vertex, and the elongation of the velocity ellipsoid. A too large value of the Oort constant A is explained by the adopted scheme of pure rotation of the Galaxy. The positive results of the test mean that the proposed method is a good basis for further improvement and application.

151.123 **Shell formation in elliptical galaxies.**
R. C. Thomson.
Mon. Not. R. Astron. Soc., Vol. 253, No. 2, p. 256 – 278 (15 Nov 1991).

Numerical simulations are described which show the formation of shell structures in elliptical/S0 galaxies. The shells are formed as density waves induced in a thick disc population of dynamically cold stars by a weak interaction with another galaxy. The elliptical galaxy is modelled as an oblate spheroidal mass distribution in which test particles are set up on short axis tube orbits with zero radial thickness. The initial particle distribution function is calculated so that the system is in equilibrium. The effect of the primary galaxy ellipticity and the mass of the secondary galaxy on the shell forming process are described. Simulated shell profiles and their radial distribution are also compared with observations. The results support the weak interaction model of shell formation in elliptical/S0 galaxies, and imply that shell galaxies are similar oblate spheroids which appear as type 1 (aligned) systems when viewed edge–on, and as type 2 (all–round) when viewed face–on. The weak interaction model of shell formation is shown to give a more accurate and robust description of the observed phenomena than does the merger model.

151.124 **An analytic inversion for anisotropic spherical galaxies.**
P. Cuddeford.
Mon. Not. R. Astron. Soc., Vol. 253, No. 3, p. 414 – 426 (1 Dec 1991).

A generalization of the Osipkov–Merritt inversion of the fundamental integral equation for anisotropic spherical galaxies is obtained. For any density profile this inversion yields a two–parameter distribution function with arbitrary degrees of both radial and trangential velocity anisotropies. Therefore, for the first time, exact analytical models can be found which generate given density profiles and possess tangential velocity anisotropies. This property disposes of the well–known radial orbit instability encountered by many models of anisotropic spherical galaxies. Moreover, the inversion provides a new opportunity to study instabilities in analytical galaxy models which have realistic density profiles. As examples of the inversion, two families of anisotropic Plummer and Jaffe models are calculated. These models are completely analytical in that the distribution functions, velocity dispersion profiles and differential energy distributions can all be expressed in terms of elementary functions. Some interesting qualitative properties of these models are also discussed.

151.125 **Extended stellar hydrodynamics for galactic discs – II.**
P. Cuddeford, P. Amendt.
Mon. Not. R. Astron. Soc., Vol. 253, No. 3, p. 427 – 444 (1 Dec 1991). With microfiche MN 253/2.

A higher–moment 3D model for axisymmetric three–integral disc galaxies is reexamined and modified slightly in order to take proper account of the possibly non–ellipsoidal character of the velocity distribution. General recurrence relations are found, relating certain successive even moments of arbitrary order, which are seen as generalizations of the original closure assumptions. Therefore the closure assumptions can be taken at any moment order, since the model equations generated by the generalized closure relations are valid at all levels of the infinite hierarchy of moment equations. By virtue of the formal equivalence between the infinite set of moments and the distribution function, this strongly suggests that the distribution function which underlies the model is a solution of the collisionless Boltzmann equation within the approximations of the model. The physical significance of the model is seen as attributing all of the vertical non–isothermality to the third isolating integral, and none to the deviations of the vertical and vertical–radial kurtoses from their Maxwellian values. It is argued that such a model is also appropriate for comparison with observations. The simple "model–free" analytic expressions derived in the previous paper (valid for all kinds of potential) for the orientation of the velocity ellipsoid at points near the plane is shown to yield the correct orientation exactly for any system in which the potential is Stäckel separable in general spheroidal coordinates. A possible functional form for the distribution function corresponding to the model is also discussed.

151.126 **Liouville field theory and the partition function for two–dimensional Newtonian gravity.**
T. Padmanabhan.
Mon. Not. R. Astron. Soc., Vol. 253, No. 3, p. 445 – 448 (1 Dec 1991).

The author shows that the grand partition function for a system of particles interacting via gravitational force can be related to the vacuum–to–vacuum amplitude of the Liouville field theory. Finite action solutions to the Liouville field equation can be expressed in closed form in two dimensions. This solution is used to construct the mean field approximation for the grand partition function. The accuracy of the mean field approximation is verified by comparing it with the exact partition function for a system with two particles.

151.127 **Stellar encounters involving red giants in globular cluster cores.**
M. B. Davies, W. Benz, J. G. Hills.
Astrophys. J., Vol. 381, No. 2, p. 449 – 461 (10 Nov 1991).

Collisions between a 0.8 $M_\odot$ red giant and impactors of mass 0.4 and 0.6 $M_\odot$ are simulated using a three–dimensional smoothed particle hydrodynamics code. The red giant is modeled as a massive core surrounded by a gaseous envelope, while the impactor, assumed to be a white dwarf or a main–sequence star, is treated as single particle. Encounters at various impact parameters and with relative velocities pertinent to globular clusters (10 km s^{-1}) are studied. The effect of the encounter on the red giant depends strongly on the impact parameter. If the impactor passes within twice the radius of the red giant ($2R_{RG}$), mass loss occurs from the red giant envelope and both objects are left bound in elliptical orbits. Calculations suggest that if the impactor has an initial distance at closest approach of approximately $1.6R_{RG}$ or less, the impactor will spiral in to the red giant envelope on subsequent periastron passages. For bound systems with a greater separation, the orbit may circularize with a radius approximately double that of the initial encounter. In both cases the system will undergo a common envelope phase: in the latter case, the red giant will eventually fill its Roche lobe as it expands up the giant branch. In the case of a sufficiently small closest approach ($R_{min} < R_{RG}$), the impactor is immediately captured by the core of the red giant. Both objects from a binary system orbiting inside a common envelope formed from the original envelope of the giant. With a white dwarf impactor, the likely outcome of the common envelope phase will be a tight white dwarf/white dwarf binary that will merge in less than a Hubble time.

151.128 **On the tidal disruption of a spherical galaxy.**
P. M. S. Namboodiri, R. K. Kochhar.
Mon. Not. R. Astron. Soc., Vol. 253, No. 4, p. 683 – 685 (15 Dec 1991).

The disruption of a satellite galaxy as it undergoes collision with a massive perturber is investigated by numerical simulations. The models are restricted by a constant pericentric distance. The disruption of the satellite is estimated in terms of the density ratio ϱ/ϱ_R where ϱ is the initial density of the satellite and ϱ_R a critical density, the Roche density. The satellite suffers considerable disruption if $\varrho/\varrho_R < 1/2$ and it survives the tidal encounter when $\varrho/\varrho_R > 1/2$. For the same initial conditions, the circular orbit models show larger tidal effects than the other models.

151.129 **Violent relaxation and the $R^{1/4}$ law.**
J. Hjorth, J. Madsen.
Mon. Not. R. Astron. Soc., Vol. 253, No. 4, p. 703 – 709 (15 Dec 1991).

Violent relaxation occurs in a finite region in space with the escape energy as no special threshold. After the escape of positive energy particles and redistribution of weakly bound particles, an isotropic equilibrium distribution function, which reproduces de Vancouleurs' empirical $R^{1/4}$ law for the surface–brightness distribution in elliptical galaxies, is predicted in a natural way. The residuals from the $R^{1/4}$ law fit the surface photometry of the standard elliptical NGC 3379. In this simple picture the $R^{1/4}$ law reflects neither initial conditions nor internal velocity distribution, but requires a deep central potential. As indicated by numerical simulations this can be achieved by cold dissipationless collapse which results in highly radial orbits but perhaps more naturally through dissipational processes for which there is auxiliary evidence.

151.130 **The velocity dispersion anisotropy and mass–to–light ratio of elliptical galaxies.**
R. P. van der Marel.
Mon. Not. R. Astron. Soc., Vol. 253, No. 4, p. 710 – 726 (15 Dec 1991).

Axisymmetric dynamical models are constructed for 37 bright elliptical galaxies for which high–quality photometrical, and both major and minor axis kinematical data are available in the literature. The models are of the type used previously by Binney, Davies and Illingworth and van der Marel, Binney and Davies. The projected kinematics are predicted from the observed surface photometry, assuming a constant mass–to–light ratio, and a velocity ellipsoid with $\sigma_r = \sigma_\theta$. For the sample as a whole it is found that these models tend to predict too much motion on the major axis. This implies that elliptical galaxies as a class must have $\sigma_r > \sigma_\theta$, which has not been demonstrated before. This result is consistent with general expectation as based on N–body simulations of dissipationless collapse. From the models accurate mass–to–light ratios are derived that are corrected for the effects of rotation and radial anisotropy. The observed velocities in the outer parts of the galaxies in the sample systematically exceed the velocities predicted by the authors' (constant mass–to–light ratio) models. From this it is argued that no axisymmetric constant mass–to–light ratio models can fit the kinematical data in the outer parts of elliptical galaxies. Either dark haloes must be present, or more detailed triaxial models must be studied.

151.131 **Post–collapse evolution of globular clusters with stochastic energy sources.**
K. Takahashi, S. Inagaki.
Publ. Astron. Soc. Jpn., Vol. 43, No. 4, p. 589 – 606 (1991).

The effects of the stochasticity of binary activity on the post–collapse evolution of globular clusters are investigated by directly integrating the Fokker–Planck equation. By comparing the present results of computations with those of continuous energy sources, it is found that the stochasticity of binary activity makes core evolution more unstable. For example, a cluster with N = 10,000 stars experiences chaotic gravothermal oscillations in the case of stochastic energy sources, whereas one with continuous energy sources exhibits regular oscillations. The stochasticity, however, does not have much influence on the overall features of gravothermal oscillations for clusters with large N.

151.132 **Influence of a dark halo on the stability of elliptical galaxies.**
M. Stiavelli, L. S. Sparke.
Astrophys. J., Vol. 382, No. 2, p. 466 – 474 (1 Dec 1991).

The authors describe N–body simulations to investigate whether the presence of dark matter in elliptical galaxies could inhibit the bar–making instability which is characteristic of systems with many stars on near–radial orbits. The calculations show that neither a fixed static halo nor a halo composed of particles with isotropic velocity dispersions is very effective in suppressing the instability; thus the presence of a dark halo is not likely to affect conclusions about how anisotropic the velocities of stars in elliptical galaxies might be. Modifications of the Jeans–type stability criterion of Polyachenko and Shukman to include the effect of the halo are in agreement with the present results for a halo of "live" particles but overestimate the stabilizing effect of a fixed halo potential. The authors also present evidence that the bar–making instability can change the radial density profile of the models, significantly reducing an initial high central concentration.

151.133 **The ordered nature of elliptical galaxies: implications for their intrinsic angular momenta and shapes.**
M. Franx, G. Illingworth, T. de Zeeuw.
Astrophys. J., Vol. 383, No. 1, p. 112 – 134 (10 Dec 1991).

The observations of rotation along the major and minor axes of 38 elliptical galaxies are analyzed to determine their intrinsic structure. Rotation along the minor axis occurs (1) as a result of projection effects in triaxial systems and/or (2) because of real misalignment of the angular momentum axis from the intrinsic short axis, as is allowed in triaxial systems. The observed misalignment of the rotation axes and the short axes ranges from 0° to 90°, with most ellipticals having misalignments that are small. This is not expected if ellipticals are generically triaxial systems with uniformly distributed angular momentum direction, and requires that particular constraints be imposed during the formation process. The analysis of the observed misalignments requires to derive the intrinsic ellipticity distribution for ellipticals. The resulting distribution, based on CCD data, shows a pronounced lack of round galaxies. The distribution of triaxiality and intrinsic misalignment of the angular momentum cannot be derived uniquely, since one observable is used to determine two internal parameters. Various assumptions regarding the intrinsic shapes and the direction of the angular momentum were used. An exploration of the solution space showed a wide range of valid distributions. All the models have a sizable fraction of galaxies ($\geqslant 35\%$) with intrinsic misalignments less than 15°. The results were compared with predictions from several groups for hierarchical formation scenarios. The predicted distributions for halos are skewed significantly toward prolate–triaxial shapes. The data show that the luminous parts of ellipticals do not have such high triaxialities. The shapes of the dark halos may well be different from the shapes of the luminous matter in ellipticals, despite their similar dynamics. This may result, for example, from the effect of gaseous dissipation.

151.134 **On core collapse.**
J. Makino, P. Hut.
Astrophys. J., Vol. 383, No. 1, p. 181 – 191 (10 Dec 1991).

Star clusters undergo core collapse on a thermal time scale. This result follows from detailed simulations in the form of N–body, Fokker–Planck, and gas sphere models. The authors present a linear stability analysis of the isothermal sphere, for different choices of heat conductivity. For some choices the core mass shrinks significantly, while for others it remains constant or even increases. Different physical systems correspond to these different choices. Gaseous systems such as contracting gas clouds with a heat transport based on radiative transfer show a heat conductivity that decreases in the high–density core. When they are isothermal, they will increase their core mass during contraction. When heat transport takes place mostly through thermal conduction by electrons, as in a moderately degenerate star, the core mass will remain nearly constant. Stellar dynamical systems such as globular star clusters have a heat conductivity which strongly peaks toward the central regions, resulting in a rapid shrinking of the core mass.

151.135 **The formation of primordial binaries in globular clusters by star–disk interactions.**
S. D. Murray, C. J. Clarke, J. E. Pringle.
Astrophys. J., Vol. 383, No. 1, p. 192 – 197 (10 Dec 1991).

The formation of primordial binaries in globular clusters is examined using simple numerical models. Clusters of protostars collapse until their velocity dispersion rises sufficiently to reverse the infall and the cluster reaches equilibrium. During the collapse, interactions between stars and protostellar disks lead to stellar capture. It is found that binary fractions of a few percent typically result. Binary formation is terminated when the velocity dispersion rises to a point at which most encounters result in disk destruction rather than capture. As a result, much gas is returned to the cluster ISM, limiting the star formation efficiency to a value significantly below 100%.

151.136 **On the effects of interstellar gas on the oscillations and the stability of galaxies.**
P. O. Vandervoort.
Astrophys. J., Vol. 383, No. 2, p. 498 – 510 (20 Dec 1991).
The modes of oscillation that would occur in galaxies that are purely stellar systems can be substantially modified in galaxies that contain interstellar gas. Thus the number of modes of a given kind is generally greater when gas is present than when it is absent. Moreover, interstellar gas can induce instability with respect to certain modes that would otherwise be stable. In the present investigation, such effects of the gas are illustrated for modes of oscillation in homogeneous spheroids composed of uniformly rotating subsystems of stars and gas. It is shown for these systems that the critical value of the ratio of rotational to gravitational energy for the onset of instability with respect to the formation of a bar is a function of the eccentricity of the spheroidal figure, the fraction of the system that is gas, and the ratio of the angular velocities of the stellar and gaseous subsystems. It is also shown that slowly rotating systems of this kind are dynamically unstable with respect to a pulsation mode if the ratio of specific heats of the gas is less than a critical value which depends on the eccentricity of the figure, the fraction of the system that is gas and the ratio of the angular velocities of the two subsystems.

151.137 **Galaxy clustering using the hierarchical Peaks approach: predictions for ROSAT, SZ, and FIR.**
S. T. Myers, J. R. Bond.
J. R. Astron. Soc. Can., Vol. 85, No. 4, p. 209 – 210 (Aug 1991). Abstract. – See Abstr. 012.091 for the main entry.

151.138 **Multidimensional analysis of non-linear gravitational clustering in a cold dark matter universe.**
G. Yepes, H. M. P. Couchman, R. Dominguez-Tenreiro.
J. R. Astron. Soc. Can., Vol. 85, No. 4, p. 210 (Aug 1991). Abstract. – See Abstr. 012.091 for the main entry.

151.139 **Stochastic orbits in triaxial elliptical galaxies.**
He Nan, R. E. Pudritz.
J. R. Astron. Soc. Can., Vol. 85, No. 4, p. 210 – 211 (Aug 1991). Abstract. – See Abstr. 012.091 for the main entry.

151.140 **A nonstationary wave and star motion in a planar system.**
A. S. Baranov.
Sov. Astron., Vol. 35, No. 6, p. 582 – 588 (Nov 1991). Current Physics Microform No.: 9205X1392. English translation of Astron. Zh., Tom 68, Vyp. 6, p. 1160 – 1173 (1991).
The influence of nonstationary effects on the dynamical evolution of noninteracting particles in a certain smoothed regular field is investigated. Different kinds of spiral-wave perturbation with slowly varying parameters are considered. Some qualitative features of the effects of a nonstationary perturbation are revealed. It is shown that there exists a category of trapped stars that move together with the resonance zone and occupy noncontiguous domains in phase space. Stars that pass through the resonance zone travel in the intervals between these domains and are not subject to perturbation. It is shown that irreversible changes build up near the separatrices as a consequence of redistribution of the phase domains with different behavior, as well as due to randomization of motion in the resonance zones. It is concluded that a temporary local decrease in density should be observed during the time interval in which a certain zone is a resonance zone.

151.141 **Parabolic orbits for one model of a galactic potential.**
E. I. Timoshkova.
Sov. Astron., Vol. 35, No. 6, p. 660 – 663 (Nov 1991). Current Physics Microform No.: 9205X1470. English translation of Astron. Zh., Tom 68, Vyp. 6, p. 1315 – 1322 (1991).
A two-parameter family of parabolic orbits in the meridional plane of a galaxy has been found in the problem of perturbed motion in a rotationally symmetric field. The relationship between the family and stationary solutions is established. Analytic expressions for star coordinates and velocities are obtained in closed form for periodic and asymptotic motion along the trajectories.

151.142 **Galaxy morphology in dissipative models.**
M. B. Mosconi, D. Garcia Lambas.
Bol. Asoc. Argent. Astron., No. 36, p. 413 – 415 (1990). Abstract. In Spanish and English. – See Abstr. 012.095 for the main entry.

151.143 **Analytical derivation of the probability for the escape of stars from colliding galaxies.**
P. M. Cincotta, J. C. Muzzio, J. A. Núñez.
Celest. Mech. Dyn. Astron., Vol. 52, No. 3, p. 263 – 275 (1991).
The authors use the impulse approximation to derive analytical formulae for the escape probability from a simple binary system that interacts with a third body. The binary system is made up of a massless body in circular orbit around a massive object, and it is assumed that the two massive bodies follow a Schuster (or Plummer) distribution. Within the ranges imposed by the impulse approximation to the parameters of the encounter, the authors find good agreement between their results and those obtained from numerical experiments.

151.144 **Distribution of stars perpendicular to the plane of the Galaxy.**
S. Chatterjee.
J. Astrophys. Astron., Vol. 12, No. 4, p. 269 – 280 (Dec 1991).
The author presents rigorous analytical solutions for the Boltzmann-Poisson equation concerning the distribution of stars above the galactic plane.

151.145 **Kinematic structures of merger remnants.**
S. K. Okumura, T. Ebisuzaki, J. Makino.
Publ. Astron. Soc. Jpn., Vol. 43, No. 6, p. 781 – 793 (1991).
The authors performed a set of N-body simulations of the merging of two identical spherical galaxies using a special-purpose computer, GRAPE-1. They found that the kinematic properties of the resulting merger remnants are consistent with those of observed elliptical galaxies. The ratios of the maximum rotation velocity (V_{max}) to the velocity dispersion (σ_0) are smaller than 0.6, even in the case of large impact parameters. This is in good agreement with the lack of elliptical galaxies whose V_{max}/σ_0 exceeds 0.6. Previous N-body calculations have contradicted the observed kinematic studies of elliptical galaxies, because relaxation was numerically enhanced due to the small number of particles (~ 500). The authors' simulations with $N = 16384$ resolve this problem. The central velocity dispersion increased by 10 – 20% through merging for a wide range of impact parameters. This increase is consistent with the Faber-Jackson relation.

151.146 **The distribution of stars in a star cluster with a Kerr black hole in its center.**
Fan Junhui, Xie Guangzhong, Tang Zhiming, Wang Yongjiu.
Publ. Yunnan Obs., No. 3, p. 6 – 13 (1991). In Chinese.
The motion of test particles in the Kerr metric and the distribution of stars in a star cluster with a Kerr black hole in its center are discussed. The result shows that there are more lower velocity stars in that kind of stable cluster.

Dynamics of disc galaxies. Proceedings.
See Abstr. 012.028.

The formation and evolution of star clusters. Proceedings.
See Abstr. 012.050.

Warped disks and inclined rings around galaxies. Proceedings.
See Abstr. 012.054.

Exact solutions for Hamiltonian systems using integer arithmetic.
See Abstr. 021.016.

Techniques for N–body codes on the connection machine.
See Abstr. 021.030.

Treecode with a special–purpose processor.
See Abstr. 021.042.

DREAM: a special–purpose computer for large scale mesh calculations.
See Abstr. 021.043.

GRAPE–1A: special–purpose computer for *N*–body simulation with a tree code.
See Abstr. 021.052.

The Connection Machine: principle and application in astronomy and astrophysics.
See Abstr. 021.056.

A modified Aarseth code for GRAPE and vector processors.
See Abstr. 021.057.

Hamiltonian dynamics of a rigid body in a central gravitational field.
See Abstr. 042.005.

On forming of binaries by close triple approaches with equal– and different–mass bodies.
See Abstr. 042.035.

The Lissajous transformation. I. Basics.
See Abstr. 042.039.

On the representativity of initial conditions of triple systems.
See Abstr. 042.048.

Nonlinear density waves in gaseous disks.
See Abstr. 062.126.

Limits to pressure in outer layers of stars and velocity dispersions in outer regions of stellar systems.
See Abstr. 062.224.

Physics of accretion by spiral shock waves.
See Abstr. 062.265.

Spiral shocks in accretion disks: a preliminary numerical study.
See Abstr. 062.266.

Mass transfer by tidally induced spiral shocks in an accretion disc.
See Abstr. 062.267.

Accretion disks and the link between an AGN and its host galaxy.
See Abstr. 062.268.

Large–scale accretion flows in AGN.
See Abstr. 062.269.

Collisions of giant stars with compact objects: hydrodynamical calculations.
See Abstr. 065.059.

Tidal disruption of stars in black hole clusters as a source of AGN luminosity.
See Abstr. 067.023.

On the evolution of gravitationally unstable protostellar disks.
See Abstr. 131.355.

On the origin of the blue stragglers in the globular cluster NGC 5053.
See Abstr. 154.065.

The tidal radii of globular clusters in M31.
See Abstr. 154.090.

Core oscillations in globular cluster evolution: recent results.
See Abstr. 154.091.

Towards the comparison of observations of globular clusters and Fokker–Planck simulations.
See Abstr. 154.092.

Two high–velocity stars shot out from the core of the globular cluster 47 Tucanae.
See Abstr. 154.155.

Large–scale oscillations of the Galaxy and the kinematics of the solar neighbourhood.
See Abstr. 155.039.

An improved model of the galactic mass distribution for orbit computations.
See Abstr. 155.132.

Time variation of ellipticity of globular clusters in the Large Magellanic Cloud.
See Abstr. 156.025.

Dynamical segregation of direct and retrograde orbits in binary disk galaxies.
See Abstr. 157.001.

The velocity dispersion gradients in spiral galaxies. I. The Galaxy and NGC 3198.
See Abstr. 157.002.

What determines the physical quantities of galaxies? A two–component gas model for protogalaxies with energy input from supernovae.
See Abstr. 157.045.

Arm classification and velocity gradients in spiral galaxies.
See Abstr. 157.097.

Abundance gradients and galaxy formation.
See Abstr. 157.106.

A numerical study of star formation in interacting disk galaxies.
See Abstr. 157.108.

The velocity dispersion of molecular cloud ensembles and the formation of GMC's in spiral galaxies.
See Abstr. 157.123.

Behavior of warps in extended disks.
See Abstr. 157.202.

The structure of cold dark matter halos.
See Abstr. 157.207.

The outer Lindblad resonance and the morphology of early type disk galaxies.
See Abstr. 157.216.

The case of NGC 1365.
See Abstr. 157.228.

Dissipationless collapse and the shape of isophotes.
See Abstr. 157.264.

A post–core–collapse model for the nucleus of M33.
See Abstr. 157.282.

Computer statistics for three–body systems and mass determination in triplets of galaxies.
See Abstr. 157.293.

Is the kinematically distinct core in NGC 5982 necessarily the remnant of an accreted companion?
See Abstr. 157.306.

External fueling and stochastic ignition of starburst in isolated galaxies.
See Abstr. 157.350.

The merging runaway.
See Abstr. 160.032.

Dynamical evolution of compact groups of galaxies.
See Abstr. 160.122.

Bound populations around cD galaxies and cD velocity offsets in clusters of galaxies.
See Abstr. 160.152.

Cosmogony and the structure of rich clusters of galaxies.
See Abstr. 160.174.

Dynamical evolution of clusters of galaxies.
See Abstr. 160.179.

Hydrodynamic simulations of the intracluster medium.
See Abstr. 160.181.

Large scale–free N–body simulations.
See Abstr. 161.087.

Precollapse scale invariance in gravitational instability.
See Abstr. 161.629.

152 Stellar Associations

152.001 **New members of the Chamaeleon I pre–main sequence association.**
T. Harrison, P. McGregor.
Bull. Am. Astron. Soc., Vol. 23, No. 2, p. 885 (1991). Abstract. – See Abstr. 012.008 for the main entry.

152.002 **UIT ultraviolet imaging and photometry of the extragalactic OB association NGC 206.**
B. B. Pfarr, R. H. Cornett, J. K. Hill, J. E. Isensee, W. B. Landsman, P. C. Chen, S. P. Maran, S. G. Neff, E. P. Smith, T. P. Stecher, M. S. Roberts.
Bull. Am. Astron. Soc., Vol. 23, No. 2, p. 909 (1991). Abstract. – See Abstr. 012.008 for the main entry.

152.003 **UIT ultraviolet color–magnitude diagrams of the stellar content of OB associations within M31.**
J. E. Isensee, R. H. Cornett, J. K. Hill, W. B. Landsman, B. B. Pfarr, P. C. Chen, S. P. Maran, S. G. Neff, E. P. Smith, T. P. Stecher, M. S. Roberts.
Bull. Am. Astron. Soc., Vol. 23, No. 2, p. 909 (1991). Abstract. – See Abstr. 012.008 for the main entry.

152.004 **Galactic OB associations in the northern Milky Way galaxy.**
C. D. Garmany, R. E. Stencel.
Bull. Am. Astron. Soc., Vol. 23, No. 2, p. 909 (1991). Abstract. – See Abstr. 012.008 for the main entry.

152.005 **Regions of active star formation in O associations. II.**
A. V. Oskanyan.
Astrophysics, Vol. 33, No. 2, p. 447 – 453 (Sep – Oct 1990). English translation of Astrofizika, Tom 33, Vyp. 2, p. 259 – 269 (Oct 1990).

Analysis of an ensemble of O associations has shown that most regions of active star formation are in O associations in which there is at least one cn cluster. The lifetime of an H II region associated with a cn cluster is estimated at $10^5 - 10^6$yr, which is much less than the lifetime of O associations. On the basis of these results, the ensemble of O associations is divided into ages. In the group of the youngest O associations, many regions of active star formation are observed. In the group of older O associations, there are oc clusters and isolated multiple Trapezium type systems. In the oldest O associations, clusters are not found. The last two groups of O associations are poor in young regions of star formation.

152.006 **A log(P) – ⟨K⟩ relation for SRC–type variables in the Per OB1 association and the LMC distance modulus.**
A. K. Dambis.
Astron. Tsirk., No. 1545, p. 19 – 20 (Aug – Sep 1990).

152.007 **Massive star clusters and OB associations.**
C. D. Garmany.
Symposium on the Formation and Evolution of Star Clusters as Part of the 102. Summer Scientific Meeting of the Astronomical Society of the Pacific, p. 23 – 34 (1991). – See Abstr. 012.050 for the main entry.

The history study of OB associations and their stellar content is reviewed. Current work on Galactic and Magellanic Cloud associations is discussed, and the relevance of these studies to related questions of star formation, the initial mass function and massive stellar evolution is pointed out.

152.008 **The Serpens OB2 association – lighting the fire under "Stockert's chimney".**
D. Forbes.
Symposium on the Formation and Evolution of Star Clusters as Part of the 102. Summer Scientific Meeting of the Astronomical Society of the Pacific, p. 148 – 150 (1991). – See Abstr. 012.050 for the main entry.

Some results are presented of an optical study of the rich OB association Ser OB2, which powers the large H II region S 54 and the unusual thermal spur or "chimney" mapped by Müller, Reif and Reich (1987).

152.009 **Massive stars in Cygnus OB2.**
A. B. Thompson, P. Massey.
Symposium on the Formation and Evolution of Star Clusters as Part of the 102. Summer Scientific Meeting of the Astronomical Society of the Pacific, p. 180 (1991). Abstract. – See Abstr. 012.050 for the main entry.

152.010 **A search for red supergiant stars in the northern Milky Way.**
R. Stencel, E. Overgard, K. Garmany, J. P. Casa, R. Wing.
Symposium in Celebration of the 50. Anniversary of McDonald Observatory of the University of Texas at Austin: Frontiers of stellar evolution, p. 621 (1991). Abstract. - See Abstr. 012.085 for the main entry.

152.011 **A stellar diamond in Virgo.**
N. Brosch.
Mon. Not. R. Astron. Soc., Vol. 253, No. 3, p. 545 - 548 (1 Dec 1991).
A system of five stars, containing objects of 12–13 mag and arranged in the form of a diamond, was discovered at a relatively high galactic latitude. Photometry, spectral classification and radial velocities were collected. The stars have similar spectral types and radial velocities, indicating that the system is probably an evaporating small cluster in the Galactic halo.

152.012 **New UBV and Hβ photometry: towards a luminosity function for the Scorpius–Centaurus OB association.**
R. W. Slawson, J. D. Landstreet.
J. R. Astron. Soc. Can., Vol. 85, No. 4, p. 194 - 195 (Aug 1991). Abstract. - See Abstr. 012.091 for the main entry.

152.013 **Hydrogen emission in the direction of h and χ Persei.**
L. V. Mirzoyan, M. Marselan, A. T. Garibdzhanyan, J. Boulesteix, A. L. Mirzoyan, Y. M. Georgelin.
Astrofizika, Tom 34, Vyp. 1, p. 67 - 71 (Feb 1991). With 1 plate. English translation in Astrophysics, Vol. 34, No. 1.
The results of Fabry–Perot observations in h and χ Persei region are presented. The intrinsic H_α emission is detected apparently comming from the OB–association situated on the distance of about 1200 pc. The intensity of the detected H_α–emission is estimated to be around 8 Rayleighs which gives in more usual units an intrinsic monochromatic brightness about 2×10^{-6} erg $cm^{-2}s^{-1}sr^{-1}$.

A model for the infrared emission from an OB star cluster environment.
See Abstr. 063.109.

S Persei: optical and water maser variability – 1984 to 1990.
See Abstr. 112.033.

Observations in the fields of Cassiopeia, Gemini and Monoceros with the Glazar space telescope.
See Abstr. 114.042.

On the physical existence of the ζ Her moving group: a detailed analysis of ϕ^2 Pavonis.
See Abstr. 114.078.

Observation of fields in Cassiopeia, Gemini, and Monoceros with the Glazar space telescope.
See Abstr. 114.088.

The observational H–R diagram and IMF of massive stars.
See Abstr. 115.027.

Evidence that the active stars HD 102077 (V838 Cen) and HD 139084 (V343 Nor) are members of the Local Association.
See Abstr. 116.079.

Spectroscopic binaries, rotation and magnetism in open clusters and associations.
See Abstr. 120.020.

The distribution of early–type stars in the direction of stellar associations Per OB1, Sco OB1 and Cyg OB1.
See Abstr. 121.048.

Distribution of stars of early types in the directions of the stellar associations Per OB1, Sco OB1, and Cyg OB1.
See Abstr. 121.084.

Slow flares in stellar aggregates. III.
See Abstr. 122.032.

The peculiar UV and optical spectrum of the LMC supergiant HD 269216.
See Abstr. 122.036.

High resolution observations of interstellar Na I and Ca II towards the Scorpio–Centaurus association.
See Abstr. 131.005.

An IRAS supershell associated with Cyg OB1.
See Abstr. 131.029.

OB associations and the fossil record of star formation.
See Abstr. 131.206.

MM–wave images of star forming clouds: structure, kinematics, and the conditions for cluster formation.
See Abstr. 131.224.

Interaction of massive stars with the interstellar medium.
See Abstr. 131.225.

OB stars and the structure of the interstellar medium.
See Abstr. 131.227.

An IUE and optical study of interstellar gas towards stars within the Cygnus OB1 and OB3 associations.
See Abstr. 131.245.

A study of the high velocity interstellar gas towards Cygnus.
See Abstr. 131.372.

A study of Sco OB1 and NGC 6231. II. A new analysis.
See Abstr. 153.005.

The role of dense cores in isolated and cluster star formation.
See Abstr. 153.040.

Rotation periods for seven stars in the Trapezium cluster.
See Abstr. 153.081.

LMC–2 as the blowout of a hot superbubble.
See Abstr. 156.029.

Toward an objective identification technique for OB associations in galaxies.
See Abstr. 157.053.

On the distribution of blue stars and gas in M33.
See Abstr. 157.094.

Stellar distribution in H II regions of M33.
See Abstr. 157.258.

153 Open Clusters

153.001 **The statistical method of open star cluster structure study and investigations of cluster dynamics.**
V. M. Danilov, A. F. Seleznev.
Bull. Inf. Cent. Données Stellaires, No. 38, p. 109 – 113 (Dec 1990).
The methods of performance and processing of star counts in the neigbourhood of open star clusters were proposed. On the basis of star counts, completed by the available photometrical data, the catalogue of structural and dynamical characteristics of 103 open clusters was obtained. These data are being used for the cluster dynamics investigation.

153.002 **Liste des étoiles Ap et Am dans les amas ouverts (Edition révisée).**
P. Renson.
Bull. Inf. Cent. Données Stellaires, No. 38, p. 151 – 164 (Dec 1990).
The author gives a revised, and more complete, version of the "Catalogue of Ap and Am Stars in Open Clusters" previously published elsewhere (46.153.037).

153.003 **The proper motions of stars in the open cluster Tr 37.**
E. V. Glushkova.
Pis'ma Astron. Zh., Tom 17, No. 6, p. 513 – 520 (Jun 1991). In Russian. English translation in Sov. Astron. Lett., Vol. 17, No. 3, p. 218 – 222 (1991).
The proper motions of 826 stars to $14^{m}.5$ in the young open cluster Tr 37 have been measured with the accuracy of $0''.0045$/year. 112 cluster members have been selected. A lack of faint stars ($M_V > +0^{m}.4$) among them has been found.

153.004 **Four-colour and Hβ photometry for low mass members in open clusters. I. The Praesepe.**
V. Reglero, J. Fabregat.
Astron. Astrophys., Suppl. Ser., Vol. 90, No. 1, p. 25 – 32 (Oct 1991).
uvby and Hβ photometry has been obtained for 32 stars in the Praesepe cluster. The stars observed were chosen from the proper motion study of Jones and Cudworth with colours ranging between 0.4 and 0.6 in (b–y) and are thus mostly G and K members of the lower main sequence. The related photometric diagrams give values of distance modulus and metal contents compatible with common accepted values. The c_1–(b–y) diagram shows the lack of the Hyades anomaly in the K star range.

153.005 **A study of Sco OB1 and NGC 6231. II. A new analysis.**
C. L. Perry, G. Hill, D. M. Christodoulou.
Astron. Astrophys., Suppl. Ser., Vol. 90, No. 2, p. 195 – 223 (Oct 1991). = Contrib. Lousiana State Univ. Obs., No. 228.
The authors have undertaken a new analysis of the Sco OB1 association, and the two open clusters NGC 6231 and Tr 24, making use of the extensive amount of published data for this region of the southern sky. The average interstellar reddening, due mostly to foreground particulate matter, is E(b–y) = 0.326 mag or E(B–V) = 0.44 mag, with a pronounced north-to-south variation across the association. The distance moduli of the three aggregates are indistinguishable from one another; their mean distance from the Sun is 2.0 kpc. A logarithmic age of 6.9 ± 0.2 years was derived for the systems. Photometric criteria were used to assign different degrees of membership to the B-type stars. Potential variable star candidates were noted. Contrary to previous suggestions, NGC 6231 retains its role as the nucleus of Sco OB1 and, finally, the physical reality of Tr 24 itself is questioned and briefly discussed.

153.006 **Cousins BVRI CCD photometry of stars in the M67 "dipper asterism".**
C. Chevalier, S. A. Ilovaisky.
Astron. Astrophys., Suppl. Ser., Vol. 90, No. 2, p. 225 – 229 (Oct 1991).
The authors present Cousins BVRI CCD photometric data for 29 stars in the M67 "dipper asterism" with a standard deviation per star $\leqslant 1\%$ for stars brighter than V = 13 and 1 – 3% for V = 14. Comparison is made with other data recently published.

153.007 **Tidal sizes of open star clusters in the combined field of the Galaxy and a gas-star complex.**
V. M. Danilov.
Sov. Astron., Vol. 35, No. 3, p. 239 – 246 (May 1991). Current Physics Microform No.: 9201X0521. English translation of Astron. Zh., Tom 68, Vyp. 3, p. 487 – 500 (1991).
The tidal radius of an open star cluster (OSC) in the combined field of the Galaxy and a gas-star complex (GSC) is estimated. The mass of the complex is $\sim 10^7 M_\odot$. Clusters moving together with the GSC and clusters escaping from the GSC have tidal sizes 1.5 – 2.5 and 1.2 – 1.25 times smaller, respectively, than do clusters in the field of the Galaxy alone. The influence of the gravitational field of the Local System on the tidal size of OSCs is discussed as one possible reason for the dependence of their linear diameter on galactocentric distance in the solar neighborhood.

153.008 **Photometric analysis of the young open cluster Hogg 16.**
R. A. Vázquez, A. Feinstein.
Astron. Astrophys., Suppl. Ser., Vol. 90, No. 2, p. 317 – 325 (Oct 1991).
With UBVRI photoelectric photometry the authors investigated the open cluster Hogg 16. The data analysis shows that the cluster is composed by a poor concentrated central star group of 4 arcmin radius around the CPD –60°4703 star even though some members may still be found at a distance of 10 arcmin from it. A mean colour excess $\langle E_{B-V} \rangle = 0.44$ and a true distance modulus $V_O - M_V = 11.65$, equivalent to a distance of d = 2130 pc, were derived. So the cluster is placed within the Sagittarius arm. Its "turn-off" point agrees with an age of $(25.8 \pm 1.5) \times 10^6$yr while from core overshooting models it results to be 23×10^6yr old. A possible relation with the open cluster Cr 272 is briefly discussed.

153.009 **Photometry of suspected low mass Hyades cluster members.**
C. Bryja, G. Lawrence.
Bull. Am. Astron. Soc., Vol. 23, No. 2, p. 908 (1991). Abstract. – See Abstr. 012.008 for the main entry.

153.010 **Preliminary CCD photometry of the open cluster NGC 2355.**
T. Colegrove, N. Moore, A. Schultz, U. Fink.
Bull. Am. Astron. Soc., Vol. 23, No. 2, p. 909 (1991). Abstract. – See Abstr. 012.008 for the main entry.

153.011 **Discovery of an open cluster containing the 22-day classical Cepheid WZ Sagittarii.**
D. G. Turner.
Bull. Am. Astron. Soc., Vol. 23, No. 2, p. 909 (1991). Abstract. – See Abstr. 012.008 for the main entry.

153.012 **VBLUW photometry of the very young open cluster NGC 2264.**
P. T. M. Feldbrugge, A. M. van Genderen.
Astron. Astrophys., Suppl. Ser., Vol. 91, No. 2, p. 209 – 215 (Dec 1991).
VBLUW photometry of 112 stars with V < 13 in the area of the very young cluster NGC 2264, is presented and discussed. Temperatures, gravities and reddening for stars hotter than 8400K are determined. The reddening and the distance are in good agreement with the results of other references. To the

existing list of presumable non-members six stars have been tentatively added as possible non-members.

153.013 **On efficiency of kinematic methods for membership determination of open clusters with small proper motions.**
G. A. Lazorenko, P. F. Lazorenko.
Kinematika Fiz. Nebesn. Tel, Tom 7, No. 4, p. 67 – 72 (Jul–Aug 1991). In Russian. English translation in Kinematics Phys. Celest. Bodies, Vol. 7, No. 4.
The membership determination errors, inherent in some known kinematic methods, used for separation of field stars and cluster members, are discussed. An error analysis has been carried out by means of simulation of real star distribution in the vicinity of two open clusters – their proper motions, angular coordinates and visual magnitudes. It has been shown that, depending on the method applied and on characteristics of star distribution, the separation can lead to losses of up to $S_c = 10 - 60\%$ of the total number of cluster members n. Simultaneously up to $S_\Phi = 0.1n - 2n$ field stars can be erroneously included in the list of cluster members. A method allowing both for kinematic data and visual magnitudes V or V_0 distribution has minimum errors $S_c = (0.1 - 0.4)n$, $S_\Phi = (0.1 - 0.2)n$, and is especially suitable for identification of bright stars. For clusters with small proper motions none of the known methods allows to obtain reliable lists of members on the error level of $S_c + S_\Phi < 0.1n$.

153.014 **The galactic cluster NGC 188: W Ursae Majoris contact binaries as a clue to two separate bursts of star formation.**
C. Maceroni, F. van't Veer.
Astron. Astrophys., Vol. 248, No. 2, p. 430 – 434 (Aug 1991).
The authors revisit the problem of the unique richness in W UMa binaries of the galactic cluster NGC 188, applying their recent results on the dynamical evolution of close binaries. They show how the presence of seven contact binaries and the alleged absence of less close ones finds a straightforward explanation in terms of a scenario developed from the properties of field solar type close binaries. Their model of NGC 188 requires: (i) that the dynamical evolution of late–type MS or solar–type close binaries is governed by the loss and transfer of angular momentum, (ii) that a second burst of star formation has taken place about 2.5 Gyr ago. This second burst can also explain some other characteristics of this peculiar cluster, for instance the presence of several blue stragglers, which may be the results of the delayed star formation.

153.015 **On the physics of lithium depletion.**
E. Schatzman, A. Baglin.
Astron. Astrophys., Vol. 249, No. 1, p. 125 – 133 (Sep 1991).
The authors first discuss the generation by stellar rotation of the turbulent flow that is supposed to generate the mixing process responsible for lithium depletion in main sequence stars as observed in the Hyades. The behaviour of the turbulent diffusion coefficient as a function of the stellar mass excludes the possibility of explaining the lithium abundance in stars of late spectral type with a process only rotation dependant. In the region of the lithium gap, the distribution of the observed velocities of rotation is incompatible with a unique relation between rotational velocity and the lithium abundance.

153.016 **Radial velocities in the open cluster Trumpler 14.**
H. Levato, S. Malaroda, B. García, N. Morrell, G. Solivella, M. Grosso.
Astrophys. Space Sci., Vol. 183, No. 1, p. 147 – 153 (Sep 1991).
The authors present radial velocity measurements of probable members of Trumpler 14, a very young open cluster associated with the Carina Nebula. They have found 4 double–line spectroscopic binaries and 2 single–line spectroscopic binaries. They have succeeded in deriving orbital elements for the SB1s. All the binaries are of potential astrophysical interest and are worth further observational efforts.

153.017 **Photometric binary stars in Praesepe and the search for globular cluster binaries.**
M. Bolte.
Astrophys. J., Vol. 376, No. 2, p. 514 – 519 (1 Aug 1991).
A radial velocity study of the stars which are located on a second sequence above the single–star zero–age main sequence at a given color in the color–magnitude diagram of the open cluster Praesepe (M44) shows that 10, and possibly 11, of 17 stars are binary systems. Of the binary systems, five have full amplitudes for their velocity variations that are greater than 50 km s^{-1}. To the extent that they can be applied to globular clusters, these results suggest that (1) observations of "second–sequence" stars in globular clusters would be an efficient way of finding main–sequence binary systems in globulars, and (2) current instrumentation on large telescopes is sufficient for establishing unambiguously the existence of main–sequence binary systems in nearby globular clusters.

153.018 **Erratum: "Infrared and optical CCD images of the young cluster associated with GM 24" [Astron. Astrophys., Vol. 242, No. 2, p. 388 – 396 (Feb 1991)].**
M. Tapia, M. Roth, J. A. López, M. Rubio, P. Persi, M. Ferrari–Toniolo.
Astron. Astrophys., Vol. 250, No. 1, p. 292 (Oct 1991). See Abstr. 53.153.027.

153.019 **Investigation of the open star cluster NGC 957.**
T. P. Gerasimenko.
Astron. Zh., Tom 68, Vyp. 5, p. 1103 – 1107 (Sep–Oct 1991). In Russian. English translation in Sov. Astron., Vol. 35, No. 5.
A photographic photometry in the Johnson and Morgan UBV system for 250 stars brighter than $V = 14^m5$ in the region of the open star cluster NGC 957 is carried out. The cluster size, extinction of light in the cluster's direction, the distance and cluster age are derived.

153.020 **Photometry and the HR diagram of the open cluster NGC 2286.**
Pan Rongshi, Zhao Junliang, Chen Li, Tian Kaiping, He Yanping.
Acta Astrophys. Sin., Vol. 11, No. 4, p. 336 – 343 (Oct 1991). In Chinese. English translation in Chin. Astron. Astrophys., Vol. 16, No. 1, p. 24 – 32 (Jan – Mar 1992).
Photographic V magnitudes and (B–V) colours in the UBV system are determined for 305 stars in the field of the open cluster NGC 2286, and the rms errors of V and (B–V) for one plate are estimated to be $\sigma_V = 0^m063$ and $\sigma_{B-V} = 0^m084$, respectively. The reddening value of the cluster is $0^m40 \pm 0^m07$. The derived distance modulus to the cluster is approximately $10^m9 \pm 0^m1$. Using the composite diagram method, the age of the cluster is estimated as $6 \cdot 10^7$yr. The HR diagram of the cluster is obtained and discussed. Furthermore, a better determination of membership of the cluster is made based on the proper motion membership probabilities and the photometric data of these 305 stars.

153.021 **Strömgren photometry of reddened B–type stars.**
A. J. Delgado, E. J. Alfaro.
Meeting on Precision Photometry: Astrophysics of the Galaxy, p. 163 (1991). – See Abstr. 012.027 for the main entry.

153.022 **Some remarks on the applicability of Sanders' method to open cluster members selection.**
E. V. Glushkova.
Astron. Zh., Tom 68, Vyp. 6, p. 1190 – 1196 (Nov–Dec 1991). In Russian. English translation in Sov. Astron., Vol. 35, No. 6.
Sanders' method of members selection in open star clusters was analysed. It was shown that this method can be used if the dispersion of proper motions in the cluster and the distance between proper motion centroids of field and cluster stars are small.

153.023 **V, R, I, Hα photometry of NGC 2264.**
C. B. Luginbuhl, F. J. Vrba, S. E. Strom, K. M. Strom.
Meeting on Precision Photometry: Astrophysics of the Galaxy, p. 323 - 326 (1991). - See Abstr. 012.027 for the main entry.

153.024 **Photometry of the young open cluster NGC 581.**
R. L. Phelps, K. A. Janes.
Meeting on Precision Photometry: Astrophysics of the Galaxy, p. 331 - 334 (1991). - See Abstr. 012.027 for the main entry.
The authors have begun a systematic photometric investigation of a group of young open clusters in the Perseus spiral arm. This study will yield information on the time scale of cluster formation in this region of the galaxy, and whether or not cluster formation occurred randomly or sequentially along a preferred direction. The authors report here the preliminary results of the investigation of one of the clusters, NGC 581.

153.025 **CCD Schmidt photometry of NGC 5822.**
W. Weller, B. J. Anthony-Twarog, B. A. Twarog.
Meeting on Precision Photometry: Astrophysics of the Galaxy, p. 345 - 348 (1991). - See Abstr. 012.027 for the main entry.
Though CCD photometry has become the technique of choice for cluster research, the small field of most systems remains unsuitable for some applications, particularly nearby open clusters. As an alternative to using a larger format CCD, the authors have been exploring the option of using a Schmidt telescope with a standard CCD to improve the field size. VBR photometry of the open cluster NGC 5822 shows that simple aperture photometry of the frames provides accuracy in uncrowded fields equal to that of more conventional CCD data.

153.026 **Detection of variable stars among blue stragglers in M67.**
M. Simoda.
Inf. Bull. Variable Stars, No. 3675, p. 1 - 4 (22 Oct 1991).

153.027 **Blue stragglers: a search for binaries in the infrared.**
M. Manteiga, C. Martinez Roger, C. Morales, L. Sabau.
Astron. Astrophys., Vol. 251, No. 1, p. 49 - 58 (Nov 1991).
The authors present infrared photometry in the *J*, *H* and *K* bands, performed on a sample of 33 blue stragglers (BS) belonging to 8 open clusters. This photometry has been combined with published Johnson *UBV* photometry to search for possible cold companions among the BS, as expected for Case B close binary evolution. Probable binaries were identified by comparison with standard mean colour relations for single stars. The number of binary detections is discussed together with the results of a previous spectroscopic search for binaries among BS (Manteiga et al. 1989), and the consequences for the hypothesis of binarity for the origin of BS are analysed.

153.028 **Collinder 223: a moderately young open cluster in the Carina spiral feature.**
J. J. Clariá, E. Lapasset.
Publ. Astron. Soc. Pac., Vol. 103, No. 667, p. 998 - 1004 (Sep 1991).
The authors have secured UBV photoelectric data for 110 stars and DDO data for seven stars in the field of the previously unstudied open cluster Collinder 223. These data show that Cr 223 is a genuine open cluster located at a distance of (2900 ± 270)pc from the Sun and 96 pc below the galactic plane. Nearly 40% of the sample stars - including one red giant, two blue stragglers, and two suspected variables - are found to be cluster members. The reddening across the cluster is not uniform, the mean value being E(B-V) = 0.26 ± 0.03(s.d.). An age of 3.6×10^7yr has been derived, which places Cr 223 within the IC 4665 age group.

153.029 **Magnetic Ap stars in the Hyades and Ursae Majoris superclusters.**
S. Hubrig, H. Schwan.
Astron. Astrophys., Vol. 251, No. 2, p. 469 - 473 (Nov 1991).
The membership of some magnetic Ap stars in the Hyades and UMa superclusters is re-investigated using new highly accurate proper motions most of which have recently become available within the work on the Fifth Fundamental Catalogue, FK5 (Fricke et al. 1988). The improved absolute magnitudes derived from individual distance moduli for probable cluster members have allowed to locate these Ap stars in the HR diagram with much more confidence than previously and to make thus more reliable statements on their evolutionary state.

153.030 **Photometric study of the old open cluster Be 32.**
J. Kałuzny, B. Mazur.
Acta Astron., Vol. 41, No. 3, p. 167 - 190 (1991).
CCD photometry on the UBV and Washington systems is presented for Be 32, a so far unstudied rich open cluster located in the Galactic anticenter. Morphology of the cluster color-magnitude diagram indicates that it is an old object with age close to ages of NGC 188 and NGC 2243 (about 6 Gyr on the scale of VandenBerg isochrones). The cluster reddening is E(B-V) = 0.16 and the estimated metallicity is [Fe/H] = -0.37 ± 0.05. The apparent distance modulus is $12.8 \leqslant (m-M)_v \leqslant 13.1$. Two gaps have been identified on the upper part of the main sequence. Three highly probable candidates for blue stragglers are present, close to the cluster center.

153.031 **The Hyades-age anticenter cluster NGC 2266.**
J. Kałuzny, B. Mazur.
Acta Astron., Vol. 41, No. 3, p. 191 - 212 (1991).
CCD photometry on the UBV and Washington systems is presented for NGC 2266, a so far unstudied open cluster located in the Galactic anticenter. The foreground reddening is calculated to be E(B-V) = 0.10. The color-magnitude diagram reveals a well defined main sequence and the populous red giant clump. The red giant clump is about 0.15 mag bluer than Hyades/Praesepe giants, indicating that NGC 2266 has a metallicity lower than these clusters. The Washington color excess $\Delta(M-T_1)$ implies [Fe/H] = -0.26 ± 0.18 while a comparison of the cluster CMD with the CMD of NGC 1817 suggests [Fe/H] ≈ -0.39. The cluster age is close to the age of the Hyades. For the assumed [Fe/H] = -0.26 the apparent distance modulus of NGC 2266 is 12.95 which corresponds to a position about 3.4 kpc near the direction of the galactic anticenter, and 600 pc above the galactic plane. The contamination of the upper part of the cluster CMD, and especially its main sequence, by field stars is negligible. This makes NGC 2266 a very useful object for testing of theoretical isochrones.

153.032 **Color-magnitude diagrams of the open cluster 092-SC18 and the globular cluster Ruprecht 106.**
M. Kubiak.
Acta Astron., Vol. 41, No. 3, p. 231 - 236 (1991).
Given are the preliminary results of a BVI CCD photometry of the open cluster 092-SC18, for which no earlier data existed. The diagram is too "shallow" to allow any firm conclusions about physical parameters of the cluster but it seems probable that it is an old open cluster, of the age about 8 Gy, lying far, about 1 kpc, from the Galactic plane. The data on Ru 106 are in agreement with recent observations by Buonanno et al. (1990).

153.033 **The Praesepe-age anticenter cluster NGC 2355.**
J. Kałuzny, B. Mazur.
Acta Astron., Vol. 41, No. 4, p. 279 - 290 (1991).
CCD photometry on the UBV system is presented for the central part of NGC 2355, a so far unstudied open cluster located in the galactic anticenter. The color-magnitude diagram (CMD) reveals a main sequence and red giant clump. The foreground reddening is estimated to be E(B-V) = 0.12. The cluster age and metallicity are close to age and metallicity of Praesepe. For the assumed [Fe/H] = 0.13 the apparent distance modulus of NGC 2355 is 12.1 which corresponds to a position about 2.2 kpc

near the direction of the galactic anticenter, and 440 pc above the galactic plane. Eleven candidates for the cluster giants were identified. Examination of the CMD for the comparison field, located 19 arcmin from the cluster center, suggest that majority of these candidates are cluster members. Red giants belonging to NGC 2355 show atypically large spread of colors as compared with populations of giants observed in other clusters with similar ages.

153.034 **Catalogue of a photographic UBV photometry of stars of the old open cluster NGC 188.**
V. N. Frolov.
Astron. Tsirk., No. 1544, p. 39 – 40 (May – Jul 1990).
The UBV magnitudes for 2172 stars within an 80′ × 80′ area centered on the old open custer NGC 188 were determined using plates taken with the Schmidt telescope of the Radio Astrophysical Observatory of the Latvian Academy of Science. The catalogue was recorded on magnetic tape and sent to the Astronomical Data Center of the Astronomical Council of USSR Academy of Sciences.

153.035 **Are galactic open clusters formed from the LMC matter?**
A. A. Suchkov.
Astron. Tsirk., No. 1545, p. 9 – 10 (Aug – Sep 1990).
An analysis of photometric and astrometric data shows that the metal-deficient open clusters of our Galaxy, most probably, have been formed from gas clouds which fell down from the large Magellanic Cloud.

153.036 **Memberships and CM diagrams of young open clusters. I. NGC 225.**
M. G. Lattanzi, G. Massone, U. Munari.
Astron. J., Vol. 102, No. 1, p. 177 – 199 (Jul 1991). Current Physics Microform No.: 9107B1873.
An astrometric, photometric, and spectroscopic investigation of the loose open cluster NGC 225 is presented. The proper motions of 289 stars (cluster + field) have been measured, with errors ranging from 0.10 to 0.25″/century, on PDS scans of astrometric plates spanning a time interval of up to 88 yr. A total of 28 members down to $B = 16.1^m$ have been identified. V magnitude and (B–V) color index have been determined for all 289 stars with mean errors of about 0.04^m and 0.07^m, respectively, while spectral classifications have been obtained for 62 stars. The cluster NGC 225 is of low mass ($60\,M_\odot \leqslant M_{TOT} \leqslant 88\,M_\odot$), young ($t = 1.2 \times 10^8$yr), nearby ($r = 525 \pm 73$ pc), moderately obscured ($\langle E_{B-V} \rangle = 0.25 \pm 0.08$), and with a marked differential reddening. The more massive stars are concentrated towards the cluster core; they seem to move more slowly than the less massive stars, which would indicate relaxation and energy equipartition. Also, a possible runaway has been identified and a real deficiency of less massive stars seems to be present.

153.037 **Reinvestigation of the binary frequency in the open cluster IC 4665.**
N. Morrell, H. A. Abt.
Astrophys. J., Vol. 378, No. 1, p. 157 – 162 (1 Sep 1991).
The authors measured the radial velocities of 15 bright B3 – A2 stars in the cluster IC 4665, using a CCD. Only four of the 15 stars are spectroscopic binaries, giving a binary frequency of 27%. Four of the six sharp-lined stars ($V \sin i \leqslant 50$ km s^{-1}) are binaries, and none of the more rapidly rotating stars are binaries with $K \geqslant 10$ km s^{-1}. Statistically, nearly all of the stars with low projected rotational velocities are slow rotators, and most are binaries; two of the four binaries may rotate synchronously. In two of the four binaries, the mass ratios are near 1.0, as one would expect for binary formation in three-body interactions after many crossing times. The measured cluster velocity dispersion is only 1.6 km s^{-1}, but most of that is still probably due to measuring errors.

153.038 **Metallicities of open clusters.**
A. Strobel.
Astron. Nachr., Vol. 312, No. 3, p. 177 – 196 (1991).
The main aim of this paper is (i) to examine the available metallicity data of open clusters, (ii) to gather a large set of homogeneous estimations of the characteristics, and (iii) to check whether open cluster metallicities are related to spatial and age distribution of these objects.

153.039 **Determination of proper motions in the Pleiades cluster.**
E. Schilbach.
Astron. Nachr., Vol. 312, No. 3, p. 197 – 200 (1991).
For 458 stars in the Pleiades field from the catalogue of Eichhorn et al. (1970) proper motions were derived on Tautenburg and CERGA Schmidt telescope plates measured with the automated measuring machine MAMA in Paris. The catalogue positions were considered as first epoch coordinates with an epoch difference of ca. 33 years to the author's observations. The results show good coincidence of proper motions derived with both Schmidt telescopes within the error bars. Comparison with proper motions determined by Vasilevskis et al. (1979) displays some significant differences but no systematic effects depending on plate coordinates or magnitudes could be found. An accuracy of 0.″3/100 a for one proper motion component was estimated. According to the criterion of common proper motion 34 new cluster members were identified.

153.040 **The role of dense cores in isolated and cluster star formation.**
P. C. Myers.
Symposium on the Formation and Evolution of Star Clusters as Part of the 102. Summer Scientific Meeting of the Astronomical Society of the Pacific, p. 73 – 77 (1991). – See Abstr. 012.050 for the main entry.
Recent observations with imaging arrays indicate that many young massive stars are accompanied by hundreds of previously unknown low-mass stars. The problem of how stars form in clusters has therefore taken on increased importance. Dense cores in embedded clusters tend to have greater size, density, and velocity dispersion than do more isolated cores. These core properties imply that cores in embedded clusters have a substantially greater rate of mass accretion than do isolated cores. This difference in dense core accretion rates should grow more pronounced as a cluster accumulates more matter and forms more stars. Thus, cores in and near a young cluster may resemble isolated, low-mass cores and produce low-mass stars, while cores in a more developed cluster may tend to have more mass, and to produce more massive stars.

153.041 **Height pattern of the formation places of open clusters.**
E. J. Alfaro, J. Cabrera-Caño, A. J. Delgado.
Symposium on the Formation and Evolution of Star Clusters as Part of the 102. Summer Scientific Meeting of the Astronomical Society of the Pacific, p. 133 – 135 (1991). – See Abstr. 012.050 for the main entry.
From the Lyngå's catalog (1981) the authors have selected a sample of 82 open clusters with determined distances and younger than 10^7 years. Their basic idea is to use this sample as a discrete set of explorations of the variable Z in the (X,Y) plane in order to estimate the function Z(X,Y) in the whole area. The dominant feature present in this view is an extended and deep elliptical depression in the third quadrant. A clear relation between large regions of star-forming activity and the Z-structure of the galactic plane is apparent from this study.

153.042 **The age distribution of galactic open clusters.**
P. Battinelli, R. Capuzzo-Dolcetta.
Symposium on the Formation and Evolution of Star Clusters as Part of the 102. Summer Scientific Meeting of the Astronomical Society of the Pacific, p. 139 – 141 (1991). – See Abstr. 012.050 for the main entry.
The fifth edition of the Lyngå "Catalogue of open cluster data" (1987) is used to obtain a number-complete sample of galactic open clusters, which represents a suitable data base to deduce

information on the evolution of the system of open clusters in our Galaxy. From the sample age distribution the authors have deduced: (1) the present-day cluster formation rate, (2) the efficiency of cluster formation; (3) the destruction function; (4) an evaluation of cluster lifetimes.

153.043 **The open cluster NGC 7419.**
A. Beauchamp, A. F. J. Moffat.
Symposium on the Formation and Evolution of Star Clusters as Part of the 102. Summer Scientific Meeting of the Astronomical Society of the Pacific, p. 142 – 144 (1991). – See Abstr. 012.050 for the main entry.

Photometry of the young open cluster NGC 7419 has been obtained leading to a distance ≈3.6 kpc behind 6.8 magnitudes of visual extinction. Radial velocities confirm the membership of 5 red supergiants. However, membership of the carbon star located within the boundary of the cluster is still problematic.

153.044 **Age calibration and age distribution of LMC clusters.**
A. K. Pandey, H. S. Mahra.
Symposium on the Formation and Evolution of Star Clusters as Part of the 102. Summer Scientific Meeting of the Astronomical Society of the Pacific, p. 167 – 169 (1991). – See Abstr. 012.050 for the main entry.

Using the model of Pandey, et al. (1989) ages of the clusters in the LMC have been obtained. The resulting age distribution of LMC clusters is quite different from the age distribution of galactic clusters and supports the conclusion of a burst in the cluster formation rate about 3 Gyr ago.

153.045 **A study of the spatial variation of mass functions in some young open clusters.**
A. K. Pandey, H. S. Mahra, R. Sagar.
Symposium on the Formation and Evolution of Star Clusters as Part of the 102. Summer Scientific Meeting of the Astronomical Society of the Pacific, p. 170 – 172 (1991). – See Abstr. 012.050 for the main entry.

The effect of mass segregation on cluster mass functions has been studied and it is found that the slope of the mass function is steeper in the outer parts of open clusters. Since dynamical relaxation time is always larger than the age of the clusters, it is inferred that the observed mass segregation might have taken place at the time of cluster formation.

153.046 **The young open cluster Be 87 and its associated Wolf–Rayet star as a possible source of γ-rays.**
R. L. Phelps, K. A. Janes.
Symposium on the Formation and Evolution of Star Clusters as Part of the 102. Summer Scientific Meeting of the Astronomical Society of the Pacific, p. 173 – 175 (1991). – See Abstr. 012.050 for the main entry.

The authors present a model to explain the observed γ-ray flux from the direction of the young open cluster Be 87. Previous photometry has shown an intervening "cloud" to be present, possibly associated with Be 87 and the ON2 star formation region. The presence of stellar winds and likely shocks from numerous early-type stars, particularly the WO class Wolf–Rayet star ST3 in or near Be 87, provides a source of cosmic rays for the production of the observed γ-ray flux by the interaction of the stellar winds with the foreground cloud. The model also supports the photometric assignment of membership of ~946 pc for the ON2 region.

153.047 **CCD photometry of the old open clusters King 11 and Be 42. Preliminary results.**
A. Aparicio, G. Bertelli, C. Chiosi, J. M. Garcia–Pelayo.
Symposium on the Formation and Evolution of Star Clusters as Part of the 102. Summer Scientific Meeting of the Astronomical Society of the Pacific, p. 230 – 232 (1991). – See Abstr. 012.050 for the main entry.

The authors present a brief resume of the main preliminary results obtained from the photometric study of the rich open clusters King 11 and Be 42. The C–M diagram of King 11 is discussed in detail with the aid of synthetic diagrams and models incorporating convective overshoot.

153.048 **CCD photometry of the galactic open cluster NGC 2243.**
A. Bonifazi, M. Tosi, F. Fusi Pecci, G. Romeo.
Symposium on the Formation and Evolution of Star Clusters as Part of the 102. Summer Scientific Meeting of the Astronomical Society of the Pacific, p. 233 (1991). Abstract. – See Abstr. 012.050 for the main entry.

153.049 **New Wolf–Rayet stars in galactic open clusters: Sher 1 and the giant H II region core Westerlund 2.**
A. F. J. Moffat, M. M. Shara, M. Potter.
Astron. J., Vol. 102, No. 2, p. 642 – 653 (Aug 1991). Current Physics Microform No.: 9108A1578.

The authors have found two new Galactic Wolf–Rayet stars in open clusters: a WN4 star in the O9 cluster Sher 1 and a WN7 star in the O7 cluster Westerlund 2. This confirms a previous trend, namely that fainter, hotter WN stars tend to be older than brighter, cooler WN stars. This may be a consequence of evolution via extreme mass loss.

153.050 **The luminosity function of the young cluster NGC 2362.**
D. J. Wilner, C. J. Lada.
Astron. J., Vol. 102, No. 3, p. 1050 – 1055 (Sep 1991). Current Physics Microform No.: 9109D1454.

The authors construct the I band luminosity function of the young open cluster NGC 2362 from CCD photometric observations which are complete to a limiting magnitude of 17.6. After empirically correcting for the presence of background field stars, they find the cluster luminosity function to be significantly deficient in low luminosity stars compared to the luminosity function derived from the IMF. Although it is impossible to entirely dismiss systematic uncertainties in the correction for field stars, artificial star tests confirm that the deficiency in the cluster luminosity function is not due to observational incompleteness at faint magnitudes. If the observed luminosity function is representative of the luminosity function of the entire cluster, then the authors' observations provide strong evidence that the IMF varies on the scale of individual open clusters. However, their data does not allow the authors to rule out the existence of an outer halo of low luminosity stars that could account for this deficit. The authors also find marginal evidence for a dip in the observed luminosity function at magnitudes corresponding to ZAMS spectral types mid-A to mid-F, a characteristic that has been observed in at least one other cluster of similar age.

153.051 **A BV photographic and CCD analysis of the intermediate-age open cluster NGC 3680.**
B. J. Anthony–Twarog, E. A. Heim, B. A. Twarog, N. Caldwell.
Astron. J., Vol. 102, No. 3, p. 1056 – 1069 (Sep 1991). Current Physics Microform No.: 9109D1460. With plate 35.

New photoelectric and CCD data on the BV system have been obtained in the open cluster, NGC 3680, and used to calibrate a photographic survey of over 490 stars down to V = 17.6 in the field of the cluster. The cluster color–magnitude diagram exhibits a well populated red giant clump, but effectively has no red giant branch. The turnoff region has a sharp cutoff at (B–V) = 0.4, with a red hook and a large amount of scatter, consistent with the explanation of a significant binary population. Comparison with model field star counts for this region of the sky demonstrates that the cluster main sequence is, at best, poorly populated below V = 14.0 (M_V = 4.0). Adopting E(B–V) = 0.05, [Fe/H] = 0.1, and (m–M) = 10.0, the standard isochrones of VandenBerg (1985) produce an age of $1.9 \pm 0.5 \times 10^9$ yr, but appear unable to match the morphology of the turnoff region. The discrepancy is removed when the comparison is made to isochrones which include convective overshoot. An approximate transformation relation between the respective age scales leads to an age of 4.5×10^9 yr for NGC 3680 on a system where the Hyades is 1.2×10^9 yr old and M67 is between 6 and 7×10^9 yr old. Comparison to the color–magnitude diagrams of clusters of similar metallicity indicates that NGC 3680 is intermediate in age to the Hyades and IC 4651, but only slightly younger than IC 4651, consistent with the color distribution of the red giants. The comparisons to both theory and observation imply that the distance modulus for the cluster needs to be increased to (m–M) = 10.3.

153.052 **On the metallicity and the turnoff temperature of M67.**
L. M. Hobbs, J. A. Thorburn.
Astron. J., Vol. 102, No. 3, p. 1070 – 1079 (Sep 1991). Current Physics Microform No.: 9109D1474.

High–dispersion echelle spectra of five stars located along the main sequence of M67 are analyzed, using the model atmospheres of Kurucz. An average iron abundance of the cluster, relative to the Sun, of [Fe/H] = -0.04 ± 0.12 is derived from the observed equivalent widths of up to four weak Fe II lines in each of the five spectra and in the solar spectrum. With this known relative iron abundance, the equivalent widths of up to six Fe I lines in these same spectra then require that the effective temperatures of the two brightest, hottest program stars average $\langle T_e \rangle_2 = 6165 \pm 60$K. This average temperature approximates satisfactorily the upper limit on the temperatures actually reached in the turnoff region. The uncertain Fe II oscillator strengths were calibrated by requiring that, for the physically very similar Sun, the identical analysis recover the correct effective temperature from the same Fe II and Fe I lines. Finally, a turnoff age of $\tau = 5.2 \pm 1.0$ Gyr is derived by comparing the limiting turnoff temperature $\langle T_e \rangle_2$ to appropriate isochrones in the theoretical H–R diagram, as calculated by VandenBerg (1990, private communication). A primary advantage of this purely spectroscopic method is that it effectively requires no estimates of the reddening or the intrinsic colors of the cluster stars or of the distance to the cluster; the attendant uncertainties in these cluster parameters are thereby circumvented entirely.

153.053 **The luminosity function of Praesepe. I. A proper motion and photometric search for candidate members.**
B. F. Jones, J. R. Stauffer.
Astron. J., Vol. 102, No. 3, p. 1080 – 1090 (Sep 1991). Current Physics Microform No.: 9109D1484.

The authors have measured proper motions for a color selected sample of stars in a square 4°x4° region centered on the Praesepe open cluster. The magnitude limit of the photometry is V ~ 19 and of the proper motions V ~ 18. The authors present a list of 765 probable and possible members from V = 9 to 18 based on the proper motions, magnitudes, and colors. The authors compare their results with other proper motion surveys, and calculate a preliminary luminosity function to $M_v = 11$.

153.054 **Evidence of an interacting stellar wind and possible high energy emission in the open cluster Berkeley 87.**
V. F. Polcaro, F. Giovannelli, R. K. Manchanda, L. Norci, C. Rossi.
Astron. Astrophys., Vol. 252, No. 2, p. 590 – 596 (Dec 1991).

The interaction of the strong (≈ 5200 km s^{-1}) stellar wind from the Wolf–Rayet star ST 3, belonging to the open cluster Berkeley 87, with the cluster stars and molecular clouds appears to produce dissipative shock waves. Optical spectroscopy and imaging of the cluster support this hypothesis. These shocks can give rise to the hard X–ray flux consistent with that detected by EXOSAT in the cluster sky area. Shock waves can accelerate the ambient cosmic ray particles, emitted by the numerous early type stars contained in the cluster, up to highly relativistic energies. This process can possibly explain the observation of high energy emission of the COS–B source 2CG 075+00 whose error box is centered on Berkeley 87.

153.055 **A Hyades distance modulus from trigonometric parallaxes from northern and southern hemisphere observatories.**
R. J. Patterson, P. A. Ianna.
Astron. J., Vol. 102, No. 3, p. 1091 – 1102 (Sep 1991). Current Physics Microform No.: 9109D1495.

New trigonometric parallaxes of high precision are presented for 10 members of the Hyades. Independent measurements of the parallax of each star were made at Leander McCormick and Mount Stromlo Observatories. No evidence for a statistically significant difference between the parallaxes obtained at the two observatories is found. The eighteen parallaxes (ten stars) with the greatest precision yield a weighted distance modulus of $3^m.18 \pm 0^m.09$.

153.056 **More radial–velocity measurements in young open clusters.**
T. Liu, K. A. Janes, T. M. Bania.
Astron. J., Vol. 102, No. 3, p. 1103 – 1107 (Sep 1991). Current Physics Microform No.: 9109D1507.

The authors report further high resolution radial–velocity measurements in 23 young open clusters using the Kitt Peak CCD coudé spectrograph on the 0.9 m feed telescope. The radial velocities for the cluster stars are derived with the technique of cross correlation. The internal precision of the velocity measurements is typically 2 km s^{-1} for early type stars. From these new data and their previously published velocities, the authors find that the observed stars in two clusters, NGC 663 and NGC 2287, showed a relatively small dispersion in the measured mean velocities. Further observations of stars in young clusters will be useful in helping to establish an early type star velocity standard system.

153.057 **Rotational velocities of low mass stars in young clusters.**
J. R. Stauffer.
NATO Advanced Research Workshop on Angular Momentum Evolution of Young Stars, p. 117 – 134 (1991). – See Abstr. 012.055 for the main entry.

Reviews of stellar rotation written prior to 1980 included essentially no information on the rotational velocities of stars later than G0 other than that they were generally slow rotators. Due to the advent of new instrumental capabilities, a large body of data on the rotational velocities of low mass stars is now available. In particular, several nearby clusters have good rotational velocity data for low mass stars, allowing us to estimate the evolution of rotation on and near the main sequence. Somewhat unexpectedly, many low mass stars have very large rotational velocities when they arrive on the main sequence. However, these stars spin down rapidly so that by the age of the Hyades essentially all of the low mass stars are relatively slow rotators. This paper provides a brief review of the "classical" rotation studies of the 1960's, an outline of the new discoveries of the 1980's and their current interpretation, and a description of the outstanding problems that remain.

153.058 **Rotational velocities of stars in open clusters: the time dependence revisited.**
M. Mayor, J.–C. Mermilliod.
NATO Advanced Research Workshop on Angular Momentum Evolution of Young Stars, p. 143 – 150 (1991). – See Abstr. 012.055 for the main entry.

Using well defined rotational velocity distributions of G–type stars in the Hyades plus Praesepe and Pleiades clusters, the authors have reexamined the age dependence of stellar rotation. The shape of the $v \sin i$ distribution is an extremely sensitive function of the α index of the braking law ($dv/dt = -Av^{\alpha}$). The comparison of Hyades plus Praesepe and Pleiades observations favours the value $\alpha = 2$. The $v \sin i$ distributions could not be reproduced with $\alpha = 3$, the value corresponding to the so–called $t^{-1/2}$ law.

153.059 **A radial velocity and photometric analysis of NGC 752.**
D. W. Latham, S. A. Daniel, B. A. Twarog, R. D. Mathieu.
Symposium on the Formation and Evolution of Star Clusters as Part of the 102. Summer Scientific Meeting of the Astronomical Society of the Pacific, p. 275 – 277 (1991). – See Abstr. 012.050 for the main entry.

153.060 **Studies of open clusters at Uttar Pradesh State Observatory.**
H. S. Mahra, A. K. Pandey.
Symposium on the Formation and Evolution of Star Clusters as Part of the 102. Summer Scientific Meeting of the Astronomical Society of the Pacific, p. 282 – 284 (1991). – See Abstr. 012.050 for the main entry.

Based on UBV observations, the parameters such as reddening, memberships, distance and age of a number of clusters have been obtained. Several studies related to integrated parameters, age distribution and lifetimes, stability and structure, star formation efficiency in open clusters and galactic structure have been carried out and some important results have been discussed.

153.061 **The main sequence gap in intermediate–age and old open clusters.**
L. Pigatto.
Symposium on the Formation and Evolution of Star Clusters as Part of the 102. Summer Scientific Meeting of the Astronomical Society of the Pacific, p. 287 – 289 (1991). – See Abstr. 012.050 for the main entry.

The main sequences of four famous intermediate–age old open clusters, i.e. Praesepe, NGC 752, M67 and NGC 188, are examined in order to interpret the gap by comparison with recent evolutionary sequences.

153.062 **CCD photometry with a Schmidt telescope.**
W. G. Weller, B. J. Anthony–Twarog, B. A. Twarog.
Symposium on the Formation and Evolution of Star Clusters as Part of the 102. Summer Scientific Meeting of the Astronomical Society of the Pacific, p. 302 – 304 (1991). – See Abstr. 012.050 for the main entry.

The authors report on preliminary results of a study of the open cluster NGC 5822 by means of a CCD photometry with a Schmidt telescope.

153.063 **Spectroscopic investigation of OB stars of open clusters NGC 2244 and NGC 2264.**
S. A. Pyatkes.
Bull. Spec. Astrophys. Obs. – North Caucasus, Vol. 32, p. 89 – 103 (1990). English translation of Astrofiz. Issled. Izv. Spets. Astrofiz. Obs., Tom 32, p. 106 – 124 (1990).

Equivalent widths for 10 stars of cluster NGC 2244 and eight stars of cluster NGC 2264 were determined from spectrograms obtained on the 6–m telescope in the $\lambda 3700 - 4900$ Å region (spectral resolution 0.5 and 0.8 Å). The quantitative spectral classes, absolute magnitudes, and rotational velocities of the investigated stars were found. The parameters T_θ and lg g were determined by the method of model atmospheres. Possible peculiarity of six objects of the program was found. The spectroscopic distance moduli obtained in work exceed all those determined earlier (V_0–M_V), which is a consequence of using the new calibration of M_V and taking into account the anomalous absorption law for NGC 2264.

153.064 **Radial velocities of turnoff stars in two intermediate–age open clusters.**
J. Andersen, B. Nordström.
Symposium on the Formation and Evolution of Star Clusters as Part of the 102. Summer Scientific Meeting of the Astronomical Society of the Pacific, p. 357 – 359 (1991). – See Abstr. 012.050 for the main entry.

The authors have studied membership and duplicity of turnoff stars in IC 4651 and NGC 3680 from CORAVEL radial velocities. Overshooting models appear to match the true cluster sequences best. However, their main conclusion is that only with proper non–member and binary identification can one reliably fit critical details of computed isochrones to observed cluster color–magnitude diagrams.

153.065 **The LkHα 101 infrared cluster.**
M. Barsony, J. M. Schombert, K. Kis–Halas.
Astrophys. J., Vol. 379, No. 1, p. 221 – 231 (20 Sep 1991). With plates 2 – 3.

An $8' \times 12'$ area, corresponding to a 1.9×2.8 pc field centered on the Herbig Ae/Be star, LkHα 101, has been mapped in the near–infrared J and K bandpasses at 1".35 pixel resolution using IRIM on the KPNO 1.3 m telescope. The authors report the discovery of an embedded cluster of more than 100 young stellar objects within this area. The completeness limits are 16.5 mag at J and 14.5 mag at K, although sources as faint as K = 16.0 and J = 17.5 were detected. The authors have identified four types of cluster members, defined by their location in the K versus J–K color magnitude diagram: (1) accreting protostars, (2) probable accreting protostars, (3) embedded, low–mass, pre–main–sequence (PMS) stars younger than 10^6yr, and (4) young ($\approx 10^6$yr), embedded, brown dwarf candidates. According to this classification scheme, the embedded cluster population of the LkHα 101 cloud core consists of 16 accreting protostars; 39 probable accreting protostars; 40 young, embedded, low–mass, PMS stars; and 46 young, embedded, brown dwarf candidates. The spatial distribution of the sources in the cloud core is highly nonuniform, such that the youngest sources are the most centrally concentrated. The star formation efficiency in the central $2' \times 2'$ of this cluster is $\approx 18\%$. In the densest parts of the cluster, the observed stellar density places a limit of 1 star per $\approx (2.7 \times 10^{17}\mathrm{cm})^3$ for the initial volume of gas which can eventually form a star. This volume corresponds to initial masses of 0.045, 0.45, and 4.5 $M_\odot$ for uniform densities of $\varrho_{H2} = 10^3, 10^4$, and $10^5\mathrm{cm}^{-3}$, respectively. An upper limit of 4×10^5yr is derived for the protostellar collapse phase in this cluster.

153.066 **Determination of proper motions in the Pleiades with Schmidt telescopes.**
E. Schilbach.
International Seminar on Problems of Astrometry and Space Geodynamics, p. 104 – 105 (1991). – See Abstr. 012.049 for the main entry.

Full text will be published in Astron. Nachr. (1991).

153.067 **Observational confirmation of mass segregation in the dynamically well–evolved open cluster Roslund 3.**
D. G. Turner.
Symposium on the Formation and Evolution of Star Clusters as Part of the 102. Summer Scientific Meeting of the Astronomical Society of the Pacific, p. 452 – 454 (1991). – See Abstr. 012.050 for the main entry.

Star densities are presented for the sparse, intermediate–age, open cluster Roslund 3, for which cluster membership has been established for stars with $B < 15.5$ from a photometric study. Mass segregation is apparent for the ~ 73 cluster members with $B < 19$.

153.068 **Image restoration of R136: the dense core of NGC 2070.**
N. Weir, S. Djorgovski, G. Meylan.
Symposium on the Formation and Evolution of Star Clusters as Part of the 102. Summer Scientific Meeting of the Astronomical Society of the Pacific, p. 455 – 457 (1991). – See Abstr. 012.050 for the main entry.

The authors present results from the application of a new seeing deconvolution technique to images of the dense core of NGC 2070 = 30 Doradus (the controversial object R136a). They utilize a new maximum entropy algorithm capable of restoring images to very high resolution, even to sub–pixel accuracy. Images of the object in different colors, as well as comparisons with previous speckle results, have been used to check the validity of the results. The images of R136 have an angular resolution approaching 0.2", better than any direct images of this object published to date, and corroborate evidence that R136a is a dense star cluster core, rather than a single ultraluminous object.

153.069 **Light elements as tracers of chemical evolution.**
A. M. Boesgaard.
Symposium on the Formation and Evolution of Star Clusters as Part of the 102. Summer Scientific Meeting of the Astronomical Society of the Pacific, p. 463 – 478 (1991). – See Abstr. 012.050 for the main entry.

The composition of galactic clusters can now reveal information about the evolution of the galactic disk because of a new level of precision in stellar abundance determinations. Many parameters such as age, metal content, stellar rotation influence the light element abundances. The significance of various parameters can be determined from the chemical composition of stars in clusters.

153.070 **Metallicities and kinematics of old open clusters.**
E. D. Friel, K. A. Janes.
Symposium on the Formation and Evolution of Star Clusters as Part of the 102. Summer Scientific Meeting of the Astronomical Society of the Pacific, p. 569 – 572 (1991). – See Abstr. 012.050 for the main entry.
The authors present preliminary results from a spectroscopic study of a sample of evolved stars in 24 open clusters with ages greater than roughly 1 billion years. Measured velocities and metallicities allow to investigate the kinematics, abundance gradients, and age–metallicity relationship of the old cluster system.

153.071 **The IC 2391 supercluster.**
O. J. Eggen.
Astron. J., Vol. 102, No. 6, p. 2028 – 2040 (Dec 1991). Current Physics Microform No.: 9112C2004.
Sixty–three field stars and the sparse cluster IC 2391 have been identified as members of the IC 2391 supercluster. The members have at least 95% of their space motion, directed toward (A,D) = ($5^h.82$, $-12°.44$). Like members of the very similar Pleiades supercluster, the IC 2391 supercluster contains objects of two ages: $t = 8x10^7$ and $2.5x10^8$yr. A prototype very active chromosphere (VAC) star and the enigmatic giant variable TZ For are supercluster members together with two A–type stars suspected of possessing particulate disks. An especially interesting member is the system of Xi Cep (HR 8417) in which the two brightest components are greatly undermassive.

153.072 **On the tidal sizes of open stellar clusters in the joint field of the Galaxy and of a gas–star complex.**
V. M. Danilov.
Astron. Tsirk., No. 1545, p. 7 – 8 (Aug – Sep 1990). In Russian.
The results of estimates of the tidal radius of open star clusters in the joint field of the Galaxy and of a gas–star complex are presented in graphic form.

153.073 **Open star cluster NGC 957.**
T. P. Gerasimenko.
Sov. Astron., Vol. 35, No. 5, p. 553 – 555 (Sep 1991). Current Physics Microform No.: 9203X1727. English translation of Astron. Zh., Tom 68, Vyp. 5, p. 1103 – 1107 (1991).
The size of the open star cluster NGC 957, the extinction in its direction, and its distance and age are determined on the basis of photographic photometry in the Morgan–Johnson UBV system for 250 nearby stars brighter than V = $14^m.5$.

153.074 **Moving cluster parallaxes and determination of the distance of Hyades cluster.**
Chen Li, Zhao Junliang.
Prog. Astron., Vol. 9, No. 3, p. 221 – 227 (Sep 1991). In Chinese.
A brief review is made on various methods of determining moving cluster parallaxes using proper motion and radial velocity data, and some characteristics of those methods are discussed. Finally, a comparison among various determinations of the distance of the Hyades cluster is made.

153.075 **A deep proper motion survey of the Pleiades for low–mass stars and brown dwarfs.**
N. C. Hambly, M. R. S. Hawkins, R. F. Jameson.
Mon. Not. R. Astron. Soc., Vol. 253, No. 1, p. 1 – 8 (1 Nov 1991).
The authors have used COSMOS measures of POSS and recent UKSTU red photographic Schmidt plates to produce a deep proper motion survey of the Pleiades open cluster. Follow–up infrared photometry of a small sample of candidate low–mass stars and brown dwarfs is presented. In addition to producing a deep luminosity function and mass function, the authors find a number of very interesting low–luminosity stars and comparison is made with recent theoretical predictions and observations of very low mass stars and brown dwarfs.

153.076 **A possible galactic supernova progenitor.**
L. Norci, C. Rossi, F. Giovannelli, V. F. Polcaro.
Vulcano Workshop 1988: Frontier objects in astrophysics and particle physics, p. 39 – 46 (1989). – See Abstr. 012.079 for the main entry.
V439 Cyg is a peculiar star which lies in the young open cluster Berkeley 87. Objective prism spectra of V439 Cyg taken in 1958 were characterized by C_2, CH and CN bands and by deep absorptions on Balmer lines, while, up–to–date, its spectrum is hot, showing H_α and H_β in strong emission, H_γ and H_δ in absorption with a narrow central emission and numerous absorption lines of He I, He II, C II, O II, Si III, N II and N III. The general appearance of the optical spectrum looks like the only recorded one of Sk –69°202 ten years before the SN 1987A explosion and the star position in the CM diagram of the cluster is absolutely anomalous. The peculiarity of the object is enhanced by the fact that V439 Cyg is identified as the strong infrared source IRAS 20198+3716.

153.077 **Precise positions of reference stars in open clusters NGC 457 and NGC 581.**
N. M. Evstigneeva, Yu. A. Shokin, A. O. Yuferov.
Tr. Gos. Astron. Inst. im. Shternberga, Tom 62, p. 234 – 242 (1991). In Russian.
Precise positions of 145 reference stars in open clusters NGC 457 and NGC 581 are determined.

153.078 **Low–mass stars in the region of the Hyades cluster.**
S. K. Leggett, M. R. S. Hawkins.
Symposium in Celebration of the 50. Anniversary of McDonald Observatory of the University of Texas at Austin: Frontiers of stellar evolution, p. 604 (1991). Abstract. – See Abstr. 012.085 for the main entry.

153.079 **Near–infrared and Strömgren photometry of the open clusters NGC 663, NGC 1502 and NGC 1893.**
M. Tapia, R. Costero, J. Echevarria, M. Roth.
Mon. Not. R. Astron. Soc., Vol. 253, No. 4, p. 649 – 661 (15 Dec 1991).
Near–infrared JHK, and Strömgren uvby and Hβ photometry of the central parts of three open clusters, NGC 663, NGC 1502 and NGC 1893, are presented. The following results were derived from the data: NGC 663 is at a distance of 2.5 kpc ($m-M = 12.03 \pm 0.20$) and is obscured by $A_V = 1.98 \pm 0.04$, its age is $9x10^6$yr; NGC 1502 is at a distance of 1.5 kpc ($m-M = 10.84 \pm 0.10$) and is obscured by $A_V = 1.83 \pm 0.03$, its age is $5x10^6$yr; NGC 1893 is at a distance of 4.3 kpc ($m-M = 13.18 \pm 0.11$) and is obscured by $A_V = 1.68 \pm 0.06$, its age is $4x10^6$yr. The total–to–selective extinction ratio of V, $R_V = A_V/E(B-V)$, is slightly smaller towards NGC 663 and 1893 than the average galactic value, but towards NGC 1502, it is found to be 2.42 ± 0.09 (root mean squared), significantly lower than that characteristic of the general interstellar medium. This result suggests that the size distribution of the dust particles in this line of sight is biased towards very small values.

153.080 **New short period variable stars in the open cluster NGC 6882–5.**
R. Peniche, J. H. Pena, R. Garrido.
ESO Workshop on Rapid Variability of OB–Stars: Nature and Diagnostic Value, p. 95 – 96 (1991). – See Abstr. 012.092 for the main entry.

153.081 **Rotation periods for seven stars in the Trapezium cluster.**
G. N. Mandel, W. Herbst.
Astrophys. J., Lett., Vol. 383, No. 2, p. L75 – L78 (20 Dec 1991).
The Trapezium cluster in the Orion Nebula has been monitored on 25 nights from November 1990 to March 1991, with a CCD attached to the Perkin telescope at Van Vleck Observatory. Photometry in the I band of more than 150 stars has been obtained, and 28 showed significant variation during this study.

A periodogram analysis reveals that seven of these have significant peaks in their power spectra, which are interpreted as the rotation periods of the stars. Periods range from 6.2 to 14.3 days. Three of them have periods longer than any previously detected for pre-main-sequence objects, and the group average, 8.7 days, is more than twice as long as the average period of previously measured T Tauri stars. Possible explanations for this result are briefly considered.

153.082 **IUE observations of solar-type stars in the Pleiades and the Hyades.**
J.-P. Caillault, O. Vilhu, J. L. Linsky.
Astrophys. J., Vol. 383, No. 2, p. 594 - 601 (20 Dec 1991).
The authors have made the first extensive set of IUE observations of solar-type stars (spectral types F5 - G5) in the Pleiades. Spectra were obtained in 1988 January and August for both the transition region (C IV 1550 Å; low-resolution) and chromospheric (Mg II 2800 Å; high-resolution) emission wavelength regions, respectively. The Pleiades solar-type stars lie at the extreme limit of what IUE can observe in very long exposures. The authors were able to detect Mg II fluxes for two out of three Pleiades stars and interesting C IV upper limits for two of these stars. Long-wavelength, high-resolution spectra were also obtained for previously unobserved solar-type stars in the Hyades. With the inclusion of spectra of additional Hyades stars obtained from the IUE archives, the authors have calculated surface fluxes and fractional luminosities for both clusters' solar-type stars. These values provide a better estimate for the Mg II saturation line for single (as opposed to contact-binary) stars. The spectra of such young stars (Pleiades $\sim 10^{7.8}$ yr; Hyades $\sim 10^{8.8}$yr) extend the main-sequence age baseline for activity - age relations over more than two orders of magnitude.

153.083 **Monte-Carlo simulation of open cluster colour-magnitude diagrams.**
D. Weltman, M. P. Vatter FitzGerald.
J. R. Astron. Soc. Can., Vol. 85, No. 4, p. 199 - 200 (Aug 1991). Abstract. - See Abstr. 012.091 for the main entry.

153.084 **Some remarks on the limits of applicability of the Sanders method of identifying cluster members.**
E. V. Glushkova.
Sov. Astron., Vol. 35, No. 6, p. 596 - 599 (Nov 1991). Current Physics Microform No.: 9205X1406. English translation of Astron. Zh., Tom 68, Vyp. 6, p. 1190 - 1196 (1991).
The Sanders method of identifying members of an open cluster from their proper motions is analyzed. This method can be used if the variance of proper motion of cluster stars is small, as is the distance between the centers of the distributions of proper motion of cluster stars and field stars.

153.085 **Membership of the WR binary V444 Cygni in the open cluster Berkeley 86.**
D. Forbes.
J. R. Astron. Soc. Can., Vol. 85, No. 4, p. 213 (Aug 1991). Abstract. - See Abstr. 012.091 for the main entry.

153.086 **Detection of a circumstellar reddening component in the extinction of late B-type stars.**
D. G. Turner.
J. R. Astron. Soc. Can., Vol. 85, No. 4, p. 216 (Aug 1991). Abstract. - See Abstr. 012.091 for the main entry.

153.087 **The open cluster NGC 5662 and its neighbouring cepheid V Centauri.**
J. J. Clariá, E. Lapasset, M. A. Bosio.
Bol. Asoc. Argent. Astron., No. 36, p. 15 - 27 (1990). - See Abstr. 012.095 for the main entry.
Based on new photometric and kinematical data the structure and stellar composition of the open cluster NGC 5662 is examined. The presence of differential reddening across the cluster field is detected and the cluster stability examined. The luminosity function, distance, age, and chemical composition of NGC 5662 are determined from photometric and spectroscopic data. The new astrophysical parameters derived support the conclusion that the cepheid V Cen (P = 5.5 days) is very likely a cluster member.

153.088 **Integrated luminosity distribution of galactic open clusters.**
B. C. Bhatt, A. K. Pandey, H. S. Mahra.
J. Astrophys. Astron., Vol. 12, No. 3, p. 179 - 185 (Sep 1991).
The integrated magnitudes of 221 Galactic open clusters have been used to derive the luminosity function. The completeness of the data has also been discussed. In the luminosity distribution the maximum frequency of clusters occurs near $I(M_v) = -3^m.5$, and some plausible reasons for a sharp cut-off at $I(M_v) = -2^m.0$ have been discussed. It is concluded that the paucity of the clusters fainter than $I(M_v) = -2^m.0$ is not purely due to selection effects.

153.089 **Study of faint young open clusters as tracers of spiral features in our galaxy. Paper 5: NGC 2236 (OCl 501).**
G. S. D. Babu.
J. Astrophys. Astron., Vol. 12, No. 3, p. 187 - 198 (Sep 1991).
Continuing the study of faint young open clusters as tracers of spiral features in our Galaxy, photoelectric and photographic photometry of 39 stars was done in the field of the faint open cluster NGC 2236 ≡ OCl 501 in the direction of Monoceros constellation. Out of these stars, a total of 22 down to $m_v \cong 15.4$ have been found to be probable members. The median age of this cluster is estimated to be 7.6×10^7years and the cluster is thereby considered as belonging to the marginally old category. Thus, it cannot be specifically used as a spiral arm tracer in the study of our Galaxy.

The formation and evolution of star clusters. Proceedings.
See Abstr. 012.050.

Study of galactic and extra galactic star clusters with a large ground based optical telescope.
See Abstr. 013.028.

Minimization of the small powers of deviations as one of the robust methods.
See Abstr. 036.009.

Photometrie mit der ST-4 Kamera.
See Abstr. 036.016.

Towards a search for a taxonomy of celestial objects. I. Open clusters.
See Abstr. 036.300.

J2000.0 positions and proper motions of 257 stars in the central part of the Praesepe astrometric standard region.
See Abstr. 041.009.

J2000.0 positions and proper motions of 257 stars in the central part of the Praesepe astrometric standard region.
See Abstr. 041.023.

A model for the infrared emission from an OB star cluster environment.
See Abstr. 063.109.

A grid of synthetic spectra for the determination of effective temperature, gravity and metallicity of F, G, and K stars. I. Description of the method.
See Abstr. 064.002.

Main-sequence mass-loss: lithium dilution in the Hyades and the Sun.
See Abstr. 064.191.

On lithium depletion: a possible discrimination of the physical processes.
See Abstr. 065.051.

Lithium depletion induced by rotation in young stars.
See Abstr. 065.079.

Remarks on the evolution of low–mass stars: more or less opaque?
See Abstr. 065.144.

Problems for the WIMP hypothesis?
See Abstr. 065.150.

New star with high proper motion in direction of NGC 6611.
See Abstr. 111.003.

Radial velocity measurements in the Pleiades.
See Abstr. 111.004.

Proper motions of stars in the open cluster NGC 7092.
See Abstr. 111.005.

Proper motions of stars in the open cluster NGC 7092.
See Abstr. 111.013.

The light curves of periodic Be stars in NGC 3766.
See Abstr. 112.096.

Reddening effects on the calibration of Strömgren photometric indices.
See Abstr. 113.031.

An improved metal abundance calibration for the Washington system.
See Abstr. 113.036.

The age evolution of Ca II emission in late type stars.
See Abstr. 114.040.

The study of lithium in stars like the Sun.
See Abstr. 114.050.

Chromospheric activity and lithium abundances in Pleiades low–mass stars.
See Abstr. 114.058.

Lithium in F dwarfs in IC 4651.
See Abstr. 114.081.

Brown dwarfs in young stellar clusters.
See Abstr. 115.001.

Tables of isochrones computed from stellar models with mass loss and overshooting.
See Abstr. 115.005.

Calibration of colour indices $(b-y)_0$ and absolute magnitudes M_V for A4–F3 supergiants from uvbyβ data.
See Abstr. 115.006.

Calibration of $(b-y)_0$ color indices and absolute magnitudes M_V for A4–F3 supergiants using uvbyβ photometry.
See Abstr. 115.008.

The luminosity of the B2 hypergiant HD 80077.
See Abstr. 115.013.

On the DDO absolute magnitudes of K giants.
See Abstr. 115.017.

A new age calibrator for red horizontal branch populations.
See Abstr. 115.029.

Ca II H and K and Hα, and Li abundances in the Pleiades late K main–sequence.
See Abstr. 116.033.

Rotation, chromospheric activity, and lithium abundances in G and K dwarfs of the Pleiades.
See Abstr. 116.068.

Light curve and radial velocity curve analyses of the eclipsing binary H235 in the open cluster NGC 752.
See Abstr. 117.022.

The dynamical evolution of the nearby multiple stellar systems ADS 48, ADS 6175 (α Geminorum = Castor), α Centauri, and ADS 9909 (ξ Scorpii).
See Abstr. 118.020.

A royal road to clusters and stellar evolution: binaries–in–clusters applications to cluster studies.
See Abstr. 119.040.

Spectroscopic binaries, rotation and magnetism in open clusters and associations.
See Abstr. 120.020.

A search for binaries in two open clusters.
See Abstr. 120.022.

Detection of the secondary spectrum in θ^2 Tauri.
See Abstr. 120.023.

A rich cluster of radio stars in the ϱ Ophiuchi cloud cores.
See Abstr. 121.064.

Slow flares in stellar aggregates. III.
See Abstr. 122.032.

A search for southern open cluster flare stars.
See Abstr. 122.120.

Some additional data on the new short–period variable in NGC 7142.
See Abstr. 123.011.

Interaction of massive stars with the interstellar medium.
See Abstr. 131.225.

IR– and mm– observations of regions of massive star formation.
See Abstr. 131.283.

First results from the Faint Object Camera: high–resolution observations of the central object R136 in the 30 Doradus nebula.
See Abstr. 132.027.

Dissolution of tidally limited star clusters in galaxies by passing giant molecular clouds or other massive objects.
See Abstr. 151.091.

On one type of models of spherical stellar clusters.
See Abstr. 151.092.

The role of binaries in cluster dynamical evolution.
See Abstr. 154.084.

Are there two disk star cluster systems in the LMC?
See Abstr. 154.097.

Progress towards an understanding of the blue straggler problem.
See Abstr. 154.098.

Spiral arms of the Galaxy from observations of interstellar extinction.
See Abstr. 155.060.

Topography of the Galactic disk: Z-structure and large-scale star formation.
See Abstr. 155.062.

A cluster of He I emission-line stars in the Galactic center.
See Abstr. 155.123.

BV CCD photometry of five young Large Magellanic Cloud star clusters.
See Abstr. 156.005.

The young Magellanic Cluster NGC 2004.
See Abstr. 156.023.

Detection of the helium flash gap in the integrated (U-B) versus (B-V) diagram for 624 Large Magellanic Cloud clusters.
See Abstr. 156.034.

154 Globular Clusters

154.001 **Carbon and nitrogen abundances in three supergiants of the young SMC globular cluster: NGC 330.**
B. Barbuy, M. Spite, F. Spite, A. Milone.
Astron. Astrophys., Vol. 247, No. 1, p. 15 - 19 (Jul 1991).

Carbon and nitrogen abundances were derived in a few supergiants of the young and metal-poor globular cluster NGC 330 of the Small Magellanic Cloud. C_2 and CN bands were used to derive carbon and nitrogen abundances of the two cool supergiants A7 and B40, and the atomic C I line at λ = 538.032 nm was used to derive the carbon abundance of the hot supergiant B38. The ratios C/Fe and N/Fe are found to be solar and it is concluded that mixing with CN processed material has probably not affected the surface abundances of the stars. These abundances should represent the initial composition of the gas from which the stars of NGC 330 formed. The abundances of the light elements in the stars of NGC 330 are compared to the abundances found in the field stars and in the H II regions of the SMC. In contrast with the overdeficiency of C and N found by Dufour et al. (1982) in the H II regions of the SMC, it is shown that in the stars of NGC 330 the deficiency of carbon and nitrogen is about the same as the deficiency of the other elements in particular oxygen and sulfur. The authors recall that the overdeficiency of carbon observed in the H II regions was not found previously in the field stars of the SMC. Additional observations are needed.

154.002 **Chemical composition of globular cluster stars.**
P. François.
Astron. Astrophys., Vol. 247, No. 1, p. 56 - 63 (Jul 1991).

High resolution spectra of six giants belonging to galactic globular clusters have been obtained. Accurate abundance determinations of various elements have been obtained by means of model atmosphere analysis. The author found that the Q_{39} index becomes extremely sensitive to a change in metallicity in objects with [Fe/H] > -1, ruling out the linearity of the relation $Q_{39} = f([Fe/H])$. The overabundance of α-elements found by previous authors is confirmed. The results are discussed in the light of models of self-enrichment of globular clusters.

154.003 **Velocity dispersions of stars and masses of globular clusters M4, M5, M10, M12 and M71.**
A. S. Rastorguev, N. N. Samus'.
Pis'ma Astron. Zh., Tom 17, No. 10, p. 915 - 931 (Oct 1991). In Russian. English translation in Sov. Astron. Lett., Vol. 17, No. 5.

The authors present the results of measured radial velocities of 159 red stars in the globular clusters M3, M4, M5, M10, M12, M13, M71 carried out with a correlation radial-velocity-meter. Single-dimensional velocity dispersions for the clusters M4, M5, M10, M12, and M71 have been calculated in the frame of King's single-component models, and the values of mass-to-luminosity ratio M/L_V and masses have been determined. $(M/L_V)_\odot$ values are in the range of 1 - 2 with uncertainties of 40 to 100 percent. The mass of the globular cluster M71 is $(1-2)\cdot10^4 M_\odot$ whereas for the other clusters the masses are in the range $(1-2)\cdot10^5 M_\odot$. For the globular cluster M5 the authors find rotation with the angular velocity 0.25 ± 0.13 km $(s\cdot pc)^{-1}$ and the position angle of the rotation axis projection 130° (in the model of rigid body rotation). The rotation axis is almost perpendicular to the major axis of central equidensities, thus evidencing for the possible connection between flattening and rotation.

154.004 **A UBVRI photoelectric sequence in the SMC globular cluster NGC 121.**
F. Alvarado, E. Wenderoth, G. Alcaino, W. Liller.
Astron. Astrophys., Suppl. Ser., Vol. 90, No. 1, p. 191 - 193 (Oct 1991).

The authors present a UBVRI photoelectric sequence of 10 stars in the magnitude range 11.8 < V < 14.8 in the Small Magellanic Cloud globular cluster NGC 121. Seven of these stars have been observed in BV by Tifft (1963) and the results are in acceptable agreement.

154.005 **The evolution of globular clusters: post-core collapse with stellar evolution.**
B. W. Murphy.
Bull. Am. Astron. Soc., Vol. 23, No. 2, p. 947 (1991). Abstract. - See Abstr. 012.008 for the main entry.

154.006 **Dynamical mass estimates in the globular clusters M13 and M71.**
P. J. T. Leonard, H. B. Richer, G. G. Fahlman.
Bull. Am. Astron. Soc., Vol. 23, No. 2, p. 947 (1991). Abstract. - See Abstr. 012.008 for the main entry.

154.007 **Brightness and color profiles for the globular cluster NGC 2419.**
D. H. Martins.
Bull. Am. Astron. Soc., Vol. 23, No. 2, p. 947 (1991). Abstract. - See Abstr. 012.008 for the main entry.

154.008 **UIT imagery of Omega Centauri.**
W. B. Landsman, R. H. Cornett, J. K. Hill, R. A. Parise, P. M. N. Hintzen, S. P. Maran, A. G. Michalitsianos, T. P. Stecher, R. W. O'Connell, M. S. Roberts.
Bull. Am. Astron. Soc., Vol. 23, No. 2, p. 947 (1991). Abstract. - See Abstr. 012.008 for the main entry.

154.009 **UIT images of the globular cluster M79.**
R. S. Hill, M. R. Greason, J. K. Hill, R. H. Cornett, W. B. Landsman, R. A. Parise, P. M. N. Hintzen, S. P. Maran, A. G. Michalitsianos, T. P. Stecher, M. S. Roberts, R. W. O'Connell.
Bull. Am. Astron. Soc., Vol. 23, No. 2, p. 947 (1991). Abstract. - See Abstr. 012.008 for the main entry.

154.010 **Ultraviolet Imaging Telescope (UIT) observations of the globular cluster NGC 1851.**
R. A. Parise, P. M. Hintzen, S. P. Maran, A. G. Michalitsianos, T. P. Stecher, M. S. Roberts, R. H. Cornett, M. R. Greason, J. K. Hill, W. B. Landsman, R. O'Connell.
Bull. Am. Astron. Soc., Vol. 23, No. 2, p. 948 (1991). Abstract. – See Abstr. 012.008 for the main entry.

154.011 **Coping with the Hubble Space Telescope's PSF: crowded field stellar photometry of M15.**
E. M. Malumuth, J. D. Neill.
Bull. Am. Astron. Soc., Vol. 23, No. 2, p. 948 (1991). Abstract. – See Abstr. 012.008 for the main entry.

154.012 **Blue stragglers in the core of 47 Tucanae.**
F. Paresce, M. Shara, G. Meylan, F. Macchetto, D. Baxter, J. C. Blades, P. Greenfield, R. Jedrzejewski, A. Nota, W. B. Sparks, R. Albrecht, C. Barbieri, A. Boksenberg, P. Crane, J. M. Deharveng, M. J. Disney, P. Jakobsen, T. M. Kamperman, I. R. King, C. D. Mackay, G. Weigelt.
Bull. Am. Astron. Soc., Vol. 23, No. 2, p. 948 (1991). Abstract. – See Abstr. 012.008 for the main entry.

154.013 **The formation of primordial binaries in globular clusters.**
S. D. Murray, C. J. Clarke, J. E. Pringle.
Bull. Am. Astron. Soc., Vol. 23, No. 2, p. 948 (1991). Abstract. – See Abstr. 012.008 for the main entry.

154.014 **Main sequence close binary stars in NGC 288.**
B. Margon, E. Wilcots, M. Bolte.
Bull. Am. Astron. Soc., Vol. 23, No. 2, p. 948 (1991). Abstract. – See Abstr. 012.008 for the main entry.

154.015 **Ca II emission from the core of M15.**
A. K. Dupree, B. Whitney.
Bull. Am. Astron. Soc., Vol. 23, No. 2, p. 949 (1991). Abstract. – See Abstr. 012.008 for the main entry.

154.016 **Abundances of the most metal poor globular clusters.**
D. Geisler, J. Claria, D. Minniti.
Bull. Am. Astron. Soc., Vol. 23, No. 2, p. 949 (1991). Abstract. – See Abstr. 012.008 for the main entry.

154.017 **The formation of globular clusters in galaxy mergers.**
K. M. Ashman, S. E. Zepf.
Bull. Am. Astron. Soc., Vol. 23, No. 2, p. 954 (1991). Abstract. – See Abstr. 012.008 for the main entry.

154.018 **Diameters of Galactic globular clusters.**
S. van den Bergh, C. Morbey, J. Pazder.
Astrophys. J., Vol. 375, No. 2, p. 594 – 599 (10 Jul 1991).
All published aperture photometry of Galactic globular clusters has been used to derive total magnitudes, V_t, and values of the diameter containing half the cluster light in projection, $D_{0.5}$. For the 98 globulars for which this information is available, the cluster diameter is found to correlate with Galactocentric distance. In this respect the Galactic globular cluster system resembles those associated with M31, NGC 5128, and the Large Magellanic Cloud, in which cluster diameters also increase with increasing Galactocentric distance. The observed correlation $D_{0.5} \sim R^{1/2}$ suggests that compact clusters formed preferentially from the dense gas clouds near the centers of galaxies, whereas more open clusters formed in the less dense halo regions of galaxies. Perhaps surprisingly, the cluster diameter does not correlate with cluster luminosity (mass). Furthermore, the cluster diameter does not appear to correlate with metallicity.

154.019 **The formation and detection of strongly magnetic white dwarf binaries in globular clusters.**
G. Chanmugam, A. Ray, K. P. Singh.
Astrophys. J., Vol. 375, No. 2, p. 600 – 607 (10 Jul 1991).
The authors discuss the mechanism for the formation of magnetic cataclysmic variables (MCVs) by the tidal capture of isolated magnetized white dwarfs and determine those globular clusters where the largest number of them are likely to be formed. By performing a statistical analysis of the X-ray emission of known MCVs, in the solar neighborhood, the authors determine the hard X-ray luminosity functions of the DQ Her and AM Her binaries and the soft X-ray luminosity function of the AM Her binaries. The DQ Her binaries have a mean hard X-ray luminosity a factor of 10 higher than that for the AM Her binaries. The authors also estimate the X-ray fluxes of globular cluster MCVs by using a solar neighborhood survey, and identify those globular clusters which are likely to contain MCVs which may be detectable by means of their X-ray emission.

154.020 **The luminosity function for the globular cluster M15.**
M. Simoda, T. Fukuoka, K. Nakajima.
Astrophys. Space Sci., Vol. 183, No. 1, p. 51 – 65 (Sep 1991).
The authors obtained the luminosity function from the Main-Sequence turnoff region to the subgiants for M15 ($V = 17.0$–19.6), using plates taken by the Kitt Peak 4 m reflector. The measured region is $r = 2\rlap{.}'5$–$8\rlap{.}'0$ zone. The authors extrapolated the star counts to the cluster center and to the tidal limit, using King's (1966) photometry, King et al.'s (1968) counts, Peterson and King's (1975) result, and Simoda and Iben's (1970) theoretical results. The comparison of the observed function with the theoretical ones by Simoda and Iben (1970) indicates that the observation is consistent with the theory for $X \sim 0.9 \pm 0.1$, thus marginally inconsistent with the Big Bang cosmology value ($X = 0.76$). The comparison with Bertelli et al. (1979) theoretical results gives contradictory results. Possible causes of this problem are the improper extrapolation of star counts to the cluster center, and/or the inaccuracies of the theoretical luminosity functions so far published, and/or the impropriety of the canonical structure and evolution theory of Population II stars. In the last two cases, the ages and the helium abundances of globular clusters may vary. The authors also obtained the luminosity function for blue horizontal branch stars from $V = 17.0$ to 19.6.

154.021 ***BVI* CCD surface photometry of the globular cluster NGC 2419.**
G. Natali, F. Pedichini, M. Righini.
Astron. Astrophys., Vol. 248, No. 2, p. 426 – 429 (Aug 1991).
The authors present the radial surface luminosity distribution of the globular cluster NGC 2419 in the *B*, *V*, and *I* photometric bands. Data are derived by single frame CCD images of (25×17) arcmin^2. The authors describe the simple methodology used to obtain measurements actually related to the diffuse emission of the globular cluster without contamination due to bright stars, field stars or other kinds of singularities. Surface brightness is compared to the theoretical one proposed by King (1965) for homogeneous self-gravitating systems. Data, extending to half the tidal radius, show good agreement with the model. The core radius ($R_c = 0.34 \pm 0.01$ arcmin) returns less than that calculated by Peterson and King (1975) with data of Kron and Mayall (1960): $R_c = 0.42$ arcmin.

154.022 **Double globular clusters in the Galaxy?**
V. G. Surdin.
Astrophys. Space Sci., Vol. 183, No. 1, p. 129 – 134 (Sep 1991).
Four pairs of "visual double" star clusters with the lowest mass-to-luminosity ratio have been selected among globular clusters of our Galaxy. By taking into accounts the effects of dynamical friction and compressive gravitational shocks, the author concludes that the probability for the pairs to be gravitationally bound is very low.

154.023 **The central dynamics of 47 Tucanae.**
I. R. King.
Workshop on the First Year of HST Observations, p. 6 – 9 (1991). – See Abstr. 012.017 for the main entry.
The blue stragglers recently discovered in 47 Tucanae have an excess concentration in the very center of the cluster that is statistically significant at a marginal level. Studies of the distribution of the ordinary stars (mainly near the main-sequence

turnoff) may imply the presence of a population of massive remnants near the center of the cluster, although more extensive observations will be needed to confirm such a hypothesis.

154.024 **Inhomogeneous halo collapse and early Galactic chemical evolution: globular cluster metallicities.**
G. Malinie, D. H. Hartmann, G. J. Mathews.
Astrophys. J., Vol. 376, No. 2, p. 520 – 530 (1 Aug 1991).
The authors propose a new solution to the observed Galactic globular cluster metallicity distribution by considering inhomogeneous collapse of a centrally condensed proto–Galaxy. In contrast to the standard one–zone model, it is shown that for inhomogeneous models the metallicity distribution can be reproduced without the need to decrease the nucleosynthetic yield for metal–poor stars. Chemical evolution in free fall is calculated analytically, and a newly developed hybrid hydro–particle code is used to study the effect of supernova induced pressure.

154.025 **Two distinct supergiant branches in the young Large Magellanic Cloud star cluster NGC 2214.**
R. Sagar, T. Richtler, K. S. de Boer.
Astron. Astrophys., Vol. 249, No. 2, p. L5 – L8 (Sep 1991). Letter–to–the–editor.
The authors present compelling evidence that the LMC star cluster NGC 2214 actually consists of two stellar age groups, clearly visible in the form of two well–defined supergiant branches in the CCD *V*, (*B–V*) colour–magnitude diagram. They are separated by ~2 mag in luminosity and have respective ages ~60 and 170 Myr or both smaller or larger depending upon the theoretical evolutionary tracks used. The spatial distribution of the stars of the two sequences indicates that the older population is more concentrated towards the cluster centre while the younger shows a more extended distribution.

154.026 **Oxygen and iron in Omega Centauri giants.**
J. A. Brown, G. Wallerstein, K. Cunha, V. V. Smith.
Astron. Astrophys., Vol. 249, No. 2, p. L13 – L15 (Sep 1991). Letter–to–the–editor.
The authors have derived oxygen and iron abundances in 18 red giants in ω Cen, a globular cluster with a substantial spread in metallicity. Over the range in metallicity from –1.0 to –2.0 dex the [O/Fe] value remains sensibly constant near [O/Fe] = +0.4. The relationship of [O/Fe] and [Fe/H] appears to be the same in ω Cen as in the halo of the Galaxy. The authors discuss this similarity in connection with the retention of supernova ejecta by ω Cen as [Fe/H] increased from –2.0 to –1.0 and [O/H] increased from –1.6 to –0.6.

154.027 **Metal–rich globular clusters: giant branch morphology as a metallicity criterion.**
S. Ortolani, B. Barbuy, E. Bica.
Astron. Astrophys., Vol. 249, No. 2, p. L31 – L33 (Sep 1991). Letter–to–the–editor.
The authors have obtained *VI* CCD photometry of the very metal–rich inner bulge globular clusters NGC 6528, NGC 6553 and Terzan 1, and of 47 Tuc and NGC 6356, which are usually called metal–rich clusters. They discuss the RGB morphology in the *V* vs. (*V–I*) and *I* vs. (*V–I*) diagrams for ranking the metallicities of these clusters. They used the arc–like structure present in the RGB, due to blanketing in the cooler giants, as a metallicity criterion, which is corroborated by line strengths in integrated spectra. The authors propose criteria of morphological parameters based on RGB inclination for increasing metallicities. They conclude that Terzan 1 is the most metal–rich globular cluster of this sample.

154.028 **Estimation of the B– and V–magnitudes of the globular cluster candidates in M81.**
Ts. B. Georgiev, N. A. Tikhonov, I. D. Karachentsev.
Pis'ma Astron. Zh., Tom 17, No. 11, p. 994 – 998 (Nov 1991). In Russian. English translation in Sov. Astron. Lett., Vol. 17, No. 6.
A photometry of 51 globular cluster candidates in M81 is carried out on the plates of 6–m and 2–m telescopes. The most bright ones of these objects, except one, have V > $19^{m}.3$ and $M_V > -8^{m}.5$. By their brightness, the globular cluster candiadate system of M81 is intermediate between the globular cluster system of M31 and M33. Some excess of more redish objects in M81 is pointed out.

154.029 **Studies of dynamical properties of globular clusters. VI. The high–concentration cluster NGC 6397.**
G. Meylan, M. Mayor.
Astron. Astrophys., Vol. 250, No. 1, p. 113 – 126 (Oct 1991).
Mean radial velocities of 127 stars, members of the metal–poor globular cluster NGC 6397, have been obtained from the ESO at La Silla (Chile) with the photoelectric spectrometer CORAVEL mounted on the 1.5 m Danish telescope. The systematic radial velocity of the cluster equals $+18.8 \pm 0.01$ km s^{-1}. The velocity dispersion, obtained in 5 boxes of about 25 stars each, decreases monotonously from 4.5 km s^{-1} in the center to 2.2 km s^{-1} in the outer parts. In order to reproduce the above observations, a multi–mass anisotropic dynamical (King–Michie) model, based on an assumed form of the phase–space distribution function $f(E,J)$, is used. The constraint in luminosity is a merger of different kinds of observations found in the literature. The recent observational determinations of the mass function for the stars still on the main sequence are also taken into account. From a grid of about 600 models, allowing an extensive exploration of the parameter space, only the best (lowest χ^2) eight models are retained. They provide the following representative values of the astrophysical parameters describing the cluster: the mean concentration $\langle c \rangle = \log r_t/r_c \sim 2.5$, with $\langle r_c \rangle = 0.22$ pc = 19″ and $\langle r_t \rangle = 66$ pc; the mean integrated magnitude $\langle V_t \rangle = 5.6$ and the mean central surface brightness $\langle \mu_0 \rangle = 7.8$ mag arcmin^{-2}; the mean total mass $\langle M_{tot} \rangle = 10^5 M_\odot$, with the mean global mass–to–light ratio $\langle M/L_V \rangle = 2.1$, whereas the mean central mass–to–light ratio $\langle (M/L_V)_0 \rangle = 1.7$. Classified as a collapsed cluster, NGC 6397 is reasonably well fitted by high–concentration King–Michie models.

154.030 **Compactness of globular cluster and its evolution.**
Zhang Mei, Ma Er.
Chin. Astron. Astrophys., Vol. 15, No. 3, p. 251 – 259 (Sep 1991). English translation of Acta Astrophys. Sin., Vol. 11, No. 2, p. 128 – 136 (Apr 1991). See Abstr. 53.154.031.
The evolution of the compactness of a globular cluster is found to be closely dependent on its mass and on its position in the Galaxy. At large galactocentric distances, the evolution is very small and it is the slower the larger the mass. A statistical study of the available material on the compactness shows that its distribution at smaller galactocentric distances agrees with the theoretical expectation. At large distances, the distribution is very different for different masses, possibly reflecting a close relation between the mass and the condensation at the stage of formation of the cluster.

154.031 **Mass functions of five young Large Magellanic Cloud star clusters.**
R. Sagar, T. Richtler.
Astron. Astrophys., Vol. 250, No. 2, p. 324 – 339 (Oct 1991).
The authors have analyzed *BV* CCD data of 5 young star clusters in the LMC in order to determine their mass functions. The clusters are NGC 1711, 2004, 2100, 2164 and 2214. They span a range in age from ~10–100 Myr. In order to convert luminosity functions into mass functions, three different sets of theoretical stellar evolutionary models have been used. The authors assumed that the mass functions can be represented by power laws and found indeed no obvious contradiction to that. They evaluate statistical errors in the mass function slope, *x*, by means of Monte–Carlo simulations. In the mass range ~2–14 $M_\odot$, the value of x in four well observed clusters is approximately the same, the average value being $x = 1.1$. It is not too different from the Salpeter value ($x = 1.35$). Neither the very flat nor the very steep mass functions found respectively by Elson et al. (1989) and Mateo (1988) could be confirmed. For three clusters, the mass function slopes in two annular rings located at ~80 and 160 pixels (corresponding to ~20 and 40 pc

respectively) from the cluster centre agree within the errors. In the case of NGC 2214, the authors do find a difference between the two slopes but NGC 2214 actually consists of two clusters of different ages (Sagar et al. 1991).

154.032 **NGC 1866 revisited: semiconvection versus overshoot.**
J. C. Lattanzio, A. Vallenari, G. Bertelli, C. Chiosi.
Astron. Astrophys., Vol. 250, No. 2, p. 340 – 350 (Oct 1991).

The authors calculate a new set of stellar tracks and related isochrones that take into account the presence of semiconvection in the He-core burning phase. The stellar masses range from 2.5 to 9 $M_\odot$ for two values of the metallicity: Z = 0.02 and Z = 0.003. These models are compared with a set calculated with convective overshoot where the metallicity, the He content and the other parameters are homogeneous. Since the treatment of the convection in the core of a real star is affected by many uncertainties, the authors look at the observations of the star clusters of the LMC to discriminate among semiconvection and overshoot. In particular, they compare the theoretical models with the colour–magnitude diagram and the luminosity function of NGC 1866 previously presented by Chiosi et al. (1989). Both models reproduce the features of the colour–magnitude diagram; however only the convective overshoot hypothesis is in agreement with the observed ratio of main sequence to evolved stars.

154.033 **Horizontal-branch models.**
A. V. Sweigart.
Meeting on Precision Photometry: Astrophysics of the Galaxy, p. 13 – 26 (1991). – See Abstr. 012.027 for the main entry.

A number of observational constraints on the properties of theoretical models for globular-cluster horizontal-branch (HB) stars are reviewed with particular emphasis on the luminosity difference between the tip of the red-giant branch (RGB) and the HB, the HB luminosity function and the relative number of HB versus RGB or asymptotic-giant-branch (AGB) stars. The author discusses the importance of such observational constraints in resolving the current theoretical uncertainty concerning the end of the HB phase.

154.034 **The age spread among galactic globular clusters.**
P. Demarque.
Meeting on Precision Photometry: Astrophysics of the Galaxy, p. 45 – 54 (1991). – See Abstr. 012.027 for the main entry.

Recent progress in estimating the relative ages of galactic globular clusters is described. The evidence for an age spread of several Gyr among halo clusters is discussed. This result has important implications for the understanding of the formation and evolution of the Galaxy. Finally, current problems and work in progress on absolute ages are reviewed, primarily within the context of the age spread.

154.035 **Globular cluster photometry near the turn-off: relative ages and the horizontal branch.**
A. Sarajedini.
Meeting on Precision Photometry: Astrophysics of the Galaxy, p. 55 – 67 (1991). – See Abstr. 012.027 for the main entry.

With the use of CCD detectors for precise stellar photometry, the morphology of the main sequence turn-off region in Galactic globular clusters has become better defined. Due to the extreme sensitivity of the turn-off region to variations in cluster age, this has allowed unprecedented precision in the determination of relative cluster ages. One method recently devised to study relative ages is based on the color difference between the turn-off and the point at which the subgiant branch turns upward. Although it is possible to define this color difference in any combination of filters, work done so far has utilized the (B–V) color difference and is thus denoted by Δ(B–V). The advantages of Δ(B–V) are that it is independent of reddening, distance modulus, and photometric zeropoint. However, perhaps the most important advantage is that Δ(B–V) is highly insensitive to cluster metallicity over a wide range of metallicities. The author has compiled Δ(B–V) measurements for approximately 20 clusters. The distribution of these values and how they vary with abundance gives information on the relative ages of the globular clusters and thus the formation history of the Galactic halo.

154.036 **On the color and population gradients in the central-cusp globular cluster M 15.**
P. B. Stetson.
Meeting on Precision Photometry: Astrophysics of the Galaxy, p. 69 – 88 (1991). – See Abstr. 012.027 for the main entry.

The author presents preliminary color–magnitude diagrams and integrated color profiles for the innermost two arcminutes of the central-cusp globular cluster M15. These data will be used to investigate the nature and the cause of the color gradients reported for this and similar clusters.

154.037 **CCD detectors applied to globular cluster research.**
H. B. Richer, G. G. Fahlman.
Meeting on Precision Photometry: Astrophysics of the Galaxy, p. 89 – 97 (1991). – See Abstr. 012.027 for the main entry.

Because of their enhanced sensitivity and digital readout, CCDs permit measurement of fainter objects with higher accuracy and more convenience than was possible with photographic plates or early versions of other electronic detectors. Further, because it is possible to carry out Monte-Carlo simulations on the data frames themselves, a better estimate of the incompleteness and the errors involved in measuring objects can be obtained. The authors illustrate the improvements possible using these detectors by looking at two areas of research interest involving globular star clusters: ages, and luminosity (and mass) functions.

154.038 **CCD four-color photometry of globular clusters stars.**
A. G. D. Philip.
Meeting on Precision Photometry: Astrophysics of the Galaxy, p. 153 – 161 (1991). – See Abstr. 012.027 for the main entry.

Four-color CCD frames have been taken in the globular clusters M4, M5, M22, M55 and M92. Three frames in each color were obtained in each field studied. A cm-diagram has been constructed for all the stars in the frame, as well as an additional cm-diagram of just the BHB stars.

154.039 **Membership of ultraviolet-bright stars in the globular cluster M12.**
M. Geffert, H.-J. Tucholke, Ts. B. Georgiev, J.-F. LeCampion.
Astron. Astrophys., Suppl. Ser., Vol. 91, No. 3, p. 487 – 495 (Dec 1991).

Positions and proper motions of 165 stars in the field of the globular cluster M12 (NGC 6218) have been determined using 20 plates covering a time interval of about 85 years. Membership probability for each star was derived from proper motions. The data show a good segregation of cluster and field stars. For ten of the 13 UV-bright stars the proper motion membership result agrees with conclusions from radial velocities as given by Harris et al. (1983).

154.040 **High precision stellar photometry in M13.**
A. Penny, B. Griffiths.
Gemini, No. 33, p. 4 – 5 (Sep 1991).

154.041 **Dynamical masses for Galactic globular clusters.**
G. Mandushev, N. Spassova (*N. M. Spasova*), A. Staneva (*A. V. Staneva*).
Astron. Astrophys., Vol. 252, No. 1, p. 94 – 99 (Dec 1991).

Dynamical masses have been determined for 147 Galactic globular clusters. First, using single-mass King models, masses were obtained for 32 clusters with reliable central velocity dispersions and the mass luminosity relation was derived. It was used to obtain the masses of the remaining 115 clusters, assuming that it is valid for the whole range of globular cluster luminosities. The authors found the masses of the Galactic globular clusters to be between $2.6 \times 10^2 M_\odot$ (AM-4) and $2.4 \times 10^6 M_\odot$ (ω Cen) with a mean of $1.9 \times 10^5 M_\odot$. The mass-to-light ratios for the sample of 32 clusters range from 0.66 (NGC 5272) to 2.9 (NGC 6093), with a mean of 1.21. The authors show that some clusters of low and intermediate mass lie outside the survival zone in the (log M – log r_h) diagram and this is considered to be evidence for rapid mass loss through stellar evaporation. It is also found that the M/L_v ratio is most likely dependent on the cluster mass, with the most massive clusters having the largest M/L_v ratios. A comparison is made between the masses obtained with single–component and multi–component King models, which shows that the mass–luminosity relations for both types of models are parallel within the uncertainties and that the multi–component masses are about twice as large as the single–component ones.

154.042 **Perspective of the search for globular clusters in the Galaxy.**
V. G. Surdin.
Astron. Tsirk., No. 1545, p. 5 – 6 (Aug – Sep 1990). In Russian.

Numerical experiments show the relation between the limiting magnitude of the all sky survey (V_{lim}) and the number of discovered globular clusters (N). The total number of globular clusters in the Galaxy $N_{tot} \approx 180$.

154.043 **A new look at the globular cluster M28.**
R. F. Rees, K. M. Cudworth.
Astron. J., Vol. 102, No. 1, p. 152 – 158 (Jul 1991). Current Physics Microform No.: 9107B1848. With plate 8.

Reductions of PDS microdensitometer scans of 20 plates from three different telescopes over an epoch range of 92 yr have yielded proper motions and photometry for more than 250 stars down to V ~ 16 in the region of the poorly studied globular cluster M28 (NGC 6626). Membership probabilities for these stars, including several variables, are derived from the proper motions, and a color–magnitude diagram of probable members is presented and discussed. The authors find strong evidence of systematic errors in previous photographic photometry. Some questions about the metal abundance of M28 are addressed; in particular, a dependence upon method of the derived metallicity. All previously known variables included in this study are at least possible members. The authors find evidence of variability of a few stars not previously known to vary. A derivation of the cluster's absolute proper motion shows the space velocity to be surprisingly disklike for a moderately metal–poor cluster.

154.044 **BVRI CCD photometry of the metal–poor globular cluster NGC 4372.**
G. Alcaino, W. Liller, F. Alvarado, E. Wenderoth.
Astron. J., Vol. 102, No. 1, p. 159 – 176 (Jul 1991). Current Physics Microform No.: 9107B1855. With plates 9 – 13.

The authors present BVRI CCD photometry in two overlapping fields in the metal–poor globular cluster NGC 4372. The observations extend approximately 2 mag below the main–sequence turnoff to V ~ 21. By comparing the color–magnitude diagram (CMD) with those of clusters with similar metallicities, the authors find that $E(B-V) = 0.50 \pm 0.03$, and $(m-M)_v = 14.75 \pm 0.06$. Comparison with theoretical isochrones leads the authors to a value $E(B-V) = 0.53 \pm 0.03$. Comparison of their CMD with that of bright stars published by other authors yields a value for $\Delta V_{TO-HB} = 3.3 \pm 0.3$. The weighted mean value of the age of the cluster, derived from the four colors, is 15 ± 4 Gyr (estimated external uncertainty).

154.045 **On self–enrichment models for the halo globular clusters.**
J. W. Truran, J. Brown, A. Burkert.
Symposium on the Formation and Evolution of Star Clusters as Part of the 102. Summer Scientific Meeting of the Astronomical Society of the Pacific, p. 78 – 87 (1991). – See Abstr. 012.050 for the main entry.

Recent spectroscopic studies of the detailed abundance patterns characterizing globular cluster stars are briefly reviewed. Guided by the current knowledge of nucleosynthesis as a function of stellar mass occurring in stars and supernovae, the authors identify some interesting constraints that these combined observational and theoretical considerations impose on the nature and timescale of the early chemical evolution of the Milky Way globular clusters. The authors are led to the conclusion that the abundance patterns observed in the halo clusters are very likely a consequence of self–enrichment. A model for such self–enrichment is presented.

154.046 **On deriving globular–cluster luminosity functions from CCD observations, with a particular application to the main–sequence turnoff/subgiant branch in metal–poor clusters.**
P. B. Stetson.
Symposium on the Formation and Evolution of Star Clusters as Part of the 102. Summer Scientific Meeting of the Astronomical Society of the Pacific, p. 88 – 111 (1991). – See Abstr. 012.050 for the main entry.

Some techniques involved in deriving and interpreting observational stellar luminosity functions from CCD observations are reviewed. Then the combined luminosity function for the three metal–poor globular clusters M68 (NGC 4590), NGC 6397, and M92 (NGC 6341) is examined. This data set contains enough stars to permit interesting tests of stellar evolutionary models. In particular, if WIMPs ("weakly interacting massive particles") exist in quantities large enough to close the Universe, their effects on stellar evolution may be visible in the main–sequence turnoff/subgiant–branch part of the luminosity function.

154.047 **The manifold of globular clusters.**
S. Djorgovski.
Symposium on the Formation and Evolution of Star Clusters as Part of the 102. Summer Scientific Meeting of the Astronomical Society of the Pacific, p. 112 – 119 (1991). – See Abstr. 012.050 for the main entry.

A multivariate statistical analysis of a large data set on the Galactic globular clusters indicates that at least 4 (possibly 5 or 6) parameters are needed in order to describe fully the global properties of these objects. Some nontrivial correlations are present in the data, and are described in the text. One of them is the scaling relation between the luminosity and the central velocity dispersion, $L \sim \sigma^{1.54}$, whose possible physical origin is briefly discussed. Many of the systematic trends found in the data suggest that dynamical effects (e.g., tidal shocks and stripping) played an imporant, or even a dominant, role in determining the present–day global properties of globular clusters.

154.048 **Main sequence mass functions in globular clusters.**
H. B. Richer, G. G. Fahlman.
Symposium on the Formation and Evolution of Star Clusters as Part of the 102. Summer Scientific Meeting of the Astronomical Society of the Pacific, p. 120 – 132 (1991). – See Abstr. 012.050 for the main entry.

The methods and associated pitfalls involved in the construction of main sequence mass functions in globular clusters are discussed. Several examples of such functions are given and their morphology is considered. Some preliminary conclusions based on this small sample are discussed, and the implication of these for star formation in the early universe, the halo dark matter, and dynamical evolution of globular clusters is mentioned.

154.049 **M/L ratios of two LMC globular clusters: NGC 1835 and NGC 1978.**
G. Meylan, P. Dubath, M. Mayor.
Symposium on the Formation and Evolution of Star Clusters as Part of the 102. Summer Scientific Meeting of the Astronomical Society of the Pacific, p. 158 – 160 (1991). – See Abstr. 012.050 for the main entry.

Projected velocity dispersions are deduced from integrated light spectra of the cores of two LMC globular clusters: NGC 1835 and NGC 1978. A numerical cross–correlation technique gives a projected velocity dispersion $\sigma_p(\text{core}) = 10.1 \pm 0.2$ km s^{-1} in the case of NGC 1835 and 5.8 ± 1.2 km s^{-1} in the case of NGC 1978. For NGC 1835 the values obtained for the total mass of the cluster range from 0.70 to $1.55 \times 10^6 M_\odot$, corresponding to mass–to–light ratios M/L_V ranging from 2.5 to 5.2 $M/L_V)_\odot$. For NGC 978, depending on the model, the values obtained for the total mass of the cluster range from 0.36 to $1.44 \times 10^6 M_\odot$, corresponding to mass–to–light ratios M/L_V ranging from 1.2 to 4.2 $(M/L_V)_\odot$. These values are typical of rich galactic globular clusters.

154.050 **The role of the galactic disk in globular cluster formation.**
S. D. Murray, D. N. C. Lin.
Symposium on the Formation and Evolution of Star Clusters as Part of the 102. Summer Scientific Meeting of the Astronomical Society of the Pacific, p. 164 – 166 (1991). – See Abstr. 012.050 for the main entry.
The authors propose a scenario which explains the lack of young globular clusters in the Galaxy.

154.051 **Magnetic fields and radiative shocks in protogalaxies and the origin of globular clusters.**
P. R. Shapiro, A. Clocchiatti, H. Kang.
Symposium on the Formation and Evolution of Star Clusters as Part of the 102. Summer Scientific Meeting of the Astronomical Society of the Pacific, p. 176 – 179 (1991). – See Abstr. 012.050 for the main entry.
The suggestion has been made that globular clusters formed within protogalaxies by gravitational instability in the compressed gas resulting from radiative shocks with velocities of the order of the virial velocity of the protogalaxy or from thermal instability. Recent calculations of the thermal and chemical evolution and radiative transfer in metal-free gas overtaken by such shocks have shown that the success of the gravitational instability in forming fragments of mass comparable to the masses of globular clusters requires that H_2 formation and cooling be suppressed by a strong UV or soft X-ray source within the protogalaxy. The authors consider there the effects of a magnetic field in the protogalactic gas.

154.052 **Properties of the globular cluster mass functions.**
G. Piotto.
Symposium on the Formation and Evolution of Star Clusters as Part of the 102. Summer Scientific Meeting of the Astronomical Society of the Pacific, p. 200 – 204 (1991). – See Abstr. 012.050 for the main entry.
The dependence of the mass function slopes on the physical parameters is critically discussed for a sample of 14 galactic globular clusters with published and unpublished deep CCD luminosity functions. The author has found a possible trend of the mass function power-law index x with the distance from the galactic center and the galactic disk.

154.053 **The horizontal-branch stars in globular clusters as tracers of galactic evolution.**
Y.-W. Lee.
Symposium on the Formation and Evolution of Star Clusters as Part of the 102. Summer Scientific Meeting of the Astronomical Society of the Pacific, p. 205 – 218 (1991). – See Abstr. 012.050 for the main entry.
Recent advances in the understanding of the horizontal-branch stars in globular clusters are discussed, with special emphasis on the Sandage period-shift effect among RR Lyrae stars, and the second parameter problem of horizontal-branch morphology. They provide important implications for the age-metallicity and the age-galactocentric distance relations among globular clusters, which have direct impacts on the chronology of the Galaxy formation. The emerging picture of Galactic evolution suggests that the formation of the halo was a prolonged and chaotic affair.

154.054 **CCD Strömgren studies in NGC 6397.**
B. J. Anthony-Twarog, B. A. Twarog, N. B. Suntzeff.
Symposium on the Formation and Evolution of Star Clusters as Part of the 102. Summer Scientific Meeting of the Astronomical Society of the Pacific, p. 227 – 229 (1991). – See Abstr. 012.050 for the main entry.

154.055 **CCD photometry of NGC 6101: another globular cluster with blue straggler stars.**
A. Sarajedini, G. S. da Costa.
Astron. J., Vol. 102, No. 2, p. 628 – 641 (Aug 1991). Current Physics Microform No.: 9108A1564.
The authors present CCD photometry for a large sample of stars in the previously little-studied southern globular cluster NGC 6101. The photometry reaches below the cluster main-sequence turnoff. The color-magnitude (C-M) diagram exhibits the morphology expected for a metal-poor globular cluster; the horizontal branch is predominantly blue and the giant branch is steep. From these and other data the authors determine V(HB) = 16.6 ± 0.2, E(B-V) = 0.06 ± 0.02 and [Fe/H] = -1.80 ± 0.1 dex for this cluster. The age of NGC 6101 is investigated through a comparison with theoretical isochrones and by the application of more precise relative dating techniques. The authors see no indication of any difference in age, at the < 2 Gyr level, between NGC 6101 and other clusters of similar abundance such as M92. The C-M diagram for NGC 6101 also reveals a significant blue straggler population. These stars are shown to be more centrally concentrated than the cluster subgiants of similar magnitude, indicative of a larger mass for the blue straggler stars. The NGC 6101 blue straggler sample is also compared to and combined with blue straggler samples from four other globular clusters. The authors find that the magnitude of the brightest blue straggler is similar in all five clusters and that all five samples occupy similar regions in the C-M diagram. Further, the luminosity function of the total sample shows a marked drop at $M_v \approx +2.6$, a magnitude which corresponds on the ZAMS to twice the turnoff mass for the average composition of the clusters considered. These results are consistent with the binary mass transfer or merger hypotheses for the origin of blue straggler stars.

154.056 **A new approach to astrophysical parameters evaluation in globular clusters: test on M15.**
G. Bono, M. Fulle, L. Pulone.
Symposium on the Formation and Evolution of Star Clusters as Part of the 102. Summer Scientific Meeting of the Astronomical Society of the Pacific, p. 234 – 236 (1991). – See Abstr. 012.050 for the main entry.
The aim of this contribution is to develop an objective method that is able to carry out a reliable comparison between the theoretical and the observational star distribution function of the M15 Horizontal Branch (HB).

154.057 **The HB width as a function of metals and helium.**
G. Bono, L. Pulone.
Symposium on the Formation and Evolution of Star Clusters as Part of the 102. Summer Scientific Meeting of the Astronomical Society of the Pacific, p. 237 – 239 (1991). – See Abstr. 012.050 for the main entry.
Theoretical HB luminosity functions (LF) have been computed using new HB tracks with evolutionary core masses. The affinities and the differences found in comparison with the observational relations are briefly discussed.

154.058 **Ruprecht 106: a young metal-poor galactic globular cluster.**
R. Buonanno, G. Buscema, F. Fusi Pecci, H. B. Richer, G. G. Fahlman.
Symposium on the Formation and Evolution of Star Clusters as Part of the 102. Summer Scientific Meeting of the Astronomical Society of the Pacific, p. 240 – 242 (1991). Also published in Astron. J., Vol. 100, No. 6, p. 1811 – 1840 (Dec 1990). – See Abstr. 012.050 for the main entry.
CCD-photometry for about 2500 stars in the Galactic globular cluster Ruprecht 106 has been performed yielding the first color-magnitude diagram (CMD) for this object. From the analysis of the CMD the following basic parameters have been determined: [Fe/H] = $-1.91(+0.2,-0.4)$, and primordial helium abundance $Y_p = 0.20 \pm 0.05$. A sizeable, highly centrally concentrated population of blue stragglers has also been detected. The location and morphology of the mean ridge lines in the CMD of Ruprecht 106 have then been differentially compared with that of well-studied metal-poor clusters finding significant differences in the positions of the TO's. These differences can be explained if Ruprecht 106 is 4 – 5 Gyr younger.

154.059 **NGC 2808: still a globular cluster with peculiar horizontal branch morphology?**
Y.-I. Byun, Y.-W. Lee.
Symposium on the Formation and Evolution of Star Clusters as Part of the 102. Summer Scientific Meeting of the Astronomical Society of the Pacific, p. 243 – 245 (1991). – See Abstr. 012.050 for the main entry.
The authors report their progress in constructing the new color–magnitude diagram based on B and V photometry obtained with a CCD.

154.060 **Further evidence for a color gradient in M15.**
S. E. Cederbloom, M. J. Moss, H. N. Cohn, P. M. Lugger, C. D. Bailyn, J. E. Grindlay, R. D. McClure.
Symposium on the Formation and Evolution of Star Clusters as Part of the 102. Summer Scientific Meeting of the Astronomical Society of the Pacific, p. 246 – 249 (1991). – See Abstr. 012.050 for the main entry.
The authors construct color maps in B–V and H_α–R of the central regions of the globular cluster M15. Comparison in B–V of the region within 6″.6 of the cluster center with a coadded sample taken from the surrounding region demonstrates a color gradient, in the sense that the central region is bluer than the surrounding region at a high level of statistical significance. Analysis of the H_α–R color map indicates that the central region is "H_α deficient" than the coadded region. This result argues against the possibility that the excess blue light from the central region is due to cataclysmics variables. It is instead suggested that the color gradient is due to a centrally concentrated population of blue subdwarfs, possibly resulting from binary–single star interactions.

154.061 **IR photometry of Terzan clusters.**
C. A. Christian, E. D. Friel.
Symposium on the Formation and Evolution of Star Clusters as Part of the 102. Summer Scientific Meeting of the Astronomical Society of the Pacific, p. 250 – 252 (1991). – See Abstr. 012.050 for the main entry.
The giant branch of galactic clusters (Terzan clusters) have been studied using the IRCAM at the United Kingdom Infrared Telescope. A preliminary color–magnitude diagram of one of these objects (Terzan 2) has been obtained to study these clusters relative to clusters with known metallicities and ages to determine independent estimates for the distances, ages and metallicities of the clusters. These clusters are particularly important as they are among the most metal–rich members of the so called disk component of the globular cluster system, and they are located in a direction very close to the galactic center, hence observations are particularly difficult due to interstellar obscuration.

154.062 **HB theory and observations: new data on M3 and M13.**
D. A. Crocker.
Symposium on the Formation and Evolution of Star Clusters as Part of the 102. Summer Scientific Meeting of the Astronomical Society of the Pacific, p. 253 – 255 (1991). – See Abstr. 012.050 for the main entry.
A project to determine the origin of the multiple populations of stars on the horizontal branch (HB) of globular clusters has been continued using new data of M3.

154.063 **The reddening gradient in M4.**
K. M. Cudworth, R. F. Rees.
Symposium on the Formation and Evolution of Star Clusters as Part of the 102. Summer Scientific Meeting of the Astronomical Society of the Pacific, p. 256 – 258 (1991). – See Abstr. 012.050 for the main entry.
The authors have found strong evidence for an east–west reddening gradient across the globular cluster M4 in the sense that the redding is larger to the west. Removal of the effects of this gradient tightens all sequences in the color–magnitude diagram of cluster members. Solution for the ratio of total to selective extinction yields $R = 3.3 \pm 0.7$, with the uncertainty apparently dominated by the intrinsic scatter in luminosity (due to stellar evolution) of HB stars.

154.064 **Solar calibration and the ages of M67, NGC 188, and NGC 6791.**
P. Demarque, D. B. Guenther, E. M. Green.
Symposium on the Formation and Evolution of Star Clusters as Part of the 102. Summer Scientific Meeting of the Astronomical Society of the Pacific, p. 259 – 261 (1991). – See Abstr. 012.050 for the main entry.
Ages are derived for the old open star clusters M67, NGC 188 and NGC 6791, using solar calibrated theoretical isochrones.

154.065 **On the origin of the blue stragglers in the globular cluster NGC 5053.**
P. J. T. Leonard, G. G. Fahlman.
Astron. J., Vol. 102, No. 3, p. 994 – 1004 (Sep 1991). Current Physics Microform No.: 9109D1398.
The hypothesis that the blue stragglers in the low–density globular cluster NGC 5053 are merged stars formed as a result of physical stellar collisions during dynamical interactions involving primordial binary stars has been investigated in detail. This was done by carrying out a large number of binary–binary and binary–single scattering experiments in which the stars were allowed to instantly merge if they touched.

154.066 **CCD photometry of globular clusters. IV. NGC 7006.**
R. Buonanno, F. Fusi Pecci, E. Cappellaro, S. Ortolani, T. Richtler, E. H. Geyer.
Astron. J., Vol. 102, No. 3, p. 1005 – 1025 (Sep 1991). Current Physics Microform No.: 9109D1409.
A new color–magnitude diagram (CMD) for the galactic globular cluster NGC 7006 is presented, based on B, V CCD photometry of ~ 3000 stars to a limiting magnitude of $V \sim 24$. From the analysis of the CMD the following conclusions have been drawn: (1) The overall morphology of the main branches reproduces in essence earlier results; in particular, the horizontal branch (HB) is unusually red for the cluster metallicity ([Fe/H] = –1.77). This confirms that NGC 7006 is a genuine member of the group of clusters affected by the so–called "second parameter" syndrome. (2) The difference in luminosity between the HB and the main sequence turn–off (TO) is $\Delta^{TO}{}_{HB} = 3.50 \pm 0.12$, totally consistent with the mean value of 3.55 ± 0.09 obtained by Buonanno et al. (1989) for 19 Galactic globulars. (3) The mean color of the HB in the inner regions ($r \leqslant 30$ arcsec) appears to be significantly shifted towards bluer colors compared to the outer regions. (4) The combination of the "normal" location of the TO with the peculiar radial variation in HB morphology leads to exclude age as the second parameter in this cluster. A strong influence of the "environmental" conditions on HB morphology seems supported by these data. (4) Adopting the approach presented by Buonanno et al. (1989), the present data yield for this cluster: $M_V(HB) = 0.63$, $(m-M)_V = 18.17$ and, finally, $d_\odot = 40 \pm 2$ kpc.

154.067 **Mass–to–light ratios for globular clusters. II. The low–concentration clusters NGC 288, NGC 5466, and M55 (NGC 6809).**
C. Pryor, R. D. McClure, J. M. Fletcher, J. E. Hesser.
Astron. J., Vol. 102, No. 3, p. 1026 – 1042 (Sep 1991). Current Physics Microform No.: 9109D1430.
The authors have measured radial velocities, accurate to ~ 1 km s^{-1}, for 20–30 stars in each of the three low–concentration globular clusters NGC 288, NGC 5466, and M55. These data are sufficient to calculate velocity dispersions and M/L's for these clusters, which the authors use to investigate their stellar populations.

154.068 **How age and [O/Fe] differences affect the location of a globular cluster's giant branch relative to its turnoff in the (V, B–V) plane.**
D. A. VandenBerg, P. B. Stetson.
Astron. J., Vol. 102, No. 3, p. 1043 – 1049 (Sep 1991). Current Physics Microform No.: 9109D1447.
On the basis of theoretical isochrones for [Fe/H] = –2.26 and –1.26, and assuming $0.0 \leqslant [O/Fe] \leqslant 1.15$ and

$0.0 \leqslant$ [O/Fe] $\leqslant 0.75$, respectively, analytic expressions are derived to quantify the B–V displacement of the lower red–giant branch from the turnoff as a function of both age and [O/Fe]. As a result, the color–difference method developed by VandenBerg et al. (1990) for the precise determination of relative globular cluster ages is now extended to allow for the possibility that [O/Fe] varies at a given [Fe/H]. To illustrate the implications of their results, the authors have reanalyzed the color–magnitude diagrams of NGC 288 and NGC 362. This comparison, within $\sim \pm 2$ mag of their turnoffs, confirms previous results by the authors and others that NGC 288 is $\sim$20% older than NGC 362, if both clusters have the same [O/Fe] value. Alternatively, for the age difference to be near zero, the calculations predict that NGC 288 would have to be more oxygen rich than NGC 362 by δ[O/Fe] $\approx$ 0.6 (assuming that no other parameter is varying). Horizontal–branch considerations indicate a preference for the former interpretation over the latter.

154.069 **New determinations of element abundances in NGC 330.**
F. Spite, T. Richtler, M. Spite.
Astron. Astrophys., Vol. 252, No. 2, p. 557 – 564 (Dec 1991).

Abundances have been determined for four stars of the young globular cluster NGC 330 of the Small Magellanic Cloud. If the interstellar reddening has the classical low value, the four stars turn out to have the same metallicity, which is slightly but significantly lower than the metallicity of the field stars of about the same age. If the reddening has the high value advocated by Bessell (1991), and if the temperature of the models are increased, following the decrease of the correct color indices, the metallicity of the two cool stars is hardly changed, the metallicity of the two hot stars is increased, and becomes similar to the metallicity of the field stars, but the spectroscopic data (profile of the Hα line of A19, equivalent widths of the lines of different excitation potential) will not be well represented by the hotter models. The pattern of the relative abundances is hardly dependent on the reddening value. It is essentially the same for the four stars. In particular, it shows essentially solar ratios for the CNO elements relative to iron. The s–process elements are not generally enhanced, but europium (r–process) is high in the four stars.

154.070 **CCD photometry of the outer–halo cluster Palomar 15.**
W. E. Harris.
Astron. J., Vol. 102, No. 4, p. 1348 – 1357 (Oct 1991). Current Physics Microform No.: 9110D1880.

CCD photometry obtained from the CFHT in (V,B–V) is presented for the sparse globular cluster Palomar 15. The data extend to a limit V $\sim$ 24, just sufficient to reach the main–sequence turnoff. Comparative matching to other moderately metal–poor globular clusters ([Fe/H] $\sim$ –1.8) suggests that it has a basically normal CMD morphology for this metallicity (blue–horizontal branch, moderately steep giant and subgiant branches), and an age $\approx$14 Gy if fitted to oxygen–enhanced isochrones. The CFHT data confirm the surprisingly high reddening of E_{B-V} = 0.40 found in a previous study by Seitzer & Carney (1990). Palomar 15 appears to contain a significant number of blue stragglers; about six are visible in a rough sequence forming a luminous upward extension of the main sequence. The cluster is $\approx$36 kpc distant from the Galactic center, thus lies near the outer edge of the "normal" halo cluster region; its structural parameters suggest that it may now be experiencing tidal truncation and dissolution into the field.

154.071 **Star counts, blue stragglers, and mass segregation in NGC 7492.**
P. Côté, H. B. Richer, G. G. Fahlman.
Astron. J., Vol. 102, No. 4, p. 1358 – 1370 (Oct 1991). Current Physics Microform No.: 9110D1890.

New B, V photometry of the sparse globular cluster NGC 7492 is used to construct the cluster color–magnitude diagram (CMD) down to about three magnitudes below the main–sequence turnoff. Adopting a reddening of E(B–V) = 0.00 mag, the authors confirm the cluster metallicity to be [Fe/H] = –1.51. The relative–age technique of VandenBerg et al. (1990) implies that NGC 7492 is approximately 1.7$\pm$1.2 Gyr older than NGC 6752. A distance modulus of $(m-M)_0$ = 17.09$\pm$0.20 mag has been derived by comparing the NGC 6752 fiducial given in VandenBerg et al. to the NGC 7492 fiducial and adopting $(m-M)_0$ = 13.17$\pm$0.17 mag for NGC 6752. The authors also report the discovery of 27 candidate blue stragglers which are found to be more centrally concentrated than the cluster subgiants of similar brightness. If mass segregation has occurred in the cluster, this result suggests that the blue stragglers are more massive than the stars at the main–sequence turnoff. A total of 3331 stars brighter than V = 24.0 mag have been counted in ten concentric annuli centered on the cluster core. There is no evidence for mass segregation in either the surface density profiles or in the luminosity functions at different radial positions. This is not unexpected since the stellar mass range sampled by the data is quite small, M $\sim$ 0.80–0.60 $M_\odot$. A comparison of the observed and theoretical luminosity functions suggests that the power–law index of the mass function is x = 0.1$\pm$0.5.

154.072 **BV CCD photometry of the low–latitude globular cluster NGC 5946.**
G. Alcaino, W. Liller, F. Alvarado, E. Wenderoth.
Astron. J., Vol. 102, No. 4, p. 1371 – 1380 (Oct 1991). Current Physics Microform No.: 9110D1903. With plates 39 – 40.

The authors present BV CCD photometry of the previously unstudied, highly concentrated, low galactic latitude globular cluster NGC 5946 observed with the 2.2 m Max–Planck–Institut telescope at ESO/La Silla. The photometry extends to V $\sim$ 21.5 reaching about 4 mag below the brighter edge of the horizontal branch. The horizontal branch extends continuously from blue to red color indices in the magnitude range 17.0 < V < 18.5. The adopted reddening of E(B–V) = 0.53$\pm$0.05 places the cluster at 10.7 kpc from the Sun and 5.8 kpc from the galactic center, assuming R_0 = 8.5 kpc. Indications are that the cluster is more metal–poor than previously believed.

154.073 **Color and population gradients in globular clusters.**
S. Djorgovski, G. Mallen–Ornelas, G. Piotto.
Symposium on the Formation and Evolution of Star Clusters as Part of the 102. Summer Scientific Meeting of the Astronomical Society of the Pacific, p. 262 – 264 (1991). – See Abstr. 012.050 for the main entry.

The authors present preliminary results from a survey for color and population gradients in globular cluster cores. Color gradients, in the sense of becoming bluer inwards, are always found in post–core–collapse clusters. They seem to be caused by the demise of red giants, and possibly an increased number of blue stragglers. This may be a consequence of stellar interactions during and after the core collapse. No gradients are seen in clusters with King–model morphology.

154.074 **The detection of the red giant luminosity function bump in a sample of galactic globular clusters.**
F. R. Ferraro, F. Fusi Pecci, D. A. Crocker, R. T. Rood, R. Buonanno.
Symposium on the Formation and Evolution of Star Clusters as Part of the 102. Summer Scientific Meeting of the Astronomical Society of the Pacific, p. 265 (1991). Abstract. – See Abstr. 012.050 for the main entry.

154.075 **A differential comparison of 47 Tuc and M71.**
J. N. Heasley, C. A. Christian.
Symposium on the Formation and Evolution of Star Clusters as Part of the 102. Summer Scientific Meeting of the Astronomical Society of the Pacific, p. 266 – 268 (1991). – See Abstr. 012.050 for the main entry.

New BV photometry of the metal–rich Galactic globular cluster M71 is presented. Comparison of photometry with that of the well–studied cluster 47 Tuc and with theoretical isochrones suggests that M71 is $\sim$3 Gyr older than the former.

154.076 **A search for faint variable stars in M71.**
P. J. C. Hodder, J. M. Nemec, H. B. Richer, G. G. Fahlman.
Symposium on the Formation and Evolution of Star Clusters as Part of the 102. Summer Scientific Meeting of the Astronomical Society of the Pacific, p. 269 – 271 (1991). Also published in "Cool stars, stellar systems, and the Sun", edited by G. Wallerstein, ISBN 0–937707–27–9, San Francisco, CA (USA), Astron. Soc. Pac. Conf. Ser., Vol. 9, p. 485 – 487 (1991). – See Abstr. 012.050 for the main entry.
A 60″ × 100″ field in the metal–rich globular cluster M71 was examined for variable blue stragglers (VBS) and other faint variable stars using 75 CCD images. One VBS was positively identified. A second variable, with a similar period, but much fainter and much bluer was also found. Two candidate eclipsing binary systems have also been discovered – these appear to be worthy of further study. No main sequence variables were discovered.

154.077 **Photometry of two metal–poor low–latitude globular clusters.**
K. A. Janes, J. N. Heasley.
Symposium on the Formation and Evolution of Star Clusters as Part of the 102. Summer Scientific Meeting of the Astronomical Society of the Pacific, p. 272 – 274 (1991). – See Abstr. 012.050 for the main entry.
The authors have undertaken a preliminary investigation of the metal–poor low–latitude globular clusters NGC 6293 and NGC 6333 using CCD images from the University of Hawaii 2.24–meter telescope, obtained on two nights when the seeing was about 0.6″. The color–magnitude (C–M) diagrams for the two clusters show well–defined giant branches and blue horizontal branches and reach just to the main–sequence turnoff. Their C–M diagram morphology resemble each other very closely as well as that of M92; a comparison of NGC 6293 and NGC 6333 with the latter yields probable reddening values of ~0.47 and ~0.44 mag and apparent distance modulii of ~16.0 and ~15.8, respectively, placing the clusters roughly 1 – 2 Kpc from the Galactic center.

154.078 **CCD four–color observations of globular clusters.**
A. G. D. Philip.
Symposium on the Formation and Evolution of Star Clusters as Part of the 102. Summer Scientific Meeting of the Astronomical Society of the Pacific, p. 285 – 286 (1991). Abstract. – See Abstr. 012.050 for the main entry.

154.079 **Cleaning up the M28 color–magnitude diagram.**
R. F. Rees, K. M. Cudworth.
Symposium on the Formation and Evolution of Star Clusters as Part of the 102. Summer Scientific Meeting of the Astronomical Society of the Pacific, p. 290 – 292 (1991). – See Abstr. 012.050 for the main entry.
Reductions of PDS microdensitometer scans of 20 plates over an epoch range of 92 years have yielded proper motions and photometry for more than 250 stars down to V ~ 16 in the region of the globular cluster M28 (NGC 6626). Membership probabilities for these stars, including several variables, are derived, and a color–magnitude diagram of probable members is presented and discussed. The absolute proper motion and space velocity of the cluster are derived.

154.080 **The horizontal branch luminosity– metallicity relation: preliminary work.**
A. Sarajedini, A. Lederman.
Symposium on the Formation and Evolution of Star Clusters as Part of the 102. Summer Scientific Meeting of the Astronomical Society of the Pacific, p. 293 – 295 (1991). – See Abstr. 012.050 for the main entry.
Using the Yale Luminosity Functions, the authors parametrize the absolute magnitude of the red giant branch (RGB) "bump" in terms of [Fe/H] and age. Using available values for these quantities in seven globular clusters all relative to NGC 104 (47 Tuc), the authors compute a slope of 0.15 ± 0.05 (s.d.) for the variation of HB luminosity with metallicity. They use the recent determinations of RR Lyrae absolute magnidutes in NGC 6121 (M4) via the Baade–Wesselink Method to set their zeropoint. This gives a value of 1.00 ± 0.08 mag. for the zeropoint. In general, a slope of 0.32 or greater is ruled out by this analysis unless a majority of the clusters considered here are younger than 47 Tuc by ≈2 Gyrs.

154.081 **Evolution versus pulsation on the horizontal branch.**
N. R. Simon.
Symposium on the Formation and Evolution of Star Clusters as Part of the 102. Summer Scientific Meeting of the Astronomical Society of the Pacific, p. 296 – 298 (1991). – See Abstr. 012.050 for the main entry.
The authors proposes a test for the horizontal branch evolutionary tracks of globular clusters by means of observable properties of RR Lyrae stars.

154.082 **Horizontal–branch evolution with time–dependent convective overshooting.**
A. V. Sweigart.
Symposium on the Formation and Evolution of Star Clusters as Part of the 102. Summer Scientific Meeting of the Astronomical Society of the Pacific, p. 299 – 301 (1991). – See Abstr. 012.050 for the main entry.
Horizontal–branch (HB) sequences have been computed for different rates of convective overshooting at the edge of the convective core. The author finds that the convective core undergoes a series of "breathing pulses" during most of the HB phase. The average behavior of these sequences closely mimics the behavior of sequences computed with canonical semiconvection.

154.083 **Compact objects in globular clusters.**
C. D. Bailyn.
Symposium on the Formation and Evolution of Star Clusters as Part of the 102. Summer Scientific Meeting of the Astronomical Society of the Pacific, p. 307 – 323 (1991). – See Abstr. 012.050 for the main entry.
The recent flood of progress, both observational and theoretical, relating to compact objects in globular clusters is reviewed. These new results pose a number of problems relating to the stellar populations and dynamics of the clusters.

154.084 **The role of binaries in cluster dynamical evolution.**
S. L. W. McMillan.
Symposium on the Formation and Evolution of Star Clusters as Part of the 102. Summer Scientific Meeting of the Astronomical Society of the Pacific, p. 324 – 338 (1991). – See Abstr. 012.050 for the main entry.
Binary star systems are believed to play a critical part in influencing the dynamical evolution of star clusters, both during and after core collapse. The mechanisms through which binaries form in globular cluster are reviewed, and the processes whereby they affect their environment are discussed, along with some of the many approaches that have been used to study them. Soon, it will become both feasible and desireable to undertake detailed modeling of globular clusters on a star–by–star basis. Some recent and proposed advances in this area are outlined.

154.085 **Observations of reexpanding cores of globular clusters.**
M. Auriere, G. Coupinot, S. Ortolani.
Symposium on the Formation and Evolution of Star Clusters as Part of the 102. Summer Scientific Meeting of the Astronomical Society of the Pacific, p. 339 – 342 (1991). – See Abstr. 012.050 for the main entry.
The authors present unpublished material concerning NGC 6397 and a new investigation, using a wavelet analysis, of the archetype of post core collapse globular clusters M15.

154.086 **The structure and evolution of rich star clusters in the Large Magellanic Cloud.**
R. A. W. Elson.
Symposium on the Formation and Evolution of Star Clusters as Part of the 102. Summer Scientific Meeting of the Astronomical Society of the Pacific, p. 350 – 356 (1991). – See Abstr. 012.050 for the main entry.

The detailed structure of the cores of rich star clusters in the Large Magellanic Cloud (LMC), with ages $\sim 10^7$–10^9, is investigated. Surface brightness profiles of the older clusters are well represented by models with a King–like core, while those of many of the younger clusters show departures from such models in the form of bumps, sharp "shoulders", and central dips. There is an upper limit to the core radii which increases with age from ≈ 1 pc for the youngest clusters, to ≈ 6 pc for the oldest ones. This trend probably reflects expansion of the cores driven by mass loss from evolving stars. Implications for the formation and early evolution of the clusters are discussed.

154.087 **Optical counterparts of globular cluster X–ray sources.**
M. Aurière, L. Koch–Miramond.
Symposium on the Formation and Evolution of Star Clusters as Part of the 102. Summer Scientific Meeting of the Astronomical Society of the Pacific, p. 360 – 362 (1991). – See Abstr. 012.050 for the main entry.

The authors present new results on some sources they are currently investigating.

154.088 **The optical counterpart of the X–ray source in NGC 6712.**
C. D. Bailyn, A. Cool, J. E. Grindlay, H. N. Cohn, P. M. Lugger, R. D. McClure.
Symposium on the Formation and Evolution of Star Clusters as Part of the 102. Summer Scientific Meeting of the Astronomical Society of the Pacific, p. 363 – 365 (1991). – See Abstr. 012.050 for the main entry.

The authors report the results of CFHT and CTIO photometry of the X–ray error box in NGC 6712. They confirm earlier reports of a faint extremely blue star, which is also consistent with the position of a newly discovered faint radio source. The authors have succeeded in resolving the blend of this star and its neighbors, and find a U–band magnitude of 20.2 ± 0.4 and $U-B < -1.0$.

154.089 **Globular clusters in the Sombrero galaxy (M104).**
T. J. Bridges, D. A. Hanes.
Symposium on the Formation and Evolution of Star Clusters as Part of the 102. Summer Scientific Meeting of the Astronomical Society of the Pacific, p. 366 – 369 (1991). – See Abstr. 012.050 for the main entry.

The authors have carried out deep CCD photometry of the globular cluster system of the Sombrero galaxy (M104) at the prime focus of the CFHT. The spatial distribution of the rich cluster system does not match that of the underlying halo light, with the former being rather less centrally concentrated. The turnover in the luminosity function (LF) is not detected, which implies either that M104 is significantly more distant than has long been believed or that the LF in M104 is unlike that in the spiral galaxies of the Local Group. The authors stress, however, that the analysis is as yet complete only for their frames taken in moderate seeing; the full reduction of the deeper best–seeing frames may lead to modify these conclusions.

154.090 **The tidal radii of globular clusters in M31.**
J. Cohen, K. Freeman.
Symposium on the Formation and Evolution of Star Clusters as Part of the 102. Summer Scientific Meeting of the Astronomical Society of the Pacific, p. 377 – 380 (1991). Also published in Astron. J., Vol. 101, No. 2, p. 483 – 489 (Feb 1991). – See Abstr. 012.050 for the main entry.

The authors have determined tidal radii for 30 globular clusters in the halo of M31 from deep direct images. Although the uncertainty of each individual measurement is large, in the mean the tidal radii suggest that the M31 globular clusters have orbital eccentricities and mass–to–light ratios that are very similar to those of Galactic globular clusters.

154.091 **Core oscillations in globular cluster evolution: recent results.**
H. N. Cohn, P. M. Lugger, R. P. Grabhorn, J. L. Breeden, N. H. Packard, B. W. Murphy, P. Hut.
Symposium on the Formation and Evolution of Star Clusters as Part of the 102. Summer Scientific Meeting of the Astronomical Society of the Pacific, p. 381 – 384 (1991). – See Abstr. 012.050 for the main entry.

The authors review recent results from studies of globular cluster dynamical evolution following core collapse that consider energy input from hard binary stars. A number of evolutionary simulations have indicated that binaries drive an unstable core reexpansion, resulting in large amplitude oscillations of the core radius and central density. Core oscillations are found to occur for all binary formation mechanisms considered: (1) the star formation process itself, (2) three–body interactions, and (3) two–body tidal capture. The authors summarize two recent studies of core oscillations.

154.092 **Towards the comparison of observations of globular clusters and Fokker–Planck simulations.**
G. A. Drukier, G. G. Fahlman, H. B. Richer.
Symposium on the Formation and Evolution of Star Clusters as Part of the 102. Summer Scientific Meeting of the Astronomical Society of the Pacific, p. 385 – 387 (1991). – See Abstr. 012.050 for the main entry.

Fokker–Planck simulations of globular cluster evolution, including the post–collapse phase, are presented. The authors include the tidal stripping of a steady galactic field and heating due to three–body binary formation and destruction in the theory. The calculations are designed to match observed number density profiles and mass functions of the globular cluster M71. Constraints placed on the models by the observations, and the limitations of this approach, are illustrated.

154.093 **The globular cluster system and the mass of M31: a progress report.**
L. Federici, B. Marano, F. Bònoli, F. Fusi Pecci, V. A. Lipovetsky (*V. A. Lipovetskij*), S. I. Neizvestny (*S. I. Neizvestnyj*), N. Spassova.
Symposium on the Formation and Evolution of Star Clusters as Part of the 102. Summer Scientific Meeting of the Astronomical Society of the Pacific, p. 388 – 391 (1991). – See Abstr. 012.050 for the main entry.

New spectroscopic observations of M31 globular cluster candidates located at large galactocentric distances have been carried out at the 4 m KPNO and 6 m BTA telescope. The estimate of the M31 mass, obtained by applying statistical mass estimators to the total sample of 162 confirmed clusters, gives a value of $M_{M31} = 4.1 \times 10^{11} M_{\odot}$ at $R \sim 16$ kpc. Cluster metallicities have also been investigated using appropriate calibrations of metallicity versus $(V-K)_0$ colors. There is no evidence for any metallicity radial gradient. However, strong selection effects are still present in the IR sample.

154.094 **Evolving, dynamical models for collapsed–core globular clusters: M15 and NGC 6624.**
R. P. Grabhorn, H. N. Cohn, P. M. Lugger, B. W. Murphy.
Symposium on the Formation and Evolution of Star Clusters as Part of the 102. Summer Scientific Meeting of the Astronomical Society of the Pacific, p. 392 – 395 (1991). – See Abstr. 012.050 for the main entry.

The authors investigate the stellar populations of the collapsed–core globular clusters M15 and NGC 6624 by fitting the observed surface brightness profiles for both clusters and the projected velocity dispersion profile for M15 with theoretical profiles computed evolving Fokker–Planck models. They find that M15 is best fit by models with a mass function slope of $x = 0.90$ (where $x = 1.35$ corresponds to the Salpeter mass function) and a maximum remnant mass of $m_u = 1.3\,M_{\odot}$. NGC 6624 is well fit by models with values of x ranging from 0 to 2 which predict central velocity dispersions ranging from $18\ \mathrm{km\,s^{-1}}$ to $5\ \mathrm{km\,s^{-1}}$, respectively. The model fits to both clusters are inconsistent with the finding that the main–sequence mass functions of globular clusters grow flatter with increasing metallicity.

154.095 **CVs and millisecond pulsar progenitors in globular clusters.**
J. E. Grindlay, A. M. Cool, C. D. Bailyn.
Symposium on the Formation and Evolution of Star Clusters as Part of the 102. Summer Scientific Meeting of the Astronomical Society of the Pacific, p. 396 – 400 (1991). – See Abstr. 012.050 for the main entry.

The recent discovery of a large population of millisecond pulsars in globular clusters, together with earlier studies of both low luminosity X–ray sources and LMCXs in globulars, suggest there should be significant numbers of CVs in globulars. Although they have been searched for without success in selected cluster X–ray source fields, systematic surveys are lacking and would constrain binary production and both stellar and dynamical evolution in globular clusters. The authors describe the beginnings of such a search, using narrow band Hα imaging, and the sensitivities it might achieve.

154.096 **Early phases of LMC star clusters?**
E. Kontizas, A. Michalitsianos, M. Kontizas.
Symposium on the Formation and Evolution of Star Clusters as Part of the 102. Summer Scientific Meeting of the Astronomical Society of the Pacific, p. 404 – 406 (1991). – See Abstr. 012.050 for the main entry.

On the basis of optical and IUE observations the authors have investigated the possibility that some of the very young stellar systems of the LMC, such as SL 360, may be regarded as globular clusters at the very early phases of their dynamical evolution after the gas cloud expulsion.

154.097 **Are there two disk star cluster systems in the LMC?**
M. Kontizas, E. Kontizas.
Symposium on the Formation and Evolution of Star Clusters as Part of the 102. Summer Scientific Meeting of the Astronomical Society of the Pacific, p. 407 – 409 (1991). – See Abstr. 012.050 for the main entry.

The surface distribution of star clusters in the LMC has revealed the existence of two elliptical systems superimposed. A small surface density ($5\ \mathrm{cl\,deg^{-2}}$) outer ellipse outlining the large system and a higher density ($15 - 25\ \mathrm{cl\,deg^{-2}}$) elliptical inner system. The major axis of the two subsystems form an angle of 60°. The central subsystem contains all youngest populous globular star clusters from both the stellar and dynamical point of view. The large outer system seems to contain all old globular clusters, showing a clumpy distribution and outlining two arms at the northeast and southwest. The small LMC clusters were found to have masses $< 10^3 M_\odot$ and are distributed all over the large elliptical system.

154.098 **Progress towards an understanding of the blue straggler problem.**
D. W. Latham, A. A. E. Milone.
Symposium on the Formation and Evolution of Star Clusters as Part of the 102. Summer Scientific Meeting of the Astronomical Society of the Pacific, p. 410 (1991). Abstract. – See Abstr. 012.050 for the main entry.

154.099 **U–band CCD surface brightness profiles for candidate collapsed–core globular clusters.**
P. M. Lugger, H. N. Cohn, J. E. Grindlay, C. D. Bailyn, P. L. Hertz.
Symposium on the Formation and Evolution of Star Clusters as Part of the 102. Summer Scientific Meeting of the Astronomical Society of the Pacific, p. 414 – 417 (1991). – See Abstr. 012.050 for the main entry.

The authors have determined U–band CCD surface brightness profiles for 15 of the 21 globular clusters that have been characterized as having "collapsed" cores. They have fitted these profiles with two types of seeing–convolved models: single–mass King models and power laws. The authors find that in many cases the characterization of a profile as being either "king" or "cusp" is somewhat subtle. Of the clusters in our sample, 10 appear to have central power–law cusps with either unresolved or marginally resolved cores, 3 appear to have resolved cores with surrounding power–law structure that is unlike a King model, and 2 are approximately described by King model fits. The observed power–law cusps are consistent with the predictions of Fokker–Planck models for core–collapse.

154.100 **Binaries among the blue stragglers in M67.**
A. A. E. Milone, D. W. Latham, R. L. Kurucz, J. A. Morse.
Symposium on the Formation and Evolution of Star Clusters as Part of the 102. Summer Scientific Meeting of the Astronomical Society of the Pacific, p. 424 – 426 (1991). – See Abstr. 012.050 for the main entry.

The authors have used calculated spectra as templates to derive radial velocities for eleven of the classical blue stragglers in M67 by cross–correlations. The mean velocities for all eleven stars are close to the cluster mean, thus confirming their membership in the cluster. For the blue straggler F190 they derive a spectroscopic orbit with period 4.1829 d, eccentricity 0.21, and mass function $1.6 \times 10^{-3} M_\odot$. Thus F190 is a good candidate for a blue straggler that resulted from mass transfer in a close binary. Several of the other blue stragglers show evidence for low–amplitude velocity variations.

154.101 **Discovery with the Sigma telescope of hard X–rays from the globular cluster Terzan 2.**
D. Barret, S. Mereghetti, J. P. Roques, P. Mandrou, L. Salotti, F. Lebrun, P. Laurent, J. Ballet, E. Churazov, M. Gilfanov (*M. R. Gil'fanov*), R. Sunyaev (*R. A. Syunyaev*), N. G. Khavenson, I. Chulkov, B. Novikov, A. V. Kuznetzov, A. Dyachkov (*A. V. D'yachkov*).
Astrophys. J., Lett., Vol. 379, No. 1, p. L21 – L24 (20 Sep 1991).

During the survey of the Galactic center performed with the Sigma hard X–ray / soft gamma–ray telescope, a new source has been discovered in the 35 – 200 keV energy range. The location of the source is consistent with the position of the globular cluster Terzan 2. Although other possbilities are discussed, it is likely that the hard X–rays emission originates in X1724–308 (4U 1724–30), a source of X–ray bursts located in the globular cluster. Under this assumption, it would be the first detection of a low–mass X–ray binary in this energy range.

154.102 **Extragalactic globular clusters. III. Metallicity comparisons and anomalies.**
J. P. Brodie, J. P. Huchra.
Astrophys. J., Vol. 379, No. 1, p. 157 – 167 (20 Sep 1991). = UCO/Lick Obs. Bull., No. 1187.

Using a method based on the strengths of six absorption line indices measured in integrated spectra, the authors have derived metallicities for 22 globular clusters associated with the Sc galaxy M33, 10 globular clusters associated with the giant elliptical galaxy M87, eight globular clusters associated with the Sb(r)I–II galaxy M81, and three globular clusters associated with the Fornax dwarf elliptical galaxy. In addition, mean metallicities have been derived for 38 bright galaxies, mostly ellipticals, 29 dwarf elliptical galaxies in the Virgo cluster, 10 dwarf elliptical galaxies in the Fornax cluster and four local group dwarf galaxies. Comparing these results with metallicities previously derived for 149 clusters in M31 and with the Milky Way cluster metallicities, the authors show that the mean metallicity of a cluster system is linearly related to the luminosity of the parent galaxy. A similar relationship is suggested between galaxy metallicity and luminosity for the bright galaxies and dwarf galaxies. The slope of the relationship between the mean metallicity of a cluster system and parent galaxy luminosity is very similar to the slope of the relationship between galaxy metallicity and luminosity, but there is an offset in the sense that the clusters are more metal poor.

154.103 **Lithium in halo stars: constraining the effects of helium diffusion on globular cluster ages and cosmology.**
C. P. Deliyannis, P. Demarque.
Astrophys. J., Vol. 379, No. 1, p. 216 – 220 (20 Sep 1991).
It has been proposed that helium diffusion can reduce age estimates for globular clusters through its effect on the main-sequence lifetime as well as on the morphology of isochrones near the turnoff. Using stellar evolutionary models with diffusion, the authors show that observations of lithium in extreme halo stars provide crucial constraints on the magnitude of the effects of helium diffusion. The flatness (and/or possible slight increasing slope) of the observed Li – T_{eff} relation (the Spite Li plateau) severely constrains diffusion Li isochrones, which tend to curve downward toward higher T_{eff}. Previously published age reductions of order 20% – 30% seem to be ruled out by the Li flatness constraint from the current observations; however, age reductions of order 10% might be possible.

154.104 **Post–core collapse scenarios of globular clusters.**
B. W. Murphy, H. N. Cohn.
Symposium on the Formation and Evolution of Star Clusters as Part of the 102. Summer Scientific Meeting of the Astronomical Society of the Pacific, p. 436 – 438 (1991). – See Abstr. 012.050 for the main entry.
The post–collapse evolution of globular clusters is examined with several different heating mechanisms. These mechanisms are (1) the inclusion of dynamically formed binaries; (2) a central black hole and stellar evolution; and (3) physical interactions of stars. The first two mechanisms are examined with a mass spectrum of stars included in the calculations. The third is examined first with a single main–sequence mass group and then with main–sequence and heavy degenerate mass groups.

154.105 **Primordial binary stars in the low density globular clusters NGC 5053 and NGC 5466.**
C. Pryor, R. A. Schommer, E. W. Olszewski.
Symposium on the Formation and Evolution of Star Clusters as Part of the 102. Summer Scientific Meeting of the Astronomical Society of the Pacific, p. 439 – 442 (1991). – See Abstr. 012.050 for the main entry.
The authors have measured ~280 radial velocities for 72 stars in the directions of the globular clusters NGC 5053 and NGC 5466 in order to search for binaries. Fourteen stars show evidence of having variable velocities and at least 5 of these are very likely to be binaries.

154.106 **Binary star fraction in star clusters.**
R. W. Romani, M. D. Weinberg.
Symposium on the Formation and Evolution of Star Clusters as Part of the 102. Summer Scientific Meeting of the Astronomical Society of the Pacific, p. 443 – 445 (1991). – See Abstr. 012.050 for the main entry.
An unresolved binary system will be systematically brighter and redder than the primary component alone. These systems may measurably widen the main sequence in the color–magnitude diagram. The authors show analytically that a "second sequence" is indeed expected for observed IMFs and standard color–magnitude relations. The relative intensity of the second sequence depends on the steepness of the mass function. For NGC 6341 (M92) and NGC 7099 (M30), the authors construct a model single–star + binary color–magnitude diagram based on the best fitting isochrones and power–law IMFs. The binary fraction f is estimated by maximum likelihood. The authors find that the highest quality M92 field contains $\lesssim 8\%$ binaries. Two fields observed in M30 have a significantly lower fraction of binary candidates, $\lesssim 3\%$.

154.107 **Speckle imaging of globular cluster cores.**
B. J. Sams III.
Symposium on the Formation and Evolution of Star Clusters as Part of the 102. Summer Scientific Meeting of the Astronomical Society of the Pacific, p. 446 – 448 (1991). – See Abstr. 012.050 for the main entry.
Speckle imaging of globular cluster cores can produce very high resolution images of the occasional bright stars which give the clusters their "granular" appearance. A recently developed predictive model of speckle imaging shows that the majority of stars, which are faint and blue, have a surface brightness too low to image clearly by speckle. Nevertheless, the speckle images are valuable in the particular case of NGC 7099, which has a blue, unresolved center. The author presents the first speckle images of a cluster core, which show that the central condensation in NGC 7099 breaks up into multiple components at scales of 0.3".

154.108 **Disk–shocking and the mass function of globular clusters.**
M. Stiavelli, G. Piotto, M. Capaccioli, S. Ortolani.
Symposium on the Formation and Evolution of Star Clusters as Part of the 102. Summer Scientific Meeting of the Astronomical Society of the Pacific, p. 449 – 451 (1991). – See Abstr. 012.050 for the main entry.
The authors consider the hypothesis that disk–shocking can induce the observed correlation between the position of a globular cluster in the Galaxy and the slope of its mass–function. They show that the effect is present in realistic, though approximate, models.

154.109 **Central velocity dispersion measurements in M30 and five other centrally concentrated GGCs.**
S. R. Zaggia, G. Piotto, M. Capaccioli.
Symposium on the Formation and Evolution of Star Clusters as Part of the 102. Summer Scientific Meeting of the Astronomical Society of the Pacific, p. 458 – 460 (1991). – See Abstr. 012.050 for the main entry.
New ESO CASPEC measurements of the central velocity dispersion in the core–collapsed galactic globular cluster NGC 7099 = M30 and of other centrally concentrated GGCs are presented and briefly discussed against the striking behaviour of M15.

154.110 **Abundance of elements of the Fe–group in globular cluster stars: trends with metal abundances.**
R. G. Gratton.
Symposium on the Formation and Evolution of Star Clusters as Part of the 102. Summer Scientific Meeting of the Astronomical Society of the Pacific, p. 479 – 484 (1991). – See Abstr. 012.050 for the main entry.
Abundance ratios for Fe group elements ($20 \leqslant Z \leqslant 28$) and Si have been determined from a reanalysis of the spectra of 30 giants in globular clusters. Full consideration of the hyperfine structure was made for lines of the odd elements. Differences between Fe abundances deduced from neutral and singly ionized lines are clearly correlated with overall metallicity. This correlation is briefly discussed in connection with uncertainties in reddening and effective temperature estimates, model atmospheres, and departures from LTE. On the whole, stars in globular clusters have the same pattern of abundances of Fe–group elements of metal poor field stars; there are marginal evidence that cluster stars have slightly larger overabundance of Si and underabundance of Mn. Rather large Si excesses have been found in M4 and M5.

154.111 **CNO abundances in globular cluster stars.**
R. A. Bell.
Symposium on the Formation and Evolution of Star Clusters as Part of the 102. Summer Scientific Meeting of the Astronomical Society of the Pacific, p. 485 – 494 (1991). – See Abstr. 012.050 for the main entry.
Some recent work on carbon, nitrogen and oxygen abundances and $^{12}C/^{13}C$ ratios in globular cluster stars is reviewed. The variation of carbon abundance with M_v in the very metal poor cluster NGC 6397 is described, with comments on the influence of the oxygen abundance and overall metal abundance on the final results. Estimates of the oxygen abundances in NGC 288 and 362 are given. Carbon and nitrogen abundances are given for stars over the range $0.5 < M_v < 5.0$ in 47 Tuc. Variations in CN band strength are seen in the turnoff region stars, with the inevitable implication that some of the variation seen in the more luminous stars is simply this earlier variation.

154.112 **Metallicities and dynamics of extragalactic globular clusters.**
J. P. Brodie, J. P. Huchra.
Symposium on the Formation and Evolution of Star Clusters as Part of the 102. Summer Scientific Meeting of the Astronomical Society of the Pacific, p. 495 – 511 (1991). – See Abstr. 012.050 for the main entry.
There is at least one property of globular clusters which is not a universal quantity. The mean metallicity of a cluster system depends on the luminosity of the parent galaxy and the dependence is roughly linear. There exist abundance anomalies. The abundances of certain elements vary from system to system and there are hints that these variations may be primordial in origin.

154.113 **The halo and old disk clusters of the Milky Way.**
R. Zinn.
Symposium on the Formation and Evolution of Star Clusters as Part of the 102. Summer Scientific Meeting of the Astronomical Society of the Pacific, p. 532 – 542 (1991). – See Abstr. 012.050 for the main entry.
The similarities between the old star clusters and the stellar populations of the Milky Way are discussed. While the clusters and stars of the halo and thick disk populations have identical properties, the globular clusters in the direction of the nuclear bulge are not representative of its stellar population. Recent developments in the dating of globular clusters are reviewed, and the implications for the evolution of the Galaxy are briefly discussed.

154.114 **Globular cluster metallicities from Ca II triplet strengths in cluster giants.**
T. E. Armandroff, G. S. Da Costa.
Symposium on the Formation and Evolution of Star Clusters as Part of the 102. Summer Scientific Meeting of the Astronomical Society of the Pacific, p. 543 (1991). Abstract. – See Abstr. 012.050 for the main entry.

154.115 **A statistical approach to the estimate of the helium abundance in galactic globular clusters.**
G. Bono, M. Fulle, L. Pulone.
Symposium on the Formation and Evolution of Star Clusters as Part of the 102. Summer Scientific Meeting of the Astronomical Society of the Pacific, p. 547 – 549 (1991). – See Abstr. 012.050 for the main entry.
A new method, based on the statistical modelling of the stellar mass distribution of the horizontal branches, has been applied to estimate the primordial helium abundance of the globular clusters M92 and M5. On the basis of this approach $Y = 0.25$ should be considered as the most probable mean value.

154.116 **Multicolor CCD photometry of globular clusters in three Virgo elliptical galaxies.**
J. Couture, W. E. Harris.
Symposium on the Formation and Evolution of Star Clusters as Part of the 102. Summer Scientific Meeting of the Astronomical Society of the Pacific, p. 559 – 561 (1991). – See Abstr. 012.050 for the main entry.
The authors present CCD photometry in B, V and I bands for globular clusters in three Virgo elliptical galaxies M87, NGC 4472 and NGC 4649. Color gradients have been detected in NGC 4472 and NGC 4649 while no gradient exists in M87. This suggests that merger processes have been more important in the giant galaxy M87 while the gradients in the fairly normal galaxies are consistent with the result of a dissipative collapse.

154.117 **Survey for LMC star clusters with 3 < age < 12 Gyrs.**
G. S. Da Costa.
Symposium on the Formation and Evolution of Star Clusters as Part of the 102. Summer Scientific Meeting of the Astronomical Society of the Pacific, p. 562 (1991). Abstract. – See Abstr. 012.050 for the main entry.

154.118 **Long-slit spectroscopy of CN absorption in the M31 globular clusters Vet 12 and Vet 282.**
T. J. Davidge.
Symposium on the Formation and Evolution of Star Clusters as Part of the 102. Summer Scientific Meeting of the Astronomical Society of the Pacific, p. 566 – 568 (1991). – See Abstr. 012.050 for the main entry.
Long-slit spectroscopic observations of the M31 globular clusters Vet 12 and Vet 282 are used to investigate the spatial behaviour of the 3883 Å and 4215 Å CN bands in these objects. Radial variations at the 2–σ level are detected in each cluster.

154.119 **C and N abundances from spectroscopy of 47 Tuc main-sequence stars.**
J. E. Hesser, M. M. Briley, R. A. Bell.
Symposium on the Formation and Evolution of Star Clusters as Part of the 102. Summer Scientific Meeting of the Astronomical Society of the Pacific, p. 572 – 574 (1991). – See Abstr. 012.050 for the main entry.

154.120 **UBVR photometry of young, massive M31 clusters.**
J. R. King, R. H. Lupton.
Symposium on the Formation and Evolution of Star Clusters as Part of the 102. Summer Scientific Meeting of the Astronomical Society of the Pacific, p. 575 – 577 (1991). – See Abstr. 012.050 for the main entry.
The authors present the results of UBVR photometry of ten bright, blue disk clusters in M31. The V band luminosities range from 3×10^4 to $4 \times 10^5 L_\odot$. The mean, (B–V) color is <0.2, comparable to the massive disk clusters found in the Magellanic Clouds but much less than Galactic globular clusters. The ages estimated from stellar population models range from 2×10^7 to 2×10^8yr. Stellar population models of mass-to-luminosity ratio as a function of age yield masses for these clusters of 3×10^3 to $5 \times 10^4 M_\odot$ – larger than Galactic open clusters, but comparable to the masses of young, rich clusters found in the LMC.

154.121 **Evidence regarding a metallicity spread in M22.**
J. B. Laird, R. J. Wilhelm, R. C. Peterson.
Symposium on the Formation and Evolution of Star Clusters as Part of the 102. Summer Scientific Meeting of the Astronomical Society of the Pacific, p. 578 – 580 (1991). – See Abstr. 012.050 for the main entry.
The authors have investigated the possibility of a star-to-star spread in heavy element abundance in M22 using low-signal-to-noise spectra of 26 cluster giants. They find $\sigma([m/H]) = 0.20$, which is probably consistent with observational scatter. Based on a plot of metallicity vs. color, it appears that the observed range in color is more likely due to variations in reddening than in abundance.

154.122 **Similarities of thick disk clusters in the Milky Way and intermediate-age SMC clusters.**
T. Richtler, W. Seggewiss, A. Vallenari, K. S. de Boer.
Symposium on the Formation and Evolution of Star Clusters as Part of the 102. Summer Scientific Meeting of the Astronomical Society of the Pacific, p. 598 – 600 (1991). – See Abstr. 012.050 for the main entry.
The authors emphasize the possibility that some Galactic globular clusters of the disk system are the counterparts of intermediate-age Magellanic Cloud globular clusters.

154.123 **Synthetic integrated globular cluster spectra.**
M. J. Tripicco, R. A. Bell.
Symposium on the Formation and Evolution of Star Clusters as Part of the 102. Summer Scientific Meeting of the Astronomical Society of the Pacific, p. 601 – 604 (1991). – See Abstr. 012.050 for the main entry.
The authors present preliminary results of their research into synthetic integrated globular cluster spectra. Synthetic stellar spectra have been used to build integrated spectra which correspond to selected points in a new grid of isochrones.

Synthetic horizontal branch stars are then added to simulate the effect of varying HB morphology.

154.124 **Space motions of globular clusters NGC 362 and NGC 6218 (M12).**
P. Brosche, H.-J. Tucholke, A. R. Klemola, S. Ninković, M. Geffert, P. Doerenkamp.
Astron. J., Vol. 102, No. 6, p. 2022 – 2027 (Dec 1991). Current Physics Microform No.: 9112C1998.
The authors measured the absolute proper motions of NGC 6218 (M12) against distant galaxies and of NGC 362 against stars of the Small Magellanic Cloud (typical accuracy $\pm 0''.13$ $cent^{-1}$). The results are combined with those of the two studied clusters NGC 4147 and NGC 5466. A variety of parameters is determined within the approximation of a simple Galactic potential, including the description of the last transits of the clusters through the main plane of the Galaxy. The path of the orbits in a meridional section (based on a more refined potential) is provided for the last 10^{10}yr. The authors draw attention to orbits with north-south asymmetry, as exemplified by the tube orbit of NGC 4147.

154.125 **The behavior of magnesium and sodium lines among giants within bimodal-CN globular clusters.**
G. H. Smith, G. D. Wirth.
Publ. Astron. Soc. Pac., Vol. 103, No. 669, p. 1158 – 1175 (Nov 1991).
Echelle spectra have been obtained of Mg I and Na I absorption lines in the wavelength range 5100 Å – 5600 Å for small numbers of CN-strong and CN-weak red giants in the globular clusters NGC 6752, 47 Tuc, and M22. In the case of the former two clusters, both of which exhibit bimodal CN band strength distributions, the CN-strong stars observed are generally found to exhibit enhanced Na lines but comparable Mg lines in comparison to CN-weak stars of similar (B-V) color and V magnitude. By contrast, a spectrum of a CN-rich giant in M22 is found to show stronger lines of Na, Mg, Fe, Ti, and Sc than that of a CN-poor giant of comparable magnitude. It is suggested on the basis of available data that the nucleosynthesis products of helium- and carbon-burning reactions are heterogeneously distributed within globular clusters at the time of their formation.

154.126 **BV photometry of RR Lyrae variables in the globular cluster M5.**
J. Storm, B. W. Carney, J. A. Beck.
Publ. Astron. Soc. Pac., Vol. 103, No. 670, p. 1264 – 1278 (Dec 1991).
The authors present BV CCD photometry for eleven RR Lyrae variables in the globular cluster M5 (NGC 5904). New ephemerides are computed, and V, B, and (B-V) light curves are given. A color-magnitude diagram based on two V and two B short-exposure frames is given, which reaches from the tip of the giant branch to below the main-sequence turnoff. The RRab and RRc variables overlap in mean (B-V) colors, but not in effective temperatures computed using periods, blue amplitudes, and the metallicity of the cluster. This supports the previous claim about the unsuitability of mean (B-V) values to estimate mean or equilibrium temperatures for RR Lyrae stars.

154.127 **Velocity dispersion of stars and the mass of the globular clusters M4, M5, M10, M12, and M71.**
A. S. Rastorguev, N. N. Samus'.
Sov. Astron. Lett., Vol. 17, No. 5, p. 388 – 396 (Sep 1991). Current Physics Microform No.: 9204X1335. English translation of Pis'ma Astron. Zh., Tom 17, No. 10, p. 915 – 931 (1991).
Measurements of radial velocities of 159 red stars in the globular clusters M3, M4, M5, M10, M12, M13, and M71 are presented. These measurements were taken using a correlation spectrometer. One-dimensional stellar velocity dispersions are calculated for M4, M5, M10, M12, and M71 within the framework of King's single-component models, and the masses of the clusters and the mass-to-light ratios (M/L_V) are determined. The mass of M71 is $(1\text{–}2)\cdot 10^4 M_\odot$, whereas the masses of the remaining clusters lie in the range $(1\text{–}2)\cdot 10^5 M_\odot$. The ratio M/L_V is 1 – 2 (in solar units, with uncertainties of 40 – 100%). M5 is shown to rotate at 0.25 ± 0.13 km/sec·pc, and the position angle of the projected rotation axis is 130° (in the model of rigid-body rotation). The rotation axis is almost perpendicular to the major axis of the apparent ellipse of central density contours, suggesting a possible link between the flattening and rotation of the cluster.

154.128 **The stellar content and structure of the globular cluster NGC 5053.**
G. G. Fahlman, H. B. Richer, J. Nemec.
Astrophys. J., Vol. 380, No. 1, p. 124 – 139 (10 Oct 1991).
Five overlapping fields in the metal poor globular cluster NGC 5053 have been observed with the CCD system at the CFH Telescope. The data have been used to obtain deep star counts, $V \leqslant 25.0$, covering an east-west strip extending through the cluster center to close to the tidal radius. From the star counts, a minimum mass-to-visual light ratio of $M/L_V = 0.68 \pm 0.07$ is determined. The stellar luminosity functions at different radial positions within the cluster were converted to the corresponding mass functions using a 16 Gyr isochrone of metallicity [Fe/H] = -2.03. Mass segregation is observed. For a mass function of the form $n(m) \sim m^{-(1+x)}$, the authors find that the core of the cluster has $x \cong 0.8$, whereas the outer region has $x \cong 2.0$. The global luminosity function for the stars with $V \leqslant 23.5$ corresponds very well to a mass function with $x = 1.5$. The projected mass and number density profiles for the cluster have been fitted with a series of representative multimass King models. A global mass-to-light ratio of $M/L_V \approx 1.2$ is derived. The authors also present a new CM diagram for NGC 5053, which reaches $V = 23.5$, well below the main-sequence turn-off. From a comparison with the main-sequence fiducial of M92, they estimate the distance modulus and reddening of NGC 5053 to be $(m\text{–}M)_V = 16.08$ and $E(B\text{–}V) = 0.06$, respectively.

154.129 **Interaction of mass lost in globular clusters with the halo medium.**
D. J. Faulkner, G. H. Smith.
Astrophys. J., Vol. 380, No. 2, p. 441 – 451 (20 Oct 1991).
This paper examines the possibility that the diffuse, soft (0.5 – 3.5 keV) X-ray features observed in three globular clusters by Hartwick, Cowley, and Grindlay arise through the interaction of intracluster gas with the Galactic halo medium. The bulk of this gas is assumed to be shed during the red giant evolution of cluster stars, although its properties might be greatly modified by other events such as novae. Three types of possible interaction with the halo medium are considered: (1) the formation of a bow shock as an intracluster gas reservoir translates with a cluster and thus impacts the medium; (2) the direct interaction of a shell ejected from a star by an energetic mass loss event accompanying the helium core flash; and (3) interactions associated with nova outbursts in a cluster. In no case does the interaction produce soft X-ray emission which is luminous enough to explain the observed diffuse sources. Thus the various theories describing the possible fate of gas shed by stars within globular clusters continue to lack observational verification.

154.130 **The Galactic globular cluster system: theoretical constraints for α-enhanced compositions.**
D. Bencivenni, F. Caputo, M. Manteiga, M. L. Quarta.
Astrophys. J., Vol. 380, No. 2, p. 484 – 494 (20 Oct 1991).
Synthetic horizontal branches are constructed from a grid of horizontal-branch (HB) evolutionary tracks with the α-elements (O, Ne, Mg, and Ca) enhanced by a factor of 5 with respect to the solar-scaled mixture ([α/Fe] = 0.7) and various surface helium [Y(HB) = 0.20, 0.25) and heavy-element ([Fe/H] = -1.27, -2.27) abundances. The relevant constraints on HB morphology and mean properties of RR Lyrae variables are presented and discussed. The comparison of α-enriched with canonical ([α/Fe] = 0) computations shows that, with the helium content and the overall metallicity Z fixed, the luminosity log L(3.83) and the mass-to-light (M/L) ratio A(3.83) of the ZAHB at the mean temperature of the RR Lyrae gap, have virtually no dependence on the ratio between α- and heavy-element abundances. This leads to general log L(3.83) versus log Z and A(3.83) versus log Z

relations which hold irrespective of the adopted enhancement of α–elements. From the models, the authors determine the mean luminosity ⟨log L⟩ and the M/L ratio ⟨A⟩ of RR Lyrae variables as functions of metallicity as well as HB morphology. With these relations, the observed relationship between M/L ratio and [Fe/H] (Sandage) is analyzed. The result is that all the observational data – except for M15 – could be fitted to within the errors with [α/Fe] in the range 0.7 – 0.4 and a constant primordial helium Y(MS) $\sim$ 0.23. For a sample of globular clusters observed by Buonanno, Corsi, and Fusi Pecci, the authors derive, at [α/Fe] = 0, a mean age of 16.9 ± 1.5 Gyr, with on the average ~17.9 Gyr at [Fe/H] = –2.3 and ~15.8 Gyr at [Fe/H] = –0.8. Concerning the effect of no scaled–solar mixture, it is found that the age is rather independent of the degree of α–enhancement.

154.131 **The pregalactic formation of globular clusters.**
E. I. Rosenblatt.
AIP Conf. Proc., No. 222, p. 210 – 214 (15 Apr 1991). Current Physics Microform No.: 9106E0210. – See Abstr. 012.069 for the main entry.

154.132 **Globular cluster mass functions.**
H. B. Richer, G. G. Fahlman, R. Buonanno, F. Fusi Pecci, L. Searle, I. B. Thompson.
Astrophys. J., Vol. 381, No. 1, p. 147 – 159 (1 Nov 1991).

Deep imaging data in the I bandpass for three globular clusters were used to construct their luminosity and mass functions. These new data for ω Cen, M5, and NGC 6752, together with existing data for M13, NGC 6397, and M71 were then employed to investigate the cluster mass function slopes and their evolution. A related question also considered was the relation between the currently observed and the IMF slopes. While definitive answers regarding the evolution of globular cluster mass function slopes are not yet possible, mainly due to the small existing sample of observed mass functions extending down to low masses ($M < 0.2\,M_\odot$) and to restrictions in the theoretical modeling, certain trends are becoming apparent. In particular, it does appear that evolution of cluster stars has been important in modifying the cluster mass function slopes for at least two and possibly three of the clusters studied. A very important result is that some (and perhaps all) clusters probably have very steep IMFs with the slope likely exceeding 2.5 (Salpeter value 1.3). This result suggests that low mass Population II stars may be an important component of the mass budget of the halo if the spheroid's mass function is similar to that seen in globular clusters.

154.133 **Carbon isotopic abundances in giant stars in the CN–bimodal globular clusters NGC 6752 and M4.**
N. B. Suntzeff, V. V. Smith.
Astrophys. J., Vol. 381, No. 1, p. 160 – 172 (1 Nov 1991).

Carbon isotopic abundances (^{12}C and ^{13}C) are determined for 45 red giants in the CN–bimodal globular clusters M4 and NGC 6752 through low resolution infrared spectroscopy of the first–overtone CO bands. All of the red giants are found to have quite low isotope ratios ($^{12}C/^{13}C \sim 3 - 10$), indicating that all of these low–mass, low metallicity giants have undergone extensive mixing of their outer envelopes with material processed through the CN cycle. It is also found that those giants with the strongest CN bands have the lowest C abundances: this implies an anticorrelation of the C and N abundances and again suggests the presence of various fractions of CN–processed material mixed with unprocessed material in the outer envelopes of these giants. In addition, low–resolution spectra of the violet CN (3883 Å) and CH (4300 Å) bands in main–sequence and subgiant stars in the globular clusters NGC 6752, M5, and 47 Tuc are found to exhibit CN and CH anticorrelations, implying that even main–sequence stars have differing (and anticorrelated) abundances of C and N. It is argued that, with all of the observations taken together, the CN–bimodal globular clusters most probably have "primordial" abundance variations of C and N, as evidenced by the main–sequence stars, while all of the low–mass giants undergo extensive mixing as they ascend the red giant branch, as evidenced by low $^{12}C/^{13}C$ ratios. It is speculated that an early generation of intermediate–mass asymptotic giant branch stars polluted the early globular cluster environment with slow stellar winds composed of CN–processed material.

154.134 **Globular clusters: New tools of the trade.**
C. Bailyn.
Nature, Vol. 352, No. 6332, p. 195 – 196 (18 Jul 1991).

The author reports on the discovery of further millisecond pulsars in globular clusters, on 47 Tuc, which hosted not one but 11 millisecond pulsars. This exacerbates a problem which has created some contention recently: the birthrate of millisecond pulsars.

154.135 **Discovery of then millisecond pulsars in the globular cluster 47 Tucanae.**
R. N. Manchester, A. G. Lyne, C. Robinson, N. D'Amico, M. Bailes, J. Lim.
Nature, Vol. 352, No. 6332, p. 219 – 221 (18 Jul 1991). Letter–to–the–editor.

In the past four years a total of 13 millisecond pulsars have been found in 12 different globular clusters. These pulsars are believed to be old neutron stars that have been spun up ('recycled') in low–mass X–ray binary systems, although some may have been formed by the accretion–induced collapse of white dwarfs in binaries. The globular cluster 47 Tucanae has an especially dense core, and is therefore a likely site for millisecond pulsar formation. Using the Parkes radiotelescope, the authors have now detected ten additional millisecond pulsars in 47 Tuc, more than half of which are members of binary systems. Almost half of the known millisecond pulsars and more than a quarter of the known binary pulsars now reside in this one cluster.

154.136 **Evacuation of gas from globular clusters by winds from millisecond pulsars.**
D. N. Spergel.
Nature, Vol. 352, No. 6332, p. 221 – 222 (18 Jul 1991). Letter–to–the–editor.

Why is so little gas observed within globular clusters? A typical globular cluster contains $\sim 10^3$ post–turnoff stars, each of which will lose $\sim 0.2\,M_\odot$ before its asymptotic giant branch phase of evolution, and $\sim 0.1\,M_\odot$ during this phase, and so should accumulate $10^2 - 10^3 M_\odot$ of gas in the 10^8–year interval between passages of the globular cluster through the galactic disk. (At each disk passage, gas ram pressure will remove the accumulated material.) But observational searches show that there is scant intracluster gas; in many clusters, there is less than $1\,M_\odot$ of gas, orders of magnitude less than theoretical predictions. The recent discovery of multiple millisecond pulsars in globular clusters may resolve this long–standing problem: the relativistic wind from these pulsars is enough to drive gas from stellar mass loss out of the globular cluster.

154.137 **Stellar evolution: Collisions and mergers.**
J. M. Nemec.
Nature, Vol. 352, No. 6333, p. 286 – 287 (25 Jul 1991).

The author reports on the origin and evolutionary history of blue stragglers in globular clusters.

154.138 **Blue stragglers in the core of the globular cluster 47 Tucanae.**
F. Paresce, M. Shara, G. Meylan, D. Baxter, P. Greenfield, R. Jedrzejewski, A. Nota, W. B. Sparks, R. Albrecht, C. Barbieri, J. C. Blades, A. Boksenberg, P. Crane, J. M. Deharveng, M. J. Disney, P. Jakobsen, T. M. Kamperman, I. R. King, F. Macchetto, C. D. Mackay, G. Weigelt.
Nature, Vol. 352, No. 6333, p. 297 – 301 (25 Jul 1991).

High–resolution observations of the core of the globular cluster 47 Tucanae with the Faint Object Camera on the Hubble Space Telescope reveal a high density of "blue straggler" stars, occupying the upper end of the main sequence from which all stars in the cluster should have long since evolved. Their presence in the dense core supports the hypothesis that they formed by stellar collision and coalescence, and, as the heaviest objects in the cluster, have drifted to the core.

154.139 **Gamma-ray emission from millisecond pulsars in globular clusters.**
K. Chen.
Nature, Vol. 352, No. 6337, p. 695 – 697 (22 Aug 1991). Letter-to-the-editor.
Recent observations indicate that a typical globular cluster may contain hundreds of millisecond pulsars. The author estimates the γ-ray luminosity of globular clusters, assuming that they indeed harbour large numbers of millisecond pulsars. An upper limit obtained by the COS-B satellite on the γ-ray emission from 47 Tucanae, which contains at least 10 millisecond pulsars, is too large to constrain the model presented here, but the estimated luminosities are high enough to be detectable by the recently launched Gamma Ray Observatory. In the near future, γ-ray observations will thus be able to test this model, and the theory of γ-ray emission by millisecond pulsars.

154.140 **Ages of globular clusters.**
P. Demarque, C. P. Deliyannis, A. Sarajedini.
NATO Advanced Research Workshop on Observational Tests of Inflation, p. 111 – 129 (1991). – See Abstr. 012.076 for the main entry.
A brief historical introduction is given. Recent advances in precision photometry of faint stars in globular clusters and in the modeling of turnoffs and synthetic horizontal branches, indicate that there is a spread in age of at least 4 Gyr among Galactic globular clusters. On the observational side, chemical abundances, [Fe/H] and [O/Fe], are the major sources of uncertainty. Two effects hitherto ignored in evolutionary calculations and currently under study, rotation and diffusion, are discussed. The diffusion of helium near the main sequence could be important in reducing the derived ages of the oldest clusters; its efficiency, which is poorly known, can however be constrained and evaluated by lithium abundance observations. Finally, an approximate chronology of the Galaxy is presented on the basis of our current best understanding of cluster dating. The lowest age derived for the cluster M92 is 14 Gyr, which puts a strong constraint on the cosmological timescale.

154.141 **Globular cluster ages and cosmology.**
A. Renzini.
NATO Advanced Research Workshop on Observational Tests of Inflation, p. 131 – 146 (1991). – See Abstr. 012.076 for the main entry.

154.142 **CCD-photometry of galactic globular clusters. III. NGC 6171.**
F. R. Ferraro, G. Clementini, F. Fusi Pecci, R. Buonanno.
Mon. Not. R. Astron. Soc., Vol. 252, No. 3, p. 357 – 377 (1 Oct 1991).
B and V CCD photometry is presented for 1565 stars in the low-concentration globular cluster NGC 6171 = M107. The main conclusions drawn from the analysis of the colour-magnitude diagram are presented and discussed in this paper.

154.143 **Ca II emission from old red giants in the globular cluster M15.**
A. K. Dupree, B. A. Whitney.
Nature, Vol. 354, No. 6351, p. 284 – 286 (28 Nov 1991). Letter-to-the-editor.
Murphy et al. recently observed characteristic emission of Ca II from the globular cluster M15, and argued that this emission came from a population of primordial binary stars. Their conclusion appears theoretically attractive in relation to the study of the dynamics and evolution of globular clusters. The authors argue here, however, that the Ca II K-line emission in various types of hard binary systems differs in strength and width from the M15 spectra, and they demonstrate that the M15 emission closely resembles that from metal-deficient red giants of the halo population in our Galaxy. Chromospheric activity, indicated by emission in Ca II, Hα and Mg II, occurs in red giants in both globular clusters and halo-population field giants. A simple detection of Ca II emission is thus not a unique signature of binaries, and in the case of M15 the authors argue that red giants are the more likely cause.

154.144 **Inhomogeneities in globular clusters.**
Chun Munsuk.
Proc. Astron. Soc. Aust., Vol. 9, No. 1, p. 37 – 40 (1991). – See Abstr. 012.078 for the main entry.
Inhomogeneities in globular clusters are reviewed with the observational evidence for chemical abundance variations from star to star in individual clusters and the large-scale structural variation of clusters. The reality of the radial colour gradient is tested in 47 Tuc (NGC 104). The result shows that the observed radial colour gradient comes from the integration of the calculated colours of individual stars. The cause of this radial colour variation is the result of the concentration of evolved stars and the reddening of the main sequence in the central region. The author proposes that the CNO abundance gradient in the early stage of a cluster's formation is the interpretation of the observed radial colour gradient.

154.145 **Bow shocks in globular clusters?**
D. J. Faulkner, G. H. Smith.
Proc. Astron. Soc. Aust., Vol. 9, No. 1, p. 132 – 134 (1991). – See Abstr. 012.078 for the main entry.
It has been suggested that diffuse X-ray sources observed in globular clusters may arise in bow shocks formed by the interaction of intracluster gas with a tenuous halo medium. This paper presents a study both of the cluster space motions and of the energetics of any possible bow shocks which would seem to rule out this hypothesis.

154.146 **Globular clusters as tests of stellar evolution.**
J. E. Hesser.
Symposium in Celebration of the 50. Anniversary of McDonald Observatory of the University of Texas at Austin: Frontiers of stellar evolution, p. 185 – 212 (1991). – See Abstr. 012.085 for the main entry.
This paper reviews recent spectroscopic tests of mixing within globular cluster stars, as well as current and future uses of color-magnitude diagrams to test stellar evolutionary models. Aspects of the conflicting evidence for mixing within stars of globular clusters in two metallicity groups divided somewhat arbitrarily at [M/H] = –1.6 are examined. Stress is placed on the CNO group of elements and on the lack of suitable [O/Fe] data. The second focus, HR diagrams and stellar evolution, is subdivided into comments on color-magnitude diagram morphology, binaries in clusters, luminosity functions, and comparisons of stellar models with observations for the purposes of deriving ages. The latter ends on an optimistic note that it should be possible, with existing techniques, to measure much more reliable relative ages throughout the Galactic halo than heretofore recognized.

154.147 **Blue-sided gaps in the horizontal branches of globular clusters.**
G. Bono, M. Fulle, L. Pulone.
Symposium in Celebration of the 50. Anniversary of McDonald Observatory of the University of Texas at Austin: Frontiers of stellar evolution, p. 593 (1991). Abstract. – See Abstr. 012.085 for the main entry.

154.148 **Carbon isotope ratios in globular cluster giants.**
J. A. Brown, G. Wallerstein.
Symposium in Celebration of the 50. Anniversary of McDonald Observatory of the University of Texas at Austin: Frontiers of stellar evolution, p. 595 (1991). Abstract. – See Abstr. 012.085 for the main entry.

154.149 **Four-color observations of BHB stars as test of stellar evolution.**
A. G. D. Philip.
Symposium in Celebration of the 50. Anniversary of McDonald Observatory of the University of Texas at Austin: Frontiers of stellar evolution, p. 613 (1991). Abstract. – See Abstr. 012.085 for the main entry.

154.150 **A new approach to aging globular clusters.**
L. Pulone, G. Bono, M. Fulle, E. Brocato.
Symposium in Celebration of the 50. Anniversary of McDonald Observatory of the University of Texas at Austin: Frontiers of stellar evolution, p. 614 (1991). Abstract. – See Abstr. 012.085 for the main entry.

154.151 **Ages of globular clusters derived from BVRI CCD photometry.**
G. Alcaino.
Liller Symposium: Asteroids to quasars, p. 127 – 139 (1991). – See Abstr. 012.087 for the main entry.

154.152 **Metal-rich giant branches of bulge clusters: application to color-magnitude diagrams of M31 and M32.**
E. Bica, B. Barbuy, S. Ortolani.
Astrophys. J., Lett., Vol. 382, No. 1, p. L15 – L17 (20 Nov 1991).
Color-magnitude diagrams (CMDs) for the metal-rich bulge clusters NGC 6528, NGC 6553, and Terzan 1 are calibrated in M_{I_o} versus $(V-I)_o$, and are compared to that of 47 Tuc. In such diagrams the blanketing effects for increasing metallicities are directly estimated. Such studies are important for the interpretation of CMDs of metal-rich populations in Local Group external galaxies. The authors compare their results to published CMDs of M31 and M32 bulges. The available CMDs are not deep enough to reveal the metal-rich populations similar to that of NGC 6553, which indicates that the determinations of average metallicities from the giant branch width in the M31 and M32 fields are lower limits.

154.153 **Deep VLA images of globular clusters.**
H. M. Johnston, S. R. Kulkarni, W. M. Goss.
Astrophys. J., Lett., Vol. 382, No. 2, p. L89 – L92 (1 Dec 1991).
The authors present sensitive high-resolution images of four globular clusters made with the VLA at 1.4 GHz. They detect a variety of objects: known pulsars, an X-ray binary, and two planetary nebulae. Surprisingly, the authors do not detect any emission from the rich cluster Liller 1. If Liller 1 is truly one of the densest clusters in the Galaxy, then this result suggests that the pulsar population in a cluster is a function not merely of the structural parameters. The authors also report the detection of seven unidentified sources in or near the cores of three clusters. Some of these may be unrelated background sources, but it is probable that some may be associated with the low-luminosity X-ray sources identified by the Einstein satellite.

154.154 **Galactic globular clusters with extremely low metallicities.**
J. J. Clariá, D. Geisler, D. Minniti.
Bol. Asoc. Argent. Astron., No. 36, p. 3 – 14 (1990). In Spanish. – See Abstr. 012.095 for the main entry.
The authors present some preliminary results obtained at Cerro Tololo Inter-American Observatory from Washington photometry of 165 red giants belonging to 15 globular clusters. The metallicities have been determined using a new empirical calibration, which accounts for the effect of decreasing abundance sensitivity of Δ(C–M) with decreasing temperature. The mean [Fe/H] values obtained for 10 of the 15 clusters range between –2.0 and –3.0. These values considerably extend the metallicity distribution of the galactic globular cluster system towards values close to –3.0, thus favouring the secondary formation theory of galactic globular clusters.

154.155 **Two high-velocity stars shot out from the core of the globular cluster 47 Tucanae.**
G. Meylan, P. Dubath, M. Mayor.
Astrophys. J., Vol. 383, No. 2, p. 587 – 593 (20 Dec 1991).
The authors report the discovery of two high-velocity stars in the core of the globular cluster 47 Tucanae. Located respectively at about 3″ and 38″ from the center, they have radial velocities relative to the cluster equal to –36.7 and +32.4 km s^{-1}, corresponding to 4.0 and 3.6 times the velocity dispersion in the core, which is 9.1 km s^{-1}. These velocities are of the order of, or larger than, the central escape velocity estimates. The long time baseline between repeated observations and the constancy of the radial velocity values indicate that neither of these two stars is a binary or a pulsating star. They have positions on the red giant branch and asymptotic giant branch in the color-magnitude diagram which favor their membership. Because of its rather high Galactic latitude (b = –44°), 47 Tuc does not suffer too strong a Galactic pollution by field stars. Therefore, the probability of finding two such stars inside the central 40″ of 47 Tuc is extremely low. There is no indication that these stars are not members of the cluster. The main mechanism which may be called upon for explaining these two interlopers is the ejection out of the core by stellar encounters between a single star and a binary, or between two binary stars. It is noted that the mechanism of ejection must have operated quite recently; and the two stars, having had only little time to evolve in the color-magnitude diagram, were ejected when they were already giants.

154.156 **Dynamics of the young LMC cluster NGC 1866.**
P. Fischer, D. L. Welch, P. Côté, M. Mateo, B. F. Madore.
J. R. Astron. Soc. Can., Vol. 85, No. 4, p. 189 (Aug 1991). Abstract. – See Abstr. 012.091 for the main entry.

154.157 **Deep CCD photometry of the globular cluster M15.**
P. R. Durrell, W. E. Harris.
J. R. Astron. Soc. Can., Vol. 85, No. 4, p. 194 (Aug 1991). Abstract. – See Abstr. 012.091 for the main entry.

154.158 **A colour-magnitude study of the distant "second parameter" globular cluster Pal 14.**
S. Holland, W. E. Harris.
J. R. Astron. Soc. Can., Vol. 85, No. 4, p. 194 (Aug 1991). Abstract. – See Abstr. 012.091 for the main entry.

154.159 **Three-colour photometry for the galactic globular cluster NGC 6517.**
J. J. Kavelaars, D. A. Hanes.
J. R. Astron. Soc. Can., Vol. 85, No. 4, p. 194 (Aug 1991). Abstract. – See Abstr. 012.091 for the main entry.

154.160 **Star counts, blue stragglers and mass segregation in NGC 7492.**
P. Côté, H. B. Richer, G. G. Fahlman.
J. R. Astron. Soc. Can., Vol. 85, No. 4, p. 213 (Aug 1991). Abstract. – See Abstr. 012.091 for the main entry.

154.161 **Ages of globular clusters derived from BVRI CCD photometry.**
G. Alcaino, W. Liller, F. Alvarado, E. Wenderoth.
Rev. Mex. Astron. Astrofis., Vol. 22, No. 2, p. 309 – 310 (Oct 1991). Abstract. – See Abstr. 012.127 for the main entry.

154.162 **Two new variables in the central region of the globular cluster M5.**
V. V. Kravtsov.
Pis'ma Astron. Zh., Tom 17, No. 12, p. 1101 – 1104 (Dec 1991). In Russian. English translation in Sov. Astron. Lett., Vol. 17, No. 6.
Two new variable stars are found in the central region ($r \lesssim 2'$) of the globular cluster M5. Their coordinates, periods and amplitudes of variability are determined. Both light curves and periods of these stars are typical of RR Lyr variables (RR_c).

154.163 **"A dust bubble within the globular cluster NGC 6624?"**
J. C. Forte, S. Cellone, M. Mendez, I. Vega.
Bol. Asoc. Argent. Astron., No. 36, p. 418 – 419 (1990). Abstract. In Spanish and English. – See Abstr. 012.095 for the main entry.

154.164 **UV–(2000 Å) observations of galactic metal–poor globular clusters.**
M. Laget, D. Burgarella, B. Milliard, J. Donas.
Adv. Space Res., Vol. 11, No. 11, p. 139 – 142 (1991). – See Abstr. 012.107 for the main entry.

Several flights of a balloon–borne 40–cm aperture ultraviolet imaging telescope have provided observational data on metal–poor, little reddened, galactic globular clusters M3, M5, M13, M92 and their surroundings. In the case of M13, a catalogue of 605 entries includes positions, cross–indentifications and ultraviolet magnitudes for objects detected outside the ~1 arcmin radius central part. The blue horizontal branch and its extension down to the region of hot subdwarfs at $M_v \sim 4$ are detected. The various classes of objects detected in the field are reviewed.

The formation and evolution of star clusters. Proceedings.
See Abstr. 012.050.

Study of galactic and extra galactic star clusters with a large ground based optical telescope.
See Abstr. 013.028.

The first year of observations with the Hubble Space Telescope.
See Abstr. 035.184.

Ultraviolet imaging of globular clusters and galaxies.
See Abstr. 036.316.

HST and dense stellar systems.
See Abstr. 051.019.

A grid of synthetic spectra for the determination of effective temperature, gravity and metallicity of F, G, and K stars. I. Description of the method.
See Abstr. 064.002.

The evolution of He–burning stars: horizontal and asymptotic branches in Galactic globulars.
See Abstr. 065.063.

The status of stellar evolution models.
See Abstr. 065.069.

Stellar models with envelope overshoot.
See Abstr. 065.070.

Implications of helium diffusion for globular cluster isochrones and luminosity functions.
See Abstr. 065.124.

Remarks on the evolution of low–mass stars: more or less opaque?
See Abstr. 065.144.

WIMP–like stellar models and the subgiant branch.
See Abstr. 065.149.

Problems for the WIMP hypothesis?
See Abstr. 065.150.

The formation of neutron stars in the Galaxy.
See Abstr. 067.111.

An improved metal abundance calibration for the Washington system.
See Abstr. 113.036.

New empirical metal abundance calibrations for Washington photometry of G and K giants.
See Abstr. 113.048.

A new abundance calibration for the Washington system and some early results.
See Abstr. 113.049.

HUT observations of ultraviolet–bright stars in globular clusters.
See Abstr. 114.010.

[O/Fe] in giants in Omega Centauri.
See Abstr. 114.014.

An abundance analysis of ROA 24, a likely post–AGB star in Omega Centauri.
See Abstr. 114.015.

Exploring the properties of stellar CO bands using synthetic photometry and spectroscopy.
See Abstr. 114.070.

CN and CH variations on the M5 subgiant branch.
See Abstr. 114.082.

C, N, and O abundances in red giants in globular clusters.
See Abstr. 114.083.

What has happened to Arcturus?
See Abstr. 114.130.

A new age calibrator for red horizontal branch populations.
See Abstr. 115.029.

Evidence for strong activity in Population II binaries.
See Abstr. 116.019.

Light curve analyses of the eclipsing binaries in the globular cluster NGC 5466.
See Abstr. 119.012.

RR Lyrae variables in the field around NGC 2210.
See Abstr. 122.009.

RR Lyrae luminosity calibrations using the Baade–Wesselink method.
See Abstr. 122.066.

The new variable stars in the central region of the globular cluster M53.
See Abstr. 123.017.

Bemerkungen zu V 133 in Kernnähe des Kugelhaufens M3.
See Abstr. 123.023.

Beobachtungen von V 81, V 82, V 83, V 84, V 85 im Kugelhaufen M3.
See Abstr. 123.024.

Beobachtungen von V 86, V 87 im Kugelhaufen M3.
See Abstr. 123.030.

Gamma–ray emission from millisecond pulsars in globular clusters.
See Abstr. 126.009.

Zehn neue Millisekunden–Pulsare in einem einzigen Kugelsternhaufen.
See Abstr. 126.010.

Do we really need α–enhanced theoretical isochrones?
See Abstr. 126.034.

Eclipses of the ablating binary pulsar PSR 1744–24A.
See Abstr. 126.080.

Star formation in globular clusters and dwarf galaxies, and implications for the early evolution of galaxies.
See Abstr. 131.226.

Stellar luminosity functions as probes of star formation history.
See Abstr. 131.229.

A grain–heated, dusty planetary nebula in M22.
See Abstr. 134.053.

X1850–086.
See Abstr. 142.038.

Galactic wind and metal enrichment in spheroidal star systems.
See Abstr. 151.020.

Merger of binary globular clusters: case of unequal masses.
See Abstr. 151.060.

The few–body problem in astrophysics.
See Abstr. 151.067.

Hydrodynamics of collisions between binary stars.
See Abstr. 151.090.

Dissolution of tidally limited star clusters in galaxies by passing giant molecular clouds or other massive objects.
See Abstr. 151.091.

The effect of dynamical friction on globular cluster motion in triaxial galaxies.
See Abstr. 151.093.

Hydrodynamic simulations of close triple encounters.
See Abstr. 151.095.

Star cluster evolution with primordial binaries.
See Abstr. 151.096.

Binaries in a medium of fast low–mass objects.
See Abstr. 151.097.

The secular instability of axisymmetric collisionless star clusters.
See Abstr. 151.102.

Dynamical evolution of globular clusters with mass spectrum.
See Abstr. 151.118.

Stellar encounters involving red giants in globular cluster cores.
See Abstr. 151.127.

Post–collapse evolution of globular clusters with stochastic energy sources.
See Abstr. 151.131.

On core collapse.
See Abstr. 151.134.

The formation of primordial binaries in globular clusters by star–disk interactions.
See Abstr. 151.135.

The distribution of stars in a star cluster with a Kerr black hole in its center.
See Abstr. 151.146.

Photometric binary stars in Praesepe and the search for globular cluster binaries.
See Abstr. 153.017.

Color–magnitude diagrams of the open cluster 092–SC18 and the globular cluster Ruprecht 106.
See Abstr. 153.032.

An observational age–metallicity relation for the Galaxy.
See Abstr. 155.003.

Detection of a Galactic color gradient for blue horizontal–branch stars of the halo field and implications for the halo age and density distributions.
See Abstr. 155.005.

Probing the Galactic halo with RR Lyr stars, globular clusters and 3C 273 IUE.
See Abstr. 155.017.

On the formation of globular clusters. I. Dynamical limits on globular cluster metallicities.
See Abstr. 155.031.

Determination of the absolute proper motion of the globular cluster M3.
See Abstr. 155.081.

Inhomogeneous halo collapse and globular cluster metallicities.
See Abstr. 155.085.

Oxygen abundances in halo giants. I. Giants in the very metal–poor globular clusters M92 and M15 and the metal–poor halo field.
See Abstr. 155.086.

BV CCD photometry of five young Large Magellanic Cloud star clusters.
See Abstr. 156.005.

FOC observations of R136a in the 30 Doradus nebula.
See Abstr. 156.011.

Time variation of ellipticity of globular clusters in the Large Magellanic Cloud.
See Abstr. 156.025.

Detection of the helium flash gap in the integrated (U–B) versus (B–V) diagram for 624 Large Magellanic Cloud clusters.
See Abstr. 156.034.

Observational tests of stellar evolution: the Magellanic Clouds.
See Abstr. 156.039.

NGC 2214; a merging binary star cluster in the Large Magellanic Cloud.
See Abstr. 156.044.

Globular cluster systems in galaxies beyond the Local Group.
See Abstr. 157.154.

The M31 globular cluster system.
See Abstr. 157.307.

The globular cluster systems of cD galaxies in poor clusters.
See Abstr. 157.333.

The globular cluster system of NGC 3842.
See Abstr. 157.334.

Globular cluster systems in central cD galaxies.
See Abstr. 157.335.

Young globular clusters?
See Abstr. 160.057.

A non–baryonic galactic halo and the real galactic age.
See Abstr. 161.011.

Thermal instabilities and the origin of globular clusters.
See Abstr. 161.221.

Globular clusters as extragalactic distance indicators.
See Abstr. 161.486.

155 Galaxy

155.001 **GRIS observations of ^{26}Al gamma-ray line emission from two points in the Galactic plane.**
B. J. Teegarden, S. D. Barthelmy, N. Gehrels, J. Tueller, M. Leventhal, C. J. MacCallum.
Astrophys. J., Lett., Vol. 375, No. 1, p. L9 – L12 (1 Jul 1991).
The GRIS (Gamma-Ray Imaging Spectrometer) experiment observed the Galactic center region during a balloon flight on 1988 October 28 – 30. Two observations of the Galactic plane were made at l = 0° and 335°, respectively. ^{26}Al gamma-ray line emission (1809 keV) was detected during both observations, although that in the latter was statistically marginal. For an assumed Galactic longitude distribution similar to high-energy gamma rays, a flux of $4.5(+1.5,-1.6)\times10^{-4}$ $cm^{-1}s^{-1}rad^{-1}$ (2.8σ) is calculated for the combined l = 0° and 335° data sets. These observations are consistent with the assumed high-energy gamma-ray distribution, but they are consistent with other distributions as well. The marginal statistics of the observations combined with the broad field of view do not permit meaningful tests of the model distributions to be made. The data do, however, suggest that the ^{26}Al emission is distributed over Galactic longitude rather than confined to a point source. The GRIS data also indicate that the 1809 keV line is broadened.

155.002 **GRIS observations of positron annihilation radiation from the Galactic center.**
N. Gehrels, S. D. Barthelmy, B. J. Teegarden, J. Tueller, M. Leventhal, C. J. MacCallum.
Astrophys. J., Lett., Vol. 375, No. 1, p. L13 – L16 (1 Jul 1991).
The authors report new observations of the Galactic center positron annihilation source in which, for the first time, the 511 keV line has been spectrally resolved and the Galactic plane and center components have been independently measured. High-resolution spectra were obtained of the Galactic center with the Gamma-Ray Imaging Spectrometer (GRIS) during May and October of 1988 balloon flights. The 511 keV line source was found to be turned on again in October of 1988, at a flux of $(11.8\pm1.6)\times10^{-4}$ photons $cm^{-2}s^{-1}$, after being observed in a quiescent state since the early 1980s. There is evidence for variability between May and October. An observation of a point in the Galactic plane 25° west of the center was made during the October flight. The 511 keV line flux for this pointing was significantly lower, less than 4.0×10^{-4} photons $cm^{-2}s^{-1}$, than that for the center pointing. The GRIS data combined with that from other narrow-field observations in the 1980s support the two-component model of the positron annihilation source. The best fit to these data sets gives a flux of $\sim8\times10^{-4}$ photons $cm^{-2}s^{-1}$ for the center point source component in its high state. A strong hard X-ray continuum was detected in the Galactic plane observation. This "diffuse" continuum component makes up the difference between the observed continuum in the GRIS Galactic center pointing and the known point source contributions.

155.003 **An observational age-metallicity relation for the Galaxy.**
A. Strobel.
Astron. Astrophys., Vol. 247, No. 1, p. 35 – 44 (Jul 1991).
An observational age-metallicity diagram for the Galaxy has been constructed on the basis of star cluster data available in the literature. This diagram exhibits two independent and simultaneous metallicity enhancement processes identified with two dynamical galactic substructures: the thin and the thick discs. Within the observational uncertainties both these processes can be satisfactory approximated by infall-inflow models of the chemical evolution with different high, although rapidly decaying with time, star formation and infall rates. The comparison of the galactic age-metallicity relations with that for the Large Magellanic Cloud shows some similarities between processes of the chemical enhancement in the Large Magellanic Cloud and the thin disc subsystem in the Galaxy.

155.004 **Probing Galactic halo dark matter with microlensing.**
R. J. Nemiroff.
Astron. Astrophys., Vol. 247, No. 1, p. 73 – 76 (Jul 1991).
A revised estimate of the probability of microlensing of resolved stars by compact objects with low stellar masses in the Galactic halo is presented using the detection volume formalism (Nemiroff 1989), and computed as a function of lens mass, source amplification, Galactic rotation speed, and extent of the Galactic halo. The probability of lensing at any specific time is calculated and shown to be a weak function of both the extent of the Galactic halo and the mass of the stellar object but proportional to the amplitude of lensing at large amplitude (for $A \geqslant 2.5$). The detection volume formalism is then extended to include relative lens motion. The probable waiting time between observable events is computed and also found to be a weak function of the extent of the Galactic halo but a strong function of the lens mass and source amplification. Practically observable events of microlensing of stars in the Large Magellanic Clouds are detectable for a Milky Way halo composed of objects with mass between $10^{-8}M_\odot$ and $1\ M_\odot$. The effects of multiple star lenses and sources are discussed.

155.005 **Detection of a Galactic color gradient for blue horizontal-branch stars of the halo field and implications for the halo age and density distributions.**
G. W. Preston, S. A. Shectman, T. C. Beers.
Astrophys. J., Vol. 375, No. 1, p. 121 – 147 (1 Jul 1991).
From analysis of a large sample of field blue horizontal-branch stars the authors find that the mean unreddened color, $\langle B-V\rangle_w$, in the color window $-0.02 < (B-V)_0 < +0.18$ increases outward in the Galaxy by ~0.025 in the range 2 kpc to 12 kpc. Calibration of the horizontal-branch (HB) parameter B/(B+R) with two theoretical, age-dependent isochrone parameters and an empirical correlation between $\langle B-V\rangle_w$ and B/(B+R) suggest that the mean age of the field horizontal branch decreases outward in the halo systematically by a few Gyr. HB models appropriate for standard red giant evolution lead to a similar conclusion. The authors calculate space densities of blue horizontal-branch (BHB) stars from the present sample and recalculate space densities for RR Lyrae stars from published data. The RR Lyrae densities are well represented by a spheroidal power-law model in which flattening decreases outward. The power-law exponent for the RR Lyrae stars is $n = -3.2\pm0.1$. BHB stars in a more restricted distance range yield $n = -3.5$. Predicted densities of BHB and RR Lyrae stars in the solar neighborhood, their ratio (6.5/1), and the abundance distribution and horizontal-branch morphologies of globular clusters are used to estimate the local density of red HB stars and the total local HB number density, 42 kpc^{-3}. Total HB populations for 44 globular clusters, derived from published photometry and the use of King models, establish that the visual luminosity of a globular cluster is closely proportional to the number of HB stars that it contains. The ratio is $L_V/N_{HB} = 540\pm40$.

155.006 **The galactic gradient of N/^{15}N.**
F. Matteucci, F. D'Antona.
Astron. Astrophys., Vol. 247, No. 2, p. L37 – L40 (Jul 1991). Letter-to-the-editor.
Predictions are given about the temporal and spatial behaviour of the isotopic ratio N/^{15}N in our Galaxy. It is assumed that N is produced in stars of all masses (but the bulk comes from low and intermediate mass stars) as a purely "secondary" element, whereas ^{15}N originates only from explosive H-burning taking place during nova outbursts. The model of galactic chemical evolution used here already reproduces many features of the solar neighbourhood and the galactic disk, such as the gradient of C/^{13}C. It is predicted that N/^{15}N decreases by a factor of $\cong 1.8$ between the epoch of formation of the solar system and now, and that its gradient along the galactic disk (14–4 kpc) is $\cong +6.8\ kpc^{-1}$.

155.007 **The determination of the radiohalo size on the basis of fluctuations of diffuse radiation.**
A. L. Lazaryan, G. V. Chibisov.
Pis'ma Astron. Zh., Tom 17, No. 6, p. 490 – 494 (Jun 1991). In Russian. English translation in Sov. Astron. Lett., Vol. 17, No. 3, p. 208 – 210 (1991).
A new method of determination of the radiohalo size is suggested based on a statistical study of synchrotron fluctuations where H II regions are used as distance indicators.

155.008 **RGU–photometric space–densities in a field close to the galactic northpole (NGC 4147) compared to the gradients predicted by the five Basle standard models.**
R. P. Fenkart, G. del Río.
Astron. Astrophys., Suppl. Ser., Vol. 89, No. 2, p. 217 – 224 (Aug 1991).
The authors present the results of the last investigation within the frame of the model comparison phase (MCP) of the Basle Halo Program. RGU photometric total (photometric population I+II) space density histograms for late–type giants and 1^m–interval main–sequence M(G)–groups ($3^m < M(G) < 8^m$) along the direction to NGC 4147 ($l = 256°$, $b = +78°$) have been determined with the help of the corresponding Basle method and compared to the gradients predicted by the standard set of five multicomponent models of the MCP, according to the method described in the appendix. The evaluation of the comparison results yields strong evidence for the existence of a thick disk component in the galactic density distribution, confirming so, again, the overall finding of the MCP.

155.009 **A three–component dynamical model of the Galaxy.**
A. A. Sumin, A. M. Fridman, U. A. Haud.
Pis'ma Astron. Zh., Tom 17, No. 8, p. 698 – 701 (Aug 1991). In Russian. English translation in Sov. Astron. Lett., Vol. 17, No. 4.
A three–component model of the Galaxy is built up. It is shown that this model is dynamically close to the six–component model, and therefore it offers the possibility to use a more simple model for investigating the problems of the Galaxy dynamics.

155.010 **Interpretation of the frequency distribution of isochrone ages of nearby F and G stars.**
H. Meusinger.
Astrophys. Space Sci., Vol. 182, No. 1, p. 19 – 34 (Aug 1991).
The frequency distribution of isochrone ages of nearby F and G stars is used in a synthetic approach to the history of the star formation rate (SFR). On the basis of stellar evolutionary sequences, age distributions are calculated for different assumptions on the SFR history. Models with a constant past SFR provide a good fit of the empirical age distributions, unless a feature at ages $\tau \approx 4$ Gyr. If real, a shoulder in the empirical frequency distribution of isochrone ages at 4–5 Gyr points toward an enhanced SFR during a more or less (1 to 6 Gyr) extended period which ceased with a peak about 4 Gyr ago. It is shown that models with such a nonmonotonous SFR history reproduce the empirical stellar age distribution and, moreover, the age–metallicity relation and the metallicity distribution of nearby stars. An era of enhanced star formation in this time interval is consistent with recently published interpretation of several age–dependent properties in different samples of nearby stars.

155.011 **On the dynamical role of the central molecular ring in the seven–component Galaxy model.**
A. A. Sumin, A. M. Fridman, U. A. Haud.
Pis'ma Astron. Zh., Tom 17, No. 9, p. 779 – 786 (Sep 1991). In Russian. English translation in Sov. Astron. Lett., Vol. 17, No. 5.
On the basis of observational data a Galaxy model with sharp kink of the gas component's surface density $\sigma_g(r)$ (two orders of magnitude) near the outer edge of the molecular ring is built up. Within this model the partial contributions of each population to the model curve of rotation of the Galaxy are calculated. It turns out that, the value of dimensionless increment of hydrodynamical instability for the gas component, being small enough with respect to 1, is very close to that in the gravitational field of the whole Galaxy. From this fact two conclusions follow: 1) the unstable gas component of the Galaxy lies near the boundary of hydrodynamical stability, which is in accordance with the Le Chatelier principle. 2) the stellar populations of the Galaxy probably do not affect dynamically the generation of the spiral structure in the gaseous component.

155.012 **Large and small scale structure in the galactic magnetic field from rotation measures of low–b^{II} extragalactic sources.**
A. W. Clegg, J. M. Cordes, J. H. Simonetti, S. R. Kulkarni.
Bull. Am. Astron. Soc., Vol. 23, No. 2, p. 889 (1991). Abstract. – See Abstr. 012.008 for the main entry.

155.013 **Neutral hydrogen in the outer Galaxy.**
A. Diplas, B. D. Savage.
Bull. Am. Astron. Soc., Vol. 23, No. 2, p. 889 (1991). Abstract. – See Abstr. 012.008 for the main entry.

155.014 **Infrared imaging polarimetry of the Galactic center.**
M. Tamura, M. W. Werner.
Bull. Am. Astron. Soc., Vol. 23, No. 2, p. 889 – 890 (1991). Abstract. – See Abstr. 012.008 for the main entry.

155.015 **The luminosity of the Galactic center.**
J. A. Davidson, M. W. Werner, X. D. Wu, D. F. Lester, P. M. Harvey, M. Joy, M. Morris.
Bull. Am. Astron. Soc., Vol. 23, No. 2, p. 890 (1991). Abstract. – See Abstr. 012.008 for the main entry.

155.016 **Structural details of the ionized gas within the light year of the Galactic center.**
F. Yusef–Zadeh.
Bull. Am. Astron. Soc., Vol. 23, No. 2, p. 890 (1991). Abstract. – See Abstr. 012.008 for the main entry.

155.017 **Probing the Galactic halo with RR Lyr stars, globular clusters and 3C 273 IUE.**
G. S. Burks.
Bull. Am. Astron. Soc., Vol. 23, No. 2, p. 895 (1991). Abstract. – See Abstr. 012.008 for the main entry.

155.018 **Compact radio sources near the galactic plane.**
S. Zoonematkermani, D. J. Helfand, R. H. Becker, R. L. White.
Bull. Am. Astron. Soc., Vol. 23, No. 2, p. 895 (1991). Abstract. – See Abstr. 012.008 for the main entry.

155.019 **A 6–cm survey of the Milky Way.**
D. J. Helfand, S. Zoonematkermani, R. H. Becker, R. L. White.
Bull. Am. Astron. Soc., Vol. 23, No. 2, p. 895 (1991). Abstract. – See Abstr. 012.008 for the main entry.

155.020 **A photometric study of the bright cloud B in Sagittarius. VII. 1165 new variable stars and 65 diffuse objects.**
A. Terzan, E. Gosset.
Astron. Astrophys., Suppl. Ser., Vol. 90, No. 3, p. 451 – 493 (Nov 1991).
As part of our on-going photometric study of the bright cloud B in Sagittarius, the authors investigated a new field situated in the direction of the galactic center: 1165 new variable stars were found on the basis of the inspection of B, V and R Schmidt plates acquired at the Mount Palomar Observatory and at the European Southern Observatory (ESO). A catalogue is presented which, for each star, gives the position (equatorial and galactic coordinates), the R magnitudes at the observed extrema, the corresponding epochs as well as the amplitude of variation, and some remarks. A result of the present work is a considerable increase of the surface density per square degree of variable stars in this direction of the central part of the galactic bulge. The amount of variables in the present field B is in fact similar to the one reported for the adjacent field A, which contrasts with the poorness of such stars in field D. The histogram of the

amplitudes derived in the present work supports the hypothesis that the variable stars of type L and/or M are most probably more numerous in the direction of the galactic center than in other regions of the Galaxy. As an additional result of the authors' investigation, new diffuse objects have been detected in the same field. A catalogue is presented of 65 objects with a tentative classification based only on their morphology as seen on the Schmidt plates.

155.021 **Westerbork observations of high–velocity clouds. The data.**
B. P. Wakker.
Astron. Astrophys., Suppl. Ser., Vol. 90, No. 3, p. 495 – 511 (Nov 1991).
The results of Westerbork observations of small–scale structure in high–velocity clouds (HVCs) at 1′ angular and 1 km s^{-1} velocity resolution are presented in the form of a table of observational parameters, maps of hydrogen column density, velocity–right ascension cuts, and histograms of the linewidth.

155.022 **A three–component dynamical model of the Galaxy.**
A. A. Sumin, A. M. Fridman, U. A. Haud.
Sov. Astron. Lett., Vol. 17, No. 4, p. 295 – 296 (Jul 1991). Current Physics Microform No.: 9202X0471. English translation of Pis'ma Astron. Zh., Tom 17, No. 8, p. 698 – 701 (Aug 1991).
A three–component dynamical model of the Galaxy is presented. It is shown that this model is dynamically similar to a six–component model, making it possible to use the simpler model for future studies of problems of Galaxy dynamics.

155.023 **A feasibility study for detection of gravitational microlensing by baryonic dark matter via CCD photometry.**
K. H. Cook, C. Alcock, T. S. Axelrod, D. P. Bennett, H.–S. Park, K. Griest, S. Perlmutter, C. W. Stubbs, K. C. Freeman, B. A. Peterson, P. J. Quinn, A. W. Rodgers.
Bull. Am. Astron. Soc., Vol. 23, No. 2, p. 952 (1991). Abstract. – See Abstr. 012.008 for the main entry.

155.024 **HEAO A–4 observations of galactic positron annihilation radiation.**
M. S. Briggs, J. L. Matteson, D. E. Gruber, L. E. Peterson.
Bull. Am. Astron. Soc., Vol. 23, No. 2, p. 959 (1991). Abstract. – See Abstr. 012.008 for the main entry.

155.025 **First VLBA images of Sgr A*.**
R. L. Brown, J. M. Benson.
Bull. Am. Astron. Soc., Vol. 23, No. 2, p. 959 (1991). Abstract. – See Abstr. 012.008 for the main entry.

155.026 **Colliding winds at the galactic center.**
L. Ozernoy, V. Usov.
Bull. Am. Astron. Soc., Vol. 23, No. 2, p. 959 (1991). Abstract. – See Abstr. 012.008 for the main entry.

155.027 **A search for enhanced lithium in the galactic center.**
D. A. Lubowich, B. E. Turner, L. M. Hobbs.
Bull. Am. Astron. Soc., Vol. 23, No. 2, p. 959 (1991). Abstract. – See Abstr. 012.008 for the main entry.

155.028 **Far–infrared emission levels in and around Baade's window.**
W. H. Waller, F. Boulanger, L. E. Tacconi–Garman, K. Okumura.
Bull. Am. Astron. Soc., Vol. 23, No. 2, p. 960 (1991). Abstract. – See Abstr. 012.008 for the main entry.

155.029 **Galactic evolution of lithium.**
F. D'Antona, F. Matteucci.
Astron. Astrophys., Vol. 248, No. 1, p. 62 – 71 (Aug 1991).
The authors investigate the galactic chemical evolution of lithium in the framework of a complete model of galactic chemical evolution. Lithium production is assumed to be a result of the evolution of intermediate mass stars during their short Asymptotic Giant Branch (AGB) lifetime, and/or of the evolution of nova systems. Production by novae is followed assuming a detailed model which relates the lithium production to the time delay from the formation of the white dwarf to the secular evolution of the system. In the same framework the authors discuss also the production of ^{13}C. As there are large uncertainties in the production factors, the numerical inputs relative to lithium production are chosen to fit the lithium abundance of the solar system at the solar age (log N(Li) $\sim$ 3.3) starting from a primoridal lithium abundance log N(Li) = 2.10. The resulting functional behaviour [Fe/H] – log N(Li) is compared with the data, in the hypothesis that the upper envelope of the observational points represents the result of galactic enrichment of lithium and not of other processes. The authors find that good agreement can be achieved if a model is adopted including production both by novae and by AGB stars. At the same time, also the observed present day $^{13}C/^{12}C$ is well reproduced, as well as, for the first time, the gradient of $^{13}C/^{12}C$ along the galactic disk. The model, however, predicts an increase of lithium abundance by factor two in the last 4.5 Gyr, a prediction which should be tested adequately.

155.030 **Kinematic characteristics of the stellar complex Loop III.**
I. A. Zenina.
Astrophysics, Vol. 33, No. 2, p. 461 – 465 (Sep – Oct 1990). English translation of Astrofizika, Tom 33, Vyp. 2, p. 283 – 290 (Oct 1990).
The kinematic characteristics of 840 B and A stars in the region of Loop III have been investigated. The existence of a stellar component of Loop III, previously detected through an enhanced star density and the spectral composition of the stars, is confirmed by kinematic methods. Seven groups of stars belonging to the complex of Loop III have been identified. The parameters of the velocity distribution of the stars have been calculated for them. There are found to be pairs of streams of stars separating in opposite directions along the envelope of Loop III.

155.031 **On the formation of globular clusters. I. Dynamical limits on globular cluster metallicities.**
J. H. Brown, A. Burkert, J. W. Truran.
Astrophys. J., Vol. 376, No. 1, p. 115 – 122 (20 Jul 1991).
The authors address the question of the efficiency with which supernova ejecta are dispersed into the ISM and its implications for the metal abundances of early stellar populations. Based upon the dynamical limits imposed by the dispersive physical processes, the authors derive a permitted range of metallicities for a closed system such as a globular cluster. The mean supernova progenitor mass, an adjustable parameter in these models, is taken to be $\sim 30\ M_{\odot}$. The authors discuss this choice in the context of observed abundance patterns and an assumed IMF. A lower metallicity limit arises from the fact that at most $\sim$8% of the input energy is available for gas motions (dispersion). The upper metallicity limit is rather a consequence of momentum conservation: the minimum energy that can be retained is $\sim$0.4%, implying very inefficient mixing and therefore a higher achieved metallicity. The authors find it to be both encouraging and suggestive that the permitted range, $\sim 10^{-2}$ to $10^{-1} Z_{\odot}$, is compatible with that observed for the Galactic halo globular clusters.

155.032 **δm_1, δc_1–$E(b–y)$ relations in a magnitude–limited complete sample of A and F stars at the North Galactic Pole.**
J. Knude.
Astron. Astrophys., Vol. 249, No. 1, p. 88 – 98 (Sep 1991).
An observing program of all A and F stars earlier than $\approx$G1 and brighter than $B \sim 11^{m}.5$ has been undertaken in 1200 square degrees centered on the North Galactic Pole. The average intrinsic stellar parameter $\langle \delta m_1 \rangle$ and $\langle \delta c_1 \rangle$ in 0.010 mag wide color excess bins are found to correlate rather strongly with the interstellar reddening, $E(b–y)$. The relations are thought to be real, not caused by observational errors. The tight bands between the average stellar chemical composition, the state of stellar

evolution and the amount of dust along the lines of sight are not a property of the Galaxy, of course, but exclusively due to bias in a magnitude limited sample. The explanations suggested for the two correlations are based on the following two points: (a) There is not an identical probability to observe any color excess all over the distance range covered by the complete sample. (b) For a given intrinsic stellar color the probability to observe less evolved stars decreases with distance. Studies of the Galaxy's chemical evolution, in the form of enrichment with time and presence of gradients perpendicular to the plane, are effected by the relations found in the present paper. Particularly point (b) may be serious because it implies the unability to observe similar age ranges for various metallicities.

155.033 **Implications of the IRAS survey for galactic gamma–ray astronomy.**
F. W. Stecker, A. K. Harding, J. Skibo.
AIP Conf. Proc., No. 220, p. 20 – 24 (Apr 1991). Current Physics Microform No.: 9106B1154. – See Abstr. 012.020 for the main entry.

Using the results of the IRAS far–infrared survey of the Galaxy, the authors have obtained the large scale radial distributions of galactic far infrared emission independently for both the northern and southern hemisphere sides of the Galaxy. They find the dominant feature in these distributions to be a broad peak coincident with the "5 kpc" molecular gas cloud ring. They also find evidence of spiral arm features. Strong correlations are evident between the large scale galactic distributions of far infrared emission, γ–ray emission and CO emission. The authors find a particularly tight correlation between the distribution of warm molecular clouds and far–infrared emission on a galactic scale.

155.034 **A self–regulated state for the interstellar medium: radial dependence in the galactic plane.**
A. Parravano, J. Mantilla Ch.
Astron. Astrophys., Vol. 250, No. 1, p. 70 – 83 (Oct 1991).

The radial dependence in the galactic plane of the star formation rate (SFR) and the value of various variables related to the state of the interstellar medium are calculated assuming that two regulating mechanisms act simultaneously; viz., the "heating regulated warm gas condensation" (HRWGC) that acts as a large scale self–regulating mechanism of star formation, and the "photoionization–regulated star formation" (PIRSF) that acts at the scale of single molecular clouds. These two mechanisms are linked via the far UV interstellar radiation field (FUV–ISRF). Since the FUV–ISRF is determined by the large scale distribution of massive stars, in this model the state of the warm gas and molecular clouds is related with the recent SFR. The simultaneous action of the two mechanisms tends to maintain the SFR near an equilibrium value due to the fact that when the SFR is higher (lower) than a critical value then both, the rate of condensation of warm gas and the star formation efficiency in molecular clouds are reduced (increased). The HRWGC controls the SFR at large scale; the PIRSF assures a fast response of the system to departures from the equilibrium. The HRWGC assumption permits to calculate the radial dependence of the FUV–ISRF, the formation rate of massive stars, the thermal pressure, and the temperature and fractional ionization of the warm gas and of the cores of the recently formed small neutral clouds. The PIRSF assumption permits to calculate the mass of the molecular gas in active star forming regions that is compatible with the equilibrium SFR.

155.035 **A high galactic latitude survey of far–ultraviolet excess objects.**
J. V. Bixler, S. Bowyer, M. Laget.
Astron. Astrophys., Vol. 250, No. 2, p. 370 – 388 (Oct 1991).

Optical spectra have been obtained for a selection of objects included in a catalog of far ultraviolet bright, high galactic latitude objects detected with a balloon–borne survey telescope. The observed objects provide a sample of subdwarf O and B stars, white dwarfs, and binary systems including a hot subluminous member. Model atmospheres analysis of the subdwarf sample is used to determine the temperature, gravity, and helium to hydrogen ratio of the individual objects. A smooth distribution of objects is found on the gravity versus temperature diagram near the theoretical location of the extended horizontal branch. A break between the helium rich and helium poor objects occurs at 40000K. Suspected binary objects were found and analyzed to determine the temperature and gravity of the hot subluminous member in each system. The number of subdwarf stars contained in binaries is determined to be from 65% to 100%. The number versus ultraviolet magnitude distribution of the subdwarf B sample is modeled to obtain a midplane density of $3.3 \times 10^{-6} pc^{-3}$ and a population scale height of 240 pc. The proportion of white dwarfs that experience the subdwarf phase of evolution is found to be 0.94%.

155.036 **Neutral hydrogen in the outer Galaxy.**
A. Diplas, B. D. Savage.
Astrophys. J., Vol. 377, No. 1, p. 126 – 140 (10 Aug 1991).

The authors have used the Bell Laboratories northern hemisphere H I survey data of Stark and coworkers to study the distribution of H I in the outer Galaxy. These measurements are well suited for investigations of faint Galactic H I emission because the sidelobe contamination of the data is low. A flat rotation curve was used to calculate the H I space density n_H as a function of Galactic position, expressed in a Galactocentric cylindrical coordinate system (R, θ, z). The authors present a series of contour maps of $N_H(R, \theta, z)$ for fixed values of R, θ, and z for $\theta < 220°$. The data set was examined in a variety of ways in order to study the various systematic problems which influence the derived results. These data are then used to probe the nature of the Galactic warp and the extension of the gas to large z in the outermost portions of the Galaxy. For 12 kpc < R < 18 kpc the z distribution of the H I is complicated and reveals the existence of a confined component and an extended component. The confined component exists only within a Galactocentric radius R $\sim$ 18 – 20 kpc, while the extended component can be followed in some directions to R $\sim$ 25 to 30 kpc. In the directions of maximum warp ($90° < \theta < 110°$), the average distance, $<z>$, of the gas away from the Galactic plane is found to reach $\sim$4 kpc at R $\sim$ 24 kpc, and the flaring of the layer as measured by z_{rms} increases to $\sim$3 kpc. In these directions the warping outer layer appears to be reliably traced over a region at least 10 kpc thick in z.

155.037 **Two–dimensional Milky Way spiral arm contrast.**
Xiang Delin.
Chin. Astron. Astrophys., Vol. 15, No. 3, p. 364 (Sep 1991). Abstract. The full paper is published in Acta Astrophys. Sin., Vol. 11, No. 2, p. 154 – 164 (Apr 1991). See Abstr. 53.155.080.

155.038 **Metal rich RR Lyrae stars and the thick disk.**
A. W. Rodgers.
Conference on Dynamics of Disc Galaxies, p. 29 – 35 (1991). – See Abstr. 012.028 for the main entry.

It is argued that the metal rich RR Lyrae stars are members of the thick disk.

155.039 **Large–scale oscillations of the Galaxy and the kinematics of the solar neighbourhood.**
K. Kuijken, S. Tremaine.
Conference on Dynamics of Disc Galaxies, p. 71 – 104 (1991). – See Abstr. 012.028 for the main entry.

Because of our location within the Galaxy, the large–scale kinematic models that we deduce for the Galaxy depend strongly on our interpretation of the small–scale kinematics of the solar neighbourhood. For example, asymmetry of the northern and southern H I distributions in velocity–longitude space could be caused by either a distortion of the outer Galaxy or radial motion of the Local Standard of Rest. The authors review the evidence for oscillations of the Galactic disk, with particular emphasis on the interplay between these oscillations and the kinematics of the solar neighbourhood.

155.040 **K giants and the amount of local dark matter.**
J. Bahcall, C. Flynn, A. Gould.
Conference on Dynamics of Disc Galaxies, p. 105 – 109 (1991). – See Abstr. 012.028 for the main entry.
The authors analyse the amount of dark matter in the galactic disk from K–giant stars observed in a cone at the South Galactic Pole. A well–mixed homogeneous sample using magnitudes, abundances and absolute magnitudes has been obtained with rigorous selection criteria. Bahcall's method was used to solve the self–consistent equations for the distribution of possible disk matter. The best–fit P–model (in which dark matter is distributed proportionally to known matter) has 53% more dark than visible matter.

155.041 **Dynamics of the inner Galactic gas disk.**
O. E. Gerhard.
Conference on Dynamics of Disc Galaxies, p. 111 – 115 (1991). – See Abstr. 012.028 for the main entry.
The author studies dynamics of cold gas at the Galactic center; the results strongly support the conjecture that the large–scale distribution of the central gas is dominated by a rapidly rotating barred potential.
Contents: Kinematical model of Galactic center gas. Dynamical friction on Galactic center clouds.

155.042 **Lopsidedness, $m = 1$ spiral, and magnetic structures at the Galactic Center.**
M. Tagger.
Conference on Dynamics of Disc Galaxies, p. 117 – 121 (1991). – See Abstr. 012.028 for the main entry.
It is shown that the poloidal magnetic field inferred from various observations of the Galactic Center on a scale of a few tens of parsecs is so strong that it can affect the spiral density waves. A consistent picture of the observed lopsidedness (asymmetry) of the molecular gas distribution, of the radio arc and of the Galactic Center lobe, could thus be provided by a density wave with $m = 1$ wavenumber (i.e. a one–armed spiral), as expected from the large velocity dispersion and the peaked rotation curve. As it propagates in the disk the perturbation should emit vertically (along field lines) Alfven waves, which might be responsible for the radio lobe.

155.043 **E(b–y) as the primary parameter for polar δm_1, δc_1 and Z(pc) relations.**
J. Knude.
Meeting on Precision Photometry: Astrophysics of the Galaxy, p. 147 – 151 (1991). – See Abstr. 012.027 for the main entry.
In a complete magnitude limited sample of A5 – (F9 – GO) stars at the NGP significant correlations are found between the extrinsic parameter E(b–y) and the intrinsic parameters δm_1 and δc_1.

155.044 **Washington CCD photometry: abundance distributions in different latitudes toward the galactic bulge.**
N. D. Tyson.
Meeting on Precision Photometry: Astrophysics of the Galaxy, p. 193 – 207 (1991). – See Abstr. 012.027 for the main entry.
The Washington photometric system permits measurement of metal abundances for individual stars in the abundance range $-3 < [Fe/H] < +0.5$. The broad–band phases of the system allow work on distant objects with small aperture telescopes. For objects such as resolvable stars in the galactic bulge or nearby galaxies, the Washington system may prove superior even to multi–object spectroscopy. As a particular example, the author presents the first results of a survey of the stellar abundance distribution in different latitudes toward the galactic bulge.

155.045 **Ultraviolet excess properties of the thick disk and halo populations.**
S. R. Majewski.
Meeting on Precision Photometry: Astrophysics of the Galaxy, p. 209 – 224 (1991). – See Abstr. 012.027 for the main entry.
First results are presented from a new, complete survey of stellar multi–color (UBVI) photometry and proper motions to B ~ 22.5. Using high quality photometry, ultraviolet excesses and photometric parallaxes have been derived for a subsample of 250 stars with $0.3 < (B–V) < 1.10$ in a field at the North Galactic Pole. Thus the chemical and kinematical distributions of late F, G and early K dwarfs have been probed to distances of up to 25 kpc above the Galactic plane, allowing for the first time an unbiased view of the properties of the thick disk and halo from unevolved stars in situ.

155.046 **uvby photometry of A–type stars at high galactic latitude.**
J. S. Drilling, A. G. D. Philip.
Meeting on Precision Photometry: Astrophysics of the Galaxy, p. 225 – 230 (1991). – See Abstr. 012.027 for the main entry.
Philip has presented a plan for a systematic study of stellar populations at high galactic latitudes. The original plan specified that 25 square–degree objective–prism plates be obtained at each 15 degrees of galactic latitude for the four galactic longitudes 0, 76, 180 and 290, and that the plates then be searched for stars of spectral class A7 and earlier down to as faint a limiting magnitude as possible. The fields for which finding lists have been published or for which uvby photometry has been obtained are listed.

155.047 **Probing the stellar population of the outer galactic disk: star clusters and "windows" through the galactic plane.**
K. A. Janes.
Meeting on Precision Photometry: Astrophysics of the Galaxy, p. 233 – 243 (1991). – See Abstr. 012.027 for the main entry.
It is likely that the most recently–formed large–scale structure of the Galaxy is the outer part of the disk; its development may even be continuing at the present time. However, because of the difficulty of observing through the disk, one knows little about the character of the stellar population of those distant regions. This review is a report on the initial progress in a study of the outer disk, based on observations of distant open clusters and a photometric survey through low–galactic latitude windows of low obscuration. There are a small number of clear windows, exactly in the galactic plane, which permit one to study the stellar population along a line of sight right through to the edge of the Galaxy. This review will include a discussion of some initial results from a photometric survey of one of these windows located at $l = 243°$.

155.048 **UBV photometry of OB^+ stars in the southern Milky Way.**
J. S. Drilling.
Meeting on Precision Photometry: Astrophysics of the Galaxy, p. 299 – 301 (1991). – See Abstr. 012.027 for the main entry.
1226 new observations are combined with previously published results of the author to yield an internally consistent set of magnitudes and colors on the international UBV system for 666 stars classified as OB in the Stephenson–Sanduleak OB star survey. The (U–B), (B–V) diagram indicates that these stars consist primarily of O–type stars and early B–type supergiants, reddened by up to E(B–V) = 2.1 mag.

155.049 **Gamma rays from the Galactic center region.**
E. P. Liang.
Publ. Astron. Soc. Pac., Vol. 103, No. 666, p. 771 – 773 (Aug 1991). – See Abstr. 012.034 for the main entry.
The author reviews the recent new gamma–ray data from the Galactic center region and reexamines the nature of the source in the context of black hole accretions. The mass of the black hole is estimated to be of the order of tens of solar masses.

155.050 **The turbulence investigation in the disc of the Galaxy.**
A. Lazarian.
Conference on Dynamics of Disc Galaxies, p. 293 – 295 (1991). – See Abstr. 012.028 for the main entry.
The method of investigating turbulence in the disc of the Galaxy is discussed. The emission of hydrogen at 21 cm wavelength is proposed to be used. The method for the

investigation random density and momentum fields through fluctuations of observable parameters is introduced.

155.051 **Neutral hydrogen filaments at high galactic latitudes. II. Existence of large-scale waves in interstellar space.**
G. L. Verschuur.
Astrophys. Space Sci., Vol. 185, No. 1, p. 137 - 152 (Nov 1991).
Large-scale morphological waves with wavelengths of order of 30° and amplitude = $\lambda/8$ exist in several long H I filaments in a 540 square degrees of sky around $l = 230°$, $b = +40°$. The extent of the longest filament is greater than 72°, the limit set by the boundaries of the area surveyed. The H I gas appears to be controlled by magnetic fields and the motion within flux tubes, determined by analysis of velocity along the filament axes, shows the presence of wave patterns with amplitude 5 to 6 km s^{-1} on an angular scale similar to that seen in the spatial structure projected on the sky.

155.052 **Metallicity and velocity distribution of giants toward the Galactic poles.**
K. U. Ratnatunga, K. M. Yoss.
Astrophys. J., Vol. 377, No. 2, p. 442 - 461 (20 Aug 1991).
A catalog of line-of-sight velocities, metallicities, and photometric distances for spectroscopically selected giants near the Galactic poles ($|b| > 80°$) is analyzed statistically. Maximum likelihood techniques are applied to study the dependence of vertical velocity dispersion σ_w on metallicity. The authors model the distribution as a sum of discrete components, each with an assumed isothermal Gaussian velocity distribution. The distribution of estimated metallicity is also assumed to be Gaussian for each component. The vertical density distribution is defined dynamically. The optimum representation has three discrete components: a solar abundance thin disk with $\sigma_W = 12 \pm 1$ km s^{-1}; an intermediate old disk with $\sigma_W = 28 \pm 2$ km s^{-1}; and a metal-weak component with $\sigma_W = 54 \pm 4$ km s^{-1}. The metallicity and vertical density distributions of the components are poorly constrained by this sample, wich is incomplete over both. The metal-weak component is not statistically separable into a halo and a "thick disk", whose existence or absence is unconstrained by this sample. The authors describe a numerical algorithm for maximum likelihood estimation, used to derive the kinematical representation, which is independent of any subjective division of the metallicity range.

155.053 **A statistical study of the correlation of galactic supernova remnants and spiral arms.**
Li Zongwei, J. C. Wheeler, F. N. Bash, W. H. Jefferys.
3. Chinese Academy of Sciences and Max-Planck Society Workshop: High energy astrophysics: compact stars and active galaxies, p. 135 - 140 (1991). - See Abstr. 012.026 for the main entry.
The distribution of observed giant H II regions and supernova remnants (SNR) in Galactic longitude is analyzed with Monte Carlo models to determine the possible correlation with spiral arms or disk populations. The H II regions are not especially well fit with any model, but show a radial scale length in the solar neighborhood consistent with the stellar distribution of ~ 3.5 kpc. The SNR distribution is consistent with half of them being in spiral arms and half not. This result is interesting because neither extreme is favored. There is little support for the notion that shell-type remnants, which dominate the sample, are a pure disk population, as would be the case if they arose solely from old-disk populations SN Ia events. Likewise, there is no substantial evidence that SNR are especially concentrated in spiral arms. The SNR seem to require a disk scale length of $\geqslant 5$ kpc which distinguishes them from the stellar disk and the H II regions. The models incorporate selection effects in a parametrized way and suggest that ~ 1000 SNR exist in the Galaxy, with ~ 800 remaining to be discovered.

155.054 **New results concerning the galactic fountain.**
P. R. Shapiro, R. A. Benjamin.
Publ. Astron. Soc. Pac., Vol. 103, No. 666, p. 923 - 927 (Aug 1991). - See Abstr. 012.034 for the main entry.
The authors have calculated the time-dependent, nonequilibrium thermal and ionization history of hot, optically thin, radiatively cooling gas in a one-dimensional, planar, steady-state flow model of the galactic fountain. They find that the calculation of such a flow which neglects photoionization cannot simultaneously reproduce the observed C IV, Si IV, and N V column densities through galactic halo gas. They show, however, that when photoionization is taken into account, a fountain flow can match the observations of both UV absorption and emission lines for a range of ionizing fluxes which depends on the characteristic size of the cooling regions within the flow. They determine what levels of external photoionizing flux can produce agreement with observational results. Remarkably, they find that the ionizing radiation emitted by the cooling gas itself is sufficient to cause the flow to match the observed column densities and line emission. Such a match implies a fountain flow with initial hydrogen density of $\sim 2 \times 10^{-2} cm^{-3}$, outflow velocity of ~ 100 km s^{-1}, and initial size for the cooling regions of ~ 1 kpc or greater.

155.055 **A cloud-Galaxy collision: observation and theory.**
H. Meyerdierks.
Astron. Astrophys., Vol. 251, No. 1, p. 269 - 275 (Nov 1991).
The author compares the observational evidence concerning the North Celestial Pole Loop (NCP Loop) and the Chain A complex of H I high velocity clouds with the theoretical description of a cloud-Galaxy collision. The main points motivating this study are that the NCP Loop is cylindrical rather than spherical, and that low, intermediate, and high velocity H I clouds in the area seem interrelated. The observations are consistent with the scenario of a halo cloud of 350 $M_\odot$ having collided with the galactic disc at a velocity of 325 km s^{-1}. The collision took place in two steps: in the collision with the warm inter-cloud medium, the infalling cloud is only moderately decelerated. This collision has already finished and we observe the shocked layer as high velocity clouds still moving downwards into the galactic disc. In a second step parts of this shocked layer encounter disc clouds thus producing the intermediate velocity clouds. The NCP Loop itself results from the lateral expansion of the cavity produced during the infall.

155.056 **Chemical evolution of the Galaxy.**
N. C. Rana.
Annu. Rev. Astron. Astrophys., Vol. 29, p. 129 - 162 (1991). - See Abstr. 003.010 for the main entry.
The author gives a brief historical overview. He surveys, and if necessary makes corrections to the observational material, namely, the G-dwarf metallicity distribution, the age-metallicity relation, gas and metallicity distributions in the Galaxy, the luminosity functions of stars and white dwarfs in the solar neighborhood, the distribution of pulsars, etc. He derives the history of star formation in the solar neighborhood from the study of chemical evolution, and applies these results to a derivation of the initial mass function of stars, the birthrates of white dwarfs and pulsars, the evolution of iron in the solar neighborhood and the metallicity gradient in the Galaxy. He concludes by highlighting the difficulties of the simple model and suggesting studies that may improve our understanding.

155.057 **Distribution of CO in the Milky Way.**
F. Combes.
Annu. Rev. Astron. Astrophys., Vol. 29, p. 195 - 237 (1991). - See Abstr. 003.010 for the main entry.
Serious questions arise when one tries to determine the H_2/CO conversion ratio. Can this ratio be measured? Once measured, can it be regarded as a universal constant? These fundamental questions are discussed as they form the basis of all interpretations of the CO observations in terms of the molecular distribution in the Milky Way. The author reviews the results of CO surveys of the Galaxy: most of these were carried out during the last decade. Although interpretation of the various surveys gave rise to some controversy in the early 1980s, because of more extensive and complete mapping a clearer picture can now be drawn. With the northern and southern hemisphere surveys combined, a large fraction ($\approx 30\%$) of the sky is covered. The main result from these surveys is that the CO and H I

distributions are very different. H_2 molecules dominate the central ISM, while atomic gas is much more important in the outer parts of the Galaxy. On small scales, the molecular component is very clumpy and so might be a better tracer of spiral structure than the atomic component. Moreover, because the interstellar medium is expected to turn into the H_2 phase at high densities and low temperatures, the molecular component provides the critical link between cloud formation and star formation. The author discuses the physical state of the molecular component: is it distributed in clouds, and what are the statistics about the mass spectrum of the clouds, their size–linewidth relations, and are they gravitationally bound or pressure supported? These statistics refer essentially to clouds in the "molecular ring", between 4 and 8 kpc. Clouds in the Galactic center are very special, and are described seperately. Then the outer parts of the Galaxy are addressed and the z–distribution of the CO emission: plane thickness and warping of the plane.

155.058 **Neutral hydrogen filaments at high galactic latitudes. III. The nature of enhanced emission features ("clouds").**
G. L. Verschuur.
Astrophys. Space Sci., Vol. 185, No. 2, p. 305 – 332 (Nov 1991).
Small angular scale twists in H I filaments identified at high galactic latitudes are associated with the location of enhanced emission features (EEFs), structures that have tradionally been referred to as "clouds". It is shown that in these directions the column density of gas is enhanced through geometrical effects because the line–of–sight intersects a greater pathlength when viewing along a segment of filament axis (flux tube) than when the line–of–sight is more normal to the filament axis. By interpreting the EEFs as isolated entities (clouds) an incorrect impression is derived as regards the properties of interstellar H I. This study leads to questions about the way observations of 21 cm, molecular, and 100 μ emission are currently being interpreted. It is concluded that much of what is observed to be "cloud" structure in the interstellar medium (except in regions directly associated with star formation) is telling us about geometry of filaments and not about the physics of "clouds". The very notion of an interstellar "cloud" may have outlived its usefulness and previous work that has attempted to account for these structures in terms of gravitational stability or pressure equilibrium has to be reconsidered in the light of the existence of complex patterns of filamentary structure, not only in the H I distribution but also of interstellar cirrus defined by 100 μ emission.

155.059 **The mass of the Galaxy.**
M. Fich, S. Tremaine.
Annu. Rev. Astron. Astrophys., Vol. 29, p. 409 – 445 (1991). – See Abstr. 003.010 for the main entry.
The authors discuss the observational constraints on the mass distribution in the outer Galaxy and compare these constraints to simple theoretical models.

155.060 **Spiral arms of the Galaxy from observations of interstellar extinction.**
L. A. Urasin, S. S. Tokhtas'ev, I. A. Dubyago, S. K. Fomin.
Izv. Astron. Obs. Ehngel'gardt., Vyp. 54, p. 3 – 16 (1989). In Russian.
A review of photometric investigations on the Engelhardt observatory from 1960 to 1985 is given. Spatial distribution of the interstellar dust matter in the galactic longitudes from 7° to 222° is obtained. A two arms model of the spiral structure of the Galaxy is proposed with the following parameters: the distance of the Sun from the galactic center is equal to 10 kpc, the characteristic (pitch) angle is 6°5.

155.061 **A statistical study of the correlation of Galactic supernova remnants and spiral arms.**
Li Zongwei, J. C. Wheeler, F. N. Bash, W. H. Jefferys.
Astrophys. J., Vol. 378, No. 1, p. 93 – 105 (1 Sep 1991).
The distribution of observed giant H II regions and supernova remnants (SNRs) in Galactic longitude is analyzed with Monte Carlo models to determine the possible correlation with spiral arms or disk populations. The novel feature of the current analysis is that it depends mainly on the angular distribution of the H II regions and SNRs and hence is independent of distance. The SNRs seem to require a long disk scale length ($\gtrsim$5 kpc). The possible correlations of the SNR distribution with the old stellar disk and the average radial distribution of the H I, H_2, and total gas density in the Galaxy is discussed. The models incorporate selection effects in a parameterized way and suggest that $\sim$1000 SNRs exist in the Galaxy, with $\sim$850 remaining to be discovered. There is no substantial evidence that SNRs are especially concentrated in spiral arms as they would be if they were selectively born in spiral arms or if the high gas density in the arms were the dominant selection effect for the production of long–lived remnants. On the basis of the current evidence, the hypothesis that the shell–type remnants that dominate the SNRs sample arise from old disk SN Ia events cannot be strongly rejected. The H II region distribution is not especially well fitted with any of the present models, but suggests a radial scale length in the solar neighborhood of $\sim$3.5 kpc.

155.062 **Topography of the Galactic disk: Z–structure and large–scale star formation.**
E. J. Alfaro, J. Cabrera–Caño, A. J. Delgado.
Astrophys. J., Vol. 378, No. 1, p. 106 – 118 (1 Sep 1991).
A sample of young open clusters with cataloged distances, treated with the Kriging techniques have been used to delineate a three–dimensional morphological description of the Galactic disk defined by the young stellar population. The basic idea entails considering the values of the positional variable Z for the cluster sample as probes of the displacement of the Galactic disk in respect to the Galactic equator in the (X, Y) plane. A view of the Galactic disk in a region of about 3 kpc around the Sun emerges from this treatment, showing a through–peak structure with four main depressions as the more striking features. The most prominent depression is apparent at a distance of $\approx$1.8 kpc from the Sun in the l $\approx$ 240° direction. This feature has an elliptical shape with axis sizes of 1.5 and 3 kpc, reaching a Z value of 200 pc below the formal Galactic plane. The four depressions are found to be related to the four young star–gas supercomplexes detected in the solar vicinity of Efremov and Sitnik. This finding leads the authors to suggest a physical link between the processes driving the large–scale star formation in the Galaxy and those being able to generate the Z–structure of the Galactic disk.

155.063 **The metallicity of RR Lyrae stars in Baade's window.**
A. R. Walker, D. M. Terndrup.
Astrophys. J., Vol. 378, No. 1, p. 119 – 126 (1 Sep 1991).
The authors derive metallicities via the ΔS method for 59 RR Lyrae variables in the Baade's window field of the Galactic nuclear bulge. Both RRab and RRc variables have $<[Fe/H]> = -1.0$. The metallicity distribution for the whole sample is sharply peaked, with a dispersion in [Fe/H] of only 0.16 dex, far smaller than suggested by previous studies. The narrow metallicity distribution does not permit the various current estimates of the magnitude of a luminosity / metallicity relation for RR Lyrae stars to be distinguished. The authors also find that the metal abundance of the globular cluster NGC 6522 in Baade's window is close to that of the bulge RR Lyraes, which sets the reddening to the Baade window to be E(B–V) = 0.50 ± 0.05 and the distance to the Galactic center to be 8.2 ± 1 kpc.

155.064 **A finger–like extension of the 20 kilometer per second cloud toward the Galactic center.**
S. K. Okumura, M. Ishiguro, E. B. Formalont, T. Hasegawa, T. Kasuga, K.–I. Morita, R. Kawabe, H. Kobayashi.
Astrophys. J., Vol. 378, No. 1, p. 127 – 130 (1 Sep 1991).
High–resolution ammonia–line images of the M –0.13–0.08 ("20 km s^{-1}") cloud in the Galactic center region have been obtained with the Nobeyama Millimeter Array. The 20 km s^{-1} cloud has a finger–like structure extending toward the Galactic center. It almost reaches to the circumnuclear ring (disk)

surrounding Sgr A West in projection. A large systematic velocity gradient along the elongation is found over the whole length of the cloud. The authors suggest that these characteristics are caused by tidal forces in the gravitational field of the Galactic center. Part of the finger–like structure might be torn away from the cloud and may fall toward the cirumnuclear ring. The 20 km s^{-1} cloud would thus supply neutral gas to the central ~10 pc of the Galactic center.

155.065 **Galactic structure from the Spacelab infrared telescope. II. Luminosity models of the Milky Way.**
S. M. Kent, T. M. Dame, G. Fazio.
Astrophys. J., Vol. 378, No. 1, p. 131 – 138 (1 Sep 1991).
The three–dimensional luminosity distribution of the Milky Way has been determined from a 2.4 μm map of the northern Galactic plane. The radial surface brightness profile of the disk has an exponential scale length of 3.0 kpc (assuming the distance of the Sun to the Galactic center to be R_0 = 8 kpc). The vertical distribution of light follows the law $\exp(-|z|/h_z)$ more closely than a canonical sec h^2 profile. In the solar neighborhood, the scale height h_z is ~247 pc. The scale height is not constant with radius but decreases to ~165 pc at R = 5 kpc. To first order, the bulge can be represented by an oblate spheroid with an axis ratio b/a = 0.61 and an exponential luminosity profile. The 2.2 μm/12 μm flux ratio for the bulge is typical of other dust–free spheroidal systems. Luminosity fluctuations along the Galactic plane are found to be caused chiefly by variations in the line–of–sight extinction. These fluctuations can be reproduced quantitatively by a detailed model for the dust distribution and allow an approximate calibration of the dust/gas ratio in regions of high optical depth.

155.066 **The space frequency and origin of the runaway O and B stars.**
R. C. Stone.
Astron. J., Vol. 102, No. 1, p. 333 – 349 (Jul 1991). Current Physics Microform No.: 9107B2029.
Determinations of space frequency for runaway O and B stars are in very poor agreement, and many possible sources of systematic errors have been put forward. This paper investigates these sources, as well as others, and presents means for determining unbiased (true) space frequencies. When corrected for major sources of systematic error, space frequency determinations for the runaway O stars are in very good agreement but less so for the B–type runaways. The true space frequencies of runaway O and B stars are, respectively, 46% and 4% as discussed in this paper. The former value is very supportive, along with other arguments presented in this paper, for runaway O stars being produced from supernova explosions in massive binaries. The origin of the B–type runaways is less clear.

155.067 **Analysis of the galactic structure by photographic surface photometry.**
S. Kimswenger.
Astron. Nachr., Vol. 312, No. 3, p. 221 (1991). Dissertation abstract.

155.068 **VLA radio recombination line observations of Sagittarius A West.**
D. A. Roberts, W. M. Goss, J. H. van Gorkom, J. P. Leahy.
Workshop on the Interpretation of Modern Synthesis Observations of Spiral Galaxies, p. 57 (1991). – See Abstr. 012.041 for the main entry.
Abstract. See also Abstr. 53.155.009.

155.069 **The large scale radio structures in the Milky Way.**
R. M. Price.
Workshop on the Interpretation of Modern Synthesis Observations of Spiral Galaxies, p. 71 – 77 (1991). – See Abstr. 012.041 for the main entry.
Our Milky Way Galaxy has a number of distinct radio components visible at decimeter wavelengths. In particular, an understanding of the base disk, spiral features, thick disk, halo, and other peculiar features such as spurs and loops give us a better chance of studying such features in external galaxies where they can be observed by synthesis techniques. A brief review of the radio structure of our Galaxy is followed by a list of features which have not yet been unambiguously observed in external galaxies and should be searched for.

155.070 **A deep survey for galactic Wolf–Rayet stars. I. Motivation, search technique, and first results.**
M. M. Shara, A. F. J. Moffat, L. F. Smith, M. Potter.
Astron. J., Vol. 102, No. 2, p. 716 – 743 (Aug 1991). Current Physics Microform No.: 9108A1652.
The authors are using direct narrowband and broadband Schmidt plates to survey large areas of the southern Milky Way for Wolf–Rayet (WR) stars to 17–18th magnitude. They expect to find more than 50 new WR stars. The newly detected stars should be among the most distant and/or reddened known in the Galaxy. The survey is also designed to test the completeness of previous bright WR star surveys. All but two of the 19 known WR stars fainter than V = 10 in the first four regions the authors have surveyed so far (in the Carina area) were identified as candidates. The exceptions are WR 43 in the core of the dense star cluster NGC 3603 and WR 30a. The survey should be very sensitive to all but the most crowded WR stars. The authors have obtained spectra of the candidate WR stars in about 40 square degrees centered near b = 0°, and stretching from l ≈ 282° to l ≈ 294° in longitude and Δb ~ 3° in latitude. Thirteen new Wolf–Rayet stars have been found. Eleven are WN and two are WC stars. The authors present and describe their spectra and subtypes, derive their distances, and discuss the significance of their locations in the Galaxy. The authors' new WR stars strongly suggest, for the first time, an increasing number ratio of WN to WC stars with increasing Galactocentric distance. The authors also consider the completeness of their survey, emphasize that previous surveys in Carina are significantly incomplete even within 4 kpc of the Sun, and discuss ways to optimize Their search technique. Two of the new objects are members of clusters and a third is only 3″ from a previously known WR star. The authors' search technique is particularly useful at finding WR stars in crowded fields missed by previous, confusion–limited objective prism surveys.

155.071 **The Milky Way: lopsided or barred?**
K. Kuijken.
Workshop on Warped Disks and Inclined Rings Around Galaxies, p. 159 – 169 (1991). – See Abstr. 012.054 for the main entry.
The asymmetries in the inner Galaxy are often attributed to spiral arms. In the outer Galaxy, where spiral arms are much weaker, large assymetries are also in evidence; these are the subject of this paper.

155.072 **Kinematics of Sandage–Fouts stars in three cardinal directions.**
K. M. Yoss, D. J. Bell, H. L. Detweiler.
Astron. J., Vol. 102, No. 3, p. 975 – 993 (Sep 1991). Current Physics Microform No.: 9109D1379.
The authors have observed V and (B–V) for 429, 442, and 427 stars in the directions of the anticenter, Cygnus, and the NGP, respectively, all within 10° of the cardinal directions (l = 180°, b = 0°), (l = 90°, b = 0°), and (b = 90°), and DDO colors for those stars G0 and later. The samples observed are ones for which Sandage & Fouts (1987) have published radial velocities (and for the NGP, radial velocities for additional stars by R. Griffin). U, V, and W components of space velocity are derived from DDO distances and available proper motions. A clear correlation exists between the velocity dispersions and [Fe/H], with the dispersions decreasing as [Fe/H] increases to ~+0.00, where ⟨V⟩ then becomes significantly negative. The "thick disk" component is clearly recognized, with σ_W ~ 36 km s^{-1}.

155.073 **A gaseous tail ablated from the supergiant IRS 7 near the Galactic center.**
E. Serabyn, J. H. Lacy, J. M. Achtermann.
Astrophys. J., Vol. 378, No. 2, p. 557 – 564 (10 Sep 1991).
A 1″ resolution map of the [Ne II] 12.8 μm emission from the M supergiant IRS 7 near the Galactic center has revealed a 5″

long ionized "plume" or "tail" of gas leaving the star, in a direction opposite to the radio point source at the Galactic center. While the stellar envelope and tail are likely ionized by the ultraviolet radiation from the Galactic center, ram pressure is more likely to be responsible for the ablation of the supergiant's mass outflow envelope. This ram pressure could arise in one of two sources: a wind from the vicinity of the Galactic center, or the drag caused by the passage of IRS 7 through an ambient medium. In the wind scenario, since the external wind and radiation which act on this star as it orbits about the Galactic center are the same agents which act on a comet orbiting about the Sun (neglecting magnetic fields), the tail of IRS 7 can be understood as the analog of a cometary tail. Numerical calculations of the velocity field expected along the plume for both wind and drag models are reasonably consistent with the observed velocity pattern, but higher resolution observations should be able to discriminate definitively between the two.

155.074 **On local stellar age distributions and the history of the stellar birthrate in the galactic disk.**
H. Meusinger.
Astron. Nachr., Vol. 312, No. 4, p. 231 – 237 (1991).
The fraction of nearby stars born at different galactocentric radii is roughly estimated using the epicyclic approximation of stellar orbits in the galactic disk. Especially, the question of a possible burst origin of features in the local stellar age distribution is discussed. If the "bump" at about 4 Gyr in the frequency distribution of stellar isochrone ages (Meusinger 1991) is explained by star formation events then they have been locally at the solar galactocentric distance or globally in the whole disk.

155.075 **Galactic bulge M giants. IV. 0.5 – 2.5 micron spectrophotometry and abundances for stars in Baade's Window.**
D. M. Terndrup, J. A. Frogel, A. E. Whitford.
Astrophys. J., Vol. 378, No. 2, p. 742 – 755 (10 Sep 1991).
The authors have obtained 0.45 – 2.45 μm spectrophotometry with a resolution $\lambda/\Delta\lambda \sim 1000$ of a representative sample of M giants in Baade's Window and the solar neighborhood. From an analysis of strong atomic lines of Na I and Ca I in the K band, the authors derive a mean metallicity of the M giants in Baade's Window of $<[M/H]> = +0.3$. It is shown that the difference in [M/H] between the bulge and field stars based on the first overtone CO bands is consistent with that derived from the atomic lines. However, when compared with local M giants of the same spectral class, bulge M giants are hotter by about 400K, and have higher gravities by about 0.6 dex. A major difference in the overall spectral energy distributions of bulge and local M giants is that the classical H band bump, attributed to the opacity minimum of the H^- ion near 1.6 μm, is considerably reduced in many of the bulge stars. This is consistent with the suggestion in earlier papers that other opacity sources may compete with or even replace H^- as the major opacity source and thus be responsible for the difference in the JHK colors of bulge and field giants. However, the new data do not support the suggestion that H_2O absorption is the main cause.

155.076 **Chemically peculiar and main sequence stars: the distribution on the galactic *z*–coordinate.**
S. V. Marchenko, T. P. Sergeeva.
Kinematika Fiz. Nebesn. Tel, Tom 7, No. 5, p. 87 – 92 (Sep–Oct 1991). In Russian. English translation in Kinematics Phys. Celest. Bodies, Vol. 7, No. 5.
Chemically peculiar (CP) and main sequence (MS) stars distribution on the galactic *z*–coordinate are obtained for three samples: 1) single stars from "Catalogue of Bright Stars" ($m \leqslant 6^m.5$); 2) suspected binaries ($m_V \leqslant 8^m.0$); 3) CP and MS stars from the $\approx 14°\cdot 14°$ zone in the Galactic anticenter direction ($m_V \leqslant 12^m.5$). In the first case CP and MS stars have identical *z*–distribution. The third case is most interesting: CP stars are much more concentrated to the Galactic plane than the MS stars. It is shown that CP stars keep the tendency to create compact ($d \lesssim 50$ pc) space clusters as it was found for the nearest ($r \leqslant 1$ kpc) CP stars. At the same time, space distribution of MS stars is more homogeneous.

155.077 **Neutron star and black hole binaries in the Galaxy.**
R. Narayan, T. Piran, A. Shemi.
Astrophys. J., Lett., Vol. 379, No. 1, p. L17 – L20 (20 Sep 1991).
The authors present a statistical analysis of the number and birthrate of double neutron star (NS – NS) binaries. They estimate that there are $\sim 10^{4.5} z_0$ such systems in the Galaxy, with a birthrate $\sim 10^{-5} z_0\,yr^{-1}$, where z_0 (kpc) is their scale height; one expects $z_0 \sim$ a few kpc. The authors find that mergers of NS – NS binaries through gravitational radiation losses should be occurring roughly once a year in galaxies within 200/h Mpc. Also, the authors estimate that one out of $\sim 10^2/z_0$ massive stars in close binaries ends up in a NS – NS system. Progenitor stars more massive than $\sim 50\,M_\odot$ probably form black holes (BHs). The authors estimate that the number and formation rate of BH – NS binaries in the Galaxy are comparable to the corresponding estimates for NS – NS binaries, and they predict that a BH – NS binary is likely to be discovered in pulsar surveys. Such an object may be an even better general relativistic laboratory than PSR 1913+16 or PSR 1534+12. BH – NS binaries may also be strong sources of gravitational radiation.

155.078 **The galactic diffuse component of soft X–rays and its source function.**
W. Hirth, U. Mebold, M. Dahlem, P. Müller.
Astrophys. Space Sci., Vol. 186, No. 2, p. 211 – 223 (Dec 1991).
A new model for the source distribution of galactic soft X–ray (B and C band) emission is presented. From the mean dependence of count rates on galactic latitude b (i.e., the brightness distribution), the authors derive the soft X–ray source function Q as function of the optical depth τ by solving the equation of radiative transfer with the aid of a Laplace transform. Q is found to increase strongly, but not abruptly, in the range $1.5 < \tau < 2.5$, indicating a noticeable emission of X–rays from beyond the H I scale height. Using standard X–ray absorption cross sections for the interstellar medium, the authors find that the B band X–ray emission coefficient is non–zero within the H I disk and has a maximum at a z–value slightly above the H I scale height. In the C band, the emission coefficient slightly decreases with increasing z, at least up to the H I scale height. A non–zero source function near the galactic plane implies that the interstellar medium within the H I scale height is not only on absorbing layer but is mixed with X–ray emitting regions. The so–called "local hot bubble" is adopted as one of these regions. The maximum of the B band emission coefficient, together with the sharp increase of Q, is strong evidence for the existence of a galactic soft X–ray "halo", and, moreover, gives rise to the assumption of a general intergalactic X–ray background.

155.079 **Molecular clouds as tracers of galactic spiral structure.**
T. M. Dame.
3. Haystack Observatory Conference on the Interstellar Medium: Skylines – atoms, ions and molecules – new results in spectral line astrophysics, p. 43 (1991). Abstract. – See Abstr. 012.051 for the main entry.

155.080 **Molecular gas surrounding the Galactic center.**
J. C. Szczepanski, P. T. P. Ho, R. Güsten.
3. Haystack Observatory Conference on the Interstellar Medium: Skylines – atoms, ions and molecules – new results in spectral line astrophysics, p. 143 – 151 (1991). – See Abstr. 012.051 for the main entry.
Observations have been made in the $CH_3 4_{-1} \rightarrow 3_0 E$ line using the Haystack 37–m telescope and in the $NH_3(3,3)$ line using the Bonn 100–m telescope and the NRAO Very Large Array (VLA). Angular resolution for the single–dish studies is comparable, ~40" in each case. There is a large amount of molecular material intimately associated with both SgrA East and SgrA West. In particular, molecular material is found to encircle the entire southeastern edge of SgrA East. Evidence of shocks and compressions is present, suggesting that SgrA East is expanding into a dense molecular cloud. Kinematics suggest that the giant

molecular cloud M–0.02–0.07 is behind SgrA East, SgrA West is in front of SgrA East, and M–0.13–0.08 is in front of SgrA East and SgrA West. Overall morphology, kinematics, and excitation suggest that supernovae are impacting on the molecular clouds. The material that is "loosened" as a result may be pulled in toward the central circumnuclear ring and the Galactic Center itself.

155.081 **Determination of the absolute proper motion of the globular cluster M3.**
R. D. Scholz.
International Seminar on Problems of Astrometry and Space Geodynamics, p. 97 – 104 (1991). – See Abstr. 012.049 for the main entry.
More details and further reductions to be published in Mon. Not. R. Astron. Soc.

155.082 **Mixture models for studying stellar populations.**
J. Nemec, A. F. Linnell Nemec.
Symposium on the Formation and Evolution of Star Clusters as Part of the 102. Summer Scientific Meeting of the Astronomical Society of the Pacific, p. 512 – 531 (1991). – See Abstr. 012.050 for the main entry.
Finite mixture models useful for decomposing the (overlapping) distribution functions of stellar populations are described. The basic properties of the models are summarized, and parameter estimation methods are briefly discussed. The theoretical results are illustrated using several variables that are relevant for studies of globular and open star clusters and solar neighbourhood stars: metal abundance [Fe/H], age, and the space velocities U, V, and W. Simulated observations are generated by using several different Gaussian mixture models, with representative parameter values. These are used to illustrate some important bivariate relationships: the W–[Fe/H] relationship; the V_{rot}–[Fe/H] relationship; and, the age–[Fe/H] relationship.

155.083 **Luminosity function and velocity distribution of late F– and early G–stars at the South Galactic Pole. IV.**
P. K. Lu.
Symposium on the Formation and Evolution of Star Clusters as Part of the 102. Summer Scientific Meeting of the Astronomical Society of the Pacific, p. 581 – 583 (1991). – See Abstr. 012.050 for the main entry.
Studies to investigate the galactic gravitational force law and the local mass density in the Galaxy are now in progress. Using faint dwarf F–stars as tracer objects to a distance of $z = 1.5$ kpc, the immediate goals of this project are to determine their velocity distribution, to search for evidence of a thick disk, and to study dynamically the missing mass and dark matter in the galactic disk. A preliminary luminosity function from a sample of 2669 late F– and early G–stars centered at the South Galactic Pole is given.

155.084 **Deuterium in the galactic center molecular clouds.**
C. M. Walmsley, T. Jacq.
3. Haystack Observatory Conference on the Interstellar Medium: Skylines – atoms, ions and molecules – new results in spectral line astrophysics, p. 305 – 312 (1991). – See Abstr. 012.051 for the main entry.
Study of deuterium fractionation in molecular clouds is in principle capable of yielding information both about the chemical history of the clouds and the abundance of interstellar deuterium. In particular, this is of interest in the hot galactic center clouds where one would naively expect the intrinsic D abundance to be smaller than in the solar neighbourhood. The current knowledge of D in the galactic center clouds is reviewed and it is concluded that in fact the D abundance in the center clouds is probably a factor 5 less than the local value but could be as large as that found in the solar neighbourhood.

155.085 **Inhomogeneous halo collapse and globular cluster metallicities.**
G. Malinie, D. H. Hartmann, G. J. Mathews.
Symposium on the Formation and Evolution of Star Clusters as Part of the 102. Summer Scientific Meeting of the Astronomical Society of the Pacific, p. 584 – 587 (1991). – See Abstr. 012.050 for the main entry.
Unique information about the early phases of Galactic evolution can be derived from the metallicity distribution of the spheroidal component of the Galaxy. Here, the authors consider the low–metallicity component ("halo clusters") and assume that it was creaed during the collpase of the protogalaxy. In particular, the authors study the effects of initial density gradients on the chemical evolution.

155.086 **Oxygen abundances in halo giants. I. Giants in the very metal–poor globular clusters M92 and M15 and the metal–poor halo field.**
C. Sneden, R. P. Kraft, C. F. Prosser, G. E. Langer.
Astron. J., Vol. 102, No. 6, p. 2001 – 2021 (Dec 1991). Current Physics Microform No.: 9112C1977.
The authors derive oxygen, iron, vanadium, and scandium abundances for very metal–poor giants in the globular clusters M92 and M15, and giants of comparable metallicity in the local halo field. They analyze the [O I] doublet ($\lambda\lambda$ 6300, 6363) and nearby metallic lines in spectra obtained with a TI CCD detector and the Hamilton echelle spectrograph at the Lick Observatory 3.0 m telescope, using line analysis and spectral synthesis codes developed at the University of Texas.

155.087 **The age and formation of the Galaxy: clues from the white dwarf luminosity function.**
M. A. Wood.
NATO Advanced Research Workshop – 7. European Workshop on White Dwarfs, p. 89 – 97 (1991). – See Abstr. 012.044 for the main entry.
Since the onset of star formation in the local disk of the Galaxy, stars have been evolving into white dwarf stars, which have cooled with time to produce the white dwarf luminosity function that is observed today. Because the cooling timescale of a typical white dwarf is longer than the age of the local disk, even those white dwarfs formed from the earliest generation of stars are still visible. Using homogeneous sets of mass–dependent cooling curves and analytical approximations to the details of galactic evolution, it is possible to synthesize theoretical white dwarf luminosity functions which can then be compared directly with the observations to probe the age and star formation history of the local disk, and the sensitivity of these to the inputs. The author has completed just such a parametric study, and reports here in brief some of the major results and their bearing on the understanding of the process of spiral–galaxy formation.

155.088 **Balloon observations of diffuse infrared emission in the galaxy.**
H. Okuda.
Infrared Phys., Vol. 32 , p. 365 – 375 (1991).
The observational conditions for extraterrestrial infrared emission is remarkably improved at the balloon altitudes comparing to those at the airplane altitudes. By utilizing this advantage effectively, a number of observations of the diffuse infrared emission in the Galaxy have been successfully done, providing unique information on new phenomena in interstellar space as well as new structures of the Galaxy. The observations of near infrared brightness surveys in the Galaxy have first given a comprehensive picture of the stellar distribution in the inner region of the Galaxy, while the far infrared brightness mappings of the galactic plane have revealed the distribution of interstellar dust condensed tightly in the galactic plane. Introduction of spectroscopic observations have opened a new scope for studies of interstellar space; for instance, the observations of the [C II]–line (157.7 μm) have shown that the emission is very strong and ubiquitously distributed in interstellar space. The [C II]–line would become a useful probe for the investigation of the ever unexplored part of the Galaxy i.e. photodissociation regions or

[CII] regions with softly ionized, warm and medium density gases.

155.089 **Direct evidence for a bar at the Galactic center.**
L. Blitz, D. N. Spergel.
Astrophys. J., Vol. 379, No. 2, p. 631 – 638 (1 Oct 1991).

The bar at the center of the Milky Way, long postulated from a number of different studies, is unambiguously detected in the 2.4 μm observations of the Galactic center of Matsumoto et al.(1982). The authors model the emission from a triaxial stellar bar and describe its distinctive signature. This signature is found in the data. The near side of the bar is in the first Galactic quadrant and the bar is tilted with respect to the Galactic plane in a sense consistent with the work of Sinha and of Liszt and Burton, who first proposed a tilted bar to explain the kinematics of the H I and CO at the Galactic center. The small extinction that is observed at latitudes greater than 3° at 2.4 μm cannot account for the asymmetries observed in the distribution of the infrared emission. The bar is distinct from the triaxial spheroid postulated by Blitz and Spergel as the source of the asymmetries in the H I distribution in the outer Galaxy.

155.090 **The dynamical role of the central molecular ring in the seven–component model of the Galaxy.**
A. A. Sumin, A. M. Fridman, U. A. Haud.
Sov. Astron. Lett., Vol. 17, No. 5, p. 329 – 332 (Sep 1991). Current Physics Microform No.: 9204X1276. English translation of Pis'ma Astron. Zh., Tom 17, No. 9, p. 779 – 786 (1991).

On the basis of observational data a model of the Galaxy was constructed in which the surface density $\sigma_g(r)$ of the gaseous component has a sharp jump (by two orders of magnitude) near the outer radius of the molecular ring. This model was used to calculate the contributions of the various subsystems of the Galaxy to the model rotation curve. The dimensionless hydrodynamic instability growth rate for the gaseous component alone, which is much less than unity, turns out to be about the same as the growth rate for the same gas in the gravitational field of the entire Galaxy. The authors can draw two conclusions from this: 1) being unstable, the gaseous component of the Galaxy is close to the limit of hydrodynamic stability, which is in accordance with Le Chatelier's principle; 2) the stellar subsystems of the Galaxy probably do not have any dynamical effect on the generation of the spiral structure in the gaseous component.

155.091 **Galactic center gasdynamics: a one–armed spiral in a Keplerian disk.**
J. H. Lacy, J. M. Achtermann, E. Serabyn.
Astrophys. J., Lett., Vol. 380, No. 2, p. L71 – L74 (20 Oct 1991). With plates L4 – L5.

The inner 3 × 4 pc of the Galaxy have been imaged in the [Ne II] (12.8 μm) line with 2″ and 30 km s^{-1} resolution. The morphology of much of the ionized gas can be explained by a one–armed linear ($r \sim \theta$) spiral. The kinematics of the gas along the spiral can be modeled by approximately circular orbits with a nearly Keplerian rotation curve, requiring a central mass of $(2.0 \pm 0.5) \times 10^6 M_\odot$. Several possible origins are suggested for the spiral.

155.092 **On the distribution of the diffuse Galactic gamma–ray line emissions at 1809 keV and 511 keV.**
P. von Ballmoos.
Astrophys. J., Vol. 380, No. 1, p. 98 – 103 (10 Oct 1991).

It is suggested that the diffuse gamma–ray line emissions from ^{26}Al and the e^+e^- annihilation are both reflected by the distribution of the Galactic ridge thin hot gas. Thermal bremsstrahlung emitted by this plasma has been mapped by X–ray telescopes on different scales. Its distribution resembles the Galactic CO distribution except for a narrow peak (~2° FWHM) at the Galactic center containing one–third of the total emission in the central radian. The proposed distribution is consistent with the present measurements of the gamma–ray line at 1809 keV and almost all the observations at 511 keV. Evidence for a connection between thin hot plasma and gamma–ray line emission is presented: injection of kinetic energy into the interstellar medium is inherent in the sites of ^{26}Al nucleosynthesis. High–velocity shock fronts are formed when the kinetic energy of supernova ejecta and stellar winds of massive stars is dissipating in the ISM. The thermal bremsstrahlung emitted by the shock–heated plasma can be used as a tracer for the sites where synthesis of intermediate and heavy nuclei takes place. In these sites the decaying radionuclides (^{56}Co, ^{26}Al) will produce positrons that can annihilate with lifetimes comparable to that of ^{26}Al. The rate of annihilation is found to be sufficient to explain all the observed 511 keV radiation.

155.093 **Population studies. X. Constraints on the mass and extent of the Galaxy's dark corona.**
J. E. Norris, M. R. S. Hawkins.
Astrophys. J., Vol. 380, No. 1, p. 104 – 115 (10 Oct 1991).

The authors report the results of a search for blue horizontal branch (BHB) stars at galactocentric distances of ~40 – 65 kpc. Heliocentric line–of–sight velocities are presented for 10 horizontal branch stars (nine BHB and one RR Lyrae object) in this distance interval and are used to place constraints on the extent and mass of the Galaxy's dark corona. The dispersion of the radial component of galactocentric velocity of this group of objects is 111 ± 25 km s^{-1}. Together with results for satellites in the range 50 – 240 kpc, the new data suggest that the mass of the dark corona continues to grow out to at least 40 kpc. The data are discussed within the framework of Little and Tremaine. If the stellar orbits are isotropic, the data show little evidence for a departure from a flat rotation curve having $v_c \sim 200$ km s^{-1}, and the mass within 50 kpc is $\sim 6 \times 10^{11} M_\odot$. If the orbits are radial, the data for the BHB stars may be interpreted in terms of a corona which extends only out to ~40 kpc, and in this case the mass of the Galaxy is $5(+2, -1) \times 10^{11} M_\odot$. The authors argue, however, that the assumptions of such a finite corona and radial orbits are inconsistent with the observed distribution of data for the 26 objects currently known to have distances greater than 40 kpc. The results obtained here are compared with recent mass estimates based on the distant satellites of the Galaxy.

155.094 **JHK imaging of a field in Baade's window.**
T. J. Davidge.
Astrophys. J., Vol. 380, No. 1, p. 116 – 123 (10 Oct 1991).

A 256x256 Hg:Cd:Te array has been used to obtain J, H, and K images of a field in Baade's Window. The data extend down to K ~ 16, which is near the base of the red giant branch at the distance of the Galactic bulge. The distribution of data points on the (K, J–K) and (J–H, H–K) diagrams is similar to that defined by previous aperture measurements. Furthermore, the redward edge of the data envelope on the (K, J–K) diagram roughly follows the giant branch of the cluster M67 to K ~ 15, indicating that the blanketing agent thought to dominate the broad–band photometric properties of super metal–rich stars must be present over a broad range of photospheric temperatures and surface gravities. The data are also used to investigate the horizontal branch content of Baade's Window, although the number of horizontal branch stars found in the current field is small. Horizontal branch stars are found over an interval spanning almost two magnitudes in K, with the bluest of these centered near K ~ 13.2. The relative frequency of horizontal branch stars, measured with respect to bright red giants, is significantly lower than that derived from I band observations, a difference which may be due to blanketing at wavelengths shortward of 1 μm.

155.095 **Population studies. XI. The extended disk, halo configuration.**
J. E. Norris, S. G. Ryan.
Astrophys. J., Vol. 380, No. 2, p. 403 – 418 (20 Oct 1991).

A comparison has been made between the predictions of various putative disk – halo configurations in the solar neighborhood, and the body of data from newly available kinematically and nonkinematically selected samples. It is argued that the continuous extended disk configuration [having a considerable tail to lower abundance (down to [Fe/H] ~ –1.5) and hotter kinematics ($\sigma_W \sim 50 - 60$ km s^{-1}) at this abundance] together with a kinematically decoupled halo, is better able to represent the available observational material than is a three discrete component model containing a thin disk, a thick disk, and a halo. In particular, only the extended disk, halo configuration has the potential of explaining the apparently discrepant results of Ratnatunga and Freeman, on the one hand, who report a rotational lag of 30 km s^{-1} for material having distances from the Galactic plane z = 2 – 5 kpc and [Fe/H] > –0.7, and of Wyse and Gilmore, on the other, who for the same distance regime but covering all abundances, present data indicating a considerably higher rotational lag of 80 km s^{-1}. This phenomenon is reconcilable in terms of a model in which the kinematics of the disk are a continuously varying function of abundance. The implications of the advocated extended disk, halo configuration for current models of Galactic cosmogony are dicussed.

155.096 **Polarimetric imaging of the Galactic center at 12.4 microns: the detailed magnetic field structure in the northern arm and east–west bar.**
D. K. Aitken, D. Gezari, C. H. Smith, M. McCaughrean, P. F. Roche.
Astrophys. J., Vol. 380, No. 2, p. 419 – 428 (20 Oct 1991).
The authors present new closely sampled (~1″) imaging polarimetry of the central 0.5 pc of the Galaxy at 12.4 μm with a 58x62 array camera system. The image shows a detailed view of the morphology of the intrinsically polarized emission (and the associated aligning magnetic field) in the ionized filaments of the Sgr A West complex. The new results show that the magnetic field in the northern arm is continuous on 1″ scale and is a property of the filamentary structure rather than the compact peaks of infrared emission and that it continues in a southwesterly direction from IRS 1 for several arcseconds into the east–west bar. In the southern and western part of the east–west bar the intrinsic polarization becomes complex, with small–scale variations of polarization fraction and position angle, probably due to superposition of different field directions along the line of sight. There is no evidence of perturbation of the field structure by the energetic outflows from the IRS 16 complex, and it is concluded that IRS 16 cannot be at the center of the mass distribution of the Sgr A West system. The authors consider that the magnetic field in Sgr A West is amplified due to shearing motions in the ionized material and that the direction of the field is that of the flow. The run of magnetic field directions in the inner rim of the northern arm and east–west bar then indicate a flow pattern which shows striking symmetry about the position of Sgr A*.

155.097 **Space density and extinction analysis at *l*, *b* = (303°, –17°).**
M. P. V. Fitzgerald, B. C. Reed.
Objective–prism and other surveys – a meeting in memory of Nicholas Sanduleak, p. 45 – 54 (1991). – See Abstr. 012.074 for the main entry.
The distribution of O – M stars and extinction in an 18.25 square degree field at (*l*, *b*) = (303°, –17°) has been determined. The results are based on objective–prism spectra of 1309 stars, UBVI photographic and UBV photoelectric photometry. Photographic and photoelectric results are in good agreement. The extinction is extremely patchy, ranging from zero to almost opaque at 1400 pc. The field includes the Thumb Print Nebula, and is immediately adjacent to several Bok bright–dark nebulae.

155.098 **Galactic Center.**
IAU Circ., No. 5323, p. 1 (13 Aug 1991).
Further information is given in Nos. 5395, 5398.

155.099 **Discovery of a non–thermal radio source associated with the Galactic centre 511–keV annihilation–line source.**
D. A. Leahy.
Mon. Not. R. Astron. Soc., Vol. 251, No. 1, p. 22P – 23P (1 Jul 1991).
The detection of a compact radio source at the position of the X–ray source 1E 1740.7-2942 is reported. This source has recently been identified with the Galactic centre point source of positron annihilation–line radiation. The radio source has a non–thermal spectrum and has yielded the following precise position for the annihilation–line source: RA $17^h40^m41^s.82$, Dec. –29°41′39″.0 (1950 coordinates) with uncertainty of 0″.9.

155.100 **The Galactic center radio source shines below the Compton limit.**
C. R. Gwinn, R. M. Danen, J. Middleditch, L. M. Ozernoy (*L. M. Ozernoj*), T. K. Tran.
Astrophys. J., Lett., Vol. 381, No. 2, p. L43 – L46 (10 Nov 1991).
Absence of refractive scintillation of Sgr A*, the Galactic center radio source, at 1.3 and 0.8 mm wavelengths places an observational limit of brightness temperature $T_b \lesssim 0.5 \times 10^{12}$K on the source. This is less than the maximum brightness of an incoherently synchrotron–emitting plasma, 10^{12}K, known as the Compton limit. The expected refractive scintillations, due to strong broadening of the source by the interstellar plasma, are observed by Zhao et al. at decimeter wavelengths. They could not be observed at 1.3 and 0.8 mm wavelengths over time spans between 0.1 s and 24 hr. Such scintillations are quenched by a source size greater than about 0.1 AU, or, equivalently, $T_b < 0.5 \times 10^{12}$K for Sgr A*, at 0.8 mm wavelength. The scintillations would also require fluctuations with scale sizes of 0.1 AU in the scattering plasma, moving across the line of sight at velocities $V_\perp \gtrsim 100$ km s^{-1}. The authors argue that the absence of scintillations is due to the size of the source and that these observation strengthens the identification of Sgr A* as a quiescent galactic nucleus.

155.101 **An alternative explanation of the evolution of element abundance ratios during the history of the Galaxy.**
M. G. Edmunds, R. M. Greenhow, D. Johnson, V. Klückers, M. B. Vila.
Mon. Not. R. Astron. Soc., Vol. 251, No. 2, p. 33P – 36P (15 Jul 1991).
It is well established that stellar element abundances show a variation of some ratios, such as oxygen to iron and calcium to iron, with the overall metallicity of the star. This has been ascribed to a time–delay effect due to the different mean lifetimes of the stars and binaries that give rise to Type II and Type I supernovae. The authors propose here that this identification may be premature, and that the observations could equally well be explained by a "pseudo–secondary" behaviour in the nucleosynthesis of iron.

155.102 **A search for massive compact halo objects in our Galaxy.**
D. P. Bennett, C. Alcock, T. Axelrod, K. Cook, H. Park, K. Griest, C. Stubbs, K. Freeman, B. Peterson, P. Quinn, A. Rodgers.
AIP Conf. Proc., No. 222, p. 446 – 450 (15 Apr 1991). Current Physics Microform No.: 9106E0446. – See Abstr. 012.069 for the main entry.
MAssive Compact Halo Objects (MACHO) such as brown dwarfs, Jupiters, and black holes are prime candidates to comprise the dark halo of our Galaxy. Paczynski noted that these objects can be detected via gravitational microlensing of stars in the Magellanic Clouds with the caveat that only about one in 10^6 stars will be lensed at any given time. The authors are involved in constucting a dedicated observing system at the Mount Stromlo Observatory for the Magellanic Clouds. During the first year of operation (1991 – 1992), they hope to detect (or rule out) objects in the mass range $0.001\ M_\odot \leqslant M \leqslant 0.1\ M_\odot$, and after five years, they hope to have covered the range $10^{-6} M_\odot < M \lesssim 100\ M_\odot$.

155.103 **Phase space distribution of halo particles and detection of WIMPS.**
Ryu Dongsu, A. Pérez.
AIP Conf. Proc., No. 222, p. 463 – 466 (15 Apr 1991). Current Physics Microform No.: 9106E0463. – See Abstr. 012.069 for the main entry.
The authors calculate the velocity distribution of halo particles using realistic models for the density profiles of the Galactic disk, spheroid, and halo components. They get the resulting velocity

distribution, which is different from the Maxwellian distribution of isothermal halo. The authors investigate the consequences concerning the time averaged count rate and the amplitude of the annual modulation in WIMPs detection experiments. Compared to the previous estimate based on the Maxwellian distribution, the time average count rate is up to 40% larger. On the other hand, the amplitude of the annual modulation is up to a factor of two larger.

155.104 **The oxygen yield and the infall history of the solar neighbourhood.**
S. A. Josey, R. J. Tayler.
Mon. Not. R. Astron. Soc., Vol. 251, No. 3, p. 474 – 481 (1 Aug 1991).

If the yield of oxygen has not changed significantly during past galactic history and its production is adequately described by the instantaneous recycling approximation, the distribution of oxygen abundance in low-mass stars in the solar neighbourhood together with the present fractional gas content can provide an estimate of the oxygen yield and of the time history of infall to the galactic disc. Use of Pagel's G-dwarf distribution gives a yield of about 80% of the present oxygen abundance and a current infall rate of about a third to a half of the rate at which mass is locked up in stars. If the infalling gas has non-zero oxygen abundance, both the yield and the current infall are reduced, but the infall remains significant. These results are in good agreement with those of a recent solar neighbourhood model by Pagel. If observations suggest that current infall is unimportant, the present yield must be approximately halved but its value must have been significantly higher in early galactic history.

155.105 **Probing the Galactic halo along the 3C 273 sight line using IUE.**
G. S. Burks, D. G. York, J. C. Blades, R. C. Bohlin, W. Wamsteker.
Astrophys. J., Vol. 381, No. 1, p. 55 – 62 (1 Nov 1991).

High-resolution spectra of the QSO 3C 273 were obtained in 1989 and added to two spectra recorded in 1982. The co-added spectra represent 600,000 s of integration at a resolution of 30 km s^{-1} from 1200 to 2000 Å. The line of sight passes through the Virgo Cluster, the entire halo of the Milky Way, at Galactic latitude b = 64°, and foreground X-ray emitting material from local disturbed gas. The observed equivalent width of Galactic C IV is greater than the equivalent width of C II, a situation that is uncommon in local gas but is often found in QSO absorption line systems. The lines of C IV, Al III, and Si IV may arise in an extended halo or may be associated with local disturbed gas in a radio loop. The remaining detected species (H I, C II, Si II) arise predominantly from the local gas. Most of the gas is probably within 5 kpc, but no firm conclusion is reached. Possible detection of interstellar absorption due to the Virgo Cluster is discussed.

155.106 **The statistics of slow interstellar accretion onto neutron stars.**
O. Blaes, M. Rajagopal.
Astrophys. J., Vol. 381, No. 1, p. 210 – 219 (1 Nov 1991).

The authors perform Monte Carlo calculations of neutron star orbits in the Galaxy adopting the birth distribution of the slow moving radio pulsar population found by Narayan and Ostriker. The authors incorporate a smooth model of the interstellar medium and calculate the distribution of accreted masses onto neutron stars in the solar neighborhood. Also included are appendices describing a general analytic approach to the accretion problem. This agrees very well with the Monte Carlo results and is valid under more general assumptions about the interstellar matter distribution and birth properties of neutron stars. These results may be used to test theories of high-energy sources which invoke interstellar accretion. In particular, the authors explore the implications for a model of gamma-ray bursts based on electron capture-induced starquakes.

155.107 **An all-sky survey of interstellar reddening in the solar neighborhood.**
C. L. Perry, J. E. Tohline, D. M. Christodoulou.
Objective-prism and other surveys – a meeting in memory of Nicholas Sanduleak, p. 71 – 74 (1991). – See Abstr. 012.074 for the main entry.

Photometric color excesses and distances, derived for 7,000 A and early F stars, were used to determine the spatial distribution of the dust component of the interstellar medium in the solar neighborhood. The reddening map traces two large cloud complexes; both are characterized by local irregularities in density and spatial extent, which attest to the variable nature of the distribution of particulate matter in the solar vicinity.

155.108 **The vertical structure of our Galaxy from digitized objective-prism spectra.**
J. A. Rose, R. Agostinho, J. Stock.
Objective-prism and other surveys – a meeting in memory of Nicholas Sanduleak, p. 77 – 83 (1991). – See Abstr. 012.074 for the main entry.

The authors are conducting a long-term survey to investigate the vertical structure of our Galaxy, with a particular emphasis on studying the intermediate (i.e., thick disk) population(s). The survey is based on digitized objective-prism, spectra obtained with the 10° prism at the CTIO Curtis Schmidt. The authors' goal is to obtain quantitative information on surface gravity, chemical composition, and kinematics for all F-K stars down to a limiting B magnitude of 12.5. Hence they are investigating the vertical structure using a complete survey over a wide range of spectral type, rather than restricting themselves to a specific tracer population.

155.109 **Understanding the kinematics of Galactic Centre gas.**
J. Binney, O. E. Gerhard, A. A. Stark, J. Bally, K. I. Uchida.
Mon. Not. R. Astron. Soc., Vol. 252, No. 2, p. 210 – 218 (15 Sep 1991).

A coherent picture is constructed of the Galaxy's H I, CO and CS emissions in the region $|l| \leqslant 10°$, $|b| \leqslant 0.5°$. The flow of gas at the Galactic Centre is dominated by a bar that has corotation at $r = 2.4 \pm 0.5$ kpc, which one views at an angle of $\theta_{incl} = 16 \pm 2°$ from its major axis. The first CO emission arises where gas is obliged to switch from x_1 orbits to x_2 orbits in the notation of Contopoulos. This gives rise to a shock and a clear signature in the (l, v) diagram. The great Galactic Centre molecular clouds such as Sgr B are on x_2 orbits. From the structure of the H I terminal velocity envelope the authors deduce that the central mass density scales as $\varrho \sim r^{-1.75}$ out to at least ~ 1.2 kpc along the bar's major axis. Consequently the circular velocity curve is rising significantly through the radius range where naive analysis of the tangent velocity leads to a falling rotation curve. The great ring of molecular material at $r \cong 3.5$ kpc is probably associated with the bar's outer Lindblad resonance, and the region of low gas densities inwards from there with corotation.

155.110 **LMC or halo stars? The Sanduleak–Fehrenbach controversy.**
W. P. Bidelman.
Objective-prism and other surveys – a meeting in memory of Nicholas Sanduleak, p. 85 – 89 (1991). – See Abstr. 012.074 for the main entry.

In 1972 Nick Sanduleak published a list of 47 A- and early F-type stars, considered by Fehrenbach to be LMC members because of their large radial velocities, that appeared to have stronger hydrogen lines than would be expected. The question raised by the existence of these stars has never been satisfactorily answered. Sanduleak did, however, obtain slit-spectrograph plates of seven of them in 1975, and examination of these by the writer strongly suggests that none are LMC members. It seems probable that a large proportion of the objects listed by Sanduleak are in fact halo stars of our own Galaxy.

155.111 **Is the bulge of our Galaxy triaxial?**
Y. Nakada, S. Deguchi, O. Hashimoto, H. Izumiura, T. Onaka, K. Sekiguchi, I. Yamamura.
Nature, Vol. 353, No. 6340, p. 140 – 141 (12 Sep 1991). Letter–to–the–editor.

As in the case of other spiral galaxies, the rotation curve of gas in our Galaxy indicates that its nuclear bulge is triaxial. Despite several studies of the distribution of specific objects in the galactic bulge, no observational evidence of asymmetry has emerged. Recently, however, Blitz and Spergel have reanalysed balloon–infrared observations at 2.4 μm, where obscuration is relatively low, to suggest that the galactic bulge is bar–like, and tilted with respect to the plane of the Galaxy. Here the authors report that the distribution of IRAS bulge stars shows an asymmetry with respect to the galactic centre, providing clear evidence for triaxiality.

155.112 **Milky Way galaxy: Annihilation near the centre.**
R. Ramaty, R. E. Lingenfelter.
Nature, Vol. 353, No. 6341, p. 215 – 216 (19 Sep 1991).

155.113 **Is the Galactic Centre gamma–ray source 1E 1740.7–2942 accreting from a molecular cloud?**
J. Bally, M. Leventhal.
Nature, Vol. 353, No. 6341, p. 234 – 237 (19 Sep 1991). Letter–to–the–editor.

For twenty years there have been detections of an intermittent source of 511–keV electron–positron annihilation radiation at the centre of our Galaxy. Until recently, none of the balloon or satellite experiments lucky enough to catch the source in its "on" state had spatial resolution better than several degrees, but on 13 – 14 October 1990 the French SIGMA experiment aboard the Soviet GRANAT spacecraft observed a day–long burst of narrow–band γ rays, probably annihilation radiation, from within 1.5 arcmin of the known X–ray source 1E 1740.7–2942, about 50 arcmin from Sgr A West, which contains the dynamical centre of the Milky Way. The authors report millimetre–wavelength observations showing that this variable X–ray source lies on a line of sight to the dense $10^5 M_{\odot}$ molecular cloud G –0.86–0.08, near the Galactic Centre. They suggest that the γ–ray source, probably a compact object, possibly a black hole, is accreting dense gas directly from the molecular cloud.

155.114 **A filamentary radio source near the Galactic Centre.**
A. D. Gray, L. E. Cram, R. D. Ekers, W. M. Goss.
Nature, Vol. 353, No. 6341, p. 237 – 239 (19 Sep 1991). Letter–to–the–editor.

Near the Galactic Centre are several filamentary radio structures that have become known as "threads" and which are thought to signify either magnetic loops or electric current paths between active regions and the galactic magnetic field. Using the Molonglo Observatory Synthesis Telescope, the authors have found a new elongated radio source, G 359.1–0.2, within 1° of the Galactic Centre. Observations with the Australia Telescope Compact Array and the Very Large Array show it to be filamentary in nature, extending more than 20 arcmin in length but only 10 arcsec wide. This object is spectrally similar to the previously known threads, but unlike them has kinks along its length, and does not appear to be obviously connected with any active region that might be responsible for its generation.

155.115 **A census of the Milky Way.**
G. Gilmore.
Nature, Vol. 353, No. 6345, p. 604 – 605 (17 Oct 1991).

155.116 **High–resolution VLA images of the Galactic Centre at 2–cm wavelength with large dynamic range.**
J.-H. Zhao, W. M. Goss, K. Y. Lo, R. D. Ekers.
Nature, Vol. 354, No. 6348, p. 46 – 48 (7 Nov 1991). Letter–to–the–editor.

The discovery of a bright compact radio source (Sgr A*) at the Galactic Centre suggested the presence of a central exciting source. Recently faint extended radio structures were reported which were interpreted as thermally emitting blobs of plasma physically associated with Sgr A*. The authors have re–observed the Galactic Centre using the Very Large Array, obtaining images at a wavelength of 2 cm with resolution of ~0.1 arcsec (0.005 pc). The r.m.s. noise of 70 μJy per beam, with a dynamic range of 22,000:1, is comparable to earlier observations. The authors discuss several ways that low–level artefacts near Sgr A* might have been generated. They also confirm the presence of a plume of emission 1 arcsec southwest of Sgr A*.

155.117 **The galactic distribution of free electrons.**
J. M. Cordes, J. M. Weisberg, D. A. Frail, S. R. Spangler, M. Ryan.
Nature, Vol. 354, No. 6349, p. 121 – 124 (14 Nov 1991).

Dispersion, distance and scattering measurements of pulsars and other radio sources are now sufficiently numerous to allow modelling of the entire galactic distribution of free electrons. A two–component axisymmetric model of local electron density, fitted on all scales from 100 km to a few parsecs, accounts for most of the data; a population of dense, discrete clouds is also needed, and there is some evidence for spiral structure.

155.118 **Equilibrium models for two samples of OH/IR stars.**
P. te Lintel Hekkert, H. Dejonghe, H. J. Habing.
Proc. Astron. Soc. Aust., Vol. 9, No. 1, p. 20 – 25 (1991). – See Abstr. 012.078 for the main entry.

The authors present a progress report on the dynamical analysis of the IRAS sample of OH/IR stars. This sample is complete within a distance of 8 kpc. The authors distinguish two groups of OH/IR stars, one with high and the other with low expansion velocity of the circumstellar shell. Using a quadratic programming method they fit a database of Galactic orbits to the observed distribution of radial velocities, longitudes and latitudes. The dynamical model yields a distribution function, based on two integrals of motion, for each group. Integration gives the density, the mean velocity and the velocity dispersions as function of R and z. The distribution functions for the two groups differ enough to suggest that they represent two different stellar populations. The authors estimate the stellar lifetime for each group by comparing the velocity dispersions and z scale–heights at the solar radius with those found for local stellar populations. It is concluded that the group with the low expansion velocity is likely to be part of the thick disc, while the high expansion velocity group has dynamical characteristics resembling an old disc population; this is in contrast with the findings for differently selected samples of OH/IR stars, which were found to be less than 1 Gyr old.

155.119 **The galactic magnetic field from pulsar rotation measures.**
M. E. Costa, P. A. Hamilton, P. M. McCulloch.
Proc. Astron. Soc. Aust., Vol. 9, No. 1, p. 131 (1991). – See Abstr. 012.078 for the main entry.

Pulsar rotation measures have been used to investigate the structure of the local Galactic magnetic field. The Galactic field is found to be concentrated in the spiral arms and hence to be a spiral field. From the pulsars within 2 kpc of the Sun, the field in the local spiral feature was modelled with Gaussian profiles in altitude and azimuth. In this model the field has a peak strength of 4.3 ± 0.2 μG directed towards Galactic longitude $l = 73° \pm 6°$.

155.120 **Stellar evolution and galactic chemical evolution.**
F. Matteucci.
Symposium in Celebration of the 50. Anniversary of McDonald Observatory of the University of Texas at Austin: Frontiers of stellar evolution, p. 539 – 570 (1991). – See Abstr. 012.085 for the main entry.

Observational constraints and theoretical ingredients necessary to compute the chemical evolution of the Galaxy are reviewed. Special attention is devoted to the discussion and interpretation of the abundance patterns observed in stars in the solar vicinity as tracers of chemical evolution. The influence of stellar evolution and nucleosynthesis on the results is discussed, and examples are given of how chemical evolution models can constrain stellar evolution theories. A specific model for galactic chemical evolution – where the evolution of the abundances of several chemical elements (He, C, ^{13}C, N, O, Ne, Mg, Si, S, Ca, and Fe) is studied in detail – is presented and its predictions compared with observational data.

155.121 **Galactic evolution.**
R. B. Larson.
Symposium in Celebration of the 50. Anniversary of McDonald Observatory of the University of Texas at Austin: Frontiers of stellar evolution, p. 571 – 587 (1991). – See Abstr. 012.085 for the main entry.

Observational evidence and theoretical notions concerning the evolution of the halo and disk components of the Galaxy are reviewed, and the implications of some recent results are discussed. Data on the ages and other properties of globular clusters suggest that the formation of the galactic halo was a prolonged and chaotic process, possibly involving the merging of galaxy–like subsystems. There appears to be approximate chronological and chemical continuity between the halo and the disk, which suggests that disk formation took up where halo formation left off some 10 to 12 Gyr ago. Theoretical considerations and direct evidence concerning the history of star formation in the local disk and in typical Sc galaxies are both consistent with a simple model in which the star formation rate is approximately proportional to the gas content. With current data, the variation of the star formation rate (SFR) in such a model is moderate, and the ratio of past average to present SFR is only about 2.

155.122 **Evidence of an older disk age derived from the white dwarf luminosity function.**
T. A. Smecker, R. F. G. Wyse.
Symposium in Celebration of the 50. Anniversary of McDonald Observatory of the University of Texas at Austin: Frontiers of stellar evolution, p. 619 (1991). Abstract. – See Abstr. 012.085 for the main entry.

155.123 **A cluster of He I emission–line stars in the Galactic center.**
A. Krabbe, R. Genzel, S. Drapatz, V. Rotaciuc.
Astrophys. J., Lett., Vol. 382, No. 1, p. L19 – L22 (20 Nov 1991). With plates L1 – L2.

The authors report 1."9 ($300\ \mathrm{km\ s^{-1}}$) resolution imaging spectroscopy of He I $n = 2^1P \rightarrow n = 2^1S$ 2.06 μm line emission in the Galactic center. There is a group of about a dozen compact, broad–line "He I emission–line stars" most of which are positionally coincident with compact 2 μm continuum sources. IRS 16 and its "broad–line region" is probably a central concentration of this 1 pc diameter He I star cluster. The He I stars may be blue supergiants with heavy mass loss. The high He I/H I (Brα, Brγ) ratios suggest an overabundance of helium in their atmospheres indicative of a post–main–sequence phase. Mass–loss rates may range between a few 10^{-5} and $10^{-4} M_{\odot} yr^{-1}$, and outflow velocities are between 500 and 1000 $\mathrm{km\ s^{-1}}$. The He I stars are members of a cluster of massive stars that can very likely account for the entire luminosity, the flux of Lyman–continuum photons, and the fairly low excitation of the central few parsecs. These data show very clearly that massive star formation has taken place in the Galactic center within the last few 10^6yr.

155.124 **SPARTAN 1 X–ray observation of diffuse emission at the galactic center.**
W. A. Snyder, R. G. Cruddace, M. P. Kowalski, G. G. Fritz, E. E. Fenimore.
Workshop on Iron Line Diagnostics in X–Ray Sources, p. 53 – 56 (1991). – See Abstr. 012.088 for the main entry.

The 1–10 keV X–ray detectors on SPARTAN 1 scanned the central 2° region of the galactic center during 20–21 June 1985. Four strong point sources and extended emission ~1° in diameter centered near the galactic center were detected (Kawal et al. 1988). The authors have re–analyzed the diffuse component data following the discovery (Koyama et al. 1989) of intense Fe line emission. The radial distribution of this component is examined by grouping the data into radial bins centered on Sgr A West, after first removing known point sources. The authors confirm that this component is an intense source of Fe line radiation, but find that Fe line flux is more sharply peaked and narrowly distributed than previously reported. A Gaussian surface brightness distribution with FWHM of 49–92 arcmin (90% confidence limits) is consistent with the SPARTAN 1 data. The spectrum can be approximated with either a single power–law or thermal form, but the high χ^2 suggests a more complex shape. The radial distribution of the Fe line and continuum emission is presented and used to address the nature of the diffuse component.

155.125 **Wide binaries in the Woolley catalog.**
I. Wasserman, M. D. Weinberg.
Astrophys. J., Vol. 382, No. 1, p. 149 – 167 (20 Nov 1991).

The authors investigate constraints on the properties of local dark matter based on the distribution of wide binary separations. Unlike previous studies, they analyze a volume–selected binary catalog (a particular subset of the Woolley catalog), which eliminates the need for cumbersome selection functions. First, the authors present a list of candidate wide binaries from a search of main–sequence stars with $M_V \leqslant 9$ in the Woolley catalog. Of these candidates, the authors identify nine pairs with $5 \leqslant M_V \leqslant 9$ and separations 0.01 pc $< s <$ 1 pc that are likely to be bound. The existence of these pairs implies that ~3% of the stars with $5 \leqslant M_V \leqslant 9$ are members of wide binary systems with separations 0.01 pc $< s <$ 1 pc. The implications of this candidate list are considered. The authors estimate that ~60 pairs of similar absolute magnitude should be found in a volume–limited astrometric data set with limiting parallax ~0."02 and limiting apparent magnitude ≈12.5. Next, the authors construct theoretical, evolutionary models for the binary distribution, which include perturbations due to giant molecular clouds and to either low–mass ($M_p \lesssim 50\ M_{\odot}$) or intermediate–mass ($50 \lesssim M_p \lesssim 10^5 M_{\odot}$) pointlike objects. The analysis suggests that the binary distribution at birth follows a law $\sim a^{-2}$ to a^{-1} (a is the semimajor axis). For models in which the dominant perturbative effects are due to ordinary stars, the authors argue that at least some wide pairs must form with semimajor axes $a \gtrsim 2000$ AU.

155.126 **H I absorption measurements toward 15 pulsars and the radial distribution of diffuse ionized gas in the Galaxy.**
D. A. Frail, J. M. Cordes, T. H. Hankins, J. M. Weisberg.
Astrophys. J., Vol. 382, No. 1, p. 168 – 181 (20 Nov 1991).

The authors report neutral hydrogen absorption measurements at 21 cm toward 15 pulsars with the Arecibo 305 m telescope and the VLA. The distances to these pulsars derived from the kinematic method rule out one potential pulsar – supernova remnant association while giving support to another. In addition, the secular change in the absorption spectrum toward one pulsar (PSR 1821+05) may provide independent confirmation of the recent discovery of AU–sized H I structures in the ISM. A study of the radial distribution of the diffuse ionized medium has been carried out using mean electron density measurements made toward pulsars. With the new measurements and those from the literature, there are a total of 18 lines of sight that sample the electron density at small galactocentric radii. The data show a clear rise in the electron density in the inner Galaxy, to about 3 times the local average. From this and other evidence, the authors argue that O stars are the principal heating agent of the warm ionized medium.

155.127 **Vertical distribution and support of Galactic H I.**
F. J. Lockman, C. S. Gehman.
Astrophys. J., Vol. 382, No. 1, p. 182 – 188 (20 Nov 1991).

The shape of high–latitude 21 cm H I emission profiles suggests that the local neutral interstellar medium is supported in the Galactic gravitational potential primarily by turbulence, that is, by the motion of individual clouds. Composite H I spectra derived from averages over large areas of the high–latitude sky are modeled by several isothermal components whose equilibrium vertical distribution in a plausible Galactic potential successfully reproduces the observed thickness of the H I layer. The analysis implies that there are several H I populations with differing central density, velocity dispersion, and scale height, although at present these populations cannot be characterized with precision. The kinetic energy in bulk vertical motions is $\gtrsim 6\times 10^{51}$ergs kpc^{-2} for a full disk. About 10% of the local H I is not part of a relaxed distribution, but is infalling or is in ordered flows.

155.128 **Young halo stars and Galactic evolution.**
C. M. Lance.
Comments Astrophys., Vol. 15, No. 6, p. 355 – 374 (Oct 1991).

Young, high-velocity, early-type stars have been observed in the Galactic halo, many kiloparsecs away from the disk. This finding is inconsistent with standard models of stellar and Galactic evolution. The author describes the studies that discovered the young halo stars and identified their characteristics, and examines the theories that have been advanced to account for them. A comparison of their properties shows that high-velocity stars of spectral types A and B have similar characteristics, but they appear to be different from those of high-velocity O stars. The significance of these anomalous stars to our understanding of the structure and evolution of the Milky Way is discussed.

155.129 **The distance to the galactic center.**
V. M. Blanco.
Bol. Asoc. Argent. Astron., No. 36, p. 87 – 106 (1990). – See Abstr. 012.095 for the main entry.

Methods used to determine the distance (R_o) to the galactic center are described and recent results are presented. These suggest that $R_o \cong 7$ Kpc. Possible improvements in the determination of R_o are suggested.

155.130 **Subdwarf luminosity function.**
S.-G. Lee.
J. Korean Astron. Soc., Vol. 24, No. 2, p. 161 – 171 (Dec 1991).

The author has derived the luminosity function for subdwarfs on the basis of the proper motion data in the LHS Catalogue, utilizing the reduced proper motion diagram for the selection of subdwarfs and the hybrid method combining the mean absolute magnitude method and V/V_m method to estimate the distance and density of subdwarfs. The luminosity function found is almost flat, showing a very slow increase up to $M_V = 9$ or $M_B = 10$, and the overall halo density is larger than those derived by Schmidt (1975), Chiu (1980), Reid (1984), Lee (1985), and Dawson (1986), but smaller than that by Eggen (1983). Comparison with 1/100 of the disk stellar luminosity funtion implies that no conclusive dip in the halo luminosity function is found.

155.131 **Disk luminosity function based on the Lowell Proper Motion Survey.**
M. Kim, S.-G. Lee.
J. Korean Astron. Soc., Vol. 24, No. 2, p. 173 – 190 (Dec 1991).

Disk stellar luminosity function has been derived with stars in the Lowell Proper Motion Survey with $\mu \geqslant 0\farcs27$ of arc/yr, $8 < m_{pg} < 17$ and with bright stars in the SAO Star Catalogue. The luminosity function has been obtained with stars within 20 pc by Luyten's mean absolute magnitudes method using Reduced Proper Motion Diagram to select disk stars. Magnitudes and colors have been transformed to the UBV system from the published UBV data. It has been found that stars which have higher proper motion than the original limit of the proper motion survey are missed, when the relation between the absolute magnitude and reduced proper motion is applied to sample stars without considering the dispersion in magnitude. Correction factors to missing stars have been estimated according to their limits of proper motion which are dependent on the absolute magnitude. Resulting luminosity function shows Wielen's dip at $M_B \sim 10$, and systematic enhancement of stars on the average of about $\Delta \log \Phi (M_B) \sim 0.2$ compared with Luyten's luminosity function.

155.132 **An improved model of the galactic mass distribution for orbit computations.**
C. Allen, A. Santillán.
Rev. Mex. Astron. Astrofis., Vol. 22, No. 2, p. 255 – 263 (Oct 1991).

A realistic, yet very simple and analytical model for the galactic mass distribution is presented, which improves upon a similar model proposed in Allen and Martos (1986). The new potential is completely analytical; the density can be obtained from it in closed form and it is positive everywhere. Extreme mathematical simplicity is retained, as well as a good representation of the observed values of both the rotation curve and the perpendicular force. The new model consists of a spherical central bulge and a disk, both of the Miyamoto–Nagai (1975) form, plus a massive, spherical, halo. The total mass of the model is 9.00×10^{11} solar masses. The model escape velocity for objects in the solar vicinity is 535.7 km s^{-1}. The values obtained for the galactic rotation constants are $A = 12.95$ km s^{-1}kpc^{-1} and $B = -12.93$ km s^{-1}kpc^{-1}; they are in good agreement with recent observational data. In contrast to other models for our galaxy, the proposed potential function is extremely simple, fully analytical, continuous, and with continuous derivatives everywhere. The suitability of this potential for efficient and accurate numerical orbit computations is demonstrated by recalculating the orbits of several nearby, high velocity stars.

155.133 **Erratum: "Distribution and studies of the infrared stellar population in the Galaxy. I. The model" [Rev. Mex. Astron. Astrofis., Vol. 22, No. 1, p. 27 – 41 (Apr 1991)].**
R. A. Ruelas-Mayorga.
Rev. Mex. Astron. Astrofis., Vol. 22, No. 2, p. 297 (Oct 1991).
See Abstr. 53.155.190.

155.134 **Erratum: "Distribution and studies of the infrared stellar population in the Galaxy. II. The data base" [Rev. Mex. Astron. Astrofis., Vol. 22, No. 1, p. 43 – 97 (Apr 1991)].**
R. A. Ruelas-Mayorga.
Rev. Mex. Astron. Astrofis., Vol. 22, No. 2, p. 297 (Oct 1991).
See Abstr.53.155.191.

A database of Ap stars.
See Abstr. 002.020.

The interstellar lines catalogue.
See Abstr. 002.027.

Determination of positions and proper motions of stars from the plates of the northern sky photographic survey program.
See Abstr. 002.046.

Catalogue of proper motions of stars in the North Galactic pole region.
See Abstr. 002.047.

Early-type emission-line stars with large infrared excesses.
See Abstr. 002.065.

Catalogue of physical and kinematical parameters for main sequence F stars within 80 parsecs of the Sun.
See Abstr. 002.083.

Surveys for cool supergiants in the Galaxy.
See Abstr. 002.096.

The space distribution of supergiants and other stars in the Michigan Spectral Catalogue, Volumes 1 – 4.
See Abstr. 002.099.

Thin prism survey for F and G stars at the south galactic pole.
See Abstr. 002.101.

Highly reddened, probable supergiants found in the recent Milky Way infrared survey at the Warner and Swasey Observatory.
See Abstr. 002.104.

Along the Milky Way.
See Abstr. 004.077.

Stellar abundance studies with a large telescope.
See Abstr. 013.026.

The programme MEGA: results and outlook.
See Abstr. 013.067.

HST and dense stellar systems.
See Abstr. 051.019.

The r–process and nucleochronology.
See Abstr. 061.035.

Direct detection of galactic axions with Rydberg atoms in an inhibited cavity regime.
See Abstr. 061.057.

Radioactive dating of the elements.
See Abstr. 061.066.

Solitary dipole vortexes in rotating gravitating systems generation of turbulent spectrum.
See Abstr. 062.127.

Model atmospheres of massive hot stars.
See Abstr. 064.126.

A new approach to gravitational microlensing: time delays and the Galactic mass distribution.
See Abstr. 066.126.

1E 1740.7–2942: a black hole in a molecular cloud?
See Abstr. 067.085.

The formation of neutron stars in the Galaxy.
See Abstr. 067.111.

EUV emission from gamma–ray–burst sources.
See Abstr. 067.155.

Compton backscattered 511 keV annihilation line emission and the 170 keV line from the Galactic center direction.
See Abstr. 067.213.

Early type high–velocity stars in the solar neighborhood. IV. Four–color and Hβ photometry.
See Abstr. 111.009.

Investigation of absolute proper motions of stars of the Special Compiled Catalogue.
See Abstr. 111.012.

Determination of absolute proper motions in the main meridional section of the Galaxy.
See Abstr. 111.015.

SiO maser emission from OH/IR stars close to the Galactic Centre.
See Abstr. 112.045.

Excitation of the hydroxyl maser in OH/IR stars.
See Abstr. 112.073.

Photoelectric UBV photometry of northern stars from the HK objective–prism survey. II.
See Abstr. 113.023.

Photoelectric UBV photometry of stars selected in the HK objective–prism survey.
See Abstr. 113.025.

UBV photometry of OB^+ stars in the southern Milky Way.
See Abstr. 113.026.

***uvby–β* photometry of high–velocity and metal–poor stars. V. Distances, kinematics and ages of halo and disk stars.**
See Abstr. 113.027.

Photometric provision for the photographic survey of the north sky and MEGA programmes.
See Abstr. 113.037.

UBV photometry of luminous stars at $l = 268$ and $l = 296$.
See Abstr. 113.041.

Surveys of high galactic latitude areas: CCD four–color photometry of population II stars.
See Abstr. 113.042.

The lithium abundance in extreme halo stars.
See Abstr. 114.002.

IUE observations of early–type stars at high Galactic latitudes.
See Abstr. 114.004.

HgMn stars, good or bad for optical soundings in the Milky Way?
See Abstr. 114.007.

A spectroscopic survey of faint compact objects.
See Abstr. 114.114.

The mass range of carbon stars.
See Abstr. 115.019.

Unresolved binaries and the stellar luminosity function.
See Abstr. 115.021.

Massive stars – from the solar neighborhood to starburst galaxies.
See Abstr. 115.024.

Optical identifications of compact galactic X–ray sources: Liller lore.
See Abstr. 117.258.

Multiplicity among solar–type stars in the solar neighbourhood. II. Distribution of the orbital elements in an unbiased sample.
See Abstr. 118.010.

Radial velocities of 44 southern Milky Way Cepheids.
See Abstr. 122.015.

The space distribution and evolutionary status of short and intermediate period oxygen–rich Miras.
See Abstr. 122.017.

Absolute magnitudes and kinematic properties of Cepheids.
See Abstr. 122.069.

Results of a nova search: 1982–1991.
See Abstr. 124.009.

A historical supernova's lower limit to the galactic stellar collapse rate.
See Abstr. 125.031.

Galactic and extragalactic supernova rates.
See Abstr. 125.040.

The distribution of supernova remnants in the Galaxy.
See Abstr. 125.092.

White dwarf space densities and birth rates reconsidered.
See Abstr. 126.043.

New limits on the population of millisecond pulsars in the galactic plane.
See Abstr. 126.078.

Trapped H_2O in SiO condensate: an explanation for the 3 micron band observed toward the Galactic center.
See Abstr. 131.001.

Molecular clouds in Vela.
See Abstr. 131.006.

Diffuse Galactic emission at millimetric wavelengths.
See Abstr. 131.007.

The study of interstellar absorption in the galactic longitude range 65 – 165°.
See Abstr. 131.011.

Star formation and the origin of the far–IR emission from giant molecular clouds in the inner Galaxy.
See Abstr. 131.045.

Westerbork observations of high–velocity clouds. Discussion.
See Abstr. 131.095.

Distribution and origin of high–velocity clouds. II. Statistical analysis of the whole–sky survey.
See Abstr. 131.096.

Distribution and origin of high–velocity clouds. III. Clouds, complexes and populations.
See Abstr. 131.097.

Redetermination of the interstellar $^{12}C/^{13}C$ ratio in the solar vicinity.
See Abstr. 131.107.

Fragments of our Galaxy at the galactic longitude 55°.
See Abstr. 131.115.

The interstellar extinction towards Selected Area 203: a distance estimation to the Chamaeleon–Musca dark–clouds complex.
See Abstr. 131.120.

Processes controlling the initial mass function: interstellar turbulence and magnetic fields.
See Abstr. 131.164.

Location of the thermal continuum source Sagittarius D.
See Abstr. 131.251.

First results from a composite large–scale CO survey at high Galactic latitudes in the second quadrant.
See Abstr. 131.262.

A molecular shell surrounding the galactic center.
See Abstr. 131.265.

The distribution of nearby H I and H II gas.
See Abstr. 131.295.

Millimeter continuum observations of Galactic center giant molecular cloud cores.
See Abstr. 131.296.

Interstellar Ca II in the galactic halo and in QSO absorption systems.
See Abstr. 131.308.

Preliminary spectral observations of the Galaxy with a 7° beam by the Cosmic Background Explorer (COBE).
See Abstr. 131.316.

Images of atomic carbon in the interstellar medium.
See Abstr. 131.326.

The nearby interstellar material in the general direction towards M22.
See Abstr. 131.327.

High–resolution Na I observations of the local interstellar medium.
See Abstr. 131.332.

Observations of Ca II absorption along the nearly parallel lines of sight toward 3C 273 and SN 1991T.
See Abstr. 131.361.

The hot interstellar medium.
See Abstr. 131.390.

How hot are the molecular clouds at the galactic centre?
See Abstr. 131.395.

The structure and kinematics of Sgr B1, a "forgotten" giant H II region near the Galactic center.
See Abstr. 132.004.

Statistical analysis of galactic H II region velocity fields: a progress report.
See Abstr. 132.045.

An extensive study of planetary nebulae in the Galactic bulge. II. Statistical properties of the nebular envelopes.
See Abstr. 134.002.

An extensive study of planetary nebulae in the galactic bulge. I. Spectrophotometric data.
See Abstr. 134.004.

An extensive study of planetary nebulae in the Galactic bulge. IV. The luminosity function as a probe of the planetary nebulae population.
See Abstr. 134.050.

Two planetary nebulae near the galactic center.
See Abstr. 134.080.

Radio sources in a field centered on BG 2107+49.
See Abstr. 141.001.

Investigation of radiosources of the second galactic quadrant from the Kallas and Reich survey at 7.6 and 31.3 cm wavelengths.
See Abstr. 141.005.

Two hard X–ray sources in 100 square degrees around the Galactic Center.
See Abstr. 142.001.

The spectrum and position of the hard galactic centre source 1E 1740.7–2942.
See Abstr. 142.025.

Vacuum ultraviolet imagery of the Virgo cluster region. III. Diffuse far–ultraviolet radiation at high Galactic latitudes.
See Abstr. 142.026.

Spectroscopic limits to an extragalactic far–ultraviolet background.
See Abstr. 142.027.

The intensity and spectrum of the diffuse X–ray background.
See Abstr. 142.028.

The diffuse soft X–ray background as seen with the Einstein Observatory.
See Abstr. 142.033.

GRS 1758–258.
See Abstr. 142.040.

GX 1+4.
See Abstr. 142.041.

A search for the 478 keV line from the decay of nucleosynthetic ^{7}Be.
See Abstr. 143.003.

The Cos–B GeV gamma–ray sources.
See Abstr. 143.018.

On the Galactic origin of gamma–ray bursts.
See Abstr. 143.033.

Leptonic origin of diffuse galactic high–energy gamma rays. Indications for a varying cosmic ray proton–to–electron ratio.
See Abstr. 143.035.

A search for the 2.223 MeV neutron capture gamma–ray line from the directions of Cygnus and the Galactic center.
See Abstr. 143.050.

SIGMA discovery of variable $e^{+}-e^{-}$ annihilation radiation from the near Galactic center variable compact source 1E 1740.7–2942.
See Abstr. 143.052.

Three spectral states of 1E 1740.7–2942: from standard Cygnus X–1 type spectrum to the evidence of electron–positron annihilation feature.
See Abstr. 143.053.

Search for diffuse cosmic gamma rays above 200 TeV with the Utah–Michigan array.
See Abstr. 144.044.

The distribution of cosmic ray nuclei in our Galaxy.
See Abstr. 144.062.

Cosmic–ray age and the interstellar medium.
See Abstr. 144.074.

The inverse Compton scattering of cosmic ray electrons in the Galaxy.
See Abstr. 144.091.

Cosmic ray positron production by gamma ray interactions on starlight.
See Abstr. 144.092.

Resonance excitation of spiral density waves in a gaseous disk. II. A nonlinear theory and application to the 3 kiloparsec arm.
See Abstr. 151.022.

The disk heating problem.
See Abstr. 151.049.

Pattern speeds of density waves.
See Abstr. 151.056.

The velocity dispersion of giant molecular clouds. II. Mathematical and numerical refinements.
See Abstr. 151.087.

Approximate third integrals in axisymmetric galaxies and the tilt of the velocity ellipsoid in the solar neighborhood.
See Abstr. 151.098.

Stellar kinematics in Kovalsky–Kapteyn figures. II.
See Abstr. 151.099.

Kinematics of stars in Kovalsky–Kapteyn figures. II.
See Abstr. 151.122.

Distribution of stars perpendicular to the plane of the Galaxy.
See Abstr. 151.144.

Height pattern of the formation places of open clusters.
See Abstr. 153.041.

Light elements as tracers of chemical evolution.
See Abstr. 153.069.

Study of faint young open clusters as tracers of spiral features in our galaxy. Paper 5: NGC 2236 (OCl 501).
See Abstr. 153.089.

Diameters of Galactic globular clusters.
See Abstr. 154.018.

Inhomogeneous halo collapse and early Galactic chemical evolution: globular cluster metallicities.
See Abstr. 154.024.

Compactness of globular cluster and its evolution.
See Abstr. 154.030.

The age spread among galactic globular clusters.
See Abstr. 154.034.

Perspective of the search for globular clusters in the Galaxy.
See Abstr. 154.042.

The role of the galactic disk in globular cluster formation.
See Abstr. 154.050.

The horizontal–branch stars in globular clusters as tracers of galactic evolution.
See Abstr. 154.053.

The halo and old disk clusters of the Milky Way.
See Abstr. 154.113.

Interaction of mass lost in globular clusters with the halo medium.
See Abstr. 154.129.

The Galactic globular cluster system: theoretical constraints for α–enhanced compositions.
See Abstr. 154.130.

Globular cluster mass functions.
See Abstr. 154.132.

Cosmic rays in the Large and Small Magellanic Clouds and the relationship to our Galaxy.
See Abstr. 156.012.

The velocity dispersion gradients in spiral galaxies. I. The Galaxy and NGC 3198.
See Abstr. 157.002.

Stellar populations in galaxies.
See Abstr. 157.076.

The role of neutral hydrogen in the evolution of spiral and irregular galaxies.
See Abstr. 157.083.

Large $^{12}CO/^{13}CO$ intensity ratios in luminous mergers.
See Abstr. 157.101.

Observational properties of disks.
See Abstr. 157.111.

The peak 21–cm surface brightness of spiral galaxies.
See Abstr. 157.178.

The interpretation of modern synthesis observations of spiral galaxies: Summary lecture.
See Abstr. 157.187.

Wolf–Rayet stars in Local Group galaxies: numbers and spectral properties.
See Abstr. 157.199.

Evolution of dust grains through a hot gaseous halo.
See Abstr. 157.270.

Abundances in galaxies.
See Abstr. 157.352.

First results from the Goddard High–Resolution Spectrograph: the Galactic halo and the Lyα forest at low redshift in 3C 273.
See Abstr. 159.044.

Dynamical models of the Local Group of galaxies.
See Abstr. 160.166.

A non–baryonic galactic halo and the real galactic age.
See Abstr. 161.011.

The search for non–baryonic dark matter.
See Abstr. 161.155.

The evolution of the galactic lithium abundance.
See Abstr. 161.424.

Chemical evolution of galaxies.
See Abstr. 161.425.

Galactic dust emission and the cosmic microwave background.
See Abstr. 161.513.

Cosmic microwave background and Galactic quadrupoles in the millimetric region.
See Abstr. 161.630.

156 Magellanic Clouds

156.001 **Photometric variability study of Wolf–Rayet binaries in the Magellanic Clouds.**
W. Seggewiss, A. F. J. Moffat, R. Lamontagne.
Astron. Astrophys., Suppl. Ser., Vol. 89, No. 1, p. 105 – 119 (Jul 1991).

Photometry in different photometric systems is presented for 14 Wolf–Rayet (WR) stars in the Large Magellanic Cloud and 4 Wolf–Rayet stars in the Small Magellanic Cloud. All but two stars are spectroscopic binaries. Nine of the 16 binaries show phase–dependent variations due to wind transparency effects or, in one case, eclipses of the stars. In the former case, the shape of the light curve of 5 stars is used to derive the orbital inclination and the WR star mass loss rate.

156.002 **Non violent star formation in the region of the 30 Doradus nebula.**
M.-C. Lortet, G. Testor.
Astron. Astrophys., Suppl. Ser., Vol. 89, No. 1, p. 185 – 213 (Jul 1991).

The authors have revisited a volume of the 30 Doradus Nebula (about 12′ in diameter) with the aim of identifying distinct sheets of ionized gas and of relating them genetically to the stellar content, taking as a basis published velocity measurements of the gas and new spectroscopic data. Apart from the rich and young cluster around R 136 (about 30″ in diameter) which is responsible for the ionization of part of the gas, young or forming stars are scattered all about and a few examples of local ionization are shown. On the other hand, a continuous range in ages of the stars is demonstrated and related to the morphology of the neighbouring gas. No simple sequential star formation is indicated, because of the intrinsic complexity of giant H II regions and confusing effects along the line of sight. The authors also draw attention to different kinds of dark features which can be used for tracing additional ionization fronts or whose existence may lead to spurious subdivisions in the ionized gas. They find marginal indication that the higher velocity gas (~ 300 km s^{-1}) is in front of the bulk of the nebula.

156.003 **A new tidal model of the Magellanic Stream.**
W. L. Shuter.
Bull. Am. Astron. Soc., Vol. 23, No. 2, p. 889 (1991). Abstract. – See Abstr. 012.008 for the main entry.

156.004 **The initial mass function for massive stars in the Magellanic Clouds.**
B. Hill.
Bull. Am. Astron. Soc., Vol. 23, No. 2, p. 899 (1991). Abstract. – See Abstr. 012.008 for the main entry.

156.005 **BV CCD photometry of five young Large Magellanic Cloud star clusters.**
R. Sagar, T. Richtler, K. S. de Boer.
Astron. Astrophys., Suppl. Ser., Vol. 90, No. 3, p. 387 – 436 (Nov 1991).

The authors present BV CCD data for the young star clusters NGC 1711, 2004, 2100, 2164 and 2214 and their nearby field regions in the Large Magellanic Cloud. The data have been calibrated with a zero point accuracy of ± 0.04 mag in both B and V. The total sample consists of over 8960 stars reaching down to V = 21 mag. The data serve as a base for the study of mass functions and for the comparison with theoretical models. The present CCD data compare very well with photoelectric observations of stars having a range in brightness ($11.8 \leqslant V \leqslant 18.1$) as well as in colour ($-0.17 \leqslant (B-V) \leqslant 2.17$) but generally shows systematically varying differences with photographic data. In NGC 2004 and 2164, e.g., the differences vary with brightness in both V and (B–V) while in NGC 2100, photographic V data agree very well with CCD ones upto $V \sim 15.5$ mag, but the difference increases rapidly for fainter stars.

156.006 **The chemical composition of the Magellanic Clouds from stellar studies.**
W. R. J. Rolleston.
Ir. Astron. J., Vol. 20, No. 2, p. 60 – 69 (Sep 1991).

Results obtained from stellar studies of the chemical composition in the Magellanic Clouds are discussed. Element abundances determined from the spectroscopic analyses of OB–type stars, F – G and red supergiants and variable stars are presented. Metallicity measurements as obtained from the photometric analyses of cepheids and red clusters are also discussed. It is evident that both clouds are metal deficient with the Small Magellanic Cloud being the more heavily depleted in metals.

156.007 **ASTRO–UIT UV images of the Magellanic Clouds: search for ultraviolet counterparts to discrete X–ray sources.**
A. G. Michalitsianos, M. Greason, R. S. Hill, P. M. N. Hintzen, J. Isensee, R. W. O'Connell, A. M. Smith, T. P. Stecher.
Bull. Am. Astron. Soc., Vol. 23, No. 2, p. 946 (1991). Abstract. – See Abstr. 012.008 for the main entry.

156.008 **The star formation history of the Large Magellanic Cloud.**
C. Chiosi, G. Bertelli, A. Bressan, M. Mateo.
Bull. Am. Astron. Soc., Vol. 23, No. 2, p. 949 (1991). Abstract. The full paper is published in "Astrophysical ages and dating methods", p. 225 – 232 (1990) (see 53.156.161). – See Abstr. 012.008 for the main entry.

156.009 **On use of the recombination echo around SN 1987A as a distance yardstick to the Large Magellanic Cloud.**
J. E. Felten, E. Dwek.
Bull. Am. Astron. Soc., Vol. 23, No. 2, p. 972 (1991). Abstract. – See Abstr. 012.008 for the main entry.

156.010 **Infrared photometry of compact objects in the Magellanic Clouds.**
F. P. Israel, J. Koornneef.
Astron. Astrophys., Vol. 248, No. 2, p. 404 – 414 (Aug 1991).
The authors present the results of *JHK* ($\dot{L}M$) photometry of 34, mostly compact, objects in the Magellanic Clouds. At *JHK*, slightly less than half are dominated by stellar emission or are in fact stars. The planetary nebula N145 (LMC) was also observed. Comparison with IRAS observations shows that almost all optically compact nebulae are also compact at longer infrared wavelengths. IRAS 12 μm "excess" emission is inversely correlated with hot and compact IR emission. Some compact objects, notably the nebula N88 (SMC) show a rather extreme blue *J–H* colour which needs further investigation. The authors suggest that N59–IRS and especially N150 (LMC) are good candidates to search for traces of recent star formation. Results on the complexes N159 (LMC), N83/N84, and N88 (SMC) are briefly discussed.

156.011 **FOC observations of R136a in the 30 Doradus nebula.**
G. Weigelt, R. Albrecht, C. Barbieri, J. C. Blades, A. Boksenberg, P. Crane, J. M. Deharveng, M. J. Disney, P. Jakobsen, T. M. Kamperman, I. R. King, F. Macchetto, C. D. Mackay, F. Paresce, D. Baxter, P. Greenfield, R. Jedrzejewski, A. Nota, W. B. Sparks.
Workshop on the First Year of HST Observations, p. 208 – 211 (1991). – See Abstr. 012.017 for the main entry.
R136a is the central object in the 30 Doradus nebula. The physical nature of R136a has been the subject of controversy over the last few years. One suggestion was that R136a might be a single object with a mass of the order of 1000 solar masses. The other suggestion was that R136a is a compact star cluster and that is consists of several O and WR stars. The HST observations described here confirm that R136a is a compact star cluster. There is good agreement with the speckle observations.

156.012 **Cosmic rays in the Large and Small Magellanic Clouds and the relationship to our Galaxy.**
C. E. Fichtel, P. Sreekumar.
AIP Conf. Proc., No. 220, p. 25 – 29 (Apr 1991). Current Physics Microform No.: 9106B1159. – See Abstr. 012.020 for the main entry.
For the LMC the intensity level of the cosmic rays in the central region is similar to that in the local region of our Galaxy. In the SMC, the cosmic ray density seems to be lower than in our Galaxy and lower than if the conditions required for dynamic equilibrium were in place for the SMC as a whole.

156.013 **The detection of the hot gaseous halo around the Small Magellanic Cloud.**
Q. Wang.
Astrophys. J., Lett., Vol. 377, No. 2, p. L85 – L88 (20 Aug 1991). With plate L4.
The author reports the detection of diffuse X-ray emission from the vicinity of the Small Magellanic Cloud. The diffuse emission constitutes about 60% of the total observed X-ray flux in a 0.16 – 3.5 keV band from a field mapped out with the Einstein IPC, and extends beyond the active main body of the galaxy, suggesting substantial emission from the galaxy's halo. The emission spectrum, getting soft away from the main body, is consistent with a picture in which hot gas, created in recent star formation regions, cools off on the way out into the halo. While the total observed luminosity of the diffuse X-rays is only $10^{38.7}$ergs s^{-1} in the 0.16 – 3.5 keV band, the low temperature ($\lesssim 10^6$K) of the X-ray emitting gas implies that its total cooling rate could be sufficient in balancing the supernova heating in the galaxy. The bulk of the radiation should be in a spectral range from EUV to extreme soft X-ray. A general scenario is described for the dynamical and thermal evolution of hot gas blowing out from the disk plane into the halo of a galaxy.

156.014 **The Magellanic Clouds: how well do we know what our closest neighbours are made of?**
S. Russell.
Vistas Astron., Vol. 34, Part 3, p. 235 – 248 (1991). – See Abstr. 012.032 for the main entry.
A series of comparisons are presented between various recent analyses of chemical abundances in the Clouds. It is shown that analyses of similar stars in the field, by different authors, yield consistent results. Similarly, analyses of both cool and hot stars yield consistent results in the field. Cluster abundances are broadly similar to the field, but show an overabundance of the heaviest elements. Finally, hot stars in general are consistently overabundant in oxygen relative to cool stars.

156.015 **Radiatively and shock excited H_2 in Magellanic H II regions.**
F. P. Isreal, J. Koornneef.
Astron. Astrophys., Vol. 250, No. 2, p. 475 – 483 (Oct 1991).
Near-IR spectroscopy of 14 small Magellanic Cloud H II regions has led to the detection of the 1–0 S(1) transition of molecular hydrogen (H_2) in 8 objects. In 5 objects (SMC: N81, N88; LMC: N83B, N159A5 and N160A2) various H_2 transitions were observed in order to identify the H_2 excitation mechanism. Early-type stars in the observed objects have stellar winds weaker than galactic O-type stars. The strength of the observed near-IR He I transitions is briefly discussed.

156.016 **The evolution of carbon stars in the Magellanic Clouds.**
B. E. Westerlund, M. Azzopardi, J. Breysacher, E. Rebeirot.
Astron. Astrophys., Suppl. Ser., Vol. 91, No. 3, p. 425 – 451 (Dec 1991).
JHK photometric data are presented for over 100 field carbon stars in the Small Magellanic Cloud and for 10 in the Large Cloud together with spectroscopic results for about half of them. These data are combined with previously published data for a discussion of the luminosity–temperature distribution of the field carbon stars in the Magellanic Clouds. In the Small Cloud carbon stars were found at higher temperatures and lower luminosities than previously observed. The faintest are below the top of the red giant branch. Their formation can thus hardly be explained by thermal pulses along the asymptotic giant branch so that other evolutionary processes may have to be considered. The medium- and low-luminosity C stars in the M–C transition zone have a low C_2 content. At these luminosities most of the J-type stars are found close to the C_2-poor stars in the HR diagram. Their C_2 content is, however, about as high as in the coolest, most evolved C stars. It appears likely that we see two scenarios of the formation of carbon stars: one being M–S–C(C_2–poor)–C, the other M–J–C. In the Large Cloud a number of high-luminosity J-type stars appear in the HR diagram between the M giant stars and the high-luminosity MS-type stars. They may represent a brief ^{13}C producing stage in the evolution of the more massive AGB stars.

156.017 **A model for Magellanic Stream.**
Liu Yongzhen.
3. Chinese Academy of Sciences and Max-Planck Society Workshop: High energy astrophysics: compact stars and active galaxies, p. 74 - 79 (1991). - See Abstr. 012.026 for the main entry.

156.018 **An Australia Telescope H I absorption survey of the Magellanic Clouds and Stream.**
U. Mebold, E. W. Greisen, W. Wilson, R. F. Haynes, U. Herbstmeier, P. M. W. Kalberla.
Astron. Astrophys., Vol. 251, No. 1, p. L1 - L4 (Nov 1991). Letter-to-the-editor.
21-cm emission and absorption line observations were carried out with the Parkes 64-m radio telescope and the Australia Telescope Compact Array, respectively, towards extragalactic radio continuum background sources in the area of the Magellanic Clouds and Stream. The H I emission indicates strong clumping on angular scale-sizes less than or equal to the 15′ beam of the Parkes radio telescope. No H I absorption was found down to an optical-depth limit of $\tau \leqslant 0.002$. Comparing average absorption properties, the Magellanic System has less absorbing material than the Galaxy and M31. If clumping is ignored, lower limits to the spin temperature of up to 600K are found for the gas in the outer halo and in the bridge region of the Magellanic Clouds.

156.019 **Reduced upper limits on the equivalent width of interstellar Li I 670.8 towards SN 1987A.**
D. Baade, S. Cristiani, T. Lanz, R. A. Malaney, K. C. Sahu, G. Vladilo.
Astron. Astrophys., Vol. 251, No. 1, p. 253 - 258 (Nov 1991).
A new search for interstellar Li I 670.8 along the line of sight to SN 1987A has been carried out using all 34 known (and previously published) high-resolution Reticon spectra obtained at the European Southern Observatory. In spite of the large number and high quality of the observations, the line remains undetected even in the carefully weighted and filtered mean spectrum. The most probable 3-σ detection limit in equivalent width is 7.3×10^{-6}nm. For the cloud at radial velocity $+286$ km s^{-1}, standard relations then imply a maximum abundance, Li/H, of $1.7\ 10^{-11}$ in the gaseous interstelllar medium of the LMC. If depletion to dust grains is the same for both lithium and potassium, this figure has to be increased by 1.1 dex. Accordingly, conventional present-day reasoning suggests that the elemental abundance of lithium does not exceed 2.7×10^{-10}. However, the cumulative uncertainties are such that a primordial abundance which is an order of magnitude higher is not firmly ruled out. Constraints on primoridal nucleosynthesis models are therefore, only weak.

156.020 **Cosmic rays in the Small Magellanic Cloud.**
P. Sreekumar, C. E. Fichtel.
Astron. Astrophys., Vol. 251, No. 2, p. 447 - 453 (Nov 1991).
A study is made of the SMC in an attempt to determine whether this galaxy is in dynamic quasi-stable equilibrium between the expansive pressure of the hot cosmic ray gas, the magnetic fields, and the kinetic motion of interstellar gas and the gravitational attraction of matter. It is concluded that the cosmic ray density level is 3 to 5 times smaller than that which would exist if the galaxy were in approximate dynamic equilibrium and the cosmic ray sources were adequate. Further, the scale of coupling between the cosmic rays and matter in the projected dimension that can be observed is $<1/2$ kpc, which is rather small to be consistent with a stable cosmic ray distribution. The conclusion that the cosmic ray density in the SMC is lower than that if the conditions of dynamic equilibrium were in place for the SMC as a whole and substantially lower than that in our galaxy is in contrast with that for the LMC, where the cosmic ray intensity has been shown to be consistent with dynamic quasi-stable equilibrium, and similar to that in the local region of our galaxy. Using the observed matter distribution and the calculated cosmic ray density distribution, the expected gamma ray intensity distribution is determined. The resulting flux integrated over the SMC region is found to be 5 to 25 times smaller than the previous estimates based on quasi-stable equilibrium or a similar cosmic ray density.

156.021 **Extinction towards 30 Doradus (LMC) derived from common upper level P_γ/H_δ and P_δ/H_ε line ratios.**
A. Greve, J. Castles, C. D. McKeith.
Astron. Astrophys., Vol. 251, No. 2, p. 575 - 580 (Nov 1991).
The authors have used the ratios P_γ/H_δ and P_δ/H_ε of corresponding hydrogen multiplet lines (which originate from the same upper levels) to derive the extinction A_V in the direction of 30 Doradus in the LMC.

156.022 **Cosmic rays and the dynamic balance in the Large Magellanic Cloud.**
C. E. Fichtel, R. G. Stone, M. E. Ozel.
IAU Colloquium No. 123: Observatories in Earth orbit and beyond, p. 537 - 541 (1990). - See Abstr. 012.038 for the main entry.
Present and future measurement of the Large Magellanic Cloud (LMC) particularly in the radio and high energy gamma ray range offer the possibility of understanding the density and distribution of the cosmic rays in a galaxy other than our own and the role that they play in galactic dynamic balance. After a study of the consistency of the measurements and interpretation of the synchrotron radiation from our own galaxy, the cosmic ray distribution for the LMC is calculated under the assumption that the cosmic ray nucleon to electron ratio is the same and the relation to the magnetic fields are the same, although the implications of alternatives are discussed.

156.023 **The young Magellanic Cluster NGC 2004.**
D. Bencivenni, E. Brocato, R. Buonanno, V. Castellani.
Astron. J., Vol. 102, No. 1, p. 137 - 151 (Jul 1991). Current Physics Microform No.: 9107B1833.
The results of a new CCD photometric investigation of stars in the field of the LMC cluster NGC 2004 are reported and discussed. B, V photometry is given for more than 1500 objects, the large majority to be ascribed to the cluster. A new color-magnitude diagram is presented. It shows a well-defined blue sequence (BS) of H-burning stars, running from $V \approx 21$ mag up to $V \approx 13$ mag, together with a group of cluster He-burning red supergiants with $V \approx 13$ mag and a clump of field red giants at about $V \approx 19$ mag. Theoretical evolutionary computations are reported covering the range of massive stars evolving in the cluster. These evolutionary tracks can match the observed CM diagram provided that a low efficiency of the surface convection (a low value of the mixing length) is adopted. Comparison of the theoretical luminosity function with the upper portion of the observed BS distribution suggests a cluster age of about 8 million years, in agreement with previous evaluations. The completeness of the sample and the contribution of field stars have been estimated. The CM distribution of cluster stars has been compared with the results of synthetic clusters based on the adopted evolutionary scenario. After a short review of the current theoretical scenario concerning massive stars, the CM diagram of field stars is discussed, reaching the conclusion that the observed clump of red giants is the natural consequence of the peculiar large lifetimes in the central He burning phase of stars with $M \approx 3\ M_\odot$.

156.024 **A log(P) - ⟨K⟩ relation for red supergiants - SRC-type variable stars in the LMC.**
A. K. Dambis.
Astron. Tsirk., No. 1545, p. 17 - 18 (Aug - Sep 1990).

156.025 **Time variation of ellipticity of globular clusters in the Large Magellanic Cloud.**
K. Akiyama.
Earth, Moon, Planets, Vol. 54, No. 3, p. 203 - 239 (Sep 1991).
Dynamical evolution of globular clusters in the LMC is investigated by means of *N*-body simulations; particular attention is paid to time evolution in the elliptical figure of globular

clusters. The simulations were started with a binary globular cluster. It merged into a single cluster with ellipticity of about 0.3. The simulations were continued until the cluster became rounder due to the effects of two body relaxation and of tidal field of LMC. It is found that the outward angular momentum transport due to the gravothermal contraction makes the inner region rounder; the ellipticity at about the initial half-mass radius (r_h) decreases with the *e*-folding time of 20 relaxation times. The outer region becomes rounder due to the stripping of stars by the tidal field; the ellipticity at about 3 r_h decreases with the *e*-folding time of 80 crossing times therein, though the time scale depends on the direction of the tidal field relative to the spin of the cluster. These two effects are comparable at about the half-mass radius. Taking account of such theoretical results the author reanalyzed observed data for the ellipticity at about the half-mass radius of LMC clusters. He estimated the relaxation time and crossing time for each of the observed clusters, from which he calculated the effective time of getting round of the cluster. He plotted the observed ellipticity of the clusters against their age normalized by the effective time. He found that observed ellipticity distribution is consistent with his picture.

156.026 **Massive star evolution in the LMC.**
Y. Tuchman, J. C. Wheeler.
10. Santa Cruz Summer Workshop in Astronomy and Astrophysics: Supernovae, p. 159 – 171 (1991). – See Abstr. 012.052 for the main entry.

156.027 **A radio continuum study of the Magellanic Clouds. I. Complete multi-frequency maps.**
R. F. Haynes, U. Klein, S. R. Wayte, R. Wielebinski, J. D. Murray, E. Bajaja, D. Meinert, U. R. Buczilowski, J. I. Harnett, A. J. Hunt, R. Wark, L. Sciacca.
Astron. Astrophys., Vol. 252, No. 2, p. 475 – 486 (Dec 1991).

New radio continuum and polarization surveys which completely cover both, the Large and the Small Magellanic Cloud (LMC and SMC), have been carried out recently at 2.45, 4.75, and 8.55 GHz, using the Parkes 64-m telescope. Here the authors present the results, along with a preliminary discussion of the radio morphology and the integrated radio continuum emission of the Clouds. It is argued that ram pressure effects caused by the motion of the LMC through the Milky Way halo are to some extent responsible for its overall radio morphology. Strong linear polarization is detected in the LMC (degrees of polarization reaching 25%), which appears to be associated with the region of highest H I and CO column densities. The observed degree of linear polarization implies a well-ordered magnetic field over kpc scales the alignment of which may also result from ram pressure effects. A preliminary analysis indicates a trailing spiral magnetic field the origin of which is as jet unclear. Only weak polarization ($\leqslant$ a few %) is seen in the SMC, possibly due to its large line-of-sight depth and complex structure. Large variations of the star formation efficiency across the LMC are disclosed by a radio-CO comparison. The integrated radio continuum spectra of the LMC and SMC are indicative of high relative amounts of thermal radio emission in these galaxies. Various impacts of the data base presented here for future studies are outlined.

156.028 **Observations of molecular clouds associated with supernova remnants in the LMC.**
J. P. Hughes, L. Bronfman, L. Nyman.
10. Santa Cruz Summer Workshop in Astronomy and Astrophysics: Supernovae, p. 679 – 682 (1991). – See Abstr. 012.052 for the main entry.

156.029 **LMC-2 as the blowout of a hot superbubble.**
Q. Wang, D. J. Helfand.
Astrophys. J., Vol. 379, No. 1, p. 327 – 337 (20 Sep 1991).

Based on the data collected with the imaging proportional counter (IPC) aboard the Einstein Observatory, the authors have discovered a bright extended X-ray emission region which is enclosed by the supershell LMC-2 identified from optical and radio observations. The diffuse X-rays appear to come from hot gas in a corresponding superbubble. The X-ray count rate of the superbubble is ~ 1.4 counts s^{-1} in the IPC broad band (0.16 – 3.5 keV), corresponding to a luminosity of $\sim 2 \times 10^{37}$ergs s^{-1} with a gas temperature $\sim 5 \times 10^6$K. The bright X-ray emission region appears as a ring with a radius of $\sim 15'$ and correlates well with a cold ISM cavity seen in both the H I and IRAS maps. The authors suggest that LMC-2 has been created by the breakout of the superbubble from the dense galactic plane of the Large Cloud. The bright X-ray emission arises near the plane, where the superbubble is energized by stellar winds and supernova explosions. The blown-out superbubble expands rapidly in the low-density region away from the plane on the far side of the galaxy; its boundary is marked by absorption of soft X-rays in the dense H I shell it creates, and by rim-brightened optical filaments at the interface between the hot interior and the cold swept-up medium. This observation represents the first clear evidence of such a hot superbubble in an external galaxy.

156.030 **A spectroscopic survey of B supergiants in the Large Magellanic Cloud.**
E. L. Fitzpatrick.
Publ. Astron. Soc. Pac., Vol. 103, No. 669, p. 1123 – 1148 (Nov 1991).

The author presents the results of a low-dispersion ($\Delta\lambda \cong 2$ Å) digital, optical, spectral survey of ~ 100 B-type supergiants in the Large Magellanic Cloud. The MK spectral classification framework for B supergiants has been transferred to the metal-weak LMC stars, and recommended classification standards have been designated. Variations among the metal line strengths are examined. The most extreme variations are found for the nitrogen lines, for which a range of a factor of 3 or more may be seen in the equivalent widths within some spectral subclasses. It is suggested that these variations indicate a range of nitrogen surface abundances among the B supergiants, resulting from contamination of some of the stellar surfaces by processed material from the original H-burning core.

156.031 **Ionized carbon in the Large Magellanic Cloud.**
R. T. Boreiko, A. L. Betz.
Astrophys. J., Lett., Vol. 380, No. 1, p. L27 – L30 (10 Oct 1991).

The authors have observed the 158 μm $^2P_{3/2} - {}^2P_{1/2}$ fine-structure transition of C^+ at selected locations in the Large Magellanic Cloud. The C II emission is most intense toward far-infrared continuum peaks and generally ist not seen in positions exhibiting strong CO J = 2–1 radiation. Where both C II and CO emission are detected, the V_{LSR} centroids are similar, but the C II line is wider. The differences in spatial distribution and spectral shape suggest a more pronounced physical separation between the predominantly neutral atomic and molecular gas regions than is the case in our Galaxy. In the LMC, the intense and extended C II emission near 30 Dor implies a total amount of C^+ several times greater than that of Galactic molecular cloud complexes. The authors also attempted to detect the 289 μm J = 9–8 transition of ^{12}CO in a few locations. The observed upper intensity limit for N159 implies that moderate density (n > 4×10^4cm^{-3}) molecular gas fills less than 5% of the antenna beam and that most of the low J CO emission comes from lower density gas.

156.032 **New field RR Lyrae variables in the outer parts of the LMC and the properties of the LMC halo.**
T. D. Kinman, L. L. Stryker, J. E. Hesser, J. A. Graham, A. R. Walker, M. L. Hazen, J. M. Nemec.
Publ. Astron. Soc. Pac., Vol. 103, No. 670, p. 1279 – 1291 (Dec 1991).

Five new type-ab RR Lyrae variables that belong to the field of the LMC were found. A King model that fits the surface densities of the known LMC field RR Lyrae stars of type ab has a limiting radius of 15 kpc. An exponential disk model that fits these densities has a scale length of 2.6 kpc. These models contain $\sim 10^4$ type-ab RR Lyrae stars and have a central surface density of ~ 200 of these stars per square degree. The LMC field halo of Population II stars has $M_v \sim -15.2$ and a mass of $\sim 1.6 \times 10^8 M_\odot$. This halo therefore makes up about 2% of the mass of the LMC. The LMC field RR Lyraes have a period–frequency distribution similar to those found in the outer halo of our Galaxy. The ratio of the surface density of these RR Lyraes to that of the older long–period variables is ~ 13 at some 3 to 4 kpc from the LMC center. The central visual surface brightness of the LMC halo is ~ 25.5 mag per square arc sec. Its average visual surface brightness is ~ 26.4 mag per square arc sec.

156.033 **Stellar populations and large–scale structure of the SMC. III. The geometry of the northern and north–western outlying regions.**
L. T. Gardiner, M. R. S. Hawkins.
Mon. Not. R. Astron. Soc., Vol. 251, No. 1, p. 174 – 191 (1 Jul 1991).
The third paper in the series of investigations on the large–scale structure of the Small Magellanic Cloud is presented here. Following the study by Hatzidimitriou and Hawkins of a large area covering nearly 50 square degrees to the north–east and south–west of the SMC main body, further results from the northern and north–western areas have been obtained. Photographic photometry derived from COSMOS analysis of sets of blue and red UK Schmidt plates provided the data for the study. Results on the projected two–dimensional appearance of the SMC and the three–dimensional geometry of these areas are presented on the basis of the properties of the red horizontal branch/clump stellar population.

156.034 **Detection of the helium flash gap in the integrated (U–B) versus (B–V) diagram for 624 Large Magellanic Cloud clusters.**
E. Bica, J. J. Clariá, H. Dottori, J. C. F. Santos Jr., A. Piatti.
Astrophys. J., Lett., Vol. 381, No. 2, p. L51 – L53 (10 Nov 1991).
The authors have increased the number of Large Magellanic Cloud clusters with integrated UBV photometry by more than a factor of 4, now totaling 624 objects. A gap in the cluster distribution through the color–color diagram appears in the region of the equivalent SWB type IV. The amplitude of the gap is ≈ 0.1 mag in both colors. The turnoff ages of a few clusters near the gap edges perfectly fit theoretical predictions of a red giant branch phase transition. This color jump is due to the first appearance of stars suffering the helium flash, which form a bright and populous red giant branch that persists through the subsequent cluster evolution. As an additional result, evidence is found that Hodge 7 (SL 735) might be a classical globular cluster.

156.035 **A new Wolf–Rayet star in the Small Magellanic Cloud.**
D. H. Morgan, E. Vassiliadis, M. A. Dopita.
Mon. Not. R. Astron. Soc., Vol. 251, No. 3, p. 51P – 53P (1 Aug 1991).
A new WR star in the Small Magellanic Cloud identified on a UK Schmidt Telescope objective prism plate is found to be similar in character to other SMC WR stars in that it has weak emission lines plus an O–star absorption spectrum. It is classified as WN2.5 + O5:.

156.036 **The OB stellar content of the Magellanic Clouds: Sanduleak's contribution in perspective.**
N. R. Walborn.
Objective–prism and other surveys – a meeting in memory of Nicholas Sanduleak, p. 157 – 161 (1991). – See Abstr. 012.074 for the main entry.
The development of our knowledge of the OB stellar content of the Magellanic Clouds is an ongoing enterprise. Nick Sanduleak provided vital contributions to a specific stage in its development. The enterprise has already moved beyond that stage, although the derivations from Nick's contributions are far from exhausted.

156.037 **The geometry of the SMC.**
L. T. Gardiner, D. Hatzidimitriou, M. R. S. Hawkins.
Proc. Astron. Soc. Aust., Vol. 9, No. 1, p. 80 – 81 (1991). – See Abstr. 012.078 for the main entry.
The authors have undertaken a study of the 3–D spatial distribution of the older ($\tau \geqslant$ 1–2 Gyr) stellar population lying beyond ~ 2 kpc in projected radial distance from the centre of the SMC. The study will eventually cover 120 square degrees including six overlapping Schmidt fields. Here the authors present the results from an area of 80 square degrees including the western, northern and north–eastern outer parts of the SMC.

156.038 **Heavy element abundances in the Magellanic Clouds.**
S. C. Russell.
Proc. Astron. Soc. Aust., Vol. 9, No. 1, p. 82 – 83 (1991). – See Abstr. 012.078 for the main entry.
This paper presents a brief discussion of the apparent underdepletion of the heavy neutron–capture elements (elements heavier than Ba), compared with Fe in the Magellanic Clouds. The s–process appears to have been only effective in forming elements in the light neutron–capture group (Sr, Y, Zr) in the Magellanic Clouds, but to have much reduced effectiveness in forming the heavy neutron–capture group. The abundances of the elements heavier than Ba have a distribution that indicates that they were produced by the r–process alone.

156.039 **Observational tests of stellar evolution: the Magellanic Clouds.**
P. R. Wood.
Symposium in Celebration of the 50. Anniversary of McDonald Observatory of the University of Texas at Austin: Frontiers of stellar evolution, p. 213 – 228 (1991). – See Abstr. 012.085 for the main entry.
The author examines some areas of stellar evolution where research on Magellanic Cloud stars is currently active. A single review could not hope to cover all areas of stellar evolution, so those areas covered reflect his personal preferences; in particular, the review will concentrate on stars less massive than $\sim 10\ M_{\odot}$.

156.040 **CCD photometry of the LMC cluster NGC 2004.**
E. Brocato, R. Buonanno, V. Castellani.
Symposium in Celebration of the 50. Anniversary of McDonald Observatory of the University of Texas at Austin: Frontiers of stellar evolution, p. 594 (1991). Abstract. – See Abstr. 012.085 for the main entry.

156.041 **CCD photometry of the LMC cluster NGC 1866.**
E. Brocato, R. Buonanno, V. Castellani, A. R. Walker.
Symposium in Celebration of the 50. Anniversary of McDonald Observatory of the University of Texas at Austin: Frontiers of stellar evolution, p. 594 (1991). Abstract. – See Abstr. 012.085 for the main entry.

156.042 **Cocoon stars in the Large Magellanic Cloud.**
N. Reid.
Astrophys. J., Vol. 382, No. 1, p. 143 – 148 (20 Nov 1991).
The author presents near–infrared (JHK) photometry of 13 of the 29 IRAS sources toward the norther LMC that were suggested as possible dust–embedded asymptotic giant branch stars by Reid, Tinney, and Mould. Two prove to be more luminous LMC red supergiants, while one appears to be a substantially more distant extragalactic object. The remaining 10 objects are identified as LMC "cocoon" stars, with most having flux distributions similar to highly reddened Galactic OH/IR sources and bolometric luminosities in the range $-5 \leqslant M_{bol} \leqslant -6$. Estimates of the mass–loss rates imply lifetimes of less than 10^5yr.

156.043 **Spatial distribution and star cluster formation in the LMC bar region.**
E. Bica, J. J. Clariá, H. Dottori.
Bol. Asoc. Argent. Astron., No. 36, p. 153 – 165 (1990). In Spanish. – See Abstr. 012.095 for the main entry.
New integrated UBV data of star clusters projected on the LMC bar region have been obtained. Most of the observed clusters appear to be physically associated to the LMC bar. An appreciable fraction of them lie within the $0.4 \leqslant (B-V) \leqslant 0.6$ range so that the extensively discussed gap in the B – V histogram of the LMC clusters disappears. An analysis of the spatial distribution and ages of the observed clusters support the

conclusion that a strong star forming event has occured in the eastern part of the bar about 100 Myr ago, which has probably propagated to the western portion at the present time.

156.044 **NGC 2214; a merging binary star cluster in the Large Magellanic Cloud.**
R. K. Bhatia, H. T. MacGillivray.
Bull. Astron. Soc. India, Vol. 19, Nos. 1 – 2, p. 91 (Mar – Jun 1991). Abstract. – See Abstr. 012.108 for the main entry.

156.045 **The period–luminosity relation for classical cepheids of the Large Magellanic Cloud.**
R. A. Vardanyan, A. V. Pogosyan.
Astrofizika, Tom 34, Vyp. 2, p. 199 – 204 (Apr 1991). In Russian. English translation in Astrophysics, Vol. 34, No. 2.
The relation period–luminosity is obtained for classical cepheids of the Large Magellanic Cloud using a new method allowing to take into account the influence of the luminosity and colour of stars. It is shown that the accuracy of the distance determination for galaxies by means of this relation depends on the interval of the cepheid periods. The interval of periods is determined giving the most accurate distances of galaxies.

156.046 **Trouble in the Magellanic Clouds! First results from the key programme on coordinated investigations of selected regions in the Magellanic Clouds.**
K. S. De Boer, F. Spite, P. François, R. Cayrel, M. Spite, B. Baschek, J. Köppen, B. Wolf, O. Stahl, A. Jüttner, W. Seggewiss, D. J. Bomans, E. K. Grebel, E. H. Geyer, T. Richtler, A. Vallenari, J. Koornneef, F. P. Israel, P. Molaro, S. Monai, G. Vladilo, S. D'Odorico, R. Leisy, M. Dennefeld, R. Ferlet, A. Vidal-Madjar, G. Stasińska, M. Azzopardi, N. Meyssonnier, G. Muratorio, E. Rebeirot, J. Lequeux.
Messenger, No. 66, p. 14 – 17 (Dec 1991).

Positional reference stars in the Magellanic Clouds.
See Abstr. 002.028.

Catalogue of blue stars between the Magellanic Clouds.
See Abstr. 002.040.

Maximum entropy deconvolution of a Wide Field Camera image of R 136.
See Abstr. 036.037.

Model atmospheres of massive hot stars.
See Abstr. 064.126.

Evolution of massive stars in the Large Magellanic Cloud: models with semiconvection.
See Abstr. 065.084.

Spectral analyses of 19 Wolf–Rayet (WN) stars in the LMC.
See Abstr. 114.020.

Quantitative spectroscopy of B supergiants in the Galaxy, the LMC and the SMC. I. The B0.5 Ia supergiants $\varkappa$ Orionis, Sk –68°41 and SK 159.
See Abstr. 114.072.

Calibration of colour indices $(b-y)_0$ and absolute magnitudes M_V for A4–F3 supergiants from uvbyβ data.
See Abstr. 115.006.

Calibration of $(b-y)_0$ color indices and absolute magnitudes M_V for A4–F3 supergiants using uvbyβ photometry.
See Abstr. 115.008.

The evolutionary status of B supergiant stars in the LMC.
See Abstr. 115.009.

Massive star evolution.
See Abstr. 115.026.

A new age calibrator for red horizontal branch populations.
See Abstr. 115.029.

Spectroscopic survey of southern symbiotic stars.
See Abstr. 117.295.

The peculiar UV and optical spectrum of the LMC supergiant HD 269216.
See Abstr. 122.036.

Cepheids in LMC clusters.
See Abstr. 122.075.

Nova in the Large Magellanic Cloud 1991.
See Abstr. 124.019.

Properties of the SN 1987A circumstellar ring and the distance to the Large Magellanic Cloud.
See Abstr. 125.235.

Observations of the LMC by the TTM instrument onboard KVANT module in 1988 – 1989.
See Abstr. 125.267.

NTT interstellar Na I observations of the two faint ($V \approx 15.5$) optical companions of SN 1987A.
See Abstr. 131.002.

LMC–2 as the blow–out of a hot superbubble.
See Abstr. 131.028.

Has interstellar [Fe X] been detected in the spectrum of SN 1987A?
See Abstr. 131.064.

Distribution and origin of high–velocity clouds. II. Statistical analysis of the whole–sky survey.
See Abstr. 131.096.

30 Doradus, starburst Rosetta.
See Abstr. 131.284.

On the nature of the SMC H II region DEM 161.
See Abstr. 132.003.

H II regions with He II emission.
See Abstr. 132.021.

A search for supernova remnants in the 30 Doradus region of the LMC.
See Abstr. 132.023.

First results from the Faint Object Camera: high–resolution observations of the central object R136 in the 30 Doradus nebula.
See Abstr. 132.027.

Abundances in four type I LMC planetary nebulae.
See Abstr. 134.015.

Photoionization modeling of Magellanic Cloud planetary nebulae. II.
See Abstr. 134.040.

Optical spectroscopy of Magellanic Cloud planetary nebulae. II.
See Abstr. 134.042.

N67 as an X–ray bright planetary nebula in the Small Magellanic Cloud.
See Abstr. 134.063.

The intensity and spectrum of the diffuse X–ray background.
See Abstr. 142.028.

A log(P) – ⟨K⟩ relation for SRC–type variables in the Per OB1 association and the LMC distance modulus.
See Abstr. 152.006.

Massive star clusters and OB associations.
See Abstr. 152.007.

Are galactic open clusters formed from the LMC matter?
See Abstr. 153.035.

Age calibration and age distribution of LMC clusters.
See Abstr. 153.044.

Carbon and nitrogen abundances in three supergiants of the young SMC globular cluster: NGC 330.
See Abstr. 154.001.

A UBVRI photoelectric sequence in the SMC globular cluster NGC 121.
See Abstr. 154.004.

Two distinct supergiant branches in the young Large Magellanic Cloud star cluster NGC 2214.
See Abstr. 154.025.

Mass functions of five young Large Magellanic Cloud star clusters.
See Abstr. 154.031.

NGC 1866 revisited: semiconvection versus overshoot.
See Abstr. 154.032.

New determinations of element abundances in NGC 330.
See Abstr. 154.069.

The structure and evolution of rich star clusters in the Large Magellanic Cloud.
See Abstr. 154.086.

Early phases of LMC star clusters?
See Abstr. 154.096.

Are there two disk star cluster systems in the LMC?
See Abstr. 154.097.

Survey for LMC star clusters with 3 < age < 12 Gyrs.
See Abstr. 154.117.

Similarities of thick disk clusters in the Milky Way and intermediate–age SMC clusters.
See Abstr. 154.122.

Dynamics of the young LMC cluster NGC 1866.
See Abstr. 154.156.

An observational age–metallicity relation for the Galaxy.
See Abstr. 155.003.

LMC or halo stars? The Sanduleak–Fehrenbach controversy.
See Abstr. 155.110.

157 Normal Galaxies (Structure, Evolution, Pairs, etc.)

157.001 **Dynamical segregation of direct and retrograde orbits in binary disk galaxies.**
W. C. Keel.
Astrophys. J., Lett., Vol. 375, No. 1, p. L5 – L7 (1 Jul 1991).

A subset of the Karachentsev galaxy–pair catalog has been identified, using geometric and velocity criteria, which must be undergoing either direct or retrograde encounters. Rotation curves have been used to define the sense of disk rotation in each case and thus the orbital sense of each companion. A significant difference in orbital properties appears between direct and retrograde encounters, such that direct companions are moving systematically faster than retrograde ones. This is interpreted as evidence of more circular orbits in retrograde cases, perhaps indicating that such orbits are less strongly modified during tidal encounters than are direct ones. The difference between direct and retrograde companions is likely to reflect differential dynamical evolution, so that the properties of companion orbits now have a nearly bimodal distribution. Any difference in energy transfer between pairs of different orbital senses requires either relatively low halo masses, so that the disk is dynamically significant, or that the halo rotates in the same direction as the disk.

157.002 **The velocity dispersion gradients in spiral galaxies. I. The Galaxy and NGC 3198.**
V. Antonuccio–Delogu.
Astron. Astrophys., Vol. 247, No. 1, p. 45 – 55 (Jul 1991).

An attempt to model the observed radial velocity dispersion profiles for our Galaxy and NGC 3198 by using exact solutions of Jeans equations is presented. No use is made of the epicylcic approximations, so the author's solutions can be applied quite generally to model the velocity dispersions in the plane of axisymmetric collisionless systems. Quite satisfactory fits are obtained by assuming that the isotropy parameter $\beta = \sigma_\varphi/\sigma_r$ of the stellar component is minimal. The author then considers the possibility of constraining models of the mass distribution of a spiral by using the velocity dispersion profile. For our Galaxy, using the data of Lewis and Freeman (1989) and Woolley et al. (1977), no firm conclusions about the best fitting mass model can be reached, mainly because the data are incomplete and affected by observational uncertainties. For NGC 3198, although the solutions fit the observations quite well, the observational errors do not allow any strong constraint on the mass models.

157.003 **The CO luminosity function of galaxies.**
F. Verter.
Astrophys. J., Vol. 375, No. 1, p. 95 – 104 (1 Jul 1991).

The CO luminosity function that was originally derived by Verter is here reevaluated for a galaxy sample that is drawn from the published portion of the FCRAO Extragalactic CO Survey satisfying $B_T{}^0 \leqslant 12$ mag. The derived CO luminosity function is approximately a power law $[L_{CO}]^\alpha$, with $\alpha \approx -1.3$. This is consistent with previous galaxy luminosity functions in the blue and infrared, and with correlations between L_{CO} and L_B or L_{IR}.

157.004 **The interstellar medium of the hot–spot galaxy NGC 2903.**
J. M. Jackson, A. Eckart, M. Cameron, W. Wild, P. T. P. Ho, R. W. Pogge, A. I. Harris.
Astrophys. J., Vol. 375, No. 1, p. 105 – 115 (1 Jul 1991).

The authors report 21 cm radio continuum, Hα, IRAS, and $J = 1 \rightarrow 0$ and $2 \rightarrow 1$ CO observations of the hot–spot galaxy NGC 2903. The radio continuum, Hα, and CO, are distributed almost identically in a central bar. The CO data are consistent with emission from giant molecular clouds with properties typical of Milky Way clouds; this is, with large optical depths, excitation temperatures near 10K, and sizes about 100 pc. If the standard CO luminosity to H_2 mass conversion factor holds in NGC 2903, then the central 770 pc contains about $3 \times 10^8 M_\odot$ in molecular gas, comparable to the central 500 pc of the Galaxy. The star formation rate per unit area is enhanced by an order of magnitude in the nuclear region. The global far-infrared-to-CO luminosity ratio, which is often interpreted as a measure of star formation efficiency, is typical of that for normal spirals. However, the local FIR/CO ratio varies considerably. In particular, 100 μm emission extends well outside the molecular bar into a region of H I emission. The authors associate this emission with cirrus, which contributes about 30% of the global 100 μm flux.

157.005 **Rotation curves for 10 thin spiral edge-on galaxies.**
I. D. Karachentsev.
Pis'ma Astron. Zh., Tom 17, No. 6, p. 485 - 489 (Jun 1991). In Russian. English translation in Sov. Astron. Lett., Vol. 17, No. 3, p. 206 - 207 (1991).

Rotational curves for 10 northern galaxies with angular diameters a > 2′ and apparent axial ratios a/b > 7 are presented as a supplement to the first list of 40 galaxies (Karachentsev, Zhou Xu, 1991). For the whole sample of the objects their spin orientation shows some signs of an anisotropy.

157.006 **Peculiar molecular clouds in NGC 3256?**
S. Aalto, J. H. Black, R. S. Booth, L. E. B. Johansson.
Astron. Astrophys., Vol. 247, No. 2, p. 291 - 302 (Jul 1991).

The authors have mapped the luminous merging galaxy system NGC 3256 in the $J = 1\text{-}0$ and $J = 2\text{-}1$ emission lines of ^{12}CO and have measured selected points in the $J = 1\text{-}0$ and $J = 2\text{-}1$ transitions of ^{13}CO. They have observed an unusually high ^{12}CO/^{13}CO intensity ratio in $J = 1\text{-}0$, ≈ 35, in the centre of the system. This would conventionally imply optically thin emission. However this intensity ratio and the other line intensities suggest that most of the molecular emission arises in small, thick clouds of high temperature, $T \approx 100\text{-}300$K. Small size ($\lesssim 1$ pc radius) is required over the full range of measurement uncertainties, although the temperature is not yet well constrained. Models of the molecular clouds are developed that explain the CO data and that predict other observable properties. The models and available data suggest a highly disrupted molecular cloud ensemble which radiates powerfully in response to the starlight and mechanical luminosity of an associated starburst.

157.007 **New blue compact dwarf galaxies with low heavy-element abundances.**
Yu. I. Izotov, V. A. Lipovetsky (*V. A. Lipovetskij*), N. G. Guseva, A. Yu. Kniazev (*A. Yu. Knyazev*), J. A. Stepanian (*D. A. Stepanyan*).
Astron. Astrophys., Vol. 247, No. 2, p. 303 - 306 (Jul 1991).

The spectrophotometric observations of five blue compact dwarf galaxies from the Second Byurakan Survey with very low heavy element abundances are present. The oxygen abundances in the studied galaxies range from 1/17 to 1/42 of the solar value.

157.008 **Dense gas in nearby galaxies. V. Multilevel studies of CH_3CCH and CH_3CN.**
R. Mauersberger, C. Henkel, C. M. Walmsley, L. J. Sage, T. Wiklind.
Astron. Astrophys., Vol. 247, No. 2, p. 307 - 314 (Jul 1991).

The first detections of methyl acetylene (CH_3CCH) and methyl cyanide (CH_3CN) emission from outside our Galaxy are reported. Five mm-wave lines of CH_3CCH have been measured toward the molecular hotspot 15″ SW of the center of M82; four transitions have been observed toward the center of the starburst galaxy NGC 253. CH_3CCH is the most complex molecule detected outside the Galaxy. Its abundances in M82 and NGC 253 are similar to those measured in Galactic clouds. CH_3CN has been detected in at least three mm-wave transitions toward the center of NGC 253, where its abundance is "normal". A sensitive search toward the center of M82 shows that CH_3CN is underabundant compared with Galactic clouds. These results confirm the different chemical compositions of the nuclear gas in M82 and NGC 253. Aside of a likely difference in kinetic temperature, there must be additional effects causing the observed discrepancies; several alternatives are discussed. For both galaxies studied, densities of a few times $10^4 cm^{-3}$ are required to match the observed CH_3CCH and CH_3CN line intensity ratios. As a byproduct, mm-emission of the ring molecule C_3H_2 was detected toward M82 and NGC 253. Toward the SW hotspot of M82, the H41α recombination line was measured.

157.009 **The core of M87: new high spatial-resolution kinematic measurements.**
B. J. Jarvis, R. F. Peletier.
Astron. Astrophys., Vol. 247, No. 2, p. 315 - 319 (Jul 1991).

The authors present high resolution, high signal-to-noise measurements of the stellar kinematics of the core of M87, taken at various position angles, and partly under subarcsecond seeing conditions. In general, the results are consistent with the recent findings of Dressler and Richstone (1990). In addition, it is found that an area of about 5″ diameter around the nucleus is blue-shifted with a velocity of $\cong 40$ km s^{-1} with respect to the main body of the galaxy. The stars at the position of the non-thermal nucleus itself move at the same velocity as the main body of the galaxy. Furthermore, the authors' spectra show that the velocity dispersion in the central regions of M87 is in general constant at $\sigma \cong 370$ km s^{-1} with a dip of about 50 km s^{-1} within $r \cong 1''$. It is proposed that the core is not in complete dynamical equilibrium and that the signature of a galaxy merging with M87 may be observed.

157.010 **Star formation, molecular clouds and ^{13}CO emission from nearby galaxies.**
L. J. Sage, D. W. Isbell.
Astron. Astrophys., Vol. 247, No. 2, p. 320 - 334 (Jul 1991).

This paper presents observations of ^{13}CO $1 \to 0$ emission from 16 nearby spiral galaxies. Through comparisons with ^{12}CO and model calculations the authors have determined that ^{12}CO $1 \to 0$ is a reliable tracer of molecular mass with, at most, a factor of two uncertainty when a standard Galactic CO to H_2 conversion ratio is applied. The ratio R of ^{12}CO to ^{13}CO emission is independent of metallicity within the range examined, $9.1 \leqslant 12 + \log([O]/[H]) \leqslant 9.6$. The authors found that R is insensitive to variations in global parameters such as inclination angle and Hubble type, although there may be a dependence on average dust temperature for the galaxies with $T_{dust} \gtrsim 40$K. R was expected to be correlated with the ratio of CO to CS emission (which traces very dense gas), but this was not seen. The authors detected ^{13}CO emission from the centers of all 16 galaxies. Central beam ratios range from 5 to 16.6, with most in the 7 to 11 range. R was also determined along the major axes of four galaxies and at four positions in the disk of M101, with detections at all but one location. The observed line ratios lie between 2.5 and $\gtrsim 17$, but most are again in the 7 to 11 range. The ^{12}CO and ^{13}CO line profiles agree very well.

157.011 **Morphology and kinematics of the interacting elliptical galaxies NGC 750 and NGC 751 = Arp 166. Velocity fields of tidally distorted elliptical galaxies.**
R. Madejsky.
Astron. Astrophys., Vol. 247, No. 2, p. 348 - 356 (Jul 1991).

New morphological and kinematical data of the elliptical galaxies NGC 750 and 751 show that they are in the process of efficient tidal interaction. The galaxy centres are separated by 21 arcsec (or 10 kpc in projection). The isophotes show drastic deviations from the elliptical shape at all radii. Pointed and drop-shaped near-central isophotes suggest that the time-varying potential of the two galaxies is evolving slowly. Moreover, both luminosity profiles show significant deviations from the $r^{1/4}$-law at all radii. Long-slit spectroscopy for five slit positions reveals peculiar kinematical properties of the two galaxies, which have almost identical central radial velocities. While NGC 750 exhibits nearly flat radial velocity curves, the radial velocity curves of NGC 751 are characterized by large variations of more than

100 km s^{-1} along the slit for all slit positions, probably reflecting initial regular rotation of NGC 751 which was disturbed in the course of the encounter. Both galaxies exhibit asymmetric and in part by large amounts radially increasing velocity dispersion curves. An encounter scenario is proposed in which the three-dimensional relative velocity of the two galaxies is very low, so that tidal coupling of a large fraction of stellar orbits is the dominant tidal effect. If this scenario applies, the galaxies rapidly will merge.

157.012 **The stellar velocity dispersion of the edge-on spiral galaxy NGC 891.**
R. Bottema, P. C. van der Kruit, E. A. Valentijn.
Astron. Astrophys., Vol. 247, No. 2, p. 357 - 367 (Jul 1991).
At three positions, along a line perpendicular to the major axis, absorption line spectra have been obtained of the edge-on spiral galaxy NGC 891. These have been used to determine the stellar kinematics at three distances from the centre. For a distance of 52″ radial velocities and velocity dispersions are presented as a function of height above the plane. At that position the average dispersion value amounted to 82 ± 15 km s^{-1}, while at distances of 5″ and 132″ average velocity dispersions of 112 ± 20 km s^{-1} and 54 ± 10 km s^{-1} were determined. The observations were translated into dispersion values as a function of radius inside the galaxy. To that aim a galaxy model was constructed including a disk, bulge and absorbing layer. When the radial velocity dispersion decreases exponentially as a function of radius with a scalelength twice the photometric scalelength, the central dispersion amounts to 120 ± 20 km s^{-1}. This is comparable to the value of 100 ± 10 km s^{-1} for our Galaxy. From the dispersion a M/L ratio (in the B-band) between 6 and 10 was determined by assuming a constant Toomre Q value of 1.7 and an $(M/L)_B$ of 5.3 ± 2.5 by assuming the old disk stars make up an isothermal disk. These mass-to-light ratios compare well with $(M/L)_B = 7$ given by measuring the thickness of the hydrogen layer. Finally, it is concluded that the dispersion is not large enough to allow a so called "maximum disk" fit to the rotation curve.

157.013 **Distance of nearby galaxies NGC 2366, IC 2574 and NGC 4236 from photographic photometry of their brightest stars.**
N. A. Tikhonov, B. I. Bilkina, I. D. Karachentsev, Ts. B. Georgiev.
Astron. Astrophys., Suppl. Ser., Vol. 89, No. 1, p. 1 - 13 (Jul 1991).
The results from B, and V stellar photographic photometry made with the 6 m telescope are presented for the three members of the M81 group. Red and blue stars are selected and the average (V_3) and (B_3) magnitudes are used as distance indicators. The corresponding true distance moduli for the galaxies NGC 2366, IC 2574 and NGC 4236 are $27^m.63$, $27^m.80$ and $27^m.36$, respectively. They are obtained without accounting for internal light absorption in the galaxies. The authors expect that the standard deviation of the differential moduli does not exceed $0^m.2$. Using the new photometric data, following distances are obtained: 3.4 Mpc (NGC 2366), 3.6 Mpc (IC 2574), and 3.0 Mpc (NGC 4236).

157.014 **Analysis of the distribution of H II regions in external galaxies. I. Position and inclination angles.**
C. Garcia-Gomez, E. Athanassoula.
Astron. Astrophys., Suppl. Ser., Vol. 89, No. 1, p. 159 - 184 (Jul 1991).
The authors have compiled a sample of galaxies with published catalogues of H II regions forming the basis for subsequent studies of the H II region distribution in spiral galaxies. As a first step the authors address the problem of the deprojection of a galaxy image as seen on the sky. They have devised two methods to do this, based on the H II region distributions themselves, and apply them to their sample. The results are compared with the ones obtained using several other methods and it is found that these new methods are capable of obtaining good values for the position and inclination angles of the galaxies.

157.015 **Calán-Tololo survey. III. H II galaxies.**
J. Maza, M. T. Ruiz, M. Peña, L. E. González, M. Wischnjewsky.
Astron. Astrophys., Suppl. Ser., Vol. 89, No. 2, p. 389 - 397 (Aug 1991).
The third list of the Calán-Tololo Survey consists of 42 H II galaxies with bright emission lines most of which were previously unknown. The authors present identification charts, equatorial coordinates, and a short description for each object. The present list was drawn from a sample of approximately 200 narrow emission line galaxies chosen because their bright emission lines make them interesting for further spectrophotometric study.

157.016 **Erratum: "Galaxy velocity dispersions using a cross-correlation method" [Astrophys. J., Vol. 366, No. 1, p. 38 - 49 (1 Jan 1991)].**
C. Dalle Ore, S. M. Faber, J. J. Gonzalez, R. Stoughton, D. Burstein.
Astrophys. J., Vol. 375, No. 1, p. 427 (1 Jul 1991). See Abstr. 53.157.040.

157.017 **The brightest stars and the distance to the dwarf galaxy Ho IX.**
Ts. B. Georgiev, N. A. Tikhonov, I. D. Karachentsev, B. I. Bilkina.
Astron. Astrophys., Suppl. Ser., Vol. 89, No. 3, p. 529 - 536 (Sep 1991).
The authors have used B and V plates obtained with the 6 m telescope to extract photometric data on the brightest stars in the dwarf galaxy Ho IX, a nearby companion of M81. They find that three brightest red supergiants have average magnitude $\langle V(3) \rangle = 19.96$ mag, and the three brightest blue stars have $\langle B(3) \rangle = 20.66$ mag. These data give a distance modulus (m-M) = 27.67 ± 0.25 mag for the Ho IX, which is nearly at the same distance as for M81. The authors also discuss the magnitude scale differences found between their data and photometric data presented by Sandage (1984) and Hopp and Schulte-Ladbeck (1987).

157.018 **Gas velocity field in the central parts of the spiral galaxies NGC 23, NGC 1134, and NGC 4536.**
V. L. Afanas'ev, A. V. Zasov, G. V. Popravko, O. K. Sil'chenko.
Pis'ma Astron. Zh., Tom 17, No. 9, p. 771 - 778 (Sep 1991). In Russian. English translation in Sov. Astron. Lett., Vol. 17, No. 5.
New results are presented of gas line-of-sight velocity measurements in the central parts ($r < 15''$) of three spiral galaxies: NGC 23, 1134, 4536 with large line-of-sight velocity gradients in the nuclei. Observations were carried out at a two-dimensional multipupil fiber spectrometer at the 6-m telescope of the Special Astrophysical Observatory. A proximity of position angles of photometric and dynamical major axes gives the evidence that gas in the nuclear regions of galaxies considered has a circular motion in the plane of stellar disks of galaxies. In NGC 23 an extended region 8x4″ of abnormally low line-of-sight velocities is found.

157.019 **Optical structure of infrared-luminous galaxies.**
U. Klaas, H. Elsässer.
Astron. Astrophys., Suppl. Ser., Vol. 90, No. 1, p. 33 - 54 (Oct 1991).
A detailed optical investigation of eight extragalactic IRAS sources which were recently identified and observed by means of CCD imaging and spectroscopy is presented. The authors found 14 individual galaxies associated with these eight sources all showing peculiar morphologies and strong spatially extended line emission. The excitation conditions for the optical radiation as well as the heating mechanisms of the thermal far-infrared emission are investigated and the evolutionary stage of these systems is discussed. In at least seven cases the observational facts can be explained as a consequence of strong tidal interaction or even merging inducing a burst of star formation. Albeit the triggering mechanism of the activity seems to be the same for all objects, different scenarios have to be applied to explain the

different strength of energy output and the variety of excitation states.

157.020 **A study of southern extreme IRAS galaxies. II. Radio continuum observations.**
W. van Driel, A. C. van den Broek, T. de Jong.
Astron. Astrophys., Suppl. Ser., Vol. 90, No. 1, p. 55 – 81 (Oct 1991).
As part of a multi–wavelength study of an infrared–complete sample of extreme IRAS galaxies with high far–infrared/blue luminosity ratios ($L_{FIR}/L_B{}^0 \gtrsim 3$) the authors have observed 76 of them at 6 cm wavelength with the VLA, in order to study the star forming regions unimpeded by dust. They also have CCD images and spectra of these objects, as well as CO(1–0) spectra of a subsample.They found that 20 radio sources are clearly resolved with intrinsic sizes [$H_0 = 75$ km s^{-1} Mpc^{-1}] ranging from 0.6 to 2.2 kpc (average 1.1 kpc). The resolved sources have a surface brightness similar to that of galactic nuclei, but considerably higher than that of galactic disks. Galaxies with a higher $L_{FIR}/L_B{}^0$ ratio appear to have more compact radio sources, implying that in objects with higher star formation activity, this activity is concentrated in smaller regions. With one exception all objects follow the same tight far–infrared/radio luminosity correlation as other IRAS and RSA galaxies, indicating star formation as the main energy source, also in those with active nuclei. However, the galaxies with AGN–type spectra have a significantly higher average far–infrared luminosity and radio power than the H II–type galaxies. The authors estimate that on average a substantial fraction, about 40%, of the total radio power emitted by the galaxies at 6 cm has a thermal origin. This ratio is consistent with published radio spectral indices of similar extreme IRAS galaxies.

157.021 **Thickness of thin stellar disks and the dark halo mass.**
A. V. Zasov, D. I. Makarov, E. A. Mikhajlova.
Pis'ma Astron. Zh., Tom 17, No. 10, p. 884 – 890 (Oct 1991). In Russian. English translation in Sov. Astron. Lett., Vol. 17, No. 5.
It is shown that the small thickness of the observed edge–on thin galaxies with visual axis ratios $b/a < 0.1$ may be well treated as possessing a massive dark halo. An analysis of the condition of stability of stellar disks and N–body numerical simulation results lead to the conclusion that the minimal relative thickness z_0/R_0 of a collisionless gravitating disk is decreasing along with relative halo mass μ. The observed values of z_0/R_0 of normal edge–on galaxies can be used to estimate the halo mass.

157.022 **Photograhic photometry and distance modulus of the dwarf galaxy "Garland" near NGC 3077.**
M. E. Sharina.
Pis'ma Astron. Zh., Tom 17, No. 10, p. 904 – 912 (Oct 1991). In Russian. English translation in Sov. Astron. Lett., Vol. 17, No. 5.
The photographic photometry of the brightest stars in the irregular dwarf galaxy "Garland" near NGC 3077 has been carried out on the plates of the 6–m telescope. A colour–magnitude diagram and luminosity function are presented. The derived distance modulus for the galaxy is equal to $(m–M)_0 = 27^m.76$.

157.023 **The global properties of supernova parent galaxies.**
D. S. Davis, O.–G. Richter.
Bull. Am. Astron. Soc., Vol. 23, No. 2, p. 882 (1991). Abstract. – See Abstr. 012.008 for the main entry.

157.024 **Multicolor surface photometry of early–type galaxies.**
J. Penereiro, V. Gorjian, S. Djorgovski,
R. de Carvalho, D. Thompson, N. Weir.
Bull. Am. Astron. Soc., Vol. 23, No. 2, p. 890 (1991). Abstract. – See Abstr. 012.008 for the main entry.

157.025 **Stellar population induced peculiar velocities in elliptical galaxies.**
M. D. Gregg.
Bull. Am. Astron. Soc., Vol. 23, No. 2, p. 891 (1991). Abstract. – See Abstr. 012.008 for the main entry.

157.026 **UIT observations of old populations in nearby galaxies.**
R. W. O'Connell, R. C. Bohlin, N. Collins,
R. H. Cornett, J. K. Hill, R. S. Hill, W. B. Landsman,
T. P. Stecher.
Bull. Am. Astron. Soc., Vol. 23, No. 2, p. 891 (1991). Abstract. – See Abstr. 012.008 for the main entry.

157.027 **Infrared bright early–type galaxies.**
J. S. Mulchaey, M. D. Lehnert, T. M. Heckman.
Bull. Am. Astron. Soc., Vol. 23, No. 2, p. 891 (1991). Abstract. – See Abstr. 012.008 for the main entry.

157.028 **Interstellar matter in early–type galaxies: gaseous components and galaxy types.**
D. E. Hogg, M. S. Roberts, J. N. Bregman.
Bull. Am. Astron. Soc., Vol. 23, No. 2, p. 891 (1991). Abstract. – See Abstr. 012.008 for the main entry.

157.029 **Dark matter in elliptical galaxies with accurate X–ray spectra.**
M. Loewenstein.
Bull. Am. Astron. Soc., Vol. 23, No. 2, p. 891 (1991). Abstract. – See Abstr. 012.008 for the main entry.

157.030 **HUT spectra of the hot stellar components in the elliptical galaxy NGC 1399 and the bulge of M31.**
H. C. Ferguson, A. F. Davidsen, G. A. Kriss, W. P. Blair,
C. W. Bowers, W. V. Dixon, S. T. Durrance, P. D. Feldman,
R. C. Henry, J. W. Kruk, H. W. Moos, O. Vancura,
R. A. Kimble, K. Long.
Bull. Am. Astron. Soc., Vol. 23, No. 2, p. 899 (1991). Abstract. – See Abstr. 012.008 for the main entry.

157.031 **Observations of extragalactic water vapor masers.**
L. J. Greenhill, J. M. Moran.
Bull. Am. Astron. Soc., Vol. 23, No. 2, p. 899 – 900 (1991). Abstract. – See Abstr. 012.008 for the main entry.

157.032 **The dusty emission filaments in M87.**
H. C. Ford, W. B. Sparks.
Bull. Am. Astron. Soc., Vol. 23, No. 2, p. 900 (1991). Abstract. – See Abstr. 012.008 for the main entry.

157.033 **The UV properties of normal galaxies. I. The IUE data.**
G. Longo, M. Capaccioli, A. Ceriello.
Astron. Astrophys., Suppl. Ser., Vol. 90, No. 2, p. 375 – 386 (Oct 1991).
As a first step toward a better understanding of the energy distribution of normal galaxies at wavelengths shorter than 3300 Å, all IUE low resolution spectra of normal galaxies have been reduced to a photometric system consisting of seven intermediate–widths bands, five of which matching the ANS photometric system. Dereddened ultraviolet magnitudes are given for 156 galaxies.

157.034 **A study of southern extreme IRAS galaxies. I. Optical observations.**
A. C. van den Broek, W. van Driel, T. de Jong, J. Lub,
M. H. K. de Grijp, P. Goudfrooij.
Astron. Astrophys., Suppl. Ser., Vol. 91, No. 1, p. 61 – 128 (Nov 1991).
In order to study the processes that (strongly) enhance the star formation rate in some galaxies the authors have selected 62 galaxies from the IRAS point source catalogue with far–infrared to blue luminosity ratios ($L_{FIR}/L_B{}^0$) of three or larger. In this paper they present optical observations and show spectra, and broad–band, colour and continuum–subtracted Hα CCD images for more than 50 galaxies. The spectra are mostly H II–types although some 25% are AGN–types. The broad–band images show, among other things, that this sample predominantly consists of barred spirals and distorted interacting system, while the Hα images show in most cases small compact sources. From the spectra were derived redshifts, line–ratios, equivalent widths, and extinction determinations from the Hβ/Hα line ratios. The

authors also derived integrated magnitudes, colours, Hα fluxes, diameters and luminosities of the galaxies from the CCD images. They determined average luminosity profiles of all early-type galaxies and all possible merger candidates in the sample, and found that all these galaxies have exponential disks except two objects that follow a de Vaucouleurs $r^{1/4}$ law.

157.035 **Wolf-Rayet stars in IC 10.**
P. Massey, T. E. Armandroff, P. S. Conti.
Bull. Am. Astron. Soc., Vol. 23, No. 2, p. 918 – 919 (1991). Abstract. – See Abstr. 012.008 for the main entry.

157.036 **A nonthermal superbubble in the irregular galaxy IC 10.**
H. Yang, E. D. Skillman.
Bull. Am. Astron. Soc., Vol. 23, No. 2, p. 919 (1991). Abstract. – See Abstr. 012.008 for the main entry.

157.037 **The use of H II region characteristic size scale for distance determination.**
N. Strobel.
Bull. Am. Astron. Soc., Vol. 23, No. 2, p. 919 (1991). Abstract. – See Abstr. 012.008 for the main entry.

157.038 **The dark matter content of the Fornax and Carina dwarf galaxies.**
M. Mateo, E. Olszewski, D. Welch, P. Fischer, W. Kunkel.
Bull. Am. Astron. Soc., Vol. 23, No. 2, p. 919 (1991). Abstract. – See Abstr. 012.008 for the main entry.

157.039 **Chemical abundances in the extremely low metallicity dwarf irregular galaxy UGC 4483.**
E. D. Skillman, R. Terlevich, E. Terlevich, R. C. Kennicutt, D. R. Garnett, R. J. Dufour, H. O. Casteñeda, J. M. Vilchez.
Bull. Am. Astron. Soc., Vol. 23, No. 2, p. 919 (1991). Abstract. – See Abstr. 012.008 for the main entry.

157.040 **Radio observations of blue compact dwarf galaxies: comparison to optical images.**
H. J. Deeg, E. Brinks, N. Duric.
Bull. Am. Astron. Soc., Vol. 23, No. 2, p. 920 (1991). Abstract. – See Abstr. 012.008 for the main entry.

157.041 **A directed search for intergalactic H I using blue compact galaxies.**
C. L. Taylor, E. D. Skillman, E. Brinks.
Bull. Am. Astron. Soc., Vol. 23, No. 2, p. 920 (1991). Abstract. – See Abstr. 012.008 for the main entry.

157.042 **Deep visual images of II Zw 40.**
C. R. A. Wilton.
Bull. Am. Astron. Soc., Vol. 23, No. 2, p. 920 (1991). Abstract. – See Abstr. 012.008 for the main entry.

157.043 **UIT observations of two interacting galaxy pairs: NGC 2992/3 and NGC 1316 (Fornax A)/7.**
N. R. Collins, R. H. Cornett, J. K. Hill, J. E. Isensee, W. B. Landsman, B. B. Pfarr, P. C. Chen, S. P. Maran, S. G. Neff, E. P. Smith, T. P. Stecher, M. S. Roberts.
Bull. Am. Astron. Soc., Vol. 23, No. 2, p. 920 (1991). Abstract. – See Abstr. 012.008 for the main entry.

157.044 **A spectrophotometric catalogue of H II galaxies.**
R. Terlevich, J. Melnick, J. Masegosa, M. Moles, M. V. F. Copetti.
Astron. Astrophys., Suppl. Ser., Vol. 91, No. 2, p. 285 – 324 (Dec 1991).

A spectrophotometric catalogue is presented of 425 emission line galaxies discovered in objective prism surveys for which the authors derived redshifts, emission line intensities, equivalent widths, and absolute fluxes. The vast majority of objects in the catalogue are H II region-like galaxies (H II galaxies). In more than 80 H II galaxies the line [O III] λ4363 was measured with accuracy good enough to permit precise electron temperature determinations. The observational parameters that define the properties of H II galaxies as a class are characterized and discussed.

157.045 **What determines the physical quantities of galaxies? A two-component gas model for protogalaxies with energy input from supernovae.**
S. Ikeuchi, C. A. Norman.
Astrophys. J., Vol. 375, No. 2, p. 479 – 483 (10 Jul 1991).

The equilibrium structure of a protogalaxy, which is assumed to be composed of cold clouds and hot ambient gas in pressure equilibrium, is examined. The underlying characteristic time scales are the free-fall time of the system and the cooling time of the hot ambient gas. The global collapse condition used here is the well-known condition that the cooling time must be less than the free-fall time. Feedback processes are incorporated into the model via an energy balance condition between the power injected from supernovae and related processes due to the evolution of newly formed massive stars and the luminosity radiated by cooling processes. These considerations allow calculation of the characteristic mass and size of these two-component protogalaxies that can be expressed in terms of fundamental constants. In addition, the pressure equilibrium condition for protogalaxies with such a two-component interstellar medium indicates a characteristic mass for clouds, which is roughly comparable to the mass scale of globular clusters depending on the relevant cooling mechanism. The physical properties derived here for these protogalactic systems could be useful in interpreting high-redshift objects that may be possibly associated with galaxy formation.

157.046 **Observing galaxies through time.**
A. Dressler.
Sky Telesc., Vol. 82, No. 2, p. 126 – 132 (Aug 1991).

Where are the ancestors of present-day galaxies, and when did they form their first generation of stars?

157.047 **The evolution of luminous infrared galaxies.**
H. E. Smith, R. T. Harmon, C. Lonsdale, R. G. McMahon.
Bull. Am. Astron. Soc., Vol. 23, No. 2, p. 931 (1991). Abstract. – See Abstr. 012.008 for the main entry.

157.048 **BBXRT observations of M82, NGC 253, and M81.**
R. Petre.
Bull. Am. Astron. Soc., Vol. 23, No. 2, p. 932 (1991). Abstract. – See Abstr. 012.008 for the main entry.

157.049 **A search for black holes in galaxy nuclei.**
J. Kormendy.
Bull. Am. Astron. Soc., Vol. 23, No. 2, p. 937 (1991). Abstract. – See Abstr. 012.008 for the main entry.

157.050 **Galaxy pairs in redshift catalogs as a probe of dark matter.**
J. C. Charlton, E. E. Salpeter.
Astrophys. J., Vol. 375, No. 2, p. 517 – 531 (10 Jul 1991).

Pairs of galaxies extracted from the CfA $m_B = 14.5$ and SSRS redshift catalogs are studied, separately in high- and low-density regions, focusing mostly on the low-density pairs. The main conclusions are: (1) In low-density regions there is an excess of pairs with radial velocity differences $\Delta v < 150$ km s^{-1}, even for pairs with projected separations r_p as large as 1 Mpc. The transit time $r_p/\Delta v$ is less than the Hubble time for at least those pairs with $r_p < 1$ Mpc; thus, any initial correlations would have been disrupted before now, and these pairs must be bound. (2) No Kepler relationship is seen between the pair properties (Δv, r_p, and L, the pair luminosity), even for pairs that are isolated. (3) The distribution of r_p for low-density pairs with $\Delta v < 150$ km s^{-1} is flat and featureless from very small separations out to 2 Mpc. These three facts indicate that large amounts of dark matter (possibly as much as $10^{13} M_\odot$ out to 1 or 2 Mpc) surround a typical galaxy. The third point suggests a model where pairs at small r_p merge rapidly by dynamical friction but are replenished by pairs that begin at larger separation and are heading toward merger. The authors suggest that many of the galaxies in low-density regions that are classified as elliptical or S0 are in fact merger remnants. Furthermore, galaxy groupings in low-density regions seem to appear in common halos of dark matter rather than each galaxy having its own large spherical isothermal halo.

157.051 **A long, hard look at the Virgo cluster.**
F. Flam.
Science, Vol. 253, No. 5016, p. 138 – 139 (12 Jul 1991).
New images reveal the stars in a distant galaxy – and heat up a long-simmering debate about the universe's age and size.

157.052 **The history of star formation rates in galaxies.**
P. Hodge.
Bull. Am. Astron. Soc., Vol. 23, No. 2, p. 949 (1991). Abstract. – See Abstr. 012.008 for the main entry.

157.053 **Toward an objective identification technique for OB associations in galaxies.**
P. Battinelli, R. Capuzzo-Dolcetta, P. W. Hodge.
Bull. Am. Astron. Soc., Vol. 23, No. 2, p. 949 (1991). Abstract. – See Abstr. 012.008 for the main entry.

157.054 **A new twist on the FIR-radio correlation.**
R. Price, N. Duric, B. Campbell.
Bull. Am. Astron. Soc., Vol. 23, No. 2, p. 950 (1991). Abstract. – See Abstr. 012.008 for the main entry.

157.055 **Star formation properties of barred spiral galaxies.**
A. C. Phillips.
Bull. Am. Astron. Soc., Vol. 23, No. 2, p. 950 (1991). Abstract. – See Abstr. 012.008 for the main entry.

157.056 **UIT imaging of the starburst galaxy M82.**
S. P. Maran, S. G. Neff, A. M. Smith, E. P. Smith, T. P. Stecher, R. W. O'Connell, W. B. Landsman, M. S. Roberts.
Bull. Am. Astron. Soc., Vol. 23, No. 2, p. 950 (1991). Abstract. – See Abstr. 012.008 for the main entry.

157.057 **Observations of X-ray starburst galaxy candidates.**
A. Fruscione, R. E. Griffiths, J. W. MacKenty.
Bull. Am. Astron. Soc., Vol. 23, No. 2, p. 950 (1991). Abstract. – See Abstr. 012.008 for the main entry.

157.058 **VLBI search for compact radio cores in starbursts.**
C. Lonsdale, C. Lonsdale, H. E. Smith.
Bull. Am. Astron. Soc., Vol. 23, No. 2, p. 950 (1991). Abstract. – See Abstr. 012.008 for the main entry.

157.059 **Evolving starburst galaxies and faint number counts.**
C. Lonsdale, R. T. Harmon.
Bull. Am. Astron. Soc., Vol. 23, No. 2, p. 950 (1991). Abstract. – See Abstr. 012.008 for the main entry.

157.060 **IUE spectra of the ring galaxy Arp 147.**
A. B. Schultz, L. D. Spight, M. Rodrigue, P. T. Colegrove, M. A. DiSanti.
Bull. Am. Astron. Soc., Vol. 23, No. 2, p. 953 (1991). Abstract. – See Abstr. 012.008 for the main entry.

157.061 **NGC 4027: an interacting one-armed spiral galaxy.**
B. Phookun, L. G. Mundy.
Bull. Am. Astron. Soc., Vol. 23, No. 2, p. 953 (1991). Abstract. – See Abstr. 012.008 for the main entry.

157.062 **Are hypernovae detectable?**
S. J. Park, E. T. Vishniac.
Astrophys. J., Vol. 375, No. 2, p. 565 – 567 (10 Jul 1991).
It has been previously suggested that galaxy mergers may lead to mergers between the extremely massive black holes found in the cores of many (perhaps most) galaxies. The burst of gravitational radiation which results from such a merger would be extremely conspicuous, even if the hypernova itself happened at a high redshift. The authors point out that plausible models for the dynamical evolution of black hole binaries include the possibility that a hypernova, set in motion by a galaxy merger at high z, may not actually take place until the present epoch. The authors calculate the expected rate of hypernovae and find that it is of order one every 10 yr. This result is only weakly dependent on the cosmological model, and is optimistic only if the binary evolution time scale is much longer than the age of the universe. This event rate constitutes a minimal estimate of events in the vicinity of 10^{-3} Hertz. If the binary relaxation time is on the order of the current age of the universe, then hypernovae will give rise to more intense, and higher frequency, bursts of gravitational radiation than previous estimates.

157.063 **The properties of NGC 2777: are companion galaxies young?**
H. Arp, J. W. Sulentic.
Astrophys. J., Vol. 375, No. 2, p. 569 – 582 (10 Jul 1991).
Observations of NGC 2777 show it to be a disturbed, high surface brightness galaxy with a very early type stellar spectrum. It has a "hot" 60/100 μm infrared flux ratio, and there is evidence for galaxy-wide expansion of gas as a result of strong current star formation. NGC 2777 contains a large amount of hydrogen, including an H I tail extending toward the larger nearby Sa galaxy, NGC 2775. Evidence for an extension of non-thermal continuum radiation from NGC 2775 toward the companion NGC 2777 is discussed. NGC 2777 is shown to be typical of a class of galaxies that are composed predominantly of recently formed stars and that are companions to larger galaxies. It is argued, however, that star formation activity is not stimulated by gravitational or collisional encounter with the main galaxy. Evidence is discussed which suggests that these kinds of companions are formed in ejection events from older galaxies and that they are more recently created galaxies. Star formation in jets is proposed as the explanation for the strong correlation between nonthermal and infrared emission in galaxies.

157.064 **Multicolor optical mosaics of M31.**
E. A. Magnier, Z. Haiman, W. H. G. Lewin, P. L. Schechter, J. van Paradijs, M. van der Klis.
Bull. Am. Astron. Soc., Vol. 23, No. 2, p. 965 (1991). Abstract. – See Abstr. 012.008 for the main entry.

157.065 **High resolution X-ray observations of the central region of M31.**
F. A. Primini, M. Garcia.
Bull. Am. Astron. Soc., Vol. 23, No. 2, p. 965 (1991). Abstract. – See Abstr. 012.008 for the main entry.

157.066 **UIT observations of the ultraviolet morphology of M33.**
M. S. Roberts, R. H. Cornett, J. K. Hill, J. E. Isensee, W. B. Landsman, B. B. Pfarr, P. C. Chen, R. C. Bohlin, S. P. Maran, S. G. Neff, E. P. Smith, T. P. Stecher.
Bull. Am. Astron. Soc., Vol. 23, No. 2, p. 965 (1991). Abstract. – See Abstr. 012.008 for the main entry.

157.067 **UIT ultraviolet imagery of M81.**
J. K. Hill, R. H. Cornett, J. E. Isensee, W. B. Landsman, B. B. Pfarr, P. C. Chen, S. P. Maran, S. G. Neff, E. P. Smith, T. P. Stecher, M. S. Roberts.
Bull. Am. Astron. Soc., Vol. 23, No. 2, p. 965 (1991). Abstract. – See Abstr. 012.008 for the main entry.

157.068 **UIT ultraviolet images of the ScI galaxy NGC 628 (M74).**
P. C. Chen, J. K. Hill, R. H. Cornett, J. E. Isensee, W. B. Landsman, B. B. Pfarr, S. P. Maran, S. G. Neff, E. P. Smith, T. P. Stecher, M. S. Roberts.
Bull. Am. Astron. Soc., Vol. 23, No. 2, p. 965 (1991). Abstract. – See Abstr. 012.008 for the main entry.

157.069 **A comparative study of the H II regions of M51 and M101.**
P. A. Scowen, R. J. Dufour, J. J. Hester.
Bull. Am. Astron. Soc., Vol. 23, No. 2, p. 966 (1991). Abstract. – See Abstr. 012.008 for the main entry.

157.070 **The structure of NGC 2403.**
K. Krienke, P. Hodge.
Bull. Am. Astron. Soc., Vol. 23, No. 2, p. 966 (1991). Abstract. – See Abstr. 012.008 for the main entry.

157.071 **Position and inclination angles of NGC 4303.**
B. Miller, P. W. Hodge.
Bull. Am. Astron. Soc., Vol. 23, No. 2, p. 966 (1991). Abstract. – See Abstr. 012.008 for the main entry.

157.072 **Testing the planetary nebula extragalactic distance method across Hubble types in NGC 891 (Sb) and NGC 1023 (SB0).**
G. Jacoby, R. Ciardullo, W. Harris.
Bull. Am. Astron. Soc., Vol. 23, No. 2, p. 972 – 973 (1991). Abstract. – See Abstr. 012.008 for the main entry.

157.073 **The flat dark matter "halo" of polar–ring galaxy NGC 4650A.**
P. D. Sackett, B. Jarvis.
Bull. Am. Astron. Soc., Vol. 23, No. 2, p. 973 (1991). Abstract. – See Abstr. 012.008 for the main entry.

157.074 **High resolution imaging of a spiral galaxy in the Virgo Cluster: the brightest stars and the extragalactic distance scale.**
M. J. Pierce, R. Racine, R. D. McClure.
Bull. Am. Astron. Soc., Vol. 23, No. 2, p. 973 (1991). Abstract. – See Abstr. 012.008 for the main entry.

157.075 **Kinematics of the extended H I disk of NGC 628.**
F. H. Briggs, J. Kamphuis.
Bull. Am. Astron. Soc., Vol. 23, No. 2, p. 973 – 974 (1991). Abstract. – See Abstr. 012.008 for the main entry.

157.076 **Stellar populations in galaxies.**
A. K. Pati.
National Large Telescope Workshop: Scientific requirements and requirement of site, p. 37 – 46 (1991). – See Abstr. 012.009 for the main entry.
The study of populations, both in the Galaxy and in other galaxies, will entail a lot of future research, especially with the large telescopes of tomorrow. The author summarizes briefly the known properties of some stellar populations which contrast with the traditional concepts.

157.077 **Cooling of gas behind the explosive shocks and consequent star–formation in the central region of a disk galaxy.**
T. Kanjilal, B. Basu.
Indian J. Phys., Part B, Vol. 65, No. 4, p. 299 – 311 (Aug 1991).
Propagation of a shock produced by a nuclear explosion in a disk galaxy through a medium of constant density has been considered for different values of the initial Mach number. Cooling times have been computed under various shock conditions using a polytropic relation between the density and temperature within the cooling layer both in the presence and absence of a magnetic field. The relative amount of the shocked material which is likely to produce the molecular clouds has also been considered under different shock conditions. Calculations indicate that weaker shocks are probably more efficient for the formation of star than stronger shocks.

157.078 **Spiral wave pattern in M31.**
Yu. N. Mishurov.
Astron. Tsirk., No. 1549, p. 5 – 6 (May – Aug 1991). In Russian.
It was shown that the galactic density wave pattern in M31 with angular velocity of rotation $\Omega p = 20\ km\cdot s^{-1}\cdot kpc^{-1}$ coincides well with the spiral structure presented by Arp from emission nebulae observations.

157.079 **On radial velocity of the elliptical galaxy A 2058 + 16.**
G. D. Polyakova, V. S. Popov.
Astron. Tsirk., No. 1549, p. 7 – 8 (May – Aug 1991).
In the sample of Fe I lines the value of radial velocity does not change with wavelength λ_0, i.e., it remains constant within the accuracy of measurements. The average value $\langle v_i \rangle = 8900 \pm 100$ km/s was adopted to be the radial velocity of the galaxy. At the same time in the sample of the alkaline and alkalescent metals a statistically reliable dependence between v and λ_0 has been found. The results obtained show that the problem of determination of the galaxies radial velocities needs more accurate methodical approach.

157.080 **The spectra of seven new galaxies with UV excess.**
V. S. Tamazyan.
Astron. Tsirk., No. 1549, p. 9 – 10 (May – Aug 1991). In Russian.
The redshifts and absolute magnitudes of seven new galaxies with UV excess from lists published by M. Kazaryan (Nos. 18, 24, 68, 96, 228, 327, 491) are presented.

157.081 **Photometric and kinematic study of VV 247 (NGC 6621/22).**
V. P. Reshetnikov, O. K. Sil'chenko.
Astrophysics, Vol. 33, No. 2, p. 385 – 393 (Sep – Oct 1990). English translation of Astrofizika, Tom 33, Vyp. 2, p. 157 – 168 (Oct 1990).
The results are given of a detailed photometric and kinematic study of the interacting double system VV 247 (NGC 6621/22). The 2.6–m telescope of the Byurakan Astrophysical Observatory of the Armenian Academy of Sciences was used in UBV surface photometry of VV 247. The standard photometric characteristics of the galaxies have been determined, and it has been shown that the members of this system are a gas–rich SB(a–b) galaxy (NGC 6621) and an E galaxy (NGC 6622). Spectroscopic observations of VV 247 were made with the 6–m telescope of the Special Astrophysical Observatory of the USSR Academy of Sciences with the long–slit spectrograph UAGS and the two–dimensional photon counter KVANT. It is shown that in the nuclear region of NGC 6621 ($\pm 4''$ –$6''$ from the nucleus) there exists a gas system which rotates in the opposite direction to the galaxy as a whole. The existence of the counterrotating gas system is attributed to the interaction of this galaxy with its companion – the orbital motion of NGC 6622 is in the opposite direction to that of NGC 6621. The example of the double system VV 247 indicates that there can be effective gravitational interaction of galaxies in the case of a retrograde encounter.

157.082 **Spectrophotometric and morphological investigation of galaxies with ultraviolet excess. VIII.**
M. A. Kazaryan, Eh. S. Kazaryan.
Astrophysics, Vol. 33, No. 2, p. 394 – 400 (Sep – Oct 1990). English translation of Astrofizika, Tom 33, Vyp. 2, p. 169 – 179 (Oct 1990).
The results are given of spectroscopic and morphological investigations of the galaxies Nos. 69, 147, 211 in the lists of Kazaryan (1979) and of galaxy No. 147a. The masses of the gas components of these galaxies and the mass of galaxy No. 211 have been determined. For objects No. 26, No. 69, No. 95, No. 147, No. 211 the FWHM of the emission lines differ in their values. The largest values correspond to the lines [O III] $\lambda\lambda$ 5007, 4959, and H_β, and the lowest to H_α, [S II] 6731, 6717. In all probability, the reason for this is that lines with different FWHM arise in regions of nuclei or formations of galaxies with ultraviolet

excess that have different dynamical properties. It is shown that object No. 147 is an Sy 2 galaxy, and that No. 69 is a LINER.

157.083 **The role of neutral hydrogen in the evolution of spiral and irregular galaxies.**
A. L. Gyul'budagyan.
Astrophysics, Vol. 33, No. 2, p. 404 – 411 (Sep – Oct 1990). English translation of Astrofizika, Tom 33, Vyp. 2, p. 187 – 198 (Oct 1990).
The abundances of neutral and molecular hydrogen in the Galaxy and also in other galaxies and in clusters and groups of galaxies are considered. The following conclusions are drawn: 1. Both in the Galaxy and in other spirals, clouds of molecular hydrogen are transformed into clouds of neutral hydrogen. 2. The older a galaxy, the higher is the percentage abundance in it of neutral hydrogen. 3. The evolutionary path for spirals is from early spirals to irregular galaxies (Sa→Sb→Sc→Irr).

157.084 **The dwarf galaxy UGC 5272 and its small companion galaxy.**
U. Hopp, R. E. Schulte-Ladbeck.
Astron. Astrophys., Vol. 248, No. 1, p. 1 – 11 (Aug 1991).
The authors present multicolor optical images and spectroscopy of the dwarf irregular galaxy UGC 5272. The CCD images reveal the presence of a closeby, small neighboring galaxy to UGC 5272 in a projected distance of 119" (3.6 kpc). The neighboring galaxy is also of irregular type and has a Holmberg diameter of 0.6 kpc. The authors discuss if these two galaxies form a physical pair. The distance to both galaxies is discussed and a value of 6 Mpc is estimated. Both galaxies are resolved into single stars. Photometry in B and R to $B_{lim} \approx 23^m.8$ is presented. From the color–magnitude diagram the presence of a young stellar population with $\leqslant 10^7$yr and a mass range of $15 \leqslant M/M_\odot \leqslant 60$ are inferred. The authors can not identify a red–supergiant population, and suggest that the blue–to–red supergiant ratio is high because of the galaxies' low metallicity. Combing the new data with radio–, IRAS– and spectrophotometric results from the literature, the present state of UGC 5272 is discussed in detail. While its extremely blue colors are suggestive for a recent star burst and several other features (OB star number, H II regions) are at least indicative for a high star formation rate, the structural parameters are surprisingly normal. The contribution of gas to the total mass is high. The best estimate of the total mass of $10^9 M_\odot$ results from the authors rotation curve.

157.085 **Radio continuum and CO (J = 2→1) observations of NGC 3628.**
H.-P. Reuter, M. Krause, R. Wielebinski, H. Lesch.
Astron. Astrophys., Vol. 248, No. 1, p. 12 – 22 (Aug 1991).
The authors have studied the radio continuum emission at λ20 cm and the $^{12}CO(2-1)$ and $^{13}CO(2-1)$ lines in the edge–on galaxy NGC 3628. The observations have an angular resolution of 12 arcsec which corresponds to ~400 pc (D ~ 6.7 Mpc). The radio continuum exhibits a spur–like feature ~3 kpc above the plane of the galaxy. Polarized intensity of the radio emission is visible up to 3.5 kpc above and below the disk of the galaxy. From the CO observations the authors derive a total molecular gas mass of $1.3 \times 10^9 M_\odot$ for the central 4×1 kpc^2 of the galaxy. A small fraction of this gas ($3.3 \times 10^7 M_\odot$) is concentrated in a small fast rotating disk coincident with the radio continuum peak. In addition, the authors could distinguish six molecular cloud complexes with masses of $(1-2) \times 10^6 M_\odot$ dominating the extended disk distribution of the gas in the central 800×200 pc^2. Only a small fraction of the molecular gas is concentrated in high density cores of the molecular clouds. The authors find an unusually low value for the ratio $I(^{12}CO)/I(^{13}CO)$ in the center (~7) increasing to moderate values in the outer parts (~15). The low value in the center indicates high optical depth fractionation of the CO gas and/or ^{13}C enhancement due to mass loss from massive stars. The authors find evidence for a spiral arm structure from both the radio continuum and the CO observations.

157.086 **The magnetic field structure in the radio halos of NGC 891 and NGC 4631.**
E. Hummel, R. Beck, M. Dahlem.
Astron. Astrophys., Vol. 248, No. 1, p. 23 – 29 (Aug 1991).
Linearly polarized radio continuum emission has been detected in the halos of NGC 891 and NGC 4631. In both galaxies the magnetic field shows uniform structures over scales of several kpc. The field lines point radially outward from the disk of NGC 4631, while in NGC 891 no overall orientation is visible. The increase in the degree of polarization with increasing distance from the plane is explained by Faraday depolarization by thermal gas with scale heights of 0.9 and 1.3 kpc for NGC 891 and NGC 4631, respectively. The value for NGC 891 agrees with recent Hα data. The magnetic field is assumed to be in equipartition with the cosmic ray energy density. In that case the scale heights of the fields are ~4 and ~8 kpc, respectively, which is consistent with the depolarization data. Hence the field lines reach far out into the halo. In the outer halo, the degree of polarization decreases again, maybe due to an increasing field turbulence. Both dynamo and galactic wind models can explain the two observed field structures. Either NGC 891 and NGC 4631 have dynamo fields of different modes (quadrupole and dipole), or the galactic wind is much stronger in NGC 4631.

157.087 **Colour change of galaxies with oscillating star–forming activities.**
Y. Yoshii, N. Arimoto.
Astron. Astrophys., Vol. 248, No. 1, p. 30 – 36 (Aug 1991).
The authors have studied the colour evolution of galaxies with oscillating star forming activities. In star burst galaxies, the pattern of colour oscillation is determined by the burst strength and the effective stellar mass which contributes most to the total galactic light; both are strongly dependent on the IMF slope. For flat IMF, the trajectory draws an open loop in the colour–colour diagram, while for steep IMF the trajectory fills in a parallelogram–shaped region and eventually degenerates into a line with tiny kinks marking the epochs of star bursts. This suggests that the slope of the IMF for burst populations is Salpeter like (x = 1.35) or more flattish (x = 0.35). Any star burst with x = 2.35, if it happens, could be hardly detected. Implications of the results are discussed in connection with colours of interacting galaxy pairs and the excess of blue galaxies in clusters of intermediate redhsifts.

157.088 **Models of the evolution of star formation regions of blue compact dwarf galaxies. III. Comparison with the observed oxygen abundances.**
L. S. Pilyugin.
Kinematika Fiz. Nebesn. Tel, Tom 7, No. 4, p. 73 – 80 (Jul–Aug 1991). In Russian. English translation in Kinematics Phys. Celest. Bodies, Vol. 7, No. 4.
The gaseous oxygen abundance evolution of star formation regions of blue compact dwarf galaxies is considered. It is shown that the gaseous oxygen abundance of star formation regions varies essentially in the process of their evolution. The computed evolutionary tracks of star formation regions agree with observed positions of star formation regions in ($O/O_\odot$, $W_{H\beta}$) and M_{SFR}, N_{LC}) planes.

157.089 **Scale–length determinations of exponential discs of spiral galaxies: a comparison.**
J. H. Knapen, P. C. van der Kruit.
Astron. Astrophys., Vol. 248, No. 1, p. 57 – 61 (Aug 1991).
Values for the scale lengths h of the exponential discs of spiral galaxies, as found by different authors, have been compared. For a given galaxy, two published values for h may show a discrepancy of almost a factor two. The average value of the discrepancy for more than 120 galaxies is 23%, with a standard deviation of 20%. No systematic trends are seen, nor correlations of the discrepancies with obvious galaxy parameters, such as morphological type, arm class or inclination. It is observed that semi–automated ways of determining the scale length introduce relatively large discrepancies. These results show the difficulty of determining reliable scale length values, and further emphasize

that one should be careful when interpreting physically the exponential nature of discs of spiral galaxies.

157.090 **Correlations between far–infrared, radio, and blue luminosities of spiral galaxies.**
J. J. Condon, M. L. Anderson, G. Helou.
Astrophys. J., Vol. 376, No. 1, p. 95 – 103 (20 Jul 1991).

Accurate far–infrared (FIR) and 1.49 GHz radio flux densities of optically selected spiral and irregular galaxies brighter than $B_T = +12$ and of infrared–selected galaxies stronger than $S = 5.24$ Jy at 60 μm were used to derive FIR–radio luminosity correlations for two statistically complete samples spanning a wide luminosity range. The observed correlations are tight but not linear, and the observed scatter is primarily intrinsic, with negligible broadening from measurement errors. Published methods to improve (linearize and tighten) the observed FIR–radio correlation by subtracting the (presumed) radio–quiet infrared cirrus or by compensating for cosmic–ray diffusion losses do not work for both samples simultaneously. An empirical correction to the radio luminosities based on the observed blue luminosities was found that makes the corrected FIR–radio correlation linear and eliminates most of the residual variance. The authors propose a two–component model for the heating luminosity within galaxies consistent with the improved correlation: (1) a radio–loud component containing young ($\lesssim 3 \times 10^7$yr) massive ($M \gtrsim 8\ M_\odot$) stars that heats the warm dust and contributes significantly to the diffuse interstellar radiation field that heats the cool cirrus dust; (2) a radio–quiet component comprised of older stars with $M < 8\ M_\odot$.

157.091 **Galactic models with massive coronae. III. Giant elliptical galaxy M87.**
P. Tenjes, J. Einasto, U. Haud.
Astron. Astrophys., Vol. 248, No. 2, p. 395 – 403 (Aug 1991).

A model of the giant elliptical galaxy M87 is calculated using the algorithm described in Paper I. The method is based on the assumption that (1) the galaxy is in a steady state, (2) galactic populations are physically homogeneous, and (3) equidensity surfaces of the galactic populations are similar concentric ellipsoids or can be represented as sums of such ellipsoids. All available observational data on surface photometry from the U to K bands, stellar velocity dispersions, distribution of globular clusters and their velocity dispersion, as well as the distribution of X–rays have been collected. These data were used to decompose the galaxy into a central nucleus, a metal–rich bulge, a metal–poor halo and a massive dark corona. The structural parameters, colour indices and mass–to–light ratios of these populations are derived. The gravitational potential and mass distribution functions of M87 and subsystems are found. Problems concerning the possible presence of a massive black hole in the centre of the galaxy and the mass of the metal–poor halo are discussed. The mass–to–luminosity ratio of optically visible parts of the galaxy is $M/L_B = 4.9 \pm 2.0\ M_\odot/L_\odot$, the ratio of the total mass to the visible mass $M_T/M_{vis} = 130$ inside the radius of 90′.

157.092 **Stellar winds and UV extinction of OB stars in M33 and M31.**
L. Bianchi, J. B. Hutchings, P. Massey.
Astron. Astrophys., Vol. 249, No. 1, p. 14 – 24 (Sep 1991).

The authors report on the continued study of the most luminous hot stars in galaxies of the Local Group, with the aim of studying their spectroscopic and wind characteristics, and comparing them with similar Milky Way stars. They present results from ground–based optical and IUE UV observations of five bright blue stars in M33 and four in M31. Four of the M33 stars are in two close pairs. The authors discuss these together with the previously published results on five stars. They discuss the limits of feature detectability in faint IUE spectra. Where possible, the authors have derived spectral classifications and fit continuum models to the data, from U–V and visible range spectra, and visible range photometry. The stars are all O, B or W–R type of high luminosity. As reported previously, the authors have continued to find that, compared with Galactic counterparts, the UV line spectrum and the stellar wind indicators are weak. It is also found that the UV extinction law in both M33 and M31 stars is steeper than the Galactic law, and similar to that found in the LMC. The authors discuss the evidence for stellar and extinction anomalies in our Galaxy, and conclude that their sample stars are disk objects in M33 and M31.

157.093 **Extragalactic $^{18}O/^{17}O$ ratios and star formation: high–mass stars preferred in starburst systems?**
L. J. Sage, R. Mauersberger, C. Henkel.
Astron. Astrophys., Vol. 249, No. 1, p. 31 – 35 (Sep 1991).

The authors have detected emission from the two lowest rotational transitions of $C^{17}O$ from NGC 253, M82, and, tentatively, IC 342. Observations of the corresponding transitions of $C^{18}O$ were used to determine $^{18}O/^{17}O$ isotope ratios, which are 8.3($\pm$1.3) toward M82, 10($\pm$3) toward NGC 253 and $\gtrsim$5.5 toward IC 342. These values are much higher than the $\sim$3.6 found in Galactic sources and two are definitely larger than the solar system ratio, 5.5. The results can be understood if, during a previous epoch of star formation, high mass stars were preferentially formed near the centers of NGC 253 and M82, if we assume that ^{18}O is mainly produced in high mass stars. This leads to an underabundance of ^{17}O relative to ^{18}O. This interpretation is also consistent with the high ratios of $^{12}CO/^{13}CO$ observed in NGC 253 and M82, which suggest an underabundance of ^{13}C.

157.094 **On the distribution of blue stars and gas in M33.**
Ts. B. Georgiev, G. R. Ivanov.
Astrofiz. Issled. Izv. Spets. Astrofiz. Obs., Tom 31, p. 29 – 35 (1991). In Russian. English translation in Bull. Spec. Astrophys. Obs. – North Caucasus.

A catalogue of 481 OB–associations with 3599 blue stars is processed. Some statistical data about the associations are given. A comparison between the radial distributions of the blue stars, neutral hydrogen and ultraviolet surface brightness of the disk is made. The appeared regularities are explained as evidences of star formation going on in the whole galaxy, the activity of which correlates with the gas concentration.

157.095 **Winds, outflows, and inflows in X–ray elliptical galaxies. I.**
L. Ciotti, A. D'Ercole, S. Pellegrini, A. Renzini.
Astrophys. J., Vol. 376, No. 2, p. 380 – 403 (1 Aug 1991).

A new class of one–dimensional hydrodynamical evolutionary sequences for the gas flows in elliptical galaxies is presented. The model galaxies are described by either King's or de Vaucouleurs's virialized density distributions, constrained to lie on the fundamental plane of elliptical galaxies, and are surrounded by variable amouts of dark matter. Two source terms operate: mass loss from evolving stars, and a secularly declining heating by Type I supernovae. The resulting models evolve through up to three consecutive evolutionary stages: the wind, outflow, and inflow phases. Agreement with the observed trends of the X–ray luminosity L_X, with the optical characteristics of the galaxies, is achieved for dark–to–visible mass ratios of $\sim$10, when a standard SN I rate is assumed. Also the large dispersion of L_X for given optical luminosity is easily accounted for. These sequences indicate that the X–ray faint galaxies are still in their wind phase, while the bulk of ellipticals are in the outflow phase, and a few of the brightest galaxies may have already experienced their transition to the inflow regime. The effects of interactions with the intracluster medium, ram pressure, encounters, and merging events are qualitatively discussed. It is argued that the iron abundance in the flows can constrain the SN I heating rate in the galaxies, while the total mass of iron in the intracluster medium provides a constraint on the total number of Type I SNs so far exploded in cluster galaxies.

157.096 **Dependence of certain rotational properties of 78 spiral galaxies on Hubble's type and luminosity.**
F. L. Tektunali, H. G. Tektunali.
Astrophys. Space Sci., Vol. 183, No. 1, p. 155 – 162 (Sep 1991).

Dependence of the central velocity gradients on Hubble's type is presented for 78 spiral galaxies with existing rotation curves.

Also the dependence of the maximum rotational velocities of 27 galaxies on both Hubble's type and the luminosity is studied. The central velocity gradient is shown to be related with Hubble's type. Maximum rotational velocities of 27 galaxies of the authors' sample depend on Hubble's type such that the mean values of maximum rotational velocity decrease from Sa through Sc. It is also determined that there is a dependence of the maximum rotational velocity on the absolute blue magnitude for each intrinsic Hubble type. For each Hubble type, maximum rotational velocity increases with increasing absolute blue magnitudes.

157.097 **Arm classification and velocity gradients in spiral galaxies.**
A. Biviano, M. Girardi, G. Giuricin, F. Mardirossian, M. Mezzetti.
Astrophys. J., Vol. 376, No. 2, p. 458 – 466 (1 Aug 1991).
On the basis of published rotation curves, the authors have compiled a list of velocity gradients for 94 galaxies. A significant correlation is found in this sample of galaxies between their gradients and arm classes (as given by Elmegreen and Elmegreen); galaxies with steeper velocity curves tend to have a flocculent arm structure, and galaxies with flatter curves tend to have a grand design morphology. This result is in agreement with previous suggestions by Whitmore and with the recent result by Elmegreen and Elmegreen; it is also consistent with the predictions of density wave theory for the formation of spiral structure.

157.098 **CO (3→2) mapping and gas excitation in the core of M82.**
R. P. J. Tilanus, L. J. Tacconi, E. C. Sutton, S. Zhou, D. B. Sanders, C. G. Wynn-Williams, K. Y. Lo, S. A. Stephens.
Astrophys. J., Vol. 376, No. 2, p. 500 – 504 (1 Aug 1991).
The authors have observed the CO (3→2) emission in the central region of M82 with a 14" FWHM beam using the Berkeley SIS receiver system at the 15 m JCM Telescope on Mauna Kea. The map clearly shows the double-peaked structure seen at other wavelengths, that has been interpreted as a circumnuclear ring. The first ^{13}CO (3→2) detections in M82 are also reported. Using a large velocity gradient model, the six observed lines (the three lowest transitions of ^{12}CO and ^{13}CO) allow a rather direct determination of the properties of the molecular gas in M82. The ratios of these lines are rather unusual and cannot be reproduced with a standard one-component model and an abundance ratio $^{12}CO/^{13}CO$ of 70 (solar), nor by simple mixtures of optically thin and thick CO components. On the other hand, using a one-component model a significantly better fit to all of the CO lines can be obtained if the abundance ratio, $^{12}CO/^{13}CO$, is lowered to values of 25 or below. Such values are similar to the abundance ratio observed in the center of the Milky Way. For an assumed abundance ratio of 20, the analysis suggests that the molecular gas in the central region of M82 is both hotter and denser on average than in the Milky Way.

157.099 **Very small grains and the infrared colors of galaxies.**
G. Helou, C. Ryter, B. T. Soifer.
Astrophys. J., Vol. 376, No. 2, p. 505 – 513 (1 Aug 1991).
The authors use IRAS data to study the contribution of very small grains with fluctuating temperature to the infrared emission of disk galaxies and to estimate the variation among galaxies of the abundance of these small grains relative to the larger "classical", thermally stable grains. It is argued that the 12 μm emission from spiral galaxies orginates in their interstellar medium. The integrated IRAS colors of galaxies are compared to the IRAS colors of local emission within Galactic nebulae. The ratio $\Gamma = \nu f_\nu(12\ \mu m)/F(40–120\ \mu m)$ is advanced as an estimator of the small-to-large grain ratio, and the interpretation of its statistical behavior in samples of galaxies is discussed. In a color-color diagram of the emission from Galactic nebulae, Γ depends only on $\Theta = f_\nu(60\ \mu m)/f_\nu(100\ \mu m)$, decreasing slowly as Θ increases. The same trend is observed for the integrated colors of galaxies and is seen in three different samples: optically selected "normal" spirals in the Virgo cluster, infrared-luminous galaxies from the IRAS bright galaxy sample and "cold" galaxies selected for their low ratios of $f_\nu(60\ \mu m)$ to $f_\nu(100\ \mu m)$. The dispersion on Γ is small (rms ≅ 40%) for $\Theta < 1/3$, and increases to about a factor of 2 at $\Theta \cong 1$. A physical interpretation of this result is presented.

157.100 **Optical and UV polarization observations of the M87 jet.**
P. E. Hodge, F. Macchetto, W. B. Sparks.
Workshop on the First Year of HST Observations, p. 196 – 199 (1991). – See Abstr. 012.017 for the main entry.

157.101 **Large $^{12}CO/^{13}CO$ intensity ratios in luminous mergers.**
S. Aalto, J. H. Black, L. E. B. Johansson, R. S. Booth.
Astron. Astrophys., Vol. 249, No. 2, p. 323 – 326 (Sep 1991).
The authors have observed the $J = 1–0$ transitions of ^{12}CO and ^{13}CO in 21 galaxies. They measure unusually large $^{12}CO/^{13}CO$ (1–0) intensity ratios, $R > 20$, in four merging galaxies NGC 3256, NGC 1614, Arp 220 (IC 4553) and Arp 299 (IC 694 and NGC 3690). All four are warm (fluxes at 60 and 100 μm, $f_{60}/f_{100} \geqslant 0.78$) and luminous in the infrared ($L_{IR} \gtrsim 3\times10^{11} L_\odot$). The ratio in the rest of the objects, which display a variety of morphologies, is well concentrated around a mean value of 11. The authors suggest that the large $^{12}CO/^{13}CO$ intensity ratios reflect a decrease in mean optical depth of molecular clouds that have been disturbed by powerful tidal forces or by a starburst triggered in the merging process. The intensity ratio $R \approx 11$ is a factor of about 2 times larger than that of giant molecular clouds in the Galactic disk, leading to a conclusion that the Galactic disk is a poor model for some extragalactic molecular emission. Since nuclear CO emission dominates in several of the sample galaxies, a comparison with the molecular clouds of the Galactic Centre may be more appropriate.

157.102 **Extragalactic dust. IV. NGC 7625.**
N. Brosch, F. Loinger.
Astron. Astrophys., Vol. 249, No. 2, p. 327 – 336 (Sep 1991).
The authors present a photometric study of the early-type galaxy with dust lanes NGC 7625. They derive the apparent extinction law of the particles in the dust lanes and find $R \sim 2.8$, explained as a dust grains that are marginally smaller than Galactic dust grains. They analyze the emission characteristics of NGC 7625 at various spectral bands and discuss star formation process therein.

157.103 **Evolutionary method of population synthesis for bulge-disc systems: optical-infrared colour-magnitude relation of spiral galaxies.**
N. Arimoto, P. Jablonka.
Astron. Astrophys., Vol. 249, No. 2, p. 374 – 388 (Sep 1991).
Based on the evolutionary method of population synthesis, the authors have built a photometric evolution model for the bulge-disc system to study star formation history in spiral galaxies. Through a confrontation of the model with observational data, they have shown that most photometric properties of spiral galaxies can be explained essentially by a single parameter, the bulge-to-disc light ratio, L_B/L_D. In particular, (1) there is a tightly correlated $(B–H)_{-0.5}$–L_B/L_D relationship that plays a fundamental role in establishing the $(B–H)_{-0.5}$ vs. $H_{-0.5}^{abs}$ colour-magnitude (CM) relation; (2) the L_B/L_D ratio is the single dominant parameter which leads to the observed trends of optical and optical-infrared colours with morphological type; the scatters in these colours at one type are due to the intrinsic scatter in the L_B/L_D ratio; (3) bulges and discs have almost the same *JHK* colours, and hence the near infrared colours do not depend significantly on the L_B/L_D ratio, thus showing no clear trend along the Hubble sequence; (4) the observed gradients in *UBV* colours along the discs of spiral galaxies can be fully explained by the local bulge-to-disc light ratio, $L_B(r)/L_D(r)$. No radial metallicity gradients across the bulge or the disc are required to explain the observed colour gradients in *UBV*.

157.104 **Erratum: "The nuclear starburst activity in M83: imaging at near–IR wavelengths" [Astron. Astrophys., Vol. 243, No. 2, p. 309 – 318 (Mar 1991)].**
P. Gallais, D. Rouan, F. Lacombe, D. Tiphène, I. Vauglin.
Astron. Astrophys., Vol. 249, No. 2, p. 594 (Sep 1991). See Abstr. 53.157.205.

157.105 **The stellar populations of M33.**
S. van den Bergh.
Publ. Astron. Soc. Pac., Vol. 103, No. 665, p. 609 – 622 (Jul 1991).
A review is given of present ideas on the evolution and stellar content of the Triangulum nebula = M33 = NGC 598. A distance modulus of $(m-M)_0 = 24.5 \pm 0.2$ (D = 795 ± 75 kpc) and a Galactic foreground reddening $E_{B-V} = 0.07$, from which $M_V = -18.87$, are adopted throughout this paper. The disk of M33 is embedded in a halo of globular clusters, metal–poor red giants, and RR Lyrae stars. Its nuclear bulge component is weak ($M_V > -14$). The ages of M33 clusters do not appear to exhibit a hiatus in their star–forming history. The integrated light of the semistellar nucleus of M33, which contains the strongest X–ray source in the Local Group, is dominated by a young metal–rich population. At optical wavelengths the disk scale length of M33 is 9′.6. The ratio of the nova rate in M33 to that in M31 is approximately equal to the ratio of their luminosities. The gas–depletion time scale in the central region of M33 is found to be $\sim 1.7 \times 10^9$yr, which is significantly shorter than a Hubble time. Available data do not yet allow an unambiguous choice between the density wave and self–propagating star–formation models for the two main spiral arms of M33.

157.106 **Abundance gradients and galaxy formation.**
M. Stiavelli, F. Matteucci.
Astrophys. J., Lett., Vol. 377, No. 2, p. L79 – L83 (20 Aug 1991).
A three–dimensional collisionless code is used to simulate the formation of elliptical galaxies with a time scale for star formation comparable to the collapse time. A one–zone chemical evolution scheme is used to compute the evolution of the abundance of several chemical elements. The initial "clumpy" system is made of gas and, in some cases, of warm dark matter. In the latter case, abundance gradients are produced. For a galaxy of $10^{11} M_\odot$ of gas and $3 \times 10^{11} M_\odot$ of warm dark matter, the predicted gradients of Fe and Mg are $d[Fe/H]/d\log r = -0.11$ and $d[Mg/H]d\log r = 0.05$, respectively. In the absence of dark matter a greater amount of dissipation is required to produce abundance gradients. The mechanism proposed here should be relevant for bright and extended objects where there is evidence for the presence of dark matter.

157.107 **A statistical analysis of cosmic ray propagation effects on the low frequency radio spectrum of spiral galaxies.**
M. Pohl, R. Schlickeiser, E. Hummel.
Astron. Astrophys., Vol. 250, No. 2, p. 302 – 311 (Oct 1991).
In the last years more and more evidence has emerged for a spectral flattening at low frequencies of the radio continuum emission from normal spiral galaxies. It is the purpose of this paper to calculate theoretical spectra possessing breaks caused by propagation effects of relativistic electrons and to present a statistically significant analysis whether these effects can account for the observed spectra. It is shown that the theoretical spectra change their slope smoothly over a range of at least two decades in frequency. The authors use the χ^2 and the maximum likelihood method to investigate the spectra of ten galaxies for which data over more than two decades of frequency are available. Six of them evidently have spectral breaks consistent with electron propagation models, three allow no definite decision, and one clearly does not show such features. The sample includes galaxies of different types such as normal galaxies, starbursters and tidally interacting ones.

157.108 **A numerical study of star formation in interacting disk galaxies.**
J. C. Mihos, D. O. Richstone, G. D. Bothun.
Astrophys. J., Vol. 377, No. 1, p. 72 – 88 (10 Aug 1991).
The authors present models of star–forming disk galaxies involved in galaxy – galaxy interactions. The models employ a restricted three–body approach to the gravitational forces, a discrete cloud model of the ISM, and a modified Schmidt law for star formation. The authors use the models to examine the star–forming properties of a wide range of encounters with differing galaxy orbits and disk inclinations. The majority of the encounters lead to galaxies which exhibit normal star–forming properties, and the triggering of very active starburst galaxies is rare. Additionally, close encounters can lead to suppressed star formation, because of the rapid loss of gas in the inner disk. However, the models are unable to account for the most active star–forming galaxies, with order–of–magnitude increases in their global SFR. This inability to produce such a large increase in star formation suggests that IR luminous galaxies are either the result of galaxy mergers which cannot be modeled by the present scheme, star formation that is governed by a different star–forming law and/or IMF, or excess luminosity caused by the triggering of an AGN. It is shown that while mass transfer can lead to centrally concentrated star formation under certain circumstances, it is not the dominant mechanism for such bursts.

157.109 **Wolf–Rayet galaxies: an introduction and a catalog.**
P. S. Conti.
Astrophys. J., Vol. 377, No. 1, p. 115 – 125 (10 Aug 1991).
The first catalog of Wolf–Rayet galaxies is presented and discussed. These are a subset of emission–line galaxies in which a broad 4686 Å He II emission feature due to Wolf–Rayet (W–R) stars is observed in the integrated spectrum. About one–quarter of these galaxies also show a broad line at 4640 Å due to N III which is prominent in the latest type WN stars. The number of Lyman continuum photons emitted by W–R galaxies scales roughly with their integrated magnitudes, similar to emission–line galaxies. Both W–R galaxies and emission–line galaxies form an overlap with an luminous extension of giant H II regions in these properties. The fraction of W–R stars to O stars in some of these galaxies appears to be large, which if confirmed by more detailed quantitative work would imply that the initial mass function was relatively "flat". There is also another small subset of emission–line galaxies in which a narow 4686 Å He II emission feature is found – this is probably a recombination nebular line whose excitation mechanism is uncertain.

157.110 **UBVR photometry and rotation of late–type galaxies: NGC 959, 1156, 1160, 6643, 7292.**
V. F. Esipov, G. A. Kyazumov, A. R. Dzhafarov.
Astron. Zh., Tom 68, Vyp. 5, p. 909 – 917 (Sep–Oct 1991). In Russian. English translation in Sov. Astron., Vol. 35, No. 5.
Brightness and colour distribution profiles have been obtained from observations with the magnetic–focusing image tube for 5 late–type galaxies. Rotation curves and density distribution models have been derived as well. Comparison of photometric and kinematic data shows that the stellar–population content changes slowly with the galactocentric distance. Density depression in the central parts of the disc is needed for explaining the rotation curves of three galaxies out of the five studied.

157.111 **Observational properties of disks.**
K. C. Freeman.
Conference on Dynamics of Disc Galaxies, p. 15 – 28 (1991). – See Abstr. 012.028 for the main entry.
Our knowledge of the structure and kinematics of the disks of spiral galaxies comes partly from our Galaxy and partly from the observed properties of other spiral galaxies. The author discusses some of the structural and kinematical properties of the thin and thick disks of spiral galaxies, and some important results about the dynamical evolution of the nearby stars.

157.112 **The deprojection of face–on disk galaxies.**
M. R. Merrifield.
Conference on Dynamics of Disc Galaxies, p. 37 – 41 (1991). – See Abstr. 012.028 for the main entry.
This paper explores an alternative approach to the deprojection problem that avoids the difficulties caused by extinction by using detailed kinematic data from a face–on galaxy to obtain the distribution of matter perpendicular to the plane of such a

system. When combined with photometric data, this method allows the full three-dimensional density distribution of a disk galaxy to be recovered.

157.113 **Dynamical properties of the Integral Sign galaxy.**
I. Márquez, A. del Olmo.
Conference on Dynamics of Disc Galaxies, p. 43 – 47 (1991). – See Abstr. 012.028 for the main entry.
The galaxy A 0705 + 7155, called the Integral Sign because of its singular aspect, was spectroscopically examined. Physical conditions of the ISM, rotation curve, mass, morphological type, warps, and the distance of A 0705 + 7155 are discussed.

157.114 **H I in disk galaxies.**
A. Bosma.
Conference on Dynamics of Disc Galaxies, p. 49 – 60 (1991). – See Abstr. 012.028 for the main entry.
The author reviews recent H I studies of disk galaxies with emphasis on the dynamical problems which they address. He reports on ongoing studies of warped and flaring H I disks: the frequency of warps seems to be diminished in early type spiral galaxies, while the flaring of the H I layer in NGC 100 indicates, in first approximation, a not too flattened dark halo. Finally, he comments on recent work on the presence of holes in the detailed H I distribution of several galaxies.

157.115 **Galaxy kinematics from ISIS spectra.**
J. Castles, C. D. McKeith, A. Greve.
Vistas Astron., Vol. 34, Part 3, p. 187 – 200 (1991). – See Abstr. 012.032 for the main entry.
The ISIS spectrograph on the 4.2 m WHT was used to observe three galaxies NGC 2146, NGC 1569, and NGC 3034 (M82). Rotation curves along the major axes of these galaxies are presented. For NGC 3034, the rotation curve derived using the relatively new technique for measuring stellar motion from one of the Ca II IR triplet (~8550 Å) absorption lines is compared with those for the ionized gas motion measured from H_α and [N II] (6583 Å) emission lines. The line splitting with slit oriented along the minor axis of NGC 3034 covering 4 arc min is examined.

157.116 **H I kinematics in NGC 4736.**
P. S. Mulder, W. Van Driel.
Conference on Dynamics of Disc Galaxies, p. 61 – 63 (1991). – See Abstr. 012.028 for the main entry.
Recent re-observation of the H I in the ringed Sab-type galaxy NGC 4736 with the Westerbork Synthesis Radio Telescope and interpretation of these data with modern software for the analysis of velocity fields has yielded a better insight into the kinematics of the H I gas.

157.117 **What about missing light?**
J. I. Davies.
Conference on Dynamics of Disc Galaxies, p. 65 – 69 (1991). – See Abstr. 012.028 for the main entry.
The author looks at a simple model to illustrate the effect optically thick discs may have on the interpretation of rotation curve data.

157.118 **Colors of IRAS galaxies.**
Wang Gang.
Acta Astrophys. Sin., Vol. 11, No. 4, p. 311 – 320 (Oct 1991). In Chinese. English translation in Chin. Astron. Astrophys., Vol. 16, No. 1, p. 8 – 16 (Jan – Mar 1992).
Based on a sample of 369 galaxies which are detected by IRAS in all four bands, the colors of IRAS galaxies are discussed. By their colors, they could be classified into four groups, i.e. normal, starburst, AGN and normal + peculiar. The colors at short wavebands are sensitive for the classification. A simple model of two black body components could be used to fit the IRAS emission of galaxies well.

157.119 **Rotation of the inner parts of spiral galaxies. NGC 497, 895, 972 and 3646.**
V. L. Afanas'ev, A. N. Burenkov, A. V. Zasov, O. K. Sil'chenko.
Astron. Zh., Tom 68, Vyp. 6, p. 1134 – 1149 (Nov–Dec 1991). In Russian. English translation in Sov. Astron., Vol. 35, No. 6.
The radial distributions of line-of-sight velocity of ionized gas at different position angles, and rotation curves are presented for several normal spiral galaxies. Observations were carried out by the long-slit spectrograph with a digital TV-detector at the prime focus of the 6-m telescope. In this paper, NGC 497, 895, 972 and 3646 are considered.

157.120 **Molecular gas in the central regions of barred galaxies.**
J. D. P. Kenney.
Conference on Dynamics of Disc Galaxies, p. 171 – 174 (1991). – See Abstr. 012.028 for the main entry.
The strongest CO emission in the central regions of many barred galaxies arises from twin peaks oriented perpendicular to the large scale stellar bar, and located where dust lanes intersect a ring at the inner Lindblad resonance. In the nuclear starburst galaxy NGC 3504, a large concentration of molecular gas is centered on the nucleus, inside the inner Lindblad resonance. The starburst is confined to the solid body part of the rotation curve by the onset of strong tidal forces where the rotation curve turns over.

157.121 **Deep CCD photometry of spiral galaxies.**
A. H. Broeils, J. H. Knapen.
Astron. Astrophys., Suppl. Ser., Vol. 91, No. 3, p. 469 – 479 (Dec 1991).
Deep optical surface photometry in the R-band is presented for the three spiral galaxies NGC 1003, NGC 5533 and NGC 6674. The luminosity profiles, derived by fitting ellipses to the isophotes, have been decomposed into bulge and disk contributions using the method proposed by Kent (1986).

157.122 **Distances of the nearby galaxies NGC 1560, NGC 2976 and DDO 165 from their brightest stars.**
I. D. Karachentsev, N. A. Tikhonov, Ts. B. Georgiev, B. I. Bilkina, M. E. Sharina.
Astron. Astrophys., Suppl. Ser., Vol. 91, No. 3, p. 503 – 512 (Dec 1991).
For three galaxies in the M81/IC 342 neighbourhood, photographic BV-photometry of the bright stellar population has been made. Using the three brightest red and three brightest blue supergiants the following distance moduli are obtained: 27.86 and 28.03 for NGC 1560, 28.36 and 28.18 for NGC 2976, and 28.43 and 28.47 for DDO 165, respectively. All the galaxies are associated with the M81 + IC 342 group.

157.123 **The velocity dispersion of molecular cloud ensembles and the formation of GMC's in spiral galaxies.**
G. Rydbeck, P. Bergman, Å. Hjalmarson, M. Thomasson, T. Wiklind.
Conference on Dynamics of Disc Galaxies, p. 235 – 239 (1991). – See Abstr. 012.028 for the main entry.
The authors have in particular investigated the difference between arm and interarm velocity dispersion of the molecular clouds. They found that in all four spiral galaxies M 51, M 83, IC 342 and NGC 2997 the molecular cloud on-arm velocity dispersion is generally considerably smaller than the off-arm dispersion.

157.124 **Dynamics and molecular clouds in the merger NGC 3256.**
S. Aalto, J. H. Black, R. S. Booth, L. E. B. Johansson, R. A. Schommer.
Conference on Dynamics of Disc Galaxies, p. 241 – 243 (1991). – See Abstr. 012.028 for the main entry.
This is a preliminary report of a study of the kinematics of both the ionised and neutral, molecular gas in NGC 3256.

157.125 **The effect of interactions on the molecular medium in galaxies.**
S. Aalto, J. H. Black, L. E. B. Johansson, R. S. Booth.
Conference on Dynamics of Disc Galaxies, p. 245 – 247 (1991). – See Abstr. 012.028 for the main entry.

In order to get a handle on how the large-scale cloud properties might vary with degree of interaction the authors have studied the $^{12}CO/^{13}CO$ $J = 1\text{–}0$ intensity ratio in galaxies in various degrees of interaction.

157.126 **Archimedes spiral and the surface distribution of dark clouds in M31. III.**
I. A. Issa.
Astrophys. Space Sci., Vol. 184, No. 2, p. 227 – 234 (Oct 1991).

The surface distribution of dark clouds along the major and minor axes were compiled to give a general formula expressing the surface distribution along both axes simultaneously. Absorption values were determined, assuming the parameters of the standard cloud model as a function of the distance from the two-dimensional distribution of dark clouds. Iso-cloud line numbers were drawn as a function of both X and Y. Two knots appeared in the new distribution function. The knots may indicate the least possible number of the dark clouds per kiloparsec beyond which spiral structure may be possible. The projected distance of the two knots on the X-axis may correspond to some spiral features found by other investigators.

157.127 **Molecular gas in the Circinus galaxy.**
L. E. B. Johansson, S. Aalto, R. S. Booth, G. Rydbeck.
Conference on Dynamics of Disc Galaxies, p. 249 – 250 (1991). – See Abstr. 012.028 for the main entry.

157.128 **CO observations of galaxies in the Grus quartet: NGC 7552 and NGC 7582.**
R. Sahai, M. J. Claussen.
Conference on Dynamics of Disc Galaxies, p. 251 – 256 (1991). – See Abstr. 012.028 for the main entry.

In order to understand the distributed molecular component of barred spiral galaxies, the authors have detected and mapped NGC 7582 and NGC 7552 in ^{12}CO $J = 1\text{–}0$ line emission at 45 arcsec resolution using the SEST.

157.129 **Triggered star formation along the spiral arms of M51.**
J. H. Knapen, J. E. Beckman, J. Cepa, T. van der Hulst, R. J. Rand, N. Nakai.
Conference on Dynamics of Disc Galaxies, p. 337 – 340 (1991). – See Abstr. 012.028 for the main entry.

The authors have combined recently published data on the distribution of ionized, molecular and atomic hydrogen in the disk of M51. Studying especially the distributions along the two main arms of the galaxy, a symmetric pattern of peaks and dips in the Hα intensity along the arms is found. The authors conclude that the observed global pattern is due to enhanced star formation at specific radii in the disk. These coincide with resonances of a density wave system suggesting that the resonance pattern locally affects the process of star formation.

157.130 **Narrow-band photometry and the evolution of galaxies. III. The intermediate – redshift sample.**
K. D. Rakos, J. M. Schombert, T. J. Kreidl.
Astrophys. J., Vol. 377, No. 2, p. 382 – 391 (20 Aug 1991).

A near-UV and blue rest frame narrow-band filter system is used to investigate the spectrophotometric evolution of elliptical galaxies. Data are presented for mean colors of ellipticals in six distant clusters ($z = 0.37 - 0.46$) for comparison with the present-day ellipticals and models of color evolution. The authors find good agreement with the UV-cold models of Guiderdoni and Rocca-Volmerange, particularly with the confirmation of a strong post-MS light contribution at $z = 0.4$. The fraction of starforming to nonstarforming objects (Butcher-Oemler effect) is calculated from rest frame colors. Mean values of $f_B = (18 \pm 3)\%$ at $z = 0.2$ and $(35 \pm 4)\%$ at $z = 0.4$ are found. Unlike near-blue colors, the 4000 Å break colors show little evolution, although this may be due to competing metallicity effects in post-MS light.

157.131 **Infrared emission regions in the interacting galaxy system Arp 299.**
C. G. Wynn-Williams, S. A. Eales, E. E. Becklin, K.-W. Hodapp, R. D. Joseph, I. S. McLean, D. A. Simons, G. S. Wright.
Astrophys. J., Vol. 377, No. 2, p. 426 – 433 (20 Aug 1991).

The authors present infrared imagery and photometry of the interacting galaxy system Arp 299 (NGC 3690/IC 694). In the central region of NGC 3690 they see two 2.2 μm sources; one of these (component B2) coincides with the apparent visible nucleus, while the other (component B1) coincides with the 6 cm and 10 μm sources. The authors see no evidence at 2.2 μm for a compact AGN in IC 694 (component A). Approximately 75% of the 12 – 25 μm power from Arp 299 comes from three regions each smaller than 600 pc in diameter centered on components A, B1, and C, making the distribution of mid-infrared emission more concentrated than the distribution of starlight. At present, it is not possible to decide whether or not source C marks the position of a further galaxy nucleus. It is more luminous than any known nonnuclear astronomical object and has physical conditions even more extreme than those in well established starburst nuclei. The main arguments against its being a nucleus are based on the implausibility of multiple simultaneous mergers.

157.132 **The molecular structure of a starburst: ^{13}CO in the nucleus of Maffei 2.**
R. L. Hurt, J. L. Turner.
Astrophys. J., Vol. 377, No. 2, p. 434 – 441 (20 Aug 1991).

The authors present high-resolution ($\sim 6''$) maps of the ^{13}CO emission in the nucleus of the nearby spiral galaxy Maffei 2 made with the Owens Valley Millimeter Interferometer. The optically thin ^{13}CO emission, a tracer of H_2 gas density, indicates that the molecular cloud distribution in the nucleus of Maffei 2 is remarkably well-correlated with the starburst as traced by radio continuum emission. Evidently the molecular density distribution determines the spatial structure of this starburst. The observed range of ^{13}CO emission implies H_2 column densities of $N_{H2} = 0.13 - 1.3 \times 10^{32} cm^{-2}$, and visual extinctions of $A_V = 14 - 140$ mag on size scales of ~ 150 pc. The molecular material forms a barlike structure 40″ (1 kpc at an assumed distance of 5 Mpc) in extent, with a total H_2 mass of $2.3 \times 10^8 M_\odot$. Two dynamical components are evident in the ^{13}CO maps: a region within ~ 120 pc of the nucleus that appears to rotate rapidly as a solid body; then a nearly Keplerian fall-off to a less rapidly rotating region extending to a radius of ~ 500 pc. The most intense star formation activity appears in the inner solid body region of the rotation curve. Within the solid body region, gas accounts for $\sim 40\%$ of the estimated dynamical mass.

157.133 **Analysis of the distribution of H II regions in external galaxies. A case study: M31.**
C. García-Gómez, L. Athanassoula.
Conference on Dynamics of Disc Galaxies, p. 365 – 368 (1991). – See Abstr. 012.028 for the main entry.

The authors studied the distribution of H II regions in spiral galaxies with the aim (a) the study of the deprojection of the galaxies using methods based only on the H II regions themselves, (b) the study of the global distribution of H II regions, (c) the study of the clustering of the H II regions, and (d) the analysis of the spiral structure. As an example they discuss here M31.

157.134 **On the evolution of disk galaxies.**
C. Firmani, A. Miranda, A. Tutukov.
Publ. Astron. Soc. Pac., Vol. 103, No. 666, p. 898 – 901 (Aug 1991). – See Abstr. 012.034 for the main entry.

A grid of models concerning the physics and the evolution of irregular galaxies is presented. A numerical code that simulates the gravitational contraction of a galaxy since the protogalactic phase has been developed. The models are based on a balance between the turbulent energy dissipation of the gas and the energy input from supernovae, the star formation being regulated by the gas conditions through Schmidt's law. The grid includes closed and open models. In the open models the gas accretion is produced from the intergalactic gas or from a surrounding gaseous halo that supplies the gas and absorbs a fraction of the yield, diluting the metallicity of the galaxy. Tests with the observed present-day gas contents and metallicities of the galaxies support the idea of the open models and particularly of the surrounding gaseous halo.

157.135 **The H II regions of the spiral galaxy M101.**
P. A. Scowen.
Publ. Astron. Soc. Pac., Vol. 103, No. 666, p. 902 - 905 (Aug 1991). - See Abstr. 012.034 for the main entry.

Employment of a new identification and integration algorithm to CCD images of extragalactic H II regions has yielded new results for radial changes in spectrophotometric parameters. The author identifies a change in the radial oxygen abundance, with an apparent discontinuity at a radius of 5'.5, as well as presenting evidence for an increase in the reddening of each region toward the center of the galaxy. A high-resolution overlay of the best radio data onto the optical imagery is included.

157.136 **Fourier analysis of structures in spiral galaxies and the link to dynamical parameters.**
I. Puerari, H. A. Dottori.
Conference on Dynamics of Disc Galaxies, p. 369 - 372 (1991). - See Abstr. 012.028 for the main entry.

The distribution of H II regions in 11 spiral galaxies is analysed through 2-D Fourier transform.

157.137 **Photoionization of high-latitude gas in NGC 891.**
J. Sokolowski, J. Bland-Hawthorn.
Publ. Astron. Soc. Pac., Vol. 103, No. 666, p. 911 - 915 (Aug 1991). - See Abstr. 012.034 for the main entry.

The authors have modeled the photoionization of diffuse, high-latitude gas in NGC 891 by the UV radiation from various in situ stellar populations. They find that although OB stars dominate the production of stellar UV photons, models using only OB stars cannot reproduce the observed [N II] $\lambda 6584/H\alpha$ ratios seen throughout this diffuse ionized medium (DIM). Models that combine OB stars with the high-temperature end of the white-dwarf distribution and the central stars of planetary nebulae are capable of reproducing the observed [N II] $\lambda 6584/H\alpha$ ratios. Estimation of ionizing photon fluxes, using free-free and infrared emission, combined with these models requires the filling factor throughout the DIM in NGC 891 to be of order 10^{-2} and that white dwarfs and planetary nebula central stars contribute $\cong 25\%$ of the ionizing photons at $|z| = 250$ pc. The model results indicate that the filling factor of the DIM in the Milky Way is approximately 10^{-1}, in agreement with previous estimates.

157.138 **Relative abundances in the low-metallicity dwarf irregular galaxy UGC 4483.**
E. D. Skillman.
Publ. Astron. Soc. Pac., Vol. 103, No. 666, p. 919 - 922 (Aug 1991). - See Abstr. 012.034 for the main entry.

UGC 4483 is a dwarf irregular galaxy in the M81 group. Narrow-band optical imaging has revealed an H II region in UGC 4483 with an $H\alpha$ flux of $\approx 2 \times 10^{-13}$ erg cm^{-2}Å^{-1}s^{-1}. Optical, ultraviolet, and near-infrared spectroscopy of this H II region yield He, C, N, O, Ne, and S abundances for the ISM in this galaxy. With an oxygen abundance of 2.1×10^{-5} ($12 + \log(O/H) = 7.3$), this galaxy is among the most metal-poor dwarf irregulars known to date. A comparison of the S/O abundance ratio in this galaxy with that in I Zw 18 supports the claim by Garnett 1990 that the S/O ratio is consistent with the solar ratio for low-metallicity dwarf irregulars. The C/O ratio in UGC 4483 is lower than that derived for I Zw 18 and, therefore, more in line with the trend seen in higher metallicity H II regions. The derived helium abundance is He/H = 0.075, which converts to a helium mass fraction of 0.23, consistent with earlier determinations of the primordial helium abundance.

157.139 **CO in mergers. III. NGC 1614 and NGC 3256.**
F. Casoli, C. Dupraz, F. Combes, I. Kazès.
Astron. Astrophys., Vol. 251, No. 1, p. 1 - 10 (Nov 1991).

The authors present $^{12}CO(1-0)$ and $^{12}CO(2-1)$ maps of two southern merger remnants, NGC 1614 and NGC 3256, obtained with the SEST 15 m telescope. The molecular masses of these galaxies are $6.5 \times 10^{9} M_{\odot}$ and $1.3 \times 10^{10} M_{\odot}$ respectively, with a standard conversion factor from CO emissivities to H_2 column densities of 2.3×10^{20}cm^{-2}. It is suggested that in mergers the bulk of the molecular material experiences physical conditions not drastically different from those found in spirals. The authors combine these observations with their results on the ongoing mergers Arp 220, NGC 2623 and NGC 6240, and on the merger remnant NGC 7252. They find that the far-infrared colour temperature $T_{60/100}$ and the ratios L_{FIF}/L_B and $L_{FIR}/M(H_2)$ evolve in the course of the merging in a correlated fashion. These indicators of the star-formation activity increase in the early stages of the collision between the parent spirals, reach a climax and decrease to attain the low values that are characteristic of elliptical galaxies.

157.140 **Dynamics, interstellar medium, and stellar activity interplay in galaxies.**
D. Pfenniger.
Conference on Dynamics of Disc Galaxies, p. 389 - 393 (1991). - See Abstr. 012.028 for the main entry.

Contents: 1. Stars and galaxies as self-gravitating systems. 2. Energetics and powers. 3. Stellar activity, dynamics, and Tully-Fisher. 4. Conclusions and remarks about the Hubble sequence.

157.141 **Evidence of recurrent star bursts in M82.**
Huang Jiehao, Su Hongjun, Deng Zugan,
G. H. Jacoby.
3. Chinese Academy of Sciences and Max-Planck Society Workshop: High energy astrophysics: compact stars and active galaxies, p. 168 - 169 (1991). Abstract. Abstract. - See Abstr. 012.026 for the main entry.

157.142 **Gaseous discs, stability and the cosmic distance scale.**
L. Staveley-Smith, W. M. Walsh, J. Bland-Hawthorn,
D. J. Axon, R. D. Davies, J. Smoker.
Conference on Dynamics of Disc Galaxies, p. 395 - 399 (1991). - See Abstr. 012.028 for the main entry.

The properties of two unusual low-luminosity gas-rich galaxies are described. Neither appears to require the presence of dark matter to explain the observed rotation curves, despite large H I-to-optical radius ratios, and both appear to be stable disc-like systems. These galaxies indicate, first, that the presence of dark matter is not a universal requirement for the kinematics of low-luminosity galaxies and, secondly, that the formation of a gaseous proto-disc may be a recognisable stage in galactic evolution.

157.143 **The Schmidt star formation law in the Andromeda galaxy.**
P. Tenjes, U. Haud.
Astron. Astrophys., Vol. 251, No. 1, p. 11 - 14 (Nov 1991).

The averaged radial surface density distribution of a young stellar component in M31 is derived, taking into account the distribution of stellar associations, open clusters and young field stars. This distribution is compared with the distribution of neutral ($H I + H_2$) hydrogen. The density of the young stellar component is found to be proportional to the gas surface density in the power of 1.30 ± 0.22, or to the space density in the power of 1.17 ± 0.25.

157.144 **Magnetic fields and interstellar gas clouds in the spiral galaxy NGC 6946.**
R. Beck.
Astron. Astrophys., Vol. 251, No. 1, p. 15 - 26 (Nov 1991).

The radio continuum emission from the spiral galaxy NGC 6946 has been observed with the VLA at 18.0 and 20.5 cm wavelengths in total power and in linear polarization with

42 arcsec (≅1.5 kpc) resolution. While the distribution of total emission is smooth, the polarized emission is lopsided due to asymmetric Faraday depolarization effects which are determined with help of Effelsberg observations at λ6.3 cm. The average Faraday dispersion is consistent with a disk of thermal electrons of 500 pc thickness and 0.03 cm^{-3} density including a random magnetic field of 12 μG strength. Corrected for Faraday depolarization, 10±3% of the total emission is polarized on average. The degree of polarization is below 1% near the center of the galaxy and exceeds 10% beyond 10 kpc radial distance. Faraday depolarization cannot account for this effect. The orientations of the uniform magnetic field lines in the upper disk closely follow the optical spiral arms. The pitch angles against the azimuthal direction vary with azimuthal angle as well as with radius. The degree of field uniformity is clearly anticorrelated with Hα, H I and CO line intensities. A model of the interaction between magnetic fields and gas clouds is presented. The south–western quadrant of NGC 6946 shows excess Faraday rotation of both signs and is also largely depolarized. A large-scale field structure perpendicular to the plane, a "galactic coronal hole", extending over ≅12 kpc in radius, can account for these observations.

157.145 **Molecular gas in the RSAB galaxy NGC 4736.**
M. Gerin, F. Casoli, F. Combes.
Astron. Astrophys., Vol. 251, No. 1, p. 32 – 42 (Nov 1991).

The authors have mapped the central region of the nearby spiral galaxy NGC 4736 with high spatial resolution (HPBW = 23″ and 14″) in the two CO lines J = 1–0 and J = 2–1. They found a total mass of molecular gas in the map of $5.1 \times 10^8 M_\odot$ for an assumed distance of 6.3 Mpc; this represents only about 4% of the dynamical mass in the mapped region (r < 80″). Molecular gas is found in the nucleus ($\approx 10^8 M_\odot$ in the central 30″), in a ring feature associated with the ring of H II regions lying at 50″ from the center and in an extended disc. The excitation of the gas is everywhere subthermal, with CO (J = 2–1) brightness temperature approaching the CO (J = 1–0) temperature only at particular positions. The neutral gas is mainly molecular in the central 2 arcmin with a gas fraction in H_2 greater than 0.7. Previous investigators have suggested that the galaxy potential is slightly barred, and that the inner ring of H II regions is associated with an ILR. The authors followed the formation of gas rings in a model galaxy represented by a massive oval with b/a = 0.8. Both an inner and an outer ring form in this potential suggesting that the main body of NGC 4736 is indeed oval.

157.146 **PB 6378 and PB 9261: two new systems of interacting galaxies with starbursts.**
C. Vanderriest, H. Reboul.
Astron. Astrophys., Vol. 251, No. 1, p. 43 – 48 (Nov 1991).

Extracted from the Berger–Fringant catalogue of faint blue objects, PB 6378 and PB 9261 are two systems of interacting galaxies with prominent starbursts. Photometry of these systems was done by electronography; bidimensional spectroscopic data with spatial resolution ≤1″ were obtained by using a fibre optics spectrograph. Spectra integrated over the different components are then obtained, as well as velocity fields and intensity maps in the continuum and in the emission lines. This allows to estimate the physical conditions in these starburst systems. The authors discuss the interest in extending the method to similar but more distant objects.

157.147 **Spectroscopy of the symbiotic carbon star C–1 in the Draco dwarf galaxy.**
U. Munari.
Astron. Astrophys., Vol. 251, No. 1, p. 103 – 105 (Nov 1991).

The author presents the results of a spectroscopic monitoring of the object C–1 in the Draco dwarf galaxy during 1987–90. It is shown that C–1 is not a planetary nebula in the process of formation but instead a symbiotic binary, composed of a carbon giant, of unusual blue colors, and of a very hot compact companion, possibly an accreting white dwarf. The nebular regions are characterized by high electron densities and the emission lines appear to form in the carbon star atmosphere energized by the white dwarf's hard radiation. Large variations in the profile and radial velocity of the emission lines are detected, which relate to orbital motion. The orbital period seems to be relatively short. Variations have been observed in the integrated flux of emission lines, as expected for changing orbital phases.

157.148 **The Cepheid distance scale.**
B. F. Madore, W. L. Freedman.
Publ. Astron. Soc. Pac., Vol. 103, No. 667, p. 933 – 957 (Sep 1991).

Developments concerning of the period–luminosity (PL) relation for classical Cepheids are discussed with particular emphasis on the fact that Cepheids continue to provide the most homogeneous and reliable set of distances to many nearby galaxies. Observational programs relating to extragalactic Cepheids are also reviewed, and new distances and reddenings to individual galaxies are determined from these data using multiwavelength PL relations, based for the first time on self-consistent calibrating data sets. Three recent promising secondary extragalactic distance indicators (Tully–Fisher, planetary nebula luminosity functions, and surface–brightness fluctuations) tie their zero points to the Cepheid distance scale and give globally consistent values for the Hubble constant of about 80 km $sec^{-1}Mpc^{-1}$. Tying the Faber–Jackson relation and the supernova distance scales to the Leo and Virgo clusters yields similar results. Such a value for the Hubble constant is marginally inconsistent with new estimates of the ages of globular clusters, and it may prove to be a serious problem for standard cold dark–matter models of the universe.

157.149 **Origins of the 12 – 200 micron flux in NGC 6946: starlight and continuum dust emission from an Sc galaxy.**
G. Engargiola.
Astrophys. J., Suppl. Ser., Vol. 76, No. 3, p. 875 – 910 (Jul 1991).

The author has mapped NGC 6946 in continuum dust emission at 100, 160, and 200 μm, in starlight in the g band and the r band, and in Hα + [N II] from ionized gas. Maps of the galaxy at 12, 25, and 60 μm, made with 1′ resolution from IRAS data are also presented. The continuum dust emission is studied in relation to spiral and radial structure, to the distribution of the interstellar medium, and to ionizing and nonionizing radiation from stars. This paper attempts to clarify the sources of dust heating, the optical and spectral characteristics of the nuclear source and disk, the properties of dust grains, and the origins of infrared luminosity in NGC 6946.

157.150 **Spectropolarimetry of the Hα emission in M82.**
N. Foley, M. Scarrott, D. Axon.
Gemini, No. 34, p. 8 – 9 (Dec 1991).

157.151 **Ultra–deep INT CCD imaging of the faintest galaxies.**
N. Metcalfe, T. Shanks, D. Fong.
Gemini, No. 34, p. 12 – 13 (Dec 1991).

157.152 **Line analysis and nebular abundances in H II galaxies.**
M. Peña, M. T. Ruiz, J. Maza.
Astron. Astrophys., Vol. 251, No. 2, p. 417 – 430 (Nov 1991).

The authors present spectrophotometry of 42 southern H II galaxies (H II Gs) recently discovered in the Calán–Tololo Spectroscopic Survey. Physical conditions and He, N, Ne and O abundances of the ionized regions of these galaxies were derived. For a sample that includes the objects in this work and those reported by Campbell et al. (1986), the authors discuss the relation between some oxygen emission line intensity ratios and the physical conditions of the ionized region by comparing the observational data with the predictions of a grid of ionization structure models. An attempt is made to study the behavior of the effective temperature of the ionizing cluster and the ionization parameter for different metallicities. It is found that the mean T_{eff} of the high O abundance H II Gs is lower than that in low abundance objects. The authors have estimated the ionization parameter for the regions in the sample and they derive a mean value of 0.005. From the H II Gs showing the highest ionization degree it is deduced that, for a given O abundance, there is an upper limit for the ionization parameter. This upper limit increases smoothly with decreasing metallicity.

157.153 **A study of southern extreme IRAS galaxies. III. CO (J = 1–0) line observations.**
W. van Driel, A. C. van den Broek.
Astron. Astrophys., Vol. 251, No. 2, p. 431 – 441 (Nov 1991).
As part of a multi-wavelength study of an infrared–complete sample of some 50 extreme IRAS galaxies with a high far-infrared/blue luminosity ratio CO (J = 1–0) line observations were made of 19 of these galaxies with the SEST, and detections of 5 galaxies from the sample were taken from the literature. The total of 21 galaxies observed includes a distance–limited ($V < 6000$ km s^{-1}) "statistical CO subsample" of 16 galaxies, 10 of which were detected, as well as 5 additional objects, all of which were detected. The 10 detected galaxies from the statistical CO subsample, which have L_{FIR} ranging from 7×10^9 to $7\times10^{10} L_{\odot}$ (H_0 = 75 km s^{-1}Mpc^{-1}), have estimated total H_2 masses ranging from 0.5 to $7.5\times10^9 M_{\odot}$, and L_{FIR}/M_{H2} ratios between 6 and 30 $L_{\odot}M_{\odot}{}^{-1}$. This ratio, which is considered to be a measure of the "star formation efficiency" is on average some 4 times higher in these galaxies than in spirals, and comparable to the ratios found for other samples of IRAS galaxies with moderate luminosities, and for "classical" starburst galaxies such as M82. Of the three distinct subclasses in which the authors can divide their infrared–complete sample (dwarfs, barred spirals, and interacting/merging systems), they tend to find the least infrared–extreme objects in their statistical CO subsample due to the distance limit. Two of the most infrared–luminous objects have OH megamasers.

157.154 **Globular cluster systems in galaxies beyond the Local Group.**
W. E. Harris.
Annu. Rev. Astron. Astrophys., Vol. 29, p. 543 – 579 (1991). – See Abstr. 003.010 for the main entry.
This review emphasizes the globular cluster systems in galaxies beyond the Local Group, their global properties, and their use as tracers for galactic structure and galaxy formation.

157.155 **On the low frequency radio spectrum of spiral galaxies.**
E. Hummel.
Astron. Astrophys., Vol. 251, No. 2, p. 442 – 446 (Nov 1991).
The author reanalyses the low frequency radio continuum data on spiral galaxies published by Israel and Mahoney (1990, IM). He concludes that most spiral galaxies do indeed show a change in spectral index with frequency but that, contrary to IM, the amount of the change is not significantly dependent on the inclination of the galaxies. The latter makes the free–free absorption model proposed by IM to explain the flattening at low frequencies less compelling. A steepening of the spectrum at high frequencies (> 1 GHz) due to propagation effects of the relativistic electrons is at least equally plausible.

157.156 **Molecular gas in galaxies.**
J. S. Young, N. Z. Scoville.
Annu. Rev. Astron. Astrophys., Vol. 29, p. 581 – 625 (1991). – See Abstr. 003.010 for the main entry.
The authors discuss CO as a tracer of H_2 mass in spiral galaxies, including new evidence that global H_2 masses for luminous spirals are generally accurate to $\pm 50\%$ (1σ), including both measurement uncertainties and variations galaxy to galaxy in the CO to H_2 conversion constant (Devereux and Young 1990b). They then review the radial distributions of gas and star formation, spiral structure, molecular properties of galaxies in the Local Group , global gas contents (Section 6), star formation rates and efficiencies (Section 7), effects of galactic interactions, and nuclear gas concentrations.

157.157 **Surface photometry of a sample of elliptical and S0 galaxies.**
R. R. de Carvalho, S. Djorgovski, L. N. da Costa.
Astrophys. J., Suppl. Ser., Vol. 76, No. 4, p. 1067 – 1083 (Aug 1991).
The authors present the results of CCD surface photometry in the B and R bands of 38 early–type galaxies, including 26 ellipticals, three E/S0's, and nine S0's. Most of the objects (23) are in the Fornax Cluster, and the remaining ones (15) belong to various smaller groups and the general field. The surface brightness profiles typically extend out to $\sim 70''$ in radius, or down to a surface brightness of ~ 24.6 mag arcsec^{-2}. Detailed comparisons with previously published work and internal and external error estimates for all quantities are provided. The authors point out some serious systematic discrepancies in the older aperture photometry of some of the galaxies in the present sample.

157.158 **^{13}CO emission in the irregular galaxy NGC 55 and the IR–luminous galaxy NGC 3256.**
R. Becker, W. Freudling.
Astron. Astrophys., Vol. 251, No. 2, p. 454 – 456 (Nov 1991).
The authors report the detection of ^{13}CO emission in the Magellanic–type irregular galaxy NGC 55 and in the IR–luminous galaxy NGC 3256. The ^{12}CO/^{13}CO intensity ratio for NGC 55 is 15 ± 3, which is still consistent with values observed in normal spiral galaxies. The ratio in NGC 3256, however, is 29 ± 6, much higher than typical values observed in spiral galaxies. The high ratio in NGC 3256 might be due to a higher gas kinetic temperature (and hence a lower ^{13}CO opacity) in the CO emitting region, as compared to normal galaxies.

157.159 **IRAS 10214+4724 – a highly polarized galaxy.**
P. Draper, C. Rolph, M. Scarrott, C. Tadhunter, A. Lawrence, R. McMahon.
Gemini, No. 34, p. 14 (Dec 1991).

157.160 **Rotation of spiral galaxy NGC 6070.**
V. F. Esipov, A. A. Strigachev.
Astron. Tsirk., No. 1544, p. 1 – 2 (May – Jul 1990). In Russian.
The rotation curve of the spiral galaxy NGC 6070 is presented. Velocity of the nucleus relative to the centroid of the Local Group is $V_0 = 1870\pm10$ km/s, which gives $D = 25\pm0.1$ Mpc (H_0 = 75 km/s/Mpc). The total mass and mass–to–luminosity ratio are estimated: $M(R_{25}) = (1.62\pm0.2)\times10^{11} M_{\odot}$, $M/L_B = 7.9\pm1$.

157.161 **Circumnuclear star formation in the barred galaxy NGC 1022.**
J. A. García-Barreto, D. Downes, F. Combes, L. Carrasco, M. Gerin, I. Cruz-Gonzalez.
Astron. Astrophys., Vol. 252, No. 1, p. 19 – 26 (Dec 1991).
Observations of ring like structures around the compact nucleus of a bar galaxy are useful for determining whether the structure is at the expected location of a resonance and also whether there is star formation in them. The authors observed the barred spiral galaxy NGC 1022 in the radio continuum at 20, 6, and 2 cm, in CO(1–0) and CO(2–1) and in the near infrared J, H and K bands. No evidence was found for a circumnuclear ring like structure either from the radio continuum or the CO observations. Radio continuum emission is only detected from the central 20″ of the galaxy, in a compact core and a weaker second component to the north. The spectral indices of the integrated fluxes at the three wavelengths are flatter than an average disk spectral index, $\alpha_{disk} \cong -0.75$, namely $\alpha_{20}{}^{6} = -0.52$, $\alpha_{20}{}^{2} = -0.23$ and $\alpha_{6}{}^{2} = -0.03$. From their mm observations the authors estimate a molecular hydrogen mass within a diameter of 3.3 kpc of $1\times10^9 M_{\odot}$. The observed near–IR colors are J–H = 0.82 and H–K = 0.35. From the observations with an aperture of 14″, together with smaller aperture observations by

Devereux et al. the authors estimate a mass within 200 pc of $2 \times 10^9 M_\odot$.

157.162 **High resolution infrared maps of IRAS galaxies.**
S. K. Ghosh, T. N. Rengarajan, R. P. Verma.
IAU Symposium No. 147: Fragmentation of molecular clouds and star formation, p. 411 – 412 (1991). – See Abstr. 012.043 for the main entry.

157.163 **Counter–rotation in the interacting starburst galaxies UGC 9922.**
S. Considère, E. Davoust.
Astron. Astrophys., Vol. 252, No. 1, p. 56 – 62 (Dec 1991).

The authors present a detailed morphological and kinematical study of the close pair of galaxies UGC 9922. The larger galaxy, which resembles a perturbed bar with three knots, shows peculiar internal motions. The ionized gas is in differential rotation, with a maximum velocity of about 120 km s^{-1}, and a very low gradient near the center, possibly due to streaming motions. The stars follow the gas in the innermost regions, rotate in the opposite sense between 0.7 and 2.2 kpc, and rotate in the same sense as the gas, but much faster (maximum rotation velocity: 180 km s^{-1}) in the outer regions. These kinematical peculiarities may be accounted for by a ring of counter–rotating stars, presumably due to a former companion spiralling toward the center of the galaxy in a retrograde orbit, and by an outer warp of the gaseous component, due to the present companion. UGC 9922 bears some resemblance to Mrk 306 and Arp 81, as all three have a bar, a small companion, a strong far–infrared luminosity, and a larger velocity width in neutral than in ionized gaseous motion.

157.164 **An 8″ resolution CO (J = 3–2) map of IC 342.**
R. Mauersberger, A. Schulz, J. W. M. Baars, H. Steppe.
IAU Symposium No. 147: Fragmentation of molecular clouds and star formation, p. 460 – 461 (1991). – See Abstr. 012.043 for the main entry.

IC342 is one of the few galaxies investigated in molecular multi–level studies. The CO (J = 3–2) measurements show the molecular hotspots and possible star–forming regions.

157.165 **Neutral hydrogen observations of NGC 4565 and NGC 891.**
M. P. Rupen.
Astron. J., Vol. 102, No. 1, p. 48 – 106 (Jul 1991). Current Physics Microform No.: 9107B1744.

The H I distributions in the edge–on spiral galaxies NGC 4565 and NGC 891 have been mapped at a resolution of ~4 arcsec with the VLA. The peak brightness temperatures are similar to those in the Milky Way, as are the comparable small– and large–scale structures. Both galaxies show large asymmetries in the distribution of the neutral gas. The warp in NGC 4565 contains roughly one–fifth of the total H I; it is very abrupt, changing by ~45° in position angle within a few hundred parsecs, and may turn back at large radii to parallel the plane of the galaxy. The velocity structure implies that the warp is continuous around the edge of the disk, but is not azimuthally symmetric. Even excluding emission from the warp, there is up to 20% more H I throughout the northern half of the galaxy, with significant asymmetries in the radial profiles as far in as R = 7 kpc. Four or five H I "rings" are detected, possibly associated with spiral structure, with the distance between rings about 2.5 kpc. In NGC 891, the H I extends beyond the main disk on both sides of the galaxy, although the southern extension is far more dramatic. Although interpretation is difficult, the gas here also appears to bend away from, and then bend back towards, the plane of the galaxy; again, the extension covers a fair range in azimuth. Unlike NGC 4565, the asymmetry in the total H I seems completely explained by the asymmetry in the gas in the extension, with the extension containing at least 6% of the total H I mass. Again, there are a number of bright H I "blobs" that may correspond to spiral arms; these features mask any general asymmetries in the H I distribution about the nucleus.

157.166 **Luminous clusters of Wolf–Rayet stars in the SBmIII galaxy NGC 4214.**
W. L. W. Sargent, A. V. Filippenko.
Astron. J., Vol. 102, No. 1, p. 107 – 112 (Jul 1991). Current Physics Microform No.: 9107B1803. With plates 4 – 5.

The authors have discovered strong, broad emission lines attributed to Wolf–Rayet (W–R) stars in spectra of several bright knots in the nearby Magellanic irregular galaxy NGC 4214. The shape of the broad blend at 4660 Å shows that WC stars must be present, in addition to the generally more prevalent WN stars. This is consistent with the relatively high metallicity of NGC 4214 in comparison with most other W–R galaxies. Broad Hα emission, probably produced by late WN stars, is visible in the brightest knot; it serves as a warning that such a feature is not always an indication of Seyfert activity. If the H II regions associated with the knots are ionization bounded, the derived number ratio of W–R stars to O stars at two locations is approximately unity. Either these knots were produced by very short–lived bursts of star formation, or they have unusually flat initial mass functions. Steep spatial gradients in the intensities of the W–R lines are observed, with the strongest W–R emission lines present in knots having the brightest stellar continuum. This correlation might indicate significant spatial variations in the initial mass function, but it is more likely to be an age effect in sudden, localized starbursts.

157.167 **A stellar content survey of NGC 2403 and M81.**
F.–J. Zickgraf, R. M. Humphreys.
Astron. J., Vol. 102, No. 1, p. 113 – 129 (Jul 1991). Current Physics Microform No.: 9107B1809. With plates 6 – 7.

The authors report on a survey of the stellar content of the nearby pair of spiral galaxies NGC 2403 (ScIII) and M81 (SbI–II) from photographic $UBVR_{ck}$ plates. The slopes of the luminosity functions they derived for these two galaxies show no significant differences. For the magnitude range $19.5m \leqslant V < 21m$ the slope of the visual luminosity function is 0.55 ± 0.10 and 0.51 ± 0.08 for NGC 2403 and M81, respectively. The slopes also agree well with previous results for M31 (SbI–II) and M33 (ScIII) obtained using similar color criteria for the selection of blue stars. The authors' new results thus confirm the proposed similarity of the slopes of the luminosity functions among Sb– and Sc–type galaxies. Stellar and nonstellar objects have been separated for a catalog of the brightest stars in the fields of NGC 2403 and M81. The catalogs include positions, magnitudes (V) and colors B–V, U–B, and $V–R_{ck}$. For NGC 2403 the authors found the two brightest blue stars to have M_v = –9.5m and –9.3m. Both are spectroscopically confirmed members. The three brightest blue candidates in M81 appear to have visual luminosities of M_v = –9.8m, –9.5m, and 9.1m. They still must be confirmed spectroscopically as single star members. The visually brightest red supergiants appear to have visual luminosities around M_v = –8m in both galaxies, similar to Local Group galaxies. Based on the results for the luminosity functions and the properties of the brightest stars it appears that the stellar populations in NGC 2403 and M81 are rather similar, unless the brightest blue candidates in M81 are not single star members. In this case, M81 would also show a lack of the brightest blue supergiants compared to Sc–type spirals like the morphologically similar galaxy M31.

157.168 **Phoenix: an intermediate dwarf galaxy in the Local Group.**
F. van de Rydt, S. Demers, W. E. Kunkel.
Astron. J., Vol. 102, No. 1, p. 130 – 136 (Jul 1991). Current Physics Microform No.: 9107B1826.

The authors present CCD BVI photometry of the Local Group dwarf galaxy Phoenix. Phoenix looks like a dwarf spheroidal with a sprinkle of young blue stars. The apparent V and I magnitudes of its red giants are used to estimate its distance at $(m–M_V)_0 = 23.1 \pm 0.1$. Phoenix is definitively a member of the Local Group. Its metal abundance, determined from the I vs (V–I) diagram, is found to be low at [Fe/H] = –2.0. The radius of Phoenix is comparable to the dwarf spheroidal Leo II while its absolute magnitude is similar to that of the Carina dwarf. With

$M_V \approx -9.9$, Phoenix is as faint as the dwarf irregular galaxies LGS3 or SagDIG. Its stellar mass is less than one million solar masses. The few bright blue stars, observed in its CMD, can be explained by a burst of star formation less than 150 million years ago. The presence of a few very red giants suggests that an intermediate age population is also present in Phoenix making its star–formation history more complex than any known dwarf spheroidal galaxy. For these reasons the authors would put Phoenix in a class of galaxies intermediate between irregulars and spheroidals.

157.169 **The discovery of a long H I plume near the peculiar galaxy NGC 2782 (Arp 215).**
B. J. Smith.
Astrophys. J., Vol. 378, No. 1, p. 39 – 46 (1 Sep 1991).

The author has mapped the peculiar starburst galaxy NGC 2782 (Arp 215) in the 21 cm H I line using the VLA and has discovered a massive H I plume extending $\sim 5'$ (~ 54 kpc) toward the northwest, containing $1.4 \times 10^9 M_\odot$ of atomic hydrogen ($\sim 40\%$ of the total H I mass of the system). There is also a shorter H I structure ($\sim 2'$) extending toward the east, which contains $6.6 \times 10^8 M_\odot$ of H I. In the Arp Atlas photograph, NGC 2782 has a disturbed body with bright arcs, possibly ripples, throughout the disk, and a pronounced stellar tail extending 2'.7 toward the east. The eastern H I structure is associated with the beginning of this stellar tail. PDS scanning of the Palomar plates in the vicinity of NGC 2782 reveals a low surface brightness counterpart to the long H I plume in the northwest. The most likely formation scenario for NGC 2782 is a merger, possibly of unequal mass, gas rich galaxies.

157.170 **Compact starbursts in ultraluminous infrared galaxies.**
J. J. Condon, Z.-P. Huang, Q. F. Yin, T. X. Thuan.
Astrophys. J., Vol. 378, No. 1, p. 65 – 76 (1 Sep 1991).

The 40 ultraluminous [$\log (L_{FIR}/L_\odot) \geqslant 11.25$] galaxies in the IRAS Bright Galaxy Sample of sources stronger than 5.24 Jy at $\lambda = 60\ \mu m$ were mapped with $\sim 0''.25$ resolution at 8.44 GHz. Twenty–five contain diffuse radio sources obeying the FIR – radio correlation; these are almost certainly starburst galaxies. Fourteen other galaxies have nearly blackbody FIR spectra with color temperatures $60K \lesssim T_c \lesssim 80K$, so their (unmeasured) FIR angular sizes must exceed $\theta \gtrsim 0''.25$, yet they contain compact (but usually resolved) radio sources smaller than this limit. The unique radio and FIR properties of these galaxies can be modeled by ultraluminous nuclear starbursts so dense that they are optically thick to free–free absorption at $\nu \sim 1.49$ GHz and dust absorption at $\lambda \sim 25\ \mu m$. Only one galaxy (Mrk 231) is dominated by a variable radio source too compact ($\lesssim 1$ pc) to be an ultraluminous starburst; it must be powered by a "monster".

157.171 **Magnetic fields in spiral galaxies.**
R. Beck.
Workshop on the Interpretation of Modern Synthesis Observations of Spiral Galaxies, p. 43 – 56 (1991). – See Abstr. 012.041 for the main entry.

The large–scale structures of the interstellar magnetic fields in spiral galaxies are of axisymmetric–spiral or bisymmetric–spiral type, but also "zero" cases without uniform direction have been found. In all cases the field lines are roughly parallel to the optical spiral arms. The uniform field is strongest in the interarm regions while the unresolved ("turbulent") field is strongest in the spiral arms. This can be understood if the connection between molecular clouds and field turbulence, as observed in a region of M31, is a general phenomenon. In regions of low molecular gas density the disk field may bend into the halo, observed as high Faraday rotation in several face–on galaxies. Exceptionally high star formation may drive a galactic wind pulling field lines radially outwards.

157.172 **Discrete sources in disks of nearby spiral galaxies.**
F. Viallefond.
Workshop on the Interpretation of Modern Synthesis Observations of Spiral Galaxies, p. 79 – 90 (1991). – See Abstr. 012.041 for the main entry.

Aperture synthesis radio continuum observations of nearby spiral galaxies reveal the presence of a number of discrete sources superposed on an extended disk component and on arms of diffuse emission. Most of these sources are H II regions; only a small fraction have been identified with supernova remnants. The properties of the population of these discrete sources in a few nearby galaxies is reviewed. Comparisons with data at other wavelengths are also reviewed.

157.173 **M81: locations of spiral tracers and latest results.**
M. Kaufman, F. N. Bash.
Workshop on the Interpretation of Modern Synthesis Observations of Spiral Galaxies, p. 103 – 112 (1991). – See Abstr. 012.041 for the main entry.

Digitized radio, optical, and ultraviolet images and CO observations are used to describe the spiral arm morphology of M81. All of the spiral tracers are intermingled across a broad spiral arm. The small diffuse H I clouds seem to be the only component to undergo a spiral shock in M81. In several respects, the spiral arms of M81 differ considerably from those of M51 but are similar to the spiral arms of M33. Much of the difference in spiral arm morphology between M81 and M51 results from the greater surface density of interstellar gas in M51.

157.174 **The relation between infrared and radio emission in spiral galaxies.**
G. Helou.
Workshop on the Interpretation of Modern Synthesis Observations of Spiral Galaxies, p. 125 – 133 (1991). – See Abstr. 012.041 for the main entry.

A remarkable correlation between the far infrared and the radio continuum emission of star forming galaxies was one of the early results on IRAS data, and has remained one of the most intriguing. Recent work has extended the correlation to early type galaxies, revealing a slightly different ratio in lenticulars. When radio and infrared maps of disk galaxies are compared, the radio disks appear systematically more diffuse. This has been interpreted as a manifestation of the diffusion of cosmic–ray electrons, and has allowed a fresh look at the behavior of magnetic fields and cosmic rays in spiral galaxies, and at their relation to the rest of the interstellar medium.

157.175 **X–ray and radio continuum correlation in late–type galaxies.**
G. Fabbiano.
Workshop on the Interpretation of Modern Synthesis Observations of Spiral Galaxies, p. 135 – 144 (1991). – See Abstr. 012.041 for the main entry.

X–ray observations give us a direct view of galaxian components that would otherwise remain unknown. These components include the end products of stellar evolution, the hot phase of the interstellar medium, and phenomena related to nuclear activity. Here comparisons of X–ray and radio continuum data are presented, which suggest the possibility of a link between the emission in these two wavebands.

157.176 **Imaging of extragalactic ammonia: hot gas in the nucleus of IC 342.**
P. T. P. Ho.
Workshop on the Interpretation of Modern Synthesis Observations of Spiral Galaxies, p. 145 – 152 (1991). – See Abstr. 012.041 for the main entry.

The authors have successfully imaged extragalactic NH_3 with 5" resolution in the nearby spiral galaxy IC 342. The NH_3 emission appears to be closely associated with the innermost portions of a nuclear bar seen in CO, apparently tracing the hotter and denser material. The NH_3 lines therefore appear to be new spectroscopic tools for VLA studies of the external galaxies.

157.177 **High–resolution observations of H I in and around M51.**
A. H. Rots.
Workshop on the Interpretation of Modern Synthesis Observations of Spiral Galaxies, p. 153 (1991). – See Abstr. 012.041 for the main entry.

157.178 **The peak 21–cm surface brightness of spiral galaxies.**
J. M. Dickey.
Workshop on the Interpretation of Modern Synthesis Observations of Spiral Galaxies, p. 163 – 173 (1991). – See Abstr. 012.041 for the main entry.

High resolution VLA observations of the H I emission from M31 by Braun (1990) show the very surprising result that the peak brightness temperature of the 21–cm line is consistently higher there than anywhere in our Galaxy (175 to 180K vs. 125 to 130K for the Milky Way). The author suggests that such differences may be common among spiral galaxies, particularly those edge–on enough to have high velocity gradients along the line of sight, caused simply by a different mixture of the warm and cool phases of H I. M31 has much less cool H I than the Milky Way, but about the same amount of total atomic hydrogen. Using this one can explain Braun's numbers with a simple radiative transfer model.

157.179 **Pattern speeds from integration of the continuity equation: the speed and origin of the spiral pattern in M81.**
D. Westpfahl.
Workshop on the Interpretation of Modern Synthesis Observations of Spiral Galaxies, p. 175 – 182 (1991). – See Abstr. 012.041 for the main entry.

By integration of the continuity equation the pattern speed of a disk galaxy may be related to the velocity field and surface brightness of a well–defined class of pattern tracers. Application to the nearby spiral M81 gives an average pattern speed of $24\ \mathrm{km\ s^{-1}kpc^{-1}}$, but the pattern is shearing at the rate $-2\ \mathrm{km\ s^{-1}kpc^{-2}}$. This is consistent with the pattern being caused by Toomre's swing amplifier due to a close passage of M82 and NGC 3077 about 5×10^{8}years ago.

157.180 **Radio continuum properties of irregular galaxies.**
D. A. Hunter.
Workshop on the Interpretation of Modern Synthesis Observations of Spiral Galaxies, p. 183 – 190 (1991). – See Abstr. 012.041 for the main entry.

157.181 **Radio emission from small galaxies.**
J. S. Gallagher III.
Workshop on the Interpretation of Modern Synthesis Observations of Spiral Galaxies, p. 191 – 196 (1991). – See Abstr. 012.041 for the main entry.

The relationship between far infrared (FIR) and radio luminosity is considered for the case of small, star forming irregular galaxies. Since the radio emission in irregulars is largely thermal (versus non–thermal in spirals), the existence of a universal FIR–radio flux–flux correlation that encompasses both spirals and irregulars is surprising. It may therefore be telling us something fundamental about how local processes, such as star formation, interact with global galactic environments.

157.182 **Active nuclei and star formation in E/S0 galaxies.**
J. M. Wrobel.
Workshop on the Interpretation of Modern Synthesis Observations of Spiral Galaxies, p. 197 – 207 (1991). – See Abstr. 012.041 for the main entry.

Recent deep VLA surveys of the nuclei of E/S0 galaxies commonly find radio continuum sources, some with powers as low as the Galactic Center region on similar linear scales. Many of these newly discovered radio sources are probably driven by an active galactic nucleus (AGN), while others are candidates for being driven by recent star formation.

157.183 **Central star formation in S0 and Sa galaxies.**
L. L. Dressel.
Workshop on the Interpretation of Modern Synthesis Observations of Spiral Galaxies, p. 209 – 214 (1991). – See Abstr. 012.041 for the main entry.

Optical spectra, radio and infrared maps, radio and far infrared flux densities, and far infrared colors show that star formation is occurring at a high rate in the central few kiloparsecs of a significant fraction of S0 and early–type spiral galaxies. Star–formation–powered radio sources have been found up to a power level of $5\times10^{22}\mathrm{WHz^{-1}}$ at 2380 MHz (for $H_0 = 50\ \mathrm{km\ s^{-1}Mpc^{-1}}$) in a magnitude–limited sample of S0 and Sa galaxies.

157.184 **Emission from the nuclei of normal spiral galaxies.**
J. M. van der Hulst.
Workshop on the Interpretation of Modern Synthesis Observations of Spiral Galaxies, p. 215 – 225 (1991). – See Abstr. 012.041 for the main entry.

The advent of modern, high resolution synthesis instruments has greatly improved our knowledge of the structure of nuclei of spiral galaxies. This review provides a summary of what we have learned in the last decade about the radio structure of galaxy nuclei covering a range of luminosities and briefly discusses the implications for the physical mechanisms which are responsible for the radio emission.

157.185 **The enigmatic jets of NGC 1097.**
A. E. Wehrle, W. C. Keel, D. L. Jones.
Workshop on the Interpretation of Modern Synthesis Observations of Spiral Galaxies, p. 251 – 261 (1991). – See Abstr. 012.041 for the main entry.

The authors use deep optical images, spectroscopy and radio observations of the optical jets in NGC 1097 to constrain models for their origin. They detect no steep–spectrum emission at 90–cm associated with fossil radio lobes. If star formation in the optical jets was triggered by ejection of relativistic matter from an active nucleus, the time of activity was at least 10 million years ago.

157.186 **VLA observations of the peculiar interacting galaxies NGC 7463–7464.**
Yin Qifeng, D. S. Heeschen, W. C. Saslaw, Huang Zhenpin.
Workshop on the Interpretation of Modern Synthesis Observations of Spiral Galaxies, p. 263 – 266 (1991). – See Abstr. 012.041 for the main entry.

The authors have mapped the NGC 7463–7464 pair of interacting galaxies with the VLA at 1465 MHz, 4885 MHz and 8415 MHz. The results show three distinctive features: a low frequency bridge connecting the two galaxies, box–like intensity contours in the outer part of NGC 7463, and a possible dependence of the peak radio intensity positions on frequency. The optical appearance of NGC 7463 is also box–like and suggests interesting tidal distortion.

157.187 **The interpretation of modern synthesis observations of spiral galaxies: Summary lecture.**
E. R. Seaquist.
Workshop on the Interpretation of Modern Synthesis Observations of Spiral Galaxies, p. 267 – 275 (1991). – See Abstr. 012.041 for the main entry.

The author discusses four main topics: (a) Origin and propagation of cosmic rays. (b) Large–scale magnetic field structure. (c) The morphology of spiral tracers. (d) Activity in galaxies.

157.188 **The neutral hydrogen content of early type disk galaxies.**
J. Eder, R. Giovanelli, M. P. Haynes.
Astron. J., Vol. 102, No. 2, p. 572 – 588 (Aug 1991). Current Physics Microform No.: 9108A1508.

This paper presents the results of a sensitive 21 cm survey of massive early type galaxies made with the Arecibo radio telescope. Of the 81 galaxies observed, the detections comprise 48% of the S0's, 73% of the S0a's, and 96% of the Sa's. Only a few of the detections would have been missed by previous surveys due to extraordinarily wide spectral line profiles, so it is unlikely that a significant population of massive galaxies have gone undetected due to this intrumental selection effect. The relatively high detection rate enabled the large range in neutral hydrogen content found for S0 galaxies to be investigated. The values of the hybrid, distance–independent H I surface densities of the S0 galaxies in the authors' sample ranged continuously from amounts comparable to the most gas–rich Sa galaxies to low estimated upper limits of the H I content. CCD images of most of the gas–rich S0's revealed either faint spiral features or patchy structure in the disks. While no firm correlation between H I content and environmental density is apparent for the galaxies in the sample, two–sample statistics suggest a difference between the highest and the lowest density bins. Early type disk galaxies within low density environments tend to have higher gas surface densities than those within high density environments.

157.189 **Neutral gas infall into NGC 628.**
J. Kamphuis, F. Briggs.
Workshop on Warped Disks and Inclined Rings Around Galaxies, p. 6 – 12 (1991). – See Abstr. 012.054 for the main entry.

H I observations of the warped, face–on Sc I galaxy NGC 628 with the VLA show the presence of two high velocity gas complexes in the outer parts of the H I distribution and a large extension to the southwest with $\sim 7 \times 10^8 M_\odot$ of H I. A possible explanation is that NGC 628 is in the process of accreting a gas–rich, dwarf irregular galaxy.

157.190 **Using gas kinematics to measure M/L in elliptical galaxies.**
T. de Zeeuw.
Workshop on Warped Disks and Inclined Rings Around Galaxies, p. 33 – 39 (1991). – See Abstr. 012.054 for the main entry.

The kinematics of cold gas in elliptical galaxies can be used to determine M/L as a function of radius, but this requires fitting the observed velocity fields with realistic triaxial models in which the gas occupies non–circular closed orbits. The derived M/L profiles may differ significantly from the data obtained on the basis of circular motion in spherical potentials.

157.191 **Modeling the atomic gas in NGC 4278.**
J. F. Lees.
Workshop on Warped Disks and Inclined Rings Around Galaxies, p. 50 – 59 (1991). – See Abstr. 012.054 for the main entry.

Using VLA data on the extended atomic gas disk in the elliptical galaxy NGC 4278, the author calculates kinematic models for the velocity distribution. Axisymmetric models cannot reproduce a number of important features in the velocity field, such as the kinking of the zero–velocity contour near the nucleus. A triaxial model is presented that fits this and other pecularities very well.

157.192 **A few statistics from the catalog of polar–ring galaxies.**
B. C. Whitmore.
Workshop on Warped Disks and Inclined Rings Around Galaxies, p. 60 – 72 (1991). – See Abstr. 012.054 for the main entry.

A recently published photographic atlas and catalog of polar–ring galaxies and related objects is described, and a statistical analysis of some properties of the galaxies are tabulated. Roughly 5% of all nearby S0 galaxies appear to have or have had in the past polar rings. Polar rings show a variety of shapes; roughly half are narrow like a wedding ring while the other half are extended annuli with the central region cut out. Roughly two–thirds of the rings show some obvious warping; of these, roughly half have an integral sign type warp, which is typical of most warped disk galaxies, while the other half show a peculiar banana–shaped warp. A fair correlation of the asymmetry of the ring with the presence of outer debris, and a weak correlation of the angle between the ring and disk with the ring asymmetry, offer the possibility of determining the approximate age of a polar–ring galaxy based on its morphological appearance.

157.193 **Kinematics of polar–ring galaxies.**
P. D. Sackett.
Workshop on Warped Disks and Inclined Rings Around Galaxies, p. 73 – 84 (1991). – See Abstr. 012.054 for the main entry.

A review of polar–ring kinematics is presented. Implications for the nature and origin of polar–ring galaxies, the shape of their potentials, and possibilities for further study are presented.

157.194 **Second–epoch VLA observations of compact radio sources in NGC 253.**
J. S. Ulvestad, R. R. J. Antonucci.
Astron. J., Vol. 102, No. 3, p. 875 – 881 (Sep 1991). Current Physics Microform No.: 9109D1279.

The nearby starburst galaxy NGC 253 has been reobserved with the highest resolution configuration of the VLA at 6 cm, and observed for the first time at 3.6 cm. No new compact radio sources stronger than 3 mJy at 6 cm appeared in the 18 months since the first–epoch observations at 6 cm. This flux–density limit corresponds to an intrinsic luminosity three times that of the galactic supernova remnant Cas A. The 95% confidence upper limit to the type II supernova rate is 3.0 yr^{-1}, assuming that 2/3 of such supernovae reaching their 6 cm maxima between epochs would have been strong enough radio sources to be detected at the second epoch. None of the 6 cm sources stronger than 1 mJy was found to undergo significant flux decreases over 18 months, a result differing from that found for M82 by Kronberg & Sramek (1985). Errors in the flux–density scales should not have caused this difference between NGC 253 and M82, implying that the sources in NGC 253 are intrinsically different from those in M82. In principle, the bulk of the sources in NGC 253 could be supernova remnants or H II regions, or they may be young supernovae of a different type from those in M82. The 3.6 cm data show a substantial reduction in confusion and support the estimate that there are ~100 compact radio sources in NGC 253; some of the sources seen at both 3.6 and 6 cm probably have flat spectra and may be H II regions. A shell supernova remnant whose intrinsic radio size and luminosity are similar to Cas A has been discovered about 330 pc from the galaxy nucleus.

157.195 **A photometric study of the inner disk of M104.**
M. S. Burkhead.
Astron. J., Vol. 102, No. 3, p. 893 – 895 (Sep 1991). Current Physics Microform No.: 9109D1297.

Nonisothermal and Hubble law models of the core of the bulge component have been subtracted from BVRI data to reveal the photometric structure of the inner disk of M104 (NGC 4594). Disk parameters, integrated magnitudes, and profiles for the transition region between the inner disk ($a < 11$ arcsec) and the prominent disk for $a > 20$ arcsec are presented.

157.196 **Absorption line gradients in the compact elliptical galaxies NGC 4478, NGC 4486B, and NGC 5846A.**
T. J. Davidge.
Astron. J., Vol. 102, No. 3, p. 896 – 903 (Sep 1991). Current Physics Microform No.: 9109D1300.

Long–slit spectra have been obtained of the compact elliptical galaxies NGC 4478, NGC 4486B, and NGC 5846A. The data are used to investigate the radial behavior of various absorption features in these galaxies and, whereas systematic radial trends appear to be abscent in NGC 5846A, systematic gradients have been detected in NGC 4478 and NGC 4486B. The Mg_2 index in NGC 4478 changes with radius at a rate which is not markedly different from what is seen in more luminous elliptical galaxies. A steeper Mg_2 gradient may be present in NGC 4486B, although the uncertainty in the results for this galaxy are large. Nevertheless, if interpreted in terms of a radial variation in chemical composition, these data suggest that the metallicity gradients in NGC 4478 and NGC 4486B are at least as steep as those in more luminous elliptical galaxies. It is also shown that the manner in which the various indices change with respect to Mg_2 in the outer regions of NGC 4478 is qualitatively similar to what is seen in other elliptical galaxies.

157.197 **A dark–halo–dominated galaxy: NGC 5585.**
S. Côté, C. Carignan, R. Sancisi.
Astron. J., Vol. 102, No. 3, p. 904 – 913 (Sep 1991). Current Physics Microform No.: 9109D1308. With plates 27 – 28.

Studies of the light distribution, the H I kinematics, and the mass distribution of NGC 5585, a late–type spiral galaxy in the M101 group, are presented. From the photometry, an extrapolated central surface brightness $B(0)_c = 21.94$ mag $arcsec^{-2}$ and a scale length $\alpha^{-1} = 1.4$ kpc are derived for the stellar disk. The

H I distribution reaches a diameter of 12 arcmin, more than two times the optical diameter. The velocity field reveals a strong warp. The position angle is shown to vary from 45° to 55° in the outer parts, while the inclination is fairly constant at 52°. The rotation curve, derived from this velocity field, reaches a maximum velocity of $\sim 90\ km\ s^{-1}$, extending out to 7 optical scale lengths. Combining the optical and kinematical data, mass models for NGC 5585 were built using both the "best-fit" and the "maximum-disk" approaches. For both models, the dark-halo component is by far the main contributor to the rotation curve and its basic parameters are fairly well constrained with a core radius $r_c = 3.0 \pm 0.2$ kpc and a central density $\varrho_0 = 0.054 \pm 0.006\ M_\odot\ pc^{-3}$. At the last observed point ($r = 9.65$ kpc), $\sim 90\%$ of the total mass is dark. Even the gaseous component is dynamically more important than the stellar disk component. In fact, despite being a fairly bright system with $M^0_T(B) = -17.5$, NGC 5585 is similar to many fainter dwarf systems where the stellar disk is clearly not self-gravitating even in the inner parts.

157.198 **A kinematic study of the Fornax dwarf spheroidal galaxy.**
M. Mateo, E. Olszewski, D. L. Welch, P. Fischer, W. Kunkel.
Astron. J., Vol. 102, No. 3, p. 914 - 926 (Sep 1991). Current Physics Microform No.: 9109D1318. With plates 29 - 31.

The primary goal of this paper is to present a reliable estimate of the central velocity dispersion and mass density of the Fornax dwarf galaxy based on precise radial velocity measurements of a sample of 44 stars and four of the globular clusters.

157.199 **Wolf-Rayet stars in Local Group galaxies: numbers and spectral properties.**
T. E. Armandroff, P. Massey.
Astron. J., Vol. 102, No. 3, p. 927 - 950 (Sep 1991). Current Physics Microform No.: 9109D1331. With plates 32 - 34.

A large number of spectra have been obtained of Wolf-Rayet (WR) candidates found by CCD surveys in regions of M31, M33, NGC 6822, and IC 1613. Using these newly acquired data as well as published work, the authors attempt to answer the questions posed at the beginning of their survey project concerning the relative population of WR stars in these galaxies, what correlation exists between metallicity and WC/WN ratio, and whether the stellar wind properties of these stars vary from system to system.

157.200 **Star formation in dwarf irregular galaxies: Sextans B.**
M. Tosi, L. Greggio, G. Marconi, P. Focardi.
Astron. J., Vol. 102, No. 3, p. 951 - 974 (Sep 1991). Current Physics Microform No.: 9109D1355.

A new method to study the star formation histories of nearby irregular galaxies is presented. The authors have obtained B, V, and R CCD photometry of 2434 objects down to $V \approx 26$ in two regions of the Local Group galaxy Sextans B. Major information on the stellar population, star formation rate, and initial mass function in the galaxy are derived from the comparison of the color-magnitude diagrams of the observed stars with synthetic diagrams generated by a numerical simulation code based on stellar evolutionary tracks. The observational luminosity function of all the objects and that of only main sequence stars are also compared with the corresponding functions derived from the synthetic diagrams. At the observational limit of reliability ($B \approx 24$), the two regions of Sextans B contain similar stellar populations with ages up to 1 Gyr or more, and metallicity $Z = 0.001$. The models in better agreement with the observational data suggest that the star formation has proceeded there in long (some 10^8yr) quiet phases of moderate activity separated by short (not longer than 1×10^8yr) quiescent intervals. Despite the low rates of star formation derived with the authors' method, a continuous activity over the whole galaxy lifetime would provide results inconsistent with the observed values of the present metallicity, gas and total mass of the galaxy. Finally, by testing several initial mass functions, the authors find that the fraction of massive stars in Sextans B is not much larger than in normal spirals.

157.201 **Warped and flaring H I disks.**
A. Bosma.
Workshop on Warped Disks and Inclined Rings Around Galaxies, p. 181 - 189 (1991). - See Abstr. 012.054 for the main entry.

The author presents statistics of warped H I disks, both for edge-on galaxies in which the warp can be seen directly, and for more face-on disks in which the warp is inferred from the kinematics. The fraction of warped H I disks is of order 50% at least. From the face-on sample there is evidence that a concentrated halo inhibits the warp. Circumstantial evidence on low surface brightness giants and very early type spirals seems to reinforce this conclusion. The problem of flaring H I disks is briefly addressed: some predictions are given, and some difficulties are outlined.

157.202 **Behavior of warps in extended disks.**
F. Briggs, J. Kamphuis.
Workshop on Warped Disks and Inclined Rings Around Galaxies, p. 190 - 199 (1991). - See Abstr. 012.054 for the main entry.

The new H I observations of the warped, face-on Scl galaxy NGC 628 serve to illustrate the kinematical behavior of large spiral galaxies whose warped disks can be detected far outside their optical dimensions. The outlying gas follows inclined orbits that appear to be differentially regressing. The model derived for NGC 628 from the H I observations can be re-oriented so that the inner, optical disk is viewed exactly edge-on. In this way, a model for an observed "kinematical" warp can be used to predict a range of appearance of "morphological" warps.

157.203 **Observational constraints for the explanation of warps.**
E. Battaner, E. Florido, M.-L. Sanchez-Saavedra, M. Prieto.
Workshop on Warped Disks and Inclined Rings Around Galaxies, p. 200 - 206 (1991). - See Abstr. 012.054 for the main entry.

Three observational constraints are discussed: a) colour gradients in the warp region, from CCD observations made a La Palma with the 2.5 m Isaac Newton Telescope, b) frequency of the phenomenon and c) possible non-random spatial distribution of warps. A theoretical model in which extragalactic magnetic fields are responsible for warps is also proposed.

157.204 **Warps in S0s: observations versus theories.**
G. Galletta.
Workshop on Warped Disks and Inclined Rings Around Galaxies, p. 207 - 214 (1991). - See Abstr. 012.054 for the main entry.

157.205 **Dark halo close to the Andromeda nebula.**
V. G. Surdin.
Priroda, No. 7, p. 102 (Jul 1991). In Russian.

157.206 **CO observations of southern S0 galaxies.**
L. J. Tacconi, L. E. Tacconi-Garman, M. Thornley, H. van Woerden.
Astron. Astrophys., Vol. 252, No. 2, p. 541 - 549 (Dec 1991).

The authors present the first attempt to measure CO distributions in a number of southern S0 galaxies. In all but two of the 11 galaxies in their sample, they searched for CO emission in several offset positions as well as in the central beam. CO was detected at the $\geqslant 3\sigma$ level in at least one position in six of the galaxies. In NGC 1326 the CO emission is found to be extended beyond the central regions, and has been detected in several positions in the disk. In all of the other galaxies, the data are consistent with the CO emission being partially resolved. However, more recent observations of a second sample of galaxies show extended, resolved CO emission in a few more S0's. To investigate the gas distributions in these systems the authors have developed a modeling routine which generates and fits synthetic spectra to the data. In NGC 1326 the best fit indicates that the molecular gas is mainly distributed in a ring which is located ~ 1 kpc from the nucleus of the galaxy. The simple ringed distribution cannot explain all of profile features which are seen along the major axis, however. A zeroth order attempt to include a CO bar in the model did not improve the model fits to the data.

157.207 **The structure of cold dark matter halos.**
J. Dubinski, R. G. Carlberg.
Astrophys. J., Vol. 378, No. 2, p. 496 – 503 (10 Sep 1991).
The density profiles and shapes of dark halos are studied using the results of N–body simulations of the gravitational collapse of density peaks. The simulations use from 3×10^4 to 3×10^5 particles, which allow density profiles and shapes to be well resolved. The core radius of a typical dark halo is found to be no greater than the softening radius, $\varepsilon = 1.4$ kpc. The density profiles can be fitted with an analytical model with an effective power law which varies between –1 in the center to –4 at large radii. The dark halos have circular velocity curves which behave like the circular velocity contribution of the dark component of spiral galaxies inferred from rotation curve decompositions. The halos are strongly triaxial and very flat, with $<c/a> = 0.50$ and $<b/a> = 0.71$. There are roughly equal numbers of dark halos with oblate and prolate forms. The distribution of ellipticities in projection for dark halos reaches a maximum at $\varepsilon = 0.5$, in contrast to the ellipticity distribution of elliptical galaxies, which peaks at $\varepsilon = 0.2$.

157.208 **A volume–limited sample of IRAS galaxies to 4000 km s^{-1}. I. Neutral hydrogen observations from Jodrell Bank.**
J. Roth, J. R. Mould, R. D. Davies.
Astron. J., Vol. 102, No. 4, p. 1303 – 1313 (Oct 1991). Current Physics Microform No.: 9110D1835.
The authors have constructed a volume–limited catalog of normal spiral galaxies within the redshift interval 0–4000 km s^{-1} and brighter than 1.9 Jy at 60 μm to study relationships between fundamental galaxy properties and to map to the peculiar velocity field. Herein the authors report observations of 57 galaxies from this sample at Jodrell Bank, 42 of which resulted in detections, and study relationships between far–infrared and neutral hydrogen properties. Gas and dynamical or halo mass are well correlated with each other and moderately so with far–infrared luminosity. A decrease in dynamical mass with increasing far–infrared color temperature suggests that less massive, later–type systems enjoy higher specific star–formation rates.

157.209 **Optical nebulosity in X–ray–selected, early type galaxies.**
J. C. Shields.
Astron. J., Vol. 102, No. 4, p. 1314 – 1334 (Oct 1991). Current Physics Microform No.: 9110D1846.
Results of an Hα + [N II] narrowband imaging survey of X–ray–selected E and S0 galaxies are presented. A new technique is described for objectively optimizing removal of stellar continuum light while providing well–defined estimates of systematic errors. The procedure has the additional benefit of eliminating sky contamination specifically in image regions occupied by galaxy light. Detailed consideration of the measured spectral energy distributions is included in the flux calibration procedure, and emission–line luminosities (or upper limits), corrected for Galactic foreground extinction, are tabulated for metric apertures. The resulting luminosities for the sample members show no tendency to correlate with X–ray luminosities. Observed Hα + [N II] luminosity ($L_{H\alpha+[N\,II]}$) does correlate with far–infrared luminosity (L_{FIR}), with typical ratios of $L_{H\alpha+[N\,II]}/L_{FIR}$ that are an order of magnitude below those for late type galaxies. Some tendency is found for $L_{H\alpha+[N\,II]}$ to correlate with nonthermal radio power, though this may be due to eccentricities of the sample selection. No connection is apparent between the "boxiness" or "diskiness" of stellar isophotes and emission–line or far–infrared luminosity. Optical nebulosity in early type galaxies probably contains a significant multiparameter dependence on active galactic nuclei behavior, accretion from the hot interstellar medium, and mass injection from external sources. Emission associated with cooling flow phenomena on a galaxy scale may be overwhelmed in many instances by other sources of mass and ionization, suggesting that analogies between nebulosity in these sources and emission in cluster cooling–flow galaxies should be treated with caution.

157.210 **Distribution of the brightnesses and colors of the bulge of galaxy M81.**
R. G. Getov, Ts. B. Georgiev.
Bull. Spec. Astrophys. Obs. – North Caucasus, Vol. 32, p. 1 – 11 (1990). English translation of Astrofiz. Issled. Izv. Spets. Astrofiz. Obs., Tom 32, p. 3 – 15 (1990).
B, V, and B–V profiles, maps, and photometric diagrams of galaxy M81 (in $4' \times 7'$ frames) based on plates of the 6–m and 2–m telescopes are given. A noticeable reddening of the northwestern region of the image, the formation of an inner ring, and formation of a minibar were detected.

157.211 **Erratum: "Multifrequency windows on spiral galaxies. I. UBV and Hα aperture photometry" [Astron. J., Vol. 101, No. 4, p. 1207 – 1230 (Apr 1991)].**
G. Gavazzi, A. Boselli, R. Kennicutt.
Astron. J., Vol. 102, No. 4, p. 1569 (Oct 1991). See Abstr. 53.157.050.

157.212 **An estimate of the distance to M100 in the Virgo Cluster via VLBI of SN 1979c: updates and prospects.**
N. Bartel.
10. Santa Cruz Summer Workshop in Astronomy and Astrophysics: Supernovae, p. 760 – 765 (1991). – See Abstr. 012.052 for the main entry.

157.213 **Emission–line galaxies in the third list of the Case Low–Dispersion Northern Sky Survey.**
D. Weistrop, R. A. Downes.
Astron. J., Vol. 102, No. 5, p. 1680 – 1690 (Nov 1991). Current Physics Microform No.: 9111D0686.
Spectra have been obtained for 47 galaxies in the third list of the Case Low–Dispersion Northern Sky Survey, which covers an area R.A. = 08^h05^m–10^h10^m, Dec. = $+33.0° - +38.0°$. Thirty–five of the galaxies in the sample were selected for the presence of emission lines on the objective prism plates. At the higher spectral dispersion of the authors' data, significant line emission was found in 46 of the 47 galaxies. Twenty–six galaxies are undergoing significant bursts of star formation. Ten additional galaxies may be starburst galaxies with low–excitation spectra. Two galaxies are probably type Seyfert 2. The most distant object, CG 200, at a redshift z = 0.144, has a strong, broad Hα emission line, and is probably a Seyfert 1. Seventeen of the galaxies have been detected by IRAS. Eight of the IRAS galaxies have H II region–type spectra and eight have low–ionization starburst spectra. The galaxies represent a mixture of types, ranging from intrinsically faint dwarf galaxies with $M_B = -16$ mag, to powerful galaxies with $M_B = -23$ mag. Galaxies CG 234, CG 235 are interacting, as are galaxies CG 269, CG 270.

157.214 **Models of the evolution of star–forming regions in blue compact dwarf galaxies. I. Photometric evolution.**
L. S. Pilyugin.
Kinematics Phys. Celest. Bodies, Vol. 7, No. 2, p. 55 – 59 (1991). English translation of Kinematika Fiz. Nebesn. Tel, Tom 7, No. 2, p. 62 – 66 (1991).
The photometric characteristics of star–forming regions in blue compact dwarf galaxies are examined. The bolometric luminosities, the quantum fluxes in the Lyman continuum, and the equivalent widths of the H_β emission line are represented as functions of the age of the star–forming region for three different initial mass functions. It is shown that the age determined for the star–forming region from the equivalent–width of the H_β emission line is sensitive to the value of the upper mass limit of the forming stars and practically independent of the value of the lower mass limit.

157.215 **The plane W(Na I) × W(Mg I): effects of interstellar Na I in a sample of southern galaxies.**
E. Bica, M. G. Pastoriza, M. Maia, L. A. L. da Silva, H. Dottori.
Astron. J., Vol. 102, No. 5, p. 1702 – 1714 (Nov 1991). Current Physics Microform No.: 9111D0708.

Galaxy spectra from a subsample of the Southern Sky Redshift Survey data bank were used to study the equivalent width plane for the lines Na I λ5893 Å vs Mg I λ5175 Å. The authors try to estimate how important the contribution of the interstellar gas for the sodium line is compared to that of the stellar population. The sample is made up of galaxies with morphological types from E to Sc and are distributed up to radial velocities of 25,000 km s^{-1}, most of them smaller than 15,000 km s^{-1}. Most early type galaxies with dust lanes, particularly nearly edge–on S0's, present an enhancement of the Na I line. Inclined spiral galaxies tend to present enhanced Na I with respect to face–on spirals. This tendency, previously found in a smaller sample of galaxies limited to V $\leqslant$ 6000 km s^{-1}, is now confirmed for more distant ones. In the large velocity sample the authors are observing the global bulge rather than the very nucleus; the persistence of the effect suggests that the scale height of the gas layer in the central disk can reach a considerable fraction of the bulge radius.

157.216 **The outer Lindblad resonance and the morphology of early type disk galaxies.**
R. Buta, D. A. Crocker.
Astron. J., Vol. 102, No. 5, p. 1715 – 1723 (Nov 1991). Current Physics Microform No.: 9111D0721. With plates 45 – 56.

The morphology of outer rings and pseudorings in SB and SAB galaxies includes a number of distinct categories that bear a remarkable resemblance to the gaseous rings which have developed near the outer Lindblad resonance in n–body models of barred spirals. This paper displays these categories using CCD images and color index maps of 22 galaxies from the Catalogue of Southern Ringed Galaxies (CSRG). The prominence of the OLR morphologies in these and many other galaxies provides an important handle on pattern speeds of bars and ovals, as well as insights into internal dynamics. It also provides an important distinction to be considered when comparing early and late–type galaxies. An interesting possible correlation between the presence of nuclear star formation and outer ring/pseudoring morphology emerges from the new images.

157.217 **A luminous X–ray binary in M82.**
J. T. Stocke, R. Wurtz, H. Kühr.
Astron. J., Vol. 102, No. 5, p. 1724 – 1728 (Nov 1991). Current Physics Microform No.: 9111D0730. With plate 57.

Optical CCD imaging at the position of a bright X–ray point source just off the visible disk of the nearby irregular galaxy M82 constrain it to be an X–ray binary with a 0.2–4 keV X–ray luminosity of $\geqslant 9 \times 10^{38}$ergs/s. Assuming that the X–ray emission is due to a single object, then by assuming an Eddington limit appropriate for pure electron scattering this luminosity requires a minimum primary mass of $\geqslant 15\ M_{\odot}$; thus, this X–ray source is a black hole candidate. If the brightest optical object detected inside the X–ray error circle at B $\approx$ 22.2 is the secondary to the X–ray emitter, it is most likely an early type star making this system the most luminous massive X–ray binary system known.

157.218 **The stellar luminosity function in the bulge of M31.**
R. L. Davies, J. A. Frogel, D. M. Terndrup.
Astron. J., Vol. 102, No. 5, p. 1729 – 1733 (Nov 1991). Current Physics Microform No.: 9111D0735.

The authors have derived the luminosity function for stars in the bulge of M31 in a region ~7.2′ (1.44 kpc) from its nucleus. It has an upper limit ~0.5 mag brighter than the steep cutoff observed in the Galactic bulge luminosity function. The shape of the authors' M31 luminosity function, that of Rich and Mould for a region 4.1′ from the nucleus, and the differences between them are all consistent with a model based on a Galactic bulge luminosity function modified by the addition of disk stars located behind M31's bulge. Kent's disk/bulge deconvolution of M31 is used to fix independently the relative amounts of disk contamination at different radii. There is no compelling reason to invoke a stellar population for the bulge of M31 that differs from that of the bulge of our own Galaxy.

157.219 **NGC 4286 and NGC 3377A: galaxies with mixed morphologies between the dwarf Im and dE/dS0 types.**
A. Sandage, G. L. Hoffman.
Astrophys. J., Lett., Vol. 379, No. 2, p. L45 – L47 (1 Oct 1991). With plates L2 – L4.

Optical and radio data for the nucleated dwarf galaxy NGC 4286 show a mixed morphology between a star–forming Im type and a gas–poor dS0, N early–type dwarf. The Im dwarf NGC 3377A shows no gas depletion, but its morphology and size are similar to the "huge, very low surface brightness" (VLSB) gas–poor galaxies recently identified in the Virgo cluster. The authors interpret the morphology of NGC 4286 as an original Im system that has transformed itself into a gas–poor dS0, N by losing most (but not all) of its original gas through an internal supergalactic wind that was generated earlier by a superluminous star cluster. The remnant of this cluster has now sunk to the center by dynamical friction to form the observed unresolved bright nucleus. Such star clusters and superwinds are known to exist in many present–day starburst Sm and amorphous galaxies such as NGC 1569 and NGC 1705.

157.220 **An H I view of the Universe.**
J. H. van Gorkom.
3. Haystack Observatory Conference on the Interstellar Medium: Skylines – atoms, ions and molecules – new results in spectral line astrophysics, p. 1 – 5 (1991). – See Abstr. 012.051 for the main entry.

The H I morphology and large scale distribution of gas rich galaxies is discussed. Isolated spirals have sharp edges in H I, possibly due to ionization of the lower H I column densities by the intergalactic UV radiation field. The prototypical low surface brightness galaxy Malin 1 has an H I disk which has a very low surface density across the entire disk. This might explain why star formation has been so inefficient in this galaxy. An optically unbiased search for gas rich dwarf galaxies or optically faint H I clouds in the Perseus–Pisces supercluster and void has been performed. Seventeen objects were detected in the supercluster, ten of them previously uncatalogued. No objects were found in the void. H I rich dwarf galaxies follow the spatial structure as defined by bright galaxies, though a moderate bias between dwarfs and bright galaxies cannot yet be ruled out. The data leave no room for a substantial population of H I dwarfs or gas rich, low surface brightness giants that could have been missed by optical catalogs.

157.221 **Molecules in external galaxies.**
K.–Y. Lo.
3. Haystack Observatory Conference on the Interstellar Medium: Skylines – atoms, ions and molecules – new results in spectral line astrophysics, p. 7 – 18 (1991). – See Abstr. 012.051 for the main entry.

A review of the current status of observations of molecules in external galaxies is given. Emphasis is placed on examining the assumptions and limitations involved. Direct measurements of extragalactic gas properties are essential to the understanding of the underlying physical processes of phenomena such as starbursts and galactic structure. Observations of multi–transitions from different molecules and their isotopes with sufficiently high resolution to resolve individual molecular clouds are needed. Millimeter–wave and submillimeter–wave interferometers are invaluable for such studies.

157.222 **Molecular gas and triggers of star formation in galaxies: spiral arms, bars and mergers.**
R. Kawabe, M. Ishiguro, S. Ishizuki, S. K. Okumura, T. Kasuga, K.–I. Morita, T. Tosaki.
3. Haystack Observatory Conference on the Interstellar Medium: Skylines – atoms, ions and molecules – new results in spectral line astrophysics, p. 19 – 27 (1991). – See Abstr. 012.051 for the main entry.

The authors present their recent results of high spatial resolution CO and ^{13}Co(J = 1–0) observations with the Nobeyama Millimeter Array. They observed the central (1 – 2 kpc) regions of nearby galaxies, IC 342, Maffei 2, and NGC 6946 with 2″ – 4″ resolution (45 – 90 pc) in CO and ^{13}CO, the 3' region of M51 with 4″ – 6″ resolution (180 – 280 pc), and seven IR–luminous galaxies with $L_{FIR} \gtrsim 10^{11} L_{\odot}$ (e.g., Mrk231, Arp220, NGC6240, and NGC828) with 2″ – 5″ resolution. They discuss the structures and kinematics of molecular gas in these galaxies, and discuss the trigger of star formation from viewpoints of galactic shock, gas fueling by bars, and merging of gas–rich spirals.

157.223 **[CII] 158 μm line images of spiral galaxies.**
G. J. Stacey, N. Geis, J. M. Jackson, A. Poglitsch.
3. Haystack Observatory Conference on the Interstellar Medium: Skylines – atoms, ions and molecules – new results in spectral line astrophysics, p. 29 – 33 (1991). – See Abstr. 012.051 for the main entry.

The authors present large scale maps of the face on spiral galaxies M51, M83, and NGC 6946 in the 158 μm [CII] fine structure line. The [CII] line is ubiquitous and bright; easily traced over the inner 5 kpc radius regions of these galaxies and accounting for ~0.5% of the total far–infrared luminosity. The [CII] line arises in the warm, dense photodissociated interface regions on the surfaces of UV exposed molecular clouds. The authors compare their [CII] maps with those obtained in the ^{12}CO (1→0) line, and interpret the [CII]/^{12}CO(1→0) line intensity ratio as a measure of global star formation activity. They conclude that the nuclear and "bar" regions of M83 are undergoing vigorous OB star formation activity, with somewhat less activity over the disks of M83 and M51. NGC 6946 displays the least amount of OB star formation activity of this galaxy sample.

157.224 **Disrupted H I environment around M82.**
M. S. Yun, P. T. P. Ho, K.–Y. Lo.
3. Haystack Observatory Conference on the Interstellar Medium: Skylines – atoms, ions and molecules – new results in spectral line astrophysics, p. 35 – 42 (1991). – See Abstr. 012.051 for the main entry.

The authors obtained 21 cm H I aperture synthesis maps of the "exploding galaxy" M82 using the D–configuration of the VLA. By studying the distribution and kinematics of the atomic gas in and around M82 with 20 times better sensitivity than any previous work, they aimed at unveiling the triggering event for the nuclear starburst in this unusual nearby galaxy. Much of the H I gas found around M82 is organized into several streamers and dense clumps with a smooth velocity gradient. Two main H I streamers, each $\gtrsim$10 kpc long, appear to be anchored to M82 dynamically as well as morphologically. The authors suggest that the H I gas around M82, particularly the two main H I steamers, are tidally disrupted outer disk material of M82 rather than tidally captured gas from M81. The deep optical imaging survey failed to defect any optical counterpart to these H I streamers or other H I features found around M82. However, a flaring is found in the outer stellar disk, which corresponds exactly with the beginning point of the northern streamer, further supporting the tidal hypothesis.

157.225 **Periodic variations of the extragalactic H_2O maser in NGC 4258.**
A. D. Haschick, W. A. Baan.
3. Haystack Observatory Conference on the Interstellar Medium: Skylines – atoms, ions and molecules – new results in spectral line astrophysics, p. 67 – 72 (1991). – See Abstr. 012.051 for the main entry.

The H_2O maser emission in the galaxy NGC 4258 has been monitored over the time period Apr 1984 to Jun 1990 and has shown rapid variability in the velocity range centered on 465 km s^{-1}. Seven distinct flares with a separation of 85 days were observed during the last period of monitoring of a year and an half. The maser was observed to skip an expected flare in 1989 Nov but has since continued to display the periodic variations. The center velocity has shown a systematic change of 10 km s^{-1} during these seven outbursts. This dependence indicates that the maser is mostly unsaturated during the bursts. Two other emission features at 495 and 525 km s^{-1} did not vary significantly during the whole monitoring period of five years. A foreground molecular region pumped by a variable star and amplifying the compact nuclear radio source is suggested as producing the periodic variations.

157.226 **The dark matter content of spiral galaxies.**
P. Salucci, K. M. Ashman, M. Persic.
Astrophys. J., Vol. 379, No. 1, p. 89 – 93 (20 Sep 1991).

The authors present a new technique for calculating the fraction of dark material within the optical radius of spiral galaxies. The method employs the well established observational result that spiral galaxies have similar central surface brightnesses, as well as published stellar synthesis evolutionary models, color – magnitude relations and optical rotation curves. No assumptions about the dark matter distribution are necessary. The authors find that the ratio of disk–to–dynamical mass within the optical radius increases roughly as $L_B^{0.4}$. This is in good agreement with the results of Persic and Salucci, which are derived from independent considerations.

157.227 **Molecular gas in elliptical galaxies.**
J. F. Lees, G. R. Knapp, M. P. Rupen, T. G. Phillips.
Astrophys. J., Vol. 379, No. 1, p. 177 – 215 (20 Sep 1991).

Observations of 24 early–type galaxies, mostly far–infrared (FIR) bright ellipticals, in the ^{12}CO (2–1) emission line resulted in 10 detections including eight out of the 20 ellipticals. Eight of these galaxies have not previously been detected. A search of the literature reveals a total of 17 elliptical galaxies with observed molecular gas, as well as 27 upper limits. The detected ellipticals have typical molecular gas masses of about 10^7– $10^8 M_{\odot}$, similar to their H I masses, and an approximately power–law distribution of M_{H2}/L_B, extending to much lower values than for the spirals. The lenticular galaxies have CO properties intermediate between the elliptical and spiral systems. The 27 early–type galaxies which are detected in both molecular and atomic gas have the same mean ratio of molecular to atomic gas mass as later type spirals, similar ratios of Hα luminosity and FIR luminosity to molecular gas mass, and identical FIR colors. Thus, these ellipticals seem to have global interstellar medium properties similar to those in late–type galaxies, but on a much smaller scale. A comparison of the properties of those ellipticals which have been detected in CO with the undetected galaxies indicates that the molecular gas may occur preferentially in the bluer, lower luminosity dwarf ellipticals, which have more than twice the detection rate of the brighter galaxies.

157.228 **The case of NGC 1365.**
J. Kopecký, J. Palouš.
Bull. Astron. Inst. Czech., Vol. 42, No. 3, p. 192 – 199 (Jun 1991).

Motions in the strong bar galaxy NGC 1365 are analyzed with the radial velocities published by Jörsäter et al. (1984). The authors found that, inside the bar, the highly elliptical streamlines are aligned with the bar major axis. Further out the less elliptical streamlines are inclined to the bar major axis at 10° to 30°. The residual velocity dispersion between 15 to 25 km s^{-1} is much less than that found earlier. Some large residuals near the bar end may be due to the chaotic orbits predicted by Pfenniger (1984). However, the conclusion suffers from many gaps in the data coverage of the galactic plane. Additional observations in some key regions of NGC 1365 are urgently needed.

157.229 **CO emission at z = 2.2867 in the galaxy IRAS F10214+4724.**
R. L. Brown, P. A. vanden Bout.
Astron. J., Vol. 102, No. 6, p. 1956 – 1959 (Dec 1991). Current Physics Microform No.: 9112C1932.

The authors have detected CO (J = 3–2) emission at a redshift of 2.2867 from the galaxy IRAS F10214+4724.

157.230 **Nearby galaxies in high energy gamma rays.**
M. E. Özel.
7. Course of the International School of Cosmic–Ray Astrophysics: Cosmic rays, supernovae, and the interstellar medium – A NATO Advanced Study Institute, p. 307 – 311 (1991). – See Abstr. 012.053 for the main entry.

157.231 **UBVR photometry and rotation of late–type galaxies: NGC 959, 1156, 1160, 6643, and 7292.**
V. F. Esipov, G. A. Kyazumov, A. R. Dzhafarov.
Sov. Astron., Vol. 35, No. 5, p. 452 – 456 (Sep 1991). Current Physics Microform No.: 9203X1626. English translation of Astron. Zh., Tom 68, Vyp. 5, p. 909 – 917 (1991).
Using a magnetic–focusing image tube to enhance brightness, UBVR profiles of brightness distribution and color have been obtained for five late–type galaxies. Rotation curves have also been obtained and models of the brightness distribution constructed for these galaxies. A comparison of photometric and kinematic data shows that the composition of the star population varies little along a radius. In three of the five galaxies, a density depression in the central part of the disk is required to explain the rotation curve.

157.232 **The stellar populations of M31.**
S. van den Bergh.
Publ. Astron. Soc. Pac., Vol. 103, No. 668, p. 1053 – 1068 (Oct 1991). Invited review paper.
Available data on stellar populations in the Andromeda nebula (M31) are reviewed. A distance modulus $(m-M)_0 = 24.3 \pm 0.1$, corresponding to a distance of 725 ± 35 kpc, and a foreground reddening $E_{B-V} = 0.08$ will be adopted in this paper. The high rotational velocity and large velocity dispersion observed in the semistellar nucleus of the Andromeda nebula demonstrate that it contains a compact mass of $\sim 10^7 M_\odot$. Spectroscopic observation of line and molecular features now rules out the possibility that the high mass–to–light ratio observed in the nucleus of M31 is due to a dwarf–enriched mass spectrum of star formation. Such observations also militate against the hypothesis that the dominant stellar population of the nucleus consists of stars that once belonged to globular clusters that were captured by tidal friction. The semistellar nucleus of M31 and its dynamical center appear to be separated by $\sim 0\farcs5$. The halo of M31 is found to be rich in RR Lyrae stars that belong to Oosterhoff Type I. The mean metallicity of stars in the halo remains controversial. The color–magnitude diagram of the halo of M31 appears to show that the blue horizontal–branch population is weak.

157.233 **Supernovae and supernova remnants in starbursts.**
R. A. Chevalier.
Massive Stars in Starbursts Meeting, p. 169 – 181 (1991). – See Abstr. 012.062 for the main entry.
Type Ib and type II supernovae are of special interest for starburst galaxies because they are thought to have massive star progenitors. Starburst galaxies are thought to have high supernova rates, but these events are difficult to observe optically because of high dust obscuration and high background. These problems are reduced at radio wavelengths and, on the basis of radio evolution, it is likely that there was a massive type II supernova in the center of M82 in the mid–1950's and type Ib event in late 1980. The high supernova rate generates a high interstellar pressure in nuclear starbursts, so that the evolution of the circumstellar medium around massive stars is modified. In particular, the red supergiant wind is stopped by the interstellar pressure at a diameter of a few 0.1's of a pc, where a dense shell is formed. The interaction of the supernovae with such dense shells may give rise to the compact radio sources that are frequently observed in starburst galaxy nuclei. Because the filling factor of cool gas is small, supernovae are most likely to deposit their energy in hot, low density gas that is easily able to escape from the galactic nucleus in a wind. There is good evidence for such a wind in M82. The high pressure in the nuclear region can drive a shock front into the surrounding galactic disk. The ring of neutral gas observed in M82 may be formed in this way; the expansion of the ring may be accompanied by a propagating starburst.

157.234 **Observations and models of blue compact dwarf galaxies.**
T. X. Thuan.
Massive Stars in Starbursts Meeting, p. 183 – 203 (1991). – See Abstr. 012.062 for the main entry.
The author discusses the evidence for starbursts in blue compact dwarf (BCD) galaxies. He discusses the star formation history in BCDs and compares the burst phenomenon in BCDs with that in larger, more massive starburst galaxies. He defines here a starburst to be an event in which, during a time short compared to the Hubble time, the star formation rate (SFR) increases to be several orders of magnitudes greater than the mean SFR, averaged over the lifetime of the galaxy.

157.235 **The initial mass function in M82.**
G. H. Rieke.
Massive Stars in Starbursts Meeting, p. 205 – 216 (1991). – See Abstr. 012.062 for the main entry.
Recent observations provide a large number of tight constraints on starburst models of M82. The most critical of these constraints are now deduced independently from a variety of measurements often using differing techniques. Comparison with starburst models using modern stellar evolutionary tracks demonstrates that these constrates cannot be met with a conventional local initial mass function; formation of low mass stars must be less common in M82 than predicted by such an IMF.

157.236 **Population synthesis models of starbursts.**
B. Rocca–Volmerange.
Massive Stars in Starbursts Meeting, p. 217 – 231 (1991). – See Abstr. 012.062 for the main entry.
Ultraluminous starbursts have been discovered in the far–IR while spectral signatures of intense star forming processes are detected from the far–UV to the near–IR, inside blue compact dwarf galaxies, faint galaxies in deep surveys and most distant radio galaxies. Are such bursts of star formation related to infrared starbursts? This is possible if a merging scenario drives the evolution of galaxies. Typical signatures are analyzed and the interpretation of various samples, more specially of high–z galaxies, makes interactions of galaxies a common scenario of galaxy evolution. This simultaneously implies luminosity and number evolution.

157.237 **H_2 and infrared in global starburst galaxies.**
N. Scoville, B. T. Soifer.
Massive Stars in Starbursts Meeting, p. 233 – 257 (1991). – See Abstr. 012.062 for the main entry.
Far infrared measurements from the IRAS survey combined with estimates of the molecular gas content provide a fundamental basis for the analysis of the starburst phenomena in galaxies. When the ratio of far infrared luminosity to molecular gas mass significantly exceeds that in normal galaxies star formation is occurring on a shorter timescale, possibly with an initial mass function biased towards high mass stars. In the highest luminosity IRAS galaxies the luminosity to H_2 mass ratio is typically 40 $L_\odot M_\odot^{-1}$, indicating star formation rates of 10 – 100 $M_\odot yr^{-1}$ and cycling times for the ISM much less than 10^9yr. In the very luminous infrared galaxies, the optical morphology almost invariably shows evidence of a strong galactic interaction and a substantial fraction of the total molecular gas content is seen at radii ≤ 1 kpc. Dense molecular gas probably plays a pivotal role in the evolution of such dynamically disturbed systems. The authors show that the shape of the high luminosity end of the infrared galaxy luminosity function can be reproduced by a model in which normal spiral galaxies, represented in the Schecter function, undergo collision–induced starbursts. Statistics from the IRAS survey are consistent with the percentage of all spiral galaxies currently undergoing a global starburst being approximately 0.2% and the lifetime of the starburst being a dynamical time, approximately 2×10^8years. The present epoch rate is therefore such that 2% of all galaxies participate in a merger every 10^9years and with standard cosmological evolution, nearly all galaxies would be undergoing such merger–induced

starbursts at z = 1. Galactic merging and starburst activity must therefore play a central role in galactic evolution.

157.238 **Stellar content of starburst galaxies.**
R. Joseph.
Massive Stars in Starbursts Meeting, p. 259 – 270 (1991). – See Abstr. 012.062 for the main entry.
The infrared spectral region provides a number of unique and powerful tools for gaining insight into the stellar content of galaxies. The author emphazised the infrared approach to investigate the stellar content of starburst galaxies.

157.239 **Models of starburst galaxies.**
C. A. Norman.
Massive Stars in Starbursts Meeting, p. 271 – 287 (1991). – See Abstr. 012.062 for the main entry.
Dynamical studies are discussed showing how gas moves to the center of a merging and interacting galaxy on timescales on the order of the dynamical timescale. The effects of the massive stars formed in a starburst on the interstellar medium of such a galaxy are outlined including the formation of superbubbles, chimneys, and galactic winds. The problem of understanding initial mass functions that are skewed towards massive stars is presented and various possible theoretical solutions that have been proposed are reviewed. Current ideas on what makes the bursts and why they are important for galaxy formation and evolution is sketched.

157.240 **The starburst–AGN connection.**
T. M. Heckman.
Massive Stars in Starbursts Meeting, p. 289 – 315 (1991). – See Abstr. 012.062 for the main entry.
The author reviews the possibility that there is a link between starbursts and active galactic nuclei (AGN's). He estimates that AGN's and starbursts are of comparable energetic significance in the local Universe and that AGN's and massive stars have produced similar amounts of total radiant energy over the history of the universe. He discusses whether and how one can discriminate between phenomena powered by massive stars and those demanding a more exotic energy source such as a supermassive black hole. He then addresses whether there might be an indirect or evolutionary connection between AGN's and starbursts. He reviews the evidence that AGN's may reside in galaxies with unusually high present or recent star–formation–rates. This involves comparing the stellar population and ISM in galaxies with and without AGN's.

157.241 **Cosmological consequences of starbursts.**
D. W. Weedman.
Massive Stars in Starbursts Meeting, p. 317 – 333 (1991). – See Abstr. 012.062 for the main entry.
Massive star formation in galaxies is the key process being sought observationally to find evidence for the first generations of primordial star formation. Starting with a local luminosity function for star forming galaxies determined from IRAS and radio data, various constraints defined by other observations in X–ray, ultraviolet, optical and far infrared astronomy are discussed. It is shown how limits from the X–ray background and from faint optical galaxy counts restrict the amount of evolution which could have occurred for starburst galaxies. Predictions are made for the far infrared background from star forming galaxies that should be observed by COBE, and it is shown that this background will provide the tightest constraint yet available on the nature of primordial formation.

157.242 **A COSMOS study of the structure and content of NGC 6822.**
P. W. Hodge, T. R. Smith, P. Eskridge, H. T. MacGillivray, S. M. Beard.
Astrophys. J., Vol. 379, No. 2, p. 621 – 630 (1 Oct 1991).
A COSMOS survey of deep Schmidt plates in four colors is used to examine the distribution of stars and unresolved light in the local group dwarf irregular galaxy NGC 6822. Global color – magnitude diagrams and luminosity functions are derived from background – substracted data. These show more clearly than previous studies the prominence of the main sequence and the comparatively small number of red giants. Maps of stars separated by magnitude and by color, as well as CMDs for various areas, show a wide range in recent histories of star formation over the image. The results are compared to the H I density distribution; most star–forming regions are offset by ~200 pc from H I peaks and some dense H I peaks have no detectable stars. The star formation rate is calculated and compared to those of other galaxies.

157.243 **Gas velocity field in the central regions of NGC 23, NGC 1134, and NGC 4536.**
V. L. Afanas'ev, A. V. Zasov, G. V. Popravko, O. K. Sil'chenko.
Sov. Astron. Lett., Vol. 17, No. 5, p. 325 – 328 (Sep 1991). Current Physics Microform No.: 9204X1272. English translation of Pis'ma Astron. Zh., Tom 17, No. 9, p. 771 – 778 (1991).
New results are presented from measurements of radial gas velocities in the central regions ($r < 15''$) of three spiral galaxies (NGC 23, NGC 1134, and NGC 4536) with large radial–velocity gradients at the center. The observations were carried out with the two–dimensional multipupil fiber optic spectrometer of the 6–m telescope of the Special Astrophysical Observatory of the USSR Academy of Sciences. The closeness of the position angles of the photometric and dynamical major axes of the galaxies in question indicates circular motions of the gas in the nuclear regions, occurring in the plane of a stellar disk. An extended region roughly $8 \times 4''$ in size with an anomalously low radial gas velocity has been detected in NGC 23.

157.244 **Thickness of thin stellar disks and the mass of the dark halo.**
A. V. Zasov, D. I. Makarov, E. A. Mikhajlova.
Sov. Astron. Lett., Vol. 17, No. 5, p. 374 – 377 (Sep 1991). Current Physics Microform No.: 9204X1321. English translation of Pis'ma Astron. Zh., Tom 17, No. 10, p. 884 – 890 (1991).
It is shown that thin galaxies with an apparent axial ratio $b/a \leqslant 0.1$ viewed edge–on have a geometrical thickness that can be explained only by the presence of a massive dark halo. Analysis of the stability condition for stellar disks and the results of numerical simulation of N–body galaxies lead to the conclusion that the minimum possible relative thickness z_0/R_0 of a collisionless disk decreases with increasing relative mass μ of the halo. Therefore, existing measurements of the ratio z_0/R_0 for disks of normal galaxies can be used to estimate the relative halo mass.

157.245 **Photographic photometry and distance modulus of the "Garland" dwarf galaxy near NGC 3077.**
M. E. Sharina.
Sov. Astron. Lett., Vol. 17, No. 5, p. 383 – 386 (Sep 1991). Current Physics Microform No.: 9204X1330. English translation of Pis'ma Astron. Zh., Tom 17, No. 10, p. 904 – 912 (1991).
The author has performed photographic photometry on the brightest stars in the "Garland" irregular dwarf galaxy near NGC 3077. He has constructed a color–magnitude diagram and a luminosity function for the Garland stars. He has determined the distance modulus to the galay: $(m-M)_0 = 27\overset{m}{.}76$.

157.246 **X–ray emission from starburst galaxies.**
Y. Rephaeli, D. Gruber, M. Persic, D. MacDonald.
Astrophys. J., Lett., Vol. 380, No. 2, p. L59 – L62 (20 Oct 1991).
The authors report the results of an investigation of X–ray emission from a sample of 53 IRAS – selected candidate starburst galaxies. Superposed soft and hard X–ray emission from these galaxies in the Einstein IPC and HEAO 1 A–2 and A–4 energy bands, which span 0.5 to 160 keV, is detected at the 99.6% confidence level, after allowing for confusion noise in the HEAO 1 data. Above 15 keV the confidence level is 97%. The combined spectrum is flat, with a (photon) power–law index of 1.0 ± 0.3. The contribution of the population of sources represented by this sample to the 3 – 50 keV residual cosmic X–ray background is estimated to be at least ~4%, assuming no

evolution. Moderate evolution, for which there is some observational evidence, increases this fractional contribution to ~26%.

157.247 **Infrared hot spots in the nucleus of NGC 253.**
D. A. Forbes, M. J. Ward, D. L. DePoy.
Astrophys. J., Lett., Vol. 380, No. 2, p. L63 – L66 (20 Oct 1991). With plate L2.
Near infrared observations of the starburst galaxy NGC 253 using a large format array and high spatial sampling show four knots of enhanced emission (hot spots) in the nuclear region. The authors find that each hot spot may correspond spatially to a compact radio knot. They discuss several possible origins of these hot spots and suggest observational tests for discriminating between these explanations.

157.248 **Molecular gas excitation in NGC 253.**
W. F. Wall, D. T. Jaffe, F. P. Israel, F. N. Bash.
Astrophys. J., Vol. 380, No. 2, p. 384 – 391 (20 Oct 1991).
The authors have observed the $J = 3 \rightarrow 2$ and $J = 2 \rightarrow 1$ lines of ^{13}CO and ^{12}CO with 22″– 24″ resolution in the central region of the spiral galaxy NGC 253. The ^{13}CO $J = 3 \rightarrow 2$ emission is pointlike, whereas the emission in the other CO lines is consistent with emission from a ~40″ × 10″ FWHM bar. The ratio of ^{13}CO $J = 3 \rightarrow 2$ to ^{13}CO $J = 2 \rightarrow 1$ line intensities toward the nucleus requires warm ($T_K \gtrsim 100K$), dense [$n(H_2) \gtrsim 10^4 – 10^5 cm^{-3}$] molecular gas. Spatial coincidence of 3 mm continuum and [Ne II] 12.8 μm emission from H II regions and mid–IR emission from hot dust with the ^{13}CO $J = 3 \rightarrow 2$ emission implies a common source of excitation, possibly related to the intense star formation that has occurred in the nucleus. The decline in the ^{13}CO $J = 3 \rightarrow 2$ intensity with distance from the nucleus reflects a decline in both density and temperature. The molecular gas mass in the central 50″, estimated from the optically thin lines of ^{13}CO, is $\sim 4 \times 10^7 M_\odot$.

157.249 **Star formation and the molecular content of IC 342.**
L. J. Sage, P. M. Solomon.
Astrophys. J., Vol. 380, No. 2, p. 392 – 402 (20 Oct 1991).
An almost fully sampled map, at 45″ resolution, of CO $J = 1 \rightarrow 0$ emission from the inner 3′ of IC 342 is presented, along with sparse observations out to 10′ from the center. The CO emission is centrally concentrated. CO is enhanced near the spiral arms but is observed over the whole map. The H_2 content can be described in terms of two components: the nuclear region and a molecular disk. The CO emission is strongest from the nuclear region, which extends 1.5 kpc from the center. The molecular disk lies between radii of 1.5 kpc and 13 kpc, across which the CO emission drops slowly. Beyond 13 kpc is an atomic disk in which H I dominates. In the nuclear region, H_2 comprises 27% of the dynamical mass and 99% of the gaseous interstellar medium. In the molecular disk, H_2 is ~8% of the total mass and 75% of the gas mass. The authors use spectra with very high S/N ratio to determine the inner rotation curve of IC 342. They find the region of solid body rotation to extend ~1.7 kpc from the center, much less than the previously reported 5 kpc. The efficiency of massive star formation in the nuclear region, determined from the ratio of far–infrared luminosity to molecular mass, is only slightly lower than in the Galactic region W49, and approaches values seen in strongly interacting and merging galaxies. Averaged over the whole disk of IC 342, the massive star formation efficiency is comparable to the average for all molecular clouds in the disk of the Milky Way.

157.250 **Extragalactic dust. V. NGC 801.**
N. Brosch.
Mon. Not. R. Astron. Soc., Vol. 251, No. 1, p. 24 – 27 (1 Jul 1991).
A deep photometric image of the edge–on spiral NGC 801 was analysed to search for predicted extragalactic dust layers. The search was negative. Other observations of spirals relevant for similar searches are reviewed.

157.251 **On the cause of metallicity gradients in spiral galaxies.**
S. Phillipps, M. G. Edmunds.
Mon. Not. R. Astron. Soc., Vol. 251, No. 1, p. 84 – 92 (1 Jul 1991).
The authors investigate particular models for the chemical evolution of galactic discs in an attempt to account for the observed behaviour of metallicity with surface density, and hence explain abundance gradients. The models include simple Schmidt star formation rate laws, models in which the formation of massive stars is suppressed below a threshold interstellar gas density, and models with self–regulation of star formation. Except for the ones with self–regulating star formation, these models are not particularly successful unless rather extreme parameter choices are made. The failure is, perhaps surprisingly, worse for gravitational instability thresholds than for some more arbitrary density thresholds. The authors suggest that the modulation of star formation rate by additional terms dependent, explicitly or implicitly, on radius may be required to generate the observed gradients.

157.252 **The radio structures of ultraluminous IRAS galaxies.**
H. M. Sopp, P. Alexander.
Mon. Not. R. Astron. Soc., Vol. 251, No. 1, p. 112 – 118 (1 Jul 1991).
The authors present multifrequency radio observations of five ultraluminous IRAS galaxies, Arp 220, Mkn 273, UGC 05101, IRAS 08572 + 3915 and IRAS 15250 + 3609, made with the VLA in three array configurations; a MERLIN map of UGC 05101 at 1.66 GHz; and integrated flux measurements at 151 MHz with the CLFST for all five galaxies. Although the radio emission from these objects tends to be concentrated within the central 1 kpc, it is not from a compact active nucleus, and there is evidence for extended radio emission in each case. Three of the galaxies show low–frequency turnovers in their spectra which may be caused by thermal absorption of their non–thermal spectra. The authors conclude that in each object almost all of the radio and far–infrared emission is associated with circum–nuclear or disc–wide star formation and not with active nuclei, although the observations do not preclude the existence of active nuclei in these galaxies.

157.253 **Erratum: "The far–infrared morphology of the double–ringed galaxy NGC 4736 (M94): a ring surrounding an extended nucleus" [Astrophys. J., Vol. 373, No. 1, p. 66 – 74 (20 May 1991)].**
B. J. Smith, D. F. Lester, P. M. Harvey, R. W. Pogge.
Astrophys. J., Vol. 380, No. 2, p. 677 (20 Oct 1991). See Abstr. 53.157.303.

157.254 **Discovery of a low–redshift ultraluminous "E + A" galaxy.**
W. R. Oegerle, J. M. Hill, J. G. Hoessel.
Astrophys. J., Lett., Vol. 381, No. 1, p. L9 – L12 (1 Nov 1991). With plate L1.
The authors report the serendipitous discovery of a relatively low redshift ($z = 0.0875$) "E + A" (poststarburst) galaxy. The galaxy is remarkable for its extremely high luminosity, strong Balmer absorption lines, and morphology, indicating that a merger precipitated the starburst ~1 Gyr ago.

157.255 **The 0.6 Mpc halo of the cD galaxy (MCG 05–33–002) in the cluster Abell 3571 revealed by the co–addition of eight Schmidt plates.**
S. N. Kemp, J. Meaburn.
Mon. Not. R. Astron. Soc., Vol. 251, No. 1, p. 10P – 13P (1 Jul 1991).
Eight IIIaJ sky–limited plates of the same field taken with the UK 1.2–m Schmidt telescope have been scanned by the APM machine and digitally co–added. The field contains the cD galaxy MCG 05–33–002, the central galaxy in the cluster A3571. A giant faint halo, of dimensions ~600x200 kpc, has been found to envelop most of the cluster. It is estimated that at least a mass of $2 \times 10^{12} M_\odot$ is contained within the halo which would account for

some of the "missing" mass in such clusters in terms of purely luminous baryonic matter.

157.256 **The H I content of the Local Group dwarf (spheroidal or irregular?) galaxy Phoenix.**
C. Carignan, S. Demers, S. Côté.
Astrophys. J., Lett., Vol. 381, No. 1, p. L13 – L16 (1 Nov 1991).
Neutral hydrogen was detected in the Local Group dwarf galaxy Phoenix. The 21 cm H I line emission, observed at a heliocentric radial velocity of 56 km s^{-1}, is clearly separated from a much larger scale component at $\sim$140 km s^{-1} which was previously associated with the Magellanic Stream. The profile width $\Delta V_{50} \cong 21$ km s^{-1}, combined with the high inclination of the system, suggests that not much rotation is present in this system which would then be mainly supported by turbulent motions. Using a newly determined distance estimate, the integrated profile gives a total H I mass of $\cong 1.0 \times 10^5 M_\odot$ for an $M_{HI}/L_V \cong 0.07 M_\odot/L_\odot$. Most of the known properties of Phoenix imply that it is intermediate between typical dwarf irregular and dwarf spheroidal galaxies. It is suggested that, in fact, the two types of galaxies may belong to the same class of objects having a different history of star formation.

157.257 **Corrugations in the discs of spiral galaxies. NGC 4244 and 5023.**
E. Florido, E. Battaner, M. Prieto, E. Mediavilla, M. L. Sanchez-Saavedra.
Mon. Not. R. Astron. Soc., Vol. 251, No. 2, p. 193 – 198 (15 Jul 1991).
Corrugations of the disc shape, which are a well known phenomenon in the Milky Way, have been observed in the edge-on galaxies NGC 4244 and 5023. The morphological characteristics are described in this paper. The bluer maps show larger and sharper corrugations. The characteristic wavelength is larger in the redder maps. The deformation seems to have a "centre", decreasing from this point not only inwards but also outwards. The possible connection between corrugations and warps is discussed.

157.258 **Stellar distribution in H II regions of M33.**
G. R. Ivanov.
Mon. Not. R. Astron. Soc., Vol. 251, No. 2, p. 281 – 283 (15 Jul 1991).
About 200 H II regions are coincident with OB stellar concentrations in M33. Their characteristic length is $\cong$80 pc and they appear to cluster in the same classical stellar associations as those of Humphreys and Sandage. These small dense groups are real OB associations because their stellar content is the same as those in the Galaxy. The radial distribution of blue stars in the H II regions shows that a strong concentration of OB stars is observed at the centre of bright H II regions. The dependence of stellar density on the distance from the centre of H II regions is proportional to $r^{-2.2}$.

157.259 **Evidence for dark matter from rotation curves: ten years later.**
V. C. Rubin.
AIP Conf. Proc., No. 222, p. 371 – 380 (15 Apr 1991). – See Abstr. 012.069 for the main entry.
During the past decade, observations have confirmed that rotation curves are generally flat or rising across the optical galaxies out to the limits of neutral hydrogen. Regrettably, parameters of the dark halos can be evaluated only in terms of assumed mass distributions. However, it seems well established that dark matter forms an increasingly dominant component with decreasing galaxy luminosity. A major advance of the decade has been the observation that in some gas-rich dwarf irregulars the mass in dark matter exceeds that in luminous matter at virtually all radii. Interior to the final measured point in DDO 154, 90% of the mass is dark, and integral $M/L \geqslant 75$.

157.260 **Dark matter, mergers, and morphology: clues from galaxy pairs.**
J. C. Charlton, E. E. Salpeter.
AIP Conf. Proc., No. 222, p. 385 – 388 (15 Apr 1991). Current Physics Microform No.: 9106E0385. – See Abstr. 012.069 for the main entry.
The authors have examined the distributions of radial velocity differences Δv and projected separations r_p of pairs in low density regions of the CfA and SSRS redshift catalogs. Bound pairs exist even at relatively large projected separations (up to 1 Mpc), indicating large amounts of dark matter around galaxies. The distribution of r_p for bound pairs is flat, with significant numbers of pairs at projected separations less than 50 kpc. Pairs at small separations merge rapidly, but the distribution is replenished by those originally at larger separation that are being drawn toward merger by dynamical friction. As a result of this merger/replenishment process, at least 20% of galaxies in low density regions are the product of a merger; thus it seems that nearly all ellipicals and S0's in these regions are created in this way.

157.261 **Ultramassive dark object in the luminous infrared galaxy NGC 6240?**
J. Bland-Hawthorn, A. S. Wilson.
AIP Conf. Proc., No. 222, p. 389 – 392 (15 Apr 1991). Current Physics Microform No.: 9106E0389. – See Abstr. 012.069 for the main entry.
Fabry-Perot Hα observations of the superluminous infrared galaxy NGC 6240 have provided the first kinematic evidence that this system comprises at least two merging disks. One of the disks appears to be strongly associated with the main body of the 2 μm continuum; the other disk displays remarkable kinematics and has no obvious counterpart in the continuum.

157.262 **Systematic properties of dark matter distribution.**
S. Casertano, J. H. van Gorkom.
AIP Conf. Proc., No. 222, p. 393 (15 Apr 1991). Abstract. Current Physics Microform No.: 9106E0393. – See Abstr. 012.069 for the main entry.

157.263 **Neutral hydrogen observations of galaxies in superclusters.**
M. P. Haynes, R. Giovanelli.
Astrophys. J., Suppl. Ser., Vol. 77, No. 3, p. 331 – 362 (Nov 1991).
As part of a continuing study of the three-dimensional distribution of galaxies in the nearby extragalactic sky, neutral hydrogen observations of spiral galaxies in the Uppsala General Catalog were made with the late 91 m telescope in Green Bank. In this paper, results of the H I observations are presented for 309 objects north of declination +38° in the right ascension range $10^h <$ R.A. $< 22^h$ and for a sample of 46 objects just south of the equator in the region around $0^h <$ R.A. $< 3^h$. Many of these objects are outlying members of superclusters and may be useful in the study of deviations from the Hubble flow on supercluster scales. Velocity widths as measured by five different algorithms are presented along with a profile quality index to aid in applications of the Tully-Fisher relation using this data set.

157.264 **Dissipationless collapse and the shape of isophotes.**
M. Stiavelli, P. Londrillo, A. Messina.
Mon. Not. R. Astron. Soc., Vol. 251, No. 3, p. 57P – 59P (1 Aug 1991).
By analysing the "photometric" properties of the end-products of dissipationless collapse the authors have shown that a system can appear either boxy or discy depending on the viewing angle. Some consequences of this result are illustrated.

157.265 **Tidally triggered galaxy formation. I. Evolution of the galaxy luminosity function.**
C. Lacey, J. Silk.
Astrophys. J., Vol. 381, No. 1, p. 14 – 32 (1 Nov 1991).
Motivated by accumulating evidence that large-scale galactic star formation is initiated and sustained by tidal interactions, the authors develop a phenomenological model for the galaxy

luminosity function, commencing from a galaxy mass function that is predicted by a hierarchical model of structure formation such as the cold dark matter dominated cosmology. The epoch of luminous galaxy formation and the galactic star–formation rate are determined by the environment. Gas cooling and star–formation feedback are incorporated, and the model is able to satisfactorily reproduce the present epoch luminosity function of bright galaxies and the distribution of galaxy colors. Biasing, via the preferential formation of luminous galaxies in denser regions associated with groups or clusters, is a natural outcome of this tidally triggered star–formation model. A large frequency is inferred of "failed" galaxies, prematurely stripped by supernova–driven winds, that populate groups and clusters in the form of low surface brightness gas–poor dwarfs, and of "retarded" galaxies, below the threshold for effective star formation, in the field, detectable as gas–rich, extremely low surface brightness objects. Predictions are presented for the evolution with redshift of the distribution of characteristic star formation time–scales, galaxy ages, and colors.

157.266 **Massive black holes in the nuclei of spiral galaxies: M31 and the problem of the initial conditions.**
M. Bradley, A. Curir, F. de Felice.
Astrophys. J., Vol. 381, No. 1, p. 72 – 78 (1 Nov 1991).
The formation of a stationary black hole is constrained by the conditions $M > 3 M_{\odot}$ and $cJ/GM^2 \equiv a/m < 1$, J and M being the total angular momentum and mass of the configuration which has collapsed to the hole. Recent observations indicate that some spiral galaxies, like M31, harbor a massive black hole in their center. In the case of M31, for example, the mass of the hole is estimated to be of the order of 10^6 - $10^7 M_{\odot}$. The authors calculate the a/m content of a fluid model for the bulge of a spiral galaxy and search for the limiting mass and angular momentum distributions which would be compatible with the formation of a massive black hole in the center. The model is then compared with the data available for the nucleus of M31 to see how far the observed properties differ from the hypothetical initial condition.

157.267 **Possible deflection of a jet by molecular clouds in NGC 4258.**
R. L. Plante, K. Y. Lo, J.–R. Roy, P. Martin, L. Noreau.
Astrophys. J., Vol. 381, No. 1, p. 110 – 117 (1 Nov 1991).
The CO ($J = 1 \rightarrow 0$) emission from the central 1′ of NGC 4258, a SABbc spiral galaxy with a pair of anomalous arms, was mapped using the Owens Valley millimeter–wave interferometer. The CO emission is found to arise from a ringlike structure, and the bulk of the detected CO appears to coincide with dust features seen in optical images. The CO distribution also appears to confine the inner part of the anomalous Hα arm in NGC 4258. To account for the unusual velocity field of the CO emission, the authors obtain a simple kinematic model for the motion of the molecular gas, where expansion away from the galaxy's center is superposed on the elliptical motion. The authors identify a feature in the CO emission whose motion is perturbed with respect to this model and suggest that this deviation is evidence for a deflection of the line–emitting jet by a dense molecular cloud.

157.268 **Orbit crowding of molecular gas at a bar–spiral arm transition zone in M83.**
J. D. P. Kenney, S. D. Lord.
Astrophys. J., Vol. 381, No. 1, p. 118 – 129 (1 Nov 1991).
High–resolution (9″ × 6″) CO ($J = 1 \rightarrow 0$) observations in the barred spiral galaxy M83 reveal a massive molecular gas complex associated with luminous H II regions at the southern bar–spiral arm transition zone. The cloud complex is resolved into two principal components, one of which is clearly linked to the bar, the other to the spiral arm. The bar component is located along a ridge of highly polarized radio continuum emission at the place where the ridge bends away from following the bar to following the spiral arm. The spiral arm component is located at the same galaxy azimuth as the bar component, but 250 pc further out in radius, just upstream from a ridge of H II regions on the spiral arm. The spiral arm component has a tangential velocity 60 km s^{-1} (in the plane of the disk) greater than the bar component. The CO morphology and kinematics are consistent with the entire molecular complex originating through orbit crowding, in which gas on highly elliptical orbits from the bar region converges with gas on more circular orbits from the region just beyond the bar. The orbit crowding is strongly enhanced by radial streaming motions due to density waves, which are in opposite directions along the bar and spiral arm, thereby producing a region of streamline convergence where the bar meets the spiral arm.

157.269 **A molecular gas ridge offset from the dust lane in a spiral arm of M83.**
S. D. Lord, J. D. P. Kenney.
Astrophys. J., Vol. 381, No. 1, p. 130 – 136 (1 Nov 1991). With plate 3.
The authors present a high–resolution (5″ × 12″) interferometric map of the CO ($J = 1 \rightarrow 0$) emission on the eastern spiral arm of M83. The detected emission originates in about five unresolved components located parallel but ~300 pc downstream from the dust lane which lies along the inner edge of the spiral arm. With one exception, the CO components in the map are all located within 130 pc of an H II region, and may represent emission from locally heated gas. While heating by H II regions may enhance the CO luminosity in the arm, the lack of CO emission on the dust lane indicates that the dense molecular gas does not pile up here in M83. The distribution of molecular emission in this spiral arm of M83 is shown to be remarkably different from the spiral arms of M51, where CO emission peaks on the dust lane. The difference between the molecular gas distributions in the two galaxies is attributed to the difference in the strength of their density waves. The observations of M83 are consistent with the model of Elmegreen, in which diffuse gas is compressed at the shock front producing the dust lane at the inner edge of the spiral arm, while dense giant molecular clouds pass through the front and form a broad distribution on the arm.

157.270 **Evolution of dust grains through a hot gaseous halo.**
A. Ferrara, F. Ferrini, J. Franco, B. Barsella.
Astrophys. J., Vol. 381, No. 1, p. 137 – 146 (1 Nov 1991).
The evolution of bare spherical dust grains in the halo of spiral galaxies is analyzed. Two different grain compounds, graphites and astronomical silicates, are considered. The detailed mass and luminosity distributions for the Milky Way and NGC 3198 (considered as representative of the Sb and Sc types, respectively) are used to evaluate the range of possible evolutionary tracks. Aside from radiative and gravitational forces, the effects of drag and sputtering from a gaseous halo are included. A simple isothermal and hydrostatic density structure, with temperatures in the range 3×10^5 – 10^6K, has been used for this gaseous halo. Most results depend on the optical properties of dust compounds, because the velocities define the final fate of the grains (i.e., destruction in the halo or expulsion from the host galaxy). Graphites move faster than silicates, but both types of grains can reach values in excess of 10^2km s^{-1}. The efficiency for grain destruction increases with (1) increasing halo gas temperature, (2) decreasing galaxy mass, and (3) decreasing radiation pressure coefficients. Thus, silicates behave as a "fragile" grain population in low mass galaxies with a hot (~10^6K) corona. For a central halo gas density of 0.1 cm^{-3}, spherical silicates with an initial radius of 10^{-5}cm lose almost 20% of their mass in time scales between 10^8 and 2×10^8yr.

157.271 **The ^{12}CO to H_2 ratio in the centre of M82.**
P. A. Smith, P. W. J. L. Brand, C. M. Mountain, P. J. Puxley, N. Nakai.
Mon. Not. R. Astron. Soc., Vol. 252, No. 1, p. 6P – 11P (1 Sep 1991).
The authors present 450–μm continuum observations of M82 at high signal–to–noise and spatial resolution. A 1′ scan along the major axis shows a main peak SW of the nucleus with a secondary peak to the NW. This morphology differs from that of ^{12}CO J = 1–0 at the same resolution, but is quite similar to that of optically thin ^{13}CO and $C^{18}O$. The authors use their 450–μm data to evaluate the CO to H_2 conversion factor, $N(H_2)/I_{CO}$, and find that it varies along the major axis and is up to a factor of 3

lower than the Milky Way value, suggesting that the galactic conversion factor may overestimate $N(H_2)$ in "starburst" galaxies. Recent multitransition observations of CO and other molecules in M82 point towards a clumpy, disturbed interstellar medium subject to intense UV fields, in which it is unlikely that the ^{12}CO reliably traces molecular hydrogen. Dust continuum observations, on the other hand, offer a far more direct and straightforward tracer of H_2, that does not depend so critically on the excitation and chemistry of the interstellar medium.

157.272 **The scattered Hα emission–line filaments in M82.**
S. M. Scarrott, N. Eaton, D. J. Axon.
Mon. Not. R. Astron. Soc., Vol. 252, No. 1, p. 12P – 16P (1 Sep 1991).

The authors present a linear polarization map of the starburst galaxy M82, obtained in the light of monochromatic Hα. Polarization levels, ~10–30%, are observed throughout the filamentary system and the centro–symmetric pattern formed by the polarization vectors demonstrates that the emission–line filaments are part of a huge bipolar reflection nebula. The circular symmetry of the polarization pattern requires a central illuminating source (≤4–arcsec diameter) which, due to light travel time arguments, must have dominated the Hα emission of the galaxy for at least 5000 yr. This source is located in a small active zone within the central starburst region very close to the strong radio source 41.9+58 and the centre of the IR emission longward of 10 μm. The observation that the filaments show up in polarized light implies that they must be due, at least in part, to illumination effects.

157.273 **Diameter functions of UGC and ESO galaxies.**
M. J. Hudson, D. Lynden–Bell.
Mon. Not. R. Astron. Soc., Vol. 252, No. 2, p. 219 – 228 (15 Sep 1991).

An analysis of the diameter function of galaxies is presented for three redshift surveys. The completeness of the UGC and ESO catalogues is investigated, the completeness of the redshift surveys is calculated and density–independent maximum likelihood estimators are applied in determining the parameters of the diameter functions. The diameter function is well described by an exponential with characteristic diameter $D^*_{UGC} = 6.38 \pm 0.15\, h^{-1}$kpc; $D^*_{ESO} = 7.56 \pm 0.19\, h^{-1}$kpc; $D^*_{25} = 5.72 \pm 0.2\, h^{-1}$kpc for diameters greater than $4.36\, h^{-1}$kpc. The ESO and UGC diameter functions are compatible with each other when they are scaled to the same depth.

157.274 **An optical and near–infrared polarization survey of nearby starburst and interacting galaxies.**
C. Brindle, J. H. Hough, J. A. Bailey, D. J. Axon, W. B. Sparks.
Mon. Not. R. Astron. Soc., Vol. 252, No. 2, p. 288 – 299 (15 Sep 1991).

The authors present broadband optical and near–infrared polarization and flux density measurements of a sample of 27 starburst and interacting galaxies. The observed polarizations are generally small ($P < 1\%$) and decrease with increasing wavelength. For those galaxies with significant polarization, the polarization E vector is usually parallel to a dust lane. The most likely polarization mechanism is the transmission of radiation through aligned dust grains lying in the plane of the galaxy. In some cases scattering by dust grains may contribute to the polarization. NGC 253 shows high polarization (4.3% in the V band), with a strong wavelength–dependence resembling a Serkowski curve having a small value of λ_{max} (0.34 μm). Modelling shows that this wavelength–dependence can also be produced by light polarized according to the Serkowski formula with a normal value of λ_{max} (0.55 μm) combined with unpolarized light with a redder spectrum.

157.275 **The common halo and tidal features of the interacting pair of galaxies NGC 5215a and b revealed by the co–addition of eight Schmidt plates.**
S. N. Kemp, J. Meaburn.
Mon. Not. R. Astron. Soc., Vol. 252, No. 2, p. 27P – 30P (15 Sep 1991).

The interacting pair of galaxies NGC 5215a and b has been investigated by means of a co–added data array of APM scans of eight Schmidt plates. The pair is found to be embedded in a common halo, and each member has a tidal tail which trails the rotation of the core region. Distortions of the isophotes of the pair confirm the tidal interaction between the two galaxies. A detached filament to the north of the pair is presumed to consist of stars tidally stripped from the pair, and the halo is considerably extended to the south, presumably due to further stars becoming unbound. The mass within the common halo is at least $\sim 8 \times 10^{10} M_\odot$. The large spiral NGC 5220, at a similar distance, forms a much disturbed chain with the components of NGC 5215 and some smaller galaxies.

157.276 **Galactic magnetic fields: Spinning a tangled web.**
E. Zweibel.
Nature, Vol. 352, No. 6338, p. 755 – 756 (29 Aug 1991).

157.277 **Correlation of magnetic and optical structure in the barred spiral galaxy M83.**
N. Neininger, U. Klein, R. Beck, R. Wielebinski.
Nature, Vol. 352, No. 6338, p. 781 – 783 (29 Aug 1991). Letter–to–the–editor.

The authors have observed the barred spiral galaxy M83 using the 100–m Effelsberg telescope at a frequency of 10.6 GHz, at which Faraday rotation should be less than 1°. There is significant polarization in the inner part of the galaxy, where they find that the magnetic field closely follows the optical structure. They see some deviations from the ordered magnetic structure that coincide with optical patchiness, and two pronounced polarization minima that correspond with enhancements of CO emission. The results suggest that the inner structure of the galaxy is more orderly than earlier work had indicated, and that any gross disorder, perhaps due to star formation, occurs only in the innermost regions. In addition, the observed distribution of Faraday rotation across the galaxy, deduced from comparison of observations at different frequencies, provides evidence that a significant component of the magnetic field extends away from the disk.

157.278 **Morphological differences between optical and infrared images of the spiral galaxy NGC 309.**
D. L. Block, R. J. Wainscoat.
Nature, Vol. 353, No. 6339, p. 48 – 50 (5 Sep 1991). Letter–to–the–editor.

The morphological classification of spiral galaxies into various types is suspected to be highly dependent on the wavelength of observation, because optical images emphasize young population I stars at the expense of other stellar types, as well as ionized gas and dust. Extension of the classification of galaxies out to wavelengths of a few micrometres has had to await the development of large–format near–infrared array cameras. The authors present images at 2.1 μm wavelength of NGC 309, one of the largest "grand design" (type ScI) spiral galaxies, obtained with the 256×256 array camera developed for the NICMOS (near–infrared camera and multiobject spectrograph) instrument, designed for installation on the Hubble Space Telescope. Optically, NGC 309 presents a classic multi–arm morphology, but at 2.1 μm one sees a two–arm spiral and the appearance of a prominent central bar; it resembles the SBa galaxy NGC 1358. These studies underscore existing indications that the disk structure of spiral galaxies may be unrelated to the Hubble type assigned from the transient population I morphology.

157.279 **Calibrating Cepheid sequences in nearby galaxies.**
N. Metcalfe, T. Shanks.
NATO Advanced Research Workshop on Observational Tests of Inflation, p. 187 – 192 (1991). – See Abstr. 012.076 for the main entry.

Using the CCD camera at prime focus of the 2.5 m Isaac Newton telescope on La Palma, the authors have checked the stellar photographic sequences on which Cepheid photometry was originally based in four local Tully–Fisher calibrating galaxies, M31, M33, NGC 2403 and M81. Correcting their distance moduli on the basis of their CCD recalibration the authors find an infra-red Tully–Fisher relation of $M_H = -10.3 (\log V_{21} - 2.5) - 21.4$, with a scatter of 0^m.44.

157.280 **Dependence of galaxy properties on viewing angle.**
D. Burstein, M. P. Haynes, S. M. Faber.
Nature, Vol. 353, No 6344, p. 515 – 521 (10 Oct 1991).
Galaxies are three-dimensional objects projected onto the sky at random angles of inclination. Deduction of their true structure from their appearance requires an understanding of the variation of their apparent diameter, luminosity and surface brightness with viewing angle. But these variations in turn depend on galactic structure. It has taken astronomers several decades to realize that only with a direct knowledge of galactic distances can these two effects be disentangled.

157.281 **Correlations of spin angular momenta of galaxies.**
M. Iye.
IUPAP Conference on Primordial Nucleosynthesis and Evolution of Early Universe, p. 403 – 408 (1991). – See Abstr. 012.071 for the main entry.
The distributions of galaxy spin in the Local Super Cluster, groups of galaxies, and pairs of galaxies were studied by using a data base of spiral winding sense of galaxies with an assumption that all the spirals are trailing. The author argues that there is a possibility that the actual distribution of galaxy spin shows a dipole or a quadrupole component depending on the scenarios of galaxy formation. No significant correlation, however, was identified in any ensemble. Implications of the present study on the origin of the spin angular momenta of galaxies are discussed.

157.282 **A post-core-collapse model for the nucleus of M33.**
L. Hernquist, P. Hut, J. Kormendy.
Nature, Vol. 354, No. 6352, p. 376 – 377 (5 Dec 1991). Letter-to-the-editor.
Recent high-resolution optical observations show that the nucleus of the nearby spiral galaxy M33 has a small core radius and an unusually low velocity dispersion. It is suggested that the nucleus of M33 has undergone core collapse. The authors propose a model for the post-collapse core of M33. By analogy with globular clusters, the formation of tight binary systems powers the re-expansion of the core, and it is expected in particular that low-mass X-ray binaries form with a rate about equal to that of all galactic globular clusters combined. About a dozen such binaries should be present, and their combined emission may explain the large (10^{39} erg s^{-1}) and enigmatic unresolved X-ray emission from the nucleus of M33. In addition, tidal formation of cataclysmic binaries may lead to a nova rate of $\gtrsim 1$ per century. Numerous blue stragglers have probably formed through mergers caused by stellar collisions, but their density is probably too low to explain the blue colour of the nucleus. Young stars are the likely cause for the colour, and their presence may complicate the simple dynamical picture presented.

157.283 **Dwarf galaxies' vanishing act.**
C. Lacey.
Nature, Vol. 354, No. 6353, p. 430 – 431 (12 Dec 1991).

157.284 **A sample of ultraluminous IRAS galaxies in southern sky.**
Z.-L. Zou, X.-Y. Xia, Z.-G. Deng.
IUPAP Conference on Primordial Nucleosynthesis and Evolution of Early Universe, p. 571 – 572 (1991). – See Abstr. 012.071 for the main entry.

157.285 **Were small galaxies once the dominant cosmological population?**
L. L. Cowie, A. Songaila, E. M. Hu.
Nature, Vol. 354, No. 6353, p. 460 – 461 (12 Dec 1991). Letter-to-the-editor.
Deep sky surveys have turned up unexpectedly large numbers of faint blue galaxies. Unless distances to these faint galaxies are known, it is difficult to distinguish between luminosity evolution (in which case, galaxies in the past were systematically brighter) and density evolution (as, for example, if modern galaxies are the results of mergers of earlier ones). The authors have obtained redshifts and K magnitudes for a small but complete sample of 22 galaxies with B magnitude down to 24. In the luminosity range B = 23 – 24, the B-band galaxy counts are dominated by a population of small blue galaxies at $z \approx 0.25$, which may collectively contain as much baryonic matter as the normal galaxies. It is possible either that these earlier galaxies have undergone merging to create the present galaxy population, or that they represent a quite different galactic population which has now faded or disappeared. Either possibility has considerable implications for our understanding of galaxy formation.

157.286 **Where galaxies collide. I. NGC 3077 and star formation in the M81 system.**
H. A. Thronson Jr., C. Wilton, A. Ksir.
Mon. Not. R. Astron. Soc., Vol. 252, No. 4, p. 543 – 549 (15 Oct 1991).
The authors present and discuss deep B_J (0.47 μm), R (0.65 μm), I (0.85 μm), and Hα images of the dwarf elliptical galaxy NGC 3077, which is suffering a period of enhanced star formation presumably as a result of its acquisition of gaseous material in an encounter with its neighbour, the giant spiral M81. There is a set of distinct partial shells in the R image, indicative of ingestion of matter by NGC 3077. The authors estimate a current star formation rate of about 0.1 $M_\odot$ yr^{-1} over a region about 700 pc across centred on the nucleus. This rate per surface area is about two orders of magnitude greater than the average over the disc of normal spirals. The derived efficiency of star formation is higher than that found for isolated spirals, although there are large uncertainties in the assumptions that go into this calculation. The authors' emission line image shows extremely sharp borders, typical of ionization-bounded H II regions. They compare the characteristics of star formation and the interstellar medium in NGC 3077 with that in the other dwarf member of the M81 system, M82. Althouth the total mass of gas in the two galaxies is roughly similar, M82 seems to be much more effective at turning it into molecules, the fuel for star formation. The authors speculate on a few reasons for this.

157.287 **The FIR properties of spiral parent galaxies of supernovae.**
A. R. Petrosian (*A. R. Petrosyan*), M. Turatto.
ESO/EIPC Workshop on Supernova 1987A and other Supernovae, p. 627 – 630 (Jun 1991). – See Abstr. 012.075 for the main entry.
The authors present the preliminary results of the study of the FIR properties of SN parent galaxies, in particular of Type II and Ib.

157.288 **Where galaxies collide. II. NGC 5195, the M51 system and the luminosity function of H II regions.**
H. A. Thronson Jr., H. Rubin, A. Ksir.
Mon. Not. R. Astron. Soc., Vol. 252, No. 4, p. 550 – 557 (15 Oct 1991).
The authors investigate the characteristics of the interstellar medium and massive star formation in the lenticular galaxy NGC 5195 and adjacent and overlying parts of the Whirlpool spiral (M51, NGC 5194) via broad-band red and Hα imaging. Although NGC 5195 has a moderate amount of interstellar gas, the authors find no visual line emission that is unambiguously associated with star formation in this galaxy. The authors use their R (0.65 μm) and I (0.8 μm) images to derive a map of the obscuring dust distribution toward this galaxy. They find good agreement between the distribution of the obscuration and published 21-cm H I maps. The estimate of A_V/N(H I) toward NGC 5195 is in fair agreement with that found for the Milky Way. From their map of the dust extinction, the authors correct their observed red image to reveal the unobscured stellar light distribution. The resultant image looks like a nearly normal S0 galaxy, but with a faint bar-like structure and ring. The Hα image also includes most of the northern half of the spiral, which allows the authors to investigate star formation where the galaxies have apparently sideswiped each other. They identify about 100 individual H II regions in their image, from which they derive the

H II region luminosity functions (LFs) for the pair of northern spiral arms. The authors find that there is no obvious variation in this function with position along or across the arms nor between this portion of M51 and the entire galaxy, although the LF for the two arms is different. They propose that the shape of the high-luminosity end of the H II region LF in galaxies is determined largely by the mass function of the molecular clouds.

157.289 **The proportion of ultraluminous IRAS galaxies in interacting systems.**
Zou Zhenlong, Xia Xiaoyang, Deng Zugan, Su Hongjun.
Mon. Not. R. Astron. Soc., Vol. 252, No. 4, p. 593 - 599 (15 Oct 1991).
The authors have examined a sample consisting of 41 ultraluminous IRAS galaxies with far-infrared luminosities LIR > $10^{12}L_\odot$ (H_0 = 50 km $s^{-1}Mpc^{-1}$) and a control sample of 41 low-luminosity IRAS galaxies with LIR < $10^{9.7}L_\odot$. All objects are taken from the southern part ($\delta < -17.^\circ5$) of the QDOT (QMW-Durham-Oxford-Toronto) survey, a random one-in-six sample with IRAS 60-μm flux density S(60) > 0.6 Jy. Their images on the film copies of the SERC Sky Survey were scanned with a pixel size of 5 μm (0.34 arcsec) over a square area of typically 67.2 arcsec to the side, using the microdensitometer PDS of the Purple Mountain Observatory. The density contour maps and the sky survey material show that the proportion of ultraluminous IRAS galaxies which are interacting or merger systems, 61±12%, is much higher than that of low-luminosity ones (10±5%), but still well short of the 100% suggested in several recent reports. This result seems to support the conclusion that galaxy interaction or merging is more common at high-IR luminosity, but it is probably not the only cause of their activity.

157.290 **Hidden mass in galaxies.**
A. V. Zasov.
Zemlya Vselennaya, No. 4, p. 3 - 13 (Jul - Aug 1991). In Russian.

157.291 **The role of SNs in the chemical evolution of galaxies.**
F. Matteucci.
ESO/EIPC Workshop on Supernova 1987A and other Supernovae, p. 703 - 708 (Jun 1991). - See Abstr. 012.075 for the main entry.
The chemical evolution of galaxies of different morphological type is studied by assuming the same prescriptions for SN (Ia, Ib and II) progenitors and their nucleosynthesis. The Hubble sequence is thought of as a sequence of decreasing collapse and star formation rate from early to late-type galaxies. In this framework, ellipticals and bulges evolve very rapidly as opposed to irregular galaxies which evolve very slowly. Differences in the IMF between galaxies are not excluded. In particular, a flatter IMF in the domain of massive stars with respect to the galactic disc, is expected in bulges and ellipticals in order to fit the majority of their properties. On the other hand, the properties of irregulars such as the Magellanic Clouds seem to be well reproduced by using the same IMF as in the galactic disc.

157.292 **A CCD Hα survey of nearby southern spiral galaxies.**
S. D. Ryder, M. A. Dopita.
Proc. Astron. Soc. Aust., Vol. 9, No. 1, p. 89 - 90 (1991). - See Abstr. 012.078 for the main entry.
Some initial results of a flux-calibrated CCD Hα imaging program of bright, nearby southern spiral galaxies are presented. Very few southern hemisphere spiral galaxies have ever been completely imaged in Hα, let alone with a CCD. This survey (which mainly uses the MSSSO 1.0-m reflector with an f/3.5 focal reducer) will, when combined with spectrophotometry of the H II regions thus revealed, allow to trace the chemical evolution of each galaxy. Furthermore, since the absolute Hα flux is a reliable measure of the high-mass star formation rate in a galaxy, such observational data will permit to test the predictions of the various star formation theories.

157.293 **Computer statistics for three-body systems and mass determination in triplets of galaxies.**
A. D. Chernin, S. Mikkola.
Mon. Not. R. Astron. Soc., Vol. 253, No. 1, p. 153 - 159 (1 Nov 1991).
The authors have simulated numerically the observations of triple galaxies to obtain a proper way to use the virial theorem for mass estimations. The results are applied to the triple galaxies catalogued by Karachentsev (1987). The dynamics was simulated using both the point-mass model and softened potential models. The results indicate that significant deviation from the point-mass model requires that the typical intergalactic distances are comparable to the half-mass radii of the galaxies. Such physical conditions would lead to rapid mergers of the galaxies. Thus the existence of the observed triplets suggests that they are at sufficient distances from each other to justify the use of the point-mass model at least for their present dynamical state. The determination of individual triple system masses is not possible: unpredictable errors can extend to orders of magnitude, but statistical estimation of a characteristic mass for observed samples of galaxy triplets seems possible in theory. The mass estimates the authors obtain for the Karachentsev triplets have almost an order of magnitude excess compared to the visible mass, when a standard mass-to-light ratio is assumed.

157.294 **Global magnetic structure in spiral galaxies.**
S. J. Spencer.
Proc. Astron. Soc. Aust., Vol. 9, No. 1, p. 95 - 97 (1991). - See Abstr. 012.078 for the main entry.
Global magnetic structure has been inferred in spiral galaxies from radio and optical observations. The author has developed a three dimensional, quasi-global kinematic dynamo model which demonstrates the possibility of maintaining large-scale magnetic fields by the $\alpha\omega$ mechanism. Analytical expressions for large-scale galactic magnetic fields near temporal steady states have been derived. Illustrations of the spatial distribution of vector and absolute large-scale magnetic field strength are presented. When coupled with observations this model may help in understanding the maintenance and morphologies inferred for disk and halo global magnetic fields in spiral galaxies.

157.295 **The Durham/UKST galaxy redshift survey.**
A. Broadbent, D. Hale-Sutton, T. Shanks, R. Fong, F. G. Watson, A. P. Oates, C. A. Collins, H. T. MacGillivray, R. Nichol, Q. A. Parker.
NATO Advanced Research Workshop on Observational Tests of Inflation, p. 447 - 448 (1991). Abstract. - See Abstr. 012.076 for the main entry.

157.296 **Mass loss in dwarf galaxies.**
G. R. Meurer.
Proc. Astron. Soc. Aust., Vol. 9, No. 1, p. 98 - 99 (1991). - See Abstr. 012.078 for the main entry.
Four examples of dwarf galaxies apparently undergoing mass loss are discussed. All have an Hα morphology suggesting a minor axis flow. This is supported by velocity field measurements in three cases. All also have at least one compact central object, probably a young star cluster, apparently powering the flow. Some implications of these results are briefly discussed.

157.297 **CCD multi-object spectroscopy of galaxies at the UKST.**
F. G. Watson, A. P. Oates, T. Shanks, D. Hale-Sutton.
Mon. Not. R. Astron. Soc., Vol. 253, No. 2, p. 222 - 234 (15 Nov 1991).
The authors have built a CCD detector system for use with the FLAIR multi-object spectrograph at the UKST. They show that using FLAIR, velocities can be obtained for ~80% of galaxies sampled uniformly to $b_J \lesssim 16.75$ mag. The velocity accuracy for absorption-line redshifts is typically ±150 km s^{-1}. These results apply to observations made in the blue (~4000-5000 Å) spectral range; the proportion of galaxies for which absorption-line redshifts can be obtained is significantly higher than in initial tests when only long wavelength (>5000 Å) observations were

possible. Using this observational configuration the authors have completed a survey of galaxies in nine UKST fields; here they have obtained redshifts for 179 galaxies sampled at a rate of one in six from catalogues complete to this limiting magnitude. These results confirm that CCD fibre spectroscopy at the UKST is now capable of major spectroscopic surveys of such astronomical targets. In particular, there is the possibility of using the UKST to make a 3D map of the large-scale distribution of galaxies in the southern hemisphere, via a galaxy redshift survey to $b_J \approx 17$ mag.

157.298 **Abundance analysis of giant H II regions in M51.**
A. I. Díaz, E. Terlevich, J. M. Vílchez, B. E. J. Pagel, M. G. Edmunds.
Mon. Not. R. Astron. Soc., Vol. 253, No. 2, p. 245 – 255 (15 Nov 1991).

The authors present spectrophotometric observations in the optical and near infrared of six giant H II regions in the high metallicity spiral galaxy M51. Five of the regions are located in the spiral arms and the sixth one is very close to the nucleus. These observations and the application of adequate photo-ionization models allow the determination of the functional parameters of the regions: ionization parameter, effective temperature of the ionizing radiation and metallicity. The results of this work confirm previously found trends of decreasing ionization parameter and radiation hardness with increasing metallicity. N/O and S/O ratios are in general higher and lower than solar respectively. The first trend is expected from the view point of stellar nucleosynthesis and chemical evolution theories. However, the existence of a slower S/O ratio in regions of high metallicity, if confirmed as real, needs to be accounted for.

157.299 **Supernovae and the spectra of high-redshift galaxies.**
M. Bithell.
Mon. Not. R. Astron. Soc., Vol. 253, No. 2, p. 320 – 340 (15 Nov 1991).

The author estimates the effect of supernova remnants (SNR) on the UV continuum of galaxies. He shows that the radiation from the remnants may lead to a very small rest frame Lyman break. The author calculates the appearance of such galaxies at high redshift, and concludes that even galaxies with negligible rest frame Lyman break will have large observed breaks because of absorption by intervening Lyα clouds. Star-forming galaxies at high redshifts may thus supply the ionizing radiation background without this radiation being observable in the spectrum. Emission from SNR also leads to a much harder galactic spectrum, especially shortward of 228 Å. This means that high-ionization lines observed in high-redshift radio galaxies or metal line systems may result from photo-ionization by hard radiation emerging from hot SNR.

157.300 **Erratum: "A redshift survey of IRAS galaxies. II. Methods for determining self-consistent velocity and density fields" [Astrophys. J., Vol. 372, No. 2, p. 380 – 393 (10 May 1991)].**
A. Yahil, M. A. Strauss, M. Davis, J. P. Huchra.
Astrophys. J., Vol. 381, No. 1, p. 348 (1 Nov 1991). See Abstr. 53.157.297.

157.301 **Advanced evolutionary phases of large-metallicity low-mass stars and the problem of the ultraviolet excess of elliptical galaxies.**
M. Castellani, A. Tornambè.
Astrophys. J., Vol. 381, No. 2, p. 393 – 408 (10 Nov 1991).

The authors present a set of evolutionary tracks of large-metallicity low-mass stars, covering the advanced evolutionary phases from the zero-age horizontal branch to the final white dwarf cooling phase, at the onset of crystallization. These star models are intended to represent the hot component of the stellar population of elliptical galaxies and are suitable for population synthesis purposes. The models include stars leaving the asymptotic giant branch (AGB) phase before the onset of thermal pulses (TP) and evolving to the blue side of the H-R diagram according to an evolutionary scheme earlier defined as post-early AGB (P-EAGB) evolution. The authors find that the maximum UV flux that post-AGB models can provide is not enough to account for the ultraviolet emission of the strong UV-emitting elliptical galaxies, whereas models following the P-EAGB evolutionary behavior produce much more UV flux. Among the star models evolving off the early AGB, before the onset of thermal pulses, those maximizing the production of UV flux give enough UV light to explain even the most UV upturned branches observed in the spectra of the most metal-rich elliptical galaxy. This is because in post-EAGB models the excursion toward the blue side of the H-R diagram is supported by the additional energy source provided by the He-burning shell.

157.302 **The QMW IRAS galaxy catalogue: a highly complete and reliable IRAS 60-μm galaxy catalogue.**
M. Rowan-Robinson, W. Saunders, A. Lawrence, K. Leech.
Mon. Not. R. Astron. Soc., Vol. 253, No. 3, p. 485 – 495 (1 Dec 1991).

A highly complete and reliable IRAS 60-μm galaxy catalogue has been constructed covering 82% of the sky. The excluded areas are the coverage gaps (3% of the sky) and areas flagged as of high source-density (15% of the sky). IRAS colour conditions have been used to exclude galactic sources, including the remaining cirrus sources. All sources flagged as extended, confused for having poor correlation coefficient with a point-source template have been examined with the raw IRAS data and accurate fluxes determined using mapping routines. The completeness, reliability and flux accuracy of the catalogue are discussed. Identifications have been made with existing optical galaxy catalogues and with galaxy redshift surveys in the literature. The authors estimate that redshifts are available for 79% of the galaxies in the catalogue with $V < 5000$ km s^{-1} and the three-dimensional distribution of such galaxies is displayed. The dipole component of the surface-brightness distribution of galaxies in the catalogue is discussed in detail. An algorithm for extending the catalogue into regions flagged as of high source-density at 100 μm (a further 11% of the sky) is discussed.

157.303 **Recent star formation in the interacting galaxy system NGC 520.**
S. A. Stanford.
Astrophys. J., Vol. 381, No. 2, p. 409 – 417 (10 Nov 1991).

New optical spectroscopy, near-infrared spectroscopy, and mid-infrared imaging of the interacting system NGC 520 are presented. The data were obtained at relatively high spatial resolution to determine the spatial distribution of young and old stellar populations. The results show that the heavily obscured primary nucleus has a current star-forming rate ~35 times the typical rate for a late-type galaxy nucleus, which may account for most of the large far-infrared luminosity of NGC 520. The secondary nucleus has little current star formation, but has recently undergone a mild starburst. Optical colors and absorption line equivalent widths suggest that the extranuclear regions in both galaxies have experienced enhanced star formation roughly at the same time as the secondary nucleus, though at a lesser intensity. The extent of the extranuclear burst is large; at least half of the projected surface area of the two disks took part to varying degrees in the episode. The age and strength of the recent star-formation episode in NGC 520 is uncertain. The best estimate for the initiation time of a few 10^8yr ago is consistent with the system-wide starburst being caused by the galaxy collision.

157.304 **Profiles of optically thick galaxies.**
S. Phillipps, R. Evans, J. I. Davies, M. J. Disney.
Mon. Not. R. Astron. Soc., Vol. 253, No. 3, p. 496 – 504 (1 Dec 1991).

If disc galaxies are optically thick in their central regions then their luminosity profiles in, say, the B band will not reflect the true underlying distributions of stars. In galaxies with known gas distributions one can attempt to correct for the associated dust absorption. By following this procedure for some well-studied objects one can gain an insight into how dramatic the optical depth effects on luminosity profiles can be. In particular the authors find that (1) in all cases looked at (which include galaxies with a wide range of ISM densities) the extrapolated central surface brightnesses of the discs are brightened by about 0.8 mag and (2) galaxies with Type II luminosity profiles (i.e. those with a dip or shoulder near the centre, relative to the inward extrapolation of the outer disc component) have "normal" bulge plus disc profiles after the absorption corrections have been applied. As a check on this "missing light", the authors also present a comparison with profiles obtained from IRAS data; in all cases there is an excellent correspondence between the light lost from the B band, as predicted by the authors' optically thick disc models, and that (re)emitted in the far infrared.

157.305 **Evidence for redshift periodicity in nearby field galaxies.**
B. N. G. Guthrie, W. M. Napier.
Mon. Not. R. Astron. Soc., Vol. 253, No. 3, p. 533 – 544 (1 Dec 1991).
The claim by Tifft and colleagues that extragalactic redshifts are periodic in ranges around 24.2, 36.3 or 72.5 km s^{-1} was investigated for an independent sample of 89 nearby spirals, in the general field, with accurately determined heliocentric redshifts. Corrections were applied for a range of solar vectors encompassing various derivations of the solar motion relative to the Galactic Centre and the centroid of the Local Group, and power spectrum analyses were carried out on the sets of corrected redshifts. A strong periodicity of ~37.2 km s^{-1} emerged, against a white noise background, for an assumed solar vector coincidental, within the uncertainties, with that corresponding to the Sun's probable motion around the Galactic Centre. Comparison with sets of synthetic data simulating the overall characteristics of the real data showed the periodicity to be present at a high confidence level. As an independent test, the sample was divided into two groups of 40 and 49 galaxies with more and less accurately determined redshifts respectively. The measured strength of the periodicity appears to be significantly greater for the group of 40 galaxies. The overall probability that the periodicity of 37.2 km s^{-1} arises by chance is in the range $3x10^{-6} \lesssim p \lesssim 3x10^{-4}$.

157.306 **Is the kinematically distinct core in NGC 5982 necessarily the remnant of an accreted companion?**
T. S. Statler.
Astrophys. J., Lett., Vol. 382, No. 1, p. L11 – L14 (20 Nov 1991).
NGC 5982 is an elliptical galaxy with a "kinematically distinct core" whose projected rotation axis is ~90° away from that of the galaxy's outer parts. Distinct cores in other ellipticals have been attributed to dynamically distinct subsystems that are the remnants of accreted companions. Here a model for NGC 5982 is presented in which the kinematic twist in the core results simply from the projection of the major families of circulating orbits in a triaxial potential. Orbits in the model are populated smoothly, and the core is not a dynamical subsystem. The model predicts a complicated overall velocity field for the galaxy, with particularly low velocity amplitudes ($\leqslant$15 km s^{-1}) along a direction ~150° counterclockwise from the major axis, a direction not yet observed spectroscopically.

157.307 **The M31 globular cluster system.**
J. P. Huchra.
Liller Symposium: Asteroids to quasars, p. 153 – 166 (1991). – See Abstr. 012.087 for the main entry.

157.308 **Spiral structure and star formation in galaxies.**
D. M. Elmegreen.
Liller Symposium: Asteroids to quasars, p. 169 – 175 (1991). – See Abstr. 012.087 for the main entry.

157.309 **The discovery of hot coronae around early type galaxies.**
W. Forman, C. Jones.
Liller Symposium: Asteroids to quasars, p. 177 – 192 (1991). – See Abstr. 012.087 for the main entry.
This review describes both the discovery and more recent observations of hot coronae around early-type galaxies. The observations show that early-type galaxies have 0.5–4.5 keV luminosities up to several 10^{42}erg s^{-1} (H_0 = 50 km $s^{-1}Mpc^{-1}$) dominated by thermal emission from $10^{10}M_\odot$ of hot (~10^7K) gas. The cooling time of the gas is sufficiently short that cooling flows as large as a few solar masses per year are common. Finally, the properties of the coronae imply that galactic winds are not now common around early-type galaxies and that the stellar matter of these galaxies is embedded in a massive dark halo extending to ~100 kpc.

157.310 **Constraints on the origin of the ultraviolet upturn in elliptical galaxies from Hopkins Ultraviolet Telescope observations of NGC 1399.**
H. C. Ferguson, A. F. Davidsen, G. A. Kriss, W. P. Blair, C. W. Bowers, W. Van Dyke Dixon, S. T. Durrance, P. D. Feldman, R. C. Henry, J. W. Kruk, H. W. Moos, O. Vancura, K. S. Long, R. A. Kimble.
Astrophys. J., Lett., Vol. 382, No. 2, p. L69 – L73 (1 Dec 1991).
The giant elliptical galaxy NGC 1399 was observed with the Hopkins Ultraviolet Telescope (HUT) aboard the Astro-1 space shuttle mission in 1990 December. These are the first data to provide measurements of the spectral energy distribution of an elliptical galaxy shortward of Lyα. The spectrum shows no emission features other than well-known geocoronal lines. Strong Lyman series absorption is evident at the galaxy redshift. Other absorption features attributable to the hot stellar population are detected, but there is no evidence of C IV λλ1548, 1551 absorption. The authors use these data to set constraints on the origin of the UV upturn in elliptical galaxies. The lack of detectable C IV absorption and the shape of the continuum spectrum exclude star formation with a standard IMF as the sole contributor to the 1550 Å flux. Postasymptotic giant branch (PAGB) stars can account for the spectrum near 1550 Å, because many have low carbon abundance in their envelopes. However, synthetic spectra produced from theoretical evolutionary tracks and stellar model atmospheres fail to fit to the observed spectral-energy distribution. If PAGB stars produce the UV upturn, then either they do not evolve along the standard tracks, or the (solar metallicity) model atmospheres used to construct the synthetic spectrum are inappropriate. The observed continuum flux decreases from 1050 Å to the Lyman limit, indicating that the light is dominated by stars with temperatures less than 25,000K.

157.311 **First observations of the CO J = 6–5 transition in starburst galaxies.**
A. I. Harris, R. E. Hills, J. Stutzki, U. U. Graf, A. P. G. Russell, R. Genzel.
Astrophys. J., Lett., Vol. 382, No. 2, p. L75 – L79 (1 Dec 1991).
The authors report the detection of carbon monoxide (CO) J = 6–5 emission from the starburst nuclei of the galaxies NGC 253, M82, and IC 342. They have used the 6–5 transition as a thermometer of warm molecular gas in starburst nuclei, unambiguously finding that the nuclear molecular gas in starburst galaxies is substantially warmer than in typical disk clouds. Nuclear gas heating could be either large scale, for example, turbulent motions or cloud-cloud collisions in noncircular bar orbits, or more local effects of radiation from high-mass star formation and supernovae, or some combination. High CO excitation temperatures can affect galactic nuclear mass estimates and consequently star formation efficiencies deduced from lower J lines of CO.

157.312 **Erratum: "Surface brightness fluctuations: a bridge from M31 to the Hubble constant" [Astrophys. J., Lett., Vol. 373, No. 1, p. L1 – L4 (20 May 1991)].**
J. L. Tonry.
Astrophys. J., Lett., Vol. 382, No. 2, p. L109 (1 Dec 1991). See Abstr. 53.157.305.

157.313 **A radio-quiet galaxy at redshift z = 3.409.**
D. A. Turnshek, F. Macchetto, M. V. Bencke, C. Hazard, W. B. Sparks, R. G. McMahon.
Astrophys. J., Vol. 382, No. 1, p. 26 – 31 (20 Nov 1991). With plates 1 – 3.
The authors report the discovery of an unresolved radio-quiet object with redshift z_{em} = 3.409 and visual magnitude V ≅

24.3 ± 0.5. The object was identified during an imaging search for Lyα emission near the redshift of a $z_{abs} = 3.390$ damped Lyα absorption-line system which appears in the spectrum of the $z_{em} = 4.11$ quasar QSO 0000–2619. For $H_0 = 50$ km s^{-1}Mpc^{-1} and $q_0 = 0.5$, the bulk of the object's Lyα and continuum emission comes from an unresolved region less than 9 kpc in diameter. A long-slit CCD spectrum at 500 km s^{-1} resolution shows that it has kinematically unresolved Lyα emission with a rest equivalent width in the range ~10 – 20 Å. The object's properties are consistent with those of a galaxy or a burst of star formation in a galaxy. This new object appears to be the highest redshift radio-quiet galaxy known. The object is observed to be ~135 kpc (19") in projection from the sight line to the QSO. Previously unidentified weak absorption lines due to low- and high-ionization metals are possibly present in the QSO spectrum at $z_{abs} = 3.409$ (i.e., at a redshift consistent with the emission-line redshift). If present, they would indicate the existence of an extended, gaseous region associated with the optical object. The presence of metals in the extended gas would suggest that it has been enriched by star formation. However, if this absorption system proves not to be real, then this places an upper limit on the size/covering factor of any extended absorbing region surrounding the galaxy.

157.314 **Projections of ellipsoidal galaxies.**
B. S. Ryden.
Mon. Not. R. Astron. Soc., Vol. 253, No. 4, p. 743 – 752 (15 Dec 1991).
The formula for the projected surface brightness of an elliptical galaxy is derived under the assumption that the isoluminosity surfaces are coaxial ellipsoids, with axis ratios varying with radius. The surface brightness for a set of five model galaxies is computed explicitly; the models have luminosity profiles which are close to a de Vaucouleurs law and have smoothly varying axis ratios. Certain models show large isophote twists (up to 90°) from some viewing angles while showing negligible twists from other viewing angles. The projected isophotes usually show a discy distortion from a purely elliptical shape, but with an amplitude which is less than 0.5% in radius. The principal axes of the projected isophotes may stray far away from the projected principal axes of the isoluminosity ellipsoids.

157.315 **The discrete X-ray source population in M31.**
G. Trinchieri, G. Fabbiano.
Astrophys. J., Vol. 382, No. 1, p. 82 – 99 (20 Nov 1991). With plate 8.
The authors report the results of the analysis of the entire set of Einstein imaging observations of M31. They detect 108 individual X-ray sources, with (0.2 – 4.0 keV) luminosities ranging from 5×10^{36}ergs s^{-1} to above 10^{38}ergs s^{-1} at the distance of M31. Fourteen of these sources are luminous enough to allow a spectral fit of the X-ray data. Fourteen sources are found to vary between Einstein observations. The authors do not find any substantial difference between the luminosity distributions of X-ray sources in the bulge and in the disk, contrary to earlier reports. The luminosity distribution of these X-ray sources is well represented by a power law with slopes between -1 and -0.66 (1σ). About 20% – 30% of the counts detected in the bulge region are not accounted for by the individual sources. This additional emission can be easily explained by the integrated contribution of sources below the detection threshold. Nevertheless, if one uses this component to set a conservative upper limit on the emission of a putative gaseous component, it is found that no more than $\sim 2 \times 10^6 M_\odot$ of hot gas can be present in the bulge. The authors also present contour maps of the X-ray emission in the central region, and compare the radial distribution of the surface brightness with the optical profiles.

157.316 **Polarized radio emission from the edge-on spiral galaxies NGC 891 and NGC 4565.**
S. Sukumar, R. J. Allen.
Astrophys. J., Vol. 382, No. 1, p. 100 – 107 (20 Nov 1991). With plates 9 – 14.
The distribution of radio continuum intensity and linear polarization has been observed with the VLA in the two nearby edge-on spiral galaxies NGC 891 and NGC 4565 at 6 and 20 cm, respectively. The authors present here the results at a resolution of 20". The linearly polarized emission has been measured at high Z-distances of up to ~2 kpc, where the degree of polarization increases to over 30%. Although the distribution of polarized emission is nearly symmetric in Z about the major axis for both galaxies, a strong asymmetry about the minor axis is evident in NGC 4565. For both galaxies, the position angle of maximum polarization is generally perpendicular to the plane of the disk. The authors present a unified model to account for the main features of the radio polarization in these two edge-on galaxies. The model geometry is determined from recent observations of face-on galaxies where the polarized emission is found to be strongest in the dark inter-arm and outer parts of the disks. A substantial Z-thickness is ascribed to this polarized emission. The asymmetries observed in the distribution of polarized emission when the model is viewed edge-on are a consequence of depolarization over the observing beam by differential Faraday rotation in an intervening plasma with randomly oriented magnetic fields.

157.317 **Properties of the redshift. III. Temporal variation.**
W. G. Tifft.
Astrophys. J., Vol. 382, No. 2, p. 396 – 415 (1 Dec 1991).
Comparisons between new 21 cm redshifts and older data, especially older 300 foot telescope data, are used to demonstrate the presence of systematic deviations, toward higher redshifts, in all the older data. The deviations frequently contain several components inconsistent with simple random spread. A linear velocity approximation, present in some reduction software, is discussed and shown not to be a factor in the deviations. The deviations are shown to relate to global quantization. Displacements correlate with global redshift phase and are interpreted as real time-dependent changes in redshift. Global quantization is rediscussed, taking time variability into account. Slight adjustments in period and solar motion are shown to produce strong time-dependent correlations for all types of galaxies. In particular, galaxies of intermediate profile width, previously not demonstratively globally quantized, now show strong periodicities. Both 21 cm profile width and shape are important parameters in the correlations.

157.318 **The relationship between the carbon monoxide intensity and the radio continuum emission in spiral galaxies.**
D. S. Adler, R. J. Allen, K. Y. Lo.
Astrophys. J., Vol. 382, No. 2, p. 475 – 482 (1 Dec 1991).
The relationship between the velocity-integrated CO emission and the nonthermal radio continuum brightness in the disks of normal spiral galaxies is examined on a variety of length scales. On a global scale, the total CO intensity correlates strongly with the total radio continuum flux density for a sample of 31 galaxies. On scales of ≳2 kpc in the disk of individual galaxies, the ratio I_{CO}/T_{20} remains fairly constant over the entire disk as well as from galaxy to galaxy. For the eight spirals in the sample, the disk-averaged values of I_{CO}/T_{20} range from 0.6 – 2.4, with the average over all eight galaxies being 1.3 ± 0.6. However, studies of giant molecular cloud complexes in the arms of nearby galaxies indicate that the relationship ceases to be constant on smaller (≲1 – 2 kpc) scales, with the CO dominating the ratio. The authors conclude that what these various length scales actually trace are differences in the primary heating mechanism of the gas in the beam. The observed relationship between CO and nonthermal radio continuum emission can be explained by assuming that molecular gas in galactic disks is heated primarily by cosmic rays. Finally, it is concluded that while in a general sense star formation is undoubtedly the primary source of energy that causes the relationship to exist, the details of the specific mechanism(s) at work are complicated by the very different scale heights and spatial distributions of the CO and nonthermal radio continuum emission.

157.319 **A complete CO survey of M31.**
E. Koper, T. M. Dame, F. P. Israel, P. Thaddeus.
Astrophys. J., Lett., Vol. 383, No. 1, p. L11 – L14 (10 Dec 1991).
M31 was fully surveyed at an angular resolution of 8.'7 in the J = 1→0 rotational transition of CO. The 153' × 45' area

observed covers the entire galaxy to a radius of at least 16 kpc, well beyond the optical disk and the Population I peak at 10 kpc. The bulk of the CO emission lies between 8 and 11 kpc from the galactic center. Compared with H I, the CO emission is more confined to optical Population I and tends to peak ~1 kpc closer to the center. Although previously few CO observations have been directed toward the NE tangent of the Population I ring, it appears as rich in molecular gas as the SW tangent. Little CO emission is found within 8 kpc of the center of M31. Using the Galactic value for the conversion of integrated CO intensity to H_2 mass, the total molecular mass of M31 is $\sim 2.5 \times 10^8 M_\odot$, less than one-fourth the molecular mass of the Milky Way computed in the same way.

157.320 **Effects of internal absorption on galaxy number count and redshift distribution.**
B. Wang.
Astrophys. J., Lett., Vol. 383, No. 2, p. L37 – L40 (20 Dec 1991).
The author examines effects of internal absorption by interstellar dust within galaxies. Observations of spiral galaxies suggest that internal absorption correlates with the luminosity of galaxies; brighter galaxies tend to be dustier. Moreover, galaxies may be more obscured by dust at earlier times because of the large amount of heavy elements available in the interstellar medium. Interstellar obscuration within galaxies thus may have profound consequences in observations of distant galaxies; here the author calculates the galaxy number–magnitude relation, N(m), and the number–redshift relation, N(z), in the B and K bands, with internal absorption taken into account. It is found that the galaxy number counts and redshift distribution can be plausibly accounted for by simple models of galaxy evolution with $\Omega = 1$; inclusion of internal absorption eliminates difficulties previously encountered in simultaneously explaining N(m) and N(z) in the B and K bands.

157.321 **H II regions in seven dwarf irregular galaxies.**
N. V. Strobel, P. Hodge, R. C. Kennicutt Jr.
Astrophys. J., Vol. 383, No. 1, p. 148 – 163 (10 Dec 1991).
CCD Hα images are used to study the characteristics of the H II regions of seven nearby dwarf irregular galaxies. Maps and coordinates identify them, and the authors use the calibrated images to measure the emission-line luminosities of all the H II regions and to determine the H II region luminosity functions for the galaxies. The luminosity functions are all normal, with a mean value for the slope of the power-law fit of -1.5 ± -0.3. H II region sizes and morphologies are examined, and values for the characteristic size scale, D_0, are given; a weak correlation is found with the absolute magnitude of the host galaxy.

157.322 **Near-infrared [Fe II] emission of M82 supernova remnants: implications for tracing the supernova content of galaxies.**
M. A. Greenhouse, C. E. Woodward, H. A. Thronson Jr., R. J. Rudy, G. S. Rossano, P. Erwin, R. C. Puetter.
Astrophys. J., Vol. 383, No. 1, p. 164 – 173 (10 Dec 1991).
The authors report near-infrared [Fe II] and Paβ spectra of two young supernova remnants and one H II region within the central 15 kpc of M82. The results, and a growing body of Galactic and extragalactic observations by others, suggest that the total [Fe II] luminosity of galaxies is uncontaminated by emission due to H II regions, their associated photodissociation regions, and other ionized regions which do not contain shock-processed grains or enriched supernova ejecta. The available data on infrared [Fe II] emission in galaxies suggests that (1) the iron depletion of the interstellar medium of galaxies is controlled by supernova activity independent of their nuclear classification; (2) the infrared [Fe II] luminosity of galaxies traces their relative supernova content; and (3) the [Fe II] 1.644 μm and $v = 1 \rightarrow 0$ S(1) H_2 emission of galaxies are correlated, suggesting that the latter is also related to supernova activity. The authors conclude that infrared [Fe II] lines can provide a new relative measure of supernova activity in galaxies. Comparison of infrared [Fe II] and H II luminosities may be useful for determining the relative age of starbursts, and comparison of [Fe II] and H_2 luminosities may provide new insight into understanding the production of global H_2 emission in galaxies.

157.323 **Planetary nebulae as standard candles. VII. A test versus Hubble type in the NGC 1023 group.**
R. Ciardullo, G. H. Jacoby, W. E. Harris.
Astrophys. J., Vol. 383, No. 2, p. 487 – 497 (20 Dec 1991).
The authors present the results of an [O III] λ5007 survey for planetary nebulae in the NGC 1023 Group. In all, they detected 143 planetary nebula candidates: 110 in the SB0 galaxy NGC 1023 and 33 in the halo of the edge-on Sb spiral NGC 891. Using a statistically complete and homogeneous sample of planetary nebulae and the procedures described in previous papers, the authors derive distances to these galaxies using the planetary nebula luminosity function (PNLF). After correcting for foreground extinction by using the estimated reddening values of Burstein and Heiles, the authors find the distance moduli for NGC 1023 and NGC 891 to be 29.97 ± 0.14 and 29.97 ± 0.16, respectively. The similarity of these distances demonstrates the insensitivity of the PNLF to galaxy Hubble types and population age, and again confirms it as one of the best extragalactic standard candles.

157.324 **Ca and Mg abundances in the Draco dwarf galaxy.**
J. E. Hesser, M. D. Lehnert, R. A. Bell, J. B. Oke.
J. R. Astron. Soc. Can., Vol. 85, No. 4, p. 181 (Aug 1991).
Abstract. – See Abstr. 012.091 for the main entry.

157.325 **Radial abundance gradients in normal and barred spiral galaxies: first results.**
P. Martin.
J. R. Astron. Soc. Can., Vol. 85, No. 4, p. 181 (Aug 1991).
Abstract. – See Abstr. 012.091 for the main entry.

157.326 **Ar/S in the blue compact dwarf galaxies I Zw 18 and II Zw 40.**
C. C. Stevenson, M. L. McCall.
J. R. Astron. Soc. Can., Vol. 85, No. 4, p. 182 – 183 (Aug 1991).
Abstract. – See Abstr. 012.091 for the main entry.

157.327 **Continuum emission at 1 mm from the elliptical galaxy NGC 205.**
M. Fich, P. Hodge.
J. R. Astron. Soc. Can., Vol. 85, No. 4, p. 185 (Aug 1991).
Abstract. – See Abstr. 012.091 for the main entry.

157.328 **Warm molecular gas in the nucleus of IC 342.**
J. A. Irwin, L. W. Avery.
J. R. Astron. Soc. Can., Vol. 85, No. 4, p. 191 – 192 (Aug 1991).
Abstract. – See Abstr. 012.091 for the main entry.

157.329 **Global O/H gradients in late-type spiral galaxies and possible bar interaction for NGC 6946.**
J. Belley, J.-R. Roy.
J. R. Astron. Soc. Can., Vol. 85, No. 4, p. 195 (Aug 1991).
Abstract. – See Abstr. 012.091 for the main entry.

157.330 **The H I content of the Local Group dwarf irregular galaxy in Phoenix.**
C. Carignan, S. Demers.
J. R. Astron. Soc. Can., Vol. 85, No. 4, p. 195 (Aug 1991).
Abstract. – See Abstr. 012.091 for the main entry.

157.331 **Spectroscopic gradients in the compact elliptical galaxies NGC 4478 and NGC 4486B.**
T. J. Davidge.
J. R. Astron. Soc. Can., Vol. 85, No. 4, p. 195 (Aug 1991).
Abstract. – See Abstr. 012.091 for the main entry.

157.332 **The near-infrared spectral gradients in elliptical galaxies and bulges, and the nature of the Na feature near 8200 Å.**
S. Delisle, E. Hardy.
J. R. Astron. Soc. Can., Vol. 85, No. 4, p. 196 (Aug 1991). Abstract. – See Abstr. 012.091 for the main entry.

157.333 **The globular cluster systems of cD galaxies in poor clusters.**
T. J. Bridges, D. A. Hanes.
J. R. Astron. Soc. Can., Vol. 85, No. 4, p. 203 (Aug 1991). Abstract. – See Abstr. 012.091 for the main entry.

157.334 **The globular cluster system of NGC 3842.**
S. T. Butterworth, W. E. Harris.
J. R. Astron. Soc. Can., Vol. 85, No. 4, p. 203 – 204 (Aug 1991). Abstract. – See Abstr. 012.091 for the main entry.

157.335 **Globular cluster systems in central cD galaxies.**
W. E. Harris, C. J. Pritchet, R. D. McClure.
J. R. Astron. Soc. Can., Vol. 85, No. 4, p. 204 (Aug 1991). Abstract. – See Abstr. 012.091 for the main entry.

157.336 **CCD photometry of multiple galaxies in rich clusters.**
J. E. Hayes.
J. R. Astron. Soc. Can., Vol. 85, No. 4, p. 210 (Aug 1991). Abstract. – See Abstr. 012.091 for the main entry.

157.337 **Rotation of the inner regions of the spiral galaxies NGC 497, 895, 972, and 3646.**
V. L. Afanas'ev, A. N. Burenkov, A. V. Zasov, O. K. Sil'chenko.
Sov. Astron., Vol. 35, No. 6, p. 569 – 576 (Nov 1991). Current Physics Microform No.: 9205X1379. English translation of Astron. Zh., Tom 68, Vyp. 6, p. 1134 – 1149 (1991).

Rotation curves and radial distributions of the radial velocity of ionized gas at different position angles are given for a number of normal spiral galaxies observed digitally with the 6-meter telescope. The galaxies NGC 497, 895, 972, and 3646 are examined.

157.338 **The variable star system of the Sculptor dwarf galaxy.**
C. G. Goldsmith.
J. R. Astron. Soc. Can., Vol. 85, No. 4, p. 220 (Aug 1991). Abstract. – See Abstr. 012.091 for the main entry.

157.339 **Planetary nebulae in NGC 3109.**
M. G. Richer, M. L. McCall.
J. R. Astron. Soc. Can., Vol. 85, No. 4, p. 220 (Aug 1991). Abstract. – See Abstr. 012.091 for the main entry.

157.340 **Extended H I halo in NGC 4449.**
E. Bajaja, W. Huchtmeier, U. Klein.
Bol. Asoc. Argent. Astron., No. 36, p. 147 – 150 (1990). In Spanish. – See Abstr. 012.095 for the main entry.

The irregular galaxy of Magellanic type NGC 4449 has been observed with the Effelsberg radiotelescope in the H I 21 cm line. The emission was detected up to a distance of 14 galactic radii. It has been confirmed that the velocities in the halo increase, towards the South, i.e. in a direction opposite to the one observed over the galaxy itself.

157.341 **Digital photographic photometry of NGC 5253.**
J. L. Sérsic, C. J. Donzelli.
Bol. Asoc. Argent. Astron., No. 36, p. 175 – 182 (1990). In Spanish. – See Abstr. 012.095 for the main entry.

The authors give the B, V, R_B total magnitudes of NGC 5253 obtained from digital surface photometry. The equivalent profile of this galaxy in the R_B-band shows a strong exponential component originated in the disk. When compared with the V-profile shows a negative gradient in the color R_B– V analogous to that found by Caldwell and Phillips in B – V, due probably to the same causes.

157.342 **Statistical properties of low surface brightness galaxies.**
M. C. Martin, E. Bajaja.
Bol. Asoc. Argent. Astron., No. 36, p. 183 – 189 (1990). In Spanish. – See Abstr. 012.095 for the main entry.

The low surface brightness galaxies have been catalogued selecting them from photographic plates applying subjective criteria. In this work the authors have tried to determine limit values for parameters, which are independent of the distance, in order to be able to classify quantitatively those objects.

157.343 **Physical considerations about the nuclear region of NGC 3256.**
E. L. Aguero, S. L. Lipari.
Bol. Asoc. Argent. Astron., No. 36, p. 408 – 409 (1990). Abstract. In Spanish and English. – See Abstr. 012.095 for the main entry.

157.344 **UV (2000Å) observations of nearby well resolved spiral galaxies.**
A. Vuillemin, V. Buat, J. Donas, B. Milliard, G. Courtes, M. Golay.
Adv. Space Res., Vol. 11, No. 11, p. 143 – 147 (1991). – See Abstr. 012.107 for the main entry.

UV (2000Å) images of nearby well resolved galaxies (M31, M33, M51, M81, M101 and NGC 4258) have been obtained during repetitive flights of a balloon-borne 40-cm, 12 – 20 arcsec resolution telescope. In the case of M51, the ultraviolet flux was derived for 36 circular regions where published gas contents exist. After corrections for dust extinction, the UV luminosity per unit area is found to correlate the HI + H2 surface density and thus confirms that the total gas density is a driving parameter for large scale star formation processes.

157.345 **Far-infrared observations of Circinus and NGC 4945 galaxies.**
R. S. Bisht, S. K. Ghosh, K. V. K. Iyengar, T. N. Rengarajan, S. N. Tandon, R. P. Verma.
Bull. Astron. Soc. India, Vol. 19, Nos. 1 – 2, p. 80 (Mar – Jun 1991). Abstract. – See Abstr. 012.108 for the main entry.

157.346 **Galaxy photometry with SEST: how big are galaxies at millimetric wavelengths?**
P. Andreani, A. Franceschini.
Messenger, No. 65, p. 30 – 33 (Sep 1991).

157.347 **Photometric properties of the bright lenticular galaxy NGC 5102.**
Y.-J. Sohn, M.-S. Chun, Y.-I. Byun.
J. Korean Astron. Soc., Vol. 24, No. 2, p. 151 – 159 (Dec 1991).

Using the photographic imagery of the S0 galaxy NGC 5102, the equivalent luminosity profile and the fractional integrated luminosity curve have been derived. Several photometric parameters of the galaxy, such as a total magnitude (B_T), equivalent radius effective surface brightness and concentration indices (C_{21}, C_{32}) were derived from the luminosity profiles. According to the decomposition method from the nonlinear least squares fitting, photometric parameters of the bulge and the exponential disk, and the bulge to total luminosity ratio (B/T) were obtained. The derived central disk surface brightness and the evaluated B/T ratio of NGC 5102 are close to the mean values of S0 galaxies. The analysis showed that a lens-like hump is embedded in the equivalent luminosity profile of NGC 5102.

157.348 **Evolution of spiral galaxies.**
A. L. Gyul'budagyan.
Sov. Phys. – Dokl., Vol. 36, No. 9, p. 607 – 608 (Sep 1991). Abstract. Current Physics Microform No.: 9206X1677.

157.349 **On superassociations (SA) at the ends of the bars of spiral Markarian galaxies.**
A. P. Magtesyan, Eh. E. Khachikyan.
Pis'ma Astron. Zh., Tom 17, No. 12, p. 1082 – 1083 (Dec 1991). In Russian. English translation in Sov. Astron. Lett., Vol. 17, No. 6.
As a result of examination of 130 spiral Markarian galaxies it is shown that there is a good tendency of superassociations (SA) forming pairs at the ends of the bars.

157.350 **External fueling and stochastic ignition of starburst in isolated galaxies.**
Y. Sofue, K.–i. Wakamatsu.
Publ. Astron. Soc. Jpn., Vol. 43, No. 5, p. L57 – L64 (1991).
Intergalactic gas clouds are captured by spiral galaxies due to the ram – pressure friction existing in gaseous halos and disks. The captured clouds are rapidly accreted toward the central region, fueling the nucleus, which may trigger starburst. This could be a triggering mechanism for starburst in isolated galaxies.

157.351 **Molecular rings in galaxies and a possible new cosmic distance indicator.**
Y. Sofue.
Publ. Astron. Soc. Jpn., Vol. 43, No. 5, p. 671 – 683 (1991).
The molecular gas distribution in the nuclear regions of galaxies can be classified into three types: a core (C) type with a high–density peak toward the nucleus; a compact–ring (CR) type with ring radius of $\sim$200 pc; and a broad–ring (BR) type with ring radius of 750 pc. The authors discuss the possibility of using the ring radius as a cosmic distance indicator for galaxies up to about 100 Mpc. They stress that this geometrical method is independent of such correction factors as extinction and reddening.

157.352 **Abundances in galaxies.**
B. E. J. Pagel.
Nuclei in the cosmos, p. 89 – 107 (1991). – See Abstr. 003.038 for the main entry.
The observed distribution of abundances of elements in the solar system, stellar populations and the interstellar medium in our own and other galaxies is affected by nucleosynthesis in the Big Bang and in former generations of stars and by many aspects of the evolution of galaxies. A rough scheme for the resulting "chemical evolution" of the Galaxy is presented here, and some observational data on H II regions and relatively unevolved stars, revealing clues about these processes, are described. Evolved stars and planetary nebulae often show modified abundances in their atmospheres (particularly for Li, C, N, O and s–process elements) resulting from nuclear reactions in their interiors followed by dredge–up due to convective mixing.

157.353 **The nuclear ring of the barred galaxy NGC 1326.**
J. A. Garcia–Barreto, R.–J. Dettmar, F. Combes, M. Gerin, B. Koribalski.
Rev. Mex. Astron. Astrofis., Vol. 22, No. 2, p. 197 – 212 (Oct 1991).
The authors have carried out observations of the radio continuum, Hα, CO(1–0) and CO(2–1) and near infrared (J, H and K) emission from the early–type barred spiral galaxy NGC 1326. The main result is that they have detected in the radio continuum and in Hα a circumnuclear structure that would correspond to a so–called nuclear ring at a mean distance of about 300 pc from the nucleus. The mean radio continuum spectral index, using the integrated total fluxes is $\alpha = -0.7$ ($S_\nu \sim \nu^\alpha$), but it is flatter when using the fluxes of individual sources suggesting a superposition of thermal and non–thermal radiation. The radio continuum and Hα emissions indicate an enhanced star formation activity in the ring giving rise to the thermal emission from H II regions and the nonthermal emission from the evolution of massive stars into supernovae and supernova remnants.

157.354 **Estado actual de la fotometría superficial de galaxias en Córdoba.**
C. J. Donzelli.
Rev. Mex. Astron. Astrofis., Vol. 22, No. 2, p. 305 (Oct 1991). Abstract. – See Abstr. 012.127 for the main entry.

157.355 **Systematics in the orientation of galaxies.**
D. García Lambas, H. Muriel.
Rev. Mex. Astron. Astrofis., Vol. 22, No. 2, p. 307 – 308 (Oct 1991). Abstract. – See Abstr. 012.127 for the main entry.

157.356 **Surface photometry of the dwarf galaxy VII Zw 009.**
G. M. Richter, K.–H. Schmidt, W. Thänert, K. Stavrev, K. Panov.
Astron. Nachr., Vol. 312, No. 5, p. 309 – 314 (1991).
Photographic photometry reveals the dwarf galaxy VII Zw 009 as an outstanding object with respect to the large colour excess of its central star burst region. It seems to be a similarity to the system IC 10, which also contains at least one star burst region with large internal extinction. The results are supported by the identification of the IRAS source 3550+6657 with VII Zw 009.

157.357 **Andromeda ha un nucleo collassato?**
A. Curir.
Orione, Vol. 11, N. 2, p. 16 – 20 (Mar – Apr 1991).

157.358 **JHK imaging of the edge–on spiral galaxy NGC 891.**
T. E. Aoki, N. Hiromoto, H. Takami, S. Okamura.
Publ. Astron. Soc. Jpn., Vol. 43, No. 6, p. 755 – 780 (1991).
New near–infrared imaging observations of the edge–on spiral galaxy NGC 891 are presented and discussed in terms of interstellar extinction and galactic structure.

The interstellar lines catalogue.
See Abstr. 002.027.

Kinematical observations of ordinary spiral galaxies: a bibliographical compilation.
See Abstr. 002.032.

A database of IUE spectra of active galactic nuclei.
See Abstr. 002.034.

An extragalactic data base. IV. Errors and misprints in catalogues of galaxies.
See Abstr. 002.070.

The Case Low–Dispersion Northern Sky Survey.
See Abstr. 002.094.

Photographic atlas for supernovae search. Vol. II: Declinations −20° to +20°.
See Abstr. 002.100.

Third Reference Catalogue of Bright Galaxies (de Vaucouleurs, de Vaucouleurs, Corwin, Buta, Paturel, and Fouqué 1991). Documentation for the machine–readable version.
See Abstr. 002.114.

The Second Byurakan Sky Survey. Spectra of galaxies. IV.
See Abstr. 002.133.

The Rosse spirals.
See Abstr. 004.082.

The interpretation of modern synthesis observations of spiral galaxies. Proceedings.
See Abstr. 012.041.

Skylines. Proceedings.
See Abstr. 012.051.

Warped disks and inclined rings around galaxies. Proceedings.
See Abstr. 012.054.

Massive stars in starbursts. Proceedings.
See Abstr. 012.062.

Galactic bulges and discs.
See Abstr. 013.020.

Star forming regions in galaxies.
See Abstr. 013.023.

Galaxies as seen by the ING.
See Abstr. 013.076.

Extragalactic research at the Instituto de Astrofisica de Canarias.
See Abstr. 013.080.

Dispersión de velocidades en galaxias elípticas.
See Abstr. 013.109.

Corrimiento al rojo de galaxias australes.
See Abstr. 013.110.

Die räumliche Verteilung von Ferninfrarotlinienstrahlung in externen Galaxien gemessen mit einem abbildenden Fabry–Pérot–Spektrometer.
See Abstr. 035.009.

The current ability of HST to reveal morphological structure in medium–redshift galaxies.
See Abstr. 035.158.

A 40–cm UV (2000 Å) balloon–borne imaging telescope: results and current work.
See Abstr. 035.222.

Structures on galaxy counts from a wavelet analysis.
See Abstr. 036.002.

Image construction from the IRAS survey and data fusion.
See Abstr. 036.032.

Low–cost 3D visualisation techniques: an application to the 3–D distribution of nearby galaxies.
See Abstr. 036.053.

Image construction from the IRAS survey.
See Abstr. 036.078.

On the derivation of selection functions from redshift survey data.
See Abstr. 036.132.

Adaptive filtering and its application.
See Abstr. 036.179.

Contrast enhancement by galaxy modelling.
See Abstr. 036.185.

Photometric decomposition of galaxies.
See Abstr. 036.198.

Comparison between different isophote–fitting programs.
See Abstr. 036.202.

A cross–correlation method for surface photometry.
See Abstr. 036.206.

Maximum entropy image construction of the galaxy M51.
See Abstr. 036.220.

The image reconstruction contest.
See Abstr. 036.221.

Comets to quasars: surface photometry from standard stars, and the morphology of the galaxy–quasar interface.
See Abstr. 036.231.

Ultraviolet imaging of globular clusters and galaxies.
See Abstr. 036.316.

Automated classification of resolved galaxies.
See Abstr. 036.336.

HST and dense stellar systems.
See Abstr. 051.019.

Infrared universe – IRAS.
See Abstr. 051.047.

Particle energization.
See Abstr. 061.067.

Dynamos in galaxies.
See Abstr. 062.100.

On the influence of turbulence of fragmentation of self–gravitating gas clouds.
See Abstr. 062.114.

Influence of turbulence on the fragmentation of self–gravitating gas clouds.
See Abstr. 062.132.

X–ray irradiation of interstellar grains in active galaxies: evaporation and infrared spectra.
See Abstr. 063.068.

The dust opacity of spiral galaxies.
See Abstr. 063.103.

The disk accretion of a tidally disrupted star onto a massive black hole.
See Abstr. 067.272.

Spectra of massive blue stars.
See Abstr. 114.089.

Massive stars – from the solar neighborhood to starburst galaxies.
See Abstr. 115.024.

Massive red stars in galaxies.
See Abstr. 115.025.

Massive star evolution.
See Abstr. 115.026.

A new age calibrator for red horizontal branch populations.
See Abstr. 115.029.

Infrared properties of the carbon symbiotic star C–1 in the Draco dwarf galaxy.
See Abstr. 117.102.

The RR Lyr stars of the Sextans Dwarf Spheroidal galaxy.
See Abstr. 122.010.

The frequency and spatial distribution of novae in M81.
See Abstr. 124.007.

Galactic and extragalactic supernova rates.
See Abstr. 125.040.

The observation and interpretation of radio supernovae in nearby galaxies.
See Abstr. 125.048.

Supernova rates, starbursts and galaxy inclinations.
See Abstr. 125.128.

Radio properties of supernova remnants in M33.
See Abstr. 125.151.

A survey of small–scale extremes in extinction and reddening at low galactic latitudes using IRAS galaxies.
See Abstr. 131.050.

Large–scale fragmentation of gas cloud rotating at the galactic center.
See Abstr. 131.141.

Physical parameters in extragalactic star forming regions.
See Abstr. 131.181.

Comparison of turbulence in H II regions and molecular clouds.
See Abstr. 131.184.

Cloud formation by combined instabilities in galactic gas layers: evidence for a Q threshold in the fragmentation of shearing wavelets.
See Abstr. 131.217.

Star formation in globular clusters and dwarf galaxies, and implications for the early evolution of galaxies.
See Abstr. 131.226.

Stellar luminosity functions as probes of star formation history.
See Abstr. 131.229.

Star formation: observations.
See Abstr. 131.341.

Observations of Ca II absorption along the nearly parallel lines of sight toward 3C 273 and SN 1991T.
See Abstr. 131.361.

Spatial inhomogeneities in giant extragalactic H II regions.
See Abstr. 132.012.

Properties of giant H II regions.
See Abstr. 132.036.

Effects of dust mixed with photoionized gas.
See Abstr. 132.042.

Cross–correlation of the X–ray background with nearby galaxies.
See Abstr. 142.005.

Iron line diagnostics in elliptical galaxies and cluster cooling flows.
See Abstr. 142.058.

X–ray detectability of gamma–ray bursts from M31.
See Abstr. 143.040.

Cosmic rays in other galaxies.
See Abstr. 144.027.

Cosmic rays in spiral galaxies.
See Abstr. 144.060.

Cosmic–ray transport away from galactic disks.
See Abstr. 144.061.

A scale invariant, general relativistic model for a galaxy in a dark matter background.
See Abstr. 151.009.

Colliding galaxies.
See Abstr. 151.010.

Global stability of self–similar Newtonian gaseous disks against axisymmetric perturbations.
See Abstr. 151.011.

Dynamics of elliptical galaxies.
See Abstr. 151.014.

Orbital decay in aspherical galaxies. II. Triaxial systems.
See Abstr. 151.015.

Ring formation in off–axis collisions of galaxies.
See Abstr. 151.017.

Galactic wind and metal enrichment in spheroidal star systems.
See Abstr. 151.020.

Resonance excitation of spiral density waves in a gaseous disk. II. A nonlinear theory and application to the 3 kiloparsec arm.
See Abstr. 151.022.

Stability of nonrotating stellar systems. II. Prolate shell–orbit models.
See Abstr. 151.024.

Galactic disk warps.
See Abstr. 151.025.

Interpreting asymmetric line profiles in merger remnants.
See Abstr. 151.026.

On the stability of gaseous disks of galaxies which have a local minimum on the rotation curve.
See Abstr. 151.027.

Modes of oscillation of a uniformly rotating, homogeneous spheroid of stars.
See Abstr. 151.029.

Bars: orbits, gas flows and observational input.
See Abstr. 151.034.

The potential of early–type barred galaxies: towards a self–consistent model for NGC 936.
See Abstr. 151.037.

Modeling the NGC 5194/95 interaction.
See Abstr. 151.043.

Molecules in disks.
See Abstr. 151.044.

Dissipational galaxy formation. I. Effects of gasdynamics.
See Abstr. 151.048.

The disk heating problem.
See Abstr. 151.049.

Warps.
See Abstr. 151.053.

Galactic disk warps.
See Abstr. 151.054.

Dynamics and evolution of acquired gaseous discs.
See Abstr. 151.055.

Pattern speeds of density waves.
See Abstr. 151.056.

Structure and dynamics of elliptical galaxies.
See Abstr. 151.068.

Dynamics of polar rings.
See Abstr. 151.073.

Evolutionary processes affecting galactic accretion disks.
See Abstr. 151.078.

Warps and bulges.
See Abstr. 151.082.

Time evolution of galactic warps.
See Abstr. 151.083.

Are warps normal modes?
See Abstr. 151.084.

Modeling edge–on elliptical galaxies: are two integrals enough?
See Abstr. 151.088.

The effect of dynamical friction on globular cluster motion in triaxial galaxies.
See Abstr. 151.093.

Triggering of repetitive starbursts in merging galaxies.
See Abstr. 151.107.

Joint evolution of a galactic nucleus and central massive black hole.
See Abstr. 151.110.

Formation and evolution of the disc in spiral galaxies: viscous models without infall.
See Abstr. 151.111.

Inner secrets of ellipticals.
See Abstr. 151.114.

Origin of kinematic subsystems in elliptical galaxies.
See Abstr. 151.115.

Merging of two galaxies with central black holes.
See Abstr. 151.116.

Collapse of a protogalactic cloud.
See Abstr. 151.121.

Shell formation in elliptical galaxies.
See Abstr. 151.123.

Violent relaxation and the $R^{1/4}$ law.
See Abstr. 151.129.

The velocity dispersion anisotropy and mass–to–light ratio of elliptical galaxies.
See Abstr. 151.130.

Influence of a dark halo on the stability of elliptical galaxies.
See Abstr. 151.132.

The ordered nature of elliptical galaxies: implications for their intrinsic angular momenta and shapes.
See Abstr. 151.133.

On the effects of interstellar gas on the oscillations and the stability of galaxies.
See Abstr. 151.136.

UIT ultraviolet imaging and photometry of the extragalactic OB association NGC 206.
See Abstr. 152.002.

UIT ultraviolet color–magnitude diagrams of the stellar content of OB associations within M31.
See Abstr. 152.003.

The formation of globular clusters in galaxy mergers.
See Abstr. 154.017.

Inhomogeneous halo collapse and early Galactic chemical evolution: globular cluster metallicities.
See Abstr. 154.024.

Magnetic fields and radiative shocks in protogalaxies and the origin of globular clusters.
See Abstr. 154.051.

Globular clusters in the Sombrero galaxy (M104).
See Abstr. 154.089.

The tidal radii of globular clusters in M31.
See Abstr. 154.090.

The globular cluster system and the mass of M31: a progress report.
See Abstr. 154.093.

Extragalactic globular clusters. III. Metallicity comparisons and anomalies.
See Abstr. 154.102.

Disk–shocking and the mass function of globular clusters.
See Abstr. 154.108.

Multicolor CCD photometry of globular clusters in three Virgo elliptical galaxies.
See Abstr. 154.116.

Long–slit spectroscopy of CN absorption in the M31 globular clusters Vet 12 and Vet 282.
See Abstr. 154.118.

UBVR photometry of young, massive M31 clusters.
See Abstr. 154.120.

Metal–rich giant branches of bulge clusters: application to color–magnitude diagrams of M31 and M32.
See Abstr. 154.152.

Large–scale oscillations of the Galaxy and the kinematics of the solar neighbourhood.
See Abstr. 155.039.

The detection of the hot gaseous halo around the Small Magellanic Cloud.
See Abstr. 156.013.

The observed relationship of X–ray and infrared emission in active and normal galaxies.
See Abstr. 158.049.

OH megamasers above $z = 0.07$.
See Abstr. 158.053.

Colour indices of the Markarian galaxies.
See Abstr. 158.068.

Radio properties of extragalactic IRAS sources.
See Abstr. 158.182.

Massive star formation and superwinds in IRAS 19254–7245 (the "Superantennae").
See Abstr. 158.185.

Very large array observations of radio–selected dumbbell galaxies.
See Abstr. 158.187.

Search for starbursts among X–ray – selected galaxies: optical spectroscopy.
See Abstr. 158.202.

IRAS 02366–3101: an accretion disk candidate among luminous IRAS galaxies.
See Abstr. 158.294.

The extraordinary radio source 3C 442.
See Abstr. 158.296.

Discovery of a Lyα galaxy near the damped Lyα system at z = 2.3 towards the QSO PHL 957.
See Abstr. 159.020.

Infrared and visible photometry of the gravitational lens system 2237+030.
See Abstr. 159.032.

Infrared K–band observations of the gravitational lens 2016+112.
See Abstr. 159.079.

A composite plot of far–infrared versus radio luminosity, and the origin of far–infrared luminosity in quasars.
See Abstr. 159.106.

Associations between galaxies and bright quasars.
See Abstr. 159.158.

Systematic effects in lens inversions: ℵ exact models for 0957+561.
See Abstr. 159.172.

Maffei 1 as an interloper of the Local Group of galaxies and the mass of the Local Group.
See Abstr. 160.002.

Fundamental parameters of brightest cluster galaxies.
See Abstr. 160.003.

Ram–pressure stripping of the multiphase interstellar medium of the Virgo Cluster elliptical galaxy M86 (NGC 4406).
See Abstr. 160.005.

Stellar populations in medium redshift clusters. II. Optical–infrared photometry and spectra.
See Abstr. 160.011.

Optical coronal emission lines from cooling flows in elliptical galaxies and galaxy clusters.
See Abstr. 160.014.

UIT observations of NGC 1275.
See Abstr. 160.018.

A catalog of galaxies in the Dorado cloud.
See Abstr. 160.022.

Hα Fabry–Perot rotation curves of a large sample of spiral galaxies in clusters.
See Abstr. 160.028.

The merging runaway.
See Abstr. 160.032.

The manifold of early–type galaxies from IR photometry in clusters of galaxies. The Virgo cluster.
See Abstr. 160.041.

Are H I–deficient galaxies of the Coma supercluster deficient in molecular gas?
See Abstr. 160.043.

Discovery of a Lyα galaxy near a damped Lyα absorber at z = 2.3.
See Abstr. 160.047.

The origin of the quasi–stellar object absorption lines.
See Abstr. 160.049.

The peculiar velocity field in the Hercules region.
See Abstr. 160.054.

On the relation between radius, luminosity, surface brightness, and surface density in elliptical galaxies.
See Abstr. 160.055.

Young globular clusters?
See Abstr. 160.057.

Abundances in Virgo spiral galaxies.
See Abstr. 160.058.

Magnetic fields in the Milky Way neighbourhood as deduced from warps in spiral galaxies.
See Abstr. 160.061.

Magnetism in the local Virgo–centered supercluster of galaxies (radius ~ 24 Mpc).
See Abstr. 160.062.

The probable extragalactic nature of some low surface brightness clouds at high galactic latitude.
See Abstr. 160.063.

A new spectroscopic survey for damped Ly_α absorption lines from high–redshift galaxies.
See Abstr. 160.064.

Dwarf galaxies in the Virgo cluster. I. The systematic photometric properties of early–type dwarfs.
See Abstr. 160.067.

The intergalactic H I supply.
See Abstr. 160.075.

The circumgalactic ring of gas in Leo.
See Abstr. 160.077.

Ultraviolet observations of galaxies in nearby clusters. II. Luminosity function and surface density distribution in the Coma cluster.
See Abstr. 160.080.

H I studies of the Sculptor group galaxies. VII. Implications on the distribution and nature of dark matter in groups.
See Abstr. 160.084.

The small–scale clustering properties of dwarf galaxies.
See Abstr. 160.090.

Galaxy correlations, counts, and moments.
See Abstr. 160.095.

The luminosity function of compact groups of galaxies.
See Abstr. 160.096.

Enrichment and heating of the intracluster medium through galactic winds.
See Abstr. 160.097.

The scale dependence of galaxy distribution functions.
See Abstr. 160.098.

Spectroscopy of the gravitational arcs in Cl 2244–02, A370 (arclet A5) and Cl 0024+1654.
See Abstr. 160.099.

The missing mass in clusters of galaxies and elliptical galaxies.
See Abstr. 160.105.

Galaxy evolution in distant, X–ray–selected clusters of galaxies. I. Cl 1358+6245.
See Abstr. 160.111.

High–resolution imaging of Virgo cluster galaxies. I. Observational techniques and first results.
See Abstr. 160.132.

New D – σ results for Coma cluster ellipticals.
See Abstr. 160.139.

Systematic bias in cluster galaxy data, affecting galaxy distances and evolutionary history.
See Abstr. 160.140.

The galaxy two–point correlation function down to V = 26 magnitudes on 0°.5 scales.
See Abstr. 160.142.

Void statistics of the CfA redshift survey.
See Abstr. 160.143.

Upper limit on interstellar gas in Coma Cluster ellipticals.
See Abstr. 160.145.

Infrared images, Virgo spirals and the Tully–Fisher law.
See Abstr. 160.150.

Bound populations around cD galaxies and cD velocity offsets in clusters of galaxies.
See Abstr. 160.152.

Optical colors of early–type galaxies in compact groups.
See Abstr. 160.158.

Modeling the evolution of galaxies in compact groups.
See Abstr. 160.159.

Dynamical models of the Local Group of galaxies.
See Abstr. 160.166.

Are field galaxies randomly distributed on scales of 100+ Mpc?
See Abstr. 160.169.

The effect of the cluster environment on galaxies.
See Abstr. 160.176.

Evidence for gas deficiency in cluster galaxies.
See Abstr. 160.177.

Properties of galaxies in groups and clusters.
See Abstr. 160.178.

Distant clusters as cosmological laboratories.
See Abstr. 160.183.

A non–baryonic galactic halo and the real galactic age.
See Abstr. 161.011.

Angular correlations of faint galaxies.
See Abstr. 161.084.

A limit on the cosmological constant.
See Abstr. 161.094.

Protogalactic mergers and galaxy number counts.
See Abstr. 161.110.

The fragmentation of "pancakes" in a dark matter–dominated universe.
See Abstr. 161.134.

The accuracy of galaxy masses from the timing argument.
See Abstr. 161.144.

Construction and analysis of a diameter–limited sample of galaxies in the northern sky.
See Abstr. 161.152.

Formation of galactic halos from seeded hot dark matter.
See Abstr. 161.154.

Analysis of the large scale structure from samples given by quasi–one–dimensional redshift survey.
See Abstr. 161.199.

Evolution of a cloudy protogalaxy interacting with an early galactic wind and galaxy formation.
See Abstr. 161.202.

Large scale peculiar velocities of galaxies.
See Abstr. 161.219.

Redshift surveys of galaxies.
See Abstr. 161.245.

Cross–correlation of the X–ray background with nearby galaxies.
See Abstr. 161.254.

Inhomogeneities in the Universe on scales of (125 h^{-1}Mpc)^3.
See Abstr. 161.265.

Galaxy formation through hierarchical clustering.
See Abstr. 161.295.

The mass distribution of groups and clusters of galaxies.
See Abstr. 161.337.

Dynamics of cooling gas in galactic dark halos.
See Abstr. 161.341.

Chemical evolution of galaxies.
See Abstr. 161.425.

The local distance scale: how reliable is it?
See Abstr. 161.480.

Gravitational lensing frequencies: galaxy cross–sections and selection effects.
See Abstr. 161.554.

Density and peculiar velocity fields in the region of Dressler's supergalactic plane survey.
See Abstr. 161.565.

Higher moments of the IRAS galaxy distribution.
See Abstr. 161.569.

Deep galactic surveys and the structure of the universe.
See Abstr. 161.636.

Large scale peculiar motions in the southern sky.
See Abstr. 161.681.

158 Active Galaxies (Seyfert Galaxies, BL Lacertae Objects, Radio Galaxies)

158.001 **Infrared broad emission lines from an obscured quasar, IRAS 23060+0505.**
T. Nakajima, N. P. Carleton, M. Nishida.
Astrophys. J., Lett., Vol. 375, No. 1, p. L1 – L4 (1 Jul 1991).
Paα and Paβ spectra have been obtained of IRAS 23060+0505, an infrared luminous galaxy at z = 0.173 with a Seyfert 2 optical spectrum. The line widths are greater at longer wavelengths. The FWHMs of Paβ (1.282 μm) and Paα (1.875 μm) are 1200 and 2050 km s^{-1}, respectively. The broad wing of Paα extends to at least 5000 km s^{-1} (FWZI). Paα is clearly dominated by the broad component. The upper limit to the narrow–line extinction and the lower limit to the broad–line extinction are $A_V = 2.3$ and 4.3, respectively. The difference between the broad– and narrow–line extinctions must be caused by dust in the inner NLR or throughout the NLR, which is seen in warm thermal emission whose temperature ranges from a few hundred to 800K.

158.002 **Broad emission lines in active galactic nuclei. II. Statistical properties of Hα and Hβ.**
G. M. Stirpe.
Astron. Astrophys., Vol. 247, No. 1, p. 3 – 10 (Jul 1991).
High quality Hα and Hβ spectra of a sample of 29 Seyfert 1 galaxies and QSOs are used to study the statistical properties of broad emission lines. Median velocities, interpercentile velocities at 10%, 20% and 50% (IPV 10, IPV 20, IPV 50) and, for Hα, the corresponding asymmetry parameters are derived. A weak dependence of the IPVs on luminosity is found, which could be due to a mass–luminosity relation. Furthermore, a correlation exists between the IPVs and the steepness of the Hβ/Hα ratio as functions of velocity. This can be explained as a projection angle effect or as a scaling of velocities with central mass. Asymmetries of up to 20% are measured, but no strong predominance of red or blue ones is present.

158.003 **Simultaneous optical and radio polarization rotation in OJ 287.**
A. Sillanpää.
Astron. Astrophys., Vol. 247, No. 1, p. 11 – 14 (Jul 1991).
Simultaneous rotation in the polarization position angles at centimeter and optical wavelengths at the rate $\approx 2°–2.5° y^{-1}$ is found between two major outbursts in 1971–72 and 1983 in OJ 287. A similar trend is visible in the optical region after the year 1983 and also in the radio in the new observations during the autumn of 1989. One new explanation for the apparent rotation is a precessing jet pointing almost directly towards us. The estimated period of this precession is as short as $\approx$150 yr. Another explanation for this phenomenon would be that the proportion of the shocked synchrotron radiation to the total flux has diminished between these years. This model can also explain the polarization behavior of OJ 287 after the 1983 outburst which is explained by a "two separate synchrotron component" model.

158.004 **The iron line and high energy bump as X–ray signatures of cold matter in Seyfert 1 galaxies.**
G. Matt, G. C. Perola, L. Piro.
Astron. Astrophys., Vol. 247, No. 1, p. 25 – 34 (Jul 1991).
Ginga observations of several bright Seyfert 1 galaxies have recently revealed the presence in their X–ray spectra of both the iron *K* fluorescence line and a high energy bump, which together are currently interpreted as due to the reprocessing of the primary radiation by cold matter located near the putative black hole. A number of quantitative model predictions on the reprocessed radiation, when the matter is assumed to be distributed in an accretion disk, are presented in this paper; they were obtained, on the basis of Monte Carlo simulations, for different geometries of the disk surface and of the primary X–ray source. The case of a flat disk and a point source placed above the disk on its symmetry axis is treated in greater detail, taking into account the general relativistic and kinematic effects, which have dramatic consequences on the shape of the *K* line and edge. These effects are less relevant for the equivalent width (W_a) of the line and for the strength of the high energy bump, and were therefore ignored for simplicity in computing these two quantities for other source geometries (an optically thin corona, which turns out to be well approximated by the point source, and an optically thick sphere), and for other configurations of the disk surface (conical and concave). Several observational implications are discussed, in particular it is noted that the relativistic and kinematic effects on the shape of the edge might lead to a misinterpretation of low resolution measurements in terms of absorption by highly ionized iron.

158.005 **Optical polarization of a complete sample of radio sources.**
C. D. Impey, C. R. Lawrence, S. Tapia.
Astrophys. J., Vol. 375, No. 1, p. 46 – 68 (1 Jul 1991).
Polarimetry of 50 strong radio sources is presented and combined with VLBI observations and optical spectroscopy in a study of the radio and optical properties of a complete 5 GHz sample. The sample is almost equally divided into radio galaxies and quasars. High polarization (p > 3%) is strongly correlated with the fraction of the total 5 GHz flux density found in a milliarcsecond core (F_c). The fraction of sources with high polarization rises from 13% for sources with $F_c < 0.1$ to 45% for sources with $F_c > 0.1$. Two–thirds of the known superluminal sources in the sample have high polarization. High optical polarization, optical power–law fraction, line–to–continuum ratio, emission lines of small equivalent width, and large amplitude flux variability are all strongly correlated with the fraction of the 5 GHz radio flux density that is unresolved on VLBI scales. The distributions of these optical properties are well matched by a model where the radio and optical radiation have the same beaming geometry, and with Doppler boosting of a factor of 10 – 100 in the optical. The authors confirm in this complete sample a striking alignment of the position angle of the VLBI structure axis and the position angle of optical polarization in highly polarized sources. The highly polarized sources have a large amount of misalignment between the VLBI structure axis and the axis of large–scale radio structure.

158.006 **Scattering–induced optical polarization in thick accretion disks.**
J. F. Kartje, A. Königl.
Astrophys. J., Vol. 375, No. 1, p. 69 – 77 (1 Jul 1991).
The authors present a general formalism for calculating the linear polarization induced by scattering within the central funnel of a thick accretion disk and show that multiple photon reflections off the funnel walls can produce polarization values of up to $\sim$10%, with the polarization position angle aligned parallel to the disk symmetry axis. The authors suggest that this process is responsible for the observed optical polarization levels in X–ray–selected BL Lac objects (XBLs), which generally show linear polarization percentages $P \lesssim 10\%$. According to this interpretation, XBLs with high optical polarization are viewed at an angle of $\lesssim 60°$ to the funnel axis and their projected polarization vectors should be preferentially aligned with the associated radio jets. The possible relevance of this model to Seyfert 1 galaxies and quasars is also discussed.

158.007 **Imaging spectrophotometry of ionized gas in NGC 1068. II. Global ionization of the inner disk.**
J. Bland–Hawthorn, J. Sokolowski, G. Cecil.
Astrophys. J., Vol. 375, No. 1, p. 78 – 84 (1 Jul 1991). With plates 3 – 5.
Spectrophotometric observations with the Hawaii Imaging Fabry–Perot Interferometer (HIFI) of the Hα and [N II] $\lambda\lambda$6548,

6584 lines has detected a diffuse ionized medium (DIM) pervading the inner ~10 kpc diameter disk of the nearby, luminous Seyfert galaxy NGC 1068. The physics and dynamics of this energetic gas are distinct from the population of adjacent disk H II regions which have line intensity ratios [N II] λ6584/Hα < 0.5, equivalent widths of Hα ≈ 15 – 100 Å, and [N II] λ6584 line widths ~100 km s^{-1} FWHM. The DIM exhibits broader lines than the H II regions, which suggests an exponential scale height of at least ~400 pc, compared with ~50 pc for the cold neutral phase. The DIM is characterized by strong [N II] λ6584 emission (3.0×10^{41} ergs s^{-1}) when compared to Hα (1.2×10^{41} ergs s^{-1}), with [N II] λ6584/Hα ~ 1 throughout and reaching 2 – 3 at regions NE and SW along the radio axis. The chemical abundances throughout the inner disk are known to be approximately solar and the authors have corrected for Balmer absorption by the underlying stellar population. Thus, these data indicate substantial excitation of the DIM. The authors discuss heating by the Seyfert nucleus and the extended starburst region.

158.008 **IUE observations of NGC 5548, 1978 – 1988: the sizes and kinematics of the broad–line region.**
A. P. Koratkar, C. M. Gaskell.
Astrophys. J., Vol. 375, No. 1, p. 85 – 94 (1 Jul 1991).
The authors have reextracted the IUE spectra of NGC 5548 taken between 1978 and 1988 and studied changes in the UV continuum flux and UV line intensities. The cross–correlation method by Gaskell and Sparke has been used to investigate the structure and kinematics of the broadline region (BLR). The authors give an extensive discussion of the errors in the cross–correlation method and of the method's applicability to nonoptimum archival data. Using the cross–correlation analysis, they find that the total flux in the C IV λ1549 line gives a peak in the cross–correlation function (CCF) corresponding to an inner radius of 8 ± 7 light–days, C III] λ1909 gives one at 11 ± 6 light–days, Lyα λ1216 one at 5 ± 5 light–days, Si IV/O IV] λ1400 one at 40 ± 11 light–days, and He II λ1640 one at 5 ± 6 light–days. The luminosity–weighted radii for the different lines were also determined from the centroid of the CCFs. Cross–correlation of the C III] and C IV lines suggests that the higher ionization line comes from a smaller radius. Cross–correlation of the wings of the C IV line shows the blue wing following the red wing variations by 2 ± 5 days (i.e., consistent with zero), implying predominantly chaotic or circular motion. Pure outflow and pure inflow are excluded at the 99% confidence level. The authors estimate the mass of the black hole to be about $2.2(+0.8, -0.9) \times 10^{8} M_{\odot}$.

158.009 **Erratum: "Turbulent velocities: a clue to the origin of extended ionized nebulosities surrounding active galactic nuclei?" [Astron. Astrophys., Vol. 238, No. 1/2, p. 15 – 27 (Nov 1990)].**
F. Durret, F. Warin.
Astron. Astrophys., Vol. 247, No. 1, p. 259 (Jul 1991). See Abstr. 52.158.052.

158.010 **On a nonthermal origin of steep far–infrared turnovers in radio–quiet active galactic nuclei.**
R. Schlickeiser, P. L. Biermann, A. Crusius–Wätzel.
Astron. Astrophys., Vol. 247, No. 2, p. 283 – 290 (Jul 1991).
A nonthermal origin of the FIR–turnover that appears to be common in Seyfert galaxies and radioweak quasars is investigated in this paper. If a double power law in the electron energy distribution accounts for turnovers steeper than a 2.5 powerlaw in the observed synchrotron emission, it is noted that the observations of the Seyfert galaxies Mrk 1014, I Zw 1 and NGC 4151 require differences in the electron powerlaw energy indices of larger than 2.4 and 4.6, respectively. One rather likely scenario in the nulcear regions of active galaxies is a combination of influences like stochastic acceleration and Coulomb interaction losses to produce the resulting electron energy distribution; this combination of processes readily leads to a concave emission spectrum as required by the observations. The authors then proceed to demonstrate that this rather general model is capable of reproducing the observations quantitatively.

158.011 **An optical study of the cD galaxy NGC 3311 and the giant elliptical galaxy NGC 3309 in the cluster Hydra I.**
A. R. Västerberg, S. Jörsäter, P. O. Lindblad.
Astron. Astrophys., Vol. 247, No. 2, p. 335 – 347 (Jul 1991).
Multicolour CCD surface photometry of the active giant elliptical (gE) galaxy NGC 3309 and the active cD galaxy NGC 3311 in the centre of the Hydra I cluster of galaxies has been performed, and has revealed a blue excess population in the centre of the cD. The *r* brightness curves of the galaxies closely follow the de Vaucouleurs law over several kpc in their interiors, but the ellipticals are less luminous in their innermost parts than predicted by this law. The $r^{1/4}$ law parameters deduced for NGC 3309 are typical for a gE while NGC 3311 has a brightness curve consistent with its classification as a cD. The authors derived a luminosity ratio L_{3311}/L_{3309} of about 3. They computed a mass of the dust feature in the centre of NGC 3311 of $3 \times 10^{4} M_{\odot}$. Also, two CCD slit spectra of the two galaxies in the region from about 620 nm to about 710 nm have been obtained. Hα and [N II] emission from ionized gas is seen in the central part of NGC 3311. The Hα emitting gas has a heliocentric radial velocity of 3828 ± 8 km s^{-1} in the centre of the cD and seems to be in ordered rotation. The emission–line spectrum of NGC 3311 seems to be H II–region like, which is strange for a bright elliptical. These lines presumably arise from photoionization by young stars. This fact and the correlation between dust absorption, blue excess and Hα emission flux suggest that the blue excess population represents a young population where formation of bright stars may still be going on.

158.012 **Centaurus A. Die nächste aktive Radiogalaxie.**
N. Junkes, R. Haynes.
Sterne Weltraum, Jahrg. 30, Nr. 11, p. 656 – 662 (Nov 1991).

158.013 **Spectroscopy of Zwicky galaxies identified with B3 radiosources.**
R. Merighi, L. Basso, M. Vigotti, J. F. Lahulla, M. Lopez–Arroyo.
Astron. Astrophys., Suppl. Ser., Vol. 89, No. 2, p. 225 – 235 (Aug 1991).
The authors present long–slit CCD spectroscopy of 33 Zwicky galaxies identified as radiosources of the B3 Survey, whose redshifts were not previously known. The radio powers of the sample range between 23.3 ⩽ log (P408) ⩽ 24.7 W/Hz (H_0 = 100). 14 galaxies of the sample are classified as E/S0 and 19 as spiral. 24 spectra show emission lines and two of them show the typical emission lines of Seyfert II galaxies. The equivalent width of [N II] λ6584 emission line has been measured down to a limit of 1.5 Å. The authors find no significant differences between the percentage of galaxies with emission lines in this sample and the samples of optically selected galaxies of similar equivalent width limit.

158.014 **Is evolution of radio sources necessary?**
G. Bigot, S. Rauzy, R. Triay.
Phys. Lett. A, Vol. 158, No. 6/7, p. 282 – 290 (9 Sep 1991).
The null–correlation approach has been used to study the distribution of the 3CR sample of radio galaxies in the Hubble diagram, with the following results: This sample is about complete at flux density limit S_{178} = 10 Jy, a relatively small number of strong sources S_{178} > 100 Jy are missing (that correspond probably to local sources). The radio luminosity distribution function is given by a power law $\sim L^{-\gamma(z)}$ with $\gamma(z) \approx 2$, which interprets the large scatter of the distribution in the Hubble diagram. Because of such a distribution in luminosity, this sample is not adapted for deriving the world model. The evolutionary effects are not substantial enough to be revealed within the range of observations.

158.015 **The connection between radio emission spectra and interaction signs in powerful radio galaxies.**
A. G. Gubanov.
Pis'ma Astron. Zh., Tom 17, No. 8, p. 684 – 690 (Aug 1991). In Russian. English translation in Sov. Astron. Lett., Vol. 17, No. 4.
The connection between radio spectra and galaxy interactions/mergers has been explored for 40 powerful radio galaxies with strong optical emission lines and with optical morphologies. Differences of radio spectra of two groups of radio galaxies agree with the suggestion that galactic mergers of interactions could be responsible for the generation of powerful radio galaxies when radio sources have to arise as a rule on late stages of mergers or tidal interactions. Analogous radio emission spectra differences have been found for samples of radio quasars and of radio galaxies from Jackson and Browne (1990) where different [O III] 5007 Å line luminosities of these samples have been discovered. The radio spectra distinction may be interpreted as a later stage of activity of radio galaxies nuclei with relation to quasars that is able to result in differences of [O II] emission line luminosities.

158.016 **Intrinsically weak active galactic nuclei.**
V. S. Artyukh, M. A. Ogannisyan.
Pis'ma Astron. Zh., Tom 17, No. 10, p. 891 – 896 (Oct 1991). In Russian. English translation in Sov. Astron. Lett., Vol. 17, No. 5.
Results of interplanetary scintillating observations of nearby galaxies at 102 MHz have been analysed. Compact (scintillating) sources have been found in spiral and lens galaxies. No compact sources have been found in elliptical galaxies. The authors believe that this result is due to different physical conditions in intrinsically weak active galactic nuclei as found earlier in classical AGNs.

158.017 **Can interstellar scintillations explain a short–time flux variability of extragalactic radio sources?**
N. Ya. Shapirovskaya.
Pis'ma Astron. Zh., Tom 17, No. 10, p. 897 – 903 (Oct 1991). In Russian. English translation in Sov. Astron. Lett., Vol. 17, No. 5.
Diffractive interstellar scintillations like pulsar scintillations must be observed for extragalactic sources with an intrinsic short–time variability. Observed parameters of short–time variability of the quasar 0917+62 are explained by interstellar scintillations for a source model consisting of two components: an extended one, which is not scintillating, and a compact one, which is scintillating.

158.018 **Simultaneous UBVRI photopolarimetric observations of blazars.**
L. O. Takalo.
Astron. Astrophys., Suppl. Ser., Vol. 90, No. 1, p. 161 – 174 (Oct 1991).
The author presents results from truly simultaneous UBVRI photopolarimetric monitoring of eleven blazars, using the Nordic Optical Telescope. Polarization is seen to vary in every object in timescales from days to weeks. The author has detected wavelength dependent polarization in all the objects, several of them showed also wavelength dependent position angle.

158.019 **Multiwavelength observations of fourteen new X–ray selected AGN from the HEAO 1 all–sky survey.**
R. J. Brissenden, D. A. Schwartz, H. V. Bradt, R. A. Remillard, I. R. Tuohy.
Bull. Am. Astron. Soc., Vol. 23, No. 2, p. 892 (1991). Abstract. – See Abstr. 012.008 for the main entry.

158.020 **The origin of [O III] linewidth in Seyfert galaxies.**
M. Whittle.
Bull. Am. Astron. Soc., Vol. 23, No. 2, p. 892 – 893 (1991). Abstract. – See Abstr. 012.008 for the main entry.

158.021 **Infrared nuclei in nearby powerful radio galaxies: evidence for hidden quasars?**
S. Djorgovski, J. Mazzarella, D. Thompson, N. Weir, J. Graham, K. Matthews.
Bull. Am. Astron. Soc., Vol. 23, No. 2, p. 893 (1991). Abstract. – See Abstr. 012.008 for the main entry.

158.022 **Near–infrared spectra and classification diagnostics of Seyfert galaxies.**
H. D. Tran, D. E. Osterbrock, S. Veilleux.
Bull. Am. Astron. Soc., Vol. 23, No. 2, p. 893 (1991). Abstract. – See Abstr. 012.008 for the main entry.

158.023 **A study of a 21–cm–selected sample of galaxies.**
P. A. Henning.
Bull. Am. Astron. Soc., Vol. 23, No. 2, p. 900 (1991). Abstract. – See Abstr. 012.008 for the main entry.

158.024 **High resolution radio and optical observations of the Seyfert 1 galaxy NGC 3516.**
T. Miyaji, A. S. Wilson, I. Pérez–Fournon.
Bull. Am. Astron. Soc., Vol. 23, No. 2, p. 921 (1991). Abstract. – See Abstr. 012.008 for the main entry.

158.025 **Multi–frequency observations of H 2106–099: a weak blue bump, variable Seyfert 1 with strong UV Fe II.**
B. Grossan, R. Remillard, H. Bradt, R. Brissenden, T. Sakao, T. Ohashi.
Bull. Am. Astron. Soc., Vol. 23, No. 2, p. 921 (1991). Abstract. – See Abstr. 012.008 for the main entry.

158.026 **Was 49: an interacting pair of Seyfert galaxies.**
E. C. Moran, J. P. Halpern, G. D. Bothun.
Bull. Am. Astron. Soc., Vol. 23, No. 2, p. 921 (1991). Abstract. – See Abstr. 012.008 for the main entry.

158.027 **Galaxien im Zusammenstoß.**
J. Barnes, L. Hernquist, F. Schweizer.
Spektrum Wiss., No. 10, p. 80 – 88 (Oct 1991).

158.028 **Ultraviolet Imaging Telescope observations of M87.**
M. R. Greason, R. W. O'Connell, S. G. Neff, E. P. Smith, P. M. N. Hintzen, A. M. Smith, T. P. Stecher, M. S. Roberts, R. H. Cornett, J. K. Hill, W. B. Landsman.
Bull. Am. Astron. Soc., Vol. 23, No. 2, p. 922 (1991). Abstract. – See Abstr. 012.008 for the main entry.

158.029 **HUT observations of the far ultraviolet spectrum of the Seyfert galaxy NGC 4151.**
G. A. Kriss, A. F. Davidsen, J. W. Kruk, W. P. Blair, C. W. Bowers, W. V. Dixon, S. T. Durrance, P. D. Feldman, R. C. Henry, H. W. Moos, O. Vancura, H. C. Ferguson, K. S. Long, R. A. Kimble.
Bull. Am. Astron. Soc., Vol. 23, No. 2, p. 922 (1991). Abstract. – See Abstr. 012.008 for the main entry.

158.030 **Ultraviolet spectroscopy of the nucleus of NGC 1068 with the Faint Object Spectrograph.**
L. Armus, S. Caganoff, H. C. Ford, I. N. Evans, R. R. J. Antonucci, G. Hartig, A. L. Kinney, G. Kriss.
Bull. Am. Astron. Soc., Vol. 23, No. 2, p. 922 (1991). Abstract. – See Abstr. 012.008 for the main entry.

158.031 **The first ultraviolet spectropolarimetry of the Seyfert galaxy NGC 1068.**
A. D. Code, K. H. Nordsieck, C. M. Anderson, B. L. Babler, K. S. Bjorkman, G. C. Clayton, A. M. Magalhães, M. R. Meade, R. E. Schulte–Ladbeck, M. Taylor, B. A. Whitney.
Bull. Am. Astron. Soc., Vol. 23, No. 2, p. 922 (1991). Abstract. – See Abstr. 012.008 for the main entry.

158.032 **Ultraviolet Imaging Telescope observations – NGC 1068.**
S. G. Neff, A. M. Smith, P. C. Chen, M. S. Roberts, R. C. Bohlin, R. W. O'Connell, J. E. Isensee, E. P. Smith, P. M. N. Hintzen, T. P. Stecher.
Bull. Am. Astron. Soc., Vol. 23, No. 2, p. 923 (1991). Abstract. – See Abstr. 012.008 for the main entry.

158.033 **Broad band X-ray telescope observations of NGC 4151.**
K. A. Weaver.
Bull. Am. Astron. Soc., Vol. 23, No. 2, p. 923 (1991). Abstract. – See Abstr. 012.008 for the main entry.

158.034 **Extreme X-ray variability of the BL Lac PKS 2155-304.**
J. Sokolowski, C. M. Urry, R. Warwick, S. Sembay, T. Ohashi.
Bull. Am. Astron. Soc., Vol. 23, No. 2, p. 923 (1991). Abstract. – See Abstr. 012.008 for the main entry.

158.035 **Optical polarimetry of PKS 2155-304 and constraints on accretion disk models for BL Lacs.**
M. L. Sitko, P. S. Smith.
Bull. Am. Astron. Soc., Vol. 23, No. 2, p. 923 (1991). Abstract. – See Abstr. 012.008 for the main entry.

158.036 **The near infrared continuum emission components in AGN.**
M. J. Ward, J. K. Kotilainen.
Bull. Am. Astron. Soc., Vol. 23, No. 2, p. 926 (1991). Abstract. – See Abstr. 012.008 for the main entry.

158.037 **Rapid variability of compact extragalactic radio sources.**
A. Quirrenbach.
Bull. Am. Astron. Soc., Vol. 23, No. 2, p. 927 (1991). Abstract. – See Abstr. 012.008 for the main entry.

158.038 **Rapid UV variability in the BL Lac PKS 2155-304.**
R. Edelson, J. Saken, G. Pike.
Bull. Am. Astron. Soc., Vol. 23, No. 2, p. 927 (1991). Abstract. – See Abstr. 012.008 for the main entry.

158.039 **BBXRT observations of BL Lac object PKS 2155-304.**
G. Madejski.
Bull. Am. Astron. Soc., Vol. 23, No. 2, p. 928 (1991). Abstract. – See Abstr. 012.008 for the main entry.

158.040 **The buried quasar in Cygnus A.**
A. S. Wilson, M. J. Ward, P. R. Blanco, M. Nishida.
Bull. Am. Astron. Soc., Vol. 23, No. 2, p. 928 (1991). Abstract. – See Abstr. 012.008 for the main entry.

158.041 **Formation of broad-line emission clouds in active galactic nuclei.**
A. S. Zentsova.
Sov. Astron., Vol. 35, No. 3, p. 320 – 322 (May 1991). Current Physics Microform No.: 9201X0602. English translation of Astron. Zh., Tom 68, Vyp. 3, p. 652 – 656 (1991).

The hypothesis that the emission clouds that produce allowed lines in the spectra of active galactic nuclei are formed in the outer regions of the accretion disk around a massive black hole is substantiated. At typical accretion rates for active galaxies, the outer regions of the disk become "self-gravitating" and are unstable against fragmentation into individual clouds. Nonlinear effects in the disk should be expected to result in slow drift of the clouds perpendicular to the disk plane, and radiation pressure from the central source should cause them to spiral away from the center.

158.042 **A connection between radio emission spectra and signs of interaction in powerful radio galaxies.**
A. G. Gubanov.
Sov. Astron. Lett., Vol. 17, No. 4, p. 289 – 292 (Jul 1991). Current Physics Microform No.: 9202X0465. English translation of Pis'ma Astron. Zh., Tom 17, No. 8, p. 684 – 690 (Jul 1991).

A connection between the radio emission spectra and signs of tidal gravitational interaction is explored for a sample of 40 powerful radio galaxies with strong optical emission lines and optical morphologies. The differences in the radio spectra of two groups of radio galaxies support the assertion that galactic mergers or strong tidal interactions could be responsible for the origin of powerful radio sources and that these radio sources must turn on in the late stages of such an encounter. Analogous differences in the radio spectra are also found between quasars and radio galaxies in the samples of Jackson and Browne (1990) where differences in the [O III] 5007 Å line luminosities were discovered. These differences in the radio spectra may be explained as a later stage of nuclear activity for radio galaxies than for quasars, which also results in a difference in [O III] emission line luminosities.

158.043 **Numerical models of extragalactic radio sources.**
J. O. Burns, M. L. Norman, D. A. Clarke.
Science, Vol. 253, No. 5019, p. 522 – 530 (2 Aug 1991).

Numerical simulations with supercomputers allow analysis of the wide range of nonlinear physics inherent in the hydrodynamic and magnetohydrodynamic equations. When applied to extragalactic radio sources, these numerical models have begun to reproduce many of the complex structures observed on telescopic images. This combination of telescopic and numerical observations provides powerful probes of the physics of radio galaxies. In this review, some of the recent results from both two-dimensional and three-dimensional numerical simulations of the formation and evolution of extended radio morphologies are described. These numerical models have allowed the exploration of important physical phenomena.

158.044 **First observations of the Lyman limit in a high redshift radio galaxy.**
M. Dickinson, H. Spinrad, J. Liebert.
Bull. Am. Astron. Soc., Vol. 23, No. 2, p. 951 (1991). Abstract. – See Abstr. 012.008 for the main entry.

158.045 **Optical identifications and spectroscopy of a subsample from the MG radio source catalog.**
H. Spinrad, M. Dickinson.
Bull. Am. Astron. Soc., Vol. 23, No. 2, p. 951 (1991). Abstract. – See Abstr. 012.008 for the main entry.

158.046 **The interaction-activity connection. II: Radio observations.**
N. E. Kassim, E. P. Smith.
Bull. Am. Astron. Soc., Vol. 23, No. 2, p. 954 (1991). Abstract. – See Abstr. 012.008 for the main entry.

158.047 **Micro-Jansky radio source counts and spectral indices at 8.4 GHz.**
R. A. Windhorst, E. B. Fomalont, R. B. Partridge, J. D. Lowenthal.
Bull. Am. Astron. Soc., Vol. 23, No. 2, p. 956 (1991). Abstract. – See Abstr. 012.008 for the main entry.

158.048 **Identifications and spectroscopy of high-redshift radio galaxies from the B3 survey.**
D. Thompson, S. Djorgovski, M. Vigotti, G. Grueff.
Bull. Am. Astron. Soc., Vol. 23, No. 2, p. 957 (1991). Abstract. – See Abstr. 012.008 for the main entry.

158.049 **The observed relationship of X-ray and infrared emission in active and normal galaxies.**
P. J. Green, S. F. Anderson, M. J. Ward.
Bull. Am. Astron. Soc., Vol. 23, No. 2, p. 957 (1991). Abstract. – See Abstr. 012.008 for the main entry.

158.050 **BBXRT observations of active galaxies.**
F. E. Marshall.
Bull. Am. Astron. Soc., Vol. 23, No. 2, p. 958 – 959 (1991). Abstract. – See Abstr. 012.008 for the main entry.

158.051 **Imaging spectrophotometry of ionized gas in NGC 1068. III. Anisotropic excitation of the large-scale disk by scattering of nuclear continuum.**
J. Sokolowski, J. Bland-Hawthorn, G. Cecil.
Astrophys. J., Vol. 375, No. 2, p. 583 – 593 (10 Jul 1991).
Imaging Fabry–Perot spectrophotometry has revealed a diffuse ionized medium (DIM) pervading the inner 10 kpc of the disk of NGC 1068. These data show that the DIM has high [N II] λ6584/Hα emission–line ratios (from 0.6 to 1.3) over much of the disk, with even higher ratios (from 1.3 to 2.5) near the nucleus and in isolated regions near the radio jet axis. Previous studies find that other low–excitation emission, such as [O II] λ3727 and [S II] λλ6716, 6731, is also strong away from the jet axis, while high–excitation emission such as [Ne V] λλ3346, 3426 and He II λ4686 is found only within two diametric sectors of opening angle ~80°, aligned along the axis. It is shown that these emission characteristics can be explained by a model, in which the intrinsic type 1 Seyfert nucleus, concealed by an obscuring medium, is visible only through scattered, polarized light. In this picture the gas excitation is anisotropic, the high–excitation gas along the jet axis being photoionized by direct nuclear continuum, while the lower excitation gas away from the axis sees only indirect nuclear emission scattered into it. The authors consider in detail a model of low optical depth, near–nuclear Thomson scattering and show that even with moderate dust extinction ($E \leqslant 0.3$ mag kpc^{-1}), this mechanism is sufficient to energize the DIM.

158.052 **The phenomenon of radio galaxies: a conceptual harvest from the synthesis of radio and optical studies.**
Gopal Krishna.
National Large Telescope Workshop: Scientific requirements and requirement of site, p. 25 – 33 (1991). – See Abstr. 012.009 for the main entry.
The author tries to present a perspective of some of the major advances in understanding the nature of radio galaxies, enabled by their observation made through radio and optical windows over the past 3 decades.

158.053 **OH megamasers above $z = 0.07$.**
I. Kazès, W. A. Baan.
Astron. Astrophys., Vol. 248, No. 1, p. L15 – L18 (Aug 1991). Letter–to–the–editor.
The Nançay radio telescope has been used to search for high redshifted OH megamasers. The authors report four new detections with redhsifts above $z = 0.07$. The properties of high redshift OH megamasers with velocities from 21000 to 51000 km/s are evaluated in terms of the dependence of the OH luminosity on the square of the far–infrared luminosity. Intrinsic linewidths of the OH line may help to distinguish between a starburst or an AGN nature for the nuclear galactic nucleus.

158.054 **Faint radio sources with ultra–steep spectra as candidates for distant radio galaxies.**
M. H. Wieringa, P. Katgert.
Astron. Astrophys., Vol. 248, No. 2, p. L31 – L34 (Aug 1991). Letter–to–the–editor.
The authors present the results of an optical identification program of a sample of 75 faint radio sources with ultra–steep radio spectra ($\alpha \gtrsim 1.1$, $S_\nu \sim \nu_{-\alpha}$), selected from Westerbork surveys at 327 MHz and with a median flux of 65 mJy.

158.055 **Cosmic evolution of linear sizes and luminosity function of powerful radio galaxies: is there a common cause?**
Gopal–Krishna.
Astron. Astrophys., Vol. 248, No. 2, p. 415 – 418 (Aug 1991).
For the population of active elliptical galaxies associated with powerful double radio sources, it is proposed that the empirically derived brightening of their radio luminosity function (RLF) and the statistical decrease in their linear size at higher redshifts (up to $z \sim 1$, or 2) could both arise if the hot gaseous halos seen to surround the ellipticals were denser at earlier epochs. The postulated density increase of the halo medium by about an order of magnitude between $z = 0$ and 1 would not just impart an extra stopping power needed to explain the smaller sizes of the double radio sources in the past but would, at the same time, enhance their efficiency for converting the beam power into radio emission by a factor which is quantitatively shown to be consistent with the empirically deduced evolution of the RLF up to $z \sim 1$. It seems thus possible to envision a unified framework for understanding the physical origin of these two well–known cosmological trends displayed by powerful radio galaxies. The author also briefly explores a possible physical link between the two basic morphological classes of extragalactic radio sources, namely the powerful, edge–brightened and the weaker, edge–darkened double radio sources.

158.056 **M82 in γ–rays.**
A. Akyüz, N. Brouillet, M. E. Özel.
Astron. Astrophys., Vol. 248, No. 2, p. 419 – 420 (Aug 1991).
In view of the reported high energy density in cosmic ray electrons from recent radio continuum observations (Klein et al. 1988), it is estimated that the nearest starburst galaxy M82 may be observable by the GRO–EGRET instrument at γ–rays above 100 MeV. Implications of a possible detection for the physics of cosmic rays, their extent and role in galactic processes are noted.

158.057 **Two BL Lacertae objects discovered through their optical variability.**
M. R. S. Hawkins, P. Véron, R. W. Hunstead, A. M. Burgess.
Astron. Astrophys., Vol. 248, No. 2, p. 421 – 425 (Aug 1991).
A survey for variable quasars over an 18 deg^2 field was started in 1977 with the UK 1.2 m Schmidt telescope. In the course of this survey, the authors have found 23 objects with a variability amplitude $\Delta B \geqslant 1$ mag. Spectroscopic observations of 14 of them have shown that two have featureless spectra; both have a high degree of optical polarization (5.9 and 9.6% respectively), and both are radio sources with a flat spectrum between 843 and 4790 MHz. There is no doubt that they are BL Lac objects. The minimum corresponding surface density of ~0.1 deg^{-2} at $B \sim 19$ is equal to the upper limit derived by Woltjer and Setti in 1982.

158.058 **A search for reflection features in the X–ray spectrum of NGC 4151.**
M. Maisack, T. Yaqoob.
Astron. Astrophys., Vol. 249, No. 1, p. 25 – 30 (Sep 1991).
The authors report results from a contemporary observation of NGC 4151 from 4–120 keV by Ginga and the HEXE instrument aboard Kvant from 1987 November 21st to 24th. This is the first spectrum of an AGN over this broad energy range since HEAO–1, but with higher sensitivities and higher energy resolution. No source variability was found during the observations in both the Ginga and the HEXE data. The authors found the source to be in a low state of emission in the X–ray range. The spectrum is well described by a single power law with an intensity of $1.44(+0.45,-0.32) \times 10^{-2}$photons/cm^{2}s keV at 1 keV and an energy index $\alpha = 0.37 \pm 0.11$ over the entire energy range, plus absorption by a hydrogen column of 6.7×10^{22}cm^{-2} and an iron emission line at 6.4 keV. The spectrum is fitted equally well by a model including some reflection from a layer of cold matter, but the reflected component is significantly smaller than would be expected if half the primary radiation were reprocessed. The authors discuss the constraints their fits can place on the importance and magnitude of a relfected component and the influence of pairs on the non–thermal power–law emission from NGC 4151 in the observed low state.

158.059 **The anomalous radio features in NGC 4388 and NGC 4438.**
E. Hummel, D. J. Saikia.
Astron. Astrophys., Vol. 249, No. 1, p. 43 – 56 (Sep 1991).
The authors present detailed, multifrequency VLA observations at 1.49 and 4.86 GHz of the edge–on spiral galaxies

NGC 4388 and NGC 4438. They are both members of the Virgo cluster of galaxies and have linear elongated radio features emerging from the nucleus and which are perpendicular to the plane of the galaxy. From these observations the authors have been able to identify the nucleus of NGC 4388 which has been reported recently to have a hidden Seyfert 1 nucleus. The larger scale structure of NGC 4388 consists of an elongated collimated structure to the south and a diffuse extended bubble to the north. The ridge of the radio emission is bent, possibly by density gradients perpendicular to the plane of the galaxy. The authors have not been able to idetify unambiguously a nuclear source in NGC 4438. The observations show a shell-like structure near the nucleus with a significant enhancement of the radio emission near the centre of the shell. The authors present the results of these observations and discuss possible explanations for the observed radio structures.

158.060 **AN 0248+43: a cold highly luminous FIR-galaxy with two nonthermal nuclei.**
W. Kollatschny, M. Dietrich, U. Borgeest, K.-J. Schramm.
Astron. Astrophys., Vol. 249, No. 1, p. 57 – 64 (Sep 1991).
The galaxy AN 0248+43, which seems to be compact on the Palomar Sky Survey, appears as a double nucleus system on CCD-images. The projected spatial separation of the two nuclei amounts to 3.8 arcsec corresponding to 3.8 kpc assuming a galaxy distance of 205 Mpc. From direct imaging and optical spectra the authors derived the physical properties of this system, showing that a spiral and an elliptical galaxy are in the merging process. This double nucleus galaxy is one of the very rare systems where both nuclei show nonthermal characteristics. The object is also a very strong FIR-source. The observed extreme excess of cold dust shows this system to be a unique object in the study of Seyfert and starburst properties.

158.061 **The beaming model and Hubble diagram of BL Lacertae objects.**
G. Z. Xie, F. K. Liu, B. F. Liu, R. W. Lu, K. H. Li, Y. Y. Zhu.
Astron. Astrophys., Vol. 249, No. 1, p. 65 – 69 (Sep 1991).
Based on the orientation of relativistic beaming of BL Lacertae objects, the authors show that the observed flux density is enhanced by a factor $\delta^{3+\alpha}$, where δ is the bulk relativistic Doppler factor. Thus, the observed apparent magnitude m_v^{ob} must be corrected for the Doppler effect. For 33 BL Lac objects given by Burbidge and Hewitt (1987), the corrected apparent magnitude m_v^{corr}– log z diagram shows a good correlation.

158.062 **The soft X-ray variability of NGC 5548.**
J. S. Kaastra.
Astron. Astrophys., Vol. 249, No. 1, p. 70 – 74 (Sep 1991).
The soft X-ray variability of NGC 5548 is studied using EXOSAT's low energy telescope in conjunction with the channel multiplyer array (CMA). For the first time, the pulse height distribution of the CMA data is used to infer information on the spectral variability. The spectrum of NGC 5548 consists of a hard power law and a soft component. During a one day observation in 1986 a considerable softening of the soft component flux was observed while the power law brightened only slightly. Significant changes in the properties of the soft X-ray emitting region or in a warm absorbing medium may explain the spectral softening.

158.063 **CCD surface photometry of the low-luminosity radio galaxies containing radio jets B2 0034+25 and B2 0206+35.**
J. I. González-Serrano, I. Pérez-Fournon.
Astron. Astrophys., Vol. 249, No. 1, p. 75 – 87 (Sep 1991).
The authors present broad-band CCD imaging of the radio galaxies B2 0034+25 and B2 0206+36. Both galaxies are of low radio power and contain radio jets at the arcsec scale. The galaxies have close companions and the optical morphology suggests that they are interacting. The surface photometry method and its application to interacting galaxies is described. An estimate of the photometric accuracy is obtained by applying this method to realistic simulations of galaxy pairs. This technique is found to be a powerful tool for studying galaxy interactions allowing the detection of photometric effects of the interaction such as non-concentric isophotes and deviations of the galaxy surface brightness profiles in the outer parts from the extrapolation of a de Vaucouleurs law fitted to the inner parts of the galaxy. In particular the two objects discussed in this paper appear to be strongly interacting with nearby companions. The authors estimate lower limits for the radio-to-optical spectral indices of the jets and conclude that the optical emission from the inner parts of the jets lies below a power-law extrapolated from radio wavelengths. The role of hot gas around elliptical galaxies on the collimation of the radio jets is briefly discussed.

158.064 **Variable soft X-ray excesses in active galactic nuclei from nonthermal electron-positron pair cascades.**
A. A. Zdziarski, P. S. Coppi.
Astrophys. J., Vol. 376, No. 2, p. 480 – 489 (1 Aug 1991).
The formation of steep soft X-ray excesses superposed on flatter hard X-ray power-law spectra in nonthermal electron-positron pair cascade sources is studied. Such soft X-ray excesses are similar to those seen below ~1 keV in AGN spectra. The soft X-ray excesses in the pair cascade AGN models are due to the Comptonization of UV bump photons by pairs decelerated in the cascade process and thermalized to a nonrelativistic equilibrium temperature. Simple formulae for the spectral form of the excess are found. The soft excess in these models appears as a steep power law superposed on the tail of the UV bump and the flat nonthermal (hard X-ray) power law. The model parameter space in which an excess is visible in soft X-rays is determined. Taking the UV photons to be a blackbody distribution, a characteristic temperature $kT_s \gtrsim 7$ eV is required for an excess to be visible at 300 eV. The time variability of soft excesses in pair cascade models is also examined. It is found that the variability patterns of the soft and hard X-rays may be distinctly different (as often seen in AGNs) despite their common origin in the nonthermal source. It is found that soft X-ray excesses can be produced by a recently developed Compton reflection model of UV and X-ray emission from the central engine of an AGN. In this model, the observed radiation comes from a nonthermal pair cascade source and a cold (~10^5K), optically thick matter distribution which intercepts and reprocesses part of the nonthermal emission.

158.065 **A tale of three jets.**
F. Macchetto.
Workshop on the First Year of HST Observations, p. 10 – 24 (1991). – See Abstr. 012.017 for the main entry.
The study of the optical counterparts to radio jets with the Faint Object Camera on board the HST has already produced new and unexpected results.

158.066 **On the origin of the optical and UV continuum in active galactic nuclei.**
S. Collin-Souffrin.
Astron. Astrophys., Vol. 249, No. 2, p. 344 – 350 (Sep 1991).
The extraordinary good correlations between the variations of lines and underlying continuum fluxes observed in several AGN, in particular in NGC 4151 and NGC 5848, and between the continuum at different wavelengths, raise a problem, as they are neither explained in the framework of current photoionization models, nor in those of emission by an accretion disc. They can be explained if the optical and UV continuum is due partly to reprocessing of X-ray continuum near the central object. Contrary to a widely held opinion, the "cold cloud" model proposed recently is not suitable for such a processing. On the contrary the reprocessing should naturally occur in the photosphere of an accretion disc irradiated by an X-ray continuum. Even if the X-ray luminosity does not dominate the bolometric luminosity, the contribution of the reprocessed luminosity in the optical-UV continuum is not negligible, and can explain the observed variations. Variations in the whole optical-UV range should be correlated with the X-ray flux, and the spectrum should harden when it brightens, as observed.

158.067 **VLBI structures of low frequency variable sources at 608 MHz.**
L. Padrielli, W. Eastman, L. Gregorini, F. Mantovani, S. Spangler.
Astron. Astrophys., Vol. 249, No. 2, p. 351 – 358 (Sep 1991).
The authors present 608 MHz VLBI maps of eleven sources from the Bologna low frequency variability survey. The sources observed are 0224+671, 0605–085, 0607–157, 1055+018, 1117+146, 1358+624, 1611+343, 1633+382, 1641+399, 1901+319, and 2255+415. These observations are of use to investigate the refractive scintillation theory for low frequency variability, as well as to provide additional information for the astrophysics of compact extragalactic radio sources. The authors find that for almost all of the sources, a close correspondence exists between the VLBI structure at 608 MHz and that visible on VLBI maps at higher frequencies. For some of the sources this VLBI structure possesses an obvious relationship to structure visible on arcsecond scales. In addition to the significance of this result for the astrophysics of these sources, it means that the observed structure for these sources is intrinsic, rather than produced by diffractive or refractive scattering phenomena.

158.068 **Colour indices of the Markarian galaxies.**
D. K. Ojha, S. C. Joshi.
Astrophys. Space Sci., Vol. 183, No. 2, p. 245 – 257 (Sep 1991).
By use of the ANS ultraviolet fluxes, optical (*UBV*) and IRAS infrared fluxes, several colour indices for a few Markarian galaxies have been investigated. A comparison of these indices with those for normal NGC galaxies has been carried out. Similarities and differences in the indices for the two groups of galaxies have been discussed and possible causes have been indicated.

158.069 **FOS spectroscopy of resolved structure in the nucleus of NGC 1068.**
S. Caganoff, R. R. J. Antonucci, H. C. Ford, G. A. Kriss, G. Hartig, L. Armus, I. N. Evans, E. Rosenblatt, R. C. Bohlin, A. L. Kinney.
Astrophys. J., Lett., Vol. 377, No. 1, p. L9 – L12 (10 Aug 1991).
The Second Special Hubble Space Telescope Issue.
The authors report the results of FOS spectroscopy of the continuum peak of NGC 1068 in the wavelength range 2200 – 7000 Å. Detailed analysis reveals the emission lines to be complex in shape and to consist of a number of different velocity components. Both the Balmer lines and the strong forbidden lines have a contribution from a component with full width at half–maximum on the order of 2200 km s^{-1}. However, the authors do not detect the broad Hβ emission observed in polarized light. This implies that the continuum peak is not the mirror which reflects light from the hidden Seyfert 1 nucleus. It could possibly contribute some of the reflected light, but the equivalent width of the Hβ BLR line, if present at all, is much weaker than expected from the ground–based nuclear polarized–flux spectrum.

158.070 **Faint Object Spectrograph observations of the low–luminosity Seyfert galaxy NGC 1566.**
G. A. Kriss, G. F. Hartig, L. Armus, W. P. Blair, S. Caganoff, L. Dressel.
Astrophys. J., Lett., Vol. 377, No. 1, p. L13 – L16 (10 Aug 1991).
The Second Special Hubble Space Telescope Issue.
FOS observations of the Seyfert galaxy NGC 1566 show it to possess the broad lines characteristic of a typical Seyfert 1 galaxy, even in states of very low luminosity. The continuum, however, is steep ($f_\nu \sim \nu^{-1.5}$) with no suggestion of the ultraviolet excess seen in higher luminosity AGNs. The broad emission lines have symmetric profiles, but are redshifted by 200 – 1000 km s^{-1} relative to the systemic velocity. The broad–line ratios indicate a low ionization parameter, and the strength of lines such as Fe II, Mg II, and the Balmer continuum relative to the UV continuum supports models for X–ray heating of the broad–line cloud interiors. The sharp core of the HST point response function gives a significant advantage over gound–based telescopes for observations of low luminosity active galaxies, despite the spherical aberration of the primary mirror.

158.071 **First results from the Goddard High–Resolution Spectrograph: ultraviolet spectra of a starburst knot in NGC 1068.**
J. B. Hutchings, F. Bruhweiler, A. Boggess, S. R. Heap, D. Ebbets, E. Beaver, E. Rosenblatt, K. Q. Truong, M. Perez, R. Westmacott.
Astrophys. J., Lett., Vol. 377, No. 1, p. L25 – L28 (10 Aug 1991).
The Second Special Hubble Space Telescope Issue.
The authors present UV spectroscopy of a circumnuclear starburst knot in the Seyfert 2 galaxy NGC 1068. The spectrum shows the presence of several thousand O and B stars which appear to have formed about 3 million years ago. The authors also discuss the extinction and interstellar medium in the line of sight.

158.072 **0735+178: the cosmic conspiracy.**
L. B. Bååth, F. J. Zhang, H. S. Chu.
Astron. Astrophys., Vol. 250, No. 1, p. 50 – 56 (Oct 1991).
This paper discusses a radio outburst in the BL Lac object 0735+178 which started in early 1988, ended during 1990 and was succeeded by a new outburst which is still in progress. The image obtained with a global VLBI network shows the source to mainly consist of a core–jet like structure ending in diffuse emission region of low surface brightness located to the east of the core. The authors identify two new components as having been produced during the late 1980's. They are moving out with apparent superluminal motion in a viewing angle of $\gtrsim 10°$ at a distance $\lesssim 2$ mas from the core. The core includes another new component, but the authors' resolution is just sufficient to resolve out this component from the core at the epoch of observation. It is suggested that the weak emission region is caused by an older component reaching the point at which the Doppler boosting is at its maximum and that the long time scale flux variations are due to successive events when evolved components reach that point.

158.073 **MG 1131+0456: discovery of the optical Einstein ring with the NTT.**
F. Hammer, O. Le Fèvre, M. C. Angonin, G. Meylan, A. Smette, J. Surdej.
Astron. Astrophys., Vol. 250, No. 2, p. L5 – L8 (Oct 1991).
Letter–to–the–editor.
Deep optical imaging and spectroscopy have been obtained for the well known radio Einstein ring MG 1131+0456 with the ESO/NTT. The direct CCD frames show a red emission excess at the exact location of the radio ring. Subsequent subtraction of the lensing galaxy, assumed to be an elliptical galaxy, results in the detection of an optical ring which is strikingly similar to the radio ring. Gravitational lensing models, including source reconstruction, predicts that the source is likely a radio galaxy for which both the optical and extended radio emissions are sufficiently distorted by the foreground lensing galaxy as to produce a nearly complete ring. The authors' spectroscopy and broad band photometry suggest that the lensing galaxy has a spectral energy distribution typical of a non evolved elliptical, tentatively at $z = 0.85$. The radio galaxy associated with the radio ring is likely to be an elliptical at larger redshift, tentatively $z = 1.13$.

158.074 **Search for fast ultraviolet variability in NGC 5548 and non–linearities in IUE's SWP camera.**
R. Walter, T. J.–L. Courvoisier.
Astron. Astrophys., Vol. 250, No. 2, p. 312 – 315 (Oct 1991).
Large observation campaigns of Seyfert galaxies give the opportunity to check the linearity of IUE short wavelength camera at low flux number and moderate background. In particular, many pairs of quasi simultaneous observations of NGC 5548 have been obtained with IUE. The authors selected pairs of spectra which include one well exposed and one poorly exposed spectrum. The ratio of the two broad band continuum fluxes measured in these paired spectra is variable and well correlated to the integrated flux of the underexposed observation indicating that this apparent variability is due to instrumental effects rather than source variability. The response of the SWP system (camera + processing) decreases by up to 25% at low flux

numbers. Taking these non linearities into account, NGC 5548 did not show variability in the ultraviolet on time scale of a few hours like in the soft X–ray.

158.075 **Radio source orientation and the angular diameter – redshift relation.**
L. I. Onuora.
Astrophys. J., Vol. 377, No. 1, p. 36 – 38 (10 Aug 1991).
The effect of a nonrandom source orientation on the angular diameter – redshift relation is considered for the 3CR sample of Laing et al., based on the "unified" scheme of Barthel. For an inhomogeneous sample of objects displaying milliarcsecond scale structure, it is found that there is no evidence for a systematic variation of orientation angle with redshift. However, if it is assumed that quasars are closer to the line of sight than powerful extended radio galaxies, then the observed angular size – redshift relation could be interpreted in terms of source orientation, rather than linear size evolution.

158.076 **Unique broad–line profile variations in the radio galaxy 3C 390.3.**
S. Veilleux, W. Zheng.
Astrophys. J., Vol. 377, No. 1, p. 89 – 99 (10 Aug 1991).
The authors report the results of a variability study of the flux and profile of Hβ in 3C 390.3. Spectra of this object were obtained at 39 epochs during the years 1974 – 1988. The fluctuations of the Hβ flux closely mimic the variations of the optical continuum during that period. The time scale for the emission line variations is shorter than a year, and perhaps as short as a month. The authors note that the blue side of the Hβ profile ("blue bump") also followed closely the continuum variations, but the red side ("red bump") did not. The ratio of the fluxes of the blue and red bumps behaved in a very regular, nearly sinusoidal fashion. It reached a maximum in mid–1975 and at the end of 1985, a time interval of 10.4 yr. The velocity of the peak of the blueshifted bump drifted by $+900\ km\ s^{-1}$ during the 14 yr interval covered by the observations. Various scenarios are proposed to explain these results. The more successful models are those involving an inhomogeneous line–emitting disk or a biconical broad–line region.

158.077 **The featureless continua and hydrogen lines of Seyfert 2 galaxies.**
A. L. Kinney, R. R. J. Antonucci, M. J. Ward, A. S. Wilson, M. Whittle.
Astrophys. J., Vol. 377, No. 1, p. 100 – 114 (10 Aug 1991).
The authors present optical ultraviolet spectra taken in similar sized large apertures for 15 Seyfert 2 galaxies. Measurements of emission–line strengths of Lyα, Hα, and Hβ are used together with measurements of the ultraviolet slopes to estimate the ratio of recombination photons to ionizing photons. The photon ratios indicate that the ionizing continuum is being emitted anisotropically for at least six of the eight objects with UV slope measurements. The median value of the UV slope is $\alpha = -1.35$ ($F_\nu \sim \nu^\alpha$), which is indistinguishable from the UV slope of Seyfert 1 galaxies. This result suggests that in the occultation / reflection picture of Seyfert nuclei, the reflecting particles are free electrons and not dust. The authors verify previous observations of variability in Mrk 477 and show that this galaxy exhibits characteristics of both Seyfert 1 and Seyfert 2 galaxies.

158.078 **Zones of high ionization of gas in nuclei of Seyfert galaxies and quasars.**
N. G. Bochkarev, L. S. Nazarova.
Astron. Zh., Tom 68, Vyp. 5, p. 918 – 935 (Sep–Oct 1991). In Russian. English translation in Sov. Astron., Vol. 35, No. 5.
On the basis of Sobolev's probability method, the authors calculate relative line intensities and level populations of hydrogen and He II for a broad range of physical parameters [$N_e(H) = 10^{9-11}cm^{-3}$, $T_e = (2-3)\cdot 10^4 K$, velocity gradients in the gas $dV/dR = 10^{-4}{}^{-}{}^{-6}s^{-1}$] under the assumption that the stationary population of levels is due to a simultaneous action of shock and radiative mechanisms of gas heating and ionization. It is accepted that the central source of radiation is nonthermal: $F_\nu = F_0(\nu/\nu_0)^{-\alpha}$ with $F_0 = 10^{-5} erg\ cm^{-2}s^{-1}Hz^{-1}$ with various spectral indices $\alpha = 0-1$. On the basis of a comparison of the calculated and observed relative intensities of hydrogen and He II lines, physical parameters are estimated for the regions of formation of high excitation lines, located above the accretion disc surface: $N_e = 10^{11}cm^{-3}$, $T_e = 3\cdot 10^4 K$, $dV/dR = 10^{-5}s^{-1}$.

158.079 **Counts and distribution of spectral indices of extra–galactic radio sources.**
A. G. Gorshkov.
Astron. Zh., Tom 68, Vyp. 6, p. 1121 – 1133 (Nov–Dec 1991). In Russian. English translation in Sov. Astron., Vol. 35, No. 6.
Assuming that at 408 MHz the counts of radio sources may be represented by $n(S) = KS^{-a-\ln Sb}$, the dependence "spectral index vs flux density" is absent and the shape of the source spectrum may be approximated by $S \sim \nu^{-\alpha}$ for all the observed range of frequencies. Analytical solutions for n(S) and P(α, S) at any frequency have been obtained. Results of the calculations for n(S) and P(α, S) are in good agreement with the observed data in the range 25 – 100000 MHz. It is shown that the counts vs frequency and spectral index vs flux density dependences are not the result of the cosmological evolution of the radio sources.

158.080 **Asymmetry of polarization properties of radio emission of the inner lobes of Centaurus A as determined from observations at the wavelengths 13 and 31 cm.**
E. N. Vinyajkin, V. I. Abramov.
Astron. Zh., Tom 68, Vyp. 6, p. 1174 – 1180 (Nov–Dec 1991). In Russian. English translation in Sov. Astron., Vol. 35, No. 6.
Linear polarization of Centaurus A inner lobes at the wavelengths 13 and 31 cm has been investigated. At these wavelengths (as at centimeter wavelengths), the NE lobe's polarized radio emission is brighter than the SW lobe's one. The degree of polarization of the integrated 31–cm radio emission of the SW lobe in the limits of measurement errors is the same as at shorter wavelengths; this indicates to the absence of considerable Faraday depolarization up to this wavelength. On the other hand, the degree of polarization of the NE lobe's integrated radio emission noticeably decreases with wavelength increasing. Taking into account the data on position angles at centimeter wavelengths, the rotation measures for both lobes have been determined: $R_{NE} = -60 \pm 2.,5$, $R_{SW} = -221 \pm 2\ rad\cdot m^{-2}$. The magnetic field is predominantly aligned with the lobes.

158.081 **Bidirectional jet–like feature in M51.**
Su Hongjun, Huang Jiehao, Deng Zugan.
3. Chinese Academy of Sciences and Max–Planck Society Workshop: High energy astrophysics: compact stars and active galaxies, p. 73 (1991). Abstract. – See Abstr. 012.026 for the main entry.

158.082 **Erratum: "Precise positions of optical counterparts of some radio objects" [Astrophys. Space Sci., Vol. 177, Nos. 1 – 2, p. 93 – 96 (Mar 1991)].**
G. Chiumiento, M. G. Lattanzi, G. Massone, R. Morbidelli, R. Pannunzio, M. Sarasso.
Astrophys. Space Sci., Vol. 184, No. 2, p. 331 (Oct 1991). See Abstr. 53.158.149.

158.083 **A model for the spectral variability of BL Lacertae objects at high frequencies.**
A. Celotti, L. Maraschi, A. Treves.
Astrophys. J., Vol. 377, No. 2, p. 403 – 416 (20 Aug 1991).
The average amplitude of variability of BL Lac objects is larger at higher frequencies, and the spectra in the X–ray range show a hardening with increasing intensity. This is shown to be a natural consequence of the relativistic jet model proposed by Ghisellini, Maraschi, and Treves, where higher frequencies are produced nearer to the jet core. Time–dependent properties are computed assuming that a perturbation travels at fixed speed down the jet, producing enhancements of constant amplitude of the relativistic particle density and of the magnetic field, in a slab or self–similar cone geometry. The time–dependent spectral intensities due to synchrotron radiation and first order Compton scattering are

computed numerically. The evolution of the spectral shape with time and the light curves at fixed frequencies are presented and discussed in detail. The results are compared with observations. Finally, the authors estimate the expected intrinsic amplitude assuming that the perturbation is due to a planar relativistic shock wave. The dependence of the observed amplitude on the viewing angle and on the shock and fluid velocities is discussed.

158.084 **X-ray observations of the bright Seyfert galaxy IC 4329A.**
K. P. Singh, A. R. Rao, M. N. Vahia.
Astrophys. J., Vol. 377, No. 2, p. 417 - 425 (20 Aug 1991).
The authors have performed an analysis of the broad-band (0.1 - 25 keV) EXOSAT observations of the nearby Seyfert type I galaxy IC 4329A. The observations were carried out on two occasions, 1984 July and 1985 July. The X-ray luminosity decreased by ~35% over one year. A significant short-term variability, where the intensity decreased by ~12% on a time scale of $\sim 2 \times 10^4$s was detected in the 1985 data. The observed X-ray spectra are all well fitted by simple power-law models with nearly 10 times larger absorption than the Galactic value, indicating significant absorbing matter local to the source. There is no evidence for low-energy excess in the spectra. The photon index Γ is measured to be 1.85 ± 0.06 in 1984 and 1.77(+0.06, -0.05) in 1985. The authors find a Fe line emission at 6.4 keV with an equivalent width of 169(+124, -113)eV in the 1984 data when the source was bright. The results are used to constrain the mass of the putative supermassive black hole in the nucleus of IC 4329A.

158.085 **Physics, structure, and fueling of active galactic nuclei.**
D. E. Osterbrock.
Publ. Astron. Soc. Pac., Vol. 103, No. 666, p. 874 - 880 (Aug 1991). - See Abstr. 012.034 for the main entry.
Some aspects of recent results on the nature of active galactic nuclei are reviewed. The importance of photoionization by a hard spectrum containing high-energy photons, extending to the X-ray region, is emphasized. Time-variation studies plus spectral line ratios show that simple models cannot satisfy all the observational constraints, and some of the suggested more sophisticated models are briefly described. Observational and theoretical results which emphasize the importance of interactions between galaxies in refueling the accretion disks about massive black holes at their centers are reviewed.

158.086 **A spectrophotometric study of Seyfert galaxies: physical conditions of the emitting gas.**
I. Cruz-González, J. Guichard, A. Serrano, L. Carrasco.
Publ. Astron. Soc. Pac., Vol. 103, No. 666, p. 888 - 890 (Aug 1991). - See Abstr. 012.034 for the main entry.
The authors present a spectrophotometric study of a sample of 52 Seyfert galaxies. The main results are: (1) The densities in the [O III] zone are high (10^5 - 10^6cm^{-3}). (2) The helium abundance is roughly solar, while the heavy element abundances (oxygen and neon) are subsolar. (3) The comparison with the photoionization models of Stasinska 1984 suggests that the photoionization mechanism for line formation is dominant in the narrow-line regions.

158.087 **Kinematics of the helical nuclear jets of NGC 4258.**
G. Cecil.
Publ. Astron. Soc. Pac., Vol. 103, No. 666, p. 894 - 897 (Aug 1991). - See Abstr. 012.034 for the main entry.
The HIFI imaging Fabry-Perot was used to synthesize 44,000 Hα and [N II] λ6583 line profiles across the disk of the nearby Seyfert galaxy NGC 4258. The spectral grid spans 5′ and resolves velocity structure across the prominent "four-branched jet". The author finds that the jets have total ionized mass $1.5 \times 10^6 M_\odot$ and [N II] λ6583/Hα ratios consistent with shock excitation. Velocity dispersions along the SE jet average 80 km s^{-1}, compared to 40 km s^{-1} in the H II regions. This broadening arises from a striking spatio-kinematic helicity of at least two plasma streams that wind through the observed phase subspace. Gaussian decomposition shows that the internal velocity dispersion of each helix is 100 km s^{-1}. A ballistic interpretation of the braiding at radii within 1 kpc and tangential velocities of 150 km s^{-1} yields a deprojected outflow velocity of 1200 km s^{-1}, compact binary source masses of $O(10^7)M_\odot$, and source separation O(10) pc. However, the detailed kinematic structure shows that some of the motion is along the helices.

158.088 **Diffuse ionized medium in NGC 1068.**
J. Bland-Hawthorn, J. Sokolowski, G. Cecil.
Publ. Astron. Soc. Pac., Vol. 103, No. 666, p. 906 - 910 (Aug 1991). - See Abstr. 012.034 for the main entry.
The authors present additional evidence for the existence of an extended (~10 kpc; ~135″), diffuse ionized medium (DIM) recently detected in the nearby, luminous Seyfert galaxy NGC 1068. The DIM phase was recognized from spectrophotometric Hα and [N II] λλ6548, 6584 line emission maps obtained with the Hawaii Imaging Fabry-Perot Interferometer (HIFI). They demonstrate that the Hα equivalent-width distribution, when combined with the [N II]λ6584/Hα line ratio distribution, permits a clean separation of the DIM from the adjacent H II region population. The authors summarize the main reasons that lead them to believe the DIM is photoexcited predominantly by scattered light from the active nucleus rather than the extended stellar population.

158.089 **High radiation brightness temperatures in active galactic nuclei.**
A. R. Crusius-Wätzel.
Astron. Astrophys., Vol. 251, No. 1, p. L5 - L6 (Nov 1991). Letter-to-the-editor.
The radiation generated in the synchrotron-self-Compton process by a peaked relativistic electron energy distribution is considered. It is found that the brightness temperature T_b of the emission can be above 10^{12}K when the self-absorption occurs well below the typical synchrotron frequency, so that the energy density in the photons is reduced by a significant amount compared to the case of complete self-absorption. The results also hold for a more general family of particle distributions with low-energy cutoffs. This is in contrast to the case of a full power-law distribution, where this value cannot be exceeded. It is argued that sources with directly measured brightness temperatures in the range $(1-4) \times 10^{12}$K possibly can be understood within the framework of this model.

158.090 **Relativistic beaming effect of BL Lac objects.**
Xie Guangzhong, Liu Bifang, Liu Fukun.
3. Chinese Academy of Sciences and Max-Planck Society Workshop: High energy astrophysics: compact stars and active galaxies, g, p. 158 - 161 (1991). - See Abstr. 012.026 for the main entry.
The authors note the jets of BL Lac objects are relativistic and the orientation of their axes to the line of sight is very small and therefore the observed flux density is enhanced by a Doppler factor of $\delta^{3+\alpha}$. Then the observed apparent magnitude m_v^{ob} or luminosity L_v^{ob} must be corrected for this effect. In this paper, the authors give two examples to illustrate the importance of relativistic effect on the Hubble diagram and the correlation of the luminosities.

158.091 **Explanation of the spectral behaviour of two BL Lac objects.**
Wang Jiangcheng, Xie Guangzhong.
3. Chinese Academy of Sciences and Max-Planck Society Workshop: High energy astrophysics: compact stars and active galaxies, p. 162 - 167 (1991). - See Abstr. 012.026 for the main entry.
An evolutionary model for relativistic electron momentum spectra is developed to explain flattening in the optical spectrum of two BL Lac objects, OI 090.4 and OJ 131 in the phases of their outburst in 1988. The model is based on the interplay of stochastic acceleration by turbulent waves, Fermi acceleration at shock waves, radiation losses due to synchrotron and inverse Compton emission, and particle escape in the turbulent plasma provided by the central object.

158.092 **The X-ray spectra and XUV excess of some AGNs.**
Ma Yuquian, Wang Tinggui, Zhou Youyuan.
3. Chinese Academy of Sciences and Max-Planck Society Workshop: High energy astrophysics: compact stars and active galaxies, p. 170 – 175 (1991). – See Abstr. 012.026 for the main entry.

Using EXOSAT ME and LE observation data and accurate galactic N_H value, the authors investigate the X-ray spectra properties of four AGNs. Associated with IUE data, the soft XUV excesses are examined. The results show some differences from the previous work.

158.093 **The proportion of ultraluminous IRAS galaxies in interacting systems.**
Zou Zhenlong, Xia Xiaoyang, Deng Zugan.
3. Chinese Academy of Sciences and Max-Planck Society Workshop: High energy astrophysics: compact stars and active galaxies, p. 176 – 182 (1991). – See Abstr. 012.026 for the main entry.

The authors have examined a sample consisting of 41 ultraluminous IRAS galaxies with far infrared luminosities LIR > $10^{12}L_{\odot}$(H_0 = 50 km/s Mpc), and a control sample of 41 low-luminosity IRAS galaxies with LIR < $10^{9.7}L_{\odot}$. All objects are taken from the southern part ($\delta < -17^{\circ}.5$) of QMC-Cambridge-Durham survey. The density contour maps and the sky survey material show that the proportion of ultraluminous IRAS galaxies which are interacting or merger systems, $61 \pm 12\%$, is much higher than that of low-luminosity ones ($10 \pm 5\%$), but still well short of the 100% suggested in several recent reports. This result seems to support the conclusion that galaxy interaction or merging is more common at high IR luminosity, but it is not the only cause of their activity.

158.094 **Analysis of infrared radiation mechanism for the Seyfert galaxies observed in IRAS survey.**
Wang Tinggui, Cheng Fuzhen, Zhang Jialu.
3. Chinese Academy of Sciences and Max-Planck Society Workshop: High energy astrophysics: compact stars and active galaxies, p. 183 (1991). Abstract. Abstract. – See Abstr. 012.026 for the main entry.

158.095 **IRAS colors of AGNs.**
Wang Gang.
3. Chinese Academy of Sciences and Max-Planck Society Workshop: High energy astrophysics: compact stars and active galaxies, p. 184 – 188 (1991). – See Abstr. 012.026 for the main entry.

93 AGNs in VCV's Catalog detected by IRAS in all of the 4 bands are discussed in the paper. The IRAS colors could be used to select AGNs from normal galaxies. The ratio of F_{60}/F_{25} is sensitive for selection of Seyferts, while the ratio of F_{25}/F_{12} is sensitive for starburst galaxies. A simple model of two blackbody components could be used to fit the IRAS emission of AGNs well.

158.096 **Observational distinction between two types of active galactic nuclei.**
Xie Guangzhong, Liu Fukun, Liu Bifang, K. H. Li, R. W. Lu, Lu Fuju.
3. Chinese Academy of Sciences and Max-Planck Society Workshop: High energy astrophysics: compact stars and active galaxies, p. 189 – 191 (1991). – See Abstr. 012.026 for the main entry.

The authors consider 69 sources including BL Lac objects (24), quasars (23), and Seyfert galaxies (22). Using two criteria, they discuss the observational distinction between relativistic beaming and pure accretion for AGNs.

158.097 **The stratification of Fe ions in shocked clouds in the narrow emission-line region of active galactic nuclei.**
M. Contini, S. M. Viegas.
Astron. Astrophys., Vol. 251, No. 1, p. 27 – 31 (Nov 1991).

The authors discuss the problems linked with the structure of shocked clouds in the narrow line region of active galactic nuclei, in the particular case of iron. Composite models are used in the calculations by the SUMA code, which accounts consistently for the ionizing flux and for the shock. The authors find that some interesting features, due to the presence of the shock, can explain several peculiarities of line emission spectra, as the absence of Fe I and of much higher ionization lines. On the other hand, results of calculation show that, theoretically, weak lines from the Fe^{+3} ion and from other intermediate ions, should be present in the observed spectra. Variabilities on time scales of some days are anticipated even for lines emitted from the narrow line region.

158.098 **Recent developments in gravitational lensing.**
P. Schneider.
3. Chinese Academy of Sciences and Max-Planck Society Workshop: High energy astrophysics: compact stars and active galaxies, p. 192 – 202 (1991). – See Abstr. 012.026 for the main entry.

The author reviews some of theoretical, but much more observational results that have been derived during the last two or three years in gravitational lensing, and which have aroused our excitement about this interesting phenomenon.

158.099 **Active galactic nuclei: a general view from optical and ultraviolet observations.**
K. Ayani.
Astron. Her., Vol. 84, No. 11, p. 352 – 358 (1991). In Japanese.

158.100 **Extended emission line region of active galactic nuclei.**
M. Yoshida.
Astron. Her., Vol. 84, No. 11, p. 359 – 363 (1991). In Japanese.

158.101 **Nuclear activity in galaxies: monster vs. starburst.**
Y. Taniguchi.
Astron. Her., Vol. 84, No. 12, p. 384 – 394 (1991). In Japanese.

158.102 **An X-ray and radio study of steep-spectrum radio sources. II. Four fields from a 22 MHz polar cap survey.**
P. E. Dewdney, C. H. Costain, I. McHardy, A. G. Willis, D. E. Harris, C. P. Stern.
Astrophys. J., Suppl. Ser., Vol. 76, No. 4, p. 1055 – 1066 (Aug 1991).

Four fields containing radio sources with spectral indices greater than unity at low frequencies have been observed over a wide range of radio wavelengths and at X-ray wavelengths. For most of the steep-spectrum sources no optical identifications have been found to the detection limit of the Palomar Schmidt prints, even with the positional precision of $\sim 1''$ from VLA maps. This implies that the associated galaxies or clusters lie at a great distance. X-ray emission was not detected from any of the steep-spectrum radio sources. In one case, an extended source appears to be a radio halo in an uncataloged cluster of galaxies at a redshift of 0.125, and the authors suggest that this object is a promising candidate for the detection of inverse Compton X-rays, since it is not a high-luminosity source of *thermal* X-rays.

158.103 **Variability and the origin of continuum emission in AGN.**
J. N. Bregman.
Georgia State University Conference on Variability of Active Galactic Nuclei, p. 1 – 12 (1991). – See Abstr. 012.040 for the main entry.

A picture that has emerged in recent years is that the continuum emission in Active Galactic Nuclei can be divided into two physically different models. Nearly all of the power emitted from radio quiet AGN in the radio through X-ray region has a thermal origin. In contrast to these "thermal" AGN are the blazars, where rapid variability, high polarization, and a lack of spectral features suggest that their continuum emission is produced by nonthermal processes such as synchrotron and inverse Compton emission. The author reports on an effort to determine how the optical and radio emitting regions, vastly different in size, are related.

158.104 **Simultaneous multifrequency observations of NGC 4051.**
C. Done.
Georgia State University Conference on Variability of Active Galactic Nuclei, p. 21 – 25 (1991). – See Abstr. 012.040 for the main entry.
A typical Active Galactic Nucleus emits roughly equal power per decade of frequency from the far infrared to X-ray frequencies. This power is either spread over the whole energy range by the primary emission process of the central engine, or is re-processed and re-distributed by secondary mechanisms at some distance from the source. If the broad spectral energy distribution is due to the primary process then variability at different frequencies is likely to be correlated. A simultaneous, multifrequency study of rapid variability can distinguish between these two possibilities and so shed light on the emission mechanism(s) responsible for the formation of the continuum.

158.105 **Variability of wavelength-dependent polarization in radio-selected BL Lacs.**
P. S. Smith.
Georgia State University Conference on Variability of Active Galactic Nuclei, p. 26 – 28 (1991). – See Abstr. 012.040 for the main entry.
During the 1980's a new generation of efficient optical and near-IR polarimeters has allowed detailed studies of the broadband linear polarization properties of radio-selected BL Lacertae objects. The author presents analyses of broadband optical and near-IR polarimetry of a few of the most well-studied BL Lacs. The emphasis here is to study the variation of polarization, both degree of polarization and position angle, with wavelength.

158.106 **Relativistic beaming and the parent population of BL Lacertae objects.**
C. M. Urry, P. Padovani.
Georgia State University Conference on Variability of Active Galactic Nuclei, p. 32 – 48 (1991). – See Abstr. 012.040 for the main entry.
A strong test of the hypothesis that BL Lacertae objects are a special subset of Fanaroff-Riley type I (low-luminosity) radio galaxies, namely those dominated by beamed emission from a relativistic jet aligned with the line of sight, is the ratio of the relative number densities. The authors show that available data in the X-ray, radio, and optical bands are consistent with this hypothesis.

158.107 **Time dependent shock acceleration of energetic electrons including synchrotron losses.**
K.-D. Fritz.
Georgia State University Conference on Variability of Active Galactic Nuclei, p. 49 – 51 (1991). – See Abstr. 012.040 for the main entry.

158.108 **Some long-term properties of a large group of active galactic nuclei.**
A. G. Smith, A. D. Nair, S. D. Clements.
Georgia State University Conference on Variability of Active Galactic Nuclei, p. 52 – 66 (1991). – See Abstr. 012.040 for the main entry.

158.109 **A search for faint variable galactic nuclei in SA57.**
M. A. Bershady, D. Trevese, R. G. Kron.
Georgia State University Conference on Variability of Active Galactic Nuclei, p. 67 – 71 (1991). – See Abstr. 012.040 for the main entry.
Nine Mayall prime focus IIIa-J plates spanning an eleven year baseline are analyzed in a uniform manner for the detection of variable objects to $B_J = 22.6$. Techniques successful in independently finding a complete sample of faint, color-selected, spectroscopically confirmed QSOs, are applied to search for nuclear variability in extended objects.

158.110 **Variability measurements of a faint, complete QSO sample.**
D. Trevese, R. G. Kron.
Georgia State University Conference on Variability of Active Galactic Nuclei, p. 72 – 74 (1991). – See Abstr. 012.040 for the main entry.
Nine 4-m Mayall prime focus blue (Kodak IIIaJ + GG385) plates in SA57 spanning 11 years and seven epochs were digitzed and processed to obtain a catalog of magnitudes normalized to a common zero point for all the stellar-like objects of the scanned area brighter than $J = 22.5$.

158.111 **Variability and the EUV spectrum of AGNs.**
P. M. Gondhalekar.
Georgia State University Conference on Variability of Active Galactic Nuclei, p. 78 – 82 (1991). – See Abstr. 012.040 for the main entry.
The observed variability of emission lines in the low luminosity Seyfert 1 galaxy 3C120 is modelled to estimate the extreme ultraviolet spectrum seen by the broad line region in this galaxy. The models are extrapolated to higher luminosity active galaxies and some speculations about the ionizing radiation and the nature of the broad line clouds in AGNs are made.

158.112 **Infrared-optical synchrotron emission from AGN.**
B. J. Wills.
Georgia State University Conference on Variability of Active Galactic Nuclei, p. 87 – 103 (1991). – See Abstr. 012.040 for the main entry.
A summary is given of results concerning the synchrotron component of luminous AGN, with emphasis on recent optical-infrared polarization studies. The author discusses the variability, polarization and spectral characteristics, and the relation between radio structure and polarization for AGN with dominant synchrotron components (the BL Lac objects).

158.113 **IR variability in GQ Comae.**
M. L. Sitko.
Georgia State University Conference on Variability of Active Galactic Nuclei, p. 104 – 107 (1991). – See Abstr. 012.040 for the main entry.
GQ Comae is a radio-quiet QSO with substantial optical/ultraviolet variability. Because of this, it is an ideal object for studying the spatial geometries of the various continuum-emitting regions. The author obtained a series of ultraviolet (IUE) spectra along with optical and infrared photometry (0.33 μm to 3.5 μm) over a three-year time span in order to determine the structure of its IR-emitting region.

158.114 **Optical variability of 6 BL Lacertae objects serendipitously discovered by EXOSAT.**
P. Grandi, P. Ciliegi, P. Giommi, G. G. C. Palumbo, G. Tagliaferri.
Georgia State University Conference on Variability of Active Galactic Nuclei, p. 108 – 110 (1991). – See Abstr. 012.040 for the main entry.
The EXOSAT Observatory Channel Multiplier Array allowed the detection of hundreds of serendipitous sources. A small fraction of these, when optically identified, turned out to be BL Lac objects. The authors have used the EXOSAT sample of new BL Lac objects to investigate the optical variability of X-ray selected BL Lacs. They present here the results of a set of observations in the BVRI bands of 6 objects serendipitously discovered by EXOSAT.

158.115 **The microvariability of selected blazars.**
H. R. Miller, M. T. Carini.
Georgia State University Conference on Variability of Active Galactic Nuclei, p. 111 – 119 (1991). – See Abstr. 012.040 for the main entry.

158.116 **Microvariability in the BL Lac Object 0716+71.**
S. J. Wagner, F. Sanchez-Pons, K. Anton, A. Quirrenbach, A. Witzel.
Georgia State University Conference on Variability of Active Galactic Nuclei, p. 120 – 122 (1991). – See Abstr. 012.040 for the main entry.

In addition to the well-known variability on time-scales of weeks to years, which have been studied in the majority of catalogued BL Lac objects, at least some of those objects vary on time scales of hours as well. Little is known about the spectrum of the rapidly variable component and possible relations with variability on longer time-scales. The former question is of special importance since the detections of intraday variability in the radio regime. The authors report on the results of observations of the BL Lac object 0716+71. It is variable in both frequency ranges. The variability has time-scales of several hours to days, amplitudes of approximately 40% in the optical and 15% in the radio regime. The behaviour was rather similar during several monitoring campaigns.

158.117 **The timescales of the optical variability of OJ287.**
M. T. Carini.
Georgia State University Conference on Variability of Active Galactic Nuclei, p. 123 – 127 (1991). – See Abstr. 012.040 for the main entry.

158.118 **The microvariability of the blazar OI 090.4.**
J. Noble, M. T. Carini, H. R. Miller.
Georgia State University Conference on Variability of Active Galactic Nuclei, p. 128 – 130 (1991). – See Abstr. 012.040 for the main entry.

158.119 **Structural variations in active galactic nuclei observed with VLBI.**
T. J. Pearson.
Georgia State University Conference on Variability of Active Galactic Nuclei, p. 134 – 150 (1991). – See Abstr. 012.040 for the main entry.

VLBI imaging allows a direct determination of the variations in the radio brightness distribution of active galactic nuclei on the parsec scale. The author reviews recent VLBI observations with particular emphasis on the work of the Caltech group. His studies have concentrated on a complete, flux-density limited sample of 65 sources, for which one now has images at three epochs. He describes the structural variations in three classes of objects: the one-sided core-jet sources, which frequently show superluminal motion; the compact steep-spectrum sources, which sometimes do; and the so-called "compact double" sources, which do not.

158.120 **Variability of nonthermal emission in compact jets.**
A. P. Marscher, J. P. Travis.
Georgia State University Conference on Variability of Active Galactic Nuclei, p. 153 – 156 (1991). – See Abstr. 012.040 for the main entry.

The class of extragalactic objects referred to as "blazars" is characterized by distinctly nonthermal continua subject to pronounced and rapid variability. At centimeter wavelenghts, VLBI images show that a blazar generally consists of a very compact core plus a knotty jet. The knots, when bright enough to follow, almost always move away from the core at apparent superluminal speeds. An outburst in total flux density is often found to coincide with the extrapolated appearance of a new superluminal knot near the core.

158.121 **On the optical polarization of BL Lacs and HPQs.**
D. J. Saikia.
Georgia State University Conference on Variability of Active Galactic Nuclei, p. 157 – 159 (1991). – See Abstr. 012.040 for the main entry.

158.122 **Radio source variability and unification schemes.**
D. J. Saikia, A. K. Singal, P. J. Wiita.
Georgia State University Conference on Variability of Active Galactic Nuclei, p. 160 – 164 (1991). – See Abstr. 012.040 for the main entry.

One of the interesting suggestions made recently by Scheuer and Barthel amongst others, is that elliptical galaxies associated with powerful radio galaxies appear as quasars when they are inclined at small angles to the line of sight. In such a scheme, the radio galaxies are believed to lie close to the plane of the sky, while the lobe- and core-dominated quasars are seen at progressively smaller angles to the line of sight. The authors discuss three sets of correlations, all of which are in accord with this simple unification scheme: linear-size vs. flux core fraction; bending angle; flux density variations.

158.123 **Intraday variability of extragalactic radio sources.**
A. Quirrenbach.
Georgia State University Conference on Variability of Active Galactic Nuclei, p. 165 – 168 (1991). – See Abstr. 012.040 for the main entry.

Intrinsic variability of extragalactic radio sources on timescales of weeks to years is a well-known phenomenon. Since due to the intensive Compton brightness limit the radio cores of active galactic nuclei are bound to be much larger than their optical counterparts, it was quite unexpected when Heeschen et al. found variations with amplitudes of $\sim 10\%$ and timescales of ~ 1 day at 11 cm wavelength in several flat spectrum sources. Here the author presents some observational results on intraday radio variability of compact radio sources and discusses some ideas to explain them.

158.124 **Radio variability of active galactic nuclei and related objects.**
H. D. Aller, P. A. Hughes, M. F. Aller.
Georgia State University Conference on Variability of Active Galactic Nuclei, p. 172 – 183 (1991). – See Abstr. 012.040 for the main entry.

This presentation summarizes some of the observational characteristics at centimeter wavelengths of the variability in flux density and linear polarization of active galactic nuclei and quasi stellar sources. In addition, the authors draw attention to the theoretical model fitting effort at Michigan which appears to be able to account for the observed variability in several sources by axial shock forming in a relativistic flow down a jet structure that is aligned near the line of sight.

158.125 **A statistical analysis of total and polarized cm-waveband flux variations.**
P. A. Hughes, M. F. Aller, H. D. Aller.
Georgia State University Conference on Variability of Active Galactic Nuclei, p. 187 – 189 (1991). – See Abstr. 012.040 for the main entry.

158.126 **Constraints on central engine models from rapid X-ray variability.**
M. C. Begelman, M. de Kool.
Georgia State University Conference on Variability of Active Galactic Nuclei, p. 198 – 212 (1991). – See Abstr. 012.040 for the main entry.

The authors discuss the implications of recently published GINGA and EXOSAT data on rapid X-ray variability in AGN. They conclude that the observed behavior of the iron Kα fluorescence line cannot be explained by the reprocessing of hard X-rays by a standard accretion disk, unless most of the X-ray luminosity is generated in local regions close to the disk. Observed power density spectra of X-ray variability can be reasonably well modelled with a reservoir mechanism.

158.127 **The complex and variable X-ray spectrum of Seyfert galaxies.**
K. A. Pounds, K. Nandra, G. C. Stewart.
Georgia State University Conference on Variability of Active Galactic Nuclei, p. 213 – 224 (1991). – See Abstr. 012.040 for the main entry.

Exosat observations (1983 – 86) showed rapid and large amplitude variability to be a common feature in the X-ray-radiation from Seyfert galaxies. Recent observations with the more sensitive Ginga LAC (1987 – 90), for many of the same bright Seyferts, have revealed additional features in the hard power law (2 – 30 keV) spectrum. An emission line is found to be common, as is a flattening of the continuum above 10 – 12 keV, both features probably arising from the interaction of the hard power law photons with nearby "cold", optically thick matter. Previous interpretations of flux–correlated, power law slope changes in the X–ray spectra of several Seyferts, in terms of a variable "compactness parameter", are now brought into question, since a similar effect can arise from the combination of intensity changes in the primary power law and variations in the reflected continuum.

158.128 **X–ray and gamma–ray variability in a nonthermal AGN model.**
A. A. Zdziarski, P. S. Coppi, R. Svensson.
Georgia State University Conference on Variability of Active Galactic Nuclei, p. 225 – 228 (1991). – See Abstr. 012.040 for the main entry.
The authors study time variability in nonthermal models of AGN. They concentrate on a particular model which fits well both the average X–ray Seyfert 1 spectra and the γ–ray constraints from the measurements of several AGN as well as the cosmic γ–ray background.

158.129 **Pulsating electromagnetic cascades and rapid time variability in active galactic nuclei.**
B. Stern, R. Svensson, M. Sikora.
Georgia State University Conference on Variability of Active Galactic Nuclei, p. 229 – 231 (1991). – See Abstr. 012.040 for the main entry.
An instability leading to limit cycle behavior of electromagnetic cascades possibly occurring in active galactic nuclei (AGN) near accreting black holes has been discovered. The power from accelerated protons gets converted by the cascade into soft radiation (X–rays and below) if the photon compactness is sufficiently large. Then the proton–photon system may develop limit cycles much like a prey–predator system with each component interchangeably dominating. This causes periodic large amplitude short time variability of the AGNs nonthermal luminosity even if the acceleration or injection process is completely steady.

158.130 **Ginga observations of X–ray variability in the broad line radio galaxy 3C 382.**
J. S. Kaastra.
Georgia State University Conference on Variability of Active Galactic Nuclei, p. 232 – 235 (1991). – See Abstr. 012.040 for the main entry.

158.131 **Non–thermal emission and acceleration of jets in AGN.**
G. Ghisellini.
Georgia State University Conference on Variability of Active Galactic Nuclei, p. 236 – 252 (1991). – See Abstr. 012.040 for the main entry.
Models for the formation of the non–thermal continuum in AGN are reviewed. It is shown that electron–positron pair production is a key process in AGN, and can be responsible for the "universal" X–ray spectral index. Different types of variability expected in simple models are briefly discussed.

158.132 **MERLIN observations of 4C 29.6 (3C 59) and 4C 35.37: two radio structures misidentified with Seyfert galaxies.**
E. J. A. Meurs, S. W. Unger.
Astron. Astrophys., Vol. 252, No. 1, p. 63 – 68 (Dec 1991).
The two radio sources 4C 29.6 (3C 59) and 4C 35.37 have been identified optically with rather distant Seyfert galaxies. Observations with the MERLIN array show both cases to be misidentifications. 4C 35.37 is an unrelated, double–lobed background source, whereas the Seyfert itself is not detected. 4C 29.6 appears mainly a superposition of two background sources (a core source with one–sided jet, and a small–scale double), while the Seyfert is also detected. The observed radio structures and the relevance of these results for the understanding of the overall radio properties of Seyferts are discussed.

158.133 **The power spectra of the short term AGN X–ray variability.**
Zhang Xiaohe.
Georgia State University Conference on Variability of Active Galactic Nuclei, p. 253 – 255 (1991). – See Abstr. 012.040 for the main entry.

158.134 **X–ray variability of AGNs.**
H. J. Lehto, I. M. McHardy, R. G. Abraham.
Georgia State University Conference on Variability of Active Galactic Nuclei, p. 256 – 259 (1991). – See Abstr. 012.040 for the main entry.
The authors have analysed X–ray variability of AGNs using the EXOSAT database. They selected only bright sources but the sample is not yet complete. They have calculated the power spectra for these observations. The power spectra were then averaged in linear space. The analysis revealed some interesting features.

158.135 **Rapid iron line variability in NGC 6814.**
T. J. Turner, H. Kunieda, R. Mushotzky, H. Awaki.
Georgia State University Conference on Variability of Active Galactic Nuclei, p. 260 – 264 (1991). – See Abstr. 012.040 for the main entry.
AGN observations with the Large Area Counter on board of the GINGA satellite have obtained the highest quality X–ray time variablility data to date, allowing examination of the detailed spectra up to 20 keV on relatively short time scales. The authors observed the rapidly variable X–ray emitting Seyfert galaxy NGC 6814 with the GINGA satellite. The most striking result of the spectral fits is the correlated change of the iron line flux with the continuum flux.

158.136 **The search for short time scale variations in Seyfert galaxies.**
E. Xanthopoulos, M. M. de Robertis.
Georgia State University Conference on Variability of Active Galactic Nuclei, p. 265 – 268 (1991). – See Abstr. 012.040 for the main entry.
The only means of getting information on the size and kinematics of the broad line region (BLR) of Seyfert galaxies, which is unresolved, is by spectral line variability. The period of the variability can set a constraint on the size of the BLR. The purpose of this study is to search for very short time scale variations (hour – day) in broad Hα in a sample of nearby Seyfert galaxies.

158.137 **Characterization of X–ray variability of Seyfert galaxies.**
G. Madejski, C. Done, J. Krolik.
Georgia State University Conference on Variability of Active Galactic Nuclei, p. 269 – 272 (1991). – See Abstr. 012.040 for the main entry.
The authors explore the Fourier phases of X–ray light curves from 5 Seyfert galaxies observed by the EXOSAT satellite. They find that the distribution of phase with Fourier frequency is not random, indicating that the light curves are not produced by a purely stochastic process. One observes phase correlation over a wide range of frequencies, which implies that there is coherence between processes occuring on a wide range of time scales.

158.138 **X–ray spectral variations in MCG–6–30–15.**
F. Fiore, M. Yamauchi, M. Matsuoka.
Georgia State University Conference on Variability of Active Galactic Nuclei, p. 273 – 276 (1991). – See Abstr. 012.040 for the main entry.
The authors have found that the overall 1.7 – 23.3 keV spectrum of the Seyfert 1 galaxy MCG–6–30–15 significantly varied during 2–day GINGA observations. Color diagrams show that the spectral variations up to and above $\sim$4 keV are not correlated. The data above $\sim$4 keV may be explained by a two component model in which the relative normalization varies. The main component, variable on all time scales down few hundred seconds, has a spectral index consistent with a constant value,

while the hard component seems to be also variable but on longer time scales.

158.139 **Spectroscopic monitoring campaigns at optical wavelengths: preliminary results for NGC 5548 and strategies for the future.**
B. M. Peterson.
Georgia State University Conference on Variability of Active Galactic Nuclei, p. 277 – 291 (1991). – See Abstr. 012.040 for the main entry.
The earliest IUE and ground-based campaigns to study spectral variability in AGNs showed that the integrated fluxes in the broad emission lines undergo rapid changes which correspond closely to observed continuum variations. This basic observation alone reveals several important facts about AGN physics.

158.140 **The NGC 5548 monitoring campaign. II. Continuum variability.**
M. A. Malkan.
Georgia State University Conference on Variability of Active Galactic Nuclei, p. 292 – 300 (1991). – See Abstr. 012.040 for the main entry.
Using ultraviolet and optical monitoring data from the NGC 5548 campaign, the author has considered the implications of these data for the continuum emission mechanism.

158.141 **Mapping the BLR in NGC 5548 and NGC 4151.**
J. Clavel.
Georgia State University Conference on Variability of Active Galactic Nuclei, p. 301 – 324 (1991). – See Abstr. 012.040 for the main entry.

158.142 **Broad line profile variability of NGC 5548.**
W. Kollatschny, M. Dietrich.
Georgia State University Conference on Variability of Active Galactic Nuclei, p. 325 – 327 (1991). – See Abstr. 012.040 for the main entry.
Parallel to the UV-monitoring program of NGC 5548 with the IUE satellite the authors have done optical slit spectroscopy (medium and high resolution) using the 2.2 m and 3.5 m telescopes at Calar Alto observatory. These observations were part of the international campaign for optical monitoring of this galaxy. The authors have obtained more than 100 spectra with good S/N ratio for 43 epochs from Dec 88 to the end of Aug 89. The main interest was to study the profile variability of the broad emission lines (Hα, Hβ, He II 4686 and He I 5876) of NGC 5548. Furthermore, it will be possible in same cases to examine the variations of Hγ with respect to the other Balmer lines. The authors want to present preliminary results of the optical emission line profile variability study.

158.143 **Gravitational infall of the BLR clouds in NGC 5548.**
D. M. Crenshaw, J. H. Blackwell Jr.
Georgia State University Conference on Variability of Active Galactic Nuclei, p. 328 – 331 (1991). – See Abstr. 012.040 for the main entry.
Observations from the international campaign to monitor the variable Seyfert 1 galaxy NGC 5548 with the IUE are used to study the response of the C IV λ1550 emission-line profile to changes in the photoionizing continuum. Near the end of the IUE campaign, the continuum flux at 1440 Å and the total C IV flux dropped by factors of 2.9 and 1.8, respectively, in 16 days. The red wing of the C IV profile responded more rapidly to the sharp continuum drop than the blue wing, indicating that clouds in the inner broad-line region (BLR) are undergoing gravitational infall. These results provide direct evidence that the central engine is a supermassive object.

158.144 **Studying the galactic central engine from space observatories.**
H. D. Greyber.
IAU Colloquium No. 123: Observatories in Earth orbit and beyond, p. 559 – 561 (1990). – See Abstr. 012.038 for the main entry.
The expanded use of space observatories in all wavelength ranges should enable to understand finally the physics of the "central engine" that powers the enormous energy output one observes from quasars and AGN. This physics will probably also explain the formation of galaxies and the variety of objects of galactic dimension that are observed.

158.145 **Echo mapping of NGC 5548.**
K. D. Horne, W. F. Welsh.
Georgia State University Conference on Variability of Active Galactic Nuclei, p. 332 – 334 (1991). – See Abstr. 012.040 for the main entry.
The broad emission-line regions close to an Active Galactic Nucleus (AGN) are too small to be resolved by direct techniques. However, one can hope to determine the size and structure of the broad-line region by observing the delayed response of the emission lines to variability of the photo-ionizing continuum of the AGN. The authors have developed a computer code MEMECHO which inverts the convolution equation to solve for the transfer function from observations of continuum and emission line fluxes. MEMECHO uses the maximum entropy method to find the "smoothest" positive transfer function that fits the data.

158.146 **Implications of variability for the standard model.**
A. P. Koratkar, C. M. Gaskell.
Georgia State University Conference on Variability of Active Galactic Nuclei, p. 339 – 342 (1991). – See Abstr. 012.040 for the main entry.
Extensive investigation of cross-correlation method advocated by Gaskell and Sparke (1986) shows that useful results with well understood uncertainties can be obtained from IUE archival data. The authors have been able to obtain estimates of the inner radii of the broad line regions, the luminosity weighted radii, and upper limits to the overall size. The sizes are all smaller than predicted by the standard photoionization models.

158.147 **The trouble with reverberation.**
M. V. Penston.
Georgia State University Conference on Variability of Active Galactic Nuclei, p. 343 – 357 (1991). – See Abstr. 012.040 for the main entry.
The possibilty of using the relationship between line and continuum variations to "map" the Broad Line Region in active galaxies has stimulated several observational and interpretational efforts. Current methods of analysis are critically reviewed in this article. Hidden assumptions are examined and it is concluded that present methods may be defective in a number of respects.

158.148 **C IV line profile correlations in NGC 4151.**
D. Kazanas, R. P. Fahey, A. G. Michalitsianos.
Georgia State University Conference on Variability of Active Galactic Nuclei, p. 375 – 377 (1991). – See Abstr. 012.040 for the main entry.

158.149 **Broad emission line variability in Seyfert 1 galaxies.**
G. M. Stirpe.
Georgia State University Conference on Variability of Active Galactic Nuclei, p. 378 – 380 (1991). – See Abstr. 012.040 for the main entry.
The two bright Seyfert 1 galaxies NGC 5548 and Markarian 279 have been monitored spectroscopically on short time scales, with the purpose of studying the response of the broad lines to the continuum variations, and therefore setting

constraints on the dynamics and kinematics of the broad line region.

158.150 **Initial results from the Colorado Active Galaxy Database: ultraviolet variability of the CfA Seyfert 1 galaxies.**
G. F. Pike, R. A. Edelson.
Georgia State University Conference on Variability of Active Galactic Nuclei, p. 381 – 383 (1991). – See Abstr. 012.040 for the main entry.
The authors have constructed a large database of ~4000 IUE spectra of Active Galactic Nuclei. The first application involves the six best-studied CfA Seyfert 1s, which are used to study the relation of luminosity to variability and continuum shape, as well as the Baldwin effect.

158.151 **Evidence for rapid narrow line variation.**
W. Zheng, E. Pérez.
Georgia State University Conference on Variability of Active Galactic Nuclei, p. 384 – 387 (1991). – See Abstr. 012.040 for the main entry.

158.152 **Discovery of two radio galaxies in the Zwicky Cluster 1437.3+6436.**
N. M. Elias II, M. F. Struble.
Astron. J., Vol. 102, No. 1, p. 30 – 33 (Jul 1991). Current Physics Microform No.: 9107B1726. With plates 2 – 3.
Two radio galaxies were serendipitously discovered in the Zwicky Cluster 1437.3+6436. One radio galaxy is a resolved point source. The other radio galaxy consists of jets that extend from a resolved core and eventually reach a large region of extended emission in a direction away from the cluster center. From the Palomar Observatory Sky Survey prints, the authors find that the optical counterparts of these radio sources are elliptical galaxies. The radio and optical properties of these galaxies are consistent with those of galaxies in other clusters.

158.153 **Short timescale broad emission–line flux variability in high–luminosity active galactic nuclei.**
K. T. Korista.
Astron. J., Vol. 102, No. 1, p. 41 – 47 (Jul 1991). Current Physics Microform No.: 9107B1737.
The standard model of AGN suggests that the broad–line regions of high–luminosity AGN should be so large that any variability in the broad emission–line fluxes should be on timescales of ~2–4 yr or more. Spectrophotometric observations of eight radio–quiet QSOs over a period of 1 yr revealed definite Hβ flux variability in one of these objects and possibly in two others. This may be an indication that the BLRs in high–luminosity AGNs are much smaller than those predicted by the standard model, as is apparently the case in Seyfert galaxies.

158.154 **Neutrino–heated stars and broad–line emission from active galactic nuclei.**
J. MacDonald, T. Stanev, P. L. Biermann.
Astrophys. J., Vol. 378, No. 1, p. 30 – 38 (1 Sep 1991).
Nonthermal radiation from active galactic nuclei indicates the presence of highly relativistic particles. The interaction of these high–energy particles with matter and photons gives rise to a flux of high–energy neutrinos. In this paper, the authors investigate the influence of the expected high neutrino fluxes on the structure and evolution of single, main–sequence stars. Sequences of models of neutrino–heated stars in thermal equilibrium are presented for masses 0.25, 0.5, 0.8, and 1.0 $M_\odot$. In addition, a set of evolutionary sequences for mass 0.5 $M_\odot$ have been computed for different assumed values for the incident neutrino energy flux. It is found that winds driven by the heating due to high–energy particles and hard electromagnetic radiation of the outer layers of neutrino–bloated stars may satisfy the requirements of the model of Kazanas for the broad–line emission clouds in active galactic nuclei.

158.155 **Multidirectional views of the active nucleus of NGC 1068.**
J. S. Miller, R. W. Goodrich, W. G. Mathews.
Astrophys. J., Vol. 378, No. 1, p. 47 – 64 (1 Sep 1991). = UCO/Lick Obs. Bull., No. 1188.
Previous spectropolarimetric work has shown that the type 2 Seyfert galaxy NGC 1068 contains a broad–line region hidden from our direct view and visible only in scattered, polarized light. New observations, including imaging polarimetry as well as spectropolarimetry, reveal off–nucleus regions of high polarization as places in the body of the galaxy where dust reflects light from the hidden broad–line region. This allows to obtain spectra of an active nucleus as it would appear from several different viewing directions. From these other directions the spectrum of NGC 1068 would appear as a normal Seyfert 1 galaxy. The authors derive a value of 1.44×10^{42} ergs s^{-1} for the Hβ luminosity that would be seen in the absence of obscuration, also typical of Seyfert 1 galaxies. The broad Hβ line as seen reflected in the external dusty regions is significantly narrower than that observed in the nuclear polarized flux, adding additional weight to the view that electron scattering is responsible for the nuclear polarization. The observational data are used to constrain a theoretical model for the electron – scattering region. The analysis suggests the surprising result that the central continuum source must be inherently anisotropic.

158.156 **The EXOSAT high Galactic latitude survey.**
P. Giommi, G. Tagliaferri, K. Beuermann, G. Branduardi–Raymont, R. Brissenden, U. Graser, K. O. Mason, J. P. D. Mittaz, P. Murdin, G. Pooley, H.–C. Thomas, I. Tuohy.
Astrophys. J., Vol. 378, No. 1, p. 77 – 92 (1 Sep 1991).
A survey of serendipitous sources performed in the very soft X–ray band (0.05 – 2.0 keV) using the EXOSAT imaging telescopes is presented. The survey covers 783 square degrees of high Galactic latitude sky and includes 210 serendipitous sources which define a complete (flux–limited) sample. An extensive program of optical and radio observations together with cross–correlations with catalogs of known objects leads to the identification of 200 of the 210 detected sources. The 10 remaining objects have been preliminarily classified on the basis of their X–ray to optical flux ratio, thus making the sample essentially fully identified. The complete sample has been used to study the log N – log S relation and the average spectral slope of active galactic nuclei (AGN). It is found that the log N – log S slope is consistent with that of the Einstein Extended Medium–Sensitivity Survey and with the "Euclidean" value of 1.5. The normalization of the log N – log S relation is a strong function of the assumed spectral slope of AGNs. Consistency with the results of the EMSS implies that the average (energy) slope of extragalactic sources in the soft X–ray band is very steep ($\alpha \approx 1.5$). An analysis of the association between AGN detection and Galactic N_H also shows that α is steep and inconsistent with the canonical value of 0.7.

158.157 **A coarse analysis of Seyfert–I narrow–line properties.**
P. Rafanelli, H. Schulz.
Astron. Nachr., Vol. 312, No. 3, p. 167 – 176 (1991).
Medium–resolution spectrophotometry of 19 Seyfert–I galaxies with M_v between -20^m and -24^m is presented. Balmer line broad and narrow components are decomposed and relative narrow–line intensities are compared with a grid of power law photoionization models. Diagrams displaying ratios of strong forbidden lines are devised which allow to read power law indices α and ionization parameters U for individual objects. On the average, α–values derived from [O I]/[O III] versus [N II]/[O III] are in good agreement with power–law slope indices α_{ux} derived from continuum measurements at 1450 Å (U) and 0.7 keV (X). However, for the same parameters, the theoretical [S II]/[O III] values appear to be 40% too large. Based on trends in current photoionization models the authors suggest that this discrepancy might be removed by including a black–body component in the ionizing continuum and a high–density component in the narrow–line region. Mkn 1044 has been isolated as an unusual

object whose strong [S II] $\lambda\lambda$4068,4076 feature suggests that its NLR emission is dominated by a high-density zone with $N_e \sim 10^7 cm^{-3}$.

158.158 **The extended radio continuum emission of NGC 4388.**
J. A. Irwin.
Workshop on the Interpretation of Modern Synthesis Observations of Spiral Galaxies, p. 59 – 66 (1991). – See Abstr. 012.041 for the main entry.
Multi-array VLA observations of the Virgo Cluster spiral, NGC 4388, show extended out-of-plane emission associated with this galaxy. The bipolar nuclear radio lobes clearly emerge out of the plane and display very different spectral index distributions on either side of the nucleus. Several disk-related kpc-scale extensions are also present, possibly originating from ram pressure stripping by the intergalactic medium.

158.159 **A synchrotron halo around M82.**
E. R. Seaquist, N. Odegard.
Workshop on the Interpretation of Modern Synthesis Observations of Spiral Galaxies, p. 67 – 70 (1991). – See Abstr. 012.041 for the main entry.
The authors report the discovery of a nonthermal radio halo surrounding the prototypical starburst galaxy M82. Some of the features of the halo are discussed along with its relationship to the outflow and supernova activity in the nucleus.

158.160 **Radio luminosity functions.**
J. J. Condon.
Workshop on the Interpretation of Modern Synthesis Observations of Spiral Galaxies, p. 113 – 123 (1991). – See Abstr. 012.041 for the main entry.
The local radio luminosity function is presented, along with separate luminosity functions for radio sources powered by the "starbursts" and "monsters" usually found in spiral and elliptical galaxies, respectively. The starburst radio luminosity function may be derived from the far-infrared luminosity function and the tight far-infrared-radio luminosity correlation. A starburst population-synthesis model can account for this correlation quantitatively. Spiral galaxies appear to evolve on cosmological time scales and may account for nearly half of the $T \sim 0.1K$ source background at $\nu = 1.4$ GHz.

158.161 **The radio emission of Seyfert galaxies.**
A. S. Wilson.
Workshop on the Interpretation of Modern Synthesis Observations of Spiral Galaxies, p. 227 – 250 (1991). – See Abstr. 012.041 for the main entry.
This paper reviews developments over the last decade in our understanding of the radio emission of Seyfert galaxies. After summarizing the characteristics and selection criteria of radio surveys, the author describes properties of the integrated radio emission – radio luminosity functions, correlations of radio power with other properties, and spectra. Radio structures are then discussed with emphasis on the different kinds of morphologies seen. "Linear" radio sources are morphologically and kinematically closely associated with the narrow line regions, and models attempting to account for this association are described. "Diffuse" radio sources are fuelled by circumnuclear star formation. The review concludes with some remarks on the implications of radio properties for "unified" models of type 1 and type 2 Seyferts.

158.162 **High redshift radio galaxies from the Third Bologna Catalogue. I.**
P. J. McCarthy.
Astron. J., Vol. 102, No. 2, p. 518 – 521 (Aug 1991). Current Physics Microform No.: 9108A1454.
The author reports on the results of optical observations of a sample of 1 Jy class radio galaxies selected from the Third Bologna Catalogue. The sources were selected to have 408 MHz flux densities greater than 1.0 Jy, 408 MHz spectral indices steeper than -0.9 and no identifications at the limit of the Palomar Sky Survey. Optical identifications and redshifts for eight sources are presented. Most of these sources are identified with faint galaxies with redshifts of order 1. Two of the identifications have strong Ly α emission at redshifts between 2 and 3. A Ly α image of one of these objects shows extended emission over roughly 50 kpc.

158.163 **High redshift radio galaxies from the Molonglo Catalogue. II.**
P. J. McCarthy, W. van Breugel, V. K. Kapahi, C. R. Subrahmanya.
Astron. J., Vol. 102, No. 2, p. 522 – 536 (Aug 1991). Current Physics Microform No.: 9108A1458.
The authors report further results from their ongoing study of radio galaxies from the Molonglo Reference Catalogue. The sources were selected to have 408 MHz flux densities greater than 0.95 Jy and low frequency spectral indices steeper than -0.9. VLA 6 cm maps, optical identifications and redshifts for 18 sources are presented. Most of these sources are identified with faint galaxies, many of which have extended optical structures with morphologies characteristic of high redshift ($z > 1$) radio galaxies. Long slit spectroscopic observations have yielded redshifts for all of the galaxies, three of which have $z > 2$. Spectrophotometry of 1 Jy and the 3CR samples are used to distinguish between an L(emission line) versus L(radio) correlation as opposed to an L(emission line) versus redshift correlation. The results strongly support an intrinsic L(emission line) versus L(radio) correlation. The structure of the sources in this, and other 1 Jy samples, are substantially different from the 3C sources, the 1 Jy galaxies having stronger cores and jets.

158.164 **VLA images of 23 extragalactic radio sources.**
J. P. Leahy, R. A. Perley.
Astron. J., Vol. 102, No. 2, p. 537 – 561 (Aug 1991). Current Physics Microform No.: 9108A1473. With plates 19 – 24.
The authors present radio images at 20 cm of 22 radio galaxies and one quasar made with National Radio Astronomy Observatory (NRAO) Very Large Array. The sources are drawn from the $z < 0.5$ subset of the complete radio sample defined by Laing, et al. (1983). The images have resolutions of 1.1–5.8 arcsec, chosen to give about 50 beamwidths across each source, and sufficient sampling of the aperture plane to represent accurately the largest-scale structure. With the current images, nearly all members of the LRL sample with $z < 0.5$ have high-sensitivity images from the VLA or Westerbork.

158.165 **An optical study of 3C 31, 3C 66B, 3C 120, and their jets.**
D. Fraix-Burnet, D. Golombek, F. D. Macchetto.
Astron. J., Vol. 102, No. 2, p. 562 – 571 (Aug 1991). Current Physics Microform No.: 9108A1498.
The authors present the results of BVRI CCD photometry of the radiogalaxies 3C 31, 3C 66B, and 3C 120, and V polarimetry of 3C 120. The photometry of the jet of 3C 66B definitively establishes the synchrotron nature of the optical emission. No optical counterpart of the radio counterjet in 3C 66B and of the radio jets in 3C 31 and 3C 120 is found. A rotating ring and an ionized region are present, respectively, in 3C 31 (NGC 383) and its companion galaxy NGC 382, but the authors find no isophotal distortions which could have revealed a gravitational interaction between the two galaxies as is the case in 3C 66B. The elliptical isophotes of 3C 120 show a slight off-centering, roughly in the direction of the radio jet, very much like 3C 66B. The authors find an upper limit of 20% for the polarization level of the condensations in 3C 120.

158.166 **Erratum: "3C 120: study of continuum-emitting condensations close to the nucleus" [Astron. Astrophys., Vol. 222, No. 1/2, p. 27 – 34 (Sep 1989)].**
A. Soubeyran, G. Wlérick, A. Bijaoui, G. Lelièvre, P. Bouchet, D. Horville, L. Renard, B. Servan.
Astron. Astrophys., Vol. 252, No. 1, p. 431 (Dec 1991). See Abstr. 50.158.084.

158.167 **VLA H I observations of the radio galaxy Centaurus A.**
J. M. van der Hulst, J. H. van Gorkom,
A. D. Haschick, A. D. Tubbs.
Workshop on Warped Disks and Inclined Rings Around Galaxies, p. 13 – 14 (1991). – See Abstr. 012.054 for the main entry.

The existence of conflicting theoretical predictions regarding the origin of the dust lane, together with the unique opportunity to study the gas kinematics of a radio galaxy in detail prompted the authors to image NGC 5128 in neutral hydrogen with the VLA. This paper summarizes the results.

158.168 **A geometric model for the dust–band of Centaurus A.**
R. A. Nicholson, K. Taylor, J. Bland.
Workshop on Warped Disks and Inclined Rings Around Galaxies, p. 15 – 24 (1991). – See Abstr. 012.054 for the main entry.

The authors demonstrated that the morphology and kinematics of the dust–band in Cen A, as revealed through the comprehensive TAURUS data of the Hα ionized gas disc, can be successfully explained with a thin warped disc geometry and a rotation curve consistent with that expected from an $r^{1/4}$ law mass distribution.

158.169 **Imprints of a cooling flow on the radio continuum spectrum of NGC 4631.**
M. Pohl, R. Schlickeiser, H. Lesch.
Astron. Astrophys., Vol. 252, No. 2, p. 493 – 497 (Dec 1991).

The radio spectrum of NGC 4631 exhibits a spectral flattening at lower frequencies, that is much too fast to be explained by common models (cf. Pohl et al., 1991). The authors propose an interpretation assuming an electron energy spectrum flattened by ionization losses to constitute the low frequency radio spectrum of NGC 4631. This model implies an effective gas density and magnetic field strength about one order of magnitude higher than previously estimated. The discrepancy may be solved by assuming a cooling flow as radial infall of the intergalactic medium. Thermal instabilities lead to the formation of dense ($10^3 cm^{-3}$) clouds with typical diameter 1 pc and a filling factor of 10^{-3}. The cooling flow may generally explain the systematical differences between the mass estimates obtained from rotation curves and those determined for pairs of galaxies. In addition, relativistic electrons are contained in shell–like structures around the clouds, and therefore the mean gas density of their environment is increased.

158.170 **VLBI observations of 24 3CR CSS radio sources at 50 cm.**
Nan Rendong, R. T. Schilizzi, C. Fanti, R. Fanti.
Astron. Astrophys., Vol. 252, No. 2, p. 513 – 527 (Dec 1991).

The authors present the results of snapshot observations carried out at 50 cm with a Global VLBI Network on 24 compact steep–spectrum sources (CSS) from the 3CR catalogue (3C 43, 48, 49, 67, 119, 138, 186, 190, 216, 237, 241, 268.3, 287, 298, 299, 303.1, 305.1, 309.1, 318, 343, 343.1, 346, 380, 454). The angular resolution ranges from about 20 mas for the global data to about 100 mas for the local networks (EVN and/or EVN+USVN) only. The minimum measurable surface brightness ranges from ≈10 to ≈40 mJy/beam. Cores have been identified or confirmed in six sources (3C 43; 186, 190, 216, 298, 346) through the spectral information contained in these observations. For radio components ≲150 mas the new data allow the radio spectra to be evaluated down to 610 MHz. In general, no evidence is found for self–absorption, except in the radio cores. Data reduction techniques have been developed to deal with the poor UV coverage of snapshot VLBI observations.

158.171 **The Coma cluster radio source 1253+275, revisited.**
G. Giovannini, L. Feretti, C. Stanghellini.
Astron. Astrophys., Vol. 252, No. 2, p. 528 – 537 (Dec 1991).

Multifrequency observations obtained with the WSRT and the VLA are presented for the extended radio source 1253+275, located at the Coma cluster periphery. This source is of astrophysical interest for its large extent (30′ × 10′) and the lack of an obvious optical identification. With the present high resolution data, the authors exclude the existence of a discrete compact radio source which could feed the extended radio emission. The structure of 1253+275 is diffuse with a central ridge, a bridge of radio emission connecting the NAT Coma cluster radio galaxy NGC 4789 and a SW leading edge. From the distribution of the spectral index and the physical properties of 1253+275, the authors suggest a connection of this puzzling source with the large scale radio emission present in the Coma cluster: the central halo source Coma C and the diffuse emission found between Coma C and 1253+275 itself. In this case the diffuse radio emission permeating the Coma cluster would have a linear size of more than 2 Mpc, in the direction of the Coma–A1367 supercluster. Alternatively, the possibility suggested in previous works cannot be excluded that 1253+275 could be the relic radio emission of the Coma cluster galaxy IC 3900.

158.172 **A turned–over optical spectrum of OJ287 during the 1989 minimum.**
M. R. Kidger, L. O. Takalo, E. Valtaoja, J. A. de Diego,
A. Sillanpää.
Astron. Astrophys., Vol. 252, No. 2, p. 538 – 540 (Dec 1991).

The authors have carried out further analysis of the optical and radio observations of the previously reported deep 1989 minimum of OJ 287 (Takalo et al. 1990). The radio spectrum shows little change until well after minimum, when it steepens considerably. The optical spectrum also shows considerable changes, with an increase of a factor of two in its slope between maximum and minimum. Calculation of the slope however loses its significance during the decline when the red end inverts and a sharp peak is seen in the continuum, before the spectrum reverts to a normal shape at minimum. The best explanation seems to be a self–absorbed synchrotron component in the accretion disk. This, however, requires strong magnetic fields and small emitting regions (30 G for a 0.1 AU emitting region, 3×10^5G for a 1 AU sized region).

158.173 **The dust enshrouded quasar in the ultraluminous galaxy Markarian 463: radio, near–infrared, and optical imaging.**
J. M. Mazzarella, R. A. Gaume, B. T. Soifer, J. R. Graham,
G. Neugebauer, K. Matthews.
Astron. J., Vol. 102, No. 4, p. 1241 – 1252 (Oct 1991). Current Physics Microform No.: 9110D1773.

New radio continuum, near–infrared, and optical images are presented for the infrared ultraluminous galaxy Mrk 463, which prior optical work has shown to be an interacting/merging system containing two Seyfert 2 nuclei separated by 4″ (4 kpc). The eastern nucleus, Mrk 463E, has a steep–spectrum ($f_\nu \sim \nu^{-1}$) radio core with a luminosity density that is characteristic of a quasar or radio galaxy. Imaging at 20 cm reveals previously unknown components 4″ north and 18″ south of Mrk 463E, and a source 10″ to the northwest which is apparently linked to the nuclear region. The western nucleus, Mrk 463W, has been detected at 6 and 20 cm ($f_\nu \sim \nu^{-0.5}$); it has a radio luminosity comparable to that of a moderately luminous Seyfert galaxy or a highly luminous starburst galaxy. Near–infrared images at wavelengths less than 2.3 μm show the two nuclei; only Mrk 463E has been detected at 3.7 μm. Mrk 463E has an extremely red V–K color of 6.8 mag. Brightness profiles of Mrk 463E are unresolved at K and L' but extended at J; this is likely due to the effects of extinction and centrally concentrated hot dust emission. Radio continuum components 4–18 kpc from the nucleus are aligned with previously known 0.05–1.5 kpc radio structure and extended, conical [O III] emission. This suggests that Mrk 463E is powering weak radio lobes and may therefore exemplify a transition between the confined linear sources in Seyfert galaxies and the extended lobes in classical quasars and radio galaxies. Mrk 463E has a 3.7 μm luminosity and K–L' color comparable to PG quasars, suggesting that dust obscures an embedded quasar.

158.174 **The micro–Jansky radio source population at 5 GHz.**
E. B. Fomalont, R. A. Windhorst, J. A. Kristian, K. I. Kellermann.
Astron. J., Vol. 102, No. 4, p. 1258 – 1277 (Oct 1991). Current Physics Microform No.: 9110D1790. With plate 38.
The authors have imaged a field of area $1.5x10^{-5}$sr with unprecedented sensitivity using the VLA at 5 GHz. The total integration time of 126 hr was spread over the D, C, and B configurations in order to obtain images of faint radio sources with angular scales between 1″ and 100″ with an rms noise of 2.5 μJy. Subsequent observations at 1.5 GHz provided spectral information about the sources. The inner part of this field, centered at $\alpha = 14^h16^m$ and $\delta = 52\overset{\circ}{.}7$ (1950.0) has also been imaged with the Palomar 200 in. 4–shooter CCD array. The authors have compiled a complete (>95%) catalog of 62 sources with a peak flux density at 5 GHz greater than 16 μJy at 8″ resolution and a typical positional accuracy of 0″.2. The properties of this radio source population are discussed in detail.

158.175 **A study of emission–line variability of the nucleus in the Seyfert galaxy NGC 4151 in 1986–1988.**
N. G. Bochkarev, A. I. Shapovalova, S. A. Zhekov.
Astron. J., Vol. 102, No. 4, p. 1278 – 1293 (Oct 1991). Current Physics Microform No.: 9110D1810.
NGC 4151 optical spectra (in the 3960–4960 Å spectral range) were obtained by the TV scanner of the 6 m telescope at the Special Astrophysical Observatory of the Academy of Sciences of the USSR in the spring of 1984, in 1986 March–April, in 1987 March and May–June, and in 1988 February–March. The dispersion is 0.5, 1.0, and 2.0 Å/channel. The signal–to–noise ratio in individual spectra is S/N = 10–25. The contribution of the absorption spectrum of the NGC 4151 nucleus stellar component to continuum is $40 \pm 10\%$ near the Hβ line. The parameters of the narrow and broad components of Hβ, Hγ, Hδ, and He II 4686 Å lines, of the narrow [O III] 4363 Å line, and of the broad Hα and He I 4471 Å emissions are analyzed. As before 1984, the NGC 4151 nuclear spectrum corresponds to the Sy 1.5 type with a significant asymmetry and a blue shift of the broad–line components. The emission–line variability is expressed more weakly as compared with the period before the minimum of 1984. An analysis of the radial velocities of faint spectral features has shown that a fraction of them can be narrow satellites of the Balmer Hβ–Hδ and He II 4686 lines whose radial velocities are −7500 and +9400 km/s. The broad components of the Hβ and Hγ lines got much weaker in 1987. The asymmetry of the broad components of these lines was enhanced in 1988.

158.176 **Host galaxy and spectral variability of the BL Lacertae object H0414+00.**
R. Falomo, E. G. Tanzi.
Astron. J., Vol. 102, No. 4, p. 1294 – 1297 (Oct 1991). Current Physics Microform No.: 9110D1826.
Surface photometry of the nebulosity surrounding the X–ray selected BL Lacertae object H0414+00 is presented. The radial profile is consistent with a giant elliptical galaxy of $M_V = -23.2$. Simultaneous optical to IR observations at three different epochs show wavelength dependent variability. The spectral flux distribution is interpreted in terms of a galaxian contribution superposed onto a flat ($\alpha_\nu \sim 0.3$), variable nonthermal component.

158.177 **The X–ray spectrum (0.1–10 keV) of the broad line radio galaxy: 3C 390.3.**
K. K. Ghosh, S. Soundararajaperumal.
Astron. J., Vol. 102, No. 4, p. 1298 – 1302 (Oct 1991). Current Physics Microform No.: 9110D1830.
The broad line radio galaxy, 3C 390.3, was observed with EXOSAT on several occassions between 1984 and 1986 using the low (LE) and medium (ME) energy detectors. Detailed results of the spectral analysis of the LE (0.1–2 keV) and ME (2–10 keV) spectra which were obtained from the EXOSAT archives, of this source are presented. A simple power law and uniform absorption model provided a good fit to the combined LE and ME spectra but the hydrogen column density (N_H) values are much smaller than the galactic N_H value. However, the above model with the fixed N_H value also provides a good fit to the datasets. The observed X–ray spectrum is best fit by a power law of a photon index, Γ, of 1.63 ± 0.07. Only the 2–10 keV flux shows a weak change (~25%) over a year and, practically, no changes were observed in the low energy flux. Weak soft excess below 2 keV was detected in this galaxy. Two power law models fit well with the data, but these models become insensitive to a change of the fit parameters. Best fit results of the broken power law model which shows a slight improvement in the χ^2_{red} values over the single power law model, suggests that two power law components are required to explain the X–ray spectrum of 3C 390.3. A highly significant (>99.9%) Gaussian line feature around 5.80 keV was detected which could be the redshifted 6.4 keV due to the fluorecsence of cold iron from an optically thick accretion disk around the central nonthermal X–ray source.

158.178 **Surface photometry and spectral analysis of Mk 50: a Seyfert 1 nucleus in an early type galaxy.**
M. G. Pastoriza, E. Bica, C. Bonatto, E. Mediavilla, E. Perez.
Astron. J., Vol. 102, No. 5, p. 1696 – 1701 (Nov 1991). Current Physics Microform No.: 9111D0702.
The authors report results obtained from spectrophotometry and surface photometry of Mk 50. Using this data the authors study the surface brightness distribution and the nature of the stellar population, as well as the properties of the emitting gas. In particular they infer on the nature of the host galaxy and the nearby point source.

158.179 **Spectrophotometric investigation of galaxy Markaryan 325.**
A. N. Burenkov, E. A. Nazarov, Eh. E. Khachikyan.
Bull. Spec. Astrophys. Obs. – North Caucasus, Vol. 32, p. 12 – 22 (1990). English translation of Astrofiz. Issled. Izv. Spets. Astrofiz. Obs., Tom 32, p. 16 – 30 (1990).
A spectral investigation of 12 H II regions in clumpy irregular galaxy (CIG) Mark 325 was carried out. The content of oxygen, nitrogen, sulfur, and neon in the H II regions was determined. The content of elements in the galaxy ([O/H] = 8.60 ± 0.07 and [N/H] = 7.29 ± 0.18) is typical for a CIG (where the mean content of three galaxies, Mark 7, Mark 297, and Mark 325, is [O/H = 8.58 ± 0.07 and [N/H] = 7.21 ± 0.17).

158.180 **Searching for supernovae in starburst galaxies.**
M. Richmond, A. V. Filippenko.
10. Santa Cruz Summer Workshop in Astronomy and Astrophysics: Supernovae, p. 731 – 736 (1991). – See Abstr. 012.052 for the main entry.

158.181 **Steep radio spectra in high–redshift radio galaxies.**
J. H. Krolik, W. Chen.
Astron. J., Vol. 102, No. 5, p. 1659 – 1662 (Nov 1991). Current Physics Microform No.: 9111D0665.
The generic spectrum of an optically thin synchrotron source steepens by 0.5 in spectral index from low frequencies to high whenever the source lifetime is greater than the energy–loss timescale for at least some of the radiating electrons. Three effects tend to decrease the frequency ν_b of this spectral bend as the source redshift increases: (1) for fixed bend frequency ν_* in the rest frame, $\nu_b = \nu_*/(1+z)$; (2) losses due to inverse Compton scattering the microwave background rise with redshift as $(1+z)^4$, so that, for fixed residence time in the radiating region, the energy of the lowest energy electron that can cool falls rapidly with increasing redshift; and (3) if the magnetic field is proportional to the equipartition field and the emitting volume is fixed or slowly varying, flux–limited samples induce a selection effect favoring low ν_* at high z because higher redshift sources require higher emissivity to be included in the sample, and hence have stronger implied fields and more rapid synchrotron losses. A combination of these effects may explain the trend observed in the 3CR sample for higher redshift radio galaxies to have steeper spectra, and the successful use of ultrasteep spectrum surveys to locate high–redshift galaxies.

58.182 **Radio properties of extragalactic IRAS sources.**
J. J. Condon, J. J. Broderick.
Astron. J., Vol. 102, No. 5, p. 1663 – 1679 (Nov 1991). Current Physics Microform No.: 9111D0669.

Extragalactic sources with $S_{60\,\mu m} \geqslant 0.2$ Jy and $\alpha(12\,\mu m, 60\,\mu m) \geqslant 0$ from the IRAS Faint Source Catalog Version 1 covering $|b| > 50°$ were identified by position coincidence with radio sources stronger than S = 25 mJy and lying north of $\delta = +5°$ on the Green Bank 4.85 GHz sky maps. There are N = 7702 far-infrared (FIR) sources and N $\sim 1.3\text{x}10^4$ radio sources in this $\Omega \sim 1.28$ sr area. Published VLA maps, new 4.86 GHz VLA maps made with 15 arcsec resolution, and accurate optical positions were used to confirm 122 of these candidate identifications. Normal and starburst spiral galaxies comprise ~97% of the FIR flux-limited sample. Their logarithmic FIR/radio flux-density ratio distribution is quite narrow ($\sigma_q = 0.14 \pm 0.02$) with mean $\langle q \rangle = 2.75 \pm 0.03$, and all are relatively cool [$\alpha(25\,\mu m, 60\,\mu m) \gtrsim +1.5$] FIR sources. Radio-loud "monsters" with $q < 2.25$ dominate the radio emission from about 2% of the FIR source sample, and radio-quiet monsters are responsible for the FIR emission from $\lesssim 1\%$ of the FIR sample. The FIR emission from the two BL Lac objects identified, B2 1308+32 and OQ 530, is nonthermal; the remaining identifications all appear to be thermal FIR sources. A significant fraction of the thermal FIR sources in radio-loud galaxies are so warm [$\alpha(25\,\mu m, 60\,\mu m) < +1.5$] that their heating sources are probably monsters, not starbursts. All of the radio-identified sources are optically identified, mostly with relatively bright nearby galaxies. There is no evidence in the authors' sample for any new populations of high-redshift FIR sources, nonthermal sources with steep FIR/optical spectra, or dust-shrouded sources visible only at FIR and radio wavelengths.

158.183 **VLBI observations of the nuclear jet in Cygnus A.**
C. L. Carilli, N. Bartel, R. P. Linfield.
Astron. J., Vol. 102, No. 5, p. 1691 – 1695 (Nov 1991). Current Physics Microform No.: 9111D0697. With plate 44.

The authors present an image of the "core" and inner jet of Cygnus A at an angular resolution of 2 mas obtained from Mark III VLBI observations at 6 cm wavelength with an array of nine antennas. The core is resolved, at least partly due to its being broadened by scattering in the Galactic interstellar medium. The jet has several condensations out to a distance of ~25 mas from the core and has a fairly constant FWHM of ~2.2 mas. The parsec-scale jet points toward the arcsec-scale jet observed with the VLA to within 4°. The ratio between the flux density of the jet and that of the counterjet is >15:1. From a comparison with an earlier VLBI image of Cygnus A, the authors find an upper limit to the apparent velocity of the brightest knot in the jet of about 2 c, where c is the speed of light (for $H_0 = 60$ km s^{-1}Mpc^{-1}, $q_0 = 0$).

158.184 **Active nuclei and molecular megamasers.**
W. A. Baan.
3. Haystack Observatory Conference on the Interstellar Medium: Skylines – atoms, ions and molecules – new results in spectral line astrophysics, p. 45 – 55 (1991). – See Abstr. 012.051 for the main entry.

Molecular megamaser activity provides new insights into the physics of the nuclear region of active galaxies. The density structure of the nuclear ISM is discussed for active/starburst galaxies. The spectral properties of the distinct class of OH megamasers are discussed. The luminosity function of the flux-limited sample of OH sources is used to study the high luminosity region of the luminosity distribution and to discuss a possible population of Gigamasers.

158.185 **Massive star formation and superwinds in IRAS 19254–7245 (the "Superantennae").**
L. Colina, S. Lipari, F. Macchetto.
Astrophys. J., Vol. 379, No. 1, p. 113 – 121 (20 Sep 1991). With plate 1.

The ultraluminous galaxy IRAS 19254–7245 is investigated by means of aperture spectroscopy and images. It shows the spectral characteristics of a Seyfert 2 galaxy with the emission lines split into four independent kinematical components separated by a velocity of 1491 km s^{-1}. A heavy internal obscuration corresponding to E(B–V) = 1.40 is measured in the circumnuclear emission line regions. The total amount of molecular gas detected in the whole system is $M_{H2} = 3.0 \times 10^{10} M_{\odot}$. The optical morphology shows a distorted main body composed of two differentiated nuclei separated by about 10 kpc, while two long tails extending over a total size of 305 kpc in projection are also detected. The measured amounts of cold gas and dust together with the extraordinarily large size of the tails indicate that the system is the merger of two almost equally massive gas-rich spiral galaxies and that the merging process did start 10^9yr ago. The line splitting is interpreted as produced in a biconical outflow, which is explained as produced by supernova explosions and stellar mass loss winds generated during a process of steady star formation with SFR = 150 $M_{\odot}$yr^{-1}. A quasar-type central nucleus, strongly obscured by dust, is assumed to produce the Seyfert 2 excitation conditions.

158.186 **Heavy, overpressured jets and the radio source Hercules A.**
D. L. Meier, A. C. Sadun, K. R. Lind.
Astrophys. J., Vol. 379, No. 1, p. 141 – 156 (20 Sep 1991).

The authors investigate the flow of jets with pressure and density initially much greater than the medium into which they are injected. These heavy, overpressured jets expand at the Mach angle and have a conical expansion shock and large convex Mach disk. The "ice cream cone" shape of these shocks should be visible in high dynamic range radio maps. Some numerical models are observed to recollimate, creating broader jets at lower density. The authors develop analytic expressions which describe the cocoon formation and recollimation processes. Under the assumption that the rings seen in high resolution images of Hercules A are shocks, the authors model the western lobe of this source as a heavy overpressured flow where, about 20" from the nucleus, the jet suddenly becomes much greater in pressure than the surrounding medium and freely expands. The authors argue that jet models which are supersonic with respect to both the internal and external speeds are applicable and find that, in spite of the prominent ring structure in Her A, the flow must be dominated by a smooth conical jet. The rings are weak spherical shocks in that flow: their size is small compared to the scale length for the sound speed to change.

158.187 **Very large array observations of radio-selected dumbbell galaxies.**
P. Parma, R. A. Cameron, H. R. de Ruiter.
Astron. J., Vol. 102, No. 6, p. 1960 – 1976 (Dec 1991). Current Physics Microform No.: 9112C1936.

An unbiased sample of radio sources associated with optical dumbbell galaxies is presented. This sample has been assembled to study the relationship between the radio and optical properties of radio-loud dumbbell galaxies. The radio sources were selected from the Parkes, B2 and 3C catalogues. High-quality radio data already exist for a number of the sources in the authors' sample, but those sources without good data have been observed with the Very Large Array at 20 or 6 cm. These new observations are described here and radio images are presented. Analysis of both the radio structure of the sources and their radio luminosity has been carried out, and a comparison is made with the properties of a complete sample of radio sources associated with single galaxies. Radio sources associated with dumbbell galaxies are found on average to have more distorted structures than sources associated with single galaxies, demonstrating the influence of the dumbbell dynamics in shaping the large-scale structure of the radio sources. It is shown that in the range 10^{24}–10^{26}W Hz^{-1} at 408 MHz the radio luminosity function of dumbbell systems is flatter than that of single-galaxy radio sources, indicating that a close companion may trigger a radio source in the main galaxy, or alternatively increase the luminosity of an existing radio source.

158.188 **Variability of the broad line spectrum of Markarian 372.**
S. A. Gregory, W. G. Tifft, W. J. Cocke.
Astron. J., Vol. 102, No. 6, p. 1977 – 1979 (Dec 1991). Current Physics Microform No.: 9112C1953.

The authors report on spectroscopic observations of the AGN galaxy Mrk 372. In 1986 the spectrum was that of a Seyfert type 1.5 with very broad permitted lines and modestly broadened forbidden lines. In 1990 the spectrum is that of a Seyfert type 1.9. At Hβ the broad line component disappeared leaving only a very weak, narrow line. Observations near Hα, however, do reveal the presence of a broad component.

158.189 **Broadband imaging of eight isolated (?) Seyfert galaxies.**
E. Xanthopoulos, M. M. de Robertis.
Astron. J., Vol. 102, No. 6, p. 1980 – 1990 (Dec 1991). Current Physics Microform No.: 9112C1956.

Tidal triggering by companion galaxies has been proposed as the main mechanism responsible for nuclear activity in Seyfert galaxies. However, although Seyfert galaxies show a tendency to occur in multiple systems, there is a small fraction of these galaxies that appear to be completely isolated. For eight interaction activity class 1 Seyfert galaxies ["completely isolated" as defined by Dahari, 1984], the authors obtained CCD V images to search for possible faint or hidden companions or other signs of distortions – evidence of a recent interaction. Two galaxies appear to have distant companions. More than 3/4 of the sample contains some type of morphological asymmetry. One galaxy (Mrk 686) shows a very prominent bar and three more (Mrk 1469, Mrk 885, and NGC 4235) show a hint of a bar. Mrk 486 appears to have a jetlike distortion. Only Mrk 1388 appears to be symmetrical.

158.190 **Redshifts of three BL Lacertae objects: PKS 0118–27, PKS 0829+04, and MC 1057+10.**
R. Falomo.
Astron. J., Vol. 102, No. 6, p. 1991 – 1993 (Dec 1991). Current Physics Microform No.: 9112C1967.

The authors present optical spectroscopy of three BL Lacertae objects which allows to derive the previously unknown redshift of the objects. For PKS 0118–27 an absorption doublet of Mg II at $z = 0.559$ is found. For PKS 0829+04 spectral signature of stellar population is detected at $z = 0.18$. Spectra of MC 1057+10 show broad emission lines which, identified with Mg II 2800 and C III] 1909, yield $z = 1.317$.

158.191 **A large radio galaxy behind the supernova remnant HB9.**
F. D. Seward, D. G. Fabricant, R. P. Kirshner, J. E. McClintock, R. C. Smith, R. A. Remillard.
Astron. J., Vol. 102, No. 6, p. 2047 – 2048 (Dec 1991). Current Physics Microform No.: 9112C2023.

HB9 is a 2° diameter radio shell with X-ray emission from a broad diffuse band in the interior. Within this band, a weak unresolved X-ray source coincides with the radio source 4C 46.09. To identify this source the authors have obtained CCD I and V band pictures and a spectrum. The source is a galaxy at redshift 0.195, and is therefore only accidentally aligned with HB9. The galaxy is large and bright and has the characteristics of a cD galaxy, often found at the center of clusters. If the galaxy is centered in a cluster, some of the surrounding diffuse X-rays may also be extragalactic.

158.192 **About ejections from the galaxy NGC 5128.**
A. L. Gyul'budagyan.
Astrofizika, Tom 33, Vyp. 3, p. 437 – 440 (Dec 1990). In Russian.

The existence of extremely young objects situated in the jet ejected from the nucleus of NGC 5128 and distributed rather far from this nucleus is a fact in favour of formation of these objects after ejection from the nucleus. In normal galaxies these objects are distributed in the spiral arms. The hypothesis of ejection decay of dense protostellar bodies was considered.

158.193 **Zones of highly ionized gas in the nuclei of Seyfert galaxies and quasars.**
N. G. Bochkarev, L. S. Nazarova.
Sov. Astron., Vol. 35, No. 5, p. 457 – 465 (Sep 1991). Current Physics Microform No.: 9203X1631. English translation of Astron. Zh., Tom 68, Vyp. 5, p. 918 – 935 (1991).

The relative line intensities and level populations for hydrogen and He II are calculated on the basis of Sobolev's probabilistic method for a wide range of physical conditions [$N_e(H) = 10^9 - 10^{11} cm^{-3}$, $T_e = (2-3)\cdot 10^4 K$, velocity gradient in the gas $dV/dR = 10^{-4} - 10^{-6} sec^{-1}$], assuming that the stationary level population is due to the simultaneous action of collisional and radiative mechanisms of gas heating and ionization. The central radiation source is assumed to be nonthermal, $F_\nu = F_0(\nu/\nu_0)^{-\alpha}$ and $F_0 = 10^{-5} erg/(cm^2 sec\ Hz)$, with a range of spectral indices: $\alpha = 0 - 1$. From a comparison of the calculated and observed intensity ratios of hydrogen and He II lines, the physical conditions in the regions of formation of high-excitation lines located above the surface of the accretion disk are estimated to be $N_e = 10^{11} cm^{-3}$, $T_e = 3\cdot 10^4 K$, and $dV/dR = 10^{-5} sec^{-1}$.

158.194 **Super-shells of M82.**
J. H. Huang, H. J. Su, Z. G. Deng, G. H. Jacoby.
3. ESO/ST-ECF Data Analysis Workshop, p. 67 – 71 (1991). – See Abstr. 012.061 for the main entry.

The H_α image of M82 has been processed with variable threshold zonal filtering technique, which illustrates multiple shell-like structures like those produced from sequential supernova explosions in the M82 nuclear region. The [SII] image of this galaxy shows that some of these supershells are shock-excited at the nuclear region.

158.195 **Echo images of broad-line regions in active galactic nuclei.**
W. F. Welsh, K. Horne.
Astrophys. J., Vol. 379, No. 2, p. 586 – 591 (1 Oct 1991).

Continuum and emission line variations can be used to form two-dimensional images of gas flows in the nuclear regions of active galaxies. The "echo image," $\Psi(v, \tau)$, is the two-dimensional transfer function that maps continuum variations to emission-line variations. Convolving a continuum light curve C(t) with $\Psi(v, \tau)$ yields, L(v, t), the line light curve at each velocity across the line. A procedure for constructing the echo image of a given BLR model is described. Echo images for several BLR models are presented and discussed. A simulated continuum light curve is generated and convolved with the echo images to produce trailed spectrograms showing variations of the line fluxes and line profiles. Because the echo image incorporates both velocity and time-delay information, it is a far more powerful probe of the structure of the BLR than one-dimensional velocity profiles, transfer functions, or cross-correlation functions.

158.196 **Modeling an accretion disk in NGC 985.**
R. M. Stanga, J. M. Rodríguez Espinosa, F. Mannucci.
Astrophys. J., Vol. 379, No. 2, p. 592 – 595 (1 Oct 1991).

An optically thick relativistic Keplerian accretion disk model is used to fit the broad component of the Hα, Hβ, and He I λ5876 emission line profiles of the active nucleus of NGC 985. A set of values of the relevant parameters is derived from independent fits of the three lines. Those values are consistent for the lines examined, and are within the range expected for accretion disks on the basis of current models.

158.197 **IUE observations of NGC 1068: the extremely luminous starburst knots.**
F. C. Bruhweiler, K. Q. Truong, B. Altner.
Astrophys. J., Vol. 379, No. 2, p. 596 – 608 (1 Oct 1991). With plate 4.

The authors present a complete study of the UV characteristics of the starburst knots in the inner 3 kpc region of NGC 1068, mapped by IUE. These knots probably represent the most luminous H II regions yet studied at optical and UV wavelengths. The brightest UV knot, knot 1, has a total luminosity $\geq 10^{43}$

..gs s^{-1}. The total observed flux of the UV knots at 1350 Å exceeds that of the nucleus of this nearby Seyfert 2 galaxy. The UV spectra clearly show that the knots have undergone recent intense star formation. The stellar population of the knots is dominated by luminous B0 – B1 supergiants with ages less than 10^7yr. The spectra of the knots exhibit with very strong emission of N V, C IV, He II, and C III]. Apparently the spectrum of a very highly ionized extended nebular emission line region is superposed upon that of the UV knots. There is no evidence of excessive extinction by dust in the vicinity of the starburst knots. The interstellar extinction deduced from the UV data for the knots corresponds to an intrinsic E(B–V) = 0.15, assuming an LMC extinction law plus an additional Galactic foreground component of 0.03 mag. The low reddening deduced from the UV may reflect the preferential destruction of small dust grains in the inner disk of NGC 1068 as a result of supernova shocks and the intense EUV radiation field of both the central active galactic nucleus and the starburst knots.

158.198 **IUE observations of NGC 1068: the extended high–ionization emission region.**
K. Q. Truong, F. C. Bruhweiler.
Astrophys. J., Vol. 379, No. 2, p. 609 – 620 (1 Oct 1991).

The UV emission morphology of the inner disk of NGC 1068 reveals extremely strong emission of the highly ionized species, N V λ1240, C IV λ1550, and He II λ1640, in regions of distances up to 25″ (2.5 kpc) from the nucleus. The strong emission from these species cannot be due to photoionization from OB stars and requires either photoionization from the nonthermal continuum of the AGN or yet some other ionization mechanism. The high–ionization emission region (HIER) is concentrated northeast of the nucleus, but is completely absent in the southwest. Flux maps for N V, C IV, and He II covering the inner disk region of NGC 1068 are presented. The data are consistent with the previously proposed dusty torus model of AGNs where the ionizing flux of the AGN is beamed in directions away from a dusty disk surrounding the central engine. The correlation of the strong HIER lines with starburst knots indicates that the cone of beamed radiation from the nucleus intersects the galactic plane. To explain the asymmetry of the HIER relative to the nucleus, the authors suggest that there is preferential clearing of the dense molecular gas, close to the nucleus, by the radio jet emanating from the nucleus toward the NE. Additional sweeping of the interstellar gas by the expanding CO ring and the rotating stellar bar provides further clearing of the gas in the inner disk at larger distance from the nucleus. This action allows EUV fluxes escaping from the central source to efficiently photoionize the gas in the northeast direction. The absence of these emission features in the southwest direction can be explained if there exists a large amount of absorbing material, SW of the nucleus, blocking the beamed UV radiation in this direction.

158.199 **On the origin of optical microvariability and X–ray flickering in active galactic nuclei.**
S. K. Chakrabarti, P. J. Wiita.
International Centre for Theoretical Physics, Trieste (Italy), IC—90/476. 6 p. (Dec 1990).

The authors suggest that non–stationary spiral shocks can produce X–ray flickering and optical microvariability. They show this by detailed non–axisymmetric numerical simulation of accretion disks in a binary system. They show that non–axisymmetric features appear and disappear in time scales of order of few local orbital periods near the inner edge of the disk and shorter than the orbital periods at a large distance. The former type may explain the X–ray flickering and the later type, the optical microvariability.

158.200 **Weakly active galactic nuclei.**
V. S. Artyukh, M. A. Ogannisyan.
Sov. Astron. Lett., Vol. 17, No. 5, p. 377 – 379 (Sep 1991). Current Physics Microform No.: 9204X1324. English translation of Pis'ma Astron. Zh., Tom 17, No. 10, p. 891 – 896 (1991).

Observations of nearby galaxies at 102 MHz by the method of interplanetary scintillation are analyzed. Compact (scintillating) radio sources are detected in spiral and lens–shaped galaxies but not in elliptical galaxies. It is suggested that this result may be explained by the different physical conditions in weakly active nuclei of galaxies of different morphological types, as has been found earlier for active galactic nuclei.

158.201 **Are short–term flux variations interstellar scintillations of extragalactic radio sources?**
N. Ya. Shapirovskaya.
Sov. Astron. Lett., Vol. 17, No. 5, p. 380 – 382 (Sep 1991). Current Physics Microform No.: 9204X1327. English translation of Pis'ma Astron. Zh., Tom 17, No. 10, p. 897 – 903 (1991).

Diffractive interstellar scintillations similar to pulsar scintillations should be observed for extragalactic sources with intrinsic variability. The parameters of the short–term variations of quasar 0917 + 62 are explained by interstellar scintillations with a model of the source as two components: an extended component that does not scintillate and a compact component that does.

158.202 **Search for starbursts among X–ray – selected galaxies: optical spectroscopy.**
A. Fruscione, R. E. Griffiths.
Astrophys. J., Lett., Vol. 380, No. 1, p. L13 – L16 (10 Oct 1991).

To establish the nature of X–ray – selected galaxies and to probe the potential contribution of "starburst" galaxies to the X–ray background, the authors have obtained low–resolution spectra, in the range 3500 – 7300 Å, of the optical counterparts of 15 X–ray sources selected from the Einstein Observatory Extended Medium–Sensitivity Survey and Imaging Proportional Counter Source Catalog. At least three X–ray sources can be classified as starburst galaxies on the basis of line–ratio diagnostics alone. This is consistent with the number expected from recent models which predict that starburst galaxies could account for at least ~15% – 20% of the extragalactic X–ray background at 2 keV.

158.203 **Some observational and theoretical progress of BL Lac objects.**
Xie Guangzhong, Li Bifang, Liu Fukun, Li Kaihua.
Prog. Astron., Vol. 9, No. 3, p. 201 – 210 (Sep 1991). In Chinese.

Recent progress in observational and theoretical research of BL Lac objects is collected. The authors review some developments in observations. Theories and models for BL Lac objects are summarized, especially the accretion disk model, the beaming model, and the criteria used for discrimination between these models on the basis of variability time–scale being discussed in more details. The cosmological implication of BL Lac objects is presented.

158.204 **A two–phase model for the X–ray emission from Seyfert galaxies.**
F. Haardt, L. Maraschi.
Astrophys. J., Lett., Vol. 380, No. 2, p. L51 – L54 (20 Oct 1991).

A two–phase accretion disk model is considered. A substantial fraction, f, of the gravitational power is assumed to be dissipated via buoyancy and reconnection of magnetic fields in a hot tenuous "corona" surrounding the main body of the disk. The main cooling mechanism of the hot layer is Comptonization of soft photons, thermally produced in the underlying cold phase. Hard Comptonized photons are backscattered into the thick phase, where they are in part reprocessed into the soft blackbody power and in part reflected. Coupled thermal balance equations for the two phases yield the temperature of the hot phase and the slope of the Comptonized component self–consistently as a function of f and τ, the optical depth of the hot phase. The authors find that for f $\cong$ 1, and $\tau < 1$, the temperature of the hot phase adjusts so as to maintain $\alpha \cong 1$. For small τ the temperature of the hot phase is sufficient for production of electron – positron pairs to be important. Pair production contributes to the optical depth and limits the maximum temperature allowed. The model accounts in a self–consistent way for the average power–law index and reflection component observed in the X–ray emission of Seyfert galaxies.

158.205 **On the nebulosity surrounding the BL Lacertae object PKS 2155–304.**
R. Falomo, E. Giraud, L. Maraschi, J. Melnick, E. G. Tanzi, A. Treves.
Astrophys. J., Lett., Vol. 380, No. 2, p. L67 – L69 (20 Oct 1991). With plate L3.

High resolution imaging of the bright BL Lacertae object PKS 2155–304 reveals a symmetric nebulosity extending up to 15″ from the nucleus and confirms the presence of a faint galaxy located ~4″ away from the nucleus. The authors show that the redshift of z = 0.117, previously attributed to the BL Lac object, pertains in fact to the companion galaxy. The radial brightness profile of the surrounding nebulosity is consistent with that of an elliptical galaxy of effectiv radius of 4.″5 which, assuming an absolute magnitude typical of luminous ellipticals, puts the object at z ~ 0.1. The field of PKS 2155–304 appears to be rich in galaxies, one of which is at z = 0.116. A scenario in which the BL Lac object is hosted by a giant elliptical (M_V ~ –22.5), which is the brightest member of a group of galaxies, is suggested by these observations.

158.206 **The luminosity function and cosmological evolution of X–ray – selected BL Lacertae objects.**
S. L. Morris, J. T. Stocke, I. M. Gioia, R. E. Schild, A. Wolter, T. Maccacaro, R. Della Ceca.
Astrophys. J., Vol. 380, No. 1, p. 49 – 65 (10 Oct 1991).

The authors present redshifts for an X–ray – selected, flux-limited sample of BL Lac objects. They derive a median X–ray luminosity of $10^{19.83}$ W Hz^{-1} and further find a correlation between the optical and radio luminosity of these objects. The complete redshift information allows a measurement of the objects' evolution with redshift and the calculation of a luminosity function at zero redshift. There is evidence for strong "negative" evolution (BL Lac objects were either less common or less luminous in the recent past). The authors discuss the differences between radio – and X–ray – selected samples of BL Lac objects, and suggest that X–ray selection picks out chiefly beamed FR I radio galaxies, while radio selection finds a mix of beamed and microlensed quasars, along with extreme examples of beamed FR I radio galaxies.

158.207 **What are the gigahertz peaked–spectrum radio sources?**
C. P. O'Dea, S. A. Baum, C. Stanghellini.
Astrophys. J., Vol. 380, No. 1, p. 66 – 77 (10 Oct 1991).

The authors discuss the astrophysical implications of recent radio and optical observations of the powerful, compact gigahertz peaked–spectrum (GPS) radio sources. The defining characteristics of these objects are (1) their peaked radio spectra with narrow spectral shape and steep high–frequency spectrum, (2) low radio polarization, (3) large radio luminosities, and (4) mostly compact radio structure. The turnover in the spectrum is probably due to synchrotron self–absorption. The fact that some GPS sources have a very narrow spectral shape is consistent with the hypothesis that there is a narrow range of size scales which dominate the radio luminosity. Based on the steep high–frequency radio spectrum and the low variability, there does not appear to be a strong beamed "jetlike" component in these objects. The highly inverted low–frequency radio spectrum is consistent with the hypothesis that the radio source is tightly confined. It is found that the radio polarization of these sources is systematically low. This is consistent either with a very tangled magnetic field or very large Faraday rotation measures. The optical identifications are a mixture of galaxies and quasars. The authors suggest that GPS radio sources are formed when the radio plasma is confined on the scale of the narrow–line region by an unusually dense and clumpy ISM. In this scenario, the high electron densities are responsible for depolarizing the radio source and in some cases producing very high Faraday rotation measures. The existing optical spectroscopic results are also consistent with the existence of a dense and dusty nuclear ISM. The GPS quasars tend to have very high redshifts (half with $z \gtrsim 3$). The hosts of these radio sources may be protogalaxies with very dense and clumpy interstellar media. The lower redshift GPS galaxies may have recently cannibalized a gas–rich companion, resulting in a very dense and clumpy nuclear ISM.

158.208 **Short–term X–ray variability of the BL Lacertae object PKS 2155–304: power spectrum and cross–correlation analysis.**
G. Tagliaferri, L. Stella, L. Maraschi, A. Treves, A. Celotti.
Astrophys. J., Vol. 380, No. 1, p. 78 – 83 (10 Oct 1991).

The authors present a detailed power spectrum and cross–correlation analysis of the X–ray light curves of the BL Lac object PKS 2155–304, obtained with EXOSAT. The power spectra show a marked red noise type variability, with the power increasing steeply toward low frequency. The variability is more pronounced in the 1 – 6 keV band than in the 0.1 – 2 keV band. It is shown that the presence of data gaps can flatten significantly the logarithmic slope of the power spectrum. A technique for bridging data gaps, which reduces the power spectrum distortions, is introduced. In this way an average power law slope of about –2.5 is obtained for the power spectrum from the 1 – 6 keV data. The average power spectrum of the light curves after the removal of a linear trend, as required in this case, is characterized by a slope of -1.9 ± 0.4 over the 10^{-4} to 3×10^{-3}Hz frequency range. The 3 – 6 keV light curves are cross–correlated with the 0.1 – 2 keV and the 1 – 2 keV light curves, separately. The short–term X–ray variability of PKS 2155–304 is compared with that of Seyfert galaxies and is briefly discussed within the framework of BL Lac emission models.

158.209 **Have all questions been asked?**
P. L. Biermann.
Workshop on Variability of Active Galaxies, p. 1 – 7 (1991). – See Abstr. 012.070 for the main entry.

The author's question is specifically whether energetic neutrinos might play a role in the broad emission line regions of active galactic nuclei. He derives a possible neutrino source from arguments about the heating of the circumnuclear dusty disk which has recently been deduced to exist in most AGN. In normal quasars, the heating of the circumnuclear dusty disk is an unsolved problem. The author argues that a flux of relativistic particles, protons and heavy nuclei, with a hard spectrum and an intensity comparable to the flux in electromagnetic radiation is plausible. These particles can easily provide disk heating. Such a flux of energetic particles also invariably leads to an increased flux of neutrinos, with a hard spectrum and extending to very high energies. The author notes that stars on the main sequence are optically thick to energetic neutrinos, with optical depths of order unity or larger. He demonstrates that a rather moderate flux of neutrinos, at a level of only a few percent to a few tens of percent of the electromagnetic flux changes the internal structure of low mass stars drastically. The stars expand, due to internal heating and begin to look like subgiants or giants. Such stars, exposed to the strong external flux of both electromagnetic radiation and relativistic particles then produce a wind, just as has been argued already for a stellar component in a former compact binary system, of which we now observe only a millisecond pulsar. These winds may contribute to the broad emission line regions in quasars. In fact, the effect of neutrinos is so strong, that it limits the lifetime of low mass stars due to mass loss; such lifetime arguments might be usuable to actually put limits on the neutrino flux in quasars.

158.210 **Line and continuum variability in NGC 4151.**
M. A. J. Snijders.
Workshop on Variability of Active Galaxies, p. 9 – 17 (1991). – See Abstr. 012.070 for the main entry.

The author presents results on the variability of the absorption lines, the emission lines and the X–ray to optical continuum of NGC 4151. The ultraviolet and optical continua vary together and the emission lines follow these continuum changes with delays depending on the line studied. The strength of the absorption lines depends on the continuum strength too but the variability pattern is rather complicated. When the ultraviolet/optical continuum is bright there is no correlation with the X–rays. When this continuum is faint the X–ray flux and the optical/ultraviolet continuum are strictly correlated. Results from Principle Component Analysis of the IUE spectra are discussed.

158.211 **The optical, ultraviolet and X-ray variations in NGC 4151.**
M.-H. Ulrich.
Workshop on Variability of Active Galaxies, p. 19 – 30 (1991). – See Abstr. 012.070 for the main entry.
The author proposes two processes to produce quasi simultaneous UV/optical variations in AGNs: instabilities in the inner part of the disk and irradiation of the disk surface by the variable X-ray source. The quasi simultaneous variations observed in the ultraviolet and optical ranges in NGC 4151 led to this proposition. Such variations are not consistent with the continuum in these wavelength ranges, being produced only by accretion flow process, and therefore other components must contribute significantly to the UV/optical flux.

158.212 **Mapping the BLR in NGC 5548 and NGC 4151.**
J. Clavel.
Workshop on Variability of Active Galaxies, p. 31 – 46 (1991). – See Abstr. 012.070 for the main entry.
The author presents new results from intensive monitoring campaigns of the Seyfert I galaxies NGC 5548 and NGC 4151 with the IUE satellite which allow a reliable determination of the BLR dimension in these sources. Most of the gas is concentrated at about 12 lt–d from the continuum source in NGC 5548, and 4 lt–d in NGC 4151. Furthermore, there is a gradient in the degree of ionization of the gas in NGC 5548, the latter increasing inward. The picture is more complex in NGC 4151 because the density of the gas also increases toward small radial distances. In NGC 4151, the line wings respond more rapidly and with a larger amplitude than the "cores" ($V \leqslant 3000\ km\ s^{-1}$) to the continuum variations, therefore implying that the velocity dispersion of the gas also increases inward. A similar though shallower velocity gradient is also present in NGC 5548. The red and blue part of the line profiles vary simultaneously in both galaxies ruling-out the possibility that radial motions play a significant role in the velocity field of the BLR. The continuum variations are simultaneous from $\cong$1300 Å to 5000 Å to within $\pm$1 day. This pattern is difficult to reconcile with models where the "big-bump" represents thermal emission from an optically thick and geometrically thin accretion disk.

158.213 **Ground-based studies of emission-line variability: recent results for NGC 5548 and future plans.**
B. M. Peterson.
Workshop on Variability of Active Galaxies, p. 47 – 56 (1991). – See Abstr. 012.070 for the main entry.

158.214 **X-ray observations of the Seyfert 1 galaxy Markarian 464.**
L. Piro, G. Matt, E. Costa, D. Dal Fiume, F. Frontera, E. Morelli.
Astrophys. J., Vol. 380, No. 2, p. 357 – 361 (20 Oct 1991).
The authors present results of EXOSAT (0.05 – 10 keV) and hard X-ray (25 – 45 keV) balloon observations of the Seyfert 1 galaxy Mrk 464. The spectrum observed with EXOSAT is well fitted by a power law with a photon spectral index 1.70 ± 0.15, in agreement with a previous measurement by HEAO 1 A-2. The absorption column density is found consistent with the galactic column density. Both the EXOSAT and the HEAO 1 upper limits ($N_H < 10^{22} cm^{-2}$) are in disagreement with the very high column density of $2.5 \times 10^{23} cm^{-2}$ observed by Ariel 5. The balloon experiment detected a bright hard X-ray source in a field of view centered on Mrk 464. This measurement is most likely explained by an outburst of this Seyfert galaxy, with a peak luminosity one order of magnitude greater than the average one. Similar variations have been observed in two other objects of this class, namely Centaurus A and NGC 4151.

158.215 **Variability of line profiles in NGC 5548.**
W. Kollatschny, M. Dietrich.
Workshop on Variability of Active Galaxies, p. 57 – 64 (1991). – See Abstr. 012.070 for the main entry.
High resolution profiles of the $H\alpha$ and $H\beta$ emission lines of NGC 5548 have been investigated as part of the international UV/optical monitoring campaign. The $H\alpha$ and $H\beta$ difference profiles show strong variations. The broad line difference profiles consist of three components. The main component is centered at the same wavelength as the narrow component. The intensity of this component is correlated with the UV continuum. It is symmetric and has constant width at zero intensity during the variations. Furthermore a blue and red component – best visible during low active states – varies independently.

158.216 **Continuum variability in NGC 5548: implications for theoretical models.**
S. Molendi, L. Maraschi, L. Stella.
Workshop on Variability of Active Galaxies, p. 65 – 69 (1991). – See Abstr. 012.070 for the main entry.
The recent campaign on NGC 5548 has shown the continuum to vary simultaneously at ultraviolet and optical frequencies, with an upper limit to the delay of less than 4 days. The authors examine this result in the context of accretion disk theory. In the first part the standard α-disk model is discussed. The effects connected with heating of the outer regions by radiation from the inner regions are also considered. Neither scenario lends itself to a simple explanation of the observed data. A more successful model involves a geometrically thin outer disk heated by a central extended source of X-ray radiation.

158.217 **Variability in Markarian 279.**
G. M. Stirpe.
Workshop on Variability of Active Galaxies, p. 71 – 78 (1991). – See Abstr. 012.070 for the main entry.
Markarian 279, a spiral galaxy with an inclination of about 40°, was immediately identifield as a Seyfert 1 galaxy. All its characteristics make it an "average" medium-bright object of its kind. Its strong broad emission lines present both a median shift and an asymmetry towards red wavelengths: neither of these characteristics is extreme within the Seyfert 1 population. The author reviews what is known about the broad line variability of Markarian 279 and presents some recent results.

158.218 **Limits on the expansion of the radio source OQ 208 in Mkn 668.**
A. G. de Bruyn.
Workshop on Variability of Active Galaxies, p. 105 – 106 (1991). – See Abstr. 012.070 for the main entry.

158.219 **BLR models and the L-M relation for AGNs.**
H. Netzer.
Workshop on Variability of Active Galaxies, p. 107 – 116 (1991). – See Abstr. 012.070 for the main entry.
This paper discusses the L-M relation for AGNs, in view of the currently available data and the present understanding of the theory of active galactic nuclei. The author reviews all data sets that are, in his opinion, reliable enough to deduce a cross correlation size for the BLR (only 5 objects) and discusses some important information about the gas distribution that can be extracted from them. An apparent contradiction between different observations of NGC 5548, and the transfer function for the NGC 4151 $H\beta$ line, are the best evidences for the existence of a thick geometry in those objects. The complications due to the non-linear response of several emission lines are also discussed, and a preliminary L-M relation is suggested.

158.220 **The discovery of a young radio galaxy at $z = 2.390$: probing initial star formation at $z \lesssim 3.0$.**
R. A. Windhorst, D. Burstein, D. F. Mathis, L. W. Neuschaefer, F. Bertola, L. M. Buson, D. C. Koo, K. Matthews, P. D. Barthel, K. C. Chambers.
Astrophys. J., Vol. 380, No. 2, p. 362 – 383 (20 Oct 1991).
The authors report the discovery of a weak radio galaxy from the Leiden Berkeley Deep Survey at a redshift of 2.390, the faint

optical and IR counterpart of the steep–spectrum, compact radio source 53W002. Its λ–dependent optical continuum morphology is compact with linear size ~10 – 35 kpc ($H_0 = 50$, $q_0 = 0$). In redshifted Lyα, the galaxy is somewhat more extended ($\lesssim 67 \times 40$ kpc). Its radio source is 7 times smaller than, and confined by, the Lyα gas. It shows alignment with the Lyα gas and the best seeing optical continuum images. The authors present nine–band photometry (Lyα, UBgriJHK) for the galaxy as well as surrounding objects. The source 53W002 is not variable on time scales of years, in either radio or optical. The data suggest that 53W002 has a Seyfert 1 – like AGN. Several independent age estimates yield a consistent value of 0.25 – 0.32 Gyr. The analysis of the data also indicates that at z = 2.390, 53W002 had processed only a few times $10^{11} M_\odot$ into stars, significantly less than most powerful radio galaxies have converted into stars at z ~ 2 – 3.8. The available data are consistent with 53W002 being a genuinely young (radio) galaxy seen at z = 2.390 during its first major starburst. It likely started forming most of its current stars at redshifts 2.5 – 3.0 (for $H_0 = 5 - 100$, $q_0 = 0.0 - 0.5$). This suggests that (radio) galaxies do not form the bulk of their stars coevally, but start doing so over a wide range of cosmic time.

158.221 **The signatures of an accretion disk in the electromagnetic spectra of quasars and AGNs.**
M.–H. Ulrich.
6. IAP Astrophysics Meeting – IAU Colloquium No. 129: Structure and emission properties of accretion disks, p. 43 – 52 (1991).
– See Abstr. 012.046 for the main entry.
The radiation from an accretion disk can plausibly contribute in direct re–radiated and re–processed forms to the continuum of quasars and AGNs from the infrared to MeV energies. Other components however can contribute to the continuum. These must first be evaluated and substrated from the observed continuum before models of accretion disk energy distribution can be applied, and M andd $\dot{M}$ determined. The author examines these questions and emphasizes the importance of modelling simultaneously the time variablility and the energy distribution in the optical, UV and X–ray ranges.

158.222 **Variability of BL Lac objects in the radio regime.**
E. Valtaoja.
Workshop on Variability of Active Galaxies, p. 117 – 129 (1991).
– See Abstr. 012.070 for the main entry.
Results from the Finnish quasar monitoring program are discussed. Both the spectra and the radio variability indicate that both the basic structure and the mechanism for variability are the same for all AGN. The basic radio properties of low and high polarization quasars are in agreement with orientation–dependent scenarios. The contradictory properties of BL Lac objects can be understood if they in reality consist of two physically different classes. One class, consisting of highly boosted, strongly variable, distant BL Lacs with quasarlike spectra, represents the extreme end of the FR II–LPQ–HPQ population, sources aligned to a few degrees with our line of sight. The other BL Lac population consists of much less boosted, moderately variable, nearby FR I radio galaxies beamed towards us.

158.223 **The high frequency properties of synchrotron radiation.**
K. Ballard.
Workshop on Variability of Active Galaxies, p. 143 – 153 (1991).
– See Abstr. 012.070 for the main entry.
The aim of this paper is to examine the relevance of models involving shocks to the high frequency near infrared and optical emission in BL–Lacs. A short study of the effect that losses can have upon non–thermal electron distributions is presented. This is combined with previous work relating to specific magnetic field geometries behind shock fronts to yield model spectra which can be compared to those actually observed. Agreement between the theory and observations is found only if the electrons are allowed to remain isotropic. Some of the limitations and implications for this model of the emission region are considered.

158.224 **Synchrotron source models and the infrared–optical variability of blazars.**
K. Meisenheimer.
Workshop on Variability of Active Galaxies, p. 155 – 162 (1991).
– See Abstr. 012.070 for the main entry.
The author explores whether the source models developed for extended radio sources can be scaled down to explain the polarized infrared–optical continua of blazars and their dramatic variations. He discusses in some detail the overall polarization and flux from a jet with a helical field containing a shock. It turns out that even very simple models of this class are able to explain the exotic behaviour observed in OJ 287 in January 1983.

158.225 **Rapid variability of BL Lac objects in the optical regime.**
S. J. Wagner.
Workshop on Variability of Active Galaxies, p. 163 – 173 (1991).
– See Abstr. 012.070 for the main entry.
Some of the results of an extensive four–week monitoring campaign of the variability of several BL Lac objects on short time–scales, simultaneously at radio and optical frequencies, are described. In some objects the variability provides strong evidence that the optical emission is beamed as has been demonstrated for the continuum emission at radio frequencies. In the BL Lac object 0716 + 714 the character and the time–scales of short period variability changed significantly within less than 48 hours during the four week campaign. The similarity of radio and optical lightcurves during both states and the simultanity of the change indicates a close correlation of the variability over more than five orders of magnitude in frequency. This correlation and the short time–scales rule out any variability introduced by propagation effects and requires an intrinsic origin. Some implications are briefly discussed.

158.226 **UV variability of blazars.**
A. Treves, E. Girardi.
Workshop on Variability of Active Galaxies, p. 175 – 178 (1991).
– See Abstr. 012.070 for the main entry.

158.227 **The connection between frequency dependent polarization (FDP) and growing radio shocks in blazars.**
L. Valtaoja.
Workshop on Variability of Active Galaxies, p. 179 – 180 (1991).
– See Abstr. 012.070 for the main entry.
The shocked jet models have had success in explaining the detailed variations associated with radio outbursts. The authors suggest that the early stages of radio shocks could cause the observed variable frequency dependence in the optical polarization.

158.228 **The connection between blazars and compact radio sources.**
C.–I. Björnsson.
Workshop on Variability of Active Galaxies, p. 187 – 191 (1991).
– See Abstr. 012.070 for the main entry.
It is argued that the unresolved VLBI–cores in flat spectrum, compact radio sources have properties very similar to the associated blazar emission region: large scale magnetic field, anisotropic pitch–angle distribution and the importance of relativistic streaming motion. Problems afflicting models invoking Compton–cooling behind a relativistic shock in a jet to explain the initial phase of out–bursts in the mm/far–infrared regime are discussed. As an alternative, a qualitative discussion is given of a model where relativistic streaming and pitch–angle scattering are the main ingredients.

158.229 **VLBI knots and superluminal motion.**
C. A. Hummel.
Workshop on Variability of Active Galaxies, p. 193 – 199 (1991).
– See Abstr. 012.070 for the main entry.
The author briefly describes high resolution radio imaging work carried out on "active galactic nuclei" (AGN). The relation between radio flux density variability and structural variability, and jet dynamics are investigated. New observations of structural characteristics of the jets close to the "central regime" are presented.

158.230 **Relativistic jet models and variable knot emission.**
M. Camenzind.
Workshop on Variability of Active Galaxies, p. 201 – 210 (1991).
– See Abstr. 012.070 for the main entry.
Variability in the radio waveband and moving features detected by VLBI observations are commonly explained in terms of shocks moving in relativistic collimated flows. The author discusses the origin of these jets with their relation to the magnetic structure of accretion discs. Dynamo action in the disk creates a rotating magentosphere which is filled with disk-plasma. Along open flux tubes, the plasma is accelerated to the relativistic speeds and collimated outside the light cylinder of the magnetosphere. The relaxation of these newly formed jets towards a minimal energy equilibrium is the origin for strong X-ray emission, ejection of knots along the jets and rapid variability in X-rays and the infrared-optical region.

158.231 **Particle acceleration and variability: magnetic reconnection in AGN.**
H. Lesch.
Workshop on Variability of Active Galaxies, p. 211 – 220 (1991).
– See Abstr. 012.070 for the main entry.
Magnetic reconnection is discussed as an important acceleration mechanism in active galactic nuclei. The dissipation of magnetic energy is directly connected with the activity in a magnetized accretion disk. Ascending magnetic flux is dissipated in the magnetosphere by reconnection. Thereby an electric field parallel to the magnetic field is induced which accelerates only the electrons to relativistic energies. This process is nonstationary, occurs on small scales, and could provide the energetics and the observed time scales of variability in AGN.

158.232 **Variable radio sources and the two-fluid models.**
G. Pelletier, J. Roland, H. Sol.
Workshop on Variability of Active Galaxies, p. 225 – 234 (1991).
– See Abstr. 012.070 for the main entry.
Superluminal radio sources associated with extended double sources can be explained by a two-fluid model that consist of i) a non-relativistic e^--p jet responsible for the formation of the extended lobes and ii) a relativistic e^--e^+ beam responsible for superluminal radio sources. When the magnetic field is greater than a critical value B_c, the beam can propagate along the magnetic field lines without dissipation. Natural consequences of the beam-jet interaction are, in the dissipation area, the acceleration of relativistic electrons by strong Langmuir turbulence, the possible formation of shocks, the emission of coherent emision and the generation of a local toroidal magnetic field. Depending on the value of the magnetic field in the beam compared to the critical value, one can observe superluminal quasars, BL Lac objects or flickering radio sources.

158.233 **Infrared variability of active nuclei.**
M. Ward.
Workshop on Variability of Active Galaxies, p. 235 – 239 (1991).
– See Abstr. 012.070 for the main entry.
AGN are known to vary in flux at near infrared wavelengths. Unfortunately because of limited accuracy and poor time coverage, the interpretation of the results is often ambiguous. Recent statistical surveys and detailed studies of individual AGN have improved the situation. The introduction of the large format infrared arrays will help to disentangle the variable and non-variable components of the infrared continuum.

158.234 **Brightness and color variations of accretion disks: implications for the parameters.**
R. Wehrse, H. Störzer.
Workshop on Variability of Active Galaxies, p. 275 – 279 (1991).
– See Abstr. 012.070 for the main entry.
For the interpretation of long-term continuum variations of AGNs detailed non-grey models for stationary accretion disks are used to estimate the brightness and color variations as functions of the accretion rate. Possible consequences of multiple solutions of the structure equations are briefly discussed.

158.235 **Lensing of BL Lac objects.**
M. Stickel.
Workshop on Variability of Active Galaxies, p. 303 – 312 (1991).
– See Abstr. 012.070 for the main entry.
Four highly variable, high redshift ($z \gtrsim 0.9$) BL Lac objects have been found to show a spatially resolved appearance on deep direct images. Because of the high redshift of the BL Lac objects, their non-stellar images are most likely due to an intervening galaxy with a much smaller redshift located nearly on the line of sight. These objects represent very good candidates for gravitationally (micro)lensed BL Lac objects. A number of properties of these objects and the BL Lac class in general are examined within the context of gravitational (micro)lensing.

158.236 **Evidence for disks in active galactic nuclei and QSOs from polarimetric observations.**
J. S. Miller.
6. IAP Astrophysics Meeting – IAU Colloquium No. 129: Structure and emission properties of accretion disks, p. 71 – 74 (1991).
– See Abstr. 012.046 for the main entry.
Direct information about the nature of the central objects in active galaxies and QSOs is very hard to obtain. It is sobering to realize that after nearly 30 years of intensive observational study of these objects we still do not know what mechanism or mechanisms produce the generally featureless continua we observe in the UV, optical, and infrared spectral regions. By a process more of elimination rather than by direct evidence it has become accepted that the ultimate source of energy is accretion of material by a massive compact object, presumably a black hole. In an attempt to find evidence for disks, recently a considerable amount of effort has been devoted to a study of the thermal emission that could be produced by the arrection disk itself. The results in some cases show excellent agreement between the model continua and what is observed, but there remain some important discrepancies. Polarization data strongly support the idea that disk-line structures are common in AGN and QSOs. It is quite reasonable and consistent with the data that these structures are in fact intimately related to central accretion disks.

158.237 **Continuum emission in AGN, can we isolate the disc component?**
T. J.-L. Courvoisier, J. Clavel.
6. IAP Astrophysics Meeting – IAU Colloquium No. 129: Structure and emission properties of accretion disks, p. 75 – 84 (1991).
– See Abstr. 012.046 for the main entry.
The emission component in Seyfert galaxies and quasars dominating in the optical-ultraviolet spectral domain, the blue bump, has often been associated with thermal emission from the surface of an optically thick and geometrically thin accretion disc. The intrinsic shape of the continuum emission in this domain is poorly measured, as it depends on a complex deconvolution, implying models of the other components emitting in the near infrared to soft X-ray domains. However, both the continuum energy distribution and the time variability observed in the optical and UV domains cannot be understood in the frame of standard geometrically thin disc models. More promising alternatives include thick accretion tori in the inner regions of the accretion disc. Such structures are required not only for extremely luminous nuclei but also for weakly active Seyfert I galaxies.

158.238 **NGC 5128.**
IAU Circ., No. 5313, p. 1 (30 Jul 1991).

158.239 **Accretion disk models for active galactic nuclei.**
M. Malkan.
6. IAP Astrophysics Meeting – IAU Colloquium No. 129: Structure and emission properties of accretion disks, p. 165 – 176 (1991). – See Abstr. 012.046 for the main entry.

158.240 **AGN disks: theory vs. observations.**
H. Netzer.
6. IAP Astrophysics Meeting – IAU Colloquium No. 129: Structure and emission properties of accretion disks, p. 177 – 186 (1991). – See Abstr. 012.046 for the main entry.

There are fundamental limitations in the application of thin accretion disk models to AGNs. The largest theoretical uncertainties are due to the incomplete treatment of the radiation transfer, the unknown viscosity, the presence of a central X-ray source and the disk corona. Many observed AGN continua can be fitted by thin disk emission, but the model parameters, such as the central mass and the accretion rate, cannot be accurately determined. There are two other ways to look for thin AGN disks, using model independent properties: (a) The continuum of classical (Newtonian) disks, with scattering atmospheres, is highly polarized. This is not the case in AGN disks, where general relativistic effects, and realistic opacity, reduce the degree of polarization. A strong polarization feature, typical of all models, should appear near the Lyman edge wavelength. It provides a characteristic signature of massive, thin disks and a direct way for discovering such objects. (b) The continuum radiation of thin accretion disks is softer in more massive objects, because of the larger central mass. In addition, the angle dependence of the radiation introduces a relation between the emission line intensities and the apparent continuum luminosity. These effects, when incorporated into photoionization calculations, predict some specific correlations between the line and continuum luminosities. A comparison of the model prediction with the observations shows a good agreement. This lends further support to the thin disk idea in AGNs.

158.241 **MERLIN observations of superluminal motion in the jet of 3C 120.**
T. W. B. Muxlow, P. N. Wilkinson.
Mon. Not. R. Astron. Soc., Vol. 251, No. 1, p. 54 – 62 (1 Jul 1991).

The authors present evidence for changes in the radio structure of the jet in 3C 120 between 1980 December and 1988 December derived from MERLIN observations at 1.67 GHz. The changes are consistent with outward motion of 19 ± 3 mas (2.4 ± 0.4 mas yr^{-1}, i.e. $3.7 \pm 0.6\ h^{-1}$ c, for $H_0 = 100\ h$ km s^{-1} Mpc^{-1}) in the inner jet, about 0.15 arcsec ($\sim 75\ h^{-1}$ pc) from the core. This confirms the suggestion by Benson et al. that the jet does not slow down quickly just outside the core region usually imaged by VLBI. However, the authors find no evidence for motion at the position of the knot 4 arcsec ($\sim 2\ h^{-1}$ kpc) from the nucleus as suggested by Walker et al. The authors' limit is < 1.25 mas yr^{-1}, i.e. $< 1.9\ h^{-1}$ c, compared with Walker's et al.'s measurement of 2.4 ± 0.8 mas yr^{-1}, i.e. $3.7 \pm 1.2\ h^{-1}$ c.

158.242 **Accretion models of some AGN and observational constraints.**
S. Tsuruta.
6. IAP Astrophysics Meeting – IAU Colloquium No. 129: Structure and emission properties of accretion disks, p. 289 – 298 (1991). – See Abstr. 012.046 for the main entry.

The author's primary aim is to explore the central engine of AGN. Therefore, his studies have been devoted to the class of Seyferts which exhibit very large amplitude, rapid time variability and interesting spectral features. This is because the effects of the matter outside the engine should be least important in these rapidly variable sources and the dominant fraction of the observed radiation should indeed be coming from the central engine. Otherwise such rapid variability would have been washed out by reprocessing in the broader region. The author adopts a type of model where the central engine is envisioned as an accreting supermassive black hole, and his attention is restricted to the innermost region, within ~ 100–$500\ R_G$ (where R_G is the gravitational radius).

158.243 **Near-infrared spectrophotometry of eight 3CR radiogalaxies: the first detections of [O III] 500.7 and [S III] 953.2 emission lines in galaxies with z > 1.**
S. Rawlings, S. Eales, M. Lacy.
Mon. Not. R. Astron. Soc., Vol. 251, No. 1, p. 17P – 21P (1 Jul 1991).

The authors present the results of near-IR spectrophotometry of eight 3CR radiogalaxies with redshifts in the range $0.256 \leqslant z \leqslant 1.575$. The narrow [O III] 500.7 emission line was detected in 3C 13 ($z = 1.351$), and the narrow [S III] 953.2 emission line in 3C 79 ($z = 0.256$), in the northern near-IR knot of 3C 356 ($z = 1.079$) and marginally in 3C 368 ($z = 1.132$). These emission lines are among the faintest to be detected with near-IR spectroscopy, and include the first measurements of [O III] 500.7 and [S III] 953.2 in any galaxy with $z > 1$. The authors show that narrow-line emission can sometimes account for a substantial fraction ($\sim 25\%$) of the broad-band near-IR flux arising from the substructures of high-z radiogalaxies, but that this fraction is strongly dependent on, and therefore a valuable probe of, physical conditions within the substructures.

158.244 **The intrinsic colour of radio galaxies and photometric determination of redshift.**
G. T. Rixon, J. V. Wall, C. R. Benn.
Mon. Not. R. Astron. Soc., Vol. 251, No. 2, p. 243 – 247 (15 Jul 1991).

Spectroscopic observations of 11 radio galaxies from the 5C12 survey are reported. The galaxies have redshifts $0.12 < z < 0.31$ and radio luminosities P_{408} (W Hz^{-1}) in the range $24.4 < \log_{10} P_{408} < 26.3$ ($H_0 = 50$ km $s^{-1} Mpc^{-1}$). The spectra show no evidence of nuclear activity or of strong star formation. V–R colours in the rest-frame of the galaxies have been determined from the spectra and show an intrinsic scatter of < 0.02 mag. There is no significant change of rest-frame colour with red-shift, consistent with no evolution of the stellar population at $z < 0.3$, and with there having been a single short burst of star formation at high redshift. The spectroscopic redshifts agree with those predicted on the basis of broad-band photographic colours, with rms difference $\sigma_z = 0.025$. Given sufficiently accurate photometry, the small intrinsic range of V–R colours may allow redshifts of similar radio sources to be predicted from broad-band colours with σ_z as small as 0.01. The median absolute magnitude of the galaxies is $M_V = -22.8$ mag, with rms scatter about this value of 0.4 mag.

158.245 **Near-infrared observations of the z ~ 2.3 IRAS source FSC 10214+4724.**
B. T. Soifer, G. Neugebauer, J. R. Graham, K. Matthews, J. M. Mazzarella, C. J. Lonsdale, M. Rowan-Robinson, T. Broadhurst, A. Lawrence, R. McMahon.
Astrophys. J., Lett., Vol. 381, No. 2, p. L55 – L57 (10 Nov 1991).

Near-infrared imaging and spectroscopy of the extremely luminous IRAS source FSC 10214+4724 have been obtained using the Cassegrain infrared camera on the 200 inch Hale Telescope. A low-resolution spectrum ($\lambda/d\lambda \sim 100$) in the 2.0 – 2.4 μm atmospheric window shows a very strong Hα line at the optically determined redshift $z = 2.286$. The observed rest-frame equivalent width of Hα is $0.07 \pm 0.02\ \mu$m, consistent with the largest values found in quasars. The images show an unresolved source, while the near-infrared colors are somewhat redder than the mean colors of quasars observed at the same redshift. The reddening inferred is $A_V \sim 1.5$ mag, with an upper limit of ~ 3.0 mag. The observations presented here cannot distinguish between starburst and quasar models for this source. Starburst models require $\sim 10^{12} M_\odot$ of stars to have been formed over 10^7– 10^8 yr. If FSC 10214+4724 is a quasar, the reddening-corrected bolometric luminosity is approximately equal to the observed infrared luminosity.

158.246 **Near IR high spatial resolution observations of the nuclear region of the active galaxies: NGC 5506, NGC 7469, NGC 7582 and IC 4329A.**
I. Cruz–González, A. Chelli.
6. IAP Astrophysics Meeting – IAU Colloquium No. 129: Structure and emission properties of accretion disks, p. 415 – 416 (1991). – See Abstr. 012.046 for the main entry.

The nuclear region of the Seyfert galaxies NGC 7582 (Sy 2), NGC 5506 (Sy 2), NGC 7469 (Sy 1) and IC 4329A (Sy 1) is studied in the near–IR (*HKL*) by means of a slow slit scanning system. The spatial profiles and visibilities for these galaxies were obtained to derive the infrared properties of the detected components: an unresolved nuclear component of a size $\leqslant 0.1$ arcsec; and for three galaxies an extended component of a size ≈ 2 arcsec. The IR colors of both components are in excess to normal stellar bulge colors.

158.247 **Quasi–periodic continuum variability of NGC 6814.**
C. Done, K. Koyama, H. Kunieda, G. Madejski, R. Mushotzky, T. J. Turner.
6. IAP Astrophysics Meeting – IAU Colloquium No. 129: Structure and emission properties of accretion disks, p. 417 – 418 (1991). – See Abstr. 012.046 for the main entry.

NGC 6814 was observed over a period of 3 days with the GINGA satellite. Continual, rapid, large amplitude variability was seen, indicating that the nucleus is extremely compact. In these GINGA data a Fourier analysis of the X–ray variability is complicated by the uneven sampling of the light curve, but a power spectrum uncontaminated by the effects of interpolating between the measured data points can be calculated and is consistent with the periodic component observed previously.

158.248 **Daily observations of compact radio sources at 2.7 and 8.1 GHz: 1979 – 1987.**
E. B. Waltman, R. L. Fiedler, K. J. Johnston, J. H. Spencer, D. R. Florkowski, F. J. Josties, D. D. McCarthy, D. N. Matsakis.
Astrophys. J., Suppl. Ser., Vol. 77, No. 3, p. 379 – 404 (Nov 1991).

The authors present 2.7 and 8.1 GHz radio light curves for 46 compact radio sources observed with the Green Bank interferometer during the period 1979 – 1987. Flux densities for the period 1979 – 1985 were reported in a previous paper. The flux densities reported here extend the daily monitoring an additional 2 years. This marks the end of the dual use of the interferometer by the US Naval Observatory to measure Earth rotation parameters and by the Naval Research Laboratory to study compact source variability. The interferometer is currently dedicated to searching for extreme scattering events (ESEs). ESEs may be found in the radio light curves of 0300+470, 0333+321, 0954+658, 1502+106, 1611+343, 1821+107, and 2352+495 during the period 1979 – 1987, including three events observed in 0300+470 and 0333+321 during the additional 2 years reported here.

158.249 **On the magnetic energy density in the X–ray emission region of Active Galactic Nuclei.**
A. Celotti, G. Ghisellini, A. C. Fabian.
Mon. Not. R. Astron. Soc., Vol. 251, No. 3, p. 529 – 540 (1 Aug 1991).

The authors estimate upper limits on the magnetic energy density in the compact X–ray emission region of the Seyfert galaxies NGC 4051 and NGC 6814 and the quasars 3C 273, 3C 279 and H1821+643 using synchrotron self–Compton theory. The size of the emission region is obtained from the X–ray variability time–scale, and the electron density from the assumption that the X–ray emission in AGN is due to nonthermal Comptonization of soft "blue bump" photons. The competition between synchrotron self–Compton emission and direct Comptonization then enables the authors to estimate the relative importance of the magnetic and radiation fields. Equipartition between magnetic and X–ray energy density is at best only marginally possible and in several cases is strongly ruled out. This poses serious problems for many theories for the acceleration and radiation of particles in the X–ray emission region. The authors discuss several alternative pictures that do allow equipartition to occur. The most promising model, suggested by the work of Rees, paradoxically requires the magnetic field to be strong. It then confines many small, dense, clouds in the region which cause the optical and UV photosphere to lie beyond the X–ray emission region. The primary radiation is then synchrotron self–Compton emission.

158.250 **High–velocity gas in powerful radio galaxies.**
C. N. Tadhunter.
Mon. Not. R. Astron. Soc., Vol. 251, No. 3, p. 46P – 50P (1 Aug 1991).

Components of high–velocity gas have been discovered in the extended emission line regions around two powerful radio galaxies. The velocity shifts – up to –1800 km s^{-1} relative to systemic in the case of 3C 405 (Cygnus A) and up to +1600 km s^{-1} in the case of 3C 265 – are among the largest yet recorded in the extended regions around radio galaxies. In both objects the shifted components appear to have relatively high ionization states – similar to, or higher than, the unshifted components of extended ionized gas. The nature of the acceleration mechanism is currently unclear, but it appears most likely that the large velocity shifts are extreme manifestations of the activity – perhaps the result of interactions between the radio plasma and the warm or hot interstellar medium.

158.251 **Statistical properties of AGN emission line spectrum in relation with illuminated accretion disc model.**
M. Joly, S. Collin–Souffrin, A. M. Dumont.
6. IAP Astrophysics Meeting – IAU Colloquium No. 129: Structure and emission properties of accretion disks, p. 447 – 448 (1991). – See Abstr. 012.046 for the main entry.

158.252 **Soft X–ray variability and the accretion disk of NGC 5548.**
J. S. Kaastra.
6. IAP Astrophysics Meeting – IAU Colloquium No. 129: Structure and emission properties of accretion disks, p. 449 – 450 (1991). – See Abstr. 012.046 for the main entry.

158.253 **Massive thin accretion discs. III. Comparison with the observations.**
A. Laor.
6. IAP Astrophysics Meeting – IAU Colloquium No. 129: Structure and emission properties of accretion disks, p. 465 – 466 (1991). – See Abstr. 012.046 for the main entry.

158.254 **Spectral/luminosity evolution of active galactic nuclei.**
D. Leiter, E. Boldt.
6. IAP Astrophysics Meeting – IAU Colloquium No. 129: Structure and emission properties of accretion disks, p. 469 – 470 (1991). – See Abstr. 012.046 for the main entry.

158.255 **Arakelian 120.**
IAU Circ., No. 5353, p. 1 (25 Sep 1991).

158.256 **The starburst ring around the Seyfert nucleus in NGC 7469.**
A. S. Wilson, T. T. Helfer, C. A. Haniff, M. J. Ward.
Astrophys. J., Vol. 381, No. 1, p. 79 – 84 (1 Nov 1991). With plate 2.

The authors present high–resolution radio continuum, optical emission line, and optical continuum images of the luminous ($\sim 3\times10^{11} L_{\odot}$) Seyfert plus circumnuclear starburst hybrid galaxy NGC 7469. The radio images reveal a compact (0.″1 – 0.″2) source associated with the Seyfert nucleus, plus a broken ring of emission centered on it with diameter 3″ ($\cong$ 1.0 kpc), both these radio components being nonthermal in origin. The distribution of the emission line fluxes over the central 3 kpc reflects a mixture of Seyfert – and H II region – like excited gas. The present images and a recent speckle masking image by Hofmann, Mauder, and Weigelt reveal optical continuum features associated with the radio ring. The authors suggest that the starburst ring in

NGC 7469 has resulted from a gas ring formed through bar-forcing at a dynamical (probably inner Lindblad) resonance. Current starburst models can account for both the far infrared and the nonthermal radio emission if the supernova rate is ~1 per year. Models in which Seyfert nuclei are fueled through inflow of gas associated with bar- or oval distortion-driven resonances, on scales ranging from galaxy wide (tens of kpc) to circumnuclear (< 500 pc), are supported by these results.

158.257 **Deconvolution of broad emission line profiles in Seyfert 1 galaxies: disk kinematics and line emissivity.**
F. Mannucci, R. M. Stanga, G. M. Stirpe.
6. IAP Astrophysics Meeting – IAU Colloquium No. 129: Structure and emission properties of accretion disks, p. 477 – 478 (1991). – See Abstr. 012.046 for the main entry.

158.258 **Einstein Observatory SSS and MPC observations of the complex X-ray spectra of Seyfert galaxies.**
T. J. Turner, K. A. Weaver, R. F. Mushotzky, S. S. Holt, G. M. Madejski.
Astrophys. J., Vol. 381, No. 1, p. 85 – 100 (1 Nov 1991).
The authors have reinvestigated the X-ray spectra of 25 Seyfert galaxies measured with the Solid State Spectrometer on the Einstein Observatory. This new investigation utilizes simultaneous data from the Monitor Proportional Counter, and automatic correction for systematic effects in the Solid State Spectrometer, which were previously handled subjectively. The authors find that the best-fit single power-law indices generally agree with those previously reported, but that soft excesses of some form are inferred for ~48% of the sources. One possible explanation of the soft excess emission is a blend of soft X-ray lines, centered around 0.8 keV. The implications of these results for accretion disk models are discussed.

158.259 **IC 4329a: a (BLR) twin of Akn 120.**
P. Marziani, M. Calvani, J. W. Sulentic, W. Zheng.
6. IAP Astrophysics Meeting – IAU Colloquium No. 129: Structure and emission properties of accretion disks, p. 479 – 480 (1991). – See Abstr. 012.046 for the main entry.

158.260 **High ionization line zone above disk in AGN.**
L. S. Nazarova, N. G. Bochkarev.
6. IAP Astrophysics Meeting – IAU Colloquium No. 129: Structure and emission properties of accretion disks, p. 491 – 492 (1991). – See Abstr. 012.046 for the main entry.
Relative intensities of hydrogen and He II lines in the broad-line regions (BLR) of AGNs have been calculated by the Sobolev's method. Physical conditions for high ionization lines of BLR were determined.

158.261 **An unusually strong Einstein ring in the radio source PKS 1830-211.**
D. L. Jauncey, J. E. Reynolds, A. K. Tzioumis, T. W. B. Muxlow, R. A. Perley, D. W. Murphy, R. A. Preston, E. A. King, A. R. Patnaik, D. L. Jones, D. L. Meier, D. J. Bird, D. G. Blair, J. D. Bunton, R. W. Clay, M. E. Costa, R. A. Duncan, R. H. Ferris, R. G. Gough, P. A. Hamilton, D. W. Hoard, A. Kemball, M. J. Kesteven, E. T. Lobdell, A. N. Luiten, P. M. McCulloch, J. D. Murray, G. D. Nicolson, A. P. Rao, A. Savage, M. W. Sinclair, L. Skjerve, L. Taaffe, R. M. Wark, G. L. White.
Nature, Vol. 352, No. 6331, p. 132 – 134 (11 Jul 1991). Letter-to-the-editor.
The authors have obtained high-resolution radio images of PKS 1830-211 from several interferometric radiotelescope networks, which show an unusual elliptical ring-like structure connecting two brighter components. The presence of the ring, and the similarity of the two brighter spots, argue strongly that this is indeed a gravitationally lensed system, specifically an Einstein ring in which lens and lensed object are closely aligned. Although the source is close to the galactic plane, it seems that both the lens and background (lensed) object are extragalactic. This object is one hundred times brighter than either of the two previously discovered radio Einstein rings, and is among the six brightest flat-spectrum sources in the sky. Its brightness makes it a peculiar object: it must involve either a chance alignment of a lensing object with an unusually bright background source, or an alignment with a less bright object but amplified to an unusual degree.

158.262 **Accretion disk model for the broad line region of three active galactic nuclei.**
E. Rokaki.
6. IAP Astrophysics Meeting – IAU Colloquium No. 129: Structure and emission properties of accretion disks, p. 511 – 512 (1991). – See Abstr. 012.046 for the main entry.

158.263 **NGC 985: new evidence for an accretion disk in a Seyfert 1 galaxy?**
R. M. Stanga, J. M. Rodríguez Espinosa, F. Mannucci.
6. IAP Astrophysics Meeting – IAU Colloquium No. 129: Structure and emission properties of accretion disks, p. 527 – 528 (1991). – See Abstr. 012.046 for the main entry.

158.264 **Active galactic nuclei: A sideways look at quasars.**
M. Urry.
Nature, Vol. 353, No. 6339, p. 18 – 19 (5 Sep 1991).

158.265 **Accretion disk modeling of the multi-wavelength spectra of Seyfert 1 galaxy Markarian 335.**
Sun Weihsin, C. R. Shrader, T. J. Turner, P. M. Hintzen, M. A. Malkan, Lin Sungnan.
6. IAP Astrophysics Meeting – IAU Colloquium No. 129: Structure and emission properties of accretion disks, p. 537 – 538 (1991). – See Abstr. 012.046 for the main entry.

158.266 **An accretion disk model for NGC 6814?**
T. J. Turner, C. Done, H. Kunieda, R. Mushotzky, H. Awaki.
6. IAP Astrophysics Meeting – IAU Colloquium No. 129: Structure and emission properties of accretion disks, p. 545 – 546 (1991). – See Abstr. 012.046 for the main entry.

158.267 **The effect of anisotropic emission from thick accretion disks on the luminosity functions of AGN.**
C. M. Urry, P. Marziani, M. Calvani.
6. IAP Astrophysics Meeting – IAU Colloquium No. 129: Structure and emission properties of accretion disks, p. 547 – 548 (1991). – See Abstr. 012.046 for the main entry.

158.268 **Formation of bipolar radio jets and lobes from accretion disk around forming blackhole at the center of protogalaxies.**
Y. Uchida, R. Matsumoto, S. Hirose, K. Shibata.
IUPAP Conference on Primordial Nucleosynthesis and Evolution of Early Universe, p. 409 – 416 (1991). – See Abstr. 012.071 for the main entry.
The authors propose that radio jets and lobes from QSO's are "magnetic bipolar jets from forming blackholes", physically analogous to those of star-formation bipolar flows, but with very much greater energy due to very much greater depth in gravitational potential. They perform 2.5 D MHD simulations. Jets from nearby AGN's may be due to a rejuvenated blackhole accretion.

158.269 **Accretion disk emission from a BL Lacertae object.**
A. Wandel, C. M. Urry.
6. IAP Astrophysics Meeting – IAU Colloquium No. 129: Structure and emission properties of accretion disks, p. 553 – 554 (1991). Cf. Astrophys. J., Vol. 367, No. 1, p. 78 – 85 (20 Jan 1991) – see Abstr. 53.158.030. – See Abstr. 012.046 for the main entry.

158.270 **Wavelength dependent polarization in active galaxies.**
W. Webb, M. Malkan, G. Schmidt, C. Impey.
6. IAP Astrophysics Meeting – IAU Colloquium No. 129: Structure and emission properties of accretion disks, p. 555 – 556 (1991). – See Abstr. 012.046 for the main entry.

158.271 **Microvariability in blazars via accretion disk instabilities.**
P. J. Wiita, H. R. Miller, M. T. Carini, A. Rosen.
6. IAP Astrophysics Meeting - IAU Colloquium No. 129: Structure and emission properties of accretion disks, p. 557 - 558 (1991). - See Abstr. 012.046 for the main entry.

158.272 **Broad-band imaging of 8 non-interacting Seyfert galaxies: are they really completely isolated?**
E. Xanthopoulos, M. M. De Robertis.
6. IAP Astrophysics Meeting - IAU Colloquium No. 129: Structure and emission properties of accretion disks, p. 563 - 564 (1991). - See Abstr. 012.046 for the main entry.

158.273 **3C 390.3: a case for supermassive binary?**
Zheng Wei.
6. IAP Astrophysics Meeting - IAU Colloquium No. 129: Structure and emission properties of accretion disks, p. 565 - 566 (1991). - See Abstr. 012.046 for the main entry.

158.274 **Two radio supernovae in the unusual galaxy Markarian 297.**
Q. F. Yin, D. S. Heeschen.
Nature, Vol. 354, No. 6349, p. 130 - 132 (14 Nov 1991). Letter-to-the-editor.

The authors report the discovery of two radio supernovae in the clumpy irregular galaxy Markarian 297 (NGC 6052), which has been estimated to have a star formation rate ten times that of a normal spiral galaxy, and supernova rate of 0.3 - 0.5 per year. It may consist of two interacting galaxies. A variable radio source found in 1982 was attributed to possible supernova activity, and the observations with the Very Large Array, combined with earlier published data, define a reasonable radio light curve for a supernova that exploded in July 1979. The variability resembles that of the young supernova SN 1986J, but at maximum the supernova in Mkn 297 would have been about six times as bright. The authors also identify a second source in the galaxy as a supernova. The presence of two extraordinarily bright supernovae in the same region may be a particular result of the interaction between two galaxies.

158.275 **High-resolution optical observations compared to radio structure of the jet in M87.**
M. Stiavelli, P. Møller, W. W. Zeilinger.
Nature, Vol. 354, No. 6349, p. 132 - 134 (14 Nov 1991). Letter-to-the-editor.

The jet in the nearby elliptical galaxy NGC 4486 (M87) is a prototype of this kind of nuclear activity. High-resolution observations over a wide wavelength range have been used to investigate the propagation of energy along the jet and the radiation mechanism, but these studies have been hampered by the much lower resolution available at optical compared with radio wavelengths. The authors present high-resolution multi-colour images of the jet, obtained using image reconstruction techniques. Morphological features in the jet known from radio observations are seen in the optical images. In particular, they find that the radio-bright "knots" are spatially extended, and that there is significant emission in the regions between the knots. There is also evidence for limb-brightening at a number of positions along the jet. Taken together, their observations suggest that the jet propagates essentially free of losses, generating radio and optical emission at its edges, where interstellar material may be encountered.

158.276 **Optical imaging of BL Lac host galaxies.**
R. G. Abraham, I. M. McHardy, C. S. Crawford.
Mon. Not. R. Astron. Soc., Vol. 252, No. 4, p. 482 - 504 (15 Oct 1991).

The authors have imaged 23 BL Lacertae objects (BL Lacs) with the William Herschel Telescope in a search for underlying host galaxies. Fourteen of these objects are resolved in the survey. The authors have determined the morphology of three of these galaxies for the first time, and have confirmed the morphology of an additional three. All the observed underlying hosts are centred on their bright cores, and two of the newly classified hosts are disc systems. These results are in disagreement with the predictions of standard beaming and lensing models for these objects. The authors have calculated absolute magnitudes for observed hosts with known redshifts and find them to be consistent with the results obtained by previous authors, despite the unexpected morphology exhibited by some of the underlying hosts.

158.277 **The relative frequency of broad-lined and narrow-lined active galactic nuclei: implications for unified schemes.**
A. Lawrence.
Mon. Not. R. Astron. Soc., Vol. 252, No. 4, p. 586 - 592 (15 Oct 1991).

The author discusses evidence concerning the relative occurrence of narrow-lined and broad-lined active galactic nuclei (AGN), in optical, IR, X-ray, and radio-selected samples. Both narrow-lined AGN and reddened broad-lined AGN occur more frequently at lower source powers. There is marginal evidence that narrow-lined AGN were more common in the past. Narrow-lined objects have weaker [O III] at a given radio power. These data are inconsistent with the simplest "unified scheme" where a similar thick molecular torus surrounds all AGN, and with the simplest modification, that the torus geometrical thickness is a function of source power. A substantial range of geometrical thicknesses must exist at all powers. Both the typical geometrical thickness and the typical optical thickness of obscuring material probably vary with source power. This may arise naturally in accretion models, as smaller black holes are more easily smothered. The author speculates that the obscuring material, rather than being an orbiting torus of cool molecular matter, may be an expelled shell of gas, only the outer parts of which are dusty.

158.278 **Optical and IR emission of a sample of IRAS galaxies.**
S. Lípari, C. Bonatto, M. G. Pastoriza.
Mon. Not. R. Astron. Soc., Vol. 253, No. 1, p. 19 - 26 (1 Nov 1991).

The authors report on observations of 12 southern warm IRAS galaxies. Using absolute fluxes of lines in the optical region in connection with diagnostic diagrams, the authors have classified these galaxies in terms of the principal excitation mechanisms and have found that all the galaxies are Seyferts. Far-infrared colours also are used to analyse these objects. Good correlations between the luminosity at 25 and 60 μm and the Hα luminosity are found. The far-infrared luminosity of Seyfert galaxies seems to depend more strongly on the number of ionizing photons than on the amount of dust.

158.279 **The emission-line filaments in Centaurus A.**
G. V. Bicknell.
Proc. Astron. Soc. Aust., Vol. 9, No. 1, p. 93 - 94 (1991). - See Abstr. 012.078 for the main entry.

This paper examines recent investigations of emission-line filaments along the edge of the radio source in Centaurus A. It shows how these studies have ramifications for the excitation of the narrow line regions and extended emission-line regions of other active galaxies.

158.280 **Radio-luminosity dependence of the IR-radio alignment effect in high-z radio galaxies.**
J. S. Dunlop, J. A. Peacock.
NATO Advanced Research Workshop on Observational Tests of Inflation, p. 463 - 466 (1991). - See Abstr. 012.076 for the main entry.

The authors present initial results from a K-band imaging study of a subsample of radio galaxies from the $S_{2.7\,GHz} > 0.1$ Jy Parkes Selected Regions (PSR), along with a comparison sample of 3CR galaxies which are a factor ~ 10 more radio luminous. The K-band images of 3CR galaxies at $z \cong 1$ show that (1) the established optical-radio alignment effect is still sufficiently apparent at K to have been easily discovered in the near-IR, and (2) the IRCAM images are certainly deep enough to reveal comparable distortions in the PSR galaxies, should they exist. Preliminary results indicate that, at least at K, radio-induced contamination of the light of the underlying galaxy may only be a significant problem for the most extreme radio luminosities. This suggests that radio galaxies may still be relevant probes of galaxy evolution in general, provided one avoids the most extreme radio luminosities found in the 3CR galaxies at $z \sim 1$.

158.281 **Demise of a galactic supernova remnant candidate, G274.68–2.84.**
J. L. Caswell, R. T. Stewart.
Proc. Astron. Soc. Aust., Vol. 9, No. 1, p. 103 – 104 (1991). – See Abstr. 012.078 for the main entry.
Young supernova remnants are poorly understood and few are known. Further examples may be present, but unrecognised, within existing catalogues of radio sources. G274.68–2.84 was recently suggested as one such candidate but the authors' the new radio map from the Australia Telescope indicates that it is not a supernova remnant. Rather, it appears to be extragalactic with a strong, variable, flat–spectrum nucleus and a bright X–ray counterpart.

158.282 **Spectroscopy of radio sources from the Parkes Selected Regions.**
J. R. Allington–Smith, J. A. Peacock, J. S. Dunlop.
Mon. Not. R. Astron. Soc., Vol. 253, No. 2, p. 287 – 294 (15 Nov 1991).
The authors have measured redshifts ($0.2 < z < 3$) for 14 radio source identifications from the Parkes Selected Regions sample and from other complete radio samples defined at 2.7 GHz. One further identification has been found to be a featureless BL Lac object for which no redshift can be determined. Using these data, the authors discuss the relationship between the strength of [O II] λ3727 emission and radio luminosity for narrow–line radio galaxies.

158.283 **The large–scale clustering of radio galaxies.**
J. A. Peacock, D. Nicholson.
Mon. Not. R. Astron. Soc., Vol. 253, No. 2, p. 307 – 319 (15 Nov 1991).
The authors use an all–sky sample of radio galaxies at redshifts $z \lesssim 0.1$ to study clustering in the Universe on scales up to several hundred Mpc. The two–point correlation function for these galaxies is consistent with their high optical luminosity and location in moderately rich environments: $\xi(r) \cong (r/11\ h^{-1}Mpc)^{-1.8}$, where $h = H_0/100\ km\ s^{-1}Mpc^{-1}$. The authors discuss direct methods for obtaining the power spectrum of the density field traced by the radio galaxies, taking into account the selection function of the sample. The results of the power–spectrum analysis indicate that the distribution of radio galaxies is more uniform on very large scales than would be predicted from an extrapolation of the power–law clustering found on small scales. There is a break to an effective spectral index $n \gtrsim -0.5$ for wavelengths $\lambda \gtrsim 200\ h^{-1}Mpc$. The variance in $\delta N/N$, σ^2, is about 0.3 for wavelengths $80 < \lambda < 200\ h^{-1}Mpc$, and there is no concentration towards the supergalactic plane for $z \gtrsim 0.02$. The authors thus do not confirm suggestions made by Broadhurst et al. and by Tully of strong clustering on scales of $\gtrsim$100 Mpc.

158.284 **The cosmological evolution and luminosity function of X–ray selected active galactic nuclei.**
R. Della Ceca, T. Maccacaro.
Workshop on Space Distribution of Quasars as Part of the 75. Anniversary Celebrations of the National Research Council of Canada (NRCC), p. 150 – 159 (1991). – See Abstr. 012.084 for the main entry.
Using a sample of more than 420 X–ray selected Active Galactic Nuclei (AGNs) the authors have derived the local ($z \leqslant 0.1$) observed X–ray Luminosity Function (XLF) and the XLF in different redshift shells up to $z \sim 2$. The AGNs XLF at different redshifts gives direct evidence of cosmological evolution. The data have been analyzed within the framework of luminosity evolution models and the two most common evolutionary forms have been considered. Luminosity–dependent evolution is required if the evolution function has the exponential form, whereas the simpler pure luminosity evolution model is still acceptable if the evolution function has the power–law form. Using the whole sample of objects a de–evolved ($z - 0$) XLF has been derived. A good fit with a broken power law has been obtained.

158.285 **Evolution of the X–ray luminosity function of AGN.**
H. L. Marshall.
Workshop on Space Distribution of Quasars as Part of the 75. Anniversary Celebrations of the National Research Council of Canada (NRCC), p. 184 – 190 (1991). – See Abstr. 012.084 for the main entry.
The HEAO A1 X–ray survey has detected a large number of active galactic nuclei and quasars, as recently identified using results from the HEAO A3 experiment. The sample shows some incompleteness and is not completely identified, so these effects are taken out in the analysis by correcting the effective coverage of the survey. Preliminary analysis indicates that significant evolution can be detected in these data and that the evolution rate appears stronger than that inferred from a sample selected at lower energies.

158.286 **The cosmological evolution of BL Lacertae objects.**
J. T. Stocke, E. S. Perlman, R. Wurtz, S. L. Morris.
Workshop on Space Distribution of Quasars as Part of the 75. Anniversary Celebrations of the National Research Council of Canada (NRCC), p. 218 – 229 (1991). – See Abstr. 012.084 for the main entry.
Using a complete, X–ray selected sample of BL Lacertae Objects (XBLs), the yuthorsn find that $<V_e/V_a> = 0.34 \pm 0.06$ implying that the cosmological evolution of this type of AGN is "negative"; i.e. fewer or less luminous BL Lacs in the past compared to the current epoch. In the process of deriving this result, the selection criteria and observational properties of XBLs are described.

158.287 **The large–scale clustering of radio galaxies.**
J. Peacock.
Workshop on Space Distribution of Quasars as Part of the 75. Anniversary Celebrations of the National Research Council of Canada (NRCC), p. 230 – 247 (1991). Also published in Mon. Not. R. Astron. Soc., Vol. 253, No. 2, p. 307 – 319 (15 Nov 1991). – See Abstr. 012.084 for the main entry.
The spatial distribution of radio galaxies within the volume $z < 0.1$ is investigated using a new all–sky redshift survey. At small separations, the correlation function is well fitted by $\xi = (r/r_0)^{-1.8}$, with $r_0 = 11 \pm 1\ h^{-1}Mpc$. On larger scales, direct power–spectrum analysis is the preferred technique for quantifying clustering. The power spectrum has a break at $\lambda \cong 200\ h^{-1}Mpc$; there is less large–scale clustering than would be implied by the power–law model for ξ. This power spectrum has the same shape as that inferred from optical and IRAS samples. For adiabatic CDM, it argues for a larger amplitude of fluctuations than is conventional, but the model then predicts excessive small–scale power. However, a CDM model seeded with textures agrees well with observation.

158.288 **3C 390.3: modeling variable profile humps.**
W. Zheng, S. Veilleux, S. A. Grandi.
Astrophys. J., Vol. 381, No. 2, p. 418 – 425 (10 Nov 1991).
A comparison of the Hα and Hβ profiles of the radio galaxy 3C 390.3 at different epochs shows that the Hβ/Hα ratio is significantly higher in the line "humps" than in the rest of the profiles. No humps are found in the Lyα profile, although Lyα is as broad as the Balmer lines. Therefore, the profile humps are probably produced in a region of higher density. A double–stream model fitted to the Hα profiles requires that the conical structure is close to the line of sight and subtends a large angle ($\geqslant 60°$) and a limited radial distance ($R_{max}/R_{min} \leqslant 10$). The pair of humps in the profile may arise from gas lumps or a narrow stream inside the cone. Fluctuations in the blue–to–red hump flux ratio may be due to variations in the relative intensity of the ionizing continuum in the two cones or to changes in the orientation of these cones. Another model is a disk with a

corotating enhanced emission region ("hot patch"). With a significant angular size ($\sim 90°$), such a region would contribute a time–dependent emission component which migrates between the two humps and changes their relative intensity.

158.289 **A near–infrared spectroscopic study of stellar populations in the starburst core of M82.**
D. Lester, J. Carr, M. Joy, N. Gaffney.
Symposium in Celebration of the 50. Anniversary of McDonald Observatory of the University of Texas at Austin: Frontiers of stellar evolution, p. 605 (1991). Abstract. – See Abstr. 012.085 for the main entry.

158.290 **The clustering of IRAS Seyfert galaxies.**
I. Georgantopoulos, T. Shanks.
Workshop on Space Distribution of Quasars as Part of the 75. Anniversary Celebrations of the National Research Council of Canada (NRCC), p. 300 – 303 (1991). – See Abstr. 012.084 for the main entry.

The authors investigate the clustering properties of 192 IRAS Seyfert galaxies distributed over almost the whole sky. They detect clear clustering (5σ) at scales $<20\,h^{-1}$Mpc. Comparing the Seyfert with the QSO clustering results at higher redshifts, the authors find that a comoving model for AGN clustering evolution is more consistent with the data, although there is some uncertainty due to the low number of pairs at small separations.

158.291 **Quasars in rich galaxy cluster environments.**
E. Ellingson, H. K. C. Yee.
Workshop on Space Distribution of Quasars as Part of the 75. Anniversary Celebrations of the National Research Council of Canada (NRCC), p. 334 – 343 (1991). – See Abstr. 012.084 for the main entry.

Extensive imaging and spectroscopic observations of the galaxy cluster environments of quasars and other AGN indicate that environment plays an important role in the evolution of AGN activity. Imaging studies provide evidence that radio–loud and radio–quiet quasars prefer significantly different environments and that radio quasars in rich environments evolve approximately six times faster than their counterparts in poor environments. This rapid evolution is consistent with radioloud quasars and radio galaxies in rich clusters forming a continuous evolutionary sequence since $z \sim 0.7$. Spectroscopic observations of galaxies in these environments indicate that this evolution is accompanied by a significant rise in the cluster velocity dispersion, in qualitative agreement with scenarios which link galaxy–galaxy interaction with the triggering of AGN activity.

158.292 **A search for primeval galaxies.**
D. J. Thompson, S. Djorgovski, J. Trauger.
Workshop on Space Distribution of Quasars as Part of the 75. Anniversary Celebrations of the National Research Council of Canada (NRCC), p. 354 – 357 (1991). – See Abstr. 012.084 for the main entry.

The authors are conducting a search for emission–line objects at large redshifts, ostensibly young and forming galaxies. The method is to search for strong line emission, in particular Ly α, using two techniques: a direct, narrow–band imaging search using a Fabry–Perot interferometer; and a serendipitous long–slit spectroscopic search. Assuming that none of the faint candidates found so far are actually primeval galaxies, one can set the strongest limits to date on the surface density of Ly α–luminous primeval galaxies on the sky. Some of the faint emission line galaxies detected in the survey at $z \sim 0.5 - 1$ may be representative of the faint blue galaxy population detected in deep continuum imaging surveys.

158.293 **Radio galaxies and galaxy formation.**
J. Peacock.
Workshop on Space Distribution of Quasars as Part of the 75. Anniversary Celebrations of the National Research Council of Canada (NRCC), p. 379 – 388 (1991). – See Abstr. 012.084 for the main entry.

Radio galaxies at large redshift constrain galaxy formation models in two ways: (1) through their comoving density; (2) through the ages of their stellar population. The observed density of radio galaxies at $z \cong 4$ sets important limits on power–spectrum normalization in hierarchical models. CDM models probably require a low degree of bias to survive. The "alignment effect" has been used to argue that observed high–z radio galaxies are young objects in which the majority of stars are generated by the jets from the AGN. However, new data show that alignments are confined to sources of the most extreme radio power. It therefore seems likely that the stellar populations in high–z radio galaxies are old – as expected in hierarchical models for galaxy formation.

158.294 **IRAS 02366–3101: an accretion disk candidate among luminous IRAS galaxies.**
L. Colina, S. Lípari, F. Macchetto.
Astrophys. J., Lett., Vol. 382, No. 2, p. L63 – L67 (1 Dec 1991). With plate L3.

The authors report the detection of double–peaked broad Balmer Hα and Hβ emission line profiles in a luminous IRAS galaxy: IRAS 02366–3101 with $\log L_{FIR} = 10.94\,L_\odot$. These double–peaked broad emission–line profiles consist of three components: (1) a narrow component with FWHM = 571 km s^{-1}, (2) a blueward broad component with FWHM = 2794 km s^{-1} and a velocity shift of the peak of –2120 km s^{-1}, (3) a redward broad component with FWHM = 2614 km s^{-1} and velocity shift of the peak of +2911 km s^{-1}. These profiles are consistent with the emission line profiles predicted by models such as those of Chen and Halpern, where the low–ionization line emission comes from a relativistic accretion disk around a supermassive black hole. This detection shows that IRAS 02366–3101 is a candidate to host an accretion disk around a supermassive black hole in its nucleus, although other alternatives cannot be ruled out.

158.295 **The correlated radio and optical asymmetries of powerful radio galaxies.**
R. Liu, G. Pooley.
Mon. Not. R. Astron. Soc., Vol. 253, No. 4, p. 669 – 674 (15 Dec 1991).

The authors present radio continuum and polarization maps of four "classical double" radio galaxies, overlaid on optical emission–line images. The sources (3C 247, 263.1, 280 and 299) are all small, powerful and distant (with $D < 200$ kpc, $P_{178} > 10^{26.8}$W Hz^{-1}sr^{-1}, $0.37 < z < 1.0$). In three cases the authors find that the emission–line gas lies on that side of the radio source for which the depolarization is stronger, the hotspot closer to the core and the radio spectrum steeper. The authors discuss the origin of these features, and suggest that a non–symmetrical distribution of gas around the host galaxy could account for the observed asymmetries.

158.296 **The extraordinary radio source 3C 442.**
N. F. Comins, F. N. Owen.
Astrophys. J., Vol. 382, No. 1, p. 108 – 114 (20 Nov 1991). With plate 15.

The authors have studied the system 3C 442 with the VLA at 6, 20, and 90 cm. This source is associated with two galaxies, NGC 7236/7237. The image of the radio source shows a mass of filaments and diffuse emission surrounding the two galaxies. A weak resolved radio core lies at the position of NGC 7237, and two arcs of steep–spectrum radio emission cross the region containing the optical galaxies, but no opaque radio core or head–tail jet structures are found, as is usually the case for radio galaxies of similar radio luminosity. The spectral index over the entire system is roughly constant. The authors propose that the unusual morphology of 3C 442 is due to a galactic merger in progress. The high ellipticity of the cluster of galaxies in which NGC 7236/7237 are imbedded suggests that they may be the central galaxies in a pair of merging clusters. It is suggested that the collimated gas was produced by a previously existing radio galaxy whose radio source was disrupted by the gravitational influence of the merger on the gas surrounding the source. The large–scale structure of 3C 442 is the result of these electrons interacting with the hot gas of the disturbed intracluster medium.

158.297 **Infrared spectroscopy of Cygnus A: implications for the obscured active nucleus.**
M. J. Ward, P. R. Blanco, A. S. Wilson, M. Nishida.
Astrophys. J., Vol. 382, No. 1, p. 115 – 120 (20 Nov 1991).
The authors present near–infrared spectroscopic observations of the central regions of the luminous radio galaxy Cygnus A (3C 405) and interpret these results in terms of an obscured quasar nucleus. The authors have detected strong emission in the molecular hydrogen lines $v = 1$–0 S(1) and $v = 1$–0 S(3), the strengths of which are plausibly accounted for through heating by the nuclear hard X–ray source. The large equivalent widths of these molecular hydrogen lines and the near–infrared narrow hydrogen recombination lines suggest that the observed nuclear continuum is strongly attenuated at 2 μm. The observed upper limit to the flux of broad Paα implies an extinction to the putative broad line region A_V(BL) of at least 24 mag, and the observed continuum intensity of the nuclear point source at 2.2 μm gives an extinction $A_V(NC) = 54 \pm 9$ mag toward the optical–infrared continuum. These estimates are consistent with the gas column density inferred from the low–energy X–ray cutoff. The authors also report strong [Si VI] 1.962 μm line emission from Cygnus A, the first such detection in a radio galaxy.

158.298 **Fe II emission and anisotropy in broad–line regions.**
W. Zheng, W. C. Keel.
Astrophys. J., Vol. 382, No. 1, p. 121 – 124 (20 Nov 1991).
The authors examine the relative strength of Fe II λ4570 in 138 active galactic nuclei in relation to the line width of Hβ. For objects with FWHM (Hβ) greater than 6000 km s^{-1}, the mean Fe II λ4570/Hβ intensity ratio is 0.21, less than half of that in the other objects. The result confirms a general tendency, seen in earlier work on smaller samples, that strong optical Fe II emission is not found in objects showing very broad emission lines. In particular, all five objects with exceptionally broad and flat profiles, such as Arp 102B, have weak or absent Fe II emission. The authors have verified that this effect is not an artifact of blending of the Fe II multiplets or of underestimation of the Hβ line width due to blending with Fe II features. The anticorrelation between the Fe II λ4570 strength and line width may be interpreted as an effect of source orientation. The Fe II region is either highly anisotropic or is not coextensive with the conventional broad–line region such as the assumed disk–shaped low–ionization zone. This is consistent with the Fe II region being associated with a small–scale jet and embedded in a thick torus, and hence being visible only in nearly face–on positions.

158.299 **IR properties of high z radio galaxies and the alignment effect.**
A. Chokshi, P. R. M. Eisenhardt.
Comments Astrophys., Vol. 15, No. 6, p. 343 – 354 (Oct 1991).
The authors summarise the status of high redshift radio galaxies (HZRGs) in which the principal axes of optical and infrared emission are often aligned with the radio axis. The red and infrared properties of HZRGs are emphasized since the broad and spectral energy distribution (SED) of morphologically and chronologically diverse objects appear to be similar at these wavelengths. This suggests a common origin for the red/infrared emission. A survey of the theoretical models poses several intriguing questions regarding the nature and the age of these objects. Whereas any star formation triggered by radio jets might suggest small ages for HZRGs, the similarity in the shape of infrared spectra from object to object suggests that a mature stellar population dominates. The authors propose observational and theoretical tests that may resolve some of these issues.

158.300 **Iron lines in AGN and black hole candidates.**
A. C. Fabian, I. M. George.
Workshop on Iron Line Diagnostics in X–Ray Sources, p. 169 – 176 (1991). – See Abstr. 012.088 for the main entry.
To put the discovery of the fluorescent iron lines in some perspective, the authors first give a brief, historical review of the observations and interpretation of the X–ray spectra of AGN and black hole candidates. They then describe how recent observations have added to the knowledge of the innermost regions of AGN.

158.301 **Iron line emission and the distribution of cold matter in Seyfert galaxies.**
K. Nandra, K. A. Pounds, G. C. Stewart.
Workshop on Iron Line Diagnostics in X–Ray Sources, p. 177 – 186 (1991). – See Abstr. 012.088 for the main entry.
The authors present the results of an analysis of 29 Ginga Seyfert spectra. In particular they address the question of how common the features reported for individual objects are, in the context of a large sample.

158.302 **MERLIN observations of superluminal motion in the jet of 3C120.**
T. W. B. Muxlow, P. N. Wilkinson.
IAU Colloquium No. 131: Radio interferometry – theory, techniques, and applications, p. 312 – 316 (1991). Also published in Mon. Not. R. Astron. Soc., Vol. 251, No. 1, p. 54 – 62 (1 Jul 1991). – See Abstr. 012.089 for the main entry.
The authors present evidence for changes in the radio structure of the jet in 3C120 between December 1980 and December 1988 derived from MERLIN observations at 1.67 GHz. This confirms the suggestion by Benson et al. (1988) that the jet does not slow down quickly just outside the core region usually imaged by VLBI.

158.303 **Spectral features in AGN: problems and questions.**
L. Piro.
Workshop on Iron Line Diagnostics in X–Ray Sources, p. 187 – 195 (1991). – See Abstr. 012.088 for the main entry.
GINGA observations of several bright Seyfert 1 galaxies have revealed the existence of a new hard X–ray component above 8 keV, the "high energy bump". This component is probably produced by reprocessing of the direct continuum (through absorption and Compton scattering) on to very thick ($>10^{24} cm^{-2}$) cold matter surrounding the central source, possibly in an accretion disk. Detection of iron fluorescent lines with equivalent width of 100 – 200 eV and energy of 6.4 keV in all these objects supports this interpretation. Some relevant predictions of a detailed accretion disk model (iron line parameters, high energy bump level) are compared with GINGA results, including the detection of iron lines with equivalent width $\cong 300 - 400$ eV observed in two other Seyfert galaxies. A partial covering and reprocessing by thick clouds offers also a good explanation to the phenomenology observed by GINGA, and is supported by the latest results on the Broad Line Region obtained by line reverberation measurements in the optical–UV bands.

158.304 **The ionization structure of iron in AGN.**
I. M. George, A. C. Fabian, R. R. Ross.
Workshop on Iron Line Diagnostics in X–Ray Sources, p. 196 – 200 (1991). – See Abstr. 012.088 for the main entry.

158.305 **Iron K fluorescence and high energy bump from accretion disks in AGNs.**
G. Matt, G. C. Perola, L. Piro.
Workshop on Iron Line Diagnostics in X–Ray Sources, p. 201 – 204 (1991). – See Abstr. 012.088 for the main entry.
Recent Ginga observations have revealed the existence of both the iron K fluorescence line and a high energy bump in the X–ray spectra of some bright Seyfert 1 galaxies, which are currently attributed to the reprocessing of the primary X–ray photons by cold matter located in the neighbourhood of the putative black hole. Assuming that the matter is distributed in an optically thick accretion disk, the authors have calculated, by means of Monte Carlo simulations, the intensities of the line and of the bump as a function of the disk inclination for different geometries of the disk surface. For the simplest case of a flat disk they have calculated the line profile taking into account also the kinematical and relativistic effects. The results obtained apply to matter which is neutral or mildly ionized ($T \lesssim 10^5$K) and with a cosmic chemical abundance (Morrison and McCammon, 1983). A full

account of this work is given in Matt et al. (1990); here some relevant results are briefly outlined.

158.306 **X-ray spectra from clouds in active galactic nuclei.**
J. Felsteiner, R. Opher.
Workshop on Iron Line Diagnostics in X-Ray Sources, p. 209 - 213 (1991). - See Abstr. 012.088 for the main entry.
The authors present a cloud model which can explain certain basic characteristics of recently observed X-ray spectra from active galactic nuclei. The model includes cold opaque clouds and hot transparent clouds. Both nonisotropic power law emission from the central engine and nonsolar abundances of iron are considered.

158.307 **Discovery of strong absorption in the X-ray spectrum of Seyfert 2 galaxy NGC 4507.**
H. Awaki, H. Kunieda, Y. Tawara, K. Koyama.
Publ. Astron. Soc. Jpn., Vol. 43, No. 4, p. L37 - L42 (1991).
The authors discovered an obscured nucleus in the Seyfert 2 galaxy NGC 4507 with the Ginga satellite. The 2–10 keV X-ray luminosity is 3.2×10^{43} erg s^{-1}, and the continuum radiation is well described by a power law model with a photon index of 1.34(+0.28/-0.25). The column density of the obscuring matter is $(4.9 \pm 0.7) \times 10^{23}$ cm^{-2}, which is sufficiently thick to hide the broad-line region and the nucleus. An iron line with equivalent width of 420(+100/-110) eV has also been detected in the X-ray spectrum. In order to explain the large equivalent width, a large covering factor for the matter around the nucleus is needed. The obscuration can be naturally explained in terms of a torus in the unified Seyfert model.

158.308 **The origin of the X-ray spectra of AGN.**
G. Ghisellini, A. A. Zdziarski, I. M. George, R. Svensson, A. C. Fabian, C. Done.
Workshop on Iron Line Diagnostics in X-Ray Sources, p. 214 - 217 (1991). - See Abstr. 012.088 for the main entry.
The presence of a spectral component due to Compton reflection in the X-ray spectra of AGN, as recently observed by Ginga, give new clues to the properties of the central regions of these sources. These observations revive models in which much of the luminosity generated in X- and γ-ray energy bands in AGN is reprocessed via electron-positron pairs. In these models the X-ray spectral index is $\alpha_x \cong 1$ for a wide range of parameters. If a substantial fraction of this radiation is incident upon cold material, the composite (directly observed + reflected) spectrum has an average spectral index of $\alpha_x \cong 0.7$ in the 1–30 keV band, and steepens above $\sim$1 MeV. Such spectra agree well with the X-ray and γ-ray observations of AGN and the constraints imposed by the cosmic γ-ray background.

158.309 **Two phase accretion disks as a model for the X-ray emission from Seyfert galaxies.**
F. Haardt, L. Maraschi.
Workshop on Iron Line Diagnostics in X-Ray Sources, p. 218 - 225 (1991). - See Abstr. 012.088 for the main entry.
A thermal two-phase accretion disk model is considered. A substantial fraction f of the gravitational power is assumed to be dissipated via buoyancy and reconnection of magnetic fields in a hot tenuous "corona" surrounding the main body of the disk. The main cooling mechanism of the hot layer is Comptonization of soft photons thermally produced in the underlying cold phase. About 50% of the hard Comptonized photons are backscattered into the thick phase where they are in part reprocessed into the soft blackbody power and in part reflected. The fraction of backscattered, absorbed and reflected photons is estimated by Monte Carlo simulations. Coupled thermal balance equations for the two phases yield the temperature of the hot phase and the slope of the Comptonized component as a function of f and τ, the optical depth of the hot phase. The authors find that for $f \geqslant 2/3$, and $\tau < 1$, the temperature of the hot phase adjusts so as to maintain $\alpha \cong 1$. For small τ the temperature of the hot phase is sufficient for pair production to be important. Monte Carlo simulations with parameters satisfying the balance equations, show that the model can account in a self-consistent way for the components present in the X-ray spectra of Seyfert galaxies, for their average properties and for the observed short term spectral variability.

158.310 **Compton broadening of the iron line in NGC 3227.**
B. Czerny, M. Zbyszewska, D. J. Raine.
Workshop on Iron Line Diagnostics in X-Ray Sources, p. 226 - 229 (1991). - See Abstr. 012.088 for the main entry.
Here the authors propose that the cool gas producing the iron line is embeded in a hot corona. The existence of such a corona seems to be required since it can effectively smear out the Lyman edge expected in the intrinsic disk spectrum. The Comptonization by a hot corona results also in a broadening of the iron line even if the disk is observed from the top and Doppler shifts are almost negligible.

158.311 **Rapid variability of iron line from NGC 6814.**
H. Kunieda, H. Awaki, K. Koyama, Y. Tawara, Y. Tsusaka, C. Done, R. Mushotzky, T. J. Turner, K. Leighly, S. Tsuruta.
Workshop on Iron Line Diagnostics in X-Ray Sources, p. 241 - 246 (1991). - See Abstr. 012.088 for the main entry.
GINGA observational results of the fastest variable Seyfert I galaxy NGC 6814 are presented. The shortest variable time scale of 50 s was found. It suggests the central mass of $10^6 M_\odot$. Strong iron line at 6.4 keV is observed with the equivalent width of up to 500 eV. The iron line flux changed together with the total X-ray flux. The delay time seems to be less than 256 s. If the line were attributed to cold iron, this result might imply the cold accretion matter comes closer than 10^{13}cm to the central source. Quasi-periodical time behavior of $\sim$12000 s is suggested with dips and sharp drops. This time scale corresponds to the Keplerian orbit at $R \sim 10^{13}$cm around the central object of $10^6 M_\odot$.

158.312 **The ultraviolet spectrum of NGC 4151 from 1978 to 1990: general characteristics and evolution.**
M.-H. Ulrich, A. Boksenberg, G. E. Bromage, J. Clavel, A. Elvius, M. V. Penston, G. C. Perola, M. A. J. Snijders.
Astrophys. J., Vol. 382, No. 2, p. 483 - 500 (1 Dec 1991).
From the examination of all the low-dispersion IUE spectra of NGC 4151 taken since the launch of IUE in 1978, including the latest data taken in 1990, the authors follow the evolution of this active nucleus on time scales ranging from days to 10 years. Here the analysis is concentrated on the campaigns of observations during which the nucleus was not continuously in a minimum state. Regarding the continuum, the authors find that the short variability time scales (a few days) in the optical and ultraviolet ranges rule out the possibility that the continuum intensity variations in these wavelength intervals are due to variations in the accretion rate of the possibly present accretion disk. Possible origins could be instabilities in the inner radiation - dominated part of the disk or irradiation of the disk by the variable X-ray flux. The authors study in detail the long- and short-term temporal behavior of the wings of the C IV and Mg II lines and the variations of the C III] line. The C IV line maintains a high degree of symmetry during the decrease and increase branches of the deep minimum of 1988 November 29 - 1989 January 30, but, in general, there are large changes in the C IV wings intensity ratio from campaign to campaign, and different responses of the blue and red wings to the continuum variations are frequently observed during a campaign. These effects could be due to the redistribution or rearrangement of the fast-moving gas on a time scale of one to a few years, probably in the form of the arrival of a new gas cloud in the inner part of the broad-line region.

158.313 **Fanaroff-Riley I galaxies as the parent population of the BL Lacertae objects. III. Radio constraints.**
C. M. Urry, P. Padovani, M. Stickel.
Astrophys. J., Vol. 382, No. 2, p. 501 - 507 (1 Dec 1991).
There is increasing support for the idea that BL Lacertae objects are dominated by beamed emission from a relativistic jet aligned with the line of sight. If so, the most likely parent (i.e., misaligned) objects are Fanaroff-Riley type I (F-R I) radio galaxies. The authors test this hypothesis by determining the ratio

of number densities of the two populations in a manner appropriate for flux-limited samples. Through a direct comparison of predicted and observed luminosity functions, it is shown that available radio samples of F-R I galaxies and BL Lac objects are fully consistent with the beaming hypothesis. This model makes specific predictions about the luminosity function of BL Lacertae objects and their radio counts below currently observed levels. A distribution of Lorentz factors is required, covering roughly the range $5 \lesssim \gamma \lesssim 35$, but skewed toward the low end. These Lorentz factors are higher than the estimate produced via a similar calculation of X-ray samples, although the different quality of the observational data in the two bands makes the significance of this increase hard to assess. Still, a relation between Lorentz factor and wavelength could be explained in conventional inhomogeneous jet models if the Lorentz factor increased along the jet.

158.314 **Multifrequency observations of BL Lacertae in 1988.**
N. Kawai, M. Matsuoka, J. N. Bregman, H. D. Aller, M. F. Aller, P. A. Hughes, S. A. Balbus, T. J. Balonek, K. C. Chambers, R. E. S. Clegg, S. D. Clements, R. J. Leacock, A. G. Smith, J. S. Miller, M. Hereld, M. G. Hoare, V. A. Hughes, G. K. Miley, G. H. Moriarty-Schieven, K. Matthews, G. Neugebauer, T. Ohashi, P. F. Roche, H. A. Thronson, E. Valtaoja, H. Terasranta, J. R. Webb, R. Goodrich, B. J. Wills, D. Wills.
Astrophys. J., Vol. 382, No. 2, p. 508 - 514 (1 Dec 1991).
Simultaneous multiwavelength observations of BL Lacertae were performed on two occasions separated by 1 month in 1988 June and July, covering the radio, submillimeter, infrared, optical, ultraviolet, and X-ray wave bands. In the wide-band photon spectra, the X-ray flux lies clearly above the extension of radio-ultraviolet continuum. The slope of the X-ray spectra is significantly flatter than that at optical - ultraviolet regimes, and its spectral index 0.7 - 1.0 corresponds to the slope in the submillimeter band. Comparison with earlier observations, in fact, indicates that the X-ray flux is correlated with the submillimeter band, and not with the others, and supports the synchrotron self-Compton (SSC) model.

158.315 **The structure and engines of AGNs.**
Lu Jufu.
Prog. Astron., Vol. 9, No. 4, p. 271 - 278 (Dec 1991). In Chinese.
First, basic observational facts of AGNs are listed. Then the author discusses the structure of the these objects, with emphasis on the physical properties of the broad-line region and the narrow-line region and the process of photoionization. He reviews the standard model of AGN engines i.e. the black hole accretion disk model, including the most recent progress on the origin and fueling of the galactic activity.

158.316 **The observations and studies of OH megamaser extragalaxies.**
Yu Zhiyao.
Prog. Astron., Vol. 9, No. 4, p. 289 - 297 (Dec 1991). In Chinese.
The author has reviewed the search, observations, and theoretical studies of extragalactic OH megamasers. So far 31 OH megamaser galaxies have been found. The OH megamaser galaxy IC 4553 was the first one discovered and a unique one observed with VLA-A, MERLIN, and VLBI. According to the present theoretical studies the mechanism for forming extragalactic OH megamasers is that the pumping of infrared radiation inverts OH maser molecules and radio continuous radiation from the core of the galaxies is amplified.

158.317 **Iron line variability and geometry in Seyfert galaxies.**
R. Walter.
Workshop on Iron Line Diagnostics in X-Ray Sources, p. 247 - 250 (1991). - See Abstr. 012.088 for the main entry.
Significant iron features were recently found in several bright active galactic nuclei (AGN) with Tenma, Exosat and Ginga. When detected, the 6.4 keV fluorescent line equivalent width (EW) is equal to a few 0.1 keV. Such large equivalent widths are the signature of optically thick matter reprocessing X-ray photons. This matter largely covers the hard X-ray source independently of geometry. The author shows here that variable iron line equivalent widths are found in four Seyfert galaxies. For these objects, the iron line equivalent widths are decreasing with increasing X-ray continuum flux. In NGC 5548 and NGC 3516, these variations are correlated with those of the ratio of hard and soft X-ray fluxes. Such behavior could be understood with geometrical variations where the soft X-ray luminosity is proportional to the square of the distance separating the X-ray source and the reprocessing matter. Such geometrical variations cannot be fitted with an extended reprocessing thin disk. Possible geometries for the soft X-ray source include the central part of a geometrically thick disk.

158.318 **X-ray spectral changes in rapidly variable Seyfert galaxies.**
F. Fiore, G. C. Perola, L. Piro, M. Yamauchi, M. Matsuoka.
Workshop on Iron Line Diagnostics in X-Ray Sources, p. 251 - 254 (1991). - See Abstr. 012.088 for the main entry.
The authors have analyzed Ginga observations of the Seyfert 1 galaxies MCG 6-30-15 and NGC 4051, which are known to be highly and rapidly variable. In this paper they present some results of the analysis of the spectral variations above 4 keV.

158.319 **Soft excess emission in Einstein SSS + MPC observations of Seyfert galaxies - a possible interpretation as a blend of soft X-ray lines.**
T. J. Turner, K. A. Weaver, R. F. Mushotzky, S. S. Holt, G. M. Madejski.
Workshop on Iron Line Diagnostics in X-Ray Sources, p. 255 - 258 (1991). - See Abstr. 012.088 for the main entry.
The authors have re-investigated the X-ray spectra of 26 Seyfert galaxies for which simultaneous Solid State Spectrometer (SSS) and Monitor Proportional Counter (MPC) data were available. This analysis has improved over the previous analysis of the SSS and MPC data by (a) using a predictive model for the amount and geometry of ice on the SSS detector (based on calibration observations), (b) use of a more sophisticated background subtraction algorithm for the MPC data, and (c) simultaneous fitting of the 2 datasets using XSPEC, giving an effective bandpass of ~0.6-20 keV.

158.320 **Relativistic electron-positron beam formation in the framework of the two-flow model for active galactic nuclei.**
G. Henri, G. Pelletier.
Astrophys. J., Lett., Vol. 383, No. 1, p. L7 - L10 (10 Dec 1991).
The authors propose a model for relativistic beam formation in active galactic nuclei, in which electrons and positrons are generated by pair cascade on the UV photons of the accretion disk. *In situ* turbulent acceleration by Alfvén waves in a pair plasma balances the Compton losses and prevents annihilation. The origin of the turbulence is in the subrelativistic magnetohydrodynamic accretion flow, which guarantees an extended energy reservoir up to distances where the Compton drag becomes inefficient. It is shown that above some threshold of the turbulence level the pair creation becomes catastrophic, and so the dramatic increase of the relativistic pressure expels the pair plasma along the widening flux tubes. Bulk Lorentz factors between 3 and 20 are obtained. Moreover, the model provides a natural explanation of the decentered location of the hard X-ray sources suggested by the strong iron line and the reflected component of this spectra. Most of the turbulence energy is transferred to the X-rays, and a much smaller fraction is converted into beam kinetic energy, which is enough to account for the power involved in VLBI knots.

158.321 **The ultraviolet spectra of nearby radio galaxies.**
W. C. Keel, R. A. Windhorst.
Astrophys. J., Vol. 383, No. 1, p. 135 - 147 (10 Dec 1991).
The authors report new and archival IUE SWP spectra for nine nearby radio galaxies (V < 15 mag), together with optical emission-line data for these galaxies as well as a number of candidates with weaker line emission. They examine both their

UV line and continuum properties, as well as their UV and UV–optical line ratios. They find that: 1. Lyα emission is common among local radio galaxies. No apparent relation is found between L(Lyα) and radio power for the nearby radio galaxies alone. 2. The Lyα/Hα ratio in low power nearby radio galaxies is 2 – 5 times lower than the prediction for case B recombination. Destruction of Lyα photons by grains during resonant scattering can explain the observed deficiency for reasonable metallicities. The role of internal reddening is likely to be small, since the continuum shapes do not allow enough reddening to affect Lyα so strongly. 3. The nearby radio galaxies have in general a small C IV/Lyα ratio (<0.1). Comparison of the C IV and Lyα strengths with those in luminous active nuclei suggests that most of the UV continuum comes from the stellar population, and not from the AGN. Most (4/5) of these nearby radio galaxies show an upturn in the continuum below 1450 Å, which is found in only 36% of the nearby radio–quiet early type galaxies, and not seen in high–redshift radio galaxies.

158.322 **Relativistic enhancement of the Compton–reflected component in active galactic nuclei.**
R. D. Rogers.
Astrophys. J., Vol. 383, No. 2, p. 550 – 553 (20 Dec 1991).
Compton reflection may be an important process in active galactic nuclei (AGNs) since it provides an improvement over power–law fits to AGN spectra observed by Ginga and can also explain the spectrum of the cosmic X–ray background. Here the author calculates the fraction of the total X–ray luminosity which is incident upon a thin steady α–disk if the X–rays are produced by inverse–Compton scattering of soft, disk photons off relativistic electrons located above the disk. The author calls this fraction the Compton reflection covering factor, f, and finds that it can range between 0.5 and 0.88. This large range in f is sufficient to explain the typical covering (≈ 0.5) observed in bright AGNs by Ginga, as well as the large covering factor (≈ 0.9) required to explain the cosmic X–ray background in the Compton reflection model.

158.323 **Multifrequency radio observations of Cygnus A: spectral aging in powerful radio galaxies.**
C. L. Carilli, R. A. Perley, J. W. Dreher, J. P. Leahy.
Astrophys. J., Vol. 383, No. 2, p. 554 – 573 (20 Dec 1991).
The authors present a detailed analysis of the radio spectrum across the lobes of Cygnus A. These observations provide the first critical test of synchrotron spectral aging theory. The results are in good agreement with the jet model for powerful radio galaxies, involving particle acceleration at the hot spots and outflow into the radio lobes, with subsequent energy loss due to synchrotron radiation. The hot spot spectra are well represented by a spectral aging model involving continuous injection of relativistic particles. Both hot spots have spectral break frequencies around 10 GHz. An injection index of –0.5 is found for both hot spots, consistent with diffusive shock acceleration at a strong, nonrelativistic shock in a Newtonian fluid. The break frequency distribution across each lobe shows a clear trend of decreasing break frequency with distance from the hot spot. The lowest break frequency in the source is 750 MHz, which implies a source age of 6 Myr in minimum–energy magnetic fields. The separation velocity assuming minimum–energy fields is 0.06c ($H_0 = 75\ \mathrm{km\ s^{-1} Mpc^{-1}}$). This is much larger than the source advance speed required for ram–pressure confinement of the heads of the lobes. A self–consistent model is possible in which the fields are a factor of 3 below minimum–energy values. This results in a source advance speed approximately equal to the separation velocity, ≈ 0.01c, and a source age of 30 Myr. A clear difference between the high–frequency behavior of hot spot spectra and that of the lobes is noted. The injection index across the lobes is 0.7, whereas the observed hot spot injection indices are –0.5. The cause for this injection index discrepancy remains a mystery.

158.324 **EXOSAT observations of the Seyfert 1 galaxy NGC 3516.**
K. K. Ghosh, S. Soundararajaperumal.
Astrophys. J., Vol. 383, No. 2, p. 574 – 579 (20 Dec 1991).
X–ray spectra (0.1 – 10 keV) of the Seyfert galaxy NGC 3516 are presented based on EXOSAT observations obtained on four epochs in 1985. The data were obtained from the EXOSAT archives. The soft (0.1 – 2 keV) X–ray flux of this galaxy varied by a factor of ~ 3 on a time scale of 2 days, and the hard (2 – 10 keV) X–ray flux varied by a factor of ~ 1.5 over a period of a month. The observed spectra are best fitted by a power–law, with uniform absorption, and a Gaussian line model. Best–fit parameters of this model suggest that NGC 3516 is a flat spectrum source. Weak soft excess and significant intrinsic absorption ($N_H \sim 10^{22} \mathrm{cm^{-2}}$) have been detected in the spectrum of this galaxy. An emission line, detected with high significance ($>99.9\%$) around 6.0 keV, could be the redshifted 6.4 keV iron line due to the fluorescence of thick cold iron present around the nucleus, or could be due to the redshifted 6.7 keV helium–like iron line produced in an optically thin hot plasma of temperature around $\sim 10^8$K. The measured equivalent width of the line ranges between 106 and 1240 eV.

158.325 **Optical polarimetry of PKS 2155–304 and constraints on accretion disk models for BL Lacertae objects.**
P. S. Smith, M. L. Sitko.
Astrophys. J., Vol. 383, No. 2, p. 580 – 586 (20 Dec 1991).
The authors present optical broad–band polarimetry and photometry of the BL Lacertae object PKS 2155–304 during late 1990. Variability in both flux and linear polarization was moderate during this period. The optical polarization ranged from 2% to 7%, while photometric variations were within 0.2 mag. Accurate multicolor measurements were made to study any wavelength dependence of the polarization. Specifically, the authors use these data to test the model of Wandel and Urry, which proposes that the UV to soft X–ray continuum of this object arises from an accretion disk. Wavelength–dependent polarization (WDP) is observed in PKS 2155–304. However, the polarization is always seen to decrease with wavelength when WDP is observed. This is opposite to the expected sense of WDP if the UV continuum is dominated by an accretion disk. The authors emphasize that there is as yet no direct observational evidence that thermal emission from accretion disks is an important contributor to the UV and optical continua of BL Lac objects.

158.326 **60 μm Peakers: morphology and nuclear activity explained by interactions.**
C. A. Heisler, J. P. Vader, J. A. Frogel.
J. R. Astron. Soc. Can., Vol. 85, No. 4, p. 181 (Aug 1991).
Abstract. – See Abstr. 012.091 for the main entry.

158.327 **Core, jet and counterjet prominence in powerful radio galaxies and quasars.**
A. H. Bridle.
J. R. Astron. Soc. Can., Vol. 85, No. 4, p. 182 (Aug 1991).
Abstract. – See Abstr. 012.091 for the main entry.

158.328 **Polarization sensitive very long baseline interferometry observations of BL Lacertae objects.**
D. C. Gabuzda, T. V. Cawthorne, D. H. Roberts, J. F. C. Wardle.
J. R. Astron. Soc. Can., Vol. 85, No. 4, p. 211 (Aug 1991).
Abstract. – See Abstr. 012.091 for the main entry.

158.329 **Counts of extragalactic radio sources and the distribution of their spectral indices.**
A. G. Gorshkov.
Sov. Astron., Vol. 35, No. 6, p. 563 – 568 (Nov 1991). Current Physics Microform No.: 9205X1373. English translation of Astron. Zh., Tom 68, Vyp. 6, p. 1121 – 1133 (1991).
Assuming that the counts of radio sources at 408 MHz may be represented in the form $n(S) = KS^{-a-\ln Sb}$, the distribution P(α) of spectral indices does not depend on flux and the spectra of radio sources may be approximated by a power law $S \sim \nu^{-\alpha}$ over the entire observed frequency range. An analytical solution is found for the counts and the distribution of spectral indices at any other frequency. The model functions n(S) and P(α,S) agree well with observations in the 25 – 10,000 MHz range. It is shown that the variation of the function n(S) with frequency and the

dependence of the spectral index on flux are not the result of cosmological evolution of radio sources.

158.330 **Asymmetry of the polarization characteristics of radio emission from the inner lobes of Centaurus A at 13 and 31 cm.**
E. N. Vinyajkin, V. I. Abramov.
Sov. Astron., Vol. 35, No. 6, p. 589 - 592 (Nov 1991). Current Physics Microform No.: 9205X1399. English translation of Astron. Zh., Tom 68, Vyp. 6, p. 1174 - 1180 (1991).
The linear polarization of emission from the inner lobes of the radio galaxy Centaurus A at 13 and 31 cm is investigated. At those wavelengths, as is the case in the centimeter range, the linearly polarized radiation from the northeast lobe is brighter than that from the southwest lobe. The degree of linear polarization of the integrated radio emission from the latter at 31 cm coincides to within the measurement errors with the values at shorter wavelengths, indicating a lack of significant Faraday depolarization of radiation from the southwest lobe up to that wavelength. On the other hand, the degree of polarization of the integrated emission from the northeast lobe decreases considerably with increasing wavelength. The rotation measure of the integrated emission from each lobe is determined with allowance for data on the position angles at centimeter wavelengths: $R_{NE} = -60 \pm 2.5$ rad·m^{-2}, $R_{SW} = -221 \pm 2$ rad·m^{-2}. The magnetic field is directed mainly along the lobes.

158.331 **Broad-band imaging of isolated (?) Seyfert galaxies.**
E. Xanthopoulos, M. M. De Robertis.
J. R. Astron. Soc. Can., Vol. 85, No. 4, p. 220 (Aug 1991). Abstract. - See Abstr. 012.091 for the main entry.

158.332 **Observations with the 6 m telescope on the emission spectrum for the nucleus of peculiar galaxy NGC 1275. I.**
I. I. Pronik, N. I. Merkulova.
Bull. Crimean Astrophys. Obs., Vol. 81, p. 113 - 118 (1990). English translation of Izv. Krym. Astrofiz. Obs., Tom 81, p. 124 - 130 (1990). See Abstr. 53.158.100.
The equivalent widths and relative intensities have been derived for emission lines from the peculiar galaxy NGC 1275 in the range from 3700 to 5200 Å from observations with a 1000-channel TV spectrophotometer working with the 6 m telescope at the Special Astrophysical Observatory on 30 November and 1 December 1984. The observations have been compared with spectral ones made with the same telescope on 12 - 15 January 1977. The relative line intensities can vary within a day. The intensities for lines with low excitation levels vary more than those for high-excitation ones.

158.333 **Emission-line variability and the dual nucleus of Seyfert galaxy NGC 1275. Part 2.**
I. I. Pronik.
Bull. Crimean Astrophys. Obs., Vol. 81, p. 119 - 125 (1990). English translation of Izv. Krym. Astrofiz. Obs., Tom 81, p. 131 - 138 (1990). See Abstr. 53.158.101.
Published equivalent widths are collected for the H_β and 4959 + 5007 Å [O III] emission lines together with UBV photometry data for the nucleus of NGC 1275 Seyfert galaxy between 1971 and 1982. The time course of the intensities for H_β and [O III] does not conflict with this nucleus being a binary system, in which both components are Seyfert-type variables. During the flare activity in the continuum in 1971-9, there were components in the gas-shell radiation corresponding to each nucleus. When the brightness in the continuum fell, as did that in a 27″ stop, V(27″) ⩾ 12.8, there was a step change in $I_{[O\ III]}/I_\beta$, which on that hypothesis is interpreted as one of the nuclei fading. The two spectra differ in excitation and ionization. For a given $I_{[O\ III]}/I_\beta$, the relative intensities $I_{[O\ I]}/I_\beta$, $I_{[O\ II]}/I_\beta$, and $I_{[S\ II]}/I_\beta$ for one component are higher. Among galaxies of Sy 2 type, there are ones with similar spectra for one or the other component of NGC 1275.

158.334 **Long-term optical-spectrum observations on the nuclei of active galaxies. Part 1. NGC 1275.**
K. K. Chuvaev.
Bull. Crimean Astrophys. Obs., Vol. 81, p. 126 - 133 (1990). English translation of Izv. Krym. Astrofiz. Obs., Tom 81, p. 138 - 148 (1990). See Abstr. 53.158.102.
A spectrograph fitted with an electron-optical converter has been used during 1972-6 with the 2.6 m Shain telescope at the Crimean Astrophysical Observatory to record spectrograms from the nucleus of NGC 1275 in the region of the H_β and H_α lines on 108 and 90 dates correspondingly. An accuracy of about ±15 - 25% has been attained in determining the equivalent widths W_λ of the following lines and blends: H_β[O III] 4959 + 5007, [O I] 6300, and H_α + [N II]. Within that accuracy, the intensities remained constant during the observation times. There is virtually no gas around the nucleus in which fairly strong broad components of the allowed emission lines could occur. The nucleus is assigned to type Sy 2. A qualitative description is given for the shapes of certain emission lines in the state of low nucleus luminosity. Evidence has been obtained for appreciable stratification in the gas that radiates the various lines.

158.335 **Millimeter-wavelength quiescent spectra and burst behavior for compact extragalactic sources.**
E. Valtaoja, S. Haarala, H. Lehto, L. Valtaoja, M. Valtonen, I. G. Moiseev, N. S. Nesterov, E. Salonen, H. Terasranta, S. Urpo, M. Tiuri.
Bull. Crimean Astrophys. Obs., Vol. 81, p. 134 - 151 (1990). English translation of Izv. Krym. Astrofiz. Obs., Tom 81, p. 148 - 169 (1990). See Abstr. 53.158.103.
Observations have been made on 27 extragalactic radio sources at the Crimean Astrophysical Observatory and at the Radio Observatory at Helsinki Technical University in 1980-5 at 12, 22, 37, and 77 GHz, which have been combined with published data. The quiescent spectra have been identified, as have ones during burst. The first indicate that even the most compact cores become optically transparent at frequencies f_m < 200 GHz. The spectra of 17 bursts recorded for 15 objects can be fitted to a curve corresponding to the canonical spectrum for a homogeneous synchrotron source, in which $\alpha = +2.5$ and $\alpha_{thin} \approx -0.2$ at the frequencies where it is optically dense. The simple burst spectra and the evolution support a model in which the bursts from these sources are due to shock waves from the nucleus propagating along the relativistic plasma jet directed toward the observer. The wave accelerates the electrons in the jet. The electron-energy distribution parameter $\gamma = 1.4$ is the same for the steady burst development and in the radio spectra for the various optical types of sources.

158.336 **Observational evidence for accretion disks in galactic nuclei.**
M.-H. Ulrich.
NATO Advanced Research Workshop on Theory of Accretion Disks, p. 3 - 18 (1989). - See Abstr. 012.121 for the main entry.
The electromagnetic spectrum of quasars and Seyfert 1 galaxies shows an excess of energy at λ > 4000 Å over the extrapolation of the infrared/optical spectrum: This "blue bump" is generally interpreted as thermal radiation by gas at T > 3×10^4K. But is it an accretion disk which emits this thermal radiation? To answer this question geometrical and kinematical information on the dense gas in the vicinity of the black hole is needed. The author examines the most important observations in favor or against the presence of an accretion disk: the photoelectric X-ray absorption, the energy budget in the broad line region, and the profiles of the Balmer lines. The optical polarization and the occurrence of Lyman edge absorptions are also reviewed.

158.337 **The low-frequency spectra of accretion disks in active galactic nuclei: the dark side of the big blue bump.**
M. Malkan.
NATO Advanced Research Workshop on Theory of Accretion Disks, p. 19 - 28 (1989). - See Abstr. 012.121 for the main entry.
Contents: 1. Previous observation and interpretation of the big blue bump (BBB). 2. Observing the long-wavelength side of the

BBB. 3. The 1 μm minimum in νf_ν and the curvature of the infrared continuum. 4. Modelling curvature in the red "power law": does it have an exponential cut-off? 5. Thermal vs. nonthermal origins for the infrared continuum. 6. Does the BBB flux turn drop at low frequencies?

158.338 **Observational constraints on viscosity in the AGN accretion discs.**
A. Siemiginowska, B. Czerny.
NATO Advanced Research Workshop on Theory of Accretion Disks, p. 29 – 33 (1989). – See Abstr. 012.121 for the main entry.
The optical/UV/soft X-ray big bump can be modelled as a thermal radiation from an accretion disc. Assuming that the observed UV variability in AGN spectra is caused by accretion disc instabilities we can set constraints on the viscosity. Simple bimodal disc approximation was used – the outer regions of the disc radiate locally as a black body whilst the inner regions locally become optically thin in a thermal time scale giving a negligible contribution to the UV radiation. Calculation were made for several models with $L/L_{Edd} = 1$ and $L/L_{Edd} = 0.01$ and various central masses. Comparison between thermal and observed variability time scales in 10 Seyfert galaxies and 16 QSOs indicate the parameter α value of the order of 0.01.

158.339 **Statistical modelling of IR/UV spectra in AGN.**
Z. Loska, B. Czerny.
NATO Advanced Research Workshop on Theory of Accretion Disks, p. 71 – 74 (1989). – See Abstr. 012.121 for the main entry.
The authors model IR/UV spectra of AGN assuming that the radiation originates in accretion discs, relativistic electrons, and dust. Within a broad range of parameters, the properties of the theoretical sample are similar to the properties of the Palomar–Green survey. The presence of a "break" frequency at $\log(\nu) = 14.5$ is particularly well reproduced. Better agreement with the data is for accretion discs radiating at their Eddington limits.

158.340 **Quasar monitoring program at Turku University Observatory.**
L. O. Takalo.
South. Stars, Vol. 34, No. 3, p. 101 – 105 (18 Jun 1991). – See Abstr. 012.120 for the main entry.
The quasar monitoring program at Turku University Observatory started in 1980. Since then 40 bright quasars and BL Lac objects have been monitored at radio and optical wavelengths. Some of the interesting results from the optical monitoring are described.

158.341 **Interesting problems in wide-angle tailed radio galaxies.**
A. O'Donoghue.
News Lett. Astron. Soc. N.Y., Vol. 3, No. 9, p. 24 (Feb 1991). Abstract. – See Abstr. 012.119 for the main entry.

158.342 **Jet models and observed jets in a sample of superluminal active galactic nuclei.**
I. Cruz-González, R. Carrillo.
Rev. Mex. Astron. Astrofis., Vol. 22, No. 2, p. 217 – 233 (Oct 1991).
The authors present a sample of 33 extragalactic sources with radio jets and superluminal motions, that includes different types of AGNs: 22 quasars, 6 blazars, 1 Seyfert galaxy, 3 radiogalaxies, and 1 empty field radio source. The observed parameters of the sources and their jets (jet velocity, redshift, fluxes and jet sizes) have been used as input to the relativistic jet model predictions to derive the following model parameters: the source luminosity, the maximum brightness temperature, the characteristic frequency of the synchrotron radiation and the magnetic field. The global parameters of the jets and central sources have been compared, to study the relations between the kinematic and radiative properties of the jets. The results are indicative of a possible relation between the emission mechanisms that are producing the radio properties of the jet sources and their FIR and X-ray emission.

158.343 **Dusty disks and the infrared emission from AGN.**
E. S. Phinney.
NATO Advanced Research Workshop on Theory of Accretion Disks, p. 457 – 470 (1989). – See Abstr. 012.121 for the main entry.
The distortions inferred in the gaseous disks of active galaxies suggest that a significant, and possibly dominant fraction of the 1 – 1000 μm radiation observed from AGN must be thermal emission from gas and dust heated by the central source. The author reports calculations of the growth and sublimation of dust grains in the outer parts of accretion disks appropriate to AGN. The far-infrared and submillimeter emission from radio-quiet quasars and Seyfert galaxies is more naturally interpreted as reradiation by dust, than as nonthermal emission from the inner accretion disk.

An atlas of normal and active galaxies from IUE archive.
See Abstr. 002.033.

A database of IUE spectra of active galactic nuclei – the results for quasars.
See Abstr. 002.035.

Spectroscopic investigations of stellar objects from the Second Byurakan Sky Survey. II. The fields centered on $\alpha = 09^h50^m$, $\delta = +55°00'$ and $\alpha = 11^h30^m$, $\delta = +59°00'$.
See Abstr. 002.048.

A search for steep-spectrum sources in the UTRAO, GB and RC catalogues.
See Abstr. 002.067.

An extragalactic data base. IV. Errors and misprints in catalogues of galaxies.
See Abstr. 002.070.

Search for steep-spectrum sources in the UTRAO, GB, and RC catalogs.
See Abstr. 002.074.

The Einstein Observatory Extended Medium-Sensitivity Survey. II. The optical identifications.
See Abstr. 002.076.

The Case Low-Dispersion Northern Sky Survey. XII. A region in southern Canes Venatici.
See Abstr. 002.077.

Spectral investigations of the Second Byurakan Sky Survey objects. Stellar objects. III. Fields $\alpha = 12^h22^m$, $\delta = +55°00'$ and $\alpha = 15^h30^m$, $\delta = +59°00'$.
See Abstr. 002.090.

Spectroscopic investigations of stellar objects from the Second Byurakan Sky Survey. III. The fields centered on $\alpha = 12^h22^m$, $\delta = +55°00'$ and $\alpha = 15^h30^m$, $\delta = +59°00'$.
See Abstr. 002.107.

PKSCAT90: the southern radio source database.
See Abstr. 002.108.

A catalogue of quasars and active nuclei. (5th edition).
See Abstr. 002.112.

The Second Byurakan Sky Survey. Spectra of galaxies. IV.
See Abstr. 002.133.

Conference on active nuclei of galaxies held at the Crimean Astrophysical Observatory on 30 October – 2 November 1990.
See Abstr. 011.007.

High energy astrophysics: compact stars and active galaxies. Proceedings.
See Abstr. 012.026.

Plasmas in astrophysics. Proceedings. 3. Texas–Mexico Conference on Astrophysics: Plasmas in astrophysics, Houston, TX (USA), 14 – 15 Feb 1991.
See Abstr. 012.034.

Variability of active galactic nuclei. Proceedings.
See Abstr. 012.040.

The interpretation of modern synthesis observations of spiral galaxies. Proceedings.
See Abstr. 012.041.

Structure and emission properties of accretion disks. Proceedings.
See Abstr. 012.046.

Warped disks and inclined rings around galaxies. Proceedings.
See Abstr. 012.054.

Variability of active galaxies. Proceedings.
See Abstr. 012.070.

Iron line diagnostics in X–ray sources. Proceedings.
See Abstr. 012.088.

Big bang, active galactic nuclei and supernovae. Proceedings.
See Abstr. 012.113.

Theory of accretion disks. Proceedings.
See Abstr. 012.121.

Polarization studies of AGN using a large telescope.
See Abstr. 013.022.

Extragalactic research at the Instituto de Astrofisica de Canarias.
See Abstr. 013.080.

The first year of observations with the Hubble Space Telescope.
See Abstr. 035.184.

An eight month monitoring campaign of NGC 5548 with IUE: GEX data reduction and measurement of emission line and continuum fluxes.
See Abstr. 036.144.

UV variability of NGC 5548: geometry of the broad line region.
See Abstr. 036.146.

Global fringe fitting applied to Mk3 VLBI data.
See Abstr. 036.269.

The determination of selected extragalactic radio source positions by photographic observations.
See Abstr. 041.020.

Optical positions of extragalactic radio sources in the FK5 system.
See Abstr. 041.021.

EUVE observations of active galactic nuclei.
See Abstr. 051.010.

EXOSAT EUV observations.
See Abstr. 051.056.

Numerical simulation of a hypersonic jet.
See Abstr. 062.038.

Numerical observations of a jet–cloud collision in 3–D.
See Abstr. 062.039.

Asymmetric morphology of radio sources: numerical simulation of jets in atmospheres.
See Abstr. 062.040.

On the structure of magnetized thin accretion disks of active galactic nuclei and their coronal radiations.
See Abstr. 062.111.

Structure of radiatively cooled jets.
See Abstr. 062.141.

Dynamical instabilities in self–gravitating annuli.
See Abstr. 062.176.

The Kelvin–Helmholtz instability for relativistic particle beams. II. Flows bounded by a simple shear layer.
See Abstr. 062.185.

Three–dimensional simulations of the jets of extragalactic radio sources.
See Abstr. 062.186.

On the structure of thin magnetized accretion disks of active galactic nuclei and their coronal radiation.
See Abstr. 062.187.

MHD forces in astrophysical disks and jets.
See Abstr. 062.203.

Jets and hotspots in extragalactic radio sources.
See Abstr. 062.227.

Particle acceleration at shock fronts in AGNs and jets.
See Abstr. 062.228.

Cosmic ray transport in AGNs and jets.
See Abstr. 062.230.

Accretion disk magnetohydrodynamics and the origin of jets.
See Abstr. 062.244.

Formation of relativistic MHD jets in the magnetosphere of accretion disks.
See Abstr. 062.246.

Accretion disks and the link between an AGN and its host galaxy.
See Abstr. 062.268.

Large–scale accretion flows in AGN.
See Abstr. 062.269.

Generation of gamma–rays in electron–photon cascades. II. Photon field of the power–law spectrum.
See Abstr. 063.011.

A novel mechanism for nonthermal X–ray emission.
See Abstr. 063.035.

Coherent radiation from energetic electron streams via collisionless bremsstrahlung in strong plasma turbulence.
See Abstr. 063.041.

X–ray irradiation of interstellar grains in active galaxies: evaporation and infrared spectra.
See Abstr. 063.068.

Meteorology of the broad line region.
See Abstr. 063.094.

An atlas of 2–dimensional transfer functions of the broad line region.
See Abstr. 063.095.

Electron–ion coupling in Compton–heated plasmas.
See Abstr. 063.096.

X–ray spectra from dynamical tuning.
See Abstr. 063.099.

X–ray continuum emission caused by relativistic protons.
See Abstr. 063.120.

Magnetically driven jets and winds.
See Abstr. 064.128.

Conference summary. Structure and emission properties of accretion disks.
See Abstr. 064.147.

The dynamics of twisted accretion discs.
See Abstr. 064.206.

Tidal disruption of stars in black hole clusters as a source of AGN luminosity.
See Abstr. 067.023.

Line profiles from a disk around a rotating black hole.
See Abstr. 067.038.

Line profile of radially flowing clouds with accretion disk.
See Abstr. 067.074.

Line profile of radially flowing clouds with accretion disk.
See Abstr. 067.077.

Polarization of radiation from AGN accretion disks.
See Abstr. 067.082.

A model for variability of AGN.
See Abstr. 067.083.

High–polarization radiation from magnetized accretion disks.
See Abstr. 067.086.

Hybrid accretion disks in active galactic nuclei. I. Structure and spectra.
See Abstr. 067.152.

Variability of active galactic nuclei: a theorist's view.
See Abstr. 067.159.

The stability of thick disks.
See Abstr. 067.160.

Fuelling accretion disks in AGN.
See Abstr. 067.161.

Accretion disk models of active galactic nuclei: a crucial role for reprocessing?
See Abstr. 067.162.

Instabilities in thick disks.
See Abstr. 067.163.

Thermal continuum from QSO disks.
See Abstr. 067.167.

Variability of active galactic nuclei.
See Abstr. 067.169.

Line emission from accretion disks: observation vs. theory.
See Abstr. 067.173.

Molecules in AGN's accretion discs.
See Abstr. 067.180.

Time dependent accretion disk model in AGN.
See Abstr. 067.191.

Smoothed particle hydrodynamics and accretion processes around a massive black hole.
See Abstr. 067.192.

Variable line profiles from accretion disks and black hole mass.
See Abstr. 067.193.

The physical structure of self–gravitating accretion disks in the inner parts of AGN's.
See Abstr. 067.194.

Stability of slim accretion discs – effects of central mass and viscosity.
See Abstr. 067.203.

The collisional acceleration of relativistic electrons in the central regions of active galactic nuclei.
See Abstr. 067.211.

Neutron stars and black holes.
See Abstr. 067.212.

Polarization properties of emission lines from relativistic accretion disks.
See Abstr. 067.225.

Line profiles from a disk around a rotating black hole.
See Abstr. 067.228.

Iron K–lines from accretion disks around Schwarzschild black holes.
See Abstr. 067.229.

Quantum–field approach to the theory of the relativistic jet activity of quasars and galactic nuclei.
See Abstr. 067.250.

Inflation of stars by TeV neutrinos.
See Abstr. 067.253.

Pair cascade model constraints on energetic particle emission from active galactic nuclei.
See Abstr. 067.257.

Collisional acceleration of electrons in central regions of active galactic nuclei.
See Abstr. 067.259.

A possible mechanism for the precession of extragalactic jets.
See Abstr. 067.263.

UBVRI photometry of stars in the fields of X–ray selected BL Lacertae objects.
See Abstr. 113.028.

Black hole candidates in binaries.
See Abstr. 117.260.

A large radio galaxy behind the supernova remnant HB 9.
See Abstr. 125.018.

Radio sources in a field centered on BG 2107+49.
See Abstr. 141.001.

Spectra and positions of three sources from the Zelenchuk survey.
See Abstr. 141.003.

A list of steep spectrum radio sources.
See Abstr. 141.004.

Investigation of radiosources of the second galactic quadrant from the Kallas and Reich survey at 7.6 and 31.3 cm wavelengths.
See Abstr. 141.005.

Spectra and coordinates of three sources from the Zelenchuk survey.
See Abstr. 141.008.

Radio surveys and large scale structure.
See Abstr. 141.014.

EUV observations as a diagnostic of possible AGN sources of the cosmic X–ray background.
See Abstr. 142.032.

What is in the X–ray sky?
See Abstr. 142.055.

Einstein deep surveys.
See Abstr. 142.056.

Iron line diagnostics in elliptical galaxies and cluster cooling flows.
See Abstr. 142.058.

Variability of active galactic nuclei and galactic QPO sources: a diagnosis.
See Abstr. 142.064.

High energy and very high energy gamma–rays from electromagnetic cascade induced by relativistic neutrons in AGN.
See Abstr. 143.056.

Collective instabilities in stellar environments of galactic nuclei as source of their activity and cause of starbursts.
See Abstr. 151.008.

Merger origin of starburst galaxies.
See Abstr. 151.081.

Collective instabilities in the stellar environment of a galactic nucleus: a source of activity and an initiator of starbursts.
See Abstr. 151.101.

Calán–Tololo survey. III. H II galaxies.
See Abstr. 157.015.

A study of southern extreme IRAS galaxies. I. Optical observations.
See Abstr. 157.034.

A spectrophotometric catalogue of H II galaxies.
See Abstr. 157.044.

Observing galaxies through time.
See Abstr. 157.046.

VLBI search for compact radio cores in starbursts.
See Abstr. 157.058.

Are hypernovae detectable?
See Abstr. 157.062.

Spectrophotometric and morphological investigation of galaxies with ultraviolet excess. VIII.
See Abstr. 157.082.

The magnetic field structure in the radio halos of NGC 891 and NGC 4631.
See Abstr. 157.086.

Galactic models with massive coronae. III. Giant elliptical galaxy M87.
See Abstr. 157.091.

Wolf–Rayet galaxies: an introduction and a catalog.
See Abstr. 157.109.

Colors of IRAS galaxies.
See Abstr. 157.118.

^{13}CO emission in the irregular galaxy NGC 55 and the IR–luminous galaxy NGC 3256.
See Abstr. 157.158.

Compact starbursts in ultraluminous infrared galaxies.
See Abstr. 157.170.

Active nuclei and star formation in E/S0 galaxies.
See Abstr. 157.182.

Second–epoch VLA observations of compact radio sources in NGC 253.
See Abstr. 157.194.

Emission–line galaxies in the third list of the Case Low–Dispersion Northern Sky Survey.
See Abstr. 157.213.

Observations and models of blue compact dwarf galaxies.
See Abstr. 157.234.

Population synthesis models of starbursts.
See Abstr. 157.236.

The starburst–AGN connection.
See Abstr. 157.240.

X–ray emission from starburst galaxies.
See Abstr. 157.246.

Infrared hot spots in the nucleus of NGC 253.
See Abstr. 157.247.

On superassociations (SA) at the ends of the bars of spiral Markarian galaxies.
See Abstr. 157.349.

Observational constraints on disc models for quasars and Seyfert galaxies.
See Abstr. 159.031.

Forbidden high–ionization lines in QSO spectra.
See Abstr. 159.047.

On the trajectories of the superluminal knots in 3C 345.
See Abstr. 159.052.

EUV radiation from active galactic nuclei and quasars.
See Abstr. 159.090.

Intraday radio variability of quasars and BL Lac objects.
See Abstr. 159.095.

Optical observations of rapid variability in the QSO PG 0117+213.
See Abstr. 159.096.

Continuum variability in quasars and Seyfert galaxies.
See Abstr. 159.097.

A composite plot of far–infrared versus radio luminosity, and the origin of far–infrared luminosity in quasars.
See Abstr. 159.106.

Microlensing and the structure of active galactic nucleus accretion disks.
See Abstr. 159.108.

Depolarization asymmetry in the quasar 3C 47.
See Abstr. 159.112.

A disrupted radio jet inside the host galaxy of the quasar 3C 48.
See Abstr. 159.116.

Superclustering at high redshifts.
See Abstr. 159.156.

Inference of the extreme–ultraviolet continuum shape of quasars.
See Abstr. 159.171.

Models of the continuum of quasars and AGNs in the EUV and adjacent energy ranges.
See Abstr. 159.179.

Quasars, Seyfert 1 (Sy1) type galaxies and observational selection.
See Abstr. 159.182.

The X–ray cooling flow in the cluster of calaxies around PKS 2354–35.
See Abstr. 160.040.

A peculiar group of galaxies in the field of the BL Lacertae object H 0414+00.
See Abstr. 160.042.

Neutral hydrogen in A 2052 and other cooling flow clusters.
See Abstr. 160.045.

The probable extragalactic nature of some low surface brightness clouds at high galactic latitude.
See Abstr. 160.063.

The environment of 3C 255.
See Abstr. 160.065.

Discovery of a ring Seyfert in a group of galaxies.
See Abstr. 160.069.

Parkes radio sources in the directions of southern rich clusters.
See Abstr. 160.089.

Powerful radio sources in clusters of galaxies.
See Abstr. 160.129.

Comments on the V/V_{max} test.
See Abstr. 161.041.

Construction and analysis of a diameter–limited sample of galaxies in the northern sky.
See Abstr. 161.152.

Kinematics of relativistic ejection with Hubble flow in subluminal motion for the determination of Hubble constant H_0.
See Abstr. 161.207.

Redshift surveys of galaxies.
See Abstr. 161.245.

Origin of cosmic X–ray background radiation.
See Abstr. 161.250.

A Compton reflection model for the cosmic X–ray and gamma–ray backgrounds.
See Abstr. 161.251.

Cross–correlation of the X–ray background with nearby galaxies.
See Abstr. 161.254.

Spectral–luminosity evolution of active galactic nuclei and the cosmic X–ray background.
See Abstr. 161.389.

The X–ray flux dipole of AGNs as an indicator of the local gravitational field.
See Abstr. 161.398.

On the cosmological significance of the apparent deficit of small interplanetary scintillation sources.
See Abstr. 161.416.

Bounds on intergalactic compact objects from observations of compact radio sources.
See Abstr. 161.417.

Aligned radio galaxies.
See Abstr. 161.518.

The confinement and cosmological evolution of extragalactic radio source components.
See Abstr. 161.638.

159 Quasars

159.001 **Discovery of a new bright southern QSO.**
L. Wisotzki, W. Wamsteker, D. Reimers.
Astron. Astrophys., Vol. 247, No. 1, p. L17 – L20 (Jul 1991). Letter–to–the–editor.

Within the newly started Hamburg/ESO survey for bright southern quasars the low–redshift QSO HE1029–1401 ($z = 0.086$) was found by automated search on objective prism plates and subsequently confirmed by slit spectroscopy. With $V = 13.86$ it is among the three brightest QSOs in the sky and the brightest ever found by optical selection. Broad band optical and NIR photometry together with IUE observations are used to construct the energy distribution between 3.5 μm and 1450 Å. The UV continuum flux at 1600 Å of $f_\lambda = 4\times10^{-14}$erg s^{-1}cm^{-2}Å^{-1} ranges it among the UV brightest AGN in total, making HE 1029–1401 a prime target for future monitoring studies. While its dereddened absolute magnitude of $M_V = -25.1$ ($H_0 = 50$ km s^{-1}Mpc^{-1}) clearly distinguishes it from Seyfert 1 galaxies, HE 1029–1401 is close enough to allow an easy detection of its host galaxy already on an ESO (R) survey print.

159.002 **Evidence for a massive, dark gravitational lensing object in Q2345+007.**
R. C. Duncan.
Astrophys. J., Lett., Vol. 375, No. 2, p. L41 – L44 (10 Jul 1991).

The author analyzes the narrow absorption line spectra of the double quasar Q2345+007. The observed pattern of Lyα forest absorption lines gives evidence that the quasar is gravitationally lensed. If Q2345+007 is lensed by an object at z = 1.49, as suggested by observations of heavy–element absorption lines, then a low–luminosity lensing mass of $\sim10^{13}M_\odot$ contained within a transverse distance $\sim40h_0{}^{-1}$kpc is implied. The Lyα

cloud diameters at $z \approx 2$ would then be $(20 \pm 10)h_0^{-1}$kpc $[q_0 = 0.5]$ or $(28.0 \pm 14)h_0^{-1}$kpc $[q_0 = 0]$. If the quasar is not lensed, then the Lyα clouds are larger by a factor of ~6.5. Several observational tests are suggested, which might verify the presence of a massive, dark, lensing object and could elucidate its properties.

159.003 **A quasar lights up the Universe.**
R. Talcott.
Astronomy, Vol. 19, No. 9, p. 42 – 45 (Sep 1991).

159.004 **The emerging picture of quasars.**
B. J. Wilkes.
Astronomy, Vol. 19, No. 12, p. 34 – 41 (Dec 1991).
Radio, infrared, ultraviolet, and X–ray radiation from quasars shows that they are supermassive black holes surrounded by a thick disk of dust and gas.

159.005 **Non–parametric tests for clustering in Lyα absorption systems.**
H. G. Bi, G. Börner, Y. Chu.
Astron. Astrophys., Vol. 247, No. 2, p. 276 – 282 (Jul 1991).
The authors extend their previous analysis of the Lyα absorption lines of Q0420–388 to 17 other Lyα samples, as well as to 100 simulated synthetic Lyα samples containing no correlation originally. The departure of the probability distribution function of line intervals from the Poisson statistics is shown to be significant for 11 samples at the 98% confidence level and for 3 at 80%; the Weibull statitics that was suggested to model the effect of lineblending with $k = 1.72\,(+0.56,-0.35)$ fits well to all the samples except one. By studying their spatial power spectra, it is found that the samples are non–renewal or non–white–noise. The simulation of synthetic samples shows that if the original Lyα clouds were randomly distributed along the lines of sight, the observed forest should form a renewal process. Thus the authors' result implies the existence of intrinsic clustering. The conclusion is assessed at the $1\text{–}10^{-5}$ confidence level when all samples are combined. The authors use another nonparametric methods to study the cross dependence between the lines of three neighbouring QSO's (Q1624+268, A and B, or Sample 11, 12 and 13, respectively) observed by Crotts (1989). The bivariate Lyα sample formed from S11 and S13 shows interdependence at the 99.5% confidence level. S11–S12 shows also a significant dependence but S12–S13 shows no dependence. If the decrease is due to the decrease of system separations, it can be deduced that the proper mean cloud radius of hypothetical Lyα absorbers, or the cloud–cluster radius, is equal to $0.60\,h^{-1}$Mpc at redshift 2.2.

159.006 **On the possible nature of X–ray radiation sources in quasars.**
V. F. Sulejmanov.
Pis'ma Astron. Zh., Tom 17, No. 7, p. 575 – 582 (Jul 1991). In Russian. English translation in Sov. Astron. Lett., Vol. 17, No. 4.
The thermal energy balance equation in accretion disks is considered. It is shown that a few different temperature states may exist in accretion disks around supermassive black holes. The possibility of high temperature solutions ($\sim 10^7$K) does not depend on a central object's mass. X–ray radiation of quasars may be explained by the existence of these states.

159.007 **IO Andromedae is a quasar, but not a cataclysmic variable.**
N. V. Borisov, N. F. Vojkhanskaya, S. N. Fabrika.
Pis'ma Astron. Zh., Tom 17, No. 7, p. 583 – 585 (Jul 1991). In Russian. English translation in Sov. Astron. Lett., Vol. 17, No. 4.
Spectral observations of the supposed cataclysmic variable IO And have shown that it is a quasar with $z = 0.134$. Some of its characteristics resemble bright Seyfert I–type nuclei. Radio– and X–luminosity estimates do not allow us to ascribe IO And to blasars, though a spectral and photometric variability in optics is possible.

159.008 **The nature of the redshift and directly observed quasar statistics.**
I. E. Segal, J. F. Nicoll, P. Wu, Z. Zhou.
Naturwissenschaften, Vol. 78, No. 7, p. 289 – 296 (Jul 1991).
The nature of the cosmic redhsift is one of the most fundamental questions in modern science. Hubble's discovery of the apparent expansion of the Universe is derived from observations on a small number of galaxies at very low redhsifts. Today, quasar redshifts have a range more than 1000 times greater than those in Hubble's sample, and represent more than 100 times as many objects. A recent comprehensive compilation of published measurements provides the basis for a study indicating that quasar observations are not in good agreement with the original predictions of the expanding Universe theory, but are well fit by the predictions of an alternative theory having fewer adjustable parameters.

159.009 **A comparison of the ultraviolet and optical emission line profiles in low redshift QSOs.**
V. T. Junkkarinen, R. D. Cohen, R. C. Puetter, H. E. Smith.
Bull. Am. Astron. Soc., Vol. 23, No. 2, p. 892 (1991). Abstract. – See Abstr. 012.008 for the main entry.

159.010 **A multicolor CCD survey for QSOs to m ~ 24.**
S. F. Anderson, P. L. Schechter, R. A. Windhorst, D. C. Koo, S. R. Majewski.
Bull. Am. Astron. Soc., Vol. 23, No. 2, p. 892 (1991). Abstract. – See Abstr. 012.008 for the main entry.

159.011 **Extinction in Mg II broad absorption line QSOs.**
D. Sprayberry, C. B. Foltz.
Bull. Am. Astron. Soc., Vol. 23, No. 2, p. 893 (1991). Abstract. – See Abstr. 012.008 for the main entry.

159.012 **Quasare unter der Lupe.**
T. Bührke.
Bild Wiss., No. 12, p. 128 – 129 (Dec 1991).

159.013 **Observations with the Faint Object Spectrograph of the low redshift QSO CSO 251.**
E. I. Rosenblatt, E. A. Beaver, E. M. Burbidge, R. D. Cohen, V. T. Junkkarinen, R. W. Lyons.
Bull. Am. Astron. Soc., Vol. 23, No. 2, p. 920 – 921 (1991). Abstract. – See Abstr. 012.008 for the main entry.

159.014 **Der Quasar 3C273.**
T. J. L. Courvoisier, E. I. Robson.
Spektrum Wiss., No. 8, p. 94 – 102 (Aug 1991). German translation of 53.159.155.

159.015 **Spectroscopic observations of 3C 273 with the Faint Object Spectrograph.**
E. M. Burbidge, E. A. Beaver, R. D. Cohen, V. T. Junkkarinen, R. W. Lyons, E. I. Rosenblatt.
Bull. Am. Astron. Soc., Vol. 23, No. 2, p. 923 (1991). Abstract. – See Abstr. 012.008 for the main entry.

159.016 **Optical variations in blue bump quasars.**
O. Kuhn, A. Siemiginowska, M. Elvis, B. J. Wilkes, J. McDowell.
Bull. Am. Astron. Soc., Vol. 23, No. 2, p. 926 (1991). Abstract. – See Abstr. 012.008 for the main entry.

159.017 **The theoretical γ–ray spectrum of quasars.**
M. E. Dixson, W. A. Stein.
Bull. Am. Astron. Soc., Vol. 23, No. 2, p. 927 (1991). Abstract. The full paper is published in AIP Conf. Proc., Vol. 220, p. 35 – 39 (5 Apr 1991). – See Abstr. 012.008 for the main entry.

159.018 **Observations of [O III] λ5007 Å and H–beta in the z = 2.1 quasar Q 1331+170.**
M. G. Smith, J. A. Bailey, P. Puxley, T. R. Geballe, G. S. Wright, R. F. Carswell, C. M. Mountain, S. K. Ramsay, B. R. Espey, M. J. Ward.
Bull. Am. Astron. Soc., Vol. 23, No. 2, p. 928 (1991). Abstract. – See Abstr. 012.008 for the main entry.

159.019 **K–band imaging of gravitational lens systems.**
J. T. Annis.
Bull. Am. Astron. Soc., Vol. 23, No. 2, p. 931 (1991). Abstract. – See Abstr. 012.008 for the main entry.

159.020 **Discovery of a Lyα galaxy near the damped Lyα system at z = 2.3 towards the QSO PHL 957.**
J. Lowenthal, C. J. Hogan, R. F. Green, A. Caulet, B. Woodgate, L. Brown, C. B. Foltz.
Bull. Am. Astron. Soc., Vol. 23, No. 2, p. 931 – 932 (1991). Abstract. – See Abstr. 012.008 for the main entry.

159.021 **On the possible nature of the X–ray source in quasars.**
V. F. Sulejmanov.
Sov. Astron. Lett., Vol. 17, No. 4, p. 245 – 248 (Jul 1991). Current Physics Microform No.: 9202X0421. English translation of Pis'ma Astron. Zh., Tom 17, No. 7, p. 575 – 582 (Jul 1991).
The equation of thermal energy balance in accretion disks is studied. It is shown that accretion disks around supermassive black holes can exist in several states with different temperatures. The possibility of a high–temperature ($\sim 10^7$K) solution does not depend on the mass of the central object. The X–ray emission of quasars can be explained by the existence of such states.

159.022 **IO Andromedae is a quasar, rather than a cataclysmic variable.**
N. V. Borisov, N. F. Vojkhanskaya, S. N. Fabrika.
Sov. Astron. Lett., Vol. 17, No. 4, p. 249 – 250 (Jul 1991). Current Physics Microform No.: 9202X0425. English translation of Pis'ma Astron. Zh., Tom 17, No. 7, p. 583 – 585 (Jul 1991).
Spectroscopic observations of the putative cataclysmic variable IO And have shown that it is a quasar with $z = 0.134$. In some respects it resembles bright type–I Seyfert nuclei. Estimates of its radio and X–ray luminosity show that it is not a blazar, but it may possess spectral variability and rapid photometric variability in the visible.

159.023 **The emission–line properties of steep radio spectrum quasars.**
M. R. Corbin.
Astrophys. J., Vol. 375, No. 2, p. 503 – 516 (10 Jul 1991).
The emission–line properties of 45 intermediate redshift ($z \sim 2$) steep spectrum quasars are presented and compared with those of a sample of flat radio spectrum sources of comparable mean luminosity and redshift. It is found that the steep–spectrum sources have, on average, smaller widths of the C IV λ1549 and C III] λ1909 line profiles, larger Mg II λ2800 equivalent widths, and smaller (He II + O III])/C IV intensity ratios than the flat–spectrum sources. The steep–spectrum sources also display C IV profiles that tend to be asymmetric to the red, whereas those in the flat–spectrum sources are found to be asymmetric to the blue. These line parameters appear to be weakly correlated with the 6 – 11 cm radio spectral index in the combined sample of steep– and flat–spectrum sources. The Baldwin effect and the correlation between the Mg II – C IV and C III] – C IV redshift differences reported elsewhere are also present both within the combined sample and among the steep–spectrum objects alone. These results support the proposed unified models of steep and flat radio spectrum quasars. Given the sensitivities of the associated line parameters to luminosity, the senses of the correlations with the radio spectral index suggest that the ionization of the broad–line region gas is anisotropic, such that it is most strongly ionized in the direction of the radio axis.

159.024 **Gravitational lenses: observational status and applications.**
J. N. Hewitt.
Bull. Am. Astron. Soc., Vol. 23, No. 2, p. 938 (1991). Abstract. – See Abstr. 012.008 for the main entry.

159.025 **Another radio–tail quasar (z = 0.54) at the center of an extremely rich cluster of galaxies.**
R. J. Oliversen, P. M. Hintzen, F. Owen, W. Romanishin, K.-P. Cheng.
Bull. Am. Astron. Soc., Vol. 23, No. 2, p. 951 – 952 (1991). Abstract. – See Abstr. 012.008 for the main entry.

159.026 **The QSOs influence on the Lyman–α forest.**
A. Dobrzycki, J. Bechtold.
Bull. Am. Astron. Soc., Vol. 23, No. 2, p. 952 (1991). Abstract. – See Abstr. 012.008 for the main entry.

159.027 **Observations of X–ray spectra of two nearby QSO's: Mrk 205 and Mrk 1148.**
K. P. Singh, A. R. Rao, M. N. Vahia.
Astron. Astrophys., Vol. 248, No. 1, p. 37 – 44 (Aug 1991).
The 0.1–10 keV X–ray spectra of two nearby quasistellar objects (QSO's) known as Markarian 205 (Mrk 205) and Mrk 1148 (PG 0049+171), are presented based on observations with the EXOSAT Observatory. Simple power law models with the line of sight absorption consistent with the values obtained from the 21 cm surveys are sufficient to explain the observed spectra of both the X–ray sources. The energy index (α) for the X–ray spectrum of Mrk 205 is measured to be 0.96 ± 0.08. The value of α for Mrk 1148 is found to be $0.61(+0.09,-0.10)$. These determinations assume that the column densities of equivalent hydrogen towards these sources are given by the accurate 21–cm measurements of Elvis et al. (1989). The present data do not require the presence of soft excess in these sources. There is no evidence for intrinsic absorption in these sources and no line features are detected in their X–ray spectra. The X–ray sources were found to be steady during the observations. The 2–10 keV X–ray luminosities of Mrk 205 and Mrk 1148 were 1.6×10^{44}erg s^{-1} and 2.1×10^{44}erg s^{-1}, respectively.

159.028 **Enhanced probability of intensity outbursts via superluminal microlensing of quasars.**
K. Subramanian, Gopal–Krishna.
Astron. Astrophys., Vol. 248, No. 1, p. 55 – 56 (Aug 1991).
The apparent faster–than–light motion of the emission knots associated with relativistic jets of quasars implies that relative to the stars of any foreground galaxy, their motion could easily be superluminal and thus a few orders–of–magnitude faster than the stellar velocities. The resulting "superluminal microlensing" of the emission knots would not only impress an ultra–rapid flux variability on even intrinsically non–varying jets, as proposed by the authors previously, but the probability of seeing microlensing events would itself be markedly enhanced, particularly when the intrinsic optical depth of the galaxy for microlensing is modest ($\sim$0.1). The probabilities are quantitatively estimated for the simple but plausible case of the dark galactic halos being made of low mass stars, or Jupiter–like bodies. For a given duration of observation, the likelihood of microlensing events could widely differ depending on whether the emission from the lensed source is dominated by a relativistic jet, or by a stationary component.

159.029 **Bounds on the enhancement of quasar counts by lenses.**
I. Kovner.
Astrophys. J., Vol. 376, No. 1, p. 70 – 89 (20 Jul 1991).
The author derives constraints on the association between QSOs and galaxies caused by gravitational lenses from the conservation of brightness, within a variety of sets of general assumptions which do not include the particular nature of the lenses. The author calculates the association enhancement, E, that is, the factor by which the QSO counts taken near galaxies are increased, as compared with the counts taken from arbitrary sites uncorrelated with galaxies. Two kinds of assumption sets are considered: in the first, the background counts are little affected

by lenses (unaffected background); in the second, they are (possibly strongly) affected (affected background). All the bounds on E are easily calculated, and some of them can be found from analytic expressions. Given an unaffected background, the application of these bounds to the QSO number-magnitude relation obtained by Boyle, Shanks, and Peterson (BSP) indicates that the number of QSO–galaxy pairs reported by Webster et al. (WHHW) is too large: neglecting the observational uncertainties, there is an indication that the background counts may be appreciably affected by lenses, at a ~90% confidence level for model independent bounds, and at a ~99% level for conventional lens models. For an affected background, the least model–dependent bound, E_{am}, is obtained if the net contribution of QSOs to the sky brightness is put into QSOs at the flux limit of the association sample. If this bound, calculated on the basis of the BSP sample, is taken as a unit, the WHHW result is $E_{WHHW}/E_{am} \approx 0.19$.

159.030 **The theory of formation of QSO spectra with broad absorption lines.**
Eh. Ya. Vil'koviskij.
Astron. Zh., Tom 68, Vyp. 6, p. 1150 – 1159 (Nov–Dec 1991). In Russian. English translation in Sov. Astron., Vol. 35, No. 6.

The theory of absorption line formation in a plasma flow is considered, basing on a joint solution of plasma dynamics and line radiation transfer equations. Model spectra of QSO with broad absorption lines are computed.

159.031 **Observational constraints on disc models for quasars and Seyfert galaxies.**
T. J.-L. Courvoisier, J. Clavel.
Astron. Astrophys., Vol. 248, No. 2, p. 389 – 394 (Aug 1991).

The emission component in Seyfert galaxies and quasars dominating in the visible–ultraviolet spectral range, the so–called "blue bump", has often been associated with thermal emission from the surface of an optically thick and geometrically thin accretion disc. Although the intrinsic shape of the continuum spectral energy distribution is poorly known, some of the spectral properties of AGN are in contradiction with expectations from the model. Furthermore, the observed simultaneity of continuum variations in UV, optical and near IR bands is in clear conflict with model expectations. Though they are still in their infancy, geometrically thick accretion disc models could potentially resolve these discrepancies and account for most of the observed properties of the continuum emission of quasars and Seyfert galaxies.

159.032 **Infrared and visible photometry of the gravitational lens system 2237+030.**
D. Nadeau, H. K. C. Yee, W. J. Forrest, J. D. Garnett, Z. Ninkov, J. L. Pipher.
Astrophys. J., Vol. 376, No. 2, p. 430 – 438 (1 Aug 1991). With plate 7.

As part of a program of high spatial resolution imaging of gravitationally lensed sources in the visible and infrared, images of 2237+030 were obtained in the Gunn r and infrared J, H, K, and 3.3 μm filters. The results of the photometry of the four bright quasar components provide evidence of extinction through the lens and a determination of the extinction law in the galaxy is made. The energy distribution shows evidence of a sharp decrease of the spectral index α ($f_\nu \propto \nu^\alpha$) at wavelengths longer than a rest wavelength of 1 μm. Assuming that microlensing amplification in the infrared is of similar strength as in the visible, the data constrain the suggested microlensing event of 1988 August – September to a time scale of 100 days.

159.033 **Early observations of gravitational lenses with the Planetary Camera of Hubble Space Telescope.**
J. Kristian, S. P. Ewald, E. J. Groth, J. J. Hester, J. A. Holtzman, T. R. Lauer, R. M. Light, D. P. Schneider, E. J. Shaya, W. A. Baum, B. G. Campbell, A. D. Code, D. G. Currie, G. E. Danielson, S. M. Faber, J. G. Hoessel, D. Hunter, T. Kelsall, C. R. Lynds, G. Mackie, D. G. Monet, E. J. O'Neil Jr., P. K. Seidelmann, B. Smith, J. A. Westphal.
Workshop on the First Year of HST Observations, p. 25 – 32 (1991). – See Abstr. 012.017 for the main entry.

Preliminary data for six "classic" gravitational lens systems were obtained.

159.034 **FOS observations of the absorption spectrum of 3C 273.**
J. N. Bahcall, B. T. Jannuzi, D. P. Schneider, G. F. Hartig, R. Bohlin, V. Junkkarinen.
Workshop on the First Year of HST Observations, p. 46 – 57 (1991). – See Abstr. 012.017 for the main entry.

The authors describe the FOS observations of the absorption line spectrum of 3C 273 and compare the results with the GHRS observations of the same quasar. Three Ly–α lines appear to be produced by gas in the Virgo cluster or by the halos of galaxies associated with the Virgo cluster. The authors identify a total of seven Ly–α absorption systems with equivalent widths greater than 0.2 Å. The evolution of the number density of Ly–α clouds cannot be determined with confidence by comparing only 3C 273 observations with those of large redshift quasars. The inferred redshift–dependence of the number density depends critically upon whether or not the Virgo–cluster lines are included or upon an uncertain extrapolation of the equivalent width distribution for strong lines found at large redshifts.

159.035 **Results and some implications of the GHRS observations of the Lyman α forest in 3C 273.**
R. J. Weymann.
Workshop on the First Year of HST Observations, p. 58 – 67 (1991). – See Abstr. 012.017 for the main entry.

The results of the first GHRS exposures of 3C 273 and related observations are summarized, and their implications for our understanding of the Ly α forest assessed. The potential for further understanding of the Ly α forest and intergalactic medium through additional HST and ground–based programs is briefly discussed. The number of low redshift Ly α lines found is substantially higher than expected based upon extrapolation from ground–based data using current estimates of the slope of the log(dN/dz) vs. log(1+z) relationship. However, the author considers the more significant results to be the small value of the slope (~0.80) between redshift 0 and 2 and the establishing of the very existence of low redshift Ly α clouds in numbers sufficient for their properties to be investigated in some detail.

159.036 **FOC images of the gravitational lens system G2237+0305.**
P. Crane, R. Albrecht, C. Barbieri, J. C. Blades, A. Boksenberg, J. M. Deharveng, M. J. Disney, P. Jakobsen, T. M. Kamperman, I. R. King, F. Macchetto, C. D. Mackay, F. Paresce, G. Weigelt, D. Baxter, P. Greenfield, R. Jedrzejewski, A. Nota, W. B. Sparks.
Workshop on the First Year of HST Observations, p. 188 – 191 (1991). – See Abstr. 012.017 for the main entry.

The gravitational lens G2237+0305 appears as a result of an extremely fortuitous alignment of a background QSO at z = 1.695 with the nucleus of a 14th magnitude foreground galaxy at z = 0.039. Although ground–based images of this lens with excellent seeing (≈0.48 arcsec FWHM) have resolved the four QSO images, clearly better resolution is required to (a) improve or confirm their positions and magnitudes, (b) better determine the galaxy's nuclear structure which has an important effect on the QSO images, and (c) search for the fifth image predicted near the nucleus by current lensing theory.

159.037 **Reduction of PG1115+080 images.**
E. J. Groth, J. A. Kristian, S. P. Ewald, J. J. Hester, J. A. Holtzman, T. R. Lauer, R. M. Light, E. J. Shaya, W. A. Baum, B. Campbell, A. Code, D. G. Currie, G. E. Danielson, S. M. Faber, J. Hoessel, D. Hunter, T. Kelsall, R. Lynds, G. Mackie, D. G. Monet, E. J. O'Neil Jr., D. P. Schneider, P. K. Seidelmann, B. Smith, J. A. Westphal.
Workshop on the First Year of HST Observations, p. 192 – 195 (1991). – See Abstr. 012.017 for the main entry.

The lensing object can be seen in the processed image. One would like to subtract the four QSO images to leave behind a clear picture of the lens.

159.038 **The non–proprietary Snapshot Survey: a search for gravitationally–lensed quasars using the HST Planetary Camera.**
D. Maoz, J. N. Bahcall, R. Doxsey, D. P. Schneider, N. A. Bahcall, O. Lahav, B. Yanny.
Workshop on the First Year of HST Observations, p. 200 – 203 (1991). – See Abstr. 012.017 for the main entry.

159.039 **3C 43: a jet deflected by dense gas?**
C. E. Akujor, R. E. Spencer, D. J. Saikia.
Astron. Astrophys., Vol. 249, No. 2, p. 337 – 340 (Sep 1991).
The compact steep spectrum quasar 3C 43 has a misaligned and asymmetric structure with a remarkable bent jet. It is shown that such an unusually distorted structure cannot easily be explained by the jet being inclined at a small angle to the line of sight. Strong interaction between the jet and dense interstellar medium is a more plausible hypothesis, and may be common in a sub–class of CSS quasars.

159.040 **Faint Object Spectrograph observations of CSO 251.**
R. D. Cohen, E. A. Beaver, E. M. Burbidge, V. T. Junkkarinen, R. W. Lyons, E. I. Rosenblatt.
Workshop on the First Year of HST Observations, p. 204 – 207 (1991). – See Abstr. 012.017 for the main entry.
The QSO CSO 251 (z = 0.0788, $F_\lambda(5500\,Å) = 2\times10^{-15}$) was observed as an early release observation (3065) with the FOS.

159.041 **The theoretical γ–ray spectrum of quasars.**
M. E. Dixson.
AIP Conf. Proc., No. 220, p. 35 – 39 (Apr 1991). Current Physics Microform No.: 9106B1169. – See Abstr. 012.020 for the main entry.
Preliminary results of calculations concerning the theoretical γ–ray spectrum of quasi–stellar objects is presented. The calculated spectra extend (in γ–ray energy) from approximately 0.1 to 1000 MeV. Numerous γ–ray producing mechanisms are presently included to contribute to the total output. These include the decay of neutral pions, electron and positron bremsstrahlung, positron annihilation and γ–ray line emission from eight different elements. Although very preliminary, the resulting model spectrum seems able to account for the observed spectrum. It is also found that the contribution to the total γ–ray spectrum by positron annihilation is significant. The source of emission is considered to be a region of trapped relativistic protons surrounding the central engine of the QSO.

159.042 **Far–ultraviolet spectroscopy of the quasar UM 675 with the Faint Object Spectrograph on the Hubble Space Telescope.**
E. A. Beaver, E. M. Burbidge, R. D. Cohen, V. T. Junkkarinen, R. W. Lyons, E. I. Rosenblatt, G. F. Hartig, B. Margon, A. F. Davidsen.
Astrophys. J., Lett., Vol. 377, No. 1, p. L1 – L4 (10 Aug 1991). The Second Special Hubble Space Telescope Issue.
Observations of UM 675 with the HST and FOS demonstrate the proper functioning of the FOS. The authors detect light in UM 675 down to 520 Å in the rest frame and set limits to He I emission, the He I Gunn – Peterson effect at z = 2.148, and Lyα absorption at z ≈ 0.5.

159.043 **The ultraviolet absorption spectrum of 3C 273.**
J. N. Bahcall, B. T. Jannuzi, D. P. Schneider, G. F. Hartig, R. Bohlin, V. Junkkarinen.
Astrophys. J., Lett., Vol. 377, No. 1, p. L5 – L8 (10 Aug 1991). The Second Special Hubble Space Telescope Issue.
The authors identify six Lyα absorption systems with equivalent widths greater than 0.2 Å in the ultraviolet spectrum of the nearby quasar 3C 273, more than expected on the basis of extrapolations of the observed number of Lyα clouds seen at large redshifts. Two Lyα lines appear to be produced by gas in the Virgo Cluster or by the halos of galaxies associated with the Virgo Cluster. The authors also detect all the interstellar absorption lines that are expected on the basis of abundance, atomic physics, and ionization considerations.

159.044 **First results from the Goddard High–Resolution Spectrograph: the Galactic halo and the Lyα forest at low redshift in 3C 273.**
S. L. Morris, R. J. Weymann, B. D. Savage, R. L. Gilliland.
Astrophys. J., Lett., Vol. 377, No. 1, p. L21 – L24 (10 Aug 1991). The Second Special Hubble Space Telescope Issue.
The authors present results of initial spectroscopy using the Goddard High–Resolution Spectrograph and the Hubble Space Telescope on the bright low–redshift quasar 3C 273. They detect absorption lines produced by gas in the disk and halo of the Milky Way from numerous neutral and weakly ionized atoms and also from highly ionized atoms (Si IV, C IV, and N V). The detection of N V provides additional strong support for the existence of gas with $T \sim 2\times10^5$K at large distances away from the disk of the Milky Way. The authors also find 10 absorption lines with equivalent widths (EW) greater than 50 mÅ, and four possible lines with EW between 25 and 50 mÅ, in the region between 1235 and 1406 Å. These cannot be identified with any lines in the Galactic interstellar medium and must therefore all be Lyα systems at very low redshifts. This number of lines is 5 – 10 times larger than expected from a simple extrapolation of the high–redshift evolution derived from ground–based observations and is consistent with no evolution in cloud properties since z = 2.

159.045 **An interstellar refraction model for the rapid radio polarization changes in the quasar 0917+624.**
J. H. Simonetti.
Astron. Astrophys., Vol. 250, No. 1, p. L1 – L4 (Oct 1991). Letter–to–the–editor.
The quasar 0917+624 is one of a set of "rapid radio variables" with significant intraday variations. A 180° polarization–position–angle swing, which occurred within one day, is commonly thought to rule out non–intrinsic models for the source variations. The author discusses a model which explains the position angle swing as due to refractive focusing from an interstellar shock moving across the line of sight. The aim is to demonstrate the possibility of an extrinsic explanation of the observations. A specific model which best reproduces the observed position angle swing assumes 0917+624 has a small component (<0.07 mas) whose polarization position angle is very different from that of the rest of the source; the shock lies at a distance of 100 pc, has a shock–front thickness of 0.04 AU, and a velocity of 178 km/s, while the life–of–sight electron column–density increases by $3\times10^{14}\mathrm{cm}^{-2}$ as the shock passes.

159.046 **Assemblage of two UVX–QSO samples to $B \sim 19$ in the vicinity of the South Galactic Pole.**
L. E. Campusano.
Astron. Astrophys., Vol. 250, No. 1, p. 9 – 22 (Oct 1991).
The author has defined two new optically selected samples of medium bright QSOs with $z < 2.2$ and $B \geqslant 19$. The sample areas are 22.9 and 2.5 deg^2 (the Large Field and Master Region respectively), the smaller one being a sub–area of the other. First, the author considered the contribution of QSOs coming from the complete spectroscopic follow–up $B = 19$ of the Cerro El Roble UVX search. 30 such QSOs with $z < 2.2$ were found in the spectroscopic survey of the 192 UVXs inside the Large Field using the 2.5 m du Pont telescope at Las Campanas Observatory. Secondly, from the ~80% complete spectroscopic follow–up of the objective–prism candidates (including the observations at Las Campanas), 11 additional QSOs with $z < 2.2$ were incorporated. Finally, 9 more QSOs with $z < 2.2$ were contributed by other five samples of QSO–candidates inside the Large Field, most of the latter having complete spectroscopic follow–up. This ensemble of 50 QSOs constitutes the medium bright sample (MBQS) in the Large Field, which is shown to be a well–defined UVX sample. The MBQS increases by ~40% the area of the Hubble diagram covered by "complete" samples in the

17.5–19 mag range. In connection with the MBQS, it is argued that the subset containing the 14 QSOs that lie inside the 2.5 deg^2 Master Region constitutes a sample on its own. The estimated completeness for the MBQS and the Master Region samples is of ~65–80% and of nearly 100%, respectively. Evidence in favour of inhomogeneities in the surface distribution of QSOs inside the Large Field is presented and discussed.

159.047 **Forbidden high–ionization lines in QSO spectra.**
I. Appenzeller, S. J. Wagner.
Astron. Astrophys., Vol. 250, No. 1, p. 57 – 61 (Oct 1991).

The authors observed [Fe VII] or [Ne V] lines in a sample of QSOs with different luminosities. Compared to the lower–luminosity Seyfert nuclei the high ionization forbidden lines were found to be enhanced relative to the lower–ionization forbidden lines, but of similar strengths relative to the nuclear continuum emission. As in lower–luminosity AGN, the forbidden high–ionization lines tend to be blueshifted relative to and broader than the low–ionization NLR features. The widths of the forbidden high–ionization lines of AGN appear correlated with the widths of the permitted BLR components. The observations are consistent with the assumption that the forbidden high–ionization lines result from photoionized gas located between the NLR and the BLR.

159.048 **Gravitational micro–lensing for large sources.**
S. Refsdal, R. Stabell.
Astron. Astrophys., Vol. 250, No. 1, p. 62 – 66 (Oct 1991).

The authors consider gravitational micro–lensing of large sources caused by a randomly distributed field of point masses (stars, black holes) in a foreground galaxy. For an angular source size larger than a few times the Einstein ring for the representative star mass, a simple analytical formula giving the typical (de)amplification due to micro–lensing is derived. Numerical test calculations agree very well with the analytical results. For some multiply imaged QSOs the authors can give tentative upper limits on the size of the broad emission line region.

159.049 **Erratum: "Against the $\Delta\ln(1+z) = 0.205$ periodicity in quasar redhifts" [Astron. Astrophys., Vol. 242, No. 1, p. 1 – 12 (Feb 1991)].**
D. Scott.
Astron. Astrophys., Vol. 250, No. 1, p. 291 (Oct 1991). See Abstr. 53.159.018.

159.050 **Optical variability of quasars: statistics and cosmological properties.**
E. Giallongo, D. Trevese, F. Vagnetti.
Astrophys. J., Vol. 377, No. 2, p. 345 – 348 (20 Aug 1991).

The long–term optical variability of a composite, complete sample of quasars is investigated. While no correlation with luminosity is detected, a positive correlation with redshift is found ($\varrho = 0.25$). These results are reconciled with previous opposite findings of other authors, by arguing that the analysis of variability can be biased by the combined effect of the structure function of light curves and of cosmological time dilation. Such bias becomes negligible if variability is measured by a magnitude difference at a fixed and large time lag (4 yr). An interpretation of this cosmological trend is given in terms of spectral variability. The effect of statistical biases on the estimate of the evolution rate for the quasar population is discussed.

159.051 **Accretion–disk coronae in QSOs.**
W. A. Stein.
Publ. Astron. Soc. Pac., Vol. 103, No. 666, p. 748 – 751 (Aug 1991).

Some potential problems for accretion–disk models are discussed. Radiation pressure on gas in the hypothesized accretion disks of QSOs will result in a corona. The coronal temperature will be very high – $T > 10^8$K. The implied mass loss rate suggests that the Compton optical depth may be large unless some physical process disrupts the mass loss. As a result there may be a problem for observation of the ultraviolet photoionizing continuum since it would be Compton scattered to X–ray energies in contradiction to observations. Accretion disk models should include schemes to disrupt the mass loss if they are to be consistent with observations.

159.052 **On the trajectories of the superluminal knots in 3C 345.**
Qian Shanjie, T. P. Krichbaum, A. Witzel,
A. Quirrenbach, C. A. Hummel, J. A. Zensus.
3. Chinese Academy of Sciences and Max–Planck Society Workshop: High energy astrophysics: compact stars and active galaxies, p. 80 – 84 (1991). – See Abstr. 012.026 for the main entry.

In the "classical" superluminal radio source 3C 345 the inner knots C4 and C5 have been moving along different curved trajectories with apparent axial symmetry. In this paper the authors discuss the possibility that the trajectories might be the projections of two helical fieldlines around the common jet–axis which is gradually curved northwards. Under this assumption, the patterns of the helical fields can be derived and the observed trajectories can be well fitted.

159.053 **QSO and CV survey with the BAO Schmidt CCD system.**
Chen Jiansheng, Jiang Zhaoji, Wei Mingzhi.
3. Chinese Academy of Sciences and Max–Planck Society Workshop: High energy astrophysics: compact stars and active galaxies, p. 155 – 157 (1991). – See Abstr. 012.026 for the main entry.

Here the authors deal with two distinct classes of objects, i.e. QSOs – the most remotest objects in the universe, and the CVs (cataclysmic variables), most of which, if not all, have been discovered within our own Galaxy, simply because the techniques developed for the survey of optically and X–ray selected QSOs can also be used to search for the CV candidates.

159.054 **Search for quasars on large regions.**
He Xiangtao.
3. Chinese Academy of Sciences and Max–Planck Society Workshop: High energy astrophysics: compact stars and active galaxies, p. 205 – 210 (1991). – See Abstr. 012.026 for the main entry.

159.055 **Quasar pair and quasar peculiar velocity.**
Chu Yaoquan.
3. Chinese Academy of Sciences and Max–Planck Society Workshop: High energy astrophysics: compact stars and active galaxies, p. 265 (1991). Abstract. Abstract. – See Abstr. 012.026 for the main entry.

159.056 **Hubble diagram for QSOs with middle and weak interplanetary scintillation.**
Xia Changli, Wang Jianmin.
3. Chinese Academy of Sciences and Max–Planck Society Workshop: High energy astrophysics: compact stars and active galaxies, p. 270 – 273 (1991). – See Abstr. 012.026 for the main entry.

The QSOs are very scattered in the Hubble diagram. As mentioned by Qu Qinyue et al. (1979), very strongly scintillating quasars should have the same physical nature and an approximate absolute magnitude because the scintillating flux comes from the compact component in radio sources. They studied the Hubble relation of QSOs with strong interplanetary scintillations, whose ratio of the flux in the scintillating component to the total flux of the quasar is larger than 0.4. They have concluded that such QSOs could be considered as "standard candle". According to the above consideration, quasars with middle and weak interplanetary scintillation, with the above ratio smaller than 0.4, should have also the same physical nature and an approximate absolute magnitude. In this paper, the authors have made a deeper analysis of the magnitude–redshift relation of QSOs. It is concluded that there is such a relation for the sample.

159.057 **3C 273: periodic optical flares and their association with superluminal motions in the milliarcsecond jet.**
E. T. Belokon'.
Astron. Tsirk., No. 1544, p. 3 – 4 (May – Jul 1990).

159.058 **Multi-frequency observation of quasar 3C279 in the outburst.**
F. Makino, T. Kii, K. Hayashida, T. Ohashi, M. J. L. Turner, A. C. Sadun, C. M. Urry, G. Neugebauer, K. Matthews, H. Teräsranta, M. F. Aller.
Georgia State University Conference on Variability of Active Galactic Nuclei, p. 13 – 20 (1991). – See Abstr. 012.040 for the main entry.
The X-ray spectrum of 3C279 was observed in 1987, 1988 and on two occasions in 1989. The intensity in 1988 was 6 times higher than that of 1987. The authors applied the homogeneous SSC model and obtained beaming factor of 14, magnetic field of 0.08 gauss and angular diameter of 0.01 mas.

159.059 **Redshift-dependent optical variability from complete QSO samples.**
E. Giallongo, D. Trevese, F. Vagnetti.
Georgia State University Conference on Variability of Active Galactic Nuclei, p. 75 – 77 (1991). – See Abstr. 012.040 for the main entry.
Using magnitude changes at large time lag, the authors perform standard regression analysis of variability vs. optical luminosity and redshift for two complete QSO samples. They find no correlation with absolute magnitude and a positive correlation with redshift, significant at 95% confidence level. Possible interpretations are suggested.

159.060 **The optical variability of Palomar-Green quasars.**
W. C. Keel, P. M. Allan.
Georgia State University Conference on Variability of Active Galactic Nuclei, p. 83 – 86 (1991). – See Abstr. 012.040 for the main entry.
To understand properly what the role of variability is in the whole AGN population, it woulkd be very desirable to know the variability properties of a well-defined, complete sample of objects. As a step in this direction, the authors have obtained high-accuracy photometry for nearly all the objects in the Palomar-Green (PG) bright quasar survey. The PG sample is an excellent source list for overall studies of quasar properties. The authors report here observations of these objects, obtained over a period of 3.5 years to study the variability of such an optically complete sample.

159.061 **Recent activity of the quasar 3C 279.**
A. C. Sadun, S. Fajardo, M. T. Carini.
Georgia State University Conference on Variability of Active Galactic Nuclei, p. 131 – 133 (1991). – See Abstr. 012.040 for the main entry.

159.062 **43-GHz VLBI imaging of the quasar 3C345.**
J. A. Zensus.
Georgia State University Conference on Variability of Active Galactic Nuclei, p. 151 – 152 (1991). – See Abstr. 012.040 for the main entry.
43 GHz observations of the quasar 3C345 yield an image with a resolution of 0.2 milliarcsec. The source structure is identified with the core and superluminal jet-components that are seen in images at lower frequencies. This improves the measurements of the superluminal speeds in core-vicinity, and provides additional evidence for different, curved trajectories of successive components in the jet.

159.063 **Is quasar 1038+064 rotating?**
C. E. Akujor, C. H. McGruder III.
Georgia State University Conference on Variability of Active Galactic Nuclei, p. 169 – 171 (1991). – See Abstr. 012.040 for the main entry.
Radio images of the flat spectrum, high redshift (z = 1.127) quasar 1038+064 at different wavelengths are presented. It has a jet which appears to be performing angular motion. The authors calculate an apparent angular velocity.

159.064 **The cm-wavelength flux and polarization of the sources in the Pearson-Readhead sample.**
M. F. Aller, H. D. Aller, P. A. Hughes.
Georgia State University Conference on Variability of Active Galactic Nuclei, p. 184 – 186 (1991). – See Abstr. 012.040 for the main entry.

159.065 **Parametric decay instability in 3C 273.**
R. T. Gangadhara, V. Krishan.
Georgia State University Conference on Variability of Active Galactic Nuclei, p. 190 – 193 (1991). – See Abstr. 012.040 for the main entry.
The parametric decay instability (PDI) causes anomalous absorption of strong electromagnetic radiation under specific conditions of energy and momentum conservation. It is shown that in the source 3C 273, the conditions are conductive to the excitation of PDI. This process also contributes towards the absorption of 21-cm radiation.

159.066 **Observations of the cores of ten IRAS-detected radio quasars at 230 GHz.**
Gopal-Krishna, H. Steppe.
Georgia State University Conference on Variability of Active Galactic Nuclei, p. 194 – 197 (1991). – See Abstr. 012.040 for the main entry.

159.067 **X-ray spectrum of the radio-loud quasar PKS 1217+023.**
K. K. Ghosh, S. Soundararajaperumal.
Astron. Astrophys., Vol. 252, No. 1, p. 53 – 55 (Dec 1991).
A study of the X-ray (0.1–10 keV) spectrum from the radio-loud quasar, PKS 1217+023 is presented. This quasar was observed with EXOSAT on two epochs in 1984 January 29 and May 14. The data were obtained from the EXOSAT archives. The soft (0.1–2 keV) and hard (2–10 keV) X-ray fluxes have displayed significant changes between the two observations. There is neither a detection of any significant low energy absorption within the quasar nor is there any evidence for the presence of soft X-ray excess. The energy indices, α, of the power law fits to the LE+ME data of January 29 and May 14 are 0.98(+0.18,-0.14) and 0.82(+0.12,-0.11) (error bars are with 90% confindence) respectively. The values of α are consistent with the values obtained from Ginga observations but larger than those obtained from Einstein data. A thermal bremsstrahlung model also provides a good fit to the X-ray data of PKS 1217+023.

159.068 **Initial light curve of Q2237+0305.**
R. T. Corrigan, M. J. Irwin, J. Arnaud, G. G. Fahlman, J. M. Fletcher, P. C. Hewett, J. N. Hewitt, O. le Fevre, R. McClure, C. J. Pritchet, D. P. Schneider, E. L. Turner, R. L. Webster, H. K. C. Yee.
Astron. J., Vol. 102, No. 1, p. 34 – 40 (Jul 1991). Current Physics Microform No.: 9107B1730.
The authors present CCD photometry for the gravitationally lensed quasar system 2237+0305, in optical passbands from B through R, taken over a time period of more than 3 yr. These data provide new information about the probable microlensing event reported in Irwin et al. (1989); the rise time of this feature is approximately 26 days. Four additional independent brightness changes in the quasar images are detected.

159.069 **Fifth image and photometric variability in 2237+0305 ("Einstein Cross").**
R. Racine.
Astron. J., Vol. 102, No. 2, p. 454 – 460 (Aug 1991). Current Physics Microform No.: 9108A1390. With plate 16.
CCD frames in R and I obtained with HRCam at the CFHT in 0".4 (FWHM) seeing on 1990 August 23.3 U.T. were combined to produce S/N > 500 images at 0".13/pixel sampling of the brighter components of the quasar gravitationally lensed by 2237+0305.

Relative magnitudes ($\pm < 0.01$ mag), positions ($\pm < 0\rlap{.}''002$) and angular sizes are derived for the four previously known quasar images and for the core of the lensing galaxy. Compared with Yee's (1988) photometry, these data show components B brighter by 0.25 mag and C fainter by 0.15 mag while the brightness of A and D and the colors of all four images are unchanged. The general variability of all four components between 1986 and 1990 is discussed. The intrinsic dispersions of the intensity profiles are $\sigma < 0\rlap{.}''08$ for the quasar images and $\sigma = 0\rlap{.}''21 \times 0\rlap{.}''15 \pm 0\rlap{.}''05$ (FWHM = $0\rlap{.}''50 \times 0\rlap{.}''35$) in P.A. = $83° \pm 2°$ for the galaxy core. Subtraction of these components leaves a fifth stellar image 4.5 ± 0.3 mag fainter than component A and $0\rlap{.}''07 \pm 0\rlap{.}''035$ SSE of the lensing galaxy center.

159.070 **The large, bright QSO survey. IV. QSOs in two equatorial fields.**
F. H. Chaffee, C. B. Foltz, P. C. Hewett, P. J. Francis, R. J. Weymann, S. L. Morris, S. F. Anderson, G. M. MacAlpine.
Astron. J., Vol. 102, No. 2, p. 461 – 487 (Aug 1991). Current Physics Microform No.: 9108A1397.

This is the fourth paper in a series aimed at selecting ~1000 QSOs brighter than $B_J \approx 18.7$ using machine–scanned direct and objective–prism plates from the UK Schmidt Telescope. The plate material is scanned at the Automated Plate Measuring facility. The candidate list is derived using a number of complementary selection algorithms ranging from "traditional" criteria, such as the presence of strong emission features, to criteria designed to select objects whose objective–prism spectra cannot be classified as normal stars, although they may possess no obvious QSO–like features. Follow–up spectroscopy at the Multiple Mirror Telescope is used to classify each candidate. In this paper the authors present results from the two final UK Schmidt Telescope equatorial fields in which 125 QSOs have been found as well as the detection of an additional 11 QSOs in the 0000+0000 field presented in the second paper in this series. The authors present coordinates, magnitudes, redshifts and spectra of moderate resolution, and signal–to–noise ratio for all 136 QSOs as well as for eight additional extragalactic objects which fail to meet the authors' absolute magnitude criterion as QSOs.

159.071 **The nature of the galaxy around 3C 48.**
A. Stockton, S. E. Ridgway.
Astron. J., Vol. 102, No. 2, p. 488 – 494 (Aug 1991). Current Physics Microform No.: 9108A1424. With plate 17.

The authors present new images of the QSO 3C 48 in line–free bands centered at 0.58, 0.77, and 2.1 μm and in a narrowband centered on the [O III] λ5007 line. Careful profile subtraction of the continuum images shows an apparent luminosity peak centered about 1″ northeast of the QSO position, which the authors tentatively identify as the nucleus of a galaxy in the process of merging with the host galaxy of 3C 48. The arclike structure extending to the northwest appears to be a tidal tail associated with the companion galaxy. While the confirmation that the host galaxy of 3C 48 is interacting is consistent with the inference that 3C 48 and similar objects are transitional between ultraluminous starburst galaxies and classical QSOs, an apparent correlation between far–infrared spectral properties and optical morphology suggests that the supposed "transitional" characteristics may persist for at least 10^8yr, so the place of objects like 3C 48 in an evolutionary scenario depends on the typical lifetime of QSO activity.

159.072 **Search of the 3–point correlation function for quasars.**
M. J. Pons Borderia, A. Iovino, S. A. Bonometto.
Astron. J., Vol. 102, No. 2, p. 495 – 501 (Aug 1991). Current Physics Microform No.: 9108A1431.

Detecting a 3–point correlation function for a set of clustered objects gives significant information on the nature of the clustering. Analyzing available quasar samples, the authors find that their 3–point correlation function does not vanish, but is likely to be smaller than for ordinary galaxies. A clustering model based on the Kirkwood superposition hypothesis, as expected for objects related to very high fluctuation peaks, seems to be excluded. Rather, the possibility that quasars are a phenomenon associated with primeval galaxy evolution seems favored. The analysis is significant also from a technical point of view, as it outlines some difficulties found when analyzing a sample consisting of many small subsamples, and how they can be possibly overcome.

159.073 **Spectroscopy of quasistellar objects belonging to the medium bright sample in the South Galactic Pole.**
L. E. Campusano.
Astron. J., Vol. 102, No. 2, p. 502 – 517 (Aug 1991). Current Physics Microform No.: 9108A1438. With plate 18.

The author presents the redshifts, energy distributions, and some of the spectral parameters of 35 quasistellar objects. Thirty–four belong to the Medium Bright QSO Sample ($B \leqslant 19$ and $z < 2.2$) in the South Galactic Pole (Campusano, 1991) and were primarily observed with the 2.5 m du Pont telescope of the Las Campanas Observatory. Thirty–one of them come from the Cerro El Roble UVX search (Campusano & Torres, 1983), three from the objective–prism search of Clowes & Savage (1983), and one from an unpublished grism search conducted with the Cerro Tololo Inter–American Observatory 4 m telescope. This work contains the discovery report for the grism–selected QSO 00539–2805 and the confirmatory spectroscopy for 28 of the 35 quasistellar objects. In the available cases, the redshift and equivalent width determinations were found to agree well with previous measurements. The equivalent width and FWHM distributions for the strong emission lines of the author's sample are examined and compared with similar previous data. Only the FWHM distribution for the C III] line seems to differ from a comparison one and is suggestive of a luminosity effect. The C IV 1549 line is found to be blueshifted with respect to Lα by ~600 ± 200 km/s. The C IV line strengths appear to follow the Baldwin relation with a slope close to Baldwin's (1977) original value.

159.074 **PC 1247+3406: an optically selected quasar with a redshift of 4.897.**
D. P. Schneider, M. Schmidt, J. E. Gunn.
Astron. J., Vol. 102, No. 3, p. 837 – 840 (Sep 1991). Current Physics Microform No.: 9109D1241.

This paper describes observations of PC 1247+3406, a quasar with a redshift of 4.897 ± 0.011. The redshift is based on three strong emission lines and the presence of substantial absorption, presumably due to "Lyman α clouds", on the blue side of the Lyman α emission line. The quasar is about the same brightness as PC 1158+4635, the 20 th magnitude $z = 4.733$ quasar found in 1989. PC 1247+3406 was discovered in a four–filter CCD survey designed to detect quasars with redshifts between 4.0 and 5.5. The properties of PC 1247+3406 seem quite similar to other $z > 3$ quasars, although the absorption due to intervening neutral hydrogen is ≈20% less than expected. The flux in the spectrum abruptly drops at a wavelength of ≈910 Å in the rest frame of the quasar. This quick encounter with a Lyman–limit system is frequently seen in other $z > 4.3$ quasars.

159.075 **Expected color variations of the gravitationally microlensed QSO 2237+0305.**
J. Wambsganss, B. Paczyński.
Astron. J., Vol. 102, No. 3, p. 864 – 868 (Sep 1991). Current Physics Microform No.: 9109D1268.

The continuum source of a quasar may have different sizes in different wavelengths. When such a source crosses a caustic due to gravitational microlensing the observed color will change. The authors show typical light curves and the corresponding color curves for their model of QSO 2237+0305A, the image that has been reported to vary. The authors present probabilities for color changes of certain amplitudes as functions of the observed magnitude changes for different time intervals between the two observational epochs, different source sizes, and different source size ratios. It may be possible to find the ratio of the continuum source sizes in the two color bands, once a color change correlated with a continuum change is found. With a few observations showing the shape of a color event one may even determine the absolute size of the quasar in the continuum.

159.076 **Very faint quasar survey: procedures and preliminary results.**
D. Schade.
Astron. J., Vol. 102, No. 3, p. 869 – 874 (Sep 1991). Current Physics Microform No.: 9109D1273.
A survey for faint emission line objects has been carried out at Canada–France–Hawaii Telescope using multiaperture spectroscopy. The limiting magnitude is V = 23.1 and the survey covers an area of 0.068 square degrees. Above an emission line signal–to–noise threshold of 6, ten very faint emission line galaxies were detected as well as five quasar candidates. The faintest quasar candidates were detected by the presence of a single strong emission line and are likely to be at $z \sim 3$ although follow–up spectroscopy is required to confirm the redshifts. The survey (which is most sensitive over the redshift range $2.4 < z < 4.7$) is well calibrated and uses emission line search procedures designed to yield reliable space densities from the sample of objects obtained.

159.077 **The synchrotron light from the jet of 3C 273.**
H.-J. Röser, K. Meisenheimer.
Astron. Astrophys., Vol. 252, No. 2, p. 458 – 474 (Dec 1991).
New optical data from polarimetry, photometry and astrometry are presented for the jet of 3C 273. They confirm the high degree of polarization at the outer end of the jet as well as the drop in polarization level in the inner region. The optical hot spot counterpart at the outer extreme of the jet reveals itself by a 90° flip in polarization angle very similar to that found in the radio data. These findings confirm the synchrotron nature of the optical radiation and – based on life time arguments for the relativistic particles – demand in situ particle acceleration throughout the visible jet. Spectral index measurements along the jet show a spectrum $f_\nu \sim \nu^{-0.9}$ in the part with low polarization steepening with increasing distance from the nucleus to $\nu^{-1.8}$. There is no pronounced radio counterpart to the peak at the outer end of the optical jet. Based on their colours the extensions at both ends of the jet are probably chance coincidences with unrelated galaxies.

159.078 **First– and second–epoch VLBI observations of the gravitational lens system 2016+112.**
M. B. Heflin, M. V. Gorenstein, C. R. Lawrence, B. F. Burke.
Astrophys. J., Vol. 378, No. 2, p. 519 – 536 (10 Sep 1991).
Two 18 cm Mark III VLBI observations of the 2016+112 gravitational lens system indicate that A and B have structure at the milliarcsecond level. Neither A nor B can be described with only one component. The detailed brightness distributions for A and B are not uniquely determined, however, because of sparse sampling and low S/N ratios. Many models are considered. One hypothesis consistent with all of the models is that images A and B are not related by linear magnification because the source is evolving on a time scale comparable to or shorter than the time delay between images. The lens is a large structure consisting of at least two galaxies, so it is unlikely that the magnification matrix would change significantly in the 5.5 years between observational epochs. Any given model shows evidence for source evolution between epochs, but the details of that source evolution are highly model–dependent. More definite conclusions could be made if the brightness distributions of A and B could be derived without ambiguity. Significantly better sampling with an order of magnitude more data are necessary, along with a better understanding of the error contributions.

159.079 **Infrared K–band observations of the gravitational lens 2016+112.**
G. I. Langston, J. Fischer, C. Aspin.
Astron. J., Vol. 102, No. 4, p. 1253 – 1257 (Oct 1991). Current Physics Microform No.: 9110D1785. With plate 37.
Infrared observations of the gravitational lens 2016+112 at the K band ($\lambda = 2.2\,\mu$m) are presented. The infrared image is dominated by the extended emission of the radio quiet foreground galaxy, which acts as the gravitational lens. Infrared flux is seen at the location of all three radio loud components of the system. Gaussian fits to the data yield K–band fluxes for all four components. To within the errors of the fit, the i–K colors of the radio loud images are similar, while the color of the foreground galaxy is closer to that expected for an elliptical galaxy with $z \sim 1$. The relative brightnesses of the 2016+112 components at K band are significantly different than the optical g–band brightnesses. The K–band observations suggest that the central galaxy makes the dominant contribution to the mass acting as the gravitational lens. A model of the gravitational lens as a single massive elliptical galaxy is presented. This model suggests that all radio loud components of 2016+112 can be explained as images of the background quasar and are not due to a second foreground galaxy, as indicated by earlier work. Radio wavelength observations at the X band ($\lambda = 3.6$ cm) are included to show the compact radio structure of 2016+112, which also supports the single galaxy lens model.

159.080 **Persistence and change in the soft X–ray spectrum of the quasar PG 1211+143.**
M. Elvis, P. Giommi, B. J. Wilkes, J. McDowell.
Astrophys. J., Vol. 378, No. 2, p. 537 – 542 (10 Sep 1991).
Two Einstein and three EXOSAT observations of PG 1211+143 over a 6 year baseline show that strong steep–spectrum low energy X–ray emission is a persistent feature of this quasar. EXOSAT observations of PG 1211+143 detected a factor of 2.3 increase in its steep soft X–ray (C–band) flux in 18 days. The bulk of the soft X–ray emission of PG 1211+143 therefore comes from a region less than 5×10^{16}cm across. In another interval of 193 days the soft X–rays decreased by a factor of 3.7. Either an intrinsic emission variation or an increase of N_H from 3.4 to 7.6×10^{20} atoms cm^{-2} could produce the same change. In the same time interval, the hard (2 – 10 keV) X–rays decreased by a factor 1.6 ± 0.05. This suggests a connection between the two energy regimes and argues against variable absorption causing the soft X–ray variations. In order to decrease in luminosity so rapidly a thermal source in PG 1211+143 would have to be optically thick to both electron scattering and free–bound absorption.

159.081 **Samples of quasars in selected areas of the sky.**
V. L. Afanas'ev, V. V. Vlasyuk, S. N. Dodonov, H. Lorenz, V. Yu. Terebizh.
Bull. Spec. Astrophys. Obs. – North Caucasus, Vol. 32, p. 51 – 56 (1990). English translation of Astrofiz. Issled. Izv. Spets. Astrofiz. Obs., Tom 32, p. 66 – 72 (1990).
Samples of quasars obtained during photometric and spectroscopic investigations of objects in fields SA68 ($00^h14^m51.03^s$, $+15°36'19.1''$), M82 ($9^h53^m28.33^s$, $+69°51'12.7''$), and SA57 ($13^h06^m03.69^s$, $+29°36'31.8''$) are examined in the work. Lists of objects, their coordinates, magnitudes, and redshifts are given. Some properties of the quasar samples are examined, the surface density of quasars is estimated, and a comparison of the values obtained with the published data is given.

159.082 **Spectral differences in the QSO pairs Q1634+267A,B and Q2345+007A,B and the gravitational lens interpretation.**
C. C. Steidel, W. L. W. Sargent.
Astron. J., Vol. 102, No. 5, p. 1610 – 1626 (Nov 1991). Current Physics Microform No.: 9111D0616.
The authors present new high signal–to–noise ratio, moderate dispersion spectra over the wavelength range 3150–7000 Å of two of the widest separation, unconfirmed gravitational lens candidates, Q1634+267A,B (separation = 3.77″) and Q2345+007A,B (separation = 7.03″). Aside from the obvious interest in establishing whether or not these pairs are gravitationally lensed images, both of these pairs are capable of setting extremely interesting limits on the sizes of intervening Lyman α and heavy–element absorption systems. For Q1634+267A,B, the detailed line profile and continuum shapes strongly support the gravitational lensing hypothesis; however, the authors find that

the Ly α, N V, C IV, Si IV emission lines of the two images exhibit as much as a 1000 km s^{-1} relative shift in velocity, whereas the C III] and Mg II lines are consistent with no velocity shift, as found by previous workers. While the emission line profiles and redshifts of Q2345+007A and B are remarkably similar ($\Delta v_{B-A} = 15 \pm 20$ km s^{-1}), the line-to-continuum ratios for various lines differ markedly; the intensity ratios for C IV, Si IV, and Mg II appear to scale roughly with the continuum ratio of the two spectra, whereas the Lyman α and especially C III] intensity ratios differ significantly from the observed continuum ratio. The authors conclude that both of the pairs in question are gravitationally lensed; however, the peculiarities of the emission line properties are puzzling, and may be providing information on the structure of QSO broad line regions. The authors discuss several possible origins for the differing emission line spectra under the gravitational lensing hypothesis.

159.083 **The large, bright QSO survey. V. QSOs in three southern fields.**
S. L. Morris, R. J. Weymann, S. F. Anderson, P. C. Hewett, C. B. Foltz, F. H. Chaffee, P. J. Francis, G. M. MacAlpine.
Astron. J., Vol. 102, No. 5, p. 1627 – 1658 (Nov 1991). Current Physics Microform No.: 9111D0633.

This is the fifth and final paper in a series reporting the results of a program aimed at selecting ~1000 QSOs brighter than $B_J \approx 18.75$ using machine-scanned direct and objective-prism plates from the UK Schmidt Telescope. The plate material is scanned at the Automated Plate Measuring facility. The candidate list is derived using a number of complementary selection algorithms ranging from "traditional" criteria, such as the presence of strong emission features, to criteria designed to select objects whose objective-prism spectra cannot be classified as normal stars, although they may possess no obvious QSO-like features. Follow-up spectroscopy at the 2.5 m duPont telescope of the Las Campanas Observatory is used to classify each candidate. In this paper the authors present results from three UK Schmidt Telescope fields in which 200 QSOs have been found, 182 of which form part of the LBQS sample. Coordinates, magnitudes, redshifts, and spectra of moderate resolution and signal-to-noise ratio are given for all 200 QSOs as well as for 3 additional extragalactic objects which fail to meet the authors' absolute magnitude criterion as QSOs.

159.084 **The visual wavelength continua of quasars: a reexamination.**
W. A. Stein.
Astrophys. J., Vol. 379, No. 1, p. 107 – 112 (20 Sep 1991).

In recent years the preferred model of origin of the ultraviolet photoionizing continuum of QSOs is that of the hot thermal accretion disk. The model spectra of these disks are very broad and extend through the visual spectral range. However, observations indicate a power-law description of the visual continuum with negative spectral index typically, and it is not clear that accretion disk spectra can satisfy this constraint as well as extend to the low-energy X-ray regime. Synchrotron spectra exhibit a power-law character, and there is no difficulty extending them to ultraviolet wavelenths for photoionization and, with an appropriate spectral break, to the low-energy X-ray regime as well. The problems associated with attributing the visual continuum to various mechanisms are discussed here.

159.085 **Structure in host galaxies of steep-spectrum radio QSOs.**
D. L. Block, A. Stockton.
Astron. J., Vol. 102, No. 6, p. 1928 – 1932 (Dec 1991). Current Physics Microform No.: 9112C1904. With plates 59 – 63.

The authors present deep imaging in two bands of five high-luminosity steep-spectrum radio QSOs. The set of broadband filters the authors have used has been specifically designed to exclude contamination from all strong emission lines found in extended emission regions around QSOs. For four of the objects, the authors also have deep R-band imaging. Because they generally do not have suitable stars for determining the point-spread function within their limited fields, the authors have developed a simple technique for detecting asymmetric structure close to the QSO by using a rotational average of the QSO profile itself. This technique allows the authors to examine the region immediately surrounding the QSO, and they have successfully detected at least one new galaxy and several other features not evident on their original images. All five objects show either close companions or substantial host-galaxy asymmetries; these results support the view that interactions are important in luminous QSOs, but the lack of a suitable control sample prevents giving quantitative statistical substance to this statement. Companion galaxies span the whole available color range, while diffuse extensions from two of the QSOs are both quite blue; this last result, however, could well be the result of observational bias.

159.086 **Ultraviolet observations of a bright southern quasar.**
S. Kirhakos, M. A. Strauss, A. Yahil, M. Griffith, K. B. Fisher.
Astron. J., Vol. 102, No. 6, p. 1933 – 1938 (Dec 1991). Current Physics Microform No.: 9112C1909.

The authors present low-resolution ultraviolet spectra of the QSO 09149-6206 ($z = 0.057$) obtained with the International Ultraviolet Explorer. The spectrum shows strong emission lines of Ly α, C IV, C III], Mg II, and Fe II. The Fe II blend centered at 2080 Å is unusually strong; this may be a consequence of overcorrection for the 2175 Å silicate feature in the reddening curve. Absorption features from the Milky Way are clearly present. There is no evidence for further absorption features due to objects beyond our Galaxy and in the foreground of the QSO. After correcting for Galactic reddening, the spectral energy distribution is very flat from ultraviolet to far-infrared wavelengths; there is no big blue bump in this object. The QSO is detected weakly at 6 cm; the authors have only upper limits at X-ray wavelengths. The upper limit to the optical-X-ray spectral index is -1.35. The authors have measured the Hα line strength from a spectrum taken under photometric conditions; the Lyα/Hα line ratio is 0.8 after correction for reddening. A spectrum of the fuzz surrounding the QSO shows narrow emission lines of Hα, [N II], [S II], and [O I].

159.087 **Interpreting the light curve of Q2237+0305.**
R. L. Webster, A. M. N. Ferguson, R. T. Corrigan, M. J. Irwin.
Astron. J., Vol. 102, No. 6, p. 1939 – 1945 (Dec 1991). Current Physics Microform No.: 9112C1915.

Theoretical modeling of the mass distribution in the lens for Q2237+0305 will not give an accurate estimate of the microlensing parameters for each of the images. It will be possible to statistically estimate microlensing parameters from the data, however the timescales over which data must be collected, may be uncomfortably long. The required timescale depends on both the mean compact object mass and the apparent transverse velocity of the source, and is $\gtrsim 130(m/M_{\odot})^{0.5} \times (5000 \text{ km s}^{-1}/v_t)$ years. The observation of the quasar continuum source crossing a caustic or cusp line gives a good estimate of the source size. The authors' best estimate from the current data is $\lesssim 4 \times 10^{-4}(v_t/5000 \text{ km s}^{-1})$ pc. General characteristics of the light curves will tell us about the nature of cluster and halo dark matter. Preliminary indications, principally from the lensed quasar 0957+561, suggest that cluster dark matter comprises objects smaller than $\sim 0.1\ M_{\odot}$. A better understanding of the mass distribution in the principle lensing galaxy for these two gravitationally lensed quasars and careful monitoring of the individual images on a regular basis (weekly for Q2237+0305) would provide the required statistical base to better determine the quasar source size, the nature of dark matter, and the masses of compact objects responsible for microlensing.

159.088 **The connection between the low polarization quasar 3C 273 and blazars.**
L. Valtaoja, E. Valtaoja, N. M. Shakhovskoy (*N. M. Shakhovskoj*), Yu. S. Efimov, L. O. Takalo, A. Sillanpää, M. Tornikoski, M. R. Kidger, J.-A. de Diego.
Astron. J., Vol. 102, No. 6, p. 1946 – 1955 (Dec 1991). Current Physics Microform No.: 9112C1922.

The authors present photopolarimetric observations in UBVRI bands and near-infrared photometric observations of the low polarization quasar (LPQ) 3C 273 obtained between February 1989 and May 1990. The degree of polarization stayed low (below 1%) for most of the observing period and generally decreased towards shorter wavelengths. In early 1990 the authors observed optical total flux and polarization flaring, during which the degree of polarization rose up to 1.5%, and the strength and the character of the frequency dependence also varied with time. The optical flare coincided with the beginnings of a 90 GHz radio outburst, suggesting the same origin for both. The authors have earlier proposed a model in which variations in the frequency dependent optical polarization (FDP) in blazars are caused by optically thin radiation from growing radio shocks. The same model can explain the behavior of 3C 273 during the 1990 flaring event, indicating that the emission mechanisms of the LPQ 3C 273 are similar to those of highly polarized quasars and BL Lacs. A comparison of the existing optical polarization data with radio monitoring shows a tendency for polarization flares and FDP events to occur in the beginnings of large radio outbursts, as predicted by the model. The polarizations and the position angles of the variable components in optical and in radio are similar, also indicating that the radiation comes from the same component.

159.089 **The origin of quasar correlations.**
N. A. Bahcall, A. Chokshi.
Astrophys. J., Lett., Vol. 380, No. 1, p. L9 – L12 (10 Oct 1991).

The authors suggest that the origin of the observed quasar correlation function, which is intermediate in strength between that of galaxies and of rich clusters, is the preferential location of quasars in small groups of galaxies. The quasars then trace the same large–scale structure of the parent galaxy groups. According to this picture, optically selected quasars are located in small groups of ~10L* galaxies on average, and have the same correlations as expected for the small groups. Radio–selected quasars are, according to the same picture, located in richer groups of ~30L* galaxies on average, having the stronger correlations expected for these richer groups. These results are consistent with observations of quasar environment and with the universal richness – dependent cluster correlation function. The results also imply that groups and clusters of galaxies, as well as the large–scale structure they trace, exist at the high redshifts observed in quasar correlations.

159.090 **EUV radiation from active galactic nuclei and quasars.**
H. L. Marshall.
1. Berkeley Colloquium on Extreme Ultraviolet Astronomy, p. 228 – 237 (1991). – See Abstr. 012.045 for the main entry.

Contents: 1. Introduction. 2. Quasar spectra. 3. Distribution of column densities. 4. Predicted quasar count rates. 5. Detectability of quasars. 6. Spectroscopy of quasars. 7. Conclusions.

159.091 **EUV excesses in quasars.**
M. Elvis, J. C. McDowell, B. J. Wilkes.
1. Berkeley Colloquium on Extreme Ultraviolet Astronomy, p. 238 – 244 (1991). – See Abstr. 012.045 for the main entry.

Contents: 1. EUV excesses – accretion disk signatures? 2. Other kinds of soft excess. 3. Common and persistent. 4. Variable, optically thick. 5. Steep spectra. 6. Connection to the rest of the quasar continuum? 7. Visible at 100 Å?

159.092 **3C 345.**
IAU Circ., No. 5308, p. 1 (17 Jul 1991).

159.093 **3C 345: is the variability of quasars nonlinear?**
R. Vio, S. Cristiani, O. Lessi, L. Salvadori.
Astrophys. J., Vol. 380, No. 2, p. 351 – 356 (20 Oct 1991).

The authors analyze the optical light curve of the OVV quasar 3C 345, showing that it is generated by a nonlinear and nonstationary stochastic process. The observed characteristics of the 3C 345 variability, such as the varying power spectrum and the strong luminosity bursts, are easily accounted for in this mathematical framework. Conventional algorithms suitable for linear analysis (power spectrum, structure function, covariance function, ...) are not useful in this case and different approaches have to be devised. The physics acting in the central regions of QSOs is probably nonlinear and dominated by a quantity subject to random variations.

159.094 **A search for variability in PKS 1302–102.**
N. Jackson.
Workshop on Variability of Active Galaxies, p. 79 – 82 (1991). – See Abstr. 012.070 for the main entry.

159.095 **Intraday radio variability of quasars and BL Lac objects.**
A. Quirrenbach.
Workshop on Variability of Active Galaxies, p. 131 – 141 (1991). – See Abstr. 012.070 for the main entry.

Light curves of 49 flat spectrum radio sources with time resolutions of 1 to 4 hours have been obtained with the Effelsberg 100 m telescope and the VLA at cm wavelengths. Intraday variations with amplitudes up to ~30% were found in several sources; in the quasar 0917+624 they are correlated with variations of the polarized flux density by a factor ~3. In the BL Lac object 0716+714, coordinated radio and optical observations have revealed correlated variations in both wavelength regimes with typical times scales between ~1 and ~7 days. The latter findings indicate that large amplitude intraday variations are probably intrinsic to the sources; a statistical analysis suggests that that additional contributions of interstellar scintillation are present at a lower level.

159.096 **Optical observations of rapid variability in the QSO PG 0117+213.**
R. Ekins, S. Plunkett, B. McBreen.
Workshop on Variability of Active Galaxies, p. 181 – 186 (1991). – See Abstr. 012.070 for the main entry.

CCD imaging photometry data has been obtained on a sample of sources to search for rapid variability. Light curves and photometric data are presented for the QSO PG 0117+213 and the BL Lacertae objects 0414+009 and 0836+182. The B and V magnitudes of the QSO PG 0117+213 decreased by 0.86 and 0.30 in a time interval of 46 hours. In the rest frame of the QSO (z = 1.493) these variations occurred in only 18.5 hours. Furthermore this QSO varied significantly over a period of only 25 minutes. No short timescale variability was detected in the two BL Lacertae objects.

159.097 **Continuum variability in quasars and Seyfert galaxies.**
T. J.-L. Courvoisier.
Workshop on Variability of Active Galaxies, p. 241 – 246 (1991). – See Abstr. 012.070 for the main entry.

Multiwavelength observations of AGN give several pieces of information which help advance our understanding of the physics of the core regions. Very good correlations at zero lag are found between the UV and visible domains, suggesting that the variability cannot be due to variations in thin accretion discs. A possible significant correlation is found between the UV and the radio emission of 3C 273, which might be due to links between accretion and non thermal activity. Although no correlation between the X–ray flux and other components can be found, analysis including spectral information suggests that the X–ray emission may be due to Comptonised EUV flux, at least in the galaxy NGC 5548.

159.098 **The variability of Ly–alpha in 3C 273.**
P. T. O'Brien, T. J. Harries.
Workshop on Variability of Active Galaxies, p. 247 (1991). Abstract. – See Abstr. 012.070 for the main entry.

159.099 **The electromagnetic spectrum of the radio–quiet quasar 1821+643 and comparison with 3C 273.**
M.-H. Ulrich.
Workshop on Variability of Active Galaxies, p. 249 – 254 (1991). – See Abstr. 012.070 for the main entry.

The quasar 1821+643 whose ecliptic latitude is 86° is a good candidate for simultaneous observations at a number of frequencies during the ROSAT survey.

159.100 **Structure and variability in broad absorption lines of quasar spectra.**
V. L. Afanasjev (*V. L. Afanas'ev*), E. A. Nazarov, H. Lorenz.
Workshop on Variability of Active Galaxies, p. 281 – 286 (1991). – See Abstr. 012.070 for the main entry.
The authors present spectroscopic observations of two broad absorption line (BAL) QSOs 0752+617 and 0932+501, representing different classes in the complexity of structures in the absorption troughs. The aim is the determination of column densities and the analysis of the absorption line fine structures. Applying a Gauss–analysis of the absorption line profiles the authors show that the troughs may be interpreted as a set of several individual components. The column densities of absorbing atoms in these proposed clouds are estimated. Physical conditions in the clouds are likely to change monotoniously depending on their velocity. The authors compare spectra of three different seasons and find no apparent changes in the C IV and Si IV absorption troughs.

159.101 **Are the broad emission lines of quasars affected by gravitational microlensing?**
J. Wambsganss, P. Schneider.
Workshop on Variability of Active Galaxies, p. 287 – 289 (1991). – See Abstr. 012.070 for the main entry.

159.102 **Gravitational microlensing and the Hamburg quasar monitoring program.**
U. Borgeest, R. Kayser, S. Refsdal, J. Schramm, T. Schramm.
Workshop on Variability of Active Galaxies, p. 291 – 302 (1991). – See Abstr. 012.070 for the main entry.
In May 1989 started an extensive optical photometric monitoring program of quasars which are promising candidates for gravitational microlensing. The main intention is to detect and analyse microlensing high amplification events. From the data obtained up to now, good evidence was found that quasars closely associated with foreground galaxies show a higher degree of variability than usual ones and interpret this result in terms of microlensing. There is, however, no evidence for a correlation between variability and the presence of strong heavy element absorption systems. An overview is given on the basic microlensing theory and its applications to observed lightcurves. Then the authors discuss the monitoring program in some detail and present first results.

159.103 **3C 279.**
IAU Circ., No. 5311, p. 1 (22 Jul 1991).

159.104 **The systemic redshift of the quasar 1331+170.**
R. F. Carswell, C. M. Mountain, D. J. Robertson, S. M. Beard, A. R. Glendinning, D. C. Laird, L. C. Lawrence, D. Montgomery, G. Pentland, D. A. Pickup, I. A. Smith, J. A. Bailey, A. Bridger, M. M. Casali, T. R. Geballe, P. Puxley, M. G. Smith, G. S. Wright, S. K. Ramsay, A. C. Baker, B. R. Espey, M. J. Ward.
Astrophys. J., Lett., Vol. 381, No. 1, p. L5 – L8 (1 Nov 1991).
Infrared spectra covering the 1.6 and 2 μm regions of the quasar 1331+170, redshift $z = 2.09$, were obtained with the CGS4 spectrometer at the UK Infrared Telescope. These spectra show redshifted Hα, Hβ, and [O III] lines. The redshifts of the forbidden lines are similar to the low–ionization lines measured at optical wavelengths, but are significantly higher than the high–ionization line redshift. Thus the low–ionization lines are likely to provide a reliable estimate for the systemic redshift in this object, and the high–ionization material is therefore flowing out from, or into, the quasar and is subject to some obscuration.

159.105 **Supernova 1991ap actually QSO.**
IAU Circ., No. 5327, p. 1 (19 Aug 1991). Further information is given in No. 5328.

159.106 **A composite plot of far–infrared versus radio luminosity, and the origin of far–infrared luminosity in quasars.**
H. M. Sopp, P. Alexander.
Mon. Not. R. Astron. Soc., Vol. 251, No. 1, p. 14P – 16P (1 Jul 1991)
The authors have constructed a composite plot of far–infrared versus radio luminosity for late–type galaxies, Seyferts, quasars and radio galaxies. The most striking result is that the radio and far–infrared luminosities of radio–quiet quasars are correlated and follow the same correlation as normal star–forming galaxies and ultra–luminous infrared galaxies, whereas the radio–loud quasars have luminosities in both bands similar to those of radio galaxies. The authors conclude that the far–infrared emission from radio–quiet quasars is from star–forming host galaxies and not from active galactic nuclei. The far–infrared radio plot may be a powerful discriminator between host galaxy type.

159.107 **CO emission from the radio–quiet quasar E 1821+64 at $z = 0.301$: implication regarding the origin of its far infrared radiation.**
D. Alloin, R. Barvainis, R. Antonucci, M. Gordon.
6. IAP Astrophysics Meeting – IAU Colloquium No. 129: Structure and emission properties of accretion disks, p. 371 – 372 (1991). – See Abstr. 012.046 for the main entry.

159.108 **Microlensing and the structure of active galactic nucleus accretion disks.**
K. P. Rauch, R. D. Blandford.
Astrophys. J., Lett., Vol. 381, No. 2, p. L39 – L42 (10 Nov 1991).
Rapid variability has been reported in two of the four gravitationally lensed images of Q2237+0305, and this is attributed to microlensing caused by the intervening stars. The associated constraints on the source size and properties are studied and compared with a variety of stationary accretion disk models. It is found that graybody disks have a half–power radius $r \sim 2\times10^{16}(\nu/\nu_L)^{-1.7}(L_{L,46})^{0.5}(\varepsilon/0.1)^{-0.5}$ cm, where ν is the frequency, $L_{L,46}$ is the luminosity per log frequency in units of 10^{46}ergs s^{-1} at the Lyman limit ν_L, and ε is the emissivity relative to a blackbody. The reported microlensing variation in Q2237+0305 requires the disk size to be over 3 times smaller than a blackbody disk of similar luminosity, implying that the optical emission is either nonthermal or optically thin. An exploration of nonstationary disk models including orbiting transient hot spots leads to a similar conclusion. Implications for models of active galactic nucleus optical continua are briefly discussed.

159.109 **The mass of the galaxy lensing Q0957+561: limits on H_0.**
G. Rhee.
AIP Conf. Proc., No. 222, p. 459 – 462 (15 Apr 1991). Current Physics Microform No.: 9106E0459. – See Abstr. 012.069 for the main entry.
The double quasar Q0957+561 A B is believed to be a gravitational lens image formed by a galaxy at $z = 0.36$. The object is a single QSO at $z = 1.41$ with a double image produced by the intervening galaxy. The author has measured the velocity dispersion of the galaxy lensing Q0957+561 and reports this measurement and the corresponding bounds on H_0.

159.110 **A spectroscopic survey of faint QSOs.**
B. J. Boyle, L. R. Jones, T. Shanks.
Mon. Not. R. Astron. Soc., Vol. 251, No. 3, p. 482 – 507 (1 Aug 1991).
Positions, magnitudes and identifications are presented for a complete sample of 236 faint ($B \leqslant 22$ mag) stellar objects based on spectroscopic observations of multicolour selected samples from deep photographic plates. Similar information is also provided for a further 98 objects not included in the complete sample. The survey covers an effective area of 0.85 deg^2 and contains 66 newly identified QSOs. In total, 93% of the complete sample has been identified and unambiguous redshifts have been obtained for 85% of the QSOs. The integral surface density of low–redshift ($z < 2.2$) QSOs at $B < 22$ mag is estimated to be

67.6 ± 8.8 deg², confirming the strong flattening previously observed in the QSO number–magnitude relation at B ≳ 19.5 mag. At the same magnitude limit, the integral surface density of QSOs in the range 2.2 < z < 2.9 is found to be 9.2 ± 3.7 deg². Broad absorption line systems are observed in 5–10% of the QSOs in which such systems should be detectable, a similar percentage to those derived from brighter samples of QSOs. Emission line equivalent widths measured in this survey are similar to those derived from optically brighter QSO samples. The remainder of the survey is principally composed of galactic subdwarf stars and compact narrow emission line galaxies.

159.111 **The ultraviolet spectrum of the bright QSO H1821 + 643 and its continuum energy distribution.**
M. Kolman, J. P. Halpern, C. R. Shrader, A. V. Filippenko.
6. IAP Astrophysics Meeting – IAU Colloquium No. 129: Structure and emission properties of accretion disks, p. 453 – 454 (1991). – See Abstr. 012.046 for the main entry.

The authors report on the first UV observations of the bright QSO H1821 + 643. A possible turnover at the high frequency side of the UV continuum constraints fits of a thin accretion disk model to a large black hole mass ($M \cong 3 \times 10^9 M_\odot$) and a high accretion rate ($\dot{M} \cong 18\ M_\odot\ yr^{-1}$), but a small outer radius. The latter can be interpreted in the two–temperature accretion flow model, which is also consistent with the observed UV and X–ray variability.

159.112 **Depolarization asymmetry in the quasar 3C 47.**
I. Fernini, J. P. Leahy, J. O. Burns, J. P. Basart.
Astrophys. J., Vol. 381, No. 1, p. 63 – 71 (1 Nov 1991).

It has been reported recently that in many extended high–luminosity extragalactic radio sources, the lobe on the jet side shows less depolarization than the lobe on the counterjet side. This depolarization asymmetry has been interpreted as being produced by an external foreground screen. In this regard, the one–sided jet quasar 3C 47 was observed with the VLA at 6 and 20 cm. The authors derived maps of the projected magnetic field, depolarization, and rotation measure. Significant differences for these parameters have been found among the various components of the source. In particular, 3C 47 shows a strong depolarization asymmetry between the NE (counterjet side) and the SW (jet side). The distribution of the depolarization ratio is quite uniform in each lobe, but quite asymmetric between the two lobes. The authors argue that this is likely to be a signature of a foreground screen. The consequences of this model for the relativistic beaming hypothesis are also explored.

159.113 **The role of microlensing in estimates of the relative time delay for the gravitational images of Q 0957 + 561.**
E. E. Falco, J. Wambsganss, P. Schneider.
Mon. Not. R. Astron. Soc., Vol. 251, No. 4, p. 698 – 706 (15 Aug 1991).

The authors present a study of the effects of individual stars in the gravitational lens that forms the images A and B of the QSO Q0957 + 561. Their main result is evidence that it is very unlikely that microlensing effects dominate the observed fluctuations of the images. Therefore, the measured (but uncertain) correlation between the light curves of the images is likely due to intrinsic luminosity variations at the source, rather than to propagation effects at the lens. The authors find, however, that the disagreement between magnification estimates at optical and radio wavelengths is a strong indicator of microlensing. Thus, estimates for the time delay for A and B, $\Delta\tau_{BA}$, should be obtained for spectral bands that are least affected by microlensing, i.e., the largest emission regions in the QSO. However, we still lack a general method to separate intrinsic and microlensing fluctuations to obtain estimates of $\Delta\tau_{BA}$. The authors also highlight several problems associated with cross–correlating the unevenly sampled light curves of images A and B. Recent estimates for $\Delta\tau_{BA}$ are different for the two available optical data sets, and they depend strongly on the method used to cross-correlate the light curves. From simulated model data, the authors single out a reliable method for cross–correlation in the present situation. They also propose a method that yields simultaneous estimates for the time delay for an image pair and for their magnification ratio at optical wavelengths.

159.114 **The large, bright QSO survey, a digital objective–prism survey for quasars.**
C. B. Foltz.
Objective–prism and other surveys – a meeting in memory of Nicholas Sanduleak, p. 61 – 70 (1991). – See Abstr. 012.074 for the main entry.

The large, bright QSO Survey has as its goal the determination of the range of optical spectral properties for a large sample of optically–selected QSOs. The basis for selection of survey candidates is the consistent application of objectively–defined selection criteria to the digitized objective–prism spectra of a magnitude–limited sample. Candidates selected by computer algorithms have been observed with the MMT and the DuPont 2.5–m telescope, yielding an impressive dataset consisting of accurate coordinates, magnitudes, redshifts and optical spectra for more than 1000 QSOs. Details of the selection and followup procedures and the general properties of the sample are discussed.

159.115 **Thermal continuum emission of GQ Comae.**
A. Siemiginowska, R. Szczerba, M. L. Sitko.
6. IAP Astrophysics Meeting – IAU Colloquium No. 129: Structure and emission properties of accretion disks, p. 521 – 522 (1991). – See Abstr. 012.046 for the main entry.

Results of the modeling the IR/Opt/UV parts of the GQ Com spectrum are presented. Three spectrum components are taken into account: IR emission from dust, UV bump from thermal emission of an accretion disk around a supermassive black hole and non–thermal radiation (power law in the whole spectrum range).

159.116 **A disrupted radio jet inside the host galaxy of the quasar 3C 48.**
P. N. Wilkinson, A. K. Tzioumis, J. M. Benson, R. C. Walker, R. S. Simon, F. D. Kahn.
Nature, Vol. 352, No. 6333, p. 313 – 315 (25 Jul 1991). Letter–to–the–editor.

The nearby quasar 3C 48 was the first to be optically identified, and its redshift, z = 0.368, was the second, after that of 3C 273, to be determined. Despite this pedigree, its detailed radio structure has remained obscure because its angular size, ~1 arcsec, is so small. The authors present a new radio image, with resolution < 10 mas, made by very–long–baseline interferometry using an array of 17 telescopes. They find a highly disrupted radio jet quite different from those found in standard high–luminosity radio sources. The jet lies well within the body of the gas–rich host galaxy of the quasar, and it is believed that its irregular shape is due to collision with a dense clump of gas in the interstellar medium of the host galaxy, rather than to any irregularity or precession of the central engine that produces the jet.

159.117 **An evolutionary unified scheme for radio–loud quasars and blazars.**
F. Vagnetti, E. Giallongo, A. Cavaliere.
IUPAP Conference on Primordial Nucleosynthesis and Evolution of Early Universe, p. 417 – 418 (1991). – See Abstr. 012.071 for the main entry.

159.118 **A deep ROSAT observation at high galactic latitude.**
I. Georgantopoulos, T. Shanks, G. Steward, K. Pounds, B. J. Boyle, R. Griffiths.
NATO Advanced Research Workshop on Observational Tests of Inflation, p. 309 – 313 (1991). – See Abstr. 012.076 for the main entry.

A deep (30,000 sec) ROSAT observation of a high galactic latitude field reveals a high surface density of sources (> 100 deg⁻²). Spectroscopic follow up observations show that the majority of these sources are QSOs with a surface density of 66 ± 14 deg⁻², at least a factor of two higher than that found by previous X-ray experiments. The discovered QSOs could contribute as much as ~50% to the 1 keV extragalactic X-ray background.

159.119 **Quasar jets: Life in the fast lane.**
T. Heckman, C. O'Dea.
Nature, Vol. 354, No. 6352, p. 355 – 356 (5 Dec 1991).

159.120 **Large-scale superluminal motion in the quasar 3C273.**
R. J. Davis, S. C. Unwin, T. W. B. Muxlow.
Nature, Vol. 354, No. 6352, p. 374 – 376 (5 Dec 1991). Letter-to-the-editor.

The quasar 3C273 is optically the brightest example of a source with a one-sided jet. It was the first object to display apparent superluminal motion on parsec scales, a phenomenon attributed to relativistic effects on the appearance of a jet moving close to the line of sight. The same explanation allows an intrinsically similar "counter-jet", moving at high speed in the opposite direction, to be dimmed to invisibility. The authors have made observations at 1.7 GHz, using very-long-baseline interferometry with a global network of 16 radiotelescopes, resulting in a high-dynamic-range map of the jet with a ratio of peak brightness to r.m.s. noise level of 16,000:1. They fail to see a counter-jet, a result which is just barely consistent with the standard model of a superluminal jet. The jet extends out to 220 pc, and some models require that the relativistic bulk flow should continue along its entire length. Comparison with an earlier image shows that superluminal motion extends out to at least 120 pc, three times farther than previously noted.

159.121 **Survey of quasars over large areas.**
He Xiangtao.
Proc. Astron. Soc. Aust., Vol. 9, No. 1, p. 45 – 48 (1991). – See Abstr. 012.078 for the main entry.

The survey techniques for finding quasars in large areas of sky have developed rapidly in recent years. One can search for quasars in all wavelengths. The area of sky surveyed may be more than 1000 square degrees. The average density of quasar candidates down to magnitude 20 is about 20 per square degree. Therefore, more than 20000 quasar candidates are waiting for further observation. The total number of confirmed quasars is currently about 5000. This paper gives a discussion of some of the physical properties of quasars, such as space distribution, luminosity function, clustering and redshift limit.

159.122 **On the nature of the redshifts of the twin quasars with discrepant redshifts.**
A. K. Sapre, V. D. Mishra.
Proc. Astron. Soc. Aust., Vol. 9, No. 1, p. 91 – 92 (1991). – See Abstr. 012.078 for the main entry.

The authors examine the nature of the redshifts of twin quasars with discrepant redshifts ($\theta < 100$ arcsec, $\Delta z \geqslant 0.5$) by considering their Hubble diagram. If such twins are physically associated one expects a large scatter in their Hubble diagram. However, the authors find a statistically significant linear correlation between log (cz) and V_c magnitudes of such quasars, consistent with the hypothesis that their redshifts are cosmological in nature.

159.123 **Intergalactic absorption in the spectra of high-redshift QSOs.**
S. Christiani, E. Giallongo.
NATO Advanced Research Workshop on Observational Tests of Inflation, p. 457 – 459 (1991). – See Abstr. 012.076 for the main entry.

Average flux decrements in the Ly α forest have been computed from the observed spectra of seven quasars. Comparing all the available data with simulations of the average absorptions in QSO spectra, the authors suggest that, within the observational and statistical uncertainties, there is evidence neither for a dramatic change in the evolution of the absorbers, nor for the presence of a new diffuse component of neutral hydrogen in the spectra of very high redshift ($z > 4$) quasars.

159.124 **A complete quasar sample at intermediate redshift.**
F. la Franca, S. Christiani, C. Barbieri, R. G. Clowes, A. Iovino.
NATO Advanced Research Workshop on Observational Tests of Inflation, p. 461 – 462 (1991). Also published in Mon. Not. R. Astron. Soc., Vol. 250, No.3, p. 531 – 540 (1 Jun 1991). – See Abstr. 012.076 for the main entry.

A search for intermediate-redshift quasars has been carried out with slitless spectroscopy in the central 21.07 deg^2 of the SA 94.

159.125 **The CFHT (MOS/PUMA) faint quasar survey.**
D. Schade.
Workshop on Space Distribution of Quasars as Part of the 75. Anniversary Celebrations of the National Research Council of Canada (NRCC), p. 17 – 21 (1991). – See Abstr. 012.084 for the main entry.

A multi-aperture spectroscopic survey for faint quasars has been carried out at Canada-France-Hawaii telescope. The survey is capable of reaching two magnitudes deeper into the luminosity function at redshifts greater than 3 than the deepest existing surveys. The technique is discussed and preliminary results are presented.

159.126 **SA 94: a test lab for QSO counts, clustering and variability.**
S. Christiani, F. La Franca, P. Andreani, R. Vio, C. Barbieri, M. Lazzarin.
Workshop on Space Distribution of Quasars as Part of the 75. Anniversary Celebrations of the National Research Council of Canada (NRCC), p. 22 – 31 (1991). – See Abstr. 012.084 for the main entry.

A search for quasars has been carried out in the field of the SA 94, using three independent complementary techniques: multicolor, slitless and variability. 178 quasars have been confirmed with slit spectroscopy. At present two complete samples have been created: (1) 99 QSOs with $z \leqslant 2.2$ and $B \leqslant 19.9$ over an area of 10.5 deg^2 and (2) 34 slitless selected QSOs with $B \leqslant 19.4$ over an area of 21 deg^2. QSO surface densities at $z \leqslant 2.2$ and $z \leqslant 3.2$ are derived, and clustering and variability properties studied.

159.127 **The faint quasar luminosity function.**
R. G. Kron, M. A. Bershady, J. A. Munn, J. J. Smetanka, S. Majewski, D. C. Koo.
Workshop on Space Distribution of Quasars as Part of the 75. Anniversary Celebrations of the National Research Council of Canada (NRCC), p. 32 – 45 (1991). – See Abstr. 012.084 for the main entry.

Preliminary results of an expanded program to determine the faint-quasar luminosity function are described. Quasars have been selected in four fields totaling 1.2 deg^2 from four-band photometry. Out of a total of 130 quasars with good spectroscopy, 37 have $J > 21.5$ and 46 have $F > 21.0$. The spectroscopic sample is representative of all of the color-selected candidates. An estimate of the luminosity function as a function of redshift is derived.

159.128 **The log N – M relation in a sample of quasars with $B_J \leqslant 22.0$.**
G. Zamorani, B. Marano, V. Zitelli, M. Mignoli, A. Cimatti, B. Boyle.
Workshop on Space Distribution of Quasars as Part of the 75. Anniversary Celebrations of the National Research Council of Canada (NRCC), p. 46 – 54 (1991). – See Abstr. 012.084 for the main entry.

The authors present and discuss the data for a spectroscopically complete sample of 52 faint optically selected quasars with $B_J \leqslant 22.0$. Using the results obtained from three different selection criteria (colors, grism and variability) applied on the same field, they first estimate the maximum level of incompleteness that they expect in the sample and then present their results on the counts of quasars as a function of magnitude for different redshift and luminosity ranges.

159.129 **A proper motion and variability QSO survey to B = 22.5.**
S. R. Majewski, J. A. Munn, R. G. Kron, M. A. Bershady, J. J. Smetanka, D. C. Koo.
Workshop on Space Distribution of Quasars as Part of the 75. Anniversary Celebrations of the National Research Council of Canada (NRCC), p. 55 – 66 (1991). – See Abstr. 012.084 for the main entry.
A survey of proper motions and variability has been conducted of all 1185 stellar objects to J = 22.5 or F = 21.5 in the North Galactic Pole field SA 57. The authors show that a search for objects which are both variable and stationary is a powerful technique for efficiently finding QSOs with no selection bias with regard to color, redshift, spectral index, or emission line equivalent widths. A large number of their QSO candidates have the colors of normal stars, and are missed by conventional multicolor searches. The authors have spectroscopically confirmed four new QSOs which were found using the new technique but not with the multicolor analyses. The authors conclude that multicolor surveys similar to those of Koo et al. are from 11 to 34% incomplete.

159.130 **The variability of a large sample of quasars.**
I. M. Hook, R. G. McMahon, B. J. Boyle, M. J. Irwin.
Workshop on Space Distribution of Quasars as Part of the 75. Anniversary Celebrations of the National Research Council of Canada (NRCC), p. 67 – 75 (1991). – See Abstr. 012.084 for the main entry.
The variability properties of a sample of ~300 quasars near the South Galactic Pole have been studied. Quasars of high luminosity show significantly less variation than those with low luminosity. A similar, though much weaker, trend with redshift was found, lower redshift quasars varying proportionally more than high redshift quasars. The observed trends are a consequence of an intrinsic dependence of quasar variability on luminosity combined with the effects of time–dilation and have strong implications for quasar samples selected solely on variability.

159.131 **A homogeneous bright quasar survey. Description of a key–programme.**
S. Cristiani, P. Andreani, A. Gemmo, R. Vio, C. Barbieri, M. Lazzarin, M. Sanvico, F. La Franca, L. Miller, C. Goldschmidt, H. MacGillivray, R. Clowes, C. Gouiffes, A. Iovino, A. Savage.
Workshop on Space Distribution of Quasars as Part of the 75. Anniversary Celebrations of the National Research Council of Canada (NRCC), p. 76 – 79 (1991). – See Abstr. 012.084 for the main entry.
The authors describe a program to study the luminosity function and the large scale clustering of quasars.

159.132 **Outline of a Liquid Mirror Telescope QSO survey.**
B. K. Gibson, P. Hickson.
Workshop on Space Distribution of Quasars as Part of the 75. Anniversary Celebrations of the National Research Council of Canada (NRCC), p. 80 – 83 (1991). – See Abstr. 012.084 for the main entry.
Potential contributions of a QSO survey to be carried out in 1992/93 by the UBC/Laval 2.7 m Liquid Mirror Telescope (LMT) are estimated. The survey area is a strip of sky ~21' wide, centered at $\delta \approx 49°04'$ encompassing an area of ~20 deg^2. Observations will be made through a set of 40 intermediate–band filters (one per photometric night until the QSO SED is constructed), chosen to ensure uniform logarithmic sampling over the 4000 – 10000 Å spectral range. The expected limiting magnitude of B ≈ 21.5 (at S/N ≈ 10) gives access to ~1000 QSOs in the survey strip. Clustering, large–scale homogeneity of the QSO distribution, and QSO/galaxy correlations will be studied.

159.133 **The Liège grism search for quasars.**
E. Gosset, J. Surdej, F. H. Cheng, J. P. Swings.
Workshop on Space Distribution of Quasars as Part of the 75. Anniversary Celebrations of the National Research Council of Canada (NRCC), p. 84 – 87 (1991). – See Abstr. 012.084 for the main entry.
The authors report on an on–going grism survey in which nine fields, one degree in diameter, have been searched for quasar–candidates. The selection techniques, the different calibrations, as well as the spectroscopic follow–up are briefly described.

159.134 **The efficiency of CCD multicolor surveys for high redshift quasars.**
A. Porter, R. Green, P. Osmer, M. Campin, P. Ogle, D. Maraziti.
Workshop on Space Distribution of Quasars as Part of the 75. Anniversary Celebrations of the National Research Council of Canada (NRCC), p. 88 – 92 (1991). – See Abstr. 012.084 for the main entry.
The authors have observed a large (n ~ 70) sample of high redshift z > 3 quasars with UBVRI CCD imaging. This is a preliminary report on efforts to distinguish an initial sample of 16 quasars from 509 stars in the same fields by multicolor photometry. Using n^{th} nearest neighbor tests, the authors can select about 4 quasars in candidate lists of about 50 objects. This is broadly consistent with the efficiencies of much larger plate surveys. Restricting the search to objects of specified colors greatly improves its efficiency: 13 of the 16 quasars (81%) can be isolated this way.

159.135 **The Edinburgh Multicolour Survey.**
P. Goldschmidt, P. S. Mitchell, L. Miller, R. S. Stobie, B. J. Boyle, R. D. Cannon, W. K. Griffiths, S. J. Lilly, R. M. Prestage.
Workshop on Space Distribution of Quasars as Part of the 75. Anniversary Celebrations of the National Research Council of Canada (NRCC), p. 93 – 95 (1991). – See Abstr. 012.084 for the main entry.
The Edinburgh Multicolour Survey is a large–area, accurately–calibrated UBVRI survey at intermediate magnitudes (R ⩽ 18.5). It consists of 2 plates in each of U, B, V, R and I in 13 UKST fields, covering a contiguous area of 330 deg^2. The plates were taken close together in time to minimise contamination and incompleteness due to variability.

159.136 **Modeling Lyman line and continuum absorption to assess completeness of multicolour quasar samples.**
P. Møller, S. Warren.
Workshop on Space Distribution of Quasars as Part of the 75. Anniversary Celebrations of the National Research Council of Canada (NRCC), p. 96 – 99 (1991). – See Abstr. 012.084 for the main entry.
To model the colours of high redshift quasars the authors have considered a single power law distribution for the column densities of the absorbers. Taking low resolution spectra published by Sargent, Steidel, and Boksenberg (1989) there is a unique solution for the slope and the normalization of the distribution, that matches simultaneously the line and continuum absorption. The slope of the distribution matches that derived from high resolution absorption studies, but the authors find a higher normalization.

159.137 **Emission–line properties of faint quasars and compact galaxies.**
J. J. Smetanka, M. A. Bershady, R. G. Kron, J. A. Munn, D. C. Koo, S. R. Majewski.
Workshop on Space Distribution of Quasars as Part of the 75. Anniversary Celebrations of the National Research Council of Canada (NRCC), p. 100 – 104 (1991). – See Abstr. 012.084 for the main entry.
Preliminary results of an analysis of the spectroscopic properties of an unbiased sample of emission–line objects to B ⩽ 22.5 are presented. The distribution of equivalent widths of the strong emission lines found in this survey (Kron et al. 1991) are

consistent with the distribution found by Zitelli et al. (1990), showing a larger mean equivalent width of Mg II relative to C IV. The authors find that the equivalent widths of the C IV and Mg II lines are anti-correlated with continuum luminosity (i.e. the "Baldwin Effect"). The narrow-emission-line objects in the survey are shown to have line ratios typical of H II regions and consistent with the line ratios observed in the resolved, narrow emission-line galaxies with extended image structure in the same fields.

159.138 **The Baldwin effect in complete optically selected samples of quasars.**
V. Zitelli, B. Marano, G. Zamorani, M. Mignoli, B. J. Boyle.
Workshop on Space Distribution of Quasars as Part of the 75. Anniversary Celebrations of the National Research Council of Canada (NRCC), p. 105 – 108 (1991). – See Abstr. 012.084 for the main entry.

From an analysis of the Baldwin effect in five complete optically selected samples of quasars, the authors find that the best fit slope of the correlation between $L_{C\,IV}$ and L_{cont} is significantly different from that obtained for a radio selected sample. Considering also the relatively large dispersion of the data they conclude that the equivalent width of the C IV line can not be used as a reliable luminosity indicator.

159.139 **Luminosity function of quasars at large redshifts from grism surveys.**
M. Schmidt, D. P. Schneider, J. E. Gunn.
Workshop on Space Distribution of Quasars as Part of the 75. Anniversary Celebrations of the National Research Council of Canada (NRCC), p. 109 – 116 (1991). – See Abstr. 012.084 for the main entry.

The authors have carried out a systematic search for emission-line objects through photometrically calibrated CCD grism surveys with the Palomar 200-inch telescope in transit mode, covering 62 square degrees. These surveys have yielded 141 quasars detected by their C IV or Ly α emission in the redshift range 2.0 – 4.7. The authors use this sample and the known flux limits to derive the slope of the luminosity function and the space density above a given line luminosity, as a function of redshift. While space densities are approximately constant at redshifts 2.0 – 3.0, they are declining steeply for redshifts larger than 3.

159.140 **APM optical surveys for high redshift quasars.**
M. Irwin, R. G. McMahon, C. Hazard.
Workshop on Space Distribution of Quasars as Part of the 75. Anniversary Celebrations of the National Research Council of Canada (NRCC), p. 117 – 128 (1991). – See Abstr. 012.084 for the main entry.

Results of a wide field ($\sim$2000 deg^2) multicolour photographic survey for bright high redshift ($z > 4$) quasars are presented. No decline in the space density of these luminous high redshift quasars compared to $z = 2$ quasars is apparent. Preliminary analysis of the emission line properties of this new sample of 20 quasars reveal a much wider dispersion in observed line strengths and profiles than is apparent in the sample of $z > 4$ quasars found by Schneider, Schmidt and Gunn. Some implications of these sample differences are discussed.

159.141 **Radio loud quasars with high redshift.**
R. McMahon.
Workshop on Space Distribution of Quasars as Part of the 75. Anniversary Celebrations of the National Research Council of Canada (NRCC), p. 129 – 138 (1991). – See Abstr. 012.084 for the main entry.

This paper describes the results of two complementary radio studies of quasars at very high redshift; (1) the results of sensitive VLA radio observations of 29 optically selected quasars with redshifts in the range $3.5 < z < 4.7$, in which 4 have been detected at 5 Ghz with $S > 1$ mJy. (2) Preliminary results of an investigation into the evolution of the space density of flat spectrum core dominated radio loud quasars from the MIT-Green Bank (MG) 5 Ghz radio survey. The two studies combined have resulted in a sample of 8 new $z > 3.5$ radio loud quasars to add to the 5, $z > 3.5$ radio loud quasars already known. In addition the problems associated with finding radio loud quasars with $z > 5$ are discussed.

159.142 **The quasar luminosity function $2.2 \leqslant z \leqslant 4.5$.**
S. J. Warren, P. C. Hewett, P. S. Osmer.
Workshop on Space Distribution of Quasars as Part of the 75. Anniversary Celebrations of the National Research Council of Canada (NRCC), p. 139 – 149 (1991). – See Abstr. 012.084 for the main entry.

In two recent papers the authors have presented the aims, methods, and spectroscopic results, and computed the completeness of a 6-band multicolour survey covering two UK Schmidt fields, to $m_{or} = 20.0$. The completeness estimate is revised here, employing a more detailed model for the effect of Lyman line and continuum absorption, and taking into account quasar variability. With this proviso the authors have computed space densities over the redshift range $2.2 < z < 4.5$, and find (1) a two-power-law PLE luminosity function, with continuing positive evolution, fits the data over the range $2.2 < z < 3.3$, (2) relative to $z = 3$, at the 95% confidence level, the integrated space density $M_B < -26.3$ at $z = 4$ is lower by a factor >2.9, $q_0 = 0.5$ (factor >2.1, $M_B < -27.9$, $q_0 = 0.0$).

159.143 **Spectroscopy of spatially extended material around high-redshift radio-loud quasars.**
T. M. Heckman, M. D. Lehnert, G. K. Miley, W. van Breugel.
Astrophys. J., Vol. 381, No. 2, p. 373 – 385 (10 Nov 1991).

The authors present the analysis of long-slit optical spectra taken of five high-redshift ($z \approx 2 - 3$) radio-loud quasars (QSRs), selected from the imaging survey of Heckman et al. These data confirm that the QSRs have Lyα emission that is spatially extended by at least several tens of kiloparsecs, with a luminosity of order 10^{44}ergs s^{-1}. The gas is very active kinematically. After subtracting the light of the broad-line region of the central QSO, the authors find that line widths in the nebulae are 1000 – 1500 km s^{-1} (FWHM). The gas does not show any globally organized velocity gradients. The line widths are close to the maximum possible for gas that is freely falling into the potential well of a very massive galaxy. The kinematic properties of the QSR nebulae are broadly similar to those of the nebulae associated with high-redshift radio galaxies. The authors confirm the suggestion of Foltz et al. that the nuclear He II λ1640 emission line in QSRs is systematically narrower than other nuclear emission lines. It is found that this narrow-lined, nuclear He II emission has a strength that correlates with the strength of the spatially extended Lyα emission, and it is suggested that the He II emission probably arises in the inner (arcsec-scale) part of the Lyα nebula. The authors report likely ($4\sigma - 5\sigma$) detections of spatially extended He II λ1640 in one QSR nebula and C IV λ1549 in another (each at 7 – 10% of Lyα). These possible detections and the upper limits on other lines are consistent with the emission-line spectra of high-z radio galaxies. The present data are not good enough either to yield strong constraints on the metallicities in the nebulae or to discriminate decisively between alternative ionization sources. The probable He II detection is inconsistent with predictions of photoionization models for normal high-mass stars. The probable C IV detection means that at least some galaxy-scale sites in the early universe have nonprimordial chemical abundances.

159.144 **Spatially resolved spectroscopy of Lyman-alpha and C IV in the gravitational lens 2237+030.**
H. K. C. Yee, M. M. De Robertis.
Astrophys. J., Vol. 381, No. 2, p. 386 – 392 (10 Nov 1991).

Spatially resolved, long-slit spectra of components A and B of the gravitational lens system 2237+030 have been obtained in Lyα and C IV at the CFHT. For Lyα, A and B are clearly resolved in both the emission line and continuum with almost no contamination from components C or D. The spectrogram of the two components shows an absorption trough situated $\sim$255 km s^{-1} redward of the broad emission-line centroid which contains a narrow emission feature lying spatially between the two components. There are, however, two distinct absorption systems in both the N V and C IV doublets which are matched in redshift. The authors propose that this partially resolved Lyα emission component, spatially distinct from the quasars, arises from the gravitational lensing of a massive compact star-forming region that is associated with either the host galaxy or a companion galaxy and which also likely produces the absorption systems. The Lyα absorption clouds have sufficiently high column density that the two profiles merge, whereas the lower column densities in N V and C IV give rise to separate absorption lines. The C IV spectrogram shows neither this narrow emission component nor the double-peaked structure so evident in Lyα. The spatial structure of the narrow emission feature within the Lyα absorption systems is consistent with a model in which the emission-line source lies on a caustic in the source plane, producing lensed features in the image plane both inside and outside the Einstein ring.

159.145 **The effects of gravitational lensing on the quasar luminosity function.**
R. Webster.
Workshop on Space Distribution of Quasars as Part of the 75. Anniversary Celebrations of the National Research Council of Canada (NRCC), p. 160 – 169 (1991). – See Abstr. 012.084 for the main entry.

It is shown that if realistic gravitational lens models are considered, it is possible that a substantial part of the bright end of the quasar luminosity function might be due to gravitational lensing. Given the new data on quasar-galaxy associations, the possibility of a gravitational lens bias must be taken seriously.

159.146 **A biassed account of biasses in QSO surveys.**
C. Hazard.
Workshop on Space Distribution of Quasars as Part of the 75. Anniversary Celebrations of the National Research Council of Canada (NRCC), p. 170 – 179 (1991). – See Abstr. 012.084 for the main entry.

This paper comments on some of the strengths and weaknesses of the techniques used to search over very large areas for bright QSOs with $z > 2$. Particular reference is made to very high redshift surveys at $z > 3.3$. and a brief description is given of some of the background to this work.

159.147 **Spectral uncertainties and the luminosity function of quasars.**
E. Giallongo, F. Vagnetti.
Workshop on Space Distribution of Quasars as Part of the 75. Anniversary Celebrations of the National Research Council of Canada (NRCC), p. 180 – 183 (1991). – See Abstr. 012.084 for the main entry.

Uncertainties in the spectral index of the UV spectrum of QSOs induce systematic biases in the estimate of the cosmological evolution of the luminosity function (LF). In a framework of pure exponential luminosity evolution the authors obtain a slower intrinsic evolution of the LF for $z < 2.2$ after correction for uncertainties in the k-correction adopted. The slower evolution helps to reduce the apparent density decline of high redshift QSOs ($z > 2.5$) observed in the red band. Indeed they find that color selected surveys are consistent with a scenario of pure luminosity evolution up to $z \cong 4$. The apparent decline in the redshift distribution is explained in terms of the sliding down along the steep side of the luminosity function for increasing redshift.

159.148 **QSO evolution and clustering at z < 2.9.**
B. J. Boyle, L. R. Jones, T. Shanks, B. Marano, V. Zitelli, G. Zamorani.
Workshop on Space Distribution of Quasars as Part of the 75. Anniversary Celebrations of the National Research Council of Canada (NRCC), p. 191 – 201 (1991). – See Abstr. 012.084 for the main entry.

Two recently completed surveys for faint ($B \lesssim 22$ mag) QSOs are used in the derivation of new results on the clustering and evolution of QSOs at $z < 2.9$. The strong luminosity evolution witnessed at low redshifts ($z<2$) is observed to "switch off" at $z \sim 2$, with the comoving space density of QSOs remaining constant in the redshift range $2 < z < 2.9$. Extrapolation of this evolution to higher redshifts implies little decrease in the comoving space density of bright ($M_B \lesssim -27$) QSOs at $z < 4.5$. Strong QSO clustering is confirmed at small separations ($<10\ h^{-1}$Mpc) but no significant evidence is found for any evolution in the strength of QSO clustering with cosmic epoch.

159.149 **The clustering of quasars and its evolution.**
A. Iovino, P. A. Shaver, S. Cristiani.
Workshop on Space Distribution of Quasars as Part of the 75. Anniversary Celebrations of the National Research Council of Canada (NRCC), p. 202 – 211 (1991). – See Abstr. 012.084 for the main entry.

Using a combined sample totalling nearly 1000 quasars, the authors measure the clustering of quasars at a very high significance level and obtain indications of evolution of the correlation function which are consistent with previous results.

159.150 **The evolution of luminous quasars.**
P. Goldschmidt, L. Miller.
Workshop on Space Distribution of Quasars as Part of the 75. Anniversary Celebrations of the National Research Council of Canada (NRCC), p. 212 – 217 (1991). – See Abstr. 012.084 for the main entry.

The authors have compared complete samples of bright quasars at different redshifts and at the same absolute magnitude, selected from the Edinburgh Multicolour Survey. They have compared the surface densities of the bright quasars in this survey to other surveys. The authors have also found that the $(1+z)$ parameterisation of the evolution of quasars is an inadequate description.

159.151 **Large groups of quasars.**
R. G. Clowes, L. E. Campusano.
Workshop on Space Distribution of Quasars as Part of the 75. Anniversary Celebrations of the National Research Council of Canada (NRCC), p. 248 – 255 (1991). – See Abstr. 012.084 for the main entry.

In a recent paper (1991) the authors presented evidence for a large group of quasars, possibly the largest structure so far observed in the universe. This was actually the third large group of quasars to be discovered. These large groups suggest the following hypothesis. Most quasars are distributed uniformly and randomly. Occasionally large groups of quasars are embedded in this otherwise uniform distribution, and they account for all clustering, even that on small scales $<10\ h^{-1}$Mpc. The authors have used published work on small-scale clustering to suggest the existence of a fourth large group of quasars.

159.152 **About a coordinated redshift survey of quasar candidates to B ~ 19 and z = 2.2 in the south galactic pole field.**
L. E. Campusano.
Workshop on Space Distribution of Quasars as Part of the 75. Anniversary Celebrations of the National Research Council of Canada (NRCC), p. 256 – 263 (1991). – See Abstr. 012.084 for the main entry.

The author describes to what extent is known the 2-dimensional distribution of quasars in the SGP field. The present situation suggests that there are inhomogeneities on their surface density which might add, if real, information about the large scale distribution of matter. Although 50 quasars are already known within the quoted limits, the analysis indicates that their total expected number is between 80 and 130. A possible redshift survey is identified which would use a collation of candidates coming from several, existing or under way, independent machine-based searches.

159.153 **Claims for periodicity in quasar redshifts.**
D. Scott.
Workshop on Space Distribution of Quasars as Part of the 75. Anniversary Celebrations of the National Research Council of Canada (NRCC), p. 264 – 271 (1991). – See Abstr. 012.084 for the main entry.

Several separate claims have been made for periodicity in the redshifts of quasars and other high z objects. The most persistent claim has been for $\Delta \ln(1+z) \cong 0.2$, which goes back to 1971, and has received more support in the recent literature. However, the scale of this effect is just where there should be most concern about selection effects. Furthermore, the characteristic peaks and troughs found in some redshift histograms are absent in the most carefully selected samples of high z objects. It is therefore reasonable to assume that the observed effect has been entirely due to selection biases. If this explanation is accepted, then a search for the periodicity in incomplete samples provides a crude but simple method of assessing the completeness in redshift space.

159.154 **The influence of quasars on the Lyman–α forest.**
A. Dobrzycki, J. Bechtold.
Workshop on Space Distribution of Quasars as Part of the 75. Anniversary Celebrations of the National Research Council of Canada (NRCC), p. 272 – 280 (1991). – See Abstr. 012.084 for the main entry.

The authors present a large (~10 Mpc) void in the Ly–α forest in the spectrum of a $z_{em} = 3.285$ quasar, Q0302–003. They discuss the possibility that it results from ionization of the clouds by the UV radiation of a bright foreground quasar, Q0301–005 ($z_{em} = 3.223$). The authors present a model of the influence of quasars on the Ly–α forest and show that in order to produce a megaparsec–sized region with no detectable Ly–α absorption the intervening quasar has to be much brighter than Q0301–005.

159.155 **Quasar superclustering.**
N. A. Bahcall, A. Chokshi.
Workshop on Space Distribution of Quasars as Part of the 75. Anniversary Celebrations of the National Research Council of Canada (NRCC), p. 281 – 289 (1991). – See Abstr. 012.084 for the main entry.

The origin of the observed quasar superclustering is suggested to be the preferential location of quasars in small groups of galaxies. The quasars then trace the same large–scale structure as traced by the parent galaxy groups. Optically selected quasars are found to be located in small groups of typically ~10 L* galaxies and have the same correlation function as expected for the small groups. Radio selected quasars are located in richer groups of ~30 L* galaxies, having the stronger correlations expected for these richer groups. These results are consistent with observations of quasar environment and with the universal richness–dependent cluster correlation function. The results suggest a consistent picture of superclustering of galaxies, clusters and quasars.

159.156 **Superclustering at high redshifts.**
M. J. West.
Workshop on Space Distribution of Quasars as Part of the 75. Anniversary Celebrations of the National Research Council of Canada (NRCC), p. 290 – 299 (1991). Also published in Astrophys. J., Vol. 379, No. 1, p. 19 – 36 (20 Sep 1991). – See Abstr. 012.084 for the main entry.

Evidence is presented which suggests that the observed radio emission from high redshift ($z \geqslant 0.5$) radio galaxies and quasars is NOT randomly directed. Rather, the radio major axes of these objects tend to be oriented in the direction toward other neighboring quasars within ~45 h^{-1}Mpc (comoving) from the radio source. This phenomenon is very similar to alignments of neighboring Abell clusters and their brightest member galaxies seen at low redshifts, which suggests that the observed alignments of radio sources at high redshifts provide tentative evidence of the existence of a well–developed pattern of superclustering in the early universe.

159.157 **Gravitational lensing of quasars.**
N. Kaiser, P. Tribble.
Workshop on Space Distribution of Quasars as Part of the 75. Anniversary Celebrations of the National Research Council of Canada (NRCC), p. 304 – 316 (1991). – See Abstr. 012.084 for the main entry.

Estimates of the lensing optical depth due to individual galaxies can account for the small $\theta \lesssim 2''$ splittings, but multi–galaxy mass concentrations are required for the larger splittings. The authors show that the abundance of these objects matches quite well a simple extrapolation of the X–ray cluster temperature distribution, provided that the mass density profiles of the clusters continue to rise roughly as $\varrho \sim r^{-2}$ well inside the X–ray core radius. Quasar–galaxy associations, if due to microlensing, provide, in principle, a very clean probe of the mass distribution around galaxies. However, the expected effect, based on conventional models for the mass distribution and intrinsic quasar luminosity function, is about a factor 20 lower than observed. Possible loopholes are considered, but for the most part rejected.

159.158 **Associations between galaxies and bright quasars.**
M. J. Drinkwater, R. L. Webster, P. A. Thomas.
Workshop on Space Distribution of Quasars as Part of the 75. Anniversary Celebrations of the National Research Council of Canada (NRCC), p. 317 – 324 (1991). – See Abstr. 012.084 for the main entry.

The results of a CCD imaging survey of bright ($V \lesssim 17.5$) quasars are presented. There is a significant overdensity of galaxies close to the lines–of–sight to these quasars. (30 ± 8)% of the quasars have closely associated galaxies in excess of those predicted by a random distribution. If this overdensity is due to amplification of the quasars by gravitational lensing, then, allowing for the probability that a galaxy is detected, at least 60% of bright quasars may be lensed.

159.159 **A search for quasar protoclusters at z > 4.**
S. Djorgovski, J. D. Smith, D. J. Thompson.
Workshop on Space Distribution of Quasars as Part of the 75. Anniversary Celebrations of the National Research Council of Canada (NRCC), p. 325 – 330 (1991). – See Abstr. 012.084 for the main entry.

A generic prediction for practically all scenarios of largescale structure and galaxy formation is that the first collapsed objects (which may be identified with the highest known redshift quasars) should be very strongly clustered. The authors started a search for quasar–marked protoclusters at $z > 4$, using a multicolor imaging technique, in the fields of known $z > 4$ quasars. Their goal is to find less luminous AGNs in these fields, which have escaped detection so far. Discovery of clustered AGNs at such redshifts would be a crucial experiment for the understanding of the formation of galaxies, large–scale structure, and the origin of the first quasars themselves.

159.160 **Cosmic evolution of quasar radio structure.**
J. B. Hutchings, S. G. Neff.
Workshop on Space Distribution of Quasars as Part of the 75. Anniversary Celebrations of the National Research Council of Canada (NRCC), p. 331 – 333 (1991). – See Abstr. 012.084 for the main entry.

The authors discuss the results of a survey of Quasar radio structures over redshifts from 0.6 to 3.7. There are clear evolutionary trends in size and luminosity, which suggest that the duty cycle of individual Quasars has increased over cosmic time. This affects source count statistics and gives clues on the evolution of Quasar environments.

159.161 **Galaxy clustering around QSOs at 0.9 < z < 1.5.**
B. J. Boyle, S. Perlmutter, G. Goldhaber, R. Muller, H. Marvin, C. Pennypacker, W. J. Couch.
Workshop on Space Distribution of Quasars as Part of the 75. Anniversary Celebrations of the National Research Council of Canada (NRCC), p. 344 – 346 (1991). – See Abstr. 012.084 for the main entry.

Deep ($R < 23.3$ mag) CCD images around 15 QSOs in the redshift range $0.9 < z < 1.5$ have been analysed to search for associations between QSOs and faint galaxies. No significant excess of galaxies is found at separations less than 15 arcsec, in contrast to the result of Hintzen et al. (1991). These results are consistent with a field or poor cluster galaxy environment for QSOs at $0.9 < z < 1.5$.

159.162 **A search for closely–spaced gravitational lenses.**
D. Crampton, R. D. McClure, J. M. Fletcher.
Workshop on Space Distribution of Quasars as Part of the 75. Anniversary Celebrations of the National Research Council of Canada (NRCC), p. 347 – 348 (1991). – See Abstr. 012.084 for the main entry.

The DAO/CFHT image stabilizing camera was used to search for closely spaced images resulting from gravitational lensing by foreground galaxies in a sample of 100 bright quasars. No convincing objects were found.

159.163 **Quasar pairs at large redshifts.**
S. Djorgovski.
Workshop on Space Distribution of Quasars as Part of the 75. Anniversary Celebrations of the National Research Council of Canada (NRCC), p. 349 – 353 (1991). – See Abstr. 012.084 for the main entry.

A sample of apparently close physical pairs of quasars at large redshifts has been compiled. There are at least a half–dozen such systems now known. They seem to divide into two groups with about equal frequencies, $\sim 10^{-3}$ pairs per catalogued quasar: a set of wide, Mpc–scale pairs (arcmin separations), and a set of close pairs, with projected separations of a few arscec, or $\lesssim 100$ kpc. A simple hierarchical clustering model matches well the observed frequency of the wide pairs, but the close pairs are about two orders of magnitude overabundant relative to the predictions of this model. This is interpreted as a signature of galaxy interactions, which trigger the nuclear activity in both members of close quasar pairs. These systems may represent birth events of quasars at the epoch where the comoving density of quasars was near its maximum.

159.164 **A radio–quiet quasar inside an Abell cluster of richness class 0?**
A. F. Meza, L. E. Campusano.
Workshop on Space Distribution of Quasars as Part of the 75. Anniversary Celebrations of the National Research Council of Canada (NRCC), p. 358 – 360 (1991). – See Abstr. 012.084 for the main entry.

Spectroscopy of the UV–excess quasar–candidate CT 2282 does not reveal a typical bona fide quasar in the sense of Schmidt and Green (1983). Its emission line redshift, 0.211 ± 0.001, implies a rather low luminosity for a quasar ($M_B \sim -19.3$). However, some characteristics do ressemble those of quasars: stellar appearance in the ESO–B (quick) Sky Survey, FWHMs of broad emission lines within the distribution found in complete samples and compatible equivalent width values (excepting for Hβ). CT 228 lies within the identified limits of the Abell Cluster A2843. The authors' spectroscopy of a probable member–galaxy of A2843 reveals a redshift of 0.215 ± 0.001, thus suggesting that CT 228 is gravitationally tied to the cluster. The significance of this finding, possibly the first radio–quiet low–luminosity quasar known to be associated to a rich Abell cluster, is discussed.

159.165 **Quasars and galaxy formation.**
E. L. Turner.
Workshop on Space Distribution of Quasars as Part of the 75. Anniversary Celebrations of the National Research Council of Canada (NRCC), p. 361 – 368 (1991). – See Abstr. 012.084 for the main entry.

Observations of high redshift quasars provide direct information on the physical state of the universe at early epochs. The available data suggest well advanced structure formation even at the highest known quasar redshifts and are most easily understood in scenarios in which galaxies primarily formed at redhsifts well above 5. Ten separate lines of arguments supporting this general conclusion are summarized.

159.166 **A spectroscopic survey for 10000 QSOs.**
B. J. Boyle.
Workshop on Space Distribution of Quasars as Part of the 75. Anniversary Celebrations of the National Research Council of Canada (NRCC), p. 389 (1991). Abstract. – See Abstr. 012.084 for the main entry.

159.167 **CFHT quasar surveys.**
D. Crampton, A. P. Cowley, F. D. A. Hartwick, D. J. Schade.
Workshop on Space Distribution of Quasars as Part of the 75. Anniversary Celebrations of the National Research Council of Canada (NRCC), p. 390 (1991). Abstract. – See Abstr. 012.084 for the main entry.

159.168 **A deep BRX survey for high redshift quasars.**
R. G. McMahon, M. Irwin.
Workshop on Space Distribution of Quasars as Part of the 75. Anniversary Celebrations of the National Research Council of Canada (NRCC), p. 391 – 392 (1991). Abstract. – See Abstr. 012.084 for the main entry.

159.169 **A deep QSO survey.**
S. R. Majewski, M. A. Bershady, R. G. Kron, J. A. Munn, J. J. Smetanka, D. C. Koo.
Workshop on Space Distribution of Quasars as Part of the 75. Anniversary Celebrations of the National Research Council of Canada (NRCC), p. 393 (1991). Abstract. – See Abstr. 012.084 for the main entry.

159.170 **High–redshift quasar surveys in progress.**
M. Schmidt, D. P. Schneider, J. E. Gunn.
Workshop on Space Distribution of Quasars as Part of the 75. Anniversary Celebrations of the National Research Council of Canada (NRCC), p. 394 (1991). Abstract. – See Abstr. 012.084 for the main entry.

159.171 **Inference of the extreme–ultraviolet continuum shape of quasars.**
Zheng Wei.
Astrophys. J., Lett., Vol. 382, No. 2, p. L55 – L57 (1 Dec 1991).

The strength of C IV emission in weak–bump QSOs is significantly larger than that in strong–bump QSOs. Equivalent widths of Lyα, Si IV, and C III show no sign of difference between objects with various UV bump strength. The finding suggests that the "big UV bump" is a source of ionization and can be scaled by emission lines. The higher average C IV EW and C IV/Lyα ratio in weak–bump QSOs represent the values intrinsic to the bare nonthermal continuum. If the bump has a turnover between 30 and 50 eV, C IV emission would not respond as effectively as other low–ionization lines, leading to the observed correlation between C IV equivalent width and bump strength. The Baldwin effect can be explained in terms of UV bumps which are stronger at higher luminosity.

159.172 **Systematic effects in lens inversions: ℵ exact models for 0957+561.**
C. S. Kochanek.
Astrophys. J., Vol. 382, No. 1, p. 58 – 70 (20 Nov 1991).

Due to the careful study of its VLBI structures, the gravitational lens 0957 + 561 is the most accurately characterized of the lens candidates. Nonetheless, the observational data supplies only five firm constraints, compared to a minimum of nine free parameters in simple models for the lens using an elliptical galaxy and an elliptical cluster both with fixed radial structure. This paper examines the systematic variations between models for the lens potential that satisfy the constraints exactly. This is done using both the four degrees of freedom left when the radial structure of the galaxy is fixed, as well as variations due to changes in the radial structure. The systematic variations between the models allow for a range of time delays from approximately $0.1h^{-1}$ to $1.0h^{-1}$ yr for a fixed Einstein – de Sitter cosmology if the galaxy and the cluster satisfy the velocity dispersion constraints of $240 \lesssim \sigma_g \lesssim 320\,\mathrm{km\,s^{-1}}$, and $500 \lesssim \sigma_c \lesssim 1500\,\mathrm{km\,s^{-1}}$, and the center of the cluster lies within 30″ of the galaxy. Given the formal value for the velocity dispersion ($303 \pm 50\,\mathrm{km\,s^{-1}}$) and the measured time delay (1.3 ± 0.1 yr), the author finds that $h \lesssim 0.6 \pm 0.2$ considering only the observational errors in the velocity dispersion. Because the images in 0957 + 561 only weakly constrain the mass distribution of the lens, the 0957 + 561 system can be used to measure the Hubble constant only to the accuracy with which observational parameters such as central velocity dispersion, ellipticities, and orientations can be translated into information on the mass distribution in the system.

159.173 **Outflows driven by cosmic–ray pressure in broad absorption line QSOs.**
M. C. Begelman, M. de Kool, M. Sikora.
Astrophys. J., Vol. 382, No. 2, p. 416 – 432 (1 Dec 1991).
The authors present a model that can explain many features of the very fast ($\sim$0.1c) outflows that are observed in broad absorption line QSOs (BALQSOs). In this model, the wind is accelerated primarily by the pressure of ultrarelativistic protons. These protons are deposited in the wind, over a range of radii, by the decay of ultrarelativistic neutrons escaping from the central engine. Recent models for nonthermal processes in the central engines of active galactic nuclei predict a neutron luminosity which can exceed a few percent of the radiation flux, which is more than adequate to accelerate the wind. Since the energy deposition is predicted to peak at distances of order 1 – 10 pc, the main acceleration will occur outside the broad emission line region, as seems to be required by observations. The absorbing gas occupies a very small fraction of the volume of the wind, and must be confined by it. The authors model the formation of BALs in the outflow, deriving line profiles and constraining the rate at which the energy of the ultrarelativistic protons has to be converted to thermal energy in order to provide pressure confinement for the BAL clouds. Further, the authors discuss the problem of acceleration and survival of the clouds, and consider some alternative confinement mechanisms. The probable geometry of the outflow is also considered.

159.174 **Emission–line and continuum properties of 92 bright QSOs: luminosity dependence and differences between radio–selected and optically selected samples.**
C. C. Steidel, W. L. W. Sargent.
Astrophys. J., Vol. 382, No. 2, p. 433 – 465 (1 Dec 1991).
The authors have analyzed the emission lines and (rest frame) UV continua in 92 QSO spectra (resolution 4 – 6 Å) which had been obtained for a recent study of Mg II absorption in the spectra of 103 low and intermediate redshift QSOs. The sample contains 42 radio–selected QSOs (of which 14 have "steep" radio spectra) and 50 which were discovered either by UV excess or by slitless spectroscopy. The radio–loud and radio–quiet subsamples have roughly the same distribution in both redshift and absolute luminosity. In general, the spectra cover the rest wavelength interval from below the C III] λ1909 emission line to just above the Mg II λ2800 emission line. For many of the objects information is also available for the C IV λ1549 line. An examination of their mean emission–line equivalent widths and continuum properties shows that there is no significant difference between the QSOs discovered by the three different methods. Further, there is no significant difference in the optical/UV spectra of radio–flat versus radio–steep QSOs. The distributions of the following quantities differ significantly ($>2\sigma$) between the radio–loud and radio–quiet subsamples: W(Al III), I(C III])/I(Mg II), I(C III])/I(C IV), and I(Al III)/I(C III]). All other measured equivalent widths and intensity ratios are statistically indistinguishable between the two subsamples. In addition, a number of relatively subtle differences in line ratios, line shifts, and line profiles between radio–quiet and radio–loud QSOs are described. On the whole, however, the continuum and emission–line properties of optically selected and radio–selected QSOs are remarkably similar. The authors note that the most significant correlations involve the C III] line, which is density sensitive. At present, it seems not possible to relate the observed regularities to any simple physical picture of the structure of QSOs.

159.175 **Quasar pairs and quasar's peculiar velocity.**
Chu Yaoquan.
Sci. China, Ser. A, Vol. 34, No. 7, p. 841 – 847 (Jul 1991).
This paper proposes a method for determining an upper limit to the peculiar velocity of quasars by means of a statistical analysis of quasar pairs. A preliminary conclusion is that the peculiar velocity in the direction along the line joining the two quasars in a pair does not exceed, on average, 1000 km/s.

159.176 **On the trajectories of the superluminal knots in the quasar 3C 345.**
Qian Shanjie, T. Krichbaum, A. Witzel, A. Quirrenbach, C. A. Hummel, J. A. Zensus.
Acta Astron. Sin., Vol. 32, No. 4, p. 369 – 379 (Dec 1991). In Chinese.
In the superluminal radio source 3C 345 it was observed that there are two inner knots (C_4 and C_5), which moved along different curved trajectories and showed accelerated motions when they were close to the core. The authors discuss their kinematics under the assumption that the knots move along helical magnetic field lines around the jet–axis as their common axis. It is shown that for a simplified geometry suitable patterns of the helical trajectories and spatial velocities can be chosen to fit both the apparent trajectories and apparent velocities for knots C_4 and C_5 quite well. The physical implications of the results are discussed.

159.177 **Searching for gravitational lenses.**
D. Crampton, R. McClure, J. M. Fletcher.
J. R. Astron. Soc. Can., Vol. 85, No. 4, p. 179 (Aug 1991). Abstract. – See Abstr. 012.091 for the main entry.

159.178 **Theory of the production of quasar spectra with broad absorption lines.**
Eh. Ya. Vil'koviskij.
Sov. Astron., Vol. 35, No. 6, p. 576 – 581 (Nov 1991). Current Physics Microform No.: 9205X1386. English translation of Astron. Zh., Tom 68, Vyp. 6, p. 1150 – 1159 (1991).
A theory of the production of absorption lines in a plasma stream is analyzed on the basis of a simultaneous solution of the equations of plasma dynamics and radiative transfer in lines. Model spectra are calculated for quasars with broad absorption lines.

159.179 **Models of the continuum of quasars and AGNs in the EUV and adjacent energy ranges.**
M.–H. Ulrich.
Adv. Space Res., Vol. 11, No. 11, p. 105 – 111 (1991). – See Abstr. 012.107 for the main entry.
It is in the EUV band (1200 Å – 1 keV) that the energy emitted by quasars and AGNs is most intense. No direct information however is yet available in this range. What constraints on the EUV continuum flux are set by the observations of the continuum in the optical, UV and X–ray ranges, and by the observations of optical/UV emission line intensity ratios? What constraints are set by plausible models of the UV/X spectrum? What are the signatures of an accretion disk in this range? The author examines these questions and concludes that direct observations of the flux variations and the spectral shape of the EUV continuum are essential to make progress in this field.

159.180 **Properties of C IV absorbers at $z > 2.3$ in the spectra of quasistellar objects.**
P. Khare, D. G. York, R. Green.
Bull. Astron. Soc. India, Vol. 19, Nos. 1 – 2, p. 75 – 76 (Mar – Jun 1991). Abstract. – See Abstr. 012.108 for the main entry.

159.181 **Symmetry of double–lobed quasars and their linear sizes.**
D. G. Banhatti.
Bull. Astron. Soc. India, Vol. 19, Nos. 1 – 2, p. 76 (Mar – Jun 1991). Abstract. – See Abstr. 012.108 for the main entry.

159.182 **Quasars, Seyfert 1 (Sy1) type galaxies and observational selection.**
R. A. Vardanyan.
Astrofizika, Tom 34, Vyp. 1, p. 41 – 50 (Feb 1991). In Russian. English translation in Astrophysics, Vol. 34, No. 1.
The effect of absolute stellar magnitude influence on the observed relations between colours and redshift of quasars and Sy1 type galaxies is taken into account. A new evidence of connection of Sy1 type galaxies and quasars is given which can be considered in favour of the cosmological nature of their redshift.

159.183 **Analysis of absorption lines in quasar spectra.**
V. Müller, J. P. Mücket, H. J. Haubold, S. Gottlöber.
3. International Workshop on Data Analysis in Astronomy, p. 179 – 185 (1989). Abstract. – See Abstr. 012.110 for the main entry.

159.184 **CCD photometry of quasars and active galactic nuclei.**
T. J. Balonek.
News Lett. Astron. Soc. N.Y., Vol. 3, No. 9, p. 22 – 23 (Feb 1991). Abstract. – See Abstr. 012.119 for the main entry.

159.185 **A search for the CO $J = 1$–0 absorption line at $z = 2.04$ toward the QSO PKS 0458–02.**
M. Tsuboi, N. Nakai.
Publ. Astron. Soc. Jpn., Vol. 43, No. 6, p. L65 – L68 (1991).
The authors have searched the CO $J = 1$–0 absorption line of a damped Lyα absorber at $z = 2.04$ toward the QSO PKS 0458–02 using the 45-m telescope of Nobeyama Radio Observatory. They obtained the upper limits of the optical thickness as 0.024 and 0.014 at velocity resolutions of 2 and 5 km s^{-1}, respectively. The upper limit of the column density of CO molecules is 2.1×10^{15}cm^{-2} for an optical thickness of 0.024 and an excitation temperature of 20K.

A database of IUE spectra of active galactic nuclei – the results for quasars.
See Abstr. 002.035.

Spectroscopic investigations of stellar objects from the Second Byurakan Sky Survey. II. The fields centered on $\alpha = 09^h50^m$, $\delta = +55°00'$ and $\alpha = 11^h30^m$, $\delta = +59°00'$.
See Abstr. 002.048.

The Einstein Observatory Extended Medium–Sensitivity Survey. II. The optical identifications.
See Abstr. 002.076.

The Case Low–Dispersion Northern Sky Survey. XII. A region in southern Canes Venatici.
See Abstr. 002.077.

Spectral investigations of the Second Byurakan Sky Survey objects. Stellar objects. III. Fields $\alpha = 12^h22^m$, $\delta = +55°00'$ and $\alpha = 15^h30^m$, $\delta = +59°00'$.
See Abstr. 002.090.

A catalog of absorption in the spectra of quasi–stellar objects.
See Abstr. 002.102.

Spectroscopic investigations of stellar objects from the Second Byurakan Sky Survey. III. The fields centered on $\alpha = 12^h22^m$, $\delta = +55°00'$ and $\alpha = 15^h30^m$, $\delta = +59°00'$.
See Abstr. 002.107.

A catalogue of quasars and active nuclei. (5th edition).
See Abstr. 002.112.

Spectral investigations of the Second Byurakan Sky Survey objects. Stellar objects. IV.
See Abstr. 002.130.

The First Byurakan Spectral Sky Survey. Blue stellar objects. IV. Zone $+41° \leqslant \delta \leqslant +45°$.
See Abstr. 002.131.

Variability of active galaxies. Proceedings.
See Abstr. 012.070.

The space distribution of quasars. Proceedings.
See Abstr. 012.084.

Some aspects of quasars.
See Abstr. 013.021.

Quasar astronomy with a large optical telescope.
See Abstr. 013.024.

Block iterative restoration of astronomical images from the Hubble Space Telescope.
See Abstr. 036.023.

Initial results from the Four College APT: pulsations, outbursts, spots and eclipses from stars to quasars.
See Abstr. 036.040.

Luminosity functions from emission line selected samples.
See Abstr. 036.207.

Comets to quasars: surface photometry from standard stars, and the morphology of the galaxy–quasar interface.
See Abstr. 036.231.

EVN as a phase stable interferometer.
See Abstr. 036.265.

Errors in high precision source position determination.
See Abstr. 036.267.

State–of–the–art VLBI imaging: 3C345.
See Abstr. 036.275.

An automated search for gravitational lenses performed on 25 grens plates.
See Abstr. 036.289.

VLBI astrometry of PSR 1937+21.
See Abstr. 041.007.

Infrared universe – IRAS.
See Abstr. 051.047.

Gamma–rays and neutrinos from accretion processes onto collapsed objects: application to 3C 273.
See Abstr. 061.082.

On the structure of magnetized thin accretion disks of active galactic nuclei and their coronal radiations.
See Abstr. 062.111.

On the structure of thin magnetized accretion disks of active galactic nuclei and their coronal radiation.
See Abstr. 062.187.

Formation of relativistic MHD jets in the magnetosphere of accretion disks.
See Abstr. 062.246.

X–ray irradiation of interstellar grains in active galaxies: evaporation and infrared spectra.
See Abstr. 063.068.

Gravitational microlensing and the superluminal illusion.
See Abstr. 066.070.

The deconvolution of the quasar structure from microlensing light curves.
See Abstr. 066.158.

Lichtablenkung, Gravitationslinsen und Einsteinsche Ringe.
See Abstr. 066.230.

Polarization of radiation from AGN accretion disks.
See Abstr. 067.082.

High–polarization radiation from magnetized accretion disks.
See Abstr. 067.086.

Hybrid accretion disks in active galactic nuclei. I. Structure and spectra.
See Abstr. 067.152.

Thermal continuum from QSO disks.
See Abstr. 067.167.

Luminosity–function evolution of quasar accretion–disks.
See Abstr. 067.172.

Molecules in AGN's accretion discs.
See Abstr. 067.180.

Inclination effects for Kerr disks.
See Abstr. 067.185.

Polarization properties of emission lines from relativistic accretion disks.
See Abstr. 067.225.

Quantum–field approach to the theory of the relativistic jet activity of quasars and galactic nuclei.
See Abstr. 067.250.

Can ultrarelativistic neutrons from the central engine drive the outflow in broad absorption line QSOs?
See Abstr. 067.252.

Enhanced radio source scintillation due to comet Austin ($1989c_1$).
See Abstr. 103.036.

The stellar component of the Hamburg Schmidt survey.
See Abstr. 114.087.

A spectroscopic survey of faint compact objects.
See Abstr. 114.114.

Interstellar Ca II in the galactic halo and in QSO absorption systems.
See Abstr. 131.308.

Observations of Ca II absorption along the nearly parallel lines of sight toward 3C 273 and SN 1991T.
See Abstr. 131.361.

Flux densities at 8400 MHz for a large sample of radio sources.
See Abstr. 141.011.

The cosmic X–ray background and QSOs to $B < 22$: a search for correlated fluctuations.
See Abstr. 142.006.

Deep ROSAT observations of quasar fields and the X–ray background.
See Abstr. 142.007.

The origin of the cosmic X–ray background.
See Abstr. 142.046.

A deep ROSAT survey.
See Abstr. 142.054.

Einstein deep surveys.
See Abstr. 142.056.

The two high latitude COS–B γ–ray sources observed by SAS–2 in the >100 MeV energy range.
See Abstr. 143.004.

Observations of TeV photons at the Whipple Observatory.
See Abstr. 143.021.

Colliding galaxies.
See Abstr. 151.010.

Probing the Galactic halo with RR Lyr stars, globular clusters and 3C 273 IUE.
See Abstr. 155.017.

Probing the Galactic halo along the 3C 273 sight line using IUE.
See Abstr. 155.105.

The starburst–AGN connection.
See Abstr. 157.240.

Cosmological consequences of starbursts.
See Abstr. 157.241.

A radio–quiet galaxy at redshift $z = 3.409$.
See Abstr. 157.313.

Infrared broad emission lines from an obscured quasar, IRAS 23060+0505.
See Abstr. 158.001.

Broad emission lines in active galactic nuclei. II. Statistical properties of Hα and Hβ.
See Abstr. 158.002.

Simultaneous optical and radio polarization rotation in OJ 287.
See Abstr. 158.003.

Optical polarization of a complete sample of radio sources.
See Abstr. 158.005.

On a nonthermal origin of steep far–infrared turnovers in radio–quiet active galactic nuclei.
See Abstr. 158.010.

The connection between radio emission spectra and interaction signs in powerful radio galaxies.
See Abstr. 158.015.

Can interstellar scintillations explain a short–time flux variability of extragalactic radio sources?
See Abstr. 158.017.

Simultaneous UBVRI photopolarimetric observations of blazars.
See Abstr. 158.018.

Infrared nuclei in nearby powerful radio galaxies: evidence for hidden quasars?
See Abstr. 158.021.

Rapid variability of compact extragalactic radio sources.
See Abstr. 158.037.

The buried quasar in Cygnus A.
See Abstr. 158.040.

A connection between radio emission spectra and signs of interaction in powerful radio galaxies.
See Abstr. 158.042.

VLBI structures of low frequency variable sources at 608 MHz.
See Abstr. 158.067.

Radio source orientation and the angular diameter – redshift relation.
See Abstr. 158.075.

Zones of high ionization of gas in nuclei of Seyfert galaxies and quasars.
See Abstr. 158.078.

High radiation brightness temperatures in active galactic nuclei.
See Abstr. 158.089.

Observational distinction between two types of active galactic nuclei.
See Abstr. 158.096.

Recent developments in gravitational lensing.
See Abstr. 158.098.

Radio variability of active galactic nuclei and related objects.
See Abstr. 158.124.

A statistical analysis of total and polarized cm–waveband flux variations.
See Abstr. 158.125.

Studying the galactic central engine from space observatories.
See Abstr. 158.144.

Short timescale broad emission–line flux variability in high–luminosity active galactic nuclei.
See Abstr. 158.153.

The EXOSAT high Galactic latitude survey.
See Abstr. 158.156.

VLA images of 23 extragalactic radio sources.
See Abstr. 158.164.

VLBI observations of 24 3CR CSS radio sources at 50 cm.
See Abstr. 158.170.

The dust enshrouded quasar in the ultraluminous galaxy Markarian 463: radio, near–infrared, and optical imaging.
See Abstr. 158.173.

Zones of highly ionized gas in the nuclei of Seyfert galaxies and quasars.
See Abstr. 158.193.

Echo images of broad–line regions in active galactic nuclei.
See Abstr. 158.195.

Are short–term flux variations interstellar scintillations of extragalactic radio sources?
See Abstr. 158.201.

The luminosity function and cosmological evolution of X–ray – selected BL Lacertae objects.
See Abstr. 158.206.

What are the gigahertz peaked–spectrum radio sources?
See Abstr. 158.207.

Have all questions been asked?
See Abstr. 158.209.

The signatures of an accretion disk in the electromagnetic spectra of quasars and AGNs.
See Abstr. 158.221.

Variability of BL Lac objects in the radio regime.
See Abstr. 158.222.

VLBI knots and superluminal motion.
See Abstr. 158.229.

Relativistic jet models and variable knot emission.
See Abstr. 158.230.

Variable radio sources and the two–fluid models.
See Abstr. 158.232.

Evidence for disks in active galactic nuclei and QSOs from polarimetric observations.
See Abstr. 158.236.

Continuum emission in AGN, can we isolate the disc component?
See Abstr. 158.237.

Near–infrared observations of the $z \sim 2.3$ IRAS source FSC 10214+4724.
See Abstr. 158.245.

On the magnetic energy density in the X–ray emission region of Active Galactic Nuclei.
See Abstr. 158.249.

Einstein Observatory SSS and MPC observations of the complex X–ray spectra of Seyfert galaxies.
See Abstr. 158.258.

High ionization line zone above disk in AGN.
See Abstr. 158.260.

Active galactic nuclei: A sideways look at quasars.
See Abstr. 158.264.

The effect of anisotropic emission from thick accretion disks on the luminosity functions of AGN.
See Abstr. 158.267.

Formation of bipolar radio jets and lobes from accretion disk around forming blackhole at the center of protogalaxies.
See Abstr. 158.268.

Wavelength dependent polarization in active galaxies.
See Abstr. 158.270.

Microvariability in blazars via accretion disk instabilities.
See Abstr. 158.271.

The cosmological evolution and luminosity function of X–ray selected active galactic nuclei.
See Abstr. 158.284.

Quasars in rich galaxy cluster environments.
See Abstr. 158.291.

Fe II emission and anisotropy in broad–line regions.
See Abstr. 158.298.

Two phase accretion disks as a model for the X–ray emission from Seyfert galaxies.
See Abstr. 158.309.

Relativistic electron–positron beam formation in the framework of the two–flow model for active galactic nuclei.
See Abstr. 158.320.

Core, jet and counterjet prominence in powerful radio galaxies and quasars.
See Abstr. 158.327.

Millimeter–wavelength quiescent spectra and burst behavior for compact extragalactic sources.
See Abstr. 158.335.

Observational evidence for accretion disks in galactic nuclei.
See Abstr. 158.336.

The low–frequency spectra of accretion disks in active galactic nuclei: the dark side of the big blue bump.
See Abstr. 158.337.

Observational constraints on viscosity in the AGN accretion discs.
See Abstr. 158.338.

Quasar monitoring program at Turku University Observatory.
See Abstr. 158.340.

Jet models and observed jets in a sample of superluminal active galactic nuclei.
See Abstr. 158.342.

Dusty disks and the infrared emission from AGN.
See Abstr. 158.343.

Evolution of high–redshift Lyman–limit absorption systems.
See Abstr. 160.001.

Gravitational lensing and the Lyman–alpha forest.
See Abstr. 160.013.

Intergalactic Lyman–alpha clouds: photoionization and mechanisms for cooling.
See Abstr. 160.016.

The Lyman alpha clouds as cosmological probes.
See Abstr. 160.027.

Lyman–alpha depression of the continuum from high–redshift quasars: a new technique applied in search of the Gunn–Peterson effect.
See Abstr. 160.035.

A forest of lines but not a forest of clouds. Diffuse IGM models for the Lyα absorption systems of quasars.
See Abstr. 160.037.

Cooling–flow modification of the Sunyaev–Zeldovich effect in clusters of galaxies.
See Abstr. 160.038.

A $\sim$10 Mpc void in the Lyα forest at $z = 3.17$.
See Abstr. 160.046.

Discovery of a Lyα galaxy near a damped Lyα absorber at $z = 2.3$.
See Abstr. 160.047.

The origin of the quasi–stellar object absorption lines.
See Abstr. 160.049.

Young globular clusters?
See Abstr. 160.057.

A new spectroscopic survey for damped Ly_α absorption lines from high–redshift galaxies.
See Abstr. 160.064.

Confirmation of dust in damped Lyman–alpha systems.
See Abstr. 160.071.

Attenuation of Lyman–alpha emission by dust in damped Lyman–alpha systems.
See Abstr. 160.082.

Clusters of galaxies associated with quasars. II. Galaxy cluster dynamics.
See Abstr. 160.083.

The carbon abundance of the Lyman–alpha clouds.
See Abstr. 160.088.

Lyman alpha evolution at high resolution: evidence for a single population?
See Abstr. 160.110.

The effect of clustering on the equivalent width distribution of QSO Lyman–alpha clouds.
See Abstr. 160.130.

New photoionization models of intergalactic clouds.
See Abstr. 160.157.

QSO absorption systems and the origin of the ionizing background at high redshift.
See Abstr. 161.135.

Redshift distribution and evolution function of quasi–stellar objects.
See Abstr. 161.165.

Void structure in the early universe: implications for a $\Lambda > 0$ cosmology.
See Abstr. 161.225.

The double magnification bias by gravitational lensing.
See Abstr. 161.230.

Gravitational lensing, time delay, and angular diameter distance.
See Abstr. 161.249.

Selection effects in optical surveys for gravitational lenses.
See Abstr. 161.302.

Statistics of gravitational lenses: the uncertainties.
See Abstr. 161.336.

The evolution of the diffuse cosmic ultraviolet background constrained by Hubble Space Telescope observations of 3C 273.
See Abstr. 161.366.

Origin of regularities in the redshift distributions of distant galaxies and quasars.
See Abstr. 161.382.

Present and future limits on compact dark matter from gravitational lensing.
See Abstr. 161.401.

The Hubble constant from VLA measurement of the time delay in the double quasar 0957+561.
See Abstr. 161.435.

Cosmological implications of H I absorption systems.
See Abstr. 161.495.

The large-scale peculiar velocity at high redshift.
See Abstr. 161.505.

Gravitational lensing frequencies: galaxy cross-sections and selection effects.
See Abstr. 161.554.

Ly α clouds at low redshift and the cosmological constant.
See Abstr. 161.570.

Some comments on the epoch of galaxy formation.
See Abstr. 161.581.

Is the Universe a fractal? Tests on quasar samples.
See Abstr. 161.614.

Quasar-modulated galaxy clustering in a cold dark matter universe.
See Abstr. 161.615.

Are quasar redshifts cosmological?
See Abstr. 161.617.

The confinement and cosmological evolution of extragalactic radio source components.
See Abstr. 161.638.

160 Galaxy Groups, Clusters of Galaxies, Superclusters, Intergalactic Matter

160.001 **Evolution of high-redshift Lyman-limit absorption systems.**
K. M. Lanzetta.
Astrophys. J., Vol. 375, No. 1, p. 1 – 14 (1 Jul 1991).

The results of a new spectroscopic survey for high-redshift Lyman-limit absorption systems are presented. These results are based on recent observations of high-redshift QSOs by Lanzetta and coworkers, from which a total of 35 Lyman-limit systems from the spectra of 52 QSOs are identified. These new data are combined with existing observations of low- and high-redsift Lyman-limit systems in order to determine the properties of the absorbers over the redshift range $0.36 \leqslant z \leqslant 4.11$. The most striking result to emerge from this new study is that on the basis of this data set the rate of incidence of the Lyman-limit systems with $z \gtrsim 2.5$ is found to evolve strongly with redshift in the sense that the product of the number density per unit comoving volume and the absorption cross section increased with increasing redshift. Over the redshift range $2.5 \leqslant z \leqslant 3.7$ the observed redshift distribution is well fitted by $n(z) = n_0(1+z)^\gamma$ with $\gamma = 5.7 \pm 1.9$. In contrast, the rate of incidence of the Lyman-limit systems is virtually constant over the redshift range $0.35 \leqslant z \lesssim 2.5$. If the observed evolution indicates intrinsic evolution of the absorbers, this result suggests that the evolution detected previously for the C IV - selected absorbers over a similar redshift range is most naturally interpreted as evolution of the ionization level of the absorbers rather than as a chemical enrichment effect, and model photoionization calculations indicate that the observed evolution of the absorption systems may be explained if the ionization parameters of the absorbers increases with decreasing redshift by a factor of ≈ 3 within the redshift interval $2.5 \lesssim z \lesssim 3.7$.

160.002 **Maffei 1 as an interloper of the Local Group of galaxies and the mass of the Local Group.**
J.-Q. Zheng, M. J. Valtonen, G. G. Byrd.
Astron. Astrophys., Vol. 247, No. 1, p. 20 – 24 (Jul 1991).

The authors study the neighbors of the Local Group of galaxies and the possibility that they could have perturbed the two-body motion. They find that the effects of Maffei 1 on the Local Group dynamics dominate assuming that Maffei 1 is at the distance of 2.1 Mpc from us as determined by Buta and McCall (1983). This distance is only about half of the redshift distance, and consequently Maffei 1 should have had a dynamical interaction with the Local Group galaxies about 6.5 billion years ago. The authors find a solution to the Local Group dynamics. In this scenario Maffei 1 could be an escaped member of the Sculptor group which has passed through the Local Group. As a result of the interaction, the Local Group has obtained a peculiar motion of about 66 km s^{-1}. This is consistent with the observational results of Richter et al. (1987). The authors' results make a better determination of Maffei 1 distance crucial. Only if Maffei 1 is at the extreme range of the uncertainty estimated by Buta and McCall, at 3.4 Mpc, will the simple M31 - Milky Way two-body assumption and the calculation of massive halos for each be correct. Finally, the authors study the fate of the companions of M31 during the Maffei encounter. As expected, many companions are dispersed. The radial velocities of the post-encounter companions are such that their spectra should show redhsifts relative to M31 in almost all cases. This is the case also with the observed companions, as has been pointed out by Arp (1970, 1985).

160.003 **Fundamental parameters of brightest cluster galaxies.**
W. R. Oegerle, J. G. Hoessel.
Astrophys. J., Vol. 375, No. 1, p. 15 – 24 (1 Jul 1991).

The authors have obtained photometry and spectroscopy of a sample of brightest cluster galaxies (BCG) in order to determine the relationships between scale length, surface brightness, velocity dispersion, and line strength, and to make comparisons with the relations determined for lower luminosity elliptical galaxies. The velocity dispersions for 32 brightest cluster galaxies have been calculated. Together with published data, this gives a sample of 43 galaxies. CCD surface photometry of these galaxies produced characteristic scale lengths and surface brightnesses. The relation between $\log r_e$, $\log \sigma$, and mean surface brightness, reported by Dressler et al. for lower luminosity ellipticals, also describes the relation for brightest cluster galaxies fairly well. There is a slight curvature in the fundamental plane at the high-luminosity end, apparently due to the fact that the velocity dispersions of BCGs increase more slowly with luminosity than those for lower luminosity ellipticals. The authors derive the fundamental plane solutions for BCGs in $r_e - \sigma - \langle\mu\rangle_e$ space, which can be used as a distance indicator with a predicted error of $\sim 21\%$ per galaxy. The $r_e - \langle\mu\rangle_e$ relation for BCGs is also a good distance indicator (distance errors of $\sim 25\%$ per galaxy) and has the advantage of not requiring knowledge of the velocity dispersion.

160.004 **Fourier spectra of three-dimensional data.**
D. J. Baumgart, J. N. Fry.
Astrophys. J., Vol. 375, No. 1, p. 25 - 34 (1 Jul 1991).
The authors evaluate the power spectrum, bispectrum, and tentatively the trispectrum, i.e., the two-, three-, and four-point galaxy correlation functions in Fourier transform space, for data from the CfA catalog and from the Perseus-Pisces redshift compilation of Giovanelli and Haynes. Working in the transform domain enables the authors to extend our knowledge of galaxy clustering to scales of order $100h^{-1}$Mpc. At these large scales, the observed power spectrum ought to reflect directly the initial conditions. The observed power spectrum agrees well with cold dark matter models. The best fit gives $\Omega \cdot h = 0.34 \pm 0.10$, while the inferred amplitude of scale-invariant horizon fluctuations is $\varepsilon \approx (2-4) \times 10^{-5}$. The observed bispectrum obeys the so-called hierarchical pattern, with a reduced three-point amplitude that is consistent with gravitational instability in perturbation theory starting from Gaussian initial conditions. The trispectrum is only marginally detected, but is again consistent with the hierarchical pattern; the reduced four-point amplitude is determined only to be of order unity. These results constitute perhaps the first direct observational suggestion that gravitational instability in the perturbative regime indeed operates on large scales.

160.005 **Ram-pressure stripping of the multiphase interstellar medium of the Virgo Cluster elliptical galaxy M86 (NGC 4406).**
D. A. White, A. C. Fabian, W. Forman, C. Jones, C. Stern.
Astrophys. J., Vol. 375, No. 1, p. 35 - 45 (1 Jul 1991). With plates 1 - 2.
The authors have analyzed the available infrared (IRAS), optical, and X-ray (Einstein) data on the elliptical galaxy M86, which is experiencing strong ram-pressure stripping in the intracluster medium of the Virgo Cluster. A significant component of infrared 60 μm emission is displaced from the center of the galaxy. The displaced 60 μm emission lies on the optical major axis of the galaxy in the direction of perturbations to the optical isophotes and is coincident with a small feature in X-ray emission southwest of the X-ray plume. The authors interpret the displaced infrared emission in terms of collisionally heated dust exposed to hot gas where ram pressure and turbulence are disrupting the stripped interstellar medium of the galaxy. The displacement of the outer optical isophotes in the direction of the stripped gas is suggested to be due to reflection of starlight on the dust grains. Ram pressure has separated and exposed the various components of the multiphase interstellar medium of M86, enabling these constituents to be identified and analyzed separately for the first time in an elliptical galaxy.

160.006 **Staub im intergalaktischen Raum?**
S. Klose, S. Marx.
Sterne Weltraum, Jahrg. 30, Nr. 11, p. 649 - 654 (Nov 1991).

160.007 **Mean-field correlations in the core of rich galaxy clusters.**
B. J. T. Jones, J. A. López, V. J. Martinez, E. Saar.
Nuovo Cimento B, Vol. 106, No. 8, p. 933 - 940 (Aug 1991).
The authors develop a theory for the contribution to the clustering correlation function from gravitational interactions of neighboring pairs of galaxies in clusters. This is based on the "Hypernetted Chain Equation", a self-consistent integral equation relating the correlation function to the interaction potential.

160.008 **Intergalactic matter.**
A. C. Fabian, X. Barcons.
Rep. Prog. Phys., Vol. 54, No. 8, p. 1069 - 1122 (Aug 1991).
The review begins with a brief discussion of the state of our own interstellar medium and follows with a detailed study of the diffuse medium in clusters of galaxies. The distribution of the intracluster medium and the role of cooling in the cores of many clusters are given special attention. The similar properties of the hot gas in massive early-type galaxies are also discussed. Clusters and massive early-type galaxies contain the best-studied bodies of diffuse gas known. Observations made over a very wide range of wavelengths, from the radio to X-ray bands, show that the intracluster medium is complex and multiphase, consisting mainly of hot diffuse gas, but with cold embedded clouds in the cluster core. The possible properties of a diffuse intergalactic medium are then reviewed, including heat sources and the allowed degree of clumping. The most crucial observations are the absence of any measurable distortion to the microwave background spectrum and the lack of an absorption trough shortward of the wavelength of Lyman α in the spectra of quasars. There is as yet no clear evidence for any diffuse intergalactic medium, although it is widely predicted to exist and could be the dominant baryonic component in the Universe. Finally, the intergalactic absorption-line clouds, both those forming the 'Lyman α forest' in the spectra of distant quasars and the metal-rich systems, are discussed in detail.

160.009 **Dynamical estimation of the mass of small groups of galaxies.**
Zh. P. Anosova, L. G. Kiseleva, V. V. Orlov, A. D. Chernin.
Sov. Astron., Vol. 35, No. 3, p. 221 - 224 (May 1991). Current Physics Microform No.: 9201X0503. English translation of Astron. Zh., Tom 68, Vyp. 3, p. 449 - 455 (1991).
On the basis of numerical simulation of the dynamics of triplets of galaxies, virial estimation of the individual masses in triplets is shown to be unreliable because of their strong nonsteadiness and projection effects. The typical error of individual estimates is about three orders of magnitude.

160.010 **New measurements of radial velocities in clusters of galaxies. III.**
H. V. Capelato, A. Mazure, D. Proust, C. Vanderriest, J. P. Lemonnier, L. Sodré Jr.
Astron. Astrophys., Suppl. Ser., Vol. 90, No. 2, p. 355 - 364 (Oct 1991).
The authors have obtained 95 redshifts in eight Abell galaxy clusters using the multi-object spectrograph SILFID. Data on individual galaxies are presented, and the accuracy of the determined velocities are discussed as well as some properties of the cluster. Near complete samples of galaxies have been obtained for the clusters A957, A2063 and A2589 in their central regions.

160.011 **Stellar populations in medium redshift clusters. II. Optical-infrared photometry and spectra.**
A. J. Pickles, P. C. van der Kruit.
Astron. Astrophys., Suppl. Ser., Vol. 91, No. 1, p. 1 - 52 (Nov 1991).
The authors present optical and infrared photometry (BVRI, JHK) and spectra of galaxies in 6 medium redshift clusters covering the redshift range $0.19 \leqslant z \leqslant 0.4$. The array photometry is used to note the radial distribution of the cluster galaxies with optical and infrared colours. By bisecting the galaxy population into red and blue halves, the tendency of the bluer galaxies to define rims or shells around the cores of the more distant clusters is noted, and its implications for delayed star formation briefly discussed. The optical photometry is also used to confirm flux calibrations of the long slit and multiple object spectra (12 Å resolution) which have been taken for a sample of central elliptical galaxies in each cluster. The infrared photometry is used to extend the broad-band flux levels for these spectra.

160.012 **Formation and distortion of optical rings around rich cluster galaxies.**
C. S. Kochanek, R. D. Blandford.
Astrophys. J., Vol. 375, No. 2, p. 492 - 498 (10 Jul 1991).
Luminous optical rings may be formed around individual galaxies in rich clusters by gravitational deflection of light from a background galaxy. Observation of these rings can provide a direct measurement of the local surface density in the cluster if the cluster redshift, the source redshift and the lens galaxy velocity dispersion are measured. If the major and minor axes of the ring are r_+ and r_-, then the harmonic mean ring radius defined by $R = 2r_+ \cdot r_-/(r_+ + r_-) = b_g/(1-\varkappa)$ where b_g is the critical radius of the unperturbed galaxy, and $\varkappa = \Sigma/\Sigma_c$ is the

ratio of the smoothly distributed local surface mass density of the cluster to the critical surface density for gravitational lensing. The ellipticity of the ring measures the local shear in the combined potential of the cluster and the galaxy. Rings will only be found in regions with $\varkappa_{min} < \varkappa < \varkappa_{max}$, where the limit $\varkappa_{min} \cong 0.4$ depends on the sensitivity of the observations, and the limit $0.5 \lesssim \varkappa_{max} \lesssim 1$ depends on the cluster density profile. Cross sections for the formation of complete and incomplete rings are given. The formation of a closed ring implies that the source is comparable in size to or larger than the tangential caustic associated with the lens galaxy. Half-illuminated rings should be at least twice as common as completely illuminated rings.

160.013 **Gravitational lensing and the Lyman-alpha forest.**
S. Ikeuchi, E. L. Turner.
Astrophys. J., Vol. 375, No. 2, p. 499 – 502 (10 Jul 1991).

Possible connections between the inhomogeneities responsible for the Lyman-α forest in quasar spectra and gravitational lensing effects are investigated. For most models of the Lyman-α forest, no significant lensing is expected. For some versions of the CDM model-based minihalo hypothesis, gravitational lensings on scales $\lesssim 0\farcs1$ would occur with a frequency approaching that with which ordinary galaxies cause arcsecond scale lensing.

160.014 **Optical coronal emission lines from cooling flows in elliptical galaxies and galaxy clusters.**
C. L. Sarazin, C. M. Graney.
Astrophys. J., Vol. 375, No. 2, p. 532 – 543 (10 Jul 1991).

The authors calculate the predicted luminosities, surface brightness profiles, and line shapes of optical coronal emission lines for models of cooling flows in galaxy clusters and in individual elliptical galaxies. It is found that the coronal line luminosities of cooling flows depend primarily on the total rate at which gas is cooling. The surface brightness profiles of the lines show where gas is cooling below X-ray emitting temperatures. In homogeneous cooling flow models in which all of the gas flows into the central regions, the coronal lines are very centrally concentrated, coming mainly from a limb-brightened shell lying just within the sonic radius of the inflow. For homogeneous models, the central surface brightnesses of the stronger coronal lines from nearby clusters with large cooling rates are expected to be $\sim 6 \times 10^{-17}$ ergs $cm^{-2}s^{-1}arcsec^{-2}$. For inhomogeneous models in which gas cools out of the flow at large radii, the surface brightness profiles of the coronal lines are more extended, and the central surface brightness is much lower. The homogeneous and inhomogeneous models also have very different line shapes. The integrated line profiles for the homogeneous models are very wide and flat, and the central line profiles are double-peaked. The inhomogeneous models have nearly Gaussian line profiles. The predicted line fluxes from nearby clusters are fairly close to the present observational limits.

160.015 **Gas in galaxies at the present epoch: a constraint on evolution of high Z systems.**
S. Rao, F. H. Briggs.
Bull. Am. Astron. Soc., Vol. 23, No. 2, p. 952 (1991). Abstract. – See Abstr. 012.008 for the main entry.

160.016 **Intergalactic Lyman-alpha clouds: photoionization and mechanisms for cooling.**
J. M. Shull, M. Donahue.
Bull. Am. Astron. Soc., Vol. 23, No. 2, p. 952 (1991). Abstract. – See Abstr. 012.008 for the main entry.

160.017 **CCD photometry of the compact galaxy group Arp 150.**
M. Rodrigue, T. Colegrove, J. Kliwer, A. Schultz.
Bull. Am. Astron. Soc., Vol. 23, No. 2, p. 953 – 954 (1991). Abstract. – See Abstr. 012.008 for the main entry.

160.018 **UIT observations of NGC 1275.**
E. P. Smith, P. M. Hintzen, S. G. Neff, A. M. Smith, T. P. Stecher, R. W. O'Connell, G. S. Hennessy, R. H. Cornett, J. K. Hill, R. S. Hill, W. B. Landsman, M. S. Roberts.
Bull. Am. Astron. Soc., Vol. 23, No. 2, p. 954 (1991). Abstract. – See Abstr. 012.008 for the main entry.

160.019 **UIT observations of galaxies in the Perseus Cluster.**
P. M. Hintzen, K.-P. Cheng, G. S. Hennessy, R. W. O'Connell, J. K. Hill, S. G. Neff, E. P. Smith, T. P. Stecher, M. S. Roberts.
Bull. Am. Astron. Soc., Vol. 23, No. 2, p. 954 (1991). Abstract. – See Abstr. 012.008 for the main entry.

160.020 **A 20 cm VLA survey of Abell clusters of galaxies: second installment.**
R. A. White, F. N. Owen, J. O. Burns.
Bull. Am. Astron. Soc., Vol. 23, No. 2, p. 955 (1991). Abstract. – See Abstr. 012.008 for the main entry.

160.021 **A deep CCD image of the core of the Coma cluster.**
M. P. Ulmer, G. Burnstein, J. A. Tyson, P. Teague.
Bull. Am. Astron. Soc., Vol. 23, No. 2, p. 955 (1991). Abstract. – See Abstr. 012.008 for the main entry.

160.022 **A catalog of galaxies in the Dorado cloud.**
P. B. Eskridge.
Bull. Am. Astron. Soc., Vol. 23, No. 2, p. 955 (1991). Abstract. – See Abstr. 012.008 for the main entry.

160.023 **Observational estimates of masses of clusters of galaxies.**
C. M. Bird.
Bull. Am. Astron. Soc., Vol. 23, No. 2, p. 955 – 956 (1991). Abstract. – See Abstr. 012.008 for the main entry.

160.024 **Empirical limits on thermal conduction in Coma.**
L. P. David, W. Tucker.
Bull. Am. Astron. Soc., Vol. 23, No. 2, p. 956 (1991). Abstract. – See Abstr. 012.008 for the main entry.

160.025 **Extragalactic distances from surface brightness fluctuations.**
J. L. Tonry.
Bull. Am. Astron. Soc., Vol. 23, No. 2, p. 956 (1991). Abstract. – See Abstr. 012.008 for the main entry.

160.026 **A new optimal unbiased galaxian distance indicator for ellipticals.**
D. H. Gudehus.
Bull. Am. Astron. Soc., Vol. 23, No. 2, p. 956 (1991). Abstract. – See Abstr. 012.008 for the main entry.

160.027 **The Lyman alpha clouds as cosmological probes.**
W. L. W. Sargent.
Bull. Am. Astron. Soc., Vol. 23, No. 2, p. 962 (1991). Abstract. – See Abstr. 012.008 for the main entry.

160.028 **Hα Fabry-Perot rotation curves of a large sample of spiral galaxies in clusters.**
W. T. Sullivan, C. Balkowski, V. Cayatte, M. Marcelin, P. Amram, E. Le Coarer.
Bull. Am. Astron. Soc., Vol. 23, No. 2, p. 973 (1991). Abstract. – See Abstr. 012.008 for the main entry.

160.029 **A spatial and kinematic analysis of the Abell 400 cluster of galaxies.**
T. C. Beers, K. Gebhardt, J. P. Huchra, W. Forman, C. Jones, G. D. Bothun.
Bull. Am. Astron. Soc., Vol. 23, No. 2, p. 974 (1991). Abstract. – See Abstr. 012.008 for the main entry.

160.030 **Alignments in the Virgo cluster?**
N. A. Sharp.
Observatory, Vol. 111, No. 1103, p. 162 – 166 (Aug 1991).
The apparent alignments of different classes of objects in the Virgo cluster are neither as strong as claimed nor conventionally inexplicable.

160.031 **Radio observations of the galaxy clusters A1187 and A1890 at 102 MHz.**
A. T. Kalloglyan, M. A. Ogannisyan.
Astrophysics, Vol. 33, No. 2, p. 401 – 404 (Sep – Oct 1990). English translation of Astrofizika, Tom 33, Vyp. 2, p. 181 – 185 (Oct 1990).
The results are given of observations of the regions of the galaxy clusters A1187 and A1890 at frequency 102 MHz using the scintillation method. In the cluster A1890, two scintillating sources have been found, and one of them appears to coincide with the brightest galaxy of the cluster. The magnetic field strength of this source is found to be in the range 120 – 300 mG.

160.032 **The merging runaway.**
A. Cavaliere, S. Colafrancesco, N. Menci.
Astrophys. J., Lett., Vol. 376, No. 2, p. L37 – L40 (1 Aug 1991).
The authors propose an analytic description of the direct mergings that involve galaxies in groups. They predict the shape and evolution of the distribution of the galaxian masses in these finite systems, including the formation of a cD–like merger that analytically appears as a critical phenomenon. These findings constitute a close counterpart of many N–body results. Similar aggregation phenomena conceivably take place in other cosmic structures.

160.033 **The distance to the Coma cluster using the B–band Tully–Fisher relation.**
M. Fukugita, S. Okamura, K. Tarusawa, H. J. Rood, B. A. Williams.
Astrophys. J., Vol. 376, No. 1, p. 8 – 22 (20 Jul 1991).
The distance to the Coma cluster was estimated using the B–band Tully–Fisher relation. The authors took all galaxies for which H I line width data are available in a circle of 4° radius centered on the Coma cluster, and carried out photographic surface photometry on them. The sample represents a substantial fraction of all spiral galaxies with $m_z < 15.5$ mag inside this circle. A careful analysis of the errors entering the establisment of the Tully–Fisher relation is presented. In particular, the effect of the cluster population incompleteness bias on the distance estimate is considered. It is found that the correction for the sample incompleteness, when combined with the dispersion of the Tully–Fisher relation, increases the distance modulus by 0.3 mag, but that it cannot be as large as the value recently suspected by Kraan–Kortweg, Cameron and Tammann. As their best estimate for the Coma cluster the authors obtained a distance modulus of $(m-M)^0 = 34.5 \pm 0.4$. This corresponds to a Hubble constant $H_0 = 92(+21, -17)$ km s^{-1}Mpc^{-1}.

160.034 **The distribution of dark matter in the Perseus cluster.**
C. J. Eyles, M. P. Watt, D. Bertram, M. J. Church, T. J. Ponman, G. K. Skinner, A. P. Willmore.
Astrophys. J., Vol. 376, No. 1, p. 23 – 32 (20 Jul 1991).
The authors report on the analysis of spatially resolved spectral images of the Perseus cluster obtained with a coded–mask X–ray telescope flown as part of the Spacelab 2 mission. The data allow the temperature and density profiles of the intracluster gas to be measured. From these the cluster potential and hence the cluster mass as a function of radius can be traced. The results show that there is significantly less dark matter in the cluster than implied by virial mass estimates based on the galaxies. The mass–to–light ratio within $1.3h_{50}$ Mpc of the cluster center is found to be ~ 100 $(M_\odot/L_\odot)$. The dark matter is more centrally condensed than the X–ray–emitting gas and probably more so than the galaxies, contrary to expectations from most cosmogonic scenarios. The authors show that the conclusions are rather robust and insensitive to different parameters and assumptions about the X–ray structure and argue that these results favor models in which the dark matter is baryonic.

160.035 **Lyman–alpha depression of the continuum from high–redshift quasars: a new technique applied in search of the Gunn–Peterson effect.**
E. B. Jenkins, J. P. Ostriker.
Astrophys. J., Vol. 376, No. 1, p. 33 – 42 (20 Jul 1991).
The authors reexamine old and new data including the high–redshift ($z > 4$) quasar spectra of Schneider, Schmidt, and Gunn in search of opacity due to smoothly distributed hydrogen absorbing at the local wavelength of Lyα (the "Gunn–Peterson" effect). Following a detailed analysis of moderate redshift, high–resolution observations of the Lyα cloud population, the authors apply two methods to the data: (1) computing and subtracting off the mean absorption due to distinct clouds; and (2) a new method of examining the distribution of intensities in comparison with a Monte Carlo simulation. It is found that at redshift $z \approx 3$ the data imply approximately 30% more opacity than in measurable lines, which could be due to lines of very small column density or to true continuous absorption. At high redshift ($z > 4$), it is found that the total absorption is larger by $20\% \pm 30\%$ than a simple extrapolation from lower redshift data. The most plausible interpretation of this is to set a limit of $\sim 15\%$ reduction in the ionizing background from $z = 3$ to $z = 4.5$.

160.036 **Dynamics of the microwave–decrement cluster Abell 665.**
W. R. Oegerle, M. J. Fitchett, J. M. Hill, P. Hintzen.
Astrophys. J., Vol. 376, No. 1, p. 46 – 50 (20 Jul 1991).
The authors have measured the radial velocities of 41 galaxies in the field of Abell 665, the richest cluster in the Abell catalog and one of only three clusters for which detection of a microwave decrement is well established. Thirty–three of the 41 galaxies observed are determined to be cluster members. The distribution of velocities is well fitted by a Gaussian, indicating that the cluster is not a superposition of groups or clusters at significantly different velocities. Nevertheless, based on the positions of a larger sample of galaxies, the authors find that subclustering is likely to be present. The mean observed redshift of A 665 is 0.18144 ± 0.00084, and the velocity dispersion is $1201(+183, -126)$ km s^{-1}, in rough agreement with predictions from recent models of the microwave decrement and X–ray–emitting intracluster gas. The brightest cluster member has a velocity which is 447 km s^{-1} larger than the sample's median velocity. This large peculiar velocity may be indicative of velocity subclustering which is not apparent in the observed sample of 33 velocities.

160.037 **A forest of lines but not a forest of clouds. Diffuse IGM models for the Lyα absorption systems of quasars.**
Bi Hongguang.
Diss. (Dr.rer.nat.), Max–Planck–Institut für Physik und Astrophysik, Garching (Germany, F.R.). Inst. für Astrophysik; München Univ. (Germany, F.R.). Fakultät für Physik. MPA—620, 94 p. (Nov 1991).
The forest–like absorption spectrum on the blue side of the Lyα emission line of high redshift quasars is often interpreted as the absorption of neutral hydrogen in isolated intergalactic clouds – the Lyα clouds. Simulations in this thesis show that transient random mass clumps in a diffusely distributed baryonic intergalactic medium (IGM) left over from the primordial nucleosynthesis can also produce the forest–like Lyα absorptions. Therefore, the Lyα forest could be totally caused by the diffuse IGM or could be seriously contaminated by it if both the cloud component and the diffuse component are found to exist. 6 IGM models from both the hierarchical clustering universe and the CDM universe are shown to be able to match the observation. The author discusses the observational methods for the IGM detection, the dynamical evolution of the IGM and the spatial distribution of 17 observed Lyα samples.

160.038 **Cooling–flow modification of the Sunyaev–Zeldovich effect in clusters of galaxies.**
R. Schlickeiser.
Astron. Astrophys., Vol. 248, No. 2, p. L23 – L26 (Aug 1991). Letter–to–the–editor.
The existence of cooling flows in the center of cluster of galaxies can influence the Sunyaev–Zeldovich diminution of the cosmic microwave background radiation temperture in the Rayleigh–Jeans part of the spectrum by its free–free emission. This may account for the positional discrepancy of the observed peak of the diminution and the expected diminution from modelling with the X–ray observations from clusters without cooling flows. Alternatively, the S–Z effect can be employed to detect cooling flows in distant clusters of galaxies.

160.039 **Extremely low surface brightness galaxies in the Fornax Cluster: properties, stability, and luminosity fluctuations.**
G. D. Bothun, C. D. Impey, D. F. Malin.
Astrophys. J., Vol. 376, No. 2, p. 404 – 423 (1 Aug 1991).
Using the method of photographic amplification, the authors have discovered additional examples of very low surface brightness (LSB) galaxies in the Fornax Cluster. In general, the properties of these galaxies are very similar to those discovered in the Virgo Cluster. Here, the following aspects of this population are studied: (1) their morphological appearance, (2) their dynamical stability, and (3) their usefulness as distance indicators via the technique of luminosity fluctuations. In addition, this sample is combined with the Virgo sample to reinvestigate the faint–end slope of the galaxy luminosity function. The authors find that the failure to account properly for surface brightness selection biases leads to artificially flatter slopes in the case where the light profile of dwarf galaxies is exponential in nature. Under this assumption, a faint–end slope of ≈ -1.55 is derived. Most of these LSB objects exhibit a ragged morphological appearance, consisting of numerous surface brightness enhancements (i.e., blobs) embedded in an overall LSB envelope. The colors of these blobs are the same as the envelope and thus they are probably not stellar clusters but instead are random piles of stars which reflect a stellar system with chaotic orbits. From a carefully selected sample of five galaxies, the authors derive a distance modulus to Fornax of (m–M) = 31.3 ± 0.3. The typical fluctuation in these LSB galaxies is $\approx 10\%$. The derived distance modulus should not be taken seriously, since we do not yet understand the dominant stellar spectral type which is responsible for the flucutations in these systems.

160.040 **The X–ray cooling flow in the cluster of calaxies around PKS 2354–35.**
D. A. Schwartz, H. V. Bradt, R. A. Remillard, I. R. Tuohy.
Astrophys. J., Vol. 376, No. 2, p. 424 – 429 (1 Aug 1991).
A position measured with the HEAO 1 Scanning Modulation Collimator suggested the group of galaxies surrounding the radio galaxy PKS 2354–35 as the identification of the bright, hard X–ray source 4U 0009–33 (= 3A 2356–341). The authors confirm this identification with measurements of the position and spatial extent using the EXOSAT CMA detector. The source shows centrally condensed X–ray emission which is distinctly asymmetric, with a core radius ~7.′3 in a roughly N–S direction and ~2.′5 in the E–W direction. A serendipitous Einstein IPC detection confirms the shape of this extent. The EXOSAT ME data are fitted by a thermal bremsstrahlung spectrum with a temperature of 3.7 ± 0.7 keV. Previously, PKS 2354–35 had been found to be a cD galaxy in a small group at redshift 0.048. At this distance the 2 – 6 keV X–ray luminosity is 1.6×10^{44}ergs s^{-1}. New optical observations reveal strong, narrow emission lines of [O II] and [N II] with luminosities $\sim 10^{41}$ergs s^{-1}. Archival radio observations from the VLA discovered a 35″ long, jetlike component, nearly aligned with the major axis of the X–ray distribution. The X–ray observations indicate a strong accretion flow. The luminosity and core radius imply a central density of at least 8×10^{-3}cm^{-3}.

160.041 **The manifold of early–type galaxies from IR photometry in clusters of galaxies. The Virgo cluster.**
E. Recillas–Cruz, L. Carrasco, A. Serrano P. G., I. Cruz–González.
Astron. Astrophys., Vol. 249, No. 2, p. 312 – 322 (Sep 1991).
In the present paper 124 galaxies of all morphological types members of the Virgo cluster are studied via accurate near infrared *JHK* photometry. Apparent fixed aperture magnitudes and colors corrected for galactic extinction and redshift for a sample of early–type galaxies are used to construct color–color (*J–H*) vs. (*H–K*) diagrams. Color–magnitude (*J–H*), (*H–K*), (*J–K*) and (*V–K*) versus K_T diagrams are established for the galaxies with measured diameters in the sample. The (*J–K*) and (*V–K*) versus K_T diagrams show a color–luminosity relationship. From published velocity dispersion data and the IR luminosities for 26 early–type galaxies the L–σ and L–σ–I_e relationships are investigated for IR wavelengths and compared with similar relationships in other bands.

160.042 **A peculiar group of galaxies in the field of the BL Lacertae object H 0414+00.**
R. Falomo, E. G. Tanzi, A. Treves.
Astron. Astrophys., Vol. 249, No. 2, p. 341 – 343 (Sep 1991).
The authors report on the serendipitous finding of a peculiar configuration of galaxies in a CCD image of the field around the BL Lacertae object H 0414+00. The galaxies appear to form four pairs, whose members have similar magnitude and colors, and similar (~5″) angular separation. They may represent a small group of binary galaxies. A more exotic possibility that the pairs are twin galaxies produced by image splitting by a cosmic string is also discussed. The field is proposed for a detailed spectroscopic and imaging study.

160.043 **Are H I–deficient galaxies of the Coma supercluster deficient in molecular gas?**
F. Casoli, P. Boissé, F. Combes, C. Dupraz.
Astron. Astrophys., Vol. 249, No. 2, p. 359 – 373 (Sep 1991).
The first ^{12}CO(1–0) and ^{12}CO(2–1) observations of spiral galaxies located in the clusters Abell 1367 and Abell 1656 are presented. Because of their strong X–ray emission and high galaxy density, these are among the best candidates to search for H_2–deficient galaxies. The sample is made of 18 galaxies selected from their far–infrared emission. 9 are strongly deficient in atomic gas, with H I contents more than 3 times lower than in isolated galaxies of the same luminosities and types, whereas the remaining 9 display a normal H I content: the latter are used as a comparison subsample. With the IRAM 30 m telescope, all these galaxies were detected except for one which is H I deficient and one which is normal. From the measured CO(1–0) intensities, the molecular mass inside the central 10 kpc of the detected galaxies is derived. In H I–deficient galaxies, most of the gas is in molecular form, with H_2/H I mass ratios $\gtrsim 2$. In order to reveal possible deficiencies in molecular matter, the data are compared to the results of the FCRAO extragalactic survey (Young et al. 1989). The molecular content of the authors' sample galaxies referred to Young's et al. blue luminosity, namely, the $M(H_2)/L_B$ ratio, is on the average similar to what has been found for this survey. It indicates that these galaxies are probably not deficient in molecular gas. As an additional clue, galaxies in both the H I–deficient and comparison subsamples have the same average values of $M(H_2)/L_B$. The star formation activity of the Coma galaxies is normal and independent of the H I deficiency.

160.044 **A determination of the 3–point spatial correlation function for clusters of galaxies.**
Y. P. Jing, R. Valdarnini.
Astron. Astrophys., Vol. 250, No. 1, p. 1 – 8 (Oct 1991).
The authors study the 3–point correlation function of clusters of galaxies, with the moment method, on 3–D samples. A fit with the so–called hierarchical model expression is found to be acceptable. Discrepancies between north and south hemispheres, previously detected in 2–D samples, are substantially reduced, and the results of north and south estimates are marginally consistent within internal errors. The mean value of the constant

Q is about 0.7, in accordance with the previous studies. The Kirkwood superposition expression, possibly indicating strong bias, cannot be excluded by this 3–D study.

160.045 **Neutral hydrogen in A 2052 and other cooling flow clusters.**
W. Jaffe.
Astron. Astrophys., Vol. 250, No. 1, p. 67 – 69 (Oct 1991).
The author reports new upper limits on H I absorption and emission near the radio source 3C 317 in the cooling flow cluster A 2052. Considering published H I data on radio galaxies in cooling flows, he notes a probable anticorrelation between radio power and observed absorption. Clusters where absorption but no emission is observed allow estimates of the spatial extent (~20 kpc) and spin temperature (~100K) of the absorbing regions.

160.046 **A ~10 Mpc void in the Lyα forest at z = 3.17.**
A. Dobrzycki, J. Bechtold.
Astrophys. J., Lett., Vol. 377, No. 2, p. L69 – L72 (20 Aug 1991).
The spectrum of the z = 3.285 quasar, Q0302–003, shows a large (~10 Mpc) void in the Lyα forest at $z \approx 3.17$. By simulating the absorption spectrum of the quasar with a range of Lyα forest cloud parameters, the authors find that the probability of such a void occurring by chance is $\sim 2\times10^{-4}$. Since the void is near (although not centered on) the redshift of a bright foreground quasar, the authors discuss the possibility that the void results from ionization of the clouds by the EUV radiation of this QSO.

160.047 **Discovery of a Lyα galaxy near a damped Lyα absorber at z = 2.3.**
J. D. Lowenthal, C. J. Hogan, R. F. Green, A. Caulet, B. E. Woodgate, L. Brown, C. B. Foltz.
Astrophys. J., Lett., Vol. 377, No. 2, p. L73 – L77 (20 Aug 1991). With plate L3.
The authors report the detection of a galaxy associated with the damped Lyα absorbing cloud seen at z = 2.309 toward the QSO PHL 957 (z_{em} = 2.681). The galaxy was discovered in deep Fabry–Perot narrow–band CCD frames and was subsequently imaged spectroscopically. In additon to a strong but narrow Lyα emission line and weaker C IV and He II lines, the object shows continuum at V ~ 24, with a slope (in F_ν) rising slightly toward the red, similar to what is seen in high–redshift radio galaxies; however, the galaxy does not correspond to any known radio source in the literature. The detected emission lines and continuum are most easily interpreted as light from hot, recently formed stars, implying not only a sizeable star formation rate (SFR $\geqslant 5\,M_\odot yr^{-1}$) but also a scarcity of dust, which readily quenches Lyα photons. The emission region appears to be marginally, resolved spatially and is located 48″ to the NW of the QSO, corresponding to a projected distance of $270h^{-1}$ kpc (for $q_0 = 0.1$); the velocity difference with respect to the damped Lyα cloud is ~ 350 km s^{-1}. The spatial correlation of the absorbing cloud and the companion galaxy supports the interpretation of damped Lyα clouds as objects fundamentally different from the lower column density Lyα forest clouds, which show weak or no clustering. The absorption trough itself shows no Lyα emission extended or unresolved, in either the Fabry–Perot frames or in deep, moderate resolution, two–dimensional spectra.

160.048 **Studies on the Coma cluster. III. The virial mass and kinematics.**
Zhao Junliang, Huang Songnian, Pan Rongshi, He Yanping.
Chin. Astron. Astrophys., Vol. 15, No. 3, p. 260 – 269 (Sep 1991). English translation of Acta Astrophys. Sin., Vol. 11, No. 2, p. 119 – 127 (Apr 1991). See Abstr. 53.160.104.
Using the radial velocity data on 414 galaxies in the Coma region of Kent and Gunn and the relevant results of statistical membership the authors determine the virial mass of the Coma cluster and discuss several issues in its kinematics. Their analysis shows that the Coma cluster has undergone some dynamical evolution and has not yet reached a sufficiently relaxed state. Judging by present available data, mass segregation is not marked. The virial mass of Coma determined from members of large membership probability is $\mu_v = 2.63\times10^{15}\mu_\odot$. The total luminosity is $1.45\times10^{13}L_\odot$, the mass–to–light ratio is 182 and the luminosity mass is $\mu_L = 3.43\times10^{14}L_\odot$.

160.049 **The origin of the quasi–stellar object absorption lines.**
S. M. Viegas, R. B. Gruenwald.
Astrophys. J., Vol. 377, No. 1, p. 39 – 48 (10 Aug 1991).
In recent years, several results concerning the QSO absorption–line systems indicate the possibility that they may be associated with star–forming regions in intervening galaxies. Here, the authors explore the hypothesis that the metal absorption lines are produced in H II regions of intervening galaxies. The method of analysis is mainly based on the C IV and Si IV absorption lines. A comparison between observed and calculated equivalent widths fix the characteristics of the absorbing H II region (ionizing star, gas density, and chemical abundances). The calculated equivalent widths are obtained from column densities given by photoionization models. The column densities are calculated for different lines of sight crossing the H II region at different distances from the star. The analysis of 20 absorption line systems shows that H II regions in intervening galaxies can be the origin of the metal–line absorption systems. This model provides promising consequences concerning star formation rates and chemical evolution of galaxies.

160.050 **CCD photometry of some southern–cluster galaxies.**
C. K. Young, M. J. Currie.
Observatory, Vol. 111, No. 1104, p. 220 – 224 (Oct 1991).
B, V, G, R_C, and I_C CCD photometry is presented for 10 galaxies in the directions of the Horologium–Reticulum complex and the southern clusters Klemola 44 and 00 03.5 –34 59. These sequences should assist further photometric studies of these rich fields.

160.051 **Radio detections of neutral hydrogen at redshift z = 3.4.**
J. M. Uson, D. S. Bagri, T. J. Cornwell.
Phys. Rev. Lett., Vol. 67, No. 24, p. 3328 – 3331 (9 Dec 1991). Current Physics Microform No.: 9201E1316.
The authors have used the VLA to search for "21 cm" narrow–band signals from neutral hydrogen at redshift z = 3.4. They detected absorption at $z = 3.3968 \pm 0.0004$ against a radio source that has been previously identified with a "normal" galaxy at $z = 3.395 \pm 0.005$. The authors detected emission at $z = 3.3970 \pm 0.0003$ separated from the absorption by 33′ (7 – 24 Mpc, depending on the cosmological model). This narrow–band emission signal is due to about $3\times10^{14}M_\odot$ of neutral hydrogen as expected from a protocluster of galaxies. The authors identify it as the first example of a Zel'dovich "pancake".

160.052 **A study of the elongation of Abell clusters. II. A sample of 107 rich cluster.**
G. F. R. N. Rhee, M. P. van Haarlem, P. Katgert.
Astron. Astrophys., Suppl. Ser., Vol. 91, No. 3, p. 513 – 554 (Dec 1991).
The authors present the results of a study of 107 Abell clusters selected according to the following criteria: Abell richness class $R \geqslant 1$, redshift, $z \leqslant 0.1$, $10^h \leqslant \alpha \leqslant 18^h$, $\delta > -25°$, $b \geqslant 30°$. Using the Leiden plate scanning machine Palomar Sky Survey Plates were scanned to produce lists of galaxies in these clusters down to approximately $m_R \sim 18.5$ magnitude. The authors present maps of the projected galaxy distribution within $2\,h_{50}^{-1}$Mpc of the cluster center, plots of position angle as a function of the number of galaxies selected, and radial profiles for each of the clusters. They have used three methods for measuring position angles; Fourier analysis of the azimuthal distribution of galaxies, moment analysis and the method of principal axes. It is found that ~50% of the clusters are significantly elongated with a position angle uncertainty $\leqslant 15°$. For significantly elongated clusters the methods produce consistent position angle estimates. The authors have used the maximum likelihood method to fit two models for the radial profiles of clusters. The method does not involve binning the data. It is found that for 10% of the clusters one can distinguish between King and de Vaucouleurs models.

160.053 **Erratum: "GINGA observations of distant galaxy clusters: temperature and iron abundance of A2507 and A483" [Astron. Astrophys., Vol. 243, No. 1, p. 65 – 66 (Mar 1991)].**
M. Arnaud, M. Lachièze-Rey, R. Rothenflug, K. Yamashita, I. Hatsukade.
Astron. Astrophys., Vol. 250, No. 2, p. 580 (Oct 1991). See Abstr. 53.160.043.

160.054 **The peculiar velocity field in the Hercules region.**
W. Freudling, H. Martel, M. P. Haynes.
Astrophys. J., Vol. 377, No. 2, p. 349 – 364 (20 Aug 1991).
The authors examine the three-dimensional distribution of galaxies in the Hercules region, which contains the Hercules supercluster and an underdense region in front of the supercluster, the Hercules void. With the aid of the Tully-Fisher relation, the authors search for deviations from a smooth Hubble flow in that region. New I band CCD images have been obtained for 86 galaxies in the Hercules region. The sample was selected to minimize any bias in derived distances. The selection criteria are discussed in detail. Magnitudes and inclinations measured from those images are presented. The total magnitudes in combination with the inclination-corrected H I line widths are used to estimate redshift-independent distances and to derive peculiar velocities of the galaxies. A much bigger sample of 2165 galaxies for which only redshifts are known is used to model the gravitationally induced deviations from a smooth Hubble flow using the linear theory of gravitational perturbations. For the galaxies with redshift-independent distances, a comparison of the predicted velocities with those measured indicates a marginal correlation or a firm upper limit on the value for the density parameter Ω of 0.7.

160.055 **On the relation between radius, luminosity, surface brightness, and surface density in elliptical galaxies.**
E. Recillas-Cruz, A. Serrano.
Astrophys. Space Sci., Vol. 185, No. 1, p. 1 – 20 (Nov 1991).
The authors have analyzed luminosity profiles of E galaxies studied by Strom and Strom in six clusters of galaxies. They have found a relationship between radius r_e, luminosity L, and surface brightness I_e for galaxies in each of the clusters. A dependence of the zero point of the relation on the local projected density ϱ of galaxies is likewise found: $r_e \sim \varrho^{-0.14} L^{0.445} I_e^{-0.413}$. This relationship implies (i) that there is not a universal luminosity profile for elliptical galaxies, (ii) the environmental variation of radius is larger than that produced by mergers of galaxies, (iii) distance to a galaxy can be estimated from apparent magnitude, surface brightness, angular size, and apparent local projected density of galaxies.

160.056 **Cosmic-ray heating of cooling flows: a critical analysis.**
M. Loewenstein, E. G. Zweibel, M. C. Begelman.
Astrophys. J., Vol. 377, No. 2, p. 392 – 402 (20 Aug 1991).
The authors present a detailed investigation of the hypothesis that a combination of magnetohydrodynamic wave mediated cosmic-ray heating and thermal conduction might serve to balance cooling in intracluster media (ICM) and substantially reduce the rate of inflow. Because of the form of the cosmic-ray heating, a nearly static ICM with a positive temperature gradient can exist. Because of the origin of the cosmic rays – presumed to be an active nucleus in the central galaxy fueled by residual inflow – a globally stable feedback mechanism is at work. Since, while cosmic rays dominate the heating at small cluster radii, thermal conduction dominates at large radii, an (undetectably) small cosmic-ray luminosity is required. The authors derive and solve the appropriate system of steady state equations that include a new, self-consistent formulation for the cosmic-ray diffusivity. Models successful in producing substantial positive temperature gradients in static configurations are indeed found, but only if conduction is reduced by a factor of 10 or more. Unfortunately, these models become cosmic-ray pressure dominated and as a result have too flat thermal pressure profiles when compared with the observations. The effect of cosmic rays on the thermal stability of the ICM is also investigated, as is the role of cosmic rays in heating emission-line filaments.

160.057 **Young globular clusters?**
Peng Wei, J. C. Weisheit.
Publ. Astron. Soc. Pac., Vol. 103, No. 666, p. 891 – 893 (Aug 1991). – See Abstr. 012.034 for the main entry.
The authors present additional arguments for an earlier claim by Peng and Weisheit 1991 that highly ionized, metal-line absorption systems observed in QSO spectra arise in the protoglobular clusters of young galaxies. If true, an important clue to the galaxy-formation process has been identified.

160.058 **Abundances in Virgo spiral galaxies.**
G. A. Shields.
Publ. Astron. Soc. Pac., Vol. 103, No. 666, p. 916 – 918 (Aug 1991). – See Abstr. 012.034 for the main entry.
Observations of H II regions in spirals in the Virgo cluster suggest oxygen abundances substantially higher than in comparable field galaxies. Possible causes include stripping of the outer hydrogen disk and curtailment of cosmological infall in the cluster environment.

160.059 **H I deficiency in the Virgo cluster – a connection with OB effect?**
M. D. Suran, G. Oprescu, N. A. Popescu.
Conference on Dynamics of Disc Galaxies, p. 409 – 412 (1991). – See Abstr. 012.028 for the main entry.
The morphological structure of the central zone ($|d| < 6°$) of the Virgo cluster is analysed with two different methods: the classical one and the new method of extreme populations seggregations. The model using the net dichotomy between Virgo A and Virgo B, realizes the connection between H I deficiency, OB effect and C-R diagrams. The model traces a possible physical evolution of galaxies in a non-relaxed cluster as the Virgo cluster.

160.060 **Neyman-Scott cluster model and diffuse IGM model for the Lyα forest.**
H. G. Bi.
3. Chinese Academy of Sciences and Max-Planck Society Workshop: High energy astrophysics: compact stars and active galaxies, p. 235 – 241 (1991). – See Abstr. 012.026 for the main entry.
To incorporate the positive cross correlation and the zero auto correlation function found before, the author proposes that the clouds are clustered as a Neyman-Scott cluster process in the line of sight. A simulation study shows that when the clusters have a mean radius of $1\,h^{-1}$Mpc and an 1-D density $\geqslant 0.2\,h\,\mathrm{Mpc}^{-1}$ corresponding to the 3-D spatial density $\geqslant 10^{-5}(h^{-1}\mathrm{Mpc})^{-3}$ at redshift 2.75, the auto-correlation functions are efficiently erased by the line-blending effect and then the models are acceptable. Another possibility is then proposed, namely that the Lyα forest-like absorption is due to a diffusely distributed IGM. A simulation shows that the resulted absorption profile is compatible to the observation. Because the positions of lines indicate the highest density regions in an 1-D random field, they will behave as a Poisson process (e.g. Cramer and Leadbetter, 1966) and show very little clustering; while the 1-D fields along two nearby lines of sight can significantly be clustered if the original 3-D field is correlated on larger scales.

160.061 **Magnetic fields in the Milky Way neighbourhood as deduced from warps in spiral galaxies.**
E. Battaner, J. L. Garrido, M. L. Sanchez-Saavedra, E. Florido.
Astron. Astrophys., Vol. 251, No. 2, p. 402 – 410 (Nov 1991).
It is shown that warps of spiral galaxies are not randomly oriented in the Milky Way neighbourhood. By adopting a previous model, in which warps are produced by intergalactic magnetic fields, and considering all northern hemisphere warped edge-on NGC spiral galaxies, an analysis of the intergalactic magnetic field in the ~ 100 Mpc neighbourhood of our Galaxy is carried out. At the 100 Mpc scale the magnetic field is still rather

homogeneous, having a direction given by ($\alpha = 289°$, $\delta = 8°$), but a characteristic scale of about 25 Mpc is found, inside which the dispersion is very low. The region containing the Virgo Cluster has a direction of the magnetic field different from the direction found in adjacent regions.

160.062 **Magnetism in the local Virgo-centered supercluster of galaxies (radius ~ 24 Mpc).**
J. P. Vallée.
Astron. Astrophys., Vol. 251, No. 2, p. 411 – 416 (Nov 1991).

An earlier search for the magnetic field in the local supercluster of galaxies, using the rotation measure data from background quasars and galaxies through the cluster and to its sides, had already revealed the presence of a net excess cluster rotation measure, attributable to a 1.5 μG magnetic field in the Virgo cluster at a redhsift $z = 0.004$, near the center of the local supercluster. A recent study of the twin warps in nearby spiral galaxies seemed to indicate an excess warping in a preferred direction, possibly attributable to an intergalactic magnetic field. Here the author examines more closely the warp excess in nearby spiral galaxies, using more data now available, and its earlier interpretation in terms of a magnetic field on a distance scale out to a redshift of 0.3 (out to 1200 Mpc, for H_0 = 75 km s^{-1}Mpc^{-1}). It is shown that such a large scale is unwarranted, from the study of the rotation measure data as a function of redshift, that any warp excess may be rather local and spurious with a small S/N = 1.7. It implies that a local intergalactic magnetic field must be less than 50 μG.

160.063 **The probable extragalactic nature of some low surface brightness clouds at high galactic latitude.**
H. Arp, J. W. Sulentic.
Astrophys. Space Sci., Vol. 185, No. 2, p. 249 – 263 (Nov 1991).

Low surface brightness clouds observed at optical, infrared, and radio wavelengths are discussed. The authors present evidence that some clouds at high galactic latitudes are associated with the Local Group, M81, and possibly even with higher redshift extragalactic objects. Low temperature clouds at high latitude must affect at some level the short wave length side of the cosmic background radiation. If some of these clouds are extragalactic there should be a further effect on the interpretation of CBR measures.

160.064 **A new spectroscopic survey for damped Ly$_\alpha$ absorption lines from high-redshift galaxies.**
K. M. Lanzetta, A. M. Wolfe, D. A. Turnshek, L. Lu, R. G. McMahon, C. Hazard.
Astrophys. J., Suppl. Ser., Vol. 77, No. 1, p. 1 – 57 (Sep 1991).

The results of a new survey for distant galaxies identified on the basis of the damped Lyα absorption lines they produce in the spectra of high-redshift background QSOs are presented. These results are based on new, low-resolution spectroscopy of a sample of 57 QSOs, which are combined with similar low-resolution spectroscopy of 53 QSOs recently acquired by Sargent, Steidel, and Boksenberg to yield a total of 101 spectra of high-redshift QSOs, which are suitable for identifying damped Lyα absorption lines. The spectra are analyzed to generate a complete sample of absorption features which, if the features are assumed to be Lyα lines, have rest-frame equivalent widths satisfying W $\geqslant$ 5 Å. A total of 84 such damped Lyα absorption-line candidates are detected. Spectroscopy of 39 of these candidates has so far revealed that 22 are in fact damped Lyα absorption lines, whereas 17 are either blends of weaker components or single Lyα lines. This sample is used to investigate in detail the statistical properties of the damped Lyα absorption systems and the evolution of these properties with redshift.

160.065 **The environment of 3C 255.**
E. Giraud.
Astron. Astrophys., Vol. 252, No. 1, p. L5 – L8 (Dec 1991). Letter-to-the-editor.

Very deep images of the field of the radio galaxy 3C 255 at $z = 1.355$ show that the object is in a region where there is an overdensity of objects with $22.5 \leqslant R \leqslant 24$. Extending the number galaxy counts – magnitude relation up to $R = 27$, the authors find that there is no clear density excess around 3C 255 at fainter magnitudes. The radio source may be in a compact group or seen through foreground blue objects. In this 62″ × 62″ field, the mean distance to the nearest object at magnitude $25.4 \leqslant R \leqslant 27$ is 2″.6 only. The spatial distribution of faint objects suggests to search for filamentary structure or clumpiness on a larger field.

160.066 **Statistical properties of X-ray clusters of galaxies.**
R. Nesci.
Astron. Astrophys., Vol. 252, No. 1, p. 13 – 18 (Dec 1991).

A flux-limited all-sky catalogue of clusters of galaxies is derived from HEAO1-A1 data. For the sake of homogeneity, only Abell clusters (for the northern hemisphere) and Abell, Corwin and Olowin clusters (for the southern hemisphere) are considered. Some statistical properties of this cluster sample, which is the widest presently available, are discussed. Taking into account contamination from noncluster sources, the flux-number distribution is found to be under-Euclidean for clusters brighter than 0.003 A1 flux units (1.0×10^{-11}erg cm^{-2}s^{-1}), with an X-ray luminosity evolution time scale of the order of 2 Gyr. The space distribution of X-ray clusters is found to be consistent with an isotropic and homogeneous one when averaged over a volume of radius $z = 0.1$, but substantial substructure is apparent at a smaller scale; in particular, two clumps of clusters (Great Attractors?) around $z = 0.011$ and $z = 0.048$ are found. The Bautz-Morgan morphological types of X-ray-selected clusters are statistically earlier than those of the optically selected ones. In a volume-limited subsample, Rood-Sastry cD clusters are, by far, the most frequent and cooling-flow clusters are about 35%.

160.067 **Dwarf galaxies in the Virgo cluster. I. The systematic photometric properties of early-type dwarfs.**
B. Binggeli, L. M. Cameron.
Astron. Astrophys., Vol. 252, No. 1, p. 27 – 52 (Dec 1991).

The authors present the first results of an extensive photographic surface photometry of dwarf galaxies in the Virgo cluster. The mean radial surface brigthness profiles of 174 dE, 20 dS0, 24 E and S0 and 7 Sa galaxies are surveyed morphologically and analyzed by fitting model profiles.

160.068 **Cl 1358 + 6245 at z = 0.32: an X-ray selected, distant, rich cluster of galaxies.**
G. A. Luppino, B. A. Cooke, I. M. McHardy, G. R. Ricker.
Astron. J., Vol. 102, No. 1, p. 1 – 13 (Jul 1991). Current Physics Microform No.: 9107B1697. With plate 1.

The authors present results of four-color (BVRI) CCD photometry of an extremely rich, X-ray selected cluster of galaxies, Cl 1358 + 6245. Spectroscopy of the optically dominant cluster galaxy indicates the cluster is at a redshift of $z = 0.32$. The photometric catalog contains 322 galaxies measured in at least two colors detected in the CCD field. The authors include the cluster luminosity function and color-magnitude diagrams. Cl 1358 + 6245 is extremely rich (at least Abell Richness Class 3), contains a compact, concentrated core of galaxies, and exhibits the Butcher-Oemler effect with a blue galaxy fraction $0.10 < f_b < 0.18$ (depending on the background galaxy correction). Cl 1358 + 6245 is X-ray luminous. Three measurements of the X-ray luminosity of the cluster range from 4.3x10^{44}ergs s^{-1} < L_x(0.5–4.5 keV) < 8.4x10^{44}ergs s^{-1} suggesting there may be a variable X-ray source in the field. The brightest cluster galaxy has an absolute visual magnitude $M_V = -22.89$ and a rest frame color of $(B-V)_{rf} = 0.94$, consistent with the magnitude and color expected for the brightest galaxy in a rich cluster.

160.069 **Discovery of a ring Seyfert in a group of galaxies.**
E. Davoust, S. Considère, P. Poulain.
Astron. Astrophys., Vol. 252, No. 1, p. 69 – 74 (Dec 1991).

The authors report the discovery of a Seyfert galaxy with a large ring in a small group of anonymous galaxies. The Seyfert is of type 2, with a heliocentric redshift of 20226 km s^{-1}. The authors provide a detailed photometric analysis of five galaxies of the group. The brightest companion, which seems perturbed and connected to the ring, has a 4057 km s^{-1} lower redshift. A comparison is made with 4 other ring Seyferts, all IRAS sources. The authors speculate that the ring was formed by an external event, possibly a high-velocity encounter, which also triggered the Seyfert activity.

160.070 **X-ray emission from poor clusters of galaxies.**
R. Price, J. O. Burns, N. Duric, M. V. Newberry.
Astron. J., Vol. 102, No. 1, p. 14 - 29 (Jul 1991). Current Physics Microform No.: 9107B1710.

The authors present X-ray observations of 25 poor clusters and compare their X-ray, optical, and radio properties with those of rich clusters. The X-ray morphology has been classified in a manner similar to that developed for rich clusters by Jones & Forman (1984) with 13 of the 18 detected clusters in the authors' sample being D-type clusters, where the peak coincides with a dominant galaxy. For the five T-type clusters, the X-ray emission appears to trace the optical distribution of galaxies. In both classes a sizeable fraction of the X-ray emission is associated with individual galaxies. The two morphological classes have a similar range of optical properties, which suggests that they are of comparable mass and dynamical age. Poor clusters are consistent with being a scaled-down version of rich clusters in essentially all characteristics and the 13 clusters in the authors' sample that are also common to the Morgan et al. (1975) sample are found to be indistinguishable from the rest of the sample. The optical properties of seven clusters that were not detected at X-ray wavelengths are also indistinguishable from those of the X-ray detected clusters. This suggests that all poor clusters have a hot intracluster medium. For clusters that have radially symmetric X-ray emission and sufficient X-ray photon statistics, analyses of their surface-brightness profiles show them to have smaller core radii than rich clusters and, in one case, a central X-ray excess which may signify a cooling flow. Radio observations show that on average T-type clusters contain more radio sources than D-type clusters.

160.071 **Confirmation of dust in damped Lyman-alpha systems.**
Y. C. Pei, S. M. Fall, J. Bechtold.
Astrophys. J., Vol. 378, No. 1, p. 6 - 16 (1 Sep 1991).

As part of a search for dust at high redshifts, the authors have acquired new spectra of quasars in the Wolfe et al. sample: 13 with damped Lyα systems along the lines of sight and 15 without. The spectra cover the region 3950 - 6900 Å at a resolution of 15 Å with S/N ratios of 10 - 50. The authors have determined spectral indices over a common range of wavelengths in the rest frames of the quasars between Lyα and C IV emission. Several internal and external checks indicate that the typical errors in the spectral indices are ± 0.1. It is found that quasars with damped Lyα systems in the foreground are stochastically redder than those without damped Lyα systems in the foreground. This difference is significant at the 3σ level. Combining the Wolfe et al. and Sargent et al. samples increases the significance of the reddening to at least 4σ. The authors compute dust-to-gas ratios from the spectral indices on the assumption that the extinction curve in the damped Lyα systems has the same shape as that in the Milky Way or the Magellanic Clouds. Galactic-type dust appears to be ruled out by the absence of strong extinction near 2175 Å in the rest frames of several damped Lyα systems. The best estimates of the typical dust-to-gas ratio in the damped Lyα systems are then 5% - 20% of that in the Milky Way. This result, together with recent data on the gas-phase abundances of Zn and Cr in three damped Lyα systems, indicates that the overall abundances of heavy elements (in both the gas and solid phases) are about one order of magnitude lower than the Galactic value.

160.072 **Investigations of clusters of galaxies. IV. Properties of the galaxy cluster Abell 1795.**
F. W. Baier, K. Fritze, H. Tiersch, R. Ziener.
Astron. Nachr., Vol. 312, No. 3, p. 149 - 160 (1991).

The structure of the centre of the galaxy cluster A 1795 and its surroundings, representing different cluster components, has been investigated in several wavelength ranges. A preferred orientation deviating about 10-20° from the north-south direction has been found for all cluster components. The orientation of the central cD galaxy is in good agreement with this direction, too. Hints seem to exist for antisegregation in luminosity as well as for colour segregation.

160.073 **On star formation in cooling flows of clusters of galaxies.**
K.-H. Schmidt.
Astron. Nachr., Vol. 312, No. 3, p. 161 - 165 (1991).

A correlation between the masses of cooling flows in clusters of galaxies and the absolute magnitudes of the central parts of the brightest galaxies of the clusters is detected. This correlation is consistent with star formation in the cooling flows according to an initial mass function of stars truncated at stellar masses corresponding to absolute magnitudes in the most probable cases between +7 and +2.

160.074 **Kinematical study of groups of galaxies.**
D. Puche, C. Carignan.
Workshop on the Interpretation of Modern Synthesis Observations of Spiral Galaxies, p. 155 - 161 (1991). - See Abstr. 012.041 for the main entry.

The results of a kinematical study of the Sculptor group are presented. It is shown that a comparison of the dynamical study of the members with the H I kinematical analysis of the individual galaxies, gives some insight on the distribution of dark matter in the group. The dark matter halos of the galaxies need only extend to ~40 kpc to include the total dynamical mass of the group. An overdensity of at least three orders of magnitude exists between the dark matter density in the galaxies and the intergalactic medium. A preliminary dynamical analysis of the M101 group shows a similar tendency of dark matter to be more concentrated around the optical galaxies.

160.075 **The intergalactic H I supply.**
F. Briggs.
Workshop on Warped Disks and Inclined Rings Around Galaxies, p. 1 - 5 (1991). - See Abstr. 012.054 for the main entry.

An inventory of the H I observed in nearby groups and clusters shows that the bulk of the neutral gas is already bound to large spiral galaxies and is not contained in gas-rich dwarfs. However, small galaxies do significantly outnumber large ones, despite the fact that they do not form a large reservoir of neutral gas. Intergalactic H I clouds with masses in the range $M_{H\,I} \approx 3\times10^8$ to $10^{10} M_{\odot}$ must be less common than normal galaxies of similar H I mass by more than a factor of one hundred. H I clouds with low H I masses ($\sim 10^7 M_{\odot}$) could be as abundant as galaxies of comparable mass.

160.076 **Evolution of clusters of galaxies and the ROSAT All Sky Survey.**
M. Pierre.
Astron. Astrophys., Vol. 252, No. 2, p. L23 - L26 (Dec 1991). Letter-to-the-editor.

The authors present predictions for the redshift distribution of clusters of galaxies in the ROSAT All Sky Survey. The cluster population is modelled according to the recent determinations of the local temperature and luminosity functions (Edge et al. 1991) and its evolution following a modified formulation of the self-similar approximation. This solution, which assumes pre-heating of the intergalactic medium, was proposed by Kaiser (1991) in order to account for the strong negative evolution revealed by the EINSTEIN and EXOSAT data. The simulations show a conspicuous difference in the redshift distribution between models characterized by indices of the initial fluctuation spectrum $n = -1$, $n = -2$ or no evolution ($n \sim -0.5$). For the most likely - as suggested by the optical data - CDM type index $n = -1$, a total of about 20 clusters should be detected between redshifts of 0.4-0.7 in a 10° radius field centered on the North ecliptic pole where the survey sensitivity is maximum (9×10^{-14} erg/s/cm^2). At a flux limit of 7×10^{-13}erg/s/cm^2, the mean sensitivity of the survey, the most distant objects detectable should have a redshift around 0.4.

160.077 **The circumgalactic ring of gas in Leo.**
S. E. Schneider.
Workshop on Warped Disks and Inclined Rings Around Galaxies, p. 25 – 32 (1991). – See Abstr. 012.054 for the main entry.

In the course of observations subsequent to its discovery, the intergalactic gas in the M96 Group in Leo finally revealed itself to be distributed in the form of an enormous ring of material surrounding the central two galaxies of the group (Schneider et al. 1989). This ring provides an unusually "clean" opportunity for studying the dynamics of a ring of gas in the dynamical environment of a few galaxies' haloes. The author reviews the properties of this giant ring and briefly discusses some searches for intergalactic gas surrounding other early-type galaxies.

160.078 **The distribution of galaxies around NGC 5846.**
M. P. Haynes, R. Giovanelli.
Astron. J., Vol. 102, No. 3, p. 841 – 863 (Sep 1991). Current Physics Microform No.: 9109D1245.

The three-dimensional distribution of galaxies in a region surrounding the chain of galaxies in the NGC 5846 group has been investigated by both subjective and objective tests for clustering. The compilation of a complete redshift sample over the area of study allows the identification of distinct galaxy aggregates and estimation of group properties. Much of the nearby structure, including the NGC 5846 group, is connected into a single prolate cloud of galaxies that points toward the Virgo cluster. Also evident is a tenuous filament of galaxies at a velocity of ≈ 7000 km s^{-1} that appears to connect at right angles with an elongated cloud that is identified as the Zw 1400+0949 cluster seen in projection. The subconcentration around NGC 5846 is itself a tight, compact configuration, unlike other density enhancements like the Virgo III or Ursa Major Clouds. A comparison of the NGC 5846 group with similar structures in the Local Supercluster confirms the qualitative impression that the highest density environments are dominated by early type systems.

160.079 **Correlation functions of galaxies in the ESO-Uppsala catalogue.**
Y. P. Jing, H. J. Mo, G. Börner.
Astron. Astrophys., Vol. 252, No. 2, p. 449 – 457 (Dec 1991).

The authors estimate the angular two- and three-point correlation functions of galaxies in the ESO-Uppsala catalog, using angular-diameter ($d1$)-limited samples. They use the diameter function given by Lahav et al. (1988) to scale the results. The two-point correlation functions for the samples with $d_{1,\rm lim} < 3\farcm0$ can all be fitted by the power law $\xi(r) = (r/r_0)^\alpha$ with $\alpha = 1.8$ and $r_0 = 4.74\,h^{-1}$Mpc; the results for samples with larger $d_{1,\rm lim}$ can be fitted by a similar power law but with smaller values of r_0. The three-point correlation function satisfies the hierarchical form with $Q = 0.66$. The authors also analyse the dependence on morphological types. Early-type galaxies are found to have significantly larger two-point correlation amplitude, but the slope does not depend strongly on morphological types. The three-point correlation functions for different morphological types can all be expressed by the hierarchical form. These results are compared to those for the corresponding UGC samples and those given by other analysis.

160.080 **Ultraviolet observations of galaxies in nearby clusters. II. Luminosity function and surface density distribution in the Coma cluster.**
J. Donas, B. Milliard, M. Laget.
Astron. Astrophys., Vol. 252, No. 2, p. 487 – 492 (Dec 1991).

Using ultraviolet (UV) observations at $\lambda \sim 2000$ Å, the authors have obtained the ultraviolet luminosity function (ULF) and surface distribution for galaxies in a circular field of 2.18 square degrees (50′ radius) centered on the Coma cluster. They found that in the UV-band, the bright end of the ULF is highly dominated by Irr-Sp type galaxies, and the surface distribution shows low central concentration. Therefore, the Coma cluster, often taken as the archetype of the elliptical-rich and concentrated cluster, has the appearance of a spiral-rich and low concentration cluster in the UV-band.

160.081 **A systematic effect in the use of planetary nebulae as standard candles.**
L. Bottinelli, L. Gouguenheim, G. Paturel, P. Teerikorpi.
Astron. Astrophys., Vol. 252, No. 2, p. 550 – 556 (Dec 1991).

The distances determined by Jacoby et al. (1990) for 6 early type galaxies members of the Virgo cluster of galaxies, using the method of the planetary nebulae luminosity function (PNLF) are correlated with the magnitude of the parent galaxy. A similar trend is found within the Leo I group of galaxies and is confirmed by a comparison between the PNLF distances and distance estimates from the surface brightness fluctuation method. This effect works in the sense that more luminous galaxies there are assigned smaller distances. Another strange systematic feature is the increase of the specific PN density from luminous to faint galaxies, clarly seen in Virgo and Leo I data. The authors show that these systematic trends can be due to an exponential part in the bright end of the PNLF. If this is the case, the short value derived for the Virgo distance by the PNLF method should be regarded uncertain. The authors analyze shortly the use of an exponential LF as a distance indicator and note that if it is applied to the planetary nebulae, one should select the calibrators to match at least the Hubble type of the measured galaxies. As a preliminary example the relative distance $\mu_{\rm Vir} - \mu_{\rm Leo\ I}$ is derived by the exponetial LF method to be 1.15 ± 0.04 (with a fixed value for the slope of the LF) instead of the original value of 0.88. The authors emphasize the need of a detailed study of the PNLF in galaxies within Virgo and closer with a large luminosity and type range.

160.082 **Attenuation of Lyman-alpha emission by dust in damped Lyman-alpha systems.**
S. Charlot, S. M. Fall.
Astrophys. J., Vol. 378, No. 2, p. 471 – 475 (10 Sep 1991).

The authors derive relations between the Lyα surface brightness, star formation rate, and abundance of dust in damped Lyα absorption systems. These are idealized as plane-parallel objects in which the sources of Lyα photons are either distributed throughout the gas or confined to a thin, central sheet. For the dust-to-gas ratio in the damped Lyα systems, the typical value inferred from the reddening of background quasars is adopted. The results of several recent searches for Lyα emission are used to set limits on, and in one case, estimate the star formation rate per unit area of the damped Lyα systems. When the H I column density exceeds 5×10^{20}cm^{-2}, the attenuation of Lyα emission is large and the star formation rate can be one or two orders of magnitude higher than that in the solar neighborhood. The large uncertainties are caused by the unknown orientations of the damped Lyα systems and the unknown distributions of stars, gas, and dust within them. However, when the H I column density is less than 5×10^{20}cm^{-2}, the attenuation is smaller and the limits on or value of the star formation rate per unit area are similar to that in the solar neighborhood.

160.083 **Clusters of galaxies associated with quasars. II. Galaxy cluster dynamics.**
E. Ellingson, R. F. Green, H. K. C. Yee.
Astrophys. J., Vol. 378, No. 2, p. 476 – 486 (10 Sep 1991).

Multiobject spectroscopy of galaxies associated with optically luminous quasars with $0.15 < z < 0.6$ is presented. The relative radial velocities of galaxies within 500 kpc of the quasars seem to be well described by a Gaussian distribution with a dispersion of 400 – 500 km s^{-1}. This result is independent of whether the quasar is radio-loud or radio-quiet, and, more surprisingly, whether it is located in a rich cluster of galaxies or in a poor group. Comparison with galaxy cluster dynamics for different environments indicates that the rich clusters of galaxies associated with quasars have anomalously low velocity dispersions, whereas the poor groups of galaxies may be dynamically normal. This result is consistent with models of quasar formation triggered by galaxy – galaxy interaction, where the efficiency of such interactions is much greater for low relative velocities.

160.084 **H I studies of the Sculptor group galaxies. VII. Implications on the distribution and nature of dark matter in groups.**
D. Puche, C. Carignan.
Astrophys. J., Vol. 378, No. 2, p. 487 – 495 (10 Sep 1991).
Results from the large-scale mapping of the H I gas in the Sculptor group are presented. From the kinematical analysis of the members, a mean value of $(M/L_B)_{global} = 9 \pm 5\ M_\odot/L_\odot$ is found for the individual galaxies. This is a factor of ~10 smaller than the $(M/L_B)_{dyn} = 90 \pm 20\ M_\odot/L_\odot$ derived from a dynamical study of the whole group. Mass models, fitted to the rotation curves, suggest a mean density of dark matter (DM) of $\sim 5 \times 10^{-3} M_\odot pc^{-3}$ in the halos of the galaxies. This is more than three orders of magnitude greater than the maximum DM density in the intergalactic medium, if all the DM were uniformly distributed in the group. This large density contrast implies that DM is more highly concentrated around the luminous galaxies than uniformly distributed in the group. Under the assumption that the Sculptor group is a virialized system and that all the mass is associated with the galaxies, an upper limit of ~40 kpc is derived for the size of the dark halos present in the five late-type spirals of the group. Implications of this result for the nature of the DM are explored.

160.085 **A dynamical analysis of twelve clusters of galaxies.**
T. C. Beers, W. Forman, J. P. Huchra, C. Jones, K. Gebhardt.
Astron. J., Vol. 102, No. 5, p. 1581 – 1609 (Nov 1991). Current Physics Microform No.: 9111D0587.
The authors report 428 new redshifts for galaxies in the vicinity of twelve rich clusters. The clusters observed – A85, A569, A957, A1069, A1185, A1750, A1991, A2063, A2440, A2589, A2593, and A2657 – exhibit a wide range of optical and X-ray properties, from extremely luminous X-ray emitters with apparently dense central regions, to clusters which exhibit a bimodal or clumpy appearance in their galaxy and X-ray surface brightness distributions. Several of the clusters (and subclusters) contain dominant galaxies with multiple nuclei embedded in an extended luminous halo. The authors discuss the velocity distributions for each of these clusters, and obtain dynamical mass estimates.

160.086 **Detection of excess rotation measure due to intracluster magnetic fields in clusters of galaxies.**
K.-T. Kim, P. C. Tribble, P. P. Kronberg.
Astrophys. J., Vol. 379, No. 1, p. 80 – 88 (20 Sep 1991).
The Faraday rotation measures of a sample of extragalactic radio sources projected within a third of an Abell radius of a galaxy cluster were compared with those of sources located further from cluster centers. The result strongly indicates that the distribution of the residual rotation measure (RRM) in the former population is broadened, at a confidence level exceeding 99%. The broadening is detectable out to $1 h_{50}^{-1}$Mpc. The best estimate of the excess Faraday rotation measure varies from 100 ± 36 rad m^{-2} in the central sixth of an Abell radius to 36 ± 15 rad m^{-2} further out. The combination of these results with electron densities determined from X-ray data for some of the clusters suggests that magnetic field strengths in cluster gas are of order 1 microgauss.

160.087 **The magnification of stars crossing a caustic. I. Lenses with smooth potentials.**
J. Miralda-Escudé.
Astrophys. J., Vol. 379, No. 1, p. 94 – 98 (20 Sep 1991).
The author studies the possibility of observing highly magnified images of individual stars within the long arcs that have been observed in rich clusters of galaxies. These images appear when a star crosses a caustic of the potential of the cluster. When the mass of the cluster is distributed smoothly, the maximum magnification reached is typically of the order of 10^7. The star image appears suddenly and reaches the maximum magnification in a few hours; then the magnification decreases slowly as the inverse of the square root of the time. This can also occur in the reverse order in time. For typical arcs, stars with bolometric luminosities $L \sim 300\ L_\odot$ reach a maximum visual apparent magnitude of ~25. If all brighter events are detected, the rate of detection is of order one every five years, for every arc intersecting a caustic. If the source galaxy is forming stars with a Salpeter IMF, the probability that a 30 $M_\odot$ star (with $10^5 L_\odot$) is magnified to a magnitude $m_V = 28$ at any time is of order 20%. However, such a star could only be detected in the presence of time variability of its magnitude, either intrinsic to the star or produced by microlensing.

160.088 **The carbon abundance of the Lyman-alpha clouds.**
L. Lu.
Astrophys. J., Vol. 379, No. 1, p. 99 – 106 (20 Sep 1991).
A search for C IV λλ1548, 1550 absorption associated with Lyα forest clouds has been attempted, using a method similar to the composite spectrum method developed by Norris, Hartwick, and Peterson. The author tentatively reports a detection of the C IV λ1548 absorption feature at ≥99.99% significance level. If confirmed, not all Lyα clouds have primordial composition. Assuming that the absorption feature is real, the author estimates the relative abundance of carbon of the Lyα forest clouds to be roughly [C/H] ~ -3.2 from photoionization models.

160.089 **Parkes radio sources in the directions of southern rich clusters.**
D. L. Brown, J. O. Burns.
Astron. J., Vol. 102, No. 6, p. 1917 – 1927 (Dec 1991). Current Physics Microform No.: 9112C1893.
The authors have compiled a listing of all Parkes radio sources coincident within 0.3 cluster radii of southern rich clusters of galaxies (Abell et al., 1989). A pronounced peak of radio sources very near the cluster centers was found. Bautz Morgan type I clusters tend to have more powerful radio sources and are radio loud ≈5 times more often than clusters without dominant central galaxies. There is no strong correlation between radio emission and richness class. The authors find that the slope of the cluster–radio source two-point angular correlation function is twice that of the galaxy–galaxy correlation function. Furthermore, the correlation amplitude for stronger sources (>0.65 Jy) is nearly twice that of the entire Parkes catalog sample suggesting a tighter correlation with cluster centers for the stronger sources. Finally, the distribution of spectral indices for Parkes sources within ACO clusters is different from that outside the clusters; stronger cluster sources have steeper spectra on average.

160.090 **The small-scale clustering properties of dwarf galaxies.**
J. P. Vader, A. Sandage.
Astrophys. J., Lett., Vol. 379, No. 1, p. L1 – L5 (20 Sep 1991).
The authors report results on the small-scale clustering properties of dwarf galaxies identified in the vicinity of early-type Shapley-Ames galaxies on high resolution photographic plates. Dwarf galaxies display the same trend of stronger clustering toward earlier morphological type on small scales as their giant counterparts on larger scales. Early-type dwarfs have a surface density as a function of projected radial distance R from E-E/S0 galaxies of the form $\Sigma(R) \sim R^{-1.22 \pm 0.05}$ for $25 < R < 400$ kpc ($H_0 = 50$) which mimicks that of the dark halos of spiral galaxies. It corresponds to a spatial two-point correlation function $\xi(r) = (r/r_0)^{-\gamma}$ with $\gamma = 2.22$ and $r_0 = 10$ Mpc. The steepness of γ and the tendency of early-type dwarf galaxies to be satellites of giant galaxies suggest that early-type dwarf galaxies are the most strongly clustered of all galaxies. These results indicate that early-type dwarfs can be used as dynamical probes of dark halos around early-type giant galaxies. The trend of increasing early-type dwarf frequency per early-type giant with environment richness, previously established for rich groups, is found to be continued in the less rich environments studied here. The data cover a range in early-type dwarf to giant ratio of a factor of 20.

160.091 **Faint galaxies in groups of bright galaxies.**
A. T. Kalloglyan, S. R. Unanyan.
Astrofizika, Tom 33, Vyp. 3, p. 403 – 408 (Dec 1990). In Russian. English translation in Astrophysics, Vol. 33, No. 3.
On the basis of galaxy counts on Palomar Sky Survey charts it has been shown that at least a part of Geller and Huchra groups

are parts of larger formations. In many groups some substructures exist mainly due to subclustering of faint galaxies.

160.092 **A measurement of the value of the Hubble constant from the X–ray properties and the Sunyaev – Zel'dovich effect of Abell 665.**
M. Birkinshaw, J. P. Hughes, K. A. Arnaud.
Astrophys. J., Vol. 379, No. 2, p. 466 – 481 (1 Oct 1991).
A comparison of the X–ray properties and the Sunyaev – Zel'dovich effect for a cluster of galaxies may be used to measure the distance of the cluster, and hence the Hubble constant. The authors have applied this method to the rich cluster Abell 665 using IPC, HRI, and MPC data from the Einstein Observatory, LAC data from the Ginga satellite, and Sunyaev – Zel'dovich data from the OVRO 40 m telescope. The X–ray and Sunyaev – Zel'dovich effect data are consistent in their description of the cluster gas and may be fitted by a simple isothermal β–model with $\beta \approx 0.66$, cluster core radius $\approx 1\farcm6$, and gas temperature ≈ 8.2 keV. The relative normalizations of the X–ray and Sunyaev – Zel'dovich effect data then lead to an estimated Hubble constant $H_0 = 40 \pm 9$ km s^{-1}Mpc^{-1} if only the random errors are included. When the possible systematic errors are added in quadrature, the range of possible values of the Hubble constants expands to $(40 \text{ to } 50) \pm 12$ km s^{-1}Mpc^{-1}. Possible improvements in the error of this estimate for the Hubble constant and the fundamental limitations of the method are considered.

160.093 **Recent progress of studies on clusters of galaxies and member galaxies in clusters. I.**
Pan Rongshi, Zhao Junliang, Liu Ruliang.
Prog. Astron., Vol. 9, No. 3, p. 211 – 220 (Sep 1991). In Chinese.
Recent progress of studies on general characteristics of clusters of galaxies and on member galaxies in clusters are described, including catalogues, membership determination, luminosity function and dynamics of clusters, and formation and evolution of cD galaxies, environment of member galaxies, etc.

160.094 **The clustering of faint galaxies.**
G. Efstathiou, G. Bernstein, N. Katz, J. A. Tyson, P. Guhathakurta.
Astrophys. J., Lett., Vol. 380, No. 2, p. L47 – L50 (20 Oct 1991).
Deep images of the sky have revealed a population of faint blue objects that may be protogalaxies at redshifts $z \gtrsim 1$. In this letter the authors show that these faint galaxies are surprisingly weakly clustered. There are several possible explanations of this result: (1) the majority of the faint blue galaxies belong to a new population that is weakly clustered and intrinsically faint at the present epoch; (2) galaxy clustering evolves much more rapidly than expected in simple models of gravitational instability; or (3) the geometry of the universe differs significantly from the Einstein – de Sitter model.

160.095 **Galaxy correlations, counts, and moments.**
P. Carruthers.
Astrophys. J., Vol. 380, No. 1, p. 24 – 29 (10 Oct 1991).
The author explores relations among galaxy correlation functions, counts, moments, and fractal dimensions. Using counts, or factorial moments in cells of variable size, one can conveniently test for possible scaling behavior. Exact relations between density moments F_p and cumulant moments K_p are given, suggesting alternative methods to investigate the texture of galaxy distributions. In particular, one can systematically extract the sequence of cumulant moments from these identities. Specializing to translation invariance and scaling, many simplifications occur. The hierarchical, or linked pair approximation expressing higher order cumulations in terms of products of two particle cumulants, leads to several interesting results. This hypothesis is shown to be closely connected to the negative binomial distribution. Remarkably, the same correlation structures (although with distinct correlation functions) seem to work for hadronic multiplicity distributions in high–energy collisions, and for galaxies.

160.096 **The luminosity function of compact groups of galaxies.**
C. Mendes de Oliveira, P. Hickson.
Astrophys. J., Vol. 380, No. 1, p. 30 – 38 (10 Oct 1991).
The authors present an analysis of the luminosity function of the galaxies in Hickson's sample of compact groups (HCG). The luminosities of compact group galaxies are consistent with their being drawn from a Schechter luminosity function with $\alpha = -0.2(+0.8, -0.9)$, $L^* = 1.1 \pm 0.2 \times 10^{10} h^{-2} L_\odot$ and $\Phi^* = 1.3(+0.2, -0.8) \times 10^{-4} h^3$ Mpc^{-3}. Individual morphological type luminosity functions were also determined. Both the total and morphological type specific luminosity functions of compact group galaxies are significantly different from those of field, loose–group, and cluster galaxies. In particular, the luminosity function of HCG elliptical galaxies has a mean magnitude which is significantly brighter than the mean magnitude of Virgo cluster elliptical galaxies. The mean luminosity density of galaxies in compact groups is estimated to be $\boldsymbol{L} = 1.3(+0.1, -0.8) \times 10^6$ h $L_\odot$Mpc^{-3}, which constitutes $\sim 0.8\%$ of the total luminosity density of galaxies in the universe. This result is consistent with the conventional scenario in which compact groups merge to form elliptical galaxies on a relatively short time scale.

160.097 **Enrichment and heating of the intracluster medium through galactic winds.**
L. P. David, W. Forman, C. Jones.
Astrophys. J., Vol. 380, No. 1, p. 39 – 48 (10 Oct 1991).
In previous papers the authors presented the results of simulations concerning the dynamic and X–ray evolution of the interstellar medium in elliptical galaxies. They found that all model galaxies developed supernova – driven winds at early epochs and that low–luminosity ellipticals maintained galactic winds for most of their lifetime. The authors examine in this paper what effect these galactic winds have on the heavy element abundance and energetics of the intracluster medium. Substantial injection of highly enriched gas occurs within 3×10^8yr after galaxy formation through galactic winds. The resulting iron abundance of the intracluster medium through galactic winds varies considerably with the assumed stellar initial mass function, Type I supernova rate, and the primordial mass fraction of the cluster gas. The combined mechanical luminosity of the winds from the galaxies in a cluster can greatly exceed the present optical luminosity of a cluster. Assuming the distribution of galaxy luminosities in a cluster can be described by a Schechter luminosity function, the calculations show that the observed iron abundance, oxygen–to–iron mass ratio, and gas temperature in clusters can all be simultaneously reproduced only if the first generation of stars formed with a relatively flat initial mass function.

160.098 **The scale dependence of galaxy distribution functions.**
W. C. Saslaw, P. Crane.
Astrophys. J., Vol. 380, No. 2, p. 315 – 319 (20 Oct 1991).
The authors analyze how the observed spatial distribution function of galaxy number counts depends on the size of the cells which are used to measure it. Areas of cells are between 0.25 and 25 deg^2. Over these scales, corresponding to about 1 – 10 Mpc in the Zwicky catalog, the distribution functions fit the form predicted by quasi–equilibrium gravitational clustering. Values of b, the ratio of gravitational correlation energy to the kinetic energy of peculiar velocities, vary with scale in the manner expected from the form of the two–point correlation function. The results are also consistent with the scale dependence found in gravitational N–body simulations. The authors further examine how the spatial scaling depends on the magnitude cutoff of the sample. The observations are consistent with random magnitude selection. They imply that on average a galaxy's magnitude does not depend strongly on the number of galaxies now around it. Both the spatial scaling and the magnitude scaling place new constraints on theories of galaxy formation and clustering.

160.099 **Spectroscopy of the gravitational arcs in Cl 2244–02, A370 (arclet A5) and Cl 0024+1654.**
Y. Mellier, B. Fort, G. Soucail, G. Mathez, M. Cailloux.
Astrophys. J., Vol. 380, No. 2, p. 334 – 343 (20 Oct 1991).
The authors analyze new spectroscopic data on the blue giant arc in Cl 2244–02, on the arclet A5 in Abell 370, and on the split

arc in Cl 0024+1654. A probable redshift z = 2.237 is derived for the source of the Cl 2244–02 arc, making this object one of the most distant "field" galaxies known today. A star formation rate is estimated from the Lyα intensity. For A5, no prominent emission line is found, but three absorption lines point to a tentative redshift z = 1.306. The spectra of the four parts of the Cl 0024+1654 arc look rather similar, but the absence of any spectral feature allows only the determination of a redshift range: 1 < z < 2.1. The systematic search for such arcs and/or arclets, and the Toulouse spectroscopic survey of arcs which follows up, seems one of the promising methods of obtaining spectra of extremely distant field galaxies and of constraining their early evolution.

160.100 **Turbulently generated magnetic fields in clusters of galaxies.**
I. Goldman, Y. Rephaeli.
Astrophys. J., Vol. 380, No. 2, p. 344 – 350 (20 Oct 1991).
The typical scale and velocity of the dominant, turbulent eddies excited by the motion of galaxies in clusters are determined from basic considerations which are valid if a steady state is attained. Hydrodynamic turbulence in the intracluster plasma presumably generates magnetic fields; the authors estimate the typical scale and mean strength of these fields in cluster cores to be about 10 kpc and at most a few 10^{-7}G. The authors discuss implications of fields with these properties on the detectability of diffuse hard X-ray emission from clusters, and on transport properties in the intracluster space.

160.101 **Evidence for evolution of the luminosity function of clusters of galaxies.**
A. C. Edge, G. C. Stewart, A. C. Fabian, K. A. Arnaud.
AIP Conf. Proc., No. 222, p. 343 – 346 (15 Apr 1991). Current Physics Microform No.: 9106E0343. – See Abstr. 012.069 for the main entry.
An all-sky, X-ray flux-limited sample of clusters of galaxies shows evidence for a significant deficit in the number density of high luminosity clusters in the redshift range z ~ 0.1 – 0.2, compared with that of nearby clusters. This indicates that the X-ray luminous clusters are undergoing strong evolution. The strength of the effect is consistent with hierarchical merging models. The implications of such strong evolution for clusters are discussed.

160.102 **The evolving intergalactic medium: the uncollapsed baryon fraction in a cold dark matter Universe.**
P. R. Shapiro, M. L. Giroux, A. Babul.
AIP Conf. Proc., No. 222, p. 347 – 351 (15 Apr 1991). Current Physics Microform No.: 9106E0347. – See Abstr. 012.069 for the main entry.
The authors have calculated the time varying density of the IGM by coupling detailed, numerical calculations of the thermal and ionization balance and radiative transfer in a uniform IGM of H and He, including the mean effect of an evolving distribution of gas clumps embedded in a smoothly distributed ambient gas, to the linearized equations for the growth of density fluctuations in both the gaseous and dark components in a cold dark matter Universe, identifying the IGM density with the uncollapsed baryon fraction.

160.103 **Diagnosing the reionization of the Universe: the absorption spectrum of the intergalactic medium and Lyman α clouds.**
M. L. Giroux, P. R. Shapiro.
AIP Conf. Proc., No. 222, p. 352 – 356 (15 Apr 1991). Current Physics Microform No.: 9106E0352. – See Abstr. 012.069 for the main entry.
The authors have calculated in detail the thermal and ionization evolution of a uniform intergalactic medium (IGM) composed of H and He undergoing "reionization", including the mean effect of gas clumps embedded in a smoothly distributed ambient gas. They have solved the rate equations for ionization and recombination together with the equations of energy conservation including the effects of cosmological expansion, radiative and Compton cooling, and the diffuse flux emitted by the gas, and radiative transfer. The authors have included the contribution to the continuum opacity of the Universe due to the observed quasar absorption line clouds. They have considered a variety of sources of photoionization, including quasars and primeval galaxies, as well as the possibility that hydrodynamical processes deposit thermal energy in the IGM. They discuss two questions. (1) Can metal line ratios for Lyman limit quasar absorption line systems diagnose the metagalactic ionizing radiation background at z ~ 3? and (2) Can the He II Gunn–Peterson effect be used to diagnose the reionization source and/or the nature of the Ly α forest clouds?

160.104 **The hydrodynamics of relict cosmological H II regions: impact on the IGM.**
A. Meiksin, P. Madau.
AIP Conf. Proc., No. 222, p. 357 – 360 (15 Apr 1991). Current Physics Microform No.: 9106E0357. – See Abstr. 012.069 for the main entry.
Discrete sources of photoionizing radiation present at the time the intergalactic medium is largely neutral will create expanding H II regions. The high pressure of the relict H II region will then drive a shock into the cold, neutral ambient medium. The authors discuss the hydrodynamical evolution of relict cosmological H II regions generated at redshifts z ~ 5 – 15. The clouds formed may account for the Ly α forest seen in the spectra of high redshift QSOs. Gravitationally bound fragments may be candidates for blue-excess galaxies at high redshift.

160.105 **The missing mass in clusters of galaxies and elliptical galaxies.**
R. F. Mushotzky.
AIP Conf. Proc., No. 222, p. 394 – 404 (15 Apr 1991). – See Abstr. 012.069 for the main entry.
The author reviews the available data for the existence of dark matter in clusters of galaxies and elliptical galaxies. While the amount of dark matter in clusters is not well determined both the X-ray and optical data show that >50% of the total mass must be dark. Their is in general fair agreement in the binding mass estimates between the X-ray and optical techniques but there is not detailed agreement on the form of the potential or the distribution of dark matter. The X-ray spectral and spatial observations of elliptical galaxies demonstrate that dark matter is also required in these objects and that it must be considerably more extended than the stellar distribution.

160.106 **Distribution of dark matter in the Perseus cluster, and mass distributions in the Coma and Ophiuchus clusters.**
C. J. Eyles, M. P. Watt, D. Bertram, M. J. Church,
P. A. Knight, T. J. Ponman, G. K. Skinner, A. P. Willmore.
AIP Conf. Proc., No. 222, p. 405 – 408 (15 Apr 1991). Current Physics Microform No.: 9106E0405. – See Abstr. 012.069 for the main entry.
Data obtained with a coded mask X-ray spectral imaging telescope constitute the first high quality spectral imaging observations of clusters of galaxies in the hard X-ray band (2.5 – 25 keV). The authors have performed a detailed analysis of the properties of the intracluster gas in Perseus, and derive the distribution of the mass components in the cluster. Work is in progress on the Coma and Ophiuchus clusters.

160.107 **Dark matter in the Hercules cluster.**
C. M. Bird, J. M. Dickey.
AIP Conf. Proc., No. 222, p. 409 – 412 (15 Apr 1991). Current Physics Microform No.: 9106E0409. – See Abstr. 012.069 for the main entry.
The authors study the distribution of dark matter in the Hercules Cluster. The dataset consists of 120 measurements of redshifts, 47 from 21-cm observations and 73 from optical work. The authors derive the dynamical, the rotational, and the total mass. The ratio of the dynamical mass to the rotational mass is in the range from 5 – 25. This means that the bulk of the mass (80 – 95%) in A2151 is not contained within the luminous

galaxies. The light in the cluster may or may not trace the mass, but most of the mass is apparently not where the light is.

160.108 **The dynamics of the rich cluster A2634.**
G. Rhee, J. Pinkney, J. Burns, J. M. Hill, W. Oegerle, P. Hintzen, D. Batuski.
AIP Conf. Proc., No. 222, p. 413 – 416 (15 Apr 1991). Current Physics Microform No.: 9106E0413. – See Abstr. 012.069 for the main entry.
The authors present results of a study of the dynamics of the rich cluster Abell 2634. They have obtained ~100 radial velocities of galaxies in this cluster with an accuracy of ~100 km/s. The authors do not measure a significant peculiar motion for the cD galaxy. They also do not find evidence of substructure in this cluster or deviations from normality of the galaxy velocity distribution.

160.109 **Superclusters are warm, not hot: the predicted flux distribution of soft X-rays.**
A. A. Klypin, R. E. Kates.
Mon. Not. R. Astron. Soc., Vol. 251, No. 2, p. 41P – 45P (15 Jul 1991).
The authors present arguments showing that gas present in superclusters should have a characteristic temperature of approximately 10^6K. In the most popular models of large-scale structure formation, including hot and cold dark matter, peculiar velocities of matter relative to the background are ≈1000 km s^{-1}. This would imply temperatures exceeding 10^7K. However, most of the peculiar motions can be attributed to large-scale, coherent velocities. Moreover, gas inside superclusters which is not concentrated in groups or clusters will cool adiabatically with the Hubble expansion. A temperature of 10^6K implies that radiation from gas in superclusters should not be seen in X-rays of energy > 500 eV. The most favourable range for observing this radiation is about 100–300 eV. The authors present the results of numerical simulations of the evolution of large-scale structure in a cold dark matter (CDM) scenario including the thermal history of the baryonic component. These simulations provide estimates of the flux, together with examples of patterns as they would appear in the data. Soft X-rays emitted from hot gas in known superclusters may be detectable at the limits of sensitivity of ROSAT by averaging counts over a sufficiently large region.

160.110 **Lyman alpha evolution at high resolution: evidence for a single population?**
E. Giallongo.
Mon. Not. R. Astron. Soc., Vol. 251, No. 3, p. 541 – 544 (1 Aug 1991).
Recent high-resolution spectroscopic data on the Lyman alpha absorption lines allow a statistical investigation of the cosmological evolution of Ly α clouds from z = 1.8 to 3.8. The maximum likelihood analysis shows that the number density evolution is differential, being progressively slower for stronger lines. There are discrepancies in the derived evolution for lines with equivalent widths W > 0.3 Å from line samples obtained at high and intermediate resolutions. A possibl explanation is given in terms of line blending. A differential evolution of this type seems to imply a unified evolutionary scheme which incorporates Ly α and metal-line systems in a single population.

160.111 **Galaxy evolution in distant, X-ray-selected clusters of galaxies. I. Cl 1358+6245.**
D. G. Fabricant, J. E. McClintock, M. W. Bautz.
Astrophys. J., Vol. 381, No. 1, p. 33 – 54 (1 Nov 1991). With plate 1.
A photometric and spectroscopic survey of a cluster of galaxies at z = 0.327 is presented. This cluster belongs to a sample of distant, high X-ray luminosity clusters, selected from the Einstein Observatory MSS Catalog. Studies of optically selected clusters at z > 0.2 have indicated that the galaxy population in such clusters is different from that in comparable, nearby clusters. The authors have obtained V, R, and I photometry of 288 stars and galaxies in a region 5'.7 × 6'.5, which is approximately centered on the dominant galaxy at the cluster center. The photometry is complete to rest band M_V = –19.5. Spectra were obtained of 70 galaxies which are located within ~2' of the cluster center, to a magnitude limit of M_V ~ –20.5; ~90% of these galaxies are cluster members. The large velocity dispersion of the cluster, 1125 km s^{-1}, is consistent with its high X-ray luminosity. The authors find evidence for evolution of the galaxy population as compared with clusters at the present epoch. Approximately 17% of the cluster members have "E + A" spectra, with strong Balmer absorption lines, the normal absorption features of E/S0 galaxies, and no emission lines. In comparable nearby clusters, only ~1% of the galaxies have E + A spectra. Approximately 10% of the cluster members have emission-line spectra, compared with ~5% in present epoch clusters. The total "active" population, defined as the sum of the E+A and emission-line populations, represents ~27% of the population in the cluster core.

160.112 **A statistically complete survey for arc-like features in images of distant rich clusters of galaxies.**
I. Smail, R. S. Ellis, M. J. Fitchett, H. U. Nørgaard-Nielsen, L. Hansen, H. E. Jørgensen.
Mon. Not. R. Astron. Soc., Vol. 252, No. 1, p. 19 – 29 (1 Sep 1991).
Using data from an earlier search for supernovae in distant clusters, the authors have constructed a homogeneous set of V images for 19 rich clusters of mean redshift $\bar{z}$=0.32. By considering the images above a fixed surface brightness limit, the authors have analysed the data for extended arc-like features which might arise from gravitational lensing of background sources. A list of 20 candidate arcs is presented. The authors examine the usefulness of such a catalogue for deriving the background source redshift distribution, N(z). Whilst the number and shape distribution of arc candidates is consistent with the lensing hypothesis, cluster velocity dispersions of very high precision would be needed to provide useful constraints on the fraction of high-redshift galaxies to faint limits. The authors show how, in principle, a likelihood ratio test based on the radial distribution of arcs in a single well-studied cluster could determine whether a significant fraction of the faint galaxy population is at high redshift. This test also provides a means of determining accurate cluster velocity dispersions at any redshift.

160.113 **The discovery of large amounts of cold, X-ray absorbing matter in cooling flows.**
D. A. White, A. C. Fabian, R. M. Johnstone, R. F. Mushotzky, K. A. Arnaud.
Mon. Not. R. Astron. Soc., Vol. 252, No. 1, p. 72 – 81 (1 Sep 1991).
The authors report the discovery of significant excess absorption in the X-ray spectra of 12 clusters of galaxies. The spectra, from the Einstein Observatory Solid State Spectrometer, also require a cooling-flow component, which confirms the results of imaging studies of the clusters showing the strongly peaked emission characteristic of cooling flows. Several other clusters, such as the Coma cluster, do not require either extra absorption of a cooling flow component. The total mass of absorbing gas is $\sim 3x10^{11}$–$10^{12}M_\odot$, on the assumption that it is distributed through the cooling flow region and has cosmic abundance. It represents a significant new component of the intracluster medium. The authors show that the gas is most likely in the form of small cold clouds. It may be difficult to detect at other wavelengths, although further studies of the 21-cm line of hydrogen should be most successful. The absorbing gas should be readily observable in spectra from the X-ray detectors on BBXRT and ROSAT.

160.114 **The X-ray subcluster in A 2256: cluster mergers, cooling flows and diffuse radio sources.**
A. C. Fabian, S. J. Daines.
Mon. Not. R. Astron. Soc., Vol. 252, No. 1, p. 17P – 19P (1 Sep 1991).
ROSAT X-ray images of the cluster A 2256 have recently resolved the cluster core into two peaks, separated by a projected

distance of about 3.5 arcmin. The second peak is interpreted as a lower temperature subcluster in the process of merging. Motivated by this observation, the authors discuss the fate of the gas in a merging subcluster and show that it may not be shocked during the merger. In particular, if either cluster or subcluster contains a cooling flow the cooling gas will not necessarily be reheated but can thermalize its bulk kinetic energy in the less dense hot gas. As clusters evolve via hierarchical mergers, cooling flows are interrupted as the intracluster medium is stirred up but are then quickly re-established. The main cluster in A 2256 has no cooling flow, but the authors show that the subcluster itself did have. A merger of this type deposits large quantities of cold gas in the core of the main cluster and may promote a cooling flow there. The X-ray subcluster in A 2256 coincides with a diffuse radio source and is close to several head-tail radio sources. The authors suggest that these sources were originally members of the subcluster and that ram-pressure stripping of the surrounding subcluster core has created the diffuse radio source from their lobes.

160.115 **Structure and formation of superclusters. XIV. Correlation functions: dependence on the intrinsic properties of galaxy samples.**
M. Einasto.
Mon. Not. R. Astron. Soc., Vol. 252, No. 2, p. 261 – 270 (15 Sep 1991).
The author studies the dependence of the correlation functions $\xi(r)$ and $1+\xi(r)$ on the intrinsic properties of galaxies and systems of galaxies, as well as on the sample size. The dependence on morphology is shown for both of these functions. The dependence on luminosity appears in the case of early-type galaxies as well as for late-type galaxies; in both cases the correlation functions show a rapid increase at a luminosity limit of about $M < -20.0$. The correlation functions increase with sample depth; this is a real effect reflecting the luminosity of galaxies and the presence of voids in the distribution of galaxies. The correlation functions $1+\xi(r)$ show the presence of self-similar distribution of galaxies in two ranges. At separations $r < 3$ Mpc the slope of this function is ~ -1.6 to -2.5, reflecting galaxy distribution in groups and clusters and depending on the intrinsic properties of galaxies. At larger distances the slopes of these functions are ~ -0.8 and do not depend on the properties of galaxies. At these distances the functions $1+\xi(r)$ reflect the intercluster distribution. This feature is seen up to separations of around 30 Mpc in the largest samples. The crossover between the two distributions takes place at separations of about 3–3.5 Mpc; this is a very stable characteristic, similar for samples of various properties.

160.116 **Large-scale periodicity: problems with cellular models.**
B. G. Williams, J. A. Peacock, A. F. Heavens.
Mon. Not. R. Astron. Soc., Vol. 252, No. 2, p. 43P – 46P (15 Sep 1991).
The authors consider recent attempts to employ cellular models of galaxy clustering in order to explain the apparent large-scale periodicity seen in pencil beam surveys. They concentrate on the Voronoi foam model, which has been proposed to provide a natural explanation for both large-scale periodicity and the clustering of Abell clusters. The authors show that this model predicts a scale length for the galaxy-galaxy correlation function which is too large: $r_0 > 10\ h^{-1}$Mpc (h is the Hubble constant in units of 100 km $s^{-1}Mpc^{-1}$). Similar problems arise for other models based on cellular structures: the observed value of the galaxy-galaxy correlation function sets a limit on how large the cells can be.

160.117 **On the use of the conditional density as a description of galaxy clustering.**
G. Lemson, R. H. Sanders.
Mon. Not. R. Astron. Soc., Vol. 252, No. 3, p. 319 – 328 (1 Oct 1991).
It has been argued that the conditional density is superior to the two-point correlation function in providing a more sample-independent description of galaxy correlations and the approach to homogeneity (Pietronero, 1987). This is demonstrated here by applying both statistics to two different synthetic distributions of points: a Lévy flight fractal with no intrinsic scale and a Voronoi tesselation with points distributed uniformly on the walls of cells having a well-defined mean size. These synthetic distributions are used to test two suggested methods of correcting for the boundaries of a finite sample, and it is shown that the standard method of normalizing pair counts by a Monte Carlo routine is the most accurate and unbiased. This method is then applied to the CfA catalogue where the edge-correction allows the authors to extend the estimate of the conditional density beyond that of previous work (Coleman et al., 1988) to scales comparable to the depths of various volume limited subsamples. The result is that galaxy distribution apparently crosses over to homogeneity on a scale of roughly 30 Mpc and that on smaller scales the conditional density is a power law with an exponent closer to -1 (as for the Voronoi tesselation) rather than the traditional -1.8.

160.118 **New D – σ results for Coma ellipticals.**
J. R. Lucey, R. Guzman, D. Carter, R. J. Terlevich.
NATO Advanced Research Workshop on Observational Tests of Inflation, p. 193 – 197 (1991). – See Abstr. 012.076 for the main entry.
The authors discuss D–σ results for a new sample of ellipticals in the Coma cluster. Unlike previous studies, their sample includes many ellipticals that lie well outside the cluster core. These data have been used to estimate the environmental dependence of the D–σ relation. The zero-point of the D–σ relation is found to vary by only $-8\% \pm 10\%$ from Coma's core to a projected distance of $\sim 4\ h^{-1}$Mpc. The projected galaxian density from the core to the outermost ellipticals decreases by a factor of over 150. Thus over a very wide range of galaxian densities the authors find no evidence for a significant zero-point variation in the D–σ relation.

160.119 **Novae and the distance scale.**
C. J. Pritchet.
NATO Advanced Research Workshop on Observational Tests of Inflation, p. 199 – 203 (1991). – See Abstr. 012.076 for the main entry.
Observation of novae in the Virgo cluster by Pritchet and van den Bergh yield a Hubble constant $H_0 = 69 \pm 14$ km $s^{-1}Mpc^{-1}$. Some of the problems in this estimate of H_0 are discussed. In the future it should prove possible to push distance determinations using novae out beyond the Local Supercluster, and hence determine a value of H_0 less affected by local velocity anomalies.

160.120 **A high resolution, ground based observation of a Virgo galaxy.**
T. Shanks, N. Tanvir, P. Doel, C. Dunlop, R. Myers, J. Major, M. Redfern, N. Devaney, P. O'Kane.
NATO Advanced Research Workshop on Observational Tests of Inflation, p. 205 – 210 (1991). – See Abstr. 012.076 for the main entry.
The authors argue that a very simple inflationary model with $\Omega_{baryon} = 1$ would be allowed, if Hubble's constant lay below its presently accepted range. As part of a programme to check the value of Hubble's constant the authors present new, high resolution observations of the Virgo galaxy, IC 3583. The authors derive a lower limit to the Virgo distance of 14 Mpc.

160.121 **X-ray iron line of cluster of galaxies.**
K. Koyama.
IUPAP Conference on Primordial Nucleosynthesis and Evolution of Early Universe, p. 387 – 391 (1991). – See Abstr. 012.071 for the main entry.
The evolution and structure of the intracluster gas are discussed based on the observation of the Ginga satellite. Large scale X-ray emission was discovered extending about 12° from the Virgo cluster. The scan profile and the temperature structure suggest that the intra-cluster gas of Virgo cluster has two components around the bright elliptical galaxies, M87 and M49. The author demonstrates that the simple evolutionary scenario

can roughly explain the relation between X-ray luminosity and gas temperature from many clusters of galaxies observed with the Ginga satellite.

160.122 **Dynamical evolution of compact groups of galaxies.**
K. Kodaira, S. Okumura, J. Makino, T. Ebisuzaki, D. Sugimoto.
IUPAP Conference on Primordial Nucleosynthesis and Evolution of Early Universe, p. 393 - 401 (1991). - See Abstr. 012.071 for the main entry.

Observations about dynamical aspects of compact groups of galaxies are reviewed, and N-body numerical simulations are applied in order to lend theoretical interpretations to the observational results. In connection with dynamical evolution time scale the possible importance of an extended massive dark matter envelope is pointed out in which a compact galaxy group is embedded.

160.123 **GINGA observations of the Shapley Supercluster.**
C. S. R. Day, A. C. Fabian, A. C. Edge, S. Raychaudhury.
Mon. Not. R. Astron. Soc., Vol. 252, No. 3, p. 394 - 402 (1 Oct 1991).

The authors present the results of scanning and pointed X-ray observations of a region of about 40 square degrees in the vicinity of the rich cluster Shapley 8, in the Shapley Supercluster of galaxies. The Shapley Supercluster is the richest known supercluster and Shapley 8 is classified optically as richness class 4. The authors find no evidence for excess diffuse emission associated with the supercluster and obtain a strong limit on the total mass of X-ray emitting gas. Several new point sources have been discovered in the region and the X-ray properties of Shapley 8 have been measured for the first time. Despite its apparent high optical richness, Shapley 8 is not exceptionally luminous in X-rays, being about one third of the luminosity of the highest luminosity clusters known. The authors suggest that its high optical richness is in part due to neighbouring clusters, four of which are now resolved in X-rays within 2° of Shapley 8.

160.124 **EXOSAT observations of clusters of galaxies. I. The X-ray data.**
A. C. Edge, G. C. Stewart.
Mon. Not. R. Astron. Soc., Vol. 252, No. 3, p. 414 - 427 (1 Oct 1991).

X-ray observations of 45 clusters of galaxies made with the European X-ray Observatory Satellite, EXOSAT, are presented. The 0.1-20 keV spectra obtained constitute the largest homogeneous spectral sample presently available. Here the analysis of the spectral and imaging data is described. A significant scatter in the correlation between the X-ray luminosity and temperature, which can be interpreted as variations in the ratio of the gas mass to total mass amongst the clusters, is found. The measured iron abundances are consistent with a "canonical" average value of 0.3 solar. The large majority of clusters are found to be cooling in their cores, indicating that cooling flows are a common phenomenon.

160.125 **EXOSAT observations of clusters of galaxies. II. X-ray to optical correlations.**
A. C. Edge, G. C. Stewart.
Mon. Not. R. Astron. Soc., Vol. 252, No. 3, p. 428 - 441 (1 Oct 1991).

The results obtained for clusters of galaxies using X-ray data from the European X-ray Observatory Satellite, EXOSAT, are compared with the optical, infrared and radio properties of the clusters taken from the literature. A number of strong correlations are found. The ratios of the intracluster medium (ICM) temperature and the cluster velocity dispersion, β, have a mean of less than one, in agreement with values obtained from the surface brightness distribution of the X-ray emission, thus resolving the "Beta problem". A particularly strong correlation exists between X-ray luminosity and spiral fraction indicating a direct link between the ICM and the structure of galaxies. The luminosity of the brightest cluster member and the X-ray properties of the cluster are found to be related, implying that its use as a standard candle should be corrected for the X-ray properties. The correlations between galaxy number density and X-ray properties imply mass-to-light ratios of 100-250 h $M_\odot/L_\odot$ in the cores of clusters and a ratio of gas mass to galactic mass of between one and two for all clusters. The implications of these results are reviewed.

160.126 **Distance to the Coma cluster and the value of H_0.**
S. Okamura, M. Fukugita.
IUPAP Conference on Primordial Nucleosynthesis and Evolution of Early Universe, p. 451 - 459 (1991). - See Abstr. 012.071 for the main entry.

The authors report the determination of the Coma cluster distance using the B-band Tully-Fisher relation. Surface photometry was carried out to estimate the total magnitude unambiguously. The authors circumvent the Virgo cluster by combining the Coma galaxies directly with the local calibrators. Coma cluster is one of the best clusters for the determination of H_0.

160.127 **ROSAT observations of clusters of galaxies.**
H. Böhringer, W. Voges, H. Ebeling, R. A. Schwarz, A. C. Edge, U. G. Briel, J. P. Henry.
NATO Advanced Research Workshop on Observational Tests of Inflation, p. 293 - 308 (1991). - See Abstr. 012.076 for the main entry.

The expectations for observations of clusters of galaxies with ROSAT in particular with respect to the All Sky Survey are discussed. A look at the first survey results suggests that about 4000 to 8000 clusters should be detected in the All Sky Survey. The final catalogue of X-ray selected clusters of galaxies will probably be quite different from the present optical catalogues. Due to the higher spatial resolution and sensitivity compared to previous instruments a more detailed investigation of the internal structure of clusters will be possible with ROSAT as is shown for Perseus and Abell 2256.

160.128 **Radio emission in a random magnetic field: radio haloes and the structure of the magnetic field in the Coma cluster.**
P. C. Tribble.
Mon. Not. R. Astron. Soc., Vol. 253, No. 1, p. 147 - 152 (1 Nov 1991).

Models of cluster radio haloes have been constructed allowing for a random magnetic field. The resulting halo is far from smooth, both in intensity and polarization. The magnetic field must therefore be tangled on sufficiently small scales to reduce intensity fluctuations to observed levels. The smoothness of the inner regions of the radio halo in the Coma cluster, with fluctuations at about the 5% level at 1-arcmin resolution, indicates that the outer scale of magnetic field fluctuations is less than 15 kpc. Similar considerations indicate that less than 10% of the total halo emission can be due to galactic wakes and demonstrate that the halo emission and the magnetic field are spread throughout the whole volume of the cluster and are not localized with a small filling factor. In the simplest models the fractional polarization should be similar to the fluctuation level, although the polarization can be substantially decreased if the field has a high degree of symmetry.

160.129 **Powerful radio sources in clusters of galaxies.**
A. M. Unewisse, R. W. Hunstead.
Proc. Astron. Soc. Aust., Vol. 9, No. 1, p. 100 - 102 (1991). - See Abstr. 012.078 for the main entry.

Using a complete sample, the authors have looked for correlations in position between radio sources with $S_{408} \geqslant 1$ Jy in the Molonglo Reference Catalogue and rich clusters of galaxies in the Abell, Corwin and Olowin catalogue. Within a projected radius of 100 kpc of the cluster centre, they find an overdensity of radio sources similar to that seen by Robertson and Roach (1990). The latter investigators attributd this feature to centrally located dominant cluster galaxies. The authors confirm this hypothesis by looking at the morphology of the clusters within this peak and by optically identifying galaxies associated with the radio emission. Due to the relatively high flux density limit imposed, the cluster radio sources in this sample are among the most powerful in the southern sky. The probability of a cluster containing a radio source of this power is found not to depend on cluster richness.

160.130 **The effect of clustering on the equivalent width distribution of QSO Lyman–alpha clouds.**
X. Barcons, J. K. Webb.
Mon. Not. R. Astron. Soc., Vol. 253, No. 2, p. 207 – 211 (15 Nov 1991).

The authors report a significant discrepancy between the observed equivalent width distribution p(W) of Lyman α absorption features in intermediate – resolution QSO spectra and that expected from high–resolution spectra. They find that, if clouds are randomly distributed, line blending cannot account for this effect and they interpret the difference as due to enhanced line blending caused by cloud–cloud clustering. Using a clustering model for the Lyman α clouds, the authors show how the apparent p(W) depends on clustering properties and derive an estimate of the clustering model parameters by comparing the simulated p(W) to that of the real data. The clustering model deduced in this way is derived independently of and is in excellent agreement with the observed two–point correlation function.

160.131 **Faint galaxies in groups of bright galaxies.**
A. T. Kalloglyan, S. R. Unanyan.
Astrophysics, Vol. 33, No. 3, p. 532 – 535 (Nov – Dec 1990). English translation of Astrofizika, Tom 33, Vyp. 3, p. 403 – 408 (Dec 1990).

Counts of galaxies in the charts of the Palomar Sky Survey have shown that many faint galaxies are associated with the bright groups of Geller and Huchra. At least some groups belong to larger formations. In many groups, there are substructural details, which are mainly due to groupings of faint galaxies.

160.132 **High–resolution imaging of Virgo cluster galaxies. I. Observational techniques and first results.**
N. R. Tanvir, T. Shanks, J. V. Major, A. P. Doel, C. N. Dunlop, R. M. Myers, R. M. Redfern, P. O'Kane, M. N. Devaney.
Mon. Not. R. Astron. Soc., Vol. 253, No. 2, p. 21P – 24P (15 Nov 1991).

The authors investigate the potential of the image sharpening technique for making high–resolution images of Virgo cluster galaxies, in order to obtain new estimates of this cluster's distance. The authors have used both the Galway Image Sharpening Camera (Redfern et al.) and the MARTINI instrument (Doel et al.) at the William Herschel Telescope to confirm Fried's $\approx 4r_0$ optimum aperture for image sharpening. The authors have further used the MARTINI image sharpening instrument to observe the Virgo galaxy IC 3583. The resulting image has 0.60–arcsec FWHM seeing but to a limiting magnitude of R = 21.7 mag the brightest stars in IC 3583 are not resolved. This tentatively suggests a lower limit to the Virgo distance modulus of m–M > 30.7 mag.

160.133 **Statistical analysis of the galaxy distribution in a diameter–limited sample in the northern sky.**
E. Zucca, S. Bardelli, A. Cappi, L. Moscardini.
Mon. Not. R. Astron. Soc., Vol. 253, No. 3, p. 401 – 413 (1 Dec 1991).

In order to understand if the differences between the Southern Sky Redshift Survey (which is diameter–limited) and the CfA survey (which is magnitude–limited) are induced by the different selection critera, the CfA survey has been directly compared with a diameter–limited sample covering the same region of the sky (UDLS, Bardelli et al.). The authors analyse the statistical properties of the CfA and the UDLS catalogues through various tests, such as the two–point correlation function, the percolation analysis and the multiplicity function. They study the filamentary properties using the Minimal Spanning Tree technique and the void distribution through the Nearest Neighbour Statistics and the Void Probability Function. From the result of these tests the authors deduce that second–order differences are not present between the CfA and the UDLS catalogues, but the UDLS galaxies are more spread out along the structures, which therefore appear more filamentary. The authors' analysis confirms the results of the comparison between the CfA and the SSRS made by Pellegrini et al. and therefore they can outline some general properties of diameter–limited samples.

160.134 **A photoionization model for the optical line emission from cooling flows.**
M. Donahue, G. M. Voit.
Astrophys. J., Vol. 381, No. 2, p. 361 – 372 (10 Nov 1991).

The authors present the detailed predictions of a photoionization model, outlined in a previous paper, to explain the optical line emission associated with cooling flows in X–ray emitting clusters of galaxies. In this model, EUV/soft X–ray radiation from condensing gas photoionizes clouds that have already cooled. The authors discuss the energetics and specific consequences of such a model, as compared to other models put forth in the literature. They also discuss the consequences of magnetic fields and cloud–cloud shielding. The model illustrates how varying the individual column densities of the ionized clouds can reproduce the range of observed line ratios. The results strongly suggest that the emission–line nebulae are self–irradiated condensing regions at the centers of cooling flows.

160.135 **On the baryon content of the Shapley Supercluster.**
A. C. Fabian.
Mon. Not. R. Astron. Soc., Vol. 253, No. 3, p. 29P – 30P (1 Dec 1991).

The Shapley Supercluster is particularly rich in baryons, since it contains many X–ray luminous clusters. Using the relationship between X–ray luminosity and gas mass withih $3\,h_{50}^{-1}$Mpc radius found for nearby clusters, it is shown that the ratio of the baryon density to the critical density in the $37\,h_{50}^{-1}$Mpc core of the supercluster $\Omega_b > 0.18\,h_{50}^{-1.3}$. This is much larger than the value obtained from published calculations of cosmic nucleosynthesis, $\Omega_b = 0.06\,h_{50}^{-2}$. The high baryon content requires that the supercluster has a total mass overdensity $\varrho/\bar{\varrho} > 2.4$, the largest yet established on large ($\sim 35\,h_{50}^{-1}$Mpc) scales.

160.136 **Scale–invariant clustering in the large–scale distribution of galaxies.**
L. Guzzo, A. Iovino, G. Chincarini, R. Giovanelli, M. P. Haynes.
Astrophys. J., Lett., Vol. 382, No. 1, p. L5 – L9 (20 Nov 1991).

The authors explore the scaling properties of galaxy clustering through an analysis of the new redshift survey of the Perseus–Pisces region and of other published data. At small separations the authors measure the "standard" scaling, with the well–known correlation dimension $D \cong 1.2$ shown by the two–point correlation analyses. On larger scales, however, there is evidence of the existence of a second range with a different scaling characterized by a correlation dimension $D \cong 2.2$. Homogeneity seems to be reached for separations $> 30h^{-1}$ Mpc. The authors discuss the possible implications of this result concerning the origin of structure. Interpreting the large–scale $D \sim 2$ range as due to fluctuations which are still in the linear regime, it provides evidence that the present structure could have developed from simple scale–free initial conditions characterized by a power spectrum $|\delta_k|^2 \sim k^{-2}$.

160.137 **The morphology of clusters of galaxies, the formation efficiency of galaxies, and the origin of the intracluster medium.**
C. Jones, W. Forman.
Liller Symposium: Asteroids to quasars, p. 195 – 209 (1991). – See Abstr. 012.087 for the main entry.

The luminous material in clusters of galaxies falls primarily into two forms – the visible galaxies and the X–ray emitting intracluster medium. The hot intracluster medium (ICM) is the major observed baryonic component of cluster with a mass equal to or greater than that of the stellar matter. In this paper the authors review the structure and morphology of rich clusters as inferred from X-ray observations of the ICM. They also discuss changes in the efficiency of galaxy formation for different clusters and the origin of the intracluster medium.

160.138 **Testing models for the dynamical evolution of clusters of galaxies.**
P. M. Lugger.
Liller Symposium: Asteroids to quasars, p. 211 – 237 (1991). – See Abstr. 012.087 for the main entry.
The author reviews theoretical and observational studies concerning the dynamical evolution of clusters of galaxies. The theoretical studies have investigated the effects of a number of gravitational interaction mechanisms including violent relaxation, two-body relaxation, dynamical friction, galaxy mergers, and tidal stripping. A central issue addressed by many of these studies is the possible production of supergiant cD galaxies by repeated galaxy mergers in dense cluster cores; this process has been termed "galactic cannibalism". The author describes observational studies designed to test the predictions of the theoretical models. These studies have investigated several properties of galaxies, the galaxy luminosity function, and the dynamics of multiply nucleated first-ranked galaxies. The observational results indicate that the effects of dynamical evolution on clusters are more subtle then originally predicted.

160.139 **New D – σ results for Coma cluster ellipticals.**
J. R. Lucey, R. Guzmán, D. Carter, R. J. Terlevich.
Mon. Not. R. Astron. Soc., Vol. 253, No. 4, p. 584 – 598 (15 Dec 1991).
The authors present velocity dispersions (σ) and photometric diameters (D) for 51 ellipticals in the Coma cluster. Unlike previous studies, the authors' sample includes many ellipticals that lie well outside the cluster core. These data have been used to estimate the environmental dependence of the D – σ relation.

160.140 **Systematic bias in cluster galaxy data, affecting galaxy distances and evolutionary history.**
D. H. Gudehus.
Astrophys. J., Vol. 382, No. 1, p. 1 – 18 (20 Nov 1991).
Plots of galaxian surface brightness parameter versus log radius parameter have, for the past 15 years, been used to estimate cluster relative distances. The appeal of this approach is twofold: the scatter about the mean line is reasonably small and the relative distance is easily interpreted as a horizontal shift of one cluster mean line relative to another. However, the small scatter is deceiving; it results from the ratio of cosmic fluctuations in surface brigthness parameter to cosmic fluctuations in log radius parameter, i.e., 5, being approximately equal to the slope of the intrinsic mean line of the brightest galaxies, i.e., 2.5. Furthermore, actual fitted slopes and zero points will depend on the range of galaxy magnitudes used to construct each plot. A systematic error thus arises which makes cluster samples lacking large luminous galaxies appear to be farther away than they really are. This bias can be substantial and is illustrated with actual and simulated data. The data are mostly derived from CCD observations of 343 cluster galaxies which have been processed to remove the effects of overlapping images and blurring from seeing. A plot of log r against total magnitude for these data exhibits a change of slope at $M'_V = -23.04$ and a' = 931 pc or r'_e = 10.20 kpc (H_0 = 50 km $s^{-1}Mpc^{-1}$, $q_0 = 0.5$). While the systematic bias would occur even without this slope change, the magnitude of the bias is enhanced by it. Reanalyses of the data of Schombert and of Dressler et al. also exhibit the change of slope at about the same magnitude. Although introduction of a third parameter such as velocity dispersion tightens up the log r – M relation, the systematic error in the S – log r relation still persists at ~69% of its former value. In order to estimate relative galaxian distances in an unbiased way, aside from the Malmquist effect, making optimal use of the available photometric, structural, and dynamical information, a new distance estimator, the reduced galaxian radius parameter, r_g, is introduced. With this estimator, relative distances good to $\pm 31\%$ for individual galaxies can be obtained.

160.141 **Large-scale peculiar velocity fields in superclusters of galaxies.**
G. Gavazzi, M. Scodeggio, A. Boselli, G. Trinchieri.
Astrophys. J., Vol. 382, No. 1, p. 19 – 25 (20 Nov 1991).
The Tully-Fisher relation is used to derive distances of 262 galaxies in seven clusters. Care is taken to reduce possible causes of error in the distance determination and to evaluate sources of scatter in the distance – velocity relation. The scatter in Virgo, A262 (and possibly in the Coma and Hercules superclusters) is real and is evidence of large-scale (50 Mpc) peculiar velocity fields. In the Virgo region the authors confirm the existence of background galaxies infalling toward the cluster, and they find evidence for infalling foreground objects. The Cancer cluster is confirmed to be an unbound collection of groups. The Hubble relation is linear from 0 to 13,000 km s^{-1} with $H_0 = 90 \pm 17$ km $s^{-1}Mpc^{-1}$.

160.142 **The galaxy two-point correlation function down to V = 26 magnitudes on 0°5 scales.**
L. W. Neuschaefer, R. A. Windhorst, A. Dressler.
Astrophys. J., Vol. 382, No. 1, p. 32 – 43 (20 Nov 1991).
The authors extend the measurement of the galaxy angular two-point correlation function $w(\theta)$ down to V ~ 26 mag on scales $\theta \lesssim 0.^{\circ}45$. They used the Palomar 200 inch Four-Shooter CCD-array to image two deep radio-optical survey regions (0.2 deg^2) with contiguous mosaics in Gunn *gri*, yielding in total ~60,000 faint galaxies to V ~ 26 mag. Here, the authors present the first results for one mosaic in Gunn *g*. In the entire magnitude range $14 \lesssim g \lesssim 26$ mag, $w(\theta)$ is consistent with a power law of the form $w(\theta) = A_\omega \cdot \theta^{1-\gamma}$ and roughly maintains a constant slope ($1-\gamma \cong -0.6$ to -0.8). The slope of $w(\theta)$ has thus apparently undergone little change over the entire redshift range sampled for $14 \lesssim g \lesssim 26$ mag. For $20 \lesssim g \lesssim 24.5$ mag, the amplitude A_ω continues to decrease toward fainter magnitudes. Beyond $g \gtrsim 25$ mag, the present catalogs become gradually incomplete, and A_ω starts to rise again while $w(\theta)$ is maintaining roughly the same slope. Possible explanations for this effect are discussed.

160.143 **Void statistics of the CfA redshift survey.**
M. S. Vogeley, M. J. Geller, J. P. Huchra.
Astrophys. J., Vol. 382, No. 1, p. 44 – 54 (20 Nov 1991).
The authors study clustering properties of two samples from the CfA redshift survey each containing ~2500 galaxies. Comparison of the velocity distributions using a K – S test reveals structure on scales comparable with the extent of the survey. The authors use the void probability function (VPF) for these samples to examine the structure and to test for scaling relations in the galaxy distribution. Using moments of galaxy counts, the authors calculate the galaxy correlation function and find that $\xi(R) = \langle(\delta\varrho/\varrho)^2\rangle = 1$ on a scale $R = 7h^{-1}$Mpc. The shape and amplitude of the correlation function roughly agree with previous determinations. The VPFs for distance-limited samples of the CfA survey do not match the scaling relation predicted by hierarchical clustering models. On scales $\lesssim 10h^{-1}$Mpc, the VPFs for these samples roughly follow the hierarchical pattern. However, on scales $\gtrsim 10h^{-1}$Mpc, large variations between the VPFs for distance-limited subsamples of each survey region are found. The authors introduce a variant of the VPF which uses nearly all the data in magnitude-limited samples. This statistic, $P_0(\bar{N})$, accounts for the variation of the sampling density with velocity in a magnitude-limited survey. It appears to be robust with respect to variations in individual structures within the survey and may provide a useful discriminant among N-body models.

160.144 **An analysis of Abell and ACO catalogs of clusters of galaxies.**
M. A. Nicotra, D. G. Lambas.
Astrophys. J., Vol. 382, No. 1, p. 55 – 57 (20 Nov 1991).
The authors analyze angular cross-correlation functions between distance classes D = 1 – 4 and D = 6 of Abell and ACO catalogs of rich clusters of galaxies, and they study the dependence of these correlations on cluster richness. They find that anticorrelations are present in the rich sample $R \geqslant 1$ of the

ACO catalog, similar to that obtained by Szalay et al. in the Abell catalog. However, the anticorrelations vanish for the sample of nearby poor clusters R = 0, D = 1 – 4 versus R ⩾ 1, D = 6 in both catalogs. The authors apply the decontamination process discussed by Dekel et al. and Olivier et al., and analyze similar cross–correlations in the decomtaminated catalogs. The results indicate that the decontamination process does not change the systematics found in the cross–correlation functions of the original catalogs. Several statistical tests are applied in order to investigate the possibility that combined effects of Abell radius, the clustering of nearby clusters, and a bias in the assignment of distances may have contributed to an enhancement of the anticorrelations.

160.145 **Upper limit on interstellar gas in Coma Cluster ellipticals.**
C. R. Canizares, P. Blizzard.
Astrophys. J., Vol. 382, No. 1, p. 79 – 81 (20 Nov 1991).

Using data from the IPC on the Einstein Observatory, the authors search for X–ray emission from 45 early–type galaxies in Coma. Some galaxies lying within 13′ of the cluster center give statistically significant signals which, however, are shown to be spurious. Data for a clean subsample of 27 galaxies more than 13′ from the cluster center yield only upper limits, which are generally larger than those of field ellipticals with similar optical luminosities. The mean galaxy X–ray and blue luminosities for the subsample are $<L_X> < 2.2 \times 10^{40}$ ergs s^{-1} and $<L_B> = 4.4 \times 10^{10} L_\odot$, respectively. This X–ray limit lies well within the range observed for field galaxies and near the value estimated for a galaxy devoid of a hot interstellar medium. Coma Cluster galaxies do not show excess emission, such as has been attributed to galaxies in A1367, and the authors conclude that environmental effects in rich clusters do not enhance the X–ray emission from individual elliptical galaxies. Conclusive evidence that the emission is strongly suppressed, e.g., by stripping of the interstellar gas, must await more sensitive studies.

160.146 **Iron line emission features in clusters of galaxies.**
K. Koyama.
Workshop on Iron Line Diagnostics in X–Ray Sources, p. 67 – 71 (1991). – See Abstr. 012.088 for the main entry.

The Ginga satellite discovered large scale X–ray emission extending about 12° from the Virgo cluster. From the scan profile and the temperature structure, the author suggests that the intra-cluster gas of Virgo has two components around the large elliptical galaxies, M87 and M49. This two–subclustering indicates that the Virgo cluster is now under strong evolution. The author demonstrates that the simple evolutionary scenario can roughly explain the relation between X–ray luminosity and gas temperature from many clusters of galaxies observed with the Ginga satellite. The iron abundance of the outer part of the Virgo cluster is only about 0.1–0.2 of cosmic value. The author therefore interprets this to mean that the major components of intra–cluster gas are primordial. However the iron abundance of the Virgo cluster shows central concentration near M87, which indicates that the central part was contaminated by the gas ejected from each galaxy. The iron abundance from individual clusters show a trend of anti–correlation with the gas temperature. One possible reason for this anti–correlation is that the galaxy formation rate is smaller in clusters of hotter intra–cluster gas.

160.147 **SPARTAN–1 X–ray observations of iron line emission in the Perseus cluster.**
M. P. Kowalski, W. A. Snyder, R. G. Cruddace, G. G. Fritz, E. E. Fenimore, M. P. Ulmer.
Workshop on Iron Line Diagnostics in X–Ray Sources, p. 72 – 75 (1991). – See Abstr. 012.088 for the main entry.

The detection of iron K line emission in the Perseus cluster of galaxies has established the fact that the dominant source of X–rays is a hot intracluster gas emitting thermal bremsstrahlung radiation. Averaged over the whole cluster, this gas has an iron abundance about half that of the solar value. However, these observations were made with instruments which had wide fields–of–view and little or no spatial resolution. Observations of iron L–lines in the core of Perseus yielded an iron abundance of ~0.4 solar. Although the Perseus cluster has been mapped in broad bands by imaging instruments, until recently no spatially resolved spectral observations have been made. Such measurements hold the key to determining the structure and evolution of the cluster gas as derived from the density, temperature, and abundance distributions.

160.148 **The distribution of iron in galaxy clusters – results from the Spacelab 2 X–ray telescope.**
T. J. Ponman, M. P. Watt, P. A. Knight, D. Bertram, M. J. Church, C. J. Eyles, G. K. Skinner, A. P. Willmore.
Workshop on Iron Line Diagnostics in X–Ray Sources, p. 76 – 79 (1991). – See Abstr. 012.088 for the main entry.

Hard X–ray spectral imaging observations of a number of galaxy clusters were obtained by the coded mask X–ray telescope on Spacelab 2. The energy band of the telescope encompasses the iron K complex, so that the spatial distribution of the iron line emission can be mapped. Here the authors present results for the distribution of iron abundance which can be inferred in the three brightest X–ray clusters: Perseus, Coma and Ophiuchus.

160.149 **The spatial distribution of iron in the Coma cluster of galaxies.**
J. P. Hughes.
Workshop on Iron Line Diagnostics in X–Ray Sources, p. 80 – 83 (1991). – See Abstr. 012.088 for the main entry.

The discovery of K–shell emission lines from highly ionized atoms of iron in the X–ray spectra of clusters of galaxies confirmed the thermal origin of the X–ray emission and led to a number of important realizations. It became clear that (1) the hot gas which was the source of this emission constituted as much matter as the optically visible galaxies, (2) this same hot gas was a sensitive probe of the total mass binding the cluster, and (3) the electrons in the hot plasma, through inverse–Compton scattering, would distort the microwave spectrum of the cosmic background radiation (the Sunyaev–Zel'dovich effect). Over the ensuing decade, considerable observational effort has gone into studying the hot gas in clusters of galaxies along just these lines of research. Recently, data with the potential to investigate the spatial distribution of iron have become available for at least two clusters of galaxies: Perseus and Coma. The author's study concentrates on the Coma cluster.

160.150 **Infrared images, Virgo spirals and the Tully–Fisher law.**
R. F. Peletier, S. P. Willner.
Astrophys. J., Vol. 382, No. 2, p. 382 – 395 (1 Dec 1991).

For a complete (but magnitude–limited) sample of Virgo Cluster spiral galaxies, the authors present 1.65 μm photometry derived from images. Both galaxy inclinations and H–magnitudes are derived from the new data. The new photometry agrees with previous measurements. The infrared inclinations from the present data are usable for galaxies having inclinations greater than 45°, but deeper images would be needed for galaxies more face–on than this. The inclinations are the major observational uncertainty in the Tully–Fisher relation. It is shown that the relation in Virgo must have an intrinsic dispersion amounting to ~0.4 mag. Sample selection is critical for determining the intrinsic dispersion.

160.151 **Thin and giant luminous arcs: a strong test of the lensing cluster mass distribution.**
F. Hammer.
Astrophys. J., Vol. 383, No. 1, p. 66 – 71 (10 Dec 1991).

New expressions are given for the gravitational magnification and the ratio between source and image sizes, which apply to the giant thin and luminous arcs found in distant cluster cores, and for any number of lenses in the cluster. The new formulae provide the first attempt to disentangle the properties of the lensing clusters from the arc source properties. Available observations are used to constrain the nature of the sources, which are probably spiral galaxies, and hence, to derive new and independent conditions on the mass density profile of the dark matter inside the lensing clusters. Most of the lensing clusters should be much more compact than inferred from their X-ray density profiles and even more compact than inferred from their visible luminosity profile, except if there was an enormous evolution of the entire luminosity function of galaxies from $z \leqslant 0.4$ to $z \approx 0.8$. Indeed lensing clusters should have mass density profiles more compact than isothermal spheres with small core radii ($r_c \leqslant 100$ kpc for $H_0 = 50$). It is also argued that the sources of the small and faint arclets found in deep images of distant clusters, could be dwarf galaxies at moderate redshift ($z \leqslant 1$). The new test presented to derive lensing mass distribution is finally compared with the method using arclets.

160.152 **Bound populations around cD galaxies and cD velocity offsets in clusters of galaxies.**
K. Gebhardt, T. C. Beers.
Astrophys. J., Vol. 383, No. 1, p. 72 – 89 (10 Dec 1991).

The authors investigate recent claims in the literature for the existence of a low velocity dispersion population of galaxies in the cores of rich clusters. The suggested population, which has been generally interpreted as gravitationally bound to the potential well of the first-ranked cluster galaxy is taken as evidence in support of the galactic cannibalism model for D/cD galaxy formation. The statistical tests of kinematic evidence which have been used to support this claim are, however, fundamentally flawed. The authors examine five clusters which have been suggested in the literature to possess bound populations of galaxies associated with the D/cD. They develop an improved statistical test, the Indicator test, which compares the number of galaxies with relative velocities less than a given fraction of the cluster scale ("dispersion") to the number expected for an assumed parent velocity distribution. They find strong supporting kinematic evidence for the existence of a bound population in only one of the five clusters investigated, Klemola 44. Marginal kinematic evidence in support of the bound population hypothesis is found for two clusters, A1991 and A2589. They find no strong supporting kinematic evidence for a bound population in two clusters, A2107 and A2593. The authors discuss the analysis of pooled samples of relative velocities. Simulations of the expected contribution from clusters with a wide range of dispersions can account for much, if not all, of the excess counts in the central bins of pooled samples, in disagreement with previous interpretations of this effect as due to the presence of a bound population of galaxies in the vicinity of the dominant galaxy. The authors apply the bootstrap resampling technique to determine confidence intervals on the central locations of cluster velocity distributions. These intervals are then compared to the measured velocity of the D/cD galaxy to assess the reality of any supposed velocity offset. They examine 14 clusters. Of these, only four cases are supported by their analysis.

160.153 **The three-point correlation function for Abell clusters: observations and simulations.**
J. R. Gott III, B. Gao, C. Park.
Astrophys. J., Vol. 383, No. 1, p. 90 – 94 (10 Dec 1991).

The authors have done a three-dimensional study of the three-point correlation function of Abell clusters and compared this with Park's 1990 biased cold dark matter (CDM) simulations, fitting the results with formulae of the form $\zeta(r_{12},r_{13},r_{23}) = Q_{11}[\xi(r_{12})\xi(r_{13}) + \xi(r_{12})\xi(r_{23}) + \xi(r_{13})\xi(r_{23})] + Q_{111}[\xi(r_{12})\xi(r_{13})\xi(r_{23})]$. For the observations the maximum likelihood fits are $Q_{11} = 0.85$ and $Q_{111} = 0.0$ in excellent agreement with the result of Toth, Hollosi, and Szalay's 1989 two-dimensional study. For Park's simulation they find maximum likelihood fits of $Q_{11} = 0.9$ and $Q_{111} = 0.0$ in excellent agreement with the observational result. Thus the standard biased CDM gravitational instability model reproduces the observed form of the three-point correlation function for Abell clusters remarkably well.

160.154 **Expectations for X-ray cluster observations by the ROSAT satellite.**
A. E. Evrard, J. P. Henry.
Astrophys. J., Vol. 383, No. 1, p. 95 – 103 (10 Dec 1991).

The authors present predictions for characteristics of the X-ray cluster population expected to be observed by the ROSAT satellite. The theoretical modeling requires an assumed fluctuation spectrum and relations describing the behavior of cluster X-ray luminosity and core radius with cluster binding mass and redshift. The most "natural" model, which assumes self-similar scaling $L_x \propto M^{4/3}$ in a cold dark matter (CDM) universe, fails to reproduce the shape of the local X-ray cluster luminosity function. An alternative model motivated by recent evidence for more large-scale power, uses an $n = -2$ initial fluctuation spectrum and fits local cluster abundances with an intermediate scaling $L_x \propto M^{11/6}$. For the north ecliptic pole region of the ROSAT all-sky survey, which will cover a 10° radius field to a limiting flux of roughly 9×10^{-14}ergs s^{-1}cm^{-2}, the successful models predict ~ 330 cluster X-ray sources visible above the flux limit if a square detect cell of side 4.′8 is employed. The clusters would have a median redshift of 0.2, and 10% of them should have $z > 0.4$ but essentially none should be seen at $z > 1$. Roughly 2500 clusters are expected with this detect geometry in the all-sky survey above the limiting flux of 7×10^{-13}ergs s^{-1}cm^{-2}. Cluster sources would be expected to contribute $\sim 10\%$ of the observed soft X-ray background. The sensitivity of cluster detectability to treatment of the core emission and detect cell geometry is the principal source of uncertainty in these predictions.

160.155 **Evolution of clusters of galaxies.**
N. Kaiser.
Astrophys. J., Vol. 383, No. 1, p. 104 – 111 (10 Dec 1991).

Optical searches for galaxy clusters have shown that the abundance of high-velocity dispersion clusters was much as it is today even at redshifts approaching unity, while X-ray surveys show very strong negative evolution. The optical results are compatible with hierarchical theories with spectral index $n \gtrsim -1$, if self-similar models are used to predict the evolution of the cluster population. The X-ray observation is much harder to reconcile with hierarchical models, at least if the present state of the gas is a result of shocks from gravitationally induced motions, but it can be better understood if the entropy of the gas we now see was the result of some early injection of heat, perhaps at the time of galaxy formation. The author constructs a simple model in which the gas is initially uniform in entropy and settles roughly adiabatically into the evolving dark matter potential wells as these evolve. The predicted correlations of gross properties of clusters (X-ray luminosity, temperature, etc.) match quite well what is observed. The model shows strong negative evolution of the X-ray luminosity function, but only for the very brightest clusters. This agrees nicely with current observations, and the form of the evolution at lower luminosities can be tested by ROSAT.

160.156 **The velocity field of clusters of galaxies within 100 megaparsecs. I. Southern clusters.**
J. R. Mould, L. Staveley-Smith, R. A. Schommer,
G. D. Bothun, P. J. Hall, Han Ming Sheng, J. P. Huchra,
J. Roth, W. Walsh, A. E. Wright.
Astrophys. J., Vol. 383, No. 2, p. 467 – 486 (20 Dec 1991).

Distances from the Tully-Fisher relation have been measured with the Parkes radio telescope for eight additional southern clusters of galaxies in the redshift range 2500 – 5000 km s^{-1}. Most of them have large (average 1000 km s^{-1}) positive peculiar velocities in a comoving frame in which the microwave background radiation shows no dipole anisotropy. In five out of eight cases these measurements have a significance at or above the 2σ level. Outflow iw seen on both sides of the Galactic plane. The source of this large-scale flow, if gravitational in origin, lies at or beyond the limit of this sample.

160.157 **New photoionization models of intergalactic clouds.**
M. Donahue, J. M. Shull.
Astrophys. J., Vol. 383, No. 2, p. 511 – 523 (20 Dec 1991).

The authors present new photoionization models of optically thin, low-density intergalactic gas at constant pressure, photoionized by QSOs. They model all ion stages of H, He, C, N, O, Si, and Fe, plus H_2 and predict the column density ratios of clouds at specified values of ionization parameter $U = n_\gamma/n_H$ and cloud metallicity. For gas photoionized by quasars, a two-phase (cloud – intercloud) medium cannot exist at $T = (1-3) \times 10^4$K, so that pressure confinement of the Lyα clouds is problematic. The authors derive new metallicity limits for absorption system at z = 3.321 (S5 0014+81) and at z = 2.9676 (PKS 2126–158). In addition, a stable molecular phase is found in Lyα clouds, with $U \approx 10^{-7}$ to 10^{-5} and $n(H_2)/n(H\ I) \approx 10^{-2}$, in which molecular cooling produces $T < 10^4$K. Photoelectron escape from low-column H I clouds can reduce the heating and explain the observed narrow line widths. If the ionizing radiation flux is $\approx 10^{-21}$ergs $cm^{-2}s^{-1}Hz^{-1}$, as inferred for several absorption systems in QSO 2206–199N, these 10^4K clouds require high densities ($\approx 0.1\ cm^{-3}$) and have significant neutral fractions. The inferred cloud thicknesses are 10^{5-9} times smaller than the lateral extents estimated by double-QSO observations. Such aspect ratios are difficult to understand, and the large pressures would make the sheet-like clouds difficult to confine by means other than ram pressure.

160.158 **Optical colors of early-type galaxies in compact groups.**
S. E. Zepf, B. C. Whitmore, H. F. Levison.
Astrophys. J., Vol. 383, No. 2, p. 524 – 541 (20 Dec 1991). With plate 11.

The authors present UBVRI colors of a sample of 55 elliptical galaxies as well as 21 S0 galaxies in Hickson compact groups. Four of the elliptical galaxies and one of the S0 galaxies in the compact groups are significantly bluer than normal for early-type galaxies of their luminosity. A population of elliptical galaxies with blue optical colors is predicted by standard models of the evolution of galaxies in compact groups because of the favorable conditions for interactions and mergers in these regions. The brightest of the blue elliptical galaxies has an extensive shell system. However, the other two blue elliptical galaxies for which sufficient spatial resolution has been obtained, show no unusual morphological features such a tails or shells. In addition to their color, the primary distinguishing feature of the blue elliptical galaxies is their color gradient, as they tend to be bluer in the center than at large radii, which is contrary to the usual trend in elliptical galaxies.

160.159 **Modeling the evolution of galaxies in compact groups.**
S. E. Zepf, B. C. Whitmore.
Astrophys. J., Vol. 383, No. 2, p. 542 – 549 (20 Dec 1991).

The authors combine models of the dynamical evolution of compact groups with models of the evolution of the stellar populations in merger products to predict the morphology and colors of galaxies in compact groups. Models in which a compact group forms out of a loose group and evolves by merging until becoming a single elliptical galaxy in several Gyr conflict with several observations. These models predict that about 15% of the elliptical galaxies in compact groups should be more than 0.10 mag bluer in $(B-V)_0$ than normal for elliptical galaxies of their luminosity, whereas the observed value is lower [5.5(+5.3, −3.0)%]. In addition, 21±3% of the galaxies in compact groups are elliptical galaxies, but these models predict that only about 13% of the galaxies in compact groups should be elliptical galaxies. The models agree much better with the observations, if compact groups form in environments with a greater fraction of early-type galaxies relative to normal loose groups. Additional constraints from the number of ongoing mergers in compact groups and the fraction of both normal and blue elliptical galaxies of compact groups require that the average time scale for the evolution of a compact group into a single elliptical galaxy is at least 4 Gyr.

160.160 **Models of the radio halo and magnetic field in the Coma cluster.**
P. C. Tribble.
J. R. Astron. Soc. Can., Vol. 85, No. 4, p. 211 (Aug 1991). Abstract. – See Abstr. 012.091 for the main entry.

160.161 **Membership probabilities of galaxies in the region of the Virgo cluster.**
Zhao Junliang, Pan Rongshi, He Yanping.
Ann. Shanghai Obs., Acad. Sin., No. 12, p. 40 – 52 (1991). In Chinese.

The detailed results for membership determination of 572 galaxies in the region of the Virgo cluster obtained from data of radial velocities based on the maximum likelihood principle are given.

160.162 **The very steep spectrum radio source 0038–096 (Galaxy cluster Abell 85) – a radio halo associated with a subcluster?**
J. Bagchi, K. P. Singh, A. R. Patnaik, M. N. Joshi.
Bull. Astron. Soc. India, Vol. 19, Nos. 1 – 2, p. 75 (Mar – Jun 1991). Abstract. – See Abstr. 012.108 for the main entry.

160.163 **Alignment effects in clusters of galaxies.**
H. Muriel, M. Nicotra, D. Garcia Lambas, L. Ruiz.
Bol. Asoc. Argent. Astron., No. 36, p. 403 – 404 (1990). Abstract. In Spanish and English. – See Abstr. 012.095 for the main entry.

160.164 **The angular cross-correlation function in clusters of galaxies.**
D. Garcia Lambas, M. Nicotra.
Bol. Asoc. Argent. Astron., No. 36, p. 404 – 406 (1990). Abstract. In Spanish and English. – See Abstr. 012.095 for the main entry.

160.165 **A survey of nearby clusters of galaxies.**
A. Mazure, E. Escalera, P. Katgert, G. Rhee, R. Dan Hartog, P. Dubath, P. Focardi, D. Gerbal, G. Giuricin, A. Biviano, B. Jones, O. Lefevre, M. Moles, J. Perea.
Messenger, No. 65, p. 7 – 10 (Sep 1991).

160.166 **Dynamical models of the Local Group of galaxies.**
S. Ninković, A. Chernin, M. Shakenov.
Astrofizika, Tom 34, Vyp. 2, p. 213 – 225 (Apr 1991). In Russian. English translation in Astrophysics, Vol. 34, No. 2.

A set of dynamical models for the Local Group is developed to analyze the amount and possible spatial distribution of dark matter in the system. Comparison with observation data shows that the mass of dark matter is as much as 4 to 13 times of the total stellar masses of the galaxies in the Group; it gives also arguments in favour to individual dark matter coronae of the two major galaxies in the Group – Andromeda and Milky Way – rather than to common dark matter corona of the system as a whole.

160.167 **Groups with apparent high M/L ratios.**
H. Quintana.
Rev. Mex. Astron. Astrofis., Vol. 22, No. 2, p. 303 (Oct 1991). Abstract. – See Abstr. 012.127 for the main entry.

160.168 **A hierarchical algorithm to disentangle nearby groups of galaxies (<6000 km s^{-1}).**
P. Fouqué, E. Gourgoulhon, P. Chamaraux, G. Paturel.
Rev. Mex. Astron. Astrofis., Vol. 22, No. 2, p. 303 – 304 (Oct 1991). Abstract. – See Abstr. 012.127 for the main entry.

160.169 **Are field galaxies randomly distributed on scales of 100+ Mpc?**
D. C. Koo.
Rev. Mex. Astron. Astrofis., Vol. 22, No. 2, p. 308 – 309 (Oct 1991). Abstract. – See Abstr. 012.127 for the main entry.

160.170 **Numerical models of Lyα clouds.**
G. M. Williger.
Rev. Mex. Astron. Astrofis., Vol. 22, No. 2, p. 309 (Oct 1991). Abstract. – See Abstr. 012.127 for the main entry.

160.171 **Analisis óptico de cúmulos de galaxias con emisión superficial doble en rayos–X: fotometría de A1750.**
A. Ramírez, H. Quintana.
Rev. Mex. Astron. Astrofis., Vol. 22, No. 2, p. 308 (Oct 1991). Abstract. – See Abstr. 012.127 for the main entry.

160.172 **Clusters of galaxies: structure, infall, and large–scale distribution.**
M. J. Geller.
Clusters of Galaxies Meeting, p. 25 – 41 (1990). – See Abstr. 012.124 for the main entry.
Clusters of galaxies are interesting both as individual dynamical "units" and as tracers of large–scale structure. Here the author reviews three aspects of cluster research: First, she discusses the controversial issue of substructure and its importance in individual systems. Then follows a discussion of the infall region in rich clusters. Finally, the problems and the promise of clusters as tracers of large–scale structure are reviewed.

160.173 **Cosmogony with clusters of galaxies.**
A. Cavaliere, S. Colafrancesco.
Clusters of Galaxies Meeting, p. 43 – 64 (1990). – See Abstr. 012.124 for the main entry.
The authors attempt to bring together N–body simulations, analytical mass distribution functions, scaling laws and numerical computations of luminosity functions, that describe from different angles the hierarchical formation and evolution of groups and clusters of galaxies. The results are discussed in the light of the existing local observations in the optical and X–ray bands, and in the perspective of preliminary evidence and expected data from X–ray surveys at medium redshifts.

160.174 **Cosmogony and the structure of rich clusters of galaxies.**
M. J. West.
Clusters of Galaxies Meeting, p. 65 – 109 (1990). – See Abstr. 012.124 for the main entry.
N–body simulations of the formation of clusters of galaxies allow a detailed, quantitative comparison of theory with observations. This comparison can be used to relate the observed properties of rich clusters of galaxies to the cosmological initial conditions. Furthermore, N–body simulations of clusters are able to test/constrain theories for the formation of the large–scale structure of the universe.

160.175 **The dark matter distribution in clusters.**
M. J. Fitchett.
Clusters of Galaxies Meeting, p. 111 – 137 (1990). – See Abstr. 012.124 for the main entry.
This article describes the current status of various methods for determining the dark matter distribution in clusters. Despite a great deal of progress recently we still do not have good mass constraints for even one cluster. The reasons for this are discussed. New observational tools and methods of analysis should, however, lead to some results in the near future.

160.176 **The effect of the cluster environment on galaxies.**
B. C. Whitmore.
Clusters of Galaxies Meeting, p. 139 – 176 (1990). – See Abstr. 012.124 for the main entry.
Various observations indicate that the cluster environment can affect the structure and dynamics of galaxies. This review concentrates on the effect the environment can have on three of the most basic properties of a galaxy; the morphological type, the size, and the distribution of mass. A reexamination of the morphology – density relation suggests that the fundamental driver may be related to some global property of the cluster, such as the distance from the cluster center, rather than some local property, such as membership in a local subclump within the cluster. There is tentative evidence that rotation curves of spiral galaxies near the centers of clusters are falling, perhaps indicating that the dark halo has been stripped off. Rotation curves for spiral galaxies in compact groups are even more bizarre, providing strong evidence that the group environment has affected the kinematics of these galaxies.

160.177 **Evidence for gas deficiency in cluster galaxies.**
M. P. Haynes.
Clusters of Galaxies Meeting, p. 177 – 199 (1990). – See Abstr. 012.124 for the main entry.
On–going removal of the low density outer interstellar H I gas occurs in galaxies passing through the central regions of clusters with moderately high X–ray luminosity. Although the galaxies currently maintain their spiral morphology, they are H I deficient by as much as a factor of ten relative to their counterparts at larger cluster radii or in the field. The H I distribution in deficient galaxies is truncated well interior to the optical radius, as the gas is removed preferentially from the outer portions. In contrast, the molecular hydrogen component, derived from observations of CO, seems undisturbed. Galaxies that are H I poor by a factor of ten may be gas poor by only a factor of three. While the intracluster medium is the likely catalyst for gas removal, the exact sweeping mechanism is unclear. Early–type objects seem to be even more H I poor than late–type ones, perhaps supporting the suggestion of a fundamental difference in the orbital anisotropy of early–type and late–type spirals. While it seems possible that after disk fading, stripped spirals would ultimately resemble S0's, it is unlikely that all S0's result from such gas sweeping events since the process seems viable only in the cores of rich clusters.

160.178 **Properties of galaxies in groups and clusters.**
A. Sandage.
Clusters of Galaxies Meeting, p. 201 – 230 (1990). – See Abstr. 012.124 for the main entry.
Data on kinematics, spatial distributions, and galaxy morphology in different density regimes within individual galaxy clusters show that many clusters are not in a stationary state but are still in the process of forming.

160.179 **Dynamical evolution of clusters of galaxies.**
D. Richstone.
Clusters of Galaxies Meeting, p. 231 – 255 (1990). – See Abstr. 012.124 for the main entry.
Recent progress in understanding four processes that play a large role in the evolution of clusters of galaxies is reviewed. These are dynamical friction, mergers, collisional tidal stripping and the cluster mean field tide. Recent estimates for the growth rate of the cD galaxy and its frequency of appearance are discussed. It appears that a major part of the cD formation must occur in subclusters. Recent work on the formation of clusters containing galaxies and dark matter suggests that considerable mass segregation occurs in small subclusters (provided clusters form in a bottom up manner). This appears to be a result of dynamical friction. It may imply that visible clusters are embedded in large dark matter halos and that cluster M/L's have been underestimated.

160.180 **Hot gas in clusters of galaxies.**
W. Forman, C. Jones.
Clusters of Galaxies Meeting, p. 257 – 285 (1990). – See Abstr. 012.124 for the main entry.
This paper reviews the X–ray properties of clusters of galaxies and includes a brief summary of the X–ray characteristics of early–type galaxies and compact, dense groups. The X–ray images show that substructure is present in at least 30% of rich (Abell) clusters. This result suggests that most cluster classification systems which utilize a property related to dynamical evolution, require a second dimension related to the dominance of the central galaxy. X–ray surveys of rich clusters show that central, dominant galaxies are twice as common as optical classifications suggest. The evidence for mass deposition ("cooling flows") around central, dominant galaxies is reviewed. Finally, the implications of X–ray gas mass and iron abundance measurements for understanding the origin of the intracluster medium are discussed.

160.181 **Hydrodynamic simulations of the intracluster medium.**
A. E. Evrard.
Clusters of Galaxies Meeting, p. 287 – 326 (1990). – See Abstr. 012.124 for the main entry.
A new, combined N–body and 3D hydrodynamic simulation algorithm is used to study the dynamics of the intracluster medium in rich clusters of galaxies. Results of a program to study an ensemble of clusters covering a range of cluster richness within the framework of a cold dark matter dominated universe are presented. Comparison with observations for both individual cluster characteristics and properties of the ensemble is emphasized. Predictions arising from the numerical models are discussed and directions for future work in this area are outlined.

160.182 **Evolution of clusters in the hierarchical scenario.**
N. Kaiser.
Clusters of Galaxies Meeting, p. 327 – 340 (1990). – See Abstr. 012.124 for the main entry.
If the universe has closure density and the spectrum of primordial density fluctuations is a power law, the lack of any preferred scale means that the clustering should evolve in a scale invariant manner. These self–similar models allow one to approximately predict the evolution of the clustering in e.g., the "standard" cold dark matter model. The author describes how these models yield predictions for the evolution of the cluster populations. Particular attention is given to the range of spectral indices for which the scaling should be valid. It is argued than the allowed range is $-3 < n < 1$, though quite what happens for spectra near the upper bound is somewhat unclear. The cold dark matter power spectrum has spectral index $n \cong -1$ on the mass scale of clusters. For this value of n, I find that the comoving density of clusters classified according to virial temperature T_v or by Abell's richness, should show weak positive density evolution $\partial \log n(T_v, z)/\partial z \cong +0.3$. Clusters classified by total X–ray luminosity should show strong positive density evolution $\partial \log n(L_X, z)/\partial z \cong +3$.

160.183 **Distant clusters as cosmological laboratories.**
J. E. Gunn.
Clusters of Galaxies Meeting, p. 341 – 357 (1990). – See Abstr. 012.124 for the main entry.
Distant clusters provide ideal samples of galaxies in more–or–less standard environments in which to study the evolution of the galaxies themselves, bound structures, the larger–scale environment, and perhaps eventually to provide data for the classical cosmological tests. The author reviews some of the observational and theoretical aspects of these topics.

160.184 **Future key optical observations of galaxy clusters.**
J. P. Huchra.
Clusters of Galaxies Meeting, p. 359 – 375 (1990). – See Abstr. 012.124 for the main entry.
A program is proposed for future optical research on clusters of galaxies. This program includes detailed studies of the internal properties of clusters, the connection between clusters and their environment, and the role of clusters in the study of large–scale structure. It is argued that a digital all–sky survey can be feasibly made with a small telescope and a CCD camera, for studies of nearby and intermediate redshift clusters.

160.185 **Cluster research with X–ray observations.**
R. Giacconi, R. Burg.
Clusters of Galaxies Meeting, p. 377 – 395 (1990). – See Abstr. 012.124 for the main entry.
Past X–ray surveys have shown that clusters of galaxies contain hot gas. Observations of this hot gas yield measurements of the fundamental properties of clusters. Results from a recent study of the X–ray luminosity function of local Abell clusters are described. Future surveys are discussed and the potential for studying the evolution of clusters is analyzed.

Compiled list of clusters of galaxies with measured redshifts.
See Abstr. 002.056.

A catalog of absorption in the spectra of quasi–stellar objects.
See Abstr. 002.102.

A compilation of redshifts and velocity dispersions for Abell clusters (Epoch 1991.2).
See Abstr. 002.103.

Clusters of galaxies. Proceedings.
See Abstr. 012.124.

Constrained realizations of Gaussian fields: a simple algorithm.
See Abstr. 021.036.

A 40–cm UV (2000 Å) balloon–borne imaging telescope: results and current work.
See Abstr. 035.222.

Application of an adaptive kernel technique for producing optical maps of clusters.
See Abstr. 036.054.

Numerical observations of a jet–cloud collision in 3–D.
See Abstr. 062.039.

Numerical simulations of thermal instabilities in stratified gases. II. Exploration of the parameter space.
See Abstr. 062.178.

Lichtablenkung, Gravitationslinsen und Einsteinsche Ringe.
See Abstr. 066.230.

Observations of Ca II absorption along the nearly parallel lines of sight toward 3C 273 and SN 1991T.
See Abstr. 131.361.

Vacuum ultraviolet imagery of the Virgo cluster region. III. Diffuse far–ultraviolet radiation at high Galactic latitudes.
See Abstr. 142.026.

Spectroscopic limits to an extragalactic far–ultraviolet background.
See Abstr. 142.027.

What is in the X–ray sky?
See Abstr. 142.055.

Einstein deep surveys.
See Abstr. 142.056.

Iron line diagnostics in elliptical galaxies and cluster cooling flows.
See Abstr. 142.058.

The few–body problem in astrophysics.
See Abstr. 151.067.

Disk warping in a slewing potential.
See Abstr. 151.085.

Galaxy clustering using the hierarchical Peaks approach: predictions for ROSAT, SZ, and FIR.
See Abstr. 151.137.

Multidimensional analysis of non–linear gravitational clustering in a cold dark matter universe.
See Abstr. 151.138.

Probing the Galactic halo along the 3C 273 sight line using IUE.
See Abstr. 155.105.

Distance of nearby galaxies NGC 2366, IC 2574 and NGC 4236 from photographic photometry of their brightest stars.
See Abstr. 157.013.

A directed search for intergalactic H I using blue compact galaxies.
See Abstr. 157.041.

Observing galaxies through time.
See Abstr. 157.046.

Galaxy pairs in redshift catalogs as a probe of dark matter.
See Abstr. 157.050.

High resolution imaging of a spiral galaxy in the Virgo Cluster: the brightest stars and the extragalactic distance scale.
See Abstr. 157.074.

Winds, outflows, and inflows in X–ray elliptical galaxies. I.
See Abstr. 157.095.

Distances of the nearby galaxies NGC 1560, NGC 2976 and DDO 165 from their brightest stars.
See Abstr. 157.122.

CO observations of galaxies in the Grus quartet: NGC 7552 and NGC 7582.
See Abstr. 157.128.

Narrow–band photometry and the evolution of galaxies. III. The intermediate – redshift sample.
See Abstr. 157.130.

Surface photometry of a sample of elliptical and S0 galaxies.
See Abstr. 157.157.

Phoenix: an intermediate dwarf galaxy in the Local Group.
See Abstr. 157.168.

Erratum: "Multifrequency windows on spiral galaxies. I. UBV and Hα aperture photometry" [Astron. J., Vol. 101, No. 4, p. 1207 – 1230 (Apr 1991)].
See Abstr. 157.211.

The 0.6 Mpc halo of the cD galaxy (MCG 05–33–002) in the cluster Abell 3571 revealed by the co–addition of eight Schmidt plates.
See Abstr. 157.255.

Neutral hydrogen observations of galaxies in superclusters.
See Abstr. 157.263.

Tidally triggered galaxy formation. I. Evolution of the galaxy luminosity function.
See Abstr. 157.265.

The Durham/UKST galaxy redshift survey.
See Abstr. 157.295.

The QMW IRAS galaxy catalogue: a highly complete and reliable IRAS 60–μm galaxy catalogue.
See Abstr. 157.302.

The globular cluster systems of cD galaxies in poor clusters.
See Abstr. 157.333.

Globular cluster systems in central cD galaxies.
See Abstr. 157.335.

External fueling and stochastic ignition of starburst in isolated galaxies.
See Abstr. 157.350.

An optical study of the cD galaxy NGC 3311 and the giant elliptical galaxy NGC 3309 in the cluster Hydra I.
See Abstr. 158.011.

An X–ray and radio study of steep–spectrum radio sources. II. Four fields from a 22 MHz polar cap survey.
See Abstr. 158.102.

Discovery of two radio galaxies in the Zwicky Cluster 1437.3 + 6436.
See Abstr. 158.152.

The extended radio continuum emission of NGC 4388.
See Abstr. 158.158.

The Coma cluster radio source 1253 + 275, revisited.
See Abstr. 158.171.

The large–scale clustering of radio galaxies.
See Abstr. 158.283.

Quasars in rich galaxy cluster environments.
See Abstr. 158.291.

Discovery of a Lyα galaxy near the damped Lyα system at $z = 2.3$ towards the QSO PHL 957.
See Abstr. 159.020.

Another radio–tail quasar ($z = 0.54$) at the center of an extremely rich cluster of galaxies.
See Abstr. 159.025.

Bounds on the enhancement of quasar counts by lenses.
See Abstr. 159.029.

FOS observations of the absorption spectrum of 3C 273.
See Abstr. 159.034.

The ultraviolet absorption spectrum of 3C 273.
See Abstr. 159.043.

First results from the Goddard High–Resolution Spectrograph: the Galactic halo and the Lyα forest at low redshift in 3C 273.
See Abstr. 159.044.

First– and second–epoch VLBI observations of the gravitational lens system 2016 + 112.
See Abstr. 159.078.

The origin of quasar correlations.
See Abstr. 159.089.

Intergalactic absorption in the spectra of high–redshift QSOs.
See Abstr. 159.123.

Spatially resolved spectroscopy of Lyman–alpha and C IV in the gravitational lens 2237 + 030.
See Abstr. 159.144.

Galaxy clustering around QSOs at $0.9 < z < 1.5$.
See Abstr. 159.161.

A radio–quiet quasar inside an Abell cluster of richness class 0?
See Abstr. 159.164.

Systematic effects in lens inversions: $\aleph$ exact models for 0957 + 561.
See Abstr. 159.172.

A general treatment of two–point angular correlation function in cellular fractal structures.
See Abstr. 161.001.

Angular correlations of faint galaxies.
See Abstr. 161.084.

Reconstructing the initial spectrum of fluctuations of the universe from the observed nonlinear clustering of galaxies.
See Abstr. 161.085.

A limit on the cosmological constant.
See Abstr. 161.094.

Massive SIMPs as dark matter: cosmological generation and the aftermath.
See Abstr. 161.102.

Arcminute fluctuations of microwave background due to hot gas in clusters of galaxies. Dependence on cosmology and cluster evolution.
See Abstr. 161.121.

Abell 2218 and the Hubble constant.
See Abstr. 161.125.

The distribution of clusters of galaxies within 300 Mpc h^{-1} and the crossover to an isotropic and homogeneous universe.
See Abstr. 161.132.

Gravitational clustering of Gaussian density fluctuations and the origin of hierarchy correlations.
See Abstr. 161.133.

The fragmentation of "pancakes" in a dark matter–dominated universe.
See Abstr. 161.134.

QSO absorption systems and the origin of the ionizing background at high redshift.
See Abstr. 161.135.

The accuracy of galaxy masses from the timing argument.
See Abstr. 161.144.

Superclusters and pencil–beam surveys: the origin of large–scale periodicity.
See Abstr. 161.146.

Mapping the universe. Statistical measures of the galaxy distribution.
See Abstr. 161.149.

Scale–invariant models of the galaxy and cluster luminosity functions.
See Abstr. 161.169.

A search for redshifted neutral hydrogen from protoclusters of galaxies.
See Abstr. 161.171.

The evolution of void–filled cosmological structures.
See Abstr. 161.175.

A quantitative measure of phase correlations in density fields.
See Abstr. 161.177.

Evolution of a cloudy protogalaxy interacting with an early galactic wind and galaxy formation.
See Abstr. 161.202.

Correlation functions of galaxies in the ESO–Uppsala catalog.
See Abstr. 161.211.

Identification of the substructures in the Universe by percolation test.
See Abstr. 161.213.

Possible gigantic structures of all–sky rich cluster.
See Abstr. 161.215.

New calibration of the $D_n - \sigma$ relation: application to modeling the large–scale peculiar velocity field.
See Abstr. 161.217.

A study of morphological segregations of galaxies in the UK Schmidt Survey Field 349.
See Abstr. 161.218.

The double magnification bias by gravitational lensing.
See Abstr. 161.230.

The existence of the structural anisotropy of the Jagiellonian field of the galaxies.
See Abstr. 161.237.

Redshift surveys of galaxies.
See Abstr. 161.245.

Scales of structures and homogeneity in the universe.
See Abstr. 161.247.

Modelling the velocity field of elliptical galaxies on large scales.
See Abstr. 161.248.

Gravitational lensing, time delay, and angular diameter distance.
See Abstr. 161.249.

Do clusters of galaxies affect the spectrum of the microwave background?
See Abstr. 161.253.

Cross–correlation of the X–ray background with nearby galaxies.
See Abstr. 161.254.

Dynamical evolution of topology of large–scale structure.
See Abstr. 161.277.

Cosmological velocity–density relation in the quasi–linear regime.
See Abstr. 161.292.

Superclustering at high redshifts.
See Abstr. 161.293.

Galaxy formation through hierarchical clustering.
See Abstr. 161.295.

Excursion set mass functions for hierarchical Gaussian fluctuations.
See Abstr. 161.298.

Power–spectrum analysis of one–dimensional redshift surveys.
See Abstr. 161.300.

Small–scale fluctuations and angular correlations of the X–ray background in the HEAO 1 A–2 energy band: constraints on clustering of X–ray sources.
See Abstr. 161.301.

Selection effects in optical surveys for gravitational lenses.
See Abstr. 161.302.

Supercluster–void structure and Cartesian tourbillons compared.
See Abstr. 161.312.

Correlations of large scale peculiar velocity field and the potential flow analysis.
See Abstr. 161.320.

The correlation function of galaxy ellipticities produced by gravitational lensing.
See Abstr. 161.335.

Statistics of gravitational lenses: the uncertainties.
See Abstr. 161.336.

The mass distribution of groups and clusters of galaxies.
See Abstr. 161.337.

The evolution of the diffuse cosmic ultraviolet background constrained by Hubble Space Telescope observations of 3C 273.
See Abstr. 161.366.

Gravitational dip profiles of the microwave background radiation temperature towards rich clusters of galaxies.
See Abstr. 161.386.

The diffuse X–ray background and the intergalactic medium.
See Abstr. 161.388.

A search for dark clusters in redshift catalogs.
See Abstr. 161.396.

Dark matter: gravitational lens studies.
See Abstr. 161.399.

Bounds on intergalactic compact objects from observations of compact radio sources.
See Abstr. 161.417.

The topology of galaxy clustering.
See Abstr. 161.476.

Distances to Virgo and beyond.
See Abstr. 161.482.

The luminosity–line–width relations and the value of H_0.
See Abstr. 161.483.

Evolution of galaxy cluster and primordial nucleosynthesis.
See Abstr. 161.508.

Dynamical estimates of Ω_0 from galaxy clustering.
See Abstr. 161.523.

Evolution and statistics from the number counts of X–ray cluster and intergalactic medium.
See Abstr. 161.531.

Density and peculiar velocity fields in the region of Dressler's supergalactic plane survey.
See Abstr. 161.565.

The power spectrum of galaxy clustering.
See Abstr. 161.568.

Higher moments of the IRAS galaxy distribution.
See Abstr. 161.569.

Ly α clouds at low redshift and the cosmological constant.
See Abstr. 161.570.

Some comments on the epoch of galaxy formation.
See Abstr. 161.581.

Statistics of primordial density perturbations from discrete seed masses.
See Abstr. 161.608.

Is the Universe a fractal? Tests on quasar samples.
See Abstr. 161.614.

Quasar–modulated galaxy clustering in a cold dark matter universe.
See Abstr. 161.615.

Are quasar redshifts cosmological?
See Abstr. 161.617.

Constraints on cosmological models from the IRAS QDOT survey results.
See Abstr. 161.619.

Microwave background anisotropies implied by large–scale galaxy correlations: the minimum of C(0) and cosmological parameters.
See Abstr. 161.633.

A hydrodynamic approach to cosmology: texture–seeded cold dark matter and hot dark matter cosmogonies.
See Abstr. 161.634.

Nonlinear matter clustering properties of a cold dark matter universe.
See Abstr. 161.635.

Deep galactic surveys and the structure of the universe.
See Abstr. 161.636.

Models of voids in elliptic universes.
See Abstr. 161.637.

Phase shifts in gravitationally evolving density fields.
See Abstr. 161.640.

Superclustering at high redshifts.
See Abstr. 161.656.

Cosmology and cluster formation.
See Abstr. 161.675.

Large scale peculiar motions in the southern sky.
See Abstr. 161.681.

Counts, colors and clustering of medium redshift galaxies.
See Abstr. 161.683.

161 Universe, Cosmology, Background Radiation

161.001 **A general treatment of two–point angular correlation function in cellular fractal structures.**
D. Calzetti, Jing Yi–Peng, R. Ruffini.
Astron. Astrophys., Vol. 247, No. 1, p. 1 - 2 (Jul 1991).
A general expression of two–point angular correlation function $\omega(\theta)$ is presented for the cellular fractal structures. The background contribution to $\omega(\theta)$ of the cellular distribution is fully considered in the derivation. A critical discussion about how to use the results presented here to test theoretical models is given.

161.002 **Robertson–Walker viscous fluid model.**
G. Mohanty, B. D. Pradhan.
Astrophys. Space Sci., Vol. 181, No. 1, p. 1 - 5 (Jul 1991).
Robertson–Walker cosmological model with bulk viscosity is investigated with equation of state $p = (\gamma\text{–}1)$. The cosmological solution of the model is obtained with the help of the special law of variation for Hubble's deceleration parameter. Some physical consequences of the solution are studied pertaining to two extreme cases of the equation of state.

161.003 **Lie group for the Jordan and Brans–Dicke cosmology.**
G. M. Covarrubias, O. Obregón.
Astrophys. Space Sci., Vol. 181, No. 1, p. 21 - 27 (Jul 1991).
A method based on the invariance under a continuous Lie group of transformations is worked out to reduce the problem of finding solutions to the cosmological equations of Jordan and Brans–Dicke theory of gravitation for the Robertson–Walker metrics and the cases of the dust universe and the vacuum universe. The reduction consists in a first–order differential equation and a quadrature for each case. Previously known cosmological solutions are re–obtained. In particular, it becomes apparent during the development of this scheme that the flat–space solutions are indeed the general solution.

161.004 **Inhomogeneous generalizations of Bianchi type VI_h models with perfect fluid.**
S. R. Roy, A. Prasad.
Astrophys. Space Sci., Vol. 181, No. 1, p. 61 - 71 (Jul 1991).
Inhomogeneous universes admitting an Abelian G_2 of isometry and filled with perfect fluid have been derived. These contain as special cases exact homogeneous universes of Bianchi type VI_h. Many of these universes asymptotically tend to homogeneous Bianchi VI_h universes. The models have been discussed for their physical and kinematical behaviours.

161.005 **Bianchi type III and Kantowski–Sachs cosmological models in Lyra geometry.**
T. Singh, G. P. Singh.
Astrophys. Space Sci., Vol. 181, No. 1, p. 89 - 101 (Jul 1991).
The cosmological models for Bianchi type III and Kantowski–Sachs space–time within the framework of Lyra geometry are obtained. The physical behaviour of the models have also been discussed.

161.006 **Flavored string–dominated universe.**
J. L. Chkareuli.
JETP Lett., Vol. 54, No. 6, p. 295 - 297 (25 Sep 1991). Current Physics Microform No.: 9112X1413.
The author argues that the flavored Z_3 string–monopole networks, which are related to the spontaneously broken symmetry $SU(3)_H$ of quarks and leptons, inevitably come to dominate the Universe in the course of evolution. It follows that there should be a strict upper limit on the family unification scale, $V_H \leqslant 10^{10}$GeV.

161.007 **An attempt to explain the mechanism of creation of matter in an expanding nondecelerative universe.**
V. Skalský, M. Súkeník.
Astrophys. Space Sci., Vol. 181, No. 1, p. 153 - 156 (Jul 1991). Letter–to–the–editor.
In the expansive nondecelerative universe creation of matter happens as a result of a fundamental relationship and mutual compatibility of the field and matter.

161.008 **Is cosmology a sometime thing?**
T. Rothman.
Astronomy, Vol. 19, No. 7, p. 38 - 43 (Jul 1991).

161.009 **Photon production in conformally flat space–times.**
A. G. Moshkin, V. I. Petukhov.
Sov. J. Nucl. Phys., Vol. 54, No. 3, p. 459 - 461 (Sep 1991). Current Physics Microform No.: 9202X1615.
The effective Lagrangian of a weakly inhomogeneous electromagnetic field in an external gravitational field is calculated in one–loop order. An expression is obtained for the number density of created photons in an arbitrary conformally flat space which is exact in ξ and the scalar curvature, and specific values are found for various models of the Universe.

161.010 **Large–scale cosmological structure and topologically stable states of a scalar field which are periodic in the radial coordinate.**
A. A. Ansel'm.
Sov. Phys. - JETP, Vol. 73, No. 1, p. 1 - 8 (Jul 1991). Current Physics Microform No.: 9111X0361. English translation of Zh. Ehksp. Teor. Fiz., Vol. 100, No. 1, p. 5 - 19 (Jul 1991).
A quasistatic stable state of a scalar field in a closed Friedmann universe with a periodic distribution of energy density along the radial coordinate is described. The classical solution is stable by virtue of the angular nature of the pseudo–Goldstone scalar field (the self–effect of this field is described by a cosinusoidal potential). As a result, there is a conserved topological number (the winding number). The solution found here is linked with a possible periodicity in the distribution of galaxies [T. J. Broadhurst et al., Nature, Vol. 343, p. 726 (1990)]. Physically, it appears that this may not be a complete solution, applying over the entire closed universe. It may instead be only a linear metastable fragment of this entire solution, which decays because of smearing at the ends after a sufficiently long time.

161.011 **A non–baryonic galactic halo and the real galactic age.**
A. Finzi.
Astron. Astrophys., Vol. 247, No. 2, p. 261 - 268 (Jul 1991).
Consideration of both the Hubble shift and radioactive dating leads to estimates of ~ 10 Gyr for t_0, the age of the Universe. The apparent ages of nearby globular clusters give instead $t_0 \gtrsim 15$ Gyr. A possible way to reconcile these conflicting estimates (and to solve the solar neutrino problem at the same time) is to assume that halons – hypothetical massive elementary components of the non–baryonic galactic halo – accrete onto stars and speed up their evolution, so that nearby globular clusters simulate older ages. If the halon hypothesis is correct, globular clusters at galactocentric distances of ~ 100 kpc, which necessarily lie on the very outskirts of the halo, must exhibit their real ages of ~ 10 Gyr. This prediction becomes consistent with astronomical observation only if one assumes that the luminosities of HB stars (in particular RR Lyrae variables) in those clusters exceed the luminosities of HB stars in nearby metal poor globular clusters by ~ 0.4 mag. The author discusses this assumption and notes that it can be tested by comparing the apparent luminosities of central and peripheral RR Lyrae variables in M31, since the average heliocentric distances of these variables are the same.

161.012 **Large–scale cosmological structure and topologically stable periodic (along radial coordinate) scalar field states.**
A. A. Ansel'm.
Zh. Ehksp. Teor. Fiz., Vol. 100, No. 1, p. 5 – 19 (Jul 1991). In Russian. English translation in Sov. Phys. – JETP, Vol. 73, No. 1, p. 1 – 8 (Jul 1991).

161.013 **On the nature of the fractal structure of the universe.**
V. G. Gurzadyan, A. A. Kocharyan.
Europhys. Lett., Vol. 15, No. 7, p. 801 – 804 (1 Aug 1991).
The possible dependence of the fractal dimension of the large–scale distribution of the universe on the dynamical characteristics of the systems (galaxies, their clusters) is considered and the corresponding formula is derived. The importance of searching for an analogous fractal distribution in momentum–coordinate 6–space is outlined.

161.014 **A potential problem in cosmology?**
G. F. R. Ellis, J. E. F. Skea, R. K. Tavakol.
Europhys. Lett., Vol. 16, No. 8, p. 767 – 772 (21 Oct 1991).
Viewing the inflationary dynamics and the associated potential $V(\Phi)$ as a table of correspondences, the authors argue, in the light of results from dynamical systems theory and with the help of some examples, that small changes in one could lead to qualitative changes in the behaviour of the other. This could be of importance, particularly in view of the unknown nature of the inflationary potential.

161.015 **Self–generating universe and many worlds.**
J. Rosen.
Found. Phys., Vol. 21, No. 8, p. 977 – 981 (Aug 1991).
A novel, compact conceptual framework for the origin of the universe is proposed, whereby the universe is identified with a baby universe born of itself. This picture indicates an intimate meshing of gravitation, space–time, and the quantum and offers a framework for the many–worlds interpretation of quantum theory.

161.016 **Construction of the field theory of (2+1)–dimensional universes filled by topological Yang–Mills fields.**
A. R. Kavalov, R. L. Mkrtchyan.
Int. J. Mod. Phys. A, Vol. 6, No. 21, p. 3797 – 3805 (10 Sep 1991).
The action for the field theory of (2+1)–dimensional universes filled by topological Yang–Mills fields is constructed.

161.017 **Highly unstable fundamental strings in inflationary cosmologies.**
M. Gasperini, N. Sánchez, G. Veneziano.
Int. J. Mod. Phys. A, Vol. 6, No. 21, p. 3853 – 3864 (10 Sep 1991).
The authors characterize a regime of extreme (Jeans–like) instability, for strings evolving in cosmological backgrounds, in which the string's proper size grows asymptotically like the scale factor of the expanding universe. They develop a new approximation scheme, based on the asymptotic proportionality of world–sheet and conformal times, for the systematic, quantitative description of such a nonlinear regime. They find that only inflationary geometries (accelerated expansion) are compatible with this instability, and they derive an equation of state for a perfect fluid of unstable strings. The effective pressure is negative, but not large enough to sustain by itself a phase of accelerated expansion.

161.018 **Generating a hot big bang as a quantum fluctuation.**
H. E. Kandrup, P. O. Mazur.
Int. J. Mod. Phys. A, Vol. 6, No. 23, p. 4041 – 4054 (30 Sep 1991).
This review provides an overview of work exploring the hypothesis that the universe "tunneled into being" as a quantum fluctuation. The basic picture is formulated within the context of a semiclassical description, in which the tunneling is interpreted as a gravitational instanton interpolating between some Planck–sized, metrically and topologically complicated, pretunneling entity and a rapidly expanding, topologically trivial, post–tunneling universe. It is argued that the tunneling hypothesis may provide a simple explanation of gross observed features of the current universe, such as the fact that it is big, nearly homogeneous and nearly flat, and that the massless particles contained therein seem characterized by a near–thermal distribution.

161.019 **The wave function of the universe and p–adic gravity.**
I. Ya. Aref'eva, I. V. Volovich, B. Dragovich, P. H. Frampton.
Int. J. Mod. Phys. A, Vol. 6, No. 24, p. 4341 – 4358 (10 Oct 1991).
A new approach to the wave function of the universe is suggested. The key idea is to take into account fluctuating number fields and present the wave function in the form of a Euler product. For this purpose the authors define a p–adic generalization of both classical and quantum gravitational theory. Elements of p–adic differential geometry are described. The action and gravitation field equations over the p–adic number field are investigated. p–adic analogs of some known solutions to the Einstein equations are presented. It follows that in quantum cosmology one should consider summation only over algebraic manifolds. The correspondence principle with the standard approach is considered.

161.020 **Superstrings und Kosmische Strings. Was die Welt im Innersten zusammenhält.**
H.–U. Keller.
Sterne Weltraum, Jahrg. 30, Nr. 8–9, p. 496 – 504 (Aug–Sep 1991).

161.021 **Relativistic fireworks.**
B. Laurent, P. Carlqvist.
Astrophys. Space Sci., Vol. 181, No. 2, p. 211 – 235 (Jul 1991).
The authors have studied a model of relativistic fireworks. In this model it is assumed that a series of explosions occur. In each explosion the fragments fly apart in arbitrary directions with a given velocity which is a parameter in the model. The authors have succeeded in obtaining an exact expression for the distribution of fragments in velocity space after n explosions. They present an exact solution also in the limiting case of small velocity steps where the process turns into a diffusion in velocity space. The development in configuration space has been obtained through Monte–Carlo numerical simulations. The model has been applied to metagalactic cosmology. Although single explosions cannot reach the highest redshifts observed in the Hubble expansion the fireworks model offers a possibility to reach these z–values in a few explosions. The model gives a density inhomogeneity of 20% over a tenth of the Hubble distance as seen from a typical position. Observations show a considerably greater irregular variation. The model gives a local velocity dispersion which is too great to comply with observations. A development of the model is suggested.

161.022 **Cosmological formation of dyon fermion bound states.**
S. K. Bose.
J. Phys. A, Vol. 24, No. 15, p. 3711 – 3717 (7 Aug 1991).
A calculation of the process of cosmological formation of dyon fermion bound states is reported. It is found that for dyon electron systems the efficiency of bound state formation is about 10% if the dyon change $Q = e/2$ and near total if $Q \geqslant e$. For dyon–proton bound states it is again near total for all $|Q| \geqslant e/2$.

161.023 **Inflation–driven string instabilities: towards a systematic large–R expansion.**
Nguyen Suan Han, G. Veneziano.
Mod. Phys. Lett. A, Vol. 6, No. 21, p. 1993 – 2001 (10 Jul 1991).
The authors present the general structure of a large–R expansion for the evolution of classical strings during a phase of fast inflation together with explicit results for several subleading terms.

161.024 **A combined test for cosmological antimatter using the new generation of γ–ray telescopes.**
Gao Yi–tian, D. B. Cline.
Mod. Phys. Lett. A, Vol. 6, No. 29, p. 2669 – 2676 (21 Sep 1991).
In line with the new generation of γ–ray telescopes, further studies of γ–ray cosmology are entering an exciting stage. Interesting models have been proposed to explain the spectrum of the extragalactic γ–ray background (EGB), while intrinsic anisotropies in the EGB can be used to pinpoint some of the theories. In this report, the authors investigate the observational link of the EGB spectrum and large–scale anisotropies. Emphasizing that spatial anisotropies should be looked for at different observational energies, they propose a combined set of the EGB observations which may turn out to be very powerful in determining the role of antimatter in the early universe.

161.025 **Macroscopic boundaries and the wave function of the universe in the c = −2 matrix model.**
J. D. Edwards, I. R. Klebanov.
Mod. Phys. Lett. A, Vol. 6, No. 31, p. 2901 – 2908 (10 Oct 1991).
Using a matrix model, the authors calculate sums over surfaces with macroscopic boundaries of fixed lengths in two–dimensional gravity coupled to a pair of anti–commuting scalar fields with c = −2. For n boundaries, the answer depends only on the sum of their lengths and is given explicitly in terms of Bessel functions to all orders of the genus expansion. For n = 1, this defines the Hartle–Hawking ground state wave function of the universe, which is shown to satisfy the minisuperspace Wheeler – De Witt equation with a boundary condition imposed at small geometries.

161.026 **Non scale–invariant isocurvature perturbations produced in the power–law inflation.**
M. Sasaki, B. L. Spokoiny.
Mod. Phys. Lett. A, Vol. 6, No. 32, p. 2935 – 2945 (20 Oct 1991).
The authors present a careful analysis of non scale–invariant insocurvature perturbations produced in the power–law inflation. As a concrete example, they consider the case when the massless scalar field φ is a Majoron and the baryon density fluctuation is produced in proportion to sin(φ/f) where f is the symmetry breaking scale.

161.027 **Quantum cosmology in $R^1 \times S^1 \times S^2$ topological space.**
S. Chakraborty.
Mod. Phys. Lett. A, Vol. 6, No. 34, p. 3123 – 3131 (10 Nov 1991).
The author has discussed classical and quantum cosmological phenomena for an anisotropic Kantowski–Sachs model. The classical Lorentzian field equations are reduced to an autonomous system and are analyzed near critical point. The minisuperspace path–integral reduces to a single ordinary integration over the lapse, and he studies this integral by lapse method. He finds that the integral has an infinite number of saddle points, each one has identical contribution to the integral. Also the matter–gravity wave function is considered for this metric.

161.028 **Galaxy counts and contributions to the background radiation from 1 μm to 1000 μm.**
A. Franceschini, L. Toffolatti, P. Mazzei, L. Danese, G. De Zotti.
Astron. Astrophys., Suppl. Ser., Vol. 89, No. 2, p. 285 – 310 (Aug 1991).
The authors present detailed predictions for the possible outcome of surveys of extragalactic sources soon to be carried out with the forthcoming generation of both ground based and satellite–borne telescope, over the full range from near–IR to mm wavelengths. Estimates of the confusion limits, due both to stars and galaxies, for several particularly interesting wavebands have been worked out. Predictions for the near– and mid–IR bands are based on up–to–date models for chemical and photometric evolution of early– and late–type galaxies, and allow for luminosity evolution of Seyfert nuclei. These models are shown to provide a good fit to the recent deep counts in the K band; they imply that E/S0 and spiral galaxies are present in roughly equal numbers for $K \leqslant 19$. At fainter magnitudes spirals dominate; ground based surveys down to K = 21 would allow to test photometric evolution models up to $z \cong 1 - 3$. Deep surveys at longer wavelengths are particularly well suited for studying the photometric evolution of E/S0 galaxies at high redshift. Relatively shallow surveys at the same λ over an area of a few square degrees would be particularly effective in providing an unbiased sample of low luminosity AGNs. Predictions for $25\,\mu m \leqslant \lambda \leqslant 1.3$ mm are obtained by extrapolating a model fitting the IRAS deep survey data. The adopted dust emission spectrum from a few microns to 1.3 mm takes into account simple but realistic size and temperature distributions of grains and of PAH molecules. Surveys at $\lambda > 100\,\mu m$ would allow to investigate dust reradiation from distant and primeval galaxies. For example, a survey at 800 μm would have a strong likelihood of discovering actively star forming galaxies at $z \cong 5$, if they exist. The measurements of the diffuse background spectrum and intensity being carried out by COBE, will provide important constraints but will require a delicate subtraction of foreground emissions.

161.029 **On the Jeans–mass in cosmology.**
A. Mészáros, Z. Molnár.
Astrophys. Space Sci., Vol. 181, No. 2, p. 323 – 325 (Jul 1991).
Letter–to–the–editor.
The matter of the Universe after recombination is considered as an expanding gas. The Jeans–mass obtained for this gas may essentially be larger than it is usually expected. This means that the discrepancy between the theoretical Jeans–mass and the observed masses of the galaxies may be mitigated.

161.030 **Cosmological implications of axinos.**
K. Rajagopal, M. S. Turner, F. Wilczek.
Nucl. Phys. B, Part. Phys., Vol. 358, No. 2, p. 447 – 470 (15 Jul 1991).
Axinos are the supersymmetric partners of axions. They arise in models incorporating both low–energy supersymmetry and the axion solution of the strong CP problem. In the present state of knowledge several of the key properties of axinos, which control their cosmological consequences, are poorly determined. But generically there are very significant cosmological consequences, and the authors attempt to survey the possibilities. In a wide variety of models the axino is the lightest R–odd particle, and destabilizes the more conventional candidates for this title (photino, higgsino ...) on cosmological time scales. While this consideration perhaps casts some models the axino itself becomes a plausible dark matter candidate. In other models the axino is heavy, and unstable. Even then axinos are of cosmological interest, because their decay can be the dominant mechanism for production of the lightest R–odd particle.

161.031 **Cosmic abundances of stable particles: improved analysis.**
P. Gondolo, G. Gelmini.
Nucl. Phys. B, Part. Phys., Vol. 360, No. 1, p. 145 – 179 (12 Aug 1991).
An exact relativistic single–integral formula for the thermal average of the annihilation cross section times velocity, the key quantity in the determination of the cosmic relic abundance of a species, is obtained. Since it does not require expansion of the cross section at low relative velocity, it can also be used when the cross section varies rapidly with energy, e.g. near the formation of a resonance or the opening of a new annihilation channel. The authors discuss approximate formulas in these cases, and they find that dips in the relic density near resonances are significantly broader and shallower than previously thought and that spurious reductions near thresholds disappear.

161.032 **A new mechanism for neutralizing the cosmological constant.**
M. J. Duncan, L. G. Jensen.
Nucl. Phys. B, Part. Phys., Vol. 361, No. 3, p. 695 – 712 (9 Sep 1991).
The authors propose a dynamical mechanism for neutralizing the cosmological constant in our universe. The analysis is wholly

lorentzian. Quantum cosmology is an effective theory below the Planck scale, and it suffices for the purpose of the demonstration. The vacuum energy from particle physics is many orders of magnitude too great to explain the measured smallness of the cosmological constant. Any generic field that allows to shift the vacuum energy has the possibility of neutralizing it. The essential result of the paper is that the dominant vacuum state for the field is indeed the one that exactly cancels the particle physics vacuum energy. This means that a large universe, such as ours, has a vanishing cosmological constant.

161.033 **Cosmology and broken discrete symmetry.**
J. Preskill, M. B. Wise, S. P. Trivedi, F. Wilczek.
Nucl. Phys. B, Part. Phys., Vol. 363, No. 1, p. 207 – 220 (30 Sep 1991).
It is argued that under various circumstances spontaneously broken discrete symmetry which is commonly thought to lead to grave cosmological problems, need not do so. A particularly interesting example is the two–doublet version of the standard model. This extended model avoids unacceptable flavor–changing neutral currents in a natural way only if an appropriate discrete symmetry is postulated. Here it is shown that this discrete symmetry is anomalous. As a result, when it is spontaneously broken it does not produce stable domain walls (which would be problematic for cosmology). Instead an interesting cosmological scenario results, which incorporates substantial deviations from thermal equilibrium at the QCD scale.

161.034 **Computer simulations of the motion and decay of global strings.**
C. Hagmann, P. Sikivie.
Nucl. Phys. B, Part. Phys., Vol. 363, No. 1, p. 247 – 280 (30 Sep 1991).
Computer simulations have been carried out of the motion and decay of global strings, including spectrum analysis of the energy stored in the scalar field which describes the global string and the radiated Nambu–Goldstone bosons. The authors simulated relaxing pieces of bent string and collapsing circular loops. They find, for the string sizes investigated, that the spectrum of field energy hardens rather than softens while the string decays into Nambu–Goldstone radiation. They argue on theoretical grounds that this is indeed the most plausible general behaviour.

161.035 **Self–sustained inflation and dimensional reduction from fundamental strings.**
M. Gasperini, N. Sánchez, G. Veneziano.
Nucl. Phys. B, Part. Phys., Vol. 364, No. 2, p. 365 – 380 (21 Oct 1991).
In a previous paper the authors have shown that an ideal gas of fundamental strings is not able to sustain, by itself, a phase of isotropic inflation of the universe. They show here that fundamental strings can sustain, instead, a phase of anisotropic inflation accompanied by the contraction of a sufficient number of internal dimensions. The conditions to be met for the existence of such a solution to the Einstein and string equations are derived, and the possibility of a successful resolution of the standard cosmological problems in the context of this model is discussed.

161.036 **Phase transitions with sub–critical bubbles.**
M. Gleiser, E. W. Kolb, R. Watkins.
Nucl. Phys. B, Part. Phys., Vol. 364, No. 2, p. 411 – 450 (21 Oct 1991).
The authors study the dynamics of cosmological phase transitions initiated from a state of thermal equilibrium. If the effective potential satisfies certain general conditions, a homogeneous phase of false vacuum will form as the universe expands, and the transition will proceed by well–known bubble nucleation processes. If such conditions do not hold, the universe may instead be filled with a two–phase emulsion. The evolution of the transition will be determined by the free energy difference between the two phases and by the expansion rate of the universe. Thermal fluctuations between the phases will determine the final distribution of regions of the universe in each phase as they freeze out. The authors develop a method to study the dynamics of such fluctuations, which they call sub–critical bubbles, and apply it to several situations of interest, including the symmetric and asymmetric double–well, and the Coleman–Weinberg scalar potentials. They show that in certain cases it is possible to avoid supercooling, with the transition being completed by sub–critical fluctuations. Possible applications to the electroweak phase transition are briefly discussed.

161.037 **The scattering of fermions from superconducting cosmic strings.**
W. B. Perkins.
Nucl. Phys. B, Part. Phys., Vol. 364, No. 2, p. 451 – 474 (21 Oct 1991).
The scattering of fermions by a superconducting cosmic string is considered. By considering the penetration of the fermionic wave function to the core radius, the author finds that the cross section for string–catalysed baryon decay is exponentially suppressed for string currents in excess of 10^5A.

161.038 **A note on harmonic oscillator in expanding universes.**
C. M. A. Dantas, I. A. Pedrosa, B. Baseia.
Nuovo Cimento B, Vol. 106, No. 7, p. 749 – 753 (Jul 1991).
It is shown that the Hamiltonian governing the dynamics of a harmonic oscillator in expanding universes belongs to the class of quadratic Hamiltonians that generates squeezed states.

161.039 **Large–scale inhomogeneities in the Universe and the spectra of fluctuations.**
B. I. Gnatyk, V. N. Lukash, B. S. Novosyadlyj.
Pis'ma Astron. Zh., Tom 17, No. 8, p. 659 – 670 (Aug 1991). In Russian. English translation in Sov. Astron. Lett., Vol. 17, No. 4.
The requirements for the density fluctuation spectra from the point of view of conformity of the observable and theoretical typical values of the large scale structure of the Universe have been analysed. An ad hoc spectrum is found, which predicts the correlation functions of galaxies and clusters, large scale bulk motions, the Great attractor events without any contradictions with the back–ground radiation anisotropy. It is similar to the CDM–spectrum at scales $k^{-1} \leqslant 3\,h^{-1}$Mpc and $k^{-1} \geqslant 10^3h^{-1}$Mpc, but has the additional power $\sigma_z = 0.2$ at scales $3\,h^{-1}\mathrm{Mpc} \leqslant k^{-1} \leqslant 10^3h^{-1}$Mpc. The amplitude of the spectrum and its slope in this region are determined by observable characteristics of the large–scale structure. It is shown that decrease of the upper limitation in the experiment "Cholod" at scales $\cong 2°$ twice will allow one to register such a spectrum.

161.040 **Flat edge–on galaxies on the Tully–Fisher diagram. I.**
I. D. Karachentsev.
Pis'ma Astron. Zh., Tom 17, No. 8, p. 671 – 683 (Aug 1991). In Russian. English translation in Sov. Astron. Lett., Vol. 17, No. 4.
For a sample of 114 northern galaxies with angular diameters $a > 2'$ and apparent axial ratios $a/b > 7$ relations are discussed between their global parameters: the inner motion amplitude W, the blue luminosity L_B, the linear diameter A_B (or A_R), and the neutral hydrogen mass M_{HI}. Taken as a distance indicator for the given W, these parameters exhibit nearly the same efficiency. Outside the Local Supercluster the linear diameter of the flat edge–on galaxy produces a standard deviation about 25 percent as a distance estimate. The total mass–to–luminosity ratio of flat galaxies increases, when their mean surface brightness decreases.

161.041 **Comments on the V/V_{max} test.**
G. Bigot, R. Triay.
Phys. Lett. A, Vol. 159, No. 4/5, p. 201 – 207 (14 Oct 1991).
Enlightenments on the theoretical basis of the V/V_{max} test (used in cosmology) are given for Λ–cosmologies. Surprisingly, the evolutionary effects in the 3CR sample of radio galaxies are not detected by the usual test, but on purpose developments of this approach confirm their presence.

161.042 **Gravitino mass bounds in a general cosmological scenario.**
P. Roy, P. S. Joshi, S. M. Chitre.
Phys. Lett. A, Vol. 160, No. 1, p. 36 – 40 (4 Nov 1991).
A stable gravitino is shown to respect an upper mass bound ~ 1 keV and an unstable one a lower mass bound ~ 10 TeV from the global hyperbolicity property alone without the assumptions of a Friedmann universe. The specific Friedmann cosmology with a critical mass density in fact reduces the upper bound for a stable gravitino to ~ 0.1 keV. Complications due to inflation and the anti–de Sitter nature of the spacetime before the breakdown of local supersymmetry are also discussed.

161.043 **Bulk viscosity in the early universe.**
Cheng Bao–lian.
Phys. Lett. A, Vol. 160, No. 4, p. 329 – 338 (25 Nov 1991).
The coefficient of bulk viscocity during the GUT phase transition is derived. Its effects of cosmological inflation, density fluctuations, and entropy production are also investigated. The author identifies conditions for which bulk viscosity can become an inflationary source as well as generate an inflation. An appropriate fluctuation spectrum caused by bulk viscosity for galaxy formation is obtained. Also, the entropy production by bulk viscosity is calculated and the result provides a possible resolution to the entropy generation problem.

161.044 **Third quantization of gravity and the cosmological constant problem.**
G. Lavrelashvili (*G. V. Lavrelashvili*), V. A. Rubakov, P. G. Tinyakov.
5. Seminar on Quantum Gravity, p. 27 – 43 (1991). – See Abstr. 012.004 for the main entry.

161.045 **Exact Bianchi type–II, VIII, and IX cosmological models with matter and electromagnetic fields in Lyra's manifold.**
R. Venkateswarlu, D. R. K. Reddy.
Astrophys. Space Sci., Vol. 182, No. 1, p. 97 – 103 (Aug 1991).
Spatially–homogeneous and anisotropic Bianchi type–II, VIII, and IX stiff–fluid cosmological models in the presence of source–free electromagnetic fields are obtained in Lyra's manifold. Some properties of the models are also discussed.

161.046 **Particle creation in slowly–rotating Robertson–Walker universes.**
K. Maniharsingh.
Astrophys. Space Sci., Vol. 182, No. 1, p. 141 – 153 (Aug 1991).
The dynamics of particle creation in the slowly rotating Robertson–Walker universe is studied by use of the Brans–Dicke theory, and four interesting solutions are obtained. Along with the discussion of the nature of the "gravitational constant" and other physical parameters in these model universes the behaviour of these solutions with respect to Dirac's hypotheses is investigated. Also it is found that the expansion law is determined by the particles created by the expansion itself. Except for the case of a flat universe, the metric rotation as well as the matter rotation is found to have a damping effect on the creation of particles.

161.047 **True vacuum bubbles and the origin of voids.**
D. La.
Phys. Lett. B, Vol. 265, No. 3/4, p. 232 – 238 (15 Aug 1991).
Extended inflation theory can provide astrophysically interesting, nearly energy–empty bubbles at the epoch of matter–radiation equality, $t = t_{eq}$. The physical size and number of such bubbles at $T = t_{eq}$ within the present physical horizon as a function of the model parameters is given. If non–baryonic cold dark matter (CDM) is not present in the universe, the filling fraction of these bubbles in the physical volume of the present universe becomes less than a few percent, so the bubbles are marginally detectable. If CDM is presented in the universe, true vacuum bubbles can grow to occupy a major fraction of the volume of the universe today.

161.048 **Neutrino degeneracy and cosmological nucleosynthesis, revisited.**
K. A. Olive, D. Thomas, D. N. Schramm, T. P. Walker.
Phys. Lett. B, Vol. 265, No. 3/4, p. 239 – 244 (15 Aug 1991).
A reexamination of the effects of non–zero neutrino degeneracies on big bang nucleosynthesis is made. As previously noted, non–trivial alterations of the standard model conclusions can only be induced if excess lepton numbers, comparable to photon number densities, are assumed. Furthermore, the required lepton number densities must be different for ν_e than for ν_μ and ν_τ. It is shown that this loophole in the standard model of nucleosynthesis is robust and will not vanish as abundances and reaction rate determinations improve.

161.049 **The formation of cosmic structure with a 17 keV neutrino.**
J. R. Bond, G. Efstathiou.
Phys. Lett. B, Vol. 265, No. 3/4, p. 245 – 250 (15 Aug 1991).
Adding a 17 keV neutrino with a lifetime of ~ 1 yr can explain the puzzling large–scale structure observed in the galaxy distribution which is difficult to understand in the 'standard' cold dark matter model with the adiabatic scale invariant density fluctuations predicted by inflation. Decay times $\gtrsim 10$ yr are incompatible with microwave background anisotropy limits and galaxy clustering observations.

161.050 **Magnetic fields from cosmological phase transitions.**
T. Vachaspati.
Phys. Lett. B, Vol. 265, No. 3/4, p. 258 – 261 (15 Aug 1991).
Cosmological phase transitions in the early universe lead to gradients in the vacuum expectation value of the Higgs field. The variation of the Higgs field results in the presence of electromagnetic fields after the completion of the phase transition. This mechanism would also produce magnetic fields of other unbroken gauge symmetrics such as $SU(3)_c$. The authors estimates the strength of the electromagnetic fields produced due to the electroweak phase transition.

161.051 **Scale factor duality for classical and quantum strings.**
G. Veneziano.
Phys. Lett. B, Vol. 265, No. 3/4, p. 287 – 294 (15 Aug 1991).
Duality under inversion of the cosmological scale factor is discussed both for the classical motion of strings in cosmological backgrounds and for genus zero, low energy effective actions. The string–modified, Einstein–Friedmann equations are then shown to possess physically–inequivalent, duality–related solutions which generically describe the 'decay' of an initial, perturbative, flat $D = 10$ superstring vacuum towards a more interesting strong–coupling state through an appealing pre–big–bang cosmological scenario.

161.052 **Electromagnetic radiation from a global cosmic string in an external magnetic field.**
D. D. Harari, F. D. Mazzitelli.
Phys. Lett. B, Vol. 266, No. 3/4, p. 269 – 274 (29 Aug 1991).
A global cosmic string may couple to electromagnetism as a consequence of anomalous triangle diagrams. In an external magnetic field, the gradients of the Nambu–Goldstone boson field around the string give rise to an extended polarization current. When the string oscillates, it radiates electromagnetic waves. The authors estimate the strength of the effect for representative values of astrophysical magnetic fields. They also discuss implications for axion strings if a primordial cosmological magnetic field existed at early times.

161.053 **Adiabatic perturbations and minimal size of the de Sitter vacuum.**
M. Gasperini, G. Ummarino.
Phys. Lett. B, Vol. 266, No. 3/4, p. 275 – 279 (29 Aug 1991).
The authors consider transitions for a higher dimensional to a lower dimensional de Sitter phase. They show that, if the transitions are adiabatic, the evolution necessarily passes through a state of maximal curvature (minimal size). This minimal size depends only on the total entropy content, but the dependence is

so weak that no fine tuning is needed in order to set the minimum at the Planck scale.

161.054 **Nucleosynthesis constraints on cosmic strings.**
M. Quirós.
Phys. Lett. B, Vol. 267, No. 1, p. 27 – 29 (5 Sep 1991).
The author reanalyzes the nucleosynthesis predictions in the presence of a background of gravitational radiation from string loop decay. The comparison with light elements abundance provides upper bounds on the dimensionless string mass per unit length $G\mu$. These bounds are stronger than present bounds coming from the millisecond–pulsar timing stability. For $N_\nu = 3$ the nucleosynthesis bounds are consistent with the values required for galaxy formation, provided that $\alpha < 10^{-3}$ (where α is the ratio of the initial size of the loop relative to the horizon size), and are stronger than present bounds coming from the angular distortions in the microwave background radiation. In the limit $\alpha \to 0$ $G\mu < [2.6 + 12.9(3-N_\nu)] \times 10^{-6}$ is found.

161.055 **Cosmological model from generalised Maxwell–Einstein system in higher dimensions.**
J. C. Fabris.
Phys. Lett. B, Vol. 267, No. 1, p. 30 – 32 (5 Sep 1991).
The author considers a generalised Maxwell–Einstein system in a higher dimensional space–time such that the energy–momentum tensor is traceless. After reduction to four dimensions and setting the gauge–fields equal to zero, he obtains the usual four dimensional Einstein–Hilbert term coupled to two scalar fields. One of them is a Brans–Dicke type scalar field with an inverted sign of the kinetic term. The obtained cosmological solutions are non–singular.

161.056 **Plausible double inflation.**
R. Holman, Wang Yun, E. W. Kolb, S. L. Vadas.
Phys. Lett. B, Vol. 269, No. 3/4, p. 252 – 256 (31 Oct 1991).
It is likely that extended inflation is followed by an epoch of slowroll inflation. Such a sequence of events may lead to a very interesting perturbation spectrum with significant power on the scale of the transition between the extended and slowroll phase, superimposed upon a power–law spectrum with deviations from the Harrison–Zel'dovich slope. Normalization of the spectra above and below the transition scale is expected to differ.

161.057 **Axially symmetric domain wall in 2+1 dimensions.**
J. Soda, Y. Yamanaka.
Phys. Lett. B, Vol. 269, No. 3/4, p. 275 – 278 (31 Oct 1991).
An axially symmetric domain wall (string) in 2+1 dimensions is investigated in the synchronous gauge. This problem is also regarded as a cylindrical symmetric domain wall (membrane) in 3+1 dimensions. Using Israel's method, the authors present a general solution. As a special case, a static solution is obtained which agrees with the previous result of Deser and Jackiw obtained by another method.

161.058 **Quantum cosmology of the superstring universe.**
Wang Jin, S. Mohanty.
Phys. Lett. B, Vol. 269, No. 3/4, p. 279 – 282 (31 Oct 1991).
The authors study the quantum cosmology of the string universe obtained.by embedding the Robertson–Walker metric in the non–linear model of Antoniadis, Bachas, Ellis and Nanopoulos (ABEN). They find that the quantum model has some unusual features not seen in the classical model. Initially, the universe exists in a series of metastable bound states with scale factor taking discrete values. Then the universe tunnels through a barrier and comes out in an inflationary state. This tunneling (or evolution in imaginary time) also has the effect of heating up the matter field so that one has the conditions of chaotic inflation. The asymptotic solutions agree with those obtained from the classical Einstein equations by ABEN.

161.059 **The arguments against "antigravity" and the gravitational acceleration of antimatter.**
M. M. Nieto, T. Goldman.
Phys. Rep., Vol. 205, No. 5, p. 221 – 281 (Jul 1991).
After reviewing the history behind the concept of antigravity, the authors discuss modern theoretical ideas and experiments which are relevant to the possibility that there exist static, non–Newtonian components of gravity. They then consider the arguments against antigravity, pointing out their regimes of applicability and inapplicability to modern quantum theories of gravity. They specifically show that, if these arguments are applied to the ongoing experiment to measure the gravitational acceleration of the antiproton, they do not rule out a large anomalous gravitational response for the antiproton. The antiproton gravity experiment is one of five types of experiments that one considers to be the most critical in testing for finite but long–range, non–Newtonian and non–Einsteinian gravity. The authors discuss how all five are related to each other and how their completion would improve our knowledge of gravitational forces.

161.060 **Scattering of Sc galaxies on the Tully–Fisher diagram and the dark matter problem.**
I. D. Karachentsev.
Pis'ma Astron. Zh., Tom 17, No. 10, p. 867 – 876 (Oct 1991). In Russian. English translation in Sov. Astron. Lett., Vol. 17, No. 5.
A sample of 130 Sc galaxies with photometric I–band profiles, measured by Freudling, and reliable data on H I line width is considered. As a distance indicator on the Tully–Fisher diagram the author tested different global parameters of the galaxies. The use of any combinations of optical and radio parameters of galaxies does not reduce scatter of their distance estimations below the minimum value $\sigma_{min}(\lg \Delta) = 0.07 - 0.06$. This limit is not affected essentialy by the galaxian environment. About 3/5 of the galaxies show significant deviations of their brightness profile with respect to a standard linear one. However, amplitude and sign of the deviation do not correlate with the galaxy location on the TF–diagram. This feature may be related with the presence of dark matter in the galaxies. The mean ratio of total mass to I–luminosity is nearly the same ($\sim 3M_\odot/L_\odot$) as well for giant, as for dwarf galaxies, but it depends on their surface brightness, SB. A small dispersion, $\sigma[\lg(M/L)|SB] = 0.090$, obtained for the sample of 130 Sc galaxies, leads to rather hard restrictions on dark matter properties.

161.061 **One–loop quantum cosmology: the general semiclassical solution of the Wheeler–DeWitt equations and its applications.**
A. O. Barvinsky (*A. O. Barvinskij*), A. Yu. Kamenshchik.
5. Seminar on Quantum Gravity, p. 407 – 422 (1991). – See Abstr. 012.004 for the main entry.
This report is devoted to the one–loop approximation in quantum cosmology. The previous attempts to study the early stage of the Universe were mainly confined to the tree–level approximation of this theory. This resulted in a number of unresolved problems including the status of unitarity in quantum cosmology, the validity of the Hartle–Hawking quantum state as a source of the inflation and so on. Thus the authors expose here several steps in clarifying such questions at the one–loop order of a semiclassical expansion. To this end they propose the general semi–classical solution of the Wheeler–DeWitt equations, show the conservation of an approapriately defined probability of inflation for the Hartle–Hawking wave function and find the conditions when the latter can be regarded as a good candidate for the quantum state of the inflationary Universe.

161.062 **Environment–induced superselection in cosmology.**
W. H. Zurek.
5. Seminar on Quantum Gravity, p. 456 – 472 (1991). – See Abstr. 012.004 for the main entry.
Interaction between a quantum system and its environment can be often regarded as a measurement, in the course of which one of the system observables influences the evolution of the external degrees of freedom and is thus "monitored" by the

environment. This causes the system to "decohere." Loss of quantum coherence erases part of the density matrix responsible for the correlations between the eigenstates of the monitored observables. This mechanism is very efficient even in the limit of weak coupling to the environment. The classical limit of quantum theory obtains in the limit in which the coupling strength and the Planck constant simultaneously tend to zero. Transition from quantum to classical in the inflationary cosmological model can be justified in the framework of environment induced susperselection.

161.063 **The self–measurement of the Universe and the concept of time in quantum cosmology.**
M. B. Mensky.
5. Seminar on Quantum Gravity, p. 473 – 493 (1991). – See Abstr. 012.004 for the main entry.
The dynamics of the quantum Universe under self–measurement is investigated in the framework of the path–integral approach to quantum theory of continuous measurements. The corresponding propagator is evaluated explicitly for a simple superspace model. It is shown that the picture of time evolution of the quantum Universe naturally emerges as a result of its self–measurement. The Wheeler–DeWitt dynamics maintains for small times when the effect of measurement can be neglected and time dependence disappears.

161.064 **Mechanical properties and cosmological implications of current carrying cosmic strings.**
B. Carter.
5. Seminar on Quantum Gravity, p. 494 – 505 (1991). – See Abstr. 012.004 for the main entry.
After a summary of the general principles governing the mechanics of macroscopic string (and membrane) models, attention is drawn to the potential cosmological importance of ring equilibrium configurations that are shown to be likely to have been formed by current carrying cosmic string loops that may have been created during an electroweak symmetry breaking phase transition.

161.065 **Quantum string theory in curved space–times.**
N. Sánchez.
5. Seminar on Quantum Gravity, p. 506 – 523 (1991). – See Abstr. 012.004 for the main entry.
A general method to quantize strings in curved space–times is exposed. It treats the space–time metric exactly and the string excitations small as compared with the energy scale of the geometry. The method is applied to cosmological (de Sitter) and black–hole (Schwarzschild) geometries. Bogolyubov transformations in the context of string theory are derived and the Bogolyubov coefficients describing elastic and inelastic scattering and excitation of modes are computed explicitely. The string–black–hole cross section is derived and a pair mode creation phenomena is found. The quantization and scattering of strings in shock–wave geometries (ultrarelativistic boosted black–holes or Aichelburg–Sexl space time) is found to be exactly solvable.

161.066 **Anomaly meltdown.**
C. Gajdzinski, R. F. Streater.
J. Math. Phys., Vol. 32, No. 8, p. 1981 – 1983 (Aug 1991). Current Physics Microform No.: 9108C1560.
It is shown that at nonzero temperature it is possible that anomalies in representations of symmetry groups and gauge groups, present at zero temperature, disappear. Several examples are given. Thus the idea that anomalies in baryon currents might have caused the baryon imbalance in the early hot universe needs reconsideration.

161.067 **Bianchi type–I cosmological models in Lyra's geometry.**
T. Singh, G. P. Singh.
J. Math. Phys., Vol. 32, No. 9, p. 2456 – 2458 (Sep 1991). Current Physics Microform No.: 9109E1014.
Exact solutions are obtained for the anisotropic Bianchi type–I model in normal gauge for Lyra's geometry. The physical behavior of the models is examined in vacuum and in the presence of perfect fluids.

161.068 **Relativistic quantum chaos in Robertson–Walker cosmologies.**
R. Tomaschitz.
J. Math. Phys., Vol. 32, No. 10, p. 2571 – 2579 (Oct 1991). Current Physics Microform No.: 9110E0763.
Open Robertson–Walker cosmologies of multiple spatial connectivity provide a challenging example for the possible influence of the global topological structure of space–time on the laws of microscopic motion. Free geodesic motion is investigated in such cosmologies in the context of first quantization. A unique localized wave field, a solution of the Klein–Gordon equation, is found as a consequence of the topological structure of the spacelike slices $t = \text{const}$ of the manifold. This solution is closely related to the collection of the bounded chaotic trajectories.

161.069 **Fluid of multidimensional objects in cosmology.**
R. A. Maszczyk.
J. Math. Phys., Vol. 32, No. 11, p. 3141 – 3147 (Nov 1991). Current Physics Microform No.: 9111D1945.
The subject of this paper is a generalized Robertson–Walker model filled with the fluid of multidimensional objects. The solutions are classified according to curvatures of the internal and external spaces, as well as to the qualities of the dynamics of the models. The majority of solutions are singular. The analysis shows, however, that there may also be nonsingular oscillatory solutions.

161.070 **Stably causal Gödel–type models.**
W. Monteiro S.
J. Math. Phys., Vol. 32, No. 11, p. 3223 – 3227 (Nov 1991). Current Physics Microform No.: 9111D2027.
The problem of finding stationary Gödel–type cosmological space–times endowed with stable causality is solved in general relativity. These models are generated by spin fluids with strong spin–vorticity coupling, and present interesting physical properties. Their intrinsic characterization is made by means of an invariant algebraic classification. A simple technique to test the stable causality is applied to new and old distinct stationary space–times.

161.071 **Killing tensors in two–dimensional space–times with applications to cosmology.**
K. Rosquist, C. Uggla.
J. Math. Phys., Vol. 32, No. 12, p. 3412 – 3422 (Dec 1991). Current Physics Microform No.: 9112E1080.
The existence of Killing vectors and Killing tensors in two–dimensional space–times is discussed. The corresponding mechanical problem having a potential with two exponential terms is analyzed in detail and a number of cases admitting Killing tensors are given. The Killing tensors give rise to new exact solutions in perfect fluid Bianchi and Kantowski–Sachs cosmologies as well as in inflationary models with a scalar field source.

161.072 **Quantum processes in the gravitational field of a cosmic string.**
D. D. Harari, V. D. Skarzhinsky.
5. Seminar on Quantum Gravity, p. 524 – 533 (1991). – See Abstr. 012.004 for the main entry.
The authors show that many elementary particle physics processes, such as pair production by a high energy photon, that take place in Minkowski space only if an external field provides for momentum non–conservation, do occur in the space–time around a straight cosmic string, even though the spaced is locally flat and there is no local gravitational potential. The authors exemplify this mechanism through the evaluation of the cross section per unit length of string for the decay of a massless scalar particle into a pair of massive particles. The cross section of this process is typically small but increases rapidly with energy and reaches a constant value. Also, it is interesting to realize how reactions of such kind occur due to topological properties of space, rather than to the action of a local field.

161.073 **Cosmological models that describe particle creation in the early universe and evolve into the "present–day" universe.**
J. P. de Leon.
J. Math. Phys., Vol. 32, No. 12, p. 3546 – 3552 (Dec 1991). Current Physics Microform No.: 9112E1214.
The disappearance of the cosmological constant can be formally treated by means of similarity solutions of general relativity that evolve from a stage with conformal symmetry to a stage with homothetic symmetry. In this work it is assumed that in this transition the universe does not "lose its memory" completely, but it does "remember" some of its past characteristics. Specifically, it is assumed that the equation of state remains the same in both stages. Then, the most general, spherically symmetric, cosmological model compatible with this assumption is developed. It is shown that it can be used to describe, classically, the birth, near the center of a spherical domain (a "bubble") of positive density and pressure from an early universe with particle production.

161.074 **Equilibrium configurations of superconducting cosmic strings in background fields.**
O. Heinrich.
5. Seminar on Quantum Gravity, p. 534 – 552 (1991). – See Abstr. 012.004 for the main entry.
The equilibrium states of a superconducting string with an arbitrary equation of state in a stationary gravitational and electromagnetic background are described. A Hamilton–Jacobi method for the equilibrium trajectory is derived and the relation between action and energy is discussed. The extension of the formalism to string states symmetric with respect to a spacelike Killing vector is given.

161.075 **Thermodynamics of the neutral string gas in electromagnetic background.**
E. J. Ferrer, E. S. Fradkin, V. de la Incera.
5. Seminar on Quantum Gravity, p. 553 – 563 (1991). – See Abstr. 012.004 for the main entry.
The one–loop partition function of the neutral open string in electromagnetic background is computed on a space–time with the time direction compactified to a circle. It is shown that the time compactification in presence of the background field produces winding modes not only in the time direction but also in the spatial one. Some thermodynamic characteristics of the system are investigated, as it is the free energy of the hot string gas and the Hagedorn temperature. The latter results to be essentially dependent on the electric field.

161.076 **On multi–inflation cosmological models.**
H.-J. Schmidt.
5. Seminar on Quantum Gravity, p. 564 – 573 (1991). – See Abstr. 012.004 for the main entry.
For a Lagrangian L, which can be interpreted as higher–loop quantum correction to the Einstein–Hilbert Lagrangian, the field equation is explicitly written down. It is conformally equivalent to Einstein's equation with many scalar fields. These fields give rise to many different inflationary stages of cosmic evolution. Furthermore, a closed–form solution for scale–invariant gravity and the spatially flat Friedmann model is given which shows directly the attractor property of power–law inflation.

161.077 **The problem of the wave function of the Universe in the radiation dominated case.**
M. A. Castagnino, E. Gunzig, P. Nardone.
5. Seminar on Quantum Gravity, p. 600 – 612 (1991). – See Abstr. 012.004 for the main entry.
Homogeneous pure radiation universe is exploited, as a theoretical laboratory, to test Hartle–Hawking's as well as Vilenkin's wave functions. A new idea is introduced in order to define the wave function: the wave packet prescription, which appears to be more satisfactory, at least in the case under study.

161.078 **Kaluza–Klein models: from quantum state to observations.**
U. Bleyer, D.-E. Liebscher, A. G. Polnarev.
5. Seminar on Quantum Gravity, p. 613 – 632 (1991). – See Abstr. 012.004 for the main entry.
Cosmological models in more than 3 + 1 space–time dimensions are considered starting from a quantum state, going through a period of compactification of internal dimensions and inflation of the outer space and ending up with the usual Friedmann stage. The description of the quantum state by the Wheeler–DeWitt equation is discussed. On the background model, for which exact solutions are discussed, the development of mixed metric perturbations described by a tensor with one index in the internal and one in the external space is worked out. The main effect is the increase of these perturbations during the compactification resulting in a possible contribution to actual dark matter density. In this way the authors find a remnant of the higher dimensional stage of the very early universe, which can be linked to observation.

161.079 **Cosmological solutions from supersymmetric Kaluza–Klein theory.**
R. Kerner, R. Balbinot, J. C. Fabris.
5. Seminar on Quantum Gravity, p. 633 – 642 (1991). – See Abstr. 012.004 for the main entry.
The theories of Kaluza–Klein type with extra dimensions provide us with scalar fields when viewed from a four–dimensional space–time; however, without any other fields, they lead to the cosmological collapse. One of the promising additions to the classical Kaluza–Klein theory is a purely geometrical super–symmetric field. The authors discuss here the possibility of avoiding the cosmological collapse, the variation of the gravitational coupling in cosmological solutions, possible Kasner–type solutions and their properties, and the stability properties of such configurations.

161.080 **Space–time dimension of the Universe.**
H. Terazawa.
5. Seminar on Quantum Gravity, p. 643 – 658 (1991). – See Abstr. 012.004 for the main entry.
Among the three paradigms in which the space–time dimension may be determined, mainly discussed is the one in which the space–time dimension can be taken as a "dynamical variable" to be determined by the "Hamilton principle" of minimizing the extended n–dimensional Einstein–Hilbert action of gravitation.

161.081 **On the origin of classical density perturbations in the inflationary Universe.**
R. Brandenberger, R. Laflamme, M. Mijić.
5. Seminar on Quantum Gravity, p. 659 – 677 (1991). – See Abstr. 012.004 for the main entry.
Quantum fluctuations in the de Sitter phase of an inflationary Universe may produce the classical primordial density perturbations required to seed galaxies. Implicit in this theory is a prescription for extracting classical quantities from a quantum theory. The authors briefly review the original prescriptions for obtaining classical density perturbations. They then solve a simple toy model and show that the original prescriptions can be justified by applying decoherence arguments similar to those used in the quantum theory of measurement and in quantum cosmology.

161.082 **COBE's FIRAS: update on refining measurements of the cosmic microwave background radiation spectrum.**
E. S. Cheng, J. C. Mather, R. A. Shafer, D. J. Fixsen, R. E. Eplee Jr., R. B. Isaacman, S. Read, S. S. Meyer, R. Weiss.
Bull. Am. Astron. Soc., Vol. 23, No. 2, p. 896 (1991). Abstract. – See Abstr. 012.008 for the main entry.

161.083 **Limits on cold dark matter models from recent anisotropy measurements of the cosmic background radiation.**
P. Lubin, P. Meinhold, T. Gaier, J. Gunderson.
Bull. Am. Astron. Soc., Vol. 23, No. 2, p. 896 – 897 (1991). Abstract. – See Abstr. 012.008 for the main entry.

161.084 **Angular correlations of faint galaxies.**
G. Bernstein, J. A. Tyson, G. Efstathiou, N. Katz, P. Guhathakurta.
Bull. Am. Astron. Soc., Vol. 23, No. 2, p. 897 (1991). Abstract. – See Abstr. 012.008 for the main entry.

161.085 **Reconstructing the initial spectrum of fluctuations of the universe from the observed nonlinear clustering of galaxies.**
A. J. S. Hamilton, P. Kumar, E. Lu, A. Matthews.
Bull. Am. Astron. Soc., Vol. 23, No. 2, p. 897 (1991). Abstract. – See Abstr. 012.008 for the main entry.

161.086 **Large scale structure from Doppler instability.**
C. J. Hogan.
Bull. Am. Astron. Soc., Vol. 23, No. 2, p. 897 (1991). Abstract. – See Abstr. 012.008 for the main entry.

161.087 **Large scale–free N–body simulations.**
J. M. Gelb.
Bull. Am. Astron. Soc., Vol. 23, No. 2, p. 897 (1991). Abstract. – See Abstr. 012.008 for the main entry.

161.088 **The evolution of void–filled cosmological structures.**
E. Regös, M. J. Geller.
Bull. Am. Astron. Soc., Vol. 23, No. 2, p. 897 – 898 (1991). Abstract. – See Abstr. 012.008 for the main entry.

161.089 **Theory of cosmological perturbations.**
L. P. Grishchuk, Yu. V. Sidorov.
5. Seminar on Quantum Gravity, p. 678 – 688 (1991). – See Abstr. 012.004 for the main entry.
It is shown that the notion of squeezed quantum states arises naturally in the theory of quantized cosmological perturbations. Gravity–wave and matter–density perturbations are studied in some detail. A new phenomenon – desqueezing of matter–density perturbations – is also considered, and its relevance to the so–called Sakharov's oscillations is clarified.

161.090 **Can the anisotropy of microwave background radiation be discovered at present?**
V. G. Gurzadyan, A. A. Kocharyan.
5. Seminar on Quantum Gravity, p. 689 – 702 (1991). – See Abstr. 012.004 for the main entry.
An effect leading to a decrease of perturbations of microwave background radiation after recombination epoch is investigated. Behavior of correlation functions for null geodesics flow enables one to evaluate the role of this effect based purely on geometrical and topological properties of the Universe. Possible anisotropy of MwB in open Friedmann Universe is shown to be decreased due to this effect to a level far below present experimental accuracy. The fractal nature of the large scale structure of the Universe is also discussed.

161.091 **Coarse–graining and backreaction in inflationary and minisuperspace cosmology.**
B. L. Hu.
7. Latin American Symposium on Relativity and Gravitation, p. 33 – 73 (1991). – See Abstr. 012.005 for the main entry.
The author discusses how the effects of coarse–graining and backreaction appear in the context of inflationary and quantum cosmology. He presents the coarse–grained effective action (CGEA) method for treating the backreaction of the averaged environmental fields on the system of interest. He introduces the idea of scaling to depict inflation, and obtains via the CGEA a set of renormalization group equations for the effective coupling constants of the system fields, both in the case of new or chaotic and of stochastic inflation. Then the coarse–graining idea is applied to explore the validity of the minisuperspace approximation commonly assumed in quantum cosmology. By regarding the scale factor of the Robertson–Walker spacetime and the homogeneous mode of an interacting scalar field as the system (the minisuperspace) and the inhomogeneous modes as the environment, the CGEA is used to compute the backreaction of the latter on the former and it is shown that the minisuperspace approximation is valid only when this backreaction is small. The overall effect of the environment is to impart to the dynamics of the system some dissipative behavior. This has important repercussions on the issue of the initial conditions and the semiclassical limit of quantum cosmology.

161.092 **Spacetime and causal sets.**
R. D. Sorkin.
7. Latin American Symposium on Relativity and Gravitation, p. 150 – 173 (1991). – See Abstr. 012.005 for the main entry.
The causal set is a possible discrete substratum for spacetime. This idea is introduced and some aspects of causal set kinematics are presented, most of them relating to the question of how the discrete order defining the causal set corresponds to the geometrical structures of continuous spacetime. In particular a statistical notion of "fractal dimension" is developed in some detail. The authors conclude with some preliminary remarks on dynamics, and a speculation on the cosmological constant.

161.093 **Thickenings and conformal gravity.**
C. LeBrun.
Commun. Math. Phys., Vol. 139, No. 1, p. 1 – 43 (Jul 1991).
A twistor correspondence is given for complex conformal space–times with vanishing Bach and Eastwood–Dighton tensors; when the Weyl curvature is algebraically general, these equations are precisely the conformal version of Einstein's vacuum equations with cosmological constant. This gives a fully curved version of the linearized correspondence of Baston and Mason.

161.094 **A limit on the cosmological constant.**
R. G. Carlberg.
Astrophys. J., Vol. 375, No. 2, p. 429 – 431 (10 Jul 1991).
The fraction of galaxies which are merging is a rapidly increasing function of redshift, which for $z \lesssim 0.5$ varies as $(1+z)^m$, where m is quite sensitive to both the density parameter, Ω, and the cosmological constant, Λ. The index m has weak dependencies on the details of the mass and structure of the merging galaxies but does depend on the rate at which the mean velocity dispersion of the universe grows with time; however, these parameters are strongly constrained by current epoch observations. The current crude estimates of m from a variety of astronomical sources give $m \cong 4.5 \pm 1.0$. As a consequence, the current merger observations favor $|\Lambda| \lesssim 1.5 \cdot H_0^2$, $\Omega = 1.0 \pm 0.5$, in a zero curvature universe. The merger data is marginally incompatible with $\Omega = 0.1$ and any Λ. Forthcoming data on merging among moderate redshift galaxies will significantly reduce the uncertainty in the observational index m.

161.095 **Cosmic rays and cosmological microwave background fluctuations.**
A. J. Banday, M. Giler, B. Szabelska, J. Szabelski, A. W. Wolfendale.
Astrophys. J., Vol. 375, No. 2, p. 432 – 438 (10 Jul 1991).
One of the most important topics in modern cosmology concerns the search for anisotropies in the cosmic microwave background radiation. Searches for these fluctuations have been made in the field of radio astronomy, and tentative claims have been made, specifically at a frequency near to 10 GHz, for fluctuations on a scale of 8°. However, the Galactic synchrotron radiation can mimic or mask a cosmological signal, and the present work examines the extent to which the fluctuations in this local component, a component due to cosmic rays, are understood. The authors conclude that at 10 GHz, although the observed fluctuations are of similar magnitude to those predicted for Galactic synchrotron radiation, the actual temperature profiles do not agree. Thus, the nature of the detected fluctuations is not understood, although what can be stated is that there is no evidence to prove the existence of cosmic microwave background anisotropies. A prognosis is made for future searches for CMB fluctuations.

161.096 **Polytropic gravitational wave production cosmological model.**
J. Cervantes–Cota, P. Chauvet, J. Klapp.
7. Latin American Symposium on Relativity and Gravitation, p. 294 – 300 (1991). See Abstr. 012.005 for the main entry.
The authors have shown that the polytropic equation of state can be interpreted to represent in the cosmological context, the production of gravitational waves with relative high frequency, presumably coming from astrophysical events taking place within the galaxies. The cosmological models resulting represent, phenomenologically, a universe with two couple fluids, matter and gravitational radiation, whose real interaction would be otherwise difficult to portray. Here, the authors present some consequences of the dynamics of these models.

161.097 **Non–Gaussian temperature fluctuations in the cosmic microwave background sky from a random Gaussian density field.**
R. Scaramella, N. Vittorio.
Astrophys. J., Vol. 375, No. 2, p. 439 – 442 (10 Jul 1991).
The authors point out that the cosmic microwave background (CMB) temperature fluctuations, observed in the microwave sky with a large antenna beam, may not be Gaussian distributed, even if the primordial matter density fluctuations constitute a random Gaussian field. This effect is due to a dominance of low–order multipoles of the CMB anisotropy pattern and is then relevant for galaxy formation models with large–scale power, which produce CMB angular anisotropies through the familiar Sachs–Wolfe effect. The authors focus on the most popular case of a flat universe with scale–invariant, Gaussian distributed density fluctuations. In this case 30% of the cosmic observers would reject the hypothesis that CMB temperature fluctuations in their own microwave sky are Gaussian distributed at least at the 99% confidence level. The authors also give numerical bounds, at the 95% confidence level, for the skewness and kurtosis of the sky pixel temperature fluctuation distribution: these bounds will possibly be useful in discriminating against scenarios with primordial, Gaussian density fluctuations.

161.098 **Resonant neutrino oscillations in the early Universe.**
J. C. D'Olivo, M. Torres.
7. Latin American Symposium on Relativity and Gravitation, p. 356 – 361 (1991). – See Abstr. 012.005 for the main entry.
The authors investigate the effect of neutrino oscillations in the epoch previous to nucleosynthesis. The effective Mikheyev–Smirnov–Wolfenstein evolution equation for the neutrino amplitudes is derived from the covariant Dirac equation. The expansion of the Universe is incorporated through the time dependence of the neutrino energy and the background density and the conditions to have resonant flavor transformations are analyzed under the assumption of a non–null lepton asymmetry.

161.099 **Perturbations of the Robertson–Walker space: multicomponent sources and generalized gravity.**
J. Hwang.
Astrophys. J., Vol. 375, No. 2, p. 443 – 462 (10 Jul 1991).
The author presents cosmological perturbation equations in the Robertson–Walker background space, considering multicomponent fluid sources with interactions between them. He generalizes the perturbation equations applicable to a class of generalized gravity theories with multicomponent fluids and fields as sources. Equations are derived from the covariant equations and presented, following Bardeen's recent suggestion, using the raw kinematic variables of the normal frame with an unspecified gauge. In this way, the equations are easily adaptable to various gauge conditions and also to various gauge–invariant formulations. Relations between different gauges and with gauge–invariant formulations are explained. Since the energy–momentum tensor of the fluid part is separately conserved even in the generalized gravity theories, the component equations follow from the conservation equations of each component. The author presents separately the vorticity and the gravitational wave perturbation equations in the generalized gravity theories with multicomponent sources. Although various apsects of the cosmological perturbations are discussed, the main concern is the new formulation of perturbation equations applicable to a broad range of generalized gravity theories, including general relativistic formulation with multicomponent sources.

161.100 **Perturbations in a singularity–free cosmological model.**
M. Israelit, N. Rosen.
Astrophys. J., Vol. 375, No. 2, p. 463 – 472 (10 Jul 1991).
This study discusses a singularity–free model of the universe previously proposed by the authors. According to this model, at the beginning of the present expansion phase the universe was filled with prematter for which $P = -\varrho$, and $\varrho \sim \varrho_p$, the Planck density, and the prematter went over to ordinary matter in the course of time. Perturbations are treated in the comoving gauge with the help of the Einstein equations in the first–order linear approximation. Two approaches are considered. In the first it is assumed that there are initial perturbations in the prematter. These grow tremendously in the course of time. In the second it is assumed that the prematter has no perturbations, but that perturbations develop as the prematter is converted to ordinary matter. The second approach seems preferable.

161.101 **Perturbations in a singularity–free cosmological model. II. The synchronous gauge.**
M. Israelit.
Astrophys. J., Vol. 375, No. 2, p. 473 – 478 (10 Jul 1991).
In the framework of a singularity–free cosmological model of the universe, density perturbations in the early periods are treated. The behavior of perturbations is considered in the synchronous gauge. The results agree with those previously obtained in the comoving gauge. It is found that the singularity–free cosmological model may serve as a background for the development of pregalactic perturbations.

161.102 **Massive SIMPs as dark matter: cosmological generation and the aftermath.**
O. G. Benvenuto, J. E. Horvath, H. Vucetich.
7. Latin American Symposium on Relativity and Gravitation, p. 362 – 370 (1991). – See Abstr. 012.005 for the main entry.
The dark component of the Universe (commonly assumed to be some kind of weakly interacting stuff) might be composed of strongly interacting massive particles (SIMPs). The authors address in the present work a cosmological scenario for the generation and early evolution of a SIMP gas of a specific type. It is shown that the emerging features are in good agreement with the "conventional wisdom" in this field, and that such a hadronic nature of the dark matter fits the recent remarkable observations tracing the dark matter distribution in galaxy clusters.

161.103 **Time variation of fundamental constants, preferred reference systems and CPT invariance.**
P. D. Sisterna, H. Vucetich.
7. Latin American Symposium on Relativity and Gravitation, p. 371 – 378 (1991). – See Abstr. 012.005 for the main entry.
The authors show that the measured parameters of the $K^0 - K^{-0}$ system impose strong constraints on the time variation of the s and d quark masses, and on the scale parameter of QCD Λ. These results provide information about the cosmological evolution of dilaton fields in superstring theories.

161.104 **Mapping the Universe.**
M. J. Geller, J. P. Huchra.
Sky Telesc., Vol. 82, No. 2, p. 134 – 139 (Aug 1991).
Astronomers grapple with the unexpected revelation that galaxies are organized into structures spanning hundreds of millions of light–years.

161.105 **Empty voids in the expanding universe.**
A. Chamorro.
7. Latin American Symposium on Relativity and Gravitation, p. 379 – 384 (1991). – See Abstr. 012.005 for the main entry.

161.106 **The origin of cosmic structure.**
D. N. Schramm.
Sky Telesc., Vol. 82, No. 2, p. 140 – 145 (Aug 1991).
How galaxies and galaxy clusters formed is turning out to be one of the most vexing mysteries of the cosmos.

161.107 **Creation of scalar and spin 1/2 particles in an expanding cosmological universe.**
V. M. Villalba, U. Percoco.
7. Latin American Symposium on Relativity and Gravitation, p. 385 – 391 (1991). – See Abstr. 012.005 for the main entry.
The authors find exact solutions to the Klein–Gordon and Dirac wave equations in an expanding Robertson–Walker universe with a positive cosmological constant associated to a universe filled with radiation. They analyze the process of particle creation for the massless solutions.

161.108 **Problems in the minisuperspace reduction of the Hilbert action.**
J. Guven, M. Rosenbaum, M. P. Ryan Jr.
7. Latin American Symposium on Relativity and Gravitation, p. 392 – 396 (1991). – See Abstr. 012.005 for the main entry.
The authors consider a perturbed Bianchi type I cosmological model as an example of a system in which a metric with a given symmetry inserted into the Hilbert action gives the correct Einstein equations. It is shown why the special case of a $k = 0$ Friedmann–Robertson–Walker model fails to give the Einstein equations.

161.109 **Decoherence and backreaction in quantum cosmology.**
J. P. Paz, S. Sinha.
7. Latin American Symposium on Relativity and Gravitation, p. 397 – 402 (1991). – See Abstr. 012.005 for the main entry.
The authors study the correlations that define the semiclassical decohering histories of some relevant cosmological variables. They consider the interaction with the environment formed by unobserved ("irrelevant") degrees of freedom and analyze under what conditions the semiclassical coarse grained histories obey the so–called semiclassical Einstein's equations. These equations are shown to be valid only as a description of adiabatic regions of maximally decoherent histories.

161.110 **Protogalactic mergers and galaxy number counts.**
M. B. Aufderheide, G. J. Mathews, D. N. Schramm.
Bull. Am. Astron. Soc., Vol. 23, No. 2, p. 931 (1991). Abstract. – See Abstr. 012.008 for the main entry.

161.111 **Anisotropy dissipation in the very early Universe.**
E. Calzetta.
7. Latin American Symposium on Relativity and Gravitation, p. 403 – 407 (1991). – See Abstr. 012.005 for the main entry.
Applying projection techniques to the Wheeler–DeWitt equation describing a Bianchi type I space time with both quantum and classical matter, the author shows that the Wave Function of the Universe will peak around the isotropic configuration for large Universes for a wide class of initial conditions on the singularity. He presents some new results on the issue of anisotropy dissipation through particle creation processes taking place in the Very Early Universe.

161.112 **The square root of the Wheeler–DeWitt equation in cosmology.**
J. Socorro, O. Obregón, A. Macías.
7. Latin American Symposium on Relativity and Gravitation, p. 408 – 415 (1991). – See Abstr. 012.005 for the main entry.
The authors use the $N = 1$ supergravity to find the square root of the Bianchi type IX cosmological models. As a result they obtain a system of four coupled Dirac–like equations for the state function that characterizes the universe, where an operator valued potential is present. In each equation the resulting component ψ_A of the state function is an eight–dimensional vector. Positive and negative energy solutions can be related by changing the sign of the potential.

161.113 **Large–scale inhomogeneities in the Universe and fluctuation spectra.**
B. I. Gnatyk, V. N. Lukash, B. S. Novosyadlyj.
Sov. Astron. Lett., Vol. 17, No. 4, p. 278 – 283 (Jul 1991). Current Physics Microform No.: 9202X0454. English translation of Pis'ma Astron. Zh., Tom 17, No. 8, p. 659 – 670 (Jul 1991).
The requirements are analyzed for the density fluctuation spectrum of matter from the standpoint of correspondence between predicted and observed characteristics of the large–scale structure of the Universe. A spectrum is found for which the computed correlation functions of galaxies and galactic clusters, bulk motion, and Great Attractor models correspond to the observed values, and which is consistent with present constraints on the anisotropy of the microwave background radiation. It is close to the cold dark matter (CDM) spectrum on scales $k^{-1} \leqslant 3h^{-1}$Mpc, $k^{-1} \geqslant 10^3h^{-1}$Mpc, but has additional power $\sigma_z \approx 0.2$ in the range $3 \leqslant k^{-1} \leqslant 10^3h^{-1}$Mpc. The amplitude of the spectrum and its slope in this range are given, to within the errors, by the observed characteristics of large–scale structure. It is shown that halving the upper limit of the Kholod simulation at scales of about 2° will make it possible to record this spectrum.

161.114 **Flat edge–on galaxies on the Tully–Fisher diagram. I.**
I. D. Karachentsev.
Sov. Astron. Lett., Vol. 17, No. 4, p. 283 – 289 (Jul 1991). Current Physics Microform No.: 9202X0459. English translation of Pis'ma Astron. Zh., Tom 17, No. 8, p. 671 – 683 (Aug 1991).
For a sample of 114 northern galaxies with angular diameters $a > 2'$ and apparent axial ratios $a/b > 7$, the author considers the relationships between the following global parameters: amplitude of internal motions W, blue luminosity L_B, linear diameter A_B (or A_R), and mass of neutral hydrogen M_{HI}. Taken as distance indicators for a given value of W, these parameters prove to be of approximately equal effectiveness. Beyond the Local Supercluster, the linear diameter of a flat galaxy yields a standard deviation of ~25% in the distance estimate. The ratio of total mass to luminosity of a flat galaxy is seen to increase with a decrease in surface brightness.

161.115 **5–dimensional scalar–tensor cosmologies as 4–dimensional GR universes.**
J. Benitez, O. Obregon.
7. Latin American Symposium on Relativity and Gravitation, p. 475 – 479 (1991). – See Abstr. 012.005 for the main entry.
The 5–dimensional scalar–tensor Jordan–Brans–Dicke cosmologies for the Bianchi type I and for FRW metric are interpreted as 4–dimensional G.R. universes. For Bianchi I effective pressure and energy density are present. The equation of state is barotropic. Two kinds of solutions are found, one is a power law, the other an exponential one. In all cases the fifth dimension decreases approaching zero and no crack of doom is present.

161.116 **Spatially homogeneous vacuum space–time cosmological solutions for the Jordan Brans–Dicke theory in Bianchi–type universes.**
P. Chauvet, J. Cervantes–Cota, H. N. Núñez–Yépez.
7. Latin American Symposium on Relativity and Gravitation, p. 480 – 486 (1991). – See Abstr. 012.005 for the main entry.
Using a time parametrization, the authors obtain and solve a general nonlinear differential equation whose solutions are the solutions for the vacuum cosmological Bianchi–type models in Jordan–Brans–Dicke theory. Several of the solutions presented here are new.

161.117 **Anisotropic degree measurement for Bianchi type models.**
P. Chauvet, J. Cervantes–Cota, H. N. Núñez–Yépez.
7. Latin American Symposium on Relativity and Gravitation, p. 487 – 491 (1991). – See Abstr. 012.005 for the main entry.
The authors study the possible isotropization of cosmological solutions with matter for the field equations of the Jordan–Brans–Dicke theory in terms of a general, arbitrary but natural, definition of isotropization in the context of spatially homogeneous universes of Bianchi type.

161.118 **Cosmic strings of scalar fields: Brans–Dicke case.**
A. Noé Morales, O. Pimentel.
7. Latin American Symposium on Relativity and Gravitation, p. 492 – 496 (1991). – See Abstr. 012.005 for the main entry.
In the theory of gravitation of Brans and Dicke, exact solutions representing infinite straight strings are found. The gravitational effects of this string solution are similar to those obtained in General Relativity.

161.119 **Limits to the cosmic background fluctuations between angular scales 10″ to 60″.**
R. B. Partridge, J. D. Lowenthal, E. B. Fomalont, R. A. Windhorst.
Bull. Am. Astron. Soc., Vol. 23, No. 2, p. 963 (1991). Abstract. – See Abstr. 012.008 for the main entry.

161.120 **The microwave background in a Voronoi model universe.**
V Icke.
Bull. Am. Astron. Soc., Vol. 23, No. 2, p. 963 – 964 (1991). Abstract. – See Abstr. 012.008 for the main entry.

161.121 **Arcminute fluctuations of microwave background due to hot gas in clusters of galaxies. Dependence on cosmology and cluster evolution.**
M. Markevitch, W. Forman, C. Jones, R. Sunyaev (*R. Syunyaev*).
Bull. Am. Astron. Soc., Vol. 23, No. 2, p. 964 (1991). Abstract. – See Abstr. 012.008 for the main entry.

161.122 **The topology of incomplete deep sky surveys.**
K. G. Dominik, A. L. Melott.
Bull. Am. Astron. Soc., Vol. 23, No. 2, p. 964 (1991). Abstract. – See Abstr. 012.008 for the main entry.

161.123 **Void statistics of the CfA redshift survey.**
M. S. Vogeley, M. J. Geller, J. P. Huchra.
Bull. Am. Astron. Soc., Vol. 23, No. 2, p. 964 (1991). Abstract. – See Abstr. 012.008 for the main entry.

161.124 **Crude geodesy on the hypersphere and its honeycomb.**
L. E. Sepulveda.
Bull. Am. Astron. Soc., Vol. 23, No. 2, p. 964 (1991). Abstract. – See Abstr. 012.008 for the main entry.

161.125 **Abell 2218 and the Hubble constant.**
M. Birkinshaw, J. P. Hughes.
Bull. Am. Astron. Soc., Vol. 23, No. 2, p. 973 (1991). Abstract. – See Abstr. 012.008 for the main entry.

161.126 **Inflation–driven string instabilities ... and the other way around.**
G. Veneziano.
Helv. Phys. Acta, Vol. 64, No. 6, p. 877 – 889 (Aug 1991). Paper presented at the Conference on Theoretical Particle Physics at the Beginning of the Nineties, Geneva (Switzerland), 9 – 10 Nov 1991.
The author reviews recent work showing that inflation can drive Jeans–like string instabilities, while it is not so easy for such strings to drive inflation.

161.127 **Gaugino condensation and the cosmological implications of the hidden sector.**
J. E. Kim, H. P. Nilles.
Phys. Lett. B, Vol. 263, No. 1, p. 79 – 85 (4 Jul 1991).
The authors investigate the cosmological implications of models in which the breakdown of supersymmetry proceeds via gaugino condensation in a hidden sector, only gravitationally coupled to the observable sector. Semistable particles in the hidden sector are forbidden in the range between 100 eV and a few times 10^{10}GeV. These constraints derived from the restrictions on energy density and nucleosynthesis are shown to have consequences for model building. As a byproduct of these investigations the authors point out a particularly attractive solution of the μ–problem and the strong CP–problem with a composite hidden sector axion.

161.128 **What kinds of perturbation spectra can be produced by inflation?**
V. F. Mukhanov, M. I. Zelnikov.
Phys. Lett. B, Vol. 263, No. 2, p. 169 – 175 (11 Jul 1991).
The problem of the generation of the nonflat perturbation spectra in inflationary models is discussed. The general limitations on the parameters of the spectra are obtained. It is shown that Sakharov's modulation of the spectra for the adiabatic entropy perturbations, and gravitational waves can occur in the double inflation models.

161.129 **Baryogenesis from collapsing topological defects.**
R. H. Brandenberger, A. C. Davis, M. Hindmarsh.
Phys. Lett. B, Vol. 263, No. 2, p. 239 – 244 (11 Jul 1991). Also published as report BROWN–HET—792, 8 p. (Feb 1991).
Topological defects in or coupled to grand unified gauge theories can trap a substantial fraction of the energy density of the universe. When the defects collapse, they release energy in the form of superheavy particles. In theories with CP and baryon number violation, this may lead to an enhanced net baryon to entropy ratio. The authors explicitly study cosmic string loop, texture and global monopole annihilation.

161.130 **Dark matter and the observability of the electroweak transition.**
M. Sher.
Phys. Lett. B, Vol. 263, No. 2, p. 255 – 258 (11 Jul 1991).
In a recent paper, Dimopoulos et al. argued that the standard calculation of the abundance of a dark matter particle would not be affected by the electroweak transition. They considered a model with a fermion whose mass arises entirely through its interaction with the Higgs boson. The author considers somewhat more complicated, but much more realistic, models, in which the dark matter particle may also be a scalar, in which a more realistic range of couplings is chosen, and in which the possibility of bare mass terms is included. It is shown that for almost all of parameter space, the electroweak transition will have a negligible effect; a (barely) significant effect arises only if the particle is a scalar with a huge coupling and a bare mass term in excess of 1300 GeV.

161.131 **Cosmological consequences of high–frequency oscillations of Newton's constant.**
F. S. Accetta, P. J. Steinhardt.
Phys. Rev. Lett., Vol. 67, No. 3, p. 298 – 301 (15 Jul 1991). Current Physics Microform No.: 9108D2136.
The authors show that high–frequency, small–amplitude oscillations of Newton's constant G can dramatically alter cosmology even if the frequency is very high compared to the expansion rate. For example, it is possible to have a spatially flat universe in which dynamical tests of Ω – tests which attempt to directly measure the mass density of the universe – obtain values less than unity ($\Omega = 0.1 - 0.3$). The cosmological effects can be obtained in a frequency – amplitude range allowed by all known constraints on G and its time derivative.

161.132 **The distribution of clusters of galaxies within 300 Mpc h^{-1} and the crossover to an isotropic and homogeneous universe.**
R. Scaramella, G. Vettolani, G. Zamorani.
Astrophys. J., Lett., Vol. 376, No. 1, p. L1 – L4 (20 Jul 1991).
The authors present for the first time firm evidence for the spatial convergence of a cosmological dipole (i.e., peculiar acceleration). The objects used in the analysis are clusters of galaxies from the Abell and ACO catalogs. The cluster dipole increases with distance up to $\sim 180h^{-1}$ Mpc, then bends over and stays flat well within the depth of the sample, while the monopole continues to grow linearly with distance up to the maximum sample depth, $300h^{-1}$ Mpc, where it reaches a level which is $\gtrsim 5$ times the dipole level. The still–growing monopole proves that the convergence of the dipole is real and is not due to depth

incompleteness of the sample. The behavior of the two quantities is expected in an asymptotically homogeneous and isotropic universe and gives three–dimensional (all–sky) support to a FRW model, besides the two–dimensional one from cosmic backgrounds. After suitable corrections for the differences in the two catalogs of clusters of galaxies (Abell and ACO) limited at $|b| > 20°$, the asymptotic dipole direction appears to be stable pointing in a direction $\approx 25°$ off the CMB dipole apex. By a posteriori pushing the limit at $|b| > 10°$ to include the effects of the Perseus cluster, the authors find that, after correction for the missing Galactic strip, the dipole direction is within only 6°– 10° of the CMB apex.

161.133 **Gravitational clustering of Gaussian density fluctuations and the origin of hierarchy correlations.**
A. Kashlinsky.
Astrophys. J., Lett., Vol. 376, No. 1, p. L5 – L9 (20 Jul 1991).
The author presents a theory of gravitational clustering based only on (1) the assumption that the primordial density fluctuations (PDF) were Gaussian and (2) identifying galaxies and groups and clusters of galaxies with regions of PDF whose turnaround time was less than the age of the universe. It is then shown: (1) properties of the hierarchy thus produced depend uniquely on the power spectrum P(k) of PDF; (2) the resultant distribution of galaxies and groups and clusters is non–Gaussian; (3) two–point and higher order correlations can be used to constrain P(k) on scales from 1 to $(50 - 100)h^{-1}$ Mpc on the basis of the available data; (4) the present theory successfully reproduces both the observed two– and three–point correlation functions for clusters of galaxies for $n = -1$ spectrum; the data are, however, inconsistent with $n > 0$ on large scales as required in the conventional cold dark matter model; (5) the hierarchy constructed here specifies unique higher order correlations for various hierarchy members.

161.134 **The fragmentation of "pancakes" in a dark matter–dominated universe.**
W. Yuan, J. M. Centrella, M. L. Norman.
Astrophys. J., Lett., Vol. 376, No. 2, p. L29 – L32 (1 Aug 1991). With plate L5.
Galaxy formation is studied within the context of the pancake model. The collapse, cooling, and fragmentation of a massive cloud composed of baryons and collisionless dark matter in an FRW universe with $\Omega_0 = 1$, $\Omega_b = 0.1$, and $H_0 = 75\ \mathrm{km\,s^{-1}Mpc^{-1}}$ is modeled using two–dimensional numerical simulations. Perturbations having a Poisson spectrum are imposed on the cloud at the start of the calculation to provide a noise source for fragmentation. The Zeus–2D hydrodynamics code with cosmological and cooling terms is coupled to a cloud–in–cell code that evolves the dark matter. A nonuniform grid is used in the direction of collapse to resolve the central thin cooling layer. The gas cloud collapses to a pancake at redshift $z \sim 5$. It cools and undergoes instabilities, causing it to fragment into protogalactic objects on a scale of ~12.5 kpc, which then merge into larger objects.

161.135 **QSO absorption systems and the origin of the ionizing background at high redshift.**
P. Madau.
Astrophys. J., Lett., Vol. 376, No. 2, p. L33 – L36 (1 Aug 1991).
The author considers two candidates for the sources of the ionizing background: primeval galaxies and AGN–type objects. The spectrum of the diffuse radiation field is computed taking into account the opacity of intervening Lyα clouds and Lyman limit systems. The author assesses the possibility that the QSO heavy–element absorption systems are photoionized by a metagalactic flux which is dominated by the UV radiation from hot stars in young galaxies. Grids of photoionization models for metal–line systems are presented, and it is argued that the relative column densities of the most abundant ions predicted by the models are compatible with the observational constraints. The inferred physical properties of the absorbing clouds are described. Given that the observed QSOs cannot provide the required number of ionizing photons at high redshift, the results obtained support a scenario in which young, star–forming galaxies are the main sources of the UV radiation background at early epochs.

161.136 **Self–consistent cosmological perturbations from thermal field theory.**
U. Kraemmer, A. Rebhan.
Phys. Rev. Lett., Vol. 67, No. 7, p. 793 – 796 (12 Aug 1991).
Current Physics Microform No.: 9109C2134.
Finite–temperature field theory is used for a self–consistent and manifestly gauge–invariant derivation of the effective field equations governing the evolution of adiabatic perturbations of a homogeneous, isotropic cosmological model dominated by an ultrarelativistic plasma of weakly interacting particles in thermal equilibrium. Solutions are obtained in the form of rapidly converging power–series expansions.

161.137 **Large–scale structure from wiggly cosmic strings.**
T. Vachaspati, A. Vilenkin.
Phys. Rev. Lett., Vol. 67, No. 9, p. 1057 – 1061 (26 Aug 1991).
Current Physics Microform No.: 9110C0125.
Recent simulations of the evolution of cosmic strings indicate the presence of small–scale structure on the strings. The authors show that wakes produced by such "wiggly" cosmic strings can result in the efficient formation of large–scale structure and large streaming velocities in the universe without significantly affecting the microwave background isotropy. It is also argued that the motion of strings will lead to the generation of a primordial magnetic field. The most promising version of this scenario appears to be the one in which the universe is dominated by light neutrinos.

161.138 **Supersymmetric Bianchi type IX cosmology.**
R. Graham.
Phys. Rev. Lett., Vol. 67, No. 11, p. 1381 – 1383 (9 Sep 1991).
Current Physics Microform No.: 9110E1805.
The homogeneous cosmology of Bianchi type IX (mixmaster cosmology) is quantized in a supersymmetric form, allowing the Dirac–type square root of its Wheeler–DeWitt equation to be taken. A supersymmetric exact solution is given in closed form.

161.139 **Bianchi type V and VI_0 cosmological models in Lyra geometry.**
T. Singh, G. P. Singh.
Astrophys. Space Sci., Vol. 182, No. 2, p. 189 – 200 (Aug 1991).
The Bianchi type V and VI_0 models have been investigated in Lyra geometry in both the cases with the gauge function β is a time dependent and β = constant. The physical behaviour of these models has been discussed.

161.140 **On slowly–rotating cosmological viscous–fluid model universes in general relativity.**
K. Maniharsingh.
Astrophys. Space Sci., Vol. 182, No. 2, p. 207 – 219 (Aug 1991).
The dynamics of a slowly–rotating cosmological viscous–fluid universe is investigated and the rotational perturbations of such models are studied in order to substantiate the possibility that the Universe is endowed with slow rotation, in the course of presentation of some new analytic solutions. Three different cases are taken up in which the nature and role of the metric rotation $\Omega(r,t)$ as well as that of the matter rotation $\omega(r,t)$ are discussed. The periods of physical validity of some of the models and the effect of viscosity on the rotational motion are also found out. Rotating models which are expanding as well are obtained, where in all the cases the rotational velocities are found to decay with the time; and these models may be taken as good examples of real astrophysical situations.

161.141 **Can Brans–Dicke universes in five dimensions be interpreted as 4–dimensional general relativity cosmologies?**
J. Benítez, A. N. Morales, O. Obregón.
Astrophys. Space Sci., Vol. 182, No. 2, p. 221 – 225 (Aug 1991).
The 5–dimensional Jordan–Brans–Dicke cosmologies in vacuum are found for the Bianchi type I metric, their relation with general relativity cosmologies is studied. Two solutions are possible, both produce effective pressure and energy density in the 4–dimensional G.R.–universes. One is a power–law relation, with two cases, the first one is for $p_{eff} = \varrho_{eff}$ and the other $p_{eff} = \alpha\, \varrho_{eff}$ $(-1 < \alpha < 1)$ has a behaviour as the open flat universe. The second solution is an exponential only valid for $p_{eff} = -\varrho_{eff}$. In all cases the three–space expansion reaches infinity as $t \to \infty$ and the fifth dimension can be made to decrease approaching zero. The scalar field can increase or decrease with time.
¢< <

161.142 **A theory for expanding, large–scale, isolated, spherically–symmetric density perturbations with null peculiar velocity field.**
E. Andriani, R. Caimmi.
Astrophys. Space Sci., Vol. 182, No. 2, p. 261 – 280 (Aug 1991).
Exact theory for expanding density perturbations under the following assumptions: (i) Friedmann universes made of dust only; (ii) spherically symmetric perturbations; (iii) isolated perturbations; (iv) null peculiar velocity field; is reviewed and extended. The overdensity equation is derived and the related solution is determined concerning both mean overdensity and local overdensity, and including both the growing and the decreasing mode, in terms of non–dimensional variables normalized to the initial configuration. A first–order and zero–order approximation to the exact theory is also performed in view of applications to specific problems, e.g., the acquisition of angular momentum by tidal torques. The idealized case presented here make a necessary reference in dealing with more general situations involving both analytical and numerical approaches.

161.143 **Some Bianchi–type cosmological models in a new scalar–tensor theory.**
T. Singh, A. K. Agrawal.
Astrophys. Space Sci., Vol. 182, No. 2, p. 289 – 312 (Aug 1991).
Bianchi type I, III, V, VI_0, and Kantowski-Sachs type models have been investigated in a scalar tensor theory developed by Saez and Ballester (1985) and Saez (1985). The dynamical behaviour of the models has also been analyzed.

161.144 **The accuracy of galaxy masses from the timing argument.**
T. L. Kroeker, R. G. Carlberg.
Astrophys. J., Vol. 376, No. 1, p. 1 – 7 (20 Jul 1991).
The accuracy of the timing argument mass estimate is examined using binary galaxies found in an N–body simulation of an $\Omega = 1$ cold dark matter (CDM) cosmology. Masses are calculated assuming the binaries are bound, for both purely radial orbits and orbits with angular momentum. The timing masses, M_t, are comparable to the total mass within two spheres, centered on each galaxy, of radii one half the separation between the two galaxies. These "two–sphere" masses, M_{2s}, are estimated with little systematic bias for the angular momentum calculation. The corresponding radial orbit timing masses on average underestimate M_{2s} by a factor of 1.7. The standard deviation of $\log_{10}(M_t/M_{2s})$ ranges from a minimum of 15% for nonradial approaching orbits to a maximum of 34% for approaching binaries assuming radial orbits and the same number of complete orbits as the best angular momentum timing mass. Eighty percent of the binary galaxies present in the simulation have completed less than one and a half orbit. The lack of binaries in advanced stages of their orbital history probably indicates that isolated binary galaxies merge quickly. This analysis suggests that the radial orbit timing mass of the M31 – Galaxy system is quite likely 50% of the two sphere mass, M_{2s} with a 1σ range of 0.2 to 1.3 about the mean value, if the system is bound. The mass within a radius of 50 kpc (assuming a Hubble constant of 50 km $s^{-1}$$Mpc^{-1}$) is of order 1/10 the timing mass for wide binaries that are similar to the M31 – Galaxy system. The authors conclude that the timing argument data for the M31 – Galaxy pair is consistent with an extensive massive halo in an $\Omega = 1$ universe.

161.145 **Time evolution of scalar field in spatially flat Robertson–Walker space–time.**
C. F. Lo.
Phys. Scr., Vol. 44, No. 3, p. 217 – 221 (Sep 1991).
The discrete lattice formalism is applied to study the time evolution of free scalar field in spatially flat Robertson–Walker space–time. Using the evolution operator method, we obtain the exact form of the time evolution operator of the system which, in turn, enables us to find the exact wave functions readily. It is found that the time evolution operator is a direct product of single–mode squeeze operators, and thus the initial vacuum state evolves into a direct product of single–mode squeezed vacuum states at a later time. In other words, the time evolution of free scalar field in spatially flat Robertson–Walker space–time is a collective single–mode "squeezing" process.

161.146 **Superclusters and pencil–beam surveys: the origin of large–scale periodicity.**
N. A. Bahcall.
Astrophys. J., Vol. 376, No. 1, p. 43 – 45 (20 Jul 1991).
It is shown that the apparent $\sim 130h^{-1}$ Mpc periodic distribution of galaxies recently observed in narrow pencil–beam surveys originates in known large–scale superclusters. The superclusters, covering the high–latitude sky to $z \leqslant 0.08$, show the same peaks that appear in the narrow–beam survey. These results suggest a network structure of large ($\sim 100h^{-1}$ Mpc) supercluster surfaces (walls) surrounding low–density regions. The apparent periodicity is expected to diminish when averaged over different directions.

161.147 **Primordial nucleosynthesis redux.**
T. P. Walker, G. Steigman, D. N. Schramm, K. A. Olive, H.–S. Kang.
Astrophys. J., Vol. 376, No. 1, p. 51 – 69 (20 Jul 1991).
The latest nuclear reaction cross sections (including the most recent determinations of the neutron lifetime) are used to recalculate the abundances of deuterium, ^3He, ^4He, and ^7Li within the framework of primordial nucleosynthesis in the standard (homogeneous and isotropic) hot, big bang model. The observational data leading to estimates of (or bounds to) the primordial abundances of the light elements is reviewed with an emphasis on ^7Li and ^4He. A comparison between theory and observation reveals the consistency of the predictions of the standard model and leads to bounds to the nucleon–to–photon ratio: $2.8 \leqslant \eta_{10} \leqslant 4.0$ ($\eta_{10} \equiv 10^{10} n_B/n_\gamma$).This constrains the baryon density parameter to $\Omega_B \cdot h_{50}^2 = 0.05 \pm 0.01$ (with the Hubble parameter $H_0 = 50h_{50}$ km $s^{-1}$$Mpc^{-1}$). These bounds imply that the bulk of the baryons in the universe are dark if $\Omega_{TOT} = 1$ and would require that the universe be dominated by nonbaryonic matter. An upper bound to the primordial mass fraction of ^4He, $Y_p \leqslant 0.240$, constrains the number of light (equivalent) neutrinos to $N_\nu \leqslant 3.3$. Alternatively, for $N_\nu = 3$, the predicted primordial abundance of ^4He is limited to: $0.236 \leqslant Y_p \leqslant 0.243$ (for a neutron lifetime $882 \leqslant \tau_n \leqslant 896$ s).

161.148 **The relativistic cosmology under the harmonic condition. I.**
Huang Chaoguang, Zhou Peiyuan.
Sci. China, Ser. A, Vol. 33, No. 8, p. 966 – 982 (Aug 1990).
The harmonic condition is applied to the motion of galaxies in relativistic cosmology. The average distribution of matter on the large scale in the universe is considered to be homogenous and isotropic, but within each small region individual galaxies can have large or small masses. Each galaxy moves according to the laws of motion determined by the field equations of gravitation for the average distribution of matter in the whole universe. Transformations between the isotropic, the co–moving, the Robertson–Walker and the standard system of coordinates can be derived by using the harmonic condition.

161.149 **Mapping the universe. Statistical measures of the galaxy distribution.**
Mo Houjun.
Diss. (Dr.rer.nat.), Max-Planck-Institut für Physik und Astrophysik, Garching (Germany, F.R.). Inst. für Astrophysik; München Univ. (Germany, F.R.). Fakultät für Physik. MPA—593, 239 p. (Jul 1991).
The author presents statistical methods and their applications to study the large-scale structures of the universe indicated by different objects. One obtains from the observational data a number of statistical results that give constraints to the current models of galaxy formation.
Contents: 1. Modelling the universe. 2. Mapping the universe. 3. Measuring the universe. 4. Correlation functions of galaxies. 5. Morphological segregation of galaxies on large scales. 6. Luminosity segregation of galaxies. 7. The distribution of clusters of galaxies.

161.150 **Cosmological models with generic particle decays.**
A. Ballesteros, R. Domínguez-Tenreiro, J. Schneider.
Astron. Astrophys., Vol. 248, No. 2, p. 345 - 348 (Aug 1991).
A complete analysis of the dynamical evolution of some cosmological models in which a population of exotic particles decay into products which remain relativistic until today is presented. The comparison of the dynamical functions describing these models with observational data determines their viability and results in limits on the parameters of the exotic particles.

161.151 **Gravitational lensing by large-scale structures.**
M. Bartelmann, P. Schneider.
Astron. Astrophys., Vol. 248, No. 2, p. 349 - 353 (Aug 1991).
The statistical gravitational lensing effects on supercluster scales are investigated. The authors follow closely the study of Jaroszyński et al. (1990, hereafter JPPG), but they introduce a completely different, semi-analytical description of the (dark) matter distribution and account explicitly for the correlation between convergence and shear, which is neglected in the above paper. They find that gravitational lensing due to large-scale inhomogeneities is fairly weak and does, by itself, not affect the image properties of distant sources significantly. The authors' results are in remarkable agreement with those of JPPG, showing that their model for the large-scale structure is as appropriate for the present purposes as the much more elaborate model of JPPG, and that the correlation between convergence and shear has no perceptible influence.

161.152 **Construction and analysis of a diameter-limited sample of galaxies in the northern sky.**
S. Bardelli, E. Zucca, G. Vettolani, G. Zamorani.
Astron. Astrophys., Vol. 248, No. 2, p. 354 - 366 (Aug 1991).
In order to understand if the differences between the Southern Sky Redshift Survey of da Costa et al. (which is diameter-limited) and the CfA survey (which is magnitude-limited) are real or induced by selection effects, a diameter-limited sample of galaxies has been extracted from the Uppsala catalogue, covering the same sky region of the CfA survey. The authors discuss the main differences between the two samples, like their morphological composition, surface brightness and redshift distribution. In analogy to the luminosity function, they have computed the diameter function for this sample testing various methods and their dependence on the large-scale inhomogeneities. The knowledge of this function has allowed to correct for the selection function of the sample and to compute an "unbiased" mean galaxy density. This density, computed through two different estimators, is then compared with that obtained from a magnitude limited sample.

161.153 **Dissipationless galaxy formation from non-isolated density perturbations.**
N. Voglis, N. Hiotelis, P. Höflich.
Astron. Astrophys., Vol. 249, No. 1, p. 5 - 13 (Sep 1991).
The non-dissipative evolution of an aspherical initially small density perturbation and the collapse of the resulting bound object (protogalaxy) is studied numerically by the *n*-body method. The investigation is based on an improved version of the model proposed by Voglis and Hiotelis (1989). This model is based on Aasheth's *n*-body code and takes into account the dynamical effects on the evolution of the protogalaxy of both the close neighbouring material and the extensive mass environment. The results show that the final radial profile of the density can be appreciably different from the prediction of the similarity solution. The difference is due to the redistribution of binding energies among the particles after the collapse, and to the asphericity of the initial density profile. The collapse factor of the various shells is found to depend on the final radius r as $r^{-\beta}$, with $\beta \approx 0.5$. The angular momentum the protogalaxy acquires in the tidal field of its environment is discussed. The values of the spin parameter λ and their distribution versus the cumulative mass from the center of the protogalaxy are also examined. In some cases the inner parts of the protogalaxy counterrotate with respect to the outer shells.

161.154 **Formation of galactic halos from seeded hot dark matter.**
A. van Dalen, J. V. Villumsen.
Astrophys. J., Vol. 376, No. 2, p. 371 - 379 (1 Aug 1991).
The authors study the formation of galactic halos in a flat universe dominated by massive neutrinos. The linear growth of a perturbation, seeded by a stationary point mass born at some arbitrarily early time in the early universe, is calculated fully relativistically. The nonlinear evolution of the perturbation is calculated using a dissipationless three-dimensional N-body code. The halos thus generated are well approximated by isothermal spheres, generating flat rotation curves, in agreement with galactic rotational velocity observations. The result also justifies the assumption of the Tremaine-Gunn limit in the calculation. However, when a massive object dominates the central region, the core radius becomes an ill-defined quantity, so with a large fraction of the inner region made up of some other material, such as baryons, it is possible to produce small galaxies with low velocity dispersions. With a system largely dominated by neutrinos, the simulations indicate the Tremaine-Gunn limit is overly conservative. In the case studied, the core radius was 50% larger and the central density about 3 times smaller than that given by the Tremaine and Gunn calculation.

161.155 **The search for non-baryonic dark matter.**
T. J. Sumner.
Nucl. Phys., B Proc. Suppl., Vol. 22B, p. 165 - 177 (Jul 1991). - See Abstr. 012.019 for the main entry.
A brief review of the observational evidence for dark matter is presented, together with some discussion of the potential candidates which may constitute such dark matter. This is followed by an overview of current experiments to detect non-baryonic Galactic dark matter, using both direct and indirect techniques.

161.156 **Are the intermediate-scale cosmic background radiation anisotropies constrained by thermal emission of a very cold dust?**
P. Andreani, G. Dall'Oglio, L. Martinis, L. Piccirillo, L. Pizzo, L. Rossi, C. Venturino.
Astron. Astrophys., Vol. 249, No. 2, p. 299 - 306 (Sep 1991).
Results from a ground-based experiment carried out at the Italian Base in Antarctica devoted to the search for Cosmic Background Radiation anisotropies at intermediate angular scale and millimetric wavelengths seem to provide evidence for a strong contamination of the data from patchy galactic emission. The millimetric emission is in fact highly correlated with the IRAS 100 μm data even in regions far from the galactic plane. Two different procedures to reduce the atmospheric contamination of the data and a possible solution to partially get rid of the galactic thermal emission in the FIR are shown here. However, since after correction for atmospheric and galactic emission the authors find still evidence for fluctuations at a level of 2×10^{-4} at 2.5° angular scale and of 2.9×10^{-4} at 5° in contrast with the strong upper limits found at similar angular scales by other groups, they argue that at the moment the poor knowledge of the

millimetric patchy galactic emission and its distribution does not allow a quantitative determination of the "local" disturbance to this kind of measurements.

161.157 **Probing pencil beams in pancake models.**
T. Buchert, H. J. Mo.
Astron. Astrophys., Vol. 249, No. 2, p. 307 - 311 (Sep 1991).
Large two-dimensional point samples extracted from a pancake model with $\Omega = 1$ are analyzed in comparison with recent deep one-dimensional pencil beam surveys. The authors show that the observed periodicity of wall peaks is reproduced well by most of the model beams which cross several walls in the simulation. Probing large Gpc-slices, however, only weakly supports the assumption of constant comoving pair separation to larger depths. The dependence on opening angles of beams, thickness of walls, "fingering of God", spatial resolution as well as sample size is analyzed and related to typical parameters of the model. The authors use these relations to obtain a spatial normalization of the model.

161.158 **Bounds on a possible HE/VHE gamma-ray line signal of galactic dark matter.**
S. Rudaz.
AIP Conf. Proc., No. 220, p. 217 - 221 (Apr 1991). Current Physics Microform No.: 9106B1351. - See Abstr. 012.020 for the main entry.
Particle physics and cosmological considerations can be used to delineate the allowed magnitude of a possible cosmic gamma ray line signal of galactic dark matter. This could in principle be observable by large, space-borne, high resolution gamma ray telescopes.

161.159 **Successful inflation in scalar-tensor theories of gravity.**
D. Walliser.
Deutsches Elektronen-Synchrotron (DESY), Hamburg (Germany), DESY—91-063. 37 p. (Jun 1991).
Among the trajectories that solve the differential equations of general scalar-tensor theories of gravity special ones are selected that successfully inflate the universe. They are well-approximated by separatrices. Furthermore a set of necessary conditions is derived which allows for successful inflation. They do not only include sufficient horizon growth but also both phase space and further cosmological constraints. These conditions are imposed on a simple Φ^n potential and on several dilaton theories to arrive at bounds for the involved parameters.

161.160 **Modern cosmology and physics beyond the standard model.**
A. D. Dolgov.
Surv. High Energ. Phys., Vol. 6, No. 2, p. 57 - 113 (Sep 1991).
What physics badly needs beyond the standard model is cosmology. In a sense, cosmology proves that fundamental physics is not exhausted by the standard model and except for the three known quark-lepton families and strong and electroweak gauge bosons there must exist new fields and particles necessary for an explanation of the astronomical data. It is the aim here to discuss the data and to advocate the last assertion. The emphasis is on a qualitative approach and rigour is sacrificed in favour of a (half-)intuitive picture. In short, the standard model seems to be incompatible with the inflationary universe model and probably also with baryogenesis. It cannot explain the universe structure formation and the phenomenon of dark matter, neither can it solve the cosmological constant problem. These topics are discussed.

161.161 **Quantum and statistical effects in superspace cosmology.**
B. L. Hu.
NATO Advanced Research Workshop on Quantum Mechanics in Curved Space-Time, p. 361 - 402 (1990). - See Abstr. 012.023 for the main entry.
The general aim of these lectures is twofold: 1) To develop a viewpoint for understanding the kinematics of quantum cosmology in the minisuperspace approximation: i.e., superspace as the space of collective excitation modes of spacetime. Emphasis is put on the spectrum of the Lifshitz operator in spacetime and that of the Wheeler-DeWitt operator in superspace. Some established theorems on the zero-mode structure and asymptotic distribution of eigenvalues of invariant operators in curved space are invoked. 2) To apply some recent results is quantum field theory in curved spacetime to problems in quantum cosmology.

161.162 **On quantum gravity for homogeneous pure radiation universes.**
M. A. Castagnino, E. Gunzig, P. Nardone.
NATO Advanced Research Workshop on Quantum Mechanics in Curved Space-Time, p. 403 - 430 (1990). - See Abstr. 012.023 for the main entry.
Quantum gravity is studied in the case of a homogeneous pure radiation universe. Several prescriptions to define the wave function of the universe are tested in this framework. A new idea to define this wave function is introduced. It is based on the final classical behaviour of the universe.

161.163 **Inflation with massive spin-2 field in curved space-time.**
C. Sivaram, V. de Sabbata.
NATO Advanced Research Workshop on Quantum Mechanics in Curved Space-Time, p. 503 - 515 (1990). - See Abstr. 012.023 for the main entry.

161.164 **Rotational perturbations of relativistic perfect-fluid model universes interacting with gravitational field.**
K. Maniharsingh.
Astrophys. Space Sci., Vol. 183, No. 2, p. 285 - 307 (Sep 1991).
Along with the presentation of several new analytic solutions, the dynamics of slowly rotating perfect-fluid model universes are investigated, and their physical and geometrical properties are discussed from all angles. The rotational perturbations of such models are examined in detail in order to substantiate the possibility that the Universe is endowed with some rotation. The nature and role of the rotational velocity $\Omega(r,t)$ which is related to the local dragging of inertial frames and that of the matter rotation $\omega(r,t)$ are studied for uniform and non-uniform motions. The authors find out the restrictions on the radii of the models for real astrophysical situations, and the periods of physical validity of them are also obtained. Rotating models which are expanding as well are obtained, in which cases the rotational velocities are found to decay with the time; and these models may be taken as examples of real astrophysical objects in this Universe.

161.165 **Redshift distribution and evolution function of quasi-stellar objects.**
Zhou Youyuan, Hu Yuedong.
Sci. China, Ser. A, Vol. 34, No. 5, p. 598 - 612 (May 1991). In Chinese.
Taking account of the selection effect in the line identification, under the cosmological redshift hypothesis, the authors calculate the redshift-distribution functions of quasi-stellar objects (QSOs) selected with the slitless technique, using three kinds of evolution functions for QSOs, i.e. the Schmidt-Green function, the Mathez function and the Heisler-Ostriker functions, and compare them with the observed distribution. The comparison shows that with regards to both the global behavior and the detailed properties, the predictions agree well with observations.

161.166 **Dynamics of phase transitions at finite temperatures in the early universe.**
H. Patzelt.
Ann. Phys. (N.Y.), Vol. 212, No. 2, p. 285 - 314 (Dec 1991).
A statistical description of phase transitions on a macroscopic, phenomenological level is used to treat the dynamics of phase transitions and the dynamics of an expanding universe simultaneously. The author considers phase transitions with non-vanishing background densities. The ratio of the typical interaction time to the typical expansion time suggests two scenarios that are discussed concerning their influence on the evolution of the universe.

161.167 **Cosmology of three–generation Calabi–Yau superstring models.**
C. Panagiotakopoulos.
Phys. Lett. B, Vol. 269, No. 1/2, p. 63 – 70 (24 Oct 1991).
The cosmology of three–generation Calabi–Yau (CY) superstring models is discussed. A model independent lower bound $\approx 10^{14}$GeV on the largest intermediate scale in models with $SU(3)^3$ gauge group is derived by requiring that the magnetic monopole flux in the Galaxy does not exceed the Parker bound. Using this bound one argues that the baryon asymmetry in three–generation CY models should necessarily be created at temperatures of ~ 1 GeV at most. A late baryogenesis scenario is presented in the context of a model with naturally suppressed proton decay which is generalizable to others with Z_3 matter parity and low $U(1)_{B-L}$ breaking scale.

161.168 **Bianchi type–II, VIII, and IX cosmologies with conformally–invariant scalar field and electromagnetic fields.**
R. Venkateswarlu, D. R. K. Reddy.
Astrophys. Space Sci., Vol. 184, No. 1, p. 153 – 156 (Oct 1991).
Exact Bianchi type–II, VIII, and IX cosmological solutions in a conformally–invariant scalar field with trace–free electromagnetic energy–momentum tensor are obtained.

161.169 **Scale–invariant models of the galaxy and cluster luminosity functions.**
F. Bernardeau, R. Schaeffer.
Astron. Astrophys., Vol. 250, No. 1, p. 23 – 42 (Oct 1991).
The authors take advantage of recent theoretical progresses made in the understanding of the matter clustering at the kpc to the Mpc scales to construct a model for the galaxy and cluster multiplicity functions. They check that the scaling law predicted by the theory, which relates the typically luminosity of a galaxy and of a rich cluster, and a similar rule relating the velocity dispersion are in agreement with the data. Explicit relations are obtained between the dark halo parameters (mass, radius) and the observed parameters within the radius where the luminosity is concentrated. It is found that the halo radius behaves as $R \sim M^{0.3}$ and that the mass multiplicity behaves as M^{-2}, being nearly divergent, if the luminosity function is assumed to behave as $L^{-1.3}$. It appears that most of the mass is contained in the smaller galaxies with $(M/L) \sim L^{0.6}$ at the faint end. These estimates are new, and obtained by a method that differs from all the previous approaches. The results turn out to be very similar to those obtained with a moderately diverging initial spectrum such as the one obtained for Cold Dark Matter initial conditions at $M \leqslant 10^{12} M_\odot$ scales. Finally, the luminosity functions for galaxies, for rich cluster, as well as for Turner–Gott groups are constructed explicitly from this scaling theory. They agree very closely with the ones obtained from observations. Scaling rules for the 2–D group multiplicity function are also obtained. These scalings provide a specific test of the authors' theory.

161.170 **Big Bounce in the very early universe.**
H. J. Blome, W. Priester.
Astron. Astrophys., Vol. 250, No. 1, p. 43 – 49 (Oct 1991).
A cosmological model is discussed which replaces the Friedmann singularity in the early universe by a de Sitter solution with a closed spherical space metric stretched by a quantum vacuum. This model contracts from past infinity to a minimum radius and then re–expands (Big Bounce). After passing the minimum, a phase transition is required which converts the quantum vacuum into the primordial relativistic matter. During this process the exponential de Sitter expansion turns into a Friedmann–Lemaitre model. This model is compared with a similar singularity–free model proposed by Israelit and Rosen (1989), which starts from a cosmic egg at the Planck time with "prematter" of Planck density.

161.171 **A search for redshifted neutral hydrogen from protoclusters of galaxies.**
J. M. Uson, D. S. Bagri, T. J. Cornwell.
Astrophys. J., Lett., Vol. 377, No. 2, p. L65 – L68 (20 Aug 1991). With plates L1 – L2.
The authors have used the VLA to search for highly redshifted (z = 3.3) "21 cm" line emission from protoclusters of galaxies. The sensitivity, ~ 3.5 mJy per synthesized beam for spectral channels of width ~ 100 kHz, was limited at this level by statistical noise. The authors rule out the existence of any Zel'dovich "pancake" in the surveyed volume ($\sim 1.6 \times 10^6 \mathrm{Mpc}^3$) with a component of neutral hydrogen with mass greater than about $10^{14} M_\odot$. This limit does not depend on the degree of fragmentation of such a structure.

161.172 **The momentum of scalar field and the wave function of the universe.**
Fan Zuhui, Yu Yunqiang.
Chin. Astron. Astrophys., Vol. 15, No. 3, p. 241 – 250 (Sep 1991). English translation of Acta Astrophys. Sin., Vol. 11, No. 2, p. 97 – 105 (Apr 1991). See Abstr. 53.161.305.
Using the minisuperspace model the authors analyse the effect of the momentum of the scalar field on the wave function of the universe. They find that the classical forbidden region is sensitive to the value of the momentum and that, under certain conditions, there exists a classical allowed region in the a $\sim$ 0 vicinity. Also, the wave function depends on the value of the operator ordering factor p. Under the Vilenkin boundary condition, the distribution probability of the scalar field ϕ is found to be independent of ϕ and different from the result of neglecting the momentum. Different implications of the initial conditions of the inflationary universe follow and are compared with the classical version.

161.173 **A new gravitational interaction of cosmological importance.**
D. F. Crawford.
Astrophys. J., Vol. 377, No. 1, p. 1 – 6 (10 Aug 1991).
It is suggested that the combination of general relativity and wave mechanics leads to a new gravitational interaction. Applied to photons traveling through uniform material, it produces a "tired light" type of redshift that is similar to a Doppler shift. It can be combined with an almost static cosmology to make definite predictions that can be compared with observations. For example, it predicts a Hubble constant of $74C^{-1/3}$ km s^{-1}Mpc^{-1}, where C is a gas clumping factor with a value near unity, by equating the energy–loss rates of the background radiation and the cosmic gas; the interaction leads to the prediction that the temperature of the background radiation is $2.90C^{-1/6}$K. The only observational input to this calculation is from the explanation the theory provides for the background X–ray radiation. It can resolve the "missing mass" problem for clusters of galaxies and possibly for galactic rotation curves. Finally, it can explain the occurrence of "voids" in the galactic distribution as being due to extra redshifts produced by cosmic gas clouds.

161.174 **Linear perturbation theory and spherical overdensities in $\Lambda \neq 0$ Friedmann models.**
H. Martel.
Astrophys. J., Vol. 377, No. 1, p. 7 – 13 (10 Aug 1991).
This paper presents various analytical results concerning the properties of large–scale structures in Friedmann models with a positive cosmological constant Λ. It shows, using linear perturbation theory, that for given density contrast δ and density parameter Ω the effect of Λ on the peculiar velocity field is at the most of order a few percent. It also shows that the number counts of galaxies inside the zero–velocity surface of spherical perturbations are nearly independent of Λ. These results support the conclusion of two previous papers, that the cosmological constant Λ leaves little observable signature in the current dynamical state of the large–scale structure of the universe.

161.175 **The evolution of void–filled cosmological structures.**
E. Regös, M. J. Geller.
Astrophys. J., Vol. 377, No. 1, p. 14 – 28 (10 Aug 1991).
The authors simulate the growth and evolution of voids in the large–scale distribution of galaxies. For arbitrary values of the cosmological density parameter Ω, the authors examine one–, two–, and three–dimensional models. In these models, void–filled cosmological structures evolve from a high number density of shallow initial voids with unspecified origin. The growth rate of a

void decreases significantly at the cosmic time corresponding to $\Omega = 0.5$. In models constructed in two and three dimensions, suitable initial conditions lead to cellular structure with faceted voids similar to those observed in redshift surveys. These structures form for $\Omega \lesssim 0.4$. The matter compressed to planes flows more rapidly toward the condensations at the intersections than one would expect for spherical infall. The rms peculiar streaming velocities for void diameters of 5000 km s^{-1} are on the order of $1000 \cdot \Omega^{0.6}$km s^{-1}. If Ω is large, the appearance of voids and walls in redshift space can be very different from their appearance in real space, when the adjacent voids differ significantly in size; voids and walls can masquerade for each other. The three-dimensional models indicate that the redshift maps are a rather good guide to the structure in real space. For the 5000 km s^{-1} maximum void size in the CfA survey, structures in redshift space on scales less than 1000 km s^{-1} are uncertain if $\Omega = 1$. For smaller Ω, the scale for such severe distortion is smaller.

161.176 **The cosmological solution of a homogeneous, isotropic, ideal fluid at local rest.**
C. Darierescu, A. Darierescu, M. Agop.
Stud. Cercet. Fiz., Vol. 42, No. 4, p. 323 - 327 (1990).
It is shown that the Robertson-Walker metric leads to an exact solution of Einstein's field equations if one considers the three-dimensional space filled with an ideal fluid, at local rest.

161.177 **A quantitative measure of phase correlations in density fields.**
R. J. Scherrer, A. L. Melott, S. F. Shandarin.
Astrophys. J., Vol. 377, No. 1, p. 29 - 35 (10 Aug 1991).
The authors present a quantitative measure of the phase correlations in a density field based on the location of the maxima of the Fourier components of that field. They find that this measure can easily detect non-Gaussian behavior either in artificially constructed density fields or those that become non-Gaussian from gravitational clustering of Gaussian initial conditions. Different initial power spectra produce somewhat distinguishable signals, and the signals are robust against sparse sampling.

161.178 **A "larger" universe.**
G. Light.
Stud. Cercet. Fiz., Vol. 43, No. 1/2, p. 3 - 4 (1991).
A model of a "larger" universe is constructed containing both matter and "invisible energy". It is suggested that the universe of "invisible energy" may be relevant, since it may be connected to the universe of matter everywhere by miniblack holes.

161.179 **Dark matter decay, reionization and microwave background anisotropies.**
D. Scott, M. J. Rees, D. W. Sciama.
Astron. Astrophys., Vol. 250, No. 2, p. 295 - 301 (Oct 1991).
The pregalactic medium may have been ionized by UV photons resulting from the decay of "dark matter" particles. The present paper explores the implications of such a hypothesis. In a specific model involving decaying neutrinos, the authors found that, after recombination at $z \cong 1000$, the fractional ionization gradually increases with time until the medium becomes fully ionized at $z_i \cong 25$, for $\Omega_b \cong 0.1$. The optical depth back to $z \cong z_1$ is less than unity; beyond z_i it increases only as $\sim(1+z)^{1/4}$, so that the "last scattering surface" is at $z_* \cong 400$. The implications for galaxy formation of this unusual thermal history are briefly described. The predicted small-angular-scale microwave background anisotropies are below current upper limits, provided that the amplitude of the initial fluctuations is adjusted to match the presently-observed clustering. The existence of non-linear structure at redshifts ~ 5 then requires the presence of non-adiabatic or non-Gaussian perturbations on galactic scales.

161.180 **Formation of pregalactic dust and the cosmic microwave background spectrum.**
B. V. Vajner, A. Yu. Glukhov, V. V. Chuvenkov.
Astron. Zh., Tom 68, Vyp. 4, p. 673 - 680 (Jul-Aug 1991). In Russian. English translation in Sov. Astron., Vol. 35, No. 4.
The process of dust-grain formation in a cosmological model with Population III stars at redshifts $z \cong 30 - 50$ is considered. By using the kinetic theory of nucleation, the ions $H^+(H_2O)$ are shown to be the initial centers for condensation of saturated water vapour. Relative share of "polluted" ice in the total mass of matter can reach the value $\Omega_d \cong 2 \cdot 10^{-5}$; that leads to the necessity of accounting for the interaction of dust and radiation in the analysis of the cosmic microwave background spectrum.

161.181 **Constraint on cosmic strings and matter in a Robertson-Walker universe.**
C. Wolf.
Hadronic J., Vol. 14, No. 4, p. 335 - 341 (Aug 1991).
By considering a homogeneous and isotropic universe the author considers the constraints imposed upon cosmic strings and matter in the early universe. His simple analysis demonstrates that the string tension must vanish and the growth of the scale factor is power-like in time and is stable to bulk viscosity perturbations for matter represented by ultra-relativistic particles, stiff matter and dust as the matter component.

161.182 **The projective d'Alembert equation.**
G. Arcidiacono, E. Capelas de Oliveira.
Hadronic J., Vol. 14, No. 4, p. 347 - 351 (Aug 1991).
The authors discuss the d'Alembert equation extended to the de Sitter universe, and they show that in only two particular cases the solution is of the same type of the classical d'Alembert solution, namely $\Phi = f(x+ct)+g(x-ct)$.

161.183 **String cosmology in Bianchi VI_0 space-time.**
S. Chakraborty.
Indian J. Pure Appl. Phys., Vol. 29, No. 1, p. 31 - 33 (Jan 1991).
A class of cosmological solutions of massive strings are obtained in a Bianchi VI_0 space-time following the techniques used by Letelier and Stachel. Some of these solutions are in the presence of a magnetic field while the others correspond to the case without magnetic field.

161.184 **Big howler of cosmology.**
P. Molnar, M. Mészáros.
Indian J. Pure Appl. Phys., Vol. 29, No. 1, p. 34 - 40 (Jan 1991).
It is shown that there is only an equilibrium solution of the cosmology problem of the general relativity theory in the case of an homogeneous-isotropic perfect fluid and the real energy density $u \neq 0$, the pressure $p \neq 0$ and the cosmological constant $\lambda \neq 0$. Furthermore, the equilibrium solution - inclusive of Einstein's one - is stable even if the pressure and the density are positive semidefinite.

161.185 **Quantum cosmology of superstring universe.**
Wang Jin, S. Mohanty.
Mod. Phys. Lett. A, Vol. 6, No. 24, p. 2181 - 2187 (10 Aug 1991).
The authors study the quantum cosmology of the string universe obtained by embedding the Robertson-Walker metric in the nonlinear σ-model. They find that the quantum model has some unusual features not seen in the classical model. Initially the universe exists in a series of metastable bound states with scale factor taking discrete values. Then the universe tunnels through a barrier and comes out in an inflationary state. This tunneling (or evolution in imaginary time) also has the effect of heating up the matter field so that one has the conditions of chaotic inflation.

161.186 **Are the cosmic strings seen as the dislocations?**
T. T. Fujishiro, M. J. Hayashi, S. Takeshita.
Mod. Phys. Lett. A, Vol. 6, No. 24, p. 2237 - 2242 (10 Aug 1991).
The effective action from the heterotic string compactification is studied on the manifolds with absolute parallelism. The cosmic string solutions resulting from the effective action are discussed

in analogy with the dislocations in 3-dimensional crystalline solid.

161.187 **Color breaking and X–, Y– and W–boson condensation on the cosmic string.**
T. M. Bibilashvili, G. R. Dvali.
Phys. Lett. B, Vol. 263, No. 3/4, p. 359 – 363 (18 Jul 1991).
The GUT symmetry breaking pattern can be changed in the core of the cosmic string. This can lead to the accumulation of the X–, Y–, $W^{\pm}$–gauge boson condensates on the phase boundary around the vortex. A similar condensation of the gauge bosons, which may take place during different phase transitions in the early universe, can play some role in the generation of the baryon asymmetry.

161.188 **What is time?**
J. M. Barnothy, M. F. Barnothy.
Comments Astrophys., Vol. 15, No. 5, p. 279 – 281 (Sep 1991).

161.189 **Supersymmetry breaking and dark matter.**
J. L. Lopez, Kajia Yuan, D. V. Nanopoulos.
Phys. Lett. B, Vol. 267, No. 2, p. 219 – 226 (12 Sep 1991).
The authors explore the possibility of neutralino dark matter in minimal SUSY GUT models. They emphasize the use of accurate formulae for the squark and slepton masses, and perform their calculations in a general supersymmetry breaking scenario. They find that the prospects for dark matter are largely determined by the ratio of the universal gaugino mass ($m_{1/2}$) to the universal scalar mass (m_0), $\xi = m_{1/2}/m_o$. For $\xi \ll 1$, both light and heavy dark matter candidates exist, while for $\xi \approx 1$ mainly the heavy candidates survive. For $\xi \gg 1$, consistency constraints of the model eliminate most dark matter candidates.

161.190 **Inflationary models with $\Omega \neq 1$.**
G. F. R. Ellis, D. H. Lyth, M. B. Mijić.
Phys. Lett. B, Vol. 271, No. 1/2, p. 52 – 60 (14 Nov 1991).
The authors describe the general features and a few specific models of inflationary cosmology that lead to $\Omega \neq 1$ today, solve all standards problems of classical cosmology, and generate density perturbations that may be responsible for the formation of structures in the universe.

161.191 **The phase space structure in minisuperspace cosmological models.**
S. V. Shabanov.
Phys. Lett. B, Vol. 272, No. 1/2, p. 11 – 16 (28 Nov 1991).
A phase space (PS) structure in minisuperspace cosmological models with gauge fields is investigated. It is shown for the SO(n), $n > 3$, gauge group that the physical PS differs from an ordinary plane. Due to this phenomenon, the path integral representation gets modified for the ground–state wave function of the universe. It is also argued that the wormhole size quantization should change due to a non–trivial physical PS structure of gauge fields.

161.192 **Is there a big–bubble problem in extended inflation?**
D. S. Goldwirth, H. W. Zaglauer.
Phys. Rev. Lett., Vol. 67, No. 26, p. 3639 – 3642 (23 Dec 1991).
Current Physics Microform No.: 9201E1851.
The authors present a generalization of the thin–wall formalism to calculate the evolution of true–vacuum bubbles in "extended" inflation. For inflation quartic couplings of the order M/M_{Pl}, gravity will lead to the recollapse of bubbles during most of the extended inflation period, thereby altering the bubble size distribution. The completion of the phase transition will proceed through tiny bubbles satisfying homogeneity constraints of the microwave background for arbitrary values of ω. Thus, it is possible to construct models that agree with experiments even for the original version of extended inflation.

161.193 **The ground–state wavefunction of a radiation–dominated universe.**
O. Bertolami, J. M. Mourão.
Classical Quantum Gravity, Vol. 8, No. 7, p. 1271 – 1282 (Jul 1991).
The authors propose a minisuperspace construction of the ground–state (no boundary) wavefunction for a radiation–dominated universe. As in the case of a free massless conformally invariant scalar field discussed by Hartle and Hawking, the functional integrations over the gravitational and gauge fields can be performed independently. The gauge part of the wavefunction is semiclassically computed with the help of the (anti–)selfdual solutions of the Euclideanized SO(4) – symmetric Einstein – Yang – Mills systems. Implications of this construction are discussed.

161.194 **Existence of finitely perturbed Friedmann models via the Cauchy problem.**
U. Brauer.
Classical Quantum Gravity, Vol. 8, No. 7, p. 1283 – 1301 (Jul 1991).
A mathematical treatment of perturbed Friedmann models is given. Local existence, convergence and analytic linearization stability for the constraints for $k = 0, \pm 1$ are shown. To prove this, special function spaces are introduced in order to invert certain elliptic differential operators. These function spaces can also be employed to formulate precisely some decomposition theorems used for the study of linearized perturbations where elliptic operators also have to be inverted. The results for the constraints cannot easily be generalized to the evolution equations.

161.195 **Finite probability of inflation from higher dimensions.**
Y. Ezawa, T. Watanabe, T. Yano.
Prog. Theor. Phys., Vol. 86, No. 1, p. 89 – 102 (Jul 1991).
The authors estimated the probability of sufficient inflation in a higher dimensional cosmological model in which there are two compact subspaces. Using a newly defined measure on the initial data surface, the authors found the probability to be finite and reasonably large for certain regions of the parameters of their model.

161.196 **A new diagram for illustrating the horizon problem.**
T. Kiang.
Acta Astrophys. Sin., Vol. 11, No. 3, p. 197 – 212 (Jul 1991).
In the r–t diagram, where t is the cosmic time and r is the radial coordinate defined as the proper radial distance divided by the scale factor, the locus of the incoming photon emitted at the big bang and reaching us now, $r_-(t)$, is a mirror image of the locus of the outgoing photon emitted at the big bang with $r = 0$, $r_+(t)$. The former graphs our present world picture while the latter traces the growth of the (particle) horizon since the big bang. The equation $r_-(t) = r_+(t)/2$ determines an epoch t_D with the property that, for all events on the world picture before t_D, the horizon problem exists to varying degrees of severity, while for events after t_D, there is no horizon problem. For the flat Friedmann universe t_D turns out to be quite recent, long after the epoch of the MBR (microwave background radiation $z \approx 10^3$); for the inflationary model of Guth, t_D is near the beginning of the inflationary era, long before the MBR epoch. For a curved Friedmann universe the time axis of the Penrose diagram can be identified with the parameter η in the usual parametric representation of the scale factor and the space axis, with the angular radial coordinate χ. Thus interpreted, the diagram readily gives $z_D = 0.86$ (3.66) for a closed (open) universe corresponding to $H_0{}^{-1} = 20 \cdot 10^9$yr and $q_0 = 1.65$ ($\Omega_0 = 0.05$).

161.197 **Singular self–adjoint quantum cosmological models in a slow–time gauge.**
N. A. Lemos.
Classical Quantum Gravity, Vol. 8, No. 7, p. 1303 – 1310 (Jul 1991).
Dust–filled Friedmann – Robertson – Walker universes are canonically quantized in a "slow–time" gauge $t = p_R$, where p_R is

the canonical momentum conjugate to the scale factor R. In the cases of flat or hyperbolic 3–space the Hamiltonian operator turns out to be time dependent and self–adjoint for all t. Both models are shown to be singular, thus disproving a conjecture put forward by Gotay and Demaret to the effect that self–adjoint quantum dynamics in a slow–time gauge is always non–singular.

161.198 **The gravitational effects of superconducting cosmic string.**
Li Shuangjiu, Wang Yinshun, Shi Zuijian, Wu Zuren.
Acta Astrophys. Sin., Vol. 11, No. 3, p. 292 – 296 (Jul 1991). In Chinese.

The magnetic field of a superconducting current in a cosmic string changes the external metric of the string. It also makes the external spacetime of the string simulating a kind of magnetic medium with permeability $\mu(r)$, and the light rays will be deflected when they pass the region nearby the string.

161.199 **Analysis of the large scale structure from samples given by quasi–one–dimensional redshift survey.**
Deng Xinfa, Deng Zugan, Liu Yongzhen.
Acta Astrophys. Sin., Vol. 11, No. 4, p. 306 – 310 (Oct 1991). In Chinese. English translation in Chin. Astron. Astrophys., Vol. 16, No. 1, p. 2 – 7 (Jan – Mar 1992).

Large scale structure in the distribution of galaxies has been investigated by analyzing the samples given by a complete redshift survey in 5 pencil–beam fields and sparse redshift survey in 9 pencil–beam fields. The fractal dimension in the distribution has been calculated with a revised method. The result shows that the distribution of galaxies has a typical scale which is about $3 - 4.5\,h^{-1}$Mpc. This result is consistent with that obtained by Shanks et al.

161.200 **In search of a new cosmic blueprint.**
F. Flam.
Science, Vol. 254, No. 5035, p. 1106 – 1108 (22 Nov 1991).

As doubts build about a once–favored model explaining how structures formed in the universe, new theories are jockeying in a cosmological free–for–all.

161.201 **Qualitative analysis of a class of inhomogeneous self–similar cosmological models. II.**
C. G. Hewitt, J. Wainwright, M. Glaum.
Classical Quantum Gravity, Vol. 8, No. 8, p. 1505 – 1518 (Aug 1991).

The authors describe the qualitative properties of a class of self–similar spatially inhomogeneous solutions of the Einstein field equations with perfect fluid source, which are of cosmological interest. It is conjectured that these solutions act as asymptotic states at late times for more general inhomogeneous cosmologies.

161.202 **Evolution of a cloudy protogalaxy interacting with an early galactic wind and galaxy formation.**
V. G. Berman, A. A. Suchkov.
Astrophys. Space Sci., Vol. 184, No. 2, p. 169 – 192 (Oct 1991).

The authors argue that a combined evidence from galactic and extragalactic studies suggests that a major star formation in giant galaxies is preceeded by an evolutionary phase at which a strong galactic wind driven by the initial burst of star formation enriches the gaseous protogalaxy with metals and heats it up, so that the latter turns over from contraction to expansion. The result is the ejection of enriched material from the outer part of the protogalaxy into the intergalactic space, while the inner part finally contracts and cools down to form the galactic major stellar component (the "hot" model of galaxy formation). This paper presents a specific mechanism to produce a hot protogalaxy according to which an early galactic wind is imparting energy and momentum into a collapsing protogalaxy whose mass is contained mainly in clouds and only a small portion is in the intercloud gas that provides pressure confinement for the clouds. The model is then capable of accounting for the nearly equal mass and iron abundance in cluster giant galaxies and the intracluster gas provided the observationally plausible input parameters for giant galaxies and early galactic winds are adopted. It also predicts the formation of long–lived X–ray coronae with characteristics similar to those observed around giant ellipticals. The model specifies a characteristic length–scale that can be very naturally interpreted as a size for a stellar system to come; the model perfectly fits a typical size of giant ellipticals.

161.203 **Entropy, inflation and the arrow of time.**
D. S. Goldwirth, T. Piran.
Classical Quantum Gravity, Vol. 8, No. 8, p. L155 – L160 (Aug 1991). Letter–to–the–editor.

One of the motivations for the introduction of inflation was to explain the large entropy of the observed universe. Penrose has recently pointed out that the entropy of the observable universe is small when compared with the maximal possible entropy. The authors show that this phenomenon is also a natural consequence of inflation. Furthermore, inflation is generic in the expanding phase but exceedingly unlikely in the collapsing phase. This introduces an asymmetry in the evolution of the universe which resembles to a large extent the asymmetry that appears in the growth of entropy and it might be related to the question of the arrow of time.

161.204 **On the critical mass of gravitational structure.**
A. Woszczyna, A. Kułak, S. Zieba.
Classical Quantum Gravity, Vol. 8, No. 8, p. L167 – L170 (Aug 1991). Letter–to–the–editor.

The authors show that in an open universe filled with matter of constant sound velocity none of the perturbations can increase indefinitely, even when their length scale exceeds the Jeans length.

161.205 **Early cosmology inspired by the heterotic string.**
N. R. Stewart.
Classical Quantum Gravity, Vol. 8, No. 9, p. 1701 – 1715 (Sep 1991).

The author obtains cosmological solutions of the low energy heterotic string assuming Calabi – Yau compactification with non–minimal curvature terms ($R^2{}_{ABCD} - 4R^2{}_{AB} + R^2$). In particular, the problem of which initial conditions to choose is discussed, since on emergence from the quantum gravity era they are as yet undeterminable. It is found that specific constraints are necessary on the initial–value phase space to be compatible with acceptable late–time behaviour.

161.206 **On the variability of physical constants and conformal transformations.**
E. G. Mychelkin (*Eh. G. Mychelkin*).
Astrophys. Space Sci., Vol. 184, No. 2, p. 235 – 245 (Oct 1991).

The problem of fundamental physical constants variability is discussed on the basis of extended conformal symmetry principles including the concept of local mass variability. It is shown, in particular, that the canonical determination of conformal weights can lead, in general, to non–invariance of the known elementary charge–to–mass ratio. The introduction of active time–like and space–like conformal transformations leads also to possibility of simultaneous variations of dimensional constants without any observable changes in the evolution of the Universe. The generally covariant method of space–time splitting for conformal renormalization of the metric is considered in terms of Lie derivatives.

161.207 **Kinematics of relativistic ejection with Hubble flow in subluminal motion for the determination of Hubble constant H_0.**
Gong(Kung) Shumo, Xia Changli.
3. Chinese Academy of Sciences and Max–Planck Society Workshop: High energy astrophysics: compact stars and active galaxies, p. 59 – 66 (1991). – See Abstr. 012.026 for the main entry.

The expected wavelengths of the respective counterjets of two subluminals 3C 84 and M87 are calculated and compared with their observed wavelengths to determine the best value of Hubble constant. The observed wavelengths of 3C 84 and M87 are, respectively, 1.3 and 18 cm. The expected wavelengths calculated in the authors' work are in the range of $1.3 < \lambda < 6$ cm and $18 < \lambda < 32$ cm respectively for 3C 84 and M87.

161.208 **The quantum mechanics of cosmology.**
J. B. Hartle.
7. Jerusalem Winter School for Theoretical Physics: Quantum cosmology and baby universes, p. 65 – 157 (1991). – See Abstr. 012.036 for the main entry.
Contents: 1. Introduction. 2. Post–Everett quantum mechanics. 3. Generalized quantum mechanics. 4. Time in quantum mechanics. 5. The quantum mechanics of spacetime. 6. Practical quantum cosmology: The semiclassical regime; The semiclassical approximation to the quantum mechanics of a non–relativistic particle; Semiclassical prediction in quantum cosmology.

161.209 **Introductory lectures on quantum cosmology.**
J. J. Halliwell.
7. Jerusalem Winter School for Theoretical Physics: Quantum cosmology and baby universes, p. 159 – 243 (1991). – See Abstr. 012.036 for the main entry.
Contents: 1. Introduction. 2. A simple example. 3. The Hamiltonian formulation of general relativity. 4. Quantization. 5. Minisuperspace – general theory. 6. Classical spacetime. 7. The WKB approximation. 8. Boundary condition proposals. 9. No–boundary versus tunneling. 10. Beyond minisuperspace. 11. Vacuum states from quantum cosmology: De Sitter – invariant vacua; The no–boundary vacuum state. 12. Summary. 13. A guide to the literature.

161.210 **Baby universes.**
A. Strominger.
7. Jerusalem Winter School for Theoretical Physics: Quantum cosmology and baby universes, p. 269 – 346 (1991). – See Abstr. 012.036 for the main entry.
Contents: 1. Introduction. 2. Topology change and third quantization in 0+1 dimensions. 3. Third quantization in 3+1 dimensions. 4. Parent and baby universes. 5. Instantons – from quantum mechanics to quantum gravity. 6. The axion model and the instanton approximation. 7. The cosmological constant: the Hawking–Baum argument; baby universes and Coleman's argument.

161.211 **Correlation functions of galaxies in the ESO–Uppsala catalog.**
H. J. Mo, Y. P. Jing, G. Börner.
3. Chinese Academy of Sciences and Max–Planck Society Workshop: High energy astrophysics: compact stars and active galaxies, p. 211 – 216 (1991). – See Abstr. 012.026 for the main entry.
The authors estimate the angular two–point and three–point correlation functions of ESO–Uppsala galaxies in angular–diameter (d1)–limited samples. They use the diameter function given by Lahav et al. (1988) to scale the results. The two–point correlation functions of the samples with $d_{1,\mathrm{lim}} < 3'.0$ can all be fitted by the power law $\xi(r) = (r/r_0)^{\alpha}$ with $\alpha = 1.8$ and $r_0 = 4.74\ h^{-1}$Mpc; the results of samples with larger $d_{1,\mathrm{lim}}$ can be fitted by a similar power law but with smaller values of r_0. The three–point correlation function satisfies the hierarchical scaling with $Q = 0.66$. Early–type galaxies are found to have larger correlation amplitude. But the slope of the two–point correlation function does not depend strongly on morphology. These results are compared to those of the corresponding UGC samples.

161.212 **Critique of Coleman's theory of the vanishing cosmological constant.**
L. Susskind.
7. Jerusalem Winter School for Theoretical Physics: Quantum cosmology and baby universes, p. 347 – 353 (1991). – See Abstr. 012.036 for the main entry.

161.213 **Identification of the substructures in the Universe by percolation test.**
Song Guoxuan.
3. Chinese Academy of Sciences and Max–Planck Society Workshop: High energy astrophysics: compact stars and active galaxies, p. 217 – 222 (1991). – See Abstr. 012.026 for the main entry.
A quantitative description of the substructures in the distribution of galaxies in the Universe is given. By percolation test the "clusters" with the longest separation L between particles and the number of particles N are identified with given separation parameter d. For three simplified substructures, filaments, layers and clusters, the quantitative relation among L, N and d is established, by which every "cluster" could be classified.

161.214 **Multi–level fractal structure in the distribution of galaxies.**
Deng Zugan, Xia Xiaoyang, Liu Yongzhen, Zou Zhenlong.
3. Chinese Academy of Sciences and Max–Planck Society Workshop: High energy astrophysics: compact stars and active galaxies, p. 223 – 228 (1991). – See Abstr. 012.026 for the main entry.
Fractal dimensions in the distribution of galaxies have been calculated for samples given by the CfA redshifts survey and the redshift surveys of IRAS galaxies in the fields F15 and NGW. Results show that the distribution of galaxies is structured even on the largest scale of analysis which is about $60h_{50}{}^{-1}$Mpc. The distribution is not simply fractal but likely to be multi–level fractal. That is, the distribution is fractal with definite fractal dimension only in a series of scale ranges. Fractal dimension of each fractal level depends on the scale of the level and increases with the scale of the fractal level. Implications of these results have been discussed briefly.

161.215 **Possible gigantic structures of all–sky rich cluster.**
Xia Xiaoyang, Deng Zugan, Liu Yongzheng.
3. Chinese Academy of Sciences and Max–Planck Society Workshop: High energy astrophysics: compact stars and active galaxies, p. 229 – 234 (1991). – See Abstr. 012.026 for the main entry.
Four possible gigantic structures in the distribution of all–sky rich clusters have been found by using quasi–three dimensional analysis method. One of those is close to the direction of local group dipole extending about $200\ h^{-1}$Mpc. The other three structures are even larger, which have the scale about $400\ h^{-1}$Mpc. This result provides a new possible evidence for the existence of super–large scale structure in the distribution of galaxies.

161.216 **Dynamical thresholding of pancake models: a promising variant of the "hot–dark–matter" picture.**
T. Buchert.
3. Chinese Academy of Sciences and Max–Planck Society Workshop: High energy astrophysics: compact stars and active galaxies, p. 242 – 252 (1991). – See Abstr. 012.026 for the main entry.
Variants of pancake models are considered which allow for the construction of a phenomenological link to the galaxy formation process. A control parameter space is introduced which defines different scenarios of galaxy formation. The sensibility of statistical measures of the small scale structure with respect to this parameter freedom is demonstrated. This property of the galaxy formation model, together with the consequences of enlarging the box size of the simulation to a "fair–sample scale", form the basis of arguments to support the possible revival of the standard "Hot–Dark–Matter" model.

161.217 **New calibration of the $D_n - \sigma$ relation: application to modeling the large–scale peculiar velocity field.**
R. E. Kates, U. V. Weigelt.
3. Chinese Academy of Sciences and Max–Planck Society Workshop: High energy astrophysics: compact stars and active galaxies, p. 253 – 258 (1991). – See Abstr. 012.026 for the main entry.
The large–scale velocity field is modeled using an improved group catalog and the authors' recent, more accurate calibration

of the log D_n – log σ relation for elliptical galaxies. The authors consider models with constant bulk streaming, pure attractor, and streaming plus attractor. In the bulk and pure attractor models the results are rather compatible with previous work. In the streaming plus attractor model, the attractor is so far away that the velocity field in the observed region is best described as bulk streaming together with a weak linear trend. In essence, no attractor is required. In this case, the velocity at the Local Group has magnitude 506 kms^{-1} in the direction ($l = 352\overset{\circ}{.}3$, $b = 2\overset{\circ}{.}5$), and the components of the field vary by about 5% in the observed region. Particular attention is paid to the reliability of the predictions by means of a careful estimate of the statistical errors and an analysis of the measured and theoretical χ^2 values.

161.218 **A study of morphological segregations of galaxies in the UK Schmidt Survey Field 349.**
B. Yulin.
3. Chinese Academy of Sciences and Max–Planck Society Workshop: High energy astrophysics: compact stars and active galaxies, p. 259 – 264 (1991). – See Abstr. 012.026 for the main entry.

The purpose of this work is to explore the possible morphological segregations of galaxies with $b_j \leqslant 19.5$ in the UK Schmidt Survey Field 349 centred on (00^h00^m, –35°00′), based on the b_j data taken from the Edinburgh/Durham Southern Galaxy Catalogue (EDSGC, Heydon–Dumbleton et al. 1989) and the R data directly from the COSMOS–scanned parameter set (Heydon–Dumbleton et al. 1988) of images on the R plate. The total number of the galaxies concerned is 6363.

161.219 **Large scale peculiar velocities of galaxies.**
Xiang Shouping.
3. Chinese Academy of Sciences and Max–Planck Society Workshop: High energy astrophysics: compact stars and active galaxies, p. 266 – 269 (1991). – See Abstr. 012.026 for the main entry.

The investigation on large scale peculiar velocities of galaxies has the same significance for cosmology as that on the problems of cosmic microwave background radiation, and primordial helium abundance, because it is closely related to cosmological models, primeval density perturbations and the nature of cosmic matter. On theoretical side, a variety of statistics such as the bulk flow in spheres of varying radii, velocity correlation function and the cosmic Mach number have been used for comparison with observations. These statistics are related to the power spectrum of initial density fluctuations, and in practice can be measured using only the radial component of peculiar velocity. The main points of these methods are presented here.

161.220 **Catastrophe and cosmology in Finsler spacetime.**
Cao Shenglin.
3. Chinese Academy of Sciences and Max–Planck Society Workshop: High energy astrophysics: compact stars and active galaxies, p. 319 – 324 (1991). – See Abstr. 012.026 for the main entry.

In the process of evolution of the universe, the catastrophe nature of the Finsler spacetime and its cosmological implication are discussed. It is shown that the nature of the universe evolution could be attributed to the geometric feature of the Finsler spacetime. The singularities and bifurcations of caustics in the 3–dimensional space and the large scale structure of the universe are also discussed.

161.221 **Thermal instabilities and the origin of globular clusters.**
Ma Er, Chen Shi, Yu Yunqiang.
3. Chinese Academy of Sciences and Max–Planck Society Workshop: High energy astrophysics: compact stars and active galaxies, p. 325 – 330 (1991). – See Abstr. 012.026 for the main entry.

The authors try to outline a globular cluster (GC) formation scenario in the framework of the cold dark matter model in which dwarf galaxies (dG) will form ahead of primeval galaxy and will provide suitable progenitors of GC in them. With the virialization of primeval galaxy, these dG will gain kinetic energy and collide with each other which will provide the necessary condition for the transformation of the progenitors to GC.

161.222 **Three lectures on cosmology.**
E. W. Kolb.
International School of Astroparticle Physics, p. 51 – 89 (1991). – See Abstr. 012.037 for the main entry.

This review consists of three lectures on cosmology. The first lecture is a basic review of the standard hot big–bang cosmology. The second lecture reviews big–bang nucleosynthesis and decoupling of particle species in the early Universe, and the third lecture concerns cosmic inflation.

161.223 **Inflation and quantum cosmology.**
A. Linde.
International School of Astroparticle Physics, p. 91 – 154 (1991). – See Abstr. 012.037 for the main entry.

Contents: 1. Introduction. 2. Chaotic inflation: The stage of inflation; Reheating; Scalar field fluctuations, perturbations of density and galaxy formation. 3. Self–reproducing inflationary Universe. 4. Inflation and the wave function of the Universe. 5. Chaotic inflation versus extended inflation. 6. Inflation and axions: Axion mass problem; Cosmological constraints on the axion mass; Axion perturbations with time–independent f_a; Peccei–Quinn symmetry behavior during inflation; Hybrid inflation; Small λ_ϕ versus small H. 7. Stochastic approach to the baby universe formation. 8. Cosmological constant problem, universe multiplication, baby universes and anthropic principle.

161.224 **Formation of pregalactic dust and the microwave background spectrum.**
B. V. Vajner, A. Yu. Glukhov, V. V. Chuvenkov.
Sov. Astron., Vol. 35, No. 4, p. 331 – 334 (Jul 1991). Current Physics Microform No.: 9203X0517. English translation of Astron. Zh., Tom 68, Vyp. 4, p. 673 – 680 (1991).

The formation of dust particles at red shifts $z \approx 30 - 50$ is investigated in a cosmological model containing Population III stars. The kinetic theory of nucleation is used to show that $H^+(H_2O)$ complexes are centers of condensation of slighly supersaturated water vapor. The relative contribution of particles of "dirty" ice to the total mass of matter may reach $\Omega_d \approx 2 \cdot 10^{-5}$, which makes it necessary to allow for the interaction between dust and radiation in analyzing the microwave background spectrum.

161.225 **Void structure in the early universe: implications for a $\Lambda > 0$ cosmology.**
J. Hoell, W. Priester.
Astron. Astrophys., Vol. 251, No. 2, p. L23 – L26 (Nov 1991). Letter–to–the–editor.

The absorption lines in the Lyα–forest of quasars reflect the distribution of hydrogen clouds along the line of sight. Here the authors used the spectra of the QSO's 0952–01, 0420–388 and 2206–199 covering the spectral range from 3850 to 6600 Å and the redshifts z from 2.17 to 4.43. Presumed that the void structure is homogeneous on a large scale then the analysis of the separation of hydrogen clouds should be compatible with the void–sizes observed at small redshifts ($z < 0.04$). Agreement was found if the analysis is based on Friedmann–Lemaitre models with $H_0 = 90$ km/(s Mpc) and with positive cosmological constant, which provides a cosmic age in the range of 30×10^9yr. The results are not compatible with a homogeneous void structure, if $H_0 = 50$ km/(s Mpc). Furthermore, there is no evidence for any systematic internal dynamical evolution of the void–structure besides the Hubble expansion after the time of Lyα absorption corresponding to $z = 4.4$. Thus, it can be concluded that the void structure was established in the early universe at $z > 4.4$.

161.226 **Astroparticle physics and superstrings.**
D. V. Nanopoulos.
International School of Astroparticle Physics, p. 183 – 218 (1991). – See Abstr. 012.037 for the main entry.

Astroparticle physics, a new discipline involving particle physics and cosmology, has had many successes during the last few years. The author discusses several classic topics including the number of neutrino species, dark matter candidates and the likes. The main emphasis though, will be on some very recent developments in superstring cosmology, where several new interesting ideas have emerged. An alternative to inflation, quantum origin of time and string black holes are some of the hallmarks of purely stringy effects discussed.

161.227 **Measurements of the primordial abundance of helium.**
G. A. Shields.
International School of Astroparticle Physics, p. 219 – 224 (1991). – See Abstr. 012.037 for the main entry.
The author briefly recapitulates the basic theory of primordial nucleosynthesis and then examines some recent result.

161.228 **Recent advances in big bang nucleosynthesis and its relation to particle physics.**
D. N. Schramm.
International School of Astroparticle Physics, p. 225 – 249 (1991). – See Abstr. 012.037 for the main entry.
Big bang nucleosynthesis is the cornerstone of the particle/cosmology connection. This paper reviews big bang nucleosynthesis and, in particular, the recent LEP (and SLC) results on the number of neutrinos are discussed as a positive laboratory test of the standard cosmology scenario. Discussion is also presented on the improved light element observational data as well as the improved neutron lifetime data. Alternate nucleosynthesis scenarios of decaying matter or of quark–hadron induced inhomogeneities are also discussed.

161.229 **Kosmische Strings.**
U. Bleyer.
Astron. Sch., Jahrg. 28, Heft 4, p. 6 – 9 (Aug 1991).

161.230 **The double magnification bias by gravitational lensing.**
U. Borgeest, J. v. Linde, S. Refsdal.
Astron. Astrophys., Vol. 251, No. 2, p. L35 – L38 (Nov 1991). Letter–to–the–editor.
It is well known that the magnification bias (MB) by gravitational lensing provides a possible explanation for the observed overdensity of foreground galaxies around quasars reported by several investigators. The mass concentrations associated with foreground objects magnify quasar images and these quasars are preferentially included in flux (or luminosity) limited samples, thus creating an overdensity of massive objects (for instance galaxies) around quasars in such samples. The authors point out that the overdensity can be much higher if two or more spectral ranges are taken into account. For example, a quasar sample limited in radio and optical luminosity is subject to two independent MBs, assuming that the two luminosities are not strongly correlated, which seems indeed to be the case for quasars. The authors discuss some deep imaging surveys of high redshift quasars undertaken for different purposes. Due to the uncompleteness of the samples with respect to radio luminosity and/or because the samples are small, the results are difficult to evaluate quantitatively, there are however some indications for a double MB. In order to prove the effect of a double MB, a deep imaging survey of a sample being complete in radio and optical luminosity should be carried out. Important informations on the dark matter associated with galaxies and clusters of galaxies would then result.

161.231 **The two–point correlation functions applied to systems with or without a mean density.**
D. Calzetti, M. Giavalisco, R. Ruffini, G. Wiedenmann.
Astron. Astrophys., Vol. 251, No. 2, p. 385 – 388 (Nov 1991).
The definition of the correlation function is not unique and depends on the choice of the normalization constraint. Such a lack of uniqueness strongly influences the correlation amplitude and, as a consequence, may change the interpretation of physical quantities in the system considered. This effect occurs in systems where it is not possible define a mean value for the probability distribution of the average densities (e.g. an infinite fractal), while in the opposite case (as, for instance, liquids, where the average density is well defined, since it does not depend on the sample geometry) there is no contradiction between the different definitions of the correlation function. In this paper the authors indicate the sample dependence of the amplitudes of the different correlation functions in sets where the mean density of the statistical population is unknown or cannot be defined.

161.232 **High spatial resolution in three dimensions: a challenge for large–scale structure formation models.**
T. Buchert, M. Bartelmann.
Astron. Astrophys., Vol. 251, No. 2, p. 389 – 392 (Nov 1991).
The generation of statistically "fair" samples of the universe requires the simulation of large spatial regions of the order of several hundreds of megaparsecs in view of the current observational situation. On the other hand, ambitious models for the formation of large–scale structure have to include, at least on a phenomenological level, the physics of the formation of a galaxy and its halo on scales of at most a few hundreds of kiloparsecs. In order to construct a link of large–scale distributions of galaxies to galactic scales, the authors propose to treat high spatial resolution calculations of semi–analytical models for the formation of large-scale structure, which allow for the realization of this link in 3D.

161.233 **Statistics of gravitational lensing: extended sources.**
S. Marchandon, L. Nottale.
Astron. Astrophys., Vol. 251, No. 2, p. 393 – 401 (Nov 1991).
An analytical formulation for the probability distribution of magnifications due to lensing by clumps on point sources is given in the low optical depth case. Then the authors study the influence of the source size on the statistics of gravitational lensing (low optical depth), and predict that the distribution of the probability of magnifications is two–peaked: (i) the first peak occurs for small magnifications, its position being independent from the source size; (ii) the second peak occurs at maximal amplification. This is at variance with the A^{-3} tail expected in the point source case. It is found that the increase of the source size progressively transfers a fraction of the probability from the first to the second peak. The authors' results are shown to agree with previous numerical simulations.

161.234 **From the expansion of supernovae to the expansion of the Universe.**
R. V. Wagoner.
International School of Astroparticle Physics, p. 251 – 265 (1991). – See Abstr. 012.037 for the main entry.
The author summarizes the methods available to measure distances in the universe, and indicates why supernovae can provide useful beacons. Attention is focused on the expanding photosphere method as applied to type II supernovae. The independent observational tests of the physical system – model atmospheres – are emphasized, and the resulting reliability is contrasted with that of "standard candle" methods. The results of the first critical test of this method, measurement of the distance to supernova 1987A in the Large Magellanic Cloud, are presented. Plans for use of the Hubble Space Telescope to take the first steps toward an independent direct determination of the Hubble Constant are summarized, and prospects for extensions to higher redshifts are discussed.

161.235 **Slowly rotating cosmological perfect fluids in Brans–Dicke theory.**
R. K. Tarachand Singh, J. K. Usham.
Astrophys. Space Sci., Vol. 185, No. 2, p. 181 – 187 (Nov 1991).
The Brans–Dicke field equations for a perfect fluid distribution representing slowly rotating fluid spheres are investigated. Exact solutions are obtained for differential rotation and perfect dragging by imposing some physical restrictions on the matter rotation $\omega(r,t)$. The physical properties are discussed for $k = \pm 1$.

161.236 **Viscous fluid universe filled with stiff fluid in general relativity.**
R. Bali, D. Raj Jain.
Astrophys. Space Sci., Vol. 185, No. 2, p. 211 – 222 (Nov 1991).
The authors have investigated two stiff–fluid models in which the material distribution is that of viscous fluid. In the first model, the coefficient of shear viscosity is assumed to be constant while in the second model the coefficient of shear viscosity is proportional to the rate of expansion in the model. The paper also discusses some physical and geometrical aspects of the model. The behaviour of the model in absence of viscosity is also discussed.

161.237 **The existence of the structural anisotropy of the Jagiellonian field of the galaxies.**
A. J. Gregul (*A. Ya. Gregul'*), A. V. Mandzhos, I. B. Vavilova.
Astrophys. Space Sci., Vol. 185, No. 2, p. 223 – 235 (Nov 1991).
The total galaxy set of the Jagiellonian catalogue ($6° \times 6°$, $N = 15650$ galaxies) was undergone to two–dimensional clusterization procedure. The position angle of the major axis and the ratio of axes of the effective ellipse of inertia are calculated for every cluster. The histograms "position angle – number of galaxies" are investigated for the different samples of clusters. The histograms at the certain radii of clusterization ($R = 2'–3'$) and galaxy population levels ($6 \leqslant N_{min} \leqslant 15$) demonstrate the pronounced and statistically significant anisotropy with the smooth trend along the position angle F and preferable orientation in the range of $F = 105°–120°$. Taking into account the systematical catalogue errors, the problem of reality of this effect has been analysed. The question about possible space scale of anisotropy is discussed. Its lower limit is evaluated by magnitude 500 Mpc. One cannot explain the result obtained by influence of systematical errors of catalogue, as in that case they would have to be enormous.

161.238 **Maps and motions of galaxies in the local Universe.**
A. Dressler.
International School of Astroparticle Physics, p. 267 – 278 (1991). – See Abstr. 012.037 for the main entry.
The view of large–scale structure in the Universe has evolved considerably over the last decade as the prevalence of large superclusters and voids has become more apparent. Highlights of the observational steps that led to this new picture are presented, including efforts to map the distribution of galaxies and, through their peculiar motions, the distribution of dark matter on large scales. Comparison of these two maps for the local volume show rough agreement, though it is still not clear if luminous galaxies trace the distribution of dark matter with any precision.

161.239 **Large–scale structure.**
J. Silk.
International School of Astroparticle Physics, p. 279 – 298 (1991). – See Abstr. 012.037 for the main entry.
The author reviews the origin and evolution of the density fluctuations that seeded large–scale structure. Linear fluctuation growth in flat and curved backgrounds is described, and the role of dark matter, both hot and cold, is discussed. Primordial gaussian density fluctuations lead to observables such as density and velocity correlations, and the resulting predictions are compared with observational results.

161.240 **Cosmic Background Explorer (COBE) observations: new sky maps of the early Universe.**
G. F. Smoot.
International School of Astroparticle Physics, p. 299 – 319 (1991). – See Abstr. 012.037 for the main entry.
This paper presents early results obtained from the first six months of measurements of the cosmic microwave background (CMB) by instruments aboard NASA's Cosmic Background Explorer (COBE) satellite and discusses the implications for cosmology. The FIRAS measurement of the CMB spectrum supports the standard big bang model. The maps made from the DMR instrument measurements show a surprisingly smooth early Universe. The maps of galactic and local emission produced by the DIRBE instrument will be needed to identify foregrounds from extragalactic emission and thus to interpret the results in terms of events in the early Universe.

161.241 **Cold dark matter cosmologies and cosmic microwave background anisotropy at small and intermediate angular scales.**
N. Vittorio, P. F. Muciaccia.
International School of Astroparticle Physics, p. 321 – 331 (1991). – See Abstr. 012.037 for the main entry.
The authors discuss cold dark matter model predictions of cosmic microwave background (CMB) anisotropy at small and intermediate angular scales and present results of a preliminary re–analysis of three CMB anisotropy experiments. The corresponding upper limits on the CMB differential temperature fluctuations set constraints on cold dark matter scenarios: in particular on the baryonic abundance and, more efficiently, on the density parameter.

161.242 **Inflation for astronomers.**
J. V. Narlikar, T. Padmanabhan.
Annu. Rev. Astron. Astrophys., Vol. 29, p. 325 – 362 (1991). – See Abstr. 003.010 for the main entry.
The concept of inflation was introduced into cosmology by Guth about a decade ago. It has generated a remarkable degree of response, both positive and negative, from physicists. By hindsight, the idea appears a natural consequence of the concept of the phase transition, which is believed to have occurred in the very early epochs of the big bang universe, when the breakdown of the so–called grand unification symmetry took place. When it was first proposed, the concept was somewhat difficult to understand, however, as it combined ideas from particle physics with those from the general theory of relativity. Even today, controversy remains about important questions, e.g.: Was there really an inflationary phase in the Universe? If yes, what was the physical mechanism behind it? Given the mode of inflation, what tangible relics should that era have left for today? The authors present the basic idea in a form that is as free of the jargon of particle physics as possible, and focus attention on the last of the three questions posed above.

161.243 **Dark matter detection.**
D. O. Caldwell.
International School of Astroparticle Physics, p. 375 – 394 (1991). – See Abstr. 012.037 for the main entry.
The discovery of the particle which probably constitutes more than 90% of the mass of the universe would be of obvious importance to cosmology and astrophysics, but also of great significance for particle physics. A very new field at the boundary of astro and particle physics is the experimental search for this particle. Even though this work has been going on for a short time, already many types of candidate particles have been ruled out or severely restricted and wide ranges of particle masses and interaction cross sections have been eliminated. There are plans to search for the most likely remaining candidates, light neutrinos, axions, and supersymmetric neutralinos. The last of these will be looked for with newly developed very low temperature detectors capable of measuring both phonons and ionization or quasiparticles and phonons.

161.244 **Dark matter: theory.**
M. Srednicki.
International School of Astroparticle Physics, p. 395 – 415 (1991). – See Abstr. 012.037 for the main entry.
The need for nonbaryonic dark matter in modern cosmological theories is reviewed. The physics of two candidate particles is emphasized: the lightest supersymmetric partice, and the axion.

161.245 **Redshift surveys of galaxies.**
R. Giovanelli, M. P. Haynes.
Annu. Rev. Astron. Astrophys., Vol. 29, p. 499 – 541 (1991). – See Abstr. 003.010 for the main entry.
The authors attempt to give a progress report on the recent advances in terms of the scientific results, technological advances,

and the sheer quantity of measurements in the midst of extraordinary growth. In order to cover the wide variety of cosmological problems explored by redshift surveys, they focus on the current horizons, on the comparison of strategies, and on the future promises of technological developments. Only cursory treatment is given to spectral characteristics other than the redshift, and reviewing quasar studies are avoided. The authors have also focused on work of the 1980s rather than providing a complete historical development.

161.246 **Pourquoi la nuit est-elle noire? Le paradoxe de Chéseaux–Olbers (3ᵉ partie).**
C. Nitschelm.
Orion, Jahrg. 49, Nr. 247, p. 230 – 235 (Dec 1991).

161.247 **Scales of structures and homogeneity in the universe.**
P. Schuecker, H.-A. Ott.
Astrophys. J., Lett., Vol. 378, No. 1, p. L1 – L4 (1 Sep 1991).
A low-resolution objective prism redshift survey of southern galaxies ($15 < m < 20$, $z \leqslant 0.3$) provides 46,000 redshifts (with errors $0.010 \leqslant \sigma_z \leqslant 0.013$) from which subsamples are drawn for studies of large-scale structures and scales of homogeneity in the universe. The derived mean number density for galaxies with $-20.8 \leqslant M \leqslant -19.2$ is $0.024 h^3 Mpc^{-3}$ in a total volume of $1.5 \times 10^6 h^{-3} Mpc^3$. Maximum density deviations, tested with galaxies $-20.8 \leqslant M \leqslant -20.2$ lie between 10% and 30% in comoving test volumes of $5 \times 10^4 h^{-3} Mpc^3$, shifted over $10^5 h^{-3} Mpc^3$. The present data appear to be incompatible with the presence of $100 h^{-1}$ Mpc voids.

161.248 **Modelling the velocity field of elliptical galaxies on large scales.**
U. V. Weigelt, R. E. Kates.
Astron. Astrophys., Vol. 252, No. 1, p. 1 – 9 (Dec 1991).
The large-scale velocity field is modeled using an improved group catalog and the authors' recent, more accurate calibration of the log D_n–log σ relation for elliptical galaxies. The algorithm produces groups that are contained within a sphere of radius R_{typ} in real space centered at the weighted center of mass. The radius $R_{typ} \approx 660$ km s$^{-1} H_0^{-1}$ is determined by a global optimization procedure which comes as close as possible both to minimizing the scatter and to maximizing the mean virial mass of the groups. The authors consider models with constant bulk streaming, pure attractor, and streaming plus attractor. In the bulk and pure attractor models the results are rather compatible with previous work. In the streaming plus attractor model, the attractor is so far away that the velocity field in the observed region is best described as bulk streaming together with a weak linear trend. In essence, no attractor is required. In this case, the velocity at the Local Group has magnitude 506 km s^{-1} in the direction ($l = 352\overset{\circ}{.}3$, $b = 2\overset{\circ}{.}5$), and the components of the field vary by about 5% in the observed region. Depending on the model, the authors obtain estimates ranging from about 640 km s^{-1} to 890 km s^{-1} for the residual dispersion of the velocity field. Particular attention is paid to the reliability of the predictions by means of a careful estimate of the statistical errors and an analysis of the measured and theoretical χ^2 values.

161.249 **Gravitational lensing, time delay, and angular diameter distance.**
R. Narayan.
Astrophys. J., Lett., Vol. 378, No. 1, p. L5 – L9 (1 Sep 1991).
Combining a model of the mass distribution in the gravitational lens Q0957+561 with measurements of the velocity dispersion of the primary lensing galaxy and the time delay between the two images, the Hubble constant has been recently estimated to be $H_0 = 37 \pm 14$ km s^{-1}Mpc^{-1}. The result depends weakly on the assumed value of q_0 and may have systematic errors due to nonuniqueness of the model. It is shown here that the quantity that is most directly measured using a gravitational lens system is not H_0 but rather the angular diameter distance d_{OL} from the observer to the lens. The determination of d_{OL} requires neither a knowledge of the distance to the source nor any cosmological assumption other than local isotropy and homogeneity transverse to the line of sight, again with possible systematic effects. In the case of Q0957+561, where the lens ist at a redshift of $z_L = 0.36$, the result is $d_{OL} = 1700 \pm 600$ Mpc. If the mass distribution of the cluster surrounding the primary lens in Q0957+561 can be determined through independent observations, then it may be possible to estimate both H_0 and q_0.

161.250 **Origin of cosmic X-ray background radiation.**
N. Terasawa.
Astrophys. J., Lett., Vol. 378, No. 1, p. L11 – L15 (1 Sep 1991).
An AGN origin model of cosmic X-ray background radiation is investigated. It is shown that AGN spectra with a bump due to Compton reflection by cold thick matter and an absorption feature caused by cold diffuse gas reproduce well the spectrum of the cosmic X-ray background. The log N – log S relation and the surface density of AGNs are also consistent with observational results under the assumption of a luminosity evolution of AGNs.

161.251 **A Compton reflection model for the cosmic X-ray and gamma-ray backgrounds.**
R. D. Rogers, G. B. Field.
Astrophys. J., Lett., Vol. 378, No. 1, p. L17 – L20 (1 Sep 1991).
The authors calculate the gamma-ray spectrum of an AGN in the Compton reflection model and consider the contribution of such sources to the cosmic gamma-ray background. The spectrum is composed of three parts: an X-ray power law, a broad hump from ≈ 10 to 120 keV due to reprocessing of the power-law X-ray spectrum by cold gas, and gamma rays with are due to the inverse-Compton scattering of X-rays. The authors compute the contribution of such sources to the cosmic gamma-ray background and find that they provide an excellent fit to the entire background spectrum from 3 keV to 10 MeV. They derive constraints on the physical conditions in these sources and consider the possibility of observing such sources with GRO.

161.252 **Megamasers as probes of galaxy mass spectrum evolution.**
V. K. Khersonskij, N. V. Voshchinnikov.
IAU Symposium No. 147: Fragmentation of molecular clouds and star formation, p. 440 (1991). – See Abstr. 012.043 for the main entry.

161.253 **Do clusters of galaxies affect the spectrum of the microwave background?**
M. Markevitch, G. R. Blumenthal, W. Forman, C. Jones, R. A. Sunyaev (*R. A. Syunyaev*).
Astrophys. J., Lett., Vol. 378, No. 2, p. L33 – L36 (10 Sep 1991).
Motivated by recent COBE limits on the spectral distortions in the cosmic microwave background radiation, the authors calculate the distortion, averaged over the sky, due to the Compton scattering of background photons with electrons in the hot gas in clusters of galaxies. Using an existing sample of X-ray clusters, various values of the density parameter Ω, and plausible models for cluster evolution (or lack thereof), the authors generate Monte Carlo realizations of the microwave sky. The spatial structure of these simulations shows a network of discrete sources whose properties can be a strong function of both evolution and Ω. The amount of spectral distortion in the models is greatest for models characterized by self-similar cluster evolution in an open universe and is within an order of magnitude of the current upper limits. Thus, improved observational sensitivity must inevitably detect some deviation from a blackbody spectrum.

161.254 **Cross-correlation of the X-ray background with nearby galaxies.**
K. Jahoda, O. Lahav, R. F. Mushotzky, E. Boldt.
Astrophys. J., Lett., Vol. 378, No. 2, p. L37 – L40 (10 Sep 1991).
The authors have cross-correlated the diffuse 2 – 10 keV X-ray surface brightness (using the HEAO 1 A-2 data) with the galaxy surface density from the diameter – limited UGC and the ESO catalog. Although the galaxies are nearby ($z < 0.03$) and sample only a small fraction of the volume believed to produce the X-ray background, a statistically significant correlation is

present. The "zero-lag" (for 17 square degree cells) cross-correlation and associated 1σ interval is $W_{xg} = (3 \pm 1) \times 10^{-3}$. The authors estimate the volume X-ray emissivity of the local universe, $\varrho_x = (2.8 \pm 1.0) \times 10^{38}$ ergs s $^{-1}$Mpc^{-3}. This implies that unevolved populations of X-ray sources correlated with present-epoch galaxies can contribute only (13 ± 5)% of the cosmic X-ray background.

161.255 **Limitations of observational cosmology.**
M. Kafatos.
IAU Colloquium No. 123: Observatories in Earth orbit and beyond, p. 543 - 550 (1990). - See Abstr. 012.038 for the main entry.

Unlike the usual situation with theoretical physics which is testable in the laboratory, in cosmological theories of the universe one faces the following problems: The observer is part of the system, the universe, and this system cannot be altered to test physical theory. The existence of horizons of knowledge in any cosmological view of the universe is tantamount to inherent observational limits imposed by acts of observation and theory itself. For example, in the big bang cosmology the universe becomes opaque to radiation early on, and the images of extended distant galaxies merge for redshifts, z, of the order of a few. Some of the problems facing big bang cosmology, like the horizon and flatness problems, may not be preconditions on theoretical models but may instead be the manifestations of the quantum nature of the universe.

161.256 **Were all white holes in the early universe converted into black holes?**
T. K. Dey, S. Banerji.
Phys. Rev. D, Vol. 44, No. 2, p. 325 - 332 (15 Jul 1991). Current Physics Microform No.: 9108F0955.

It has been claimed that in the early universe any white hole must have been converted to a black hole. But taking the simple case of an expanding homogeneous dust sphere colliding with a homogeneous spherical shell of dust which are mutually noninteracting, the authors find that the mean motion of the combined system will appear to a distant observer to be either expanding or contracting according as the combined radius at the instant of collision is less than or greater than the Schwarzschild radius.

161.257 **Quantum creation of topological defects during inflation.**
R. Basu, A. H. Guth, A. Vilenkin.
Phys. Rev. D, Vol. 44, No. 2, p. 340 - 351 (15 Jul 1991). Current Physics Microform No.: 9108F0970.

Circular loops of string and spherical domain-wall bubbles of radius equal to the horizon can spontaneously nucleate in de Sitter space. These objects are expanded by the subsequent inflation, and by the end of the inflationary era they have a spectrum of sizes extending well beyond the present Hubble length. Monopole - antimonopole pairs with an initial separation equal to the horizon can also be produced. The cosmological implications of these effects are briefly discussed.

161.258 **Expressions for linearized perturbations in a massive scalar field dominated cosmological model.**
B. Ratra.
Phys. Rev. D, Vol. 44, No. 2, p. 352 - 364 (15 Jul 1991). Current Physics Microform No.: 9108F0982.

A spatially flat cosmological model dominated by a massive scalar (or pseudoscalar) field (Φ) with the scalar field potential $V(\Phi) = m^2(\Phi - \Phi^{(2)})^2$ (where $\Phi^{(2)}$ is the value of the scalar field at which the potential vanishes), in the limit in which the length scale corresponding to the scalar field mass m is very much smaller than the length scale set by the Hubble parameter, is a tractable approximation of the macrophysics of both the late-time axion dominated scenario as well as the early-time "reheating" subscenario of the inflation picture. The author derives solutions of the synchronous gauge relativistic linear perturbation equations, which govern the evolution of spatial irregularities in a spatially homogeneous and isotropic background.

161.259 **Dependence of cosmological energy density irregularities on the shape of the scalar-field potential during inflation and "reheating".**
B. Ratra.
Phys. Rev. D, Vol. 44, No. 2, p. 365 - 375 (15 Jul 1991). Current Physics Microform No.: 9108F0995.

Estimates for the baryon dominated epoch form of the large-scale adiabatic energy density irregularities generated during an early scalar field dominated inflation epoch, in simple inflation modified hot big bang models, are compared to the widely used approximate general expression, which is proportional to the large-scale, gauge invariant part of $H^2\langle\varphi\varphi^*\rangle/(\Phi_b)^2$ evaluated at the first Hubble radius crossing (here Φ_b and φ are the spatially homogeneous and inhomogeneous parts of the scalar field, H is the Hubble parameter). In the de Sitter inflation limit, if the inflation epoch background scalar field solution is an "attractor", or if there is sufficient inflation before the scale of interest leaves the Hubble radius, the approximate general expression identically reproduces earlier results. The approximate general formula is used to restudy energy density inhomogeneities in the quartic-potential scalar field de Sitter inflation model; the difference between the standard result in this model and the author's results in related models is traced to a difference in the form of the part of the potential used to model "reheating" at the end of inflation.

161.260 **Quark- and gluon-jet emission from primordial black holes. II. The emission over the black hole lifetime.**
J. H. MacGibbon.
Phys. Rev. D, Vol. 44, No. 2, p. 376 - 392 (15 Jul 1991). Current Physics Microform No.: 9108F1006.

The emission produced over the lifetime of black holes with masses less than $M_* \approx 4-6 \times 10^{14}$g is investigated by convolving the Hawking emission formulas with a Monte Carlo QCD jet code. Such emission may be astrophysically important if, for example, holes form from initial density perturbations in the early universe. The quark and gluon decay products contribute significantly to the M_* emission and dominate the lifetime emission from holes with initial masses less than about 10^{14}g. The M_* emission shows little sensitivity to the uncertainties in particle physics models around 100 - 300 MeV and above 100 GeV. A more precise determination of the mass of a primordial black hole which just expires today is also given as a function of the cosmological matter density and Hubble constant.

161.261 **Effect of neutrino heating in the early universe on neutrino decoupling temperatures and nucleosynthesis.**
N. C. Rana, B. Mitra.
Phys. Rev. D, Vol. 44, No. 2, p. 393 - 397 (15 Jul 1991). Current Physics Microform No.: 9108F1023.

The neutrino energy loss rates during pair annihilation in the early universe are calculated using the formulation of Herrera and Hacyan and assuming neutrinos to be massless. The work is extended to include the neutrinos of the muon and tauon families, and a method to calculate the evolution of neutrino temperatures is suggested. It is found that neutrinos of the electron family decouple at 1.5×10^{10}K, and those of the other families at 2.5×10^{10}K. The former component is heated by about 0.36%, resulting in a reduction of the primordial abundance of helium (Y_p) in the standard hot big bang model by about 0.003.

161.262 **Strange matter lumps formed in the early universe.**
C. H. Lee, H. K. Lee.
Phys. Rev. D, Vol. 44, No. 2, p. 398 - 402 (15 Jul 1991). Current Physics Microform No.: 9108F1028.

Strange matter lumps are supposed to be formed during the quark-hadron coexistence period at the critical temperature T_c. An important mechanism by which these lumps may be dissolved as the temperature goes down is boiling. However, boiling can occur only if the quark chemical potentials in the strange matter lumps become smaller than some critical values at which quark and hadron matter are in equilibrium. The authors calculate how the equilibrium conditions vary as the temperature drops from T_c to almost zero and discuss how the chemical potential and/or pressure within the strange matter lumps may be maintained above the equilibrium value to avoid boiling. The calculation shows that a gravitationally stable system consisting of a strange matter lump and a hadron shell surrounding it has to contain an unreasonably large number of baryons for the strange matter lump to avoid boiling.

161.263 **Fluctuations of the microwave background radiation on large and intermediate angular scales.**
F. Atrio–Barandela, A. G. Doroshkevich, A. A. Klypin.
Astrophys. J., Vol. 378, No. 1, p. 1 – 5 (1 Sep 1991).
The microwave background anisotropies are estimated for different types of spectra of initial perturbations. Results are presented for multipoles from $l = 2$ (the quadrupole) to $l = 100$. The authors take into account the Sachs – Wolfe and Doppler effects, but treat the recombination as instantaneous, which is a reasonable approximation for multipoles $l < 100$ ($\theta > 2°$). Measurements of $\Delta T/T$ anisotropies on angular scales 2° – 3° and larger are necessary for discrimination between different inflationary models, which predict a variety of shapes for the spectrum of initial fluctuations.

161.264 **First detection of beryllium in a very metal poor star: a test of the standard big bang model.**
G. Gilmore, B. Edvardsson, P. E. Nissen.
Astrophys. J., Vol. 378, No. 1, p. 17 – 21 (1 Sep 1991).
The primordial abundances of ^{9}Be and ^{11}B are the most sensitive indicators of the importance of inhomogeneity in the early universe which have not yet been measured. Additionally, their abundances in the oldest stars are a sensitive measure of the star formation and cosmic–ray spallation history of the early Milky Way. The authors report here the first determination of the abundance of ^{9}Be in a very metal poor star, showing that such measurements are possible, and opening the way to a new investigation of the structure of the very early universe and the evolution of the early Galaxy. The star HD 140283, which has [Fe/H] = –2.6, has a beryllium abundance of $\log(^9\mathrm{Be/H}) = -12.8 \pm 0.3$, a factor of ≈ 1000 greater than the primordial value predicted in the standard model of light element nucleosynthesis.

161.265 **Inhomogeneities in the Universe on scales of $(125\,h^{-1}\mathrm{Mpc})^3$.**
A. Picard.
Astron. J., Vol. 102, No. 2, p. 445 – 453 (Aug 1991). Current Physics Microform No.: 9108A1381.
Number counts of galaxies from a recently completed catalog of galaxies based on 16 plates obtained for the second epoch Palomar Observatory Sky Survey (POSS–II) are presented. A difference of 30% in the galaxy surface density is observed in magnitudes between the range of 16.5–19.0 in r. Because the entire survey has been brought to a common magnitude system using CCD photometry, variations in the magnitude zero point of the individual plates cannot account for this effect. Several tests have been conducted to verify the homogeneity of the catalog. The angular extent and depth of the survey suggests that structures of size 50^{-1}Mpc could be responsible for this effect.

161.266 **Dynamics of the D–dimensional FRW – cosmological models within the superstring – generated gravity model.**
M. Demiański, Z. Golda, W. Puszkarz.
Gen. Relativ. Gravitation, Vol. 23, No. 8, p. 917 – 939 (Aug 1991).
The authors study the dynamics of the generalized D–dimensional (D = 1 + 3 + d) Friedmann – Robertson – Walker (FRW) cosmological models in the framework of an extended gravity theory obtained by adding the Gauss – Bonnet term to the standard Einstein – Hilbert action. The authors extensively use methods of dynamical systems. They consider models filled in with a perfect fluid obeying the equation of state $p = (\gamma-1)\varrho$ and vacuum but non–flat models. The authors present a detailed analysis of the ten dimensional model and study, in particular, the vacuum case. Several phase portraits show how the evolution of this model depends on the parameter γ.

161.267 **Remarks about late stage homogeneous cosmological dynamics.**
C. Uggla, R. T. Jantzen, K. Rosquist, H. von Zur–Mühlen.
Gen. Relativ. Gravitation, Vol. 23, No. 8, p. 947 – 966 (Aug 1991).
Some simple observations are made to clear up misconceptions regarding the behavior of spatially homogeneous monotonically expanding general relativistic cosmological models at very late times, as well as to emphasize some simple features of the dynamics in the direction away from an initial singularity even in the case of recollapse. A few examples illustrating these ideas are discussed, including the derivation of some approximate solutions of the late stage field equations.

161.268 **General vacuum solution for Brans – Dicke – Bianchi type V.**
E. Guzmán.
Gen. Relativ. Gravitation, Vol. 23, No. 9, p. 1007 – 1010 (Sep 1991).
The author has obtained the general vacuum solution for Bianchi type V in the Brans – Dicke theory (BDT). It is shown that for the special case $\gamma = 0$, the sourceless scalar field Φ is dynamically an essential factor, which determines the cosmological expansion parameters R_i, and the singularity does not occur when $t = 0$. For this solution there is no "antigravity" ($\Phi > 0$), which disagrees with other solutions for BDT–Bianchi type V.

161.269 **Inflation in the Einstein – Cartan cosmological model.**
M. S. Berman.
Gen. Relativ. Gravitation, Vol. 23, No. 10, p. 1083 – 1088 (Oct 1991).
All the three primordial phases of the universe, required by inflation, plus what is believed to be the present phase, have been studied in the Einstein – Cartan cosmological model in a Bianchi type I metric.

161.270 **Gauge – invariant variations in the redshift and cosmic background radiation anisotropies.**
W. R. Stoeger, G. F. R. Ellis, B. G. Schmidt.
Gen. Relativ. Gravitation, Vol. 23, No. 10, p. 1169 – 1194 (Oct 1991).
The gauge – invariance of the calculations determining anisotropies in the cosmic microwave background radiation (the Sachs – Wolfe effect) is reexamined. It is shown that the results obtained are gauge – invariant only if a physically – based definition of the surface of last scattering is implemented, in a context where perturbations of the surface of last scattering as well as of the space–time are taken into account. Any physical interpretation of the results based on their splitting into "scalar", "vector", and "tensor" parts, is unique only if non–local (unverifiable) conditions are imposed; locally, any such interpretation is non–unique. The physical meaning of the Sachs – Wolfe potential B and the associated redshift formula depends on implementing a very specific gauge, without a clear physical or geometric meaning; its implications do not extend to other, more usual, gauges.

161.271 **Dynamical reduction in multidimensional cosmological models.**
H. Makaruk.
Gen. Relativ. Gravitation, Vol. 23, No. 11, p. 1205 – 1230 (Nov 1991).
The author presents conditions under which there occurs a dynamical dimensional reduction of cosmological models in the form of a Bianchi I × (N–3) – dimensional torus filled with matter of the ideal fluid type.

161.272 **The age problem in inhomogeneous universes.**
S. Bildhauer, T. Futamase.
Gen. Relativ. Gravitation, Vol. 23, No. 11, p. 1251 – 1264 (Nov 1991).
Recent determinations of globular cluster ages impose severe constraints for standard cosmological models. Before introducing a cosmological constant, other possibilities of solving the

problem should be investigated carefully. In this paper, the backreactions of inhomogeneities in realistic, clumpy universes and their influence on age determinations are studied. Applying a recently developed approximation scheme, it is shown that the backreactions lead to an underestimation of the age of the universe as inferred from a measurement of today's Hubble parameter. Observations of the cosmic microwave background radiation impose constraints on the parameters. For a simple model within the framework of pancake theory for structure formation on a flat expanding background, it is shown that the age problem may be solved by taking into account the backreactions of inhomogeneities in an averaged sense. No cosmological constant is needed.

161.273 **Cosmology with G and Λ coupling scalars.**
R. F. Sisteró.
Gen. Relativ. Gravitation, Vol. 23, No. 11, p. 1265 – 1278 (Nov 1991).
Cosmology with the gravitational and cosmological constants generalized as coupling scalars in Einstein's theory is considered. A general method of solving the field equations is given. Fifteen exact solutions for zero pressure models satisfying $G = G_0(R/R_0)^n$ are given in the appendix; they are briefly discussed.

161.274 **Comments on the computation of Liapunov exponents for the Mixmaster universe.**
B. K. Berger.
Gen. Relativ. Gravitation, Vol. 23, No. 12, p. 1385 – 1402 (Dec 1991).
To explain the discrepancy between recently computed vanishing Liapunov exponents for the evolution of Mixmaster universes and the positive Liapunov exponent for the associated 1–dimensional map first discussed by Belinskii, Khalatnikov, and Lifshitz, the Liapunov exponents computed from a numerical universe evolution are compared using several time variables. Previous numerical results of vanishing Liapunov exponents were obtained with time variables which incrased roughly exponentially in each epoch. Here it is found that minisuperspace proper time, which increases by a fixed amount during each epoch, yields nonvanishing Liapunov exponents within the limited number of epochs numerically accessible. The map parameter u as measured along the trajectory attains the values predicted by the map to very high accuracy (except near the maximum of expansion) even though the metric coefficients deviate in some cases from idealized Mixmaster behavior.

161.275 **The faint galaxy counts revisited in an $\Omega_0 = 1$ universe.**
B. Guiderdoni, B. Rocca–Volmerange.
Astron. Astrophys., Vol. 252, No. 2, p. 435 – 448 (Dec 1991).
The purpose of this paper is to address the following question: is an $\Omega_0 = 1$ universe consistent with the increasing number of constraints given by the faint galaxy counts? In a previous paper (Guiderdoni and Rocca–Volmerange 1990), the authors showed that the various constraints analysed under the assumption of pure luminosity evolution ruled out $\Omega_0 = 1$ and were consistent with $\Omega_0 \approx 0.10$ and redshifts of star formation turn–on $z_{\rm for} \geqslant 10$. Rocca–Volmerange and Guiderdoni (1990) introduced number evolution in addition to the normal luminosity evolution, as the most natural way to save $\Omega_0 = 1$. The authors extend this analysis and show that these results are valid under general assumptions on mass conservation. They develop a more refined modelling of the observations which takes into account the evolution of surface brigthness. Under the assumption that the high–redshift building blocks which are to merge to form the current giant galaxies have the surface brightnesses of nearby galaxies (giants or dwarfs) with similar masses, it turns out that these blocks are visible in the current deep surveys and appear as faint, blue objects. This phenomenological model seems to give nice fits of all the constraints deduced from faint counts, in an $\Omega_0 = 1$ universe with $z_{\rm for}$ as low as 2 to 5. Dynamical friction could be the process regulating merging and number evolution. Some of the consequences of this scenario are investigated.

161.276 **Black holes, "mole burrows" and the "time machine".**
V. P. Frolov.
Priroda, No. 8, p. 10 – 16 (Aug 1991). In Russian.
The possibility of mole burrows (where space has three–dimensional topology) will probably allow to realize the dream of all science fiction writers to travel through space with a velocity greater than the velocity of light.

161.277 **Dynamical evolution of topology of large–scale structure.**
C. Park, J. R. Gott III.
Astrophys. J., Vol. 378, No. 2, p. 457 – 460 (10 Sep 1991).
The nonlinear effects of statistical biasing and gravitational evolution on the genus are studied. The biased galaxy subset is picked for the first time by actually identifying galaxy–sized peaks above a fixed threshold in the initial conditions, and their subsequent evolution is followed. It is found that in the standard cold dark matter (CDM) model the statistical biasing in the locations of galaxies produces asymmetry in the genus curve, and coupling with gravitational evolution gives rise to a shift in the genus curve to the left in moderately nonlinear regimes. Gravitational evolution alone reduces the amplitude of the genus curve due to strong phase correlations in the density field and also produces asymmetry in the curve. Results on the genus of the mass density field for both CDM and hot dark matter models are consistent with previous work by Melott, Weinberg, and Gott.

161.278 **Models of voids in the expanding universe. II.**
W. B. Bonnor, A. Chamorro.
Astrophys. J., Vol. 378, No. 2, p. 461 – 465 (10 Sep 1991).
Spherical voids in a Friedmann universe are modeled as underdense regions represented by the Tolman dust metric. A formula is obtained for the average density of the void in terms of the time that has elapsed since the big bang. In the case of the elliptic and parabolic Friedmann universes, the formula does not allow significant voids unless (1) the void region is born before the rest of the universe and/or (2) the matter in the void is in hyperbolic motion.

161.279 **Cosmological constraints on the zero–point electromagnetic field.**
P. S. Wesson.
Astrophys. J., Vol. 378, No. 2, p. 466 – 470 (10 Sep 1991).
It has been argued by Boyer, McCrea, Puthoff, and others that the zero–point electromagnetic radiation of quantum theory really exists. This idea is examined from the cosmological viewpoint. If the field is regenerated on a cosmic scale, its energy density requires an unlikely combination of cosmological parameters. And if the field energy interacts gravitationally in the conventional way, it leads to some awkward cosmological consequences. General relativity appears to imply either that zero–point electromagnetic energy does not exist, or that it exists but does not gravitate. In either event, major revisions may be necessary in quantum mechanics and/or gravitation.

161.280 **Primordial nucleosynthesis without a computer.**
R. Esmailzadeh, G. D. Starkman, S. Dimopoulos.
Astrophys. J., Vol. 378, No. 2, p. 504 – 518 (10 Sep 1991).
Approximate results for the standard theory of big bang nucleosynthesis are derived analytically, i.e., without the use of the Wagoner code, using ideas of fixed points and freezeout.

161.281 **The brightness temperature of the south celestial pole and the temperature of the cosmic background radiation measured at 36.6 and 12 centimeter wavelength.**
G. Sironi, G. Bonelli, M. Limon.
Astrophys. J., Vol. 378, No. 2, p. 550 – 556 (10 Sep 1991).
The absolute temperature of the sky around the south celestial pole has been measured at 0.82 GHz (λ = 36.6 cm) and 2.5 GHz (λ = 12 cm) during an observing campaign from the Amundsen Scott Base at the South Pole in 1989 December. The antennas had an angular resolution of 18° × 23°. The measured temperatures of the sky have been used to determine the temperature of the cosmic background radiation in a frequency region where

large deviations from a pure Planckian spectrum are expected. The results at 2.5 GHz (T_{sky} = 2.58±0.34K, $T_{CBR}{}^{th}$ = 2.50±0.34K) are fully consistent with the results of previous observations at the same frequency. At 0.82 GHz an undesired level of the system attenuation gave large error bars (T_{sky} = 5.7±1.6K, $T_{CBR}{}^{th}$ = 2.7±1.6K). In spite of that limitation, the authors report the 0.82 GHz results, because no data at the same or nearby frequencies are presently available.

161.282 **A map of the Universe.**
M. Łysik.
Urania, Rok 62, Nr. 11, p. 290 – 294 (Nov 1991). In Polish.

161.283 **Föppl's negative gravity masses and the repulsive cosmos.**
H.-J. Treder.
Astron. Nachr., Vol. 312, No. 4, p. 229 – 230 (1991).
Föppl (1897) distincts between the positive inertial masses and gravity masses which may have positive and negative signes. Föppl has proved the masses with opposite signes of gravity repel each other. For our normal matter the principle of equivalence is valid. Our "world island" is built by masses with positive gravity only and repels each body with negative gravity.

161.284 **Cosmological singularity, inflationary evolution and problems of structure formation.**
V. Müller.
Astron. Nachr., Vol. 312, No. 4, p. 274 (1991). Dissertation abstract.

161.285 **The value of the Hubble constant through novae and supernovae.**
M. Della Valle, M. Capaccioli, E. Cappellaro, M. Turatto.
10. Santa Cruz Summer Workshop in Astronomy and Astrophysics: Supernovae, p. 737 – 740 (1991). – See Abstr. 012.052 for the main entry.

161.286 **The role of expansion for the entropy of the Universe.**
E. Ulrych, P. Voráček.
Astrophys. Space Sci., Vol. 186, No. 1, p. 157 – 158 (Dec 1991). Letter–to–the–editor.
Under assumption of the closed FRW–universe, the idea is presented that the cosmological expansion/contraction on its own, has an entropy balancing effectively the changing entropy of the cosmic fluid in such a way that at every epoch the total entropy of the Universe remains constant.

161.287 **Initial condition for the minimal isocurvature scenario.**
M. Sasaki, J. Yokoyama.
Phys. Rev. D, Vol. 44, No. 4, p. 970 – 979 (15 Aug 1991). Current Physics Microform No.: 9110D1024.
The authors derive a proper expression for the power spectrum of baryon–number fluctuations arising from decay of a heavy Majorana lepton with space dependent CP violation due to a spatially varying Majoron field. The spectrum is explicitly calculated in a power–law inflation model and found to be almost scale invariant on small scales and white noise on large scales. Under reasonable assumptions, the authors present a natural particle–physics model, which provides an appropriate initial condition for the minimal isocurvature scenario of large–scale structure formation.

161.288 **Constraints on generalized extended inflationary models.**
Wang Yun.
Phys. Rev. D, Vol. 44, No. 4, p. 991 – 998 (15 Aug 1991). Current Physics Microform No.: 9110D1045.
The author studies the physical constraints on generalized extended inflationary theories, in which the Brans–Dicke field Φ couples to the inflaton sector in the form of two arbitrary power–law functions of Φ. The most stringent constraints arise from the need to suppress large bubbles and from considerations of the density perturbations generated during extended inflation. There are sizable portions of the parameter spaces in which all the constraints are satisfied.

161.289 **Anisotropic and inhomogeneous solutions of the Einstein–Boltzmann equations for Friedmann–Robertson–Walker spacetimes.**
A. Coley, E. O'Neill.
Astrophys. Space Sci., Vol. 186, No. 2, p. 205 – 210 (Dec 1991).
It is shown how the techniques that yield anisotropic and inhomogeneous solutions of the Einstein–Liouville equations for Friedmann–Robertson–Walker (FRW) spacetimes can be used to demonstrate the existence of anisotropic and inhomogeneous FRW solutions of the Einstein–Boltzmann equations, thereby providing a possible relativistic kinetic theory basis (and consequently a more self–consistent basis) for tilting imperfect fluid FRW cosmological models (based on a set of phenomenological laws of relativistic thermodynamics).

161.290 **The large–scale structure of the Universe, great attractor and the spectra of fluctuations.**
B. I. Gnatik, V. N. Lukash, B. S. Novosyadlij.
Kinematika Fiz. Nebesn. Tel, Tom 7, No. 6, p. 48 – 61 (Nov–Dec 1991). In Russian. English translation in Kinematics Phys. Celest. Bodies, Vol. 7, No. 6.
In the frame of cosmological models with random Gaussian density fluctuations, the relations between predicted and observational characteristics of the large–scale structure of the Universe are analysed. The calculations of the galaxy and cluster correlation functions, large–scale galaxy streamings, great attractor event as large–scale peak and anisotropy of cosmic microwave radiation are performed for the models of the Universe with one and three sorts of massive neutrinos, with cold particles, with hot and cold ones, with double inflation and with additional power in the density fluctuation spectrum. The requirements to the spectrum of density fluctuations are analysed and the spectrum approximation is proposed. Such spectrum is similar to the CDM spectrum at scales $k^{-1} \leqslant 3h^{-1}$Mpc and $k^{-1} \geqslant 10^3h^{-1}$Mpc but has the additional power $\sigma_z = 0.2$ at scales $3h^{-1}\text{Mpc} \leqslant k^{-1} \leqslant 10^3h^{-1}$Mpc. The spectrum amplitude and its slope in this region are determined by the observable characteristics of the large–scale structure. It is shown that two–fold decrease of the upper limitation for temperature anisotropy in the experiment "Kholod" at scales $\theta \approx 2°$ would allow discovering such a spectrum.

161.291 **Constraints on the thermal history of the universe from the cosmic microwave background spectrum.**
C. Burigana, G. De Zotti, L. Danese.
Astrophys. J., Vol. 379, No. 1, p. 1 – 5 (20 Sep 1991).
The authors have carried out an extensive numerical analysis of distortions of the microwave background spectrum that can arise over a broad interval of cosmic times. The range of validity of existing analytic approximation formulae, holding for large and very small values of the Comptonization parameter y, has been assessed and the shape of distortions that can arise for intermediate values of y has been determined. From a comparison with existing spectral data the authors have derived upper limits to the amount of energy that could have been dissipated at each cosmic time. The effectiveness of future measurements in tightening such constraints is discussed.

161.292 **Cosmological velocity–density relation in the quasi–linear regime.**
A. Nusser, A. Dekel, E. Bertschinger, G. R. Blumenthal.
Astrophys. J., Vol. 379, No. 1, p. 6 – 18 (20 Sep 1991).
The authors develop practical methods for extracting the mass–density fluctuation field of a cosmological system, $\delta(x)$, from the peculiar velocity field smoothed on scales of a few Mpc. The methods are local. They are based on quasi–linear approximations to the gravitational equations of motion of a pressureless fluid and address density fluctuations in the range $-0.7 \leqslant \delta \leqslant 4.5$ They are tested against exact solutions in special configurations and against cosmological N–body simulations

with $\Omega = 1$ and with $\Omega = 0.2$. Two approximations are considered under the assumption of Zel'dovich displacements of particles with a universal time dependence: the exact solution of the continuity equation, δ_c, and the exact solution of the dynamical Euler–Poisson equation, δ_d, which turns out to coincide with the linear approximation, δ_0. A key new result is the derivation of the density fluctuation field, based on the Lagrangian Zel'dovich approximation, in terms of the partial derivatives of the Eulerian velocity field. Both δ_d and δ_c are tightly related to the true density, with a standard deviation $\leqslant 0.1$. The continuity density scheme is now being used in the POTENT analysis of the large-scale velocities.

161.293 **Superclustering at high redshifts.**
M. J. West.
Astrophys. J., Vol. 379, No. 1, p. 19 – 36 (20 Sep 1991).

Evidence is presented which suggests that the radio emission from distant quasars and radio galaxies is not randomly directed, but rather has preferred orientations with respect to surrounding large-scale structures. Specifically, the radio major axes of high redshift sources exhibit a significant tendency to point in the direction of neighboring quasars and radio galaxies, for separations up to $\sim 45h^{-1}$Mpc (comoving). This alignment phenomenon is very similar to that seen at low redshifts between neighboring rich clusters and their brightest member galaxies, which suggests that what is being seen here is tentative evidence of the existence of a filamentary pattern of superclustering at redshifts $z \sim 1$ or more. If the alignment effect found here is confirmed with larger and more homogeneous samples, then the orientations of extragalactic radio sources may provide a very useful probe of large-scale structure at high redshifts. These results may also help shed some light on the recently discovered tendency for the extended radio and optical emissions from high redshift radio sources to be aligned.

161.294 **Constraints on models for the high- and low-energy X-ray backgrounds.**
R. A. Daly.
Astrophys. J., Vol. 379, No. 1, p. 37 – 51 (20 Sep 1991).

The high-energy X-ray background most likely results from unresolved sources, since the model in which the background results from a smooth hot intergalactic medium is ruled out by the COBE measurements. Models in which the high- and low-energy backgrounds are produced by sources are constrained by general arguments based on energetics. In addition, the properties of the sources which produce the low-energy background, and of those producing the high-energy background if the spectrum of these sources extrapolates to low energies, are constrained by the flux and isotropy limits of the *Einstein* deep survey. The application of the energy, flux, and isotropy constraints depends on the cosmological model, that is, on whether the universe is flat and matter dominated, open, or flat and dominated by a cosmological constant. The flux and isotropy arguments depend directly on the cosmological model, and the energy arguments are indirectly tied to the cosmological model through the Hubble constant. These constraints are discussed and applied to particular models to determine which models are viable sources of the high- and low-energy backgrounds in different cosmological models. Predictions of the models are discussed, as are the constraints which will result from the ROSAT deep survey.

161.295 **Galaxy formation through hierarchical clustering.**
S. D. M. White, C. S. Frenk.
Astrophys. J., Vol. 379, No. 1, p. 52 – 79 (20 Sep 1991).

The authors develop analytic methods for studying the formation of galaxies by gas condensation within massive dark halos. In these models, structure grows through hierarchical clustering of a mixture of gas and dissipationless dark matter. The authors adopt the simplest models consistent with our current understanding of N-body work on dissipationless clustering, and of numerical and analytic work on gas evolution and cooling. They also employ standard models for the evolution of stellar populations, and construct new models for the way star formation heats and enriches the surrounding gas. The models predict the following properties of the galaxy population: current star formation rates and halo X-ray luminosities; current luminosity functions both for galaxies and for virialized systems; relations between present luminosity, circular velocity, metallicity, and stellar or total M/L ratio; the history of the OB star contribution to the metagalactic ionizing flux; and the distribution of faint blue galaxies in both apparent magnitude and redshift. In this paper the authors give detailed results only for a cold dark matter universe with $\Omega = 1$ and $H_0 = 50\ \mathrm{km\ s^{-1} Mpc^{-1}}$. Predictions depend strongly on the mean baryon density, on the fluctuation amplitude, on the models for heating and metal enrichment by massive stars, and on the initial mass function with which stars form. The most successful models require a large baryon fraction ($\Omega_b/\Omega \gtrsim 0.1$) and efficient heating and enrichment of halo gas. Successful models form their stars late, typically more than half of them since $z = 1$, making the epoch of galaxy formation easily accessible to observation.

161.296 **Contribution of the Weyl tensor to R^2 inflation.**
A. L. Berkin.
Phys. Rev. D, Vol. 44, No. 4, p. 1020 – 1027 (15 Aug 1991).
Current Physics Microform No.: 9110D1074.

To study the generality of inflation from Lagrangians containing higher powers of curvature, the effects of the square of the Weyl tensor, C^2, on the R^2 inflationary scenario is considered. By examining the stability of radiation - dominated Robertson-Walker universes for small perturbations, the C^2 term is determined to enter with a negative coupling constant. Both analytical arguments and numerical methods are used to investigate the influence of the various derivatives of the anisotropy in a Bianchi type I universe. While a small anisotropy only slightly modifies the standard R^2 scenario, for sufficiently large anisotropy the universe cannot isotropize and recollapse subsequently ensues. The critical values at which this behavior occurs are estimated. Thus, the inclusion of C^2 terms restricts the success of R^2 inflation.

161.297 **Baryon isocurvature scenario in inflationary cosmology: a particle physics model and its astrophysical implications.**
J. Yokoyama, Y. Suto.
Astrophys. J., Vol. 379, No. 2, p. 427 – 439 (1 Oct 1991).

A phenomenological model to produce isocurvature baryon-number fluctuations is proposed in the framework of inflationary cosmology. The resulting spectrum of density fluctuations is very different from the conventional Harrison - Zel'dovich shape. The model, with the parameters satisfying several requirements from particle physics and cosmology, provides an appropriate initial condition for the minimal baryon isocurvature scenario of galaxy formation discussed by Peebles.

161.298 **Excursion set mass functions for hierarchical Gaussian fluctuations.**
J. R. Bond, S. Cole, G. Efstathiou, N. Kaiser.
Astrophys. J., Vol. 379, No. 2, p. 440 – 460 (1 Oct 1991).

Most schemes for determining the mass function of virialized objects from the statistics of the initial density perturbation field suffer from the "cloud-in-cloud" problem of miscounting the number of low-mass clumps, many of which would have been subsumed into larger objects. The authors propose a solution based on the theory of the excursion sets of $F(\mathbf{r}, R_f)$, the four-dimensional initial density perturbation field smoothed with a continuous hierarchy of filters of radii R_f. The authors identify the mass fraction of matter in virialized objects with mass greater than M with the fraction of space in which the initial density contrast lies above a critical overdensity, when smoothed on some filter of radius greater than or equal to $R_f(M)$. The differential mass function is then given by the rate of first upcrossings of the critical overdensity level as one decreases R_f at constant position $\mathbf{r}$. The shape of the mass function depends on the choice of filter function. The simplest case is "sharp k-space" filtering, in which the field performs a Brownian random walk as the resolution changes. General filters (e.g., Gaussian or "top hat") are also applied. The authors compare the results of these calculations with group catalogs found from N-body simulations. Over the limited mass range probed by the N-body simulations, the differences in the predicted abundances for different filters are less than a factor of 2 and span the values found in the simulations. Thus the mass functions for sharp k-space and more general filtering all fit the N-body results reasonably well. None of the filter functions is particularly successful in identifying the particles which form low-mass groups in the N-body simulations. These calculations are extended to compute the evolution of the mass function in regions that are constrained to lie within clusters or underdensities at the present epoch. These predictions agree well with N-body results.

161.299 **Collapse of Population III objects induced by cold collisionless dark matter.**
J. C. N. de Araujo, R. Opher.
Astrophys. J., Vol. 379, No. 2, p. 461 - 465 (1 Oct 1991).

The authors study the formation of Population III objects that can be formed by cold collisionless dark matter. It is assumed that at the beginning of the recombination era there exist perturbations only in the dark matter. The authors take into account photon drag, photon cooling, recombination, photoionization, collisional ionization, Lyman-α cooling, and the formation and cooling of hydrogen molecules in the baryonic matter. It is found that for a universe that contains a greater quantity of dark matter (i.e., a larger Ω_d, the ratio of the present density of dark matter to the critical density), the collapse occurs faster for a given perturbed dark matter mass, M_d, but a smaller quantity of baryonic mass is carried with it. For a given universe (i.e., Ω_b and Ω_d given), it is found that for larger M_d a greater fraction of baryonic matter follows the dark matter collapse. The authors previously showed that the minimum mass to collapse from the recombination era without dark matter was $\sim 10^4 M_\odot$; the present calculations indicate that with dark matter the minimum mass is reduced to $\sim 10^3 M_\odot$ if virialization occurs at $\sim 0.01 r_{ta}$; and is $\sim 10^4 M_\odot$ if virialization occurs at $\sim 0.5 r_{ta}$, where r_{ta} is the turnaround radius.

161.300 **Power-spectrum analysis of one-dimensional redshift surveys.**
N. Kaiser, J. A. Peacock.
Astrophys. J., Vol. 379, No. 2, p. 482 - 506 (1 Oct 1991).

The authors consider the statistical properties of power spectra obtained from "pencil-beam" redshift surveys, with particular attention to the recent study by Broadhurst et al. (BEKS). They discuss tests of the null hypothesis that the long-wavelength one-dimensional power spectrum found in such surveys is generated by the known small-scale clustering of galaxies. To a good approximation, the Fourier transform of the one-dimensional galaxy counts is a complex Gaussian random field, and the statistical properties of the low-k power spectrum are calculable in terms of the survey selection functions and the power spectrum of small-scale galaxy clustering. This provides a prediction for the one-dimensional power level, which allows to calculate the frequency of peaks in the power spectrum, as well as such quantities as the uncertainty in the power estimated from a finite bandwidth. The authors generate realizations of power spectra and compare these with the surveys used by BEKS. They find that the power spectra for the two deep surveys alone show little evidence for periodicity or extra large-scale power. With the inclusion of the shallower samples the significance of the spike at $k \cong 0.05h\ \mathrm{Mpc}^{-1}$ increases but is still marginal (such peaks are expected to occur $\cong 6\%$ of the time). The texture of the simulations matches very well with observation and shows that (often quite isolated) spikes are common features. The limited sensitivity of these surveys means that the results are expected to be consistent with a wide range of models for large-scale structure. Most importantly, they are certainly consistent with a galaxy distribution which is very nearly uniform on 100 Mpc scales.

161.301 **Small-scale fluctuations and angular correlations of the X-ray background in the HEAO 1 A-2 energy band: constraints on clustering of X-ray sources.**
J. M. Martín-Mirones, G. De Zotti, E. A. Boldt, F. E. Marshall, L. Danese, A. Franceschini, M. Persic.
Astrophys. J., Vol. 379, No. 2, p. 507 - 516 (1 Oct 1991).

Using HEAO 1 A-2 all-sky survey data, the authors have determined the amplitude of intensity fluctuations of the extragalactic 2 - 20 keV X-ray background (XRB) over an effective solid angle of 1.84 deg^2 and their angular correlation function on angular scales of less than 3°. The observed fluctuations are $\cong 3\sigma$ above those expected if the counts keep an Euclidean slope well below the limit of the Piccinotti survey. A good empirical fit to the data is obtained assuming that the integral counts in the A-2 band have a slope of 1.65(+0.06, -0.07), fully consistent with the best-fit slope of 1.72 obtained by Piccinotti et al. directly from source counts with the same experiment. Alternatively, the data may imply a significant clustering of extragalactic X-ray sources. However, the non-Poisson constribution of rich clusters of galaxies with clustering radii consistent with optical data ($r_0 \leqslant 50 - 60$ Mpc, $H_0 = 50$) do not fully account for the observed signal. Clustering of AGNs could do the job; the required amplitude of the two-point correlation function depends primarily on the evolution of their luminosity and/or space density and of their clustering properties.

161.302 **Selection effects in optical surveys for gravitational lenses.**
C. S. Kochanek.
Astrophys. J., Vol. 379, No. 2, p. 517 - 531 (1 Oct 1991).

The selection effects present in surveys for gravitational lenses are examined in detail. The use of existing quasar catalogs to search for lenses may lead to the loss of 10% - 30% of the expected lenses, because the lensed quasar system did not meet the survey criterion for being a quasar. This occurs if the quasar survey used a stringent criterion for stellar appearance, or a limit on the apparent color of candidates. The APM survey for gravitational lenses among highly elliptical quasar candidates is an efficient scheme for finding lenses, as the elliptical subsample should have 4 times more lensed quasars per object then the sample as a whole. The elliptical subsample is also very incomplete, containing only 10% - 20% of the lenses present in the full sample. The imaging searches for lensed quasars are strongly limited by the average resolution of the images. Qualitative estimates for different seeing conditions are presented. Errors in the measured galaxy luminosities, number counts, and velocity dispersions produce only 50% statistical uncertainties, and the remainder is due to systematic problems in the model for the lens. Magnification bias strongly affects all existing optical surveys, increasing the incidence in bright quasar samples with $m_B < 18.5$ mag by a factor of 15, and in fainter samples with $m_b < 20$ mag by a factor of 4. The uncertainties in the estimates of the magnification bias are another factor of 2.

161.303 **Primordial nucleosynthesis: the next significant figure.**
G. Steigman.
Workshop on Primordial Nucleosynthesis, p. 1 - 14 (1990). - See Abstr. 012.060 for the main entry.

The author reviews the current status of the comparison between the predictions of primordial nucleosynthesis in the "standard" hot big bang model and the abundances of the light elements derived from observational data. The goals include testing the consistency of the standard model and probing - to two significant figures - the constraints on cosmology and particle physics which follow from the agreements between theory and observation. Current data are consistent with the standard hot big bang model and demand that baryons contribute between 2 and 12 percent of the critical density of the Universe. Thus, although baryons can not "close" the Universe, most of the baryons in the Universe are "dark". The author constrains the number of (equivalent) light neutrino flavors to $N_\nu \leqslant 3.4$, in excellent agreement with very recent results from $e^\pm$ colliders.

161.304 **Inhomogeneous nucleosynthesis: extending the standard model.**
C. J. Hogan.
Workshop on Primordial Nucleosynthesis, p. 15 – 27 (1990). – See Abstr. 012.060 for the main entry.
A brief review is presented to nucleosynthesis models in which matter is not uniform. The generation and evolution of inhomogeneities in the big bang is described, including the concentration of baryons in the QCD transition, the effects of heat conduction by neutrinos and photons, the diffusion of baryons through the plasma, and gravity. The effects of a lumpy matter distribution on cosmic abundances, especially effects connected with segregation between neutrons and protons, are summarized.

161.305 **Nuclear reaction uncertainties in standard and non-standard cosmologies.**
R. A. Malaney.
Workshop on Primordial Nucleosynthesis, p. 49 – 62 (1990). – See Abstr. 012.060 for the main entry.
The author discusses the uncertainties in the nuclear input data relevant for calculations of standard and non-standard primordial nucleosynthesis. He shows how these uncertainties can affect the predictive power of such calculations, and he identifies those key nuclear reactions for which improved experimental data on the associated reaction rates is most needed. Such experimental data can lead to more accurate discriminatory tests between the differing primordial nucleosynthesis scenarios.

161.306 **Perspectives: nucleosynthesis and nuclear reactions in standard and inhomogeneous big bang models and beyond.**
F.-K. Thielemann, M. Wiescher.
Workshop on Primordial Nucleosynthesis, p. 92 – 115 (1990). – See Abstr. 012.060 for the main entry.
The authors summarize the topics presented at this workshop on primordial nucleosynthesis, ranging from the standard big bang, over models with inhomogeneous density distributions, their nucleosynthesis, observational constraints on primordial abundances, to the methods and tools used in determining important thermonuclear reaction rates. This summary is complemented by an additional discussion, regarding the possibility of an r-process in inhomogeneous big bang scenarios and by a short overview over nuclear reactions which play a key role in other fields of astrophysics.

161.307 **Motion of test particles and photons in the gravitational field of a cosmic string.**
A. Banerjee, N. Banerjee.
Phys. Lett. A, Vol. 160, No. 2, p. 119 – 122 (11 Nov 1991).
The motion of light rays and test particles are studied in the gravitational field of a cosmic string. Using Hiscock's semiclassical cosmic string metric, it is found that some of the outgoing particles will traverse an infinite distance but some particles may be trapped by the string at a finite distance from the axis. For incoming particles also, some will reach the boundary of the string but some may face a potential barrier and cannot penetrate beyond a certain distance from the string.

161.308 **Mixmaster cosmological models as disturbed Toda lattices.**
M. Biesiada, M. Szydłowski.
Phys. Lett. A, Vol. 160, No. 2, p. 123 – 130 (11 Nov 1991).
The authors show that Mixmaster cosmological models belong to the class of disturbed Toda lattices. The original method of investigating chaos in Hamiltonian systems is applied to Mixmaster models by truncation at orders $i \leqslant 7$. The third order system is equivalent to the generalized Henon–Heiles system. They also show that already the third order approximation exhibits all the qualitative features of Mixmaster models.

161.309 **The density perturbation in the chaotic inflation with non-minimal coupling.**
N. Makino, M. Sasaki.
Prog. Theor. Phys., Vol. 86, No. 1, p. 103 – 118 (Jul 1991).
The authors formulate a method to rigorously evaluate the density perturbation amplitude in inflation models with non-canonical gravity. As a specific example but which contains essential features of non-canonical models, the density perturbation in the $\lambda\Phi^4$ chaotic inflation model with non-minimal coupling $\xi R\Phi^2$ is carefully investigated in terms of the Hamiltonian formalism for constrained systems. They utilize a particular conformal transformation under which the original system with a non-minimally coupled scalar field is transformed into one with a minimal coupling and derive a reduced Hamiltonian for scalar-type perturbations in which only the dynamical degrees of freedom are contained. They then evaluate the amplitude of density perturbations arising from vacuum fluctuations of the system. They find the amplitude in the conformally transformed metric exactly agrees with that in the original metric. Thus the result obtained in previous papers is confirmed in a more rigorous manner.

161.310 **Quantum cosmology and the creation of the universe.**
J. J. Halliwell.
Sci. Am., Vol. 265, No. 6, p. 28 – 35 (Dec 1991).
Einstein's general relativity enable cosmologists to describe the formation of matter and its coalescence into galaxies, stars and planets. But that theory cannot explain the events before the instant of creation. During the past decade, a group of cosmologists turned to the theories of quantum mechanics to fill the gap. Still missing is an observation, such as gravity waves, to verify their ideas.

161.311 **The relation between peculiar velocity and density parameter for a class of solutions in Newtonian cosmology.**
S. Bildhauer, T. Buchert.
Prog. Theor. Phys., Vol. 86, No. 3, p. 653 – 658 (Sep 1991).
The authors derive a relation between the peculiar velocity and the density parameter for a class of inhomogeneous three-dimensional solutions for the motion of dust in the framework of Newtonian cosmology. The domain of validity of the solutions is confined to the early non-linear regime of the formation of large-scale structure. They find that, despite of the solution's fully non-linear property, the $\Omega^{0.6}$-relation obtained in the linear perturbation theory holds. This is also true for the approximate ansatz of the pancake theory. This paper forms a generalization of a recent paper and shows that the non-linear corrections of the $\Omega^{0.6}$-relation depend very much on the special geometrical situation.

161.312 **Supercluster-void structure and Cartesian tourbillons compared.**
R. E. M. Moore.
Prog. Theor. Phys., Vol. 86, No. 4, p. 765 – 766 (Oct 1991). Letter-to-the-editor.
The cellular structure of supercluster-voids is currently regarded as a Voronoi tessellation. Descartes' theory of celestial tourbillons can also be regarded in terms of a Voronoi tessellation.

161.313 **Scatter of Sc galaxies in the Tully–Fisher diagram and the dark matter problem.**
I. D. Karachentsev.
Sov. Astron. Lett., Vol. 17, No. 5, p. 367 – 371 (Sep 1991). Current Physics Microform No.: 9204X1314. English translation of Pis'ma Astron. Zh., Tom 17, No. 10, p. 867 – 876 (1991).
The author considers a sample of 130 Sc galaxies with photometric profiles in the I-band [measured by Freudling (1990)] and reliable data on the widths of H I lines. He has tested various global parameters of the galaxies as distance indicators in the Tully–Fisher diagram. Although he tried various combinations of optical and radio parameters, he could not find any that would reduce the scatter in the galaxy distance estimates below the minimum value $\sigma_{min}(\log \Delta) = 0.07 - 0.06$. The environment of

the galaxies has no serious effect on this limit. About 3/5 of the galaxies have brightness profiles which deviate significantly from the standard line. However, neither the amplitude nor the sign of this deviation is correlated with the position of the galaxy in the TF–diagram. This property can be explained if the galaxy contains a dark component. Whether the galaxies are giants or dwarfs, the mean ratio of total mass to I–luminosity is the same (about $3M_{\odot}/L_{\odot}$). However, this ratio is a function of the surface brightness (SB). For the sample of 130 galaxies, the author finds that the standard deviation log σ[log (M/L)|SB] is 0.090, a small value which imposes severe constraints on the parameters of the dark matter.

161.314 **Radiation from transcendent matter.**
R. A. Treumann.
Europhys. Lett., Vol. 16, No. 1, p. 121 – 126 (1 Sep 1991).
Radiation from hypothetical charged particles having velocities faster than the velocity of light is examined. It is found that such particles may emit radiation as other particles too when accelerated. For longitudinal acceleration the radiation received in the observer's frame is real and positive. For transverse acceleration, however, it has negative energy. Hence either transcendent matter does not radiate at all, or it cannot be electrically charged. In the case one accepts the possibility of radiation at negative energy, transcendent matter could become a source of negative (absorbing) radiation postulated to exist in some quantum cosmologies.

161.315 **The SU(11) family unified model and the thermal history of the very early universe.**
Lu Jizong, Li Xinzhou, Zhang Guangyen.
Nuovo Cimento A, Vol. 104, No. 12, p. 1687 – 1695 (Dec 1991).
Ellis and Steigman have shown that in the SU(5) grand unified theory (GUT) the thermal history of the very early universe is trivial, i.e. in thermal equilibrium. Ogino pointed out that if torsion particles were introduced in the SU(5) GUT, the thermal history of the very early universe would become complex. One of more realistic grand unified models is the SU(11) family unified model suggested by Georgi. In this paper, the authors discuss the thermal history of the very early universe in the frame of Georgi's SU(11) model. Because a lot of superheavy fermions will appear in the SU(11) model, Ogino's results must be modified. These quantitative modifications have been calculated.

161.316 **Scalar fields and matter cosmologies.**
R. de Ritis, G. Platania, C. Rubano, R. Sabatino.
Phys. Lett. A, Vol. 161, No. 3, p. 230 – 235 (30 Dec 1991).
The authors have extended the FRW scalar field cosmology including matter. They have discussed exact solutions for potentials like $V(\varphi) = 0$ and (only in presence of dust) $V(\varphi) = \alpha e^{\eta\varphi} + e^{-\eta\varphi}$.

161.317 **The upper bound to the primordial abundance of helium and the consistency of the hot big bang model.**
K. A. Olive, G. Steigman, T. P. Walker.
Astrophys. J., Lett., Vol. 380, No. 1, p. L1 – L4 (10 Oct 1991).
Consistency of the primordial abundance of ^{4}He with the predictions of big bang nucleosynthesis provides a key test of the standard model of cosmology. Recent calculations predict that the primordial mass fraction of ^{4}He, Y_p, should exceed 0.236. The helium abundances determined from the observations of 38 metal–poor extragalactic H II regions are used to infer a 95% confidence limit statistical upper bound to Y_p which, depending on the specific method employed to extrapolate the data, and/or the subsets of data used in 1, 2, 3, or 4 parameter fits, yields $Y_p \leqslant$ 0.237 – 0.243.

161.318 **Motion of a spherical domain wall and the large–scale structure formation.**
K. Yamamoto, K. Tomita.
Prog. Theor. Phys., Vol. 86, No. 5, p. 1003 – 1020 (Nov 1991).
The evolution of a wall–like structure in the universe is investigated by assuming a simplified model of a domain wall. The domain wall is approximated as a thin spherical shell with domain wall–like matter, which is assumed to interact with dust–like dark matter in an entirely inelastic manner, and its motion in an expanding universe is numerically studied in the general–relativistic treatment. The authors evaluate the lifetime of the wall, which is defined as the characteristic time for the wall to shrink due to its own tension. It is necessary that this time is not smaller than the cosmic age, in order that the walls avoid the collapse to the present time and play an important role in the structure formation of the universe. It is shown that, in spite of the above interaction, the strong restriction is imposed on the surface density of the domain walls and the allowed values are too small to have any influences on the background model.

161.319 **Effect of Rayleigh scattering on cosmic microwave background anisotropies.**
F. Takahara, S. Sasaki.
Prog. Theor. Phys., Vol. 86, No. 5, p. 1021 – 1030 (Nov 1991).
The authors study the effect of Rayleigh scattering off neutral hydrogens on CMB anisotropies. They numerically solved the evolution of frequency dependent radiation perturbations coupled with matter perturbations for a baryon dominated $\Omega = 1$ universe in the synchronous gauge. It is found that at wavelengths shorter than 0.05 cm radiation perturbations are significantly damped and that small scale CMB anisotropies are smaller by 20% than those obtained when only Thomson scattering is taken into account. Thus the effect of Rayleigh scattering is milder than suggested by La. Rayleigh scattering has little effect on matter perturbations and radiation perturbations at longer wavelengths.

161.320 **Correlations of large scale peculiar velocity field and the potential flow analysis.**
Xiang Shouping.
Prog. Astron., Vol. 9, No. 2, p. 107 – 114 (Jun 1991). In Chinese.
The author reports on the progress which is made in the past year on theoretical analysis of large scale peculiar motions of galaxies. The correlation analysis on the large scale peculiar velocity field shows that three clusters of elliptical galaxies in the Centaurus region may dominate the velocity correlation function of elliptical galaxy samples. If the primordial perturbation had a form of a Gaussian spectrum with $n < 0$, then the velocity correlation function of the massive members of hierarchical clustering would be amplified leading to an apparently larger gravitational field, and one would expect to find larger average peculiar velocities when measured from a sample of more massive members of the hierarchy. The velocity potential analysis shows that there is a massive density perturbation peak (Great Attractor) existing in the Hydra–Centaurus region. However, there are still some uncertainties in the analysis of the large scale peculiar velocity field of galaxies due to the sparsity of galaxy samples with known distances and due to the errors in distance measurements which seriously affect the description on large scale peculiar motions of galaxies.

161.321 **New progress in cosmology.**
Chu Yaoquan, Zhu Xingfen.
Prog. Astron., Vol. 9, No. 3, p. 171 – 181 (Sep 1991). In Chinese.
Recent progress in cosmology, in particular new spectacular observational results, is reviewed. The new measurements of the cosmic microwave background radiation by the COBE satellite and the LEP experimental results support strongly the hot big bang model. On the other hand, recent discoveries of large structures in the universe, such as the Great Wall, the Great Attractor, and the periodic distribution of galaxy redshifts, seriously challenge the standard view of the cosmic model.

161.322 **Quantum cosmology, baby universe, and wormhole.**
Shen Yougen.
Prog. Astron., Vol. 9, No. 3, p. 182 – 191 (Sep 1991). In Chinese.
The author describes the methods of the Euclidean path integral (by Hartle–Hawking) and of the Vilenkin tunneling wave function; and their corresponding boundary conditions in quantum cosmology. For the above two kinds of the wave function calculations are made in the de Sitter space–time. The

different kinds of the minisuperspace models are introduced. The baby universe and the wormhole are introduced too.

161.323 **Vector inflation and vortices.**
C. M. Lewis.
Phys. Rev. D, Vol. 44, No. 6, p. 1661 – 1669 (15 Sep 1991). Current Physics Microform No.: 9111B1889.

A vector field A_μ is coupled to the Einstein equations with a linearly perturbed Friedmann – Robertson – Walker metric, constructed to generate first-order vector perturbations. A working classical chaotic vector inflation is demonstrated and then quantum fluctuations of the field are used to constrain the cosmological perturbations. In particular, the vector momentum flux T_{0i} is tracked to the epoch where radiation – dominated matter exists. Matching conditions using observational constraints of the cosmic microwave background radiation give rise to a peculiar cosmological velocity of the order of 10^{-100}c. Amplification of this number, e.g., by breaking the conformal invariance of the field, could be used to generate cosmic magnetic fields using a dynamo mechanism.

161.324 **Stochastic inflation in a simple two-field model.**
S. Mollerach, S. Matarrese, A. Ortolan, F. Lucchin.
Phys. Rev. D, Vol. 44, No. 6, p. 1670 – 1679 (15 Sep 1991). Current Physics Microform No.: 9111B1898.

The dynamics of a nondominating scalar field during inflation is considered in the framework of the stochastic approach where its motion and that of the inflaton are described by two coupled Langevin equations. Curvature perturbations induced by the inflaton make the problem that of a Brownian motion in a random medium. The associated Fokker-Planck equation is solved for a free massless field in a power-law inflation driven by an inflaton with an exponential potential: this simple model could describe the dynamics of the axion, or any other pseudo – Goldstone boson, during inflation. In spite of being free, the field shows a highly non-Gaussian behavior on scales much larger than the present horizon; on observable scales it gives rise to isocurvature perturbations which are both essentially Gaussian and scale-free.

161.325 **Inflationary cosmology based on Weyl-scaled induced gravity.**
D. La.
Phys. Rev. D, Vol. 44, No. 6, p. 1680 – 1684 (15 Sep 1991). Current Physics Microform No.: 9111B1908.

It is shown that successful extended inflation can occur in (Weyl-scaled) minimal theories of gravity, such as a certain superstring theory, Kaluza-Klein theory, N-dimensional gravity, scrR^2 gravity, and induced – gravity models. In these models, during the matter – dominated epoch, the time – varying inertia of cosmic matter acts as an effective dark matter. Therefore, the present density parameter of visible matter is $\Omega_\sigma < 1$, even if the universe is geometrically flat (k = 0).

161.326 **Cosmic superfluid phase.**
B. A. Gradwohl.
Phys. Rev. D, Vol. 44, No. 6, p. 1685 – 1690 (15 Sep 1991). Current Physics Microform No.: 9111B1913.

The universe may have undergone a superfluid – like phase during its evolution, resulting from the injection of nontopological charge into the spontaneously broken vacuum. In the presence of vortices, this charge is identified with angular momentum. This leads to turbulent domains on the scale of the correlation length. By restoring the symmetry at low temperatures the vortices dissociate and push the charges to the boundaries of these domains. If one scales this model to very low energies, one can incorporate it in a late-time phase transition and form large-scale structure in the boundary layers of the correlation volumes. The novel feature of this model lies in the fact that the dark matter is endowed with coherent motion. The author discusses the possibility of identifying this flow around superfluid vortices with the observed large-scale bulk motion. This identification would imply that a more extended map of peculiar velocities in the universe would have to reveal large-scale circulations in the flow pattern.

161.327 **Inflation in generalized Einstein theories.**
A. L. Berkin, K. Maeda.
Phys. Rev. D, Vol. 44, No. 6, p. 1691 – 1704 (15 Sep 1991). Current Physics Microform No.: 9111B1919.

The authors analyze the details of soft inflationary models, which have two scalar fields: one is the standard inflaton, whose potential is exponentially coupled to the other field. Such models are derived from both fundamental theories, and in the conformal frame of generalized Einstein theories. In the latter case, a nonstandard exponential coupling to the inflaton kinetic term may arise. The authors list and discuss the various theories which give soft inflation. They then specialize to new and chaotic inflation potentials, with both standard and nonstandard kinetic terms. The density perturbations are reduced sufficiently so that new inflation works well, with the coupling constant near the values allowed by grand unified theories. For chaotic inflation with a massive inflaton, successful inflation is obtained without any fine-tuning of the coupling constant or initial data.

161.328 **Acceleration-free spherically symmetric inhomogeneous cosmological model with shear viscosity.**
Deng Yaobing, P. D. Mannheim.
Phys. Rev. D, Vol. 44, No. 6, p. 1722 – 1730 (15 Sep 1991). Current Physics Microform No.: 9111B1950.

Some new exact solutions to the Einstein equations with an acceleration-free imperfect fluid source are obtained. Some physical restrictions on the solutions are discussed. Cosmological models built out of these solutions are found to have increasing entropy per baryon and do not possess any flatness problem.

161.329 **The cosmological quark-hadron transition.**
S. A. Bonometto.
Nucl. Phys. B, Proc. Suppl., Vol. 24B, p. 54 – 58 (Dec 1991). – See Abstr. 012.068 for the main entry.

Current results on homogeneous primeval nucleosynthesis are hardly consistent with a baryonic density large enough to allow cooling down primeval fluctuations (over galactic scale) and the growing consensus that $H \gtrsim 75\,\mathrm{km\,s^{-1}Mpc^{-1}}$. The quark-hadron transition may result in inhomogeneous nucleosynthesis. Computations of supercooling during the transition are a step towards connecting microphysical parameters and inhomogeneity features. Recent results on supercooling are given.

161.330 **Hydrodynamical computations of the cosmological quark-hadron transition.**
J. Miller, O. Pantano.
Nucl. Phys. B, Proc. Suppl., Vol. 24B, p. 59 – 64 (Dec 1991). – See Abstr. 012.068 for the main entry.

A description is given of some computations which have been made to study bubble dynamics during the cosmological quark-hadron transition. Results are presented which are relevant for the initial stages of growth of a hadron bubble immediately following its nucleation.

161.331 **Baryon inhomogeneity from the cosmic quark-hadron phase transition.**
H. Kurki-Suonio.
Nucl. Phys. B, Proc. Suppl., Vol. 24B, p. 67 – 73 (Dec 1991). – See Abstr. 012.068 for the main entry.

The authors discuss the generation of inhomogeneity in the baryon-number density during the cosmic quark-hadron phase transition. They use a simple model with thin-wall phase boundaries and ideal-gas equations of state. The nucleation of the phase transition introduces a new distance scale into the universe which will be the scale of the generated inhomogeneity. They review the estimate of this scale. During the transition baryon number is likely to collect onto a layer at the phase boundary. These layers may in the end be deposited as small regions of very high baryon density.

161.332 **A phenomenological QCD approach to the formation of primordial strange quark matter.**
T. Kajino.
Nucl. Phys. B, Proc. Suppl., Vol. 24B, p. 74 – 79 (Dec 1991). – See Abstr. 012.068 for the main entry.

A microscopic calculation using the chromoelectric flux tube model provides a low baryon penetrability at the surface of the quark–gluon plasma. Applying the low penetrability to the time evolution of universal baryon number density during the cosmic QCD phase transition, the author finds the strange quark matter can be formed if the QCD phase transition is weakly first order and satisfies $\sigma \leqslant 5 \times 10^5 \mathrm{MeV}^3$ and 80 MeV $\leqslant T_C \leqslant$ 130 MeV.

161.333 **Survival condition of primordial strange quark matter with low baryon penetrability.**
K. Sumiyoshi, T. Kajino.
Nucl. Phys. B, Proc. Suppl., Vol. 24B, p. 80 – 83 (Dec 1991). – See Abstr. 012.068 for the main entry.

The chromoelectric flux tube model, which is an effective model of the quark confinement, provides a low baryon penetrability at the surface of the strange quark matter. Applying the low penetrability to the evaporation process of the primordial strange quark nuggets, the authors find that the nuggets having a baryon number greater than 10^{39} could survive evaporation into nucleons, suggesting a possible detection of such objects or their fragments in the present Universe.

161.334 **Survival of quark nuggets and their present–day abundance.**
J. Madsen.
Nucl. Phys. B, Proc. Suppl., Vol. 24B, p. 84 – 90 (Dec 1991). – See Abstr. 012.068 for the main entry.

Evaporation and boiling of quark nuggests in the early universe is described, and survivability criteria given. Similarities and differences to heavy–ion collisions are pointed out. Limits on the present cosmic abundance and galactic flux of quark nuggets derived from big bang nucleosynthesis and the pulsar glitch phenomenon are discussed.

161.335 **The correlation function of galaxy ellipticities produced by gravitational lensing.**
J. Miralda–Escudé.
Astrophys. J., Vol. 380, No. 1, p. 1 – 8 (10 Oct 1991).

The gravitational potential of mass fluctuations in the universe deflects the light rays of background sources, distorting their shapes. This distortion introduces a correlation of the ellipticities of the sources lying within a given area of the sky. In this paper, the author calculates the correlation of ellipticities produced by gravitational lensing, as a function of the power spectrum of density fluctuations in the universe, by generalizing an analytical method developed by Gunn. This method is applied to a model where identical objects with spherically symmetric density profiles are randomly laid down in space, and to the cold dark matter model. The possibility of detecting this correlation is discussed. An ellipticity correlation can also be caused by an intrinsic alignment of the axes of galaxies belonging to a cluster or a supercluster; however, a method is suggested by which one can distinguish between the two types of correlation. This ellipticity correlation is one of the few astronomical observations that can directly probe large–scale mass fluctuations in the universe.

161.336 **Statistics of gravitational lenses: the uncertainties.**
S. Mao.
Astrophys. J., Vol. 380, No. 1, p. 9 – 14 (10 Oct 1991).

The assumptions in the analysis of gravitational lensing statistics are examined. Special emphasis is given to the uncertainties in the theoretical predictions. It is shown that a simple redshift cutoff model, which may result from galaxy evolution, can significantly reduce the lensing probability and explain the large mean separation of images in observed gravitational lenses. This effect may affect the constraint on the contribution of the cosmological constant to producing a flat universe from the number counts of the observed lenses. For the $\Omega_0 = 1$ model, the lensing probability of early–type galaxies with finite core radii is reduced roughly by a factor of 2 for high–redshift quasars as compared with the corresponding singular isothermal sphere model. The finite core radius effect is $\sim 20\%$ for a λ–dominated flat universe. It is also shown that the most recent galaxy luminosity function gives lensing probabilities that are smaller than previously estimated roughly by a factor of 3.

161.337 **The mass distribution of groups and clusters of galaxies.**
A. Cavaliere, S. Colafrancesco, R. Scaramella.
Astrophys. J., Vol. 380, No. 1, p. 15 – 23 (10 Oct 1991).

The authors reformulate the collapse theories for the mass distribution of cosmic structures – of which the theory of Press and Schechter constitutes the archetype – in terms of a rate equation with finite collapse and survival times. The authors then derive the mass distribution and the optical luminosity function expected for groups and clusters of galaxies in these hierarchical clustering scenarios, considering both the collapse of high peaks in the initial overdensity field and the subsequent mass infall. Finally, the transition from direct collapse to merging and aggregation is discussed.

161.338 **Large–scale evolution of a perturbed universe.**
J. Hwang.
Astrophys. J., Vol. 380, No. 2, p. 307 – 314 (20 Oct 1991).

The author presents a straightforward derivation of the large–scale evolution of a perturbed Robertson – Walker space. The origin of the Friedmann – like behavior of the perturbed model is clarified in a comoving gauge, and thus in a corresponding gauge–invariant formalism, using the covariant equations. Thus, when the imperfect fluid contributions are negligible, the large–scale perturbations in a nearly flat background evolve like separate Friedmann models. However, there exists a preferable gauge–invariant quantity ζ, which is conserved on large scales even considering the background spatial curvature, the cosmological constant, and the anisotropic pressure. Its independence from the anisotropic pressure allows to generalize ζ to a class of generalized gravity theories. The notation used here allows simple derivations of both the integral form solution in the large scale and also Bardeen's second–order evolution equation in the general case.

161.339 **False–vacuum decay in time–dependent and two–field models.**
L. M. Widrow.
Phys. Rev. D, Vol. 44, No. 8, p. 2306 – 2313 (15 Oct 1991). Current Physics Microform No.: 9112A1230.

The author studies the false vacuum decay of a scalar field making use of the functional Schrödinger equation. The wave function describing the nucleation and expansion of true vacuum bubbles is derived to $O(\hbar)$, and the problem of matching at the classical turning point is discussed. The author also studies the vacuum decay of a scalar field coupled to a time–dependent external field and derives the traversal time for bubble nucleation. It is shown that a quasistatic approximation is valid as long as the time scale for the external field to change is long compared to the light–travel time across the bubble. Implications for recently proposed scenarios of first–order inflation are discussed.

161.340 **Chaotic inflation driven by a minimally coupled scalar field in Brans–Dicke models.**
J. McDonald.
Phys. Rev. D, Vol. 44, No. 8, p. 2314 – 2324 (15 Oct 1991). Current Physics Microform No.: 9112A1238.

The author considers inflation driven by the energy density of a conventional scalar field minimally coupled to gravity, which has an initial vacuum expectation value which is large compared with the value at the minimum of its potential ("chaotic inflation"), in the context of Brans–Dicke models with a time–dependent Newton's constant. The equations of motion for the scalar field driving inflation and the Brans–Dicke scalar are solved in the slow–rolling approximation for the case of a $\lambda\sigma^4$ potential driving inflation, and the magnitude and spectrum of density perturbations produced during inflation are calculated. Sufficient inflation to account for the observed homogeneity and isotropy of the universe is found to occur only if the coupling ε between the Brans–Dicke scalar and the Ricci scalar is smaller than 1. It is shown that the spectrum of density perturbations is sufficiently flat to explain galaxy formation only if $\varepsilon < 0.04$.

161.341 **Dynamics of cooling gas in galactic dark halos.**
J. F. Navarro, W. Benz.
Astrophys. J., Vol. 380, No. 2, p. 320 – 329 (20 Oct 1991).
The authors simulate the dynamical evolution of both collisionless "dark" particles and a dissipative gaseous ("baryonic") component in a flat universe in order to investigate the formation process of the luminous components of galaxies at the center of galactic dark halos. It is assumed that the initial density fluctuations are Gaussian with a power spectrum of the form $P_{(k)} \sim k^{-1}$, that the gas amounts to 10% of the mass of the universe, and that both the gas and the dark matter are identically distributed in phase space at high redshifts. The gas is allowed to dissipate energy according to a cooling function corresponding to an H–He mixture with primordial abundances. The effects of star formation and supernova explosions are neglected. The results confirm previous suggestions that the merging history of the surrounding halo is a key factor in the determination of the morphological type of a galaxy. The baryonic component is found to lose more angular momentum than could be predicted by dissipationless simulations. The authors observe the formation of dense, slowly rotating baryonic cores at the center of galactic dark halos. On the other hand, the loss of angular momentum by the gaseous component is so important that prominent, rotationally supported disks like those of bright spirals apparently are unable to form in the present simulations.

161.342 **Magnetohydrodynamic cosmologies.**
R. Portugal, I. D. Soares.
Astrophys. J., Vol. 380, No. 2, p. 330 – 333 (20 Oct 1991).
The authors analyze a class of cosmological models in a magnetohydrodynamic regime extending and completing the results of a previous paper. The material content of the models is a perfect fluid plus electromagnetic fields. The fluid is neutral on average, but admits an electrical current which satisfies Ohm's law. All models fulfill the physical requirements of near-equilibrium thermodynamics and can be favorably used as a more realistic description of the interior of a collapsing star in a magnetohydrodynamic regime with or without a magnetic field.

161.343 **After the first three minutes: General introduction.**
P. J. E. Peebles.
AIP Conf. Proc., No. 222, p. 3 – 11 (15 Apr 1991). – See Abstr. 012.069 for the main entry.
What will be the character of research in physical cosmology in the 1990s? We may find a greater emphasis on semiempirical models whose development is informed by a more nearly equal balance of theory and observation than has been feasible up to now. The author presents an example of how this might go in a model for galaxy formation in a low density Universe.

161.344 **The first three minutes: 1990 version.**
D. N. Schramm.
AIP Conf. Proc., No. 222, p. 12 – 40 (15 Apr 1991). – See Abstr. 012.069 for the main entry.
Early Universe cosmology has come a long way since S. Weinberg's famous book, *The first three minutes*. The author summarizes some of the key areas: inflation, phase transitions, dark matter, and nucleosynthesis. The summary focusses what comes out of the Universe at the end of three minutes and thus serves as the initial conditions for later studies of the Universe. Particular emphasis is made on those aspects of the early Universe that are confirmed by experiment or observation and those aspects that are relatively insensitive to the model-dependent assumptions.

161.345 **Rocket measurement of the spectrum of the CBR.**
M. Halpern, H. P. Gush, E. H. Wishnow.
AIP Conf. Proc., No. 222, p. 53 – 61 (15 Apr 1991). Current Physics Microform No.: 9106E0053. – See Abstr. 012.069 for the main entry.
The spectrum of the cosmic background radiation (CBR) has been measured using a liquid helium cooled spectrometer carried to high altitude by a rocket. The spectrum in the frequency band 2 to 30 cm^{-1} is very well fitted by a Planck function of temperature 2.736K. The scatter of the equivalent temperature in the band 3 to 16 cm^{-1} is $\pm$10mK, about 1/3% of the mean, whereas the estimated overall experimental accuracy of the mean temperature is $\pm$17mK. These results are inconsistent with previously reported infrared excess in CBR, but are in good agreement with the COBE results. They form the most precise constraints reported to date on the spectrum of the CBR.

161.346 **A decade of long-wavelength CMB measurements.**
A. Kogut, M. Bensadoun, M. Bersanelli, G. De Amici, S. Levin, M. Limon, G. F. Smoot.
AIP Conf. Proc., No. 222, p. 62 – 72 (15 Apr 1991). Current Physics Microform No.: 9106E0062. – See Abstr. 012.069 for the main entry.
The authors have continued the efforts of an international collaboration begun in 1979 to measure the long-wavelength (Rayleigh–Jeans) spectrum of the cosmic microwave background (CMB). They have repeated measurements at a number of wavelengths over many years, using different equipment and techniques to investigate the possibility of unsuspected systematic errors in the measurements. They conclude that their results show no evidence for significant undiagnosed systematic effects. The results have improved in precision over ten years and yield significant limits on physical processes in the early Universe. The CMB spectrum is consistent with a blackbody over >3 decades of wavelength.

161.347 **COBE limits on explosive galaxy formation scenarios.**
J. J. Levin, K. Freese, D. N. Spergel.
AIP Conf. Proc., No. 222, p. 73 – 76 (15 Apr 1991). Current Physics Microform No.: 9106E0073. – See Abstr. 012.069 for the main entry.
Hydrodynamic models have been proposed to generate large-scale structure from small-scale initial density perturbations. These models invoke high energy explosions to catalyze the collapse of material on large mass scales. Through inverse Compton scattering, a distribution of explosions in the early Universe produces a distortion in the spectrum of the primordial photon background. The authors calculate the overall spectral distortion and compare their calculations to the recent measurements of the COBE satellite. They find that existing upper limits on the spectral distortion of the background radiation are beginning to challenge this theory of galaxy formation.

161.348 **Constraints imposed by the COBE spectral measurements.**
R. A. Daly.
AIP Conf. Proc., No. 222, p. 77 – 80 (15 Apr 1991). Current Physics Microform No.: 9106E0077. – See Abstr. 012.069 for the main entry.
Upper bounds on the magnitudes of spectral distortions of the microwave background radiation are used to constrain the amplitude of perturbations present prior to recombination. These constraints are further used to constrain the index of the initial power spectrum of density perturbations in the isocurvature and adiabatic models. The current and projected limits imposed by the COBE spectral measurements are discussed.

161.349 **Early FIRAS results and superconducting cosmic strings.**
M. Signore, N. Sanchez.
AIP Conf. Proc., No. 222, p. 81 – 84 (15 Apr 1991). Current Physics Microform No.: 9106E0081. – See Abstr. 012.069 for the main entry.
The preliminary results from FIRAS/COBE (Far Infrared Absolute Spectrometer/Cosmic Background Explorer) and the

available anisotropy in the intensity of the Cosmic Microwave Background (CMB) for an angular separation of 1° place new constraints on Superconducting Cosmic Strings (SCS) and their cosmological evolution independent of numerical simulations. The early FIRAS results and the latest PSR 1937+21 timing measurements converge to the same upper limits on $G\mu$.

161.350 **Post-COBE measurements of the cosmic microwave background spectrum.**
S. Staggs, D. Wilkinson.
AIP Conf. Proc., No. 222, p. 85 – 92 (15 Apr 1991). – See Abstr. 012.069 for the main entry.

This paper reviews experiments for $\lambda > 1$ cm and points out some ways that those experiments can be improved to achieve higher accuracy.

161.351 **Preliminary DMR measurements of the CMB isotropy.**
G. F. Smoot, C. L. Bennett, A. Kogut, J. Aymon, C. Backus, G. De Amici, K. Galuk, P. D. Jackson, P. Keegstra, L. Rokke, L. Tenorio, S. Gulkis, M. G. Hauser, M. Janssen, J. C. Mather, R. Weiss, D. T. Wilkinson, E. L. Wright, N. W. Boggess, E. S. Cheng, T. Kelsall, P. Lubin, S. Meyer, S. H. Moseley, T. L. Murdock, R. A. Shafer, R. F. Silverberg.
AIP Conf. Proc., No. 222, p. 95 – 106 (15 Apr 1991). Current Physics Microform No.: 9106E0095. – See Abstr. 012.069 for the main entry.

The COBE Differential Microwave Radiometers (DMR) instrument has produced preliminary full-sky maps at frequencies 31.5, 53, and 90 GHz. The redundant channels and matched beams at three frequencies distinguish the DMR from previous large-scale surveys. Galactic emission is seen unambiguously at all three frequencies. The only large-scale anisotropy detected in the cosmic microwave background is the dipole anisotropy. There is no clear evidence for any other large-angular-scale feature in the maps. Without correcting for any systematic effects, the authors are able to place limits $\Delta T/T_0 < 3\times10^{-5}$ for the rms quadrupole amplitude, $\Delta T/T_0 < 4\times10^{-5}$ for monochromatic fluctuations, and $\Delta T/T_0 < 4\times10^{-5}$ for Gaussian fluctuations. The data limit $\Delta T/T_0 < 10^{-4}$ for any feature larger than 7°. The authors briefly review the DMR and discuss some implications of these results for cosmology.

161.352 **Limits on Gaussian fluctuations in the cosmic microwave background at 19.2 GHz.**
S. P. Boughn, E. S. Cheng, D. A. Cottingham, D. J. Fixsen.
AIP Conf. Proc., No. 222, p. 107 – 112 (15 Apr 1991). Current Physics Microform No.: 9106E0107. – See Abstr. 012.069 for the main entry.

The northern hemisphere data from the 19.2 GHz full sky survey are analyzed to place limits on the magnitude of Gaussian fluctuations in the cosmic microwave background implied by a variety of correlation functions. Included among the models tested are the monochromatic and Gaussian-shaped families, and those with power law spectra for $-2 \leqslant n \leqslant 1$. The authors place an upper bound on the quadrupole anisotropy of $\Delta T/T < 3.2\times10^{-5}$rms, and an upper bound on scale-invariant (n = 1) fluctuations of $a_2 < 4.5\times10^{-5}$ (95% confidence level). There is significant contamination of these data from Galactic emission, and improvement of modelling of the Galaxy could yield a significant reduction of these upper bounds.

161.353 **Early results from the MIT millimeter and sub-millimeter balloon-borne anisotropy measurement.**
S. S. Meyer, E. S. Cheng, L. A. Page.
AIP Conf. Proc., No. 222, p. 113 – 118 (15 Apr 1991). Current Physics Microform No.: 9106E0113. – See Abstr. 012.069 for the main entry.

The MIT balloon-borne bolometric search for Cosmic Microwave Background Radiation (CMBR) anisotropies places the most stringent constraints to date on the fluctuations in the CMBR. Four maps of half of the Northern Hemisphere at 1.8, 1.1, 0.63 and 0.44 mm wavelength, have a beam size of 3.8° with a 1 sigma sensitivity of less than 0.1 mK (thermodynamic) per FOV in each of the first two channels. Analysis of the sky map at 1.8 mm wavelength using a likelihood ratio test for galactic latitudes $|b| \geqslant 15°$ yields a 95% confidence level (CL) upper limit on fluctuations of the CMBR at $\Delta T/T \leqslant 1.6\times10^{-5}$ with a statistical power of 92% for Gaussian fluctuations at a correlation angle of 13°. Between 3° and 22°, the upper limit of fluctuations is $\Delta T/T \leqslant 4.0\times10^{-5}$ (95% CL). An anisotropy is detected in the map but it cannot yet be attributed to primordial sources. The ultimate sensitivity for this experiment is $\Delta T/T \sim 7\times10^{-6}$ (95% CL) over this angular range for Gaussian fluctuations.

161.354 **Dynamical effects of the cosmological constant.**
O. Lahav, P. B. Lilje, J. R. Primack, M. J. Rees.
Mon. Not. R. Astron. Soc., Vol. 251, No. 1, p. 128 – 136 (1 Jul 1991).

The possibility of measuring the density parameter Ω_0 and the cosmological constant $\lambda_0 \equiv \Lambda/(3H_0^2)$ using dynamical tests is explored in linear and non-linear theory. In linear theory the authors find that the rate of growth of the perturbations at the present epoch is approximated by $f(z=0) \approx \Omega_0^{0.6} + 1/70\, \lambda_0(1 + 1/2\, \Omega_0)$. Therefore, dynamical tests such as infall around clusters and dipoles at the present epoch do not distinguish well between universes with and without a cosmological constant. At higher redshifts, the perturbations also depend mainly on the matter density at a particular epoch, $f(z) \approx \Omega^{0.6}(z)$, which has a strong dependence on λ_0 at $z \approx 0.5$–2.0. Therefore, information on both parameters can be obtained by looking at clustering at different redshifts. In practice, however, the other observables also depend on the cosmology, and in some cases conspire to give a weak dependence on λ_0. By using the non-linear spherical infall model for a family of Cold Dark Matter (CDM) power-spectra the authors also find that dynamics at $z=0$ does not tell much about λ_0. At higher redshifts there is unfortunately another conspiracy between conventional observables, which hides information about λ_0. The authors find radius of a virialized cluster (relative to the turn-around radius) is approximated by $R_f/R_{ta} \approx (1-\eta/2)/(2-\eta/2)$, where η is the ratio of Λ to the density at turn-around. Therefore a repulsive Λ gives a smaller find radius than a vanishing Λ.

161.355 **Inflation, large scale structure, and COBE.**
R. K. Schaefer.
AIP Conf. Proc., No. 222, p. 119 – 122 (15 Apr 1991). Current Physics Microform No.: 9106E0119. – See Abstr. 012.069 for the main entry.

The predictions of the inflationary cosmology are compared with the expected sensitivity of the COBE satellite. It is noted that detection of large angle temperature fluctuations in the usual biased cold dark matter version of inflation requires COBE to reduce systematic errors to the level of the detector sensitivity. However, the difficulty the usual biased cold dark matter scenario has in fitting recent large scale structure observations leads us to consider a variety of possible modifications of inflation. These changes generally increase the amplitude of the temperature fluctuations to a more detectable level, although the amount of increase depends on the specific modification. Thus we would expect that COBE will easily see temperature fluctuations in a realistic model of large scale structure formation based on inflation.

161.356 **A millimeter-wave anisotropy experiment (MAX) to search for anisotropy in the cosmic background radiation on medium angular scales.**
M. L. Fischer, D. C. Alsop, A. C. Clapp, D. A. Cottingham, A. E. Lange, P. L. Richards, G. F. Smoot, J. O. Gunderson, T. C. Koch, P. R. Meinhold, P. M. Lubin, E. S. Cheng, E. Kreysa.
AIP Conf. Proc., No. 222, p. 123 – 128 (15 Apr 1991). Current Physics Microform No.: 9106E0123. – See Abstr. 012.069 for the main entry.

The authors report preliminary results from two balloon flights of a millimeter-wave telescope designed to measure anisotropy in the cosmic microwave background (CMB) on angular scales from 0.3 to 3 degrees. During the first flight, in

November 1989, the authors measured the spectrum of the brightness of the galactic plane at a galactic longitude of $l_{II} = 24$ degrees and searched for CMB anisotropy at nine points centered near the north celestial pole. The second flight, in July 1990, employed improved bolometric detectors. Preliminary analysis of the data gives a factor of 3 improvement in sensitivity over the first flight in the $6\,cm^{-1}$ band. Future plans include a new receiver to provide an additional factor of 5 improvement in sensitivity and a new balloon-borne telescope which is optimized for low-background bolometric detectors at millimeter wavelengths.

161.357 **CMBR anisotropies: small scale measurements and the OVRO experiments.**
S. T. Myers, A. C. Readhead, C. R. Lawrence.
AIP Conf. Proc., No. 222, p. 129 - 139 (15 Apr 1991). - See Abstr. 012.069 for the main entry.

This is a short overview of current CMBR anisotropy experiments on small angular sizes. Ongoing efforts and future prospects in relation to the testing of cosmological theories are summarized. Details of previously published as well as new observations from Owens Valley 20 GHz (RING) experiment are presented.

161.358 **Fine-scale microwave fluctuations at $\lambda = 3.6$ cm.**
E. B. Fomalont, J. Lowenthal, R. B. Partridge, R. A. Windhorst.
AIP Conf. Proc., No. 222, p. 140 (15 Apr 1991). Abstract. Current Physics Microform No.: 9106E0140. - See Abstr. 012.069 for the main entry.

161.359 **Observational bounds on the cosmic microwave background anisotropy at small and intermediate angular scale.**
N. Vittorio, P. F. Muciaccia.
AIP Conf. Proc., No. 222, p. 141 - 150 (15 Apr 1991). Current Physics Microform No.: 9106E0141. - See Abstr. 012.069 for the main entry.

The authors re-analyzed the data of three CMB anisotropy experiments, performed on small and intermediate angular scales. The analysis is based on a likelihood ratio test, and considers flat and open CDM models. The authors discuss: i) the method of data analysis; ii) the statistical significance of the obtained upper limits; iii) the constraints one can pose on the density parameter and the baryonic abundance of these models.

161.360 **The biasing scheme in N-body simulations.**
C. Park.
Mon. Not. R. Astron. Soc., Vol. 251, No. 1, p. 167 - 173 (1 Jul 1991).

A popular prescription for the galaxy formation, i.e. identifying peaks in the linear density field as sites of galaxy formation, is numerically justified under the Cold Dark Matter cosmogony. In particular, the peak-background scheme, originally developed to simplify analytic calculations of peak properties in a Gaussian random density field, is proven to be an excellent presciption for assigning peak tracers in low-resolution N-body simulations through many statistical comparisons. It is found that statistical properties of peak tracers allocated by the peak-background scheme closely resemble those of true peaks when the background smoothing length is about three times the galaxy scale and the correspondence improves as they evolve gravitationally. It is also shown that these peak particles are indeed strongly associated with non-linear objects that are presently collapsed.

161.361 **In an optically thick universe, the cosmic microwave background can appear smoother than the Universe actually was.**
D. W. Marks.
AIP Conf. Proc., No. 222, p. 151 - 154 (15 Apr 1991). Current Physics Microform No.: 9106E0151. - See Abstr. 012.069 for the main entry.

The relativistically invariant equation of radiative transfer with arbitrary velocity flows, temperature and density gradients, and gravitational fields can be expanded in projected, symmetric, trace-free tensors. In the optically-thick limit, the monopole moment of the radiation field is Planckian plus a small term that depends on the proper velocity dotted with the temperature gradient (if any) divided by the absorption (but not the scattering) cross-section. The dipole moment contains a term proportional to the temperature gradient and another term proportional to the proper acceleration, both divided by the sum of the absorption cross-section and the scattering cross-section. The quadrupole term is proportional to the rate of strain (if any), divided by the sum of the absorption cross-section and nine-tenths the scattering cross-section. All higher-order moments vanish identically. Thus the cosmic microwave background should not show higher-order multipole moments, in agreement with COBE observations. Moreover, the multipole expansion shows terms proportional to the temperature gradient, but not to the density gradient. The absence of temperature fluctuations in the CMB does not imply the absence of density fluctuations in the early Universe. The smoothness of the cosmic microwave background may be a weaker constraint on theories of galaxy formation than currently thought.

161.362 **State reduction and cosmic anisotropy.**
C. H. Woo.
AIP Conf. Proc., No. 222, p. 155 - 158 (15 Apr 1991). Current Physics Microform No.: 9106E0155. - See Abstr. 012.069 for the main entry.

Between the quantum domain and the classical world of facts there is an intermediate domain for which a "classical" probabilistic description is appropriate. As a result of decoherence, such a description obeys the classical additivity of probabilities. In contrast to the usual interpretation in classical probability theory, however, the "classical" probability which emerges from an underlying quantum theory is meant to be applicable to a single system, such as the Universe. It is argued that small inhomogeneities and anisotropies in the very early Universe should be described probabilistically. If, furthermore, the transition between this probabilistic description and a later deterministic description happens as late as after the decoupling of photons, then the smoothness of the microwave background receives a different interpretation: it is not because the fluctuations at decoupling were extremely small, but because the contributions from different fluctuations add to smooth things out.

161.363 **The diffuse infrared background: COBE and other observations.**
M. G. Hauser, T. Kelsall, S. H. Moseley Jr., R. F. Silverberg, T. Murdock, G. Toller, W. Spiesman, J. Weiland.
AIP Conf. Proc., No. 222, p. 161 - 178 (15 Apr 1991). - See Abstr. 012.069 for the main entry.

The authors summarize the status of diffuse infrared background observations from the DIRBE, and compare preliminary results with those of recent rocket and satellite instruments.

161.364 **Erratum: "Merging-driven evolution of high-redshift galaxies in a universe with $\Omega_0 = 1$" [Mon. Not. R. Astron. Soc., Vol. 247, No. 1, p. 166 - 172 (1 Nov 1990)].**
B. Rocca-Volmerange, B. Guiderdoni.
Mon. Not. R. Astron. Soc., Vol. 251, No. 1, p. 192 (1 Jul 1991). See Abstr. 52.161.288.

161.365 **Cosmological implications of COBE.**
B. J. Carr, J. R. Bond, C. J. Hogan.
AIP Conf. Proc., No. 222, p. 187 - 201 (15 Apr 1991). Current Physics Microform No.: 9106E0187. - See Abstr. 012.069 for the main entry.

The authors examine the constraints placed by FIRAS and DIRBE on radiation production in the early Universe. IRAS galaxies must produce a submillimetre background at some level and, if evolutionary effects are large, one can already place an interesting upper limit on the epoch of galaxy formation. A variety of pregalactic sources could have produced radiation, in particular, the VMOs whose black hole remnants are sometimes invoked to explain the dark mass in galactic halos. In some circumstances such pregalactic radiation would have been reprocessed into the far–IR or submillimetre bands by cosmological dust. The main results of this problem are summarized. The VMO–plus–dust scenario is still consistent with FIRAS data but it may soon be excluded, in which case the radiation would need to be in the near–IR.

161.366 **The evolution of the diffuse cosmic ultraviolet background constrained by Hubble Space Telescope observations of 3C 273.**
S. Ikeuchi, E. L. Turner.
Astrophys. J., Lett., Vol. 381, No. 1, p. L1 – L4 (1 Nov 1991).
Recent Hubble Space Telescope UV spectroscopy of 3C 273 has revealed more low–redshift Lyman–α absorption lines (IGM clouds) than expected from the extrapolation from high–redshift ($z \geqslant 1.6$) observations. On the basis of the standard pressure confined cloud model of the Lyman–α forest, the authors show that this result indicates a sharp drop in the diffuse cosmic UV background from $z = 2$ to $z = 0$. If one fits a simple power law in $1+z$, the UV background flux must evolve as $(1+z)^{2.7 \pm 1.0}$. Applying this conclusion to a simple model for a photoionized diffuse component of the IGM, the authors predict that the H I optical depth will also only drop slowly or perhaps even increase with decreasing redshift at $z < 2$. The implied constraints on the density and pressure of the diffuse IGM at $z = 0$ are also derived. It is noted that the inferred evolution of the diffuse UV flux bears a striking, and perhaps not coincidental, resemblance to the most recent direct determinations of the volume emissivity of the quasar population.

161.367 **Dark matter from an oscillating Brans–Dicke scalar in the extended inflation model.**
J. McDonald.
Phys. Rev. D, Vol. 44, No. 8, p. 2325 – 2334 (15 Oct 1991).
Current Physics Microform No.: 9112A1249.
The author considers the possibility that the discrepancy between the dynamical estimate of the density of matter in the universe $\Omega \approx 0.2$ and the prediction of inflation $\Omega = 1$ can be accounted for by oscillation of the Brans–Dicke scalar of the simplest extended inflation model. It is shown that such an explanation requires an extremely small self–coupling for the Brans–Dicke scalar ($\lambda < 10^{-60}$) as well as an apparent fine–tuning of the initial amplitude of the Brans–Dicke scalar at the end of inflation, which may however be understood by the anthropic principle. The requirement that the anisotropies of the cosmic microwave background radiation induced by quantum fluctuations of the Brans–Dicke field during inflation are acceptably small constrains the coupling of the Brans–Dicke scalar to the Ricci scalar to be not much greater than ~ 0.01, which is close to the lower limit allowed in the simplest extended inflation model. It is also shown that if one imposes the constraint that the spatial fluctuations of the Brans–Dicke scalar induced by bubble growth during the first–order phase transition do not give too large a contribution to the energy density at present, then the frequency of the oscillations has an upper bound, which is typically $\sim 10^6$GHz.

161.368 **Quantum field – theoretic analysis of inflation dynamics in a (2+1)–dimensional universe.**
M. Samiullah, O. Éboli, Pi Soyoung.
Phys. Rev. D, Vol. 44, No. 8, p. 2335 – 2355 (15 Oct 1991).
Current Physics Microform No.: 9112A1259.
The authors reexamine inflationary scenarios based on slow–rollover transitions, which occur under various initial conditions of the inflation – driving scalar field. Inflation dynamics is examined using a recently developed calculational technique for studying a quantum field – theoretic system in an external environment that is itself changing with time. This method, based on the functional Schrödinger picture, uses a self–consistent Gaussian approximation that, unlike ordinary perturbation theory, reflects some of the nonlinearities of the full quantum theory. The treatment is confined to planar universes, where the approximation techniques do not suffer from well–known problems associated with scalar field self–interactions in four–dimensional space–time. However, for these toy models the authors can present concrete and explicit results.

161.369 **Gravity–wave insights to Bianchi type–IX universes.**
D. H. King.
Phys. Rev. D, Vol. 44, No. 8, p. 2356 – 2368 (15 Oct 1991).
Current Physics Microform No.: 9112A1280.
Every Bianchi type–IX universe can be interpreted as a closed Friedmann universe on which are superimposed circularly polarized gravitational waves with the longest wavelength that will fit into a closed universe. In this paper, the author gives a new derivation of this result based on the concept of homogeneous tensor fields on the three–sphere. Every homogeneous symmetric traceless tensor field is shown to be a longest wavelength three–sphere harmonic. Contrary to previous authors, it is shown that the wavelength of these gravitational waves is one–half the circumference of the universe. In order to maintain homogeneity, the gravitational waves must all have the same polarization. There are five longest wavelength modes for each polarization. This interpretation is an exact description, which is valid for every Bianchi type–IX universe. It is in no way limited to first–order perturbations of a Friedmann universe.

161.370 **Chaos in mixmaster models.**
M. Szydlowski, M. Biesiada.
Phys. Rev. D, Vol. 44, No. 8, p. 2369 – 2374 (15 Oct 1991).
Current Physics Microform No.: 9112A1293.
The authors apply their methods developed in previous papers to investigate chaos in mixmaster cosmological models. It is shown that these models have an infinite relaxation time and zero Kolmogorov entropy near the singularity. Therefore they exhibit a power–law instability instead of an exponential one. From the point of view of ergodic theory these systems should be classified as the so–called K systems.

161.371 **Amplitude–real–phase exact solutions for quantum mixmaster universes.**
V. Moncrief, M. P. Ryan.
Phys. Rev. D, Vol. 44, No. 8, p. 2375 – 2379 (15 Oct 1991).
Current Physics Microform No.: 9112A1299.
The authors present a set of exact solutions for quantum Bianchi type–IX anisotropic cosmological models (including the Taub model) of the form $\Psi = We^{-S}$. These solutions are spread over all values of anisotropy near the singularity, but at larger values of the radius of the universe they are strongly peaked around the $k = +1$ Friedmann – Robertson – Walker model.

161.372 **Lorentz Chern–Simons terms in Bianchi cosmologies and the cosmic no–hair conjecture.**
N. Kaloper.
Phys. Rev. D, Vol. 44, No. 8, p. 2380 – 2387 (15 Oct 1991).
Current Physics Microform No.: 9112A1304.
The theory describing gravity and the Kalb–Ramond axion field with topological Lorentz Chern–Simons coupling is considered. The topological coupling induces an additional term in the energy–momentum tensor, which in general can lead to violations of the cosmic no–hair conjecture. These violations are shown explicitly by analyzing the solutions in a diagonal Bianchi type–II geometry.

161.373 **Gravitational lensing properties of curved cosmic strings.**
D. N. Vollick, W. G. Unruh.
Phys. Rev. D, Vol. 44, No. 8, p. 2388 – 2396 (15 Oct 1991).
Current Physics Microform No.: 9112A1312.
The authors examine the effect of a Gaussian wave pulse, propagating along a cosmic string, on both the observed position and redshift of the double images of an object behind the string. The authors also calculate the effect of the string on a rectangular grid of stars which are located infinitely far behind the string. The calculations are done numerically, using the exact metric for a traveling wave recently discovered by Garfinkle.

161.374 **Background radiation from galaxies and COBE.**
Wang Boqi.
AIP Conf. Proc., No. 222, p. 202 – 205 (15 Apr 1991). Current Physics Microform No.: 9106E0202. – See Abstr. 012.069 for the main entry.

The authors assess the general possibility of detecting the integrated background radiation from galaxies at far-infrared wavelengths. They model the evolution of interstellar dust in disk galaxies, and obtain analytical solutions for the dust abundance as a function of the galaxy age. The total emission from dust at far-infrared wavelengths in a galaxy is calculated. Consistent with IRAS observations, they find that the integrated far-infrared background from galaxies peaks at around 100 – 130 μm, with total radiation density 0.5 to 6% that of the cosmic microwave background. The authors calculate deviations due to galaxies at submillimeter wavelengths from the blackbody spectrum of the cosmic microwave background.

161.375 **The submillimetre background expected from normal galaxies.**
S. Eales.
AIP Conf. Proc., No. 222, p. 206 – 209 (15 Apr 1991). Current Physics Microform No.: 9106E0206. – See Abstr. 012.069 for the main entry.

161.376 **Post-decoupling evolution of primordial clouds: The role of Rayleigh scattering.**
L. M. Ozernoy.
AIP Conf. Proc., No. 222, p. 215 (15 Apr 1991). Abstract. Current Physics Microform No.: 9106E0215. – See Abstr. 012.069 for the main entry.

161.377 **Maps and motions of galaxies in the local Universe.**
A. Dressler.
AIP Conf. Proc., No. 222, p. 227 – 238 (15 Apr 1991). – See Abstr. 012.069 for the main entry.

The view of large-scale structure in the Universe has evolved considerably over the last decade as the prevalence of large superclusters and voids has become more apparent. Highlights of the observational steps that led to this new picture are presented, including efforts to map the distribution of galaxies and, through their peculiar motions, the distribution of dark matter on large scales. Comparision of these two maps for the local volume show rough agreement, though it is still not clear if luminous galaxies trace the distribution of dark matter with any precision.

161.378 **Redshift surveys.**
M. J. Geller, J. P. Huchra.
AIP Conf. Proc., No. 222, p. 239 – 247 (15 Apr 1991). – See Abstr. 012.069 for the main entry.

Contents: 1. Introduction. 2. The large-scale galaxy distribution. 3. Statistics of the CfA redshift survey. 4. For the future.

161.379 **Local large-scale structure: dynamics and statistics.**
N. Kaiser.
AIP Conf. Proc., No. 222, p. 248 – 260 (15 Apr 1991). Current Physics Microform No.: 9106E0248. – See Abstr. 012.069 for the main entry.

161.380 **Analysis of the large scale structure with deep pencil beam surveys.**
A. S. Szalay, R. S. Ellis, D. C. Koo, T. J. Broadhurst.
AIP Conf. Proc., No. 222, p. 261 – 275 (15 Apr 1991). Also published in "Primordial nucleosynthesis and evolution of early Universe" edited by K. Sato and J. Audouze, ISBN 0-7923-1193-0, Dordrecht (Netherlands), Kluwer Academic Publ., p. 435 – 450 (1991) in Astrophys. Space Sci. Libr., Vol.169. Current Physics Microform No.: 9106E0261. – See Abstr. 012.069 for the main entry.

Recent observations of the large scale structure of the Universe suggest inhomogeneities on scales larger than previously thought. A deep redshift survey with a "pencil beam" geometry of galaxies at the Galactic Poles indicates strong clustering. The authors examine in detail the role of such deep "pencil beam" surveys in studies of the large scale structure using the 1-D power spectrum of the distribution to illustrate the statistical significance of the feature claimed by Broadhurst et al. (1990). They demonstrate how deep probes complement sparse-sampled wide angle local surveys providing new information on large scales.

161.381 **The origin of large-scale periodicity.**
N. A. Bahcall.
AIP Conf. Proc., No. 222, p. 276 – 282 (15 Apr 1991). Current Physics Microform No.: 9106E0276. – See Abstr. 012.069 for the main entry.

The apparent $\sim 130\,h^{-1}$Mpc periodic distribution of galaxies recently observed in pencil-beam surveys is shown to originate in known large-scale superclusters. The superclusters, covering the high-latitude sky to $z \leqslant 0.08$, show the same peaks that appear in the pencil-beam survey, indicating that the galaxy peaks are drawn from these superclusters. Predictions for future pencil-beam surveys in different directions are given. The results suggest a network structure of large ($\sim 100\,h^{-1}$Mpc) supercluster surfaces surrounding low density regions.

161.382 **Origin of regularities in the redshift distributions of distant galaxies and quasars.**
L. M. Ozernoy.
AIP Conf. Proc., No. 222, p. 283 – 287 (15 Apr 1991). Current Physics Microform No.: 9106E0283. – See Abstr. 012.069 for the main entry.

A model in which inhomogeneities originated during inflation and transformed eventually into density peaks with maximum wavelength equal to the horizon size at the decoupling epoch ($z_{dec} \approx 10^3$) would give at the present for very large scale structure a periodic distribution: $\ln(1+z) \approx 0.04\,n + 0.04$, $n = 0, 1, 2, \ldots$ This formula in which the true phase is replaced by its characteristic value fits the data on redshift distribution of distant galaxies fairly well. Some relevant implications are discussed, such as (i) difference between the constants of periodicity $C \equiv \Delta \ln(1+z)$; and (ii) a recent finding that quasars close on the sky to low redshifted galaxies show a higher concentration towards peaks in the quasar redshift distribution than do quasars in general.

161.383 **Cosmology with oscillating Hubble parameter.**
M. Morikawa.
AIP Conf. Proc., No. 222, p. 288 – 296 (15 Apr 1991). Current Physics Microform No.: 9106E0288. – See Abstr. 012.069 for the main entry.

The author proposes a cosmological model in which resonant energy exchange between gravity and matter is realized. The Hubble parameter becomes oscillatory and causes an apparent density fluctuation without conflicting with the isotropy of the microwave background radiation. The author studies a non-conformally coupled scalar field with a very small mass about 10^{-31}eV. Several new observational consequences are derived: a large scale coherent periodicity in the redshift distribution of galaxies with the periodicity about 100 Mpc, a QSO redshift clustering, a changing Hubble parameter on a scale $z \approx 0.01$, a dark matter density that increases with distance, a luminosity distance-redshift relation, and a changing gravitational constant.

161.384 **Galaxy correlation functions in redshift space from large N-body simulations.**
J. M. Gelb.
AIP Conf. Proc., No. 222, p. 312 – 315 (15 Apr 1991). Current Physics Microform No.: 9106E0312. – See Abstr. 012.069 for the main entry.

The author computed galaxy-galaxy correlation functions in redshift space with data generated from a set of large cold dark matter N-body ($\Omega = 1$) simulations. Galaxy halos were identified as local density maxima in the dark matter distribution and artificial magnitude limited galaxy catalogs were made.

161.385 **Galaxy correlation function down to V = 26 on 0.5° scales.**
R. A. Windhorst, L. W. Neuschaefer.
AIP Conf. Proc., No. 222, p. 316 – 321 (15 Apr 1991). Current Physics Microform No.: 9106E0316. – See Abstr. 012.069 for the main entry.
The authors extend the galaxy 2-point correlation function $w(\theta)$ down to $V \sim 26$ mag on scales $\theta \lesssim 0.5°$. For $14 \lesssim g \lesssim 26$ mag, $w(\theta)$ roughly maintains a constant slope ($1-\gamma \approx -0.6$ to -0.8), which has apparently undergone little change over the entire redshift range sampled. For $20 \lesssim g \lesssim 24.5$ mag, $w(\theta)$'s amplitude continues to decrease with fainter magnitudes. Beyond $g \gtrsim 25$ mag, our catalogs become gradually incomplete, and $w(\theta)$'s amplitude starts to rise again while maintaining roughly the same slope. The authors give several possible explanations for this effect.

161.386 **Gravitational dip profiles of the microwave background radiation temperature towards rich clusters of galaxies.**
M. Chodorowski.
Mon. Not. R. Astron. Soc., Vol. 251, No. 2, p. 248 – 252 (15 Jul 1991).
The anisotropies of the temperature of the microwave background radiation due to gravitational effects of clusters of galaxies are studied in the frame of the potential approximation method. To describe a "typical" cluster, the author applies the general spherical model of a pure infall of a compensated lump and normalize resulting density profiles to the Coma cluster density field. The predicted values of the temperature dip amplitude towards the cluster turn out to be well below the accuracy of present measurements.

161.387 **The nature of faint blue galaxies.**
P. Guhathakurta.
AIP Conf. Proc., No. 222, p. 322 – 325 (15 Apr 1991). Current Physics Microform No.: 9106E0322. – See Abstr. 012.069 for the main entry.
Deep CCD images in U (3600 Å), B_J (4500 Å), and R (6500 Å) are used to define and analyse a population of faint, blue galaxies that approach the confusion limit at $B_J = 28$. Indirect methods are used to place broad constraints on the redshift distribution of faint blue galaxies. The galaxies appear to be weakly clustered on the sky, and this suggests that they are progenitors of systems that are intrinsically faint at the present epoch. The evolution of the present–day galaxy luminosity function is considered.

161.388 **The diffuse X–ray background and the intergalactic medium.**
A. Loeb.
AIP Conf. Proc., No. 222, p. 329 – 338 (15 Apr 1991). Current Physics Microform No.: 9106E0329. – See Abstr. 012.069 for the main entry.
The possible origin of the X–ray background (XRB) from a hot intergalactic medium (IGM), as opposed to discrete sources, is critically analysed. The Comptonization of the microwave background radiation by the IGM is discussed for an arbitrary electron distribution function, using a multiple–scattering approach for relativistic Comptonization. The recent spectral results from COBE can be used to rule–out the existence of a homogeneous hot IGM as the dominant source of the XRB. The alternative possibility of bremsstrahlung emission by a clumped IGM is considered.

161.389 **Spectral–luminosity evolution of active galactic nuclei and the cosmic X–ray background.**
D. Leiter, E. Boldt.
AIP Conf. Proc., No. 222, p. 339 – 342 (15 Apr 1991). Current Physics Microform No.: 9106E0339. – See Abstr. 012.069 for the main entry.
Black hole accretion disk dynamo processes are generally regarded as the central power source for AGN. If the precursor active galaxies (PAG) for such AGN are formed at redshift $z \gtrsim 4$ and contain initial central seed black holes of $\sim 10^6$ solar masses, then the Eddington limited X–ray radiation emitted during their lifetime will undergo the phenomenon of "spectral–luminosity evolution". A superposition of thermal, Comptonized PAG sources can account for the residual cosmic X–ray background and can act as a source of X–ray heating of the intergalactic medium for $z \gtrsim 4$.

161.390 **Great Attractors and the value of Ω.**
A. F. Heavens.
Mon. Not. R. Astron. Soc., Vol. 251, No. 2, p. 267 – 280 (15 Jul 1991).
Recent determinations of the velocity field of galaxies indicate that the Great Attractor is a local minimum of the peculiar velocity potential. This paper develops the techniques for analysing velocity potential minima in general, and applies the method to the Great Attractor located in the direction of the Hydra–Centaurus supercluster. It is shown that the presence of a Great Attractor with its known distance and large potential depth puts a powerful constraint on the value of the density parameter Ω.

161.391 **Bubbles in McVittie spacetime.**
P. Haines, J. D. Harris.
AIP Conf. Proc., No. 222, p. 361 – 363 (15 Apr 1991). Current Physics Microform No.: 9106E0361. – See Abstr. 012.069 for the main entry.
The authors present some results concerning vacuum–filled voids embedded in the McVittie metric.

161.392 **Integral maxima of scale invariant structure.**
P. von Gronefeld.
AIP Conf. Proc., No. 222, p. 364 (15 Apr 1991). Abstract. Current Physics Microform No.: 9106E0364. – See Abstr. 012.069 for the main entry.

161.393 **Primordial magnetic field and density fluctuation from early Universe phase transition.**
J. Wang.
AIP Conf. Proc., No. 222, p. 365 – 368 (15 Apr 1991). Current Physics Microform No.: 9106E0365. – See Abstr. 012.069 for the main entry.
This paper considers a mechanism to generate primordial magnetic field and density fluctuation. The author shows that bulk viscosity being generated during the early Universe phase transition might be the origin of the two problems mentioned above.

161.394 **Ultrarelativistic Bose–Einstein condensation in the Einstein universe and energy conditions.**
L. Parker, Zhang Yang.
Phys. Rev. D, Vol. 44, No. 8, p. 2421 – 2431 (15 Oct 1991). Current Physics Microform No.: 9112A1345.
The authors study a relativistic boson gas of particles and antiparticles in the Einstein universe at high temperatures and densities. The thermodynamic potential of the boson system in curved spacetime is derived. The result shows that ultrarelativistic Bose–Einstein condensation can occur at very high densities in the Einstein universe, and by implication in the early stages of a dynamically changing universe. The authors show that in a slowly changing universe with a weak self–interaction this ultrarelativistic Bose–Einstein condensate would violate the strong energy condition, which enters into the cosmological singularity theorems.

161.395 **Constraints on properties of dark halos.**
K. M. Ashman.
AIP Conf. Proc., No. 222, p. 381 – 384 (15 Apr 1991). Current Physics Microform No.: 9106E0381. – See Abstr. 012.069 for the main entry.
The requirement that dark galactic halos virialize by the present epoch gives a lower limit to the mean mass density within such objects. In conjunction with observations of H I rotation curves and the virial theorem, this constraint can be used to limit the radial extent and the mass of dark halos. One can use this constraint to calculate the maximum contribution to the density

of the Universe made by such halos. Reversing the argument allows to obtain an upper limit on the formation epoch of dark halos. The author found that the halos of spirals formed more recently than a redshift of around 2 to 4 and contribute less than a tenth of the critical density required to close the Universe. Stronger limits are obtained by using estimates of the degree of dissipational gas collapse within dark halos along with observed disk scale–lengths.

161.396 **A search for dark clusters in redshift catalogs.**
J. M. Dickey.
AIP Conf. Proc., No. 222, p. 417 – 420 (15 Apr 1991). Current Physics Microform No.: 9106E0417. – See Abstr. 012.069 for the main entry.

The author attempts to set limits on the abundance of "dark clusters", i.e., deep gravitational potential wells similar to those in clusters of galaxies but without the galaxies. Two methods based on statistics of galaxy velocities in redshift surveys are discussed. Both methods turn up some objects which look like groups but with large velocity dispersions (more than 1000 km s^{-1}). Some of these may be larger scale structures seen in projection, while others are compact groups which seem to be bound.

161.397 **Observational evidence for dark matter from peculiar velocities in the local Universe.**
O. Lahav.
AIP Conf. Proc., No. 222, p. 421 – 430 (15 Apr 1991). – See Abstr. 012.069 for the main entry.

Deviations from the Hubble flow in the local Universe provide a direct measure of the underlying total mass distribution, assuming the gravitational instability picture. A comparison of the mass distribution deduced from the peculiar velocities with the galaxy distribution could tell us the amount of dark matter on large scales. The author discusses the origin of motion of the Local Group with respect to the Cosmic Background Radiation and reviews the peculiar velocity field deduced from distances to hundreds of elliptical and spiral galaxies, including new results for the Shapley Supercluster. The density parameter Ω_0 deduced from the relation between the peculiar velocity and acceleration (as inferred from galaxies) suggests an open Universe if light traces mass. However, it is still possible that there is enough matter to close the Universe if the galaxy distribution is more strongly clustered ("biased") than the mass distribution. The author also considers the possibility of a non–zero cosmological constant Λ and "anti–bias".

161.398 **The X–ray flux dipole of AGNs as an indicator of the local gravitational field.**
T. Miyaji, K. Jahoda, E. Boldt.
AIP Conf. Proc., No. 222, p. 431 – 434 (15 Apr 1991). Current Physics Microform No.: 9106E0431. – See Abstr. 012.069 for the main entry.

The authors have investigated the dipole anisotropy of the sky X–ray flux from nearby AGNs to compare with the peculiar motion of the Local Group. The sample shows a significant dipole anisotropy whose apex is only about 30° away from the direction of the Local Group's peculiar motion for the objects with $z < 0.015$. The present sample strengthens the conclusion that X–ray emission from AGNs traces the underlying mass distribution as strongly as optical and IR emission from galaxies.

161.399 **Dark matter: gravitational lens studies.**
J. A. Tyson.
AIP Conf. Proc., No. 222, p. 437 – 445 (15 Apr 1991). – See Abstr. 012.069 for the main entry.

Studies of dark matter using gravitational lenses are reviewed, with emphasis on recent maps of dark matter in galaxy clusters. Although there are several QSOs which are multiply imaged by isolated intervening galaxies, distant galaxies are 4 million times more numerous than QSOs and thus can sample a lens at multiple positions. To a surface brightness of 29 B mag $arcsec^{-2}$ there are about 100 background galaxies per square arcmin, which is sufficient to map statistically the dark matter distribution in a foreground galaxy cluster. Automated pattern recognition software generates a 2–d mass density map. There is evidence for a soft core of about 100 kpc radius for the dark matter in the rich clusters studied so far. These lens studies confirm the large mass in rich galaxy clusters which were implied by virial calculations using velocity dispersion. The dark matter distribution in a cluster is found to be similar to the distribution of the bright galaxies.

161.400 **Local approximations to non–linear evolution of density fluctuations.**
J. Betancort–Rijo.
Mon. Not. R. Astron. Soc., Vol. 251, No. 3, p. 399 – 407 (1 Aug 1991).

The author shows how the evaluation of any moment of the mass distribution, as given by the Zeldovich approximation, may be reduced to the computation of an integral. In this way the author computes the present rms density fluctuation in a cold dark matter model and finds it close to its linear value when this value is of order one. The arguments presented here are valid for a set of approximations that the author calls local approximations, and the expediency of some of them is discussed.

161.401 **Present and future limits on compact dark matter from gravitational lensing.**
R. J. Nemiroff.
AIP Conf. Proc., No. 222, p. 451 – 454 (15 Apr 1991). – See Abstr. 012.069 for the main entry.

161.402 **The distortion of distant galaxy images by large scale structure.**
R. D. Blandford, A. B. Saust, T. G. Brainerd, J. V. Villumsen.
AIP Conf. Proc., No. 222, p. 455 – 458 (15 Apr 1991). Current Physics Microform No.: 9106E0455. – See Abstr. 012.069 for the main entry.

Inhomogeneity in the large scale distribution of mass in the Universe on scales $\sim 100\ h^{-1}$Mpc can induce a coherent shear distortion in the images of background galaxies. This distortion may be measurable over patches of the sky a few degrees in size. Simulations of giant voids, great attractors and walls as randomly located structures, expanding with the Universe, are discussed.

161.403 **Type Ia supernovae: constraints on baryonic dark matter.**
T. A. Smecker, R. F. G. Wyse.
AIP Conf. Proc., No. 222, p. 467 – 470 (15 Apr 1991). See also Abstr. 53.161.344. Current Physics Microform No.: 9106E0467. – See Abstr. 012.069 for the main entry.

The authors examine models in which the dark matter of galaxy halos is in the form of white dwarfs (WDs). Major constraints, which have been previously neglected, result from the occurrence of type Ia supernovae, which are the products of merging WDs in binary systems whose orbits decay due to the emission of gravitational wave radiation. The lack of observed supernovae in the outer regions of galaxy halos limits the fraction of binaries to $\ll 1\%$ in extant models of WD dominated halos, for standard type Ia parameters. This binary fraction is much smaller than that believed typical for luminous stars, necessitating a further peculiarity in the star–formation process during formation of the dark halo.

161.404 **Astronomical limits on the nature of the dark matter.**
D. N. Spergel.
AIP Conf. Proc., No. 222, p. 481 – 490 (15 Apr 1991). – See Abstr. 012.069 for the main entry.

161.405 **Dark matter theory: constraints from particle physics.**
L. M. Krauss.
AIP Conf. Proc., No. 222, p. 491 – 496 (15 Apr 1991). – See Abstr. 012.069 for the main entry.

The author briefly reviews recent results from particle theory and experiment which are relevant to the question of what might make up galactic halo dark matter, and more generally what might contribute a closure density in dark matter. The author talks in turn about cold dark matter candidates, baryons, and hot dark matter (i. e. light neutrinos).

161.406 **The cosmic neutrino background.**
A. Dar.
AIP Conf. Proc., No. 222, p. 497 – 511 (15 Apr 1991). – See Abstr. 012.069 for the main entry.
The cosmic neutrino background is expected to consist of relic neutrinos from the Big Bang, of neutrinos produced during nuclear burning in stars, of neutrinos released by gravitational stellar collapse and of neutrinos produced by cosmic ray interactions with matter and radiation in the interstellar and intergalactic medium. Formation of baryonic dark matter in the early Universe, matter–antimatter annihilation in a baryonic symmetric Universe and dark matter annihilation could have also contributed significantly to the cosmic neutrino background. The purpose of this paper is to review the properties of these cosmic neutrino backgrounds, the indirect evidence for their existence and the prospects for their detection in view of the dramatic progress in neutrino physics and astronomy.

161.407 **The new discussion of a neutrino mass and issues in the formation of large–scale structure.**
A. L. Melott.
AIP Conf. Proc., No. 222, p. 521 – 524 (15 Apr 1991). Current Physics Microform No.: 9106E0521. – See Abstr. 012.069 for the main entry.

161.408 **Cosmic strings, neutrinos and the large–scale structure of the Universe.**
R. H. Brandenberger.
AIP Conf. Proc., No. 222, p. 525 – 529 (15 Apr 1991). Current Physics Microform No.: 9106E0525. – See Abstr. 012.069 for the main entry.
A cosmological model in which the energy density perturbations are seeded by cosmic strings and in which the dark matter is hot (e.g. neutrinos with a mass of 25 $h_{50}{}^2$eV) produces large–scale structure dominated by planar overdensities with mean radius and separation determined by the comoving horizon at the time of equal matter and radiation (about 40 $h_{50}{}^{-2}$Mpc). The planar overdensities are the consequence of the velocity perturbations in the wake of long rapidly moving strings. As a consequence of dark matter being hot, clustering about cosmic string loops is suppressed, leading to little nonlinear structure between the wakes.

161.409 **Cosmic strings in the Sun?**
J. Wang.
AIP Conf. Proc., No. 222, p. 530 (15 Apr 1991). Abstract. Current Physics Microform No.: 9106E0530. – See Abstr. 012.069 for the main entry.

161.410 **Dark matter or new physics?**
D. Kazanas, P. D. Mannheim.
AIP Conf. Proc., No. 222, p. 531 – 540 (15 Apr 1991). Current Physics Microform No.: 9106E0531. – See Abstr. 012.069 for the main entry.
The deviation of galactic rotation curves from their expected Keplerian falloff with distance as based on the observed distribution of light, has prompted the suggestion, by a small number of authors, that this phenomenon signifies a drastic departure of certain laws of physics at large scales from their smaller scale forms. A number of proposals is reviewed with particular attention being given to a locally scale invariant theory of gravity proposed by the authors. This latter theory and several of its consequences and comparisons to observation are presented.

161.411 **Current status of conformal Weyl gravity.**
P. D. Mannheim, D. Kazanas.
AIP Conf. Proc., No. 222, p. 541 – 544 (15 Apr 1991). Current Physics Microform No.: 9106E0541. – See Abstr. 012.069 for the main entry.
Fourth order conformal Weyl gravity has been advanced by Mannheim and Kazanas as a candidate alternative theory of gravity to the standard second order Einstein theory. The authors review its current status, discussing what has been achieved and what open questions remain.

161.412 **Weyl gravity, astrophysical observations, and laboratory measurements.**
K. S. Wood, R. J. Nemiroff.
AIP Conf. Proc., No. 222, p. 545 – 548 (15 Apr 1991). Current Physics Microform No.: 9106E0545. – See Abstr. 012.069 for the main entry.
Observational aspects of Weyl gravity are investigated. The authors find the limits that current laboratory tests and ranging on the moon and satellites in low earth orbit place on the γ parameter. These results rule out a constant γ equal to the reciprocal of the Hubble constant. On galactic scales, discrepancies exist between Weyl gravity predictions and galactic rotation curves.

161.413 **In their own words: extracts from the first fifty years.**
V. Trimble.
AIP Conf. Proc., No. 222, p. 551 – 558 (15 Apr 1991). – See Abstr. 012.069 for the main entry.
Many of the topics on the program of this meeting have been discussed in nearly their present form for more than 40 years. These include dark matter; gravitational lensing; very large scale structure and streaming (scale, amplitude, topology, origin); galaxy formation; and the microwave background and cosmological nucleosynthesis. Collected here are a number of extracts from early papers on these subjects, most of them by very well–known astronomers, for comparison, if the reader wishes, with the versions he carries in his memory.

161.414 **Summary of "After the first three minutes".**
J. R. Bond.
AIP Conf. Proc., No. 222, p. 561 – 582 (15 Apr 1991). – See Abstr. 012.069 for the main entry.
Contents: 1. Introduction. 2. Global parameters. 3. Fluctuations and large–scale structure. 4. Fluctuations and CMB anisotropy. 5. Cosmological energy injection. 6. A requiem for CDM?

161.415 **After the first three minutes: workshop epilog.**
C. L. Bennett.
AIP Conf. Proc., No. 222, p. 583 – 589 (15 Apr 1991). – See Abstr. 012.069 for the main entry.

161.416 **On the cosmological significance of the apparent deficit of small interplanetary scintillation sources.**
C. A. Hajivassiliou.
Astrophys. J., Vol. 381, No. 1, p. 3 – 5 (1 Nov 1991).
The deficit of scintillating sources with small angular sizes, revealed by interplanetary scintillation (IPS) surveys, has led to suggestions that the universe is closed. More recent developments in IPS theory show that both the magnitude and the astronomical implications of the deficit have been exaggerated.

161.417 **Bounds on intergalactic compact objects from observations of compact radio sources.**
A. Kassiola, I. Kovner, R. D. Blandford.
Astrophys. J., Vol. 381, No. 1, p. 6 – 13 (1 Nov 1991).
The authors investigate the constraints that can be imposed on the cosmic density of uniformly distributed intergalactic compact objects in the mass range $10^7 \lesssim M/M_\odot \lesssim 10^9$ using VLBI observations of compact radio sources. A compact object at a sufficiently small angular separation from a compact radio source produces a deamplified and inverted secondary image. No

such cases are known. Using published maps of compact radio sources and assuming an Einstein – de Sitter cosmology, the authors exclude a density parameter for mass in this range $\Omega_M > 0.4$ at the 99.7% confidence level. Observations using the VLBA should allow one to exclude or discover $\Omega_M \gtrsim 10^{-3}$ for compact objects in the mass range $10^5 \lesssim M/M_\odot \lesssim 10^8$.

161.418 **Statistical properties of cosmological N–body simulations.**
R. Valdarnini, S. Borgani.
Mon. Not. R. Astron. Soc., Vol. 251, No. 4, p. 575 – 583 (15 Aug 1991).

The authors follow the non–linear evolution of N–body simulations for different primordial spectra, by means of a P^3M numerical code. The spectra they choose are that of cold dark matter and two scale–free spectra with spectral indices $n = 1$ and $n = -2$. In order to study the effect of a biasing in the background distribution, the authors select for each spectrum different particle populations, which are associated with different density thresholds. A comparison of evolved power spectra and 2–point functions with previous results shows excellent agreement. An enhanced clustering for higher biasing levels is observed only for spectra with sufficient power on large scales, while no significant effects are found for the scale–free model with $n = 1$. Moreover, the authors investigate correlation functions, up to the fourth order, by means of the moment method, in order to test the predictions of the gravitational instability picture in the non–linear regime. They find that non–linear evolution leads to a shape for the 3– and 4–point functions which approaches the hierarchical expression as dynamical evolution proceeds and the biasing level increases, while the Kirkwood model seems to be ruled out. A comparison between the authors' results on higher order functions and predictions of the BBGKY equations in the stability regime shows good agreement.

161.419 **The distortion of distant galaxy images by large–scale structure.**
R. D. Blandford, A. B. Saust, T. G. Brainerd, J. V. Villumsen.
Mon. Not. R. Astron. Soc., Vol. 251, No. 4, p. 600 – 627 (15 Aug 1991).

Inhomogeneity in the distribution of mass in the Universe on scales $\lesssim 100$ Mpc can generate a coherent shear distortion or polarization of the images of background galaxies. This distortion may be measurable over patches of the sky up to a few square degrees in size. If this distortion is measured, or conversely, if its magnitude is limited, it should help us understand the degree to which luminosity traces the underlying mass over cosmological scales. A prescription is given for quantifying the galaxy distortion and a propagation equation for its evolution in an inhomogeneous universe is derived. The creation of shear by inhomogeneity is illustrated using model kinematic universes comprising random distributions of point masses, spheres and circular discs designed to simulate the superclusters, voids and "walls" reported in galaxy velocity surveys.

161.420 **The large–scale distribution of IRAS galaxies and the predicted peculiar velocity field.**
N. Kaiser, G. Efstathiou, R. Ellis, C. Frenk, A. Lawrence, M. Rowan–Robinson, W. Saunders.
Mon. Not. R. Astron. Soc., Vol. 252, No. 1, p. 1 – 12 (1 Sep 1991).

Using the QDOT IRAS galaxy redshift survey, the authors construct a smoothed density field, corrected for peculiar velocity distortion, of a region of the local universe out to $\sim 150\ h^{-1}$Mpc. Predicted peculiar velocities are calculated from this density field, assuming gravitational instability theory, and compared to observed velocities derived from Tully–Fisher and D_n–σ techniques. A good correlation is seen, from which the authors estimate that $b_{IRAS}\Omega^{-0.6} \cong 1.16 \pm 0.21$, in general agreement with earlier results from analysis of the Local Group motion. If IRAS galaxies trace the mass on large scales then the Universe must have a density very close to the critical value.

161.421 **Standard model of primordial nucleosynthesis: a few general remarks.**
J. Audouze.
IUPAP Conference on Primordial Nucleosynthesis and Evolution of Early Universe, p. 1 – 2 (1991). Abstract. – See Abstr. 012.071 for the main entry.

161.422 **The abundances of deuterium, helium and lithium test and constrain the standard model of cosmology.**
G. Steigman.
IUPAP Conference on Primordial Nucleosynthesis and Evolution of Early Universe, p. 3 – 21 (1991). – See Abstr. 012.071 for the main entry.

The standard hot big bang model with input from the experimentally tested standard model of particle physics (e.g., three flavors of neutrinos) is remarkably successful in accounting for the abundances of the light elements D, ^{3}He, ^{4}He and ^{7}Li. Accurate observations of the abundances of these light elements, along with the means to extrapolate them to their primordial values, provide an invaluable tool for testing the predictions of the standard cosmological model and for constraining the baryon density of the Universe. The current data on the abundances of the light elements is reviewed and is supplemented by model independent and some model dependent (i.e. galactic chemical evolution models) estimates to derive their primordial values. The inferred primordial abundances are compared with the predictions of nucleosynthesis in the standard cosmological model to test the consistency of the model and to constrain the allowed range of the baryon–to–photon ratio (η). The preferred value of $\eta = 4 \times 10^{-10}$ points to a very low baryon density ($\Omega_B \lesssim 0.1$) and strengthens the case for nonbaryonic dark matter.

161.423 **Lithium, beryllium and boron; observational constraints on primordial nucleosynthesis.**
J. E. Beckman.
IUPAP Conference on Primordial Nucleosynthesis and Evolution of Early Universe, p. 23 – 37 (1991). – See Abstr. 012.071 for the main entry.

The author summarizes the evidence bearing on the derivation of the primordial ^{7}Li abundance from observations of extremely metal deficient stars in the galactic halo. Arguments, primarily observational, but supported theoretically, are presented to show that the ^{7}Li observed in warm halo subdwarfs has neither been supplemented by ^{7}Li production in the early galaxy, nor significantly depleted during the lifetime of the stars themselves, and gives a good approximation to the primordial value. The absence of ^{9}Be or ^{10}B + ^{11}B in these objects at present limits of observational accuracy tends to support the primordiality of their ^{7}Li. Observing ^{9}Be, and heavier nuclides in the most metal deficient stars, does offer a possible route to distinguish between standard (SBB) and inhomogeneous (IBB) primordial cosmological models.

161.424 **The evolution of the galactic lithium abundance.**
L. M. Hobbs.
IUPAP Conference on Primordial Nucleosynthesis and Evolution of Early Universe, p. 39 – 44 (1991). – See Abstr. 012.071 for the main entry.

Studies of the lithium abundances in population I and population II stars have been used to determine the Li/H fraction produced in Big Bang nucleosynthesis. A summary of the principal results derived from the 6707 Å line measured in the spectra of various stars is presented. The value Li/H $\approx 1 \times 10^{-10}$ uniformly found for about 40 relatively hot, extreme halo stars appears to have survived, nearly unaltered, from that early epoch.

161.425 **Chemical evolution of galaxies.**
B. E. J. Pagel.
IUPAP Conference on Primordial Nucleosynthesis and Evolution of Early Universe, p. 45 – 54 (1991). Also published as report NORDITA—90/73-A(prepr.), 11. p., (1990). – See Abstr. 012.071 for the main entry.

Initial conditions are probably set by results of Big Bang nucleosynthesis (BBNS) without intervening complications affecting the composition of visible matter so that extrapolation of observed abundances to BBNS products seems fairly secure. Primordial helium and deuterium abundances deduced in this way place upper and lower limits on baryonic density implying that both baryonic and non-baryonic dark matter exist and predicting no more than 3 neutrino flavours as recently confirmed in accelerator experiments. The validity of simple galactic chemical evolution models assumed in extrapolating back to the Big Bang is examined in the light of the frequency distribution of iron or oxygen abundances in the Galactic halo, bulge and disk.

161.426 **The effect of some nonequilibrium processes on the primordial nucleosynthesis.**
A. D. Dolgov, D. P. Kirilova.
IUPAP Conference on Primordial Nucleosynthesis and Evolution of Early Universe, p. 55 – 59 (1991). – See Abstr. 012.071 for the main entry.

The effect of nonequilibrium processes, namely nonequilibrium decays of light particles and nonequilibrium neutrino oscillations, on primordial nucleosynthesis is investigated. In case of primordial nucleosynthesis with neutrino oscillations a concrete model of oscillations between active and nonthermalized sterile neutrinos is investigated. The effect on the production of helium-4 has been proved considerable for a certain range of oscillation parameters. Thus, stringent constraints on the oscillation parameters were obtained.

161.427 **Analysis of the reaction ^{7}Li(d,n)^{8}Be at subcoulomb energies.**
T. Rauscher, K. Grün, H. Krauss, H. Oberhummer.
IUPAP Conference on Primordial Nucleosynthesis and Evolution of Early Universe, p. 61 – 62 (1991). – See Abstr. 012.071 for the main entry.

The reaction ^{7}Li(d,n)^{8}Be is analyzed in the energy range 50 keV to 2 MeV using the DWBA formalism. The astrophysical S-factor for this reaction is calculated and compared to experimental data.

161.428 **Experimental study of the key reaction to the nucleosynthesis in the inhomogeneous Big Bang models.**
S. Kubono, R. Boyd, R. J. Peterson, N. Ikeda, M. H. Tanaka, M. Yasue, S. Yun, H. Toyokawa, M. Yosoi, H. Ohnuma, I. Tanihata, T. Kajino.
IUPAP Conference on Primordial Nucleosynthesis and Evolution of Early Universe, p. 63 – 64 (1991). – See Abstr. 012.071 for the main entry.

161.429 **Primordial black holes and Big Bang nucleosynthesis.**
K. Arai, M. Hashimoto, T. Futamase.
IUPAP Conference on Primordial Nucleosynthesis and Evolution of Early Universe, p. 65 – 66 (1991). – See Abstr. 012.071 for the main entry.

Primordial nucleosynthesis is examined with evaporating black holes. A low density model of the Universe is preferable because of the heating by the evaporation.

161.430 **Perturbation of coherent large-scale structure.**
S. Bildhauer, O. Aso, M. Kasai, T. Futamase.
Mon. Not. R. Astron. Soc., Vol. 252, No. 1, p. 132 – 137 (1 Sep 1991).

Linear perturbation theory on the background of a one-dimensional, non-linear inhomogeneous universe, described by the Zeldovich solution, is considered. Special attention is paid to perturbations inside the "wall"-like structures, for which the exact solution is given for an Einstein-de Sitter behaviour of the scale factor. The nonlinear effect of the density inhomogeneity on the evolution of the peculiar velocity is written in the form of a relation between the peculiar velocity and the density parameter Ω including the correction due to the density contrast δ_0 of the Zeldovich solution. It is shown that the component parallel to the walls is characterized by $\Omega^{0.6}(1+\delta_0)^{-0.57}$, whereas the background component perpendicular to the walls obeys the same relation as that in the homogeneous and isotropic background in spite of its fully non-linear property. The authors also give an exact solution for one-dimensional perturbations perpendicular to the walls.

161.431 **Particle creation, inflation, and cosmic isotropy.**
E. Calzetta.
Phys. Rev. D, Vol. 44, No. 10, p. 3043 – 3051 (15 Nov 1991). Current Physics Microform No.: 9201B0771.

Within the framework of homogeneous models of the universe, inflation provides the simplest explanation for the present cosmic isotropy, and a Bianchi type-IX (mixmaster) model is the least prejudiced guess one can make about the state of the universe before the inflationary phase. However, a mixmaster model would not inflate, unless either shear or the radiation energy density are large enough. Particle creation enhances the radiation energy density and therefore enlarges the set of inflating initial conditions for the universe.

161.432 **Baryogenesis, sphalerons, and the cogeneration of dark matter.**
S. M. Barr.
Phys. Rev. D, Vol. 44, No. 10, p. 3062 – 3066 (15 Nov 1991). Current Physics Microform No.: 9201B0790.

Processes involving the electroweak anomaly can erase completely a primordial baryon and lepton asymmetry if B-L = 0. This has led to the search for plausible mechanisms for weak-scale baryogenesis, or for the generation of a primordial B-L asymmetry. Here it is emphasized that if another quantum number conserved up to anomalies is present, electroweak anomaly processes would not necessarily erase a primordial baryon asymmetry even if B-L = 0. Moreover, an asymmetry in the new quantum number that is comparable to the baryon asymmetry is generated concomitantly due to the electroweak anomaly. This asymmetry could be the origin of dark matter.

161.433 **Prospects for observing subhorizon preinflation fluctuations in the cosmic microwave background.**
H. Kurki-Suonio, F. Graziani, G. J. Mathews.
Phys. Rev. D, Vol. 44, No. 10, p. 3072 – 3076 (15 Nov 1991). Current Physics Microform No.: 9201B0800.

The authors consider constraints on preinflation subhorizon fluctuations in a chaotic inflationary model. The requirement that the energy density in the fluctuating field does not exceed the Planck scale limits the amplitude of the fluctuations. The requirement that the preinflation fluctuations have a presently observable amplitude that is at least comparable to the quantum fluctuations generated during inflation, constrains the length scale of the fluctuations. The authors show that in order for preinflation fluctuations to be observable above the background due to quantum fluctuations generated during inflation, they must have a scale at least comparable to the preinflation Hubble length. Unless $\Omega_0 > 0.94$, such fluctuations could be observable in the present microwave background with a wavelength less than the present observed horizon.

161.434 **Inhomogeneous inflation: the initial-value problem.**
P. Laguna, H. Kurki-Suonio, R. A. Matzner.
Phys. Rev. D, Vol. 44, No. 10, p. 3077 – 3086 (15 Nov 1991). Current Physics Microform No.: 9201B0805.

The authors present a spatially three-dimensional study for solving the initial-value problem in general relativity for inhomogeneous cosmologies. They use York's conformal approach to solve the constraint equations of Einstein's field equations for scalar field sources and find the initial data which will be used in the evolution. This work constitutes the first stage in the development of a code to analyze the effects of matter and spacetime inhomogeneities on inflation.

161.435 **The Hubble constant from VLA measurement of the time delay in the double quasar 0957+561.**
D. H. Roberts, J. Lehár, J. N. Hewitt, B. F. Burke.
Nature, Vol. 352, No. 6330, p. 43 – 45 (4 Jul 1991).

In the double quasar 0957 + 561, the first detected example of a gravitational lens, a quasar at redshift z = 1.41 appears as two images separated by ~6 arcsec and accompanied by an extended radio source. The Hubble constant, H_0, can be estimated from the angular separation and the time delay between the appearance of the same brightness variations in the two images – a method that avoids all the usual uncertainties that attend the estimation of distance by the use of standard candles. From 11 years of Very Large Array (VLA) observations the authors have found the time delay to be 1.40 ± 0.10 yr, a significantly larger and more robust value than has been obtained from optical measurements. Using a standard lens model and taking the observed velocity dispersion of luminous matter in the lensing galaxy to measure the true gravitational potential, they find $H_0 = 46 \pm 14$ (42 ± 14) km $s^{-1}Mpc^{-1}$ for cosmic density parameter $\Omega_0 = 0$ (1). But if, as is likely, there is dark matter in the lensing galaxy, H_0 could be as high as 69 ± 21 (63 ± 21) km $s^{-1}Mpc^{-1}$. The agreement of several completely independent estimates of H_0 gives some confidence that the scale of the Universe is indeed known to about a factor of two.

161.436 **Constraints from primordial nucleosynthesis on neutrino degeneracy.**
H.-S. Kang.
IUPAP Conference on Primordial Nucleosynthesis and Evolution of Early Universe, p. 67 – 68 (1991). – See Abstr. 012.071 for the main entry.

The author presents the permissible ranges of, and most stringent constraints on, the lepton asymmetry (degeneracy), particularly the electron–neutrino asymmetry, for different baryon asymmetries by comparing calculated primordial abundances of the light elements (D, ^{3}He, ^{4}He, ^{7}Li) in Big Bang nucleosynthesis with those abundances inferred from current observational data. Furthermore, it is shown that the Universe can be closed by baryons only neutrinos are sufficiently degenerate. Finally, the behaviour of the abundances of the light elements for different baryon and lepton asymmetries is studied.

161.437 **Primordial nucleosynthesis in inhomogeneous Universe.**
K. Sato.
IUPAP Conference on Primordial Nucleosynthesis and Evolution of Early Universe, p. 79 – 90 (1991). – See Abstr. 012.071 for the main entry.

From recent studies of nucleosynthesis in inhomogeneous Universe, it has been pointed out that 1) $\Omega_B = 1$ model becomes consistent with light element observations, 2) heavy elements including r–process elements are abundantly synthesized, and 3) ^{9}Be plays a role of the indicator of inhomogeneity since ^{9}Be is also abundantly synthesized. These suggestions were, however, based on a simple two–zone model in which back diffusion of neutrons are neglected during the nucleosynthesis. In this short review, it is shown that 1) the upper limit of Ω_B which is consistent with light element observations is almost the same as that of homogeneous case, 2) the ^{9}Be abundance in the simple two–zone model is overestimated at least two orders of magnitude in plausible parameter space and 3) the abundances of CNO elements are very small in the proper range of baryon/photon ratio consistent with the abundances of light elements.

161.438 **Sterile neutrinos in the early Universe.**
R. A. Malaney, G. M. Fuller.
IUPAP Conference on Primordial Nucleosynthesis and Evolution of Early Universe, p. 91 – 94 (1991). – See Abstr. 012.071 for the main entry.

The authors discuss the role played by right–handed sterile neutrinos in the early Universe. They show that the well known ^{4}He constraint on the number of relativistic degrees of freedom at early times limits the equilibration of the right–handed neutrino sea with the background plasma. The authors discuss how this allows interesting constraints to be placed on neutrino properties. In particular, a new limit on the Dirac mass of the neutrino is presented.

161.439 **Multi–zone calculation of nucleosynthesis in inhomogeneous Universe and ^{9}Be abundance.**
N. Terasawa.
IUPAP Conference on Primordial Nucleosynthesis and Evolution of Early Universe, p. 101 – 106 (1991). – See Abstr. 012.071 for the main entry.

While ^{9}Be is expected to provide a test of inhomogeneous models of primordial nucleosynthesis, the author shows that the previous estimate by the simplified two–zone model of neutron diffusion overestimated the ^{9}Be abundance at least two orders of magnitude. The abundances of CNO elements expected to be the seeds of cosmic r–process are very small in the proper range of parameters consistent with the abundances of light elements.

161.440 **Signatures of inhomogeneity in the early Universe.**
W. A. Fowler, L. H. Kawano, R. A. Malaney, R. W. Kavanagh.
IUPAP Conference on Primordial Nucleosynthesis and Evolution of Early Universe, p. 107 – 116 (1991). See also Abstr. 53.161.328. – See Abstr. 012.071 for the main entry.

The authors have made a systematic study of the production of elemental CNO in inhomogeneous nucleosynthesis, investigating a much larger region of parameter space than previously studied. They have determined abundances of CNO elements and ascertained the main channels to their production.

161.441 **Diffusion coefficients of nucleons in the inhomogeneous Big Bang model.**
B. Banerjee, S. M. Chitre.
IUPAP Conference on Primordial Nucleosynthesis and Evolution of Early Universe, p. 117 – 119 (1991). – See Abstr. 012.071 for the main entry.

161.442 **Survival of strange matter lumps formed in the early Universe.**
C. H. Lee, H. K. Lee.
IUPAP Conference on Primordial Nucleosynthesis and Evolution of Early Universe, p. 123 – 124 (1991). – See Abstr. 012.071 for the main entry.

Strange matter lumps are supposed to form during the quark–hadron coexistence period at T_c. The authors discuss the conditions in which the lumps may avoid dissolution into hadron matter through boiling and survive until the present time.

161.443 **Inhomogeneous universes in the framework of lattice QCD.**
M. Hackel, M. Faber, H. Markum, H. Oberhummer.
IUPAP Conference on Primordial Nucleosynthesis and Evolution of Early Universe, p. 127 – 128 (1991). – See Abstr. 012.071 for the main entry.

There exist two calculations for the interface tension α of coexisting quark and hadron matter with two different methods. The authors perform an independent computation of α for time extension $N_t = 2$ and 4 and find that (i) the interface tension vanishes for our volume within error bars and (ii) there are systematic deviations between the differential and integral method.

161.444 **The beginning of the Universe.**
S. W. Hawking.
IUPAP Conference on Primordial Nucleosynthesis and Evolution of Early Universe, p. 129 – 139 (1991). – See Abstr. 012.071 for the main entry.

The author discusses whether the Universe has a beginning in time, and whether it will have an end. All the evidence seems to indicate, that the Universe has not existed for ever, but that it had a beginning about 15 billion years ago. This is probably the most remarkable discovery of modern cosmology. We are not yet certain whether the Universe will have an end, but if: the end will not come for at least ten billion years.

161.445 **Extended inflationary cosmology: a primer.**
P. J. Steinhardt.
IUPAP Conference on Primordial Nucleosynthesis and Evolution of Early Universe, p. 141 – 152 (1991). – See Abstr. 012.071 for the main entry.
New approaches to inflationary cosmology have been developed which may ultimately solve the fine–tuning problem that plagues previous inflationary models. The distinctive feature is that the inflationary phase transition is completed by bubble nucleation. The bubbles may lead to some interesting effects in the post–inflationary Universe, including new seeds for galaxy formation and a unique gravitational wave signature.

161.446 **The inflation sector of extended inflation.**
E. W. Kolb.
IUPAP Conference on Primordial Nucleosynthesis and Evolution of Early Universe, p. 153 – 165 (1991). – See Abstr. 012.071 for the main entry.
In extended inflation the inflationary era is brought to a close by the process of percolation of true vacuum bubbles produced in a first–order phase transition. The author discusses several effects that might obtain if the Universe undergoes an inflationary first–order phase transition.

161.447 **Inflation in generalized Einstein theories.**
K. Maeda, A. L. Berkin, J. Yokoyama.
IUPAP Conference on Primordial Nucleosynthesis and Evolution of Early Universe, p. 167 – 172 (1991). – See Abstr. 012.071 for the main entry.
The authors classify inflationary models into three types: (1) the Einstein gravity theory with inflaton scalar field, (2) modified gravity theories with ordinary matter fluid and (3) the hybrid of these two models. Extended inflation and soft inflation belong to the third class. The authors analyze the details of soft inflationary models, which have two scalar fields: one the standard inflaton, whose potential is exponentially coupled to the other field. They consider satisfaction of the inflationary constraints. New inflation works well, with the coupling constant near the values allowed by grand unified theories. For chaotic inflation with a massive inflaton, the authors find successful inflation without fine–tuning of the coupling constant or initial data.

161.448 **Baryogenesis in the Universe.**
A. D. Dolgov.
IUPAP Conference on Primordial Nucleosynthesis and Evolution of Early Universe, p. 173 – 186 (1991). – See Abstr. 012.071 for the main entry.
Contents: 1. Introduction and history. 2. Three principles of baryogenesis. 3. Heavy particle decays. 4. Baryogenesis with conserved baryonic charge. 5. Baryogenesis, inflation, and CP–violation. 6. Baryogenesis and large scale Universe structure. 7. Baryonic charge condensate. 8. Baryogenesis and CPT. 9. Baryosynthesis by topological defects. 10. Baryosynthesis on electroweak scale. 11. Conclusion.

161.449 **Formation of topological defects in the inflationary Universe.**
J. Yokoyama, M. Nagasawa.
IUPAP Conference on Primordial Nucleosynthesis and Evolution of Early Universe, p. 187 – 192 (1991). – See Abstr. 012.071 for the main entry.
It is argued that topological defects of typical grand unification scales are produced by the curvature–induced phase transition more likely than by the Kibble mechanism. Properties of the new phase transition mechanism is clarified and its cosmological consequences are discussed.

161.450 **Production of primordial magnetic field in protogalactic tori.**
S. K. Chakrabarti.
Mon. Not. R. Astron. Soc., Vol. 252, No. 2, p. 246 – 248 (15 Sep 1991).
The author shows that in the early epoch of galaxy formation, non–electrical forces in radiation pressure supported tori can drive a poloidal electric current. This can generate a toroidal magnetic field which the author interprets as the primordial "seed" field.

161.451 **Non–Zel'dovich fluctuations from inflation.**
J. R. Primack.
IUPAP Conference on Primordial Nucleosynthesis and Evolution of Early Universe, p. 193 – 202 (1991). – See Abstr. 012.071 for the main entry.
The author reviews the recent work of the Santa Cruz group on the generation of non–Zel'dovich (including non–Gaussian) fluctuations in chaotic inflation. Contents: 1. Introduction. 2. Non flat P(k) from a single–inflation model with a polynomial potential. 3. "Designer inflation" with a single inflation. 4. Trouble with double inflation. 5. Non–Gaussian fluctuations. 6. Conclusions.

161.452 **Chaotic inflation and the omega problem.**
G. J. Mathews, F. Graziani, H. Kurki–Suonio.
IUPAP Conference on Primordial Nucleosynthesis and Evolution of Early Universe, p. 213 – 223 (1991). – See Abstr. 012.071 for the main entry.
An overview of the omega problem is presented along with discussions of several ways to resolve this problem in the context of chaotic and stochastic inflation.

161.453 **Late–time cosmological phase transitions.**
D. N. Schramm.
IUPAP Conference on Primordial Nucleosynthesis and Evolution of Early Universe, p. 225 – 242 (1991). – See Abstr. 012.071 for the main entry.
It is shown that the potential galaxy formation and large–scale structure problems of (1) objects existing at high redshifts ($Z \gtrsim 5$), (2) structures existing on scales of 100 Mpc as well as velocity flows on such scales, and (3) minimal microwave anisotropies $\Delta T/T \lesssim 10^{-5}$ can be solved if the seeds needed to generate structure form in a vacuum phase transition *after* decoupling. It is argued that the basic physics of such a phase transition is no more exotic than that utilized in the more traditional GUT scale phase transitions, and that, just as in the GUT case, significant random gaussian fluctuations and/or topological defects can form. Scale lengths of ~100 Mpc for large–scale structure as well as ~1 Mpc for galaxy formation occur naturally. Possible support for new physics that might be associated with such a late–time transition comes from the preliminary results of the SAGE solar neutrino experiment, implying neutrino flavor mixing with values similar to those required for a late–time transition. It is also noted that a see–saw model for the neutrino masses might also imply a tau neutrino mass that is an ideal hot dark matter candidate. However, in general either hot or cold dark matter can be consistent with a late–time transition.

161.454 **False–vacuum decay in generalized extended inflation.**
R. Holman.
IUPAP Conference on Primordial Nucleosynthesis and Evolution of Early Universe, p. 243 – 247 (1991). – See Abstr. 012.071 for the main entry.
The author studies false–vacuum decay in context of generalized extended inflationary theories, and computes the bubble nucleation rates for these theories in the limit of $G_N \to 0$. He finds that the time dependence of the nucleation rate can be exponentially strong through the time dependence of the Jordan–Brans–Dicke field. This can have a pronounced effect on whether extended inflation can be successfully implemented.

161.455 **Reconciling a small density parameter to inflation.**
Y. Fujii, T. Nishioka.
IUPAP Conference on Primordial Nucleosynthesis and Evolution of Early Universe, p. 249 – 250 (1991). – See Abstr. 012.071 for the main entry.
A theoretical model is proposed in which a decaying cosmological constant fills up the gap between $\Omega = 1$ as expected from the inflationary scenario and the observed small value $\Omega \lesssim 0.1$.

161.456 **Soft inflation: a model for easing constraints.**
A. L. Berkin, K.–i. Maeda, J. Yokoyama.
IUPAP Conference on Primordial Nucleosynthesis and Evolution of Early Universe, p. 251 – 252 (1991). – See Abstr. 012.071 for the main entry.
The cosmology resulting from two coupled scalar fields, one which is either a new or chaotic type inflaton and the other which has an exponentially decaying potential, is studied. Such a potential arises in superstring theories and in the conformally transformed frame of generalized Einstein theories like the Jordan–Brans–Dicke theory. The constraints necessary for successful inflation are examined. Conventional GUT models such as SU(5) are compatible with new inflation.

161.457 **Stochastic inflation lattice simulations: ultra–large scale structure of the Universe.**
D. S. Salopek.
IUPAP Conference on Primordial Nucleosynthesis and Evolution of Early Universe, p. 253 – 264 (1991). – See Abstr. 012.071 for the main entry.
Non–Gaussian fluctuations for structure formation may arise in inflation from the nonlinear interaction of long wavelength gravitational and scalar fields. Long wavelength fields have spatial gradients $\alpha^{-1}\nabla$ small compared to the Hubble radius, and they are described in terms of classical random fields that are fed by short wavelength quantum noise. Lattice Langevin calculations are given for a "toy model" with a scalar field interacting with an exponential potential where one can obtain exact analytic solutions of the Fokker–Planck equation. For single scalar field models that are consistent with current microwave background fluctuations, the fluctuations are Gaussian. However, for scales much larger than our observable Universe, one expects large metric fluctuations that are non–Gaussian. This example illuminates non–Gaussian models involving multiple scalar fields which are consistent with current microwave background limits.

161.458 **Purely quantum derivation of density fluctuations in the inflationary Universe.**
Y. Nambu, M. Sasaki.
IUPAP Conference on Primordial Nucleosynthesis and Evolution of Early Universe, p. 265 – 266 (1991). – See Abstr. 012.071 for the main entry.

161.459 **Formation and evolution of domain–wall–networks.**
H. Kubotani, H. Ishihara, Y. Nambu.
IUPAP Conference on Primordial Nucleosynthesis and Evolution of Early Universe, p. 269 – 270 (1991). – See Abstr. 012.071 for the main entry.
The authors investigate dynamical evolution of global strings and domain walls produced by the phase transition of a scalar field. They found that the system evolves in a quasi–equilibrium state.

161.460 **Catastrophe of spacetime in the early Universe.**
Cao Shenglin.
IUPAP Conference on Primordial Nucleosynthesis and Evolution of Early Universe, p. 271 – 272 (1991). – See Abstr. 012.071 for the main entry.
The author discussed the catastrophe nature of the Finsler spacetime and its cosmological meaning. It is shown that the nature of evolution of the Universe could be attributed to the geometric feature of the Finsler spacetime.

161.461 **(2 + 1)–dimensional quantum gravity.**
A. Hosoya.
IUPAP Conference on Primordial Nucleosynthesis and Evolution of Early Universe, p. 273 – 274 (1991). – See Abstr. 012.071 for the main entry.
The author considers the (2 + 1)–dimensional quantum gravity as a toy model. It is shown that the torus universe in the (2 + 1)–dimensional quantum gravity is a quantum chaos in a rigorous sense.

161.462 **A stringy Universe scenario.**
M. Mashino, H. Minakata.
IUPAP Conference on Primordial Nucleosynthesis and Evolution of Early Universe, p. 275 – 276 (1991). – See Abstr. 012.071 for the main entry.
A scenario of the evolution of the early universe based on the fundamental string theory is described.

161.463 **The constant–mean–curvature slicing of the Schwarzschild–de Sitter space–time.**
K.–i. Nakao.
IUPAP Conference on Primordial Nucleosynthesis and Evolution of Early Universe, p. 277 – 278 (1991). – See Abstr. 012.071 for the main entry.

161.464 **Cosmology: More Hubble trouble?**
J. Peacock.
Nature, Vol. 352, No. 6334, p. 378 – 379 (1 Aug 1991).

161.465 **Schwarzschild–de Sitter type wormhole and cosmological constant.**
Kim Sungwon, S. K. Kim, Yang Jongmann.
IUPAP Conference on Primordial Nucleosynthesis and Evolution of Early Universe, p. 279 – 280 (1991). – See Abstr. 012.071 for the main entry.
The two Schwarzschild–de Sitter spaces connected by a wormhole following the method of Visser is considered in order to find the effect of the cosmological constant on the Minkowskian wormhole. The dynamics of the wormhole is also compared with those of the Euclidean wormhole.

161.466 **COBE: new sky maps of the early Universe.**
G. F. Smoot.
IUPAP Conference on Primordial Nucleosynthesis and Evolution of Early Universe, p. 281 – 294 (1991). – See Abstr. 012.071 for the main entry.
This paper presents early results obtained from the first six months of measurements of the cosmic microwave background (CMB) by instruments aboard NASA's Cosmic Background Explorer (COBE) satellite and discusses the implications for cosmology. The three instruments: FIRAS, DMR, and DIRBE have operated well and produced significant new results. The FIRAS measurement of the CMB spectrum supports the standard Big Bang nucleosynthesis model. The maps made from the DMR instrument measurements show a surprisingly smooth early Universe. The measurements are sufficiently precise that we must pay careful attention to potential systematic errors. The maps of galactic and local emission produced by the DIRBE instrument will be needed to identify foregrounds from extragalactic emission and thus to interpret the results in terms of events in the early Universe.

161.467 **Large scale cosmic instability.**
C. J. Hogan.
IUPAP Conference on Primordial Nucleosynthesis and Evolution of Early Universe, p. 295 – 304 (1991). – See Abstr. 012.071 for the main entry.
Spectral distortions in cosmic background radiation can create instabilities in matter, whose possible role in generating large scale structure is outlined here. A promising physical mechanism is sketched, based on Ly α resonance scattering, which exponentially amplifies perturbations in neutral hydrogen density on large scales after recombination.

161.468 **Gas–induced primary and secondary CMB anisotropies.**
J. R. Bond, S. T. Myers.
IUPAP Conference on Primordial Nucleosynthesis and Evolution of Early Universe, p. 305 – 323 (1991). – See Abstr. 012.071 for the main entry.
On scales less than a few degrees the primary CMB anisotropies (from linear processes occurring at photon decoupling) and the secondary CMB anisotropies (from nonlinear processes occurring later) arise mainly from gas–driven mechanisms such as Thomson scattering, inhomogeneous Compton cooling, and radiant emission by dust. The authors describe some of the

current experiments which probe these anisotropies and the constraints on models that can be derived from them. They pay special attention to hierarchical Gaussian models of structure formation such as the $\Omega = 1$, $H_0 = 50$, adiabatic CDM model. The authors also present their methods for generating large catalogues of galaxy groups and clusters using a hierarchical Gaussian peaks formulation. These are used to generate simulated maps of secondary CMB anisotropies due to the Sunyaev–Zeldovich effect and the associated soft X–ray emission viewable with ROSAT.

161.469 **The case for the relativistic hot Big Bang cosmology.**
P. J. E. Peebles, D. N. Schramm, E. L. Turner, R. G. Kron.
Nature, Vol. 352, No. 6338, p. 769 – 776 (29 Aug 1991).
The relativistic hot Big Bang model for the expanding Universe has yielded a set of interpretations and successful predictions that substantially outnumber the elements used in devising the theory, with no well–established empirical contradictions. It is reasonable to conclude that this standard cosmology has developed into a mature and believable physical model.

161.470 **Fundamental arguments for inflation.**
A. H. Guth.
NATO Advanced Research Workshop on Observational Tests of Inflation, p. 1 – 21 (1991). Also published as report CTP–1963, 10 p. (Apr. 1991). – See Abstr. 012.076 for the main entry.
The mechanism of inflation is described, and the fundamental arguments in favor of inflation are summarized. It is claimed that the inflationary model provides a very plausible explanation for (1) the large number of particles in the universe, (2) the Hubble expansion, (3) the large–scale uniformity of the universe, (4) the nearness of the universe to a critical density, and (5) the absence of magnetic monopoles.

161.471 **Predictions of inflation.**
A. R. Liddle.
NATO Advanced Research Workshop on Observational Tests of Inflation, p. 23 – 38 (1991). – See Abstr. 012.076 for the main entry.
Many models exist based upon the inflationary universe paradigm. The predictions of these models are outlined, and it is emphasised that the differing models can have substantially different implications. Consequently, inflation seems more likely to fall from favour through us being restricted to unappealing models rather than being explicitly ruled out. Whether or not the universe is at the critical density remains the most solid testable prediction of inflation.

161.472 **Classicality of density perturbations in the early Universe.**
R. Brandenberger, R. Laflamme, M. Mijić.
NATO Advanced Research Workshop on Observational Tests of Inflation, p. 39 – 45 (1991). – See Abstr. 012.076 for the main entry.
Quantum fluctuations in the de Sitter phase of an inflationary Universe may produce the classical primordial density perturbations required to seed galaxies. Implicit in this theory is a prescription for extracting classical quantities from a quantum theory. The author briefly reviews the original prescriptions for obtaining classical density perturbations. He then solves a simple toy model and shows that the original prescriptions can be justified by applying decoherence arguments similar to those used in quantum cosmology.

161.473 **The influence of non–linear density fluctuations on the micowave sky.**
J. L. Sanz, E. Martinez–González.
NATO Advanced Research Workshop on Observational Tests of Inflation, p. 47 – 52 (1991). – See Abstr. 012.076 for the main entry.
Secondary anisotropies are interesting because they can erase/enhance the primary ones at different scales. The authors emphasize on the integrated gravitational effect due to the presence of non–linear density fluctuations as a possible competitive effect with other linear and non–linear contributions at least on certain angular scales. In particular, they comment on the level of $\Delta T/T$ fluctuations due to cosmic voids, great attractors and second order perturbations related to theories of galaxy formation.

161.474 **Quantum cosmology and the cosmological constant.**
I. Moss.
NATO Advanced Research Workshop on Observational Tests of Inflation, p. 53 – 61 (1991). – See Abstr. 012.076 for the main entry.
The origin of our universe and the small scale structure of spacetime lie deep in the realm of quantum theory, yet they seem to relate to the large scale features of the universe. The value of the cosmological constant is probably zero, but a small value related to a light particle mass is possible.

161.475 **Lessons from inflation and cold dark matter.**
P. J. E. Peebles.
2. International Workshop on Plasma Physics: Plasma Theoretical Problems in Astro– and Fusion Physics, p. 63 – 73 (1991). – See Abstr. 012.067 for the main entry.
The progress of research in cosmology has been greatly stimulated by the flow of ideas from the inflation scenario and its simplest implementation in physical cosmology, the cold dark matter theory. The author presents his list of the major lessons we have learned from the long running debate over the observational viability of this theory.

161.476 **The topology of galaxy clustering.**
P. Coles, M. Plionis.
2. International Workshop on Plasma Physics: Plasma Theoretical Problems in Astro– and Fusion Physics, p. 75 – 79 (1991). – See Abstr. 012.067 for the main entry.
The authors discuss an objective method for quantifying the topology of the galaxy distribution using only projected galaxy counts. The method is a useful complement to fully three–dimensional studies of topology based on the genus by virtue of the enormous projected data sets available. Applying the method to the Lick counts they find no evidence for large–scale non–gaussian behaviour, whereas the small–scale distribution is strongly non–gaussian, with a shift in the meatball direction.

161.477 **Can non–Gaussian fluctuations for structure formation arise from inflation?**
D. S. Salopek.
2. International Workshop on Plasma Physics: Plasma Theoretical Problems in Astro– and Fusion Physics, p. 81 – 89 (1991). – See Abstr. 012.067 for the main entry.
Non–Gaussian fluctuations for structure formation may be generated during the inflationary epoch from the nonlinear interaction of two scalar fields with gravity. Semi–analytical calculations are given describing nonlinear long wavelength evolution in 3+1 dimensions. Typically, non–Gaussian fluctuations arise when one passes over several sharp ridges in the potential surface. One can input this richer class of initial conditions into N–body codes to see the effects on the large scale structure in the Universe. The cleanest test of non–Gaussian fluctuations will hopefully occur in the near future from large angle microwave background anisotropy experiments.

161.478 **Non–baryonic dark matter.**
S. Sarkar.
NATO Advanced Research Workshop on Observational Tests of Inflation, p. 91 – 102 (1991). – See Abstr. 012.076 for the main entry.
The author discusses the theoretical motivation for new particles which may constitute the cosmological dark matter and reviews the progress of experimental efforts at their detection.

161.479 **Are galactic halos made of brown dwarfs or black holes?**
B. J. Carr.
NATO Advanced Research Workshop on Observational Tests of Inflation, p. 103 – 109 (1991). – See Abstr. 012.076 for the main entry.
If galactic halos are baryonic, they comprise either brown dwarfs or black holes. The first possibility can be tested by

microlensing observations, the second by searches for an infrared background.

161.480 **The local distance scale: how reliable is it?**
M. W. Feast.
NATO Advanced Research Workshop on Observational Tests of Inflation, p. 147 – 160 (1991). – See Abstr. 012.076 for the main entry.
A review of local calibrators and data for the Magellanic Clouds, M31 and M33 suggests an increase in the local distance scale of ~10% over that recently adopted by several workers. The uncertainty in this local scale is also ~10% and this sets a lower limit to the uncertainty in the Hubble constant.

161.481 **Large scale anisotropy of the CMB in an open Universe and constraints on the models of galaxy formation.**
N. Gouda, N. Sugiyama, M. Sasaki.
IUPAP Conference on Primordial Nucleosynthesis and Evolution of Early Universe, p. 335 – 336 (1991). – See Abstr. 012.071 for the main entry.
The authors derived a formula by which one can calculate any multipole moment of cosmic microwave background anisotropy in an open Universe, and calculate (by using this formula) the rms quadrupole moment expected in some cosmological models. They restrict the models by comparing the calculated quadrupole moments with the observed upper limit of the quadrupole moment of the cosmic microwave background anisotropy.

161.482 **Distances to Virgo and beyond.**
M. Rowan-Robinson.
NATO Advanced Research Workshop on Observational Tests of Inflation, p. 161 – 171 (1991). – See Abstr. 012.076 for the main entry.
New distance measurements to Virgo and beyond are reviewed. The discrepancy between the Type Ia supernova and infrared Tully–Fisher methods remains unresolved. Two new distance methods based on the luminosity function for planetary nebulae and on surface brightness fluctuations in ellipticals and lenticulars are briefly discussed. Analysis of IRAS maps of 100 μm emission from interstellar dust appear to resolve the controversy about extinction towards the Galactic poles. A new model for the local flow based on the QDOT IRAS galaxy redshift survey is applied to the analysis of the Hubble diagram. The implications of the current best estimates of H_0 and Ω_0 for inflation are discussed.

161.483 **The luminosity–line–width relations and the value of H_0.**
M. J. Pierce.
NATO Advanced Research Workshop on Observational Tests of Inflation, p. 173 – 178 (1991). – See Abstr. 012.076 for the main entry.
An extensive examination of the luminosity–line–width relations as an indicator of extragalactic distances is described. Evidence for significant background contamination in the Virgo Cluster is presented. It is shown that this contamination is responsible for much of the controversy regarding the distance to the Virgo Cluster. The absolute calibration is examined and it appears that the technique is now limited by the uncertainty in the Galactic distance scale. The relation is applied to samples from the Virgo and Ursa Major clusters and the value of H_0 is determined. The cosmological time–scale test is used to examine the inflationary hypothesis.

161.484 **Observational status of H_0.**
G. A. Tammann.
NATO Advanced Research Workshop on Observational Tests of Inflation, p. 179 – 185 (1991). – See Abstr. 012.076 for the main entry.
The distance of the Virgo cluster (21.2 ± 0.6 Mpc) from seven independent methods is presently the best known extragalactic distance. On the other hand the cosmic recession velocity of the cluster (v = 1144 ± 18 km s^{-1}) within a nearly Machian frame can be determined using relative distances to more distant clusters. The resulting value of the Hubble constant – freed from all velocity perturbations – is $H_0 = 52$ km $s^{-1}Mpc^{-1}$. The formal error of this determination is very small.

161.485 **The best–fit Universe.**
M. S. Turner.
IUPAP Conference on Primordial Nucleosynthesis and Evolution of Early Universe, p. 337 – 350 (1991). – See Abstr. 012.071 for the main entry.
Inflation provides very strong motivation for a flat Universe, Harrison–Zel'dovich (constant–curvature) density perturbations, and cold dark matter. However, there are a number of cosmological observations that conflict with the predictions of the simplest such model – one with zero cosmological constant. They include the age of the Universe, dynamical determinations of Ω, galaxy–number counts, and the apparent abundance of large–scale structure in the Universe. While the discrepancies are not yet serious enough to rule out the simplest and "most well motivated" model, the current data point to a "best–fit model" with the following parameters: $\Omega_B \cong 0.03$, $\Omega_{CDM} \cong 0.17$, $\Omega_\Lambda \cong 0.8$, and $H_0 \cong 70$ km $sec^{-1}Mpc^{-1}$, which improves significantly the concordance with observations. While there is no good reason to expect such a value for the cosmological constant, there is no physical principle that would rule such out.

161.486 **Globular clusters as extragalactic distance indicators.**
D. Hanes.
NATO Advanced Research Workshop on Observational Tests of Inflation, p. 211 – 215 (1991). – See Abstr. 012.076 for the main entry.
The author describes some recent developments in the use of globular clusters as extragalactic distance indicators. He warns: globular cluster systems differ in perplexing ways in various galaxies and environments, and caution may be needed in their use as distance indicators, despite their so far exemplary behaviour.

161.487 **LEP physics and the early Universe.**
L. M. Krauss.
IUPAP Conference on Primordial Nucleosynthesis and Evolution of Early Universe, p. 351 – 361 (1991). – See Abstr. 012.071 for the main entry.
The author reviews the implications of the Z decay measurements at LEP (and SLC) for the early Universe: (a) The Z width measurements, when combined with non–accelerator data rule out GeV range Dirac neutrinos, Majorana neutrinos, and sneutrinos as WIMP dark matter, rule out most explicit "cosmions", and strongly constrain neutralinos. When combined with data from other experiments, this implies that any WIMP is likely to be heavier than 10 – 20 GeV, and could easily be in the O(100 GeV – 1 TeV) region; (b) The limit on N_V, when combined with big–bang nucleosynthesis (BBN) estimates, is consistent with both baryonic and non–baryonic dark matter and allows one to probe the consistency of BBN itself; (c) direct searches for WIMPs will probably require new detectors, sensitive to spin dependent interactions on possibly heavy targets, with event rates which could be 3 – 4 orders of magnitude smaller than those expected at present detectors; (d) limits on the Higgs and the top quark also have an impact on possible new processes in the early Universe.

161.488 **Baryonic dark matter.**
J. Sil.
IUPAP Conference on Primordial Nucleosynthesis and Evolution of Early Universe, p. 371 – 379 (1991). – See Abstr. 012.071 for the main entry.
Both canonical primordial nucleosynthesis constraints and large–scale structure measurements, as well as observations of the fundamental cosmological parameters, appear to be consistent with the hypothesis that the Universe predominantly consists of baryonic dark matter. The author reviews the arguments for BDM to consist of compact objects that are either stellar relics or substellar objects. He describes several techniques for searching for halo BDM.

161.489 **Cosmology: Texture and cosmic structure.**
J. Silk, R. Juszkiewicz.
Nature, Vol. 353, No. 6343, p. 386 – 388 (3 Oct 1991).

161.490 **Phenomenological dark matter detection rate –from WIMP to SIMP–.**
H. Sato.
IUPAP Conference on Primordial Nucleosynthesis and Evolution of Early Universe, p. 381 – 385 (1991). – See Abstr. 012.071 for the main entry.

Aim of this paper is to search for loopholes of the dark matter particles which exhibit detectable evidences.

161.491 **Spherical symmetric model for calculating large peculiar velocities of galaxies.**
Xiang Shouping, Cheng Fuzhen, Liu Jianmin.
IUPAP Conference on Primordial Nucleosynthesis and Evolution of Early Universe, p. 421 – 422 (1991). – See Abstr. 012.071 for the main entry.

161.492 **On the origin of cosmological magnetic fields.**
T. Tajima, S. Cable, K. Shibata.
IUPAP Conference on Primordial Nucleosynthesis and Evolution of Early Universe, p. 423 – 424 (1991). – See Abstr. 012.071 for the main entry.

161.493 **The Hawaii deep survey – implications for cosmology and galaxy formation.**
L. L. Cowie.
IUPAP Conference on Primordial Nucleosynthesis and Evolution of Early Universe, p. 425 – 434 (1991). – See Abstr. 012.071 for the main entry.

The Hawaii deep survey has now established 2.2 micron galaxy counts to a K magnitude of 23. The low number density of these counts at the faint end favors a $q_0 = 0.5$, $\Lambda = 0$ geometry which minimizes the available cosmological volume. However these models provide no explanation for the blue galaxies which are responsible for the large number of faint blue counts. Combining spectroscopic data with K magnitudes for a blue selected sample shows that the large number of galaxies at B = 23 – 24 appears to be a population of blue dwarf galaxies lying at typical redshifts of 0.3 to 0.4. There are no obvious descendants of these galaxies at the present epoch. The dwarf galaxies contain as much K light as the normal galaxy population and may dominate the baryonic and dark matter content of the Universe.

161.494 **Analysis of the large scale structure with deep pencil beam surveys.**
A. S. Szalay, R. S. Ellis, D. C. Koo, T. J. Broadhurst.
IUPAP Conference on Primordial Nucleosynthesis and Evolution of Early Universe, p. 435 – 450 (1991). – See Abstr. 012.071 for the main entry.

Recent observations of the large scale structure of the Universe suggest inhomogeneities on scales larger than previously thought. A deep redshift survey with a "pencil beam" geometry of galaxies at the Galactic Poles indicates strong clustering, with a provocative regularity on a scale exceeding $100\,h^{-1}$Mpc. The authors examine in detail the role of such deep "pencil beam" surveys in studies of the large scale structure using the 1–D power spectrum of the distribution to illustrate the statistical significance of the feature claimed by Broadhurst et al. (1990).

161.495 **Cosmological implications of H I absorption systems.**
S. Ikeuchi.
IUPAP Conference on Primordial Nucleosynthesis and Evolution of Early Universe, p. 461 – 468 (1991). – See Abstr. 012.071 for the main entry.

As a unified model for H I absorption systems of quasars such as the Lyman α forest, Lyman limit systems and damped Lyman α line systems, the author examines the evolution of intergalactic clouds confined by the gravity of cold dark matter, the so called minihalos. For reproducing three independent observed results, the H I column density distribution, the number density evolution and the continuum depression of high redshift quasars, the minihalos have to expand due to the increase of diffuse UV flux with decreasing redshift. From this result, the author discusses the cosmological implications of absorption systems of quasars in relation to galaxy formation.

161.496 **Gravitational lens effect on cosmological distance.**
M. Sasaki.
IUPAP Conference on Primordial Nucleosynthesis and Evolution of Early Universe, p. 469 – 474 (1991). – See Abstr. 012.071 for the main entry.

Based on the cosmological post Newtonian expansion, a general formula for the distance–redshift relation in a realistic inhomogeneous Universe is presented. Then, focusing on the gravitational lens effect, the validity of using the Friedmann formula for the distance–redshift relation in an inhomogeneous Universe is discussed.

161.497 **Apparent and biased effects in oscillating Universe.**
M. Morikawa.
IUPAP Conference on Primordial Nucleosynthesis and Evolution of Early Universe, p. 475 – 481 (1991). – See Abstr. 012.071 for the main entry.

The author studies a cosmological model in which the expansion rate H and the gravitational constant G are oscillating. This model is proposed to explain the recently observed periodicity in the redshift distribution of galaxies with a characteristic scale of $128\,h^{-1}$Mpc. Apparent and biased enhancement of the galaxy distribution in the concentric spherical shells with periodically spaced radii centered on the Milky Way is derived. The apparent effect comes from the distorsion of the ordinary distance–redshift relation due to the H change. The biased effect comes from glow of a galaxy due to the G change. The main subjects in this paper are: 1) Doublet peak structure of the periodic distribution of galaxies. 2) Difference in the strength of quasar and Lyman–α cloud clusterings. 3) Bending of the magnitude–redshift curve in the Hubble diagram.

161.498 **Oscillating physics and periodic Universes.**
P. J. Steinhardt.
IUPAP Conference on Primordial Nucleosynthesis and Evolution of Early Universe, p. 483 – 485 (1991). – See Abstr. 012.071 for the main entry.

The periodicity in the pencil–beam redshift survey of galaxies reported by Broadhurst, et al. may be an illusion caused by the periodic oscillations of physical constants. The author analyzes scenarios entailing gravitational, electromagnetic, and other interactions.

161.499 **Probes of the high–redshift Universe.**
M. J. Rees.
IUPAP Conference on Primordial Nucleosynthesis and Evolution of Early Universe, p. 487 – 494 (1991). – See Abstr. 012.071 for the main entry.

Some lines of evidence on physical conditions at redshifts $z \gtrsim 3$ are briefly reviewed. The 21 cm line emission from inhomogeneous neutral gas could yield information on redshifts preceeding the epoch of galaxy formation and reheating.

161.500 **Large scale coherent structures in the Universe.**
L. Kofman.
IUPAP Conference on Primordial Nucleosynthesis and Evolution of Early Universe, p. 495 – 508 (1991). – See Abstr. 012.071 for the main entry.

The non–linear dynamics of the large–scale structure of the Universe due to gravitational instability is considered in the framework of the adhesion model. The author argues that pancaking should occur on the all scales, forming large scale coherent structures – "superpancakes", due to large–scale coherent motions of clumps of matter formed before. Then the author considers a simple statistics of nonlinear matter distribution: the one–point probability distribution functions of the smoothed peculiar velocity field and density field. In the Zel'dovich approximation, the velocity distribution function remains constant even through nonlinear evolution, while the density distribution becomes non Gaussian very rapidly, and depends on the linear rms with the given filtered scale only. One can expect a different shape of distribution function in other cosmological scenarious besides gravitational instability.

161.501 **Non-Gaussian density field in nonlinear gravitational clustering.**
Y. Suto.
IUPAP Conference on Primordial Nucleosynthesis and Evolution of Early Universe, p. 509 – 516 (1991). – See Abstr. 012.071 for the main entry.

Nonlinear gravitational clustering is one of the most important and fundamental processes to link the theoretical models with the observed cosmic structures. The author addresses two aspects of the nonlinear gravitational effects, mainly on the basis of the recent cosmological N-body simulations; one is the growth rate of density fluctuations in nonlinear regime, and the other is the generation of the non-Gaussian density field.

161.502 **United ways of inflation in generalized gravities.**
L. Amendola, S. Capozziello, M. Litterio, F. Occhionero.
IUPAP Conference on Primordial Nucleosynthesis and Evolution of Early Universe, p. 517 – 523 (1991). – See Abstr. 012.071 for the main entry.

Generalized theories of gravity can be viewed in a unified way when the trajectories of cosmological models can be projected onto a two dimensional phase space. This tool is applied here to the case of a non-minimally coupled scalar field, to the compactification of multidimensional cosmology, and to fourth order gravity. In conclusion, as a natural follow-up of these models, the authors outline the possibility of a scalar coupling with the curvature squared term. In all cases, they have in mind a spatially flat Friedmann-Robertson-Walker metric with scale factor a(t).

161.503 **Reviving massive neutrinos for large-scale structure.**
R. J. Scherrer.
IUPAP Conference on Primordial Nucleosynthesis and Evolution of Early Universe, p. 525 – 531 (1991). – See Abstr. 012.071 for the main entry.

Although cold dark matter is currently favored for the formation of large-scale structure, neutrinos are the only presently observed particle which could serve as dark matter, and massive neutrinos act as hot dark matter, rather than cold dark matter. The author presents results of a seeded hot dark matter model, in which the Universe is dominated by massive neutrinos ($m_\nu \sim 25$ eV), and relic compact objects produce the density perturbations. The simplest such model, with randomly-distributed static point seed masses, produces results in good agreement with the observations, suggesting that massive neutrinos can be revived as a candidate for dark matter. When combined with inflation, seeded hot dark matter produces a model in which perturbations on large and small scales are completely decoupled.

161.504 **Cellular distribution of the Universe and dark matter.**
D. Calzetti, M. Giavalisco, R. Ruffini.
IUPAP Conference on Primordial Nucleosynthesis and Evolution of Early Universe, p. 533 – 542 (1991). – See Abstr. 012.071 for the main entry.

The medium-scale inhomogeneities in the distribution of structures in the Universe can be explained within a massive "inos" dark matter model. In this environment a cell-fractal distribution arises, giving at present time a general structure of fractal cells randomly distributed in space. In order to determine the compatibility of the available data with the model proposed and the average size of the cells, three different analyses have been performed on the Zwicky and CfAI catalogues, respectively. The result is that the data are in agreement with a cell-fractal structure of cell diameter $\cong 60 - 70\,h^{-1}$Mpc ($H_0 =$ 100 km/s/Mpc).

161.505 **The large-scale peculiar velocity at high redshift.**
Zhu Xingfen, Chu Yaoquan.
IUPAP Conference on Primordial Nucleosynthesis and Evolution of Early Universe, p. 543 – 544 (1991). – See Abstr. 012.071 for the main entry.

161.506 **An evidence of multiply-connected topology model of the Universe.**
Zhu Xingfen, Chu Yaoquan.
IUPAP Conference on Primordial Nucleosynthesis and Evolution of Early Universe, p. 545 – 546 (1991). – See Abstr. 012.071 for the main entry.

161.507 **Non-Gaussian initial conditions in CDM model simulations.**
F. Lucchin.
IUPAP Conference on Primordial Nucleosynthesis and Evolution of Early Universe, p. 547 – 548 (1991). – See Abstr. 012.071 for the main entry.

Results are reported on N-body simulations of the large-scale structure of the Universe starting from non-Gaussian initial conditions in a cold dark matter scenario.

161.508 **Evolution of galaxy cluster and primordial nucleosynthesis.**
V. V. Chuvenkov, A. Yu. Glukhov, B. V. Vainer.
IUPAP Conference on Primordial Nucleosynthesis and Evolution of Early Universe, p. 549 – 551 (1991). – See Abstr. 012.071 for the main entry.

The model of galaxy cluster evolution is presented, matter exchange between galaxies and intergalactic medium and time dependence of initial mass function being taken into account. In all the model versions corresponding to available observed data primordial deuterium abundance in a galaxy decreases by a factor 2 to the present time, that enables to explain the results of primordial nucleosynthesis in the standard Big Bang model with baryon density parameter $\Omega_b = 0.1$.

161.509 **Constraints on Universe models with cosmological constant from cosmic microwave background anisotropy.**
N. Sugiyama, N. Gouda, M. Sasaki.
IUPAP Conference on Primordial Nucleosynthesis and Evolution of Early Universe, p. 553 – 554 (1991). – See Abstr. 012.071 for the main entry.

Thorough numerical calculations of the fluctuations in the cosmic microwave background radiation using the gauge invariant formalism are carried out for various cosmological models with the cosmological constant.

161.510 **Gravitational lens effect and distance measure.**
K. Watanabe, M. Sasaki, K. Tomita.
IUPAP Conference on Primordial Nucleosynthesis and Evolution of Early Universe, p. 555 – 556 (1991). – See Abstr. 012.071 for the main entry.

The distance measure and time delays of images of a strongly gravitational lensed source are numerically calculated in a simple cosmological model.

161.511 **Constraints on the cosmic string scenario from the large-scale distribution of galaxies.**
S. Miyoshi, T. Hara.
IUPAP Conference on Primordial Nucleosynthesis and Evolution of Early Universe, p. 557 – 558 (1991). – See Abstr. 012.071 for the main entry.

161.512 **Clustering and velocity field of galaxies in cold dark matter scenario.**
T. Suginohara, Y. Suto.
IUPAP Conference on Primordial Nucleosynthesis and Evolution of Early Universe, p. 559 – 560 (1991). – See Abstr. 012.071 for the main entry.

161.513 **Galactic dust emission and the cosmic microwave background.**
A. J. Banday, A. W. Wolfendale.
Mon. Not. R. Astron. Soc., Vol. 252, No. 3, p. 462 – 472 (1 Oct 1991).
The authors examine the constraints imposed on observations of the Cosmic Microwave Background by the presence of galactic dust emission. In particular, they consider the nature of the dust emission, in terms of grain properties and environment, and the overall galactic emission (especially at high latitudes). The IRAS 100–μm survey should provide useful data on which to base estimates of emission at longer wavelengths. However, variations in the dust heating from place to place in the Galaxy, and, should they be found, variations in the dust/gas ratio and grain properties within the Galaxy imply that multi–wavelength coverage and detailed point to point modelling of the galactic dust emission is required if genuine CMB anisotropies are to be observed at far–infrared frequencies. If a new dust component is found which is not correlated with the H I associated dust seen by IRAS, the 100–μm data will no longer accurately trace the dust emission at longer wavelengths. Estimates are given of the likely dust contribution to observations by COBE at 90 GHz.

161.514 **Cosmology: Quark soup – do not boil.**
M. Fukugita, C. J. Hogan.
Nature, Vol. 354, No. 6348, p. 17 – 18 (7 Nov 1991).

161.515 **High–redshift tests of Ω_0.**
B. Guiderdoni.
NATO Advanced Research Workshop on Observational Tests of Inflation, p. 217 – 231 (1991). – See Abstr. 012.076 for the main entry.
This paper briefly reviews the present status of the classical cosmological tests involving high–redshift galaxies: the Hubble diagram, the Loh–Spillar test, the faint galaxy counts and the age of distant galaxies. The review focuses on the crucial problem of the correction of galaxy evolution. As a matter of fact, the simplest analysis of the data leads to a low–density open universe. Nevertheless the observations could be reconciled with the flat geometry predicted by inflation if the cosmological constant was non–zero. As an alternative to this desperate solution, the author shows that a scenario in which the evolution of galaxies is driven by merging in an $\Omega_0 = 1$ flat universe could accomodate most of the data.

161.516 **Cosmology with galaxies at high redshifts.**
S. J. Lilly.
NATO Advanced Research Workshop on Observational Tests of Inflation, p. 233 – 241 (1991). – See Abstr. 012.076 for the main entry.
Following some remarks on the methodology of the classical cosmological tests, recent observations of galaxies at high redshift are examined to see whether they place any constraints on the cosmological models. It is concluded that, at least at present, they do not. The principle difficulty is that the luminosity evolution of galaxies is poorly constrained.

161.517 **Prospects for measuring the deceleration parameter.**
R. S. Ellis.
NATO Advanced Research Workshop on Observational Tests of Inflation, p. 243 – 249 (1991). – See Abstr. 012.076 for the main entry.
The author revives q_0 (geometrical deceleration parameter) tests with a fresh approach. q_0 is important to measure independently of local Ω_0 estimates for several reasons. Foremost, inflation requires the Universe to be spatially flat, i.e. $k \equiv 0$ which can be reconciled with low estimates of the baryonic mass density, $0.009 < \Omega_{b0} h^2 < 0.04$, if the cosmological constant, Λ, is $3(1-\Omega_0)H_0{}^2$. Although somewhat contrived, this might resolve the globular cluster age conflict, recent galaxy count/redshift data and large scale structure problems. Furthermore, regardless of whether $\Lambda \neq 0$, deceleration tests make no assumptions as to the expected bias in the sources used. The test is a geometric one involving relative distances and is usually independent of H_0.

161.518 **Aligned radio galaxies.**
K. C. Chambers.
NATO Advanced Research Workshop on Observational Tests of Inflation, p. 251 – 256 (1991). – See Abstr. 012.076 for the main entry.
The discovery of high redshift radio galaxies at $z \leqslant 4$ has unleashed a host of questions about their nature. Many new objects are being found, and their unusual properties are being uncovered. They generally exhibit the "alignment effect", wherein their highly elongated lumpy morphologies are lined up along the axes of their powerful radio sources. Evidence for both large star formation rates and scattered quasar light are found in the aligned blue continuum. The aligned infrared continuum is more likely to be dominated by starlight. The K–band Hubble diagram for radio galaxies is now extended to $z \sim 4$, but it will be difficult to use it to test different cosmological world models.

161.519 **K band galaxy counts and the cosmological geometry.**
L. L. Cowie.
NATO Advanced Research Workshop on Observational Tests of Inflation, p. 257 – 265 (1991). – See Abstr. 012.076 for the main entry.
The Hawaii deep survey has now established galaxy counts to a K magnitude of 23. The low number density of these counts at the faint end favors a $q_0 = 0.5$, $\Lambda = 0$ geometry which minimizes the available cosmological volume. However these models provide no explanation for the large number of faint blue galaxies. Combining spectroscopic data with K magnitudes for a blue selected sample shows that the large number of galaxies $B = 23 - 24$ appears to be a population of blue dwarf galaxies lying at typical redshifts of 0.3 to 0.4. These dwarf galaxies contain as much K light as the normal galaxy population and may dominate the baryonic and dark matter content of the universe.

161.520 **Selection effects in redshift surveys.**
Y. Yoshii, M. Fukugita.
NATO Advanced Research Workshop on Observational Tests of Inflation, p. 267 – 271 (1991). – See Abstr. 012.076 for the main entry.
The recent Durham redshift survey to $B_J = 22.5$ mag shows a lack of galaxies with a redshift higher than $z \cong 0.7$ for their survey magnitude, while cosmological models with the Tinsley–Bruzual type luminosity evolution of galaxies predict a substantial high–redshift tail in the redshift distribution. The authors point out that the suppression of high–redshift galaxies is due to the selection effects; when the effects of using the isophotal magnitude and finite seeing are taken into account, they find that a significant fraction of high–redshift galaxies should not be observed in agreement with the current observation. This means that we do not need to modify the current models of galaxy evolution. The authors also find that the observed redshift distribution favours a low–density universe in agreement with the conclusion from the number count of faint galaxies.

161.521 **Nucleosynthesis: Beryllium and the Big Bang.**
B. E. J. Pagel.
Nature, Vol. 354, No. 6351, p. 267 – 268 (28 Nov 1991).

161.522 **An inflationary alternative to the big–bang.**
F. Hoyle.
NATO Advanced Research Workshop on Observational Tests of Inflation, p. 273 – 278 (1991). – See Abstr. 012.076 for the main entry.
The model proposed is inflationary but with inflationary units of the scale of superclusters of galaxies rather than the much larger scale of more standard inflationary theories. The physical conditions in an inflationary unit are the same as in a standard big–bang model with a baryon–to–photon ratio of $\sim 3 \cdot 10^{-10}$. Because the microwave background was smoothed after galaxies were formed, fluctuations in the background on angular scales of arc minutes to degrees are very small. The creation phase of the inflationary units is at a lookback redshift of ~ 30 and is potentially observable. A universe with repeated inflationary

phases can have relevance to basic physics, affecting cut-off processes as in quantum electrodynamics.

161.523 **Dynamical estimates of Ω_0 from galaxy clustering.**
S. D. M. White.
NATO Advanced Research Workshop on Observational Tests of Inflation, p. 279 - 291 (1991). - See Abstr. 012.076 for the main entry.
The author reviews dynamical methods for obtaining masses on scales larger than galaxies, and the assumptions and uncertainties involves in using these to estimate the density of the Universe.

161.524 **Large scale structure and inflation.**
J. P. Huchra.
NATO Advanced Research Workshop on Observational Tests of Inflation, p. 315 - 326 (1991). - See Abstr. 012.076 for the main entry.
Recent observations of the large scale distribution of galaxies and clusters of galaxies are contrasted with predictions of models for the formation of such structures (primarily n-body simulations) based on the inflationary paradigm of Ω = unity. The combined set of observations are not easily matched by any single model and many of the observations favor low Ω universes. While the failure to match the observations of galaxies and clusters is not conclusive, it strongly suggests that the simplest, $\Omega = 1$, $\Lambda = 0$ models are incorrect.

161.525 **The structure of the Universe on large scales.**
A. G. Doroshkevich.
NATO Advanced Research Workshop on Observational Tests of Inflation, p. 327 - 330 (1991). - See Abstr. 012.076 for the main entry.
A possible explanation of the results from a quantitative analysis of the deep galactic surveys is suggested. The theoretical model taking into account the influence of the potential on the galaxy distribution is given. This model allows explaining the galaxy distribution on superlarge scales and the formation of large voids. It is shown that high density of the structure elements between voids contradicts to standard spectra of the cold and hot dark matter models and to simple power spectra. This effect can be explained if one assumes a special shape of the initial spectrum in the shortwave region.

161.526 **Testing the Zeldovich spectrum.**
W. Sutherland.
NATO Advanced Research Workshop on Observational Tests of Inflation, p. 331 - 335 (1991). - See Abstr. 012.076 for the main entry.
A variety of theoretical models are confronted with the correlation function results from the APM Galaxy Survey. One can make rather good estimates of the post-recombination power spectrum out to $k \sim 1/(100\,h^{-1}Mpc)$; its slope is $n \sim -1$ on cluster to supercluster scales, and rolls over to $n \gtrsim 0$ at larger scales, but cannot reach $n \sim 1$ until $k \lesssim 1/(100\,h^{-1}Mpc)$; thus the microwave background appears to offer the most direct test.

161.527 **Ω on the scale of 3 Mm/s.**
D. Lynden-Bell.
NATO Advanced Research Workshop on Observational Tests of Inflation, p. 337 - 353 (1991). - See Abstr. 012.076 for the main entry.
The finding and naming of the Great Attractor is briefly reviewed. Although originally found as a major feature of the optical galaxy distribution, it is shown to be the most prominent feature in the all-sky distribution of IRAS galaxies. A map of the surface density of the Supergalactic Plane towards Hydra and Centaurus is presented. Optical and IRAS determinations of $\Omega^{0.6}/b$ are compared. It is evident that the bias factor b is different for optical and IRAS galaxies. It would then be surprising if the b of IRAS galaxies is precisely unity.

161.528 **Formation of galactic halos from seeded hot dark matter.**
A. van Dalen.
IUPAP Conference on Primordial Nucleosynthesis and Evolution of Early Universe, p. 561 - 562 (1991). The full paper is published in Astrophys. J., Vol. 376, No. 2, p. 371 - 379 (1 Aug 1991). - See Abstr. 012.071 for the main entry.
The author summarizes work done with J. V. Villumsen on the formation of galactic halos in a flat universe dominated by massive neutrinos. He finds that the halos generated are well approximated by isothermal spheres, generating flat rotation curves, in agreement with galactic rotational velocity observations. Further, it affirms the assumption used in the calculation of the Tremaine-Gunn limit. However, it is found that with a massive object dominating the central region the core radius becomes an ill-defined quantity, so with a large fraction of the inner region made up of some other material, such as baryons, it is possible to produce small galaxies with low velocity dispersions.

161.529 **Gravitational clustering of galaxies: comparison between thermodynamic theory and N-body simulations.**
M. Itoh.
IUPAP Conference on Primordial Nucleosynthesis and Evolution of Early Universe, p. 563 - 564 (1991). - See Abstr. 012.071 for the main entry.
f(N) statistics are very useful for describing the distribution of galaxies. Thermodynamic theory gives the functional form of f(N) and agrees with the results of N-body simulations very well. It can also give the good measure of clustering and initial conditions.

161.530 **Velocity field around superclusters: the effect of asphericity.**
T. Watanabe.
IUPAP Conference on Primordial Nucleosynthesis and Evolution of Early Universe, p. 567 - 568 (1991). - See Abstr. 012.071 for the main entry.
Velocity fields in some aspherical configurations are investigated. They tend to deviate from that of a spherical configuration due to initial shapes of systems. The value of the cosmological density parameter may be affected by the flatness of the Local Supercluster.

161.531 **Evolution and statistics from the number counts of X-ray cluster and intergalactic medium.**
H. Hanami.
IUPAP Conference on Primordial Nucleosynthesis and Evolution of Early Universe, p. 569 - 570 (1991). - See Abstr. 012.071 for the main entry.
The author discusses the distribution and the evolution of X-ray clusters of galaxies. Hot intra-cluster plasma is the source of X-ray emission. The author calculates the mass function and the temperature distribution by galaxy formation scenarios and compares the computed results with observational data.

161.532 **Backreactions of inhomogeneities on expanding universes.**
S. Bildhauer, T. Futamase.
IUPAP Conference on Primordial Nucleosynthesis and Evolution of Early Universe, p. 573 - 574 (1991). - See Abstr. 012.071 for the main entry.
In a post-Newtonian approximation scheme in general relativity, an inhomogeneous metric is constructed for the non-linear regime. After spatially averaging, the universe is nearly homogeneous and isotropic. For simple solutions of the local Newtonian equations, it is shown that the expansion in the direction where structure preferentially forms is less rapid than the average expansion. The resulting anisotropy of the CMBR depends strongly on the time at which perturbations get non-linear. A special model is constructed in which the observed dipole component may be due to large-scale inhomogeneities.

161.533 **Multi-level feature in the large scale distribution of galaxies.**
Z.-G. Deng, X.-Y. Xia, Y.-Z. Liu, Z.-L. Zou.
IUPAP Conference on Primordial Nucleosynthesis and Evolution of Early Universe, p. 575 – 576 (1991). – See Abstr. 012.071 for the main entry.

161.534 **Tests of inflation using the QDOT redshift survey.**
C. S. Frenk.
NATO Advanced Research Workshop on Observational Tests of Inflation, p. 355 – 364 (1991). – See Abstr. 012.076 for the main entry.
The QDOT survey of 2163 redshifts of IRAS galaxies was designed to study clustering on scales $\gtrsim 10\ h^{-1}$Mpc. It is ideal to test two fundamental predictions of generic models of inflation: the value of the mean cosmic density and the nature of primordial density fluctuations. The survey supports the predictions that $\Omega = 1$ and that the primordial density field is a random Gaussian process, but is inconsistent with a constant curvature power spectrum on scales $\lesssim 50\ h^{-1}$Mpc.

161.535 **Fragmentation and clustering in the wake formed by cosmic string.**
T. Hara, S. Miyoshi.
IUPAP Conference on Primordial Nucleosynthesis and Evolution of Early Universe, p. 577 – 578 (1991). – See Abstr. 012.071 for the main entry.
Numerical calculations are performed for the fragmentation, clustering and merging of dark matter under the open cosmic string model. Both the fragmentation in one dimensional collapse of pancake scenario and clustering theory from small scale to large scale within a wake are realized.

161.536 **Perturbations of the one-dimensional structure in the Universe.**
S. Bildhauer, O. Aso, M. Kasai, T. Futamase.
IUPAP Conference on Primordial Nucleosynthesis and Evolution of Early Universe, p. 579 – 580 (1991). – See Abstr. 012.071 for the main entry.
Linear perturbation theory on the background of a one-dimensional, non-linear inhomogeneous Universe is considered. For an Einstein-de Sitter behaviour of the scale factor, the exact solution of the perturbation equations is given. The non-linear effect of the density inhomogeneity on the evolution of the peculiar velocity is written in form of a relation between the peculiar velocity and the density parameter Ω including the correction due to the density contrast δ_0 of the Zeldovich solution. It is shown that the component parallel to the walls is characterized by $\Omega^{0.6}(1+\delta_0)^{-0.57}$.

161.537 **Testing inflation with peculiar velocities.**
A. Dekel.
NATO Advanced Research Workshop on Observational Tests of Inflation, p. 365 – 373 (1991). – See Abstr. 012.076 for the main entry.
The author presents three different results which indicate that the velocity and mass-density fields extracted directly from the observed peculiar velocities agree with the predictions of Inflation. (1) A comparison with the IRAS galaxy density yields $\Omega^{0.6}/b_I = 1.35 \pm 0.25$, where b_I is the biasing factor for IRAS galaxies, and non-linear analysis $\Omega \cong b_I \cong 1$. (2) The shape of the fluctuation power spectrum in the wavelength range $10 - 100\ h^{-1}$Mpc is consistent with an initially scale-invariant spectrum modified by cold dark matter. (3) The one-point probability distribution of the velocity fields by POTENT and by IRAS are consistent with Gaussian initial fluctuations. The author concludes that there is no dynamical evidence against the standard model of the formation of large-scale structure; the indicated large-scale excess of power in the galaxy distribution could be attributed to non-trivial biasing of galaxies relative to the mass distribution.

161.538 **Three points spatial correlation function on the self similar observer homogeneous structure.**
Lee Hyungwon, R. Ruffini.
IUPAP Conference on Primordial Nucleosynthesis and Evolution of Early Universe, p. 581 – 582 (1991). – See Abstr. 012.071 for the main entry.
The authors present an explicit form of 2 and 3 points spatial correlation functions for observer homogeneous and self similar structure. Crucial difference from the usual hierarchical form is the fact that it increases with separation after a certain value. This behaviour seems to be consistent with observation.

161.539 **The invisible cosmological constant.**
O. Lahav, P. B. Lilje, J. R. Primack, M. J. Rees.
NATO Advanced Research Workshop on Observational Tests of Inflation, p. 375 – 378 (1991). – See Abstr. 012.076 for the main entry.
The authors discuss the possibility of measuring the density parameter Ω_0 and the cosmological constant $\lambda_0 \equiv \Lambda/(3H_0^2)$ from dynamical tests using linear and non-linear spherical models. While information on both parameters can be obtained by looking at clustering at different redshifts, in practice the other observables also depend on the cosmology, and in some cases conspire to give a weak dependence on λ_0. The authors also show that a repulsive Λ gives a smaller final radius of a virialized cluster than a vanishing Λ.

161.540 **Limits from cosmological nucleosynthesis on the leptonic numbers of the Universe.**
A. Bianconi, H. W. Lee, R. Ruffini.
IUPAP Conference on Primordial Nucleosynthesis and Evolution of Early Universe, p. 583 – 585 (1991). – See Abstr. 012.071 for the main entry.
Constraints on chemical potentials and masses of "inos" are calculated using cosmological standard nucleosynthesis processes.

161.541 **Support for inflation from the Great Attractor.**
A. Heavens.
NATO Advanced Research Workshop on Observational Tests of Inflation, p. 379 – 383 (1991). – See Abstr. 012.076 for the main entry.
The depth of the velocity potential at the Great Attractor is used to constrain the density parameter, the spectrum of fluctuations, and the degree of bias of IRAS galaxies, using only the following assumptions: on large scales, galaxies trace the velocity field (of any dark matter); the velocity perturbations are gravitational in origin; the perturbations are Gaussian on large scales. No further assumptions are required. From the absence of fluctuations in the microwave background radiation, adiabatic CDM models must have $\Omega > 0.5$. From the distribution of IRAS galaxies, their bias parameter on large scales must satisfy $\Omega/b^{5/3} > 0.5$. For spectra which have more large-scale power, the lower limits on $\Omega/b^{5/3}$ are relaxed, but the maximum likelihood values for power-law spectra exceed 0.6.

161.542 **Inhomogeneities and non-Friedmann light propagation.**
E. V. Linder.
IUPAP Conference on Primordial Nucleosynthesis and Evolution of Early Universe, p. 587 – 588 (1991). – See Abstr. 012.071 for the main entry.
Driven by the need to reconcile large scale inhomogeneity with global isotropy, the author investigates the rigorous development of a metric realistically describing our cosmology. He then estimates the order of magnitudes for various observational effects due to the presence of inhomogeneities affecting light propagation.

161.543 **The angular large scale structure.**
Y. Hoffman.
NATO Advanced Research Workshop on Observational Tests of Inflation, p. 385 – 388 (1991). – See Abstr. 012.076 for the main entry.
The angular distribution of the IRAS galaxies has been expanded in spherical harmonics and its angular power spectrum

has been calculated. This distribution is dominated by the first few multipoles and the amplitude is a decreasing function of l. The author finds the IRAS data to have more power on the scale of a few 10 h^{-1}Mpc compared with the standard CDM model, and to be compatible with an open universe CDM model of $\Omega_0 h = 0.2$.

161.544 **Is there any observational evidence for non–Gaussian primordial density fluctuations?**
A. L. Melott.
NATO Advanced Research Workshop on Observational Tests of Inflation, p. 389 – 393 (1991). – See Abstr. 012.076 for the main entry.

A brief review of tests of observational data and a report on recent numerical experiments is used to argue that there is little motivation from observational data for non–Gaussian primordial density fluctuations. Any such motivation must come at this time from the wish to consider a specific theoretical model.

161.545 **On the generation of non–Gaussian fluctuations during chaotic inflation.**
I. Yi, E. T. Vishniac, S. Mineshige.
IUPAP Conference on Primordial Nucleosynthesis and Evolution of Early Universe, p. 589 – 590 (1991). – See Abstr. 012.071 for the main entry.

The authors solve the stochastic evolution equations for the inflaton field during the slow–roll period of chaotic inflation. We find exact analytic solutions for the $V(\Phi) = \lambda\Phi^4/4$ potential. The resulting probability distributions are sharply peaked around the classical deterministic trajectory and are approximately Gaussian, but the exact probability distributions show that skewness is very sensitively dependent on λ and also weakly dependent on initial and final conditions. The non–Gaussian statistics induced by the nonlinear stochasticity effectively measures the expected level of non–Gaussian deviation in the inflation–generated density fluctuations.

161.546 **A unified picture of H I absorption–line systems by mini–halos.**
I. Murakami, S. Ikeuchi.
IUPAP Conference on Primordial Nucleosynthesis and Evolution of Early Universe, p. 591 – 592 (1991). – See Abstr. 012.071 for the main entry.

The authors consider an evolution of intergalactic clouds confined by the gravity of cold dark matter, the so called minihalo, to reproduce the observed results of neutral hydrogen absorption line systems of quasars. Assuming a simplified evolution law of diffuse UV flux and mass function of gas clouds, one can reproduce several observational properties by expanding minihalos which experience an expansion caused by the increase of diffuse UV flux at $z > 2$.

161.547 **Gravitational lens effect on the images of high redshift objects.**
K. Tomita, K. Watanabe.
IUPAP Conference on Primordial Nucleosynthesis and Evolution of Early Universe, p. 593 – 594 (1991). – See Abstr. 012.071 for the main entry.

In order to study the gravitational lens effect of galaxies on the images of high redshift objects, the authors solved numerically the null geodesic equations in an inhomogeneous model Universe consisting of galaxies. Morphological studies were done as well as quantitative analyses for the image deformation.

161.548 **Gravitational lens determinations of cosmological parameters.**
E. L. Turner.
IUPAP Conference on Primordial Nucleosynthesis and Evolution of Early Universe, p. 595 – 598 (1991). – See Abstr. 012.071 for the main entry.

Currently available attempts to determine the three classical cosmological parameters (H_0, Ω_0, and Λ) using gravitational lens techniques and arguments are reviewed. These suggest that $H_0 \gtrsim 75$ km/s/Mpc, that contributions to Ω_0 by certain hypothetical types of dark matter are <1, and that Λ is considerably smaller than $3\,H_0^2$ if a flat cosmological model is assumed. Unfortunately, none of these three conclusions is satisfactorily free of model dependence, possible systematic error, caveats, etc.

161.549 **Global cosmological parameters.**
M. Fukugita.
IUPAP Conference on Primordial Nucleosynthesis and Evolution of Early Universe, p. 599 – 614 (1991). – See Abstr. 012.071 for the main entry.

The determination of the global cosmological parameters is discussed. It is argued that the Hubble constant is likely to lie between 75 and 100 km $s^{-1}Mpc^{-1}$, and a low value ($H_0 \approx 50$ km $s^{-1}Mpc^{-1}$) is quite unlikely. It is also shown that the current observational evidence supports a low density (typically $\Omega \approx 0.1$) Universe. Combining these parameters with the accepted value for the age of the Universe, it is concluded that a sizable amount of the cosmological constant is probably necessary. The author discusses that the same conclusion is also inferred from the number count of faint galaxies.

161.550 **Gravitational lens effects on the redshift–volume test of Ω_0.**
M. Omote, H. Yoshida.
IUPAP Conference on Primordial Nucleosynthesis and Evolution of Early Universe, p. 615 – 616 (1991). – See Abstr. 012.071 for the main entry.

Gravitational lens effects on the redshift–volume test are investigated. The authors reanalyze Loh–Spillar's data of this test by taking into account the gravitational lens effects caused by inhomogeneities of the Universe, and show that the effects decrease the value of the density parameter Ω_0 to be 0.38 ± 0.44.

161.551 **Summary of the conference "Primordial nucleosynthesis and evolution of early Universe".**
S. Hayakawa.
IUPAP Conference on Primordial Nucleosynthesis and Evolution of Early Universe, p. 617 – 621 (1991). – See Abstr. 012.071 for the main entry.

161.552 **Objective determination of the large–scale distribution of galaxies.**
R. J. Dodd, H. T. MacGillivray.
Proc. Astron. Soc. Aust., Vol. 9, No. 1, p. 49 – 51 (1991). – See Abstr. 012.078 for the main entry.

Observations of the large–scale organisation of matter in the Universe are of great importance in present day astronomy. A description is given of a southern sky catalogue containing 10^9objects recently completed and an outline of some of the extragalactic projects underway using this large body of data.

161.553 **Non–Gaussian initial conditions in cosmological N–body simulations. III. Groups in Cold Dark Matter models.**
S. Matarrese, F. Lucchin, A. Messina, L. Moscardini.
Mon. Not. R. Astron. Soc., Vol. 253, No. 1, p. 35 – 46 (1 Nov 1991).

Group catalogues are obtained in N–body simulations of the large–scale structure of the Universe with non–Gaussian initial conditions in a Cold Dark Matter cosmogony. Groups of particles are defined by the standard percolation algorithm with linking parameter $b = 0.25$ in units of the mean interparticle distance. Grouping properties are found to strongly depend on the statistics of the underlying mass distribution. By applying different tests, the clustering and the topology of the group distribution, the environmental properties of the groups and their internal geometric and dynamical features are analysed. Groups in negative models, i.e. models characterized by a predominance of primordial underdense regions, are preferentially loose systems, slightly biased over the background, with high velocity dispersion and clear signal of alignment with neighbouring groups up to $\sim 30\,h^{-2}$Mpc. Groups in positive models, i.e. models with a predominance of primordial overdense regions, are highly biased compact systems with lower velocity dispersion.

The group density field in the authors' non–Gaussian models always shifts towards the meatball topology, regardless of the topology of the underlying mass distribution.

161.554 **Gravitational lensing frequencies: galaxy cross–sections and selection effects.**
M. Fukugita, E. L. Turner.
Mon. Not. R. Astron. Soc., Vol. 253, No. 1, p. 99 – 106 (1 Nov 1991).
In order to allow more realistic predictions of multiple image galaxy–quasar lensing frequencies, the authors consider in detail four issues: the best currently available data on the galaxy velocity dispersion distribution, the effects of finite core radii potential ellipticity on lensing cross–sections, the predicted distribution of lens image separations compared to observational angular resolutions, and the preferential inclusion ("amplification bias") of lens systems in flux limited samples. The authors conclude from their considerations that flat Universe models excessively dominated by the cosmological constant are not favoured by the small observed galaxy–quasar lensing rate. Models with a dimensionless cosmological constant $\lambda_0 = 0.9$ may be marginally consistent with the available data; all $\lambda_0 = 0$ models give a good fit.

161.555 **Geometry of the Metagalaxy.**
I. L. Rozental'.
Zemlya Vselennaya, No. 6, p. 10 – 14 (Nov – Dec 1991). In Russian.

161.556 **Large–scale structure and H_0 from super–nova envelope.**
S. Pistinner, G. Shaviv.
ESO/EIPC Workshop on Supernova 1987A and other Supernovae, p. 731 – 737 (Jun 1991). – See Abstr. 012.075 for the main entry.
A method to examine perturbations to the Robertson–Walker metric on the supernova distance scale is suggested. The method is based on the determination of the redshift of a supernova envelope in two independent ways. If the two values of redshift so found are in agreement, one can (1) proceed and compare the redshift with galactic redshift as measured from line shifts and then (2) check if the redshift of the SNe is indeed due to the expansion of the universe.

161.557 **COBE DMR observations of CMB anisotropy.**
G. F. Smoot.
NATO Advanced Research Workshop on Observational Tests of Inflation, p. 395 – 412 (1991). – See Abstr. 012.076 for the main entry.
The COBE Differential Microwave Radiometers (DMR) have produced preliminary full–sky maps at frequencies 31.5, 53, and 90 GHz. Galactic emission is seen unambiguously at all three frequencies. The only large–scale anisotropy detected in the cosmic microwave background is the dipole anisotropy. There is no clear evidence for any other large–angular–scale feature in the maps. The author briefly reviews the DMR experiment and discusses some implications of these results for cosmology in general, and inflationary models in particular.

161.558 **Observations of microwave background anisotropy at Tenerife and Cambridge.**
A. N. Lasenby, R. D. Davies, R. A. Watson, R. Rebolo, C. Gutierrez, J. E. Beckman.
NATO Advanced Research Workshop on Observational Tests of Inflation, p. 413 – 418 (1991). – See Abstr. 012.076 for the main entry.
A brief summary of the status of microwave background anisotropy observations at Tenerife and Cambridge is given. New 15 GHz Tenerife results give a preliminary limit of $\Delta T/T \lesssim 2.5 \times 10^{-5}$ on an angular scale of 4°. The features reported previously at 10 GHz appear to have a galactic spectrum, and can be predicted, though only approximately, from lower frequency sky surveys. At Cambridge, the newly completed Ryle Telescope is making its first observations of clusters of galaxies in order to measure the Sunyaev–Zel'dovich effect, and a dedicated interferometer to detect primordial perturbations is under construction.

161.559 **Foreground effects and the search for fluctuations in the CMB radiation.**
A. J. Banday, M. Giler, B. Szabelska, J. Szabelski, A. W. Wolfendale.
NATO Advanced Research Workshop on Observational Tests of Inflation, p. 419 – 423 (1991). – See Abstr. 012.076 for the main entry.
Precise measurements of anisotropies in the Cosmic Microwave Background radiation are of great importance to modern cosmology. Of special interest here are the recent measurements at ~ 10 GHz where some evidence for a detection of cosmological fluctuations has been claimed. However, background effects, particularly these due to the Galactic Synchrotron Radiation and infra–red emission from dust, can contaminate the genuine signal. The authors examine these local fields in some detail and conclude that predicted fluctuations in the GSR are of the same magnitude as the observed ones at 10 GHz. Future searches for CMB fluctuations will be possible at frequencies greater than 20 GHz provided that the nature of the Galactic signal is better understood.

161.560 **Microwave background anisotropies and large scale structure in the Universe.**
G. Efstathiou.
NATO Advanced Research Workshop on Observational Tests of Inflation, p. 425 – 436 (1991). – See Abstr. 012.076 for the main entry.
Recent observations show that the Universe contains more structure on scales $\gtrsim 10\,h^{-1}$Mpc than expected in the standard cold dark matter (CDM) theory of structure formation. This may indicate that the initial fluctuations were not scale–invariant. The author investigates how this would affect microwave background anisotropies if the Universe is spatially flat with $\Omega = 1$, as seems most natural if the Universe went through an inflationary phase. If we are allowed arbitrary freedom in choosing the shape of the initial fluctuation spectrum, then it is possible to construct models which can account for observed structure in the galaxy distribution but which result in small ($\Delta T/T \lesssim 10^{-6}$) anisotropies in the microwave background radiation on angular sclaes $\gtrsim 5°$.

161.561 **Discovery of the small scale sky anisotropy at 2.7 cm: radio sources or relic emission?**
Yu. N. Parijskij, B. L. Erukhimov, M. G. Mingaliev, A. B. Berlin, N. N. Bursov, N. A. Nizhelskij, M. N. Naugolnaja, V. N. Chernenkov, O. V. Verkhodanov, A. V. Chepurnov, A. A. Starobinsky (*A. A. Starobinskij*).
NATO Advanced Research Workshop on Observational Tests of Inflation, p. 437 – 441 (1991). – See Abstr. 012.076 for the main entry.
The authors present new limits on the small scale 3K emission anisotropy. At the scales of 1° – 3°.3 all 7 harmonics have at $\lambda = 7.6$ cm amplitudes below 10^{-5} in $\Delta T/T$. At l from 6.1×10^3 to 17.5×10^3 (rad^{-1}) scale this limit is about 2.8×10^{-4} at a wavelength of 2.7 cm and may be fully explained by a discrete source contribution. The low baryonic content of the Universe and present limit on $\Delta T/T$ at the horizon scale suggest that $(\Delta M/M)_{\mathrm{gravitating}} \ll (\Delta M/M)_{\mathrm{baryonic}}$. The authors speculate that 3K anisotropy measurements demonstrate a) the existence of the dark matter, b) the nonbaryonic nature of this dark matter and c) that the spatial distribution of this matter is much more homogeneous than that of the visible one.

161.562 **Balloon–borne observations of CMB anisotropies at intermediate angular scales, at submm and mm wavelengths.**
P. de Bernardis, S. Masi, B. Melchiorri, F. Melchiorri.
NATO Advanced Research Workshop on Observational Tests of Inflation, p. 443 – 446 (1991). – See Abstr. 012.076 for the main entry.
The authors have performed extensive observations of the cosmic microwave background (CMB) at angular scales of

$6-17°$ in the wavelength range 500 – 3000 micron, by balloon-borne instrumentation. They have found an upper limit for the square root of the amplitude of the correlation function ranging from $\leqslant 2\times10^{-5}$ for primordial CDM perturbations with spectral index $n = 0$ down to $\leqslant 0.8\times10^{-5}$ for $n = 2.9$, at 95% of confidence level, after correction for residual atmospheric fluctuations and galactic dust emission.

161.563 **Time evolution of lensed image separations.**
T. Broadhurst, S. Oliver.
NATO Advanced Research Workshop on Observational Tests of Inflation, p. 449 – 451 (1991). – See Abstr. 012.076 for the main entry.

Cosmic expansion results in a slow decrease in the angular separation of gravitationally lensed images, which, if monitored, may provide a measure of the global geometry. The authors calculate the time dependence of this effect for the simple case of a double image produced by a point mass in FRW universes. They note that peculiar velocities of source and lens have a negligible effect on the separation and they briefly discuss the feasibility of observing this effect.

161.564 **Deep galactic surveys as probes of the large scale structure of the Universe.**
O. E. Buryak, M. Demianski, A. G. Doroshkevich.
NATO Advanced Research Workshop on Observational Tests of Inflation, p. 453 – 455 (1991). – See Abstr. 012.076 for the main entry.

The authors report a new method of analysis of deep galactic surveys and apply it to the available surveys. They consider a few model distributions which allows to obtain detailed information on the one dimensional characteristics of voids. The authors show that the one dimensional distribution of the elements of large scale structure (LSS) is almost Poissonian and they obtain an estimate of the typical size of voids. The authors confirm the reality of super LSS (of an approximate size of 80 h^{-1}Mpc) the observational evidences of which perhaps are explained as the existence of a population of large voids. This result is also confirmed by analysis of the C IV doublet distribution in the absorption spectra of quasars.

161.565 **Density and peculiar velocity fields in the region of Dressler's supergalactic plane survey.**
M. J. Hudson.
NATO Advanced Research Workshop on Observational Tests of Inflation, p. 467 – 468 (1991). Abstract. – See Abstr. 012.076 for the main entry.

161.566 **Scale invariance induced by non–linear growth of density fluctuations.**
F. Moutarde, J.-M. Alimi, F. R. Bouchet, R. Pellat.
NATO Advanced Research Workshop on Observational Tests of Inflation, p. 469 – 470 (1991). – See Abstr. 012.076 for the main entry.

The authors investigate analytically and numerically the formation of smooth dark matter halos in an $\Omega = 1$ universe. They have proved analytically in the spherical case that a power–law density profile is expected before shell–crossing and give a general formula relating the slope to the initial overdensity. Performing high resolution particle–mesh numerical simulations (up to 128^3 particles in a 128^3 grid) the authors have shown that a self–similar regime takes place soon after shell–crossing in various 3D and 2D collapses (spherical, cylindrical or superposition of 2 or 3 caustics). The authors have compared this regime in some details with the semi–analytic approximate model of Fillmore and Goldreich (1984). They suggest that the afore–mentioned scale invariance at collapse provides a natural starting point for the self–similar solutions, which was the missing link in the semi–analytic studies.

161.567 **The power spectrum of galaxy clustering.**
J. A. Peacock.
NATO Advanced Research Workshop on Observational Tests of Inflation, p. 471 – 474 (1991). Also published in Mon. Not. R. Astron. Soc., Vol. 253, No. 1, p. 1P – 5P (1 Nov 1991). – See Abstr. 012.076 for the main entry.

There now exist several datasets with the depth necessary to determine the power spectrum of galaxy clustering on scales $\gtrsim$100 Mpc. Results from redshift surveys and angular clustering all agree on the form of the function, which may be well approximated for optically–selected galaxies by $\Delta^2(k) = 0.08\, y^4/(1+y^{2.4})$, where $y \equiv k/0.034\, h\, \mathrm{Mpc}^{-1}$. At small k, $h = 0.5$ CDM with J_3 normalization gives an amplitude a factor 2.0 smaller. Since normalization should aim to match spectra on the largest scales, where linear theory is most nearly applicable, it seems that investigations of the CDM model should therefore use a higher amplitude of fluctuations; an increase by a factor 2 is probably allowed by present limits on fluctuations in the microwave background. N–body simulations have not studied such a model in much detail; it may be premature to reject the $\Omega = 1$ adiabatic CDM universe as a viable picture for galaxy formation.

161.568 **The power spectrum of galaxy clustering.**
J. A. Peacock.
Mon. Not. R. Astron. Soc., Vol. 253, No. 1, p. 1P – 5P (1 Nov 1991).

There now exist several data sets with the depth necessary to determine the power spectrum of galaxy clustering on scales $\gtrsim$100 Mpc. Results from redshift surveys and angular clustering all agree on the form of the function, which shows a break towards greater uniformity at very large wavelengths, consistent with scale–invariant primordial fluctuations. The amplitude of these large–wavelength fluctuations implies that the Cold Dark Matter model should use a higher normalization; an increase by a factor 2 is allowed by present microwave–background limits. Such a revised amplitude predicts too much small–scale power in linear theory, but this discrepancy is on a scale where astrophysical effects are expected to alter the shape of the power spectrum. Nevertheless, if a better match between data and linear theory is desired, models containing global texture perform extremely well.

161.569 **Higher moments of the IRAS galaxy distribution.**
C. A. Scharf.
NATO Advanced Research Workshop on Observational Tests of Inflation, p. 475 – 476 (1991). Abstract. – See Abstr. 012.076 for the main entry.

161.570 **Ly α clouds at low redshift and the cosmological constant.**
M. Fukugita, O. Lahav.
Mon. Not. R. Astron. Soc., Vol. 253, No. 1, p. 17P – 20P (1 Nov 1991).

The authors point out that the distribution of Lyα clouds at low redshift ($z < 2$) is very sensitive to the cosmological constant. Recent spectroscopic observations with the Hubble Space Telescope (HST) towards 3C 273 show that the number of Lyα absorption systems found is higher than expected from the numbers at higher redshifts. If evolution is positive, in that it acts to decrease the number density of clouds at lower redshifts, these observations suggest that a universe highly dominated by the cosmological constant is not favoured, although the possibility of a negative evolution cannot be ruled out. The authors demonstrate these aspects in the framework of the pressure–confined evolutionary model, and emphasize that any discussion of evolution at low redshift should take into account the cosmological model.

161.571 **Exact solution in a scale invariant model with vanishing cosmological constant.**
W. F. Kao.
Chin. J. Phys., Vol. 29, No. 1, p. 1 – 6 (Feb 1991).

The author analyzes a special scale invariant effective theory with vanishing cosmological constant. He shows that, in this theory, the spatial–independence of the inflaton is implied by the equation of motion incorporated with the Robertson–Walker metric. Furthermore, he finds that this theory can be solved exactly for k = 0 Robertson–Walker spaces. He also finds almost exact solution in k ≠ 0 spaces.

161.572 **The scalar field as dynamical variable in inflation.**
J. E. Lidsey.
Phys. Lett. B, Vol. 273, No. 1/2, p. 42 – 46 (12 Dec 1991).
The scalar field driving inflation in a general Robertson–Walker cosmology is used as a dynamical variable to express the field equations as a first–order differential system. This allows a study of super-inflation to be made. It is found that super-inflation is likely to occur in closed cosmologies when the slow-roll approximation is valid and a new super–inflationary solution is presented. The possibility of extending this technique to the Bianchi models is examined.

161.573 **Can particle production in de Sitter spacetime lead to a matter–dominated universe?**
C. C. Bernido.
International Centre for Theoretical Physics, Trieste (Italy), IC—90/382. 8 p. (Oct 1990).
Using path integrals, it is shown that particle–antiparticle production in the early universe can lead to an observed matter–antimatter asymmetry at later times. This result is obtained from an integrability condition of the Green function for a relativistic scalar particle in de Sitter spacetime.

161.574 **Big–bang nucleosynthesis – observational aspects.**
B. E. J. Pagel.
Phys. Scr. T, Vol. T36, p. 7 – 15 (1991). – See Abstr. 012.080 for the main entry.
Extrapolation of observational data on the abundances of D, ^{3}He, ^{4}He and ^{7}Li in various astrophysical objects to derive their primordial values leads to results in good accordance with calculations from Standard Big Bang nucleosynthesis theory over 9 orders of magnitude in abundance and has led to the following predictions: there are not more than 3 light neutrino species or other particles contributing relativistic degrees of freedom at temperatures of a few MeV; the neutron half–life is less than 10.4 min; and baryonic dark matter exists, but not in sufficient quantities to close the universe. Searches for a primordial component in the abundance of any other element heavier than hydrogen – such as might have resulted from inhomogeneities due to phase transitions in the early universe, notably the quark–hadron transition – have so far proved completely negative.

161.575 **The cosmic microwave background.**
J. Silk.
Phys. Scr. T, Vol. T36, p. 16 – 21 (1991). – See Abstr. 012.080 for the main entry.
Recent limits on spectral distortions and angular anisotropies in the cosmic microwave background are reviewed. The various backgrounds are described, and the theoretical implications are assessed. Constraints on inflationary cosmology dominated by cold dark matter and on open cosmological models dominated by baryonic dark matter, with, respectively, primordial random phase scale–invariant curvature fluctuations or non–gaussian isocurvature fluctuations are described. More exotic theories are addressed, and the author concludes with the "bottom line": what theories expect experimentalists to be measuring within the next two to three years without having to abandon their most cherished theorists.

161.576 **Big bang nucleosynthesis: the standard model and alternatives.**
D. N. Schramm.
Phys. Scr. T, Vol. T36, p. 22 – 29 (1991). – See Abstr. 012.080 for the main entry.
Big bang nucleosynthesis provides (with the microwave background radiation) one of the two quantitative experimental tests of the big bang cosmological model. This paper reviews the standard homogeneous–isotropic calculation and shows how it fits the light element abundances ranging from ^{4}He at 24% by mass through ^{2}H and ^{3}He at parts in 10^5 down to ^{7}Li at parts in 10^{10}. Furthermore, the recent LEP (and SLC) results on the number of neutrinos are discussed as a positive laboratory test of the standard scenario. Discussion is presented on the improved observational data as well as the improved neutron lifetime data. Alternate scenarios of decaying matter or of quark–hadron induced inhomogeneities are discussed.

161.577 **Inflation and quantum cosmology.**
A. Linde.
Phys. Scr. T, Vol. T36, p. 30 – 54 (1991). Also published in "General relativity and gravitation, 1989. Proceedings" edited by N. Ashby, D. F. Bartlett and W. Wyss, p. 177 – 204, ISBN 0–521–38428–1, Cambridge (1990). – See Abstr. 012.080 for the main entry.
In this article a review of the present status of inflationary cosmology is given. The author starts with a discussion of the simplest version of the chaotic inflation scenario. Then he discusses some recent developments in inflationary cosmology, including the theory of a self–reproducing inflationary universe (eternal chaotic inflation). He does it with the help of a stochastic approach to inflation. The results obtained within this approach are compared with the results obtained in the context of Euclidean quantum cosmology.

161.578 **Initial conditions for chaotic inflation.**
R. Brandenberger, J. Kung, H. Feldman.
Phys. Scr. T, Vol. T36, p. 64 – 69 (1991). – See Abstr. 012.080 for the main entry.
In contrast to many other inflationary Universe models, chaotic inflation does not depend on fine tuning initial conditions. Within the context of linear perturbation theory, it is shown that chaotic inflation is stable towards both metric and matter perturbations. Neglecting gravitational perturbations, it is shown that chaotic inflation is an attractor in initial condition space.

161.579 **Models of large scale structure.**
C. S. Frenk.
Phys. Scr. T, Vol. T36, p. 70 – 87 (1991). – See Abstr. 012.080 for the main entry.
The ingredients required to construct models of the cosmic large scale structure are discussed. Empirical evidence provided by an analysis of a redshift survey of IRAS galaxies suggests that gravity is the main agent shaping the large–scale structure. In addition, this survey implies large values of the mean cosmic density, $\Omega \gtrsim 0.5$, and is consistent with a flat geometry if IRAS galaxies are somewhat more clustered than the underlying mass. Together with current limits on the density of baryons from Big Bang nucleosynthesis, this lends support to the idea of a universe dominated by non–baryonic dark matter. Results from cosmological N–body simulations evolved from a variety of initial conditions are reviewed. In particular, neutrino dominated and cold dark matter dominated universes are discussed in detail.

161.580 **Large–scale structure in the Universe.**
G. Efstathiou.
Phys. Scr. T, Vol. T36, p. 88 – 96 (1991). – See Abstr. 012.080 for the main entry.
The authors presents the latest results from three new surveys of the galaxy distribution which have been specially designed to detect large scale structure. The surveys provide strong evidence that our Universe contains more structure on scales $\gtrsim 10h^{-1}$Mpc than would have been expected in simple versions of the Cold Dark Matter model and pose a challenge for proponents of scale–invariant fluctuations and inflation. The author discusses whether these results require fundamentally new ideas for the origin of cosmic structure or whether the basic features of the CDM model can be retained.

161.581 **Some comments on the epoch of galaxy formation.**
M. J. Rees.
Phys. Scr. T, Vol. T36, p. 97 – 101 (1991). – See Abstr. 012.080 for the main entry.
Some aspects of galaxy formation are considered, with special reference to the epoch $z \gtrsim 4$ probed by the most distant quasars and by the state of the intergalactic medium.

161.582 **Galaxy formation and evolution.**
L. L. Cowie.
Phys. Scr. T, Vol. T36, p. 102 – 107 (1991). – See Abstr. 012.080 for the main entry.
The presence of high z quasars and radio galaxies tells us that galaxy formation began at $z > 5$, but leaves unanswered the question of when the bulk of galaxies formed. Recent near infrared number counts of galaxies strongly favor a cosmological geometry with $q_0 = 0.5$ and $\Lambda = 0$. Such a model grossly underpredicts blue galaxy counts. Spectroscopy shows that the excess blue galaxies at $B = 24$ are dwarfs at $z \sim 0.4$ which are no longer seen at the present time. These dwarfs must contain a large amount of baryonic matter which is not included in current estimates of baryonic Ω.

161.583 **Primordial perturbations from inflation.**
L. Kofman.
Phys. Scr. T, Vol. T36, p. 108 – 113 (1991). – See Abstr. 012.080 for the main entry.
The generation of initial perturbations from cosmic inflation is considered. Several different mechanisms of non–standard origin – i.e., nonflat or nongaussian inhomogeneities – are reviewed. Nonstandard isocurvature and adiabatic perturbations, as well as nonperturbative inhomogeneities can be produced along with the standard adiabatic perturbations during inflations. The departure of standard perturbations crucially depends on inflationary scenario. The nongaussianity of fluctuations in the context of cosmology is discussed.

161.584 **Cosmic strings and the large–scale structure of the Universe.**
R. H. Brandenberger.
Phys. Scr. T, Vol. T36, p. 114 – 126 (1991). – See Abstr. 012.080 for the main entry.
A review of the cosmic string scenario of structure formation is given, emphasizing the basic points and the current status and problems. The main results are that in the context of a cosmic string model, hot dark matter is an acceptable dark matter candidate. The cosmic string model predicts planar density enhancements, the most numerous and prominent of which have planar dimensions comparable to the comoving horizon at t_{eq}, the time of equal matter and radiation. Cosmic strings also give rise to a distinguished signal in the microwave background.

161.585 **Cosmic global strings.**
P. Sikivie.
Phys. Scr. T, Vol. T36, p. 127 – 134 (1991). – See Abstr. 012.080 for the main entry.
The topics discussed are: Global strings. The gravitational field of a straight global string. How do global strings behave? The axion cosmological energy density. Computer simulations of the motion and decay of global strings. Electromagnetic radiation from the conversion of Nambu–Goldstone bosons in astrophysical magnetic fields.

161.586 **Population III objects and lithium abundance.**
S. Ramadurai.
Proc. Astron. Soc. Aust., Vol. 9, No. 1, p. 138 (1991). – See Abstr. 012.078 for the main entry.
The earlier work on the production of deuterium in the accretion tori around Population III objects has been criticised on the basis of relative production ratios of deuterium, helium–3 and ^{7}Li. It is shown here that the criticism on the basis of helium–3 over–production is unjustified, as no measurement of helium–3 and lithium abundances for the same stars is available. It is predicted that indeed the helium–3 abundance determination will be the best means to verify the original hypothesis.

161.587 **Primordial fluctuations and non–linear structure.**
B. Little, D. H. Weinberg, C. Park.
Mon. Not. R. Astron. Soc., Vol. 253, No. 2, p. 295 – 306 (15 Nov 1991).
The authors explore several aspects of non–linear gravitational instability using two numerical experiments, each of which employs a series of 3D cosmological N–body simulations. In the first experiment they truncate all of the initial high–frequency power above some critical wavenumber k_c; in the second experiment, they replace initial Fourier components with wavenumbers $k > k_c$ by waves taken from an independent realization with the same power spectrum but with unrelated phases. The authors evolve these initial conditions for different values of k_c to see how the progressive elimination or substitution of initial high–frequency components affects the final, non–linear structure. They find that such structure is determined mainly by initial waves with $k \lesssim k_1 \sim R_1^{-1}$, where R_1 is the comoving scale on which rms fluctuations are just becoming non–linear. Changes in high–frequency components of the initial conditions do not disturb the evolution of longer waves, and even on small scales the influence of waves with $k > k_1$ is erased by gravitational collapse. Since fluctuations with $k \ll k_1$ are weak, the authors conclude that the appearance of non–linear structure in the universe is determined by fluctuations from a fairly small range in wavenumber. The authors' first experiment also confirms that a truncated initial power spectrum leads to coherent, large–scale filamentary structures when $k_c \sim k_1$.

161.588 **Global texture as the origin of cosmic structure.**
N. Turok.
Phys. Scr. T, Vol. T36, p. 135 – 141 (1991). – See Abstr. 012.080 for the main entry.
A pedagogical account of global texture is given. The generic occurrence of texture in unified theories is discussed. The dynamics of global texture are reviewed, and the "scaling solution" in an expanding universe described. The seeding of large scale structure through texture induced perturbations is described, as well as new numerical results on the predictions for the matter distribution today. Texture is shown to have significant advantages over the "standard" inflationary fluctuations in (1) having more power on large scales, and (2) producing nonlinear structures at high redshift. The nongaussian nature of the perturbations is emphasised. Finally predictions for the cosmic microwave anisotropy are described.

161.589 **Experimental constraints on dark matter.**
J. Ellis.
Phys. Scr. T, Vol. T36, p. 142 – 152 (1991). – See Abstr. 012.080 for the main entry.
The author discusses some of the particle candidates for dark matter. For each one, he reviews the constraints on it that come from both accelerator and non–accelerator experiments. Among the new topics reviewed are the constraints on dark matter candidates given by LEP, and a recent resurgence of interest in metastable particles from a hidden sector of string theory as possible candidates for dark matter.

161.590 **Evolution of small–scale structure on cosmic strings.**
T. W. B. Kibble, E. Copeland.
Phys. Scr. T, Vol. T36, p. 153 – 166 (1991). – See Abstr. 012.080 for the main entry.
An analytic approach to cosmic string evolution is developed to complement the computer simulations, with the aim of helping to resolve the apparent conflict between the simulations of various authors. Equations are derived governing the evolution of small–scale structure on cosmic strings. It is shown that the question of whether or not the loop distribution, as well as the long–string distribution, will approach scaling depends critically on the value of a parameter q which describes the relative kinkiness of a loop as compared to a section of long string of the

same length. Several predictions are made that can readily be checked against the simulations.

161.591 **Dark matter in the Universe.**
M. S. Turner.
Phys. Scr. T, Vol. T36, p. 167 – 182 (1991). – See Abstr. 012.080 for the main entry.
What is the quantity and composition of material in the Universe? Answering this question could lead to the discovery of new particles. At present, only a partial answer is at hand: Most of the material in the Universe does not give off detectable radiation, i.e., is "dark". There are three well motivated particle dark-matter candidates: an axion of mass 10^{-6}eV to 10^{-4}eV; a neutralino of mass 10 GeV to about 3 TeV; or a neutrino of mass 20 eV to 90 eV. All three possibilities can be tested by experiments that are either being planned or are underway.

161.592 **CP–violation and electroweak baryogenesis.**
M. E. Shaposhnikov.
Phys. Scr. T, Vol. T36, p. 183 – 193 (1991). – See Abstr. 012.008 for the main entry.
The author reviews the recent results on anomalous electroweak fermion number non–conservation at high temperatures and their relevance to the early universe. The scenario of low temperature electroweak baryogenesis and cosmological prediction of the Higgs boson mass are discussed.

161.593 **Electroweak baryon number violation at high energies.**
V. A. Rubakov.
Phys. Scr. T, Vol. T36, p. 194 – 198 (1991). – See Abstr. 012.080 for the main entry.
The electroweak baryon number violation is expected to be rapid at high temperatures. This implies that at least some multiparticle processes with B–violation are rapid at high energies. The author discusses recent attempts to study the two–particle cross sections of instanton–like processes, which show that at energies below 10 TeV, the cross sections increase exponentially, and might even become large at about 10 TeV.

161.594 **First–order inflation.**
E. W. Kolb.
Phys. Scr. T, Vol. T36, p. 199 – 217 (1991). – See Abstr. 012.080 for the main entry.
In the original proposal, inflation occurred in the process of a strongly first–order phase transition. This model was soon demonstrated to be fatally flawed. Subsequent models for inflation involved phase transitions that were second–order, or perhaps weakly first–order; some even involved no phase transition at all. Recently the possibility of inflation during a strongly first–order phase transition has been reviewed. The author discusses some models for first–order inflation, and emphasizes unique signatures that result if inflation is realized in a first–order transition. Before discussing first–order inflation, he briefly reviews some of the history of inflation to demonstrate how first–order inflation differs from other models.

161.595 **The alpha parameters of wormholes.**
S. W. Hawking.
Phys. Scr. T, Vol. T36, p. 222 – 227 (1991). – See Abstr. 012.080 for the main entry.
Quantum gravity allows the possibility of wormholes or throats that connect one region of four dimensional Euclidian spacetime with another. The author discusses some features of the effective coupling constants α which are involved.

161.596 **Classical physics and Hamiltonian quantum mechanics as relics of the big bang.**
J. B. Hartle.
Phys. Scr. T, Vol. T36, p. 228 – 236 (1991). – See Abstr. 012.080 for the main entry.
In a fundamental formulation of the quantum mechanics of a closed system, such as the universe as a whole, three forms of information are needed to make predictions of the probabilities of alternative histories. These are the action functional of the elementary particles, the quantum initial condition of the universe, and the information about our specific history. The author discusses the origin of the "quasiclassical domain" of familiar experience and Hamiltonian quantum mechanics with its preferred time in such a formulation of quantum cosmology. It is argued that these features of the universe are not general properties of quantum theory, but rather approximate features emergent after the Planck time because of the character of the specific initial condition and dynamics of our universe.

161.597 **Can a man–made universe be created by quantum tunneling without an initial singularity?**
A. H. Guth.
Phys. Scr. T, Vol. T36, p. 237 – 246 (1991). – See Abstr. 012.080 for the main entry.
Essentially all modern particle theories suggest the possible existence of a false vacuum state – a metastable state with an energy density that cannot be lowered except by means of a very slow phase transition. Inflationary cosmology makes use of such a state to drive the expansion of the big bang, allowing the entire observed universe to evolve from a very small initial mass. A sphere of false vacuum in our present universe, if larger than a certain critical mass, could inflate to form a new universe which would rapidly detach from its parent. A false vacuum bubble of this size, however, cannot be produced classically unless an initial singularity is present from the outset. The author therefore explores the possibility that a bubble of subcritical size, which classically would evolve to a maximum size and collapse, might instead tunnel through a barrier to produce a new universe. He estimates the tunneling rate using semiclassical quantum gravity, and discovers some interesting ambiguities in the formalism.

161.598 **The black hole horizon as a quantum surface.**
G. t' Hooft.
Phys. Scr. T, Vol. T36, p. 247 – 252 (1991). – See Abstr. 012.080 for the main entry.
In previous work it was argued why it is practically inevitable to assume the existence of an S–matrix that describes particle absorption and production by a black hole, and the relationship between this S–matrix and string theory was derived. The physical interpretation of the corresponding mathematical expressions however is quite different from string theory. The author has an algebra of operators now defined on a two–dimensional Euclidean "world sheet". The two dimensional functional integrals correspond to a rather unusual field theory. The long distance structure of this theory follows directly from the long distance structure of the standard model at the GeV scale. The author emphasizes the rather delicate physical interpretation of this approach.

161.599 **Classical perturbations from decoherence of quantum fluctuations in the inflationary Universe.**
R. Brandenberger, R. Laflamme, M. Mijić.
Phys. Scr. T, Vol. T36, p. 265 – 268 (1991). Also published in Mod. Phys. Lett. A, Vol. 5, No. 28, p. 2311 – 2317 (10 Nov 1990). – See Abstr. 012.080 for the main entry.
The evolution of a scalar field interacting with an environment in the de Sitter phase of an inflationary Universe is studied. The environment is taken to be a second scalar field. It is shown that the coherence length of the quantum fluctuations rapidly decreases after the wavelength of the perturbation crosses the Hubble radius. Hence, the fluctuations can be interpreted as classical. This lends support to the usual derivation of the spectrum of density perturbations in inflationary Universe models.

161.600 **Unstable spacetime in the Jordan–Brans–Dicke gravity.**
M. Yoshimura.
Phys. Scr. T, Vol. T36, p. 276 – 280 (1991). – See Abstr. 012.080 for the main entry.
A Jordan–Brans–Dicke (JBD) type theory is an essential ingredient in a successful scenario of new inflation, extended inflation, and naturally appears in many unified theories including gravity. Strong field aspect of this theory is analyzed by

using a complete set of exact solutions. It is pointed out that the problem of naked singularity and instability inherent in these theories may be avoided due to a dilaton coupling to gauge fields.

161.601 **Perspectives on particle physics and cosmology.**
F. Wilczek.
Phys. Scr. T, Vol. T36 , p. 281 – 296 (1991). – See Abstr. 012.080 for the main entry.

161.602 **Fractals and the large-scale galaxy distribution.**
A. Provenzale.
Applying fractals in astronomy, p. 97 – 118 (1991). – See Abstr. 003.023 for the main entry.

The author considers the fractal behavior of the large scale galaxy distribution and introduces a family of fractal cascading models based on appropriate modifications of the β model of turbulence. These models generate distributions with scale-dependent fractal properties, as observed in the analysis of real galaxy catalogs. In particular, a smooth transition from small-scale self-similarity to large-scale homogeneity is considered. The scale dependence of the fractal properties generates a "dressing" of the asymptotic value of the fractal dimension to a large effective value D_{ef}. The physical motivation for the use of a scale-dependent β model in cosmology are discussed. The multifractal properties of the distributions are considered and an extended version of the β model for the field of density perturbations $\delta\varrho/\varrho$ is introduced.

161.603 **Is the spatial distribution of galaxies a bounded self-similar structure? Observational evidences.**
D. Calzetti, M. Giavalisco.
Applying fractals in astronomy, p. 119 – 134 (1991). – See Abstr. 003.023 for the main entry.

The authors want to show here that the distribution of galaxies in space may be described in terms of a self-similar arrangement of the clustering on scales of distances ranging from some Mpc to (30 – 40) h^{-1}Mpc. To this end, the classical two-point correlation function has been evaluated both in the angular and the spatial case on the Zwicky and CfA 1 catalogues, respectively. In contrast to previous works, the quantity $1+\omega(\theta)$ and its spatial equivalent $1+\xi(r)$ have been analysed with respect to their scaling properties with the sample depth. The authors have not studied the local scaling properties of the distribution (multifractality) and have approximated it to a simple fractal set. The resulting estimate for the fractal dimension (or better, the correlation dimension) is $D_2 \approx 1.7$ from angular data, while $D_2 \approx 2.3$ from the spatial ones. If this value is confirmed, the clustering in the galaxy distribution results to be less strong, but more extended in space, than previously found.

161.604 **Fractal aspects of galaxy clustering.**
V. J. Martínez.
Applying fractals in astronomy, p. 135 – 159 (1991). – See Abstr. 003.023 for the main entry.

In the past decade, the mathematical concept of fractal has exerted a great influence in a large variety of scientific disciplines. Cosmology has been one of the first fields where fractals have been applied. Fractal models for the cosmic structure are discussed.

161.605 **The temperature of the cosmic microwave background radiation at 3.8 GHz: results of a measurement from the South Pole site.**
G. De Amici, M. Bersanelli, A. Kogut, S. Levin, M. Limon, G. F. Smoot.
Astrophys. J., Vol. 381, No. 1, p. 341 – 347 (1 Nov 1991).

As part of an international collaboration to measure the low-frequency spectrum of the cosmic microwave background (CMB) radiation, the authors have measured its temperature at a frequency of 3.8 GHz (7.9 cm wavelength), during the austral spring of 1989, obtaining a brightness temperature, T_{CMB}, of 2.64 ± 0.07K (68% confidence level). The new result is in agreement with previous meaasurements at the same frequency obtained in 1986 – 88 from a very different site and has comparable error bars. Combining measurements from all years, the authors obtain $T_{CMB} = 2.64 \pm 0.06$K.

161.606 **Quadratic Lagrangians and horizons.**
T. Rothman, P. Anninos.
Phys. Rev. D, Vol. 44, No. 10, p. 3087 – 3095 (15 Nov 1991).
Current Physics Microform No.: 9201B0815.

It has been known for some time that the standard isotropic cosmological model with an R^2 Lagrangian has horizon-breaking solutions in the $t \to 0$ singularity limit. The authors search for horizonless power-law solutions in the vacuum Bianchi type I and orthogonal type V cosmological models. They find that type V with an R^2 Lagrangian does contain such solutions. At first glance this would suggest that quadratic theories might provide an alternative resolution to the horizon problem. However, for both the horizonless isotropic and type V solutions the authors show that the eigenvalues of the linear vector field near the stationary points in phase space have opposite signs, demonstrating that such solutions are not Lyapunov stable. The addition of free quantum fields may stabilize the solutions.

161.607 **Fractal properties in the simulations of a one-dimensional spherically expanding Universe.**
J. L. Rouet, E. Jamin, M. R. Feix.
Applying fractals in astronomy, p. 161 – 179 (1991). – See Abstr. 003.023 for the main entry.

A model Universe, expanding according to the Hubble law, is numerically simulated. The model consists of a thin central part of a spherically symmetric and homogeneous Universe. By appropriate rescaling, the dynamics becomes, in the case of critical expansion, analogous to the dynamics of a one-dimensional one-component plasma with the addition of a friction term. The motion is most appropriately solved in phase space. Physical properties of the model are examined. The results of two simulations corresponding to different initial conditions are presented; they show a hierarchical clustering process of the density with clumping of the velocities in each cluster. The configuration space density and the phase space distribution function are found to have prominent fractal structures, with for the former a fractal dimension independent of initial conditions of $0.575 \pm .035$. The importance of phase space in the study of the fractal structure of large dynamic systems is stressed.

161.608 **Statistics of primordial density perturbations from discrete seed masses.**
R. J. Scherrer, E. Bertschinger.
Astrophys. J., Vol. 381, No. 2, p. 349 – 360 (10 Nov 1991).

The authors examine the statistics of density perturbations for general distributions of seed masses with arbitrary matter accretion. Formal expressions for the power spectrum, the N-point correlation functions, and the density distribution function are derived. These results are applied to the case of uncorrelated seed masses, and power spectra are derived for accretion of both hot and cold dark matter plus baryons. The authors compute the reduced moments (cumulants) of the density distribution and use them to obtain a series expansion for the density distribution function. Analytic results are obtained for the density distribution function in the case of a distribution of seed masses with a spherical tophat accretion pattern. More generally, the formalism is able to give a complete characterization of the statistical properties of any random field generated from a discrete linear superposition of kernels. In particular, it can be applied to density fields derived by smoothing a discrete set of points with a window function.

161.609 **Finite-temperature gauge field propagator in the early universe.**
N. Banerjee, S. Mallik.
Phys. Rev. D, Vol. 44, No. 12, p. 3730 – 3736 (15 Dec 1991).
Current Physics Microform No.: 9101C0012.

The authors consider a model of the early universe where a Higgs-scalar electrodynamics describes its contents. The finite-temperature gauge field propagator is obtained in the real-time formalism during an interval of time when thermal equilibrium is maintained making effective masses large compared to curvature terms.

161.610 **Tilted universe and other remnants of the preinflationary universe.**
M. S. Turner.
Phys. Rev. D, Vol. 44, No. 12, p. 3737 – 3748 (15 Dec 1991). Current Physics Microform No.: 9101C0019.
If the inflationary epoch lasted only ~ 10 e–foldings longer than required to solve the horizon problem, observable "remnants" of the preinflationary universe may exist. They include dipole and quadrupole anisotropies of the cosmic microwave background radiation (CMBR). These "remnants" arise due to preinflationary fluctuations in scalar fields such as the inflaton, the axion, and the ilion, or due to preinflationary density perturbations. The dipole anisotropy can lead to the illusion of a "tilted universe": Viewed from the rest frame of the CMBR, galaxies throughout the entire observable universe would have a uniform streaming velocity. A dipole CMBR anisotropy could provide a very unconventional explanation for the large peculiar velocities measured for galaxies in our $\sim 50h^{-1}$ Mpc neighborhood. Among things very unlikely to be a "remnant" of inflation is a value of Ω today that is significantly different from unity.

161.611 **Erratum: "Seeded hot dark matter models for large–scale structure" [Astrophys. J., Vol. 367, No. 1, p. 37 – 44 (20 Jan 1991)].**
J. V. Villumsen, R. J. Scherrer, E. Bertschinger.
Astrophys. J., Vol. 381, No. 2, p. 601 (10 Nov 1991). See Abstr. 53.161.096.

161.612 **Monte Carlo simulation of texture formation.**
R. A. Leese, T. Prokopec.
Phys. Rev. D, Vol. 44, No. 12, p. 3749 – 3759 (15 Dec 1991). Current Physics Microform No.: 9101C0031.
The formation of global texture in both two and three spatial dimensions is studied using Monte Carlo techniques. The authors calculate the probability of texture formation for critical windings in the range [1/2, 1], finding for three dimensions average values per correlation volume in the range [0.017, 0.045], which is in good agreement with previous dynamical simulations. The authors discuss cosmological structures seeded by texture collapse.

161.613 **Gravitino–induced baryogenesis, primordial nucleosynthesis, and the Tremaine–Gunn limit.**
R. J. Scherrer, J. Cline, S. Raby, D. Seckel.
Phys. Rev. D, Vol. 44, No. 12, p. 3760 – 3766 (15 Dec 1991). Current Physics Microform No.: 9101C0042.
The authors examine constraints on the gravitino–induced baryogenesis model which arise from primordial nucleosynthesis.

161.614 **Is the Universe a fractal? Tests on quasar samples.**
P. Andreani, S. Cristiani, F. La Franca.
Mon. Not. R. Astron. Soc., Vol. 253, No. 3, p. 527 – 532 (1 Dec 1991).
The authors have used several complete samples of quasars (450 objects in total) to gain information on the large–scale structure of the Universe at redshifts $0.4 < z < 2.2$. They have evaluated the conditional density and the average density functions to establish that on scales larger than 12.5 h^{-1}Mpc the quasar distribution becomes homogeneous. The use of a more conventional and time–honoured technique such as the correlation function is therefore appropriate in the case of the present samples and equivalent to the conditional and average density functions. These latter functions allow the authors, however, to put slightly stronger constraints on the amplitude of clustering at large scales. Possible observational biases which could lead to incorrect conclusions on quasar clustering are discussed.

161.615 **Quasar–modulated galaxy clustering in a cold dark matter universe.**
A. Babul, S. D. M. White.
Mon. Not. R. Astron. Soc., Vol. 253, No. 3, p. 31P – 34P (1 Dec 1991).
The authors explore the possibility that large–scale structure in the galaxy distribution is the result of quasars modulating galaxy formation in the standard cold dark matter (CDM) model. Specifically, the authors consider the case where radiation from high–redshift quasars is assumed to suppress galaxy formation in their neighbourhood. The observed quasar population can, in principle, influence large enough regions of space to reconcile the CDM model with recent measurements of large–scale structure. In such a model the observed voids should contain failed galaxies and galaxy clusters.

161.616 **Gauge invariant perturbations in multi–component fluid cosmologies.**
P. K. S. Dunsby.
Classical Quantum Gravity, Vol. 8, No. 10, p. 1785 – 1806 (Oct 1991).
In a series of recent papers, an alternative approach to the study of cosmological density perturbations in cosmology has been presented. Covariant and gauge invariant quantities were defined that characterize density inhomogeneities in an almost–uniform model universe and propagation equations were derived for these quantities in the case of a general perfect fluid. In this paper, the author extends this work to the case of an imperfect fluid and derives generalizations of these equations. The theory is then applied to a multi–component fluid, in particular to the case where the background is described by a spatially flat Friedmann – Lemaitre – Robertson – Walker universe model containing a mixture of non–interacting dust and radiation. Solutions are obtained for the comoving fractional density gradient, the energy flux and the relative velocity in the centre of mass frame which include new modes linking the density gradient to the vorticity.

161.617 **Are quasar redshifts cosmological?**
A. Dar.
Astrophys. J., Lett., Vol. 382, No. 1, p. L1 – L4 (20 Nov 1991).
Recent observations on gravitational lensing of quasars by intervening galaxies strongly support the cosmological interpretation of quasar redshifts against the far more exotic proposals. The lensing model predicts correctly the angular separation and time delay between different images and the frequency of quasars that are lensed by galaxies only if the quasars lie at the large cosmological distances that are inferred from their redshifts.

161.618 **Structure formation in a low bias CDM Universe.**
H. M. P. Couchman, R. G. Carlberg.
Workshop on Space Distribution of Quasars as Part of the 75. Anniversary Celebrations of the National Research Council of Canada (NRCC), p. 369 – 378 (1991). – See Abstr. 012.084 for the main entry.
The authors show that a highly evolved – or low bias – Cold Dark Matter model is consistent with the current clustering and dynamical data for galaxies. Using a numerical simulation the CDM model has been evolved to the limit allowed by recent measurements of Microwave Background isotropy. This corresponds to a factor of two increase in linear amplitude over typical high bias models. Implications for the epoch of structure formation are discussed.

161.619 **Constraints on cosmological models from the IRAS QDOT survey results.**
C. Park.
Astrophys. J., Lett., Vol. 382, No. 2, p. L59 – L62 (1 Dec 1991).
Constraints on the nature of the galaxy density field are explored using the results of the recent IRAS QDOT survey. Analytic modeling of the power spectrum gives a best fit to observed density variances when $P(k) \sim k^n$ with $n = -1.26(+0.39,-0.35)$ [68% confidence limits] or $P(k) \sim P_{CDM}(\Omega h)$ with $(\Omega h) = 0.17(+0.17,-0.09)$ over the scales covered by the

survey ($\sim 100h^{-1}$ Mpc). Realistic simulations of the survey made from N–body simulations of the n = –1 power–law power spectrum and the $\Omega = 1$ CDM models show that the power–law model is basically consistent with all QDOT variance results, while the CDM model has a smaller variance than the observation at the $20h^{-1}$ Mpc Gaussian smoothing scale. It is concluded that the QDOT results are best explained if the power spectrum of the galaxy density field has a slope of ~ -1 near $\lambda \sim 100h^{-1}$ Mpc, while the CDM model has a slope of ~ 0 at that scale.

161.620 **The probability density function of the temperature autocorrelation of the cosmic microwave background for a Gaussian cosmological model.**
L. Cayón, E. Martínez–González, J. L. Sanz.
Mon. Not. R. Astron. Soc., Vol. 253, No. 4, p. 599 – 604 (15 Dec 1991).
The probability density function of the temperature autocorrelation is given in a semi–analytical form for the case of Gaussian temperature fluctuations. Based on this result the authors obtain confidence intervals for the anisotropies expected in several cosmological scenarios dominated by baryonic (B), hot dark matter (HDM) and cold dark matter (CDM) with adiabatic fluctuations. The authors also confront these expectations with the upper limits given by the present experiments at large angular scales.

161.621 **On shell crossing in the Tolman metric.**
A. Mészáros.
Mon. Not. R. Astron. Soc., Vol. 253, No. 4, p. 619 – 624 (15 Dec 1991).
The initial conditions of the Tolman metric are studied. In the hyperbolic case an exact condition leading to shell crossing is found. It is conjectured that shell crossing is a good way to form galaxies and voids in a hyperbolic universe dominated by baryons. A comparison with the standard gauge–invariant theory of small perturbations is also presented.

161.622 **Microwave background constraints on extended inflation voids.**
A. R. Liddle, D. Wands.
Mon. Not. R. Astron. Soc., Vol. 253, No. 4, p. 637 – 648 (15 Dec 1991).
Extended inflation ends via the nucleation of bubbles of true vacuum, and in principle the earliest bubbles to nucleate can be expanded by the last stages of inflation to astrophysically interesting sizes, giving large voids empty of any matter. Several authors have suggested a link with the observed bubbly large–scale structure in the Universe. The authors examine in detail the constraints on the bubble distribution from the microwave background observations, particularly COBE, for different types of dark matter. In a baryon–dominated universe, they obtain upper limits on the extended inflation parameters similar to those previously given, but these become somewhat stronger when one includes dark matter. In any scenario, extended inflation cannot by itself be responsible for the bubbly large–scale structure.

161.623 **The evolution of black holes in an expanding universe.**
J. D. Barrow, E. J. Copeland, A. R. Liddle.
Mon. Not. R. Astron. Soc., Vol. 253, No. 4, p. 675 – 682 (15 Dec 1991).
The authors examine the cosmological evolution of black holes in an expanding universe, taking into account the process of Hawking evaporation and including the influence of the black holes on the expansion dynamics. The black holes are assumed to have a mass spectrum of power–law form. The authors find analytic solutions in which the energy density in black holes is in equilibrium with radiation (though no individual black hole is), and demonstrate that they are attractors for the system. The authors study the approach to this attractor. For a matter–dominated era, they find the general solution for black hole evolution, and again demonstrate an attractor property. Some consequences of this analysis for the study of the early universe are discussed.

161.624 **Skewness and large–scale structure.**
P. Coles, C. S. Frenk.
Mon. Not. R. Astron. Soc., Vol. 253, No. 4, p. 727 – 737 (15 Dec 1991).
The evolution of the skewness, γ, of the cosmological mass distribution into the nonlinear regime is discussed. Various analytical estimates of the skewness of initially Gaussian (i.e. unskewed) density perturbations are compared with each other, with N–body results and with real clustering data. The authors find that, in general, the skewness is simply related to the rms mass fluctuation on the scale in question: $\gamma = A \times \sigma^4$ where $A \cong 3$ is a constant. This relationship can be used to construct a simple but potentially powerful test of the hypothesis that large–scale structure in the Universe evolved from Gaussian initial conditions. In particular, the authors find that the measured skewness of the QDOT redshift survey density distribution is compatible with the gravitational evolution of sufficiently high–amplitude Gaussian fluctuations. There is therefore no need to invoke non–Gaussian initial fluctuations to explain these data. The authors also discuss in detail the role of discreteness (i.e. shot noise) effects upon the interpretation of cell–count statistics.

161.625 **Anisotropies of the cosmic background radiation in a "hot" dark matter universe.**
P. Anninos, R. A. Matzner, R. Tuluie, J. Centrella.
Astrophys. J., Vol. 382, No. 1, p. 71 – 78 (20 Nov 1991). With plates 4 – 7.
The authors investigate microwave photons propagating through the evolving density inhomogeneities in a model universe filled with "hot" dark matter. The evolution takes place in a comoving cube now 165 Mpc on a side, beginning at z = 1900 before the epoch of decoupling and continuing to the present time z = 0. The authors take the density parameter $\Omega_0 = 1$ and the Hubble constant $H_0 = 100$ km $s^{-1}Mpc^{-1}$. The calculation is fully three–dimensional and the density inhomogeneities are followed into the nonlinear regime. The effect of Thomson scattering is included throughout the model history by using a Monte Carlo scattering algorithm. The optical depth for this scattering is a function of time determined by the free electron density, which is obtained by the numerical integration. The results are displayed in the form of sky maps; the maps are of $4° \times 4°$ and $2° \times 2°$ regions. The authors find structure on these scales with amplitudes $7 \times 10^{-6} \leqslant (\Delta T/T)_{rms} \leqslant 1.4 \times 10^{-5}$ with peak excursions $\Delta T/T \sim \pm 3 \times 10^{-5}$, which is consistent with the best current observational limit of 4×10^{-5} set by the COBE data at 7° and 3.5×10^{-5} set by the South Pole experiment of Meinhold and Lubin at $\sim 1°$.

161.626 **Erratum: "Signatures of inhomogeneity in the early universe" [Astrophys. J., Vol. 372, No. 1, p. 1 – 7 (1 May 1991)].**
L. H. Kawano, W. A. Fowler, R. W. Kavanagh, R. A. Malaney.
Astrophys. J., Vol. 382, No. 1, p. 358 (20 Nov 1991). See Abstr. 53.161.328.

161.627 **Gauge–invariant joining conditions for cosmological perturbations.**
J. Hwang, E. T. Vishniac.
Astrophys. J., Vol. 382, No. 2, p. 363 – 368 (1 Dec 1991).
The authors derive two independent gauge–invariant variables which are continuous at any transition where there is a discontinuous change in pressure. This is valid on any scale and for a general background, subject only to the assumption that the contents of the universe are a perfect fluid. Separately, the authors show the continuous variables for a transition from one scalar field to another. They discuss the physical significance of these variables and compare them with other, previously suggested, joining conditions. The authors then examine the evolution of large–scale cosmological perturbations, assuming that the pressure is a function only of energy density, together with these joining conditions. It is noted that in inflationary models it is not necessary to consider the evolution of the universe between the time a perturbation expands beyond the Hubble horizon, H^{-1}, and the time it comes within the horizon

again, to predict the final amplitude of the fluctuations. The results of this work are applied to power–law–type inflationary models.

161.628 **Physically permitted cosmological models with nonzero cosmological constant.**
M. Moles.
Astrophys. J., Vol. 382, No. 2, p. 369 – 376 (1 Dec 1991).
Cosmological models with nonzero cosmological constant are reconsidered. It is found, in agreement with earlier results by Glanfield, that there are two particular values of the cosmological constant, Λ_M (being $\leqslant 0$ for $\Omega_0 \leqslant 1$ and positive otherwise) and Λ_C (with $\Lambda_M < \Lambda_C$). Exploding models with $\Lambda \leqslant \Lambda_M$ or $\Lambda \geqslant \Lambda_C$ are oscillating. But the latter are not physically permitted because they never reach the present situation of the universe. Nonexploding models with $\Lambda > \Lambda_C$ are also theoretically possible, but, in general, they are physically improbable because, other than losing the explanatory capabilities of the models with a big bang, they started to expand from a nonsingular situation too short a time ago. Finally, models with $\Lambda_M < \Lambda < \Lambda_C$ are of the inflexional type, with an endless expansion. Thus, the value Λ_C plays in fact the role of an upper limit to the value of the cosmological constant. Analytical expressions for the critical values of Λ are given. A lower limit to Λ could result from time scale considerations. Given the current lowest estimates of the age of the oldest globular clusters, the condition $H_0 t_0 \geqslant 1$ has to hold for any $H_0 > 65\ \mathrm{km\ s^{-1} Mpc^{-1}}$. On that assumption it is shown that the range of permitted values of Λ is very small.

161.629 **Precollapse scale invariance in gravitational instability.**
F. Moutarde, J.–M. Alimi, F. R. Bouchet, R. Pellat, A. Ramani.
Astrophys. J., Vol. 382, No. 2, p. 377 – 381 (1 Dec 1991).
The authors numerically and analytically investigate the transition to the very nonlinear regime during the gravitational collapse of collisionless matter in an $\Omega = 1$ expanding universe. It is found numerically that the formation of halos from isolated and smooth initial overdensities leads to the progressive establishment, complete at the collapse time, of a power–law density profile. The evolutions from various types of initial perturbations are shown to produce such power laws. The slopes are different, but the scale invariance itself appears quite generic. The authors then derive an analytical approximation based on a perturbative expansion whose first–order term is the well–known "Zeldovich approximation". This new approximation is found to be in perfect agreement with the numerical simulations up to density contrasts $\delta \lesssim 50$. The establishment of a power–law density profile at collapse is thus confirmed analytically. It is crucial to note that the scale invariance pointed out is obtained *before* shell–crossing; it prepares the system for a self–similar evolution after the collapse.

161.630 **Cosmic microwave background and Galactic quadrupoles in the millimetric region.**
P. de Bernardis, S. Masi, N. Vittorio.
Astrophys. J., Vol. 382, No. 2, p. 515 – 522 (1 Dec 1991).
The authors discuss the constraints that a COBE–DMR–like experiment can set on the CMB (cosmic microwave background) quadrupole anisotropy, once that Galactic emission and receiver noise are taken into account. The authors use a simple model for describing the diffuse emission of our Galaxy at millimeter wavelengths, and they estimate that the minimum Galactic quadrupole occurs at $\lambda \sim 3 - 4$ mm, with an amplitude $\approx 50\ \mu$K. Multifrequency experiments, such as COBE–DMR, can easily disentangle the Galactic and CMB quadrupoles. The authors discuss several methods for isolating the CMB quadrupole: its measurement, according to the simulations, turns out to be limited in sensitivity mostly by detector noise and much less by Galactic contamination. At the expected level of sensitivity, COBE should be able to set a very conservative upper limit to the rms CMB quadrupole of $\cong 80\ \mu$K. This upper limit would be fatal for open ($\Omega = 0.1 - 0.4$) pure baryonic, fully ionized universes, with a primordial spectral index $n = -1$. In flat, biased, cold dark matter models, a fraction $\lesssim 50\%$ of the cosmic observers should detect a CMB quadrupole anisotropy.

161.631 **Recent developments in non–standard cosmological models.**
Bian Yulin.
Prog. Astron., Vol. 9, No. 4, p. 253 – 262 (Dec 1991). In Chinese.
Recently, some cosmologists who disagree with the standard model gave the Big Bang model a severe criticism and suggested their own model as an "alternative view". A brief review on it is presented in this article, in which among the issues discussed are those concerning the cosmic microwave background radiation, noncosmological redshift, evolution of discrete celestial objects, and some points about the alternative hypothesis itself.

161.632 **Quantum cosmology with conformal coupling scalar field in the Vilenkin theory.**
Huang Weihong, Shen Yougen.
Acta Astron. Sin., Vol. 32, No. 4, p. 380 – 388 (Dec 1991). In Chinese.
Using the Vilenkin method, the authors discuss the quantum cosmology with conformal coupling scalar field. The corresponding Wheeler–DeWitt equation has been derived, and the wave function of the universe calculated.

161.633 **Microwave background anisotropies implied by large–scale galaxy correlations: the minimum of C(0) and cosmological parameters.**
A. Kashlinsky.
Astrophys. J., Lett., Vol. 383, No. 1, p. L1 – L5 (10 Dec 1991).
The author uses data from the recently completed APM survey to show that large–scale galaxy correlations found there imply the existence of measurable microwave background radiation (MBR) anisotropies on scales greater than several degrees. It is shown that $[C(0)]^{1/2} \geqslant 3.5 \times 10^{-5}$ if the APM data are used at $\Theta_0 = 20°$, or $[C(0)]^{1/2} \geqslant 2.8 \times 10^{-5}$ if one uses $\Theta_0 = 10°$; these numbers are almost independent of the cosmological parameters Ω, Λ, and z_{LS}, the redshift of the last scattering surface. However, for finite–beamwidth experiments the minimal fluctuations depend on the cosmological parameters and are within the reach of the currently planned or conducted experiments. The minimal anisotropies are smaller in a low–Ω universe. For the experiment with FWHM = 1°.5 now being conducted by P. Lubin and colleagues, the minimal anisotropies must be $\delta T/T \geqslant 3 \times 10^{-5}$ on scales of 2°– 3° or greater if $\Omega = 1$. For other models the minimal anisotropies for such a beam would also be measurable with techniques developed for this experiment, and, hence, the upper limits on, or detection of, MBR anisotropies can constrain Ω, z_{LS}, and Λ. For COBE parameters (FWHM = 7°) the anisotropies would be $\delta T/T \geqslant 7 \times 10^{-6}$ if $\Omega = 1$ and would exceed COBE's ultimate sensitivity limit for other cosmological parameters.

161.634 **A hydrodynamic approach to cosmology: texture–seeded cold dark matter and hot dark matter cosmogonies.**
R. Y. Cen, J. P. Ostriker, D. N. Spergel, N. Turok.
Astrophys. J., Vol. 383, No. 1, p. 1 – 18 (10 Dec 1991).
The authors present hydrodynamical simulations of galaxy formation in a texture–seeded cosmology, investigating both hot dark matter (HDM) and cold dark matter (CDM) dominated $\Omega = 1$ universes. Texture induces a scale–invariant spectrum of non–Gaussian density fluctuations. The simulations include both gravitational and hydrodynamical physics with a detailed treatment of collisional and radiative thermal processes, and use a cooling criterion to estimate galaxy formation. Background radiation fields and Zel'dovich–Sunyaev fluctuations are explicitly computed. In both HDM and CDM texture scenarios, the "galaxies" and "clusters" are significantly more strongly correlated than the dark matter, due, the authors believe, to physical bias processes. The slope of the correlation function in both cases is consistent with observation. In contrast to Gaussian models, peaks in the dark matter density distribution are less correlated than average.

161.635 **Nonlinear matter clustering properties of a cold dark matter universe.**
F. R. Bouchet, R. Schaeffer, M. Davis.
Astrophys. J., Vol. 383, No. 1, p. 19 – 40 (10 Dec 1991).
The authors study the matter distribution of strongly nonlinear scales in a CDM model. To that effect, they determine the count probabilities in an unbiased catalog generated with a P^3M simulation code and compare the results with the predictions of the scale–invariant model introduced by Balian and Schaeffer and those of the thermodynamical model introduced by Saslaw. They find that the probability P_N of finding a given amount of matter, i.e., N particles in a specified volume, is a power law with a lower and an upper cutoff. These cuts may be described for all volume sizes by two universal functions that they determine, in the proper limit of a continuous distribution. These results show that such a CDM simulation indeed presents all the count probability properties expected from a scale–invariant system. On the other hand, the authors find that all the specific scale–invariant models proposed by various authors can be confidently rejected. They also find that the thermodynamical model of Saslaw reproduces the qualitative features of the results of the simulations, provided they adjust to each case the free parameter b of this model, which ought to be unique. No overall fit could be obtained for small and large N simultaneously, the discrepancy being worse at the small, more nonlinear scales which should be the most virialized. The main reason for this failure is that the thermodynamic approach does not provide for the scale–invariant properties that are seen to hold in the numerical simulations. This precludes the use of this approach to describe the matter distribution in a CDM universe.

161.636 **Deep galactic surveys and the structure of the universe.**
O. E. Buryak, M. Demiański, A. G. Doroshkevich.
Astrophys. J., Vol. 383, No. 1, p. 41 – 50 (10 Dec 1991).
The authors present a new method of analysis of deep galactic surveys and they apply it to the available data. They consider a few model distributions what allows to show that the one–dimensional distribution of elements of the large–scale structure (LSS) in the regions between voids is almost Poissonian and to estimate its parameters. The authors also estimate the typical size of voids. They confirm reality of the super LSS (of an approximate size of $80\,h^{-1}$Mpc) which probably is connected with the existence of population of large voids. This result is also supported by analysis of distribution of the C IV doublets in absorption spectra of quasars.

161.637 **Models of voids in elliptic universes.**
A. Chamorro.
Astrophys. J., Vol. 383, No. 1, p. 51 – 55 (10 Dec 1991).
Models for nonempty spherical voids in elliptic universes are constructed. The models consist of an underdense Friedmann region – the void – an intervening Tolman region, and an exterior Friedmann universe. In order to achieve a significant underdensity, in fact as small as wanted but nonzero, it is necessary to allow for a prematurely born void. It is then possible to choose the arbitrary functions in the Tolman metric of the intervening zone in such a way that the models are exempt from shell–crossings during their whole history. In that case there is a first region in the intervening Tolman zone containing the oldest matter in the universe, and a second region with the youngest.

161.638 **The confinement and cosmological evolution of extragalactic radio source components.**
A. A. Ubachukwu, S. E. Okoye, L. I. Onuora.
Astrophys. J., Vol. 383, No. 1, p. 56 – 59 (10 Dec 1991).
Simple analytical models for the cosmological evolution of the linear sizes of extragalactic radio sources have been developed based on ram and thermal pressures of the intergalactic medium. It was found that for suitable assumptions about the source magnetic field and energies, the evolution was of the form $D \sim p^q(1+z)^{-n}$ with $n \approx 2.0$ for ram pressure confinement and $n \approx 3.0$ for thermal pressure confinement. In all cases, the value of q was 0.33. It was suggested that quasars may be confined by ram pressure, while galaxies are subject to thermal confinement.

161.639 **Hubble spheres and particle horizons.**
E. Harrison.
Astrophys. J., Vol. 383, No. 1, p. 60 – 65 (10 Dec 1991).
Isotropy of the cosmic microwave background radiation highlights the horizon problem. Analysis of the horizon problem and its inflationary solution requires that one studies the properties of the Hubble sphere and distinguishes between the "Hubble surface" and the particle horizon. All regions in the visible universe become causally connected when inflation increased the distance of the particle horizon by a factor of 3. If exponential inflation occurs because of a phase transition at temperature $\sim 10^{15}$GeV, the number of e–foldings to effect this threefold increase in the particle horizon distance is $N \cong 60$. The properties of the Hubble sphere provide the answer to the question: how, in a universe of age t, can regions separated by distances much greater than ct be causally connected?

161.640 **Phase shifts in gravitationally evolving density fields.**
B. S. Ryden, M. Gramann.
Astrophys. J., Lett., Vol. 383, No. 2, p. L33 – L36 (20 Dec 1991).
The authors examine the phases of the Fourier components of an initially Gaussian density field to see how the phases are shifted away from their initial values as the field evolves gravitationally. The analytic expression for the phase shift, in second–order perturbation theory, is presented. To investigate the fully nonlinear regime, the authors ran two–dimensional N–body simulations. The power spectra used were of the form $P(k) \sim k^n$, with $n = -1$ and $n = 0$. The numerical results show that second–order perturbation theory soon breaks down when computing the phase shifts. As a function of the expansion factor, the authors find the wavenumber k_Φ at which the mean magnitude of the phase shift is equal to $\pi/4$. The wavenumber k_Φ is proportional to the wavenumber k_w at which the peculiar velocity goes nonlinear. For $n = -1$, $k_\Phi \approx 0.2k_w$; for $n = 0$, $k_\Phi \approx 0.3k_w$. The phases of the Fourier components are significantly shifted from their initial values at a scale on which their amplitudes are still in the linear stage of growth.

161.641 **Euclideanized Einstein–Yang–Mills equations, wormholes and the ground–state wave function of a radiation dominated Universe.**
O. Bertolami, J. M. Mourão.
12. Autumn School of Physics. The physical universe – the interface between cosmology, astrophysics and particle physics, p. 21 – 38 (1991). – See Abstr. 012.097 for the main entry.
SO(4)–symmetric (anti–)selfdual solutions of euclideanized Yang–Mills equations are used for constructing the ground–state wave function of the Universe. The authors discuss also SO(4)–symmetric wormhole solutions of Einstein–Yang–Mills systems and some physical implications of our minisuperspace ground–state wave function of the Universe.

161.642 **Topological defects in the Early Universe.**
E. Copeland.
12. Autumn School of Physics. The physical universe – the interface between cosmology, astrophysics and particle physics, p. 47 – 101 (1991). – See Abstr. 012.097 for the main entry.

161.643 **Understanding large–scale cosmic structure.**
B. J. T. Jones, R. van de Weygaert.
12. Autumn School of Physics. The physical universe – the interface between cosmology, astrophysics and particle physics, p. 134 – 158 (1991). – See Abstr. 012.097 for the main entry.
The purpose of this lecture is mainly to discuss new techniques that have been developed in order to best interpret various kinds of galaxy catalogues. With these techniques we leave, for the first time, the idea that we are forever confined to knowing only about the universe as projected onto the sky. One is able to construct three dimensional models of the universe. There are indeed some crucial assumptions that go into this, but they are clearly understood and appear to be entirely reasonable. It remains an important problem for the future to validate these assumptions.

161.644 **Galaxy formation – an update.**
M. S. Longair.
12. Autumn School of Physics. The physical universe – the interface between cosmology, astrophysics and particle physics, p. 174 – 196 (1991). – See Abstr. 012.097 for the main entry.

161.645 **The hypothesis of the expansion of the Universe and the global tests.**
M. Moles.
12. Autumn School of Physics. The physical universe – the interface between cosmology, astrophysics and particle physics, p. 197 – 226 (1991). – See Abstr. 012.097 for the main entry.
The author reviews and discusses the status of the standard cosmology when it is confronted to the existing observational results on the global tests. He considers first what the theoretical predictions are and how are they fitted by the data. Then the author considers how the consideration of non zero cosmological constant models or the introduction of evolutionary effects could change the situation.

161.646 **The Early Universe behaviour with non–minimal coupling.**
P. Moniz, P. Crawford, A. Barroso.
12. Autumn School of Physics. The physical universe – the interface between cosmology, astrophysics and particle physics, p. 227 – 236 (1991). – See Abstr. 012.097 for the main entry.
The authors comment on the stability problem of the equilibrium states of the Universe. With a specific ansatz for the ground state they re–examine the physics of spontaneous symmetry breaking in curved space–time. Introducing a non–minimal coupling between a scalar field (a multiplet under the action of a gauge group) and gravity they show that spontaneous symmetry breaking may occur without the need of a negative value for m^2. Within this framework the authors calculate the Higgs mass, the value of the cosmological constant and of the effective gravitational constant.

161.647 **Stability of compactification in Einstein–Yang–Mills theories.**
O. Bertolami, Yu. A. Kubyshin, J. M. Mourão.
12. Autumn School of Physics. The physical universe – the interface between cosmology, astrophysics and particle physics, p. 237 – 252 (1991). – See Abstr. 012.097 for the main entry.
The authors study the dynamics of multidimensional Einstein–Yang–Mills cosmologies. The stability of compactification after the period of inflationary expansion of the external dimensions is analysed.

161.648 **The cosmological constant, third quantization and all that.**
G. Lavrelashvili.
12. Autumn School of Physics. The physical universe – the interface between cosmology, astrophysics and particle physics, p. 292 – 304 (1991). – See Abstr. 012.097 for the main entry.
The paper consists of two main parts. The first part contains a short review of recent works on the application of the third quantization method to the cosmological constant problem. In the second part critical consideration is given to Dirac's scheme of gravity quantization. The author supposes that Dirac's scheme may turn out inadequate for gravity quantization because of anomalies arising in the algebra of constraints.

161.649 **Structure of the inflationary Universe.**
M. I. Zelnikov.
12. Autumn School of Physics. The physical universe – the interface between cosmology, astrophysics and particle physics, p. 305 – 312 (1991). – See Abstr. 012.097 for the main entry.
The present report is organized as follows. In the first section a brief discussion of scalar field behaviour at the inflationary stage is presented. The second section is devoted to an analysis of the scalar field perturbation spectrum which determines the local structure. In the third section the author considers a model in which the inflationary Universe has a nontrivial global dimensional structure.

161.650 **Forming cosmic structure with a 17 keV neutrino.**
J. R. Bond.
J. R. Astron. Soc. Can., Vol. 85, No. 4, p. 182 (Aug 1991). Abstract. – See Abstr. 012.091 for the main entry.

161.651 **Large–scale structure in a low bias universe.**
H. M. P. Couchman, R. G. Carlberg.
J. R. Astron. Soc. Can., Vol. 85, No. 4, p. 183 – 184 (Aug 1991). Abstract. – See Abstr. 012.091 for the main entry.

161.652 **Baryogenesis and the nature of dark matter.**
L. M. Widrow.
J. R. Astron. Soc. Can., Vol. 85, No. 4, p. 188 (Aug 1991). Abstract. – See Abstr. 012.091 for the main entry.

161.653 **Erratum: "Gravitational microlensing of high–redshift supernovae by compact objects" [Astrophys. J., Vol. 374, No. 1, p. 83 – 90 (10 Jun 1991)].**
K. P. Rauch.
Astrophys. J., Vol. 383, No. 1, p. 466 (10 Dec 1991). See Abstr. 53.161.371.

161.654 **Rationalist programmes in early modern cosmology.**
G. Gale.
Astron. Q., Vol. 8, No. 4, p. 193 – 218 (1991).

161.655 **Density distribution function at the non–linear stage of evolution.**
A. Klypin, L. Kofman.
J. R. Astron. Soc. Can., Vol. 85, No. 4, p. 209 (Aug 1991). Abstract. – See Abstr. 012.091 for the main entry.

161.656 **Superclustering at high redshifts.**
M. J. West.
J. R. Astron. Soc. Can., Vol. 85, No. 4, p. 211 – 212 (Aug 1991). Abstract. – See Abstr. 012.091 for the main entry.

161.657 **Cosmology a priori.**
A. J. Rutgers.
Proc. K. Ned. Akad. Wet. (1990), Vol. 94, No. 3, p. 415 – 417 (30 Sep 1991).

161.658 **Matter and antimatter in the universe.**
A. Morselli, P. G. Picozza.
5. Winter School on Hadronic Physics: Problems of fundamental modern physics II, p. 281 – 299 (1991). – See Abstr. 012.101 for the main entry.
Cosmic rays provide a unqiue probe of the most energetic processes in the universe. The authors show how the use of the Space Station could change our knowledge about the asymmetry (or symmetry) of matter and antimatter in our universe and help our investigation of the origin and evolution of matter in the Galaxy.

161.659 **Introductory notes on particle physics and cosmology.**
F. Wilczek.
NATO Advanced Study Institute on Techniques and Concepts of High Energy Physics, p. 49 – 69 (1991). – See Abstr. 012.102 for the main entry.
The author gives notes on two and a half of the four lectures he gave at the St. Croix Summer School of 1990. The purpose of these lectures was to introduce students in experimental physics to some of the central concepts in cosmology, and perhaps inspire them to consider taking up some of the experimental challenges it poses.

161.660 **The search for valid theories of the Universe.**
J.–C. Pecker.
Mem. Soc. Astron. Ital., Vol. 62, No. 3, p. 499 – 511 (1991). – See Abstr. 012.093 for the main entry.
One attempts to reply to the questions: what is the Universe? What is observed, what is theoretical, what is inferred? What is a valid theory? And, finally, how to improve the methodology used in cosmology? This discussion leads the author to outline

different aspects of the cosmological research and, in particular, to discuss the status of the so-called "cosmological facts".

161.661 **The role of quantum phenomena in our understanding of the Universe.**
J. V. Narlikar.
Mem. Soc. Astron. Ital., Vol. 62, No. 3, p. 519 – 530 (1991). – See Abstr. 012.093 for the main entry.
This discussion concerns the intimate connection between the quantum phenomena and the large scale structure of the Universe. Examples are given to illustrate not only how quantum theory has advanced our perception of the Universe but also that the interaction with the Universe may be neccessary to understand how a microscopic system operates under the laws of quantum mechanics.

161.662 **Major themes in the relation between philosophy and cosmology.**
G. Ellis.
Mem. Soc. Astron. Ital., Vol. 62, No. 3, p. 553 – 605 (1991). – See Abstr. 012.094 for the main entry.
Modern cosmology (the study of the large-scale structure of the physical universe) has made tremendous progress in recent decades. However there are substantially different approaches to the subject, based on different philosophical understandings; and the degree to which models of the universe are satisfactory depends to a considerable degree on the approach taken (indeed even the concept of what is a "satisfactory model" depends on the philosophical stance adopted). The aim of this paper is to give an overview of some of the main issues that arise and the different approaches to them; the hope is that such a perspective (representing old issues and questions, but in a new setting) helps our understanding.

161.663 **What does the big bang tell us about quantum gravity?**
R. Penrose.
Mem. Soc. Astron. Ital., Vol. 62, No. 3, p. 607 – 614 (1991). – See Abstr. 012.094 for the main entry.
Contents: 1. The big bang. 2. The need for quantum gravity. 3. Evidence from the second law of thermodynamics. 4. Space-time curvature near singularities. 5. How special was the big bang? 6. State-vector reduction as a quantum gravity effect?

161.664 **Lectures on particle physics and cosmology.**
K. A. Olive.
1990 Summer School in High Energy Physics and Cosmology, p. 421 – 494 (1991). – See Abstr. 012.105 for the main entry.
The standard big bang model including baryosynthesis and inflation is briefly reviewed. The present status of big bang nucleosynthesis and particle physics candidates for dark matter are discussed in detail.

161.665 **Cosmological and astrophysical aspects of the quark-hadron phase transition.**
K. A. Olive.
1990 Summer School in High Energy Physics and Cosmology, p. 495 – 514 (1991). – See Abstr. 012.105 for the main entry.
A transition from normal hadronic matter (e.g. protons and neutrons) to quark-gluon matter is expected at both high temperatures and densities. In physical situations, this transition may occur in heavy ion collisions, the early Universe, and in the cores of neutron stars. In this article, the author discusses the transition in the latter two of these environments. With regard to the early Universe, he focuses attention on the effect of inhomogeneities produced by the transition on big bang nucleosynthesis, the theory describing the primordial origin of the light elements. A transition to quark matter in the interior of neutron stars, further enhances the uncertainties regarding the equation of state of dense nuclear matter and neutron star properties such as the maximum mass and rotation frequencies.

161.666 **Quantum cosmology.**
J. J. Halliwell.
1990 Summer School in High Energy Physics and Cosmology, p. 515 – 599 (1991). – See Abstr. 012.105 for the main entry.
Contents: 1. Introduction. 2. A simple example. 3. The Hamiltonian formulation of general relativity. 4. Quantization. 5. Minisuperspace – general theory. 6. Classical spacetime. 7. The WKB approximation. 8. Boundary condition proposals. 9. No-boundary vs. tunneling. 10. Beyond minisuperspace. 11. Vacuum states from quantum cosmology. 12. Summary. 13. A guide to the literature.

161.667 **Wormholes and cosmological constant.**
J. Maharana.
1990 Summer School in High Energy Physics and Cosmology, p. 708 – 719 (1991). – See Abstr. 012.105 for the main entry.
The author has considered the effects of wormholes and found that they introduce indeterminacy in the constants of Nature.

161.668 **The Zeldovich effect of the universe reionized by VMOs scattering off the 3K background radiation.**
Wu Jianxin, Xie Guangzhong.
Publ. Yunnan Obs., No. 3, p. 1 – 5 (1991). In Chinese.
The authors assume that the population III stars are composed of very massive objects (VMOs). The high luminosity of VMOs may make the background universe reionized. The authors discussed the inverse Compton effect of the hot electrons of the universe in scattering off the microwave background photons. The result shows, the reionized universe in the period of VMOs make the 3K background radiation produce a small distortion and in the Rayleigh-Jeans frequency region, the temperature deviation from the black body is $10^{-3} - 10^{-5}$.

161.669 **Cosmological numerical simulation.**
M. Abadi, D. Garcia Lambas.
Bol. Asoc. Argent. Astron., No. 36, p. 433 – 434 (1990). Abstract. In Spanish and English. – See Abstr. 012.095 for the main entry.

161.670 **A modified Robertson-Walker cosmological model.**
P. C. Vaidya, R. P. Sarkar.
Bull. Astron. Soc. India, Vol. 19, Nos. 1 – 2, p. 77 (Mar – Jun 1991). Abstract. – See Abstr. 012.108 for the main entry.

161.671 **Gravitational production of massive scalars in the context of inflation.**
U. A. Yajnik.
Bull. Astron. Soc. India, Vol. 19, Nos. 1 – 2, p. 92 (Mar – Jun 1991). Abstract. – See Abstr. 012.108 for the main entry.

161.672 **Hydrogen recombiantion in the expanding Universe.**
S. I. Grachev, V. K. Dubrovich.
Astrofizika, Tom 34, Vyp. 2, p. 249 – 264 (Apr 1991). In Russian. English translation in Astrophysics, Vol. 34, No. 2.
Fractional electron density and electron temperature in the expanding Universe are calculated as functions of redshift $z(0 < z < 1800)$ by means of combined numerical solutions of statistical equilibrium equations (for hydrogen atoms) and energy equation for different values of the Hubble constant and cosmological density parameters. It is shown that for $z < 900$ there is an agreement with previous investigations (see AJB, Band 68, Ref. 44.122, AAA Vol. 32, Abstr. 066.002, AAA Vol. 40, Abstr. 161.041), in which simpler treatments of the problem are used. It is found also that the strict description of radiation transfer in the hydrogen Lyα line has no significant influence on the recombination kynetics in constrast to previous statements (see AAA, Vol. 49, Abstr. 161.178).

161.673 **Cosmic background radiations.**
G. F. Smoot.
3. International Workshop on Data Analysis in Astronomy, p. 291 – 318 (1989). – See Abstr. 012.110 for the main entry.
The photon is the best probe of the distant regions of the universe. There are two regions in the electromagnetic spectrum

where one can expect to see far back into the universe. In both of these windows there are backgrounds which are now being explored and studied. These observations have led to surprisingly strong statements about the nature of our universe.

161.674 **The role of boron in cosmology.**
V. E. Viola.
AIP Conf. Proc., No. 231, p. 1 – 12 (1 Jul 1991). Current Physics Microform No.: 9110A0001. – See Abstr. 012.123 for the main entry.

The role of the boron nucleus in nucleosynthesis is discussed. Because of the fragile nature of boron nuclei, they are destroyed rather than created in cosmological settings such as the big bang or stellar burning cycles. This instability also severely constrains the buildup of heavy elements in both processes. However, ^{8}B is an important intermediary in stellar burning, as its beta–decay neutrinos are primary objects of searches for solar neutrinos in experiments on Earth. Interactions of galactic cosmic rays with the interstellar medium are believed to form ^{10}B and ^{11}B, as well as ^{6}Li and ^{9}Be. The success of this model leads to speculations concerning the expanding universe.

161.675 **Cosmology and cluster formation.**
P. J. E. Peebles.
Clusters of Galaxies Meeting, p. 1 – 23 (1990). – See Abstr. 012.124 for the main entry.

The author discusses some issues that arise in the attempt to understand what rich clusters of galaxies might teach us about cosmology. First, the mean mass per galaxy in a cluster, if applied to all bright galaxies, yields a mean mass density ~ 30 percent of the critical Einstein – de Sitter value. Is this because the mass per galaxy is biased low in clusters, or must we learn to live in a low density universe? Second, what is the sequence of creation? There are theories in which protoclusters form before galaxies, or after, or the two are more or less coeval. Third, can we imagine that clusters formed by gravitational instability out of Gaussian primeval density fluctuations? Or do the observations point to the non–Gaussian perturbations to be expected from cosmic strings, or explosions, or even some variants of inflation? These issues depend on a fourth: do we know the gross physical properties of clusters well enough to use them as constraints on cosmology ?

161.676 **On the validity of the quasi–linear theory for the cosmological three–point correlation function.**
S. Inagaki.
Publ. Astron. Soc. Jpn., Vol. 43, No. 5, p. 661 – 669 (1991).

Quasi–linear theory for the cosmological two–point correlation function has been found to be successful. One might think that this would also be true for the three–point correlation function. In this paper it is shown that this is false. Another important conclusion obtained is that the initial correlations equal to or higher than three–point do not affect the higher–order correlations after a sufficiently long time.

161.677 **Primordial nucleosynthesis: beyond the standard model.**
R. A. Malaney.
Nuclei in the cosmos, p. 127 – 145 (1991). – See Abstr. 003.038 for the main entry.

Contents: 1. Introduction. 2. Standard primordial nucleosynthesis. 3. Observational tests of the standard model: helium; lithium. 4. The QCD phase transition. 5. Neutrinos in the early universe: sterile neutrinos; neutrino oscillations. 6. Cosmic strings. 7. Conclusions.

161.678 **Production of heavy elements in inhomogeneous cosmologies.**
F.–K. Thielemann, J. H. Applegate, J. J. Cowan, M. Wiescher.
Nuclei in the cosmos, p. 147 – 177 (1991). – See Abstr. 003.038 for the main entry.

The incorporation of quantum chromodynamics into big bang cosmology predicts that the universe shifted from a quark–gluon plasma to a meson gas when it was roughly 10 μs old. If this change is due to a first order phase transition, baryon density inhomogeneities can be produced and survive until nucleosynthesis. At that point they can affect the primordial abundances resulting from element formation in the early universe. Here, the authors undertake a detailed study of one of the proposed signatures of inhomogeneous nucleosynthesis – the production of heavy elements by neutron capture – and find that heavy elements are indeed produced, but that the level of production is probably too low to be observable.

161.679 **Dirac equation in the field of a charged cosmic string and in the field of a point charge with $Z > 137$.**
T. J. Allen, L. J. Tassie.
Aust. J. Phys., Vol. 44, No. 6, p. 585 – 590 (1991).

In both spherical and cylindrical coordinates, the radial Dirac equation can be written in the form of a Schrödinger equation with an effective potential. It is shown that the difficulties at $r \to 0$ for the Dirac equation in the field of a point charge for $Z > 137$ are the same as those for the Schrödinger equation with a $1/r^2$ potential. The effective potential is used to show that similar difficulties do not arise for the field of a line charge, so allowing the consideration of the motion of electrons in the field of a charged superconducting cosmic string without considering the internal structure of the string.

161.680 **Kinetic equation with neutrino oscillations in the early universe.**
H. Granek, B. H. J. McKellar.
Aust. J. Phys., Vol. 44, No. 6, p. 591 – 617 (1991).

This paper considers the kinetic equation for interacting neutrino gas in the context of an expanding early universe. It is suggested that if neutrino oscillations are present and CP variations occur prior to the decoupling of the neutrino gas from the rest of the universe, then lepton number may not be conserved and, in principle, significant permanent neutrino chemical potentials may develop and survive until the present day.

161.681 **Large scale peculiar motions in the southern sky.**
R. A. Schommer, G. Bothun, J. R. Mould.
Rev. Mex. Astron. Astrofis., Vol. 22, No. 2, p. 304 (Oct 1991). Abstract. – See Abstr. 012.127 for the main entry.

161.682 **Statistical properties of CDM galactic halos.**
M. Abadi, P. Tissera, D. Garcia Lambas.
Rev. Mex. Astron. Astrofis., Vol. 22, No. 2, p. 305 (Oct 1991). Abstract. – See Abstr. 012.127 for the main entry.

161.683 **Counts, colors and clustering of medium redshift galaxies.**
L. Infante, C. Pritchet.
Rev. Mex. Astron. Astrofis., Vol. 22, No. 2, p. 305 (Oct 1991). Abstract. – See Abstr. 012.127 for the main entry.

161.684 **Generalized inflation from R^3 and $R\Box R$ terms.**
S. Gottlöber, V. Müller, H.-J. Schmidt.
Astron. Nachr., Vol. 312, No. 5, p. 291 – 297 (1991).

Generalized inflationary stages as power–law solutions result from the addition of R^3 and $R\Box R$ terms to the quadratic gravitational Lagrangian. Both contributions lead to new attractor solutions in the high–curvature limit. The authors emphasize the properties of critical values of the curvature scalar in the R^3 theory. Contrarily to models with two scalar fields, the higher order curvature terms allow double inflation for special initial conditions only.

161.685 **Föppl's repulsive cosmos and hyperbolic Friedmannian matter cosmos.**
H.-J. Treder.
Astron. Nachr., Vol. 312, No. 5, p. 307 – 308 (1991).

Föppl's repulsive cosmos and the hyperbolic Friedmannian matter cosmos are compared. Both worlds have positive energy integrals. In the repulsive cosmos no cosmic singularity exists.

161.686 **Formation of galaxies with dissipation.**
M. B. Mosconi, D. García Lambas.
Rev. Mex. Astron. Astrofis., Vol. 22, No. 2, p. 309 (Oct 1991). Abstract. – See Abstr. 012.127 for the main entry.

161.687 **Fisica dell'universo primordiale.**
G. Ellis.
G. Astron., Vol. 17, N. 1 – 2, p. 6 – 14 (Mar – Jun 1991). Paper presented at "La cosmologia nella cultura dell '900". Conferenza Internazionale nel centenario della nascita di Hubble, Torino (Italy), 21 – 23 Nov 1990.

161.688 **Le cosmologie di Einstein e di De Sitter: la distinzione–identificazione fra geometrico e fisico.**
S. d'Agostino.
G. Astron., Vol. 17, N. 1 – 2, p. 24 – 31 (Mar – Jun 1991). Paper presented at "La cosmologia nella cultura dell '900". Conferenza Internazionale nel centenario della nascita di Hubble, Torino (Italy), 21 – 23 Nov 1990.

161.689 **L'energetica dell'Universo.**
A. Ferrari.
G. Astron., Vol. 17, N. 1 – 2, p. 55 – 59 (Mar – Jun 1991). Paper presented at "La cosmologia nella cultura dell '900". Conferenza Internazionale nel centenario della nascita di Hubble, Torino (Italy), 21 – 23 Nov 1990.

161.690 **But what a whimper!**
C. Kitchin.
Astron. Now, Vol. 5, No. 9, p. 18 – 21 (Sep 1991).

Gauge field theories. An introduction with applications.
See Abstr. 003.003.

Our universes.
See Abstr. 003.004.

Elementary particles.
See Abstr. 003.011.

The almighty chance.
See Abstr. 003.029.

Bausteine des Universums. Quarks und Leptonen.
See Abstr. 003.031.

The hidden universe.
See Abstr. 003.034.

Pourquoi la nuit est–elle noire? Le paradoxe de Chéseaux–Olbers (1re partie).
See Abstr. 004.028.

Pourquoi la nuit est–elle noire? Le paradoxe de Chéseaux–Olbers (2e partie).
See Abstr. 004.029.

Historical note on the discovery of cosmological background radiation.
See Abstr. 004.085.

Il principio di Weyl, passo essenziale verso la precisazione della nozione di modello di Universo.
See Abstr. 004.091.

On the occasion of the 100th anniversary of A. A. Friedmann's (1888 – 1925) birth.
See Abstr. 005.024.

Universums begynnelse bland de jämtländska fjällen.
See Abstr. 011.026.

Quantum gravity. Proceedings.
See Abstr. 012.004.

Relativity and gravitation: classical and quantum. Proceedings.
See Abstr. 012.005.

Elementary particles and the universe. Essays in honor of Murray Gell–Mann.
See Abstr. 012.007.

Current problems in particle physics and cosmology. Proceedings. Workshop on Current Problems in Particle Physics and Cosmology, Ioannina (Greece), 25 – 29 Jun 1990.
See Abstr. 012.022.

High energy astrophysics: compact stars and active galaxies. Proceedings.
See Abstr. 012.026.

Quantum cosmology and baby universes. Proceedings.
See Abstr. 012.036.

Astroparticle physics. Proceedings.
See Abstr. 012.037.

Primordial nucleosynthesis. Proceedings.
See Abstr. 012.060.

Strange quark matter in physics and astrophysics. Proceedings. International Workshop on Strange Quark Matter in Physics and Astrophysics, Aarhus (Denmark), 20 – 24 May 1991.
See Abstr. 012.068.

After the first three minutes. Proceedings.
See Abstr. 012.069.

Primordial nucleosynthesis and evolution of early universe. Proceedings.
See Abstr. 012.071.

Observational tests of cosmological inflation. Proceedings.
See Abstr. 012.076.

The birth and early evolution of our universe. Proceedings. Nobel Symposium 79, Gräftåvallen (Sweden), 11 – 16 Jun 1990.
See Abstr. 012.080.

The space distribution of quasars. Proceedings.
See Abstr. 012.084.

Kosmos – cosmology today between science and philosophy. Proceedings. 1. Venice Conference on Cosmology and Philosophy, Venice (Italy), 8 – 9 May 1987
See Abstr. 012.093.

Origin and evolution of the Universe. A Hubble Centenary Conference. Proceedings. 3. Venice Conference on Cosmology and Philosophy, Venice (Italy), 15 – 16 Dec 1989.
See Abstr. 012.094.

The physical universe: the interface between cosmology, astrophysics and particle physics. Proceedings.
See Abstr. 012.097.

High energy physics and cosmology. Proceedings.
See Abstr. 012.105.

Particles, strings and supernovae. Volumes 1, 2. Proceedings.
See Abstr. 012.109.

The Fifth Marcel Grossmann Meeting on Recent Developments in Theoretical and Experimental General Relativity, Gravitation and Relativistic Field Theories. Parts A, B. Proceedings.
See Abstr. 012.112.

Big bang, active galactic nuclei and supernovae. Proceedings.
See Abstr. 012.113.

A. A. Friedmann: centenary volume. Proceedings.
See Abstr. 012.114.

General relativity and gravitational physics. Proceedings.
See Abstr. 012.115.

On doing cosmology with a large optical telescope.
See Abstr. 013.019.

Extragalactic research at the Instituto de Astrofisica de Canarias.
See Abstr. 013.080.

Glorious Universe.
See Abstr. 014.003.

1980's cosmology and bioastronomy: an Odyssean perspective.
See Abstr. 015.030.

Will the Universe expand for ever?
See Abstr. 015.035.

The concept of the Universe and the problem of reality.
See Abstr. 015.036.

Origin and evolution of the Universe. Final panel discussions I – III.
See Abstr. 015.037.

Determining isometries of space–times using computer algebra.
See Abstr. 021.007.

Constrained realizations of Gaussian fields: a simple algorithm.
See Abstr. 021.036.

Microwave background interferometry in Cambridge.
See Abstr. 033.041.

Neutrino and dark–matter detection at low temperature.
See Abstr. 034.022.

Development of 3–D UV and soft X–ray detectors based on energy sensitive superconducting elements.
See Abstr. 034.119.

Detecting halo weakly interacting massive particles with a cryogenic phonon detector.
See Abstr. 034.120.

Early results from the Cosmic Background Explorer (COBE).
See Abstr. 035.103.

The Space Telescope and the problems of cosmology.
See Abstr. 035.212.

The wavelet transform.
See Abstr. 036.175.

Cosmic Background Explorer (COBE) satellite anisotropy experiment data analysis techniques.
See Abstr. 036.354.

COBE attitude refinement.
See Abstr. 051.066.

Flavored cosmic strings.
See Abstr. 061.001.

Particle production by the formation of a global monopole.
See Abstr. 061.002.

Vacuum instability of spin–zero particles around a charged superconducting cosmic string.
See Abstr. 061.003.

Electroweak baryogenesis in the two–doublet model.
See Abstr. 061.007.

Goldstone bosons versus domain walls bounded by cosmic strings.
See Abstr. 061.012.

Inequalities which constrain negative energy in quantum field theory.
See Abstr. 061.022.

Experimental constraints on dark matter.
See Abstr. 061.025.

Soliton operators in two and three dimensions.
See Abstr. 061.028.

Second quantised Hamiltonian for light cone gauge heterotic superstrings.
See Abstr. 061.030.

BRST effective actions on phase space.
See Abstr. 061.031.

The minimal 17 keV neutrino model.
See Abstr. 061.034.

Astrophysics and cosmology confront the 17 keV neutrino.
See Abstr. 061.037.

Constraints from primordial nucleosynthesis on the mass of the τ neutrino.
See Abstr. 061.040.

$^{17}O(n, \alpha)^{14}C$ cross section from 25 meV to approximately 1 MeV.
See Abstr. 061.049.

Cosmic neutrino radiation in the far ultraviolet and the E_6 model.
See Abstr. 061.055.

Cosmology, astrophysics, and a 17 keV neutrino.
See Abstr. 061.056.

Dark matter and the equivalence principle.
See Abstr. 061.060.

Radioactive dating of the elements.
See Abstr. 061.066.

Generalized Tremaine – Gunn limits for bosons and fermions.
See Abstr. 061.078.

Perturbations on domain walls and strings: a covariant theory.
See Abstr. 061.079.

Sourceless Abelian gauge string in an expanding universe.
See Abstr. 061.080.

High–resolution computations on inhomogeneous nucleosynthesis.
See Abstr. 061.086.

Astrophysical measurements: current and future.
See Abstr. 061.087.

Rate of ^{3}H(^{7}Li, n$_0$)^{9}Be and big–bang nucleosynthesis.
See Abstr. 061.094.

Limits to the radiativc decay of the axion.
See Abstr. 061.101.

Energetic neutrinos from heavy–neutralino annihilation in the Sun.
See Abstr. 061.102.

Semilocal cosmic strings.
See Abstr. 061.103.

Strange quark matter in physics and astrophysics.
See Abstr. 061.104.

Could cosmic QCD phase transition produce strange quark matter which survives until the present time?
See Abstr. 061.105.

Measurement of the cross section of the ^{12}C(n,γ)^{13}C reaction at stellar energy.
See Abstr. 061.107.

Big bang nucleosynthesis comes of age.
See Abstr. 061.110.

Primordial nucleosynthesis in an inhomogeneous universe.
See Abstr. 061.111.

Particle physics footprints of the invisible axion.
See Abstr. 061.112.

Time variation of fundamental constants. II. Quark masses as time–dependent parameters.
See Abstr. 061.113.

Conical bremsstrahlung in a cosmic–string spacetime.
See Abstr. 061.114.

Nucleosynthesis in big–bang models.
See Abstr. 061.122.

Concocting a cosmic repice for matter.
See Abstr. 061.123.

Physicists close in on a weighty quarry.
See Abstr. 061.124.

Beta decay far from stability and double beta decay, and consequences for astrophysics.
See Abstr. 061.130.

Effects of negatively charged massive particles on primordial stars.
See Abstr. 065.136.

An introduction to curved space–times.
See Abstr. 066.002.

On the gravity theories and cosmology from strings.
See Abstr. 066.007.

Two–dimensional quantum cosmology.
See Abstr. 066.008.

On some problems of quantum and classical gravitation.
See Abstr. 066.009.

Extended inflation and generalised scalar–tensor theories.
See Abstr. 066.011.

Minisuperspace wavefunctions for cosmology and wormholes.
See Abstr. 066.014.

Uncertainty and equivalence principles in quantum general relativity.
See Abstr. 066.019.

Quantum theory of particles with Planckean masses.
See Abstr. 066.022.

Quantum string gravity near the Planck scale.
See Abstr. 066.023.

Einstein universe as a moving fluid cosmology.
See Abstr. 066.027.

Spherically symmetric density perturbation in relativistic cosmology.
See Abstr. 066.028.

O(3)–invariant tunneling at false vacuum decay in general relativity.
See Abstr. 066.032.

Five lectures on planar gravity.
See Abstr. 066.038.

Numerical relativity.
See Abstr. 066.039.

The Yamabe constant.
See Abstr. 066.040.

On the classification of group–invariant connections.
See Abstr. 066.041.

Non–local variables in general relativity.
See Abstr. 066.042.

Time and chaos in general relativity.
See Abstr. 066.043.

Lightlike contractions – singular limits of spacetimes.
See Abstr. 066.044.

Ghost fields in principal bundles.
See Abstr. 066.045.

A Clifford bundle approach to gravitational theory.
See Abstr. 066.048.

Canonical formalism, the initial value problem and the maximum principle.
See Abstr. 066.049.

A class of interior solutions of the Kerr–Newman field.
See Abstr. 066.050.

Thermodynamics of spacetimes admitting a conformal symmetry parallel to the 4–velocity.
See Abstr. 066.051.

On the existence of affine collineations.
See Abstr. 066.052.

Collineations in Gödel–type space–times.
See Abstr. 066.053.

New solutions of the semiclassical Einstein equations.
See Abstr. 066.054.

Time shifts and non–linear interference in cylindrically symmetric Einstein–Maxwell solitons.
See Abstr. 066.055.

Cylindrically symmetric gravitational waves with rotational properties.
See Abstr. 066.056.

2+1 gravity as an exactly solved system.
See Abstr. 066.062.

The algebra of traces in some Chern–Simons theories.
See Abstr. 066.064.

Non perfect fluids in Kaluza–Klein cosmological models.
See Abstr. 066.068.

Vanishing null energy solutions and Hamiltonian approach for membranes in the intrinsic light front gauge.
See Abstr. 066.071.

Exact solutions from one dimensional subspaces in Kaluza–Klein theory.
See Abstr. 066.072.

A mass–radius relation for astrophysical objects and Weyl gravity.
See Abstr. 066.074.

Gravity from superstrings.
See Abstr. 066.083.

Mutually interacting quantum fields in curved space–times: the outcome of physical processes.
See Abstr. 066.084.

Astrophysical constraints on the gravitational constant.
See Abstr. 066.088.

Measurements of a Machian transient mass fluctuation.
See Abstr. 066.090.

A new description for a realistic inhomogeneous universe in general relativity.
See Abstr. 066.091.

Physical models of white and grey holes and their astrophysical applications.
See Abstr. 066.092.

The semidegenerate configurations of a selfgravitating system of fermions.
See Abstr. 066.105.

Radiation and false–vacuum bubble dynamics.
See Abstr. 066.121.

A new approach to gravitational microlensing: time delays and the Galactic mass distribution.
See Abstr. 066.126.

Solitonic solutions on a Bianchi II background generated by SHEEP algebraic manipulation.
See Abstr. 066.140.

Anisotropic solutions in the 5D space – time – mass theory.
See Abstr. 066.147.

Cosmological solutions in the D = 5 Einstein – Maxwell theory coupled to matter.
See Abstr. 066.151.

Algebraic invariant curves in cosmological dynamical systems and exact solutions.
See Abstr. 066.155.

Decoherence and back reaction: the origin of the semiclassical Einstein equations.
See Abstr. 066.166.

Spectral boundary conditions in one–loop quantum cosmology.
See Abstr. 066.180.

Nucleation of a universe in (2+1)–dimensional gravity with a negative cosmological constant.
See Abstr. 066.182.

Repulsive gravity in tension stars.
See Abstr. 066.189.

Conformal gravity with torsion.
See Abstr. 066.193.

Scalar field, nonminimal coupling, and cosmology.
See Abstr. 066.197.

Gravitational field of a hedgehog and the evolution of vacuum bubbles.
See Abstr. 066.199.

The fractal structure of the quantum space–time.
See Abstr. 066.203.

Microscopic anisotropy in a Robertson–Walker cosmology.
See Abstr. 066.209.

Gravitation and hot big–bang cosmology.
See Abstr. 066.217.

Bianchi cosmologies: the role of spatial topology.
See Abstr. 066.219.

Tension shells and tension stars.
See Abstr. 066.221.

Oscillating and inflating universe in SO(3)–gravitation theory.
See Abstr. 066.231.

More on the black hole S–matrix.
See Abstr. 067.018.

Absorption cross section of a mini black hole.
See Abstr. 067.050.

The entropy change of imperfect fluid collapsing system in which the fluids' own gravitational field plays a role.
See Abstr. 067.073.

Gamma rays and energetic particles from primordial black holes.
See Abstr. 067.197.

O(3)–invariant tunnelling at false vacuum decay in general relativity.
See Abstr. 067.207.

Self–gravitating magnetic monopoles, global monopoles and black holes.
See Abstr. 067.240.

Neutrino properties from SN 1987A and from future observations of GSC.
See Abstr. 080.111.

Isotopically anomalous Ti in presolar SiC from the Murchison meteorite.
See Abstr. 105.012.

Reconciling ZIP and IRAS observations of the zodiacal background.
See Abstr. 106.039.

The lithium abundance in extreme halo stars.
See Abstr. 114.002.

Exploding stars and the expanding Universe.
See Abstr. 125.025.

From the expansion of SN 1987A to the expansion of the Universe.
See Abstr. 125.088.

Supernovae Ia as standard candles.
See Abstr. 125.089.

C I/CO ratio in interstellar dense clouds and decay photons of dark matter.
See Abstr. 131.325.

The cosmic D/H ratio.
See Abstr. 131.391.

The all–sky extragalactic X–ray foreground.
See Abstr. 142.024.

Spectroscopic limits to an extragalactic far–ultraviolet background.
See Abstr. 142.027.

The intensity and spectrum of the diffuse X–ray background.
See Abstr. 142.028.

Faint X–ray source counts and the origin of the X–ray background.
See Abstr. 142.029.

The diffuse soft X–ray background as seen with the Einstein Observatory.
See Abstr. 142.033.

Cosmic X–ray background.
See Abstr. 142.045.

The origin of the cosmic X–ray background.
See Abstr. 142.046.

Einstein deep surveys.
See Abstr. 142.056.

The lightest supersymmetric particle and the extragalactic γ–ray background.
See Abstr. 143.014.

Cosmic rays at the highest energies.
See Abstr. 144.078.

Dynamics of "clouds" of protogalactic medium in the gravitational potential of dark matter.
See Abstr. 151.004.

Dynamics of "clouds" of the protogalactic medium in the gravitational potential of dark matter.
See Abstr. 151.012.

Flattening of tilted protogalactic disks.
See Abstr. 151.047.

Dissipational galaxy formation. I. Effects of gasdynamics.
See Abstr. 151.048.

The shape of the invisible halo: N–body simulations on parallel supercomputers.
See Abstr. 151.103.

N–body simulations: from galaxies to cosmology.
See Abstr. 151.104.

The hydrodynamical interaction of galaxies.
See Abstr. 151.117.

Galaxy clustering using the hierarchical Peaks approach: predictions for ROSAT, SZ, and FIR.
See Abstr. 151.137.

Multidimensional analysis of non–linear gravitational clustering in a cold dark matter universe.
See Abstr. 151.138.

Lithium in halo stars: constraining the effects of helium diffusion on globular cluster ages and cosmology.
See Abstr. 154.103.

The pregalactic formation of globular clusters.
See Abstr. 154.131.

Ages of globular clusters.
See Abstr. 154.140.

Globular cluster ages and cosmology.
See Abstr. 154.141.

Galactic evolution of lithium.
See Abstr. 155.029.

The age and formation of the Galaxy: clues from the white dwarf luminosity function.
See Abstr. 155.087.

What determines the physical quantities of galaxies? A two–component gas model for protogalaxies with energy input from supernovae.
See Abstr. 157.045.

Galaxy pairs in redshift catalogs as a probe of dark matter.
See Abstr. 157.050.

A long, hard look at the Virgo cluster.
See Abstr. 157.051.

Dynamics, interstellar medium, and stellar activity interplay in galaxies.
See Abstr. 157.140.

The Cepheid distance scale.
See Abstr. 157.148.

The structure of cold dark matter halos.
See Abstr. 157.207.

CO emission at $z = 2.2867$ in the galaxy IRAS F10214+4724.
See Abstr. 157.229.

Cosmological consequences of starbursts.
See Abstr. 157.241.

Evidence for dark matter from rotation curves: ten years later.
See Abstr. 157.259.

Dark matter, mergers, and morphology: clues from galaxy pairs.
See Abstr. 157.260.

Tidally triggered galaxy formation. I. Evolution of the galaxy luminosity function.
See Abstr. 157.265.

Diameter functions of UGC and ESO galaxies.
See Abstr. 157.273.

Correlations of spin angular momenta of galaxies.
See Abstr. 157.281.

Dwarf galaxies' vanishing act.
See Abstr. 157.283.

Were small galaxies once the dominant cosmological population?
See Abstr. 157.285.

Hidden mass in galaxies.
See Abstr. 157.290.

The Durham/UKST galaxy redshift survey.
See Abstr. 157.295.

A radio–quiet galaxy at redshift $z = 3.409$.
See Abstr. 157.313.

Properties of the redshift. III. Temporal variation.
See Abstr. 157.317.

Effects of internal absorption on galaxy number count and redshift distribution.
See Abstr. 157.320.

Planetary nebulae as standard candles. VII. A test versus Hubble type in the NGC 1023 group.
See Abstr. 157.323.

Molecular rings in galaxies and a possible new cosmic distance indicator.
See Abstr. 157.351.

Cosmic evolution of linear sizes and luminosity function of powerful radio galaxies: is there a common cause?
See Abstr. 158.055.

MG 1131+0456: discovery of the optical Einstein ring with the NTT.
See Abstr. 158.073.

Counts and distribution of spectral indices of extra–galactic radio sources.
See Abstr. 158.079.

Relativistic beaming effect of BL Lac objects.
See Abstr. 158.090.

Recent developments in gravitational lensing.
See Abstr. 158.098.

Radio luminosity functions.
See Abstr. 158.160.

The luminosity function and cosmological evolution of X–ray – selected BL Lacertae objects.
See Abstr. 158.206.

The cosmological evolution of BL Lacertae objects.
See Abstr. 158.286.

The clustering of IRAS Seyfert galaxies.
See Abstr. 158.290.

Radio galaxies and galaxy formation.
See Abstr. 158.293.

IR properties of high z radio galaxies and the alignment effect.
See Abstr. 158.299.

Counts of extragalactic radio sources and the distribution of their spectral indices.
See Abstr. 158.329.

Evidence for a massive, dark gravitational lensing object in Q2345+007.
See Abstr. 159.002.

The nature of the redshift and directly observed quasar statistics.
See Abstr. 159.008.

Bounds on the enhancement of quasar counts by lenses.
See Abstr. 159.029.

Assemblage of two UVX–QSO samples to $B \sim 19$ in the vicinity of the South Galactic Pole.
See Abstr. 159.046.

Optical variability of quasars: statistics and cosmological properties.
See Abstr. 159.050.

Search for quasars on large regions.
See Abstr. 159.054.

Hubble diagram for QSOs with middle and weak interplanetary scintillation.
See Abstr. 159.056.

The origin of quasar correlations.
See Abstr. 159.089.

Survey of quasars over large areas.
See Abstr. 159.121.

On the nature of the redshifts of the twin quasars with discrepant redshifts.
See Abstr. 159.122.

Intergalactic absorption in the spectra of high–redshift QSOs.
See Abstr. 159.123.

Large groups of quasars.
See Abstr. 159.151.

Quasar superclustering.
See Abstr. 159.155.

Superclustering at high redshifts.
See Abstr. 159.156.

A search for quasar protoclusters at $z > 4$.
See Abstr. 159.159.

Quasars and galaxy formation.
See Abstr. 159.165.

Systematic effects in lens inversions: $\aleph$ exact models for 0957+561.
See Abstr. 159.172.

Quasar pairs and quasar's peculiar velocity.
See Abstr. 159.175.

Evolution of high–redshift Lyman–limit absorption systems.
See Abstr. 160.001.

Fourier spectra of three–dimensional data.
See Abstr. 160.004.

The merging runaway.
See Abstr. 160.032.

The distance to the Coma cluster using the B–band Tully–Fisher relation.
See Abstr. 160.033.

The distribution of dark matter in the Perseus cluster.
See Abstr. 160.034.

Lyman–alpha depression of the continuum from high–redshift quasars: a new technique applied in search of the Gunn–Peterson effect.
See Abstr. 160.035.

Cooling–flow modification of the Sunyaev–Zeldovich effect in clusters of galaxies.
See Abstr. 160.038.

A peculiar group of galaxies in the field of the BL Lacertae object H 0414+00.
See Abstr. 160.042.

A determination of the 3–point spatial correlation function for clusters of galaxies.
See Abstr. 160.044.

Radio detections of neutral hydrogen at redshift $z = 3.4$.
See Abstr. 160.051.

The peculiar velocity field in the Hercules region.
See Abstr. 160.054.

Neyman–Scott cluster model and diffuse IGM model for the Lyα forest.
See Abstr. 160.060.

Magnetic fields in the Milky Way neighbourhood as deduced from warps in spiral galaxies.
See Abstr. 160.061.

Magnetism in the local Virgo–centered supercluster of galaxies (radius ~ 24 Mpc).
See Abstr. 160.062.

The probable extragalactic nature of some low surface brightness clouds at high galactic latitude.
See Abstr. 160.063.

A new spectroscopic survey for damped Ly_α absorption lines from high–redshift galaxies.
See Abstr. 160.064.

Statistical properties of X–ray clusters of galaxies.
See Abstr. 160.066.

Confirmation of dust in damped Lyman–alpha systems.
See Abstr. 160.071.

Evolution of clusters of galaxies and the ROSAT All Sky Survey.
See Abstr. 160.076.

Correlation functions of galaxies in the ESO–Uppsala catalogue.
See Abstr. 160.079.

A systematic effect in the use of planetary nebulae as standard candles.
See Abstr. 160.081.

Attenuation of Lyman–alpha emission by dust in damped Lyman–alpha systems.
See Abstr. 160.082.

H I studies of the Sculptor group galaxies. VII. Implications on the distribution and nature of dark matter in groups.
See Abstr. 160.084.

The magnification of stars crossing a caustic. I. Lenses with smooth potentials.
See Abstr. 160.087.

A measurement of the value of the Hubble constant from the X–ray properties and the Sunyaev – Zel'dovich effect of Abell 665.
See Abstr. 160.092.

The clustering of faint galaxies.
See Abstr. 160.094.

The scale dependence of galaxy distribution functions.
See Abstr. 160.098.

Diagnosing the reionization of the Universe: the absorption spectrum of the intergalactic medium and Lyman α clouds.
See Abstr. 160.103.

The hydrodynamics of relict cosmological H II regions: impact on the IGM.
See Abstr. 160.104.

The missing mass in clusters of galaxies and elliptical galaxies.
See Abstr. 160.105.

Superclusters are warm, not hot: the predicted flux distribution of soft X–rays.
See Abstr. 160.109.

Galaxy evolution in distant, X–ray–selected clusters of galaxies. I. Cl 1358+6245.
See Abstr. 160.111.

A statistically complete survey for arc–like features in images of distant rich clusters of galaxies.
See Abstr. 160.112.

Structure and formation of superclusters. XIV. Correlation functions: dependence on the intrinsic properties of galaxy samples.
See Abstr. 160.115.

Large–scale periodicity: problems with cellular models.
See Abstr. 160.116.

On the use of the conditional density as a description of galaxy clustering.
See Abstr. 160.117.

New D – σ results for Coma ellipticals.
See Abstr. 160.118.

Novae and the distance scale.
See Abstr. 160.119.

A high resolution, ground based observation of a Virgo galaxy.
See Abstr. 160.120.

Dynamical evolution of compact groups of galaxies.
See Abstr. 160.122.

Distance to the Coma cluster and the value of H_0.
See Abstr. 160.126.

The effect of clustering on the equivalent width distribution of QSO Lyman–alpha clouds.
See Abstr. 160.130.

Statistical analysis of the galaxy distribution in a diameter–limited sample in the northern sky.
See Abstr. 160.133.

On the baryon content of the Shapley Supercluster.
See Abstr. 160.135.

Scale–invariant clustering in the large–scale distribution of galaxies.
See Abstr. 160.136.

The morphology of clusters of galaxies, the formation efficiency of galaxies, and the origin of the intracluster medium.
See Abstr. 160.137.

Testing models for the dynamical evolution of clusters of galaxies.
See Abstr. 160.138.

Systematic bias in cluster galaxy data, affecting galaxy distances and evolutionary history.
See Abstr. 160.140.

Large–scale peculiar velocity fields in superclusters of galaxies.
See Abstr. 160.141.

The galaxy two–point correlation function down to V = 26 magnitudes on 0°.5 scales.
See Abstr. 160.142.

Void statistics of the CfA redshift survey.
See Abstr. 160.143.

Expectations for X–ray cluster observations by the ROSAT satellite.
See Abstr. 160.154.

The velocity field of clusters of galaxies within 100 megaparsecs. I. Southern clusters.
See Abstr. 160.156.

A hierarchical algorithm to disentangle nearby groups of galaxies (<6000 km s^{-1}).
See Abstr. 160.168.

Are field galaxies randomly distributed on scales of 100+ Mpc?
See Abstr. 160.169.

Clusters of galaxies: structure, infall, and large–scale distribution.
See Abstr. 160.172.

Cosmogony with clusters of galaxies.
See Abstr. 160.173.

Cosmogony and the structure of rich clusters of galaxies.
See Abstr. 160.174.

Dynamical evolution of clusters of galaxies.
See Abstr. 160.179.

Evolution of clusters in the hierarchical scenario.
See Abstr. 160.182.

Distant clusters as cosmological laboratories.
See Abstr. 160.183.

Author Index

The authors are listed in alphabetical order according to the initial letter following the first names. Author names which are not transliterated according to our scheme are indicated by †.

Subject Index

The Subject Index provides an alphabetical subject key to the papers abstracted in *Astronomy and Astrophysics Abstracts*. In order to explain the structure of this Index some general remarks on the construction principles of our key words will follow.

Whenever possible a key word was formed in such a way as to combine two different descriptors characterizing an abstracted article. Such a combination represents the contents of a paper more precisely than both components regarded as single entries. As an example we take the two terms Minor Planets and Orbits and the combination

Minor Planets
  Orbits

Efforts were made to choose combinations which can be inverted in order to increase the usefulness of the Index. That means the two components represent terms of equal status which both serve as primary entries. In the given example we find the two entries

Minor Planets
  Orbits

and

Orbits
  Minor Planets

But there are also two-term key words which cannot be reversed. The conditions for non-inversibility are: the secondary descriptor is either a very specific term (e.g. LTE Analyses) or a very general one (e.g. Models). Thus we distinguish two kinds of descriptors, the primary ones, serving as direct entries and the secondary ones as auxiliary entries to the Index. Our intention is to choose comprehensive as well as significant formulations.

Some further aids to the use of this Subject Index are given by the following concordances:

Asteroids	see Minor Planets
Earth: Magnetic Field	see Geomagnetic Field
Galactic...	see also Galaxies, Galaxy
Lunar...	see also Moon
Planetary...	see also Planets
Solar...	see also Sun
Stellar...	see also Stars

Efforts were made to standardize the key words. This idea led to the development of our vocabulary of astronomical and astrophysical terms. From time to time we enlarged our vocabulary and at present version 29.92 consists of 2,438 items.

The list of key words was examined in order to standardize the key words and to confine some of the synonymous entries. For example, according to our vocabulary the key words Catalogues, Positions, Star Catalogues, and Star Positions were admissible terms, likewise the combinations

Catalogues
  Star Positions

Star Catalogues
  Positions

Star Catalogues
  Star Positions

The user is – in any case – kindly requested to look for synonymous entries, because further references to the topic looked for might exist elsewhere in the Index under another current astronomical term.

Some key words concerning comets, novae, and solar eclipses are sorted by a special routine. Comets are listed according to their Roman numeral designation or preliminary designation, novae according to their stellar constellation and solar eclipses according to the date of the event.

Due to the introduction of a special object index containing a list of single objects, special object designations (e.g. SS433, BL Lacertae, Algol, etc) have been removed from the subject index and transferred to the object index (see SS433, BL Lac, β Per).

Exceptions have been made where major planets and their moons, comets, novae, and supernovae are concerned, which are given in the subject index only. Some objects (name or catalog designation) are contained in both indexes

Andromeda Galaxy	M31, NGC 224
Magellanic Clouds	LMC, SMC
Pleiades	M45
Hyades	Melotte 25
Crab Nebula	M1

In order to obtain the most comprehensive information possible the use of both indexes is recommended.

Starting with Volume 30 of *Astronomy and Astrophysics Abstracts*, this Index is given in a four-column arrangement. The overall size further has been optimized by suppression of a repetition of the first key word in a two-term combination. Thus, the primary term is printed only once (in semi-boldface) for all the following secondary key words which belong to this header.

The following List of Key Words might serve as a further guidance to the use of this Subject Index. This list, a subset of our vocabulary 29.92, consists of 1,336 entries and comprises all one-term key words and all primary terms of the two-term ones actually used in this Subject Index.

List of Key Words

A Dwarfs
A Giants
A Stars
A Supergiants
Aberration
Absolute Magnitudes
Absorption
Absorption Coefficient
Absorption Lines
Absorption Spectra
Absorption Systems
Accretion
Accretion Disk
Accretion Disks
Achondrites
Acoustic Waves
Active Galactic Nuclei
Active Galaxies
Active Optics
Adaptive Optics
Ae Stars
Aerosols
Air Showers
Airglow
Albedo
Alfven Waves
Algol Systems
Almanacs
AM Herculis Stars
Am Stars
Amplifiers
Andromeda Galaxy
Angular Diameters
Angular Momentum
Antennas
Antimatter
Antineutrinos
Ap Stars
Aperture Synthesis
Apodization
Apollo Objects
Apparent Magnitudes
Apsidal Motion
Aquarids
Archaeoastronomy
Artificial Satellites
Asteroidal Belt
Asteroseismology
Astrodynamics
Astrographic Catalogues
Astrographic Plates
Astrographs
Astrolabes
Astrometric Binaries
Astrometric Reflectors
Astrometry
Astronomical Constants
Astronomical Geodesy
Astronomical Instruments
Astronomical Optics
Astronomical Unit
Astronomy
Astrophysics
Atlases
Atmospheres
Atmospheric Optics
Atomic Clocks
Atomic Parameters
Atomic Processes
Atomic Spectra
Atomic Time
Aurorae
Auroral Arcs
Auxiliary Instrumentation

B Stars
B Subdwarfs
B Supergiants
Background Radiation
Balloons
Balmer Lines
Barium Stars
Barred Spirals
Baryogenesis
Baryons
Be Stars
Beta Cephei Stars
Beta Lyrae Stars
Binaries
Binary Pulsars
Bipolar Nebulae
Bipolar Outflows
BL Lacertae Objects
Black Holes
Black-Body Radiation
Blazars
Blue Galaxies
Blue Objects
Blue Stars
Blue Stragglers
Bolometric Corrections
Bolometric Magnitudes
Boson Stars
Bp Stars
Bremsstrahlung
Bright Galaxies
Bright Stars
Brightness Temperature
Brightness Temperatures
Broad-Band Photometry
Brown Dwarfs
BV Photometry
BY Draconis Stars

C-M Diagrams
Calendars
Callisto
Callisto Craters
Cameras
Carbon Burning
Carbon Stars
Cassegrain Spectrographs
Cataclysmic Binaries
Catalogues
CCD Cameras
CCD Detectors
CCD Observations
CCD Photometry
Celestial Bodies
Celestial Mechanics
Celestial Objects
Central Stars
Cepheids
CH Stars
Chandler Wobble
Chaos
Charon
Chondrites
Chondrules
Circumplanetary Matter
Circumstellar Clouds
Circumstellar Disks
Circumstellar Dust
Circumstellar Grains
Circumstellar Matter
Circumstellar Shell
Circumstellar Shells
Clocks
Close Binaries
Clusters of Galaxies
CNO Anomalies
Cocoon Stars
Collapse
Collapsed Stars
Collapsing Clouds
Collapsing Stars
Collimators
Color Excesses
Color Indices
Colors
Cometary Atmospheres
Cometary Comae
Cometary Dust
Cometary Ionospheres
Cometary Nebulae
Cometary Nuclei
Cometary Plasma
Cometary Tails
Comets
Compact Galaxies
Compact Objects
Compton Effect
Compton Scattering
Contact Binaries
Convection
Convective Zones
Cool Giants
Cool Stars
Cooling Flows
Coordinate Systems
Coronographs
Correlation Analysis
Cosmic Abundance
Cosmic Density Parameter
Cosmic Dust
Cosmic Microwave Background
Cosmic Rays
Cosmic Strings
Cosmochemistry
Cosmochronology
Cosmogony
Cosmological Constant
Cosmological Models
Cosmology
Coude Spectrographs
CP Stars
Crab Nebula
Curves-of-Growth
Cyclotron Radiation
Cygnus Loop

Dark Clouds
Dark Matter
Dark Nebulae
Data Bases
Data Centers
Data Networks
Data Processing
DDO Photometry
Degenerate Matter
Degenerate Stars
Deimos
Deimos Craters
Deimos Surface
Delta Delphini Stars
Delta Scuti Stars
Dense Matter
Densities
Density
Density Waves
Detached Systems
Detector Arrays
Detectors
Diameter
Diameters
Differential Rotation
Diffuse Nebulae
Disk Galaxies
Disk Population
Disk Stars
Distance
Distance Indicators
Distance Moduli
Distance Scale
Distances
Documentation
Domes
Doppler Imaging
Doppler Velocities
DQ Herculis Stars
Dust Clouds
Dwarf Galaxies
Dwarf Novae
Dwarfs
Dynamo Theory

Early Universe
Early-Stage Stars
Early-Type Galaxies
Early-Type Stars
Early-Type Supergiants
Earth
Earth Atmosphere
Earth Core
Earth Interior
Earth Ionosphere
Earth Magnetosphere
Earth Mantle
Earth Mesosphere
Earth Plasmasphere
Earth Radiation Belts
Earth Rotation
Earth Stratosphere
Earth Thermosphere
Earth Troposphere
Earth-Moon System
Echelle Spectra
Echelle Spectrographs
Echelle Spectrometers
Eclipses
Eclipsing Binaries
Effective Temperature
Effective Temperatures
Electron Beams
Electron Densities
Electron Temperatures
Elementary Particles
Elliptical Galaxies
Emission Lines
Emission Nebulae
Emission Spectra
Emission-Line Galaxies
Emission-Line Stars
Energetic Particles
Ephemerides
Ephemeris
Ephemeris Time
Equidensitometry
Equilibrium Figures
Eruptive Variables
Europa
Extended Objects
Extinction
Extragalactic Objects
Extraterrestrial Intelligence
Extraterrestrial Life
Extreme UV
Extreme UV Astronomy
Extreme UV Background
Extreme UV Sources
Extreme UV Spectra
Eyepieces

F Dwarfs
F Giants
F Stars
F Supergiants
Fabry-Perot Interferometers
Fabry-Perot Interferometry
Fabry-Perot Spectrometers
Faint Galaxies
Faint Stars
Fiber Optics
Field Galaxies
Field Stars
Field Theories
Filters
FIR Radiation
Fireballs
FK Comae Stars

Object Index

The usual bibliographic information accompanying the papers in the fields of astronomy and astrophysics certainly allows for their exact identification but the selection process of information retrieval is facilitated by supplementary information. Phenomena of all kinds, methods of observation and reduction as well as properties of classes of objects are referred to in the Subject Index.

Beginning with Vol. 39 *Astronomy and Astrophysics Abstracts* will present an additional index of astronomical objects dealt with in papers compiled. Basically all individual objects appearing in the title and the author's abstract are listed, but objects treated in the paper proper have also been included into the index. The number of single objects listed usually has been limited to 30. Papers discussing more than 30 individual objects have not been evaluated in the Object Index.

It is recommended to use the Object Index together with the key words of the Subject Index. Individual comets, novae, supernovae, major planets and their satellites are exclusively referred to in the Subject Index. A number of familiar astronomical objects will continue to appear in the Subject Index with their common designation. These objects are listed in the Object Index with their standard acronyms:

M31, NGC 224	Andromeda Galaxy
LMC, SMC	Magellanic Clouds
M45	Pleiades
Melotte 25	Hyades
M1	Crab Nebula

Novae identified with variables are referred to in the Object Index with their variable name:

GK Per	Nova Persei 1901

Astronomy and Astrophysics Abstracts supports the attempts of the IAU in standardizing the nomenclature of astronomical objects. Acronyms used follow closely the recommendations of "The First Dictionary of the Nomenclature of Celestial Objects" (A. Fernandez, M.-C. Lortet, F. Spite; Astron. Astrophys., Suppl. Ser., Vol. 52, No. 4 (1983)), and its "First Supplement" (M.-C. Lortet, F, Spite; Astron. Astrophys., Suppl. Ser., Vol. 64, No. 2, p. 329–389 (1986)).

In sorting astronomical objects deviations from the lexicographic sequence are unavoidable in many cases. Sorting of the Object Index is based on the IBM standard routine SORT, which sorts blanks followed by special characters, Latin small and capital letters, and finally numbers. The three-letter mnemonic code of the 88 stellar constellations adopted by the IAU defines the highest rank within the sorting hierarchy. This implies that all stellar and non-stellar objects, whose designations explicitely refer to a constellation, are listed under the constellation's abbreviation. Stars within a constellation are sorted according to their Greek letter, followed by the standard variable nomenclature, and finally the Flamsteed numbers as well as other catalogue designations:

Cyg A
Cyg OB2
Cyg X–3
α Cyg
χ Cyg
BF Cyg
NML Cyg
Z Cyg
V367 Cyg
V1727 Cyg
6 Cyg

Designations without reference to stellar constellations are treated – as a rule – by sorting letters before numbers. An exception has been made for the minor planets. These objects have been retained in a particular section following after the letter 'Z'. The internal sequence is (1) permanently numbered planets, (2) objects with provisional designations, and (3) Palomar-Leiden objects:

(9) Metis
(29) Amphitrite
1979 VA
1984 HA_1
6344 P–L

Efforts were made to sort objects from individual catalogues according to the internal order of these catalogues. In Durchmusterungen the stars are arranged from the equator to the pole, taking the northward direction first. Object designations which contain coordinates are sorted first according to the usual catalogue abbreviation (B2, PKS, ZwCl, etc.) or object class designator (GX, PSR, SNR, etc.) and then – if present – according first to equatorial and then to galactic coordinates. Objects with catalogue designations beginning with a number (1E, 3C, 4U, etc.) are listed under the respective number. If only a position is given, the object is referred to by the number sequence, subdivided in equatorial and galactic coordinates, at the end of the Index. A consequence of the applied sorting routine is, e.g., that Messier objects will be found at the end of the list of objects beginning with the letter 'M'.

Attempts have been made to avoid ambiguous acronyms. For instance, in the astronomical literature the acronym VB is used to designate (1) planetary nebula listed by Van den Bergh, (2) Bologna variables, (3) Van Bueren's list of Hyades stars, (4) Van Biesbroeck's list of faint stars. This ambiguity is partially resolved by using 'Van Bueren ...' for objects in (3), and 'Van Biesbroeck ...' for those in (4). With respect to the problem of multiple designations for a single object, a concordance relation is beyond the scope of the present Object Index. The user is therefore advised to search for possible alternative designations of a particular object. For example, the supergiant α Ori may also appear under BD +7°1055 or HD 39801, or the radio source Virgo A may be referred under 1218+1240, 3C 274, NGC 4486 or M87.

ASTRONOMY AND ASTROPHYSICS ABSTRACTS

A Publication of the Astronomisches Rechen-Institut Heidelberg

Editors: G. Burkhardt, U. Esser, H. Hefele, I. Heinrich, W. Hofmann, D. Krahn, V. R. Matas, L. D. Schmadel, R. Wielen, G. Zech

Volume 1	Literature 1969, Part 1	X + 435 pp. (1969)
Volume 2	Literature 1969, Part 2	X + 516 pp. (1970)
Volume 3	Literature 1970, Part 1	X + 490 pp. (1970)
Volume 4	Literature 1970, Part 2	X + 562 pp. (1971)
Volume 5	Literature 1971, Part 1	X + 505 pp. (1971)
Volume 6	Literature 1971, Part 2	X + 560 pp. (1972)
Volume 7	Literature 1972, Part 1	X + 526 pp. (1972)
Volume 8	Literature 1972, Part 2	X + 594 pp. (1973)
Volume 9	Literature 1973, Part 1	X + 610 pp. (1973)
Volume 10	Literature 1973, Part 2	X + 661 pp. (1974)
Volume 11	Literature 1974, Part 1	X + 579 pp. (1974)
Volume 12	Literature 1974, Part 2	X + 699 pp. (1975)
Volume 13	Literature 1975, Part 1	X + 632 pp. (1975)
Volume 14	Literature 1975, Part 2	X + 747 pp. (1976)
Volume 15/16	Author and Subject Indexes to Volumes 1–10	
	Literature 1969–1973	VII + 655 pp. (1976)
Volume 17	Literature 1976, Part 1	XII + 645 pp. (1976)
Volume 18	Literature 1976, Part 2	X + 859 pp. (1977)
Volume 19	Literature 1977, Part 1	X + 732 pp. (1977)
Volume 20	Literature 1977, Part 2	X + 786 pp. (1978)
Volume 21	Literature 1978, Part 1	X + 834 pp. (1978)
Volume 22	Literature 1978, Part 2	X + 849 pp. (1979)
Volume 23/24	Author and Subject Indexes to Volumes 11–14 and 17–22	
	Literature 1974–1978	VIII + 1127 pp. (1979)
Volume 25	Literature 1979, Part 1	X + 872 pp. (1979)
Volume 26	Literature 1979, Part 2	X + 794 pp. (1980)
Volume 27	Literature 1980, Part 1	X + 939 pp. (1980)
Volume 28	Literature 1980, Part 2	X + 841 pp. (1981)
Volume 29	Literature 1981, Part 1	X + 853 pp. (1981)
Volume 30	Literature 1981, Part 2	X + 792 pp. (1982)
Volume 31	Literature 1982, Part 1	X + 776 pp. (1982)
Volume 32	Literature 1982, Part 2	X + 848 pp. (1983)
Volume 33	Literature 1983, Part 1	X + 815 pp. (1983)
Volume 34	Literature 1983, Part 2	X + 921 pp. (1984)
Volume 35/36	Author and Subject Indexes to Volumes 25–34	
	Literature 1979–1983	XIII + 891 pp. (1984)

Published for Astronomisches Rechen-Institut by
Springer-Verlag Berlin Heidelberg New York

ASTRONOMY AND ASTROPHYSICS ABSTRACTS

A Publication of the Astronomisches Rechen-Institut Heidelberg

Editors: G. Burkhardt, U. Esser, H. Hefele, I. Heinrich, W. Hofmann, D. Krahn, V. R. Matas, L. D. Schmadel, R. Wielen, G. Zech

Volume 37	Literature 1984, Part 1	X + 937 pp. (1984)
Volume 38	Literature 1984, Part 2	X + 919 pp. (1985)
Volume 39	Literature 1985, Part 1	X + 1149 pp. (1985)
Volume 40	Literature 1985, Part 2	XIV + 1096 pp. (1986)
Volume 41	Literature 1986, Part 1	X + 1114 pp. (1986)
Volume 42	Literature 1986, Part 2	X + 1172 pp. (1987)
Volume 43	Literature 1987, Part 1	X + 1291 pp. (1987)
Volume 44	Literature 1987, Part 2	XII + 1100 pp. (1988)
Volume 45	Literature 1988, Part 1	X + 1259 pp. (1988)
Volume 46	Literature 1988, Part 2	X + 1298 pp. (1989)
Volume 47/48	Author, Subject, and Object Indexes to Volumes 37–46	
	Literature 1984–1988	VII + 1498 pp. (1989)
Volume 49	Literature 1989, Part 1	XII + 1423 pp. (1990)
Volume 50	Literature 1989, Part 2	X + 1429 pp. (1990)
Volume 51	Literature 1990, Part 1	X + 1526 pp. (1991)
Volume 52	Literature 1990, Part 2	X + 1553 pp. (1991)
Volume 53	Literature 1991, Part 1	X + 1625 pp. (1992)
Volume 54	Literature 1991, Part 2	X + 1582 pp. (1992)

Published for Astronomisches Rechen-Institut by
Springer-Verlag Berlin Heidelberg New York

9783662123782VOL03